What Is an "Integrated Approach"?

One of the key themes in this book is that the body is an integrated set of systems. One of your tasks as you study will be to construct for yourself this global view of the body, its systems, and the many processes that keep the systems working.

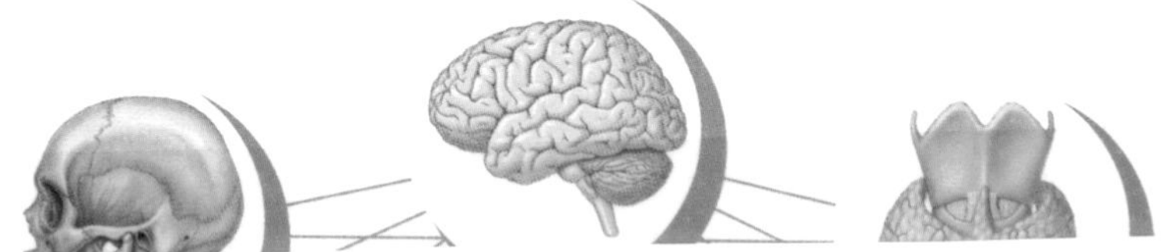

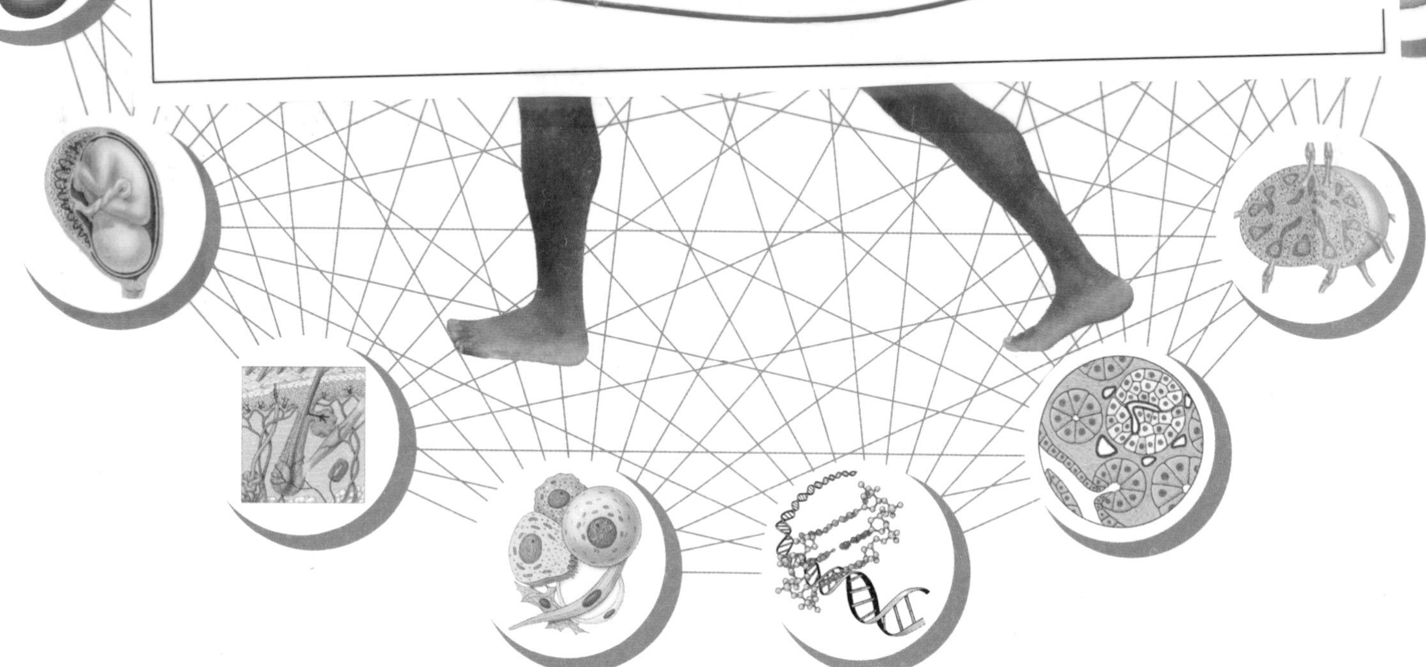

Top Ten Ways to Succeed in Classes that Use Active Learning

By Marilla Svinicki, Ph.D., Director
University of Texas Center for Teaching Effectiveness

1 Make the switch from an authority-based conception of learning to a self-regulated conception of learning. Recognize and accept your own responsibility for learning.

2 Be willing to take risks and go beyond what is presented in class or the text.

3 Be able to tolerate ambiguity and frustration in the interest of understanding.

4 See errors as opportunities to learn rather than failures. Be willing to make mistakes in class or in study groups so that you can learn from them.

5 Engage in active listening to what's happening in class.

6 Trust the instructor's experience in designing class activities and participate willingly if not enthusiastically.

7 Be willing to express an opinion or hazard a guess.

8 Accept feedback in the spirit of learning rather than as a reflection of you as a person.

9 Prepare for class physically, mentally, and materially (do the reading, work the problems, etc.).

10 Provide support for your classmate's attempts to learn. The best way to learn something well is to teach it to someone who doesn't understand.

Dr. Dee's Eleventh Rule:
DON'T PANIC! Pushing yourself beyond the comfort zone is scary, but you have to do it in order to improve.

Word Roots for Physiology

a- or **an-** without, absence
anti- against
-ase signifies an enzyme
auto self
bi-two
brady-slow
cardio- heart
cephalo- head
cerebro- brain
contra- against
-crine a secretion
crypt- hidden
cutan- skin
-cyte or **cyto-** cell
de- without, lacking
di- two
dys- difficult, faulty
-elle small
-emia blood
endo- inside or within
epi- over
erythro- red
exo- outside
extra- outside
gastro- stomach
-gen, -genic produce
gluco-, glyco- sugar or sweet
hemi- half
hemo- blood
hepato- liver
homo- same
hydro- water
hyper- above or excess
hypo- beneath or deficient
inter- between
intra- within
-itis inflammation of
kali- potassium
leuko- white
lipo- fat
lumen inside of a hollow tube
-lysis split apart or rupture
macro- large
micro- small
mono- one
multi- many
myo- muscle
oligo- little, few
para- near, close
patho-, -pathy related to disease
peri- around
poly- many
post- after
pre- before
pro- before
pseudo- false
re- again
retro- backward or behind
semi- half
sub- below
super- above, beyond
supra- above, on top of
tachy- rapid
trans- across, through

HUMAN PHYSIOLOGY

AN INTEGRATED APPROACH

SIXTH EDITION

Dee Unglaub Silverthorn, Ph.D.
University of Texas, Austin

WITH CONTRIBUTIONS BY

Bruce R. Johnson, Ph.D.
Cornell University

AND

William C. Ober, M.D.
Illustration Coordinator

Claire W. Garrison, R.N.
Illustrator

Andrew C. Silverthorn, M.D.
Clinical Consultant

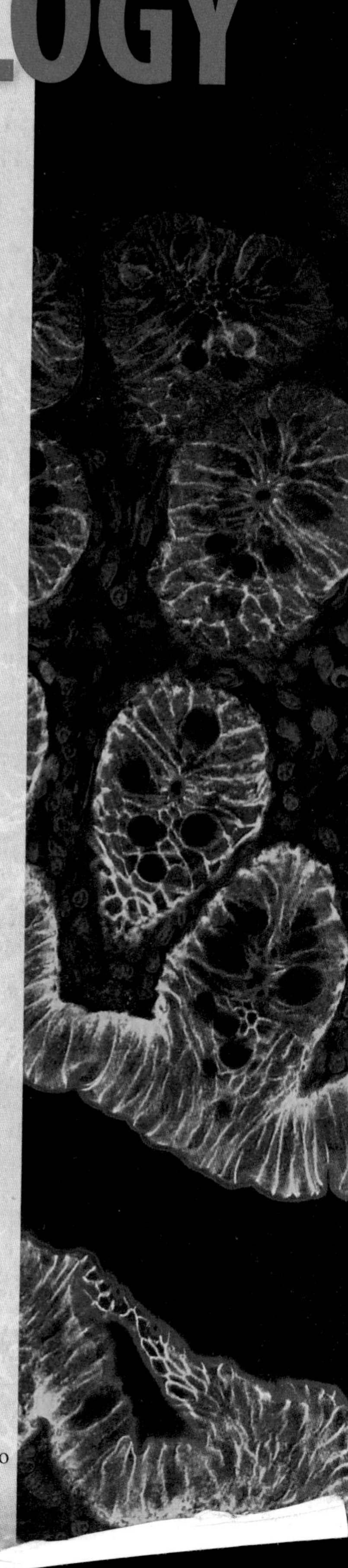

PEARSON

Boston Columbus Indianapolis New York San Francisco Upper Saddle River
Amsterdam Cape Town Dubai London Madrid Milan Munich Paris Montréal Toronto
Delhi Mexico City São Paulo Sydney Hong Kong Seoul Singapore Taipei Tokyo

Acquisitions Editor: *Kelsey K. Volker*
Project Editor: *Anne Scanlan-Rohrer*
Director of Development: *Barbara Yien*
Development Editor: *Anne A. Reid*
Assistant Editor: *Ashley Williams*
Senior Managing Editor: *Deborah Cogan*
Assistant Managing Editor: *Nancy Tabor*
Copyeditor: *Antonio Padial*
Production Management and Composition: *PreMediaGlobal*
Production Manager: *Jared Sterzer*
Interior Designer: *Jim Gibson*
Cover Designer: *Riezebos Halzbaur Design Group*
Illustrators: *William C. Ober* and *Claire W. Garrison*
Art House: *Imagineering*
Photo Researcher: *Kristin Piljay*
Photo Lead: *Donna Kalal*
Assistant Media Producer: *Annie Wang*
Senior Manufacturing Buyer: *Stacey Weinberger*
Marketing Manager: *Derek Perrigo*

Cover Image: Confocal immunofluorescent analysis of rat colon using Phospho-β-Catenin (Ser675) (D2F1) XP® Rabbit mAb (green). Actin filaments have been labeled with DY-554 Phalloidin (red). Blue pseudocolor = DRAQ5® #4084 (fluorescent DNA dye). Credit: © Cell Signaling Technology. Image reproduced courtesy of Cell Signaling Technology, Inc. (www.cellsignal.com)

Library of Congress Cataloging-in-Publication Data
Silverthorn, Dee Unglaub, 1948-
Human physiology : an integrated approach / Dee Unglaub Silverthorn ; with contributions by Bruce R. Johnson and William C. Ober, illustration coordinator ; Claire W. Garrison, illustrator ; Andrew C Silverthorn, clinical consultant. -- 6th ed.
p. cm.
Includes index.
ISBN-13: 978-0-321-75007-5 (student ed.)
ISBN-10: 0-321-75007-1 (student ed.)
ISBN-13: 978-0-321-81082-3 (instructor's review copy)
ISBN-10: 0-321-81082-1 (instructor's review copy)
1. Human physiology. I. Johnson, Bruce R. II. Ober, William C. III. Garrison, Claire W. IV. Silverthorn, Andrew C. V. Title.
QP34.5.S55 2013
612--dc23

2011044995

ISBN 10: 0-321-75007-1; ISBN 13: 978-0-321-75007-5 (Student edition)
ISBN 10: 0-321-81082-1; ISBN 13: 978-0-321-81082-3 (Instructor's Review Copy)

www.pearsonhighered.com

1 2 3 4 5 6 7 8 9 10—DOW—15 14 13 12 11

About the Author

Dee Unglaub Silverthorn studied biology as an undergraduate at Newcomb College of Tulane University, where she did research on cockroaches. For graduate school she switched to studying crabs and received a Ph.D. in marine science from the Belle W. Baruch Institute for Marine and Coastal Sciences at the University of South Carolina. Her research interest is epithelial transport, and recent work in her laboratory has focused on transport properties of the chick allantoic membrane. Her teaching career started in the Physiology Department at the Medical University of South Carolina but over the years she has taught a wide range of students, from medical and college students to those still preparing for higher education. At the University of Texas-Austin she teaches physiology in both lecture and laboratory settings, and instructs graduate students on developing teaching skills in the life sciences. She has received numerous teaching awards and honors, including a 2011 UT System Regents' Outstanding Teaching Award, the 2009 Outstanding Undergraduate Science Teacher Award from the Society for College Science Teachers, the American Physiological Society's Claude Bernard Distinguished Lecturer and Arthur C. Guyton Physiology Educator of the Year, and multiple awards from UT-Austin, including the Burnt Orange Apple Award. The first edition of her textbook won the 1998 Robert W. Hamilton Author Award for best textbook published in 1997–98 by a University of Texas faculty member. Dee recently completed six years as editor-in-chief of *Advances in Physiology Education,* and she works with members of the International Union of Physiological Sciences to improve physiology education in developing countries. She will be the president of the Human Anatomy and Physiology Society in 2012–13. She is also an active member of the American Physiological Society, the Society for Comparative and Integrative Biology, the Association for Biology Laboratory Education, and the Society for College Science Teachers. Her free time is spent creating multimedia fiber art and enjoying the Texas hill country with her husband, Andy, and their dogs.

About the Illustrators

William C. Ober, M.D. (art coordinator and illustrator) received his undergraduate degree from Washington and Lee University and his M.D. from the University of Virginia. He also studied in the Department of Art as Applied to Medicine at Johns Hopkins University. After graduation, Dr. Ober completed a residency in Family Practice and later was on the faculty at the University of Virginia in the Department of Family Medicine and in the Department of Sports Medicine. He also served as Chief of Medicine of Martha Jefferson Hospital in Charlottesville, VA. He is currently a visiting Professor of Biology at Washington & Lee University, where he has taught several courses and led student trips to the Galapagos Islands. He is part of the Core Faculty at Shoals Marine Laboratory, where he teaches Biological Illustration every summer. The textbooks illustrated by Medical & Scientific Illustration have won numerous design and illustration awards.

Claire W. Garrison, R.N. (illustrator) practiced pediatric and obstetric nursing before turning to medical illustration as a full-time career. She returned to school at Mary Baldwin College where she received her degree with distinction in studio art. Following a 5-year apprenticeship, she has worked as Dr. Ober's partner in Medical and Scientific Illustration since 1986. She is on the Core Faculty at Shoals Marine Laboratory and co-teaches the Biological Illustration course.

About the Clinical Consultant

Andrew C. Silverthorn, M.D. is a graduate of the United States Military Academy (West Point). He served in the infantry in Vietnam, and upon his return entered medical school at the Medical University of South Carolina in Charleston. He was chief resident in family medicine at the University of Texas Medical Branch, Galveston, and is currently a family physician in solo practice in Austin, Texas. When Andrew is not busy seeing patients, he may be found on the golf course or playing with his dogs, Lady Godiva (a chocolate lab), and Molly.

About the Contributor

Bruce Johnson is a Senior Research Associate in the Department of Neurobiology and Behavior at Cornell University. He earned biology degrees at Florida State University (B.A.), Florida Atlantic University (M.S.), and at the Marine Biological Laboratory in Woods Hole (Ph.D.) through the Boston University Marine Program. At Cornell he teaches an undergraduate laboratory course entitled Principles of Neurophysiology. He is a co-author of *Crawdad: a CD-ROM lab manual for Neurophysiology* and the *Laboratory Manual for Physiology,* and he continues development of model preparations for student neuroscience laboratories. Bruce has taught in faculty workshops sponsored by NSF (Crawdad) and the Faculty for Undergraduate Neuroscience (FUN)/Project Kaleidoscope, and in graduate and undergraduate neuroscience laboratory courses at the University of Copenhagen, the Marine Biological Laboratory, and the Shoals Marine Laboratory. He has received outstanding educator and distinguished teaching awards at Cornell University, and the FUN Educator of the Year Award. He is a past president of FUN. Bruce's research addresses the cellular and synaptic mechanisms of motor network plasticity.

DEDICATION

In memory of
Walter George Unglaub, M.D.
1918–1970

Contents in Brief

Contents

Chapter 11

377 Efferent Division: Autonomic and Somatic Motor Control

Chapter 12

398 Muscles

Chapter 13

441 Integrative Physiology I: Control of Body Movement

Unit 3

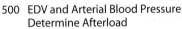

Integration of Function

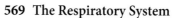

Unit 4

Metabolism, Growth, and Aging

Owner's Manual: How to Use This Book

Welcome to Human Physiology!

As you begin your study of the human body, you should be prepared to make maximum use of the resources available to you, including your instructor, the library, the Internet, and your textbook. One of my goals in this book is to provide you not only with information about how the human body functions but also with tips for studying and problem solving. Many of these study aids have been developed with the input of my students, so I think you may find them particularly helpful. On the following pages, I have put together a brief tour of the special features of the book, especially those that you may not have encountered previously in textbooks. Please take a few minutes to read about them so that you can make optimum use of the book as you study.

One of your tasks as you study will be to construct for yourself a global view of the body, its systems, and the many processes that keep the systems working. This "big picture" is what physiologists call the integration of systems, and it is a key theme in the book. To integrate information, however, you must do more than simply memorize it. You need to truly understand it and be able to use it to solve problems that you have never encountered before. If you are headed for a career in the health professions, you will do this in the clinics. If you plan a career in biology, you will solve problems in the laboratory, field, or classroom. Analyzing, synthesizing, and evaluating information are skills you need to develop while you are in school, and I hope that the features of this book will help you with this goal.

In this edition we have continued to update and focus on basic themes and concepts of physiology. Chapter 1 introduces you to the key concepts in physiology that you encounter repeatedly as you study different organ systems. It also includes several special features: one on mapping, a useful study skill that is also used for decision-making in the clinics, and one on constructing and interpreting graphs. The new Chapter 1 Running Problem introduces you to effective ways to find information on the Internet.

A new element in this edition's art program is the Essentials and Review figures. These figures distill the basics about a topic onto one or two pages, much as the Anatomy Summaries do. My students tell me they find them particularly useful for review when there isn't time to go back and read all the text.

We have also retained the four approaches to learning physiology that proved so popular since this book was first published in 1998.

1. Cellular and Molecular Physiology

Most physiological research today is being done at the cellular and molecular level, and there have been many exciting developments in molecular medicine and physiology in the ten years since the first edition. For example, now scientists are paying more attention to primary cilia, the single cilium that occurs on most cells of the body. Primary cilia are thought to play a role in some kidney and other diseases. Look for similar links between molecular and cellular biology, physiology, and medicine throughout the book.

2. Physiology as a Dynamic Field

Physiology is a dynamic discipline, with numerous unanswered questions that merit further investigation and research. Many of the "facts" presented in this text are really only our current theories, so you should be prepared to change your mental models as new information emerges from scientific research.

3. An Emphasis on Integration

The organ systems of the body do not work in isolation, although we study them one at a time. To emphasize the integrative nature of physiology, three chapters (Chapters 13, 20, and 25) focus on how the physiological processes of multiple organ systems coordinate with each other, especially when homeostasis is challenged.

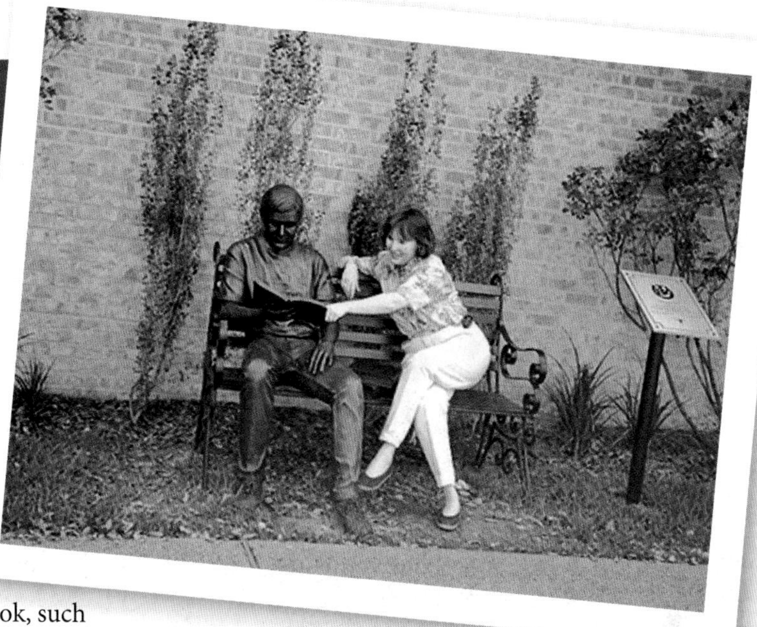

4. A Focus on Problem Solving

One of the most valuable life skills students should acquire is the ability to think critically and use information to solve problems. As you study physiology, you should be prepared to practice these skills. You will find a number of features in this book, such as the Concept Check questions and figure and graph questions, that are designed to challenge your critical thinking and analysis skills. In each chapter, read the Running Problem as you work through the text and see if you can apply what you're reading to the clinical scenario described in the problem.

Also, be sure to look at the back of the text, where we have combined the index and glossary to save time when you are looking up unfamiliar words. The appendices have the answers to the end-of-chapter questions, as well as reviews of physics, logarithms, and basic genetics. The back end papers include a periodic table of the elements, diagrams of anatomical positions of the body, and tables with conversions and normal values of blood components. Take a few minutes to look at all these features so that you can make optimum use of them.

It is my hope that by reading this book, you will develop an integrated view of physiology that allows you to enter your chosen profession with respect for the complexity of the human body and a clear vision of the potential of physiological and biomedical research. May you find physiology as fun and exciting I do. Good luck with your studies!

Warmest regards,

Dr. Dee (as my students call me)
silverthorn@mail.utexas.edu

Engaging Art Helps You Visualize Processes and Concepts—

NEW REVIEW ART FEATURES visually pull together foundational concepts so you can quickly review key topics.

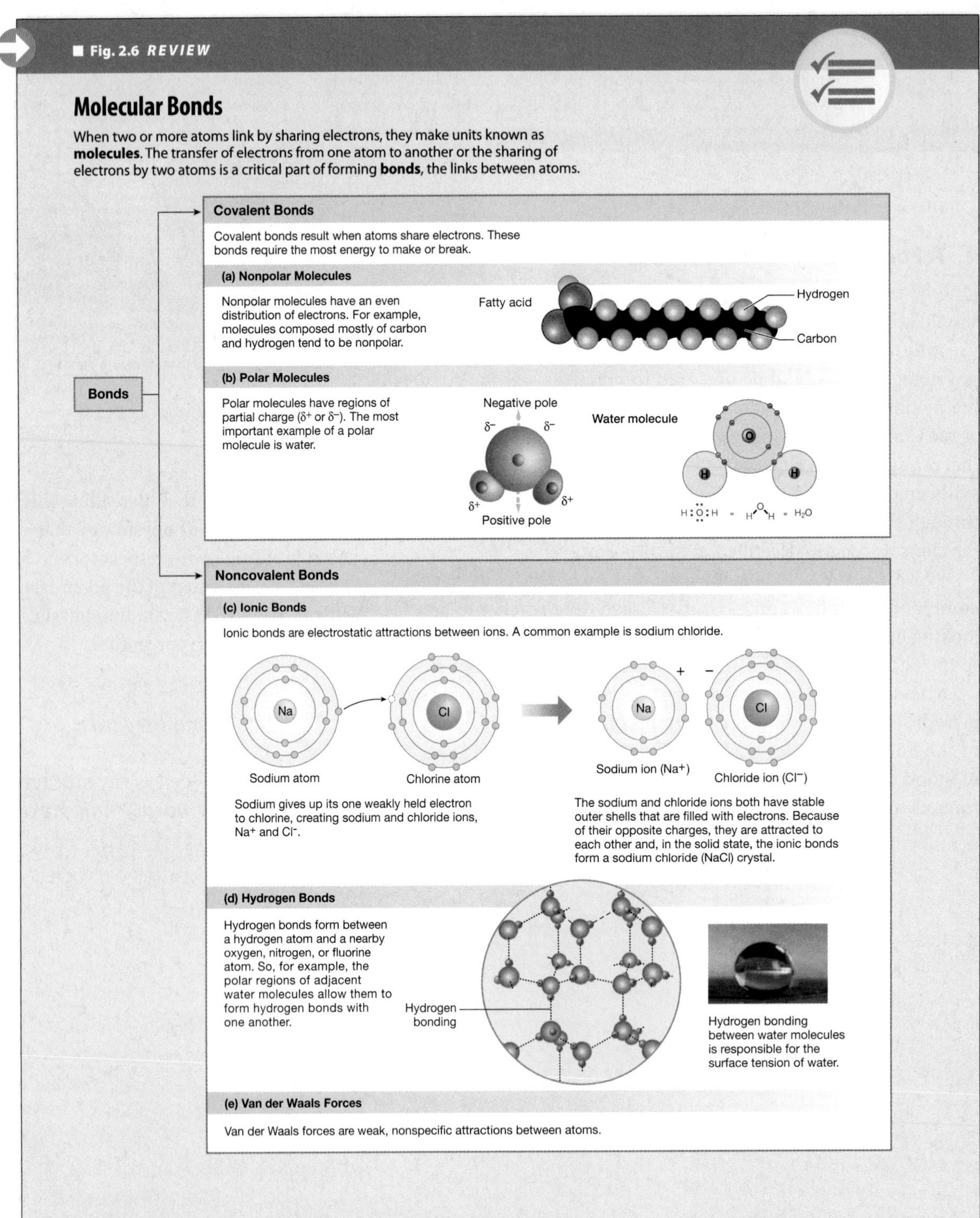

Fig. 2.6 REVIEW
Molecular Bonds
When two or more atoms link by sharing electrons, they make units known as molecules. The transfer of electrons from one atom to another or the sharing of electrons by two atoms is a critical part of forming bonds, the links between atoms.
Bonds
Covalent Bonds
Covalent bonds result when atoms share electrons. These bonds require the most energy to make or break.
(a) Nonpolar Molecules
Nonpolar molecules have an even distribution of electrons. For example, molecules composed mostly of carbon and hydrogen tend to be nonpolar.
Fatty acid
Hydrogen
Carbon
(b) Polar Molecules
Polar molecules have regions of partial charge (δ+ or δ−). The most important example of a polar molecule is water.
Negative pole
δ−
δ−
δ+
δ+
Positive pole
Water molecule
O
H
H
H:Ö:H = H-O-H = H₂O
Noncovalent Bonds
(c) Ionic Bonds
Ionic bonds are electrostatic attractions between ions. A common example is sodium chloride.
Na
Cl
Sodium atom
Chlorine atom
+
−
Sodium ion (Na+)
Chloride ion (Cl−)
Sodium gives up its one weakly held electron to chlorine, creating sodium and chloride ions, Na+ and Cl-.
The sodium and chloride ions both have stable outer shells that are filled with electrons. Because of their opposite charges, they are attracted to each other and, in the solid state, the ionic bonds form a sodium chloride (NaCl) crystal.
(d) Hydrogen Bonds
Hydrogen bonds form between a hydrogen atom and a nearby oxygen, nitrogen, or fluorine atom. So, for example, the polar regions of adjacent water molecules allow them to form hydrogen bonds with one another.
Hydrogen bonding
Hydrogen bonding between water molecules is responsible for the surface tension of water.
(e) Van der Waals Forces
Van der Waals forces are weak, nonspecific attractions between atoms.

and Focus on the Key Ideas

NEW ESSENTIALS ART FEATURES show each chapter's core concepts, helping you connect ideas visually and see the big picture of human physiology.

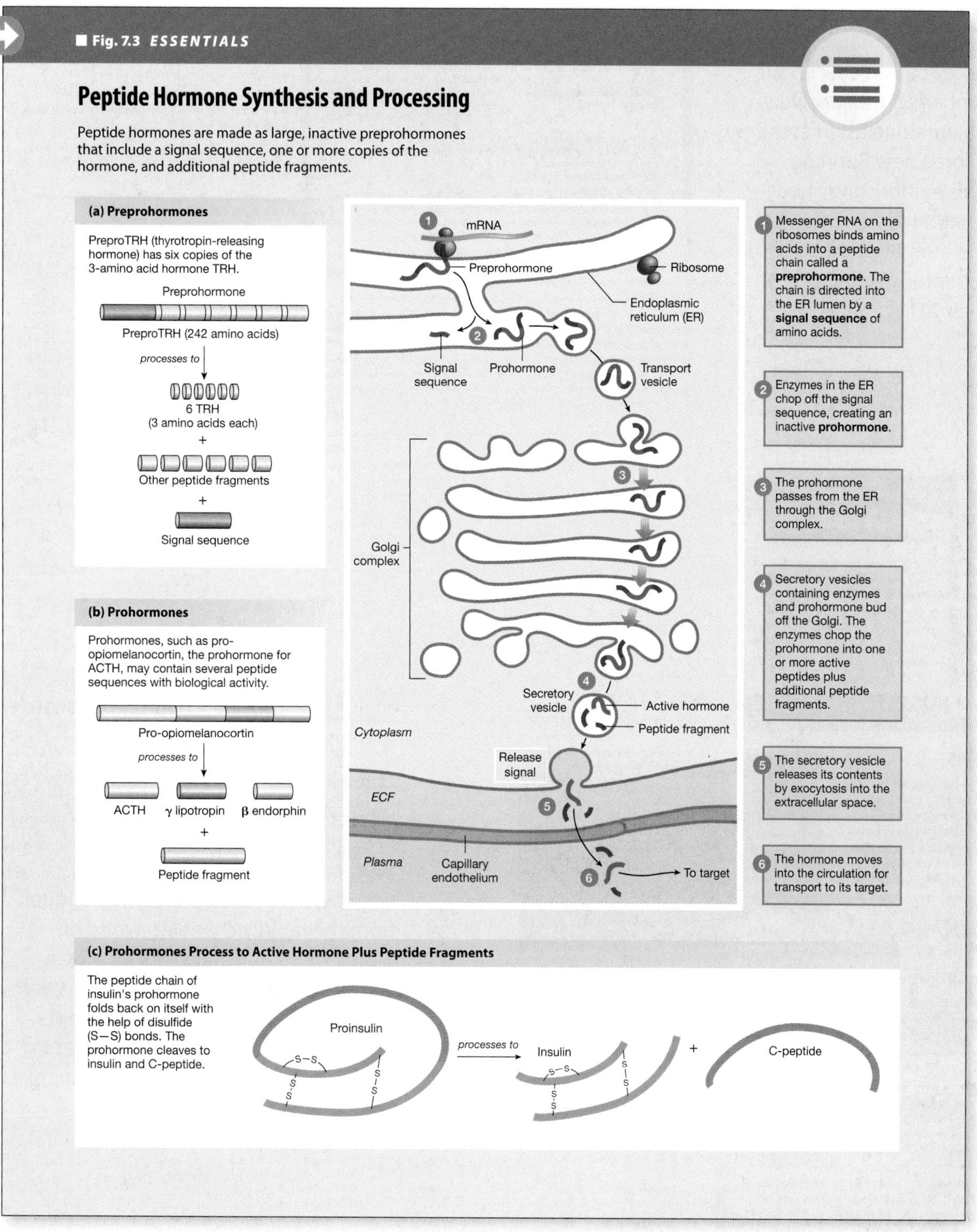

Fig. 7.3 ESSENTIALS
Peptide Hormone Synthesis and Processing
Peptide hormones are made as large, inactive preprohormones that include a signal sequence, one or more copies of the hormone, and additional peptide fragments.
(a) Preprohormones
PreproTRH (thyrotropin-releasing hormone) has six copies of the 3-amino acid hormone TRH.
Preprohormone
PreproTRH (242 amino acids)
processes to
6 TRH
(3 amino acids each)
+
Other peptide fragments
+
Signal sequence
(b) Prohormones
Prohormones, such as pro-opiomelanocortin, the prohormone for ACTH, may contain several peptide sequences with biological activity.
Pro-opiomelanocortin
processes to
ACTH
γ lipotropin
β endorphin
+
Peptide fragment
1
mRNA
Preprohormone
Ribosome
Endoplasmic reticulum (ER)
2
Signal sequence
Prohormone
Transport vesicle
3
Golgi complex
4
Secretory vesicle
Active hormone
Peptide fragment
Cytoplasm
Release signal
ECF
5
Plasma
Capillary endothelium
6
To target
1 Messenger RNA on the ribosomes binds amino acids into a peptide chain called a **preprohormone**. The chain is directed into the ER lumen by a **signal sequence** of amino acids.
2 Enzymes in the ER chop off the signal sequence, creating an inactive **prohormone**.
3 The prohormone passes from the ER through the Golgi complex.
4 Secretory vesicles containing enzymes and prohormone bud off the Golgi. The enzymes chop the prohormone into one or more active peptides plus additional peptide fragments.
5 The secretory vesicle releases its contents by exocytosis into the extracellular space.
6 The hormone moves into the circulation for transport to its target.
(c) Prohormones Process to Active Hormone Plus Peptide Fragments
The peptide chain of insulin's prohormone folds back on itself with the help of disulfide (S—S) bonds. The prohormone cleaves to insulin and C-peptide.
Proinsulin
S—S
S
S
S
S
processes to
Insulin
S—S
S
S
S
S
+
C-peptide

Everyday Applications Encourage Critical Thinking

RUNNING PROBLEMS begin each chapter with a problem involving a disease or disorder that unfolds in segments on subsequent pages. The questions in each segment ask you to apply information you have learned in the text. You can check your understanding by comparing your answers with those in the Problem Conclusion at the end of each chapter. Three new Running Problems appear in this edition: one about finding reliable information on the internet, (chapter 1), one on the human papillomavirus and cervical cancer (chapter 24), and one on cholera in Haiti (chapter 21).

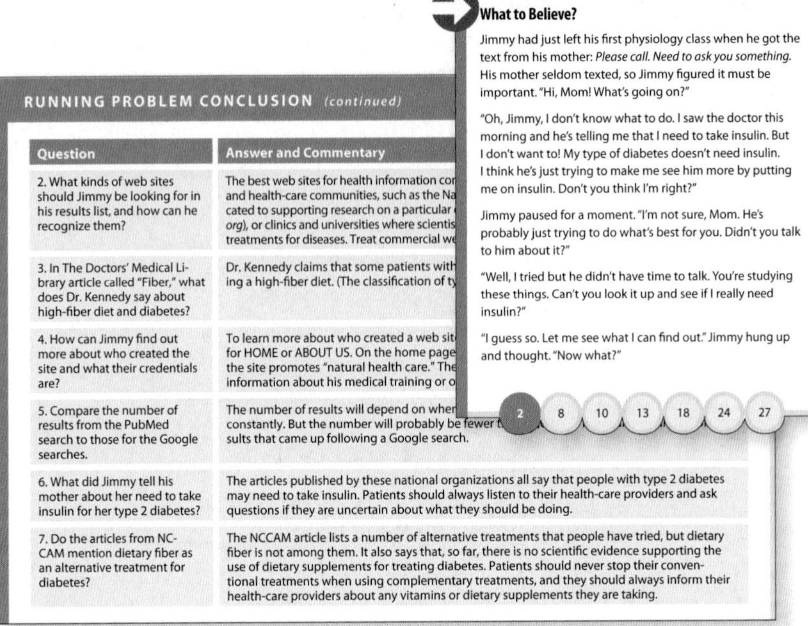

RUNNING PROBLEM

What to Believe?

Jimmy had just left his first physiology class when he got the text from his mother: *Please call. Need to ask you something.* His mother seldom texted, so Jimmy figured it must be important. "Hi, Mom! What's going on?"

"Oh, Jimmy, I don't know what to do. I saw the doctor this morning and he's telling me that I need to take insulin. But I don't want to! My type of diabetes doesn't need insulin. I think he's just trying to make me see him more by putting me on insulin. Don't you think I'm right?"

Jimmy paused for a moment. "I'm not sure, Mom. He's probably just trying to do what's best for you. Didn't you talk to him about it?"

"Well, I tried but he didn't have time to talk. You're studying these things. Can't you look it up and see if I really need insulin?"

"I guess so. Let me see what I can find out." Jimmy hung up and thought. "Now what?"

2 8 10 13 18 24 27

RUNNING PROBLEM CONCLUSION *(continued)*

Question	Answer and Commentary
2. What kinds of web sites should Jimmy be looking for in his results list, and how can he recognize them?	The best web sites for health information co… and health-care communities, such as the Na… cated to supporting research on a particular… *org*), or clinics and universities where scientis… treatments for diseases. Treat commercial w…
3. In The Doctors' Medical Library article called "Fiber," what does Dr. Kennedy say about high-fiber diet and diabetes?	Dr. Kennedy claims that some patients wit… ing a high-fiber diet. (The classification of t…
4. How can Jimmy find out more about who created the site and what their credentials are?	To learn more about who created a web sit… for HOME or ABOUT US. On the home pag… the site promotes "natural health care." Th… information about his medical training or o…
5. Compare the number of results from the PubMed search to those for the Google searches.	The number of results will depend on whe… constantly. But the number will probably be fewer t… sults that came up following a Google search.
6. What did Jimmy tell his mother about her need to take insulin for her type 2 diabetes?	The articles published by these national organizations all say that people with type 2 diabetes may need to take insulin. Patients should always listen to their health-care providers and ask questions if they are uncertain about what they should be doing.
7. Do the articles from NCCAM mention dietary fiber as an alternative treatment for diabetes?	The NCCAM article lists a number of alternative treatments that people have tried, but dietary fiber is not among them. It also says that, so far, there is no scientific evidence supporting the use of dietary supplements for treating diabetes. Patients should never stop their conventional treatments when using complementary treatments, and they should always inform their health-care providers about any vitamins or dietary supplements they are taking.

EMERGING CONCEPTS

Transporter Gene Families

One outcome of the Human Genome project has been the recognition that many proteins are closely related to each other, both within and across species. As a result, scientists have discovered that most membrane transporters for organic solutes belong to one of two gene "superfamilies": the ATP-binding cassette (ABC) superfamily or the solute carrier (SLC) superfamily. The ABC family transporters use ATP's energy to transport …

BIOTECHNOLOGY

Calcium Signals Glow in the Dark

If you've ever run your hand through a tropical ocean at night and seen the glow of bioluminescent jellyfish, you've seen a calcium signal. Aequorin, a protein complex isolated from jellyfish, is one of the molecules that scientists use to monitor the presence of calcium ions during a cellular response. When aequorin combines with calcium, it releases light that can be measured by electronic detection systems. Since the first use of aequorin in 1967, researchers have been designing increasingly sophisticated indicators that allow them to follow calcium si… called fura, Oreg… now watch calci… flow out of intra…

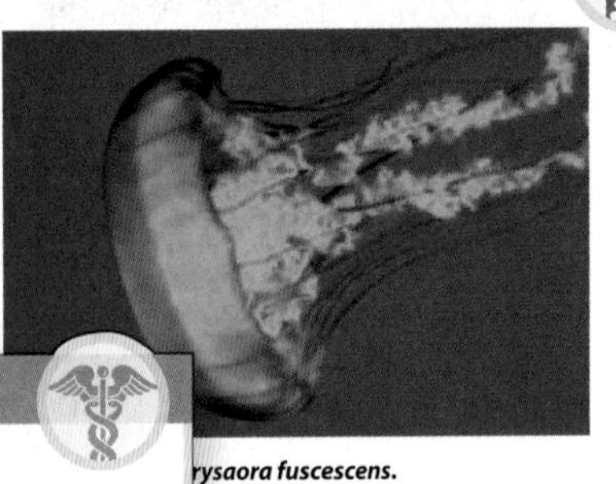

…rysaora fuscescens.

CLINICAL FOCUS

LDL: The Lethal Lipoprotein

"Limit the amount of cholesterol in your diet!" has been the recommendation for many years. So why is too much cholesterol bad for you? After all, cholesterol molecules are essential for membrane structure and for making steroid hormones (such as the sex hormones). But elevated cholesterol levels in the blood also lead to heart disease. One reason some people have too much cholesterol in their blood (*hypercholesterolemia*) is not diet but the failure of cells to take up the cholesterol. In the blood, hydrophobic cholesterol is bound to a lipoprotein carrier molecule to make it water soluble. The most common form of carrier is *low-density lipoprotein (LDL)*. When the LDL-cholesterol complex (LDL-C) binds to LDL receptors in caveolae, then it can then enter the cell in a vesicle. When people do not have adequate numbers of LDL receptors on their cell membranes, LDL-C remains in the blood. Hypercholesterolemia due to high levels of LDL-C predisposes these people to the development of **atherosclerosis,** also known as hardening of the arteries {*atheroma*, a tumor + *skleros*, hard + *-sis*, condition}. In this condition, the accumulation of cholesterol in blood vessels blocks blood flow and contributes to heart attacks.

FOCUS BOXES highlight research in physiology and medicine.

Three kinds of focus boxes help you understand the role of physiology in science and medicine today: **Biotechnology** boxes discuss applications and laboratory techniques from the rapidly changing world of biotechnology; **Clinical Focus** boxes concentrate on clinical applications and pathologies; and **Emerging Concepts** boxes describe upcoming advances in physiological research.

Insightful Pedagogy Helps You Make Connections

REFLEX PATHWAYS & CONCEPT MAPS organize physiological processes and details into a logical, visual format. These figures use consistent colors and shapes to represent processes and will guide you to a better understanding of coordinated physiological function.

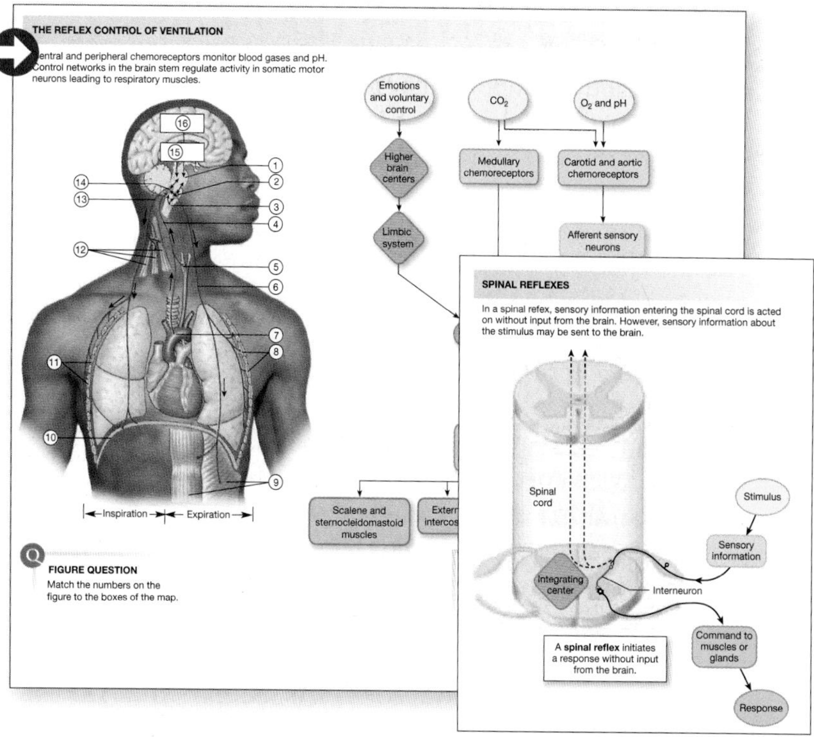

BACKGROUND BASICS, found on the chapter opening page, lists topics you will need to master for understanding the material that follows. Page references save study time, making the textbook an easy-to-use resource.

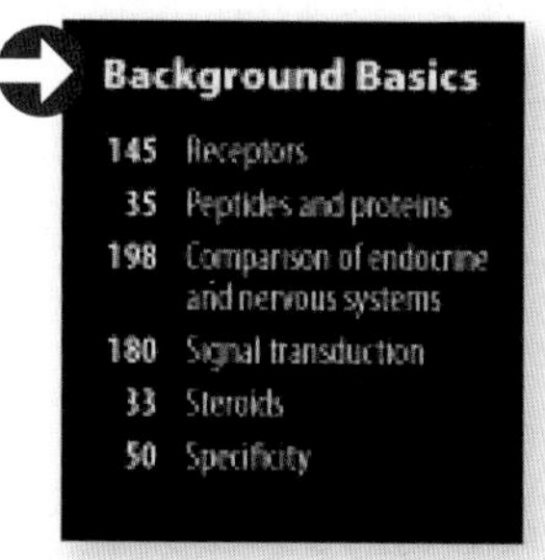
Background Basics

145 Receptors
35 Peptides and proteins
198 Comparison of endocrine and nervous systems
180 Signal transduction
33 Steroids
50 Specificity

CONCEPT LINKS are blue page numbers [p.321] embedded in the text. They connect the concepts you are reading about to topics discussed earlier in the book. Concept Links help you find material that you may have forgotten or that may be helpful in understanding new information.

response by means of a *signal transduction* system (■ Fig. 7.4). Many peptide hormones work through cAMP second messenger systems [p. 183]. A few peptide hormone receptors, such as that of insulin, have tyrosine kinase activity [p. 183] or

FIGURE AND GRAPH QUESTIONS promote analytical skills by encouraging you to interpret and apply information presented in the art and graphs. Answers to these questions appear at the end of each chapter.

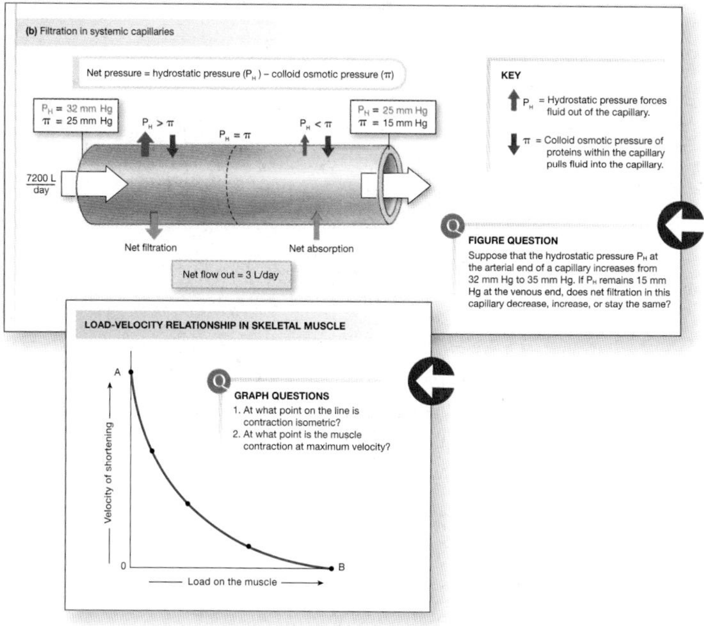

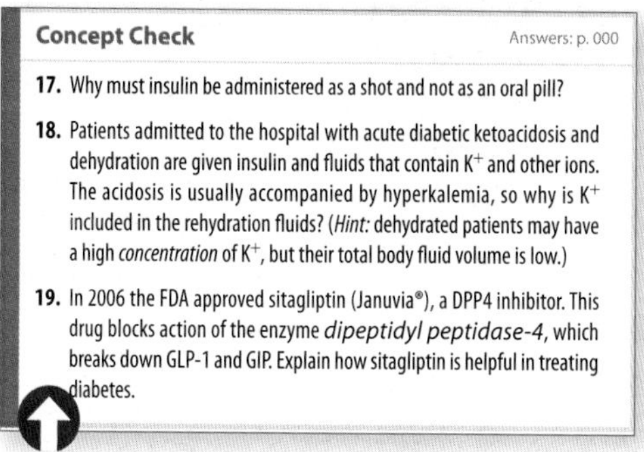
Concept Check Answers: p. 000

17. Why must insulin be administered as a shot and not as an oral pill?
18. Patients admitted to the hospital with acute diabetic ketoacidosis and dehydration are given insulin and fluids that contain K^+ and other ions. The acidosis is usually accompanied by hyperkalemia, so why is K^+ included in the rehydration fluids? (*Hint:* dehydrated patients may have a high *concentration* of K^+, but their total body fluid volume is low.)
19. In 2006 the FDA approved sitagliptin (Januvia®), a DPP4 inhibitor. This drug blocks action of the enzyme *dipeptidyl peptidase-4*, which breaks down GLP-1 and GIP. Explain how sitagliptin is helpful in treating diabetes.

CONCEPT CHECK BOXES are placed at intervals throughout the chapters, helping to test your understanding before continuing to the next topic. You can check your own answers using the key at the end of every chapter.

FOCUS ON... figures highlight the anatomy and physiology of important organs that are often overlooked in physiology texts, including the skin, the liver, the spleen, and the pineal gland.

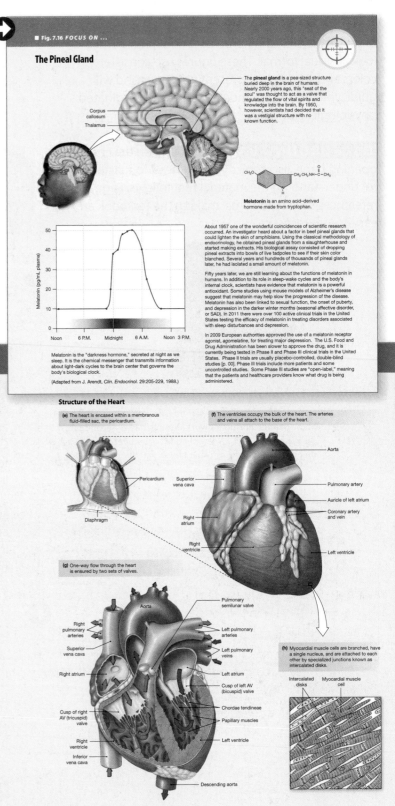

■ Fig. 7.16 *FOCUS ON ...*

The Pineal Gland

Corpus callosum

Thalamus

The **pineal gland** is a pea-sized structure buried deep in the brain of humans. Nearly 2000 years ago, this "seat of the soul" was thought to act as a valve that regulated the flow of vital spirits and knowledge into the brain. By 1950, however, scientists had decided that it was a vestigial structure with no known function.

Melatonin is an amino acid-derived hormone made from tryptophan.

Melatonin (pg/mL plasma)

Noon 6 P.M. Midnight 6 A.M. Noon 3 P.M.

Melatonin is the "darkness hormone," secreted at night as we sleep. It is the chemical messenger that transmits information about light-dark cycles to the brain center that governs the body's biological clock.

(Adapted from J. Arendt, *Clin. Endocrinol.* 29:205-229, 1988.)

About 1957 one of the wonderful coincidences of scientific research occurred. An investigator heard about a factor in beef pineal glands that could lighten the skin of amphibians. Using the classical methodology of endocrinology, he obtained pineal glands from a slaughterhouse and started making extracts. His biological assay consisted of dropping pineal extracts into bowls of live tadpoles to see if their skin color blanched. Several years and hundreds of thousands of pineal glands later, he had isolated a small amount of melatonin.

Fifty years later, we are still learning about the functions of melatonin in humans. In addition to its role in sleep-wake cycles and the body's internal clock, scientists have evidence that melatonin is a powerful antioxidant. Some studies using mouse models of Alzheimer's disease suggest that melatonin may help slow the progression of the disease. Melatonin has also been linked to sexual function, the onset of puberty, and depression in the darker winter months (seasonal affective disorder, or SAD). In 2011 there were over 100 active clinical trials in the United States testing the efficacy of melatonin in treating disorders associated with sleep disturbances and depression.

In 2009 European authorities approved the use of a melatonin receptor agonist, *agomelatine*, for treating major depression. The U.S. Food and Drug Administration has been slower to approve the drug, and it is currently being tested in Phase II and Phase III clinical trials in the United States. Phase II trials are usually placebo-controlled, double-blind studies [p. 00]. Phase III trials include more patients and some uncontrolled studies. Some Phase III studies are "open-label," meaning that the patients and healthcare providers know what drug is being administered.

ANATOMY SUMMARIES provide succinct visual overviews of a physiological system from a macro to micro perspective. Whether you are learning the anatomy for the first time or refreshing your memory, these summaries show you the essential features of each system in a single figure.

■ Fig. 14.5 *ANATOMY SUMMARY*

The Heart

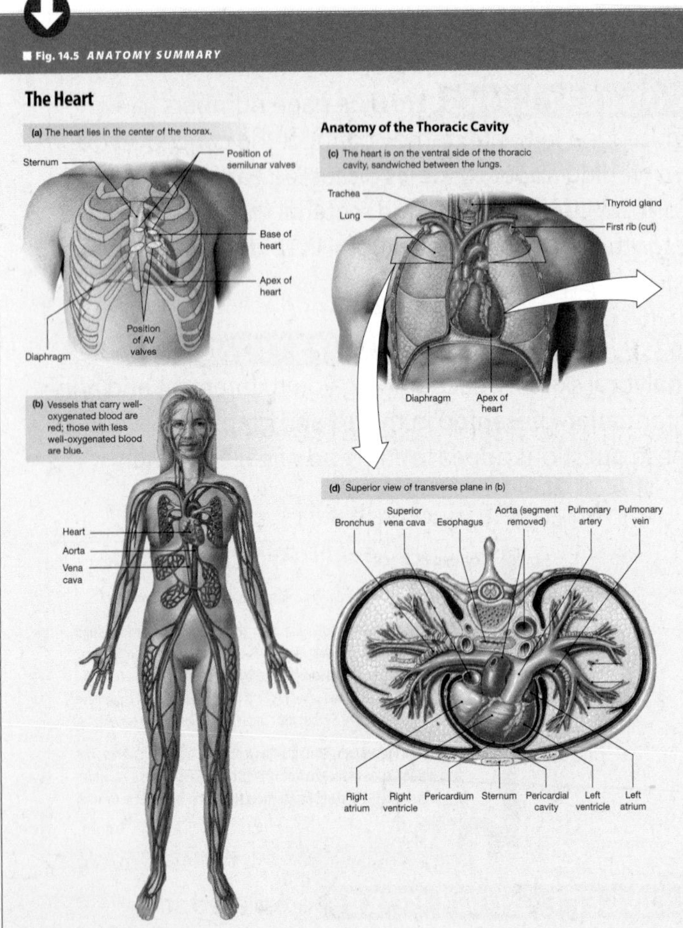

Structure of the Heart

(e) The heart is encased within a membranous fluid-filled sac, the pericardium.

Pericardium

Diaphragm

(f) The ventricles occupy the bulk of the heart. The arteries and veins all attach to the base of the heart.

Superior vena cava

Aorta

Pulmonary artery

Auricle of left atrium

Coronary artery and vein

Right atrium

Right ventricle

Left ventricle

(g) One-way flow through the heart is ensured by two sets of valves.

Aorta

Pulmonary semilunar valve

Right pulmonary arteries

Left pulmonary arteries

Superior vena cava

Left pulmonary veins

Right atrium

Left atrium

Cusp of left AV (bicuspid) valve

Cusp of right AV (tricuspid) valve

Chordae tendineae

Papillary muscles

Right ventricle

Left ventricle

Inferior vena cava

Descending aorta

(h) Myocardial muscle cells are branched, have a single nucleus, and are attached to each other by specialized junctions known as intercalated disks.

Intercalated disks

Myocardial muscle cell

Bring Physiology to Life with Mastering A&P®!

MASTERING A&P Used by over a million science students, the Mastering platform is the most effective and widely used online tutorial, homework, and assessment system for the sciences.

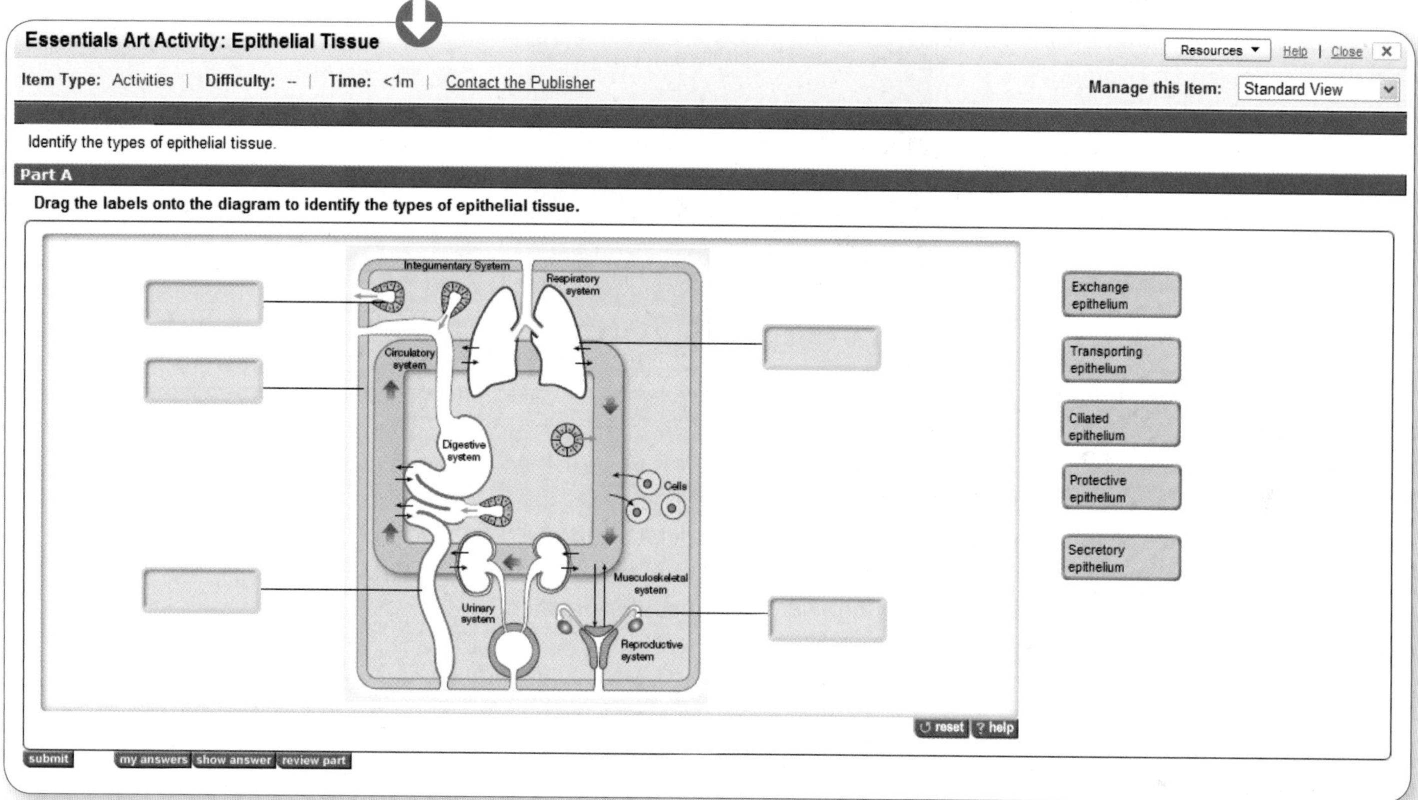

- **Study with proven A&P media:** Interactive Physiology®, PhysioEx 9.0™, A&PFlix™ Coaching Activities, and a multitude of additional question types help students learn and allow instructors to assign homework and assess their students' understanding.
- **Benefit from Personalized Coaching and Feedback:** Many MasteringA&P questions provide answer-specific hints and feedback, helping students when they need it most.
- **Come to Class Ready to Learn:** Reading Quizzes help students understand basic concepts before lecture while interactive concept maps help students check their understanding and make sense of physiological processes.
- **Understand the Essentials of Physiology:** Selected Essentials art features from the text are included as assignable activities in MasteringA&P, ensuring that students visualize and understand foundational topics in physiology.
- **Your course in one place:** All assessments, instructor media, and student self-study media from The Physiology Place are now available in MasteringA&P.

A&P FLIX™ animations provide carefully developed, step-by-step explanations with dramatic 3D representations of structures that show action and movement of processes, bringing difficult physiology concepts to life. Each animation includes a quiz with detailed hints and feedback.

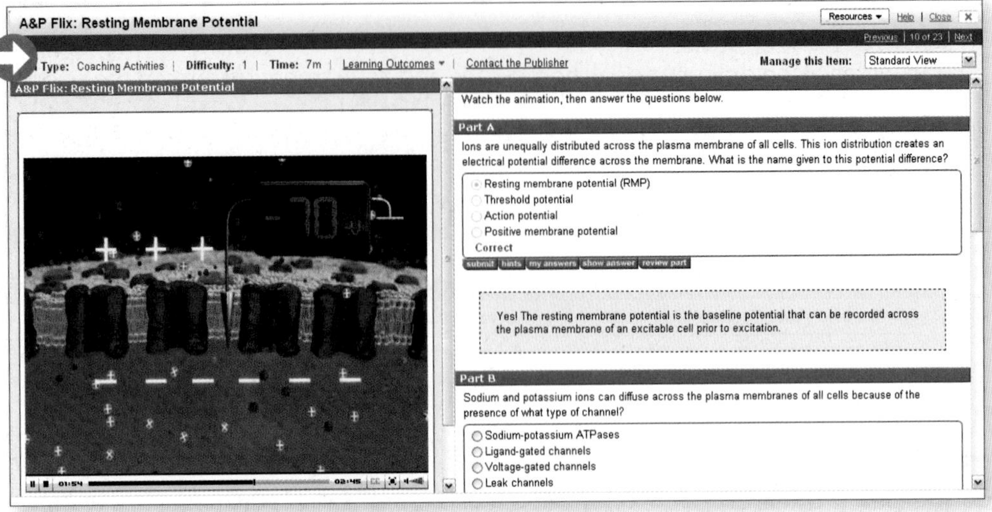

INTERACTIVE PHYSIOLOGY® 10-SYSTEM SUITE (IP-10) helps students advance beyond memorization to a genuine understanding of complex physiological processes through a series of engaging, interactive tutorials. Available on a CD and assignable within MasteringA&P®, IP-10 includes practice quizzes and printable exercise sheets and assignments.

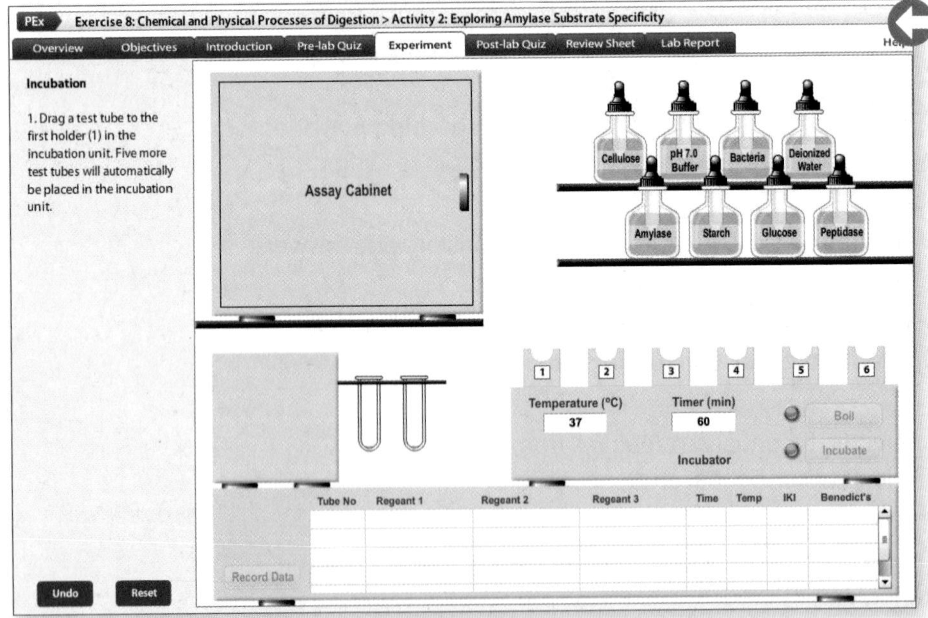

PHYSIOEX™ 9.0: LABORATORY SIMULATIONS IN PHYSIOLOGY is an easy-to-use laboratory simulation software and lab manual consisting of 12 exercises containing 63 physiology lab activities. PhysioEx allows students to repeat labs as often as they like, perform experiments without harming live animals, and conduct experiments that are difficult to perform in a wet lab environment because of time, cost, or safety concerns. Now completely online with step-by-step instructions and assessments, new Pre-lab and Post-lab quizzes, Stop & Think and Predict questions and the ability to save and print PDF lab results, PhysioEx 9.0 is the best supplement to, or replacement of, a traditional wet lab.

Dynamic and Flexible Instructor Resources for Every Course

STUDENT WORKBOOK Co-authored by Dee Silverthorn and adapted from materials developed for her own class, this workbook integrates a wide range of material, including try-it-yourself activities, lab exercises, quantitative and application-level review questions, background information with references, vocabulary lists, chapter summaries, and more!

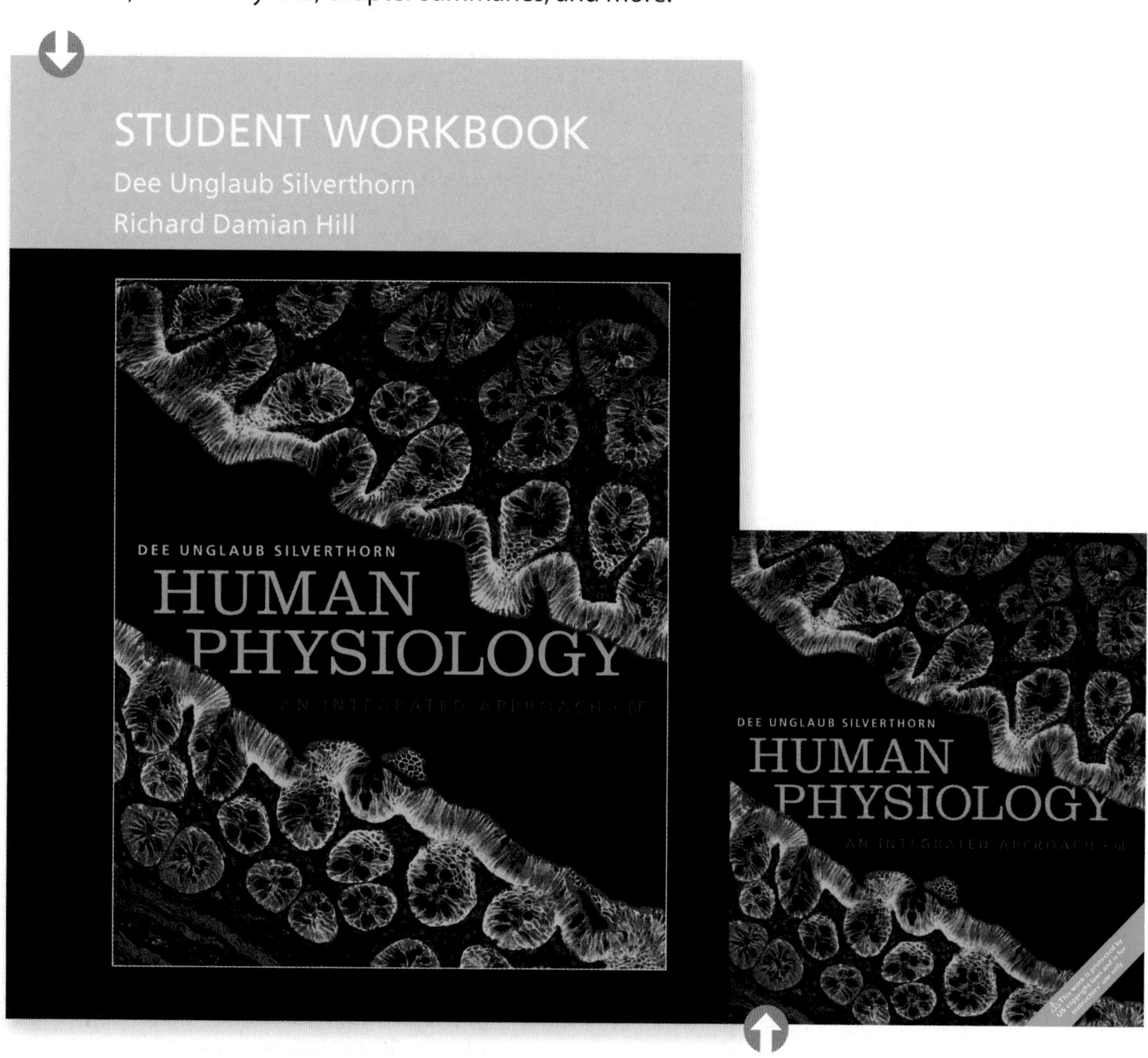

THE INSTRUCTOR RESOURCE DVD includes:

- JPEG files of all illustrations, photos, and tables from the book, plus selected illustrations from past editions.
- PowerPoint® slides with all images (labeled and unlabeled) with select items in customizable Label Edit PowerPoint and Step Edit PowerPoint; Lecture Outlines; Active Lecture Questions; Physiology Review Quiz Show Games; and a Quiz Show Game template.
- A&P Flix 3D animations cover core physiology concepts, helping students visualize tough-to-teach topics such as the cross-bridge cycle, excitation-contraction coupling of skeletal muscle, and generation of an action potential. Each animation will be available in PowerPoint with quiz questions for use with personal response systems (clickers).

New to This Edition

The sixth edition of *Human Physiology: An Integrated Approach* builds upon the thorough coverage of integrative and molecular physiology topics that have been the foundation of this book since its first publication. The text has been revised with extensive content updates, particularly in the areas of intravenous fluid therapy, taste transduction, and smooth muscle physiology.

Some sections of the text have been reorganized, particularly in the first unit (Chapters 1–6), to allow those introductory chapters to focus more on the themes of molecular interactions, compartmentation, biological energy, membrane dynamics, and communication and integration. The discussion of homeostasis is consolidated in Chapter 1. Anabolic metabolism pathways and catabolism of fats and proteins moved to Chapter 22 as part of the discussion of fed- and fasted-state metabolism. Chapter 4 now focuses on the theme of biological energy.

In a substantial revision of the art, we created Review figures for material that most students will have learned in previous courses and Essentials figures that can serve as capsule summaries of a topic for quick review. In most art, the title and caption have become part of the art to simplify extracting information about the figure. Figures from last edition that were significantly modified or eliminated are still available to instructors on the Instructor's DVD.

Chapter-by-Chapter Content Updates

Chapter 1

- New Running Problem on searching the literature
- Consolidated discussion of homeostasis
- Expanded introduction of themes includes information about core concepts highlighted in reports on biology education from the National Science Foundation, Howard Hughes Medical Institute and Association of American Medical Colleges, and the College Board
- New figures on internal and external environments; steady-state disequilibrium
- New Emerging Concepts box on *The Changing World of Omics*

Chapter 2

- Chapter now focuses on molecular interactions, with more emphasis on noncovalent interactions and molecular shape.
- Most of the basic chemistry is incorporated into review figures.
- A new Review Quiz allows students to test their recall of basic chemistry.
- Expanded discussion on protein interactions
- Review figures: Lipids; Carbohydrates; Proteins; Nucleotides; Atoms and Molecules; Molecular Bonds; Solutions; Molecular Interactions; pH
- Essentials figures: Protein Activation-Inhibition; Factors Affecting Protein Binding

Chapter 3

- Condensed discussion of cell organelles
- GPI anchors for membrane proteins
- Protein synthesis as a functional example of how compartmentation works inside cells
- New/updated info on:
  - primary cilia
  - brown fat in adults
  - Pap smears
  - autologous cultured chondrocytes
- New Emerging Concepts box: *Single Cilia Are Sensors*
- Review figure: Cell Structure
- Essentials figures: Levels of Organization: Body Compartments; The Cell Membrane; Cell Junctions; Epithelial Tissue; Types of Epithelia; Connective Tissue; Types of Connective Tissue

Chapter 4

- Moved anabolic metabolism pathways and catabolism of fats and proteins to Chapter 22 and consolidated pathways for glycolysis, citric acid cycle, and electron transport system into Essentials figures so that the text discussion focuses on energy production
- Updated discussion and terminology for protein sorting
- Essentials figures: ATP Production; Glycolysis; Pyruvate, Acetyl CoA, and the Citric Acid Cycle; The Electron Transport System; Protein Synthesis

Chapter 5

- Chapter now focuses on compartments and equilibrium/disequilibrium.
- New opening section on using coconut water for IVs
- Expanded discussion of osmolarity and tonicity using body compartments instead of single cell for the examples. Examples of calculations for IV fluid therapy.
- New figure shows diffusion in agar gel.
- Essentials figures: Body Fluid Compartments; Osmolarity and Tonicity; Membrane Transporters; Endocytosis, Exocytosis, and Membrane Recycling

Chapter 6

- Moved homeostasis to Chapter 1
- Consolidated figures
- Essentials figures: Communication in the Body; Signal Transduction Cascades and Amplification; Signal Transduction; Simple and Complex Reflex Pathways

Chapter 7

- New/updated information on:
  - paraventricular and supraoptic nuclei
  - ultra-short-loop feedback
  - melatonin
- Essentials figures: Peptide Hormone Synthesis and Processing; Steroid Hormones; The Pituitary Gland; Hypothalamic-Anterior Pituitary Hormones

Chapter 8

- Consolidated discussion of development and repair
- New/updated information on:
  - resistance, Ohm's law, length constant
  - neurotransmitter transporters
  - stem cell transplants
  - neurotransmitter synthesis
- New box on cable properties, capacitance, and time constant
- Five new Concept Check questions
- New figure question on Nernst equation
- Essentials figures: Organization of the Nervous System; Neuron Anatomy; Glial Cells; Graded Potentials; The Action Potential; Synaptic Communication; Divergence and Convergence; Fast and Slow Postsynaptic Responses; Summation

Chapter 9

- New/updated information on:
  - antidepressant drugs and neuron growth
  - sleep and memory
  - Alzheimer's disease
  - Brainbow mice
- New Level Three end-of-chapter question
- Two new figure questions
- Essentials figure: Nervous System Development

Chapter 10

- New/updated information on:
  - taste transduction
  - somatosensory aspects of taste mediated by cranial nerves and TRP receptors
  - melanopsin and melanopsin-containing retinal ganglion cells (mRGC)
  - retinal bipolar cell signal processing
  - treatment for Ménière's disease
  - pain research
  - epithelial sodium channel (ENaC)
  - transient receptor potential (TRP) channels
  - T1R (sweet and umami) and T2R receptors for bitter taste
  - taste buds: types I, II, III and basal cells
  - retinal bipolar cells, ON and OFF subtypes
  - mGluR6: metabotropic glutamate receptor in the retina
- Essentials figures: Taste; Equilibrium; Optics of the Eye

Chapter 11

- New/updated information on:
  - smoking cessation
  - statistics on student smoking
- Revised art for autonomic pathways
- Essentials figures: Efferent Divisions; Somatic Motor Division

Chapter 12

- New/updated information on:
  - satellite cells and skeletal muscle stem cells
  - Ca^{2+} entry through DHP receptor: excitation-coupled Ca^{2+} entry
  - myosin structure
- Revised section on smooth muscle physiology:
  - tonic and phasic muscles
  - revised model of filament organization
  - calcium sensitivity
  - calcium-induced calcium release
- New figure for skeletal muscle relaxation
- Essentials figures: The Sarcomere; Excitation-Contraction Coupling and Relaxation; Smooth Muscle

Chapter 13

- Essentials figure: Neural Reflexes

Chapter 14

- New/expanded discussion of:
  - ECGs
  - autonomic control of cardiac output
  - length-force relationships
- Three new figure questions
- New Concept Check question
- Essentials figures: Physics of Fluid Flow; The ECG

Chapter 15

- New/expanded discussion of:
  - blood pressure control
  - selective control of targets by the cardiovascular control centers
- New figure on integration of cardiac output and resistance
- New Concept Check question
- Four new figure questions
- New end-of-chapter quantitative problem
- Essentials figures: Mean Arterial Pressure; Cardiovascular Control

Chapter 16

- Emerging Concepts box on platelet-rich plasma therapy

Chapter 17

- New/updated information on:
  - fluid secretion in airways
  - submucosal glands in airways

  - visceral and parietal pleura
  - equations for compliance and elastance
- New figure on fluid secretion and submucosal glands in airways
- Four new figure questions
- New end-of-chapter quantitative problem
- Essentials figures: Gas Laws; Ventilation

Chapter 18

- New/updated information on:
  - Fick equation as an example of mass balance
- New figures for alveolar gas exchange and oxygen binding to hemoglobin, Fick equation and mass balance
- Essentials figure: Oxygen-Hemoglobin Saturation Curve

Chapter 19

- New/updated information on:
  - Organic anion transporter (OAT)
  - Tertiary active transport
- New figures for reabsorption in peritubular capillaries, organic anion secretion
- Five new figure questions
- New end-of-chapter quantitative problem
- Essentials figures: Nephron Function; GFR; Renal Clearance

Chapter 20

- New/updated information on:
  - natriuretic peptides
  - Henderson-Hasselbalch equation
  - ANG II effect on proximal tubule
- Added endocrine tables for aldosterone, RAS, AVP, and ANP
- New map for aldosterone
- New figure on acid-base disturbance compensations
- New Concept Check question
- Three new figure questions
- Essentials figures: Aldosterone; RAS; AVP; Natriuretic Peptides

Chapter 21

- New running problem on cholera in Haiti
- New/updated information on:
  - salivary glands and secretion of saliva
  - the swallowing reflex
  - orlistat
  - heme transport
  - iron and calcium absorption
- New figure question
- New Concept Check question
- New figures for iron and calcium absorption, GI regulation
- Essentials figures: Secretion; Digestion and Absorption—Carbohydrates; Digestion and Absorption—Fats

Chapter 22

- New/updated information on:
  - psychological influences on eating
  - cholesterol recommendations
  - ketogenic diets
  - diabetes statistics
  - revised criteria for diagnosis of diabetes
  - pre-diabetes
  - hyperglycemic hyperosmolar state
  - GLUT4 in exercising muscle
  - active cutaneous vasodilation
  - beta-3 receptors
  - brown fat in humans
- Moved protein and fat metabolism here from Chapter 4
- New boxes on BMI and on antioxidants and free radicals
- New graph of insulin and glucagon interaction
- New figure on factors involved in insulin secretion
- New figure question
- New Level Three question on thermoregulatory response experiments
- Essentials figures: Biochemical Pathways for Energy Production and Fat Synthesis

Chapter 23

- New/updated information on:
  - thyroid transporters
  - sex steroids and pubertal bone growth
  - TRPV6 and ECaC
  - POMC neurons in the hypothalamus
- New figure on acid secretion by osteoclasts
- Essentials figure: Bone

Chapter 24

- New Running Problem on HPV, cervical cancer, and HPV vaccines
- New figures of bacterial structure and first-line defenses
- Two new figure questions
- Essentials figures: Bacterial Infections; Viral Infections

Chapter 25

- New quantitative problem on cardiac output and exercise

Chapter 26

- New/updated information on:
  - Male circumcision and HIV
  - Finasteride and prostate cancer
  - AMH as an indicator of ovarian function
- New figure on follicular development
- Essentials figure: Fertilization

Acknowledgments

Writing, editing, and publishing a textbook is a group project that requires the talent and expertise of many people. No scientist has the detailed background needed in all areas to write a book of this scope, and I am indebted to all my colleagues who have so generously shared their expertise in each edition. I particularly want to acknowledge Bruce Johnson, Cornell University, Department of Neurobiology and Behavior, a superb neurobiologist and educator, who once again ensured that the chapters on neurobiology are accurate and reflect the latest developments in that rapidly changing field.

Many other people devoted time and energy to making this book a reality, and I would like to thank them all, collectively and individually. I apologize in advance to anyone whose name I have omitted.

Reviewers

I am particularly grateful to the instructors who reviewed one or more chapters of the last edition. There were many suggestions in their thoughtful reviews that I was unable to include in the text, but I appreciate the time and thought that went into their comments. The reviewers for this edition include:

Lisa Bonneau, Metropolitan Community College-Blue River
Erick Bourassa, University of Sint Eustatius School of Medicine
Betsy Brantley, Valencia College
Michael Buratovich, Spring Arbor University
Tom Davis, Loras College
Cathy Davison, SUNY Empire State College
Joseph Esdin, Antelope Valley College and University of California, Los Angeles
Suzanna Gribble, Grove City College
Steve Henderson, California State University, Chico
William Jurney, Normandale Community College
Dean Lauritzen, City College of San Francisco
David Mallory, Marshall University
Byron Noordewier, Northwestern College
Ryan Paruch, Tulsa Community College
Mark Smith, Santiago Canyon College
Jason Strandberg, Century College
Jill Tall, Youngstown State University
Wendy Vermillion, Columbus State Community College

Many other instructors and students took time to write or email queries or suggestions for clarification. I am always delighted to have input, and I apologize that I do not have room to acknowledge them all individually. Of particular help in this edition were

Ana Maria Barral, Cuyamaca College
Catherine Loudon, University of California, Irvine
Anita Woods, University of Western Ontario
Erik Swenson, University of Washington

Specialty Reviews

No one can be an expert in every area of physiology, and I am deeply thankful for my friends and colleagues who reviewed entire chapters or answered specific questions. Even with their help, there may be errors, for which I take full responsibility. The specialty reviewers for this edition were:

Susan M. Barman, Michigan State University
James Bryant, University of Texas, Austin
Jeffrey Pommerville, Maricopa Community Colleges
Marion J. Siegman, Jefferson Medical College

Student Contributions

The following students suggested topics that appear in this edition:

Vitali Azouz—coconut water IVs
Chun-Yu "Jimmy" Lee—story for the Chapter 1 Running problem
Katie Zimmerman—Ménière's disease
Claire Conroy—revision of Figure 11.5 and Running Problems on cholera and HPV

Another group of students served as a review panel as we developed the new art program:

Ryan Aghabozorg
Fadi Al-Asadi
Kayla Broadhead
Hilary Edelman
Meherin Huque
Catherine O'Krafka
Zuleikha Tyebjee
Tiana Won

Photographs

I would like to thank the following colleagues who generously provided micrographs from their research:

Kristen Harris, University of Texas
Flora M. Love, University of Texas
Jane Lubisher, University of Texas
Young-Jin Son, University of Texas

Supplements

Damian Hill once again worked with me to revise and improve the Instructor Resource Manual and Student Workbook that accompany the book. I believe that these supplements should reflect the style and approach of the text, so I am grateful that

Damian has continued to be my alter-ego for them through so many editions.

I would also like to thank my colleagues who helped with the test bank and media supplements for this edition: Steven Bassett, Lawrence Brewer, Heidi Bustamante, Michael T. Griffin, William Jurney, Robert Kemm, Peggy LePage, Chelsea Loafman, Catherine Loudon, Alice Martin, Rennee Moore, Cheryl Neudauer, Jason Strandberg, Chad Wayne, Scott Zimmerman.

The Development and Production Team

Writing a manuscript is only a first step in the long and complicated process that results in a bound book with all its ancillaries. The team that works with me on book development deserves a lot of credit for the finished product. In this edition Bill Ober and Claire Garrison, my art coauthors, along with talented designer Jim Gibson created the new *Essentials* features that appear in each chapter. Once again, Yvo Riezebos designed a striking cover that reflects how science is really art. Anne A. Reid, my long-time developmental editor, and Antonio Padial, my copy editor, were a delight to work with, as usual. Anne Scanlan-Rohrer, my project editor, is a master at getting deadlines met without creating too much stress.

The team at Pearson Education worked tirelessly to see this edition move from manuscript to bound book. I had two acquisitions editors who worked with me on the vision for this edition: Deirdre Espinoza, who retired to take care of her family, and my new editor Kelsey Volker, who jumped right in to take up the reins. Barbara Yien, Director of Development for Applied Sciences, ably filled in during the transition period and tried her best to keep me on track. Ashley Williams was the assistant editor who coordinated the print supplements and assisted whenever needed.

The task of coordinating production fell to Pearson Assistant Managing Editor Nancy Tabor. Senior Project Manager Jared Sterzer at PreMediaGlobal handled composition and project management, and Project Manager Winnie Luong at the art house, Imagineering, managed the team that prepared Bill and Claire's art for production. Kristin Piljay was the photo researcher who found the wonderful new photos that appear in this edition.

Annie Wang was the assistant media producer who kept my supplements authors on task and on schedule. Christy Lesko is the director of marketing who works with the excellent sales teams at Pearson Education and Pearson International, and Derek Perrigo is the marketing manager for the anatomy and physiology list.

Special Thanks

As always, I would like to thank my students and colleagues who looked for errors and areas that needed improvement. I've learned that awarding one point of extra credit for being the first student to report a typo works really well. My graduate teaching assistants have played a huge role in my teaching ever since I arrived at the University of Texas, and their input has helped shape how I teach. Many of them are now faculty members themselves. I would particularly like to thank:

Lynn Cialdella, M.S., M.B.A.
Patti Thorn, Ph.D.
Karina Loyo-Garcia, Ph.D.
Jan M. Machart, Ph.D.
Ari Berman, Ph.D.
Kurt Venator, Ph.D.
Peter English, Ph.D.
Kira Wenstrom, Ph.D.
Lawrence Brewer, Ph.D.
Carol C. Linder, Ph.D.
Sarah Davies, M.S.

Finally, special thanks to my colleagues in the American Physiological Society and the Human Anatomy & Physiology Society, whose experiences in the classroom have enriched my own understanding of how to teach physiology. I would also like to recognize a special group of friends for their continuing support: Penelope Hansen (Memorial University, St. John's), Mary Anne Rokitka (SUNY Buffalo), Rob Carroll (East Carolina University School of Medicine), Cindy Gill (Hampshire College), and Joel Michael (Rush Medical College), as well as Ruth Buskirk, Jeanne Lagowski, Jan M. Machart and Marilla Svinicki (University of Texas).

As always, I thank my family and friends for their patience, understanding, and support during the chaos that seems inevitable with book revisions. The biggest thank you goes to my husband Andy, whose love, support, and willingness to forgo home-cooked meals on occasion help me meet my deadlines.

A Work in Progress

One of the most rewarding aspects of writing a textbook is the opportunity it has given me to meet or communicate with other instructors and students. In the years since the first edition was published, I have heard from people around the world, and have had the pleasure of hearing how the book has been incorporated into their teaching and learning.

Because science textbooks are revised every three or four years, they are always works in progress. I invite you to contact me or my publisher with any suggestions, corrections, or comments about this edition. I am most reachable through e-mail at *silverthorn@mail.utexas.edu*. You can reach my editor at the following address:

Applied Sciences
Pearson Education
1301 Sansome Street
San Francisco, CA 94111

Dee U. Silverthorn
Integrative Biology
University of Texas
Austin, Texas

Introduction to Physiology

The current tendency of physiological thought is clearly toward an increasing emphasis upon the unity of operation of the Human Body.

—Ernest G. Martin, preface to *The Human Body* 10th edition, 1917

Thermography of the human body. Warmer areas are red, cooler are blue.

Welcome to the fascinating study of the human body! For most of recorded history, humans have been interested in how their bodies work. Early Egyptian, Indian, and Chinese writings describe attempts by physicians to treat various diseases and to restore health. Although some ancient remedies, such as camel dung and powdered sheep horn, may seem bizarre, we are still using others, such as blood-sucking leeches and chemicals derived from medicinal plants. The way we use these treatments has changed through the centuries as we learned more about the human body.

There has never been a more exciting time in human physiology. **Physiology** is the study of the normal functioning of a living organism and its component parts, including all its chemical and physical processes. The term *physiology* literally means "knowledge of nature." Aristotle (384–322 B.C.E.) used the word in this broad sense to describe the functioning of all living organisms, not just of the human body. However, Hippocrates (ca. 460–377 B.C.E.), considered the father of medicine, used the word *physiology* to mean "the healing power of nature," and thereafter the field became closely associated with medicine. By the sixteenth century in Europe, physiology had been formalized as the study of the vital functions of the human body. Today the term is again used to refer to the study of animals and plants.

Today we benefit from centuries of work by physiologists who constructed a foundation of knowledge about how the human body functions. Since the 1970s, rapid advances in the fields of cellular and molecular biology have supplemented this work. A few decades ago we thought that we would find the key to the secret of life by sequencing the human *genome,* which is the collective term for all the genetic information contained in the DNA of a species. However, this deconstructionist view of biology has proved to have its limitations, because living organisms are much more than the simple sum of their parts.

RUNNING PROBLEM

What to Believe?

Jimmy had just left his first physiology class when he got the text from his mother: *Please call. Need to ask you something.* His mother seldom texted, so Jimmy figured it must be important. "Hi, Mom! What's going on?"

"Oh, Jimmy, I don't know what to do. I saw the doctor this morning and he's telling me that I need to take insulin. But I don't want to! My type of diabetes doesn't need insulin. I think he's just trying to make me see him more by putting me on insulin. Don't you think I'm right?"

Jimmy paused for a moment. "I'm not sure, Mom. He's probably just trying to do what's best for you. Didn't you talk to him about it?"

"Well, I tried but he didn't have time to talk. You're studying these things. Can't you look it up and see if I really need insulin?"

"I guess so. Let me see what I can find out." Jimmy hung up and thought. "Now what?"

2 8 10 13 18 24 27

Physiology Is an Integrative Science

Many complex systems—including those of the human body—possess **emergent properties,** which are properties that cannot be predicted to exist based only on knowledge of the system's individual components. An emergent property is not a property of any single component of the system, and it is greater than the simple sum of the system's individual parts. Emergent properties result from complex, nonlinear interactions of the different components. For example, suppose someone broke down a car into its nuts and bolts and pieces and laid them out on a floor. Could you predict that, properly assembled, these bits of metal and plastic would become a vehicle capable of converting the energy in gasoline into movement? Who could predict that the right combination of elements into molecules and assemblages of molecules would result in a living organism? Among the most complex emergent properties in humans are emotion, intelligence, and other aspects of brain function. None of these properties can be predicted from knowing the individual properties of nerve cells.

When the Human Genome Project (*www.genome.gov*) began in 1990, scientists thought that by identifying and sequencing all the genes in human DNA, they would understand how the body worked. However, as research advanced, scientists had to revise their original idea that a given segment of DNA contained one gene that coded for one protein. It became clear that one gene may code for many proteins. The Human Genome Project ended in 2003, but before then researchers had moved beyond genomics to *proteomics,* the study of proteins in living organisms. Now scientists have realized that knowing that a protein is made in a particular cell does not always tell us the significance of that protein to the cell, the tissue, or the functioning organism. The exciting new areas in biological research are called functional genomics, systems biology, and integrative biology, but fundamentally these are all fields of physiology. The **integration** of function across many **levels of organization** is a special focus of physiology. (To *integrate* means to bring varied elements together to create a unified whole.)

■ Figure 1.1 illustrates levels of organization ranging from the molecular level all the way up to populations of different species living together in *ecosystems* and in the *biosphere.* The levels of organization are shown along with the various subdisciplines of chemistry and biology related to the study of each organizational level. There is considerable overlap between the different fields of study, and these artificial divisions vary

EMERGING CONCEPTS

The Changing World of Omics

If you read the scientific literature, it appears that contemporary research, using the tools of molecular biology, has exploded into an era of "omes" and "omics." What is an "ome"? The term apparently derives from the Latin word for a mass or tumor, and it is now used to refer to a collection of items that make up a whole, such as a genome. One of the earliest uses of the "ome" suffix in biology is the term *biome,* meaning the entire community of organisms living in a major ecological region, such as the marine biome or the desert biome. A genome is all the genes in an organism, and a proteome includes all the proteins in that organism.

The related adjective "omics" describes the research related to studying an "ome." Adding "omics" to a root word has become the cutting-edge way to describe a research field. For example, *proteomics,* the study of proteins in living organisms, is now as important as *genomics,* the sequencing of DNA (the genome). The traditional study of biochemistry includes *metabolomics* (study of metabolic pathways) and *interactomics* (the study of protein-protein interactions). If you search the Internet, you will find numerous listings for the *transcriptome* (RNA), *lipidome* (lipids), and *pharmacogenomics* (the influence of genetics on the body's response to drugs). There is even a journal named *OMICS*!

The Physiome Project (*www.physiome.org*) is an organized international effort to coordinate molecular, cellular, and physiological information about living organisms into an Internet database. Scientists around the world can access this information and apply it in their own research efforts to create better drugs or genetic therapies for curing and preventing disease. Some scientists are using the data to create mathematical models that explain how the body functions. The Physiome Project is an ambitious undertaking that promises to integrate information from diverse areas of research so that we can improve our understanding of the complex processes we call life.

according to who is defining them. Notice, however, that physiology includes multiple levels, from molecular and cellular biology to the ecological physiology of populations.

At the level of the organism, physiology is closely tied to anatomy. The structure of a tissue or organ must provide an efficient physical base for its function. For this reason, it is nearly impossible to study the physiology of a body system without understanding the underlying anatomy. Because of the interrelationship of anatomy and physiology, you will find Anatomy Summaries throughout the book. These special review features illustrate the anatomy of the physiological systems at different levels of organization.

At the simplest level of organization shown in Figure 1.1, atoms of elements link together to form molecules. Collections of molecules in living organisms form **cells,** the smallest unit of structure capable of carrying out all life processes. A lipid and protein barrier called the **cell membrane** (also called the *plasma membrane*) separates cells from their external environment. Simple organisms are composed of only one cell, but complex organisms have many cells with different structural and functional specializations. Collections of cells that carry out related functions are called **tissues** {*texere,* to weave}. Tissues form structural and functional units known as **organs** {*organon,* tool}, and groups of organs integrate their functions to create **organ systems.** [Chapter 3 reviews the anatomy of cells, tissues and organs.]

The 10 physiological organ systems in the human body are illustrated in ■ Table 1.1. Several of the systems have alternate names, given in parentheses, that are based on the organs of the system rather than the function of the system. The **integumentary system** {*integumentum,* covering}, composed of the skin, forms a protective boundary that separates the body's internal environment from the external environment (the outside world). The **musculoskeletal system** provides support and body movement.

Four systems exchange materials between the internal and external environments. The **respiratory (pulmonary) system** exchanges gases; the **digestive (gastrointestinal) system** takes up nutrients and water and eliminates wastes; the **urinary (renal) system** removes excess water and waste material; and the **reproductive system** produces eggs or sperm.

■ **Fig. 1.1** ***Levels of organization and the related fields of study***

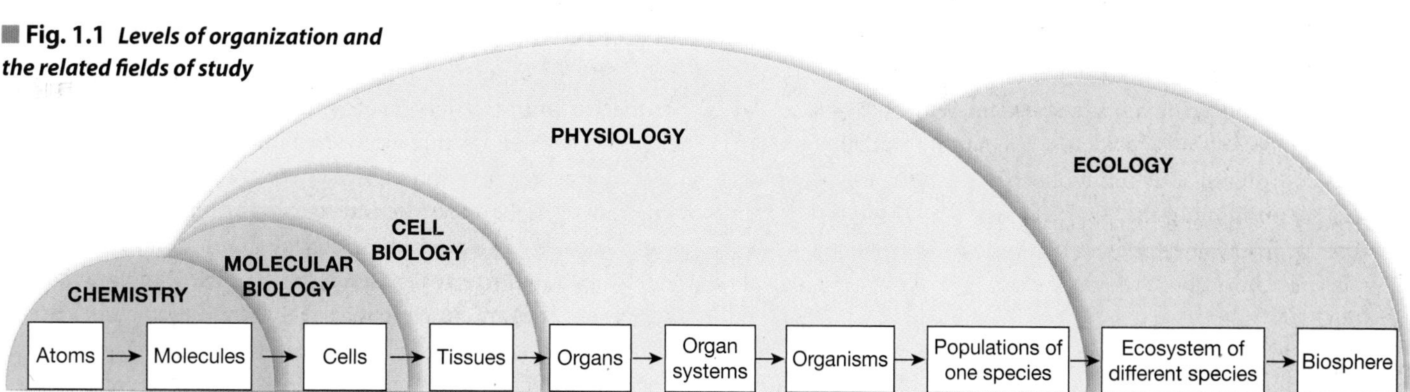

Organ Systems of the Human Body and their Integration

Table 1.1

System Name	Includes	Representative Functions	The Integration Between Systems of the Body
Circulatory	Heart, blood vessels, blood	Transport of materials between all cells of the body	
Digestive	Stomach, intestine, liver, pancreas	Conversion of food into particles that can be transported into the body; elimination of some wastes	
Endocrine	Thyroid gland, adrenal gland	Coordination of body function through synthesis and release of regulatory molecules	
Immune	Thymus, spleen, lymph nodes	Defense against foreign invaders	
Integumentary	Skin	Protection from external environment	
Musculoskeletal	Skeletal muscles, bone	Support and movement	
Nervous	Brain, spinal cord	Coordination of body function through electrical signals and release of regulatory molecules	
Reproductive	Ovaries and uterus, testes	Perpetuation of the species	
Respiratory	Lungs, airways	Exchange of oxygen and carbon dioxide between the internal and external environments	
Urinary	Kidneys, bladder	Maintenance of water and solutes in the internal environment; waste removal	

This schematic figure indicates relationships between systems of the human body. The interiors of some hollow organs (shown in white) are part of the external environment.

The remaining four systems extend throughout the body. The **circulatory (cardiovascular) system** distributes materials by pumping blood through vessels. The **nervous** and **endocrine systems** coordinate body functions. Note that the figure shows them as a continuum rather than as two distinct systems. Why? Because the lines between these two systems have blurred as we have learned more about the integrative nature of physiological function.

The one system not illustrated in Table 1.1 is the diffuse **immune system,** which includes but is not limited to the anatomical structures known as the *lymphatic system.* The specialized cells of the immune system are scattered throughout the body. They protect the internal environment from foreign substances by intercepting material that enters through the intestines and lungs or through a break in the skin. In addition, immune tissues are closely associated with the circulatory system.

Traditionally, physiology courses and books are organized by organ system. Students study cardiovascular physiology and regulation of blood pressure in one chapter, and then study the kidneys and control of body fluid volume in a different chapter. In the functioning human, however, the cardiovascular and renal systems communicate with each other, so that a change in one is likely to cause a reaction in the other. For example, body fluid volume influences blood pressure, while changes in blood pressure alter kidney function because the kidneys regulate fluid volume. In this book you will find several chapters devoted to topics of integrated function.

Developing skills to help you understand how the different organ systems work together is just as important as memorizing facts. One way physiologists integrate information is by using visual representations of physiological processes called maps. The Focus on Mapping feature in this chapter helps you learn how to make maps. The first type of map, shown in ■ Figure 1.2a, is a schematic representation of structure or function. The second type of map, shown in Figure 1.2b, diagrams a physiological process as it proceeds through time. These maps are called *flow charts* or process maps. The end-of-chapter questions throughout the book feature lists of selected terms that you can use to practice mapping.

Function and Mechanism

We define physiology as the normal functioning of the body, but physiologists are careful to distinguish between function and mechanism. The **function** of a physiological system or event is the "why" of the system or event: why does a certain response help an animal survive in a particular situation? In other words, what is the *adaptive significance* of this event for this animal?

For example, humans are large, mobile, terrestrial animals, and our bodies maintain relatively constant water content despite living in a dry, highly variable external environment. Dehydration is a constant threat to our well-being. What processes have evolved in our anatomy and physiology that enable us to survive in this hostile environment? One is the production of highly concentrated urine by the kidney, which allows the body to conserve water. This statement tells us *why* we produce concentrated urine but does not tell us *how* the kidney accomplishes that task.

Thinking about a physiological event in terms of its adaptive significance is the **teleological approach** to science. For example, the teleological answer to the question of why red blood cells transport oxygen is "because cells need oxygen and red blood cells bring it to them." This answer explains *why* red blood

■ **Fig. 1.2** ***Types of maps***

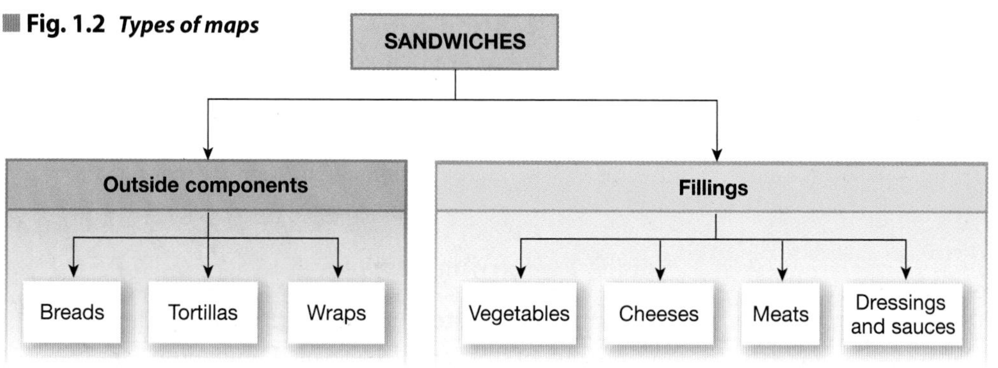

(a) A map showing structure/function relationships

FIGURE QUESTIONS

1. Can you add more details and links to map **(a)**?

2. Here is an alphabetical list of terms for a map of the body. Use the steps on the next page to create a map with them. Add additional terms to the map if you like. A sample answer appears at the end of the chapter.

- bladder
- blood vessels
- brain
- cardiovascular system
- digestive system
- endocrine system
- heart
- immune system
- integumentary system
- intestine
- kidneys
- lungs
- lymph nodes
- mouth
- musculoskeletal system
- nervous system
- ovaries
- reproductive system
- respiratory system
- stomach
- testes
- the body
- thyroid gland
- urinary system
- uterus

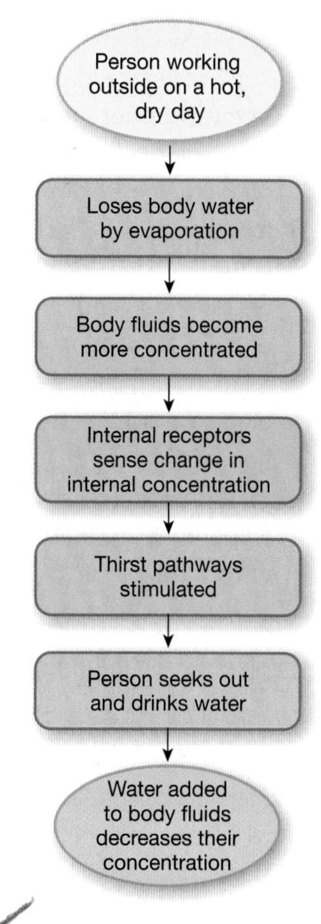

(b) A process map, or flow chart

FOCUS ON . . .

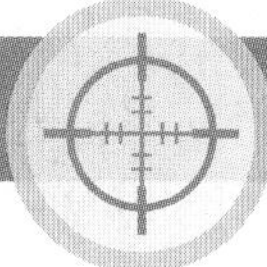

Mapping

Mapping is a nonlinear way of organizing material that is based on the theory that individuals have their own unique ways of integrating new material with what they already know. Mapping is a useful study tool because creating a map requires thinking about the importance of and relationships between various pieces of information. Studies have shown that when people interact with information by organizing it in their own way *before* they load it into memory, their understanding and retention of information improves.

Mapping is not just a study technique. Experts in a field make maps when they are trying to integrate newly acquired information into their knowledge base, and they may create two or three versions of a map before they are satisfied that it represents their understanding. Scientists map out the steps in their experiments. Health-care professionals create maps to guide them while diagnosing and treating patients.

A map can take a variety of forms but usually consists of terms (words or short phrases) linked by arrows to indicate associations. You can label the connecting arrows to describe the type of linkage between the terms (structure/function, cause/effect) or with explanatory phrases ("is composed of"). You may also choose to use different colors for arrows and terms to represent different categories of ideas. Maps in physiology usually focus either on the relationships between anatomical structures and physiological processes (*structure/function maps*) or on normal homeostatic control pathways and responses to abnormal (pathophysiological) events (*process maps,* or *flow charts*). If appropriate, a map may also include graphs, diagrams, or pictures.

Figure 1.2a is a structure map. [You can find other examples of structure/function maps in Chapter 3.] Figure 1.2b is an example of a flow chart. [Look at the end of Chapter 15 for additional examples of flow charts.]

Many maps appear in this textbook, and they may serve as the starting point for your own maps. However, the real benefit of mapping comes from preparing maps yourself. By mapping information on your own, you think about the relationships between terms, organize concepts into a hierarchical structure, and look for similarities and differences between items. Interaction with the material in this way helps you process it into long-term memory instead of simply memorizing bits of information and forgetting them.

Some people do not like the messiness of hand-drawn maps. There are several electronic ways of making maps, including PowerPoint and free and commercial software programs.

PowerPoint

1. Select the completely blank slide from FORMAT - SLIDE LAYOUT.
2. Use AUTOSHAPES to create boxes/ovals and arrows. To format the autoshape, right-click on it after you have drawn it. You can change fill color and line color.
3. Use INSERT - TEXT BOX to label your arrows and put terms inside your shapes.

Software Free concept mapping software is available from IHMC CmapTools at *http://cmap.ihmc.us.* Or search for the term *free concept map* to find other resources on the Web. A popular commercial program for mapping is *Inspiration* (*www.inspiration.com*).

Getting Started with Mapping

1. First, **select the terms or concepts to map.** (In every chapter of this text, the end-of-chapter questions include at least one list of terms to map.) Sometimes it is helpful to write the terms on individual slips of paper or on 1-by-2-inch sticky notes so that you can rearrange the map more easily.
2. Usually the most difficult part of mapping is **deciding where to begin.** Start by grouping related terms in an organized fashion. You may find that you want to put some terms into more than one group. Make a note of these terms, as they will probably have several arrows pointing to them or leading away from them.
3. Now try to **create some hierarchy with your terms.** You may arrange the terms on a piece of paper, on a table, or on the floor. In a structure/function map, start at the top with the most general, most important, or overriding concept—the one from which all the

others stem. In a process map, start with the first event to occur. Next, either break down the key idea into progressively more specific parts using the other concepts or follow the event through its time course. Use arrows to point the direction of linkages and include horizontal links to tie related concepts together. The downward development of the map will generally mean either an increase in complexity or the passage of time.

You may find that some of your arrows cross each other. Sometimes you can avoid this by rearranging the terms on the map. Labeling the linking arrows with explanatory words may be useful. For example,

channel proteins $\xrightarrow{\textit{form}}$ open channels

Color can be very effective on maps. You can use colors for different types of links or for different categories of terms. You may also add pictures and graphs that are associated with specific terms in your map.

4. Once you have created your map, **sit back and think about it.** Are all the items in the right place? You may want to move them around once you see the big picture. Revise your map to fill in the picture with new concepts or to correct wrong links. Review by recalling the main concept and then moving to the more specific details. Ask yourself questions like, What is the cause? effect? What parts are involved? What are the main characteristics?
5. Science is a collaborative field. A useful way to study with a map is to **trade maps with a classmate** and try to understand each other's maps. Your maps will almost certainly not look the same! It's OK if they are different. Remember that your map reflects the way you think about the subject, which may be different from the way someone else thinks about it. Did one of you put in something the other forgot? Did one of you have an incorrect link between two items?
6. **Practice making maps.** The study questions in each chapter will give you some ideas of what you should be mapping. Your instructor can help you get started.

cells transport oxygen—their function—but says nothing about *how* the cells transport oxygen.

In contrast, most physiologists study physiological processes, or **mechanisms**—the "how" of a system. The **mechanistic approach** to physiology examines process. The mechanistic answer to the question "How do red blood cells transport oxygen?" is "Oxygen binds to hemoglobin molecules in the red blood cells." This very concrete answer explains exactly how oxygen transport occurs but says nothing about the significance of oxygen transport to the animal.

Students often confuse these two approaches to thinking about physiology. Studies have shown that even medical students tend to answer questions with teleological explanations when the more appropriate response would be a mechanistic explanation.[1] Often they do so because instructors ask why a physiological event occurs when they really want to know how it occurs. Staying aware of the two approaches will help prevent confusion.

Although function and mechanism seem to be two sides of the same coin, it is possible to study mechanisms, particularly at the cellular and subcellular level, without understanding their function in the life of the organism. As biological knowledge becomes more complex, scientists sometimes become so involved in studying complex processes that they fail to step back and look at the significance of those processes to cells, organ systems, or the animal. Conversely, it is possible to use teleological thinking incorrectly by saying, "Oh, in this situation the body needs to do this." *This* may be a good solution, but if a mechanism for doing *this* doesn't exist, the organism is out of luck.

Applying the concept of integrated functions and mechanisms is the underlying principle in **translational research**, an approach sometimes described as "bench to bedside." Translational research uses the insights and results gained from basic biomedical research on mechanisms to develop treatments and strategies for preventing human diseases. For example, researchers working on rats found that a chemical from the pancreas named *amylin* reduced the rats' food intake. These findings led directly to a translational research study in which human volunteers injected a synthetic form of amylin and recorded their subsequent food intake, but without intentionally modifying their lifestyle.[2] The drug suppressed food intake in humans, although it has not been approved by the Food and Drug Administration for that use.

[1]D. R. Richardson. A survey of students' notions of body function as teleologic or mechanistic. *Advan Physiol Educ* 258: 8–10, Jun 1990. (*http://advan.physiology.org*)

[2]S. R. Smith *et al.* Pramlintide treatment reduces 24-h caloric intake and meal sizes and improves control of eating in obese subjects: a 6-wk translational research study. *Am J Physiol Endocrinol Metab* 293: E620–E627, 2007.

RUNNING PROBLEM

When he got back to his room Jimmy sat down at his computer and went to the Internet. He typed *diabetes* in his search box—and came up with 66.7 million results. "That's not going to work. What about *insulin*?" 22.5 million results. "How in the world am I going to get any answers?" He clicked on the first sponsored ad, for a site called *type2-diabetes-info.com*. That might be good. His mother had type 2 diabetes. But it was for a pharmaceutical company trying to sell him a drug. "What about this? *WhyInsulin.com*—That might give some answers." But it, too, was trying to sell something. "Maybe my physiology prof can help me. I'll ask tomorrow."

Q1: What search terms could Jimmy have used to get fewer results?

At the systems level, we know about most of the mechanics of body function from centuries of research. The unanswered questions today mostly involve integration and control of these mechanisms, particularly at the cellular and molecular levels. Nevertheless, explaining what happens in test tubes or isolated cells can only partially answer questions about function. For this reason, animal and human trials are essential steps in the process of applying basic research to treating or curing diseases.

Themes in Physiology

"Physiology is not a science or a profession but a point of view."[3] Physiologists pride themselves on relating the mechanisms they study to the functioning of the organism as a whole. For students, being able to think about how multiple body systems integrate their function is one of the more difficult aspects of learning physiology. To develop expertise in physiology, you must do more than simply memorize facts and learn new terminology. Researchers have found that the ability to solve problems requires a conceptual framework, or "big picture," of the field.

This book will help you build a conceptual framework for physiology by explicitly emphasizing the basic biological concepts, or themes, that are common to all living organisms. These concepts represent patterns that repeat over and over, and you will begin to recognize them when you encounter them in specific contexts. Pattern recognition is an important skill that will help simplify learning physiology.

In the last few years, three different organizations issued reports to encourage the teaching of biology using these fundamental concepts. Although the descriptions vary in the three reports, five major themes emerge:

1. structure and function across all levels of organization
2. energy transfer, storage, and use
3. information flow, storage, and use within single organisms and within a species of organism
4. homeostasis and the control systems that maintain it
5. evolution

In addition, all three reports emphasize the importance of understanding how science is done and of the quantitative nature of biology. ■ Table 1.2 lists the core concepts in biology from the three reports.

In this book we focus on the four themes most related to physiology: structure-function relationships, biological energy use, information flow within an organism, and homeostasis and the control systems that maintain it. The first six chapters introduce the fundamentals of these themes. You may already be familiar with some of them from earlier biology and chemistry classes. The themes and their associated concepts, with variations, then appear over and over in subsequent chapters of this book. Look for them in the summary material at the end of the chapters and in the end-of-chapter questions as well.

Theme 1: Structure and Function Are Closely Related

The integration of structure and function extends across all levels of organization, from the molecular level to the intact body. This theme subdivides into two major ideas: molecular interactions and compartmentation.

Molecular Interactions The ability of individual molecules to bind to or react with other molecules is essential for biological function. A molecule's function depends on its structure and shape, and even a small change to the structure or shape may have significant effects on the function. The classic example of this phenomenon is the change in one amino acid of the hemoglobin protein. (Hemoglobin is the oxygen-carrying pigment of the blood.) This one small change in the protein converts normal hemoglobin to the form associated with sickle cell disease.

Many physiologically significant molecular interactions that you will learn about in this book involve the class of biological molecules called *proteins*. Functional groups of proteins include *enzymes* that speed up chemical reactions, *signal molecules* and the *receptor proteins* that bind signal molecules, and specialized proteins that function as biological pumps, filters,

[3] R. W. Gerard. *Mirror to Physiology: A Self-Survey of Physiological Science*. Washington, D.C.: American Physiology Society, 1958.

1

Biology Concepts — Table 1.2

Scientific Foundations for Future Physicians (HHMI and AAMC)[1]	Vision and Change (NSF and AAAS)[2]	The 2010 Advanced Placement Biology Curriculum (College Board)[3]
Structure/function from molecules to organisms	Structure and function (anatomy and physiology)	Relationship of structure to function
Physical principles applied to living systems Chemical principles applied to living systems	Pathways and transformations of energy and matter	Energy transfer
Biomolecules and their functions	Information flow, exchange, and storage	Continuity and change
Organisms sense and control their internal environment and respond to external change	Systems	Regulation ("a state of dynamic balance")
Evolution as an organizing principle	Evolution	Evolution

[1] *Scientific Foundations for Future Physicians.* Howard Hughes Medical Institute (HHMI) and the Association of American Medical Colleges (AAMC), 2009. *www.aamc.org/scientificfoundations.*

[2] *Vision and Change: A Call to Action.* National Science Foundation (NSF) and American Association for the Advancement of Science (AAAS). 2011. *http://visionandchange.org/finalreport.* The report mentioned the integration of science and society as well.

[3] *College Board AP Biology Course Description,* The College Board, 2010. *http://apcentral.collegeboard.com/apc/public/repository/ap-biology-course-description.pdf.* The AP report also included "Interdependence in Nature" and "Science, Technology and Society" as two of their eight themes.

motors, or transporters. [Chapter 2 describes molecular interactions involving proteins in more detail.]

Interactions between proteins, water, and other molecules influence cell structure and the mechanical properties of cells and tissues. Mechanical properties you will encounter in your study of physiology include *compliance* (ability to stretch), *elastance* (stiffness or the ability to return to the unstretched state), strength, flexibility, and fluidity (*viscosity*).

Compartmentation Compartmentation is the division of space into separate compartments. Compartments allow a cell, a tissue, or an organ to specialize and isolate functions. Each level of organization is associated with different types of compartments. At the macroscopic level, the tissues and organs of the body form discrete functional compartments, such as body cavities or the insides of hollow organs. At the microscopic level, cell membranes separate cells from the fluid surrounding them and also create tiny compartments within the cell called organelles. [Compartmentation is the theme of Chapter 3.]

Theme 2: Living Organisms Need Energy

Growth, reproduction, movement, homeostasis—these and all other processes that take place in an organism require the continuous input of energy. Where does this energy come from, and how is it stored? We'll answer those questions and describe some of the ways that energy in the body is used for building and breaking down molecules [see Chapter 4]. In subsequent chapters you learn how energy is used to transport molecules across cell membranes and to create movement.

Theme 3: Information Flow Coordinates Body Functions

Information flow in living systems ranges from the transfer of information stored in DNA from generation to generation (genetics) to the flow of information within the body of a single organism. At the organismal level, information flow includes translation of the genetic code into proteins that are responsible for cell structure and function as well as the many forms of cell-to-cell communication that coordinate the functioning of a complex organism.

In the human body, information flow between cells takes the form of either chemical signals or electrical signals. Information may go from one cell to its neighbors (local communication) or from one part of the body to another (long-distance communication). When chemical signals reach their target cells, they must transfer their information from outside the cell to the inside of the cell.

RUNNING PROBLEM

After his second physiology class Jimmy introduced himself to his professor and explained his problem. The professor's first suggestion was simple: try to narrow the search. "One of the best ways to search is to combine terms using the connector AND. If you remember set theory from your math class, the connector AND will give you the intersection of the sets. In other words, you'll get only the results that occur in both sets."

Seemed simple enough. Jimmy went back to the Internet and tried *diabetes and insulin.* That search still had 12.4 million results but some of the results on the first page after the ads looked pretty good: *mayoclinic.com* and *diabetes.org.* Now he was getting somewhere.

Q2: What kinds of web sites should Jimmy be looking for in his results list, and how can he recognize them?

2 8 10 13 18 24 27

Some molecules are able to pass through the barrier of the cell membrane, but other molecules are unable to cross. Signal molecules that cannot enter the cell must pass their message across the cell membrane. [How molecules cross biological membranes is the topic of Chapter 5. Chapter 6 discusses chemical communication in the body.]

Theme 4: Homeostasis Maintains Internal Stability

Organisms that survive in challenging habitats cope with external variability by keeping their internal environment relatively stable, an ability known as **homeostasis** {*homeo-*, similar + *-stasis,* condition}. Homeostasis and the regulation of the internal environment are key principles of physiology and underlying themes in each chapter of this book. The next section looks in detail at the key elements of this important theme.

Homeostasis

The concept of a relatively stable internal environment is attributed to the French physician Claude Bernard in the mid-1800s. During his studies of experimental medicine, Bernard noted the stability of various physiological functions, such as body temperature, heart rate, and blood pressure. As the chair of physiology at the University of Paris, he wrote "C'est la fixité du milieu intérieur qui est la condition d'une vie libre et indépendante." (It is the constancy of the internal environment that is the condition for a free and independent life.)[4] This idea was applied to many of the experimental observations of his day, and it became the subject of discussion among physiologists and physicians.

In 1929, an American physiologist named Walter B. Cannon wrote a review for the American Physiological Society.[5] Using observations made by numerous physiologists and physicians during the nineteenth and early twentieth centuries, Cannon proposed a list of variables that are under homeostatic control. We now know that his list was both accurate and complete. Cannon divided his variables into what he described as environmental factors that affect cells (osmolarity, temperature, and pH) and "materials for cell needs" (nutrients, water, sodium, calcium, other inorganic ions, oxygen, as well as "internal secretions having general and continuous effects"). Cannon's "internal secretions" are the hormones and other chemicals that our cells use to communicate with one another.

In his essay, Cannon created the word *homeostasis* to describe the regulation of the body's internal environment. He explained that he selected the prefix *homeo-* (meaning *like* or *similar*) rather than the prefix *homo-* (meaning *same*) because the internal environment is maintained within a range of values rather than at an exact fixed value. He also pointed out that the suffix *-stasis* in this instance means a *condition,* not a state that is static and unchanging. Cannon's homeostasis therefore is a state of maintaining "a similar condition," similar to Claude Bernard's relatively constant internal environment.

Some physiologists contend that a literal interpretation of *stasis* {a state of standing} in the word *homeostasis* implies a static, unchanging state. They argue that we should use the word *homeodynamics* instead, to reflect the small changes constantly taking place in our internal environment {*dynamikos,* force or power}. Whether the process is called homeostasis or homeodynamics, the important concept to remember is that the body monitors its internal state and takes action to correct disruptions that threaten its normal function.

If the body fails to maintain homeostasis of the critical variables listed by Walter Cannon, then normal function is disrupted and a disease state, or **pathological** condition {*pathos,* suffering}, may result. Diseases fall into two general groups according to their origin: those in which the problem arises from internal failure of some normal physiological process, and those that originate from some outside source. Internal causes of disease include the abnormal growth of cells, which may cause cancer or benign tumors; the production of antibodies by the body against its own tissues (autoimmune diseases); and the premature death of cells or the failure of cell processes. Inherited disorders are also considered to have internal causes. External causes of disease include toxic chemicals, physical trauma, and foreign invaders such as viruses and bacteria.

---

[4]C. Bernard. *Introduction á l'étude de la medicine,* Paris: J.-B. Baillière, 1865. (*www.gutenberg.org/ebooks/16234*).

[5]W. B. Cannon. Organization for physiological homeostasis. *Physiol Rev* 9: 399–443, 1929.

In both internally and externally caused diseases, when homeostasis is disturbed, the body attempts to compensate (■ Fig. 1.3). If the compensation is successful, homeostasis is restored. If compensation fails, illness or disease may result. The study of body functions in a disease state is known as **pathophysiology.** You will encounter many examples of pathophysiology as we study the various systems of the body.

One very common pathological condition in the United States is **diabetes mellitus,** a metabolic disorder characterized by abnormally high blood glucose concentrations. Although we speak of diabetes as if it were a single disease, it is actually a whole family of diseases with various causes and manifestations. You will learn more about diabetes in the focus boxes scattered throughout the chapters of this book. The influence of this one disorder on many systems of the body makes it an excellent example of the integrative nature of physiology.

What Is the Body's Internal Environment?

Claude Bernard wrote of the "constancy of the internal environment," but why is constancy so essential? As it turns out, most cells in our bodies are not very tolerant of changes in their surroundings. In this way they are similar to early organisms that lived in tropical seas, a stable environment where salinity, oxygen content, and pH vary little and where light and temperature cycle in predictable ways. The internal composition of these ancient creatures was almost identical to that of seawater. If environmental conditions changed, conditions inside the primitive organisms changed as well. Even today, marine invertebrates cannot tolerate significant changes in salinity and pH, as you know if you have ever maintained a saltwater aquarium.

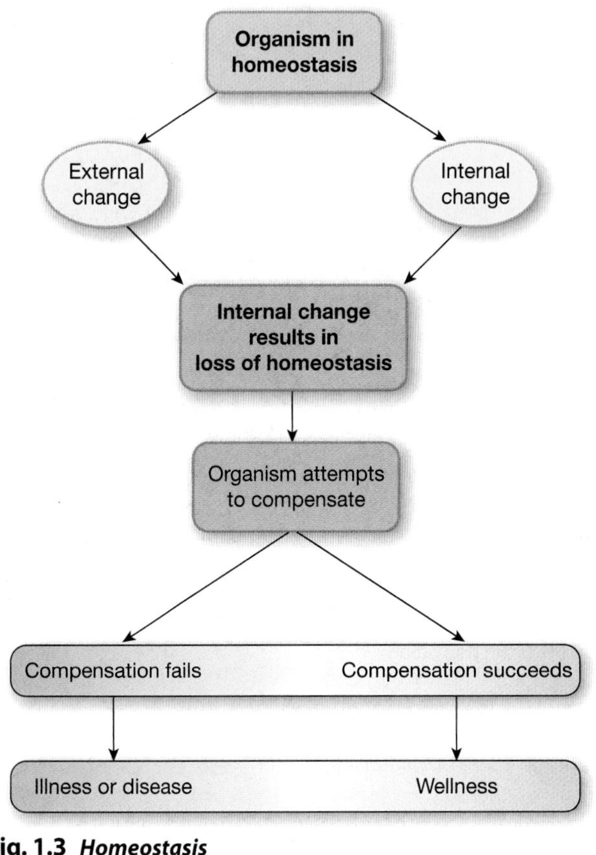

■ **Fig. 1.3** ***Homeostasis***

In both ancient and modern times, many marine organisms relied on the constancy of their external environment to keep their internal environment in balance. In contrast, as organisms evolved and migrated from the ancient seas into estuaries, then into freshwater environments and onto the land, they encountered highly variable external environments. Rains dilute the salty water of estuaries, and organisms that live there must cope with the influx of water into their body fluids. Terrestrial organisms, including humans, face the challenge of dehydration—constantly losing internal water to the dry air around them. Keeping the internal environment stable means balancing that water loss with appropriate water intake.

But what exactly is the internal environment of the body? For multicellular animals, it is the watery internal environment that surrounds the cells, a "sea within" the body called the **extracellular fluid** {*extra*-, outside of} (■ Fig. 1.4). Extracellular fluid (ECF) serves as the transition between an organism's external environment and the **intracellular fluid** (ICF) inside cells {*intra*-, within}. Because extracellular fluid is a buffer zone between cells and the outside world, elaborate physiological processes have evolved to keep its composition relatively stable.

When the extracellular fluid composition varies outside its normal range of values, compensatory mechanisms activate and try to return the fluid to the normal state. For example, when you drink a large volume of water, the dilution of your extracellular fluid triggers a mechanism that causes your kidneys to remove excess water and protect your cells from swelling. Most cells of multicellular animals do not tolerate much change. They depend on the constancy of extracellular fluid to maintain normal function.

Homeostasis Depends on Mass Balance

In the 1960s, a group of conspiracy theorists obtained a lock of Napoleon Bonaparte's hair and sent it for chemical analysis in an attempt to show that he died from arsenic poisoning. Today, a group of students sharing a pizza joke about the garlic odor on their breath. At first glance these two scenarios appear to have little in common, but in fact Napoleon's hair and "garlic breath" both demonstrate how the human body works to maintain the balance that we call homeostasis.

The human body is an open system that exchanges heat and materials with the outside environment. To maintain homeostasis, the body must maintain mass balance. The **law of mass balance** says that if the amount of a substance in the body is to remain constant, any gain must be offset by an equal loss (■ Fig. 1.5a). The amount of a substance in the body is also called the body's **load,** as in "sodium load."

For example, water loss to the external environment (output) in sweat and urine must be balanced by water intake from the external environment plus metabolic water production (input). The concentrations of other substances, such as oxygen and carbon dioxide, salts, and hydrogen ions (pH), are also maintained through mass balance. The following equation summarizes the law of mass balance:

$$\text{Total amount of substance } x \text{ in the body} = \text{intake} + \text{production} - \text{excretion} - \text{metabolism}$$

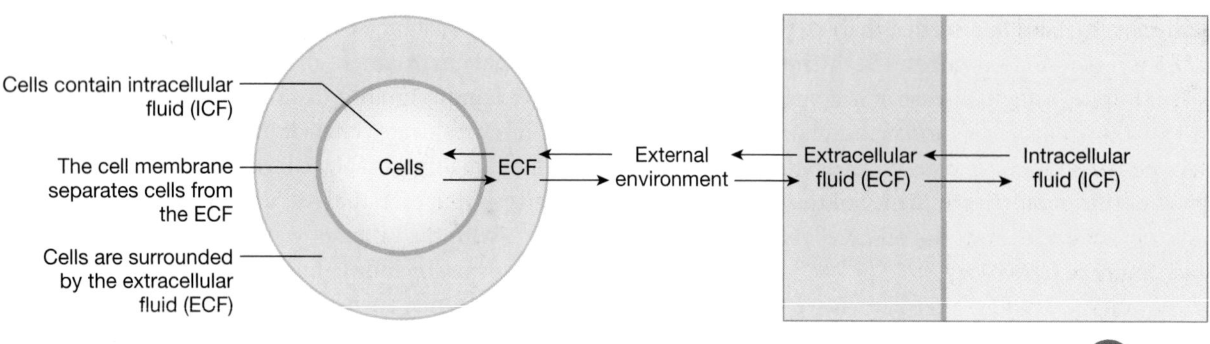

FIGURE QUESTION

Put a ★ on the cell membrane of the box diagram.

■ **Fig. 1.4** ***The body's internal and external environments.*** (a) Most cells are completely surrounded by the body's internal environment, composed of the extracellular fluid (ECF). (b) A box diagram of the body represents the body's fluid compartments, the extracellular fluid that exchanges material with the outside world and the fluid inside cells, or intracellular fluid (ICF).

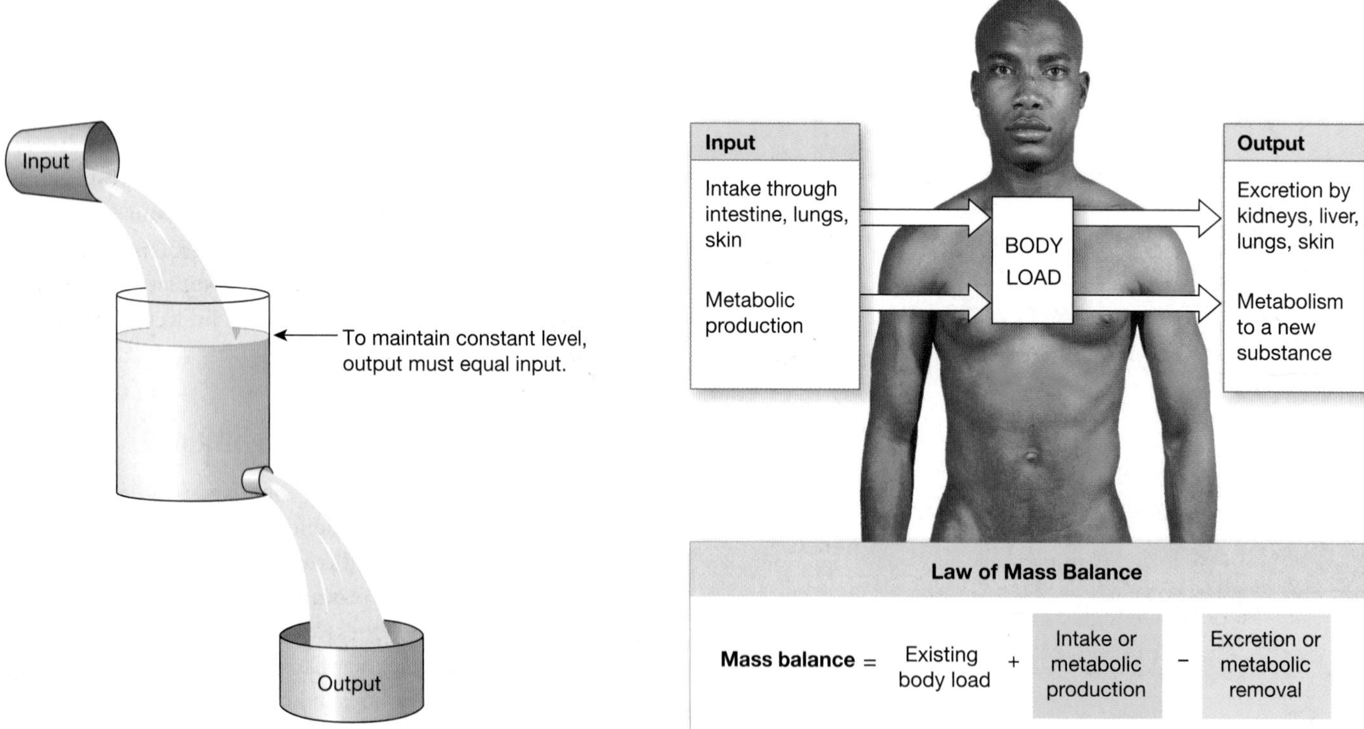

(a) Mass balance in an open system

(b) Mass balance in the body

■ **Fig. 1.5** ***Mass balance in an open system.*** (a) To maintain balance, input into the system must match output. (b) Materials enter the body primarily by ingestion or breathing or may be produced through metabolism. Materials leave the body by excretion or by metabolism.

Most substances enter the body from the outside environment, but some (such as carbon dioxide) are produced internally through metabolism (Fig. 1.5b). In general, water and nutrients enter the body as food and drink absorbed through the intestine. Oxygen and other gases and volatile molecules enter through the lungs. A few lipid-soluble chemicals make their way to the internal environment by penetrating the barrier of the skin.

To maintain mass balance, the body has two options for output. The simplest option is simply to excrete the material. **Excretion** is defined as the elimination of material from the body, usually through the urine, feces, lungs, or skin. For example, carbon dioxide produced during metabolism is excreted by the lungs. Many foreign substances that enter the body, such as drugs or artificial food additives, are excreted by the liver and kidneys. (Any foreign substance in the body is called a *xenobiotic,* from the Greek word *xenos,* a stranger.)

A second output option for maintaining mass balance is to convert the substance to a different substance through metabolism. Nutrients that enter the body become the starting point for metabolic pathways that convert the original nutrient to a different molecule. However, converting the original nutrient to something different then creates a new mass balance disturbance by adding more of the new substance, or *metabolite,* to the body. (*Metabolite* is the general term for any product created in a metabolic pathway.)

Scientists use **mass flow** to follow material throughout the body. For substance *x,* the equation for mass flow is

$$\text{Mass flow} = \text{concentration of } x \times \text{volume flow}$$
$$(\text{amount } x/\text{min}) = (\text{amount } x/\text{vol}) \times (\text{vol}/\text{min})$$

Mass flow can be used to determine the rate of intake, output, or production of *x.*

For example, suppose a person is given an intravenous (IV) infusion of glucose solution that has a concentration of 50 grams of glucose per liter of solution. If the infusion is given at a rate of 2 milliliters per minute, the mass flow of glucose into the body is:

$$\frac{50 \text{ g glucose}}{1000 \text{ mL solution}} \times 2 \text{ mL solution/min} = 0.1 \text{ g glucose/min}$$

The rate of glucose input into the body is 0.1 g glucose/min.

Mass flow applies not only to the entry, production, and removal of substances but also to the movement of substances from one compartment in the body to another. When materials enter the body, they first become part of the extracellular fluid. Where a substance goes after that depends on whether or not it can cross the barrier of the cell membrane and enter the cells.

Excretion Clears Substances from the Body

It is relatively easy to monitor how much of a substance enters the body from the outside world, but it is more difficult to track molecules inside the body to monitor their excretion or metabolism. Instead of directly measuring the substance, we can follow the rate at which the substance disappears from the blood, a concept called **clearance.** Clearance is usually expressed as a volume of blood *cleared* of substance *x* per unit of time. For this reason clearance is only an indirect measure of how substance *x* is eliminated.

The kidney and the liver are the two primary organs that clear solutes from the body. *Hepatocytes* {*hepaticus,* pertaining to the liver + *cyte,* cell}, or liver cells, metabolize many different types of molecules, especially xenobiotics such as drugs. The resulting metabolites may be secreted into the intestine for excretion in the feces or released into the blood for excretion by the kidneys. Pharmaceutical companies testing chemicals for their potential use as therapeutic drugs must know the clearance of the chemical before they can develop the proper dosing schedule.

Clearance also takes place in tissues other than the liver and kidneys. Saliva, sweat, breast milk, and hair all contain substances that have been cleared from the body. Salivary secretion of the hormone *cortisol* provides a simple noninvasive source of hormone for monitoring chronic stress. Drugs and alcohol

RUNNING PROBLEM

Jimmy called his mother with the news that he'd found some good information on the Mayo Clinic and American Diabetes Association web sites. According to both those organizations, someone with type 2 diabetes might begin to require insulin as the disease progressed. But his mother was still not convinced that she needed to start insulin injections.

"My friend Ahn read that some doctors say that if you eat a high-fiber diet, you won't need any other treatment for diabetes."

"Mom, that doesn't sound right to me."

"But it must be," Jimmy's mother replied. "It says so in The Doctors' Medical Library."

Q3: Go to The Doctors' Medical Library at *www.medical-library.net* and search for the article called "Fiber" by typing the word into the Search box or by using the alphabetical listing of Library Articles. What does Dr. Kennedy, the author of the article, say about high-fiber diet and diabetes?

Q4: Should Jimmy's mother believe what it says on this web site? How can Jimmy find out more about who created the site and what their credentials are?

secreted into breast milk are important because a breast-feeding infant will ingest these substances.

The 1960s analysis of Napoleon Bonaparte's hair tested it for arsenic because hair follicles help clear some compounds from the body. The test results showed significant concentrations of the poison in his hair, but the question remains whether Napoleon was murdered, poisoned accidentally, or died from stomach cancer. Eating garlic causes "garlic breath" because the lungs clear volatile lipid-soluble materials from the blood when these substances pass into the airways and are exhaled. The lungs also clear ethanol in the blood, and exhaled alcohol is the basis of the "breathalyzer" test used by law enforcement agencies.

Concept Check Answers: p. 30

1. If a person eats 12 milligrams (mg) of salt in a day and excretes 11 mg of it in the urine, what happened to the remaining 1 mg?
2. Glucose is metabolized to CO_2 and water. Explain the effect of glucose metabolism on mass balance in the body.

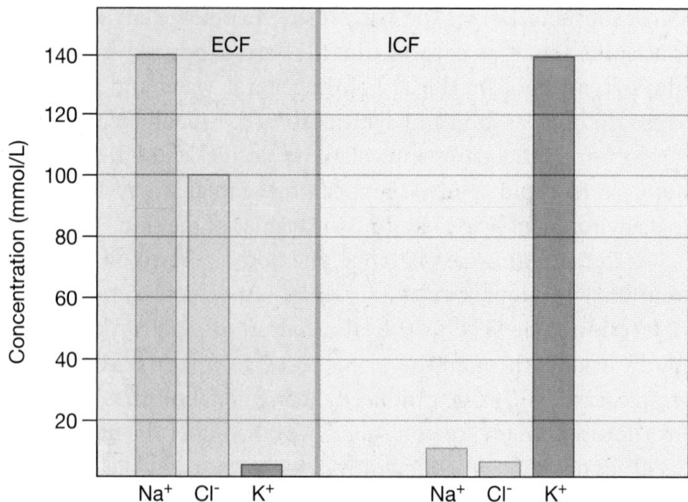

■ **Fig. 1.6** ***The body compartments are in a dynamic steady state but are not at equilibrium.*** The concentrations of sodium, potassium, and chloride are not the same in the extracellular fluid compartment (ECF) and intracellular fluid compartment (ICF).

Homeostasis Does Not Mean Equilibrium

When physiologists talk about homeostasis, they are speaking of the stability of the body's *internal environment*—in other words, the stability of the extracellular fluid compartment (ECF). One reason for focusing on extracellular fluid homeostasis is that it is relatively easy to monitor by taking a blood sample. When you centrifuge blood, it separates into two parts: **plasma,** the fluid component, and the heavier blood cells. Plasma is part of the extracellular fluid compartment, and its composition can be easily analyzed. It is much more difficult to follow what is taking place in the intracellular fluid compartment (ICF), although cells do maintain *cellular homeostasis.*

In a state of homeostasis, the composition of both body compartments is relatively stable. This condition is a dynamic **steady state.** The modifier *dynamic* indicates that materials are constantly moving back and forth between the two compartments. In a steady state, there is no *net* movement of materials between the compartments.

Steady state is not the same as **equilibrium** {*aequus*, equal + *libra*, balance}, however. Equilibrium implies that the composition of the body compartments is identical. If we examine the composition of the ECF and ICF, we find that the concentrations of many substances are different in the two compartments (■ Fig. 1.6). For example, sodium (Na^+) and chloride (Cl^-) are far more concentrated in the ECF than in the ICF, while potassium (K^+) is most concentrated in the ICF. Because of these concentration differences, the two fluid compartments are not at equilibrium. Instead the ECF and ICF exist in a state of **disequilibrium** {*dis-* is a negative prefix indicating the opposite of the base noun}. For living organisms, the goal of homeostasis is to maintain the dynamic steady states of the body's compartments, not to make the compartments the same.

Control Systems and Homeostasis

To maintain homeostasis, the human body monitors certain key functions, such as blood pressure and blood glucose concentration, that must stay within a particular operating range if the body is to remain healthy. These important **regulated variables** are kept within their acceptable (normal) range by physiological control mechanisms that kick in if the variable ever strays too far from its **setpoint**, or optimum value. There are two basic patterns of control mechanisms: local control and long-distance reflex control.

In their simplest form, all **control systems** have three components: (1) an input signal; (2) a controller, or **integrating center** {*integrare*, to restore}, that integrates incoming information and initiates an appropriate response; and (3) an output signal (■ Fig. 1.7) that creates a response. Long-distance reflex control systems are more complex than this simple model, however, as they may include input from multiple sources and have output that acts on multiple targets.

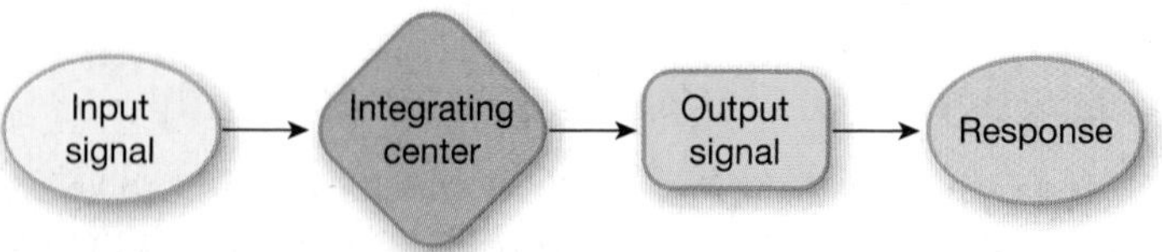

■ **Fig. 1.7** ***A simple control system***

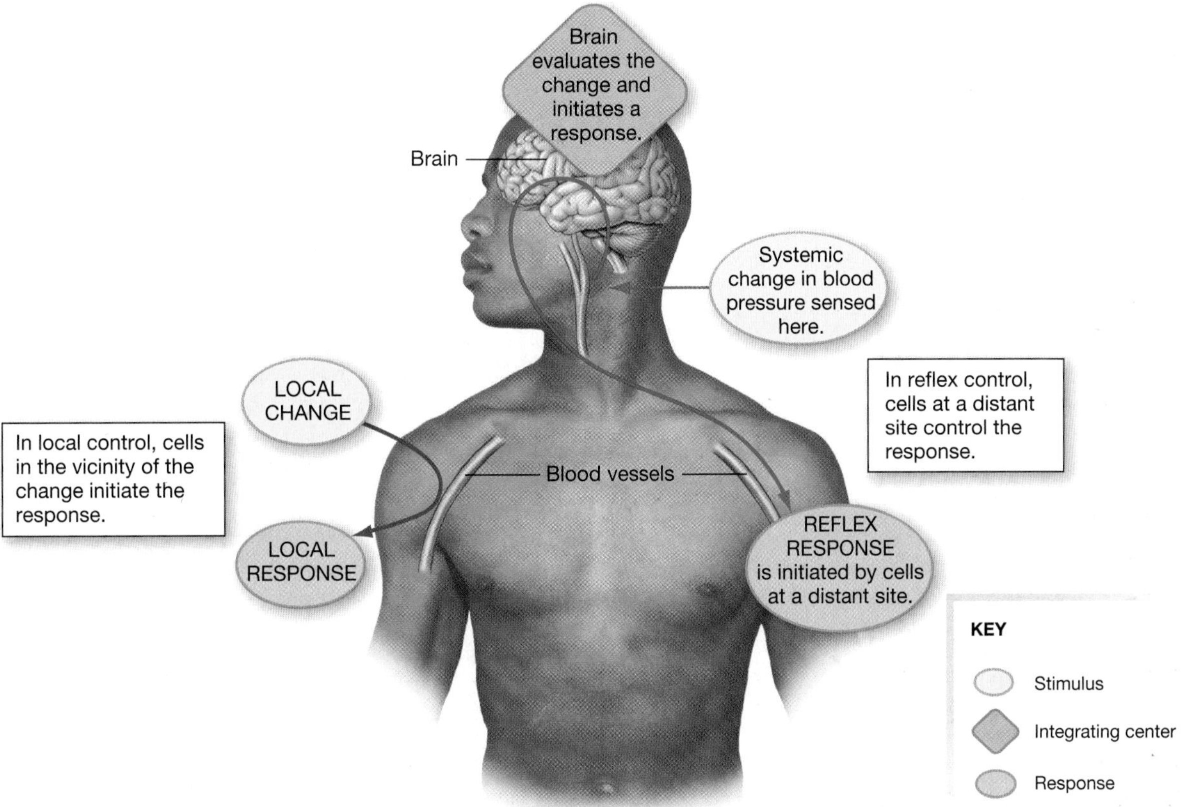
Brain evaluates the change and initiates a response.
Brain
Systemic change in blood pressure sensed here.
In reflex control, cells at a distant site control the response.
LOCAL CHANGE
In local control, cells in the vicinity of the change initiate the response.
Blood vessels
LOCAL RESPONSE
REFLEX RESPONSE is initiated by cells at a distant site.
KEY
Stimulus
Integrating center
Response

■ **Fig. 1.8** ***A comparison of local and reflex control***

Local Control Is Restricted to a Tissue

The simplest form of control is **local control**, which is restricted to the tissue or cell involved (■ Fig. 1.8). In local control, a relatively isolated change occurs in a tissue. A nearby cell or group of cells senses the change in their immediate vicinity and responds, usually by releasing a chemical. The response is restricted to the region where the change took place—hence the term *local control.*

One example of local control can be observed when oxygen concentration in a tissue decreases. Cells lining the small blood vessels that bring blood to the area sense the lower oxygen concentration and respond by secreting a chemical signal. The signal molecule diffuses to nearby muscles in the blood vessel wall, bringing them a message to relax. Relaxation of the muscles widens (*dilates*) the blood vessel, which increases blood flow into the tissue and brings more oxygen to the area.

Reflex Control Uses Long-Distance Signaling

Changes that are widespread throughout the body, or *systemic* in nature, require more complex control systems to maintain homeostasis. For example, maintaining blood pressure to drive blood flow throughout the body is a systemic issue rather than a local one. Because blood pressure is body-wide, maintaining it requires long-distance communication and coordination.

We will use the term ***reflex control*** to mean any long-distance pathway that uses the nervous system, endocrine system, or both.

A physiological reflex can be broken down into two parts: a response loop and a feedback loop (■ Fig. 1.9a). As with the simple control system just described, a **response loop** has three primary components: an *input signal,* an *integrating center* to integrate the signal, and an *output signal.* These three components can be expanded into the following sequence of seven steps to form a pattern that is found with slight variations in all reflex pathways:

Stimulus ⟶ sensor ⟶ input signal ⟶
integrating center ⟶
output signal ⟶ target ⟶ response

The input side of the response loop starts with a *stimulus*—the change that occurs when the regulated variable moves out of its desirable range. A specialized **sensor** monitors the variable. If the sensor is activated by the stimulus, it sends an input signal to the integrating center. The integrating center evaluates the information coming from the sensor and initiates an output signal. The output signal directs a target to carry out a response. If successful, the response brings the regulated variable back into the desired range.

In mammals, integrating centers are usually part of the nervous system or endocrine system. Output signals may be chemical signals, electrical signals, or a combination of both. The targets activated by output signals can be any cell of the body.

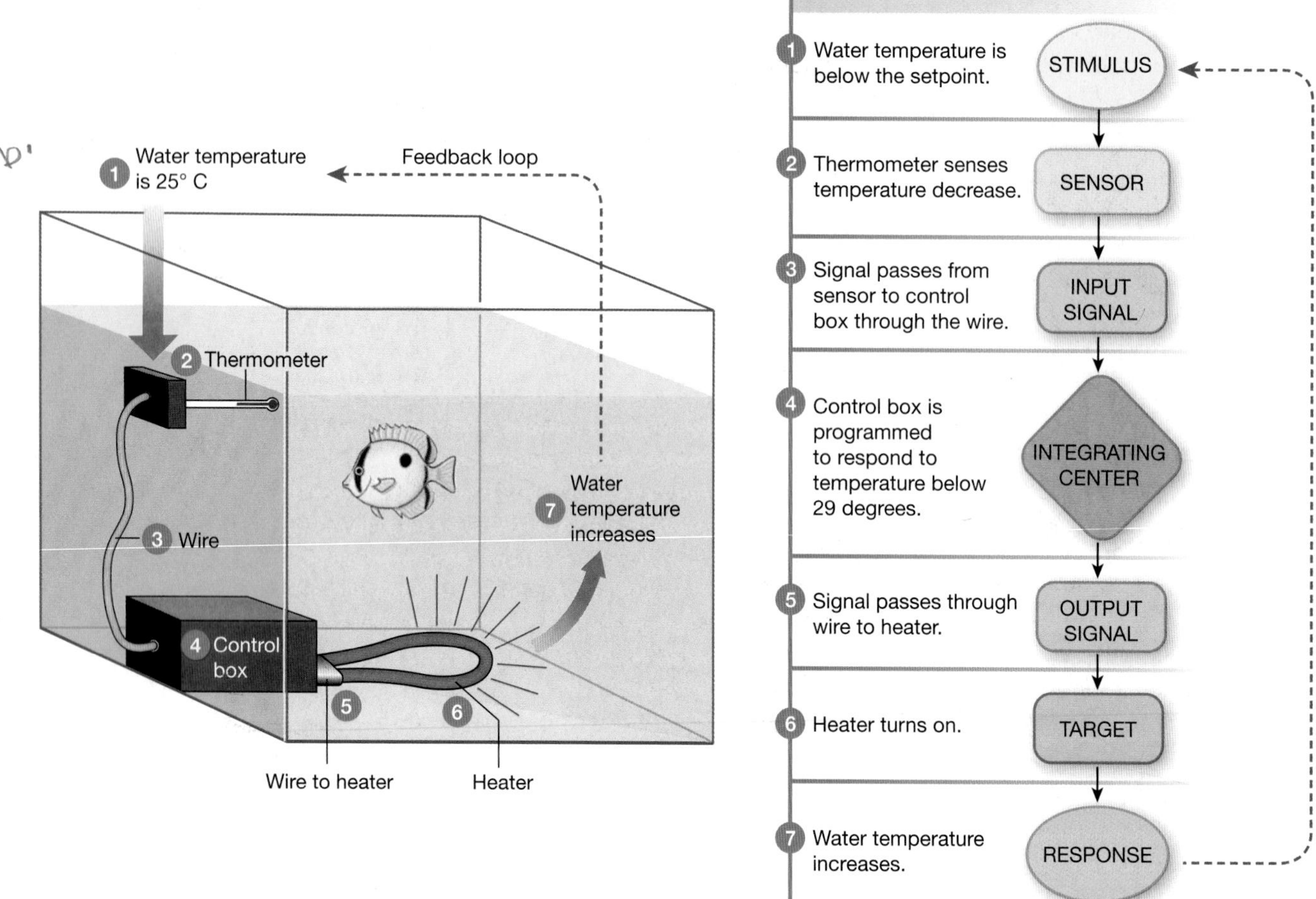
Water temperature is 25° C
Feedback loop
Thermometer
Wire
Control box
Water temperature increases
Wire to heater
Heater
Reflex steps
Water temperature is below the setpoint.
STIMULUS
Thermometer senses temperature decrease.
SENSOR
Signal passes from sensor to control box through the wire.
INPUT SIGNAL
Control box is programmed to respond to temperature below 29 degrees.
INTEGRATING CENTER
Signal passes through wire to heater.
OUTPUT SIGNAL
Heater turns on.
TARGET
Water temperature increases.
RESPONSE
Feedback loop

Fig. 1.9 ***Steps in the response loop of a reflex control pathway.*** In the aquarium example shown, the control box is set to maintain a water temperature of 30 ± 1 °C.

Response Loops Begin with a Stimulus

To illustrate response loops, let's apply the concept to a simple nonbiological example. Think about an aquarium whose heater is programmed to maintain the water temperature (the regulated variable) at 30 °C (Fig. 1.9b). The room temperature is 25 °C. The desired water temperature (30 °C) is the *setpoint* for the regulated variable.

Assume that initially the aquarium water is at room temperature, 25 °C. When you turn the control box on, you set the response loop in motion. The thermometer (sensor) registers a temperature of 25 °C. It sends this information through a wire (input signal) to the control box (integrating center). The control box is programmed to evaluate the incoming temperature signal, compare it with the setpoint for the system (30 °C), and "decide" whether a response is needed to bring the water temperature up to the setpoint. The control box sends a signal through another wire (output signal) to the heater (the target), which turns on and starts heating the water (response). This sequence—from stimulus to response—is the response loop.

This aquarium example involves a variable (temperature) controlled by a single control system (the heater). We can also describe a system that is under dual control. For example, think of a house that has both heating and air conditioning. The owner would like the house to remain at 70 °F (about 21 °C). On chilly autumn mornings, when the temperature in the house falls, the heater turns on to warm the house. Then, as the day warms up, the heater is no longer needed and turns off. When the sun heats the house above the setpoint, the air conditioner turns on to cool the house back to 70 °F. The heater and air conditioner have *antagonistic control* over house temperature because they work in opposition to each other. Similar situations occur in the human body when two branches of the nervous system or two different hormones have opposing effects on a single target.

Concept Check Answer: p. 30

3. What is the drawback of having only a single control system (a heater) for maintaining aquarium water temperature in some desired range?

Feedback Loops Modulate the Response Loop

The response loop is only the first part of a reflex. For example, in the aquarium just described, the sensor sends temperature information to the control box, which recognizes that the water is too cold. The control box responds by turning on the heater to warm the water. Once the response starts, what keeps the heater from sending the temperature up to, say, 50 °C?

The answer is a **feedback loop,** where the response "feeds back" to influence the input portion of the pathway. In the aquarium example, turning on the heater increases the temperature of the water. The sensor continuously monitors the temperature and sends that information to the control box. When the temperature warms up to the maximum acceptable value, the control box shuts off the heater, thus ending the reflex response.

Negative Feedback Loops Are Homeostatic

For most reflexes, feedback loops are homeostatic—that is, designed to keep the system at or near a setpoint so that the regulated variable is relatively stable. How well an integrating center succeeds in maintaining stability depends on the *sensitivity* of the system. In the case of our aquarium, the control box is programmed to have a sensitivity of ± 1 °C. If the water temperature drops from 30 °C to 29.5 °C, it is still within the acceptable range, and no response occurs. If the water temperature drops below 29 °C (30 – 1), the control box turns the heater on (■ Fig. 1.10). As the water heats up, the control box constantly receives information about the water temperature from the sensor. When the water reaches 31 °C (30 ± 1), the upper limit for the acceptable range, the feedback loop causes the control box to turn the heater off. The water then gradually cools off until the cycle starts all over again. The end result is a regulated variable that *oscillates* {*oscillare,* to swing} around the setpoint.

In physiological systems, some sensors are more sensitive than others. For example, the sensors that trigger reflexes to conserve water activate when blood concentration increases only 3% above normal, but the sensors for low oxygen in the blood will not respond until oxygen has decreased by 40%.

A pathway in which the response opposes or removes the signal is known as **negative feedback** (■ Fig. 1.11a). Negative feedback loops *stabilize* the regulated variable and thus aid the system in maintaining homeostasis. In the aquarium example, the heater warms the water (the response) and removes the stimulus (low water temperature). With loss of the stimulus for the pathway, the response loop shuts off. *Negative feedback loops can restore the normal state but cannot prevent the initial disturbance.*

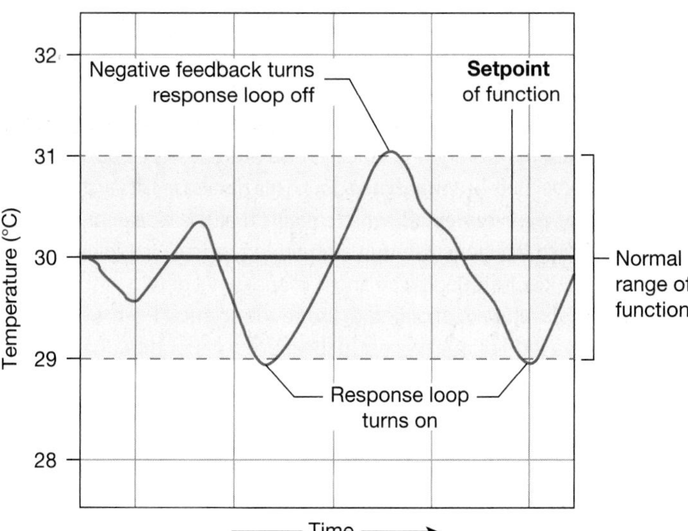

■ **Fig. 1.10** ***Oscillation around the setpoint.*** Most functions that maintain homeostasis have a setpoint, or normal value. The response loop that controls the function activates when the function moves outside a predetermined normal range.

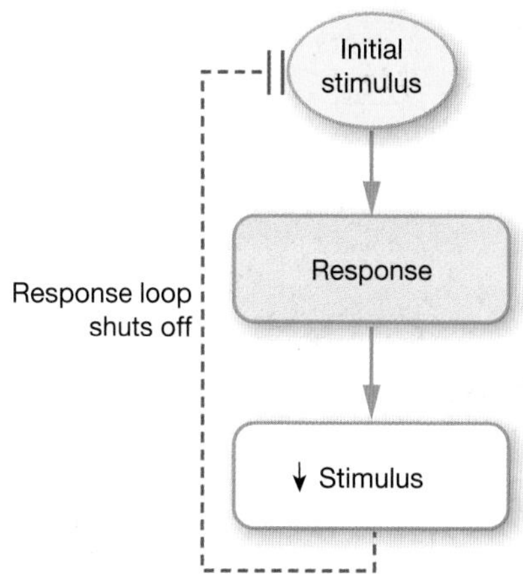

(a) Negative feedback: the response counteracts the stimulus, shutting off the response loop.

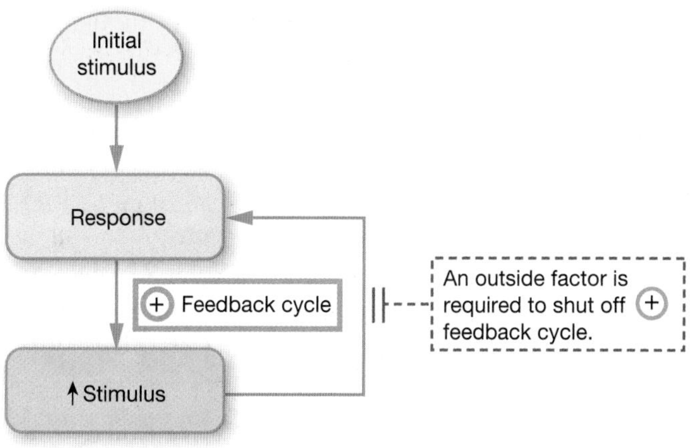

(b) Positive feedback: the response reinforces the stimulus, sending the variable farther from the setpoint.

■ **Fig. 1.11** ***Negative and positive feedback***

Positive Feedback Loops Are Not Homeostatic

A few reflex pathways are not homeostatic. In a **positive feedback loop,** the response *reinforces* the stimulus rather than decreasing or removing it. In positive feedback, the response sends the regulated variable even farther from its normal value. This initiates a vicious cycle of ever-increasing response and sends the system temporarily out of control (Fig. 1.11b). Because positive feedback escalates the response, this type of feedback requires some intervention or event outside the loop to stop the response.

One example of a positive feedback loop involves the hormonal control of uterine contractions during childbirth (■ Fig. 1.12). When the baby is ready to be delivered, it drops lower in the uterus and begins to put pressure on the *cervix,* the opening of the uterus. Sensory signals from the cervix to the brain cause release of the hormone *oxytocin,* which causes the uterus to contract and push the baby's head even harder against the cervix, further stretching it. The increased stretch causes more oxytocin release, which causes more contractions that push the baby harder against the cervix. This cycle continues until finally the baby is delivered, releasing the stretch on the cervix and stopping the positive feedback loop.

Baby drops lower in uterus to initiate labor
Cervical stretch
stimulates
Oxytocin release
causes
Uterine contractions
causing
Push baby against cervix
Positive feedback loop
Delivery of baby stops the cycle

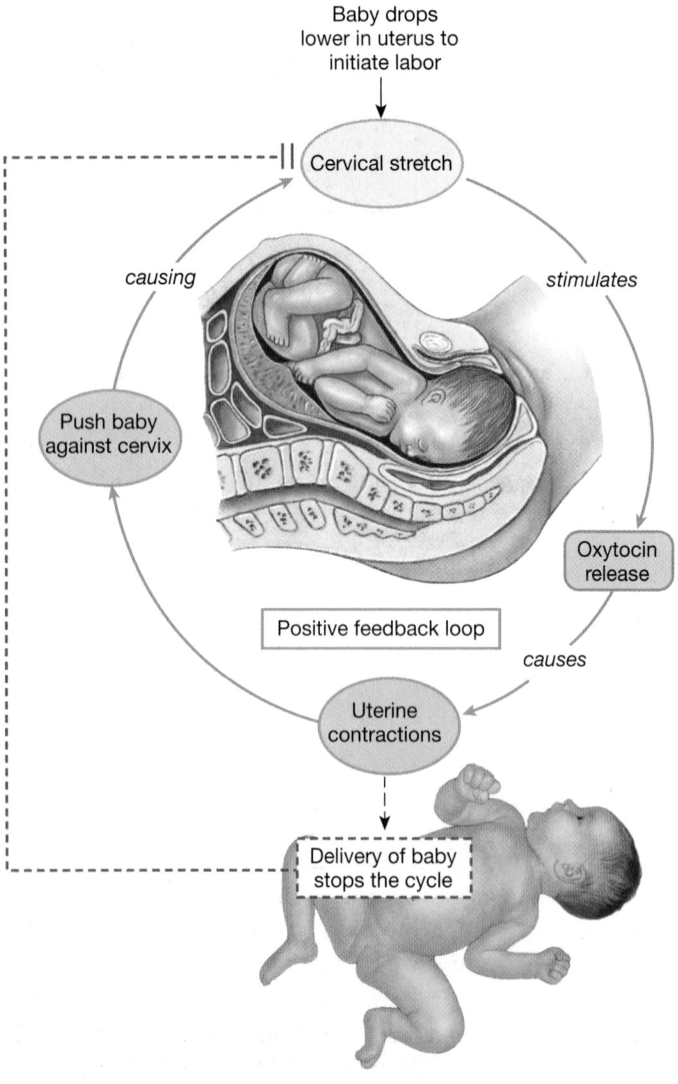

■ **Fig. 1.12** ***A positive feedback loop***

Concept Check

Answer: p. 30

4. Does the aquarium heating system in Figure 1.9 operate using positive feedback or negative feedback?

Feedforward Control Allows the Body to Anticipate Change

Negative feedback loops can stabilize a function and maintain it within a normal range but are unable to prevent the change that triggered the reflex in the first place. A few reflexes have evolved that enable the body to predict that a change is about to occur and start the response loop in anticipation of the change. These anticipatory responses are called **feedforward control.**

An easily understood physiological example of feedforward control is the salivation reflex. The sight, smell, or even the thought of food is enough to start our mouths watering in expectation of eating the food. This reflex extends even further, because the same stimuli can start the secretion of hydrochloric acid as the stomach anticipates food on the way. One of the most complex feedforward reflexes appears to be the body's response to exercise [Ch. 25].

RUNNING PROBLEM

After reading the article on fiber, Jimmy decided to go back to his professor for help. "How can I figure out who to believe on the Internet? Isn't there a better way to get health information?"

"Well, the sites you found from the Mayo Clinic and the American Diabetes Association are fine for general information aimed at the lay public. But if you want to find the same information that scientists and physicians read, you should search using MEDLINE, the database published by the U.S. National Library of Medicine. PubMed is the free public-access version (*www.pubmed.gov*). This database lists articles that are **peer-reviewed,** which means that the research described has gone through a screening process in which the work is critiqued by an anonymous panel of two or three scientists who are qualified to judge the science. Peer review acts as a kind of quality control because a paper that does not meet the standards of the reviewers will be rejected by the editor of the journal."

Q5: Jimmy went to PubMed and typed in his search terms: *type 2 diabetes and insulin therapy.* Repeat his search. Compare the number of results to the Google searches.

2 8 10 13 18 24 27

Biological Rhythms Result from Changes in a Setpoint

As discussed above, each regulated variable has a normal range within which it can vary without triggering a correction. In physiological systems, the setpoints for many regulated variables are different from person to person, or may change for the same individual over a period of time. Factors that influence an individual's setpoint for a given variable include normal biological rhythms, inheritance, and the conditions to which the person has become accustomed.

Regulated variables that change predictably and create repeating patterns or cycles of change are known as biological rhythms, or *biorhythms*. The timing of many biorhythms coincides with a predictable environmental change, such as daily light-dark cycles or the seasons. Biological rhythms reflect changes in the setpoint of the regulated variable.

For example, all animals exhibit some form of daily biological rhythm, called a **circadian rhythm** {*circa*, about + *dies*, day}. Humans have circadian rhythms for many body functions, including blood pressure, body temperature, and metabolic processes. Body temperature peaks in the late afternoon and declines dramatically in the early hours of the morning (■ Fig. 1.13a). Have you ever been studying late at night and noticed that you feel cold? This is not because of a drop in environmental temperature but because your thermoregulatory reflex has turned down your internal thermostat.

One of the interesting correlations between circadian rhythms and behavior involves body temperature. Researchers found that self-described "morning people" have temperature rhythms that cause body temperature to climb before they awaken in the morning, so that they get out of bed prepared to face the world. On the other hand, "night people" may be forced by school and work schedules to get out of bed while their body temperature is still at its lowest point, before their bodies are prepared for activity. These night people are still going strong and working productively in the early hours of the morning, when the morning people's body temperatures are dropping and they are fast asleep.

Many hormones in humans have blood concentrations that fluctuate predictably in a 24-hour cycle. Cortisol, growth hormone, and the sex hormones are among the most noted examples. A cortisol concentration in a 9:00 A.M. sample might be nearly twice as high as one taken in the early afternoon (Fig. 1.13b).

If a patient has a suspected abnormality in hormone secretion, it is therefore important to know when hormone levels are measured. A concentration that is normal at 9:00 A.M. is high at 2:00 P.M. One strategy for avoiding errors due to circadian fluctuations is to collect information for a full day and calculate an average value over 24 hours. For example, cortisol secretion is estimated indirectly by measuring all urinary cortisol metabolites excreted in 24 hours.

What is the adaptive significance of functions that vary with a circadian rhythm? Our best answer is that biological rhythms create an anticipatory response to a predictable environmental variable. There are seasonal rhythms of reproduction in many organisms. These rhythms are timed so that the offspring have food and other favorable conditions to maximize survival.

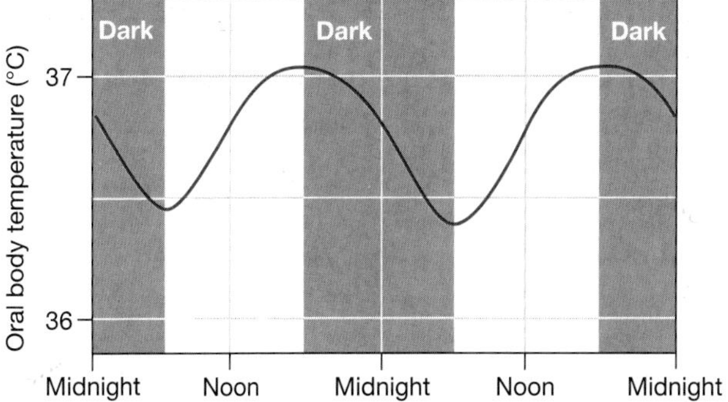

(a) Body temperature is lowest in the early morning and peaks in the late afternoon and early evening.

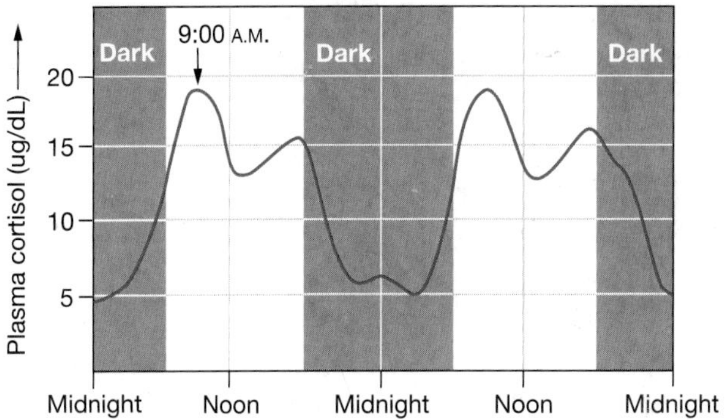

(b) Plasma cortisol is lowest during sleep and peaks shortly after awakening.

■ **Fig. 1.13** ***Circadian rhythms in humans.*** Data in (a) from W. E. Scales *et al.*, *J Appl Physiol* 65(4): 1840–1846, 1988. Data in (b) from L. Weibel *et al.*, *Am J Physiol Endocrinol Metab* 270: E608–E613, 1996.

Circadian rhythms cued by the light-dark cycle may correspond to rest-activity cycles. These rhythms allow our bodies to anticipate behavior and coordinate body processes accordingly. You may hear people who are accustomed to eating dinner at 6:00 P.M. say that they cannot digest their food if they wait until 10:00 P.M. to eat because their digestive system has "shut down" in anticipation of going to bed.

Some variability in setpoints is associated with changing environmental conditions rather than biological rhythms. The adaptation of physiological processes to a given set of environmental conditions is known as **acclimatization** when it occurs naturally. If the process takes place artificially in a laboratory setting, it is called **acclimation**. Each winter, people in the upper latitudes of the northern hemisphere go south in February, hoping to escape the bitter subzero temperatures and snows of the northern climate. As the northerners walk around in 40 °F (about 4 °C) weather in short-sleeve shirts, the southerners, all bundled up in coats and gloves, think that the northerners are crazy: the weather is cold! The difference in behavior is due to different temperature acclimatization, a difference in the setpoint for body temperature regulation that is a result of prior conditioning.

Biorhythms and acclimatization are complex processes that scientists still do not completely understand. Some rhythms arise from special groups of cells in the brain and are reinforced by information about the light-dark cycle that comes in through the eyes. Some cells outside the nervous system generate their own rhythms. Research in simpler animals such as flies is beginning to explain the molecular basis for biological rhythms. [We discuss the cellular and molecular basis for circadian rhythms in Chapter 9.]

The Science of Physiology

How do we know what we know about the physiology of the human body? The first descriptions of physiology came from simple observations. But physiology is an experimental science, one in which researchers generate **hypotheses** {*hypotithenai,* to assume; singular *hypothesis*}, or logical guesses, about how events take place. They test their hypotheses by designing experiments to collect evidence that supports or disproves their hypothesis, and they publish the results of their experiments in the scientific literature. Health-care providers look in the scientific literature for evidence from these experiments to help guide their clinical decision-making. Critically evaluating the scientific evidence in this manner is a practice known as *evidence-based medicine.* Observation and experimentation are the key elements of **scientific inquiry.**

Good Scientific Experiments Must Be Carefully Designed

A common type of biological experiment either removes or alters some variable that the investigator thinks is an essential part of an observed phenomenon. That altered variable is the **independent variable.** For example, a biologist notices that birds at a feeder seem to eat more in the winter than in the summer. She generates a hypothesis that cold temperatures cause birds to increase their food intake. To test her hypothesis, she designs an experiment in which she will keep birds at different temperatures and monitor how much food they eat. In her experiment, temperature, the manipulated element, is the independent variable. Food intake, which is hypothesized to be dependent on temperature, becomes the **dependent variable.**

Concept Check Answer: p. 30

5. Students in the laboratory run an experiment in which they drink different volumes of water and measure their urine output in the hour following drinking. What are the independent and dependent variables in this experiment?

An essential feature of any experiment is an experimental **control.** A control group is usually a duplicate of the experimental group in every respect except that the independent variable is not changed from its initial value. For example, in the bird-feeding experiment, the control group would be a set of birds maintained at a warm summer temperature but otherwise treated exactly like the birds held at cold temperatures. The purpose of the control is to ensure that any observed changes are due to the manipulated variable and not to changes in some other variable. For example, suppose that in the bird-feeding experiment food intake increased after the investigator changed to a different food. Unless she had a control group that was also fed the new food, the investigator could not determine whether the increased food intake was due to temperature or to the fact that the new food was more palatable.

During an experiment, the investigator carefully collects information, or **data** {plural; singular *datum,* a thing given}, about the effect that the manipulated (independent) variable has on the observed (dependent) variable. Once the investigator feels that she has sufficient information to draw a conclusion, she begins to analyze the data. Analysis can take many forms and usually includes statistical analysis to determine if apparent differences are statistically significant. A common format for presenting data is a graph (see Fig. 1. 14 in Focus on Graphs).

If one experiment supports the hypothesis that cold causes birds to eat more, then the experiment should be repeated to ensure that the results were not an unusual one-time event. This step is called **replication.** When the data support a hypothesis in multiple experiments, the hypothesis may become a working **model.** A model with substantial evidence from multiple investigators supporting it may become a **scientific theory.**

Most information presented in textbooks like this one is based on models that scientists have developed from the best available experimental evidence. On occasion, investigators publish new experimental evidence that does not support a current model. In that case, the model must be revised to fit the available evidence. For this reason, you may learn a physiological "fact" while using this textbook, but in 10 years that "fact" may be inaccurate because of what scientists have discovered in the interval.

For example, in 1970 students learned that the cell membrane was a "butter sandwich," a structure composed of a layer of fats sandwiched between two layers of proteins. In 1972, however, scientists presented a very different model of the membrane, in which globules of proteins float within a double layer of fats. As a result, textbook writers had to revise their descriptions of cell membranes, and students who had learned the butter sandwich model had to revise their mental model of the membrane.

Where do our scientific models for human physiology come from? We have learned much of what we know from experiments on animals ranging from squid to rats. In many instances, the physiological processes in such animals are either identical to those taking place in humans or else similar enough that we can extrapolate from the animal model to humans. It is important to use nonhuman models because experiments using human subjects can be difficult to perform.

However, not all studies done on animals can be applied to humans. For example, an antidepressant that Europeans had used safely for years was undergoing stringent testing required by the U.S. Food and Drug Administration before it could be sold in this country. When beagles took the drug for a period of months, the dogs started dying from heart problems. Scientists were alarmed until further research showed that beagles have a unique genetic makeup that causes them to break down the drug into a more toxic substance. The drug was perfectly safe in other breeds of dogs and in humans, and it was subsequently approved for human use.

The Results of Human Experiments Can Be Difficult to Interpret

There are many reasons it is difficult to carry out physiological experiments in humans, including variability, psychological factors, and ethical considerations.

Variability Human populations have tremendous genetic and environmental **variability.** Although physiology books usually present *average* values for many physiological variables, such as blood pressure, these average values simply represent a number that falls somewhere near the middle of a wide range of values. Thus, to show significant differences between experimental and control groups in a human experiment, an investigator would have to include a large number of identical subjects.

However, getting two groups of people who are *identical* in every respect is impossible. Instead, the researcher must attempt to recruit subjects who are *similar* in as many aspects as possible. You may have seen newspaper advertisements requesting research volunteers: "Healthy males between 18 and 25, nonsmokers, within 10% of ideal body weight, to participate in a study. . . ." Researchers must take into account the variability inherent in even a select group of humans when doing experiments with human subjects. This variability may affect the researcher's ability to interpret the significance of data collected on that group.

One way to reduce variability within a test population, whether human or animal, is to do a **crossover study.** In a crossover study, each individual acts both as experimental subject and as control. Thus, each individual's response to the treatment can be compared with his or her own control value. This method is particularly effective when there is wide variability within a population.

For example, in a test of blood pressure medication, investigators might divide subjects into two groups. Group A takes an inactive substance called a **placebo** (from the Latin for "I shall be pleasing") for the first half of the experiment, then changes to the experimental drug for the second half. Group B starts with the experimental drug, and then changes to the placebo. This scheme enables the researcher to assess the effect of the drug on each individual. In other words, each subject acts as his or her own control. Statistically, the data analysis can use methods that look at the changes within each individual rather than at changes in the collective group data.

Psychological Factors Another significant variable in human studies is the psychological aspect of administering a treatment. If you give someone a pill and tell the person that it will help alleviate some problem, there is a strong possibility that the pill will have exactly that effect, even if it contains only sugar or an inert substance. This well-documented phenomenon is called the **placebo effect.** Similarly, if you warn people that a drug they are taking may have specific adverse side effects, those people will report a higher incidence of the side effects than a similar group of people who were not warned. This phenomenon is called the **nocebo effect,** from the Latin *nocere,* to do harm. The placebo and nocebo effects show the ability of our minds to alter the physiological functioning of our bodies.

In setting up an experiment with human subjects, we must try to control for the placebo and nocebo effects. The simplest way to do this is with a **blind study,** in which the subjects do not know whether they are receiving the treatment or the placebo. Even this precaution can fail, however, if the researchers assessing the subjects know which type of treatment each subject is receiving. The researchers' expectations of what the treatment will or will not do may color their measurements or interpretations.

To avoid this outcome, researchers often use **double-blind studies.** A third party, not involved in the experiment, is the only one who knows which group is receiving the experimental treatment and which group is receiving the control treatment. The most sophisticated experimental design for minimizing psychological effects is the **double-blind crossover study.** In this type of study the control group in the first half of the experiment becomes the experimental group in the second half, and vice versa, but no one involved knows who is taking the active treatment.

Ethical Considerations Ethical questions arise when humans are used as experimental subjects, particularly when the subjects are people suffering from a disease or other illness. Is it ethical to withhold a new and promising treatment from the control group? A noteworthy example occurred some years ago when researchers were testing the efficacy of a treatment for dissolving blood clots in heart attack victims. The survival rate among

■ Fig. 1.14 *FOCUS ON . . . Graphs*

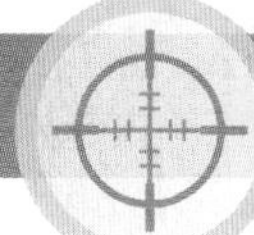

Graphs

Graphs are pictorial representations of the relationship between two (or more) variables, plotted in a rectangular region (■ Fig. 1.14a). We use graphs to present a large amount of numerical data in a small space, to emphasize comparisons between variables, or to show trends over time. A viewer can extract information much more rapidly from a graph than from a table of numbers or from a written description. A well-constructed graph should contain (in very abbreviated form) everything the reader needs to know about the data, including what the purpose of the experiment was, how it was conducted, and what the results were.

All scientific graphs have common features. The independent variable (the variable manipulated by the experimenter) is graphed on the horizontal *x*-axis. The dependent variable (the variable measured by the experimenter) is plotted on the vertical *y*-axis. If the experimental design is valid and the hypothesis is correct, changes in the independent variable (*x*-axis) will cause changes in the dependent variable (*y*-axis). In other words, *y* is a function of *x*. This relationship can be expressed mathematically as $y = f(x)$. Another way to describe the relationship between the axes is to think of it as "the effect of *x* on *y*."

Each axis of a graph is divided into units represented by evenly spaced tick marks on the axis. A label tells what variable the axis represents (time, temperature, amount of food consumed) and in what units it is marked (days, degrees Celsius, grams per day). The *origin* is the intersection of the two axes. The origin usually, but not always, has a value of zero for both axes. A graph should have a title or legend that describes what the graph represents. If multiple groups are shown on one graph, the lines or bars representing the groups may have labels, or a key may show what group each symbol or color represents.

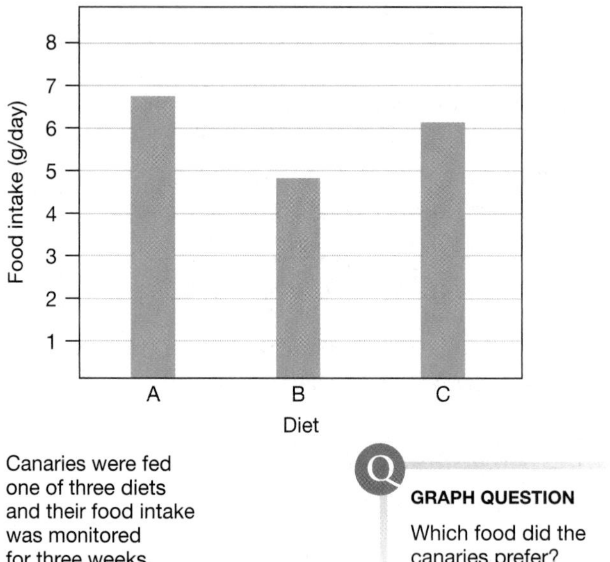

(b) Bar graph. Each bar shows a distinct variable. The bars are lined up side by side along one axis so that they can be easily compared with one another. Scientific bar graphs traditionally have the bars running vertically.

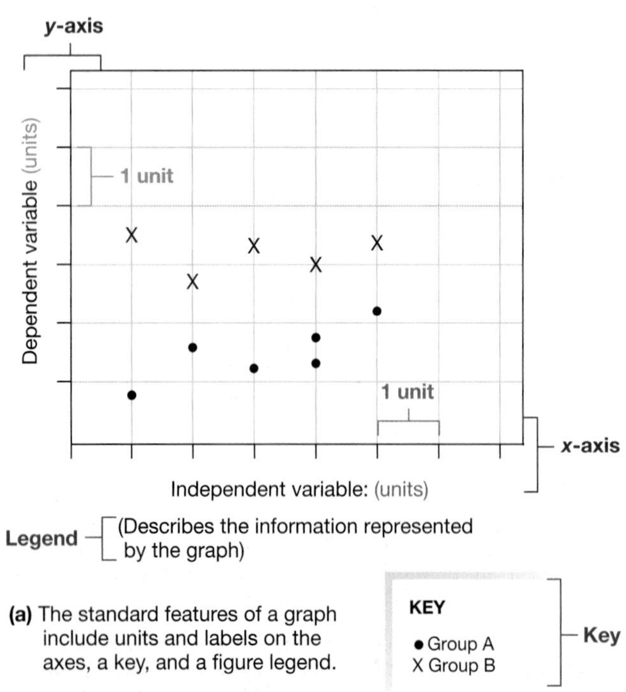

(a) The standard features of a graph include units and labels on the axes, a key, and a figure legend.

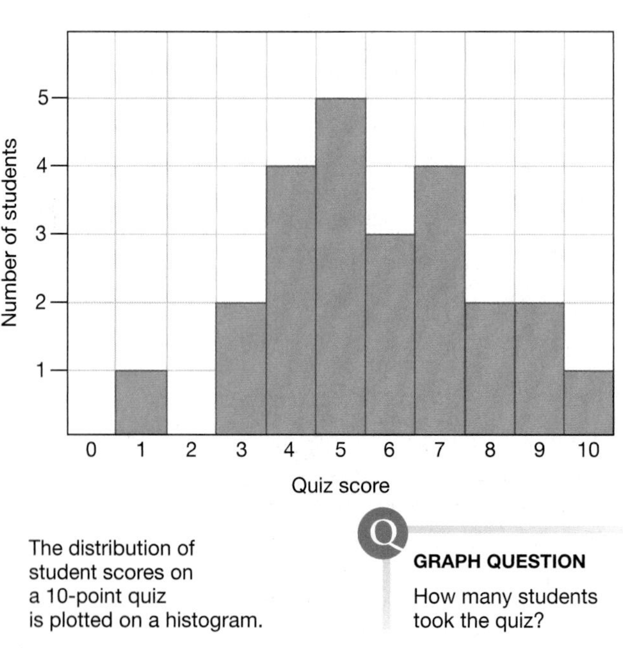

(c) Histogram. A histogram quantifies the distribution of one variable over a range of values.

■ **Fig. 1.14**

Most graphs you will encounter in physiology display data either as bars (bar graphs or histograms), as lines (line graphs), or as dots (scatter plots). Four typical types of graphs are shown in Figure 1.14b–e. **Bar graphs** (Fig. 1.14b) are used when the independent variables are distinct entities. A **histogram** (Fig. 1.14c) is a specialized bar graph that shows the distribution of one variable over a range. The *x*-axis is divided into units (called "bins" in some computer graphing programs), and the *y*-axis indicates how many pieces of data are associated with each bin.

Line graphs (Fig. 1.14d) are appropriate when the independent variable on the *x*-axis is a continuous phenomenon, such as time, temperature, or weight. Each point on the graph may represent the average of a set of observations. Because the independent variable is a continuous function, the points on the graph can be connected with a line (point-to-point connections or a mathematically calculated "best-fit" line or curve). Connecting the points allows the reader to **interpolate,** or estimate values between the measured values.

Scatter plots (Fig. 1.14e) show the relationship between two variables, such as time spent studying for an exam and performance on that exam. Usually each point on the plot represents one member of a test population. Individual points on a scatter plot are never connected by a line, but a "best-fit" line or curve may indicate a trend in the data.

Here are some questions to ask when you are trying to extract information from a graph:

1. What variable does each axis represent?
2. What is the relationship between the variables represented by the axes? This relationship can usually be expressed by substituting the labels on the axes into the following statement: "the effect of *x* on *y*." For example, graph (b) shows the effect of diet type on the canaries' daily food intake.
3. Are any trends apparent in the graph? For line graphs and scatter plots, is the line horizontal (no change in the dependent variable when the independent variable changes), or does it have a slope? Is the line straight or curved? For bar graphs, are the bars the same height or different heights? If different heights, is there a trend in the direction of height change?

Concept Check

Answers: p. 30

6. Students in a physiology laboratory collected heart-rate data on one another. In each case, heart rate was measured first for the subject at rest and again after the subject had exercised using a step test. Two findings from the experiment were (1) that heart rate was greater with exercise than at rest, and (2) that female subjects had higher resting heart rates than male subjects.
   (a) What was the independent variable in this experiment? What was the dependent variable?
   (b) Draw a graph and label each axis with the correct variable. Draw trend lines or bars that might approximate the data collected.

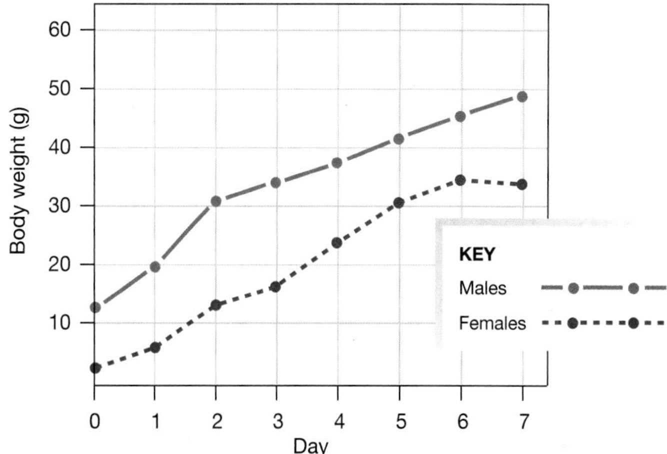

Male and female mice were fed a standard diet and weighed daily.

GRAPH QUESTION

When did male mice increase their body weight the fastest?

(d) Line graph. The *x*-axis frequently represents time; the points represent averaged observations. The points may be connected by lines, in which case the slope of the line between two points shows the rate at which the variable changed.

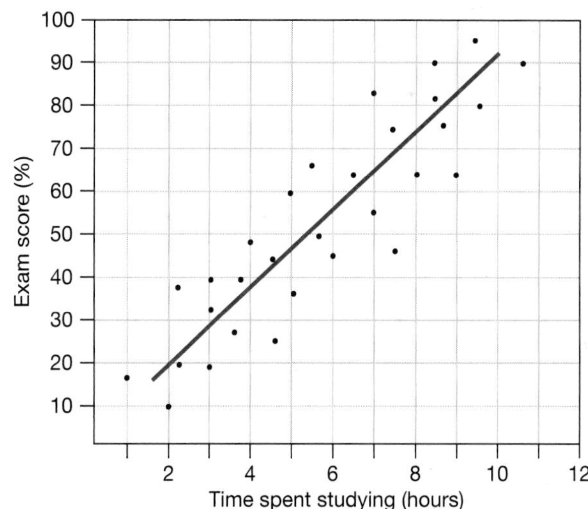

Student scores were directly related to the amount of time they spent studying.

GRAPH QUESTIONS

For graphs (d) and (e), answer the following:
- What was the investigator trying to determine?
- What are the independent and dependent variables?
- What are the results or trends indicated by the data?

(e) Scatter plot. Each point represents one member of a test population. The individual points of a scatter plot are never connected by lines, but a best fit line may be estimated to show a trend in the data, or better yet, the line may be calculated by a mathematical equation.

the treated patients was so much higher that testing was halted so that members of the control group could also be given the experimental drug.

In contrast, tests on some anticancer agents have shown that the experimental treatments were less effective in stopping the spread of cancer than were the standard treatments used by the controls. Was it ethical to undertreat patients in the experimental group by depriving them of the more effective current medical practice? Most studies now are evaluated continually over the course of the study to minimize the possibility that subjects will be harmed by their participation.

In 2002 a trial on hormone replacement therapy in postmenopausal women was halted early when investigators realized that women taking a pill containing two hormones were developing cardiovascular disease and breast cancer at a higher rate than women on placebo pills. On the other hand, the women receiving hormones also had *lower* rates of colon cancer and bone fractures. The investigators decided that the risks associated with taking the hormones exceeded the potential benefits, and they stopped the study. To learn more about this clinical trial and the pros and cons of hormone replacement therapy, go to *www.nlm.nih.gov/medlineplus/hormonereplacementtherapy.html*, the web site of the U.S. National Library of Medicine.

RUNNING PROBLEM

"Hi, professor. I'm back again." Most of the articles Jimmy found in PubMed were too focused on single experiments. And he didn't really understand the technical terms the authors used. "Is there any way to find papers that are not so complicated?"

"Yes, there are several ways. Many journals publish **review articles** that contain a synopsis of recent research on a particular topic. When you are just beginning to learn about a topic, it's best to begin with review articles. PubMed will have a link on the Results page that takes you directly to the review articles in your results. A different place you can look for basic information is MedlinePlus, another resource from the National Library of Medicine (*www.medlineplus.gov*)."

Jimmy decided to try MedlinePlus because the PubMed results seemed too technical for his simple question. On the MedlinePlus site he entered *type 2 diabetes and insulin therapy* into the search box. After reading a few of the articles he found linked there, he called his mother. "Hey, Mom! I found the answer to your question!"

Q6: Repeat Jimmy's search in MedlinePlus and look for links to articles on type 2 diabetes published by the National Institutes of Health (NIH), National Library of Medicine (NLM), or the Centers for Disease Control and Prevention (CDC). Based on what you read in those articles, what did Jimmy tell his mother about her need to take insulin for her type 2 diabetes?

Q7: What about the article that said eating a high-fiber diet could help? On the MedlinePlus results pages, look for articles on alternative treatments for diabetes published by the National Center for Complementary and Alternative Medicine. Do these articles mention dietary fiber?

2 8 10 13 18 24 27

Human Studies Can Take Many Forms

Almost daily, the newspapers carry articles about clinical trials studying the efficacy of drugs or other medical treatments. Many different aspects of experimental design can affect the validity and applicability of the results of these trials. For example, some trials are carried out for only a limited time on a limited number of people, such as studies conducted for the U.S. Food and Drug Administration's drug-approval process. In several instances in the past few years, drugs approved as a result of such studies have later been withdrawn from the market when extended use of the drug uncovered adverse side effects, including deaths.

Longitudinal studies are designed to be carried out for a long period of time. One of the most famous longitudinal studies is the Framingham Heart Study (*www.framingham.com/heart*), started in 1948 and still ongoing. Framingham is a **prospective study** {*prospectus,* outlook, looking forward} that recruited healthy people and has been following them for years to identify factors that contribute to the development of cardiovascular disease. This study has already made important contributions to health care, and it continues today with the adult children and grandchildren of the original participants.

Additional study designs you may encounter in the literature include cross-sectional and retrospective studies. **Cross-sectional studies** survey a population for the prevalence of a disease or condition. Data from cross-sectional studies identify trends to be investigated further, such as whether age group or socioeconomic status is associated with a higher risk of developing the condition being surveyed. **Retrospective studies** {*retro,* backward + *spectare,* to look} match groups of people who all have a particular disease to a similar but healthy control group. The goal of these studies is to determine whether development of the disease can be associated with a particular variable.

Often, the results of one or more published studies do not agree with the conclusions of other published studies. In some cases, the reason for the disagreement turns out to be a limitation of the experimental design, such as a small number of subjects who may not be representative of larger populations. In other cases, the disagreement may be due to small but potentially significant differences in the experimental designs of the different studies.

One way scientists attempt to resolve contradictory results is to perform a **meta-analysis** of the data {*meta-*, at a higher level}. A meta-analysis combines all the data from a group of similar studies and uses sophisticated statistical techniques to extract significant trends or findings from the combined data. For example, multiple studies have been done to assess whether glucosamine and chondroitin, two dietary supplements, can improve degenerative joint disease. However, the individual studies had small numbers of subjects (<50) and used different dosing regimens. A meta-analysis using statistical methods is one way to compare the results from these studies.[6]

The difficulty of using human subjects in experiments is one of the reasons scientists use animals to develop many of our scientific models. Since the 1970s, physiological research has increasingly augmented animal experimentation with techniques developed by cellular biologists and molecular geneticists. As we have come to understand the fundamentals of chemical signaling and communication in the body, we have unlocked the mysteries of many processes. In doing so, we also have come closer to being able to treat many diseases by correcting their cause rather than simply treating their symptoms.

More and more, medicine is turning to therapies based on interventions at the molecular level. A classic example is the treatment of cystic fibrosis, an inherited disease in which the mucus of the lungs and digestive tract is unusually thick. For many years, patients with this condition had few treatment options, and most died at a young age. However, basic research into the mechanisms by which salt and water move across cell membranes provided clues to the underlying cause of cystic fibrosis: a defective protein in the membrane of certain cells. Once molecular geneticists found the gene that coded for that protein, the possibility of replacing the defective gene in cystic fibrosis patients with the gene for the normal protein became a reality. Without the basic research into how cells and tissues carry out their normal function, however, this treatment would never have been developed.

As you read this book and learn what we currently know about how the human body works, keep in mind that many of the ideas presented in it reflect models that represent our current understanding and are subject to change. There still are many questions in physiology waiting for investigators to find the answers.

[6]See, for example, S. Wandel *et al.* Effects of glucosamine, chondroitin, or placebo in patients with osteoarthritis of hip or knee: network meta-analysis. *Br Med J* 341: c4675–c4676, 2010.

RUNNING PROBLEM CONCLUSION

What to Believe?

One skill all physiology students should acquire is the ability to find information in the scientific literature. In today's world, the scientific literature can be found both in print, in the form of books and periodicals, and on the Web. However, unless a book has a recent publication date, it may not be the most up-to-date source of information.

Many students begin their quest for information on a subject by searching the Internet. Be cautious! Anyone can create a web page and publish information on the Web. There is no screening process comparable to peer review in journals, and the reader of a web page must decide how valid the information is. Web sites published by recognized universities and not-for-profit organizations are likely to have good information, but you should view an article about vitamins on the web page of a health food store with a skeptical eye unless the article cites published peer-reviewed research.

Question	Answer and Commentary
1. What search terms could Jimmy have used to get fewer results?	Including more words in a web search is the best way to narrow the results list. For example, Jimmy could have searched for *insulin therapy diabetes.* That search would have produced about 800,000 results. Being more specific about his mother's type of diabetes would help. A search for *insulin therapy for type 2 diabetes* has only about 450,000 results. But that's still a lot of web pages to look at!

RUNNING PROBLEM CONCLUSION *(continued)*

Question	Answer and Commentary
2. What kinds of web sites should Jimmy be looking for in his results list, and how can he recognize them?	The best web sites for health information come from organizations that are part of the scientific and health-care communities, such as the National Institutes of Health (NIH), nonprofit groups dedicated to supporting research on a particular disease (The American Diabetes Association, *diabetes.org*), or clinics and universities where scientists and physicians are actively investigating causes and treatments for diseases. Treat commercial websites that end in **.com* with extra caution.
3. In The Doctors' Medical Library article called "Fiber," what does Dr. Kennedy say about high-fiber diet and diabetes?	Dr. Kennedy claims that some patients with type 2 diabetes can be "successfully treated" by eating a high-fiber diet. (The classification of type 2 diabetes as "adult onset" is obsolete.)
4. How can Jimmy find out more about who created the site and what their credentials are?	To learn more about who created a web site and why, look for links at the bottom of the page for HOME or ABOUT US. On the home page for The Doctors' Medical Library you will learn that the site promotes "natural health care." The link on Ron Kennedy, M.D., does not give you any information about his medical training or other credentials.
5. Compare the number of results from the PubMed search to those for the Google searches.	The number of results will depend on when you do the search because new articles are added constantly. But the number will probably be fewer than 50,000, far less than the millions of results that came up following a Google search.
6. What did Jimmy tell his mother about her need to take insulin for her type 2 diabetes?	The articles published by these national organizations all say that people with type 2 diabetes may need to take insulin. Patients should always listen to their health-care providers and ask questions if they are uncertain about what they should be doing.
7. Do the articles from NCCAM mention dietary fiber as an alternative treatment for diabetes?	The NCCAM article lists a number of alternative treatments that people have tried, but dietary fiber is not among them. It also says that, so far, there is no scientific evidence supporting the use of dietary supplements for treating diabetes. Patients should never stop their conventional treatments when using complementary treatments, and they should always inform their health-care providers about any vitamins or dietary supplements they are taking.

Citation Formats

When you find an article either in print or on the Web, you should write down the full citation. Citation formats for papers vary slightly from source to source but will usually include the following elements (with the punctuation shown):

Author(s). Article title. *Journal Name* volume (issue): inclusive pages, year of publication.

For example:
Echevarria M and Ilundain AA. Aquaporins. *J Physiol Biochem* 54(2): 107–118, 1998.

In many citations, the journal name is abbreviated using standard abbreviations. For example, the *American Journal of Physiology* is abbreviated as *Am J Physiol.* (One-word titles, such as *Science,* are never abbreviated.) For each calendar year, the issues of a given journal are assigned a **volume** number. The first issue of a given volume is designated **issue** 1, the second is issue 2, and so on.

Citing sources from the Web requires a different format. Here is one suggested format:

Author/Editor (if known). Revision or copyright date (if available). Title of web page [Publication medium]. Publisher of web page. *URL* [Date accessed].

For example:
Patton G (editor). 2005. Biological Journals and Abbreviations. [Online]. National Cancer Institute. *http://home.ncifcrf.gov/research/bja* [accessed April 10, 2005].

Unlike print resources, web pages are not permanent and frequently disappear or move. If you access a print journal on the Web, you should give the print citation, not the URL.

1

RUNNING PROBLEM CONCLUSION *(continued)*

Citing Other's Work

Copying or paraphrasing material from another source without acknowledging that source is academic dishonesty. Word-for-word quotations placed within quotation marks are rarely used in scientific writing. Instead, we summarize the contents of the source paper and acknowledge the source, as follows:

Some rare forms of epilepsy are known to be caused by mutations in ion channels (Mulley *et al.,* 2003).

When a paper has three or more authors, we use the abbreviation *et al.*—from the Latin *et alii,* meaning "and others"—to save space in the body of the text. All authors' names are given in the full citation, which is usually included within a References section at the end of the paper. Reference lists are often arranged alphabetically by the last name of the paper's first author.

Chapter Summary

1. **Physiology** is the study of the normal functioning of a living organism and its component parts. (p. 2)

Physiology Is an Integrative Science

2. Many complex functions are **emergent properties** that cannot be predicted from the properties of the individual component parts. (p. 2)
3. Physiologists study the many **levels of organization** in living organisms, from molecules to populations of one species. (p. 2; Fig. 1.1)
4. The **cell** is the smallest unit of structure capable of carrying out all life processes. (p. 3)
5. Collections of cells that carry out related functions make up **tissues** and **organs.** (p. 3)
6. The human body has 10 physiological organ systems: **integumentary, musculoskeletal, respiratory, digestive, urinary, immune, circulatory, nervous, endocrine,** and **reproductive.** (p. 3; Tbl. 1.1)

Function and Mechanism

7. The **function** of a physiological system or event is the "why" of the system. The **mechanism** by which events occur is the "how" of a system. The **teleological approach** to physiology explains why events happen; the **mechanistic approach** explains how they happen. (p. 5)
8. Translational research applies the results of basic physiological research to medical problems. (p. 7)

Themes in Physiology

9. The four key themes in physiology are structure/function relationships, such as **molecular interactions** and **compartmentation;** biological energy use; information flow within the body; and homeostasis. (p. 8)

Homeostasis

10. **Homeostasis** is the maintenance of a relatively constant internal environment. Variables that are regulated to maintain homeostasis include temperature, pH, ion concentrations, oxygen, and water. (p. 10)
11. Failure to maintain homeostasis may result in illness or disease. (p. 10; Fig. 1.3)
12. The body's internal environment is the **extracellular fluid.** (p. 11; Fig. 1.4)
13. The human body as a whole is adapted to cope with a variable external environment, but most cells of the body can tolerate much less change. (p. 11)
14. The **law of mass balance** says that if the amount of a substance in the body is to remain constant, any input must be offset by an equal loss. (p. 11; Fig. 1.5)
15. Input of a substance into the body comes from metabolism or from the outside environment. Output occurs through metabolism or **excretion.** (p. 13; Fig. 1.5)
16. The rate of intake, production, or output of a substance x is expressed as **mass flow,** where mass flow = concentration × volume flow. (p. 13)

17. **Clearance** is the rate at which a material is removed from the blood by excretion, metabolism, or both. The liver, kidneys, lungs, and skin all clear substances from the blood. (p. 13)
18. Cells and the extracellular fluid both maintain homeostasis, but they are not identical in composition. Their stable condition is a dynamic **steady state.** (p. 14)
19. Most solutes are concentrated in either one compartment or the other, creating a state of **disequilibrium.** (p. 14; Fig. 1.6)

Control Systems and Homeostasis

20. **Regulated variables** have a **setpoint** and a normal range. (p. 14; Fig. 1.10)
21. The simplest homeostatic control takes place at the tissue or cell level and is known as **local control.** (p. 15; Fig. 1.8)
22. **Control systems** have three components: an input signal, an **integrating center,** and an output signal. (p. 15; Fig. 1.7)
23. Reflex pathways can be broken down into **response loops** and **feedback loops.** A response loop begins when a **stimulus** is sensed by a **sensor.** The sensor is linked by the input signal to the **integrating center** that decides on an appropriate response. The output signal travels from the integrating center to a **target** that carries out the appropriate **response.** (p. 16; Fig. 1.9)
24. In **negative feedback,** the response opposes or removes the original stimulus, which in turn stops the response loop. (p. 17; Fig. 1.11a)
25. In **positive feedback** loops, the response reinforces the stimulus rather than decreasing or removing it. This destabilizes the system until some intervention or event outside the loop stops the response. (p. 18; Fig. 1.11b; Fig. 1.12)
26. **Feedforward control** allows the body to predict that a change is about to occur and start the response loop in anticipation of the change. (p. 18)
27. Regulated variables that change in a predictable manner are called biological rhythms. Those that coincide with light-dark cycles are called **circadian rhythms.** (p. 19; Fig. 1.13)

The Science of Physiology

28. Observation and experimentation are the key elements of **scientific inquiry.** A **hypothesis** is a logical guess about how an event takes place. (p. 20)
29. In scientific experimentation, the factor manipulated by the investigator is the **independent variable,** and the observed factor is the **dependent variable.** All well-designed experiments have a **control** to ensure that observed changes are due to the experimental manipulation and not to some outside factor. (p. 20)
30. **Data,** the information collected during an experiment, are analyzed and presented, often as a graph. (p. 20; Fig. 1.14)
31. A **scientific theory** is a hypothesis that has been supported by data on multiple occasions. When new experimental evidence does not support a theory or a model, then the theory or model must be revised. (p. 20)
32. Animal experimentation is an important part of learning about human physiology because of the tremendous amount of **variability** within human populations, and because it is difficult to control human experiments. In addition, ethical questions arise when using humans as experimental animals. (p. 21)
33. As the control in many experiments, some subjects may take an inactive substance known as a **placebo.** One difficulty in human experiments arises from the **placebo** and **nocebo effects,** in which changes take place even if the treatment is inactive. (p. 21)
34. In a **blind study,** the subjects do not know whether they are receiving the experimental treatment or a placebo. In a **double-blind study,** a third party removed from the experiment is the only one who knows which group is the experimental group and which is the control. In a **crossover study,** the control group in the first half of the experiment becomes the experimental group in the second half, and vice versa. (p. 21)
35. **Meta-analysis** of data combines data from many studies to look for trends. (p. 25)

Questions

Answers: p. A-1

Level One Reviewing Facts and Terms

1. Define physiology. Describe the relationship between physiology and anatomy.
2. Name the different levels of organization in the biosphere.
3. Name the 10 systems of the body and give their major function(s).
4. What does "Physiology is an integrative science" mean?
5. Define homeostasis. Name some regulated variables that are maintained through homeostasis.
6. Name four major themes in physiology.
7. Put the following parts of a reflex in the correct order for a physiological response loop: input signal, integrating center, output signal, response, sensor, stimulus, target.
8. The name for daily fluctuations of body functions such as blood pressure, temperature, and metabolic processes is a(n).

Level Two Reviewing Concepts

9. **Mapping exercise:** Make a large map showing the organization of the human body. Show all levels of organization in the body (see Fig. 1.1) and all 10 organ systems. Try to include functions of all components on the map and remember that some structures may share functions. (Hint: Start with the human body as the most important term. You may also draw the outline of a body and make your map using it as the basis.)
10. Distinguish between the items in each group of terms.
    (a) tissues and organs
    (b) *x*-axis and *y*-axis on a graph
    (c) dependent and independent variables
    (d) teleological and mechanistic approaches
    (e) the internal and external environments for a human
    (f) blind, double-blind, and crossover studies
    (g) the target and the sensor in a control system
11. Name as many organs or body structures that connect directly with the external environment as you can.

12. Which organ systems are responsible for coordinating body function? For protecting the body from outside invaders? Which systems exchange material with the external environment, and what do they exchange?
13. Explain the differences among positive feedback, negative feedback, and feedforward mechanisms. Under what circumstances would each be advantageous?

Level Three Problem Solving

14. A group of biology majors went to a mall and asked passersby, "Why does blood flow?" These are some of the answers they received. Which answers are teleological and which are mechanistic? (Not all answers are correct, but they can still be classified.)
    (a) Because of gravity
    (b) To bring oxygen and food to the cells
    (c) Because if it didn't flow, we would die
    (d) Because of the pumping action of the heart
15. Although dehydration is one of the most serious physiological obstacles that land animals must overcome, there are others. Think of as many as you can, and think of various strategies that different terrestrial animals have to overcome these obstacles. (Hint: Think of humans, insects, and amphibians; also think of as many different terrestrial habitats as you can.)

Level Four Quantitative Problems

16. A group of students wanted to see what effect a diet deficient in vitamin D would have on the growth of baby guppies. They fed the guppies a diet low in vitamin D and measured fish body length every third day for three weeks. Their data looked like this:

Day	0	3	6	9	12	15	18	21
Average body length (mm)	6	7	9	12	14	16	18	21

    (a) What was the dependent variable and what was the independent variable in this experiment?
    (b) What was the control in this experiment?
    (c) Make a fully labeled graph with a legend, using the data in the table.
    (d) During what time period was growth slowest? Most rapid? (Use your graph to answer this question.)
17. You performed an experiment in which you measured the volumes of nine slices of potato, then soaked the slices in solutions of different salinities for 30 minutes. At the end of 30 minutes, you again measured the volumes of the nine slices. The changes you found were:

% Change in volume after 30 minutes			
Solution	**Sample 1**	**Sample 2**	**Sample 3**
Distilled water	10%	8%	11%
1% salt (NaCl)	0%	−0.5%	1%
9% salt (NaCl)	−8%	−12%	−11%

    (a) What was the independent variable in this experiment? What was the dependent variable?
    (b) Can you tell from the information given whether or not there was a control in this experiment? If there was a control, what was it?
    (c) Graph the results of the experiment using the most appropriate type of graph.
18. At the end of the semester, researchers measured an intermediate-level class of 25 male weight lifters for aerobic fitness and midarm muscle circumference. The relationship between those two variables is graphed here.

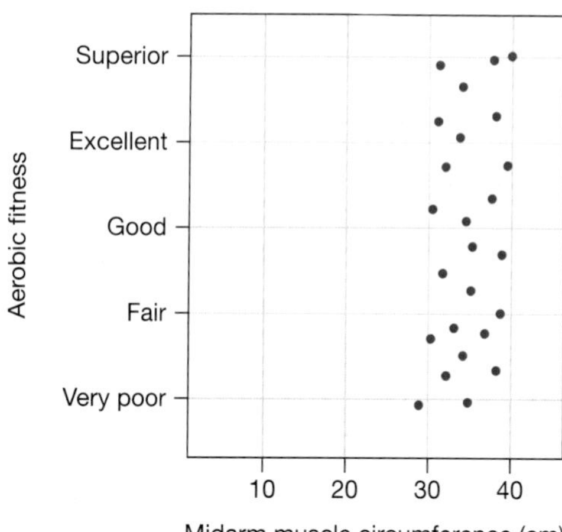

    (a) What kind of graph is this?
    (b) What question were the investigators asking?
    (c) In one sentence, summarize the relationship between the two variables plotted on the graph.
19. Answer the questions after the following article summary.

    A study[7] was carried out on human volunteers to see whether two procedures performed during arthroscopic surgery {*arthro-*, joint + *scopium,* to look at} are effective in relieving knee pain associated with osteoarthritis, or degenerative joint disease {*osteon,* bone + *arthro-*, joint + *-itis,* inflammation}. The volunteers were up to 75 years old and were recruited from a Veterans Affairs Medical Center. They were 93% male and 60% white. One-third of the subjects had placebo operations; that is, they were given anesthesia and their knees were cut open, but the remainder of the treatment procedure was not done. The other two-thirds of the subjects had one of the two treatment procedures performed. Subjects were followed for 2 years. They answered questions about their knee pain and function and were given an objective walking and stair-climbing test. At the end of the study the results showed no significant difference in knee function or perception of pain between subjects getting one of the standard treatments and those getting the placebo operation.

    (a) Do you think it is ethical to perform placebo surgeries on humans who are suffering from a painful condition, even if the subjects are informed that they might receive the placebo operation and not the standard treatment?
    (b) Give two possible explanations for the decreased pain reported by the placebo operation subjects.
    (c) Analyze and critique the experimental design of this study. Are the results of this study applicable to everyone with knee pain?
    (d) Was this study a blind, double-blind, or double-blind crossover design?
    (e) Why do you think the investigators felt it was necessary to include a placebo operation in this study?

[7] J. B. Moseley *et al.* A controlled trial of arthroscopic surgery for osteoarthritis of the knee. *N Eng J Med* 347(2): 81–88, 2002.

Answers

Answers to Concept Check Questions

Page 14

1. The remaining 1 mg of salt remains in the body.
2. Glucose metabolism adds CO_2 and water to the body, disturbing the mass balance of these two substances. To maintain mass balance, both metabolites must be either excreted or further metabolized.

Page 16

3. If the aquarium water became overheated, there is no control mechanism for bringing it back into the desired range.

Page 18

4. Negative feedback shuts off the heater.

Page 20

5. The independent variable is the amount of water the students drink. The dependent variable is their urine output.

Page 23

6. (a) Activity level was the independent variable (x-axis), and heart rate was the dependent variable (y-axis). (b) A line graph would be appropriate for the data, but a bar graph could also be used if exercise intensity was the same in all subjects. To show the difference between males and females, the graphs should have either separate lines or separate bars for males and females.

Answers to Figure and Graph Questions

Page 25

Figure 1.2: (a) You might include different kinds of breads, meats, and so on, or add a category of sandwich characteristics, such as temperature (hot, cold) or layers (single, club).

The sample map shown here is one possible way to map the terms. Notice how some items are linked to more than one other item. The map does not include many terms that could be included.

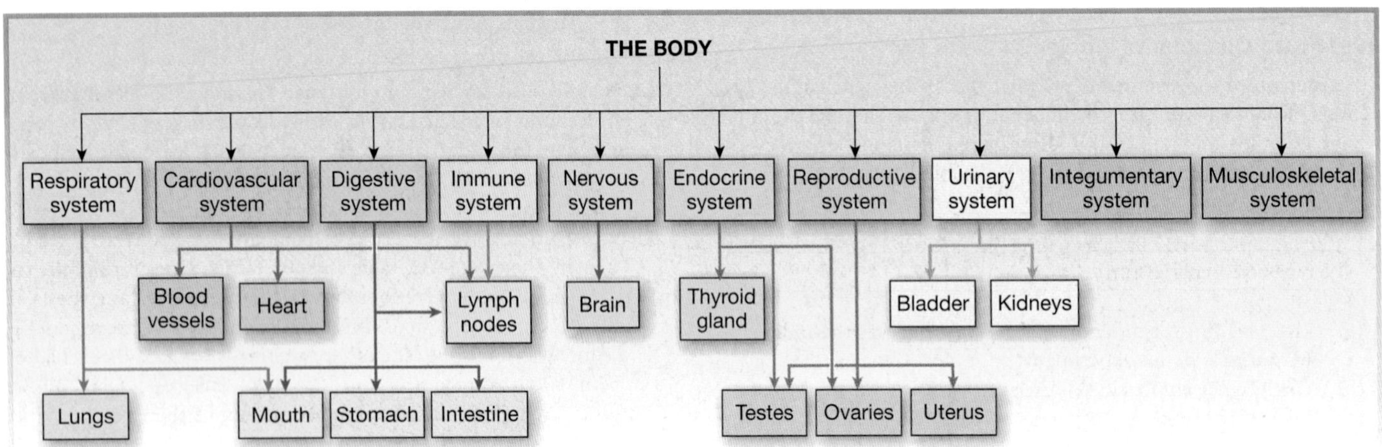

Pages 22–23

Figure 1.14b: The birds preferred diet A.

Figure 1.14c: Twenty-four students took the quiz.

Figure 1.14d: Male mice (blue line) increased their weight most between day 1 and day 2.

Figure 1.14e: (a) In graph 1.14d the investigator was looking at changes of body weight over time in male and female mice. In graph 1.14e the investigator was trying to determine if there was a relationship between the amount of time spent studying for an examination and the student's score on that examination. (b) In graph 1.14d the independent variable was time and the dependent variable was body weight. In graph 1.14e the independent variable was number of hours spent studying and the dependent variable was the student score. (c) Graph 1.14d shows that male mice weigh more than female mice from the start of the experiment and that body weight increases with time. The rate of increase is about the same in males and females, indicated by the nearly parallel lines. Graph 1.14e shows that more hours spent studying resulted in higher exam scores.

2 Molecular Interactions

It's only when life appears that you begin to get organisation on a larger scale. Life takes the atoms and molecules and . . . combines them into new and more elaborate patterns of its own.

—Aldous L. Huxley, 1945. In *Time Must Have a Stop.*

Crystals of ATP

Nearly 100 years ago two scientists, Aleksander Oparin in Russia and John Haldane in England, speculated on how life might have arisen on a primitive Earth whose atmosphere consisted mainly of hydrogen, water, ammonia, and methane. Their theories were put to the test in 1953, when a 23-year-old scientist named Stanley Miller combined these molecules in a closed flask and boiled them for a week while periodically discharging flashes of electricity through them, simulating lightning. At the end of his test, Miller found amino acids had formed in the flask. With this simple experiment, he had shown that it was possible to create organic molecules, usually associated with living creatures, from nonliving inorganic precursors.

Miller's experiments were an early attempt to solve one of the biggest mysteries of biology: how did a collection of chemicals first acquire the complex properties that we associate with living creatures? We still do not have an answer to this question. Numerous scientific theories have been proposed, ranging from life arriving by meteor from outer space to molecules forming in deep ocean hydrothermal vents. No matter what their origin, the molecules associated with living organisms have the ability to organize themselves into compartments, replicate themselves, and act as *catalysts* to speed up reactions that would otherwise proceed too slowly to be useful.

The human body is far removed from the earliest life forms, but we are still a collection of chemicals—dilute solutions of dissolved and suspended molecules enclosed in compartments with lipid-protein walls. Strong links between atoms, known as chemical bonds, store and transfer energy to support life functions. Weaker interactions between and within molecules create distinctive molecular shapes and allow biological molecules to interact reversibly with each other.

This chapter introduces some of the fundamental principles of molecular interactions that you will encounter repeatedly in your study of physiology. The human body is more than 50% water, and because most of its molecules are dissolved in this water, we will review the properties of aqueous solutions. If you would like to refresh your understanding of the key features of atoms, chemical bonds, and biomolecules, you will find a series of one- and two-page review features that encapsulate biochemistry as it pertains to physiology. You can test your knowledge of basic chemistry and biochemistry with a special review quiz at the end of the chapter.

RUNNING PROBLEM

Chromium Supplements

"Lose weight while gaining muscle," the ads promise. "Prevent heart disease." "Stabilize blood sugar." What is this miracle substance? It's chromium picolinate, a nutritional supplement being marketed to consumers looking for a quick fix. Does it work, though, and is it safe? Some athletes, like Stan—the star halfback on the college football team—swear by it. Stan takes 500 micrograms of chromium picolinate daily. Many researchers, however, are skeptical and feel that the necessity for and safety of chromium supplements have not been established.

32 39 43 47 49 52 57

Molecules and Bonds

There are more than 100 known elements on earth, but only 3—oxygen, carbon, and hydrogen—make up more than 90% of the body's mass. These 3 plus 8 additional elements are considered *major essential elements*. An additional 19 *minor essential elements* are required in trace amounts. A periodic table showing major and minor essential elements can be found inside the back cover of the book.

Most Biomolecules Contain Carbon, Hydrogen, and Oxygen

Molecules that contain carbon are known as **organic molecules,** because it was once thought that they all existed in or were derived from plants and animals. Organic molecules associated with living organisms are also called **biomolecules.** There are four major groups of biomolecules: carbohydrates, lipids, proteins, and nucleotides. The body uses the first three groups for energy and as the building blocks of cellular components. The fourth group, the nucleotides, includes DNA, RNA, ATP, and cyclic AMP. DNA and RNA are the structural components of genetic material. ATP (adenosine triphosphate) and related molecules carry energy, while cyclic AMP (adenosine monophosphate; cAMP) and related compounds regulate metabolism.

Each group of biomolecules has a characteristic composition and molecular structure. Lipids are mostly carbon and hydrogen (■ Fig. 2.1). Carbohydrates are primarily carbon, hydrogen, and oxygen, in the ratio CH_2O (■ Fig. 2.2). Proteins and nucleotides contain nitrogen in addition to carbon, hydrogen, and oxygen (■ Fig. 2.3 and 2.4). Two amino acids, the building blocks of proteins, also contain sulfur.

Not all biomolecules are pure protein, pure carbohydrate, or pure lipid, however. **Conjugated proteins** are protein molecules combined with another kind of biomolecule. For example, proteins combine with lipids to form **lipoproteins.** Lipoproteins are found in cell membranes and in the blood, where they act as carriers for less soluble molecules, such as cholesterol.

Glycosylated molecules are molecules to which a carbohydrate has been attached. Proteins combined with carbohydrates form **glycoproteins.** Lipids bound to carbohydrates become **glycolipids.** Glycoproteins and glycolipids, like lipoproteins, are important components of cell membranes [see Chapter 3].

Biochemistry of Lipids

Lipids are biomolecules made mostly of carbon and hydrogen. Most lipids have a backbone of **glycerol** and 1–3 **fatty acids**. An important characteristic of lipids is that they are nonpolar and therefore not very soluble in water. Lipids can be divided into two broad categories.

- **Fats** are solid at room temperature. Most fats are derived from animal sources.
- **Oils** are liquid at room temperature. Most plant lipids are oils.

Fatty Acids

Fatty acids are long chains of carbon atoms bound to hydrogens, with a carboxyl (–COOH) or "acid" group at one end of the chain.

Palmitic acid, a saturated fatty acid

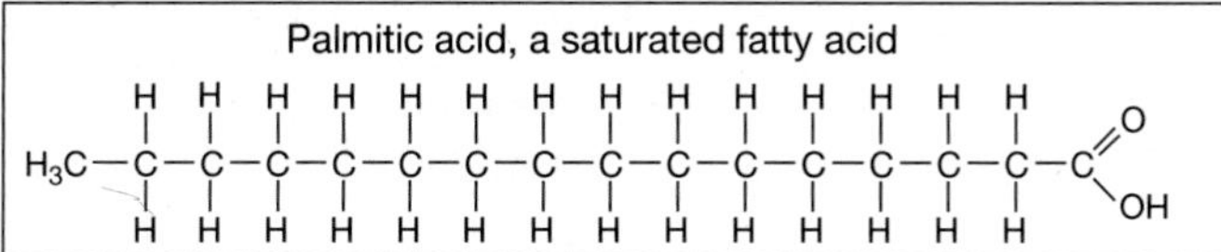

Saturated fatty acids have no double bonds between carbons, so they are "saturated" with hydrogens. The more saturated a fatty acid is, the more likely it is to be solid at room temperature.

Oleic acid, a monounsaturated fatty acid

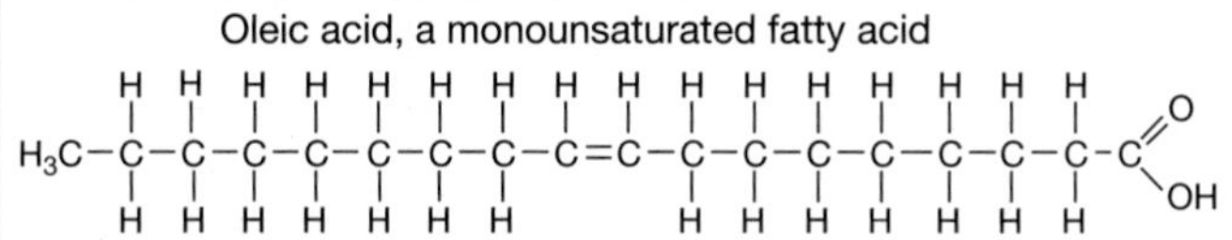

Monounsaturated fatty acids have one double bond between two of the carbons in the chain. For each double bond, the molecule has two fewer hydrogen atoms attached to the carbon chain.

Linolenic acid, a polyunsaturated fatty acid

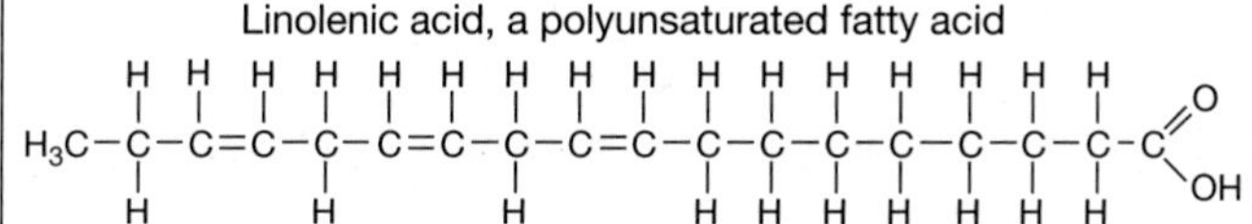

Polyunsaturated fatty acids have two or more double bonds between carbons in the chain.

Formation of Lipids

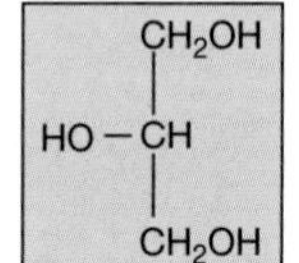

Glycerol is a simple 3-carbon molecule that makes up the backbone of most lipids.

Glycerol plus one fatty acid produces a **monoglyceride**.

Glycerol plus two fatty acids produces a **diglyceride**.

Glycerol plus three fatty acids produces a **triglyceride** (triacylglycerol). More than 90% of lipids are in the form of triglycerides.

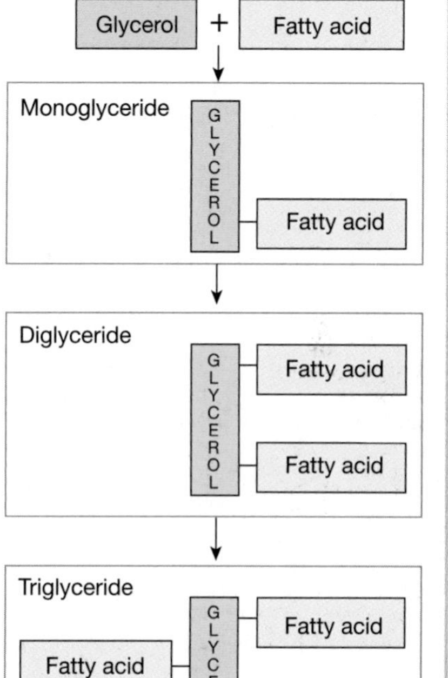

Lipid-Related Molecules

In addition to true lipids, this category includes three types of lipid-related molecules.

Eicosanoids

Eicosanoids {*eikosi*, twenty} are modified 20-carbon fatty acids with a complete or partial carbon ring at one end and two long carbon chain "tails."

O
COOH
OH OH

Prostaglandin E_2 (PGE_2)

Eicosanoids, such as thromboxanes, leukotrienes, and prostaglandins, act as regulators of physiological functions.

Steroids

Steroids are lipid-related molecules whose structure includes four linked carbon rings.

CH_3 H
H – C – CH_2 – CH_2 – CH_2 – C – CH_3
H_3C CH_3
H_3C
HO

Cholesterol is the primary source of steroids in the human body.

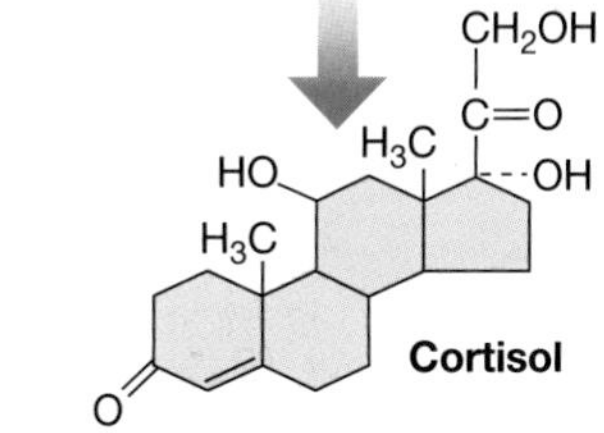

Phospholipids

Phospholipids have 2 fatty acids and a phosphate group (–H_2PO_4). Cholesterol and phospholipids are important components of animal cell membranes.

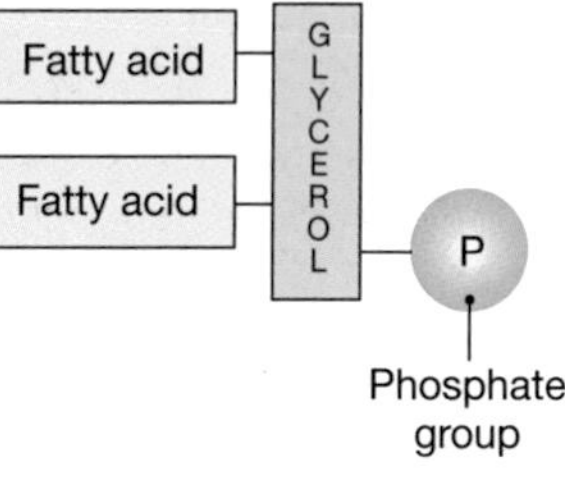

■ Fig. 2.2 *REVIEW*

Biochemistry of Carbohydrates

Carbohydrates are the most abundant biomolecule. They get their name from their structure, literally carbon {*carbo-*} with water {*hydro-*}. The general formula for a carbohydrate is $(CH_2O)_n$ or $C_nH_{2n}O_n$, showing that for each carbon there are two hydrogens and one oxygen. Carbohydrates can be divided into three categories: **monosaccharides**, **disaccharides**, and complex glucose polymers called **polysaccharides**.

Monosaccharides

Monosaccharides are simple sugars. The most common monosaccharides are the building blocks of complex carbohydrates and have either five carbons, like ribose, or six carbons, like glucose.

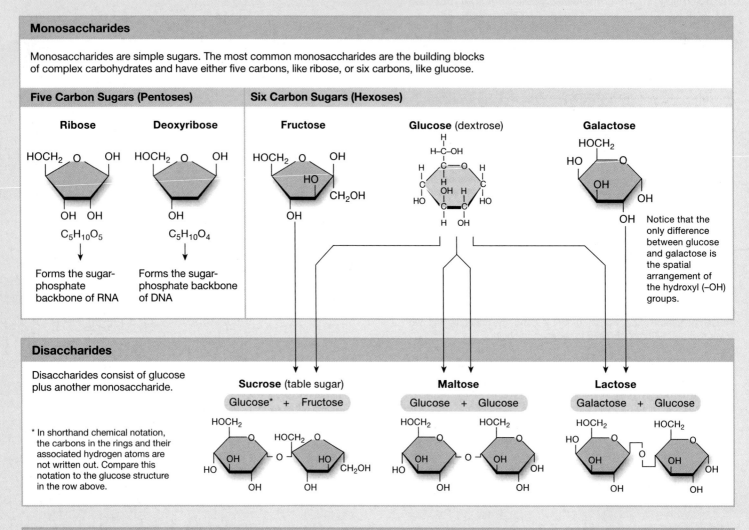

Five Carbon Sugars (Pentoses)

Ribose

HOCH$_2$ O OH

OH OH

$C_5H_{10}O_5$

↓

Forms the sugar-phosphate backbone of RNA

Deoxyribose

HOCH$_2$ O OH

OH

$C_5H_{10}O_4$

↓

Forms the sugar-phosphate backbone of DNA

Six Carbon Sugars (Hexoses)

Fructose

HOCH$_2$ O OH

HO

CH$_2$OH

OH

Glucose (dextrose)

H

H–C–OH

C—O

H H

C H C

OH H

HO C—C HO

H OH

Galactose

HOCH$_2$

HO O

OH

OH

OH

Notice that the only difference between glucose and galactose is the spatial arrangement of the hydroxyl (–OH) groups.

Disaccharides

Disaccharides consist of glucose plus another monosaccharide.

* In shorthand chemical notation, the carbons in the rings and their associated hydrogen atoms are not written out. Compare this notation to the glucose structure in the row above.

Sucrose (table sugar)

Glucose* + Fructose

HOCH$_2$ O HOCH$_2$ O

OH O HO CH$_2$OH

HO OH OH

Maltose

Glucose + Glucose

HOCH$_2$ O HOCH$_2$ O

OH O OH OH

HO OH OH

Lactose

Galactose + Glucose

HOCH$_2$ HOCH$_2$

HO O O O

OH OH OH

OH OH

Polysaccharides

Polysaccharides are glucose polymers. All living cells store glucose for energy in the form of a polysaccharide.

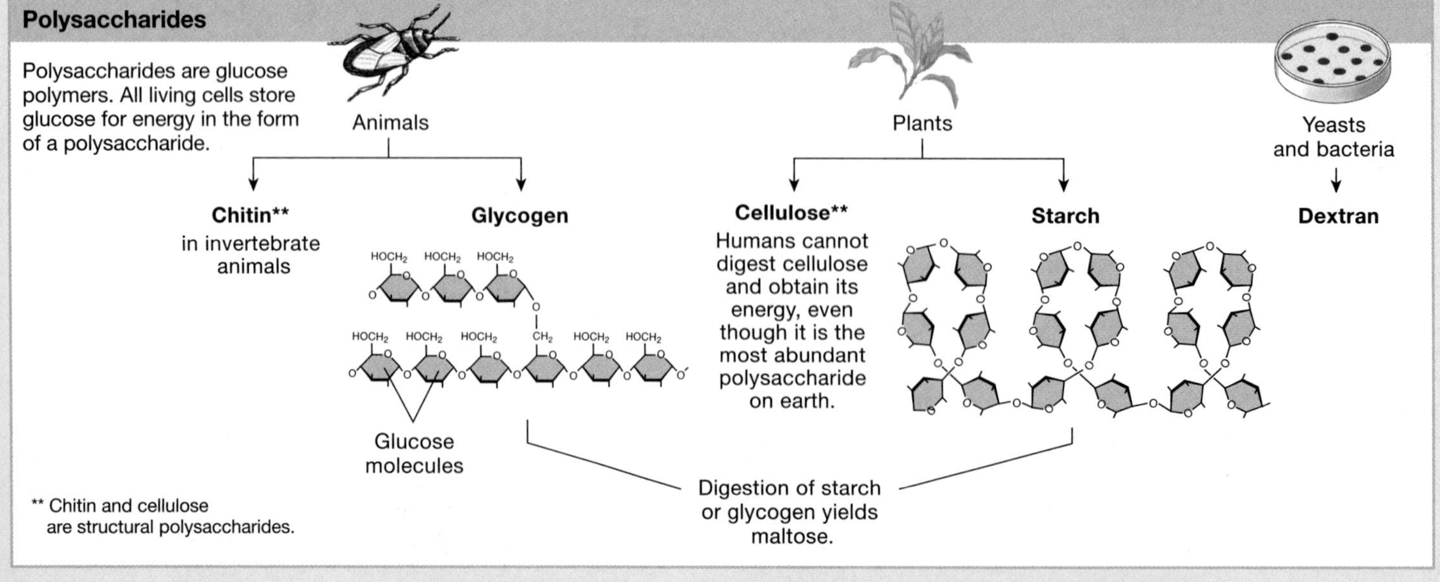

Biochemistry of Proteins

Proteins are polymers of smaller building-block molecules called **amino acids**.

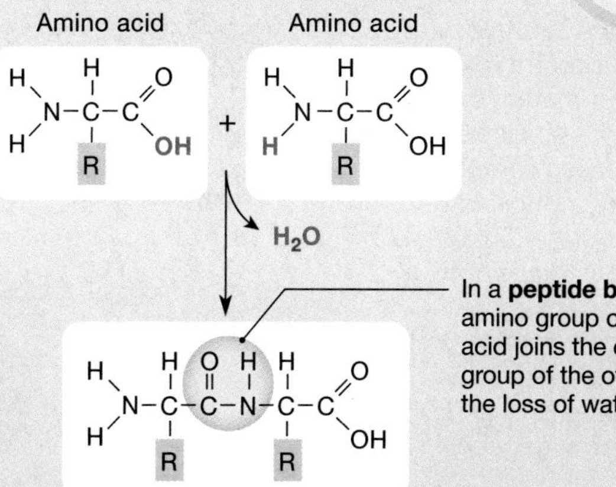

Amino Acids

All amino acids have a carboxyl group (–COOH), an amino group ($–NH_2$), and a hydrogen attached to the same carbon. The fourth bond of the carbon attaches to a variable "R" group.

The nitrogen (N) in the amino group makes proteins our major dietary source of nitrogen.

The R groups differ in their size, shape, and ability to form hydrogen bonds or ions. Because of the different R groups, each amino acid reacts with other molecules in a unique way.

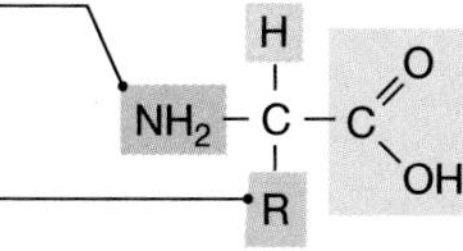

Amino Acids in Natural Proteins

Twenty different amino acids commonly occur in natural proteins. The human body can synthesize most of them, but at different stages of life some amino acids must be obtained from diet and are therefore considered *essential amino acids*.

Amino Acid	Three-Letter Abbreviation	One-Letter Symbol
Alanine	Ala	A
Arginine	Arg	R
Asparagine	Asn	N
Asparagine or aspartic acid	Asx	B
Aspartic acid	Asp	D
Cysteine	Cys	C
Glutamic acid	Glu	E
Glutamine	Gln	Q
Glutamine or glutamic acid	Glx	Z
Glycine	Gly	G
Histidine	His	H
Isoleucine	Ile	I
Leucine	Leu	L
Lysine	Lys	K
Methionine	Met	M
Phenylalanine	Phe	F
Proline	Pro	P
Serine	Ser	S
Threonine	Thr	T
Tryptophan	Trp	W
Tyrosine	Tyr	Y
Valine	Val	V

Note:

A few amino acids do not occur in proteins but have important physiological functions.

- *Homocysteine*: a sulfur-containing amino acid that in excess is associated with heart disease
- *γ-amino butyric acid* (gamma-amino butyric acid) or *GABA*: a chemical made by nerve cells
- *Creatine*: a molecule that stores energy when it binds to a phosphate group

Structure of Peptides and Proteins

Primary Structure

The 20 protein-forming amino acids assemble into polymers called peptides. The sequence of amino acids in a peptide chain is called the **primary structure**. Just as the 26 letters of our alphabet combine to create different words, the 20 amino acids can create an almost infinite number of combinations.

Peptides range in length from two to two million amino acids:

- **Oligopeptide** {*oligo*-, few}: 2–9 amino acids
- **Polypeptide**: 10–100 amino acids
- **Proteins**: >100 amino acids

Sequence of amino acids

Secondary Structure

α-helix

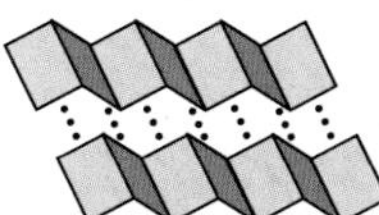

β-pleated sheets

Tertiary Structure

Chains fold.

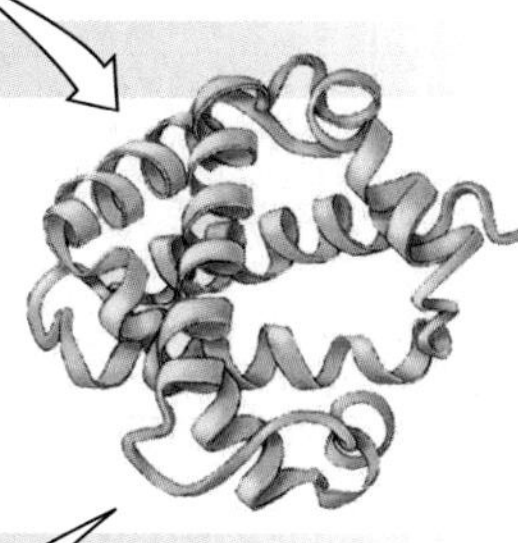

Quaternary Structure

Multiple subunits combine.

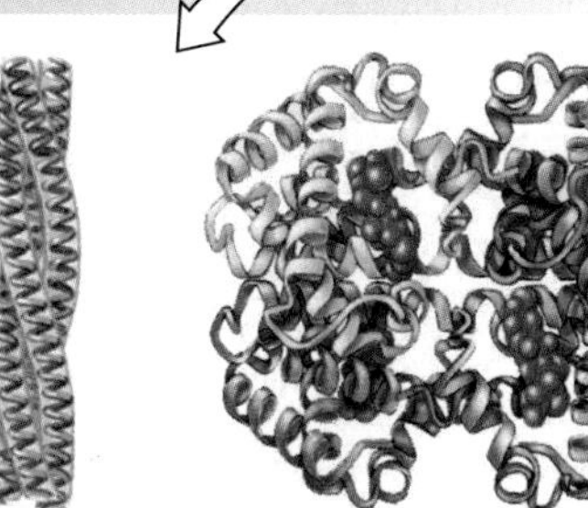

Fibrous proteins Collagen

Globular proteins Hemoglobin

Nucleotides and Nucleic Acids

Nucleotides are biomolecules that play an important role in energy and information transfer. Single nucleotides include the energy-transferring compounds **ATP** (adenosine triphosphate) and **ADP** (adenosine diphosphate), as well as **cyclic AMP**, a molecule important in the transfer of signals between cells. **Nucleic acids** (or nucleotide polymers) such as **RNA** and **DNA** store and transmit genetic information.

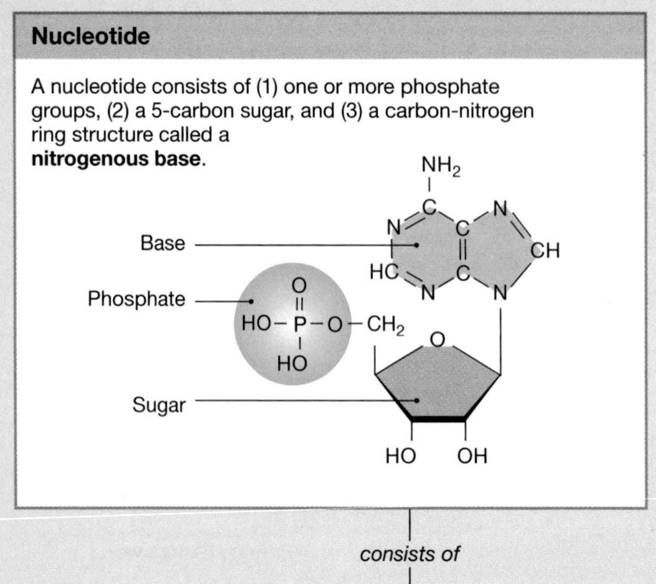

Nucleotide
A nucleotide consists of (1) one or more phosphate groups, (2) a 5-carbon sugar, and (3) a carbon-nitrogen ring structure called a **nitrogenous base**.
Base
Phosphate
Sugar
NH2
HO–P–O–CH2
HO
OH

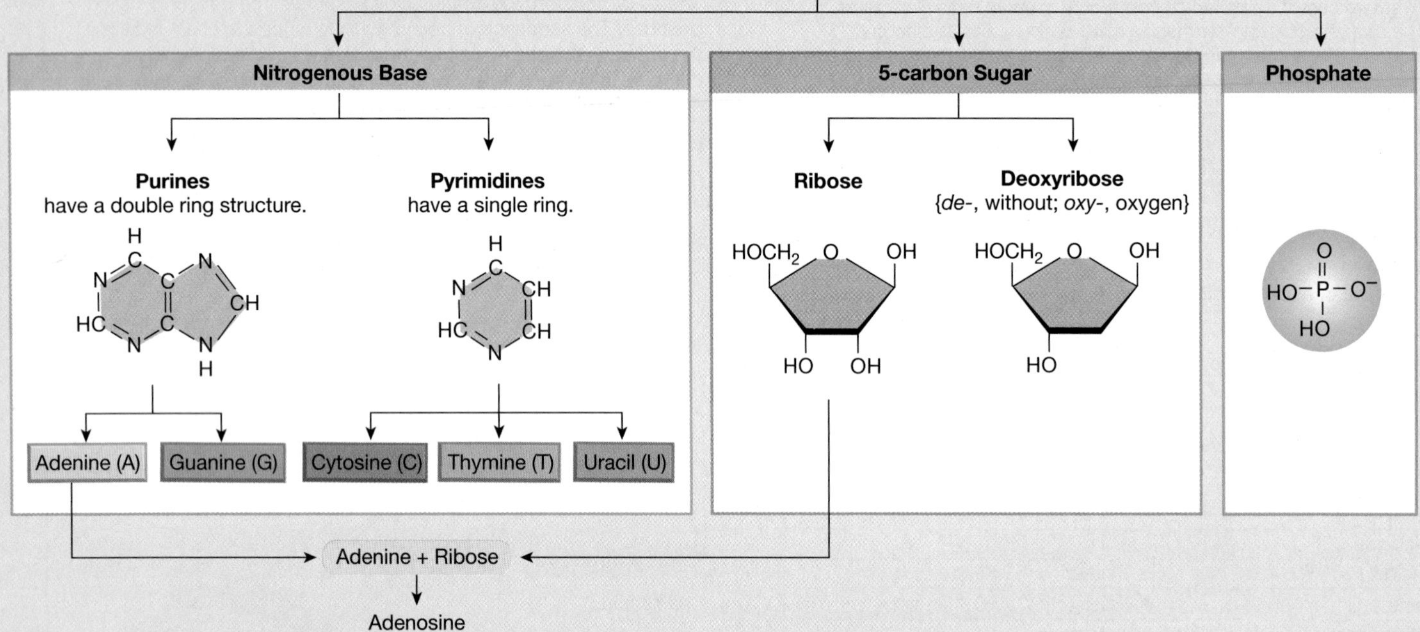

Single Nucleotide Molecules

Single nucleotide molecules have two critical functions in the human body: (1) Capture and transfer energy in high-energy electrons or phosphate bonds, and (2) aid in cell-cell communication.

Nucleotide	*consists of*	Base	+	Sugar	+	Phosphate Groups	+	Other Component	Function
ATP	=	Adenine	+	Ribose	+	3 phosphate groups			Energy capture and transfer
ADP	=	Adenine	+	Ribose	+	2 phosphate groups			Energy capture and transfer
NAD	=	Adenine	+	2 Ribose	+	2 phosphate groups	+	Nicotinamide	Energy capture and transfer
FAD	=	Adenine	+	Ribose	+	2 phosphate groups	+	Riboflavin	Energy capture and transfer
cAMP	=	Adenine	+	Ribose	+	1 phosphate group			Cell-to-cell communication

Nucleic acids (nucleotide polymers) function in information storage and transmission. The sugar of one nucleotide links to the phosphate of the next, creating a chain of alternating sugar–phosphate groups. The sugar–phosphate chains, or backbone, are the same for every nucleic acid molecule.

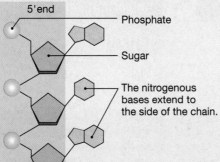

5'end

Phosphate

Sugar

The nitrogenous bases extend to the side of the chain.

3' end

Nucleotide chain. The end of the polymer that has an unbound sugar is called the 3' ("three prime") end. The end of the polymer with the unbound phosphate is called the 5' end.

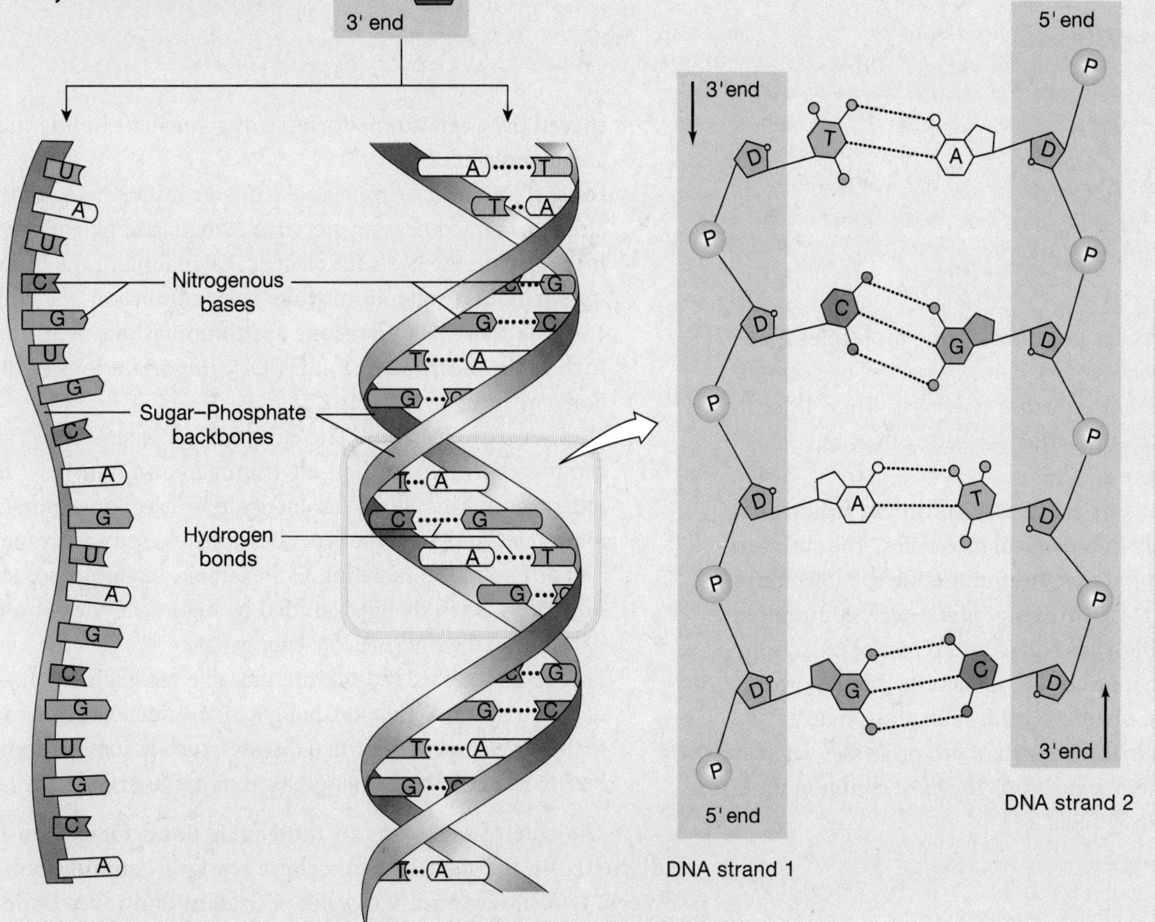

Antiparallel orientation: the 3' end of one strand is bound to the 5' end of the second strand.

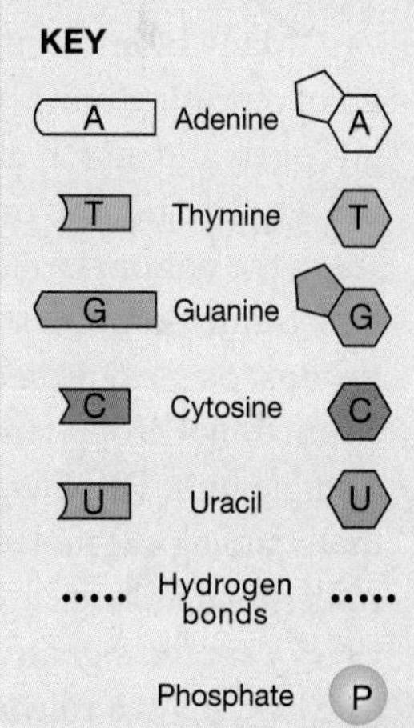

RNA (ribonucleic acid) is a single–chain nucleic acid with ribose as the sugar in the backbone, and four bases–adenine, guanine, cytosine, and *uracil*.

DNA (deoxyribonucleic acid) is a double helix, a three-dimensional structure that forms when two DNA chains link through hydrogen bonds between complementary base pairs. Deoxyribose is the sugar in the backbone, and the four bases are adenine, guanine, cytosine, and *thymine*.

Base-Pairing

Bases on one strand form hydrogen bonds with bases on the adjoining strand. This bonding follows very specific rules:

- Because purines are larger than pyrimidines, space limitations always pair a purine with a pyrimidine.
- Guanine (G) forms three hydrogen bonds with cytosine (C).
- Adenine (A) forms two hydrogen bonds with thymine (T) or uracil (U).

Guanine-Cytosine Base Pair

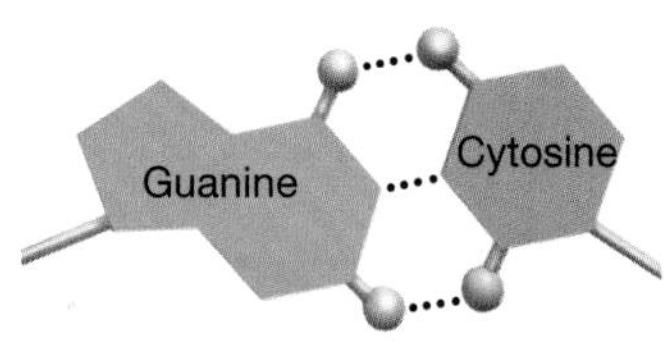

More energy is required to break the triple hydrogen bonds of G:::C than the double bonds of A::::T or A::::U.

Adenine-Thymine Base Pair

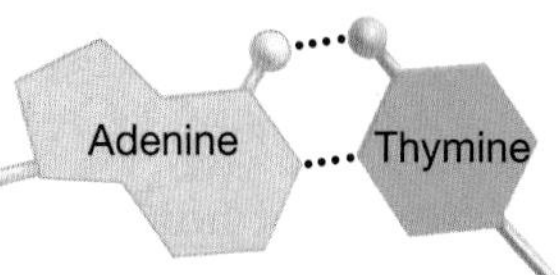

Common Functional Groups — Table 2.1

Notice that oxygen, with two electrons to share, sometimes forms a double bond with another atom.

	Shorthand	Bond Structure
Amino	$-NH_2$	$-N(H)H$
Carboxyl (acid)	$-COOH$	$-C(=O)OH$
Hydroxyl	$-OH$	$-O-H$
Phosphate	$-H_2PO_4$	$-O-P(=O)(OH)OH$

Important Ions of the Body — Table 2.2

Cations		Anions	
Na^+	Sodium	Cl^-	Chloride
K^+	Potassium	HCO_3^-	Bicarbonate
Ca^{2+}	Calcium	HPO_4^{2-}	Phosphate
H^+	Hydrogen	SO_4^{2-}	Sulfate
Mg^{2+}	Magnesium		

Many biomolecules are **polymers,** large molecules made up of repeating units {*poly*-, many + -*mer,* a part}. For example, glycogen and starch are both glucose polymers. They differ in the way the glucose molecules attach to each other, as you can see at the bottom of Figure 2.2.

Some combinations of elements, known as **functional groups,** occur repeatedly in biological molecules. The atoms in a functional group tend to move from molecule to molecule as a single unit. For example, *hydroxyl groups,* $-OH$, common in many biological molecules, are added and removed as a group rather than as single hydrogen or oxygen atoms. Amino groups, $-NH_2$, are the signature of amino acids. The phosphate group, $-H_2PO_4$, plays a role in many important cell processes, such as energy transfer and protein regulation. The most common functional groups are listed in ■ Table 2.1.

Concept Check Answers: p. 61

1. List three major essential elements found in the human body.
2. What is the general formula of a carbohydrate?
3. What is the chemical formula of an amino group? Of a carboxyl group?

Electrons Have Four Important Biological Roles

An atom of any element has a unique combination of protons and electrons that determines the element's properties (■ Fig. 2.5). We are particularly interested in the electrons because they play four important roles in physiology:

1. **Covalent bonds.** The arrangement of electrons in the outer energy level (*shell*) of an atom determines an element's ability to bind with other elements. Electrons shared between atoms form strong covalent bonds that create molecules.
2. **Ions.** If an atom or molecule gains or loses one or more electrons, it acquires an electrical charge and becomes an **ion.** Ions are the basis for electrical signaling in the body. Ions may be single atoms, like the sodium ion Na^+ and chloride ion Cl^-. Other ions are combinations of atoms, such as the bicarbonate ion HCO_3^-. Important ions of the body are listed in ■ Table 2.2.
3. **High-energy electrons.** The electrons in certain atoms can capture energy from their environment and transfer it to other atoms. This allows the energy to be used for synthesis, movement, and other life processes. The released energy may also be emitted as radiation. For example, bioluminescence in fireflies is visible light emitted by high-energy electrons returning to their normal low-energy state.
4. **Free radicals.** Free radicals are unstable molecules with an unpaired electron. They are thought to contribute to aging and to the development of certain diseases, such as some cancers. Free radicals and high-energy electrons are discussed later.

The role of electrons in molecular bond formation is discussed in the next section. There are four common bond types, two strong and two weak. Covalent and ionic bonds are strong bonds because they require significant amounts of energy to make or break. Hydrogen bonds and Van der Waals forces are weaker bonds that require much less energy to break. Interactions between molecules with different bond types are responsible for energy use and transfer in metabolic reactions as well as a variety of other reversible interactions.

Covalent Bonds Between Atoms Create Molecules

Molecules form when atoms share pairs of electrons, one electron from each atom, to create **covalent bonds.** These strong bonds require the input of energy to break them apart. It is possible to predict how many covalent bonds an atom can form by knowing how many unpaired electrons are in its outer shell, because an atom is most stable when all of its electrons are paired (■ Fig. 2.6).

RUNNING PROBLEM

What is chromium picolinate? Chromium (Cr) is an essential element that has been linked to normal glucose metabolism. In the diet, chromium is found in brewer's yeast, broccoli, mushrooms, and apples. Because chromium in food and in chromium chloride supplements is poorly absorbed from the digestive tract, a scientist developed and patented the compound chromium picolinate. Picolinate, derived from amino acids, enhances chromium uptake at the intestine. As of 2001 the recommended adequate intake (AI) of chromium for men age 19–50 is 35 μg/day. (For women it is 25 μg/day.) As we've seen, Stan takes more than 10 times this amount.

Q1: Locate chromium on the periodic table of the elements. What is chromium's atomic number? Atomic mass? How many electrons does one atom of chromium have? Which elements close to chromium are also essential elements?

32 39 43 47 49 52 57

For example, a hydrogen atom has one unpaired electron and one empty electron place in its outer shell. Because hydrogen has only one electron to share, it always forms one covalent bond. Oxygen has six electrons in an outer shell that can hold eight. That means oxygen can form two covalent bonds and fill its outer shell with electrons. If adjacent atoms share two pairs of electrons rather than just one pair, a **double bond,** represented by a double line (=), results. If two atoms share three pairs of electrons, they form a triple bond.

Polar and Nonpolar Molecules Some molecules develop regions of partial positive and negative charge when the electron pairs in their covalent bonds are not evenly shared between the linked atoms. When electrons are shared unevenly, the atom(s) with the stronger attraction for electrons develops a slight negative charge (indicated by δ^-), and the atom(s) with the weaker attraction for electrons develops a slight positive charge (δ^+). These molecules are called **polar molecules** because they can be said to have positive and negative ends, or poles. Certain elements, particularly nitrogen and oxygen, have a strong attraction for electrons and are often found in polar molecules.

A good example of a polar molecule is water (H_2O). The larger and stronger oxygen atom pulls the hydrogen electrons toward itself. This pull leaves the two hydrogen atoms of the molecule with a partial positive charge, and the single oxygen atom with a partial negative charge from the unevenly shared electrons (Fig. 2.6b). Note that the net charge for the entire water molecule is zero. The polarity of water makes it a good solvent, and all life as we know it is based on watery, or *aqueous*, solutions.

A **nonpolar molecule** is one whose shared electrons are distributed so evenly that there are no regions of partial positive or negative charge. For example, molecules composed mostly of carbon and hydrogen, such as the fatty acid shown in Figure 2.6a, tend to be nonpolar. This is because carbon does not attract electrons as strongly as oxygen does. As a result, the carbons and hydrogens share electrons evenly, and the molecule has no regions of partial charge.

Noncovalent Bonds Facilitate Reversible Interactions

Ionic bonds, hydrogen bonds, and van der Waals forces are noncovalent bonds. They play important roles in many physiological processes, including pH, molecular shape, and the reversible binding of molecules to each other.

Ionic Bonds Ions form when one atom has such a strong attraction for electrons that it pulls one or more electrons completely away from another atom. For example, a chlorine atom needs only one electron to fill the last of eight places in its outer shell, so it pulls an electron from a sodium atom, which has only one weakly held electron in its outer shell (Fig. 2.6c). The atom that gains electrons acquires one negative charge (−1) for each electron added, so the chlorine atom becomes a chloride ion Cl^-. Negatively charged ions are called **anions.**

An atom that gives up electrons has one positive charge (+1) for each electron lost. For example, the sodium atom becomes a sodium ion Na^+. Positively charged ions are called **cations.**

Ionic bonds, also known as *electrostatic attractions*, result from the attraction between ions with opposite charges. (Remember the basic principle of electricity that says that opposite charges attract and like charges repel.) In a crystal of table salt, the solid form of ionized NaCl, ionic bonds between alternating Na^+ and Cl^- ions hold the ions in a neatly ordered structure.

Hydrogen Bonds A **hydrogen bond** is a weak attractive force between a hydrogen atom and a nearby oxygen, nitrogen, or fluorine atom. No electrons are gained, lost, or shared in a hydrogen bond. Instead, the oppositely charged regions in polar molecules are attracted to each other. Hydrogen bonds may occur between atoms in neighboring molecules or between atoms in different parts of the same molecule. For example, one water molecule may hydrogen-bond with as many as four other water molecules. As a result, the molecules line up with their neighbors in a somewhat ordered fashion (Fig. 2.6d).

Hydrogen bonding between molecules is responsible for the **surface tension** of water. Surface tension is the attractive force between water molecules that causes water to form spherical droplets when falling or to bead up when spilled onto a nonabsorbent surface (Fig. 2.6d). The high cohesiveness {*cohaesus*, to cling together} of water makes it difficult to stretch or deform, as you may have noticed in trying to pick up a wet glass that is "stuck" to a slick table top by a thin film of water. The surface tension of water influences lung function [described in Chapter 17].

Atoms and Molecules

Elements are the simplest type of matter. There are over 100 known elements,* but only 3—oxygen, carbon, and hydrogen—make up more than 90% of the body's mass. These 3 plus 8 additional elements are *major essential elements*. An additional 19 *minor essential elements* are required in trace amounts. The smallest particle of any element is an **atom** {*atomos*, indivisible}. Atoms link by sharing electrons to form molecules.

* A periodic table of the elements can be found inside the back cover of the book.

Major Essential Elements	Minor Essential Elements
H, C, O, N, Na, Mg, K, Ca, P, S, Cl	Li, F, Cr, Mn, Fe, Co, Ni, Cu, Zn, Se, Y, I, Zr, Nb, Mo, Tc, Ru, Rh, La

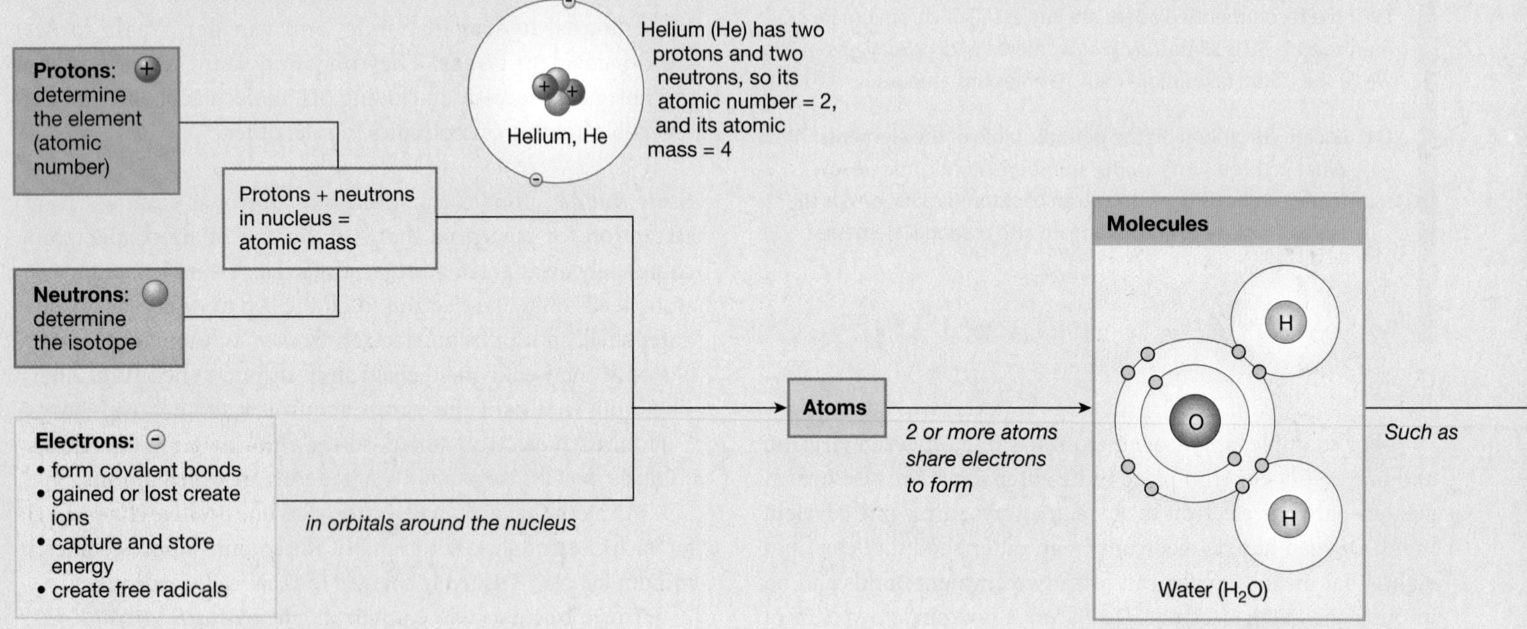

Protons: determine the element (atomic number)
Neutrons: determine the isotope
Electrons:
• form covalent bonds
• gained or lost create ions
• capture and store energy
• create free radicals
Protons + neutrons in nucleus = atomic mass
in orbitals around the nucleus
Helium, He
Helium (He) has two protons and two neutrons, so its atomic number = 2, and its atomic mass = 4
Atoms
2 or more atoms share electrons to form
Molecules
H
O
H
Water (H_2O)
Such as

Isotopes and Ions

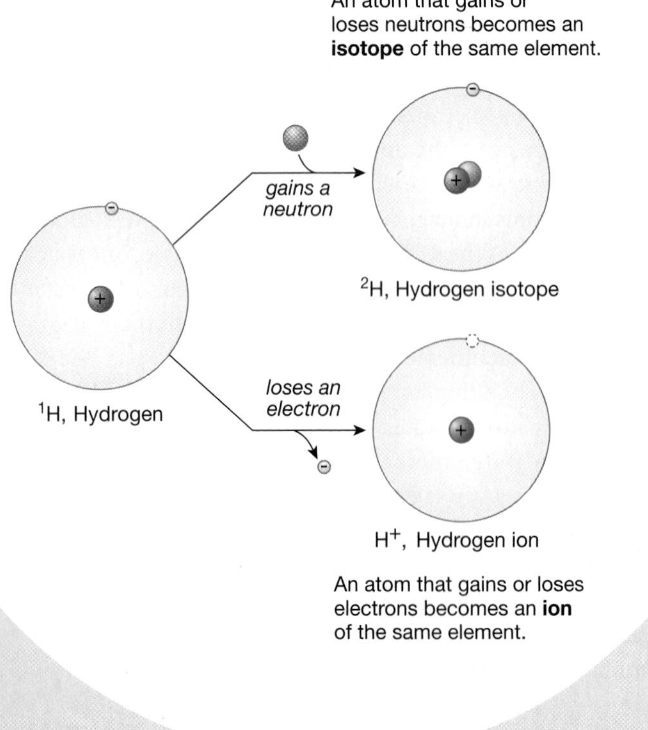

An atom that gains or loses neutrons becomes an **isotope** of the same element.
gains a neutron
^2H, Hydrogen isotope
^1H, Hydrogen
loses an electron
H^+, Hydrogen ion
An atom that gains or loses electrons becomes an **ion** of the same element.

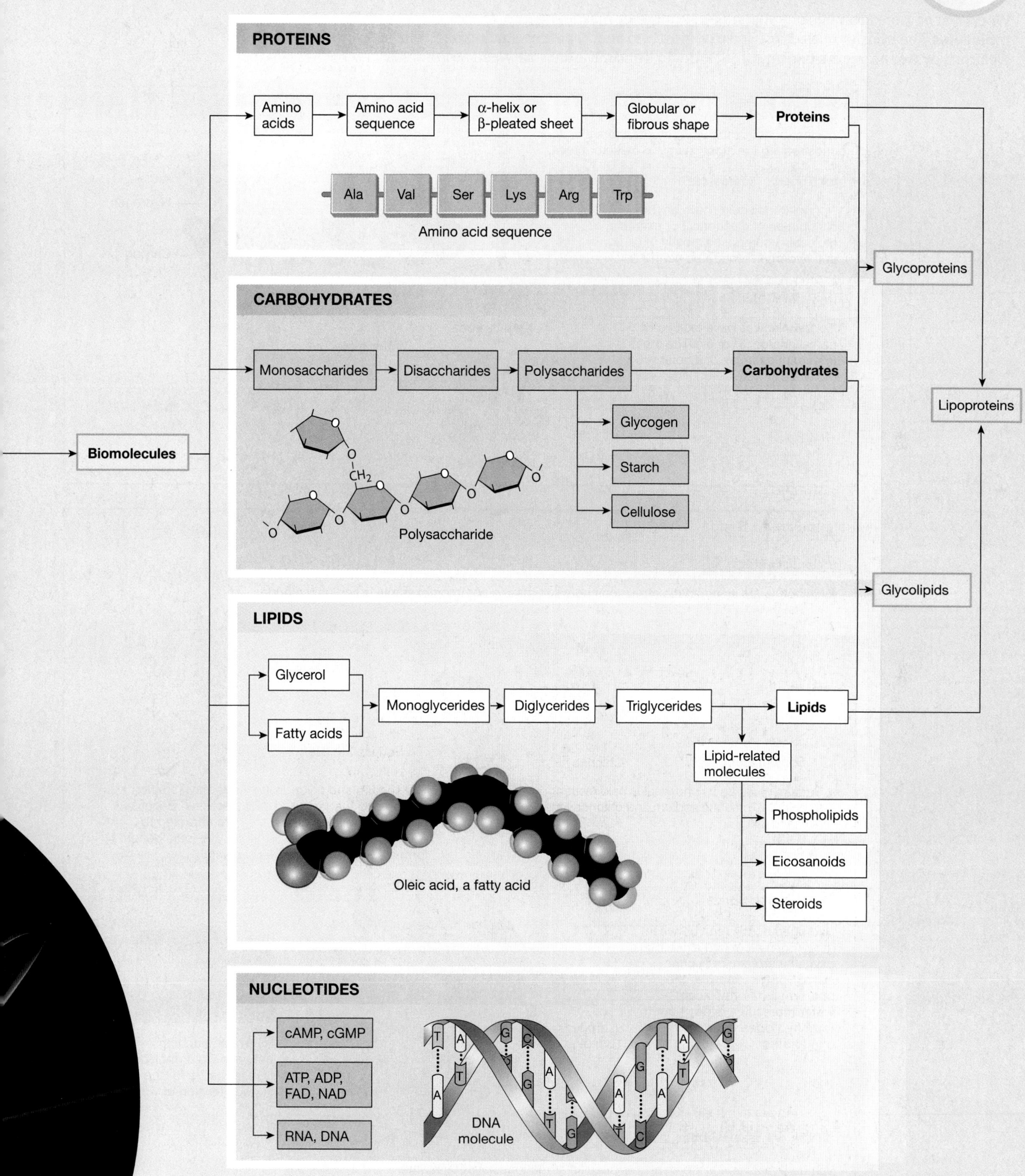

Biomolecules

PROTEINS

Amino acids → Amino acid sequence → α-helix or β-pleated sheet → Globular or fibrous shape → **Proteins**

Ala – Val – Ser – Lys – Arg – Trp

Amino acid sequence

CARBOHYDRATES

Monosaccharides → Disaccharides → Polysaccharides → **Carbohydrates**

Polysaccharides → Glycogen, Starch, Cellulose

CH_2

Polysaccharide

Proteins + Carbohydrates → Glycoproteins

Proteins + Lipids → Lipoproteins

Carbohydrates + Lipids → Glycolipids

LIPIDS

Glycerol, Fatty acids → Monoglycerides → Diglycerides → Triglycerides → **Lipids**

Lipid-related molecules → Phospholipids, Eicosanoids, Steroids

Oleic acid, a fatty acid

NUCLEOTIDES

cAMP, cGMP

ATP, ADP, FAD, NAD

RNA, DNA

DNA molecule

Molecular Bonds

When two or more atoms link by sharing electrons, they make units known as **molecules**. The transfer of electrons from one atom to another or the sharing of electrons by two atoms is a critical part of forming **bonds**, the links between atoms.

Bonds

Covalent Bonds

Covalent bonds result when atoms share electrons. These bonds require the most energy to make or break.

(a) Nonpolar Molecules

Nonpolar molecules have an even distribution of electrons. For example, molecules composed mostly of carbon and hydrogen tend to be nonpolar.

(b) Polar Molecules

Polar molecules have regions of partial charge (δ+ or δ−). The most important example of a polar molecule is water.

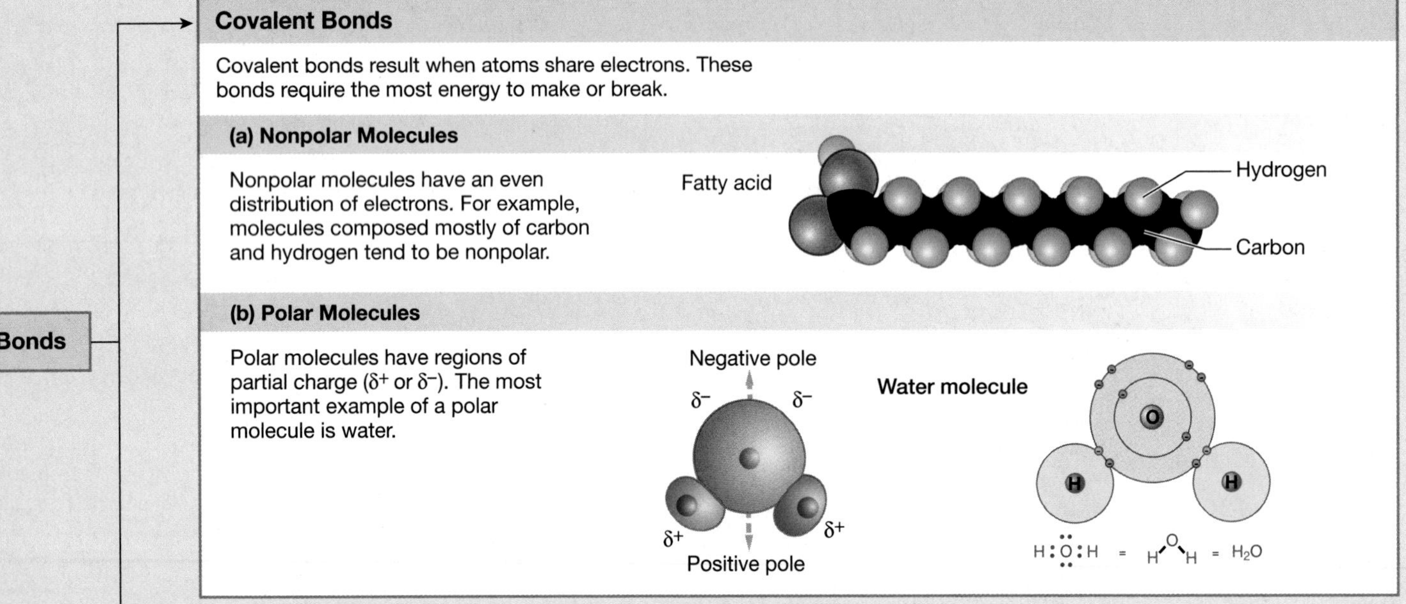

Noncovalent Bonds

(c) Ionic Bonds

Ionic bonds are electrostatic attractions between ions. A common example is sodium chloride.

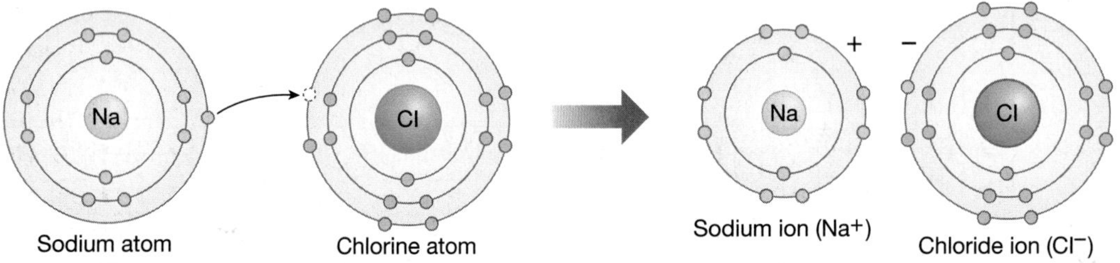

Sodium gives up its one weakly held electron to chlorine, creating sodium and chloride ions, Na^+ and Cl^-.

The sodium and chloride ions both have stable outer shells that are filled with electrons. Because of their opposite charges, they are attracted to each other and, in the solid state, the ionic bonds form a sodium chloride (NaCl) crystal.

(d) Hydrogen Bonds

Hydrogen bonds form between a hydrogen atom and a nearby oxygen, nitrogen, or fluorine atom. So, for example, the polar regions of adjacent water molecules allow them to form hydrogen bonds with one another.

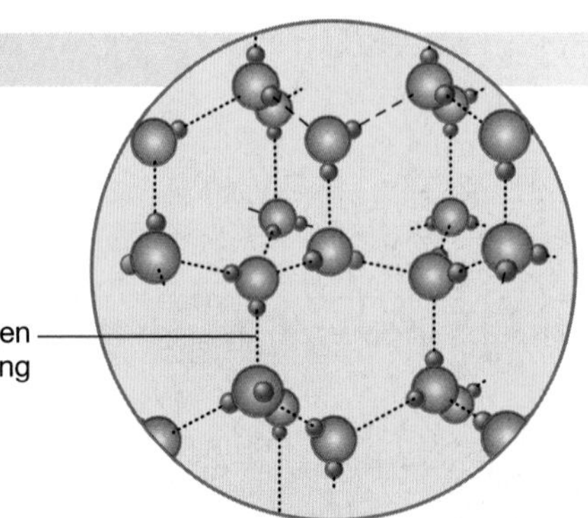

Hydrogen bonding between water molecules is responsible for the surface tension of water.

(e) Van der Waals Forces

Van der Waals forces are weak, nonspecific attractions between atoms.

Van der Waals Forces **Van der Waals forces** are weak, nonspecific attractions between the nucleus of any atom and the electrons of nearby atoms. Two atoms that are weakly attracted to each other by van der Waals forces move closer together until they are so close that their electrons begin to repel one another. Consequently, van der Waals forces allow atoms to pack closely together and occupy a minimum amount of space. A single van der Waals attraction between atoms is very weak.

Concept Check Answers: p. 61

4. Are electrons in an atom or molecule most stable when they are paired or unpaired?
5. When an atom of an element gains or loses one or more electrons, it is called a(n) ________ of that element.
6. Match each type of bond with its description:

(a) covalent bond	1. weak attractive force between hydrogen and oxygen or nitrogen
(b) ionic bond	2. formed when two atoms share one or more pairs of electrons
(c) hydrogen bond	3. weak attractive force between atoms
(d) van der Waals force	4. formed when one atom loses one or more electrons to a second atom

RUNNING PROBLEM

One advertising claim for chromium is that it improves the transfer of glucose—the simple sugar that cells use to fuel all their activities—from the bloodstream into cells. In people with diabetes mellitus, cells are unable to take up glucose from the blood efficiently. It seemed logical, therefore, to test whether the addition of chromium to the diet would enhance glucose uptake in people with diabetes. In one Chinese study, diabetic patients receiving 500 micrograms of chromium picolinate twice a day showed significant improvement in their diabetes, but patients receiving 100 micrograms or a placebo did not.

Q2: If people have a chromium deficiency, would you predict that their blood glucose level would be lower or higher than normal? From the results of the Chinese study, can you conclude that all people with diabetes suffer from a chromium deficiency?

32 39 43 47 49 52 57

Noncovalent Interactions

Many different kinds of noncovalent interactions can take place between and within molecules as a result of the four different types of bonds. For example, the charged, uncharged, or partially charged nature of a molecule determines whether that molecule can dissolve in water. Covalent and noncovalent bonds determine molecular shape and function. Finally, noncovalent interactions mediate the reversible association of proteins with other molecules, creating functional pairings such as enzymes and substrates, or signal receptors and molecules.

Hydrophilic Interactions Create Biological Solutions

Life as we know it is established on water-based, or *aqueous,* solutions that resemble dilute seawater in their ionic composition. The adult human body is about 60% water. Na^+, K^+, and Cl^- are the main ions in body fluids, with other ions making up a lesser proportion. All molecules and cell components are either dissolved or suspended in these solutions. For these reasons it is useful to understand the properties of solutions, which are reviewed in ■ Figure 2.7.

The degree to which a molecule is able to dissolve in a solvent is the molecule's **solubility:** the more easily a molecule dissolves, the higher its solubility. Water, the biological solvent, is polar, so molecules that dissolve readily in water are polar or ionic molecules whose positive and negative regions readily interact with water. For example, if NaCl crystals are placed in water, polar regions of the water molecules disrupt the ionic bonds between sodium and chloride, which causes the crystals to dissolve (■ Fig. 2.8a). Molecules that are soluble in water are said to be **hydrophilic** {*hydro-*, water + *-philic,* loving}.

In contrast, molecules such as oils that do not dissolve well in water are said to be **hydrophobic** {*-phobic,* hating}. Hydrophobic substances are usually nonpolar molecules that cannot form hydrogen bonds with water molecules. The lipids (fats and oils) are the most hydrophobic group of biological molecules.

When placed in an aqueous solution, lipids do not dissolve. Instead they separate into distinct layers. One familiar example is salad oil floating on vinegar in a bottle of salad dressing. Before hydrophobic molecules can dissolve in body fluids, they must combine with a hydrophilic molecule that will carry them into solution.

For example, cholesterol, a common animal fat, is a hydrophobic molecule. Fat from a piece of meat dropped into a glass of warm water will float to the top, undissolved. In the blood, cholesterol will not dissolve unless it binds to special water-soluble carrier molecules. You may know the combination of cholesterol with its hydrophilic carriers as HDL-cholesterol and LDL-cholesterol, the "good" and "bad" forms of cholesterol associated with heart disease.

Some molecules, such as the phospholipids, have both polar and nonpolar regions (Fig. 2.8b). This dual nature allows them to associate both with each other (hydrophobic interactions) and with polar water molecules (hydrophilic interactions). Phospholipids are the primary component of biological membranes.

■ Fig. 2.7 *REVIEW*

Solutions

Life as we know it is established on water-based, or aqueous, solutions that resemble dilute seawater in their ionic composition. The human body is 60% water. Sodium, potassium, and chloride are the main ions in body fluids. All molecules and cell components are either dissolved or suspended in these saline solutions. For these reasons, the properties of solutions play a key role in the functioning of the human body.

TERMINOLOGY

A **solute** is any substance that dissolves in a liquid. The degree to which a molecule is able to dissolve in a solvent is the molecule's **solubility**. The more easily a solute dissolves, the higher its solubility.

A **solvent** is the liquid into which solutes dissolve. In biological solutions, water is the universal solvent.

A **solution** is the combination of solutes dissolved in a solvent. The **concentration** of a solution is the amount of solute per unit volume of solution.

Concentration = solute amount/volume of solution

EXPRESSIONS OF SOLUTE AMOUNT

- **Mass** (weight) of the solute before it dissolves. Usually given in grams (g) or milligrams (mg).
- **Molecular mass** is calculated from the chemical formula of a molecule. This is the mass of one molecule, expressed in atomic mass units (amu) or, more often, in daltons (Da), where 1 amu = 1 Da.

Molecular mass = SUM [atomic mass of each element × the number of atoms of each element]

Example

What is the molecular mass of glucose, $C_6H_{12}O_6$?

Answer

Element	# of Atoms	Atomic Mass of Element
Carbon	6	12.0 amu × 6 = 72
Hydrogen	12	1.0 amu × 12 = 12
Oxygen	6	16.0 amu × 6 = 96

Molecular mass of glucose = 180 amu (or Da)

- **Moles** (mol) are an expression of the number of solute molecules, without regard for their weight. One mole = 6.02×10^{23} atoms, ions, or molecules of a substance. One mole of a substance has the same number of particles as one mole of any other substance, just as a dozen eggs has the same number of items as a dozen roses.
- **Gram molecular weight.** In the laboratory, we use the molecular mass of a substance to measure out moles. For example, one mole of glucose (with 6.02×10^{23} glucose molecules) has a molecular mass of 180 Da and weighs 180 grams. The molecular mass of a substance expressed in grams is called the gram molecular weight.
- **Equivalents** (eq) are a unit used for ions, where 1 equivalent = molarity of the ion × the number of charges the ion carries. The sodium ion, with its charge of +1, has one equivalent per mole. The hydrogen phosphate ion (HPO_4^{2-}) has two equivalents per mole. Concentrations of ions in the blood are often reported in milliequivalents per liter (meq/L).

FIGURE QUESTIONS

1. What are the two components of a solution?
2. The concentration of a solution is expressed as:
   (a) amount of solvent/volume of solute
   (b) amount of solute/volume of solvent
   (c) amount of solvent/volume of solution
   (d) amount of solute/volume of solution
3. Calculate the molecular mass of water, H_2O.
4. How much does a mole of KCl weigh?

EXPRESSIONS OF VOLUME

Volume is usually expressed as liters (L) or milliliters (mL) {*milli-*, 1/1000}. A volume convention common in medicine is the deciliter (dL), which is 1/10 of a liter, or 100 mL.

Prefixes

deci- (d)	1/10	1×10^{-1}
milli- (m)	1/1000	1×10^{-3}
micro- (μ)	1/1,000,000	1×10^{-6}
nana- (n)	1/1,000,000,000	1×10^{-9}
pico- (p)	1/1,000,000,000,000	1×10^{-12}

Useful Conversions

- 1 liter of water weighs 1 kilogram (kg) {*kilo*-, 1000}
- 1 kilogram = 2.2 pounds

EXPRESSIONS OF CONCENTRATION

- **Percent solutions.** In a laboratory or pharmacy, scientists cannot measure out solutes by the mole. Instead, they use the more conventional measurement of weight. The solute concentration may then be expressed as a percentage of the total solution, or **percent solution**. A 10% solution means 10 parts of a solute per 100 parts of total solution. Weight/volume solutions, used for solutes that are solids, are usually expressed as g/100 mL solution or mg/dL. An out-of-date way of expressing mg/dL is mg% where % means per 100 parts or 100 mL. A concentration of 20 mg/dL could also be expressed as 20 mg%.

Example

Solutions used for intravenous (IV) infusions are often expressed as percent solutions. How would you make 500 mL of a 5% dextrose (glucose) solution?	**Answer** 5% solution = 5 g glucose dissolved in water to make a final volume of 100 mL solution. 5 g glucose/100 mL = ? g/500 mL 25 g glucose with water added to give a final volume of 500 mL

- **Molarity** is the number of moles of solute in a liter of solution, and is abbreviated as either mol/L or M. A one molar solution of glucose (1 mol/L, 1 M) contains 6.02×10^{23} molecules of glucose per liter of solution. It is made by dissolving one mole (180 grams) of glucose in enough water to make one liter of solution. Typical biological solutions are so dilute that solute concentrations are usually expressed as **millimoles** per liter (mmol/L or mM).

Example

What is the molarity of a 5% dextrose solution?	**Answer** 5 g glucose/100 mL = 50 g glucose/1000 mL (or 1 L) 1 mole glucose = 180 g glucose 50 g/L × 1 mole/180 g = 0.278 moles/L or 278 mM

FIGURE QUESTIONS

5. Which solution is more concentrated: a 100 mM solution of glucose or a 0.1 M solution of glucose?
6. When making a 5% solution of glucose, why don't you measure out 5 grams of glucose and add it to 100 mL of water?

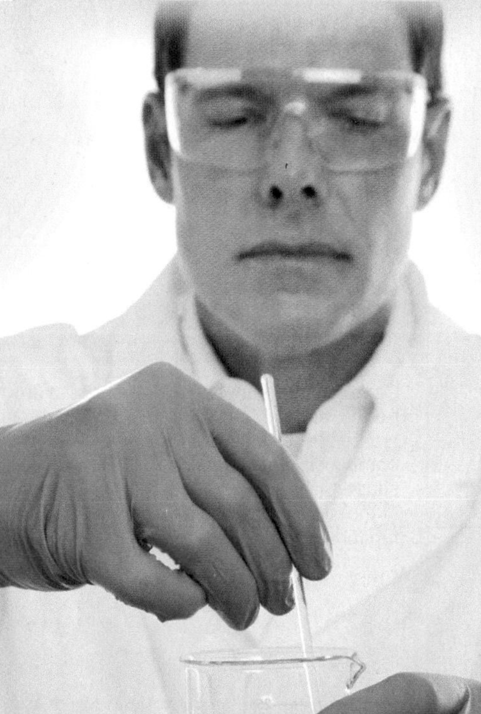

Molecular Interactions

(a) Hydrophilic Interactions

Molecules that have polar regions or ionic bonds readily interact with the polar regions of water. This enables them to dissolve easily in water. Molecules that dissolve readily in water are said to be **hydrophilic** {*hydro-*, water + *philos*, loving}.

Water molecules interact with ions or other polar molecules to form hydration shells around the ions. This disrupts the hydrogen bonding between water molecules, thereby lowering the freezing temperature of water (freezing point depression).

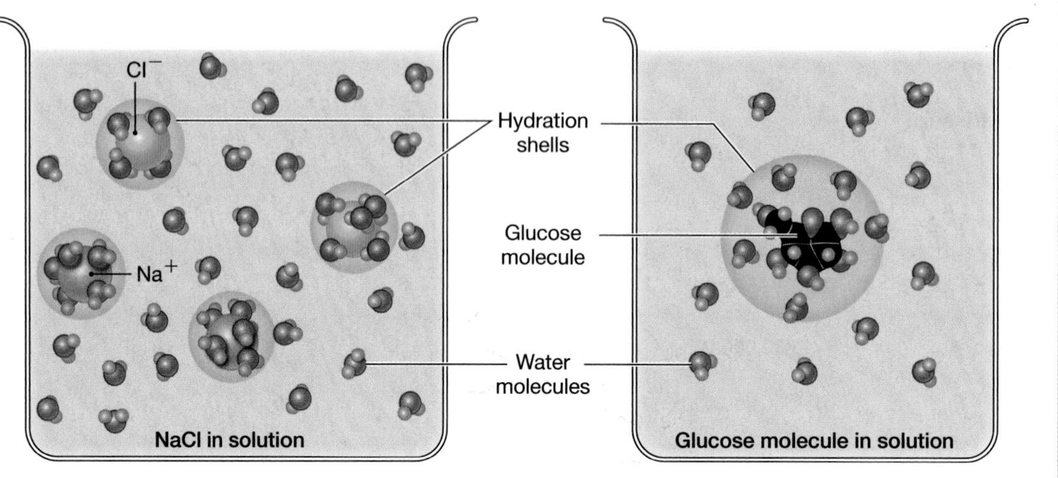

(b) Hydrophobic Interactions

Because they have an even distribution of electrons and no positive or negative poles, nonpolar molecules have no regions of partial charge, and therefore tend to repel water molecules. Molecules like these do not dissolve readily in water and are said to be **hydrophobic** {*hydro-*, water + *phobos*, fear}. Molecules such as phospholipids have both polar and nonpolar regions that play critical roles in biological systems and in the formation of biological membranes.

Phospholipid molecules have polar heads and nonpolar tails.

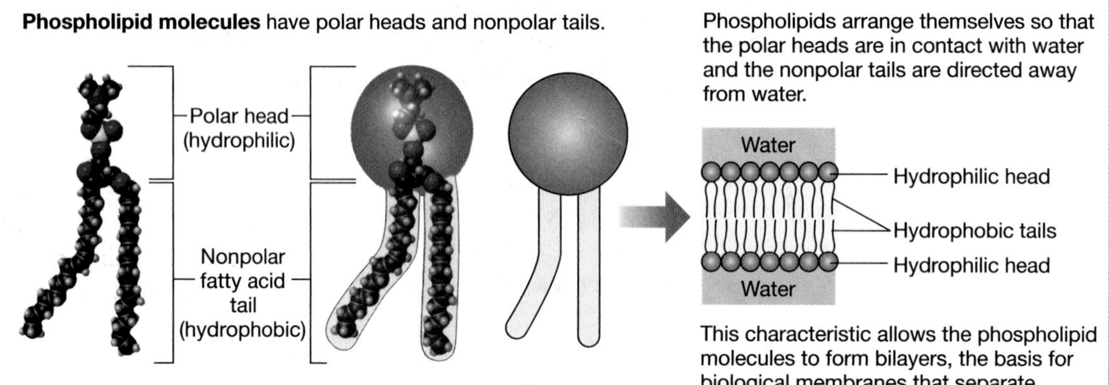

Phospholipids arrange themselves so that the polar heads are in contact with water and the nonpolar tails are directed away from water.

This characteristic allows the phospholipid molecules to form bilayers, the basis for biological membranes that separate compartments.

(c) Molecular Shape

Covalent bond angles, ionic bonds, hydrogen bonds, and van der Waals forces all interact to create the distinctive shape of a complex biomolecule. This shape plays a critical role in the molecule's function.

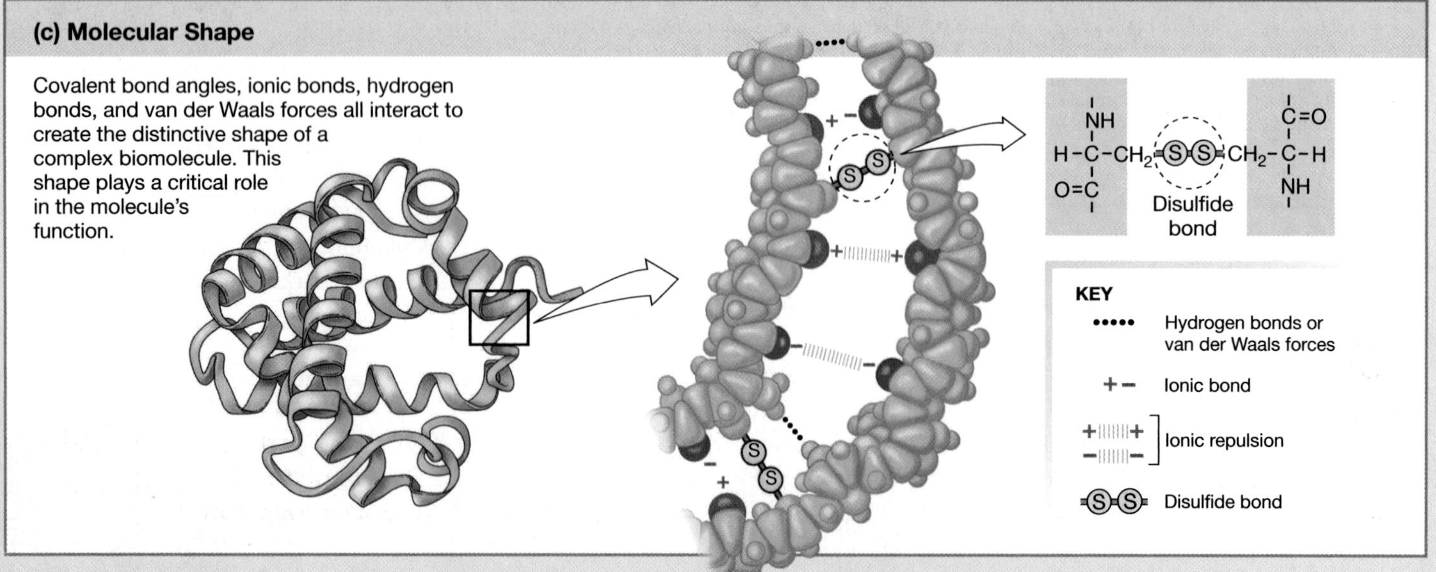

Concept Check Answers: p. 61

7. Which dissolve more easily in water, polar molecules or nonpolar molecules?
8. A molecule that dissolves easily is said to be hydro ________ ic.
9. Why does table salt (NaCl) dissolve in water?

RUNNING PROBLEM

Chromium is found in several ionic forms. The chromium usually found in biological systems and in dietary supplements is the cation Cr^{3+}. This ion is called trivalent because it has a net charge of +3. The hexavalent cation, Cr^{6+}, with a charge of +6, is used in industry, such as in the manufacturing of stainless steel and the chrome plating of metal parts.

Q3: How many electrons have been lost from the hexavalent ion of chromium? From the trivalent ion?

32 39 43 47 49 52 57

2

Molecular Shape Is Related to Molecular Function

A molecule's shape is closely related to its function. Molecular bonds—both covalent bonds and weak bonds—play a critical role in determining molecular shape. The three-dimensional shape of a molecule is difficult to show on paper, but many molecules have characteristic shapes due to the angles of covalent bonds between the atoms. For example, the two hydrogen atoms of the water molecule shown in Figure 2.6b are attached to the oxygen with a bond angle of 104.5°. Double bonds in long carbon chain fatty acids cause the chains to kink or bend, as shown by the three-dimensional model of oleic acid in Figure 2.5.

Weak noncovalent bonds also contribute to molecular shape. The complex double helix of a DNA molecule (Fig. 2.4) results both from covalent bonds between adjacent bases in each strand and the hydrogen bonds connecting the two strands of the helix.

Proteins have the most complex and varied shapes of all the biomolecules. The two most common **secondary structures** for polypeptide chains are the **α-helix** (alpha-helix) spiral and **β-pleated** (beta-pleated) **sheets** (Fig. 2.3). The covalent bond angles between amino acids create the spiral of the α-helix or the zigzag shape of the individual *β-strands* in the pleated sheet. Adjacent β-strands are stabilized by hydrogen bonding, shown as . . . in Figure 2.3. The pleated sheet configuration is very stable and occurs in many proteins destined for structural uses. Proteins with other functions may have a mix of β-pleated sheets and α-helices.

Proteins are categorized into two large groups based on their three-dimensional shape: fibrous and globular (see Fig. 2.3). **Fibrous proteins** may be β-pleated sheets or long chains of α-helices. The fibrous proteins are insoluble in water and form important structural components of cells and tissues. Examples include *collagen,* a fibrous protein found in many types of connective tissue, such as skin, and *keratin,* a fibrous protein found in hair and nails.

Globular proteins have amino acid chains that fold back on themselves to create a complex **tertiary structure** containing pockets, channels, or protruding knobs. The tertiary structure of globular proteins arises partly from the angles of covalent bonds between amino acids and partly from hydrogen bonds, van der Waals forces, and ionic bonds that stabilize the molecule's shape.

In addition to noncovalent bonds, covalent **disulfide** (S-S) **bonds** play an important role in the shape of many globular proteins (Fig. 2.8c). The amino acid *cysteine* contains sulfur as part of a *sulfhydryl group* (—SH). Two cysteines in different parts of the polypeptide chain can bond to each other with a disulfide bond that pulls the sections of chain together.

Hydrogen Ions in Solution Can Alter Molecular Shape

Hydrogen bonding is an important part of molecular shape. However, free hydrogen ions, H^+, in solution can also participate in hydrogen bonding and van der Waals forces. If free H^+ disrupts a molecule's noncovalent bonds, the molecule's shape, or *conformation*, can change. A change in shape may alter or destroy the molecule's ability to function.

The concentration of free H^+ in body fluids, or *acidity*, is measured in terms of **pH.** ■ Figure 2.9 reviews the chemistry of pH and shows a pH scale with the pH values of various substances. The normal pH of blood in the human body is 7.40, slightly alkaline. Regulation of the body's pH within a narrow range is critical because a blood pH more acidic than 7.00 ($pH < 7.00$) or more alkaline than 7.70 ($pH > 7.70$) is incompatible with life.

Where do hydrogen ions in body fluids come from? Some of them come from the separation of water molecules (H_2O) into H^+ and OH^- ions. Others come from **acids,** molecules that release H^+ when they dissolve in water (Fig. 2.9). Many of the molecules made during normal metabolism are acids. For example, carbonic acid is made in the body from CO_2 (carbon dioxide) and water. In solution, carbonic acid separates into a bicarbonate ion and a hydrogen ion:

$$CO_2 + H_2O \rightleftharpoons H_2CO_3 \text{ (carbonic acid)} \rightleftharpoons H^+ + HCO_3^-$$

Note that when the hydrogen is part of the intact carbonic acid molecule, it does not contribute to acidity. *Only free H^+ contributes to the hydrogen ion concentration.*

■ Fig. 2.9 *REVIEW*

pH

ACIDS AND BASES

An **acid** is a molecule that contributes H^+ to a solution.

- The carboxyl group, –COOH, is an acid because in solution it tends to lose its H^+:

$$R–COOH \rightarrow R–COO^- + H^+$$

A **base** is a molecule that decreases the H^+ concentration of a solution by combining with free H^+.

- Molecules that produce hydroxide ions, OH^-, in solution are bases because the hydroxide combines with H^+ to form water:

$$R–OH \rightarrow R^+ + OH^- \rightarrow OH^- + H^+ \rightarrow H_2O$$

- Another molecule that acts as a base is ammonia, NH_3. It reacts with a free H^+ to form an ammonium ion:

$$NH_3 + H^+ \rightarrow NH_4^+$$

pH

The concentration of H^+ in body fluids is measured in terms of **pH.**

- The expression pH stands for "power of hydrogen."

1 $$pH = -\log [H^+]$$

This equation is read as "pH is equal to the negative log of the hydrogen ion concentration." Square brackets are shorthand notation for "concentration" and by convention, concentration is expressed in meq/L.

- Using the rule of logarithms that says $-\log x = \log(1/x)$, pH equation (1) can be rewritten as:

2 $$pH = \log (1/[H^+])$$

This equation shows that pH is inversely related to H^+ concentration. In other words, as the H^+ concentration goes up, the pH goes down.

Example

What is the pH of a solution whose hydrogen ion concentration $[H^+]$ is 10^{-7} meq/L?

Answer

$$pH = -\log [H^+]$$
$$pH = -\log [10^{-7}]$$

Using the rule of logs, this can be rewritten as

$$pH = \log (1/10^{-7})$$

Using the rule of exponents that says $1/10^x = 10^{-x}$

$$pH = \log 10^7$$

the log of 10^7 is 7, so the solution has a pH of 7.

Pure water has a pH value of 7.0, meaning its H^+ concentration is 1×10^{-7} M.

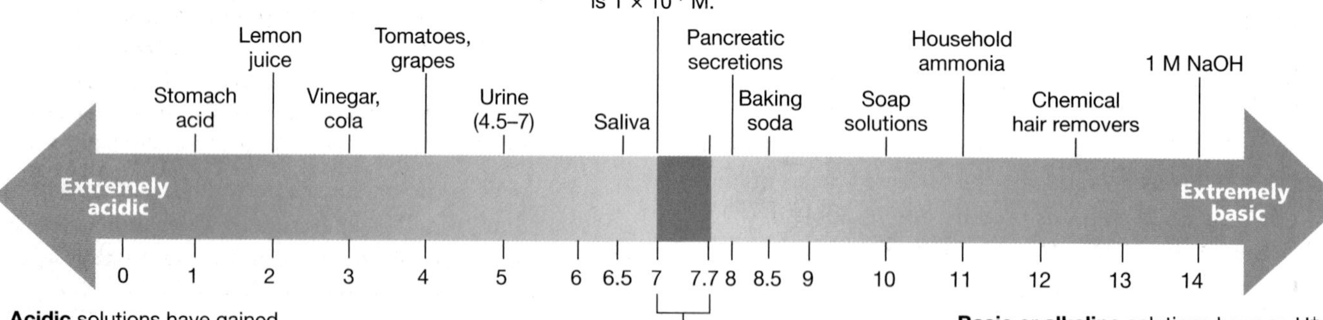

Acidic solutions have gained H^+ from an acid and have a pH less than 7.

Basic or **alkaline** solutions have an H^+ concentration lower than that of pure water and have a pH value greater than 7.

The normal pH of blood in the human body is 7.40. Homeostatic regulation is critical because blood pH less than 7.00 or greater than 7.70 is incompatible with life.

The pH of a solution is measured on a numeric scale between 0 and 14. The pH scale is logarithmic, meaning that a change in pH value of 1 unit indicates a 10-fold change in $[H^+]$. For example, if a solution changes from pH 8 to pH 6, there has been a 100-fold (10^2 or 10×10) increase in $[H^+]$.

FIGURE QUESTIONS

1. When the body becomes more acidic, does pH increase or decrease?
2. How can urine, stomach acid, and saliva have pH values outside the pH range that is compatible with life and yet be part of the living body?

We are constantly adding acid to the body through metabolism, so how does the body maintain a normal pH? One answer is buffers. A **buffer** is any substance that moderates changes in pH. Many buffers contain anions that have a strong affinity for H^+ molecules. When free H^+ is added to a buffer solution, the buffer's anions bond to the H^+, thereby minimizing any change in pH.

The bicarbonate anion, HCO_3^-, is an important buffer in the human body. The following equation shows how a sodium bicarbonate solution acts as a buffer when hydrochloric acid (HCl) is added. When placed in plain water, hydrochloric acid separates, or dissociates, into H^+ and Cl^- and creates a high H^+ concentration (low pH). When HCl dissociates in a sodium bicarbonate solution, however, some of the bicarbonate ions combine with some of the H^+ to form undissociated carbonic acid. "Tying up" the added H^+ in this way keeps the H^+ concentration of the solution from changing significantly and minimizes the pH change.

$$\underbrace{H^+ + Cl^-}_{\text{Hydrochloric acid}} + \underbrace{HCO_3^- + Na^+}_{\text{Sodium bicarbonate}} \rightleftharpoons \underbrace{H_2CO_3}_{\text{Carbonic acid}} + \underbrace{Cl^- + Na^+}_{\text{Sodium chloride (table salt)}}$$

Concept Check Answers: p. 61

10. To be classified as an acid, a molecule must do what when dissolved in water?
11. pH is an expression of the concentration of what in a solution?
12. When pH goes up, acidity goes ________.

Protein Interactions

Noncovalent molecular interactions occur between many different biomolecules and often involve proteins. For example, biological membranes are formed by the noncovalent associations of phospholipids and proteins. Glycosylated proteins and glycosylated lipids in cell membranes create a "sugar coat" on cell surfaces, where they assist cell aggregation {*aggregare*, to join together} and adhesion {*adhaerere*, to stick}.

Proteins play important roles in so many cell functions that they can be considered the "workhorses" of the body. Most soluble proteins fall into one of seven broad categories:

1 **Enzymes.** Some proteins act as **enzymes,** biological catalysts that speed up chemical reactions. Enzymes play an important role in metabolism [discussed in Chapters 4 and 22].

2 **Membrane transporters.** Proteins in cell membranes help move substances back and forth between the intracellular and extracellular compartments. These proteins may form channels in the cell membrane, or they may bind to molecules and carry them through the membrane. [Membrane transporters are discussed in detail in Chapter 5.]

3 **Signal molecules.** Some proteins and smaller peptides act as hormones and other signal molecules. [Different types of signal molecules are described in Chapters 6 and 7.]

4 **Receptors.** Proteins that bind signal molecules and initiate cellular responses are called *receptors.* [Receptors are discussed along with signal molecules in Chapter 6.]

5 **Binding proteins.** These proteins, found mostly in the extracellular fluid, bind and transport molecules throughout the body. Examples you have already encountered include the oxygen-transporting protein *hemoglobin* and the cholesterol-binding proteins such as LDL (low-density lipoprotein).

6 **Immunoglobulins.** These extracellular immune proteins, also called *antibodies,* help protect the body. [Immune functions are discussed in Chapter 24.]

7 **Regulatory proteins.** Regulatory proteins turn cell processes on and off or up and down. For example, the regulatory proteins known as *transcription factors* bind to DNA and alter gene expression and protein synthesis. The details of regulatory proteins can be found in cell biology textbooks.

RUNNING PROBLEM

The hexavalent form of chromium used in industry is known to be toxic to humans. In 1992, officials at California's Hazard Evaluation System and Information Service warned that inhaling chromium dust, mist, or fumes placed chrome and stainless steel workers at increased risk for lung cancer. Officials found no risk to the public from normal contact with chrome surfaces or stainless steel. In 1995 and 2003, a possible link between the biological trivalent form of chromium (Cr^{3+}) and cancer came from *in vitro* studies {*vitrum,* glass; that is, a test tube} in which mammalian cells were kept alive in tissue culture. In these experiments, cells exposed to moderately high levels of chromium picolinate developed potentially cancerous changes.*

Q4: From this information, can you conclude that hexavalent and trivalent chromium are equally toxic?

*D. M. Stearns *et al.* Chromium(III) picolinate produces chromosome damage in Chinese hamster ovary cells. *FASEB J* 9: 1643–1648, 1995.

D. M. Stearns *et al.* Chromium(III) tris(picolinate) is mutagenic at the hypoxanthine (guanine) phosphoribosyltransferase locus in Chinese hamster ovary cells. *Mutation Res Genet Toxicol Environ Mutagen* 513: 135–142, 2002.

32 39 43 47 49 52 57

Although soluble proteins are quite diverse, they do share some common features. They all bind to other molecules through noncovalent interactions. The binding, which takes place at a location on the protein molecule called a **binding site,** exhibits properties that will be discussed shortly: specificity, affinity, competition, and saturation. If binding of a molecule to the protein initiates a process, as occurs with enzymes, membrane transporters, and receptors, we can describe the activity rate of the process and the factors that modulate, or alter, the rate.

Any molecule or ion that binds to another molecule is called a **ligand** {*ligare,* to bind or tie}. Ligands that bind to enzymes and membrane transporters are also called **substrates** {*sub-*, below + *stratum,* a layer}. Protein signal molecules and protein transcription factors are ligands. Immunoglobulins bind ligands, but the immunoglobulin-ligand complex itself then becomes a ligand [for details, see Chapter 24].

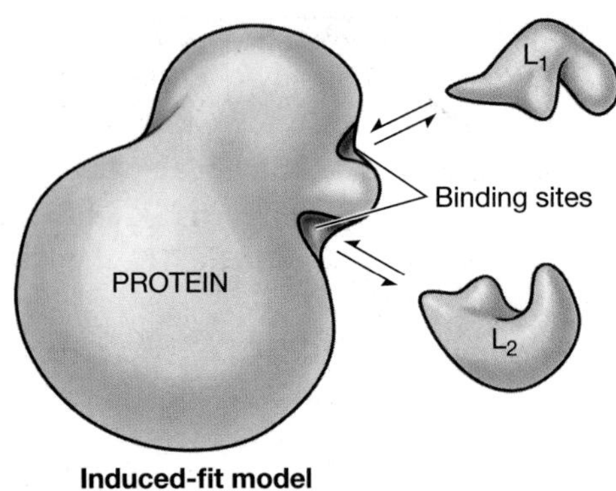

■ **Fig. 2.10** ***The induced-fit model of protein–ligand (L) binding***

Proteins Are Selective About the Molecules They Bind

The ability of a protein to bind to a certain ligand or a group of related ligands is called **specificity.** Some proteins are very specific about the ligands they bind, while others bind to whole groups of molecules. For example, the enzymes known as *peptidases* bind polypeptide ligands and break apart peptide bonds, no matter which two amino acids are joined by those bonds. For this reason peptidases are not considered to be very specific in their action. In contrast, *aminopeptidases* also break peptide bonds but are more specific. They will bind only to one end of a protein chain (the end with an unbound amino group) and can act only on the terminal peptide bond.

Ligand binding requires *molecular complementarity*. In other words, the ligand and the protein binding site must be complementary, or compatible. In protein binding, when the ligand and protein come close to each other, noncovalent interactions between the ligand and the protein's binding site allow the two molecules to bind. From studies of enzymes and other binding proteins, scientists have discovered that a protein's binding site and the shape of its ligand do not need to fit one another exactly. When the binding site and the ligand come close to each other, they begin to interact through hydrogen and ionic bonds and van der Waals forces. The protein's binding site then changes shape (*conformation*) to fit more closely to the ligand. This **induced-fit model** of protein-ligand interaction is shown in ■ Figure 2.10.

Protein-Binding Reactions Are Reversible

The degree to which a protein is attracted to a ligand is called the protein's **affinity** for the ligand. If a protein has a high affinity for a given ligand, the protein is more likely to bind to that ligand than to a ligand for which the protein has a lower affinity.

Protein binding to a ligand can be written using the same notation that we use to represent chemical reactions:

$$P + L \rightleftharpoons PL$$

where P is the protein, L is the ligand, and PL is the bound protein-ligand complex. The double arrow indicates that binding is reversible.

Reversible binding reactions go to a state of **equilibrium,** where the rate of binding ($P + L \rightarrow PL$) is exactly equal to the rate of unbinding, or *dissociation* ($P + L \leftarrow PL$). When a reaction is at equilibrium, the ratio of the product concentration, or protein-ligand complex [PL], to the reactant concentrations [P][L] is always the same. This ratio is called the *equilibrium constant* K_{eq}, and it applies to all reversible chemical reactions:

$$K_{eq} = \frac{[PL]}{[P][L]}$$

The square brackets [ ] around the letters indicate concentrations of the protein, ligand, and protein-ligand complex.

Binding Reactions Obey the Law of Mass Action

Equilibrium is a dynamic state. In the living body, concentrations of protein or ligand change constantly through synthesis, breakdown, or movement from one compartment to another. What happens to equilibrium when the concentration of P or L changes? The answer to this question is shown in ■ Figure 2.11, which begins with a reaction at equilibrium (Figure 2.11a).

In Figure 2.11b, the equilibrium is disturbed when more protein or ligand is added to the system. Now the ratio of [PL] to [P][L] differs from the K_{eq}. In response, the rate of the binding reaction increases to convert some of the added P or L into the bound protein-ligand complex (Fig. 2.11c). As the ratio approaches its equilibrium value again, the rate of the forward reaction slows down until finally the system reaches the equilibrium ratio once more (Fig. 2.11d).

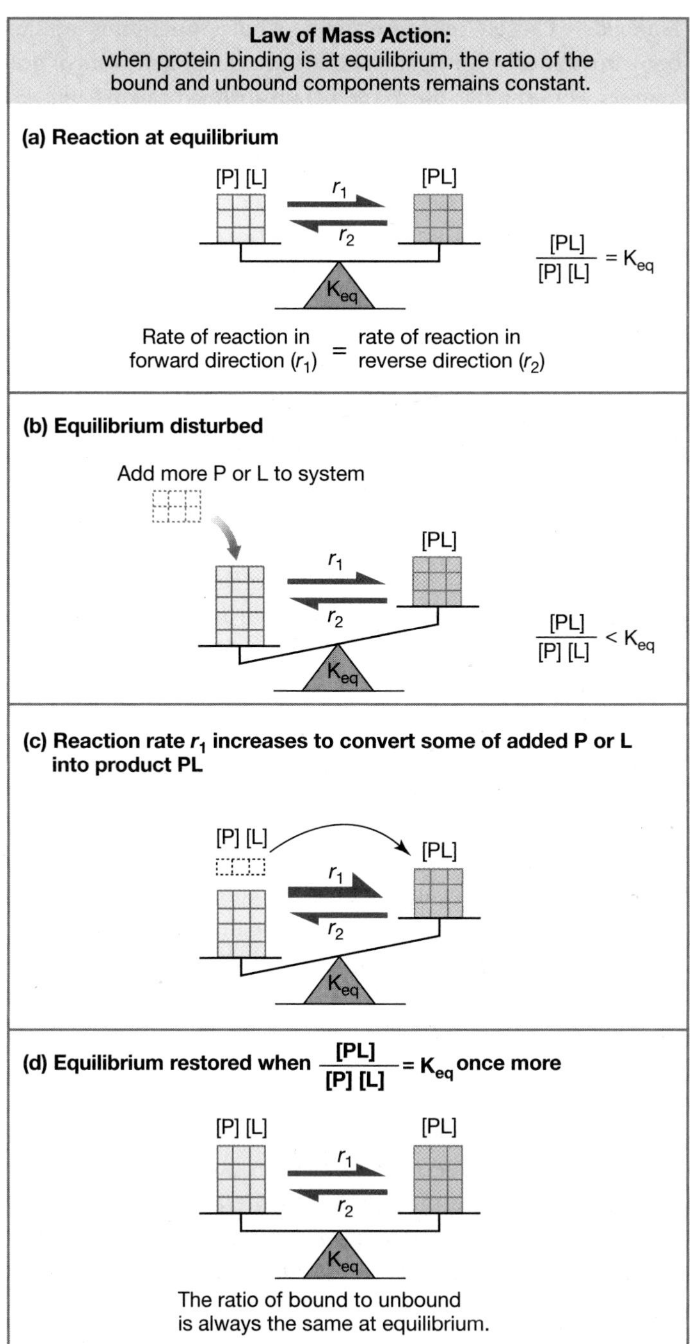

Law of Mass Action:
when protein binding is at equilibrium, the ratio of the bound and unbound components remains constant.
(a) Reaction at equilibrium
[P] [L]
[PL]
r1
r2
Keq
[PL] / [P] [L] = Keq
Rate of reaction in forward direction (r1) = rate of reaction in reverse direction (r2)
(b) Equilibrium disturbed
Add more P or L to system
[PL]
r1
r2
Keq
[PL] / [P] [L] < Keq
(c) Reaction rate r1 increases to convert some of added P or L into product PL
[P] [L]
[PL]
r1
r2
Keq
(d) Equilibrium restored when [PL] / [P] [L] = Keq once more
[P] [L]
[PL]
r1
r2
Keq
The ratio of bound to unbound is always the same at equilibrium.

Fig. 2.11 ***The law of mass action***

[P], [L], and [PL] have all increased over their initial values, but the equilibrium ratio has been restored.

The situation just described is an example of a reversible reaction obeying the **law of mass action,** a simple relationship that holds for chemical reactions whether in a test tube or in a cell. You may have learned this law in chemistry as *LeChâtelier's principle.* In very general terms, the law of mass action says that when a reaction is at equilibrium, the ratio of the products to the substrates is always the same. If the ratio is disturbed by adding or removing one of the participants, the reaction equation will shift direction to restore the equilibrium condition. (Note that the law of mass action is not the same as mass balance [see Chapter 1, p. 11].)

One example of this principle at work is the transport of steroid hormones in the blood. Steroids are hydrophobic, so more than 99% of hormone in the blood is bound to carrier proteins. The equilibrium ratio [PL]/[P][L] is 99% bound/1% unbound hormone. However, only the unbound or "free" hormone can cross the cell membrane and enter cells. As unbound hormone leaves the blood, the equilibrium ratio is disturbed. The binding proteins then release some of the bound hormone until the 99/1 ratio is again restored. The same principle applies to enzymes and metabolic reactions. Changing the concentration of one participant in a chemical reaction has a chain-reaction effect that alters the concentrations of other participants in the reaction.

Concept Check Answer: p. 61

13. Consider the carbonic acid reaction, which is reversible:

$$CO_2 + H_2O \rightleftharpoons H_2CO_3 \text{ (carbonic acid)} \rightleftharpoons H^+ + HCO_3^-$$

If the carbon dioxide concentration in the body increases, what happens to the concentration of carbonic acid (H_2CO_3)? What happens to the pH?

The Dissociation Constant Indicates Affinity

For protein-binding reactions, where the equilibrium equation is a quantitative representation of the protein's binding affinity for the ligand, the reciprocal of the equilibrium constant is called the **dissociation constant** (K_d):

$$K_d = \frac{[P][L]}{[PL]}$$

Using algebra to rearrange the equation, this can also be expressed as

$$[PL] = \frac{[P][L]}{K_d}$$

From the rearranged equation you should be able to see that when K_d is large, the value of [PL] is small. In other words, a large dissociation constant K_d means little binding of protein and ligand, and we can say the protein has a low affinity for the ligand. Conversely, a small K_d is a lower dissociation constant and means a higher value for [PL], indicating a higher affinity of the protein for the ligand.

If a protein binds to several related ligands, a comparison of their K_d values can tell us which ligand is more likely to bind to the protein. The related ligands compete for the binding sites and are said to be **competitors.** Competition between ligands is a universal property of protein binding.

Competing ligands that mimic each other's actions are called **agonists** {*agonist*, contestant}. Agonists may occur in nature, such as *nicotine,* the chemical found in tobacco, which mimics the activity of the neurotransmitter *acetylcholine* by binding to the same receptor protein. Agonists can also be synthesized using what scientists learn from the study of protein-ligand binding sites. The ability of agonist molecules to mimic the activity of naturally occurring ligands has led to the development of many drugs.

Concept Check Answer: p. 61

14. A researcher is trying to design a drug to bind to a particular cell receptor protein. Candidate molecule A has a K_d of 4.9 for the receptor. Molecule B has a K_d of 0.3. Which molecule has the most potential to be successful as the drug?

Multiple Factors Alter Protein Binding

A protein's affinity for a ligand is not always constant. Chemical and physical factors can alter, or *modulate,* binding affinity or can even totally eliminate it. Some proteins must be activated before they have a functional binding site. In this section we discuss some of the processes that have evolved to allow activation, modulation, and inactivation of protein binding.

RUNNING PROBLEM

Stan has been taking chromium picolinate because he heard that it would increase his strength and muscle mass. Then a friend told him that the Food and Drug Administration (FDA) said there was no evidence to show that chromium would help build muscle. In one study*, a group of researchers gave high daily doses of chromium picolinate to football players during a two-month training period. By the end of the study, the players who took chromium supplements had not increased muscle mass or strength any more than players who did not take the supplement.

Use Google Scholar (*http://scholar.google.com*) and search for *FDA review chromium picolinate.* Look at the articles you find listed there before you answer the next question.

Q5: Based on the FDA review, the Hallmark *et al.* study (which did not support enhanced muscle development from chromium supplements), and the studies that suggest that chromium picolinate might cause cancer, do you think that Stan should continue taking chromium picolinate?

*M. A. Hallmark *et al.* Effects of chromium and resistive training on muscle strength and body composition. *Med Sci Sports Exercise* 28(1): 139–144, 1996.

32 39 43 47 49 52 57

Isoforms Closely related proteins whose function is similar but whose affinity for ligands differs are called **isoforms** of one another. For example, the oxygen-transporting protein *hemoglobin* has multiple isoforms. One hemoglobin molecule has a quaternary structure consisting of four subunits (see Fig. 2.3). In the developing fetus, the hemoglobin isoform has two α (alpha) chains and two γ (gamma) chains that make up the four subunits. Shortly after birth, fetal hemoglobin molecules are broken down and replaced by adult hemoglobin. The adult hemoglobin isoform retains the two α chain isoforms but has two β (beta) chains in place of the γ chains. Both adult and fetal isoforms of hemoglobin bind oxygen, but the fetal isoform has a higher affinity for oxygen. This makes it more efficient at picking up oxygen across the placenta.

Activation Some proteins are inactive when they are synthesized in the cell. Before such a protein can become active, enzymes must chop off one or more portions of the molecule (■ Fig. 2.12a). Protein hormones (a type of signal molecule) and enzymes are two groups that commonly undergo such *proteolytic activation* {*lysis,* to release}. The inactive forms of these proteins are often identified with the prefix *pro-* {before}: prohormone, proenzyme, proinsulin, for example. Some inactive enzymes have the suffix *-ogen* added to the name of the active enzyme instead, as in *trypsinogen,* the inactive form of trypsin.

The activation of some proteins requires the presence of a **cofactor,** which is an ion or small organic functional group. Cofactors must attach to the protein before the binding site will activate and bind to ligand (Fig. 2.12c). Ionic cofactors include Ca^{2+}, Mg^{2+}, and Fe^{2+}. Many enzymes will not function without their cofactors.

Modulation The ability of a protein to bind a ligand and initiate a response can be altered by various factors, including temperature, pH, and molecules that interact with the protein. A factor that influences either protein binding or protein activity is called a **modulator.** There are two basic mechanisms by which modulation takes place. The modulator either (1) changes the protein's ability to bind the ligand or it (2) changes the protein's activity or its ability to create a response. ■ Table 2.3 (on page 54) summarizes the different types of modulation.

Chemical modulators are molecules that bind covalently or noncovalently to proteins and alter their binding ability or their activity. Chemical modulators may activate or enhance ligand binding, decrease binding ability, or completely inactivate the protein so that it is unable to bind any ligand. Inactivation may be either reversible or irreversible.

Antagonists, also called *inhibitors,* are chemical modulators that bind to a protein and decrease its activity. Many are simply molecules that bind to the protein and block the binding site without causing a response. They are like the guy who slips into the front of the movie ticket line to chat with his girlfriend, the cashier. He has no interest in buying a ticket, but he prevents

■ Fig. 2.12 *ESSENTIALS*

Protein Activation and Inhibition

ACTIVATION

(a) Proteolytic activation: Protein is inactive until peptide fragments are removed.

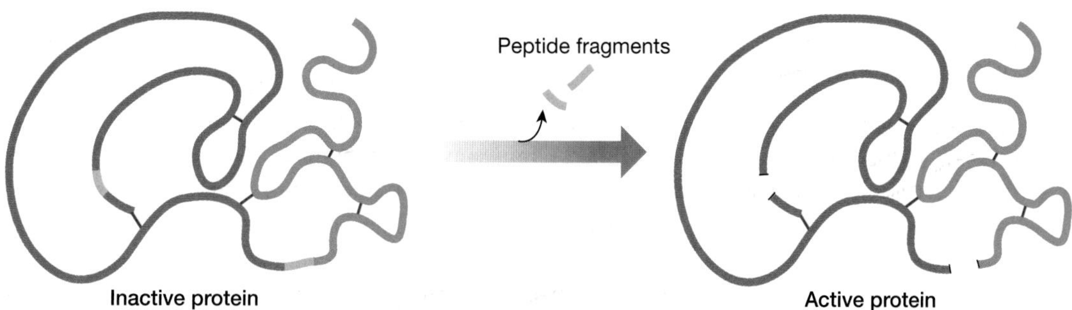

(b) Allosteric activator is a modulator that binds to protein away from binding site and turns it on.

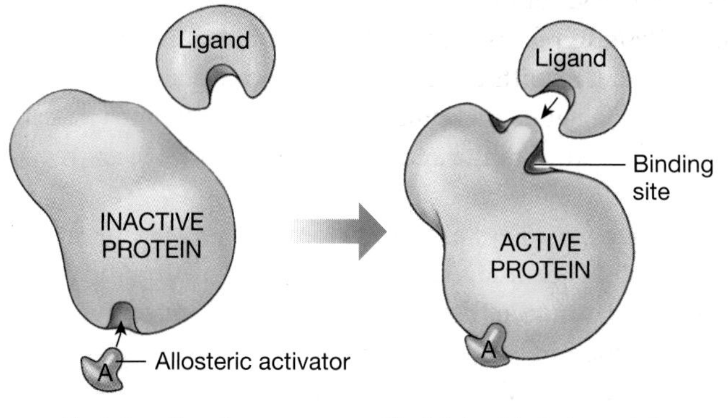

Protein without modulator is inactive.

Modulator binds to protein away from binding site.

(c) Cofactors are required for an active binding site.

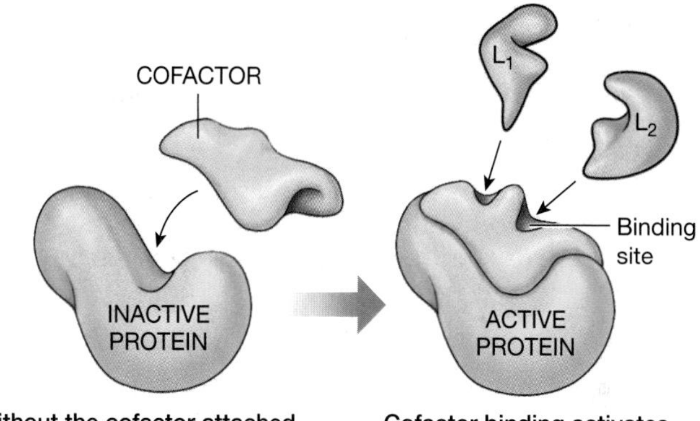

Without the cofactor attached, the protein is not active.

Cofactor binding activates the protein.

INHIBITION

(d) A **competitive inhibitor** blocks ligand binding at the binding site.

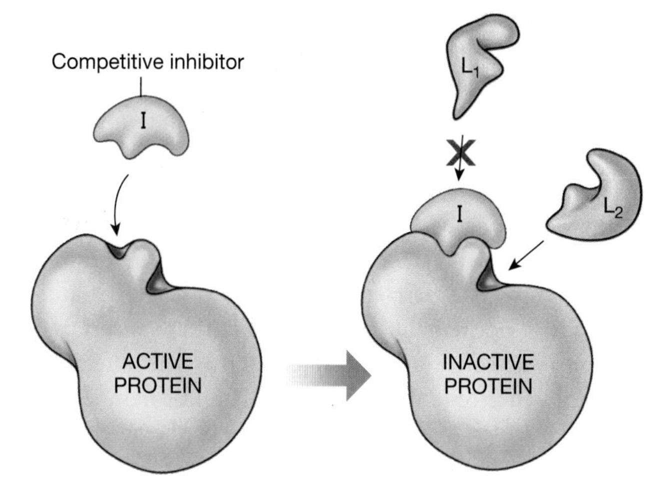

(e) Allosteric inhibitor is a modulator that binds to protein away from binding site and inactivates the binding site.

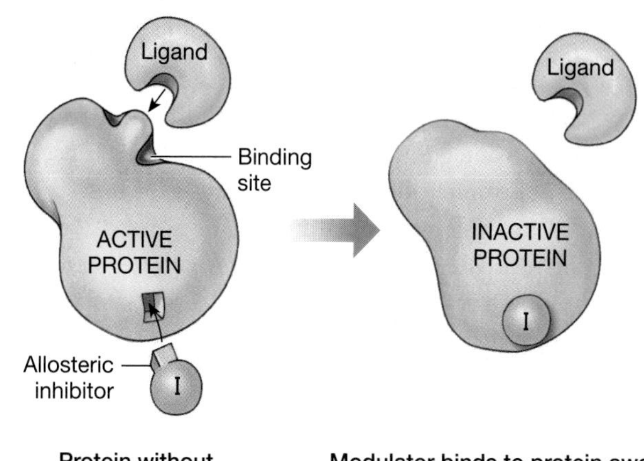

Protein without modulator is active.

Modulator binds to protein away from binding site and inactivates the binding site.

Table 2.3 Factors That Affect Protein Binding

Essential for Binding Activity	
Cofactors	Required for ligand binding at binding site
Proteolytic activation	Converts inactive to active form by removing part of molecule. Examples: digestive enzymes, protein hormones
Modulators and Factors That Alter Binding or Activity	
Competitive inhibitor	Competes directly with ligand by binding reversibly to active site
Irreversible inhibitor	Binds to binding site and cannot be displaced
Allosteric modulator	Binds to protein away from binding site and changes activity; may be inhibitors or activators
Covalent modulator	Binds covalently to protein and changes its activity. Example: phosphate groups
pH and temperature	Alter three-dimensional shape of protein by disrupting hydrogen or S—S bonds; may be irreversible if protein denatures

the people in line behind him from getting their tickets for the movie.

Competitive inhibitors are reversible antagonists that compete with the customary ligand for the binding site (Fig. 2.12d). The degree of inhibition depends on the relative concentrations of the competitive inhibitor and the customary ligand, as well as on the protein's affinities for the two. The binding of competitive inhibitors is reversible: increasing the concentration of the customary ligand can displace the competitive inhibitor and decrease the inhibition.

Irreversible antagonists, on the other hand, bind tightly to the protein and cannot be displaced by competition. Antagonist drugs have proven useful for treating many conditions. For example, tamoxifen, an antagonist to the estrogen receptor, is used in the treatment of hormone-dependent cancers of the breast.

Allosteric and covalent modulators may be either antagonists or activators. **Allosteric modulators** {*allos,* other + *stereos,* solid (as a shape)} bind reversibly to a protein at a regulatory site away from the binding site, and by doing so change the shape of the binding site. *Allosteric inhibitors* are antagonists that decrease the affinity of the binding site for the ligand and inhibit protein activity (Fig. 2.12e). *Allosteric activators* increase the probability of protein-ligand binding and enhance protein activity (Fig. 2.12b). For example, the oxygen-binding ability of hemoglobin changes with allosteric modulation by carbon dioxide, H^+, and several other factors [see Chapter 18].

Covalent modulators are atoms or functional groups that bind covalently to proteins and alter the proteins' properties. Like allosteric modulators, covalent modulators may either increase or decrease a protein's binding ability or its activity. One of the most common covalent modulators is the phosphate group. Many proteins in the cell can be activated or inactivated when a phosphate group forms a covalent bond with them, a process known as *phosphorylation.*

One of the best known chemical modulators is the antibiotic penicillin. Alexander Fleming discovered this compound in 1928, when he noticed that *Penicillium* mold inhibited bacterial growth. By 1938, researchers had extracted the active ingredient penicillin from the mold and used it to treat infections in humans. Yet it was not until 1965 that researchers figured out exactly how the antibiotic works. Penicillin is an antagonist that binds to a key bacterial protein by mimicking the normal ligand. Because penicillin forms unbreakable bonds with the protein, the protein is irreversibly inhibited. Without the protein, the bacterium is unable to make a rigid cell wall. With no rigid cell wall, the bacterium swells, ruptures, and dies.

Physical Factors Physical conditions such as temperature and pH (acidity) can have dramatic effects on protein structure and function. Small changes in pH or temperature act as modulators to increase or decrease activity (■ Fig. 2.13a). However, once these factors exceed some critical value, they disrupt the noncovalent bonds holding the protein in its tertiary conformation. The protein loses its shape and, along with that, its activity. When the protein loses its conformation, it is said to be *denatured.*

If you have ever fried an egg, you have watched this transformation happen to the egg white protein *albumin* as it changes from a slithery clear state to a firm white state. Hydrogen ions in high enough concentration to be called acids have a similar effect on protein structure. During preparation of ceviche, the national dish of Ecuador, raw fish is marinated in lime juice. The acidic lime juice contains hydrogen ions that disrupt hydrogen bonds in the muscle proteins of the fish, causing the proteins to denature. As a result, the meat becomes firmer and opaque, just as it would if it were cooked with heat.

In a few cases, activity can be restored if the original temperature or pH returns. The protein then resumes its original shape as if nothing had happened. Usually, however, denaturation produces a permanent loss of activity. There is certainly no way to unfry an egg or uncook a piece of fish. The potentially disastrous influence of temperature and pH on proteins is one reason these variables are so closely regulated by the body.

Fig. 2.13 *ESSENTIALS*

Factors That Influence Protein Activity

(a) Temperature and pH

Temperature and pH changes may disrupt protein structure and cause loss of function.

Active protein in normal tertiary conformation

GRAPH QUESTION

Is the protein more active at 30 °C or at 48 °C?

Rate of protein activity

20 30 40 50 60

Temperature (°C)

Denatured protein

This protein denatures around 50 °C.

(b) Amount of Protein

Reaction rate depends on the amount of protein. The more protein present, the faster the rate.

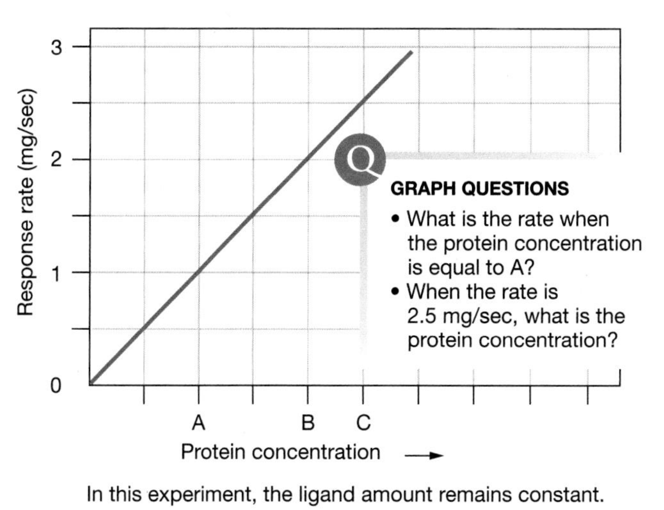

In this experiment, the ligand amount remains constant.

(c) Amount of Ligand

If the amount of binding protein is held constant, the reaction rate depends on the amount of ligand, up to the saturation point.

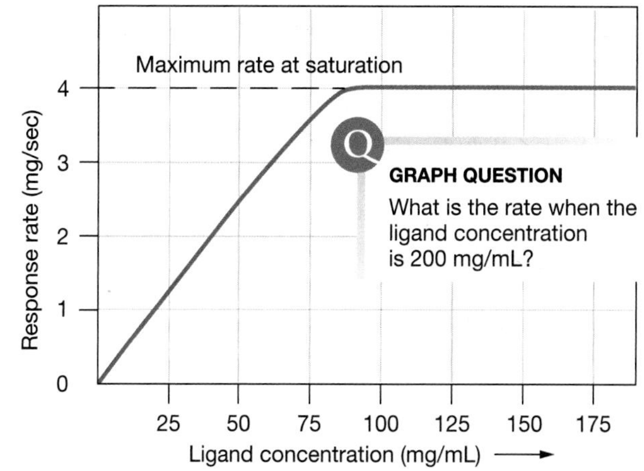

In this experiment, the amount of binding protein was constant. At the maximum rate, the protein is said to be saturated.

Concept Check Answers: p. 61

15. Match each chemical to its action(s).

(a) Allosteric modulator
(b) Competitive inhibitor
(c) Covalent modulator

1. Bind away from the binding site
2. Bind to the binding site
3. Inhibit activity only
4. Inhibit or enhance activity

The Body Regulates the Amount of Protein in Cells

The final characteristic of proteins in the human body is that the amount of a given protein varies over time, often in a regulated fashion. The body has mechanisms that enable it to monitor whether it needs more or less of particular proteins. Complex signaling pathways, many of which themselves involve proteins, direct particular cells to make new proteins or to break down (*degrade*) existing proteins. The programmed production of new proteins (receptors, enzymes, and membrane transporters, in particular) is called **up-regulation.** Conversely, the programmed removal of proteins is called **down-regulation.** In both instances, the cell is directed to make or remove proteins to alter its response.

The amount of protein present in a cell has a direct influence on the magnitude of the cell's response. For example, the graph in Figure 2.13b shows the results of an experiment in which the amount of ligand is held constant while the amount of protein is varied. As the graph shows, an increase in the amount of protein present causes an increase in the response.

As an analogy, think of the checkout lines in a supermarket. Imagine that each cashier is an enzyme, the waiting customers are ligand molecules, and people leaving the store with their purchases are products. One hundred customers can be checked out faster when there are 25 lines open than when there are only 10 lines. Likewise, in an enzymatic reaction, the presence of more protein molecules (enzyme) means that more binding sites are available to interact with the ligand molecules. As a result, the ligands are converted to products more rapidly.

Regulating protein concentration is an important strategy that cells use to control their physiological processes. Cells alter the amount of a protein by influencing both its synthesis and its breakdown. If protein synthesis exceeds breakdown, protein accumulates and the reaction rate increases. If protein breakdown exceeds synthesis, the amount of protein decreases, as does the reaction rate. Even when the amount of protein is constant, there is still a steady turnover of protein molecules.

Reaction Rate Can Reach a Maximum

If the concentration of a protein in a cell is constant, then the concentration of the ligand determines the magnitude of the response. Fewer ligands activate fewer proteins, and the response is low. As ligand concentrations increase, so does the magnitude of the response, up to a maximum where all protein binding sites are occupied.

Figure 2.13c shows the results of a typical experiment in which the protein concentration is constant but the concentration of ligand varies. At low ligand concentrations, the response rate is directly proportional to the ligand concentration. Once the concentration of ligand molecules exceeds a certain level, the protein molecules have no more free binding sites. The proteins are fully occupied, and the rate reaches a maximum value. This condition is known as **saturation.** Saturation applies to enzymes, membrane transporters, receptors, binding proteins, and immunoglobulins.

An analogy to saturation appeared in the early days of television on the *I Love Lucy* show. Lucille Ball was working at the conveyor belt of a candy factory, loading chocolates into the little paper cups of a candy box. Initially, the belt moved slowly, and she had no difficulty picking up the candy and putting it into the box. Gradually, the belt brought candy to her more rapidly, and she had to increase her packing speed to keep up. Finally, the belt brought candy to her so fast that she could not pack it all in the boxes because she was working at her maximum rate. That was Lucy's saturation point. (Her solution was to stuff the candy into her mouth as well as into the box!)

In conclusion, you have now learned about the important and nearly universal properties of soluble proteins. You will revisit these concepts many times as you work through the organ systems of the body. The body's insoluble proteins, which are key structural components of cells and tissues, are covered in later chapters.

Concept Check Answers: p. 61

16. What happens to the rate of an enzymatic reaction as the amount of enzyme present decreases?

17. What happens to the rate of an enzymatic reaction when the enzyme has reached saturation?

RUNNING PROBLEM CONCLUSION

Chromium Supplements

In this running problem, you learned that claims of chromium picolinate's ability to enhance muscle mass have not been supported by evidence from controlled scientific experiments. You also learned that studies suggest that some forms of the biological trivalent form of chromium may be toxic. To learn more about current research, go to PubMed (*www.pubmed.gov*) and search for "*chromium picolinate*" (use the quotation marks). Compare what you find there with the results of a similar Google search. Should you believe everything you read on the Web? Now compare your answers with those in the summary table.

Question	Facts	Integration and Analysis
1. Locate chromium on the periodic table of elements.	The periodic table organizes the elements according to atomic number.	N/A*
What is chromium's atomic number? Atomic mass?	Reading from the table, chromium (Cr) has an atomic number of 24 and an average atomic mass of 52.	N/A
How many electrons does one atom of chromium have?	Atomic number of an element = number of protons in one atom. One atom has equal numbers of protons and electrons.	The atomic number of chromium is 24; therefore, one atom of chromium has 24 protons and 24 electrons.
Which elements close to chromium are also essential elements?	Molybdenum, manganese, and iron.	N/A
2. If people have chromium deficiency, would you predict that their blood glucose level would be lower or higher than normal?	Chromium helps move glucose from blood into cells.	If chromium is absent or lacking, less glucose would leave the blood and blood glucose would be higher than normal.
From the result of the Chinese study, can you conclude that all people with diabetes suffer from chromium deficiency?	Higher doses of chromium supplements lowered elevated blood glucose levels, but lower doses have no effect. This is only one study, and no information is given about similar studies elsewhere.	We have insufficient evidence from the information presented to draw a conclusion about the role of chromium deficiency in diabetes.
3. How many electrons have been lost from the hexavalent ion of chromium? From the trivalent ion?	For each electron lost from an ion, a positively charged proton is left behind in the nucleus of the ion.	The hexavalent ion of chromium, Cr^{6+}, has six unmatched protons and therefore has lost six electrons. The trivalent ion, Cr^{3+}, has lost three electrons.
4. From this information, can you conclude that hexavalent and trivalent chromium are equally toxic?	The hexavalent form is used in industry and, when inhaled, has been linked to an increased risk of lung cancer. Enough studies have shown an association that California's Hazard Evaluation System and Information Service has issued warnings to chromium workers. Evidence to date for toxicity of trivalent chromium in chromium picolinate comes from studies done on isolated cells in tissue culture.	Although the toxicity of Cr^{6+} is well established, the toxicity of Cr^{3+} has not been conclusively determined. Studies performed on cells *in vitro* may not be applicable to humans. Additional studies need to be performed in which animals are given reasonable doses of chromium picolinate for an extended period of time.
5. Based on the study that did not support enhanced muscle development from chromium supplements and the studies that suggest that chromium picolinate might cause cancer, do you think Stan should continue taking picolinate?	No research evidence supports a role for chromium picolinate in increasing muscle mass or strength. Other research suggests that chromium picolinate may cause cancerous changes in isolated cells.	The evidence presented suggests that for Stan, there is no benefit from taking chromium picolinate, and there may be risks. Using risk-benefit analysis, the evidence supports stopping the supplements. However, the decision is Stan's personal responsibility. He should keep himself informed of new developments that would change the risk-benefit analysis.

32 39 43 47 49 52 57

Chemistry Review Quiz

Use this quiz to see what areas of chemistry and basic biochemistry you might need to review. Answers are at the end of the chapter. The title above each set of questions refers to a review figure on this topic.

Atoms and Molecules (Fig. 2.5)

Match each subatomic particle in the left column with all the phrases in the right column that describe it. A phrase may be used more than once.

1. electron
2. neutron
3. proton

(a) one has atomic mass of 1 amu
(b) found in the nucleus
(c) negatively charged
(d) changing the number of these in an atom creates a new element
(e) adding or losing these makes an atom into an ion
(f) gain or loss of these makes an isotope of the same element
(g) determine(s) an element's atomic number
(h) contribute(s) to an element's atomic mass

4. Isotopes of an element have the same number of ________ and ________, but differ in their number of ________. Unstable isotopes emit energy called ________.
5. Name the element associated with each of these symbols: C, O, N, and H.
6. Write the one- or two-letter symbol for each of these elements: phosphorus, potassium, sodium, sulfur, calcium, and chlorine.
7. Use the periodic table of the elements on the inside back cover to answer the following questions:
   (a) Which element has 30 protons?
   (b) How many electrons are in one atom of calcium?
   (c) Find the atomic number and average atomic mass of iodine. What is the letter symbol for iodine?
8. A magnesium ion, Mg^{2+}, has (*gained/lost*) two (*protons/neutrons/electrons*).
9. H^+ is also called a proton. Why is it given that name?
10. Use the periodic table of the elements on the inside back cover to answer the following questions about an atom of sodium.
   (a) How many electrons does the atom have?
   (b) What is the electrical charge of the atom?
   (c) How many neutrons does the average atom have?
   (d) If this atom loses one electron, it would be called a(n) *anion/cation*.
   (e) What would be the electrical charge of the substance formed in (d)?
   (f) Write the chemical symbol for the ion referred to in (d).
   (g) What does the sodium atom become if it loses a proton from its nucleus?
   (h) Write the chemical symbol for the atom referred to in (g).
11. Write the chemical formulas for each molecule depicted. Calculate the molecular mass of each molecule.

(a) $HOCH_2$, O, HO, OH, OH, OH

(b) O=C=O

(c)

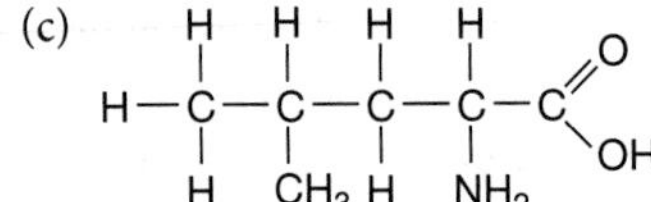

(d) COOH, H_2N—C—H, CH_3

Lipids (Fig. 2.1)

12. Match each lipid with its best description:

(a) triglyceride	1. most common form of lipid in the body
(b) eicosanoid	2. liquid at room temperature, usually from plants
(c) steroid	3. important component of cell membrane
(d) oil	4. structure composed of carbon rings
(e) phospholipids	5. modified 20-carbon fatty acid

13. Use the chemical formulas given to decide which of the following fatty acids is most unsaturated: (a) $C_{18}H_{36}O_2$ (b) $C_{18}H_{34}O_2$ (c) $C_{18}H_{30}O_2$

Carbohydrates (Fig. 2.2)

14. Match each carbohydrate with its description:

(a) starch	1. monosaccharide
(b) chitin	2. disaccharide, found in milk
(c) glucose	3. storage form of glucose for animals
(d) lactose	4. storage form of glucose for plants
(e) glycogen	5. structural polysaccharide of invertebrates

Proteins (Fig. 2.3)

15. Match these terms pertaining to proteins and amino acids:

(a) the building blocks	1. essential amino acids
(b) must be included in our diet	2. primary structure
(c) protein catalysts that speed the rate of chemical reactions	3. amino acids
(d) sequence of amino acids in a protein	4. globular proteins
(e) protein chains folded into a ball-shaped structure	5. enzymes
	6. tertiary structure
	7. fibrous proteins

16. What aspect of protein structure allows proteins to have more versatility than lipids or carbohydrates?
17. Peptide bonds form when the ________ group of one amino acid joins the ________ of another amino acid.

Nucleotides (Fig. 2.4)

18. List the three components of a nucleotide.
19. Compare the structure of DNA with that of RNA.
20. Distinguish between purines and pyrimidines.

Chapter Summary

This chapter introduces the *molecular interactions* between biomolecules, water, and ions that underlie many of the key themes in physiology. These interactions are an integral part of *information flow, energy storage and transfer*, and the *mechanical properties* of cells and tissues in the body.

Molecules and Bonds

1. The four major groups of **biomolecules** are carbohydrates, lipids, proteins, and nucleotides. They all contain carbon, hydrogen, and oxygen. (p. 33; Figs. 2.1, 2.2, 2.3, and 2.4)
2. Proteins, lipids, and carbohydrates combine to form glycoproteins, glycolipids, or lipoproteins. (p. 40; Fig. 2.5)
3. Electrons are important for covalent and ionic bonds, energy capture and transfer, and formation of free radicals. (p. 42)
4. **Covalent bonds** form when adjacent atoms share one or more pairs of electrons. (p. 42; Fig. 2.6)
5. **Polar molecules** have atoms that share electrons unevenly. When atoms share electrons evenly, the molecule is **nonpolar**. (p. 42; Fig. 2.6)
6. An atom that gains or loses electrons acquires an electrical charge and is called an **ion**. (p. 42; Fig. 2.6)
7. **Ionic bonds** are strong bonds formed when oppositely charged ions are attracted to each other. (p. 42)
8. Weak **hydrogen bonds** form when hydrogen atoms in polar molecules are attracted to oxygen, nitrogen, or fluorine atoms. Hydrogen bonding among water molecules is responsible for the surface tension of water. (p. 42; Fig. 2.6)
9. **Van der Waals forces** are weak bonds that form when atoms are attracted to each other. (p. 42)

Noncovalent Interactions

iP Fluids and Electrolytes: Acid-Base Homeostasis

10. The universal solvent for biological solutions is water. (p. 46; Fig. 2.8)
11. The ease with which a molecule dissolves in a solvent is called its **solubility** in that solvent. **Hydrophilic** molecules dissolve easily in water, but **hydrophobic** molecules do not. (p. 46)
12. Molecular shape is created by covalent bond angles and weak noncovalent interactions within a molecule. (p. 46; Fig. 2.8)
13. Free H^+ in solution can disrupt a molecule's noncovalent bonds and alter its ability to function. (p. 48)
14. The **pH** of a solution is a measure of its hydrogen ion concentration. The more acidic the solution, the lower its pH. (p. 48; Fig. 2.9)
15. **Buffers** are solutions that moderate pH changes. (p. 49)

Protein Interactions

16. Most water-soluble proteins serve as enzymes, membrane transporters, signal molecules, receptors, binding proteins, immunoglobulins, or transcription factors. (p. 49)
17. **Ligands** bind to proteins at a binding site. According to the **induced-fit model** of protein binding, the shapes of the ligand and binding site do not have to match exactly. (p. 50; Fig. 2.10)
18. Proteins are specific about the ligands they will bind. The attraction of a protein to its ligand is called the protein's **affinity** for the ligand. The **dissociation constant** (K_d) is a quantitative measure of a protein's affinity for a given ligand. (p. 50)
19. Reversible binding reactions go to equilibrium. If equilibrium is disturbed, the reaction follows the **law of mass action** and shifts in the direction that restores the equilibrium ratio. (p. 51; Fig. 2.11)
20. Ligands may compete for a protein's binding site. If competing ligands mimic each other's activity, they are **agonists**. (p. 51)
21. Closely related proteins having similar function but different affinities for ligands are called **isoforms** of one another. (p. 52)
22. Some proteins must be activated, either by **proteolytic activation** or by addition of **cofactors**. (p. 53; Fig. 2.12)
23. **Competitive inhibitors** can be displaced from the binding site, but irreversible **antagonists** cannot. (p. 53; Fig. 2.12)
24. **Allosteric modulators** bind to proteins at a location other than the binding site. **Covalent modulators** bind with covalent bonds. Both types of modulators may activate or inhibit the protein. (p. 53; Fig. 2.12)
25. Extremes of temperature or pH will **denature** proteins. (p. 55; Fig. 2.13)
26. Cells regulate their proteins by **up-regulation** or **down-regulation** of protein synthesis and destruction. The amount of protein directly influences the magnitude of the cell's response. (p. 55; Fig. 2.13)
27. If the amount of protein (such as an enzyme) is constant, the amount of ligand determines the cell's response. If all binding proteins (such as enzymes) become **saturated** with ligand, the response reaches its maximum. (p. 55; Fig. 2.13)

Questions

Answers: p. A-1

Level One Reviewing Facts and Terms

1. List the four kinds of biomolecules. Give an example of each kind that is relevant to physiology.
2. True or false? All organic molecules are biomolecules.
3. When atoms bind tightly to one another, such as H_2O or O_2, one unit is called a(n) ________.
4. An atom of carbon has four unpaired electrons in an outer shell with space for eight electrons. How many covalent bonds will one carbon atom form with other atoms?

5. Fill in the blanks with the correct bond type.
   In a(n) ________ bond, electrons are shared between atoms. If the electrons are attracted more strongly to one atom than to the other, the molecule is said to be a(n) ________ molecule. If the electrons are evenly shared, the molecule is said to be a(n) ________ molecule.
6. Name two elements whose presence contributes to a molecule becoming a polar molecule.
7. Based on what you know from experience about the tendency of the following substances to dissolve in water, predict whether they are polar or nonpolar molecules: table sugar, vegetable oil.
8. A negatively charged ion is called a(n) ________, and a positively charged ion is called a(n) ________.
9. Define the pH of a solution. If pH less than 7, the solution is ________; if pH is greater than 7, the solution is ________.
10. A molecule that moderates changes in pH is called a ________.
11. Proteins combined with fats are called ________, and proteins combined with carbohydrates are called ________.
12. A molecule that binds to another molecule is called a(n) ________.
13. Match these definitions with their terms (not all terms are used):

(a) the ability of a protein to bind one molecule but not another	1. irreversible inhibition
(b) the part of a protein molecule that binds the ligand	2. induced fit
(c) the ability of a protein to alter shape as it binds a ligand	3. binding site
	4. specificity
	5. saturation

14. An ion, such as Ca^{2+} or Mg^{2+}, that must be present in order for an enzyme to work is called a(n) ________.
15. A protein whose structure is altered to the point that its activity is destroyed is said to be ________.

Level Two Reviewing Concepts

16. Mapping exercise: Make the list of terms into a map describing solutions.

• concentration	• nonpolar molecule
• equivalent	• polar molecule
• hydrogen bond	• solubility
• hydrophilic	• solute
• hydrophobic	• solvent
• molarity	• water
• mole	

17. A solution in which $[H^+] = 10^{-3}$ M is ________ (acidic/basic), whereas a solution in which $[H^+] = 10^{-10}$ M is ________ (acidic/basic). Give the pH for each of these solutions.
18. Name three nucleotides or nucleic acids, and tell why each one is important.
19. You know that two soluble proteins are isoforms of each other. What can you predict about their structures, functions, and affinities for ligands?
20. You have been asked to design some drugs for the purposes described below. Choose the desirable characteristic(s) for each drug from the numbered list.

(a) Drug A must bind to an enzyme and enhance its activity.	1. antagonist
(b) Drug B should mimic the activity of a normal nervous system signal molecule.	2. competitive inhibitor
(c) Drug C should block the activity of a membrane receptor protein.	3. agonist
	4. allosteric activator
	5. covalent modulator

Level Three Problem Solving

21. You have been summoned to assist with the autopsy of an alien being whose remains have been brought to your lab. The chemical analysis returns with 33% C, 40% O, 4% H, 14% N, and 9% P. From this information you conclude that the cells contain nucleotides, possibly even DNA or RNA. Your assistant is demanding that you tell him how you knew this. What do you tell him?
22. The harder a cell works, the more CO_2 it produces. CO_2 is carried in the blood according to the following equation:

$$CO_2 + H_2O \rightleftharpoons H_2CO_3 \rightleftharpoons H^+ + HCO_3^-$$

What effect does hard work by your muscle cells have on the pH of the blood?

Level Four Quantitative Problems

23. Calculate the amount of NaCl you would weigh out to make one liter of 0.9% NaCl. Explain how you would make a liter of this solution.
24. A 1.0 M NaCl solution contains 58.5 g of salt per liter. (a) How many molecules of NaCl are present in this solution? (b) How many millimoles of NaCl are present? (c) How many equivalents of Na^+ are present? (d) Express 58.5 g of NaCl per liter as a percent solution.
25. How would you make 200 mL of a 10% glucose solution? Calculate the molarity of this solution. How many millimoles of glucose are present in 500 mL of this solution? (Hint: What is the molecular mass of glucose?)
26. The graph shown below represents the binding of oxygen molecules (O_2) to two different proteins, myoglobin and hemoglobin, over a range of oxygen concentrations. Based on the graph, which protein has the higher affinity for oxygen? Explain your reasoning.

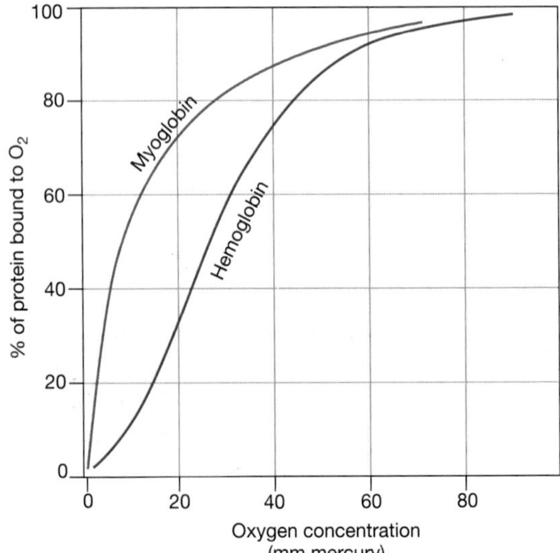

Answers

Answers to Concept Check Questions

Page 38

1. The major essential elements are O, C, H, N, P, Na, K, Ca, Mg, S, Cl.
2. $C_nH_{2n}O_n$ or $(CH_2O)_n$
3. An amino group is $-NH_2$. A carboxyl group is $-COOH$.

Page 42

4. paired
5. ion
6. (a) 2, (b) 4, (c) 1, (d) 3

Page 47

7. polar
8. hydrophilic
9. Na^+ and Cl^- ions form hydrogen bonds with the polar water molecules. This disrupts the ionic bonds that hold the NaCl crystal together.

Page 49

10. An acid dissociates into one or more H^+ plus anions.
11. pH is the concentration of H^+.
12. down

Page 51

13. Carbonic acid increases and pH decreases.

Page 52

14. Molecule B is a better candidate because its lower K_d means higher binding affinity.

Page 56

15. (a) 1, 4. (b) 2, 3. (c) 4 (can bind anywhere)
16. As the amount of protein decreases, the reaction rate decreases.
17. If a protein has reached saturation, the rate is at its maximum.

Answers to Figure and Graph Questions

Page 44

Figure 2.7:

1. A solution is composed of solute and solvent.
2. (d)
3. 18 amu = 18 Da
4. 74.6 g

Page 45

Figure 2.7:

5. A 0.1 M solution is the same as a 100 mM solution, which means the concentrations are equal.
6. The 5 g of glucose add volume to the solution, so if you begin with 100 mL of the solvent, you end up with more than 100 mL of solution.

Page 48

Figure 2.9:

1. Increased acidity means H^+ concentration increases and pH decreases. 2. Urine, stomach acid, and saliva are all inside the lumens of hollow organs, where they are not part of the body's internal environment [see Table 1.1 on p. 4].

Page 55

Figure 2.13:

(a) The protein is more active at 30 °C. (b) At protein concentration A, the rate is 1 mg/sec. Protein concentration C has a rate of 2.5 mg/sec. (c) When the ligand concentration is 200 mg/mL, the rate is 4 mg/sec.

Answers to Chemistry Review Quiz

1. c, e
2. a, b, f, h
3. a, b, d, g, h
4. *protons* and *electrons; neutrons.* Radiation.
5. carbon, oxygen, nitrogen, and hydrogen
6. P, K, Na, S, Ca, Cl
7. (a) zinc, (b) 20, (c) atomic number 53; average atomic mass = 126.9. Iodine = I.
8. Mg^{2+} has *lost* two *electrons.*
9. Loss of hydrogen's one electron leaves behind one proton.
10. (a) 11, (b) zero, (c) 12, (d) cation, (e) +1, (f) Na^+, (g) neon, (h) Ne
11. (a) $C_6H_{12}O_6$ (glucose); m.w. 180, (b) CO_2; m.w. 44, (c) leucine, $C_6H_{13}NO_2$, m.w. 131, (d) $C_3H_7NO_2$ (alanine); m.w. 89
12. (a) 1, (b) 5, (c) 4, (d) 2, (e) 3
13. Unsaturated fatty acids have double bonds between carbons. Each double bond removes two hydrogens from the molecule, therefore (c) $C_{18}H_{30}O_2$ is the most unsaturated because it has the fewest hydrogens.
14. (a) 4, (b) 5, (c) 1, (d) 2, (e) 3
15. (a) 3, (b) 1, (c) 5, (d) 2, (e) 4, 6
16. Proteins are composed of 20 different amino acids that can be linked in different numbers and an almost infinite number of sequences.
17. *amino; carboxyl* (or vice versa)
18. one or more phosphate groups, a 5-carbon sugar, and a base.
19. DNA: a double-stranded molecule with adenine, guanine, cytosine, and thymine linked in an α-helix; sugar is deoxyribose. RNA: a single-stranded molecule with uracil instead of thymine plus the sugar ribose.
20. Purines have two carbon rings. Pyrimidines have one carbon ring.

3 Compartmentation: Cells and Tissues

Cells are organisms, and entire animals and plants are aggregates of these organisms.

—Theodor Schwann, *1839*

Background Basics

Pancreas cell

What makes a compartment? We may think of something totally enclosed, like a room or a box with a lid. But not all compartments are totally enclosed . . . think of the modular cubicles that make up many modern workplaces. And not all functional compartments have walls . . . think of a giant hotel lobby divided into conversational groupings by careful placement of rugs and furniture. Biological compartments come with the same type of anatomic variability, ranging from totally enclosed structures such as cells to functional compartments without visible walls.

The first living compartment was probably a simple cell whose intracellular fluid was separated from the external environment by a wall made of phospholipids and proteins—the cell membrane. Cells are the basic functional unit of living organisms, and an individual cell can carry out all the processes of life.

As cells evolved, they acquired intracellular compartments separated from the intracellular fluid by membranes. Over time, groups of single-celled organisms began to cooperate and specialize their functions, eventually giving rise to multicellular organisms. As multicellular organisms evolved to become larger and more complex, their bodies became divided into various functional compartments.

Compartments are both an advantage and a disadvantage for organisms. On the advantage side, compartments separate biochemical processes that might otherwise conflict with one another. For example, protein synthesis takes place in one subcellular compartment while protein degradation is taking place in another. Barriers between compartments, whether inside a cell or inside a body, allow the contents of one compartment to differ from the contents of adjacent compartments. An extreme example is the intracellular compartment called the *lysosome*, with an internal pH of 5 [Fig. 2.9, p. 48]. This pH is so acidic that if the lysosome ruptures, it severely damages or kills the cell that contains it.

The disadvantage to compartments is that barriers between them can make it difficult to move needed materials from one compartment to another. Living organisms overcome this problem with specialized mechanisms that transport selected substances across membranes. [Membrane transport is the subject of Chapter 5.]

In this chapter we explore the theme of compartmentation by first looking at the various compartments that subdivide the human body, from body cavities to the subcellular compartments called organelles. We then examine how groups of cells with similar functions unite to form the tissues and organs of the body. Continuing the theme of molecular interactions, we also look at how different molecules and fibers in cells and tissues give rise to their *mechanical properties*: their shape, strength, flexibility, and the connections that hold tissues together.

RUNNING PROBLEM

Pap Tests Save Lives

Dr. George Papanicolaou has saved the lives of millions of women by popularizing the Pap test, a screening method that detects the early signs of cervical cancer. In the past 50 years, deaths from cervical cancer have dropped dramatically in countries that routinely use the Pap test. In contrast, cervical cancer is a leading cause of death in regions where Pap test screening is not routine, such as Africa and Central America. If detected early, cervical cancer is one of the most treatable forms of cancer. Today, Jan Melton, who had an abnormal Pap test several months ago, returns to Dr. Baird, her family physician, for a repeat test. The results may determine whether she should undergo treatment for cervical cancer.

63 65 72 84 90 92

Functional Compartments of the Body

The human body is a complex compartment separated from the outside world by layers of cells. Anatomically, the body is divided into three major body **cavities:** the *cranial cavity* (commonly referred to as the *skull*), the *thoracic cavity* (also called the *thorax*), and the *abdominopelvic cavity* (■ Fig. 3.1a). The cavities are separated from one another by bones and tissues, and they are lined with *tissue membranes*.

The cranial cavity {*cranium*, skull} contains the brain, our primary control center. The thoracic cavity is bounded by the spine and ribs on top and sides, with the muscular *diaphragm* forming the floor. The thorax contains the heart, which is enclosed in a membranous *pericardial sac* {*peri-*, around + *cardium*, heart}, and the two lungs, enclosed in separate *pleural sacs*.

The *abdomen* and *pelvis* form one continuous cavity, the *abdominopelvic cavity*. A tissue lining called the *peritoneum* lines the abdomen and surrounds the organs within it (stomach, intestines, liver, pancreas, gallbladder, and spleen). The kidneys lie outside the abdominal cavity, between the peritoneum and the muscles and bones of the back, just above waist level. The pelvis contains reproductive organs, the urinary bladder, and the terminal portion of the large intestine.

The Lumens of Some Organs Are Outside the Body

The hollow organs, such as heart, lungs, blood vessels, and intestines, create another set of compartments within the body. The interior of any hollow organ is called its **lumen** {*lumin*, window}. A lumen may be wholly or partially filled with air or fluid.

3

Levels of Organization: Body Compartments

BODY COMPARTMENTS

(a) ANATOMICAL: The Body Cavities

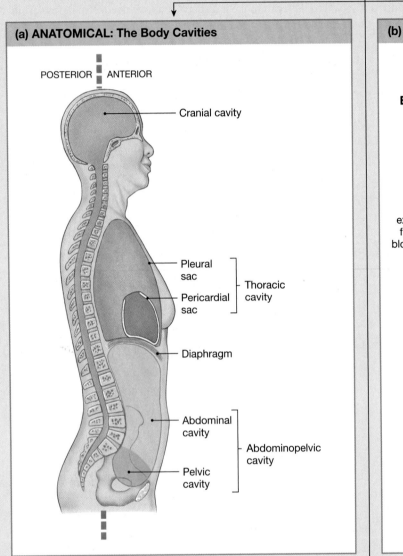

POSTERIOR | ANTERIOR

Cranial cavity

Pleural sac

Pericardial sac

Thoracic cavity

Diaphragm

Abdominal cavity

Pelvic cavity

Abdominopelvic cavity

(b) FUNCTIONAL: Body Fluid Compartments

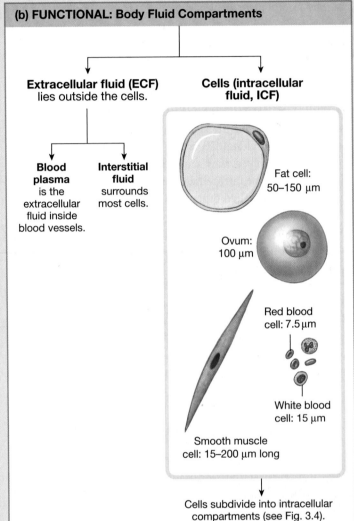

Extracellular fluid (ECF) lies outside the cells.

Blood plasma is the extracellular fluid inside blood vessels.

Interstitial fluid surrounds most cells.

Cells (intracellular fluid, ICF)

Fat cell: 50–150 μm

Ovum: 100 μm

Red blood cell: 7.5 μm

White blood cell: 15 μm

Smooth muscle cell: 15–200 μm long

Cells subdivide into intracellular compartments (see Fig. 3.4).

(c) Compartments Are Separated by Membranes

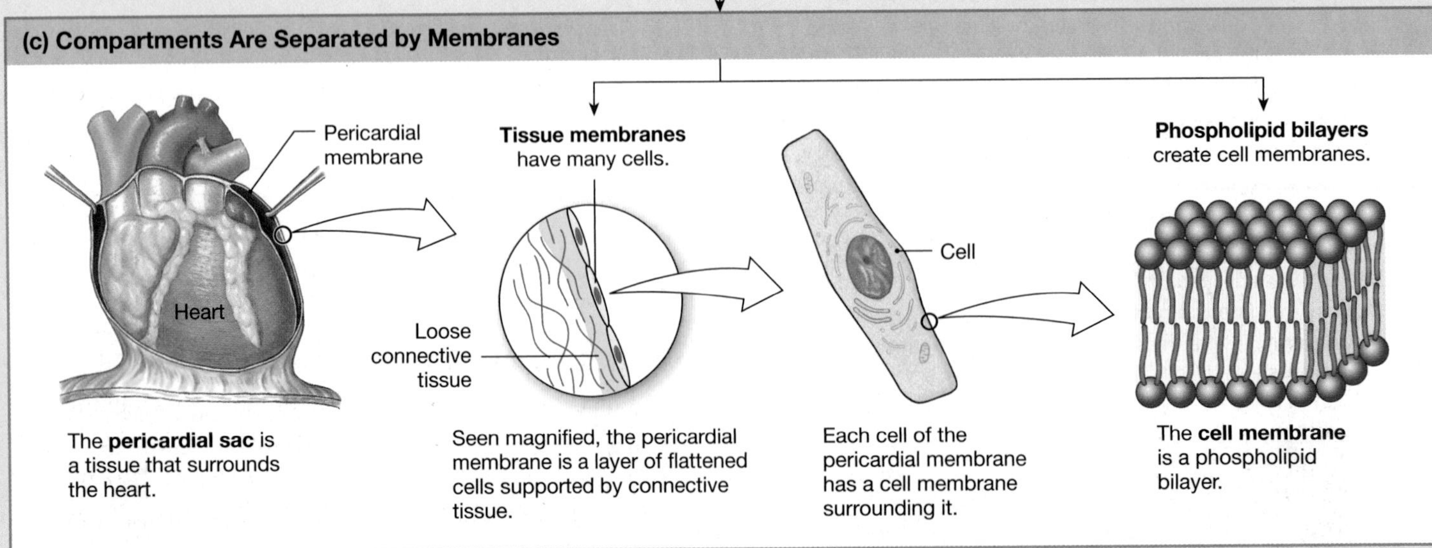

The **pericardial sac** is a tissue that surrounds the heart.

Seen magnified, the pericardial membrane is a layer of flattened cells supported by connective tissue.

Each cell of the pericardial membrane has a cell membrane surrounding it.

The **cell membrane** is a phospholipid bilayer.

For example, the lumens of blood vessels are filled with the fluid we call blood.

For some organs, the lumen is essentially an extension of the external environment, and material in the lumen is not truly part of the body's internal environment until it crosses the wall of the organ. For example, we think of our digestive tract as being "inside" our body, but in reality its lumen is part of the body's external environment [see Tbl. 1.1, p. 4]. An analogy would be the hole through a bead. The hole passes through the bead but is not actually inside the bead.

An interesting illustration of this distinction between the internal environment and the external environment in a lumen involves the bacterium *Escherichia coli.* This organism normally lives and reproduces inside the large intestine, an internalized compartment whose lumen is continuous with the external environment. When *E. coli* is residing in this location, it does not harm the host. However, if the intestinal wall is punctured by disease or accident and *E. coli* enters the body's internal environment, a serious infection can result.

Functionally, the Body Has Three Fluid Compartments

In physiology we are often more interested in functional compartments than in anatomical compartments. Most cells of the body are not in direct contact with the outside world. Instead their external environment is the extracellular fluid [Fig. 1.3, p. 11]. If we think of all the cells of the body together as one unit, we can then divide the body into two main fluid compartments: (1) the *extracellular fluid* (ECF) outside the cells and (2) the *intracellular fluid* (ICF) within the cells (Fig. 3.1b). The extracellular fluid subdivides further into **plasma**, the fluid portion of the blood, and **interstitial fluid** {*inter-*, between + *stare,* to stand}, which surrounds most cells of the body. The dividing wall between ECF and ICF is the cell membrane.

RUNNING PROBLEM

Cancer is a condition in which a small group of cells starts to divide uncontrollably and fails to differentiate into specialized cell types. Cancerous cells that originate in one tissue can escape from that tissue and spread to other organs through the circulatory system and the lymph vessels, a process known as *metastasis*.

Q1: Why does the treatment of cancer focus on killing the cancerous cells?

63 65 72 84 90 92

Biological Membranes

The word *membrane* {*membrana,* a skin} has two meanings in biology. Before the invention of microscopes in the sixteenth century, a membrane always described a tissue that lined a cavity or separated two compartments. Even today, we speak of *mucous membranes* in the mouth and vagina, the *peritoneal membrane* that lines the inside of the abdomen, the *pleural membrane* that covers the surface of the lungs, and the *pericardial membrane* that surrounds the heart. These visible membranes are tissues: thin, translucent layers of cells.

Once scientists observed cells with a microscope, the nature of the barrier between a cell's intracellular fluid and its external environment became a matter of great interest. By the 1890s, scientists had concluded that the outer surface of cells, the **cell membrane,** was a thin layer of lipids that separated the aqueous fluids of the interior and outside environment. We now know that cell membranes consist of microscopic double layers (*bilayers*) of phospholipids with protein molecules inserted in them.

Thus, the word *membrane* may apply either to a tissue or to a phospholipid-protein boundary layer (Fig. 3.1c). To add to the confusion, tissue membranes are often depicted in book illustrations as a single line, leading students to think of them as if they were similar in structure to the cell membrane. In this section you will learn more about these phospholipid membranes that create compartments for cells.

The Cell Membrane Separates Cell from Environment

There are two synonyms for the term *cell membrane: plasma membrane* and *plasmalemma*. We will use the term *cell membrane* in this book rather than *plasma membrane* or *plasmalemma* to avoid confusion with the term *blood plasma*. The general functions of the cell membrane include:

1. **Physical isolation.** The cell membrane is a physical barrier that separates intracellular fluid inside the cell from the surrounding extracellular fluid.
2. **Regulation of exchange with the environment.** The cell membrane controls the entry of ions and nutrients into the cell, the elimination of cellular wastes, and the release of products from the cell.
3. **Communication between the cell and its environment.** The cell membrane contains proteins that enable the cell to recognize and respond to molecules or to changes in its external environment. Any alteration in the cell membrane may affect the cell's activities.
4. **Structural support.** Proteins in the cell membrane hold the *cytoskeleton,* the cell's interior structural scaffolding, in place to maintain cell shape. Membrane proteins also create specialized junctions between adjacent cells or between

cells and the *extracellular matrix* {*extra-*, outside}, which is extracellular material that is synthesized and secreted by the cells. (**Secretion** is the process by which a cell releases a substance into the extracellular space.) Cell-cell and cell-matrix junctions stabilize the structure of tissues.

How can the cell membrane carry out such diverse functions? The answer lies in our current model of cell membrane structure.

Membranes Are Mostly Lipid and Protein

In the early decades of the twentieth century, researchers trying to decipher membrane structure ground up cells and analyzed their composition. They discovered that all biological membranes consist of a combination of lipids and proteins plus a small amount of carbohydrate. However, a simple and uniform structure did not account for the highly variable properties of membranes found in different types of cells. How could water cross the cell membrane to enter a red blood cell but not be able to enter certain cells of the kidney tubule? The explanation had to lie in the molecular arrangement of the proteins and lipids in the various membranes.

The ratio of protein to lipid varies widely, depending on the source of the membrane (■ Tbl. 3.1). Generally, the more metabolically active a membrane is, the more proteins it contains. For example, the inner membrane of a mitochondrion, which contains enzymes for ATP production, is three-quarters protein.

This chemical analysis of membranes was useful, but it did not explain the structural arrangement of lipids and proteins in a membrane. Studies in the 1920s suggested that there was enough lipid in a given area of membrane to create a double layer. The bilayer model was further modified in the 1930s to account for the presence of proteins. With the introduction of electron microscopy, scientists saw the cell membrane for the first time. The 1960s model of the membrane, as seen in in electron micrographs, was a "butter sandwich"—a clear layer of lipids sandwiched between two dark layers of protein.

By the early 1970s, freeze-fracture electron micrographs had revealed the actual three-dimensional arrangement of lipids and proteins within cell membranes. Because of what scientists learned from looking at freeze-fractured membranes, S. J. Singer and G. L. Nicolson in 1972 proposed the **fluid mosaic model** of the membrane. ■ Figure 3.2 highlights the major features of this contemporary model of membrane structure.

The lipids of biological membranes are mostly phospholipids arranged in a bilayer so that the phosphate heads are on the membrane surfaces and the lipid tails are hidden in the center of the membrane (Fig. 3.2b). The cell membrane is studded with protein molecules, like raisins in a slice of bread, and the extracellular surface has glycoproteins and glycolipids. All cell membranes are of relatively uniform thickness, about 8 nm.

Membrane Lipids Create a Hydrophobic Barrier

Three main types of lipids make up the cell membrane: phospholipids, sphingolipids, and cholesterol. Phospholipids are made of a glycerol backbone with two fatty acid chains extending to one side and a phosphate group extending to the other [p. 33]. The glycerol-phosphate head of the molecule is polar and thus hydrophilic. The fatty acid "tail" is nonpolar and thus hydrophobic.

When placed in an aqueous solution, phospholipids orient themselves so that the polar heads of the molecules interact with the water molecules while the nonpolar fatty acid tails "hide" by putting the polar heads between themselves and the water. This arrangement can be seen in three structures: the micelle, the liposome, and the phospholipid bilayer of the cell membrane (Fig. 3.2a). **Micelles** are small droplets with a single layer of phospholipids arranged so that the interior of the micelle is filled with hydrophobic fatty acid tails. Micelles are important in the digestion and absorption of fats in the digestive tract.

Liposomes are larger spheres with bilayer phospholipid walls. This arrangement leaves a hollow center with an aqueous core that can be filled with water-soluble molecules. Biologists think that a liposome-like structure was the precursor of the first living cell. Today, liposomes are being used as a medium to deliver drugs and cosmetics through the skin.

Phospholipids are the major lipid of membranes, but some membranes also have significant amounts of **sphingolipids.** Sphingolipids also have fatty acid tails, but their heads may be either phospholipids or glycolipids. Sphingolipids are slightly longer than phospholipids.

Table 3.1 Composition of Selected Membranes

Membrane	Protein	Lipid	Carbohydrate
Red blood cell membrane	49%	43%	8%
Myelin membrane around nerve cells	18%	79%	3%
Inner mitochondrial membrane	76%	24%	0%

The Cell Membrane

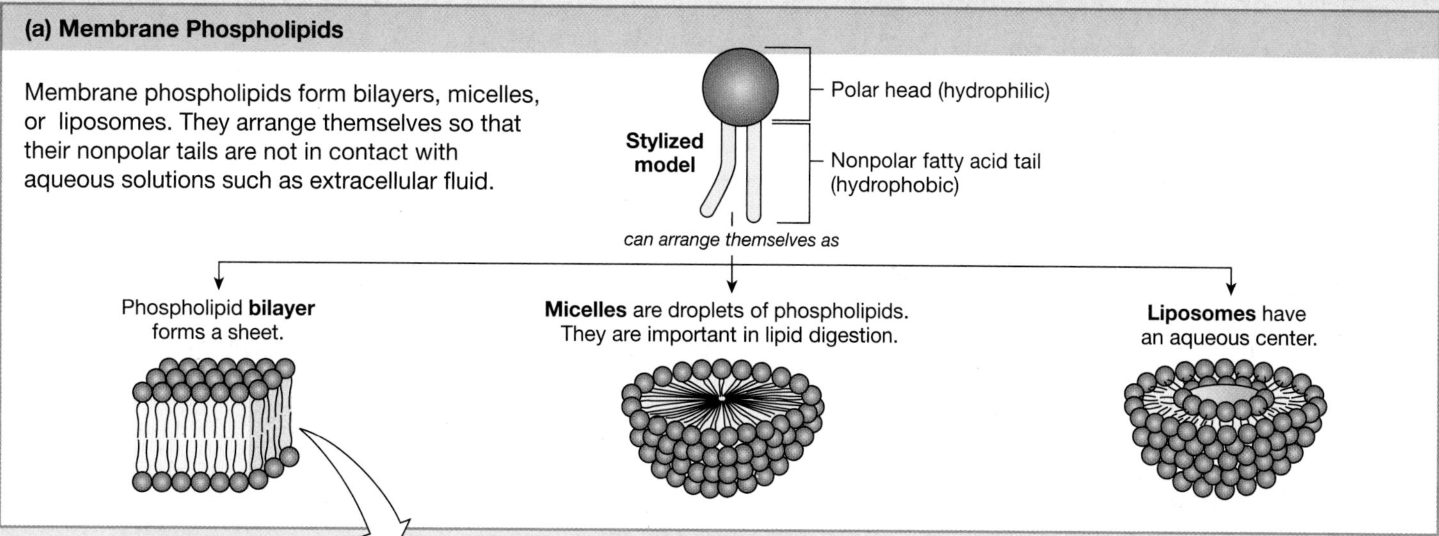
(a) Membrane Phospholipids
Membrane phospholipids form bilayers, micelles, or liposomes. They arrange themselves so that their nonpolar tails are not in contact with aqueous solutions such as extracellular fluid.
Stylized model
Polar head (hydrophilic)
Nonpolar fatty acid tail (hydrophobic)
can arrange themselves as
Phospholipid bilayer forms a sheet.
Micelles are droplets of phospholipids. They are important in lipid digestion.
Liposomes have an aqueous center.

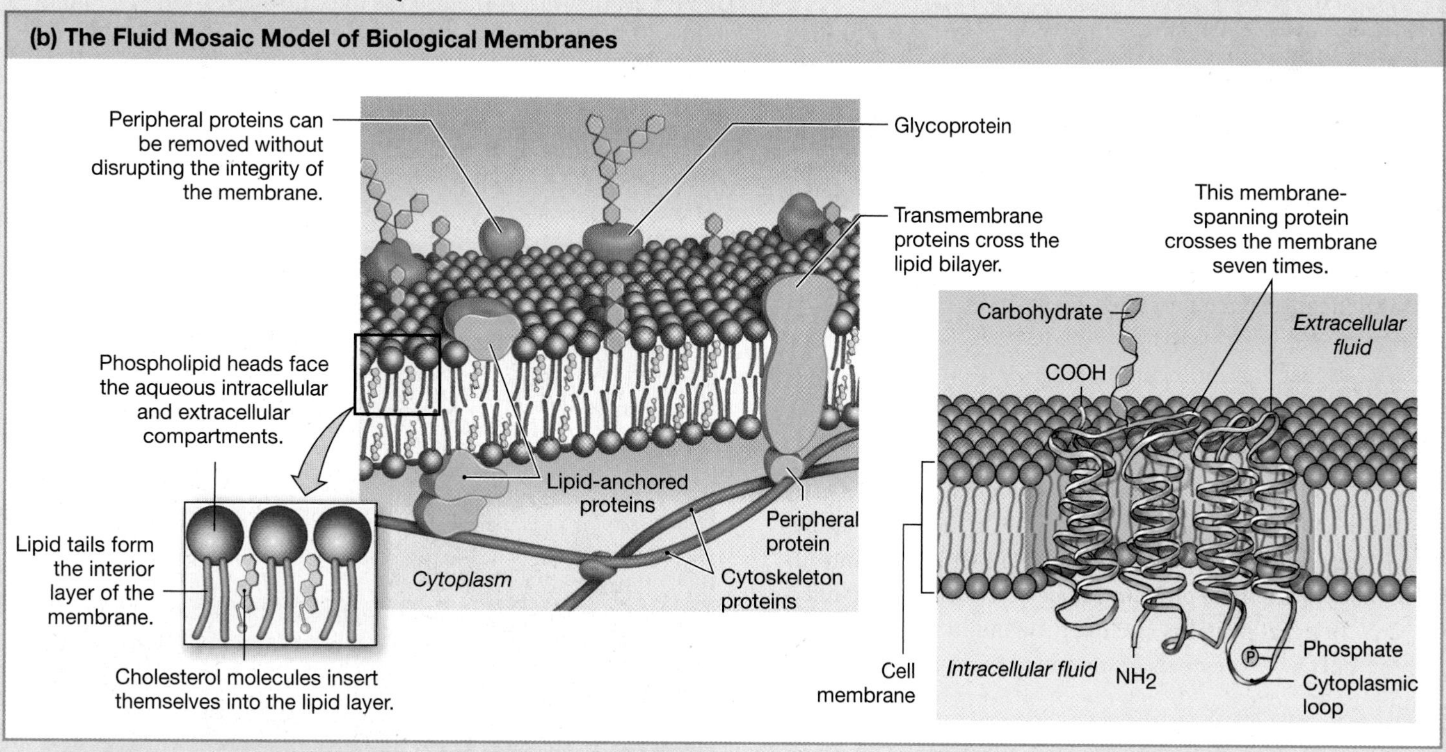
(b) The Fluid Mosaic Model of Biological Membranes
Peripheral proteins can be removed without disrupting the integrity of the membrane.
Glycoprotein
Transmembrane proteins cross the lipid bilayer.
This membrane-spanning protein crosses the membrane seven times.
Carbohydrate
Extracellular fluid
COOH
Phospholipid heads face the aqueous intracellular and extracellular compartments.
Lipid-anchored proteins
Peripheral protein
Lipid tails form the interior layer of the membrane.
Cytoplasm
Cytoskeleton proteins
Cell membrane
Intracellular fluid
NH2
Phosphate
Cytoplasmic loop
Cholesterol molecules insert themselves into the lipid layer.

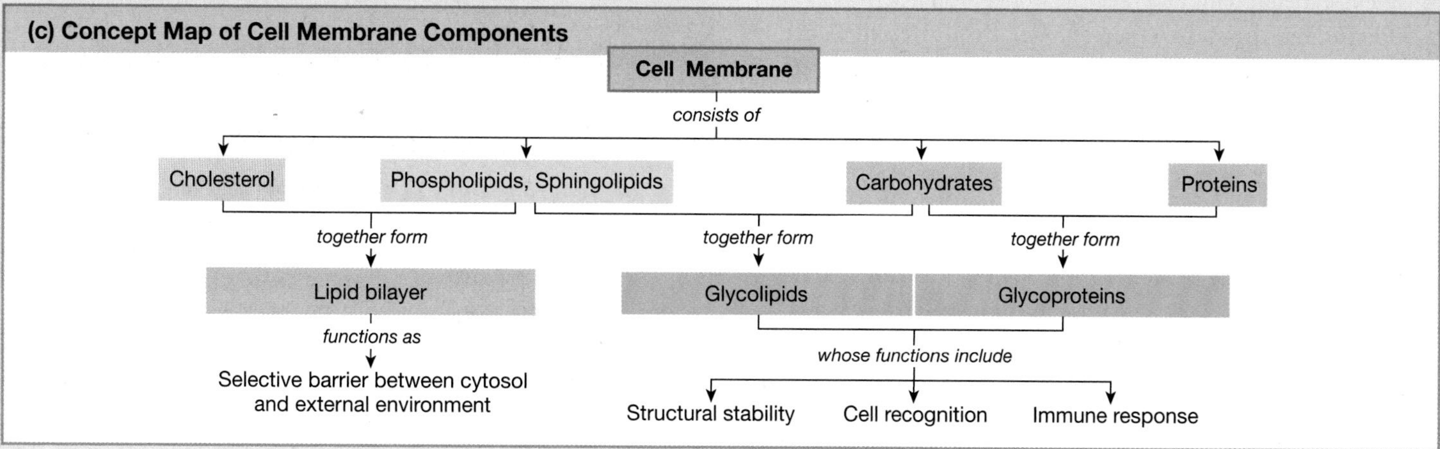
(c) Concept Map of Cell Membrane Components
Cell Membrane
consists of
Cholesterol
Phospholipids, Sphingolipids
Carbohydrates
Proteins
together form
together form
together form
Lipid bilayer
Glycolipids
Glycoproteins
functions as
Selective barrier between cytosol and external environment
whose functions include
Structural stability
Cell recognition
Immune response

BIOTECHNOLOGY

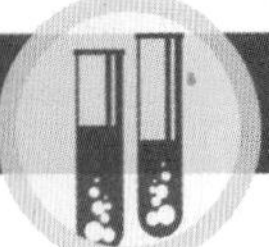

Liposomes for Beauty and Health

Many people first hear the term *liposome* in connection with cosmetic skin creams that promise to deliver ingredients to the cells that need them. This is not the only use for these tiny structures, however. Cosmetic manufacturers have simply adopted a medical technique developed to enhance the delivery of drugs. In medicine, the centers of liposomes are filled with drugs or with fragments of DNA for gene therapy. Then the liposomes are applied to the skin or injected into the bloodstream. To make drug delivery more specific, researchers now can make *immunoliposomes* that use antibodies to recognize specific types of cancer cells. By targeting the drugs to the cells they are treating, researchers hope to increase the effectiveness of the drugs and decrease unwanted side effects. To learn more about this topic, Google *liposome drug delivery* or *immunoliposomes.*

Cholesterol is also a significant part of many cell membranes. Cholesterol molecules, which are mostly hydrophobic, insert themselves between the hydrophilic heads of phospholipids (Fig. 3.2b). Cholesterol helps make membranes impermeable to small water-soluble molecules and keeps membranes flexible over a wide range of temperatures.

Membrane Proteins May Be Loosely or Tightly Bound to the Membrane

According to some estimates, membrane proteins may be nearly one-third of all proteins coded in our DNA. Each cell has between 10 and 50 different types of proteins inserted into its membranes. Membrane proteins can be described several different ways. **Integral proteins** are tightly bound to the membrane, and the only way they can be removed is by disrupting the membrane structure with detergents or other harsh methods that destroy the membrane's integrity. Integral proteins include transmembrane proteins and lipid-anchored proteins.

Peripheral proteins {*peripheria*, circumference} are attached to other membrane proteins by noncovalent interactions [p. 39] and can be separated from the membrane by chemical methods that do not disrupt the integrity of the membrane. Peripheral proteins include enzymes and some structural binding proteins that anchor the *cytoskeleton* (the cell's internal "skeleton") to the membrane (Fig. 3.2b).

Transmembrane proteins {*trans-* across} are also called *membrane-spanning* proteins because the protein's chains extend all the way across the cell membrane (Fig. 3.2c). When a protein crosses the membrane more than once, loops of the amino acid chain protrude into the cytoplasm and the extracellular fluid. Carbohydrates may attach to the extracellular loops, and phosphate groups may attach to the intracellular loops. Phosphorylation of proteins is one regulatory method cells use to alter protein function [p. 54].

Transmembrane proteins are classified into families according to how many transmembrane segments they have. Many physiologically important membrane proteins have seven transmembrane segments, as shown in Figure 3.2c. Others cross the membrane only once or up to as many as 12 times.

Membrane-spanning proteins are integral proteins, tightly but not covalently bound to the membrane. The 20–25 amino acids in the protein chain segments that pass through the bilayer are nonpolar. This allows those amino acids to create strong noncovalent interactions with the lipid tails of the membrane phospholipids, holding them tightly in place.

Some membrane proteins that were previously thought to be peripheral proteins are now known to be **lipid-anchored proteins** (Fig. 3.2b). Some of these proteins are covalently bound to lipid tails that insert themselves into the bilayer. Others, found only on the external surface of the cell, are held by a **GPI anchor** that consists of a membrane lipid plus a sugar-phosphate chain. (GPI stands for *glycosylphosphatidylinositol.*) Many lipid-anchored proteins are found in association with membrane sphingolipids, leading to the formation of specialized patches of membrane called *lipid rafts* (■ Fig. 3.3). The longer tails of the sphingolipids elevate the lipid rafts over their phospholipid neighbors.

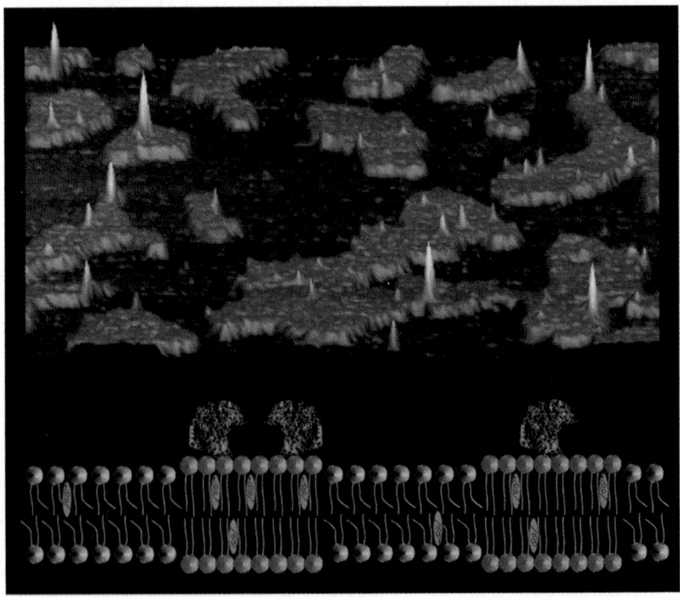

■ **Fig. 3.3** ***Lipid rafts are made of sphingolipids.*** Sphingolipids (orange) are longer than phospholipids and stick up above the phospholipids of the membrane (black). A lipid-anchored enzyme, placental alkaline phosphatase (yellow), is almost always associated with a lipid raft. Image courtesy of D. E. Saslowsky, J. Lawrence, X. Ren, D. A. Brown, R. M. Henderson, and J. M. Edwardson. Placental alkaline phosphatase is efficiently targeted to rafts in supported lipid bilayers. *J. Biol. Chem.* 277: 26966–26970, 2002.

According to the original fluid mosaic model of the cell membrane, membrane proteins could move laterally from location to location, directed by protein fibers that run just under the membrane surface. However, researchers have learned that this is not true of all membrane proteins. Some integral proteins are anchored to cytoskeleton proteins (Fig. 3.2b) and are therefore immobile. The ability of the cytoskeleton to restrict the movement of integral proteins allows cells to develop *polarity,* in which different faces of the cell have different proteins and therefore different properties. This is particularly important in the cells of the transporting epithelia, as you will see in multiple tissues in the body.

Membrane Carbohydrates Attach to Both Lipids and Proteins

Most membrane carbohydrates are sugars attached either to membrane proteins (glycoproteins) or to membrane lipids (glycolipids). They are found exclusively on the external surface of the cell, where they form a protective layer known as the **glycocalyx** {*glyco-*, sweet + *kalyx,* husk or pod}. Glycoproteins on the cell surface play a key role in the body's immune response. For example, the ABO blood groups are determined by the number and composition of sugars attached to membrane sphingolipids.

Figure 3.2c is a summary map organizing the structure of the cell membrane.

Concept Check Answers: p. 96

1. Name three types of lipids found in cell membranes.
2. Describe three types of membrane proteins and how they are associated with the cell membrane.
3. Why do phospholipids in cell membranes form a bilayer instead of a single layer?
4. How many phospholipid bilayers will a substance cross passing into a cell?

Intracellular Compartments

Much of what we know about cells comes from studies of simple organisms that consist of one cell. But humans are much more complex, with trillions of cells in their bodies. It has been estimated that there are more than 200 different types of cells in the human body, each cell type with its own characteristic structure and function.

During development, cells specialize and take specific shapes and functions. Each cell in the body inherits identical genetic information in its DNA, but no one cell uses all this information. During **differentiation,** only selected genes activate, transforming the cell into a specialized unit. In most cases, the final shape and size of a cell and its contents reflect its function. Figure 3.1b shows some of the different cells in the human body. Although these mature cells look very different from one another, they all started out alike in the early embryo, and they retain many features in common.

Cells Are Divided into Compartments

The structural organization of a cell can be compared to that of a medieval walled city. The city is separated from the surrounding countryside by a high wall, with entry and exit strictly controlled through gates that can be opened and closed. The city inside the walls is divided into streets and a diverse collection of houses and shops with varied functions. Within the city, a ruler in the castle oversees the everyday comings and goings of the city's inhabitants. Because the city depends on food and raw material from outside the walls, the ruler negotiates with the farmers in the countryside. Foreign invaders are always a threat, so the city ruler communicates and cooperates with the rulers of neighboring cities.

In the cell, the outer boundary is the cell membrane. Like the city wall, it controls the movement of material between the cell interior and the outside by opening and closing "gates" made of protein. The inside of the cell is divided into compartments rather than into shops and houses. Each of these compartments has a specific purpose that contributes to the function of the cell as a whole. In the cell, DNA in the nucleus is the "ruler in the castle," controlling both the internal workings of the cell and its interaction with other cells. Like the city, the cell depends on supplies from its external environment. It must also communicate and cooperate with other cells to keep the body functioning in a coordinated fashion.

■ Figure 3.4a is an overview map of cell structure. The cells of the body are surrounded by the dilute salt solution of the extracellular fluid. The cell membrane separates the inside environment of the cell (the intracellular fluid) from the extracellular fluid.

Internally the cell is divided into the *cytoplasm* and the *nucleus*. The cytoplasm consists of a fluid portion, called *cytosol;* insoluble particles called *inclusions;* insoluble protein fibers; and membrane-bound structures collectively known as *organelles*. Figure 3.4 shows a typical cell from the lining of the small intestine. It has most of the structures found in animal cells.

The Cytoplasm Includes Cytosol, Inclusions, Fibers, and Organelles

The **cytoplasm** includes all material inside the cell membrane except for the nucleus. The cytoplasm has four components:

1. **Cytosol** {*cyto-*, cell + sol(uble)}, or intracellular fluid: The cytosol is a semi-gelatinous fluid separated from the extracellular fluid by the cell membrane. The cytosol contains

Cell Structure

(a) This is an overview map of cell structure. The *cell membrane* separates the inside environment of the cell (the intracellular fluid) from the extracellular fluid. Internally the cell is divided into the *cytoplasm* and the *nucleus*. The cytoplasm consists of a fluid portion, called the *cytosol*; membrane-bound structures called *organelles*; insoluble particles called *inclusions*; and protein fibers that create the *cytoskeleton*.

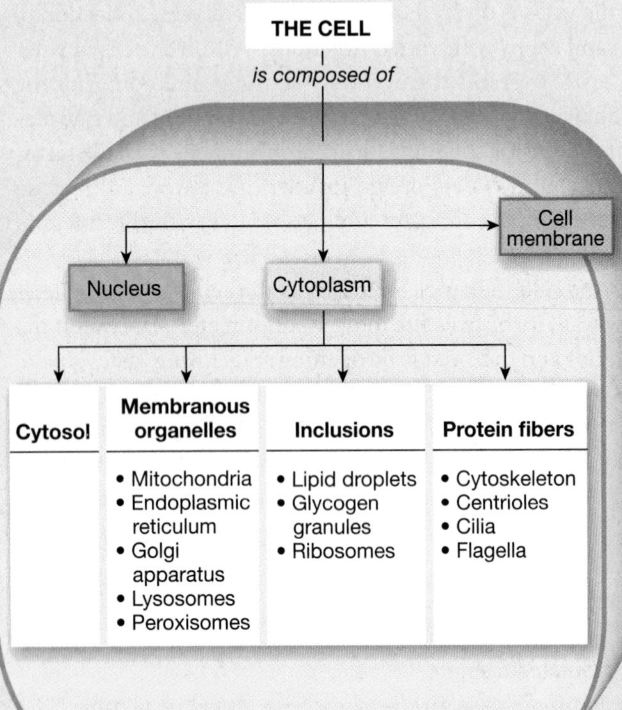

Extracellular fluid

(b) Cytoskeleton

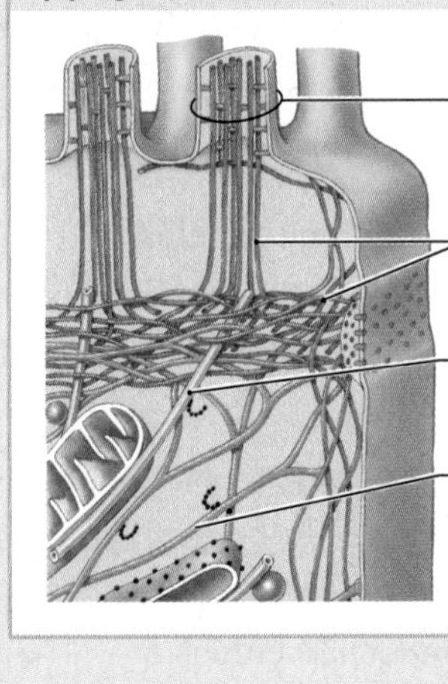

Microvilli increase cell surface area. They are supported by microfilaments.

Microfilaments form a network just inside the cell membrane.

Microtubules are the largest cytoskeleton fiber.

Intermediate filaments include myosin and keratin.

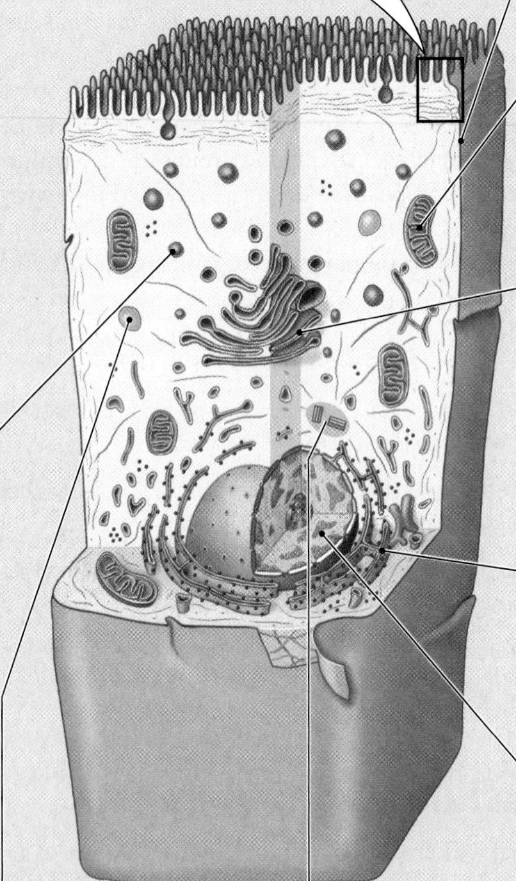

(c) Peroxisomes

Peroxisomes contain enzymes that break down fatty acids and some foreign materials.

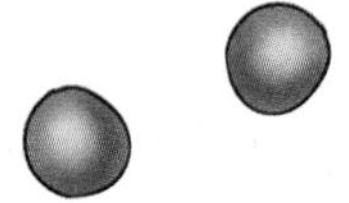

(d) Lysosomes

Lysosomes are small, spherical storage vesicles that contain powerful digestive enzymes.

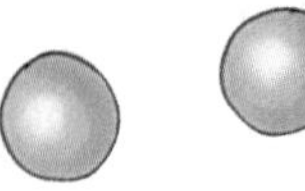

(e) Centrioles

Centrioles are made from microtubules and direct DNA movement during cell division.

Centrioles

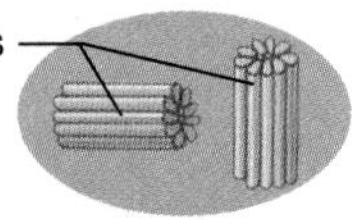

(f) Cell Membrane

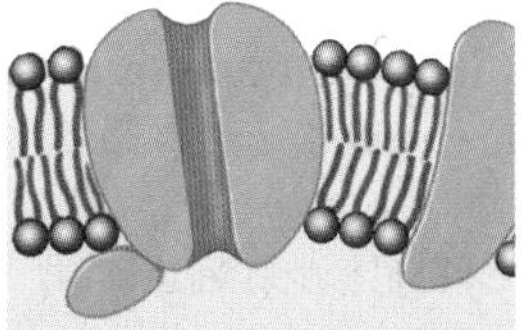

The **cell membrane** is a phospholipid bilayer studded with proteins that act as structural anchors, transporters, enzymes, or signal receptors. Glycolipids and glycoproteins occur only on the extracellular surface of the membrane. The cell membrane acts as both a gateway and a barrier between the cytoplasm and the extracellular fluid.

(g) Mitochondria

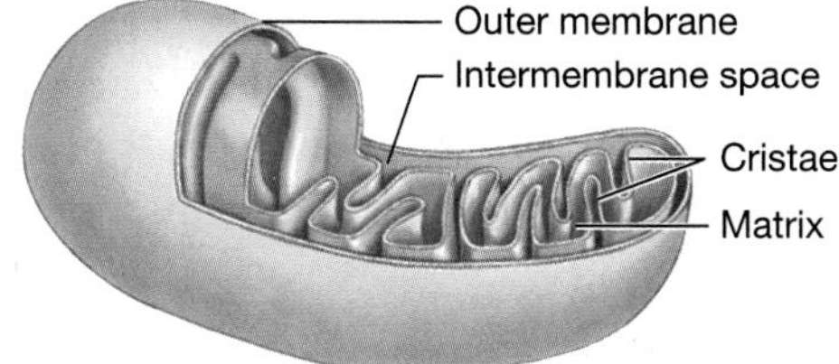

Mitchondria are spherical to elliptical organelles with a double wall that creates two separate compartments within the organelle. The inner **matrix** is surrounded by a membrane that folds into leaflets called **cristae**. The **intermembrane space**, which lies between the two membranes, plays an important role in ATP production. Mitochondria are the site of most ATP synthesis in the cell.

(h) Golgi Apparatus

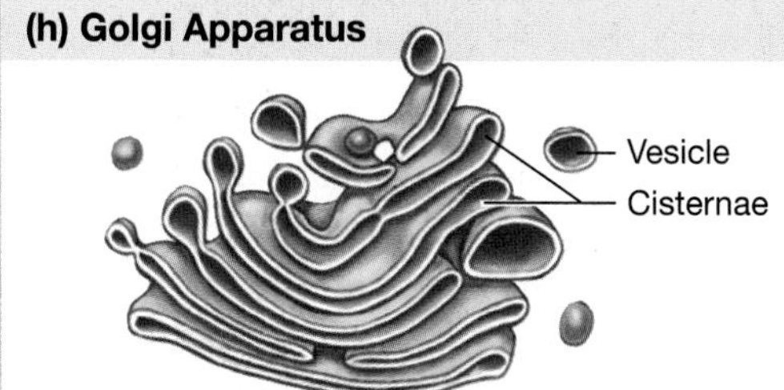

The **Golgi apparatus** consists of a series of hollow curved sacs called **cisternae** stacked on top of one another and surrounded by vesicles. The Golgi apparatus participates in protein modification and packaging.

(i) Endoplasmic Reticulum (ER)

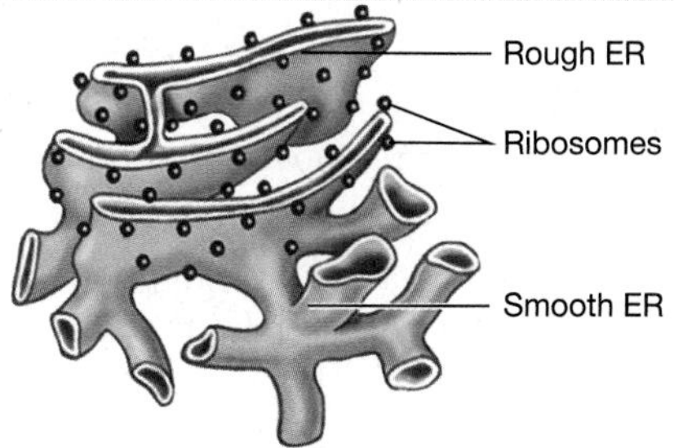

The **endoplasmic reticulum (ER)** is a network of interconnected membrane tubes that are a continuation of the outer nuclear membrane. **Rough endoplasmic reticulum** has a granular appearance due to rows of ribosomes dotting its cytoplasmic surface. **Smooth endoplasmic reticulum** lacks ribosomes and appears as smooth membrane tubes. The rough ER is the main site of protein synthesis. The smooth ER synthesizes lipids and, in some cells, concentrates and stores calcium ions.

(j) Nucleus

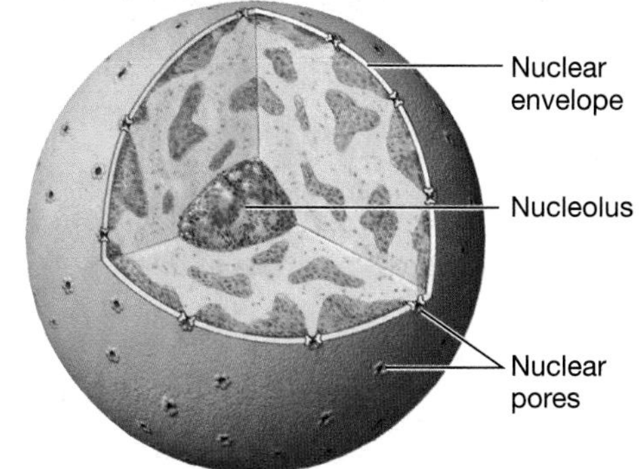

The **nucleus** is surrounded by a double-membrane **nuclear envelope**. Both membranes of the envelope are pierced here and there by **pores** to allow communication with the cytoplasm. The outer membrane of the nuclear envelope connects to the endoplasmic reticulum membrane. In cells that are not dividing, the nucleus appears filled with randomly scattered granular material composed of DNA and proteins. Usually a nucleus also contains from one to four larger dark-staining bodies of DNA, RNA, and protein called **nucleoli**.

RUNNING PROBLEM

During a Pap test for cervical cancer, tissue is sampled from the cervix (neck) of the uterus with a collection device that resembles a tiny brush. The cells are rinsed off the brush into preservative fluid that is sent to a laboratory. There the sample is processed onto a glass slide that will be examined first by a computer, then by a trained cytologist. The computer and cytologist look for *dysplasia* {*dys*-, abnormal + *-plasia*, growth or cell multiplication}, a change in the size and shape of cells that is suggestive of cancerous changes. Cancer cells can usually be recognized by a large nucleus surrounded by a relatively small amount of cytoplasm. Jan's first Pap test showed all the hallmarks of dysplasia.

Q2: What is happening in cancer cells that explains the large size of their nucleus and the relatively small amount of cytoplasm?

dissolved nutrients and proteins, ions, and waste products. The other components of the cytoplasm—inclusions, fibers, and organelles—are suspended in the cytosol.

2 **Inclusions** are particles of insoluble materials. Some are stored nutrients. Others are responsible for specific cell functions. These structures are sometimes called the *nonmembranous organelles*.

3 Insoluble **protein fibers** form the cell's internal support system, or **cytoskeleton**.

4 **Organelles**—"little organs"—are membrane-bounded compartments that play specific roles in the overall function of the cell. For example, the organelles called mitochondria (singular, *mitochondrion*) generate most of the cell's ATP, and the organelles called lysosomes act as the digestive system of the cell. The organelles work in an integrated manner, each organelle taking on one or more of the cell's functions.

Inclusions Are in Direct Contact with the Cytosol

The inclusions of cells do not have boundary membranes and so are in direct contact with the cytosol. Movement of material between inclusions and the cytosol does not require transport across a membrane. Nutrients are stored as glycogen granules and lipid droplets. Most inclusions with functions other than nutrient storage are made from protein or combinations of RNA and protein.

Ribosomes (Fig. 3.4i) are small, dense granules of RNA and protein that manufacture proteins under the direction of the cell's DNA [see Chapter 4 for details]. Ribosomes attached to the cytosolic surface of organelles are called **fixed ribosomes.** Those suspended free in the cytosol are **free ribosomes.** Some free ribosomes form groups of 10 to 20 known as **polyribosomes.** A ribosome that is fixed one minute may release and become a free ribosome the next. Ribosomes are most numerous in cells that synthesize proteins for export out of the cell.

Cytoplasmic Protein Fibers Come in Three Sizes

The three families of cytoplasmic protein fibers are classified by diameter and protein composition (■ Tbl. 3.2). All fibers are polymers of smaller proteins. The thinnest are **actin fibers,** also called *microfilaments*. Somewhat larger **intermediate filaments** may be made of different types of protein, including *keratin* in hair and skin, and *neurofilament* in nerve cells. The largest protein fibers are the hollow **microtubules,** made of a protein called **tubulin.** A large number of *accessory proteins* are associated with the cell's protein fibers.

The insoluble protein fibers of the cell have two general purposes: structural support and movement. Structural support comes primarily from the cytoskeleton. Movement of the cell or of elements within the cell takes place with the aid of protein fibers and a group of specialized enzymes called motor proteins. These functions are discussed in more detail in the sections that follow.

Table 3.2 Diameter of Protein Fibers in the Cytoplasm

	Diameter	Type of Protein	Functions
Microfilaments	7 nm	Actin (globular)	Cytoskeleton; associates with myosin for muscle contraction
Intermediate filaments	10 nm	Keratin, neurofilament protein (filaments)	Cytoskeleton, hair and nails, protective barrier of skin
Microtubules	25 nm	Tubulin (globular)	Movement of cilia, flagella, and chromosomes; intracellular transport of organelles; cytoskeleton

Microtubules Form Centrioles, Cilia, and Flagella

The largest cytoplasmic protein fibers, the microtubules, create the complex structures of centrioles, cilia, and flagella, which are all involved in some form of cell movement. The cell's *microtubule-organizing center,* the **centrosome,** assembles tubulin monomers into microtubules. The centrosome appears as a region of darkly staining material close to the cell nucleus. In most animal cells, the centrosome contains two **centrioles,** shown in the typical cell of Figure 3.4e. Each centriole is a cylindrical bundle of 27 microtubules, arranged in nine triplets. In cell division, the centrioles direct the movement of DNA strands. Cells that have lost their ability to undergo cell division, such as mature nerve cells, lack centrioles.

Cilia are short, hairlike structures projecting from the cell surface like the bristles of a brush {singular, *cilium,* Latin for eyelash}. Most cells have a single short cilium, but cells lining the upper airways and part of the female reproductive tract are covered with cilia. In these tissues, ciliary movement, like a waving field of grain, creates currents that sweep fluids or secretions across the cell surface.

The surface of a cilium is a continuation of the cell membrane. The core of *motile,* or moving, cilia contains nine pairs of microtubules surrounding a central pair (■ Fig. 3.5b). The microtubules terminate just inside the cell at the *basal body.* These cilia beat rhythmically back and forth when the microtubule pairs in their core slide past each other with the help of the motor protein *dynein.*

Flagella have the same microtubule arrangement as cilia but are considerably longer {singular, *flagellum,* Latin for whip}. Flagella are found on free-floating single cells, and in humans the only flagellated cell is the male sperm cell. A sperm cell has only one flagellum, in contrast to ciliated cells, which may have one surface almost totally covered with cilia (Fig. 3.5a). The wavelike movements of the flagellum push the sperm through fluid, just as undulating contractions of a snake's body push it headfirst through its environment. Flagella bend and move by the same basic mechanism as cilia.

EMERGING CONCEPTS

Single Cilia Are Sensors

Cilia in the body are not limited to the airways and the female reproductive tract. Scientists have known for years that most cells of the body contain a single, stationary, or *non-motile*, cilium, but they thought that these solitary **primary cilia** were mostly evolutionary remnants and of little significance. Primary cilia differ structurally from motile cilia because they lack the central pair of microtubules found in motile cilia (a 9 + 0 arrangement instead of 9 + 2; see Fig. 3.5). Researchers in recent years have learned that primary cilia actually serve a function. They can act as sensors of the external environment, passing information into the cell. For example, primary cilia in photoreceptors of the eye help with light sensing, and primary cilia in the kidney sense fluid flow. Using molecular techniques, scientists have found that these small, insignificant hairs play critical roles during embryonic development as well. Mutations to ciliary proteins cause disorders (*ciliopathies*) ranging from polycystic kidney disease and loss of vision to cancer. The role of primary cilia in other disorders, including obesity, is currently a hot topic in research.

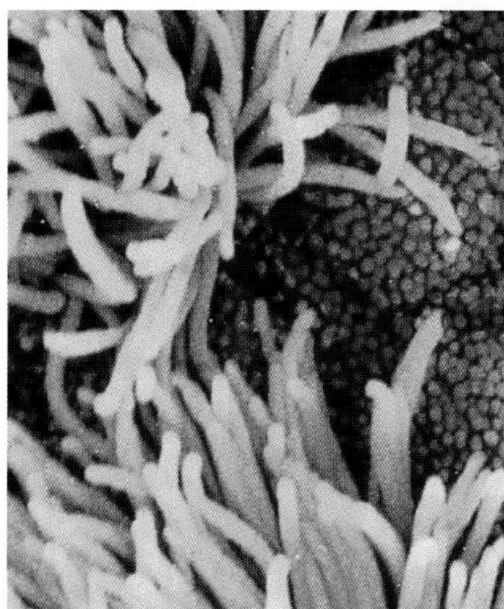

(a) Cilia

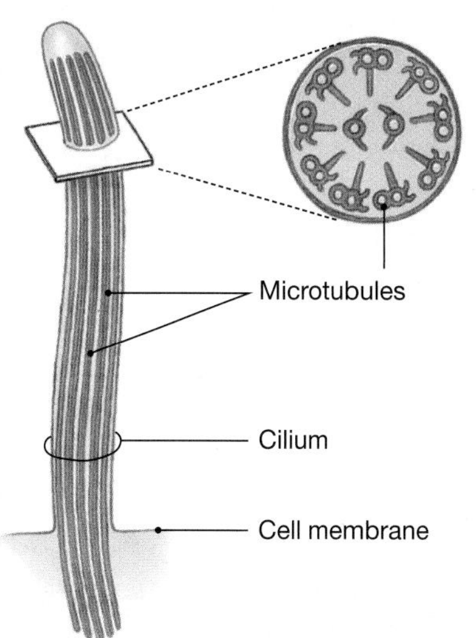

(b) Cilia and flagella have 9 pairs of microtubules surrounding a central pair.

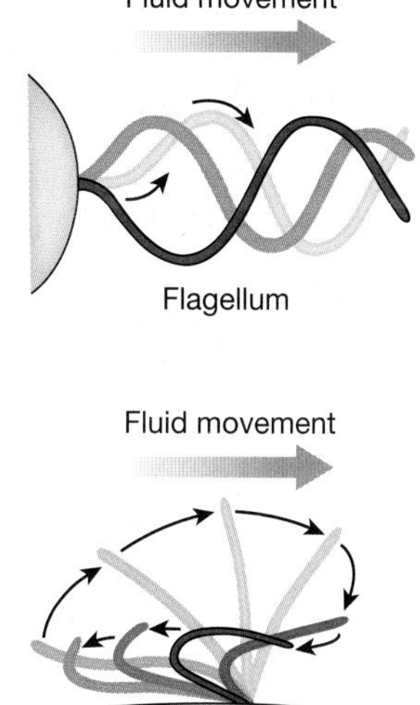

(c) The beating of cilia and flagella creates fluid movement.

■ **Fig. 3.5** ***Centrioles, cilia, and flagella.*** All three structures are formed from microtubules.

The Cytoskeleton Is a Changeable Scaffold

The cytoskeleton is a flexible, changeable three-dimensional scaffolding of actin microfilaments, intermediate filaments, and microtubules that extends throughout the cytoplasm. Some cytoskeleton protein fibers are permanent, but most are synthesized or disassembled according to the cell's needs. Because of the cytoskeleton's changeable nature, its organizational details are complex and are not discussed in detail here.

The cytoskeleton has at least five important functions.

1. **Cell shape.** The protein scaffolding of the cytoskeleton provides mechanical strength to the cell and in some cells plays an important role in determining the shape of the cell. Figure 3.4b shows how cytoskeletal fibers help support **microvilli** {*micro-*, small + *villus,* tuft of hair}, finger-like extensions of the cell membrane that increase the surface area for absorption of materials.
2. **Internal organization.** Cytoskeletal fibers stabilize the positions of organelles. Figure 3.4b illustrates organelles held in place by the cytoskeleton. Note, however, that this figure is only a snapshot of one moment in the cell's life. The interior arrangement and composition of a cell are dynamic, changing from minute to minute in response to the needs of the cell, just as the inside of the walled city is always in motion. One disadvantage of the static illustrations in textbooks is that they are unable to represent movement and the dynamic nature of many physiological processes accurately.
3. **Intracellular transport.** The cytoskeleton helps transport materials into the cell and within the cytoplasm by serving as an intracellular "railroad track" for moving organelles. This function is particularly important in cells of the nervous system, where material must be transported over intracellular distances as long as a meter.
4. **Assembly of cells into tissues.** Protein fibers of the cytoskeleton connect with protein fibers in the extracellular space, linking cells to one another and to supporting material outside the cells. In addition to providing mechanical strength to the tissue, these linkages allow the transfer of information from one cell to another.
5. **Movement.** The cytoskeleton helps cells move. For example, the cytoskeleton helps white blood cells squeeze out of blood vessels and growing nerve cells send out long extensions as they elongate. Cilia and flagella on the cell membrane are able to move because of their microtubule cytoskeleton. Special motor proteins facilitate movement and intracellular transport by using energy from ATP to slide or step along cytoskeletal fibers.

Motor Proteins Create Movement

Motor proteins are proteins that are able to convert stored energy into directed movement. Three groups of motor proteins are associated with the cytoskeleton: myosins, kinesins, and dyneins. All three groups use energy stored in ATP to propel themselves along cytoskeleton fibers.

Myosins bind to actin fibers and are best known for their role in muscle contraction [see details in Chapter 12]. **Kinesins** and **dyneins** assist the movement of vesicles along microtubules. Dyneins also associate with the microtubule bundles of cilia and flagella to help create their whiplike motion.

Most motor proteins are made of multiple protein chains arranged into three parts: two heads that bind to the cytoskeleton fiber, a neck, and a tail region that is able to bind "cargo," such as an organelle that needs to be transported through the cytoplasm (■ Fig. 3.6). The heads alternately bind to the cytoskeleton fiber, then release and "step" forward using the energy stored in ATP.

Concept Check Answers: p. 96

5. Name the three sizes of cytoplasmic protein fibers.
6. How would the absence of a flagellum affect a sperm cell?
7. What is the difference between cytoplasm and cytosol?
8. What is the difference between a cilium and a flagellum?
9. What is the function of motor proteins?

Organelles Create Compartments for Specialized Functions

Organelles are subcellular compartments separated from the cytosol by one or more phospholipid membranes similar in structure to the cell membrane. The compartments created

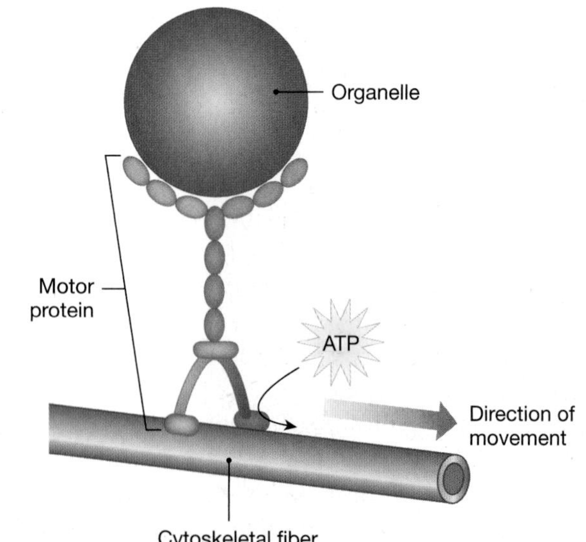

■ **Fig. 3.6** ***Motor proteins.*** Motor proteins have multiple protein chains that form two heads, a neck, and a tail that can bind to organelles or other cargo. The heads "walk" along cytoskeletal fibers with the help of energy from ATP.

by organelles allow the cell to isolate substances and segregate functions. For example, an organelle might contain substances that could be harmful to the cell, such as digestive enzymes. Figures 3.4g, 3.4h, and 3.4i show the four major groups of organelles: mitochondria, the Golgi apparatus, the endoplasmic reticulum, and membrane-bound spheres called **vesicles** {*vesicula,* bladder}.

Mitochondria **Mitochondria** {singular, *mitochondrion; mitos,* thread + *chondros,* granule} are unique organelles in several ways. First, they have an unusual double wall that creates two separate compartments within the mitochondrion (Fig. 3.4g). In the center, inside the inner membrane, is a compartment called the **mitochondrial matrix** {*matrix,* female animal for breeding}. The matrix contains enzymes, ribosomes, granules, and surprisingly, its own unique DNA. This **mitochondrial DNA** has a different nucleotide sequence from that found in the nucleus. Because mitochondria have their own DNA, they can manufacture some of their own proteins.

Why do mitochondria contain DNA when other organelles do not? This question has been the subject of intense scrutiny. According to the *prokaryotic endosymbiont theory,* mitochondria are the descendants of bacteria that invaded cells millions of years ago. The bacteria developed a mutually beneficial relationship with their hosts and soon became an integral part of the host cells. Supporting evidence for this theory is the fact that our mitochondrial DNA, RNA, and enzymes are similar to those in bacteria but unlike those in our own cell nuclei.

The second compartment inside a mitochondrion is the **intermembrane space**, which lies between the outer and inner mitochondrial membranes. This compartment plays an important role in mitochondrial ATP production, and the number of mitochondria in a cell is directly related to the cell's energy needs. For example, skeletal muscle cells, which use a lot of energy, have many more mitochondria than less active cells, such as adipose (fat) cells.

Another unusual characteristic of mitochondria is their ability to replicate themselves even when the cell to which they belong is not undergoing cell division. This process is aided by the mitochondrial DNA, which allows the organelles to direct their own duplication. Mitochondrial replication takes place by budding, during which small daughter mitochondria pinch off from an enlarged parent. For instance, exercising muscle cells that experience increased energy demands over a period of time may meet the demand for more ATP by increasing the number of mitochondria in their cytoplasm.

The Endoplasmic Reticulum The **endoplasmic reticulum,** or **ER,** is a network of interconnected membrane tubes with three major functions: synthesis, storage, and transport of biomolecules (Fig. 3.4i). The name *reticulum* comes from the Latin word for *net* and refers to the netlike arrangement of the tubules. Electron micrographs reveal that there are two forms of endoplasmic reticulum: **rough endoplasmic reticulum** (RER) and **smooth endoplasmic reticulum** (SER).

The rough endoplasmic reticulum is the main site of protein synthesis. Proteins are assembled on ribosomes attached to the cytoplasmic surface of the rough ER, then inserted into the rough ER lumen, where they undergo chemical modification.

The smooth endoplasmic reticulum lacks attached ribosomes and is the main site for the synthesis of fatty acids, steroids, and lipids [p. 33]. Phospholipids for the cell membrane are produced here, and cholesterol is modified into steroid hormones, such as the sex hormones estrogen and testosterone. The smooth ER of liver and kidney cells detoxifies or inactivates drugs. In skeletal muscle cells, a modified form of smooth ER stores calcium ions (Ca^{2+}) to be used in muscle contraction.

The Golgi Apparatus The **Golgi apparatus** (also known as the Golgi complex) was first described by Camillo Golgi in 1898 (Fig. 3.4h). For years, some investigators thought that this organelle was just a result of the fixation process needed to prepare tissues for viewing under the light microscope. However, we now know from electron microscope studies that the Golgi apparatus is indeed a discrete organelle. It consists of a series of hollow curved sacs, called *cisternae,* stacked on top of one another like a series of hot water bottles and surrounded by vesicles. The Golgi apparatus receives proteins made on the rough ER, modifies them, and packages them into the vesicles.

Cytoplasmic Vesicles Membrane-bound cytoplasmic vesicles are of two kinds: secretory and storage. **Secretory vesicles** contain proteins that will be released from the cell. The contents of most **storage vesicles**, however, never leave the cytoplasm.

Lysosomes {*lysis,* dissolution + *soma,* body} are small storage vesicles that appear as membrane-bound granules in the cytoplasm (Fig. 3.4d). Lysosomes act as the digestive system of the cell. They use powerful enzymes to break down bacteria or old organelles, such as mitochondria, into their component molecules. Those molecules that can be reused are reabsorbed into the cytosol, while the rest are dumped out of the cell. As many as 50 types of enzymes have been identified from lysosomes of different cell types.

Because lysosomal enzymes are so powerful, early workers puzzled over the question of why these enzymes do not normally destroy the cell that contains them. What scientists discovered was that lysosomal enzymes are activated only by very acidic conditions, 100 times more acidic than the normal acidity level in the cytoplasm. When lysosomes first pinch off from the Golgi apparatus, their interior pH is about the same as that of the cytosol, 7.0–7.3. The enzymes are inactive at this pH. Their inactivity serves as a form of insurance. If the lysosome breaks or accidentally releases enzymes, they will not harm the cell.

However, as the lysosome sits in the cytoplasm, it accumulates H^+ in a process that uses energy. Increasing concentrations of H^+ decrease the pH inside the vesicle to 4.8–5.0, and the enzymes activate. Once activated, lysosomal enzymes can break down biomolecules inside the vesicle. The lysosomal membrane is not affected by the enzymes.

The digestive enzymes of lysosomes are not always kept isolated within the organelle. Occasionally, lysosomes release their enzymes outside the cell to dissolve extracellular support material, such as the hard calcium carbonate portion of bone. In other instances, cells allow their lysosomal enzymes to come in contact with the cytoplasm, leading to self-digestion of all or part of the cell. When muscles *atrophy* (shrink) from lack of use or the uterus diminishes in size after pregnancy, the loss of cell mass is due to the action of lysosomes.

The inappropriate release of lysosomal enzymes has been implicated in certain disease states, such as the inflammation and destruction of joint tissue in *rheumatoid arthritis.* In the inherited conditions known as *lysosomal storage diseases,* lysosomes are ineffective because they lack specific enzymes. One of the best-known lysosomal storage diseases is the fatal inherited condition known as *Tay-Sachs disease.* Infants with Tay-Sachs disease have defective lysosomes that fail to break down glycolipids. Accumulation of glycolipids in nerve cells causes nervous system dysfunction, including blindness and loss of coordination. Most infants afflicted with Tay-Sachs disease die in early childhood.

Peroxisomes are storage vesicles that are even smaller than lysosomes (Fig. 3.4c). For years, they were thought to be a kind of lysosome, but we now know that they contain a different set of enzymes. Their main function appears to be to degrade long-chain fatty acids and potentially toxic foreign molecules.

Peroxisomes get their name from the fact that the reactions that take place inside them generate hydrogen peroxide (H_2O_2), a toxic molecule. The peroxisomes rapidly convert this peroxide to oxygen and water using the enzyme *catalase.* Peroxisomal disorders disrupt the normal processing of lipids and can severely disrupt neural function by altering the structure of nerve cell membranes.

Concept Check Answers: p. 96

10. What distinguishes organelles from inclusions?
11. What is the anatomical difference between rough endoplasmic reticulum and smooth endoplasmic reticulum? What is the functional difference?
12. How do lysosomes differ from peroxisomes?
13. Apply the physiological theme of compartmentation to organelles in general and to mitochondria in particular.
14. Microscopic examination of a cell reveals many mitochondria. What does this observation imply about the cell's energy requirements?
15. Examining tissue from a previously unknown species of fish, you discover a tissue containing large amounts of smooth endoplasmic reticulum in its cells. What is one possible function of these cells?

The Nucleus Is the Cell's Control Center

The nucleus of the cell contains DNA, the genetic material that ultimately controls all cell processes. Figure 3.4j illustrates the structure of a typical nucleus. Its boundary, or **nuclear envelope,** is a two-membrane structure that separates the nucleus from the cytoplasmic compartment. Both membranes of the envelope are pierced here and there by round holes, or **pores.**

Communication between the nucleus and cytosol occurs through the **nuclear pore complexes,** large protein complexes with a central channel. Ions and small molecules move freely through this channel when it is open, but transport of large molecules such as proteins and RNA is a process that requires energy. Specificity of the transport process allows the cell to restrict DNA to the nucleus and various enzymes to either the cytoplasm or the nucleus.

In electron micrographs of cells that are not dividing, the nucleus appears filled with randomly scattered granular material, or **chromatin,** composed of DNA and associated proteins. Usually a nucleus also contains from one to four larger dark-staining bodies of DNA, RNA, and protein called **nucleoli** {singular, *nucleolus,* little nucleus}. Nucleoli contain the genes and proteins that control the synthesis of RNA for ribosomes.

The process of protein synthesis, modification, and packaging in different parts of the cell is an excellent example of how compartmentation allows separation of function, as shown in ■ Figure 3.7. RNA for protein synthesis is made from DNA templates in the nucleus (1), then transported to the cytoplasm through the nuclear pores (2). In the cytoplasm, proteins are synthesized on ribosomes that may be free inclusions (3) or attached to the rough endoplasmic reticulum (4). The newly made protein is compartmentalized in the lumen of the rough ER (5), where it is modified before being packaged into a vesicle (6). The vesicles fuse with the Golgi apparatus, allowing additional modification of the protein in the Golgi lumen (7). The modified proteins leave the Golgi packaged in either storage vesicles (9) or secretory vesicles whose contents will be released into the extracellular fluid (10). The molecular details of protein synthesis are discussed elsewhere [see Chapter 4].

Tissues of the Body

Despite the amazing variety of intracellular structures, no single cell can carry out all the processes of the mature human body. Instead, cells assemble into the larger units we call tissues. The cells in tissues are held together by specialized connections called *cell junctions* and by other support structures. Tissues range in complexity from simple tissues containing only one cell type, such as the lining of blood vessels, to complex tissues containing many cell types and extensive extracellular material, such as connective tissue. The cells of most tissues work together to achieve a common purpose.

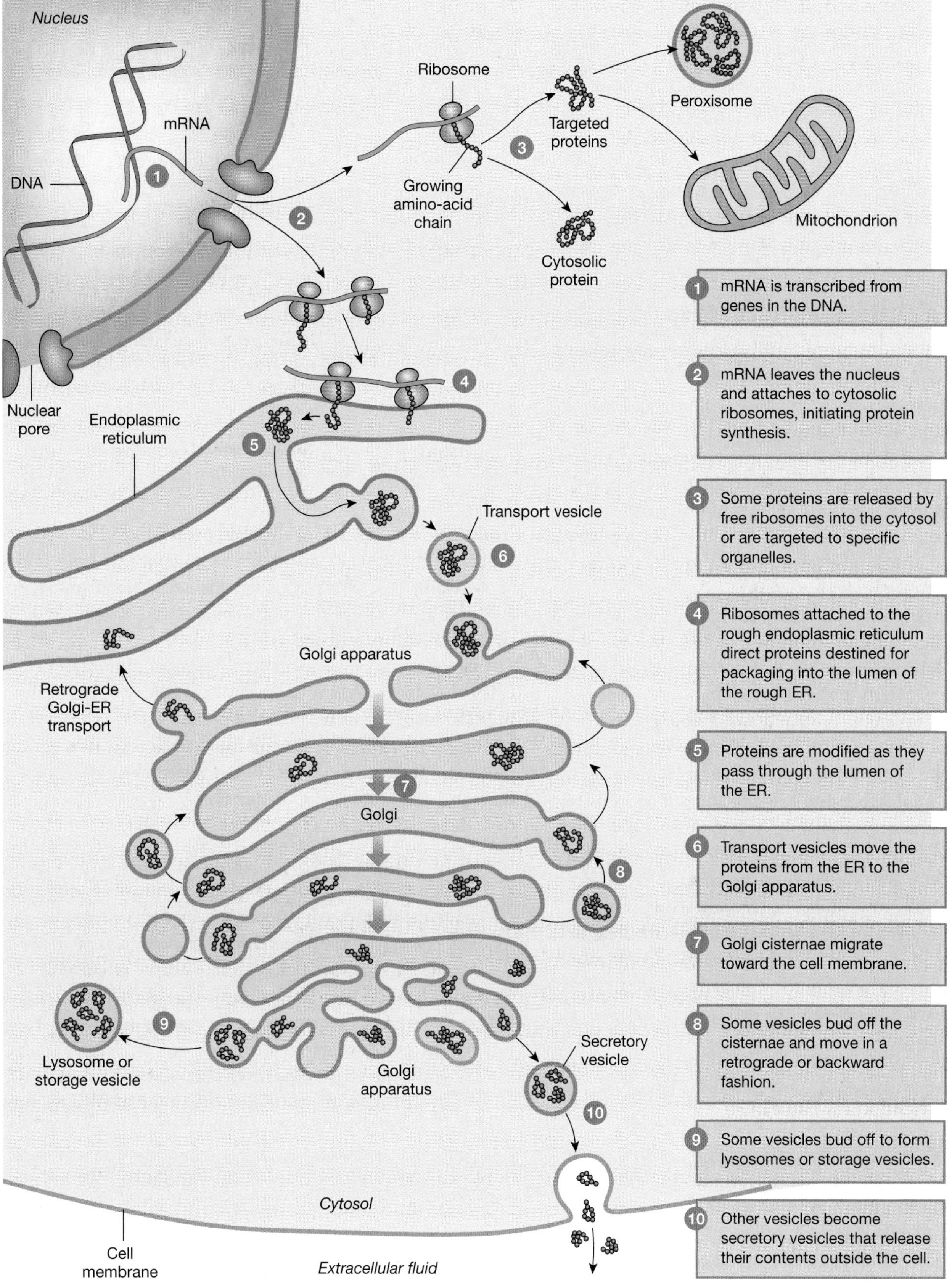
Nucleus
DNA
mRNA
Ribosome
Growing amino-acid chain
Targeted proteins
Peroxisome
Mitochondrion
Cytosolic protein
Nuclear pore
Endoplasmic reticulum
Transport vesicle
Golgi apparatus
Retrograde Golgi-ER transport
Golgi
Lysosome or storage vesicle
Golgi apparatus
Secretory vesicle
Cytosol
Cell membrane
Extracellular fluid
1 mRNA is transcribed from genes in the DNA.
2 mRNA leaves the nucleus and attaches to cytosolic ribosomes, initiating protein synthesis.
3 Some proteins are released by free ribosomes into the cytosol or are targeted to specific organelles.
4 Ribosomes attached to the rough endoplasmic reticulum direct proteins destined for packaging into the lumen of the rough ER.
5 Proteins are modified as they pass through the lumen of the ER.
6 Transport vesicles move the proteins from the ER to the Golgi apparatus.
7 Golgi cisternae migrate toward the cell membrane.
8 Some vesicles bud off the cisternae and move in a retrograde or backward fashion.
9 Some vesicles bud off to form lysosomes or storage vesicles.
10 Other vesicles become secretory vesicles that release their contents outside the cell.

Fig. 3.7 ***Protein synthesis*** shows the importance of subcellular compartmentation.

3

The study of tissue structure and function is known as **histology** {*histos,* tissue}. Histologists describe tissues by their physical features: (1) the shape and size of the cells, (2) the arrangement of the cells in the tissue (in layers, scattered, and so on), (3) the way cells are connected to one another, and (4) the amount of extracellular material present in the tissue. There are four primary tissue types in the human body: epithelial, connective, muscle, and *neural,* or nerve. Before we consider each tissue type specifically, let's examine how cells link together to form tissues.

Extracellular Matrix Has Many Functions

Extracellular matrix (usually just called *matrix*) is extracellular material that is synthesized and secreted by the cells of a tissue. For years, scientists believed that matrix was an inert substance whose only function was to hold cells together. However, experimental evidence now shows that the extracellular matrix plays a vital role in many physiological processes, ranging from growth and development to cell death. A number of disease states are associated with overproduction or disruption of extracellular matrix, including chronic heart failure and the spread of cancerous cells throughout the body (*metastasis*).

The composition of extracellular matrix varies from tissue to tissue, and the mechanical properties, such as elasticity and flexibility, of a tissue depend on the amount and consistency of the tissue's matrix. Matrix always has two basic components: proteoglycans and insoluble protein fibers. **Proteoglycans** are glycoproteins, which are proteins covalently bound to polysaccharide chains [p. 32]. Insoluble protein fibers such as *collagen, fibronectin,* and *laminin* provide strength and anchor cells to the matrix. Attachments between the extracellular matrix and proteins in the cell membrane or the cytoskeleton are ways cells communication with their external environment.

The amount of extracellular matrix in a tissue is highly variable. Nerve and muscle tissue have very little matrix, but the connective tissues, such as cartilage, bone, and blood, have extensive matrix that occupies as much volume as their cells. The consistency of extracellular matrix can vary from watery (blood and lymph) to rigid (bone).

Cell Junctions Hold Cells Together to Form Tissues

During growth and development, cells form *cell-cell adhesions* that may be transient or that may develop into more permanent **cell junctions. Cell adhesion molecules,** or **CAMs,** are membrane-spanning proteins responsible both for cell junctions and for transient cell adhesions (■ Tbl. 3.3). Cell-cell or cell-matrix adhesions mediated by CAMs are essential for normal growth and development. For example, growing nerve cells creep across the extracellular matrix with the help of *nerve-cell adhesion molecules,* or *NCAMs.* Cell adhesion helps white blood cells escape from the circulation and move into infected tissues, and it allows clumps of platelets to cling to damaged blood vessels. Because cell adhesions are not permanent, the bond between those CAMs and matrix is weak.

Table 3.3 Major Cell Adhesion Molecules (CAMs)

Name	Examples
Cadherins	Cell-cell junctions such as adherens junctions and desmosomes. Calcium-dependent.
Integrins	Primarily found in cell-matrix junctions. These also function in cell signaling.
Immunoglobulin superfamily CAMs	NCAMs (nerve-cell adhesion molecules). Responsible for nerve cell growth during nervous system development.
Selectins	Temporary cell-cell adhesions.

Stronger cell junctions can be grouped into three broad categories by function: communicating junctions, occluding junctions {*occludere,* to close up}, and anchoring junctions (■ Fig. 3.8). In animals, the communicating junctions are gap junctions. The occluding junctions of vertebrates are tight junctions that limit movement of materials between cells. Animals have three major types of junctions, described below.

1 **Gap junctions** are the simplest cell-cell junctions (Fig. 3.8b). They allow direct and rapid cell-to-cell communication through cytoplasmic bridges between adjoining cells. Cylindrical proteins called *connexins* interlock to create passageways that look like hollow rivets with narrow channels through their centers. The channels are able to open and close, regulating the movement of small molecules and ions through them.

   Gap junctions allow both chemical and electrical signals to pass rapidly from one cell to the next. They were once thought to occur only in certain muscle and nerve cells, but we now know they are important in cell-to-cell communication in many tissues, including the liver, pancreas, ovary, and thyroid gland.

2 **Tight junctions** are occluding junctions that restrict the movement of material between the cells they link (Fig. 3.8c). In tight junctions, the cell membranes of adjacent cells partly fuse together with the help of proteins called *claudins* and *occludins,* thereby making a barrier. As in many physiological processes, the barrier properties of

Cell Junctions

(a) Cell junctions connect one cell with another cell (or to surrounding matrix) with membrane-spanning proteins called **cell adhesion molecules**, or **CAMs**. This map shows the many ways cell junctions can be categorized.

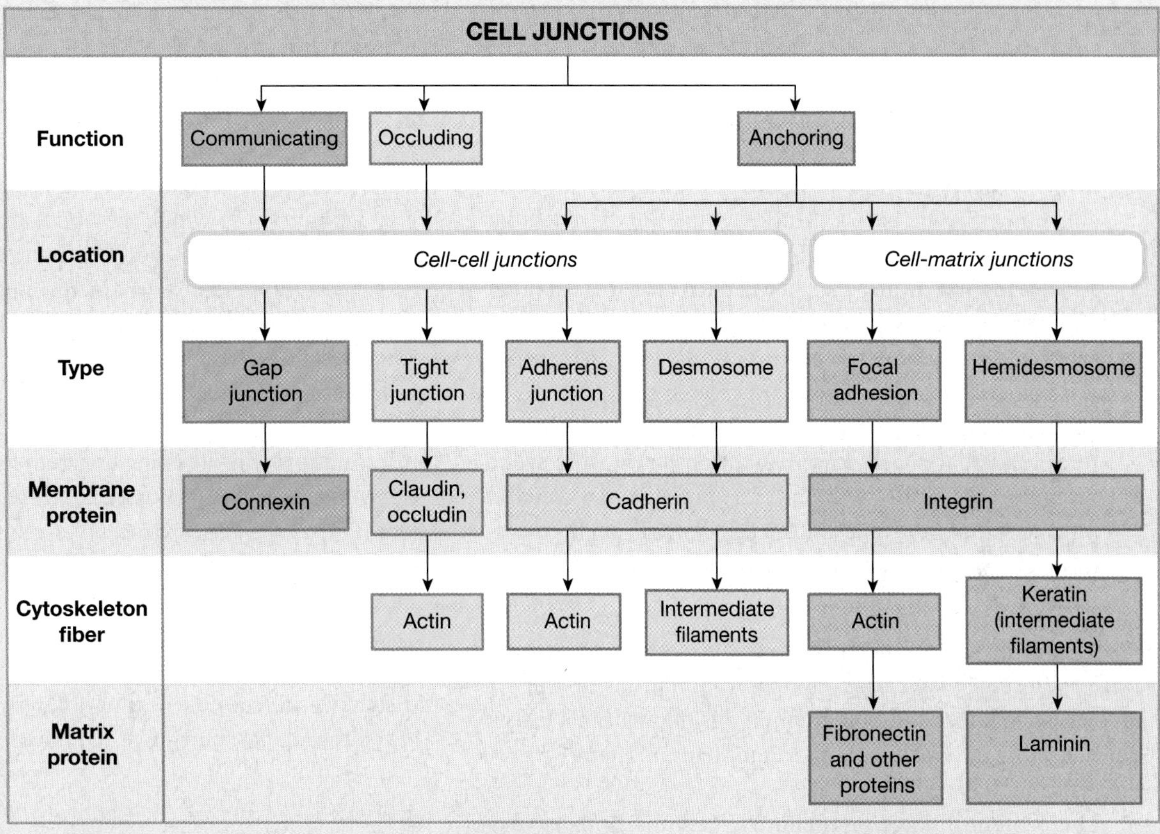

Cell junctions can be grouped into three categories: (b) **Gap junctions** which allow direct cell to cell communication, (c) **tight junctions** that block movement of material between cells, and (d) **anchoring junctions** that hold cells to one another and to the extracellular matrix.

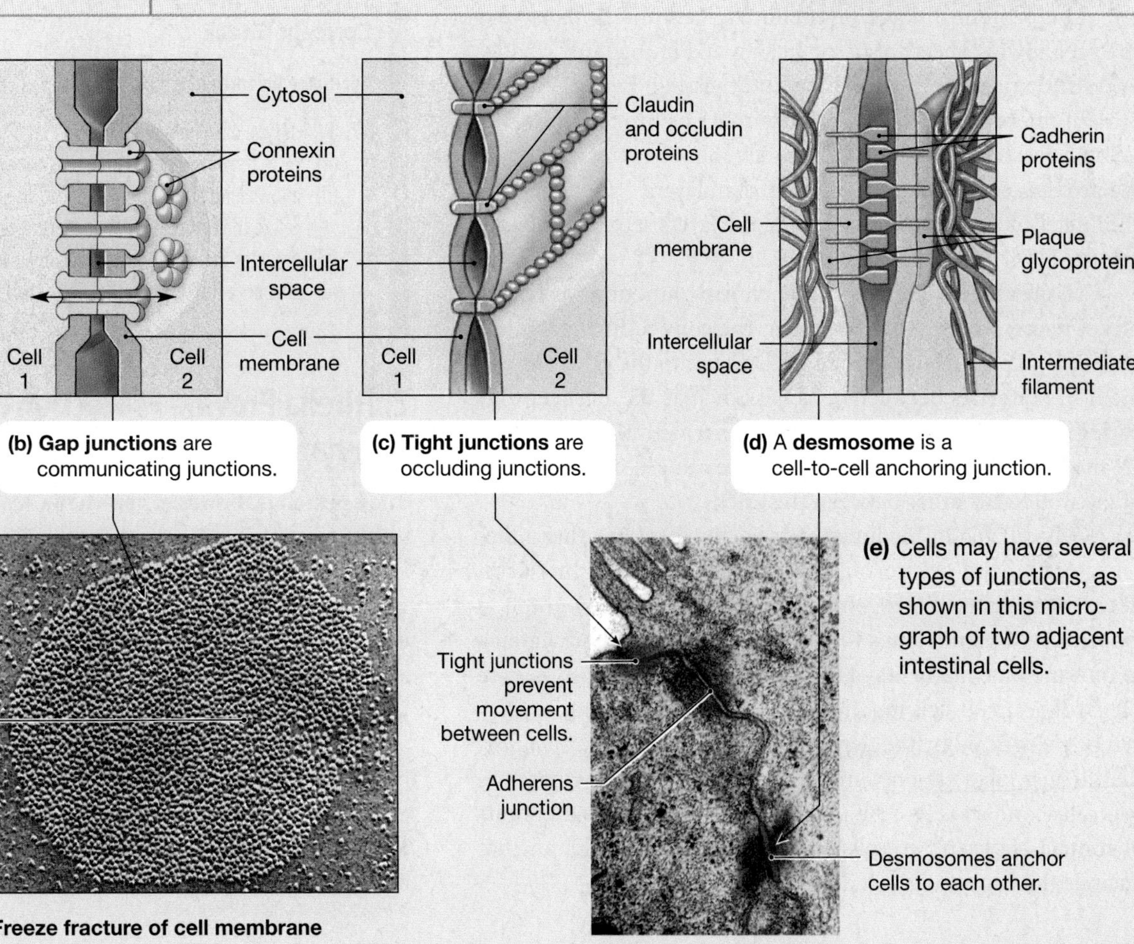

(b) Gap junctions are communicating junctions.

(c) Tight junctions are occluding junctions.

(d) A **desmosome** is a cell-to-cell anchoring junction.

Heart muscle has gap junctions that allow chemical and electrical signals to pass rapidly from one cell to the next.

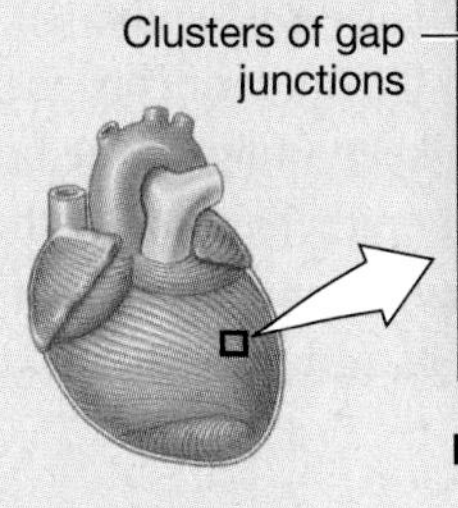

Freeze fracture of cell membrane

(e) Cells may have several types of junctions, as shown in this micrograph of two adjacent intestinal cells.

tight junctions are dynamic and can be altered depending on the body's needs. Tight junctions may have varying degrees of "leakiness."

Tight junctions in the intestinal tract and kidney prevent most substances from moving freely between the external and internal environments. In this way, they enable cells to regulate what enters and leaves the body. Tight junctions also create the so-called *blood-brain barrier* that prevents many potentially harmful substances in the blood from reaching the extracellular fluid of the brain.

3 **Anchoring junctions** (Fig. 3.8d) attach cells to each other (cell-cell anchoring junctions) or to the extracellular matrix (cell-matrix anchoring junctions). In vertebrates, cell-cell anchoring junctions are created by CAMs called **cadherins,** which connect with one another across the intercellular space. Cell-matrix junctions use CAMs called **integrins.** Integrins are membrane proteins that can also bind to signal molecules in the cell's environment, transferring information carried by the signal across the cell membrane into the cytoplasm.

Anchoring junctions contribute to the mechanical strength of the tissue. They have been compared to buttons or zippers that tie cells together and hold them in position within a tissue. Notice how the interlocking cadherin proteins in Figure 3.8c resemble the teeth of a zipper.

The protein linkage of anchoring cell junctions is very strong, allowing sheets of tissue in skin and lining body cavities to resist damage from stretching and twisting. Even the tough protein fibers of anchoring junctions can be broken, however. If you have shoes that rub against your skin, the stress can shear the proteins connecting the different skin layers. When fluid accumulates in the resulting space and the layers separate, a *blister* results.

Tissues held together with anchoring junctions are like a picket fence, where spaces between the connecting bars allow materials to pass from one side of the fence to the other. Movement of materials between cells is known as the **paracellular** pathway. In contrast, tissues held together with tight junctions are more like a solid brick wall: very little can pass from one side of the wall to the other between the bricks.

Cell-cell anchoring junctions take the form of either adherens junctions or desmosomes. **Adherens junctions** link actin fibers in adjacent cells together, as shown in the micrograph in Figure 3.8e. **Desmosomes** {*desmos*, band + *soma*, body} attach to intermediate filaments of the cytoskeleton. Desmosomes are the strongest cell-cell junctions. In electron micrographs they can be recognized by the dense glycoprotein bodies, or *plaques,* that lie just inside the cell membranes in the region where the two cells connect (Fig. 3.8e). Desmosomes may be small points of contact between two cells (spot desmosomes) or bands that encircle the entire cell (belt desmosomes).

There are also two types of cell-matrix anchoring junctions. **Hemidesmosomes** {*hemi-*, half} are strong junctions that anchor intermediate fibers of the cytoskeleton to fibrous matrix proteins such as laminin. **Focal adhesions** tie intracellular actin fibers to different matrix proteins, such as fibronectin.

The loss of normal cell junctions plays a role in a number of diseases and in metastasis. Diseases in which cell junctions are destroyed or fail to form can have disfiguring and painful symptoms, such as blistering skin. One such disease is *pemphigus,* a condition in which the body attacks some of its own cell junction proteins (*www.pemphigus.org*).

The disappearance of anchoring junctions probably contributes to the metastasis of cancer cells throughout the body. Cancer cells lose their anchoring junctions because they have fewer cadherin molecules and are not bound as tightly to neighboring cells. Once a cancer cell is released from its moorings, it secretes protein-digesting enzymes known as *proteases.* These enzymes, especially those called *matrix metalloproteinases* (MMPs), dissolve the extracellular matrix so that escaping cancer cells can invade adjacent tissues or enter the bloodstream. Researchers are investigating ways of blocking MMP enzymes to see if they can prevent metastasis.

Now that you understand how cells are held together into tissues, we will look at the four different tissue types in the body: (1) epithelial, (2) connective, (3) muscle, and (4) neural.

Concept Check Answers: p. 97

16. Name the three functional categories of cell junctions.

17. Which type of cell junction:
  (a) restricts movement of materials between cells?
  (b) allows direct movement of substances from the cytoplasm of one cell to the cytoplasm of an adjacent cell?
  (c) provides the strongest cell-cell junction?
  (d) anchors actin fibers in the cell to the extracellular matrix?

Epithelia Provide Protection and Regulate Exchange

The **epithelial tissues,** or **epithelia** {*epi-*, upon + *thele-*, nipple; singular *epithelium*}, protect the internal environment of the body and regulate the exchange of materials between the internal and external environments (■ Fig. 3.9). These tissues cover exposed surfaces, such as the skin, and line internal passageways, such as the digestive tract. *Any substance that enters or leaves the internal environment of the body must cross an epithelium.*

Some epithelia, such as those of the skin and mucous membranes of the mouth, act as a barrier to keep water in the body and invaders such as bacteria out. Other epithelia, such as those in the kidney and intestinal tract, control the movement of materials between the external environment and the extracellular

Epithelial Tissue

(a) Five Functional Categories of Epithelia

	Exchange	Transporting	Ciliated	Protective	Secretory
NUMBER OF CELL LAYERS	One	One	One	Many	One to many
CELL SHAPE	Flattened	Columnar or cuboidal	Columnar or cuboidal	Flattened in surface layers; polygonal in deeper layers	Columnar or polygonal
SPECIAL FEATURES	Pores between cells permit easy passage of molecules	Tight junctions prevent movement between cells; surface area increased by folding of cell membrane into fingerlike microvilli	One side covered with cilia to move fluid across surface	Cells tightly connected by many desmosomes	Protein-secreting cells filled with membrane-bound secretory ganules and extensive RER; steroid-secreting cells contain lipid droplets and extensive SER
WHERE FOUND	Lungs, lining of blood vessels	Intestine, kidney, some exocrine glands	Nose, trachea, and upper airways; female reproductive tract	Skin and lining of cavities (such as the mouth) that open to the environment	Exocrine glands, including pancreas, sweat glands, and salivary glands; endocrine glands, such as thyroid and gonads
KEY	exchange epithelium	transporting epithelium	ciliated epithelium	protective epithelium	 secretory epithelium

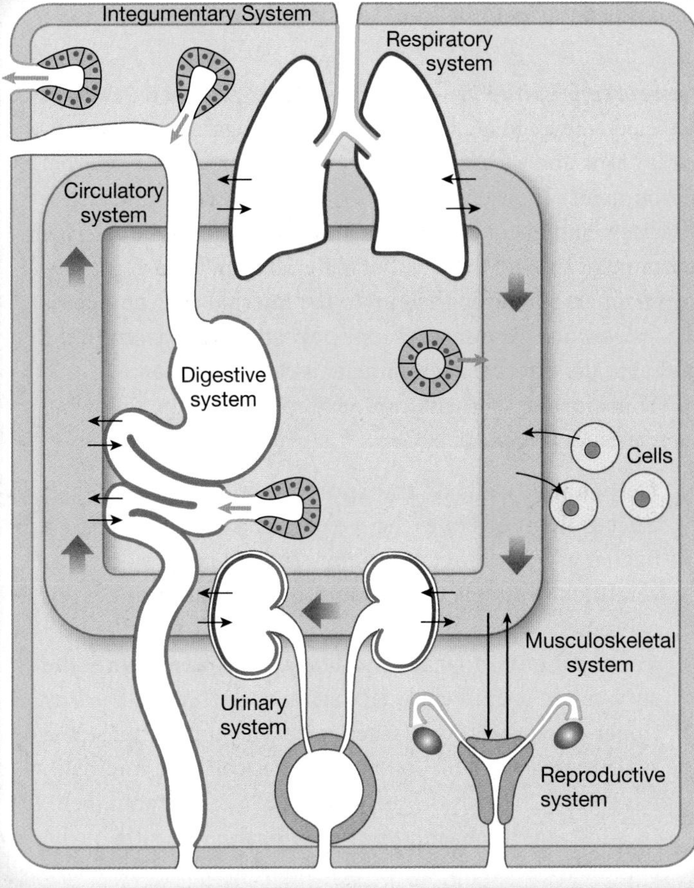

(b) This diagram shows the distribution of the five kinds of epithelia in the body outlined in the table above.

FIGURE QUESTIONS

- Where do secretions from endocrine glands go?
- Where do secretions from exocrine glands go?

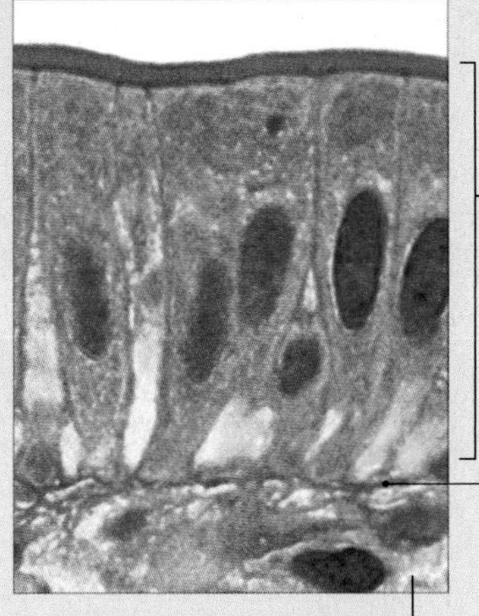

Epithelial cells attach to the basal lamina using cell adhesion molecules.
Basal lamina (basement membrane) is an acellular matrix layer that is secreted by the epithelial cells.
Underlying tissue

(c) Most epithelia attach to an underlying matrix layer called the **basal lamina** or **basement membrane**.

fluid of the body. Nutrients, gases, and wastes often must cross several different epithelia in their passage between cells and the outside world.

Another type of epithelium is specialized to manufacture and secrete chemicals into the blood or into the external environment. Sweat and saliva are examples of substances secreted by epithelia into the environment. Hormones are secreted into the blood.

Structure of Epithelia Epithelia typically consist of one or more layers of cells connected to one another, with a thin layer of extracellular matrix lying between the epithelial cells and their underlying tissues (Fig. 3.9c). This matrix layer, called the **basal lamina** {*bassus,* low; *lamina,* a thin plate}, or **basement membrane,** is composed of a network of collagen and laminin filaments embedded in proteoglycans. The protein filaments hold the epithelial cells to the underlying cell layers, just as cell junctions hold the individual cells in the epithelium to one another.

The cell junctions in epithelia are variable. Physiologists classify epithelia either as "leaky" or "tight," depending on how easily substances pass from one side of the epithelial layer to the other. In a leaky epithelium, anchoring junctions allow molecules to cross the epithelium by passing through the gap between two adjacent epithelial cells. A typical leaky epithelium is the wall of capillaries (the smallest blood vessels), where all dissolved molecules except for large proteins can pass from the blood to the interstitial fluid by traveling through gaps between adjacent epithelial cells.

In a tight epithelium, such as that in the kidney, adjacent cells are bound to each other by tight junctions that create a barrier, preventing substances from traveling between adjacent cells. To cross a tight epithelium, most substances must enter the epithelial cells and go *through* them. The tightness of an epithelium is directly related to how selective it is about what can move across it. Some epithelia, such as those of the intestine, have the ability to alter the tightness of their junctions according to the body's needs.

Types of Epithelia Structurally, epithelial tissues can be divided into two general types: (1) sheets of tissue that lie on the surface of the body or that line the inside of tubes and hollow organs and (2) secretory epithelia that synthesize and release substances into the extracellular space. Histologists classify sheet epithelia by the number of cell layers in the tissue and by the shape of the cells in the surface layer. This classification scheme recognizes two types of layering—**simple** (one cell thick) and **stratified** (multiple cell layers) {*stratum,* layer + *facere,* to make}—and three cell shapes—**squamous** {*squama,* flattened plate or scale}, **cuboidal,** and **columnar.** However, physiologists are more concerned with the functions of these tissues, so instead of using the histological descriptions, we will divide epithelia into five groups according to their function.

There are five functional types of epithelia: exchange, transporting, ciliated, protective, and secretory (■ Fig. 3.10). *Exchange epithelia* permit rapid exchange of materials, especially gases. *Transporting epithelia* are selective about what can cross them and are found primarily in the intestinal tract and the kidney. *Ciliated epithelia* are located primarily in the airways of the respiratory system and in the female reproductive tract. *Protective epithelia* are found on the surface of the body and just inside the openings of body cavities. *Secretory epithelia* synthesize and release secretory products into the external environment or into the blood.

Figure 3.9b shows the distribution of these epithelia in the systems of the body. Notice that most epithelia face the external environment on one surface and the extracellular fluid on the other. The only exception is the endocrine glands.

Exchange Epithelia The **exchange epithelia** are composed of very thin, flattened cells that allow gases (CO_2 and O_2) to pass rapidly across the epithelium. This type of epithelium lines the blood vessels and the lungs, the two major sites of gas exchange in the body. In capillaries, gaps or pores in the epithelium also allow molecules smaller than proteins to pass *between* two adjacent epithelial cells, making this a leaky epithelium (Fig. 3.10a). Histologists classify thin exchange tissue as *simple squamous epithelium* because it is a single layer of thin, flattened cells. The simple squamous epithelium lining the heart and blood vessels is also called the **endothelium.**

Transporting Epithelia The **transporting epithelia** actively and selectively regulate the exchange of nongaseous materials, such as ions and nutrients, between the internal and external environments. These epithelia line the hollow tubes of the digestive system and the kidney, where lumens open into the external environment [p. 4]. Movement of material from the external environment across the epithelium to the internal environment is called *absorption.* Movement in the opposite direction, from the internal to the external environment, is called *secretion.*

Transporting epithelia can be identified by the following characteristics (Fig. 3.10e):

1. **Cell shape.** Cells of transporting epithelia are much thicker than cells of exchange epithelia, and they act as a barrier as well as an entry point. The cell layer is only one cell thick (a simple epithelium), but cells are cuboidal or columnar.
2. **Membrane modifications.** The **apical membrane,** the surface of the epithelial cell that faces the lumen, has tiny finger-like projections called *microvilli* that increase the surface area available for transport. A cell with microvilli has at least 20 times the surface area of a cell without them. In addition, the **basolateral membrane,** the side of the epithelial cell facing the extracellular fluid, may also have folds that increase the cell's surface area.

■ Fig. 3.10 *ESSENTIALS*

Types of Epithelia

(a) Exchange Epithelium

The thin, flat cells of exchange epithelium allow movement through and between the cells.

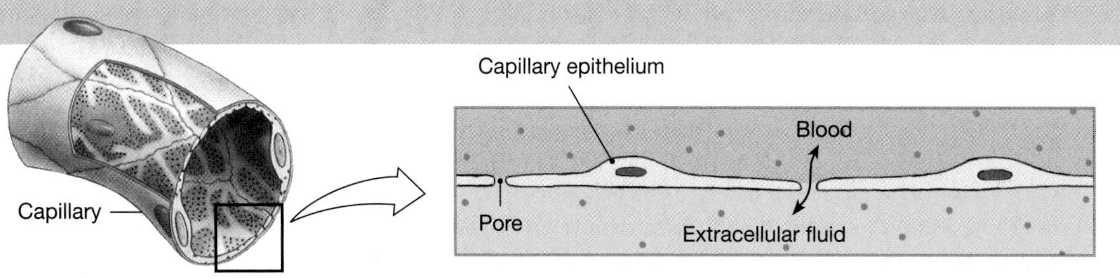

(b) Protective Epithelium

Protective epithelia have many stacked layers of cells that are constantly being replaced. This figure shows layers in skin (see Focus on Skin, p. 91).

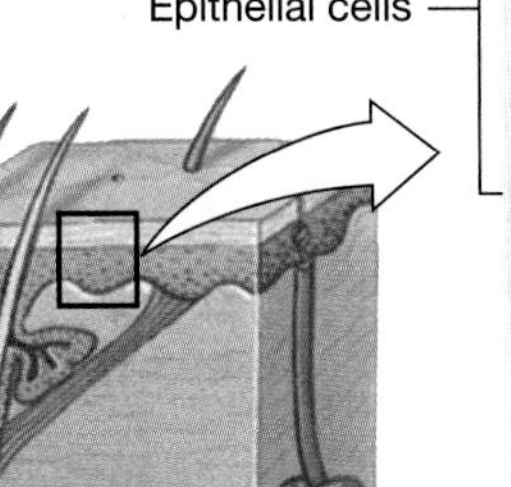

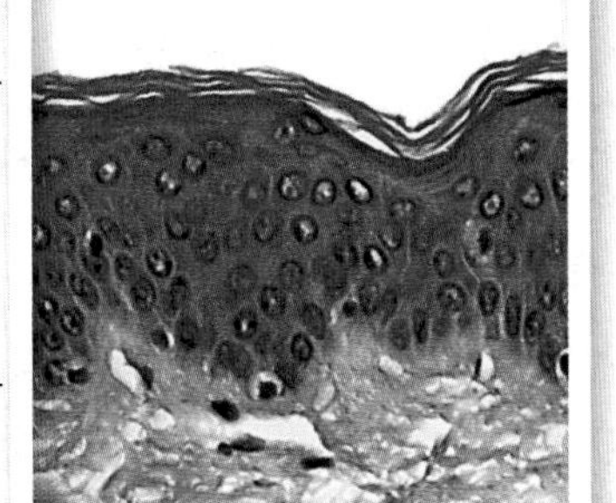

Section of skin showing cell layers.

(c) Ciliated Epithelium

Beating cilia create fluid currents that sweep across the epithelial surface.

SEM of the epithelial surface of an airway

(d) Secretory Epithelium

Secretory epithelial cells make and release a product. Exocrine secretions, such as the mucus shown here, are secreted outside the body. The secretions of endocrine cells (hormones) are released into the blood.

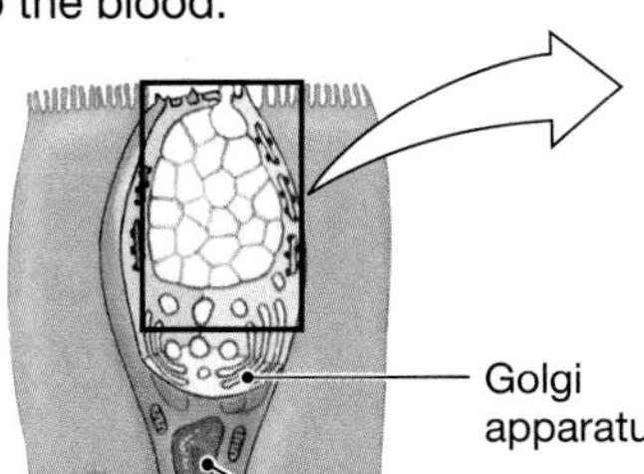

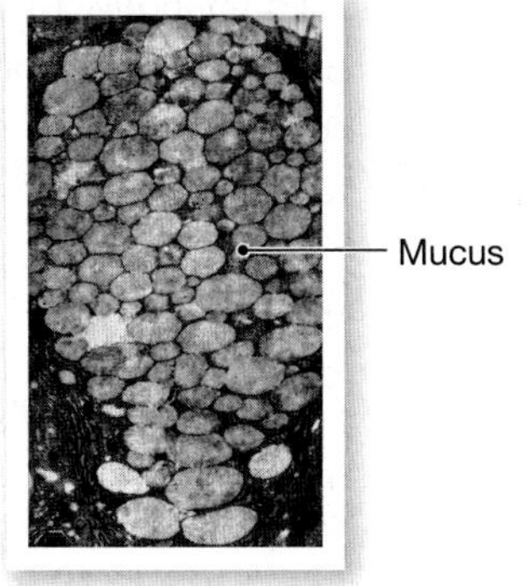

SEM of goblet cell

Goblet cells secrete mucus into the lumen of hollow organs such as the intestine.

(e) Transporting Epithelium

Transporting epithelia selectively move substances between a lumen and the ECF.

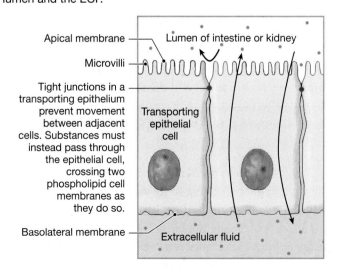

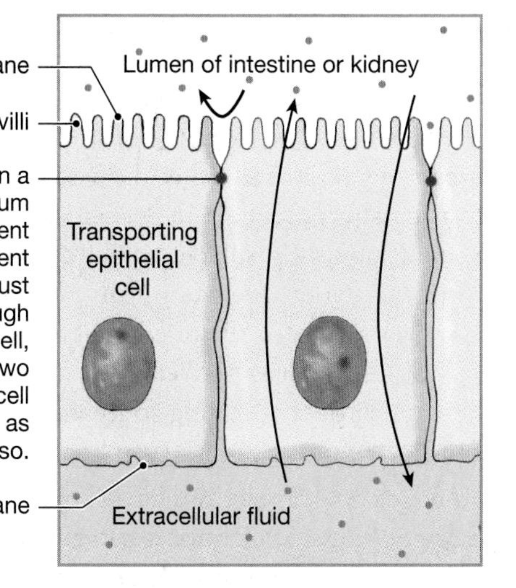

3 **Cell junctions.** The cells of transporting epithelia are firmly attached to adjacent cells by moderately tight to very tight junctions. This means that to cross the epithelium, material must move into an epithelial cell on one side of the tissue and out of the cell on the other side.

4 **Cell organelles.** Most cells that transport materials have numerous mitochondria to provide energy for transport processes [discussed further in Chapter 5]. The properties of transporting epithelia differ depending on where in the body the epithelia are located. For example, glucose can cross the epithelium of the small intestine and enter the extracellular fluid but cannot cross the epithelium of the large intestine.

The transport properties of an epithelium can be regulated and modified in response to various stimuli. Hormones, for example, affect the transport of ions by kidney epithelium. You will learn more about transporting epithelia when you study the kidney and digestive systems.

Ciliated Epithelia **Ciliated epithelia** are nontransporting tissues that line the respiratory system and parts of the female reproductive tract. The surface of the tissue facing the lumen is covered with cilia that beat in a coordinated, rhythmic fashion, moving fluid and particles across the surface of the tissue (Fig. 3.10c). Injury to the cilia or to their epithelial cells can stop ciliary movement. For example, smoking paralyzes the ciliated epithelium lining the respiratory tract. Loss of ciliary function contributes to the higher incidence of respiratory infection in smokers, when the mucus that traps bacteria can no longer be swept out of the lungs by the cilia.

Protective Epithelia The **protective epithelia** prevent exchange between the internal and external environments and protect areas subject to mechanical or chemical stresses. These epithelia are stratified tissues, composed of many stacked layers of cells (Fig. 3.10b). Protective epithelia are toughened by the secretion of *keratin* {*keras,* horn}, the same insoluble protein abundant in hair and nails. The *epidermis* {*epi,* upon + *derma,* skin} and linings of the mouth, pharynx, esophagus, urethra, and vagina are all protective epithelia.

Because protective epithelia are subjected to irritating chemicals, bacteria, and other destructive forces, the cells in them have a short life span. In deeper layers, new cells are produced continuously, displacing older cells at the surface. Each time you wash your face, you scrub off dead cells on the surface layer. As skin ages, the rate of cell turnover declines. Tretinoin (Retin-A®), a drug derived from vitamin A, speeds up cell division and surface shedding so treated skin develops a more youthful appearance.

Secretory Epithelia **Secretory epithelia** are composed of cells that produce a substance and then secrete it into the extracellular space. Secretory cells may be scattered among other epithelial cells, or they may group together to form a multicellular **gland.** There are two types of secretory glands: exocrine and endocrine.

Exocrine glands release their secretions to the body's external environment {*exo-,* outside + *krinein,* to secrete}. This may be onto the surface of the skin or onto an epithelium lining one of the internal passageways, such as the airways of the lung or the lumen of the intestine (Fig. 3.10d). In effect, an exocrine secretion leaves the body. This explains how some exocrine secretions, like stomach acid, can have a pH that is incompatible with life [Fig. 2.9, p. 48].

Most exocrine glands release their products through open tubes known as **ducts.** Sweat glands, mammary glands in the breast, salivary glands, the liver, and the pancreas are all exocrine glands.

Exocrine gland cells produce two types of secretions. **Serous secretions** are watery solutions, and many of them contain enzymes. Tears, sweat, and digestive enzyme solutions are all serous exocrine secretions. **Mucous secretions** (also called **mucus**) are sticky solutions containing glycoproteins and proteoglycans.

Goblet cells, shown in Figure 3.10d, are single exocrine cells that produce mucus. Mucus acts as a lubricant for food to be swallowed, as a trap for foreign particles and microorganisms inhaled or ingested, and as a protective barrier between the epithelium and the environment.

RUNNING PROBLEM

Many kinds of cancer develop in epithelial cells that are subject to damage or trauma. The cervix consists of two types of epithelia. Secretory epithelium with mucus-secreting glands lines the inside of the cervix. A protective epithelium covers the outside of the cervix. At the opening of the cervix, these two types of epithelia come together. In many cases, infections caused by the human papillomavirus (HPV) cause the cervical cells to develop dysplasia. Dr. Baird ran an HPV test on Jan's first Pap smear, and it was positive for the virus. Today she is repeating the tests to see if Jan's dysplasia and HPV infection have persisted.

Q3: What other kinds of damage or trauma are cervical epithelial cells normally subjected to? Which of the two types of cervical epithelia is more likely to be affected by physical trauma?

Q4: The results of Jan's first Pap test showed atypical squamous cells of unknown significance (ASCUS). Were these cells more likely to come from the secretory portion of the cervix or from the protective epithelium?

Some exocrine glands contain more than one type of secretory cell, and they produce both serous and mucous secretions. For example, the salivary glands release mixed secretions.

Unlike exocrine glands, **endocrine glands** are ductless and release their secretions, called **hormones,** into the body's extracellular compartment (Fig. 3.9d). Hormones enter the blood for distribution to other parts of the body, where they regulate or coordinate the activities of various tissues, organs, and organ systems. Some of the best-known endocrine glands are the pancreas, the thyroid gland, the gonads, and the pituitary gland. For years, it was thought that all hormones were produced by cells grouped together into endocrine glands. We now know that isolated endocrine cells occur scattered in the epithelial lining of the digestive tract, in the tubules of the kidney, and in the walls of the heart.

■ Figure 3.11 shows the epithelial origin of endocrine and exocrine glands. During development, epithelial cells grow downward into the supporting connective tissue. Exocrine glands remain connected to the parent epithelium by a duct that transports the secretion to its destination (the external environment). Endocrine glands lose the connecting cells and secrete their hormones into the bloodstream.

During development, the region of epithelium destined to become glandular tissue divides downward into the underlying connective tissue.
Epithelium
Connective tissue
Exocrine
Endocrine
Duct
Connecting cells disappear
Exocrine secretory cells
Endocrine secretory cells
Blood vessel
A hollow center, or lumen, forms in exocrine glands, creating a duct that provides a passageway for secretions to move to the surface of the epithelium.
Endocrine glands lose the connecting bridge of cells that links them to the parent epithelium. Their secretions go directly into the bloodstream.

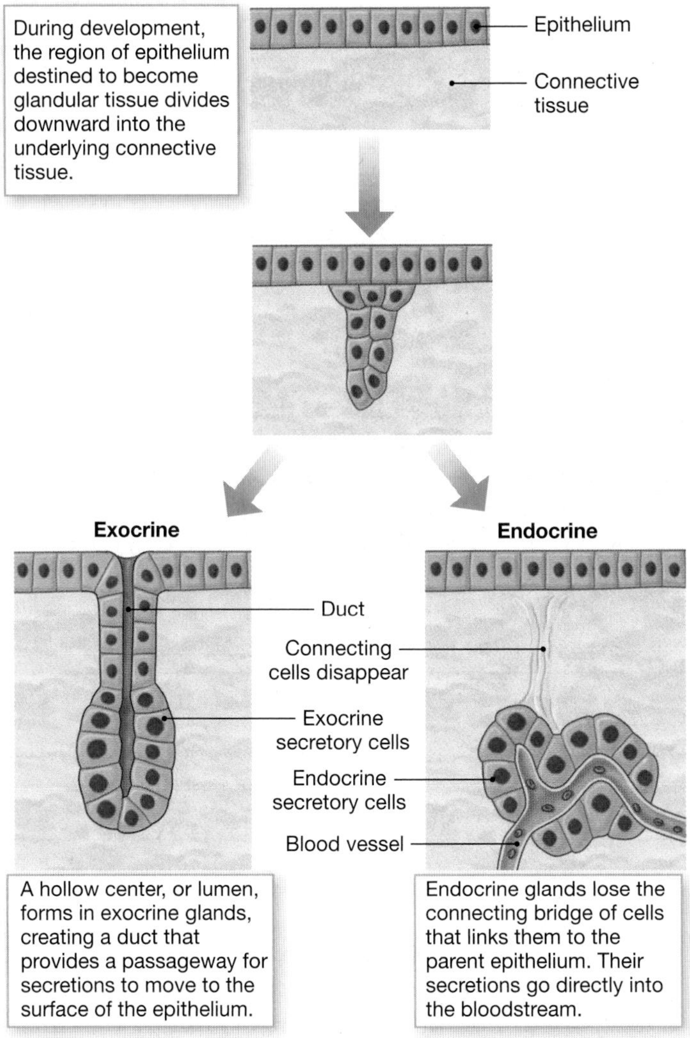

■ **Fig. 3.11** ***Development of endocrine and exocrine glands from epithelium.***

Concept Check Answers: p. 97

18. List the five functional types of epithelia.
19. Define secretion.
20. Name two properties that distinguish endocrine glands from exocrine glands.
21. The basal lamina of epithelium contains the protein fiber laminin. Are the overlying cells attached by focal adhesions or hemidesmosomes?
22. You look at a tissue under a microscope and see a simple squamous epithelium. Can it be a sample of the skin surface? Explain.
23. A cell of the intestinal epithelium secretes a substance into the extracellular fluid, where it is picked up by the blood and carried to the pancreas. Is the intestinal epithelium cell an endocrine or an exocrine cell?

Connective Tissues Provide Support and Barriers

Connective tissues, the second major tissue type, provide structural support and sometimes a physical barrier that, along with specialized cells, helps defend the body from foreign invaders such as bacteria. The distinguishing characteristic of connective tissues is the presence of extensive extracellular matrix containing widely scattered cells that secrete and modify the matrix (■ Fig. 3.12). Connective tissues include blood, the support tissues for the skin and internal organs, and cartilage and bone.

Structure of Connective Tissue The extracellular matrix of connective tissue is a **ground substance** of proteoglycans and water in which insoluble protein fibers are arranged, much like pieces of fruit suspended in a gelatin salad. The consistency of ground substance is highly variable, depending on the type of connective tissue (Fig. 3.12a). At one extreme is the watery matrix of blood, and at the other extreme is the hardened matrix of bone. In between are solutions of proteoglycans that vary in consistency from syrupy to gelatinous. The term *ground substance* is sometimes used interchangeably with *matrix.*

Connective tissue cells lie embedded in the extracellular matrix. These cells are described as *fixed* if they remain in one place and as *mobile* if they can move from place to place. **Fixed cells** are responsible for local maintenance, tissue repair, and energy storage. **Mobile cells** are responsible mainly for defense. The distinction between fixed and mobile cells is not absolute, because at least one cell type is found in both fixed and mobile forms.

Although extracellular matrix is nonliving, the connective tissue cells constantly modify it by adding, deleting, or rearranging molecules. The suffix *-blast* {*blastos,* sprout} on a connective

Fig. 3.12 *ESSENTIALS*

Connective Tissue

(a) Map of connective tissue components

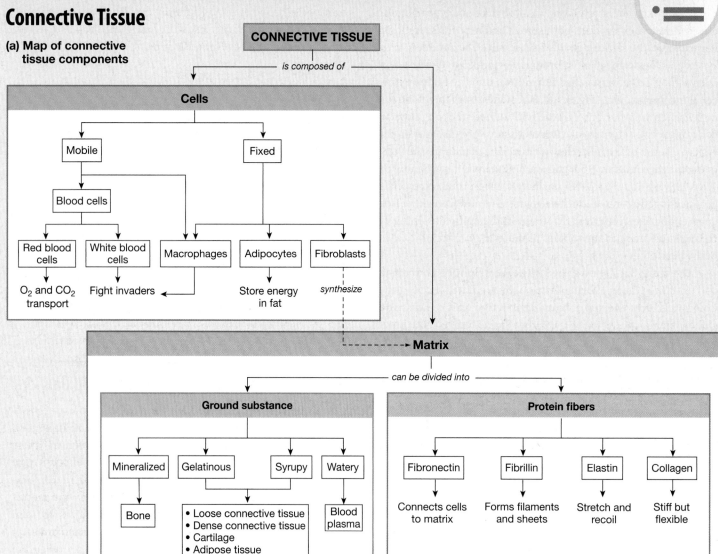

(b) Types of Connective Tissue

Tissue Name	Ground Substance	Fiber Type and Arrangement	Main Cell Types	Where Found
Loose connective tissue	Gel; more ground substance than fibers or cells	Collagen, elastic, reticular; random	Fibroblasts	Skin, around blood vessels and organs, under epithelia
Dense, irregular connective tissue	More fibers than ground substance	Mostly collagen; random	Fibroblasts	Muscle and nerve sheaths
Dense, regular connective tissue	More fibers than ground substance	Collagen; parallel	Fibroblasts	Tendons and ligaments
Adipose tissue	Very little ground substance	None	Brown fat and white fat	Depends on age and sex
Blood	Aqueous	None	Blood cells	In blood and lymph vessels
Cartilage	Firm but flexible; hyaluronic acid	Collagen	Chondroblasts	Joint surfaces, spine, ear, nose, larynx
Bone	Rigid due to calcium salts	Collagen	Osteoblasts and osteocytes	Bones

tissue cell name often indicates a cell that is either growing or actively secreting extracellular matrix. **Fibroblasts,** for example, are connective tissue cells that secrete collagen-rich matrix. Cells that are actively breaking down matrix are identified by the suffix *-clast* {*klastos,* broken}. Cells that are neither growing, secreting matrix components, nor breaking down matrix may be given the suffix *-cyte,* meaning "cell." Remembering these suffixes should help you remember the functional differences between cells with similar names, such as the osteoblast, osteocyte, and osteoclast, three cell types found in bone.

In addition to secreting proteoglycan ground substance, connective tissue cells produce matrix fibers. Four types of fiber proteins are found in matrix, aggregated into insoluble fibers. **Collagen** {*kolla,* glue + *-genes,* produced} is the most abundant protein in the human body, almost one-third of the body's dry weight. Collagen is also the most diverse of the four protein types, with at least 12 variations. It is found almost everywhere connective tissue is found, from the skin to muscles and bones. Individual collagen molecules pack together to form collagen fibers, flexible but inelastic fibers whose strength per unit weight exceeds that of steel. The amount and arrangement of collagen fibers help determine the mechanical properties of different types of connective tissues.

Three other protein fibers in connective tissue are elastin, fibrillin, and fibronectin. **Elastin** is a coiled, wavy protein that returns to its original length after being stretched. This property is known as *elastance.* Elastin combines with the very thin, straight fibers of **fibrillin** to form filaments and sheets of elastic fibers. These two fibers are important in elastic tissues such as the lungs, blood vessels, and skin. As mentioned earlier, **fibronectin** connects cells to extracellular matrix at focal adhesions. Fibronectins also play an important role in wound healing and in blood clotting.

Types of Connective Tissue Figure 3.12b compares the properties of different types of connective tissue. The most common types are loose and dense connective tissue, adipose tissue, blood, cartilage, and bone. By many estimates, connective tissues are the most abundant of the tissue types as they are a component of most organs.

Loose connective tissues (■ Fig. 3.13a) are the elastic tissues that underlie skin and provide support for small glands. **Dense connective tissues** provide strength or flexibility. Examples are tendons, ligaments, and the sheaths that surround muscles and nerves. In these dense tissues, collagen fibers are the dominant type. **Tendons** (Fig. 3.13c) attach skeletal muscles to bones. **Ligaments** connect one bone to another. Because ligaments contain elastic fibers in addition to collagen fibers, they have a limited ability to stretch. Tendons lack elastic fibers and so cannot stretch.

Cartilage and bone together are considered supporting connective tissues. These tissues have a dense ground substance that contains closely packed fibers. **Cartilage** is found in structures such as the nose, ears, knee, and windpipe. It is solid, flexible, and notable for its lack of blood supply. Without a blood supply, nutrients and oxygen must reach the cells of cartilage by diffusion. This is a slow process, which means that damaged cartilage heals slowly.

BIOTECHNOLOGY

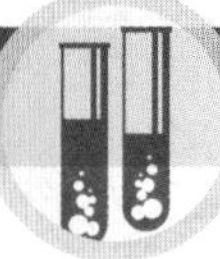

Grow Your Own Cartilage

Have you torn the cartilage in your knee playing basketball or some other sport? Maybe you won't need surgery to repair it. Replacing lost or damaged cartilage is moving from the realm of science fiction to the realm of reality. Researchers have developed a process in which they take a cartilage sample from a patient and put it into a tissue culture medium to reproduce. Once the culture has grown enough *chondrocytes*—the cells that synthesize the extracellular matrix of cartilage—the mixture is sent back to a physician, who surgically places the cells in the patient's knee at the site of cartilage damage. Once returned to the body, the chondrocytes secrete matrix and help repair the damaged cartilage. Because the person's own cells are grown and reimplanted, there is no tissue rejection. A different method for cartilage repair being used outside the United States is treatment with stem cells derived from bone marrow. Both therapies have proved to be effective treatments for selected cartilage problems.

The fibrous extracellular matrix of **bone** is said to be *calcified* because it contains mineral deposits, primarily calcium salts, such as calcium phosphate (Fig. 3.13b). These minerals give the bone strength and rigidity. We examine the structure and formation of bone along with calcium metabolism later [Chapter 23].

Adipose tissue is made up of **adipocytes,** or fat cells. An adipocyte of **white fat** typically contains a single enormous lipid droplet that occupies most of the volume of the cell (Fig. 3.13e). This is the most common form of adipose tissue in adults.

Brown fat is composed of adipose cells that contain multiple lipid droplets rather than a single large droplet. This type of fat has been known for many years to play an important role in temperature regulation in infants. Until recently it was thought to be almost completely absent in adults. However, modern imaging techniques such as combined CT and PET scans have revealed that adults do have brown fat [discussed in more detail in Chapter 22].

Blood is an unusual connective tissue that is characterized by its watery extracellular matrix called *plasma.* Plasma consists of a dilute solution of ions and dissolved organic molecules, including a large variety of soluble proteins. Blood cells and cell fragments are suspended in the plasma (Fig. 3.13d), but the insoluble protein fibers typical of other connective tissues are absent. [We discuss blood in Chapter 16.]

Fig. 3.13 *ESSENTIALS*

Types of Connective Tissue

(a) Loose Connective Tissue

Loose connective tissue is very flexible, with multiple cell types and fibers.

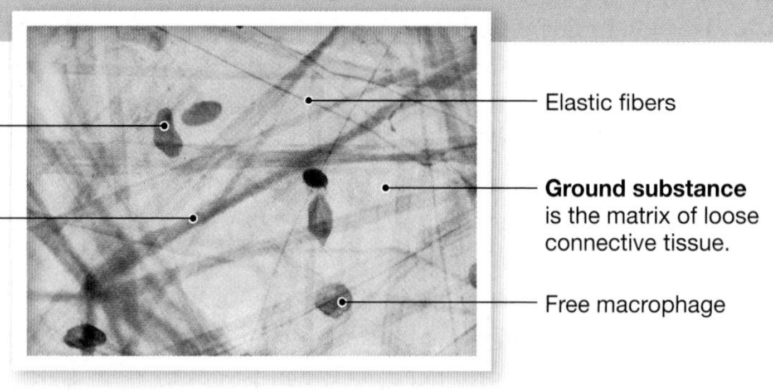

Light micrograph of loose connective tissue

(b) Bone and Cartilage

Hard bone forms when osteoblasts deposit calcium phosphate crystals in the matrix. Cartilage has firm but flexible matrix secreted by cells called chondrocytes.

Matrix

Light micrograph of bone

Chondrocytes

Matrix

Light micrograph of hyaline cartilage

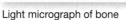

(c) Dense Regular Connective Tissue

Collagen fibers of tendon are densely packed into parallel bundles.

Collagen fibers

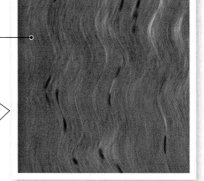

Light micrograph of tendon

(d) Blood

Blood consists of liquid matrix (*plasma*) plus red and white blood cells and the cell fragments called platelets.

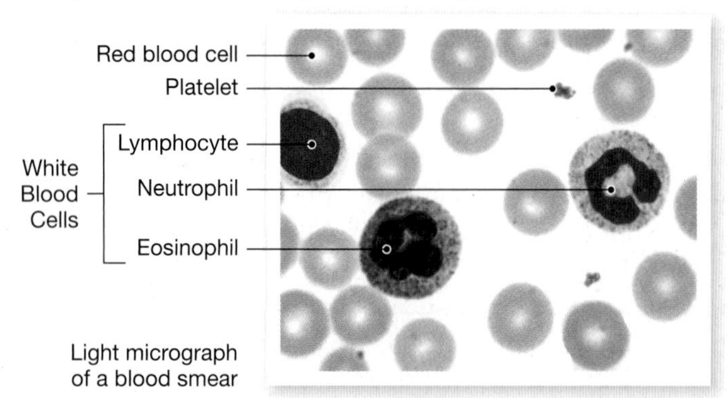

Light micrograph of a blood smear

(e) Adipose Tissue

In white fat, the cell cytoplasm is almost entirely filled with lipid droplets.

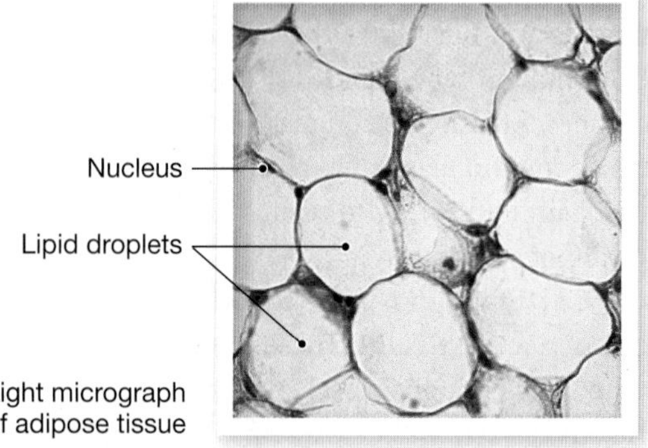

Light micrograph of adipose tissue

Concept Check Answers: p. 97

24. What is the distinguishing characteristic of connective tissues?
25. Name four types of protein fibers found in connective tissue matrix and give the characteristics of each.
26. Name six types of connective tissues.
27. Blood is a connective tissue with two components: plasma and cells. Which of these is the matrix in this connective tissue?
28. Why does torn cartilage heal more slowly than a cut in the skin?

Muscle and Neural Tissues Are Excitable

The third and fourth of the body's four tissue types—muscle and neural—are collectively called the *excitable tissues* because of their ability to generate and propagate electrical signals called *action potentials*. Both of these tissue types have minimal extracellular matrix, usually limited to a supportive layer called the *external lamina*. Some types of muscle and nerve cells are also notable for their gap junctions, which allow the direct and rapid conduction of electrical signals from cell to cell.

Muscle tissue has the ability to contract and produce force and movement. The body contains three types of muscle tissue: cardiac muscle in the heart; smooth muscle, which makes up most internal organs; and skeletal muscle. Most skeletal muscles attach to bones and are responsible for gross movement of the body. [We discuss muscle tissue in more detail in Chapter 12.]

Neural tissue has two types of cells. **Neurons,** or nerve cells, carry information in the form of chemical and electrical signals from one part of the body to another. They are concentrated in the brain and spinal cord but also include a network of cells that extends to virtually every part of the body. **Glial cells,** or neuroglia, are the support cells for neurons. [We discuss the anatomy of neural tissue in Chapter 8.] A summary of the characteristics of the four tissue types can be found in ■ Table 3.4.

Tissue Remodeling

Most people associate growth with the period from birth to adulthood. However, cell birth, growth, and death continue throughout a person's life. The tissues of the body are constantly remodeled as cells die and are replaced.

Characteristics of the Four Tissue Types — Table 3.4

	Epithelial	Connective	Muscle	Nerve
Matrix amount	Minimal	Extensive	Minimal	Minimal
Matrix type	Basal lamina	Varied—protein fibers in ground substance that ranges from liquid to gelatinous to firm to calcified	External lamina	External lamina
Unique features	No direct blood supply	Cartilage has no blood supply	Able to generate electrical signals, force, and movement	Able to generate electrical signals
Surface features of cells	Microvilli, cilia	N/A	N/A	N/A
Locations	Covers body surface; lines cavities and hollow organs, and tubes; secretory glands	Supports skin and other organs; cartilage, bone, and blood	Makes up skeletal muscles, hollow organs, and tubes	Throughout body; concentrated in brain and spinal cord
Cell arrangement and shapes	Variable number of layers, from one to many; cells flattened, cuboidal, or columnar	Cells not in layers; usually randomly scattered in matrix; cell shape irregular to round	Cells linked in sheets or elongated bundles; cells shaped in elongated, thin cylinders; heart muscle cells may be branched	Cells isolated or networked; cell appendages highly branched and/or elongated

3

Apoptosis Is a Tidy Form of Cell Death

Cell death occurs two ways, one messy and one tidy. In **necrosis,** cells die from physical trauma, toxins, or lack of oxygen when their blood supply is cut off. Necrotic cells swell, their organelles deteriorate, and finally the cells rupture. The cell contents released this way include digestive enzymes that damage adjacent cells and trigger an inflammatory response. You see necrosis when you have a red area of skin surrounding a scab.

In contrast, cells that undergo *programmed cell death,* or **apoptosis** {ap-oh-TOE-sis or a-pop-TOE-sis; *apo-*, apart, away + *ptosis,* falling}, do not disrupt their neighbors when they die. Apoptosis, also called cell suicide, is a complex process regulated by multiple chemical signals. Some signals keep apoptosis from occurring, while other signals tell the cell to self-destruct. When the suicide signal wins out, chromatin in the nucleus condenses, and the cell pulls away from its neighbors. It shrinks, then breaks up into tidy membrane-bound *blebs* that are gobbled up by neighboring cells or by wandering cells of the immune system.

Apoptosis is a normal event in the life of an organism. During fetal development, apoptosis removes unneeded cells, such as half the cells in the developing brain and the webs of skin between fingers and toes. In adults, cells that are subject to wear and tear from exposure to the outside environment may live only a day or two before undergoing apoptosis. For example, it has been estimated that the intestinal epithelium is completely replaced with new cells every two to five days.

Concept Check Answer: p. 97

29. What are some features of apoptosis that distinguish it from cell death due to injury?

RUNNING PROBLEM

The day after Jan's visit, the computerized cytology analysis system rapidly scans the cells on the slide of Jan's cervical tissue, looking for abnormal cell size or shape. The computer is programmed to find multiple views for the cytologist to evaluate. The results of Jan's two Pap tests are shown in Figure 3.14.

Q5: Has Jan's dysplasia improved or worsened? What evidence do you have to support your answer?

Q6: Use your answer to question 5 to predict whether Jan's HPV infection has persisted or been cleared by her immune system.

Stem Cells Can Create New Specialized Cells

If cells in the adult body are constantly dying, where do their replacements come from? This question is still being answered and is one of the hottest topics in biological research today. The following paragraphs describe what we currently know.

All cells in the body are derived from the single cell formed at conception. That cell and those that follow reproduce themselves by undergoing the cell division process known as **mitosis** [see Appendix C]. The very earliest cells in the life of a human being are said to be **totipotent** {*totus,* entire} because they have the ability to develop into any and all types of specialized cells. Any totipotent cell has the potential to become a functioning organism.

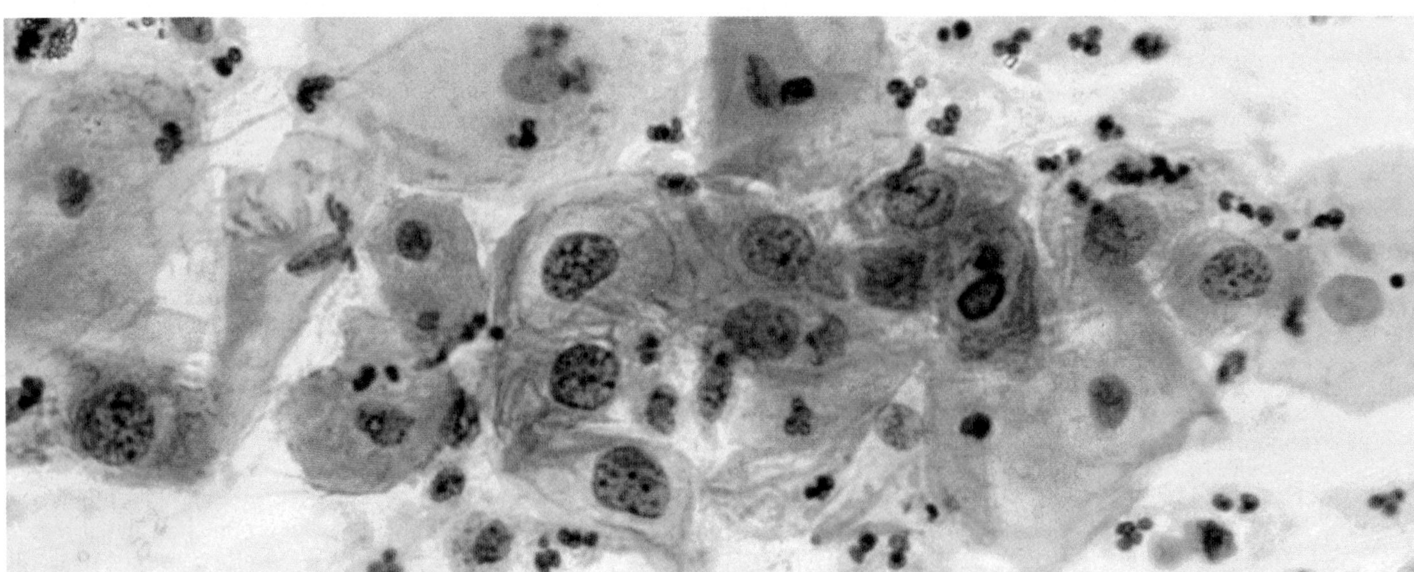

Fig. 3.14 ***Pap smears of cervical cells.*** The darker-staining structures inside the cells are the nuclei, surrounded by lighter-staining cytoplasm.

■ Fig. 3.15 *FOCUS ON ...*

The Skin

The layers of the skin

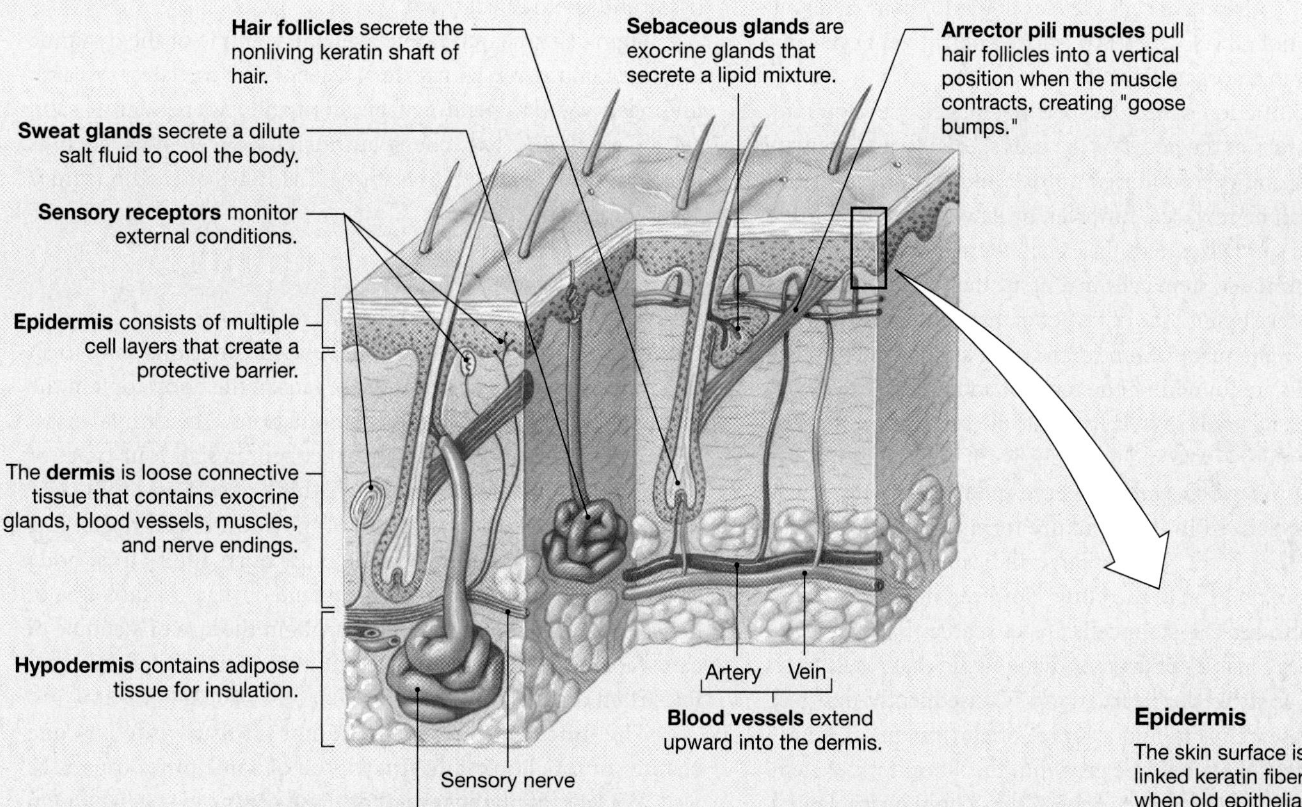

Hair follicles secrete the nonliving keratin shaft of hair.
Sebaceous glands are exocrine glands that secrete a lipid mixture.
Arrector pili muscles pull hair follicles into a vertical position when the muscle contracts, creating "goose bumps."
Sweat glands secrete a dilute salt fluid to cool the body.
Sensory receptors monitor external conditions.
Epidermis consists of multiple cell layers that create a protective barrier.
The **dermis** is loose connective tissue that contains exocrine glands, blood vessels, muscles, and nerve endings.
Hypodermis contains adipose tissue for insulation.
Artery
Vein
Blood vessels extend upward into the dermis.
Sensory nerve
Apocrine glands in the genitalia, anus, axillae (*axilla*, armpit), and eyelids release waxy or viscous milky secretions in response to fear or sexual excitement.

CLINICAL FOCUS

Melanoma is a serious form of skin cancer

Melanoma occurs when melanocytes become malignant, often following repeated exposure to UV light. One study found that people who used tanning beds were 24% more likely to develop melanoma.

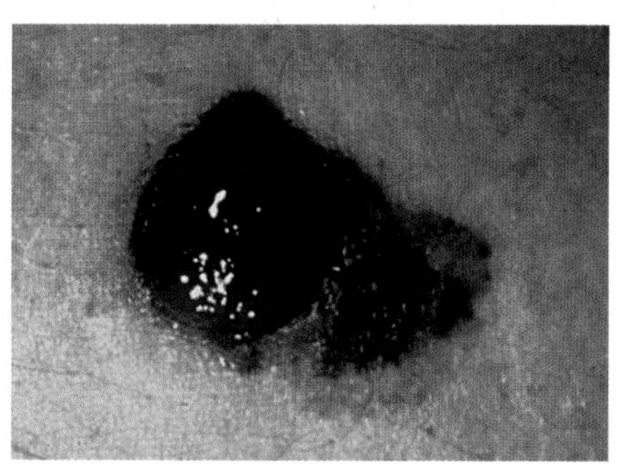

Epidermis

The skin surface is a mat of linked keratin fibers left behind when old epithelial cells die.

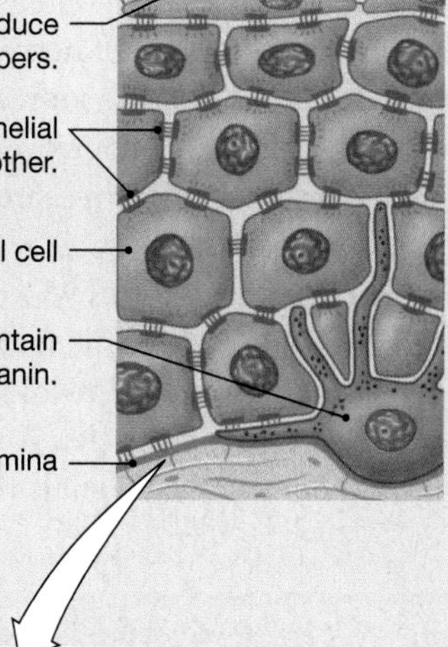

Phospholipid matrix acts as the skin's main waterproofing agent.
Surface keratinocytes produce keratin fibers.
Desmosomes anchor epithelial cells to each other.
Epidermal cell
Melanocytes contain the pigment melanin.
Basal lamina

Connection between epidermis and dermis

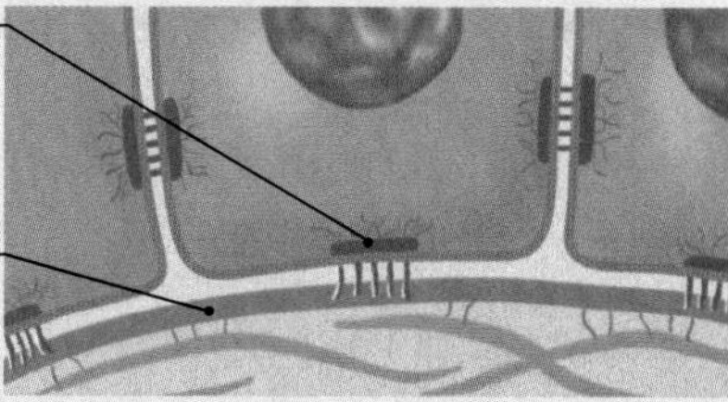

Hemidesmosomes tie epidermal cells to fibers of the basal lamina.
Basal lamina or basement membrane is an acellular layer between epidermis and dermis.

After about day 4 of development, the totipotent cells of the embryo begin to specialize, or *differentiate*. As they do so, they narrow their potential fates and become **pluripotent** {*plures,* many}. Pluripotent cells can develop into many different cell types but not all cell types. An isolated pluripotent cell cannot develop into an organism.

As differentiation continues, pluripotent cells develop into the various tissues of the body. As the cells specialize and mature, many lose the ability to undergo mitosis and reproduce themselves. They can be replaced, however, by new cells created from **stem cells,** less specialized cells that retain the ability to divide.

Undifferentiated stem cells in a tissue that retain the ability to divide and develop into the cell types of that tissue are said to be **multipotent** {*multi,* many}. Some of the most-studied multipotent adult stem cells are found in bone marrow and give rise to blood cells. However, all adult stem cells occur in very small numbers. They are difficult to isolate and do not thrive in the laboratory.

Biologists once believed that nerve and muscle cells, which are highly specialized in their mature forms, could not be replaced when they died. Now research indicates that stem cells for these tissues do exist in the body. However, naturally occurring neural and muscle stem cells are so scarce that they cannot replace large masses of dead or dying tissue that result from diseases such as strokes or heart attacks. Consequently, one goal of stem cell research is to find a source of pluripotent or multipotent stem cells that could be grown in the laboratory. If stem cells could be gown in larger numbers, they could be implanted to treat damaged tissues and *degenerative* diseases, those in which cells degenerate and die.

One example of a degenerative disease is Parkinson's disease, in which certain types of nerve cells in the brain die. Embryos and fetal tissue are rich sources of stem cells, but the use of embryonic stem cells is controversial and poses many legal and ethical questions. Some researchers hope that adult stem cells will show **plasticity,** the ability to specialize into a cell of a type different from the type for which they were destined.

There are still many challenges facing us before stem cell therapy becomes a standard medical treatment. One is finding a good source of stem cells. A second major challenge is determining the chemical signals that tell stem cells when to differentiate and what type of cell to become. And even once these two challenges are overcome and donor stem cells are implanted, the body may recognize that the new cells are foreign tissue and try to reject them.

Stem cell research is an excellent example of the dynamic and often controversial nature of science. For the latest research findings, as well as pending legislation and laws regulating stem cell research and use, check authoritative web sites, such as that sponsored by the U.S. National Institutes of Health (*http://stemcells.nih.gov*).

Organs

Groups of tissues that carry out related functions may form structures known as **organs.** The organs of the body contain the four types of tissue in various combinations. The skin is an excellent example of an organ that incorporates all four types of tissue into an integrated whole. We think of skin as a thin layer that covers the external surfaces of the body, but in reality it is the heaviest single organ, at about 16% of an adult's total body weight! If it were flattened out, it would cover a surface area of between 1.2 and 2.3 square meters, about the size of a couple of card-table tops. Its size and weight make skin one of the most important organs of the body.

The functions of the skin do not fit neatly into any one chapter of this book, and this is true of some other organs as well. We will highlight several of these organs in special *Organ Focus Features* throughout the book. These illustrated boxes discuss the structure and functions of these versatile organs so that you can gain an appreciation for the way different tissues combine for a united purpose. The first of these features, *Focus on the Skin,* appears on page 91.

As we consider the systems of the body in the succeeding chapters, you will see how diverse cells, tissues, and organs carry out the processes of the living body. Although the body's cells have different structures and different functions, they have one need in common: a continuous supply of energy. Without energy, cells cannot survive, let alone carry out all the other processes of daily living. Next we look at energy in living organisms and how cells capture and use the energy released by chemical reactions.

RUNNING PROBLEM CONCLUSION

The Pap Test, Cervical Cancer, and HPV

In this running problem, you learned that the Pap test can detect the early cell changes that precede cervical cancer. The diagnosis is not always simple because the change in cell cytology from normal to cancerous occurs along a continuum and can be subject to individual interpretation. In addition, not all cell changes are cancerous. The human papillomavirus (HPV), a common sexually transmitted infection, can also cause cervical dysplasia. In most cases, the woman's immune system overcomes the virus within two years, and the cervical cells revert to normal. A small number of women with persistent HPV infections have a higher risk of developing cervical cancer, however.

RUNNING PROBLEM CONCLUSION *(continued)*

Studies indicate that 98% of cervical cancers are associated with HPV infection. To learn more about the association between HPV and cervical cancer, go to the National Cancer Institute homepage (*www.cancer.gov*) and search for HPV. This site also contains information about cervical cancer. To check your understanding of the running problem, compare your answers with the information in the following summary table.

Question	Facts	Integration and Analysis
1. Why does the treatment of cancer focus on killing the cancerous cells?	Cancerous cells divide uncontrollably and fail to coordinate with normal cells. Cancerous cells fail to differentiate into specialized cells.	Unless removed, cancerous cells will displace normal cells. This may cause destruction of normal tissues. In addition, because cancerous cells do not become specialized, they cannot carry out the same functions as the specialized cells they displace.
2. What is happening in cancer cells that explains the large size of their nucleus and the relatively small amount of cytoplasm?	Cancerous cells divide uncontrollably. Dividing cells must duplicate their DNA prior to cell division, and this DNA duplication takes place in the nucleus, leading to the large size of that organelle [Appendix C].	Actively reproducing cells are likely to have more DNA in their nucleus as they prepare to divide, so their nuclei tend to be larger. Each cell division splits the cytoplasm between two daughter cells. If division is occurring rapidly, the daughter cells may not have time to synthesize new cytoplasm, so the amount of cytoplasm is less than in a normal cell.
3. What other kinds of damage or trauma are cervical epithelial cells normally subjected to?	The cervix is the passageway between the uterus and vagina.	The cervix is subject to trauma or damage, such as might occur during sexual intercourse and childbirth.
4. Which of its two types of epithelia is more likely to be affected by trauma?	The cervix consists of secretory epithelium with mucus-secreting glands lining the inside and protective epithelium covering the outside.	Protective epithelium is composed of multiple layers of cells and is designed to protect areas from mechanical and chemical stress [p. 84]. Therefore, the secretory epithelium with its single-cell layer is more easily damaged.
5. Jan's first Pap test showed atypical squamous cells of unknown significance (ASCUS). Were these cells more likely to come from the secretory portion of the cervix or from the protective epithelium?	Secretory cells are columnar epithelium. Protective epithelium is composed of multiple cell layers.	Protective epithelium with multiple cell layers has cells that are flat (stratified squamous epithelium). The designation ASC refers to these protective epithelial cells.
6. Has Jan's dysplasia improved or worsened? What evidence do you have to support your answer?	The slide from Jan's first Pap test shows abnormal cells with large nuclei and little cytoplasm. These abnormal cells do not appear in the second test.	The disappearance of the abnormal cells indicates that Jan's dysplasia has resolved. She will return in six months for a repeat Pap test. If it shows no dysplasia, her cervical cells have reverted to normal.
7. Use your answer to question 5 to predict whether Jan's HPV infection has persisted or been cleared by her immune system.	The cells in the second Pap test appear normal.	Once Jan's body fights off the HPV infection, her cervical cells should revert to normal. Her second HPV test showed no evidence of HPV DNA.

63 65 72 84 90 92

Chapter Summary

Cell biology and histology illustrate one of the major themes in physiology: *compartmentation*. In this chapter you learned how a cell is subdivided into two main compartments, the nucleus and the cytoplasm. You also learned how cells form tissues that create larger compartments within the body. A second theme in this chapter is the *molecular interactions* that create the *mechanical properties* of cells and tissues. Protein fibers of the cytoskeleton and cell junctions, along with the molecules that make up the extracellular matrix, form the "glue" that holds tissues together.

Functional Compartments of the Body

iP Fluids and Electrolytes: Introduction to Body Fluids

1. The **cell** is the functional unit of living organisms. (p. 63)
2. The major human body cavities are the cranial cavity (skull), thoracic cavity (thorax), and abdominopelvic cavity. (p. 64; Fig. 3.1a)
3. The **lumens** of some hollow organs are part of the body's external environment. (p. 63)
4. The body fluid compartments are the extracellular fluid (ECF) outside the cells and the intracellular fluid (ICF) inside the cells. The ECF can be subdivided into **interstitial fluid** bathing the cells and **plasma,** the fluid portion of the blood. (p. 64; Fig. 3.1b)

Biological Membranes

5. The word **membrane** is used both for cell membranes and for tissue membranes that line a cavity or separate two compartments. (p. 64; Fig. 3.1c)
6. The **cell membrane** acts as a barrier between the intracellular and extracellular fluids, provides structural support, and regulates exchange and communication between the cell and its environment. (p. 65)
7. The **fluid mosaic model** of a biological membrane shows it as a **phospholipid bilayer** with proteins inserted into the bilayer. (p. 66; Fig. 3.2b)
8. Membrane lipids include phospholipids, **sphingolipids,** and cholesterol. **Lipid-anchored proteins** attach to membrane lipids. (p. 68)
9. **Transmembrane proteins** are **integral proteins** tightly bound to the phospholipid bilayer. **Peripheral proteins** attach less tightly to either side of the membrane. (pp. 67; Fig. 3.2b, c)
10. Carbohydrates attach to the extracellular surface of cell membranes. (pp. 67)

Intracellular Compartments

11. The cytoplasm consists of semi-gelatinous **cytosol** with dissolved nutrients, ions, and waste products. Suspended in the cytosol are the other components of the cytoplasm: insoluble **inclusions** and fibers, which have no enclosing membrane, and **organelles,** which are membrane-enclosed bodies that carry out specific functions. (p. 70; Fig. 3.4a)
12. **Ribosomes** are inclusions that take part in protein synthesis.
13. Insoluble protein fibers come in three sizes: **actin fibers** (also called **microfilaments**), **intermediate filaments,** and **microtubules.** (p. 70; Table 3.2)
14. **Centrioles** that aid the movement of chromosomes during cell division, **cilia** that move fluid or secretions across the cell surface, and **flagella** that propel sperm through body fluids are made of microtubules. (p. 70; Figures 3.4e, 3.5)
15. The changeable **cytoskeleton** provides strength, support, and internal organization; aids transport of materials within the cell; links cells together; and enables motility in certain cells. (p. 70; Fig. 3.4b)
16. **Motor proteins** such as **myosins, kinesins,** and **dyneins** associate with cytoskeleton fibers to create movement. (p. 74; Fig. 3.6)
17. Membranes around organelles create compartments that separate functions. (p. 74)
18. **Mitochondria** generate most of the cell's ATP. (p. 71; Fig. 3.4g)
19. The **smooth endoplasmic reticulum** is the primary site of lipid synthesis. The **rough endoplasmic reticulum** is the primary site of protein synthesis. (p. 71; Fig. 3.4i)
20. The **Golgi apparatus** packages proteins into vesicles. **Secretory vesicles** release their contents into the extracellular fluid. (p. 71; Fig. 3.4h)
21. **Lysosomes** and **peroxisomes** are small **storage vesicles** that contain digestive enzymes. (p. 70; Fig. 3.4c and d)
22. The **nucleus** contains DNA, the genetic material that ultimately controls all cell processes, in the form of **chromatin.** The double-membrane **nuclear envelope** surrounding the nucleus has **nuclear pore complexes** that allow controlled chemical communication between the nucleus and cytosol. **Nucleoli** are nuclear areas that control the synthesis of RNA for ribosomes. (p. 71; Fig. 3.4j)
23. Protein synthesis is an example of how the cell separates functions by isolating them to separate compartments within the cell (p. 77; Fig. 3.7)

Tissues of the Body

iP Muscular: Anatomy Review—Skeletal Muscle Tissue

24. There are four **primary tissue types** in the human body: epithelial, connective, muscle, and neural. (p. 78)

25. **Extracellular matrix** secreted by cells provides support and a means of cell-cell communication. It is composed of proteoglycans and insoluble protein fibers. (p. 78)
26. Animal cell junctions fall into three categories. **Gap junctions** allow chemical and electrical signals to pass directly from cell to cell. **Tight junctions** restrict the movement of material between cells. **Anchoring junctions** hold cells to each other or to the extracellular matrix. (p. 79; Fig. 3.8)
27. Membrane proteins called **cell adhesion molecules** (CAMs) are essential in cell adhesion and in anchoring junctions. (p. 78; Table 3.3)
28. **Desmosomes** and **adherens junctions** anchor cells to each other. **Focal adhesions** and **hemidesmosomes** anchor cells to matrix. (p. 79; Fig. 3.8)
29. **Epithelial tissues** protect the internal environment, regulate the exchange of material, or manufacture and secrete chemicals. There are five functional types found in the body: exchange, transporting, ciliated, protective, and secretory. (p. 81; Fig. 3.9)
30. **Exchange epithelia** permit rapid exchange of materials, particularly gases. **Transporting epithelia** actively regulate the selective exchange of nongaseous materials between the internal and external environments. **Ciliated epithelia** move fluid and particles across the surface of the tissue. **Protective epithelia** help prevent exchange between the internal and external environments. The **secretory epithelia** release secretory products into the external environment or the blood. (p. 83; Fig. 3.10)
31. **Exocrine glands** release their secretions into the external environment through **ducts. Endocrine glands** are ductless glands that release their secretions, called **hormones,** directly into the extracellular fluid. (p. 81; Fig. 3.9b)
32. **Connective tissues** have extensive extracellular matrix that provides structural support and forms a physical barrier. (p. 86; Fig. 3.12)
33. **Loose connective tissues** are the elastic tissues that underlie skin. **Dense connective tissues,** including **tendons** and **ligaments,** have strength or flexibility because they are made of collagen. **Adipose tissue** stores fat. The connective tissue we call **blood** is characterized by a watery matrix. **Cartilage** is solid and flexible and has no blood supply. The fibrous matrix of **bone** is hardened by deposits of calcium salts. (p. 88; Fig. 3.13)
34. Muscle and neural tissues are called excitable tissues because of their ability to generate and propagate electrical signals called action potentials. **Muscle tissue** has the ability to contract and produce force and movement. There are three types of muscle: cardiac, smooth, and skeletal. (p. 89)
35. **Neural tissue** includes **neurons,** which use electrical and chemical signals to transmit information from one part of the body to another, and support cells known as **glial cells** (neuroglia). (p. 89)

3

Tissue Remodeling

36. Cell death occurs by **necrosis,** which adversely affects neighboring cells, and by **apoptosis,** programmed cell death that does not disturb the tissue. (p. 90)
37. **Stem cells** are cells that are able to reproduce themselves and differentiate into specialized cells. Stem cells are most plentiful in embryos but are also found in the adult body. (p. 92)

Organs

38. **Organs** are formed by groups of tissues that carry out related functions. The organs of the body contain the four types of tissues in various ratios. For example, skin is largely connective tissue. (p. 92)

Questions

Answers: p. A-1

Level One Reviewing Facts and Terms

1. List the four general functions of the cell membrane.
2. In 1972, Singer and Nicolson proposed the fluid mosaic model of the cell membrane. According to this model, the membrane is composed of a bilayer of ________ and a variety of embedded ________, with ________ on the extracellular surface.
3. What are the two primary types of biomolecules found in the cell membrane?
4. Define and distinguish between inclusions and organelles. Give an example of each.
5. Define cytoskeleton. List five functions of the cytoskeleton.
6. Match each term with the description that fits it best:

(a) cilia	1. in human cells, appears as single, long, whiplike tail
(b) centriole	2. short, hairlike structures that beat to produce currents in fluids
(c) flagellum	3. a bundle of microtubules that aids in mitosis
(d) centrosome	4. the microtubule-organizing center

7. Exocrine glands produce watery secretions (such as tears or sweat) called ________, or stickier solutions called ________.
8. Match each organelle with its function:

(a) endoplasmic reticulum	1. powerhouse of the cell where most ATP is produced
(b) Golgi apparatus	2. degrades long-chain fatty acids and toxic foreign molecules
(c) lysosome	3. network of membranous tubules that synthesize biomolecules
(d) mitochondrion	4. digestive system of cell, degrading or recycling components
(e) peroxisome	5. modifies and packages proteins into vesicles

9. What process activates the enzymes inside lysosomes?
10. ________ glands release hormones, which enter the blood and regulate the activities of organs or systems.
11. List the four major tissue types. Give an example and location of each.
12. The largest and heaviest organ in the body is the ________.

13. Match each protein to its function. Functions in the list may be used more than once.

(a) cadherin (b) CAM (c) collagen (d) connexin (e) elastin (f) fibrillin (g) fibronectin (h) integrin (i) occludin	1. membrane protein used to form cell junctions 2. matrix glycoprotein used to anchor cells 3. protein found in gap junctions 4. matrix protein found in connective tissue

14. What types of glands can be found within the skin? Name the secretion of each type.
15. The term *matrix* can be used in reference to an organelle or to tissues. Compare the meanings of the term in these two contexts.

Level Two Reviewing Concepts

16. List, compare, and contrast the three types of cell junctions and their subtypes. Give an example of where each type can be found in the body and describe its function in that location.
17. Which would have more rough endoplasmic reticulum: pancreatic cells that manufacture the protein hormone insulin, or adrenal cortex cells that synthesize the steroid hormone cortisol?
18. A number of organelles can be considered vesicles. Define *vesicle* and describe at least three examples.
19. Explain why a stratified epithelium offers more protection than a simple epithelium.
20. **Mapping exercise:** Transform this list of terms into a map of cell structure. Add functions where appropriate.

• actin • cell membrane • centriole • cilia • cytoplasm • cytoskeleton • cytosol • extracellular matrix • flagella • Golgi apparatus • intermediate filament • keratin • lysosome	• microfilament • microtubule • mitochondria • nonmembranous organelle • nucleus • organelle • peroxisome • ribosome • rough ER • secretory vesicle • smooth ER • storage vesicle • tubulin

21. Sketch a short series of columnar epithelial cells. Label the apical and basolateral borders of the cells. Briefly explain the different kinds of junctions found on these cells.
22. Arrange the following compartments in the order a glucose molecule entering the body at the intestine would encounter them: interstitial fluid, plasma, intracellular fluid. Which of these fluid compartments is/are considered extracellular fluid(s)?
23. Explain how inserting cholesterol into the phospholipid bilayer of the cell membrane decreases membrane permeability.
24. Compare and contrast the structure, locations, and functions of bone and cartilage.
25. Differentiate between the terms in each set below:
    (a) lumen and wall
    (b) cytoplasm and cytosol
    (c) myosin and keratin
26. When a tadpole turns into a frog, its tail shrinks and is reabsorbed. Is this an example of necrosis or apoptosis? Defend your answer.
27. Match the structures from the chapter to the basic physiological themes in the right column and give an example or explanation for each match. A structure may match with more than one theme.

(a) cell junctions (b) cell membrane (c) cytoskeleton (d) organelles (e) cilia	1. communication 2. molecular interactions 3. compartmentation 4. mechanical properties 5. biological energy use

28. In some instances, the extracellular matrix can be quite rigid. How might developing and expanding tissues cope with a rigid matrix to make space for themselves?

Level Three Problem Solving

29. One result of cigarette smoking is paralysis of the cilia that line the respiratory passageways. What function do these cilia serve? Based on what you have read in this chapter, why is it harmful when they no longer beat? What health problems would you expect to arise? How does this explain the hacking cough common among smokers?
30. Cancer is abnormal, uncontrolled cell division. What property of epithelial tissues might (and does) make them more prone to developing cancer?
31. What might happen to normal physiological function if matrix metalloproteinases are inhibited by drugs?

Answers

Answers to Concept Check Questions

Page 69

1. Membrane lipids are phospholipid, sphingolipid, and cholesterol.
2. Integral proteins are tightly bound to the membrane. Peripheral proteins are loosely bound to membrane components. Proteins may be transmembrane, lipid-anchored, or loosely bound to other proteins.
3. The tails of phospholipids are hydrophobic, and a single layer would put the tails in direct contact with aqueous body fluids.
4. A substance crosses one phospholipid bilayer to enter a cell.

Page 74

5. Cytoplasmic fibers are actin fibers (microfilaments), intermediate filaments, and microtubules.
6. Without a flagellum, a sperm would be unable to swim to find an egg to fertilize.
7. Cytoplasm is everything inside the cell membrane except the nucleus. Cytosol is the semi-gelatinous substance in which organelles and inclusions are suspended.
8. Cilia are short, usually are very numerous on a cell, and move fluid or substances across the cell surface. Flagella are longer, usually occur singly on human sperm, and are used to propel a cell through a fluid.
9. Motor proteins use energy to create movement.

Page 76

10. A membrane separates organelles from the cytosol; inclusions are suspended in the cytosol.
11. Rough ER has ribosomes attached to the cytoplasmic side of its membrane; smooth ER lacks ribosomes. Rough ER synthesizes proteins; smooth ER synthesizes lipids.
12. Lysosomes contain enzymes that break down bacteria and old organelles. Peroxisomes contain enzymes that break down fatty acids and foreign molecules.
13. The membranes of organelles create compartments that physically isolate their lumens from the cytosol. The double membrane of mitochondria creates two different compartments inside the organelle.
14. A large number of mitochondria suggests that the cell has a high energy requirement because mitochondria are the site of greatest energy production in the cell.
15. Large amounts of smooth endoplasmic reticulum suggest that the tissue synthesizes large amounts of lipids, fatty acids, or steroids, or that it detoxifies foreign molecules.

Page 80

16. Cell junctions are gap (communicating), tight (occluding), and anchoring.
17. (a) tight, (b) gap, (c) anchoring (specifically, desmosome), (d) anchoring (specifically, focal adhesion)

Page 85

18. The five functional types of epithelia are protective, secretory, transporting, ciliated, and exchange.
19. Secretion is the process by which a cell releases a substance into its environment.
20. Endocrine glands do not have ducts, and they secrete into the blood. Exocrine glands have ducts and secrete into the external environment.
21. Hemidesmosomes attach to laminin (see Fig. 3.8).
22. No, skin has many layers of cells in order to protect the internal environment. A simple squamous epithelium (which is one cell thick with flattened cells) would not be a protective epithelium.
23. The cell is an endocrine cell because it secretes its product into the extracellular space for distribution in the blood.

Page 89

24. Connective tissues have extensive matrix.
25. Collagen provides strength and flexibility; elastin and fibrillin provide elastance; fibronectin helps anchor cells to matrix.
26. Connective tissues include bone, cartilage, blood, dense connective tissues (ligaments and tendons), loose connective tissue, and adipose tissue.
27. The plasma, or liquid portion of blood, surrounds the blood cells and is therefore the extracellular matrix.
28. Cartilage lacks a blood supply, so oxygen and nutrients needed for repair must reach the cells by diffusion, a slow process.

Page 90

29. Apoptosis is a tidy form of cell death that removes cells without disrupting their neighbors. By contrast, necrosis releases digestive enzymes that damage neighboring cells.

Answers to Figure Questions

Page 81

Figure 3.9: Endocrine glands (without ducts) secrete their hormones into the blood. Exocrine glands, with ducts, secrete their products outside the body—onto the surface of the skin or into the lumen of an organ that opens into the environment outside the body.

4 Energy and Cellular Metabolism

There is no good evidence that . . . life evades the second law of thermodynamics, but in the downward course of the energy-flow it interposes a barrier and dams up a reservoir which provides potential for its own remarkable activities.

—F. G. Hopkins, 1933

Background Basics

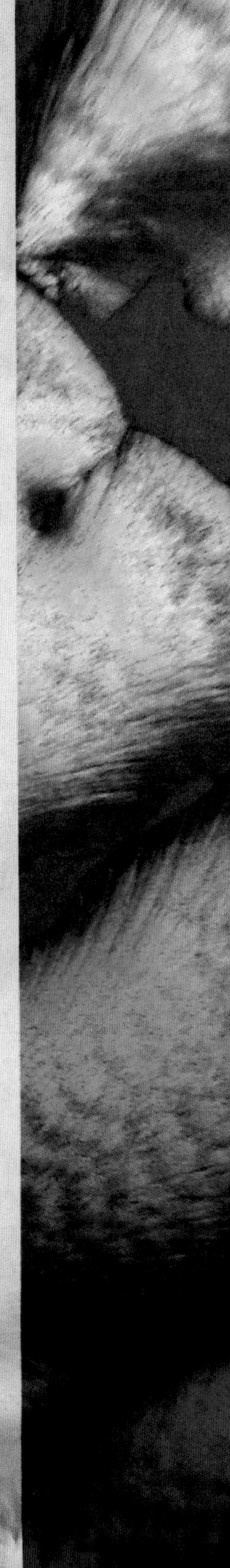

Glucose crystals

Christine Schmidt, Ph.D., and her graduate students seed isolated endothelial cells onto an engineered matrix and watch them grow. They know that if their work is successful, the tissue that results might someday help replace a blood vessel in the body. Just as a child playing with building blocks assembles them into a house, the bioengineer and her students create tissue from cells. In both cases someone familiar with the starting components, building blocks or cells, can predict what the final product will be: blocks make buildings; cells make tissues.

Why then can't biologists, knowing the characteristics of nucleic acids, proteins, lipids, and carbohydrates, explain how combinations of these molecules acquire the remarkable attributes of a living cell? How can living cells carry out processes that far exceed what we would predict from understanding their individual components? The answer is *emergent properties* [p. 2], those distinctive traits that cannot be predicted from the simple sum of the component parts. For example, if you came across a collection of metal pieces and bolts from a disassembled car motor, could you predict (without prior knowledge) that, given an energy source and properly arranged, this collection could create the power to move thousands of pounds?

The emergent properties of biological systems are of tremendous interest to scientists trying to explain how a simple compartment, such as a phospholipid liposome [p. 66], could have evolved into the first living cell. Pause for a moment and see if you can list the properties of life that characterize all living creatures. If you were a scientist looking at pictures and samples sent back from Mars, what would you look for to determine whether life exists there?

RUNNING PROBLEM

Tay-Sachs Disease: A Deadly Inheritance

In many American ultra-orthodox Jewish communities—in which arranged marriages are the norm—the rabbi is entrusted with an important, life-saving task. He keeps a confidential record of individuals known to carry the gene for Tay-Sachs disease, a fatal, inherited condition that strikes one in 3600 American Jews of Eastern European descent. Babies born with this disease rarely live beyond age 4, and there is no cure. Based on the family trees he constructs, the rabbi can avoid pairing two individuals who carry the deadly gene.

Sarah and David, who met while working on their college newspaper, are not orthodox Jews. Both are aware, however, that their Jewish ancestry might put any children they have at risk for Tay-Sachs disease. Six months before their wedding, they decide to see a genetic counselor to determine whether they are carriers of the gene for Tay-Sachs disease.

99 104 107 110 118 124

Table 4.1 Properties of Living Organisms

Properties of Living Organisms
1. Have a complex structure whose basic unit of organization is the cell
2. Acquire, transform, store, and use energy
3. Sense and respond to internal and external environments
4. Maintain homeostasis through internal control systems with feedback
5. Store, use, and transmit information
6. Reproduce, develop, grow, and die
7. Have emergent properties that cannot be predicted from the simple sum of the parts
8. Individuals adapt and species evolve

Now compare your list with the one in ■ Table 4.1. Living organisms are highly organized and complex entities. Even a one-celled bacterium, although it appears simple under a microscope, has incredible complexity at the chemical level of organization. It uses intricately interconnected biochemical reactions to acquire, transform, store, and use energy and information. It senses and responds to changes in its internal and external environments and adapts so that it can maintain homeostasis. It reproduces, develops, grows, and dies; and over time, its species evolves.

Energy is essential for these processes we associate with living things. Without energy for growth, repair, and maintenance of the internal environment, a cell is like a ghost town filled with buildings that are slowly crumbling into ruin. Cells need energy to import raw materials, make new molecules, and repair or recycle aging parts. The ability of cells to extract energy from the external environment and use that energy to maintain themselves as organized, functioning units is one of their most outstanding characteristics. In this chapter, we look at the cell processes through which the human body obtains energy and maintains its ordered systems. You will learn how protein interactions [p. 50] apply to enzyme activity and how the subcellular compartments [p. 72] separate various steps of energy metabolism.

Energy in Biological Systems

Energy cycling between the environment and living organisms is one of the fundamental concepts of biology. All cells use energy from their environment to grow, make new parts, and reproduce. Plants trap radiant energy from the sun and store it as chemical-bond energy through the process of

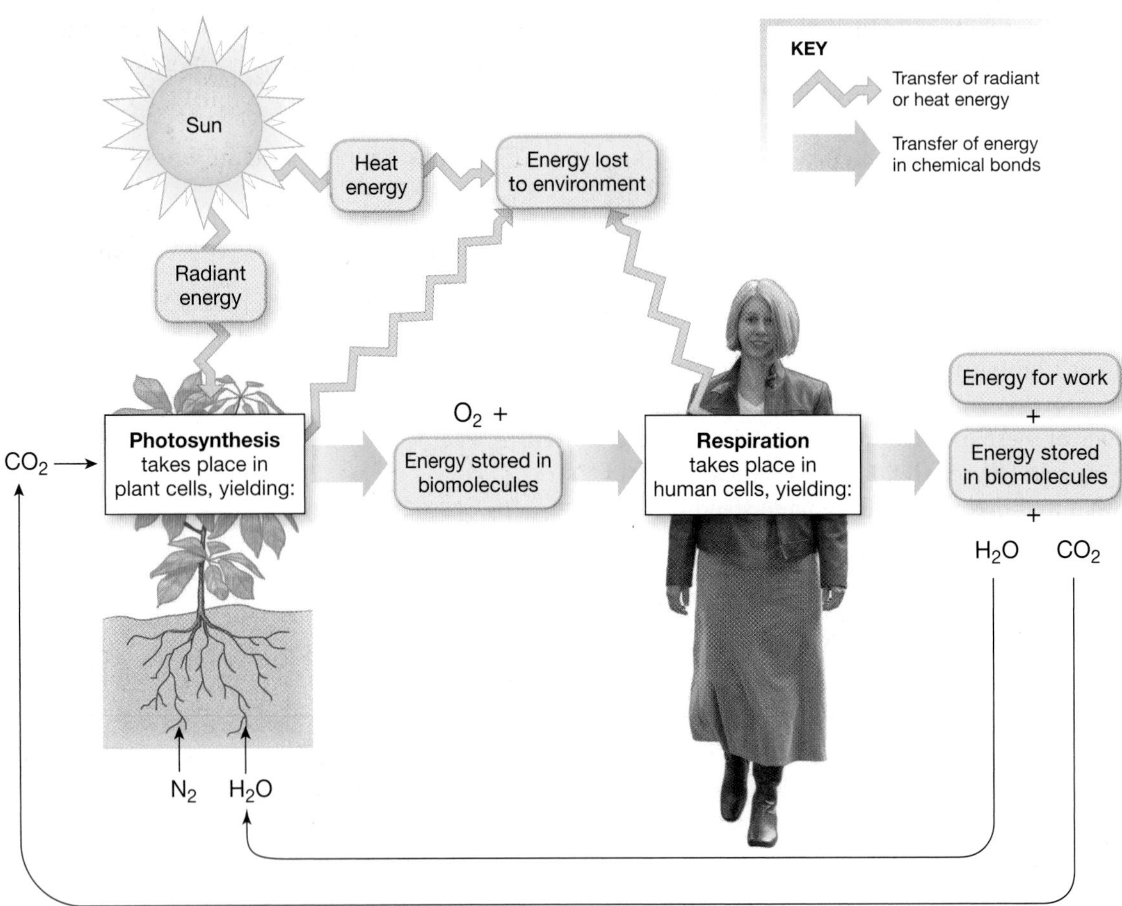

Fig. 4.1 ***Energy transfer in the environment.*** Plants trap radiant energy from the sun and use it to store energy in the chemical bonds of biomolecules. Animals eat the plants and either use the energy or store it.

photosynthesis (Fig. 4.1). They extract carbon and oxygen from carbon dioxide, nitrogen from the soil, and hydrogen and oxygen from water to make biomolecules such as glucose and amino acids.

Animals, on the other hand, cannot trap energy from the sun or use carbon and nitrogen from the air and soil to synthesize biomolecules. They must import chemical-bond energy by ingesting the biomolecules of plants or other animals. Ultimately, however, energy trapped by photosynthesis is the energy source for all animals, including humans.

Animals extract energy from biomolecules through the process of *respiration,* which consumes oxygen and produces carbon dioxide and water. If animals ingest more energy than they need for immediate use, the excess energy is stored in chemical bonds, just as it is in plants. Glycogen (a glucose polymer) and lipid molecules are the main energy stores in animals [p. 34]. These storage molecules are available for use at times when an animal's energy needs exceed its food intake.

Concept Check Answer: p. 128

1. Which biomolecules always include nitrogen in their chemical makeup?

Energy Is Used to Perform Work

All living organisms obtain, store, and use energy to fuel their activities. **Energy** can be defined as the capacity to do work, but what is *work*? We use this word in everyday life to mean various things, from hammering a nail to sitting at a desk writing a paper. In biological systems, however, the word means one of three specific things: chemical work, transport work, or mechanical work.

Chemical work is the making and breaking of chemical bonds. It enables cells and organisms to grow, maintain a suitable internal environment, and store information needed for reproduction and other activities. Forming the chemical bonds of a protein is an example of chemical work.

Transport work enables cells to move ions, molecules, and larger particles through the cell membrane and through the membranes of organelles in the cell. Transport work is particularly useful for creating **concentration gradients,** distributions of molecules in which the concentration is higher on one side of a membrane than on the other. For example, certain types of endoplasmic reticulum [p. 75] use energy to import calcium ions from the cytosol. This ion transport creates a high calcium concentration inside the organelle and a low concentration in the cytosol. If calcium is then released back into the cytosol, it creates a "calcium signal" that causes the cell to perform some action, such as muscle contraction.

Mechanical work in animals is used for movement. At the cellular level, movement includes organelles moving around in a cell, cells changing shape, and cilia and flagella beating [p. 73]. At the macroscopic level in animals, movement usually involves muscle contraction. Most mechanical work is mediated by motor proteins that make up certain intracellular fibers and filaments of the cytoskeleton [p. 70].

Energy Comes in Two Forms: Kinetic and Potential

Energy can be classified in various ways. We often think of energy in terms we deal with daily: thermal energy, electrical energy, mechanical energy. We speak of energy stored in chemical bonds. Each type of energy has its own characteristics. However, all types of energy share an ability to appear in two forms: as kinetic energy or as potential energy.

Kinetic energy is the energy of motion {*kinetikos*, motion}. A ball rolling down a hill, perfume molecules spreading through the air, electric charge flowing through power lines, heat warming a frying pan, and molecules moving across biological membranes are all examples of bodies that have kinetic energy.

Potential energy is stored energy. A ball poised at the top of a hill has potential energy because it has the potential to start moving down the hill. A molecule positioned on the high-concentration side of a concentration gradient stores potential energy because it has the potential energy to move down the gradient. In chemical bonds, potential energy is stored in the position of the electrons that form the bond [p. 38]. [To learn more about kinetic and potential energy, see Appendix B.]

A key feature of all types of energy is the ability of potential energy to become kinetic energy and vice versa.

Energy Can Be Converted from One Form to Another

Recall that a general definition of energy is the capacity to do work. Work always involves movement and therefore is associated with kinetic energy. Potential energy also can be used to perform work, but the potential energy must first be converted to kinetic energy. The conversion from potential energy to kinetic energy is never 100% efficient, and a certain amount of energy is lost to the environment, usually as heat.

The amount of energy lost in the transformation depends on the *efficiency* of the process. Many physiological processes in the human body are not very efficient. For example, 70% of the energy used in physical exercise is lost as heat rather than transformed into the work of muscle contraction.

■ Figure 4.2 summarizes the relationship of kinetic energy and potential energy:

1. Kinetic energy of the moving ball is transformed into potential energy as work is used to push the ball up the ramp (Fig. 4.2a).
2. Potential energy is stored in the stationary ball at the top of the ramp (Fig. 4.2b). No work is being performed, but the capacity to do work is stored in the position of the ball.
3. The potential energy of the ball becomes kinetic energy when the ball rolls down the ramp (Fig. 4.2c). Some kinetic energy is lost to the environment as heat due to friction between the ball and the air and ramp.

In biological systems, potential energy is stored in concentration gradients and chemical bonds. It is transformed into kinetic energy when needed to do chemical, transport, or mechanical work.

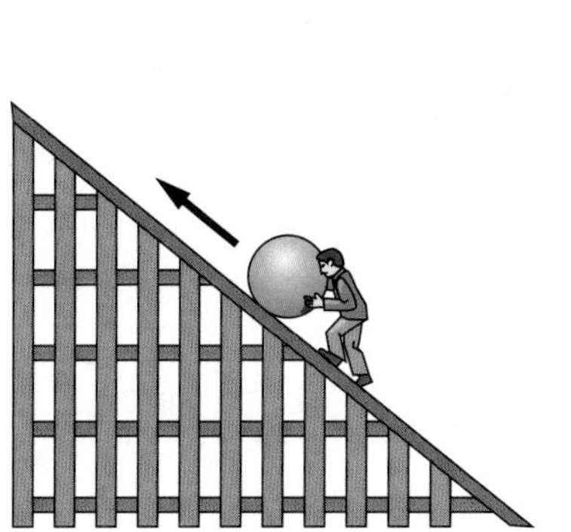

(a) Work is used to push a ball up a ramp. Kinetic energy of movement up the ramp is being stored in the potential energy of the ball's position.

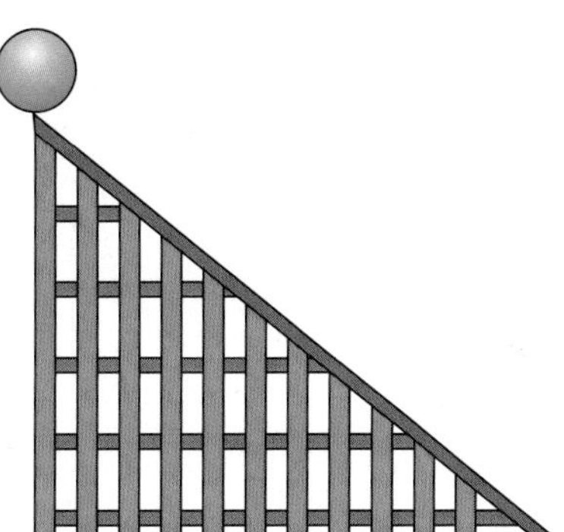

(b) The ball sitting at the top of the ramp has potential energy, the potential to do work.

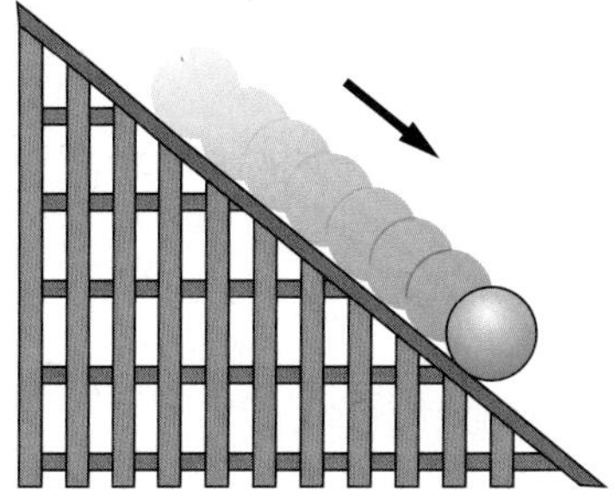

(c) The ball rolling down the ramp is converting the potential energy to kinetic energy. However, the conversion is not totally efficient, and some energy is lost as heat due to friction between the ball, ramp, and air.

■ **Fig. 4.2** ***The relationship between kinetic energy and potential energy***

Thermodynamics Is the Study of Energy Use

Two basic rules govern the transfer of energy in biological systems and in the universe as a whole. The **first law of thermodynamics,** also known as the *law of conservation of energy,* states that the total amount of energy in the universe is constant. The universe is considered to be a *closed system*—nothing enters and nothing leaves. Energy can be converted from one type to another, but the total amount of energy in a closed system never changes.

The human body is not a closed system, however. As an *open system,* it exchanges materials and energy with its surroundings. Because our bodies cannot create energy, they import it from outside in the form of food. By the same token, our bodies lose energy, especially in the form of heat, to the environment. Energy that stays within the body can be changed from one type to another or can be used to do work.

The **second law of thermodynamics** states that natural spontaneous processes move from a state of order (nonrandomness) to a condition of randomness or disorder, also known as **entropy.** Creating and maintaining order in an open system such as the body requires the input of energy. Disorder occurs when open systems lose energy to their surroundings without regaining it. When this happens, we say that the entropy of the open system has increased.

The ghost-town analogy mentioned earlier illustrates the second law. When people put all their energy into activities away from town, the town slowly falls into disrepair and becomes less organized (its entropy increases). Similarly, without continual input of energy, a cell is unable to maintain its ordered internal environment. As the cell loses organization, its ability to carry out normal functions disappears, and it dies.

In the remainder of this chapter, you will learn how cells obtain energy from and store energy in the chemical bonds of biomolecules. Using chemical reactions, cells transform the potential energy of chemical bonds into kinetic energy for growth, maintenance, reproduction, and movement.

Concept Check Answers: p. 128

2. Name two ways animals store energy in their bodies.
3. What is the difference between potential energy and kinetic energy?
4. What is entropy?

Chemical Reactions

Living organisms are characterized by their ability to extract energy from the environment and use it to support life processes. The study of energy flow through biological systems is a field known as **bioenergetics** {*bios,* life + *en-,* in + *ergon,* work}. In a biological system, chemical reactions are a critical means of transferring energy from one part of the system to another.

Energy Is Transferred Between Molecules During Reactions

In a **chemical reaction,** a substance becomes a different substance, usually by the breaking and/or making of covalent bonds. A reaction begins with one or more molecules called **reactants** and ends with one or more molecules called **products** (■ Tbl. 4.2). In this discussion, we consider a reaction that begins with two reactants and ends with two products:

$$A + B \rightarrow C + D$$

The speed with which a reaction takes place, the **reaction rate,** is the disappearance rate of the reactants (A and B) or the appearance rate of the products (C and D). Reaction rate is measured as change in concentration during a certain time period and is often expressed as molarity per second (M/sec).

The purpose of chemical reactions in cells is either to transfer energy from one molecule to another or to use energy stored in reactant molecules to do work. The potential energy stored in the chemical bonds of a molecule is known as the **free energy** of the molecule. Generally, complex molecules have more chemical bonds and therefore higher free energies. For example, a large glycogen molecule has more free energy than a single glucose molecule, which in turn has more free energy than the carbon dioxide and water from which it was synthesized. The high free energy of complex molecules such as glycogen is the reason that these molecules are used to store energy in cells.

To understand how chemical reactions transfer energy between molecules, we should answer two questions. First, how do reactions get started? The energy required to initiate a reaction is known as the *activation energy* for the reaction. Second, what happens to the free energy of the products and reactants during a reaction? The difference in free energy between reactants and products is the *net free energy change of the reaction.*

Table 4.2 Chemical Reactions

Reaction Type	Reactants (Substrates)		Products
Combination	A + B	→	C
Decomposition	C	→	A + B
Single displacement*	L + MX	→	LX + M
Double displacement*	LX + MY	→	LY + MX

*X and Y represent atoms, ions, or chemical groups.

Activation Energy Gets Reactions Started

Activation energy is the initial input of energy required to bring reactants into a position that allows them to react with one another. This "push" needed to start the reaction is shown in ■ Figure 4.3a as the little hill up which the ball must be pushed before it can roll by itself down the slope. A reaction with low activation energy proceeds spontaneously when the reactants are brought together. You can demonstrate a *spontaneous reaction* by pouring a little vinegar onto some baking soda and watching the two react to form carbon dioxide. Reactions with high activation energies either do not proceed spontaneously or else proceed too slowly to be useful. For example, if you pour vinegar over a pat of butter, no observable reaction takes place.

Energy Is Trapped or Released during Reactions

One characteristic property of any chemical reaction is the free energy change that occurs as the reaction proceeds. The products of a reaction have either a lower free energy than the reactants or a higher free energy than the reactants. A change in free energy level means that the reaction has either released or trapped energy.

If the free energy of the products is lower than the free energy of the reactants, as in Figure 4.3b, the reaction releases energy and is called an **exergonic reaction** {*ex-*, out + *ergon,* work}. The energy released by an exergonic, or *energy-producing,* reaction may be used by other molecules to do work or may be given off as heat. In a few cases, the energy released in an exergonic reaction is stored as potential energy in a concentration gradient.

An important biological example of an exergonic reaction is the combination of ATP and water to form ADP, inorganic phosphate (P_i), and H^+. Energy is released during this reaction when the high-energy phosphate bond of the ATP molecule is broken:

$$ATP + H_2O \rightarrow ADP + P_i + H^+ + \text{energy}$$

Now contrast the exergonic reaction of Figure 4.3b with the reaction represented in Figure 4.3c. In the latter, products retain part of the activation energy that was added, making their free energy greater than that of the reactants. These reactions that require a net input of energy are said to be **endergonic** {*end(o),* within + *ergon,* work}, or *energy-utilizing,* reactions.

Some of the energy added to an endergonic reaction remains trapped in the chemical bonds of the products. These energy-consuming reactions are often *synthesis* reactions, in which complex molecules are made from smaller molecules. For example, an endergonic reaction links many glucose molecules together to create the glucose polymer glycogen. The complex glycogen molecule has more free energy than the simple glucose molecules used to make it.

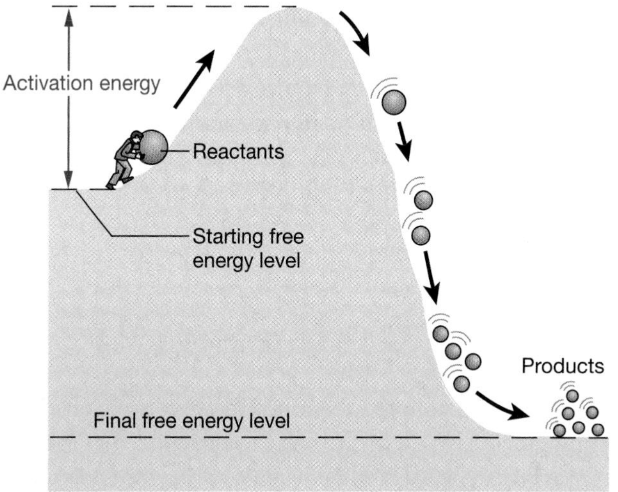

(a) Activation energy is the "push" needed to start a reaction.

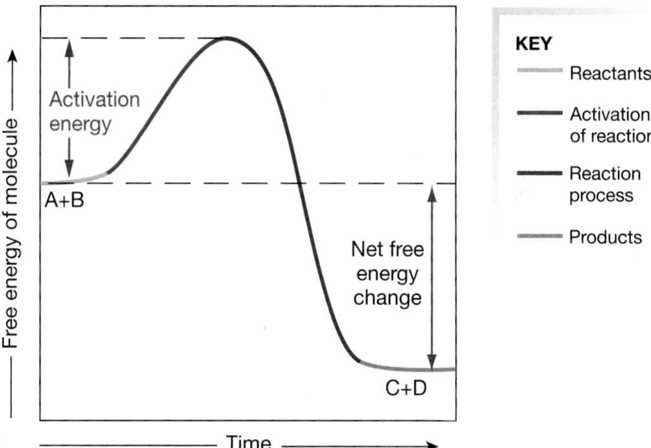

b) Exergonic reactions release energy because the products have less energy than the reactants.

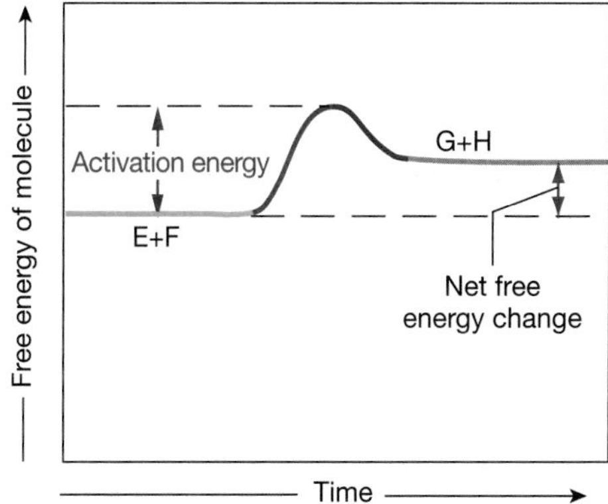

(c) Endergonic reactions trap some activation energy in the products, which then have more free energy than the reactants.

■ **Fig. 4.3** ***Activation energy and exergonic and endergonic reactions***

If a reaction traps energy as it proceeds in one direction ($A + B \rightarrow C + D$), it releases energy as it proceeds in the reverse direction ($C + D \rightarrow A + B$). (The naming of forward and reverse directions is arbitrary.) For example, the energy trapped in the bonds of glycogen during its synthesis is released when glycogen is broken back down into glucose.

Coupling Endergonic and Exergonic Reactions Where does the activation energy for metabolic reactions come from? The simplest way for a cell to acquire activation energy is to couple an exergonic reaction to an endergonic reaction. Some of the most familiar coupled reactions are those that use the energy released by breaking the high-energy bond of ATP to drive an endergonic reaction:

$$E + F \xrightarrow{\text{ATP} \;\rightarrow\; \text{ADP} + P_i} G + H$$

In this type of coupled reaction, the two reactions take place simultaneously and in the same location, so that the energy from ATP can be used immediately to drive the endergonic reaction between reactants E and F.

However, it is not always practical for reactions to be directly coupled like this. Consequently, living cells have developed ways to trap the energy released by exergonic reactions and save it for later use. The most common method is to trap the energy in the form of high-energy electrons carried on nucleotides [p. 36]. The nucleotide molecules NADH, $FADH_2$, and NADPH all capture energy in the electrons of their hydrogen atoms (■ Fig. 4.4). NADH and $FADH_2$ usually transfer most of this energy to ATP, which can then be used to drive endergonic reactions.

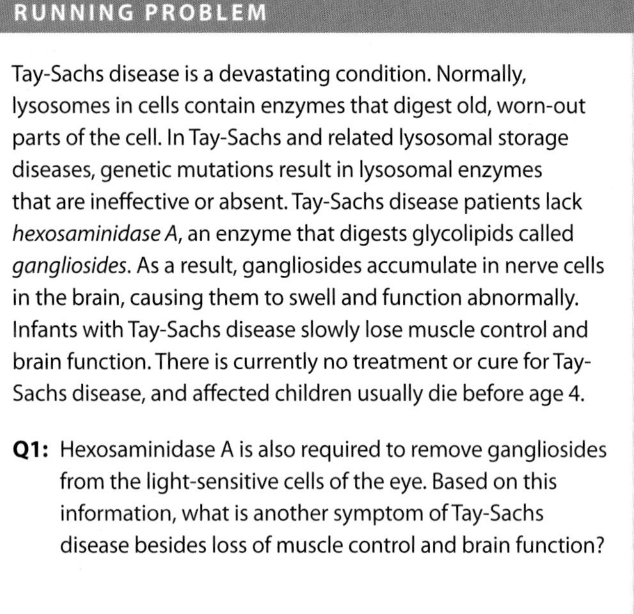
RUNNING PROBLEM

Tay-Sachs disease is a devastating condition. Normally, lysosomes in cells contain enzymes that digest old, worn-out parts of the cell. In Tay-Sachs and related lysosomal storage diseases, genetic mutations result in lysosomal enzymes that are ineffective or absent. Tay-Sachs disease patients lack *hexosaminidase A*, an enzyme that digests glycolipids called *gangliosides*. As a result, gangliosides accumulate in nerve cells in the brain, causing them to swell and function abnormally. Infants with Tay-Sachs disease slowly lose muscle control and brain function. There is currently no treatment or cure for Tay-Sachs disease, and affected children usually die before age 4.

Q1: Hexosaminidase A is also required to remove gangliosides from the light-sensitive cells of the eye. Based on this information, what is another symptom of Tay-Sachs disease besides loss of muscle control and brain function?

99 104 107 110 118 124

Net Free Energy Change Determines Reaction Reversibility

The net free energy change of a reaction plays an important role in determining whether that reaction can be reversed, because the net free energy change of the forward reaction contributes to the activation energy of the reverse reaction. A chemical reaction that can proceed in both directions is called a **reversible reaction.** In a reversible reaction, the forward reaction $A + B \rightarrow C + D$ and its reverse reaction $C + D \rightarrow A + B$ are both likely to take place. If a reaction proceeds in one direction but not the other, it is an **irreversible reaction.**

For example, look at the activation energy of the reaction $C + D \rightarrow A + B$ in ■ Figure 4.5. This reaction is the reverse of the reaction shown in Figure 4.3b. Because a lot of energy was released in the forward reaction $A + B \rightarrow C + D$, the activation energy of the reverse reaction is substantial (Fig. 4.5). As you will recall, the larger the activation energy, the less likely it is that the reaction will proceed spontaneously. Theoretically, all reactions can be reversed with enough energy input, but some reactions release so much energy that they are essentially irreversible.

In your study of physiology, you will encounter a few irreversible reactions. However, most biological reactions are reversible: if the reaction $A + B \rightarrow C + D$ is possible, then so is the reaction $C + D \rightarrow A + B$. Reversible reactions are shown with arrows that point in both directions: $A + B \rightleftharpoons C + D$. One of the main reasons that many biological reactions are reversible is that they are aided by the specialized proteins known as enzymes.

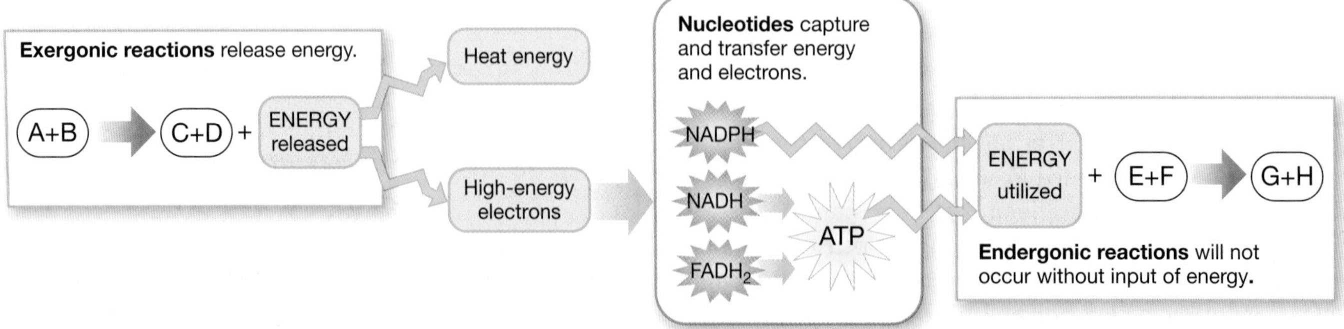

■ **Fig. 4.4** ***Energy transfer and storage in biological reactions.*** Energy released by exergonic reactions can be trapped in the high-energy electrons of NADH, $FADH_2$, or NADPH. Energy that is not trapped is given off as heat.

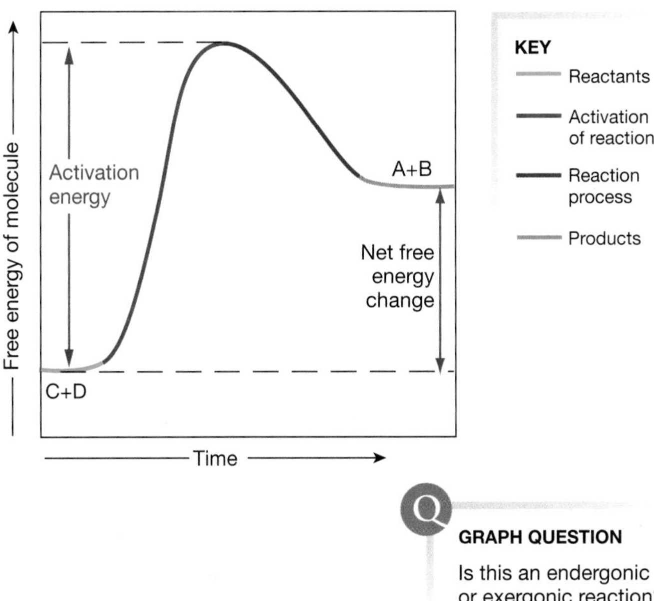

Fig. 4.5 ***Some reactions have large activation energies***

Concept Check

Answers: p. 128

5. What is the difference between endergonic and exergonic reactions?
6. If you mix baking soda and vinegar together in a bowl, the mixture reacts and foams up, releasing carbon dioxide gas. Name the reactant(s) and product(s) in this reaction.
7. Do you think the reaction of question 6 is endergonic or exergonic? Do you think it is reversible? Defend your answers.

Enzymes

Enzymes are proteins that speed up the rate of chemical reactions. During these reactions, the enzyme molecules are not changed in any way, meaning they are biological *catalysts*. Without enzymes, most chemical reactions in a cell would go so slowly that the cell would be unable to live. Because an enzyme is not permanently changed or used up in the reaction it catalyzes, we might write it in a reaction equation this way:

$$A + B + \text{enzyme} \rightarrow C + D + \text{enzyme}$$

This way of writing the reaction shows that the enzyme participates with reactants A and B but is unchanged at the end of the reaction. A more common shorthand for enzymatic reactions shows the name of the enzyme above the reaction arrow, like this:

$$A + B \xrightarrow{\text{enzyme}} C + D$$

In enzymatically catalyzed reactions, the reactants are called **substrates.**

Enzymes Are Proteins

Most enzymes are large proteins with complex three-dimensional shapes, although recently researchers discovered that RNA can sometimes act as a catalyst. Like other proteins that bind to substrates, protein enzymes exhibit specificity, competition, and saturation [p. 50 ff].

A few enzymes come in a variety of related forms (isoforms) and are known as **isozymes** {*iso-*, equal} of one another. Isozymes are enzymes that catalyze the same reaction but under different conditions or in different tissues. The structures of related isozymes are slightly different from one another, which causes the variability in their activity. Many isozymes have complex structures with multiple protein chains.

For example, the enzyme *lactate dehydrogenase* (LDH) has two kinds of subunits, named H and M, that are assembled into *tetramers*—groups of four. LDH isozymes include H_4, H_2M_2, and M_4. The different LDH isozymes are tissue specific, including one found primarily in the heart and a second found in skeletal muscle and the liver.

Isozymes have an important role in the diagnosis of certain medical conditions. For example, in the hours following a heart attack, damaged heart muscle cells release enzymes into the blood. One way to determine whether a person's chest pain was indeed due to a heart attack is to look for elevated levels of heart isozymes in the blood. Some diagnostically important enzymes and the diseases of which they are suggestive are listed in Table 4.3.

Table 4.3 Diagnostically Important Enzymes

Elevated blood levels of these enzymes are suggestive of the pathologies listed.

Enzyme	Related Diseases
Acid phosphatase*	Cancer of the prostate
Alkaline phosphatase	Diseases of bone or liver
Amylase	Pancreatic disease
Creatine kinase (CK)	Myocardial infarction (heart attack), muscle disease
Glutamate dehydrogenase (GDH)	Liver disease
Lactate dehydrogenase (LDH)	Tissue damage to heart, liver, skeletal muscle, red blood cells

*A newer test for a molecule called prostate specific antigen (PSA) has replaced the test for acid phosphatase in the diagnosis of prostate cancer.

BIOTECHNOLOGY

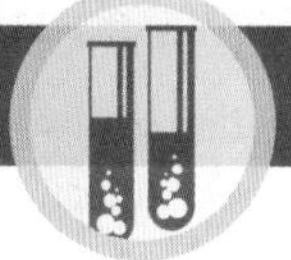

Seeing Isozymes

One way to determine which isozymes are present in a tissue sample is to use a technique known as **electrophoresis.** In this technique, a solution derived from the tissue sample is placed at one end of a container filled with a polyacrylamide polymer gel. An electric current passed through the gel causes the negatively charged proteins to move toward the positively charged end of the gel. The rate at which the proteins move depends on their size, their shape, and the electrical charge on their amino acids. As proteins move along the gel at different rates, they separate from one another and appear as individual bands of color when stained with a dye called Coomassie blue or with silver. Electrophoresis can separate mixtures of charged macromolecules, such as proteins and DNA.

Reaction Rates Are Variable

We measure the rate of an enzymatic reaction by monitoring either how fast the products are synthesized or how fast the substrates are consumed. Reaction rate can be altered by a number of factors, including changes in temperature, the amount of enzyme present, and substrate concentrations [p. 55]. In mammals we consider temperature to be essentially constant. This leaves enzyme amount and substrate concentration as the two main variables that affect reaction rate.

In protein-binding interactions, if the amount of protein (in this case, enzyme) is constant, the reaction rate is proportional to the substrate concentration. One strategy cells use to control reaction rates is to regulate the amount of enzyme in the cell. In the absence of appropriate enzyme, many biological reactions go very slowly or not at all. If enzyme is present, the rate of the reaction is proportional to the amount of enzyme and the amount of substrate, unless there is so much substrate that all enzyme binding sites are saturated and working at maximum capacity [p. 56].

This seems simple until you consider a reversible reaction that can go in both directions. In that case, what determines in which direction the reaction goes? The answer is that reversible reactions go to a state of *equilibrium,* where the rate of the reaction in the forward direction ($A + B \rightarrow C + D$) is equal to the rate of the reverse reaction ($C + D \rightarrow A + B$). At equilibrium, there is no net change in the amount of substrate or product, and the ratio $[C][D]/[A][B]$ is equal to the reaction's equilibrium constant, K_{eq} [p. 50].

If substrates or products are added or removed by other reactions in a pathway, the reaction rate increases in the forward or reverse direction as needed to restore the ratio [C][D]/[A][B]. According to the law of mass action, the ratio of [C] and [D] to [A] and [B] is always the same at equilibrium.

Enzymes May Be Activated, Inactivated, or Modulated

Enzyme activity, like the activity of other soluble proteins, can be altered by various factors. Some enzymes are synthesized as inactive molecules (*proenzymes* or *zymogens*) and activated on demand by proteolytic activation [p. 52]. Others require the binding of inorganic cofactors, such as Ca^{2+} or Mg^{2+}, before they become active.

Organic cofactors for enzymes are called **coenzymes.** Coenzymes do not alter the enzyme's binding site as inorganic cofactors do. Instead, coenzymes act as receptors and carriers for atoms or functional groups that are removed from the substrates during the reaction. Although coenzymes are needed for some metabolic reactions to take place, they are not required in large amounts.

Many of the substances that we call **vitamins** are the precursors of coenzymes. The water-soluble vitamins, such as the B vitamins, vitamin C, folic acid, biotin, and pantothenic acid, become coenzymes required for various metabolic reactions. For example, vitamin C is needed for adequate collagen synthesis.

Enzymes may be inactivated by inhibitors or by becoming denatured. Enzyme activity can be modulated by chemical factors or by changes in temperature and pH. ■ Figure 4.6 shows how enzyme activity can vary over a range of pH values. By turning reactions on and off or by increasing and decreasing the rate at which reactions take place, a cell can regulate the flow of biomolecules through different synthetic and energy-producing pathways.

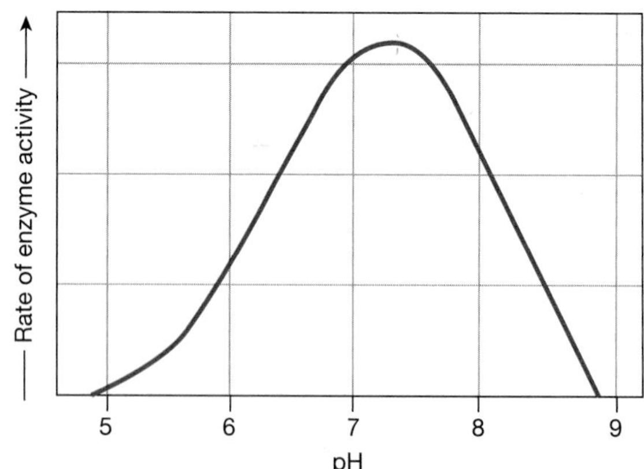

Most enzymes in humans have optimal activity near the body's internal pH of 7.4.

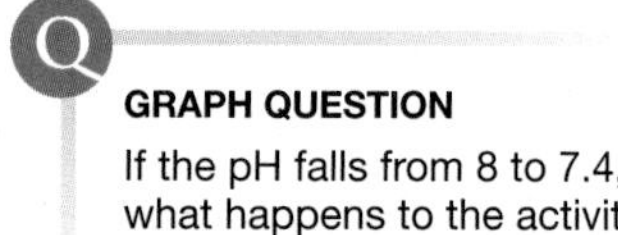

GRAPH QUESTION

If the pH falls from 8 to 7.4, what happens to the activity of the enzyme?

■ **Fig. 4.6** ***Effect of pH on enzyme activity***

Concept Check Answers: p. 128

8. What is a biological advantage of having multiple isozymes for a given reaction rather than only one form of the enzyme?
9. The four protein chains of an LDH isozyme are an example of what level of protein structure? (a) primary (b) secondary (c) tertiary (d) quaternary

Enzymes Lower Activation Energy of Reactions

How does an enzyme increase the rate of a reaction? In thermodynamic terms, it lowers the activation energy, making it more likely that the reaction will start (■ Fig. 4.7). Enzymes accomplish this by binding to their substrates and bringing them into the best position for reacting with each other. Without enzymes, the reaction would depend on random collisions between substrate molecules to bring them into alignment.

The rate of a reaction catalyzed by an enzyme is much more rapid than the rate of the same reaction taking place without the enzyme. For example, consider *carbonic anhydrase,* which facilitates conversion of CO_2 and water to carbonic acid. This enzyme plays a critical role in the transport of waste CO_2 from cells to lungs. Each molecule of carbonic anhydrase takes one second to catalyze the conversion of 1 million molecules of CO_2 and water to carbonic acid. In the absence of enzyme, it takes more than a minute for one molecule of CO_2 and water to be converted to carbonic acid. Without carbonic anhydrase and other enzymes in the body, biological reactions would go so slowly that cells would be unable to live.

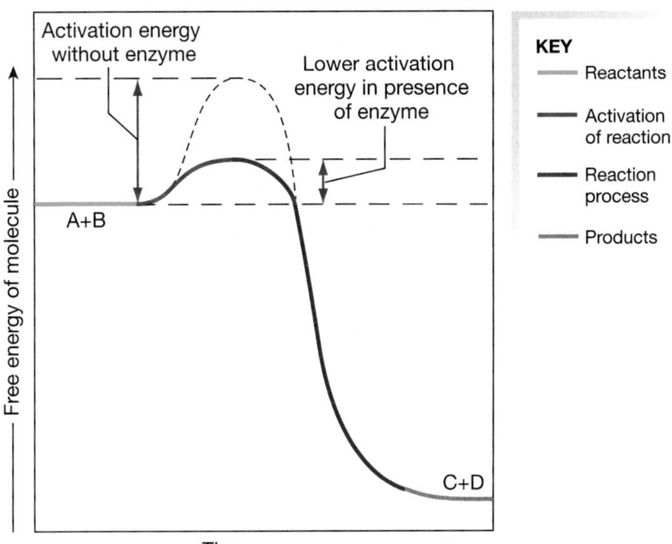

■ **Fig. 4.7** ***Enzymes lower the activation energy of reactions.*** In the absence of enzyme, the reaction (curved dashed line) would have much greater activation energy.

Enzymatic Reactions Can Be Categorized

Most reactions catalyzed by enzymes can be classified into four categories: oxidation-reduction, hydrolysis-dehydration, exchange-addition-subtraction, and ligation reactions. ■ Table 4.4 summarizes these categories and gives common enzymes for different types of reactions.

An enzyme's name can provide important clues to the type of reaction the enzyme catalyzes. Most enzymes are instantly recognizable by the suffix *-ase.* The first part of the enzyme's name (everything that precedes the suffix) usually refers to the type of reaction, to the substrate upon which the enzyme acts, or to both. For example, *glucokinase* has glucose as its substrate, and as a *kinase* it will add a phosphate group [p. 38] to the substrate. Addition of a phosphate group is called **phosphorylation.**

A few enzymes have two names. These enzymes were discovered before 1972, when the current standards for naming enzymes were first adopted. As a result, they have both a new name and a commonly used older name. Pepsin and trypsin, two digestive enzymes, are examples of older enzyme names.

Oxidation-Reduction Reactions **Oxidation-reduction reactions** are the most important reactions in energy extraction and transfer in cells. These reactions transfer electrons from one molecule to another. A molecule that gains electrons is said to be **reduced.** One way to think of this is to remember that adding negatively charged electrons *reduces* the electric charge on the molecule. Conversely, molecules that lose electrons are said to be **oxidized.** Use the mnemonic OIL RIG to remember what happens: Oxidation Is Loss (of electrons), Reduction Is Gain.

RUNNING PROBLEM

Tay-Sachs disease is a *recessive* genetic disorder caused by a defect in the gene that directs synthesis of hexosaminidase A. *Recessive* means that for a baby to be born with Tay-Sachs disease, it must inherit two defective genes, one from each parent. People with one Tay-Sachs gene and one normal gene are called *carriers* of the disease. Carriers do not develop the disease but can pass the defective gene on to their children. People who have two normal genes have normal amounts of hexosaminidase A in their blood. Carriers have lower-than-normal levels of the enzyme, but this amount is enough to prevent excessive accumulation of gangliosides in cells.

Q2: How could you test whether Sarah and David are carriers of the Tay-Sachs gene?

99 104 107 110 118 124

Classification of Enzymatic Reactions — Table 4.4

Reaction Type	What Happens	Representative Enzymes
1. Oxidation-reduction	Add or subtract electrons	**Class:*** oxidoreductase
(a) Oxidation	Transfer electrons from donor to oxygen	Oxidase
	Remove electrons and H^+	Dehydrogenase
(b) Reduction	Gain electrons	Reductase
2. Hydrolysis-dehydration	Add a water molecule	**Class:*** hydrolase
	Subtract a water molecule	
(a) Hydrolysis	Split large molecules by adding water	Peptidases, saccharidases, lipases
(b) Dehydration	Remove water to make one large molecule from several smaller ones	Dehydratases
3. Transfer chemical groups	Exchange groups between molecules	**Class:*** transferases
	Add or subtract groups	**Class:*** lyases
(a) Exchange reaction	Phosphate	Kinase
	Amino group	Transaminase
(b) Addition	Phosphate	Phosphorylase
	Amino group	Aminase
(c) Subtraction	Phosphate	Phosphatase
	Amino group	Deaminase
4. Ligation	Join two substrates using energy from ATP	**Class:*** ligases
		Synthetase

* Enzyme classes as defined by the Nomenclature Committee of the International Union of Biochemistry and Molecular Biology, *www.chem.qmul.ac.uk/iubmb/enzyme*

Hydrolysis-Dehydration Reactions Hydrolysis and dehydration reactions are important in the breakdown and synthesis of large biomolecules. In **dehydration reactions** {*de-*, out + *hydr-*, water}, a water molecule is one of the products. In many dehydration reactions, two molecules combine into one, losing water in the process. For example, the monosaccharides glucose and fructose join to make one sucrose molecule [p. 34]. In the process, one substrate molecule loses a hydroxyl group (–OH), and the other substrate molecule loses a hydrogen to create water, H_2O. When a dehydration reaction results in the synthesis of a new molecule, the process is known as *dehydration synthesis.*

In a **hydrolysis reaction** {*hydro,* water + *lysis,* to loosen or dissolve}, a substrate changes into one or more products through the addition of water. In these reactions, the covalent bonds of the water molecule are broken ("lysed") so that the water reacts as a hydroxyl group (–OH) and a hydrogen (–H). For example, an amino acid can be removed from the end of a peptide through a hydrolysis reaction.

When an enzyme name consists of the substrate name plus the suffix *-ase*, the enzyme causes a hydrolysis reaction. One example is *lipase,* an enzyme that breaks up large lipids into smaller lipids by hydrolysis. A *peptidase* is an enzyme that removes an amino acid from a peptide.

Addition-Subtraction-Exchange Reactions An **addition reaction** adds a functional group to one or more of the substrates. A **subtraction reaction** removes a functional group from one or more of the substrates. Functional groups are exchanged between or among substrates during **exchange reactions.**

For example, phosphate groups may be transferred from one molecule to another during addition, subtraction, or exchange reactions. The transfer of phosphate groups is an important means of covalent modulation [p. 54], turning reactions on or off or increasing or decreasing their rates. Several types of enzymes catalyze reactions that transfer phosphate groups. **Kinases** transfer a phosphate group from a substrate to an ADP molecule to create ATP, or from an ATP molecule to a substrate. For example, creatine kinase transfers a phosphate group from creatine phosphate to ADP, forming ATP and leaving behind creatine.

The addition, subtraction, and exchange of amino groups [p. 35] are also important in the body's use of amino acids. Removal of an amino group from an amino acid or peptide is a **deamination** reaction. Addition of an amino group is **amination,** and the transfer of an amino group from one molecule to another is **transamination.**

Ligation Reactions Ligation reactions join two molecules together using enzymes known as *synthetases* and energy from ATP. An example of a ligation reaction is the synthesis of *acetyl coenzyme A* (acetyl CoA) from fatty acids and coenzyme A. Acetyl CoA is an important molecule in the body, as you will learn in the next section.

> **Concept Check** Answers: p. 128
>
> 10. Name the substrates for the enzymes lactase, peptidase, lipase, and sucrase.
> 11. Match the reaction type or enzyme in the left column to the group or particle involved.
>
> | | |
> |---|---|
> | (a) kinase | 1. amino group |
> | (b) oxidation | 2. electrons |
> | (c) hydrolysis | 3. phosphate group |
> | (d) transaminase | 4. water |

Metabolism

Metabolism refers to all chemical reactions that take place in an organism. These reactions (1) extract energy from nutrient biomolecules (such as proteins, carbohydrates, and lipids) and (2) either synthesize or break down molecules. Metabolism is often divided into **catabolism,** reactions that release energy through the breakdown of large biomolecules, and **anabolism,** energy-utilizing reactions that result in the synthesis of large biomolecules. Anabolic and catabolic reactions take place simultaneously in cells throughout the body, so that at any given moment, some biomolecules are being synthesized while others are being broken down.

The energy released from or stored in the chemical bonds of biomolecules during metabolism is commonly measured in kilocalories (kcal). A **kilocalorie** is the amount of energy needed to raise the temperature of 1 liter of water by 1 degree Celsius. One kilocalorie is the same as a Calorie, with a capital C, used for quantifying the energy content of food. One kilocalorie is also equal to 1000 calories (small c).

Much of the energy released during catabolism is trapped in the high-energy phosphate bonds of ATP or in the high-energy electrons of NADH, $FADH_2$, or NADPH. Anabolic reactions then transfer energy from these temporary carriers to the covalent bonds of biomolecules.

Metabolism is a network of highly coordinated chemical reactions in which the activities taking place in a cell at any given moment are matched to the needs of the cell. Each step in a metabolic pathway is a different enzymatic reaction, and the reactions of a pathway proceed in sequence. Substrate A is changed into product B, which then becomes the substrate for the next reaction in the pathway. B is changed into C, and so forth:

$$A \rightarrow B \rightarrow C \rightarrow D$$

We call the molecules of the pathway **intermediates** because the products of one reaction become the substrates for the next. You will sometimes hear metabolic pathways called *intermediary metabolism*. Certain intermediates, called *key intermediates,* participate in more than one pathway and act as the branch points for channeling substrate in one direction or another. Glucose, for instance, is a key intermediate in several metabolic pathways.

In many ways, a group of metabolic pathways is similar to a detailed road map (■ Fig. 4.8). Just as a map shows a network of roads that connect various cities and towns, you can think

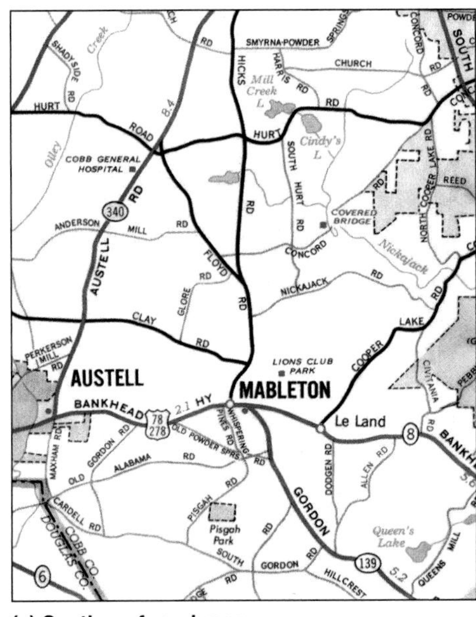

(a) Section of road map

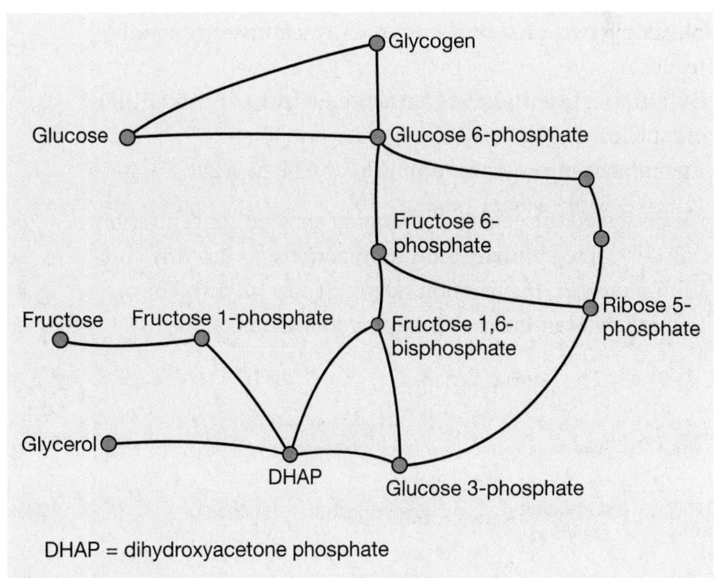

(b) Metabolic pathways drawn like a road map

■ **Fig. 4.8** ***A group of metabolic pathways resembles a road map.*** Cities on the map are equivalent to intermediates in metabolism. In metabolism, there may be more than one way to go from one intermediate to another, just as on the map there may be many ways to get from one city to another.

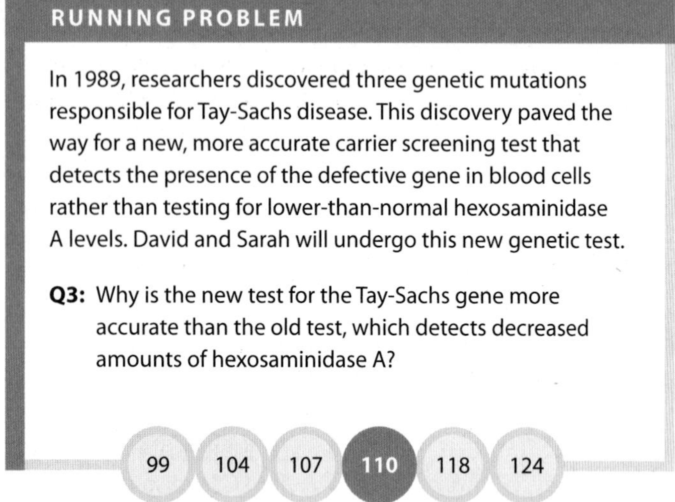

RUNNING PROBLEM

In 1989, researchers discovered three genetic mutations responsible for Tay-Sachs disease. This discovery paved the way for a new, more accurate carrier screening test that detects the presence of the defective gene in blood cells rather than testing for lower-than-normal hexosaminidase A levels. David and Sarah will undergo this new genetic test.

Q3: Why is the new test for the Tay-Sachs gene more accurate than the old test, which detects decreased amounts of hexosaminidase A?

99 104 107 110 118 124

of metabolism as a network of chemical reactions connecting various intermediate products. Each city or town is a different chemical intermediate. One-way roads are irreversible reactions, and big cities with roads to several destinations are key intermediates. Just as there may be more than one way to get from one place to another, there can be several pathways between any given pair of chemical intermediates.

Cells Regulate Their Metabolic Pathways

How do cells regulate the flow of molecules through their metabolic pathways? They do so in five basic ways:

1. By controlling enzyme concentrations
2. By producing modulators that change reaction rates
3. By using two different enzymes to catalyze reversible reactions
4. By compartmentalizing enzymes within intracellular organelles
5. By maintaining an optimum ratio of ATP to ADP

We discussed the effects of changing enzyme concentration in the discussion of protein-binding reactions: as enzyme concentration increases, the reaction rate increases [p. 55]. The sections that follow examine the remaining four items on the list.

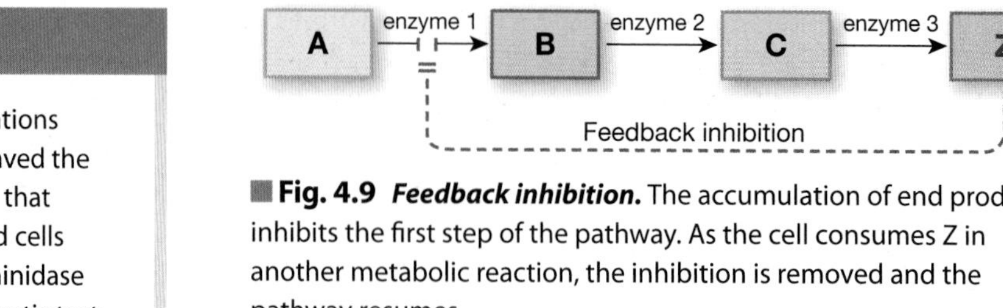

Fig. 4.9 ***Feedback inhibition.*** The accumulation of end product Z inhibits the first step of the pathway. As the cell consumes Z in another metabolic reaction, the inhibition is removed and the pathway resumes.

Enzyme Modulation Modulators, which alter the activity of a protein, were introduced in the discussion of protein binding [p. 52]. For enzymes, the production of modulators is frequently controlled by hormones and other signals coming from outside the cell. This type of outside regulation is a key element in the integrated control of the body's metabolism following a meal or during periods of fasting between meals.

In addition, some metabolic pathways have their own built-in form of modulation, called **feedback inhibition**. In this form of modulation, the end product of a pathway, shown as Z in Figure 4.9, acts as an inhibitory modulator of the pathway. As the pathway proceeds and Z accumulates, the enzyme catalyzing the conversion of A to B is inhibited. Inhibition of the enzyme slows down production of Z until the cell can use it up. Once the levels of Z fall, feedback inhibition on enzyme 1 is removed and the pathway starts to run again. Because Z is the end product of the pathway, this type of feedback inhibition is sometimes called *end-product inhibition.*

Reversible Reactions Cells can use reversible reactions to regulate the rate and direction of metabolism. If a single enzyme can catalyze the reaction in either direction, the reaction will go to a state of equilibrium, as determined by the law of mass action (Fig. 4.10a). Such a reaction therefore cannot be closely regulated except by modulators and by controlling the amount of enzyme.

However, if a reversible reaction requires two different enzymes, one for the forward reaction and one for the reverse reaction, the cell can regulate the reaction more closely (Fig. 4.10b). If no enzyme for the reverse reaction is present in the cell, the reaction is irreversible (Fig. 4.10c).

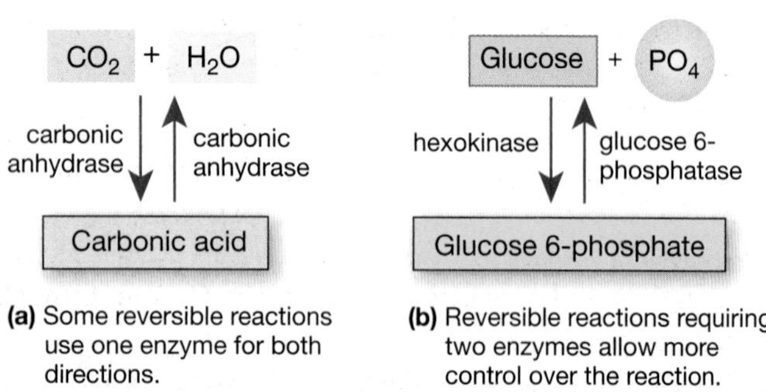

(a) Some reversible reactions use one enzyme for both directions.

(b) Reversible reactions requiring two enzymes allow more control over the reaction.

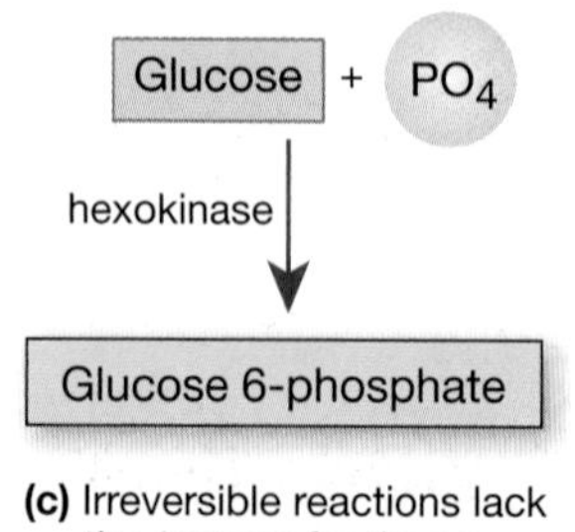

(c) Irreversible reactions lack the enzyme for the reverse direction.

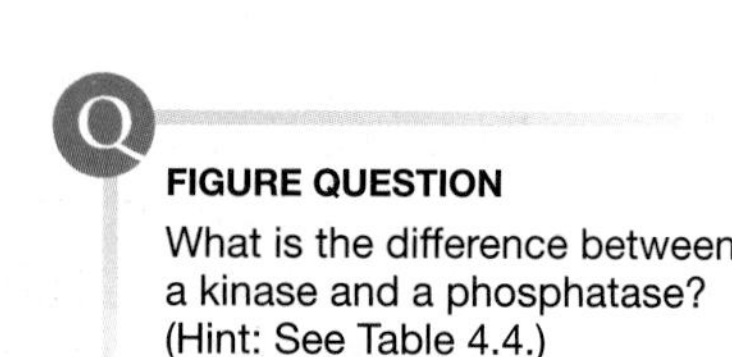

FIGURE QUESTION

What is the difference between a kinase and a phosphatase? (Hint: See Table 4.4.)

Fig. 4.10 ***The reversibility of metabolic reactions is controlled by enzymes***

Compartmentalizing Enzymes in the Cell Many enzymes of metabolism are isolated in specific subcellular compartments. Some, like the enzymes of carbohydrate metabolism, are dissolved in the cytosol, whereas others are isolated within specific organelles. Mitochondria, endoplasmic reticulum, Golgi apparatus, and lysosomes all contain enzymes that are not found in the cytosol. This separation of enzymes means that the pathways controlled by the enzymes are also separated. That allows the cell to control metabolism by regulating the movement of substrate from one cellular compartment to another. The isolation of enzymes within organelles is an important example of structural and functional compartmentation [p. 9].

Ratio of ATP to ADP The energy status of the cell is one final mechanism that can influence metabolic pathways. Through complex regulation, the ratio of ATP to ADP in the cell determines whether pathways that result in ATP synthesis are turned on or off. When ATP levels are high, production of ATP decreases. When ATP levels are low, the cell sends substrates through pathways that result in more ATP synthesis. In the next section, we look further into the role of ATP in cellular metabolism.

ATP Transfers Energy Between Reactions

The usefulness of metabolic pathways as suppliers of energy is often measured in terms of the net amount of ATP the pathways can yield. ATP is a nucleotide containing three phosphate groups [p. 36]. One of the three phosphate groups is attached to ADP by a covalent bond in an energy-requiring reaction. Energy is stored in this **high-energy phosphate bond** and then released when the bond is broken during removal of the phosphate group. This relationship is shown by the following reaction:

$$\text{ADP} + \text{P}_i + \text{energy} \rightleftharpoons \text{ADP}{\sim}\text{P}(=\text{ATP})$$

The squiggle $\sim$ indicates a high-energy bond, and P_i is the abbreviation for an inorganic phosphate group. Estimates of the amount of free energy released when a high-energy phosphate bond is broken range from 7 to 12 kcal per mole of ATP.

ATP is more important as a carrier of energy than as an energy-storage molecule. For one thing, cells can contain only a limited amount of ATP. A resting adult human needs 40 kg (88 pounds) of ATP to supply the energy required to support one day's worth of metabolic activity, far more than our cells could store. Instead, the body acquires most of its daily energy requirement from the chemical bonds of complex biomolecules. Metabolic reactions transfer that chemical bond energy to the high-energy bonds of ATP, or in a few cases, to the high-energy bonds of the related nucleotide *guanosine triphosphate,* **GTP.**

The metabolic pathways that yield the most ATP molecules are those that require oxygen—the **aerobic,** or *oxidative,* pathways. **Anaerobic** {*an-*, without + *aer,* air} pathways, which are those that can proceed without oxygen, also produce ATP molecules but in much smaller quantities. The lower ATP yield of anaerobic pathways means that most animals (including humans) are unable to survive for extended periods on anaerobic metabolism alone. In the next section we consider how biomolecules are metabolized to transfer energy to ATP.

Concept Check Answers: p. 128

12. Name five ways in which cells regulate the movement of substrates through metabolic pathways.
13. In which part of an ATP molecule is energy trapped and stored? In which part of a NADH molecule is energy stored?
14. What is the difference between aerobic and anaerobic pathways?

Catabolic Pathways Produce ATP

■ Figure 4.11 summarizes the catabolic pathways that extract energy from biomolecules and transfer it to ATP. Aerobic production of ATP from glucose commonly follows two pathways: **glycolysis** {*glyco-*, sweet + *lysis,* dissolve} and the **citric acid cycle** (also known as the tricarboxylic acid cycle). The citric acid cycle was first described by Hans A. Krebs, so it is sometimes called the *Krebs cycle.* Because Dr. Krebs described other metabolic cycles, we will avoid confusion by using the term *citric acid cycle.*

Carbohydrates enter glycolysis in the form of glucose (top of Fig. 4.11). Lipids are broken down into glycerol and fatty acids [p. 33], which enter the pathway at different points: glycerol feeds into glycolysis, and fatty acids are metabolized to acetyl CoA. Proteins are broken down into amino acids, which also enter at various points. Carbons from glycolysis and other nutrients enter the citric acid cycle, which makes a never-ending circle. At each turn, the cycle adds carbons and produces ATP, high-energy electrons, and carbon dioxide.

Both glycolysis and the citric acid cycle produce small amounts of ATP directly, but their most important contribution to ATP synthesis is trapping energy in electrons carried by NADH and $FADH_2$ to the **electron transport system** (ETS) in the mitochondria. The electron transport system, in turn, transfers energy from those electrons to the high-energy phosphate bond of ATP. At various points, the process produces carbon dioxide and water. Cells can use the water, but carbon dioxide is a waste product and must be removed from the body.

Because glucose is the only molecule that follows both pathways in their entirety, in this chapter we look at only glucose catabolism.

- ■ Figure 4.12 on page 113 summarizes the key steps of glycolysis, the conversion of glucose to pyruvate.
- ■ Figure 4.13 on page 114 shows how pyruvate is converted to acetyl CoA and how carbons from acetyl CoA go through the citric acid cycle.
- ■ Figure 4.14 on page 115 illustrates the energy-transferring pathway of the electron transport system.

ATP Production

The catabolic pathways that extract energy from biomolecules and transfer it to ATP are summarized in this overview figure of aerobic pathways. Aerobic production of ATP from glucose commonly follows two pathways: **glycolysis** and the **citric acid cycle**. Each of these pathways produces small amounts of ATP directly, but their most important contributions to ATP synthesis are high-energy electrons carried by NADH and $FADH_2$ to the electron transport system in the mitochondria.

The energy production from one glucose molecule can be summarized in the following two equations.

Aerobic Metabolism of Glucose

$$\text{Glucose} + O_2 + \text{ADP} + P_i \longrightarrow CO_2 + H_2O + \text{ATP}$$

$$C_6H_{12}O_6 + 6\,O_2 \xrightarrow{30\text{-}32\ \text{ADP} + P_i \;\to\; 30\text{-}32\ \text{ATP}} 6\,CO_2 + 6\,H_2O$$

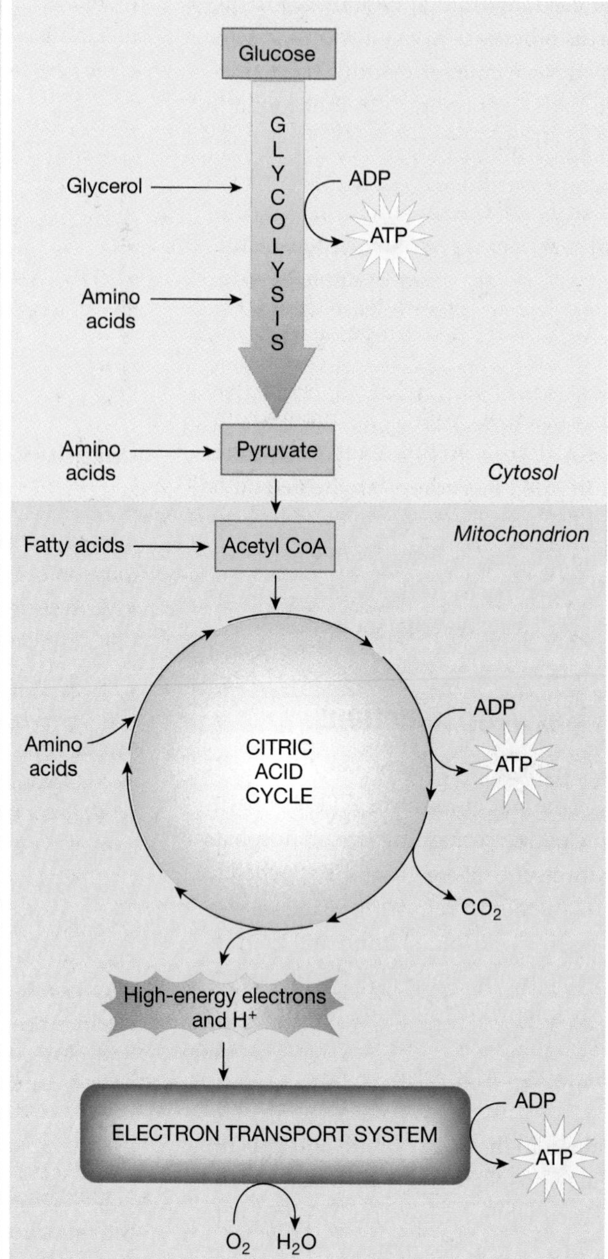

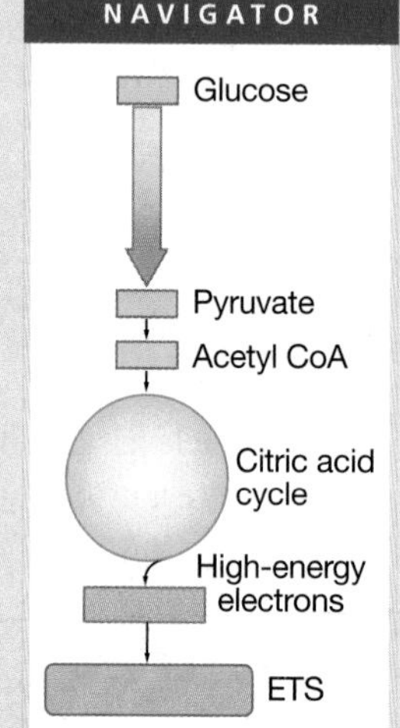

This icon represents the different steps in the metabolic summary figure. Look for it in the figures that follow to help you navigate your way through metabolism.

We will examine protein and lipid catabolism and synthetic pathways for lipids and glucose when we look at the fate of the nutrients we eat [Chapter 22].

The aerobic pathways for ATP production are a good example of compartmentation within cells. The enzymes of glycolysis are located in the cytosol, and the enzymes of the citric acid cycle are in the mitochondria. Within mitochondria, concentration of H^+ in the intermembrane compartment stores the energy needed to make the high-energy bond of ATP.

Concept Check Answers: p. 128

15. Match each component on the left to the molecule(s) it is part of:

(a) amino acids
(b) fatty acids
(c) glycerol
(d) glucose

1. carbohydrates
2. lipids
3. polysaccharides
4. proteins
5. triglycerides

16. Do endergonic reactions release energy or trap it in the products?

Glycolysis

During glycolysis, one molecule of glucose is converted by a series of enzymatically catalyzed reactions into two pyruvate molecules, producing a net release of energy.

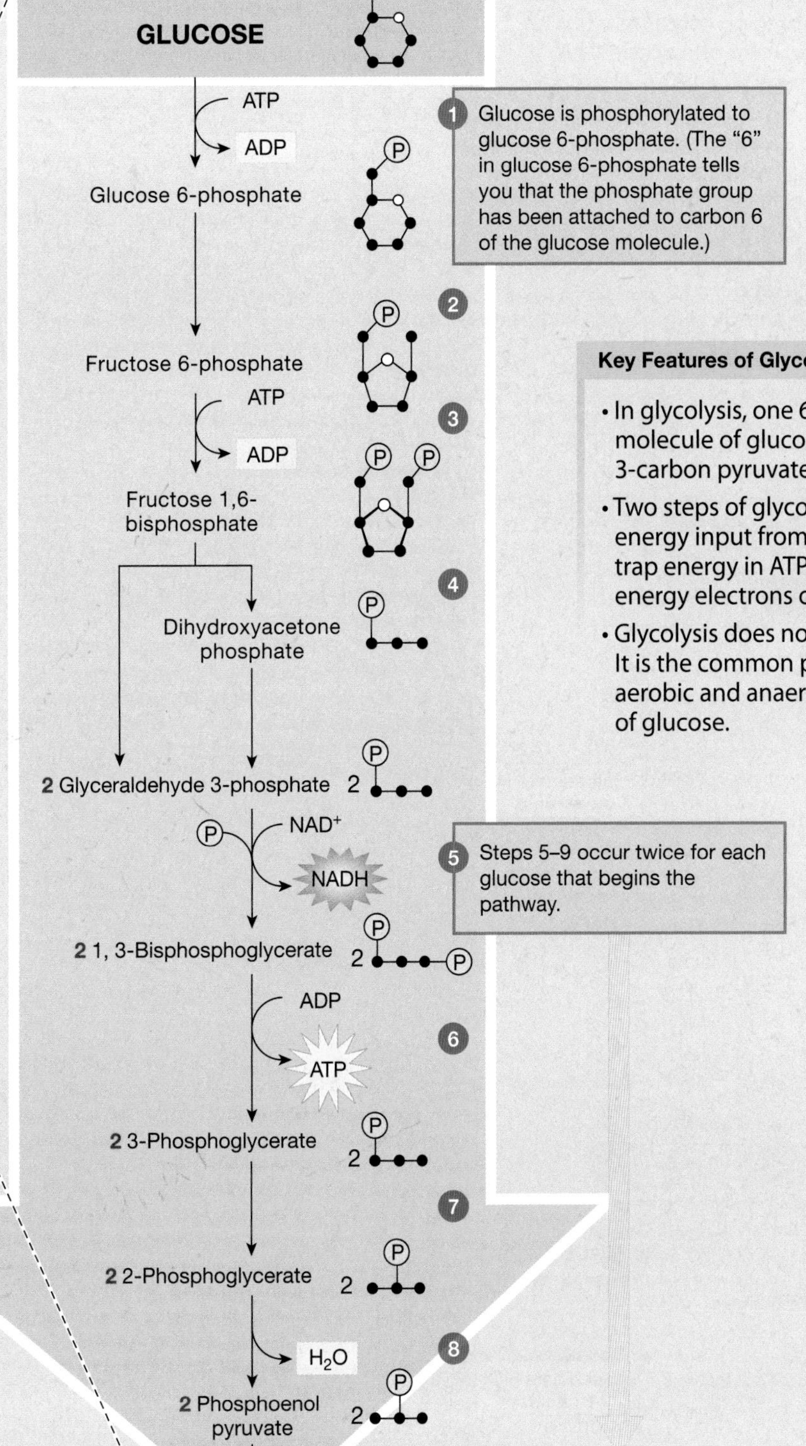

NAVIGATOR

Glucose

Pyruvate

Key Features of Glycolysis

- In glycolysis, one 6-carbon molecule of glucose becomes two 3-carbon pyruvate molecules.
- Two steps of glycolysis require energy input from ATP. Other steps trap energy in ATP and the high-energy electrons of NADH.
- Glycolysis does not require oxygen. It is the common pathway for aerobic and anaerobic catabolism of glucose.

KEY

● = Carbon
○ = Oxygen
Ⓟ = Phosphate group
(side groups not shown)

FIGURE QUESTIONS

1. Overall, is glycolysis an endergonic or exergonic pathway?
2. Which steps of glycolysis
   (a) use ATP?
   (b) make ATP or NADH?
   (c) are catalyzed by kinases?
   (d) are catalyzed by dehydrogenases? (Hint: See Table 4.4.)
3. What is the net energy yield (ATP and NADH) for one glucose?

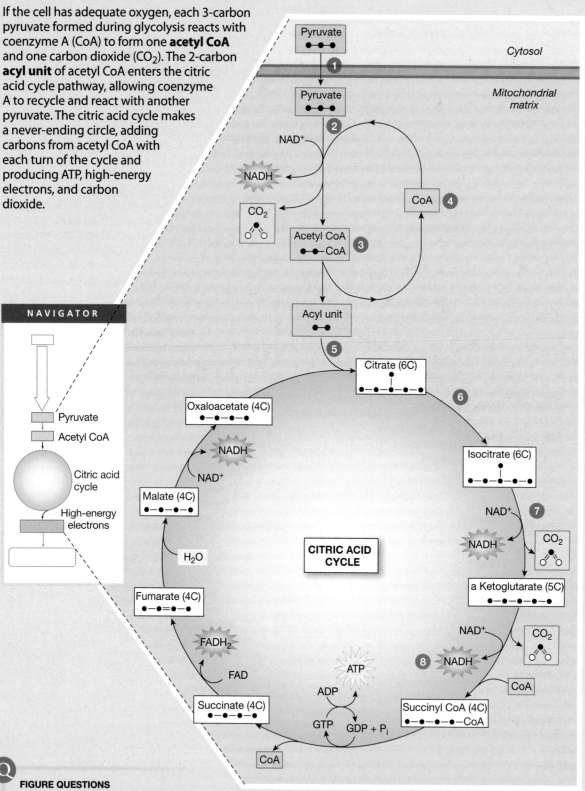

Pyruvate, Acetyl CoA, and the Citric Acid Cycle

If the cell has adequate oxygen, each 3-carbon pyruvate formed during glycolysis reacts with coenzyme A (CoA) to form one **acetyl CoA** and one carbon dioxide (CO_2). The 2-carbon **acyl unit** of acetyl CoA enters the citric acid cycle pathway, allowing coenzyme A to recycle and react with another pyruvate. The citric acid cycle makes a never-ending circle, adding carbons from acetyl CoA with each turn of the cycle and producing ATP, high-energy electrons, and carbon dioxide.

Cytosol

Mitochondrial matrix

Pyruvate → (1) → Pyruvate → (2) NAD^+ → NADH; CO_2 → Acetyl CoA (3) → CoA (4) → Acyl unit → (5)

Citrate (6C) → (6) → Isocitrate (6C) → (7) NAD^+ → NADH; CO_2 → a Ketoglutarate (5C) → NAD^+ → NADH (8); CO_2; CoA → Succinyl CoA (4C) → GTP → GDP + P_i; ADP → ATP; CoA → Succinate (4C) → FAD → $FADH_2$ → Fumarate (4C) → H_2O → Malate (4C) → NAD^+ → NADH → Oxaloacetate (4C)

CITRIC ACID CYCLE

NAVIGATOR

Pyruvate

Acetyl CoA

Citric acid cycle

High-energy electrons

1. If the cell has adequate oxygen, pyruvate is transported into the mitochondria.

2. Pyruvate reacts with coenzyme A to produce acetyl CoA, one NADH, and one CO_2.

3. Acetyl CoA has two parts: a 2-carbon acyl unit, derived from pyruvate, and coenzyme A.

4. Coenzyme A is made from the vitamin *pantothenic acid*. Coenzymes, like enzymes, are not changed during reactions and can be reused.

5. The 2-carbon acyl unit enters the cycle by combining with a 4-carbon oxaloacetate molecule.

6. The 6-carbon citrate molecule goes through a series of reactions until it completes the cycle as another oxaloacetate molecule.

7. Two carbons are removed in the form of CO_2.

8. Most of the energy released is captured as high-energy electrons on three NADH and one $FADH_2$. Some energy goes into the high-energy phosphate bond of one ATP. The remaining energy is given off as heat.

FIGURE QUESTIONS

1. Overall, is the citric acid cycle an endergonic or exergonic pathway?
2. What is the net energy yield (ATP, $FADH_2$, and NADH) for one pyruvate completing the cycle?
3. How many CO_2 are formed from one pyruvate? Compare the number of carbon atoms in the pyruvate and CO_2s.

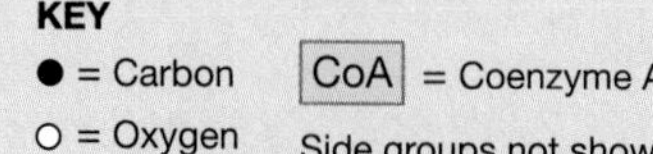

KEY

● = Carbon

○ = Oxygen

The Electron Transport System

The final step in aerobic ATP production is energy transfer from high-energy electrons of NADH and $FADH_2$ to ATP. This energy transfer requires mitochondrial proteins known as the **electron transport system** (**ETS**), located in the inner mitochondrial membrane. ETS proteins include enzymes and iron-containing **cytochromes**. The synthesis of ATP using the ETS is called **oxidative phosphorylation** because the system requires oxygen to act as the final acceptor of electrons and H^+. The **chemiosmotic theory** says that potential energy stored by concentrating H^+ in the intermembrane space is used to make the high-energy bond of ATP.

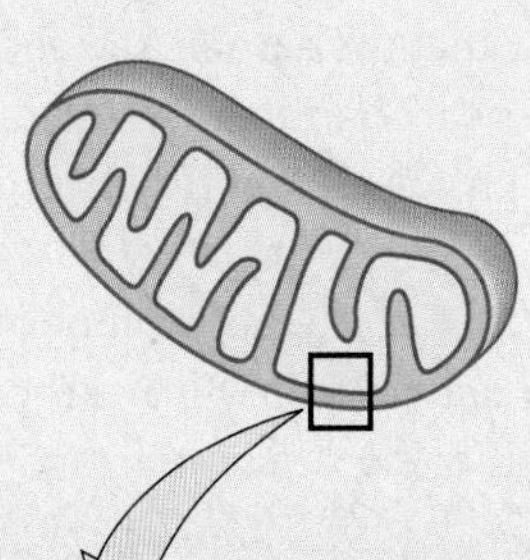

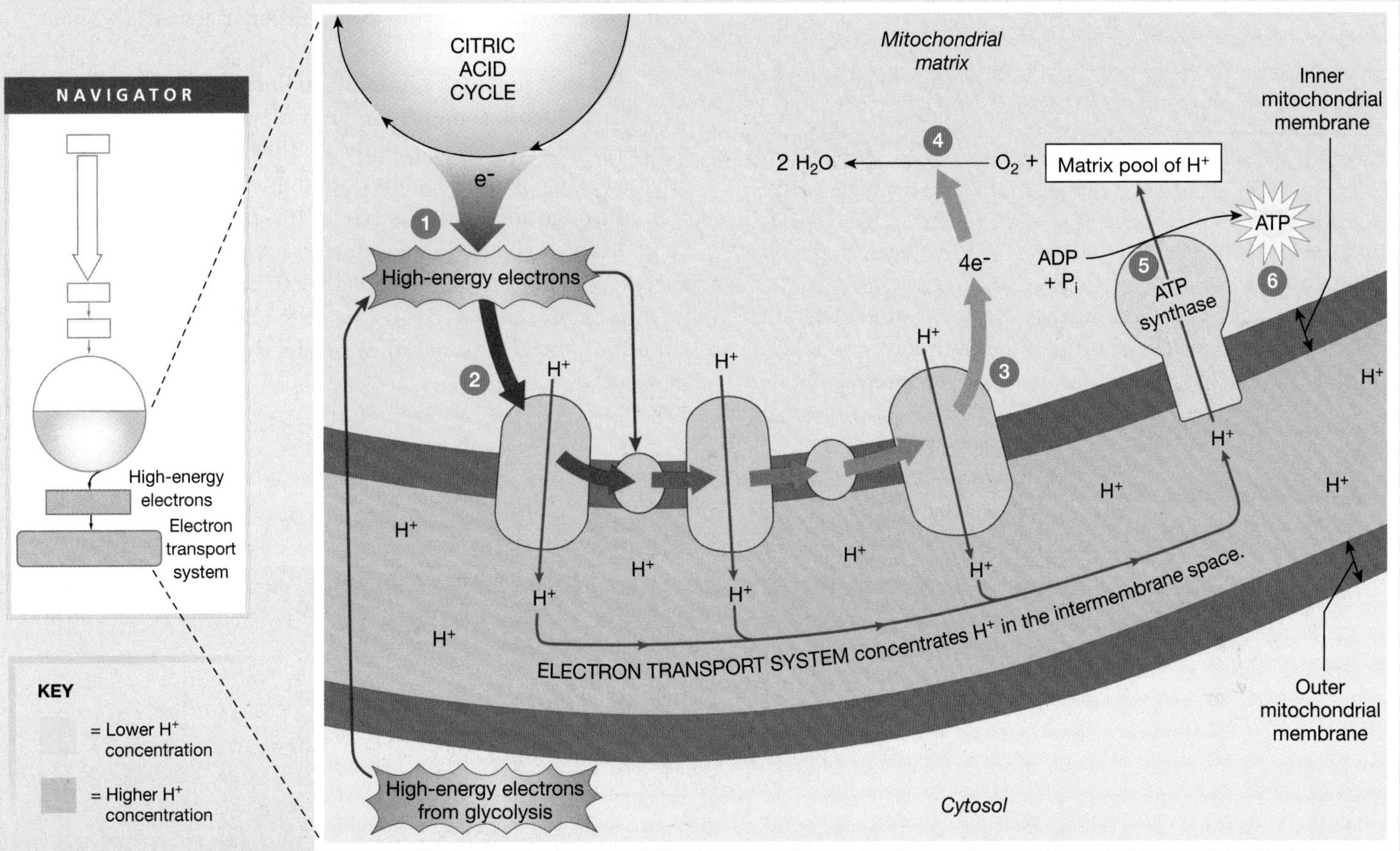

NAVIGATOR

High-energy electrons

Electron transport system

KEY

= Lower H^+ concentration

= Higher H^+ concentration

CITRIC ACID CYCLE

Mitochondrial matrix

Inner mitochondrial membrane

e^-

1 High-energy electrons

2 H^+

4 $2\ H_2O$ ← O_2 + Matrix pool of H^+

$4e^-$

3

ADP + P_i

5 ATP synthase

6 ATP

ELECTRON TRANSPORT SYSTEM concentrates H^+ in the intermembrane space.

Outer mitochondrial membrane

High-energy electrons from glycolysis

Cytosol

1. NADH and $FADH_2$ release high-energy electrons and H^+ to the ETS. NAD^+ and FAD are coenzymes that recycle.
2. Energy released when pairs of high-energy electrons pass along the transport system is used to concentrate H^+ from the mitochondrial matrix in the intermembrane space. The H^+ concentration gradient is a source of potential energy.
3. By the end of the ETS, the electrons have given up their stored energy.
4. Each pair of electrons released by the ETS combines with two H^+ and an oxygen atom, creating a molecule of water, H_2O.
5. H^+ flow back into the matrix through a protein known as **ATP synthase**. As the H^+ move down their concentration gradient, the synthase transfers their kinetic energy to the high-energy phosphate bond of ATP. Because energy conversions are never completely efficient, a portion of the energy is released as heat.
6. Each three H^+ that shuttle through the ATP synthase make a maximum of one ATP.

FIGURE QUESTIONS

1. What is phosphorylation? What is phosphorylated in oxidative phosphorylation?
2. Is the movement of electrons through the electron transport system endergonic or exergonic?
3. What is the role of oxygen in oxidative phosphorylation?

One Glucose Molecule Can Yield 30–32 ATP

Recall from Figure 4.11 that the aerobic metabolism of one glucose molecule produces carbon dioxide, water, and 30–32 ATP. Let's review the role of glycolysis and the citric acid cycle in that ATP production.

In glycolysis (Fig. 4.12), metabolism of one glucose molecule ($C_6H_{12}O_6$) has a net yield of two 3-carbon pyruvate molecules, 2 ATPs, and high-energy electrons carried on 2 NADH:

$$\text{Glucose} + 2\,\text{NAD}^+ + 2\,\text{ADP} + 2\,\text{P}_\text{i} \rightarrow$$
$$2\,\text{Pyruvate} + 2\,\text{ATP} + 2\,\text{NADH} + 2\,\text{H}^+ + 2\,\text{H}_2\text{O}$$

In the next phase, the conversion of pyruvate to acetyl CoA produces one NADH (Fig. 4.13). Carbons from one acetyl CoA going through the citric acid cycle trap energy in three NADH molecules, one $FADH_2$, and one ATP. These steps happen twice for each glucose, giving a total yield of 8 NADH, 2 $FADH_2$, and 2 ATP for the pyruvate-citric acid cycle phase of glucose metabolism.

In the final step, high-energy electrons of NADH and $FADH_2$ passing along the proteins of the electron transport system use their energy to concentrate H^+ in the intermembrane compartment of the mitochondria (Fig. 4.14). When the H^+ move down their concentration gradient through a channel in the ATP synthase, the energy released is transferred to the high-energy phosphate bond of ATP. On average, the NADH and $FADH_2$ from one glucose produce 26–28 ATPs.

When we tally the maximum potential energy yield for the catabolism of one glucose molecule through aerobic pathways, the total comes to 30–32 ATP (■ Fig. 4.15b). These numbers are the *potential* maximum because often the mitochondria do not work up to capacity. There are various reasons for this, including the fact that a certain number of H^+ ions leak from the intermembrane space back into the mitochondrial matrix without producing an ATP.

A second source of variability in the number of ATP produced per glucose comes from the two cytosolic NADH molecules produced during glycolysis. These NADH molecules are unable to enter mitochondria and must transfer their electrons through membrane carriers. Inside a mitochondrion, some of these electrons go to $FADH_2$, which has a potential average yield of only 1.5 ATP rather than the 2.5 ATP made by mitochondrial NADH. If cytosolic electrons go to mitochondrial NADH instead, they produce two additional ATP molecules.

(a) Anaerobic Metabolism

$C_6H_{12}O_6 \rightarrow 2\,C_3H_5O_3^- + 2\,H^+$

1 Glucose
GLYCOLYSIS
2 Pyruvate
2 Lactate

	NADH	$FADH_2$	ATP	CO_2
Glycolysis	2		4 −2	
2 Pyruvate → 2 Lactate	−2			
TOTALS	0 NADH		2 ATP	

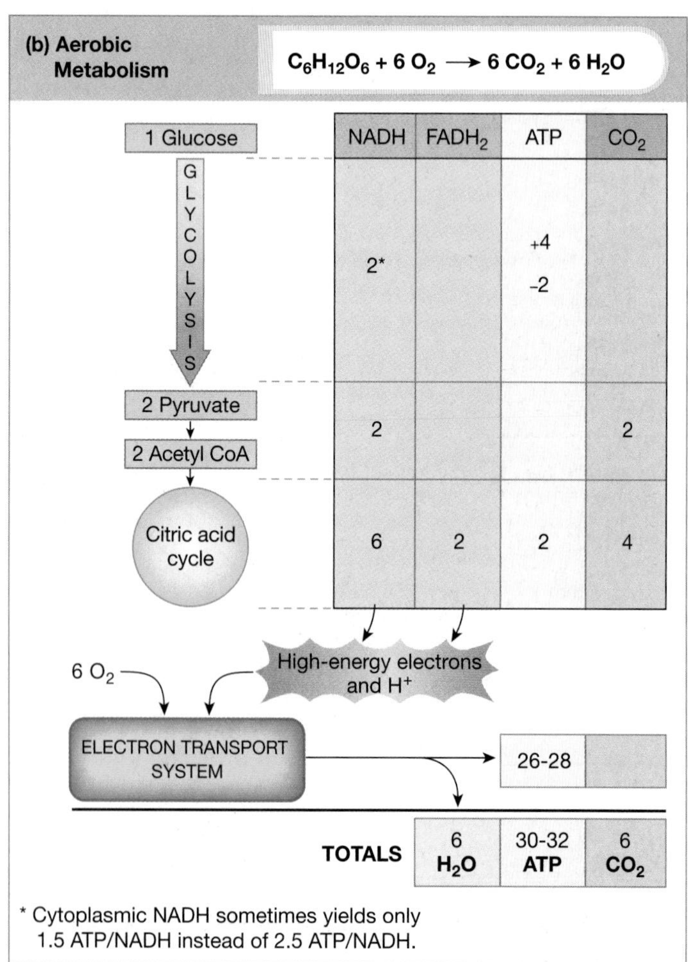

■ **Fig. 4.15** ***Summary of energy yields from catabolism of one glucose molecule.*** One glucose metabolized aerobically through the citric acid cycle yields 30–32 ATP. One glucose metabolized anaerobically yields only 2 ATP.

Anaerobic Metabolism Makes 2 ATP

The metabolism of glucose just described assumes that the cells have adequate oxygen to keep the electron transport system functioning. But what happens to a cell whose oxygen supply cannot keep pace with its ATP demand, such as often happens during strenuous exercise? In that case, the metabolism of glucose shifts from aerobic to anaerobic metabolism, starting at pyruvate (■ Fig. 4.16).

In anaerobic glucose metabolism, pyruvate is converted to lactate instead of being transported into the mitochondria:

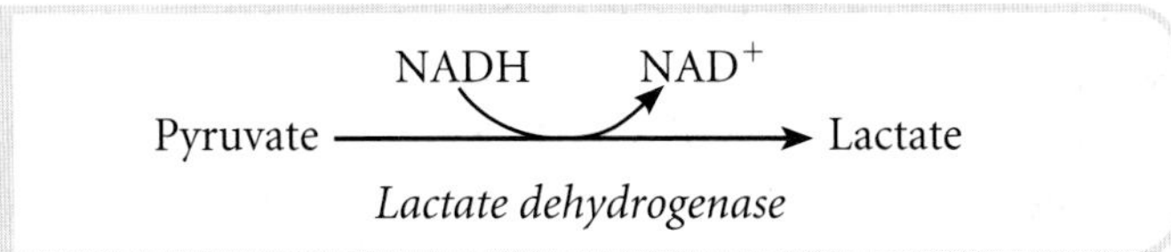

Pyruvate is a branch point for metabolic pathways, like a hub city on a road map. Depending on a cell's needs and oxygen content, pyruvate can be shuttled into the citric acid cycle or diverted into lactate production until oxygen supply improves.

The conversion of pyruvate to lactate changes one NADH back to NAD^+ when a hydrogen atom and an electron are transferred to the lactate molecule. As a result, the net energy yield for the anaerobic metabolism of one glucose molecule is 2 ATP and no NADH (Fig. 14.15a), a very puny yield when compared to the 30–32 ATP/glucose that result from aerobic metabolism (Fig. 4.15b). The low efficiency of anaerobic metabolism severely limits its usefulness in most vertebrate cells, whose metabolic energy demand is greater than anaerobic metabolism can provide. Some cells, such as exercising muscle cells, can tolerate anaerobic metabolism for a limited period of time. Eventually, however, they must shift back to aerobic metabolism. [Aerobic and anaerobic metabolism in muscle are discussed further in Chapters 12 and 25.]

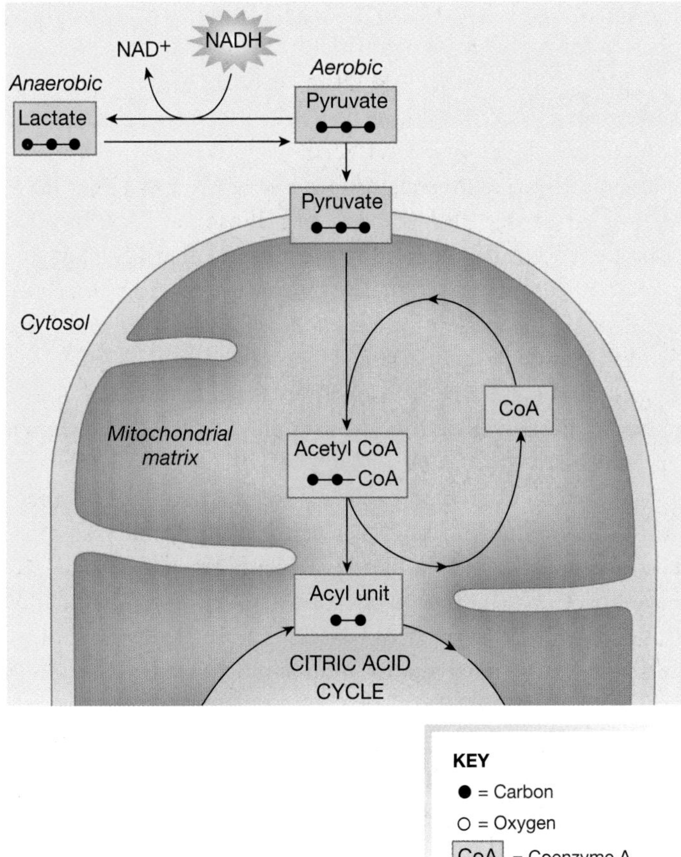

■ **Fig. 4.16** ***Pyruvate is the branch point between aerobic and anaerobic metabolism of glucose***

Concept Check Answers: p. 128

17. How is the separation of mitochondria into two compartments essential to ATP synthesis?
18. Lactate dehydrogenase acts on lactate by ________ (adding or removing?) a(n) ________ and a(n) ________. This process is called ________ (oxidation or reduction?).
19. Describe two differences between aerobic and anaerobic metabolism of glucose.

Proteins Are the Key to Cell Function

As you have seen, proteins are the molecules that run a cell from day to day. Protein enzymes control the synthesis and breakdown of carbohydrates, lipids, structural proteins, and signal molecules. Protein transporters and pores in the cell membrane and in organelle membranes regulate the movement of molecules into and out of compartments. Other proteins form the structural skeleton of cells and tissues. In these and other ways, protein synthesis is critical to cell function.

The power of proteins arises from their tremendous variability and specificity. Protein synthesis using 20 amino acids can be compared to creating a language with an alphabet of 20 letters. The "words" vary in length from three letters to hundreds of letters, spelling out the structure of thousands of different proteins with different functions. A change in one amino acid during protein synthesis can alter the protein's function, just as changing one letter turns the word "foot" into "food."

The classic example of an amino acid change causing a problem is sickle cell disease. In this inherited condition, when the amino acid valine replaces one glutamic acid in the protein chain, the change alters the shape of hemoglobin. As a result, red blood cells containing the abnormal hemoglobin take on a crescent (sickle) shape, which causes them to get tangled up and block small blood vessels.

The Protein "Alphabet" One of the mysteries of biology until the 1960s was the question of how only four nitrogenous bases in the DNA molecule—adenine (A), guanine (G), cytosine (C), and thymine (T)—could code for more than 20 different

RUNNING PROBLEM

David and Sarah had their blood drawn for the genetic test several weeks ago and have been anxiously awaiting the results. Today, they returned to the hospital to hear the news. The tests show that Sarah carries the gene for Tay-Sachs disease but David does not. This means that although some of their children may be carriers of the Tay-Sachs gene like Sarah, none of the children will develop the disease.

Q4: The Tay-Sachs gene is a recessive gene (t). If Sarah is a carrier of the gene (Tt) but David is not (TT), what is the chance that any child of theirs will be a carrier? (Consult a general biology or genetics text if you need help solving this problem.)

amino acids. If each base controlled the synthesis of one amino acid, a cell could make only four different amino acids. If pairs of bases represented different amino acids, the cell could make 4^2 or 16 different amino acids. Because we have 20 amino acids, this is still not satisfactory. If triplets of bases were the codes for different molecules, however, DNA could create 4^3 or 64 different amino acids. These triplets, called **codons**, are indeed the way information is encoded in DNA and RNA. ■ Figure 4.17 shows the genetic code as it appears in one form of RNA. Remember that RNA substitutes the base uracil (U) for the DNA base thymine [p. 37].

Second base of codon

First base of codon	U	C	A	G	Third base of codon
U	UUU Phe	UCU Ser	UAU Tyr	UGU Cys	U
	UUC	UCC	UAC	UGC	C
	UUA Leu	UCA	UAA Stop	UGA Stop	A
	UUG	UCG	UAG	UGG Trp	G
C	CUU Leu	CCU Pro	CAU His	CGU Arg	U
	CUC	CCC	CAC	CGC	C
	CUA	CCA	CAA Gln	CGA	A
	CUG	CCG	CAG	CGG	G
A	AUU Ile	ACU Thr	AAU Asn	AGU Ser	U
	AUC	ACC	AAC	AGC	C
	AUA	ACA	AAA Lys	AGA Arg	A
	AUG Met Start	ACG	AAG	AGG	G
G	GUU Val	GCU Ala	GAU Asp	GGU Gly	U
	GUC	GCC	GAC	GGC	C
	GUA	GCA	GAA Glu	GGA	A
	GUG	GCG	GAG	GGG	G

■ **Fig. 4.17** ***The genetic code as it appears in the codons of mRNA.*** The three-letter abbreviations to the right of the brackets indicate the amino acid each codon represents. The start and stop codons are also marked. [The 3-letter abbreviations are explained in Fig. 2.3, p. 35.]

Of the 64 possible triplet combinations, one DNA codon (TAC) acts as the initiator or "start codon" that signifies the beginning of a coding sequence. Three codons serve as terminator or "stop codons" that show where the sequence ends. The remaining 60 triplets all code for amino acids. Methionine and tryptophan have only one codon each, but the other amino acids have between two and six different codons each. Thus, like letters spelling words, the DNA base sequence determines the amino acid sequence of proteins.

Unlocking DNA's Code How does a cell know which of the thousands of bases present in its DNA sequence to use in making a protein? It turns out that the information a cell needs to make a particular protein is contained in a segment of DNA known as a gene. What exactly is a gene? The definition keeps changing, but for this text we will say that a **gene** is a region of DNA that contains the information needed to make a functional piece of RNA, which in turn can make a protein.

■ Figure 4.18 shows the five major steps from gene to RNA to functional protein. First, a section of DNA containing a gene must be activated so that its code can be read ①. Genes that are continuously being read and converted to RNA messages are said to be *constitutively active*. Usually these genes code for proteins that are essential to ongoing cell functions. Other genes are *regulated*; that is, their activity can be turned on (*induced*) or turned off (*repressed*) by regulatory proteins.

Once a gene is activated, the DNA base sequence of the gene is used to create a piece of RNA in the process known as **transcription** {*trans*, over + *scribe*, to write} (Fig. 4.18 ②). Human cells have three major forms of RNA: **messenger RNA** (mRNA), **transfer RNA** (tRNA), and **ribosomal RNA** (rRNA). Messenger RNA is processed in the nucleus after it is made ③. It may either undergo *alternative splicing* (discussed shortly) before leaving the nucleus or be "silenced" and destroyed by enzymes through *RNA interference*. Processed mRNA leaves the nucleus and enters the cytosol. There it works with tRNA and rRNA to direct **translation**, the assembly of amino acids into a protein chain ④.

Newly synthesized proteins are then subject to **post-translational modification** (Fig. 4.18 ⑤). They fold into complex shapes, may be split by enzymes into smaller peptides, or have various chemical groups added to them. The remainder of this chapter looks at transcription, RNA processing, translation, and post-translational modification in more detail.

DNA Guides the Synthesis of RNA

The first steps in protein synthesis are compartmentalized within the nucleus because DNA is a very large molecule that cannot pass through the nuclear envelope. Transcription uses DNA as a template to create a small single strand of RNA that can leave the nucleus (■ Fig. 4.19 on page 120). The synthesis of RNA from the double-stranded DNA template requires an

Overview of Protein Synthesis

The major steps required to convert the genetic code of DNA into a functional protein.

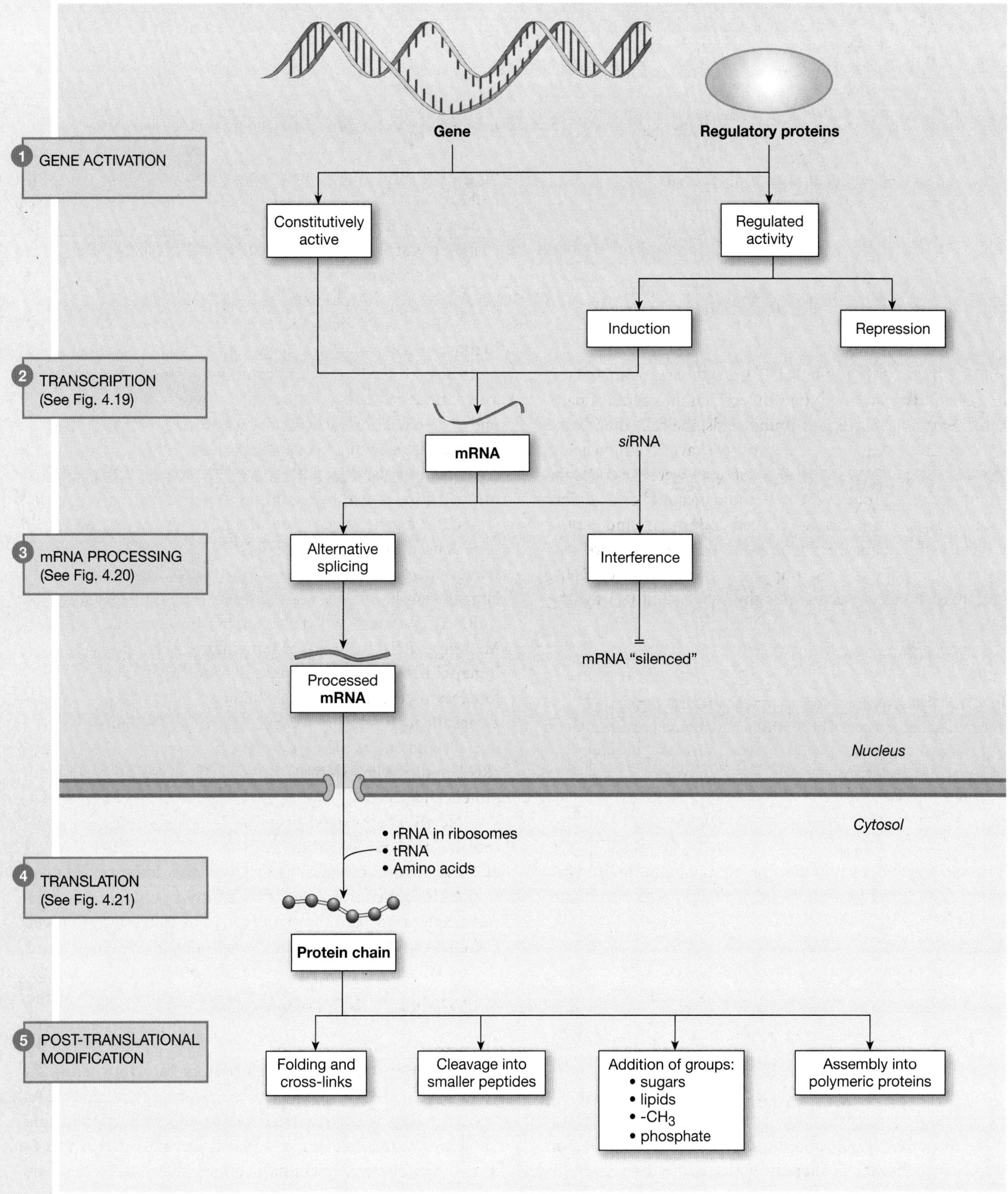

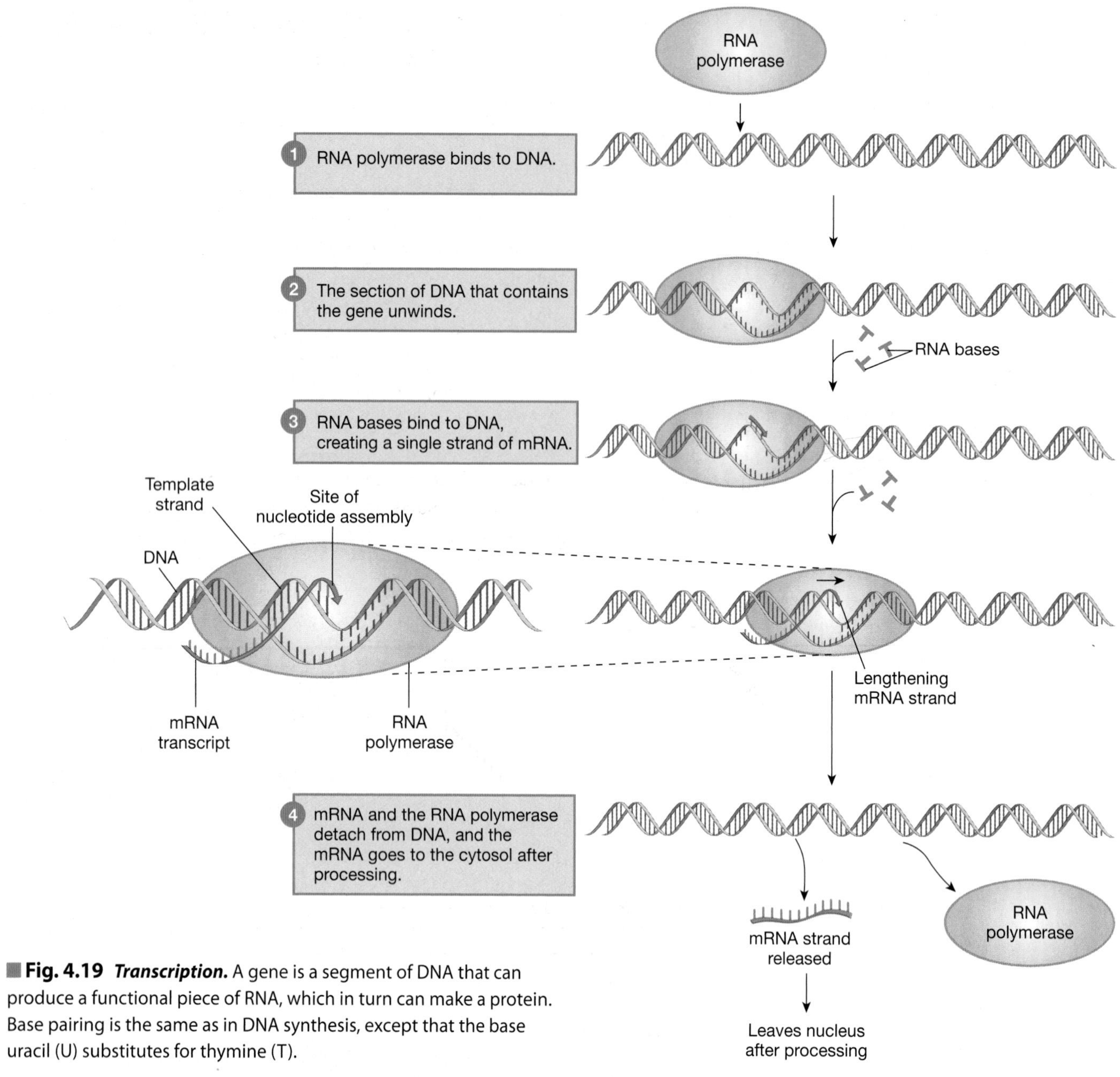

Fig. 4.19 ***Transcription.*** A gene is a segment of DNA that can produce a functional piece of RNA, which in turn can make a protein. Base pairing is the same as in DNA synthesis, except that the base uracil (U) substitutes for thymine (T).

enzyme known as **RNA polymerase**, plus magnesium or manganese ions and energy in the form of high-energy phosphate bonds:

$$\text{DNA template} + \text{nucleotides A, U, C, G} \xrightarrow[\text{and energy}]{\text{RNA polymerase, } Mg^{2+} \text{ or } Mn^{2+},} \text{DNA template} + \text{mRNA}$$

A **promoter** region that precedes the gene must be activated before transcription can begin. Regulatory-protein **transcription factors** bind to DNA and activate the promoter. The active promoter tells the RNA polymerase where to bind to the DNA (Fig. 4.19, ①). The polymerase moves along the DNA molecule and "unwinds" the double strand by breaking the hydrogen bonds between paired bases ②. One strand of DNA, called the *template strand*, serves as the guide for RNA synthesis, ③. The promoter region is not transcribed into RNA.

During transcription, each base in the DNA template strand pairs with the complementary RNA base (G-C, C-G, T-A, A-U). This pairing of complementary bases is similar to the process by which a double strand of DNA forms [see Appendix C for a review of DNA synthesis]. For example, a DNA segment containing the base sequence AGTAC is transcribed into the RNA sequence UCAUG.

As the RNA bases bind to the DNA template strand, they also bond with one another to create a single strand of RNA. During transcription, bases are linked at an average rate of 40 per second. In humans, the largest RNAs may contain as many as 5000 bases, and their transcription may take more than a minute—a long time for a cellular process. When RNA polymerase reaches the stop codon, it stops adding bases to the growing RNA strand and releases the strand (Fig. 4.19, ④).

Concept Check Answers: p. 128

20. Use the genetic code in Figure 4.17 to write the DNA codons that correspond to the three mRNA stop codons.

21. What does the name RNA polymerase tell you about the function of this enzyme?

Alternative Splicing Creates Multiple Proteins from One DNA Sequence

The next step in the process of protein synthesis is **mRNA processing**, which takes two forms (Fig. 4.18, ③). In *RNA interference*, newly synthesized mRNA is inactivated or destroyed before it can be translated into proteins (see the Emerging Concepts box). In **alternative splicing**, enzymes clip segments out of the middle or off the ends of the mRNA strand. Other enzymes then splice the remaining pieces of the strand back together.

Alternative splicing is necessary because a gene contains both segments that encode proteins (**exons**) and noncoding segments called **introns** (■ Fig. 4.20). That means the mRNA initially made from the gene's DNA contains noncoding segments that must be removed before the mRNA leaves the nucleus. The result of alternative splicing is a smaller piece of mRNA that now contains only the coding sequence for a specific protein.

One advantage of alternative splicing is that it allows a single base sequence on DNA to code for more than one protein.

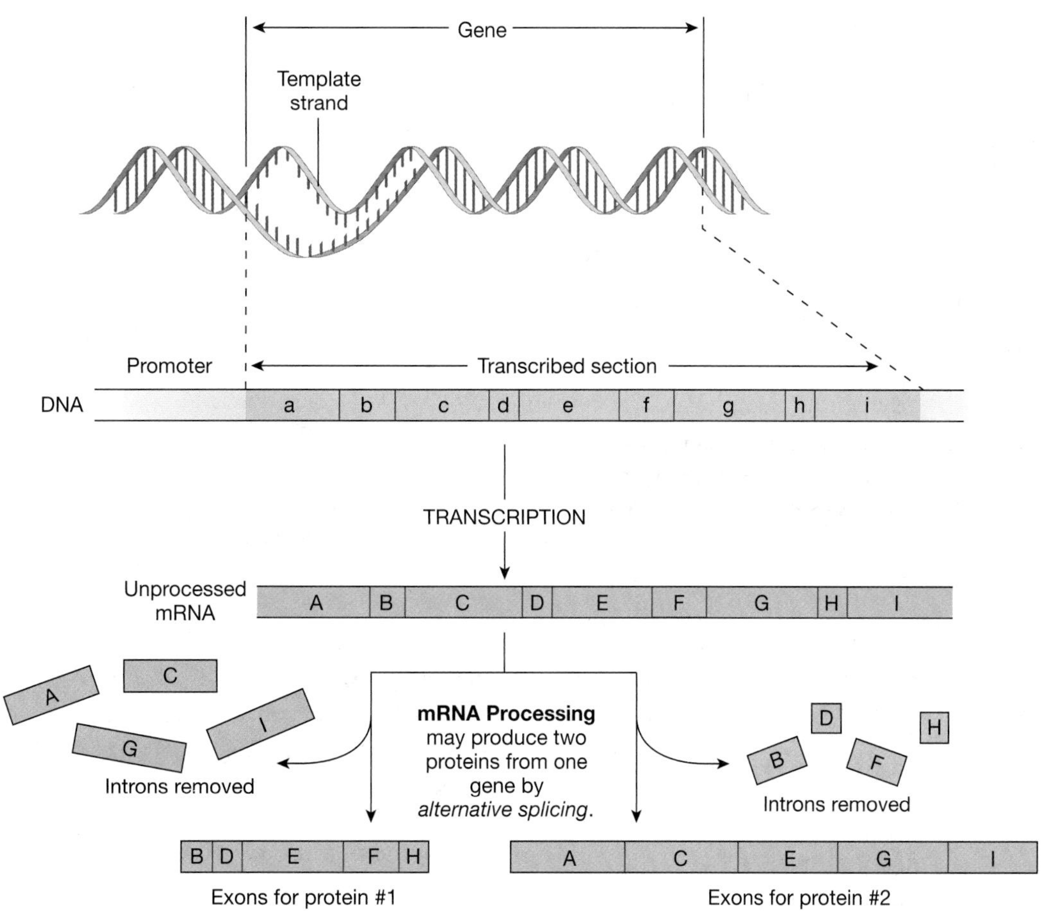

■ **Fig. 4.20** ***mRNA processing.*** The promoter segment of DNA is not transcribed into RNA. In mRNA processing, segments of the newly created mRNA strand called introns are removed. The remaining exons are spliced back together to form the mRNA that codes for a functional protein. Differential removal of introns from mRNA allows a single gene to code for multiple proteins. For protein #1, introns A, C, G, and I were removed. For protein #2, segments B, D, F, and H became the introns.

4

EMERGING CONCEPTS

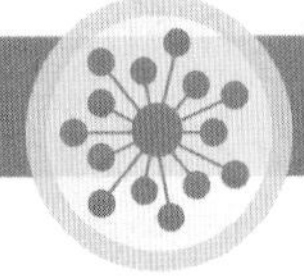

Purple Petunias and RNAi

Who could have guessed that research to develop a deep purple petunia would lead the way to one of the most exciting new areas of molecular biology research? **RNA interference** (*RNAi*) was first observed in 1990, when botanists who introduced purple pigment genes into petunias ended up with plants that were white or striped with white instead of the deeper purple color they expected. This observation did not attract attention until 1998, when scientists doing research in animal biology and medicine had similar problems in experiments on a nematode worm. Now RNAi is one of the hottest areas of biotechnology research.

Many details of RNAi are still not well understood. In very simple terms, RNA "silencing" of mRNA is a naturally occurring event accomplished through the production or introduction of short RNA molecules, such as small interfering RNA (*siRNA*) or microRNA (*miRNA*). These short pieces of RNA bind to mRNA and prevent it from being translated. They may even target the mRNA for destruction.

siRNA begins as double-stranded pieces of RNA that are chopped by an enzyme called *Dicer* into smaller (20–25 nucleotide) pieces. The double-stranded pieces combine with proteins and lose one RNA strand, becoming *RNA-induced silencing complex*, or RISC. RISC and the single-stranded miRNA bind to complementary pieces of mRNA made by the cell.

RNAi is a naturally occurring RNA processing mechanism that may have evolved as a means of blocking the replication of RNA viruses. Now, however, researchers are hoping to learn how to use it to selectively block the production of single proteins within a cell. The scientists' ultimate goal is to create technologies that can be used for the diagnosis and treatment of disease.

The designation of segments as coding or noncoding is not fixed for a given gene. Segments of mRNA that are removed one time can be left in the next time, producing a finished mRNA with a different sequence. The closely related forms of a single enzyme known as *isozymes* are probably made by alternative splicing of a single gene.

After mRNA has been processed, it exits the nucleus through nuclear pores and goes to ribosomes in the cytosol. There mRNA directs the construction of protein.

Concept Check Answer: p. 128

22. Explain in one or two sentences the relationship of mRNA, nitrogenous bases, introns, exons, mRNA processing, and proteins.

mRNA Translation Links Amino Acids

Protein synthesis requires cooperation and coordination among all three types of RNA: mRNA, rRNA, and tRNA. Upon arrival in the cytosol, processed mRNA binds to ribosomes, which are small particles of protein and several types of rRNA [p. 77]. Each ribosome has two subunits, one large and one small, that come together when protein synthesis begins (■ Fig. 4.21, ③). The small ribosomal subunit binds the mRNA, then adds the large subunit so that the mRNA is sandwiched in the middle. Now the ribosome-mRNA complex is ready to begin translation.

During translation, the mRNA codons are matched to the proper amino acid. This matching is done with the assistance of a tRNA molecule (Fig. 4.21, ④). One region of each tRNA contains a three-base sequence called an **anticodon** that is complementary to an mRNA codon. A different region of the tRNA molecule binds to a specific amino acid.

As translation begins, the anticodons of tRNAs carrying amino acids attach to the complementary codons of ribosomal mRNA. For example, a tRNA with anticodon sequence UUU carries the amino acid lysine. The UUU anticodon pairs with an AAA codon, one of two codons for lysine, on mRNA. The pairing between mRNA and tRNA puts newly arrived amino acids into the correct orientation to link to the growing peptide chain.

Dehydration synthesis links amino acids by creating a *peptide bond* between the amino group ($-NH_2$) of the newly arrived amino acid and the carboxyl end ($-COOH$) of the peptide chain [p. 35]. Once this happens, the "empty" tRNA releases from mRNA. This tRNA can then attach to another amino acid molecule with the aid of a cytosolic enzyme and ATP.

When the last amino acid has been joined to the newly synthesized peptide chain, the termination stage has been reached (Fig. 4.21, ⑤). The mRNA, the peptide, and the ribosomal subunits separate. The ribosomes are ready for a new round of protein synthesis, but the mRNA is broken down by enzymes known as *ribonucleases*. Some forms of mRNA are broken down quite rapidly, while others may linger in the cytosol and be translated many times.

Protein Sorting Directs Proteins to Their Destination

One of the amazing aspects of protein synthesis is the way specific proteins go from the ribosomes directly to where they are needed in the cell, a process called *protein sorting*. Many newly made proteins carry a *sorting signal,* an address label that tells the cell where the protein should go. Some proteins that are synthesized on cytosolic ribosomes do not have sorting signals. Without a "delivery tag," they remain in the cytosol when they are released from the ribosome [Fig. 3.7 ③, p. 77].

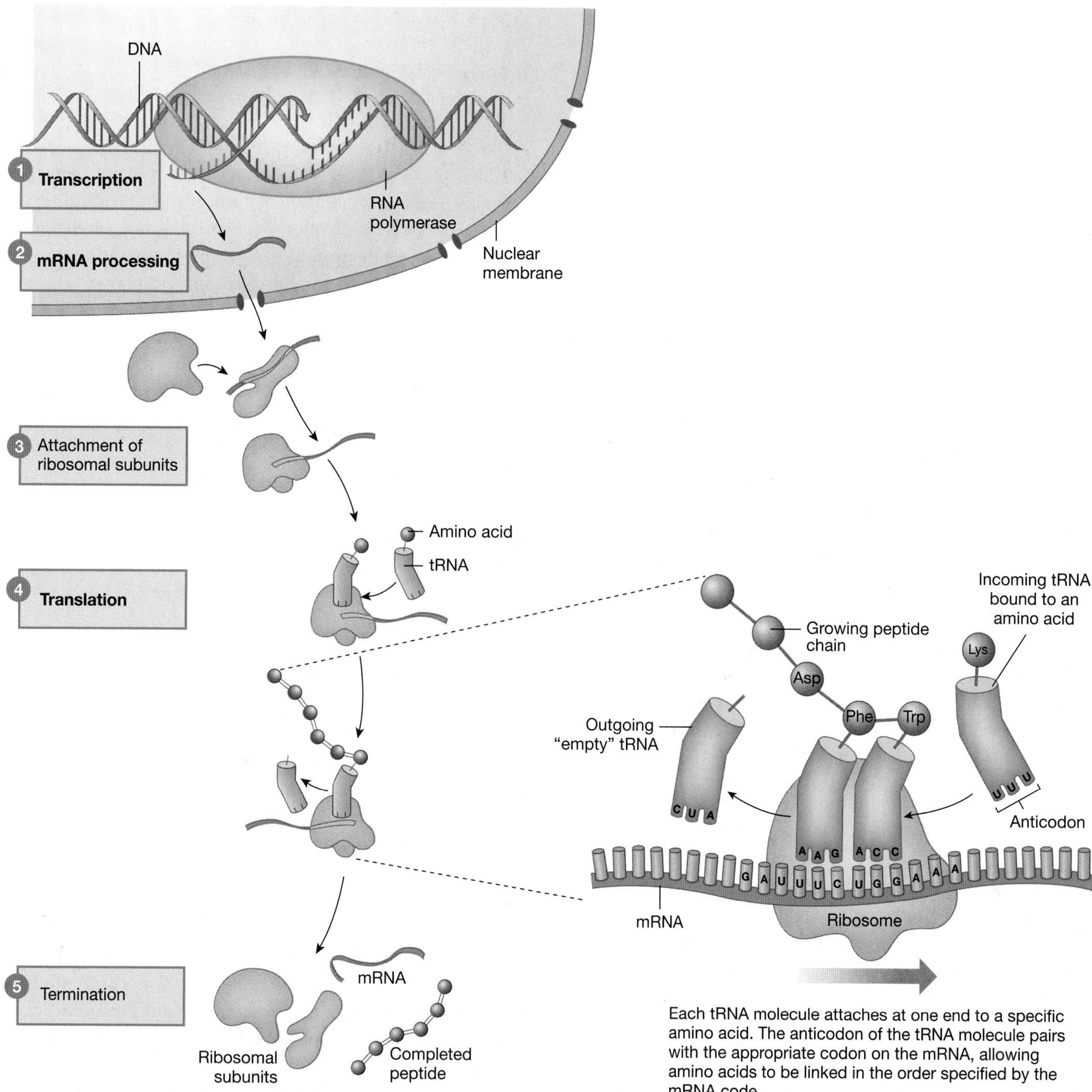

Fig. 4.21 ***Translation.*** Processed mRNA leaves the nucleus and associates with ribosomes. Translation matches the codons of RNA with amino acids to create a protein.

The sorting signal is a special segment of amino acids known as a **signal sequence**. The signal sequence tag directs the protein to the proper organelle, such as the mitochondria or peroxisomes, and allows it to be transported through the organelle membrane. Peptides synthesized on ribosomes attached to the rough endoplasmic reticulum have a signal sequence directs them through the membrane of the rough ER and into the lumen of this organelle. Once a protein enters the ER lumen, enzymes remove the signal sequence.

Proteins Undergo Post-Translational Modification

The amino acid sequence that comes off a ribosome is the primary structure of a newly synthesized protein [p. 35], but not the final form. The newly made protein can now form different types of covalent and noncovalent bonds, a process known as **post-translational modification**. Cleavage of the amino acid chain, attachment of molecules or groups, and cross-linkages are three general types of post-translational modification.

More than 100 different types of post-translational modification have been described so far.

In some common forms of post-translational modification, the amino acid chain can:

1. fold into various three-dimensional shapes. Protein folding creates the tertiary structure of the protein.
2. create cross-links between different regions of its amino acid chain
3. be cleaved (split) into fragments
4. add other molecules or groups
5. assemble with other amino acid chains into a polymeric (many-part) protein. Assembly of proteins into polymers creates the quaternary structure of the protein.

Protein Folding Peptides released from ribosomes are free to take on their final three-dimensional shape. Each peptide first forms its secondary structure, which may be an α-helix or a β-pleated sheet [p. 35]. The molecule then folds into its final shape when hydrogen bonds, covalent bonds, and ionic bonds form between amino acids in the chain. Studies show that some protein folding takes place spontaneously, but it is often facilitated by helper proteins called *molecular chaperones*.

The three-dimensional shape of proteins is often essential for proper function. Misfolded proteins, along with other proteins the cell wishes to destroy, are tagged with a protein called *ubiquitin* and sent to *proteasomes*, cylindrical cytoplasmic enzyme complexes that break down proteins.

Cross-linkage Some protein folding is held in place by relatively weak hydrogen bonds and ionic bonds. However, other proteins form strong covalent bonds between different parts of the amino acid chain. These bonds are often disulfide bonds (S–S) between two cysteine amino acids, which contain sulfur atoms. For example, the three chains of the digestive enzyme chymotrypsin are held together by disulfide bonds.

Cleavage Some biologically active proteins, such as enzymes and hormones, are synthesized initially as inactive molecules that must have segments removed before they become active. The enzyme chymotrypsin must have two small peptide fragments removed before it can catalyze a reaction [Fig. 2.12a, p. 53]. Post-translational processing also activates some peptide hormones.

Addition of Other Molecules or Groups Proteins can be modified by the addition of sugars (glycosylation) to create glycoproteins, or by combination with lipids to make lipoproteins [p. 32]. The two most common chemical groups added to proteins are phosphate groups, PO_4^{2-} and methyl groups, $-CH_3$. (Addition of a methyl group is called *methylation*.)

Assembly into Polymeric Proteins Many complex proteins have a quaternary structure with multiple subunits, in which protein chains assemble into dimers, trimers, or tetramers. One example is the enzyme lactate dehydrogenase (described on page 105). Another example is the hemoglobin molecule, with four protein chains [Fig. 2.3, p. 35].

Concept Check

Answers: p. 128

23. What is the removal of a phosphate group called?
24. List three general types of post-translational modification of proteins.
25. Is hemoglobin a monomer, dimer, trimer, or tetramer?

The many ways that proteins can be modified after synthesis add to the complexity of the human body. We must know not only the sequence of a protein but also how it is processed, where the protein occurs in or outside the cell, and what it does. Scientists working on the Human Genome Project initially predicted that our DNA would code for about 30,000 proteins, but they were not taking into account alternative splicing or post-translational modifications. Scientists working on the Human Proteomics Initiative are now predicting that we will find more than a million different proteins. The magnitude of this project means that it will continue for many years into the future.

RUNNING PROBLEM CONCLUSION

Tay-Sachs Disease

In this running problem you learned that Tay-Sachs disease is an incurable, recessive genetic disorder in which the enzyme that breaks down gangliosides in cells is missing. One in 27 Americans of Eastern European Jewish descent in the United States carries the gene for this disorder. Other high-risk populations include French Canadians, Louisiana "Cajuns," and Irish Americans. By one estimate, about one person in every 250 in the general American population is a carrier of the Tay-Sachs gene.

You have also learned that a blood test can detect the presence of this deadly gene. Check your understanding of this running problem by comparing your answers to those in the summary table. To read more on Tay-Sachs disease, see the NIH reference page (*www.ninds.nih.gov/disorders/taysachs/taysachs.htm*) or the web site of the National Tay-Sachs & Allied Diseases Association (*www.ntsad.org*).

RUNNING PROBLEM CONCLUSION *(continued)*

Question	Facts	Integration and Analysis
1. What is another symptom of Tay-Sachs disease besides loss of muscle control and brain function?	Hexosaminidase A breaks down gangliosides. In Tay-Sachs disease, this enzyme is absent, and gangliosides accumulate in cells, including light-sensitive cells of the eye, and cause them to function abnormally.	Damage to light-sensitive cells of the eye could cause vision problems and even blindness.
2. How could you test whether Sarah and David are carriers of the Tay-Sachs gene?	Carriers of the gene have lower-than-normal levels of hexosaminidase A.	Run tests to determine the average enzyme levels in known carriers of the disease (i.e., people who are parents of children with Tay-Sachs disease) and in people who have little likelihood of being carriers. Compare the enzyme levels of suspected carriers such as Sarah and David with the averages for the known carriers and noncarriers.
3. Why is the new test for the Tay-Sachs disease more accurate than the old test?	The new test detects the defective gene. The old test analyzed levels of the enzyme produced by the gene.	The new test is a direct way to test if a person is a carrier. The old test was an indirect way. It is possible for factors other than a defective gene to alter a person's enzyme level. Can you think of some? (See answer at the end of this chapter.)
4. The Tay-Sachs gene is a recessive gene (t). What is the chance that any child of a carrier (Tt) and a noncarrier (TT) will be a carrier? What are the chances that a child of two carriers will have the disease or be a carrier?	Mating of Tt × TT results in the following offspring: TT, Tt, TT, Tt. Mating of Tt × Tt results in the following offspring: TT, Tt, Tt, tt.	If only one parent is a carrier, each child has a 50% chance of being a carrier (Tt). If both parents are carriers, there is a 25% chance that a child will have Tay-Sachs disease and a 50% chance a child will be a carrier.

99 104 107 110 118 124

Chapter Summary

The major theme of this chapter is *energy in biological systems* and how it is acquired, transferred, and used to do biological work. Energy is stored in large biomolecules such as fats and glycogen and is extracted from them through the processes of metabolism. Extracted energy is often stored temporarily in the high-energy phosphate bonds of ATP. Reactions and processes that require energy often use ATP as the energy source. This is a pattern you will see repeated as you learn more about the organ systems of the body.

Other themes in the chapter involve two kinds of *structure-function relationships:* molecular interactions and compartmentation. *Molecular interactions* are important in enzymes, where the ability of an enzyme to bind to its substrate influences the enzyme's activity. *Compartmentation* of enzymes allows cells to direct energy flow by separating functions. Glycolysis takes place in the cytosol of the cell, but the citric acid cycle is isolated within mitochondria, requiring transport of substrates across the mitochondrial membrane. Modulation of enzyme activity and the separation of pathways into subcellular compartments are essential for organizing and separating metabolic processes.

Energy in Biological Systems

1. **Energy** is the capacity to do work. **Chemical work** enables cells and organisms to grow, reproduce, and carry out normal activities. **Transport work** enables cells to move molecules to create concentration gradients. **Mechanical work** is used for movement. (pp. 100–101)
2. **Kinetic energy** is the energy of motion. **Potential energy** is stored energy. (p. 101; Fig. 4.2)

Chemical Reactions

3. A **chemical reaction** begins with one or more **reactants** and ends with one or more **products** (Tbl. 4.2). **Reaction rate** is measured as the change in concentration of products with time. (p. 102)
4. The energy stored in the chemical bonds of a molecule and available to perform work is the **free energy** of the molecule. (p. 102)
5. **Activation energy** is the initial input of energy required to begin a reaction. (p. 103; Fig. 4.3)
6. **Exergonic reactions** are energy-producing. **Endergonic reactions** are energy-utilizing. (p. 103; Fig. 4.3)
7. Metabolic pathways couple exergonic reactions to endergonic reactions. (p. 104; Fig. 4.4)
8. Energy for driving endergonic reactions is stored in ATP. (p. 104)
9. A reaction that can proceed in both directions is called a **reversible reaction.** If a reaction can proceed in one direction but not the other, it is an **irreversible reaction.** The net free energy change of a reaction determines whether that reaction can be reversed. (p. 104)

Enzymes

10. **Enzymes** are biological catalysts that speed up the rate of chemical reactions without themselves being changed. In reactions catalyzed by enzymes, the reactants are called **substrates.** (p. 105)
11. Like other proteins that bind ligands, enzymes exhibit saturation, specificity, and competition. Related isozymes may have different activities. (p. 105)
12. Some enzymes are produced as inactive precursors and must be activated. This may require the presence of a **cofactor.** Organic cofactors are called **coenzymes.** (p. 106)
13. Enzyme activity is altered by temperature, pH, and modulator molecules. (p. 106)
14. Enzymes work by lowering the activation energy of a reaction. (p. 107; Fig. 4.7)
15. Most reactions can be classified as **oxidation-reduction, hydrolysis-dehydration, addition-subtraction-exchange,** or **ligation** reactions. (p. 108; Tbl. 4.4)

Metabolism

16. All the chemical reactions in the body are known collectively as **metabolism. Catabolic reactions** release energy and break down large biomolecules. **Anabolic reactions** require a net input of energy and synthesize large biomolecules. (p. 109)
17. Cells regulate the flow of molecules through their metabolic pathways by (1) controlling enzyme concentrations, (2) producing allosteric and covalent modulators, (3) using different enzymes to catalyze reversible reactions, (4) isolating enzymes in intracellular organelles, or (5) maintaining an optimum ratio of ATP to ADP. (p. 110–111)
18. **Aerobic pathways** require oxygen and yield the most ATP. **Anaerobic pathways** can proceed without oxygen but produce ATP in much smaller quantities. (p. 111)

ATP Production

iP IP Muscular: Muscle Metabolism

19. Through **glycolysis,** one molecule of glucose is converted into two pyruvate molecules, and yields 2 ATP, 2 NADH, and 2 H^+. Glycolysis does not require the presence of oxygen. (p. 113; Fig. 4.12)
20. **Aerobic metabolism** of pyruvate through the **citric acid cycle** yields ATP, carbon dioxide, and high-energy electrons captured by NADH and $FADH_2$. (p. 114; Fig. 4.13)
21. **High-energy electrons** from NADH and $FADH_2$ give up their energy as they pass through the **electron transport system.** Their energy is trapped in the high-energy bonds of ATP. (p. 115; Fig. 4.14)
22. Maximum energy yield for aerobic metabolism of one glucose molecule is 30–32 ATP.
23. In **anaerobic metabolism,** pyruvate is converted into lactate, with a net yield of 2 ATP for each glucose molecule. (p. 116; Fig. 4.15)
24. Protein synthesis is controlled by nuclear **genes** made of DNA. The code represented by the base sequence in a gene is transcribed into a complementary base code on **RNA. Alternative splicing** of mRNA in the nucleus allows one gene to code for multiple proteins. (pp. 119–121; Figs. 4.18, 4.19, 4.20)
25. mRNA leaves the nucleus and goes to the cytosol where, with the assistance of **transfer RNA** and **ribosomal RNA,** it assembles amino acids into a designated sequence. This process is called **translation.** (pp. 123; Fig. 4.21)
26. **Post-translational modification** converts the newly synthesized protein to its finished form. (p. 123–124)

Questions

Answers: p. A-1

Level One Reviewing Facts and Terms

1. List the three basic forms of work and give a physiological example of each.
2. Explain the difference between potential energy and kinetic energy.
3. State the two laws of thermodynamics in your own words.
4. The sum of all chemical processes through which cells obtain and store energy is called ________.
5. In the reaction $CO_2 + H_2O \rightarrow H_2CO_3$, water and carbon dioxide are the reactants, and H_2CO_3 is the product. Because this reaction is catalyzed by an enzyme, it is also appropriate to call water and carbon dioxide ________. The speed at which this reaction occurs is called the reaction ________, often expressed as molarity/second.
6. ________ are protein molecules that speed up chemical reactions by (increasing or decreasing?) the activation energy of the reaction.

7. Match each definition in the left column with the correct term from the right column (you will not use all the terms):

1. reaction that can run either direction	(a) exergonic
2. reaction that releases energy	(b) endergonic
3. ability of an enzyme to catalyze one reaction but not another	(c) activation energy
4. boost of energy needed to get a reaction started	(d) reversible
	(e) irreversible
	(f) specificity
	(g) free energy
	(h) saturation

8. Since 1972, enzymes have been designated by adding the suffix ________ to their name.
9. Organic molecules that must be present in order for an enzyme to function are called ________. The precursors of these organic molecules come from ________ in our diet.
10. In an oxidation-reduction reaction, in which electrons are moved between molecules, the molecule that gains an electron is said to be________, and the one that loses an electron is said to be ________.
11. The removal of H_2O from reacting molecules is called ________. Using H_2O to break down polymers, such as starch, is called ________.
12. The removal of an amino group ($-NH_2$) from a molecule (such as an amino acid) is called ________. Transfer of an amino group from one molecule to the carbon skeleton of another molecule (to form a different amino acid) is called________.
13. In metabolism, ________ reactions release energy and result in the breakdown of large biomolecules, and ________ reactions require a net input of energy and result in the synthesis of large biomolecules. In what units do we measure the energy of metabolism?
14. Metabolic regulation in which the last product of a metabolic pathway (the end product) accumulates and slows or stops reactions earlier in the pathway is called ________.
15. Explain how H^+ movement across the inner mitochondrial membrane results in ATP synthesis.
16. List the two carrier molecules that deliver high-energy electrons to the electron transport system.

Level Two Reviewing Concepts

17. Create maps using the following terms.

Map 1: Metabolism

- acetyl CoA
- ATP
- citric acid cycle
- CO_2
- cytosol
- electron transport system
- $FADH_2$
- glucose
- glycolysis
- high-energy electrons
- lactate
- mitochondria
- NADH
- oxygen
- pyruvate
- water

Map 2: Protein synthesis

- alternative splicing
- base pairing
- bases (A, C, G, T, U)
- DNA
- exon
- gene
- intron
- promoter
- mRNA
- tRNA
- ribosome
- RNA polymerase
- RNA processing
- start codon
- stop codon
- template strand
- transcription
- transcription factors
- translation

18. When bonds are broken during a chemical reaction, what are the three possible fates for the potential energy found in those bonds?
19. Match the metabolic processes with the letter of the biological theme that best describes the process:

(a) Biological energy use	1. Glycolysis takes place in the cytosol; oxidative phosphorylation takes place in mitochondria.
(b) Compartmentation	2. The electron transport system traps energy in a hydrogen ion concentration gradient.
(c) Molecular interactions	3. Proteins are modified in the endoplasmic reticulum.
	4. Metabolic reactions are often coupled to the reaction $ATP \rightarrow ADP + P_i$.
	5. Some proteins have S–S bonds between nonadjacent amino acids.
	6. Enzymes catalyze biological reactions.

20. Explain why it is advantageous for a cell to store or secrete an enzyme in an inactive form.
21. Compare the following: (a) the energy yield from the aerobic breakdown of one glucose to CO_2 and H_2O and (b) the energy yield from one glucose going through anaerobic glycolysis ending with lactate. What are the advantages of each pathway?
22. Briefly describe the processes of transcription and translation. Which organelles are involved in each process?
23. On what molecule does the anticodon appear? Explain the role of this molecule in protein synthesis.
24. Is the energy of ATP's phosphate bond an example of potential or kinetic energy?
25. If ATP releases energy to drive a chemical reaction, would you suspect the activation energy of that reaction to be large or small? Explain.

Level Three Problem Solving

26. Given the following strand of DNA, list the entire sequence of bases that would appear in the matching mRNA. For the underlined triplets, underline the corresponding mRNA codon. Then give the appropriate amino acids for which those four triplets are the code.

DNA: CGC<u>TAC</u><u>AAG</u>T<u>CAC</u>GTACC<u>GTA</u>ACG

mRNA:

Amino acids:

Level Four Quantitative Problems

27. The graph shows the free energy change for the reaction $A + B \rightarrow D$. Is this an endergonic or exergonic reaction?

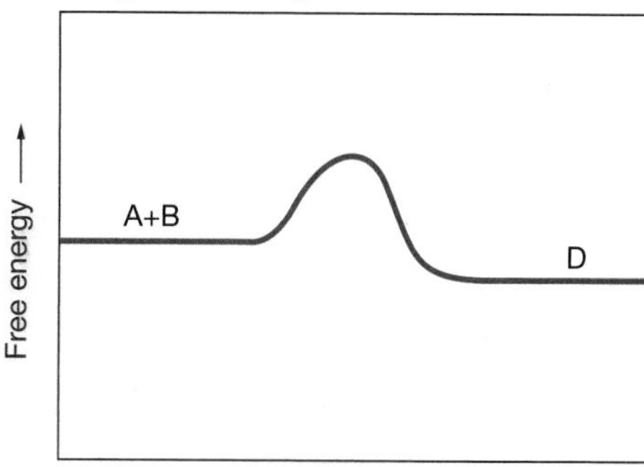

28. If the protein-coding portion of a piece of processed mRNA is 450 bases long, how many amino acids will be in the corresponding polypeptide? (*Hint:* The start codon is translated into an amino acid, but the stop codon is not.)

4

Answers

Answers to Concept Check Questions

Page 100

1. Amino acids and nucleotides always contain nitrogen.

Page 102

2. Energy in the body is stored in chemical bonds and in concentration gradients.
3. Kinetic energy is the energy of motion: something is happening. Potential energy is stored energy: something is waiting to happen.
4. Entropy is a state of randomness or disorder.

Page 105

5. Endergonic reactions consume energy; exergonic reactions release energy.
6. The reactants are baking soda and vinegar; the product is carbon dioxide.
7. The foaming indicates that energy is being released, and so this is an exergonic reaction. The large amount of energy released indicates that the reaction is not readily reversible.

Page 107

8. The presence of isozymes enables one reaction to be catalyzed under a variety of conditions.
9. The four protein chains represent the quaternary level of protein structure.

Page 109

10. The substrates are lactose (lactase), peptides (peptidase), lipids (lipase), and sucrose (sucrase).
11. (a) 3 (b) 2 (c) 4 (d) 1

Page 111

12. Cells regulate substrate movement by (1) controlling the amount of enzyme, (2) producing allosteric and covalent modulators, (3) using two different enzymes to catalyze reversible reactions, (4) isolating enzymes within intracellular organelles, and (5) altering the ratio of ADP to ATP in the cell.
13. In ATP, energy is trapped and stored in one of the three phosphate bonds. In NADH, energy is stored in high-energy electrons.
14. Aerobic pathways require sufficient quantities of molecular oxygen (O_2) in the cell. Anaerobic pathways can proceed without oxygen.

Page 112

15. (a) 4 (b) 2, 5 (c) 2, 5 (d) 1, 3
16. Endergonic reactions trap energy in the products.

Page 117

17. When H^+ ions are pumped into the intermembrane space, energy is trapped in the concentration gradient of H^+. Release of this energy as H^+ pass through the ATP synthase is coupled to ATP synthesis.
18. Lactate dehydrogenase acts on lactate by *removing* an *electron* and a *hydrogen atom*. This process is called *oxidation*.
19. Anaerobic metabolism of glucose can proceed in the absence of oxygen; aerobic metabolism requires oxygen. Anaerobic metabolism produces much less ATP per glucose than aerobic metabolism.

Page 121

20. The DNA triplets are ATT, ATC, and ACT.
21. RNA polymerase makes polymers of RNA.

Page 122

22. During mRNA processing, base sequences called introns are cut out of the mRNA. The remaining segments, the exons, are spliced back together and provide the code for a protein.

Page 124

23. Removal of a phosphate group is dephosphorylation.
24. Three types of post-translational modification are cleavage, addition of groups, and cross-linking.
25. Hemoglobin is a tetramer because it contains four protein chains.

Answers to Figure and Graph Questions

Page 105

Figure 4.5: The graph shows an endergonic reaction.

Page 106

Figure 4.6: When pH decreases from 8 to 7.4, enzyme activity increases.

Page 110

Figure 4.10: A kinase moves a phosphate group from one molecule to another. A phosphatase removes a phosphate group.

Page 113

Figure 4.12: 1. Glycolysis is an exergonic pathway. 2. (a) ①, ③. (b) ⑤, ⑥, ⑨. (c) Kinases add a phosphate group. ①, ③, ⑤. (d) Dehydrogenases remove an electron and a hydrogen atom. In step ⑤, NAD^+ acquires an electron and an H, suggesting that this step is catalyzed by a dehydrogenase. 3. Net yield for 1 glucose is 2 ATP and 2 NADH.

Page 114

Figure 4.13: 1. Citric acid cycle is exergonic. 2. Net yield for one pyruvate is 4 NADH, 1 $FADH_2$, and 1 ATP. 3. 3 CO_2 form from one pyruvate. This is the same as the 3 carbons in pyruvate.

Page 115

Figure 4.14: 1. Phosphorylation is the addition of a phosphate group. ADP is phosphorylated in oxidative phosphorylation. 2. Electron transfer is exergonic. 3. Oxygen acts as an acceptor of electrons and hydrogen ions at the end of the ETS.

Answer to Running Problem Conclusion

Page 124

Question 3. Factors other than a defective gene that could alter enzyme levels include decreased protein synthesis or increased protein breakdown in the cell. Such changes could occur even though the gene was normal.

5 Membrane Dynamics

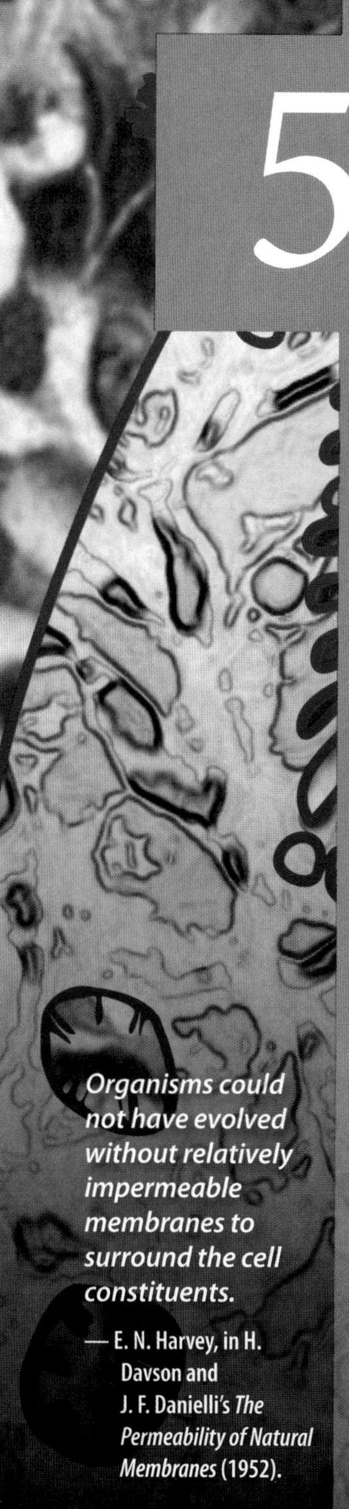

Organisms could not have evolved without relatively impermeable membranes to surround the cell constituents.

— E. N. Harvey, in H. Davson and J. F. Danielli's *The Permeability of Natural Membranes* (1952).

Background Basics

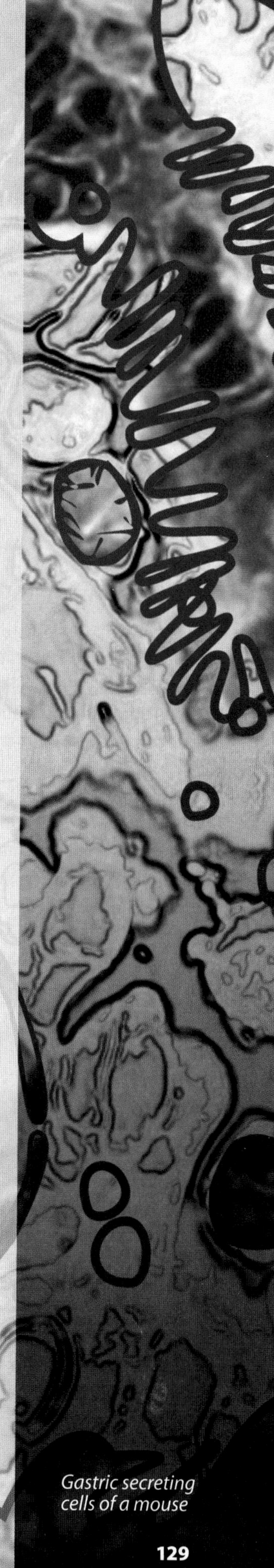

Gastric secreting cells of a mouse

In 1992 the medical personnel at isolated Atoifi Hospital in the Solomon Islands of the South Pacific were faced with a dilemma. A patient was vomiting and needed intravenous (IV) fluids, but the hospital's supply had run out, and it would be several days before a plane could bring more. Their solution was to try something they had only heard about—make an IV of coconut water, the sterile solution that forms in the hollow center of developing coconuts. For two days the patient received a slow drip of fluid into his veins directly from young coconuts suspended next to his bed. He soon recovered and was well enough to go home.*

No one knows who first tried coconut water as an IV solution, although stories have been passed down that both the Japanese and the British used it in the Pacific Theater of Operations during World War II. Choosing the appropriate IV solution is more than a matter of luck, however. It requires a solid understanding of the body's compartments and of the ways different solutes pass between them.

Homeostasis Does Not Mean Equilibrium

The body has two distinct fluid compartments: the cells and the fluid that surrounds the cells (■ Fig. 5.1). The extracellular fluid (ECF) outside the cells is the buffer between the cells and the environment outside the body. Everything that enters or leaves most cells passes through the ECF.

Water is essentially the only molecule that moves freely between cells and the extracellular fluid. Because of this free movement of water, the extracellular and intracellular compartments reach a state of **osmotic equilibrium** {*osmos*, push or thrust}, in which the fluid concentrations are equal on the two sides of the cell membrane. (Concentration is expressed as amount of solute per volume [Fig. 2.7, p. 44].) Although the overall concentrations of the ECF and intracellular fluid (ICF) are equal, some solutes are more concentrated in one of the two body compartments than in the other (Fig. 5.1d). This means the body is in a state of **chemical disequilibrium.**

Figure 5.1d shows the uneven distribution of major solutes among the body fluid compartments. For example, sodium, chloride, and bicarbonate (HCO_3^-) ions are more concentrated in extracellular fluid than in intracellular fluid. Potassium ions are more concentrated inside the cell. Calcium (not shown in the figure) is more concentrated in the extracellular fluid than in the cytosol, although many cells store Ca^{2+} inside organelles such as the endoplasmic reticulum and mitochondria.

Even the extracellular fluid is not at equilibrium between its two subcompartments, the plasma and the interstitial fluid (IF) [p. 65]. Plasma is the liquid matrix of blood and is found inside the circulatory system. Proteins and other large anions are concentrated in the plasma but cannot cross the leaky exchange epithelium of blood vessels [p. 82], so they are mostly absent from the interstitial fluid. On the other hand, smaller molecules and ions such as Na^+ and Cl^- are small enough to pass freely between the endothelial cells and therefore have the same concentrations in plasma and interstitial fluid.

The concentration differences of chemical disequilibrium are a hallmark of a living organism, as only the continual input of energy keeps the body in this state. If solutes leak across the cell membranes dividing the intracellular and extracellular compartments, energy is required to return them to the compartment they left. For example, K^+ ions that leak out of the cell and Na^+ ions that leak into the cell are returned to their original compartments by an energy-utilizing enzyme known as the *Na^+-K^+-ATPase*, or the sodium-potassium pump. When cells die and cannot use energy, they obey the second law of thermodynamics [p. 102] and return to a state of randomness that is marked by loss of chemical disequilibrium.

Many body solutes mentioned so far are ions, and for this reason we must also consider the distribution of electrical charge between the intracellular and extracellular compartments. The body as a whole is electrically neutral, but a few extra negative ions are found in the intracellular fluid, while their matching positive ions are located in the extracellular fluid. As a result, the inside of cells is slightly negative relative to the extracellular fluid. This ionic imbalance results in a state of **electrical disequilibrium.** Changes in this disequilibrium create electrical signals. We discuss this topic in more detail later in this chapter.

In summary, note that homeostasis is not the same as equilibrium. The intracellular and extracellular compartments of the body are in osmotic equilibrium, but in chemical and electrical disequilibrium. Furthermore, osmotic equilibrium and the two disequilibria are dynamic *steady states*. The goal of homeostasis is to maintain the dynamic steady states of the body's compartments.

In the remainder of this chapter, we discuss these three steady states, and the role transport mechanisms and the selective permeability of cell membranes play in maintaining these states.

RUNNING PROBLEM

Cystic Fibrosis

Over 100 years ago, midwives performed an unusual test on the infants they delivered: the midwife would lick the infant's forehead. A salty taste meant that the child was destined to die of a mysterious disease that withered the flesh and robbed the breath. Today, a similar "sweat test" will be performed in a major hospital—this time with state-of-the-art techniques—on Daniel Biller, a 2-year-old with a history of weight loss and respiratory problems. The name of the mysterious disease? Cystic fibrosis.

130 141 148 158 161 168

*D. Campbell-Falck *et al.* The intravenous use of coconut water. *Am J Emerg Med* 18: 108–111, 2000.

Body Fluid Compartments

(a) The body fluids are in two compartments: the extracellular fluid (ECF) and intracellular fluid (ICF). The ECF and ICF are in osmotic equilibrium but have very different chemical composition.

Intracellular fluid is 2/3 of the total body water volume. Material moving into and out of the ICF must cross the cell membrane.

Extracellular fluid includes all fluid outside the cells. The ECF is 1/3 of the body fluid volume. The ECF consists of:

- **Interstitial fluid** (IF), which lies between the circulatory system and the cells, is 75% of the ECF volume.
- **Plasma**, the liquid matrix of blood, is 25% of the ECF volume. Substances moving between the plasma and interstitiial fluid must cross the leaky exchange epithelium of the capillary wall.

KEY

- Intracellular fluid
- Interstitial fluid
- Plasma

(b) This figure shows the compartment volumes for the "standard" 70-kg man.

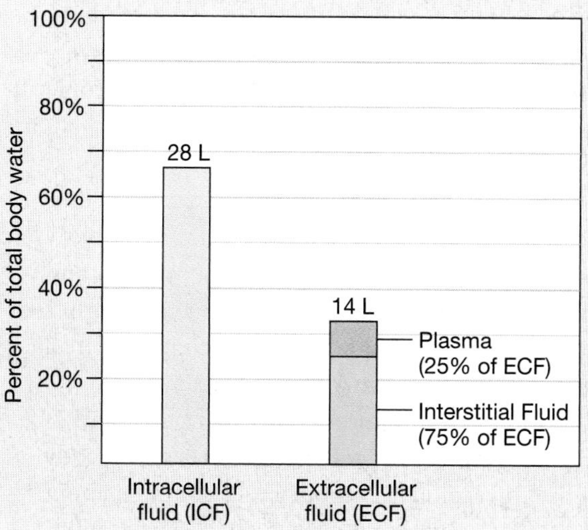

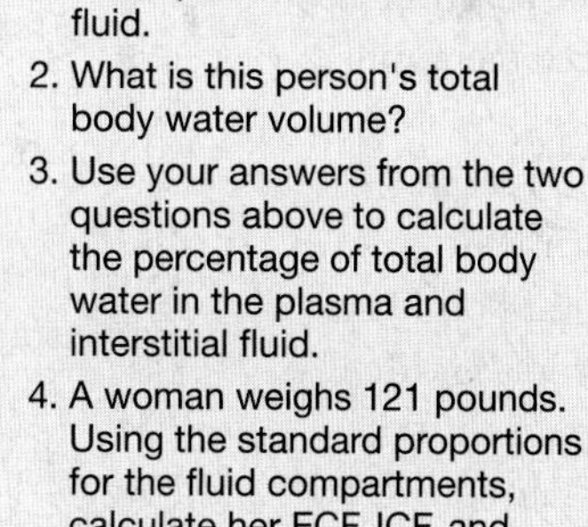

GRAPH QUESTIONS

1. Using the ECF volume shown in (b), calculate the volumes of the plasma and interstitial fluid.
2. What is this person's total body water volume?
3. Use your answers from the two questions above to calculate the percentage of total body water in the plasma and interstitial fluid.
4. A woman weighs 121 pounds. Using the standard proportions for the fluid compartments, calculate her ECF, ICF, and plasma volumes. (2.2 lb = 1 kg. 1 kg water = 1 L)

(c) Fluid compartments are often illustrated with box diagrams like this one.

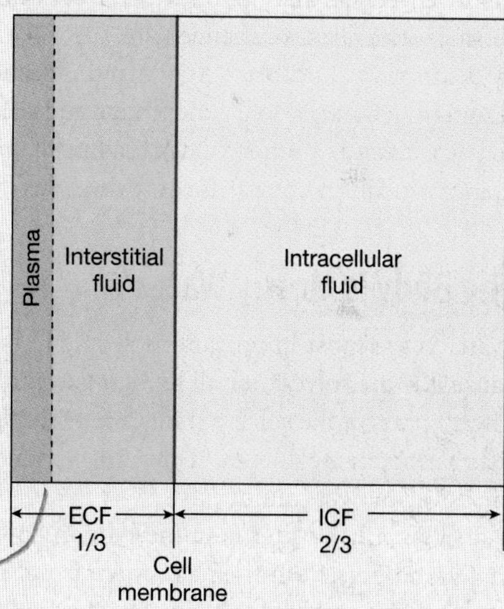

(d) The body compartments are in a state of chemical disequilibrium. The cell membrane is a selectively permeable barrier between the ECF and ICF.

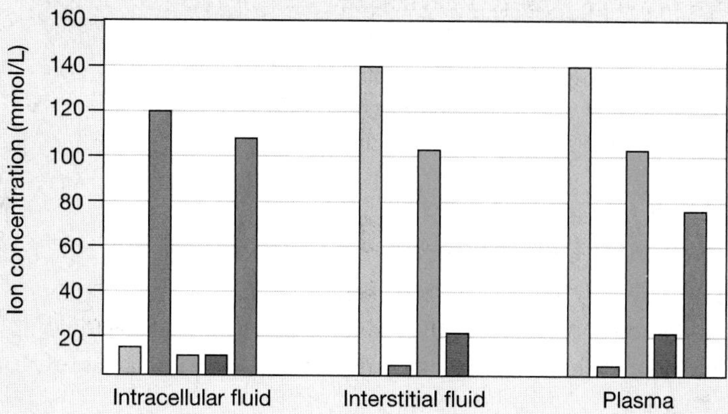

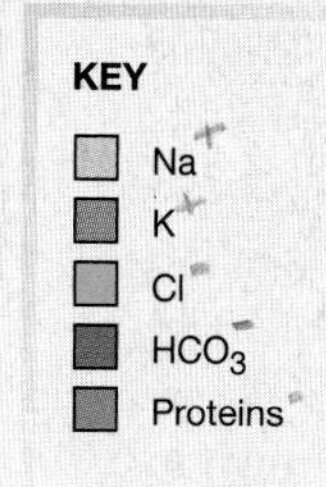

GRAPH QUESTIONS

5. How does the ion composition of plasma differ from that of the IF?
6. What ions are concentrated in the ECF? In the ICF?

Concept Check Answers: p. 172

1. Using what you learned about the naming conventions for enzymes [p. 107], explain what the name *Na^+-K^+-ATPase* tells you about this enzyme's actions.
2. The intracellular fluid can be distinguished from the extracellular fluid by the ICF's high concentration of ________ ions and low concentration of ________, ________, and ________ ions.
3. Clinically we monitor homeostasis of various dissolved solutes such as ions, blood gases, and organic solutes by taking a blood sample and analyzing its plasma. For each of the following substances, predict whether knowing its plasma concentration also tells you its concentration in the ECF and the ICF. Defend your answer.
   (a) Na^+
   (b) K^+
   (c) water
   (d) proteins

Table 5.1 Water Content as Percentage of Total Body Weight by Age and Sex

Age	Male	Female
Infant	65%	65%
1–9	62%	62%
10–16	59%	57%
17–39	61%	51%
40–59	55%	47%
60+	52%	46%

Adapted from Edelman and Leibman, *Am J Med 27*; 256–277, 1959.

Osmosis and Tonicity

The distribution of solutes in the body depends on whether a substance can cross cell membranes. Water, on the other hand, is able to move freely in and out of nearly every cell in the body by traversing water-filled ion channels and special water channels created by the protein *aquaporin* (AQP). In this section, we examine the relationship between solute movement and water movement across cell membranes. A sound understanding of this topic provides the foundation for the clinical use of intravenous (IV) fluid therapy.

The Body Is Mostly Water

universal solvent

Water is the most important molecule in the human body because it is the solvent for all living matter. As we look for life in distant parts of the solar system, one of the first questions scientists ask about a planet is, "Does it have water?" Without water, life as we know it cannot exist.

How much water is in the human body? Because one individual differs from the next, there is no single answer. However, in human physiology we often speak of standard values for physiological functions based on "the 70-kg man." These standard values are derived from data published in the mid-20th century by The International Commission on Radiological Protection. The ICRP was setting guidelines for permissible radiation exposure, and they selected a young (age 20–30) white European male who weighed 70 kilograms (kg) or 154 pounds as their "reference man," or "standard man." In 1984 Reference Man was joined by Reference Woman, a young 58 kg (127.6 lb) female. The U.S. population is getting larger and heavier, however, and in 1990 the equivalent Reference Man had grown to 77.5 kg and was 8 cm taller.

The 70-kilogram Reference Man has 60% of his total body weight, or 42 kg (92.4 lb), in the form of water. Each kilogram of water has a volume of 1 liter, so his **total body water** is 42 liters. This is the equivalent of 21 two-liter soft drink bottles!

Adult women have less water per kilogram of body mass than men because women have more adipose tissue. Large fat droplets in adipose tissue occupy most of the cell's volume, displacing the more aqueous cytoplasm [see Fig. 3.13e, p. 88]. Age also influences body water content. Infants have relatively more water than adults, and water content decreases as people grow older than 60.

■ Table 5.1 shows water content as a percentage of total body weight in people of various ages and both sexes. In clinical practice, it is necessary to allow for the variability of body water content when prescribing drugs. Because women and older people have less body water, they will have a higher concentration of a drug in the plasma than will young men if all are given an equal dose per kilogram of body mass.

The distribution of water among body compartments is less variable. When we look at the relative volumes of the body compartments, the intracellular compartment contains about two-thirds (67%) of the body's water (Fig. 5.1b, c). The remaining third (33%) is split between the interstitial fluid (which contains about 75% of the extracellular water) and the plasma (which contains about 25% of the extracellular water).

Concept Check Answer: p. 172

4. If the 58 kg Reference Woman has total body water equivalent to 50% of her body weight, what is (a) her total body water volume, (b) her ECF and ICF volumes, and (c) her plasma volume?

The Body Is in Osmotic Equilibrium

Water is able to move freely between cells and the extracellular fluid and distributes itself until water concentrations are equal throughout the body—in other words, until the body is in a state of osmotic equilibrium. The movement of water across a membrane in response to a solute concentration gradient is called **osmosis.** In osmosis, water moves to dilute the more concentrated solution. Once concentrations are equal, net movement of water stops.

Look at the example shown in ■ Figure 5.2 in which a selectively permeable membrane separates two compartments of equal volume. The membrane is permeable to water but does not allow glucose to cross. In (1), compartments A and B contain equal volumes of glucose solution. Compartment B has more solute (glucose) per volume of solution and therefore is the more concentrated solution. A concentration gradient across the membrane exists for glucose. However, because the membrane is not permeable to glucose, glucose cannot move to equalize its distribution.

Water, by contrast, can cross the membrane freely. It will move by osmosis from compartment A, which contains the dilute glucose solution, to compartment B, which contains the more concentrated glucose solution. Thus, water moves to dilute the more concentrated solution (Fig. 5.2 (2)).

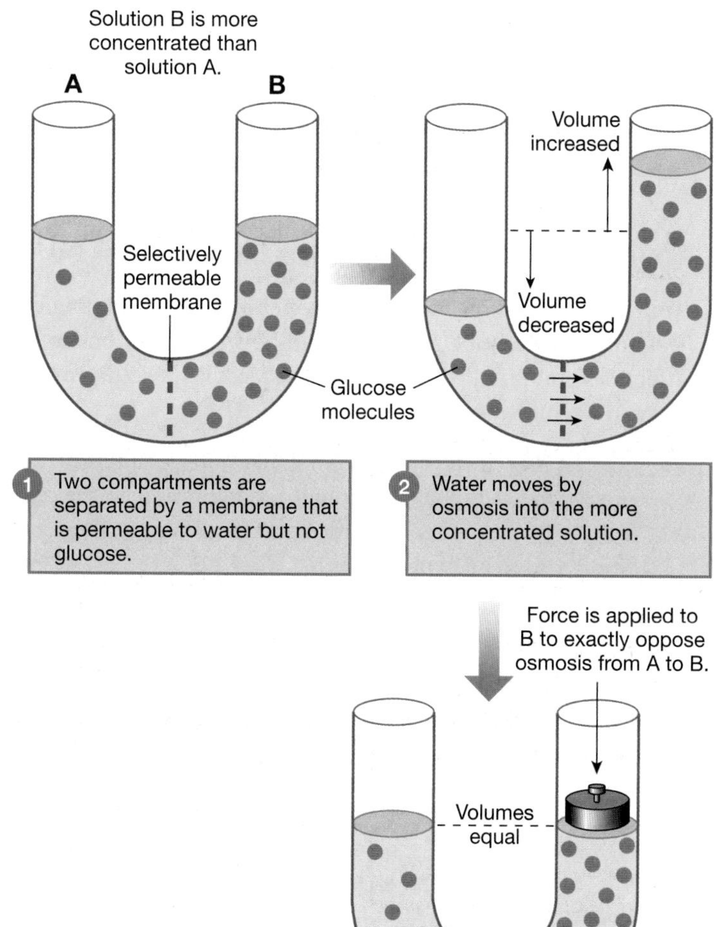
Solution B is more concentrated than solution A.
A
B
Selectively permeable membrane
Glucose molecules
1 Two compartments are separated by a membrane that is permeable to water but not glucose.
Volume increased
Volume decreased
2 Water moves by osmosis into the more concentrated solution.
Force is applied to B to exactly oppose osmosis from A to B.
Volumes equal
3 Osmotic pressure is the pressure that must be applied to B to oppose osmosis.

■ **Fig. 5.2** ***Osmosis and osmotic pressure***

How can we make quantitative measurements of osmosis? One method is shown in Figure 5.2 (3). Place a piston into compartment B, which has a higher solute concentration than compartment A. By pushing down on the piston, you can keep water from flowing from A to B. The pressure that must be applied to the piston to exactly oppose the osmotic movement of water into compartment B is known as the **osmotic pressure** of solution B. The units for osmotic pressure, just as with other pressures in physiology, are *atmospheres* (atm) or *millimeters of mercury* (mm Hg). A pressure of 1 mm Hg is equivalent to the pressure exerted on a 1-cm^2 area by a 1-mm-high column of mercury.

Osmolarity Describes the Number of Particles in Solution

Another way to predict the osmotic movement of water quantitatively is to know the concentrations of the solutions with which we are dealing. In chemistry, concentrations are usually expressed as *molarity* (*M*), which is defined as number of moles of dissolved solute per liter of solution (mol/L). Recall that one *mole* is 6.02×10^{23} molecules [Fig. 2.7, p. 45].

However, using molarity to describe biological concentrations can be misleading. The important factor for osmosis is the number of osmotically active *particles* in a given volume of solution, not the number of molecules. Because some molecules dissociate into ions when they dissolve in a solution, the number of particles in solution is not always the same as the number of molecules.

For example, one glucose molecule dissolved in water yields one particle, but one NaCl dissolved in water theoretically yields two ions (particles): Na^+ and Cl^-. Water moves by osmosis in response to the total concentration of all *particles* in the solution. The particles may be ions, uncharged molecules, or a mixture of both.

Consequently, for biological solutions we express the concentration as **osmolarity**, the number of osmotically active particles (ions or intact molecules) per liter of solution. Osmolarity is expressed in *osmoles* per liter (osmol/L or OsM) or, for very dilute physiological solutions, milliosmoles/liter (mOsM). To convert between molarity and osmolarity, use the following equation:

$$\text{molarity (mol/L)} \times \text{number of particles/molecule (osmol/mol)} = \text{osmolarity (osmol/L)}$$

Let us look at two examples, glucose and sodium chloride, and compare their molarity with their osmolarity.

One mole of glucose molecules dissolved in enough water to create 1 liter of solution yields a 1 molar solution (1 M). Because glucose does not dissociate in solution, the solution has only one mole of osmotically active particles:

$$1 \text{ M glucose} \times 1 \text{ particle per glucose molecule} = 1 \text{ OsM glucose}$$

Unlike glucose, sodium chloride dissociates when placed in solution. At body temperature, a few NaCl ions fail to separate, so instead of 2 ions per NaCl, the *dissociation factor* is about 1.8.

Thus, one mole of NaCl dissociates in solution to yield 1.8 moles of particles (Na^+, Cl^-, and NaCl). The result is a 1.8 OsM solution:

$$1 \text{ mole NaCl/L} \times 1.8 \text{ osmoles/mole NaCl} = 1.8 \text{ osmol NaCl/L}$$

Osmolarity describes only the number of particles in the solution. It says nothing about the composition of the particles. A 1 OsM solution could be composed of pure glucose or pure Na^+ and Cl^- or a mixture of all three solutes.

The normal osmolarity of the human body ranges from 280 to 296 milliosmoles per liter (mOsM). In this book, to simplify calculations we will round that number up slightly to 300 mOsM.

A term related to osmolarity is osmolality. **Osmolality** is concentration expressed as osmoles of solute per kilogram of water. Because biological solutions are dilute and little of their weight comes from solute, physiologists often use the terms *osmolarity* and *osmolality* interchangeably. Osmolality is usually used in clinical situations because it is easy to estimate people's body water content by weighing them.

Clinicians estimate a person's fluid loss in dehydration by equating weight loss to fluid loss. Because 1 liter of pure water weighs 1 kilogram, a decrease in body weight of 1 kilogram (or 2.2 lb) is considered equivalent to the loss of 1 liter of body fluid. A baby with diarrhea can easily be weighed to estimate its fluid loss. A decrease of 1.1 pounds (0.5 kg) of body weight is assumed to mean the loss of 500 mL of fluid. This calculation provides a quick estimate of how much fluid needs to be replaced.

Concept Check Answer: p. 172

5. A mother brings her baby to the emergency room because he has lost fluid through diarrhea and vomiting for two days. The staff weighs the baby and finds that he has lost 2 pounds. If you assume that the reduction in weight is due to water loss, what volume of water has the baby lost (2.2 pounds = 1 kilogram)?

Comparing Osmolarities of Two Solutions Osmolarity is a property of every solution. You can compare the osmolarities of different solutions as long as the concentrations are expressed in the same units—for example, as milliosmoles per liter. If two solutions contain the same number of solute particles per unit volume, we say that the solutions are **isosmotic** {*iso*-, equal}. If solution A has a higher osmolarity (contains more particles per unit volume, is more concentrated) than solution B, we say that solution A is **hyperosmotic** to solution B. In the same example, solution B, with fewer osmoles per unit volume, is **hyposmotic** to solution A. ■ Table 5.2 shows some examples of comparative osmolarities.

Osmolarity is a *colligative* property of solutions, meaning it depends strictly on the *number* of particles per liter of solution.

Comparing Osmolarities — Table 5.2

Solution A = 1 OsM Glucose	Solution B = 2 OsM Glucose	Solution C = 1 OsM NaCl
A is hyposmotic to B	B is hyperosmotic to A	C is isosmotic to A
A is isosmotic to C	B is hyperosmotic to C	C is hyposmotic to B

Osmolarity says nothing about what the particles are or how they behave. Before we can predict whether osmosis will take place between any two solutions divided by a membrane, we must know the properties of the membrane and of the solutes on each side of it.

If the membrane is permeable only to water and not to any solutes, water will move by osmosis from a less concentrated (hyposmotic) solution into a more concentrated (hyperosmotic) solution, as illustrated in Figure 5.2. Most biological systems are not this simple, however. Biological membranes are selectively permeable and allow some solutes to cross in addition to water. To predict the movement of water into and out of cells, you must know the *tonicity* of the solution, explained in the next section.

Tonicity Describes the Volume Change of a Cell

Tonicity {*tonikos*, pertaining to stretching} is a physiological term used to describe a solution and how that solution would affect cell volume if the cell were placed in the solution and allowed to come to equilibrium (■ Tbl. 5.3).

- If a cell placed in the solution gains water at equilibrium and swells, we say that the solution is **hypotonic** to the cell.
- If the cell loses water and shrinks at equilibrium, the solution is said to be **hypertonic**.
- If the cell in the solution does not change size at equilibrium, the solution is **isotonic**.

Tonicity of Solutions — Table 5.3

Solution	Cell Behavior When Placed in the Solution	Description of the Solution Relative to the Cell
A	Cell swells	Solution A is hypotonic.
B	Cell doesn't change size	Solution B is isotonic.
C	Cell shrinks	Solution C is hypertonic.

By convention, we always describe the tonicity of the solution relative to the cell. How, then, does tonicity differ from osmolarity?

1. Osmolarity describes the number of solute particles dissolved in a volume of solution. It has units, such as osmoles/liter. The osmolarity of a solution can be measured by a machine called an *osmometer*. Tonicity has no units; it is only a comparative term.
2. Osmolarity can be used to compare any two solutions, and the relationship is reciprocal (solution A is hyperosmotic to solution B; therefore, solution B is hyposmotic to solution A). Tonicity always compares a solution and a cell, and by convention, tonicity is used to describe only the solution—for example, "Solution A is hypotonic to red blood cells."
3. Osmolarity alone does not tell you what happens to a cell placed in a solution. Tonicity by definition tells you what happens to cell volume at equilibrium when the cell is placed in the solution.

This third point is the one that is most confusing to students. Why can't osmolarity be used to predict tonicity? The reason is that the tonicity of a solution depends not only on its concentration (osmolarity) but also on the *nature* of the solutes in the solution.

By nature of the solutes, we mean whether the solute particles can cross the cell membrane. If the solute particles (ions or molecules) can enter the cell, we call them **penetrating solutes.** We call particles that cannot cross the cell membrane **nonpenetrating solutes.** Tonicity depends on the concentration of nonpenetrating solutes only. Let's see why this is true.

First, some preliminary information. The most important nonpenetrating solute in physiology is NaCl. If a cell is placed in a solution of NaCl, the Na^+ and Cl^- ions do not enter the cell. This makes NaCl a nonpenetrating solute. (In reality, a few Na^+ ions may leak across, but they are immediately transported back to the extracellular fluid by the Na^+-K^+-ATPase. For this reason NaCl is considered a *functionally* nonpenetrating solute.)

By convention, we assume that cells are filled with other types of nonpenetrating solutes. In other words, the solutes inside the cell are unable to leave so long as the cell membrane remains intact. Now we are ready to see why osmolarity alone cannot be used to predict tonicity.

Suppose you know the composition and osmolarity of a solution. How can you figure out the tonicity of the solution without actually putting a cell in it? The key lies in knowing *the relative concentrations of nonpenetrating solutes in the cell and in the solution.* Water will always move until the concentrations of nonpenetrating solutes in the cell and the solution are equal.

Here are the rules for predicting tonicity:

1. *If the cell has a higher concentration of nonpenetrating solutes than the solution,* there will be net movement of water into the cell. The cell swells, and the solution is *hypotonic.*
2. *If the cell has a lower concentration of nonpenetrating solutes than the solution,* there will be net movement of water out of the cell. The cell shrinks, and the solution is *hypertonic.*
3. *If the concentrations of nonpenetrating solutes are the same in the cell and the solution,* there will be no net movement of water at equilibrium. The solution is *isotonic* to the cell.

How does tonicity relate osmolarity? ■ Figure 5.3 shows the possible combinations of osmolarity and tonicity. A hyposmotic solution is always hypotonic, no matter what its composition. The cell will always have a higher concentration of nonpenetrating solutes than the solution, and water will move into the cell (rule 1 above).

An isosmotic solution may be isotonic or hypotonic, but can never be hypertonic because it can never have a higher concentration of nonpenetrating solutes than the cell. If all solutes in the isosmotic solution are nonpenetrating, then the solution is also isotonic. If there are any penetrating solutes in the isosmotic solution, the solution will be hypotonic.

Hyperosmotic solutions may be hypertonic, isotonic, or hypotonic. Their tonicity depends on the relative concentration of nonpenetrating solutes in the solution compared to the cell, as described in the list above.

Normally tonicity is explained using a single cell that is placed into a solution, but here we will use a more physiologically appropriate system: a two-compartment box model that represents the total body divided into ECF and ICF (see Fig. 5.1c). To simplify the calculations, we will use a 3-liter body, with 2 liters in the ICF and 1 liter in the ECF. We assume that the starting osmolarity is 300 mOsM (0.3 OsM) and that solutes in each compartment are nonpenetrating (NP) and cannot move into the other compartment. By defining volumes and concentrations, we can use the equation *solute/volume = concentration* ($S/V = C$) to mathematically determine changes to volumes and osmolarity. *Concentration* is osmolarity.

Always begin by defining the starting conditions. This may be the person's normal state or it may be the altered state that you are trying to return to normal. An example of this would be trying to restore normal volume and osmolarity in a person who has become dehydrated through sweat loss.

■ Figure 5.4 shows the starting conditions for the 3-liter body both as a box diagram and in a table. The table format allows you to deal with an example mathematically if you know the volumes of the body and of the solution added or lost.

	OSMOLARITY		
TONICITY	**Hyposmotic**	**Isosmotic**	**Hyperosmotic**
Hypotonic	√	√	√
Isotonic		√	√
Hypertonic			√

■ **Fig. 5.3** ***The relationship between osmolarity and tonicity***

5

Osmolarity and Tonicity

For all problems, define your starting conditions. Assume that all initial body solutes are nonpenetrating (NP) and will remain in either the ECF or ICF.

Use the equation **Solute / volume = concentration (S/V = C)** to solve the problems. You will know two of the three variables and can calculate the third.

Remember that body compartments are in osmotic equilibrium. Once you know the total body's osmolarity (concentration), you also know the ECF and ICF osmolarity because they are the same.

ECF	ICF
300 mosmol NP	600 mosmol NP
1 L	2 L

Starting Condition

We have a 3-liter body that is 300 mOsM. The ECF is 1 liter and the ICF is 2 liters.

Use **S/V = C** to find out how much solute is in each of the two compartments. Rearrange the equation to solve for S: **S = CV.**

1 S_{ICF} = 300 mosmol/L × 2 L = 600 mosmol NP solute in the ICF

2 S_{ECF} = 300 mosmol/L × 1 L = 300 mosmol NP solute in the ECF

We can also do these calculations using the following table format. This table has been filled in with the values for the starting body. Remember that the ECF + ICF must always equal the total body values, and that once you know the total body osmolarity, you know the ECF and ICF osmolarity.

	Total Body	ECF	ICF
Solute (mosmoles)	900 mosmol	300 mosmol	600 mosmol
Volume (L)	3 L	1 L	2 L
Osmolarity (mOsM)	300 mOsM	300 mOsM	300 mOsM

To see the effect of adding a solution or losing fluid, start with this table and add or subtract volume and solute as appropriate. *You cannot add and subtract concentrations. You must use volumes and solute amounts.*

- Work the total body column first, adding or subtracting solutes and volume. Once you calculate the new total body osmolarity, carry that number across the bottom row to the ECF and ICF columns. (The compartments are in osmotic equilibrium.)
- Distribute nonpenetrating solutes to the appropriate compartment. NaCl stays in the ECF. Glucose goes into the cells. Use **V = S/C** to calculate the new compartment volumes.

In the tables below and on the following page, the yellow boxes indicate the unknowns that must be calculated.

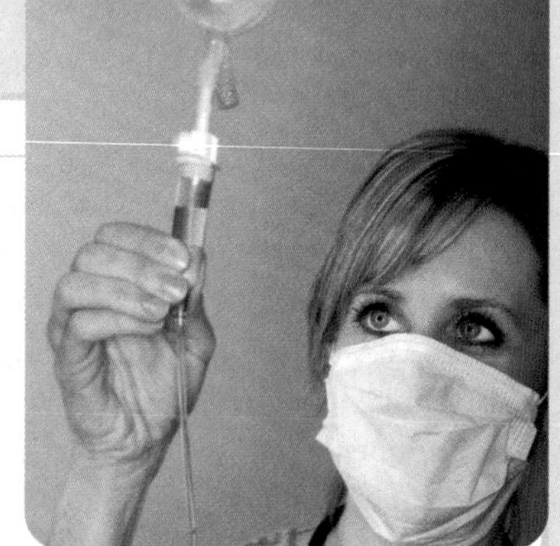

Example 1

Add an IV solution of 1 liter of 300 mOsM NaCl to this body. This solution adds 1 liter of volume and 300 mosmoles of NaCl.

Answer

Work total body first. Add solute and volume, then calculate new osmolarity (yellow box).

	Total Body
Solute (mosmoles)	900 + 300 = 1200 mosmol
Volume (L)	3 + 1 = 4 L
Osmolarity (mOsM)	1200 / 4 = 300 mOsM

Carry the new osmolarity across to the ECF and ICF boxes (arrows). All of the added NaCl will stay in the ECF, so add that solute amount to the ECF box. ICF solute amount is unchanged. Use V = S/C to calculate the new ECF and ICF volumes (yellow boxes).

	Total Body	ECF	ICF
Solute (mosmoles)	1200 mosmol	300 + 300 = 600	600 mosmol
Volume (L)	4 L	2 L	2 L
Osmolarity (mOsM)	300 mOsM →	300 mOsM →	300 mOsM

The added solution was isosmotic (300 mOsM), and its nonpenetrating concentration was the same as that of the body's (300 mOsM NP). You would predict that the solution was isotonic. That is confirmed with these calculations, which show no water entering or leaving the cells (no change in ICF volume).

Example 2

Add 2 liters of a 500 mOsM solution. The solution is equal parts NaCl (nonpenetrating) and urea (penetrating), so it has 250 mosmol/L NaCl and 250 mosmol/L urea.

Answer

This solution has both penetrating and nonpenetrating solutes, but only nonpenetrating solutes contribute to tonicity and cause water to shift between compartments.

Before working this problem, answer the following questions:

(a) This solution is __________ osmotic to the 300 mOsM body.
(b) What is the concentration of nonpenetrating solutes [NP] in the solution? ________________
(c) What is the [NP] in the body? ________
(d) Using the rules for tonicity in Table 5.4, will there be water movement into or out of the cells? If so, in what direction?
(e) Based on your answer in (d), this solution is ________ tonic to this body's cells.

Now work the problem using the starting conditions table as your starting point.
What did you add? 2 L of (250 mosmol/L urea and 250 mosmol/L NaCl) = 2 liters of volume + 500 mosmol urea + 500 mosmol NaCl.
Urea does not contribute to tonicity, so we will set the 500 mosmol of urea aside and add only the volume and NaCl in the first step:

Step 1: Add 2 liters and 500 mosmoles NaCl. Do total body column first.

	Total Body
Solute (mosmoles)	900 + 500 = 1400 mosmol
Volume (L)	3 + 2 = 5 L
Osmolarity (mOsM)	1400/5 = 280 mOsM

Step 2: Carry the new osmolarity across to ECF and ICF. NaCl all remains in the ECF, so add that solute to the ECF column. Calculate new ECF and ICF volumes.

- Notice that IFC volume + ECF volume = total body volume.

	Total Body	ECF	ICF
Solute (mosmoles)	1400 mosmol	300 + 500 = 800	600
Volume (L)	5 L	2.857 L	2.143 L
Osmolarity (mOsM)	280 mOsM →	280 mOsM →	280 mOsM

Step 3: Now add the reserved urea solute to the whole body solute to get the final osmolarity. That osmolarity carries over to the ECF and ICF compartments. Urea will distribute itself throughout the body until its concentration everywhere is equal, but it will not cause any water shift between ECF and ICF. So the ECF and ICF volumes remain as they were in step 2.

	Total Body	ECF	ICF
Solute (mosmoles)	1400 + 500 = 1900		
Volume (L)	5 L	2.857 L	2.143 L
Osmolarity (mOsM)	1900/5 = 380 mOsM →	380 mOsM →	380 mOsM

Answer the following questions from the values in the table:

(f) What happened to the body osmolarity after adding the solution? ____________
This result means the added solution was __________osmotic to the body's starting osmolarity.

(g) What happened to the ICF volume? ________________________
This means the added solution was ___________tonic to the cells.

Compare your answers in (f) and (g) to your answers for (a)–(e). Do they match? They should.

If you know the starting conditions of the body and you know the composition of a solution you are adding, you should be able to describe the solution's osmolarity and tonicity relative to the body by asking the questions in (a)–(e). Now test yourself by working Concept Check questions 8 and 9.

The body's volumes and concentration will change as the result of adding or losing solutes, water, or both—the law of mass balance [p. 11]. Additions to the body normally come through the ingestion of food and drink, but in medical situations solutions can be added directly to the ECF through intravenous (IV) infusions. Significant solute and water loss may occur with sweating, vomiting and diarrhea, or blood loss.

Once you have defined the starting conditions, you add or subtract volume and solutes to find the body's new osmolarity. The final step is to determine whether the ECF and ICF volumes will change as a result of the water and solute gain or loss. In this last step, you must separate the added solutes into penetrating solutes and nonpenetrating solutes.

In our examples, we use three solutes: NaCl, urea, and glucose. NaCl is considered nonpenetrating. Any NaCl added to the body remains in the ECF. Urea is freely penetrating and behaves as if the cell membranes dividing the ECF and ICF do not exist. An added load of urea distributes itself until the urea concentration is the same throughout the body.

Glucose (also called *dextrose*) is an unusual solute. Like all solutes, it first goes into the ECF. Over time, however, 100% of added glucose will enter the cells. When glucose enters the cells, it is phosphorylated to glucose 6-phosphate (G-6-P) and cannot leave the cell again. So although glucose enters cells, it is not freely penetrating because it stays in the cell and adds to the cell's nonpenetrating solutes.

Giving someone a glucose solution is the same as giving them a slow infusion of pure water because glucose 6-phosphate is the first step in the aerobic metabolism of glucose [p. 113]. The end products of aerobic glucose metabolism are CO_2 and water.

The examples shown in Figure 5.4 walk you through the process of adding and subtracting solutions to the body. Ask the following questions when you are evaluating the effects of a solution on the body:

1. What is the osmolarity of this solution relative to the body? (Tbl. 5.2)
2. What is the tonicity of this solution? (Use Fig. 5.3 to help eliminate possibilities.) To determine tonicity, compare the concentration of the nonpenetrating solutes in the solution to the body concentration. (All body solutes are considered to be nonpenetrating.)

For example, consider a solution that is 300 mOsM—isosmotic to a body that is 300 mOsM. The solution's tonicity depends on the concentration of nonpenetrating solutes in the solution. If the solution is 300 mOsM NaCl, the solution's nonpenetrating solute concentration is equal to that of the body. When the solution mixes with the ECF, the ECF nonpenetrating concentration and osmolarity do not change. No water will enter or leave the cells (the ICF compartment), and the solution is isotonic. You can calculate this for yourself by working through Example 1 in Figure 5.4.

Now suppose the 300 mOsM solution has urea as its only solute. Urea is a penetrating solute, so this solution has no

nonpenetrating solutes. When the 300 mOsM urea solution mixes with the ECF, the added volume of the urea solution dilutes the nonpenetrating solutes of the ECF. ($S/V = C$. The same amount of NP solute in a larger volume means a lower NP concentration.)

Now the nonpenetrating concentration of the ECF is less than 300 mOsM. The cells still have a nonpenetrating solute concentration of 300 mOsM, so water moves into the cells to equalize the nonpenetrating concentrations. (Rule: water moves into the compartment with the higher concentration of NP solutes.) The cells gain water and volume. This means the urea solution is hypotonic, even though it is isosmotic.

Example 2 in Figure 5.4 shows how combining penetrating and nonpenetrating solutes can complicate the situation. This example asks you to describe the solution's osmolarity and tonicity based on its composition before you do the mathematical calculations. This skill is important for clinical situations, when you will not know exact body fluid volumes for the person needing an IV. ■ Table 5.4 lists some rules to help you distinguish between osmolarity and tonicity.

Understanding the difference between osmolarity and tonicity is critical to making good clinical decisions about intravenous (IV) fluid therapy. The choice of IV fluid depends on how the clinician wants the solutes and water to distribute between the extracellular and intracellular fluid compartments. If the problem is dehydrated cells, the appropriate IV solution is hypotonic because the cells need fluid. If the situation requires fluid that remains in the extracellular fluid to replace blood loss, an isotonic IV solution is used. In medicine, the tonicity of a solution is usually the important consideration.

■ Table 5.5 lists some common IV solutions and their approximate osmolarity and tonicity relative to the normal human cell. What about the coconut water described at the start of the chapter? Chemical analysis shows that it is not an ideal IV solution, although it is useful for emergencies. It is isosmotic to human plasma but is hypotonic, with Na^+ concentrations much lower than normal ECF and high concentrations of glucose and fructose, along with amino acids.

Table 5.4 Rules for Osmolarity and Tonicity

1. Assume that all intracellular solutes are nonpenetrating.
2. Compare osmolarities before the cell is exposed to the solution. (At equilibrium, the cell and solution are always isosmotic.)
3. Tonicity of a solution describes the volume change of a cell at equilibrium (Tbl. 5.6).
4. Determine tonicity by comparing nonpenetrating solute concentrations in the cell and the solution. Net water movement is into the compartment with the higher concentration of nonpenetrating solutes.
5. Hyposmotic solutions are always hypotonic.

Table 5.5 Intravenous Solutions

Solution	Also Known As	Osmolarity	Tonicity
0.9% saline*	Normal saline	Isosmotic	Isotonic
5% dextrose† in 0.9% saline	D5–normal saline	Hyperosmotic	Isotonic
5% dextrose in water	D5W	Isosmotic	Hypotonic
0.45% saline	Half-normal saline	Hyposmotic	Hypotonic
5% dextrose in 0.45% saline	D5–half-normal saline	Hyperosmotic	Hypotonic

*Saline = NaCl.
†Dextrose = glucose.

Concept Check

Answers: p. 172

6. Which of the following solutions has/have the most water per unit volume: 1 M glucose, 1 M NaCl, or 1 OsM NaCl?
7. Two compartments are separated by a membrane that is permeable to water and urea but not to NaCl. Which way will water move when the following solutions are placed in the two compartments? (Hint: Watch the units!)

Compartment A	Membrane	Compartment B
(a) 1 M NaCl	|	1 OsM NaCl
(b) 1 M urea	|	2 M urea
(c) 1 OsM NaCl	|	1 OsM urea

8. Use the same 3-liter, 300 mOsM body as in Figure 5.4 for this problem.

   Add 1 liter of 260 mOsM glucose to the body and calculate the new body volumes and osmolarity once all the glucose has entered the cells and been phosphorylated. Before you do the calculations, make the following predictions: This solution is __________ osmotic to the body and is __________ tonic to the body's cells.
9. Use the same 3-liter, 300 mOsM body as in Figure 5.4 for this problem. A 3-liter person working in the hot sun loses 500 mL of sweat that is equivalent to a 130 mOsM NaCl solution. Assume all NaCl loss comes from the ECF.
   (a) The sweat lost is __________ osmotic to the body. This means that the osmolarity of the body after the sweat loss will (*increase/decrease/not change?*).
   (b) As a result of this sweat loss, the body's cell volume will (*increase/decrease/not change?*).
   (c) Using the table, calculate what happens to volume and osmolarity as a result of this sweat loss. Do the results of your calculations match your answers in (a) and (b)?
10. You have a patient who lost 1 liter of blood, and you need to restore volume quickly while waiting for a blood transfusion to arrive from the blood bank.
    (a) Which would be better to administer: 5% dextrose in water or 0.9% NaCl in water? (*Hint:* Think about how these solutes distribute in the body.) Defend your choice.
    (b) How much of your solution of choice would you have to administer to return blood volume to normal?

Transport Processes

Humans are large complex organisms, and the movement of material within and between body compartments is necessary for communication. This movement requires a variety of transport mechanisms. Some require an outside source of energy, such as that stored in the high-energy bond of ATP [p. 111], while other transport processes use only the kinetic or potential energy already in the system [p. 101]. Movement between compartments usually means a molecule must cross one or more cell membranes. Movement within a compartment is less restricted. For this reason, biological transport is another theme that you will encounter repeatedly as you study the organ systems.

The most general form of biological transport is the **bulk flow** of fluids within a compartment. Although many people equate **fluids** with liquids, in physics both gases and liquids are considered fluids because they flow. The main difference between the two fluids is that gases are compressible because their molecules are so far apart in space. Liquids, especially water, are not compressible. (Think of squeezing on a water balloon.)

In bulk flow, a *pressure gradient* causes fluid to flow from regions of higher pressure to regions of lower pressure. As the fluid flows, it carries with it all of its component parts, including substances dissolved or suspended in it. Blood moving through the circulatory system is an excellent example of bulk flow. The heart acts as a pump that creates a region of high pressure, pushing plasma with its dissolved solutes and the suspended blood cells through the blood vessels. Air flow in the lungs is another example of bulk flow that you will encounter as you study physiology.

Other forms of transport are more specific than bulk flow. When we discuss them, we must name the molecule or molecules that are moving. Transport mechanisms you will learn about in the following sections include diffusion, protein-mediated transport, and vesicular transport.

Cell Membranes Are Selectively Permeable

Although many materials move freely within a body compartment, exchange between the intracellular and extracellular compartments is restricted by the cell membrane. Whether or not a substance enters a cell depends on the properties of the cell membrane and those of the substance. Cell membranes are **selectively permeable**, which means that some molecules can cross them but others cannot.

The lipid and protein composition of a given cell membrane determines which molecules will enter the cell and which will leave [p. 66]. If a membrane allows a substance to pass through it, the membrane is said to be **permeable** to that substance {*permeare,* to pass through}. If a membrane does not allow a substance to pass, the membrane is said to be **impermeable** {*im*-, not} to that substance.

Membrane permeability is variable and can be changed by altering the proteins or lipids of the membrane. Some molecules, such as oxygen, carbon dioxide, and lipids, move easily across most cell membranes. On the other hand, ions, most polar molecules, and very large molecules (such as proteins), enter cells with more difficulty or may not enter at all.

Two properties of a molecule influence its movement across cell membranes: the size of the molecule and its lipid solubility. Very small molecules and those that are lipid soluble can cross directly through the phospholipid bilayer. Larger and less lipid-soluble molecules usually do not enter or leave a cell unless the cell has specific membrane proteins to transport these molecules across the lipid bilayer. Very large lipophobic molecules cannot be transported on proteins and must enter and leave cells in vesicles [p. 75].

There are two ways to categorize how molecules move across membranes. One scheme, just described, separates movement according to physical requirements: whether it moves directly through the phospholipid bilayer, crosses with the aid of a membrane protein, or enters the cell in a vesicle (bottom of ■ Fig. 5.5). A second scheme classifies movement according to its energy requirements (top of Fig. 5.5). **Passive transport** does not require the input of energy. **Active transport** requires the input of energy from some outside source, such as the high-energy phosphate bond of ATP.

The following sections look at how cells move material across their membranes. The principles discussed here also apply to movement across intracellular membranes, when substances move between organelles.

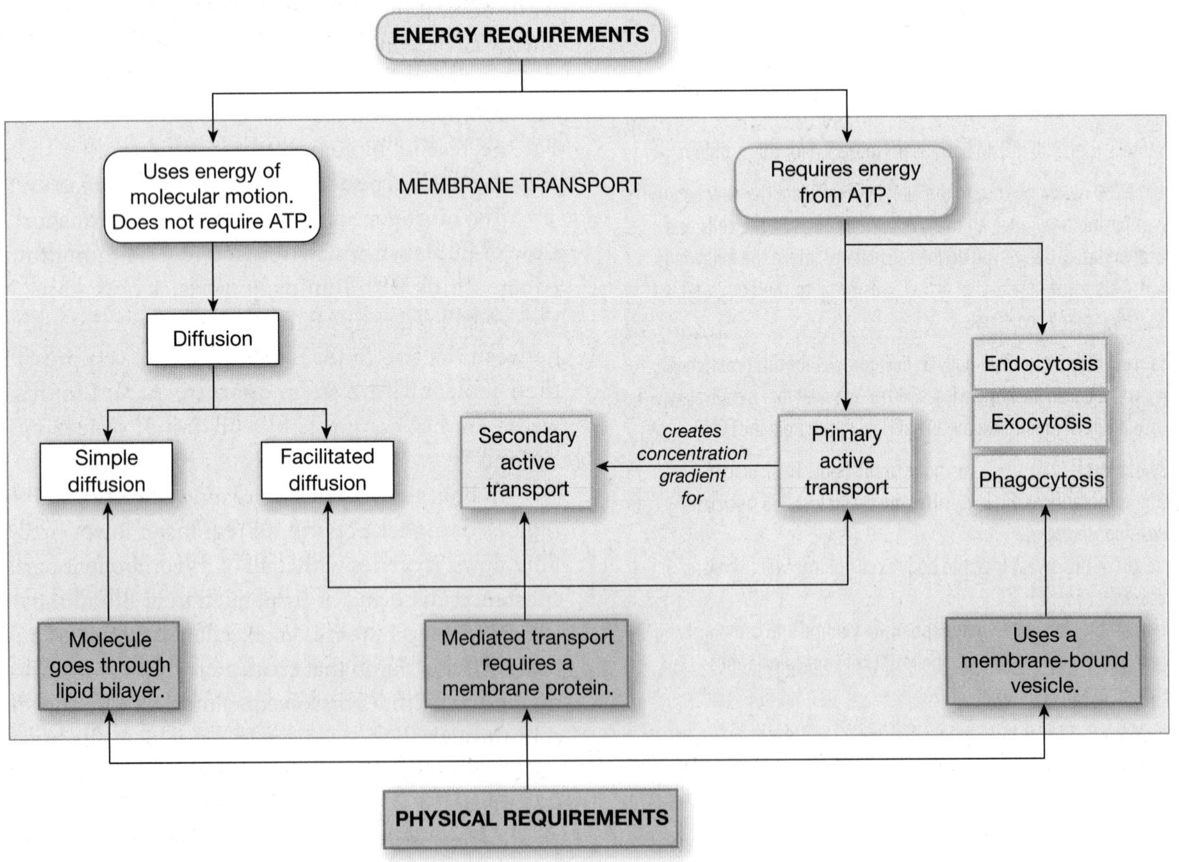

■ **Fig. 5.5** ***Map of membrane transport.*** Movement of substances across cell membranes can be classified either by the energy requirements of transport (top part of map) or according to whether transport occurs by diffusion, a membrane protein, or a vesicle (bottom part of map).

RUNNING PROBLEM

Daniel's medical history tells a frightening story of almost constant medical problems since birth: recurring bouts of respiratory infections, digestive ailments, and, for the past six months, a history of weight loss. Then, last week, when Daniel began having trouble breathing, his mother rushed him to the hospital. A culture taken from Daniel's lungs raised a red flag for cystic fibrosis: the mucus from his airways was unusually thick and dehydrated. In cystic fibrosis, this thick mucus causes life-threatening respiratory congestion and provides a perfect breeding ground for infection-causing bacteria.

Q1: In people with cystic fibrosis, movement of sodium chloride into the lumen of the airways is impaired. Why would failure to move NaCl into the airways cause the secreted mucus to be thick? (*Hint:* Remember that water moves into hyperosmotic regions.)

130 141 148 158 161 168

Diffusion

Passive transport across membranes uses the kinetic energy [p. 101] inherent in molecules. Gas molecules and molecules in solution constantly move from one place to another, bouncing off other molecules or off the sides of any container holding them. When molecules start out concentrated in one area of an enclosed space, their motion causes them to spread out gradually until they distribute evenly throughout the available space. This process is known as diffusion.

Diffusion {*diffundere,* to pour out} may be defined as the movement of molecules from an area of higher concentration of the molecules to an area of lower concentration of the molecules.* If you leave a bottle of cologne open and later notice its fragrance across the room, it is because the aromatic molecules in the cologne have diffused from where they are more concentrated (in the bottle) to where they are less concentrated (across the room).

Diffusion has the following seven properties:

1 *Diffusion is a passive process.* By *passive,* we mean that diffusion does not require the input of energy from some outside source. Diffusion uses only the kinetic energy possessed by all molecules.

2 *Molecules move from an area of higher concentration to an area of lower concentration.* A difference in the concentration of a substance between two places is called a concentration gradient, also known as a **chemical gradient**. We say that molecules diffuse *down the gradient,* from higher concentration to lower concentration.

The rate of diffusion depends on the magnitude of the concentration gradient. The larger the concentration gradient, the faster diffusion takes place. For example, when you open a bottle of cologne, the rate of diffusion is most rapid as the molecules first escape from the bottle into the air. Later, when the cologne has spread evenly throughout the room, the rate of diffusion has dropped to zero because there is no longer a concentration gradient.

3 *Net movement of molecules occurs until the concentration is equal everywhere.* Once molecules of a given substance have distributed themselves evenly, the system reaches equilibrium and diffusion stops. Individual molecules are still moving at equilibrium, but for each molecule that exits an area, another one enters. The *dynamic equilibrium* state in diffusion means that the concentration has equalized throughout the system but molecules continue to move.

4 *Diffusion is rapid over short distances but much slower over long distances.* Albert Einstein studied the diffusion of molecules in solution and found that the time required for a molecule to diffuse from point A to point B is proportional to the square of the distance from A to B. In other words, if the distance doubles from 1 to 2, the time needed for diffusion increases from 1^2 to 2^2 (from 1 to 4).

What does the slow rate of diffusion over long distances mean for biological systems? In humans, nutrients take five seconds to diffuse from the blood to a cell that is 100 μm from the nearest capillary. At that rate, it would take years for nutrients to diffuse from the small intestine to cells in the big toe, and the cells would starve to death.

To overcome the limitations of diffusion over distance, organisms use various transport mechanisms that speed up the movement of molecules. Most multicellular organisms have some form of circulatory system to bring oxygen and nutrients rapidly from the point at which they enter the body to the cells.

5 *Diffusion is directly related to temperature.* At higher temperatures, molecules move faster. Because diffusion results from molecular movement, the rate of diffusion increases as temperature increases. Generally, changes in temperature do not significantly affect diffusion rates in humans because we maintain a relatively constant body temperature.

6 *Diffusion rate is inversely related to molecular weight and size.* Smaller molecules require less energy to move over a distance and therefore diffuse faster. Einstein showed that friction between the surface of a particle and the medium through which it diffuses is a source of resistance to movement. He calculated that diffusion is inversely proportional to the radius of the molecule: the larger the molecule, the slower its diffusion through a given medium. The experiment in ■ Figure 5.6 shows that the smaller and lighter potassium iodide (KI)

*Some texts use the term *diffusion* to mean any random movement of molecules, and they call molecular movement along a concentration gradient *net diffusion.* To simplify matters, we will use the term *diffusion* to mean movement down a concentration gradient.

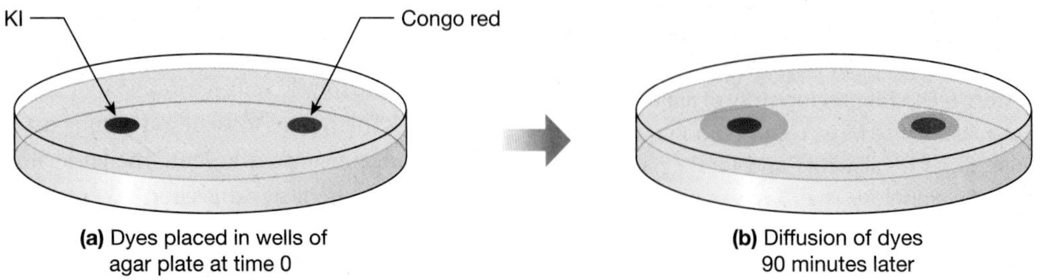

KI
Congo red

(a) Dyes placed in wells of agar plate at time 0

(b) Diffusion of dyes 90 minutes later

Fig. 5.6 ***Diffusion experiment.*** (a) Wells in an agar gel plate are filled with two dyes of equal concentration: potassium iodide (KI, 166 daltons) and Congo red (697 daltons). (b) Ninety minutes later, the smaller and lighter KI has diffused through the gel to stain a larger area.

molecules diffuse more rapidly through the agar gel than the larger and heavier Congo red molecules.

7 *Diffusion can take place in an open system or across a partition that separates two compartments.* Diffusion of cologne within a room is an example of diffusion taking place in an open system. There are no barriers to molecular movement, and the molecules spread out to fill the entire system. Diffusion can also take place between two compartments, such as the intracellular and extracellular compartments, but only if the partition dividing the two compartments allows the diffusing molecules to cross.

For example, if you close the top of an open bottle of cologne, the molecules cannot diffuse out into the room because neither the bottle nor the cap is permeable to the cologne. However, if you replace the metal cap with a plastic bag that has tiny holes in it, you will begin to smell the cologne in the room because the bag is permeable to the molecules. Similarly, if a cell membrane is permeable to a molecule, that molecule can enter or leave the cell by diffusion. If the membrane is not permeable to that particular molecule, the molecule cannot cross.

Table 5.6 summarizes these points.

An important point to note: ions do not move by diffusion, even though you will read and hear about ions *diffusing across membranes*. Diffusion is random molecular motion down a *concentration* gradient. Ion movement is influenced by *electrical* gradients because of the attraction of opposite charges and repulsion of like charges. For this reason, ions move in response to combined electrical and concentration gradients, or *electrochemical gradients*. This electrochemical movement is a more complex process than diffusion resulting solely from a concentration gradient, and the two processes should not be confused. We discuss ions and electrochemical gradients in more detail at the end of this chapter.

In summary, diffusion is the passive movement of uncharged molecules down their concentration gradient due to random molecular movement. Diffusion is slower over long distances and slower for large molecules. When the concentration of the diffusing molecules is the same throughout a system, the system has come to chemical equilibrium, although the random movement of molecules continues.

Concept Check Answer: p. 172

11. If the distance over which a molecule must diffuse triples from 1 to 3, diffusion takes how many times as long?

Lipophilic Molecules Cross Membranes by Simple Diffusion

Diffusion across membranes is a little more complicated than diffusion in an open system. Only lipid-soluble (lipophilic) molecules can pass through the phospholipid bilayer. Water and the many vital nutrients, ions, and other molecules that dissolve in water are lipo*phobic* as a rule: they do not readily dissolve in lipids. For these substances, the hydrophobic lipid core of the cell membrane acts as a barrier that prevents them from crossing.

Lipophilic substances that can pass through the lipid center of a membrane move by diffusion. Diffusion directly across the phospholipid bilayer of a membrane is called **simple diffusion** and has the following properties in addition to the properties of diffusion listed earlier.

1 *The rate of diffusion depends on the ability of the diffusing molecule to dissolve in the lipid layer of the membrane.* Another way to say this is that the diffusion rate depends on how permeable the membrane is to the diffusing molecules. Most molecules in solution can mingle with the polar phosphate-glycerol heads of the bilayer [p. 33], but only nonpolar molecules that are lipid-soluble (lipophilic) can dissolve in the central lipid core of the membrane. As a rule, only lipids, steroids, and small lipophilic molecules can move across membranes by simple diffusion.

One important exception to this statement concerns water. Water, although a polar molecule, may diffuse slowly across some phospholipid membranes. For years it was thought that the polar nature of the water molecule prevented it from moving through the lipid center of the bilayer, but experiments done with artificial membranes have shown that the small size of the water molecule allows it to slip between the lipid tails in some membranes.

Table 5.6 Rules for Diffusion of Uncharged Molecules

General Properties of Diffusion
1. Diffusion uses the kinetic energy of molecular movement and does not require an outside energy source.
2. Molecules diffuse from an area of higher concentration to an area of lower concentration.
3. Diffusion continues until concentrations come to equilibrium. Molecular movement continues, however, after equilibrium has been reached.
4. Diffusion is faster —along higher concentration gradients. —over shorter distances. —at higher temperatures. —for smaller molecules.
5. Diffusion can take place in an open system or across a partition that separates two systems.
Simple Diffusion Across a Membrane
6. The rate of diffusion through a membrane is faster if —the membrane's surface area is larger. —the membrane is thinner. —the concentration gradient is larger. —the membrane is more permeable to the molecule.
7. Membrane permeability to a molecule depends on —the molecule's lipid solubility. —the molecule's size. —the lipid composition of the membrane.

How readily water passes through the membrane depends on the composition of the phospholipid bilayer. Membranes with high cholesterol content are less permeable to water than those with low cholesterol content, presumably because the lipid-soluble cholesterol molecules fill spaces between the fatty acid tails of the lipid bilayer and thus exclude water. For example, the cell membranes of some sections of the kidney are essentially impermeable to water unless the cells insert special water channel proteins into the phospholipid bilayer. Most water movement across membranes takes place through protein channels.

2 *The rate of diffusion across a membrane is directly proportional to the surface area of the membrane.* In other words, the larger the membrane's surface area, the more molecules can diffuse across per unit time. This fact may seem obvious, but it has important implications in physiology. One striking example of how a change in surface area affects diffusion is the lung disease emphysema. As lung tissue breaks down and is destroyed, the surface area available for diffusion of oxygen decreases. Consequently, less oxygen can move into the body. In severe cases, the oxygen that reaches the cells is not enough to sustain any muscular activity and the patient is confined to bed.

The rules for simple diffusion across membranes are summarized in Table 5.6. They can be combined mathematically into an equation known as **Fick's law of diffusion**, a relationship that involves the factors just mentioned for diffusion across membranes plus the factor of concentration gradient. In an abbreviated form, Fick's law says that the diffusion rate increases when surface area, the concentration gradient, or the membrane permeability increase:

$$\text{rate of diffusion} \propto \text{surface area} \times \text{concentration gradient} \times \text{membrane permeability}$$

■ Figure 5.7 illustrates the principles of Fick's law.

Membrane permeability is the most complex of the terms in Fick's law because several factors influence it:

1 the size (and shape, for large molecules) of the diffusing molecule. As molecular size increases, membrane permeability decreases.

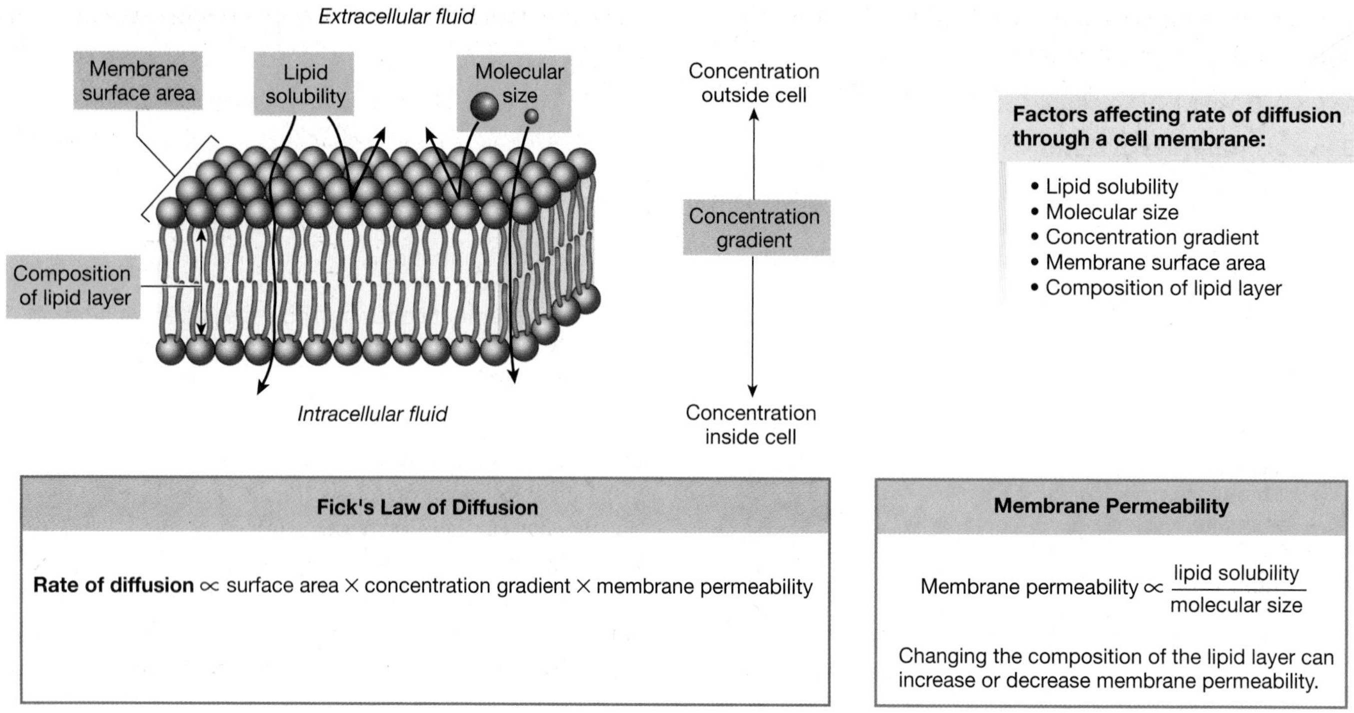

■ **Fig. 5.7** ***Fick's law of diffusion.*** This law mathematically relates factors that influence the rate of simple diffusion across a membrane.

2 the lipid-solubility of the molecule. As lipid solubility of the diffusing molecule increases, membrane permeability to the molecule increases.

3 the composition of the lipid bilayer across which it is diffusing. Alterations in lipid composition of the membrane change how easily diffusing molecules can slip between the individual phospholipids. For example, cholesterol molecules in membranes pack themselves into the spaces between the fatty acids tails and retard passage of molecules through those spaces [Fig. 3.2, p. 67], making the membrane less permeable.

We can rearrange the Fick equation to read:

$$\frac{\text{diffusion rate}}{\text{surface area}} = \text{concentration gradient} \times \text{membrane permeability}$$

This equation now describes the flux of a molecule across the membrane, because **flux** is defined as the diffusion rate per unit surface area of membrane:

$$\text{flux} = \text{concentration gradient} \times \text{membrane permeability}$$

In other words, the flux of a molecule across a membrane depends on the concentration gradient and the membrane's permeability to the molecule.

Remember that the principles of diffusion apply to all biological membranes, not just to the cell membrane. Diffusion of materials in and out of organelles follows the same rules.

Concept Check Answers: pp. 172–173

12. Where does the energy for diffusion come from?
13. Which is more likely to cross a cell membrane by simple diffusion: a fatty acid molecule or a glucose molecule?
14. What happens to the flux of molecules in each of the following cases?
    (a) Molecular size increases.
    (b) Concentration gradient increases.
    (c) Surface area of membrane decreases.
15. Two compartments are separated by a membrane that is permeable only to water and to yellow dye molecules. Compartment A is filled with an aqueous solution of yellow dye, and compartment B is filled with an aqueous solution of an equal concentration of blue dye. If the system is left undisturbed for a long time, what color will compartment A be: yellow, blue, or green? (Remember, yellow plus blue makes green.) What color will compartment B be?
16. What keeps atmospheric oxygen from diffusing into our bodies across the skin? (*Hint:* What kind of epithelium is skin?)

Protein-Mediated Transport

In the body, simple diffusion across membranes is limited to lipophilic molecules. The majority of molecules in the body are either lipophobic or electrically charged and therefore cannot cross membranes by simple diffusion. Instead, the vast majority of solutes cross membranes with the help of membrane proteins, a process we call **mediated transport**.

If mediated transport is passive and moves molecules down their concentration gradient, and if net transport stops when concentrations are equal on both sides of the membrane, the process is known as **facilitated diffusion** (Fig. 5.5). If protein-mediated transport requires energy from ATP or another outside source and moves a substance against its concentration gradient, the process is known as **active transport**.

Membrane Proteins Have Four Major Functions

Protein-mediated transport across a membrane is carried out by membrane-spanning proteins known as *transporters*. For physiologists, classifying membrane proteins by their function is more useful than classifying them by their structure. Our functional classification scheme recognizes four broad categories of membrane proteins: (1) structural proteins, (2) enzymes, (3) receptors, and (4) transporters. ■ Figure 5.8 is a map comparing the structural and functional classifications of membrane proteins. These groupings are not completely distinct, and as you will learn, some membrane proteins have more than one function, such as receptor-channels and receptor-enzymes.

Structural Proteins The **structural proteins** have three major roles.

1. They connect the membrane to the cytoskeleton to maintain the shape of the cell [Fig. 3.4, p. 70]. The microvilli of transporting epithelia are one example of membrane shaping by the cytoskeleton [Fig. 3.10e, p. 83].
2. They create cell junctions that hold tissues together, such as tight junctions and gap junctions [Fig. 3.8, p. 79].
3. They attach cells to the extracellular matrix by linking cytoskeleton fibers to extracellular collagen and other protein fibers [p. 78].

Enzymes **Membrane enzymes** catalyze chemical reactions that take place either on the cell's external surface or just inside the cell. For example, enzymes on the external surface of cells lining the small intestine are responsible for digesting peptides and carbohydrates. Enzymes attached to the intracellular surface of many cell membranes play an important role in transferring signals from the extracellular environment to the cytoplasm [see Chapter 6].

Receptors **Membrane receptor proteins** are part of the body's chemical signaling system. The binding of a receptor with its ligand usually triggers another event at the membrane

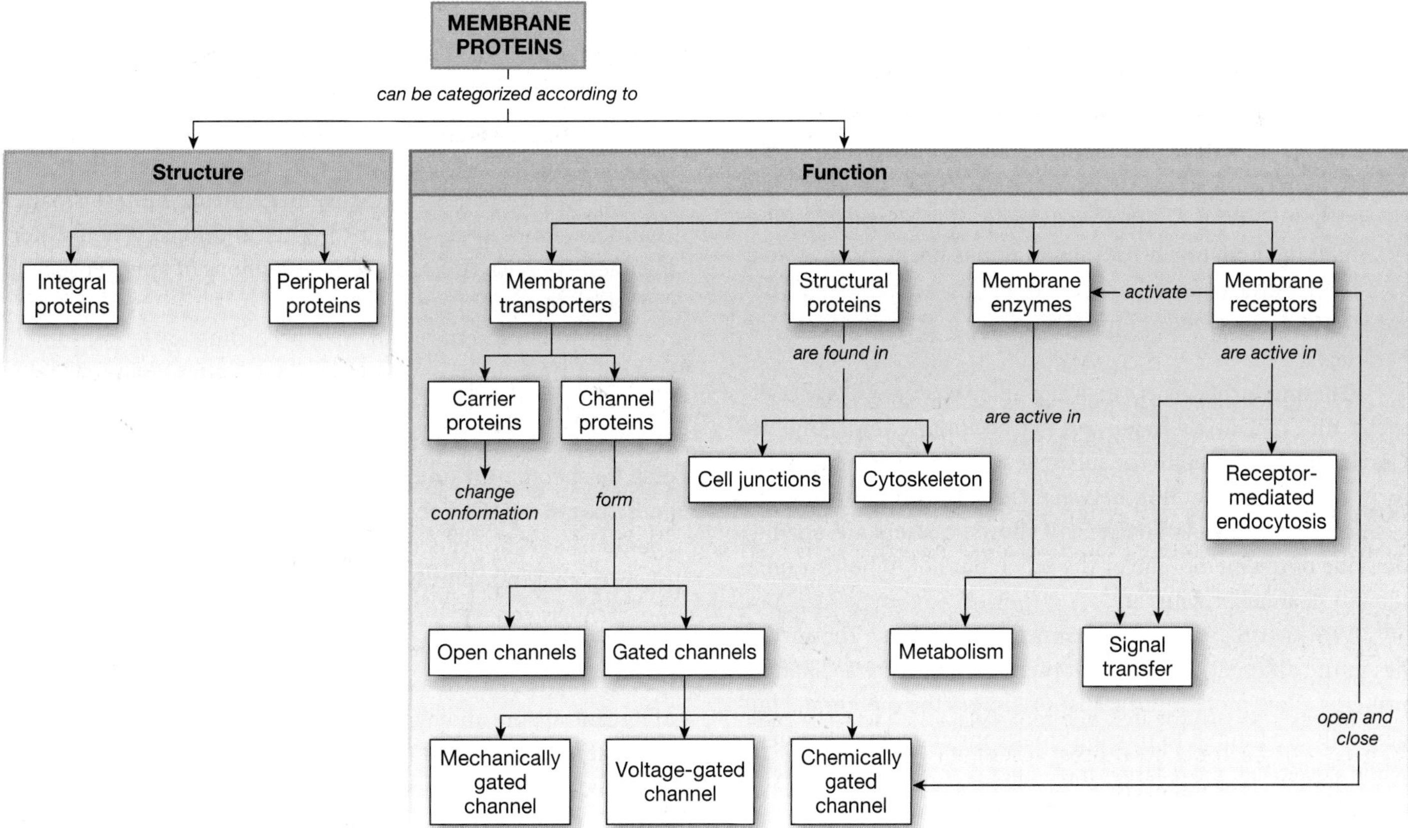

■ **Fig. 5.8** ***Map of membrane proteins.*** Functional categories of membrane proteins include transporters, structural proteins, enzymes, and receptors.

5

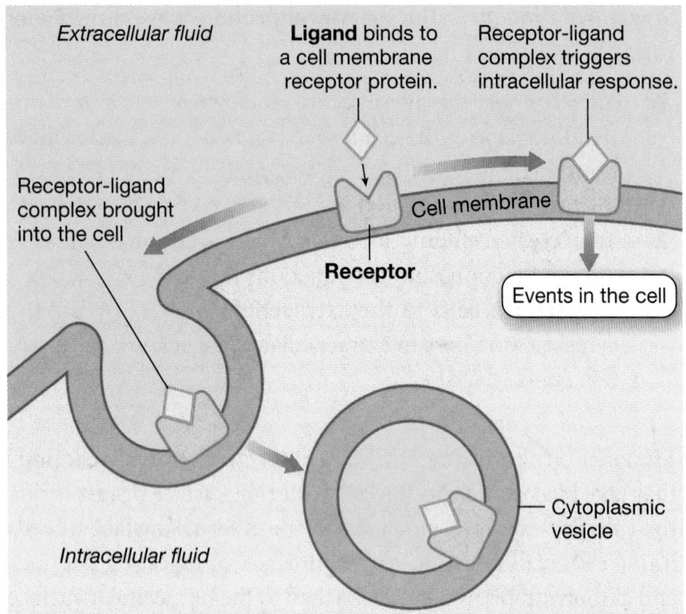

Fig. 5.9 ***Membrane receptor proteins bind extracellular ligands***

(Fig. 5.9). Sometimes the ligand remains on the cell surface, and the receptor-ligand complex triggers an intracellular response. In other instances, the receptor-ligand complex is brought into the cell in a vesicle [p. 75]. Membrane receptors also play an important role in some forms of vesicular transport, as you will learn later in this chapter.

Transporters The fourth group of membrane proteins—**transporters**—moves molecules across membranes (Fig. 5.10). The current classification scheme* recognizes two main types of transporters: channels and carriers. There is overlap between the two types, both structurally and functionally. For example, the aquaporin protein AQP has been shown to act as both a water channel and a carrier for certain small organic molecules (see Emerging Concepts: Transporter Gene Families).

Channel proteins create water-filled passageways that directly link the intracellular and extracellular compartments. **Carrier proteins** bind to the substrates that they carry but never form a direct connection between the intracellular fluid and extracellular fluid. As Figure 5.10 shows, carriers are open to one side of the membrane or the other, but not to both at once the way channel proteins are.

Why do cells need both channels and carriers? The answer lies in the different properties of the two transporters. Channel proteins allow more rapid transport across the membrane but generally are limited to moving small ions and water. Carriers, while slower, can move larger molecules than channels can.

*The Transporter Classification System, *www.tcdb.org*

EMERGING CONCEPTS

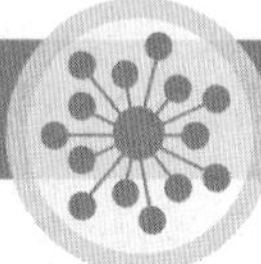

Transporter Gene Families

One outcome of the Human Genome project has been the recognition that many proteins are closely related to each other, both within and across species. As a result, scientists have discovered that most membrane transporters for organic solutes belong to one of two gene "superfamilies": the ATP-binding cassette (ABC) superfamily or the solute carrier (SLC) superfamily. The ABC family transporters use ATP's energy to transport small organic molecules across membranes. Interestingly, the CFTR chloride channel, which causes cystic fibrosis when absent or defective, is also a member of the ABC family and is the only known ion channel in that superfamily. The 43 families of the SLC superfamily family include most facilitated diffusion carriers, such as the GLUT sugar transporters, as well as the secondary active transporters shown in Table 5.8.

Channel Proteins Form Open, Water-Filled Passageways

Channel proteins are made of membrane-spanning protein subunits that create a cluster of cylinders with a tunnel or *pore* through the center. Nuclear pore complexes [p. 76] and gap junctions [Fig. 3.8b, p. 79] can be considered very large forms of channels. In this book we restrict use of the term "channel" to smaller channels whose centers are narrow, water-filled pores (Fig. 5.11 on page 148). Movement through these smaller channels is mostly restricted to water and ions. When water-filled ion channels are open, tens of millions of ions per second can whisk through them unimpeded.

Channel proteins are named according to the substances that they allow to pass. Most cells have **water channels** made from a protein called *aquaporin*. In addition, more than 100 types of **ion channels** have been identified. Ion channels may be specific for one ion or may allow ions of similar size and charge to pass. For example, there are Na^+ channels, K^+ channels, and nonspecific *monovalent* ("one-charge") cation channels that transport Na^+, K^+, and lithium ions (Li^+). Other ion channels you will encounter frequently in this text are Ca^{2+} channels and Cl^- channels.

The selectivity of a channel is determined by the diameter of its central pore and by the electrical charge of the amino acids that line the channel. If the channel amino acids are positively charged, positive ions are repelled and negative ions can pass through the channel. On the other hand, a cation channel must have a negative charge that attracts cations but prevents the passage of Cl^- or other anions.

■ Fig. 5.10 *ESSENTIALS*

Membrane Transporters

Membrane transporters are membrane-spanning proteins that help move lipophobic molecules across membranes.

MEMBRANE TRANSPORTERS
(a) Channel proteins create a water-filled pore.
ECF
Cell membrane
ICF
can be classified
Gated channels open and close in response to signals.
Open
Closed
Open channels or pores are usually open.
(b) Carrier proteins never form an open channel between the two sides of the membrane.
Carrier open to ICF
Same carrier open to ECF
can be classified
Cotransporters
Uniport carriers transport only one kind of substrate.
Symport carriers move two or more substrates in the same direction across the membrane.
Antiport carriers move substrates in opposite directions.
Glu
Na+
Glu
Na+
ATP
K+
ATP

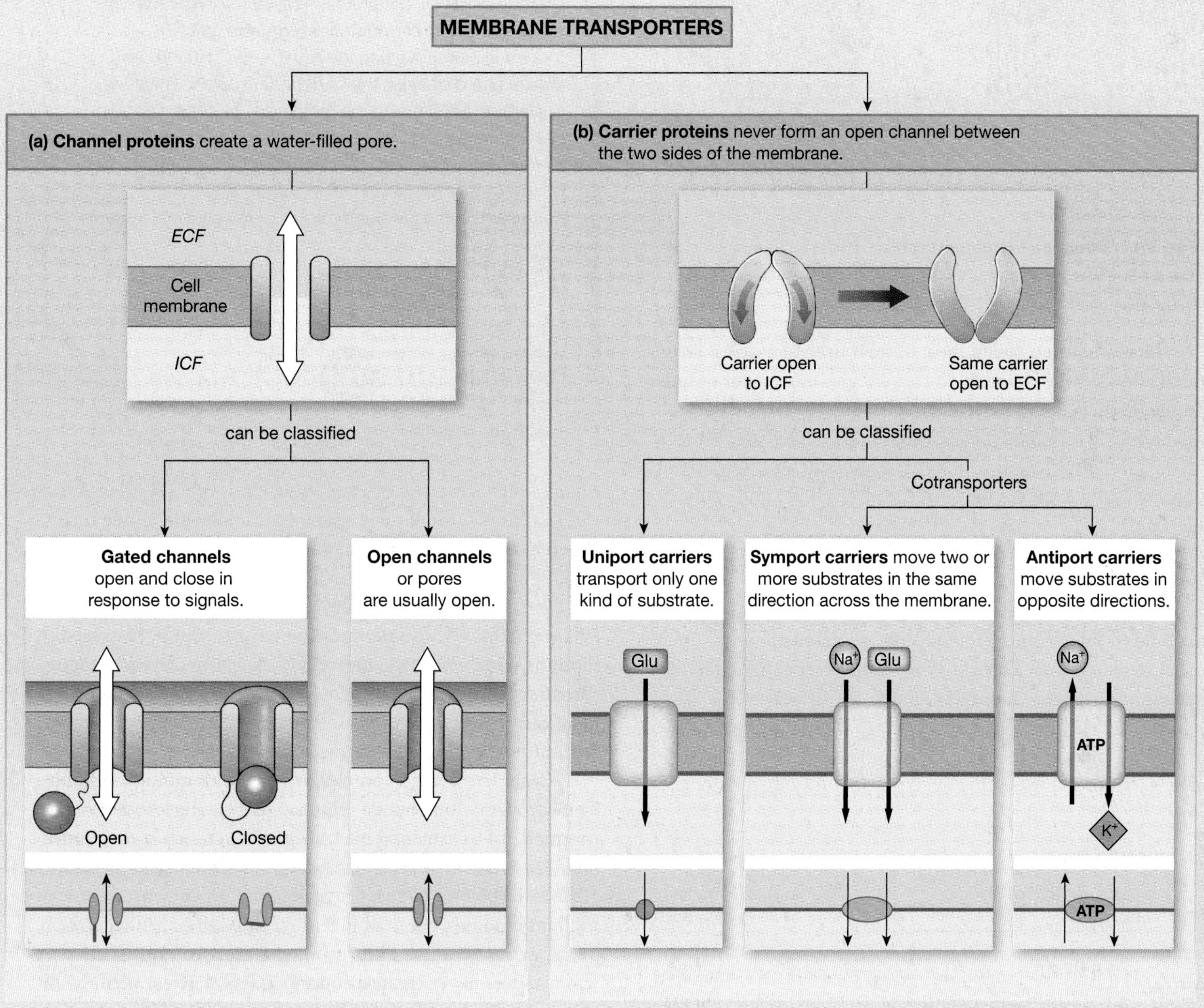

Close-up views of transporters are shown in the top two rows and distant views in the bottom row. Primary active transport is indicated by *ATP* on the protein.

Channel proteins are like narrow doorways into the cell. If the door is closed, nothing can go through. If the door is open, there is a continuous passage between the two rooms connected by the doorway. The open or closed state of a channel is determined by regions of the protein molecule that act like swinging "gates."

According to current models, channel "gates" take several forms. Some channel proteins have gates in the middle of the protein's pore. Other gates are part of the cytoplasmic side of the membrane protein. Such a gate can be envisioned as a ball on a chain that swings up and blocks the mouth of the channel (Fig. 5.10a). One type of channel in neurons has two different gates.

Channels can be classified according to whether their gates are usually open or usually closed. **Open channels** spend most of their time with their gate open, allowing ions to move back and forth across the membrane without regulation. These gates may occasionally flicker closed, but for the most part these channels behave as if they have no gates. Open channels are sometimes called either *leak channels* or pores, as in *water pores*.

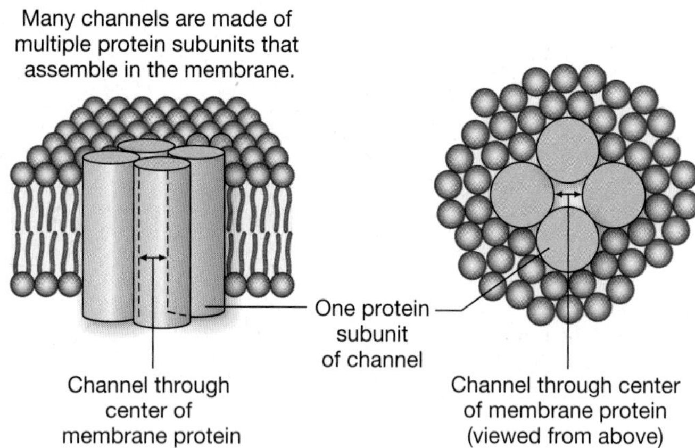

■ **Fig. 5.11** ***Structure of channel proteins.*** Hydrophilic amino acids in the protein line the channel, creating a water-filled passage that allows ions and very small molecules, such as water, to pass through.

Gated channels spend most of their time in a closed state, which allows these channels to regulate the movement of ions through them. When a gated channel opens, ions move through the channel just as they move through open channels. When a gated channel is closed, which it may be much of the time, it allows no ion movement between the intracellular and extracellular fluid.

What controls the opening and closing of gated channels? For **chemically gated channels,** the gating is controlled by intracellular messenger molecules or extracellular ligands that bind to the channel protein. **Voltage-gated channels** open and close when the electrical state of the cell changes. Finally, **mechanically gated channels** respond to physical forces, such as increased temperature or pressure that puts tension on the membrane and pops the channel gate open. You will encounter many variations of these channel types as you study physiology.

Concept Check Answer: p. 173

17. Positively charged ions are called ________, and negatively charged ions are called ________.

Carrier Proteins Change Conformation to Move Molecules

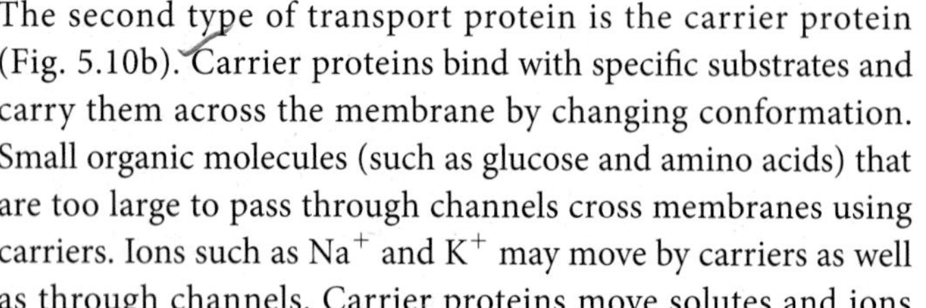

The second type of transport protein is the carrier protein (Fig. 5.10b). Carrier proteins bind with specific substrates and carry them across the membrane by changing conformation. Small organic molecules (such as glucose and amino acids) that are too large to pass through channels cross membranes using carriers. Ions such as Na^+ and K^+ may move by carriers as well as through channels. Carrier proteins move solutes and ions into and out of cells as well as into and out of intracellular organelles, such as the mitochondria.

Some carrier proteins move only one kind of molecule and are known as **uniport carriers**. However, it is common to find carriers that move two or even three kinds of molecules. A carrier that moves more than one kind of molecule at one time is called a **cotransporter**. If the molecules being transported are moving in the same direction, whether into or out of the cell, the carrier proteins are **symport carriers** {*sym*-, together + *portare,* to carry}. (Sometimes the term *cotransport* is used in place of *symport.*) If the molecules are being carried in opposite directions, the carrier proteins are **antiport carriers** {*anti,* opposite + *portare,* to carry}, also called *exchangers.* Symport and antiport carriers are shown in Figure 5.10b.

Carriers are large, complex proteins with multiple subunits. The conformation change required of a carrier protein makes this mode of transmembrane transport much slower than movement through channel proteins. A carrier protein can move only 1000 to 1,000,000 molecules per second, in contrast to tens of millions of ions per second that move through a channel protein.

Carrier proteins differ from channel proteins in another way: carriers never create a continuous passage between the inside and outside of the cell. If channels are like doorways, then carriers are like revolving doors that allow movement between inside and outside without ever creating an open hole. Carrier proteins can transport molecules across a membrane in both directions, like a revolving door at a hotel, or they can restrict their transport to one direction, like the turnstile at an amusement park that allows you out of the park but not back in.

One side of the carrier protein always creates a barrier that prevents free exchange across the membrane. In this respect, carrier proteins function like the Panama Canal (■ Fig. 5.12). Picture the canal with only two gates, one on the Atlantic side and one on the Pacific side. Only one gate at a time is open.

When the Atlantic gate is closed, the canal opens into the Pacific. A ship enters the canal from the Pacific, and the gate

RUNNING PROBLEM

Cystic fibrosis is a debilitating disease caused by a defect in a membrane channel protein that normally transports chloride ions (Cl^-). The protein—called the **cystic fibrosis transmembrane conductance regulator,** or CFTR—is located in epithelia lining the airways, sweat glands, and pancreas. A channel in the CFTR protein opens when the nucleotide ATP binds to the protein. In the lungs, this open channel transports Cl^- out of the epithelial cells and into the airways. In people with cystic fibrosis, CFTR is nonfunctional or absent. As a result, chloride transport across the epithelium is impaired, and thickened mucus is the result.

Q2: Is the CFTR a chemically gated, a voltage-gated, or a mechanically gated channel protein?

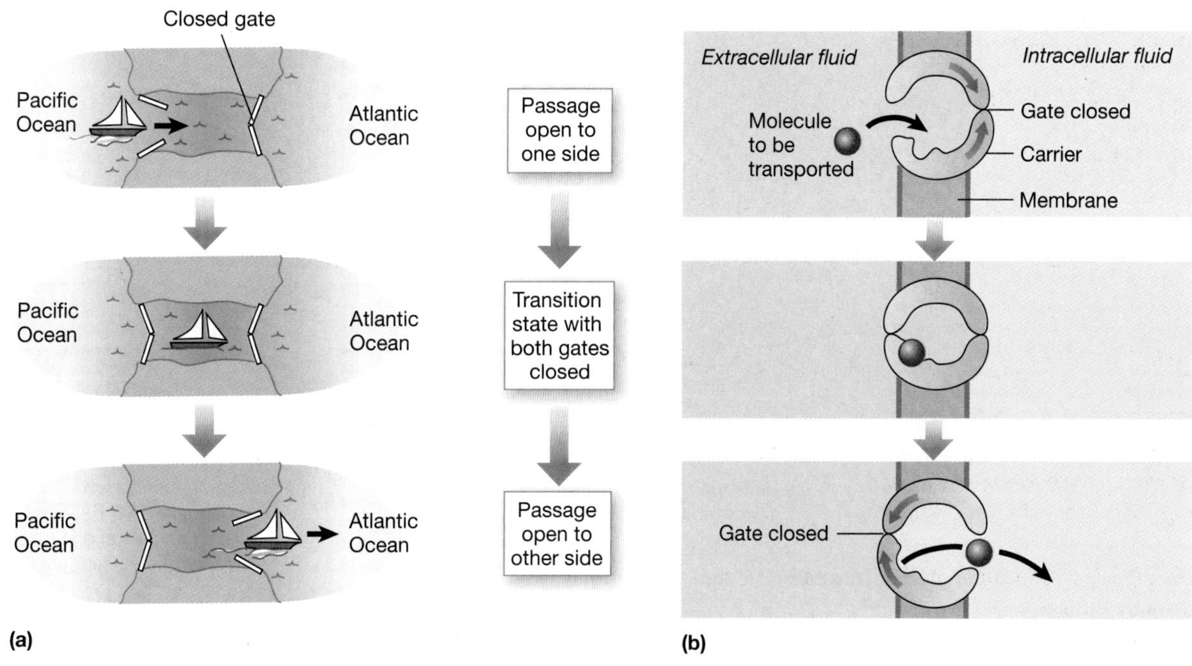

Fig. 5.12 ***Facilitated diffusion by means of a carrier protein.*** Carrier proteins, like the canal illustrated, never form a continuous passageway between the extracellular and intracellular fluids.

closes behind it. Now the canal is isolated from both oceans with the ship trapped in the middle. Then the Atlantic gate opens, making the canal continuous with the Atlantic Ocean. The ship sails out of the gate and off into the Atlantic, having crossed the barrier of the land without the canal ever forming a continuous connection between the two oceans.

Movement across the membrane through a carrier protein is similar (Fig. 5.12b). The molecule being transported binds to the carrier on one side of the membrane (the extracellular side in our example). This binding changes the conformation of the carrier protein so that the opening closes. After a brief transition in which both sides are closed, the opposite side of the carrier opens to the other side of the membrane. The carrier then releases the transported molecule into the opposite compartment, having brought it through the membrane without creating a continuous connection between the extracellular and intracellular compartments.

Concept Check

Answers: p. 173

18. Name four functions of membrane proteins.
19. Which kinds of particles pass through open channels?
20. Name two ways channels differ from carriers.
21. If a channel is lined with amino acids that have a net positive charge, which of the following ions is/are likely to move freely through the channel? Na^+, Cl^-, K^+, Ca^{2+}
22. Why can't glucose cross the cell membrane through open channels?

Carrier proteins can be divided into two categories according to the energy source that powers the transport. As noted earlier, facilitated diffusion is protein-mediated transport in which no outside source of energy is needed to move molecules across the cell membrane. Active transport is protein-mediated transport that requires an outside energy source, either ATP or the potential energy stored in a concentration gradient. We will look first at facilitated diffusion.

Facilitated Diffusion Uses Carrier Proteins

Some polar molecules appear to move into and out of cells by diffusion, even though we know from their chemical properties that they are unable to pass easily through the lipid core of the cell membrane. The solution to this seeming contradiction is that these polar molecules cross the cell membrane by facilitated diffusion, with the aid of specific carriers. Sugars and amino acids are examples of molecules that enter or leave cells using facilitated diffusion. For example, the family of carrier proteins known as **GLUT transporters** move glucose and related hexose sugars across membranes.

Facilitated diffusion has the same properties as simple diffusion (see Tbl. 5.6). The transported molecules move down their concentration gradient, the process requires no input of energy, and net movement stops at equilibrium, when the concentration inside the cell equals the concentration outside the cell (Fig. 5.13):

$$[\text{glucose}]_{\text{ECF}} = [\text{glucose}]_{\text{ICF}}^*$$

*In this book, the presence of brackets around a solute's name indicates concentration.

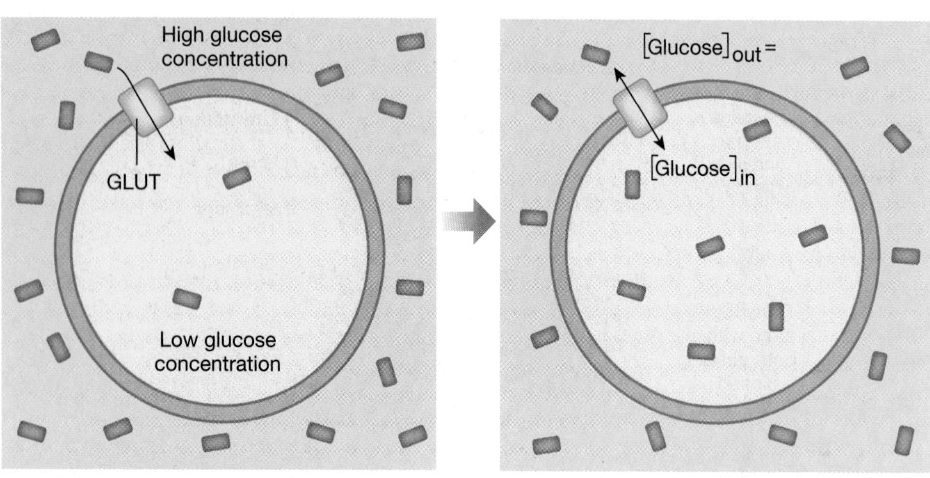

(a) Facilitated diffusion brings glucose into the cell down its concentration gradient.

(b) Diffusion reaches equilibrium when the glucose concentrations inside and outside the cell are equal.

(c) Conversion of imported glucose into glucose 6-phosphate (G-6-P) keeps intracellular glucose concentrations low so that diffusion never reaches equilibrium.

Fig. 5.13 ***Facilitated diffusion of glucose into cells.*** This figure shows glucose transport using a GLUT carrier protein.

Facilitated diffusion carriers always transport molecules down their concentration gradient. If the gradient reverses, so does the direction of transport.

Cells in which facilitated diffusion takes place can avoid reaching equilibrium by keeping the concentration of substrate in the cell low. With glucose, for example, this is accomplished by phosphorylation (Fig. 5.13c). As soon as a glucose molecule enters the cell on the GLUT carrier, it is phosphorylated to glucose 6-phosphate, the first step of glycolysis [p. 113]. Addition of the phosphate group prevents build-up of glucose inside the cell and also prevents glucose from leaving the cell.

Concept Check Answer: p. 173

23. Liver cells are able to convert glycogen to glucose, thereby making the intracellular glucose concentration higher than the extracellular glucose concentration. What do you think happens to facilitated diffusion of glucose when this occurs?

Active Transport Moves Substances against Their Concentration Gradients

Active transport is a process that moves molecules *against* their concentration gradient—that is, from areas of lower concentration to areas of higher concentration. Rather than creating an equilibrium state, where the concentration of the molecule is equal throughout the system, active transport creates a state of *dis*equilibrium by making concentration differences more pronounced. Moving molecules against their concentration gradient requires the input of outside energy, just as pushing a ball up a hill requires energy [see Fig. 4.2, p. 101]. The energy for active transport comes either directly or indirectly from the high-energy phosphate bond of ATP.

Active transport can be divided into two types. In **primary (direct) active transport**, the energy to push molecules against their concentration gradient comes directly from the high-energy phosphate bond of ATP. **Secondary (indirect) active transport** uses potential energy [p. 101] stored in the concentration gradient of one molecule to push other molecules against their concentration gradient. All secondary active transport ultimately depends on primary active transport because the concentration gradients that drive secondary transport are created using energy from ATP.

The mechanism for both types of active transport appears to be similar to that for facilitated diffusion. A substrate to be transported binds to a membrane carrier and the carrier then changes conformation, releasing the substrate into the opposite compartment. Active transport differs from facilitated diffusion because the conformation change in the carrier protein requires energy input.

Primary Active Transport Because primary active transport uses ATP as its energy source, many primary active transporters are known as **ATPases**. You may recall that the suffix *-ase* signifies an enzyme, and the stem (ATP) is the substrate upon which the enzyme is acting [p. 107]. These enzymes hydrolyze ATP to ADP and inorganic phosphate (P_i), releasing usable energy in the process. Most of the ATPases you will encounter in your study of physiology are listed in Table 5.7. ATPases are sometimes called *pumps*, as in the sodium-potassium pump, or Na^+-K^+-ATPase, mentioned earlier in this chapter.

The sodium-potassium pump is probably the single most important transport protein in animal cells because it maintains the concentration gradients of Na^+ and K^+ across the cell membrane (Fig. 5.14). The transporter is arranged in the cell membrane so that it pumps 3 Na^+ out of the cell and 2 K^+ into the cell for each ATP consumed. In some cells, the energy

Table 5.7 Primary Active Transporters

Names	Type of Transport
Na^+-K^+-ATPase or sodium-potassium pump	Antiport
Ca^{2+}-ATPase	Uniport
H^+-ATPase or proton pump	Uniport
H^+-K^+-ATPase	Antiport

The Na^+-K^+-ATPase uses energy from ATP to pump Na^+ out of the cell and K^+ into the cell.

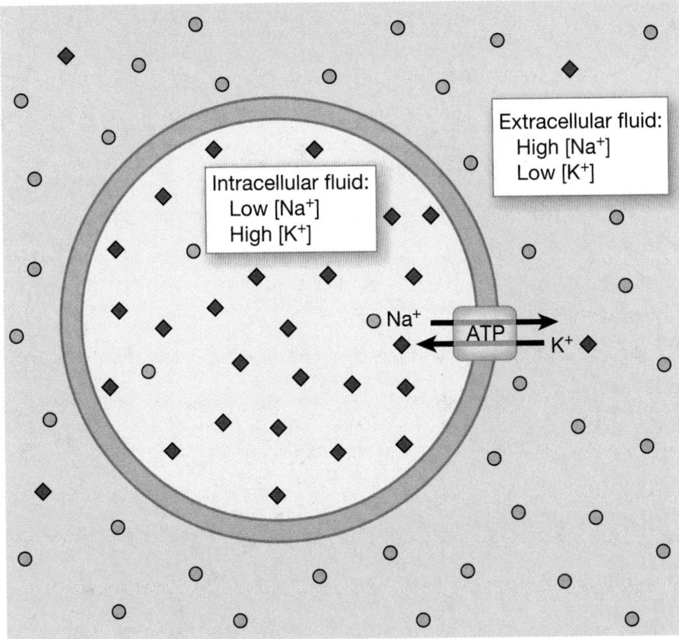

Fig. 5.14 ***The sodium-potassium pump, Na^+-K^+-ATPase.*** In this book, carrier proteins that hydrolyze ATP have the letters *ATP* written on the membrane protein.

needed to move these ions uses 30% of all the ATP produced by the cell. Figure 5.15 illustrates the current model of how the Na^+-K^+-ATPase works.

Secondary Active Transport The sodium concentration gradient, with Na^+ concentration high in the extracellular fluid and low inside the cell, is a source of potential energy that the cell can harness for other functions. For example, nerve cells use the sodium gradient to transmit electrical signals, and epithelial cells use it to drive the uptake of nutrients, ions, and water. Membrane transporters that use potential energy stored in concentration gradients to move molecules are called *secondary active transporters.*

Secondary active transport uses the kinetic energy of one molecule moving down its concentration gradient to push other molecules against their concentration gradient. The cotransported molecules may go in the same direction across the membrane (symport) or in opposite directions (antiport). The most common secondary active transport systems are driven by the sodium concentration gradient.

As a Na^+ moves into the cell, it either brings one or more molecules with it or trades places with molecules exiting the cell. The major Na^+-dependent transporters are listed in Table 5.8. Notice that the cotransported substances may be either other ions or uncharged molecules, such as glucose. As you study the different systems of the body, you will find these secondary active transporters taking part in many physiological processes.

The mechanism of the **Na^+-glucose secondary active transporter** (SGLT) is illustrated in Figure 5.16 on page 153. Both Na^+ and glucose bind to the SGLT protein on the extracellular fluid side. Sodium binds first and causes a conformational change in the protein that creates a high-affinity binding site for glucose (1). When glucose binds to SGLT (2), the protein changes conformation again and opens its channel to the intracellular fluid side (3). Sodium is released to the ICF as it

Table 5.8 Examples of Secondary Active Transporters

Symport Carriers	Antiport Carriers
Sodium-dependent transporters	
Na^+-K^+-$2Cl^-$ (NKCC)	Na^+-H^+ (NHE)
Na^+-glucose (SGLT)	Na^+-Ca^{2+} (NCX)
Na^+-Cl^-	
Na^+-HCO_3^-	
Na^+-amino acids (several types)	
Na^+-bile salts (small intestine)	
Na^+-choline uptake (nerve cells)	
Na^+-neurotransmitter uptake (nerve cells)	
Nonsodium-dependent transporters	
H^+-peptide symporter (pepT)	HCO_3^--Cl^-

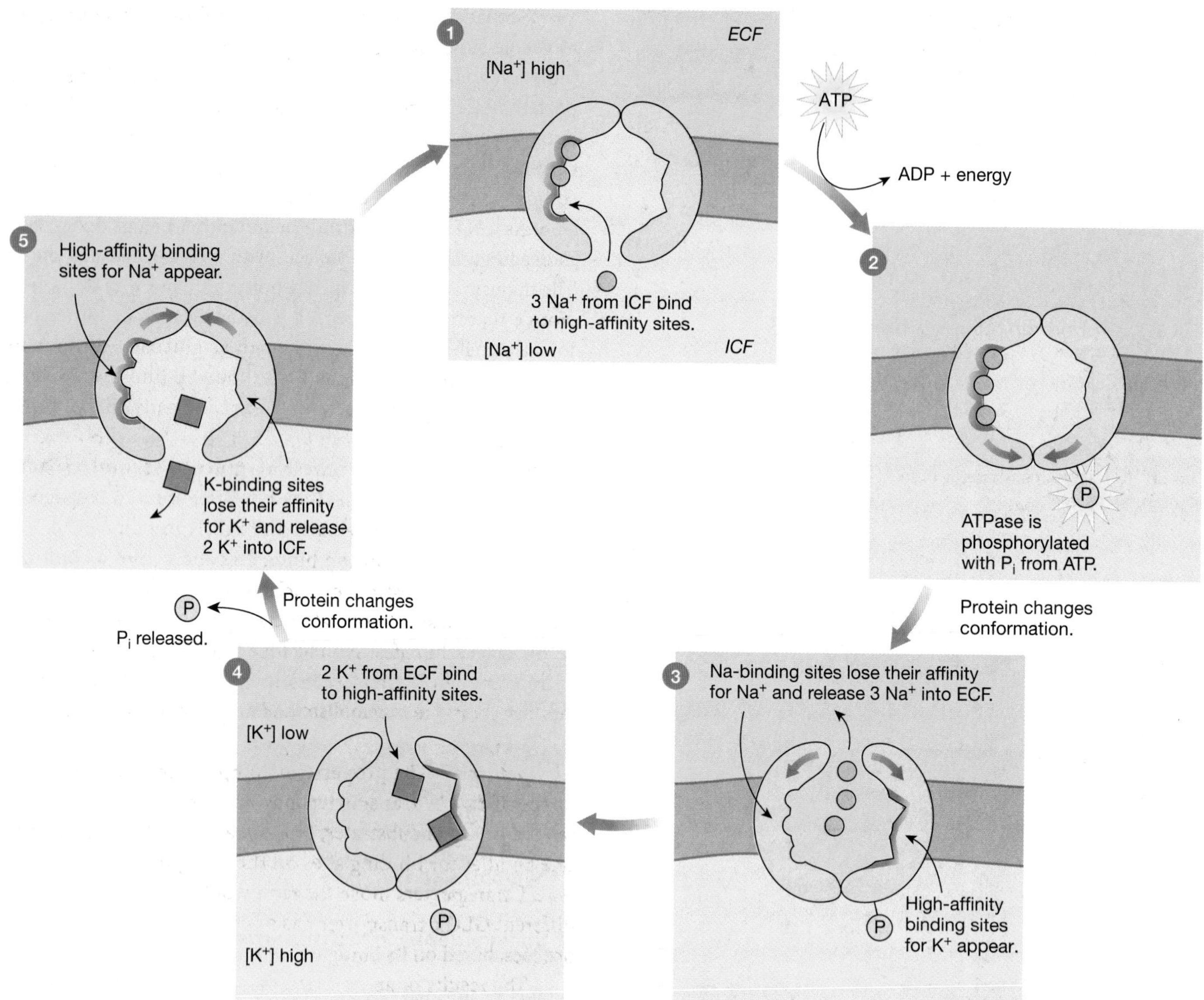

Fig. 5.15 ***Mechanism of the Na^+-K^+-ATPase.*** This figure presents one model of how the Na^+-K^+-ATPase uses energy and inorganic phosphate (P_i) from ATP to move ions across a membrane.

moves down its concentration gradient. The loss of Na^+ from the protein changes the binding site for glucose back to a low-affinity site, so glucose is released and follows Na^+ into the cytoplasm (4). The net result is the entry of glucose into the cell against its concentration gradient, coupled to the movement of Na^+ into the cell down its concentration gradient. The SGLT transporter can move glucose only into cells because glucose must follow the Na^+ gradient.

In contrast, GLUT transporters are reversible and can move glucose into or out of cell the depending on the concentration gradient. For example, when blood glucose levels are high, GLUT transporters on liver cells bring glucose into those cells. During times of fasting, when blood glucose levels fall, liver cells convert their glycogen stores to glucose. When the glucose concentration inside the liver cells builds up and exceeds the glucose concentration in the plasma, glucose leaves the cells on the reversible GLUT transporters. GLUT transporters are found on all cells of the body.

If GLUT transporters are everywhere, then why does the body need the SGLT Na^+-glucose symporter? The simple answer is that both SGLT and GLUT are needed to move glucose from one side of an epithelium to the other. Consequently, SGLT transporters are found on certain epithelial cells, such as intestinal and kidney cells, that bring glucose into the body from the external environment. We discuss the process of transepithelial transport of glucose later in this chapter.

Concept Check Answer: p. 173

24. Name two ways active transport by the Na^+-K^+-ATPase (Fig. 5.15) differs from secondary transport by the SGLT (Fig. 5.16).

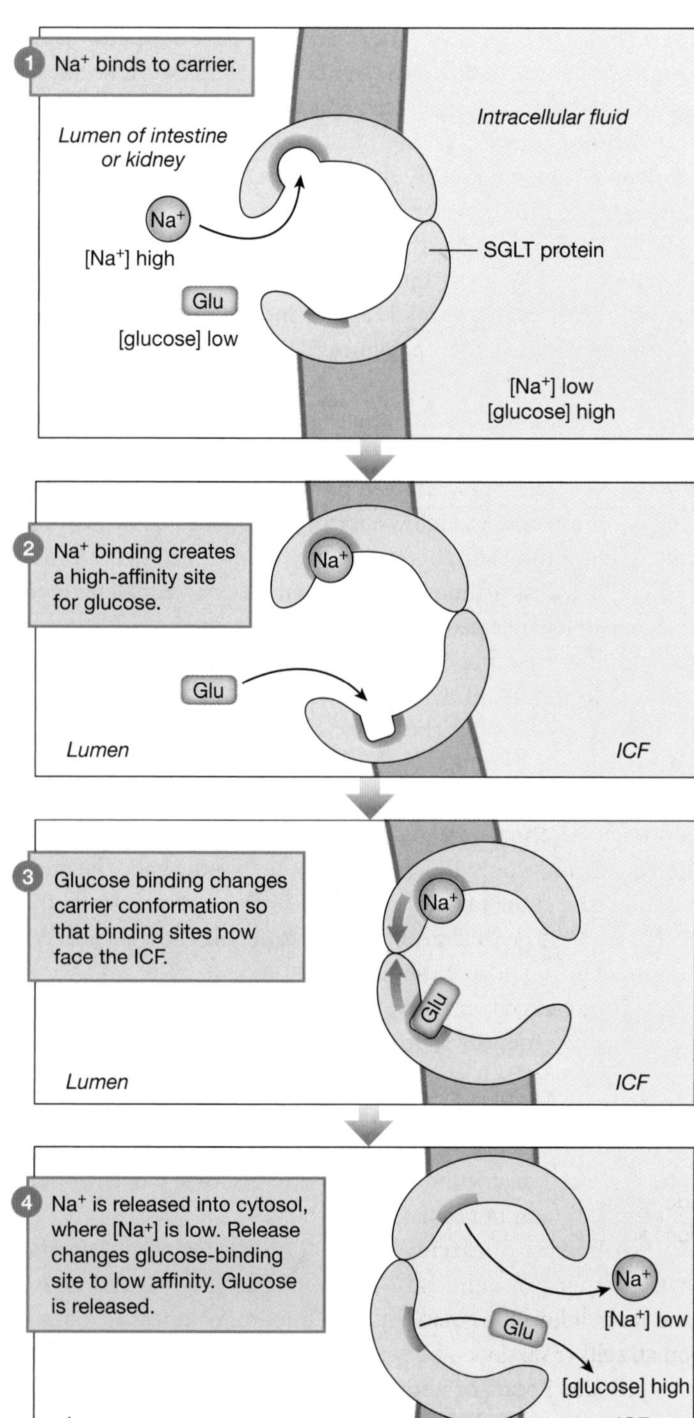
1 Na+ binds to carrier.
Lumen of intestine or kidney
Intracellular fluid
Na+
[Na+] high
SGLT protein
Glu
[glucose] low
[Na+] low
[glucose] high
2 Na+ binding creates a high-affinity site for glucose.
Na+
Glu
Lumen
ICF
3 Glucose binding changes carrier conformation so that binding sites now face the ICF.
Na+
Glu
Lumen
ICF
4 Na+ is released into cytosol, where [Na+] is low. Release changes glucose-binding site to low affinity. Glucose is released.
Na+
[Na+] low
Glu
[glucose] high
Lumen
ICF

Fig. 5.16 ***Mechanism of the SGLT transporter.*** This transporter uses the potential energy stored in the Na^+ concentration gradient to move glucose against its concentration gradient.

Carrier-Mediated Transport Exhibits Specificity, Competition, and Saturation

Both passive and active forms of carrier-mediated transport demonstrate specificity, competition, and saturation—three properties that result from the binding of a substrate to a protein [p. 56].

Specificity Specificity refers to the ability of a transporter to move only one molecule or only a group of closely related molecules [p. 56]. One example of specificity is found in the GLUT family of transporters, which move 6-carbon sugars (*hexoses*), such as glucose, mannose, galactose, and fructose [p. 34], across cell membranes. GLUT transporters have binding sites that recognize and transport hexoses, but they will not transport the disaccharide maltose or any form of glucose that is not found in nature (■ Fig. 5.17b). For this reason we can say that GLUT transporters are specific for naturally occurring 6-carbon monosaccharides.

For many years, scientists assumed that there must be different isoforms of the glucose-facilitated diffusion carrier because they had observed that glucose transport was regulated by hormones in some cells but not in others. However, it was not until the 1980s that the first glucose transporter was isolated. To date, about 12 GLUT genes have been identified. The important GLUT proteins you will encounter in this book include GLUT1, found in most cells of the body; GLUT2, found in liver and in kidney and intestinal epithelia; GLUT3, found in neurons; GLUT4, the insulin-regulated transporter of skeletal muscle; and GLUT5, the intestinal fructose transporter. The restriction of different GLUT transporters to different tissues is an important feature in the metabolism and homeostasis of glucose.

Competition The property of competition is closely related to specificity. A transporter may move several members of a related group of substrates, but those substrates compete with one another for binding sites on the transporter. For example, GLUT transporters move the family of hexose sugars, but each different GLUT transporter has a "preference" for one or more hexoses, based on its binding affinity.

The results of an experiment demonstrating competition are shown in Figure 5.17c. The graph shows glucose transport rate as a function of glucose concentration. The top line (red) shows transport when only glucose is present. The lower line (black) shows that glucose transport decreases if galactose is also present. Galactose competes for binding sites on the GLUT transporters and displaces some glucose molecules. With fewer glucose able to bind to the GLUT protein, the rate of glucose transport into the cell decreases.

Sometimes the competing molecule is not transported but merely blocks the transport of another substrate. In this case, the competing molecule is a *competitive inhibitor* [p. 54]. In the glucose transport system, the disaccharide maltose is a competitive inhibitor (Fig. 5.17b). It competes with glucose for the binding site, but once bound, it is too large to be moved across the membrane.

Competition between transported substrates has been put to good use in medicine. An example involves gout, a disease caused by elevated levels of uric acid in the plasma. One method of decreasing uric acid in plasma is to enhance its excretion in the urine. Normally, the kidney's *organic anion transporter*

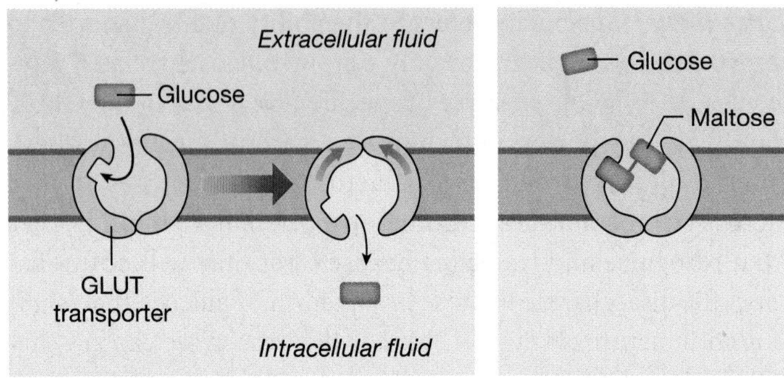

(a) The **GLUT transporter** brings glucose across cell membranes.

(b) Maltose is a competitive inhibitor that binds to the GLUT transporter but is not itself carried across the membrane.

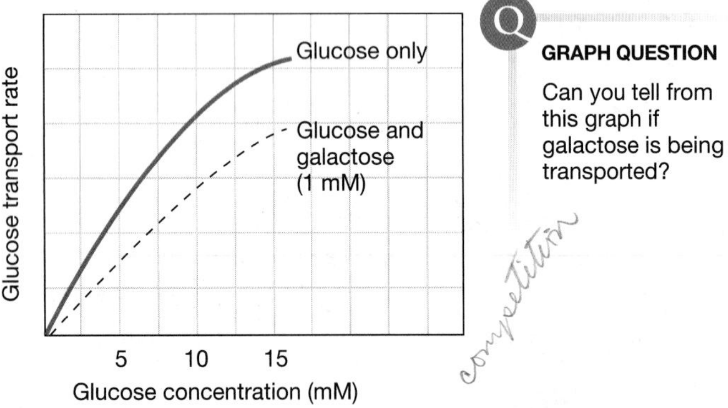

GRAPH QUESTION

Can you tell from this graph if galactose is being transported?

(c) This graph shows glucose transport rate as a function of glucose concentration. In one experiment, only glucose was present. In the second experiment, a constant concentration of galactose was present.

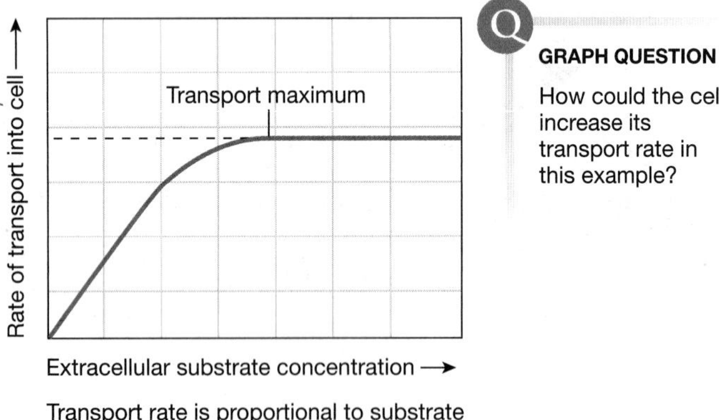

Transport rate is proportional to substrate concentration until the carriers are saturated.

GRAPH QUESTION

How could the cell increase its transport rate in this example?

(d) This graph shows that transport can reach a maximum rate when all the carrier binding sites are filled with substrate.

Fig. 5.17 ***Saturation and competition***

(OAT) reclaims urate (the anion form of uric acid) from the urine and returns the acid to the plasma. However, if an organic acid called probenecid is administered to the patient, OAT binds to probenecid instead of to uric acid, preventing the reabsorption of urate. As a result, more urate leaves the body in the urine, lowering the uric acid concentration in the plasma.

Saturation The rate of substrate transport depends on the substrate concentration and the number of carrier molecules, a property that is shared by enzymes and other binding proteins [p. 56]. For a fixed number of carriers, however, as substrate concentration increases, the transport rate increases up to a maximum, the point at which all carrier binding sites are filled with substrate. At this point, the carriers are said to have reached saturation. At saturation, the carriers are working at their maximum rate, and a further increase in substrate concentration has no effect. Figure 5.17d represents saturation graphically.

As an analogy, think of the carriers as doors into a concert hall. Each door has a maximum number of people that it can allow to enter the hall in a given period of time. Suppose that all the doors together can allow a maximum of 100 people per minute to enter the hall. This is the maximum transport rate, also called the **transport maximum**. When the concert hall is empty, three maintenance people enter the doors every hour. The transport rate is 3 people/60 minutes, or 0.05 people/minute, well under the maximum. For a local dance recital, about 50 people per minute go through the doors, still well under the maximum. When the most popular rock group of the day appears in concert, however, thousands of people gather outside. When the doors open, thousands of people are clamoring to get in, but the doors allow only 100 people/minute into the hall. The doors are working at the maximum rate, so it does not matter whether there are 1000 or 3000 people trying to get in. The transport rate saturates at 100 people/minute.

How can cells increase their transport capacity and avoid saturation? One way is to increase the number of carriers in the membrane. This would be like opening more doors into the concert hall. Under some circumstances, cells are able to insert additional carriers into their membranes. Under other circumstances, a cell may withdraw carriers to decrease movement of a molecule into or out of the cell.

All forms of carrier-mediated transport show specificity, competition, and saturation, but as you learned earlier in the chapter, they also differ in one important way: passive mediated transport—better known as facilitated diffusion—requires no input of energy from an outside source. Active transport requires energy input from ATP, either directly or indirectly.

Concept Check Answers: p. 173

25. What would you call a carrier that moves two substrates in opposite directions across a membrane?

26. In the concert-hall door analogy, we described how the maximum transport rate might be increased by increasing the number of doors leading into the hall. Using the same analogy, can you think of another way a cell might increase its maximum transport rate?

Vesicular Transport

What happens to the many macromolecules that are too large to enter or leave cells through protein channels or carriers? They move in and out of the cell with the aid of bubble-like *vesicles* [p. 75] created from the cell membrane. Cells use two basic processes to import large molecules and particles: phagocytosis and endocytosis. Some scientists consider phagocytosis to be a type of endocytosis, but mechanistically the two processes are different. Material leaves cells by the process known as exocytosis, a process that is similar to endocytosis run in reverse.

Phagocytosis Creates Vesicles Using the Cytoskeleton

If you studied *Amoeba* in your biology laboratory, you may have watched these one-cell creatures ingest their food by surrounding it and enclosing it within a vesicle that is brought into the cytoplasm. **Phagocytosis** {*phagein,* to eat + *cyte,* cell + *-sis,* process} is the actin-mediated process by which a cell engulfs a bacterium or other particle into a large membrane-bound vesicle called a **phagosome** {*soma,* body}. The phagosome pinches off from the cell membrane and moves to the interior of the cell, where it fuses with a lysosome [p. 75], whose digestive enzymes destroy the bacterium. Phagocytosis requires energy from ATP for the movement of the cytoskeleton and for the intracellular transport of the vesicles. In humans, phagocytosis occurs in certain types of white blood cells called *phagocytes,* which specialize in "eating" bacteria and other foreign particles (■ Fig. 5.18).

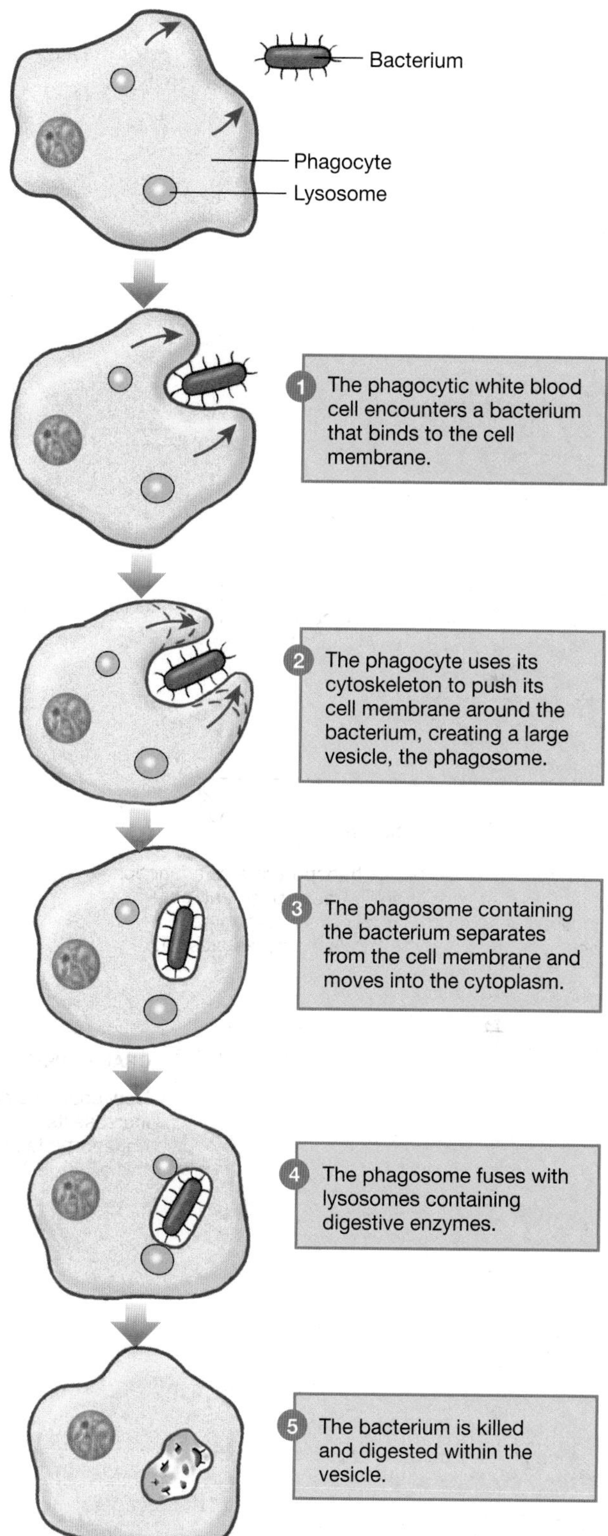

Bacterium
Phagocyte
Lysosome
1 The phagocytic white blood cell encounters a bacterium that binds to the cell membrane.
2 The phagocyte uses its cytoskeleton to push its cell membrane around the bacterium, creating a large vesicle, the phagosome.
3 The phagosome containing the bacterium separates from the cell membrane and moves into the cytoplasm.
4 The phagosome fuses with lysosomes containing digestive enzymes.
5 The bacterium is killed and digested within the vesicle.

■ **Fig. 5.18** ***Phagocytosis***

Endocytosis Creates Smaller Vesicles

Endocytosis, the second process by which large molecules or particles move into cells, differs from phagocytosis in two important ways. First, in endocytosis the membrane surface indents rather than pushes out. Second, the vesicles formed from endocytosis are much smaller. In addition, some endocytosis is *constitutive;* that is, it is an essential function that is always taking place. In contrast, phagocytosis must be triggered by the presence of a substance to be ingested.

Endocytosis is an active process that requires energy from ATP. It can be nonselective, allowing extracellular fluid to enter the cell—a process called **pinocytosis** {*pino-*, drink}—or it can be highly selective, allowing only specific molecules to enter the cell. In receptor-mediated endocytosis, a ligand binds to a membrane receptor protein to activate the process.

Receptor-Mediated Endocytosis **Receptor-mediated endocytosis** takes place in regions of the cell membrane known as **coated pits**, indentations where the cytoplasmic side of the membrane has high concentrations of protein. The most common protein found in coated pits is *clathrin,* illustrated in ■ Figure 5.19. In the first step of the process, extracellular ligands that will be brought into the cell bind to their membrane receptors (1). The receptor-ligand complex migrates along the cell surface until it encounters a coated pit (2). Once the receptor-ligand complex is in the coated pit, the membrane draws inward, or *invaginates* (3),

■ Fig. 5.19 *ESSENTIALS*

Endocytosis, Exocytosis, and Membrane Recycling

Membrane removed from the cell surface by endocytosis is recycled back to the cell surface by exocytosis.

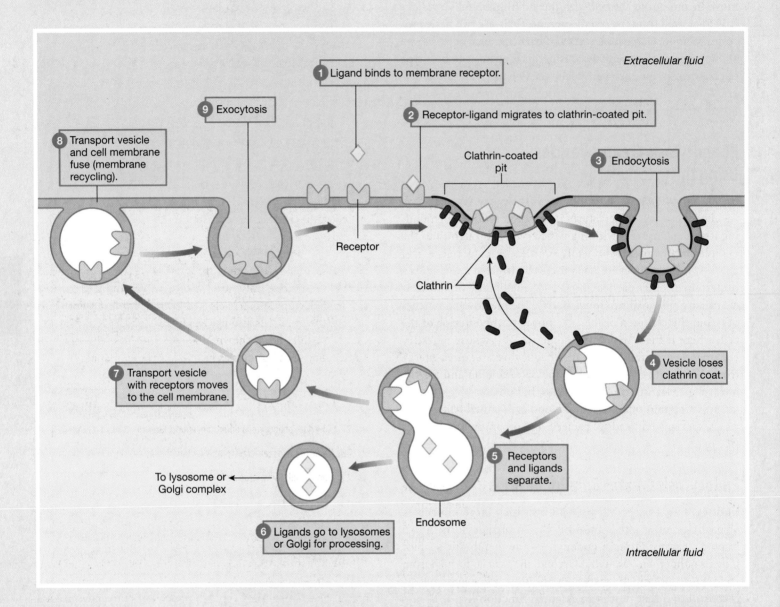

then pinches off from the cell membrane and becomes a cytoplasmic vesicle. The clathrin molecules are released and recycle back to the membrane (4). In the vesicle, the receptor and ligand separate, leaving the ligand inside an *endosome* (5). The endosome moves to a lysosome if the ligand is to be destroyed, or to the Golgi complex if the ligand is to be processed (6).

Meanwhile, the ligand's membrane-bound receptors may be reused in a process known as **membrane recycling**. The vesicle with the receptors moves to the cell membrane (7) and fuses with it (8). The vesicle membrane then is incorporated back into the cell membrane by exocytosis (9). Notice in Figure 5.19 that the cytoplasmic face of the membrane remains the same throughout endocytosis and recycling. The extracellular surface of the cell membrane becomes the inside face of the vesicle membrane.

Receptor-mediated endocytosis transports a variety of substances into the cell, including protein hormones, growth factors, antibodies, and plasma proteins that serve as carriers for iron and cholesterol. Elevated plasma cholesterol levels and cardiovascular disease are associated with abnormalities in receptor-mediated removal of cholesterol from the blood (see Clinical Focus box on LDL: The Lethal Lipoprotein).

CLINICAL FOCUS

LDL: The Lethal Lipoprotein

"Limit the amount of cholesterol in your diet!" has been the recommendation for many years. So why is too much cholesterol bad for you? After all, cholesterol molecules are essential for membrane structure and for making steroid hormones (such as the sex hormones). But elevated cholesterol levels in the blood also lead to heart disease. One reason some people have too much cholesterol in their blood (*hypercholesterolemia*) is not diet but the failure of cells to take up the cholesterol. In the blood, hydrophobic cholesterol is bound to a lipoprotein carrier molecule to make it water soluble. The most common form of carrier is *low-density lipoprotein (LDL)*. When the LDL-cholesterol complex (LDL-C) binds to LDL receptors in caveolae, then it can then enter the cell in a vesicle. When people do not have adequate numbers of LDL receptors on their cell membranes, LDL-C remains in the blood. Hypercholesterolemia due to high levels of LDL-C predisposes these people to the development of **atherosclerosis,** also known as hardening of the arteries {*atheroma,* a tumor + *skleros,* hard + *-sis,* condition}. In this condition, the accumulation of cholesterol in blood vessels blocks blood flow and contributes to heart attacks.

Caveolae Some endocytosis uses small flask-shaped indentations called **caveolae** ("little caves") rather than clathrin-coated pits to concentrate and bring receptor-bound molecules into the cell. Caveolae are membrane regions with lipid rafts [p. 68], membrane receptor proteins, and usually a coat of membrane proteins named *caveolins*. The receptors in caveolae are lipid-anchored proteins [p. 68]. In many cells, caveolae appear as small indented pockets on the cell membrane, which is how they acquired their name.

Caveolae have several functions: to concentrate and internalize small molecules, to help in the transfer of macromolecules across the capillary endothelium, and to participate in cell signaling. Caveolae appear to be involved in some disease processes, including viral and parasitic infections. Two forms of the disease *muscular dystrophy* are associated with abnormalities in the protein caveolin. Scientists are currently trying to discover more details about the role of caveolae in normal physiology and pathophysiology.

Exocytosis Releases Molecules Too Large for Transport Proteins

Exocytosis is the opposite of endocytosis. In exocytosis, intracellular vesicles move to the cell membrane, fuse with it (Fig. 5.19 (8)), and then release their contents to the extracellular fluid (9). Cells use exocytosis to export large lipophobic molecules, such as proteins synthesized in the cell, and to get rid of wastes left in lysosomes from intracellular digestion.

The process by which the cell and vesicle membranes fuse is similar in a variety of cell types, from neurons to endocrine cells. Exocytosis involves two families of proteins: *Rabs,* which help vesicles dock onto the membrane, and *SNAREs,* which facilitate membrane fusion. In regulated exocytosis, the process usually begins with an increase in intracellular Ca^{2+} concentration that acts as a signal. The Ca^{2+} interacts with a calcium-sensing protein, which in turn initiates secretory vesicle docking and fusion. When the fused area of membrane opens, the vesicle contents diffuse into the extracellular fluid while the vesicle membrane stays behind and becomes part of the cell membrane. Exocytosis, like endocytosis, requires energy in the form of ATP.

Exocytosis takes place continuously in some cells, making it a *constitutive* process. For example, goblet cells [p. 84] in the intestine continuously release mucus by exocytosis, and fibroblasts in connective tissue release collagen [p. 87]. In other cell types, exocytosis is an intermittent process that is initiated by a signal. In many endocrine cells, hormones are stored in secretory vesicles in the cytoplasm and released in response to a signal from outside the cell. Cells also use exocytosis to insert proteins into the cell membrane, as shown in Figure 5.19. You will encounter many examples of exocytosis in your study of physiology.

Concept Check Answers: p. 173

27. How does phagocytosis differ from endocytosis?
28. Name the two membrane protein families associated with endocytosis.
29. How do cells move large proteins into the cell? Out of the cell?

Epithelial Transport

All the transport processes described in the previous sections deal with the movement of molecules across a single membrane, that of the cell. However, molecules entering and leaving the body or moving between certain compartments within the body must cross a layer of epithelial cells [p. 82] that are connected to one another by adhesive junctions and tight junctions [p. 82].

The tight junctions of epithelia separate the cell membrane into two regions, or poles. The surface of the epithelial cell that faces the lumen of an organ is called the *apical* {*apex,* the highest point} membrane (■ Fig. 5.20). It is often folded into microvilli that increase its surface area. Below the tight junctions the three surfaces of the cell that face the extracellular fluid are collectively called the *basolateral* membrane {*basal,* base + *latus,* side}. The apical membrane is also called the *mucosal* membrane. The corresponding term for the basolateral membrane is *serosal* membrane.

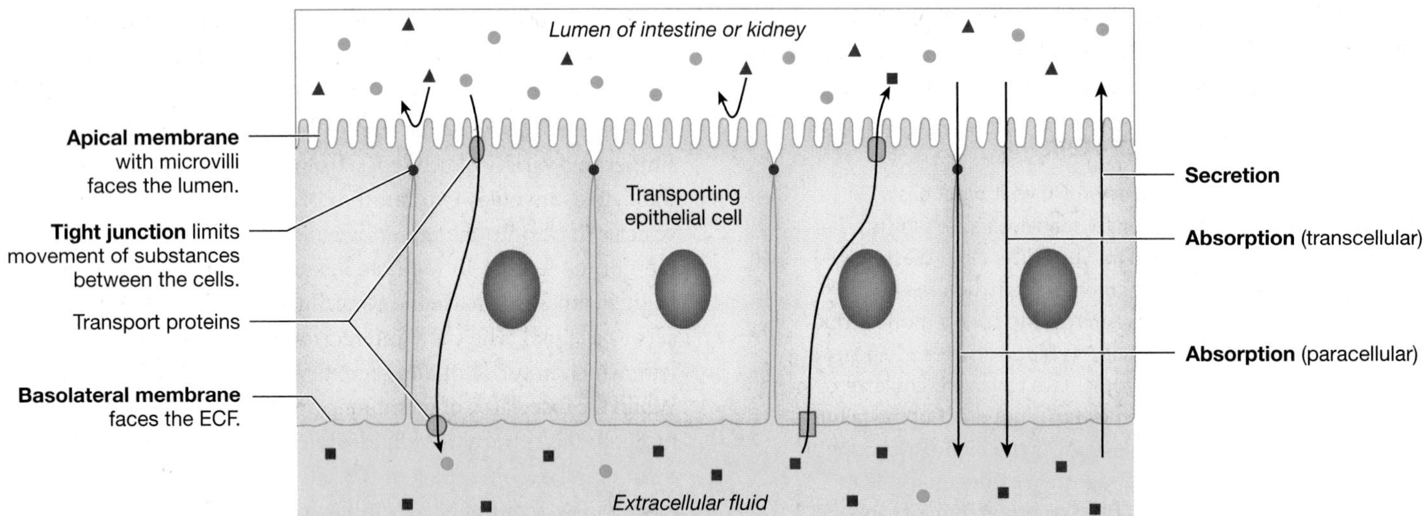
Lumen of intestine or kidney
Apical membrane with microvilli faces the lumen.
Tight junction limits movement of substances between the cells.
Transport proteins
Basolateral membrane faces the ECF.
Transporting epithelial cell
Secretion
Absorption (transcellular)
Absorption (paracellular)
Extracellular fluid
Polarized epithelia have different transport proteins on apical and basolateral membranes. This allows selective directional transport across the epithelium. Transport from lumen to ECF is called absorption. Transport from ECF to lumen is called secretion.

Fig. 5.20 ***Polarized cells of transporting epithelia.*** The apical membrane and the basolateral membrane are the two poles of the cell.

Transporting epithelial cells are said to be *polarized* because their apical and basolateral membranes have very different properties. Certain transport proteins, such as the Na^+-K^+-ATPase, are usually found only on the basolateral membrane. Others, like the Na^+-glucose symporter SGLT, are restricted to the apical membrane. This polarized distribution of transporters allows the one-way movement of certain molecules across the epithelium.

Transport of material from the lumen of an organ to the extracellular fluid is called **absorption** (Fig. 5.20). For example, the intestinal epithelium absorbs digested nutrients. When material moves from the ECF to the lumen, the process is called *secretion*. For example, the salivary glands secrete saliva to help moisten the food you eat. Note that the term *secretion* is also used more broadly to mean the release of a substance from a cell.

Epithelial Transport May Be Paracellular or Transcellular

Movement across an epithelium, or **epithelial transport**, may take place either as **paracellular transport** {*para-*, beside} through the junctions between adjacent cells or as **transcellular transport** through the epithelial cells themselves (Fig. 5.20). In "tight" epithelia, the cell-cell junctions act as barriers to minimize the unregulated diffusion of material between the cells, so there is very little paracellular transport. In recent years, however, scientists have learned that some epithelia have the ability to change the "tightness" of their junctions. It appears that some junctional proteins such as *claudins* can form large holes or pores that allow water, ions, and a few small uncharged solutes to move by the paracellular pathway. In certain pathological states, increased movement through the paracellular route is a hallmark of the disease.

RUNNING PROBLEM

The sweat test that Daniel will undergo analyzes levels of the salt NaCl in sweat. Sweat—a mixture of ions and water—is secreted into sweat ducts by the epithelial cells of sweat glands. As sweat moves toward the skin's surface through the sweat ducts, CFTR and Na^+ channels move Cl^- and Na^+ out of the sweat and back into the body. This portion of the sweat gland epithelium is not permeable to water, and so normal reabsorption of NaCl creates sweat with a low salt content. However, in the absence of CFTR in the epithelium, salt is not reabsorbed. "Normally, sweat contains about 120 millimoles of salt per liter," says Beryl Rosenstein, M.D., of the Cystic Fibrosis Center at the Johns Hopkins Medical Institutions. "In cystic fibrosis, salt concentrations in the sweat can be four times the normal amount."

Q3: Based on the information given, is CFTR protein on the apical or basolateral surface of the sweat gland epithelium?

130 141 148 158 161 168

In contrast, substances moving by the transcellular route must cross two cell membranes. Molecules cross the first membrane when they move into the epithelial cell from one compartment. They cross the second membrane when they leave the epithelial cell to enter the second compartment. Transcellular transport uses a combination of active and passive transport mechanisms.

Protein-mediated transcellular transport is usually a two-step process, with one "uphill" step that requires energy and one "downhill" step in which the molecule moves passively down its gradient. You will see these steps in the example of glucose transport that follows. Molecules that are too large to be moved by membrane proteins can be transported across the cell in vesicles.

The cells of transporting epithelia can alter their permeability by selectively inserting or withdrawing membrane proteins. Transporters pulled out of the membrane may be destroyed in lysosomes, or they may be stored in vesicles inside the cell, ready to be reinserted into the membrane in response to a signal (another example of membrane recycling). Most epithelial transport you will study in this book involves the transporting epithelia of intestine and kidney, which are specialized to selectively transport molecules into and out of the body.

Transcellular Transport of Glucose Uses Membrane Proteins

The absorption of glucose from the lumen of the kidney tubule or intestine to the extracellular fluid is an important example of directional movement across a transporting epithelium. Transepithelial movement of glucose involves three transport systems: the SGLT-mediated secondary active transport of glucose with Na^+ from the lumen into the epithelial cell at the apical membrane, followed by the movement of Na^+ and glucose out of the cell and into the extracellular fluid on separate transporters. Sodium moves out by primary active transport via a Na^+-K^+-ATPase, and glucose leaves the cell by facilitated diffusion on GLUT carriers.

■ Figure 5.21 shows the process in detail. The glucose concentration in the transporting epithelial cell is higher than the glucose concentration in either the extracellular fluid or the lumen of the kidney or intestine. For this reason, moving glucose from the lumen into the cell requires the input of energy—in this case, energy stored in the Na^+ concentration gradient. Sodium ions in the lumen bind to the SGLT carrier, as previously described (see Fig. 5.16), and bring glucose with them into the cell. The energy needed to move glucose against its concentration gradient comes from the kinetic energy of Na^+ moving down its concentration gradient (Fig. 5.21 ①).

Once glucose is in the epithelial cell, it leaves by moving down its concentration gradient on the facilitated diffusion GLUT transporter in the basolateral membrane (Fig. 5.21 ②). Na^+ is pumped out of the cell on the basolateral side using

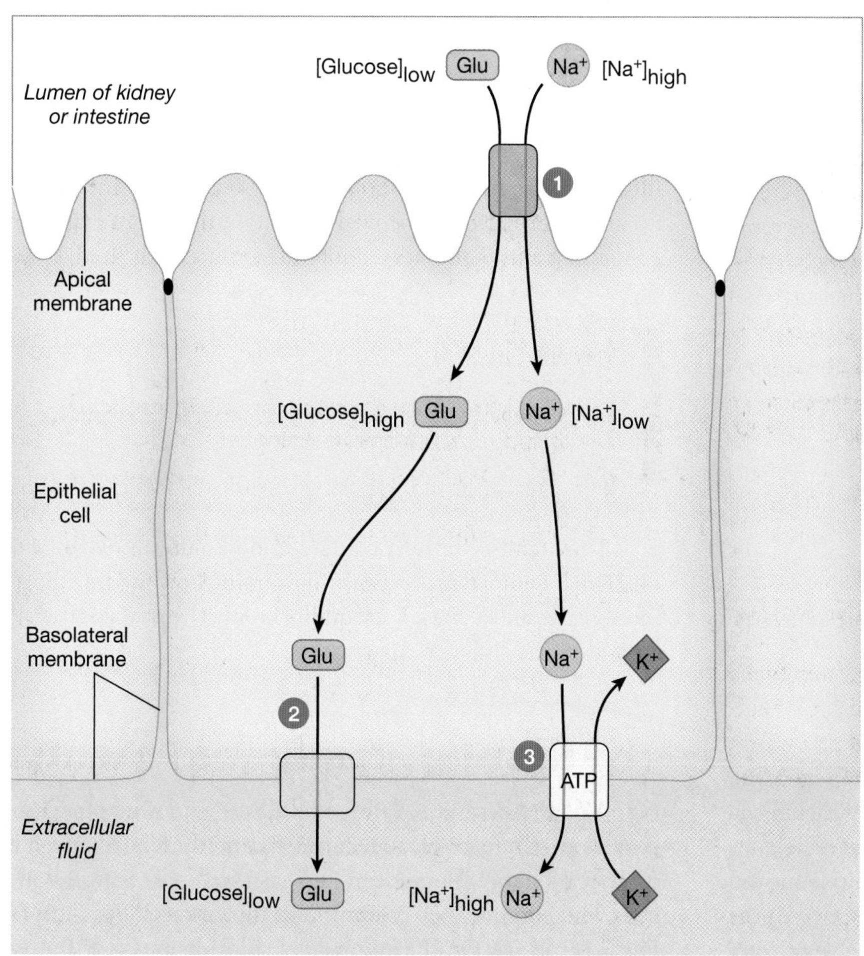

① **Na⁺- glucose symporter** brings glucose into cell against its gradient using energy stored in the Na⁺ concentration gradient.

② **GLUT transporter** transfers glucose to ECF by facilitated diffusion.

③ **Na⁺-K⁺-ATPase** pumps Na⁺ out of the cell, keeping ICF Na⁺ concentration low.

Q **FIGURE QUESTIONS**

- Match each transporter to its location.
  1. GLUT (a) apical membrane
  2. Na⁺-glucose symporter (b) basolateral membrane
  3. Na⁺-K⁺-ATPase
- Is glucose movement across the basolateral membrane active or passive? Explain.
- Why doesn't Na⁺ movement at the apical membrane require ATP?

■ **Fig. 5.21** ***Transepithelial transport of glucose.*** This process involves indirect (secondary) active transport of glucose across the apical membrane and glucose diffusion across the basolateral membrane.

Na^+-K^+-ATPase (3). This step requires energy provided by ATP because sodium is more concentrated in the extracellular fluid than in the cell.

The removal of Na^+ from the cell is essential if glucose is to continue to be absorbed from the lumen. The potential energy to run the SGLT symporter comes from the sodium concentration gradient, which depends on low intracellular concentrations of Na^+. If the basolateral Na^+-K^+-ATPase is poisoned with *ouabain* (pronounced wah-bane—a compound related to the heart drug digitalis), Na^+ that enters the cell cannot be pumped out. The Na^+ concentration inside the cell gradually increases until it is equal to that in the lumen. Without a sodium gradient, there is no energy source to run the SGLT symporter, and the movement of glucose across the epithelium stops.

Transepithelial transport can use ion movement through channels in addition to carrier-mediated transport. For example, the apical membrane of a transporting epithelium may use the Na^+-K^+-$2Cl^-$ (NKCC) symporter to bring K^+ into the cell against its concentration gradient, using energy from the Na^+ gradient. Because the K^+ concentration inside the cell is higher than in the extracellular fluid, K^+ can move out of the cell on the basolateral side through open K^+ leak channels. Na^+ must be pumped out by Na^+-K^+-ATPase. By this simple mechanism the body can absorb Na^+ and K^+ at the same time from the lumen of the intestine or the kidney.

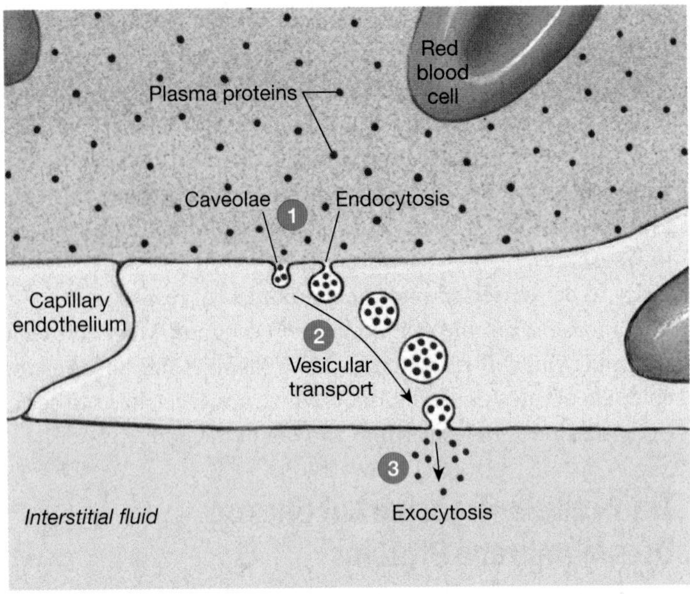

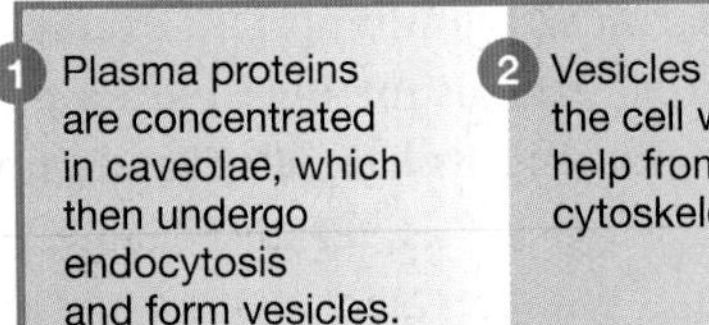

1 Plasma proteins are concentrated in caveolae, which then undergo endocytosis and form vesicles.
2 Vesicles cross the cell with help from the cytoskeleton.
3 Vesicle contents are released into interstitial fluid by exocytosis.

Fig. 5.22 ***Transcytosis across the capillary endothelium***

Concept Check Answers: p. 173

30. Why does Na^+ movement from the cytoplasm to the extracellular fluid require energy?
31. Ouabain, an inhibitor of the Na^+-K^+-ATPase, cannot pass through cell membranes. What would happen to the transepithelial glucose transport shown in Figure 5.21 if ouabain were applied to the apical side of the epithelium? To the basolateral side of the epithelium?
32. Which GLUT transporter is illustrated in Figure 5.21?

Transcytosis Uses Vesicles to Cross an Epithelium

Some molecules, such as proteins, are too large to cross epithelia on membrane transporters. Instead they are moved across epithelia by **transcytosis**, which is a combination of endocytosis, vesicular transport across the cell, and exocytosis (Fig. 5.22). In this process, the molecule is brought into the epithelial cell via receptor-mediated endocytosis. The resulting vesicle attaches to microtubules in the cell's cytoskeleton and is moved across the cell by a process known as **vesicular transport**. At the opposite side of the epithelium, the contents of the vesicle are expelled into the interstitial fluid by exocytosis.

Transcytosis makes it possible for large proteins to move across an epithelium and remain intact. It is the means by which infants absorb maternal antibodies in breast milk. The antibodies are absorbed on the apical surface of the infant's intestinal epithelium and then released into the extracellular fluid.

Concept Check Answer: p. 173

33. If you apply a poison that disassembles microtubules to a capillary endothelial cell, what happens to transcytosis?

Now that we have considered how solutes move between the body's compartments, we will examine how the transport of ions creates an electrical disequilibrium between the intracellular and extracellular compartments.

The Resting Membrane Potential

Many of the body's solutes, including organic compounds such as pyruvate and lactate, are ions and therefore carry a net electrical charge. Potassium (K^+) is the major cation within cells, and sodium (Na^+) dominates the extracellular fluid (see Fig. 5.1, p. 131). On the anion side, chloride ions (Cl^-) mostly remain with Na^+ in the extracellular fluid. Phosphate ions and negatively charged proteins are the major anions of the intracellular fluid.

RUNNING PROBLEM

Three days after Daniel's sweat test, the lab returns the grim results: salt levels in his sweat are more than twice the normal concentration. Daniel is diagnosed with cystic fibrosis. Now, along with antibiotics to prevent lung infections and therapy to loosen the mucus in his airways, Daniel must begin a regimen of pancreatic enzymes to be taken whenever he eats, for the rest of his life. In cystic fibrosis, thick mucus in the pancreatic ducts blocks the secretion of digestive enzymes into the intestine. Without artificial enzymes, he would starve.

Q4: Why will Daniel starve if he does not take artificial pancreatic enzymes?

130 141 148 158 161 168

However, the intracellular compartment is not electrically neutral: there are some protein anions inside cells that do not have matching cations, giving the cells a net negative charge. At the same time, the extracellular compartment has a net positive charge: some cations in the extracellular fluid do not have matching anions. One consequence of this uneven distribution of ions is that the intracellular and extracellular compartments are not in electrical equilibrium. Instead, the two compartments exist in a state of electrical disequilibrium [p. 130].

The concept of electrical disequilibrium has traditionally been taught in chapters on nerve and muscle function because those tissues generate electrical signals known as action potentials. Yet one of the most exciting recent discoveries in physiology is the realization that other kinds of cells also use electrical signals for communication. In fact, all living organisms, including plants, use electrical signals! This section reviews the basic principles of electricity and discusses what creates electrical disequilibrium in the body. The chapter ends with a look at how beta cells of the pancreas use changes in the distribution of ions across cell membranes to trigger insulin secretion.

Electricity Review

Atoms are electrically neutral [p. 40]. They are composed of positively charged protons, negatively charged electrons, and uncharged neutrons, but in balanced proportions, so that an atom is neither positive nor negative. The removal or addition of electrons to an atom creates the charged particles we know as ions. We have discussed several ions that are important in the human body, such as Na^+, K^+, and H^+. For each of these positive ions, somewhere in the body there is a matching electron, usually found as part of a negative ion. For example, when Na^+ in the body enters in the form of NaCl, the "missing" electron from Na^+ can be found on the Cl^-.

Remember the following important principles when you deal with electricity in physiological systems:

1. The **law of conservation of electrical charge** states that the net amount of electrical charge produced in any process is zero. This means that for every positive charge on an ion, there is an electron on another ion. Overall, the human body is electrically neutral.
2. Opposite charges (+ and −) are attracted to each other, but two charges of the same type (+ and +, or − and −) repel each other. The protons and electrons in an atom exhibit this attraction.
3. Separating positive charges from negative charges requires energy. For example, energy is needed to separate the protons and electrons of an atom.
4. When separated positive and negative charges can move freely toward each other, the material through which they move is called a **conductor**. Water is a good conductor of electrical charge. When separated charges cannot move through the material that separates them, the material is known as an **insulator**. The phospholipid bilayer of the cell membrane is a good insulator, as is the plastic coating on electrical wires.

The word *electricity* comes from the Greek word *elektron*, meaning "amber," the fossilized resin of trees. The Greeks discovered that if they rubbed a rod of amber with cloth, the amber acquired the ability to attract hair and dust. This attraction (called static electricity) arises from the separation of electrical charge that occurs when electrons move from the amber atoms to the cloth. To separate these charged particles, energy (work) must be put into the system. In the case of the amber, work was done by rubbing the rod. In the case of biological systems, the work is usually done by energy stored in ATP and other chemical bonds.

The Cell Membrane Enables Separation of Electrical Charge in the Body

In the body, separation of electrical charge takes place across the cell membrane. This process is shown in ■ Figure 5.23. The diagram shows an artificial cell filled with molecules that dissociate into positive and negative ions, represented by the plus and minus signs. Because the molecules were electrically neutral to begin with, there are equal numbers of positive and negative ions inside the cell. The cell is placed in an aqueous solution, also electrically neutral, that contains the same types of cations and anions. The phospholipid bilayer of the artificial cell, like the bilayer of a real cell, is not permeable to ions. Water can freely cross this cell membrane, making the extracellular and intracellular ion concentrations equal. In Figure 5.23a, the system is at chemical and electrical equilibrium.

In Figure 5.23b, an active transporter is inserted into the membrane. This carrier uses energy to move positive ions out

(a) Cell and solution are electrically and chemically at equilibrium.

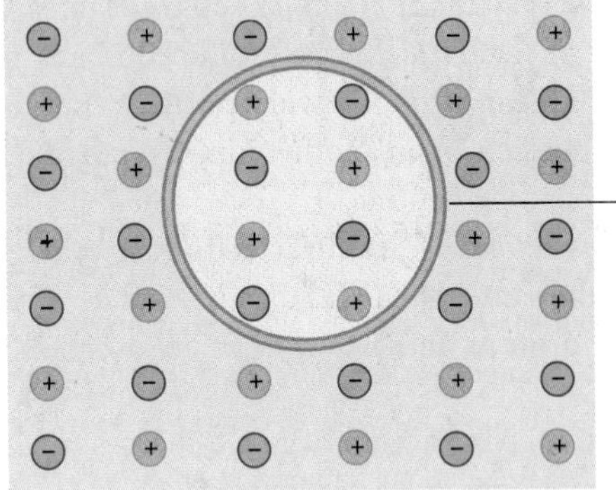

The cell membrane acts as an insulator to prevent free movement of ions between the intracellular and extracellular compartments.

(b) Energy is used to pump one cation out of the cell, leaving a net charge of -1 in the cell and + 1 outside the cell. Cell and solution are now in chemical and electrical disequilbrium.

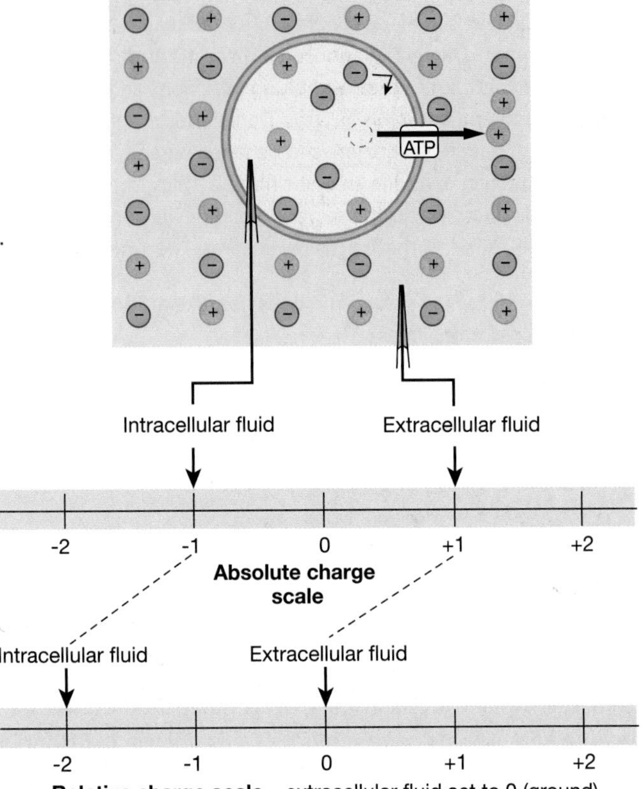

(c) On an absolute charge scale, the intracellular fluid (ICF) would be at -1 and the extracellular fluid (ECF) at +1.

Physiological measurements, however, are always on a relative scale, on which the extracellular fluid is assigned a value of 0. This shifts the scale to the left and gives the inside of the cell a relative charge of -2.

(d) In the laboratory, a cells's membrane potential is measured by placing one electrode inside the cell and a second in the extracellular bath.

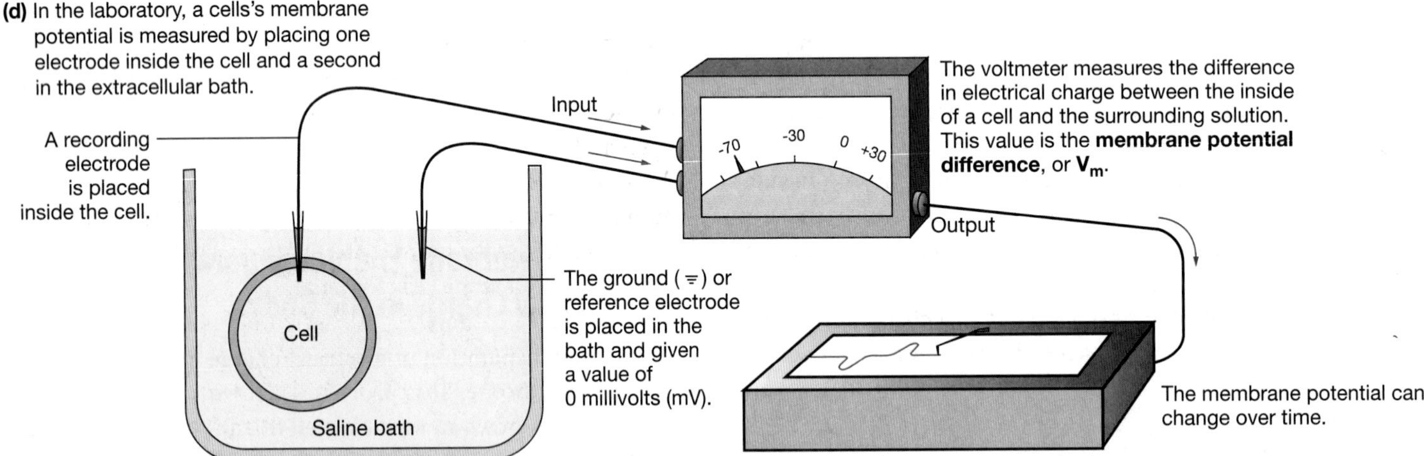

■ **Fig. 5.23** ***Separation of electrical charge: the membrane potential difference***

of the cell against their concentration gradient. The negative ions in the cell attempt to follow the positive ions because of the attraction of positive and negative charges. But because the membrane is impermeable to negative ions, they remain trapped in the cell. Positive ions outside the cell might try to move into the cell, attracted by the net negative charge of the intracellular fluid, but the membrane does not allow these cations to leak across it.

As soon as the first positive ion leaves the cell, the electrical equilibrium between the extracellular fluid and intracellular fluid is disrupted: the cell's interior has developed a net charge of −1 while the cell's exterior has a net charge of +1. The input of energy to transport ions across the membrane has created an **electrical gradient**—that is, a difference in the net charge between two regions. In this example, the inside of the cell becomes negative relative to the outside.

The active transport of positive ions out of the cell also creates a concentration gradient: there are now more positive ions outside the cell than inside. The combination of electrical and concentration gradients is called an **electrochemical gradient**. The cell remains in osmotic equilibrium because water can move freely across the membrane in response to solute movement.

An electrical gradient between the extracellular fluid and the intracellular fluid is known as the **resting membrane potential difference,** or **membrane potential** for short. Although the name sounds intimidating, we can break it apart to see what it means.

1. The *resting* part of the name comes from the fact that this electrical gradient is seen in all living cells, even those that appear to be without electrical activity. In these "resting" cells, the membrane potential has reached a steady state and is not changing.
2. The *potential* part of the name comes from the fact that the electrical gradient created by active transport of ions across the cell membrane is a form of stored, or potential, energy, just as concentration gradients are a form of potential energy. When oppositely charged molecules come back together, they release energy that can be used to do work, in the same way that molecules moving down their concentration gradient can do work. The work done by electrical energy includes opening voltage-gated membrane channels and sending electrical signals.
3. The *difference* part of the name is to remind you that the membrane potential represents a difference in the amount of electrical charge inside and outside the cell. The word *difference* is usually dropped from the name, as noted earlier.

In living systems, we measure electrical gradients on a relative scale rather than an absolute scale. Figure 5.23c compares the two scales. On the absolute scale, the extracellular fluid in our simple example has a net charge of +1 from the positive ion it gained, and the intracellular fluid has a net charge of −1 from the now unbalanced negative ion that was left behind.

However, in real life we cannot measure electrical charge as numbers of electrons gained or lost. Instead we use a device that measures the *difference* in electrical charge between two points. This device artificially sets the net electrical charge of one side of the membrane to 0 and measures the net charge of the second side relative to the first. In our example, resetting the extracellular fluid net charge to 0 gives the intracellular fluid a net charge of −2, and we call this value the cell's resting membrane potential.

The equipment for measuring a cell's membrane potential is depicted in Figure 5.23. *Electrodes* are created from hollow glass tubes drawn to very fine points. These *micropipettes* are filled with a liquid that conducts electricity and then connected to a *voltmeter,* which measures the electrical difference between two points in units of either volts (V) or millivolts (mV). A *recording electrode* is inserted through the cell membrane into the cytoplasm of the cell. A *reference electrode* is placed in the external bath, which represents the extracellular fluid.

In living systems, by convention, the extracellular fluid is designated as the *ground* and assigned a charge of 0 mV (Fig. 5.23c). When the recording electrode is placed inside a living cell, the voltmeter measures the membrane potential—in other words, the electrical difference between the intracellular fluid and the extracellular fluid. A recorder connected to the voltmeter can make a recording of the membrane potential versus time.

For nerve and muscle cells, the voltmeter will record a resting membrane potential between −40 and −90 mV, indicating that the intracellular fluid is negative relative to the extracellular fluid (0 mV). (Throughout this discussion, remember that the extracellular fluid is not really neutral because it has excess positive charges that exactly balance the excess negative charges inside the cell, as shown in Figure 5.23c. The total body remains electrically neutral at all times.)

The Resting Membrane Potential Is Due Mostly to Potassium

Which ions create the resting membrane potential in animal cells? The artificial cell shown in Figure 5.23b used an active transport protein to move an unspecified positive ion across a membrane that was otherwise impermeable to ions. But what processes go on in living cells to create an electrical gradient?

Real cells are not completely impermeable to all ions. They have open channels and protein transporters that allow ions to move between the cytoplasm and the extracellular fluid. We can use a different artificial cell to show how the resting membrane potential arises in a typical living cell.

The artificial cell in ■ Figure 5.24 has a membrane that is impermeable to ions. The cell contains K^+ and large negatively charged proteins, represented by Pr^-. The cell is placed in a solution of Na^+ and Cl^-. Both the cell and the solution are electrically neutral, and the system is in electrical equilibrium. However, it is not in chemical equilibrium. There are concentration gradients for all four types of ions in the system, and they would all diffuse down their respective concentration gradients if they could cross the cell membrane.

In Figure 5.24b, a K^+ leak channel is inserted into the membrane, making it permeable only to K^+. Because there is no K^+ in the extracellular fluid initially, some K^+ ions leak out of the cell, moving down their concentration gradient. As K^+ leaves the cell, the negatively charged proteins, Pr^-, are unable to follow because the cell membrane is not permeable to them. The proteins gradually build up a negative charge inside the cell as more and more K^+ diffuses out of the cell.

If the only force acting on K^+ were the concentration gradient, K^+ would leak out of the cell until the K^+ concentration inside the cell equaled the K^+ concentration outside. The loss of positive ions from the cell creates an electrical gradient, however. Because opposite charges attract each other, the negative proteins inside the cell try to pull K^+ back into the cell. At some point in this process, the electrical force attracting K^+ into the cell becomes equal in magnitude to the chemical concentration

(a) An artificial cell whose membrane is impermeable to ions is filled with K^+ and large protein anions. It is placed in a solution of Na^+ and Cl^-. Both cell and solution are electrically neutral.

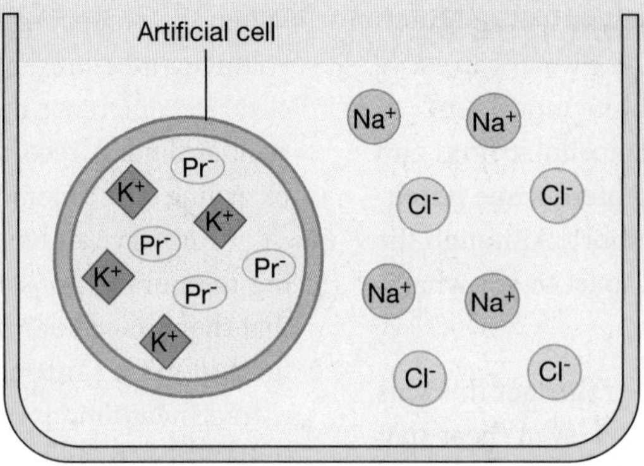

A K^+ leak channel is inserted into the membrane.

(b) K^+ leaks out of the cell because there is a K^+ concentration gradient.

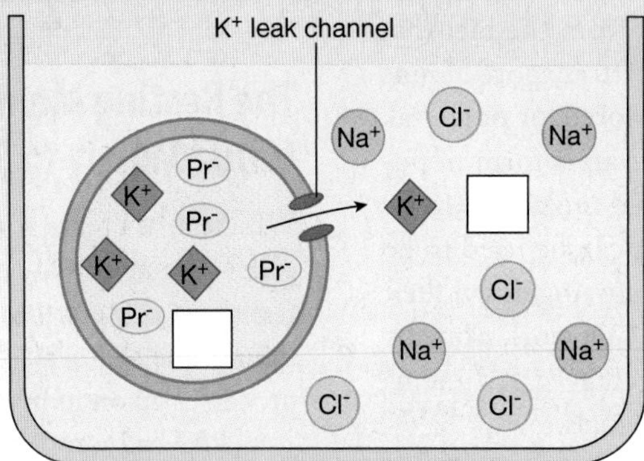

FIGURE QUESTION

In the white boxes write the net electrical charge of the intracellular and extracellular compartments as shown.

(c) The negative membrane potential attracts K^+ back into the cell. When the electrical gradient exactly opposes the K^+ concentration gradient, the resting membrane potential is the **equilibrium potential** for K^+.

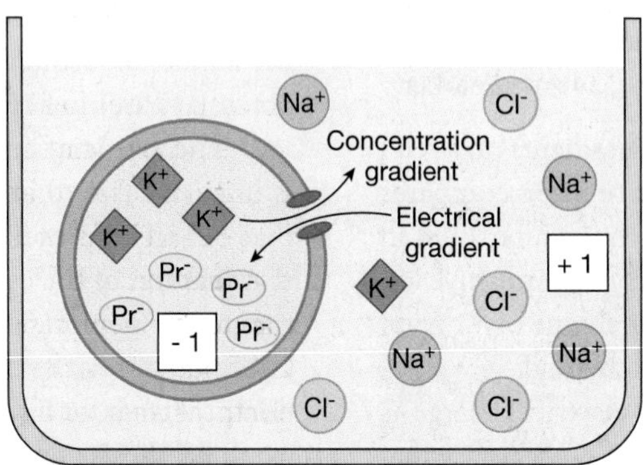

FIGURE QUESTION

Why don't Na^+, Cl^-, and the proteins (Pr^-) cross the membrane?

Now a Na^+ leak channel replaces the K^+ channel.

(d) The Na^+ concentration gradient in this artificial cell is exactly opposed by a membrane potential of +60 mV.

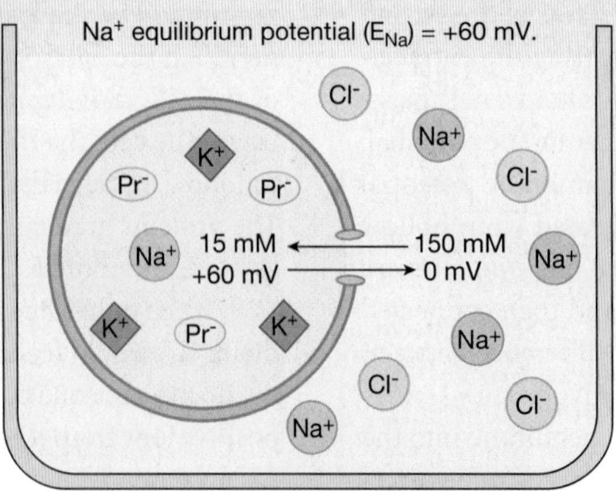

■ **Fig. 5.24** ***Equilibrium potentials.*** An ion's equilibrium potential, or E_{ion}, is the membrane potential at which the ion's chemical and electrical gradients are equal in magnitude and opposite in direction. A cell at the E_{ion} has no net movement of the ion across the membrane.

gradient driving K^+ out of the cell. At that point, net movement of K^+ across the membrane stops (Fig. 5.24c). The rate at which K^+ ions move out of the cell down the concentration gradient is exactly equal to the rate at which K^+ ions move into the cell down the electrical gradient.

In a cell that is permeable to only one ion, such as the artificial cell just described, the membrane potential that exactly opposes the concentration gradient of the ion is known as the **equilibrium potential**, or E_{ion} (where the subscript *ion* is replaced by the symbol for whichever ion we are looking at). For example, when the concentration gradient is 150 mM K^+ inside and 5 mM K^+ outside the cell, the equilibrium potential for potassium, or E_K, is −90 mV.

The equilibrium potential for any ion at 37 °C (human body temperature) can be calculated using the **Nernst equation:**

$$E_{ion} = \frac{61}{z} \log \frac{[\text{ion}]_{out}}{[\text{ion}]_{in}}$$

where 61 is 2.303 RT/F at 37 °C*

z is the electrical charge on the ion (+1 for K^+),

$[\text{ion}]_{out}$ and $[\text{ion}]_{in}$ are the ion concentrations outside and inside the cell, and

E_{ion} is measured in mV.

Now we will use the same artificial cell (K^+ and Pr^- inside, Na^+ and Cl^- outside), but this time we will make the membrane permeable only to Na^+ (Fig. 5.24d). Because Na^+ is more concentrated outside the cell, some Na^+ moves into the cell and accumulates there. Meanwhile, Cl^- left behind in the extracellular fluid gives that compartment a net negative charge. This imbalance creates an electrical gradient that tends to drive Na^+ back out of the cell. When the Na^+ concentration is 150 mM outside and 15 mM inside, the equilibrium potential for Na^+ (E_{Na}) is +60 mV. In other words, the concentration gradient moving Na^+ into the cell (150 mM outside, 15 mM inside) is exactly opposed by a positive membrane potential of +60 mV.

In reality, living cells are not permeable to only one ion. The situation in real cells is similar to a combination of the two artificial systems just described. If a cell is permeable to several ions, we cannot use the Nernst equation to calculate membrane potential. Instead we must use a related equation called the *Goldman equation* that considers concentration gradients of the permeable ions and the relative permeability of the cell to each ion. [For more detail on the Goldman equation, see Chapter 8.]

The real cell illustrated in Figure 5.25 has a resting membrane potential of −70 mV. Most cells are about 40 times more permeable to K^+ than to Na^+. As a result, a cell's resting membrane potential is closer to the E_K of −90 mV than to the E_{Na} of +60 mV.

*R is the ideal gas constant, T is absolute temperature, and F is the Faraday constant. For additional information, see Appendix B.

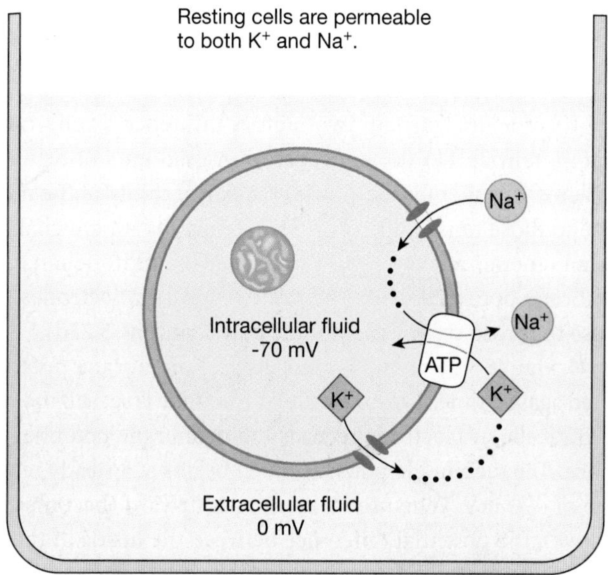

FIGURE QUESTIONS

- What force(s) promote(s) Na^+ leak into the cell?
- What force(s) promote(s) K^+ leak out of the cell?

Fig. 5.25 ***Resting membrane potential in an actual cell.*** Most cells in the human body are about 40 times more permeable to K^+ than to Na^+, and the resting membrane potential is about −70 mV.

A small amount of Na^+ leaks into the cell, making the inside of the cell less negative than it would be if Na^+ were totally excluded. Additional Na^+ that leaks in is promptly pumped out by the Na^+-K^+-ATPase, as described earlier. At the same time, K^+ ions that leak out of the cell are pumped back in. The pump contributes to the membrane potential by pumping 3 Na^+ out for every 2 K^+ pumped in. Because the Na^+-K^+-ATPase helps maintain the electrical gradient, it is called an *electrogenic* pump.

Not all ion transport creates an electrical gradient. Many transporters, like the Na^+-K^+-$2Cl^-$ (NKCC) symporter, are electrically neutral. Some make an even exchange: for each charge that enters the cell, the same charge leaves. An example is the HCO_3^--Cl^- antiporter of red blood cells, which transports these ions in a one-for-one, electrically neutral exchange. Electrically neutral transporters have little effect on the resting membrane potential of the cell.

Concept Check Answers: p. 173

34. Add a Cl^- leak channel to the artificial cell shown in Figure 5.24a, and then figure out which way Cl^- will move along the concentration and electrical gradients. Will the Cl^- equilibrium potential be positive or negative?

35. What would happen to the resting membrane potential of a cell poisoned with ouabain (an inhibitor of the Na^+-K^+-ATPase)?

Changes in Ion Permeability Change the Membrane Potential

As you have just learned, two factors influence a cell's membrane potential: (1) the concentration gradients of different ions across the membrane and (2) the permeability of the membrane to those ions. If the cell's permeability to an ion changes, the cell's membrane potential changes. We monitor changes in membrane potential using the same recording electrodes that we use to record resting membrane potential (Fig. 5.23d).

■ Figure 5.26 shows a recording of membrane potential plotted against time. The extracellular electrode is set at 0 mV, and the intracellular electrode records the membrane potential difference. The membrane potential (V_m) begins at a steady resting value of −70 mV. When the trace moves upward (becomes less negative), the potential difference between the inside of the cell and the outside (0 mV) decreases, and the cell is said to have *depolarized*. A return to the resting membrane potential is termed *repolarization*. If the resting potential becomes more negative, the potential difference has increased, and the cell has *hyperpolarized*.

A major point of confusion when we talk about changes in membrane potential is the use of the phrases "the membrane potential decreased" or "the membrane potential increased." Normally, we associate "increase" with becoming more positive and "decrease" with becoming more negative—the opposite of what is happening in our cell discussion. One way to avoid confusion is to add the word *difference* after *membrane potential*. If the membrane potential *difference* is *increasing*, the value of V_m must be moving away from the ground value of 0 and becoming *more negative*. If the membrane potential *difference* is *decreasing*, the value of V_m is moving closer to the ground value of 0 mV and is becoming *less negative*.

What causes changes in membrane potential? In most cases, membrane potential changes in response to movement of one of four ions: Na^+, Ca^{2+}, Cl^-, and K^+. The first three are more concentrated in the extracellular fluid than in the cytosol, and the resting cell is minimally permeable to them. If a cell suddenly becomes more permeable to any one of these ions, then those ions will move down their electrochemical gradient into the cell. Entry of Ca^{2+} or Na^+ depolarizes the cell (makes the membrane potential more positive). Entry of Cl^- hyperpolarizes the cell (makes the membrane potential more negative).

Most resting cells are fairly permeable to K^+, but making them more permeable allows even more K^+ to leak out. The cell hyperpolarizes until it reaches the equilibrium potential for K^+. Making the cell *less* permeable to K^+ allows fewer K^+ to leak out of the cell. When the cell retains K^+, it becomes more positive and depolarizes. You will encounter instances of all these permeability changes as you study physiology.

It is important to learn that a significant change in membrane potential requires the movement of very few ions. *The concentration gradient does not have to reverse to change the membrane potential.* For example, to change the membrane potential by 100 mV (the size of a typical electrical signal passing down a neuron), only one of every 100,000 K^+ must enter or leave the cell. This is such a tiny fraction of the total number of K^+ ions in the cell that the concentration gradient for K^+ remains essentially unchanged.

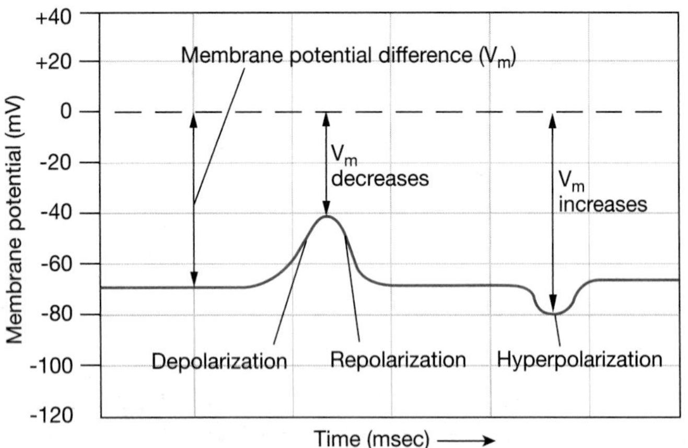

■ **Fig. 5.26** ***Membrane potential terminology.*** If the membrane potential becomes less negative than the resting potential, the cell depolarizes. If the membrane potential becomes more negative, the cell hyperpolarizes.

Integrated Membrane Processes: Insulin Secretion

The movement of Na^+ and K^+ across cell membranes has been known to play a role in generating electrical signals in excitable tissues for many years. You will study these processes in detail when you learn about the nervous and muscular systems. Recently, however, we have come to understand that small changes in membrane potential act as signals in nonexcitable tissues, such as endocrine cells. One of the best-studied examples of this process involves the beta cell of the pancreas. Release of the hormone insulin by beta cells demonstrates how membrane processes—such as facilitated diffusion, exocytosis, and the opening and closing of ion channels by ligands and membrane potential—work together to regulate cell function.

The beta cells of the pancreas synthesize the protein hormone insulin and store it in cytoplasmic secretory vesicles [p. 77]. When blood glucose levels increase, such as after a meal, the beta cells release insulin by exocytosis. Insulin then directs other cells of the body to take up and use glucose, bringing blood concentrations down to pre-meal levels.

A key question about the process that went unanswered until recently was, "How does a beta cell 'know' that glucose levels have gone up and that it needs to release insulin?" The answer, we have now learned, links the beta cell's metabolism to its electrical activity.

■ Figure 5.27a shows a beta cell at rest. Recall from earlier sections in this chapter that gated membrane channels can be opened or closed by chemical or electrical signals. The beta cell has

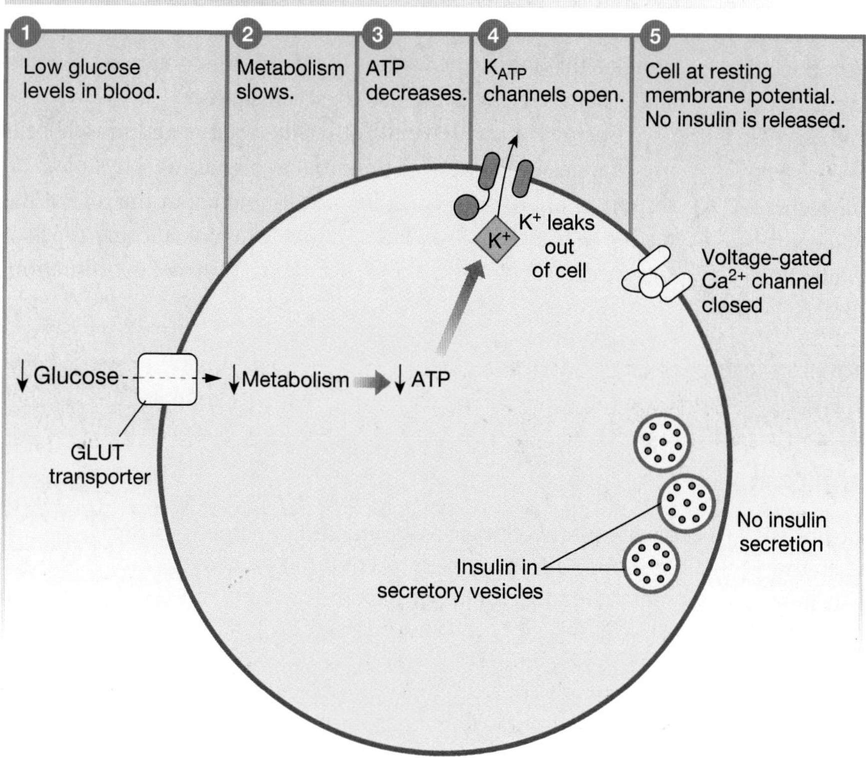
(a) Beta cell at rest. The K_{ATP} channel is open, and the cell is at its resting membrane potential.
1
Low glucose levels in blood.
2
Metabolism slows.
3
ATP decreases.
4
K_{ATP} channels open.
5
Cell at resting membrane potential. No insulin is released.
K^+ leaks out of cell
K^+
Voltage-gated Ca^{2+} channel closed
↓Glucose
↓Metabolism
↓ATP
GLUT transporter
No insulin secretion
Insulin in secretory vesicles

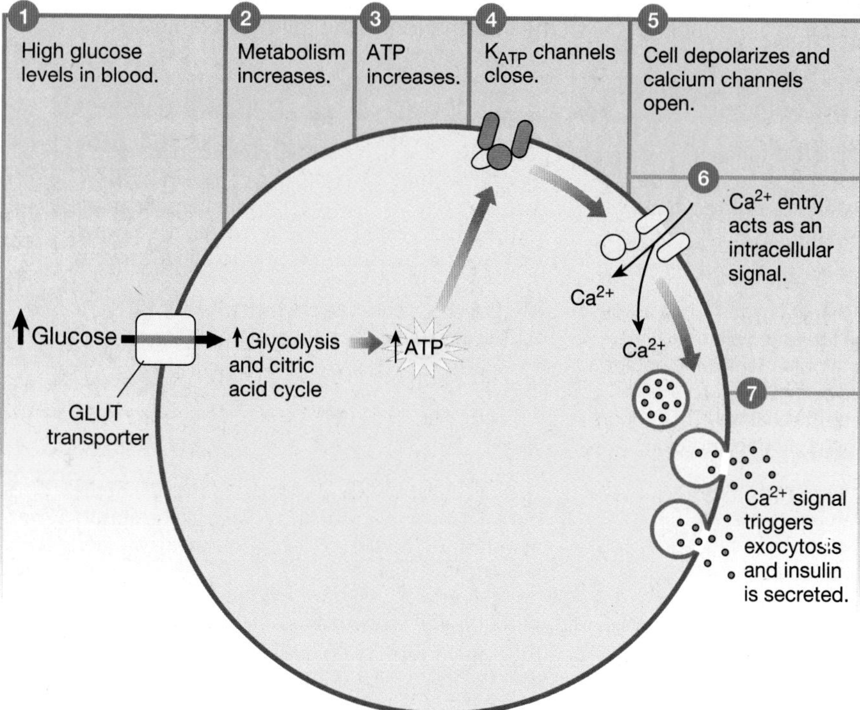
(b) Beta cell secretes insulin. Closure of K_{ATP} channel depolarizes cell, triggering exocytosis of insulin.
1
High glucose levels in blood.
2
Metabolism increases.
3
ATP increases.
4
K_{ATP} channels close.
5
Cell depolarizes and calcium channels open.
6
Ca^{2+} entry acts as an intracellular signal.
Ca^{2+}
Ca^{2+}
↑Glucose
↑Glycolysis and citric acid cycle
↑ATP
GLUT transporter
7
Ca^{2+} signal triggers exocytosis and insulin is secreted.

■ **Fig. 5.27** ***Insulin secretion and membrane transport processes***

two such channels that help control insulin release. One is a **voltage-gated Ca^{2+} channel**. This channel is closed at the cell's resting membrane potential (5 in Fig. 5.27a). The other is a K^+ leak channel (that is, the channel is usually open) that closes when ATP binds to it. It is called an **ATP-gated K^+ channel,** or **K_{ATP} channel**. In the resting cell, when glucose concentrations are low, the cell makes less ATP 1–3. There is little ATP to bind to the K_{ATP} channel, and the channel remains open, allowing K^+ to leak out of the cell 4. At the resting membrane potential, the voltage-gated Ca^{2+} channels are closed, and there is no insulin secretion 5.

Figure 5.27b shows a beta cell secreting insulin. Following a meal, plasma glucose levels increase as glucose is absorbed from the intestine (1). Glucose reaching the beta cell diffuses into the cell with the aid of a GLUT transporter. Increased glucose in the cell stimulates the metabolic pathways of glycolysis and the citric acid cycle [p. 112], and ATP production increases (2), (3). When ATP binds to the K_{ATP} channel, the gate to the channel closes, preventing K^+ from leaking out of the cell (4). Retention of K^+ depolarizes the cell (5), which then causes the voltage-sensitive Ca^{2+} channels to open (6). Calcium ions enter the cell from the extracellular fluid, moving down their electrochemical gradient. The Ca^{2+} binds to proteins that initiate exocytosis of the insulin-containing vesicles, and insulin is released into the extracellular space (7).

The discovery that cells other than nerve and muscle cells use changes in membrane potential as signals for physiological responses altered our traditional thinking about the role of the resting membrane potential. Next we will look at other types of signals that the body uses for communication and coordination.

RUNNING PROBLEM CONCLUSION

Cystic Fibrosis

In this running problem, you learned about cystic fibrosis (CF), one of the most common inherited diseases in the United States. By some estimates, more than 10 million people are symptomless carriers of the CF gene. A person must inherit two copies of the gene, one from each parent, before he or she will develop CF. Although there is no cure for this disease, treatments have become better, and the life span of CF patients continues to improve. Today the median survival age is around 37.

Cystic fibrosis is caused by a defect in the CFTR channel protein, which regulates the transport of Cl^- into and out of epithelial cells. Because CFTR channels are found in the epithelial cell membranes of several organs—the sweat glands, lungs, and pancreas—cystic fibrosis may affect many different body processes. Some of the most interesting animal research on cystic fibrosis uses genetically altered mice, called CF mice. These model animals can be bred to have either totally nonfunctional CFTR or CFTR channels with altered functions corresponding to the mutations of the CFTR gene in humans.

To learn more about current research in this disease, go to the Cystic Fibrosis Foundation web site (*www.cff.org*) and click the *Research Overview* tab. To check your understanding of the running problem, compare your answers with the information in the following table.

Question	Facts	Integration and Analysis
1. Why would failure to transport NaCl into the airways cause the secreted mucus to be thick?	If NaCl is secreted into the lumen of the airways, the solute concentration of the airway fluid increases. Water moves into compartments with higher osmolarity.	Normally, movement of NaCl creates an osmotic gradient so that water also enters the airway lumen, creating a saline solution that thins the thick mucus. If NaCl cannot be secreted into the airways, there will be no fluid movement to thin the mucus.
2. Is the CFTR a chemically gated, a voltage-gated, or a mechanically gated channel protein?	Chemically gated channels open when a ligand binds to them. Voltage-gated channels open with a change in the cell's membrane potential. Mechanically gated channels open when a physical force opens the channel. CFTRs open when ATP binds to the channel protein.	ATP is a chemical ligand, which means CFTRs are chemically gated channel proteins.
3. Based on the information given, is the CFTR protein on the apical or basolateral surface of the sweat gland epithelium?	In normal people, the CFTR channels move Cl^- from sweat into epithelial cells.	The epithelial surface that faces the lumen of the sweat gland, which contains sweat, is the apical membrane. Therefore, the CFTR proteins are on the apical surface.
4. Why will Daniel starve if he does not take artificial pancreatic enzymes?	The pancreas secretes mucus and digestive enzymes into ducts that empty into the small intestine. In cystic fibrosis, mucus in the ducts is thick because of lack of Cl^- and fluid secretion. This thick mucus blocks the ducts and prevents digestive enzymes from reaching the small intestine.	Without digestive enzymes, Daniel cannot digest the food he eats. His weight loss over the past six months suggests that this has already become a problem. Taking artificial enzymes will enable him to digest his food.

133 145 156 161 165 171

Chapter Summary

Several key themes come together in this chapter. You learned how the cell membrane acts as a barrier to create distinct intracellular and extracellular compartments, illustrating the theme of *compartmentation*. Although the contents of the intracellular and extracellular compartments differ, *homeostasis* keeps them in a dynamic steady state. Movement of materials between and within compartments is necessary for *communication* and is accomplished by *bulk flow* and *biological transport*. Flow of solutes and water across cell membranes occurs in response to osmotic, chemical (concentration), or electrical gradients. The selectively permeable cell membrane creates resistance to flow that can be overcome by changing the composition of the membrane lipids or by inserting membrane proteins that act as channels or transporters. Biological transport in the body requires *energy* from different sources: molecular motion, concentration gradients, or chemical bonds. Finally, the binding of substrates to transporters demonstrates the theme of *protein interactions*.

5

Osmosis and Tonicity

iP Fluids and Electrolytes: Introduction to Body Fluids

1. Most solutes are concentrated in either one compartment or the other, creating a state of **chemical disequilibrium.** (p. 131; Fig. 5.1)
2. Cations and anions are not distributed equally between the body compartments, creating a state of **electrical disequilibrium.** (p. 130)
3. Water moves freely between the cells and extracellular fluid, resulting in a state of **osmotic equilibrium.** (p. 130)
4. The movement of water across a membrane in response to a concentration gradient is called **osmosis.** (p. 132)
5. To compare solution concentrations, we express the concentration in terms of **osmolarity,** the number of particles (ions or intact molecules) per liter of solution, expressed as milliosmoles per liter (mOsM). (p. 133)
6. **Tonicity** of a solution describes the cell volume change that occurs at equilibrium if the cell is placed in that solution. Cells swell in **hypotonic solutions** and shrink in **hypertonic solutions.** If the cell does not change size at equilibrium, the solution is **isotonic.** (p. 134)
7. The osmolarity of a solution cannot be used to determine the tonicity of the solution. The relative concentrations of **nonpenetrating solutes** in the cell and in the solution determine tonicity. **Penetrating solutes** contribute to the osmolarity of a solution but not to its tonicity. (p. 136; Fig. 5.4)

Diffusion

8. In **bulk flow** a pressure gradient moves a fluid along with its dissolved and suspended materials. (p. 139)
9. The cell membrane is a selectively permeable barrier that restricts free exchange between the cell and the interstitial fluid. The movement of a substance across the membrane depends on the permeability of the membrane to that substance. (p. 140)
10. Movement of molecules across membranes can be classified either by energy requirements or by the physical means the molecule uses to cross the membrane. (p. 140; Fig. 5.5)
11. Lipid-soluble substances can diffuse through the phospholipid bilayer. Less lipid-soluble molecules require the assistance of a membrane protein to cross the membrane. (p. 140)
12. **Passive transport** does not require the input of energy. (p. 140)
13. **Diffusion** is the passive movement of molecules down a chemical (concentration) gradient from an area of higher concentration to an area of lower concentration. Net movement stops when the system reaches **equilibrium,** although molecular movement continues. (p. 143; Tbl. 5.6)
14. Diffusion rate depends on the magnitude of the concentration gradient. Diffusion is slow over long distances, is directly related to temperature, and is inversely related to molecular size. (p. 144)
15. **Simple diffusion** across a membrane is directly proportional to membrane surface area, concentration gradient, and membrane permeability, and inversely proportional to membrane thickness. (p. 144; Fig. 5.7)

Protein-Mediated Transport

16. Most molecules cross membranes with the aid of membrane proteins. (p. 145)
17. Membrane proteins have four functional roles: **structural proteins** maintain cell shape and form cell junctions; **membrane-associated enzymes** catalyze chemical reactions and help transfer signals across the membrane; **receptor proteins** are part of the body's signaling system; and **transport proteins** move many molecules into or out of the cell. (p. 145; Fig. 5.8)
18. **Channel proteins** form water-filled channels that link the intracellular and extracellular compartments. **Gated channels** regulate movement of substances through them by opening and closing. Gated channels may be regulated by ligands, by the electrical state of the cell, or by physical changes such as pressure. (p. 147; Fig. 5.10)

19. **Carrier proteins** never form a continuous connection between the intracellular and extracellular fluid. They bind to substrates, then change conformation. (p. 149; Fig. 5.12)
20. Protein-mediated diffusion is called **facilitated diffusion.** It has the same properties as simple diffusion. (p. 143, 150; Tbl. 5.6; Fig. 5.13)
21. **Active transport** moves molecules against their concentration gradient and requires an outside source of energy. In **primary (direct) active transport**, the energy comes directly from ATP. **Secondary (indirect) active transport** uses the potential energy stored in a concentration gradient and is indirectly driven by energy from ATP. (pp. 150–151)
22. The most important primary active transporter is the **sodium-potassium-ATPase** (Na^+-K^+-ATPase), which pumps Na^+ out of the cell and K^+ into the cell. (p. 151; Fig. 5.14)
23. Most secondary active transport systems are driven by the sodium concentration gradient. (p. 151, 153; Tbl. 5.8; Fig. 5.16)
24. All carrier-mediated transport demonstrates **specificity, competition,** and **saturation.** Specificity refers to the ability of a transporter to move only one molecule or a group of closely related molecules. Related molecules may compete for a single transporter. Saturation occurs when a group of membrane transporters are working at their maximum rate. (p. 154; Fig. 5.17)

Vesicular Transport

25. Large macromolecules and particles are brought into cells by phagocytosis or endocytosis. Material leaves cells by exocytosis. When vesicles that come into the cytoplasm by endocytosis are returned to the cell membrane, the process is called **membrane recycling.** (pp. 155–156; Figs. 5.18, 5.19)
26. In **receptor-mediated endocytosis,** ligands bind to membrane receptors that concentrate in **coated pits** or **caveolae.** (p. 156; Fig. 5.19)
27. In exocytosis, the vesicle membrane fuses with the cell membrane before releasing its contents into the extracellular space. Exocytosis requires ATP. (p. 157)

Epithelial Transport

28. Transporting epithelia have different membrane proteins on their **apical** and **basolateral** surfaces. This polarization allows one-way movement of molecules across the epithelium. (pp. 158–159; Figs. 5.20, 5.21)
29. Molecules cross epithelia by moving between the cells by the **paracellular** route or through the cells by the **transcellular** route. (pp. 158–159; Fig. 5.20)
30. Larger molecules cross epithelia by **transcytosis,** which includes **vesicular transport.** (p. 160; Fig. 5.22)

The Resting Membrane Potential

iP Nervous I: The Membrane Potential

31. Although the total body is electrically neutral, diffusion and active transport of ions across the cell membrane create an **electrical gradient,** with the inside of cells negative relative to the extracellular fluid. (p. 162; Fig. 5.23)
32. The electrical gradient between the extracellular fluid and the intracellular fluid is known as the **resting membrane potential difference.** (p. 163)
33. The movement of an ion across the cell membrane is influenced by the **electrochemical gradient** for that ion. (pp. 162–163)
34. The membrane potential that exactly opposes the concentration gradient of an ion is known as the **equilibrium potential** (E_{ion}). The equilibrium potential for any ion can be calculated using the Nernst equation. (p. 164; Fig. 5.24)
35. In most living cells, K^+ is the primary ion that determines the resting membrane potential. (pp. 163–165)
36. Changes in membrane permeability to ions such as K^+, Na^+, Ca^{2+}, or Cl^- alter membrane potential and create electrical signals. (p. 166)

Integrated Membrane Processes: Insulin Secretion

37. The use of electrical signals to initiate a cellular response is a universal property of living cells. Pancreatic beta cells release insulin in response to a change in membrane potential. (p. 167; Fig. 5.27)

Questions

Answers A1.

Level One Reviewing Facts and Terms

1. List the four functions of membrane proteins, and give an example of each.
2. Distinguish between active transport and passive transport.
3. Which of the following processes are examples of active transport, and which are examples of passive transport? Simple diffusion, phagocytosis, facilitated diffusion, exocytosis, osmosis, endocytosis.
4. List four factors that increase the rate of diffusion in air.
5. List the three physical methods by which materials enter cells.
6. A cotransporter is a protein that moves more than one molecule at a time. If the molecules are moved in the same direction, the transporters are called ________ carriers; if the molecules are transported in opposite directions, the transporters are called ________ carriers. A transport protein that moves only one substrate is called a(n) ________ carrier.
7. The two types of active transport are ________, which derives energy directly from ATP, and ________, which couples the kinetic energy of one molecule moving down its concentration gradient to the movement of another molecule against its concentration gradient.
8. A molecule that moves freely between the intracellular and extracellular compartments is said to be a(n) ________ solute. A molecule that is not able to enter cells is called a(n) ________ solute.
9. Rank the following individuals in order of how much body water they contain, from highest to lowest: (a) a 25-year-old, 70-kg male; (b) a 25-year-old, 50-kg female; (c) a 65-year-old, 50-kg female; and (d) a 1-year-old male toddler.
10. What determines the osmolarity of a solution? In what units is body osmolarity usually expressed?
11. What does it mean if we say that a solution is hypotonic to a cell? Hypertonic to the same cell? What determines the tonicity of a solution relative to a cell?

12. Match the membrane channels with the appropriate descriptions. Answers may be used once, more than once, or not at all.

(a) chemically gated channel (b) open pore (c) voltage-gated channel (d) mechanically gated channel	1. channel that spends most of its time in the open state 2. channel that spends most of its time in a closed state 3. channel that opens when resting membrane potential changes 4. channel that opens when a ligand binds to it 5. channel that opens in response to membrane stretch 6. channel through which water can pass

13. In your own words, state the four principles of electricity important in physiology.
14. Match each of the following items with its primary role in cellular activity.

(a) Na^+-K^+-ATPase	1. ion channel
(b) protein	2. extracellular cation
(c) unit of measurement for membrane potential	3. source of energy
	4. intracellular anion
(d) K^+	5. intracellular cation
(e) Cl^-	6. millivolts
(f) ATP	7. electrogenic pump
(g) Na^+	8. extracellular anion
	9. milliosmoles

15. The membrane potential at which the electrical gradient exactly opposes the concentration gradient for an ion is known as the ________.
16. A material that allows free movement of electrical charges is called a(n) ________, whereas one that prevents this movement is called a(n) ________.

Level Two Reviewing Concepts

17. Create a map of transport across cell membranes using the following terms. You may add additional terms if you wish.

• active transport	• ligand
• carrier	• Na^+-K^+-ATPase
• caveolae	• osmosis
• channel	• passive transport
• clathrin-coated pit	• phospholipid bilayer
• concentration gradient	• receptor-mediated endocytosis
• electrochemical gradient	• secondary active transport
• exocytosis	• simple diffusion
• facilitated diffusion	• small polar molecule
• glucose	• transcytosis
• GLUT transporter	• vesicle
• ion	• vesicular transport
• large polar molecule	• water

18. Draw a large rectangle to represent the total body volume. Using the information in Figure 5.1b, divide the box proportionately into compartments to represent the different body compartments. Use the information in Figure 5.1d and add solutes to the compartments. Use large letters for solutes with higher concentrations, and small letters for solutes with low concentrations. Label the cell membranes and the endothelial membrane.
19. What factors influence the rate of diffusion across a membrane? Briefly explain each one.
20. Define the following terms and explain how they differ from one another: specificity, competition, saturation. Apply these terms in a short explanation of facilitated diffusion of glucose.
21. Red blood cells are suspended in a solution of NaCl. The cells have an osmolarity of 300 mOsM, and the solution has an osmolarity of 250 mOsM. (a) The solution is (hypertonic, isotonic, or hypotonic) to the cells. (b) Water would move (into the cells, out of the cells, or not at all).
22. Two compartments are separated by a membrane that is permeable to glucose. Each compartment is filled with 1 M glucose. After 6 hours, compartment A contains 1.5 M glucose and compartment B contains 0.5 M glucose. What kind of transport occurred? Explain.
23. A 2 M NaCl solution is placed in compartment A and a 2 M glucose solution is placed in compartment B. The compartments are separated by a membrane that is permeable to water but not to NaCl or glucose. Complete the following statements. Defend your answers.
    (a) The salt solution is ________ osmotic to the glucose solution.
    (b) True or false? Water will move from one compartment to another. If water moves, it will move from compartment ________ to compartment ________.
24. Explain the differences between a chemical gradient, an electrical gradient, and an electrochemical gradient.

Level Three Problem Solving

25. Sweat glands secrete into their lumen a fluid that is identical to interstitial fluid. As the fluid moves through the lumen on its way to the surface of the skin, the cells of the sweat gland's epithelium make the fluid hypotonic by removing Na^+ and leaving water behind. Design an epithelial cell that will reabsorb Na^+ but not water. You may place water pores, Na^+ leak channels, K^+ leak channels, and the Na^+-K^+-ATPase in the apical membrane, basolateral membrane, or both.
26. Insulin is a hormone that promotes the movement of glucose into many types of cells, thereby lowering blood glucose concentration. Propose a mechanism that explains how this occurs, using your knowledge of cell membrane transport.
27. The following terms have been applied to membrane carriers: specificity, competition, saturation. Why can these terms also be applied to enzymes? What is the major difference in how enzymes and carriers carry out their work?
28. NaCl is a nonpenetrating solute and urea is a penetrating solute for cells. Red blood cells are placed in each of the solutions below. The intracellular concentration of nonpenetrating solute is 300 mOsM. What will happen to the cell volume in each solution? Label each solution with all the terms that apply: hypertonic, isotonic, hypotonic, hyperosmotic, hyposmotic, isosmotic. Watch units! Assume 1 M NaCl = 2 OsM for simplicity.
    (a) 150 mM NaCl plus 150 mM urea
    (b) 100 mM NaCl plus 50 mM urea
    (c) 100 mM NaCl plus 100 mM urea
    (d) 150 mM NaCl plus 100 mM urea
    (e) 100 mM NaCl plus 150 mM urea

29. Integral membrane glycoproteins have sugars added as the proteins pass through the lumen of the endoplasmic reticulum and Golgi complex [p. 77]. Based on this information, where would you predict finding the sugar "tails" of the proteins: on the cytoplasmic side of the membrane, the extracellular side, or both? Explain your reasoning.

Level Four Quantitative Problems

30. The addition of dissolved solutes to water lowers the freezing point of water. A 1 OsM solution depresses the freezing point of water by 1.86 °C. If a patient's plasma shows a freezing-point depression of 0.550 °C, what is her plasma osmolarity? (Assume that 1 kg water = 1 L.)
31. The patient in the previous question is found to have total body water volume of 42 L, ECF volume of 12.5 L, and plasma volume of 2.7 L.
    (a) What is her intracellular fluid (ICF) volume? Her interstitial fluid volume?
    (b) How much solute (osmoles) exists in her whole body? ECF? ICF? plasma? (*Hint:* concentration = solute amount/volume of solution)
32. What is the osmolarity of half-normal saline (= 0.45% NaCl)? [p. 139] Assume that all NaCl molecules dissociate into two ions.
33. If you give 1 L of half-normal saline (see question 32) to the patient in question 31, what happens to each of the following at equilibrium? (*Hint:* NaCl is a nonpenetrating solute.)
    (a) her total body volume
    (b) her total body osmolarity
    (c) her ECF and ICF volumes
    (d) her ECF and ICF osmolarities
34. The following graph shows the results of an experiment in which a cell was placed in a solution of glucose. The cell had no glucose in it at the beginning, and its membrane can transport glucose. Which of the following processes is/are illustrated by this experiment?

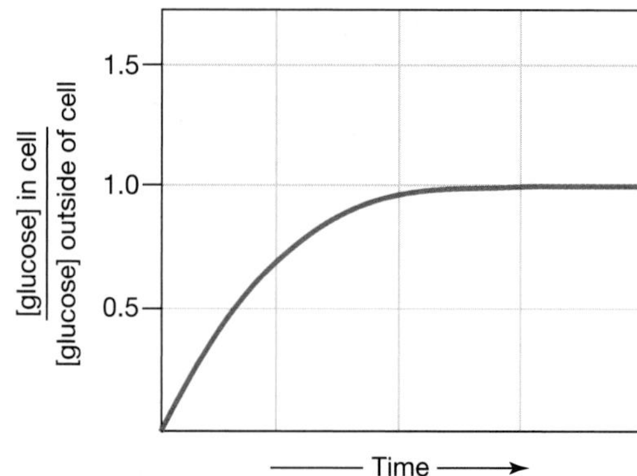

    (a) simple diffusion
    (b) saturation
    (c) competition
    (d) active transport

Answers

Answers to Concept Check Questions

Page 132

1. The stem preceding the suffix *-ase* is the name of the substrate on which the enzyme acts; therefore ATP is a substrate for this enzyme.
2. Intracellular fluid has a high K^+ concentration and low Na^+, Cl^-, and Ca^{2+} concentrations.
3. Plasma is part of the ECF, so knowing the plasma concentration of (a) Na^+ and (b) K^+ tells you the ECF concentration of these ions. The ICF ion concentrations are different from the ECF concentrations, so plasma concentration will not tell you the ICF values. (c) The body is in osmotic equilibrium, so water concentration is the same in all compartments. (d) Proteins are in the plasma but not in the interstitial fluid, so plasma protein concentration will not tell you ECF or ICF protein concentrations.
4. (a) total body water = 29 L; (b) ECF = 19.1, ICF = 38.9 L; (c) plasma = 4.8 L

Page 134

5. The baby has lost 0.91 kg of water, which is 0.91 liter.

Page 139

6. 1 M NaCl = 2 OsM NaCl. The 1 M (= 1 OsM) glucose and 1 OsM NaCl have the most water.
7. (a) Water moves into A because A is 2 OsM; (b) no net movement occurs because urea will diffuse across the membrane until it reaches equilibrium; (c) water moves into A, because A has a higher concentration of nonpenetrating solutes.
8. Prediction: 260 mOsM glucose is hyposmotic and hypotonic. Adding 260 mosmoles glucose and 1 L volume:

	Total Body	ECF	ICF
S (mosmol)	900 + 260 = 1160	300	600 + 260 = 860
V (L)	3 + 1 = 4	1.034	2.966
C (mOsM)	1160/4 = 290 mOsM →	290 →	290

9. (a) Sweat loss is hyposmotic. Body osmolarity will increase. (b) Cell volume will decrease. (c) Losing 0.5 L and 65 mosmoles NaCl. Yes, cell volume decreased and osmolarity increased.

	Total Body	ECF	ICF
S (mosmol)	900 − 65 = 835	300 − 65 = 235	600
V (L)	3 − 0.5 = 2.5	0.704	1.796
C (mOsM)	334 mOsM →	334 →	334

10. (a) The NaCl solution is better, even though both solutions are isosmotic to the body (Tbl. 5.8). Because blood is lost from the extracellular compartment, the best replacement solution would remain in the ECF. For this reason glucose is not as good a choice because it slowly enters cells, taking water with it. (b) If 1 L has been lost, you should replace at least 1 L.

Page 142

11. If distance triples, diffusion takes nine times as long.

Page 144

12. Energy for diffusion comes from molecular motion.
13. Because it is lipophilic, the fatty acid is more likely to cross by simple diffusion.
14. The flux (a) decreases, (b) increases, (c) decreases.
15. Compartment A remains yellow, and compartment B turns green.
16. The skin's thick extracellular matrix is generally impermeable to oxygen. Also, oxygen needs a moist surface to effectively diffuse across a tissue membrane.

Page 148

17. Positive ions are cations, and negative ions are anions.

Page 149

18. Membrane proteins serve as structural proteins, receptors, enzymes, and transporters.
19. Ions and water molecules move through open channels.
20. Channel proteins form continuous connections between the two sides of a membrane and transport molecules more quickly.
21. A channel lined with positive charges attracts anions, which in this instance means Cl^-.
22. Glucose is too large to pass through a channel.

Page 150

23. The direction of facilitated diffusion of glucose reverses, and glucose leaves the cell.

Page 152

24. The ATPase is an antiporter, but the SGLT is a symporter. The ATPase requires energy from ATP to change conformation, whereas the SGLT uses energy stored in the Na^+ concentration gradient.

Page 154

25. An antiporter moves substrates in opposite directions.
26. Larger doors could move more people. This would be analogous to a cell's synthesizing a new isoform of the transporter that would let the transporter move more substrate per second.

Page 157

27. In phagocytosis, the cytoskeleton pushes the membrane out to engulf a particle in a large vesicle. In endocytosis, the membrane surface indents and the vesicle is much smaller.
28. The proteins associated with endocytosis are clathrin and caveolin.
29. Proteins move into cells by endocytosis and out of cells by exocytosis.

Page 160

30. Sodium movement out of the cell requires energy because the direction of ion flow is against the concentration gradient.
31. Ouabain applied to the apical side would have no effect because there are no Na^+-K^+-ATPase molecules on that side. Ouabain applied to the basolateral side would stop the pump. Glucose transport would continue for a time until the Na^+ gradient between the cell and the lumen disappears because Na^+ has entered the cell.
32. The GLUT2 transporter is illustrated.
33. Transcytosis will stop because vesicular transport by the cytoskeleton depends on functioning microtubules.

Page 166

34. Cl^- will move into the cell down its concentration gradient, which would make the inside of the cell negative. The positive charges left outside would attract Cl^- back outside. The equilibrium potential would be negative.
35. Over time, Na^+ would leak into the cell, and the resting membrane potential would become more positive.

Answers to Figure and Graph Questions

Page 131

Figure 5.1: 1. Plasma is 25% of 14 L = 3.5 L. Interstitial fluid is 75% = 10.5 L. 2. Total body water = 42 L. 3. Plasma is 3.5 L/42 L = 8.3% of total body water; interstitial volume is 10.5 L/42 L = 25%. 4. Total body weight is 121 lb × 1 kg/2.2 lb = 55 kg. If body water = 50% of body weight, total body water = 27.5 L. ICF is 67% of 27.5 L = 18.425 L. ECF is 33% of 27.5 L = 9.075 L. Plasma is 25% of ECF = 2.269 L. 5. Plasma contains proteins and large anions not present in interstitial fluid. 6. The extracellular compartment contains more Na^+, Cl^-, and bicarbonate than the intracellular compartment, and fewer K^+.

Page 137

Figure 5.4: Example 2: (a) hyperosmotic; (b) 250 mOsM; (c) 300 mOsM; (d) Yes, into the cells; (e) hypotonic; (f) increased; hyperosmotic; (g) increased; hypotonic

Page 154

Figure 5.17: (c) You cannot tell if galactose is being transported because the curves only show glucose transport. (d) The cell could increase transport by adding more transporters to the membrane.

Page 159

Figure 5.21: 1 = b; 2 = a; 3 = b. Basolateral glucose transport is passive because the glucose moves down its concentration gradient. Na^+ movement across the apical membrane does not require ATP because Na^+ is moving down its concentration gradient.

Page 164

Figure 5.24: (b) ICF = −1 and ECF = +1. (c) The cell membrane is not permeable to Na^+, Cl^-, and proteins.

Page 165

Figure 5.25: Na^+ leak into the cell is promoted by concentration and electrical gradients. K^+ leak out of the cell is promoted by the concentration gradient.

6 Communication, Integration, and Homeostasis

Future progress in medicine will require a quantitative understanding of the many interconnected networks of molecules that comprise our cells and tissues, their interactions, and their regulation.

—Overview of the NIH Roadmap, 2003

Background Basics

Microarray

In 2003 the United States National Institutes of Health embarked on an ambitious project to promote translation of basic research into new medical treatments and strategies for disease prevention. Contributors to the NIH Common Fund Programs (*http://commonfund.nih.gov*) are compiling information on biological pathways in an effort to understand how cells communicate with one another and maintain the body in a healthy state. In this chapter, we examine the basic patterns of cell-to-cell communication and see how the coordination of function resides in chemical and electrical signals. Each cell in the body can communicate with most other cells. To maintain homeostasis, the body uses a combination of simple diffusion across small distances; widespread distribution of molecules through the circulatory system; and rapid, specific delivery of messages by the nervous system.

Cell-to-Cell Communication

In recent years the amount of information available about cell-to-cell communication has mushroomed as a result of advances in research technology. Signal pathways that once seemed fairly simple and direct are now known to be incredibly complex networks and webs of information transfer. In the sections that follow, we distill what is known about cell-to-cell communication into some basic patterns that you can recognize when you encounter them again in your study of physiology.

By most estimates the human body is composed of about 75 *trillion* cells. Those cells face a daunting task—to communicate with one another in a manner that is rapid and yet conveys a tremendous amount of information. Surprisingly, there are only two basic types of physiological signals: electrical and chemical. **Electrical signals** are changes in a cell's membrane potential [p. 163]. **Chemical signals** are molecules secreted by cells into the extracellular fluid. Chemical signals are responsible for most communication within the body. The cells that respond to electrical or chemical signals are called **target cells,** or **targets** for short.

Our bodies use four basic methods of cell-to-cell communication (■ Fig. 6.1). **Local communication** includes (1) **gap junctions,** which allow direct cytoplasmic transfer of electrical and chemical signals between adjacent cells; (2) **contact-dependent signals,** which occur when surface molecules on one cell membrane bind to surface molecules on another cell's membrane; and (3) chemicals that diffuse through the extracellular fluid to act on cells close by. **Long-distance communication** (4) uses a combination of chemical and electrical signals carried by nerve cells and chemical signals transported in the blood. A given molecule can function as a signal by more than one method. For example, a molecule can act close to the cell that released it (local communication) as well as in distant parts of the body (long-distance communication).

RUNNING PROBLEM

Diabetes Mellitus: A Growing Epidemic

It is 8:00 A.M. and Marvin Garcia, age 20, is hungry. He came to his family physician's office before breakfast to have a fasting blood glucose test as part of a routine physical examination. In this test, blood is drawn after an overnight fast, and the glucose concentration in the blood is measured. Because he knows he is in good condition, Marvin isn't worried about the results. He is surprised, then, when the nurse practitioner in the doctor's office calls two days later. "Your fasting blood sugar is a bit elevated, Marvin. It is 150 milligrams per deciliter, and normal is 110 or less. Does anyone in your family have diabetes?" "Well, yeah—my dad has it. What exactly is diabetes?"

175 178 191 194 196 198 202

Gap Junctions Create Cytoplasmic Bridges

The simplest form of cell-to-cell communication is the direct transfer of electrical and chemical signals through *gap junctions,* protein channels that create cytoplasmic bridges between adjacent cells (Fig. 6.1a). A gap junction forms from the union of membrane-spanning proteins, called *connexins,* on two adjacent cells [p. 78]. The united connexins create a protein channel, or *connexon,* that can open and close. When the channel is open, the connected cells function like a single cell that contains multiple nuclei (a *syncytium*).

When gap junctions are open, ions and small molecules such as amino acids, ATP, and cyclic AMP diffuse directly from the cytoplasm of one cell to the cytoplasm of the next. Larger molecules cannot pass through gap junctions. In addition, gap junctions are the only means by which electrical signals can pass *directly* from cell to cell. Movement of molecules and electrical signals through gap junctions can be modulated or shut off completely.

Gap junctions are not all alike. Scientists have discovered more than 20 different isoforms of connexins that may mix or match to form gap junctions. The variety of connexin isoforms allows gap junction selectivity to vary from tissue to tissue. In mammals, gap junctions are found in almost every cell type, including heart muscle, some types of smooth muscle, lung, liver, and neurons of the brain.

Contact-Dependent Signals Require Cell-to-Cell Contact

Some cell-to-cell communication requires that surface molecules on one cell membrane bind to a membrane protein of another cell (Fig. 6.1b). Such *contact-dependent signaling* occurs in the immune system and during growth and development, such as when nerve cells send out long extensions that must grow from the central axis of the body to the *distal* (distant) ends of

Communication in the Body

Cell-to-cell communication uses chemical and electrical signaling to coordinate function and maintain homeostasis.

LOCAL COMMUNICATION

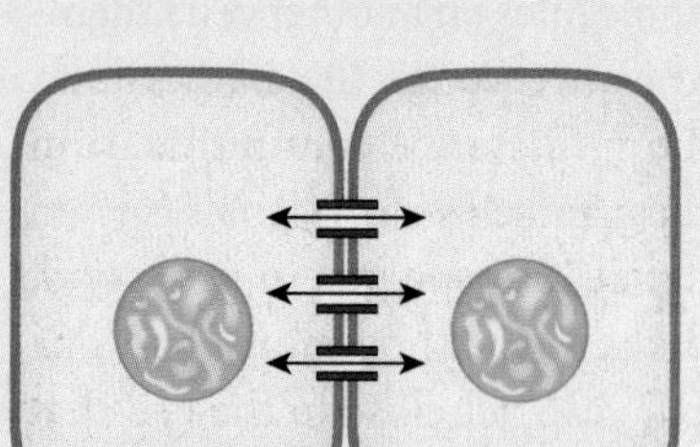

(a) Gap junctions form direct cytoplasmic connections between adjacent cells.

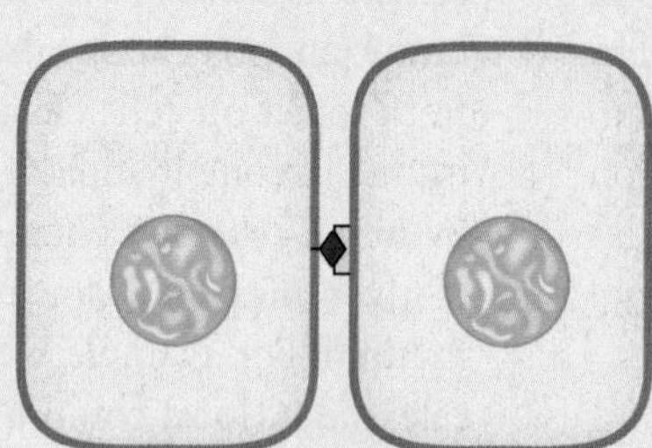

(b) Contact-dependent signals require interaction between membrane molecules on two cells.

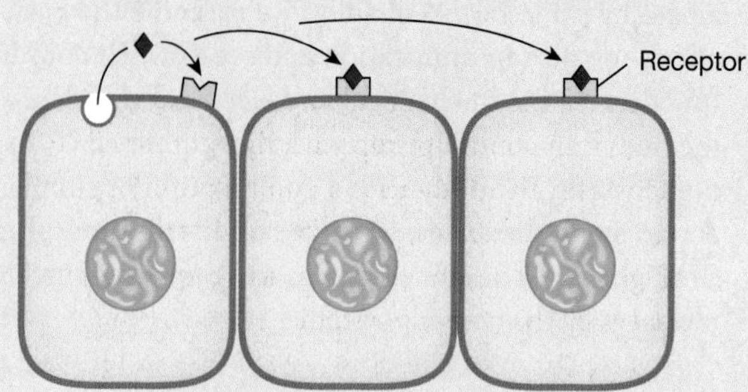

(c) Autocrine signals act on the same cell that secreted them. **Paracrine signals** are secreted by one cell and diffuse to adjacent cells.

LONG-DISTANCE COMMUNICATION

Long-distance signaling may be electrical signals passing along neurons or chemical signals that travel through the circulatory system.

Endocrine System

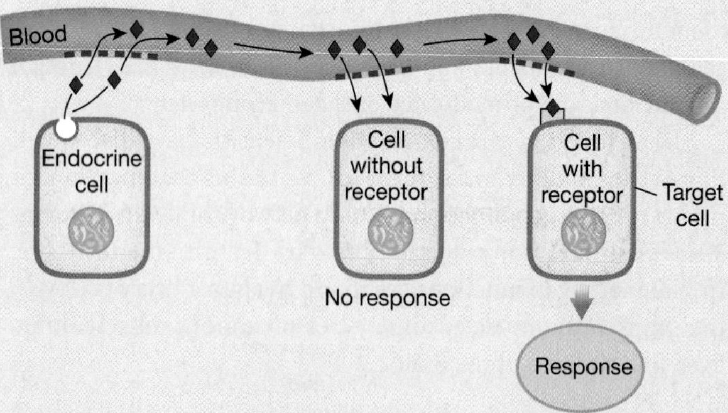

(d) Hormones are secreted by endocrine glands or cells into the blood. Only target cells with receptors for the hormone respond to the signal.

Nervous System

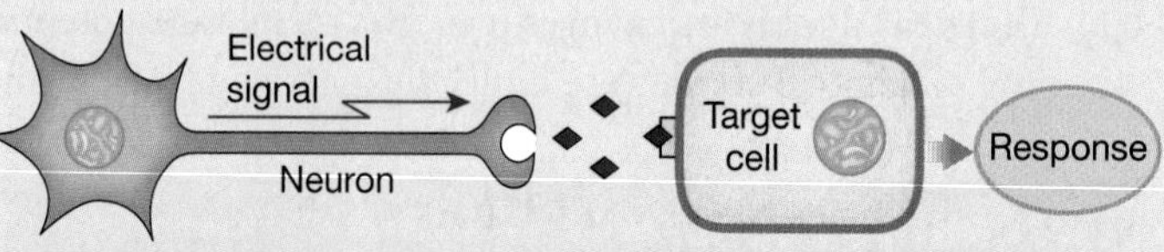

(e) Neurotransmitters are chemicals secreted by neurons that diffuse across a small gap to the target cell.

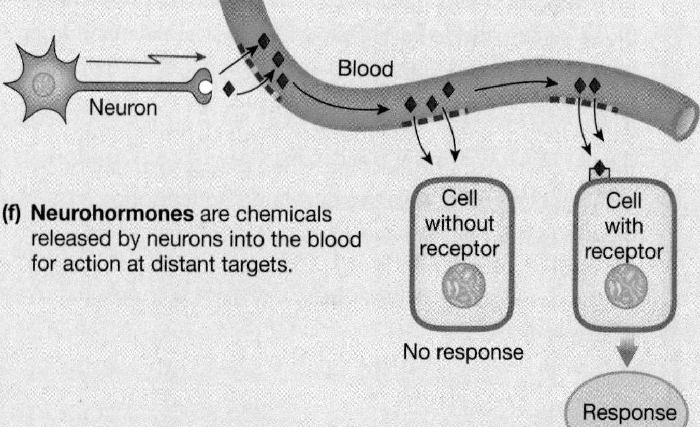

(f) Neurohormones are chemicals released by neurons into the blood for action at distant targets.

the developing limbs. **Cell adhesion molecules (CAMs)** first known for their role in cell-to-cell adhesion [p. 78], have now been shown to act as receptors in cell-to-cell signaling. CAMs are linked to the cytoskeleton and to intracellular enzymes. Through these linkages, CAMs transfer signals in both directions across cell membranes.

Paracrine and Autocrine Signals Carry Out Local Communication

Local communication takes place through paracrine and autocrine signaling. A **paracrine signal** {*para–*, beside + *krinen*, to secrete} is a chemical that acts on cells in the immediate vicinity of the cell that secreted the signal. A chemical signal that acts on the cell that secreted it is called an **autocrine signal** {*auto–*, self}. In some cases a molecule may act as both an autocrine signal and a paracrine signal.

Paracrine and autocrine signal molecules reach their target cells by diffusing through the interstitial fluid (Fig. 6.1c). Because distance is a limiting factor for diffusion, the effective range of paracrine signals is restricted to adjacent cells. A good example of a paracrine molecule is *histamine,* a chemical released from damaged cells. When you scratch yourself with a pin, the red, raised *wheal* that results is due in part to the local release of histamine from the injured tissue. The histamine acts as a paracrine signal, diffusing to capillaries in the immediate area of the injury and making them more permeable to white blood cells and antibodies in the plasma. Fluid also leaves the blood vessels and collects in the interstitial space, causing swelling around the area of injury.

Several important classes of molecules act as local signals. *Cytokines* are regulatory peptides, and *eicosanoids* [p. 33] are lipid-derived paracrine and autocrine signal molecules. We discuss cytokines and eicosanoids in more detail below.

Long-Distance Communication May Be Electrical or Chemical

All cells in the body can release paracrine signals, but most long-distance communication between cells takes place through the nervous and endocrine systems. The endocrine system communicates by using **hormones** {*hormon,* to excite}, chemical signals that are secreted into the blood and distributed all over the body by the circulation. Hormones come in contact with most cells of the body, but only those cells with receptors for the hormone are target cells (Fig. 6.1d).

The nervous system uses a combination of chemical signals and electrical signals to communicate over long distances. An electrical signal travels along a nerve cell (*neuron*) until it reaches the very end of the cell, where it is translated into a chemical signal secreted by the neuron. Such a chemical signal is called a **neurocrine.**

If a neurocrine molecule diffuses from the neuron across a narrow extracellular space to a target cell and has a rapid effect, it is called a **neurotransmitter** (Fig. 6.1f). If a neurocrine acts more slowly as an autocrine or paracrine signal, it is called a **neuromodulator**. If a neurocrine released by a neuron diffuses into the blood for distribution, it is called a **neurohormone** (Fig. 6.1e). The similarities between neurohormones and classic hormones secreted by the endocrine system blur the distinction between the nervous and endocrine systems, making them a functional continuum rather than two distinct systems.

Cytokines May Act as Both Local and Long-Distance Signals

Cytokines are among the most recently identified communication molecules. Initially the term *cytokine* referred only to proteins that modulate immune responses, but in recent years the definition has been broadened to include a variety of regulatory peptides. All nucleated cells synthesize and secrete cytokines in response to stimuli. Cytokines control cell development, cell differentiation, and the immune response. In development and differentiation, cytokines usually function as autocrine or paracrine signals. In stress and inflammation, some cytokines may act on relatively distant targets and may be transported through the circulation just as hormones are.

How do cytokines differ from hormones? In general, cytokines act on a broader spectrum of target cells. In addition, cytokines are not produced by specialized cells the way hormones are, and they are made on demand. In contrast, most protein or peptide hormones are made in advance and stored in the endocrine cell until needed. Also, the signal pathways for cytokines are usually different from those for hormones. However, the distinction between cytokines and hormones is sometimes blurry. For example, erythropoietin, the molecule that controls synthesis of red blood cells, is by tradition considered a hormone but functionally fits the definition of a cytokine.

Concept Check

Answers: p. 204

1. Match the communication method on the left with its property on the right.

(a) autocrine
(b) cytokine
(c) gap junction
(d) hormone
(e) neurohormone
(f) neurotransmitter
(g) paracrine

Communication occurs by:

1. electrical signals
2. chemical signals
3. both electrical and chemical signals

2. Which signal molecules listed in the previous question are transported through the circulatory system? Which are released by neurons?

3. A cat sees a mouse and pounces on it. Do you think the internal signal to pounce could have been transmitted by a paracrine signal? Give two reasons to explain why or why not.

Signal Pathways

Chemical signals in the form of paracrine and autocrine molecules and hormones are released from cells into the extracellular compartment. This is not a very specific way for these signals to find their targets because substances that travel through the blood reach nearly every cell in the body. Yet cells do not respond to every signal that reaches them.

Why do some cells respond to a chemical signal while other cells ignore it? The answer lies in the target-cell **receptor proteins** to which chemical signals bind [p. 145]. *A cell can respond to a chemical signal only if the cell has the appropriate receptor proteins for that signal* (Fig. 6.1).

If a target cell has a receptor for a signal molecule, binding of the signal to the receptor protein initiates a response. All signal pathways share the following features (■ Fig. 6.2):

1. The signal molecule is a *ligand* that binds to a protein receptor. The ligand is also known as a *first messenger* because it brings information to the target cell.
2. Ligand-receptor binding activates the receptor.
3. The receptor in turn activates one or more intracellular signal molecules.
4. The last signal molecule in the pathway initiates synthesis of target proteins or modifies existing target proteins to create a response.

In the following sections, we describe some basic signal pathways. They may seem complex at first, but they follow patterns that you will encounter over and over as you study the systems of the body. Most physiological processes, from the beating of your heart to learning and memory, use some variation of these pathways. One of the wonders of physiology is the fundamental importance of these signal pathways and the way they have been conserved in animals ranging from worms to humans.

Signal Pathways

Most signal pathways consist of the 5 steps shown. Use the shapes and colors of the steps shown here to identify the pattern in later illustrations.

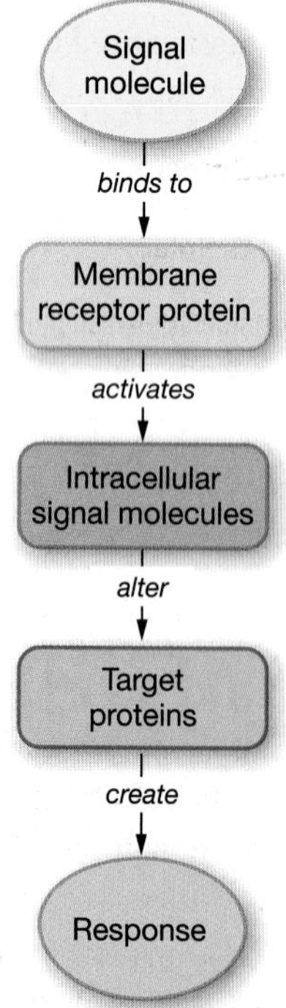

■ **Fig. 6.2**

Receptor Proteins Are Located Inside the Cell or on the Cell Membrane

Protein receptors for signal molecules play an important role in physiology and medicine. About half of all drugs currently in use act on receptor proteins. Target-cell receptor proteins may be found in the nucleus, in the cytosol, or on the cell membrane as integral proteins. Where a chemical signal binds to its receptor largely depends on whether that signal molecule is lipophilic or lipophobic (■ Fig. 6.3).

Lipophilic signal molecules can diffuse through the phospholipid bilayer of the cell membrane [p. 66] and bind to *cytosolic receptors* or *nuclear receptors* (Fig. 6.3a). In these cases, receptor activation often turns on a gene and directs the nucleus to make new mRNA (transcription, [p. 118]). The mRNA then provides a template for synthesis of new proteins (translation, [p. 118]). This process is relatively slow and the cell's response may not be noticeable for an hour or longer. In some instances the activated receptor can also turn off, or *repress*, gene activity. Many lipophilic signal molecules that follow this pattern are hormones.

RUNNING PROBLEM

Later that day in the physician's office, the nurse practitioner explains diabetes to Marvin. Diabetes mellitus is a family of metabolic disorders caused by defects in the homeostatic pathways that regulate glucose metabolism. Several forms of diabetes exist, and some can be inherited. One form, called *type 1 diabetes mellitus*, occurs when endocrine cells of the pancreas are destroyed and stop making insulin, a protein hormone involved in blood glucose homeostasis. In another form, type 2 diabetes mellitus, insulin may be present in normal or above-normal levels, but the insulin-sensitive cells of the body do not respond normally to the hormone.

Q1: In which type of diabetes is the target cell signal pathway for insulin more likely to be defective?

Q2: Insulin is a protein hormone. Would you expect to find its receptor on the cell surface or in the cytoplasm of the target cells?

175 178 191 194 196 198 202

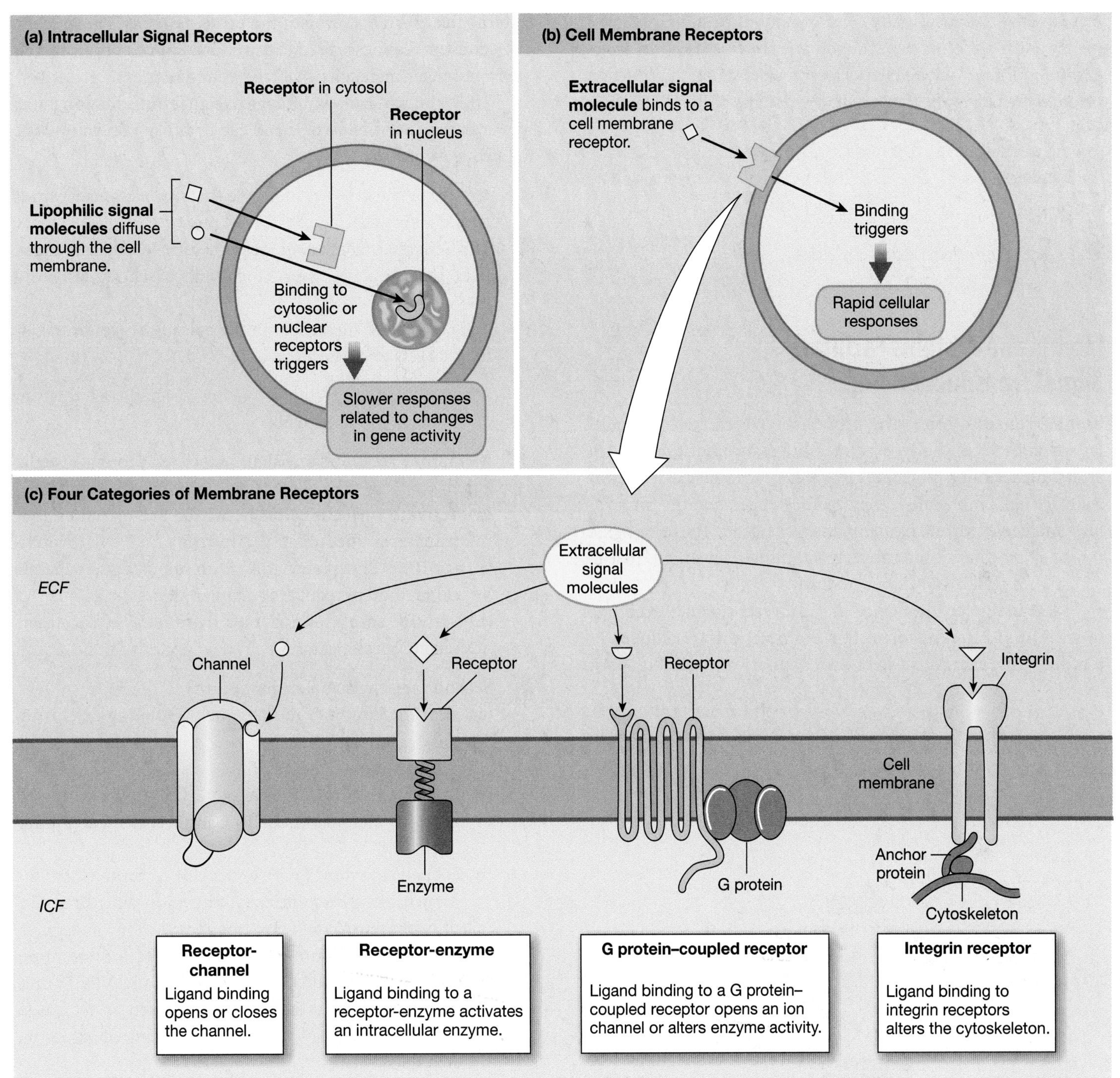
(a) Intracellular Signal Receptors
Receptor in cytosol
Receptor in nucleus
Lipophilic signal molecules diffuse through the cell membrane.
Binding to cytosolic or nuclear receptors triggers
Slower responses related to changes in gene activity
(b) Cell Membrane Receptors
Extracellular signal molecule binds to a cell membrane receptor.
Binding triggers
Rapid cellular responses
(c) Four Categories of Membrane Receptors
ECF
Extracellular signal molecules
Channel
Receptor
Receptor
Integrin
Cell membrane
Enzyme
G protein
Anchor protein
Cytoskeleton
ICF
Receptor-channel
Ligand binding opens or closes the channel.
Receptor-enzyme
Ligand binding to a receptor-enzyme activates an intracellular enzyme.
G protein–coupled receptor
Ligand binding to a G protein–coupled receptor opens an ion channel or alters enzyme activity.
Integrin receptor
Ligand binding to integrin receptors alters the cytoskeleton.

Fig. 6.3 ***Target cell receptors may be located on the cell surface or inside the cell***

Lipophobic signal molecules are unable to diffuse through the phospholipid bilayer of the cell membrane. Instead, these signal molecules remain in the extracellular fluid and bind to receptor proteins on the cell membrane (Fig. 6.3b). (Some lipophilic signal molecules also bind to cell membrane receptors in addition to their intracellular receptors.) In general, the response time for pathways linked to membrane receptor proteins is very rapid: responses can be seen within milliseconds to minutes.

We can group membrane receptors into four major categories, illustrated in Figure 6.3c. The simplest receptors are chemically gated (*ligand-gated*) ion channels called *receptor-channels* [p. 148]. Ligand binding opens or closes the channel and alters ion flow across the membrane.

Three other receptor types are shown in Figure 6.3c: *receptor-enzymes, G protein–coupled receptors,* and *integrin receptors*. For all three, information from the signal molecule must be passed across the membrane to initiate an intracellular response. This transmission of information from one side of a membrane to the other using membrane proteins is known as *signal transduction*. We will take a closer look at signal transduction before returning to the four receptor types that participate in it.

Concept Check Answers: p. 205

4. List four components of signal pathways.
5. Name three cellular locations of receptors.

Membrane Proteins Facilitate Signal Transduction

Signal transduction is the process by which an extracellular signal molecule activates a membrane receptor that in turn alters intracellular molecules to create a response. The extracellular signal molecule is the first messenger, and the intracellular molecules form a *second messenger system*. The term *signal transduction* comes from the verb *to transduce,* meaning "to lead across" {*trans,* across + *ducere,* to lead}.

A **transducer** is a device that converts a signal from one form into a different form. For example, the transducer in a radio converts radio waves into sound waves (■ Fig. 6.4).

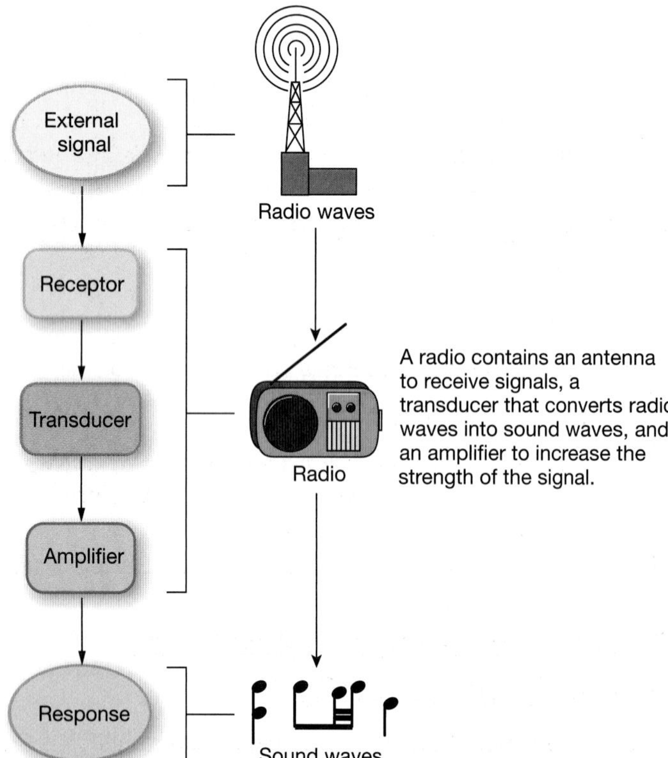
Signal transduction converts one form of signal into a different form.
External signal
Radio waves
Receptor
Transducer
Amplifier
Radio
A radio contains an antenna to receive signals, a transducer that converts radio waves into sound waves, and an amplifier to increase the strength of the signal.
Response
Sound waves

■ **Fig. 6.4** ***Signal transduction***

In biological systems, membrane proteins act as transducers. They convert the message of extracellular signals into intracellular messenger molecules that trigger a response.

The basic pattern of a biological signal transduction pathway is shown in ■ Figure 6.5a and can be broken down into the following events.

1. An extracellular signal molecule (the *first messenger*) binds to and activates a membrane receptor.
2. The activated membrane receptor turns on its associated proteins and starts an intracellular cascade of **second messengers.**
3. The last second messenger in the cascade acts on intracellular targets to create a response.

A more detailed version of the basic signal transduction pathway is shown in Figure 6.5b.

1. Membrane receptors and their associated proteins usually either
   (a) activate **protein kinases,** which are enzymes that transfer a phosphate group from ATP to a protein [p. 108]. Phosphorylation is an important biochemical method of regulating cellular processes.
   (b) activate amplifier enzymes that create intracellular second messengers.
2. Second messenger molecules in turn
   (a) alter the gating of ion channels. Opening or closing ion channels creates electrical signals by altering the cell's membrane potential [p. 163].
   (b) increase intracellular calcium. Calcium binding to proteins changes their function, creating a cellular response.
   (c) change enzyme activity, especially of protein kinases or **protein phosphatases,** enzymes that remove a phosphate group. The phosphorylation or *dephosphorylation* of a protein can change its configuration and create a response. Examples of changes that occur with phosphorylation include increased or decreased enzyme activity and opening or closing of gated ion channels.
3. The proteins modified by calcium binding and phosphorylation control one or more of the following:
   (a) metabolic enzymes
   (b) motor proteins for muscle contraction and cytoskeletal movement
   (c) proteins that regulate gene activity and protein synthesis
   (d) membrane transport and receptor proteins

If you think this list includes almost everything a cell does, you're right!

■ Figure 6.6a on page 182 shows how the steps of a signal transduction pathway form a **cascade.** A signaling cascade

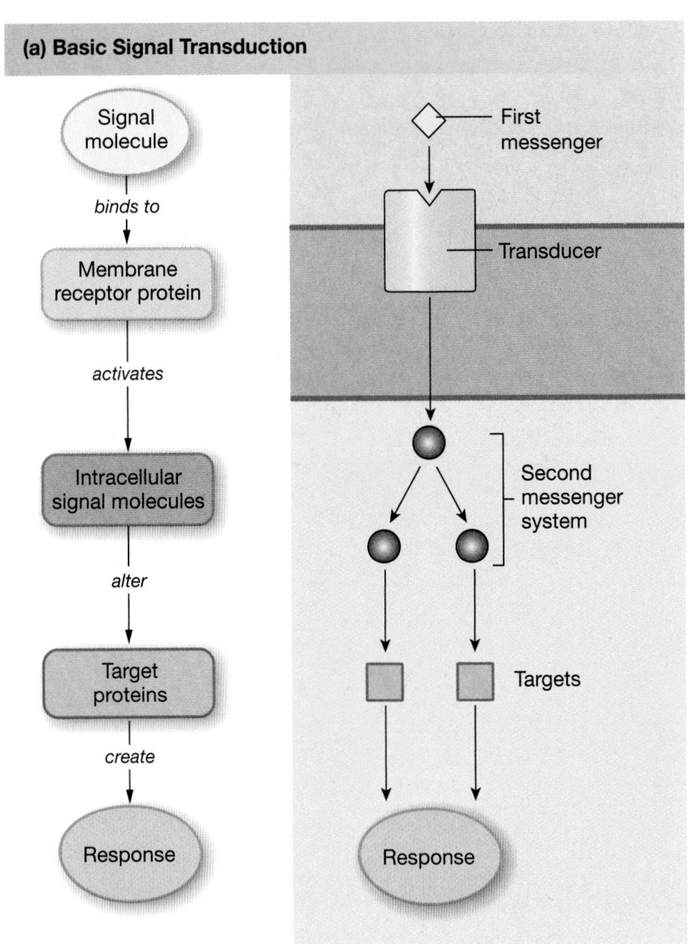

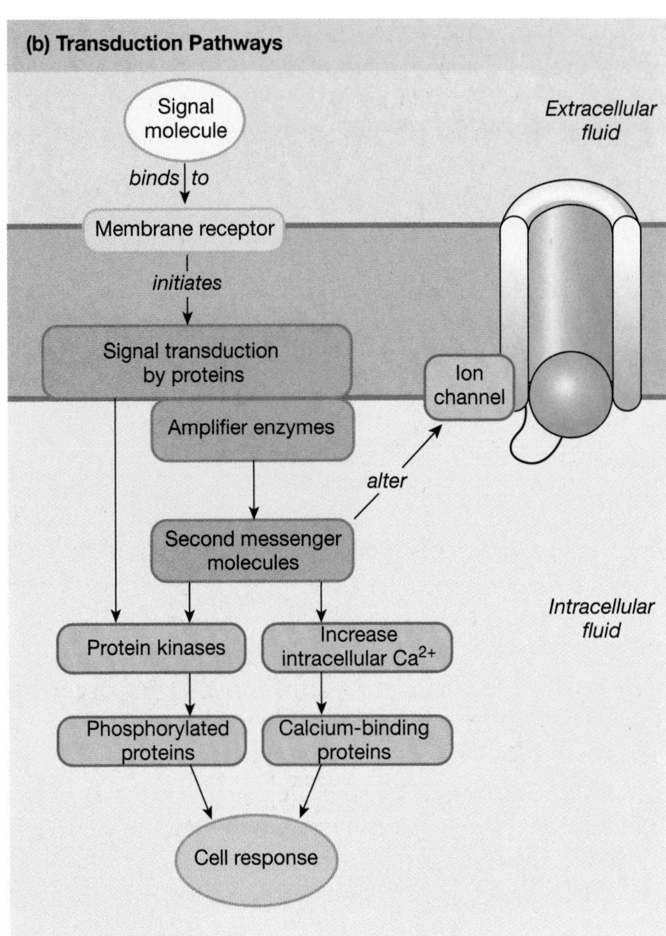

Fig. 6.5 ***Biological signal transduction***

6

starts when a stimulus (the signal molecule) converts inactive molecule A (the receptor) to an active form. Active A then converts inactive molecule B into active B, active molecule B in turn converts inactive molecule C into active C, and so on, until at the final step a substrate is converted into a product. Many intracellular signal pathways are cascades. Blood clotting is an important example of an extracellular cascade.

In signal transduction pathways, the original signal is not only transformed but also amplified {*amplificare,* to make larger}. In a radio, the radio wave signal is also amplified. In cells, **signal amplification** turns one signal molecule into multiple second messenger molecules (Fig. 6.6b). The process begins when the first messenger ligand combines with its receptor. The receptor-ligand complex turns on an **amplifier enzyme.** The amplifier enzyme activates several molecules, which in turn each activate several more molecules as the cascade proceeds. By the end of the process, the effects of the ligand have been amplified much more than if there were a 1:1 ratio between each step. Amplification gives the body "more bang for the buck" by enabling a small amount of ligand to create a large effect. The most common amplifier enzymes and second messengers are listed in the table in Figure 6.6c.

In the sections that follow, we will examine in more detail the four major types of membrane receptors (see Fig. 6.3c). Keep in mind that these receptors may be responding to any of the different kinds of signal molecules—hormones, neurohormones, neurotransmitters, cytokines, paracrines, or autocrines.

Concept Check

Answers: p. 205

6. What are the four steps of signal transduction?
7. What happens during amplification?
8. Why do steroid hormones not require signal transduction and second messengers to exert their action? (*Hint:* Are steroids lipophobic or lipophilic? [p. 142])

Receptor-Enzymes Have Protein Kinase or Guanylyl Cyclase Activity

Receptor-enzymes have two regions: a receptor region on the extracellular side of the cell membrane, and an enzyme region on the cytoplasmic side (see Fig. 6.3c). In some instances,

Signal Transduction

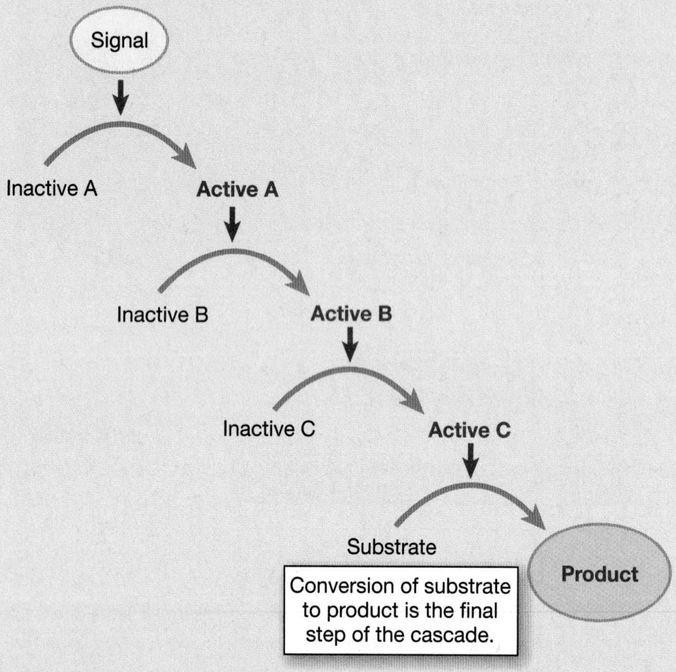
(a) Signal transduction pathways form a cascade.
Signal
Inactive A
Active A
Inactive B
Active B
Inactive C
Active C
Substrate
Product
Conversion of substrate to product is the final step of the cascade.

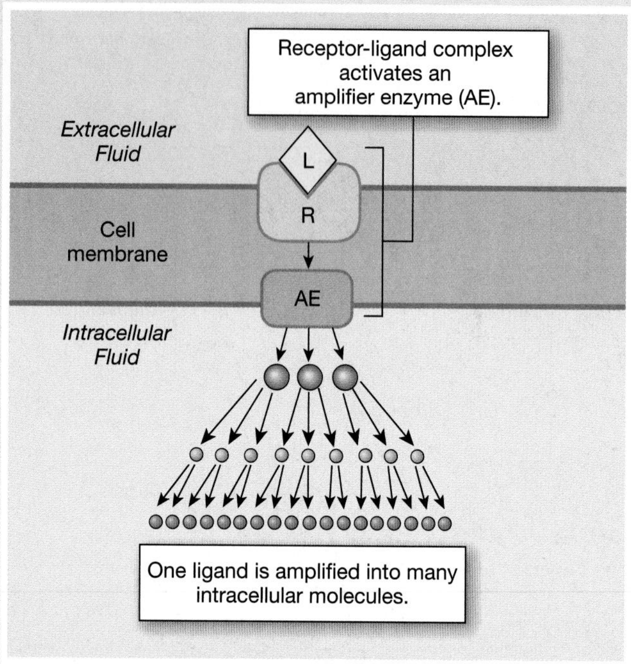
(b) Signal amplification allows a small amount of signal to have a large effect.
Receptor-ligand complex activates an amplifier enzyme (AE).
Extracellular Fluid
Cell membrane
Intracellular Fluid
L
R
AE
One ligand is amplified into many intracellular molecules.

(c) Second messenger pathways

SECOND MESSENGER	MADE FROM	AMPLIFIER ENZYME	LINKED TO	ACTION	EFFECTS
Nucleotides					
cAMP	ATP	Adenylyl cyclase (membrane)	GPCR*	Activates protein kinases, especially PKA. Binds to ion channels.	Phosphorylates proteins. Alters channel opening.
cGMP	GTP	Guanylyl cyclase (membrane)	Receptor-enzyme	Activates protein kinases, especially PKG.	Phosphorylates proteins.
		Guanylyl cyclase (cytosol)	Nitric oxide (NO)	Binds to ion channels.	Alters channel opening.
Lipid-derived*					
IP_3	Membrane phospholipids	Phospholipase C (membrane)	GPCR	Releases Ca^{2+} from intracellular stores.	See Ca^{2+} effects below.
DAG				Activates protein kinase C.	Phosphorylates proteins.
Ions					
Ca^{2+}				Binds to calmodulin. Binds to other proteins.	Alters enzyme activity. Exocytosis, muscle contraction, cytoskeleton movement, channel opening.

*GPCR = G protein–coupled receptor. IP_3 = Inositol trisphosphate. DAG = diacylglycerol

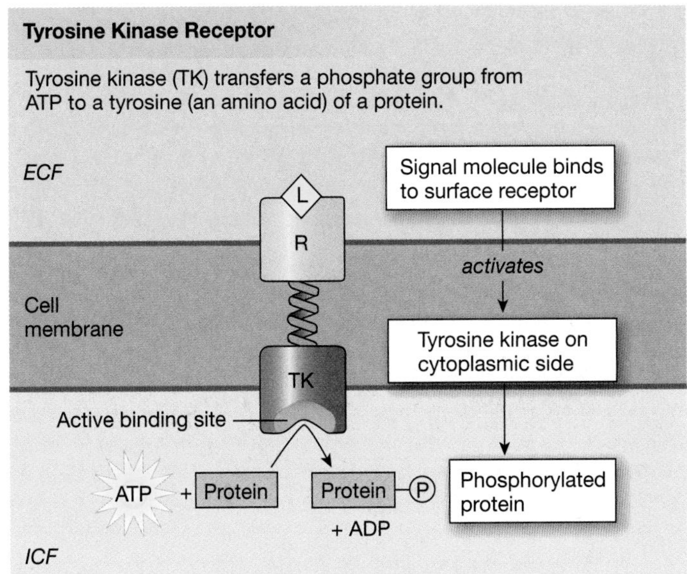

■ **Fig. 6.7**

the receptor region and enzyme region are parts of the same protein molecule. In other cases, the enzyme region is a separate protein. Ligand binding to the receptor activates the enzyme. The enzymes of receptor-enzymes are either protein kinases, such as *tyrosine kinase* (■ Fig. 6.7), or *guanylyl cyclase,* the amplifier enzyme that converts GTP to **cyclic GMP (cGMP)** [p. 36].

Ligands for receptor-enzymes include the hormone insulin as well as many cytokines and growth factors. The insulin receptor protein has its own intrinsic tyrosine kinase activity. In contrast, most cytokine receptor proteins do not have intrinsic enzyme activity. Instead, cytokine binding activates a cytosolic enzyme called *Janus family tyrosine kinase,* usually abbreviated as *JAK kinase.*

Most Signal Transduction Uses G Proteins

The **G protein–coupled receptors (GPCR)** are a large and complex family of membrane-spanning proteins that cross the phospholipid bilayer seven times (see Fig. 6.3c). The cytoplasmic tail of the receptor protein is linked to a three-part membrane transducer molecule known as a **G protein.** Hundreds of G protein–coupled receptors have been identified, and the list continues to grow. The types of ligands that bind to G protein–coupled receptors include hormones, growth factors, olfactory molecules, visual pigments, and neurotransmitters. In 1994, Alfred G. Gilman and Martin Rodbell received a Nobel prize for the discovery of G proteins and their role in cell signaling (see *http://nobelprize.org/nobel_prizes/medicine*).

G proteins get their name from the fact that they bind guanosine nucleotides [p. 36]. Inactive G proteins are bound to guanosine diphosphate (GDP). Exchanging the GDP for guanosine triphosphate (GTP) activates the G protein. When G proteins are activated, they either (1) open an ion channel in the membrane or (2) alter enzyme activity on the cytoplasmic side of the membrane.

G proteins linked to amplifier enzymes make up the bulk of all known signal transduction mechanisms. The two most common amplifier enzymes for G protein–coupled receptors are adenylyl cyclase and phospholipase C. The pathways for these amplifier enzymes are described next.

Many Lipophobic Hormones Use GPCR-cAMP Pathways

The **G protein–coupled adenylyl cyclase-cAMP system** was the first identified signal transduction pathway (■ Fig. 6.8a). It was discovered in the 1950s by Earl Sutherland when he was studying the effects of hormones on carbohydrate metabolism. This discovery proved so significant to our understanding of signal transduction that in 1971 Sutherland was awarded a Nobel prize for his work.

The G protein–coupled adenylyl cyclase-cAMP system is the signal transduction system for many protein hormones. In this system, *adenylyl cyclase* is the amplifier enzyme that converts ATP to the second messenger molecule *cyclic AMP* (cAMP). Cyclic AMP then activates *protein kinase A* (PKA), which in turn phosphorylates other intracellular proteins as part of the signal cascade.

G Protein–Coupled Receptors Also Use Lipid-Derived Second Messengers

Some G protein–coupled receptors are linked to a different amplifier enzyme: phospholipase C (Fig. 6.8b). When a signal molecule activates this G protein–coupled pathway, **phospholipase C (PLC)** converts a membrane phospholipid (*phosphatidylinositol bisphosphate*) into two lipid-derived second messenger molecules: diacylglycerol and inositol trisphosphate.

Diacylglycerol (DAG) is a nonpolar diglyceride that remains in the lipid portion of the membrane and interacts with **protein kinase C (PKC),** a Ca^{2+}-activated enzyme associated with the cytoplasmic face of the cell membrane. PKC phosphorylates cytosolic proteins that continue the signal cascade.

Inositol trisphosphate (IP_3) is a water-soluble messenger molecule that leaves the membrane and enters the cytoplasm. There it binds to a calcium channel on the endoplasmic reticulum (ER). IP_3 binding opens the Ca^{2+} channel, allowing Ca^{2+} to diffuse out of the ER and into the cytosol. Calcium is itself an important signal molecule, as discussed below.

G Protein–Coupled Signal Transduction

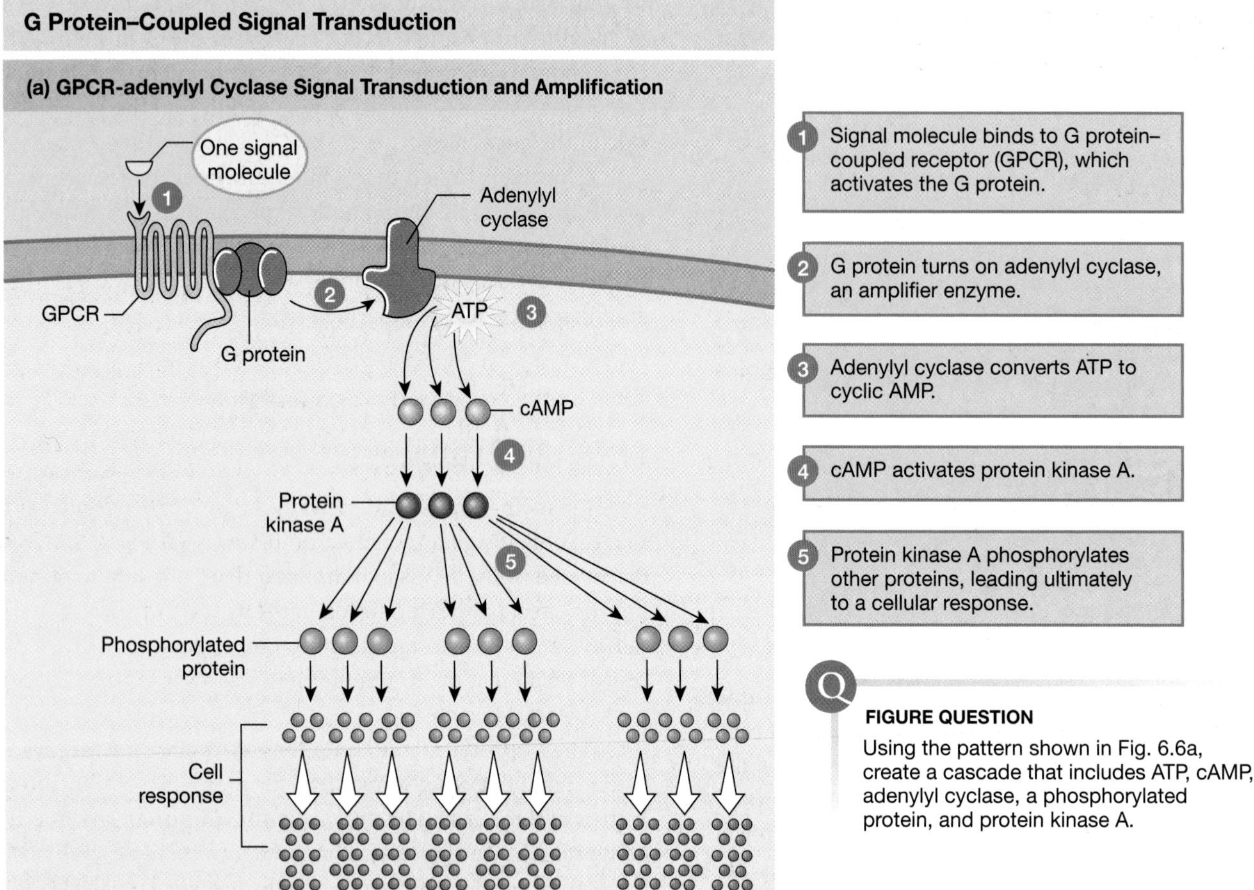

1 Signal molecule binds to G protein–coupled receptor (GPCR), which activates the G protein.

2 G protein turns on adenylyl cyclase, an amplifier enzyme.

3 Adenylyl cyclase converts ATP to cyclic AMP.

4 cAMP activates protein kinase A.

5 Protein kinase A phosphorylates other proteins, leading ultimately to a cellular response.

FIGURE QUESTION

Using the pattern shown in Fig. 6.6a, create a cascade that includes ATP, cAMP, adenylyl cyclase, a phosphorylated protein, and protein kinase A.

(b) GPCR-phospholipase C Signal Transduction

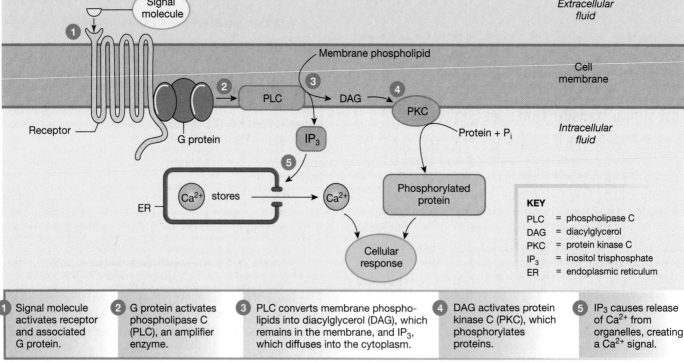

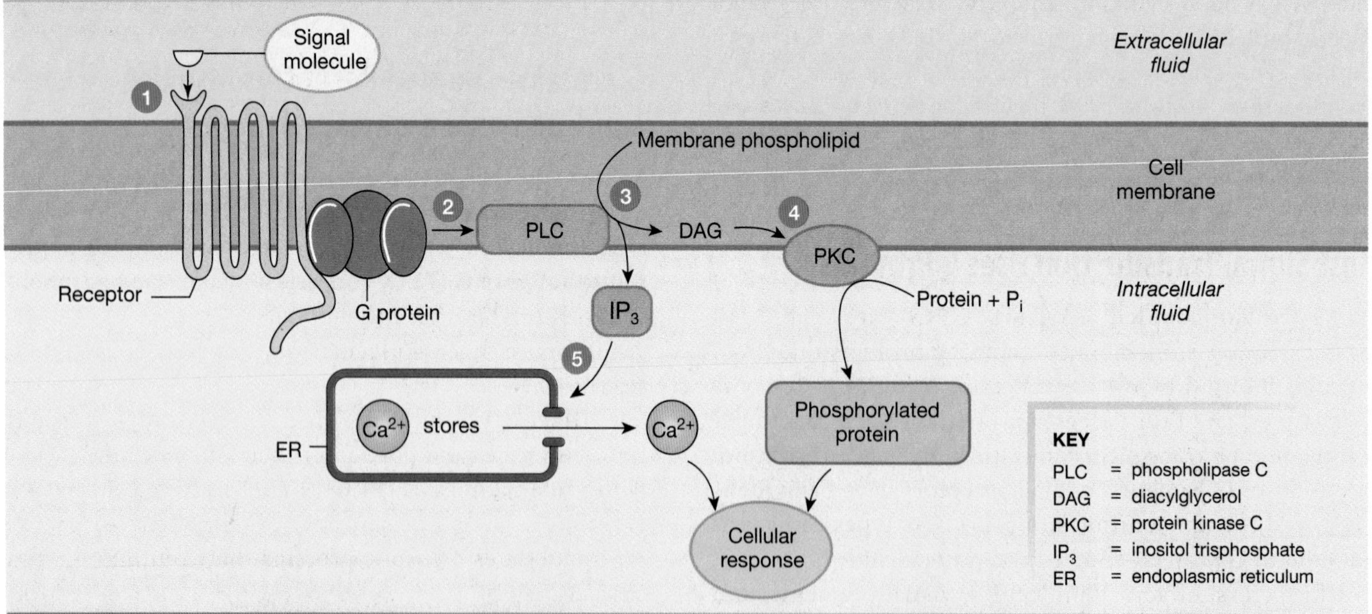

Signal molecule

Extracellular fluid

Membrane phospholipid

Cell membrane

Receptor

G protein

PLC

DAG

PKC

IP_3

Protein + P_i

Intracellular fluid

ER

Ca^{2+} stores

Ca^{2+}

Phosphorylated protein

Cellular response

KEY

PLC = phospholipase C
DAG = diacylglycerol
PKC = protein kinase C
IP_3 = inositol trisphosphate
ER = endoplasmic reticulum

1 Signal molecule activates receptor and associated G protein.

2 G protein activates phospholipase C (PLC), an amplifier enzyme.

3 PLC converts membrane phospholipids into diacylglycerol (DAG), which remains in the membrane, and IP_3, which diffuses into the cytoplasm.

4 DAG activates protein kinase C (PKC), which phosphorylates proteins.

5 IP_3 causes release of Ca^{2+} from organelles, creating a Ca^{2+} signal.

Fig. 6.8

Integrin Receptors Transfer Information from the Extracellular Matrix

The membrane-spanning proteins called *integrins* [p. 30] mediate blood clotting, wound repair, cell adhesion and recognition in the immune response, and cell movement during development. On the extracellular side of the membrane, integrin receptors bind either to proteins of the extracellular matrix [p. 78] or to ligands such as antibodies and molecules involved in blood clotting. Inside the cell, integrins attach to the cytoskeleton via *anchor proteins* (Fig. 6.3c). Ligand binding to the receptor causes integrins to activate intracellular enzymes or alter the organization of the cytoskeleton.

The importance of integrin receptors is illustrated by inherited conditions in which the receptor is absent. In one condition, platelets—cell fragments that play a key role in blood clotting—lack an integrin receptor. As a result, blood clotting is defective in these individuals.

The Most Rapid Signal Pathways Change Ion Flow Through Channels

The simplest receptors are ligand-gated ion channels, and most of them are neurotransmitter receptors found in nerve and muscle. The activation of **receptor-channels** initiates the most rapid intracellular responses of all receptors. When an extracellular ligand binds to the receptor-channel protein, a channel gate opens or closes, altering the cell's permeability to an ion. Increasing or decreasing ion permeability rapidly changes the cell's membrane potential [p. 163], creating an electrical signal that alters voltage-sensitive proteins (■ Fig. 6.9).

One example of a receptor-channel is the acetylcholine-gated monovalent ("one-charge") cation channel of skeletal muscle. The neurotransmitter *acetylcholine* released from an adjacent neuron binds to the acetylcholine receptor and opens the channel. Both Na^+ and K^+ flow through the open channel, K^+ leaving the cell and Na^+ entering the cell along

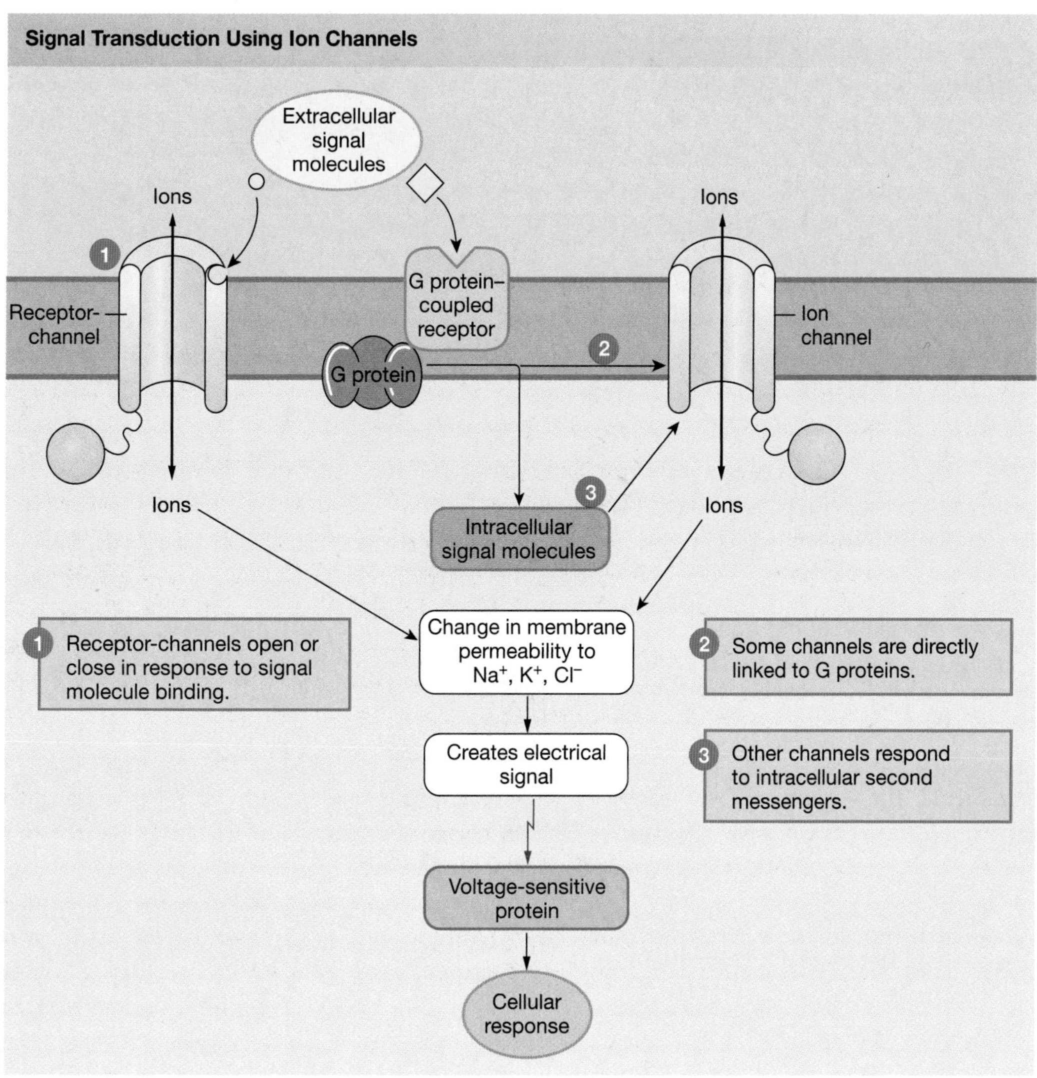

■ **Fig. 6.9** ***Signal transduction using ion channels***

■ Fig. 6.10 *ESSENTIALS*

Summary Map of Signal Transduction

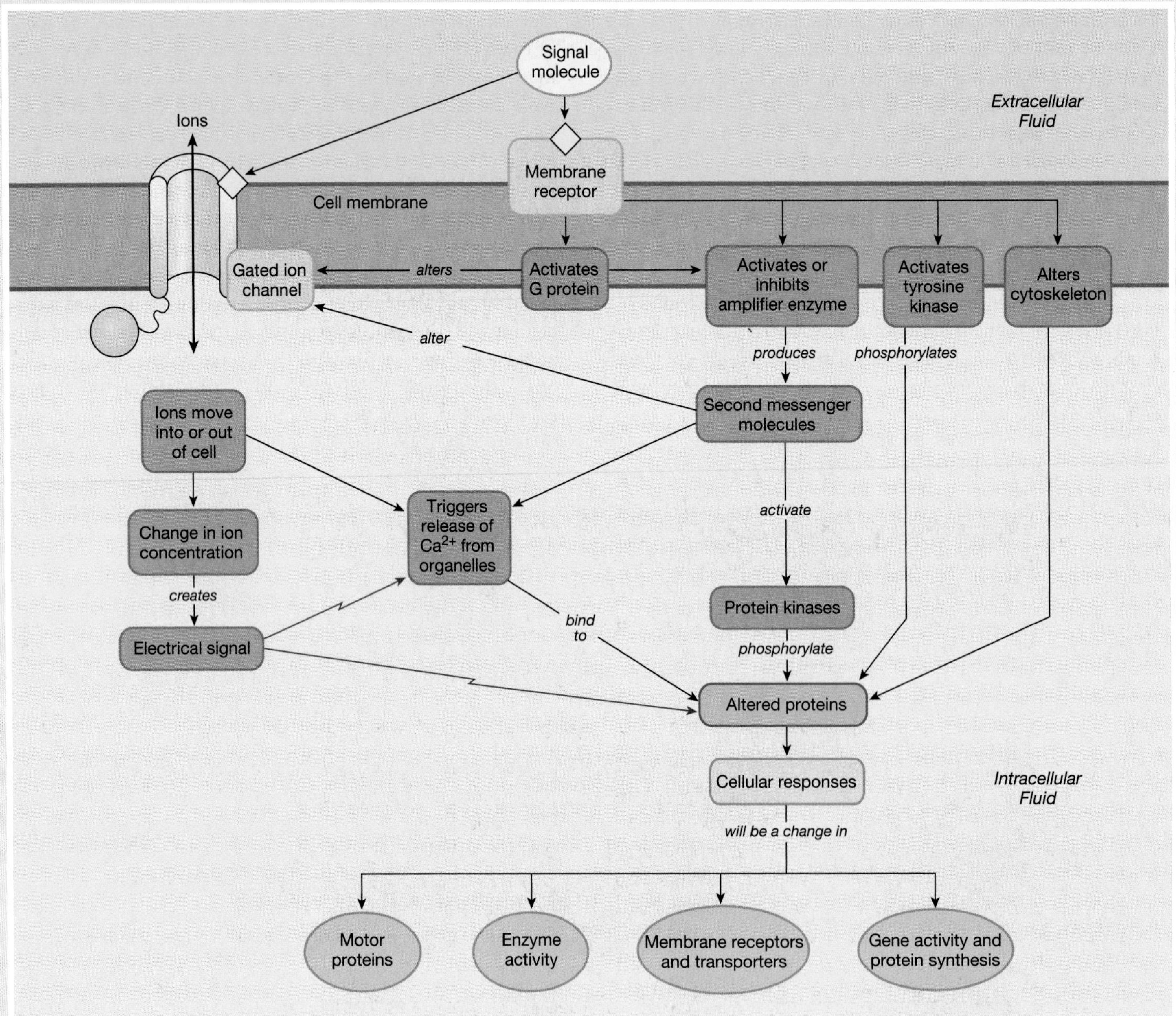

their electrochemical gradients. The sodium gradient is stronger, however, so net entry of positively charged Na^+ depolarizes the cell. In skeletal muscle, this cascade of intracellular events results in muscle contraction.

Not all ion channel signal transduction is mediated by receptor-channels. Some ligand-gated ion channels are controlled by intracellular second messengers, such as cAMP or ATP. The ATP-gated K^+ channels of the pancreatic beta cell are an example [Fig. 5.27, p. 167]. Other ion channels open or close in response to extracellular signals, but the signal ligand does not bind to the channel protein. Instead it binds to a G protein–coupled receptor that is linked to the ion channel.

■ Figure 6.10 is a summary map of basic signal transduction, showing the general relationships among first messengers, membrane receptors, second messengers, and cell responses.

Concept Check Answers: p. 205

9. Name the four categories of membrane receptors.
10. What is the difference between a first messenger and a second messenger?
11. Place the following terms in the correct order for a signal transduction pathway:
    (a) cell response, receptor, second messenger, ligand
    (b) amplifier enzyme, cell response, phosphorylated protein, protein kinase, second messenger
12. In each of the following situations, will a cell depolarize or hyperpolarize?
    (a) Cl^- channel opens
    (b) K^+ channel opens
    (c) Na^+ channel opens

Novel Signal Molecules

The following sections introduce you to some unusual signal molecules that are important in physiology and medicine. They include an ion (Ca^{2+}), three gases, and a family of lipid-derived messengers. The processes controlled by these signal molecules have been known for years, but the control signals themselves were discovered only relatively recently.

Calcium Is an Important Intracellular Signal

Calcium ions are the most versatile ionic messengers (■ Fig. 6.11). Calcium enters the cell either through voltage-gated Ca^{2+} channels or through ligand-gated or mechanically gated channels. Calcium can also be released from intracellular compartments by second messengers, such as IP_3. Most intracellular Ca^{2+} is stored in the endoplasmic reticulum [p. 75], where it is concentrated by active transport.

Release of Ca^{2+} into the cytosol (from any of the sources just mentioned) creates a Ca^{2+} signal, or Ca^{2+} "spark," that can be recorded using special Ca^{2+}-imaging techniques (see Biotechnology box on calcium signals). The calcium ions combine with cytoplasmic calcium-binding proteins to exert various effects. Several types of calcium-dependent events occur in the cell:

1. Ca^{2+} binds to the protein **calmodulin,** found in all cells. Calcium binding alters enzyme or transporter activity or the gating of ion channels.
2. Calcium binds to other regulatory proteins and alters movement of contractile or cytoskeletal proteins such as microtubules. For example, Ca^{2+} binding to the regulatory protein *troponin* initiates muscle contraction in a skeletal muscle cell.

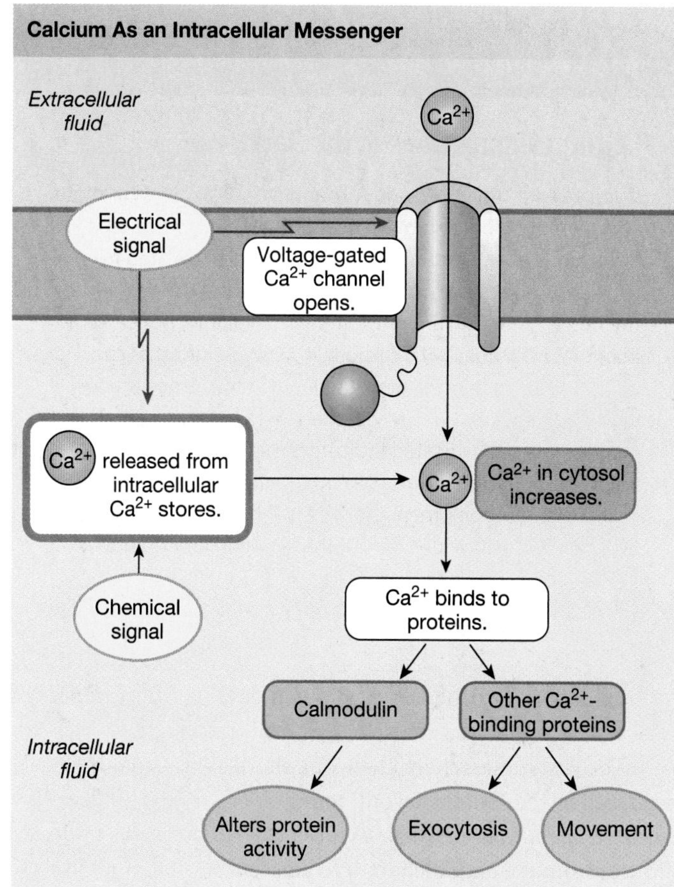

■ **Fig. 6.11**

3. Ca^{2+} binds to regulatory proteins to trigger exocytosis of secretory vesicles [p. 157]. For example, the release of insulin from pancreatic beta cells occurs in response to a calcium signal.
4. Ca^{2+} binds directly to ion channels to alter their gating state. An example of this target is a Ca^{2+}-activated K^+ channel found in nerve cells.
5. Ca^{2+} entry into a fertilized egg initiates development of the embryo.

Concept Check Answer: p. 205

13. The extracellular fluid Ca^{2+} concentration averages 2.5 mmol/L. Free cytosolic Ca^{2+} concentration is about 0.001 mmol/L. If a cell is going to move calcium ions from its cytosol to the extracellular fluid, will it use passive or active transport? Explain.

Gases Are Ephemeral Signal Molecules

Soluble gases are short-acting paracrine/autocrine signal molecules that act close to where they are produced. The best-known gaseous signal molecule is **nitric oxide** (NO), but carbon

6

BIOTECHNOLOGY

Calcium Signals Glow in the Dark

If you've ever run your hand through a tropical ocean at night and seen the glow of bioluminescent jellyfish, you've seen a calcium signal. Aequorin, a protein complex isolated from jellyfish, is one of the molecules that scientists use to monitor the presence of calcium ions during a cellular response. When aequorin combines with calcium, it releases light that can be measured by electronic detection systems. Since the first use of aequorin in 1967, researchers have been designing increasingly sophisticated indicators that allow them to follow calcium signals in cells. With the help of molecules called fura, Oregon green, BAPTA, and chameleons, we can now watch calcium ions diffuse through gap junctions and flow out of intracellular organelles.

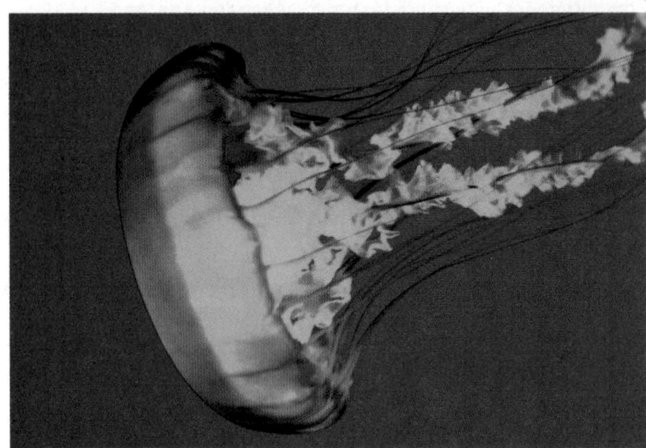

The sea nettle *Chrysaora fuscescens*.

monoxide and hydrogen sulfide, two gases better known for their noxious effects, can also act as local signals.

For years researchers knew of a short-lived signal molecule produced by the endothelial cells lining blood vessels. They initially named it *endothelial-derived relaxing factor* (EDRF). This molecule diffuses from the endothelium into adjacent smooth muscle cells, causing the muscle to relax and dilate the blood vessel. Scientists took years to identify EDRF as nitric oxide because it is rapidly broken down, with a half-life of only 2 to 30 seconds. (*Half-life* is the time required for the signal to lose half of its activity.) As a result of this difficult work on NO in the cardiovascular system, Robert Furchgott, Louis Ignarro, and Ferid Murad received the 1998 Nobel prize for physiology and medicine.

In tissues, NO is synthesized by the action of the enzyme *nitric oxide synthase* (NOS) on the amino acid arginine:

$$\text{Arginine} + O_2 \xrightarrow{\textit{nitric oxide synthase}} \text{NO} + \text{citrulline (an amino acid)}$$

The NO produced in this reaction diffuses into target cells, where it binds to a receptor that activates the cytosolic form of guanylyl cyclase and causes formation of the second messenger cGMP. In addition to relaxing blood vessels, NO in the brain acts as a neurotransmitter and a neuromodulator.

Carbon monoxide (CO), a gas known mostly for its toxic effects, is also a signal molecule produced in minute amounts by certain cells. Like NO, CO activates guanylyl cyclase and cGMP, but it may also work independently to exert its effects. Carbon monoxide targets smooth muscle and neural tissue.

The newest gaseous signal molecule to be described is **hydrogen sulfide (H_2S)**. Hydrogen sulfide also acts in the cardiovascular system to relax blood vessels. Garlic is a major dietary source of the sulfur-containing precursors, which may explain studies suggesting that eating garlic has protective effects on the heart.

CLINICAL FOCUS

From Dynamite to Medicine

Who would have thought that a component of smog and a derivative of dynamite would turn out to be a biological messenger? Certainly not the peer reviewers who initially rejected Louis Ignarro's attempts to publish his research findings on the elusive gas. However, the ability of nitrate-containing compounds to relax blood vessels has been known for more than 100 years, ever since workers in Alfred Nobel's dynamite factory complained of headaches caused by nitrate-induced vasodilation. Since the 1860s, physicians have used nitroglycerin to relieve *angina,* heart pain that results from constricted blood vessels. Even today heart patients carry little nitroglycerin tablets to slide under their tongues when angina strikes. Still, it took years of work to isolate nitric oxide (NO), the short-lived gas that is the biologically active molecule derived from nitroglycerin. Despite all our twenty-first-century technology, direct research on NO is still difficult. Many studies look at its influence indirectly by studying the location and activity of nitric oxide synthase (NOS), the enzyme that produces NO.

Some Lipids Are Important Paracrine Signals

One of the interesting developments from sequencing the human genome and using genes to find proteins has been the identification of *orphan receptors,* receptors that have no known ligand.

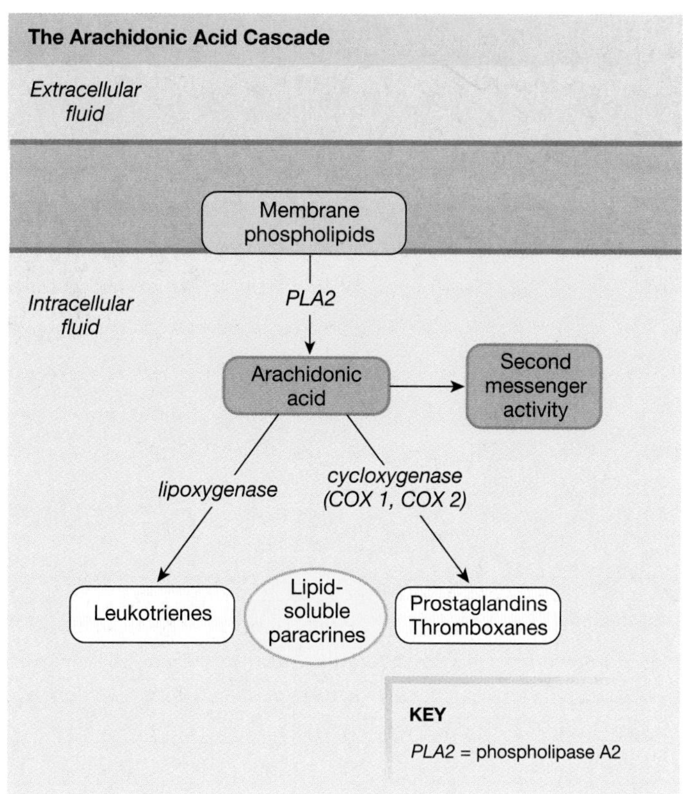

■ **Fig. 6.12**

Scientists are trying to work backward through signal pathways to find the ligands that bind to these orphan receptors. It was from this type of research that investigators recognized the importance and universality of *eicosanoids,* lipid-derived paracrine signals that play important roles in many physiological processes.

All eicosanoid signal molecules are derived from arachidonic acid, a 20-carbon fatty acid. The synthesis process is a network called the *arachidonic acid cascade* (■ Fig. 6.12). For simplicity, we will break the cascade into steps.

Arachidonic acid is produced from membrane phospholipids by the action of an enzyme, **phospholipase A2 (PLA2).** The activity of PLA2 is controlled by hormones and other signals. Arachidonic acid itself may act directly as a second messenger, altering ion channel activity and intracellular enzymes. It may also be converted into one of several classes of eicosanoid paracrines. These lipid-soluble molecules can diffuse out of the cell and combine with receptors on neighboring cells to exert their action.

There are two major groups of arachidonic acid-derived paracrines to be aware of:

1. **Leukotrienes** are molecules produced by the action of the enzyme *lipoxygenase* on arachidonic acid {*leuko–*, white + *triene,* a molecule with three double bonds between carbon atoms}. Leukotrienes are secreted by certain types of white blood cells. They play a significant role in asthma, a lung condition in which the smooth muscle of the airways constricts, making it difficult to breathe, and in the severe allergic reaction known as *anaphylaxis*. For this reason, pharmaceutical companies have been actively developing drugs to block leukotriene synthesis or action.
2. **Prostanoids** are molecules produced when the enzyme **cyclooxygenase (COX)** acts on arachidonic acid. Prostanoids include **prostaglandins** and **thromboxanes.** These eicosanoids act on many tissues of the body, including smooth muscle in various organs, platelets, kidney, and bone. In addition, prostaglandins are involved in sleep, inflammation, pain, and fever.

The nonsteroidal anti-inflammatory drugs (NSAIDs), such as aspirin and ibuprofen, help prevent inflammation by inhibiting COX enzymes and decreasing prostaglandin synthesis. However, NSAIDs are not specific and may have serious unwanted side effects, such as bleeding in the stomach. The discovery of two COX isozymes, COX1 and COX2, enabled the design of drugs that target a specific COX isozyme. By inhibiting only COX2, the enzyme that produces inflammatory prostaglandins, physicians hoped to treat inflammation with fewer side effects. However, studies have shown that some patients who take COX2 inhibitors and other NSAIDs have increased risk of heart attacks and strokes, so these drugs are not recommended for long-term use.

Eicosanoids are not the only known lipid signal molecules. Lipids called *sphingolipids* also act as extracellular signals to help regulate inflammation, cell adhesion and migration, and cell growth and death. Like the eicosanoids, sphingolipids combine with G protein–coupled receptors in the membranes of their target cells.

6

Concept Check Answer: p. 205

14. One drug blocks leukotriene action in its target cells. A different drug blocks leukotriene synthesis. Use what you have learned about leukotrienes, signal molecules, and signal transduction to predict what these drugs are doing to have those effects.

Modulation of Signal Pathways

As you have just learned, signal pathways in the cell can be very complex. To complicate matters, different cells may respond differently to one kind of signal molecule. How can one chemical trigger response A in tissue 1 and response B in tissue 2? *For most signal molecules, the target cell response depends on its receptor or its associated intracellular pathways, not on the ligand.*

One Ligand May Have Multiple Receptors

For many years physiologists were unable to explain the observation that a single signal molecule could have different effects in different tissues. For example, the neurohormone epinephrine dilates blood vessels in skeletal muscle but constricts blood vessels in the intestine. How can one chemical have opposite effects? The answer became clear when scientists discovered that receptors, like other proteins, may come as families of related *isoforms* [p. 52].

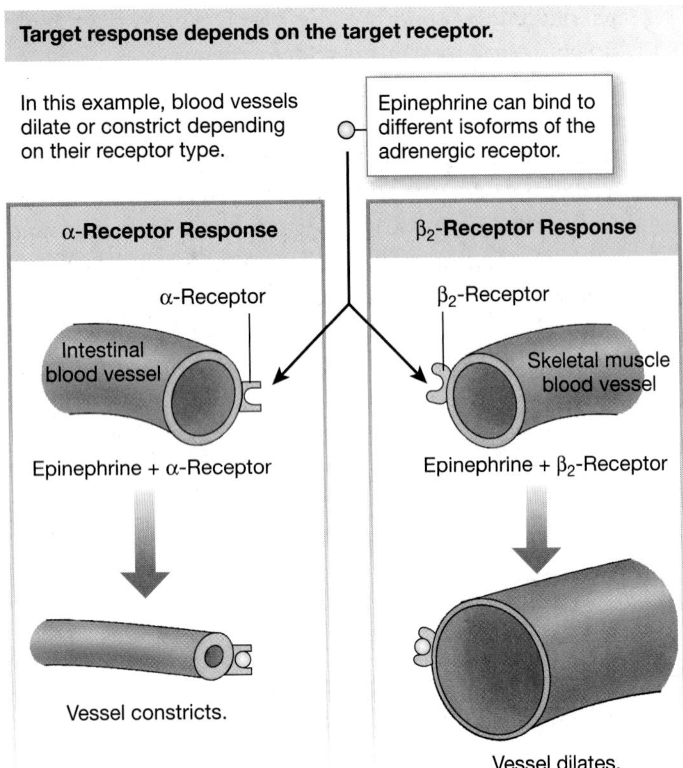
Target response depends on the target receptor.
In this example, blood vessels dilate or constrict depending on their receptor type.
Epinephrine can bind to different isoforms of the adrenergic receptor.
α-Receptor Response
β2-Receptor Response
α-Receptor
β2-Receptor
Intestinal blood vessel
Skeletal muscle blood vessel
Epinephrine + α-Receptor
Epinephrine + β2-Receptor
Vessel constricts.
Vessel dilates.

■ **Fig. 6.13**

The cellular response that follows binding of a signal molecule to a receptor depends on which isoform of the receptor is involved. For example, the α- and β_2-adrenergic receptors for epinephrine are isoforms of each other. When epinephrine binds to α-receptors on smooth muscle in intestinal blood vessels, signal pathways begin that cause the vessels to constrict (■ Fig. 6.13). When epinephrine binds to β_2-receptors on certain skeletal muscle blood vessels, the associated signal transduction pathways cause the vessels to dilate. In other words, the response of the blood vessel to epinephrine depends on the receptor isoform and its signal transduction pathway, not on the ligand that activates the receptor. Many drugs now are designed to be specific for only one receptor isoform.

Receptors Exhibit Saturation, Specificity, and Competition

Because receptors are proteins, receptor-ligand binding exhibits the general protein-binding characteristics of specificity, competition, and saturation [discussed in Chapter 2, p. 50]. Similar protein-binding reactions occur in enzymes [Chapter 4, p. 105] and transporters [Chapter 5, p. 153].

Specificity and Competition: Multiple Ligands for One Receptor Receptors have binding sites for their ligands, just as enzymes and transporters do. As a result, different ligand molecules with similar structures may be able to bind to the same receptor. A classic example of this principle involves two neurocrines responsible for the fight-or-flight response: the neurotransmitter

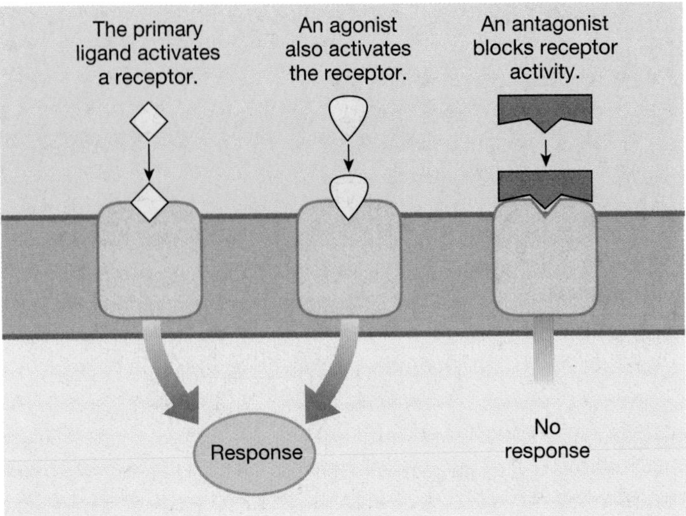
The primary ligand activates a receptor.
An agonist also activates the receptor.
An antagonist blocks receptor activity.
Response
No response

■ **Fig. 6.14**

norepinephrine and its cousin the neurohormone *epinephrine* (also called *adrenaline*). Both molecules bind to a class of receptors called *adrenergic receptors*. (*Adrenergic* is the adjective relating to adrenaline.) The ability of adrenergic receptors to bind these neurocrines, but not others, demonstrates the specificity of the receptors.

Epinephrine and norepinephrine also compete for a single receptor type. Both neurocrines bind to subtypes of adrenergic receptors designated alpha (α) and beta (β). However, α-receptors have a higher binding affinity for norepinephrine, and the β_2-receptor subtype has a higher affinity for epinephrine.

Agonists and Antagonists When a ligand combines with a receptor, one of two events follows. Either the ligand activates the receptor and elicits a response, or the ligand occupies the binding site and prevents the receptor from responding (■ Fig. 6.14). Ligands that turn receptors on are known as **agonists.** Ligands that block receptor activity are called **antagonists.**

Pharmacologists use the principle of competing agonists [p. 51] to design drugs that are longer-acting and more resistant to enzymatic degradation than the *endogenous* ligand produced by the body {*endo-*, within + *-genous*, developing}. One example is the family of modified estrogens (female sex hormones) in birth control pills. These drugs are agonists of naturally occurring estrogens but have chemical groups added to protect them from breakdown and extend their active life.

Concept Check Answers: p. 205

15. What do receptors, enzymes, and transporters have in common that explains why they all exhibit saturation, specificity, and competition?
16. Insulin increases the number of glucose transporters on a skeletal muscle cell but not on the membrane of a liver cell. List two possible mechanisms that could explain how this one hormone can have these two different effects.

Up- and Down-Regulation Enable Cells to Modulate Responses

Saturation of proteins refers to the fact that protein activity reaches a maximum rate because cells contain limited numbers of protein molecules [p. 56]. Saturation can be observed with enzymes, transporters, and receptors. A cell's ability to respond to a chemical signal therefore can be limited by the number of receptors for that signal.

A single cell contains between 500 and 100,000 receptors on the surface of its cell membrane, with additional receptors in the cytosol and nucleus. In any given cell, the number of receptors changes over time. Old receptors are withdrawn from the membrane by endocytosis and are broken down in lysosomes. New receptors are inserted into the membrane by exocytosis. Intracellular receptors are also made and broken down. This flexibility permits a cell to vary its responses to chemical signals depending on the extracellular conditions and the internal needs of the cell.

What happens when a signal molecule is present in the body in abnormally high concentrations for a sustained period of time? Initially the increased signal level creates an enhanced response. As this enhanced response continues, the target cells may attempt to bring their response back to normal by either down-regulation or desensitization of the receptors for the signal [p. 56].

Down-regulation is a decrease in receptor number. The cell can physically remove receptors from the membrane through endocytosis [Fig. 5.19, p. 156]. A quicker and more easily reversible way to decrease cell response is *desensitization,* which can be achieved by binding a chemical modulator to the receptor protein. For example, the β-adrenergic receptors described in the previous section can be desensitized by phosphorylation of the receptor. The result of decreased receptor number or desensitization is a diminished response of the target cell even though the concentration of the signal molecule remains high. Down-regulation and desensitization are one explanation for the development of *drug tolerance,* a condition in which the response to a given dose decreases despite continuous exposure to the drug.

In the opposite situation, when the concentration of a ligand decreases, the target cell may use up-regulation in an attempt to keep its response at a normal level. In **up-regulation,** the target cell inserts more receptors into its membrane. For example, if a neuron is damaged and unable to release normal amounts of neurotransmitter, the target cell may up-regulate its receptors. More receptors make the target cell more responsive to whatever neurotransmitters are present. Up-regulation is also programmed during development as a mechanism that allows cells to vary their responsiveness to growth factors and other signal molecules.

RUNNING PROBLEM

"My dad takes insulin shots for his diabetes," Marvin says. "What does insulin do?" The nurse practitioner replies that normally insulin helps most cells take up and use glucose. In both types of diabetes, however, fasting blood glucose concentrations are elevated because the cells are not taking up and using glucose normally. If people with type 1 diabetes are given shots of insulin, their blood glucose levels decline. If people with type 2 diabetes are given insulin, blood glucose levels may change very little.

Q3: In which form of diabetes are the insulin receptors more likely to be up-regulated?

175 178 191 194 196 198 202

Concept Check Answer: p. 205

17. To decrease a receptor's binding affinity, a cell might (select all that apply):

(a) synthesize a new isoform of the receptor
(b) withdraw receptors from the membrane
(c) insert new receptors into the membrane
(d) use a covalent modulator [*Hint:* p. 54]

Cells Must Be Able to Terminate Signal Pathways

In the body, signals turn on and off, so cells must be able to tell when a signal is over. This requires that signaling processes have built-in termination mechanisms. For example, to stop the response to a calcium signal, a cell removes Ca^{2+} from the cytosol by pumping it either back into the endoplasmic reticulum or out into the extracellular fluid.

Receptor activity can be stopped in a variety of ways. The extracellular ligand can be degraded by enzymes in the extracellular space. An example is the breakdown of the neurotransmitter acetylcholine. Other chemical messengers, particularly neurotransmitters, can be removed from the extracellular fluid through transport into neighboring cells. A widely used class of antidepressant drugs called *selective serotonin reuptake inhibitors,* or SSRIs, extends the active life of the neurotransmitter serotonin by slowing its removal from the extracellular fluid.

Once a ligand is bound to its receptor, activity can also be terminated by endocytosis of the receptor-ligand complex [illustrated in Fig. 5.19, p. 156]. After the vesicle is in the cell, the ligand is removed, and the receptors can be returned to the membrane by exocytosis.

Many Diseases and Drugs Target the Proteins of Signal Transduction

As researchers learn more about cell signaling, they are realizing how many diseases are linked to problems with signal pathways. Diseases can be caused by alterations in receptors or by problems with G proteins or second messenger pathways

Table 6.1 Some Diseases or Conditions Linked to Abnormal Signaling Mechanisms

Genetically Inherited Abnormal Receptors

Receptor	Physiological Alteration	Disease or Condition That Results
Vasopressin receptor (X-linked defect)	Shortens half-life of the receptor	Congenital diabetes insipidus
Calcium sensor in parathyroid gland	Fails to respond to increase in plasma Ca^{2+}	Familial hypercalcemia
Rhodopsin receptor in retina of eye	Improper protein folding	Retinitis pigmentosa

Toxins Affecting Signal Pathways

Toxin	Physiological Effect	Condition That Results
Bordetella pertussis toxin	Blocks inhibition of adenylyl cyclase (i.e., keeps it active)	Whooping cough
Cholera toxin	Blocks enzyme activity of G proteins; cell keeps making cAMP	Ions secreted into lumen of intestine, causing massive diarrhea

(see Tbl. 6.1 for some examples). A single change in the amino acid sequence of a receptor protein can alter the shape of the receptor's binding site, thereby either destroying or modifying its activity.

Pharmacologists are using information about signaling mechanisms to design drugs to treat disease. Some of the alphabet soup of drugs in widespread use are ARBs (angiotensin receptor blockers), "beta blockers" (β-adrenergic receptor blockers), and calcium-channel blockers for treating high blood pressure; SERMs (selective estrogen receptor modulators) for treating estrogen-dependent cancers; and H_2 (histamine type 2) receptor antagonists for decreasing acid secretion in the stomach. You may encounter many of these drugs again if you study the systems in which they are effective.

Homeostatic Reflex Pathways

The cellular signal mechanisms just described are often just one small component of the body's signaling systems that maintain homeostasis. For local control mechanisms, a relatively isolated change occurs in a cell or tissue, and the chemical paracrine or autocrine signals released there are the entire pathway. But in more complicated *reflex control pathways* [p. 15], information must be transmitted throughout the body using chemical signals or a combination of electrical and chemical signaling. In the last section of this chapter we look at some of the patterns of reflex control pathways you will encounter as you study the various organ systems of the body.

Cannon's Postulates Describe Regulated Variables and Control Systems

Walter Cannon, the father of American physiology, described a number of properties of homeostatic control systems in the 1920s based on his observations of the body in health and disease states.*

Cannon's four postulates are:

1. **The nervous system has a role in preserving the "fitness" of the internal environment.** *Fitness* in this instance means conditions that are compatible with normal function. The nervous system coordinates and integrates blood volume, blood osmolarity, blood pressure, and body temperature, among other regulated variables. (In physiology, a regulated variable is also known as a **parameter** {*para-*, beside + *meter,* measure}).
2. **Some systems of the body are under tonic control {*tonos,* tone}.** To quote Cannon, "An agent may exist which has a moderate activity which can be varied up and down." Tonic control is like the volume control on a radio. The radio is always on, but by turning a single knob, you can make the sound level louder or softer.

   A physiological example of a tonically controlled system is the neural regulation of diameter in certain blood vessels, in which increased input from the nervous system decreases diameter, and decreased input from the nervous system increases diameter (Fig. 6.15a). *Tonic control* is one of the more difficult concepts in physiology because we have a tendency to think of responses stopping and starting when a controller turns off or on rather than as a response that is always on but can increase or decrease.
3. **Some systems of the body are under antagonistic control.** Cannon wrote, "When a factor is known which can shift a homeostatic state in one direction, it is reasonable to look for a factor or factors having an opposing effect." Systems that are not under tonic control are usually under *antagonistic control,* either by hormones or the nervous system.

   In pathways controlled by the nervous system, the sympathetic and parasympathetic divisions often have opposing effects. For example, chemical signals from a

*W. B. Cannon. Organization for physiological homeostasis. *Physiological Reviews* 9: 399–443, 1929.

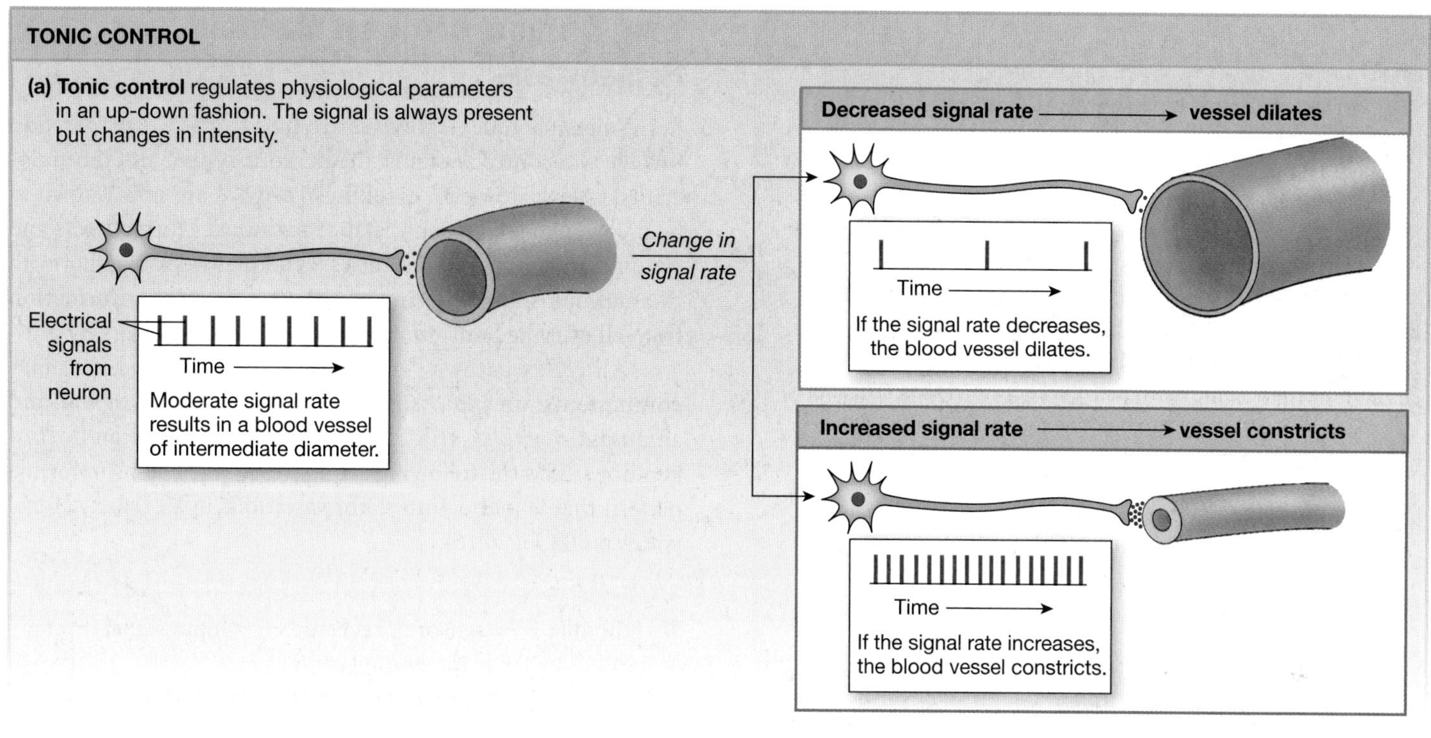

ANTAGONISTIC CONTROL

(b) Antagonistic control uses different signals to send a parameter in opposite directions. In this example, antagonistic neurons control heart rate: some speed it up, while others slow it down.

Sympathetic neuron

Parasympathetic neuron

Stimulation by sympathetic nerves increases heart rate.

Heartbeats

0 1 2 3

Time (sec) →

Stimulation by parasympathetic nerves decreases heart rate.

Heartbeats

0 1 2 3

Time (sec) →

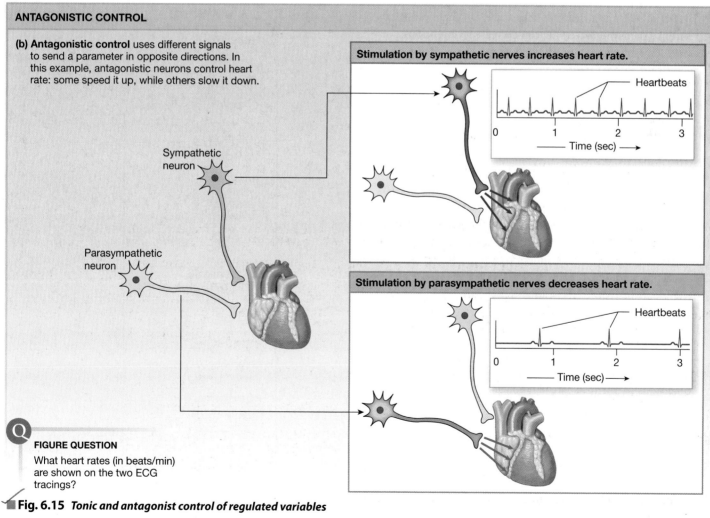

FIGURE QUESTION

What heart rates (in beats/min) are shown on the two ECG tracings?

Fig. 6.15 ***Tonic and antagonist control of regulated variables***

6

RUNNING PROBLEM

"Why is an elevated blood glucose concentration bad?" Marvin asks. "The elevated blood glucose itself is not bad," says the nurse practitioner, "but when it is high after an overnight fast, it suggests that there is something wrong with the way your body is regulating its glucose metabolism." When a normal person absorbs a meal containing carbohydrates, blood glucose levels increase and stimulate insulin release. When cells have taken up the glucose from the meal and blood glucose levels fall, secretion of another pancreatic hormone, glucagon, increases. Glucagon raises blood glucose and helps keep the level within the homeostatic range.

Q4: The homeostatic regulation of blood glucose levels by the hormones insulin and glucagon is an example of which of Cannon's postulates?

sympathetic neuron increase heart rate, but chemical signals from a parasympathetic neuron decrease it (Fig. 6.15b).

When chemical signals have opposing effects, they are said to be antagonistic to each other. For example, insulin and glucagon are antagonistic hormones. Insulin decreases the glucose concentration in the blood and glucagon increases it.

4 **One chemical signal can have different effects in different tissues.** Cannon observed correctly that "homeostatic agents antagonistic in one region of the body may be cooperative in another region." However, it was not until scientists learned about cell receptors that the basis for the seemingly contradictory actions of some hormones or nerves became clear. As you learned in the first part of this chapter, a single chemical signal can have different effects depending on the receptor and signal pathway of the target cell. For example, epinephrine constricts or dilates blood vessels, depending on whether the vessel has α- or β_2-adrenergic receptors (Fig. 6.13).

The remarkable accuracy of Cannon's postulates, now confirmed with cellular and molecular data, is a tribute to the observational skills of scientists in the nineteenth and early twentieth centuries.

Concept Check Answers: p. 205

18. What is the difference between tonic control and antagonistic control?

19. How can one chemical signal have opposite effects in two different tissues?

Long-Distance Pathways Maintain Homeostasis

Long-distance reflex pathways are traditionally considered to involve two control systems: the nervous system and the endocrine system. However, cytokines [p. 177] are now known to be involved in some long-distance pathways. During stress and systemic inflammatory responses, cytokines work together with the nervous and endocrine systems to integrate information from all over the body into coordinated responses.

All reflex pathway **response loops** have three primary components: an *input signal, integration of the signal,* and an *output signal* [p. 16]. These three components can be broken down into the following sequence of seven steps to form a pattern that is found with slight variations in all reflex control pathways (■ Fig. 6.16):

> Stimulus ⟶ sensor or receptor ⟶ input signal ⟶
> integrating center ⟶
> output signal ⟶ target ⟶ response

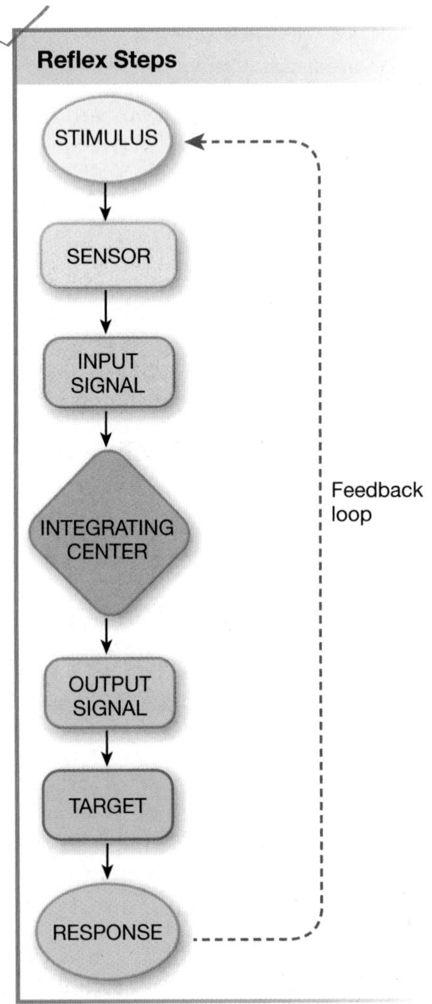

■ **Fig. 6.16** ***Steps of a reflex pathway***

- A **stimulus** is the disturbance or change that sets the pathway in motion. The stimulus may be a change in temperature, oxygen content, blood pressure, or any one of a myriad of other regulated variables.
- A **sensor** or sensory receptor continuously monitors its environment for a particular variable.
- When activated by a change, the sensor sends an **input** (*afferent*) **signal** to the integrating center for the reflex.
- The **integrating center** compares the input signal with the **setpoint,** or desired value of the variable.
- If the variable has moved out of the acceptable range, the integrating center initiates an output signal.
- The **output** (*efferent*) **signal** is an electrical and/or chemical signal that travels to the target.
- The **target,** or **effector** {*effectus,* the carrying out of a task} is the cell or tissue that carries out the appropriate **response** to bring the variable back within normal limits.

Let's look in more detail at these components of a reflex.

Sensors In the first step in a physiological response loop, a stimulus activates a sensor or receptor. Notice that this is a new and different application of the word *receptor.* Like many other terms in physiology, *receptor* can have different meanings (■ Fig. 6.17). The sensory receptors of a neural reflex are not protein receptors that bind to signal molecules, like those involved in signal transduction. Rather, neural receptors are specialized cells, parts of cells, or complex multicellular receptors (such as the eye) that respond to changes in their environment.

There are many sensory receptors in the body, each located in the best position to monitor the variable it detects. The eyes, ears, and nose are receptors that sense light, sound and motion, and odors, respectively. Your skin is covered with less complex receptors that sense touch, temperature, vibration, and pain. Other sensors are internal: receptors in the joints of the skeleton that send information to the brain about body position, or blood pressure and oxygen receptors in blood vessels that monitor conditions in the circulatory system. The sensory receptors involved in neural reflexes are divided into central receptors and peripheral receptors. *Central receptors* are located in the brain or are closely linked to the brain. An example is the brain's chemoreceptor for carbon dioxide. *Peripheral receptors* reside elsewhere in the body and include the skin receptors and internal receptors just described.

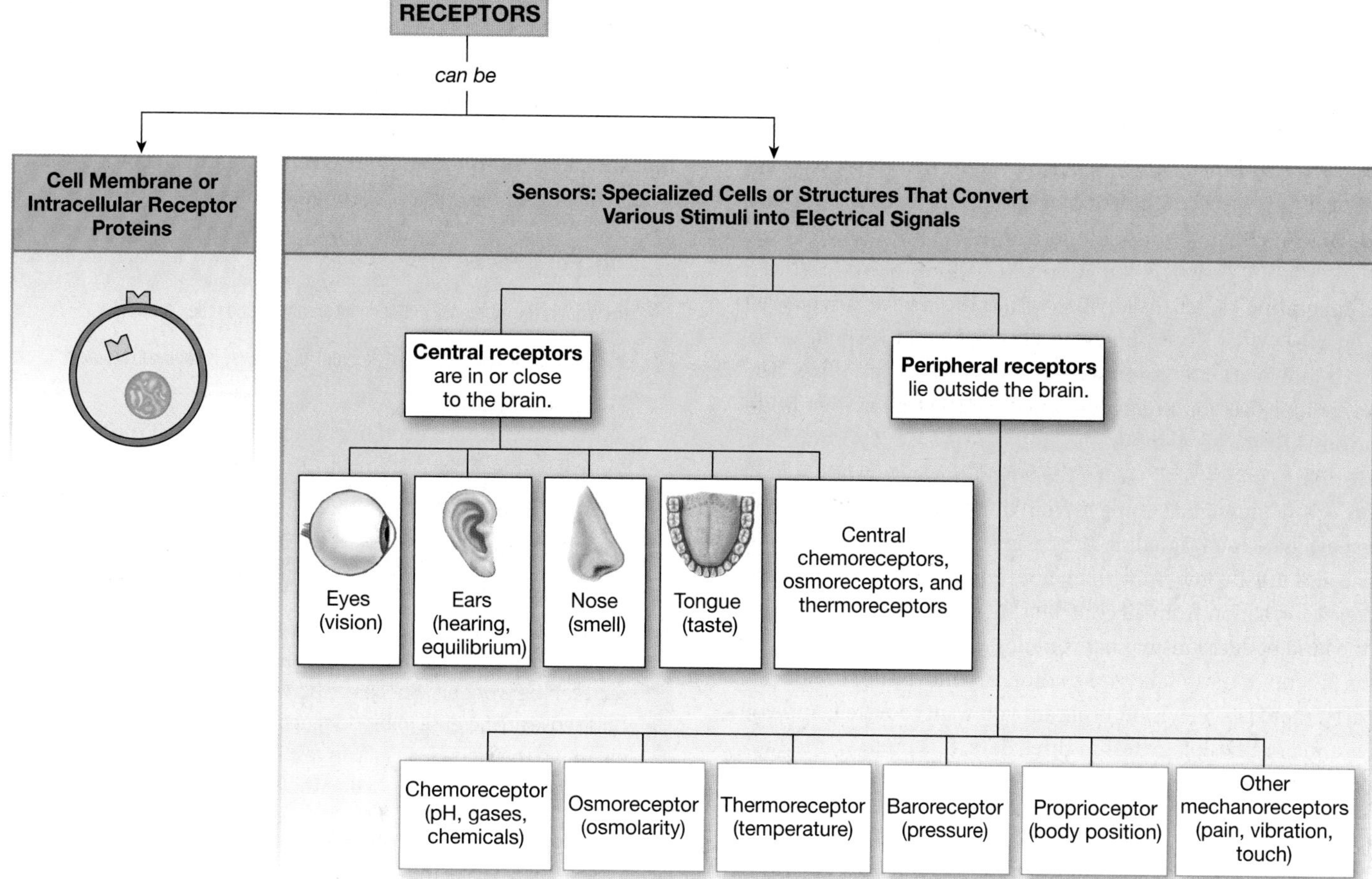

■ **Fig. 6.17** ***Multiple meanings of the word receptor.*** The word *receptor* may mean a protein that binds to a ligand. Receptor can also mean a specialized cell or structure for transduction of stimuli into electrical signals (a *sensory receptor* or *sensor*). Sensory receptors are classified as central or peripheral, depending on whether they are found in the brain or outside the brain.

6

All sensors have a **threshold,** the minimum stimulus needed to set the reflex response in motion. If a stimulus is below the threshold, no response loop is initiated.

You can demonstrate threshold in a sensory receptor easily by touching the back of your hand with a sharp, pointed object, such as a pin. If you touch the point to your skin lightly enough, you can see the contact between the point and your skin even though you do not feel anything. In this case, the stimulus (pressure from the point of the pin) is below threshold, and the pressure receptors of the skin are not responding. As you press harder, the stimulus reaches threshold, and the receptors respond by sending a signal to the brain, causing you to feel the pin.

Endocrine reflexes that are not associated with the nervous system do not use sensory receptors to initiate their pathways. Instead, endocrine cells act both as sensor and integrating center for the reflex. For example, a pancreatic beta cell sensing and responding directly to changes in blood glucose concentrations is an endocrine cell that is both sensor and integrating center [Fig. 5.27, p. 167].

Input Signal The input signal in a reflex varies depending on the type of reflex. In a neural pathway, such as the pin touch above, the input signal is electrical and chemical information transmitted by a sensory neuron. In an endocrine reflex, there is no input pathway because the stimulus acts directly on the endocrine cell, which serves as both sensor and integrating center.

Integrating Center The integrating center in a reflex pathway is the cell that receives information about the regulated variable and can initiate an appropriate response. In endocrine reflexes, the integrating center is the endocrine cell. In neural reflexes, the integrating center usually lies within the *central nervous system* (CNS), which is composed of the brain and the spinal cord.

If information is coming from a single stimulus, it is a relatively simple task for an integrating center to compare that information with the setpoint and initiate a response if appropriate. Integrating centers really "earn their pay," however, when two or more conflicting signals come in from different sources. The center must evaluate each signal on the basis of its strength and importance and must come up with an appropriate response that integrates information from all contributing receptors. This is similar to the kind of decision-making you must do when on one evening your parents want to take you to dinner, your friends are having a party, there is a television program you want to watch, and you have a major physiology test in three days. It is up to you to rank those items in order of importance and decide what you will do.

Output Signals Output signal pathways are relatively simple. In the nervous system, the output signal is always the electrical and chemical signals transmitted by an efferent neuron. Because all electrical signals traveling through the nervous system are identical, the distinguishing characteristic of the signal is the anatomical pathway of the neuron—the route through which the neuron delivers its signal. For example, the vagus nerve carries neural signals to the heart, and the phrenic nerve carries neural signals to the diaphragm. Output pathways in the nervous system are given the anatomical name of the nerve that carries the signal. For example, we speak of the vagal control of heart rate (*vagal* is the adjective for *vagus*).

In the endocrine system, the anatomical routing of the output signal is always the same—all hormones travel in the blood to their target. Hormonal output pathways are distinguished by the chemical nature of the signal and are therefore named for the hormone that carries the message. For example, the output signal for a reflex integrated through the endocrine pancreas will be either the hormone insulin or the hormone glucagon, depending on the stimulus and the appropriate response.

Targets The targets of reflex control pathways are the cells or tissues that carry out the response. The targets of neural pathways may be any type of muscle, endocrine or exocrine glands, or adipose tissue. Targets of an endocrine pathway are the cells that have the proper receptor for the hormone.

Responses There are two levels of response for any reflex control pathway. One is the very specific *cellular response* that takes place in the target cell. The more general *systemic response* describes what those specific cellular events mean to the organism as a whole. For example, when the hormone epinephrine combines with β_2-adrenergic receptors on the walls of certain blood vessels, the cellular response is relaxation of the smooth muscle. The systemic response to relaxation of the blood vessel wall is increased blood flow through the vessel.

Concept Check Answers: p. 205

20. What is the difference between local control and reflex control?
21. Name the seven steps in a reflex control pathway in their correct order.

Now that you have been introduced to the basic parts of a reflex control pathway, we can turn to an analysis of the two primary control systems, the nervous system and the endocrine system.

RUNNING PROBLEM

Marvin is fascinated by the ability of the body to keep track of glucose. "How does the pancreas know which hormone to secrete?" he wonders. Special cells in the pancreas called beta cells sense an increase in blood glucose concentrations after a meal, and they release insulin in response. Insulin then acts on many tissues of the body so that they take up and use glucose.

Q5: In the insulin reflex pathway that regulates blood glucose levels, what are the stimulus, the sensor, the integrating center, the output signal, the target(s), and the response(s)?

175 178 191 194 196 198 202

Control Systems Vary in Their Speed and Specificity

Physiological reflex control pathways are mediated by the nervous system, the endocrine system, or a combination of the two (■ Fig. 6.18). A reflex mediated solely by the nervous system or solely by the endocrine system is relatively simple, but combination reflex pathways can be quite complex. In the most complex pathways, signals pass through three different integrating centers before finally reaching the target tissue. With so much overlap between pathways controlled by the nervous and endocrine systems, it makes sense to consider these systems as parts of a continuum rather than as two discrete systems.

Why does the body need different types of control systems? To answer that question, let us compare endocrine control with neural control to see what the differences are. Five major differences are summarized in ■ Table 6.2 and discussed next.

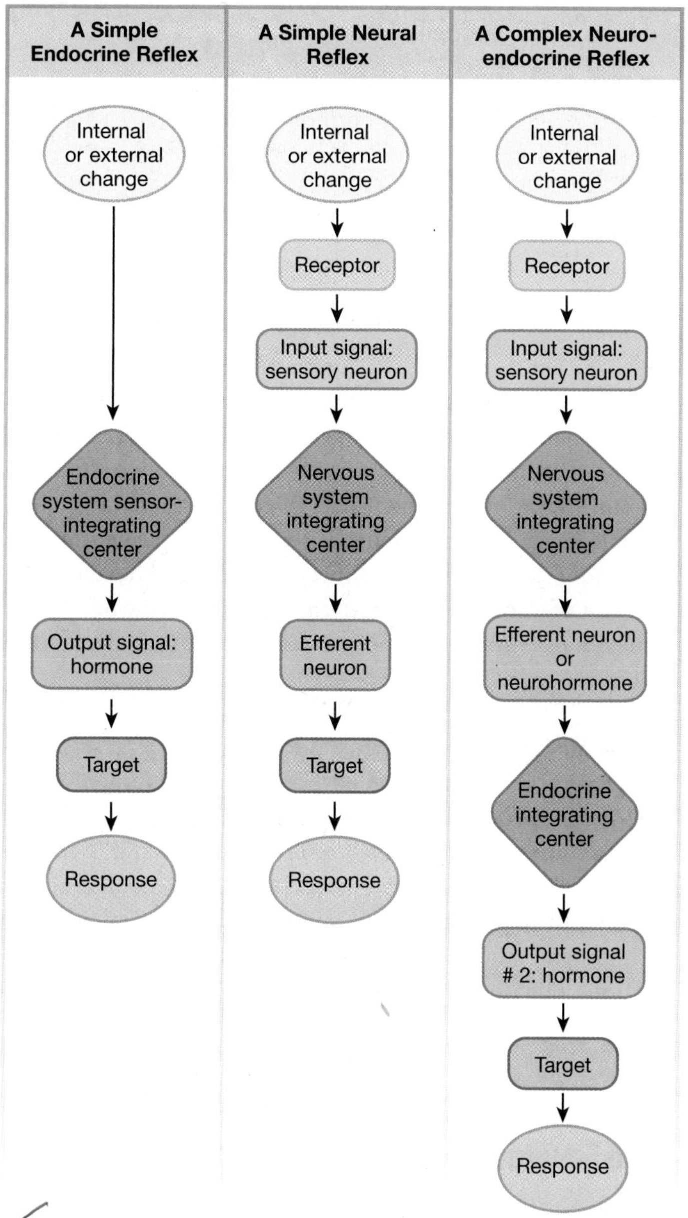

A Simple Endocrine Reflex: Internal or external change → Endocrine system sensor-integrating center → Output signal: hormone → Target → Response

A Simple Neural Reflex: Internal or external change → Receptor → Input signal: sensory neuron → Nervous system integrating center → Efferent neuron → Target → Response

A Complex Neuro-endocrine Reflex: Internal or external change → Receptor → Input signal: sensory neuron → Nervous system integrating center → Efferent neuron or neurohormone → Endocrine integrating center → Output signal # 2: hormone → Target → Response

■ **Fig. 6.18** ***Endocrine, neural, and neuroendocrine control pathways***

Specificity Neural control is very specific because each neuron has a specific target cell or cells to which it sends its message. Anatomically, we can isolate a neuron and trace it from its origin to where it terminates on its target.

Endocrine control is more general because the chemical messenger is released into the blood and can reach virtually every cell in the body. As you learned in the first half of this chapter, the body's response to a specific hormone depends on which cells have receptors for that hormone and which receptor type they have. Multiple tissues in the body can respond to a hormone simultaneously.

Nature of the Signal The nervous system uses both electrical and chemical signals to send information throughout the body. Electrical signals travel long distances through neurons, releasing chemical signals (neurotransmitters) that diffuse across the small gap between the neuron and its target. In a limited number of instances, electrical signals pass directly from cell to cell through gap junctions.

The endocrine system uses only chemical signals: hormones secreted into the blood by endocrine glands or cells. Neurohormone pathways represent a hybrid of the neural and endocrine reflexes. In a neurohormone pathway, a neuron creates an electrical signal, but the chemical released by the neuron is a neurohormone that goes into the blood for general distribution.

Concept Check Answer: p. 205

22. (a) In the simple neural reflex shown in Figure 6.18, which box or boxes represent the brain and spinal cord? (b) Which box or boxes represent the central and peripheral sense organs? (c) In the simple neural reflex, add a dashed line connecting boxes to show how a negative feedback loop would shut off the reflex [p. 17].

Speed Neural reflexes are much faster than endocrine reflexes. The electrical signals of the nervous system cover great distances very rapidly, with speeds of up to 120 m/sec. Neurotransmitters also create very rapid responses, on the order of milliseconds.

Hormones are much slower than neural reflexes. Their distribution through the circulatory system and diffusion from capillary to receptors take considerably longer than signals through neurons. In addition, hormones have a slower onset of action. In target tissues, the response may take minutes to hours before it can be measured.

Why do we need the speedy reflexes of the nervous system? Consider this example. A mouse ventures out of his hole and sees a cat ready to pounce on him and eat him. A signal must go from the mouse's eyes and brain down to his feet, telling him to run back into the hole. If his brain and feet

Comparison of Neural and Endocrine Control — Table 6.2

Property	Neural Reflex	Endocrine Reflex
Specificity	Each neuron terminates on a single target cell or on a limited number of adjacent target cells.	Most cells of the body are exposed to a hormone. The response depends on which cells have receptors for the hormone.
Nature of the signal	Electrical signal that passes through neuron, then chemical neurotransmitters that carry the signal from cell to cell. In a few cases, signals pass from cell to cell through gap junctions.	Chemical signals secreted in the blood for distribution throughout the body.
Speed	Very rapid.	Distribution of the signal and onset of action are much slower than in neural responses.
Duration of action	Usually very short. Responses of longer duration are mediated by neuromodulators.	Responses usually last longer than neural responses.
Coding for stimulus intensity	Each signal is identical in strength. Stimulus intensity is correlated with increased frequency of signaling.	Stimulus intensity is correlated with amount of hormone secreted.

were only 5 micrometers (5 μm = 1/200 millimeter) apart, it would take a chemical signal 20 milliseconds (msec) to diffuse across the space and the mouse could escape. If the brain and feet were 50 μm (1/20 millimeter) apart, diffusion would take 2 seconds and the mouse might get caught. But because the head and tail of a mouse are *centimeters* apart, it would take a chemical signal *three weeks* to diffuse from the mouse's head to his feet. Poor mouse!

Even if the distribution of the chemical signal were accelerated by help from the circulatory system, the chemical message would still take 10 seconds to get to the feet, and the mouse would become cat food. The moral of this tale is that reflexes requiring a speedy response are mediated by the nervous system because they are so much more rapid.

RUNNING PROBLEM

"OK, just one more question," says Marvin. "You said that people with diabetes have high blood glucose levels. If glucose is so high, why can't it just leak into the cells?"

Q6: Why can't glucose always diffuse into cells when the blood glucose concentration is higher than the intracellular glucose concentration?

Q7: What do you think happens to the rate of insulin secretion when blood glucose levels fall? What kind of feedback loop is operating?

175 178 191 194 196 198 202

Duration of Action Neural control is of shorter duration than endocrine control. The neurotransmitter released by a neuron combines with a receptor on the target cell and initiates a response. The response is usually very brief, however, because the neurotransmitter is rapidly removed from the vicinity of the receptor by various mechanisms. To get a sustained response, multiple repeating signals must be sent through the neuron.

Endocrine reflexes are slower to start, but they last longer. Most of the ongoing, long-term functions of the body, such as metabolism and reproduction, fall under the control of the endocrine system.

Coding for Stimulus Intensity As a stimulus increases in intensity, control systems must have a mechanism for conveying this information to the integrating center. The signal strength from any one neuron is constant in magnitude and therefore cannot reflect stimulus intensity. Instead, the frequency of signaling through the afferent neuron increases. In the endocrine system, stimulus intensity is reflected by the amount of hormone released: the stronger the stimulus, the more hormone is released.

Complex Reflex Control Pathways Have Several Integrating Centers

■ Figure 6.19 summarizes variations in the neural, neuroendocrine, and endocrine reflex control pathways.

Fig. 6.19 *ESSENTIALS*

Simple and Complex Reflex Pathways

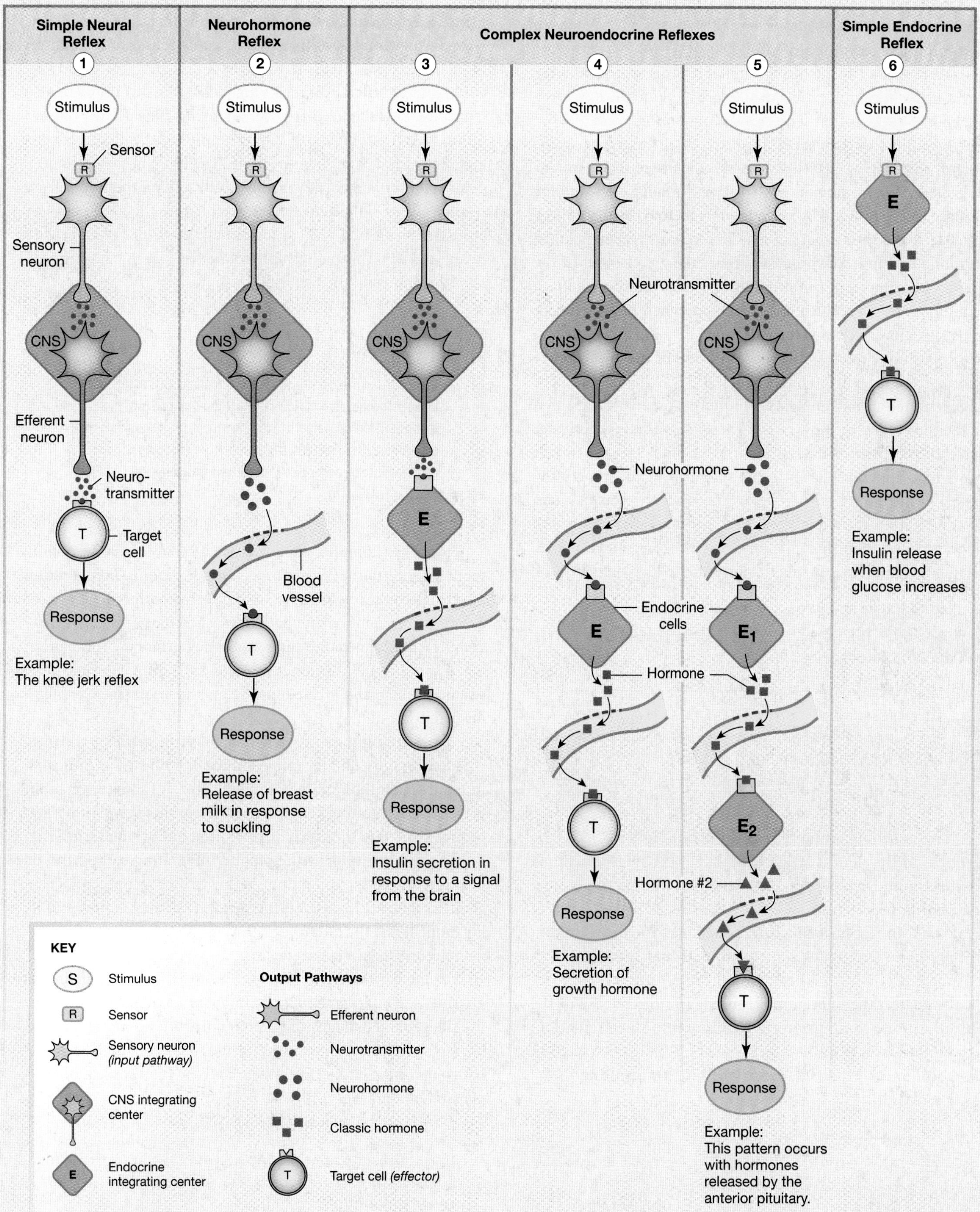

In a simple endocrine reflex pathway (Fig. 6.19 ⑥), the endocrine cell monitors the regulated variable and acts as both sensor and integrating center; there is no input pathway. The output pathway is the hormone, and the target is any cell having the appropriate receptor protein.

An example of a simple endocrine reflex is secretion of the hormone insulin in response to changes in blood glucose level. The endocrine cells that secrete insulin monitor blood glucose concentrations by using ATP production in the cell as an indicator [Fig. 5.27, p. 167]. When blood glucose increases, intracellular ATP production exceeds the threshold level, and the endocrine cells respond by secreting insulin into the blood. Any target cell in the body that has insulin receptors responds to the hormone and initiates processes that take glucose out of the blood. The removal of the stimulus acts in a negative feedback manner, and the response loop shuts off when blood glucose levels fall below a certain concentration.

In a simple neural reflex, all the steps of the pathway are present, from receptor to target (Fig. 6.19 ①). The neural reflex is represented in its simplest form by the knee jerk (or patellar tendon) reflex. A blow to the knee (the stimulus) activates a stretch receptor. An electrical and chemical signal travels through an afferent neuron to the spinal cord (the integrating center). If the blow is strong enough (exceeds threshold), a signal travels from the spinal cord through an efferent neuron to the muscles of the thigh (the target or effector). In response, the muscles contract, causing the lower leg to kick outward (the knee jerk).

Concept Check

Answer: p. 205

23. Match the following terms for parts of the knee jerk reflex to the parts of the simple neural reflex shown in Figure 6.19 ①: blow to knee, leg muscles, neuron to leg muscles, sensory neuron, brain and spinal cord, stretch receptor, muscle contraction.

The neurohormone reflex, shown in Figure 6.19 ②, is identical to the neural reflex except that the neurohormone released by the neuron travels in the blood to its target, just like a hormone. A simple neurohormone reflex is the release of breast milk in response to a baby's suckling. The baby's mouth on the nipple stimulates sensory signals that travel through sensory neurons to the brain (integrating center). An electrical signal in the efferent neuron triggers the release of the neurohormone oxytocin from the brain into the circulation. Oxytocin is carried to the breast, where it causes contraction of smooth muscles in the breast (the target), resulting in the ejection of milk.

In complex pathways, there may be more than one integrating center. Figure 6.19 shows three examples of complex neuroendocrine pathways. The simplest of these, Figure 6.19 ③, combines a neural reflex with a classic endocrine reflex. An example of this pattern can be found in the control of insulin release. The pancreatic beta cells monitor blood glucose concentrations directly (Fig. 6.19 ⑥), but they are also controlled by the nervous system. During a meal, the presence of food in the stomach stretches the wall of the digestive tract and sends input signals to the brain. The brain in turn sends excitatory output signals to the beta cells, telling them to release insulin. These signals take place even before the food has been absorbed and blood glucose levels have gone up (a *feedforward reflex* [p. 18]). This pathway therefore has two integrating centers (the brain and the beta cells).

Concept Check

Answer: p. 205

24. Match the following terms with the appropriate parts of the simple neuroendocrine reflex in Fig. 6.19 ③ (terms may be used more than once): food in stomach following a meal, brain and spinal cord, endocrine cells of pancreas, stretch receptors, efferent neuron to pancreas, insulin, adipose cell, blood, sensory neuron.

Another complex reflex (Fig. 6.19 ④) uses a neurohormone to control the release of a classic hormone. The secretion of growth hormone is an example of this pathway. The most complex neuroendocrine pathways, shown as Figure 6.19 ⑤, include a neurohormone and two classic hormones. This pattern is typical of some hormones released by the anterior pituitary, an endocrine gland located just below the brain [see Chapter 7 for details].

In describing complex neuroendocrine reflex pathways, we identify only one receptor and input pathway, as indicated in Figure 6.19. In the three complex pathways shown, the brain is the first integrating center and the neurohormone is the first output pathway. In Figure 6.19 ⑤ the endocrine target (E_1) of the neurohormone is the second integrating center, and its hormone is the second output pathway. The second endocrine gland in the pathway (E_2) is the third integrating center, and its hormone is the third output pathway. The target of the last signal in the sequence is the effector.

■ Table 6.3 compares the various steps in neural, neuroendocrine, and endocrine reflexes. In the remainder of the text, we use the general patterns shown in Figure 6.19 as a tool for classifying complex reflex pathways. Next we look in detail at some pathways of the endocrine system and the roles these pathways play in homeostasis.

Comparison of Neural, Neuroendocrine, and Endocrine Reflexes

Table 6.3

	Neural	Neuroendocrine	Endocrine
Sensor	Special and somatic sensory receptors	Special and somatic sensory receptors	Endocrine cell
Input signal	Sensory neuron	Sensory neuron	None
Integrating center	Brain or spinal cord	Brain or spinal cord	Endocrine cell
Output signal	Efferent neuron (electrical signal and neurotransmitter)	Efferent neuron (electrical signal and neurohormone)	Hormone
Target(s)	Muscles and glands, some adipose tissue	Most cells of the body	Most cells of the body
Response	Contraction and secretion primarily; may have some metabolic effects	Change in enzymatic reactions, membrane transport, or cell proteins	Change in enzymatic reactions, membrane transport, or cell proteins

RUNNING PROBLEM CONCLUSION

Diabetes Mellitus

Marvin underwent further tests and was diagnosed with early type 2 diabetes. With careful attention to his diet and with a regular exercise program, he has managed to keep his blood glucose levels under control. Diabetes is a growing epidemic in the United States, with more than 25 million diabetics in the United States in 2011 (about 8% of the population). Even scarier is the estimate that another 79 million people are considered "prediabetic"—at significant risk of becoming diabetic. You will learn more about diabetes as you work through the chapters in this book. To learn more about diabetes now, see the American Diabetes Association web site (*www.diabetes.org*) or the Centers for Disease Control and Prevention (*www.cdc.gov/diabetes*).

In this running problem, you learned about glucose homeostasis and how it is maintained by insulin and glucagon. The disease diabetes mellitus is an indication that glucose homeostasis has been disrupted. Check your understanding of this running problem by comparing your answers to the information in the summary table.

Question	Facts	Integration and Analysis
1. In which type of diabetes is the signal pathway for insulin more likely to be defective?	Insulin is a peptide hormone that uses membrane receptors linked to second messengers to transmit its signal to cells. People with type 1 diabetes lack insulin; people with type 2 diabetes have normal-to-elevated insulin levels.	Normal or high insulin levels suggest that the problem is not with amount of insulin but with the action of the insulin at the cell. The problem in type 2 diabetes could be a defective signal transduction mechanism.
2. Insulin is a protein hormone. Would you expect to find its receptor on the cell surface or in the cytoplasm of the target cells?	Lipophilic signal molecules have intracellular receptors. Lipophobic molecules have cell membrane receptors.	Proteins are lipophobic so protein hormones have cell surface receptors.

RUNNING PROBLEM CONCLUSION *(continued)*

Question	Facts	Integration and Analysis
3. In which form of diabetes are the insulin receptors more likely to be up-regulated?	Up-regulation of receptors usually occurs if a signal molecule is present in unusually low concentrations [p. 191]. In type 1 diabetes, insulin is not secreted by the pancreas.	In type 1 diabetes, insulin levels are low. Therefore, type 1 is more likely to cause up-regulation of the insulin receptors.
4. The homeostatic regulation of blood glucose levels by the hormones insulin and glucagon is an example of which of Cannon's postulates?	Cannon's postulates describe the role of the nervous system in maintaining homeostasis, and the concepts of tonic activity, antagonistic control, and different effects of signals in different tissues.	Insulin decreases blood glucose levels, and glucagon increases them. Therefore, the two hormones are an example of an antagonistic control.
5. In the insulin pathway that regulates blood glucose levels, what are the stimulus, the sensor, the integrating center, the output signal, the target(s), and the response(s)?	See the steps of reflex pathways [p. 194].	*Stimulus:* increase in blood glucose levels; *sensor:* beta cells of the pancreas that sense the change; *integrating center:* beta cells; *output signal*: insulin; *targets:* any tissues of the body that respond to insulin; *responses:* cellular uptake and use of glucose.
6. Why can't glucose always diffuse into cells when the blood glucose concentration is higher than the intracellular glucose concentration?	Glucose is lipophobic. Simple diffusion goes across the phospholipid bilayer. Facilitated diffusion uses protein carriers [p. 149].	Because glucose is lipophobic, it cannot cross the membrane by simple diffusion. It must go by facilitated diffusion. If a cell lacks the necessary carriers, facilitated diffusion cannot take place.
7. What do you think happens to the rate of insulin secretion when blood glucose levels fall? What kind of feedback loop is operating?	The stimulus for insulin release is an increase in blood glucose levels. In negative feedback, the response offsets the stimulus. In positive feedback, the response enhances the stimulus.	An increase in blood glucose concentration stimulates insulin release; therefore, a decrease in blood glucose should decrease insulin release. In this example, the response (lower blood glucose) offsets the stimulus (increased blood glucose), so a negative feedback loop is operating.

175 178 191 194 196 198 202

Chapter Summary

Two of the major themes in physiology stand out in this chapter: *the control of homeostasis* and *communication*. The sensors, integrating centers, and targets of physiological control systems are described in the context of reflex control pathways, which vary from simple to complex. Functional control systems require efficient communication that uses various combinations of chemical and electrical signals. Those signals that cannot cross cell membranes must use membrane receptor proteins and signal transduction to transfer their information into the cell. The interaction of signal molecules with protein receptors illustrates another fundamental theme of physiology, *molecular interactions*.

Cell-to-Cell Communication

1. There are two basic types of physiological signals: chemical and electrical. Chemical signals are the basis for most communication within the body. (p. 175)
2. There are four methods of cell-to-cell communication: (1) direct cytoplasmic transfer through gap junctions, (2) contact-dependent signaling, (3) local chemical communication, and (4) long-distance communication. (p. 176; Fig. 6.1)
3. **Gap junctions** are protein channels that connect two adjacent cells. When they are open, chemical and electrical signals pass directly from one cell to the next. (p. 175)
4. **Contact-dependent signals** require direct contact between surface molecules of two cells. (p. 175)
5. Local communication uses **paracrine signals**, chemicals that act on cells close to the cell that secreted the paracrine. A chemical that acts on the cell that secreted it is called an **autocrine signal**. The activity of paracrine and autocrine signal molecules is limited by diffusion distance. (p. 177)
6. Long-distance communication uses **neurocrine molecules** and electrical signals in the nervous system, and **hormones** in the endocrine system. Only cells that possess receptors for a hormone will be **target cells.** (p. 177)
7. **Cytokines** are regulatory peptides that control cell development, differentiation, and the immune response. They serve as both local and long-distance signals. (p. 177)

Signal Pathways

8. Chemical signals bind to **receptors** and change intracellular signal molecules that direct the response. (p. 178)
9. Lipophilic signal molecules enter the cell and combine with cytoplasmic or nuclear receptors. Lipophobic signal molecules and some lipophilic molecules combine with membrane receptors. (p. 179; Fig. 6.3)
10. **Signal transduction** pathways use membrane receptor proteins and intracellular second messenger molecules to translate signal information into an intracellular response. (p. 180; Fig. 6.4)
11. Some signal transduction pathways activate **protein kinases**. Others activate **amplifier enzymes** that create **second messenger** molecules. (p. 181; Fig. 6.5)
12. Signal pathways create intracellular **cascades** that amplify the original signal. (p. 182; Fig. 6.6)
13. **Receptor-enzymes** activate protein kinases, such as **tyrosine kinase** (Fig. 6.7), or the amplifier enzyme **guanylyl cyclase,** which produces the second messenger **cGMP**. (p. 183)
14. **G proteins** linked to amplifier enzymes are the most prevalent signal transduction system. **G protein–coupled receptors** also alter ion channels. (p. 184; Fig. 6.8)
15. The **G protein–coupled adenylyl cyclase-cAMP-protein kinase A** pathway is the most common pathway for protein and peptide hormones. (p. 184; Fig. 6.8a)
16. In the **G protein–coupled phospholipase C** pathway, the amplifier enzyme **phospholipase C** creates two second messengers: **IP_3** and **diacylglycerol** (DAG). IP_3 causes Ca^{2+} release from intracellular stores. Diacylglycerol activates **protein kinase** C. (p. 184; Fig. 6.8b)
17. **Integrin** receptors link the extracellular matrix to the cytoskeleton. (p. 179; Fig. 6.3c)
18. **Ligand-gated ion channels** open or close to create electrical signals. (p. 185; Fig. 6.9)

Novel Signal Molecules

19. Calcium is an important signal molecule that binds to **calmodulin** to alter enzyme activity. It also binds to other cell proteins to alter movement and initiate exocytosis. (p. 187; Fig. 6.11)
20. **Nitric oxide (NO), carbon monoxide (CO),** and **hydrogen sulfide (H_2S)** are short-lived gaseous signal molecules. NO activates guanylyl cyclase directly. (pp. 187–188)
21. The arachidonic acid cascade creates lipid signal molecules, such as **leukotrienes**, **prostaglandins**, and **thromboxanes**. (p. 189; Fig. 6.12)

Modulation of Signal Pathways

22. The response of a cell to a signal molecule is determined by the cell's receptor for the signal. (p. 189)
23. Receptors come in related forms called **isoforms**. One ligand may have different effects when binding to different isoforms. (p. 190; Fig. 6.13)
24. A receptor may have multiple ligands. **Agonists** mimic the action of a signal molecule. **Antagonists** block the signal pathway. (p. 190; Fig. 6.14)
25. Receptor proteins exhibit specificity, competition, and saturation. (p. 190)
26. Cells exposed to abnormally high concentrations of a signal for a sustained period of time attempt to bring their response back to normal through down-regulation or by desensitization. In **down-regulation**, the cell decreases the number of receptors. In *desensitization*, the cell decreases the binding affinity of the receptors. **Up-regulation** is the opposite of down-regulation and involves increasing the number of receptors for a signal. (p. 191)
27. Cells have mechanisms for terminating signal pathways, such as removing the signal molecule or breaking down the receptor-ligand complex. (p. 191)
28. Many diseases have been linked to defects in various aspects of signal pathways, such as missing or defective receptors. (p. 192; Tbl. 6.1)

6

Homeostatic Reflex Pathways

29. Walter Cannon first stated four basic postulates of homeostasis: (1) The nervous system plays an important role in maintaining homeostasis. (2) Some parameters are under **tonic control**, which allows the parameter to be increased or decreased by a single signal (Fig. 6.14a). (3) Other parameters are under **antagonistic control**, in which one hormone or neuron increases the parameter while another decreases it (Fig. 6.14b). (4) Chemical signals can have different effects in different tissues of the body, depending on the type of receptor present at the target cell. (Fig. 6.15) (p. 195)
30. In **reflex** control pathways, the decision to respond to a change is made by an integrating center. A chemical or electrical signal to the target cell or tissue then initiates the response. Long-distance reflex pathways involve the nervous and endocrine systems and cytokines. (p. 194)
31. Neural control is faster and more specific than endocrine control but is usually of shorter duration. Endocrine control is less specific and slower to start but is longer lasting and is usually amplified. (p. 198; Tbl. 6.2)
32. Many reflex pathways are complex combinations of neural and endocrine control mechanisms. (p. 199; Fig. 6.19)

Questions

Answers: p. A-1

Level One Reviewing Facts and Terms

1. What are the two routes for long-distance signal delivery in the body?
2. Which two body systems maintain homeostasis by monitoring and responding to changes in the environment?
3. What two types of physiological signals does the body use to send messages? Of these two types, which is available to all cells?
4. In a signal pathway, the signal ligand, also called the first messenger, binds to a(n) ________, which activates and changes intracellular ________ .
5. The three main amplifier enzymes are (a) ________, which forms cAMP; (b) ________, which forms cGMP; and (c) ________, which converts a phospholipid from the cell's membrane into two different second messenger molecules.
6. An enzyme known as protein kinase adds the functional group ________ to its substrate, by transferring it from a(n) ________ molecule.
7. Distinguish between central and peripheral receptors.
8. Receptors for signal pathways may be found in the ________, ________, or ________ of the cell.
9. Down-regulation results in a(n) ________ (increased or decreased?) number of receptors in response to a prolonged signal.
10. List two ways a cell may decrease its response to a signal.
11. In a negative feedback loop, the response moves the system in the ________ (same/opposite) direction as the stimulus moves it.

Level Two Reviewing Concepts

12. Explain the relationships of the terms in each of the following sets. Give a physiological example or location if applicable.
    (a) gap junctions, connexins, connexon
    (b) autocrine, paracrine, cytokine, neurocrine, hormone
    (c) agonist, antagonist
    (d) transduction, amplification, cascade
13. List and compare the four classes of membrane receptors for signal pathways. Give an example of each.
14. Who was Walter Cannon? Restate his four postulates in your own words.
15. Briefly define the following terms and give an anatomical example when applicable: input signal, integrating center, output signal, response, sensor, stimulus, target.
16. Compare and contrast the advantages and disadvantages of neural versus endocrine control mechanisms.
17. Would the following reflexes have positive or negative feedback?
    (a) glucagon secretion in response to declining blood glucose
    (b) increasing milk release and secretion in response to baby's suckling
    (c) urgency in emptying one's urinary bladder
    (d) sweating in response to rising body temperature
18. Identify the target tissue or organ for each example in question 17.
19. Now identify the integrating center for examples (a), (c), and (d) in question 17.

Level Three Problem Solving

20. In each of the following situations, identify the components of the reflex.
    (a) You are sitting quietly at your desk, studying, when you become aware of the bitterly cold winds blowing outside at 30 mph, and you begin to feel a little chilly. You start to turn up the thermostat, remember last month's heating bill, and reach for an afghan to pull around you instead. Pretty soon you are toasty warm again.
    (b) While you are strolling through the shopping district, the aroma of cinnamon sticky buns reaches you. You inhale appreciatively, but remind yourself that you're not hungry, because you had lunch just an hour ago. You go about your business, but 20 minutes later you're back at the bakery, sticky bun in hand, ravenously devouring its sweetness, saliva moistening your mouth.
21. A researcher is studying the smooth muscle of the respiratory system airways. When she exposes the airways to the neurotransmitter acetylcholine, the smooth muscle contracts. When she exposes the airways to the neurohormone epinephrine, the airways relax.
    (a) The phenomenon just described is an example of ________ control.
    (b) What distinguishes a neurotransmitter from a neurohormone?
    (c) Which chemical messenger is secreted in higher concentrations: acetylcholine or epinephrine? Defend your answer.

Level Four Quantitative Problems

22. In a signal cascade for rhodopsin, a photoreceptor molecule, one rhodopsin activates 1000 molecules of transducin, the next molecule in the signal cascade. Each transducin activates one phosphodiesterase, and each phosphodiesterase converts 4000 cGMP to GMP.
    (a) What is the name of the phenomenon described in this paragraph?
    (b) Activation of one rhodopsin will result in the production of how many GMP molecules?

Answers

Answers to Concept Check Questions

Page 177

1. All methods listed are chemical signals except for (c) gap junctions, which transfer both chemical and electrical signals. Neurohormones (e) and neurotransmitters (f) are associated with electrical signaling in neurons but are themselves chemicals.
2. Cytokines, hormones, and neurohormones travel through the blood. Cytokines, neurohormones, and neurotransmitters are released by neurons.
3. The signal to pounce could not have been a paracrine signal because the eyes are too far away from the legs and because the response was too rapid for it to have taken place by diffusion.

Page 180

4. The components of signal pathways are signal molecule, receptor, intracellular signal molecule(s), and target proteins.
5. The cellular locations of receptors are cell membrane, cytosol, and nucleus.

Page 181

6. The steps of signal transduction are (1) signal molecule binds to receptor that (2) activates a protein that (3) creates second messengers that (4) create a response.
7. Amplification turns one signal molecule (first messenger) into multiple second messenger molecules.
8. Steroids are lipophilic, so they can enter cells and bind to intracellular receptors.

Page 187

9. Receptors are either ligand-gated ion channels, receptor-enzymes, G protein–coupled receptors, or integrins.
10. First messengers are extracellular; second messengers are intracellular.
11. (a) ligand, receptor, second messenger, cell response; (b) amplifier enzyme, second messenger, protein kinase, phosphorylated protein, cell response
12. (a) Cl^- channel opens: cell hyperpolarizes; (b) K^+ channel opens: cell hyperpolarizes; (c) Na^+ channel opens: cell depolarizes.

Page 187

13. The cell must use active transport to move Ca^{2+} against its concentration gradient.

Page 189

14. A drug that blocks leukotriene action could act at the target cell receptor or at any step downstream. A drug that blocks leukotriene synthesis might inhibit lipoxygenase.

Page 190

15. Receptors, enzymes, and transporters are all proteins.
16. Insulin could be using one receptor isoform with different second messenger systems in different cells or could be binding to different receptor isoforms.

Page 191

17. Choices (a) and (d) could decrease binding affinity. Changing receptor number would not affect binding affinity.

Page 194

18. Tonic control usually involves one control system, but antagonistic control uses two.
19. A signal can have opposite effects by using different receptors or different signal pathways.

Page 196

20. In local control, the stimulus, integration of the signal, and response all take place in or very close to the target cell. With reflex control, integration of the input signal and initiation of a response may take place far from the location where the change occurred. In addition, the reflex response is often systemic and not localized.
21. Stimulus, sensor or sensory receptor, input signal (afferent pathway), integrating center, output signal (efferent pathway), target or effector, response (tissue and systemic)

Page 197

22. (a) The "neural system integrating center" is the brain and spinal cord. (b) "Receptor" represents the sense organs. (c) The dashed line indicating negative feedback runs from "Response" back to "Internal or external change."

Page 200

23. blow to knee = internal or external change; leg muscles = targets; neuron to leg muscles = efferent neuron; sensory neuron = input signal; brain and spinal cord = CNS integrating center; stretch receptor = sensor or receptor; muscle contraction = response.

Page 200

24. food in stomach = stimulus; brain and spinal cord = CNS integrating center; endocrine cells of pancreas = E (integrating center); stretch receptors = receptor; efferent neuron to pancreas = efferent neuron; insulin = classic hormone; adipose cell = target cell; sensory neuron = afferent neuron. Blood is the anatomical route that hormones use to reach their target but is not part of the reflex pathway.

Answers to Figure Questions

Page 184

Figure 6.8: A (inactive and active) = adenylyl cyclase; inactive B = ATP; active B = cAMP; C (inactive and active) = protein kinase A; product = phosphorylated protein.

Page 193

Figure 6.15: 180 beats/min for the top ECG and 60 beats/min for the bottom ECG.

7

Introduction to the Endocrine System

The separation of the endocrine system into isolated subsystems must be recognized as an artificial one, convenient from a pedagogical point of view but not accurately reflecting the interrelated nature of all these systems.

—Howard Rasmussen, *in* Williams' Textbook of Endocrinology, 1974

Background Basics

Gamma scan of a goiter of the thyroid gland

David was seven years old when the symptoms first appeared. His appetite at meals increased, and he always seemed to be in the kitchen looking for food. Despite eating more, however, he was losing weight. When he started asking for water instead of soft drinks, David's mother became concerned, and when he wet the bed three nights in a row, she knew something was wrong. The doctor confirmed the suspected diagnosis after running tests to determine the concentration of glucose in David's blood and urine. David had diabetes mellitus. In his case, the disease was due to lack of insulin, a hormone produced by the pancreas. David was placed on insulin injections, a treatment he would continue for the rest of his life.

One hundred years ago, David would have died not long after the onset of symptoms. The field of **endocrinology,** the study of hormones, was then in its infancy. Most hormones had not been discovered, and the functions of known hormones were not well understood. There was no treatment for diabetes, no birth control pill for contraception. Babies born with inadequate secretion of thyroid hormone did not grow or develop normally.

Today, all that has changed. We have identified a long and growing list of hormones. The endocrine diseases that once killed or maimed can now be controlled by synthetic hormones and sophisticated medical procedures. Although physicians do not hesitate to use these treatments, we are still learning exactly how hormones act on their target cells. This chapter provides an introduction to the basic principles of hormone structure and function. You will learn more about individual hormones as you encounter them in your study of the various systems.

Hormones

As you have learned, hormones are chemical messengers secreted into the blood by specialized cells. Hormones are responsible for many functions that we think of as long-term, ongoing functions of the body. Processes that fall mostly under hormonal control include growth and development, metabolism, regulation of the internal environment (temperature, water balance, ions), and reproduction. Hormones act on their target cells in one of three basic ways: (1) by controlling the rates of enzymatic reactions, (2) by controlling the transport of ions or molecules across cell membranes, or (3) by controlling gene expression and the synthesis of proteins.

RUNNING PROBLEM

Graves' Disease

The ball slid by the hole and trickled off the green: another bogey. Ben Crenshaw's golf game was falling apart. The 33-year-old professional had won the Masters Tournament only a year ago, but now something was not right. He was tired and weak, had been losing weight, and felt hot all the time. He attributed his symptoms to stress, but his family thought otherwise. At their urging, he finally saw a physician. The diagnosis? Graves' disease, which results in an overactive thyroid gland.

207 216 224 226 229 230 232

Hormones Have Been Known Since Ancient Times

Although the scientific field of endocrinology is relatively young, diseases of the endocrine system have been documented for more than a thousand years. Evidence of endocrine abnormalities can even be seen in ancient art. For example, one pre-Colombian statue of a woman shows a mass on the front of her neck (■ Fig. 7.1). The mass is an enlarged thyroid gland, or *goiter,* a common condition high in the Andes, where the dietary iodine needed to make thyroid hormones was lacking.

The first association of endocrine structure and function was probably the link between the testes and male sexuality. Castration of animals and men was a common practice in both Eastern and Western cultures because it decreased the sex drive and rendered males infertile.

■ **Fig. 7.1** ***An endocrine disorder in ancient art.*** This pre-Colombian stone carving of a woman shows a mass at her neck. This mass is an enlarged thyroid gland, a condition known as goiter. It was considered a sign of beauty among the people who lived high in the Andes mountains.

In 1849, A. A. Berthold performed the first classic experiment in endocrinology. He removed the testes from roosters and observed that the castrated birds had smaller combs, less aggressiveness, and less sex drive than uncastrated roosters. If the testes were surgically placed back into the donor rooster or into another castrated bird, normal male behavior and comb development resumed. Because the reimplanted testes were not connected to nerves, Berthold concluded that the glands must be secreting something into the blood that affected the entire body.

Experimental endocrinology did not receive much attention, however, until 1889, when the 72-year-old French physician Charles Brown-Séquard made a dramatic announcement of his sexual rejuvenation after injecting himself with extracts made from bull testes ground up in water. An international uproar followed, and physicians on both sides of the Atlantic began to inject their patients with extracts of many different endocrine organs, a practice known as *organotherapy*.

We now know that the increased virility Brown-Séquard reported was most likely a placebo effect because testosterone is a hydrophobic steroid that cannot be extracted by an aqueous preparation. His research opened the door to hormone therapy, however, and in 1891 organotherapy had its first true success: a woman was treated for low thyroid hormone levels with glycerin extracts of sheep thyroid glands.

As the study of "internal secretions" grew, Berthold's experiments became a template for endocrine research. Once a gland or structure was suspected of secreting hormones, the classic steps for identifying an endocrine gland became:

1. **Remove the suspected gland.** This is equivalent to inducing a state of *hormone deficiency*. If the gland does produce hormones, the animal should start to exhibit anatomical, behavioral, or physiological abnormalities.
2. **Replace the hormone**. This can be done by placing the gland back in the animal or administering an extract of the gland. This *replacement therapy* should eliminate the symptoms of hormone deficiency.
3. **Create a state of hormone excess**. Take a normal animal and implant an extra gland or administer extract from the gland to see if symptoms characteristic of *hormone excess* appear.

Once a gland is identified as a potential source of hormones, scientists purify extracts of the gland to isolate the active substance. They test for hormone activity by injecting animals with the purified extract and monitoring for a response.

Hormones identified by this technique are sometimes called *classic hormones*. They include hormones of the pancreas, thyroid, adrenal glands, pituitary, and gonads, all discrete endocrine glands that could be easily identified and surgically removed. Not all hormones come from identifiable glands, however, and we have been slower to discover them. For example, many hormones involved in digestion are secreted by endocrine cells scattered throughout the wall of the stomach or intestine, which has made them difficult to identify and isolate.

CLINICAL FOCUS: DIABETES

The Discovery of Insulin

Diabetes mellitus, the metabolic condition associated with pathologies of insulin function, has been known since ancient times. Detailed clinical descriptions of insulin-deficient diabetes were available to physicians, but they had no means of treating the disease. Patients invariably died. However, in a series of classic experiments in endocrine physiology, Oscar Minkowski at the University of Strasbourg (Germany) pinpointed the relationship between diabetes and the pancreas. In 1889, Minkowski surgically removed the pancreas from dogs (*pancreatectomy*) and noticed that they developed symptoms that mimicked those of diabetes. He also found that implanting pieces of pancreas under the dogs' skin would prevent development of diabetes. Subsequently, in 1921 Fredrick G. Banting and Charles H. Best (Toronto, Canada) identified an antidiabetic substance in pancreas extracts. Banting and Best and others injected pancreatic extracts into diabetic animals and found that the extracts reversed the elevated blood glucose levels of diabetes. From there, it was a relatively short process until, in 1922, purified insulin was used in the first clinical trials. Science had found a treatment for a once-fatal disease.

The Anatomy Summary in ■ Figure 7.2 (on page 210) lists the major hormones of the body and the glands or cells that secrete them, along with the major effects of each hormone.

What Makes a Chemical a Hormone?

In 1905, the term *hormone* was coined from the Greek verb meaning "to excite or arouse." The traditional definition of a **hormone** is a chemical secreted by a cell or group of cells into the blood for transport to a distant target, where it exerts its effect at very low concentrations. However, as scientists learn more about chemical communication in the body, this definition is continually being challenged.

Hormones Are Secreted by a Cell or Group of Cells Traditionally, the field of endocrinology has focused on chemical messengers secreted by endocrine *glands*, the discrete and readily identifiable tissues derived from epithelial tissue [p. 80]. However, we now know that molecules that act as hormones are secreted not only by classic endocrine glands but also by isolated endocrine cells (hormones of the *diffuse endocrine system*), by neurons (*neurohormones*), and by cells of the immune system (*cytokines*).

Hormones Are Secreted into the Blood **Secretion** is the movement of a substance from inside a cell to the extracellular fluid or directly into the external environment [p. 66]. According to the traditional definition of a hormone, hormones are secreted into the blood. However, the term *ectohormone* {*ektos,* outside} has been given to signal molecules secreted into the external environment.

Pheromones {*pherein,* to bring} are specialized ectohormones that act on other organisms of the same species to elicit a physiological or behavioral response. For example, sea anemones secrete alarm pheromones when danger threatens, and ants release trail pheromones to attract fellow workers to food sources. Pheromones are also used to attract members of the opposite sex for mating. Sex pheromones are found throughout the animal kingdom, in animals from fruit flies to dogs.

But do humans have pheromones? This question is still a matter of debate. Some studies have shown that human *axillary* (armpit) sweat glands secrete volatile steroids related to sex hormones that may serve as human sex pheromones. In one study, when female students were asked to rate the odors of T-shirts worn by male students, each woman preferred the odor of men who were genetically dissimilar from her. In another study, female axillary secretions rubbed on the upper lip of young women altered the timing of their menstrual cycles. The selling of putative human pheromones as perfume is becoming the latest fad in the mating game, as you will see if you do a Google search for *human pheromone.* How humans may sense pheromones is discussed later [see Chapter 10].

Hormones Are Transported to a Distant Target By the traditional definition, a hormone must be transported by the blood to a distant target cell. Experimentally, this property is sometimes difficult to demonstrate. Molecules that are suspected of being hormones but not fully accepted as such are called *candidate hormones.* They are usually identified by the word *factor.* For example, in the early 1970s, the hypothalamic regulating hormones were known as "releasing factors" and "inhibiting factors" rather than releasing and inhibiting hormones.

Currently, **growth factors,** a large group of substances that influence cell growth and division, are being studied to determine if they meet all the criteria for hormones. Although many growth factors have been shown to act locally as *autocrines* or *paracrines* [p. 177], most do not seem to be distributed widely in the circulation. A similar situation exists with the lipid-derived signal molecules called *eicosanoids* [p. 33].

Complicating the classification of signal molecules is the fact that a molecule may act as a hormone when secreted from one location but as a paracrine or autocrine signal when secreted from a different location. For example, in the 1920s scientists discovered that *cholecystokinin* (CCK) in extracts of intestine caused contraction of the gallbladder. For many years thereafter, CCK was known only as an intestinal hormone. Then in the mid-1970s, CCK was found in neurons of the brain, where it acts as a neurotransmitter or neuromodulator. In recent years, CCK has gained attention because of its possible role in controlling hunger.

Hormones Exert Their Effect at Very Low Concentrations One hallmark of a hormone is its ability to act at concentrations in the nanomolar (10^{-9} M) to picomolar (10^{-12} M) range. Some chemical signals transported in the blood to distant targets are not considered hormones because they must be present in relatively high concentrations before an effect is noticed. For example, histamine released during severe allergic reactions may act on cells throughout the body, but its concentration exceeds the accepted range for a hormone.

As researchers discover new signal molecules and new receptors, the boundary between hormones and nonhormonal signal molecules continues to be challenged, just as the distinction between the nervous and endocrine systems has blurred. Many *cytokines* [p. 177] seem to meet the previously stated definition of a hormone. However, experts in cytokine research do not consider cytokines to be hormones because peptide cytokines are synthesized and released on demand, in contrast to classic peptide hormones, which are made in advance and stored in the parent endocrine cell. A few cytokines—for example, *erythropoietin,* the molecule that controls red blood cell production—were classified as hormones before the term *cytokine* was coined, contributing to the overlap between the two groups of signal molecules.

Hormones Act by Binding to Receptors

All hormones bind to target cell receptors and initiate biochemical responses. These responses are the **cellular mechanism of action** of the hormone. As you can see from the table in Figure 7.2, one hormone may act on multiple tissues. To complicate matters, the effects may vary in different tissues or at different stages of development. Or a hormone may have no effect at all in a particular cell. Insulin is an example of a hormone with varied effects. In muscle and adipose tissues, insulin alters glucose transport proteins and enzymes for glucose metabolism. In the liver, it modulates enzyme activity but has no direct effect on glucose transport proteins. In the brain and certain other tissues, glucose metabolism is totally independent of insulin.

Concept Check Answer: p. 236

1. Name the membrane transport process by which glucose moves from the extracellular fluid into cells.

The variable responsiveness of a cell to a hormone depends primarily on the cell's receptor and signal transduction pathways [p. 180]. If there are no hormone receptors in a tissue, its cells cannot respond. If tissues have different receptors and receptor-linked pathways for the same hormone, they will respond differently.

Hormones

Location	Hormone	Primary Target(s)
Pineal gland	Melatonin [A]	Brain, other tissues
Hypothalamus (N)	Trophic hormones [P] (see Fig. 7.8)	Anterior pituitary
Posterior pituitary (N)	Oxytocin [P] Vasopressin (ADH) [P]	Breast and uterus Kidney
Anterior pituitary (G)	Prolactin [P] Growth hormone (somatotropin) [P] Corticotropin (ACTH) [P] Thyrotropin (TSH) [P] Follicle-stimulating hormone [P] Luteinizing hormone [P]	Breast Liver Many tissues Adrenal cortex Thyroid gland Gonads Gonads
Thyroid gland	Triiodothyronine and thyroxine [A] Calcitonin [P]	Many tissues Bone
Parathyroid gland	Parathyroid hormone [P]	Bone, kidney
Thymus gland	Thymosin, thymopoietin [P]	Lymphocytes
Heart (C)	Atrial natriuretic peptide [P]	Kidneys
Liver (C)	Angiotensinogen [P] Insulin-like growth factors [P]	Adrenal cortex, blood vessels Many tissues
Stomach and small intestine (C)	Gastrin, cholecystokinin, secretin, and others [P]	GI tract and pancreas
Pancreas (G)	Insulin, glucagon, somatostatin, pancreatic polypeptide [P]	Many tissues
Adrenal cortex (G)	Aldosterone [S] Cortisol [S] Androgens [S]	Kidney Many tissues Many tissues
Adrenal medulla (N)	Epinephrine, norepinephrine [A]	Many tissues
Kidney (C)	Erythropoietin [P] 1,25 Dihydroxy-vitamin D_3 (calciferol) [S]	Bone marrow Intestine
Skin (C)	Vitamin D_3 [S]	Intermediate form of hormone
Testes (male) (G)	Androgens [S] Inhibin [P]	Many tissues Anterior pituitary
Ovaries (female) (G)	Estrogen, progesterone [S] Inhibin [P] Relaxin (pregnancy) [P]	Many tissues Anterior pituitary Uterine muscle
Adipose tissue (C)	Leptin, adiponectin, resistin	Hypothalamus, other tissues
Placenta (pregnant females only) (C)	Estrogen, progesterone [S] Chorionic somatomammotropin [P] Chorionic gonadotropin [P]	Many tissues Many tissues Corpus luteum

KEY

G = gland
C = endocrine cells
N = neurons

P = peptide
S = steroid
A = amino acid–derived

Main Effect(s)
Circadian rhythms; immune function; antioxidant
Release or inhibit pituitary hormones
Milk ejection; labor and delivery; behavior
Water reabsorption
Milk production
Growth factor secretion
Growth and metabolism
Cortisol release
Thyroid hormone synthesis
Egg or sperm production; sex hormone production
Sex hormone production; egg or sperm production
Metabolism, growth, and development
Plasma calcium levels (minimal effect in humans)
Regulates plasma Ca^{2+} and phosphate levels
Lymphocyte development
Increases Na^+ excretion
Aldosterone secretion; increases blood pressure
Growth
Assist digestion and absorption of nutrients
Metabolism of glucose and other nutrients
Na^+ and K^+ homeostasis
Stress response
Sex drive in females
Fight-or-flight response
Red blood cell production
Increases calcium absorption
Precursor of 1,25 dihydroxy-vitamin D_3
Sperm production, secondary sex characteristics
Inhibits FSH secretion
Egg production, secondary sex characteristics
Inhibits FSH secretion
Relaxes muscle
Food intake, metabolism, reproduction
Fetal, maternal development
Metabolism
Hormone secretion

Hormone Action Must Be Terminated

Signal activity by hormones and other chemical signals must be of limited duration if the body is to respond to changes in its internal state. For example, insulin is secreted when blood glucose concentrations increase following a meal. As long as insulin is present, glucose leaves the blood and enters cells. However, if insulin activity continues too long, blood glucose levels can fall so low that the nervous system becomes unable to function properly—a potentially fatal situation. Normally the body avoids this situation in several ways: by limiting insulin secretion, by removing or inactivating insulin circulating in the blood, and by terminating insulin activity in target cells.

In general, hormones in the bloodstream are *degraded* (broken down) into inactive metabolites by enzymes found primarily in the liver and kidneys. The metabolites are then excreted in either the bile or the urine. The rate of hormone breakdown is indicated by a hormone's **half-life** in the circulation, the amount of time required to reduce the concentration of hormone by one-half. Half-life is one indicator of how long a hormone is active in the body.

Hormones bound to target membrane receptors have their activity terminated in several ways. Enzymes that are always present in the plasma can degrade peptide hormones bound to cell membrane receptors. In some cases, the receptor-hormone complex is brought into the cell by endocytosis, and the hormone is then digested in lysosomes [Figure 5.19, p. 156]. Intracellular enzymes metabolize hormones that enter cells.

Concept Check Answer: p. 236

2. What is the suffix in a chemical name that tells you a molecule is an enzyme? [*Hint:* p. 107] Use that suffix to name an enzyme that digests peptides.

The Classification of Hormones

Hormones can be classified according to different schemes. The scheme used in Figure 7.2 groups them according to their source. A different scheme divides hormones into those whose release is controlled by the brain and those whose release is not controlled by the brain. Another scheme groups hormones according to whether they bind to G protein–coupled receptors, tyrosine kinase–linked receptors, or intracellular receptors, and so on.

A final scheme divides hormones into three main chemical classes: peptide/protein hormones, steroid hormones, and amino acid-derived, or amine, hormones (■ Tbl. 7.1). The peptide/protein hormones are composed of linked amino acids. The steroid hormones are all derived from cholesterol [p. 33]. The amino acid-derived hormones, also called *amine hormones*, are modifications of single amino acids, either tryptophan or tyrosine.

Comparison of Peptide, Steroid, and Amino Acid-Derived Hormones Table 7.1

	Peptide Hormones	Steroid Hormones	Amine Hormones (Tyrosine Derivatives)	
			Catecholamines	Thyroid Hormones
Synthesis and storage	Made in advance; stored in secretory vesicles	Synthesized on demand from precursors	Made in advance; stored in secretory vesicles	Made in advance; precursor stored in secretory vesicles
Release from parent cell	Exocytosis	Simple diffusion	Exocytosis	Simple diffusion
Transport in blood	Dissolved in plasma	Bound to carrier proteins	Dissolved in plasma	Bound to carrier proteins
Half-life	Short	Long	Short	Long
Location of receptor	Cell membrane	Cytoplasm or nucleus; some have membrane receptors also	Cell membrane	Nucleus
Response to receptor-ligand binding	Activation of second messenger systems; may activate genes	Activation of genes for transcription and translation; may have nongenomic actions	Activation of second messenger systems	Activation of genes for transcription and translation
General target response	Modification of existing proteins and induction of new protein synthesis	Induction of new protein synthesis	Modification of existing proteins	Induction of new protein synthesis
Examples	Insulin, parathyroid hormone	Estrogen, androgens, cortisol	Epinephrine, norepinephrine	Thyroxine (T_4)

Concept Check Answers: p. 234

3. What is the classic definition of a hormone?
4. Based on what you know about the organelles involved in protein and steroid synthesis [p. 72], what would be the major differences between the organelle composition of a steroid-producing cell and that of a protein-producing cell?

Most Hormones Are Peptides or Proteins

The peptide/protein hormones range from small peptides of only three amino acids to larger proteins and glycoproteins. Despite the size variability among hormones in this group, they are usually called peptide hormones for the sake of simplicity. You can remember which hormones fall into this category by exclusion: if a hormone is not a steroid hormone and not an amino-acid derivative, then it must be a peptide or protein.

Peptide Hormone Synthesis, Storage, and Release The synthesis and packaging of peptide hormones into membrane-bound secretory vesicles is similar to that of other proteins. The initial peptide that comes off the ribosome is a large inactive protein known as a preprohormone (■ Fig. 7.3 ①). **Preprohormones** contain one or more copies of a peptide hormone, a *signal sequence* that directs the protein into the lumen of the rough endoplasmic reticulum, and other peptide sequences that may or may not have biological activity.

As an inactive preprohormone moves through the endoplasmic reticulum and Golgi complex, the signal sequence is removed, creating a smaller, still-inactive molecule called a **prohormone** (Fig. 7.3 ④). In the Golgi complex, the prohormone is packaged into secretory vesicles along with *proteolytic* {*proteo-*, protein + *lysis*, rupture} enzymes that chop the prohormone into active hormone and other fragments. This process is called *post-translational modification* [p. 123].

The secretory vesicles containing peptides are stored in the cytoplasm of the endocrine cell until the cell receives a signal for secretion. At that time, the vesicles move to the cell membrane and

■ Fig. 7.3 *ESSENTIALS*

Peptide Hormone Synthesis and Processing

Peptide hormones are made as large, inactive preprohormones that include a signal sequence, one or more copies of the hormone, and additional peptide fragments.

(a) Preprohormones

PreproTRH (thyrotropin-releasing hormone) has six copies of the 3-amino acid hormone TRH.

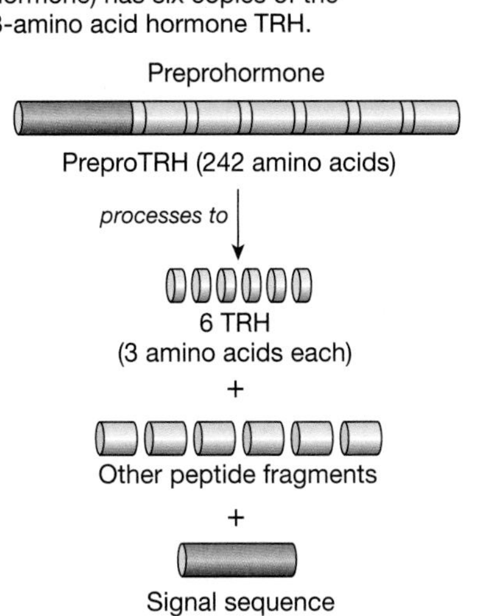

(b) Prohormones

Prohormones, such as pro-opiomelanocortin, the prohormone for ACTH, may contain several peptide sequences with biological activity.

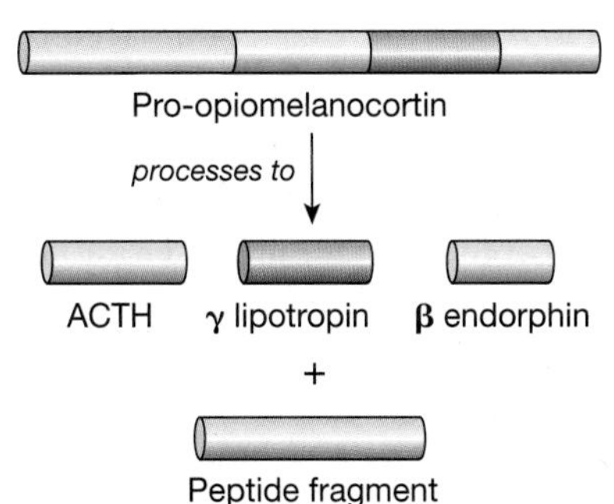

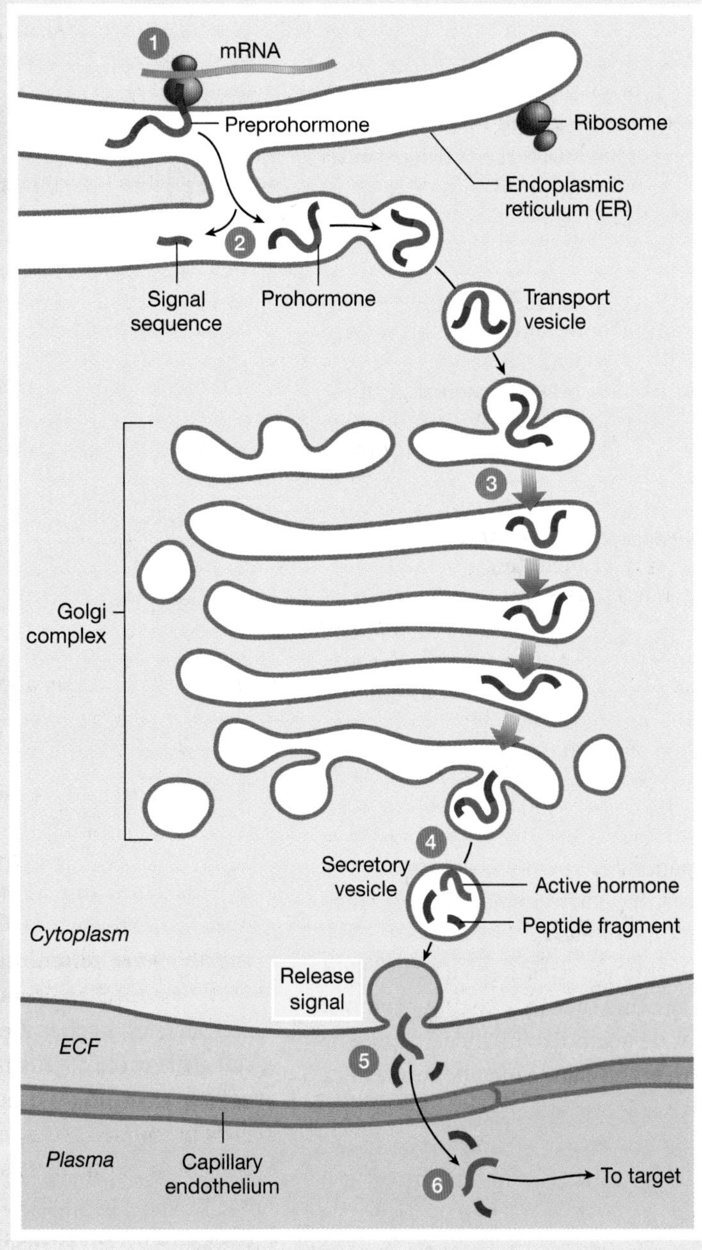

1 Messenger RNA on the ribosomes binds amino acids into a peptide chain called a **preprohormone**. The chain is directed into the ER lumen by a **signal sequence** of amino acids.

2 Enzymes in the ER chop off the signal sequence, creating an inactive **prohormone**.

3 The prohormone passes from the ER through the Golgi complex.

4 Secretory vesicles containing enzymes and prohormone bud off the Golgi. The enzymes chop the prohormone into one or more active peptides plus additional peptide fragments.

5 The secretory vesicle releases its contents by exocytosis into the extracellular space.

6 The hormone moves into the circulation for transport to its target.

(c) Prohormones Process to Active Hormone Plus Peptide Fragments

The peptide chain of insulin's prohormone folds back on itself with the help of disulfide (S—S) bonds. The prohormone cleaves to insulin and C-peptide.

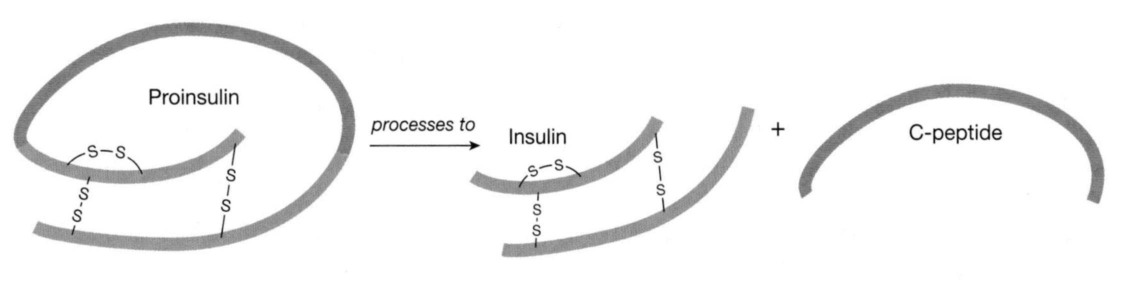

release their contents by calcium-dependent exocytosis [p. 157]. All of the peptide fragments created from the prohormone are released together into the extracellular fluid, in a process known as *co-secretion* (Fig. 7.3 ⑤).

Post-Translational Modification of Prohormones Studies of prohormone processing have led to some interesting discoveries. Some prohormones, such as that for *thyrotropin-releasing hormone* (TRH), contain multiple copies of the hormone (Figure 7.3a). Another interesting prohormone is *pro-opiomelanocortin* (Figure 7.3b). This prohormone splits into three active peptides plus an inactive fragment. In some instances, even the fragments are clinically useful. For example, proinsulin is cleaved into active insulin and an inactive fragment known as *C-peptide* (Figure 7.3c). Clinicians measure the levels of C-peptide in the blood of diabetics to monitor how much insulin the patient's pancreas is producing.

Transport in the Blood and Half-Life of Peptide Hormones Peptide hormones are water soluble and therefore generally dissolve easily in the extracellular fluid for transport throughout the body. The half-life for peptide hormones is usually quite short, in the range of several minutes. If the response to a peptide hormone must be sustained for an extended period of time, the hormone must be secreted continually.

Cellular Mechanism of Action of Peptide Hormones Because peptide hormones are lipophobic, they are usually unable to enter the target cell. Instead, they bind to surface membrane receptors. The hormone-receptor complex initiates the cellular response by means of a *signal transduction* system (■ Fig. 7.4). Many peptide hormones work through cAMP second messenger systems [p. 183]. A few peptide hormone receptors, such as that of insulin, have tyrosine kinase activity [p. 183] or work through other signal transduction pathways.

The response of cells to peptide hormones is usually rapid because second messenger systems modify existing proteins. The changes triggered by peptide hormones include opening or closing membrane channels and modulating metabolic enzymes or transport proteins. Researchers have recently discovered that some peptide hormones also have longer-lasting effects when their second messenger systems activate genes and direct the synthesis of new proteins.

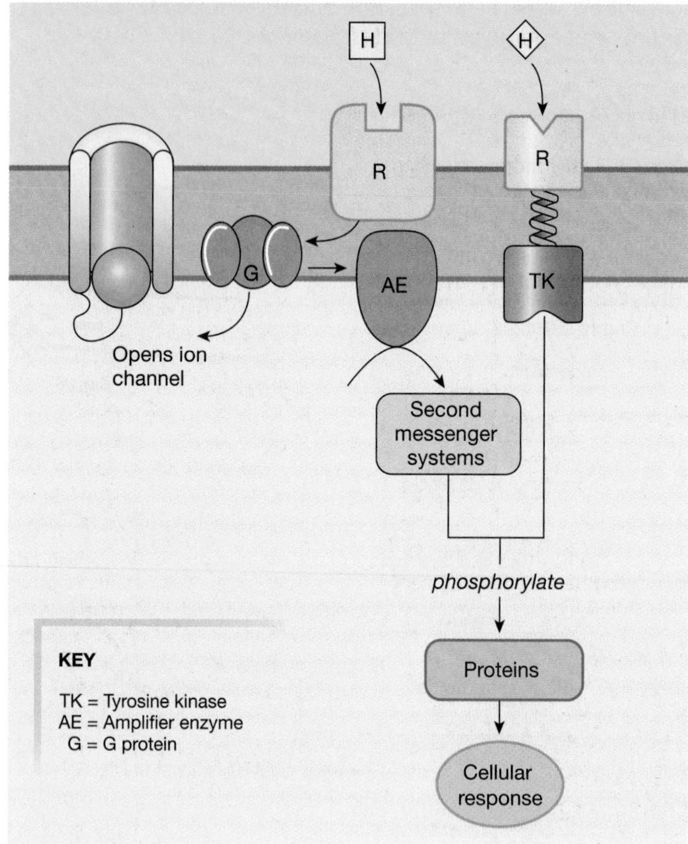

■ **Fig. 7.4** ***Membrane receptors and signal transduction for peptide hormones***

Steroid Hormones Are Derived from Cholesterol

Steroid hormones have a similar chemical structure because they are all derived from cholesterol (■ Fig. 7.5a). Unlike peptide hormones, which are made in tissues all over the body, steroid hormones are made in only a few organs. The **adrenal cortex**, the outer portion of the adrenal glands {*cortex*, bark}, makes several types of steroid hormones. One **adrenal gland** sits atop each kidney {*ad-*, upon + *renal*, kidney}. The gonads produce the sex steroids (estrogens, progesterone, and androgens). In pregnant women, the placenta is also a source of steroid hormones.

Steroid Hormone Synthesis and Release Cells that secrete steroid hormones have unusually large amounts of smooth endoplasmic reticulum, the organelle in which steroids are synthesized. Steroids are lipophilic and diffuse easily across membranes, both out of their parent cell and into their target cell. This property also means that steroid-secreting cells cannot store hormones in secretory vesicles. Instead, they synthesize their hormone as it is needed. When a stimulus activates the endocrine cell, precursors in the cytoplasm are rapidly converted to active hormone. The hormone concentration in the cytoplasm rises, and the hormones move out of the cell by simple diffusion.

Transport in the Blood and Half-Life of Steroid Hormones Like their parent cholesterol, steroid hormones are not very soluble in plasma and other body fluids. For this reason, most of the steroid hormone molecules found in the blood are bound to protein carrier molecules (Fig. 7.5b ①). Some hormones have specific carriers, such as *corticosteroid-binding globulin*. Others simply bind to general plasma proteins, such as *albumin*.

The binding of a steroid hormone to a carrier protein protects the hormone from enzymatic degradation and results in an extended half-life. For example, **cortisol,** a hormone produced

■ Fig. 7.5 *ESSENTIALS*

Steroid Hormones

Most steroid hormones are made in the adrenal cortex or gonads (ovaries and testes). Steroid hormones are not stored in the endocrine cell because of their lipophilic nature. They are made on demand and diffuse out of the endocrine cell.

(a) Cholesterol is the parent compound for all steroid hormones.

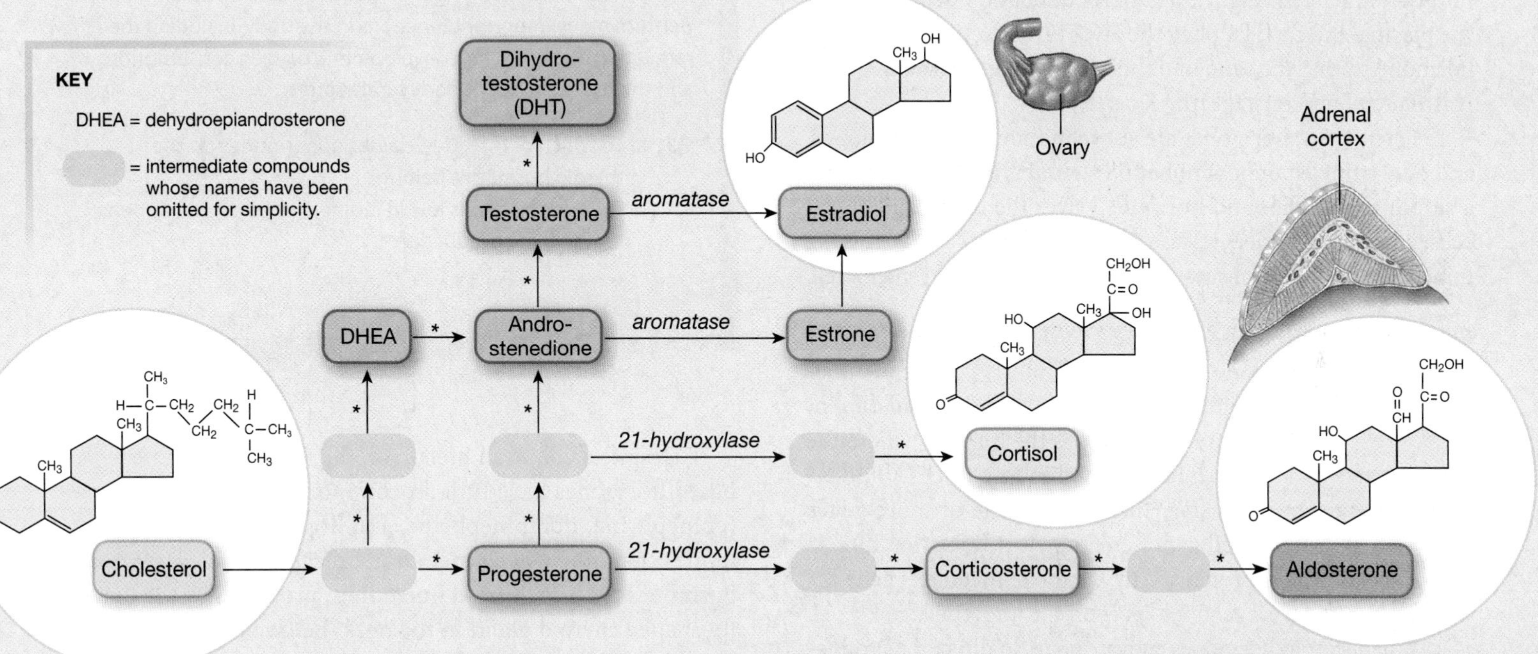

* Each step is catalyzed by an enzyme, but only two enzymes are shown in this figure.

(b) Steroid hormones act primarily on intracellular receptors.

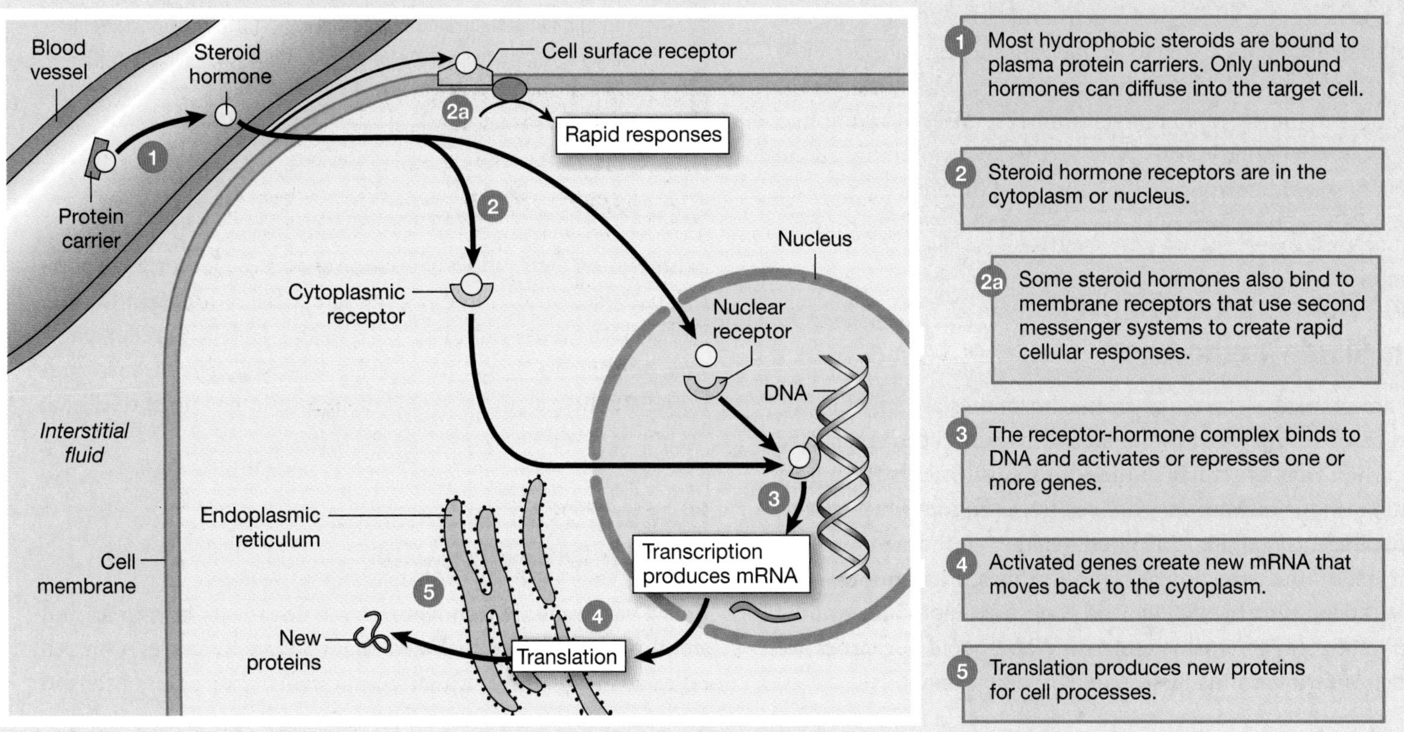

1. Most hydrophobic steroids are bound to plasma protein carriers. Only unbound hormones can diffuse into the target cell.

2. Steroid hormone receptors are in the cytoplasm or nucleus.

2a. Some steroid hormones also bind to membrane receptors that use second messenger systems to create rapid cellular responses.

3. The receptor-hormone complex binds to DNA and activates or represses one or more genes.

4. Activated genes create new mRNA that moves back to the cytoplasm.

5. Translation produces new proteins for cell processes.

by the adrenal cortex, has a half-life of 60–90 minutes. (Compare this with epinephrine, an amino acid–derived hormone whose half-life is measured in seconds.)

Although binding steroid hormones to protein carriers extends their half-life, it also blocks their entry into target cells. The carrier-steroid complex remains outside the cell because the carrier proteins are lipophobic and cannot diffuse through the membrane. Only an unbound hormone molecule can diffuse into the target cell (Fig. 7.5b ②). As unbound hormone leaves the plasma, the carriers obey the law of mass action and release hormone so that the ratio of unbound to bound hormone in the plasma remains constant [the K_d; p. 51].

Fortunately, hormones are active in minute concentrations, and only a tiny amount of unbound steroid is enough to produce a response. As unbound hormone leaves the blood and enters cells, additional carriers release their bound steroid so that some unbound hormone is always in the blood and ready to enter a cell.

Cellular Mechanism of Action of Steroid Hormones The best-studied steroid hormone receptors are found within cells, either in the cytoplasm or in the nucleus. The ultimate destination of steroid receptor-hormone complexes is the nucleus, where the complex acts as a *transcription factor,* binding to DNA and either activating or *repressing* (turning off) one or more genes (Fig. 7.5b ③). Activated genes create new mRNA that directs the synthesis of new proteins. Any hormone that alters gene activity is said to have a *genomic effect* on the target cell.

When steroid hormones activate genes to direct the production of new proteins, there is usually a lag time between hormone-receptor binding and the first measurable biological effects. This lag can be as much as 90 minutes. Consequently, steroid hormones do not mediate reflex pathways that require rapid responses.

In recent years researchers have discovered that several steroid hormones, including estrogens and aldosterone, have cell membrane receptors linked to signal transduction pathways, just as peptide hormones do. These receptors enable those steroid hormones to initiate rapid **nongenomic responses** in addition to their slower genomic effects. With the discovery of nongenomic effects of steroid hormones, the functional differences between steroid and peptide hormones seem almost to have disappeared.

Some Hormones Are Derived from Single Amino Acids

The amino acid–derived, or amine, hormones are small molecules created from either tryptophan or tyrosine, both notable for the carbon ring structures in their R-groups [p. 35]. The pineal gland hormone **melatonin** is derived from tryptophan (see *Focus on the Pineal Gland,* Fig. 7.16) but the other amino acid–derived hormones—the catecholamines and thyroid hormones—are derived from tyrosine (■ Fig. 7.6). Catecholamines are a modification of a single tyrosine molecule. The thyroid hormones are made from two tyrosine molecules plus iodine atoms.

RUNNING PROBLEM

Shaped like a butterfly, the thyroid gland straddles the trachea just below the Adam's apple. Responding to hormonal signals from the hypothalamus and anterior pituitary, the thyroid gland concentrates iodine, an element found in food (most notably as an ingredient added to salt), and combines it with the amino acid tyrosine to make two thyroid hormones, thyroxine and triiodothyronine. These thyroid hormones perform many important functions in the body, including the regulation of growth and development, oxygen consumption, and the maintenance of body temperature.

Q1: a. To which of the three classes of hormones do the thyroid hormones belong?
b. If a person's diet is low in iodine, predict what happens to thyroxine production.

Despite a common precursor, the two groups of tyrosine-based hormones have little in common. The **catecholamines** (epinephrine, norepinephrine, and dopamine) are neurohormones that bind to cell membrane receptors the way peptide hormones do. The **thyroid hormones,** produced by the butterfly-shaped thyroid gland in the neck, behave more like steroid hormones, with intracellular receptors that activate genes.

Concept Check Answers: p. 236

5. What are the three chemical classes of hormones?
6. The steroid hormone aldosterone has a short half-life for a steroid hormone—only about 20 minutes. What would you predict about the degree to which aldosterone is bound to blood proteins?

Control of Hormone Release

A fundamental principle of homeostasis is the importance of reflex pathways in maintaining the internal environment [p. 10]. The sections that follow apply the basic patterns of reflex pathways to the control pathways for hormones. This discussion is not all-inclusive, and you will encounter a few hormones that do not fit exactly into these patterns.

Hormones Can Be Classified by Their Reflex Pathways

Reflex pathways are a convenient way to classify hormones and simplify learning the pathways that regulate their secretion. All reflex pathways have similar components: a stimulus, a sensor,

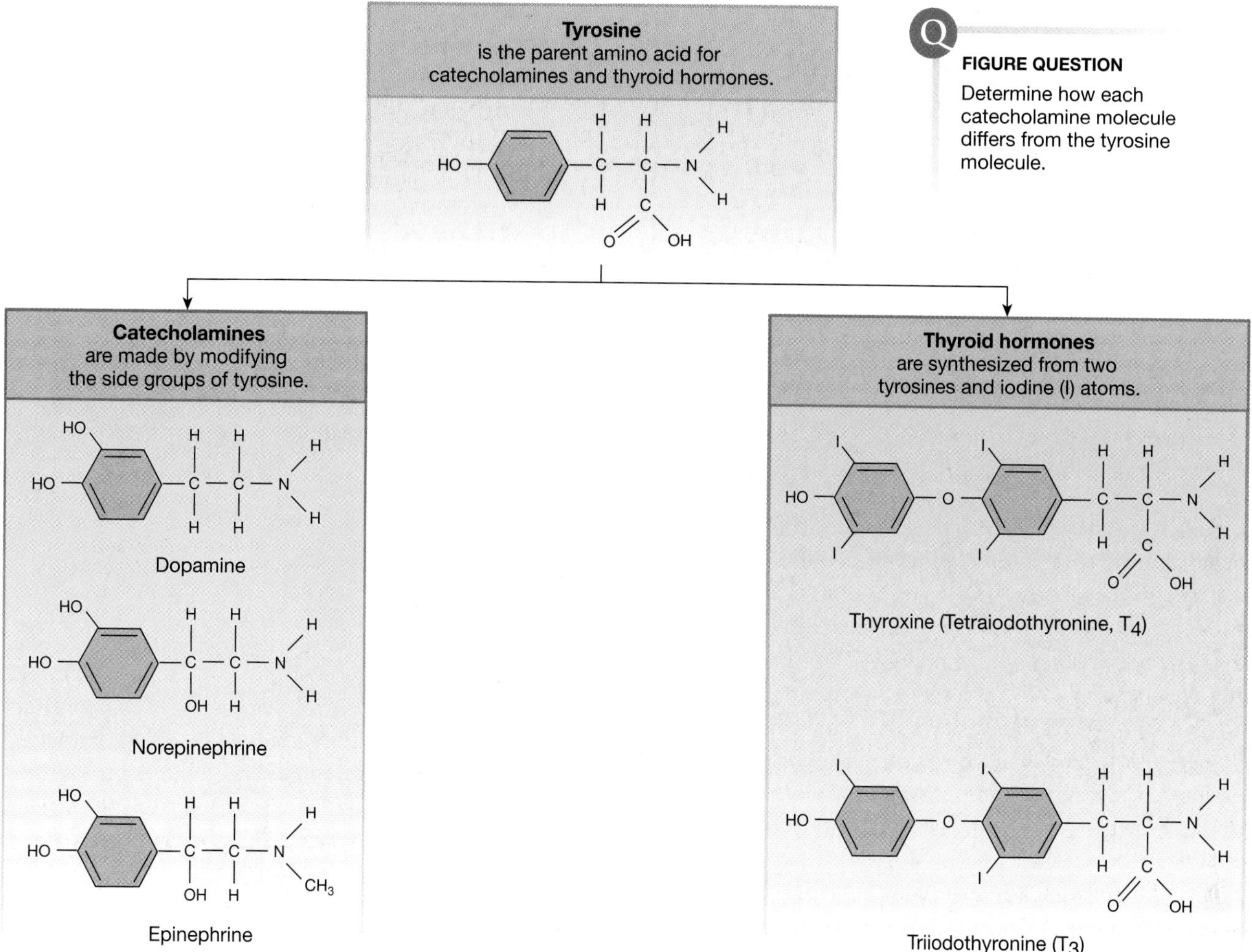

■ Fig. 7.6

an input signal, integration of the signal, an output signal, one or more targets, and a response [Fig. 6.16, p. 194]. In endocrine and neuroendocrine reflexes, the output signal is a hormone or a neurohormone. Some hormones have clear stimuli that initiate their release, such as insulin secreted in response to increasing blood glucose concentrations. Other hormones have less obvious stimuli or are secreted continuously, often with a circadian rhythm [p. 19].

The Endocrine Cell Is the Sensor in the Simplest Endocrine Reflexes

The simplest reflex control pathways in the endocrine system are those in which an endocrine cell directly senses a stimulus and responds by secreting its hormone [Fig. 6.19, pathway 6, p. 199]. In this type of pathway, the endocrine cell acts as both sensor and integrating center. The hormone is the output signal, and the response usually serves as a *negative feedback* signal that turns off the reflex [Fig. 1.11a, p. 17].

Parathyroid hormone (PTH) is an example of a hormone that operates via this simple endocrine reflex. PTH is secreted by four small parathyroid glands that lie behind the thyroid gland. The parathyroid endocrine cells monitor plasma Ca^{2+} concentration with the aid of G protein–coupled Ca^{2+} receptors on their cell membranes. When a certain number of receptors are bound to Ca^{2+}, PTH secretion is inhibited. If the plasma Ca^{2+} concentration falls below a certain level and fewer Ca^{2+} receptors are bound, inhibition ceases and the parathyroid cells secrete PTH (■ Fig. 7.7a). Parathyroid hormone travels through the blood to act on bone, kidney, and intestine, initiating responses that increase the concentration of Ca^{2+} in the plasma. The increase in plasma Ca^{2+} is a negative feedback signal that turns off the reflex, ending the release of parathyroid hormone.

Other hormones that follow a simple endocrine reflex pattern include the classic hormones insulin and glucagon, as well as some hormones of the diffuse endocrine system. For example, pancreatic beta cells are sensors that monitor blood

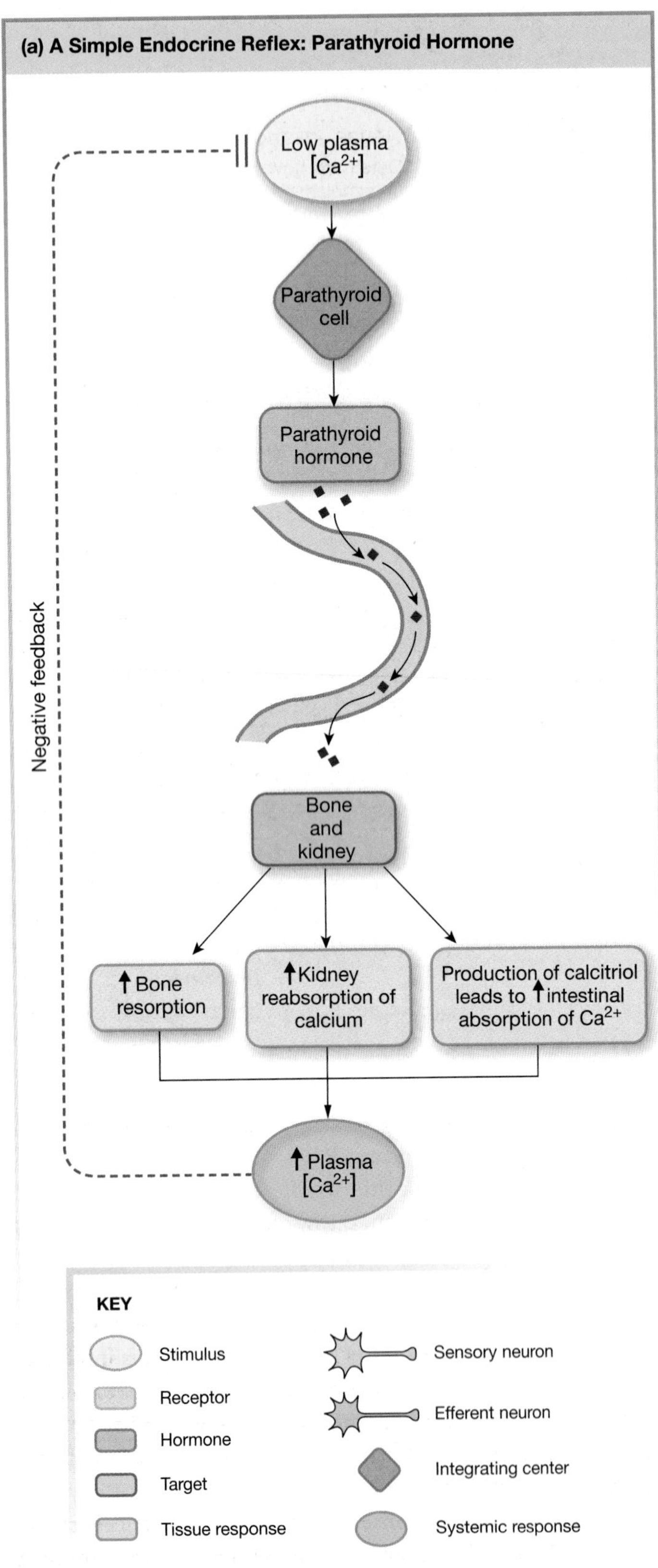

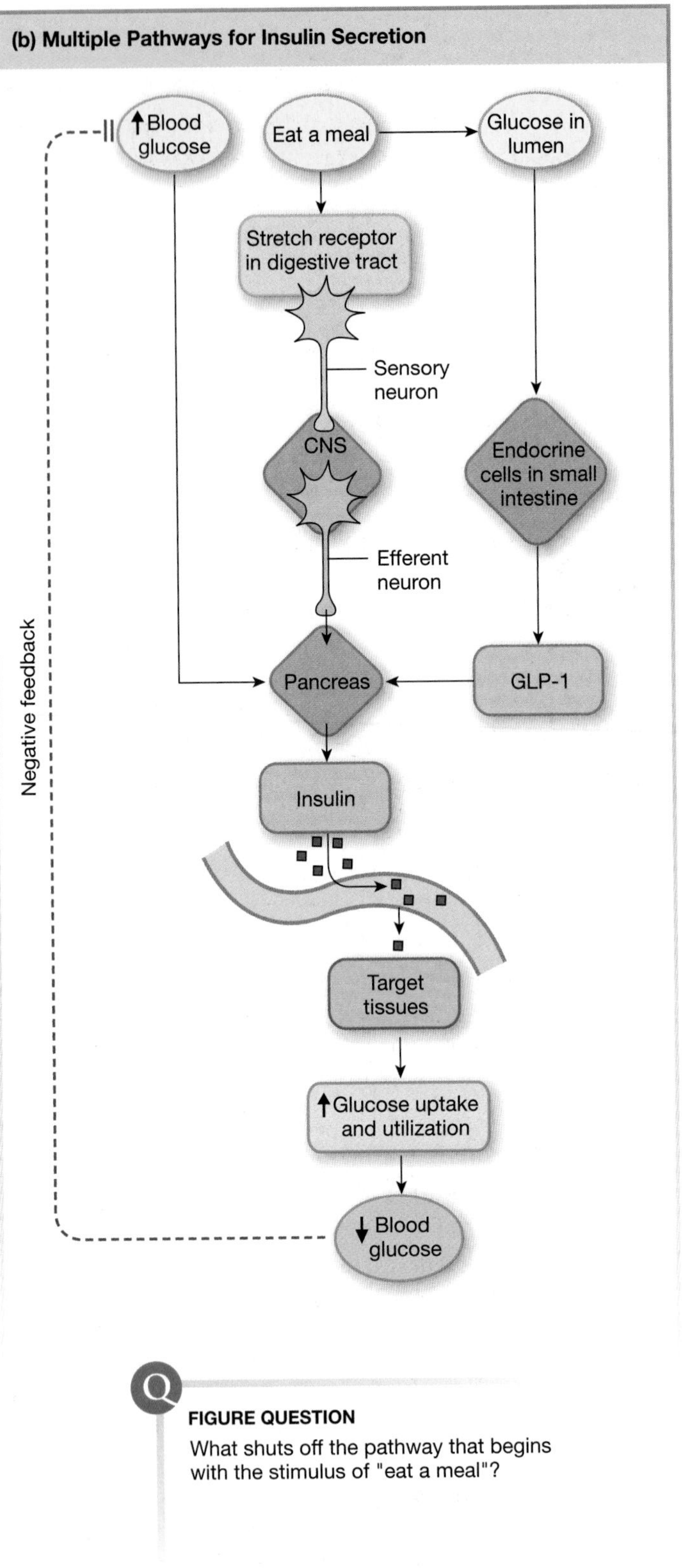

FIGURE QUESTION

What shuts off the pathway that begins with the stimulus of "eat a meal"?

Fig. 7.7 *Examples of simple endocrine pathways*

glucose concentration [p. 166]. If blood glucose increases, they respond by secreting insulin (Fig. 7.7b). Insulin travels through the blood to its target tissues, which increase their glucose uptake and metabolism. Glucose moving into cells decreases the blood concentration, which acts as a negative feedback signal that turns off the reflex, ending release of insulin.

Hormones can be released by more than one pathway, however. For example, insulin secretion can also be triggered by signals from the nervous system or by a hormone secreted from the digestive tract after a meal is eaten (Fig. 7.7b). The pancreatic endocrine cells—the integrating center for these reflex pathways—therefore must evaluate input signals from multiple sources when "deciding" whether to secrete insulin.

Concept Check Answers: p. 236

7. In the blood glucose example, the increase in blood glucose corresponds to which step of a reflex pathway? Insulin secretion and the decrease in blood glucose correspond to which steps?
8. Which insulin release pathway in Figure 7.7b is a simple endocrine reflex? Which is a complex endocrine reflex? Which is a combination neural-endocrine reflex?
9. Glucagon is released from the endocrine pancreas when blood glucose levels decrease and it acts on multiple target tissues to increase blood glucose. Draw a reflex pathway to match this description.

Many Endocrine Reflexes Involve the Nervous System

The nervous system and the endocrine system overlap in both structure and function [see Fig. 6.19, pathways 3–5, p. 199]. Stimuli integrated by the central nervous system influence the release of many hormones through efferent neurons, as previously described for insulin. In addition, specialized groups of neurons secrete neurohormones, and two endocrine structures are incorporated in the anatomy of the brain: the pineal gland (see Fig. 7.16, p. 231) and the pituitary gland.

One of the most fascinating links between the brain and the endocrine system is the influence of emotions over hormone secretion and function. Physicians for centuries have recorded instances in which emotional state has influenced health or normal physiological processes. Women today know that the timing of their menstrual periods may be altered by stressors such as travel or final exams. The condition known as "failure to thrive" in infants can often be linked to environmental or emotional stress that increases secretion of some pituitary hormones and decreases production of others. The interactions among stress, the endocrine system, and the immune system are receiving intense study by scientists [Chapter 24.].

Neurohormones Are Secreted into the Blood by Neurons

As noted previously, neurohormones are chemical signals released into the blood by a neuron [p. 177]. The human nervous system produces three major groups of neurohormones: (1) catecholamines, (described earlier) made by modified neurons in the adrenal medulla, (2) hypothalamic neurohormones secreted from the posterior pituitary, and (3) hypothalamic neurohormones that control hormone release from the anterior pituitary. Because the latter two groups of neurohormones are associated with the pituitary gland, we describe that important endocrine structure next.

Concept Check Answer: p. 236

10. Catecholamines belong to which chemical class of hormone?

The Pituitary Gland Is Actually Two Fused Glands

The **pituitary gland** is a lima bean–sized structure that extends downward from the brain, connected to it by a thin stalk and cradled in a protective pocket of bone (■ Fig. 7.8a). The first accurate description of the function of the pituitary gland came from Richard Lower (1631–1691), an experimental physiologist at Oxford University. Using observations and some experiments, he theorized that substances produced in the brain passed down the stalk into the gland and from there into the blood.

Lower did not realize that the pituitary gland is actually two different tissue types that merged during embryonic development. The **anterior pituitary** is a true endocrine gland of epithelial origin, derived from embryonic tissue that formed the roof of the mouth [Fig. 3.11, p. 85]. It is also called the *adenohypophysis* {*adeno-*, gland + *hypo-*, beneath + *phyein*, to grow}, and its hormones are *adenohypophyseal* secretions. The **posterior pituitary,** or *neurohypophysis,* is an extension of the neural tissue of the brain. It secretes neurohormones made in the *hypothalamus*, a region of the brain that controls many homeostatic functions.

The Posterior Pituitary Stores and Releases Two Neurohormones

The posterior pituitary is the storage and release site for two neurohormones: oxytocin and vasopressin (Fig. 7.8c). The neurons producing oxytocin and vasopressin are clustered together in areas of the hypothalamus known as the the *paraventricular* and *supraoptic nuclei.* (A cluster of nerve cell bodies in the central nervous system is called a nucleus.) Each neurohormone is made in a separate cell type, and the synthesis and processing follow the standard pattern for peptide hormones described earlier in this chapter.

Once the neurohormones are packaged into secretory vesicles, the vesicles are transported to the posterior pituitary through long extensions of the neurons called *axons.* After vesicles reach the axon terminals, they are stored there, waiting for the release signal.

When a stimulus reaches the hypothalamus, an electrical signal passes from the neuron cell body in the hypothalamus to the *distal* (distant) end of the cell in the posterior pituitary.

■ Fig. 7.8 *ESSENTIALS*

The Pituitary Gland

The pituitary is actually two glands with different embryological origins that fused during development.

(a) The pituitary gland sits in a protected pocket of bone, connected to the brain by a thin stalk.

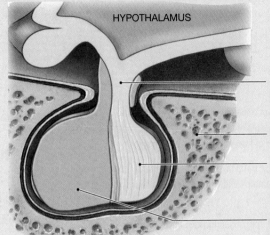

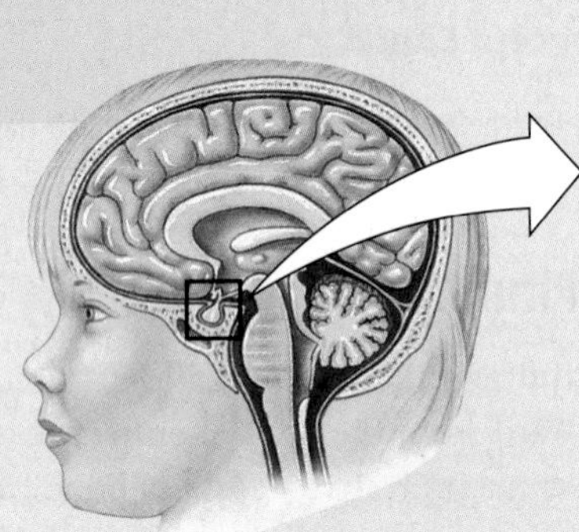

(b) The **anterior pituitary** is a true endocrine gland that secretes six classic hormones. Neurohormones from the hypothalamus control release of the anterior pituitary hormones. The hypothalamic hormones reach the anterior pituitary through a specialized region of the circulation called a portal system.

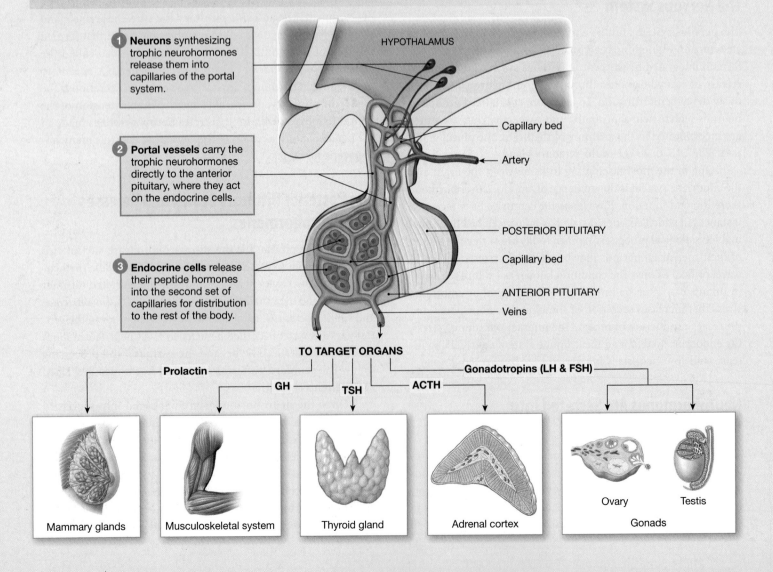

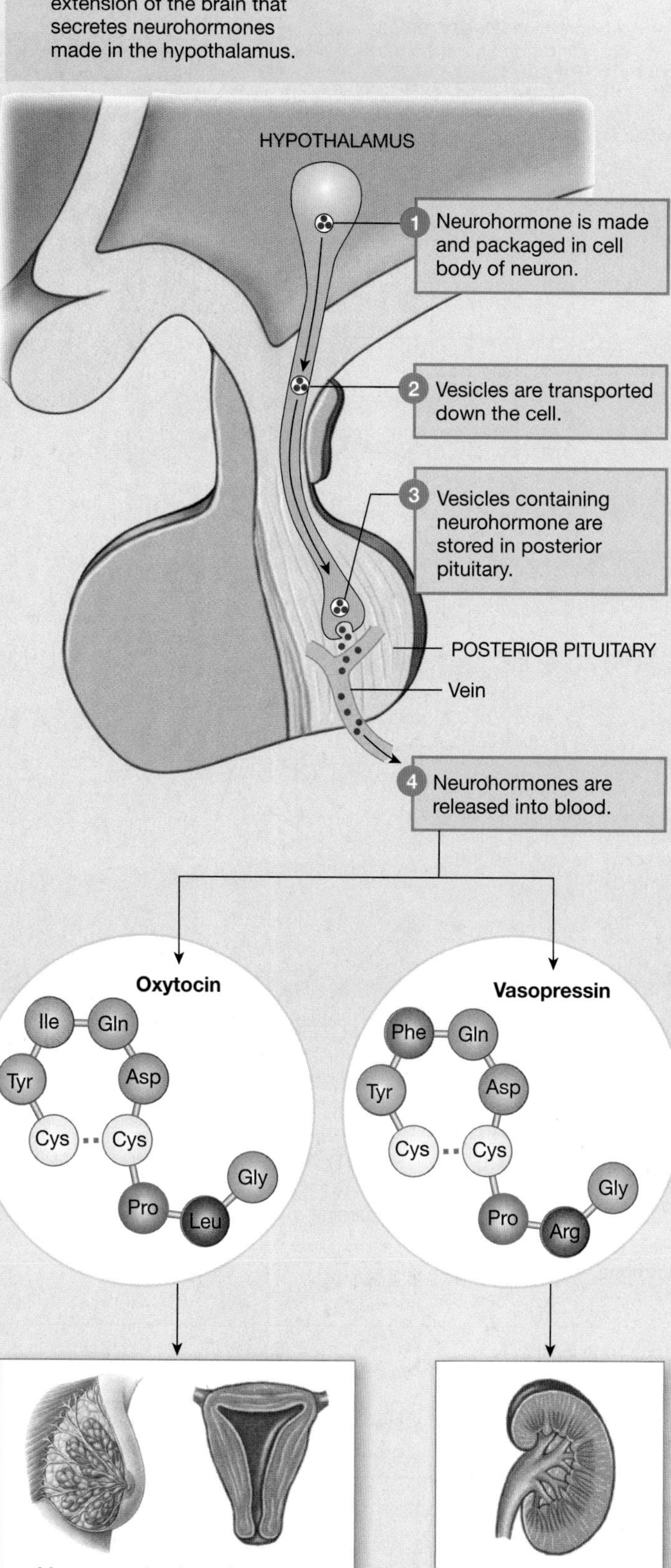
(c) The **posterior pituitary** is an extension of the brain that secretes neurohormones made in the hypothalamus.
HYPOTHALAMUS
1 Neurohormone is made and packaged in cell body of neuron.
2 Vesicles are transported down the cell.
3 Vesicles containing neurohormone are stored in posterior pituitary.
POSTERIOR PITUITARY
Vein
4 Neurohormones are released into blood.
Oxytocin
Ile
Gln
Tyr
Asp
Cys
Cys
Gly
Pro
Leu
Vasopressin
Phe
Gln
Tyr
Asp
Cys
Cys
Gly
Pro
Arg
Mammary glands and uterus
Kidneys

Depolarization of the axon terminal opens voltage-gated Ca^{2+} channels, and Ca^{2+} enters the cell. Calcium entry triggers exocytosis and the vesicle contents are released into the circulation. [Compare to insulin release, Fig. 5.27, p. 167.] Once in the blood, the neurohormones travel to their targets.

The two posterior pituitary neurohormones are composed of nine amino acids each. **Vasopressin,** also known as *antidiuretic hormone* or *ADH*, acts on the kidneys to regulate water balance in the body. In women, **oxytocin** released from the posterior pituitary controls the ejection of milk during breastfeeding and contractions of the uterus during labor and delivery.

A few neurons release oxytocin as a neurotransmitter or neuromodulator onto neurons in other parts of the brain. A number of animal experiments plus a few human experiments suggest that oxytocin plays an important role in social, sexual, and maternal behaviors. Some investigators postulate that *autism,* a developmental disorder in which patients are unable to form normal social relationships, may be related to defects in the normal oxytocin-modulated pathways of the brain.

Concept Check Answers: p. 236

11. What intracellular structure is used for transport of secretory vesicles within the cell?
12. Name the membrane process by which the contents of secretory vesicles are released into the extracellular fluid.

The Anterior Pituitary Secretes Six Hormones

As late as 1889, it was being said in reviews of physiological function that the pituitary was of little or no use to higher vertebrates! By the early 1900s, however, researchers had discovered that animals with their anterior pituitary glands surgically removed were unable to survive more than a day or two. This observation, combined with the clinical signs associated with pituitary tumors, made scientists realize that the anterior pituitary is a major endocrine gland that secretes not one but six physiologically significant hormones: prolactin (PRL), thyrotropin (TSH), adrenocorticotropin (ACTH), growth hormone (GH), follicle-stimulating hormone (FSH), and luteinizing hormone (LH) (Fig. 7.8b). Secretion of all the anterior pituitary hormones is controlled by hypothalamic neurohormones.

The anterior pituitary hormones, their associated hypothalamic neurohormones, and their targets are illustrated in ■ Fig 7.9. Notice that all the anterior pituitary hormones except prolactin have another endocrine gland or cell as one of their targets. A hormone that controls the secretion of another hormone is known as a **trophic hormone.**

The adjective *trophic* comes from the Greek word *trophikós,* which means "pertaining to food or nourishment" and refers to the manner in which the trophic hormone "nourishes" the target cell. Trophic hormones often have names that end with the

Hormones of the Hypothalamic–Anterior Pituitary Pathway

The hypothalamus secretes releasing hormones (-RH) and inhibiting hormones (-IH) that act on endocrine cells of the anterior pituitary to influence secretion of their hormones.

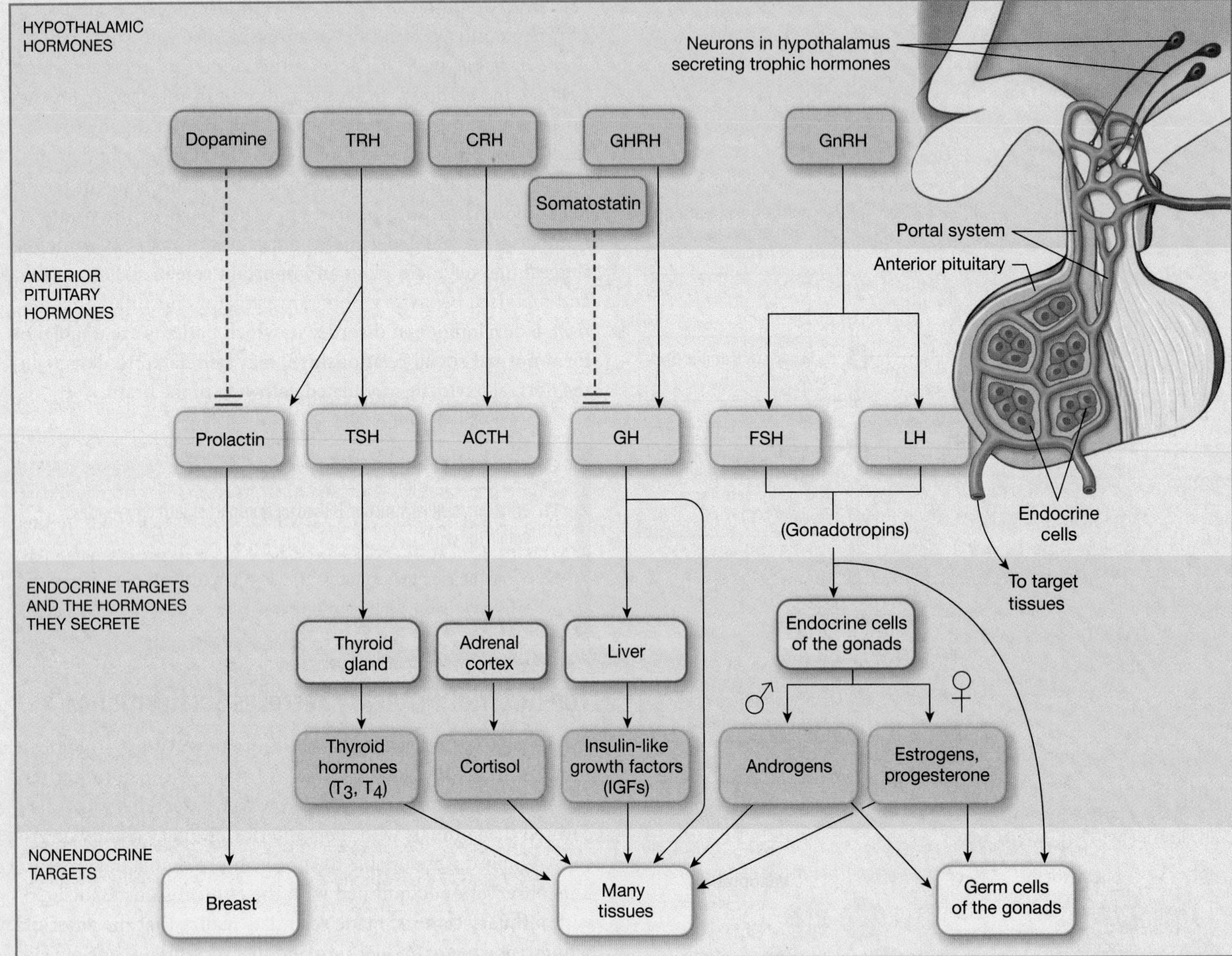

ANTERIOR PITUITARY HORMONE	HYPOTHALAMIC RELEASING HORMONE	HYPOTHALAMIC INHIBITING HORMONE
Prolactin (PRL)	Oxytocin ?	Dopamine
Thyrotropin, Thyroid-stimulating hormone (TSH)	Thyrotropin-releasing hormone (TRH)	
Adrenocorticotropin, Adrenocorticotrophic hormone (ACTH)	Corticotropin-releasing hormone (CRH)	
Growth hormone (GH), Somatotropin	GHRH (dominant)	Somatostatin (SS), also called growth hormone–inhibiting hormone (GHIH)
Gonadotropins: Follicle-stimulating hormone (FSH) Luteinizing hormone (LH)	Gonadotropin-releasing hormone (GnRH)	

suffix *-tropin,* as in *gonadotropin.** The root word to which the suffix is attached is the target tissue: the gonadotropins are hormones that are trophic to the gonads. The hypothalamic neurohormones that control release of the anterior pituitary hormones are also trophic hormones, but for historical reasons they are described as either *releasing hormones* (e.g., thyrotropin-releasing hormone) or *inhibiting hormones* (e.g., growth hormone–inhibiting hormone).

You should be aware that many of the hypothalamic and anterior pituitary hormones have multiple names as well as standardized abbreviations. For example, hypothalamic **somatostatin** (SS) is also called *growth hormone–inhibiting hormone* (GHIH), or in older scientific papers, *somatotropin release-inhibiting hormone* (SRIH). The table in Figure 7.9 lists the hypothalamic and anterior pituitary abbreviations and current alternate names.

A Portal System Delivers Hormones from Hypothalamus to Anterior Pituitary

The signals that regulate secretion of the anterior pituitary hormones come from the brain in the form of neurohormones. These hypothalamic releasing and inhibiting hormones are secreted into the circulation in the hypothalamus. They go directly from the hypothalamus to the pituitary through a special set of blood vessels known as the **hypothalamic-hypophyseal portal system** (Fig. 7.8b).

A **portal system** is a specialized region of the circulation consisting of two sets of capillaries connected in series (one after the other) by a set of larger blood vessels. There are three portal systems in the body: one in the kidneys, one in the digestive system, and this one in the brain.

Hormones secreted into a portal system have a distinct advantage over hormones secreted into the general circulation because, with a portal system, a much smaller amount of hormone can be secreted to elicit a given level of response. A dose of hormone secreted into the general circulation is rapidly diluted by the total blood volume, which is typically more than 5 L. The same dose secreted into the tiny volume of blood flowing through the portal system remains concentrated while it goes directly to its target. In this way, a small number of neurosecretory neurons in the hypothalamus can effectively control the anterior pituitary.

The minute amounts of hormone secreted into the hypothalamic-hypophyseal portal system posed a great challenge to the researchers who first isolated these hormones. Because such tiny quantities of hypothalamic-releasing hormones are secreted, Roger Guillemin and Andrew Shalley had to work with huge amounts of tissue to obtain enough hormone to analyze. Guillemin and his colleagues processed more than 50 tons of sheep hypothalami, and a major meat packer donated more than 1 million pig hypothalami to Shalley and his associates. For the final analysis, they needed 25,000 hypothalami to isolate and identify the amino acid sequence of just 1 mg of thyrotropin-releasing hormone (TRH), a tiny peptide made of three amino acids (see Fig. 7.3a). For their discovery, Guillemin and Shalley shared a Nobel prize in 1977 (see *http://nobelprize.org*).

Anterior Pituitary Hormones Control Growth, Metabolism, and Reproduction

The hormones of the anterior pituitary control so many vital functions that the pituitary is often called the master gland of the body. In general, we can say that the anterior pituitary hormones control metabolism, growth, and reproduction, all very complex processes.

One anterior pituitary hormone, **prolactin** (PRL), controls milk production in the female breast, along with other effects. In both sexes, prolactin appears to play a role in regulation of the immune system. **Growth hormone** (GH; also called *somatotropin*) affects metabolism of many tissues in addition to stimulating hormone production by the liver (■ Fig. 7.10). Prolactin and growth hormone are the only two anterior pituitary hormones whose secretion is controlled by both releasing hormones and inhibiting hormones, as you can see in Figure 7.9. We discuss prolactin and growth hormone in detail later [Chapters 26 and 23, respectively].

The remaining four anterior pituitary hormones all have another endocrine gland as their primary target. **Follicle-stimulating hormone** (FSH) and **luteinizing hormone** (LH), known collectively as the **gonadotropins,** were originally named for their effects on the ovaries, but both hormones are trophic on testes as well. **Thyroid-stimulating hormone** (TSH, or *thyrotropin*) controls hormone synthesis and secretion in the thyroid gland. **Adrenocorticotrophic hormone** (ACTH, or *adrenocorticotropin*) acts on certain cells of the adrenal cortex to control synthesis and release of the steroid hormone cortisol.

Concept Check

Answers: p. 236

13. Map the pathways for:

(a) the hypothalamic releasing hormone—prolactin—breast pattern just described

(b) the growth hormone pathway shown in Figure 7.10 [*Hint:* Figure 6.19, p. 199]

14. What is the target tissue of a hypothalamic neurohormone secreted into the hypothalamic-hypophyseal portal system?

Feedback Loops Are Different in the Hypothalamic-Pituitary Pathway

The pathways in which anterior pituitary hormones act as trophic hormones are among the most complex endocrine reflexes because they involve three integrating centers: the

*A few hormones whose names end in *-tropin* do not have endocrine cells as their targets. For example, melanotropin acts on pigment-containing cells in many animals.

The Growth Hormone Pathway

Hypothalamic growth hormone–releasing hormone (GHRH) stimulates growth hormone (GH) secretion. Growth hormone acts directly on many body tissues but also influences liver production of insulin-like growth factors (IGFs or somatomedins), another group of hormones that regulate growth.

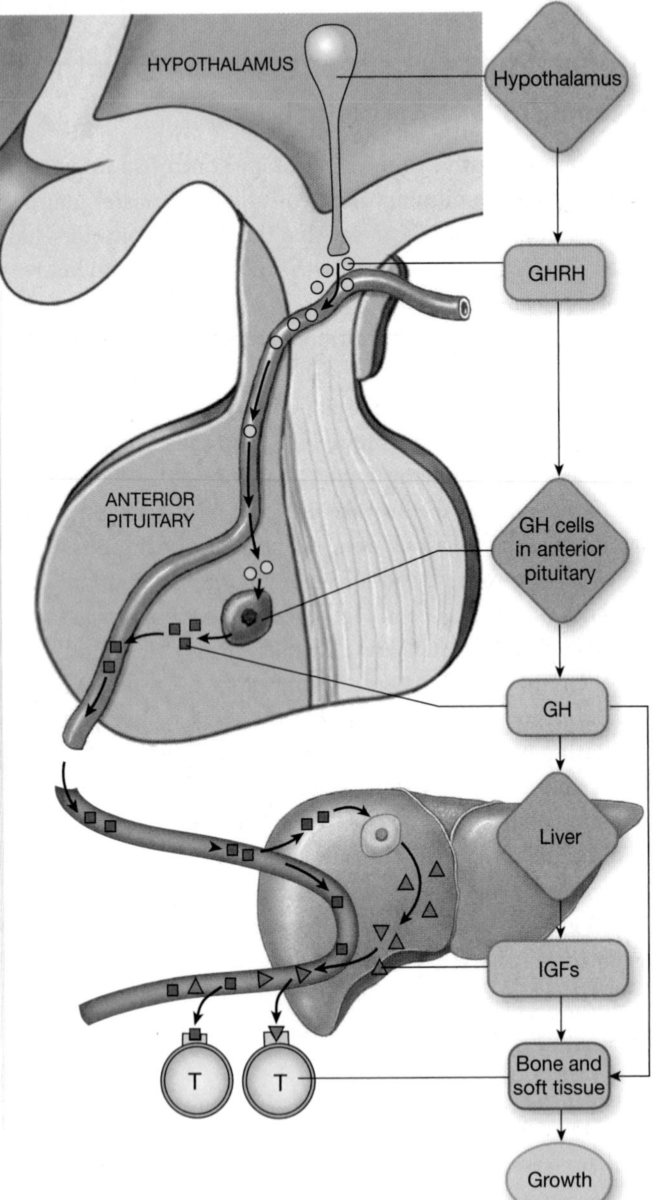

■ **Fig. 7.10**

hypothalamus, the anterior pituitary, and the endocrine target of the pituitary hormone (■ Fig. 7.11a). Feedback in these complex pathways follows a pattern that is different from the pattern described previously. Instead of the response acting as the negative feedback signal, the hormones themselves are the feedback signal. One reason this is necessary is that for most anterior pituitary hormone pathways, there is no single response that the body can easily monitor. The hormones of these pathways act on multiple tissues and have different, often subtle, effects in different tissues. There is no single parameter, such as blood glucose concentration, that can be the signal for negative feedback.

In hypothalamic-pituitary pathways, each hormone in the pathway feeds back to suppress hormone secretion by integrating centers earlier in the pathway. When secretion of one hormone changes, the secretion of other hormones also changes because of the feedback loops that link the hormones. In pathways with two or three hormones in sequence, the "downstream" hormone usually feeds back to suppress the hormone(s) that controlled its secretion. A major exception to this is feedback by ovarian hormones [Chapter 26], where feedback alternates between positive and negative.

The hormones of the hypothalamic-pituitary-adrenal (HPA) pathway provide a good example of feedback loops. Cortisol secreted from the adrenal cortex feeds back to suppress secretion of hypothalamic corticotropin-releasing hormone (CRH) and adrenocorticotropin (ACTH) from the anterior pituitary (Fig. 7.11b). When the last hormone in a pathway feeds back to suppress secretion of its trophic hormones, the relationship is called **long-loop negative feedback**.

In **short-loop negative feedback**, pituitary hormones feed back to decrease hormone secretion by the hypothalamus. We see this type of feedback in cortisol secretion in Fig. 7.11b, where ACTH exerts short-loop negative feedback on the secretion of CRH. There can also be *ultra-short-loop feedback*, in which a hormone acts as an autocrine to influence the cell that secreted it.

With this hormone-based system of negative feedback, the hormones in a pathway normally stay within the range needed for an appropriate response. Feedback patterns are important in the diagnosis of endocrine pathologies, discussed later in the chapter.

RUNNING PROBLEM

Thyroid hormone production is regulated by thyroid-stimulating hormone (TSH), a hormone secreted by the anterior pituitary. The production of TSH is in turn regulated by the neurohormone thyrotropin-releasing hormone (TRH) from the hypothalamus.

Q2: a. In a normal person, when thyroid hormone levels in the blood increase, will negative feedback increase or decrease the secretion of TSH?

b. In a person with a hyperactive gland that is producing too much thyroid hormone, would you expect the level of TSH to be higher or lower than in a normal person?

207 216 224 226 229 230 232

7

Negative Feedback in Complex Endocrine Pathways

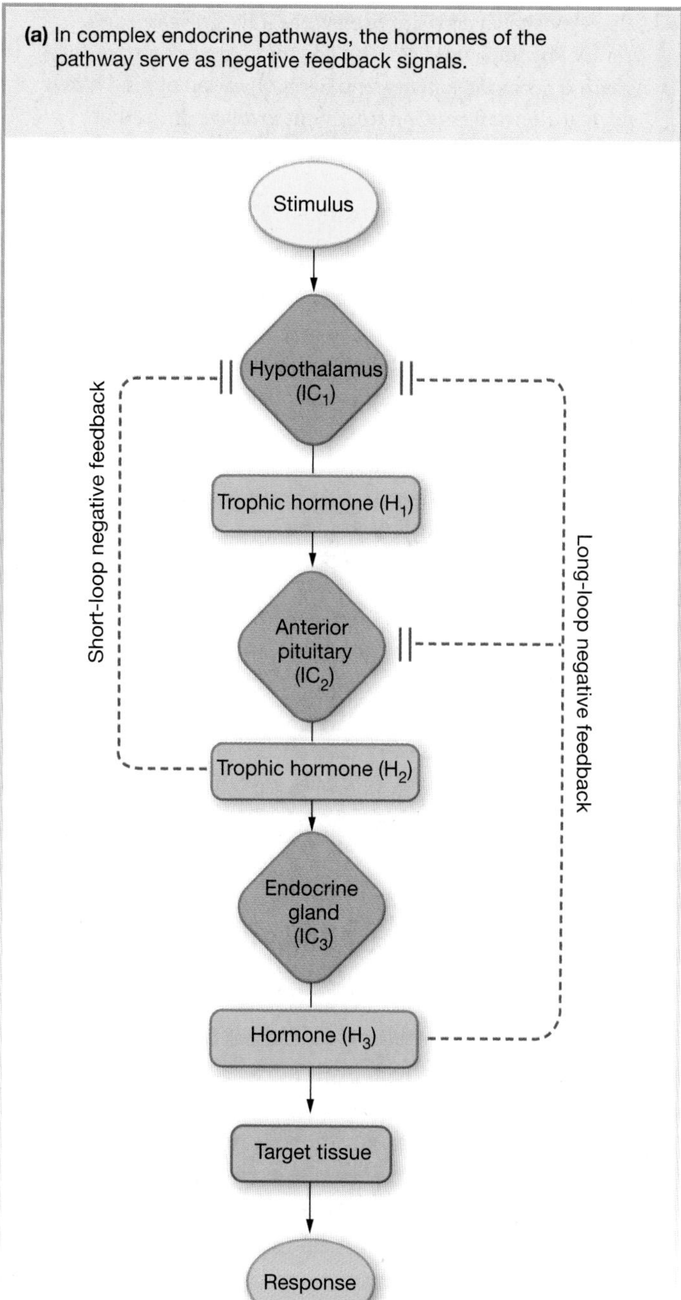

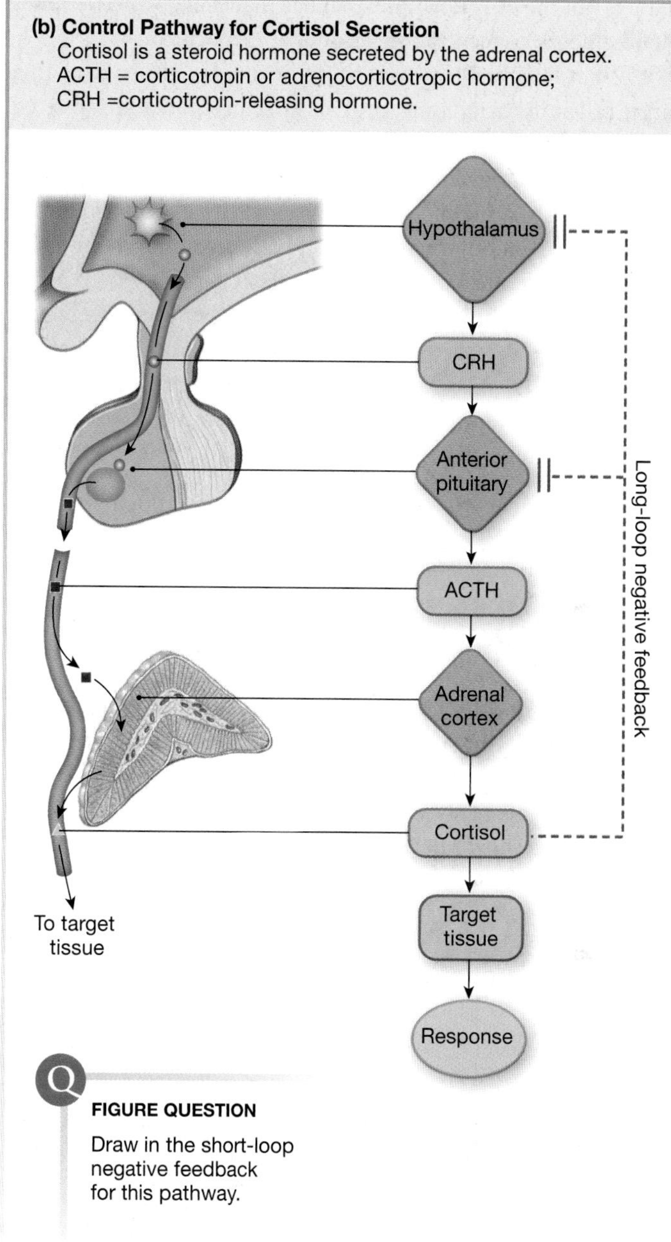

Fig. 7.11

Hormone Interactions

One of the most complicated and confusing aspects of endocrinology is the way hormones interact at their target cells. It would be simple if each endocrine reflex were a separate entity and if each cell were under the influence of only a single hormone. In many instances, however, cells and tissues are controlled by multiple hormones that may be present at the same time. Complicating the picture is the fact that multiple hormones acting on a single cell can interact in ways that cannot be predicted by knowing the individual effects of the hormone. In this section, we examine three types of hormone interaction: synergism, permissiveness, and antagonism.

In Synergism, the Effect of Interacting Hormones Is More Than Additive

Sometimes different hormones have the same effect on the body, although they may accomplish that effect through different cellular mechanisms. One example is the hormonal control of blood glucose levels. Glucagon from the pancreas is the hormone primarily responsible for elevating blood glucose levels, but it is not the only hormone that has that effect. Cortisol raises blood glucose concentration, as does epinephrine.

What happens if two of these hormones are present in a target cell at the same time, or if all three hormones are secreted at the same time? You may expect their effects to be additive. In other words, if a given amount of epinephrine elevates blood glucose 5 mg/100 mL blood, and glucagon elevates blood glucose 10 mg/100 mL blood, you may expect both hormones acting at the same time to elevate blood glucose 15 mg/100 mL blood (5 + 10).

Frequently, however, two (or more) hormones interact at their targets so that the combination yields a result that is greater than additive (1 + 2 > 3). This type of interaction is called **synergism**. For our epinephrine/glucagon example, a synergistic reaction would be:

epinephrine	elevates blood glucose	5 mg/100 mL blood
glucagon	elevates blood glucose	10 mg/100 mL blood
epinephrine + glucagon	elevate blood glucose	22 mg/100 mL blood

In other words, the combined effect of the two hormones is greater than the sum of the effects of the two hormones individually.

An example of synergism involving epinephrine, glucagon, and cortisol is shown in ■ Figure 7.12. The cellular mechanisms that underlie synergistic effects are not always clear, but with peptide hormones, synergism is often linked to overlapping effects on second messenger systems.

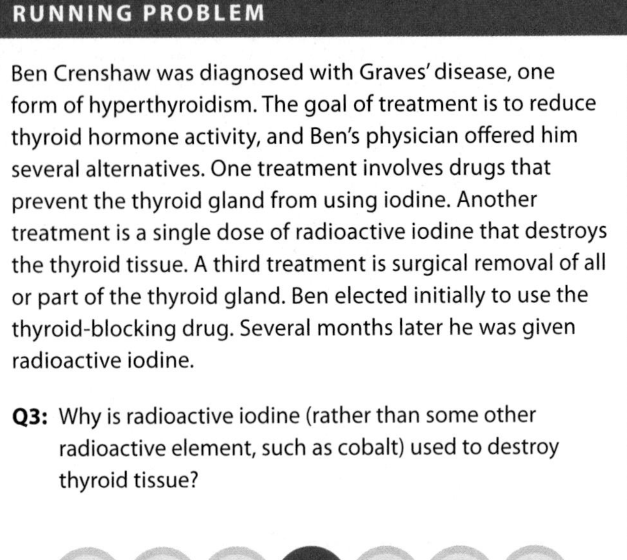

RUNNING PROBLEM

Ben Crenshaw was diagnosed with Graves' disease, one form of hyperthyroidism. The goal of treatment is to reduce thyroid hormone activity, and Ben's physician offered him several alternatives. One treatment involves drugs that prevent the thyroid gland from using iodine. Another treatment is a single dose of radioactive iodine that destroys the thyroid tissue. A third treatment is surgical removal of all or part of the thyroid gland. Ben elected initially to use the thyroid-blocking drug. Several months later he was given radioactive iodine.

Q3: Why is radioactive iodine (rather than some other radioactive element, such as cobalt) used to destroy thyroid tissue?

207 216 224 **226** 229 230 232

Synergism is not limited to hormones. It can occur with any two (or more) chemicals in the body. Pharmacologists have developed drugs with synergistic components. For example, the effectiveness of the antibiotic penicillin is enhanced by the presence of clavulanic acid in the same pill.

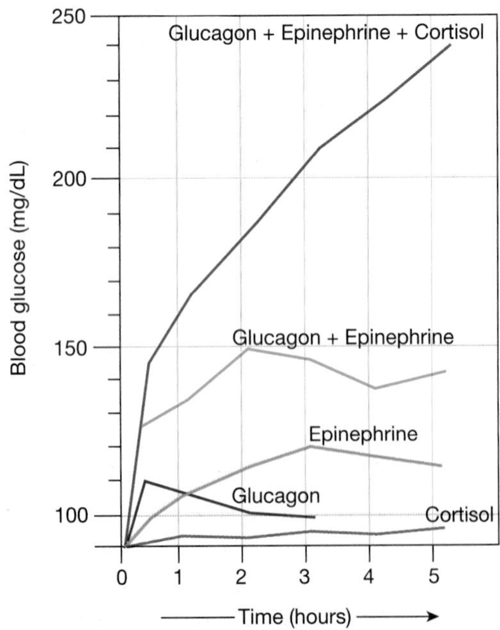

■ **Fig. 7.12** ***Synergism.*** This graph shows the effect of hormone infusions on blood glucose levels. The effects of combined hormones are greater than the summed effects of the individual hormones, indicating synergistic relationships. (Data adapted from Eigler *et al., J. Clin. Invest.* 63: 114, 1979.)

A Permissive Hormone Allows Another Hormone to Exert Its Full Effect

In **permissiveness,** one hormone cannot fully exert its effects unless a second hormone is present (0 + 2 > 2). For example, maturation of the reproductive system is controlled by gonadotropin-releasing hormone from the hypothalamus, gonadotropins from the anterior pituitary, and steroid hormones from the gonads. However, if thyroid hormone is not present in sufficient amounts, maturation of the reproductive system is delayed. Because thyroid hormone by itself cannot stimulate maturation of the reproductive system, thyroid hormone is considered to have a permissive effect on sexual maturation.

The results of this interaction can be summarized as follows:

thyroid hormone alone	no development of reproductive system
reproductive hormones alone	delayed development of reproductive system
reproductive hormones with adequate thyroid hormone	normal development of reproductive system

The molecular mechanisms responsible for permissiveness are not well understood in most instances.

Antagonistic Hormones Have Opposing Effects

In some situations, two molecules work against each other, one diminishing the effectiveness of the other. This tendency of one substance to oppose the action of another is called *antagonism*. Antagonism may result when two molecules compete for the same receptor [p. 52]. When one molecule binds to the receptor but does not activate it, that molecule acts as a *competitive inhibitor*, or antagonist, to the other molecule. This type of receptor antagonism has been put to use in the development of pharmaceutical compounds, such as the estrogen receptor antagonist *tamoxifen*, which is used to treat breast cancers that are stimulated by estrogen.

In endocrinology, two hormones are considered *functional antagonists* if they have opposing physiological actions. For example, both glucagon and growth hormone raise the concentration of glucose in the blood, and both are antagonistic to insulin, which lowers the concentration of glucose in the blood. Hormones with antagonistic actions do not necessarily compete for the same receptor. Instead, they may act through different metabolic pathways, or one hormone may decrease the number of receptors for the opposing hormone. For example, evidence suggests that growth hormone decreases the number of insulin receptors, providing part of its functional antagonistic effects on blood glucose concentration.

The synergistic, permissive, and antagonistic interactions of hormones make the study of endocrinology both challenging and intriguing. With this brief survey of hormone interactions, you have built a solid foundation for learning more about hormone interactions.

Endocrine Pathologies

As one endocrinologist said, "There are no good or bad hormones. A balance of hormones is important for a healthy life. . . . Unbalance leads to diseases."* We can learn much about the normal functions of a hormone by studying the diseases caused by hormone imbalances. There are three basic patterns of endocrine pathology: hormone excess, hormone deficiency, and abnormal responsiveness of target tissues to a hormone.

To illustrate endocrine pathologies, we will use a single example, that of cortisol production by the adrenal cortex (see Fig. 7.11b). This is a complex reflex pathway that starts with the secretion of corticotropin-releasing hormone (CRH) from the hypothalamus. CRH stimulates release of adrenocorticotropin (ACTH) from the anterior pituitary. ACTH in turn controls the synthesis and release of cortisol from the adrenal cortex. As in other homeostatic reflex pathways, negative feedback shuts off the pathway. As cortisol increases, it acts as a negative feedback signal, causing the pituitary and hypothalamus to decrease their output of ACTH and CRH, respectively.

*W. König, preface to *Peptide and Protein Hormones,* New York: VCH Publishers, 1993.

Hypersecretion Exaggerates a Hormone's Effects

If a hormone is present in excessive amounts, the normal effects of the hormone are exaggerated. Most instances of hormone excess are due to **hypersecretion.** There are numerous causes of hypersecretion, including benign tumors (*adenomas*) and cancerous tumors of the endocrine glands. Occasionally, nonendocrine tumors secrete hormones.

Any substance coming from outside the body is referred to as *exogenous* {*exo-*, outside}, and sometimes a patient may exhibit signs of hypersecretion as the result of medical treatment with an exogenous hormone or agonist. In this case, the condition is said to be *iatrogenic,* or physician-caused {*iatros,* healer + *-gen,* to be born}. It seems simple enough to correct the hormone imbalance by stopping treatment with the exogenous hormone, but this is not always the case.

In our example, exogenous cortisol in the body acts as a negative feedback signal, just as cortisol produced within the body would, shutting off the production of CRH and ACTH (■ Fig. 7.13). Without the trophic "nourishing" influence of ACTH, the body's own cortisol production shuts down. If the pituitary remains suppressed and the adrenal cortex is deprived of ACTH long enough, the cells of both glands shrink and lose their ability to manufacture ACTH and cortisol. The loss of cell mass is known as **atrophy** {*a-*, without + *trophikós,* nourishment}.

If the cells of an endocrine gland atrophy because of exogenous hormone administration, they may be very slow or totally unable to regain normal function when the treatment with exogenous hormone is stopped. As you may know, steroid hormones can be used to treat poison ivy and severe allergies. However, when treatment is complete, the dosage must be tapered off gradually to allow the pituitary and adrenal gland to

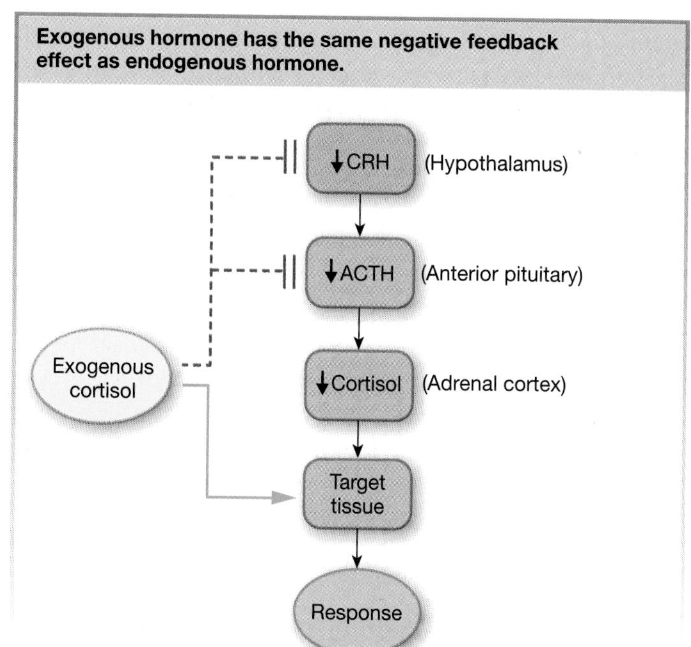

■ **Fig. 7.13**

work back up to normal hormone production. As a result, packages of steroid pills direct patients ending treatment to take six pills one day, five the day after that, and so on. Low-dose, over-the-counter steroid creams usually do not pose a risk of feedback suppression when used as directed.

Hyposecretion Diminishes or Eliminates a Hormone's Effects

Symptoms of hormone deficiency occur when too little hormone is secreted (**hyposecretion**). Hyposecretion may occur anywhere along the endocrine control pathway, in the hypothalamus, pituitary, or other endocrine glands. For example, hyposecretion of thyroid hormone may occur if there is insufficient dietary iodine for the thyroid gland to manufacture the iodinated hormone. The most common cause of hyposecretion pathologies is atrophy of the gland due to some disease process.

Negative feedback pathways are affected in hyposecretion, but in the opposite direction from hypersecretion. The absence of negative feedback causes trophic hormone levels to rise as the trophic hormones attempt to make the defective gland increase its hormone output. For example, if the adrenal cortex atrophies as a result of tuberculosis, cortisol production diminishes. The hypothalamus and anterior pituitary sense that cortisol levels are below normal, so they increase secretion of CRH and ACTH, respectively, in an attempt to stimulate the adrenal gland into making more cortisol.

Receptor or Second Messenger Problems Cause Abnormal Tissue Responsiveness

Endocrine diseases do not always arise from problems with endocrine glands. They may also be triggered by changes in the responsiveness of target tissues to the hormones. In these situations, the target tissues show abnormal responses even though the hormone levels may be within the normal range. Changes in the target tissue response are usually caused by abnormal interactions between the hormone and its receptor or by alterations in signal transduction pathways.

Down-Regulation If hormone secretion is abnormally high for an extended period of time, target cells may *down-regulate* (decrease the number of) their receptors in an effort to diminish their responsiveness to excess hormone. **Hyperinsulinemia** {*hyper-*, elevated + insulin + *-emia*, in the blood} is a classic example of down-regulation in the endocrine system. In this disorder, sustained high levels of insulin in the blood cause target cells to remove insulin receptors from the cell membrane. Patients suffering from hyperinsulinemia may show signs of diabetes despite their high blood insulin levels.

Receptor and Signal Transduction Abnormalities Many forms of inherited endocrine pathologies can be traced to problems with hormone action in the target cell. Endocrinologists once believed that these problems were rare, but they are being recognized more frequently as scientists increase their understanding of receptors and signal transduction mechanisms.

Some pathologies are due to problems with the hormone receptor. If a mutation alters the protein sequence of the receptor, the cellular response to receptor-hormone binding may be altered. In other mutations, the receptors may be absent or completely nonfunctional. For example, in *testicular feminizing syndrome,* androgen receptors are nonfunctional in the male fetus because of a genetic mutation. As a result, androgens produced by the developing fetus are unable to influence development of the genitalia. The result is a child who appears to be female but lacks a uterus and ovaries.

Genetic alterations in signal transduction pathways can lead to symptoms of hormone excess or deficiency. In the disease called *pseudohypoparathyroidism* {*pseudo-*, false + *hypo-*, decreased + parathyroid + *-ism,* condition or state of being}, patients show signs of low parathyroid hormone even though blood levels of the hormone are normal or elevated. These patients have inherited a defect in the G protein that links the hormone receptor to the cAMP amplifier enzyme, adenylyl cyclase. Because the signal transduction pathway does not function, target cells are unable to respond to parathyroid hormone, and signs of hormone deficiency appear.

Diagnosis of Endocrine Pathologies Depends on the Complexity of the Reflex

Diagnosis of endocrine pathologies may be simple or complicated, depending on the complexity of the reflex. For example, consider a simple endocrine reflex, such as that for parathyroid hormone. If there is too much or too little hormone, the problem can arise in only one location: the parathyroid glands (see Figure 7.10). However, with complex hypothalamic-pituitary-endocrine gland reflexes, the diagnosis can be much more difficult.

If a pathology (deficiency or excess) arises in the last endocrine gland in a reflex, the problem is considered to be a **primary pathology**. For example, if a tumor in the adrenal cortex begins to produce excessive amounts of cortisol, the resulting condition is called *primary hypersecretion*. If dysfunction occurs in one of the tissues producing trophic hormones, the problem is a **secondary pathology**. For example, if the pituitary is damaged because of head trauma and ACTH secretion diminishes, the resulting cortisol deficiency is considered to be *secondary hyposecretion* of cortisol.

The diagnosis of pathologies in complex endocrine pathways depends on understanding negative feedback in the control pathway. ■ Figure 7.14 shows three possible causes of excess cortisol secretion. To determine which is the correct *etiology* (cause) of the disease in a particular patient, the clinician must assess the levels of the three hormones in the control pathway.

If cortisol levels are high but levels of both trophic hormones are low, the problem must be a primary disorder (Figure 7.14a). There are two possible explanations: endogenous

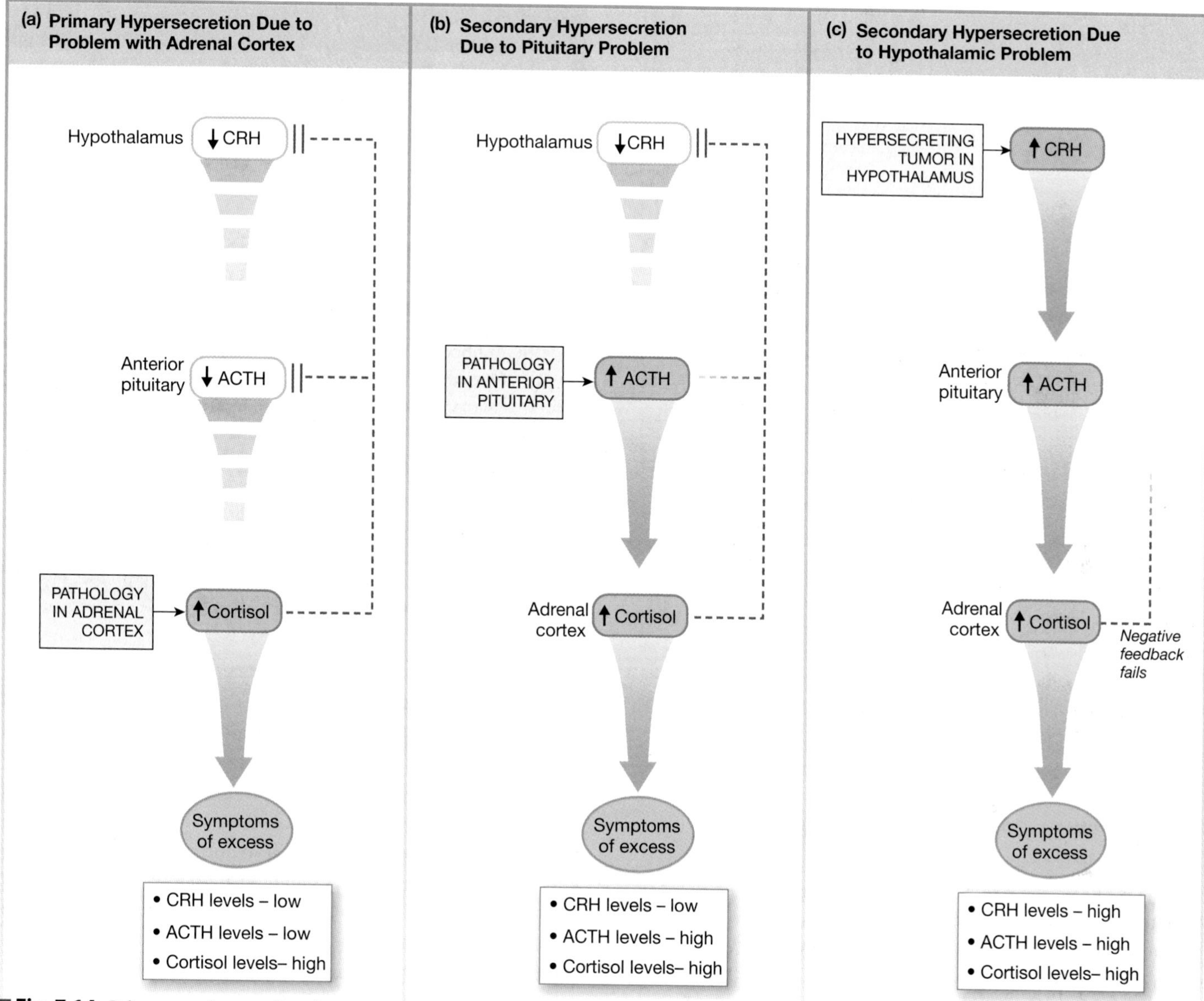

■ **Fig. 7.14** ***Primary and secondary hypersecretion of cortisol.*** When there is a pathology in an endocrine gland, negative feedback fails.

cortisol hypersecretion or the exogenous administration of cortisol for therapeutic reasons (see Figure 7.13). In either case, high levels of cortisol act as a negative feedback signal that shuts off production of CRH and ACTH. The pattern of high cortisol with low trophic hormone levels points to a primary disorder.

When the problem is endogenous—an adrenal tumor that is secreting cortisol in an unregulated fashion—the normal control pathways are totally ineffective. Although negative feedback shuts off production of the trophic hormones, the tumor is not dependent on them for cortisol production, so cortisol secretion continues in their absence. The tumor must be removed or suppressed before cortisol secretion can be controlled.

Figure 7.14b shows a secondary hypersecretion of cortisol due to an ACTH-secreting tumor of the pituitary. The high levels of ACTH cause high cortisol production, but in this example the high

RUNNING PROBLEM

Graves' disease is one form of thyroid gland hyperactivity. For this reason, people with Graves' disease have elevated thyroxine levels in the blood. Their TSH levels are very low.

Q4: If levels of TSH are low and thyroxine levels are high, is Graves' disease a primary disorder or a secondary disorder (one that arises as a result of a problem with the anterior pituitary or the hypothalamus)? Explain your answer.

207 216 224 226 229 230 232

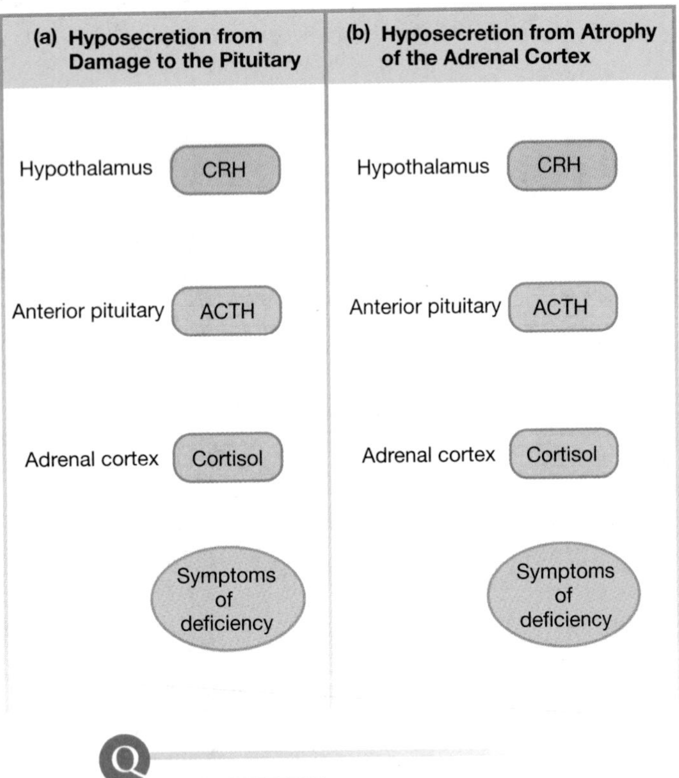

Fig. 7.15 ***Patterns of hormone secretion in hypocortisolism***

cortisol level has a negative feedback effect on the hypothalamus, decreasing production of CRH. The combination of low CRH and high ACTH isolates the problem to the pituitary. This pathology is responsible for about two-thirds of cortisol hypersecretion *syndromes* {*syn*-, together + *-drome*, running; a combination of symptoms characteristic of a particular pathology}.

If the problem is overproduction of CRH by the hypothalamus (Figure 7.14c) CRH levels are higher than normal. High CRH in turn causes high ACTH, which in turn causes high cortisol. This is therefore secondary hypersecretion arising from a problem in the hypothalamus. In clinical practice, hypothalamic hypersecretion pathologies are rare.

Figure 7.15 shows two possible etiologies for hyposecretion of cortisol. You can apply your understanding of negative feedback in the hypothalamic-pituitary control pathway to predict whether the levels of CRH, ACTH, and cortisol will be high or low in each case.

Hormone Evolution

Chemical signaling is an ancient method for communication and the maintenance of homeostasis. As scientists sequence the genomes of diverse species, they are discovering that in many cases hormone structure and function have changed amazingly little from the most primitive vertebrates through the mammals. In fact, hormone signaling pathways that were once considered exclusive to vertebrates, such as those for thyroid hormones and insulin, have now been shown to play physiological or developmental roles in invertebrates such as echinoderms and insects. This *evolutionary conservation* of hormone function is also demonstrated by the fact that some hormones from other organisms have biological activity when administered to humans. By studying which portions of a hormone molecule do not change from species to species, scientists have acquired important clues to aid in the design of agonist and antagonist drugs.

The ability of nonhuman hormones to work in humans was a critical factor in the birth of endocrinology. When Best and Banting discovered insulin in 1921 and the first diabetic patients were treated with the hormone, the insulin was extracted from cow, pig, or sheep pancreases. Before the mid-1980s slaughterhouses were the major source of insulin for the medical profession. Now, with genetic engineering, the human gene for insulin has been inserted into bacteria, which then synthesize the hormone, providing us with a plentiful source of human insulin.

Although many hormones have the same function in most vertebrates, a few hormones that play a significant role in the physiology of lower vertebrates seem to be evolutionarily "on their way out" in humans. Calcitonin is a good example of such a hormone. Although it plays a role in calcium metabolism in fish, calcitonin apparently has no significant influence on daily calcium balance in adult humans. Neither calcitonin deficiency nor calcitonin excess is associated with any pathological condition or symptom.

Although calcitonin is not a significant hormone in humans, the calcitonin gene does code for a biologically active protein. In the brain, cells process mRNA from the calcitonin gene to make

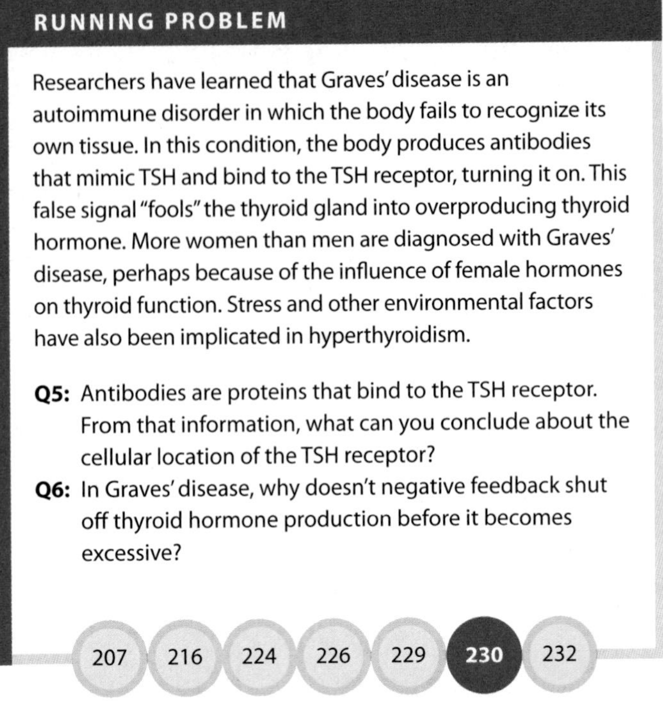

RUNNING PROBLEM
Researchers have learned that Graves' disease is an autoimmune disorder in which the body fails to recognize its own tissue. In this condition, the body produces antibodies that mimic TSH and bind to the TSH receptor, turning it on. This false signal "fools" the thyroid gland into overproducing thyroid hormone. More women than men are diagnosed with Graves' disease, perhaps because of the influence of female hormones on thyroid function. Stress and other environmental factors have also been implicated in hyperthyroidism.
Q5: Antibodies are proteins that bind to the TSH receptor. From that information, what can you conclude about the cellular location of the TSH receptor?
Q6: In Graves' disease, why doesn't negative feedback shut off thyroid hormone production before it becomes excessive?
207 216 224 226 229 230 232

■ Fig. 7.16 *FOCUS ON ...*

The Pineal Gland

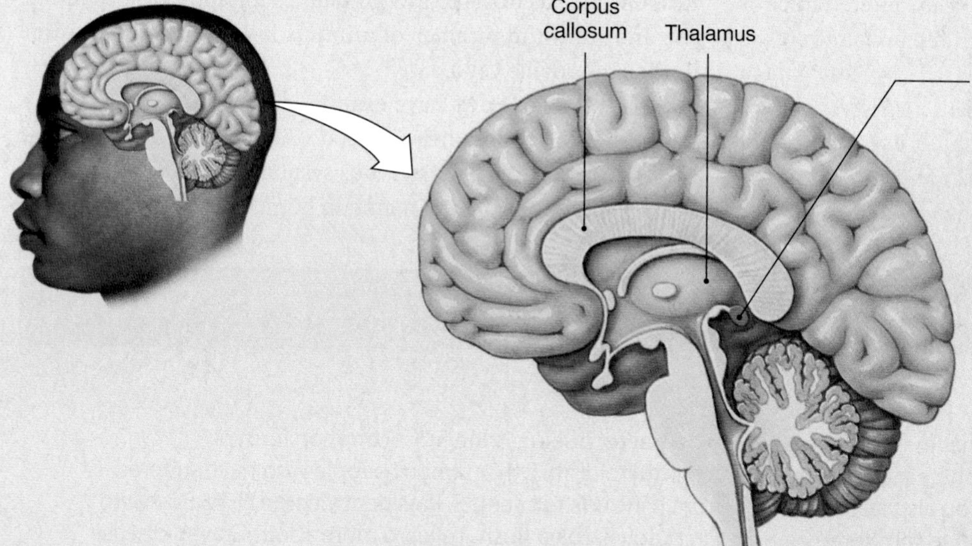

The **pineal gland** is a pea-sized structure buried deep in the brain of humans. Nearly 2000 years ago, this "seat of the soul" was thought to act as a valve that regulated the flow of vital spirits and knowledge into the brain. By 1950, however, scientists had decided that it was a vestigial structure with no known function.

CH_3O—[indole ring with N–H]—$CH_2CH_2NH-\overset{O}{\overset{\|}{C}}-CH_3$

Melatonin is an amino acid–derived hormone made from tryptophan.

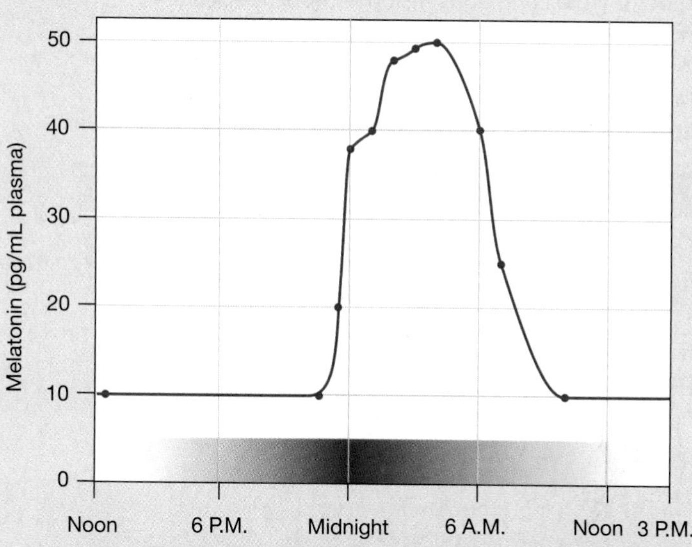

Melatonin is the "darkness hormone," secreted at night as we sleep. It is the chemical messenger that transmits information about light-dark cycles to the brain center that governs the body's biological clock.

(Adapted from J. Arendt, *Clin. Endocrinol.* 29:205–229, 1988.)

About 1957 one of the wonderful coincidences of scientific research occurred. An investigator heard about a factor in beef pineal glands that could lighten the skin of amphibians. Using the classical methodology of endocrinology, he obtained pineal glands from a slaughterhouse and started making extracts. His biological assay consisted of dropping pineal extracts into bowls of live tadpoles to see if their skin color blanched. Several years and hundreds of thousands of pineal glands later, he had isolated a small amount of melatonin.

Fifty years later, we are still learning about the functions of melatonin in humans. In addition to its role in sleep-wake cycles and the body's internal clock, scientists have evidence that melatonin is a powerful antioxidant. Some studies using mouse models of Alzheimer's disease suggest that melatonin may help slow the progression of the disease. Melatonin has also been linked to sexual function, the onset of puberty, and depression in the darker winter months (seasonal affective disorder, or SAD). In 2011 there were over 100 active clinical trials in the United States testing the efficacy of melatonin in treating disorders associated with sleep disturbances and depression.

In 2009 European authorities approved the use of a melatonin receptor agonist, *agomelatine*, for treating major depression. The U.S. Food and Drug Administration has been slower to approve the drug, and it is currently being tested in Phase II and Phase III clinical trials in the United States. Phase II trials are usually placebo-controlled, double-blind studies. Phase III trials include more patients and some uncontrolled studies. Some Phase III studies are "open-label," meaning that the patients and healthcare providers know what drug is being administered.

a peptide known as *calcitonin gene-related peptide* (CGRP), which acts as a neurotransmitter. CGRP can act as a powerful dilator of blood vessels, and one recent study found that a CGRP receptor antagonist effectively treated migraine headaches, which occur when cerebral blood vessels dilate (vasodilation). The ability of one gene to produce multiple peptides is one reason research is shifting from genomics to physiology and *proteomics* (the study of the role of proteins in physiological function).

Some endocrine structures that are important in lower vertebrates are *vestigial* {*vestigium,* trace} in humans, meaning that in humans these structures are present as minimally functional glands. For example, *melanocyte-stimulating hormone* (MSH) from the intermediate lobe of the pituitary controls pigmentation in reptiles and amphibians. However, adult humans have only a vestigial intermediate lobe and normally do not have measurable levels of MSH in their blood.

In the research arena, *comparative endocrinology*—the study of endocrinology in nonhuman organisms—has made significant contributions to our quest to understand the human body. Many of our models of human physiology are based on research carried out in fish or frogs or rats, to name a few. For example, the pineal gland hormone *melatonin* (■ Fig. 7.16) was discovered through research using tadpoles. Many small nonhuman vertebrates have short life cycles that facilitate studying aging or reproductive physiology. Genetically altered mice (transgenic or knockout mice) have provided researchers valuable information about proteomics.

Opponents of animal research argue that scientists should not experiment with animals at all and should use only cell cultures and computer models. Cell cultures and models are valuable tools and can be helpful in the initial stages of medical research, but at some point new drugs and procedures must be tested on intact organisms prior to clinical trials in humans. Responsible scientists follow guidelines for appropriate animal use and limit the number of animals killed to the minimum needed to provide valid data.

In this chapter we have examined how the endocrine system with its hormones helps regulate the slower processes in the body. As you will see, the nervous system takes care of the more rapid responses needed to maintain homeostasis.

RUNNING PROBLEM CONCLUSION

Graves' Disease

In this running problem, you learned that in Graves' disease, thyroid hormone levels are high because an immune-system protein mimics TSH. You also learned that the thyroid gland concentrates iodine for synthesis of thyroid hormones and that radioactive iodine can concentrate in the gland and destroy the thyroid cells. Ben Crenshaw's treatment for Graves' disease was successful. He went on to win the Masters Tournament for a second time in 1995 and he still plays golf professionally today.

Graves' disease is the most common form of hyperthyroidism. Other famous people who have suffered from it include former U.S. President George H. W. Bush and First Lady Barbara Bush. To learn more about Graves' disease and other thyroid conditions, visit the Endocrine Society's Hormone Foundation web site at *www.hormone.org* or the American Thyroid Association at *www.thyroid.org*. Check your answers to the problem questions by comparing them to the information in the summary table below.

Question	Facts	Integration and Analysis
1a. To which of the three classes of hormones do thyroid hormones belong?	The three classes of hormones are peptides, steroids, and amino-acid derivatives.	Thyroid hormones are made from the amino acid tyrosine; therefore, they are amino-acid derivatives.
1b. If a person's diet is low in iodine, predict what happens to thyroxine production.	The thyroid gland concentrates iodine and combines it with the amino acid tyrosine to make thyroid hormones.	If iodine is lacking in the diet, a person is unable to make thyroid hormones.
2a. In a normal person, when thyroid hormone levels in the blood increase, will negative feedback increase or decrease the secretion of TSH?	Negative feedback shuts off response loops.	Normally negative feedback decreases TSH secretion.
2b. In a person with a hyperactive gland that is producing too much thyroid hormone, would you expect the level of TSH to be higher or lower than in a normal person?		If thyroid hormone is high, you would expect strong negative feedback and even lower levels of TSH.
3. Why is radioactive iodine (rather than some other radioactive element, such as cobalt) used to destroy thyroid tissue?	The thyroid gland concentrates iodine to make thyroid hormones.	Radioactive iodine is concentrated in the thyroid gland and therefore selectively destroys that tissue. Other radioactive elements distribute more widely throughout the body and may harm normal tissues.
4. If levels of TSH are low and thyroxine levels are high, is Graves' disease a primary disorder or a secondary disorder (one that arises as a result of a problem with the anterior pituitary or the hypothalamus)? Explain your answer.	In secondary hypersecretion disorders, you would expect the levels of the hypothalamic and/or anterior pituitary trophic hormones to be elevated.	In Graves' disease, TSH from the anterior pituitary is very low. Therefore, the oversecretion of thyroid hormones is not the result of elevated TSH. This means that Graves' disease is a primary disorder that is caused by a problem in the thyroid gland itself.

RUNNING PROBLEM CONCLUSION *(continued)*

Question	Facts	Integration and Analysis
5. Antibodies are proteins that bind to the TSH receptor. From that information, what can you conclude about the cellular location of the TSH receptor?	Receptors may be membrane receptors or intracellular receptors. Proteins cannot cross the cell membrane.	The TSH receptor is a membrane receptor. It uses the cAMP second messenger pathway for signal transduction.
6. In Graves' disease, why doesn't negative feedback shut off thyroid hormone production before it becomes excessive?	In normal negative feedback, increasing levels of thyroid hormone shut off TSH secretion. Without TSH stimulation, the thyroid stops producing thyroid hormone.	In Graves' disease, high levels of thyroid hormone have shut off endogenous TSH production. However, the thyroid gland still produces hormone in response to the binding of antibody to the TSH receptor. In this situation, negative feedback fails to correct the problem.

207 216 224 226 229 230 232

Chapter Summary

This chapter introduced you to the endocrine system and the role it plays in *communication* and *control* of physiological processes. As you've seen before, the *compartmentalization of the body* into intracellular and extracellular compartments means that special mechanisms are required to enable signals to pass from one compartment to the other. The chapter also presented basic patterns that you will encounter again as you study various organ systems: differences among the three chemical classes of hormones, reflex pathways for hormones, types of hormone interactions, and endocrine pathologies.

Hormones

iP Endocrine System: Endocrine System Review

1. The specificity of a hormone depends on its receptors and their associated signal transduction pathways. (p. 208)
2. A **hormone** is a chemical secreted by a cell or group of cells into the blood for transport to a distant target, where it is effective at very low concentrations. (p. 208)
3. **Pheromones** are chemical signals secreted into the external environment. (p. 209)
4. Hormones bind to receptors to initiate responses known as the **cellular mechanism of action.** (p. 209)
5. Hormone activity is limited by terminating secretion, removing hormone from the blood, or terminating activity at the target cell. (p. 209)
6. The rate of hormone breakdown is indicated by a hormone's **half-life.** (p. 211)

The Classification of Hormones

iP Endocrine System: Biochemistry, Secretion and Transport of Hormones, and the Actions of Hormones on Target Cells

7. There are three types of hormones: **peptide/protein hormones,** composed of three or more amino acids; **steroid hormones,** derived from cholesterol; and **amino acid–derived hormones,** derived from either tyrosine (e.g., catecholamines and thyroid hormones) or tryptophan (e.g., melatonin). (p. 212; Tbl. 7.1)
8. Peptide hormones are made as inactive **preprohormones** and processed to **prohormones.** Prohormones are chopped into active hormone and peptide fragments that are co-secreted. (p. 213; Fig. 7.3)

9. Peptide hormones dissolve in the plasma and have a short half-life. They bind to surface receptors on their target cells and initiate rapid cellular responses through signal transduction. In some instances, peptide hormones also initiate synthesis of new proteins. (p. 214; Fig. 7.4)
10. Steroid hormones are synthesized as they are needed. They are hydrophobic, and most steroid hormones in the blood are bound to protein carriers. Steroids have an extended half-life. (p. 215; Fig. 7.5)
11. Traditional steroid receptors are inside the target cell, where they turn genes on or off and direct the synthesis of new proteins. Cell response is slower than with peptide hormones. Steroid hormones may bind to membrane receptors and have nongenomic effects. (p. 215; Fig. 7.5)
12. Amine hormones may behave like typical peptide hormones or like a combination of a steroid hormone and a peptide hormone. (p. 217; Fig. 7.6)

Control of Hormone Release

iP Endocrine System: The Hypothalamic-Pituitary Axis

13. Classic endocrine cells act as both sensor and integrating center in the simple reflex pathway. (p. 218; Fig. 7.7)
14. Many endocrine reflexes involve the nervous system, either through **neurohormones** or through neurons that influence hormone release. (p. 216)
15. The pituitary gland is composed of the anterior pituitary (a true endocrine gland) and the posterior pituitary (an extension of the brain). (p. 220; Fig. 7.8a)
16. The posterior pituitary releases two neurohormones, oxytocin and vasopressin, that are made in the hypothalamus. (p. 220; Fig. 7.8c)
17. **Trophic hormones** control the secretion of other hormones. (p. 221)
18. Hypothalamic releasing hormones and inhibiting hormones control the secretion of anterior pituitary hormones. (p. 222; Fig. 7.9)
19. The hypothalamic trophic hormones reach the pituitary through the **hypothalamic-hypophyseal portal system.** (p. 222; Fig. 7.9)
20. There are six anterior pituitary hormones: prolactin, growth hormone, follicle-stimulating hormone, luteinizing hormone, thyroid-stimulating hormone, and adrenocorticotrophic hormone. (p. 222; Fig. 7.9)
21. In complex endocrine reflexes, hormones of the pathway act as negative feedback signals. (p. 225; Fig. 7.11)

Hormone Interactions

22. If the combination of two or more hormones yields a result that is greater than additive, the interaction is **synergism.** (p. 226; Fig. 7.12)
23. If one hormone cannot exert its effects fully unless a second hormone is present, the second hormone is said to be **permissive** to the first. (p. 226)
24. If one hormone opposes the action of another, the two are **antagonistic** to each other. (p. 227)

Endocrine Pathologies

25. Diseases of hormone excess are usually due to **hypersecretion.** Symptoms of hormone deficiency occur when too little hormone is secreted (**hyposecretion**). **Abnormal tissue responsiveness** may result from problems with hormone receptors or signal transduction pathways. (pp. 227–228)
26. **Primary pathologies** arise in the last endocrine gland in a reflex. A **secondary pathology** is a problem with one of the tissues producing trophic hormones. (p. 229; Fig. 7.14)

Hormone Evolution

27. Many human hormones are similar to hormones found in other vertebrate animals. (pp. 230–232)

Questions

Answers: p. A-1

Level One Reviewing Facts and Terms

1. The study of hormones is called ________.
2. List the three basic ways hormones act on their target cells.
3. List five endocrine glands, and name one hormone secreted by each. Give one effect of each hormone you listed.
4. Match the following researchers with their experiments:

(a) Lower (b) Berthold (c) Guillemin and Shalley (d) Brown-Séquard (e) Banting and Best	1. isolated trophic hormones from the hypothalami of pigs and sheep 2. claimed sexual rejuvenation after injections of testicular extracts 3. isolated insulin 4. accurately described the function of the pituitary gland 5. studied comb development in castrated roosters

5. Put the following steps for identifying an endocrine gland in order:
   (a) Purify the extracts and separate the active substances.
   (b) Perform replacement therapy with the gland or its extracts and see if the abnormalities disappear.
   (c) Implant the gland or administer the extract from the gland to a normal animal and see if symptoms characteristic of hormone excess appear.
   (d) Put the subject into a state of hormone deficiency by removing the suspected gland, and monitor the development of abnormalities.
6. For a chemical to be defined as a hormone, it must be secreted into the ________ for transport to a(n) ________ and take effect at ________ concentrations.
7. What is meant by the term *half-life* in connection with the activity of hormone molecules?
8. Metabolites are inactivated hormone molecules, broken down by enzymes found primarily in the ________ and ________, to be excreted in the ________ and ________, respectively.

9. Candidate hormones often have the word ________ as part of their name.
10. List and define the three chemical classes of hormones. Name one hormone in each class.
11. Decide if each of the following characteristics applies best to peptide hormones, steroid hormones, both classes, or neither class.
    (a) are lipophobic and must use a signal transduction system
    (b) have a short half-life, measured in minutes
    (c) often have a lag time of 90 minutes before effects are noticeable
    (d) are water-soluble, and thus easily dissolve in the extracellular fluid for transport
    (e) most hormones belong to this class
    (f) are all derived from cholesterol
    (g) consist of three or more amino acids linked together
    (h) are released into the blood to travel to a distant target organ
    (i) are transported in the blood bound to protein carrier molecules
    (j) are all lipophilic, so diffuse easily across membranes
12. Why do steroid hormones usually take so much longer to act than peptide hormones?
13. When steroid hormones act on a cell nucleus, the hormone-receptor complex acts as a(n) ________ factor, binds to DNA, and activates one or more ________, which create mRNA to direct the synthesis of new ________.
14. Researchers have discovered that some cells have additional steroid hormone receptors on their ________, enabling a faster response.
15. Melatonin is made from the amino acid ________, and the catecholamines and thyroid hormones are made from the amino acid ________.
16. A hormone that controls the secretion of another hormone is known as a(n) ________ hormone.
17. In reflex control pathways involving trophic hormones and multiple integrating centers, the hormones themselves act as ________ ________ signals, suppressing trophic hormone secretion earlier in the reflex.
18. What characteristic defines neurohormones?
19. List the two hormones secreted by the posterior pituitary gland. To what chemical class do they belong?
20. What is the hypothalamic-hypophyseal portal system? Why is it important?
21. List the six hormones of the anterior pituitary gland; give an action of each. Which ones are trophic hormones?
22. How do long-loop negative feedback and short-loop negative feedback differ? Give an example of each type in the body's endocrine system.
23. When two hormones work together to create a result that is greater than additive, that interaction is called ________. When two hormones must both be present to achieve full expression of an effect, that interaction is called ________. When hormone activities oppose each other, that effect is called ________.

Level Two Reviewing Concepts

24. Compare and contrast the terms in each of the following sets:
    (a) paracrine, hormone, cytokine
    (b) primary and secondary endocrine pathologies
    (c) hypersecretion and hyposecretion
    (d) anterior and posterior pituitary
25. Compare and contrast the three chemical classes of hormones.
26. Map the following groups of terms. Add terms if you like.

List 1	List 2
co-secretion	ACTH
endoplasmic reticulum	anterior pituitary
exocytosis	blood
Golgi complex	endocrine cell
hormone receptor	gonadotropins
peptide hormone	growth hormone
preprohormone	hypothalamus
prohormone	inhibiting hormone
secretory vesicle	neurohormone
signal sequence	neuron
synthesis	oxytocin
target cell response	peptide/protein
	portal system
	posterior pituitary
	prolactin
	releasing hormone
	trophic hormone
	TSH
	vasopressin

Level Three Problem Solving

27. The terms *specificity, receptors,* and *down-regulation* can be applied to many physiological situations. Do their meanings change when applied to the endocrine system? What chemical and physical characteristics do hormones, enzymes, transport proteins, and receptors have in common that makes specificity important?
28. Dexamethasone is a drug used to suppress the secretion of adrenocorticotrophic hormone (ACTH) from the anterior pituitary. Two patients with hypersecretion of cortisol are given dexamethasone. Patient A's cortisol secretion falls to normal as a result, but patient B's cortisol secretion remains elevated. Draw maps of the reflex pathways for these two patients (see Fig. 7.11b for a template) and use the maps to determine which patient has primary hypercortisolism. Explain your reasoning.
29. Some early experiments for male birth control pills used drugs that suppressed gonadotropin (FSH and LH) release. However, men given these drugs stopped taking them because the drugs decreased testosterone secretion, which decreased the men's sex drive and caused impotence.
    (a) Use the information given in Figure 7.9 to draw the GnRH-FSH/LH-testosterone reflex pathway. Use the pathway to show how suppressing gonadotropins decreases sperm production and testosterone secretion.
    (b) Researchers subsequently suggested that a better treatment would be to give men extra testosterone. Draw another copy of the reflex pathway to show how testosterone could suppress sperm production without the side effect of impotence.

Level Four Quantitative Problems

30. The following graph represents the disappearance of a drug from the blood as the drug is metabolized and excreted. Based on the graph, what is the half-life of the drug?

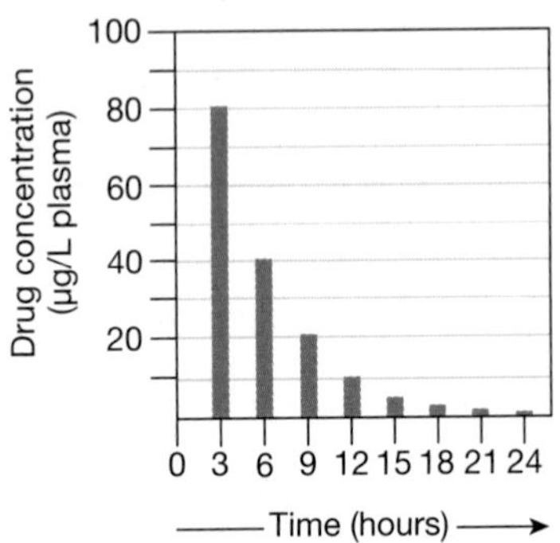

31. The following graph shows plasma TSH concentration in three groups of subjects. Which pattern would be consistent with the following pathologies? Explain your reasoning.
    (a) primary hypothyroidism
    (b) primary hyperthyroidism
    (c) secondary hyperthyroidism

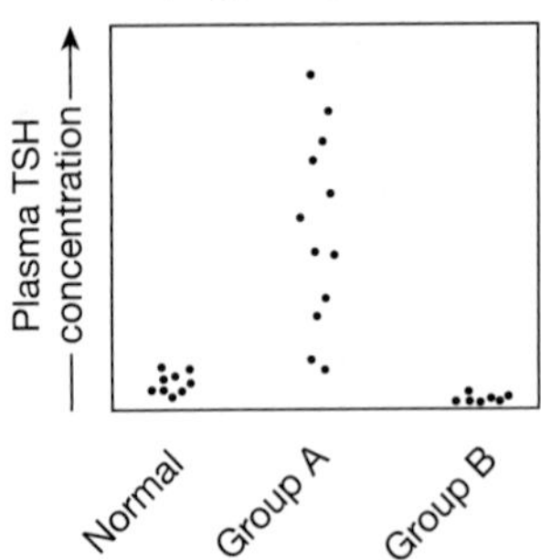

32. Based on what you have learned about the pathway for insulin secretion, draw and label a graph showing the effect of plasma glucose concentration on insulin secretion.

Answers

Answers to Concept Check Questions

Page 209

1. Glucose enters cells by facilitated diffusion (GLUT transporters).

Page 211

2. The suffix *-ase* indicates an enzyme. A *peptidase* digests peptides.

Page 212

3. A hormone is a chemical that is secreted into the blood and acts on a distant target in very low concentrations.
4. A steroid-producing cell would have extensive smooth endoplasmic reticulum; a protein-producing cell would have lots of rough endoplasmic reticulum and secretory vesicles.

Page 216

5. The three chemical classes of hormones are peptide, steroid, and amino acid–derived.
6. The short half-life suggests that aldosterone is not bound to plasma proteins as much as other steroid hormones are.

Page 219

7. Increased blood glucose is the stimulus. Insulin secretion is the efferent pathway; decrease in blood glucose is the response.
8. Insulin release by blood glucose is a simple endocrine reflex. Insulin release in response to a digestive hormone is the complex endocrine reflex. Insulin release triggered by a neural signal following a meal is the neural-endocrine reflex.
9. Stimulus: decreased blood glucose; sensor/integrating center: pancreatic endocrine cells; efferent path: glucagon; target: multiple target tissues; response: increased blood glucose.

Page 219

10. Catecholamines are amino acid–derived hormones.

Page 221

11. Microtubules of the cytoskeleton move secretory vesicles.
12. Contents of secretory vesicles are released by exocytosis.

Page 223

13. (a) pathway 4; (b) pathway 4 for GH acting directly on targets, and pathway 5 for GH acting on the liver.
14. The target is endocrine cells of the anterior pituitary.

Answers to Figure Questions

Page 217

Figure 7.6: The conversion of tyrosine to dopamine adds a hydroxyl (—OH) group to the 6-carbon ring and changes the carboxyl (—COOH) group to a hydrogen. Norepinephrine is made from dopamine by changing one hydrogen to a hydroxyl group. Epinephrine is made from norepinephrine by changing one hydrogen attached to the nitrogen to a methyl (—CH_3) group.

Page 218

Figure 7.7: The pathway begun by eating a meal shuts off when the stretch stimulus disappears as the meal is digested and absorbed from the digestive tract.

Page 225

Figure 7.11: In short-loop negative feedback, ACTH feeds back to inhibit hypothalamic release of CRH.

Page 230

Figure 7.15: (a) CRH high, ACTH low, cortisol low. No negative feedback loops are functioning. (b) CRH normal/high, ACTH high, cortisol low. Absence of negative feedback by cortisol increases trophic hormones. Short-loop negative feedback from ACTH may keep CRH within the normal range.

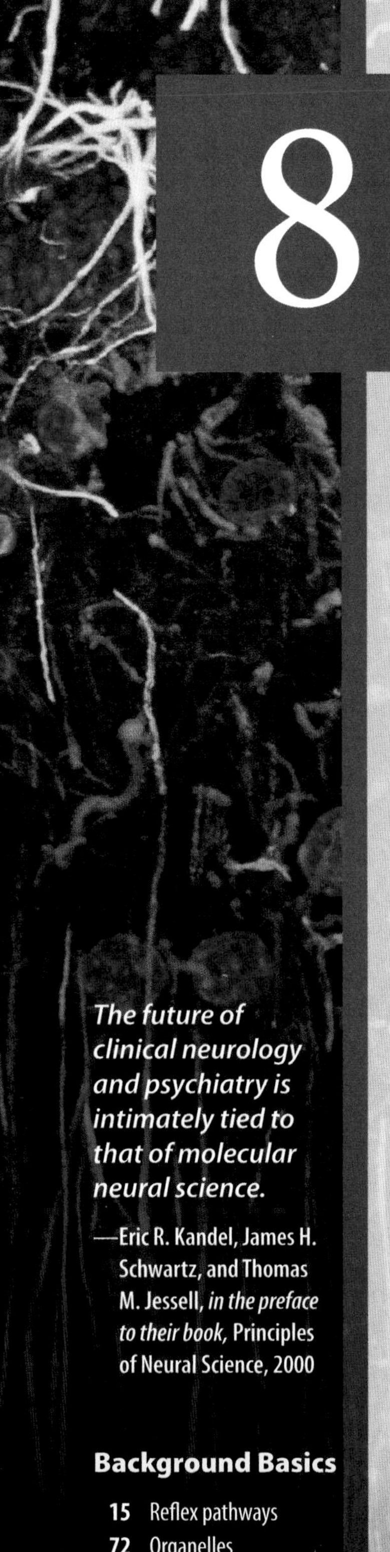

8 Neurons: Cellular and Network Properties

The future of clinical neurology and psychiatry is intimately tied to that of molecular neural science.

—Eric R. Kandel, James H. Schwartz, and Thomas M. Jessell, *in the preface to their book,* Principles of Neural Science, 2000

Background Basics

Purkinje cells (red) and glial cells (green) in the cerebellum

In an eerie scene from a science fiction movie, white-coated technicians move quietly through a room filled with bubbling cylindrical fish tanks. As the camera zooms in on one tank, no fish can be seen darting through aquatic plants. The lone occupant of the tank is a gray mass with a convoluted surface like a walnut and a long tail that appears to be edged with beads. Floating off the beads are hundreds of fine fibers, waving softly as the oxygen bubbles weave through them. This is no sea creature. . . . It is a brain and spinal cord, removed from its original owner and awaiting transplantation into another body. Can this be real? Is this scenario possible? Or is it just the creation of an imaginative movie screenwriter?

The brain is regarded as the seat of the soul, the mysterious source of those traits that we think of as setting humans apart from other animals. The brain and spinal cord are also integrating centers for homeostasis, movement, and many other body functions. They are the control center of the **nervous system**, a network of billions or trillions of nerve cells linked together in a highly organized manner to form the rapid control system of the body.

Nerve cells, or **neurons**, carry electrical signals rapidly and, in some cases, over long distances. They are uniquely shaped cells, and most have long, thin extensions, or **processes**, that can extend up to a meter in length. In most pathways, neurons release chemical signals, called **neurotransmitters,** into the extracellular fluid to communicate with neighboring cells. In a few pathways, neurons are linked by *gap junctions* [p. 175], allowing electrical signals to pass directly from cell to cell.

Using electrical signals to release chemicals from a cell is not unique to neurons. For example, pancreatic beta cells generate an electrical signal to initiate exocytosis of insulin-containing storage vesicles [p. 166]. Single-celled protozoa and plants also employ electrical signaling mechanisms, in many cases using the same types of ion channels as vertebrates do.

RUNNING PROBLEM

Mysterious Paralysis

"Like a polio ward from the 1950s" is how Guy McKhann, M.D., a neurology specialist at the Johns Hopkins School of Medicine, describes a ward of Beijing Hospital that he visited on a trip to China in 1986. Dozens of paralyzed children—some attached to respirators to assist their breathing—filled the ward to overflowing. The Chinese doctors thought the children had Guillain-Barré syndrome (GBS), a rare paralytic condition, but Dr. McKhann wasn't convinced. There were simply too many stricken children for the illness to be the rare Guillain-Barré syndrome. Was it polio—as some of the Beijing staff feared? Or was it another illness, perhaps one that had not yet been discovered?

238 239 242 264 266 270 281

Scientists sequencing ion channel proteins have found that many of these channel proteins have been highly conserved during evolution, indicating their fundamental importance.

Although electrical signaling is universal, sophisticated neural networks are unique to animal nervous systems. Reflex pathways in the nervous system do not necessarily follow a straight line from one neuron to the next. One neuron may influence multiple neurons, or many neurons may affect the function of a single neuron. The intricacy of neural networks and their neuronal components underlies the emergent properties of the nervous system. **Emergent properties** are complex processes, such as consciousness, intelligence, and emotion, that cannot be predicted from what we know about the properties of individual nerve cells and their specific connections. The search to explain emergent properties makes neuroscience one of the most active research areas in physiology today.

Neuroscience, like many other areas of science, has its own specialized language. In many instances, multiple terms describe a single structure or function, which potentially can lead to confusion. ■ Table 8.1 lists some neuroscience terms used in this book, along with their common synonyms, which you may encounter in other publications.

Table 8.1 Synonyms in Neuroscience

Term Used in This Book	Synonym(s)
Action potential	AP, spike, nerve impulse, conduction signal
Autonomic nervous system	Visceral nervous system
Axon	Nerve fiber
Axonal transport	Axoplasmic flow
Axon terminal	Synaptic knob, synaptic bouton, presynaptic terminal
Axoplasm	Cytoplasm of an axon
Cell body	Cell soma, nerve cell body
Cell membrane of an axon	Axolemma
Glial cells	Neuroglia, glia
Interneuron	Association neuron
Rough endoplasmic reticulum	Nissl substance, Nissl body
Sensory neuron	Afferent neuron, afferent

Organization of the Nervous System

The nervous system can be divided into two parts (■ Fig. 8.1). The **central nervous system (CNS)** consists of the **brain** and the **spinal cord**. The **peripheral nervous system (PNS)** consists of **sensory (afferent) neurons** and **efferent neurons**. Information flow through the nervous system follows the basic pattern of a reflex: stimulus → sensor → input signal → integrating center → output signal → target → response [p. 15].

Sensory receptors throughout the body continuously monitor conditions in the internal and external environments. These sensors send information along sensory neurons to the CNS, which is the integrating center for neural reflexes. CNS neurons integrate information that arrives from the sensory division of the PNS and determine whether a response is needed.

If a response is needed, the CNS sends output signals that travel through efferent neurons to their targets, which are mostly muscles and glands. Efferent neurons subdivide into the **somatic motor division**, which controls skeletal muscles, and the **autonomic division**, which controls smooth and cardiac muscles, exocrine glands, some endocrine glands, and some types of adipose tissue. Terminology used to describe efferent neurons can be confusing. The expression *motor neuron* is sometimes used to refer to all efferent neurons. However, clinically, the term *motor neuron* (or *motoneuron*) is often used to describe somatic motor neurons that control skeletal muscles.

The autonomic division of the PNS is also called the *visceral nervous system* because it controls contraction and secretion in the various internal organs {*viscera*, internal organs}. Autonomic neurons are further divided into **sympathetic** and **parasympathetic branches** which can be distinguished by their anatomical organization and by the chemicals they use to communicate with their target cells. Many internal organs receive innervation from both types of autonomic neurons, and it is common for the two divisions to exert *antagonistic control* over a single target [p. 192].

In recent years, a third division of the nervous system has received considerable attention. The **enteric nervous system** is a network of neurons in the walls of the digestive tract. It is frequently controlled by the autonomic division of the nervous system, but it is also able to function autonomously as its own integrating center. You will learn more about the enteric nervous system when you study the digestive system.

It is important to note that the CNS can initiate activity without sensory input, such as when you decide to text a friend. Also, the CNS need not create any measurable output to the efferent divisions. For example, thinking and dreaming are complex higher-brain functions that can take place totally within the CNS.

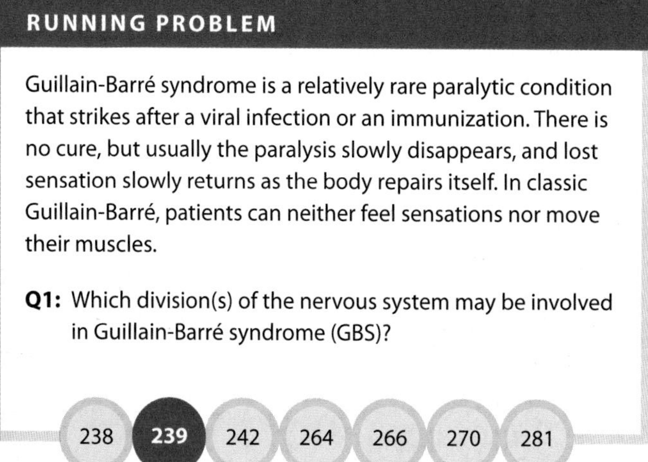
RUNNING PROBLEM
Guillain-Barré syndrome is a relatively rare paralytic condition that strikes after a viral infection or an immunization. There is no cure, but usually the paralysis slowly disappears, and lost sensation slowly returns as the body repairs itself. In classic Guillain-Barré, patients can neither feel sensations nor move their muscles.
Q1: Which division(s) of the nervous system may be involved in Guillain-Barré syndrome (GBS)?
238
239
242
264
266
270
281

Concept Check Answer: p. 286

1. Organize the following terms describing functional types of neurons into a map or outline: afferent, autonomic, brain, central, efferent, enteric, parasympathetic, peripheral, sensory, somatic motor, spinal, sympathetic.

Cells of the Nervous System

The nervous system is composed primarily of two cell types: neurons—the basic signaling units of the nervous system—and support cells known as *glial cells* (or glia or neuroglia).

Neurons Carry Electrical Signals

The neuron, or nerve cell, is the functional unit of the nervous system. (A *functional unit* is the smallest structure that can carry out the functions of a system.) Neurons are uniquely shaped cells with long processes that extend outward from the *nerve cell body*. These processes are usually classified as either **dendrites,** which receive incoming signals, or **axons,** which carry outgoing information. The shape, number, and length of axons and dendrites vary from one neuron to the next, but these structures are an essential feature that allows neurons to communicate with one another and with other cells. Neurons may be classified either structurally or functionally (■ Fig. 8.2).

Structurally, neurons are classified by the number of processes that originate from the cell body. The model neuron that is commonly used to teach how a neuron functions is *multipolar*, with many dendrites and branched axons (Fig. 8.2e). Multipolar neurons in the CNS look different from multipolar efferent neurons (Fig. 8.2d). In other structural neuron types, the axons and dendrites may be missing or modified. *Pseudounipolar* neurons have the cell body located off one side of one long process that is called the axon (Fig. 8.2a). (During development, the dendrites fused and became part of the axon.) *Bipolar* neurons

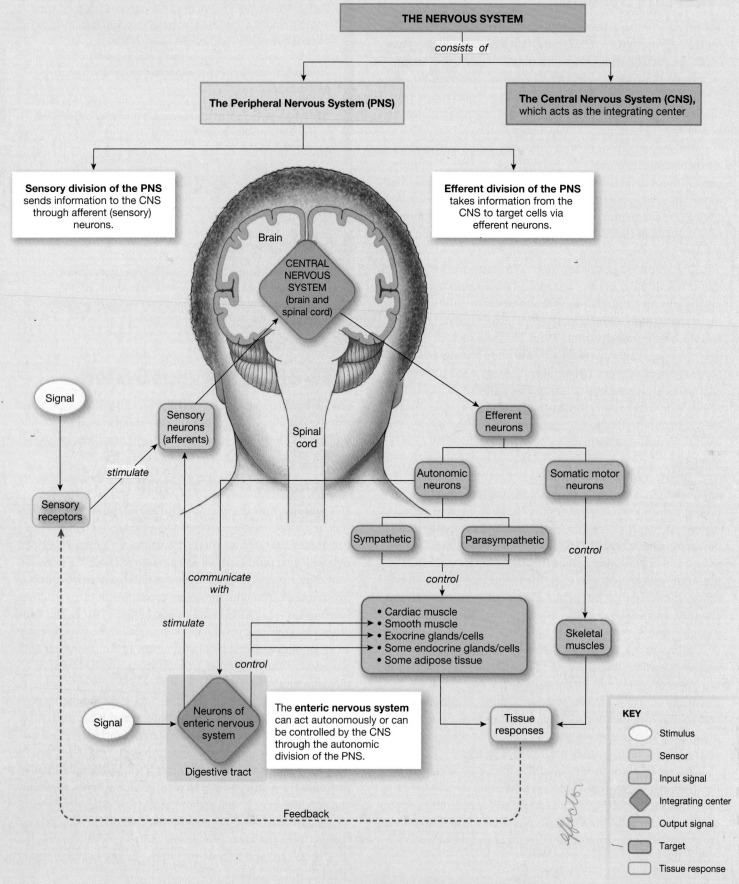

The Organization of the Nervous System

■ Fig. 8.2 *ESSENTIALS*

Neuron Anatomy

Functional Categories

Sensory Neurons

Somatic senses

Neurons for smell and vision

Dendrites

Schwann cell

Axon

Interneurons of CNS

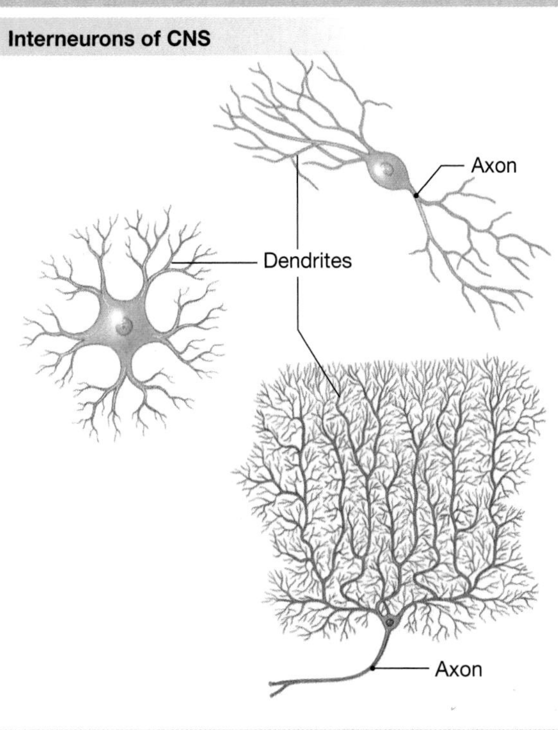

Efferent Neurons

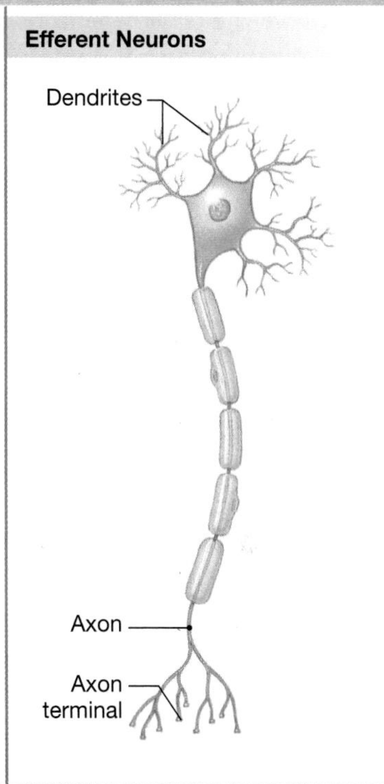

Structural Categories

Pseudounipolar

(a) Pseudounipolar neurons have a single process called the axon. During development, the dendrite fused with the axon.

Bipolar

(b) Bipolar neurons have two relatively equal fibers extending off the central cell body.

Anaxonic

(c) Anaxonic CNS interneurons have no apparent axon.

Multipolar

(d) Multipolar CNS interneurons are highly branched but lack long extensions.

(e) A typical multipolar efferent neuron has five to seven dendrites, each branching four to six times. A single long axon may branch several times and end at enlarged axon terminals.

(f) Parts of a Neuron

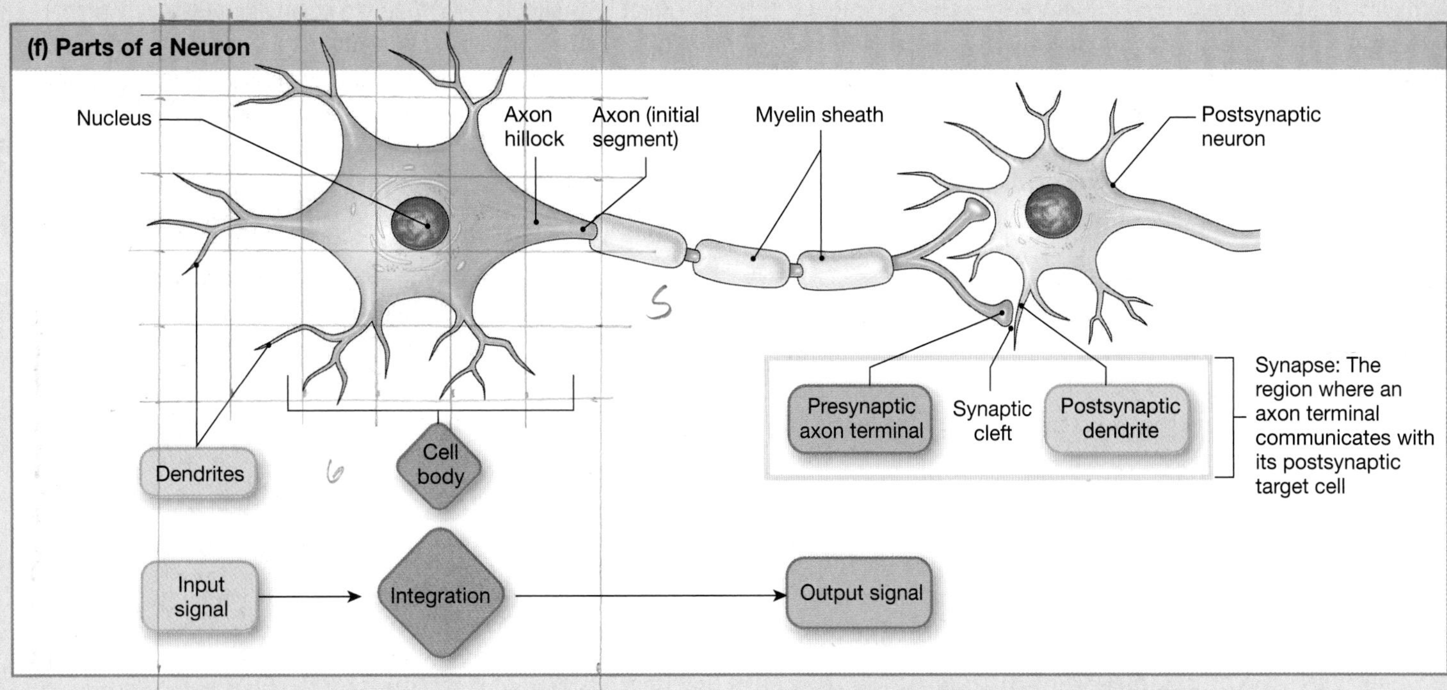

have a single axon and single dendrite coming off the cell body (Fig. 8.2b). *Anaxonic* neurons lack an identifiable axon but have numerous branched dendrites (Fig. 8.2c).

Because physiology is concerned chiefly with function, however, we will classify neurons according to their functions: sensory (afferent) neurons, interneurons, and efferent (somatic motor and autonomic) neurons. Sensory neurons carry information about temperature, pressure, light, and other stimuli from sensory receptors to the CNS. Peripheral sensory neurons are pseudounipolar, with cell bodies located close to the CNS and very long processes that extend out to receptors in the limbs and internal organs. In these sensory neurons, the cell body is out of the direct path of signals passing along the axon (Fig. 8.2a). In contrast, sensory neurons in the nose and eye are much smaller bipolar neurons. Signals that begin at the dendrites travel through the cell body to the axon (Fig. 8.2b).

Neurons that lie entirely within the CNS are known as **interneurons** (short for *interconnecting neurons*). They come in a variety of forms but often have quite complex branching processes that allow them to communicate with many other neurons (Fig. 8.2c, d). Some interneurons are quite small compared to the model neuron.

Efferent neurons, both somatic motor and autonomic, are generally very similar to the neuron in Figure 8.2e. Efferent neurons have enlarged axon terminals. Many autonomic neurons also have enlarged regions along the axon called **varicosities** [see Fig. 11.7, p. 384]. Both axon terminals and varicosities store and release neurotransmitter.

The long axons of both afferent and efferent peripheral neurons are bundled together with connective tissue into cord-like fibers called **nerves** that extend from the CNS to the targets of the component neurons. Nerves that carry afferent signals only are called **sensory nerves**, and those that carry efferent signals only are called **motor nerves.** Nerves that carry signals in both directions are **mixed nerves**. Many nerves are large enough to be seen with the naked eye and have been given anatomical names. For example, the *phrenic nerve* runs from the spinal cord to the muscles of the diaphragm.

The Cell Body Is the Control Center The **cell body** (*cell soma*) of a neuron resembles a typical cell, with a nucleus and all organelles needed to direct cellular activity [p. 70]. An extensive cytoskeleton extends outward into the axon and dendrites. The position of the cell body varies in different types of neurons, but in most neurons the cell body is small, generally making up one-tenth or less of the total cell volume. Despite its small size, the cell body with its nucleus is essential to the well-being of the cell because it contains DNA that is the template for protein synthesis [p. 117].

Dendrites Receive Incoming Signals Dendrites {*dendron*, tree} are thin, branched processes that receive incoming information from neighboring cells (Fig. 8.2f). Dendrites increase the surface area of a neuron, allowing it to communicate with multiple other neurons. The simplest neurons have only a single dendrite. At the other extreme, neurons in the brain may have multiple dendrites with incredibly complex branching (Fig. 8.2d). A dendrite's surface area can be expanded even more by the presence of **dendritic spines** that vary from thin spikes to mushroom-shaped knobs (see Fig. 8.25, p. 279).

The primary function of dendrites in the peripheral nervous system is to receive incoming information and transfer it to an integrating region within the neuron. Within the CNS, dendrite function is more complex. Dendritic spines can function as independent compartments, sending signals back and forth with other neurons in the brain. Many dendritic spines contain polyribosomes and can make their own proteins.

Dendritic spines can change their size and shape in response to input from neighboring cells. Changes in spine morphology are associated with learning and memory as well as with various pathologies, including genetic disorders that cause mental retardation and degenerative diseases such as Alzheimer's disease. Because of these associations, dendritic spines are a hot topic in neuroscience research.

Axons Carry Outgoing Signals Most peripheral neurons have a single axon that originates from a specialized region of the cell body called the **axon hillock** (Fig. 8.2f). Axons vary in length from more than a meter to only a few micrometers. They often branch sparsely along their length, forming **collaterals** {*col-*, with + *lateral*, something on the side}. In our model neuron, each collateral ends in a swelling called an **axon terminal**. The axon terminal contains mitochondria and membrane-bound vesicles filled with *neurocrine* molecules [p. 177].

The primary function of an axon is to transmit outgoing electrical signals from the integrating center of the neuron to the end of the axon. At the distal end of the axon, the electrical signal is usually translated into a chemical message by secretion of a neurotransmitter, neuromodulator, or neurohormone. Neurons that secrete neurotransmitters and neuromodulators terminate near their target cells, which are usually other neurons, muscles, or glands.

RUNNING PROBLEM

In classic Guillain-Barré syndrome, the disease affects both sensory and somatic motor neurons. Dr. McKhann observed that although the Beijing children could not move their muscles, they could feel a pin prick.

Q2: Do you think the paralysis found in the Chinese children affected both sensory (afferent) and somatic motor neurons? Why or why not?

238 239 **242** 264 266 270 281

Concept Check Answers: p. 286

2. Where do neurons that secrete neurohormones terminate?
3. What is the difference between a nerve and a neuron?

Axons are specialized to convey chemical and electrical signals. Their cytoplasm is filled with many types of fibers and filaments but lacks ribosomes and endoplasmic reticulum. For this reason, any proteins destined for the axon or the axon terminal must be synthesized on the rough endoplasmic reticulum in the cell body. The proteins are then moved down the axon by a process known as **axonal transport**.

Slow axonal transport moves material by **axoplasmic** (cytoplasmic) **flow** from the cell body to the axon terminal. Material moves at a rate of only 0.2–2.5 mm/day, which means that slow transport can be used only for components that are not consumed rapidly by the cell, such as enzymes and cytoskeleton proteins.

Fast axonal transport moves organelles at rates of up to 400 mm (about 15.75 in.) per day (■ Fig. 8.3). The neuron uses stationary microtubules as tracks along which transported vesicles and mitochondria "walk" with the aid of attached footlike motor proteins [p. 74]. These motor proteins alternately bind and unbind to the microtubules with the help of ATP, stepping their organelles along the axon in a stop-and-go fashion. The role of motor proteins in axonal transport is similar to their role in muscle contraction and in the movement of chromosomes during cell division.

Fast axonal transport goes in two directions. Forward (or *anterograde*) transport moves synaptic and secretory vesicles and mitochondria from the cell body to the axon terminal. Backward (or *retrograde*) transport returns old cellular components from the axon terminal to the cell body for recycling. There is evidence that nerve growth factors and some viruses also reach the cell body by fast retrograde transport.

8

Establishing Synapses Depends on Chemical Signals

The region where an axon terminal meets its target cell is called a **synapse** {*syn-*, together + *hapsis*, to join}. The neuron that delivers a signal to the synapse is known as the **presynaptic cell**, and the cell that receives the signal is called the **postsynaptic cell** (Fig. 8.2f). The narrow space between the two cells is called the **synaptic cleft.** Although illustrations make the synaptic cleft look like an empty gap, it is filled with extracellular matrix whose fibers hold the presynaptic and postsynaptic cells in position.

During embryonic development, how can more than 100 billion neurons in the brain find their correct targets and make synapses among more than 10 times that many glial cells? How can a somatic motor neuron in the spinal cord find the correct

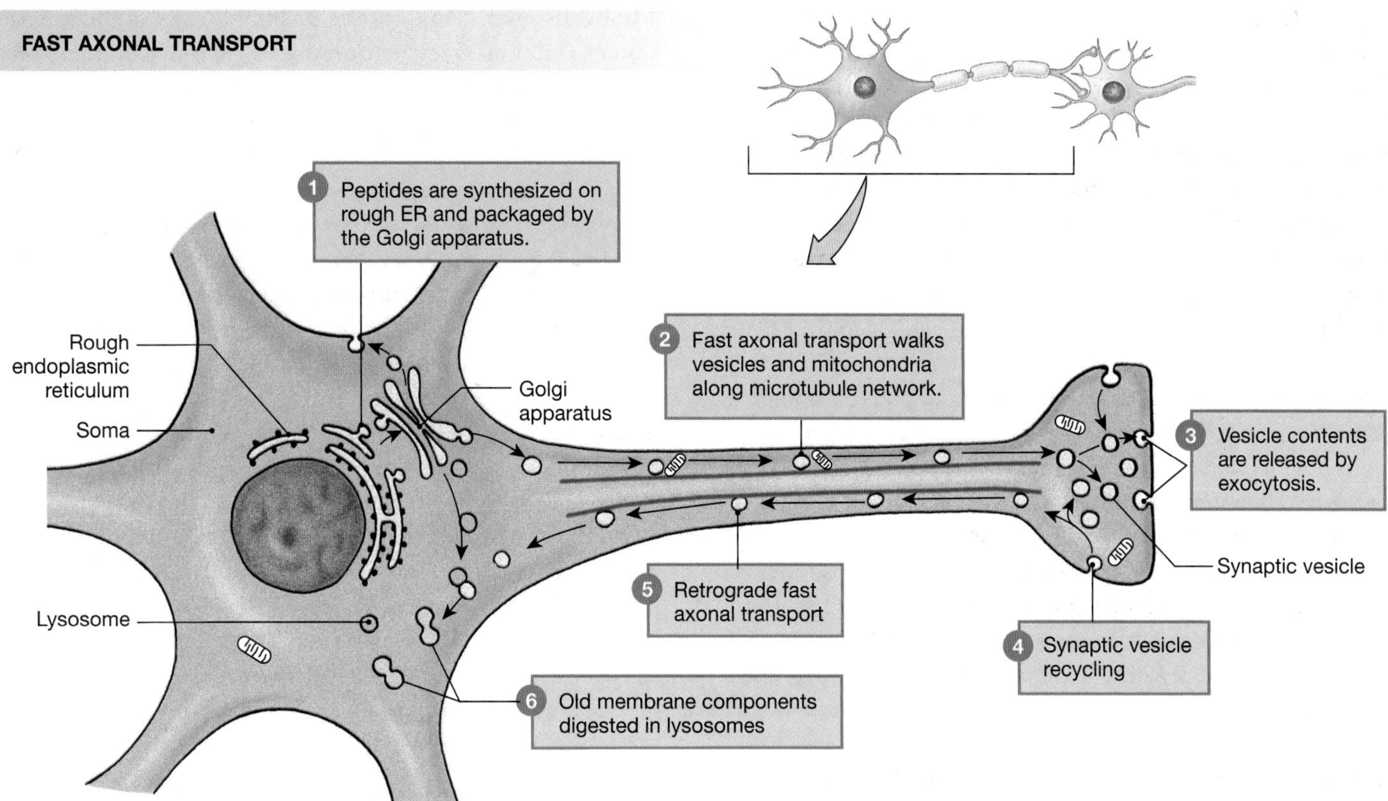

FAST AXONAL TRANSPORT
1 Peptides are synthesized on rough ER and packaged by the Golgi apparatus.
Rough endoplasmic reticulum
Soma
Golgi apparatus
2 Fast axonal transport walks vesicles and mitochondria along microtubule network.
3 Vesicle contents are released by exocytosis.
Synaptic vesicle
4 Synaptic vesicle recycling
5 Retrograde fast axonal transport
6 Old membrane components digested in lysosomes
Lysosome

■ **Fig. 8.3**

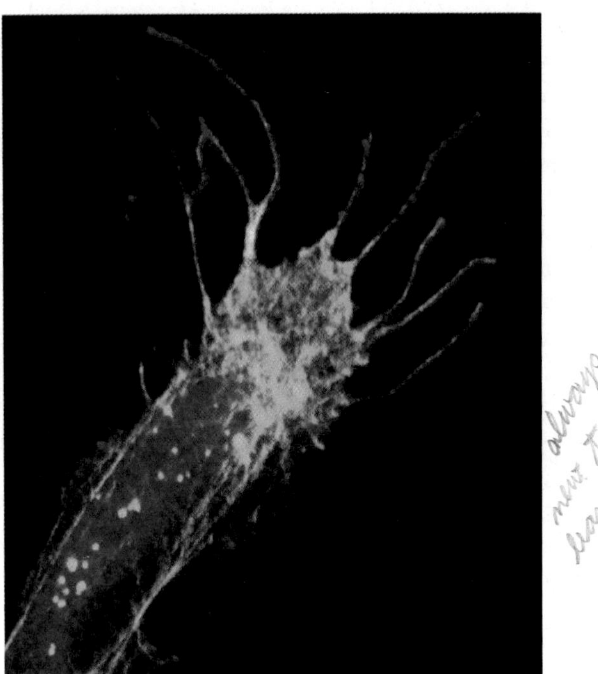

■ **Fig. 8.4** ***The growth cone of a developing axon.***
The growing tip of a developing axon (blue) is a flattened region filled with microtubules (green) and actin filaments (red and yellow) that continuously assemble at their distal ends, extending the tip of the axon as it seeks its target.

pathway to form a synapse with its target muscle in the big toe? The answer lies with the chemical signals used by the developing embryo, ranging from factors that control differentiation of stem cells into neurons and glia to those that direct an elongating axon to its target.

The axons of embryonic nerve cells send out special tips called **growth cones** that extend through the extracellular compartment until they find their target cell (■ Fig. 8.4). In experiments where target cells are moved to an unusual location in the embryo, the axons in many instances are still able to find their targets by "sniffing out" the target's chemical scent. Growth cones depend on many different types of signals to find their way: growth factors, molecules in the extracellular matrix, and membrane proteins on the growth cones and on cells along the path. For example, *integrins* [p. 80] on the growth cone membrane bind to *laminins*, protein fibers in the extracellular matrix. *Nerve-cell adhesion molecules* (NCAMs) [p. 78] interact with membrane proteins of other cells.

Once an axon reaches its target cell, a synapse forms. However, synapse formation must be followed by electrical and chemical activity, or the synapse will disappear. The survival of neuronal pathways depends on **neurotrophic factors** {*trophikos,* nourishment} secreted by neurons and glial cells. There is still much we have to learn about this complicated process, and it is an active area of physiological research.

This "use it or lose it" scenario is most dramatically reflected by the fact that the infant brain is only about one-fourth the size of the adult brain. Further brain growth is due not to an increase in cell number but to an increase in size and number of axons, dendrites, and synapses. This development depends on electrical signaling between sensory pathways, interneurons, and efferent neurons.

Babies who are neglected or deprived of sensory input may experience delayed development ("failure to thrive") because of the lack of nervous system stimulation. On the other hand, there is no evidence that extra stimulation in infancy enhances intellectual development, despite a popular movement to expose babies to art, music, and foreign languages before they can even walk. Once synapses form, they are not fixed for life. Variations in electrical activity can cause rearrangement of the synaptic connections, a process that continues throughout life. Maintaining synapses is one reason that older adults are urged to keep learning new skills and information.

always something new to learn

Concept Check Answer: p. 286

4. Draw a chain of three neurons that synapse on one another in sequence. Label the presynaptic and postsynaptic ends of each neuron, the cell bodies, dendrites, axons, and axon terminals.

Glial Cells Provide Support for Neurons

Glial cells {*glia,* glue} are the unsung heroes of the nervous system, outnumbering neurons by 10–50 to 1. For many years scientists thought that the primary function of glial cells was physical support, and that glial cells had little influence on information processing. That view has changed. Although glial cells do not participate directly in the transmission of electrical signals over long distances, they do communicate with and provide important biochemical support to neurons. The peripheral nervous system has two types of glial cells—Schwann cells and satellite cells—and the CNS has four types: oligodendrocytes, microglia, astrocytes, and ependymal cells (■ Fig. 8.5a).

Myelin-Forming Glia Neural tissue secretes very little extracellular matrix [p. 78], and glial cells provide structural stability to neurons by wrapping around them. **Schwann cells** in the PNS and **oligodendrocytes** in the CNS support and insulate axons by forming **myelin,** a substance composed of multiple concentric layers of phospholipid membrane (Fig. 8.5c). In addition to providing support, the myelin acts as insulation around axons and speeds up their signal transmission.

Myelin forms when the glial cells wrap around an axon, squeezing out the glial cytoplasm so that each wrap becomes two membrane layers (Fig. 8.5d). As an analogy, think of wrapping a deflated balloon tightly around a pencil. Some neurons have as many as 150 wraps (300 membrane layers) in the myelin sheath that surrounds their axons. Gap junctions connect the membrane layers and allow the flow of nutrients and information from layer to layer.

One difference between oligodendrocytes and Schwann cells is the number of axons each cell wraps around. In the CNS, one oligodendrocyte branches and forms myelin around portions of several axons (Fig. 8.5b). In the peripheral nervous system, one Schwann cell associates with one axon.

Schwann Cells A single axon may have as many as 500 different Schwann cells along its length. Each Schwann cell wraps around a 1–1.5 mm segment of the axon, leaving tiny gaps, called the **nodes of Ranvier**, between the myelin-insulated areas (Fig. 8.5c). At each node, a tiny section of axon membrane remains in direct contact with the extracellular fluid. The nodes play an important role in the transmission of electrical signals along the axon.

Satellite Cells The second type of PNS glial cell, the **satellite cell,** is a nonmyelinating Schwann cell (Fig. 8.5a). Satellite cells form supportive capsules around nerve cell bodies located in ganglia. A **ganglion** {cluster or knot} is a collection of nerve cell bodies found outside the CNS. Ganglia appear as knots or swellings along a nerve. (A cluster of nerve cell bodies inside the CNS, the equivalent of a peripheral ganglion, is called a **nucleus** {plural, *nuclei*}.)

Astrocytes **Astrocytes** {*astron*, a star} are highly branched glial cells that by some estimates make up about half of all cells in the brain (Fig. 8.5a, b). They come in several subtypes and form a functional network by communicating with one another through gap junctions. Astrocytes have multiple roles. The terminals of some astrocyte processes are closely associated with synapses, where they take up and release chemicals. Astrocytes also provide neurons with substrates for ATP production, and they help maintain homeostasis in the CNS extracellular fluid by taking up K^+ and water. Finally, the terminals of some astrocyte processes surround blood vessels and become part of the so-called *blood-brain barrier* that regulates the movement of materials between blood and extracellular fluid.

Microglia The glial cells known as **microglia** are specialized immune cells that reside permanently in the CNS (Fig. 8.5a, b). When activated, they remove damaged cells and foreign invaders. However, it now appears that microglia are not always helpful. Activated microglia sometimes release damaging *reactive oxygen species (ROS)* that form free radicals. The *oxidative stress* caused by ROS is believed to contribute to neurodegenerative diseases such as *amyotrophic lateral sclerosis* (*ALS*, also known as Lou Gehrig's disease).

Ependymal Cells The final class of glial cells is the **ependymal cells,** specialized cells that create a selectively permeable epithelial layer, the *ependyma,* that separates the fluid compartments of the CNS (Fig. 8.5a, b). The ependyma is one source of **neural stem cells** [p. 92], immature cells that can differentiate into neurons and glial cells.

8

All glial cells communicate with neurons and with one another primarily through chemical signals. Glial-derived growth and *trophic* (nourishing) factors help maintain neurons and guide them during repair and development. Glial cells in turn respond to neurotransmitters and neuromodulators secreted by neurons. Glial cell function is an active area of neuroscience research, and scientists are still exploring the roles these important cells play in the nervous system.

Concept Check Answers: p. 286

5. What is the primary function of each of the following: myelin, microglia, ependymal cells?
6. Name the two glial cell types that form myelin. How do they differ from each other?

Can Stem Cells Repair Damaged Neurons?

Neurons grow when we are young, but what happens when adult neurons are injured? The responses of mature neurons to injury are similar in many ways to the growth of neurons during development. Both processes rely on a combination of chemical and electrical signals.

When a neuron is damaged, if the cell body dies, the entire neuron dies. If the cell body is intact and only the axon is severed, the cell body and attached segment of axon survive (■ Fig. 8.6). The section of axon separated from the cell body usually degenerates slowly and dies because axons lack the cellular organelles to make essential proteins.

What are the cellular events that follow damage to a neuron? First, the axon cytoplasm leaks out at the injury site until membrane is recruited to seal the opening. The segment of axon still attached to the cell body swells as organelles and filaments brought in by axonal transport accumulate. Schwann cells near the injury site send chemical signals to the cell body to tell it that an injury has occurred.

In the distal segment of the axon, synaptic transmission ceases almost immediately. The axon, deprived of its protein source, slowly begins to collapse. The myelin sheath around the distal axon also begins to unravel. Scavenger microglia or

■ Fig. 8.5 *ESSENTIALS*

Glial Cells

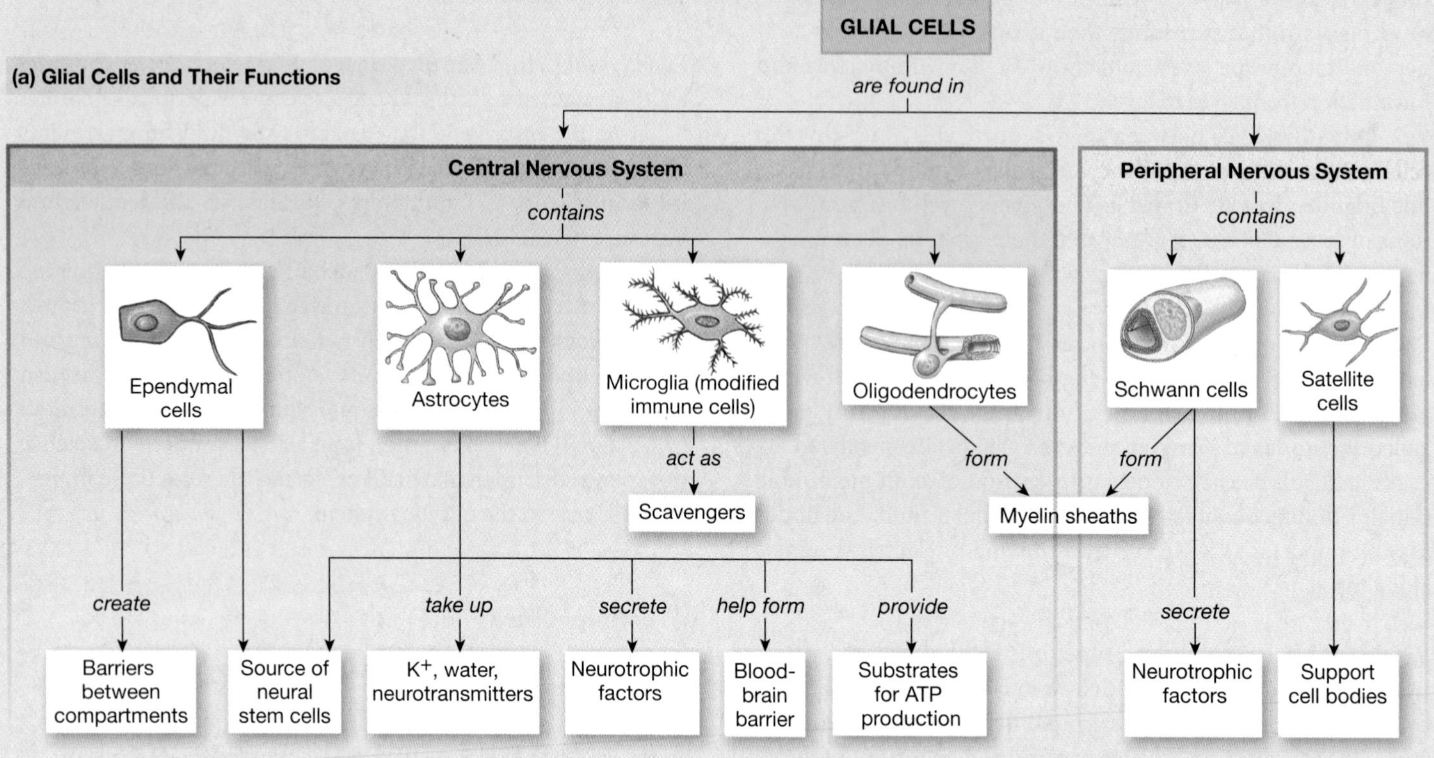

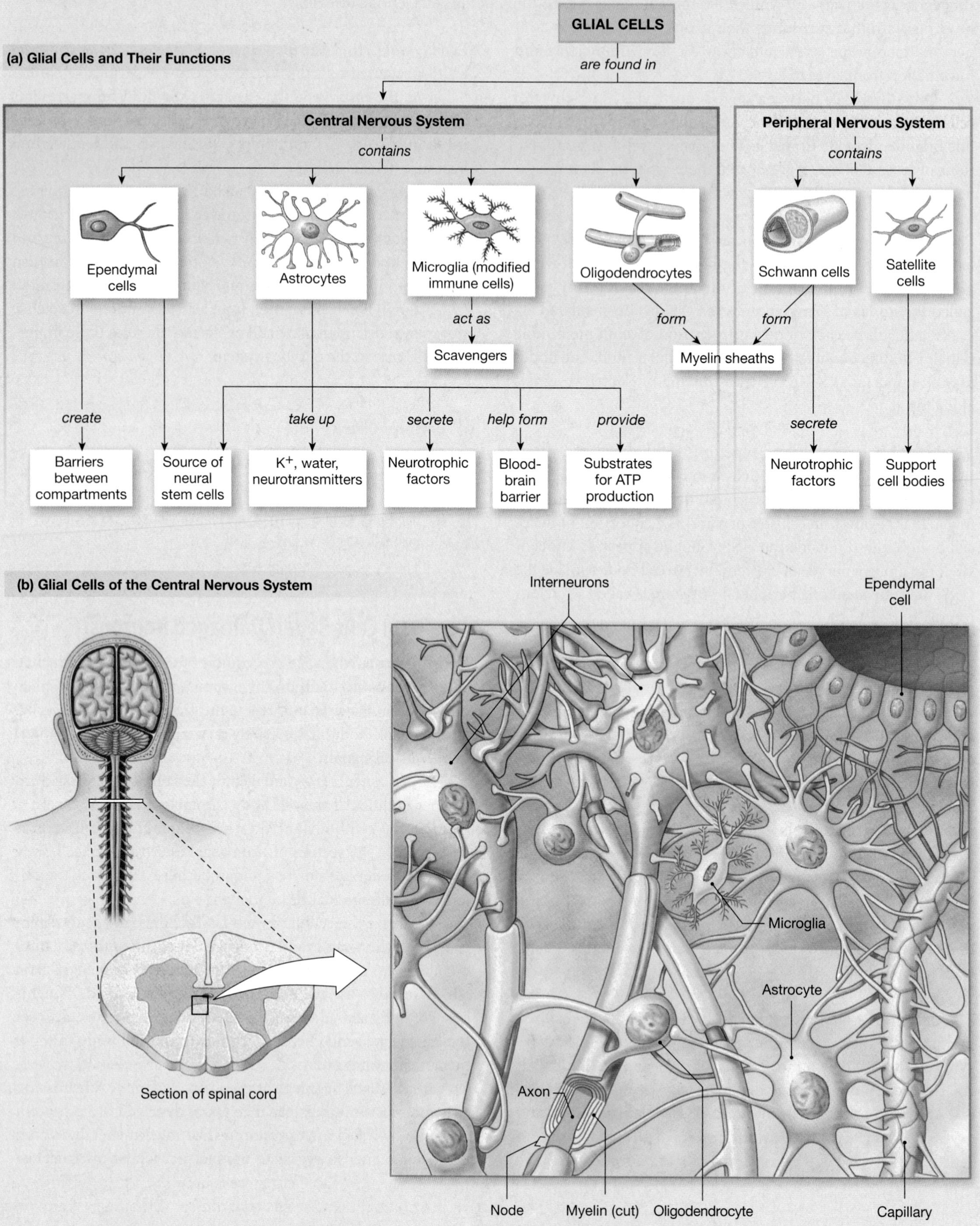

(b) Glial Cells of the Central Nervous System

Section of spinal cord

Interneurons

Ependymal cell

Microglia

Astrocyte

Axon

Node

Myelin (cut)

Oligodendrocyte

Capillary

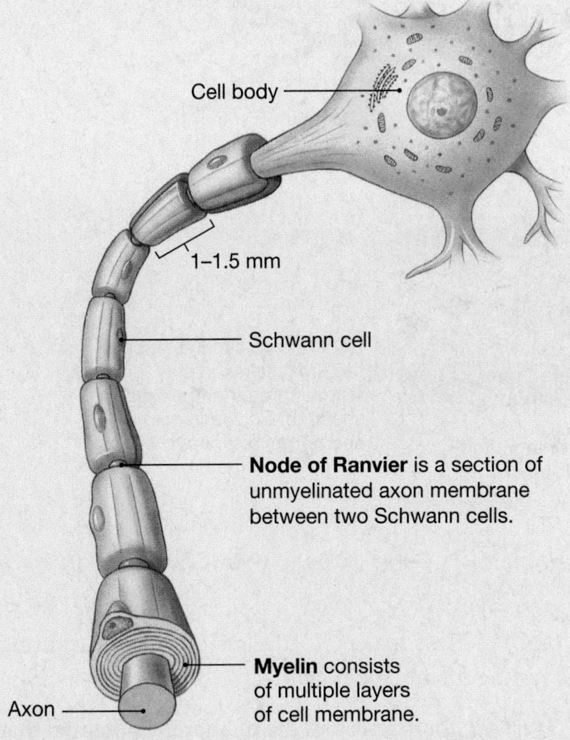
(c) Each Schwann Cell Forms Myelin Around a Small Segment of One Axon.
Cell body
1–1.5 mm
Schwann cell
Node of Ranvier is a section of unmyelinated axon membrane between two Schwann cells.
Myelin consists of multiple layers of cell membrane.
Axon

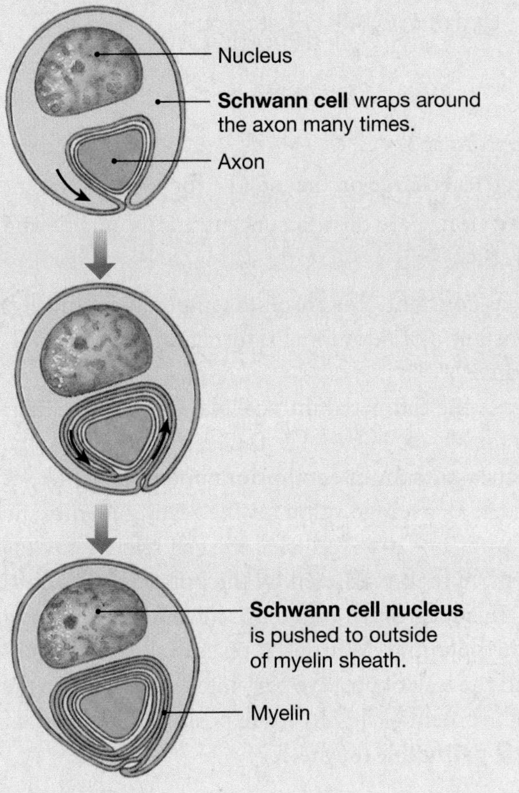
(d) Myelin Formation in the Peripheral Nervous System
Nucleus
Schwann cell wraps around the axon many times.
Axon
Schwann cell nucleus is pushed to outside of myelin sheath.
Myelin

phagocytes ingest and clear away the debris. This process may take a month or longer.

If the severed axon belongs to a somatic motor neuron, death of the distal {distant} axon results in permanent paralysis of the skeletal muscles *innervated* by the neuron. (The term *innervated* means "controlled by a neuron.") If the damaged neuron is a sensory neuron, the person may experience loss of sensation (numbness or tingling) in the region previously innervated by the neuron.

Under some conditions, axons in the peripheral nervous system can regenerate and re-establish their synaptic connections. Schwann cells secrete neurotrophic factors that keep the cell body alive and stimulate regrowth of the axon. The growing tip of a regenerating axon behaves much like the growth cone of a developing axon, following chemical signals in the extracellular matrix along its former path until the axon forms a new synapse with its target cell. Sometimes the loss of the distal axon is permanent, however, and the pathway is destroyed.

Regeneration of axons in the central nervous system is less likely to occur naturally. CNS glial cells tend to seal off and scar the damaged region, and damaged CNS cells secrete factors that inhibit axon regrowth. Many scientists are studying the mechanisms of axon growth and inhibition in the hopes of finding treatments that can restore function to victims of spinal cord injury and degenerative neurological disorders.

Scientists once believed that if a neuron died, it could never be replaced. The discovery of neural stem cells changed that view. During early development, an undifferentiated cell layer called *neuroepithelium* lines the lumen of the neural tube, a structure that will later become the brain and spinal cord. As development proceeds, some cells migrate out of the neuroepithelium and differentiate into neurons. Other cells bordering the lumen of the neural tube specialize into the epithelium of the ependyma. However, among the ependymal cells and in the subependymal layer, some neural stem cells remain unspecialized, waiting until they are called upon to replace damaged cells. Neural stem cells have also been found in other parts of the body, including the *hippocampus* of the brain and the enteric nervous system of the gut.

When neural stem cells receive the correct signals, they transform into neurons and glial cells. Scientists are working intensely to learn how to control this transformation, in the hope that stem cell transplants can reverse the loss of function that comes with degenerative neurological diseases. Most of these studies are being done with mice and rats, but in late 2006 a stem cell transplant into a human brain took place. The patient was a child suffering from Batten disease, a fatal lysosomal enzyme disorder similar to Tay-Sachs [p. 99]. Physicians hoped the transplanted neural stem cells would produce the missing enzymes and slow or stop progression of the disease. Within the year, another five patients with Batten disease also received transplants. Although the stem cells in this first trial did nothing to treat Batten disease, this trial did show that large numbers

PERIPHERAL NEURON INJURY

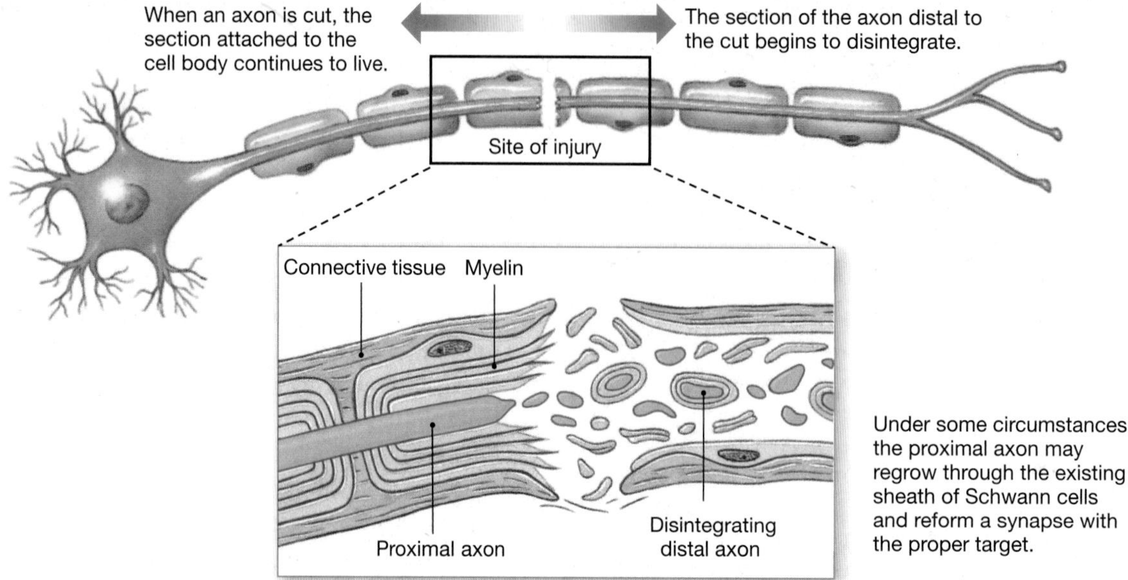

When an axon is cut, the section attached to the cell body continues to live.
The section of the axon distal to the cut begins to disintegrate.
Site of injury
Connective tissue
Myelin
Proximal axon
Disintegrating distal axon
Under some circumstances, the proximal axon may regrow through the existing sheath of Schwann cells and reform a synapse with the proper target.

■ **Fig. 8.6**

of stem cells can be transplanted into the brain without causing adverse effects. Scientists also hope to use neural stem cells to treat Parkinson's disease, a degenerative condition in which dopamine-secreting neurons in the brain die.

Electrical Signals in Neurons

Nerve and muscle cells are described as *excitable tissues* because of their ability to propagate electrical signals rapidly in response to a stimulus. We now know that many other cell types generate electrical signals to initiate intracellular processes [see insulin secretion, p. 166], but the ability of nerve and muscle cells to send a constant electrical signal over long distance is characteristic of electrical signaling in these tissues.

The Nernst Equation Predicts Membrane Potential for a Single Ion

Recall that all living cells have a resting membrane potential difference (V_m) [p. 163] that represents the separation of electrical charge across the cell membrane. Two factors influence the membrane potential:

1. *The uneven distribution of ions* across the cell membrane. Normally, sodium (Na^+), chloride (Cl^-), and calcium (Ca^{2+}) are more concentrated in the extracellular fluid than in the cytosol. Potassium (K^+) is more concentrated in the cytosol than in the extracellular fluid.
2. *Differing membrane permeability to those ions.* The resting cell membrane is much more permeable to K^+ than to Na^+ or Ca^{2+}. This makes K^+ the major ion contributing to the resting membrane potential.

The *Nernst equation* describes the membrane potential that would result if the membrane were permeable to only one ion [p. 165]. For any given ion, this membrane potential is called the *equilibrium potential* of the ion (E_{ion}):

$$E_{ion}(\text{in mV}) = \frac{61}{z} \log \frac{[\text{ion}]_{out}}{[\text{ion}]_{in}}$$

where:

61 is 2.303 *RT/F* at 37 °C

z is the electrical charge on the ion (+1 for K^+), and

$[\text{ion}]_{out}$ and $[\text{ion}]_{in}$ are the ion concentrations outside and inside the cell.

(*R* is the ideal gas constant, *T* is absolute temperature, and *F* is the Faraday constant. For additional information on these values, see Appendix B.)

When we use the estimated intracellular and extracellular concentrations for K^+ (■ Tbl. 8.2) in the Nernst equation, the equation predicts a potassium equilibrium potential, or E_K, of −90 mV. However, an average value for the resting membrane potential of neurons is −70 mV (inside the cell relative to outside), more positive than predicted by the potassium equilibrium potential. This means that other ions must be contributing to the membrane potential. Neurons at rest are slightly permeable to Na^+, and the leak of positive Na^+ into the cell makes the resting membrane potential slightly more positive than it would be if the cell were permeable only to K^+.

8

Ion Concentrations and Equilibrium Potentials

Table 8.2

Ion	Extracellular Fluid (mM)	Intracellular Fluid (mM)	E_{ion} at 37 °C
K^+	5 mM (normal: 3.5–5)	150 mM	−90 mV
Na^+	145 mM (normal: 135–145)	15 mM	+60 mV
Cl^-	108 mM (normal: 100–108)	10 mM (normal: 5–15)	−63 mV
Ca^{2+}	1 mM	0.0001 mM	See Concept Check question 7

Concept Check Answer: p. 286

7. Given the values in Table 8.2, use the Nernst equation to calculate the equilibrium potential for Ca^{2+}. Express the concentrations as powers of 10 and use your knowledge of logarithms [Appendix B] to try the calculations without a calculator.

The GHK Equation Predicts Membrane Potential Using Multiple Ions

In living systems, several different ions contribute to the membrane potential of cells. The **Goldman-Hodgkin-Katz (GHK) equation** calculates the membrane potential that results from the contribution of all ions that can cross the membrane. The GHK equation includes membrane permeability values because the permeability of an ion influences its contribution to the membrane potential. If the membrane is not permeable to an ion, that ion does not affect the membrane potential.

For mammalian cells, we assume that Na^+, K^+, and Cl^- are the three ions that influence membrane potential in resting cells. Each ion's contribution to the membrane potential is proportional to its ability to cross the membrane. The GHK equation for cells that are permeable to Na^+, K^+, and Cl^- is

$$V_m = 61 \log \frac{P_K[K^+]_{out} + P_{Na}[Na^+]_{out} + P_{Cl}[Cl^-]_{in}}{P_K[K^+]_{in} + P_{Na}[Na^+]_{in} + P_{Cl}[Cl^-]_{out}}$$

where:

V_m is the resting membrane potential in mV at 37 °C

61 is 2.303 *RT/F* at 37 °C

P is the relative permeability of the membrane to the ion shown in the subscript, and

$[ion]_{out}$ and $[ion]_{in}$ are the ion concentrations outside and inside the cell.

Although this equation looks quite intimidating, it can be simplified into words to say: Resting membrane potential is determined by the combined contributions of the (concentration gradient × membrane permeability) for each ion.

If the membrane is not permeable to an ion, the permeability value for that ion is zero, and the ion drops out of the equation. For example, cells at rest normally are not permeable to Ca^{2+}, and therefore Ca^{2+} is not part of the GHK equation.

The GHK equation predicts resting membrane potentials based on given ion concentrations and membrane permeabilities. Notice that if permeabilities for Na^+ and Cl^- are zero, the equation reverts back to the Nernst equation for K^+. The GHK equation explains how the cell's slight permeability to Na^+ makes the resting membrane potential more positive than the E_K determined with the Nernst equation. The GHK equation can also be used to predict what happens to membrane potential when ion concentrations or membrane permeabilities change.

Ion Movement Creates Electrical Signals

The resting membrane potential of living cells is determined primarily by the K^+ concentration gradient and the cell's resting permeability to K^+, Na^+, and Cl^-. A change in either the K^+ concentration gradient or ion permeabilities changes the membrane potential. If you know numerical values for ion concentrations and permeabilities, you can use the GHK equation to calculate the new membrane potential.

In medicine you usually will not have numerical values, however, so it is important to be able to think conceptually about the relationship between ion concentrations, permeabilities, and membrane potential. For example, at rest, the cell membrane of a neuron is only slightly permeable to Na^+. If the membrane suddenly increases its Na^+ permeability, Na^+ enters the cell, moving down its electrochemical gradient [p. 163].

The addition of positive Na^+ to the intracellular fluid *depolarizes* the cell membrane and creates an electrical signal.

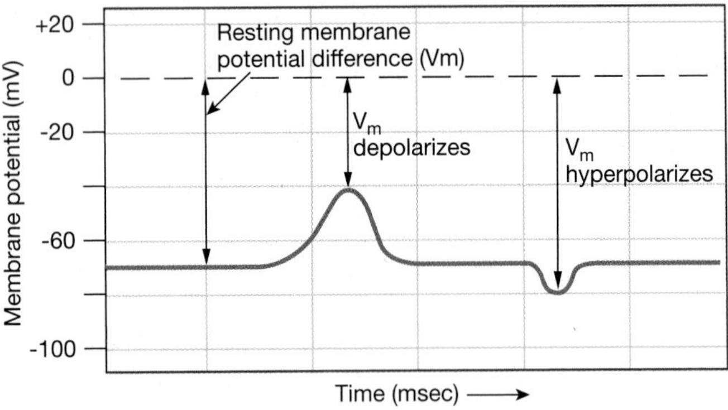

The movement of ions across the membrane can also *hyperpolarize* a cell. If the cell membrane suddenly becomes more permeable to K^+, positive charge is lost from inside the cell, and the cell becomes more negative (hyperpolarizes). A cell may also hyperpolarize if negatively charged ions, such as Cl^-, enter the cell from the extracellular fluid.

Concept Check Answers: p. 287

8. Would a cell with a resting membrane potential of −70 mV depolarize or hyperpolarize in the following cases? (You must consider both the concentration gradient and the electrical gradient of the ion to determine net ion movement.)
   (a) Cell becomes more permeable to Ca^{2+}.
   (b) Cell becomes less permeable to K^+.
9. Would the cell membrane depolarize or hyperpolarize if a small amount of Na^+ leaked into the cell?

It is important to understand that a change in membrane potential from −70 mV to a positive value, such as +30 mV, *does not mean that the ion concentration gradients have reversed!* A significant change in membrane potential occurs with the movement of very few ions. For example, to change the membrane potential by 100 mV, only 1 of every 100,000 K^+ must enter or leave the cell. This is such a tiny fraction of the total number of K^+ in the cell that the intracellular concentration of K^+ remains essentially unchanged even though the membrane potential has changed by 100 mV.

To appreciate how a tiny change can have a large effect, think of getting one grain of beach sand into your eye. There are so many grains of sand on the beach that the loss of one grain is not significant, just as the movement of one K^+ across the cell membrane does not significantly alter the concentration of K^+. However, the electrical signal created by moving a few K^+ across the membrane has a significant effect on the cell's membrane potential, just as getting that one grain of sand in your eye creates significant discomfort.

Gated Channels Control the Ion Permeability of the Neuron

How does a cell change its ion permeability? The simplest way is to open or close existing channels in the membrane. Neurons contain a variety of gated ion channels that alternate between open and closed states, depending on the intracellular and extracellular conditions [p. 148]. A slower method for changing membrane permeability is for the cell to insert new channels into the membrane or remove some existing channels.

Ion channels are usually named according to the primary ion(s) they allow to pass through them. There are four major types of selective ion channels in the neuron: (1) Na^+ channels, (2) K^+ channels, (3) Ca^{2+} channels, and (4) Cl^- channels. Other channels are less selective, such as the *monovalent cation channels* that allow both Na^+ and K^+ to pass.

The ease with which ions flow through a channel is called the channel's **conductance** (G) {*conductus,* escort}. Channel conductance varies with the gating state of the channel and with the channel protein isoform. Some ion channels, such as the *K^+ leak channels* that are the major determinant of resting membrane potential, spend most of their time in an open state. Other channels have gates that open or close in response to particular stimuli. Most gated channels fall into one of three categories:

1. **Mechanically gated ion channels** are found in sensory neurons and open in response to physical forces such as pressure or stretch.
2. **Chemically gated ion channels** in most neurons respond to a variety of ligands, such as extracellular neurotransmitters and neuromodulators or intracellular signal molecules.
3. **Voltage-gated ion channels** respond to changes in the cell's membrane potential. Voltage-gated Na^+ and K^+ channels play an important role in the initiation and conduction of electrical signals along the axon.

Not all voltage-gated channels behave in exactly the same way. The **threshold voltage** for channel opening varies from one channel type to another. For example, some channels we think of as leak channels are actually voltage-gated channels that remain open in the voltage range of the resting membrane potential.

The speed with which a gated channel opens and closes differs among different types of channels. Channel opening to allow ion flow is called channel *activation*. For example, Na^+ channels and K^+ channels of axons are both activated by cell depolarization. The Na^+ channels open very rapidly, but the K^+ channels are slower to open. The result is an initial flow of Na^+ across the membrane, followed later by a flow of K^+.

Many channels that open in response to depolarization close only when the cell repolarizes. The gating portion of the channel protein has an electrical charge that moves the gate between open and closed positions as membrane potential changes. This is like a spring-loaded door that opens when you push on it, then closes when you release it.

CLINICAL FOCUS

Mutant Channels

Ion channels are proteins, and like other proteins they may lose or change function if their amino acid sequence is altered. **Channelopathies** {*pathos*, suffering} are inherited diseases caused by mutations in ion channel proteins. The most common channelopathy is cystic fibrosis, which results from defects in chloride channel function [see Chapter 5 Running Problem]. Because ion channels are so intimately linked to the electrical activity of cells, many channelopathies manifest themselves as disorders of the excitable tissues (nerve and muscle). One significant contribution of molecular biology to medicine was the discovery that what the medical community considers to be one disease can actually be a family of related diseases with different causes but similar symptoms. For example, the condition known as *long Q-T syndrome* (LQTS; named for changes in the electrocardiogram test) is a cardiac problem characterized by an irregular heartbeat {*arrhythmia*; *a-*, without}, fainting, and sometimes sudden death. Scientists have identified eight different gene mutations in K^+, Na^+, or Ca^{2+} channels that result in various subtypes of LQTS. Other well-known channelopathies include some forms of epilepsy and malignant hyperthermia.

But some channels that open with a stimulus close even though the activating stimulus continues, a process known as *inactivation*. This is similar to doors with an automatic timed open-close mechanism. The door opens when you hit the button, then after a certain period of time, it closes itself, whether you are still standing in the doorway or not. An inactivated channel returns to its normal closed state shortly after the membrane repolarizes. The specific mechanisms underlying channel inactivation vary with different channel types.

Each major channel type has several to many subtypes with varying properties, and the list of subtypes gets longer each year. Within each subtype there may be multiple isoforms that express different opening and closing *kinetics* {*kinetikos*, moving} and associated proteins that modify channel properties. In addition, channel activity can be modulated by chemical factors that bind to the channel protein, such as phosphate groups.

Current Flow Obeys Ohm's Law

When ion channels open, ions may move into or out of the cell. The flow of electrical charge carried by an ion is called the ion's **current,** abbreviated $\mathbf{I_{ion}}$. The direction of ion movement depends on the *electrochemical* (combined electrical and concentration) gradient of the ion. Potassium ions usually move out of the cell. Na^+, Cl^-, and Ca^{2+} usually flow into the cell. The net flow of ions across the membrane depolarizes or hyperpolarizes the cell, creating an electrical signal.

Current flow, whether across a membrane or inside a cell, obeys a rule known as **Ohm's Law**. Ohm's Law says that current flow (I) is directly proportional to the electrical potential difference (in volts, V) between two points and inversely proportional to the resistance (R) of the system to current flow: $I = V \times 1/R$ or $I = V/R$. In other words, as resistance increases, current flow decreases. (You will encounter a variant of Ohm's Law when you study fluid flow in the cardiovascular and respiratory systems.)

Resistance in biological flow is the same as resistance in everyday life: it is a force that opposes flow. Electricity is a form of energy and, like other forms of energy, it dissipates as it encounters resistance. As an analogy, think of rolling a ball along the floor. A ball rolled across a smooth wood floor encounters less resistance than a ball rolled across a carpeted floor. If you throw both balls with the same amount of energy, the ball that encounters less resistance retains energy longer and travels farther along the floor.

In biological electricity, resistance to current flow comes from two sources: the resistance of the cell membrane (R_m) and the internal resistance of the cytoplasm (R_i). The phospholipid bilayer of the cell membrane is normally an excellent insulator, and a membrane with no open ion channels has very high resistance and low conductance. If ion channels open, ions (current) flow across the membrane if there is an electrochemical gradient for them. Opening ion channels therefore decreases the membrane resistance.

The internal resistance of most neurons is determined by the composition of the cytoplasm and is relatively constant. The membrane and internal resistances together determine how far current will flow through a cell before the energy is dissipated and the current dies. The combination of the two resistances is called the *length constant* for a given neuron.

Voltage changes across the membrane can be classified into two basic types of electrical signals: graded potentials and action potentials (■ Tbl. 8.3). **Graded potentials** are variable-strength signals that travel over short distances and lose strength as they travel through the cell. They are used for short-distance communication. If a depolarizing graded potential is strong enough when it reaches an integrating region within a neuron, the graded potential initiates an action potential. **Action potentials** are very brief, large depolarizations that travel for long distances through a neuron without losing strength. Their function is rapid signaling over long distances, such as from your toe to your brain.

Graded Potentials Reflect Stimulus Strength

Graded potentials in neurons are depolarizations or hyperpolarizations that occur in the dendrites and cell body or, less frequently, near the axon terminals. These changes in membrane

Comparison of Graded Potential and Action Potential in Neurons

Table 8.3

	Graded Potential	Action Potential
Type of signal	Input signal	Regenerating conduction signal
Occurs where?	Usually dendrites and cell body	Trigger zone through axon
Types of gated ion channels involved	Mechanically, chemically, or voltage-gated channels	Voltage-gated channels
Ions involved	Usually Na^+, Cl^-, Ca^{2+}	Na^+ and K^+
Type of signal	Depolarizing (e.g., Na^+) or hyperpolarizing (e.g., Cl^-)	Depolarizing
Strength of signal	Depends on initial stimulus; can be summed	All-or-none phenomenon; cannot be summed
What initiates the signal?	Entry of ions through gated channels	Above-threshold graded potential at the trigger zone opens ion channels
Unique characteristics	No minimum level required to initiate	Threshold stimulus required to initiate
	Two signals coming close together in time will sum	Refractory period: two signals too close together in time cannot sum
	Initial stimulus strength is indicated by frequency of a series of action potentials	

potential are called "graded" because their size, or *amplitude* {*amplitudo*, large}, is directly proportional to the strength of the triggering event. A large stimulus causes a strong graded potential, and a small stimulus results in a weak graded potential.

In neurons of the CNS and the efferent division, graded potentials occur when chemical signals from other neurons open chemically gated ion channels, allowing ions to enter or leave the neuron. Mechanical stimuli (such as stretch) or chemical stimuli open ion channels in some sensory neurons. Graded potentials may also occur when an open channel closes, decreasing the movement of ions through the cell membrane. For example, if K^+ leak channels close, fewer K^+ leave the cell. The retention of K^+ depolarizes the cell.

Concept Check Answer: p. 287

10. Match each ion's movement with the type of graded potential it creates.

(a) Na^+ entry
(b) Cl^- entry
(c) K^+ exit
(d) Ca^{2+} entry

1. depolarizing
2. hyperpolarizing

■ Figure 8.7a shows a graded potential that begins when a stimulus opens monovalent cation channels on the cell body of a neuron. Sodium ions move into the neuron, bringing in electrical energy. The positive charge carried in by the Na^+ spreads as a wave of depolarization through the cytoplasm, just as a stone thrown into water creates ripples or waves that spread outward from the point of entry. The wave of depolarization that moves through the cell is known as **local current flow**. By convention, current in biological systems is the net movement of *positive* electrical charge.

The strength of the initial depolarization in a graded potential is determined by how much charge enters the cell, just as the size of waves caused by a stone tossed in water is determined by the size of the stone. If more Na^+ channels open, more Na^+ enters, and the graded potential has higher initial amplitude. The stronger the initial amplitude, the farther the graded potential can spread through the neuron before it dies out.

Why do graded potentials lose strength as they move through the cytoplasm? Two factors play a role:

1 *Current leak*. The membrane of the neuron cell body has open leak channels that allow positive charge to leak out into the extracellular fluid. Some positive ions leak out of

Graded Potentials

(a) Graded potentials decrease in strength as they spread out from the point of origin.

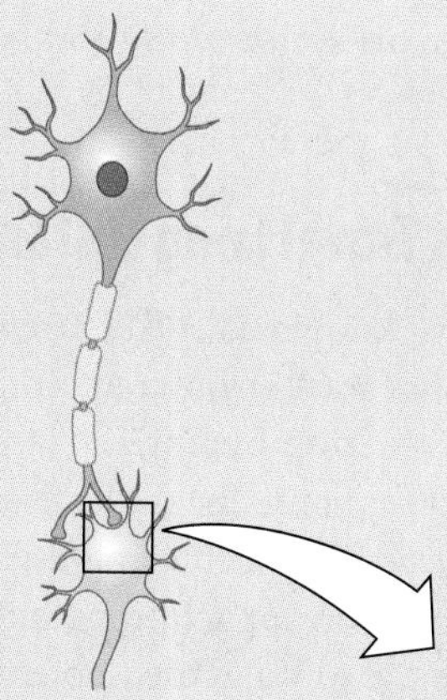

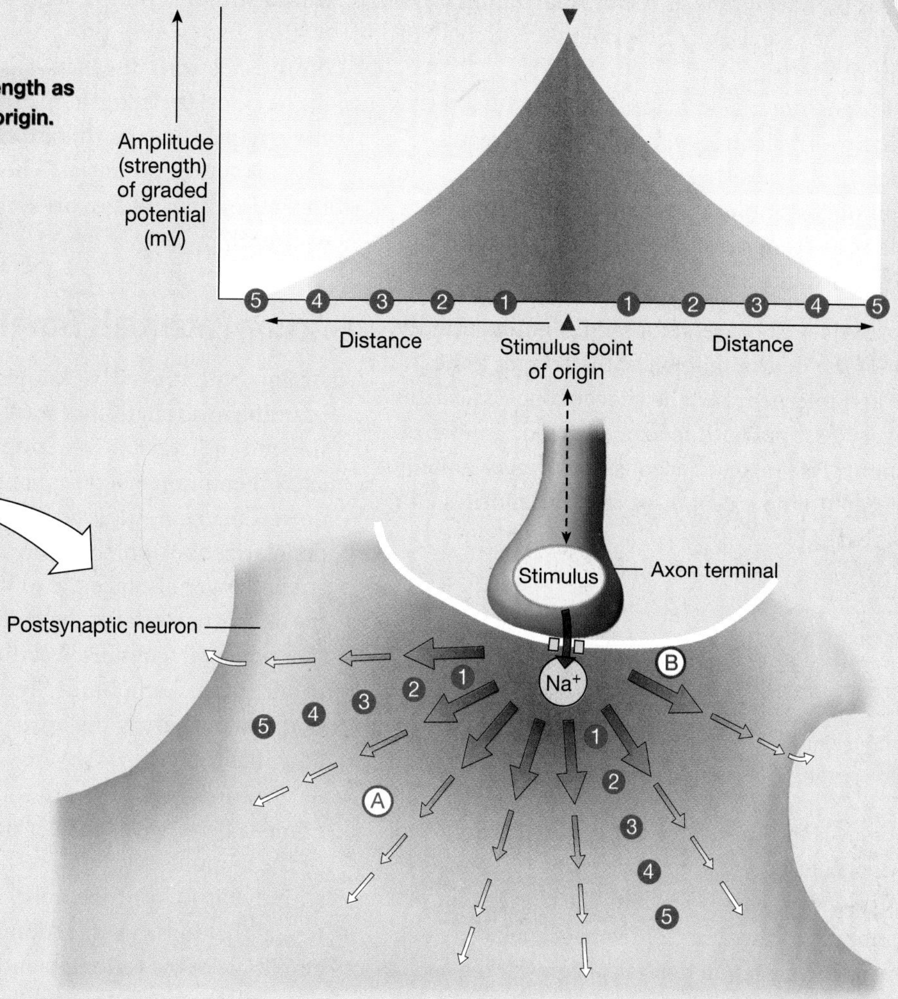

FIGURE QUESTION

At which point of the neuron will the graded potential be stronger, (A) or (B)? On the curve of the graph above, mark and label the approximate locations of (A) and (B).

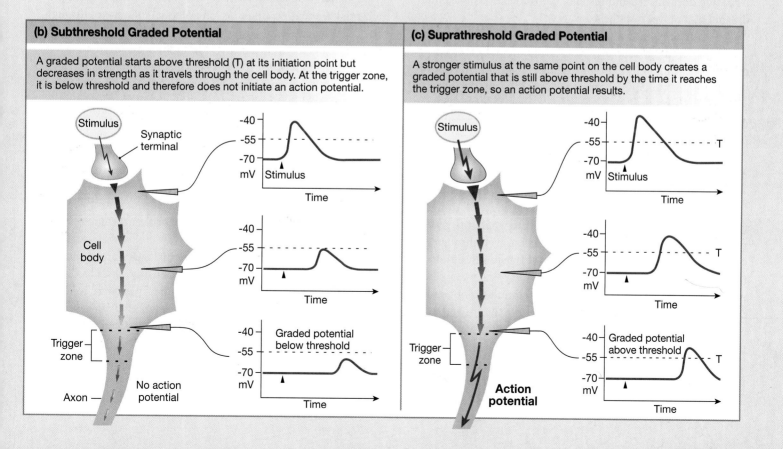

(b) Subthreshold Graded Potential

A graded potential starts above threshold (T) at its initiation point but decreases in strength as it travels through the cell body. At the trigger zone, it is below threshold and therefore does not initiate an action potential.

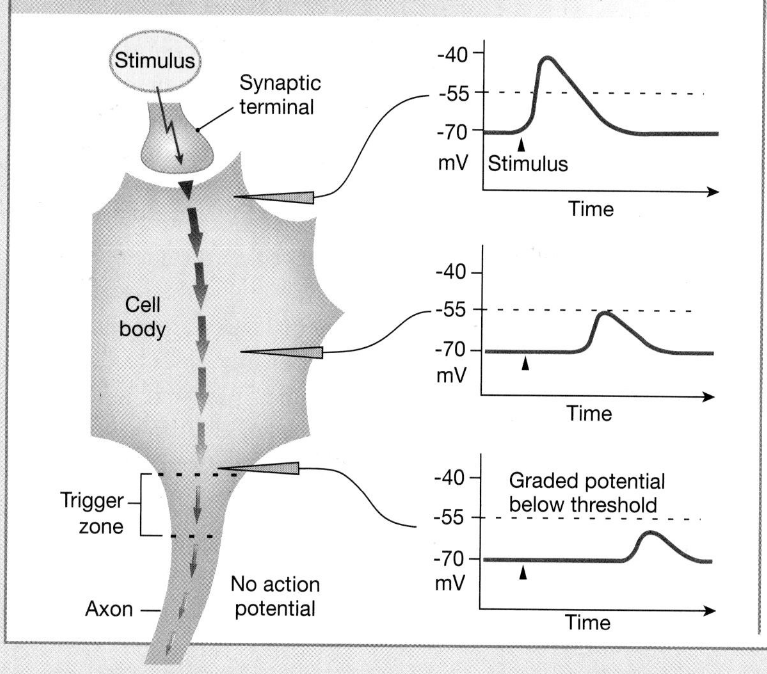

(c) Suprathreshold Graded Potential

A stronger stimulus at the same point on the cell body creates a graded potential that is still above threshold by the time it reaches the trigger zone, so an action potential results.

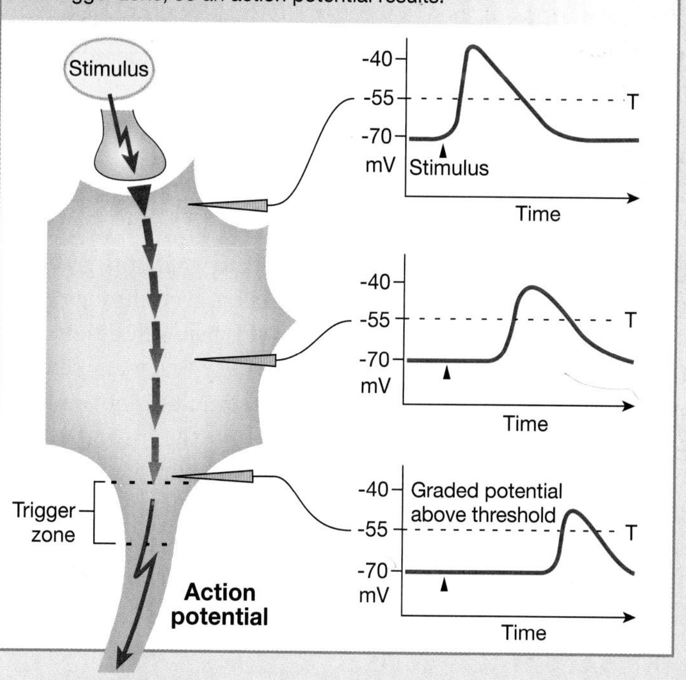

the cell across the membrane as the depolarization wave moves through the cytoplasm, diminishing the strength of the signal inside the cell.

2 *Cytoplasmic resistance.* The cytoplasm provides resistance to the flow of electricity, just as water creates resistance that diminishes the waves from the stone. The combination of current leak and cytoplasmic resistance means that the strength of the signal inside the cell decreases over distance.

Graded potentials that are strong enough eventually reach the region of the neuron known as the **trigger zone**. In efferent neurons and interneurons, the trigger zone is the *axon hillock* and the very first part of the axon, a region known as the **initial segment**. In sensory neurons, the trigger zone is immediately adjacent to the receptor, where the dendrites join the axon (see Fig. 8.2).

Concept Check Answer: p. 287

11. Identify the trigger zones of the neurons illustrated in Figure 8.2, if possible.

The trigger zone is the integrating center of the neuron and contains a high concentration of voltage-gated Na^+ channels in its membrane. If graded potentials reaching the trigger zone depolarize the membrane to the threshold voltage, voltage-gated Na^+ channels open, and an action potential begins. If the depolarization does not reach threshold, the graded potential simply dies out as it moves into the axon.

Because depolarization makes a neuron more likely to fire an action potential, depolarizing graded potentials are considered to be *excitatory*. A hyperpolarizing graded potential moves the membrane potential farther from the threshold value and makes the neuron less likely to fire an action potential. Consequently, hyperpolarizing graded potentials are considered to be *inhibitory*.

Figure 8.7b shows a neuron with three recording electrodes placed at intervals along the cell body and trigger zone. A single stimulus triggers a *subthreshold* graded potential, one that is below threshold by the time it reaches the trigger zone. Although the cell is depolarized to −40 mV at the site where the graded potential begins, the current decreases as it travels through the cell body. As a result, the graded potential is below threshold by the time it reaches the trigger zone. (For the typical mammalian neuron, threshold is about −55 mV.) The stimulus is not strong enough to depolarize the cell to threshold at the trigger zone, and the graded potential dies out without triggering an action potential.

Figure 8.7c shows *suprathreshold* graded potential, one that is strong enough to cause an action potential. A stronger initial stimulus on the cell body initiates a stronger depolarization and current flow. Although this graded potential also diminishes with distance as it travels through the neuron, its higher initial strength ensures that it is above threshold at the trigger zone. In this example, the graded potential triggers an action potential. The ability of a neuron to respond to a stimulus and fire an action potential is called the cell's **excitability**.

Action Potentials Travel Long Distances

Action potentials, also known as *spikes,* are electrical signals of uniform strength that travel from a neuron's trigger zone to the end of its axon. In action potentials, ion channels in the axon membrane open sequentially as electrical current passes down the axon. Additional Na^+ entering the cell reinforce the depolarization, which is why an action potential does not lose strength over distance the way a graded potential does. Instead, the action potential at the end of an axon is identical to the action potential that started at the trigger zone: a depolarization of about 100 mV amplitude. The high-speed movement of an action potential along the axon is called **conduction** of the action potential.

Action potentials are sometimes called **all-or-none** phenomena because they either occur as a maximal depolarization (if the stimulus reaches threshold) or do not occur at all (if the stimulus is below threshold). The strength of the graded potential that initiates an action potential has no influence on the amplitude of the action potential.

When we talk about action potentials, it is important to realize that there is no single action potential that moves through the cell. The action potential that occurs at the trigger zone is like the movement in the first domino of a series of dominos standing on end (■ Fig. 8.8a). As the first domino falls, it strikes the next, passing on its kinetic energy. As the second domino falls, it passes kinetic energy to the third domino, and so on. If you could take a snapshot of the line of falling dominos, you would see that as the first domino is coming to rest in the fallen position, the next one is almost down, the third one most of the way down, and so forth, until you reach the domino that has just been hit and is starting to fall.

In an action potential, a wave of electrical energy moves down the axon. Instead of getting weaker over distance, action potentials are replenished along the way so that they maintain constant amplitude. As the action potential passes from one part of the axon to the next, the membrane's energy state is reflected in the membrane potential of each region. If we were to insert a series of recording electrodes along the length of an axon and start an action potential at the trigger zone, we would see a series of overlapping action potentials, each in a different part of the waveform, just like the dominos that are frozen in different positions (Fig. 8.8b).

8

CONDUCTION OF AN ACTION POTENTIAL

(a) The conduction of an action potential down an axon is similar to energy passed along a series of falling dominos. In this snapshot, each domino is in a different phase of falling. In the axon, each section of membrane is in a different phase of the action potential.

(b) A wave of electrical current passes down the axon.

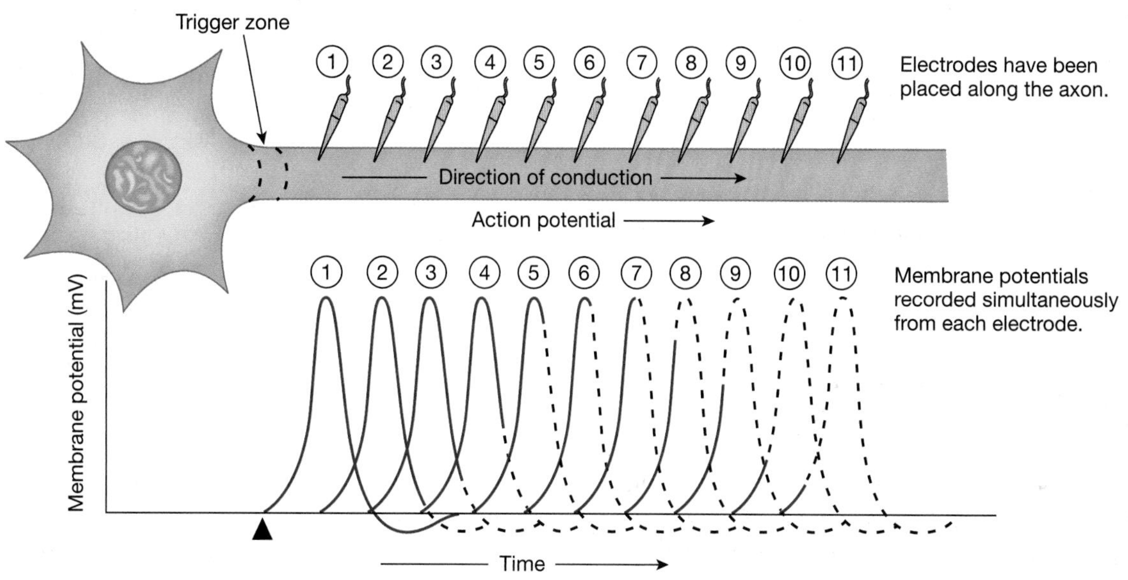

Simultaneous recordings show that each section of axon is experiencing a different phase of the action potential.

■ **Fig. 8.8**

Concept Check Answer: p. 287

12. What is the difference between conductance and conduction in neurons?

Na^+ and K^+ Move Across the Membrane During Action Potentials

What is happening to the neuron membrane when an action potential takes place? The explanation of action potential generation that follows is typical of an unmyelinated PNS neuron. In this simple but elegant mechanism, for whose description A. L. Hodgkin and A. F. Huxley won a 1963 Nobel prize, a suprathreshold (above-threshold) stimulus at the trigger zone initiates an action potential, as you saw in Figure 8.8b. Conduction of an action potential requires only a few types of ion channels: a voltage-gated Na^+ channel and a voltage-gated K^+ channel, plus some leak channels that help set the resting membrane potential.

Action potentials begin when voltage-gated ion channels open, altering membrane permeability (P) to Na^+ (P_{Na+}) and K^+ (P_{K+}). ■ Figure 8.9 shows the voltage and ion permeability changes that take place in one section of membrane during an action potential. Before and after the action potential, at (1) and (9), the neuron is at its resting membrane potential of −70 mV. The action potential itself can be divided into three phases: a rising phase, a falling phase, and the after-hyperpolarization phase.

Rising Phase of the Action Potential The rising phase is due to a sudden temporary increase in the cell's permeability to Na^+. An action potential begins when a graded potential reaching the trigger zone depolarizes the membrane to threshold (−55 mV) (3). As the cell depolarizes, voltage-gated

Fig. 8.9 *ESSENTIALS*

The Action Potential

Changes in ion permeability (P_{ion}) along the axon create ion flow and voltage changes.

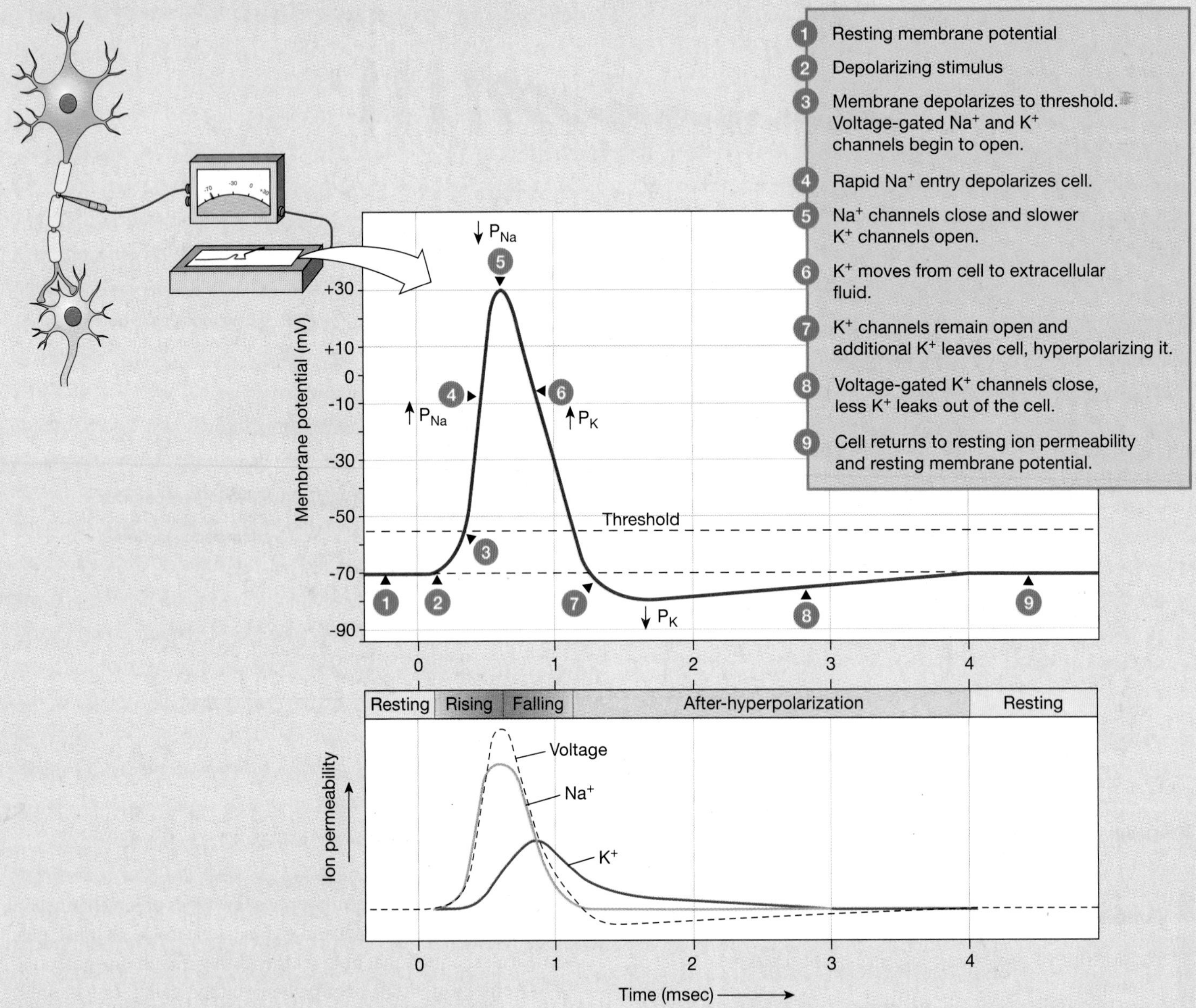

1. Resting membrane potential
2. Depolarizing stimulus
3. Membrane depolarizes to threshold. Voltage-gated Na^+ and K^+ channels begin to open.
4. Rapid Na^+ entry depolarizes cell.
5. Na^+ channels close and slower K^+ channels open.
6. K^+ moves from cell to extracellular fluid.
7. K^+ channels remain open and additional K^+ leaves cell, hyperpolarizing it.
8. Voltage-gated K^+ channels close, less K^+ leaks out of the cell.
9. Cell returns to resting ion permeability and resting membrane potential.

Na^+ channels open, making the membrane much more permeable to Na^+. Because Na^+ is more concentrated outside the cell and because the negative membrane potential inside the cell attracts these positively charged ions, Na^+ flows into the cell.

The addition of positive charge to the intracellular fluid depolarizes the cell membrane, making it progressively more positive (shown by the steep rising phase on the graph ④). In the top third of the rising phase, the inside of the cell has become more positive than the outside, and the membrane potential has reversed polarity. This reversal is represented on the graph by the *overshoot,* that portion of the action potential above 0 mV.

As soon as the cell membrane potential becomes positive, the electrical driving force moving Na^+ into the cell disappears. However, the Na^+ concentration gradient remains, so Na^+ continues to move into the cell. As long as Na^+ permeability remains high, the membrane potential moves toward the Na^+ *equilibrium potential* (E_{Na}) of +60 mV. (Recall that E_{Na} is the membrane potential at which the movement of Na^+ into the cell down its concentration gradient is exactly opposed by the positive membrane potential [p. 165]). The action potential peaks at +30 mV, when Na^+ channels in the axon close and potassium channels open (5).

Falling Phase of the Action Potential The falling phase corresponds to an increase in K^+ permeability. Voltage-gated K^+ channels, like Na^+ channels, open in response to depolarization. The K^+ channel gates are much slower to open, however, and peak K^+ permeability occurs later than peak Na^+ permeability (Fig. 8.9, lower graph). By the time the K^+ channels finally open, the membrane potential of the cell has reached +30 mV because of Na^+ influx through faster-opening Na^+ channels.

When the Na^+ channels close at the peak of the action potential, the K^+ channels have just finished opening, making the membrane very permeable to K^+. At a positive membrane potential, the concentration and electrical gradients for K^+ favor movement of K^+ out of the cell. As K^+ moves out of the cell, the membrane potential rapidly becomes more negative, creating the falling phase of the action potential (6) and sending the cell toward its resting potential.

When the falling membrane potential reaches −70 mV, the K^+ permeability has not returned to its resting state. Potassium continues to leave the cell through both voltage-gated and K^+ leak channels, and the membrane hyperpolarizes, approaching the E_K of −90 mV. This after-hyperpolarization (7) is also called the *undershoot*.

Finally the slow voltage-gated K^+ channels close, and some of the outward K^+ leak stops (8). Retention of K^+ and leak of Na^+ into the axon bring the membrane potential back to −70 mV (9), the value that reflects the cell's resting permeability to K^+, Cl^-, and Na^+.

To summarize, the action potential is a change in membrane potential that occurs when voltage-gated ion channels in the membrane open, increasing the cell's permeability first to Na^+ (which enters) and then to K^+ (which leaves). The *influx* (movement into the cell) of Na^+ depolarizes the cell. This depolarization is followed by K^+ *efflux* (movement out of the cell), which restores the cell to the resting membrane potential.

One Action Potential Does Not Alter Ion Concentration Gradients

As you just learned, an action potential results from ion movements across the neuron membrane. First Na^+ moves into the cell, and then K^+ moves out. However, it is important to understand that very few ions move across the membrane in a single action potential, so that *the relative Na^+ and K^+ concentrations inside and outside the cell remain essentially unchanged.* For example, only 1 in every 100,000 K^+ must leave the cell to shift the membrane potential from +30 to −70 mV, equivalent to the falling phase of the action potential. The tiny number of ions that cross the membrane during an action potential does not disrupt the Na^+ and K^+ concentration gradients.

Normally, the ions that do move into or out of the cell during action potentials are rapidly restored to their original compartments by Na^+-K^+-ATPase (also known as the Na^+-K^+ pump). The pump uses energy from ATP to exchange Na^+ that enters the cell for K^+ that leaked out of it [p. 150]. *This exchange does not need to happen before the next action potential fires, however, because the ion concentration gradient was not significantly altered by one action potential!* A neuron without a functional Na^+-K^+ pump could fire a thousand or more action potentials before a significant change in the ion gradients occurred.

8

Concept Check Answer: p. 287

13. If you put ouabain, an inhibitor of the Na^+-K^+ pump, on a neuron and then stimulate the neuron repeatedly, what do you expect to happen to action potentials generated by that neuron?

(a) They cease immediately.
(b) There is no immediate effect, but they diminish with repeated stimulation and eventually disappear.
(c) They get smaller immediately, then stabilize with smaller amplitude.
(d) Ouabain has no effect on action potentials.

Axonal Na^+ Channels Have Two Gates

One question that puzzled scientists for many years was how the voltage-gated Na^+ channels could close at the peak of the action potential, when the cell was depolarized. Why should these channels *close* when depolarization was the stimulus for Na^+ channel *opening*? After many years of study, they found the answer. These voltage-gated Na^+ channels have two gates to regulate ion movement rather than a single gate. The two gates, known as **activation** and **inactivation gates**, flip-flop back and forth to open and close the Na^+ channel.

When a neuron is at its resting membrane potential, the activation gate of the Na^+ channel closes and no Na^+ can move through the channel (■ Fig. 8.10a). The inactivation gate, an amino acid sequence behaving like a ball and chain on the cytoplasmic side of the channel, is open. When the cell membrane near the channel depolarizes, the activation gate swings open (Fig. 8.10b). This opens the channel and allows Na^+ to move into the cell down its electrochemical gradient (Fig. 8.10c).

THE VOLTAGE-GATED Na^+ CHANNEL

The distinguishing feature of this channel is the presence of two gates: an activation gate that opens rapidly and an inactivation gate that is slower to close.

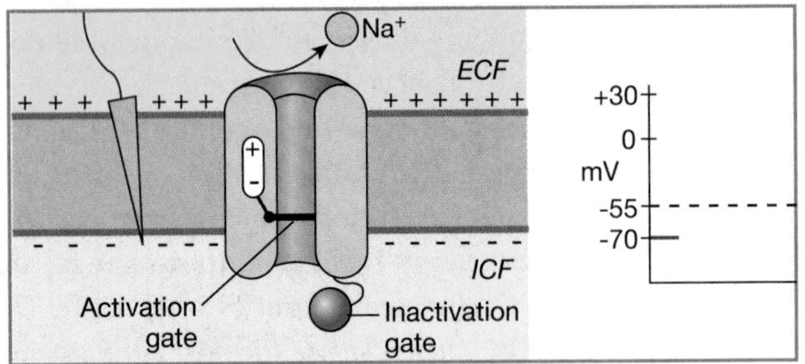

(a) At the resting membrane potential, the activation gate closes the channel.

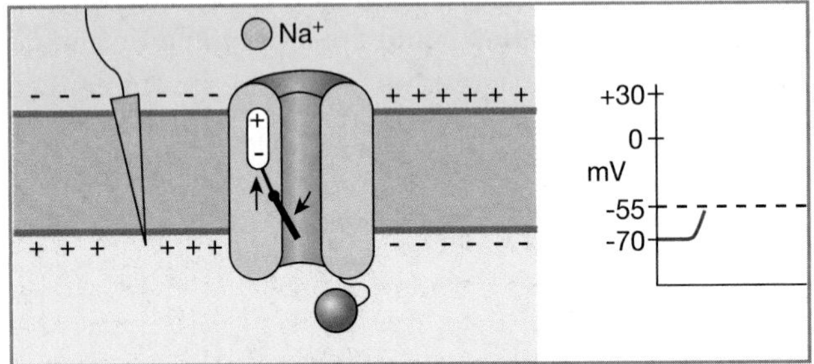

(b) Depolarizing stimulus arrives at the channel. Activation gate opens.

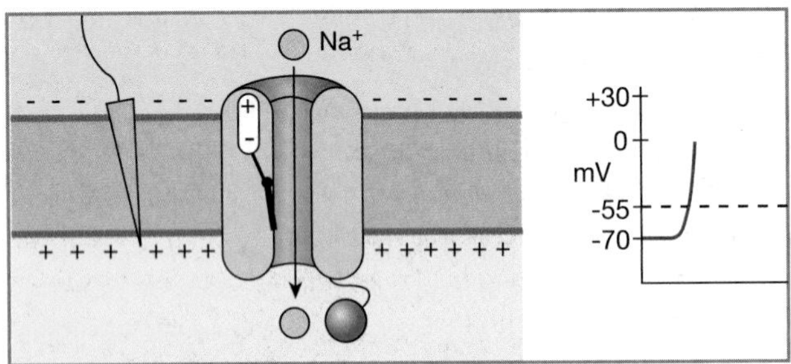

(c) With activation gate open, Na^+ enters the cell.

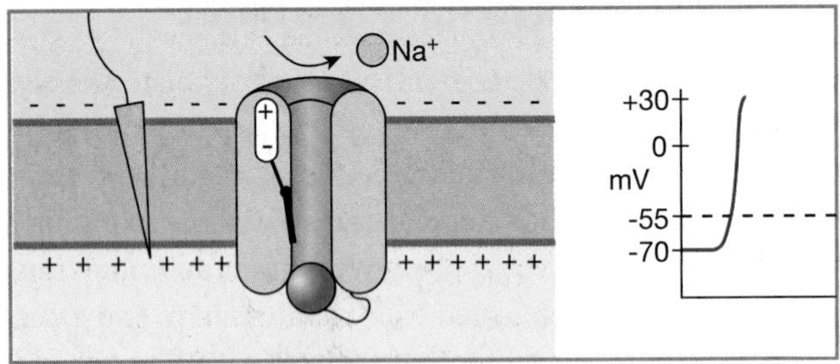

(d) Inactivation gate closes and Na^+ entry stops.

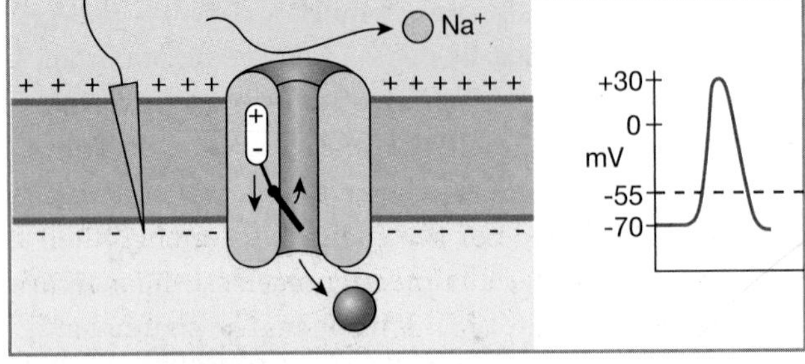

(e) During repolarization caused by K^+ leaving the cell, the two gates reset to their original positions.

■ Fig. 8.10

The addition of positive charge further depolarizes the inside of the cell and starts a *positive feedback loop* [p. 18] (■ Fig. 8.11). More Na^+ channels open, and more Na^+ enters, further depolarizing the cell. As long as the cell remains depolarized, activation gates in Na^+ channels remain open.

Positive feedback loops require outside intervention to stop them. In axons, the inactivation gates in the Na^+ channels are the outside intervention that stops the escalating depolarization of the cell. Both activation and inactivation gates move in response to depolarization, but the inactivation gate delays its movement for 0.5 msec. During that delay, the Na^+ channel is open, allowing enough Na^+ influx to create the rising phase of the action potential. When the slower inactivation gate finally closes, Na^+ influx stops, and the action potential peaks (Fig. 8.10d).

While the neuron repolarizes during K^+ efflux, the Na^+ channel gates reset to their original positions so they can respond to the next depolarization (Fig. 8.10e). The double-gating mechanism found in axonal voltage-gated Na^+ channels allows electrical signals to be conducted in only one direction, as you will see in the next section.

Concept Check Answers: p. 287

14. The pyrethrin insecticides, derived from chrysanthemums, disable inactivation gates of Na^+ channels so that the channels remain open. In neurons poisoned with pyrethrins, what happens to the membrane potential? Explain your answer.

15. When Na^+ channel gates are resetting, is the activation gate opening or closing? Is the inactivation gate opening or closing?

Action Potentials Will Not Fire During the Absolute Refractory Period

The double gating of Na^+ channels plays a major role in the phenomenon known as the **refractory period**. The adjective *refractory* comes from a Latin word meaning "stubborn." The "stubbornness" of the neuron refers to the fact that once an action potential has begun, a second action potential cannot be triggered for about 1–2 msec, no matter how large the stimulus. This delay, which represents the time required for the Na^+ channel gates to reset to their resting positions, is called the **absolute refractory period** (■ Fig. 8.12). Because of the absolute refractory period, a second action potential cannot occur before the first has finished. Consequently, *action potentials moving from trigger zone to axon terminal cannot overlap and cannot travel backward.*

A **relative refractory period** follows the absolute refractory period. During the relative refractory period, some but not

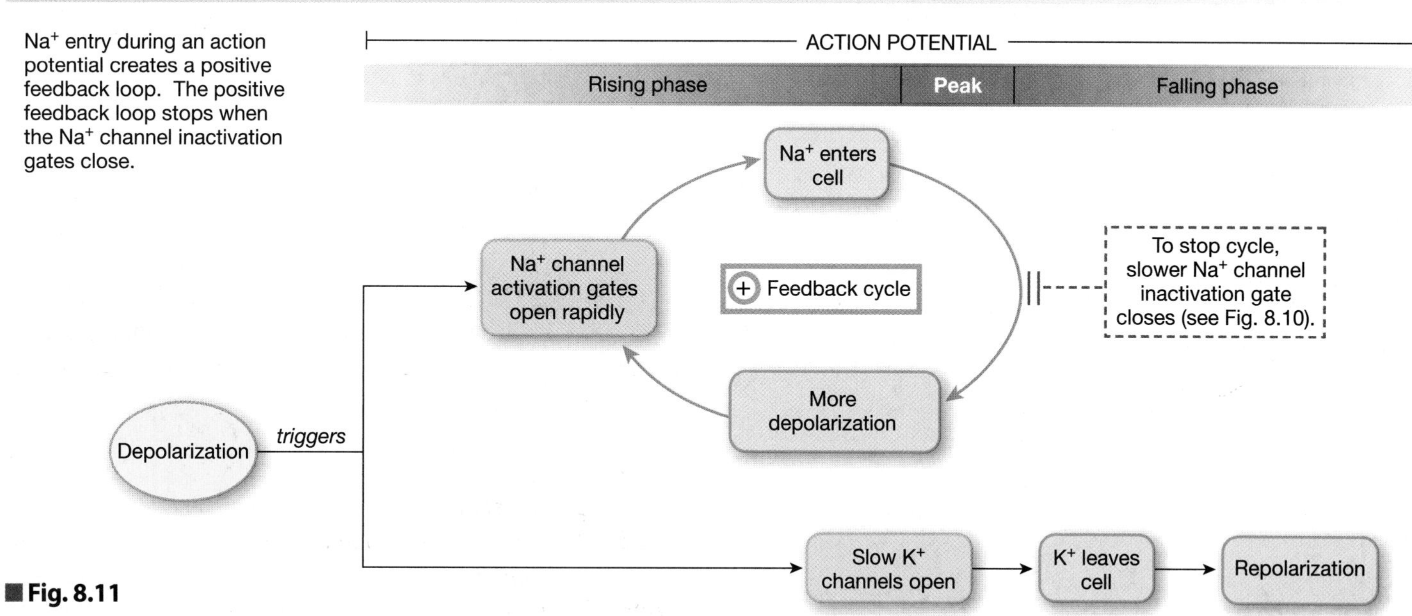
POSITIVE FEEDBACK
Na+ entry during an action potential creates a positive feedback loop. The positive feedback loop stops when the Na+ channel inactivation gates close.
ACTION POTENTIAL
Rising phase
Peak
Falling phase
Na+ enters cell
Na+ channel activation gates open rapidly
+ Feedback cycle
More depolarization
To stop cycle, slower Na+ channel inactivation gate closes (see Fig. 8.10).
Depolarization
triggers
Slow K+ channels open
K+ leaves cell
Repolarization

■ Fig. 8.11

all Na^+ channel gates have reset to their original positions. In addition, during the relative refractory period, K^+ channels are still open.

The Na^+ channels that have not quite returned to their resting position can be reopened by a stronger-than-normal graded potential. In other words, the threshold value has temporarily moved closer to zero, which requires a stronger depolarization to reach it. Although Na^+ enters through newly reopened Na^+ channels, depolarization due to Na^+ entry is offset by K^+ loss through still-open K^+ channels. As a result, any action potentials that fire during the relative refractory period will be of smaller amplitude than normal.

The refractory period is a key characteristic that distinguishes action potentials from graded potentials. If two stimuli reach the dendrites of a neuron within a short time, the successive graded potentials created by those stimuli can be added to one another. If, however, two suprathreshold graded potentials reach the action potential trigger zone within the absolute refractory period, the second graded potential has no effect because the Na^+ channels are inactivated and cannot open again so soon.

Refractory periods limit the rate at which signals can be transmitted down a neuron. The absolute refractory period also ensures one-way travel of an action potential from cell body to axon terminal by preventing the action potential from traveling backward.

Action Potentials Are Conducted

A distinguishing characteristic of action potentials is that they can travel over long distances of a meter or more without losing energy. The action potential that reaches the end of an axon is identical to the action potential that started at the trigger zone. To see how this happens, we must examine the conduction of action potentials at the cellular level.

The depolarization of a section of axon causes positive current to spread through the cytoplasm in all directions by local current flow (■ Fig. 8.13). Simultaneously, on the outside of the axon membrane, current flows back toward the depolarized section. The local current flow in the cytoplasm diminishes over distance as energy dissipates. Forward flow down the axon would eventually die out were it not for voltage-gated channels.

The axon is well supplied with voltage-gated Na^+ channels. Whenever a depolarization reaches those channels, they open, allowing more Na^+ to enter the cell and reinforcing the depolarization—the positive feedback loop shown in Figure 8.11. Let's see how this works when an action potential begins at the axon's trigger zone.

First, a graded potential above threshold enters the trigger zone (■ Fig. 8.14 ①). Its depolarization opens voltage-gated Na^+ channels, Na^+ enters the axon, and the initial segment of axon depolarizes ②. Positive charge from the depolarized trigger zone spreads by local current flow to adjacent sections of membrane ③, repelled by the Na^+ that entered the cytoplasm and attracted by the negative charge of the resting membrane potential.

The flow of local current toward the axon terminal (to the right in Figure 8.14) begins conduction of the action potential. When the membrane distal to the trigger zone depolarizes from local current flow, its Na^+ channels open, allowing Na^+ into the cell ④. This starts the positive feedback loop: depolarization opens Na^+ channels, Na^+ enters, causing more depolarization and opening more Na^+ channels in the adjacent membrane.

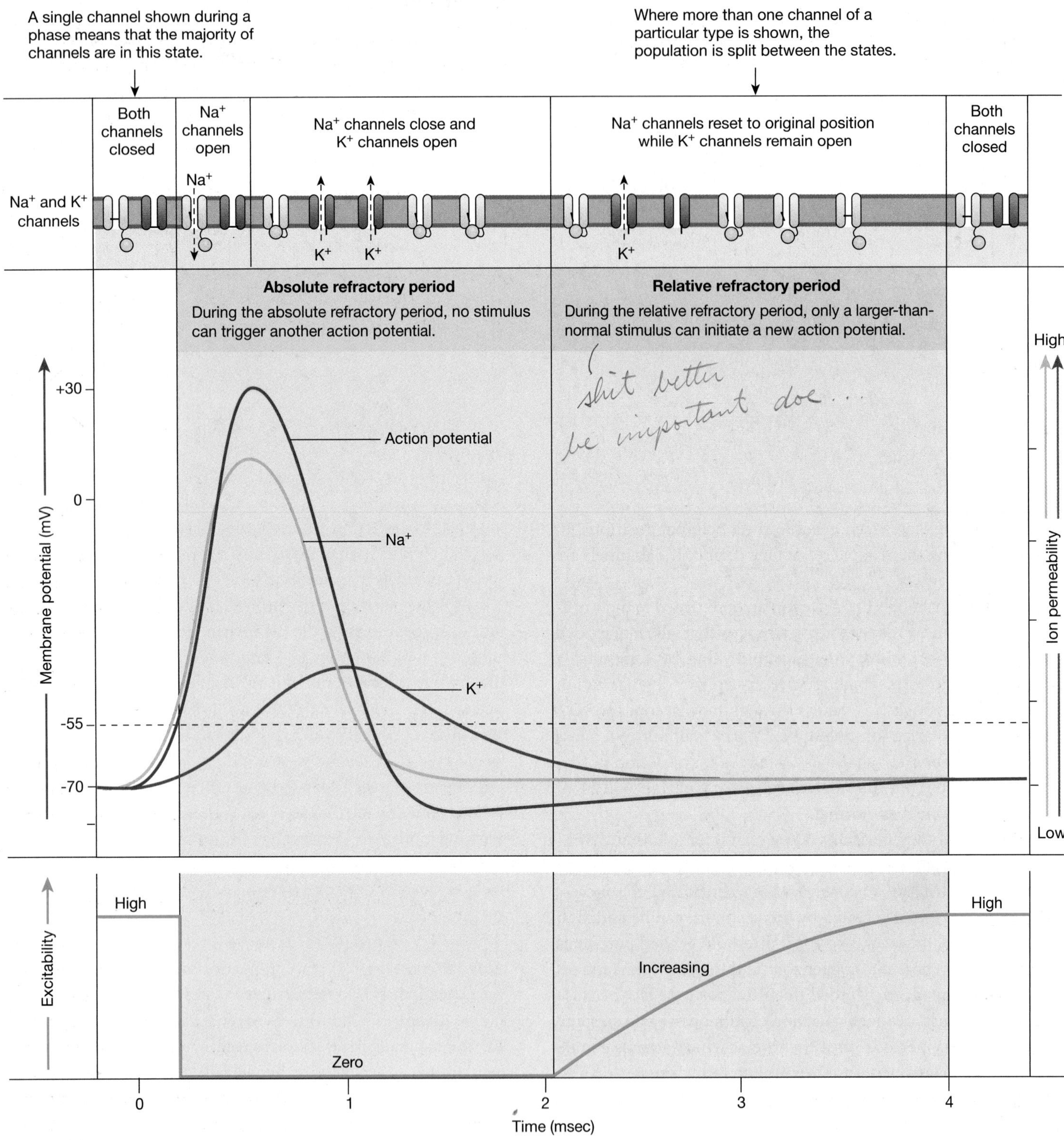
REFRACTORY PERIODS FOLLOWING AN ACTION POTENTIAL
A single channel shown during a phase means that the majority of channels are in this state.
Where more than one channel of a particular type is shown, the population is split between the states.
Both channels closed
Na+ channels open
Na+ channels close and K+ channels open
Na+ channels reset to original position while K+ channels remain open
Both channels closed
Na+ and K+ channels
Na+
K+
K+
K+
Absolute refractory period
During the absolute refractory period, no stimulus can trigger another action potential.
Relative refractory period
During the relative refractory period, only a larger-than-normal stimulus can initiate a new action potential.
Membrane potential (mV)
+30
0
-55
-70
Action potential
Na+
K+
Ion permeability
High
Low
Excitability
High
Zero
Increasing
High
0
1
2
3
4
Time (msec)

■ **Fig. 8.12**

LOW CURRENT FLOW

When a section of axon depolarizes, positive charges move by local current flow into adjacent sections of the cytoplasm. On the extracellular surface, current flows toward the depolarized region.

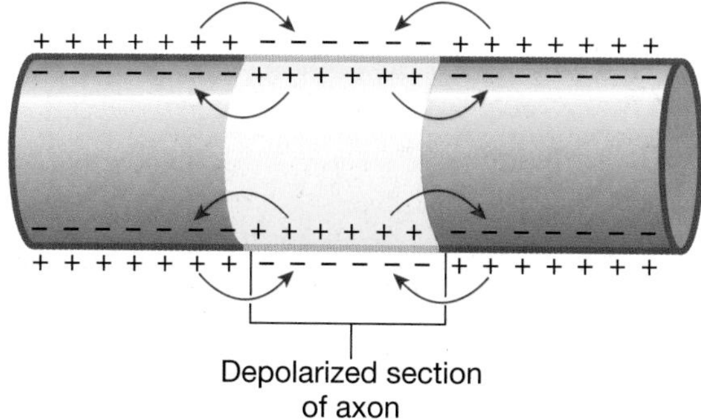

■ Fig. 8.13

The continuous entry of Na^+ as Na^+ channels open along the axon means that the strength of the signal does not diminish as the action potential propagates itself. (Contrast this with graded potentials in Figure 8.7, in which Na^+ enters only at the point of stimulus, resulting in a membrane potential change that loses strength over distance.)

As each segment of axon reaches the peak of the action potential, its Na^+ channels inactivate. During the action potential's falling phase, K^+ channels are open, allowing K^+ to leave the cytoplasm. Finally, the K^+ channels close, and the membrane in that segment of axon returns to its resting potential.

Although positive charge from a depolarized segment of membrane may flow backward toward the trigger zone ⑤, depolarization in that direction has no effect on the axon. The section of axon that has just completed an action potential is in its absolute refractory period, with its Na^+ channels inactivated. For this reason, the action potential cannot move backward.

What happens to current flow backward from the trigger zone into the cell body? Scientists used to believe that there were few voltage-gated ion channels in the cell body, so that retrograde current flow could be ignored. However, they now know that the cell body and dendrites do have voltage-gated ion channels and may respond to local current flow from the trigger zone. These retrograde signals are able to influence and modify the next signal that reaches the cell. For example, depolarization flowing backward from the axon could open voltage-gated channels in the dendrites, strengthening an externally initiated graded potential.

Concept Check Answer: p. 287

16. If you place an electrode in the middle of an axon and artificially depolarize the cell above threshold, in which direction will an action potential travel: to the axon terminal, to the cell body, or to both? Explain your answer.

Larger Neurons Conduct Action Potentials Faster

Two key physical parameters influence the speed of action potential conduction in a mammalian neuron: (1) the diameter of the axon and (2) the resistance of the axon membrane to ion leakage out of the cell. The larger the diameter of the axon or the more leak-resistant the membrane, the faster an action potential will move.

To understand the relationship between diameter and conduction, think of a water pipe with water flowing through it. The water that touches the walls of the pipe encounters resistance due to friction between the flowing water molecules and the stationary walls. The water in the center of the pipe meets no direct resistance from the walls and therefore flows faster. In a large-diameter pipe, a smaller fraction of the water flowing through the pipe is in contact with the walls, making the total resistance lower.

In the same way, charges flowing inside an axon meet resistance from the membrane. Thus, the larger the diameter of the axon, the lower its resistance to ion flow. The connection between axon diameter and speed of conduction is especially evident in the giant axons that certain organisms, such as squid, earthworms, and fish, use for rapid escape responses. These giant axons may be up to 1 mm in diameter. Because of their large diameter, they can easily be punctured with electrodes (■ Fig. 8.15). As a result, these species have been very important in research on electrical signaling.

If you compare a cross section of a squid giant axon with a cross section of a mammalian nerve, you find that the mammalian nerve contains about 200 axons in the same cross-sectional area. Complex nervous systems pack more axons into a small nerve by using smaller-diameter axons wrapped in insulating membranes of myelin instead of large-diameter unmyelinated axons.

Conduction Is Faster in Myelinated Axons

The conduction of action potentials down an axon is faster in high-resistance axons, in which current leak out of the cell is minimized. The unmyelinated axon depicted in Figure 8.14 has low resistance to current leak because the entire axon membrane is in contact with the extracellular fluid and has ion channels through which current can leak.

In contrast, myelinated axons limit the amount of membrane in contact with the extracellular fluid. In these axons, small sections of bare membrane—the nodes of Ranvier—alternate with longer segments wrapped in multiple layers of membrane (the myelin sheath). The myelin sheath creates a high-resistance wall that prevents ion flow out of the cytoplasm. The myelin membranes are analogous to heavy coats of plastic surrounding electrical wires, as they increase the effective thickness of the axon membrane by as much as 100-fold.

CONDUCTION OF ACTION POTENTIALS

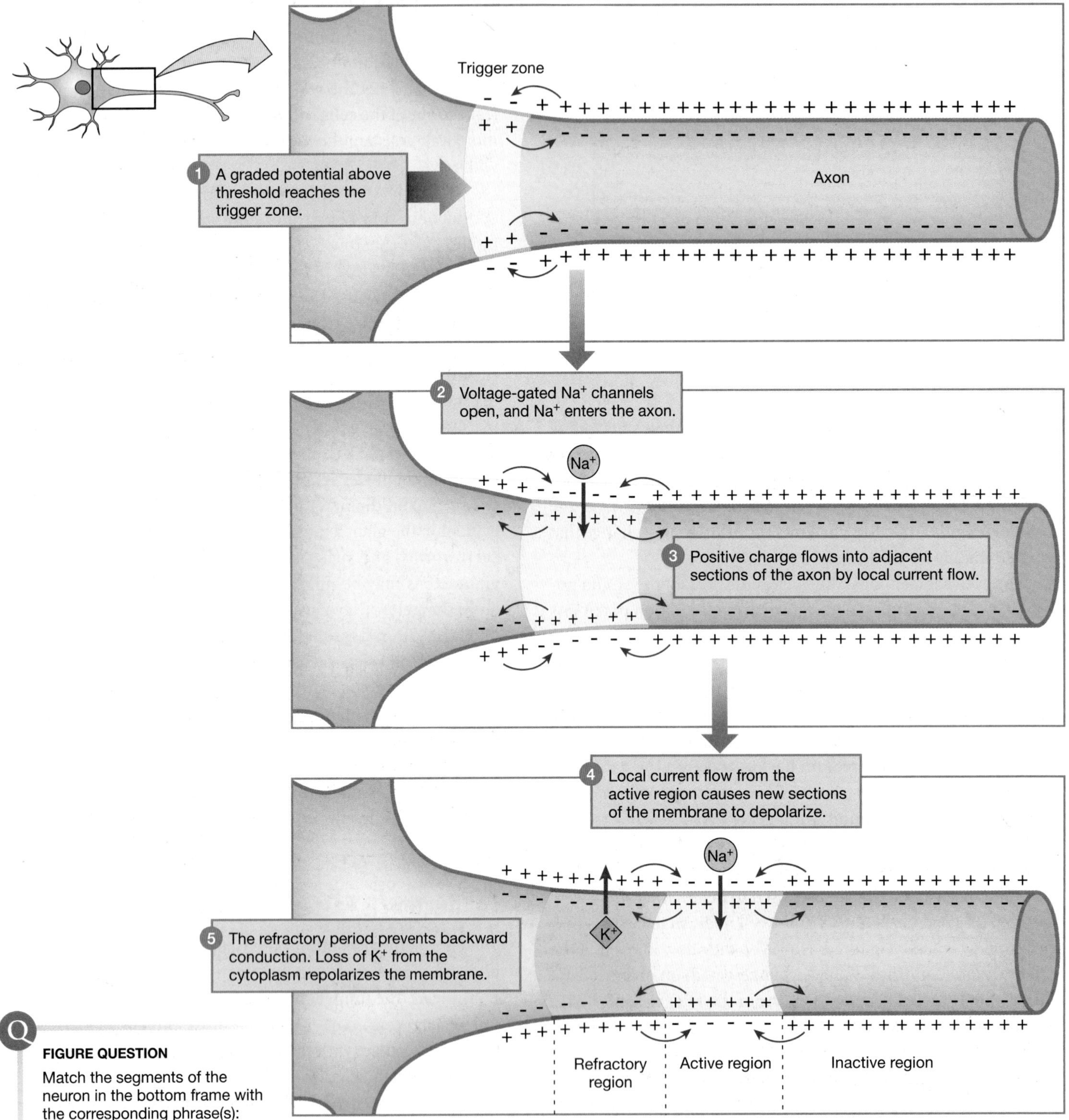
Trigger zone
Axon
1 A graded potential above threshold reaches the trigger zone.
2 Voltage-gated Na+ channels open, and Na+ enters the axon.
Na+
3 Positive charge flows into adjacent sections of the axon by local current flow.
4 Local current flow from the active region causes new sections of the membrane to depolarize.
K+
5 The refractory period prevents backward conduction. Loss of K+ from the cytoplasm repolarizes the membrane.
Refractory region
Active region
Inactive region

FIGURE QUESTION

Match the segments of the neuron in the bottom frame with the corresponding phrase(s):

(a) proximal axon (blue)
(b) absolute refractory period (pink)
(c) active region (yellow)
(d) relative refractory period (purple)
(e) distal inactive region (blue)

1. rising phase of action potential
2. falling phase of action potential
3. after-hyperpolarization
4. resting potential

Fig. 8.14

LARGE AXONS OFFER LESS RESISTANCE.

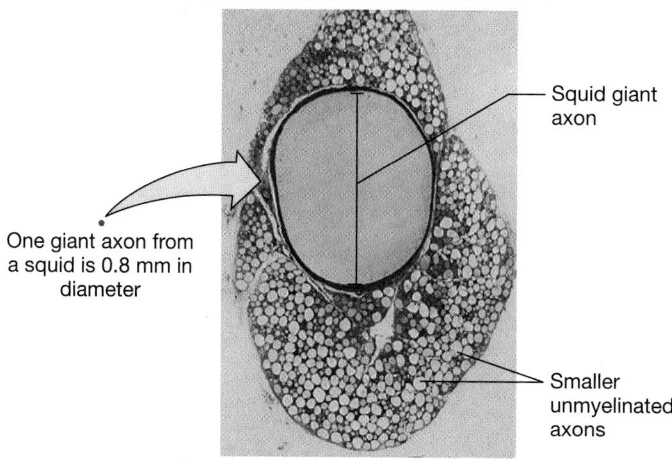

FIGURE QUESTION

A squid giant axon is 0.8 mm in diameter. A myelinated mammalian axon is 0.002 mm in diameter. What would be the diameter of a mammalian nerve if it contained 100 axons that were each the size of a squid giant axon? (Hint: The area of a circle is π x radius^2, and π = 3.1459.)

■ **Fig. 8.15**

As an action potential passes down the axon from trigger zone to axon terminal, it passes through alternating regions of myelinated axon and nodes of Ranvier (■ Fig. 8.16a). The conduction process is similar to that described previously for the unmyelinated axon, except that it occurs only at the nodes in myelinated axons. Each node has a high concentration of voltage-gated Na^+ channels, which open with depolarization and allow Na^+ into the axon. Sodium ions entering at a node reinforce the depolarization and restore the amplitude of the action potential constant as it passes from node to node. The apparent jump of the action potential from node to node is called **saltatory conduction**, from the Latin word *saltare*, meaning "to leap."

What makes conduction more rapid in myelinated axons? Part of the answer lies with the *cable properties* of neurons (see Biotechnology box). Also, channel opening slows conduction slightly. In unmyelinated axons, channels must open sequentially all the way down the axon membrane to maintain the amplitude of the action potential. One clever student compared this process to moving the cursor across a computer screen by repeatedly pressing the space bar.

In myelinated axons, however, only the nodes need Na^+ channels because of the insulating properties of the myelin membrane. As the action potential passes along myelinated segments, conduction is not slowed by channel opening. In the student's analogy, this is like zipping across the screen by using the Tab key.

BIOTECHNOLOGY

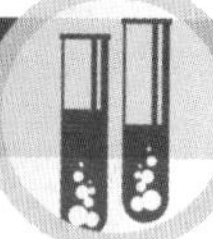

The Body's Wiring

Many aspects of electrical signaling in the body have their parallels in the physical world. The flow of electricity along an axon or through a muscle fiber is similar to the flow of electricity through wires. In both cells and wires, the flow of electrical current is influenced by the physical properties of the material, also known as the *cable properties*. In cells, two factors alter current flow: resistance (discussed in the text) and capacitance.

Capacitance refers to the ability of the cell membrane to store charge (like a battery). A system with high capacitance requires more energy for current flow because some of the energy is diverted to "storage" in the system's *capacitor*. In physics, a capacitor is two plates of conducting material separated by a layer of insulator. In the body, the extracellular and intracellular fluids are the conducting materials, and the phospholipid cell membrane is the insulator.

So what does this have to do with electrical signaling in the body? A simple answer is that the cable properties of cell membranes determine how fast voltage can change across a section of membrane (the *time constant*). For example, cable properties influence how fast a neuron depolarizes to initiate an action potential. The time constant τ (tau) is directly proportional to the resistance of the cell membrane R_m and the capacitance of the membrane C_m: $\tau = R_m \times C_m$. Before current can flow across the membrane to change the voltage, the capacitor must be fully charged. Time spent charging or discharging the capacitor slows voltage changes across the membrane.

Membrane capacitance is normally a constant for biological membranes. However, capacitance becomes important when comparing electrical signaling in myelinated and unmyelinated axons. Capacitance is inversely related to distance: as distance between the conducting compartments increases, capacitance decreases. The stacked membrane layers of the myelin sheath increase the distance between the ECF and ICF and therefore decrease capacitance in that region of the axon. Decreasing membrane capacitance makes voltage changes across the membrane faster—part of the reason conduction of action potentials is faster in myelinated axons. When myelin is lost in demyelinating diseases, the membrane capacitance increases and voltage changes across the membrane take longer. This contributes to slower action potential conduction in diseases such as multiple sclerosis.

Saltatory conduction thus is an effective alternative to large-diameter axons and allows rapid action potentials through small axons. A myelinated frog axon 10-μm in diameter conducts action potentials at the same speed as an unmyelinated 500-μm squid axon. A myelinated 8.6-μm mammalian neuron

SALTATORY CONDUCTION

(a) Action potentials appear to jump from one node of Ranvier to the next. Only the nodes have voltage-gated Na^+ channels.

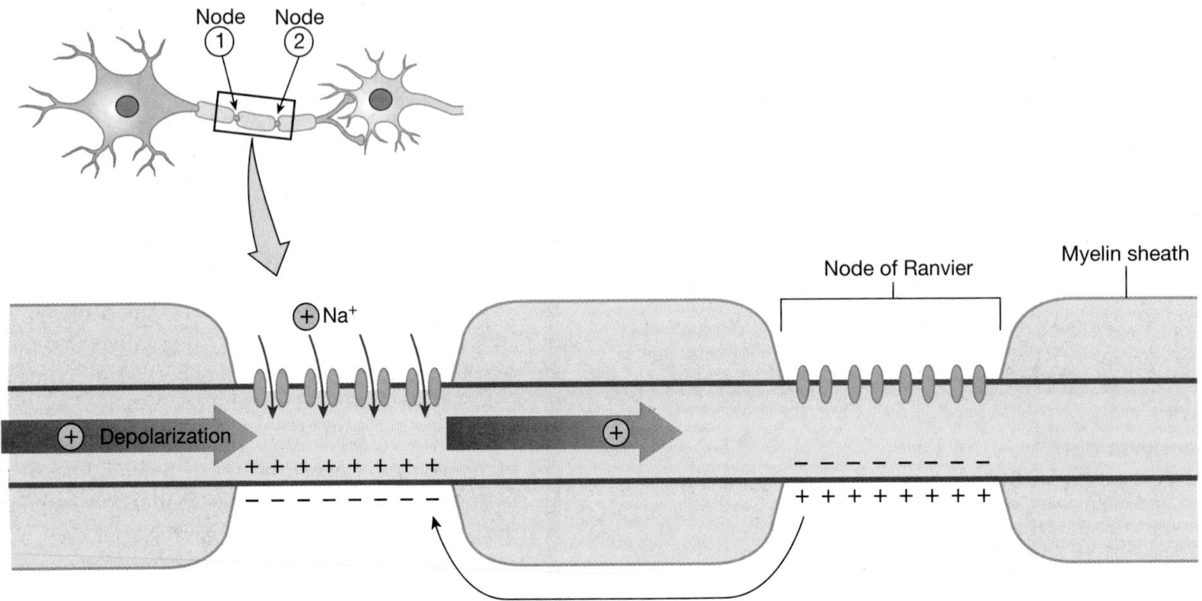

(b) In demyelinating diseases, conduction slows when current leaks out of the previously insulated regions between the nodes.

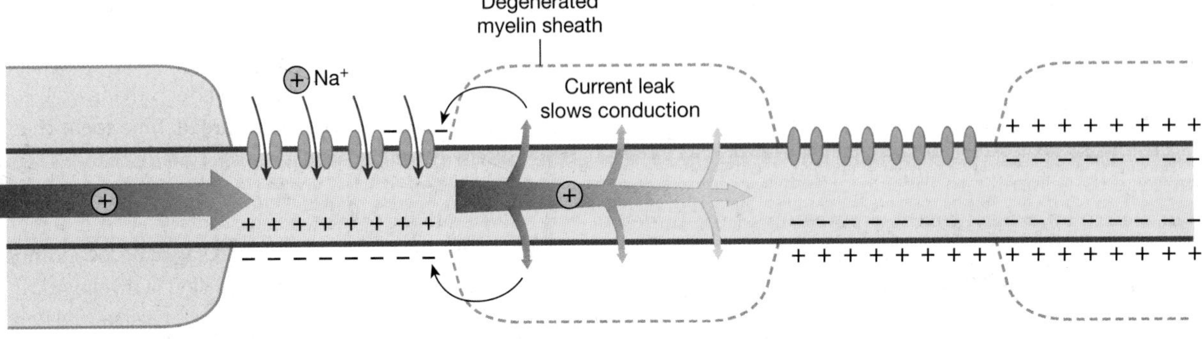

Fig. 8.16

conducts action potentials at 120 m/sec (432 km/hr or 268 miles per hour), while action potentials in a smaller, unmyelinated 1.5-μm pain fiber travel only 2 m/s (7.2 km/hr or 4.5 mph). In summary, action potentials travel through different axons at different rates, depending on the two parameters of axon diameter and myelination.

Concept Check Answer: p. 287

17. Place the following neurons in order of their speed of conduction, from fastest to slowest:
    (a) myelinated axon, diameter 20-μm
    (b) unmyelinated axon, diameter 20-μm
    (c) unmyelinated axon, diameter 200-μm

RUNNING PROBLEM

The classic form of Guillain-Barré syndrome found in Europe and North America is an illness in which the myelin that insulates axons is destroyed. One way that GBS, multiple sclerosis, and other demyelinating illnesses are diagnosed is through the use of a nerve conduction test. This test measures the combined strength of action potentials from many neurons and the rate at which these action potentials are conducted as they travel down axons.

Q3: In GBS, what would you expect the results of a nerve conduction test to be?

238 239 242 264 266 270 281

In *demyelinating diseases,* the loss of myelin from vertebrate neurons can have devastating effects on neural signaling. In the central and peripheral nervous systems, the loss of myelin slows the conduction of action potentials. In addition, when current leaks out of the now-uninsulated regions of membrane between the channel-rich nodes of Ranvier, the depolarization that reaches a node may no longer be above threshold, and conduction may fail (Fig. 8.16b).

Multiple sclerosis is the most common and best-known demyelinating disease. It is characterized by a variety of neurological complaints, including fatigue, muscle weakness, difficulty walking, and loss of vision. Guillain-Barré syndrome, described in this chapter's Running Problem, is also characterized by the destruction of myelin. At this time, we can treat some of the symptoms but not the causes of demyelinating diseases, which are mostly either inherited or autoimmune disorders. Currently, researchers are using recombinant DNA technology to study demyelinating disorders in mice.

Chemical Factors Alter Electrical Activity

A large variety of chemicals alter the conduction of action potentials by binding to Na^+, K^+, or Ca^{2+} channels in the neuron membrane. For example, some *neurotoxins* bind to and block Na^+ channels. Local anesthetics such as procaine, which block sensation, function the same way. If Na^+ channels are not functional, Na^+ cannot enter the axon. A depolarization that begins at the trigger zone then cannot be replenished as it travels; it loses strength as it moves down the axon, much like a normal graded potential. If the wave of depolarization manages to reach the axon terminal, it may be too weak to release neurotransmitter. As a result, the message of the presynaptic neuron is not passed on to the postsynaptic cell, and electrical signaling fails.

Alterations in the extracellular fluid concentrations of K^+ and Ca^{2+} are also associated with abnormal electrical activity in the nervous system. The relationship between extracellular fluid K^+ levels and the conduction of action potentials is the most straightforward and easiest to understand, as well as one of the most clinically significant.

The concentration of K^+ in the blood and interstitial fluid is the major determinant of the resting potential of all cells. If K^+ concentration in the blood moves out of the normal range of 3.5–5 mmol/L, the result is a change in the resting membrane potential of cells (■ Fig. 8.17). This change is not important to most cells, but it can have serious consequences to the body as a whole because of the relationship between resting potential and the excitability of nervous and muscle tissue.

At normal K^+ levels, subthreshold graded potentials do not trigger action potentials, and suprathreshold graded potentials do (Fig. 8.17a, b). An increase in blood K^+ concentration—**hyperkalemia** {*hyper-*, above + *kalium,* potassium + *-emia,* in the blood}—shifts the resting membrane potential of a neuron

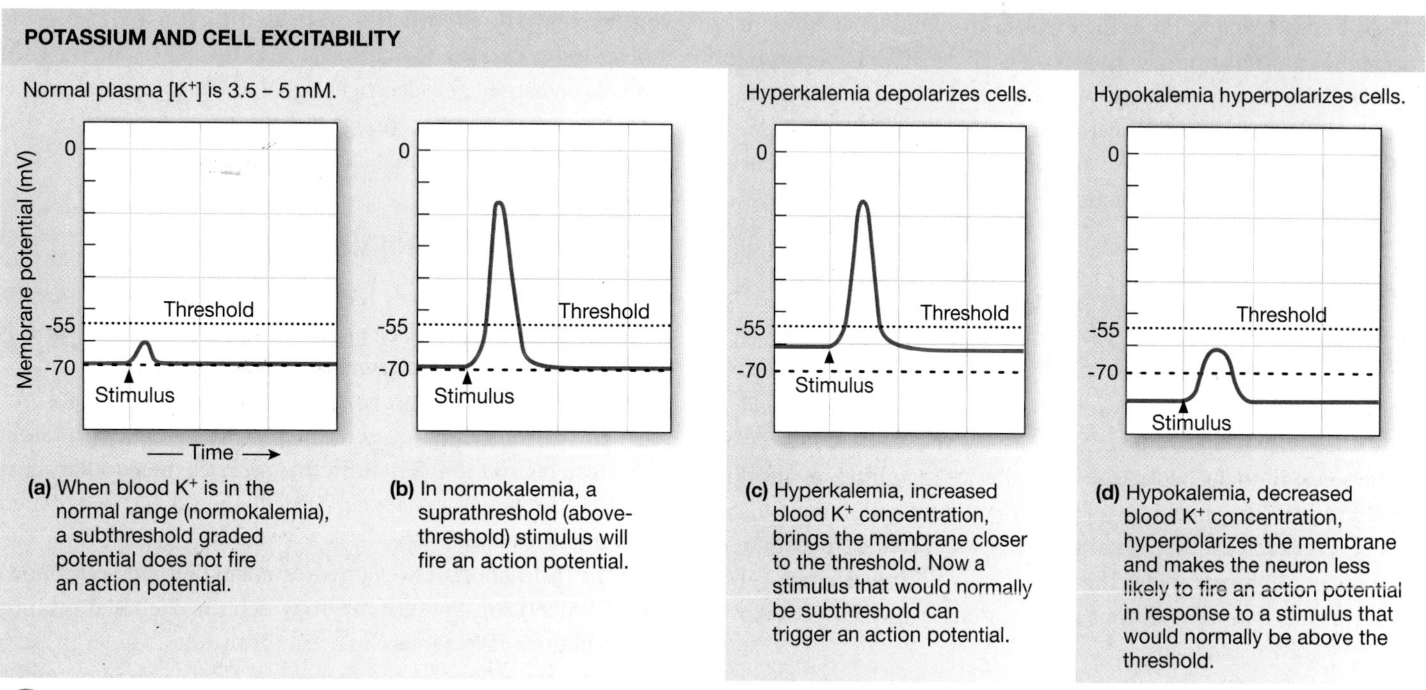

(a) When blood K^+ is in the normal range (normokalemia), a subthreshold graded potential does not fire an action potential.

(b) In normokalemia, a suprathreshold (above-threshold) stimulus will fire an action potential.

(c) Hyperkalemia, increased blood K^+ concentration, brings the membrane closer to the threshold. Now a stimulus that would normally be subthreshold can trigger an action potential.

(d) Hypokalemia, decreased blood K^+ concentration, hyperpolarizes the membrane and makes the neuron less likely to fire an action potential in response to a stimulus that would normally be above the threshold.

FIGURE QUESTION

The E_K of -90 mV is based on ECF [K^+] = 5 mM and ICF [K^+] = 150 mM.
Use the Nernst equation to calculate E_K when ECF [K^+] is (a) 2.5 mM and (b) 6 mM.

■ **Fig. 8.17**

RUNNING PROBLEM

Dr. McKhann decided to perform nerve conduction tests on some of the paralyzed children in Beijing Hospital. He found that although the rate of conduction along the children's nerves was normal, the strength of the summed action potentials traveling down the nerve was greatly diminished.

Q4: Is the paralytic illness that affected the Chinese children a demyelinating condition? Why or why not?

238 239 242 264 **266** 270 281

closer to threshold and causes the cells to fire action potentials in response to smaller graded potentials (Fig. 8.17c).

If blood K^+ concentration falls too low—a condition known as **hypokalemia**—the resting membrane potential of the cells hyperpolarizes, moving farther from threshold. In this case, a stimulus strong enough to trigger an action potential when the resting potential is the normal −70 mV does not reach the threshold value (Fig. 8.17d). This condition shows up as muscle weakness because the neurons that control skeletal muscles are not firing normally.

Hypokalemia and its resultant muscle weakness are one reason that sport drinks supplemented with Na^+ and K^+, such as Gatorade™, were developed. When people sweat excessively, they lose both salts and water. If they replace this fluid loss with pure water, the K^+ remaining in the blood is diluted, causing hypokalemia. By replacing sweat loss with a dilute salt solution, a person can prevent potentially dangerous drops in blood K^+ levels. Because of the importance of K^+ to normal function of the nervous system, the body regulates blood K^+ levels within a narrow range.

Cell-to-Cell Communication in the Nervous System

Information flow through the nervous system using electrical and chemical signals is one of the most active areas of neuroscience research today because so many devastating diseases affect this process. The specificity of neural communication depends on several factors: the signal molecules secreted by neurons, the target cell receptors for these chemicals, and the anatomical connections between neurons and their targets, which occur in regions known as synapses.

Neurons Communicate at Synapses

Each synapse has two parts: (1) the axon terminal of the *presynaptic cell* and (2) the membrane of the *postsynaptic cell* (see Fig. 8.2f, p. 241). In a neural reflex, information moves from presynaptic cell to postsynaptic cell. The postsynaptic cells may be neurons or non-neuronal cells. In most neuron-to-neuron synapses, the presynaptic axon terminals are next to either the dendrites or the cell body of the postsynaptic neuron.

In general, postsynaptic neurons with many dendrites also have many synapses. A moderate number of synapses is 10,000, but some cells in the brain are estimated to have 150,000 or more synapses on their dendrites! Synapses can also occur on the axon and even at the axon terminal of the postsynaptic cell. Synapses are classified as electrical or chemical depending on the type of signal that passes from the presynaptic cell to the postsynaptic one.

Electrical Synapses **Electrical synapses** pass an electrical signal, or current, directly from the cytoplasm of one cell to another through the pores of gap junction proteins [p. 78]. Information can flow in both directions through most gap junctions, but in some current can flow in only one direction (a *rectifying synapse*).

Electrical synapses occur mainly in neurons of the CNS. They are also found in glial cells, in cardiac and smooth muscle, and in nonexcitable cells that use electrical signals, such as the pancreatic beta cell. The primary advantage of electrical synapses is rapid conduction of signals from cell to cell that synchronizes activity within a network of cells. Gap junctions also allow chemical signal molecules to diffuse between adjacent cells.

Chemical Synapses The vast majority of synapses in the nervous system are **chemical synapses**, which use neurocrine molecules to carry information from one cell to the next. At chemical synapses, the electrical signal of the presynaptic cell is converted into a neurocrine signal that crosses the synaptic cleft and binds to a receptor on its target cell.

Neurons Secrete Chemical Signals

The number of molecules identified as neurocrine signals is large and growing daily. Neurocrine chemical composition is varied, and these molecules may function as neurotransmitters, neuromodulators, or neurohormones [p. 177]. Neurotransmitters and neuromodulators act as *paracrine signals*, with target cells located close to the neuron that secretes them. Neurohormones, in contrast, are secreted into the blood and distributed throughout the body.

The distinction between neurotransmitter and neuromodulator depends on the receptor to which the chemical is binding, as many neurocrines can act in both roles. Generally, if a molecule primarily acts at a synapse and elicits a rapid response, we call it a neurotransmitter, even if it can also act as a neuromodulator. Neuromodulators act at both synaptic and nonsynaptic sites and are slower acting. Some neuromodulators and neurotransmitters also act on the cell that secretes them, making them *autocrine* signals as well as paracrines.

Neurocrine Receptors The neurocrine receptors found in chemical synapses can be divided into two categories: receptor-channels, which are ligand-gated ion channels, and G protein–coupled receptors (GPCR) [p. 183]. Receptor-channels mediate rapid responses by altering ion flow across the membrane, so they are also called **ionotropic receptors**. Some ionotropic receptors are specific for a single ion, such as Cl^-, but others are less specific, such as the nonspecific monovalent cation channel.

G protein–coupled receptors mediate slower responses because the signal must be transduced through a second messenger system. GPCRs for neuromodulators are described as **metabotropic receptors**. Some metabotropic GPCRs regulate the opening or closing of ion channels.

All neurotransmitters except nitric oxide bind to one or more receptor types. Each receptor type may have multiple subtypes, allowing one neurotransmitter to have different effects in different tissues. Receptor subtypes are distinguished by combinations of letter and number subscripts. For example, serotonin (5-HT) has at least 20 receptor subtypes that have been identified, including 5-HT_{1A} and 5-HT_4.

The study of neurotransmitters and their receptors has been greatly simplified by two advances in molecular biology. The genes for many receptor subtypes have been cloned, allowing researchers to create mutant receptors and study their properties. In addition, researchers have discovered or synthesized a variety of agonist and antagonist molecules that mimic or inhibit neurotransmitter activity by binding to the receptors [p. 51].

Neurotransmitters Are Highly Varied

The array of neurocrines in the body and their many receptor types is truly staggering (■ Tbl. 8.4). Neurocrine molecules can be informally grouped into seven classes according to their structure: (1) acetylcholine, (2) amines, (3) amino acids, (4) peptides, (5) purines, (6) gases, and (7) lipids. CNS neurons release many different neurocrines, including some polypeptides known mostly for their hormonal activity, such as the hypothalamic releasing hormones and oxytocin and vasopressin [p. 221]. In contrast, the PNS secretes only three major neurocrines: the neurotransmitters acetylcholine and norepinephrine, and the neurohormone epinephrine. Some PNS neurons co-secrete additional neurocrines, such as ATP, which we will mention when they are functionally important.

Acetylcholine **Acetylcholine** (ACh), in a chemical class by itself, is synthesized from choline and acetyl coenzyme A (acetyl CoA). Choline is a small molecule also found in membrane phospholipids. Acetyl CoA is the metabolic intermediate that links glycolysis to the citric acid cycle [p. 114]. The synthesis of ACh from these two precursors is a simple enzymatic reaction that takes place in the axon terminal. Neurons that secrete ACh and receptors that bind ACh are described as **cholinergic**.

Cholinergic receptors come in two main subtypes: **nicotinic**, named because *nicotine* is an agonist, and **muscarinic**, for which *muscarine*, a compound found in some fungi, is an agonist. Cholinergic nicotinic receptors are receptor-channels found on skeletal muscle, in the autonomic division of the PNS, and in the CNS. Nicotinic receptors are monovalent cation channels through which both Na^+ and K^+ can pass. Sodium entry into cells exceeds K^+ exit because the electrochemical gradient for Na^+ is stronger. As a result, net Na^+ entry depolarizes the postsynaptic cell and makes it more likely to fire an action potential.

Cholinergic muscarinic receptors come in five related subtypes. They are all G protein–coupled receptors linked to second messenger systems. The tissue response to activation of a muscarinic receptor varies with the receptor subtype. These receptors occur in the CNS and in the autonomic parasympathetic division of the PNS.

Amines The amine neurotransmitters are all active in the CNS. Like the amine hormones [p. 216], these neurotransmitters are derived from single amino acids. **Serotonin**, also called *5-hydroxytryptamine* or 5-HT, is made from the amino acid tryptophan. *Histamine*, made from histadine, plays a role in allergic responses in addition to serving as a neurotransmitter.

The amino acid tyrosine is converted to **dopamine**, **norepinephrine**, and **epinephrine**. Norepinephrine is the major neurotransmitter of the PNS autonomic sympathetic division. All three tyrosine-derived neurocrines can also function as neurohormones.

Neurons that secrete norepinephrine are called **adrenergic neurons**, or, more properly, **noradrenergic neurons**. The adjective *adrenergic* does not have the same obvious link to its neurotransmitter as *cholinergic* does to *acetylcholine*. Instead, the adjective derives from the British name for epinephrine, *adrenaline*. In the early part of the twentieth century, British researchers thought that sympathetic neurons secreted adrenaline (epinephrine), hence the modifier *adrenergic*. Although our understanding has changed, the name persists. Whenever you see reference to "adrenergic control" of a function, you must make the connection to a neuron secreting norepinephrine.

Adrenergic receptors are divided into two classes: α (alpha) and β (beta), with multiple subtypes of each. Like cholinergic muscarinic receptors, adrenergic receptors are linked to G proteins. The two subtypes of adrenergic receptors work through different second messenger pathways. The action of epinephrine on β-receptors in dog liver led E. W. Sutherland to the discovery of cyclic AMP and the concept of second messenger systems as transducers of extracellular messengers [p. 183].

Major Neurocrines*

Table 8.4

Chemical	Receptor	Type	Receptor Location	Key Agonists, Antagonists, and Potentiators†
Acetylcholine (ACh)	Cholinergic			
	Nicotinic	ICR‡ (Na^+, K^+)	Skeletal muscles, autonomic neurons, CNS	**Agonist**: nicotine **Antagonists**: curare, α-bungarotoxin
	Muscarinic	GPCR	Smooth and cardiac muscle, endocrine and exocrine glands, CNS	**Agonist**: muscarine **Antagonist**: atropine
Amines				
Norepinephrine (NE)	Adrenergic (α, β)	GPCR	Smooth and cardiac muscle, glands, CNS	**Antagonists**: α-receptors: ergotamine, phentolamine β-receptors: propranolol
Dopamine (DA)	Dopamine (D)	GPCR	CNS	**Agonist**: bromocriptine **Antagonists**: antipsychotic drugs
Serotonin (5-hydroxytryptamine, 5-HT)	Serotonergic (5-HT)	ICR (Na^+, K^+), GPCR	CNS	**Agonist**: sumatriptan **Antagonist**: LSD
Histamine	Histamine (H)	GPCR	CNS	**Antagonists**: ranitidine (Zantac®) and cimetidine (Tagamet®)
Amino acids				
Glutamate	Glutaminergic ionotropic (iGluR)			
	AMPA	ICR (Na^+, K^+)	CNS	**Agonist**: quisqualate
	NMDA	ICR (Na^+, K^+, Ca^{2+})	CNS	**Potentiator**: serine
	Glutaminergic metabotropic (mGluR)	GPCR	CNS	**Potentiator**: glycine
GABA (γ-aminobutyric acid)	GABA	ICR (Cl^-), GPCR	CNS	**Antagonist**: picrotoxin **Potentiators**: alcohol, barbiturates
Glycine	Glycine	ICR (Cl^-)	CNS	**Antagonist**: strychnine
Purines				
Adenosine	Purine (P)	GPCR	CNS	
Gases				
Nitric oxide (NO)	None	N/A	N/A	

*This table does not include the numerous peptides that can act as neurocrines.

†This list does not include many chemicals that are used as agonists and antagonists in physiological research.

‡ICR = ion channel-receptor; GPCR = G protein–coupled receptor; AMPA = α-amino-3-hydroxy-5-methyl-4 isoxazole proprionic acid; NMDA = N-methyl-D-aspartate; LSD = lysergic acid diethylamine; N/A = not applicable.

8

Concept Check Answer: p. 287

18. When pharmaceutical companies design drugs, they try to make a given drug as specific as possible for the particular receptor subtype they are targeting. For example, a drug might target adrenergic β_1- receptors rather than all adrenergic α- and β-receptors. What is the advantage of this specificity?

Amino Acids Several amino acids function as neurotransmitters in the CNS. **Glutamate** is the primary excitatory neurotransmitter of the CNS, and **aspartate** is an excitatory neurotransmitter in selected regions of the brain. *Excitatory neurotransmitters* depolarize their target cells, usually by opening ion channels that allow flow of positive ions into the cell.

The main inhibitory neurotransmitter in the brain is **gamma-aminobutyric acid (GABA)**. The primary inhibitory neurotransmitter of the spinal cord is the amino acid **glycine**. These *inhibitory neurotransmitters* hyperpolarize their target cells by opening Cl^- channels and allowing Cl^- to enter the cell.

Glutamate also acts as a neuromodulator. The action of glutamate at a particular synapse depends on which of its receptor types occurs on the target cell. Metabotropic glutaminergic receptors act through GPCRs. Two ionotropic glutamate receptors are receptor-channels.

AMPA receptors are ligand-gated monovalent cation channels similar to nicotinic acetylcholine channels. Glutamate binding opens the channel, and the cell depolarizes because of net Na^+ influx. AMPA receptors are named for their agonist *α-amino-3-hydroxy-5-methylisoxazole-4-proprionic acid.*

NMDA receptors are named for the glutamate agonist N-*methyl*-D-*aspartate*. They are unusual for several reasons. First, they are nonselective cation channels that allow Na^+, K^+, and Ca^{2+} to pass through the channel. Second, channel opening requires both glutamate binding and a change in membrane potential. The NMDA receptor-channel's action is described in the section on long-term potentiation later in this chapter.

Glycine and the amino acid *D-serine* potentiate, or enhance, the excitatory effects of glutamate at one type of glutamate receptor. D-serine is made and released by glial cells as well as neurons, which illustrates the role that glial cells can play in altering electrical communication.

Peptides The nervous system secretes a variety of peptides that act as neurotransmitters and neuromodulators in addition to functioning as neurohormones. These peptides include **substance P**, involved in some pain pathways, and the **opioid peptides** (**enkephalins** and **endorphins**) that mediate pain relief, or *analgesia* {*an*-, without + *algos*, pain}. Peptides that function as both neurohormones and neurotransmitters include *cholecystokinin (CCK), vasopressin,* and *atrial natriuretic peptide.* Many peptide neurotransmitters are co-secreted with other neurotransmitters.

CLINICAL FOCUS

Myasthenia Gravis

What would you think was wrong if suddenly your eyelids started drooping, you had difficulty watching moving objects, and it became difficult to chew, swallow, and talk? What disease attacks these skeletal muscles but leaves the larger muscles of the arms and legs alone? The answer is **myasthenia gravis** {*myo*-, muscle + *asthenes*, weak + *gravis*, severe}, an autoimmune disease in which the body fails to recognize the acetylcholine (ACh) receptors on skeletal muscle as part of "self." The immune system then produces antibodies to attack the receptors. The antibodies bind to the ACh receptor protein and change it in some way that causes the muscle cell to pull the receptors out of the membrane and destroy them. This destruction leaves the muscle with fewer ACh receptors in the membrane. Even though neurotransmitter release is normal, the muscle target has a diminished response that is exhibited as muscle weakness. Currently medical science does not have a cure for myasthenia gravis, although various drugs can help control its symptoms. To learn more about this disease, visit the web site for the Myasthenia Gravis Foundation of America at *www.myasthenia.org*.

Purines *Adenosine, adenosine monophosphate* (AMP), and *adenosine triphosphate* (ATP) can all act as neurotransmitters. These molecules, known collectively as *purines* [p. 36], bind to *purinergic* receptors in the CNS and on other excitable tissues such as the heart. The purines all bind to G protein–coupled receptors.

Gases One of the most interesting neurotransmitters is *nitric oxide* (NO), an unstable gas synthesized from oxygen and the amino acid arginine. Nitric oxide acting as a neurotransmitter diffuses freely into a target cell rather than binding to a membrane receptor [p. 187]. Once inside the target cell, nitric oxide binds to proteins. With a half-life of only 2–30 seconds, nitric oxide is elusive and difficult to study. It is also released from cells other than neurons and often acts as a paracrine.

Recent work suggests that *carbon monoxide* (CO) and hydrogen sulfide (H_2S), both known as toxic gases, are produced by the body in tiny amounts to serve as neurotransmitters.

Lipids Lipid neurocrines include several eicosanoids [p. 33] that are the endogenous ligands for *cannabinoid receptors*. The CB_1 cannabinoid receptor is found in the brain, and the CB_2 receptor is found on immune cells. The receptors were named for one of their exogenous ligands, Δ^9-*tetrahydrocannabinoid* (THC), which comes from the plant *Cannabis sativa*, more

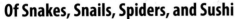

BIOTECHNOLOGY

Of Snakes, Snails, Spiders, and Sushi

What do snakes, marine snails, and spiders have to do with neurophysiology? They all provide neuroscientists with compounds for studying synaptic transmission, extracted from the neurotoxic venoms these creatures use to kill their prey. The Asian snake *Bungarus multicinctus* provides us with α-bungarotoxin, a long-lasting poison that binds tightly to nicotinic acetylcholine receptors. The fish-hunting cone snail, *Conus geographus*, and the funnel web spider, *Agelenopsis aperta*, use toxins that block different types of voltage-gated Ca^{2+} channels. One of the most potent poisons known, however, comes from the Japanese puffer fish, a highly prized delicacy whose flesh is consumed as sushi. The puffer has tetrodotoxin (TTX) in its gonads. This neurotoxin blocks Na^+ channels on axons and prevents the transmission of action potentials, so ingestion of only a tiny amount can be fatal. The Japanese chefs who prepare the puffer fish, or *fugu*, for consumption are carefully trained to avoid contaminating the fish's flesh as they remove the toxic gonads. There's always some risk involved in eating *fugu,* though—one reason that traditionally the youngest person at the table is the first to sample the dish.

commonly known as marijuana. Lipid neurocrines all bind to G protein–coupled receptors.

Neurotransmitters Are Released from Vesicles

When we examine the axon terminal of a presynaptic cell with an electron microscope, we find many small **synaptic vesicles** filled with neurotransmitter that is released on demand (■ Fig. 8.18).

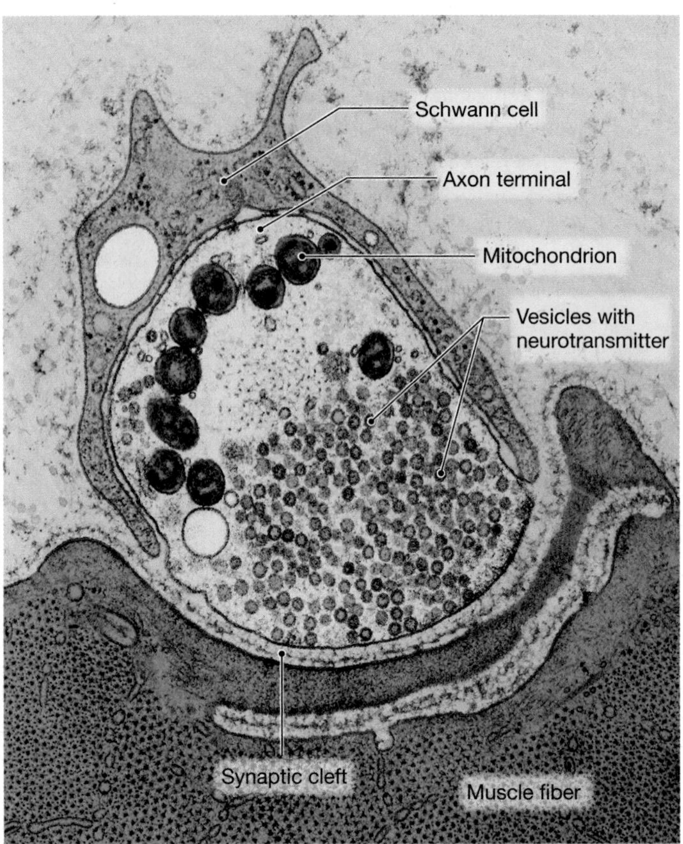

■ **Fig. 8.18** ***A chemical synapse.*** The axon terminal contains mitochondria and synaptic vesicles filled with neurotransmitter. The postsynaptic membrane has receptors for neurotransmitter that diffuses across the synaptic cleft.

RUNNING PROBLEM

Dr. McKhann then asked to see autopsy reports on some of the children who had died of their paralysis at Beijing Hospital. In the reports, pathologists noted that the patients had normal myelin but damaged axons. In some cases, the axon had been completely destroyed, leaving only a hollow shell of myelin.

Q5: Do the results of Dr. McKhann's investigation suggest that the Chinese children had classic Guillain-Barré syndrome? Why or why not?

238 239 242 264 266 270 281

Some vesicles are "docked" at active zones along the membrane closest to the synaptic cleft, waiting for a signal to release their contents. Other vesicles act as a reserve pool, clustering close to the docking sites. Axon terminals also contain mitochondria to produce ATP for metabolism and transport. In this section we discuss general patterns of neurotransmitter synthesis, storage, release, and termination of action.

Neurotransmitter Synthesis Neurotransmitter synthesis takes place both in the nerve cell body and in the axon terminal. Polypeptide neurotransmitters must be made in the cell body because axon terminals do not have the organelles needed for protein synthesis. Protein synthesis follows the usual pathway [p. 77]. The large *propeptide* that results is packaged into vesicles along with the enzymes needed to modify it. The vesicles then move from the cell body to the axon terminal by fast axonal transport. Inside the vesicle, the propeptide is broken down into smaller active peptides—a pattern similar to the preprohormone-prohormone-active hormone process in endocrine cells [p. 212]. For example, one propeptide contains the amino acid sequences for three active peptides that are co-secreted: ACTH, gamma(γ)-lipotropin, and beta(β)-endorphin.

Smaller neurotransmitters, such as acetylcholine, amines, and purines, are synthesized and packaged into vesicles in the axon terminal. The enzymes needed for their synthesis are made in the cell body and released into the cytosol. The dissolved enzymes are then brought to axon terminals by slow axonal transport.

Concept Check Answers: p. 287

19. Which organelles are needed to synthesize proteins and package them into vesicles?

20. What is the function of mitochondria in a cell?

21. How do mitochondria get to the axon terminals?

Neurotransmitter Release Neurotransmitters in the axon terminal are stored in vesicles, so their release into the synaptic cleft takes place by exocytosis [p. 157]. From what we can tell, exocytosis in neurons is similar to exocytosis in other types of cells, but much faster. Neurotoxins that block neurotransmitter release, including tetanus and botulinum toxins, exert their action by inhibiting specific proteins of the cell's exocytotic apparatus.

■ Figure 8.19a shows how neurotransmitters are released by exocytosis. When the depolarization of an action potential reaches the axon terminal, the change in membrane potential sets off a sequence of events (1). The axon terminal membrane has voltage-gated Ca^{2+} channels that open in response to depolarization (2). Calcium ions are more concentrated in the extracellular fluid than in the cytosol, and so they move into the cell.

Ca^{2+} entering the cell binds to regulatory proteins and initiates exocytosis (3). The membrane of the synaptic vesicle fuses with the cell membrane, aided by multiple membrane proteins. The fused area opens, and neurotransmitter inside the synaptic vesicle moves into the synaptic cleft (4). The neurotransmitter molecules diffuse across the gap to bind with membrane receptors on the postsynaptic cell. When neurotransmitters bind to their receptors, a response is initiated in the postsynaptic cell (5). Each synaptic vesicle contains the same amount of neurotransmitter, so measuring the magnitude of the target cell response is an indication of how many vesicles released their content.

In the classic model of exocytosis, the membrane of the vesicle becomes part of the axon terminal membrane [Fig. 5.19, p. 156]. To prevent a large increase in membrane surface area, the membrane is recycled by endocytosis of vesicles at regions away from the active sites (see Fig. 8.3, p. 243). The recycled vesicles are then refilled with newly made neurotransmitter.

The transporters that concentrate neurotransmitter into vesicles are H^+-dependent antiporters [p. 148]. The vesicles use H^+-ATPases to concentrate H^+ inside the vesicle, then exchange the H^+ for the neurotransmitter.

Recently, a second model of secretion has emerged. In this model, called the **kiss-and-run pathway**, synaptic vesicles fuse to the presynaptic membrane at a complex called the **fusion pore**. This fusion opens a small channel that is just large enough for neurotransmitter to pass through. Then, instead of opening the fused area wider and incorporating the vesicle membrane into the cell membrane, the vesicle pulls back from the fusion pore and returns to the pool of vesicles in the cytoplasm.

Concept Check Answers: p. 287

22. In an experiment on synaptic transmission, a synapse was bathed in a Ca^{2+}-free medium that was otherwise equivalent to extracellular fluid. An action potential was triggered in the presynaptic neuron. Although the action potential reached the axon terminal at the synapse, the usual response of the postsynaptic cell did not occur. What conclusion did the researchers draw from these results?

23. Classify the H^+-neurotransmitter exchange as facilitated diffusion, primary active transport, or secondary active transport. Explain your reasoning.

Termination of Neurotransmitter Activity A key feature of neural signaling is its short duration, due to the rapid removal or inactivation of neurotransmitter in the synaptic cleft. Recall that ligand binding to a protein is reversible and goes to a state of equilibrium, with a constant ratio of unbound to bound ligand [p. 165]. If unbound neurotransmitter is removed from the synapse, the receptors release bound neurotransmitter, terminating its activity, to keep the ratio of unbound/bound transmitter constant.

Removal of unbound neurotransmitter from the synaptic cleft can be accomplished in various ways (Fig. 8.19b). Some neurotransmitter molecules simply diffuse away from the synapse, becoming separated from their receptors. Other neurotransmitters are inactivated by enzymes in the synaptic cleft. For example, acetylcholine (ACh) in the extracellular fluid is rapidly broken down into choline and acetyl CoA by the enzyme **acetylcholinesterase** (AChE) in the extracellular matrix and in the membrane of the postsynaptic cell (■ Fig. 8.20). Choline from degraded ACh is transported back into the presynaptic axon terminal on a Na^+-dependent cotransporter. Once back in the axon terminal, it can be used to make new acetylcholine.

Many neurotransmitters are removed from the extracellular fluid by transport either back into the presynaptic cell or into adjacent neurons or glial cells. For example, norepinephrine action is terminated when the intact neurotransmitter is transported back into the presynaptic axon terminal. Norepinephrine uptake uses a Na^+-dependent cotransporter. Once back in the axon terminal, norepinephrine is either transported back into

■ Fig. 8.19 *ESSENTIALS*

Synaptic Communication

Cell-to-cell communication uses chemical and electrical signaling to coordinate function and maintain homeostasis.

(a) Neurotransmitter Release

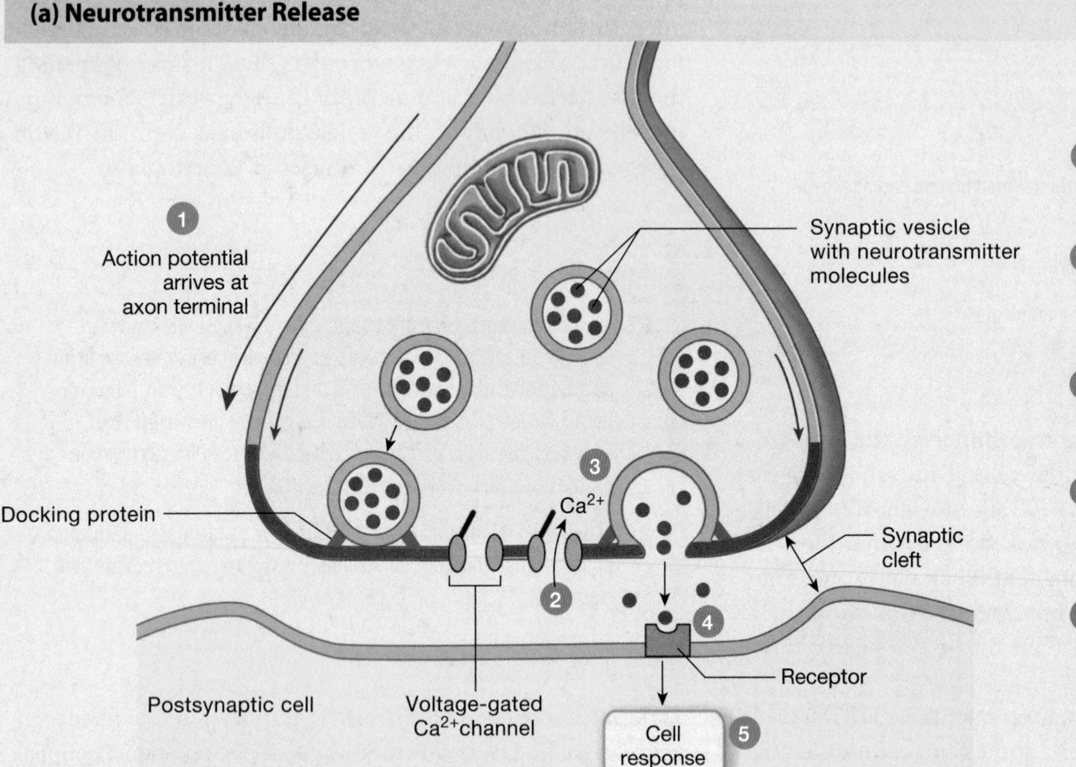

1. An action potential depolarizes the axon terminal.
2. The depolarization opens voltage-gated Ca^{2+} channels, and Ca^{2+} enters the cell.
3. Calcium entry triggers exocytosis of synaptic vesicle contents.
4. Neurotransmitter diffuses across the synaptic cleft and binds with receptors on the postsynaptic cell.
5. Neurotransmitter binding initiates a response in the postsynaptic cell.

(b) Neurotransmitter Termination

Neurotransmitter action terminates when the chemicals are broken down, are taken up into cells, or diffuse away from the synapse.

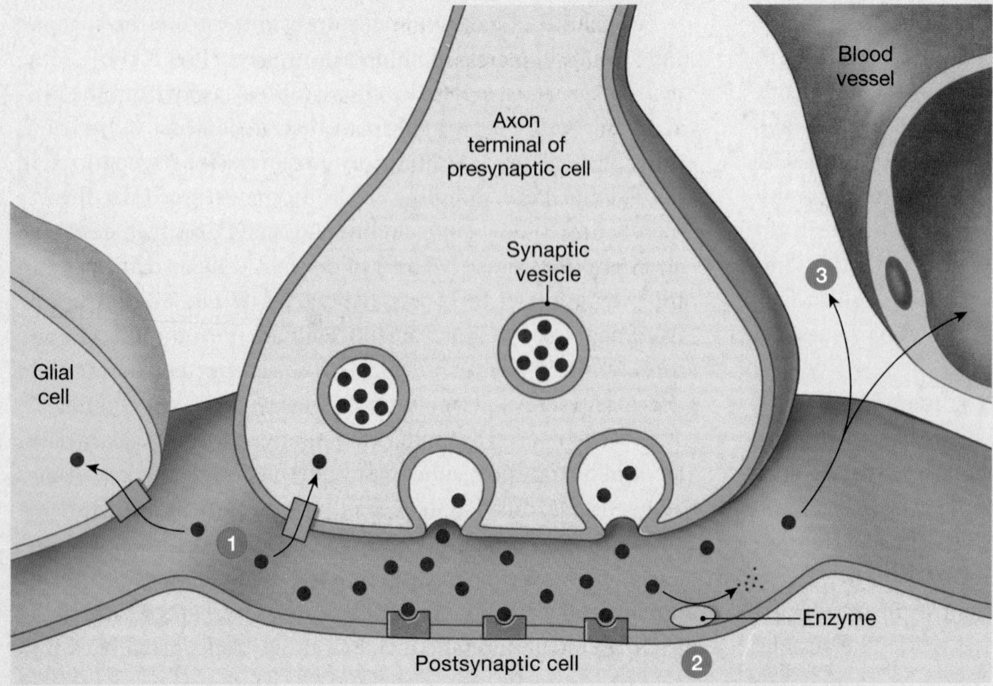

1. Neurotransmitters can be returned to axon terminals for reuse or transported into glial cells.
2. Enzymes inactivate neurotransmitters.
3. Neurotransmitters can diffuse out of the synaptic cleft.

8

SYNTHESIS AND RECYCLING OF ACETYLCHOLINE

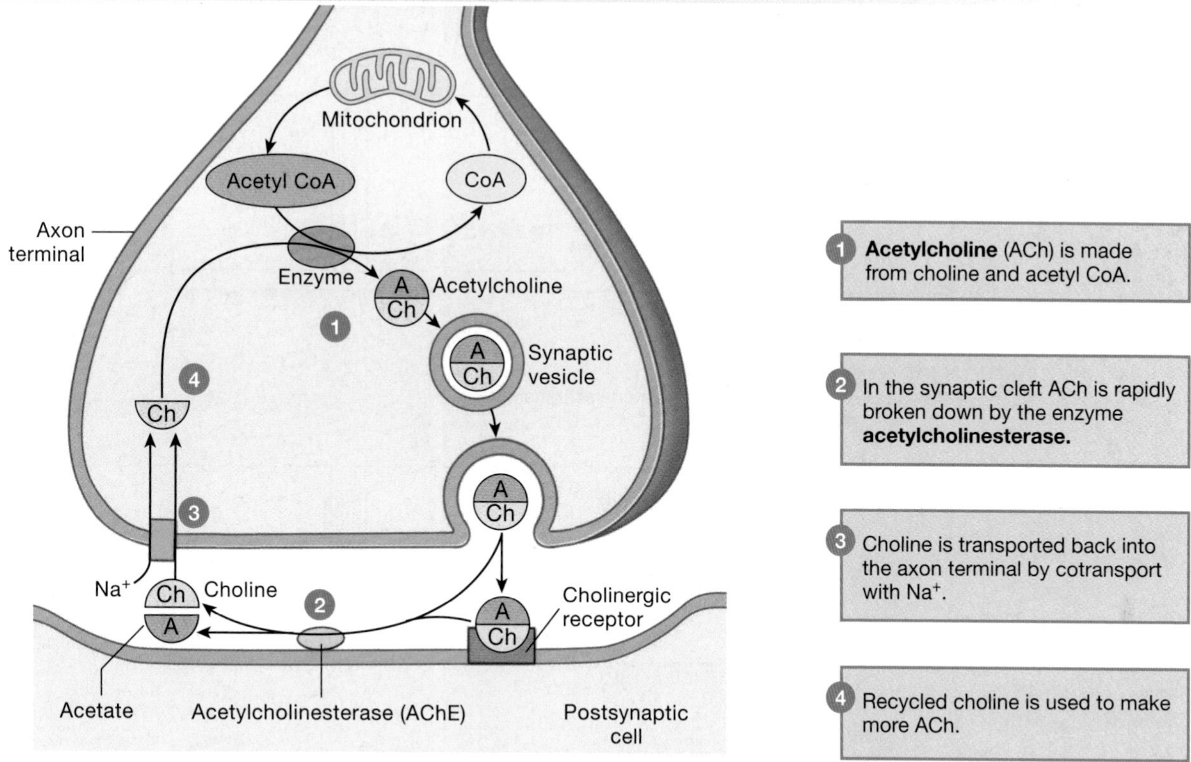
Mitochondrion
Acetyl CoA
CoA
Axon terminal
Enzyme
Acetylcholine
Synaptic vesicle
Na+
Choline
Cholinergic receptor
Acetate
Acetylcholinesterase (AChE)
Postsynaptic cell
1 **Acetylcholine** (ACh) is made from choline and acetyl CoA.
2 In the synaptic cleft ACh is rapidly broken down by the enzyme **acetylcholinesterase.**
3 Choline is transported back into the axon terminal by cotransport with Na+.
4 Recycled choline is used to make more ACh.

■ Fig. 8.20

vesicles or broken down by intracellular enzymes such as *monoamine oxidase (MAO)*, found in mitochondria. Neurotransmitters and their components can be recycled to refill empty synaptic vesicles.

Concept Check Answers: p. 287

24. One class of antidepressant drugs is called selective serotonin reuptake inhibitors (SSRIs). What do these drugs do to serotonin activity at the synapse?

25. How does the axon terminal make acetyl CoA for acetylcholine synthesis? (Hint: p. 267)

26. Is Na^+-dependent neurotransmitter reuptake facilitated diffusion, primary active transport, or secondary active transport? Explain your reasoning.

Stronger Stimuli Release More Neurotransmitter

A single action potential arriving at the axon terminal releases a constant amount of neurotransmitter. Neurons therefore can use the frequency of action potentials to transmit information about the strength and duration of the stimuli that activated them. A stronger stimulus causes more action potentials per second to arrive at the axon terminal, which in turn results in more neurotransmitter release.

For example, let's consider how a sensory neuron tells the CNS the intensity of an incoming stimulus. An above-threshold graded potential reaching the trigger zone of the sensory neuron does not trigger just one action potential. Instead, even a small graded potential that is above threshold triggers a burst of action potentials (■ Fig. 8.21a). As graded potentials increase in strength (amplitude), they trigger more frequent action potentials (Fig. 8.21b).

Usually a burst of action potentials arriving at the axon terminal results in increased neurotransmitter release, as shown in Figure 8.21b. However, in some cases of sustained activity, neurotransmitter release may decrease over time because the axon cannot replenish its neurotransmitter supply rapidly enough.

Electrical signaling patterns in the CNS are more variable. Brain neurons show different electrical personalities by firing action potentials in a variety of patterns, sometimes spontaneously, without an external stimulus to bring them to threshold. For example, some neurons are *tonically active* [p. 192], firing regular trains of action potentials (beating pacemakers). Other neurons exhibit *bursting*, bursts of action potentials rhythmically alternating with intervals of quiet (rhythmic pacemakers).

These different firing patterns in CNS neurons are created by ion channel variants that differ in their activation and inactivation voltages, opening and closing speeds, and sensitivity to neuromodulators. This variability makes brain neurons more dynamic and complicated than the simple somatic motor neuron we use as our model.

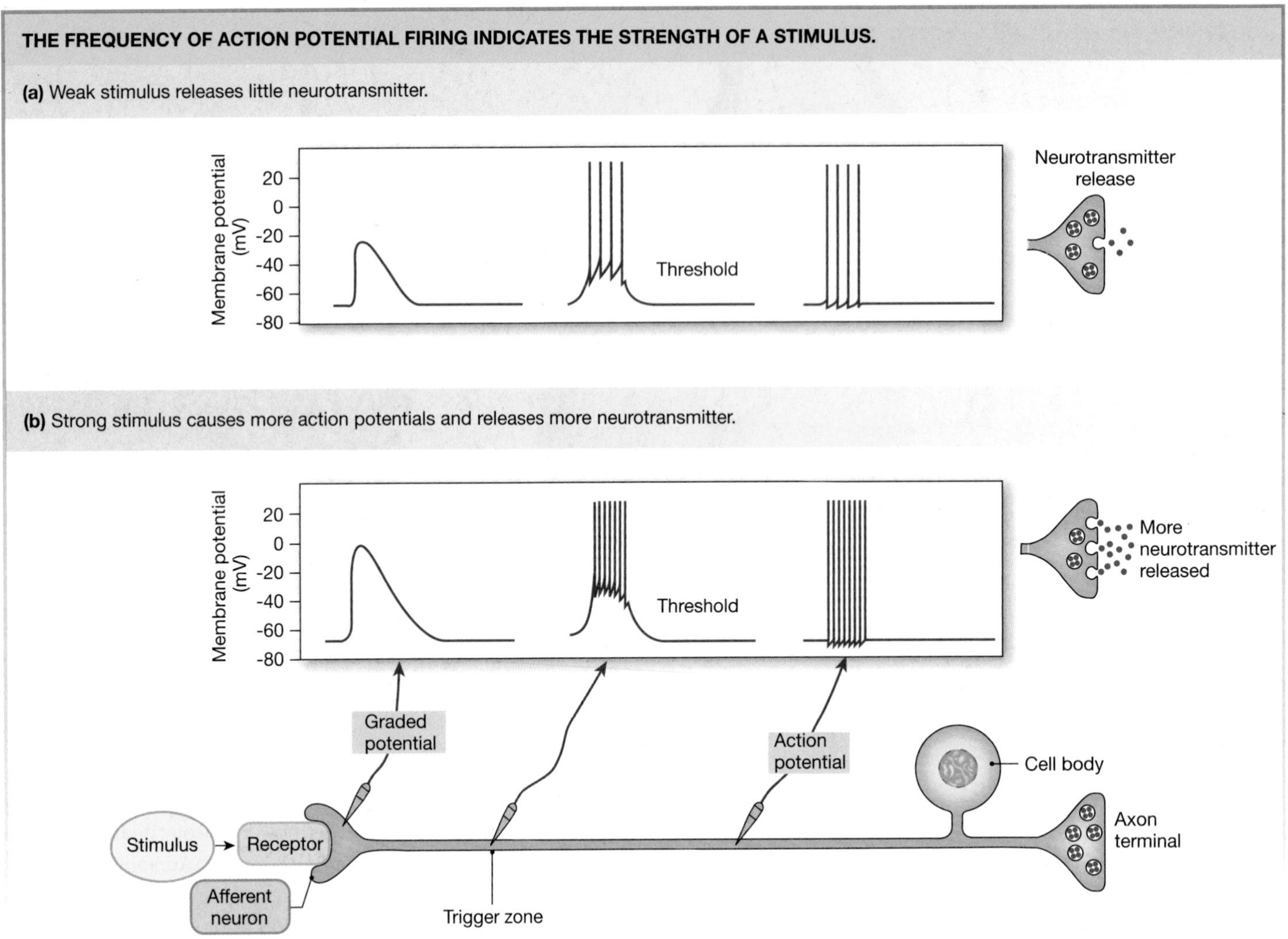

■ **Fig. 8.21**

Integration of Neural Information Transfer

Communication between neurons is not always a one-to-one event as we have been describing. Frequently, a single presynaptic neuron branches, and its collaterals synapse on multiple target neurons. This pattern is known as **divergence** (■ Fig. 8.22a). On the other hand, when a group of presynaptic neurons provide input to a smaller number of postsynaptic neurons, the pattern is known as **convergence** (Fig. 8.22b).

Combination of convergence and divergence in the CNS may result in one postsynaptic neuron with synapses from as many as 10,000 presynaptic neurons (Fig. 8.22c). For example, the Purkinje neurons of the CNS have highly branched dendrites so that they can receive information from many neurons (Fig. 8.22d).

In addition, we now know that the traditional view of chemical synapses as sites of one-way communication, with all messages moving from presynaptic cell to postsynaptic cell, is not always correct. In the brain, there are some synapses where cells on both sides of the synaptic cleft release neurotransmitters that act on the opposite cell. Perhaps more importantly, we have learned that many postsynaptic cells "talk back" to their presynaptic neurons by sending neuromodulators that bind to presynaptic receptors. Variations in synaptic activity play a major role in determining how communication takes place in the nervous system.

The ability of the nervous system to change activity at synapses is called **synaptic plasticity** {*plasticus*, that which may be molded}. Short term plasticity may enhance activity at the synapse (facilitation) or decrease it (depression). Sometimes changes at the synapse persist for significant periods of time (long-term depression or long-term potentiation). In the sections that follow we examine some of the ways that communication at synapses can be modified.

Postsynaptic Responses May Be Slow or Fast

A neurotransmitter combining with its receptor sets in motion a series of responses in the postsynaptic cell (■ Fig. 8.23). Neurotransmitters that bind to G protein–coupled receptors linked to second messenger systems initiate slow postsynaptic responses.

Divergence and Convergence

(a) In a **divergent pathway**, one presynaptic neuron branches to affect a larger number of postsynaptic neurons.

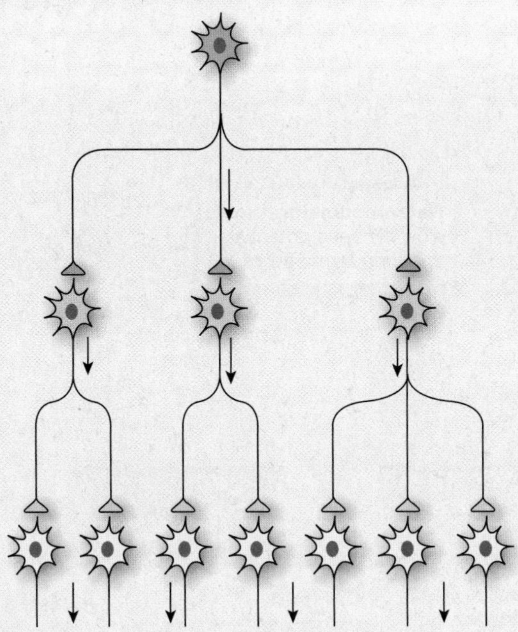

(b) In a **convergent pathway**, many presynaptic neurons provide input to influence a smaller number of postsynaptic neurons.

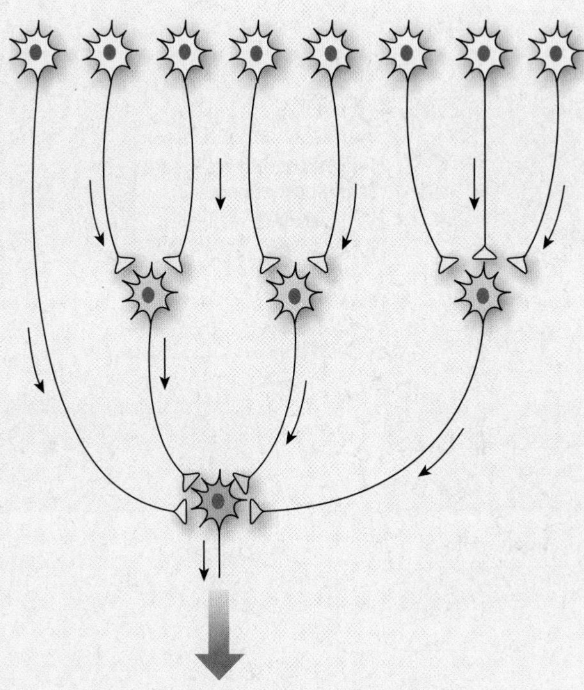

FIGURE QUESTION

The pattern of divergence in (a) is similar to ______________ in a second messenger system.

(c) The cell body of a somatic motor neuron is nearly covered with synapses providing input from other neurons.

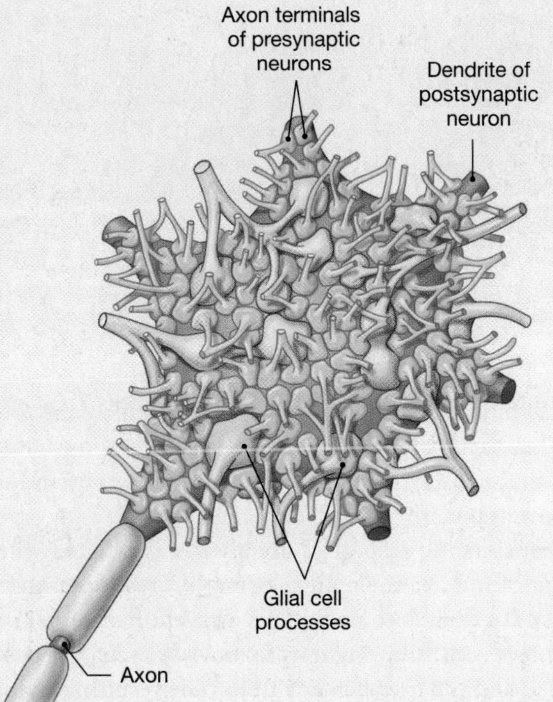

(d) The highly branched dendrites of a Purkinje cell (neuron) demonstrate convergence of signals from many synapses onto a cell body.

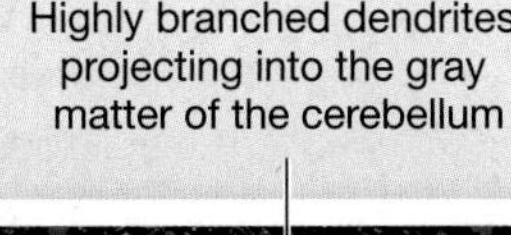

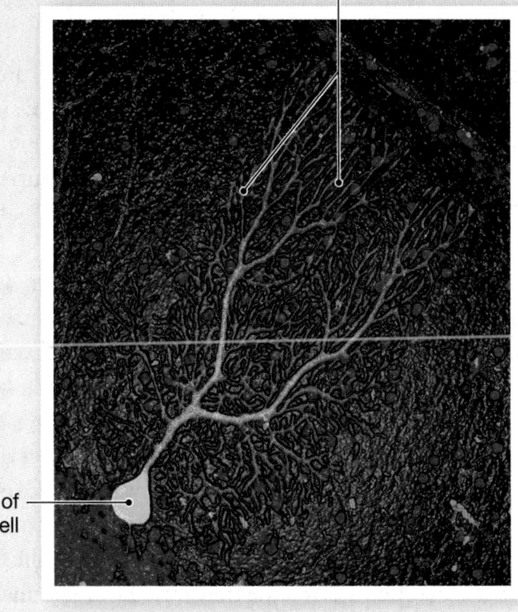

Light micrograph of Purkinje cells in cerebellum

■ Fig. 8.23 *ESSENTIALS*

Fast and Slow Postsynaptic Responses

Fast responses are mediated by ion channels.

Slow responses are mediated by G protein–coupled receptors.

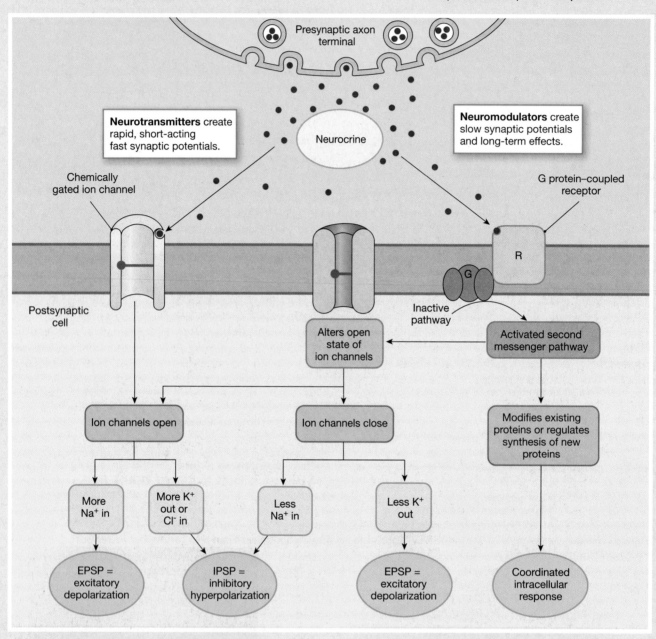

Presynaptic axon terminal

Neurotransmitters create rapid, short-acting fast synaptic potentials.

Neurocrine

Neuromodulators create slow synaptic potentials and long-term effects.

Chemically gated ion channel

G protein–coupled receptor

R

G

Postsynaptic cell

Inactive pathway

Alters open state of ion channels

Activated second messenger pathway

Ion channels open

Ion channels close

Modifies existing proteins or regulates synthesis of new proteins

More Na^+ in

More K^+ out or Cl^- in

Less Na^+ in

Less K^+ out

EPSP = excitatory depolarization

IPSP = inhibitory hyperpolarization

EPSP = excitatory depolarization

Coordinated intracellular response

Some second messengers act from the cytoplasmic side of the cell membrane to open or close ion channels. Changes in membrane potential resulting from these alterations in ion flow are called **slow synaptic potentials** because the response of the second messenger pathway takes longer than the direct opening or closing of a channel. In addition, the response itself lasts longer, usually seconds to minutes.

Slow postsynaptic responses are not limited to altering the open state of ion channels. Neurotransmitters acting on GPCRs may also modify existing cell proteins or regulate the production of new cell proteins. These types of slow response have been linked to the growth and development of neurons and to the mechanisms underlying long-term memory.

Fast synaptic responses are always associated with the opening of ion channels. In the simplest response, the neurotransmitter binds to and opens a receptor-channel on the postsynaptic cell, allowing ions to move between the postsynaptic cell and the extracellular fluid. The resulting change in membrane potential is called a **fast synaptic potential** because it begins quickly and lasts only a few milliseconds.

If the synaptic potential is depolarizing, it is called an **excitatory postsynaptic potential (EPSP)** because it makes the cell more likely to fire an action potential. If the synaptic potential is hyperpolarizing, it is called an **inhibitory postsynaptic potential (IPSP)** because hyperpolarization moves the membrane potential farther from threshold and makes the cell less likely to fire an action potential.

Neural Pathways May Involve Many Neurons

When two or more presynaptic neurons converge on the dendrites or cell body of a single postsynaptic cell, the response of the postsynaptic cell is determined by the summed input from the presynaptic neurons. If the stimuli all create subthreshold excitatory postsynaptic potentials (EPSPs), those EPSPs can sum to create a suprathreshold potential at the trigger zone.

The initiation of an action potential from several nearly simultaneous graded potentials is an example of **spatial summation**. The word *spatial* {*spatium,* space} refers to the fact that the graded potentials originate at different locations (spaces) on the neuron.

■ Figure 8.24a illustrates spatial summation when three presynaptic neurons releasing excitatory neurotransmitters ("excitatory neurons") converge on one postsynaptic neuron. Each neuron's EPSP is too weak to trigger an action potential by itself, but if the three presynaptic neurons fire simultaneously, the sum of the three EPSPs is suprathreshold and creates an action potential.

Postsynaptic inhibition may occur when a presynaptic neuron releases an inhibitory neurotransmitter onto a postsynaptic cell and alters its response. Figure 8.24b shows three neurons, two excitatory and one inhibitory, converging on a postsynaptic cell. The neurons fire, creating one inhibitory postsynaptic potential (IPSP) and two excitatory graded potentials that sum as they reach the trigger zone. The IPSP counteracts the two EPSPs, creating an integrated signal that is below threshold. As a result, no action potential leaves the trigger zone.

Summation of graded potentials does not always require input from more than one presynaptic neuron. Two subthreshold graded potentials from the same presynaptic neuron can be summed if they arrive at the trigger zone close enough together in time. Summation that occurs from graded potentials overlapping in time is called **temporal summation** {*tempus,* time}. Let's see how this can happen.

Figure 8.24c shows recordings from an electrode placed in the trigger zone of a neuron. A stimulus (X1) starts a subthreshold graded potential on the cell body at the time marked on the x-axis. The graded potential reaches the trigger zone and depolarizes it, as shown on the graph (A1), but not enough to trigger an action potential. A second stimulus (X2) occurs later, and its subthreshold graded potential (A2) reaches the trigger zone sometime after the first. The interval between the two stimuli is so long that the two graded potentials do not overlap. Neither potential by itself is above threshold, so no action potential is triggered.

In Figure 8.24d, the two stimuli occur closer together in time. As a result, the two subthreshold graded potentials arrive at the trigger zone at almost the same time. The second graded potential adds its depolarization to that of the first, causing the trigger zone to depolarize to threshold.

In many situations, graded potentials in a neuron incorporate both temporal and spatial summation. The summation of graded potentials demonstrates a key property of neurons: *postsynaptic integration.* When multiple signals reach a neuron, postsynaptic integration creates a signal based on the relative strengths and durations of the signals. If the integrated signal is above threshold, the neuron fires an action potential. If the integrated signal is below threshold, the neuron does not fire.

■ Figure 8.25 shows the distribution of excitatory and inhibitory synapses on a three-dimensional reconstruction of dendritic spines of various shapes and sizes. The summed input from these synapses determines the activity of the postsynaptic neuron.

Concept Check

Answers: p. 287

27. In Figure 8.24b, assume the postsynaptic neuron has a resting membrane potential of −70 mV and a threshold of −55 mV. If the inhibitory presynaptic neuron creates an IPSP of −55 mV, and the two excitatory presynaptic neurons have EPSPs of 10 and 12 mV, will the postsynaptic neuron fire an action potential?
28. In the graphs of Figure 8.24c and 8.24d, why doesn't the membrane potential change at the same time as the stimulus?

Synaptic Activity Can Be Modified

The examples of synaptic integration we just discussed all took place on the postsynaptic side of a synapse, but the activity of presynaptic cells can also be altered. When an inhibitory or excitatory neuron terminates on or close to an axon terminal of a presynaptic cell, its IPSP or EPSP can alter the action potential reaching the terminal and alter neurotransmitter release by the presynaptic cell.

If activity in an inhibitory neuron decreases neurotransmitter release, the modulation is called *presynaptic inhibition* (■ Fig. 8.26a). Presynaptic inhibition allows selective modulation of collaterals and their targets. One collateral can be inhibited while others remain unaffected. In *presynaptic facilitation,* input from an excitatory neuron increases neurotransmitter release by the presynaptic cell.

Presynaptic alteration of neurotransmitter release provides a more precise means of control than postsynaptic modulation. In postsynaptic modulation, if a neuron synapses on the dendrites and cell body of a neuron, the responsiveness of the entire postsynaptic neuron is altered. In that case, all target cells of the postsynaptic neuron are affected equally (Fig. 8.26b).

Summation

Spatial Summation

Spatial summation occurs when the currents from nearly simultaneous graded potentials combine.

(a) Summation of several subthreshold signals results in an action potential.

(b) Postsynaptic inhibition. An inhibitory presynaptic neuron prevents an action potential from firing.

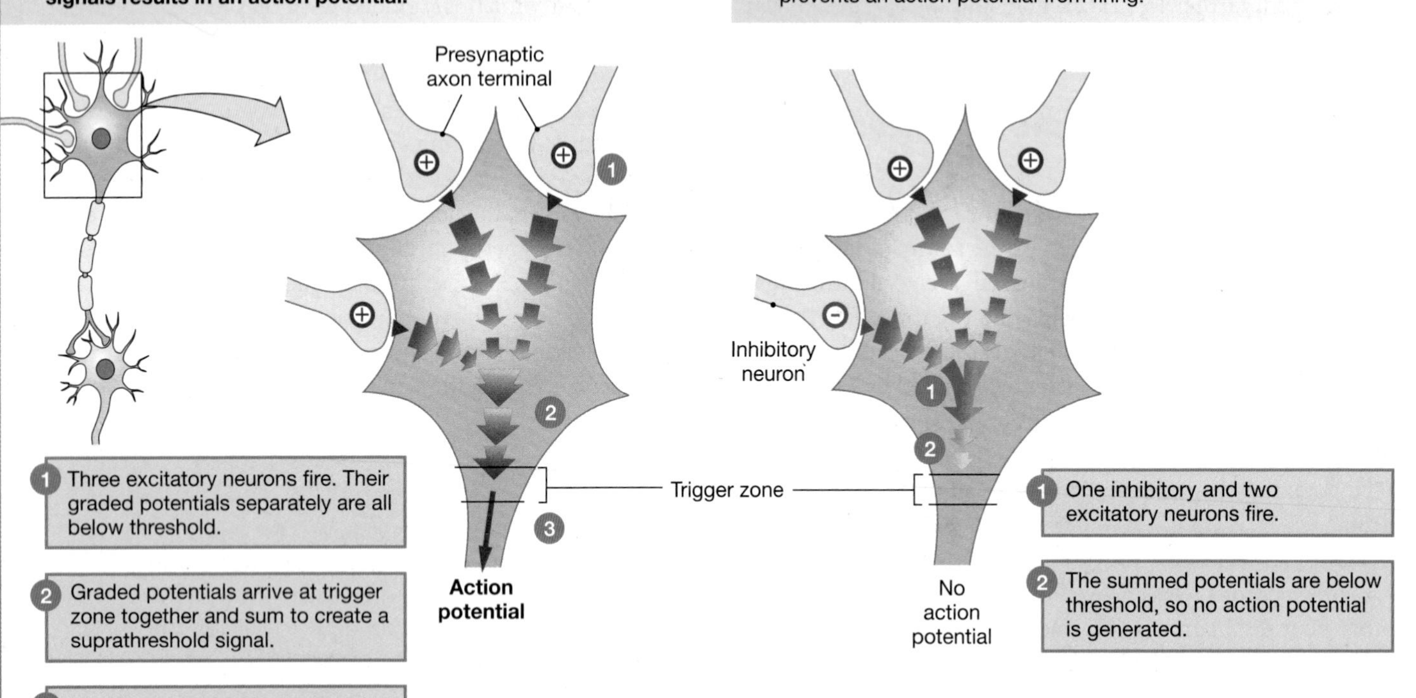

Temporal Summation

Temporal summation occurs when two graded potentials from one presynaptic neuron occur close together in time.

(c) No summation. Two subthreshold graded potentials will not initiate an action potential if they are far apart in time.

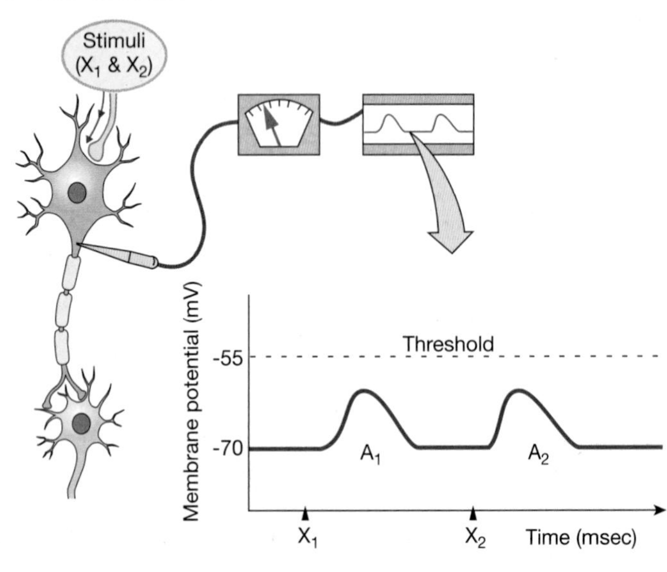

(d) Summation causing action potential. If two subthreshold potentials arrive at the trigger zone within a short period of time, they may sum and initiate an action potential.

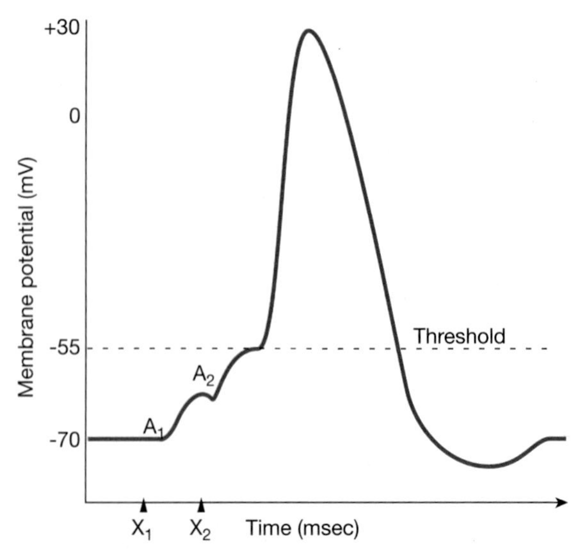

A THREE-DIMENSIONAL RECONSTRUCTION OF DENDRITIC SPINES AND THEIR SYNAPSES

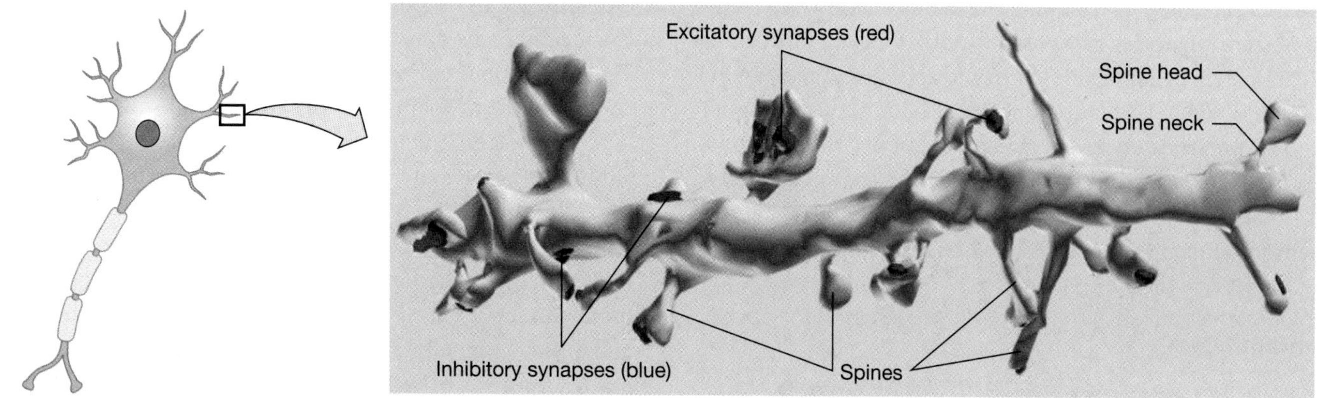

Fig. 8.25

PRESYNAPTIC AND POSTSYNAPTIC INHIBITION

(a) In **presynaptic inhibition**, an inhibitory neuron synapses on one collateral of the presynaptic neuron and selectively inhibits one target.

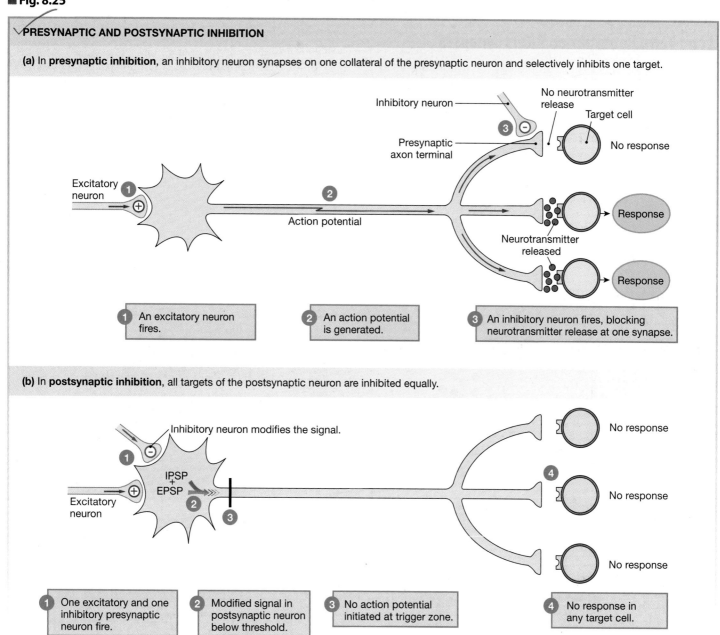

Inhibitory neuron

No neurotransmitter release

Target cell

Presynaptic axon terminal

No response

Excitatory neuron

Action potential

Response

Neurotransmitter released

Response

1 An excitatory neuron fires.

2 An action potential is generated.

3 An inhibitory neuron fires, blocking neurotransmitter release at one synapse.

(b) In **postsynaptic inhibition**, all targets of the postsynaptic neuron are inhibited equally.

Inhibitory neuron modifies the signal.

IPSP + EPSP

Excitatory neuron

No response

No response

No response

1 One excitatory and one inhibitory presynaptic neuron fire.

2 Modified signal in postsynaptic neuron below threshold.

3 No action potential initiated at trigger zone.

4 No response in any target cell.

Fig. 8.26

Synaptic activity can also be altered by changing the target (postsynaptic) cell's responsiveness to neurotransmitter. This may be accomplished by changing the identity, affinity, or number of neurotransmitter receptors. Modulators can alter all of these parameters by influencing the synthesis of enzymes, membrane transporters, and receptors. Most neuromodulators act through second messenger systems that alter existing proteins, and their effects last much longer than do those of neurotransmitters. One signal molecule can act as either a neurotransmitter or a neuromodulator, depending upon its receptor (Fig. 8.23).

Concept Check Answer: p. 287

29. Why are axon terminals sometimes called "biological transducers"?

Long-Term Potentiation Alters Synapses

Two of the "hot topics" in neurobiology today are **long-term potentiation** (LTP) {*potentia*, power} and *long-term depression* (LTD), processes in which activity at a synapse brings about sustained changes in the quality or quantity of synaptic connections. Many times changes in synaptic transmission, such as the facilitation and inhibition we just discussed, are of limited duration. However, if synaptic activity persists for longer periods, the neurons may adapt through LTP and LTD. Our understanding of LTP and LTD is changing rapidly, and the mechanisms may not be the same in different brain areas. The descriptions below reflect some of what we currently know about long-term adaptations of synaptic transmission.

A key element in long-term changes in the CNS is the amino acid glutamate, the main excitatory neurotransmitter in the CNS. As you learned previously, glutamate has two types of receptor-channels: AMPA receptors and NMDA receptors. The NMDA receptor has an unusual property. First, at resting membrane potentials, the NMDA channel is blocked by both a gate and a Mg^{2+} ion. Glutamate binding opens the ligand-activated gate, but ions cannot flow past the Mg^{2+}. However, if the cell depolarizes, the Mg^{2+} blocking the channel is expelled, and then ions flow through the channel. Thus, the NMDA channel opens only when the receptor is bound to glutamate and the cell is depolarized.

In long-term potentiation, when presynaptic neurons release glutamate, the neurotransmitter binds to both AMPA and NMDA receptors on the postsynaptic cell (■ Fig. 8.27 ①). Binding to the AMPA receptor opens a cation channel, and net Na^+ entry depolarizes the cell ②. Simultaneously, glutamate binding to the NMDA receptor opens the channel gate, and

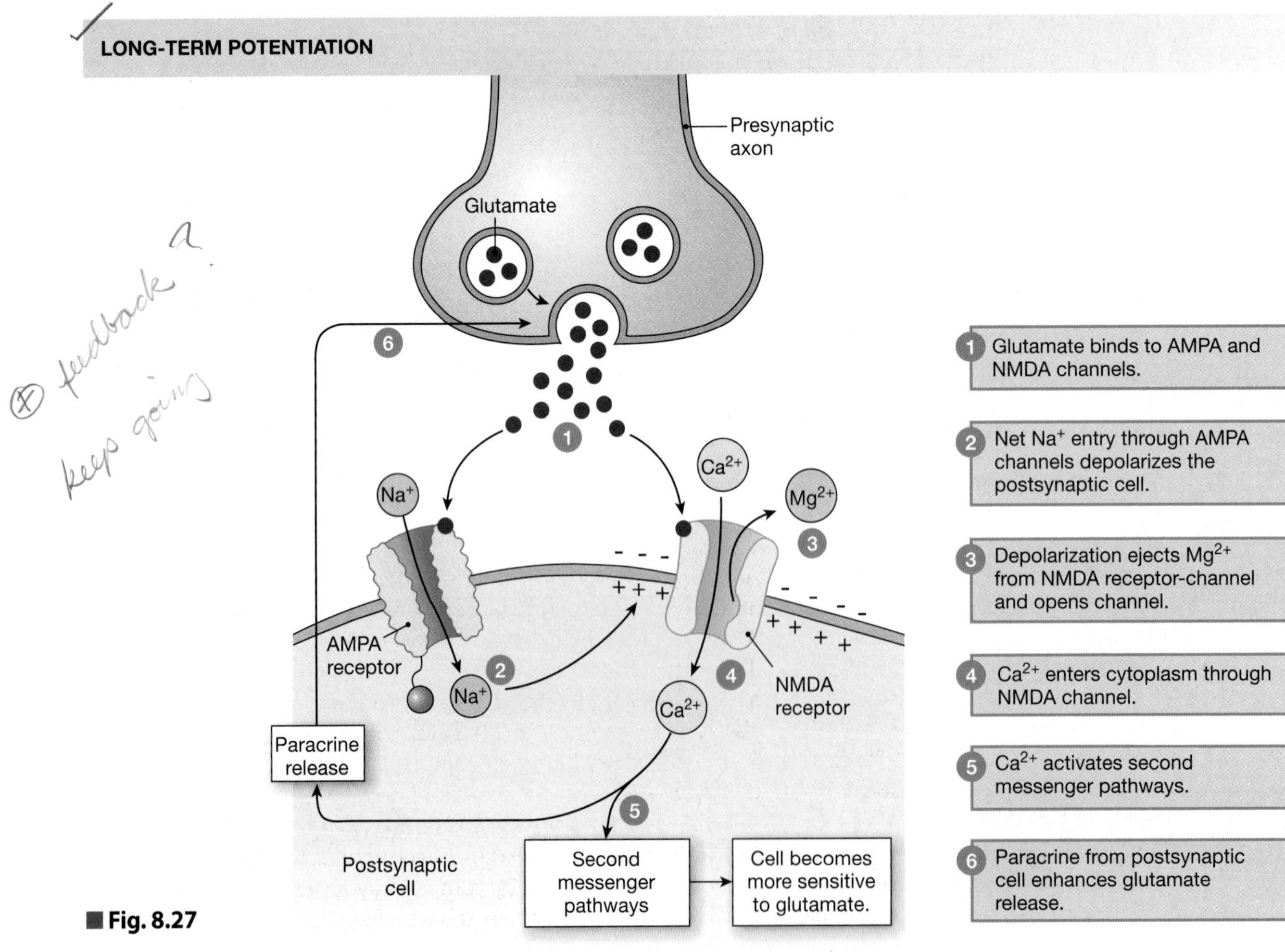
LONG-TERM POTENTIATION
Presynaptic axon
Glutamate
6
1
Na^+
Ca^{2+}
Mg^{2+}
3
AMPA receptor
2
Na^+
4
Ca^{2+}
NMDA receptor
Paracrine release
5
Postsynaptic cell
Second messenger pathways
Cell becomes more sensitive to glutamate.
1 Glutamate binds to AMPA and NMDA channels.
2 Net Na^+ entry through AMPA channels depolarizes the postsynaptic cell.
3 Depolarization ejects Mg^{2+} from NMDA receptor-channel and opens channel.
4 Ca^{2+} enters cytoplasm through NMDA channel.
5 Ca^{2+} activates second messenger pathways.
6 Paracrine from postsynaptic cell enhances glutamate release.

■ **Fig. 8.27**

depolarization of the cell creates electrical repulsion that knocks the Mg^{2+} out of the NMDA channel (3). Once the NMDA channel is open, Ca^{2+} enters the cytosol (4).

The Ca^{2+} signal initiates second messenger pathways (5). As a result of these intracellular pathways, the postsynaptic cell becomes more sensitive to glutamate, possibly by inserting more glutamate receptors in the postsynaptic membrane [up-regulation, p. 191]. In addition the postsynaptic cell releases a paracrine that acts on the presynaptic cell to enhance glutamate release (6).

Long-term depression seems to have two components: a change in the number of postsynaptic receptors and a change in the isoforms of the receptor proteins. In the face of continued neurotransmitter release from presynaptic neurons, the postsynaptic neurons withdraw AMPA receptors from the cell membrane by endocytosis [p. 155], a process similar to down-regulation of receptors in the endocrine system [p. 191]. In addition, different subunits are inserted into the AMPA receptors, changing current flow through the ion channels.

Researchers believe that long-term potentiation and depression are related to the neural processes for learning and memory, and to changes in the brain that occur with clinical depression and other mental illnesses. The clinical link makes LTP and LTD hot topics in neuroscience research.

Concept Check Answer: p. 287

30. Why would depolarization of the membrane drive Mg^{2+} from the channel into the extracellular fluid?

Disorders of Synaptic Transmission Are Responsible for Many Diseases

Synaptic transmission is the most vulnerable step in the process of signaling through the nervous system. It is the point at which many things go wrong, leading to disruption of normal function. Yet, at the same time, the receptors at synapses are exposed to the extracellular fluid, making them more accessible to drugs than intracellular receptors are. In recent years scientists have linked a variety of nervous system disorders to problems with synaptic transmission. These disorders include Parkinson's disease, schizophrenia, and depression. The best understood diseases of the synapse are those that involve the neuromuscular junction, such as myasthenia gravis. Diseases resulting from synaptic transmission problems within the CNS have proved more difficult to study because they are more difficult to isolate anatomically.

Drugs that act on synaptic activity, particularly synapses in the CNS, are the oldest known and most widely used of all pharmacological agents. Caffeine, nicotine, and alcohol are common drugs in many cultures. Some of the drugs we use to treat conditions such as schizophrenia, depression, anxiety, and epilepsy act by influencing events at the synapse. In many disorders arising in the CNS, we do not yet fully understand either the cause of the disorder or the drug's mechanism of action. This subject is one major area of pharmacological research, and new classes of drugs are being formulated and approved every year.

RUNNING PROBLEM

Dr. McKhann suspected that the disease afflicting the Chinese children—which he named acute motor axonal polyneuropathy (AMAN)—might be triggered by a bacterial infection. He also thought that the disease initiates its damage of axons at the neuromuscular junctions.

Q6: Based on information provided in this chapter, name other diseases involving altered synaptic transmission.

238 239 242 264 266 270 **281**

RUNNING PROBLEM CONCLUSION

Mysterious Paralysis

In this running problem you learned about acute motor axonal polyneuropathy (AMAN), a baffling paralytic illness that physicians thought might be a new disease. Although its symptoms resemble those of classic Guillain-Barré syndrome, AMAN is not a demyelinating disease. It affects only motor neurons. However, in both classic GBS and AMAN, the body's immune system makes antibodies against nervous system components. This similarity led experts eventually to conclude that AMAN is a subtype of GBS. The classic form of GBS has been renamed acute inflammatory demyelinating polyneuropathy, or AIDP. AIDP is more common in Europe and North America, while AMAN is the predominant form of GBS in China, Japan, and South America. A significant number of patients with AMAN develop their disease following a gastrointestinal illness caused by the bacterium *Campylobacter jejuni*, and experts suspect that antibodies to the bacterium also attack glycolipids called gangliosides in the axonal membrane. To learn more about the link between *Campylobacter* and GBS, see "*Campylobacter* Species and Guillain-Barré Syndrome," *Clin Microbiol Rev* 11: 555–567, July 1998 (*http://cmr.asm.org*). Check your understanding of this running problem by comparing your answers to the information in the summary table below.

8

RUNNING PROBLEM CONCLUSION *(continued)*

Question	Facts	Integration and Analysis
1. Which division(s) of the nervous system may be involved in Guillain-Barré syndrome (GBS)?	The nervous system is divided into the central nervous system (CNS) and the afferent (sensory) and efferent subdivisions of the peripheral nervous system. Efferent neurons are either somatic motor neurons, which control skeletal muscles, or autonomic neurons, which control glands and smooth and cardiac muscle.	Patients with GBS can neither feel sensations nor move their muscles. This suggests a problem in both afferent and somatic motor neurons. However, it is also possible that there is a problem in the CNS integrating center. You do not have enough information to determine which division is affected.
2. Do you think the paralysis found in the Chinese children affected both sensory (afferent) and somatic motor neurons? Why or why not?	The Chinese children can feel a pin prick but cannot move their muscles.	Sensory (afferent) function is normal if they can feel the pin prick. Paralysis of the muscles suggests a problem with somatic motor neurons, with the CNS centers controlling movement, or with the muscles themselves.
3. In GBS, what would you expect the results of a nerve conduction test to be?	Nerve conduction tests measure conduction speed and strength of conduction action potentials. In GBS, myelin around neurons is destroyed.	Myelin insulates axons and increases speed. Without myelin, ions leak out of the axon. Thus, in GBS you would expect decreased conduction speed and decreased strength of action potentials.
4. Is the paralytic illness that affected the Chinese children a demyelinating condition? Why or why not?	Nerve conduction tests showed normal conduction speed but decreased strength of the summed action potentials.	Myelin loss should decrease conduction speed as well as action potential strength. Therefore, this illness is probably not a demyelinating disease.
5. Do the results of Dr. McKhann's investigation suggest that the Chinese children had classic Guillain-Barré syndrome? Why or why not?	Autopsy reports on children who died from the disease showed that the axons were damaged but the myelin was normal.	Classic GBS is a demyelinating disease that affects both sensory and motor neurons. The Chinese children had normal sensory function, and nerve conduction tests and histological studies indicated normal myelin. Therefore, it was reasonable to conclude that the disease was not GBS.
6. Based on information provided in this chapter, name other diseases involving altered synaptic transmission.	Synaptic transmission can be altered by blocking neurotransmitter release from the presynaptic cell, by interfering with the action of neurotransmitter on the target cell, or by removing neurotransmitter from the synapse.	Parkinson's disease, depression, schizophrenia, and myasthenia gravis are related to problems with synaptic transmission.

238 239 242 264 266 270 281

Chapter Summary

This chapter introduces the nervous system, one of the major control systems responsible for maintaining *homeostasis*. The divisions of the nervous system correlate with the steps in a reflex pathway. Sensory receptors monitor regulated variables and send input signals to the central nervous system through sensory (afferent) neurons. Output signals, both electrical and chemical, travel through the efferent divisions (somatic motor and autonomic) to their targets throughout the body. Information transfer and *communication* depend on electrical signals that pass along neurons, on *molecular interactions* between signal molecules and their receptors, and on signal transduction in the target cells.

1. The **nervous system** is a complex network of neurons that form the rapid control system of the body. (p. 238)
2. **Emergent properties** of the nervous system include consciousness, intelligence, and emotion. (p. 238)

Organization of the Nervous System

3. The nervous system is divided into the **central nervous system (CNS)**, composed of the **brain** and **spinal cord**, and the **peripheral nervous system (PNS)**. (p. 240; Fig. 8.1)
4. The peripheral nervous system has **sensory** (afferent) **neurons** that bring information into the CNS, and **efferent neurons** that carry information away from the CNS back to various parts of the body. (p. 240)
5. The efferent neurons include **somatic motor neurons,** which control skeletal muscles, and **autonomic neurons,** which control smooth and cardiac muscles, glands, and some adipose tissue. (p. 240)
6. Autonomic neurons are subdivided into **sympathetic** and **parasympathetic** branches. (p. 240)

Cells of the Nervous System

iP Nervous System I: Anatomy Review

7. Neurons have a **cell body** with a nucleus and organelles to direct cellular activity, **dendrites** to receive incoming signals, and an **axon** to transmit electrical signals from the cell body to the **axon terminal**. (p. 241; Fig. 8.2)
8. **Interneurons** are neurons that lie entirely within the CNS. (p. 241; Fig. 8.2c,d)
9. Material is transported between the cell body and axon terminal by **axonal transport**. (p. 243; Fig. 8.3)
10. The region where an axon terminal meets its target cell is called a **synapse**. The target cell is called the **postsynaptic cell**, and the neuron that releases the chemical signal is known as the **presynaptic cell**. The region between these two cells is the **synaptic cleft**. (p. 241; Fig. 8.2f)
11. Developing neurons find their way to their targets by using chemical signals. (p. 243)
12. **Glial cells** provide physical support and communicate with neurons. **Schwann cells** and **satellite cells** are glial cells associated with the peripheral nervous system. **Oligodendrocytes**, **astrocytes**, **microglia**, and **ependymal cells** are glial cells found in the CNS. Microglia are modified immune cells that act as scavengers. (p. 246; Fig. 8.5)
13. Schwann cells and oligodendrocytes form insulating **myelin sheaths** around neurons. The **nodes of Ranvier** are sections of uninsulated membrane occurring at intervals along the length of an axon. (p. 246; Fig. 8.5c)
14. **Neural stem cells** that can develop into new neurons and glia are found in the ependymal layer as well as in other parts of the nervous system. (p. 245)

Electrical Signals in Neurons

iP Nervous System I: The Membrane Potential; Ion Channels; The Action Potential

15. The **Nernst equation** describes the membrane potential of a cell that is permeable to only one ion. (p. 248)
16. Membrane potential is influenced by the concentration gradients of ions across the membrane and by the permeability of the membrane to those ions. (p. 248)
17. The **Goldman-Hodgkin-Katz (GHK) equation** predicts membrane potential based on ion concentration gradients and membrane permeability for multiple ions. (p. 249)
18. The permeability of a cell to ions changes when ion channels in the membrane open and close. Movement of only a few ions significantly changes the membrane potential. (p. 249)
19. Gated ion channels in neurons open or close in response to chemical or mechanical signals or in response to depolarization of the cell membrane. Channels also close through inactivation. (p. 250)
20. Current flow (I) obeys **Ohm's Law**: $I = \text{voltage/resistance}$. **Resistance** to current flow comes from the cell membrane, which is a good insulator, and from the cytoplasm. **Conductance** (G) is the reciprocal of resistance: $G = 1/R$. (p. 251)
21. **Graded potentials** are depolarizations or hyperpolarizations whose strength is directly proportional to the strength of the triggering event. Graded potentials lose strength as they move through the cell. (p. 252–253; Tbl. 8.3; Fig. 8.7)
22. The wave of depolarization that moves through a cell is known as **local current flow**. (p. 259)
23. **Action potentials** are rapid electrical signals that travel undiminished in amplitude (strength) down the axon from the cell body to the axon terminals. (p. 251)
24. Action potentials begin in the **trigger zone** if a single graded potential or the sum of multiple graded potentials exceeds the **threshold** voltage. (p. 253; Fig. 8.7c)
25. Depolarizing graded potentials make a neuron more likely to fire an action potential. Hyperpolarizing graded potentials make a neuron less likely to fire an action potential. (p. 251)
26. Action potentials are uniform, **all-or-none** depolarizations that can travel undiminished over long distances. (p. 252)
27. The rising phase of the action potential is due to increased Na^+ permeability. The falling phase of the action potential is due to increased K^+ permeability. (p. 256; Fig. 8.9)
28. The voltage-gated Na^+ channels of the axon have a fast **activation gate** and a slower **inactivation gate**. (p. 258; Fig. 8.10)
29. Very few ions cross the membrane during an action potential. The Na^+-K^+-ATPase eventually restores Na^+ and K^+ to their original compartments. (p. 255)

30. Once an action potential has begun, there is a brief period of time known as the **absolute refractory period** during which a second action potential cannot be triggered, no matter how large the stimulus. Because of this, action potentials cannot be summed. (p. 260; Fig. 8.12)
31. During the **relative refractory period**, a higher-than-normal graded potential is required to trigger an action potential. (p. 258)
32. The myelin sheath around an axon speeds up conduction by increasing membrane resistance and decreasing current leakage. Larger-diameter axons conduct action potentials faster than smaller-diameter axons do. (p. 261)
33. The apparent jumping of action potentials from node to node is called **saltatory conduction**. (p. 264; Fig. 8.16)
34. Changes in blood K^+ concentration affect resting membrane potential and the conduction of action potentials. (p. 265; Fig. 8.17)

Cell-to-Cell Communication in the Nervous System

iP Nervous System II: Anatomy Review; Synaptic Transmission; Ion Channels

35. In **electrical synapses**, an electrical signal passes directly from the cytoplasm of one cell to another through gap junctions. **Chemical synapses** use neurotransmitters to carry information from one cell to the next, with the neurotransmitters diffusing across the synaptic cleft to bind with receptors on target cells. (p. 266)
36. Neurotransmitters come in a variety of forms. **Cholinergic** neurons secrete **acetylcholine. Adrenergic** neurons secrete **norepinephrine. Glutamate, GABA, serotonin, adenosine,** and **nitric oxide** are other major neurotransmitters. (p. 268; Tbl. 8.4)
37. Neurotransmitter receptors are either ligand-gated ion channels (ionotropic receptors) or G protein–coupled receptors (metabotropic receptors). (p. 269)
38. Neurotransmitters are synthesized in the cell body or in the axon terminal. They are stored in **synaptic vesicles** and are released by exocytosis when an action potential reaches the axon terminal. (p. 272; Fig. 8.19a)
39. Neurotransmitter action is rapidly terminated by reuptake into cells, diffusion away from the synapse, or enzymatic breakdown. (p. 272; Fig. 8.19b)
40. Information about the strength and duration of a stimulus is conveyed by the amount of neurotransmitter released. Increased frequency of action potentials releases more neurotransmitter. (p. 274; Fig. 8.21)

Integration of Neural Information Transfer

iP Nervous System II: Synaptic Potentials & Cellular Integration

41. When a presynaptic neuron synapses on a larger number of postsynaptic neurons, the pattern is known as **divergence**. When several presynaptic neurons provide input to a smaller number of postsynaptic neurons, the pattern is known as **convergence**. (p. 275; Fig. 8.22)
42. Synaptic transmission can be modified in response to activity at the synapse, a process known as **synaptic plasticity**. (p. 274)
43. G protein–coupled receptors either create **slow synaptic potentials** or modify cell metabolism. Ion channels create **fast synaptic potentials**. (p. 276; Fig. 8.23)
44. The summation of simultaneous graded potentials from different neurons is known as **spatial summation**. The summation of graded potentials that closely follow each other sequentially is called **temporal summation**. (p. 278; Fig. 8.24)
45. **Presynaptic modulation** of an axon terminal allows selective modulation of collaterals and their targets. **Postsynaptic modulation** occurs when a modulatory neuron synapses on a postsynaptic cell body or dendrites. (p. 279; Fig. 8.26)
46. **Long-term potentiation** and **long-term depression** are mechanisms by which neurons change the strength of their synaptic connections. (p. 280; Fig. 8.27)

Questions

Answers: p. A-1

Level One Reviewing Facts and Terms

1. List the three functional classes of neurons, and explain how they differ structurally and functionally.
2. Somatic motor neurons control ________, and ________ neurons control smooth and cardiac muscles, glands, and some adipose tissue.
3. Autonomic neurons are classified as either ________ or ________ neurons.
4. Match each term with its description:

(a) axon (b) dendrite (c) afferent (d) efferent (e) trigger zone	1. process of a neuron that receives incoming signals 2. sensory neuron, transmits information to CNS 3. long process that transmits signals to the target cell 4. region of neuron where action potential begins 5. neuron that transmits information from CNS to the rest of the body

5. Name the two primary cell types found in the nervous system.
6. Draw a typical neuron and label the cell body, axon, dendrites, nucleus, trigger zone, axon hillock, collaterals, and axon terminals. Draw mitochondria, rough endoplasmic reticulum, Golgi complex, and vesicles in the appropriate sections of the neuron.
7. Axonal transport refers to the
   (a) release of neurotransmitters into the synaptic cleft.
   (b) use of microtubules to send secretions from the cell body to the axon terminal.
   (c) movement of organelles and cytoplasm up and down the axon.
   (d) movement of the axon terminal to synapse with a new postsynaptic cell.
   (e) none of these
8. Match the numbers of the appropriate characteristics with the two types of potentials. Characteristics may apply to one or both types.

(a) action potential (b) graded potential	1. all-or-none 2. can be summed 3. amplitude decreases with distance 4. exhibits a refractory period 5. amplitude depends on strength of stimulus 6. has no threshold

9. Match the glial cell(s) on the right to the functions on the left. There may be more than one correct answer for each function.

(a) modified immune cells	1. astrocytes
(b) help form the blood-brain barrier	2. ependymal cells
(c) form myelin	3. microglia
(d) separate CNS fluid compartments	4. oligodendrocytes
(e) found in peripheral nervous system	5. satellite cells
(f) found in ganglia	6. Schwann cells

10. List the four major types of ion channels found in neurons. Are they chemically gated, mechanically gated, or voltage-gated?
11. Arrange the following events in the proper sequence:
    (a) Efferent neuron reaches threshold and fires an action potential.
    (b) Afferent neuron reaches threshold and fires an action potential.
    (c) Effector organ responds by performing output.
    (d) Integrating center reaches decision about response.
    (e) Sensory organ detects change in the environment.
12. An action potential is (circle all correct answers)
    (a) a reversal of the Na^+ and K^+ concentrations inside and outside the neuron.
    (b) the same size and shape at the beginning and end of the axon.
    (c) initiated by inhibitory postsynaptic graded potentials.
    (d) transmitted to the distal end of a neuron and causes release of neurotransmitter.
13. Choose from the following ions to fill in the blanks correctly: Na^+, K^+, Ca^{2+}, Cl^-.
    (a) The resting cell membrane is more permeable to ________ than to ________. Although ________ contribute little to the resting membrane potential, they play a key role in generating electrical signals in excitable tissues.
    (b) The concentration of ________ is 12 times greater outside the cell than inside.
    (c) The concentration of ________ is 30 times greater inside the cell than outside.
    (d) An action potential occurs when ________ enter the cell.
    (e) The resting membrane potential is due to the high ________ permeability of the cell.
14. What is the myelin sheath?
15. List two factors that enhance conduction speed.
16. List three ways neurotransmitters are removed from the synapse.
17. Draw and label a graph of an action potential. Below the graph, draw the positioning of the K^+ and Na^+ channel gates during each phase.

Level Two Reviewing Concepts

18. Create a map showing the organization of the nervous system using the following terms, plus any terms you choose to add:

• afferent signals	• neuron
• astrocyte	• neurotransmitter
• autonomic division	• oligodendrocyte
• brain	• parasympathetic division
• CNS	• peripheral division
• efferent neuron	• satellite cell
• ependymal cell	• Schwann cell
• glands	• sensory division
• glial cells	• somatic motor division
• integration	• spinal cord
• interneuron	• stimulus
• microglia	• sympathetic division
• muscles	• target

19. What causes the depolarization phase of an action potential? (Circle all that apply.)
    (a) K^+ leaving the cell through voltage-gated channels
    (b) K^+ being pumped into the cell by the Na^+-K^+-ATPase
    (c) Na^+ being pumped into the cell by the Na^+-K^+-ATPase
    (d) Na^+ entering the cell through voltage-gated channels
    (e) opening of the Na^+ channel inactivation gate
20. Name any four neurotransmitters, their receptor(s), and tell whether the receptor is an ion channel or a GPCR.
21. Arrange the following terms to describe the sequence of events after a neurotransmitter binds to a receptor on a postsynaptic neuron. Terms may be used more than once or not at all.
    (a) action potential fires at axon hillock
    (b) trigger zone reaches threshold
    (c) cell depolarizes
    (d) exocytosis
    (e) graded potential occurs
    (f) ligand-gated ion channel opens
    (g) local current flow occurs
    (h) saltatory conduction occurs
    (i) voltage-gated Ca^{2+} channels open
    (j) voltage-gated K^+ channels open
    (k) voltage-gated Na^+ channels open
22. Match the best term (hyperpolarize, depolarize, repolarize) to the following events. The cell in question has a resting membrane potential of -70 mV.
    (a) membrane potential changes from -70 mV to -50 mV
    (b) membrane potential changes from -70 mV to -90 mV
    (c) membrane potential changes from $+20$ mV to -60 mV
    (d) membrane potential changes from -80 mV to -70 mV
23. A neuron has a resting membrane potential of -70 mV. Will the neuron hyperpolarize or depolarize when each of the following events occurs? (More than one answer may apply; list all those that are correct.)
    (a) Na^+ enters the cell
    (b) K^+ leaves the cell
    (c) Cl^- enters the cell
    (d) Ca^{2+} enters the cell
24. Define, compare, and contrast the following concepts:
    (a) threshold, subthreshold, suprathreshold, all-or-none, overshoot, undershoot
    (b) graded potential, EPSP, IPSP
    (c) absolute refractory period, relative refractory period
    (d) afferent neuron, efferent neuron, interneuron
    (e) sensory neuron, somatic motor neuron, sympathetic neuron, autonomic neuron, parasympathetic neuron
    (f) fast synaptic potential, slow synaptic potential
    (g) temporal summation, spatial summation
    (h) convergence, divergence
25. If all action potentials within a given neuron are identical, how does the neuron transmit information about the strength and duration of the stimulus?
26. The presence of myelin allows an axon to
    (a) produce more frequent action potentials.
    (b) conduct impulses more rapidly.
    (c) produce action potentials of larger amplitude.
    (d) produce action potentials of longer duration.

8

Level Three Problem Solving

27. If human babies' muscles and neurons are fully developed and functional at birth, why can't they focus their eyes, sit up, or learn to crawl within hours of being born? (*Hint:* Muscle strength is not the problem.)
28. The voltage-gated Na^+ channels of a neuron open when the neuron depolarizes. If depolarization opens the channels, what makes them close when the neuron is maximally depolarized?
29. One of the pills that Jim takes for high blood pressure caused his blood K^+ level to decrease from 4.5 mM to 2.5 mM. What happens to the resting membrane potential of his liver cells? (Circle all that are correct.)
   (a) decreases
   (b) increases
   (c) does not change
   (d) becomes more negative
   (e) becomes less negative
   (f) fires an action potential
   (g) depolarizes
   (h) hyperpolarizes
   (i) repolarizes
30. Characterize each of the following stimuli as being mechanical, chemical, or thermal:
   (a) bath water at 106 °F
   (b) acetylcholine
   (c) a hint of perfume
   (d) epinephrine
   (e) lemon juice
   (f) a punch on the arm
31. An unmyelinated axon has a much greater requirement for ATP than a myelinated axon of the same diameter and length. Can you explain why?

Level Four Quantitative Problems

32. The GHK equation is sometimes abbreviated to exclude chloride, which plays a minimal role in membrane potential for most cells. In addition, because it is difficult to determine absolute membrane permeability values for Na^+ and K^+, the equation is revised to use the ratio of the two ion permeabilities, expressed as $\alpha = P_{Na}/P_K$:

$$V_m = 61 \log \frac{[K^+]_{out} + \alpha[Na^+]_{out}}{[K^+]_{in} + \alpha[Na^+]_{in}}$$

Thus, if you know the relative membrane permeabilities of the two ions and their intracellular (ICF) and extracellular (ECF) concentrations, you can predict the membrane potential for a cell.
   (a) A resting cell has an alpha value of 0.025 and the following ion concentrations:

$$Na^+: ICF = 5\text{ mM}, ECF = 135\text{ mM}$$
$$K^+: ICF = 150\text{ mM}, ECF = 4\text{ mM}$$

   What is the cell's membrane potential?
   (b) The Na^+ permeability of the cell in (a) suddenly increases so that $\alpha = 20$. Now what is the cell's membrane potential?
   (c) Mrs. Nguyen has high blood pressure, and her physician puts her on a drug whose side effect decreases her plasma (ECF) K^+ from 4 mM to 2.5 mM. Using the other values in (a), now what is the membrane potential?
   (d) The physician prescribes a potassium supplement for Mrs. Nguyen, who decides that if two pills are good, four must be better. Her plasma (ECF) K^+ now goes to 6 mM. What happens to the membrane potential?
33. In each of the following scenarios, will an action potential be produced? The postsynaptic neuron has a resting membrane potential of -70 mV.
   (a) Fifteen neurons synapse on one postsynaptic neuron. At the trigger zone, 12 of the neurons produce EPSPs of 2 mV each, and the other three produce IPSPs of 3 mV each. The threshold for the postsynaptic cell is -50 mV.
   (b) Fourteen neurons synapse on one postsynaptic neuron. At the trigger zone, 11 of the neurons produce EPSPs of 2 mV each, and the other three produce IPSPs of 3 mV each. The threshold for the postsynaptic cell is -60 mV.
   (c) Fifteen neurons synapse on one postsynaptic neuron. At the trigger zone, 14 of the neurons produce EPSPs of 2 mV each, and the other one produces an IPSP of 9 mV. The threshold for the postsynaptic cell is -50 mV.

Answers

Answers to Concept Check Questions

Page 239

1. Compare your answer to the map in Figure 8.1, p. 240.

Page 243

2. Neurons that secrete neurohormones terminate close to blood vessels so that the neurohormones can enter the circulation.
3. A neuron is a single nerve cell. A nerve is a bundle of axons from many neurons.

Page 244

4. See Figure 8.2.

Page 245

5. Myelin insulates axon membranes. Microglia are scavenger cells in the CNS. Ependymal cells form epithelial barriers between fluid compartments of the CNS.
6. Schwann cells are in the PNS, and each Schwann cell forms myelin around a small portion of one axon. Oligodendrocytes are in the CNS, and one oligodendrocyte forms myelin around axons of several neurons.

Page 249

7. For Ca^{2+}, the electrical charge z is $+2$; the ratio of ion concentrations is $1/0.0001 = 10{,}000$ or 10^4. Log of 10^4 is 4 (see Appendix B). Thus E_{ion} (in mV) $= (61 \times 4)/(+2) = 122$ mV.

Page 250

8. (a) depolarize (b) depolarize
9. depolarize

Page 252

10. (a) 1, (b) 2, (c) 2, (d) 1

Page 254

11. The trigger zone for the sensory neurons is close to where the dendrites converge. You cannot tell where the trigger zone is for the anaxonic neuron. For multipolar neurons, the trigger zone is at the junction of the cell body and the axon.

Page 255

12. Conductance refers to the movement of ions across a cell membrane. Conduction is the rapid, undiminished movement of an electrical signal down the axon of a neuron.

Page 257

13. (b)

Page 258

14. The membrane potential depolarizes and remains depolarized.
15. During resetting, the activation gate is closing, and the inactivation gate is opening.

Page 261

16. The action potential will go in both directions because the Na^+ channels around the stimulation site have not been inactivated by a previous depolarization. See discussion of refractory periods.

Page 264

17. (a), (c), (b)

Page 269

18. Because different receptor subtypes work through different signal transduction pathways, targeting drugs to specific receptor subtypes decreases the likelihood of unwanted side effects.

Page 271

19. Proteins are synthesized on the ribosomes of the rough endoplasmic reticulum; then the proteins are directed into the Golgi apparatus to be packaged into vesicles.
20. Mitochondria are the primary sites of ATP synthesis.
21. Mitochondria reach the axon terminal by fast axonal transport along microtubules.

Page 271

22. The researchers concluded that some event between arrival of the action potential at the axon terminal and depolarization of the postsynaptic cell is dependent on extracellular Ca^{2+}. We now know that this event is neurotransmitter release.
23. The exchange is secondary active transport because it uses energy stored in the H^+ concentration gradient to concentrate neurotransmitter inside the vesicles.

Page 273

24. SSRIs decrease reuptake of serotonin into the axon terminal, thereby increasing the time serotonin is active in the synapse.
25. Acetyl CoA is made from pyruvate, the end product of glycolysis, and CoA.
26. Neurotransmitter uptake is secondary active transport because it uses energy stored in the Na^+ concentration gradient to concentrate neurotransmitter inside the axon terminal.

Page 277

27. The postsynaptic neuron will fire an action potential, because the net effect would be a 17 mV depolarization to $-70 - (-17) = -53$ mV, which is just above the threshold of -55 mV.
28. The membrane potential does not change at the same time as the stimulus because the depolarization must travel from the point of the stimulus to the recording point.

Page 280

29. Axon terminals convert (transduce) the electrical action potential signal into a chemical neurotransmitter signal.

Page 281

30. Membrane depolarization makes the inside of the membrane more positive than the outside. Like charges repel one another, so the more positive membrane potential tends to repel Mg^{2+}.

Answers to Figure Questions

Page 253

Figure 8.7: The graded potential is stronger at B. On the graph, A is between 3 and 4, and B is about at 1.

Page 262

Figure 8.14: (a) 4; (b) 2, 3; (c) 1; (d) 3; (e) 4

Page 263

Figure 8.15: Area of 100 giant axons is 50.3344 mm^2, $r^2 = 16$ mm, $r = 4$ mm, so diameter $= 8$ mm.

Page 265

Figure 8.17: (a) −108 mV; (b) −85 mV

Page 275

Figure 8.22: Amplification

9 The Central Nervous System

Neuronal assemblies have important properties that cannot be explained by the additive qualities of individual neurons.

—O. Hechter, *in* Biology and Medicine into the 21st Century, 1991

Background Basics

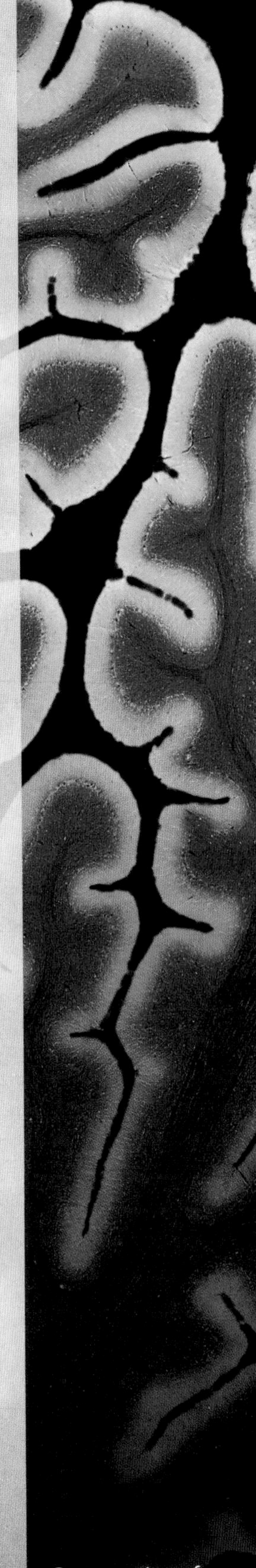

Cross-section of the cerebellum

Matt Nagle, paralyzed from the neck down and breathing with the aid of a ventilator, sat immobile in his wheelchair, a small box perched on top of his skull. But this was no ordinary box: it was part of a brain-computer interface (BCI) with 96 recording electrodes implanted in Matt's brain. On the computer screen in front of Matt was a cursor that moved across the screen as Matt thought about where he wanted it to go. By the end of the revolutionary one-year experiment in 2005, Matt could open email, play a game, and open and close a robotic hand using only his thoughts. This story may sound like science fiction, but the BrainGate® BCI and the experiment are real. The scientists who developed BrainGate were using what we know about the human brain to harness its electrical signals and create wireless bridges to external machines. In this chapter we take a look at the structure and organization of our main integrating center, the central nervous system.

Emergent Properties of Neural Networks

Neurons in the nervous system link together to form circuits that have specific functions. The most complex circuits are those of the brain, in which billions of neurons are linked into intricate networks that converge and diverge, creating an infinite number of possible pathways. Signaling within these pathways creates thinking, language, feeling, learning, and memory—the complex behaviors that make us human. Some neuroscientists have proposed that the functional unit of the nervous system be changed from the individual neuron to neural networks because even the most basic functions require circuits of neurons.

How is it that combinations of neurons linked together into chains or networks collectively possess emergent properties not found in any single neuron? We do not yet have an answer to this question. Some scientists seek to answer it by looking for parallels between the nervous system and the integrated circuits of computers.

Computer programs have been written that attempt to mimic the thought processes of humans. This field of study, called *artificial intelligence*, has created some interesting programs, such as the "psychiatrist" programmed to respond to typed complaints with appropriate comments and suggestions. We are nowhere near creating a brain as complex as that of a human, however, or even one as complex as that of Hal, the computer in the classic movie *2001: A Space Odyssey.*

Probably one reason computers cannot yet accurately model brain function is that computers lack *plasticity*, the ability to change circuit connections and function in response to sensory input and past experience [p. 274]. Although some computer programs can change their output under specialized conditions, they cannot begin to approximate the plasticity of human brain networks, which easily restructure themselves as the result of sensory input, learning, emotion, and creativity. In addition, we now know that the brain can add new connections when neural stem cells differentiate. Computers cannot add new circuits to themselves.

How can simply linking neurons together create **affective behaviors**, which are related to feeling and emotion, and **cognitive behaviors** {*cognoscere,* to get to know} related to thinking? In their search for the organizational principles that lead to these behaviors, scientists seek clues in the simplest animal nervous systems.

Evolution of Nervous Systems

All animals have the ability to sense and respond to changes in their environment. Even single-cell organisms such as *Paramecium* are able to carry out the basic tasks of life: finding food, avoiding becoming food, finding a mate. Yet these unicellular organisms have no obvious brain or integrating center. They use the resting membrane potential that exists in living cells and many of the same ion channels as more complex animals to coordinate their daily activities.

The first multicellular animals to develop neurons were members of the phylum *Cnidaria*, the jellyfish and sea anemones. Their nervous system is a *nerve net* composed of sensory neurons, connective interneurons, and motor neurons that innervate muscles and glands (■ Fig. 9.1a). These animals respond to stimuli with complex behaviors, yet without input from an identifiable control center. If you watch a jellyfish swim or a sea anemone maneuver a piece of shrimp into its mouth, it is hard to imagine how a diffuse network of neurons can create such complex coordinated movements. However, the same basic principles of neural communication apply to jellyfish and humans. Electrical signals in the form of action potentials, and chemical signals passing across synapses, are the same in all animals. It is only in the number and organization of the neurons that one species differs from another.

RUNNING PROBLEM

Infantile Spasms

At four months of age, Ben could roll over, hold up his head, and reach for things. At seven months, he was nearly paralyzed and lay listlessly in his crib. He had lost his abilities so gradually that it was hard to remember when each one had slipped away, but his mother could remember exactly when it began. She was preparing to feed him lunch one day when she heard a cry from the highchair where Ben was sitting. As she watched, Ben's head dropped to his chest, came back up, then went hurtling toward his lap, smacking into his highchair tray. Ben's mother snatched him up into her arms, and she could feel him still convulsing against her shoulder. This was the first of many such spells that came with increasing frequency and duration.

289 297 312 314 316 320

EVOLUTION OF THE NERVOUS SYSTEM

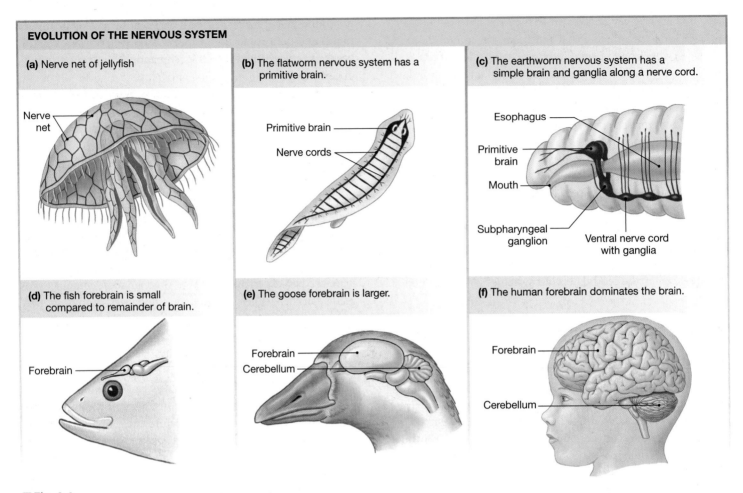

(a) Nerve net of jellyfish

Nerve net

(b) The flatworm nervous system has a primitive brain.

Primitive brain

Nerve cords

(c) The earthworm nervous system has a simple brain and ganglia along a nerve cord.

Esophagus

Primitive brain

Mouth

Subpharyngeal ganglion

Ventral nerve cord with ganglia

(d) The fish forebrain is small compared to remainder of brain.

Forebrain

(e) The goose forebrain is larger.

Forebrain

Cerebellum

(f) The human forebrain dominates the brain.

Forebrain

Cerebellum

Fig. 9.1

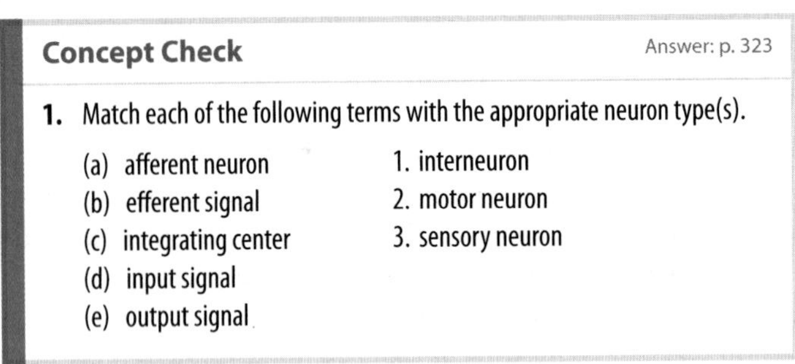

Concept Check Answer: p. 323

1. Match each of the following terms with the appropriate neuron type(s).

(a) afferent neuron	1. interneuron
(b) efferent signal	2. motor neuron
(c) integrating center	3. sensory neuron
(d) input signal	
(e) output signal	

In the primitive flatworms, we see the beginnings of a nervous system as we know it in higher animals, although in flatworms the distinction between central nervous system and peripheral nervous system is not clear. Flatworms have a rudimentary brain consisting of a cluster of nerve cell bodies concentrated in the head (*cephalic*) region. Two large nerves called *nerve cords* come off the primitive brain and lead to a nerve network that innervates distal regions of the flatworm body (Fig. 9.1b).

The segmented worms, or annelids, such as the earthworm, have a more advanced central nervous system (Fig. 9.1c). Clusters of cell bodies are no longer restricted to the head region, as they are in flatworms, but also occur in fused pairs, called *ganglia* (singular *ganglion*) [p. 245], along a nerve cord. Because each segment of the worm contains a ganglion, simple reflexes can be integrated within a segment without input from the brain. Reflexes that do not require integration in the brain also occur in higher animals and are called **spinal reflexes** in humans and other vertebrates.

Annelids and higher invertebrates have complex reflexes controlled through neural networks. Researchers use leeches (a type of annelid) and *Aplysia*, a type of shell-less mollusk, to study neural networks and synapse formation because the neurons in these species are 10 times larger than human brain neurons, and because the networks have the same organization of neurons from animal to animal. The neural function of these invertebrates provides a simple model that we can apply to more complex vertebrate networks.

Nerve cell bodies clustered into brains persist throughout the more advanced phyla and become increasingly more complex. One advantage to cephalic brains is that in most animals, the head is the part of the body that first contacts the environment as the animal moves. For this reason, as brains evolved, they became associated with specialized cephalic receptors, such as eyes for vision and chemoreceptors for smell and taste.

In the higher arthropods, such as insects, specific regions of the brain are associated with particular functions. More complex brains are associated with complex behaviors, such as the ability of social insects like ants and bees to organize themselves into colonies, divide labor, and communicate with one

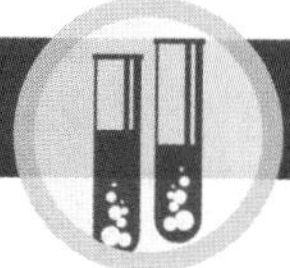

BIOTECHNOLOGY

Tracing Neurons in a Network

One of the challenges facing neurophysiologists and neuroanatomists is tracing the networks of neurons that control specific functions, a task that can be likened to following one tiny thread through a tangled mass the size of a beach ball. In 1971, K. Kristensson and colleagues introduced the use of *horseradish peroxidase* (HRP), an enzyme that acts in the presence of its substrate to produce a visible product (colored or fluorescent). When injected into the extracellular fluid near axon terminals, HRP is brought into the neuron by endocytosis. The HRP vesicles are then transported by fast retrograde axonal transport to the cell body and dendrites [p. 243]. Once the enzyme-substrate reaction is completed, the neuron becomes visible, allowing the researcher to trace the entire neuron from its target back to its origin. Now, with the help of fluorescent antibodies, scientists have used genetic modification in mice to create neurons that light up in a rainbow of colors: the Brainbow mouse.

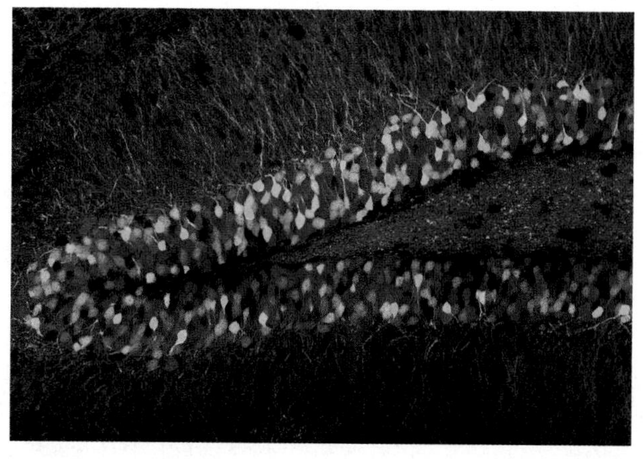

another. The octopus (a cephalopod mollusk) has the most sophisticated brain development among the invertebrates, as well as the most sophisticated behavior.

In vertebrate brain evolution, the most dramatic change is seen in the *forebrain* region {*fore*, in front}, which includes the **cerebrum** {*cerebrum*, brain; adjective *cerebral*}. In fish, the forebrain is a small bulge dedicated mainly to processing olfactory information about odors in the environment (Fig. 9.1d). In birds and rodents, part of the forebrain has enlarged into a cerebrum with a smooth surface (Fig. 9.1e).

In humans, the cerebrum is the largest and most distinctive part of the brain, with deep grooves and folds (Fig. 9.1f). More than anything else, the cerebrum is what makes us human. All evidence indicates that it is the part of the brain that allows reasoning and cognition.

The other brain structure whose evolution is obvious in the vertebrates is the *cerebellum*, a region of the *hindbrain* devoted to coordinating movement and balance. Birds (Fig. 9.1e) and humans (Fig. 9.1f) both have well-developed cerebellar structures. The cerebellum, like the cerebrum, is readily identifiable in these animals by its grooves and folds.

In this chapter we begin with an overview of CNS anatomy and functions. We then look at how neural networks create the higher brain functions of thought and emotion.

Anatomy of the Central Nervous System

The vertebrate central nervous system (CNS) consists of the brain and the spinal cord. As you learned in the previous section, brains increase in complexity and degree of specialization as we move up the phylogenetic tree from fish to humans. However, if we look at the vertebrate nervous system during development, a basic anatomical pattern emerges. In all vertebrates, the CNS consists of layers of neural tissue surrounding a fluid-filled central cavity lined with epithelium.

The CNS Develops from a Hollow Tube

In the very early embryo, cells that will become the nervous system lie in a flattened region called the **neural plate**. As development proceeds (at about day 20 of human development), neural plate cells along the edge migrate toward the midline (■ Fig. 9.2a).

By about day 23 of human development, the neural plate cells have fused with each other, creating a **neural tube** (Fig. 9.2b). *Neural crest cells* from the lateral edges of the neural plate now lie dorsal to the neural tube. The lumen of the neural tube will remain hollow and become the central cavity of the CNS.

The cells lining the neural tube will either differentiate into the epithelial *ependyma* [p. 245] or remain as undifferentiated *neural stem cells*. The outer cell layers of the neural tube will become the neurons and glia of the CNS. Neural crest cells will become the sensory and motor neurons of the peripheral nervous system.

By week 4 of human development, the anterior portion of the neural tube has begun to specialize into the regions of the brain (Fig. 9.2c). Three divisions are obvious: a **forebrain**, a **midbrain**, and a **hindbrain**. The tube posterior to the hindbrain will become the spinal cord. At this stage, the portion of the forebrain that will become the cerebrum is not much larger than the other regions of the brain.

As development proceeds, the growth of the cerebrum begins to outpace that of the other regions (Fig. 9.2d). By week 6, the CNS has formed the seven major divisions that are present at birth. Six of these regions are in the brain—(1) the cerebrum, (2) the *diencephalon*, (3) the midbrain, (4) and (5) the cerebellum and *pons*, (6) the *medulla oblongata*—and the seventh is the spinal cord. The cerebrum and diencephalon develop from the forebrain. The cerebellum, pons, and medulla oblongata are divisions of the hindbrain.

By week 6 the central cavity (lumen) of the neural tube has begun to enlarge into the hollow **ventricles** {*ventriculus*, belly} of

Development of the Human Nervous System

(a) Day 20

In the 20-day embryo (dorsal view), neural plate cells (purple) migrate toward the midline. Neural crest cells migrate with the neural plate cells.

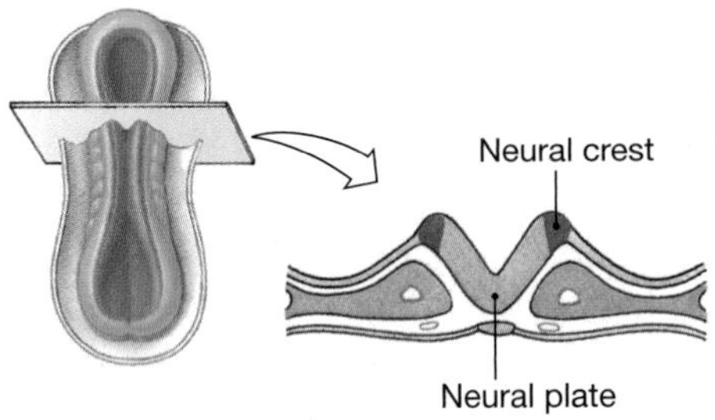

(b) Day 23

By day 23 of embryonic development, neural tube formation is almost complete.

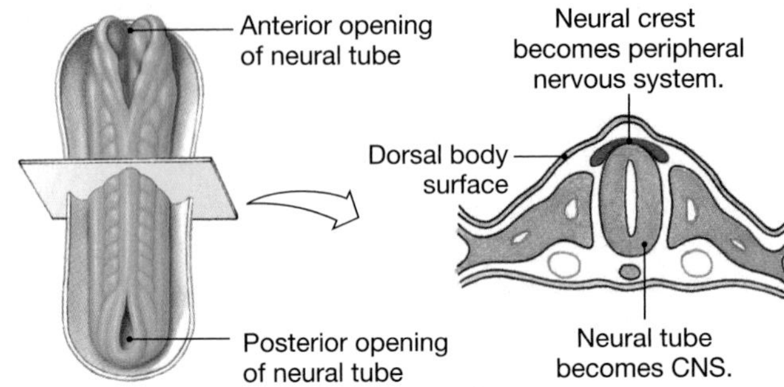

(c) 4 Weeks

A 4-week human embryo showing the anterior end of the neural tube which has specialized into three brain regions.

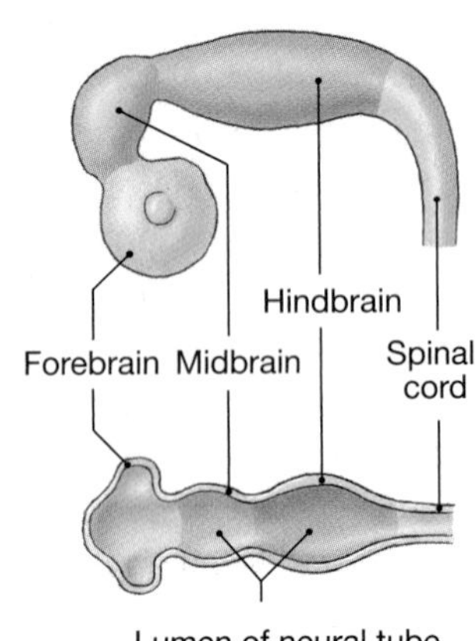

(d) 6 Weeks

At 6 weeks, the neural tube has differentiated into the brain regions present at birth. The central cavity (lumen) shown in the cross section will become the ventricles of the brain (see Fig. 9.4).

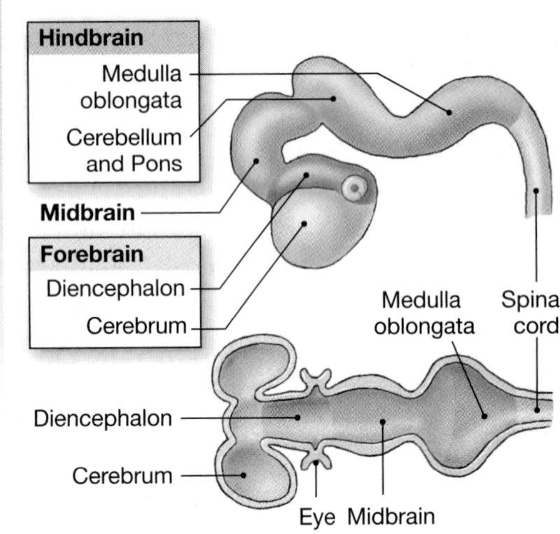

(e) 11 Weeks

By 11 weeks of embryonic development, the growth of the cerebrum is noticeably more rapid than that of the other divisions of the brain.

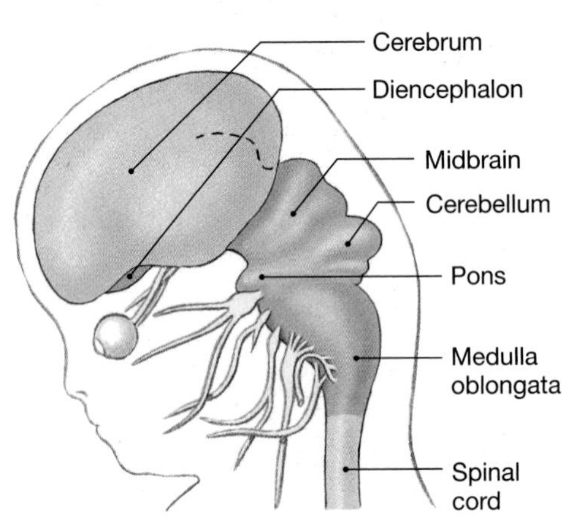

(f) 40 Weeks

At birth, the cerebrum has covered most of the other brain regions. Its rapid growth within the rigid confines of the cranium forces it to develop a convoluted, furrowed surface.

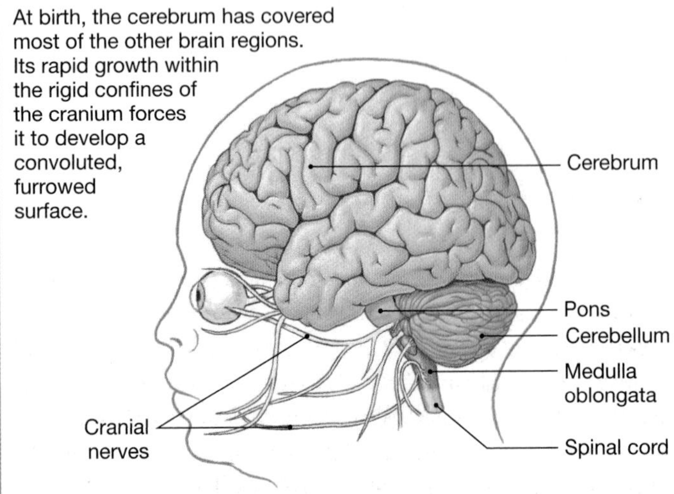

(g) Child

The directions "dorsal" and "ventral" are different in the brain because of flexion in the neural tube during development.

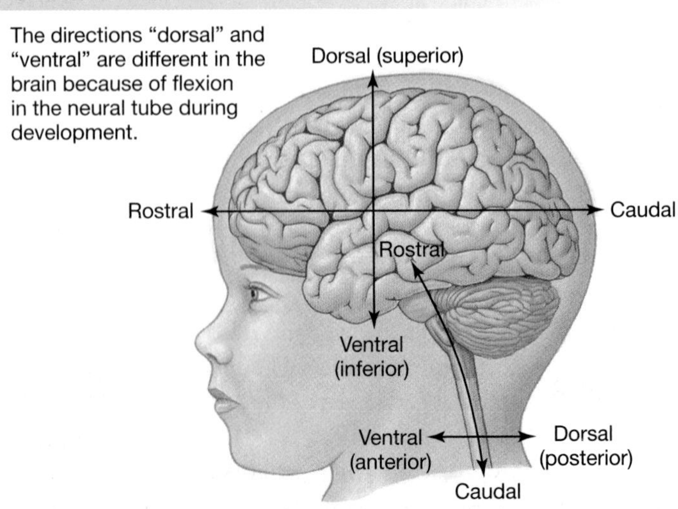

the brain. There are two *lateral ventricles* (the first and second) and two *descending ventricles* (the third and fourth). The central cavity of the neural tube also becomes the *central canal* of the spinal cord.

By week 11 the cerebrum is noticeably enlarged (Fig. 9.2e), and at birth the cerebrum is the largest and most obvious structure we see when looking at a human brain (Fig. 9.2f). The fully developed cerebrum surrounds the diencephalon, midbrain, and pons, leaving only the cerebellum and medulla oblongata visible below it. Because of the flexion (bending) of the neural tube early in development (see Fig. 9.2c), some directional terms have different meanings when applied to the brain (Fig. 9.2g).

The CNS Is Divided into Gray Matter and White Matter

The central nervous system, like the peripheral nervous system, is composed of neurons and supportive glial cells. Interneurons are those neurons completely contained within the CNS. Sensory (afferent) and efferent neurons link interneurons to peripheral receptors and effectors.

When viewed on a macroscopic level, the tissues of the CNS are divided into gray matter and white matter (■ Fig. 9.3c). **Gray matter** consists of unmyelinated nerve cell bodies, dendrites, and axon terminals. The cell bodies are assembled in an organized fashion in both the brain and the spinal cord. They form layers in some parts of the brain and in other parts cluster into groups of neurons that have similar functions. Clusters of cell bodies in the brain and spinal cord are known as *nuclei*. Nuclei are usually identified by specific names—for example, the *lateral geniculate nucleus,* where visual information is processed.

White matter is mostly myelinated axons and contains very few cell bodies. Its pale color comes from the myelin sheaths that surround the axons. Bundles of axons that connect different regions of the CNS are known as **tracts.** Tracts in the central nervous system are equivalent to nerves in the peripheral nervous system.

The consistency of the brain and spinal cord is soft and jellylike. Although individual neurons and glial cells have highly organized internal cytoskeletons that maintain cell shape and orientation, neural tissue has minimal extracellular matrix and must rely on external support for protection from trauma. This support comes in the form of an outer casing of bone, three layers of connective tissue membrane, and fluid between the membranes (Fig. 9.3b, c).

Concept Check Answer: p. 323

2. Name the four kinds of glial cells found in the CNS, and describe the function(s) of each [p. 244].

Bone and Connective Tissue Support the CNS

In vertebrates, the brain is encased in a bony **skull,** or **cranium** (Fig. 9.3a), and the spinal cord runs through a canal in the **vertebral column.** The body segmentation that is characteristic of many invertebrates can still be seen in the bony **vertebrae** (singular *vertebra*), which are stacked on top of one another and separated by disks of connective tissue. Nerves of the peripheral nervous system enter and leave the spinal cord by passing through notches between the stacked vertebrae (Fig. 9.3c).

Three layers of membrane, collectively called the **meninges** {singular *meninx,* membrane}, lie between the bones and tissues of the central nervous system. These membranes help stabilize the neural tissue and protect it from bruising against the bones of the skeleton. Starting from the bones and moving toward the neural tissue, the membranes are (1) the dura mater, (2) the arachnoid membrane, and (3) the pia mater (Fig. 9.3b, c).

The **dura mater** {*durare,* to last + *mater,* mother} is the thickest of the three membranes (think *durable*). It is associated with veins that drain blood from the brain through vessels or cavities called *sinuses*. The middle layer, the **arachnoid** {*arachnoides,* cobweblike} **membrane,** is loosely tied to the inner membrane, leaving a *subarachnoid space* between the two layers. The inner membrane, the **pia mater** {*pius,* pious + *mater,* mother}, is a thin membrane that adheres to the surface of the brain and spinal cord. Arteries that supply blood to the brain are associated with this layer.

The final protective component of the CNS is extracellular fluid, which helps cushion the delicate neural tissue. The cranium has an internal volume of 1.4 L, of which about 1 L is occupied by the cells. The remaining volume is divided into two distinct extracellular compartments: the blood (100–150 mL), and the *cerebrospinal fluid* and interstitial fluid (250–300 mL). The cerebrospinal fluid and interstitial fluid together form the extracellular environment for neurons. Interstitial fluid lies inside the pia mater. Cerebrospinal fluid is found in the ventricles and in the space between the pia mater and the arachnoid membrane. The cerebrospinal and interstitial fluid compartments communicate with each other across the leaky junctions of the pial membrane and the ependymal cell layer lining the ventricles.

Concept Check Answers: p. 323

3. What is a ganglion? What is the equivalent structure in the CNS?
4. Peripheral nerves are equivalent to what organizational structure in the CNS?

The Brain Floats in Cerebrospinal Fluid

Cerebrospinal fluid, or **CSF,** is a salty solution that is continuously secreted by the **choroid plexus,** a specialized region on the walls of the ventricles (■ Fig. 9.4b). The choroid plexus is remarkably similar to kidney tissue and consists of capillaries and a transporting epithelium [p. 157] derived from the ependyma. The choroid plexus cells selectively pump sodium and other solutes from plasma into the ventricles, creating an osmotic gradient that draws water along with the solutes (Fig. 9.4c).

9

The Central Nervous System

(a) Posterior View of the Central Nervous System

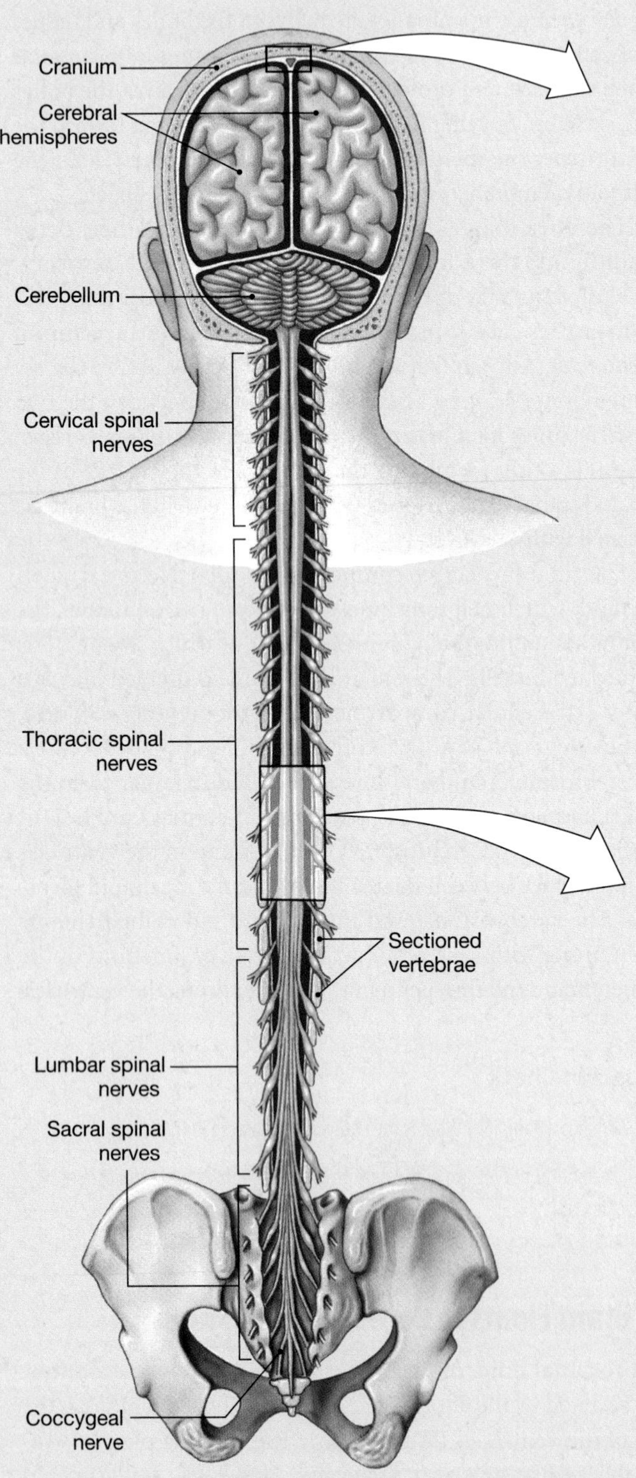

(b) Sectional View of the Meninges of the Brain, showing how they cushion and protect delicate neural tissue

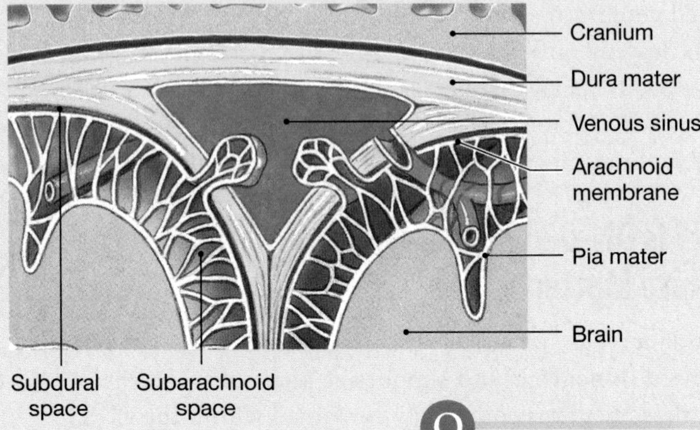

FIGURE QUESTION

Moving from the cranium in, name the meninges that form the boundaries of the venous sinus and the subdural and subarachnoid spaces.

(c) Posterior View of Spinal Cord and Vertebra

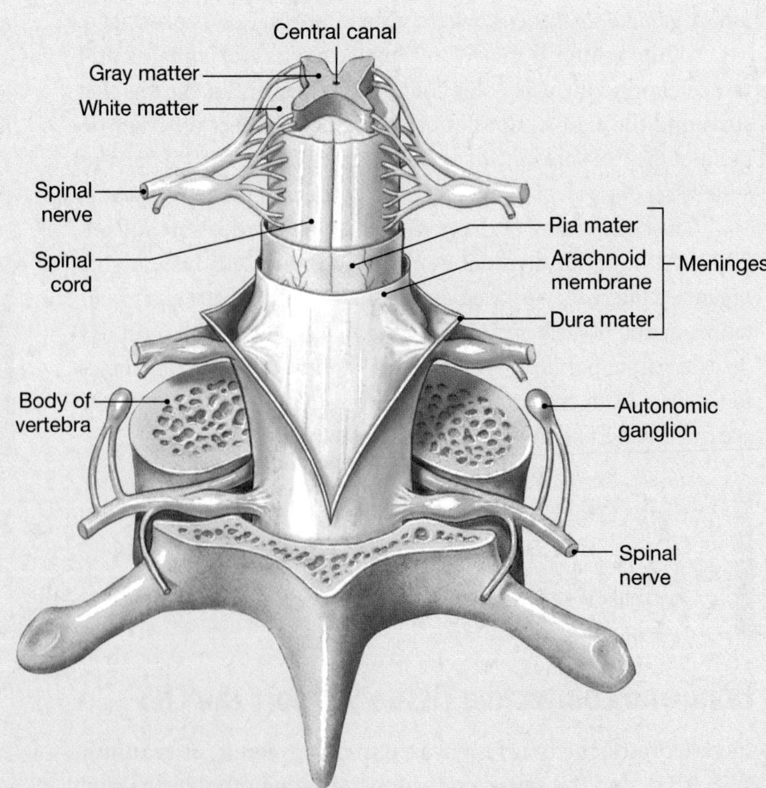

Cerebrospinal Fluid

(a) The Ventricles of the Brain

The lateral ventricles consist of the first and second ventricles. The third and fourth ventricles extend through the brain stem and connect to the central canal that runs through the spinal cord. Compare the frontal view to the cross section in Fig. 9.10a.

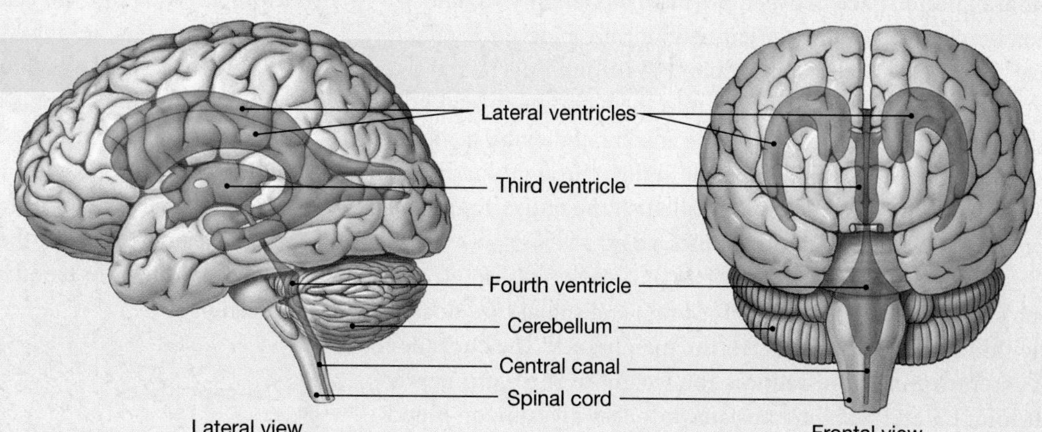

(b) Cerebrospinal Fluid Secretion

Cerebrospinal fluid is secreted into the ventricles and flows throughout the subarachnoid space, where it cushions the central nervous system.

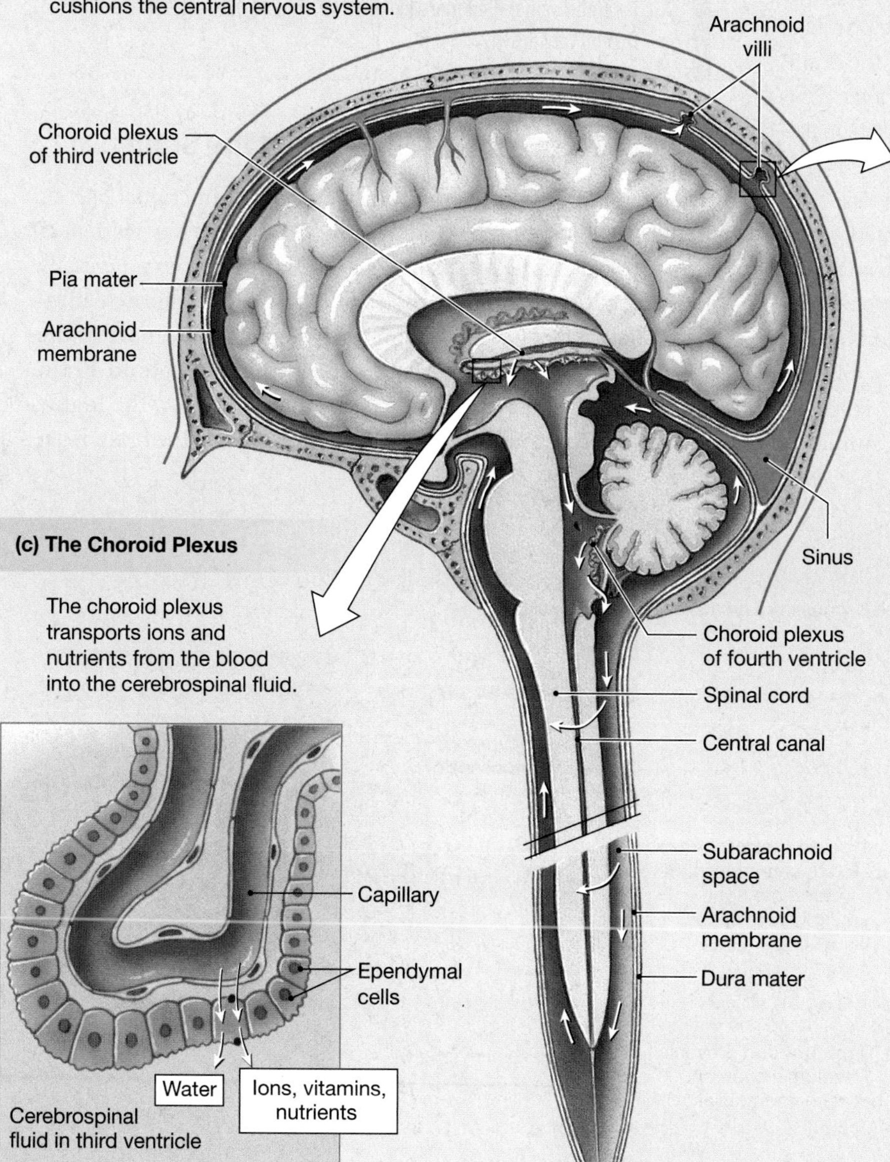

(c) The Choroid Plexus

The choroid plexus transports ions and nutrients from the blood into the cerebrospinal fluid.

(d) Cerebrospinal Fluid Reabsorption

Cerebrospinal fluid is reabsorbed into the blood at fingerlike projections of the arachnoid membrane called villi.

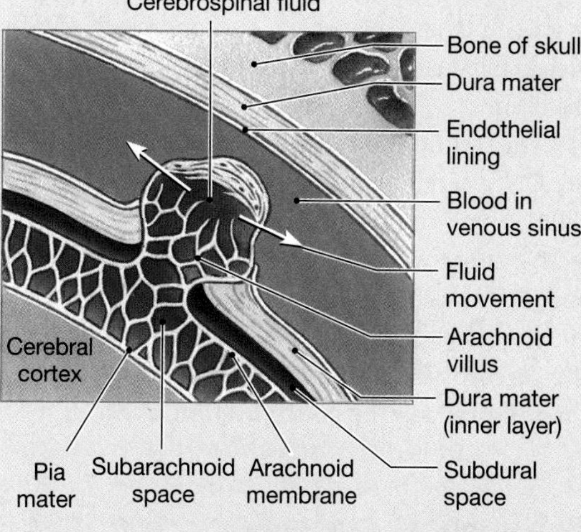

FIGURE QUESTIONS

1. Physicians may extract a sample of cerebrospinal fluid when they suspect an infection in the brain. Where is the least risky and least difficult place for them to insert a needle through the meninges? (See Fig. 9.4b.)
2. The aqueduct of Sylvius is the narrow passageway between the third and fourth ventricles. What happens to CSF flow if the aqueduct becomes blocked by infection or tumor, a condition known as aqueductal stenosis {*stenos*, narrow}? On a three-dimensional imaging study of the brain, how would you distinguish aqueductal stenosis from a blockage of CSF flow in the subarachnoid space near the frontal lobe?

From the ventricles, cerebrospinal fluid flows into the **subarachnoid space** between the pia mater and the arachnoid membrane, surrounding the entire brain and spinal cord in fluid (Fig. 9.4b). The cerebrospinal fluid flows around the neural tissue and is finally absorbed back into the blood by special **villi** {singular *villus,* shaggy hair} on the arachnoid membrane in the cranium (Fig. 9.4d). The rate of fluid flow through the central nervous system is sufficient to replenish the entire volume of cerebrospinal fluid about three times a day.

Cerebrospinal fluid serves two purposes: physical protection and chemical protection. The brain and spinal cord float in the thin layer of fluid between the membranes. The buoyancy of cerebrospinal fluid reduces the weight of the brain nearly 30-fold. Lighter weight translates into less pressure on blood vessels and nerves attached to the CNS.

The cerebrospinal fluid also provides protective padding. When there is a blow to the head, the CSF must be compressed before the brain can hit the inside of the cranium. However, water is minimally compressible, which helps CSF cushion the brain. For a dramatic demonstration of the protective power of cerebrospinal fluid, shake a block of tofu (representing the brain) in an empty jar. Then shake a second block of tofu in a jar completely filled with water to see how cerebrospinal fluid safeguards the brain.

In addition to physically protecting the delicate tissues of the CNS, cerebrospinal fluid creates a closely regulated extracellular environment for the neurons. The choroid plexus is selective about which substances it transports into the ventricles, and, as a result, the composition of cerebrospinal fluid is different from that of the plasma. The concentration of K^+ is lower in the cerebrospinal fluid, and the concentration of H^+ is higher than in plasma. The concentration of Na^+ in CSF is similar to that in the blood. Cerebrospinal fluid normally contains very little protein and no blood cells.

Cerebrospinal fluid exchanges solutes with the interstitial fluid of the CNS and provides a route by which wastes can be removed. Clinically, a sample of cerebrospinal fluid is presumed to be an indicator of the chemical environment in the brain. This sampling procedure, known as a *spinal tap* or *lumbar puncture,* is generally done by withdrawing fluid from the subarachnoid space between vertebrae at the lower end of the spinal cord. The presence of proteins or blood cells in cerebrospinal fluid suggests an infection.

Concept Check Answers: p. 323

5. If the concentration of H^+ in cerebrospinal fluid is higher than that in the blood, what can you say about the pH of the CSF?
6. Why is rupturing a blood vessel running between the meninges ruptures potentially a surgical emergency?
7. Is cerebrospinal fluid more like plasma or more like interstitial fluid? Defend your answer.

The Blood-Brain Barrier Protects the Brain

The final layer of protection for the brain is a functional barrier between the interstitial fluid and the blood. This barrier is necessary to isolate the body's main control center from potentially harmful substances in the blood and from blood-borne pathogens such as bacteria. To achieve this protection, most of the 400 miles of brain capillaries create a functional **blood-brain barrier** (■ Fig. 9.5). Although not a literal barrier, the highly selective permeability of brain capillaries shelters the brain from

THE BLOOD-BRAIN BARRIER

Neurons are protected from harmful substances in the blood because brain capillaries are not leaky.

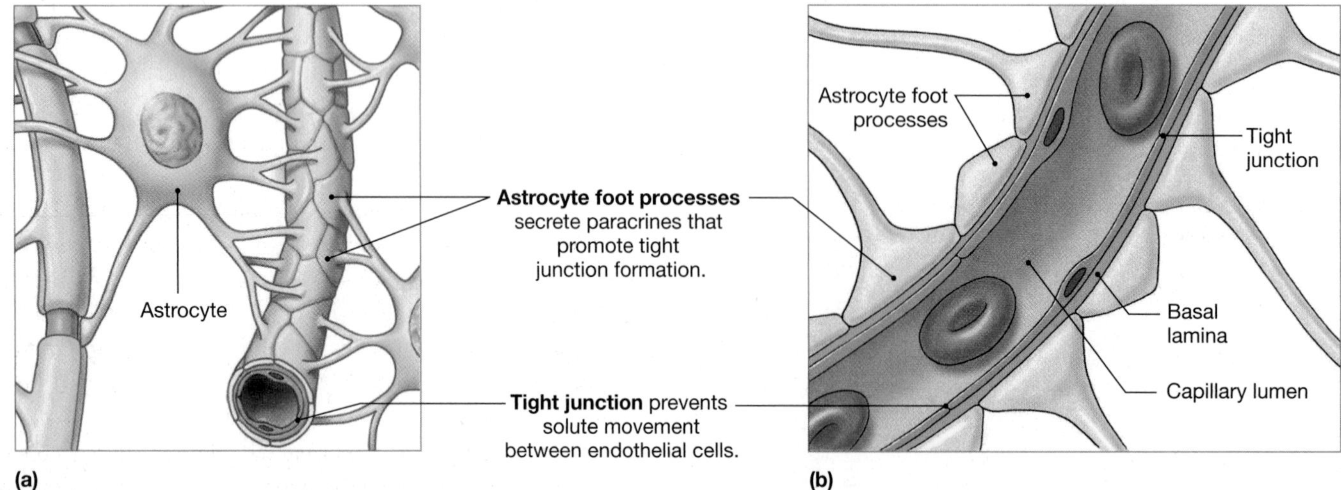
Astrocyte
Astrocyte foot processes secrete paracrines that promote tight junction formation.
Tight junction prevents solute movement between endothelial cells.
Astrocyte foot processes
Tight junction
Basal lamina
Capillary lumen
(a)
(b)

■ **Fig. 9.5**

toxins and from fluctuations in hormones, ions, and neuroactive substances such as neurotransmitters in the blood.

Why are brain capillaries so much less permeable than other capillaries? In most capillaries, leaky cell-cell junctions and pores allow free exchange of solutes between the plasma and interstitial fluid [p. 78]. In brain capillaries, however, the endothelial cells form tight junctions with one another, junctions that prevent solute movement between the cells. Tight junction formation apparently is induced by paracrine signals from adjacent astrocytes whose foot processes surround the capillary. As a result, it is the brain tissue itself that creates the blood-brain barrier.

The selective permeability of the blood-brain barrier can be attributed to its transport properties. The capillary endothelium uses selected membrane carriers and channels to move nutrients and other useful materials from the blood into the brain interstitial fluid. Other transporters move wastes from the interstitial fluid into the plasma. Any water-soluble molecule that is not transported on one of these carriers cannot cross the blood-brain barrier.

One interesting illustration of how the blood-brain barrier works is seen in *Parkinson's disease,* a neurological disorder in which brain levels of the neurotransmitter dopamine are too low because dopaminergic neurons are either damaged or dead. Dopamine administered in a pill or injection is ineffective because it is unable to cross the blood-brain barrier. The dopamine precursor *L-dopa*, however, is transported across the cells of the blood-brain barrier on an amino acid transporter [p. 144]. Once neurons have access to L-dopa in the interstitial fluid, they metabolize it to dopamine, thereby allowing the deficiency to be treated.

The blood-brain barrier effectively excludes many water-soluble substances, but smaller lipid-soluble molecules can diffuse through the cell membranes [p. 142]. This is one reason some antihistamines make you sleepy but others do not. Older antihistamines were lipid-soluble amines that readily crossed the blood-brain barrier and acted on brain centers controlling alertness. The newer drugs are much less lipid soluble and as a result do not have the same sedative effect.

A few areas of the brain lack a functional blood-brain barrier, and their capillaries have leaky endothelium like most of the rest of the body. In these areas of the brain, the function of adjacent neurons depends in some way on direct contact with the blood. For instance, the hypothalamus releases neurosecretory hormones that must pass into the capillaries of the *hypothalamic-hypophyseal portal system* for distribution to the anterior pituitary [p. 223].

Another region that lacks the blood-brain barrier is the vomiting center in the medulla oblongata. These neurons monitor the blood for possibly toxic foreign substances, such as drugs. If they sense something harmful, they initiate a vomiting reflex. Vomiting removes the contents of the digestive system and helps eliminate ingested toxins.

RUNNING PROBLEM

Ben was diagnosed with *infantile spasms*, a form of epilepsy characterized by the onset of head-drop seizures at four to seven months and by arrested or deteriorating mental development. Ben was started on a month-long regimen of adrenocorticotropin (ACTH) [p. 223] shots to control the seizures. Scientists are unsure why this hormone is so effective in controlling this type of seizure. They have found that, among its effects, it increases myelin formation, increases blood-brain barrier integrity, and enhances binding of the neurotransmitter GABA at synapses. As expected, Ben's seizures disappeared completely before the month of treatment ended, and his development began to return to a normal level.

Q1: How might a leaky blood-brain barrier lead to a cascade of action potentials that trigger a seizure?

Q2: GABA opens Cl^- channels on the postsynaptic cell. What does this do to the cell's membrane potential? Does GABA make the cell more or less likely to fire action potentials?

Q3: Why is it important to limit the duration of ACTH therapy, particularly in very young patients? [p. 227]

289 297 312 314 316 320

Neural Tissue Has Special Metabolic Requirements

A unique property of the central nervous system is its specialized metabolism. Neurons require a constant supply of oxygen and glucose to make ATP for active transport of ions and neurotransmitters. Oxygen passes freely across the blood-brain barrier, and membrane transporters move glucose from the plasma to the brain's interstitial fluid. Unusually low levels of either substrate can have devastating results on brain function.

Because of its high demand for oxygen, the brain receives about 15% of the blood pumped by the heart. If blood flow to the brain is interrupted, brain damage occurs after only a few minutes without oxygen. Neurons are equally sensitive to lack of glucose. Under normal circumstances the only energy source for neurons is glucose.

By some estimates, the brain is responsible for about half of the body's glucose consumption. Consequently, the body uses several homeostatic pathways to ensure that glucose concentrations in the blood always remain adequate to meet the brain's demand. If homeostasis fails, progressive **hypoglycemia** (low blood glucose levels) leads to confusion, unconsciousness, and eventually death.

Now that you have a broad overview of the central nervous system, we will examine the structure and function of the spinal cord and brain in more detail.

CLINICAL FOCUS: DIABETES

Hypoglycemia and the Brain

Neurons are picky about their food. Under most circumstances, the only biomolecule that neurons use for energy is glucose. Surprisingly, this can present a problem for diabetic patients, whose problem is too much glucose in the blood. In the face of sustained hyperglycemia (elevated blood glucose), the cells of the blood-brain barrier down-regulate [p. 191] their glucose transporters. Then, if the patient's blood glucose level falls below normal because of excess insulin or failing to eat, the neurons of the brain may not be able to take up glucose fast enough to sustain their electrical activity. The individual may exhibit confusion, irritability, and slurred speech as brain function begins to fail. Prompt administration of sugar, either by mouth or intravenous infusion is necessary to prevent permanent damage. In extreme cases, hypoglycemia can cause coma or even death.

Concept Check

Answers: p. 323

8. Oxidative phosphorylation takes place in which organelle?
9. Name the two metabolic pathways for aerobic metabolism of glucose. What happens to NADH produced in these pathways?
10. In the late 1800s the scientist Paul Ehrlich injected blue dye into the bloodstream of animals. He noticed that all tissues except the brain stained blue. He was not aware of the blood-brain barrier, so what conclusion do you think he drew from his results?
11. In a subsequent experiment, a student of Ehrlich's injected the dye into the cerebrospinal fluid of the same animals. What do you think he observed about staining in the brain and in other body tissues?

The Spinal Cord

The spinal cord is the major pathway for information flowing back and forth between the brain and the skin, joints, and muscles of the body. In addition, the spinal cord contains neural networks responsible for locomotion. If the spinal cord is severed, there is loss of sensation from the skin and muscles as well as *paralysis,* loss of the ability to voluntarily control muscles.

The spinal cord is divided into four regions (*cervical, thoracic, lumbar,* and *sacral*), named to correspond to the adjacent vertebrae (see Fig. 9.3a). Each spinal region is subdivided into segments, and each segment gives rise to a bilateral pair of **spinal nerves**. Just before a spinal nerve joins the spinal cord, it divides into two branches called **roots** (■ Fig. 9.6a).

SPINAL CORD ORGANIZATION

The spinal cord contains nuclei with cell bodies of efferent neurons and tracts of axons going to and from the brain.

(a) One segment of spinal cord, ventral view, showing its pair of nerves.

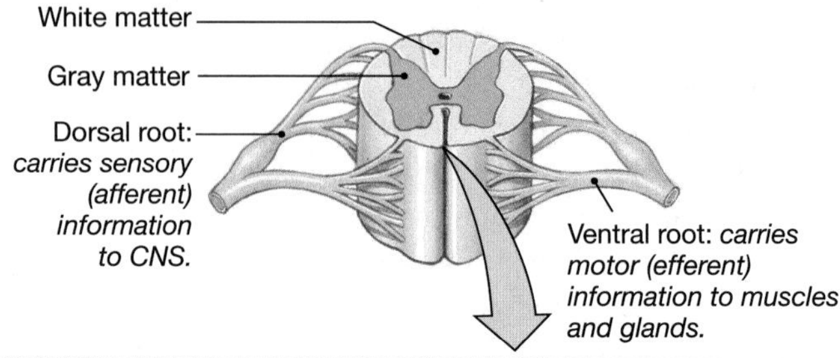

(b) Gray matter consists of sensory and motor nuclei.

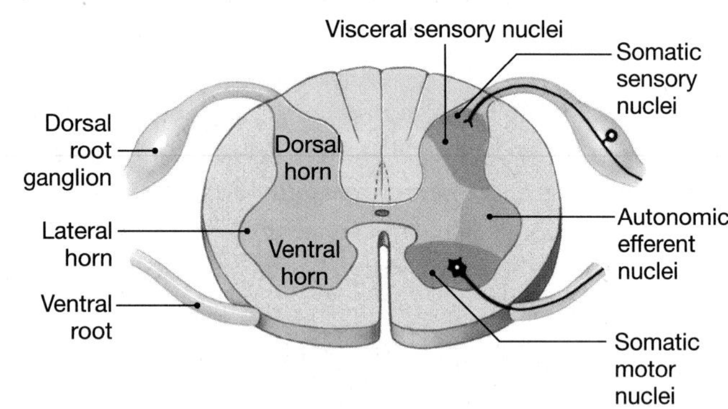

(c) White matter in the spinal cord consists of tracts of axons carrying information to and from the brain.

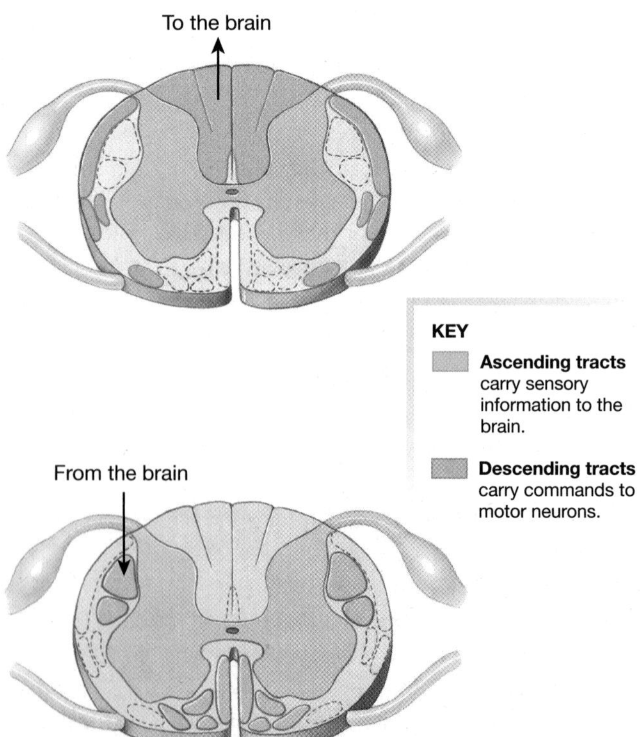

■ **Fig. 9.6**

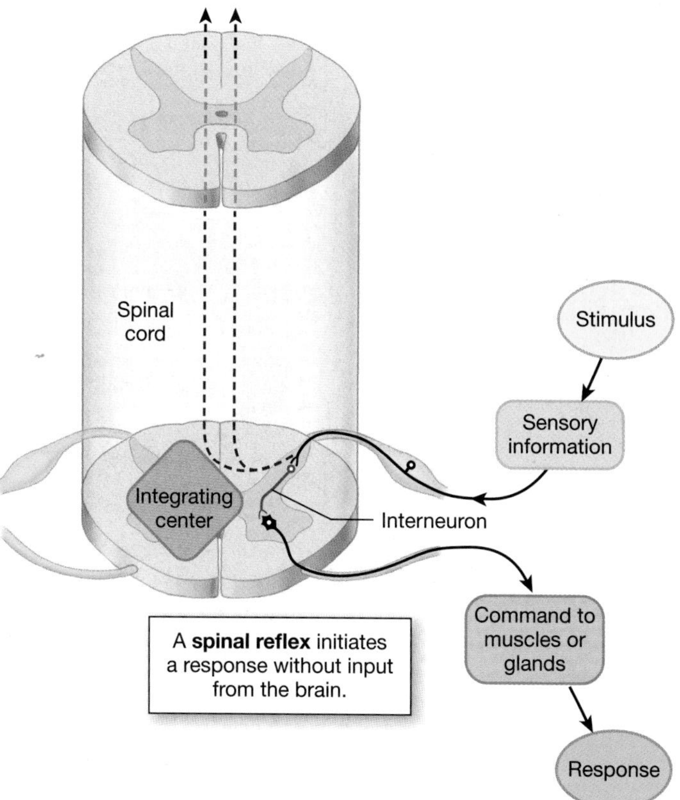
SPINAL REFLEXES
In a spinal refex, sensory information entering the spinal cord is acted on without input from the brain. However, sensory information about the stimulus may be sent to the brain.
Spinal cord
Stimulus
Sensory information
Integrating center
Interneuron
Command to muscles or glands
A spinal reflex initiates a response without input from the brain.
Response

■ **Fig. 9.7**

The **dorsal root** of each spinal nerve is specialized to carry incoming sensory information. The **dorsal root ganglia**, swellings found on the dorsal roots just before they enter the cord (Fig. 9.6b), contain cell bodies of sensory neurons. The **ventral root** carries information from the CNS to muscles and glands.

In cross section, the spinal cord has a butterfly- or H-shaped core of gray matter and a surrounding rim of white matter. Sensory fibers from the dorsal roots synapse with interneurons in the **dorsal horns** of the gray matter. The dorsal horn cell bodies are organized into two distinct nuclei, one for somatic information and one for visceral information (Fig. 9.6b).

The **ventral horns** of the gray matter contain cell bodies of motor neurons that carry efferent signals to muscles and glands. The ventral horns are organized into somatic motor and autonomic nuclei. Efferent fibers leave the spinal cord via the ventral root.

The white matter of the spinal cord is the biological equivalent of fiber-optic cables that telephone companies use to carry our communications systems. White matter can be divided into a number of **columns** composed of tracts of axons that transfer information up and down the cord. **Ascending tracts** take sensory information to the brain. They occupy the dorsal and external lateral portions of the spinal cord (Fig. 9.6c). **Descending tracts** carry mostly efferent (motor) signals from the brain to the cord. They occupy the ventral and interior lateral portions of the white matter. **Propriospinal tracts** {*proprius*, one's own} are those that remain within the cord.

The spinal cord can function as a self-contained integrating center for simple *spinal reflexes*, with signals passing from a sensory neuron through the gray matter to an efferent neuron (■ Fig. 9.7). In addition, spinal interneurons may route sensory information to the brain through ascending tracts or bring commands from the brain to motor neurons. In many cases, the interneurons also modify information as it passes through them. Reflexes play a critical role in the coordination of body movement.

Concept Check Answers: p. 324

12. What are the differences between horns, roots, tracts, and columns of the spinal cord?
13. If a dorsal root of the spinal cord is cut, what function will be disrupted?

The Brain

Thousands of years ago, Aristotle declared that the heart was the seat of the soul. However, most people now agree that the brain is the organ that gives the human species its unique attributes. The challenge facing today's scientists is to understand how circuits formed by millions of neurons result in complex behaviors such as speaking, writing a symphony, or creating imaginary worlds for an interactive computer game. Brain function may be the ultimate emergent property [p. 2]. The question remains whether we will ever be able to decipher how emotions such as happiness and love arise from the chemical and electrical signals passing along circuits of neurons.

It is possible to study the brain at many levels of organization. The most reductionist view looks at the individual neurons and at what happens to them in response to chemical or electrical signals. A more integrative study might look at groups of neurons and how they interact with one another in *circuits, pathways*, or networks. The most complicated approach starts with a behavior or physiological response and works backward to dissect the neural circuits that create the behavior or response.

For centuries, studies of brain function were restricted to anatomical descriptions. However, when we study the brain we see no tidy 1:1 relationship between structure and function. An adult human brain has a mass of about 1400 g and contains an estimated 10^{12} neurons. When you consider that each one

Central Nervous System

(a) Lateral View of the CNS

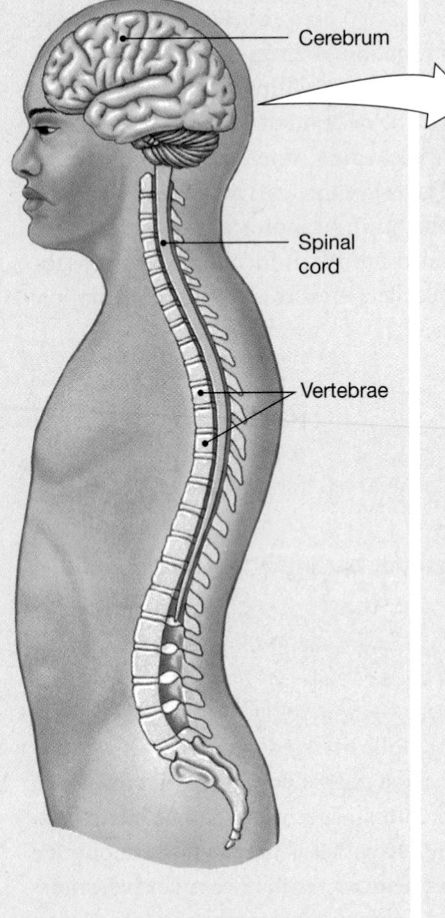

Anatomy of the Brain

(b) Lateral View of Brain

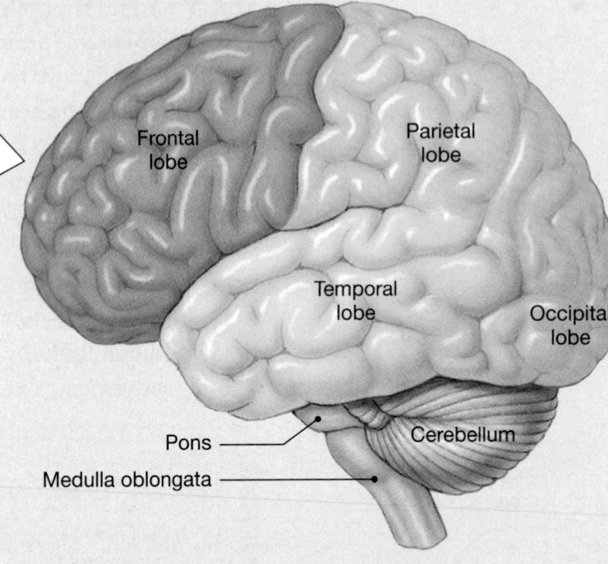

(c) Mid-Sagittal View of Brain

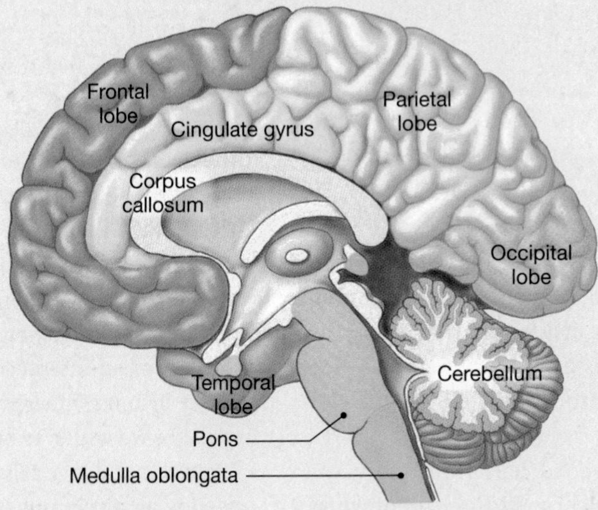

(d) Diencephalon

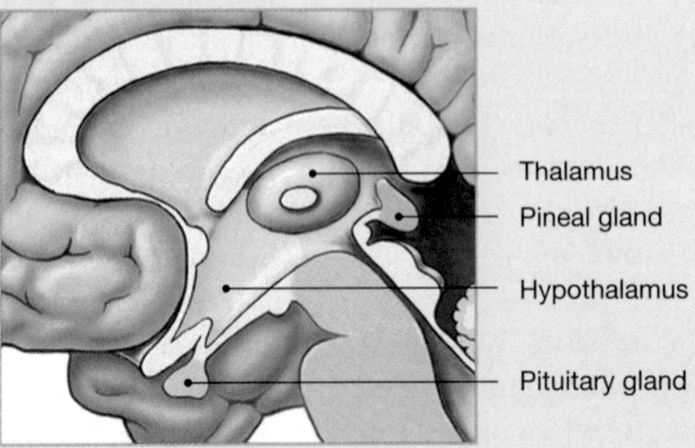

Functions of the Cerebrum

Cerebral cortex
See Figure 9.13

Sensory areas
- Perception

Motor areas
- Skeletal muscle movement

Association areas
- Integration of information and direction of voluntary movement

Basal ganglia (not shown)
See Figure 9.10
- Movement

Limbic System (not shown)
See Figure 9.11

Amygdala
- Emotion
- Memory

Hippocampus
- Learning
- Memory

Functions of the Cerebellum
- Movement coordination

Functions of the Diencephalon

Thalamus
- Integrating center and relay station for sensory and motor information

Pineal gland
- Melatonin secretion

Hypothalamus
See Table 9.2
- Homeostasis
- Behavioral drives

Pituitary gland
- Hormone secretion

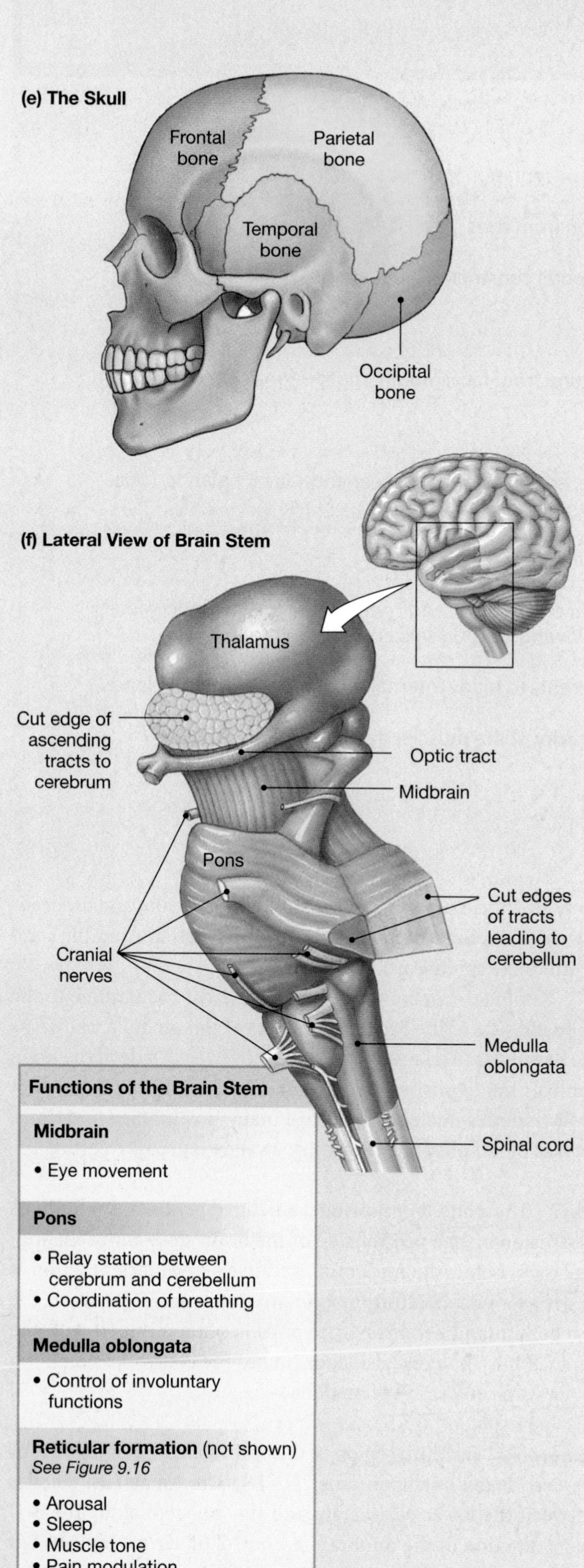

of these millions of neurons may receive as many as 200,000 synapses, the number of possible neuronal connections is mind boggling. To complicate matters even more, those synapses are not fixed and are constantly changing.

A basic principle to remember when studying the brain is that one function, even an apparently simple one such as bending your finger, will involve multiple brain regions (as well as the spinal cord). Conversely, one brain region may be involved in several functions at the same time. In other words, understanding the brain is not simple and straightforward.

■ Figure 9.8 is an anatomy summary to follow as we discuss major brain regions, moving from the most primitive to the most complex. Of the six major divisions of the brain present at birth (see Fig. 9.2e), only the medulla, cerebellum, and cerebrum are visible when the intact brain is viewed in profile. The remaining three divisions (diencephalon, midbrain, and pons) are covered by the cerebrum.

The Brain Stem Is the Oldest Part of the Brain

The **brain stem** is the oldest and most primitive region of the brain and consists of structures that derive from the embryonic midbrain and hindbrain. The brain stem can be divided into white matter and gray matter, and in some ways its anatomy is similar to that of the spinal cord. Some ascending tracts from the spinal cord pass through the brain stem, while other ascending tracts synapse there. Descending tracts from higher brain centers also travel through the brain stem on their way to the spinal cord.

Pairs of peripheral nerves branch off the brain stem, similar to spinal nerves along the spinal cord (Fig. 9.8f). Eleven of the 12 **cranial nerves** (numbers II–XII) originate along the brain stem. (The first cranial nerve, the olfactory nerve, enters the forebrain.) Cranial nerves carry sensory and motor information for the head and neck (■ Tbl. 9.1).

The cranial nerves are described according to whether they include sensory fibers, efferent fibers, or both (mixed nerves). For example, cranial nerve X, the **vagus nerve** {*vagus*, wandering}, is a mixed nerve that carries both sensory and motor fibers for many internal organs. An important component of a clinical neurological examination is testing the functions controlled by these nerves.

The brain stem contains numerous discrete groups of nerve cell bodies (*nuclei*). Many of these nuclei are associated with the **reticular formation**, a diffuse collection of neurons that extends throughout the brain stem. The name *reticular* means "network" and comes from the crisscrossed axons that branch profusely up into superior sections of the brain and down into the spinal cord. Nuclei in the brain stem are involved in many basic processes, including arousal and sleep, muscle tone and stretch reflexes, coordination of breathing, blood pressure regulation, and modulation of pain.

Table 9.1 The Cranial Nerves

Number	Name	Type	Function
I	Olfactory	Sensory	Olfactory (smell) information from nose
II	Optic	Sensory	Visual information from eyes
III	Oculomotor	Motor	Eye movement, pupil constriction, lens shape
IV	Trochlear	Motor	Eye movement
V	Trigeminal	Mixed	Sensory information from face, mouth; motor signals for chewing
VI	Abducens	Motor	Eye movement
VII	Facial	Mixed	Sensory for taste; efferent signals for tear and salivary glands, facial expression
VIII	Vestibulocochlear	Sensory	Hearing and equilibrium
IX	Glossopharyngeal	Mixed	Sensory from oral cavity, baro- and chemoreceptors in blood vessels; efferent for swallowing, parotid salivary gland secretion
X	Vagus	Mixed	Sensory and efferents to many internal organs, muscles, and glands
XI	Spinal accessory	Motor	Muscles of oral cavity, some muscles in neck and shoulder
XII	Hypoglossal	Motor	Tongue muscles

Concept Check Answers: p. 324

14. Are the following white matter or gray matter? (a) ascending tracts, (b) reticular formation, (c) descending tracts.
15. Using the information from Table 9.1, describe the types of activities you might ask a patient to perform if you wished to test the function of each cranial nerve.
16. In anatomical directional terminology, the cerebrum, which is located next to the top of the skull, is said to be ________ to the brain stem.

Starting at the spinal cord and moving toward the top of the skull, the brain stem consists of the medulla oblongata, the pons, and the midbrain (Fig. 9.8f). Some authorities include the cerebellum as part of the brain stem. The diamond-shaped fourth ventricle runs through the interior of the brain stem and connects to the central canal of the spinal cord (see Fig. 9.4a).

Medulla The **medulla oblongata**, frequently just called the *medulla* {*medulla*, marrow; adjective *medullary*}, is the transition from the spinal cord into the brain proper (Fig. 9.8f). Its white matter includes ascending **somatosensory tracts** {*soma*, body} that bring sensory information to the brain, and descending **corticospinal tracts** that convey information from the cerebrum to the spinal cord.

About 90% of corticospinal tracts cross the midline to the opposite side of the body in a region of the medulla known as the **pyramids**. As a result of this crossover, each side of the brain controls the opposite side of the body. Gray matter in the medulla includes nuclei that control many involuntary functions, such as blood pressure, breathing, swallowing, and vomiting.

Pons The **pons** {*pons*, bridge; adjective *pontine*} is a bulbous protrusion on the ventral side of the brain stem above the medulla and below the midbrain. Because its primary function is to act as a relay station for information transfer between the cerebellum and cerebrum, the pons is often grouped with the cerebellum. The pons also coordinates the control of breathing along with centers in the medulla.

Midbrain The third region of the brain stem, the **midbrain**, or *mesencephalon* {*mesos*, middle}, is a relatively small area that lies between the lower brain stem and the diencephalon. The primary function of the midbrain is control of eye movement, but it also relays signals for auditory and visual reflexes.

The Cerebellum Coordinates Movement

The **cerebellum** is the second largest structure in the brain (Fig. 9.8a–c). It is located inside the base of the skull, just above the nape of the neck. The name *cerebellum* {adjective *cerebellar*} means "little brain," and, indeed, most of the nerve cells in the brain are in the cerebellum. The specialized function of the cerebellum is to process sensory information and coordinate the execution of movement. Sensory input into the cerebellum comes from somatic receptors in the periphery of the body and from receptors for equilibrium and balance located in the inner ear. The cerebellum also receives motor input from neurons in the cerebrum.

The Diencephalon Contains the Centers for Homeostasis

The **diencephalon**, or "between-brain," lies between the brain stem and the cerebrum. It is composed of two main sections, the thalamus and the hypothalamus, and two endocrine structures, the pituitary and pineal glands (■ Fig. 9.9).

Most of the diencephalon is occupied by many small nuclei that make up the **thalamus** {*thalamus*, bedroom; adjective *thalamic*}. The thalamus receives sensory fibers from the optic tract, ears, and spinal cord as well as motor information from the cerebellum. It projects fibers to the cerebrum, where the information is processed.

The thalamus is often described as a relay station because almost all sensory information from lower parts of the CNS passes through it. Like the spinal cord, the thalamus can modify information passing through it, making it an integrating center as well as a relay station.

The **hypothalamus** lies beneath the thalamus. Although the hypothalamus occupies less than 1% of total brain volume, it is the center for homeostasis and contains centers for various behavioral drives, such as hunger and thirst. Output from the hypothalamus also influences many functions of the autonomic division of the nervous system, as well as a variety of endocrine functions (■ Tbl. 9.2).

The hypothalamus receives input from multiple sources, including the cerebrum, the reticular formation, and various

THE DIENCEPHALON

The diencephalon lies between the brain stem and the cerebrum. It consists of thalamus, hypothalamus, pineal gland, and pituitary gland.

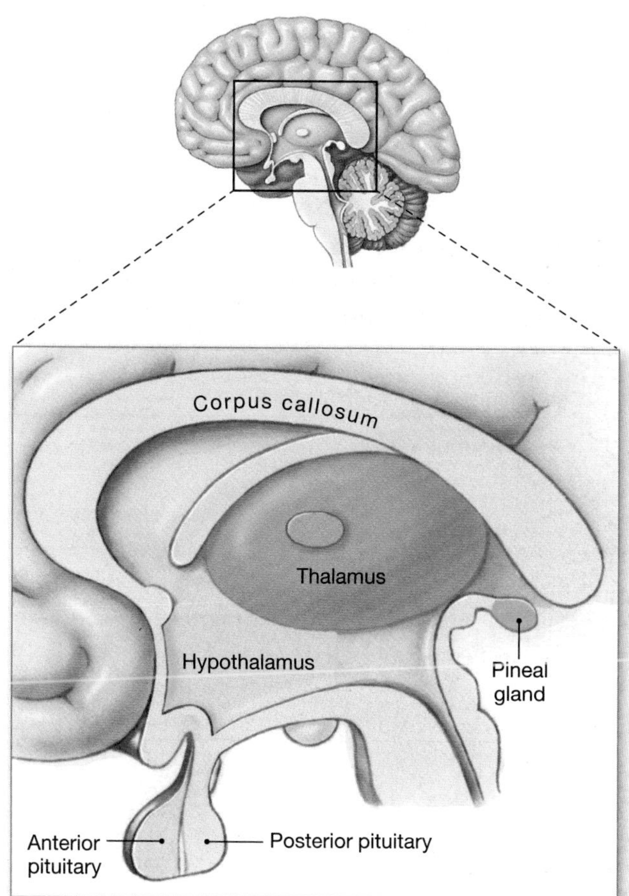

■ **Fig. 9.9**

Table 9.2 Functions of the Hypothalamus

1. Activates sympathetic nervous system
   - Controls catecholamine release from adrenal medulla (as in fight-or-flight reaction)
   - Helps maintain blood glucose concentrations through effects on endocrine pancreas
2. Maintains body temperature
   - Stimulates shivering and sweating
3. Controls body osmolarity
   - Motivates thirst and drinking behavior
   - Stimulates secretion of vasopressin [p. 219]
4. Controls reproductive functions
   - Directs secretion of oxytocin (for uterine contractions and milk release)
   - Directs trophic hormone control of anterior pituitary hormones FSH and LH [p. 223]
5. Controls food intake
   - Stimulates satiety center
   - Stimulates feeding center
6. Interacts with limbic system to influence behavior and emotions
7. Influences cardiovascular control center in medulla oblongata
8. Secretes trophic hormones that control release of hormones from anterior pituitary gland

sensory receptors. Output from the hypothalamus goes first to the thalamus and eventually to multiple effector pathways.

Two important endocrine structures are located in the diencephalon: the pituitary gland and the pineal gland [p. 231]. The posterior pituitary (*neurohypophysis*) is a down-growth of the hypothalamus and secretes neurohormones that are synthesized in hypothalamic nuclei. The anterior pituitary (*adenohypophysis*) is a true endocrine gland. Its hormones are regulated by hypothalamic neurohormones secreted into the hypothalamic-hypophyseal portal system. The pineal gland, which secretes the hormone melatonin [p. 231], is discussed later in this chapter.

Concept Check Answers: p. 324

17. Starting at the spinal cord and moving up, name the subdivisions of the brain stem.
18. What are the four primary structures of the diencephalon?

The Cerebrum Is the Site of Higher Brain Functions

As noted earlier in the chapter, the cerebrum is the largest and most distinctive part of the human brain and fills most of the cranial cavity. It is composed of two hemispheres connected primarily at the **corpus callosum** (Figs. 9.8c and 9.9), a distinct structure formed by axons passing from one side of the brain to the other. This connection ensures that the two hemispheres communicate and cooperate with each other. Each cerebral hemisphere is divided into four lobes, named for the bones of the skull under which they are located: *frontal, parietal, temporal,* and *occipital* (Fig. 9.8b, c, e).

The surface of the cerebrum in humans and other primates has a furrowed, walnut-like appearance, with grooves called *sulci* {singular *sulcus,* a furrow} dividing convolutions called *gyri* {singular *gyrus,* a ring or circle}. During development, the cerebrum grows faster than the surrounding cranium, causing the tissue to fold back on itself to fit into a smaller volume. The degree of folding is directly related to the level of processing of which the brain is capable. Less-advanced mammals, such as rodents, have brains with a relatively smooth surface. The human brain, on the other hand, is so convoluted that if it were inflated enough to smooth the surfaces, it would be three times as large and would need a head the size of a beach ball.

Gray Matter and White Matter Cerebral gray matter can be divided into three major regions: the cerebral cortex, the basal ganglia, and the limbic system. The **cerebral cortex** {*cortex,* bark or rind; adjective *cortical,* plural *cortices*} is the outer layer of the cerebrum, only a few millimeters thick (■ Fig. 9.10a). Neurons of the cerebral cortex are arranged in anatomically distinct vertical columns and horizontal layers (Fig. 9.10b). It is within these layers that our higher brain functions arise.

The second region of cerebral gray matter consists of the **basal ganglia** (Fig. 9.10a), which are involved in the control of

GRAY MATTER OF THE CEREBRUM

The cerebral cortex and basal ganglia are two of the three regions of gray matter in the cerebrum. The third region, the limbic system, is detailed in Figure 9.11. The frontal view shown here is similar to the sectional view obtained using modern diagnostic imaging techniques.

FIGURE QUESTION

The section through this brain is a section through the________ plane.

(a) coronal
(b) lateral
(c) frontal
(d) transverse
(e) sagittal

(a) Section through the brain showing the basal ganglia

Basal ganglia
Corpus callosum
Lateral ventricle
Tracts of white matter
Tip of lateral ventricle
Gray matter of cerebral cortex

(b) Cell bodies in the cerebral cortex form distinct layers and columns.

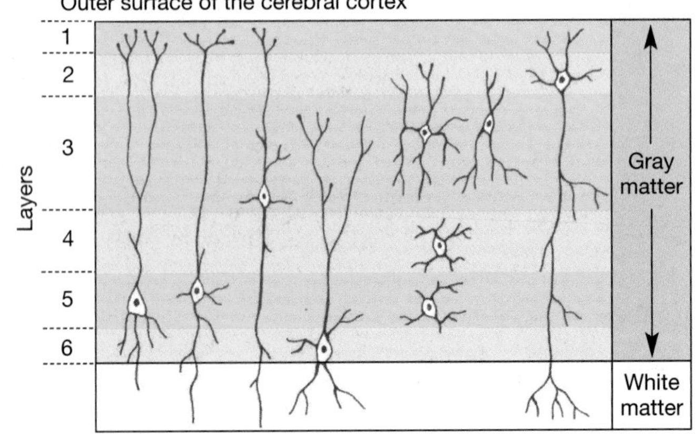

■ **Fig. 9.10**

movement. The basal ganglia are also called the *basal nuclei.* Neuroanatomists prefer to reserve the term ganglia for clusters of nerve cell bodies outside the CNS, but the term basal ganglia is commonly used in clinical settings.

The third region of the cerebrum is the **limbic system** {*limbus,* a border}, which surrounds the brain stem (■ Fig. 9.11). The limbic system represents probably the most primitive region of the cerebrum. It acts as the link between higher cognitive functions, such as reasoning, and more primitive emotional responses, such as fear. The major areas of the limbic system are the **amygdala** and **cingulate gyrus**, which are linked to emotion and memory, and the **hippocampus**, which is associated with learning and memory.

White matter in the cerebrum is found mostly in the interior (Fig. 9.10a). Bundles of fibers allow different regions of the cortex to communicate with one another and transfer information from one hemisphere to the other, primarily through the corpus callosum. According to some estimates, the corpus callosum may have as many as 200 million axons passing through it! Information entering and leaving the cerebrum goes along tracts that pass through the thalamus (with the exception of olfactory information, which goes directly from olfactory receptors to the cerebrum).

THE LIMBIC SYSTEM

The limbic system includes the amygdala, hippocampus, and cingulate gyrus. Anatomically, the limbic system is part of the gray matter of the cerebrum. The thalamus is shown for orientation purposes and is not part of the limbic system.

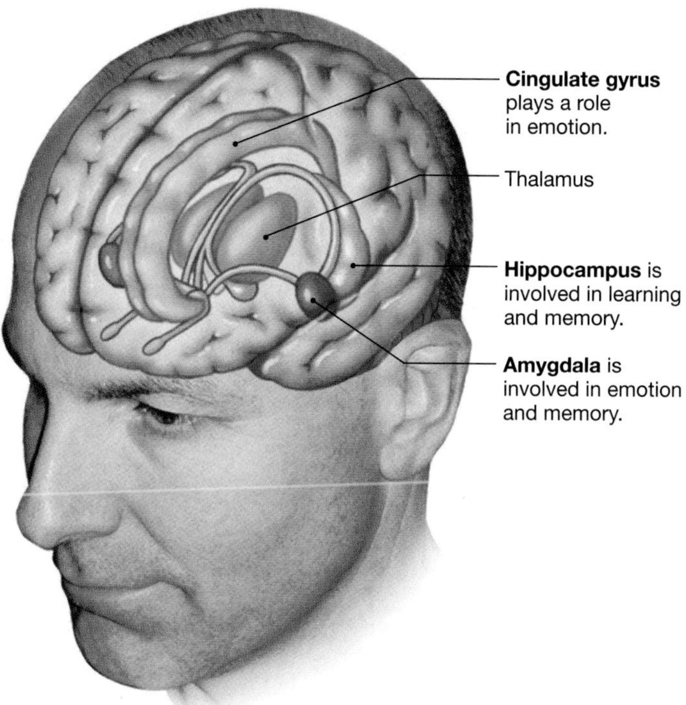

■ **Fig. 9.11**

> **Concept Check** Answers: p. 324
>
> **19.** Name the anatomical location in the brain where neurons from one side of the body cross to the opposite side.
>
> **20.** Name the divisions of the brain in anatomical order, starting from the spinal cord.

Brain Function

From a simplistic view, the CNS is an information processor much like a computer. For many functions it follows a basic reflex pathway [p. 15]. The brain receives sensory input from the internal and external environments, integrates and processes the information, and, if appropriate, creates a response (■ Fig. 9.12a). What makes the brain more complicated than this simple reflex pathway, however, is its ability to generate information and output signals *in the absence of external input.* Modeling this intrinsic input requires a more complex diagram.

Larry Swanson of the University of Southern California presents one approach to modeling brain function in his book *Brain Architecture: Understanding the Basic Plan* (Oxford University Press, 2003). He describes three systems that influence output by the motor systems of the body: (1) the **sensory system,** which monitors the internal and external environments and initiates reflex responses; (2) a **cognitive system** that

SIMPLE AND COMPLEX PATHWAYS IN THE BRAIN

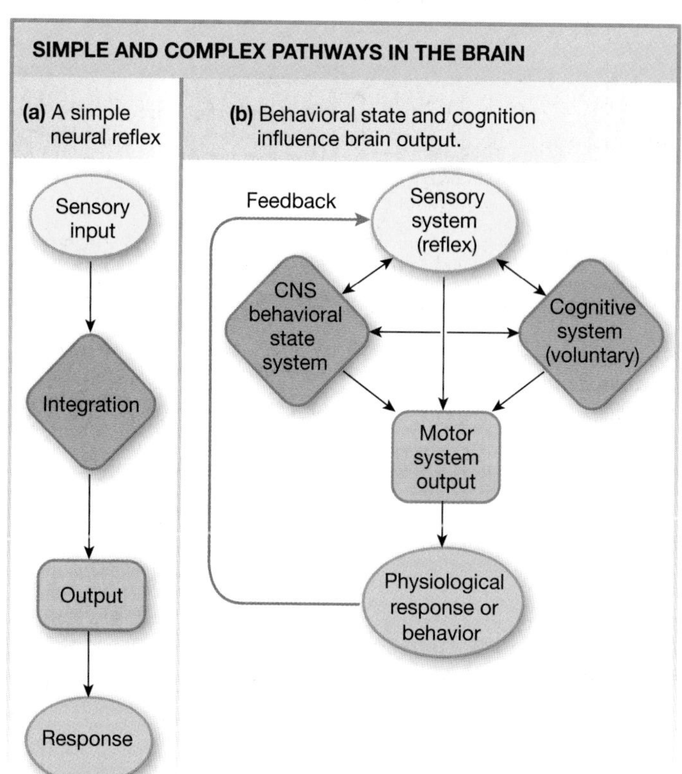

■ **Fig. 9.12**

resides in the cerebral cortex and is able to initiate voluntary responses; and (3) a **behavioral state system**, which also resides in the brain and governs sleep-wake cycles and other intrinsic behaviors. Information about the physiological or behavioral responses created by motor output feeds back to the sensory system, which in turn communicates with the cognitive and behavioral state systems (Fig. 9.12b).

In most of the physiological organ systems of the body that you will study, simple reflex pathways initiated through the sensory system and executed by motor output are adequate to explain homeostatic control mechanisms. However, the cognitive and behavioral state systems remain potential sources of influence. At its simplest, this influence may take the form of voluntary behaviors, such as breath-holding, that override automatic functions. More subtle and complicated interactions include the effect of emotions on normal physiology, such as stress-induced heart palpitations, and the role of circadian rhythms in jet lag and shift work.

In the sections that follow we take a brief look at sensory and motor systems in the brain. We conclude this chapter with a discussion of some aspects of the behavioral state system and the cognitive system, such as circadian rhythms, sleep-wake cycles, emotion, learning, and memory.

The Cerebral Cortex Is Organized into Functional Areas

The cerebral cortex serves as an integrating center for sensory information and a decision-making region for many types of motor output. If we examine the cortex from a functional viewpoint, it can be divided into three specializations: (1) **sensory areas** (also called sensory fields), which receive sensory input and translate it into perception (awareness); (2) **motor areas,** which direct skeletal muscle movement; and (3) **association areas** (association cortices), which integrate information from sensory and motor areas and can direct voluntary behaviors (■ Fig. 9.13). Information passing along a pathway is usually processed in more than one of these areas.

The functional areas of the cerebral cortex do not necessarily correspond to the anatomical lobes of the brain. For one thing, functional specialization is not symmetrical across the cerebral cortex: each lobe has special functions not shared by the matching lobe on the opposite side. This **cerebral lateralization** of function is sometimes referred to as *cerebral dominance,* more popularly known as left brain–right brain dominance (■ Fig. 9.14). Language and verbal skills tend to be concentrated on the left side of the brain, with spatial skills concentrated on

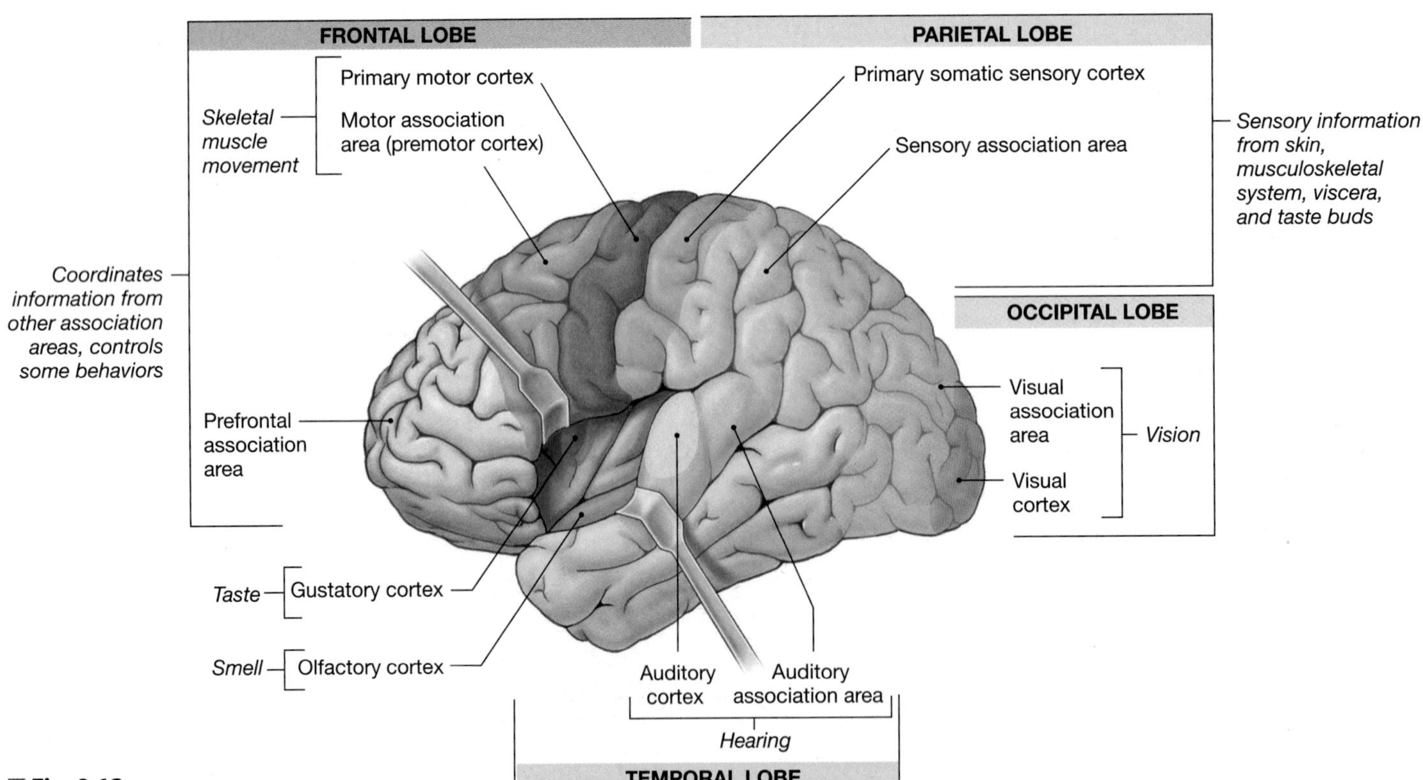

■ **Fig. 9.13**

the right side. The left brain is the dominant hemisphere for right-handed people, and it appears that the right brain is the dominant hemisphere for many left-handed people.

Even these generalizations are subject to change, however. Neural connections in the cerebrum, like those in other parts of the nervous system, exhibit a certain degree of plasticity. For example, if a person loses a finger, the regions of motor and sensory cortex previously devoted to control of the finger do not go dormant. Instead, adjacent regions of the cortex extend their functional fields and take over the parts of the cortex that are no longer used by the absent finger. Similarly, skills normally associated with one side of the cerebral cortex can be developed in the other hemisphere, as when a right-handed person with a broken hand learns to write with the left hand.

Much of what we know about functional areas of the cerebral cortex comes from study of patients who have either inherited neurological defects or suffered wounds in accidents or war. In some instances, surgical lesions made to treat some medical condition, such as uncontrollable epilepsy, have revealed functional relationships in particular brain regions. Imaging techniques such as *positron emission tomography* (PET) scans and *functional magnetic resonance imaging* (fMRI) provide noninvasive ways for us to watch the human brain at work.

The Spinal Cord and Brain Integrate Sensory Information

The sensory system monitors the internal and external environments and sends information to neural integrating centers, which in turn initiate appropriate responses. In its simplest form, this pathway is the classic reflex, illustrated in

CEREBRAL LATERALIZATION

The distribution of functional areas in the two cerebral hemispheres is not symmetrical.

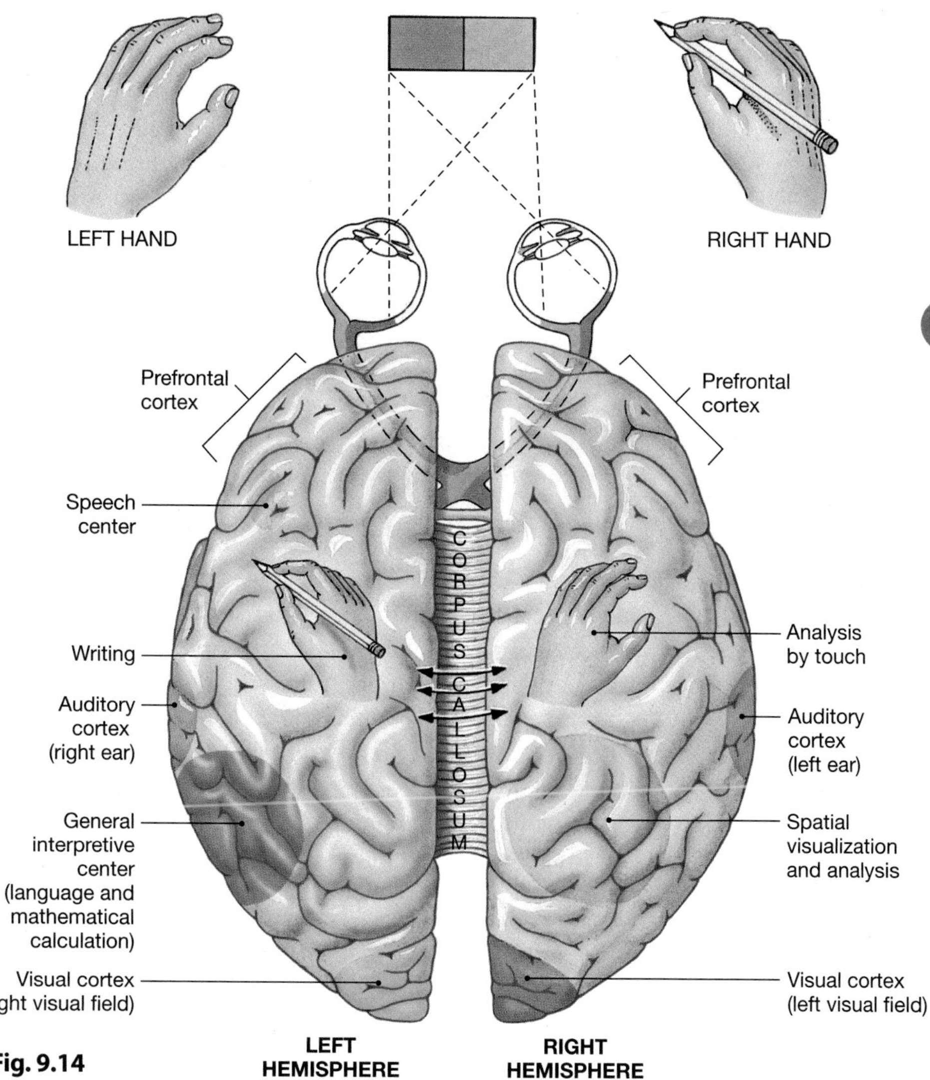

Fig. 9.14

FIGURE QUESTIONS

1. What would a person see if a stroke destroyed all function in the right visual cortex?
2. What is the function of the corpus callosum?
3. Many famous artists, including Leonardo da Vinci and Michelangelo, were left-handed. How is this related to cerebral lateralization?

Figure 9.12a. The simplest reflexes can be integrated in the spinal cord, without input from higher brain centers (see Fig. 9.7). However, even simple spinal reflexes usually send sensory information to the brain, creating perception of the stimulus. Brain functions dealing with perception are the most difficult to study because they require communication between the subject and the investigator—the subject must be able to tell the investigator what he or she is seeing, hearing, or feeling.

Sensory information from the body travels in ascending pathways to the brain. Information about muscle and joint position and movement goes to the cerebellum as well as to the cerebral cortex, allowing the cerebellum to assist with automatic subconscious coordination of movement. Most sensory information continues on to the cerebral cortex, where five sensory areas process information.

The **primary somatic sensory cortex** (also called the *somatosensory cortex*) in the parietal lobe is the termination point of pathways from the skin, musculoskeletal system, and viscera (see Fig. 9.13). The somatosensory pathways carry information about touch, temperature, pain, itch, and body position. Damage to this part of the brain leads to reduced sensitivity of the skin on the opposite side of the body because sensory fibers cross to the opposite side of the midline as they ascend through the spine or medulla.

The special senses of vision, hearing, taste, and olfaction (smell) each have different brain regions devoted to processing their sensory input (see Fig. 9.13). The **visual cortex**, located in the occipital lobe, receives information from the eyes. The **auditory cortex**, located in the temporal lobe, receives information from the ears. The **olfactory cortex**, a small region in the temporal lobe, receives input from chemoreceptors in the nose. The **gustatory cortex,** deeper in the brain near the edge of the frontal lobe, receives sensory information from the taste buds.

Sensory Information Is Processed into Perception

Once sensory information reaches the appropriate cortical area, information processing has just begun. Neural pathways extend from sensory areas to appropriate association areas, which integrate somatic, visual, auditory, and other stimuli into *perception*, the brain's interpretation of sensory stimuli.

Often the perceived stimulus is very different from the actual stimulus. For instance, photoreceptors in the eye receive light waves of different frequencies, but we perceive the different wave energies as different colors. Similarly, the brain translates pressure waves hitting the ear into sound and interprets chemicals binding to chemoreceptors as taste or smell.

One interesting aspect of perception is the way our brain fills in missing information to create a complete picture, or translates a two-dimensional drawing into a three-dimensional shape (■ Fig. 9.15). Thus, we sometimes perceive what our brains expect to perceive. Our perceptual translation of sensory stimuli allows the information to be acted upon and used in voluntary motor control or in complex cognitive functions such as language.

PERCEPTION

The brain has the ability to interpret sensory information to create the perception of (a) shapes or (b) three-dimensional objects.

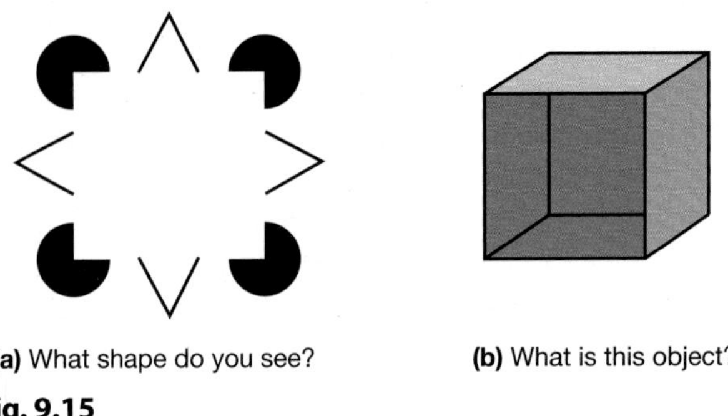

(a) What shape do you see? **(b)** What is this object?

■ **Fig. 9.15**

The Motor System Governs Output from the CNS

The motor output component of the nervous system is associated with the efferent division of the nervous system [Fig. 8.1, p. 240]. Motor output can be divided into three major types: (1) skeletal muscle movement, controlled by the somatic motor division; (2) neuroendocrine signals, neurohormones secreted into the blood by neurons located primarily in the hypothalamus and adrenal medulla; and (3) *visceral* responses, the actions of smooth and cardiac muscle or endocrine and exocrine glands. Visceral responses are governed by the autonomic division of the nervous system.

Information about skeletal muscle movement is processed in several regions of the CNS. Simple stimulus-response pathways, such as the knee jerk reflex, are processed either in the spinal cord or in the brain stem. Although these reflexes do not require integration in the cerebral cortex, they can be modified or overridden by input from the cognitive system.

Voluntary movements are initiated by the cognitive system and originate in the **primary motor cortex** and **motor association area** in the frontal lobes of the cerebrum (see Fig. 9.13). These regions receive input from sensory areas as well as from the cerebellum and basal ganglia. Long output neurons called *pyramidal cells* project axons from the motor areas through the brain stem to the spinal cord. Other pathways go from the cortex to the basal ganglia and lower brain regions. Descending motor pathways cross to the opposite side of the body, which means that damage to a motor area manifests as paralysis or loss of function on the opposite side of the body.

Neuroendocrine and visceral responses are coordinated primarily in the hypothalamus and medulla. The brain stem contains the control centers for many of the automatic life functions we take for granted, such as breathing and blood pressure.

It receives sensory information from the body and relays motor commands to peripheral muscles and glands.

The hypothalamus contains centers for temperature regulation, eating, and control of body osmolarity, among others. The responses to stimulation of these centers may be neural or hormonal reflexes or a behavioral response. Stress, reproduction, and growth are also mediated by the hypothalamus by way of multiple hormones. You will learn more about these reflexes in later chapters as we discuss the various systems of the body.

Sensory input is not the only factor determining motor output by the brain. The behavioral state system can modulate reflex pathways, and the cognitive system exerts both voluntary and involuntary control over motor functions.

The Behavioral State System Modulates Motor Output

The behavioral state system is an important modulator of sensory and cognitive processing. Many neurons in the behavioral state system are found in regions of the brain outside the cerebral cortex, including parts of the reticular formation in the brain stem, the hypothalamus, and the limbic system.

The neurons collectively known as the **diffuse modulatory systems** originate in the reticular formation in the brain stem and project their axons to large areas of the brain (■ Fig. 9.16). There are four modulatory systems that are generally classified according to the neurotransmitter they secrete: *noradrenergic* (norepinephrine), *serotonergic* (serotonin), *dopaminergic* (dopamine), and *cholinergic* (acetylcholine). The diffuse modulatory systems regulate brain function by influencing attention, motivation, wakefulness, memory, motor control, mood, and metabolic homeostasis.

One function of the behavioral state system is control of levels of consciousness and sleep-wake cycles. **Consciousness** is the body's state of arousal or awareness of self and environment. Experimental evidence shows that the **reticular activating system**, a diffuse collection of neurons in the reticular formation, plays an essential role in keeping the "conscious brain" awake.

If connections between the reticular formation and the cerebral cortex are disrupted surgically, an animal becomes comatose. Other evidence for the importance of the reticular formation in states of arousal comes from studies showing that general anesthetics depress synaptic transmission in that region of the brain. Presumably, blocking ascending pathways between the reticular formation and the cerebral cortex creates a state of unconsciousness.

Why Do We Sleep?

Physiologically, what distinguishes being awake from various stages of sleep? One way to define arousal states is by the pattern of electrical activity created by the cortical neurons. The measurement of brain activity is recorded by a procedure known as **electroencephalography**. Surface electrodes placed on or in the scalp detect depolarizations of the cortical neurons in the region just under the electrode.

In awake states, many neurons are firing but not in a coordinated fashion (■ Fig. 9.17a). Presumably the desynchronization of electrical activity in waking states is produced by ascending signals from the reticular formation. An *electroencephalogram,* or **EEG**, of the waking-alert (eyes open) state shows a rapid, irregular pattern with no dominant waves.

In awake-resting (eyes closed) states, sleep, or coma, electrical activity of the neurons begins to synchronize into waves with characteristic patterns. As the person's state of arousal lessens, the frequency of the waves decreases. The more synchronous the firing of cortical neurons, the larger the amplitude of the waves. Accordingly, the awake-resting state is characterized by low-amplitude, high-frequency waves, and deep sleep is marked by high-amplitude, low-frequency waves (Fig. 9.17a). The complete cessation of brain waves is one of the clinical criteria for determining death.

In humans, our major rest period is marked by a behavior known as **sleep,** defined as an easily reversible state of inactivity characterized by lack of interaction with the external environment. Most mammals and birds show the same stages of sleep as humans, telling us that sleep is a very ancient property of vertebrate brains.

Why we need to sleep is one of the unsolved mysteries in neurophysiology. Some explanations that have been proposed include to conserve energy, to avoid predators, to allow the body to repair itself, and to process memories. A number of recent studies have shown that sleep deprivation impairs our performance on tasks and tests, but also that 20–30 minute "power naps" can help make up a sleep deficit.

Until the 1960s, sleep was thought to be a passive state that resulted from withdrawal of stimuli to the brain. Then experiments showed that neuronal activity in ascending tracts from the brain stem to the cerebral cortex was required for sleep. From other studies, we know that the sleeping brain consumes as much oxygen as the awake brain, and sometimes even more. As a result, we now consider sleep to be an active state.

Sleep is divided into four stages, each marked by identifiable, predictable events associated with characteristic somatic changes and EEG patterns (Fig. 9.17a). The two major sleep phases are **slow-wave sleep** (also called **deep sleep** or **non-REM sleep**, stage 4) and **REM (rapid eye movement) sleep** (stage 1). Slow-wave sleep is apparent on the EEG by the presence of *delta waves,* high-amplitude, low-frequency waves of long duration that sweep across the cerebral cortex. During this phase of the sleep cycle, sleepers adjust body position without conscious commands from the brain to do so.

In contrast, REM sleep is marked by an EEG pattern closer to that of an awake person, with low-amplitude, high-frequency waves. During REM sleep, brain activity inhibits motor neurons to skeletal muscles, paralyzing them. Exceptions to this pattern

DIFFUSE MODULATORY SYSTEMS

The neurons collectively known as the diffuse modulatory systems originate in the reticular formation of the brain stem and project their axons to large areas of the brain. The four systems are named for their neurotransmitters.

(a) Noradrenergic (Norepinephrine)

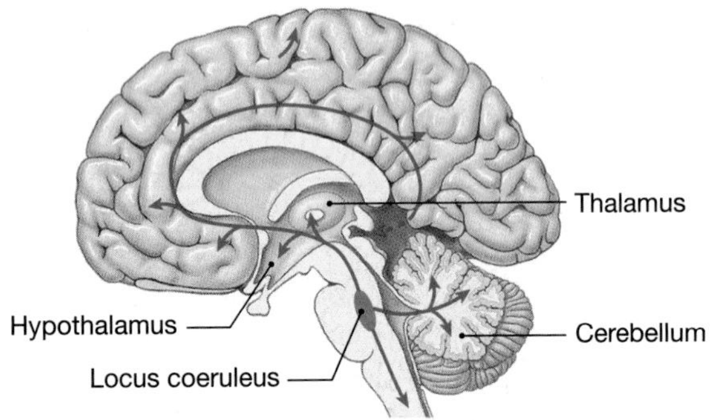

Functions:	Attention, arousal, sleep-wake cycles, learning, memory, anxiety, pain, and mood
Neurons originate:	Locus coeruleus of the pons
Neurons terminate:	Cerebral cortex, thalamus, hypothalamus, olfactory bulb, cerebellum, midbrain, spinal cord

(b) Serotonergic (Serotonin)

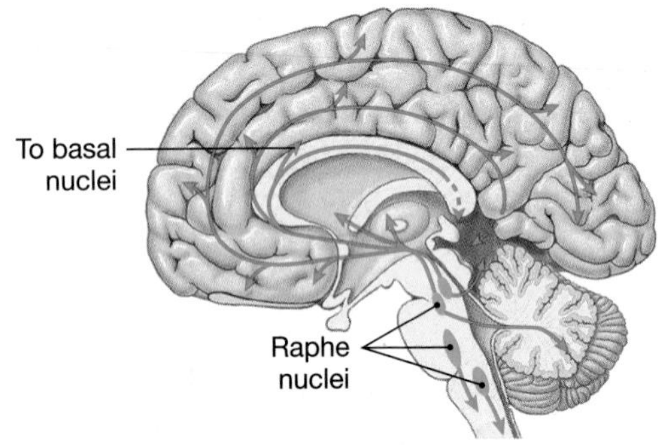

Functions:	1. Lower nuclei: Pain, locomotion 2. Upper nuclei: Sleep-wake cycle; mood and emotional behaviors, such as aggression and depression
Neurons originate:	Raphe nuclei along brain stem midline
Neurons terminate:	1. Lower nuclei project to spinal cord 2. Upper nuclei project to most of brain

(c) Dopaminergic (Dopamine)

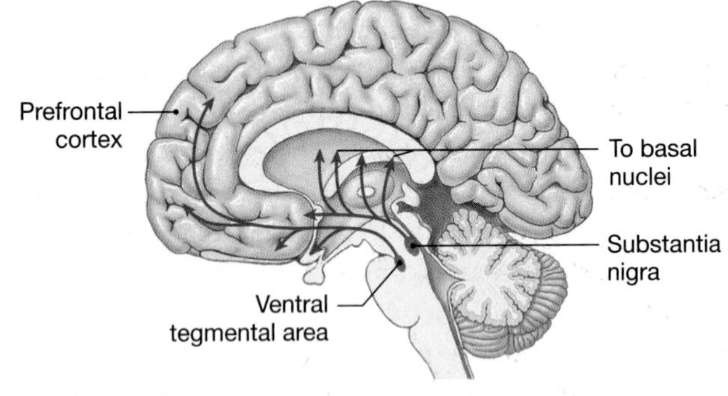

Functions:	1. Motor control 2. "Reward" centers linked to addictive behaviors
Neurons originate:	1. Substantia nigra in midbrain 2. Ventral tegmentum in midbrain
Neurons terminate:	1. Cortex 2. Cortex and parts of limbic system

(d) Cholinergic (Acetylcholine)

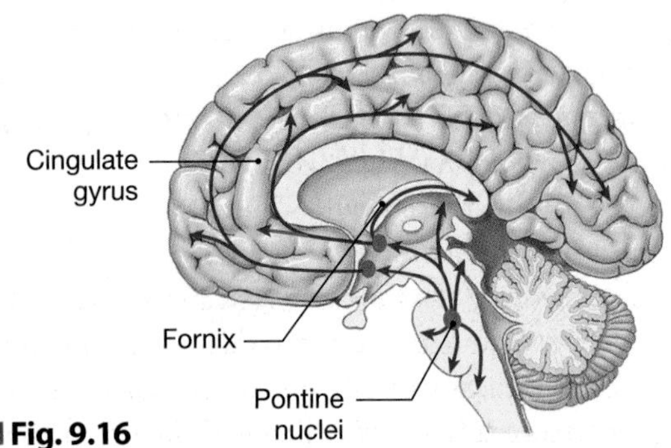

Functions:	Sleep-wake cycles, arousal, learning, memory, sensory information passing through thalamus
Neurons originate:	Base of cerebrum; pons and midbrain
Neurons terminate:	Cerebrum, hippocampus, thalamus

■ **Fig. 9.16**

ELECTROENCEPHALOGRAMS (EEGs) AND THE SLEEP CYCLE

(a) Recordings of electrical activity in the brain during awake-resting and sleep periods show characteristic patterns.

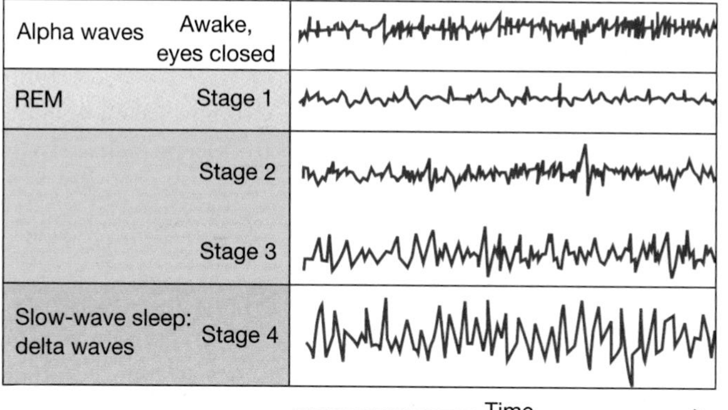

(b) The deepest sleep occurs in the first three hours.

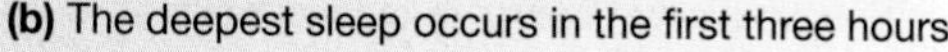

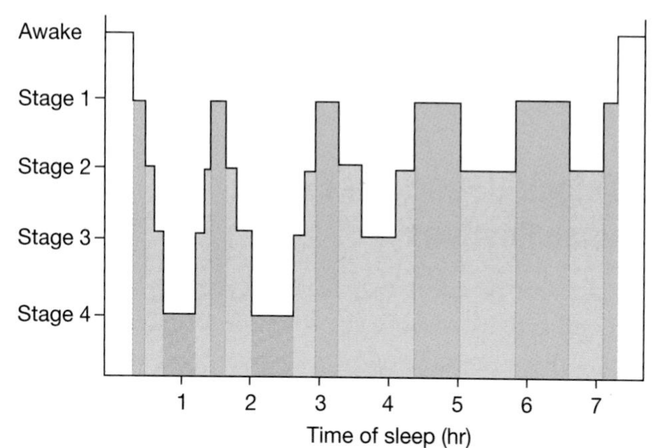

FIGURE QUESTIONS

1. Which EEG pattern has the fastest frequency? The greatest amplitude?
2. In a 20–30 minute "power nap," what sleep stages will the napper experience?

KEY

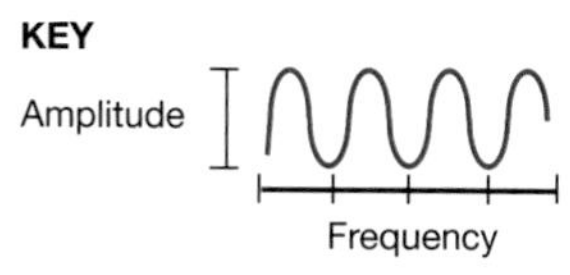

Fig. 9.17

are the muscles that move the eyes and those that control breathing. The control of homeostatic functions is depressed during REM sleep, and body temperature falls toward ambient temperature.

REM sleep is the period during which most dreaming takes place. The eyes move behind closed lids, as if following the action of the dream. Sleepers are most likely to wake up spontaneously from periods of REM sleep.

A typical eight-hour sleep consists of repeating cycles, as shown in Figure 9.17b. In the first hour, the person moves from wakefulness into a deep sleep (stage 4; first blue area in Figure 9.17b). The sleeper then cycles between deep sleep and REM sleep (stage 1), with stages 2–3 occurring in between. Near the end of an eight-hour sleep period a sleeper spends the most time in stage 2 and REM sleep, until finally awakening for the day.

If sleep is a neurologically active process, what is it that makes us sleepy? The possibility of a sleep-inducing factor was first proposed in 1913, when scientists found that cerebrospinal fluid from sleep-deprived dogs could induce sleep in normal animals. Since then, a variety of sleep-inducing factors have been identified. Curiously, many of them are also substances that enhance the immune response, such as interleukin-1, interferon, serotonin, and tumor necrosis factor. As a result of this finding, some investigators have suggested that one answer to the puzzle of the biological reason for sleep is that we need to sleep to enhance our immune response. Whether or not that is a reason for why we sleep, the link between the immune system and sleep induction may help explain why we tend to sleep more when we are sick.

Sleep disorders are relatively common, as you can tell by looking at the variety of sleep-promoting agents available over the counter in drugstores. Among the more common sleep disorders are *insomnia* (the inability to go to sleep or remain asleep long enough to awake refreshed), sleep apnea, and sleepwalking. *Sleep apnea* {*apnoos*, breathless} is a condition in which the sleeper awakes when the airway muscles relax to the point of obstructing normal breathing.

Sleepwalking, or *somnambulism* {*somnus*, sleep + *ambulare*, to walk}, is a sleep behavior disorder that for many years was thought to represent the acting out of dreams. However, most dreaming occurs during REM sleep (stage 1), while sleepwalking takes place during deep sleep (stage 4). During sleepwalking episodes, which may last from 30 seconds to 30 minutes, the subject's eyes are open and registering the surroundings. The subject is able to avoid bumping into objects, can negotiate stairs, and in some cases is reported to perform such tasks as preparing food or folding clothes. The subject usually has little if any conscious recall of the sleepwalking episode upon awakening.

Sleepwalking is most common in children, and the frequency of episodes declines with age. There is also a genetic component, as the tendency to sleepwalk runs in families. To learn more about the different sleep disorders, see the U.S. National Institutes of Health web site for the National Center for Sleep Disorder Research (*www.nhlbi.nih.gov/about/ncsdr*).

Concept Check Answer: p. 324

21. During sleep, relay neurons in the thalamus reduce information reaching the cerebrum by altering their membrane potential. Are these neurons more likely to have depolarized or hyperpolarized? Explain your reasoning.

Physiological Functions Exhibit Circadian Rhythms

All organisms (even plants) have alternating daily patterns of rest and activity. Sleep-wake rhythms, like many other biological cycles, generally follow a 24-hour light-dark cycle and are known as *circadian rhythms* [p. 19]. When an organism is placed in conditions of constant light or darkness, these activity rhythms persist, apparently cued by an internal clock.

In mammals, the primary "clock" resides in networks of neurons located in the **suprachiasmatic nucleus** of the hypothalamus. A very simple interpretation of recent experiments on the molecular basis of the clock is that clock cycling is the result of a complex feedback loop in which specific genes turn on and direct protein synthesis. The proteins accumulate, turn off the genes, and then are themselves degraded. In the absence of the proteins, the genes turn back on and the cycle begins again. The clock has intrinsic activity that is synchronized with the external environment by sensory information about light cycles received through the eyes.

Circadian rhythms in humans can be found in most physiological functions and usually correspond to the phases of our sleep-wake cycles. For example, body temperature and cortisol secretion both cycle on a daily basis [Fig. 1.13, p. 19]. Melatonin secretion by the pineal gland also is strongly linked to light-dark cycling and appears to feed back to the suprachiasmatic nucleus to modulate clock cycling.

Emotion and Motivation Involve Complex Neural Pathways

Emotion and motivation are two aspects of brain function that probably represent an overlap of the behavioral state system and cognitive system. The pathways involved are complex and form closed circuits that cycle information among various parts of the brain, including the hypothalamus, limbic system, and cerebral cortex. We still do not understand the underlying neural mechanisms, and this is a large and active area of neuroscience research.

Emotions are difficult to define. We know what they are and can name them, but in many ways they defy description. One characteristic of emotions is that they are difficult to voluntarily turn on or off. The most commonly described emotions, which arise in different parts of the brain, are anger, aggression, sexual feelings, fear, pleasure, contentment, and happiness.

EMERGING CONCEPTS

Adenosine and That "Java Jolt"

Caffeine and its methylxanthine cousins *theobromine* and *theophylline* (found in chocolate and tea) are probably the most widely consumed psychoactive drugs, known since ancient times for their stimulant effect. Molecular research has revealed that the methylxanthines are receptor antagonists for *adenosine,* a molecule composed of the nitrogenous base adenine plus the sugar ribose [p. 36]. Adenosine acts as an important neuromodulator in the central nervous system. Four subtypes of adenosine receptor have been identified, and they are all G protein–coupled, cAMP-dependent membrane proteins. The discovery that the stimulant effect of caffeine comes from its blockade of adenosine receptors has led scientists to investigate adenosine's role in sleep-wake cycles. Evidence suggests that adenosine accumulates in the extracellular fluid during waking hours, increasingly suppressing activity of the neurons that promote wakefulness. Other roles for adenosine in the brain include its possible involvement in the addiction/reward system and in the development of depression.

RUNNING PROBLEM

About six months after the start of ACTH treatment, Ben's head-drop seizures returned, and his development began to decline once again. An EEG following Ben's relapse did not demonstrate the erratic wave patterns specific to infantile spasms but did show abnormal activity in the right cortex. A neurologist ordered a positron emission tomography (PET) scan to determine the focus of Ben's seizure activity.

Ben received an injection of radioactively labeled glucose. He was then placed in the center of a PET machine lined with radiation detectors that created a map of his brain showing areas of high and low radioactivity. Those parts of his brain that were more active absorbed more glucose and thus emitted more radiation when the radioactive compound began to decay.

Q4: What is the rationale for using radioactively labeled glucose (and not some other nutrient) for the PET scan?

289 297 312 314 316 320

The limbic system, particularly the region known as the *amygdala,* is the center of emotion in the human brain. Scientists have learned about the role of this brain region through experiments in humans and animals. When the amygdala is artificially stimulated in humans, as it might be during surgery for epilepsy, patients report experiencing feelings of fear and anxiety. Experimental lesions that destroy the amygdala in animals cause the animals to become tamer and to display hypersexuality. As a result, neurobiologists believe that the amygdala is the center for basic instincts such as fear and aggression.

The pathways for emotions are complex (■ Fig. 9.18). Sensory stimuli feeding into the cerebral cortex are constructed in the brain to create a representation (perception) of the world.

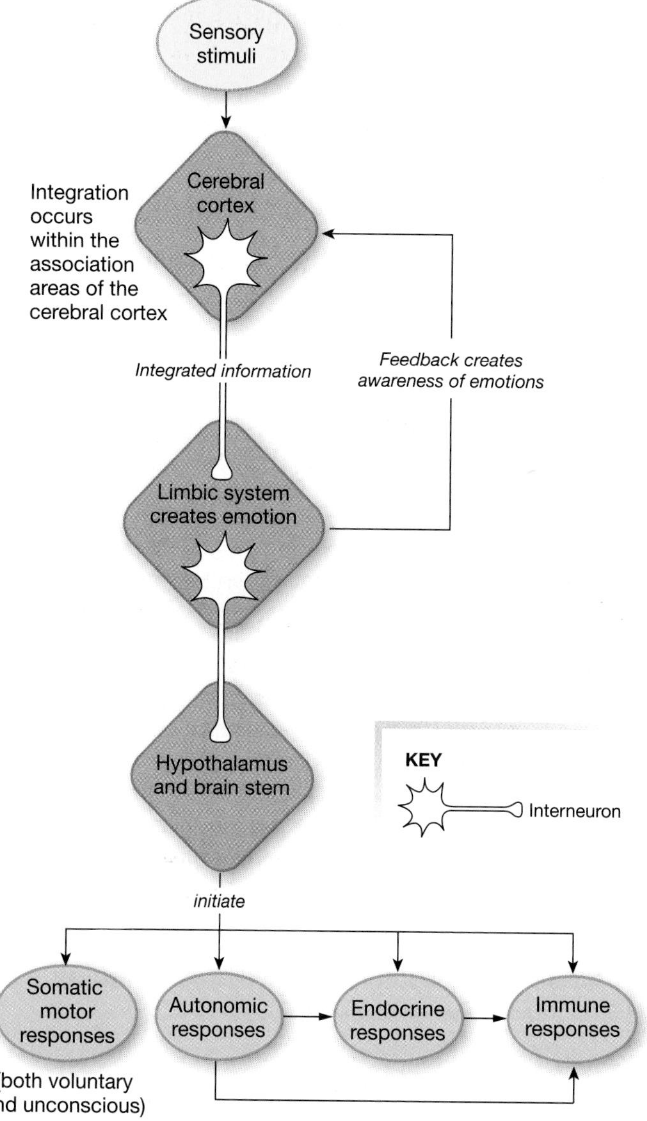

■ **Fig. 9.18**

After information is integrated by the association areas, it is passed on to the limbic system. Feedback from the limbic system to the cerebral cortex creates awareness of the emotion, while descending pathways to the hypothalamus and brain stem initiate voluntary behaviors and unconscious responses mediated by autonomic, endocrine, immune, and somatic motor systems.

The physical result of emotions can be as dramatic as the pounding heart of a fight-or-flight reaction or as insidious as the development of an irregular heartbeat. The links between mind and body are difficult to study and will take many years of research to understand.

Motivation is defined as internal signals that shape voluntary behaviors. Some of these behaviors, such as eating, drinking, and having sex, are related to survival. Others, such as curiosity and having sex (again), are linked to emotions. Some motivational states are known as **drives** and generally have three properties in common: (1) they create an increased state of CNS arousal or alertness, (2) they create goal-oriented behavior, and (3) they are capable of coordinating disparate behaviors to achieve that goal.

Motivated behaviors often work in parallel with autonomic and endocrine responses in the body, as you might expect with behaviors originating in the hypothalamus. For example, if you eat salty popcorn, your body osmolarity increases. This stimulus acts on the thirst center of the hypothalamus, motivating you to seek something to drink. Increased osmolarity also acts on an endocrine center in the hypothalamus, releasing a hormone that increases water retention by the kidneys. In this way one stimulus triggers both a motivated behavior and a homeostatic endocrine response.

Some motivated behaviors can be activated by internal stimuli that may not be obvious even to the person in whom they are occurring. Eating, curiosity, and sex drive are three examples of behaviors with complex stimuli underlying their onset. We may eat, for example, because we are hungry or because the food looks good or because we do not want to hurt someone's feelings. Many motivated behaviors stop when the person has reached a certain level of satisfaction, or **satiety**, but they may also continue *despite* feeling satiated.

Pleasure is a motivational state that is being intensely studied because of its relationship to *addictive behaviors,* such as drug use. Animal studies have shown that pleasure is a physiological state that is accompanied by increased activity of the neurotransmitter dopamine in certain parts of the brain. Drugs that are addictive, such as cocaine and nicotine, act by enhancing the effectiveness of dopamine, thereby increasing the pleasurable sensations perceived by the brain. As a result, use of these drugs rapidly becomes a learned behavior.

Interestingly, not all behaviors that are addictive are pleasurable. For example, there are a variety of compulsive behaviors that involve self-mutilation, such as pulling out hair by the roots. Fortunately, many behaviors can be modulated, given motivation.

Moods Are Long-Lasting Emotional States

Moods are similar to emotions but are longer-lasting, relatively stable subjective feelings related to one's sense of well-being. Moods are difficult to define at a neurobiological level, but evidence obtained in studying and treating mood disorders suggests that mood disturbances reflect changes in CNS function, such as abnormal neurotransmitter release or reception in different brain regions.

Mood disorders are estimated to be the fourth leading cause of illness in the world today. **Depression** is a mood disturbance that affects nearly 10% of the United States population each year. It is characterized by sleep and appetite disturbances and alterations of mood and libido that may seriously affect the person's ability to function at school or work or in personal relationships. Many people do not realize that depression is not a sign of mental or moral weakness, or that it can be treated successfully with drugs and psychotherapy. (For detailed information about depression, go to *www.nlm.nih.gov/medlineplus/depression.html*).

The drug therapy for depression has changed in recent years, but all the major categories of antidepressant drugs alter some aspect of synaptic transmission. The older *tricyclic antidepressants,* such as amitriptyline, block reuptake of norepinephrine into the presynaptic neuron, thus extending the active life of the neurotransmitter. The antidepressants known as *selective serotonin reuptake inhibitors,* or SSRIs, slow down the removal of serotonin (and possibly also norepinephrine) from the synapse. As a result of uptake inhibition, serotonin lingers in the synaptic cleft longer than usual, increasing serotonin-dependent activity in the postsynaptic neuron. Other antidepressant drugs alter brain levels of dopamine. The effectiveness of these different classes of antidepressant drugs suggests that norepinephrine, serotonin, and dopamine are all involved in brain pathways for mood and emotion.

Interestingly, patients need to take antidepressant drugs for several weeks before they experience their full effect. This delay suggests that the changes taking place in the brain are long-term modulation of pathways rather than simply enhanced fast synaptic responses. One recent study in mice* found evidence that antidepressants promote the growth of new neurons, which would also explain the delayed onset of full action.

The causes of major depression are complex and probably involve a combination of genetic factors, the serotonergic and noradrenergic diffuse modulatory systems, trophic factors such as *brain-derived neurotrophic factor* (BDNF), and stress. The search to uncover the biological basis of disturbed brain function is a major focus of neuroscience research today.

*A. Surget *et al.* Antidepressants recruit new neurons to improve stress response regulation. *Mol Psychiatry* 3 May 2011, *www.nature.com/mp/journal/vaop/ncurrent/full/mp201148a.html.* (free access)

Some research into brain function has become quite controversial, particularly that dealing with sexuality and the degree to which behavior in general is genetically determined in humans. We will not delve deeply into any of these subjects because they are complex and would require lengthy explanations to do them justice. Instead, we will look briefly at some of the recent models proposed to explain the mechanisms that are the basis for higher cognitive functions.

Learning and Memory Change Synaptic Connections in the Brain

For many years, motivation, learning, and memory (all of which are aspects of the cognitive state) were considered to be in the realm of psychology rather than biology. Neurobiologists in decades past were more concerned with the network and cellular aspects of neuronal function. In recent years, however, the two fields have overlapped more and more. Scientists have discovered that the underlying basis for cognitive function seems to be explainable in terms of cellular events that influence plasticity—events such as long-term potentiation [p. 280]. The ability of neuronal connections to change with experience is fundamental to the two cognitive processes of learning and memory.

Learning Is the Acquisition of Knowledge

How do you know when you have learned something? Learning can be demonstrated by behavioral changes, but behavioral changes are not required in order for learning to occur. Learning can be internalized and is not always reflected by overt behavior while the learning is taking place. Would someone watching you read your textbook or listen to a professor's lecture be able to tell whether you had learned anything?

RUNNING PROBLEM

Ben's halted development is a feature unique to infantile spasms. The abnormal portions of the brain send out continuous action potentials during frequent seizures and ultimately change the interconnections of brain neurons. The damaged portions of the brain harm normal portions to such an extent that medication or surgery should be started as soon as possible. If intervention is not begun early, the brain can be permanently damaged and development will never recover.

Q5: The brain's ability to change its synaptic connections as a result of neuronal activity is called ________.

289 297 312 **314** 316 320

Learning can be classified into two broad types: associative and nonassociative. **Associative learning** occurs when two stimuli are associated with each other, such as Pavlov's classic experiment in which he simultaneously presented dogs with food and rang a bell. After a period of time, the dogs came to associate the sound of the bell with food and began to salivate in anticipation of food whenever the bell was rung. Another form of associative learning occurs when an animal associates a stimulus with a given behavior. An example would be a mouse that gets a shock each time it touches a certain part of its cage. It soon associates that part of the cage with an unpleasant experience and avoids the area.

Nonassociative learning is a change in behavior that takes place after repeated exposure to a single stimulus. This type of learning includes habituation and sensitization, two adaptive behaviors that allow us to filter out and ignore background stimuli while responding more sensitively to potentially disruptive stimuli. In **habituation**, an animal shows a decreased response to an irrelevant stimulus that is repeated over and over. For example, a sudden loud noise may startle you, but if the noise is repeated over and over again, your brain begins to ignore it. Habituated responses allow us to filter out stimuli that we have evaluated and found to be insignificant.

Sensitization is the opposite of habituation, and the two behaviors combined help increase an organism's chances for survival. In sensitization learning, exposure to a noxious or intense stimulus causes an enhanced response upon subsequent exposure. For example, people who become ill while eating a certain food may find that they lose their desire to eat that food again. Sensitization is adaptive because it helps us avoid potentially harmful stimuli.

Memory Is the Ability to Retain and Recall Information

Memory is the ability to retain and recall information. Memory is a very complex function, but scientists have tried to classify it in different ways. We think of several types of memory: short-term and long-term, reflexive and declarative. Processing for different types of memory appears to take place through different pathways. With noninvasive imaging techniques such as MRI and PET scans, researchers have been able to track brain activity as individuals learned to perform tasks.

Memories are stored throughout the cerebral cortex in pathways known as **memory traces**. Some components of memories are stored in the sensory cortices where they are processed. For example, pictures are stored in the visual cortex, and sounds in the auditory cortex.

Learning a task or recalling a task already learned may involve multiple brain circuits that work in parallel. This *parallel processing* helps provide backup in case one of the circuits is damaged. It is also believed to be the means by which specific memories are generalized, allowing new information to be matched to stored information. For example, a person who has never seen a volleyball will recognize it as a ball because the volleyball has the same general characteristics as all other balls the person has seen.

In humans, the hippocampus seems to be an important structure in both learning and memory. Patients who have part of the hippocampus destroyed to relieve a certain type of epilepsy also have trouble remembering new information. When given a list of words to repeat, they can remember the words as long as their attention stays focused on the task. If they are distracted, however, the memory of the words disappears, and they must learn the list again. Information stored in long-term memory before the operation is not affected. This inability to remember newly acquired information is a defect known as **anterograde amnesia** {*amnesia,* oblivion}.

Memory has multiple levels of storage, and our memory bank is constantly changing (■ Fig. 9.19). When a stimulus comes into the CNS, it first goes into **short-term memory**, a limited storage area that can hold only about 7 to 12 pieces of information at a time. Items in short-term memory disappear unless an effort, such as repetition, is made to put them into a more permanent form.

Working memory is a special form of short-term memory processed in the prefrontal lobes. This region of the cerebral cortex is devoted to keeping track of bits of information long enough to put them to use in a task that takes place after the information has been acquired. Working memory in these regions is linked to long-term memory stores, so that newly acquired information can be integrated with stored information and acted on.

For example, suppose you are trying to cross a busy road. You look to the left and see that there are no cars coming for

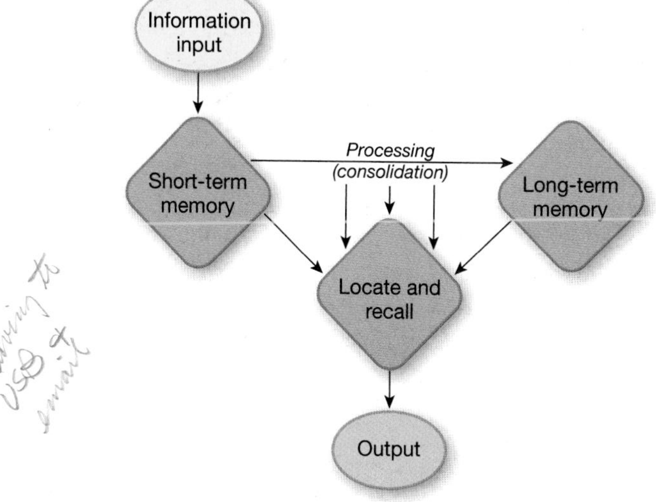

■ **Fig. 9.19**

9

several blocks. You then look to the right and see that there are no cars coming from that direction either. Working memory has stored the information that the road to the left is clear, and so using this stored knowledge about safety, you are able to conclude that there is no traffic from either direction and it is safe to cross the road.

In people with damage to the prefrontal lobes of the brain, this task becomes more difficult because they are unable to recall whether the road is clear from the left once they have looked away to assess traffic coming from the right. Working memory allows us to collect a series of facts from short- and long-term memory and connect them in a logical order to solve problems or plan actions.

Long-term memory is a storage area capable of holding vast amounts of information. The processing of information that converts short-term memory into long-term memory is known as **consolidation** (Fig. 9.19). Consolidation can take varying periods of time, from seconds to minutes. Information passes through many intermediate levels of memory during consolidation, and in each of these stages, the information can be located and recalled.

As scientists studied the consolidation of short-term memory into long-term memory, they discovered that the process involves changes in synaptic connections of the circuits involved in learning. In some cases, new synapses form; in others, the effectiveness of synaptic transmission is altered either through long-term potentiation or through long-term depression. These changes are evidence of plasticity and show us that the brain is not "hard-wired."

Long-term memory has been divided into two types that are consolidated and stored using different neuronal pathways (■ Tbl. 9.3). **Reflexive (implicit) memory,** which is automatic and does not require conscious processes for either creation or recall, involves the amygdala and the cerebellum. Information stored in reflexive memory is acquired slowly through repetition. Motor skills fall into this category, as do procedures and rules.

For example, you do not need to think about putting a period at the end of each sentence or about how to pick up a fork. Reflexive memory has also been called *procedural memory* because it generally concerns how to do things. Reflexive memories can be acquired through either associative or nonassociative learning processes, and these memories are stored.

Declarative (explicit) memory, on the other hand, requires conscious attention for its recall. Its creation generally depends on the use of higher-level cognitive skills such as inference, comparison, and evaluation. The neuronal pathways involved in this type of memory are in the temporal lobes. Declarative memories deal with knowledge about ourselves and the world around us that can be reported or described verbally.

Sometimes information can be transferred from declarative memory to reflexive memory. The quarterback on a football team is a good example. When he learned to throw the football as a small boy, he had to pay close attention to gripping the ball and coordinating his muscles to throw the ball accurately. At that point of learning to throw the ball, the process was in declarative memory and required conscious effort as the boy analyzed his movements.

Table 9.3

Types of Long-Term Memory

Reflexive (Implicit) Memory	Declarative (Explicit) Memory
Recall is automatic and does not require conscious attention	Recall requires conscious attention
Acquired slowly through repetition	Depends on higher-level thinking skills such as inference, comparison, and evaluation
Includes motor skills and rules and procedures	Memories can be reported verbally
Procedural memories can be demonstrated	

RUNNING PROBLEM

The PET scan revealed two abnormal spots, or *loci* (plural of *locus*), on Ben's right hemisphere, one on the parietal lobe and one overlapping a portion of the primary motor cortex. Because the loci triggering Ben's seizures were located on the same hemisphere and were in the cortex, Ben was a candidate for a *hemispherectomy*, removal of the cortex of the affected hemisphere. Surgeons removed 80 percent of his right cerebral cortex, sparing areas crucial to vision, hearing, and sensory processing. Normally the motor cortex would be spared as well, but in Ben's case a seizure locus overlapped much of the region.

Q6: In which lobes are the centers for vision, hearing, and sensory processing located?

Q7: Which of Ben's abilities might have suffered if his left hemisphere had been removed instead?

Q8: By taking only the cortex of the right hemisphere, what parts of the cerebrum did surgeons leave behind?

Q9: Why were the surgeons careful to spare Ben's right lateral ventricle?

289 297 312 314 316 320

With repetition, however, the mechanics of throwing the ball were transferred to reflexive memory: they became a reflex that could be executed without conscious thought. That transfer allowed the quarterback to use his conscious mind to analyze the path and timing of the pass while the mechanics of the pass became automatic. Athletes often refer to this automaticity of learned body movements as *muscle memory*.

Memory is an individual thing. We process information on the basis of our experiences and perception of the world. Because people have widely different experiences throughout their lives, it follows that no two people will process a given piece of information in the same way. If you ask a group of people about what happened during a particular event such as a lecture or an automobile accident, no two descriptions will be identical. Each person processed the event according to her or his own perceptions and experiences. Experiential processing is important to remember when studying in a group situation, because it is unlikely that all group members learn or recall information the same way.

Memory loss and the inability to process and store new memories are devastating medical conditions. In younger people, memory problems are usually associated with trauma to the brain from accidents. In older people, strokes and progressive *dementia* {*demens*, out of one's mind} are the main causes of memory loss. **Alzheimer's disease** is a progressive neurodegenerative disease of cognitive impairment that accounts for about half the cases of dementia in the elderly. Alzheimer's is characterized by memory loss that progresses to a point where the patient does not recognize family members. Over time, even the personality changes, and in the final stages, other cognitive functions fail so that patients cannot communicate with caregivers.

Diagnosis of Alzheimer's is usually made through the patient's declining performance on cognitive function examinations. Scientists are testing whether certain blood tests or imaging studies can reveal if a person has the disease, but these studies are in their initial stages. The only definitive diagnosis of Alzheimer's comes after death, when brain tissue can be examined for neuronal degeneration, extracellular plaques made of *β-amyloid protein,* and intracellular tangles of *tau,* a protein that is normally associated with microtubules.

The presence of amyloid plaques and tau tangles is diagnostic, but the underlying cause of Alzheimer's is unclear. There is a known genetic component, and other theories include oxidative stress and chronic inflammation. Currently there is no proven prevention or treatment, although drugs that are acetylcholine agonists or acetylcholinesterase inhibitors slow the progression of the disease.

By one estimate, Alzheimer's affects about 5.4 million Americans, with the number expected to rise as Baby Boomers age. The forecast of 14 million people with Alzheimer's by the year 2050 has put this disease in the forefront of neurobiological research.

Language Is the Most Elaborate Cognitive Behavior

One of the hallmarks of an advanced nervous system is the ability of one member of a species to exchange complex information with other members of the same species. Although found predominantly in birds and mammals, this ability also occurs in certain insects that convey amazingly detailed information by means of sound (crickets), touch and sight (bees), and odor (ants). In humans, the exchange of complex information takes place primarily through spoken and written language. Because language is considered the most elaborate cognitive behavior, it has received considerable attention from neurobiologists.

Language skills require the input of sensory information (primarily from hearing and vision), processing in various centers in the cerebral cortex, and the coordination of motor output for vocalization and writing. In most people the centers for language ability are found in the left hemisphere of the cerebrum. Even 70% of people who are either left-handed (right-brain dominant) or ambidextrous use their left brain for speech. The ability to communicate through speech has been divided into two processes: the combination of different sounds to form words (vocalization) and the combination of words into grammatically correct and meaningful sentences.

The integration of spoken language in the human brain involves two regions in the cerebral cortex: **Wernicke's area** in the temporal lobe and **Broca's area** in the frontal lobe close to the motor cortex (■ Fig. 9.20). Most of what we know about these areas comes from studies of people with brain lesions (because nonhuman animals are not capable of speech). Even primates that communicate on the level of a small child through sign language and other visual means do not have the physical ability to vocalize the sounds of human language.

Input into the language areas comes from either the visual cortex (reading) or the auditory cortex (listening). Sensory input from either cortex goes first to Wernicke's area, then to Broca's area. After integration and processing, output from Broca's area to the motor cortex initiates a spoken or written action. If damage occurs to Wernicke's area, a person is unable to understand any spoken or visual information. The person's own speech, as a result, is nonsense, because the person is unaware of his or her own errors. This condition is known as **receptive aphasia** {*a-*, not + *phatos,* spoken} because the person is unable to understand sensory input.

On the other hand, people with damage to Broca's area understand spoken and written language but are unable to speak or write in normal syntax. Their response to a question often consists of appropriate words strung together in random order. These patients may have a difficult time dealing with their disability because they are aware of their mistakes but are powerless to correct them. Damage to Broca's area causes an **expressive aphasia**.

9

LANGUAGE PROCESSING

People with damage to Wernicke's area do not understand spoken or written communication. Those with damage to Broca's area understand but are unable to respond appropriately.

(a) Speaking a written word

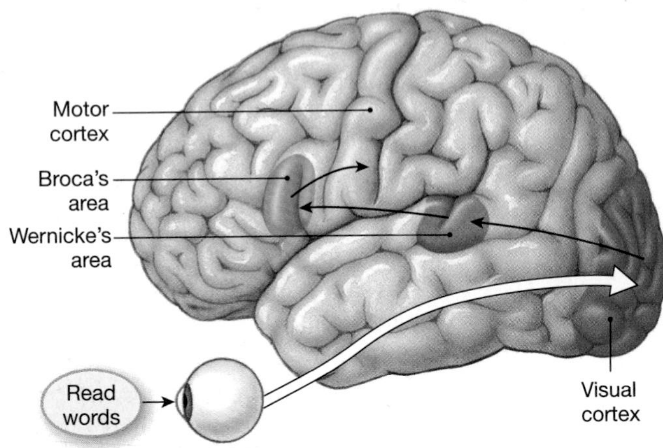

(b) Speaking a heard word

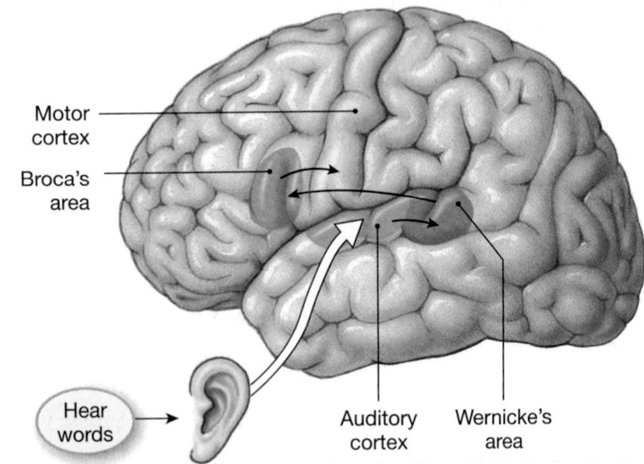

(c) PET scan of the brain at work

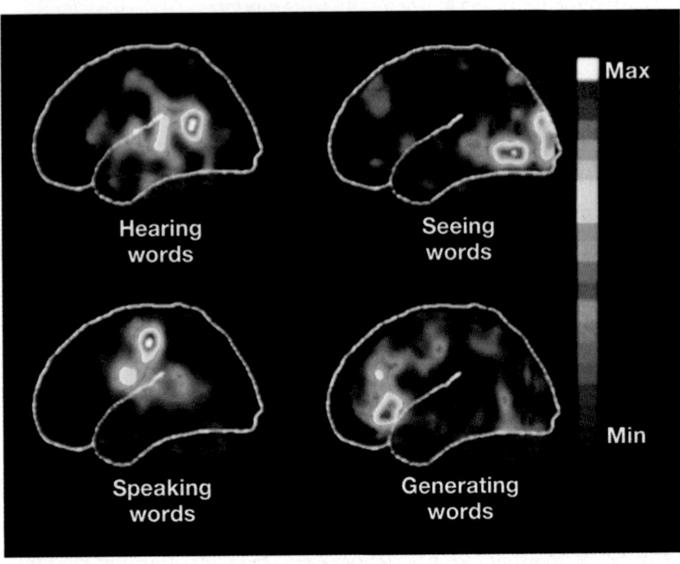

FIGURE QUESTION

In the image above, the brain area active in seeing words is in the ________ lobe, and the brain area active during word generation is in the ________ lobe.

■ **Fig. 9.20**

Mechanical forms of aphasia occur as a result of damage to the motor cortex. Patients with this type of damage find themselves unable to physically shape the sounds that make up words, or unable to coordinate the muscles of their arm and hand to write.

Personality Is a Combination of Experience and Inheritance

One of the most difficult aspects of brain function to translate from the abstract realm of psychology into the physical circuits of neurobiology is the combination of attributes we call **personality**. What is it that makes us individuals? The parents of more than one child will tell you that their offspring were different from birth, and even in the womb. If we all have the same brain structure, what makes us different?

This question fascinates many people. The answer that is evolving from neurobiology research is that we are a combination of our experiences and the genetic constraints we inherit. One complicating factor is the developmental aspect of "experience," as scientists are showing that exposure of developing embryos to hormones while still in the womb can alter brain pathways.

What we learn or experience and what we store in memory create a unique pattern of neuronal connections in our brains. Sometimes these circuits malfunction, creating depression, schizophrenia, or any number of other personality disturbances. Psychiatrists for many years attempted to treat these disorders as if they were due solely to events in the person's life, but now we know that there is a genetic component to many of these disorders.

Schizophrenia {*schizein,* to split + *phren,* the mind} is an example of a brain disorder that has both a genetic and an environmental basis. In the American population as a whole, the risk of developing schizophrenia is about 1%. However, if one parent has schizophrenia, the risk increases to 10%, indicating that people can inherit a susceptibility to developing the disease. The cause of schizophrenia is not currently known. However, as with many other conditions involving altered mental states, schizophrenia can be treated with drugs that influence neurotransmitter release and activity in the brain. To learn more about diagnosis and treatment of schizophrenia, see the National Institutes of Health web site *www.nlm.nih.gov/medlineplus/schizophrenia.html.*

We still have much to learn about repairing damage to the CNS. One of the biggest tragedies in life are the intellectual and personality changes that sometimes accompany head trauma. Physical damage to the delicate circuits of the brain, particularly to the frontal lobe, can create a whole new personality. The person who exists after the injury may not be the same personality who inhabited that body before the injury. Although the change may not be noticeable to the injured person, it can be devastating to the victim's family and friends. Perhaps as we learn more about how neurons link to one another, we will be able to find a means of restoring damaged networks and preventing the lasting effects of head trauma and brain disorders.

RUNNING PROBLEM CONCLUSION

Infantile Spasms

Ben has remained seizure-free since the surgery and shows normal development in all areas except motor skills. He remains somewhat weaker and less coordinated on his left side, the side opposite (*contralateral*) to the surgery, but over time the weakness should subside with the aid of physical therapy. Ben's recovery stands as a testament to the incredible plasticity of the brain. Apart from the physical damage caused to the brain, a number of children with epilepsy have developmental delays that stem from the social aspects of their disorder. Young children with frequent seizures often have difficulty socializing with their peers because of overprotective parents, missed school days, and the fear of people who do not understand epilepsy. Their problems can extend into adulthood, when people with epilepsy may have difficulty finding employment or driving if their seizures are not controlled. There are numerous examples of adults who undergo successful epilepsy surgery but are still unable to fully enter society because they lack social and employment skills. Not surprisingly, the rate of depression is much higher among people with epilepsy. To learn more about this disease, start with the Epilepsy Foundation (*www.epilepsyfoundation.org*).

This Running Problem was written by Susan E. Johnson while she was an undergraduate student at the University of Texas at Austin studying for a career in the biomedical sciences.

Question	Facts	Integration and Analysis
1. How might a leaky blood-brain barrier lead to action potentials that trigger a seizure?	Neurotransmitters and other chemicals circulating freely in the blood are normally separated from brain tissue by the blood-brain barrier.	Ions and neurotransmitters entering the brain might depolarize neurons and trigger action potentials.
2. What does GABA do to the cell's membrane potential? Does GABA make the cell more or less likely to fire action potentials?	GABA opens Cl^- channels.	Cl^- entering a neuron hyperpolarizes the cell and makes it less likely to fire action potentials.
3. Why is it important to limit the duration of ACTH therapy?	Exogenous ACTH acts in a short negative feedback loop, decreasing the output of CRH from the hypothalamus and ACTH production by the anterior pituitary. [See Figure 7.13, p. 227.]	Long-term suppression of endogenous hormone secretion by ACTH can cause CRH- and ACTH-secreting neurons to atrophy, resulting in a lifelong cortisol deficiency.
4. What is the rationale for using radioactively labeled glucose (and not some other nutrient) for the PET scan?	Glucose is the primary energy source for the brain.	Glucose usage is more closely correlated to brain activity than any other nutrient in the body. Areas of abnormally high glucose usage are suggestive of overactive cells.

RUNNING PROBLEM CONCLUSION *(continued)*

Question	Facts	Integration and Analysis
5. The brain's ability to change its synaptic connections as a result of neuronal activity is called ________.	Changes in synaptic connections as a result of neuronal activity are an example of plasticity.	N/A
6. In which lobes are the centers for vision, hearing, and sensory processing located?	Vision is processed in the occipital lobe, hearing in the temporal lobe, and sensory information in the parietal lobe.	N/A
7. Which of Ben's abilities might have suffered if his left hemisphere had been removed instead?	In most people, the left hemisphere contains Wernicke's area and Broca's area, two centers vital to speech. The left brain controls right-sided sensory and motor functions.	Patients who have undergone left hemispherectomies have difficulty with speech (abstract words, grammar, and phonetics). They show loss of right-side sensory and motor functions.
8. By taking only the cortex of the right hemisphere, what parts of the cerebrum did surgeons leave behind?	The cerebrum consists of gray matter in the cortex and interior nuclei, white matter, and the ventricles.	The surgeons left behind the white matter, interior nuclei, and ventricles.
9. Why were the surgeons careful to spare Ben's right lateral ventricle?	The walls of the ventricles contain the choroid plexus, which secretes cerebrospinal fluid (CSF). CSF plays a vital protective role by cushioning the brain.	CSF protection is particularly important following removal of portions of brain tissue because the potential damage from jarring of the head is much greater.

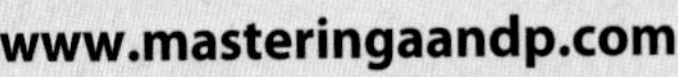

Chapter Summary

The brain is the primary control center of the body, and (as you will learn in later chapters), homeostatic responses in many organ systems are designed to maintain brain function. The ability of the brain to create complex thoughts and emotions in the absence of external stimuli is one of its *emergent properties*.

Emergent Properties of Neural Networks

1. Neural networks create **affective** and **cognitive behaviors**. (p. 289)
2. The brain exhibits **plasticity**, the ability to change connections as a result of experience. (p. 289)

Evolution of Nervous Systems

3. Nervous systems evolved from a simple network of neurons to complex brains. (p. 290; Fig. 9.1)
4. The **cerebrum** is responsible for thought and emotion. (p. 291)

Anatomy of the Central Nervous System

5. The central nervous system consists of layers of cells around a fluid-filled central cavity and develops from the **neural tube** of the embryo. (p. 292; Fig. 9.2)
6. The **gray matter** of the CNS consists of unmyelinated nerve cell bodies, dendrites, and axon terminals. The cell bodies either form layers in parts of the brain or else cluster into groups known as **nuclei**. (p. 293)
7. Myelinated axons form the **white matter** of the CNS and run in bundles called **tracts**. (p. 293)

8. The brain and spinal cord are encased in the **meninges** and the bones of the **cranium** and vertebrae. The meninges are the **pia mater**, the **arachnoid membrane**, and the **dura mater**. (p. 294; Fig. 9.3)
9. The **choroid plexus** secretes **cerebrospinal fluid** (CSF) into the **ventricles** of the brain. Cerebrospinal fluid cushions the tissue and creates a controlled chemical environment. (p. 295; Fig. 9.4)
10. Tight junctions in brain capillaries create a **blood-brain barrier** that prevents possibly harmful substances in the blood from entering the interstitial fluid. (p. 296; Fig. 9.5)
11. The normal fuel source for neurons is glucose, which is why the body closely regulates blood glucose concentrations. (p. 297)

The Spinal Cord

12. Each segment of the spinal cord is associated with a pair of **spinal nerves**. (p. 298)
13. The **dorsal root** of each spinal nerve carries incoming sensory information. The **dorsal root ganglia** contain the nerve cell bodies of sensory neurons. (p. 298; Fig. 9.6)
14. The **ventral roots** carry information from the central nervous system to muscles and glands. (p. 299)
15. **Ascending tracts** of white matter carry sensory information to the brain, and **descending tracts** carry efferent signals from the brain. **Propriospinal tracts** remain within the spinal cord. (p. 299)
16. **Spinal reflexes** are integrated in the spinal cord. (p. 299; Fig. 9.7)

The Brain

17. The brain has six major divisions: cerebrum, diencephalon, midbrain, cerebellum, pons, and medulla oblongata. (p. 300; Fig. 9.8)
18. The **brain stem** is divided into medulla oblongata, pons, and midbrain (mesencephalon). **Cranial nerves** II to XII originate here. (p. 302; Tbl. 9.1)
19. The **reticular formation** is a diffuse collection of neurons that play a role in many basic processes. (p. 301)
20. The **medulla oblongata** contains **somatosensory** and **corticospinal tracts** that convey information between the cerebrum and spinal cord. Most tracts cross the midline in the **pyramid** region. The medulla contains control centers for many involuntary functions. (p. 302)
21. The **pons** acts as a relay station for information between the cerebellum and cerebrum. (p. 302)
22. The **midbrain** controls eye movement and relays signals for auditory and visual reflexes. (p. 302)
23. The **cerebellum** processes sensory information and coordinates the execution of movement. (p. 303)
24. The **diencephalon** is composed of the thalamus and hypothalamus. The **thalamus** relays and modifies sensory and motor information going to and from the cerebral cortex. (p. 303; Fig. 9.9)
25. The **hypothalamus** contains centers for behavioral drives and plays a key role in homeostasis by its control over endocrine and autonomic function. (p. 303; Tbl. 9.2)
26. The **pituitary gland** and **pineal gland** are endocrine glands located in the diencephalon. (p. 304)
27. The cerebrum is composed of two hemispheres connected at the **corpus callosum**. Each cerebral hemisphere is divided into **frontal**, **parietal**, **temporal**, and **occipital lobes**. (p. 304)
28. Cerebral gray matter includes the **cerebral cortex**, basal ganglia, and limbic system. (p. 304; Fig. 9.10)
29. The **basal ganglia** help control movement. (p. 304)
30. The **limbic system** acts as the link between cognitive functions and emotional responses. It includes the **amygdala** and **cingulate gyrus**, linked to emotion and memory, and the **hippocampus**, associated with learning and memory. (p. 305; Fig. 9.11)

Brain Function

31. Three brain systems influence motor output: a **sensory system**, a **cognitive system**, and a **behavioral state system**. (p. 305; Fig. 9.12)
32. Higher brain functions, such as reasoning, arise in the cerebral cortex. The cerebral cortex contains three functional specializations: **sensory areas**, **motor areas**, and **association areas**. (p. 306; Fig. 9.13)
33. Each hemisphere of the cerebrum has developed functions not shared by the other hemisphere, a specialization known as **cerebral lateralization**. (p. 307; Fig. 9.14)
34. Sensory areas receive information from sensory receptors. The **primary somatic sensory cortex** processes information about touch, temperature, and other somatic senses. The **visual cortex**, **auditory cortex**, **gustatory cortex**, and **olfactory cortex** receive information about vision, sound, taste, and odors, respectively. (p. 308)
35. **Association areas** integrate sensory information into perception. **Perception** is the brain's interpretation of sensory stimuli. (p. 308)
36. Motor output includes skeletal muscle movement, neuroendocrine secretion, and visceral responses. (p. 308)
37. Motor areas direct skeletal muscle movement. Each cerebral hemisphere contains a **primary motor cortex** and **motor association area**. (p. 308)
38. The **behavioral state system** controls states of arousal and modulates the sensory and cognitive systems. (p. 309)
39. The **diffuse modulatory systems** of the reticular formation influence attention, motivation, wakefulness, memory, motor control, mood, and metabolic homeostasis. (p. 310; Fig. 9.16)
40. The **reticular activating system** keeps the brain **conscious**, or aware of self and environment. Electrical activity in the brain varies with levels of arousal and can be recorded by **electroencephalography**. (p. 311; Fig. 9.17)
41. **Circadian rhythms** are controlled by an internal clock in the **suprachiasmatic nucleus** of the hypothalamus. (p. 312)
42. **Sleep** is an easily reversible state of inactivity with characteristic stages. The two major phases of sleep are **REM (rapid eye movement) sleep** and **slow-wave sleep** (non-REM sleep). The physiological reason for sleep is uncertain. (p. 309)
43. The limbic system is the center of **emotion** in the human brain. Emotional events influence physiological functions. (p. 313; Fig. 9.18)
44. **Motivation** arises from internal signals that shape voluntary behaviors related to survival or emotions. Motivational **drives** create goal-oriented behaviors. (p. 313)
45. **Moods** are long-lasting emotional states. Many mood disorders can be treated by altering neurotransmission in the brain. (p. 314)
46. **Learning** is the acquisition of knowledge about the world around us. **Associative learning** occurs when two stimuli are associated with each other. **Nonassociative learning** includes imitative behaviors, such as learning a language. (p. 314)
47. In **habituation**, an animal shows a decreased response to a stimulus that is repeated over and over. In **sensitization**, exposure to a noxious or intense stimulus creates an enhanced response on subsequent exposure. (p. 315)

48. **Memory** has multiple levels of storage and is constantly changing. Information is first stored in **short-term memory** but disappears unless consolidated into long-term memory. (p. 315; Fig. 9.19)
49. **Long-term memory** includes **reflexive memory**, which does not require conscious processes for its creation or recall, and **declarative memory**, which uses higher-level cognitive skills for formation and requires conscious attention for its recall. (p. 316; Tbl. 9.3)
50. The **consolidation** of short-term memory into long-term memory appears to involve changes in the synaptic connections of the circuits involved in learning. (p. 316)
51. Language is considered the most elaborate cognitive behavior. The integration of spoken language in the human brain involves information processing in **Wernicke's area** and **Broca's area**. (p. 318; Fig. 9.20)

Questions

Answers: p. A-1

Level One Reviewing Facts and Terms

1. The ability of human brains to change circuit connections and function in response to sensory input and past experience is known as __________.
2. __________ behaviors are related to feeling and emotion. __________ behaviors are related to thinking.
3. The part of the brain called the __________ is what makes us human, allowing human reasoning and cognition.
4. In vertebrates, the central nervous system is protected by the bones of the __________ and __________.
5. Name the meninges, beginning with the layer next to the bones.
6. List and explain the purposes of cerebrospinal fluid (CSF). Where is CSF made?
7. Compare the CSF concentration of each of the following substances with its concentration in the blood plasma.
   (a) H^+
   (b) Na^+
   (c) K^+
8. The only fuel source for neurons under normal circumstances is __________. Low concentration of this fuel in the blood is termed __________. To synthesize enough ATP to continually transport ions, the neurons consume large quantities of __________. To supply these needs, about __________% of the blood pumped by the heart goes to the brain.
9. Match each of the following areas with its function.

(a) medulla oblongata	1. coordinates execution of movement
(b) pons	2. is composed of the thalamus and hypothalamus
(c) midbrain	3. controls arousal and sleep
(d) reticular formation	4. fills most of the cranium
(e) cerebellum	5. contains control centers for blood pressure and breathing
(f) diencephalon	6. relays and modifies information going to and from the cerebrum
(g) thalamus	7. transfers information to the cerebellum
(h) hypothalamus	8. contains integrating centers for homeostasis
(i) cerebrum	9. relays signals and visual reflexes, plus eye movement

10. What is the blood-brain barrier, and what is its function?
11. How are gray matter and white matter different from each other, both anatomically and functionally?
12. Name the cerebral cortex areas that (a) direct perception, (b) direct movement, and (c) integrate information and direct voluntary behaviors.
13. What does *cerebral lateralization* refer to? What functions tend to be centered in each hemisphere?
14. Name the 12 cranial nerves in numerical order and their major functions.
15. Name and define the two major phases of sleep. How are they different from each other?
16. List several homeostatic reflexes and behaviors influenced by output from the hypothalamus. What is the source of emotional input into this area?
17. The __________ region of the limbic system is believed to be the center for basic instincts (such as fear) and learned emotional states.
18. What are the broad categories of learning? Define habituation and sensitization. What anatomical structure of the cerebrum is important in both learning and memory?
19. What two centers of the cortex are involved in integrating spoken language?

Level Two Reviewing Concepts

20. Map the following terms describing CNS anatomy. You may draw pictures or add terms if you wish.

• arachnoid membrane	• ependyma
• ascending tracts	• gray matter
• blood-brain barrier	• lumbar nerves
• brain	• meninges
• capillaries	• nuclei
• cell bodies	• pia mater
• cerebrospinal fluid	• propriospinal tracts
• cervical nerves	• sacral nerves
• choroid plexus	• spinal cord
• cranial nerves	• thoracic nerves
• descending tracts	• ventral root
• dorsal root	• ventricles
• dorsal root ganglion	• vertebral column
• dura mater	• white matter

21. Trace the pathway that the cerebrospinal fluid follows through the nervous system.
22. What are the three brain systems that regulate motor output by the CNS?
23. Explain the role of Wernicke's and Broca's areas in language.

24. Compare and contrast the following concepts:
    (a) diffuse modulatory systems, reticular formation, limbic system, and reticular activating system
    (b) different forms of memory
    (c) nuclei and ganglia
    (d) tracts, nerves, horns, nerve fibers, and roots
25. Replace each question mark in the following table with the appropriate word(s):

Cerebral Area	Lobe	Functions
Primary somatic sensory cortex	?	Receives sensory information from peripheral receptors
?	Occipital	Processes information from the eyes
Auditory cortex	Temporal	?
?	Temporal	Receives input from chemoreceptors in the nose
Motor cortices	?	?
Association areas	NA	?

26. Given the wave shown below, draw (a) a wave having a lower frequency, (b) a wave having a larger amplitude, (c) a wave having a higher frequency. (*Hint:* See Figure 9.17, p. 311.)

27. What properties do motivational states have in common?
28. What changes occur at synapses as memories are formed?

Level Three Problem Solving

29. Mr. Andersen, a stroke patient, experiences expressive aphasia. His savvy therapist, Cheryl, teaches him to sing to communicate his needs. What signs did he exhibit before therapy? How do you know he did not have receptive aphasia? Using what you have learned about cerebral lateralization, hypothesize why singing worked for him.
30. A study was done in which 40 adults were taught about the importance of using seat belts in their cars. At the end of the presentation, all participants scored at least 90% on a comprehensive test covering the material taught. The people were also secretly videotaped entering and leaving the parking lot of the class site. Twenty subjects entered wearing their seat belts; 22 left wearing them. Did learning occur? What is the relationship between learning and actually buckling the seat belts?
31. In 1913, Henri Pieron kept a group of dogs awake for several days. Before allowing them to sleep, he withdrew cerebrospinal fluid from the sleep-deprived animals. He then injected this CSF into normal, rested dogs. The recipient dogs promptly went to sleep for periods ranging from two hours to six hours. What conclusion can you draw about the possible source of a sleep-inducing factor? What controls should Pieron have included?
32. A 2002 study* presented the results of a prospective study [p. 24] done in Utah. The study began in 1995 with cognitive assessment of 1889 women whose mean age was 74.5 years. Investigators asked about the women's history of taking calcium, multivitamin supplements, and postmenopausal hormone replacement therapy (estrogen or estrogen/progesterone). Follow-up interviews in 1998 looked for the development of Alzheimer's disease in the study population. Data showed that 58 of 800 women who had not used hormone replacement therapy developed Alzheimer's, compared with 26 of 1066 women who had used hormones.
    (a) Can the researchers conclude from the data given that hormone replacement therapy decreases the risk of developing Alzheimer's? Should other information be factored into the data analysis?
    (b) How applicable are these findings to American women as a whole? What other information might you want to know about the study subjects before you draw any conclusions?
33. A young woman having a seizure was brought into the emergency room. Her roommate said that the woman had taken the street drug Ecstasy the night before and that she had been drinking a lot of water. A blood test showed that her plasma Na^+ was very low: 120 mM (normal 135–145), and her plasma osmolality was 250 mOsmol/kg (normal 280–296). Why would her low osmolality and low Na^+ concentration disrupt her brain function and cause seizures?

*Hormone replacement therapy and incidence of Alzheimer disease in older women: The Cache County study. *JAMA* 288: 2123-2129, 2002 Nov. 6.

9

Answers

Answers to Concept Check Questions

Page 290

1. (a) 3; (b) 2; (c) 1; (d) 3; (e) 2

Page 293

2. Glial cells in the CNS are astrocytes, oligodendrocytes, microglia, and ependyma. See Figure 8.5, p. 245, for functions.

Page 293

3. A ganglion is a cluster of nerve cell bodies outside the CNS. The CNS equivalent is a nucleus.
4. Tracts are the CNS equivalent of peripheral nerves.

Page 296

5. When H^+ concentration increases, pH decreases, which means CSF pH must be lower than blood pH.
6. Blood will collect in the space between the membranes, pushing on the soft brain tissue under the skull. (This is called a *subdural hematoma.*)
7. Cerebrospinal fluid is more like interstitial fluid because both these fluids contain little protein and no blood cells.

Page 298

8. Oxidative phosphorylation takes place in mitochondria.
9. The two pathways are glycolysis and the citric acid cycle. Glucose is metabolized to pyruvate through glycolysis and then enters the

citric acid cycle (also called the tricarboxylic acid cycle). $NADH_2$ passes high-energy electrons to the electron transport system for ATP synthesis.

10. Ehrlich concluded that some property of brain tissue made it resistant to staining by the dye.
11. The brain stained blue this time, but none of the other body tissues were stained because the dye was unable to cross the blood-brain barrier and enter the bloodstream.

Page 299

12. Horns are areas of gray matter in the spinal cord. Roots are sections of spinal nerves just before they enter the spinal cord. Tracts are long projections of white matter (axons) that extend up and down the spinal cord. Columns are groups of tracts carrying similar information.
13. Cutting a dorsal root disrupts sensory function.

Page 302

14. (a) and (c) are white matter, (b) is gray matter.
15. Activities would include moving the eyes, jaw, or tongue and testing taste, smell, and hearing.
16. The cerebrum is dorsal or superior to the brain stem.

Page 304

17. The three subdivisions of the brain stem are medulla oblongata, pons, and midbrain.
18. The diencephalon is composed of thalamus, hypothalamus, pituitary gland, and pineal gland.

Page 305

19. Neurons cross from one side of the body to the other at the pyramids in the medulla.
20. The divisions of the brain, starting at the spinal cord, are medulla, pons, cerebellum, midbrain, diencephalon, and cerebrum.

Page 312

21. Neurons that are sending fewer signals have probably hyperpolarized because they would then require a larger stimulus to initiate an action potential.

Answers to Figure and Graph Questions

Page 294

Figure 9.3: Dura mater completely surrounds the venous sinus and forms one boundary of the subdural space. The arachnoid membrane separates the subdural and subarachnoid spaces. Pia mater forms the other boundary of the subarachnoid space.

Page 295

Figure 9.4: 1. The easiest access is into the subarachnoid space below the bottom of the spinal cord, where there is less risk of damaging the cord. This is called a spinal tap or lumbar puncture. 2. Blockage of the aqueduct will cause CSF to accumulate in the first, second, and third ventricles. Blockage near the frontal lobe will cause fluid build-up in all the ventricles. You would look for enlargement of the fourth ventricle to help localize the site of the blockage.

Page 304

Figure 9.10: (a) and (c) are both correct.

Page 307

Figure 9.14: 1. Losing function in the right visual cortex would mean that the person could see nothing in the left visual field, indicated by the red box at the top of the figure. 2. The tracts of the corpus callosum exchange information between the two sides of the cerebrum. 3. Left-handed people are right-brain dominant, and the right brain is responsible for spatial visualization, which is important in art.

Page 311

Figure 9.17: 1. Alpha waves have the highest frequency, and delta waves have the greatest amplitude. 2. Stages 1 and 2.

Page 318

Figure 9.20: occipital, frontal

10 Sensory Physiology

Nature does not communicate with man by sending encoded messages.

—Oscar Hechter, in *Biology and Medicine into the 21st Century,* 1991

Background Basics

Vestibular hair cells

Imagine floating in the dark in an indoor tank of buoyant salt water: there is no sound, no light, and no breeze. The air and water are the same temperature as your body. You are in a sensory deprivation chamber, and the only sensations you are aware of come from your own body. Your limbs are weightless, your breath moves in and out effortlessly, and you feel your heart beating. In the absence of external stimuli, you turn your awareness inward to hear what your body has to say.

In decades past, flotation tanks for sensory deprivation were a popular way to counter the stress of a busy world. These facilities are hard to find now, but they illustrate the role of the afferent division of the nervous system: to provide us with information about the environment outside and inside our bodies. Sometimes we perceive sensory signals when they reach a level of conscious awareness, but other times they are processed completely at the subconscious level (■ Tbl. 10.1). Stimuli that usually do not reach conscious awareness include changes in muscle stretch and tension as well as a variety of internal parameters that the body monitors to maintain homeostasis, such as blood pressure and pH. The responses to these stimuli constitute many of the subconscious reflexes of the body, and you will encounter them in later chapters as we explore the processes that maintain physiological homeostasis.

In this chapter we are concerned primarily with sensory stimuli whose processing reaches the conscious level of perception. These stimuli are those associated with the **special senses** of vision, hearing, taste, smell, and equilibrium, and the **somatic senses** of touch, temperature, pain, itch, and proprioception. **Proprioception**, which is defined as the awareness of body movement and position in space, is mediated by muscle and joint sensory receptors called **proprioceptors** and may be either unconscious or conscious. If you close your eyes and raise your arm above your head, you are aware of its position because of the activation of proprioceptors.

We first consider general properties of sensory pathways. We then look at the unique receptors and pathways that distinguish the different sensory systems from one another.

RUNNING PROBLEM

Ménière's Disease

On December 23, 1888, Vincent Van Gogh, the legendary French painter, returned to his room in a boardinghouse in Arles, France, picked up a knife, and cut off his own ear. A local physician, Dr. Felix Ray, examined Van Gogh that night and wrote that the painter had been "assailed by auditory hallucinations" and in an effort to relieve them, "mutilated himself by cutting off his ear." A few months later, Van Gogh committed himself to a lunatic asylum. By 1890, Van Gogh was dead by his own hand. Historians have postulated that Van Gogh suffered from epilepsy, but some American neurologists disagree. They concluded that the painter's strange attacks of dizziness, nausea, and overwhelming tinnitus (ringing or other sounds in the ears), which he described in desperate letters to his relatives, are more consistent with Ménière's disease, a condition that affects the inner ear. Today, Anant, a 20-year-old college student, will be examined by an otolaryngologist (ear-nose-throat specialist) to see if his periodic attacks of severe dizziness and nausea are caused by the same condition that might have driven Van Gogh to suicide.

326 331 348 356 361 365 371

Table 10.1 Information Processing by the Sensory Division

Stimulus Processing Usually Conscious	
Special Senses	**Somatic Senses**
Vision	Touch
Hearing	Temperature
Taste	Pain
Smell	Itch
Equilibrium	Proprioception
Stimulus Processing Usually Subconscious	
Somatic Stimuli	**Visceral Stimuli**
Muscle length and tension	Blood pressure
Proprioception	Distension of gastrointestinal tract
	Blood glucose concentration
	Internal body temperature
	Osmolarity of body fluids
	Lung inflation
	pH of cerebrospinal fluid
	pH and oxygen content of blood

General Properties of Sensory Systems

All sensory pathways have certain elements in common. They begin with a stimulus, in the form of physical energy that acts on a sensory receptor. The receptor is a *transducer* that converts the stimulus into an intracellular signal, usually a change in membrane potential. If the stimulus is above threshold, action potentials pass along a sensory neuron to the central nervous system, where incoming signals are integrated. Some stimuli pass upward to the cerebral cortex, where they reach conscious perception, but others are acted on subconsciously, without our awareness. At each synapse along the pathway, the nervous system can modulate and shape the sensory information.

Sensory systems in the human body vary widely in complexity. The simplest systems are single sensory neurons with branched dendrites that function as receptors, such as pain and itch receptors. The most complex systems include multicellular **sense organs**, such as the ear and the eye. The cochlea of the ear contains about 16,000 sensory receptors and more than a million associated parts, and the human eye has about 126 million sensory receptors.

Receptors Are Sensitive to Particular Forms of Energy

Receptors in the sensory system vary widely in complexity, ranging from the branched endings of a single sensory neuron to complex, highly organized cells such as photoreceptors. The simplest receptors consist of a neuron with naked ("free") nerve endings (■ Fig. 10.1a). In more complex receptors, the nerve endings are encased in connective tissue capsules (Fig. 10.1b). The axons of both simple and complex receptors may be myelinated or unmyelinated.

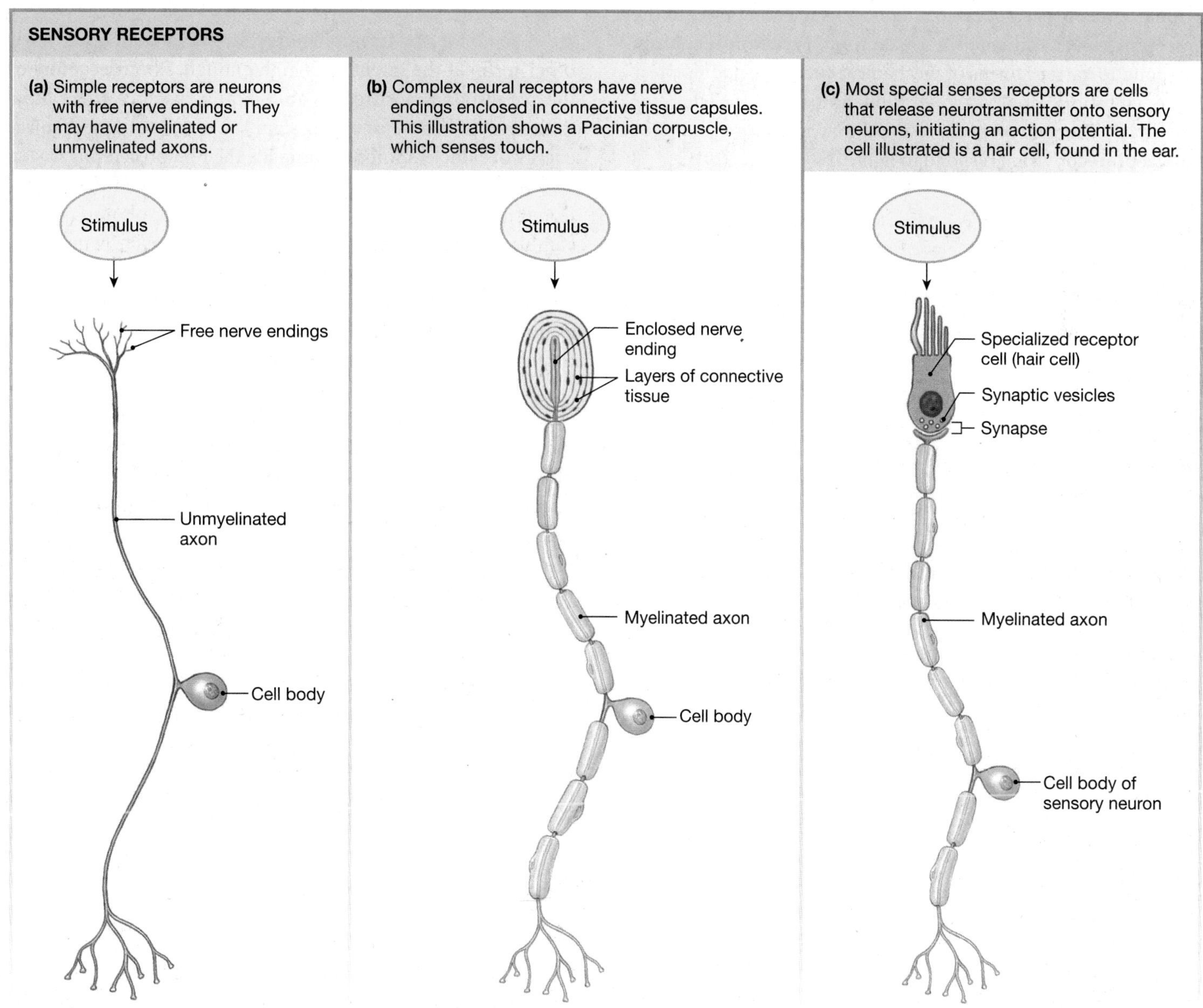
SENSORY RECEPTORS
(a) Simple receptors are neurons with free nerve endings. They may have myelinated or unmyelinated axons.
(b) Complex neural receptors have nerve endings enclosed in connective tissue capsules. This illustration shows a Pacinian corpuscle, which senses touch.
(c) Most special senses receptors are cells that release neurotransmitter onto sensory neurons, initiating an action potential. The cell illustrated is a hair cell, found in the ear.
Stimulus
Free nerve endings
Unmyelinated axon
Cell body
Stimulus
Enclosed nerve ending
Layers of connective tissue
Myelinated axon
Cell body
Stimulus
Specialized receptor cell (hair cell)
Synaptic vesicles
Synapse
Myelinated axon
Cell body of sensory neuron

■ **Fig. 10.1**

The special senses have the most highly specialized receptors. The receptors for smell are neurons, but the other four special senses use non-neural receptor cells that synapse onto sensory neurons. The *hair cell* of the ear, shown in Figure 10.1c, is an example of a non-neural receptor. When activated, the hair cell releases a neurotransmitter that initiates an action potential in the associated sensory neuron. Both neural and non-neural receptors develop from the same embryonic tissue.

Non-neural *accessory structures* are critical to the operation of many sensory systems. For example, the lens and cornea of the eye help focus incoming light onto photoreceptors. The hairs on our arms help **somatosensory receptors** sense movement in the air millimeters above the skin surface. Accessory structures often enhance the information-gathering capability of the sensory system.

Receptors are divided into four major groups, based on the type of stimulus to which they are most sensitive (■ Tbl. 10.2). **Chemoreceptors** respond to chemical ligands that bind to the receptor (taste and smell, for example). **Mechanoreceptors** respond to various forms of mechanical energy, including pressure, vibration, gravity, acceleration, and sound (hearing, for example). **Thermoreceptors** respond to temperature, and **photoreceptors** for vision respond to light.

Table 10.2 Types of Sensory Receptors

Type of Receptor	Examples of Stimuli
Chemoreceptors	Oxygen, pH, various organic molecules such as glucose
Mechanoreceptors	Pressure (baroreceptors), cell stretch (osmoreceptors), vibration, acceleration, sound
Photoreceptors	Photons of light
Thermoreceptors	Varying degrees of heat

Concept Check

Answers: p. 375

1. What advantage do myelinated axons provide?
2. What accessory role does the outer ear (the pinna) play in the auditory system?
3. For each of the somatic and visceral stimuli listed in Table 10.1, which of the following receptor types is the appropriate transducer: mechano-, chemo-, photo-, or thermoreceptors?

Sensory Transduction Converts Stimuli into Graded Potentials

How do receptors convert diverse physical stimuli, such as light or heat, into electrical signals? The first step is **transduction**, the conversion of stimulus energy into information that can be processed by the nervous system [p. 180]. In many receptors, the opening or closing of ion channels converts mechanical, chemical, thermal, or light energy directly into a change in membrane potential. Some sensory transduction mechanisms include signal transduction and second messenger systems that initiate the change in membrane potential.

Each sensory receptor has an **adequate stimulus**, a particular form of energy to which it is most responsive. For example, thermoreceptors are more sensitive to temperature changes than to pressure, and mechanoreceptors respond preferentially to stimuli that deform the cell membrane. Although receptors are specific for one form of energy, they can respond to most other forms if the intensity is high enough. Photoreceptors of the eye respond most readily to light, for instance, but a blow to the eye may cause us to "see stars," an example of mechanical energy of sufficient force to stimulate the photoreceptors.

Sensory receptors can be incredibly sensitive to their preferred form of stimulus. For example, a single photon of light stimulates certain photoreceptors, and a single *odorant* molecule may activate the chemoreceptors involved in the sense of smell. The minimum stimulus required to activate a receptor is known as the **threshold**, just as the minimum depolarization required to trigger an action potential is called the threshold [p. 254].

How is a physical or chemical stimulus converted into a change in membrane potential? The stimulus opens or closes ion channels in the receptor membrane, either directly or indirectly (through a second messenger). In most cases, channel opening results in net influx of Na^+ or other cations into the receptor, depolarizing the membrane. In a few cases, the response to the stimulus is hyperpolarization when K^+ leaves the cell. In the case of vision, the stimulus (light) closes cation channels to hyperpolarize the receptor.

The change in sensory receptor membrane potential is a graded potential [p. 251] called a **receptor potential**. In some cells, the receptor potential initiates an action potential that travels along the sensory fiber to the CNS. In other cells, receptor potentials influence neurotransmitter secretion by the receptor cell, which in turn alters electrical activity in an associated sensory neuron.

A Sensory Neuron Has a Receptive Field

Somatic sensory and visual neurons are activated by stimuli that fall within a specific physical area known as the neuron's **receptive field**. For example, a touch-sensitive neuron in the skin responds to pressure that falls within its receptive field. In the

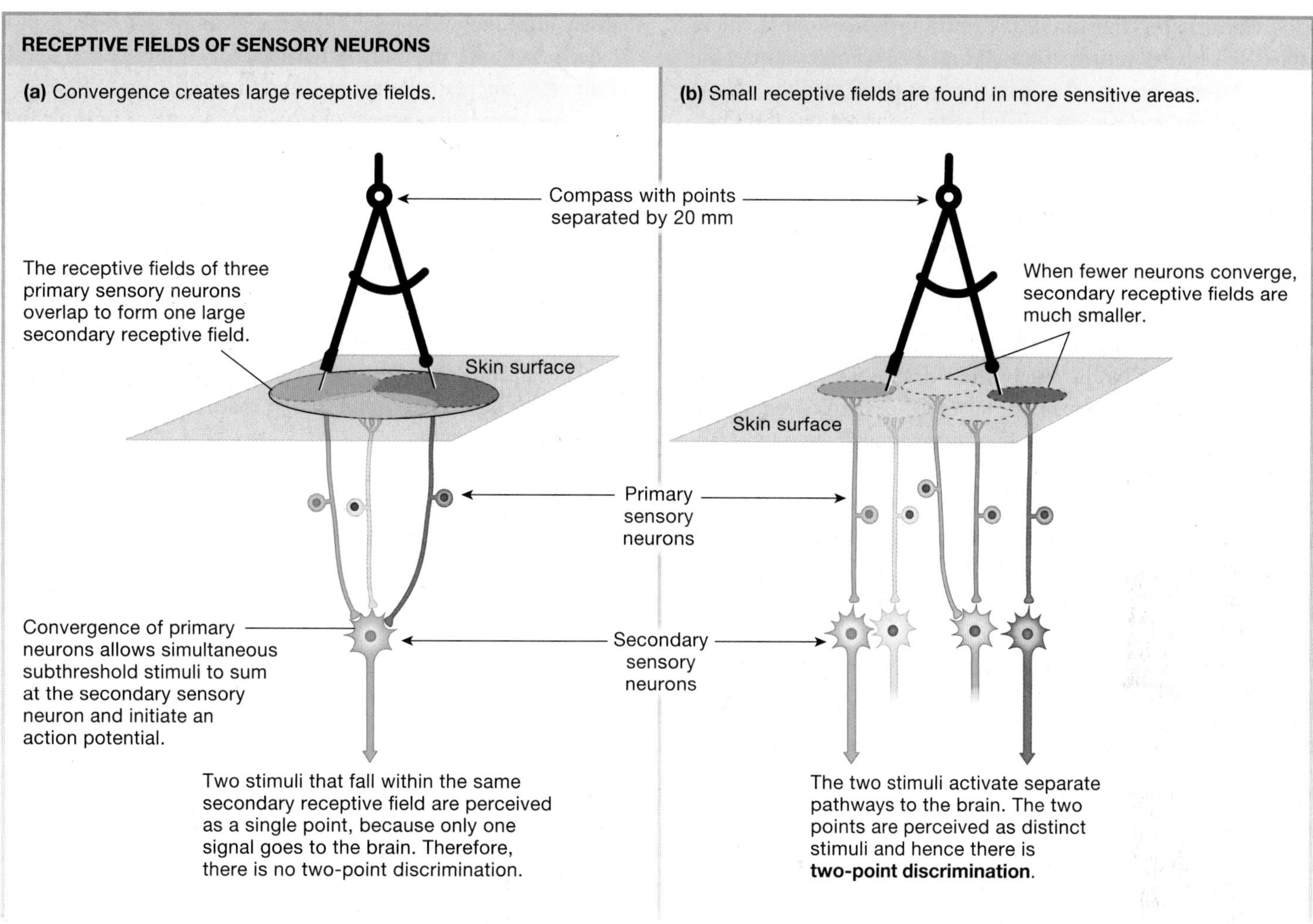
RECEPTIVE FIELDS OF SENSORY NEURONS
(a) Convergence creates large receptive fields.
(b) Small receptive fields are found in more sensitive areas.
Compass with points separated by 20 mm
The receptive fields of three primary sensory neurons overlap to form one large secondary receptive field.
When fewer neurons converge, secondary receptive fields are much smaller.
Skin surface
Skin surface
Primary sensory neurons
Convergence of primary neurons allows simultaneous subthreshold stimuli to sum at the secondary sensory neuron and initiate an action potential.
Secondary sensory neurons
Two stimuli that fall within the same secondary receptive field are perceived as a single point, because only one signal goes to the brain. Therefore, there is no two-point discrimination.
The two stimuli activate separate pathways to the brain. The two points are perceived as distinct stimuli and hence there is **two-point discrimination**.

■ **Fig. 10.2**

10

simplest case, one receptive field is associated with one sensory neuron (the **primary sensory neuron** in the pathway), which in turn synapses on one CNS neuron (the **secondary sensory neuron**). (Primary and secondary sensory neurons are also known as *first-order* and *second-order neurons.*) Receptive fields frequently overlap with neighboring receptive fields.

In addition, sensory neurons of neighboring receptive fields may exhibit *convergence* [p. 274], in which multiple presynaptic neurons provide input to a smaller number of postsynaptic neurons (■ Fig. 10.2). Convergence allows multiple simultaneous subthreshold stimuli to sum at the postsynaptic (secondary) neuron. When multiple primary sensory neurons converge on a single secondary sensory neuron, their individual receptive fields merge into a single, large *secondary receptive field,* as shown in Figure 10.2a.

The size of secondary receptive fields determines how sensitive a given area is to a stimulus. For example, sensitivity to touch is demonstrated by a **two-point discrimination test**. In some regions of skin, such as that on the arms and legs, two pins placed within 20 mm of each other are interpreted by the brain as a single pinprick. In these areas, many primary neurons converge on a single secondary neuron, so the secondary receptive field is very large (Fig. 10.2a).

In contrast, more sensitive areas of skin, such as the fingertips, have smaller receptive fields, with as little as a 1:1 relationship between primary and secondary sensory neurons (Fig. 10.2b). In these regions, two pins separated by as little as 2 mm can be perceived as two separate touches.

The CNS Integrates Sensory Information

Sensory information from much of the body enters the spinal cord and travels through ascending pathways to the brain. Some sensory information goes directly into the brain stem via the cranial nerves [p. 301]. Sensory information that initiates visceral reflexes is integrated in the brain stem or spinal cord and usually does not reach conscious perception. An example of an unconscious visceral reflex is the control of blood pressure by centers in the brain stem.

Each major division of the brain processes one or more types of sensory information (■ Fig. 10.3). For example, the midbrain receives visual information, and the medulla oblongata receives input for sound and taste. Information about balance and equilibrium is processed primarily in the cerebellum. These pathways, along with those carrying somatosensory information, project to the thalamus, which acts as a relay and processing station before passing the information on to the cerebrum.

Only *olfactory* {*olfacere,* to sniff} information is not routed through the thalamus. The sense of smell, a type of chemoreception, is considered to be one of the oldest senses, and even the most primitive vertebrate brains have well-developed regions for processing olfactory information. Information about odors travels from the nose through the first cranial nerve [p. 302] and *olfactory bulb* to the olfactory cortex in the cerebrum. Perhaps it is because of this direct input to the cerebrum that odors are so closely linked to memory and emotion. Most people have experienced encountering a smell that suddenly brings back a flood of memories of places or people from the past.

One interesting aspect of CNS processing of sensory information is the **perceptual threshold**, the level of stimulus intensity necessary for you to be aware of a particular sensation. Stimuli bombard your sensory receptors constantly, but your brain can filter out and "turn off" some stimuli. You experience a change in perceptual threshold when you "tune out" the radio while studying or when you "zone out" during a lecture. In both cases, the noise is adequate to stimulate sensory neurons in the ear, but neurons higher in the pathway dampen the perceived signal so that it does not reach the conscious brain.

Decreased perception of a stimulus, or *habituation*, is accomplished by *inhibitory modulation* [p. 277]. Inhibitory modulation diminishes a suprathreshold stimulus until it is below

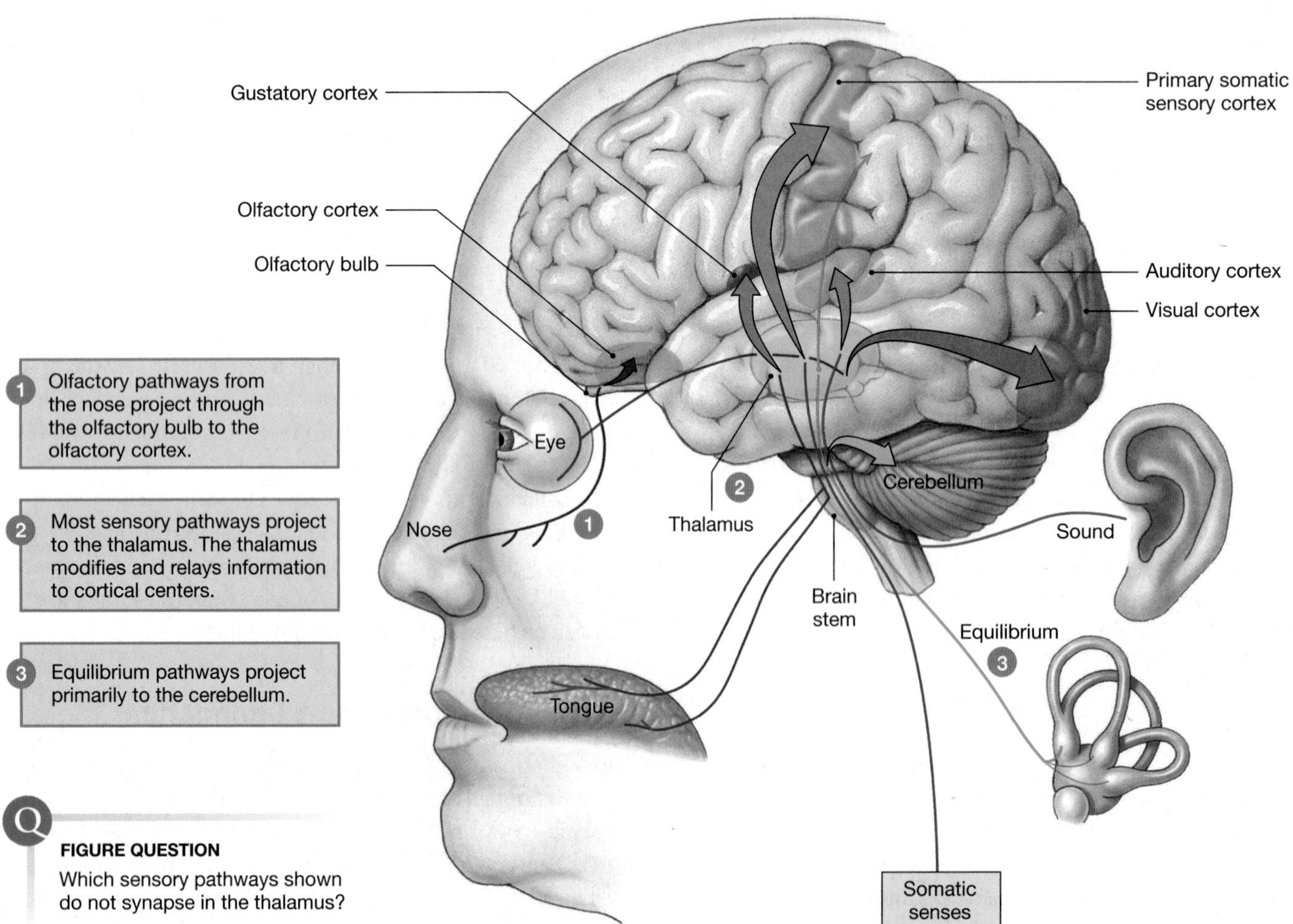
SENSORY PATHWAYS IN THE BRAIN
Most pathways pass through the thalamus on their way to the cerebral cortex.
Gustatory cortex
Olfactory cortex
Olfactory bulb
Primary somatic sensory cortex
Auditory cortex
Visual cortex
1 Olfactory pathways from the nose project through the olfactory bulb to the olfactory cortex.
2 Most sensory pathways project to the thalamus. The thalamus modifies and relays information to cortical centers.
3 Equilibrium pathways project primarily to the cerebellum.
Eye
Nose
1
2
Thalamus
Cerebellum
Sound
Brain stem
Equilibrium
3
Tongue
Somatic senses
FIGURE QUESTION
Which sensory pathways shown do not synapse in the thalamus?

■ **Fig. 10.3**

the perceptual threshold. It often occurs in the secondary and higher neurons of a sensory pathway. If the modulated stimulus suddenly becomes important, such as when the professor asks you a question, you can consciously focus your attention and overcome the inhibitory modulation. At that point, your conscious brain seeks to retrieve and recall recent sound input from your subconscious so that you can answer the question.

Coding and Processing Distinguish Stimulus Properties

If all stimuli are converted to action potentials in sensory neurons and all action potentials are identical, how can the central nervous system tell the difference between, say, heat and pressure, or between a pinprick to the toe and one to the hand? The attributes of the stimulus must somehow be preserved once the stimulus enters the nervous system for processing. This means that the CNS must distinguish four properties of a stimulus: (1) its nature, or **modality**, (2) its location, (3) its intensity, and (4) its duration.

Sensory Modality The modality of a stimulus is indicated by which sensory neurons are activated and by where the pathways of the activated neurons terminate in the brain. Each receptor type is most sensitive to a particular modality of stimulus. For example, some neurons respond most strongly to touch; others respond to changes in temperature. Each sensory modality can be subdivided into qualities. For instance, color vision is divided into red, blue, and green according to the wavelengths that most strongly stimulate the different visual receptors.

In addition, the brain associates a signal coming from a specific group of receptors with a specific modality. This 1:1 association of a receptor with a sensation is called **labeled line coding**. Stimulation of a cold receptor is always perceived as cold, whether the actual stimulus was cold or an artificial depolarization of the receptor. The blow to the eye that causes us to "see" a flash of light is another example of labeled line coding.

Location of the Stimulus The location of a stimulus is also coded according to which receptive fields are activated. The sensory regions of the cerebrum are highly organized with respect to incoming signals, and input from adjacent sensory receptors is processed in adjacent regions of the cortex. This arrangement preserves the topographical organization of receptors on the skin, eye, or other regions in the processing centers of the brain.

For example, touch receptors in the hand project to a specific area of the cerebral cortex. Experimental stimulation of that area of the cortex during brain surgery is interpreted as a touch to the hand, even though there is no contact. Similarly, one type of the *phantom limb pain* reported by amputees occurs when secondary sensory neurons in the spinal cord become hyperactive, resulting in the sensation of pain in a limb that is no longer there.

Auditory information is an exception to the localization rule, however. Neurons in the ears are sensitive to different frequencies of sound, but they have no receptive fields and their activation provides no information about the location of the sound. Instead, the brain uses the timing of receptor activation to compute a location, as shown in ■ Figure 10.4.

A sound originating directly in front of a person reaches both ears simultaneously. A sound originating on one side

RUNNING PROBLEM

Ménière's disease—named for its discoverer, the nineteenth-century French physician Prosper Ménière—is associated with a build-up of fluid in the inner ear and is also known as *endolymphatic hydrops* {*hydro-*, water}. Symptoms of Ménière's disease include episodic attacks of vertigo, nausea, and tinnitus, accompanied by hearing loss and a feeling of fullness in the ears. *Vertigo* is a false sensation of spinning movement that patients often describe as dizziness.

Q1: In which part of the brain is sensory information about equilibrium processed?

326 331 348 356 361 365 371

The Brain Uses Timing Differences to Localize Sound

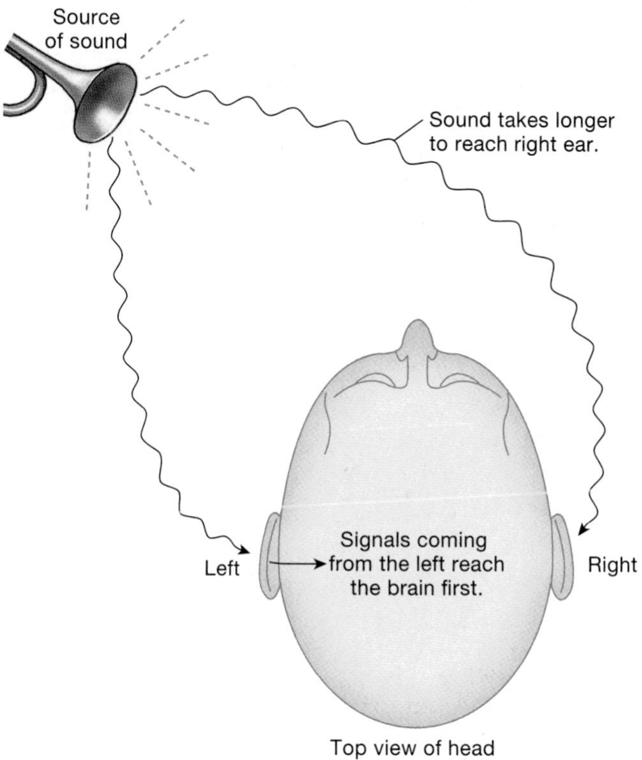
Source of sound
Sound takes longer to reach right ear.
Left
Signals coming from the left reach the brain first.
Right
Top view of head

■ **Fig. 10.4**

reaches the closer ear several milliseconds before it reaches the other ear. The brain registers the difference in the time it takes for the sound stimuli to reach the two sides of the auditory cortex and uses that information to compute the sound's source.

Lateral inhibition, which increases the contrast between activated receptive fields and their inactive neighbors, is another way of isolating the location of a stimulus. ■ Figure 10.5 shows this process for a pressure stimulus to the skin. A pin pushing on the skin activates three primary sensory neurons, each of which releases neurotransmitters onto its corresponding secondary neuron.

However, the three secondary neurons do not all respond in the same fashion. The secondary neuron closest to the stimulus (neuron B) suppresses the response of the secondary neurons lateral to it (that is, on either side), where the stimulus is weaker, and simultaneously allows its own pathway to proceed without interference. The inhibition of neurons farther from the stimulus enhances the contrast between the center and the sides of the receptive field, making the sensation more easily localized. In the visual system, lateral inhibition sharpens our perception of visual edges.

The pathway in Figure 10.5 also is an example of **population coding**, the way multiple receptors function together to send the CNS more information than would be possible from a single receptor. By comparing the input from multiple receptors, the CNS can make complex calculations about the quality and spatial and temporal characteristics of a stimulus.

Concept Check Answer: p. 375

4. In Figure 10.5, what kind(s) of ion channel might open in neurons A and C that would depress their responsiveness: Na^+, K^+, Cl^-, or Ca^{2+}?

Intensity and Duration of the Stimulus The intensity of a stimulus cannot be directly calculated from a single sensory neuron action potential because a single action potential is "all-or-none." Instead, stimulus intensity is coded in two types of

LATERAL INHIBITION

Lateral inhibition enhances contrast and makes a stimulus easier to perceive. The responses of primary sensory neurons A, B, and C are proportional to the intensity of the stimulus in each receptor field. Secondary sensory neuron B inhibits secondary neurons A and C, creating greater contrast between B and its neighbors.

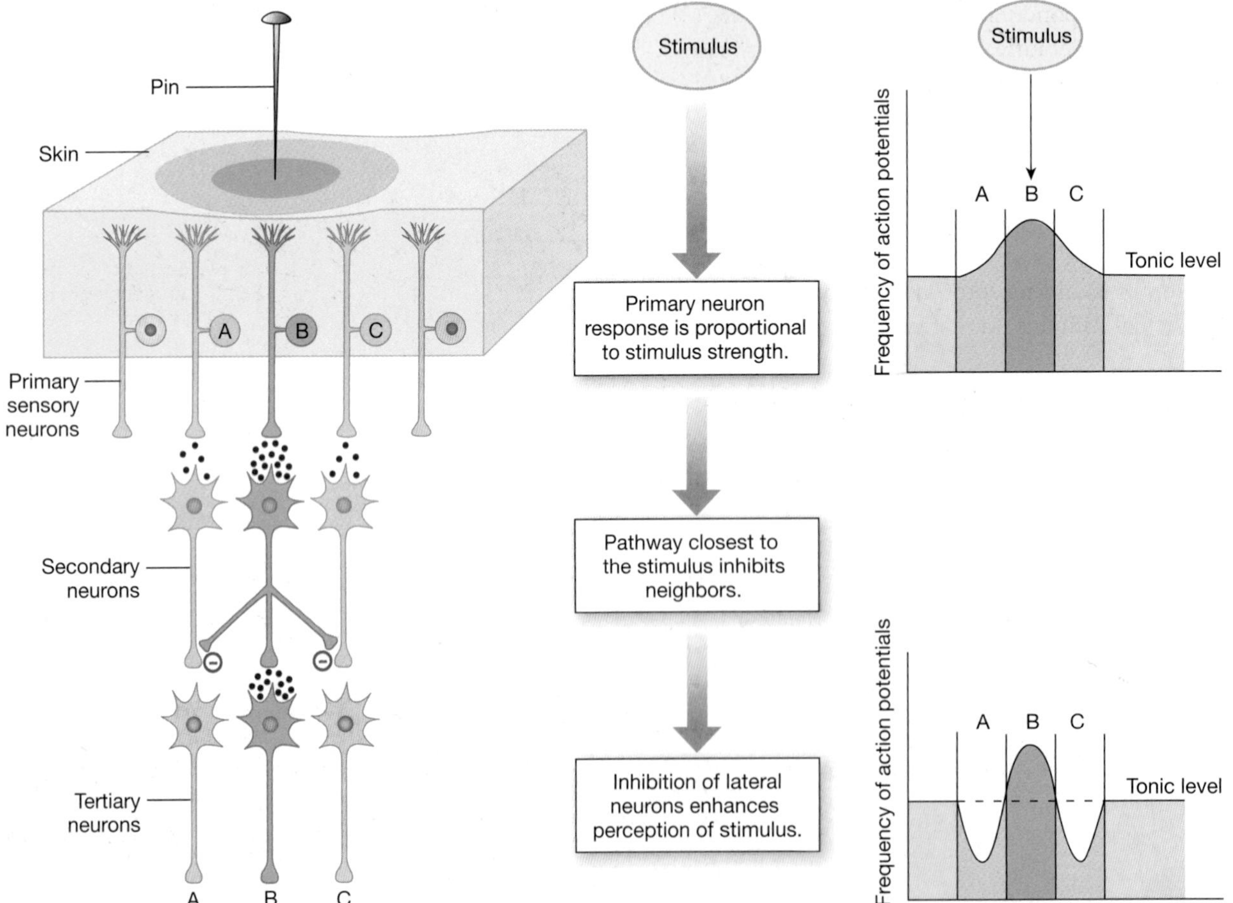

■ **Fig. 10.5**

information: the number of receptors activated (another example of population coding) and the frequency of action potentials coming from those receptors, called *frequency coding*.

Population coding for intensity occurs because the threshold for the preferred stimulus is not the same for all receptors. Only the most sensitive receptors (those with the lowest thresholds) respond to a low-intensity stimulus. As a stimulus increases in intensity, additional receptors are activated. The CNS then translates the number of active receptors into a measure of stimulus intensity.

For individual sensory neurons, intensity discrimination begins at the receptor. If a stimulus is below threshold, the primary sensory neuron does not respond. Once stimulus intensity exceeds threshold, the primary sensory neuron begins to fire action potentials. As stimulus intensity increases, the receptor potential amplitude (strength) increases in proportion, and the frequency of action potentials in the primary sensory neuron increases, up to a maximum rate (■ Fig. 10.6).

Similarly, the duration of a stimulus is coded by the duration of action potentials in the sensory neuron. In general, a longer stimulus generates a longer series of action potentials in the primary sensory neuron. However, if a stimulus persists, some receptors **adapt**, or cease to respond. Receptors fall into one of two classes, depending on how they adapt to continuous stimulation.

Tonic receptors are slowly adapting receptors that fire rapidly when first activated, then slow and maintain their firing as long as the stimulus is present (■ Fig. 10.7a). Pressure-sensitive baroreceptors, irritant receptors, and some tactile receptors and proprioceptors fall into this category. In general, the stimuli that activate tonic receptors are parameters that must be monitored continuously by the body.

In contrast, **phasic receptors** are rapidly adapting receptors that fire when they first receive a stimulus but cease firing if the strength of the stimulus remains constant (Fig. 10.7b). Phasic receptors are attuned specifically to *changes* in a parameter.

10

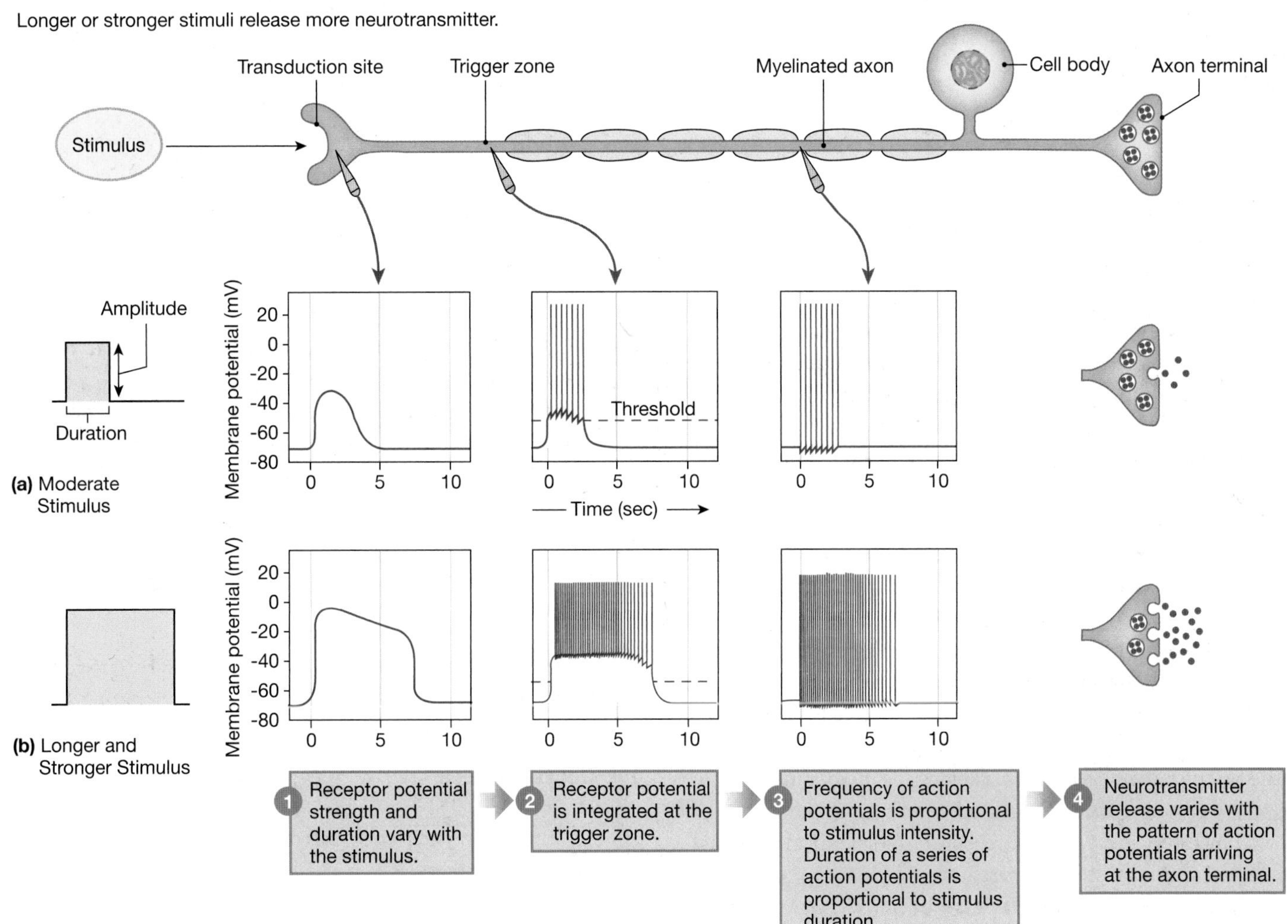
CODING FOR STIMULUS INTENSITY AND DURATION
Longer or stronger stimuli release more neurotransmitter.
Transduction site
Trigger zone
Myelinated axon
Cell body
Axon terminal
Stimulus
Amplitude
Duration
(a) Moderate Stimulus
(b) Longer and Stronger Stimulus
Membrane potential (mV)
20
0
-20
-40
-60
-80
0
5
10
Threshold
Time (sec)
1 Receptor potential strength and duration vary with the stimulus.
2 Receptor potential is integrated at the trigger zone.
3 Frequency of action potentials is proportional to stimulus intensity. Duration of a series of action potentials is proportional to stimulus duration.
4 Neurotransmitter release varies with the pattern of action potentials arriving at the axon terminal.

■ **Fig. 10.6**

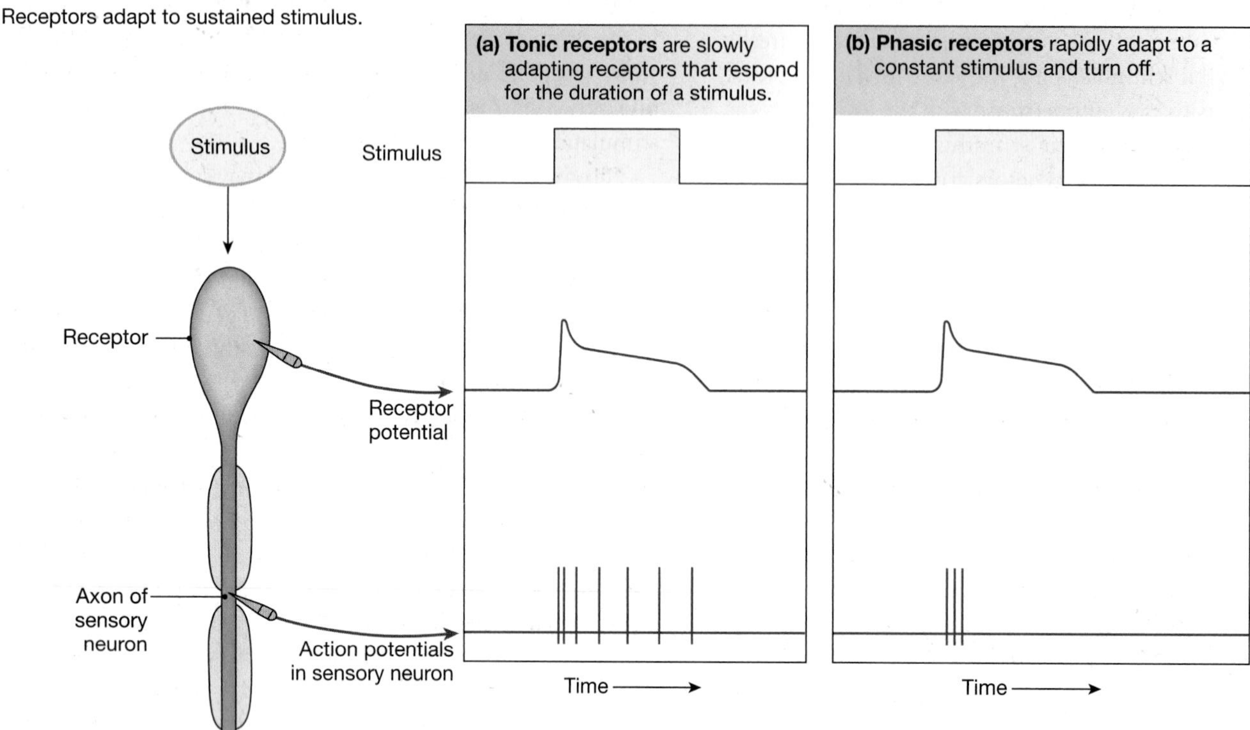
RECEPTOR ADAPTATION
Receptors adapt to sustained stimulus.
(a) Tonic receptors are slowly adapting receptors that respond for the duration of a stimulus.
(b) Phasic receptors rapidly adapt to a constant stimulus and turn off.
Stimulus
Stimulus
Receptor
Receptor potential
Axon of sensory neuron
Action potentials in sensory neuron
Time
Time

Fig. 10.7

Once a stimulus reaches a steady intensity, phasic receptors adapt to the new steady state and turn off. This type of response allows the body to ignore information that has been evaluated and found not to threaten homeostasis or well-being.

Our sense of smell is an example of a sense that uses phasic receptors. For example, you can smell your cologne when you put it on in the morning, but as the day goes on your olfactory receptors adapt and are no longer stimulated by the cologne molecules. You no longer smell the fragrance, yet others may comment on it.

Adaptation of phasic receptors allows us to filter out extraneous sensory information and concentrate on what is new, different, or essential. In general, once adaptation of a phasic receptor has occurred, the only way to create a new signal is to either increase the intensity of the excitatory stimulus or remove the stimulus entirely and allow the receptor to reset.

The molecular mechanism for sensory receptor adaptation depends on the receptor type. In some receptors, K^+ channels in the receptor membrane open, causing the membrane to repolarize and stopping the signal. In other receptors, Na^+ channels quickly inactivate. In yet other receptors, biochemical pathways alter the receptor's responsiveness.

Accessory structures may also decrease the amount of stimulus reaching the receptor. In the ear, for example, tiny muscles contract and dampen the vibration of small bones in response to loud noises, thus decreasing the sound signal before it reaches auditory receptors.

To summarize, the specificity of sensory pathways is established in several ways:

1. Each receptor is most sensitive to a particular type of stimulus.
2. A stimulus above threshold initiates action potentials in a sensory neuron that projects to the CNS.
3. Stimulus intensity and duration are coded in the pattern of action potentials reaching the CNS.
4. Stimulus location and modality are coded according to which receptors are activated or (in the case of sound) by the timing of receptor activation.
5. Each sensory pathway projects to a specific region of the cerebral cortex dedicated to a particular receptive field. The brain can then tell the origin of each incoming signal.

Concept Check Answers: p. 375

5. How do sensory receptors communicate the intensity of a stimulus to the CNS?
6. What is the adaptive significance of irritant receptors that are tonic instead of phasic?

Somatic Senses

There are four somatosensory modalities: touch, proprioception, temperature, and *nociception*, which includes pain and itch. [We discuss details of proprioception in Chapter 13.]

Pathways for Somatic Perception Project to the Cortex and Cerebellum

Receptors for the somatic senses are found both in the skin and in the viscera. Receptor activation triggers action potentials in the associated primary sensory neuron. In the spinal cord, many primary sensory neurons synapse onto interneurons that serve as the secondary sensory neurons. The location of the synapse between a primary neuron and a secondary neuron varies according to the type of receptor (■ Fig. 10.8).

Neurons associated with receptors for nociception, temperature, and coarse touch synapse onto their secondary neurons shortly after entering the spinal cord. In contrast, most fine touch, vibration, and proprioceptive neurons have very long axons that project up the spinal cord all the way to the medulla.

All secondary sensory neurons cross the midline of the body at some point, so that sensations from the left side of the body are processed in the right hemisphere of the brain and vice versa. The secondary neurons for nociception, temperature, and coarse touch cross the midline in the spinal cord, then ascend to the brain. Fine touch, vibration, and proprioceptive neurons cross the midline in the medulla.

In the thalamus, all secondary sensory neurons synapse onto **tertiary sensory neurons**, which in turn project to the somatosensory region of the cerebral cortex. In addition, many sensory pathways send branches to the cerebellum so that it can use the information to coordinate balance and movement.

The **somatosensory cortex** [p. 306] is the part of the brain that recognizes where ascending sensory tracts originate. Each sensory tract has a corresponding region of the cortex, so that all sensory pathways for the left hand terminate in one area, all pathways for the left foot terminate in another area, and so on (■ Fig. 10.9). Within the cortical region for a particular body part, columns of neurons are devoted to particular types of receptors.

For example, a cortical column activated by cold receptors in the left hand may be found next to a column activated by pressure receptors in the skin of the left hand. This columnar arrangement creates a highly organized structure that maintains the association between specific receptors and the sensory modality they transmit.

Some of the most interesting research about the somatosensory cortex has been done on patients during brain surgery for epilepsy. Because brain tissue has no pain fibers, this type of surgery can be performed with the patient awake under local anesthesia. The surgeon stimulates a particular region of the brain and asks the patient about sensations that occur. The ability of the patient to communicate with the surgeon during this process has expanded our knowledge of brain regions tremendously.

Experiments can also be done on nonhuman animals by stimulating peripheral receptors and monitoring electrical activity in the cortex. We have learned from these experiments that the more sensitive a region of the body is to touch and other stimuli, the larger the corresponding region in the cortex. Interestingly, the size of the regions is not fixed. If a particular body part is used more extensively, its topographical region in the cortex will expand. For example, people who are visually handicapped and learn to read Braille with their fingertips develop an enlarged region of the somatosensory cortex devoted to the fingertips.

In contrast, if a person loses a finger or limb, the portion of the somatosensory cortex devoted to the missing structure begins to be taken over by sensory fields of adjacent structures. Reorganization of the somatosensory cortex "map" is an example of the remarkable plasticity [p. 274] of the brain. Unfortunately, sometimes the reorganization is not perfect and can result in sensory sensations, including pain, that the brain interprets as being located in the missing limb (phantom limb pain).

Contemporary research in this field now uses noninvasive imaging techniques, such as *functional magnetic resonance imaging* (fMRI) and *positive emission tomography* (PET) scans to watch brains at work. Both techniques measure the metabolic activity of neurons, so that more active areas of neuronal activity become highlighted and can be associated with their location. [See Fig. 9.20c, p. 318, for PET scans of the brain.]

Touch Receptors Respond to Many Different Stimuli

Touch receptors are among the most common receptors in the body. These receptors respond to many forms of physical contact, such as stretch, steady pressure, fluttering or stroking movement, vibration, and texture. They are found both in the skin (■ Fig. 10.10 on page 338) and in deeper regions of the body.

Touch receptors in the skin come in many forms. Some are free nerve endings, such as those that respond to noxious stimuli. Others are more complex. Most touch receptors are difficult to study because of their small size. However, **Pacinian corpuscles**, which respond to vibration, are some of the largest receptors in the body, and much of what we know about somatosensory receptors comes from studies on these structures.

Pacinian corpuscles are composed of nerve endings encapsulated in layers of connective tissue (see Fig. 10.1b). They are found in the subcutaneous layers of skin and in muscles, joints, and internal organs. The concentric layers of connective tissue in the corpuscles create large receptive fields.

Pacinian corpuscles respond best to high-frequency vibrations, whose energy is transferred through the connective tissue

SOMATOSENSORY PATHWAYS

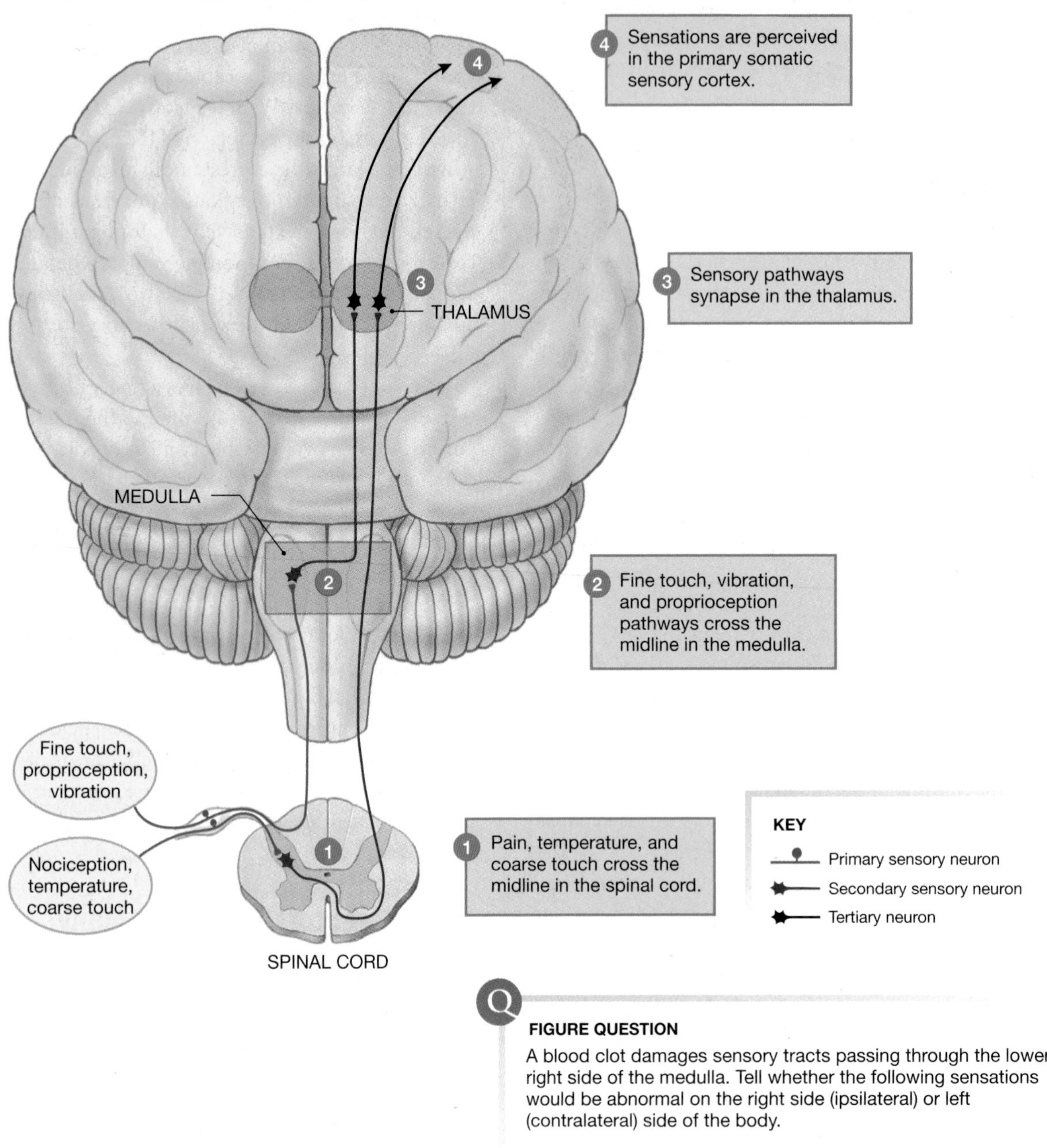
4 Sensations are perceived in the primary somatic sensory cortex.
THALAMUS
3 Sensory pathways synapse in the thalamus.
MEDULLA
2 Fine touch, vibration, and proprioception pathways cross the midline in the medulla.
Fine touch, proprioception, vibration
Nociception, temperature, coarse touch
1 Pain, temperature, and coarse touch cross the midline in the spinal cord.
SPINAL CORD
KEY
Primary sensory neuron
Secondary sensory neuron
Tertiary neuron

FIGURE QUESTION

A blood clot damages sensory tracts passing through the lower right side of the medulla. Tell whether the following sensations would be abnormal on the right side (ipsilateral) or left (contralateral) side of the body.

(a) pain (b) proprioception (c) temperature

	PRIMARY SENSORY	SECONDARY SENSORY	SYNAPSE WITH	TERTIARY SENSORY
Fine touch, proprioception, vibration	Primary sensory neuron synapses in the medulla.	Secondary sensory neuron crosses midline of body in medulla.	Synapse with tertiary sensory neuron in the thalamus.	Tertiary sensory neuron terminates in somatosensory cortex.
Irritants, temperature, coarse touch	Primary sensory neuron synapses in dorsal horn of spinal cord.	Secondary sensory neuron crosses midline of body in spinal cord.		

■ Fig. 10.8

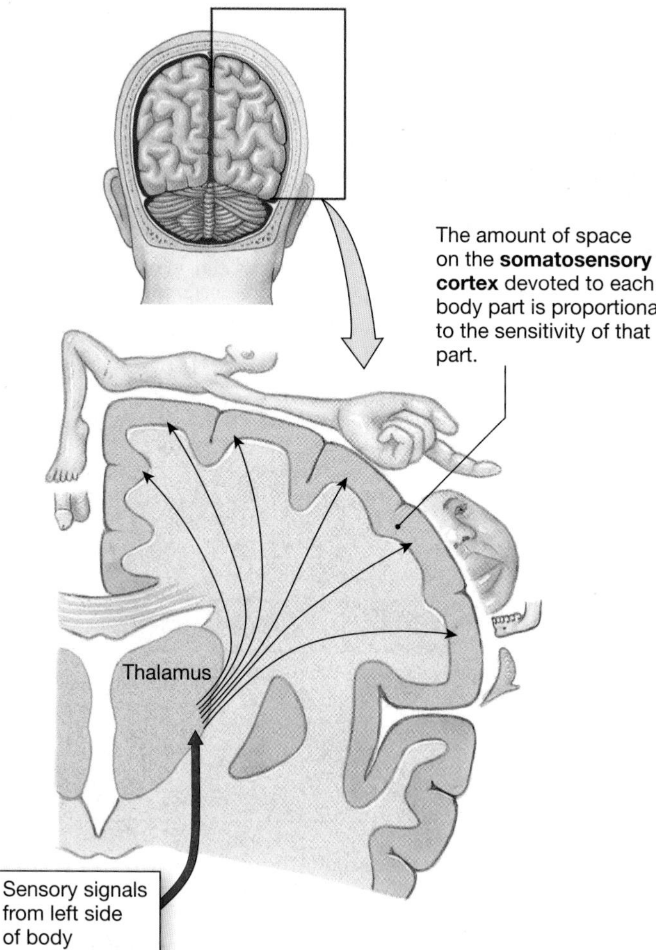

THE SOMATOSENSORY CORTEX
Each body part is represented next to the area of the sensory cortex that processes stimuli for that body part. This mapping was created by two neurosurgeons, W. Penfield and T. Rasmussen, in 1950 and is called a homunculus (little man).
The amount of space on the somatosensory cortex devoted to each body part is proportional to the sensitivity of that part.
Thalamus
Sensory signals from left side of body
Cross section of the right cerebral hemisphere and sensory areas of the cerebral cortex

Fig. 10.9

capsule to the nerve ending, where the energy opens mechanically gated ion channels [p. 148]. Pacinian corpuscles are rapidly adapting phasic receptors, and this property allows them to respond to a change in touch but then ignore it. For example, you notice your shirt when you first put it on, but the touch receptors soon adapt. Properties of the remaining touch receptors depicted in Figure 10.10—Meissner's corpuscles, Ruffini corpuscles, and Merkel receptors—are summarized in the table of that figure.

Temperature Receptors Are Free Nerve Endings

Temperature receptors are free nerve endings that terminate in the subcutaneous layers of the skin. **Cold receptors** are sensitive primarily to temperatures lower than body temperature. **Warm receptors** are stimulated by temperatures in the range extending from normal body temperature (37 °C) to about 45 °C. Above that temperature, pain receptors are activated, creating a sensation of painful heat. Thermoreceptors in the brain play an important role in thermoregulation.

The receptive field for a thermoreceptor is about 1 mm in diameter, and the receptors are scattered across the body. There are considerably more cold receptors than warm ones. Temperature receptors slowly adapt between 20 and 40 °C. Their initial response tells us that the temperature is changing, and their sustained response tells us about the ambient temperature. Outside the 20–40 °C range, where the likelihood of tissue damage is greater, the receptors do not adapt. Thermoreceptors use a family of cation channels called *transient receptor potential* or TRP channels.

Nociceptors Initiate Protective Responses

Nociceptors {*nocere*, to injure} are receptors that respond to a variety of strong noxious stimuli (chemical, mechanical, or thermal) that cause or have the potential to cause tissue damage. Activation of nociceptors initiates adaptive, protective responses, such as the reflexive withdrawal of a hand from a hot stove touched accidentally. Nociceptors are not limited to the skin. Discomfort from overuse of muscles and joints helps us avoid additional damage to these structures. Two sensations may be perceived when nociceptors are activated: pain and itch.

Nociceptors are sometimes called pain receptors, even though pain is a perceived sensation rather than a stimulus. **Nociceptive pain** is mediated by free nerve endings whose ion channels are sensitive to a variety of chemical, mechanical, and thermal stimuli. For example, the membrane channels called *vanilloid receptors* respond to damaging heat from a stove or other source, as well as to *capsaicin*, the chemical that makes hot chili peppers burn your mouth. (Vanilloid receptors are also called *transient receptor potential* V_1 or $TRPV_1$ channels, in the same family as the thermoreceptor channels.) At the opposite end of the temperature spectrum, researchers recently identified a membrane protein that responds both to cold and to menthol, one reason mint-flavored foods feel cool.

Nociceptor activation is modulated by local chemicals that are released upon tissue injury, including K^+, histamine, and prostaglandins released from damaged cells; serotonin released from platelets activated by tissue damage; and the peptide **substance P**, which is secreted by primary sensory neurons. These chemicals, which also mediate the inflammatory response at the site of injury, either activate nociceptors or sensitize them by lowering their activation threshold. Increased sensitivity to pain at sites of tissue damage is called **inflammatory pain**.

Nociceptors may activate two pathways: (1) reflexive protective responses that are integrated at the level of the spinal cord and (2) ascending pathways to the cerebral cortex that become conscious sensation (pain or itch). Primary sensory neurons from

SENSORY RECEPTORS IN THE SKIN

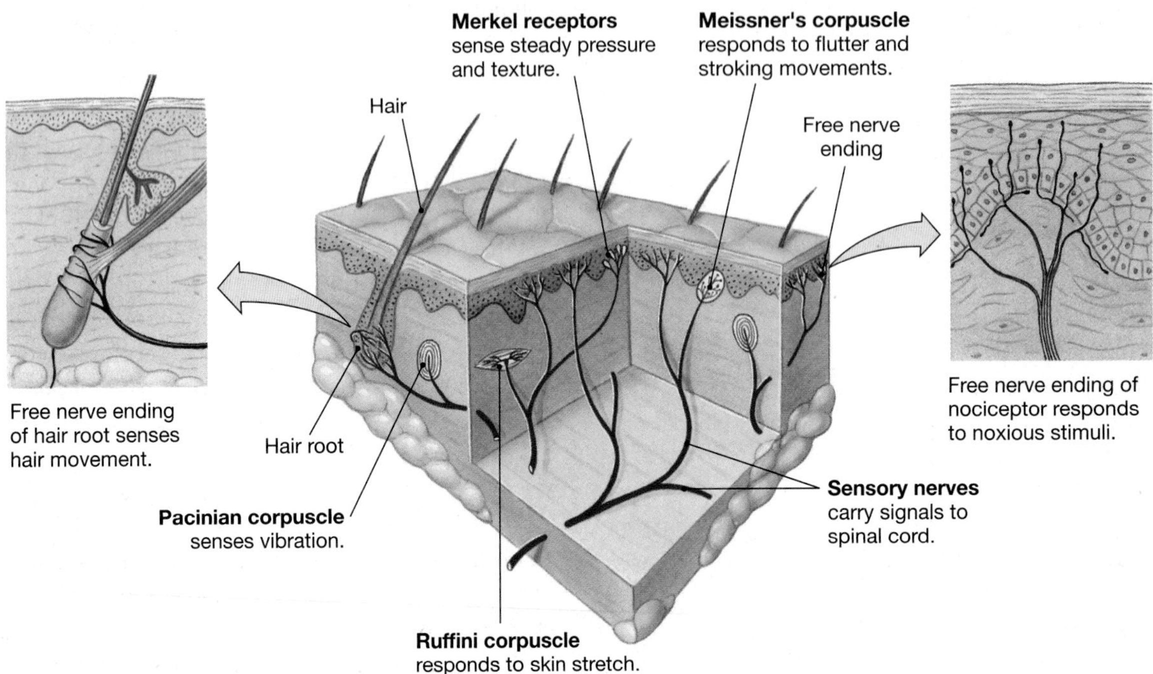
Merkel receptors sense steady pressure and texture.
Meissner's corpuscle responds to flutter and stroking movements.
Hair
Free nerve ending
Free nerve ending of hair root senses hair movement.
Hair root
Free nerve ending of nociceptor responds to noxious stimuli.
Sensory nerves carry signals to spinal cord.
Pacinian corpuscle senses vibration.
Ruffini corpuscle responds to skin stretch.

RECEPTOR	STIMULUS	LOCATION	STRUCTURE	ADAPTATION
Free nerve endings	Temperature, noxious stimuli, hair movement	Around the hair roots and under surface of skin	Unmyelinated nerve endings	Variable
Meissner's corpuscles	Flutter, stroking	Superficial layers of skin	Encapsulated in connective tissue	Rapid
Pacinian corpuscles	Vibration	Deep layers of skin	Encapsulated in connective tissue	Rapid
Ruffini corpuscles	Stretch of skin	Deep layers of skin	Enlarged nerve endings	Slow
Merkel receptors	Steady pressure, texture	Superficial layers of skin	Enlarged nerve endings	Slow

■ **Fig. 10.10**

nociceptors terminate in the dorsal horn of the spinal cord (see Fig. 10.8). There they synapse onto secondary sensory neurons that project to the brain or onto interneurons for local circuits.

Irritant responses that are integrated in the spinal cord initiate rapid unconscious protective reflexes that automatically remove a stimulated area from the source of the stimulus. For example, if you accidentally touch a hot stove, an automatic **withdrawal reflex** causes you to pull back your hand even before you are aware of the heat. This is one example of a spinal reflex [p. 290].

The lack of brain involvement in many protective reflexes has been demonstrated in the classic "spinal frog" preparation, in which the animal's brain has been destroyed. If the frog's foot is placed in a beaker of hot water or strong acid, the withdrawal reflex causes the leg to contract and move the foot away from the stimulus. The frog is unable to feel pain because the brain, which translates sensory input into perception, is not functional, but its protective spinal reflexes are intact.

Pain and Itching Are Mediated by Nociceptors

Afferent signals from nociceptors are carried to the CNS in two types of primary sensory fibers: *Aδ* (A-delta) *fibers,* and *C fibers* (■ Tbl. 10.3). The most common sensation carried by these pathways is perceived as pain, but when histamine or some other stimulus activates a subtype of C fiber, we perceive the sensation we call **itch** (*pruritus*). Itch comes only from nociceptors in the skin and is characteristic of many rashes and

Classes of Somatosensory Nerve Fibers

Table 10.3

Fiber Type	Fiber Characteristics	Speed of Conduction	Associated With
$A\beta$ (beta)	Large, myelinated	30–70 m/sec	Mechanical stimuli
$A\delta$ (delta)	Small, myelinated	12–30 m/sec	Cold, fast pain, mechanical stimuli
C	Small, unmyelinated	0.5–2 m/sec	Slow pain, heat, cold, mechanical stimuli

other skin conditions. However, itch can also be a symptom of a number of systemic diseases, including multiple sclerosis, hyperparathyroidism, and diabetes mellitus.

The higher pathways for itch are not as well understood as the pathways for pain, but there is an antagonistic interaction between the two sensations. When something itches, we scratch it, creating a mildly painful sensation that seems to interrupt the itch sensation. And many of the opioid painkillers, such as morphine, relieve pain but in some people they also induce the side effect of itching.

Pain is a subjective perception, the brain's interpretation of sensory information transmitted along pathways that begin at nociceptors. Pain is also highly individual and may vary with a person's emotional state. The discussion here is limited to the sensory experience of pain.

Fast pain, described as sharp and localized, is rapidly transmitted to the CNS by small, myelinated $A\delta$ fibers. **Slow pain**, described as duller and more diffuse, is carried on small, unmyelinated C fibers. The timing distinction between the two is most obvious when the stimulus originates far from the CNS, such as when you stub your toe. You first experience a quick stabbing sensation (fast pain), followed shortly by a dull throbbing (slow pain).

The ascending pathways for nociception cross the body's midline in the spinal cord and ascend to the thalamus and sensory areas of the cortex (see Fig. 10.8). The pathways also send branches to the limbic system and hypothalamus. As a result, pain may be accompanied by emotional distress (suffering) and a variety of autonomic reactions, such as nausea, vomiting, or sweating.

Our perception of pain is subject to modulation at several levels in the nervous system. It can be magnified by past experiences or suppressed in emergencies when survival depends on ignoring injury. In such emergencies, descending pathways that travel through the thalamus inhibit nociceptor neurons in the spinal cord. Artificial stimulation of these inhibitory pathways is one of the newer techniques being used to control chronic pain.

Pain can also be suppressed in the dorsal horn of the spinal cord, before the stimuli are sent to ascending spinal tracts. Normally, tonically active inhibitory interneurons in the spinal cord inhibit ascending pathways for pain (■ Fig. 10.11a). C fibers from nociceptors synapse on these inhibitory interneurons. When activated by a noxious stimulus, the C fibers simultaneously excite the ascending path and block the tonic inhibition (Fig. 10.11b). This action allows the pain signal from the C fiber to travel unimpeded to the brain.

In the **gate control theory** of pain modulation, $A\beta$ fibers carrying sensory information about mechanical stimuli help block pain transmission (Fig. 10.11c). The $A\beta$ fibers synapse on the inhibitory interneurons and *enhance* the interneuron's inhibitory activity. If simultaneous stimuli reach the inhibitory neuron from the $A\beta$ and C fibers, the integrated response is partial inhibition of the ascending pain pathway so that pain perceived by the brain is lessened. The gate control theory explains why rubbing a bumped elbow or shin lessens your pain: the tactile stimulus of rubbing activates $A\beta$ fibers and helps decrease the sensation of pain.

Pain can be felt in skeletal muscles (*deep somatic pain*) as well as in the skin. Muscle pain during exercise is associated with the onset of anaerobic metabolism and is often perceived as a burning sensation in the muscle (as in "go for the burn!"). Some investigators have suggested that the exercise-induced metabolite responsible for the burning sensation is K^+, known to enhance the pain response. Muscle pain from **ischemia** (lack of adequate blood flow that reduces oxygen supply) also occurs during *myocardial infarction* (heart attack).

Pain in the heart and other internal organs (*visceral pain*) is often poorly localized and may be felt in areas far removed from the site of the stimulus (■ Fig. 10.12a). For example, the pain of cardiac ischemia may be felt in the neck and down the left shoulder and arm. This **referred pain** apparently occurs because multiple primary sensory neurons converge on a single ascending tract (Fig. 10.12b). According to this model, when painful stimuli arise in visceral receptors, the brain is unable to distinguish visceral signals from the more common signals arising from somatic receptors. As a result, it interprets the pain as coming from the somatic regions rather than the viscera.

Chronic pain of one sort or another affects millions of people in this country every year. This type of pain is often much greater than nociceptor activation would indicate and reflects damage to or long-term changes in the nervous system. Chronic pain is a **pathological pain** and is also called *neuropathic pain*.

THE GATE CONTROL MODEL

In the gate control model of pain modulation, nonpainful stimuli can diminish the pain signal.

(a) In absence of input from C fibers, a tonically active inhibitory interneuron suppresses pain pathway.

(b) With strong pain, C fiber stops inhibition of the pathway, allowing a strong signal to be sent to the brain.

(c) Pain can be modulated by simultaneous somatosensory input.

■ **Fig. 10.11**

CLINICAL FOCUS

Natural Painkillers

Many drugs we use today for pain relief are derivatives of plant or animal molecules. One of the newest painkillers in this group is *ziconotide,* a synthetic compound related to the poison that South Pacific cone snails use to kill fish. This drug works by blocking calcium channels on nociceptive neurons. Ziconotide, approved in 2004 for the treatment of severe chronic pain, is highly toxic. To minimize systemic side effects, it must be injected directly into the cerebrospinal fluid surrounding the spinal cord. Ziconotide relieves pain but may also cause hallucinations and other psychiatric symptoms, so it is a last-resort treatment. Other painkilling drugs from biological sources include aspirin, derived from the bark of willow trees (genus *Salix*), and opiate drugs such as morphine and codeine that come from the opium poppy, *Papaver somniferum*. These drugs have been used in Western and Chinese medicine for centuries, and even today you can purchase willow bark as an herbal remedy.

One of the most common forms of neuropathic pain is *diabetic neuropathy,* which develops as a consequence of chronically elevated blood glucose concentrations. Scientists do not yet fully understand what causes glucose neurotoxicity or neuropathic pain, which makes its treatment difficult.

The alleviation of pain is of considerable interest to health professionals. **Analgesic drugs** {*analgesia,* painlessness} range from aspirin to potent opioids such as morphine. Aspirin inhibits prostaglandins, decreases inflammation, and presumably slows the transmission of pain signals from the site of injury. The opioid drugs act directly on CNS *opioid receptors* that are part of an analgesic system that responds to endogenous opioid molecules [p. 269]. Activation of opioid receptors blocks pain perception by decreasing neurotransmitter release from primary sensory neurons and by postsynaptic inhibition of the secondary sensory neurons.

The endogenous opioids include three families: endorphins, enkephalins, and dynorphins. **Enkephalins** and **dynorphins** are secreted by neurons associated with pain pathways. The endogenous opioid **β-endorphin** is produced from the same prohormone as ACTH (adrenocorticotropin) in neuroendocrine cells of the hypothalamus [Fig. 7.3b, p. 213]. Although opioid drugs are effective at relieving pain, a person taking them for long periods of time will develop tolerance and need larger and larger doses.

As a result, scientists are exploring alternative drugs and strategies for pain relief. Some chronic pain may be caused by sensitization of nociceptive nerve endings near a site of injury when the body releases chemicals in response to the damage. Non-narcotic anti-inflammatory drugs such as aspirin and COX2 inhibitors can often relieve pain, but even over-the-counter doses may have adverse side effects. New research is focused on blocking receptors in the sensitized nociceptor nerve endings.

For people with severe chronic pain, possible treatments include electrically stimulating inhibitory pain pathways to the

REFERRED PAIN

(a) Pain in internal organs is often sensed on the surface of the body, a sensation known as **referred pain**.

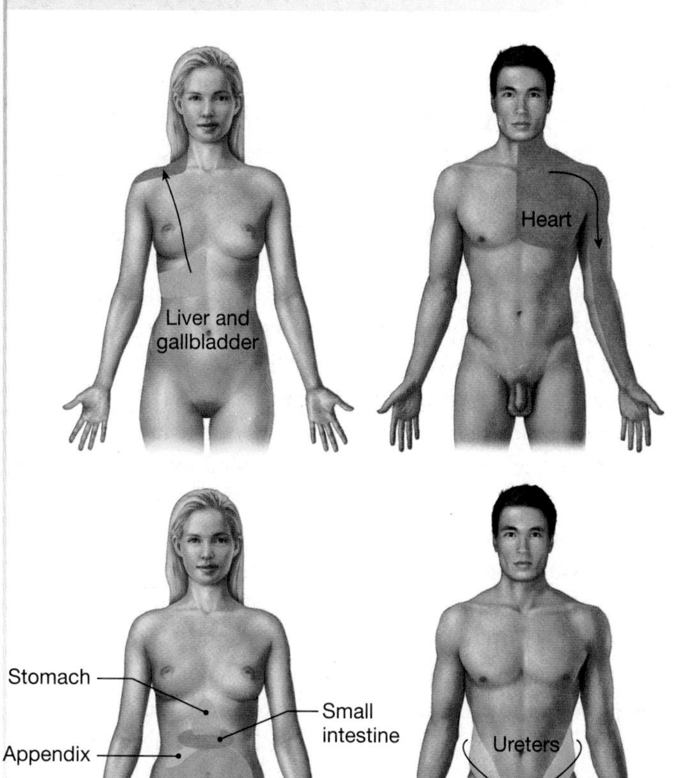

(b) One theory of referred pain says that nociceptors from several locations converge on a single ascending tract in the spinal cord. Pain signals from the skin are more common than pain from internal organs, and the brain associates activation of the pathway with pain in the skin. Adapted from H.L. Fields, *Pain* (McGraw Hill, 1987).

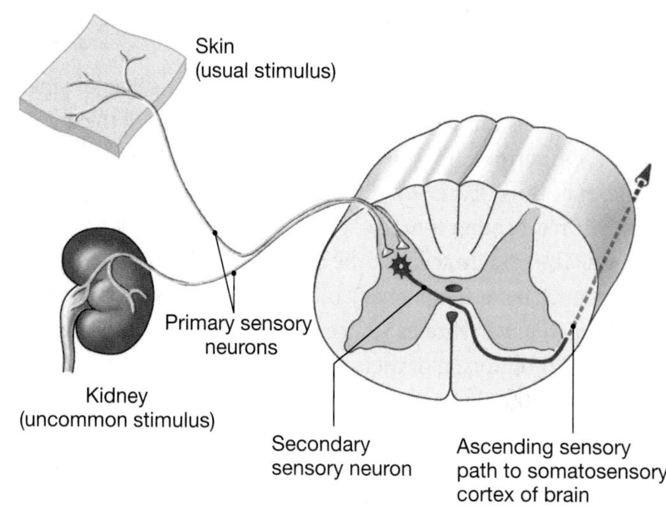

FIGURE QUESTION

A man goes to his physician and complains of pain that radiates down his left arm. This suggests to the physician that the man may have a problem with what organ?

Fig. 10.12

brain, or in extreme cases, surgically severing sensory nerves at the dorsal root. Acupuncture can also be effective, although the physiological reason for its effectiveness is not clear. The leading theory on how acupuncture works proposes that properly placed acupuncture needles trigger the release of endorphins by the brain.

Concept Check Answers: p. 375

7. What is the adaptive advantage of a spinal reflex?
8. Rank the speed of signal transmission through the following fiber types, from fastest to slowest: (a) small diameter, myelinated fiber; (b) large diameter, myelinated fiber; (c) small diameter, unmyelinated fiber.
9. Your sense of smell uses phasic receptors. What other receptors (senses) adapt to ongoing stimuli?

Chemoreception: Smell and Taste

The five special senses—smell, taste, hearing, equilibrium, and vision—are concentrated in the head region. Like somatic senses, the special senses rely on receptors to transform information about the environment into patterns of action potentials that can be interpreted by the brain. Smell and taste are both forms of *chemoreception,* one of the oldest senses from an evolutionary perspective. Unicellular bacteria use chemoreception to sense their environment, and primitive animals without formalized nervous systems use chemoreception to locate food and mates. It has been hypothesized that chemoreception evolved into chemical synaptic communication in animals.

Olfaction Is One of the Oldest Senses

Imagine waking up one morning and discovering a whole new world around you, a world filled with odors that you never dreamed existed—scents that told you more about your surroundings than you ever imagined from looking at them. This is exactly what happened to a young patient of Dr. Oliver Sacks (an account is in *The Man Who Mistook His Wife for a Hat and Other Clinical Tales*). Or imagine skating along the sidewalk without a helmet, only to fall and hit your head. When you regain consciousness, the world has lost all odor: no smell of grass or perfume or garbage. Even your food has lost much of its taste, and you now eat only to survive because eating has lost its pleasure.

We do not realize the essential role that our sense of smell plays in our lives until a head cold or injury robs us of the ability to smell. **Olfaction** allows us to discriminate among thousands of different odors. Even so, our noses are not nearly as sensitive as those of many other animals whose survival depends on olfactory cues. The **olfactory bulb**, the extension of the forebrain that receives input from the primary olfactory neurons, is much better developed in vertebrates whose survival is more closely linked to chemical monitoring of their environment (■ Fig. 10.13a).

The human olfactory system consists of primary sensory neurons (**olfactory sensory neurons**) whose axons form the *olfactory nerve* (cranial nerve I [p. 302]). The olfactory nerve synapses with secondary sensory neurons in the olfactory bulb, which then processes the incoming information (Fig. 10.13b). Secondary and higher-order neurons project from the olfactory bulb through the *olfactory tract* to the *olfactory cortex* (Fig. 10.13a). The olfactory tract, unlike most other sensory pathways, bypasses the thalamus.

This arrangement seems quite simple, but complex processing takes place in the olfactory bulb before signals pass on to the cortex. Some descending modulatory pathways from the cortex terminate in the olfactory bulb, and there are reciprocal modulatory connections within and between the two branches of the olfactory bulb. In addition, olfactory pathways lead to the amygdala and hippocampus, parts of the limbic system involved with emotion and memory.

The link between smell, memory, and emotion is one amazing aspect of olfaction. A special cologne or the aroma of food can trigger memories and create a wave of nostalgia for the time, place, or people with whom the aroma is associated. In some way that we do not understand, the processing of odors through the limbic system creates deeply buried olfactory memories. Particular combinations of olfactory receptors become linked to other patterns of sensory experience so that stimulating one pathway stimulates them all.

In rodents, an accessory olfactory structure in the nasal cavity, the **vomeronasal organ** (VNO), is known to be involved in behavioral responses to sex pheromones [p. 209]. Anatomical studies in humans have not provided clear evidence for or against a functional VNO, but experiments with compounds believed to act as human pheromones suggest that humans do communicate with chemical signals.

Olfactory sensory neurons in humans are concentrated in a 3-cm^2 patch of **olfactory epithelium** high in the nasal cavity (Fig. 10.13a). Olfactory sensory neurons have a single dendrite that extends down from the cell body to the surface of the olfactory epithelium, and a single axon that extends up to the olfactory bulb, located on the underside of the frontal lobe. Olfactory sensory neurons, unlike other neurons in the body, have very short lives, with a turnover time of about two months (Fig. 10.13c). Stem cells in the basal layer of the olfactory epithelium are continuously dividing to create new neurons. The axon of each newly formed neuron must then find its way to the olfactory bulb and make the proper synaptic connections. Scientists are studying how these neurons manage to repeat the same connection each time to give us insight into how developing neurons find their targets.

The surface of the olfactory epithelium is composed of the knobby terminals of the olfactory sensory neurons, each knob sprouting multiple nonmobile cilia that function as dendrites (Fig. 10.13c). The cilia are embedded in a layer of mucus, and odorant molecules must first dissolve in and penetrate the mucus before they can bind to an **odorant receptor** protein. Each odorant receptor is sensitive to a variety of substances.

Odorant receptors are G protein–linked membrane receptors [p. 183]. Odorant receptor genes form the largest known gene family in vertebrates (about 1000 genes, or 3–5% of the genome), but only about 400 odorant receptor proteins are expressed in humans. The combination of an odorant molecule with its odorant receptor activates a special G protein, $\mathbf{G_{olf}}$, which in turn increases intracellular cAMP. The increase in cAMP concentration opens cAMP-gated cation channels, depolarizing the cell and triggering a signal that travels along the olfactory sensory neuron axon to the olfactory bulb.

What is occurring at the cellular and molecular levels that allows us to discriminate between thousands of different odors? Current research suggests that each individual olfactory sensory neuron contains a single type of odorant receptor. The axons of cells with the same receptors converge on a few secondary neurons in the olfactory bulb, which then can modify the information before sending it on to the olfactory cortex. The brain uses information from hundreds of olfactory sensory neurons in different combinations to create the perception of many different smells, just as combinations of letters create different words. This is another example of population coding in the nervous system [p. 332].

Concept Check

Answers: p. 375

10. Create a map or diagram of the olfactory pathway from an olfactory sensory neuron to the olfactory cortex.
11. Create a map or diagram that starts with a molecule from the environment binding to its odorant receptor in the nose and ends with neurotransmitter release from the primary olfactory neuron.
12. The dendrites are which part of an olfactory sensory neuron?
13. Are olfactory neurons pseudounipolar, bipolar, or multipolar? [*Hint:* See Fig. 8.2, p. 241.]

The Olfactory System

(a) Olfactory Pathways
The olfactory epithelium lies high within the nasal cavity, and its olfactory neurons project to the olfactory bulb. Sensory input at the receptors is carried through the olfactory cortex to the cerebral cortex and the limbic system.

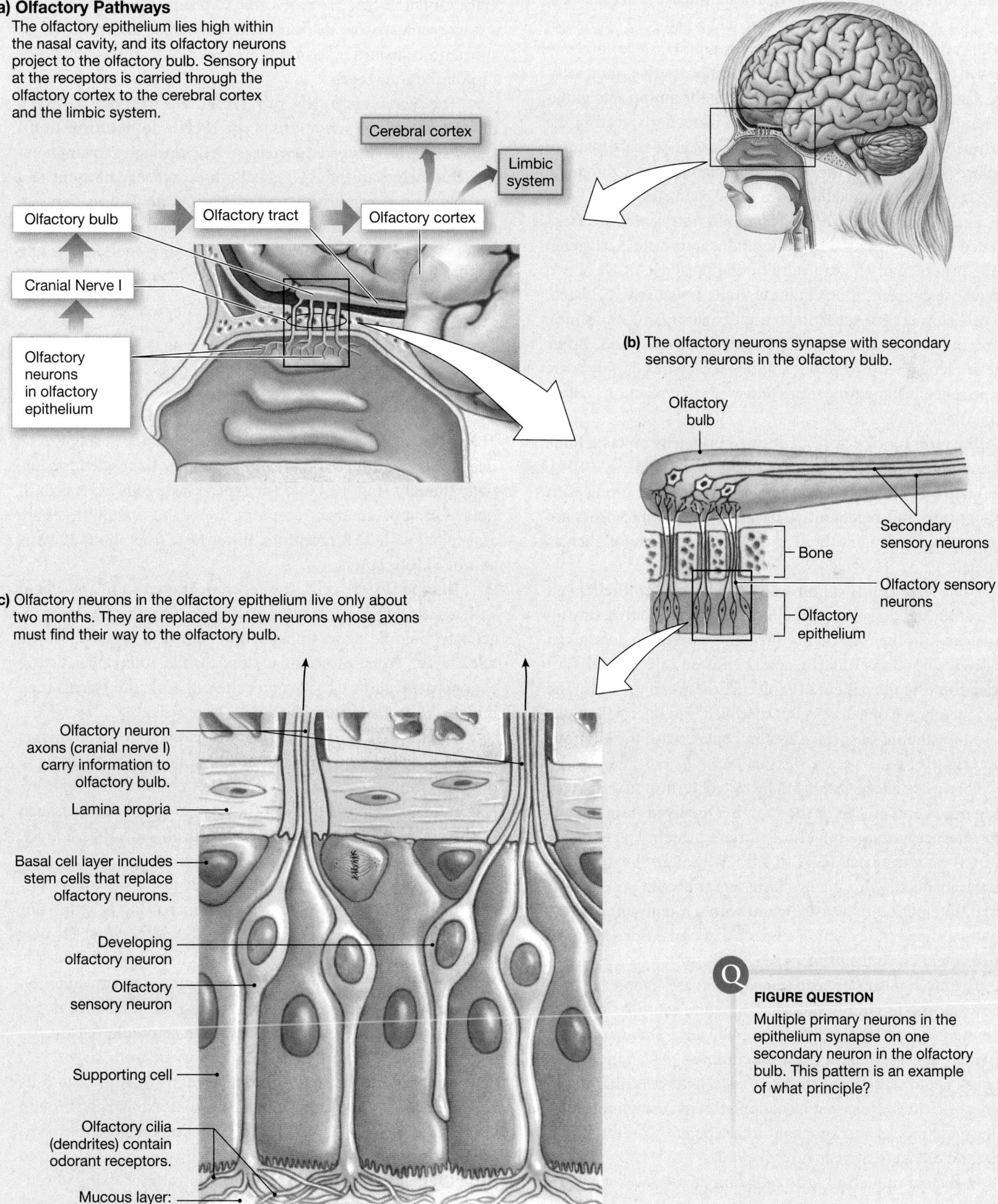

(b) The olfactory neurons synapse with secondary sensory neurons in the olfactory bulb.

(c) Olfactory neurons in the olfactory epithelium live only about two months. They are replaced by new neurons whose axons must find their way to the olfactory bulb.

FIGURE QUESTION

Multiple primary neurons in the epithelium synapse on one secondary neuron in the olfactory bulb. This pattern is an example of what principle?

Taste Is a Combination of Five Basic Sensations

Our sense of taste, or **gustation,** is closely linked to olfaction. Indeed, much of what we call the taste of food is actually the aroma, as you know if you have ever had a bad cold. Although smell is sensed by hundreds of receptor types, taste is currently believed to be a combination of five sensations: sweet, sour, salty, bitter, and **umami**, a taste associated with the amino acid glutamate and some nucleotides. Umami, a name derived from the Japanese word for "deliciousness," is a basic taste that enhances the flavor of foods. It is the reason that monosodium glutamate (MSG) is used as a food additive in some countries.

Each of the five currently recognized taste sensations is associated with an essential body function. Sour taste is triggered by the presence of H^+ and salty by the presence of Na^+, two ions whose concentrations in body fluids are closely regulated. The other three taste sensations result from organic molecules. Sweet and umami are associated with nutritious food. Bitter taste is recognized by the body as a warning of possibly toxic components. If something tastes bitter, our first reaction is often to spit it out.

The receptors for taste are located primarily on **taste buds** clustered together on the surface of the tongue (■ Fig. 10.14). One taste bud is composed of 50–150 **taste cells**, along with support cells and regenerative basal cells. Taste receptors are also scattered through other regions of the oral cavity, such as the palate.

Each taste cell is a non-neural polarized epithelial cell [p. 158] tucked down into the epithelium so that only a tiny tip protrudes into the oral cavity through a *taste pore*. In a given bud, tight junctions link the apical ends of adjacent cells together, limiting movement of molecules between the cells. The apical membrane of a taste cell is modified into microvilli to increase the amount of surface area in contact with the environment (Fig. 10.14a).

For a substance (*tastant*) to be tasted, it must first dissolve in the saliva and mucus of the mouth. Dissolved taste ligands then interact with an apical membrane protein (receptor or channel) on a taste cell (Fig. 10.14b). Although the details of signal transduction for the five taste sensations are still controversial, interaction of a taste ligand with a membrane protein initiates a signal transduction cascade that ends with a series of action potentials in the primary sensory neuron.

The mechanisms of taste transduction are a good example of how our models of physiological function must periodically be revised as new research data are published. For many years the widely held view of taste transduction was that an individual taste cell could sense more than one taste, with cells differing in their sensitivities. However, gustation research using molecular biology techniques and knockout mice currently indicates that each taste cell is sensitive to only one taste.

In the old model, all taste cells formed synapses with primary sensory neurons (*gustatory neurons*). Now it has been shown that there are at least two different types of taste cells. Taste buds contain four morphologically distinct cell types designated I, II, and III, plus *basal cells* that may be the taste stem cells. Only the *type III* taste cells, also called *presynaptic cells,* synapse with sensory neurons. The presynaptic taste cells release the neurotransmitter serotonin by exocytosis. Presynaptic cells respond to sour tastes.

The *type II* taste cells, or *receptor cells,* respond to sweet, bitter, and umami sensations. Type II cells do not form traditional synapses. Instead they release ATP through gap junction–like channels, and the ATP acts both on sensory neurons and on neighboring presynaptic cells. This communication between neighboring taste cells creates complex interactions. Currently it is not clear which cell type is responsible for responding to salt. Some evidence suggests that the glia-like type I, or *support cells,* may be the salt sensors.

Taste Transduction Uses Receptors and Channels

The details of taste cell signal transduction, once thought to be relatively straightforward, are more complex than scientists initially thought (Fig. 10.14b). The type II taste cells express multiple G protein–coupled receptors. Sweet and umami tastes are associated with T1R receptors. Bitter taste uses about 30 variants of T2R receptors.

In type II taste cells, the receptor proteins are associated with a special G protein called **gustducin**. Gustducin appears to activate multiple signal transduction pathways. Some pathways release Ca^{2+} from intracellular stores, while others open cation channels and allow Ca^{2+} to enter the cell. Calcium signals then initiate ATP release from the type II taste cells.

In contrast, salty and sour transduction mechanisms both appear to be mediated by ion channels rather than by G protein–coupled receptors. In the current model for salty taste, Na^+ enters the presynaptic cell through an apical channel, such as the *epithelial* Na^+ *channel (ENaC,* pronounced *ēē-knack).* Sodium entry depolarizes the taste cell.

Transduction mechanisms for sour tastes are more controversial, complicated by the fact that increasing H^+, the sour taste signal, also changes pH. There is evidence that H^+ acts on ion channels of the presynaptic cell from both extracellular and intracellular sides of the membrane. The transduction mechanisms remain uncertain. Ultimately, H^+-mediated depolarization of the presynaptic cell results in serotonin release. Serotonin in turn excites the primary sensory neuron.

Neurotransmitters (ATP and serotonin) from taste cells activate primary gustatory neurons whose axons run through cranial nerves VII, IX, and X to the medulla, where they synapse. Sensory information then passes through the thalamus to the gustatory cortex (see Fig. 10.3). Central processing of sensory information compares the input from multiple taste cells

■ Fig. 10.14 *ESSENTIALS*

Taste

(a) Taste buds. Each taste bud is composed of taste cells joined near the apical surface with tight junctions.

Taste buds are located on the dorsal surface of the tongue.
Taste pore
Light micrograph of a taste bud
Taste ligands create Ca^{2+} signals that release serotonin or ATP.
Sweet
Umami
Bitter
Sour
Tight junction
Type I support cells may sense salt when Na^+ enters through Na^+ channels.
Salt?
Presynaptic cell (III)
ATP
Serotonin
Receptor cells (type II)
Primary sensory neurons
(Adapted from Tomchik *et al.*, *J Neurosci* 27(40): 10840–10848, 2007.)

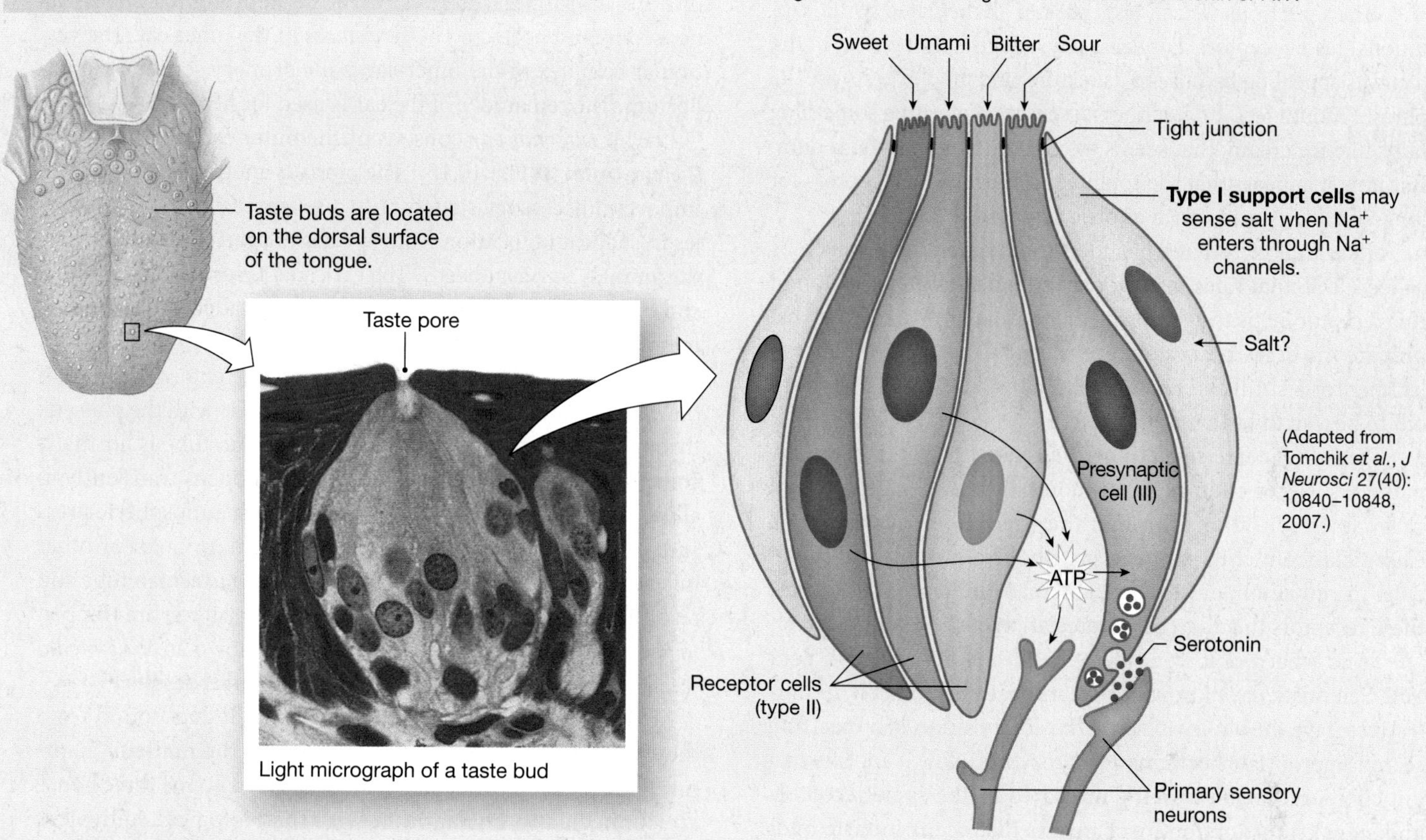

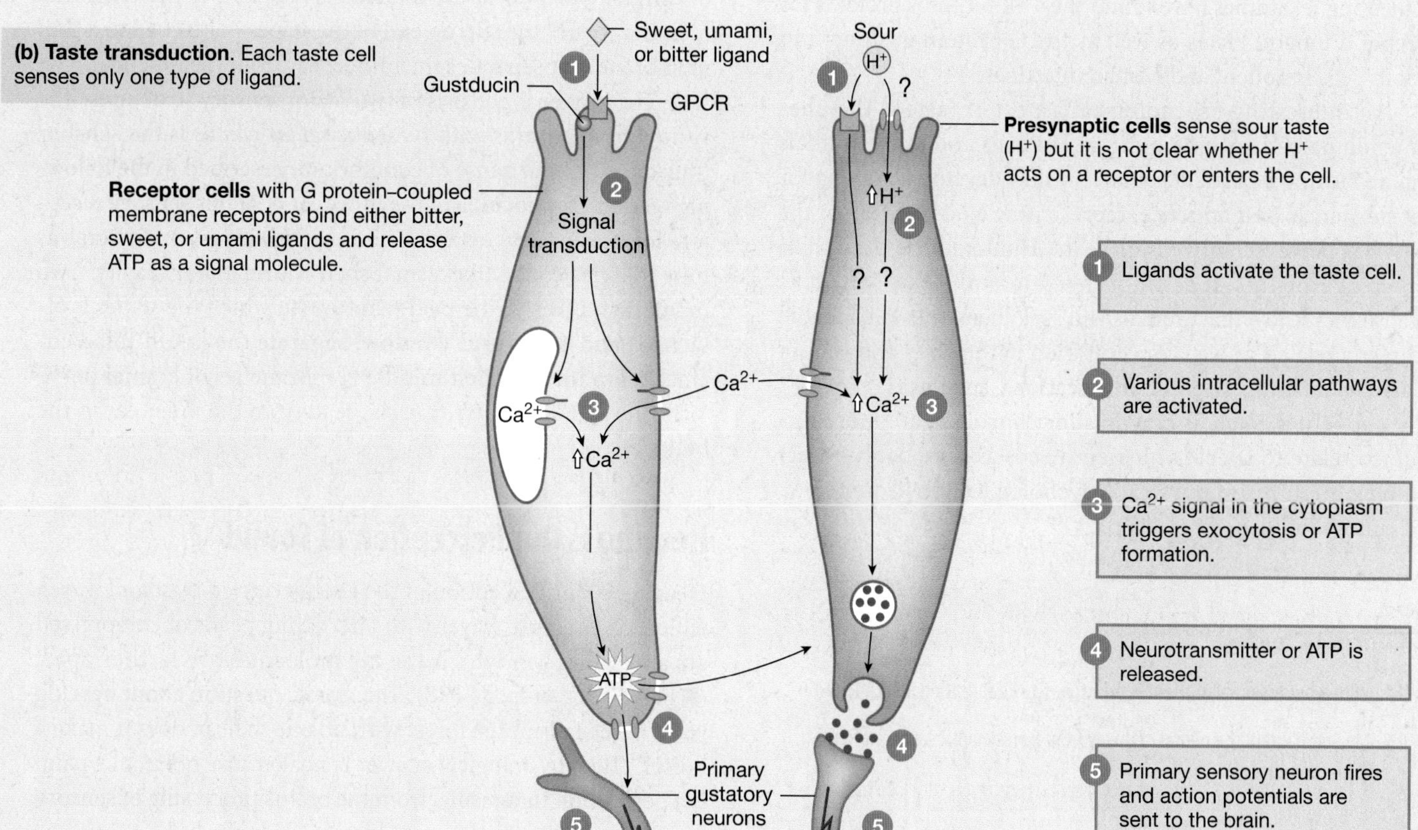

(b) Taste transduction. Each taste cell senses only one type of ligand.
Sweet, umami, or bitter ligand
Gustducin
GPCR
1
2
Signal transduction
Receptor cells with G protein–coupled membrane receptors bind either bitter, sweet, or umami ligands and release ATP as a signal molecule.
Ca^{2+}
3
⇧Ca^{2+}
ATP
4
Primary gustatory neurons
5
Sour
H^+
?
Presynaptic cells sense sour taste (H^+) but it is not clear whether H^+ acts on a receptor or enters the cell.
⇧H^+
? ?
⇧Ca^{2+}
1 Ligands activate the taste cell.
2 Various intracellular pathways are activated.
3 Ca^{2+} signal in the cytoplasm triggers exocytosis or ATP formation.
4 Neurotransmitter or ATP is released.
5 Primary sensory neuron fires and action potentials are sent to the brain.

and interprets the taste sensation based on which populations of neurons are responding most strongly. Signals from the sensory neurons also initiate behavioral responses, such as feeding, and feedforward responses [p. 18] that activate the digestive system.

The sensations we call taste are not all mediated by the traditional taste receptors. For years physiologists thought fat in the diet was appealing because of its texture, and food experts use the phrase "mouth feel" to describe the sensation of eating something fatty, like ice cream, that seems to coat the inside of the mouth. But now it appears that the tongue may have taste receptors for long-chain fatty acids, such as oleic acid [Fig. 2.1, p. 33].

Research in rodents has identified a membrane receptor called *CD36* that lines taste pores and binds fats. Activation of this receptor helps trigger the feedforward digestive reflexes that prepare the digestive system for a meal. Currently evidence is lacking for a similar receptor in humans, but "fatty" may turn out to be a sixth taste sensation.

Yet other taste sensations are related to somatosensory pathways. Nerve endings in the mouth have TRP receptors and carry spicy sensations through the *trigeminal nerve* (CN V). *Capsaicin* from chili peppers, menthol from mint, and molecules in cinnamon, mustard oil, and many Indian spices activate these receptors to add to our appreciation of the food we eat.

And what would you say to the idea of taste buds in your gut? Scientists have known for years that the stomach and intestines have the ability to sense the composition of a meal and secrete appropriate hormones and enzymes. Now it appears that gut chemoreception is being mediated by the same receptors and signal transduction mechanisms that occur in taste buds on the tongue. Studies have found the T1R receptor proteins for sweet and umami tastes as well as the G protein gustducin in various cells in rodent and human intestines.

An interesting psychological aspect of taste is the phenomenon named **specific hunger**. Humans and other animals that are lacking a particular nutrient may develop a craving for that substance. **Salt appetite**, representing a lack of Na^+ in the body, has been recognized for years. Hunters have used their knowledge of this specific hunger to stake out salt licks because they know that animals will seek them out. Salt appetite is directly related to Na^+ concentration in the body and cannot be assuaged by ingestion of other cations, such as Ca^{2+} or K^+. Other appetites, such as cravings for chocolate, are more difficult to relate to specific nutrient needs and probably reflect complex mixtures of physical, psychological, environmental, and cultural influences.

Concept Check Answers: p. 375

14. With what essential nutrient is the umami taste sensation associated?
15. Map or diagram the neural pathway from a presynaptic taste cell to the gustatory cortex.

The Ear: Hearing

The ear is a sense organ that is specialized for two distinct functions: hearing and equilibrium. It can be divided into external, middle, and inner sections, with the neurological elements housed in and protected by structures in the inner ear. The vestibular complex of the inner ear is the primary sensor for equilibrium. The remainder of the ear is used for hearing.

The *external ear* consists of the outer ear, or **pinna**, and the **ear canal** (■ Fig. 10.15). The pinna is another example of an important accessory structure to a sensory system, and it varies in shape and location from species to species, depending on the animals' survival needs. The ear canal is sealed at its internal end by a thin membranous sheet of tissue called the **tympanic membrane**, or *eardrum*.

The tympanic membrane separates the external ear from the *middle ear*, an air-filled cavity that connects with the pharynx through the **Eustachian tube**. The Eustachian tube is normally collapsed, sealing off the middle ear, but it opens transiently to allow middle ear pressure to equilibrate with atmospheric pressure during chewing, swallowing, and yawning. Colds or other infections that cause swelling can block the Eustachian tube and result in fluid buildup in the middle ear. If bacteria are trapped in the middle ear fluid, the ear infection known as *otitis media* {*oto-*, ear + *-itis*, inflammation + *media*, middle} results.

Three small bones of the middle ear conduct sound from the external environment to the inner ear: the **malleus** {hammer}, the **incus** {anvil}, and the **stapes** {stirrup}. The three bones are connected to one another with the biological equivalent of hinges. One end of the malleus is attached to the tympanic membrane, and the stirrup end of the stapes is attached to a thin membrane that separates the middle ear from the inner ear.

The inner ear consists of two major sensory structures. The *vestibular apparatus* with its *semicircular canals* is the sensory transducer for our sense of equilibrium, described in the following section. The **cochlea** of the inner ear contains sensory receptors for hearing. On external view the cochlea is a membranous tube that lies coiled like a snail shell within a bony cavity. Two membranous disks, the **oval window** (to which the stapes is attached) and the **round window**, separate the liquid-filled cochlea from the air-filled middle ear. Branches of cranial nerve VIII, the *vestibulocochlear nerve*, lead from the inner ear to the brain.

Hearing Is Our Perception of Sound

Hearing is our perception of the energy carried by *sound waves*, which are pressure waves with alternating peaks of compressed air and valleys in which the air molecules are farther apart (■ Fig. 10.16a on page 348). The classic question about hearing is, "If a tree falls in the forest with no one to hear, does it make a noise?" The physiological answer is no, because noise, like pain, is a perception that results from the brain's processing of sensory

■ Fig. 10.15 *ANATOMY SUMMARY*

The Ear

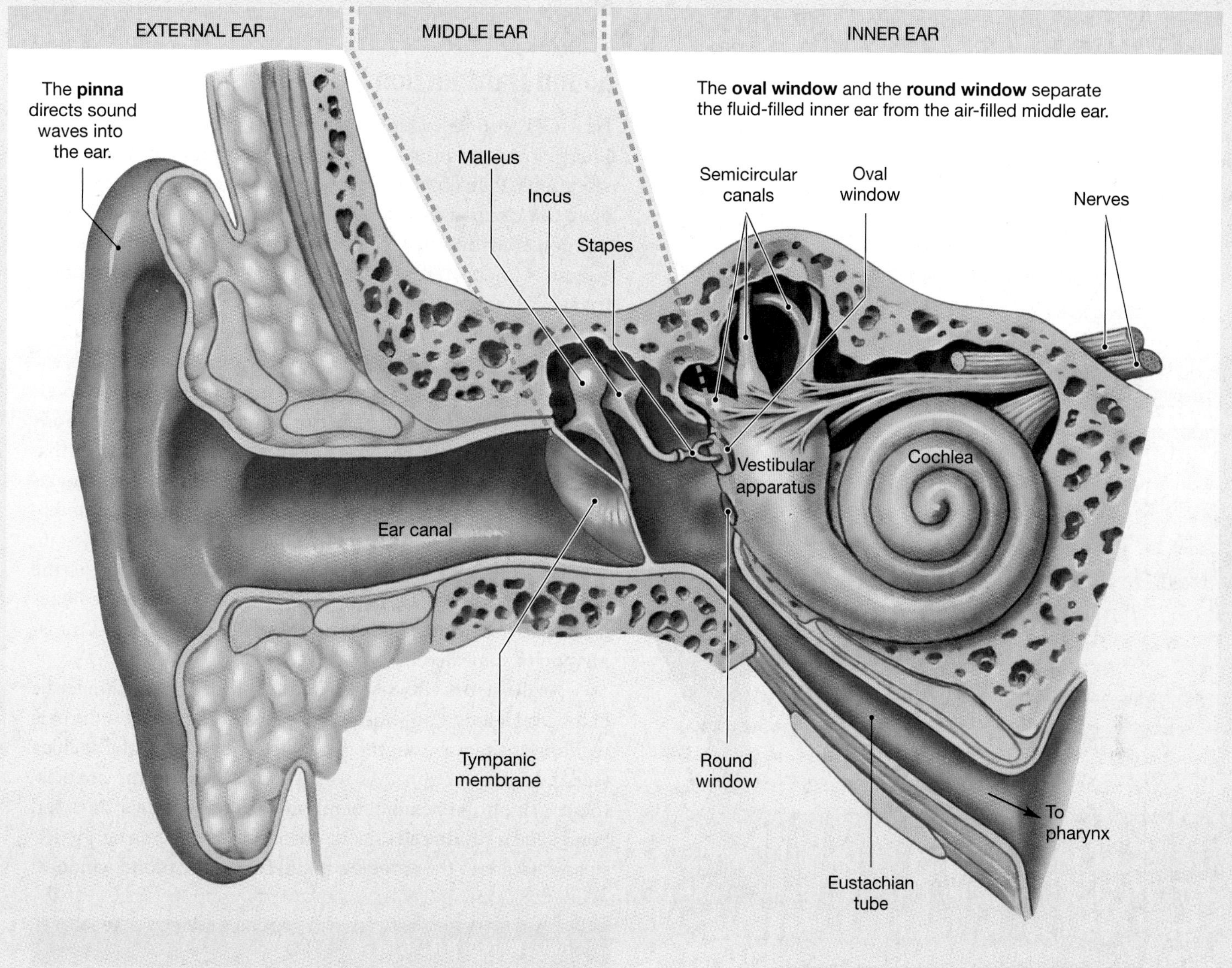

information. A falling tree emits sound waves, but there is no noise unless someone or something is present to process and perceive the wave energy as sound.

Sound is the brain's interpretation of the frequency, amplitude, and duration of sound waves that reach our ears. Our brains translate **frequency** of sound waves (the number of wave peaks that pass a given point each second) into the **pitch** of a sound. Low-frequency waves are perceived as low-pitched sounds, such as the rumble of distant thunder. High-frequency waves create high-pitched sounds, such as the screech of fingernails on a blackboard.

Sound wave frequency (Fig. 10.16b) is measured in waves per second, or **hertz (Hz)**. The average human ear can hear sounds over the frequency range of 20–20,000 Hz, with the most acute hearing between 1000–3000 Hz. Our hearing is not as acute as that of many other animals, just as our sense of smell is less acute. Bats listen for ultra-high-frequency sound waves (in the kilohertz range) that bounce off objects in the dark. Elephants and some birds can hear sounds in the infrasound (very low frequency) range.

Loudness is our interpretation of sound intensity and is influenced by the sensitivity of an individual's ear. The intensity of a sound wave is a function of the wave **amplitude** (Fig. 10.16b). Intensity is measured on a logarithmic scale in units called **decibels (dB)**. Each 10 dB increase represents a 10-fold increase in intensity.

SOUND WAVES

(a) Sound waves alternate peaks of compressed air and valleys where the air is less compressed.

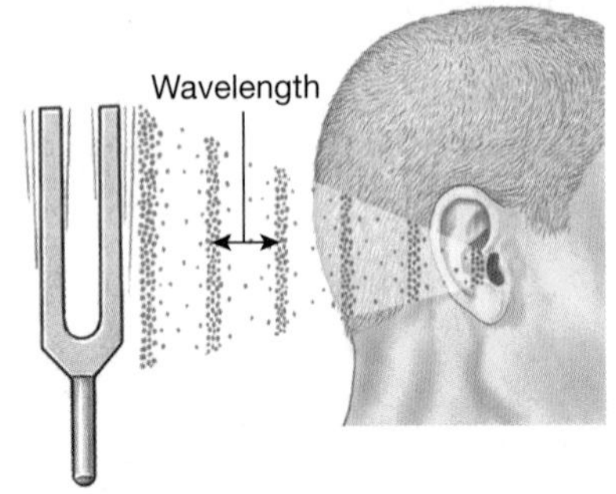

Wavelength

Tuning fork

(b) Sound waves are distinguished by their amplitude, measured in decibels (dB), and frequency, measured in hertz (Hz).

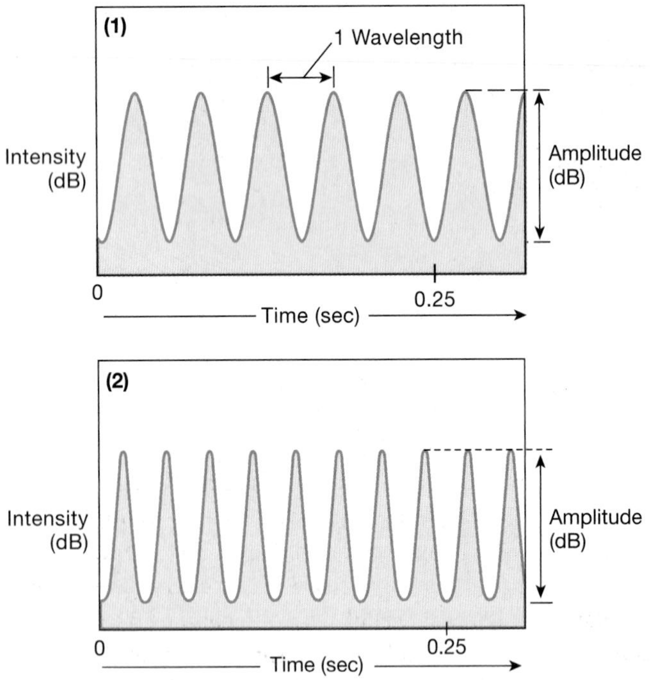

(1) 1 Wavelength; Intensity (dB); Amplitude (dB); Time (sec) 0 0.25

(2) Intensity (dB); Amplitude (dB); Time (sec) 0 0.25

FIGURE QUESTIONS

1. What are the frequencies of the sound waves in graphs (1) and (2) in Hz (waves/second)?
2. Which set of sound waves would be interpreted as having lower pitch?

■ Fig. 10.16

Normal conversation has a typical noise level of about 60 dB. Sounds of 80 dB or more can damage the sensitive hearing receptors of the ear, resulting in hearing loss. A typical heavy metal rock concert has noise levels around 120 dB, an intensity that puts listeners in immediate danger of damage to their hearing. The amount of damage depends on the duration and frequency of the noise as well as its intensity.

Concept Check Answer: p. 375

16. What is a kilohertz?

Sound Transduction Is a Multistep Process

Hearing is a complex sense that involves multiple transductions. Energy from sound waves in the air becomes (1) mechanical vibrations, then (2) fluid waves in the cochlea. The fluid waves open ion channels in *hair cells*, the sensory receptors for hearing. Ion flow into hair cells creates (3) electrical signals that release (4) neurotransmitter (chemical signal), which in turn triggers (5) action potentials in the primary auditory neurons.

These transduction steps are shown in ■ Figure 10.17. Sound waves striking the outer ear are directed down the ear canal until they hit the tympanic membrane and cause it to vibrate (first transduction). The tympanic membrane vibrations are transferred to the malleus, the incus, and the stapes, in that order. The arrangement of the three connected middle ear bones creates a "lever" that multiplies the force of the vibration (*amplification*) so that very little sound energy is lost due to friction. If noise levels are so high that there is danger of damage to the inner ear, small muscles in the middle ear can pull on the bones to decrease their movement and thereby dampen sound transmission to some degree.

As the stapes vibrates, it pulls and pushes on the thin tissue of the oval window, to which it is attached. Vibrations at the oval window create waves in the fluid-filled channels of the cochlea (second transduction). As waves move through the cochlea, they push on the flexible membranes of the *cochlear duct* and bend sensory **hair cells** inside the duct. The wave energy dissipates back into the air of the middle ear at the round window.

RUNNING PROBLEM

Anant reports to the otolaryngologist that he never knows when his attacks of dizziness will strike and that they last from 10 minutes to an hour. They often cause him to vomit. He also reports that he has a persistent low buzzing sound in one ear and that he does not seem to hear low tones as well as he could before the attacks started. The buzzing sound (tinnitus) often gets worse during his dizzy attacks.

Q2: Subjective tinnitus occurs when an abnormality somewhere along the anatomical pathway for hearing causes the brain to perceive a sound that does not exist outside the auditory system. Starting from the ear canal, name the auditory structures in which problems may arise.

SOUND TRANSMISSION THROUGH THE EAR

1. Sound waves strike the tympanic membrane and become vibrations.
2. The sound wave energy is transferred to the three bones of the middle ear, which vibrate.
3. The stapes is attached to the membrane of the oval window. Vibrations of the oval window create fluid waves within the cochlea.
4. The fluid waves push on the flexible membranes of the cochlear duct. Hair cells bend and ion channels open, creating an electrical signal that alters neurotransmitter release.
5. Neurotransmitter release onto sensory neurons creates action potentials that travel through the cochlear nerve to the brain.
6. Energy from the waves transfers across the cochlear duct into the tympanic duct and is dissipated back into the middle ear at the round window.

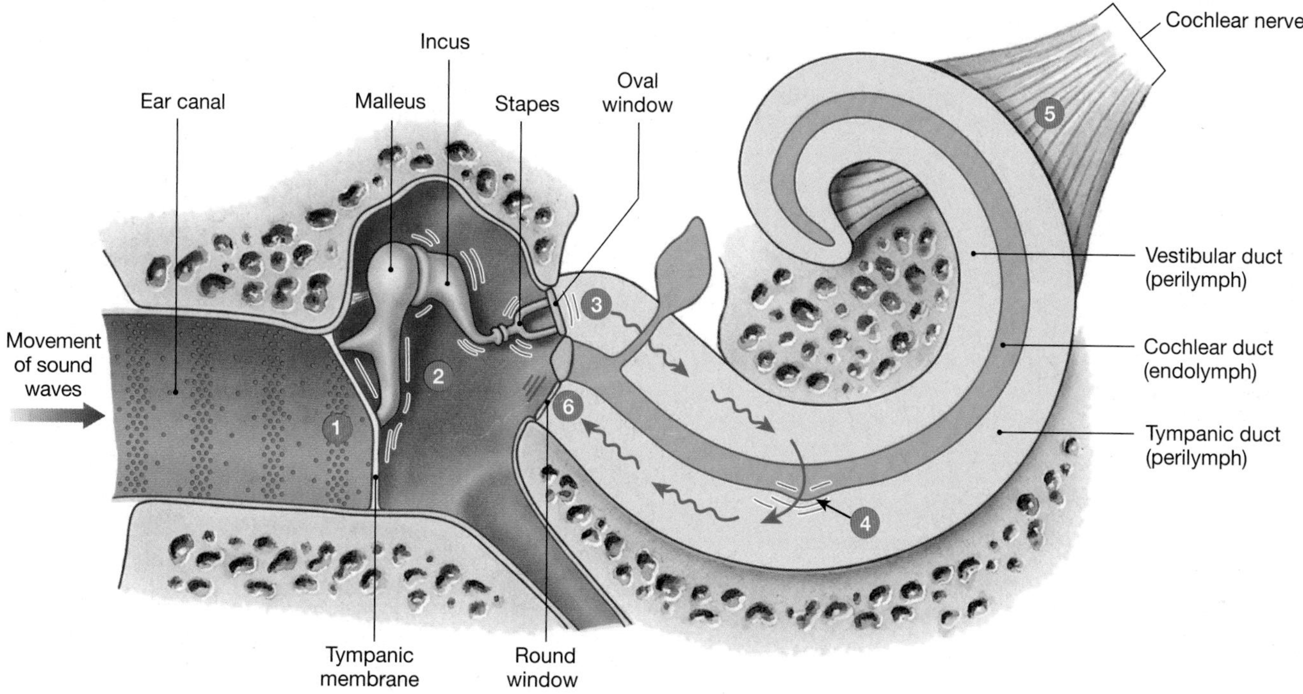

■ **Fig. 10.17**

Movement of the cochlear duct opens or closes ion channels on hair cell membranes, creating electrical signals (third transduction). These electrical signals alter neurotransmitter release (fourth transduction). Neurotransmitter binding to the primary auditory neurons initiates action potentials (fifth transduction) that send coded information about sound through the *cochlear branch of the vestibulocochlear nerve* (cranial nerve VIII) and the brain.

The Cochlea Is Filled with Fluid

The transduction of wave energy into action potentials takes place in the cochlea of the inner ear. Uncoiled, the cochlea can be seen to be composed of three parallel, fluid-filled channels: (1) the **vestibular duct**, or *scala vestibuli* {*scala*, stairway; *vestibulum*, entrance}; (2) the central **cochlear duct**, or *scala media* {*media*, middle}; and (3) the **tympanic duct**, or *scala tympani* {*tympanon*, drum} (■ Fig. 10.18). The vestibular and tympanic ducts are continuous with each other, and they connect at the tip of the cochlea through a small opening known as the **helicotrema** {*helix*, a spiral + *trema*, hole}. The cochlear duct is a dead-end tube, but it connects to the vestibular apparatus through a small opening.

The fluid in the vestibular and tympanic ducts is similar in ion composition to plasma and is known as **perilymph**. The cochlear duct is filled with **endolymph** secreted by epithelial cells in the duct. Endolymph is unusual because it is more like intracellular fluid than extracellular fluid in composition, with high concentrations of K^+ and low concentrations of Na^+.

The cochlear duct contains the **organ of Corti**, composed of hair cell receptors and support cells. The organ of Corti sits on the **basilar membrane** and is partially covered by the **tectorial membrane** {*tectorium*, a cover}, both flexible tissues that move in response to fluid waves passing through the vestibular duct (Fig. 10.18). As the waves travel through the cochlea, they displace basilar and tectorial membranes, creating up-and-down oscillations that bend the hair cells.

Hair cells, like taste cells, are non-neural receptor cells. The apical surface of each hair cell is modified into 50–100 stiffened cilia known as **stereocilia**, arranged in ascending height (■ Fig. 10.19a). The stereocilia of the hair cells are embedded in the overlying tectorial membrane. If the tectorial membrane moves, the underlying cilia do also.

When hair cells move in response to sound waves, their stereocilia flex, first one way, then the other. The stereocilia are

The Cochlea

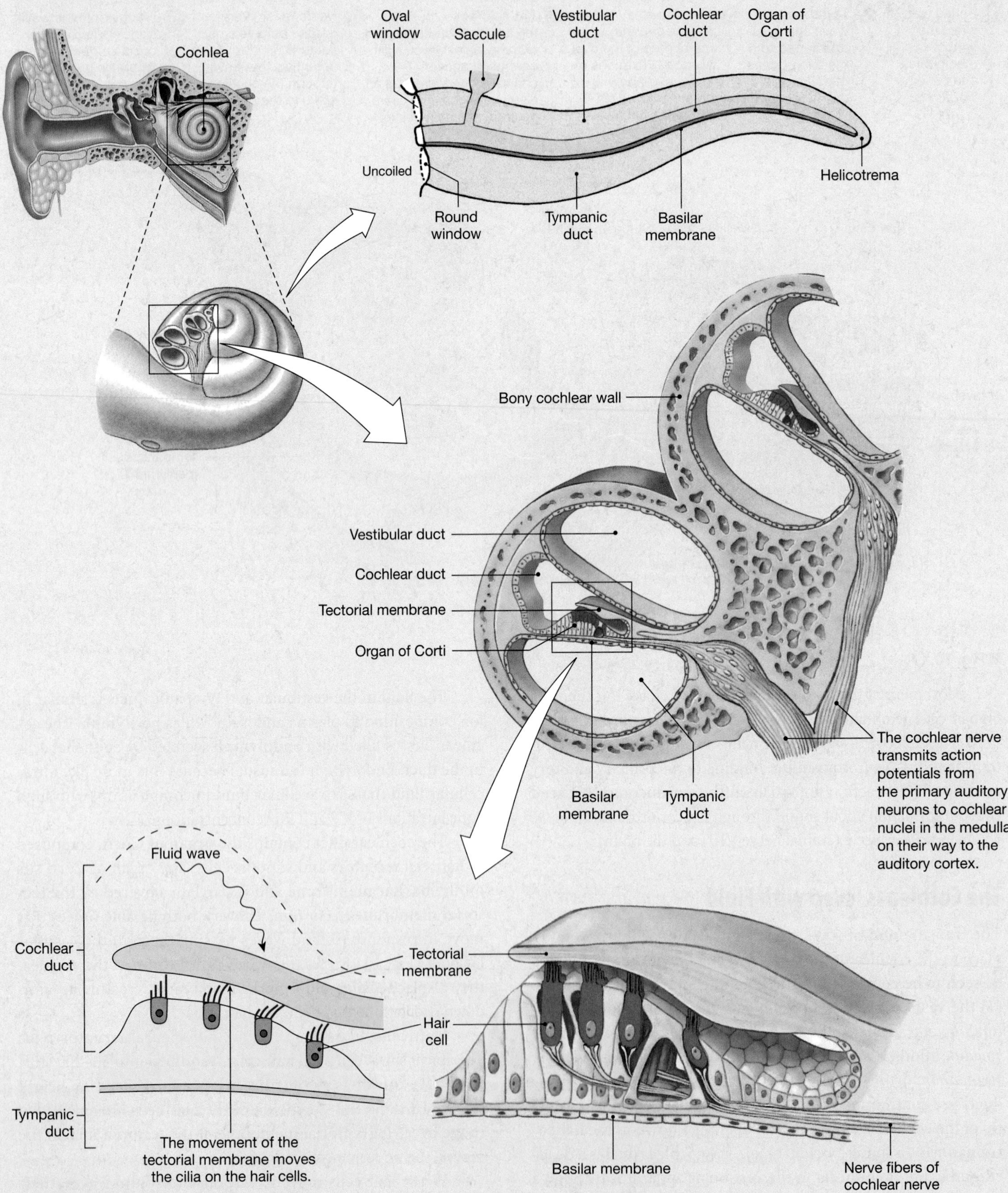
Cochlea
Oval window
Saccule
Vestibular duct
Cochlear duct
Organ of Corti
Helicotrema
Uncoiled
Round window
Tympanic duct
Basilar membrane
Bony cochlear wall
Vestibular duct
Cochlear duct
Tectorial membrane
Organ of Corti
Basilar membrane
Tympanic duct
The cochlear nerve transmits action potentials from the primary auditory neurons to cochlear nuclei in the medulla, on their way to the auditory cortex.
Fluid wave
Cochlear duct
Tectorial membrane
Hair cell
Tympanic duct
The movement of the tectorial membrane moves the cilia on the hair cells.
Basilar membrane
Nerve fibers of cochlear nerve

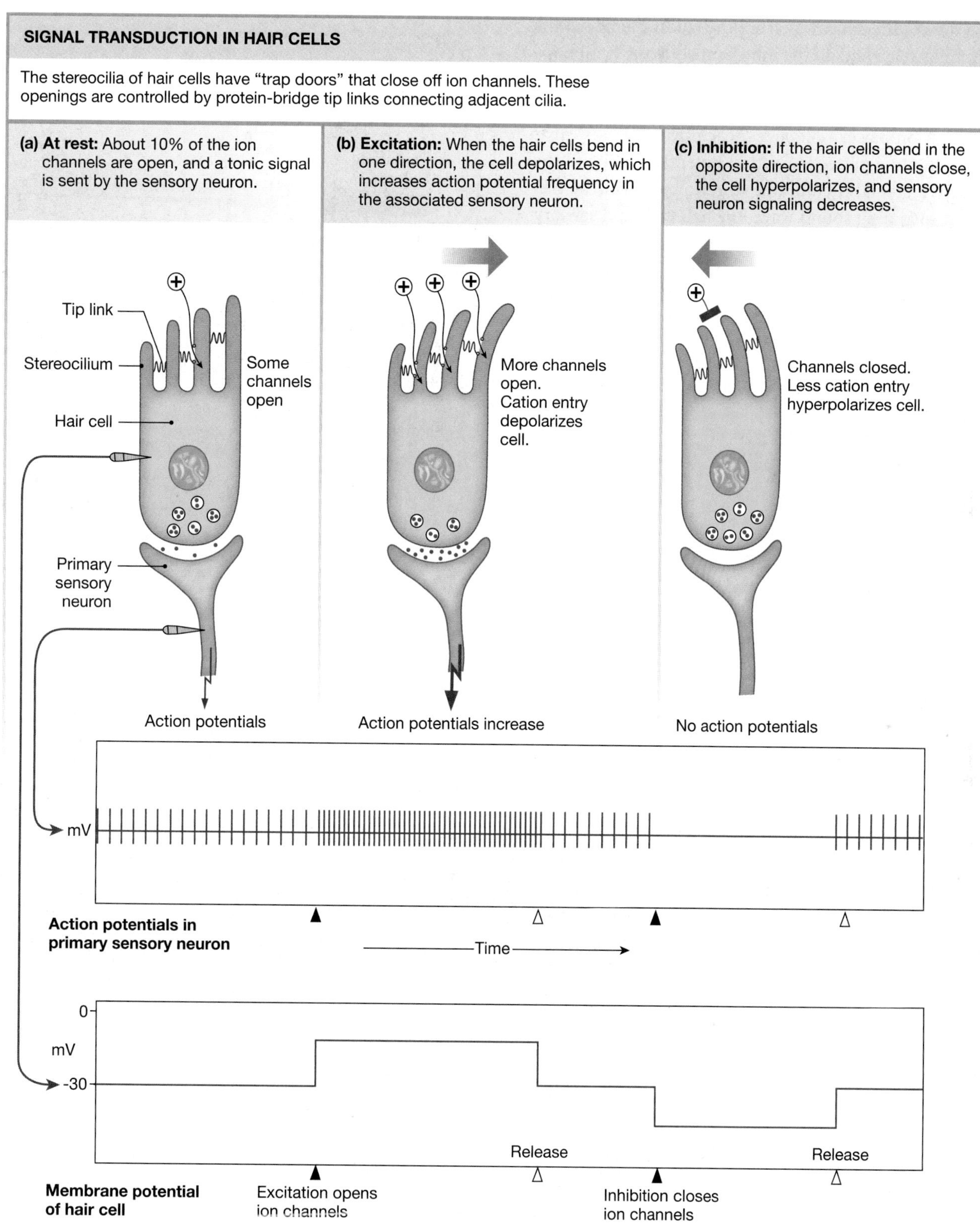

SIGNAL TRANSDUCTION IN HAIR CELLS
The stereocilia of hair cells have "trap doors" that close off ion channels. These openings are controlled by protein-bridge tip links connecting adjacent cilia.
(a) At rest: About 10% of the ion channels are open, and a tonic signal is sent by the sensory neuron.
(b) Excitation: When the hair cells bend in one direction, the cell depolarizes, which increases action potential frequency in the associated sensory neuron.
(c) Inhibition: If the hair cells bend in the opposite direction, ion channels close, the cell hyperpolarizes, and sensory neuron signaling decreases.
Tip link
Stereocilium
Hair cell
Some channels open
Primary sensory neuron
Action potentials
More channels open.
Cation entry depolarizes cell.
Action potentials increase
Channels closed.
Less cation entry hyperpolarizes cell.
No action potentials
mV
Action potentials in primary sensory neuron
Time
0
mV
-30
Release
Release
Membrane potential of hair cell
Excitation opens ion channels
Inhibition closes ion channels

Fig. 10.19

attached to each other by protein bridges called *tip links*. The tip links act like little springs and are connected to gates that open and close ion channels in the cilia membrane. When the hair cells and cilia are in a neutral position, about 10% of the ion channels are open, and there is a low tonic level of neurotransmitter released onto the primary sensory neuron.

When waves deflect the tectorial membrane so that cilia bend toward the tallest members of a bundle, the tip links pop more channels open, so cations (primarily K^+ and Ca^{2+}) enter the cell, which then depolarizes (Fig. 10.19b). Voltage-gated Ca^{2+} channels open, neurotransmitter release increases, and the sensory neuron increases its firing rate. When the tectorial membrane pushes the cilia away from the tallest members, the springy tip links relax and all the ion channels close. Cation influx slows, the membrane hyperpolarizes, less transmitter is released, and sensory neuron firing decreases (Fig. 10.19c).

The vibration pattern of waves reaching the inner ear is thus converted into a pattern of action potentials going to the CNS. Because tectorial membrane vibrations reflect the frequency of the incoming sound wave, the hair cells and sensory neurons must be able to respond to sounds of nearly 20,000 waves per second, the highest frequency audible by a human ear.

Concept Check Answer: p. 376

17. Normally when cation channels on a cell open, either Na^+ or Ca^{2+} enters the cell. Why does K^+ rather than Na^+ enter hair cells when their cation channels open?

Sounds Are Processed First in the Cochlea

The auditory system processes sound waves so that they can be discriminated by location, pitch, and loudness. Localization of sound is a complex process that requires sensory input from both ears coupled with sophisticated computation by the brain (see Fig. 10.4). In contrast, the initial processing for pitch and loudness takes place in the cochlea of each ear.

Coding sound for pitch is primarily a function of the basilar membrane. This membrane is stiff and narrow near its attachment between the round and oval windows but widens and becomes more flexible near its distal end (■ Fig. 10.20a). High-frequency waves entering the vestibular duct create maximum displacement of the basilar membrane close to the oval window and consequently are not transmitted very far along the cochlea. Low-frequency waves travel along the length of the basilar membrane and create their maximum displacement near the flexible distal end.

This differential response to frequency transforms the temporal aspect of frequency (number of sound waves per second) into spatial coding for pitch by location along the basilar membrane (Fig. 10.20b). A good analogy is a piano keyboard, where the location of a key tells you its pitch. The spatial coding of the basilar membrane is preserved in the auditory cortex as neurons project from hair cells to corresponding regions in the brain. Loudness is coded by the ear in the same way that signal strength is coded in somatic receptors. The louder the noise, the more rapidly action potentials fire in the sensory neuron.

Auditory Pathways Project to the Auditory Cortex

Once the cochlea transforms sound waves into electrical signals, sensory neurons transfer this information to the brain. The cochlear (auditory) nerve is a branch of cranial nerve VIII, the *vestibulocochlear nerve* [p. 302]. Primary auditory neurons project from the cochlea to *cochlear nuclei* in the medulla oblongata (■ Fig. 10.21). Some of these neurons carry information that is processed into the timing of sound, and others carry information that is processed into the sound quality.

From the medulla, secondary sensory neurons project to two higher nuclei, one *ipsilateral* (on the same side of the body) and one *contralateral* (on the opposite side). Splitting sound signals between two ascending tracts means that each side of the brain gets information from both ears. These ascending tracts then synapse in nuclei in the midbrain and thalamus before projecting

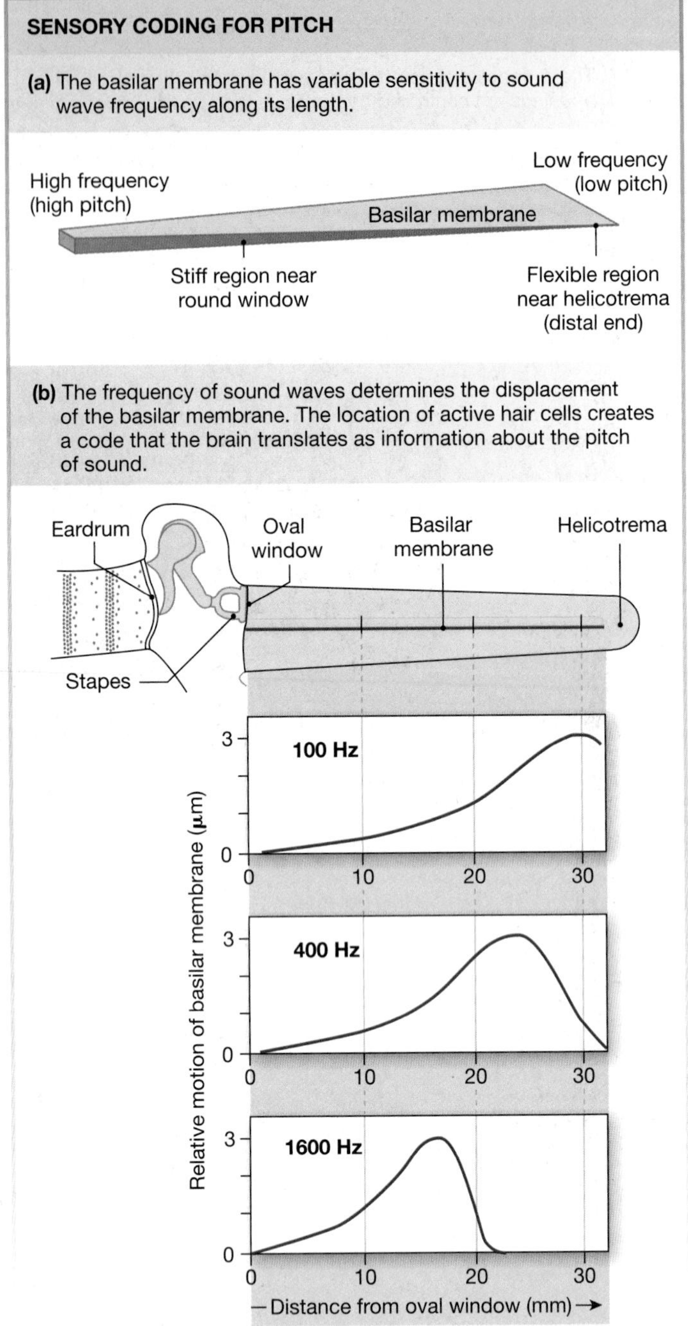

■ **Fig. 10.20**

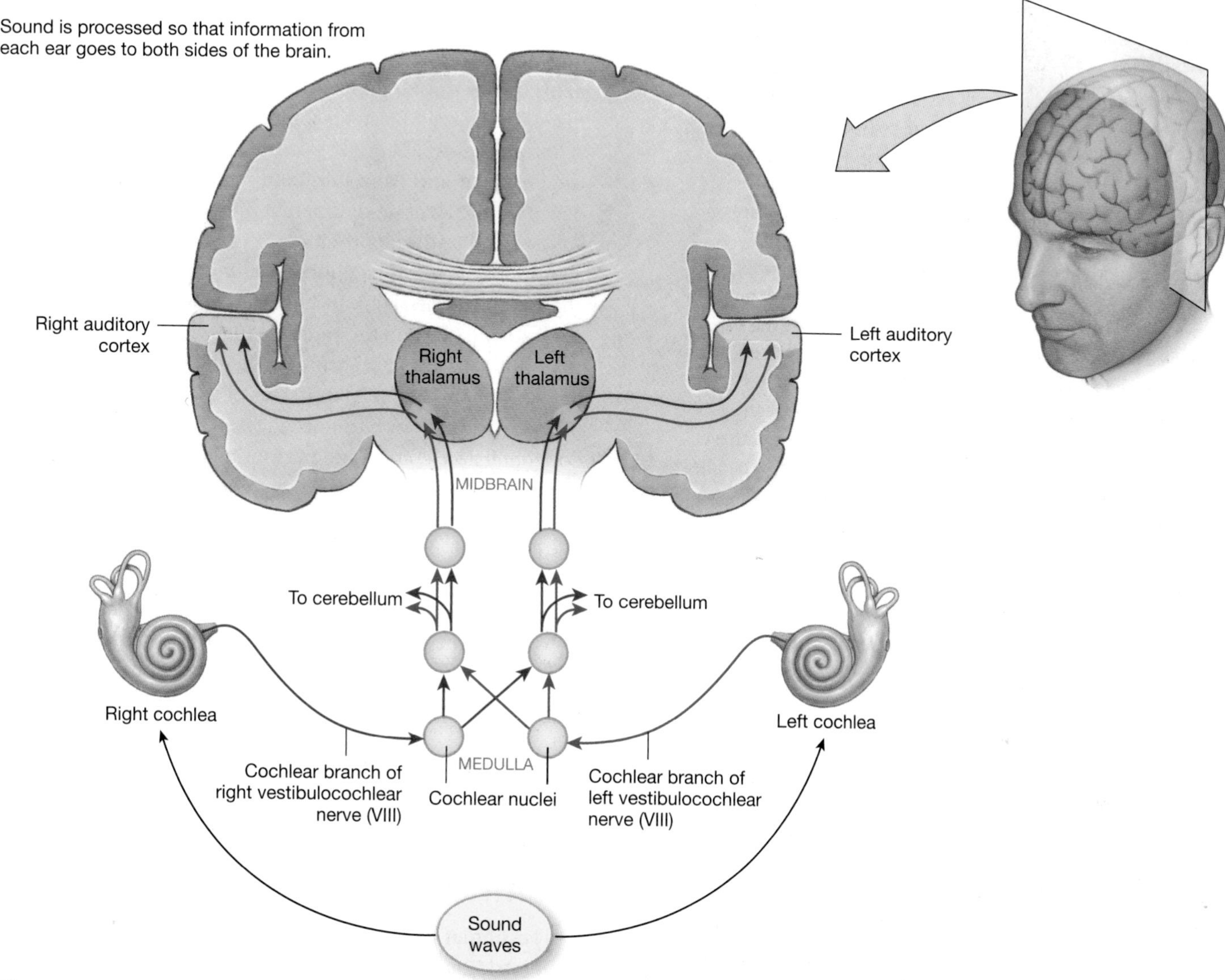

Fig. 10.21

to the auditory cortex (see Fig. 10.3). Collateral pathways take information to the reticular formation and the cerebellum.

The localization of a sound source is an integrative task that requires simultaneous input from both ears. Unless sound is coming from directly in front of a person, it will not reach both ears at the same time (see Fig. 10.4). The brain records the time differential for sound arriving at the ears and uses complex computation to create a three-dimensional representation of the sound source.

Hearing Loss May Result from Mechanical or Neural Damage

There are three forms of hearing loss: conductive, central, and sensorineural. In *conductive hearing loss,* sound cannot be transmitted through either the external ear or the middle ear. The causes of conductive hearing loss range from an ear canal plugged with earwax (*cerumen*), to fluid in the middle ear from an infection, to diseases or trauma that impede vibration of the malleus, incus, or stapes. Correction of conductive hearing loss includes microsurgical techniques in which the bones of the middle ear can be reconstructed.

Central hearing loss results either from damage to the neural pathways between the ear and cerebral cortex or from damage to the cortex itself, as might occur from a stroke. This form of hearing loss is relatively uncommon.

Sensorineural hearing loss arises from damage to the structures of the inner ear, including death of hair cells as a result of loud noises. The loss of hair cells in mammals is currently irreversible. Birds and lower vertebrates, however, are able to regenerate hair cells to replace those that die. This discovery has researchers exploring strategies to duplicate the process in mammals, including transplantation of neural stem cells and

BIOTECHNOLOGY

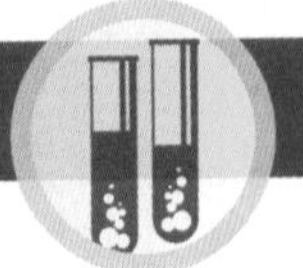

Artificial Ears

One technique used to treat sensorineural hearing loss is the cochlear implant. The newest cochlear implants have multiple components. Externally, a microphone, tiny computerized speech processor, and transmitter fit behind the ear like a conventional hearing aid. The speech processor is a transducer that converts sound into electrical impulses. The transmitter converts the processor's electrical impulses into radio waves and sends these signals to a receiver and 8–24 electrodes, which are surgically placed under the skin. The electrodes take electrical signals directly into the cochlea or to the auditory nerve, bypassing any damaged areas. After surgery, recipients go through therapy so that they can learn to understand the sounds they hear. Cochlear implants have been remarkably successful for many profoundly deaf people, allowing them to hear loud noises and modulate their own voices. In the most successful cases, individuals can even use the telephone. To learn more about cochlear implants, visit the web site of the National Institute for Deafness and Other Communication Disorders (*www.nidcd.nih.gov/health/hearing*).

gene therapy to induce nonsensory cells to differentiate into hair cells.

Therapy that replaces hair cells would be an important advance. The incidence of hearing loss in younger people is increasing because of prolonged exposure to rock music and environmental noises. Ninety percent of hearing loss in the elderly—called *presbycusis* {*presbys*, old man + *akoustikos*, able to be heard}—is sensorineural. Currently the primary treatment for sensorineural hearing loss is the use of hearing aids, but amazing results have been obtained with cochlear implants attached to tiny computers (see Biotechnology box).

Hearing is probably our most important social sense. Suicide rates are higher among deaf people than among those who have lost their sight. More than any other sense, hearing connects us to other people and to the world around us.

Concept Check

Answers: p. 376

18. Map or diagram the pathways followed by a sound wave entering the ear, starting in the air at the outer ear and ending on the auditory cortex.

19. Why is somatosensory information projected to only one hemisphere of the brain but auditory information is projected to both hemispheres? (*Hint:* See Figs. 10.4 and 10.8.)

20. Would a cochlear implant help a person who suffers from nerve deafness? From conductive hearing loss?

The Ear: Equilibrium

Equilibrium is a state of balance, whether the word is used to describe ion concentrations in body fluids or the position of the body in space. The special sense of equilibrium has two components: a dynamic component that tells us about our movement through space, and a static component that tells us if our head is not in its normal upright position. Sensory information from the inner ear and from joint and muscle proprioceptors tells our brain the location of different body parts in relation to one another and to the environment. Visual information also plays an important role in equilibrium, as you know if you have ever gone to one of the 360° movie theaters where the scene tilts suddenly to one side and the audience tilts with it!

Our sense of equilibrium is mediated by hair cells lining the fluid-filled vestibular apparatus of the inner ear. These nonneural receptors respond to changes in rotational, vertical, and horizontal acceleration and positioning. The hair cells function just like those of the cochlea, but gravity and acceleration rather than sound waves provide the force that moves the stereocilia. Vestibular hair cells have a single long cilium called a **kinocilium** {*kinein*, to move} located at one side of the ciliary bundle. The kinocilium creates a reference point for the direction of bending.

When the cilia bend, tip links between them open and close ion channels. Movement in one direction causes the hair cells to depolarize; with movement in the opposite direction, they hyperpolarize. This is similar to what happens in cochlear hair cells (see Fig. 10.19).

The Vestibular Apparatus Provides Information about Movement and Position

The **vestibular apparatus**, also called the *membranous labyrinth*, is an intricate series of interconnected fluid-filled chambers. (In Greek mythology the labyrinth was a maze that housed a monster called the Minotaur.) In humans, the vestibular apparatus consists of two saclike **otolith organs**—the **saccule** and the **utricle**—along with three **semicircular canals** that connect to the utricle at their bases (■ Fig. 10.22a). The otolith organs tell us about *linear acceleration* and head position. The three semicircular canals sense *rotational acceleration* in various directions.

The vestibular apparatus, like the cochlear duct, is filled with high-K^+, low-Na^+ endolymph secreted by epithelial cells. Like cerebrospinal fluid, endolymph is secreted continuously and drains from the inner ear into the venous sinus in the dura mater of the brain.

If endolymph production exceeds the drainage rate, buildup of fluid in the inner ear may increase fluid pressure within the vestibular apparatus. Excessive accumulation of endolymph is believed to contribute to *Ménière's disease*, a condition marked by episodes of dizziness and nausea. If the organ of Corti in the cochlear duct is damaged by fluid pressure within the vestibular apparatus, hearing loss may result.

Equilibrium

The vestibular apparatus of the inner ear responds to changes in the body's position in space. The cristae are sensory receptors for rotational acceleration. The maculae are sensory receptors for linear acceleration and head position.

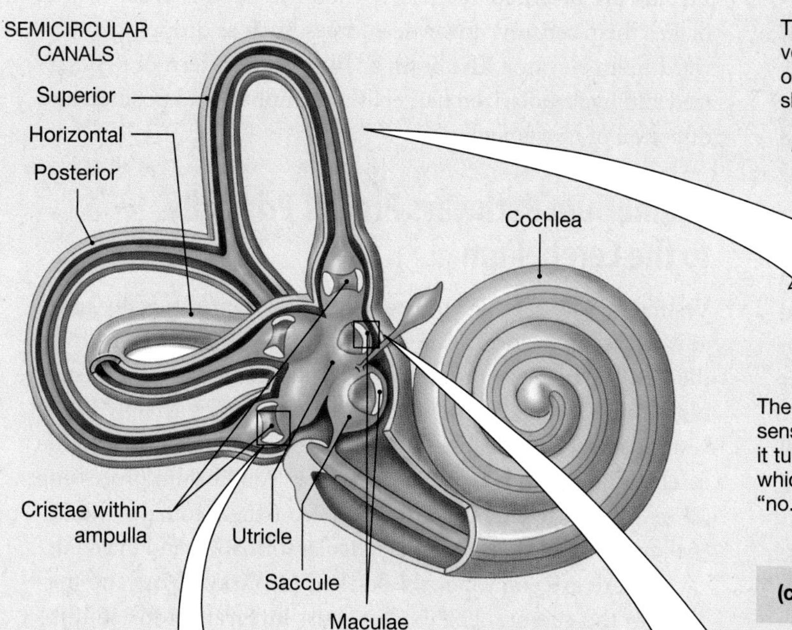

(a) Semicircular Canals

The posterior canal of the vestibular apparatus senses the tilt of the head toward the right or left shoulder.

Left ⟷ *right*

The superior canal senses rotation of the head from front to back, such as that which occurs when nodding "yes."

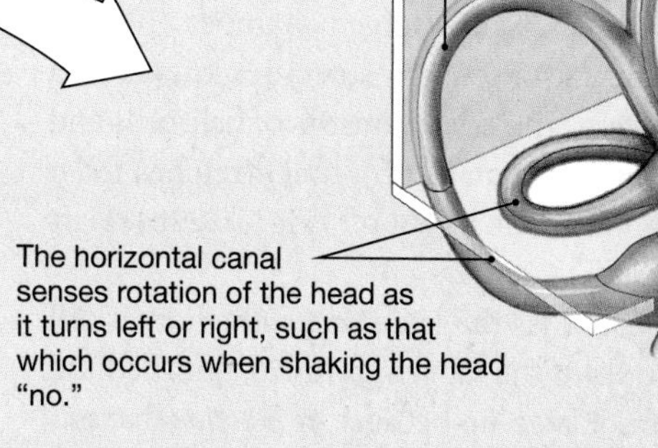

The horizontal canal senses rotation of the head as it turns left or right, such as that which occurs when shaking the head "no."

(b) Crista

Movement of the endolymph pushes on the gelatinous cupula and activates the hair cells.

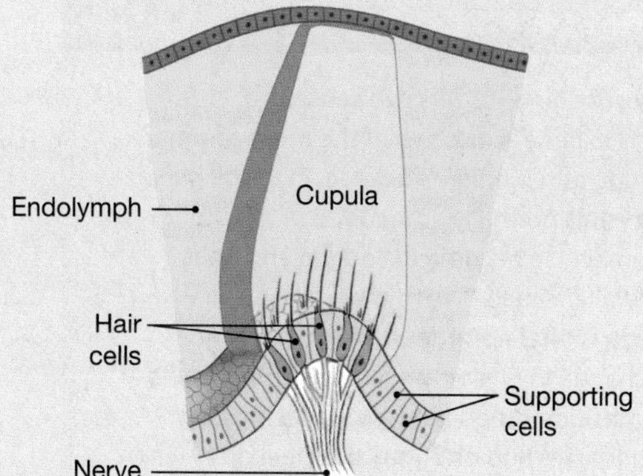

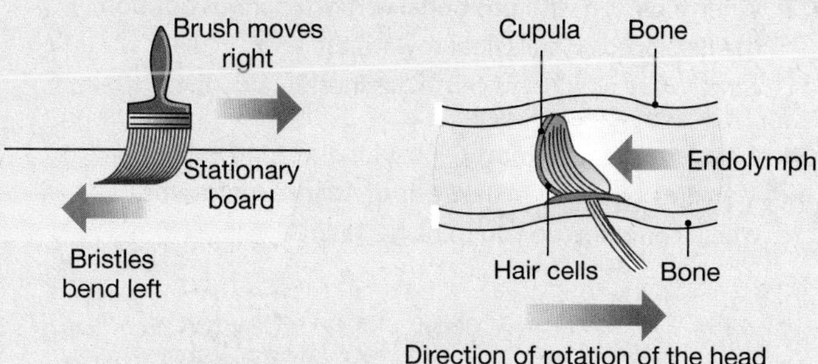

When the head turns right, endolymph pushes the cupula to the left.

(c) Macula

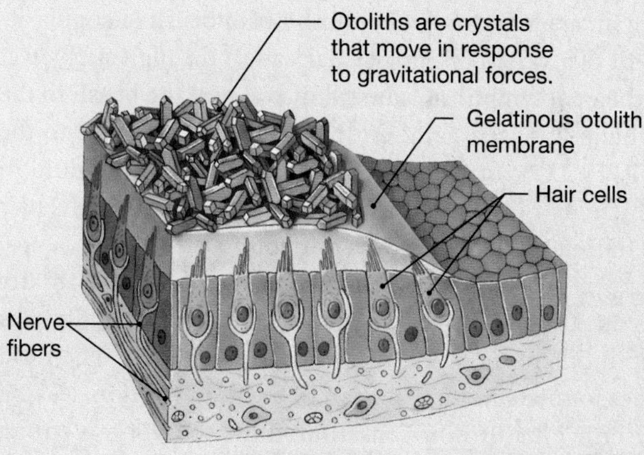

Head in neutral position

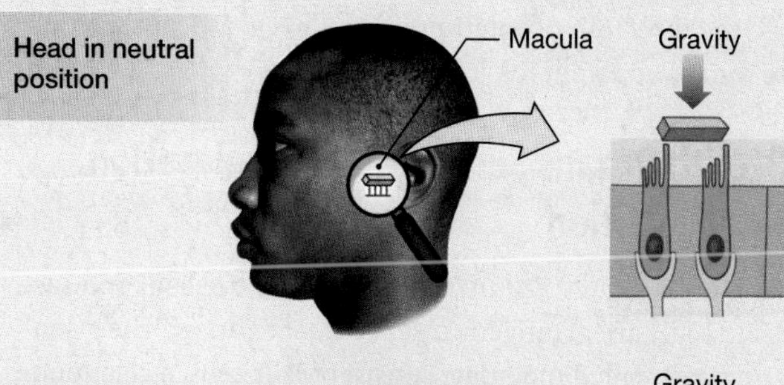

Head tilted posteriorly

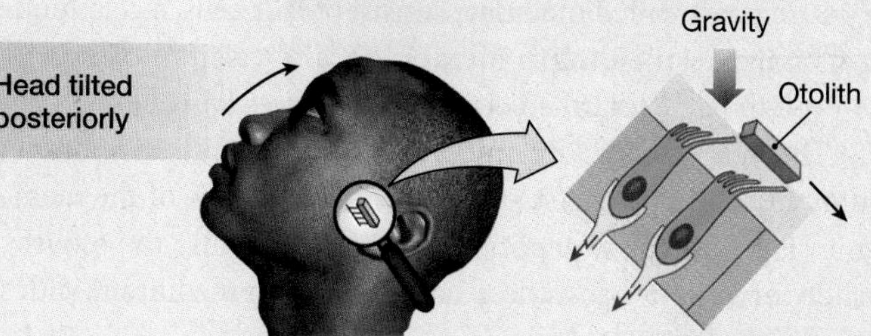

The Semicircular Canals Sense Rotational Acceleration

The three semicircular canals of the vestibular apparatus monitor rotational acceleration. They are oriented at right angles to one another, like three planes that come together to form the corner of a box (Fig. 10.22a). The horizontal canal monitors rotations that we associate with turning, such as an ice skater's spin or shaking your head left and right to say "no." The posterior canal monitors left-to-right rotation, such as the rotation when you tilt your head toward your shoulders or perform a cartwheel. The superior canal is sensitive to forward and back rotation, such as nodding your head front to back or doing a somersault.

At one end of each canal is an enlarged chamber, the **ampulla** {bottle}, which contains a sensory structure known as a **crista** {a crest; plural *cristae*}. The crista consists of hair cells and a gelatinous mass, the **cupula** {small tub}, that stretches from floor to ceiling of the ampulla, closing it off (Fig. 10.22b). Hair cell cilia are embedded in the cupula.

How is rotation sensed? As the head turns, the bony skull and the membranous walls of the labyrinth move, but the fluid within the labyrinth cannot keep up because of *inertia* (the tendency of a body at rest to remain at rest). In the ampullae, the drag of endolymph bends the cupula and its hair cells in the direction *opposite* to the direction in which the head is turning.

For an analogy, think of pulling a paintbrush (a cupula attached to the wall of a semicircular canal) through sticky wet paint (the endolymph) on a board. If you pull the brush to the right, the drag of the paint on the bristles bends them to the left (Fig. 10.22b). In the same way, the inertia of the fluid in the semicircular canal pulls the cupula and the cilia of the hair cells to the left when the head turns right.

If rotation continues, the moving endolymph finally catches up. Then if head rotation stops suddenly, the fluid has built up momentum and cannot stop immediately. The fluid continues to rotate in the direction of the head rotation, leaving the person with a turning sensation. If the sensation is strong enough, the person may throw his or her body in the direction opposite the direction of rotation in a reflexive attempt to compensate for the apparent loss of equilibrium.

The Otolith Organs Sense Linear Acceleration and Head Position

The two otolith organs, the utricle {*utriculus*, little bag} and saccule {little sac}, are arranged to sense linear forces. Their sensory structures, called **maculae**, consist of hair cells, a gelatinous mass known as the **otolith membrane**, and calcium carbonate and protein particles called **otoliths** {*oto*, ear + *lithos*, stone}.

The hair cell cilia are embedded in the otolith membrane, and otoliths bind to matrix proteins on the surface of the membrane (Fig. 10.22c). If gravity or acceleration cause the otoliths to slide forward or back, the gelatinous otolith membrane slides with them, bending the hair cell cilia and setting off a signal. For example, the maculae are horizontal when the head is in its normal upright position. If the head tips back, gravity displaces the otoliths, and the hair cells are activated.

The maculae of the utricle sense forward acceleration or deceleration as well as head tilt. In contrast, the maculae of the saccule are oriented vertically when the head is erect, which makes them sensitive to vertical forces, such as dropping downward in an elevator. The brain analyzes the pattern of depolarized and hyperpolarized hair cells to compute head position and direction of movement.

Equilibrium Pathways Project Primarily to the Cerebellum

Vestibular hair cells, like those of the cochlea, are tonically active and release neurotransmitter onto primary sensory neurons of the **vestibular nerve** (a branch of cranial nerve VIII, the vestibulocochlear nerve). Those sensory neurons either synapse in the *vestibular nuclei* of the medulla or run without synapsing to the cerebellum, which is the primary site for equilibrium processing (■ Fig. 10.23). Collateral pathways run from the medulla to the cerebellum or upward through the reticular formation and thalamus.

There are some poorly defined pathways from the medulla to the cerebral cortex, but most integration for equilibrium occurs in the cerebellum. Descending pathways from the vestibular nuclei go to certain motor neurons involved in eye movement. These pathways help keep the eyes locked on an object as the head turns.

RUNNING PROBLEM

Although many vestibular disorders can cause the symptoms Anant is experiencing, two of the most common are positional vertigo and Ménière's disease. In *positional vertigo,* calcium crystals normally embedded in the otolith membrane of the maculae become dislodged and float toward the semicircular canals. The primary symptom of positional vertigo is brief episodes of severe dizziness brought on by a change in position, such as moving to the head-down yoga position called "downward-facing dog." People with positional vertigo often say they feel dizzy when they lie down or turn over in bed.

Q3: When a person with positional vertigo changes position, the displaced crystals float toward the semicircular canals. Why would this cause dizziness?

Q4: Compare the symptoms of positional vertigo and Ménière's disease. On the basis of Anant's symptoms, which condition do you think he has?

326 331 348 356 361 365 371

EQUILIBRIUM PATHWAYS

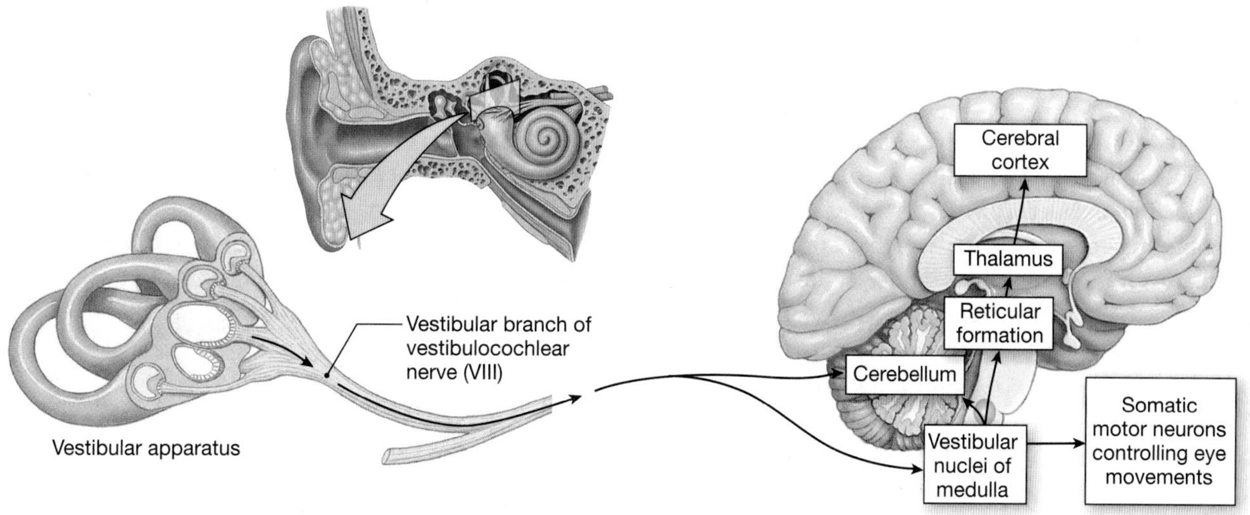

■ **Fig. 10.23**

> **Concept Check** Answers: p. 376
>
> **21.** The stereocilia of hair cells are bathed in endolymph, which has a very high concentration of K^+ and a low concentration of Na^+. When ion channels in the stereocilia open, which ions move in which direction to cause depolarization?
>
> **22.** Why does hearing decrease if an ear infection causes fluid buildup in the middle ear?
>
> **23.** When dancers perform multiple turns, they try to keep their vision fixed on a single point ("spotting"). How does spotting keep a dancer from getting dizzy?

The Eye and Vision

The eye is a sensory organ that functions much like a camera. It focuses light on a light-sensitive surface (the retina) using a lens and an aperture or opening (the pupil) whose size can be adjusted to change the amount of entering light. **Vision** is the process through which light reflected from objects in our environment is translated into a mental image. This process can be divided into three steps:

1. Light enters the eye, and the lens focuses the light on the retina.
2. Photoreceptors of the retina transduce light energy into an electrical signal.
3. Neural pathways from retina to brain process electrical signals into visual images.

The Skull Protects the Eye

The external anatomy of the eye is shown in ■ Figure 10.24. Like sensory elements of the ears, the eyes are protected by a bony cavity, the *orbit,* which is formed by facial bones of the skull.

Accessory structures associated with the eye include six *extrinsic eye muscles,* skeletal muscles that attach to the outer surface of the eyeball and control eye movements. Cranial nerves III, IV, and VI innervate these muscles.

The upper and lower *eyelids* close over the anterior surface of the eye, and the *lacrimal apparatus,* a system of glands and ducts, keeps a continuous flow of tears washing across the exposed surface so that it remains moist and free of debris. Tear secretion is stimulated by parasympathetic neurons from cranial nerve VII.

EXTERNAL ANATOMY OF THE EYE

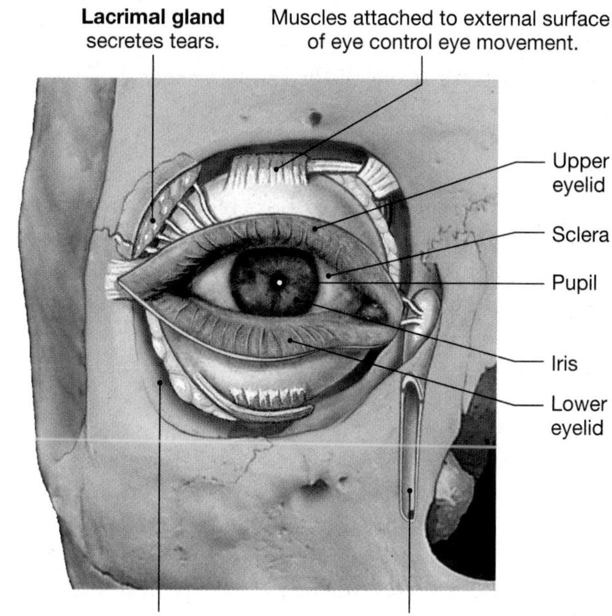

■ **Fig. 10.24**

The **pupil** is an opening through which light can pass into the interior of the eye. Pupil size varies with the contraction and relaxation of a ring of smooth *pupillary muscle*. The pupil appears as the black spot inside the colored ring of pigment known as the **iris.** The pigments and other components of the iris determine eye color.

The eye itself is a hollow sphere divided into two compartments (chambers) separated by a lens (■ Fig. 10.25). The **lens,** suspended by ligaments called **zonules,** is a transparent disk that focuses light. The anterior chamber in front of the lens is filled with **aqueous humor** {*humidus*, moist}, a low-protein, plasma-like fluid secreted by the *ciliary epithelium* supporting the lens. Behind the lens is a much larger chamber, the **vitreous chamber**, filled mostly with the **vitreous body** {*vitrum*, glass; also called the *vitreous humor*}, a clear, gelatinous matrix that helps maintain the shape of the eyeball. The outer wall of the eyeball, the **sclera**, is composed of connective tissue.

Light enters the anterior surface of the eye through the **cornea**, a transparent disk of tissue that is a continuation of the sclera. After passing through the opening of the pupil, light strikes the lens, which has two convex surfaces. The cornea and lens together bend incoming light rays so that they focus on the **retina**, the light-sensitive lining of the eye that contains the photoreceptors.

When viewed through the pupil with an ophthalmoscope {*ophthalmos*, eye}, the retina is seen to be crisscrossed with small arteries and veins that radiate out from one spot, the **optic disk** (Fig. 10.25b). The optic disk is the location where neurons of the visual pathway form the **optic nerve** (cranial nerve II) and exit the eye. Lateral to the optic disk is a small dark spot, the *fovea*. The fovea and a narrow ring of tissue surrounding it, the *macula*, are the regions of the retina with the most acute vision.

Neural pathways for the eyes are illustrated in ■ Figure 10.26. The optic nerves from the eyes go to the **optic chiasm** in the brain, where some of the fibers cross to the opposite side. After synapsing in the **lateral geniculate body** (*lateral geniculate nucleus*) of the thalamus, the vision neurons of the tract terminate in the occipital lobe at the **visual cortex**. Collateral pathways go from the thalamus to the midbrain, where they synapse with efferent neurons of cranial nerve III that control the diameter of the pupils.

CLINICAL FOCUS

Glaucoma

The eye disease *glaucoma,* characterized by degeneration of the optic nerve, is the leading cause of blindness worldwide. Many people associate glaucoma with increased *intraocular* (within the eyeball) pressure, but scientists have discovered that increased pressure is only one risk factor for the disease. A significant number of people with glaucoma have normal intraocular pressure, and not everyone with elevated pressure develops glaucoma. Many cases of elevated eye pressure are associated with excess aqueous humor, a fluid that is secreted by the ciliary epithelium near the lens. Normally the fluid drains out through the canal of Schlemm in the anterior chamber of the eye, but if outflow is blocked, the aqueous humor accumulates, causing pressure to build up inside the eye. Treatments to decrease intraocular pressure include drugs that inhibit aqueous humor production and surgery to reopen the canal of Schlemm. Research suggests that the optic nerve degeneration in glaucoma may be due to nitric oxide or apoptosis-inducing factors, and studies in these areas are underway.

Concept Check Answer: p. 376

24. What functions does the aqueous humor serve?

Light Enters the Eye through the Pupil

In the first step of the visual pathway, light from the environment enters the eye. Before it strikes the retina, however, the light is modified two ways. First, the amount of light that reaches photoreceptors is modulated by changes in the size of the pupil. Second, the light is focused by changes in the shape of the lens.

The human eye functions over a 100,000-fold range of light intensity. Most of this ability comes from the sensitivity of the photoreceptors, but the pupils assist by regulating the amount of light that falls on the retina. In bright sunlight, the pupils narrow to about 1.5 mm in diameter when a parasympathetic pathway constricts the circular pupillary muscles. In the dark, the opening of the pupil dilates to 8 mm, a 28-fold increase in pupil area. Dilation occurs when radial muscles lying perpendicular to the circular muscles contract under the influence of sympathetic neurons.

Testing **pupillary reflexes** is a standard part of a neurological examination. Light hitting the retina in one eye activates the reflex. Signals travel through the optic nerve to the thalamus, then to the midbrain, where efferent neurons constrict the pupils in *both* eyes (Fig. 10.26c). This response is known as the **consensual reflex** and is mediated by parasympathetic fibers running through cranial nerve III.

Concept Check Answers: p. 376

25. Use the neural pathways in Figure 10.26 to answer the following questions.
  (a) Why does shining light into one eye cause pupillary constriction in both eyes?
  (b) If you shine a light in the left eye and get pupillary constriction in the right eye but not in the left eye, what can you conclude about the afferent path from the left eye to the brain? About the efferent pathways to the pupils?

26. Parasympathetic fibers constrict the pupils, and sympathetic fibers dilate them. The two autonomic divisions can be said to have __________ effects on pupil diameter.

■ Fig. 10.25 *ANATOMY SUMMARY*

The Eye

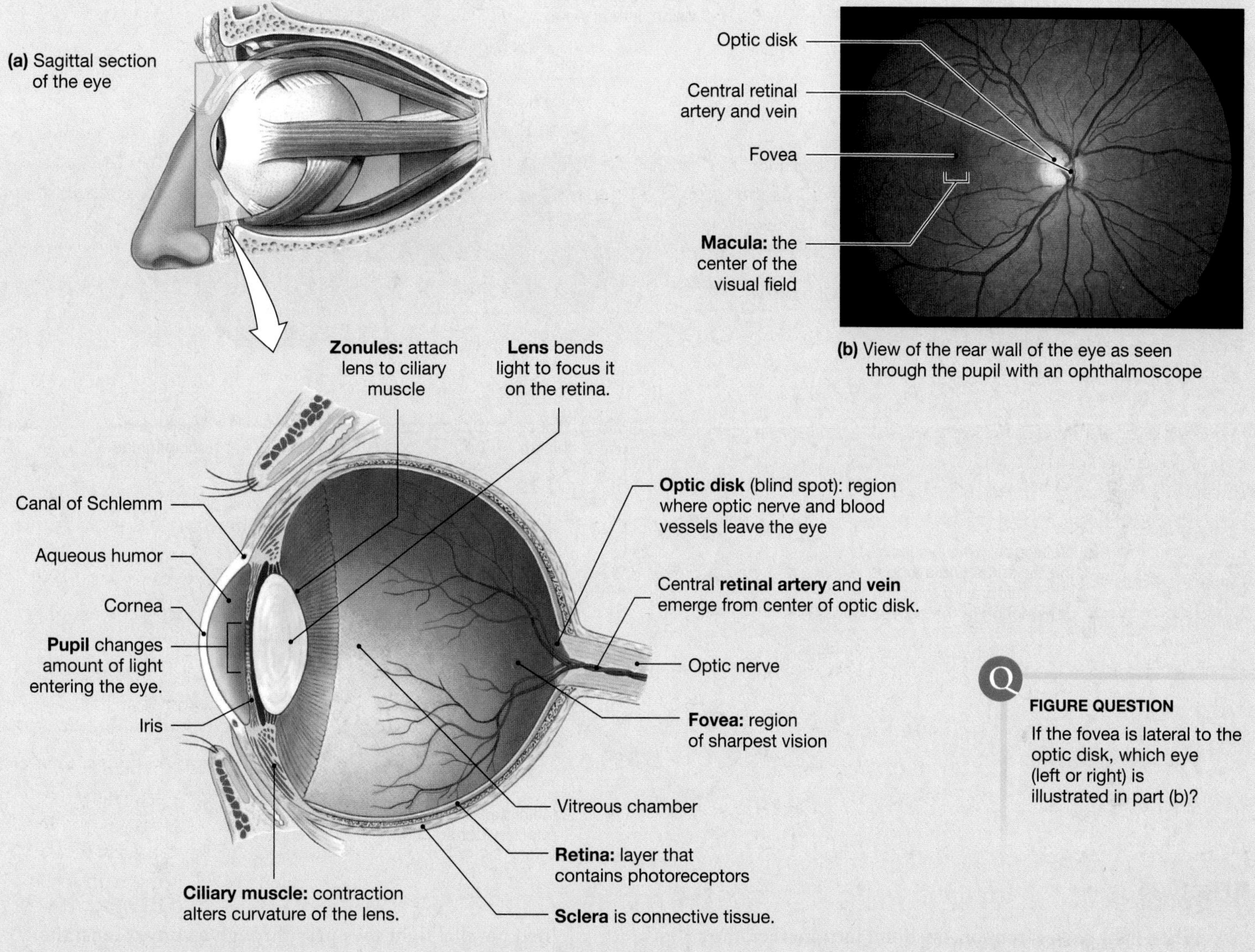

FIGURE QUESTION

If the fovea is lateral to the optic disk, which eye (left or right) is illustrated in part (b)?

In addition to regulating the amount of light that hits the retina, the pupils create what is known as **depth of field**. A simple example comes from photography. Imagine a picture of a puppy sitting in the foreground amid a field of wildflowers. If only the puppy and the flowers immediately around her are in focus, the picture is said to have a shallow depth of field. If the puppy and the wildflowers all the way back to the horizon are in focus, the picture has full depth of field. Full depth of field is created by constricting the pupil (or the diaphragm on a camera) so that only a narrow beam of light enters the eye. In this way, a greater depth of the image is focused on the retina.

The Lens Focuses Light on the Retina

The physics that describes the behavior and properties of light is a field known as **optics**. When light rays pass from air into a medium of different density, such as glass or water, they bend, or **refract**. Light entering the eye is refracted twice: first when it passes through the cornea, and again when it passes through the lens. About two-thirds of the total refraction (bending) occurs at the cornea and the remaining one-third occurs at the lens. Here we consider only the refraction that occurs as light passes through the lens because the lens is capable of changing its shape to focus light.

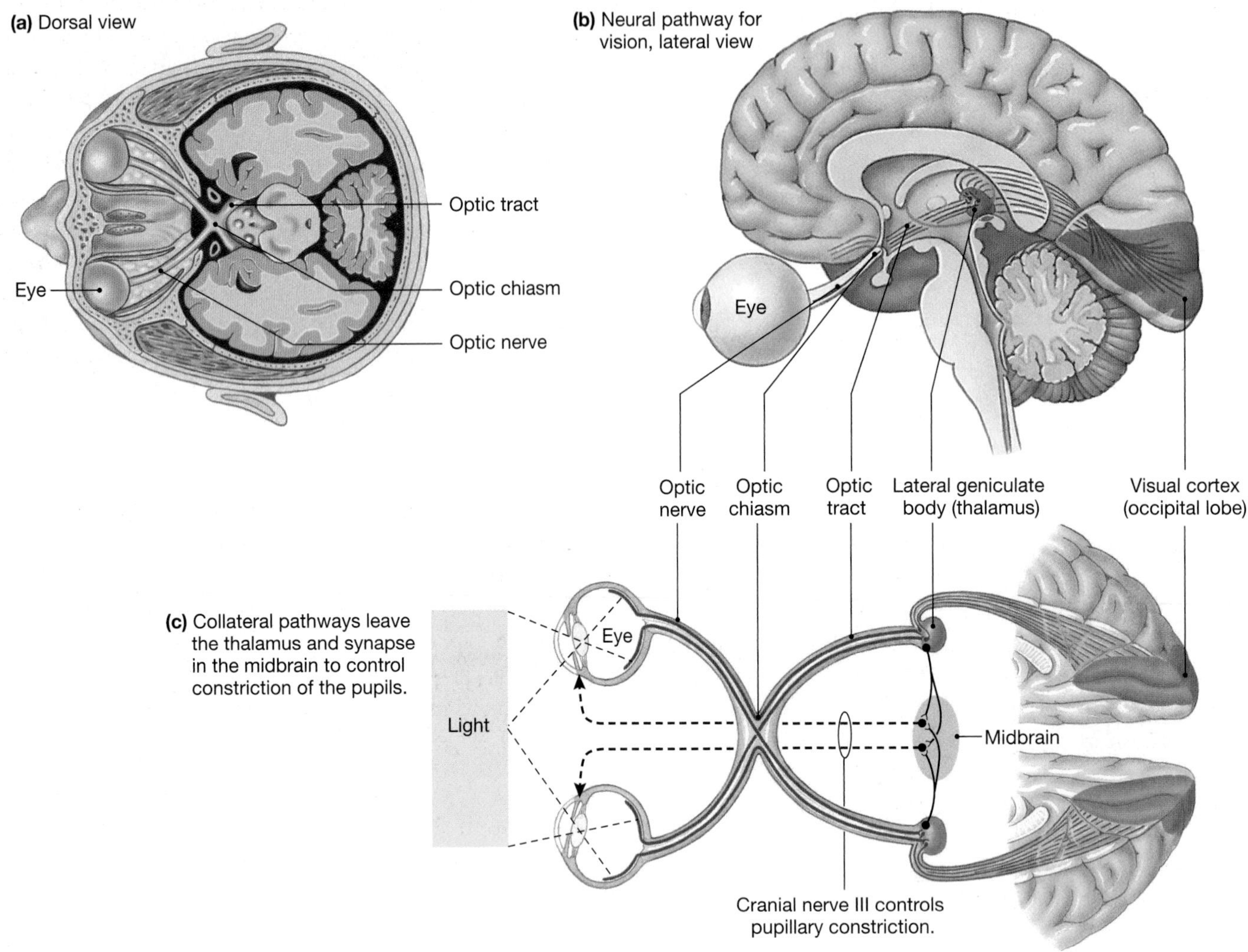

■ **Fig. 10.26**

When light passes from one medium into another, the angle of refraction (how much the light rays bend) is influenced by two factors: (1) the difference in density of the two media and (2) the angle at which the light rays meet the surface of the medium into which it is passing. For light passing through the lens of the eye, we assume that the density of the lens is the same as the density of the air and thus ignore this factor. The angle at which light meets the face of the lens depends on the curvature of the lens surface and the direction of the light beam.

Imagine parallel light rays striking the surface of a transparent lens. If the lens surface is perpendicular to the rays, the light passes through without bending. If the surface is not perpendicular, however, the light rays bend. Parallel light rays striking a *concave* lens, such as that shown in ■ Figure 10.27a, are refracted into a wider beam. Parallel rays striking a *convex* lens bend inward and focus to a point—*convex lenses converge* light waves (Fig. 10.27b). You can demonstrate the properties of a convex lens by using a magnifying glass to focus sunlight onto a piece of paper or other surface.

When parallel light rays pass through a convex lens, the single point where the rays converge is called the **focal point** (Fig. 10.27b). The distance from the center of a lens to its focal point is known as the **focal length** (or *focal distance*) of the lens. For any given lens, the focal length is fixed. For the focal length to change, the shape of the lens must change.

When light from an object passes through the lens of the eye, the focal point and object image must fall precisely on the retina if the object is to be seen in focus. In Figure 10.27c, parallel light rays strike a lens whose surface is relatively flat. For this lens, the focal point falls on the retina. The object is therefore in focus. For the normal human eye, any object that is 20 feet or more from the eye creates parallel light rays and will be in focus when the lens is flatter.

What happens, though, when an object is closer than 20 feet to the lens? In that case, the light rays from the object are not parallel and strike the lens at an oblique angle that changes the distance from the lens to the object's image (Fig. 10.27d). The focal point now lies behind the retina, and the object image becomes fuzzy and out of focus.

RUNNING PROBLEM

The otolaryngologist strongly suspects that Anant has Ménière's disease, with excessive endolymph in the vestibular apparatus and cochlea. Many treatments are available, beginning with simple dietary changes. For now, the physician suggests that Anant limit his salt intake and take diuretics, drugs that cause the kidneys to remove excess fluid from the body.

Q5: Why is limiting salt (NaCl) intake suggested as a treatment for Ménière's disease? (Hint: What is the relationship between salt, osmolarity, and fluid volume?)

To keep a near object in focus, the lens must become more rounded to increase the angle of refraction (Fig. 10.27e). Making a lens more convex shortens its focal length. In this example, rounding the lens causes light rays to converge on the retina instead of behind it, and the object comes into focus.

The process by which the eye adjusts the shape of the lens to keep objects in focus is known as **accommodation**, and the closest distance at which it can focus an object is known as the **near point of accommodation**. You can demonstrate changing focus with the *accommodation reflex* easily by closing one eye and holding your hand up about 8 inches in front of your open eye, fingers spread apart.

Focus your eye on some object in the distance that is visible between your fingers. Notice that when you do so, your fingers remain visible but out of focus. Your lens is flattened for distance vision, so the focal point for near objects falls behind the retina. Those objects appear out of focus. Now shift your gaze to your fingers and notice that they come into focus. The light rays reflecting off your fingers have not changed their angle, but your lens has become more rounded, and the light rays now converge on the retina.

How can the lens, which is clear and does not have any muscle fibers in it, change shape? The answer lies in the **ciliary muscle**, a ring of smooth muscle that surrounds the lens and is attached to it by the inelastic ligaments called zonules (Fig. 10.27f). If no tension is placed on the lens by the ligaments, the lens assumes its natural rounded shape because of the elasticity of its capsule. If the ligaments pull on the lens, it flattens out and assumes the shape required for distance vision.

Tension on the ligaments is controlled by the ciliary muscle. When the ciliary muscle is relaxed, the ring is more open and the lens is pulled into a flatter shape (Fig. 10.27g). When this circular muscle contracts, the muscle ring gets smaller, releasing tension on the ligaments so that the lens rounds (Fig. 10.27h).

Young people can focus on items as close as 8 cm, but the accommodation reflex diminishes from the age of 10 on. By age 40, accommodation is only about half of what it was at age 10. By age 60, many people lose the reflex completely because the lens has lost flexibility and remains in its flatter shape for distance vision. The loss of accommodation, **presbyopia**, is the reason most people begin to wear reading glasses in their 40s.

Two other common vision problems are near-sightedness and far-sightedness. Near-sightedness, or **myopia,** occurs when the focal point falls in front of the retina (Fig. 10.27j). Far-sightedness, or **hyperopia**, occurs when the focal point falls behind the retina (Fig. 10.27i). These vision problems are caused by abnormally curved or flattened corneas or by eyeballs that are too long or too short. Placing a lens with the appropriate curvature in front of the eye changes the refraction of light entering the eye and corrects the problem. A third common vision problem, **astigmatism**, is usually caused by a cornea that is not a perfectly shaped dome, resulting in distorted images.

Concept Check Answers: p. 376

27. If a person's cornea, which helps focus light, is more rounded than normal (has a greater curvature), is this person more likely to be hyperopic or myopic? (*Hint:* See Fig. 10.27.)

28. The relationship between the focal length of a lens (F), the distance between an object and the lens (P), and the distance from the lens to the object's image (Q) is expressed as $1/F = 1/P + 1/Q$.

(a) If the focal length of a lens does not change but an object moves closer to the lens, what happens to the image distance Q?

(b) If an object moves closer to the lens and the image distance Q must stay the same for the image to fall on the retina, what must happen to the focal length F of the lens? For this change in F to occur, should the lens become flatter or more rounded?

29. (a) Explain how convex and concave corrective lenses change the refraction of light.

(b) Which type of corrective lens should be used for myopia, and why? For hyperopia?

Phototransduction Occurs at the Retina

In the second step of the visual pathway, photoreceptors of the retina convert light energy into electrical signals. Light energy is part of the electromagnetic spectrum, which ranges from high-energy, very-short-wavelength waves such as X-rays and gamma rays to low-energy, lower-frequency microwaves and radio waves (■ Fig. 10.28). However, our brains can perceive only a small portion of this broad energy spectrum. For humans, **visible light** is limited to electromagnetic energy with waves that have a frequency of $4.0–7.5 \times 10^{14}$ cycles per second (hertz, Hz) and a wavelength of 400–750 nanometers (nm). Electromagnetic energy is measured in units called *photons.*

Our unaided eyes see visible light but do not respond to ultraviolet and infrared light, whose wavelengths border the ends of our visible light spectrum. On the other hand, the eyes of some other animals can see these wavelengths. For example, bees use ultraviolet "runways" on flowers to guide them to pollen and nectar.

Optics of the Eye

Light passing through a curved surface will bend or refract.

(a) A **concave lens** scatters light rays.

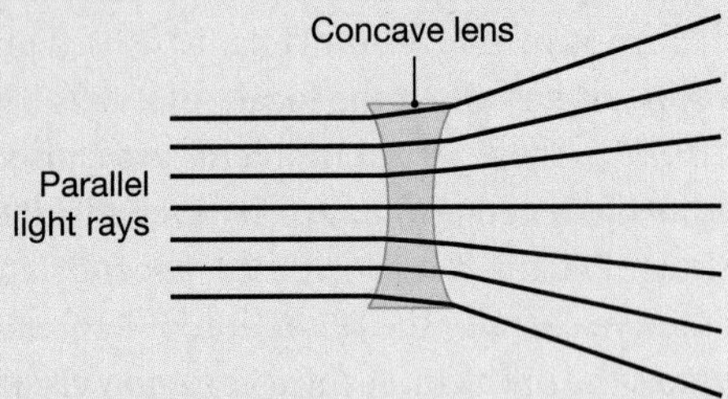

(b) A **convex lens** causes light rays to converge.

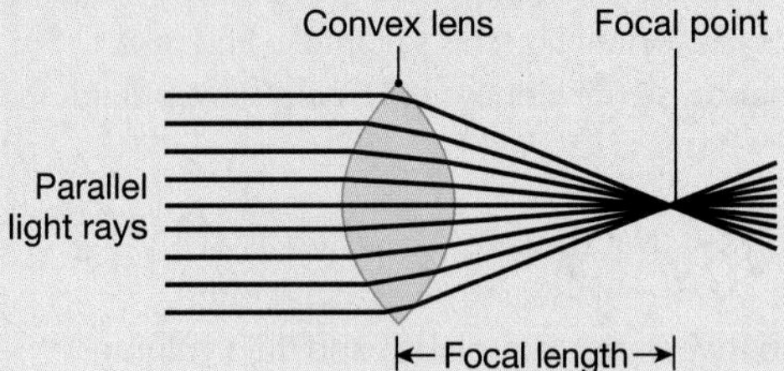

The **focal length** of the lens is the distance from the center of the lens to the **focal point.**

For clear vision, the focal point must fall on the retina.

(c) Parallel light rays pass through a flattened lens, and the focal point falls on the retina.

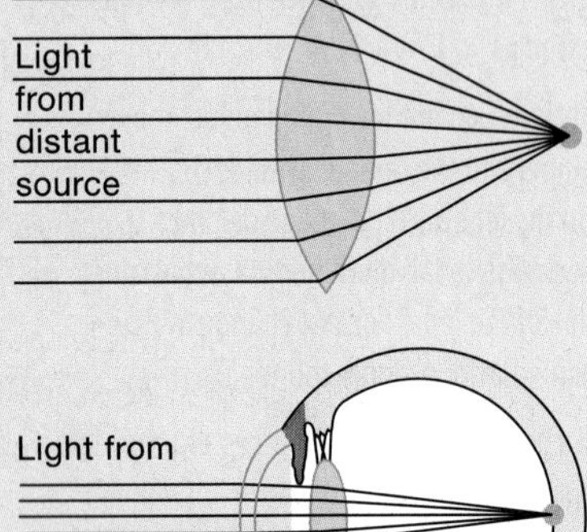

(d) For close objects, the light rays are no longer parallel. The lens and its focal length have not changed, but the object is seen out of focus because the light beam is not focused on the retina.

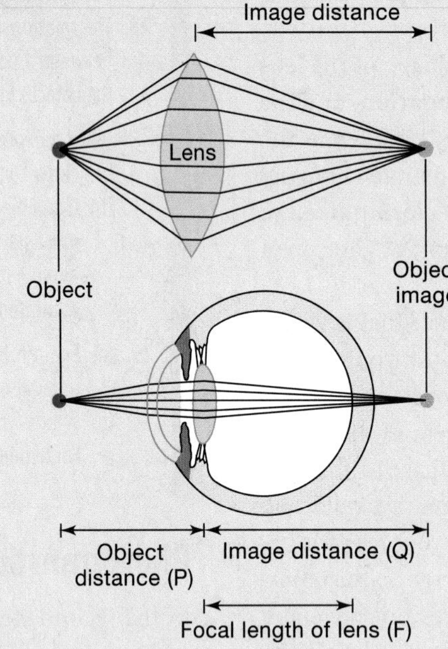

(e) To keep an object in focus as it moves closer, the lens becomes more rounded.

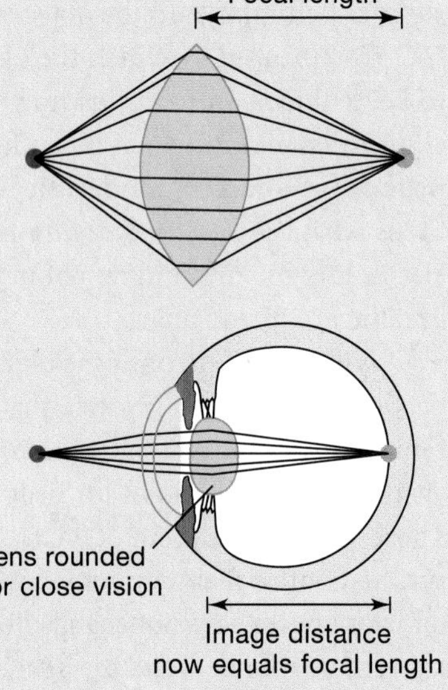

Changes in lens shape are controlled by the ciliary muscle.

(f) The lens is attached to the ciliary muscle by inelastic ligaments (zonules).

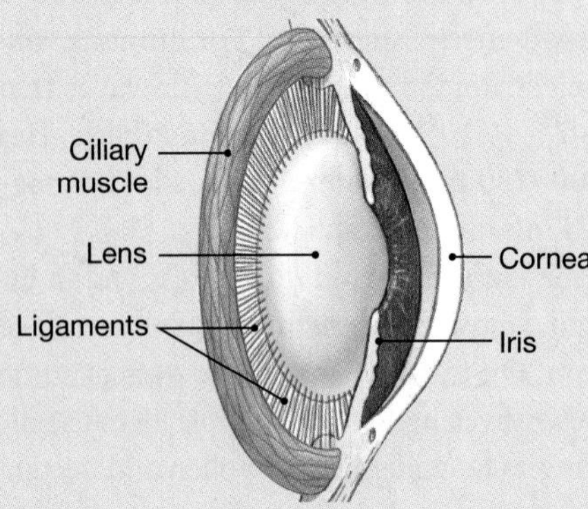

(g) When ciliary muscle is relaxed, the ligaments pull on and flatten the lens.

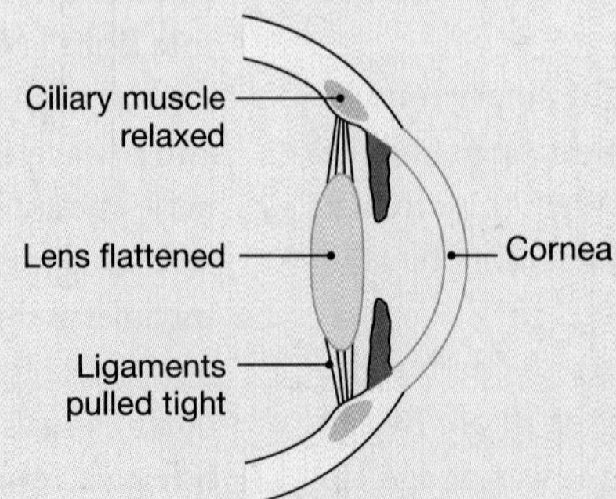

(h) When ciliary muscle contracts, it releases tension on the ligaments and the lens becomes more rounded.

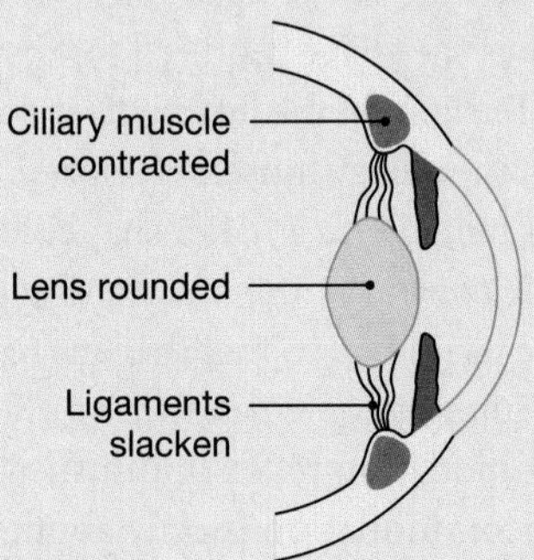

Common vision defects can be corrected with external lenses.

(i) Hyperopia, or far-sightedness, occurs when the focal point falls behind the retina.

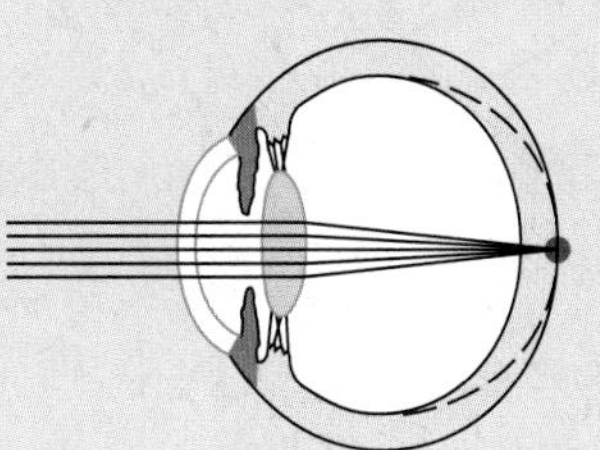

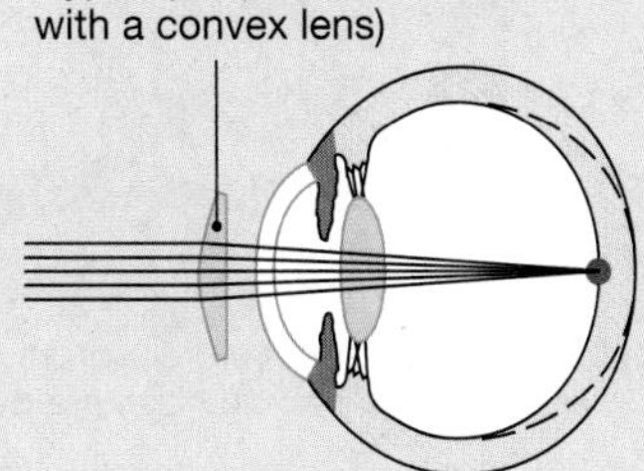

(j) Myopia, or near-sightedness, occurs when the focal point falls in front of the retina.

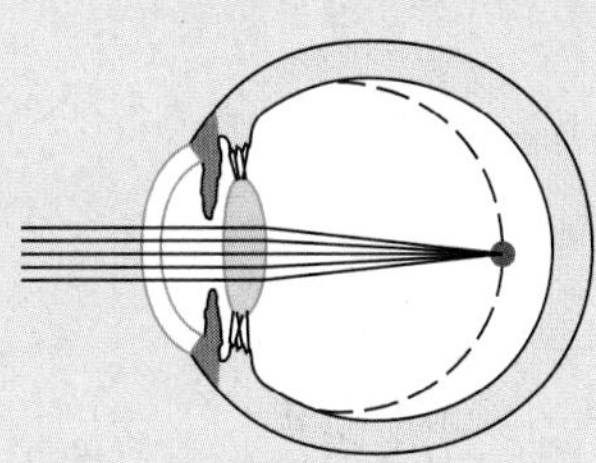

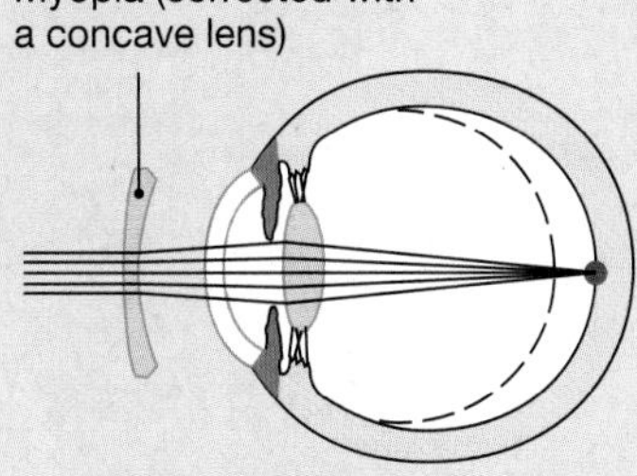

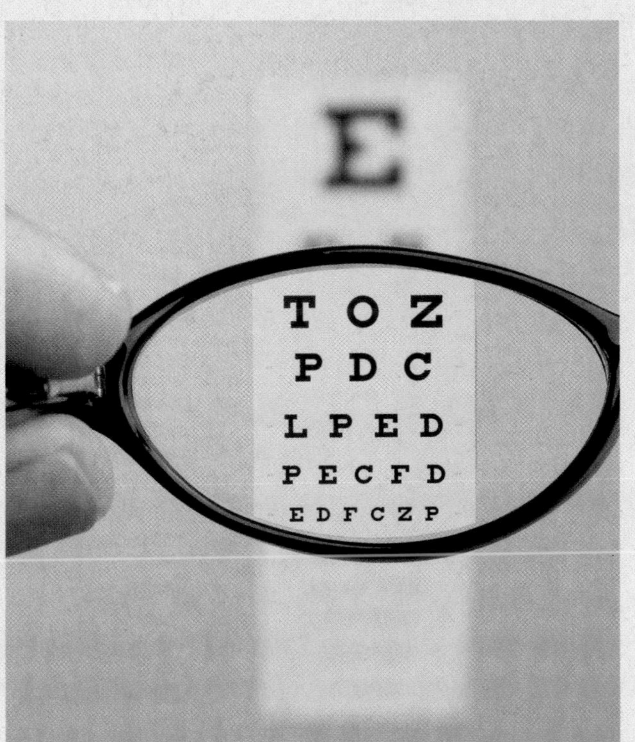

Phototransduction is the process by which animals convert light energy into electrical signals. In humans, phototransduction takes place when light hits the retina, the sensory organ of the eye (■ Fig. 10.29). The retina develops from the same embryonic tissue as the brain, and (as in the cortex of the brain) neurons in the retina are organized into layers. There are five types of neurons in the retinal layers: photoreceptors, bipolar cells, ganglion cells, amacrine cells, and horizontal cells (Fig. 10.29f).

Backing the photosensitive portion of the human retina is a dark **pigment epithelium** layer. Its function is to absorb any light rays that escape the photoreceptors, preventing distracting light from reflecting inside the eye and distorting the visual image. The black color of these epithelial cells comes from granules of the pigment *melanin*.

Photoreceptors are the neurons that convert light energy into electrical signals. There are two main types of photoreceptors, rods and cones, as well as a recently discovered photoreceptor that is a modified ganglion cell (see Emerging Concepts Box: Melanopsin). You might expect photoreceptors to be on the surface of the retina facing the vitreous chamber, where light will strike them first, but the retinal layers are actually in reverse order. The photoreceptors are the bottom layer, with their photosensitive tips against the pigment epithelium. Most light entering the eye must pass through several relatively transparent layers of neurons before striking the photoreceptors.

One exception to this organizational pattern occurs in a small region of the retina known as the **fovea** {pit}. This area is free of neurons and blood vessels that would block light reception, so photoreceptors receive light directly, with minimal scattering. (Fig. 10.29d). As noted earlier, the fovea and the **macula** immediately surrounding it are the areas of most acute vision, and they form the center of the visual field.

THE ELECTROMAGNETIC SPECTRUM

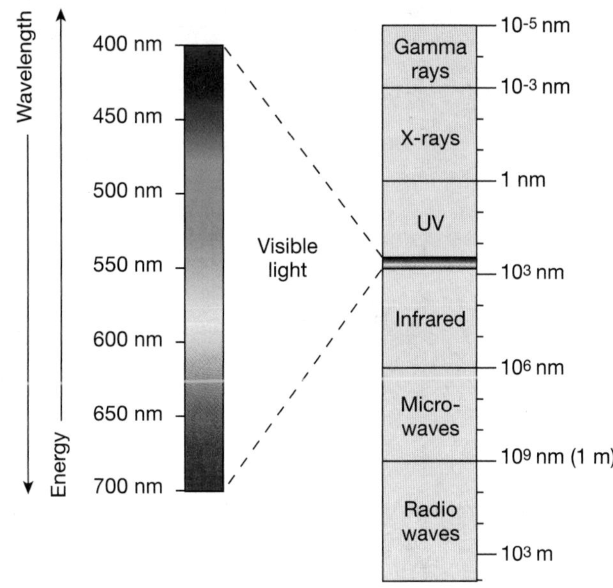

■ **Fig. 10.28**

The Retina

(a) Dorsal view of a section of the right eye

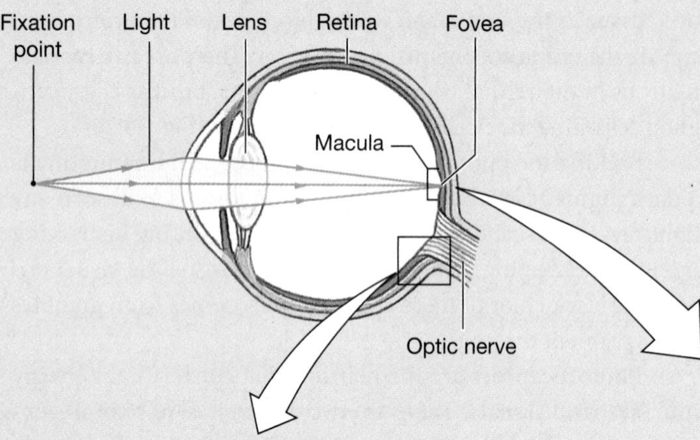

(c) Axons from the retina exit via the optic nerve.

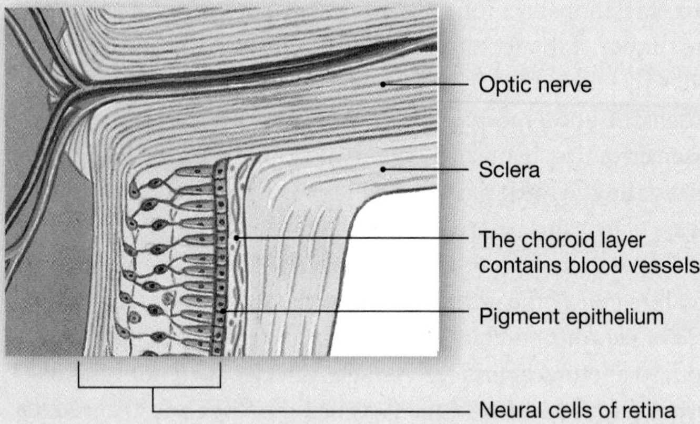

(e) Convergence in the retina

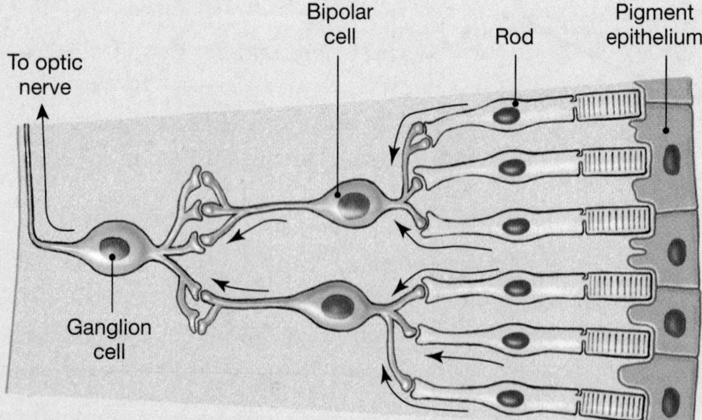

FIGURE QUESTION

How many rods converge on the ganglion cell in (e)?

(b) The projected image is upside down on the retina. Visual processing in the brain reverses the image.

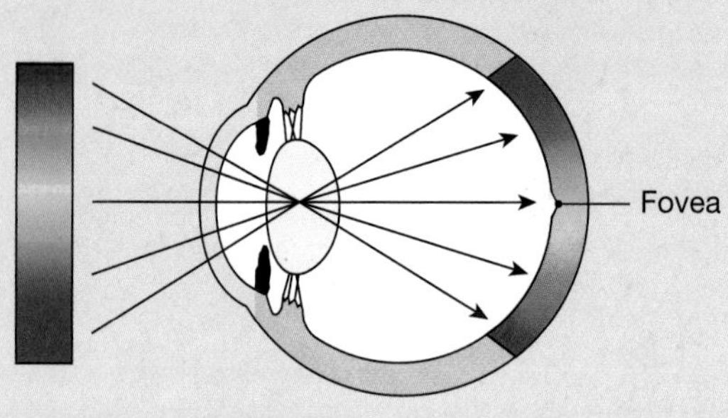

(d) Light strikes the photoreceptors in the fovea directly because overlying neurons are pushed aside.

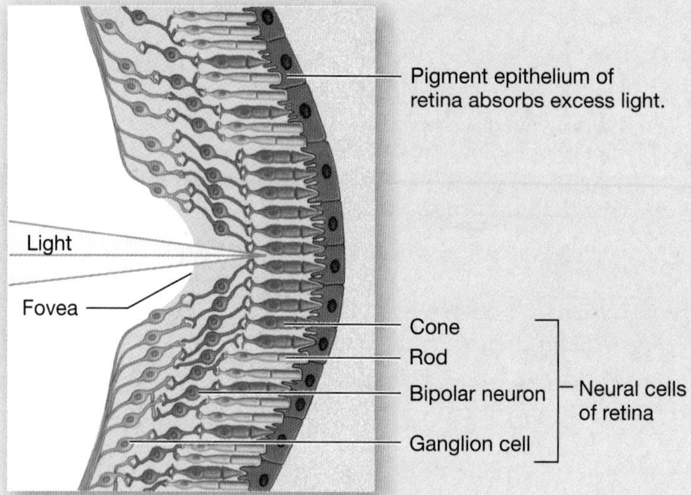

(f) Retinal photoreceptors are organized into layers.

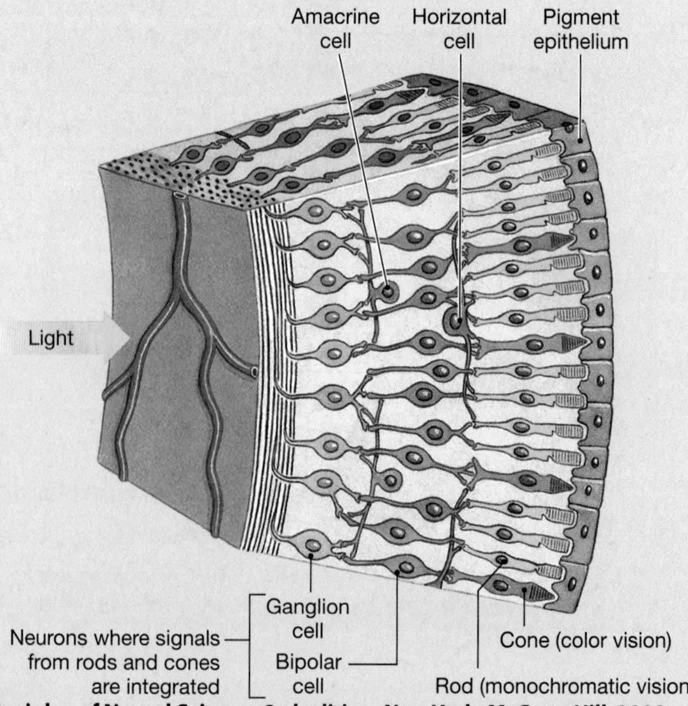

Drawing of photoreceptors in the fovea adapted from E. R. Kandel et al.,* Principles of Neural Science, *3rd edition. New York: McGraw Hill, 2000.

EMERGING CONCEPTS

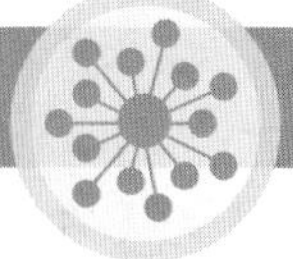

Melanopsin

Circadian rhythms in mammals are cued by light entering the eyes. For many years, scientists believed that rods and cones of the retina were the primary photoreceptors linked to the *suprachiasmatic nucleus* (SCN), the brain center for circadian rhythms. However, in 1999 researchers found that transgenic mice lacking both rods and cones still had the ability to respond to changing light cues, suggesting that an additional photoreceptor must exist in the retina. Now scientists believe they have found it: a subset of retinal ganglion cells that contain an opsin-like pigment called *melanopsin* (mRGCs). Axons from these mRGC ganglion cells project to the SCN as well as to brain areas that process visual information. It appears that these newly identified photoreceptors join rods and cones as the light-sensing cells of the mammalian retina, and scientists may have to revise the traditional models of visual processing. To learn more, see C. Sedwick, Melanopsin ganglion cells: a different way of seeing things. *PLoS Biol* 8(12): e1001003, 2010 (*www.plosbiology.org*).

When you look at an object, the lens focuses the object image on the fovea. For example, in Figure 10.29b, the eye is focused on the green-yellow border of the color bar. Light from that section of the visual field falls on the fovea and is in sharp focus. Notice also that the image falling on the retina is upside down. Subsequent visual processing by the brain reverses the image again so that we perceive it in the correct orientation.

Sensory information about light passes from the photoreceptors to **bipolar neurons**, then to a layer of **ganglion cells** (Fig. 10.29e). The axons of ganglion cells form the optic nerve, which leaves the eye at the optic disk. Because the optic disk has no photoreceptors, images projected onto this region cannot be seen, creating what is called the eye's **blind spot**.

Concept Check

Answers: p. 376

30. Animals that see well in very low light, such as cats and owls, lack a pigment epithelium and instead have a layer called the *tapetum lucidum* behind the retina. What property might this layer have that would enhance vision in low light?
31. How is the difference in visual acuity between the fovea and the edge of the visual field similar to the difference in touch discrimination between the fingertips and the skin of the arm?
32. Macular degeneration is the leading cause of blindness in Americans over the age of 55. Impaired function of the macula causes vision loss in which part of the visual field?

RUNNING PROBLEM

Anant's condition does not improve with the low-salt diet and diuretics, and he continues to suffer from disabling attacks of vertigo with vomiting. In severe cases of Ménière's disease, surgery is sometimes performed when less invasive treatments have failed. In one surgical procedure for the disease, a drain is inserted to relieve pressure in the endolymph by removing some of the fluid. If that fails to provide relief, as a last resort the vestibular nerve can be severed. This surgery is difficult to perform, as the vestibular nerve lies near many other important nerves, including facial nerves and the auditory nerve. Patients who undergo this procedure are advised that the surgery can result in deafness if the cochlear nerve is inadvertently severed.

Q6: Why would severing the vestibular nerve alleviate Ménière's disease?

326 331 348 356 361 365 371

Photoreceptors Transduce Light into Electrical Signals

There are two main types of photoreceptors in the eye: rods and cones. **Rods** function well in low light and are used in night vision, when objects are seen in black and white rather than in color. They outnumber cones by a 20:1 ratio, except in the fovea, which contains only cones.

Cones are responsible for *high-acuity* vision and color vision during the daytime, when light levels are higher. *Acuity* means keenness and is derived from the Latin *acuere,* meaning "to sharpen." The fovea, which is the region of sharpest vision, has a very high density of cones.

The two types of photoreceptors have the same basic structure (■ Fig. 10.30): (1) an outer segment whose tip touches the pigment epithelium of the retina, (2) an inner segment that contains the cell nucleus and organelles for ATP and protein synthesis, and (3) a basal segment with a synaptic terminal that releases glutamate onto bipolar cells.

In the outer segment, the cell membrane has deep folds that form disk-like layers. Toward the tip of the outer segments in rods, these layers actually separate from the cell membrane and form free-floating membrane disks. In the cones, the disks stay attached.

Light-sensitive **visual pigments** are bound to the disk membranes in outer segments of photoreceptors. These visual pigments are transducers that convert light energy into a change in membrane potential. Rods have one type of visual pigment, **rhodopsin**. Cones have three different pigments that are closely related to rhodopsin.

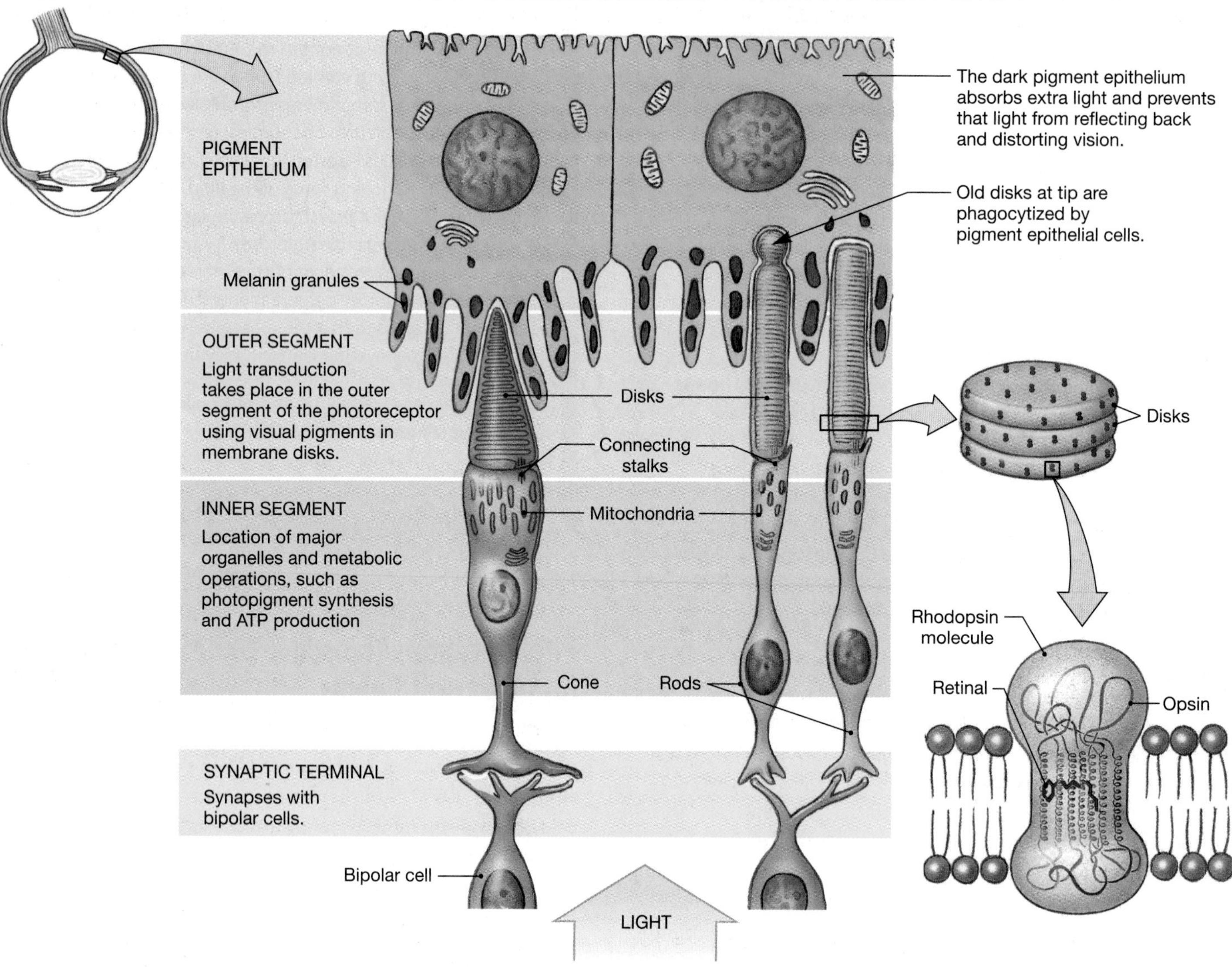
PHOTORECEPTORS: RODS AND CONES
PIGMENT EPITHELIUM
Melanin granules
The dark pigment epithelium absorbs extra light and prevents that light from reflecting back and distorting vision.
Old disks at tip are phagocytized by pigment epithelial cells.
OUTER SEGMENT
Light transduction takes place in the outer segment of the photoreceptor using visual pigments in membrane disks.
Disks
Connecting stalks
Disks
INNER SEGMENT
Location of major organelles and metabolic operations, such as photopigment synthesis and ATP production
Mitochondria
Rhodopsin molecule
Retinal
Opsin
Cone
Rods
SYNAPTIC TERMINAL
Synapses with bipolar cells.
Bipolar cell
LIGHT

■ Fig. 10.30

The visual pigments of cones are excited by different wavelengths of light, allowing us to see in color. White light is a combination of colors, as demonstrated when you separate white light by passing it through a prism. The eye contains cones for red, green, and blue light. Each cone type is stimulated by a range of light wavelengths but is most sensitive to a particular wavelength (■ Fig. 10.31). Red, green, and blue are the three primary colors that make the colors of visible light, just as red, blue, and yellow are the three primary colors that make different colors of paint.

The color of any object we are looking at depends on the wavelengths of light reflected by the object. Green leaves reflect green light, and bananas reflect yellow light. White objects reflect most wavelengths. Black objects absorb most wavelengths, which is one reason they heat up in sunlight while white objects stay cool.

Our brain recognizes the color of an object by interpreting the combination of signals coming to it from the three different color cones. The details of color vision are still not fully understood, and there is some controversy about how color is processed in the cerebral cortex. **Color-blindness** is a condition in which a person inherits a defect in one or more of the three types of cones and has difficulty distinguishing certain colors. Probably the best-known form of color-blindness is red-green, in which people have trouble telling red and green apart.

Concept Check Answer: p. 376

33. Why is our vision in the dark in black and white rather than in color?

LIGHT ABSORPTION BY VISUAL PIGMENTS

There are three types of cone pigment, each with a characteristic light absorption spectrum. Rods are for black and white vision in low light.

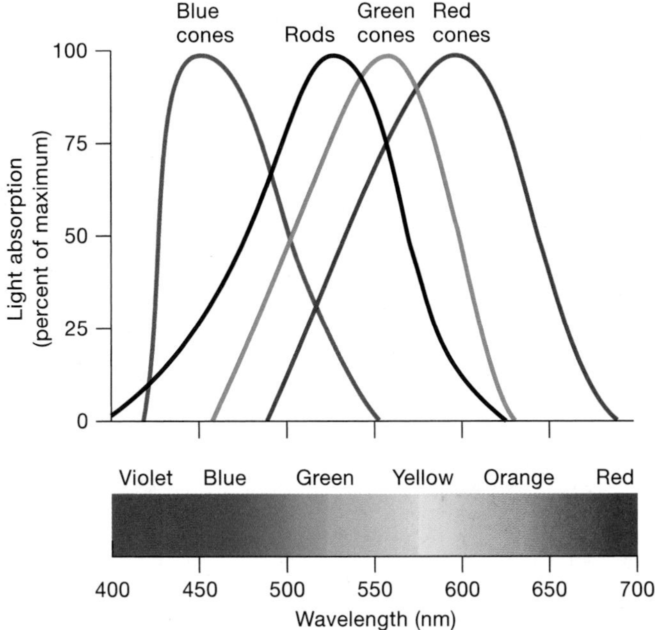

GRAPH QUESTIONS

1. Which pigment absorbs light over the broadest spectrum of wavelengths?
2. Over the narrowest?
3. Which pigment absorbs the most light at 500 nm?

■ **Fig. 10.31**

Phototransduction The process of phototransduction is similar for rhodopsin (in rods) and the three color pigments (in cones). Rhodopsin is composed of two molecules: **opsin**, a protein embedded in the membrane of the rod disks, and **retinal**, a vitamin A derivative that is the light-absorbing portion of the pigment (see Fig. 10.30). In the absence of light, retinal binds snugly into a binding site on the opsin (■ Fig. 10.32). When activated by as little as one photon of light, retinal changes shape to a new configuration. The activated retinal no longer binds to opsin and is released from the pigment in the process known as **bleaching**.

How does rhodopsin bleaching lead to action potentials traveling through the optical pathway? To understand the pathway, we must look at other properties of the rods. Electrical signals in cells occur as a result of ion movement between the intracellular and extracellular compartments. Rods contain three main types of cation channels: **cyclic nucleotide-gated channels** (CNG channels) that allow Na^+ and Ca^{2+} to enter the rod, K^+ channels that allow K^+ to leak out of the rod, and voltage-gated Ca^{2+} channels in the synaptic terminal that help regulate exocytosis of neurotransmitter.

When a rod is in darkness and rhodopsin is not active, cyclic GMP (cGMP) levels in the rod are high, and both CNG and K^+ channels are open (Fig. 10.32 ①). Sodium and Ca^{2+} ion influx is greater than K^+ efflux, so the rod stays depolarized to an average membrane potential of −40 mV (instead of the more usual −70 mV). At this slightly depolarized membrane potential, the voltage-gated Ca^{2+} channels are open and there is tonic (continuous) release of the neurotransmitter glutamate from the synaptic portion of the rod onto the adjacent bipolar cell.

When light activates rhodopsin, a second-messenger cascade is initiated through the G protein **transducin** (Fig. 10.32 ②). (Transducin is closely related to gustducin, the G protein found in bitter taste receptors.) The transducin second-messenger cascade decreases the concentration of cGMP, which closes the CNG channels. As a result, cation influx slows or stops.

With decreased cation influx and continued K^+ efflux, the inside of the rod hyperpolarizes, and glutamate release onto the bipolar neurons decreases. Bright light closes all CNG channels and stops all neurotransmitter release. Dimmer light causes a response that is graded in proportion to the light intensity.

After activation, retinal diffuses out of the rod and is transported into the pigment epithelium. There it reverts to its inactive form before moving back into the rod and being reunited with opsin (Fig. 10.32 ③). The recovery of rhodopsin from bleaching can take some time and is a major factor in the slow adaptation of the eyes when moving from bright light into the dark.

Concept Check Answer: p. 376

34. Draw a map or diagram to explain phototransduction. Start with bleaching and end with release of neurotransmitter.

Signal Processing Begins in the Retina

We now move from the cellular mechanism of light transduction to the processing of light signals by the retina and brain, the third and final step in our vision pathway. Signal processing in the retina is an excellent example of convergence [p. 274], in which multiple neurons synapse onto a single postsynaptic cell (■ Fig. 10.33a). Depending on location in the retina, as many as 15 to 45 photoreceptors may converge on one bipolar neuron.

Multiple bipolar neurons in turn innervate a single ganglion cell, so that the information from hundreds of millions of retinal photoreceptors is condensed down to a mere 1 million axons leaving the eye in each optic nerve. Convergence is minimal in the fovea, where some photoreceptors have a 1:1 relationship with their bipolar neurons, and greatest at the outer edges of the retina.

Signal processing in the retina is modulated by input from two additional sets of cells that we will not discuss (Fig. 10.29f). **Horizontal cells** synapse with photoreceptors and bipolar cells. **Amacrine cells** modulate information flowing between bipolar cells and ganglion cells.

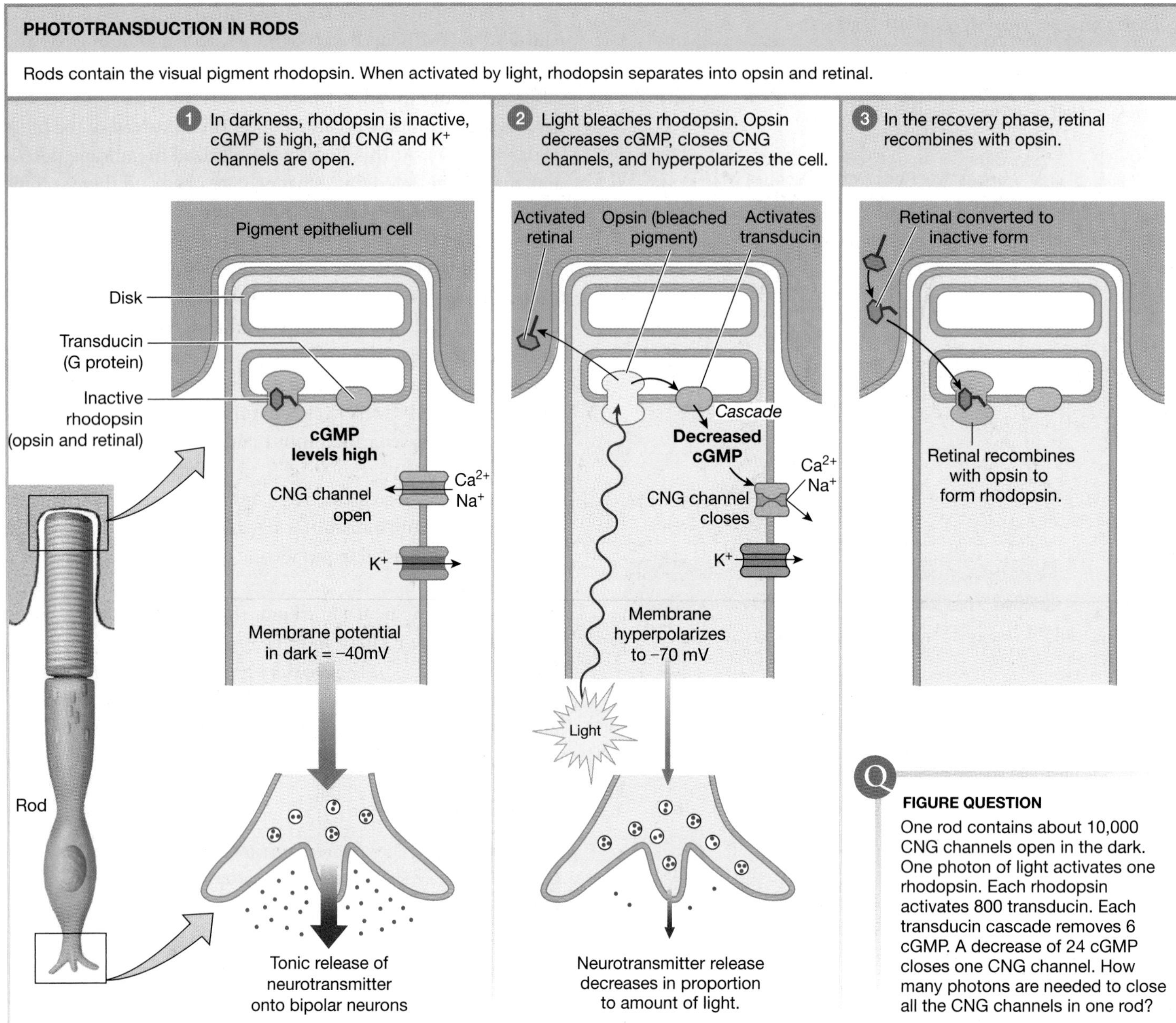
PHOTOTRANSDUCTION IN RODS
Rods contain the visual pigment rhodopsin. When activated by light, rhodopsin separates into opsin and retinal.
1 In darkness, rhodopsin is inactive, cGMP is high, and CNG and K+ channels are open.
2 Light bleaches rhodopsin. Opsin decreases cGMP, closes CNG channels, and hyperpolarizes the cell.
3 In the recovery phase, retinal recombines with opsin.
Pigment epithelium cell
Disk
Transducin (G protein)
Inactive rhodopsin (opsin and retinal)
cGMP levels high
CNG channel open
Ca2+
Na+
K+
Membrane potential in dark = -40mV
Rod
Tonic release of neurotransmitter onto bipolar neurons
Activated retinal
Opsin (bleached pigment)
Activates transducin
Cascade
Decreased cGMP
CNG channel closes
Membrane hyperpolarizes to -70 mV
Light
Neurotransmitter release decreases in proportion to amount of light.
Retinal converted to inactive form
Retinal recombines with opsin to form rhodopsin.
FIGURE QUESTION
One rod contains about 10,000 CNG channels open in the dark. One photon of light activates one rhodopsin. Each rhodopsin activates 800 transducin. Each transducin cascade removes 6 cGMP. A decrease of 24 cGMP closes one CNG channel. How many photons are needed to close all the CNG channels in one rod?

Fig. 10.32

Bipolar Cells Glutamate release from photoreceptors onto bipolar neurons begins signal processing. There are two types of bipolar cells, *light-on* (ON bipolar cells) and *light-off* (OFF bipolar cells). ON bipolar cells are activated in the light when glutamate secretion by photoreceptors decreases. In the dark, ON bipolar cells are inhibited by glutamate release. OFF bipolar cells are excited by glutamate release in the dark. In the light, with less glutamate, OFF bipolar cells are inhibited. By using different glutamate receptors, one stimulus (light) creates two different responses with a single neurotransmitter.

Whether glutamate is excitatory or inhibitory depends on the type of glutamate receptor on the bipolar neuron. ON bipolar cells have a metabotropic glutamate receptor called *mGluR6* that hyperpolarizes the cell when the receptor binds glutamate in the dark. When mGluR6 is not activated, the ON bipolar cell depolarizes. OFF bipolar cells have an ionotropic glutamate receptor that opens ion channels and depolarizes the OFF bipolar cell in the dark. Bipolar cell signal processing is also modified by input from the horizontal and amacrine cells.

Ganglion Cells Bipolar cells synapse with ganglion cells, the next neurons in the pathway. We know more about ganglion cells because they lie on the surface of the retina, where their axons are the most accessible to researchers. Extensive studies have been done in which researchers stimulated the retina with carefully placed light and evaluated the response of the ganglion cells.

Each ganglion cell receives information from a particular area of the retina. These areas, known as **visual fields**,

VISUAL FIELDS

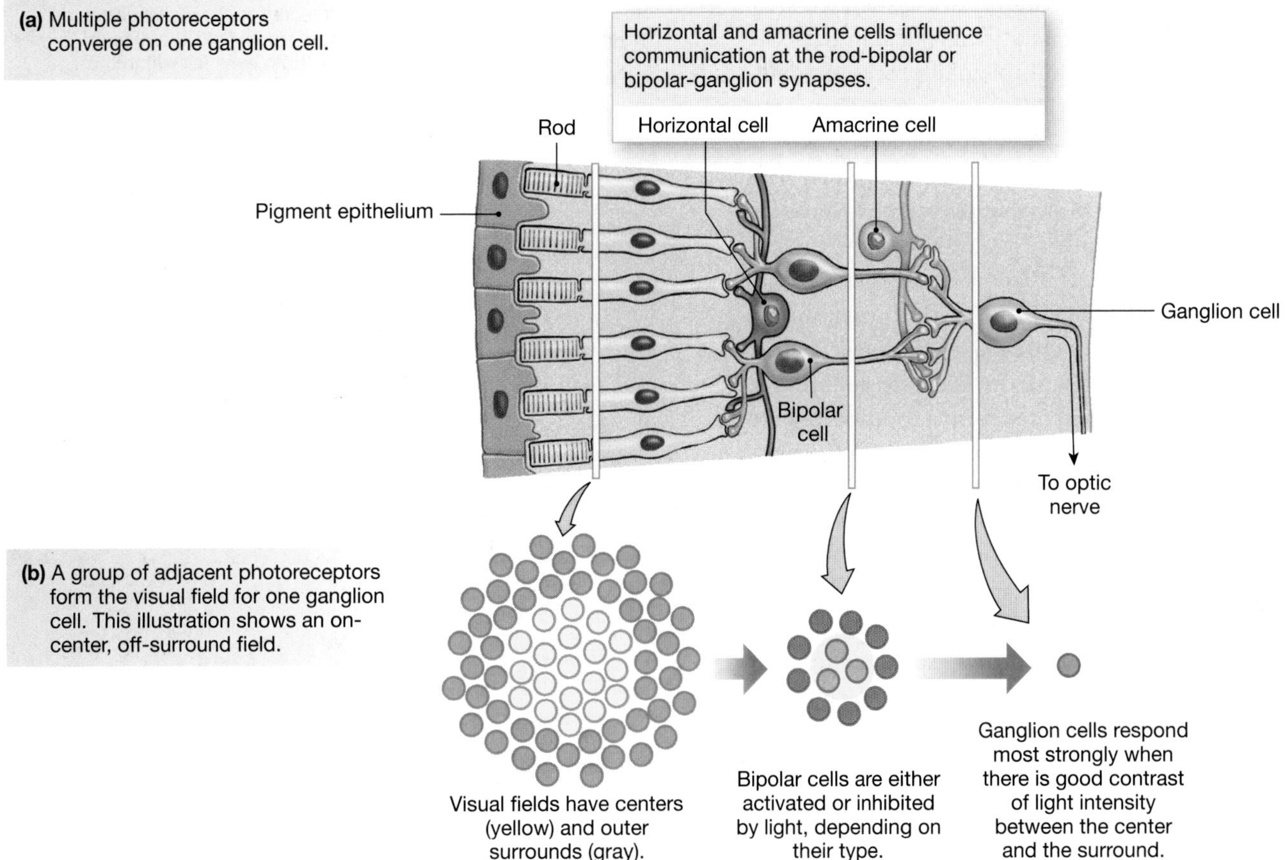
Horizontal and amacrine cells influence communication at the rod-bipolar or bipolar-ganglion synapses.
Horizontal cell
Amacrine cell
Rod
Pigment epithelium
Ganglion cell
Bipolar cell
To optic nerve
Visual fields have centers (yellow) and outer surrounds (gray).
Bipolar cells are either activated or inhibited by light, depending on their type.
Ganglion cells respond most strongly when there is good contrast of light intensity between the center and the surround.

(a) Multiple photoreceptors converge on one ganglion cell.

(b) A group of adjacent photoreceptors form the visual field for one ganglion cell. This illustration shows an on-center, off-surround field.

(c) The retina uses contrast rather than absolute light intensity for better detection of weak stimuli.

Visual field type	Field is on-center/off-surround	Field is off-center/on-surround
On-center, off-surround Bright light onto center	Ganglion cell is excited by light in the center of the visual field.	Ganglion cell is inhibited by light in the center of the visual field.
Off-center, on-surround Bright light onto surround Bright light onto surround	Ganglion cell is inhibited by light on the surround of the visual field.	Ganglion cell is excited by light on the surround of the visual field.
Both field types Diffuse light on both center and surround	Ganglion cell responds weakly.	Ganglion cell responds weakly.

Fig. 10.33

10

are similar to receptive fields in the somatic sensory system [p. 328]. The visual field of a ganglion cell near the fovea is quite small. Only a few photoreceptors are associated with each ganglion cell, and so visual acuity is greatest in these areas. At the edge of the retina, multiple photoreceptors converging onto a single ganglion cell results in vision that is not as sharp (Fig. 10.33a).

An analogy for this arrangement is to think of pixels on your computer screen. Assume that two screens have the same number of "photoreceptors," as indicated by a maximal screen resolution of 1280 × 1024 pixels. If screen A has one photoreceptor becoming one "ganglion cell" pixel, the actual screen resolution is 1280 × 1024, and the image is very clear. If eight photoreceptors on screen B converge into one ganglion cell pixel, then the actual screen resolution falls to 160 × 128, resulting in a very blurry and perhaps indistinguishable image.

Visual fields of ganglion cells are roughly circular (unlike the irregular shape of somatic sensory receptive fields) and are divided into sections: a round center and its doughnut-shaped **surround** (Fig. 10.33b). This organization allows each ganglion cell to use contrast between the center and its surround to interpret visual information. Strong contrast between the center and surround elicits a strong excitatory response (a series of action potentials) or a strong inhibitory response (no action potentials) from the ganglion cell. Weak contrast between center and surround gets an intermediate response.

There are two types of ganglion cell visual fields. In an *on-center/off-surround field,* the associated ganglion cell responds most strongly when light is brightest in the center of the field (Fig. 10.33c). If light is brightest in the off-surround region of the field, the *on*-center/off-surround field ganglion cell is inhibited and stops firing action potentials. The reverse happens with *off-center/on-surround fields.*

What happens if light is uniform across a visual field? In that case, the ganglion cell responds weakly. Thus, the retina uses *contrast* rather than absolute light intensity to recognize objects in the environment. One advantage of using contrast is that it allows better detection of weak stimuli.

Scientists have now identified multiple types of ganglion cells in the primate retina. The two predominant types, which account for 80% of retinal ganglion cells, are M cells and P cells. Large *magnocellular* ganglion cells, or **M cells**, are more sensitive to information about movement. Smaller *parvocellular* ganglion cells, or **P cells**, are more sensitive to signals that pertain to form and fine detail, such as the texture of objects in the visual field. A recently discovered subtype of ganglion cell, the *melanopsin retinal ganglion cell,* apparently also acts as a photoreceptor.

Processing Beyond the Retina Once action potentials leave ganglion cell bodies, they travel along the optic nerves to the CNS for further processing. As noted earlier, the optic nerves enter the brain at the optic chiasm. At this point, some nerve fibers from each eye cross to the other side of the brain for processing. ■ Figure 10.34 shows how information from the right side of each eye's visual field is processed on the left side of the

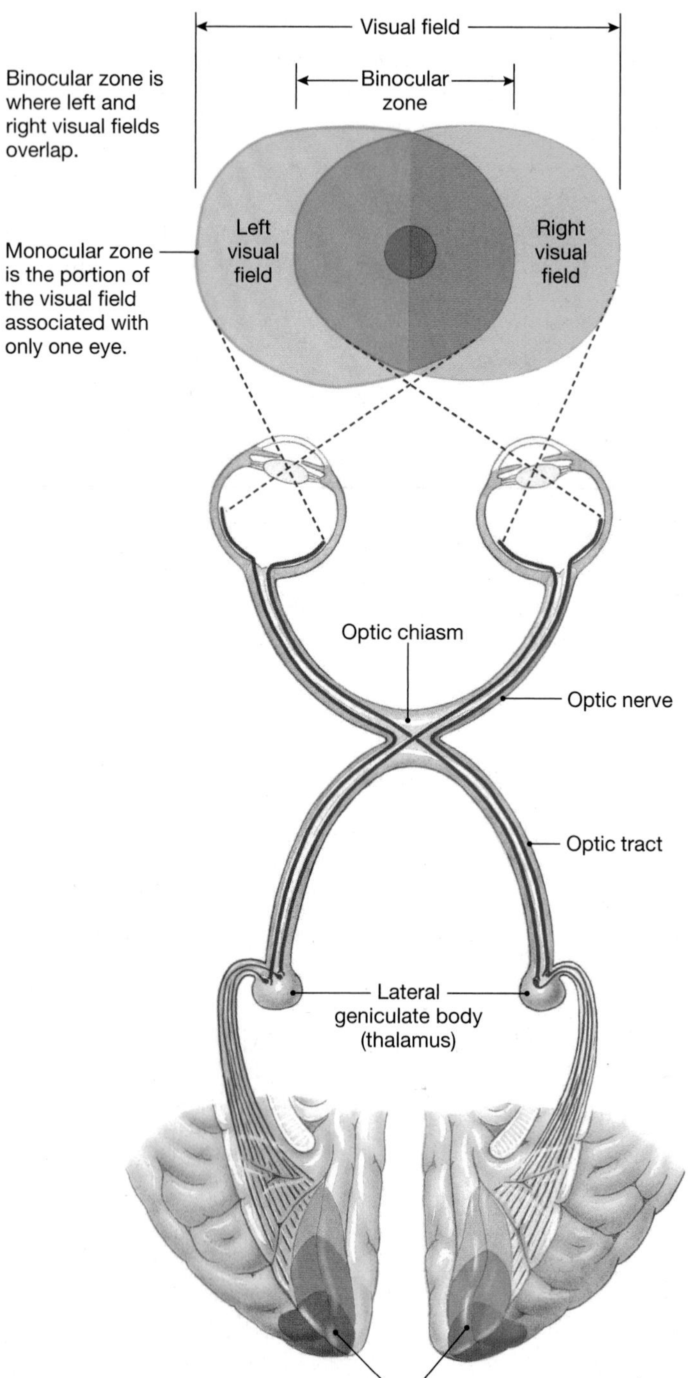

BINOCULAR VISION
The left visual field of each eye is projected to the visual cortex on the right side of the brain, and the right visual field is projected to the left visual cortex. Objects seen by both eyes fall within the binocular zone and are perceived in three dimensions. Objects seen with only one eye fall outside the binocular zone and are perceived in only two dimensions.
Visual field
Binocular zone
Binocular zone is where left and right visual fields overlap.
Monocular zone is the portion of the visual field associated with only one eye.
Left visual field
Right visual field
Optic chiasm
Optic nerve
Optic tract
Lateral geniculate body (thalamus)
Visual cortex

■ **Fig. 10.34**

brain, and information from the left side of the field is processed on the right side of the brain.

The central portion of the visual field, where left and right sides of each eye's visual field overlap, is the **binocular zone**. The two eyes have slightly different views of objects in this region, and the brain processes and integrates the two views to create three-dimensional representations of the objects. Our sense of depth perception—that is, whether one object is in front of or behind another—depends on binocular vision. Objects that fall within the visual field of only one eye are in the **monocular zone** and are viewed in two dimensions.

Once axons leave the optic chiasm, some fibers project to the midbrain, where they participate in control of eye movement or coordinate with somatosensory and auditory information for balance and movement (see Fig. 10.26). Most axons, however, project to the lateral geniculate body of the thalamus, where the optic fibers synapse onto neurons leading to the visual cortex in the occipital lobe.

The lateral geniculate body is organized in layers that correspond to the different parts of the visual field, which means that information from adjacent objects is processed together. This **topographical organization** is maintained in the visual cortex, with the six layers of neurons grouped into vertical columns. Within each portion of the visual field, information is further sorted by form, color, and movement.

The cortex merges monocular information from the two eyes to give us a binocular view of our surroundings. Information from on/off combinations of ganglion cells is translated into sensitivity to line orientation in the simplest pathways, or into color, movement, and detailed structure in the most complex. Each of these attributes of visual stimuli is processed through a separate pathway, creating a network whose complexity we are just beginning to unravel.

RUNNING PROBLEM CONCLUSION

Ménière's Disease

Anant was told about the surgical options but elected to continue medical treatment for a little longer. Over the next two months, his Ménière's disease gradually resolved. The cause of Ménière's disease is still unknown, which makes treatment difficult. To learn more about treatments that are available to alleviate Ménière's disease, do a Google search. Now check your understanding of this running problem by comparing your answers to those in the summary table.

Question	Facts	Integration and Analysis
1. In which part of the brain is sensory information about equilibrium processed?	The major equilibrium pathways project to the cerebellum. Some information is also processed in the cerebrum.	[Not applicable]
2. Subjective tinnitus occurs when an abnormality somewhere along the anatomical pathway for hearing causes the brain to perceive a sound that does not exist outside the auditory system. Starting from the ear canal, name the auditory structures in which problems may arise.	The middle ear consists of malleus, incus, and stapes, bones that vibrate with sound. The hearing portion of the inner ear consists of hair cells in the fluid-filled cochlea. The cochlear (auditory) nerve leads to the brain.	Subjective tinnitus could arise from a problem with any of the structures named. Abnormal bone growth can affect the middle ear bones. Excessive fluid accumulation in the inner ear will affect the hair cells. Neural defects may cause the cochlear nerve to fire spontaneously, creating the perception of sound.
3. When a person with positional vertigo changes position, the displaced crystals float toward the semicircular canals. Why would this cause dizziness?	The ends of the semicircular canals contain sensory cristae, each crista consisting of a cupula with embedded hair cells. Displacement of the cupula creates a sensation of rotational movement.	If the floating crystals displace the cupula, the brain will perceive movement that is not matched to sensory information coming from the eyes. The result is vertigo, an illusion of movement.
4. Compare the symptoms of positional vertigo and Ménière's disease. On the basis of Anant's symptoms, which condition do you think he has?	The primary symptom of positional vertigo is brief dizziness following a change in position. Ménière's disease combines vertigo with tinnitus and hearing loss.	Anant complains of dizzy attacks typically lasting up to an hour that come on without warning, making it more likely that Anant has Ménière's disease.

RUNNING PROBLEM CONCLUSION *(continued)*

Question	Facts	Integration and Analysis
5. Why is limiting salt (NaCl) intake suggested as a treatment for Ménière's disease?	Ménière's disease is characterized by too much endolymph in the inner ear. Endolymph is an extracellular fluid.	Reducing salt intake should also reduce the amount of fluid in the extracellular compartment because the body will retain less water. Reduction of ECF volume may decrease fluid accumulation in the inner ear.
6. Why would severing the vestibular nerve alleviate Ménière's disease?	The vestibular nerve transmits information about balance and rotational movement from the vestibular apparatus to the brain.	Severing the vestibular nerve prevents false information about body rotation from reaching the brain, thus alleviating the vertigo of Ménière's disease.

326 331 348 356 361 365 371

Chapter Summary

We all live in the same world, but different animals perceive the world differently. Dogs hear sounds we can't, for instance, and nocturnal animals have better night vision than we do. An animal can perceive only those stimuli for which it has sensory receptors. In this chapter, you explored sensory receptors in the human body and learned how each type is designed to enable us to perceive different aspects of the world around us.

Despite the unique characteristics of each sense, basic patterns emerge for sensory transduction and perception. *Molecular interactions* between signal molecules and ion channels or G protein–coupled receptors initiate many sensory pathways. Neural and non-neural sensory receptors convert chemical, mechanical, thermal, and light *energy* into electrical signals that pass along sensory neurons to CNS *control centers*. The brain processes and filters incoming signals, sometimes acting on sensory *information* without that information ever reaching conscious awareness. Many of the visceral reflexes you will study are unconscious responses to sensory input.

General Properties of Sensory Systems

1. Sensory stimuli are divided into the **special senses** of vision, hearing, taste, smell, and equilibrium, and the **somatic senses** of touch, temperature, pain, itch, and proprioception. (p. 326)
2. Sensory pathways begin with a stimulus that is converted by a receptor into an electrical potential. (p. 326)
3. If the stimulus is above threshold, action potentials pass along a sensory neuron to the central nervous system. We become aware of some stimuli but are never conscious of others. (p. 326; Tbl. 10.1)
4. Sensory receptors vary from free nerve endings to encapsulated nerve endings to specialized receptor cells. (p. 327; Fig. 10.1)
5. There are four types of sensory receptors, based on the stimulus to which they are most sensitive: **chemoreceptors, mechanoreceptors, thermoreceptors**, and **photoreceptors**. (p. 328; Tbl. 10.2)
6. Each receptor type has an **adequate stimulus**, a particular form of energy to which it is most responsive. (p. 328)
7. A stimulus that is above **threshold** creates a graded potential in the receptor. (p. 328)
8. Multiple sensory neurons may converge on one secondary neuron and create a single large **receptive field**. (p. 329; Fig. 10.2)
9. Sensory information from the spinal cord projects to the thalamus, then on to the sensory areas of the cerebral cortex. Olfactory information does not pass through the thalamus. (p. 330; Fig. 10.3)

10. The central nervous system is able to modify our level of awareness of sensory input. The **perceptual threshold** is the level of stimulus intensity necessary for us to be aware of a particular sensation. (p. 330)
11. The **modality** of a signal and its location are indicated by which sensory neurons are activated. The association of a receptor with a specific sensation is called **labeled line coding.** (p. 331)
12. Localization of auditory information depends on the timing of receptor activation in each ear. (p. 331; Fig. 10.4)
13. **Lateral inhibition** enhances the contrast between the center of the receptive field and the edges of the field. In **population coding,** the brain uses input from multiple receptors to calculate location and timing of a stimulus. (p. 332; Fig. 10.5)
14. Stimulus intensity is coded by the number of receptors activated and by the frequency of their action potentials. (p. 333; Fig. 10.6)
15. For **tonic receptors,** the sensory neuron fires action potentials as long as the **receptor potential** is above threshold. **Phasic receptors** respond to a change in stimulus intensity but adapt if the strength of the stimulus remains constant. (p. 334; Fig. 10.7)

Somatic Senses

16. There are four somatosensory modalities: touch, proprioception, temperature, and nociception. (p. 335)
17. **Secondary sensory neurons** cross the midline so that one side of the brain processes information from the opposite side of the body. Ascending sensory tracts terminate in the **somatosensory cortex.** (p. 336; Fig. 10.8)
18. Touch receptors come in many varieties. Temperature receptors sense heat and cold. (p. 338; Fig. 10.10)
19. **Nociceptors** are free nerve endings that respond to chemical, mechanical, or thermal stimuli. Their activation is perceived as pain and itch. (p. 338)
20. Some responses to irritants, such as the withdrawal reflex, are protective **spinal reflexes.** (p. 338)
21. **Fast pain** is transmitted rapidly by small, myelinated fibers. **Slow pain** is carried by small, unmyelinated fibers. Pain may be modulated either by descending pathways from the brain or by **gating** mechanisms in the spinal cord. (pp. 339, 340; Fig. 10.11, Tbl. 10.3)
22. **Referred pain** from internal organs occurs when multiple primary sensory neurons converge onto a single ascending tract. (p. 341; Fig. 10.12)

Chemoreception: Smell and Taste

23. Chemoreception is divided into the special senses of smell (**olfaction**) and taste (**gustation**). (pp. 342–344)
24. **Olfactory sensory neurons** in the nasal cavity are bipolar neurons whose pathways project directly to the olfactory cortex. (p. 343; Fig. 10.13)
25. **Odorant receptors** are G protein–coupled membrane proteins. (p. 342)
26. Taste is a combination of five sensations: sweet, sour, salty, bitter, and **umami.** (p. 344)
27. **Taste cells** are non-neural cells with membrane channels or receptors that interact with taste ligands. This interaction creates an intracellular Ca^{2+} signal that ultimately activates the primary sensory neuron. (p. 345; Fig. 10.14)

The Ear: Hearing

28. **Hearing** is our perception of the energy carried by sound waves. Sound transduction turns air waves into mechanical vibrations, then fluid waves, chemical signals, and finally action potentials. (p. 349; Fig. 10.17)
29. The **cochlea** of the inner ear contains three parallel, fluid-filled ducts. The **cochlear duct** contains the **organ of Corti,** which contains **hair cell** receptors. (p. 350; Fig. 10.18)
30. When sound bends hair cell cilia, the hair cell membrane potential changes and alters release of neurotransmitter onto sensory neurons. (p. 351; Fig. 10.19)
31. The initial processing for pitch, loudness, and duration of sound takes place in the cochlea. Localization of sound is a higher function that requires sensory input from both ears and sophisticated computation by the brain. (p. 352; Fig. 10.20, 10.4)
32. The auditory pathway goes from cochlear nerve to medulla, pons, midbrain, and thalamus before terminating in the auditory cortex. Information from both ears goes to both sides of the brain. (p. 353; Fig. 10.21)

The Ear: Equilibrium

10

33. **Equilibrium** is mediated through hair cells in the **vestibular apparatus** and **semicircular canals** of the inner ear. Gravity and acceleration provide the force that moves the cilia. (p. 355; Fig. 10.22)

The Eye and Vision

34. **Vision** is the translation of reflected light into a mental image. Photoreceptors of the **retina** transduce light energy into an electrical signal that passes to the visual cortex for processing. (p. 357)
35. The amount of light entering the eye is altered by changing the size of the pupil. (p. 357)
36. Light waves are focused by the lens, whose shape is adjusted by contracting or relaxing the **ciliary muscle.** (p. 362; Fig. 10.27)
37. Light is converted into electrical energy by the **photoreceptors** of the retina. Signals pass through bipolar neurons to ganglion cells, whose axons form the optic nerve. (p. 364; Fig. 10.29)
38. The **fovea** has the most acute vision because it has the smallest receptive fields. (p. 363)
39. **Rods** are responsible for monochromatic nighttime vision. **Cones** are responsible for high-acuity vision and color vision during the daytime. (p. 366; Fig. 10.30)
40. Light-sensitive **visual pigments** in photoreceptors convert light energy into a change in membrane potential. The visual pigment in rods is **rhodopsin.** Cones have three different visual pigments. (p. 367; Fig. 10.31)
41. Rhodopsin is composed of **opsin** and **retinal.** In the absence of light, retinal binds snugly to opsin. (p. 368; Fig. 10.32)
42. When light bleaches rhodopsin, **retinal** is released and **transducin** begins a second-messenger cascade that hyperpolarizes the rod and releases less glutamate onto the bipolar neurons. (p. 367)
43. Signals pass from photoreceptors through bipolar neurons to ganglion cells, with modulation by horizontal and amacrine cells. (p. 369; Fig. 10.33)
44. Ganglion cells called **M cells** convey information about movement. Ganglionic **P cells** transmit signals that pertain to the form and texture of objects in the visual field. (p. 370)
45. Information from one side of the visual field is processed on the opposite side of the brain. Objects must be seen by both eyes to appear three-dimensional. (p. 370; Fig. 10.34)

Questions

(Answers: p. A-1)

Level One Reviewing Facts and Terms

1. What is the role of the afferent division of the nervous system?
2. Define proprioception.
3. What are the common elements of all sensory pathways?
4. List and briefly describe the four major types of somatic receptors based on the type of stimulus to which they are most sensitive.
5. The receptors of each primary sensory neuron pick up information from a specific area, known as the ________.
6. Match the brain area with the sensory information processed there:

(a) sounds	1. midbrain
(b) odors	2. cerebrum
(c) visual information	3. medulla
(d) taste	4. cerebellum
(e) equilibrium	5. none of the above

7. The conversion of stimulus energy into a change in membrane potential is called ________. The form of energy to which a receptor responds is called its ________. The minimum stimulus required to activate a receptor is known as the ________.
8. When a sensory receptor membrane depolarizes (or hyperpolarizes in a few cases), the change in membrane potential is called the ________ potential. Is this a graded potential or an all-or-none potential?
9. Explain what is meant by adequate stimulus to a receptor.
10. The organization of sensory regions in the ________ of the brain preserves the topographical organization of receptors on the skin, eye, or other regions. However, there are exceptions to this rule. In which two senses does the brain rely on the timing of receptor activation to determine the location of the initial stimulus?
11. What is lateral inhibition?
12. Define tonic receptors and list some examples. Define phasic receptors and give some examples. Which type adapts?
13. Heart pain perceived as coming from the neck and down the left arm is an example of ________ pain.
14. What are the five basic tastes? What is the adaptive significance of each taste sensation?
15. The unit of sound wave measurement is ________, which is a measure of the frequency of sound waves per second. The loudness, or intensity, of a sound is a function of the ________ of the sound waves and is measured in ________. The range of hearing for the average human ear is from ________ to ________ [units], with the most acute hearing in the range of ________ to ________ [units].
16. Which structure of the inner ear codes sound for pitch? Define spatial coding.
17. Loud noises cause action potentials to: (choose all correct answers)
    (a) fire more frequently.
    (b) have higher amplitudes.
    (c) have longer refractory periods.
18. Once sound waves have been transformed into electrical signals in the cochlea, sensory neurons transfer information to the ________, from which collaterals then take the information to the ________ and ________. The main auditory pathway synapses in the ________ and ________ before finally projecting to the ________ in the ________.
19. The parts of the vestibular apparatus that tell our brain about our movements through space are the ________, which sense rotation, and the ________ organs, which respond to linear forces.
20. List the following structures in the sequence in which a beam of light entering the eye will encounter them: (a) aqueous humor, (b) cornea, (c) lens, (d) pupil, (e) retina.
21. The three primary colors of vision are ________, ________, and ________. White light containing these colors stimulates photoreceptors called ________. Lack of the ability to distinguish some colors is called ________.
22. List six types of cells found in the retina, and briefly describe their functions.

Level Two Reviewing Concepts

23. Compare and contrast the following:
    (a) the special senses with the somatic senses
    (b) different types of touch receptors with respect to structure, size, and location
    (c) transmission of sharp localized pain with transmission of dull and diffuse pain (include the particular fiber types involved as well as the presence or absence of myelin in your discussion)
    (d) the forms of hearing loss
    (e) convergence of retinal neurons with convergence of primary somatic sensory neurons
24. Draw three touch receptors having overlapping receptive fields (see Fig. 10.2) and number the fields 1–3. Draw a primary and secondary sensory neuron for each receptor so that they have separate ascending pathways to the cortex. Use the information in your drawing to answer this question: How many different regions of the skin can the brain distinguish using input from these three receptors?
25. Describe the neural pathways that link pain with emotional distress, nausea, and vomiting.
26. Trace the neural pathways involved in olfaction. What is G_{olf}?
27. Compare the current models of signal transduction in taste buds for salty/sour ligands and sweet/bitter/umami ligands.
28. Put the following structures in the order in which a sound wave would encounter them: (a) pinna, (b) cochlear duct, (c) stapes, (d) ion channels, (e) oval window, (f) hair cells/stereocilia, (g) tympanic membrane, (h) incus, (i) vestibular duct, (j) malleus
29. Sketch the structures and receptors of the vestibular apparatus for equilibrium. Label the components. Briefly describe how they function to notify the brain of movement.
30. Explain how accommodation by the eye occurs. What is the loss of accommodation called?
31. List four common visual problems, and explain how they occur.
32. Explain how the intensity and duration of a stimulus are coded so that the stimulus can be interpreted by the brain. (Remember, action potentials are all-or-none phenomena.)
33. Make a table of the special senses. In the first row, write these stimuli: sound, standing on the deck of a rocking boat, light, a taste, an aroma. In row 2, describe the location of the receptor for each sense. In row 3, describe the structure or properties of each receptor. In a final row, name the cranial nerve(s) that convey(s) each sensation to the brain. [p. 302]

34. Map the following terms related to vision. Add terms if you wish.

Map 1

• accommodation reflex	• depth of field	• lens
• binocular vision	• field of vision	• macula
• blind spot	• focal point	• optic chiasm
• ciliary muscle	• fovea	• optic disk
• cornea	• iris	• optic nerve
• cranial nerve III	• lateral geniculate	• phototransduction
• pupillary reflex	• visual cortex	• zonules
• retina	• visual field	

Map 2: the retina

• amacrine cells	• ganglion cells	• pigment epithelium
• bipolar cells	• horizontal cells	• retinal
• bleaching	• melanin	• rhodopsin
• cGMP	• melanopsin	• rods
• cones	• opsin	• transducin

Level Three Problem Solving

35. You are prodding your blindfolded lab partner's arm with two needle probes (with her permission). Sometimes she can tell you are using two probes. But when you probe less sensitive areas, she thinks there is just one probe. Which sense are you testing? Which receptors are being stimulated? Explain why she sometimes feels only one probe.
36. Consuming alcohol depresses the nervous system and vestibular apparatus. In a sobriety check, police officers use this information to determine if an individual is intoxicated. What kinds of tests can you suggest that would show evidence of this inhibition?
37. Often, children are brought to medical attention because of speech difficulties. If you were a clinician, which sense would you test first in such patients, and why?
38. A clinician shines a light into a patient's left eye, and neither pupil constricts. Shining the light into the right eye elicits a normal consensual reflex. What problem in the reflex pathway could explain these observations?
39. An optometrist wishes to examine a patient's retina. Which of the following classes of drugs might dilate the pupil? Explain why you did or did not select each choice.
    (a) a sympathomimetic [*mimicus*, imitate]
    (b) a muscarinic antagonist
    (c) a cholinergic agonist
    (d) an anticholinesterase
    (e) a nicotinic agonist
40. The iris of the eye has two sets of antagonistic muscles, one for dilation and one for constriction. One set of muscles is radial (radiating from the center of the pupil), and the other set is circular. Draw an iris and pupil, and arrange the muscles so that contraction of one set causes pupillary constriction and contraction of the other set causes dilation.
41. As people age, their ability to see at night decreases. What changes in the retina might explain this?

Level Four Quantitative Problems

42. The relationship between focal length (F) of a lens, object distance (P), and image distance or focal point (Q) is $1/F = 1/P + 1/Q$. Assume the distance from lens to retina is 20 mm.
    (a) For a distant object, P = infinity (∞) and $1/\infty = 0$. If Pavi sees a distant object in focus, what is the focal length of her lens in meters?
    (b) If the object moves to 1 foot in front of Pavi's lens and the lens does not change shape, what is the image distance (1 in = 2.54 cm)? What must happen to Pavi's lens for the closer image to come into focus?

Answers

Answers to Concept Check Questions

Page 328

1. Myelinated axons have a faster conduction velocity than unmyelinated axons.
2. The pinna funnels sound into the ear canal.
3. Muscle length/tension, proprioception = mechanoreception. Pressure, inflation, distension = mechanoreception. Osmolarity = mechanoreception. Temperature = thermoreception. Oxygen, glucose, pH = chemoreception.

Page 332

4. K^+ and Cl^- channels in neurons A and C are probably opening and causing hyperpolarization.

Page 334

5. Sensory neurons signal intensity of a stimulus by the rate at which they fire action potentials.
6. Irritant receptors warn the body of danger. If possible, the body should respond in some way that stops the harmful stimulus. Therefore, it is important that signals continue as long as the stimulus is present, meaning the receptors should be tonic rather than phasic.

Page 341

7. The adaptive advantage of a spinal reflex is a rapid reaction.
8. b, a, c (see Tbl. 10.3).
9. There are many examples, including receptors for taste and touch.

Page 342

10. Olfactory sensory neuron (primary neuron) → cranial nerve I → secondary neuron in olfactory bulb → olfactory tract → olfactory cortex in temporal lobe.
11. If you need help, use Figure 10.13 as the basic pattern for creating this map.
12. The knobby terminals of olfactory sensory neurons function as dendrites.
13. Olfactory neurons are bipolar neurons.

Page 346

14. Umami is associated with ingestion of the amino acid glutamate.
15. Presynaptic taste cell → primary sensory neuron through cranial nerves VII, IX, or X → medulla (synapse with secondary neuron) → thalamus → gustatory cortex in parietal lobe.

Page 348

16. A kilohertz is 1000 Hz, which means 1000 waves per second.

Page 352

17. Endolymph has high [K^+] and low [Na^+] so the electrochemical gradient favors K^+ movement into the cell.

Page 354

18. Use Figures 10.15, 10.17, and 10.21 to create your map.
19. Somatosensory information projects to the hemisphere of the brain opposite to the side of the body on which the signal originates. The location of sound is coded by the time a stimulus arrives in each hemisphere, so a signal to both hemispheres is necessary.
20. A cochlear implant would not help people with nerve deafness or conductive hearing loss. It can help only those people with sensorineural hearing loss.

Page 357

21. K^+ entry into hair cells causes depolarization.
22. When fluid builds up in the middle ear, the eardrum is unable to move freely and cannot transmit sound through the bones of the middle ear as efficiently.
23. When a dancer spots, the endolymph in the ampulla moves with each head rotation but then stops as the dancer holds the head still. This results in less inertia than if the head were continuously turning.

Page 358

24. The aqueous humor supports the cornea and lens. It also brings nutrients to and removes wastes from the epithelial layer of the cornea, which has no blood supply.

Page 358

25. (a) The sensory pathway from one eye diverges to activate motor pathways for both pupils. (b) The afferent path and its integration must be functioning because there is an appropriate response on the right side. The motor (efferent) path to the left eye must not be functioning.
26. *antagonistic*

Page 361

27. A more curved cornea causes light rays to converge more sharply. This causes the focal point to fall in front of the retina, so the person will be myopic.
28. (a) Image distance gets longer. (b) Focal length must decrease, which is accomplished by the lens becoming more rounded.
29. (a) Convex lenses focus a beam of light, and concave lenses scatter a beam of light passing through them. (b) In myopia, the focal point lies in front of the retina so a concave corrective lens increases the focal length and moves the focal point onto the retina. In hyperopia, the focal point lies behind the retina so a convex corrective lens shortens the focal length. This moves the focal point onto the retina.

Page 365

30. The tapetum lucidum reflects light, which enhances the amount of light hitting the photoreceptors.
31. In both the retina and skin, the finest discrimination occurs in the region with the smallest visual or receptive fields.
32. Damage to the macula, which surrounds the fovea, results in vision loss in the central portion of the visual field. Peripheral vision remains unaffected.

Page 366

33. Our dark vision is in black and white because only rods (black and white vision), not cones (color vision), are sensitive enough to be stimulated by such low levels of light.

Page 367

34. Use the information in Figures 10.30 and 10.32 to create your map.

Answers to Figure and Graph Questions

Page 330

Figure 10.3: The olfactory and some equilibrium pathways do not synapse in the thalamus.

Page 336

Figure 10.8: Sensations affected would be contralateral pain and temperature, and ipsilateral proprioception.

Page 341

Figure 10.12: His heart

Page 343

Figure 10.13: Multiple neurons synapsing on a single neuron is an example of convergence.

Page 348

Figure 10.16: Graph (1) shows 20 Hz waves (5 waves in the 0.25-sec interval shown means 20 waves in 1 minute). Graph (2) shows 32 Hz waves. The waves in (1) have the lower pitch because they have the lower frequency.

Page 359

Figure 10.25: The right eye is shown in this photograph.

Page 364

Figure 10.29: Six rods converge on the ganglion cell.

Page 367

Figure 10.31: The pigment in red cones absorbs light over the broadest spectrum, and blue cones absorb over the narrowest range. At 500 nm, the pigments in blue and green cones absorb light equally.

Page 368

Figure 10.32: (10,000 CNG channels $\times$ 24 cGMP/channel) $\times$ 1 transducin/6 cGMP $\times$ 1 rhodopsin/800 transducin $\times$ 1 photon/rhodopsin = 50 photons needed

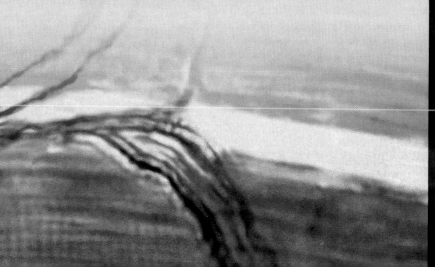

11 Efferent Division: Autonomic and Somatic Motor Control

Because a number of cells in the autonomic nervous system act in conjunction, they have relinquished their independence to function as a coherent whole.

—Otto Appenzeller and Emilio Oribe, in *The Autonomic Nervous System*, 1997

Background Basics

Muscle fibers with motor neurons terminating at neuromuscular junctions.

The picnic lunch was wonderful. You are now dozing on the grass in the warm spring sunlight as you let the meal digest. Suddenly you feel something moving across your lower leg. You open your eyes, and as they adjust to the bright light, you see a four-foot-long snake slithering over your foot. More by instinct than reason, you fling the snake into the grass while scrambling to a safe perch on top of the nearby picnic table. You are breathing heavily, and your heart is pounding.

In less than a second, your body has gone from a state of quiet rest and digestion to a state of panic and frantic activity. This reflex reaction is integrated and coordinated through the central nervous system (CNS), then carried out by the efferent division of the peripheral nervous system (PNS). The fibers of efferent neurons are bundled together into nerves that carry commands from the CNS to the muscles and glands of the body. Some nerves, called *mixed nerves,* also carry sensory information through afferent fibers [p. 242].

The efferent division of the peripheral nervous system can be subdivided into **somatic motor neurons**, which control skeletal muscles, and **autonomic neurons**, which control smooth muscle, cardiac muscle, many glands, and some adipose tissue. The somatic and autonomic divisions are sometimes called the voluntary and involuntary divisions of the nervous system, respectively. However, this distinction does not always hold true. Although most movement controlled by somatic pathways requires conscious thought, some skeletal muscle reflexes, such as swallowing and the knee jerk reflex, are involuntary. And although autonomic reflexes are mainly involuntary, a person can use biofeedback training to learn to modulate some autonomic functions, such as heart rate and blood pressure.

RUNNING PROBLEM

A Powerful Addiction

Every day, more than 1.3 billion people around the world intentionally absorb a chemical that kills about 5 million people a year. Why would people knowingly poison themselves? If you've guessed that the chemical is nicotine, you already know part of the answer. One of more than 4000 chemicals found in tobacco, nicotine is highly addictive. So powerful is this addiction that fewer than 20% of tobacco users are able to quit smoking the first time they try. Shanika, a smoker for six years, is attempting for the second time to stop smoking. The odds are in her favor this time, however, because she has made an appointment with her physician to discuss all the options available to help her break her addiction to nicotine and smoking.

378 380 381 387 389 393

We begin our study of the efferent division of the PNS by looking at the autonomic division. Then we consider the somatic motor division, as preparation for learning about muscles [Chapter 12].

The Autonomic Division

The autonomic division of the efferent nervous system (or *autonomic nervous system* for short) is also known in older writings as the *vegetative nervous system*, reflecting the observation that its functions are not under voluntary control. The word *autonomic* comes from the same roots as *autonomous*, meaning *self-governing*. Another name for the autonomic division is *visceral nervous system* because of its control over internal organs.

The autonomic division is subdivided into **sympathetic** and **parasympathetic branches** (often called the *sympathetic* and *parasympathetic nervous systems*). Some parts of the sympathetic branch were first described by the Greek physician Claudius Galen (ca. C.E. 130–200), who is famous for his compilation of anatomy, physiology, and medicine as they were known during his time. As a result of his dissections, Galen proposed that "animal spirits" flowed from the brain to the tissues through hollow nerves, creating "sympathy" between the different parts of the body. Galen's "sympathy" later gave rise to the name for the sympathetic branch. The prefix *para-*, for the parasympathetic branch, means *beside* or *alongside*.

The sympathetic and parasympathetic branches can be distinguished anatomically, but there is no simple way to separate the actions of the two branches on their targets. They are distinguished best by the type of situation in which they are most active. The picnic scene that began the chapter illustrates the two extremes at which the sympathetic and parasympathetic branches function. If you are resting quietly after a meal, the parasympathetic branch is dominant, taking command of the routine, quiet activities of day-to-day living, such as digestion. Consequently, parasympathetic neurons are sometimes said to control "rest and digest" functions.

In contrast, the sympathetic branch is dominant in stressful situations, such as the potential threat from the snake. One of the most dramatic examples of sympathetic action is the **fight-or-flight** response, in which the brain triggers massive simultaneous sympathetic discharge throughout the body. As the body prepares to fight or flee, the heart speeds up; blood vessels to muscles of the arms, legs, and heart dilate; and the liver starts to produce glucose to provide energy for muscle contraction. Digestion becomes a low priority when life and limb are threatened, and so blood is diverted from the gastrointestinal tract to skeletal muscles.

The massive sympathetic discharge that occurs in fight-or-flight situations is mediated through the hypothalamus and is a total-body response to a crisis. If you have ever been scared by the squealing of brakes or a sudden sound in the

THE AUTONOMIC DIVISION

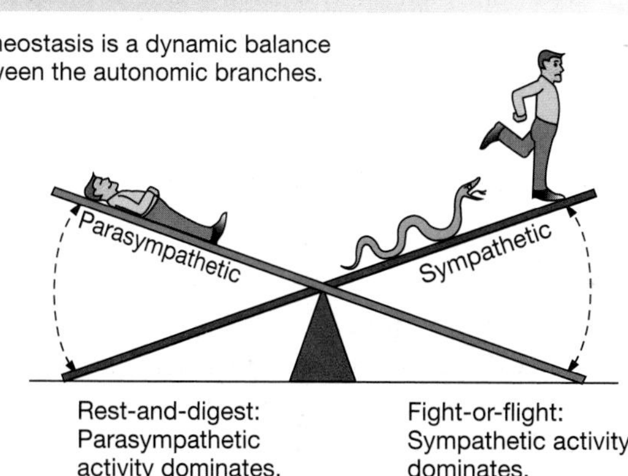

Fig. 11.1

dark, you know how rapidly the nervous system can influence multiple body systems. Most sympathetic responses are not the all-out response of a fight-or-flight reflex, however, and more importantly, activating one sympathetic pathway does not automatically activate them all.

The role of the sympathetic nervous system in mundane daily activities is as important as a fight-or-flight response. For example, one key function of the sympathetic branch is control of blood flow to the tissues. Most of the time, autonomic control of body function "seesaws" back and forth between the sympathetic and parasympathetic branches as they cooperate to fine-tune various processes (■ Fig. 11.1). Only occasionally, as in the fight-or-flight example, does the seesaw move to one extreme or the other.

Concept Check Answers: p. 396

1. The afferent division of the nervous system has what two components?
2. The central nervous system consists of the ________ and the ________.

Autonomic Reflexes Are Important for Homeostasis

The autonomic nervous system works closely with the endocrine system and the behavioral state system [p. 306] to maintain homeostasis in the body. Sensory information from somatosensory and visceral receptors goes to homeostatic control centers in the hypothalamus, pons, and medulla (■ Fig. 11.2). These centers monitor and regulate important functions such as blood pressure, temperature control, and water balance (■ Fig. 11.3).

The hypothalamus also contains neurons that act as sensors, such as *osmoreceptors*, which monitor osmolarity, and thermoreceptors, which monitor body temperature.

The hypothalamus, pons, and medulla initiate autonomic, endocrine, and behavioral responses.

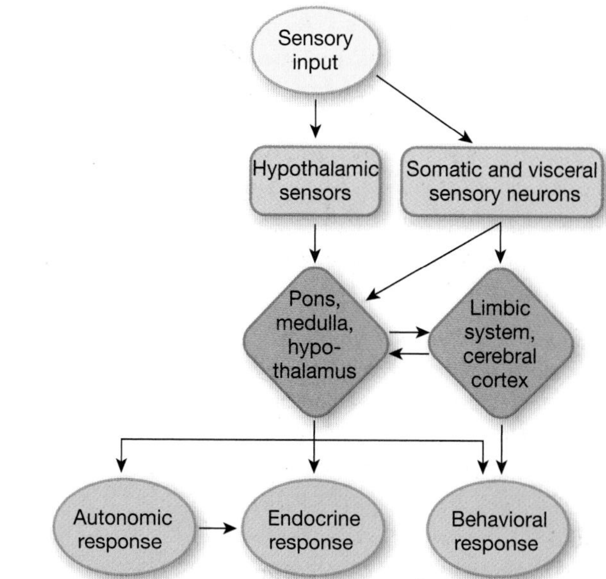

Fig. 11.2

AUTONOMIC CONTROL CENTERS

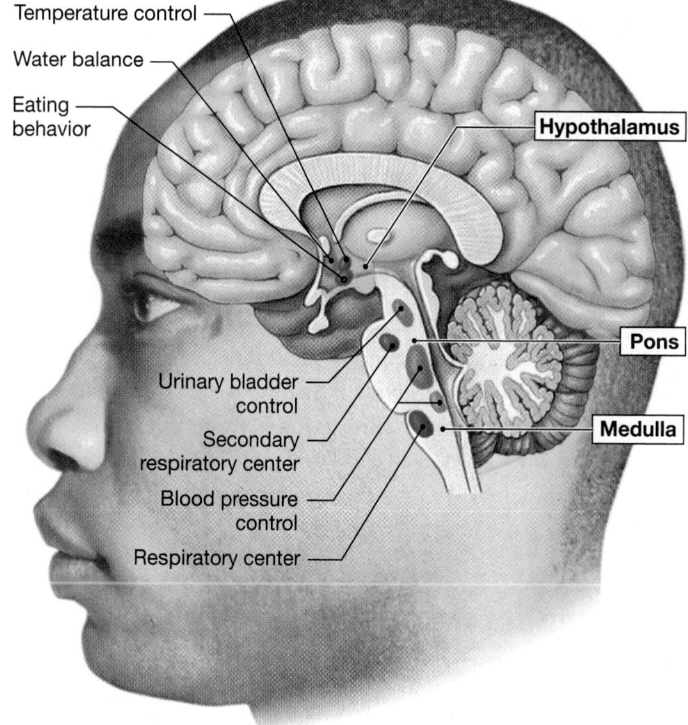

Fig. 11.3

11

RUNNING PROBLEM

Neuroscientists have learned that addictive behaviors develop because certain chemicals act as *positive reinforcers* in the brain, creating physical and psychological dependence. Nicotine is an addictive drug that enhances dopamine release in the brain's reward centers and creates pleasurable sensations. Over time, the brain also begins to associate the social aspects of cigarette smoking with pleasure, a conditioned response that makes quitting difficult. If smokers do stop smoking, they may suffer from unpleasant physical withdrawal symptoms, including lethargy, hunger, and irritability.

Q1: To avoid withdrawal symptoms, people continue to smoke, resulting in chronically elevated nicotine levels in their blood. Nicotine binds to nicotinic acetylcholine receptors (nAChR). What is the usual response of cells that are chronically exposed to elevated concentrations of a signal molecule? [Hint: p. 191]

Motor output from the hypothalamus and brain stem creates autonomic responses, endocrine responses, and behavioral responses such as drinking, food-seeking, and temperature regulation (getting out of the heat, putting on a sweater). These behavioral responses are integrated in brain centers responsible for motivated behaviors and control of movement.

In addition, sensory information integrated in the cerebral cortex and limbic system can create emotions that influence autonomic output, as Figure 11.2 illustrates. Blushing, fainting at the sight of a hypodermic needle, and "butterflies in the stomach" are all examples of emotional influences on autonomic functions. Understanding the autonomic and hormonal control of organ systems is the key to understanding the maintenance of homeostasis in virtually every system of the body.

Some autonomic reflexes are capable of taking place without input from the brain. These *spinal reflexes* [Figure 9.7, p. 299] include urination, defecation, and penile erection—body functions that can be influenced by descending pathways from the brain but do not require this input. For example, people with spinal cord injuries that disrupt communication between the brain and spinal cord may retain some spinal reflexes but lose the ability to sense or control them.

Antagonistic Control Is a Hallmark of the Autonomic Division

The sympathetic and parasympathetic branches of the autonomic nervous system display all four of Walter Cannon's properties of homeostasis: (1) preservation of the fitness of the internal environment, (2) up-down regulation by tonic control, (3) antagonistic control, and (4) chemical signals with different effects in different tissues [p. 192].

Most internal organs are under *antagonistic control,* in which one autonomic branch is excitatory and the other branch is inhibitory (see the table in Figure 11.5). For example, sympathetic innervation increases heart rate, while parasympathetic stimulation decreases it. Consequently, heart rate can be regulated by altering the relative proportions of sympathetic and parasympathetic control.

Exceptions to dual antagonistic innervation include the sweat glands and the smooth muscle in most blood vessels. These tissues are innervated only by the sympathetic branch and rely strictly on tonic (up-down) control.

Although the two autonomic branches are usually antagonistic in their control of a given target tissue, they sometimes work cooperatively on different tissues to achieve a common goal. For example, blood flow for penile erection is under control of the parasympathetic branch, and muscle contraction for sperm ejaculation is directed by the sympathetic branch.

In some autonomic pathways, the neurotransmitter receptor determines the response of the target tissue. For instance, most blood vessels contain one type of *adrenergic receptor* [p. 267] that causes smooth muscle contraction (vasoconstriction). However, some blood vessels also contain a second type of adrenergic receptor that causes smooth muscle relaxation (vasodilation). Both receptors are activated by catecholamines [p. 216]. In this example the receptor, not the chemical signal, determines the response.

Concept Check Answer: p. 396

3. Define homeostasis.

Autonomic Pathways Have Two Efferent Neurons in Series

All autonomic pathways (sympathetic and parasympathetic) consist of two neurons in series (■ Fig. 11.4). The first neuron, called the **preganglionic neuron**, originates in the central

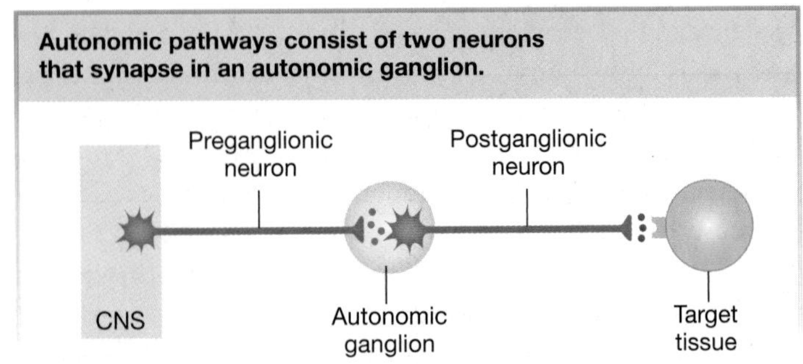

■ **Fig. 11.4**

nervous system and projects to an **autonomic ganglion** outside the CNS. There the preganglionic neuron synapses with the second neuron in the pathway, the **postganglionic neuron**. This neuron has its cell body in the ganglion and projects its axon to the target tissue. (A *ganglion* is a cluster of nerve cell bodies that lie outside the CNS. The equivalent in the CNS is a *nucleus* [p. 245].)

Divergence [p. 274] is an important feature of autonomic pathways. On average, one preganglionic neuron entering a ganglion synapses with 8 or 9 postganglionic neurons. Some synapse on as many as 32 neurons! Each postganglionic neuron may then innervate a different target, meaning that a single signal from the CNS can affect a large number of target cells simultaneously.

In the traditional view of the autonomic division, autonomic ganglia were simply a way station for the transfer of signals from preganglionic neurons to postganglionic neurons. We now know, however, that ganglia are more than a simple collection of axon terminals and nerve cell bodies: they also contain neurons that lie completely within them. These neurons enable the autonomic ganglia to act as mini-integrating centers, receiving sensory input from the periphery of the body and modulating outgoing autonomic signals to target tissues. Presumably this arrangement means that a reflex could be integrated totally within a ganglion, with no involvement of the CNS. That pattern of control is known to exist in the enteric nervous system [p. 239], which is discussed with the digestive system [Chapter 21].

RUNNING PROBLEM

Shanika's doctor congratulates her for trying once more to stop smoking. He explains that quitting is most likely to be successful if the smoker uses a combination of behavioral modification strategies and drug therapy. Currently there are three types of pharmacological treatments used for nicotine addiction: nicotine replacement, bupropion (Zyban®), and varenicline (Chantix®). Bupropion inhibits reuptake of the monoamines (dopamine, serotonin, and norepinephrine) by neurons, mimicking the effects of nicotine. Varenicline binds to nicotinic cholinergic receptors (nAChR). Nicotinic receptors are found throughout the nervous system, and evidence suggests that activation of nAChR by nicotine in certain regions of the brain plays a key role in nicotine addiction.

Q2: Cholinergic receptors are classified as either nicotinic or muscarinic, on the basis of the agonist molecules that bind to them. What happens to a postsynaptic cell when nicotine rather than ACh binds to a nicotinic cholinergic receptor?

378 380 **381** 387 389 393

Sympathetic and Parasympathetic Branches Originate in Different Regions

How, then, do the two autonomic branches differ anatomically? The main anatomical differences are (1) the pathways' point of origin in the CNS and (2) the location of the autonomic ganglia. As ■ Figure 11.5 shows, most sympathetic pathways (red) originate in the thoracic and lumbar regions of the spinal cord. *Sympathetic ganglia* are found primarily in two ganglion chains that run along either side of the bony vertebral column, with additional ganglia along the descending aorta. Long nerves (axons of postganglionic neurons) project from the ganglia to the target tissues. Because most sympathetic ganglia lie close to the spinal cord, sympathetic pathways generally have short preganglionic neurons and long postganglionic neurons.

Many parasympathetic pathways (shown in blue in Figure 11.5) originate in the brain stem, and their axons leave the brain in several cranial nerves [p. 301]. Other parasympathetic pathways originate in the sacral region (near the lower end of the spinal cord) and control pelvic organs. In general, parasympathetic ganglia are located either on or near their target organs. Consequently, parasympathetic preganglionic neurons have long axons, and parasympathetic postganglionic neurons have short axons.

Parasympathetic innervation goes primarily to the head, neck, and internal organs. The major parasympathetic tract is the **vagus nerve** (cranial nerve X), which contains about 75% of all parasympathetic fibers. This nerve carries both sensory information from internal organs to the brain and parasympathetic output from the brain to organs (■ Fig. 11.6).

Vagotomy, a procedure in which the vagus nerve is surgically cut, was an experimental technique used in the nineteenth and early twentieth centuries to study the effects of the autonomic nervous system on various organs. For a time, vagotomy was the preferred treatment for stomach ulcers because removal of parasympathetic innervation to the stomach decreased the secretion of stomach acid. However, this procedure had many unwanted side effects and has been abandoned in favor of drug therapies that treat the problem more specifically.

Concept Check Answers: p. 396

4. A nerve that carries both sensory and motor information is called a(n) ________ nerve.
5. Name the four regions of the spinal cord in order, starting from the brain stem.

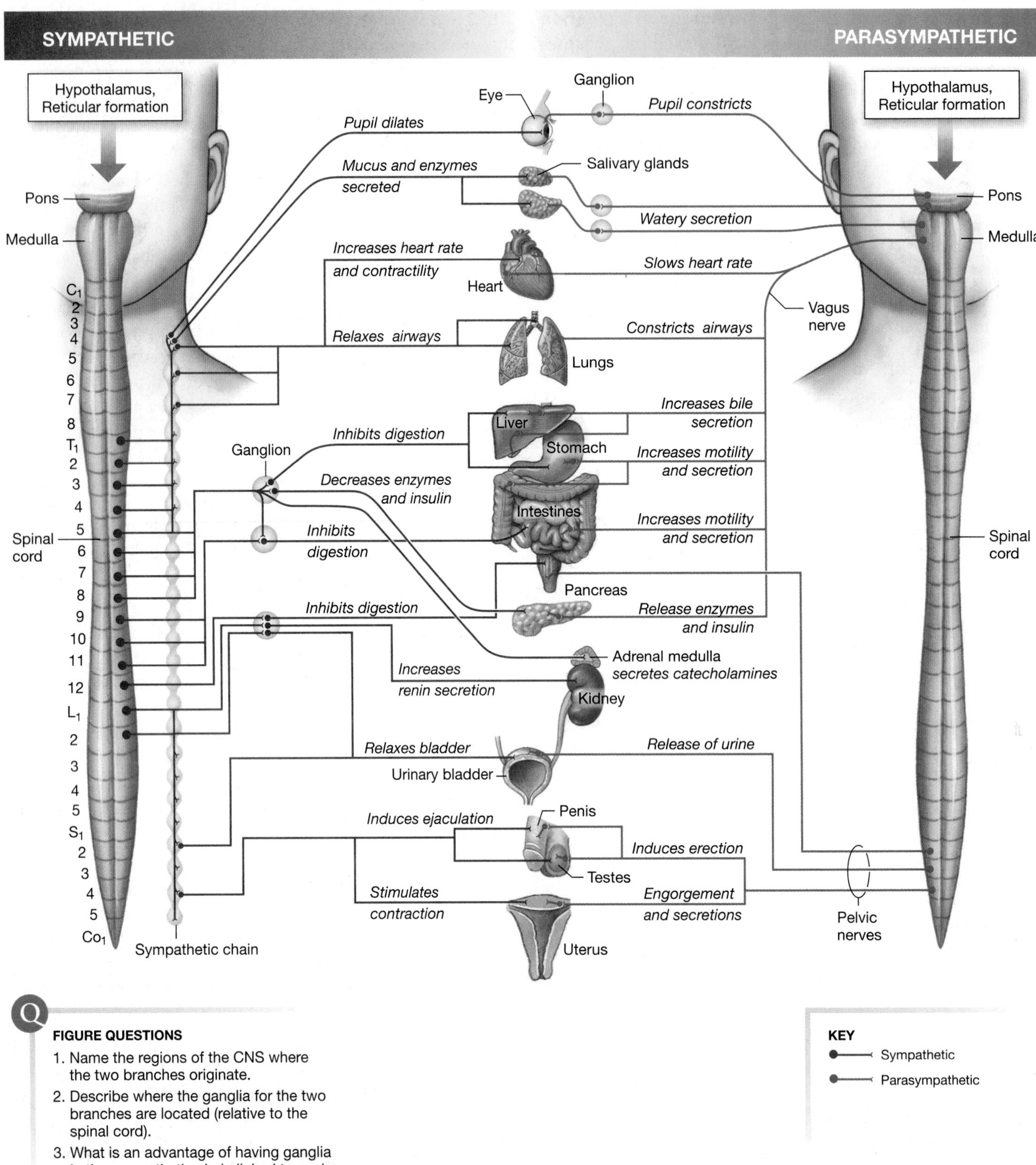

FIGURE QUESTIONS

1. Name the regions of the CNS where the two branches originate.
2. Describe where the ganglia for the two branches are located (relative to the spinal cord).
3. What is an advantage of having ganglia in the sympathetic chain linked to each other?

Fig. 11.5

EFFECTOR ORGAN	SYMPATHETIC RESPONSE	ADRENERGIC RECEPTOR	PARASYMPATHETIC RESPONSE **
Pupil of eye	Dilates	α	Constricts
Salivary glands	Mucus, enzymes	α and β_2	Watery secretion
Heart	Increases rate and force of contraction	β_1	Slows rate
Arterioles and veins	Constricts Dilates	α β_2	– –
Lungs	Bronchioles dilate	β_2*	Bronchioles constrict
Digestive tract	Decreases motility and secretion	α, β_2	Increases motility and secretion
Exocrine pancreas	Decreases enzyme secretion	α	Increases enzyme secretion
Endocrine pancreas	Inhibits insulin secretion	α	Stimulates insulin secretion
Adrenal medulla	Secretes catecholamines	–	– –
Kidney	Increases renin secretion	β_1	– –
Urinary bladder	Urinary retention	α, β_2	Release of urine
Adipose tissue	Fat breakdown	β	– –
Male and female sex organs	Ejaculation (male)	α	Erection
Uterus	Depends on stage of cycle	α, β_2	Depends on stage of cycle
Lymphoid tissue	Generally inhibitory	α, β_2	– –
	*Hormonal epinephrine only		**All parasympathetic responses are mediated by muscarinic receptors.

The Autonomic Nervous System Uses a Variety of Chemical Signals

Chemically, the sympathetic and parasympathetic branches can be distinguished by their neurotransmitters and receptors, using the following rules and ■ Figure 11.6:

1. Both sympathetic and parasympathetic preganglionic neurons release acetylcholine (ACh) onto *nicotinic cholinergic receptors* (AChR) on the postganglionic cell [p. 267].

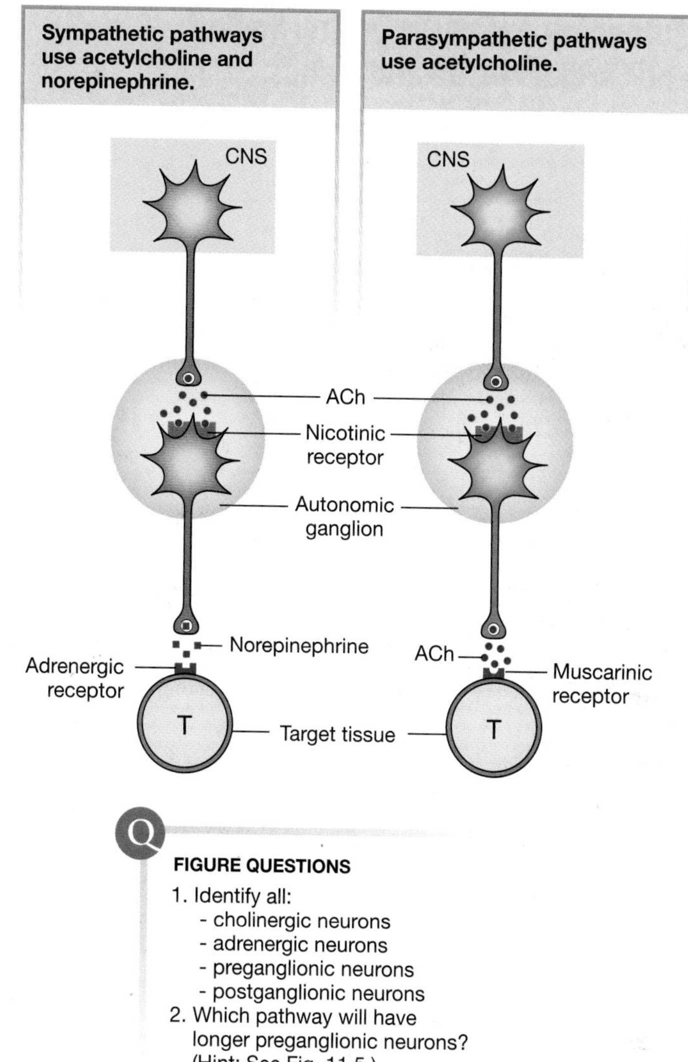
Sympathetic pathways use acetylcholine and norepinephrine.
Parasympathetic pathways use acetylcholine.
CNS
CNS
ACh
Nicotinic receptor
Autonomic ganglion
Norepinephrine
ACh
Adrenergic receptor
Muscarinic receptor
T
Target tissue
T
Q
FIGURE QUESTIONS
1. Identify all:
- cholinergic neurons
- adrenergic neurons
- preganglionic neurons
- postganglionic neurons
2. Which pathway will have longer preganglionic neurons? (Hint: See Fig. 11.5.)

■ **Fig. 11.6** ***Sympathetic and parasympathetic neurotransmitters and receptors***

2. Most postganglionic sympathetic neurons secrete norepinephrine (NE) onto *adrenergic receptors* on the target cell.
3. Most postganglionic parasympathetic neurons secrete acetylcholine onto *muscarinic cholinergic receptors* on the target cell.

However, there are some exceptions to these rules. A few sympathetic postganglionic neurons, such as those that terminate on sweat glands, secrete ACh rather than norepinephrine. These neurons are therefore called *sympathetic cholinergic neurons.*

A small number of autonomic neurons secrete neither norepinephrine nor acetylcholine and are known as *nonadrenergic, noncholinergic neurons.* Some of the chemicals they use as neurotransmitters include substance P, somatostatin, vasoactive intestinal peptide (VIP), adenosine, nitric oxide, and ATP. The nonadrenergic, noncholinergic neurons are assigned to either the sympathetic or parasympathetic branch according to where their preganglionic fibers leave the nerve cord.

Autonomic Pathways Control Smooth and Cardiac Muscle and Glands

The targets of autonomic neurons are smooth muscle, cardiac muscle, many exocrine glands, a few endocrine glands, lymphoid tissues, and some adipose tissue. The synapse between a postganglionic autonomic neuron and its target cell is called the **neuroeffector junction** (recall that targets are also called effectors).

The structure of an autonomic synapse differs from the model synapse [Figure 8.2f, p. 241]. Autonomic postganglionic axons end with a series of swollen areas at their distal ends, like beads spaced out along a string (■ Fig. 11.7a). Each of these swellings, known as a **varicosity** {*varicosus*, abnormally enlarged or swollen}, contains vesicles filled with neurotransmitter.

The branched ends of the axon lie across the surface of the target tissue, but the underlying target cell membrane does not

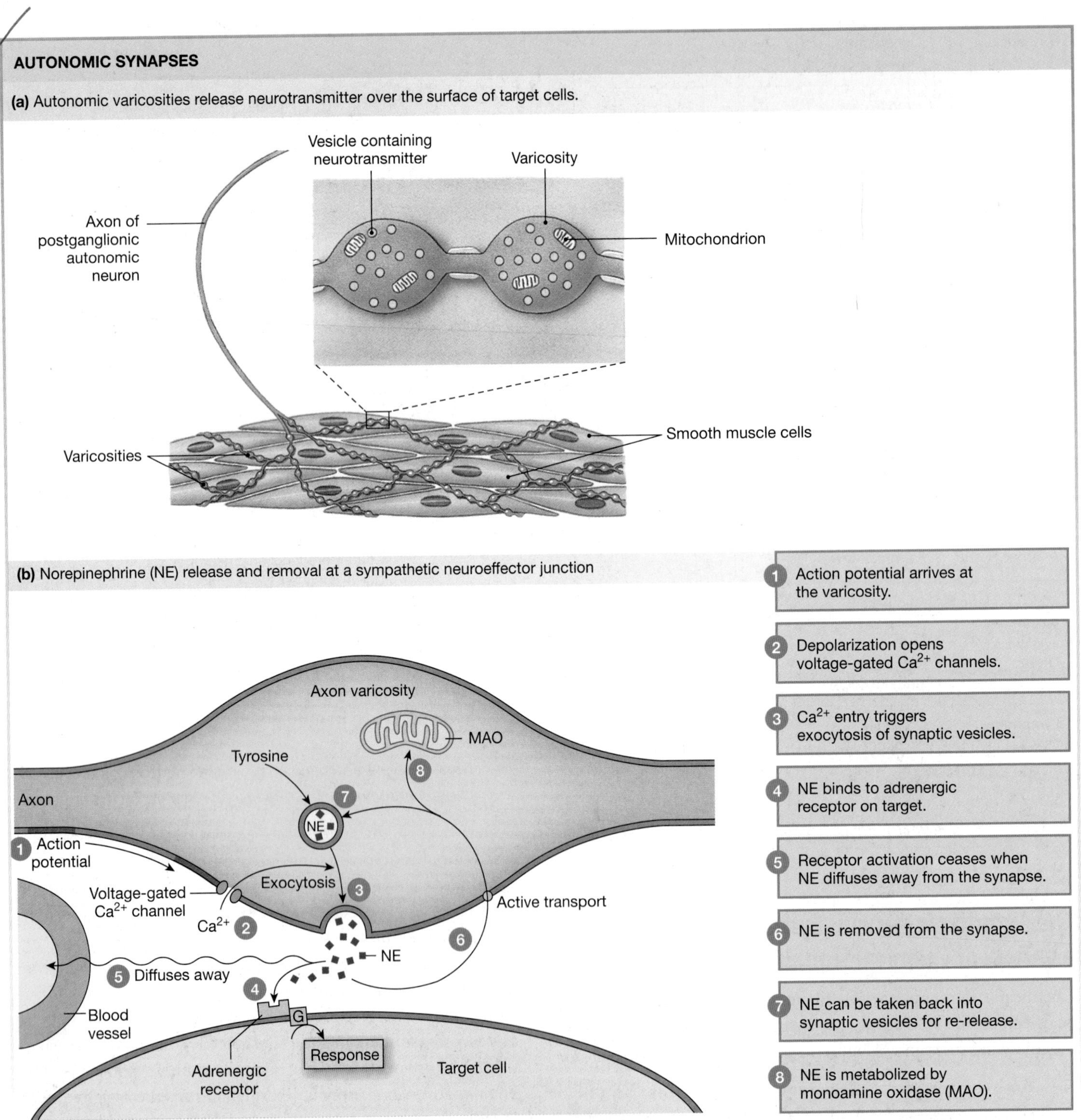
AUTONOMIC SYNAPSES
(a) Autonomic varicosities release neurotransmitter over the surface of target cells.
Vesicle containing neurotransmitter
Varicosity
Axon of postganglionic autonomic neuron
Mitochondrion
Varicosities
Smooth muscle cells
(b) Norepinephrine (NE) release and removal at a sympathetic neuroeffector junction
Axon varicosity
MAO
Tyrosine
Axon
NE
1 Action potential
Voltage-gated Ca^{2+} channel
Exocytosis 3
Ca^{2+} 2
Active transport
6
NE
5 Diffuses away
4
Blood vessel
G
Response
Adrenergic receptor
Target cell
7
8
1 Action potential arrives at the varicosity.
2 Depolarization opens voltage-gated Ca^{2+} channels.
3 Ca^{2+} entry triggers exocytosis of synaptic vesicles.
4 NE binds to adrenergic receptor on target.
5 Receptor activation ceases when NE diffuses away from the synapse.
6 NE is removed from the synapse.
7 NE can be taken back into synaptic vesicles for re-release.
8 NE is metabolized by monoamine oxidase (MAO).

■ **Fig. 11.7**

possess clusters of neurotransmitter receptors in specific sites. Instead, the neurotransmitter is simply released into the interstitial fluid to diffuse to wherever the receptors are located. The result is a less-directed form of communication than that which occurs between a somatic motor neuron and a skeletal muscle. The diffuse release of autonomic neurotransmitter means that a single postganglionic neuron can affect a large area of target tissue.

The release of autonomic neurotransmitters is subject to modulation from a variety of sources. For example, sympathetic varicosities contain receptors for hormones and for paracrines such as histamine. These modulators may either facilitate or inhibit neurotransmitter release. Some preganglionic neurons co-secrete neuropeptides along with acetylcholine. The peptides act as neuromodulators, producing slow synaptic potentials that modify the activity of postganglionic neurons [p. 276].

Autonomic Neurotransmitters Are Synthesized in the Axon

In the autonomic division, neurotransmitters are synthesized in the axon varicosities (Fig. 11.7b). The primary autonomic neurotransmitters are acetylcholine (ACh) and norepinephrine, both small molecules easily synthesized by cytoplasmic enzymes. Neurotransmitter made in the varicosities is packaged into synaptic vesicles for storage.

Neurotransmitter release follows the pattern found in other cells: depolarization—calcium signal—exocytosis [p. 166]. When an action potential arrives at the varicosity, voltage-gated Ca^{2+} channels open, Ca^{2+} enters the neuron, and the synaptic vesicle contents are released by exocytosis. Once neurotransmitters are released into the synapse, they either diffuse through the interstitial fluid until they encounter a receptor on the target cell or drift away from the synapse.

The concentration of neurotransmitter in the synapse is a major factor in the control that an autonomic neuron exerts on its target: more neurotransmitter means a longer or stronger response. The concentration of neurotransmitter in a synapse is influenced by its rate of breakdown or removal (Fig. 11.7b). Neurotransmitter activation of its receptor terminates when the neurotransmitter (1) diffuses away, (2) is metabolized by enzymes in the extracellular fluid, or (3) is actively transported into cells around the synapse. The uptake of neurotransmitter by varicosities allows neurons to reuse the chemicals.

These steps are shown for norepinephrine in Figure 11.7b. Norepinephrine is synthesized in the varicosity from the amino acid tyrosine. Once released into the synapse, norepinephrine may combine with an adrenergic receptor on the target cell, diffuse away, or be transported back into the varicosity. Inside the neuron, recycled norepinephrine is either repackaged into vesicles or broken down by **monoamine oxidase** (MAO), the main enzyme responsible for degradation of catecholamines. [See Figure 8.20, p. 273 for a similar figure on acetylcholine.]

■ Table 11.1 compares the characteristics of the two primary autonomic neurotransmitters.

Autonomic Receptors Have Multiple Subtypes

11

Sympathetic Pathways Sympathetic pathways secrete catecholamines that bind to adrenergic receptors on their target cells. Adrenergic receptors come in two varieties: α (alpha) and β (beta), with several subtypes of each. **Alpha receptors**—the most common sympathetic receptor—respond strongly to norepinephrine and only weakly to epinephrine (■ Tbl. 11.2).

The three main subtypes of beta receptors differ in their affinity for catecholamines. **β_1-receptors** respond equally strongly to norepinephrine and epinephrine. **β_2-receptors** are more sensitive to epinephrine than to norepinephrine. Interestingly, the β_2-receptors are not innervated (no sympathetic neurons terminate near them), which limits their exposure to the neurotransmitter norepinephrine. **β_3-receptors,** which are found primarily

Postganglionic Autonomic Neurotransmitters — Table 11.1

	Sympathetic Division	Parasympathetic Division
Neurotransmitter	Norepinephrine (NE)	Acetylcholine (ACh)
Receptor types	α- and β-adrenergic	Nicotinic and muscarinic cholinergic
Synthesized from	Tyrosine	Acetyl CoA + choline
Inactivation enzyme	Monoamine oxidase (MAO) in mitochondria of varicosity	Acetylcholinesterase (AChE) in synaptic cleft
Varicosity membrane transporters for	Norepinephrine	Choline

Properties of Adrenergic Receptors — Table 11.2

Receptor	Found in	Sensitivity	Effect on Second Messenger
α_1	Most sympathetic target tissues	NE > E*	Activates phospholipase C
α_2	Gastrointestinal tract and pancreas	NE > E	Decreases cAMP
β_1	Heart muscle, kidney	NE = E	Increases cAMP
β_2	Certain blood vessels and smooth muscle of some organs	E > NE	Increases cAMP
β_3	Adipose tissue	NE > E	Increases cAMP

*NE = norepinephrine, E = epinephrine.

on adipose tissue, are innervated and more sensitive to norepinephrine than to epinephrine.

All adrenergic receptors are G protein–coupled receptors rather than ion channels [p. 183]. This means that the target cell response is slower to start and usually lasts longer. The different adrenergic receptor subtypes use different second messenger pathways (Tbl. 11.2). Catecholamine binding to β-receptors increases cyclic AMP and triggers the phosphorylation of intracellular proteins. The target cell response then depends on the specific downstream pathway. For example, activation of β_1-receptors enhances cardiac muscle contraction, but activation of β_2-receptors relaxes smooth muscle in many tissues.

α_1-receptors activate phospholipase C, creating inositol trisphosphate (IP_3) and diacylglycerol (DAG) [Fig. 6.8b, p. 184]. DAG initiates a cascade that phosphorylates proteins. IP_3 opens Ca^{2+} channels, creating intracellular Ca^{2+} signals. In general, activation of α_1-receptors causes muscle contraction or secretion by exocytosis. **α_2-receptors** decrease intracellular cyclic AMP and cause smooth muscle relaxation (gastrointestinal tract) or decreased secretion (pancreas).

For all adrenergic receptors, second messenger activity in the target tissue can persist for a longer time than is usually associated with the rapid action of the nervous system. The long-lasting metabolic effects of some autonomic pathways result from modification of existing proteins or from the synthesis of new proteins. We discuss the specific effects of catecholamines on various tissues in later chapters.

Parasympathetic Pathways As a rule, parasympathetic neurons release ACh at their targets. As noted earlier, the neuroeffector junctions of the parasympathetic branch have muscarinic cholinergic receptors [p. 267]. Muscarinic receptors are all G protein–coupled receptors. Receptor activation initiates second messenger pathways, some of which open K^+ or Ca^{2+} channels. The tissue response to activation of a muscarinic receptor varies with the receptor subtype, of which there are at least five.

Concept Check — Answers: p. 396

6. In what organelle is most intracellular Ca^{2+} stored?
7. What enzyme (a) converts ATP to cAMP? (b) does cAMP activate? [Fig. 6.8a, p. 184]

The Adrenal Medulla Secretes Catecholamines

The **adrenal medulla** {*ad–*, upon + *renal*, kidney; *medulla*, marrow} is a specialized neuroendocrine tissue associated with the sympathetic nervous system. During development, the neural tissue destined to secrete the catecholamines norepinephrine and epinephrine splits into two functional entities: the sympathetic branch of the nervous system, which secretes norepinephrine, and the adrenal medulla, which secretes epinephrine primarily.

The adrenal medulla forms the core of the *adrenal glands,* which sit atop the kidneys (■ Fig. 11.8a). Like the pituitary gland, each adrenal gland is actually two glands of different embryological origin that fused during development (Fig. 11.8b). The outer portion, the *adrenal cortex,* is a true endocrine gland of epidermal origin that secretes steroid hormones [p. 214]. The adrenal medulla, which forms the small core of the gland, develops from the same embryonic tissue as sympathetic neurons and is a neurosecretory structure.

The adrenal medulla is often described as a *modified sympathetic ganglion*. Preganglionic sympathetic neurons project from the spinal cord to the adrenal medulla, where they synapse (Fig. 11.8c). However, the postganglionic neurons lack the axons that would normally project to target cells. Instead, the axonless cell bodies, called *chromaffin cells,* secrete the neurohormone epinephrine directly into the blood. In response to alarm signals from the CNS, the adrenal medulla releases large amounts of epinephrine for general distribution throughout the body as part of a fight-or-flight response.

THE ADRENAL MEDULLA SECRETES EPINEPHRINE INTO THE BLOOD.

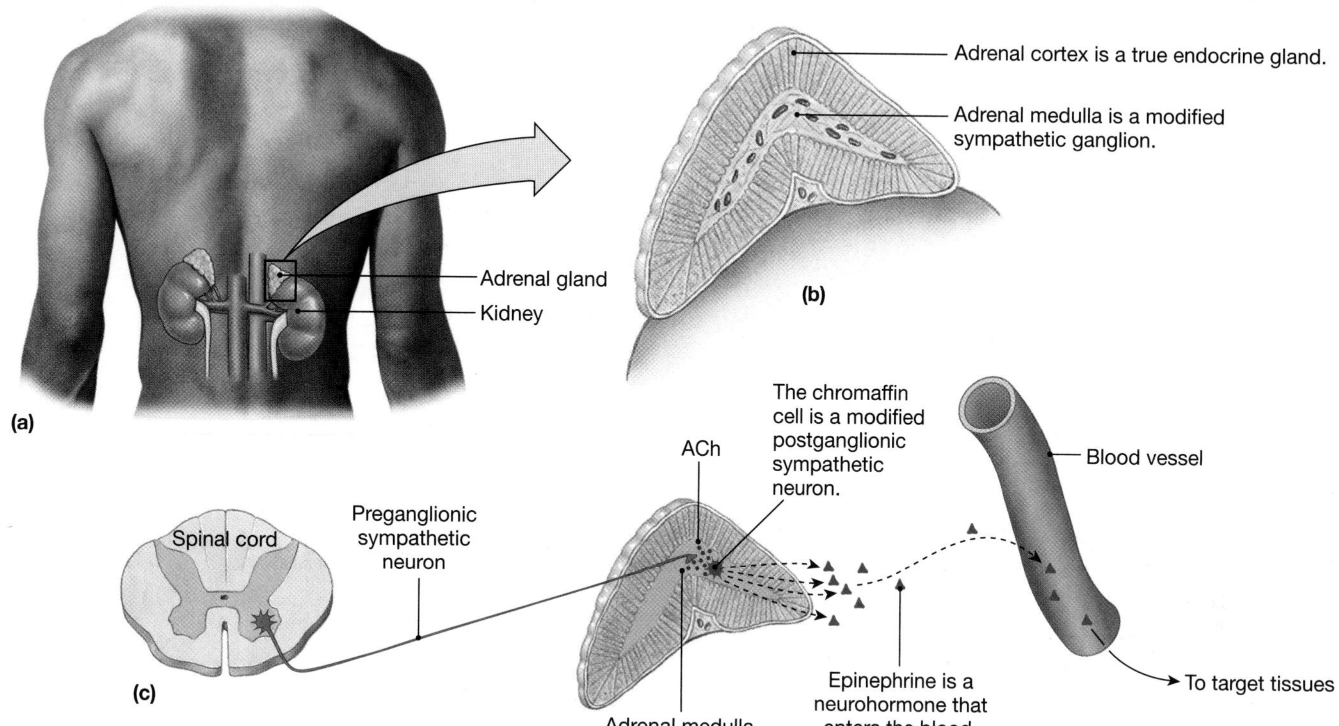
Adrenal gland
Kidney
(a)
Adrenal cortex is a true endocrine gland.
Adrenal medulla is a modified sympathetic ganglion.
(b)
Spinal cord
Preganglionic sympathetic neuron
(c)
ACh
The chromaffin cell is a modified postganglionic sympathetic neuron.
Blood vessel
Adrenal medulla
Epinephrine is a neurohormone that enters the blood.
To target tissues

■ **Fig. 11.8**

Concept Check Answers: p. 396

8. Is the adrenal medulla most like the anterior pituitary or the posterior pituitary? Explain.
9. Predict which type of ACh receptors (nicotinic or muscarinic) chromaffin cells have?

RUNNING PROBLEM

The action of nicotine on nAChR is complicated. Normally, chronic exposure of cells to a receptor agonist such as ACh or nicotine causes the cells to down-regulate their receptors. However, one research study that examined brains at autopsy found that smokers have more nAChR receptors on their cell membranes than nonsmokers do. This increase in receptor numbers, or up-regulation [p. 191], usually occurs when cells are chronically exposed to receptor *antagonists*.

Q3: Although ACh and nicotine have been shown in short-term studies to be nAChR agonists, continued exposure of the receptors to ACh has been shown to close, or inactivate, the channel. Speculate why this could explain the up-regulation of nAChR observed in smokers.

Q4: Name another ion channel you have studied that opens in response to a stimulus but inactivates and closes shortly thereafter [p. 257].

378
380
381
387
389
393

Autonomic Agonists and Antagonists Are Important Tools in Research and Medicine

The study of the two autonomic branches has been greatly simplified by advances in molecular biology. The genes for many autonomic receptors and their subtypes have been cloned, allowing researchers to create mutant receptors and study their properties. In addition, researchers have either discovered or synthesized a variety of agonist and antagonist molecules (■ Tbl. 11.3). Direct agonists and antagonists combine with the target receptor to mimic or block neurotransmitter action. Indirect agonists and antagonists act by altering secretion, reuptake, or degradation of neurotransmitters.

For example, cocaine is an indirect agonist that blocks the reuptake of norepinephrine into adrenergic nerve terminals, thereby extending norepinephrine's excitatory effect on the target. This is demonstrated by the toxic effect of cocaine on the heart, where sympathetic-induced vasoconstriction of the heart's blood vessels can result in a heart attack. *Anticholinesterases* (cholinesterase inhibitors) are indirect agonists that block ACh degradation and extend the active life of each ACh molecule. The toxic *organophosphate insecticides,* such as parathion and malathion,

Agonists and Antagonists of Neurotransmitter Receptors Table 11.3

Receptor Type	Neurotransmitter	Agonist	Antagonists	Indirect Agonists / Antagonists
Cholinergic	Acetylcholine			AChE* *inhibitors:* neostigmine
Muscarinic		Muscarine	Atropine, scopolamine	
Nicotinic		Nicotine	α-bungarotoxin (muscle only), TEA (tetraethylammonium; ganglia only), curare	
Adrenergic	Norepinephrine (NE), epinephrine			*Stimulate NE release:* ephedrine, amphetamines; *Prevents NE uptake:* cocaine
Alpha (α)		Phenylephrine	"Alpha-blockers"	
Beta (β)		Isoproterenol	"Beta-blockers": propranolol (β_1 and β_2), metoprolol (β_1 only)	

*AChE = acetylcholinesterase

CLINICAL FOCUS

Diabetes: Autonomic Neuropathy

Primary disorders of the autonomic division are rare, but the secondary condition known as **diabetic autonomic neuropathy** is quite common. This complication of diabetes often begins as a sensory neuropathy, with tingling and loss of sensation in the hands and feet. In some patients, pain is the primary symptom. About 30% of diabetic patients go on to develop autonomic neuropathies, manifested by dysfunction of the cardiovascular, gastrointestinal, urinary, and reproductive systems (abnormal heart rate, constipation, incontinence, impotence). The cause of diabetic neuropathies is controversial. Patients who have chronically elevated blood glucose levels are more likely to develop neuropathies, but the underlying metabolic pathway has not been identified. Other contributing factors for neuropathy include oxidative stress and autoimmune reactions. Currently there is no prevention for diabetic neuropathies other than controlling blood glucose levels, and no cure. The only recourse for patients is taking drugs that treat the symptoms.

are anticholinesterases. They kill insects by causing sustained contraction of their breathing muscles so that they suffocate.

Many drugs used to treat depression are indirect agonists that act either on membrane transporters for neurotransmitters (tricyclic antidepressants and selective serotonin reuptake inhibitors) or on their metabolism (monoamine oxidase inhibitors). The older antidepressant drugs that act on norepinephrine transport and metabolism (tricyclics and MAO inhibitors) may have side effects related to their actions in the autonomic nervous system, including cardiovascular problems, constipation, urinary difficulty, and sexual dysfunction {*dys*-, abnormal or ill}. The serotonin reuptake inhibitors have fewer autonomic side effects. Some of the newest drugs influence the action of both norepinephrine and serotonin.

Many new drugs have been developed from studies of agonists and antagonists. The discovery of α- and β-adrenergic receptors led to the development of drugs that block only one of the two receptor types. The drugs known as beta-blockers have given physicians a powerful tool for treating high blood pressure, one of the most common disorders in the United States today. Early α-adrenergic receptor antagonists had many unwanted side effects, but now pharmacologists can design drugs to target specific receptor subtypes. For example, tamsulosin (Flomax®) blocks alpha-1A adrenergic receptors (ADRA1A) found largely on smooth muscle of the prostate gland and bladder. Relaxing these muscles helps relieve the urinary symptoms of prostatic enlargement.

Primary Disorders of the Autonomic Nervous System Are Relatively Uncommon

Diseases and malfunction of the autonomic nervous system are relatively rare. Direct damage (trauma) to hypothalamic control centers may disrupt the body's ability to regulate water balance

RUNNING PROBLEM

After discussing her options with her doctor, Shanika decides to try the nicotine patch, one form of nicotine replacement therapy. These adhesive patches allow the former smoker to gradually decrease nicotine levels in the body, preventing withdrawal symptoms during the time the cells are down-regulating their receptors back to the normal number. When Shanika reads the package insert prior to applying her first nicotine patch, she notices a warning to keep the patches away from children. An overdose of nicotine (highly unlikely when the patch is used as directed) could result in complete paralysis of the respiratory muscles (the diaphragm and the skeletal muscles of the chest wall).

Q5: Why might excessive levels of nicotine cause respiratory paralysis?

or temperature. Generalized sympathetic dysfunction may result from systemic diseases such as cancer and diabetes mellitus. There are also some conditions, such as *multiple system atrophy*, in which the CNS control centers for autonomic functions degenerate.

In many cases of sympathetic dysfunction, the symptoms are manifested most strongly in the cardiovascular system, when diminished sympathetic input to blood vessels results in abnormally low blood pressure. Other prominent symptoms of sympathetic pathology include urinary *incontinence* {*in-*, unable + *continere*, to contain}, which is the loss of bladder control, and *impotence*, which is the inability to achieve or sustain a penile erection.

Occasionally, patients suffer from primary autonomic failure when sympathetic neurons degenerate. In the face of continuing diminished sympathetic input, target tissues up-regulate [p. 191], putting more receptors into the cell membrane to maximize the cell's response to available norepinephrine. This increase in receptor abundance leads to *denervation hypersensitivity*, a state in which the administration of exogenous adrenergic agonists causes a greater-than-expected response.

Summary of Sympathetic and Parasympathetic Branches

As you have seen in this discussion, the branches of the autonomic nervous system share some features but are distinguished by others. Many of these features are summarized in ■ Figure 11.9 and compared in ■ Table 11.4.

1. Both sympathetic and parasympathetic pathways consist of two neurons (preganglionic and postganglionic) in series. One exception to this rule is the adrenal medulla, in which postganglionic sympathetic neurons have been modified into a neuroendocrine organ.

Comparison of Sympathetic and Parasympathetic Branches — Table 11.4

	Sympathetic	Parasympathetic
Point of CNS origin	1st thoracic to 2nd lumbar segments	Midbrain, medulla, and 2nd–4th sacral segments
Location of peripheral ganglia	Primarily in paravertebral sympathetic chain; 3 outlying ganglia located alongside descending aorta	On or near target organs
Structure of region from which neurotransmitter is released	Varicosities	Varicosities
Neurotransmitter at target synapse	Norepinephrine (adrenergic neurons)	ACh (cholinergic neurons)
Inactivation of neurotransmitter at synapse	Uptake into varicosity, diffusion	Enzymatic breakdown, diffusion
Neurotransmitter receptors on target cells	Adrenergic	Muscarinic
Ganglionic synapse	ACh on nicotinic receptor	ACh on nicotinic receptor
Neuron-target synapse	NE on α- or β-adrenergic receptor	ACh on muscarinic receptor

11

■ Fig. 11.9 *ESSENTIALS*

Efferent Divisions of the Nervous System

SOMATIC MOTOR PATHWAY

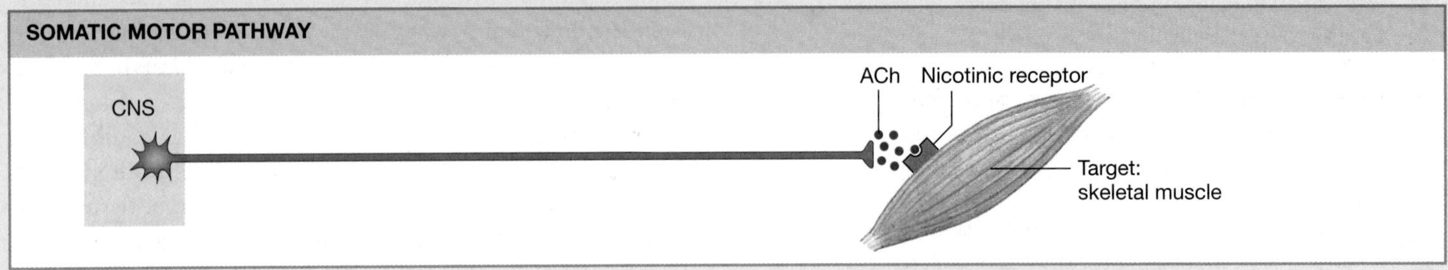

AUTONOMIC PATHWAYS

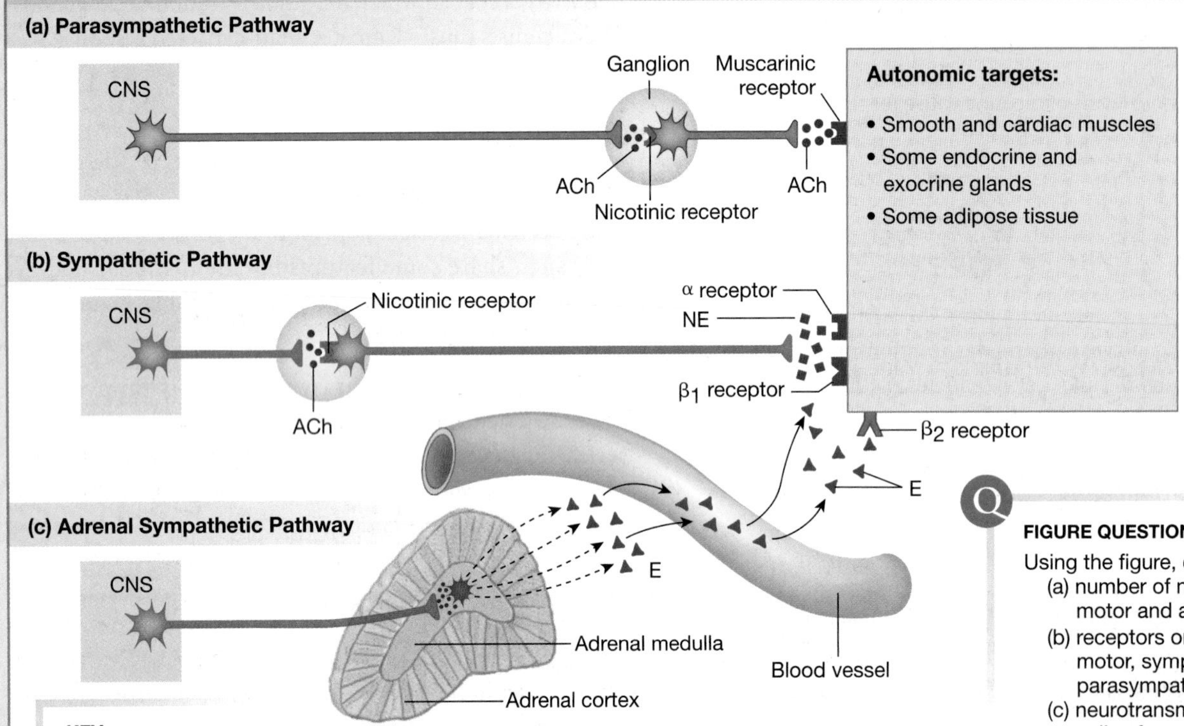

KEY
- ● ACh = acetylcholine
- ▲ E = epinephrine
- ■ NE = norepinephrine

FIGURE QUESTIONS

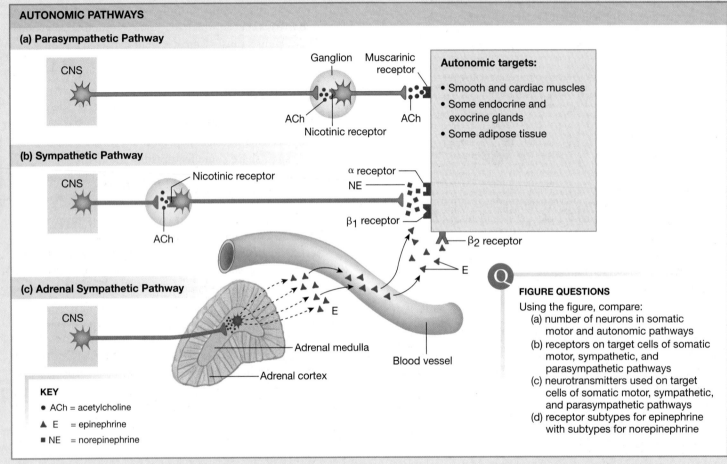

Using the figure, compare:
(a) number of neurons in somatic motor and autonomic pathways
(b) receptors on target cells of somatic motor, sympathetic, and parasympathetic pathways
(c) neurotransmitters used on target cells of somatic motor, sympathetic, and parasympathetic pathways
(d) receptor subtypes for epinephrine with subtypes for norepinephrine

Comparison of Somatic Motor and Autonomic Divisions

	SOMATIC MOTOR	AUTONOMIC
Number of neurons in efferent path	1	2
Neurotransmitter/receptor at neuron-target synapse	ACh/nicotinic	ACh/muscarinic or NE/α- or β-adrenergic
Target tissue	Skeletal muscle	Smooth and cardiac muscle; some endocrine and exocrine glands; some adipose tissue
Neurotransmitter released from	Axon terminals	Varicosities and axon terminals
Effects on target tissue	Excitatory only: muscle contracts	Excitatory or inhibitory
Peripheral components found outside the CNS	Axons only	Preganglionic axons, ganglia, postganglionic neurons
Summary of function	Posture and movement	Visceral function, including movement in internal organs and secretion; control of metabolism

2 All preganglionic autonomic neurons secrete acetylcholine onto nicotinic receptors. Most sympathetic neurons secrete norepinephrine onto adrenergic receptors. Most parasympathetic neurons secrete acetylcholine onto muscarinic receptors.
3 Sympathetic pathways originate in the thoracic and lumbar regions of the spinal cord. Parasympathetic pathways leave the CNS at the brain stem and in the sacral region of the spinal cord.
4 Most sympathetic ganglia are located close to the spinal cord (are *paravertebral*). Parasympathetic ganglia are located close to or in the target tissue.
5 The sympathetic branch controls functions that are useful in stress or emergencies (fight-or-flight). The parasympathetic branch is dominant during rest-and-digest activities.

The Somatic Motor Division

Somatic motor pathways, which control skeletal muscles, differ from autonomic pathways both anatomically and functionally (see the table in Figure 11.9). Somatic motor pathways have a single neuron that originates in the CNS and projects its axon to the target tissue, which is always a skeletal muscle. Somatic pathways are always excitatory, unlike autonomic pathways, which may be either excitatory or inhibitory.

A Somatic Motor Pathway Consists of One Neuron

The cell bodies of somatic motor neurons are located either in the ventral horn of the spinal cord [p. 299] or in the brain, with a long single axon projecting to the skeletal muscle target (Fig. 11.9). These myelinated axons may be a meter or more in length, such as the somatic motor neurons that innervate the muscles of the foot and hand.

Somatic motor neurons branch close to their targets. Each branch divides into a cluster of enlarged axon terminals that lie on the surface of the skeletal muscle fiber (■ Fig. 11.10a). This branching structure allows a single motor neuron to control many muscle fibers at one time.

The synapse of a somatic motor neuron on a muscle fiber is called the **neuromuscular junction**, or NMJ (Fig. 11.10b). Like all other synapses, the NMJ has three components: (1) the motor neuron's presynaptic axon terminal filled with synaptic vesicles and mitochondria, (2) the synaptic cleft, and (3) the postsynaptic membrane of the skeletal muscle fiber.

In addition, the neuromuscular junction includes extensions of Schwann cells that form a thin layer covering the top of the axon terminals. For years it was thought that this cell layer simply provided insulation to speed up the conduction of the action potential, but we now know that Schwann cells secrete a variety of chemical signal molecules. These signal molecules play a critical role in the formation and maintenance of neuromuscular junctions.

On the postsynaptic side of the neuromuscular junction, the muscle cell membrane that lies opposite the axon terminal is modified into a **motor end plate**, a series of folds that look like shallow gutters (Fig. 11.10c). Along the upper edge of each gutter, nicotinic ACh receptor (nAChR) channels cluster together in an active zone. Between the axon and the muscle, the synaptic cleft is filled with a fibrous matrix whose collagen fibers hold the axon terminal and the motor end plate in the proper alignment. The matrix also contains **acetylcholinesterase** (AChE), the enzyme that rapidly deactivates ACh by degrading it into acetyl and choline [p. 271].

Concept Check Answer: p. 396

10. Is the ventral horn of the spinal cord, which contains the cell bodies of somatic motor neurons, gray matter or white matter?

11

The Neuromuscular Junction Contains Nicotinic Receptors

As in all neurons, action potentials arriving at the axon terminal open voltage-gated Ca^{2+} channels in the membrane. Calcium diffuses into the cell down its electrochemical gradient, triggering the release of ACh-containing synaptic vesicles. Acetylcholine diffuses across the synaptic cleft and combines with nicotinic receptor channels (nAChR) on the skeletal muscle membrane (Fig. 11.10d).

The nAChR channels of skeletal muscle are similar but not identical to the nicotinic ACh receptors found on neurons. This difference is illustrated by the fact that the snake toxin *α-bungarotoxin* binds to nicotinic skeletal muscle receptors but not to those in autonomic ganglia. Both muscle and neuronal nAChR proteins have five subunits encircling the central pore. However, skeletal muscle has α, β, δ, and ε subunit isoforms, while neuronal nAChR has only the α and β isoforms. These isoforms of nAChR are inactivated [p. 251] with extended exposure to ACh or other agonists.

Nicotinic cholinergic receptors are chemically gated ion channels with two binding sites for ACh (Fig. 11.10d). When ACh binds to the receptor, the channel gate opens and allows monovalent cations to flow through. In skeletal muscle, net Na^+ entry into the muscle fiber depolarizes it, triggering an action potential that causes contraction of the skeletal muscle cell.

Acetylcholine acting on a skeletal muscle's motor end plate is always excitatory and creates muscle contraction. There is no antagonistic innervation to relax skeletal muscles. Instead, relaxation occurs when the somatic motor neurons are

Somatic Motor Neurons and the Neuromuscular Junction

(a) The **neuromuscular junction** consists of axon terminals, motor end plates on the muscle membrane, and Schwann cell sheaths.
Somatic motor neuron branches at its distal end.
Skeletal muscle fiber
Motor end plate
(b) The **motor end plate** is a region of muscle membrane that contains high concentrations of ACh receptors.
Schwann cell sheath
Axon terminal
Mitochondrion
Motor end plate
(c) The neuromuscular junction
Synaptic vesicle (ACh)
Presynaptic membrane
Synaptic cleft
Nicotinic ACh receptors
Postsynaptic membrane is modified into a motor end plate.
(d) An action potential arrives at the axon terminal, causing voltage-gated Ca^{2+} channels to open. Calcium entry causes synaptic vesicles to fuse with the presynaptic membrane and release ACh into the synaptic cleft.
Synaptic vesicle (ACh)
Ca^{2+}
Ca^{2+}
ACh
Voltage-gated Ca^{2+} channel
Acetyl + choline
AChE
Nicotinic receptor
Skeletal muscle fiber
Acetylcholine (ACh) is metabolized by acetylcholinesterase (AChE).
(e) The nicotinic cholinergic receptor binds two ACh molecules, opening a nonspecific monovalent cation channel. The open channel allows Na^+ and K^+ to pass. Net Na^+ influx depolarizes the muscle fiber.
Na^+
K^+
ACh
K^+
Na^+
Closed channel
Open channel

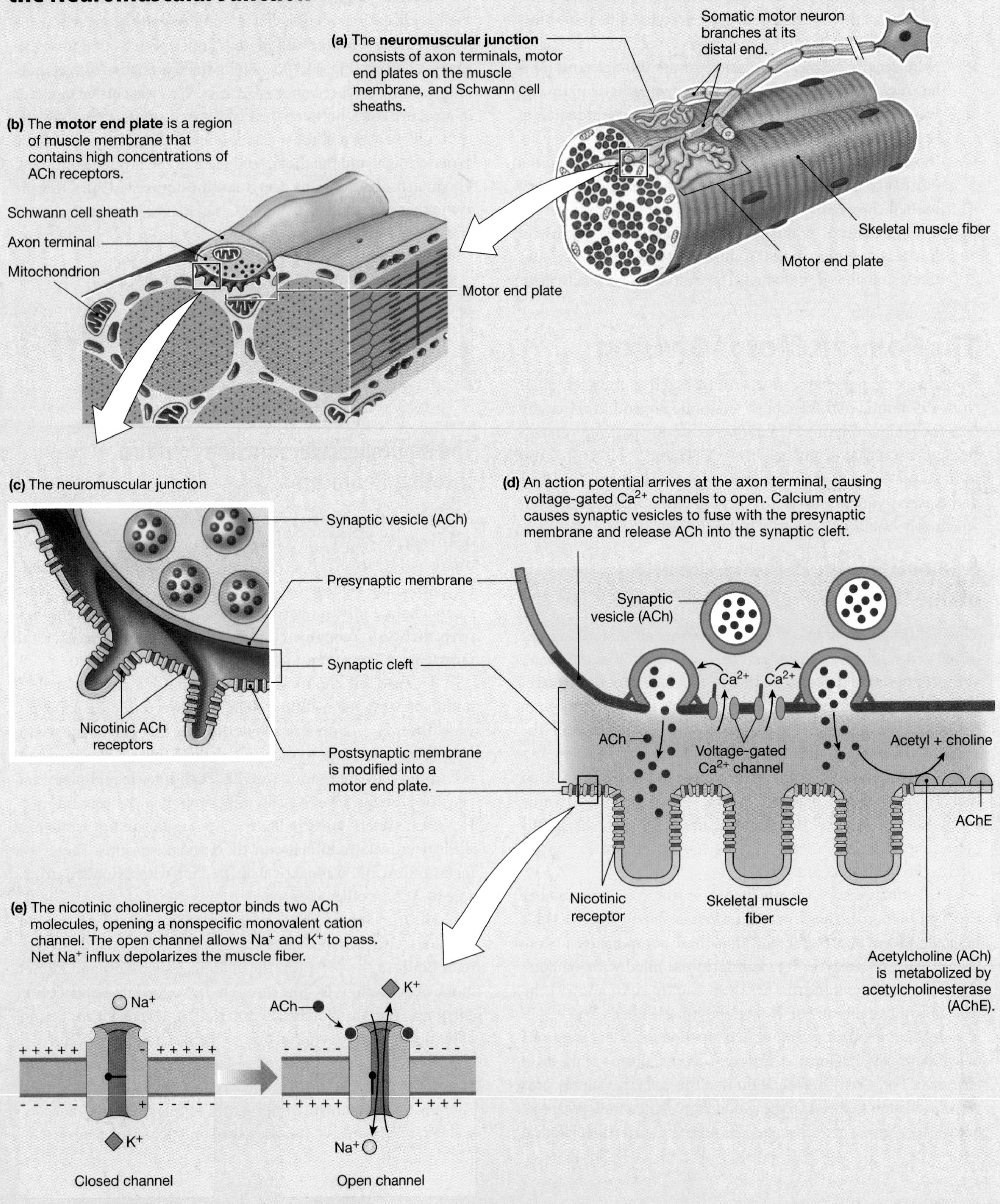

inhibited in the CNS, preventing ACh release. You will learn later about how inhibition of somatic motor pathways controls body movement.

Somatic motor neurons do more than simply create contractions: they are necessary for muscle health. "Use it or lose it" is a cliché that is very appropriate to the dynamics of muscle mass because disrupting synaptic transmission at the neuromuscular junction has devastating effects on the entire body. Without communication between the motor neuron and the muscle, the skeletal muscles for movement and posture weaken, as do the skeletal muscles for breathing. In the severest cases, loss of respiratory function can be fatal unless the patient is placed on artificial ventilation. Myasthenia gravis, a disease characterized by loss of ACh receptors, is the most common disorder of the neuromuscular junction.

Concept Check Answers: p. 396

11. Compare gating and ion selectivity of acetylcholine receptor-channels in the motor end plate with that of ion channels along the axon of a somatic motor neuron.
12. A nonsmoker who chews nicotine-containing gum might notice an increase in heart rate, a function controlled by sympathetic neurons. Postganglionic sympathetic neurons secrete norepinephrine, not ACh, so how could nicotine affect heart rate?
13. Patients with myasthenia gravis have a deficiency of ACh receptors on their skeletal muscles and have weak muscle function as a result. Why would administration of an anticholinesterase drug (one that inhibits acetylcholinesterase) improve muscle function in these patients?

RUNNING PROBLEM CONCLUSION

A Powerful Addiction

Shanika is determined to stop smoking this time because her grandfather, a smoker for many years, was just diagnosed with lung cancer. Finding that the patch alone does not stop her craving for a cigarette, she attends behavioral modification classes. In these classes, she learns to avoid situations that make her likely to smoke and to substitute other activities, such as chewing gum, for smoking. After six months, Shanika proudly informs her family that she thinks she has kicked the habit.

Nicotine replacement may not be the ideal treatment for smoking cessation because although the former smoker is no longer exposed to cigarette smoke, the nicotine addiction may remain. Varenicline (Chantix®) acts as a partial nAChR agonist and may help break the addiction. However, unwanted side effects, such as nightmares and psychological disturbances, have been reported with its use. Some smokers have quit with the help of bupropion (Zyban®), a drug that is also used as an antidepressant. Two drugs that act on cannabinoid receptors [p. 269] were effective in clinical trials but were withdrawn from the market after people taking them exhibited serious psychological side effects. A vaccine against nicotine is currently in the last stages of clinical trials in the United States. To learn more about nicotine addiction and smoking cessation programs, see Medline Plus (*www.nlm.nih.gov/medlineplus*). Check your understanding of this running problem by comparing your answers to the information in the following summary table.

Question	Facts	Integration and Analysis
1. What is the usual response of cells that are chronically exposed to elevated concentrations of a signal molecule?	A cell exposed to elevated concentrations of a signal molecule will decrease (down-regulate) its receptors for that molecule.	Down-regulation of receptors allows a cell to respond normally even if the concen-tration of ligand is elevated.
2. What happens to a postsynaptic cell when nicotine rather than ACh binds to a nicotinic cholinergic receptor?	Nicotine is an agonist of ACh. Agonists mimic the activity of a ligand.	Nicotine binding to a nAChR will open ion channels in the postsynaptic cell, and the cell will depolarize. This is the same effect that ACh binding creates.
3. Although ACh and nicotine have been shown in short-term studies to be nAChR agonists, continued exposure of the receptors to ACh has been shown to close, or inactivate, the channel. Speculate why this could explain the up-regulation of nAChR observed in smokers.	Chronic exposure to an agonist usually causes down-regulation. Chronic exposure to an antagonist usually causes up-regulation. nAChR channels open with initial exposure to agonists but close with continued exposure.	Although nicotine is a short-term agonist, it appears to be having the same effect as an antagonist during long-term exposure. With both antagonism and the inactivation described here, the cell's activity decreases. The cell subsequently up-regulates the number of receptors in an attempt to restore activity.

11

RUNNING PROBLEM CONCLUSION *(continued)*

Question	Facts	Integration and Analysis
4. Name another ion channel you have studied that opens in response to a stimulus but inactivates and closes shortly thereafter.	The voltage-gated Na^+ channel of the axon first opens, then closes when the inactivation gate shuts.	N/A
5. Why might excessive levels of nicotine cause respiratory paralysis?	Nicotinic receptors are found at the neuromuscular junction that controls skeletal muscle contraction. The diaphragm and chest wall muscles that regulate breathing are skeletal muscles.	The nicotinic receptors of the neuromuscular junction are not as sensitive to nicotine as are those of the CNS and autonomic ganglia. However, excessively high amounts of nicotine will activate the nAChR of the motor end plate, causing the muscle fiber to depolarize and contract. The continued presence of nicotine keeps these ion channels open, and the muscle remains depolarized. In this state, the muscle is unable to contract again, resulting in paralysis.

378 380 381 387 389 393

Chapter Summary

The autonomic and somatic motor divisions are the output pathways of the peripheral nervous system. *Communication* among the sensory and efferent divisions and the CNS depends primarily on chemical signaling and *molecular interactions* between neurotransmitters and their receptors. *Homeostasis* requires constant surveillance of body parameters by the nervous system, working in conjunction with the endocrine and immune systems. As you learn about the function of other body systems, you will continue to revisit the principles of communication and coordination.

The Autonomic Division

1. The efferent division of the peripheral nervous system consists of **somatic motor neurons,** which control skeletal muscles, and **autonomic neurons,** which control smooth muscle, cardiac muscle, many glands, lymphoid tissue, and some adipose tissue. (p. 378)
2. The autonomic division is subdivided into a **sympathetic branch** and a **parasympathetic branch**. (p. 389; Tbl. 11.4)
3. The maintenance of homeostasis within the body is a balance of autonomic control, endocrine control, and behavioral responses. (p. 379; Fig. 11.2)
4. The autonomic division is controlled by centers in the hypothalamus, pons, and medulla. Some autonomic reflexes are spinal reflexes. Many of these can be modulated by input from the brain. (p. 379; Fig. 11.3)
5. The two autonomic branches demonstrate Cannon's properties of homeostasis: maintenance of the internal environment, tonic control, antagonistic control, and variable tissue responses. (p. 380)
6. All autonomic pathways are composed of a **preganglionic neuron** from the CNS that synapses with a **postganglionic neuron** in an **autonomic ganglion**. Autonomic ganglia can modulate and integrate information passing through them. (p. 380; Fig. 11.4)
7. Most sympathetic pathways originate in the thoracic and lumbar regions of the spinal cord. Most sympathetic ganglia lie either close to the spinal cord or along the descending aorta. (p. 382; Fig. 11.5)
8. Parasympathetic pathways originate in the brain stem or the sacral region of the spinal cord. Parasympathetic ganglia are located on or near their target organs. (p. 382; Fig. 11.5)

9. The primary autonomic neurotransmitters are **acetylcholine** and **norepinephrine**. All preganglionic neurons secrete ACh onto **nicotinic cholinergic receptors**. As a rule, postganglionic sympathetic neurons secrete norepinephrine onto **adrenergic receptors**, and postganglionic parasympathetic neurons secrete ACh onto **muscarinic cholinergic receptors**. (p. 383, 385; Fig. 11.6, Tbl. 11.1)
10. The synapse between an autonomic neuron and its target cells is called the **neuroeffector junction**. (p. 384)
11. Postganglionic autonomic axons end with **varicosities** from which neurotransmitter is released. (p. 384, 386; Figs. 11.7, 11.8)
12. The **adrenal medulla** secretes epinephrine and is controlled by sympathetic preganglionic neurons. (p. 386; Fig. 11.8)
13. Adrenergic receptors are G protein–coupled receptors. Alpha receptors respond most strongly to norepinephrine. **β_1-receptors** respond equally to norepinephrine and epinephrine. **β_2-receptors** are not associated with sympathetic neurons and respond most strongly to epinephrine. **β_3-receptors** respond most strongly to norepinephrine. (p. 386, 390; Fig. 11.9, Tbl. 11.2)
14. Cholinergic muscarinic receptors are also G protein–coupled receptors. (p. 386)

The Somatic Motor Division

15. Somatic motor pathways, which control skeletal muscles, have a single neuron that originates in the CNS and terminates on a skeletal muscle. Somatic motor neurons are always excitatory and cause muscle contraction. (p. 390; Fig. 11.9)
16. A single **somatic motor neuron** controls many muscle fibers at one time. (p. 391)
17. The synapse of a somatic motor neuron on a muscle fiber is called the **neuromuscular junction.** The muscle cell membrane is modified into a **motor end plate** that contains a high concentration of nicotinic ACh receptors. (p. 392; Fig. 11.10)
18. ACh binding to nicotinic receptor opens cation channels. Net Na^+ entry into the muscle fiber depolarizes the fiber. Acetylcholine in the synapse is broken down by the enzyme **acetylcholinesterase.** (p. 392; Fig. 11.10)

Questions

Answers: p. A1

Level One Reviewing Facts and Terms

1. Name the two efferent divisions of the peripheral nervous system. What type of effectors does each control?
2. The autonomic nervous system is sometimes called the ________ nervous system. Why is this an appropriate name? List some functions controlled by the autonomic nervous system.
3. What are the two branches of the autonomic nervous system? How are these branches distinguished from each other anatomically and physiologically?
4. Which neurosecretory endocrine gland is closely allied to the sympathetic branch?
5. Neurons that secrete acetylcholine are described as ________ neurons, whereas those that secrete norepinephrine are called either ________ or ________ neurons.
6. List four things that can happen to autonomic neurotransmitters after they are released into a synapse.
7. The main enzyme responsible for catecholamine degradation is ________, abbreviated as ________.
8. Somatic motor pathways
   (a) are excitatory or inhibitory?
   (b) are composed of a single neuron or a preganglionic and a postganglionic neuron?
   (c) synapse with glands or with smooth, cardiac, or skeletal muscle?
9. What is acetylcholinesterase? Describe its action.
10. What kind of receptor is found on the postsynaptic cell in a neuromuscular junction?

Level Two Reviewing Concepts

11. What is the advantage of divergence of neural pathways in the autonomic nervous system?
12. Compare and contrast
    (a) neuroeffector junctions and neuromuscular junctions.
    (b) alpha, beta, muscarinic, and nicotinic receptors. Describe where each is found and the ligands that bind to them.
13. **Concept map:** Use the following terms to make a map comparing the somatic motor division and the sympathetic and parasympathetic branches of the autonomic division.
    You may add terms.

• acetylcholine	• muscarinic receptor
• adipose tissue	• nicotinic receptor
• alpha receptor	• norepinephrine
• autonomic division	• one-neuron pathway
• beta receptor	• parasympathetic branch
• cardiac muscle	• skeletal muscle
• cholinergic receptor	• smooth muscle
• efferent division	• somatic motor division
• endocrine gland	• sympathetic branch
• exocrine gland	• two-neuron pathway
• ganglion	

14. Compare and contrast
    (a) autonomic ganglia and CNS nuclei.
    (b) the adrenal medulla and the posterior pituitary gland.
    (c) axon terminals and varicosities.
15. If a target cell's receptor is ________ (use items in left column), the neuron(s) releasing neurotransmitter onto the receptor must be ________ (use all appropriate items from the right column).

(a) nicotinic cholinergic	1. somatic motor neuron
(b) adrenergic α	2. autonomic preganglionic neuron
(c) muscarinic cholinergic	3. sympathetic postganglionic neuron
(d) adrenergic β	4. parasympathetic postganglionic neuron

16. Ganglia contain the cell bodies of (choose all that apply)
    (a) somatic motor neurons
    (b) preganglionic autonomic neurons
    (c) interneurons

(d) postganglionic autonomic neurons
(e) sensory neurons

Level Three Problem Solving

17. If nicotinic receptor channels allow both Na^+ and K^+ to flow through, why does Na^+ influx exceed K^+ efflux? [*Hint:* p. 163]
18. You have discovered a neuron that innervates an endocrine cell in the intestine. To learn more about this neuron, you place a marker substance at the endocrine cell synapse. The marker is taken into the neuron and transported in a vesicle by retrograde axonal transport to the nerve cell body.
    (a) By what process is the marker probably taken into the axon terminal?
    (b) The nerve cell body is found in a ganglion very close to the endocrine cell. To which branch of the peripheral nervous system does the neuron probably belong? (Be as specific as you can.)
    (c) Which neurotransmitter do you predict will be secreted by the neuron onto the endocrine cell?

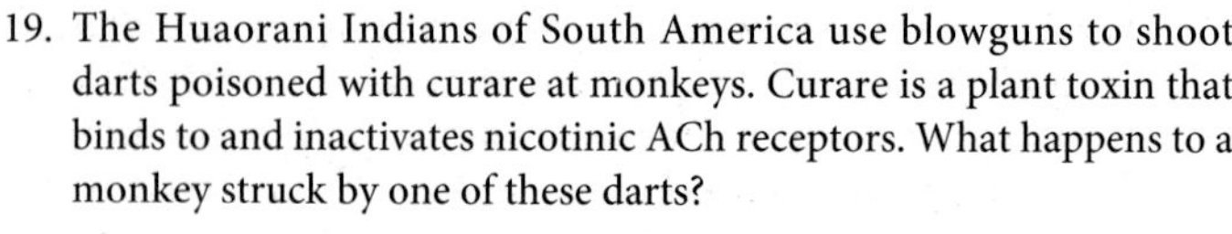

19. The Huaorani Indians of South America use blowguns to shoot darts poisoned with curare at monkeys. Curare is a plant toxin that binds to and inactivates nicotinic ACh receptors. What happens to a monkey struck by one of these darts?

Level Four Quantitative Problems

20. The U.S. Centers for Disease Control and Prevention (CDC) conduct biennial Youth Risk Behavior Surveys (YRBS) in which they ask high school students to self-report risky behaviors such as alcohol consumption and smoking. The graphs that follow were created from data in the latest report on cigarette smoking among American high school students. *Current smoking* is defined as smoking cigarettes on at least one day in the 30 days preceding the survey. *(http://www.cdc.gov/mmwr/pdf/ss/ss5905.pdf)*
    (a) What can you say about cigarette smoking among high school students in the period from 1991 to 2009?
    (b) Which high school students are most likely to be smokers? Least likely to be smokers?

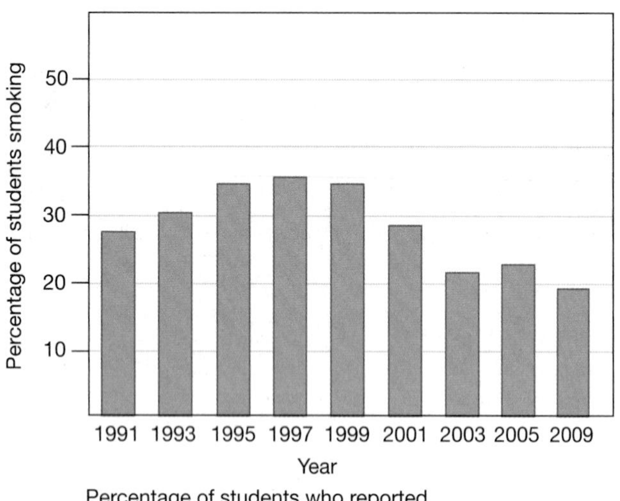

Percentage of students who reported current smoking (1991-2009)

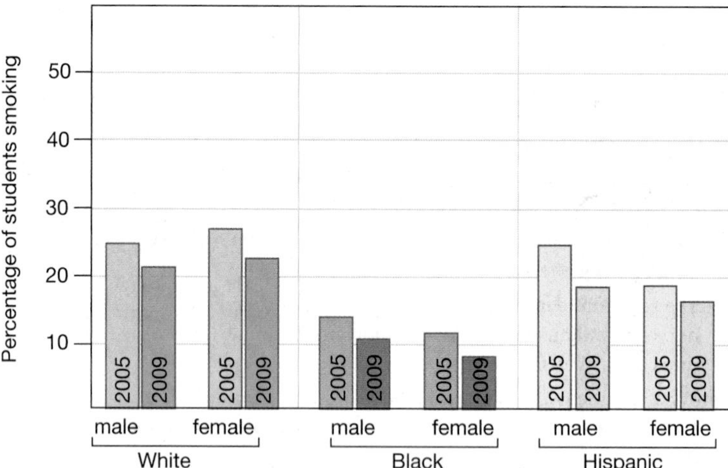

Percentage of students in 2005 and 2009 who reported current smoking, separated by sex and race/ethnicity*

* Other race/ethnic groups are not shown because the numbers were too small for meaningful statistical analysis.

Answers

Answers to Concept Check Questions

Page 379

1. The afferent division consists of sensory receptors and sensory neurons.
2. The CNS consists of the brain and spinal cord.

Page 380

3. Homeostasis is the maintenance of a relatively stable internal environment.

Page 381

4. Mixed nerves carry sensory and motor signals.
5. The regions of the spinal cord are cervical, thoracic, lumbar, and sacral.

Page 386

6. Ca^{2+} is stored in the endoplasmic reticulum.
7. (a) Adenylyl cyclase converts ATP to cAMP. (b) cAMP activates protein kinase A.

Page 387

8. The adrenal medulla is neurosecretory and therefore like the posterior pituitary.
9. Chromaffin cells are modified postganglionic neurons, so they have nicotinic receptors.

Page 391

10. The ventral horn is gray matter.

Page 393

11. The nAChR of the motor end plate is a ligand-gated monovalent cation (Na^+ and K^+) channel. The axon contains voltage-gated channels, with separate channels for Na^+ and K^+ [p. 257].
12. Postganglionic sympathetic neurons are activated by ACh acting on nicotinic receptors. This means that nicotine also excites sympathetic neurons, such as those that increase heart rate.

13. Anticholinesterase drugs decrease the rate at which ACh is broken down at the motor end plate. Slower breakdown rate allows ACh to remain active at the motor end plate for a longer time and helps offset the decrease in active receptors.

Answers to Figure Questions

Page 382

Figure 11.5: 1. Sympathetic pathways originate in the thoracic and lumbar regions of the spinal cord; parasympathetic pathways originate in the brain stem or sacral region. 2. Sympathetic ganglia are located close to the spinal column or along the descending aorta (not shown); parasympathetic ganglia are located on or near their target organs. 3. Connections between the sympathetic ganglia allow rapid communication within the sympathetic branch.

Page 383

Figure 11.6: 1. The three neurons that secrete ACh are cholinergic. The one neuron that secretes norepinephrine is adrenergic. The cell bodies of preganglionic neurons are in the CNS; the cell bodies of postganglionic neurons are in a ganglion. 2. Parasympathetic pathways have the longer preganglionic neurons.

Page 387

Figure 11.9: (a) Somatic has one neuron, autonomic has two. (b) Somatic motor targets have nicotinic ACh receptors, parasympathetic targets have muscarinic ACh receptors, and sympathetic targets have adrenergic receptors. (c) Somatic motor and parasympathetic pathways use ACh; sympathetic uses norepinephrine. (d) Epinephrine is most active on β_1- and β_2-receptors; norepinephrine is most active on β_1- and α-receptors. (e) Sympathetic ganglia are close to the CNS; parasympathetic ganglia are closer to their target tissues.

11

12 Muscles

A muscle is . . . an engine, capable of converting chemical energy into mechanical energy. It is quite unique in nature, for there has been no artificial engine devised with the great versatility of living muscle.

—Ralph W. Stacy and John A. Santolucito, in *Modern College Physiology,* 1966

Background Basics

Striated muscle sarcomeres and sarcoplasmic reticulum between the fibers

It was his first time to be the starting pitcher. As he ran from the bullpen onto the field, his heart was pounding and his stomach felt as if it were tied in knots. He stepped onto the mound and gathered his thoughts before throwing his first practice pitch. Gradually, as he went through the familiar routine of throwing and catching the baseball, his heart slowed and his stomach relaxed. It was going to be a good game.

The pitcher's pounding heart, queasy stomach, and movements as he runs and throws all result from muscle contraction. Our muscles have two common functions: to generate motion and to generate force. Our skeletal muscles also generate heat and contribute significantly to the homeostasis of body temperature. When cold conditions threaten homeostasis, the brain may direct our muscles to shiver, creating additional heat.

The human body has three types of muscle tissue: skeletal muscle, cardiac muscle, and smooth muscle. Most **skeletal muscles** are attached to the bones of the skeleton, enabling these muscles to control body movement. **Cardiac muscle** {*kardia,* heart} is found only in the heart and moves blood through the circulatory system. Skeletal and cardiac muscles are classified as **striated muscles** {*stria,* groove} because of their alternating light and dark bands seen under the light microscope (■ Fig. 12.1a, b).

Smooth muscle is the primary muscle of internal organs and tubes, such as the stomach, urinary bladder, and blood vessels. Its primary function is to influence the movement of material into, out of, and within the body. An example is the passage of food through the gastrointestinal tract. Viewed under the microscope, smooth muscle lacks the obvious cross-bands of striated muscles (Fig. 12.1c). Its lack of banding results from the less organized arrangement of contractile fibers within the muscle cells.

Skeletal muscles are often described as voluntary muscles, and smooth and cardiac muscle as involuntary. However, this is not a precise classification. Skeletal muscles can contract without conscious direction, and we can learn a certain degree of conscious control over some smooth and cardiac muscle.

Skeletal muscles are unique in that they contract only in response to a signal from a somatic motor neuron. They cannot initiate their own contraction, and their contraction is not influenced directly by hormones.

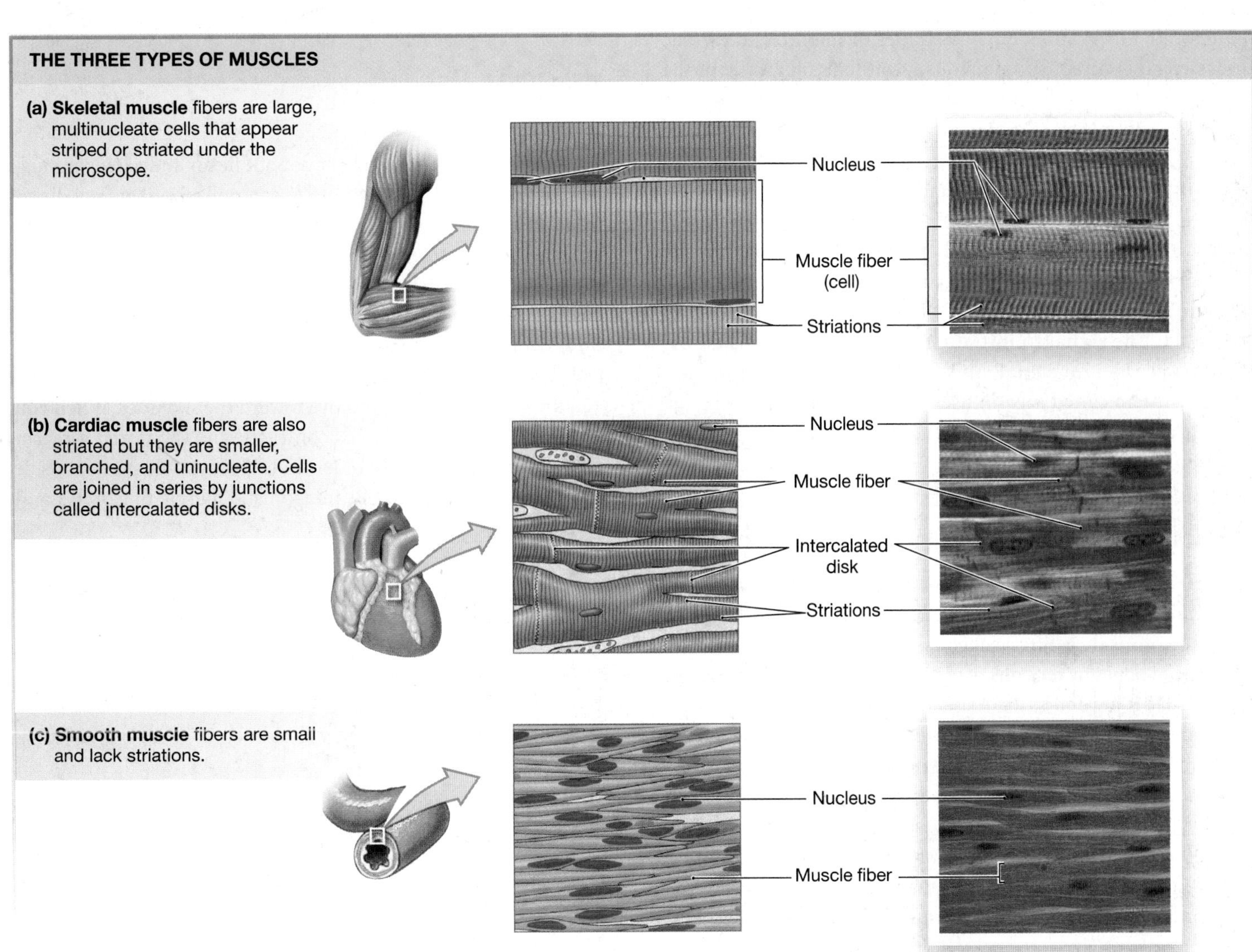

THE THREE TYPES OF MUSCLES
(a) Skeletal muscle fibers are large, multinucleate cells that appear striped or striated under the microscope.
Nucleus
Muscle fiber (cell)
Striations
(b) Cardiac muscle fibers are also striated but they are smaller, branched, and uninucleate. Cells are joined in series by junctions called intercalated disks.
Nucleus
Muscle fiber
Intercalated disk
Striations
(c) Smooth muscle fibers are small and lack striations.
Nucleus
Muscle fiber

■ **Fig. 12.1**

12

RUNNING PROBLEM

Periodic Paralysis

This morning, Paul, age 6, gave his mother the fright of her life. One minute he was happily playing in the backyard with his new beagle puppy. The next minute, after sitting down to rest, he could not move his legs. In answer to his screams, his mother came running and found her little boy unable to walk. Panic-stricken, she scooped him up, brought him into the house, and dialed 9-1-1. But as she hung up the phone and prepared to wait for the paramedics, Paul got to his feet and walked over to her. "I'm OK now, Mom," he announced. "I'm going outside."

400 413 415 425 428 435

In contrast, cardiac and smooth muscle have multiple levels of control. Their primary extrinsic control arises through autonomic innervation, but some types of smooth and cardiac muscle can contract spontaneously, without signals from the central nervous system. In addition, the activity of cardiac and some smooth muscle is subject to modulation by the endocrine system. Despite these differences, smooth and cardiac muscle share many properties with skeletal muscle.

In this chapter we discuss skeletal and smooth muscle anatomy and contraction, and conclude by comparing the properties of skeletal muscle, smooth muscle, and cardiac muscle. All three muscle types have certain properties in common. The signal to initiate muscle contraction is an intracellular calcium signal, and movement is created when a motor protein called *myosin* uses energy from ATP to change its conformation. The details of these processes vary with the different muscle types.

Skeletal Muscle

Skeletal muscles make up the bulk of muscle in the body and constitute about 40% of total body weight. They position and move the skeleton, as their name suggests. Skeletal muscles are usually attached to bones by **tendons** made of collagen [p. 87]. The **origin** of a muscle is the end of the muscle that is attached closest to the trunk or to the more stationary bone. The **insertion** of the muscle is the more *distal* {*distantia,* distant} or more mobile attachment.

When the bones attached to a muscle are connected by a flexible joint, contraction of the muscle moves the skeleton. The muscle is called a **flexor** if the centers of the connected bones are brought closer together when the muscle contracts, and the movement is called *flexion*. The muscle is called an **extensor** if the bones move away from each other when the muscle contracts, and the movement is called *extension*.

ANTAGONISTIC MUSCLES

Antagonistic muscle groups move bones in opposite directions. Muscle contraction can pull on a bone but cannot push a bone away.

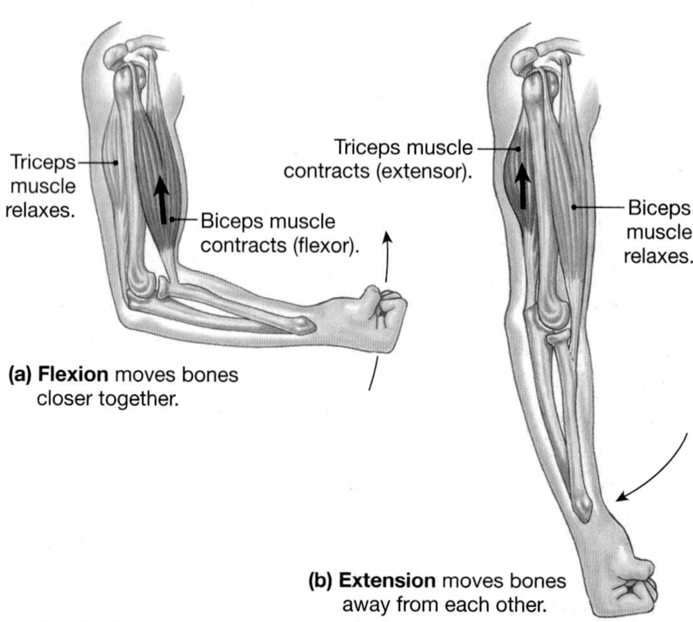
Triceps muscle relaxes.
Biceps muscle contracts (flexor).
Triceps muscle contracts (extensor).
Biceps muscle relaxes.
(a) Flexion moves bones closer together.
(b) Extension moves bones away from each other.

Fig. 12.2

Most joints in the body have both flexor and extensor muscles, because a contracting muscle can pull a bone in one direction but cannot push it back. Flexor-extensor pairs are called **antagonistic muscle groups** because they exert opposite effects. Figure 12.2 shows a pair of antagonistic muscles in the arm: the *biceps brachii* {*brachion,* arm}, which acts as the flexor, and the *triceps brachii,* which acts as the extensor. When you do a "dumbbell curl" with a weight in your hand, the biceps muscle contracts and the hand and forearm move toward the shoulder. When you lower the weight, the triceps contracts, and the flexed forearm moves away from the shoulder. In each case, when one muscle contracts and shortens, the antagonistic muscle must relax and lengthen.

Concept Check Answer: p. 439

1. Identify as many pairs of antagonistic muscle groups in the body as you can. If you cannot name them, point out the probable location of the flexor and extensor of each group.

Skeletal Muscles Are Composed of Muscle Fibers

Muscles function together as a unit. A skeletal muscle is a collection of muscle cells, or **muscle fibers**, just as a nerve is a collection of neurons. Each skeletal muscle fiber is a long, cylindrical cell with up to several hundred nuclei near the surface of the fiber (see Anatomy Summary, ■ Fig. 12.3a). Skeletal muscle fibers are the largest cells in the body, created by the fusion of many individual embryonic muscle cells. Committed stem cells called **satellite cells** lie just outside the muscle fiber membrane. Satellite cells activate and differentiate into muscle when needed for muscle growth and repair.

The fibers in a given muscle are arranged with their long axes in parallel (Fig. 12.3a). Each skeletal muscle fiber is sheathed in connective tissue, with groups of adjacent muscle fibers bundled together into units called **fascicles**. Collagen, elastic fibers, nerves, and blood vessels are found between the fascicles. The entire muscle is enclosed in a connective tissue sheath that is continuous with the connective tissue around the muscle fibers and fascicles and with the tendons holding the muscle to underlying bones.

Muscle Fiber Anatomy Muscle physiologists, like neurobiologists, use specialized vocabulary (■ Tbl. 12.1). The cell membrane of a muscle fiber is called the **sarcolemma** {*sarkos,* flesh + *lemma,* shell}, and the cytoplasm is called the **sarcoplasm**. The main intracellular structures in striated muscles are **myofibrils** {*myo-,* muscle}, highly organized bundles of contractile and elastic proteins that carry out the work of contraction.

Skeletal muscles also contain extensive **sarcoplasmic reticulum** (SR), a form of modified endoplasmic reticulum that wraps around each myofibril like a piece of lace (Figs. 12.3b, ■ 12.4). The sarcoplasmic reticulum consists of longitudinal tubules with enlarged end regions called the **terminal cisternae** {*cisterna,* a reservoir}. The sarcoplasmic reticulum concentrates and sequesters Ca^{2+} {*sequestrare,* to put in the hands of a trustee} with the help of a *Ca^{2+}-ATPase* in the SR membrane. Calcium release from the SR creates calcium signals that play a key role in contraction in all types of muscle.

The terminal cisternae are adjacent to and closely associated with a branching network of **transverse tubules**, also known as **t-tubules** (Fig. 12.4). One t-tubule and its two flanking terminal cisternae are called a *triad*. The membranes of t-tubules are a continuation of the muscle fiber membrane, which makes the lumen of t-tubules continuous with the extracellular fluid.

Table 12.1 Muscle Terminology

General Term	Muscle Equivalent
Muscle cell	Muscle fiber
Cell membrane	Sarcolemma
Cytoplasm	Sarcoplasm
Modified endoplasmic reticulum	Sarcoplasmic reticulum

To understand how this network of t-tubules deep inside the muscle fiber communicates with the outside, take a lump of soft clay and poke your finger into the middle of it. Notice how the outside surface of the clay (analogous to the surface membrane of the muscle fiber) is now continuous with the sides of the hole that you poked in the clay (the membrane of the t-tubule).

T-tubules allow action potentials to move rapidly from the cell surface into the interior of the fiber so that they reach the terminal cisternae nearly simultaneously. Without t-tubules, the action potential would reach the center of the fiber only by conduction of the action potential through the cytosol, a slower and less direct process that would delay the response time of the muscle fiber.

The cytosol between the myofibrils contains many glycogen granules and mitochondria. Glycogen, the storage form of glucose found in animals, is a reserve source of energy. Mitochondria provide much of the ATP for muscle contraction through oxidative phosphorylation of glucose and other biomolecules.

muscles need energy source

Myofibrils Are Muscle Fiber Contractile Structures

One muscle fiber contains a thousand or more myofibrils that occupy most of the intracellular volume, leaving little space for cytosol and organelles (Fig. 12.3b). Each myofibril is composed of several types of proteins: the contractile proteins *myosin* and *actin,* the regulatory proteins *tropomyosin* and *troponin,* and the giant accessory proteins *titin* and *nebulin.*

12

Myosin {*myo-,* muscle} is a motor protein with the ability to create movement [p. 74]. Various isoforms of myosin occur in different types of muscle and help determine the muscle's speed of contraction. Each myosin molecule is composed of protein chains that intertwine to form a long tail and a pair of tadpole-like heads (Fig. 12.3e). The rodlike tail is stiff, but the protruding myosin heads have an elastic hinge region where the heads join the rods. This hinge region allows the heads to swivel around their point of attachment.

Each myosin head has two protein chains: a heavy chain and a smaller *light chain*. The heavy chain is the *motor domain* that binds ATP and uses the energy from ATP's high-energy phosphate bond to create movement. Because the motor domain acts as an enzyme, it is considered a **myosin ATPase**. The heavy chain also contains a binding site for actin. In skeletal muscle, about 250 myosin molecules join to create a **thick filament**. Each thick filament is arranged so that the myosin heads are clustered at each end of the filament, and the central region of the filament is a bundle of myosin tails.

Actin {*actum,* to do} is a protein that makes up the **thin filaments** of the muscle fiber. One actin molecule is a globular protein (*G-actin*), represented in Figure 12.3f by a round ball. Usually, multiple G-actin molecules polymerize to form long

Skeletal Muscles

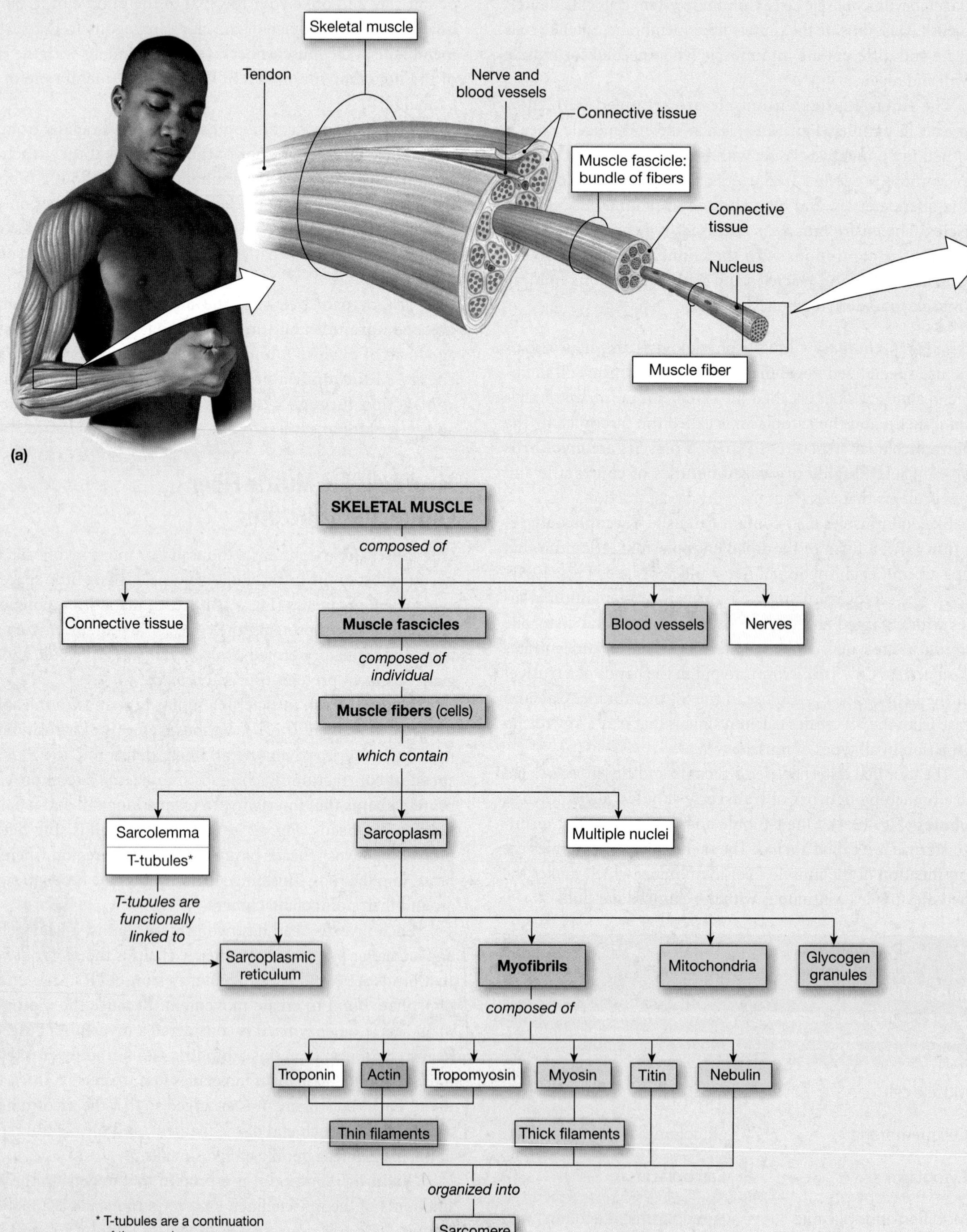

Ultrastructure of Muscle

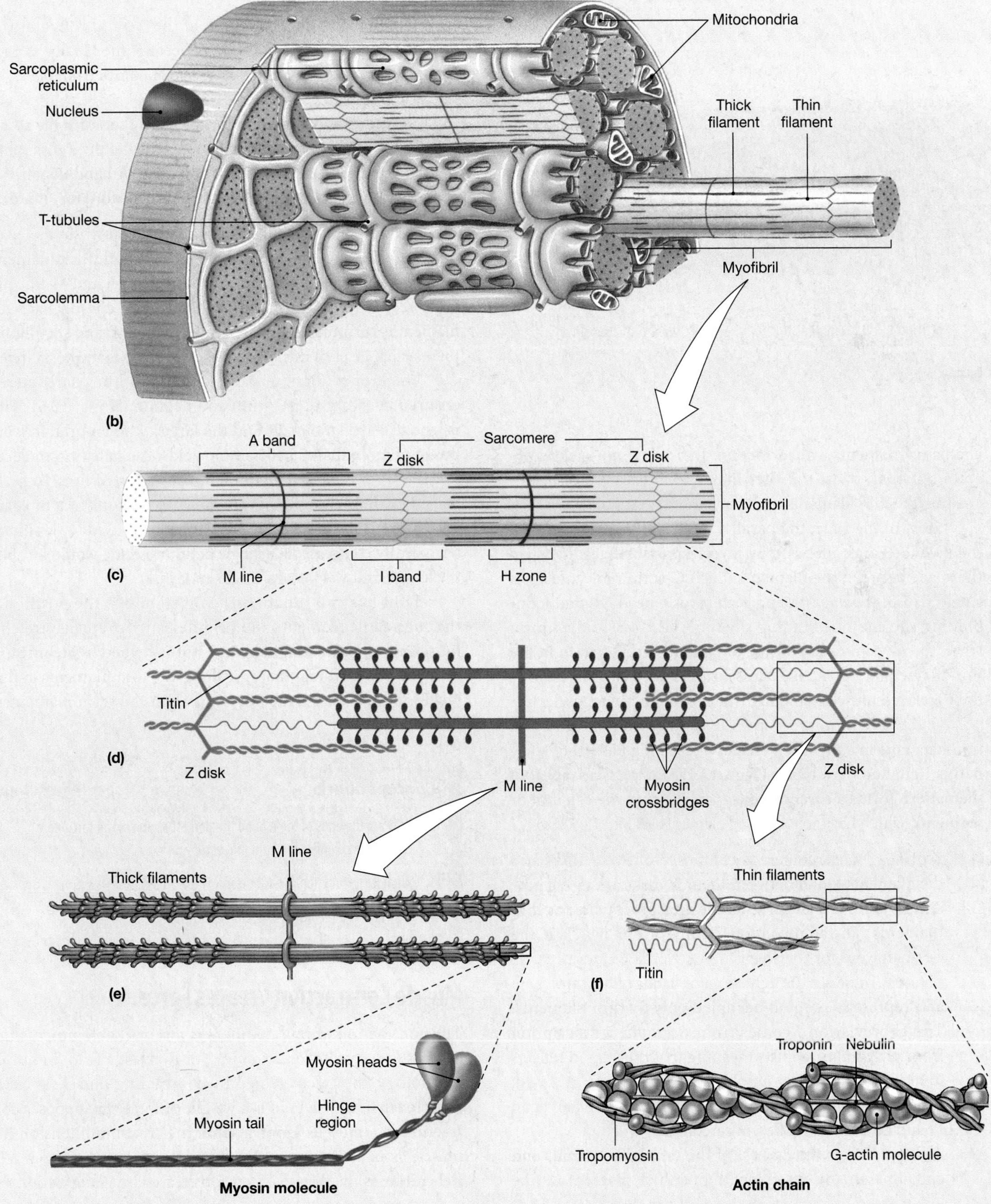

Mitochondria
Sarcoplasmic reticulum
Nucleus
Thick filament
Thin filament
T-tubules
Myofibril
Sarcolemma
(b)
A band
Sarcomere
Z disk
Z disk
Myofibril
(c)
M line
I band
H zone
Titin
(d)
Z disk
M line
Myosin crossbridges
Z disk
M line
Thick filaments
Thin filaments
Titin
(e)
(f)
Myosin heads
Hinge region
Myosin tail
Myosin molecule
Troponin
Nebulin
Tropomyosin
G-actin molecule
Actin chain

T-TUBULES

T-tubules are extensions of the cell membrane (sarcolemma) that associate with the ends (terminal cisternae) of the sarcoplasmic reticulum.

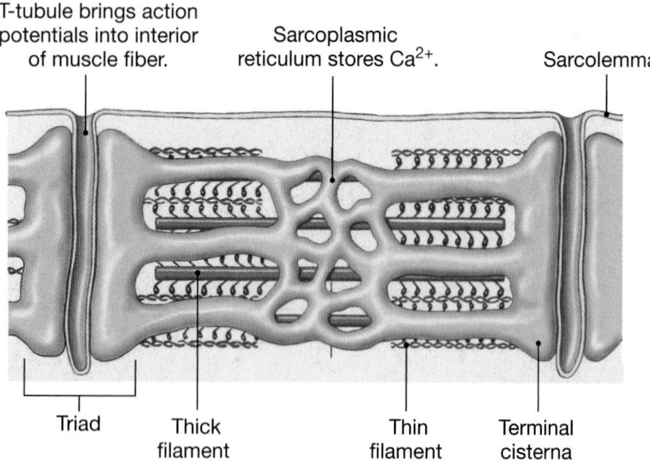
T-tubule brings action potentials into interior of muscle fiber.
Sarcoplasmic reticulum stores Ca^{2+}.
Sarcolemma
Triad
Thick filament
Thin filament
Terminal cisterna

■ **Fig. 12.4**

chains or filaments, called *F-actin*. In skeletal muscle, two F-actin polymers twist together like a double strand of beads, creating the thin filaments of the myofibril.

Most of the time, the parallel thick and thin filaments of the myofibril are connected by myosin **crossbridges** that span the space between the filaments. Each G-actin molecule has a single *myosin-binding site*, and each myosin head has one actin-binding site and one binding site for ATP. Crossbridges form when the myosin heads of thick filaments bind to actin in the thin filaments (Fig. 12.3d). Crossbridges have two states: low-force (relaxed muscles) and high-force (contracting muscles).

Under a light microscope, the arrangement of thick and thin filaments in a myofibril creates a repeating pattern of alternating light and dark bands (Figs. 12.1a, 12.3c). One repeat of the pattern forms a **sarcomere** {*sarkos,* flesh + *-mere,* a unit or segment}, which has the following elements (■ Fig. 12.5):

1. **Z disks**. One sarcomere is composed of two Z disks and the filaments found between them. Z disks are zigzag protein structures that serve as the attachment site for thin filaments. The abbreviation *Z* comes from *zwischen,* the German word for "between."
2. **I bands**. These are the lightest color bands of the sarcomere and represent a region occupied only by thin filaments. The abbreviation *I* comes from *isotropic,* a description from early microscopists meaning that this region reflects light uniformly under a polarizing microscope. A Z disk runs through the middle of every I band, so each half of an I band belongs to a different sarcomere.
3. **A band**. This is the darkest of the sarcomere's bands and encompasses the entire length of a thick filament. At the outer edges of the A band, the thick and thin filaments overlap. The center of the A band is occupied by thick filaments only. The abbreviation *A* comes from *anisotropic* {*an-*, not}, meaning that the protein fibers in this region scatter light unevenly.
4. **H zone**. This central region of the A band is lighter than the outer edges of the A band because the H zone is occupied by thick filaments only. The *H* comes from *helles,* the German word for "clear."
5. **M line**. This band represents proteins that form the attachment site for thick filaments, equivalent to the Z disk for the thin filaments. Each M line divides an A band in half. *M* is the abbreviation for *mittel,* the German word for "middle."

In three-dimensional array, the actin and myosin molecules form a lattice of parallel, overlapping thin and thick filaments, held in place by their attachments to the Z-disk and M-line proteins, respectively (Fig. 12.5b). When viewed end-on, each thin filament is surrounded by three thick filaments, and six thin filaments encircle each thick filament (Fig. 12.5c, rightmost circle).

The proper alignment of filaments within a sarcomere is ensured by two proteins: titin and nebulin (■ Fig. 12.6). **Titin** is a huge elastic molecule and the largest known protein, composed of more than 25,000 amino acids. A single titin molecule stretches from one Z disk to the neighboring M line. To get an idea of the immense size of titin, imagine that one titin molecule is an 8-foot-long piece of the very thick rope used to tie ships to a wharf. By comparison, a single actin molecule would be about the length and weight of a single eyelash.

Titin has two functions: (1) it stabilizes the position of the contractile filaments and (2) its elasticity returns stretched muscles to their resting length. Titin is helped by **nebulin**, an inelastic giant protein that lies alongside thin filaments and attaches to the Z disk. Nebulin helps align the actin filaments of the sarcomere.

Concept Check Answers: p. 439

2. Why are the ends of the A band the darkest region of the sarcomere when viewed under the light microscope?
3. What is the function of t-tubules?
4. Why are skeletal muscles described as striated?

Muscle Contraction Creates Force

The contraction of muscle fibers is a remarkable process that enables us to create force to move or to resist a load. In muscle physiology, the force created by contracting muscle is called **muscle tension**. The **load** is a weight or force that opposes contraction of a muscle. **Contraction**, the creation of tension in a muscle, is an active process that requires energy input from ATP. **Relaxation** is the release of tension created by a contraction.

■ Fig. 12.5 *ESSENTIALS*

The Sarcomere

Organization of a Sarcomere

The Z disk (not shown in part (c)) has accessory proteins that link the thin filaments together, similar to the accessory proteins shown for the M line. Myosin heads are omitted for simplicity.

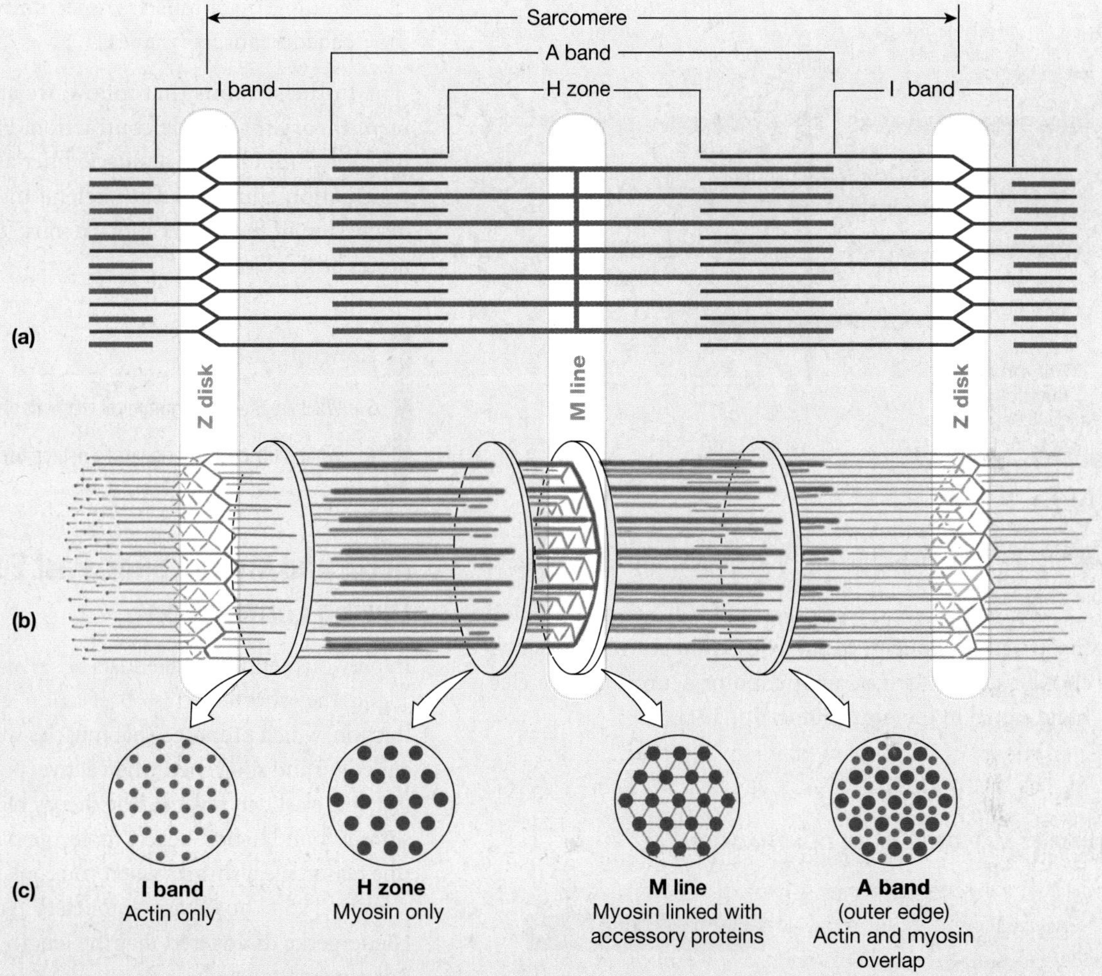

Sarcomere
A band
I band
H zone
I band
(a)
Z disk
M line
Z disk
(b)

KEY
Actin
Myosin

(c)

I band
Actin only

H zone
Myosin only

M line
Myosin linked with accessory proteins

A band
(outer edge)
Actin and myosin overlap

The Sarcomere Shortens During Contraction. As contraction takes place, actin and myosin do not change length but instead slide past one another.

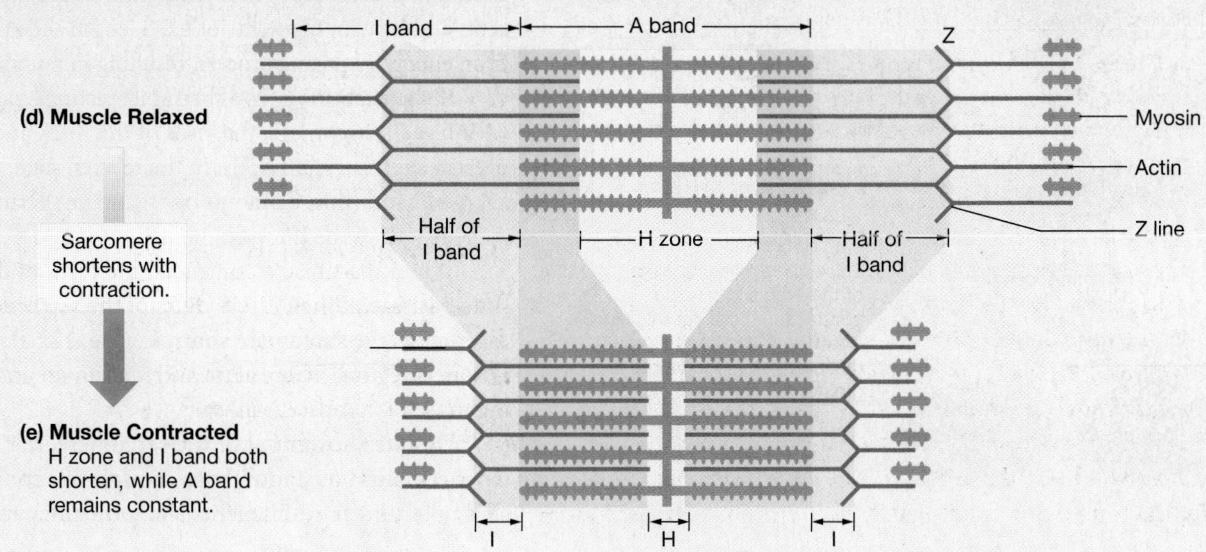

TITIN AND NEBULIN

Titin and nebulin are giant accessory proteins. Titin spans the distance from one Z disk to the neighboring M line. Nebulin, lying along the thin filaments, attaches to a Z disk but does not extend to the M line.

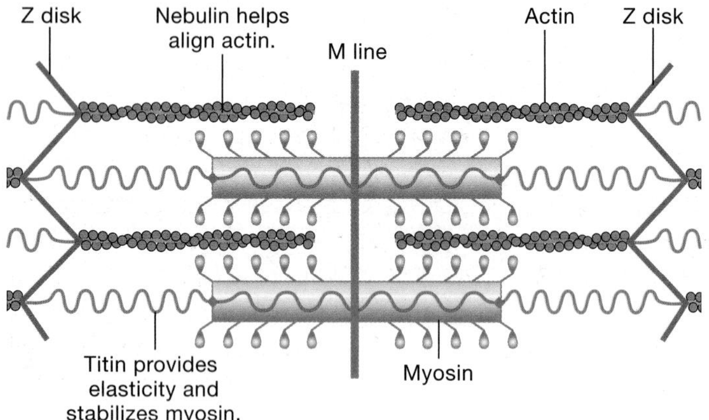

■ **Fig. 12.6**

■ Figure 12.7 maps the major steps leading up to skeletal muscle contraction.

1 **Events at the neuromuscular junction** convert an acetylcholine signal from a somatic motor neuron into an electrical signal in the muscle fiber [p. 391].

SUMMARY MAP OF MUSCLE CONTRACTION

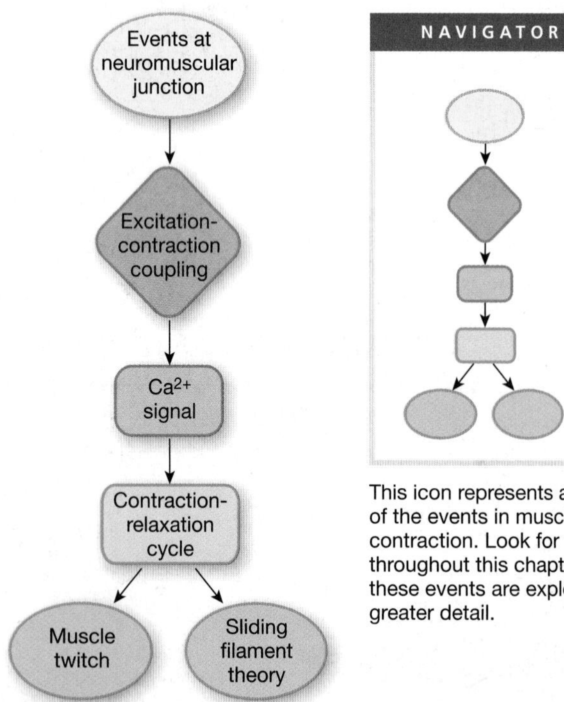

NAVIGATOR

This icon represents a map of the events in muscle contraction. Look for this icon throughout this chapter as these events are explored in greater detail.

■ **Fig. 12.7**

2 **Excitation-contraction (E-C) coupling** is the process in which muscle action potentials initiate calcium signals that in turn activate a contraction-relaxation cycle.

3 At the molecular level, a **contraction-relaxation cycle** can be explained by the *sliding filament theory of contraction.* In intact muscles, one contraction-relaxation cycle is called a muscle *twitch.*

In the sections that follow, we start with the sliding filament theory for muscle contraction. From there, we look at the integrated function of a muscle fiber as it undergoes excitation-contraction coupling. The skeletal muscle section ends with a discussion of the innervation of muscles and how muscles move bones around joints.

Concept Check Answers: p. 439

5. What are the three anatomical elements of a neuromuscular junction?
6. What is the chemical signal at a neuromuscular junction?

Actin and Myosin Slide Past Each Other During Contraction

In previous centuries, scientists observed that when muscles move a load, they shorten. This observation led to early theories of contraction, which proposed that muscles were made of molecules that curled up and shortened when active, then relaxed and stretched at rest, like elastic in reverse. The theory received support when myosin was found to be a helical molecule that shortened upon heating (the reason meat shrinks when you cook it).

In 1954, however, scientists Andrew Huxley and Rolf Niedergerke discovered that the length of the A band of a myofibril remains constant during contraction. Because the A band represents the myosin filament, Huxley and Niedergerke realized that shortening of the myosin molecule could not be responsible for contraction. Subsequently, they proposed an alternative model, the **sliding filament theory of contraction**. In this model, overlapping actin and myosin filaments of fixed length slide past one another in an energy-requiring process, resulting in muscle contraction.

If you examine a myofibril at its resting length, you see that within each sarcomere, the ends of the thick and thin filaments overlap slightly (Fig. 12.5d). In the relaxed state, a sarcomere has a large I band (thin filaments only) and an A band whose length is the length of the thick filament.

When the muscle contracts, the thick and thin filaments slide past each other. The Z disks of the sarcomere move closer together as the sarcomere shortens (Fig. 12.5e). The I band and H zone—regions where actin and myosin do not overlap in resting muscle—almost disappear.

Despite shortening of the sarcomere, the length of the A band remains constant. These changes are consistent with the sliding of thin actin filaments along the thick myosin filaments as the actin filaments move toward the M line in the center of the sarcomere. It is from this process that the sliding filament theory of contraction derives its name.

The sliding filament theory explains how a muscle can contract and create force without creating movement. For example, if you push on a wall, you are creating tension in many muscles of your body without moving the wall. According to the sliding filament theory, tension generated in a muscle fiber is directly proportional to the number of high-force crossbridges between the thick and thin filaments.

Myosin Crossbridges Move Actin Filaments

The movement of myosin crossbridges provides force that pushes the actin filament during contraction. The process can be compared to a competitive sailing team, with many people holding the rope that raises a heavy mainsail. When the order to raise the mainsail comes, each person on the team begins pulling on the rope, hand over hand, grabbing, pulling, and releasing repeatedly as the rope moves past.

In muscle, myosin heads bind to actin molecules, which are the "rope." A calcium signal initiates the **power stroke**, when myosin crossbridges swivel and push the actin filaments toward the center of the sarcomere. At the end of a power stroke, each myosin head releases actin, then swivels back and binds to a new actin molecule, ready to start another contractile cycle. During contraction, the heads do not all release at the same time or the fibers would slide back to their starting position, just as the mainsail would fall if the sailors all released the rope at the same time.

The power stroke repeats many times as a muscle fiber contracts. The myosin heads bind, push, and release actin molecules over and over as the thin filaments move toward the center of the sarcomere.

Myosin ATPase Where does energy for the power stroke come from? The answer is ATP. Myosin converts the chemical bond energy of ATP into the mechanical energy of crossbridge motion.

Myosin is an ATPase (*myosin ATPase*) that hydrolyzes ATP to ADP and inorganic phosphate (P_i). The energy released by ATP hydrolysis is trapped by myosin and stored as potential energy in the angle between the myosin head and the long axis of the myosin filament. Myosin heads in this position are said to be "cocked," or ready to rotate. The potential energy of the cocked heads becomes kinetic energy in the power stroke that moves actin.

Calcium Signals Initiate Contraction

How does a calcium signal turn muscle contraction on and off? The answer is found in **troponin** (TN), a calcium-binding complex of three proteins. Troponin controls the positioning of an elongated protein polymer, **tropomyosin** {*tropos,* to turn}.

In resting skeletal muscle, tropomyosin wraps around actin filaments and partially covers actin's myosin-binding sites (■ Fig. 12.8a). This is tropomyosin's blocking or "off" position.

(a) Relaxed state. Myosin head cocked. Tropomyosin partially blocks binding site on actin. Myosin is weakly bound to actin.

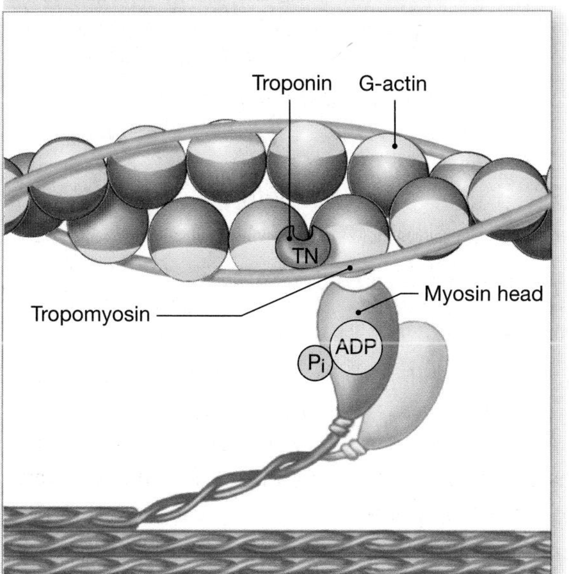

(b) Initation of contraction. A calcium signal initiates contraction.

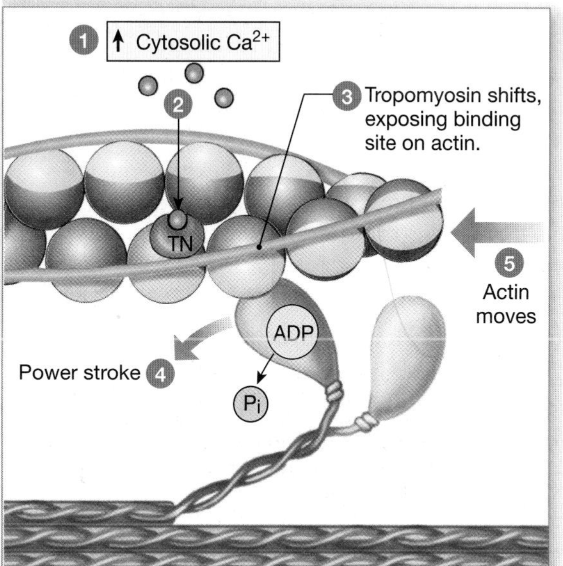

1. Ca^{2+} levels increase in cytosol.
2. Ca^{2+} binds to troponin (TN).
3. Troponin-Ca^{2+} complex pulls tropomyosin away from actin's myosin-binding site.
4. Myosin binds strongly to actin and completes power stroke.
5. Actin filament moves.

■ **Fig. 12.8**

Weak, low-force actin-myosin binding can still take place, but myosin is blocked from completing its power stroke, much as the safety latch on a gun keeps the cocked trigger from being pulled. Before contraction can occur, tropomyosin must be shifted to an "on" position that uncovers the remainder of actin's myosin-binding site.

The off-on positioning of tropomyosin is regulated by troponin. When contraction begins in response to a calcium signal ((1) in Fig. 12.8b), one protein of the complex—**troponin C**—binds reversibly to Ca^{2+} (2). The calcium-troponin C complex pulls tropomyosin completely away from actin's myosin-binding sites (3). This "on" position enables the myosin heads to form strong, high-force crossbridges and carry out their power strokes (4), moving the actin filament (5). Contractile cycles repeat as long as the binding sites are uncovered.

For muscle relaxation to occur, Ca^{2+} concentrations in the cytosol must decrease. By the law of mass action [p. 51], when cytosolic calcium decreases, Ca^{2+} unbinds from troponin. In the absence of Ca^{2+}, troponin allows tropomyosin to return to the "off" position, covering most of actin's myosin-binding sites. During the brief portion of the relaxation phase when actin and myosin are not bound to each other, the filaments of the sarcomere slide back to their original positions with the aid of titin and elastic connective tissues within the muscle.

The discovery that Ca^{2+}, not the action potential, is the signal for muscle contraction was the first piece of evidence suggesting that calcium acts as a messenger inside cells. Initially scientists thought that calcium signals occurred only in muscles, but we now know that calcium is an almost universal second messenger [p. 187].

Myosin Heads Step Along Actin Filaments

■ Figure 12.9 shows the molecular events of a contractile cycle in skeletal muscle. We will start a cycle with the **rigor state** {*rigere,* to be stiff}, where the myosin heads are tightly bound to G-actin molecules. No nucleotide (ATP or ADP) is bound to myosin. In living muscle, the rigor state occurs for only a very brief period. Then:

1. **ATP binds and myosin detaches.** An ATP molecule binds to the myosin head. ATP-binding decreases the actin-binding affinity of myosin, and myosin releases from actin.
2. **ATP hydrolysis provides energy for the myosin head to rotate and reattach to actin.** The ATP-binding site on the myosin head closes around ATP and hydrolyzes it to ADP and inorganic phosphate (P_i). Both ADP and P_i remain bound to myosin as energy released by ATP hydrolysis rotates the myosin head until it forms a 90° angle with the long axis of the filaments. In this cocked position, myosin binds to a new actin that is 1–3 molecules away from where it started.

   The newly formed actin-myosin crossbridge is weak and low-force because tropomyosin is partially blocking actin's binding site. However, in this rotated position myosin has stored potential energy, like a stretched spring. The head is cocked, just as someone preparing to fire a gun pulls back or cocks the spring-loaded hammer before firing. Most resting muscle fibers are in this state, cocked and prepared to contract, and just waiting for a calcium signal.
3. **The power stroke.** The power stroke (*crossbridge tilting*) begins after Ca^{2+} binds to troponin to uncover the rest of the myosin-binding site. The crossbridges transform into strong, high-force bonds as myosin releases P_i. Release of P_i allows the myosin head to swivel. The heads swing toward the M line, sliding the attached actin filament along with them. The power stroke is also called crossbridge tilting because the myosin head and hinge region tilt from a 90° angle to a 45° angle.
4. **Myosin releases ADP.** At the end of the power stroke, myosin releases ADP, the second product of ATP hydrolysis. With ADP gone, the myosin head is again tightly bound to actin in the rigor state. The cycle is ready to begin once more as a new ATP binds to myosin.

The Rigor State Although the contractile cycle began with the rigor state in which no ATP or ADP was bound to myosin, relaxed muscle fibers remain mostly in step 2. The rigor state in living muscle is normally brief because the muscle fiber has a sufficient supply of ATP that quickly binds to myosin once ADP is released in step 4.

After death, however, when metabolism stops and ATP supplies are exhausted, muscles are unable to bind more ATP, so they remain in the tightly bound rigor state. In the condition known as *rigor mortis,* the muscles "freeze" owing to immovable crossbridges. The tight binding of actin and myosin persists for a day or so after death, until enzymes within the decaying fiber begin to break down the muscle proteins.

Concept Check Answers: p. 439

7. Each myosin molecule has binding sites for what molecules?
8. What is the difference between F-actin and G-actin?
9. Myosin hydrolyzes ATP to ADP and P_i. Enzymes that hydrolyze ATP are collectively known as ________.

Although the preceding discussion sounds as if we know everything there is to know about the molecular basis of muscle contraction, in reality this is simply our current model. The process is more complex than presented here, and it now appears that myosin can influence Ca^{2+}-troponin binding, depending

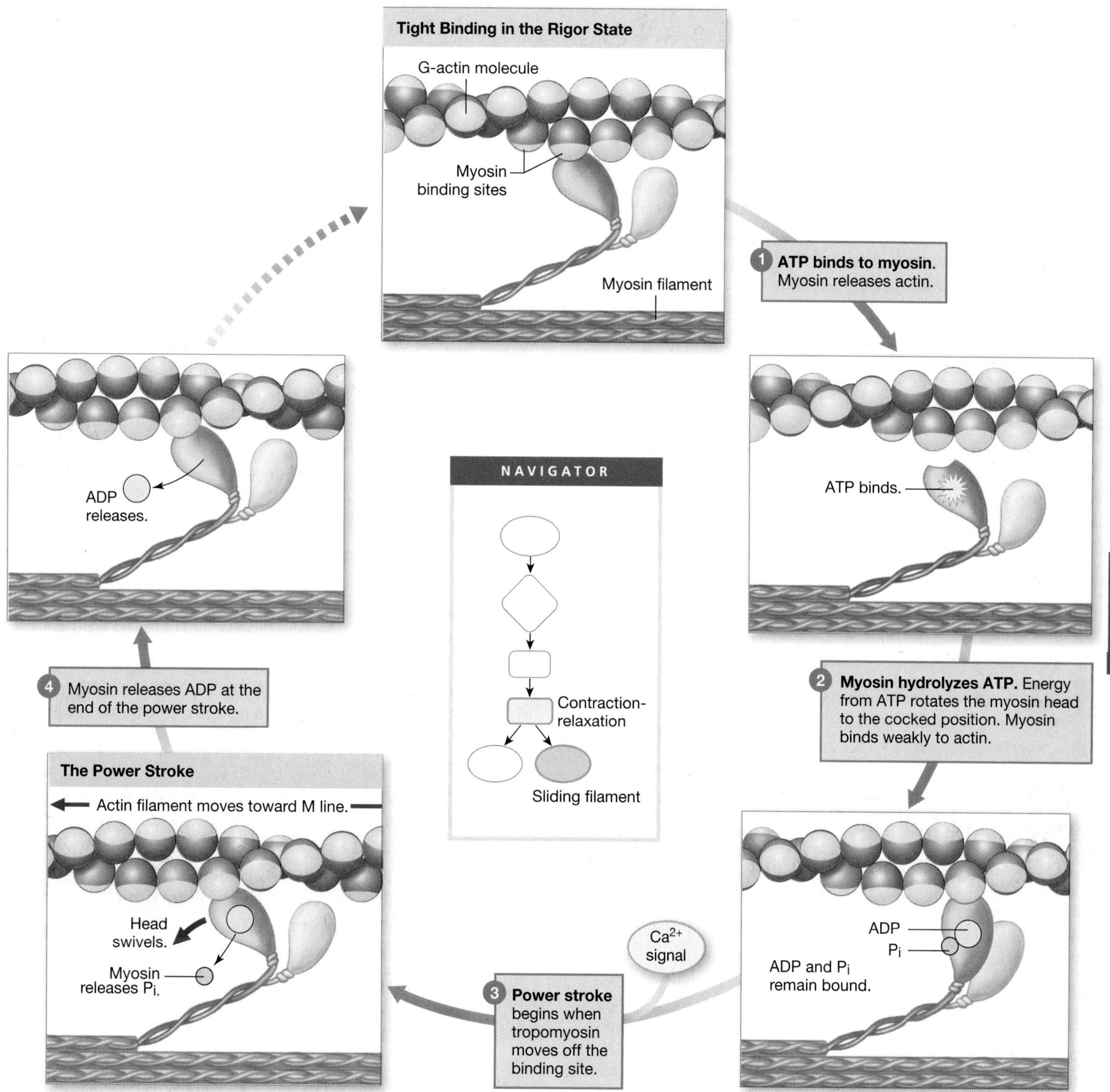
THE CONTRACTION CYCLE
Tight Binding in the Rigor State
G-actin molecule
Myosin binding sites
Myosin filament
1 ATP binds to myosin. Myosin releases actin.
ATP binds.
2 Myosin hydrolyzes ATP. Energy from ATP rotates the myosin head to the cocked position. Myosin binds weakly to actin.
ADP
Pi
ADP and Pi remain bound.
Ca2+ signal
3 Power stroke begins when tropomyosin moves off the binding site.
The Power Stroke
Actin filament moves toward M line.
Head swivels.
Myosin releases Pi.
4 Myosin releases ADP at the end of the power stroke.
ADP releases.
NAVIGATOR
Contraction-relaxation
Sliding filament

Fig. 12.9

on whether the myosin is bound to actin in a strong (rigor) state, bound to actin in a weak state, or not bound at all. The details of this influence are still being worked out.

Studying contraction and the movement of molecules in a myofibril has proved very difficult. Many research techniques rely on crystallized molecules, electron microscopy, and other tools that cannot be used with living tissues. Often we can see the thick and thin filaments only at the beginning and end of contraction. Progress is being made, however, and perhaps in the next decade you will see a "movie" of muscle contraction, constructed from photographs of sliding filaments.

BIOTECHNOLOGY

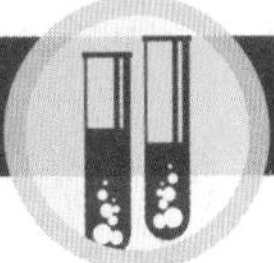

Watching Myosin Work

One big step forward in understanding the power stroke of myosin was the development of the *in vitro* motility assay in the 1980s. In this assay, isolated myosin molecules are randomly bonded to a specially coated glass coverslip. A fluorescently labeled actin molecule is placed on top of the myosin molecules. With ATP as a source of energy, the myosin heads bind to the actin and move it across the coverslip, marked by a fluorescent trail as it goes. In even more ingenious experiments, developed in 1995, a single myosin molecule is bound to a tiny bead that elevates it above the surface of the cover slip. An actin molecule is placed on top of the myosin molecule, like the balancing pole of a tightrope walker. As the myosin "motor" moves the actin molecule, lasers measure the nanometer movements and piconewton forces created with each cycle of the myosin head. Because of this technique, researchers can now measure the mechanical work being done by a single myosin molecule! For an animation and movie of the process, visit *http://physiology.med.uvm.edu/warshaw/TechspglnVitro.html.*

Concept Check Answers: p. 439

10. Name an elastic fiber in the sarcomere that aids relaxation.
11. In the sliding filament theory of contraction, what prevents the filaments from sliding back to their original position each time a myosin head releases to bind to the next actin binding site?

Acetylcholine Initiates Excitation-Contraction Coupling

Now let's start at the neuromuscular junction and follow the events leading up to contraction. As you learned earlier in the chapter, this combination of electrical and mechanical events in a muscle fiber is called *excitation-contraction coupling*. E-C coupling has four major events:

1. Acetylcholine (ACh) is released from the somatic motor neuron.
2. ACh initiates an action potential in the muscle fiber.
3. The muscle action potential triggers calcium release from the sarcoplasmic reticulum.
4. Calcium combines with troponin and initiates contraction.

Now let's look at these steps in detail. Acetylcholine released into the synapse at a neuromuscular junction binds to ACh receptor-channels on the motor end plate of the muscle fiber (■ Fig. 12.10a (1)) [p. 391]. When the ACh-gated channels open, they allow both Na^+ and K^+ to cross the membrane. However, Na^+ influx exceeds K^+ efflux because the electrochemical driving force is greater for Na^+ [p. 163]. The addition of net positive charge to the muscle fiber depolarizes the membrane, creating an **end-plate potential (EPP)**. Normally, end-plate potentials always reach threshold and initiate a muscle action potential (Fig. 12.10a (2)).

The action potential travels across the surface of the muscle fiber and into the t-tubules by the sequential opening of voltage-gated Na^+ channels. The process is similar to the conduction of action potentials in axons, although action potentials in skeletal muscle are conducted more slowly than action potentials in myelinated axons [p. 259].

The action potential that moves down the t-tubules causes Ca^{2+} release from the sarcoplasmic reticulum (Fig. 12.10b (3), (4)). Free cytosolic Ca^{2+} levels in a resting muscle are normally quite low, but after an action potential, they increase about 100-fold. As you've learned, when cytosolic Ca^{2+} levels are high, Ca^{2+} binds to troponin, tropomyosin moves to the "on" position (5), and contraction occurs (6).

At the molecular level, transduction of the electrical signal into a calcium signal requires two key membrane proteins. The t-tubule membrane contains a voltage-sensing **L-type calcium channel** called a **dihydropyridine (DHP) receptor** (Fig. 12.10b (3)). These specialized ion channels are mechanically linked to **Ca^{2+} release channels** in the adjacent sarcoplasmic reticulum. The SR Ca^{2+} release channels are also known as **ryanodine receptors**, or **RyR**.

When the depolarization of an action potential reaches a DHP receptor, the receptor changes conformation. The conformation change opens the RyR Ca^{2+} release channels in the sarcoplasmic reticulum (Fig. 12.10 (4)). Stored Ca^{2+} then flows down its electrochemical gradient into the cytosol, where it initiates contraction.

Scientists used to believe that the calcium channel we call the DHP receptor did not form an open channel for calcium entry from the ECF. However, in recent years it has become apparent that there is some limited Ca^{2+} entry through the DHP receptor, described as *excitation-coupled* Ca^{2+} *entry*. Skeletal muscle contraction can still take place if there is no ECF Ca^{2+}, however, so the physiological role of excitation-coupled Ca^{2+} entry is still being investigated.

Relaxation To end a contraction, calcium must be removed from the cytosol. The sarcoplasmic reticulum pumps Ca^{2+} back into its lumen using a **Ca^{2+}-ATPase** [p. 151]. As the free cytosolic Ca^{2+} concentration decreases, the equilibrium between bound and unbound Ca^{2+} is disturbed. Calcium releases from troponin, which allows tropomyosin to slide back and block actin's myosin-binding site. As crossbridges release, the muscle fiber relaxes with the help of elastic fibers in the sarcomere and in the connective tissue of the muscle.

Fig. 12.10 ***ESSENTIALS***

Excitation-Contraction Coupling and Relaxation

(a) Initiation of Muscle Action Potential

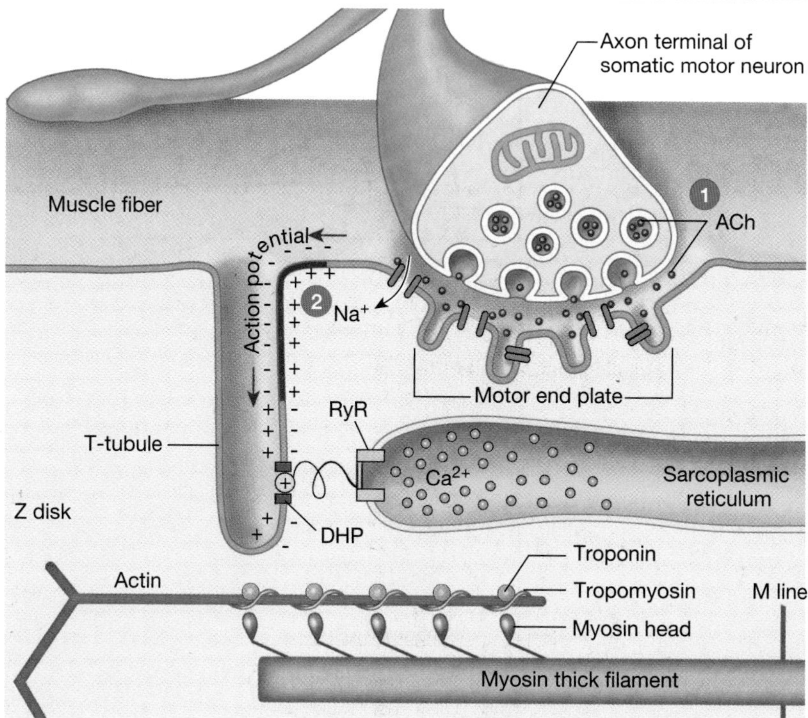

KEY

DHP = dihydropyridine L-type calcium channel

RyR = ryanodine receptor-channel

1. Somatic motor neuron releases ACh at neuromuscular junction.
2. Net entry of Na^+ through ACh receptor-channel initiates a muscle action potential.

(b) Excitation-Contraction Coupling

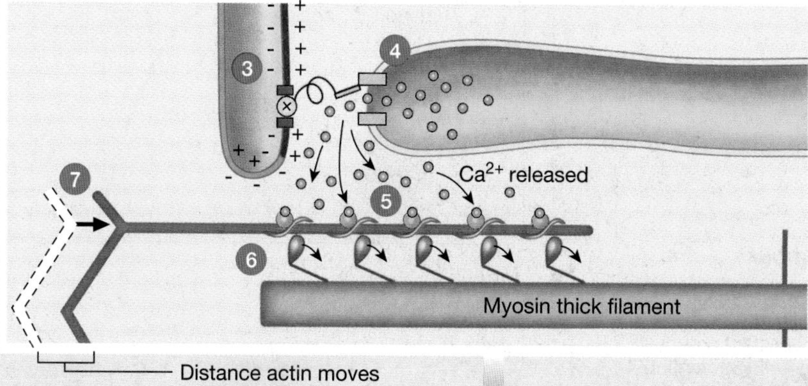

3. Action potential in t-tubule alters conformation of DHP receptor.
4. DHP receptor opens RyR Ca^{2+} release channels in sarcoplasmic reticulum, and Ca^{2+} enters cytoplasm.
5. Ca^{2+} binds to troponin, allowing actin-myosin binding.
6. Myosin heads execute power stroke.
7. Actin filament slides toward center of sarcomere.

(c) Relaxation Phase

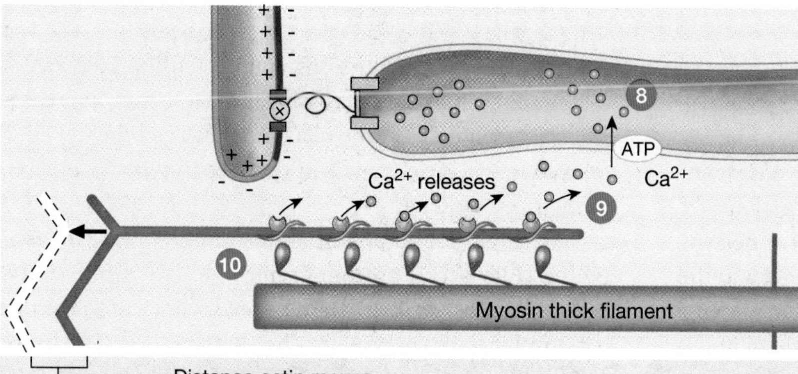

8. Sarcoplasmic Ca^{2+}-ATPase pumps Ca^{2+} back into SR.
9. Decrease in free cytosolic $[Ca^{2+}]$ causes Ca^{2+} to unbind from troponin.
10. Tropomyosin re-covers binding site. When myosin heads release, elastic elements pull filaments back to their relaxed position.

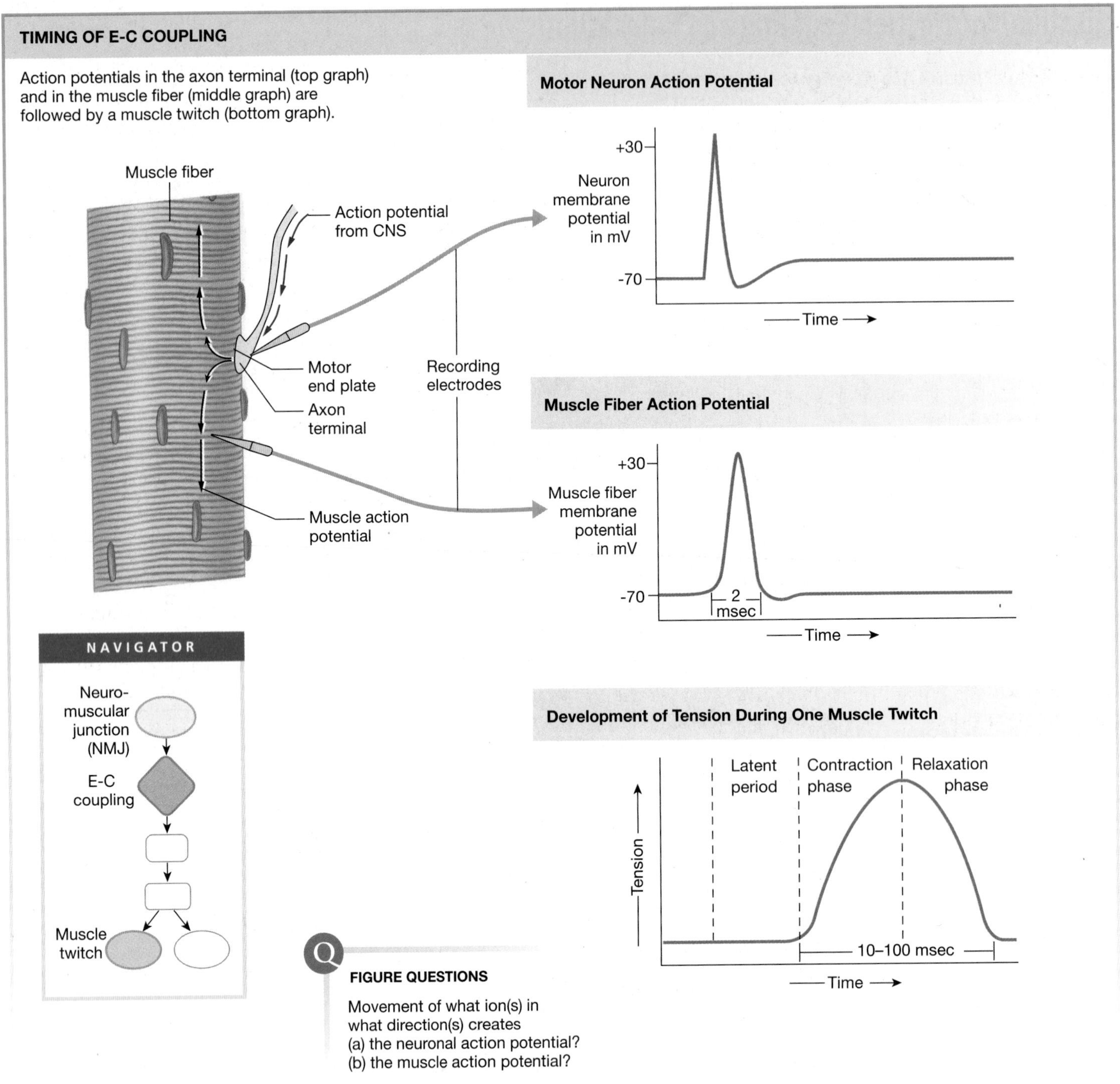

Fig. 12.11

Timing of E-C Coupling The graphs in ■ Figure 12.11 show the timing of electrical and mechanical events during E-C coupling. The somatic motor neuron action potential is followed by the skeletal muscle action potential, which in turn is followed by contraction. A single contraction-relaxation cycle in a skeletal muscle fiber is known as a **twitch**. Notice that there is a short delay—the **latent period**—between the muscle action potential and the beginning of muscle tension development. This delay represents the time required for calcium release and binding to troponin.

Once contraction begins, muscle tension increases steadily to a maximum value as crossbridge interaction increases. Tension then decreases in the relaxation phase of the twitch. During relaxation, elastic elements of the muscle return the sarcomeres to their resting length.

A single action potential in a muscle fiber evokes a single twitch (Fig. 12.11, bottom graph). However, muscle twitches vary from fiber to fiber in the speed with which they develop tension (the rising slope of the twitch curve), the maximum

RUNNING PROBLEM

Paul had experienced mild attacks of muscle weakness in his legs before, usually in the morning. Twice the weakness had come on after exposure to cold. Each attack had disappeared within minutes, and Paul seemed to suffer no lasting effects. On the advice of Paul's family doctor, Mrs. Leong takes her son to see a specialist in muscle disorders, who suspects a condition called periodic paralysis. The periodic paralyses are a family of disorders caused by Na^+ or Ca^{2+} ion channel mutations in the membranes of skeletal muscle fibers. The specialist believes that Paul has a condition in which defective voltage-gated Na^+ channels fail to inactivate after they open.

Q1: When Na^+ channels on the muscle membrane open, which way does Na^+ move?

Q2: What effect would continued movement of Na^+ have on the membrane potential of muscle fibers?

tension they achieve (the height of the twitch curve), and the duration of the twitch (the width of the twitch curve). You will learn about factors that affect these parameters in upcoming sections. First we discuss how muscles produce ATP to provide energy for contraction and relaxation.

Concept Check Answers: p. 439

12. Which part of contraction requires ATP? Does relaxation require ATP?

13. What events are taking place during the latent period before contraction begins?

Skeletal Muscle Contraction Requires a Steady Supply of ATP

The muscle fiber's use of ATP is a key feature of muscle physiology. Muscles require energy constantly: during contraction for crossbridge movement and release, during relaxation to pump Ca^{2+} back into the sarcoplasmic reticulum, and after E-C coupling to restore Na^+ and K^+ to the extracellular and intracellular compartments, respectively. Where do muscles get the ATP they need for this work?

The amount of ATP in a muscle fiber at any one time is sufficient for only about eight twitches. As a backup energy source, muscles contain **phosphocreatine**, a molecule whose high-energy phosphate bonds are created from creatine and ATP when muscles are at rest (■ Fig. 12.12). When muscles become active, such as during exercise, the high-energy phosphate group of phosphocreatine is transferred to ADP, creating more ATP to power the muscles.

The enzyme that transfers the phosphate group from phosphocreatine to ADP is **creatine kinase** (CK), also known as *creatine phosphokinase* (CPK). Muscle cells contain large amounts of this enzyme. Consequently, elevated blood levels of creatine kinase usually indicate damage to skeletal or cardiac muscle. Because the two muscle types contain different isozymes [p. 105], clinicians can distinguish cardiac tissue damage during a heart attack from skeletal muscle damage.

Energy stored in high-energy phosphate bonds is very limited, so muscle fibers must use metabolism to transfer energy from the chemical bonds of nutrients to ATP. Carbohydrates, particularly glucose, are the most rapid and efficient source of energy for ATP production. Glucose is metabolized through glycolysis to pyruvate [p. 113]. In the presence of adequate oxygen, pyruvate goes into the citric acid cycle, producing about 30 ATP for each molecule of glucose.

When oxygen concentrations fall during strenuous exercise, muscle fiber metabolism relies more on *anaerobic glycolysis*. In this pathway, glucose is metabolized to lactate with a yield of only 2 ATP per glucose [p. 117]. Anaerobic metabolism of

Phosphocreatine

Resting muscle stores energy from ATP in the high-energy bonds of phosphocreatine. Working muscle then uses that stored energy.

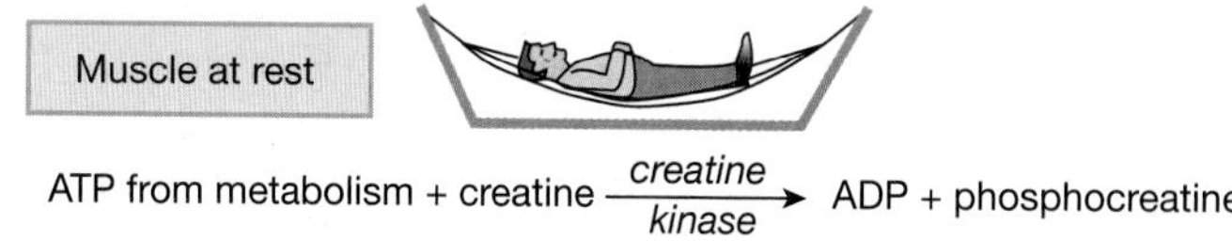

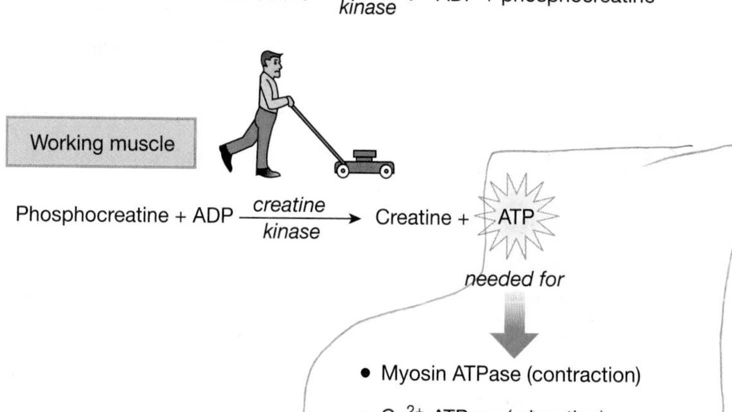

■ **Fig. 12.12**

12

glucose is a quicker source of ATP but produces many fewer ATP per glucose. When muscle energy demands outpace the amount of ATP that can be produced through anaerobic metabolism of glucose, muscles can function for only a short time without fatiguing.

Muscle fibers also obtain energy from fatty acids, although this process always requires oxygen. During rest and light exercise, skeletal muscles burn fatty acids along with glucose, one reason that modest exercise programs of brisk walking are an effective way to reduce body fat. However, the metabolic process by which fatty acids are converted to acetyl CoA is relatively slow and cannot produce ATP rapidly enough to meet the energy needs of muscle fibers during heavy exercise. Under these conditions, muscle fibers rely more on glucose.

Proteins normally are not a source of energy for muscle contraction. Most amino acids found in muscle fibers are used to synthesize proteins rather than to produce ATP.

Do muscles ever run out of ATP? You might think so if you have ever exercised to the point of fatigue, the point at which you feel that you cannot continue or your limbs refuse to obey commands from your brain. Most studies show, however, that even intense exercise uses only 30% of the ATP in a muscle fiber. The condition we call fatigue must come from other changes in the exercising muscle.

Concept Check Answers: p. 439

14. According to the convention for naming enzymes, what does the name creatine kinase tell you about this enzyme's function? [*Hint:* p. 107]

15. The reactions in Figure 12.12 show that creatine kinase catalyzes the creatine-phosphocreatine reaction in both directions. What then determines the direction that the reaction goes at any given moment? [*Hint:* p. 51]

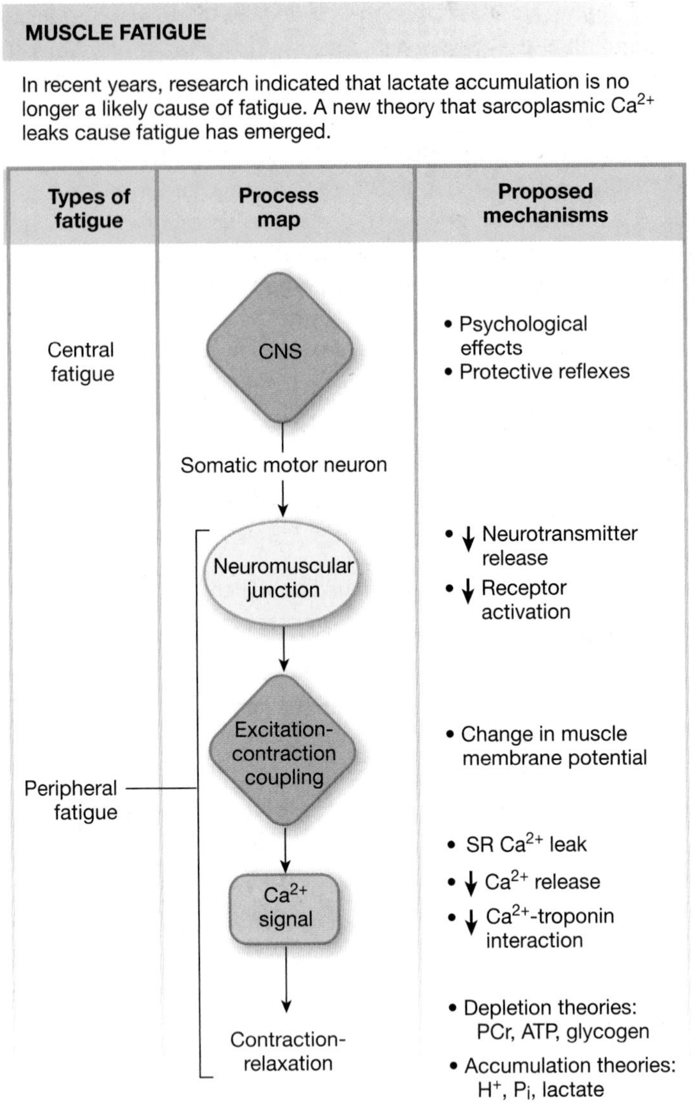

MUSCLE FATIGUE

In recent years, research indicated that lactate accumulation is no longer a likely cause of fatigue. A new theory that sarcoplasmic Ca^{2+} leaks cause fatigue has emerged.

Types of fatigue	Process map	Proposed mechanisms
Central fatigue	CNS	• Psychological effects • Protective reflexes
	Somatic motor neuron ↓	
Peripheral fatigue	Neuromuscular junction ↓	• ↓ Neurotransmitter release • ↓ Receptor activation
	Excitation-contraction coupling ↓	• Change in muscle membrane potential
	Ca^{2+} signal ↓	• SR Ca^{2+} leak • ↓ Ca^{2+} release • ↓ Ca^{2+}-troponin interaction
	Contraction-relaxation	• Depletion theories: PCr, ATP, glycogen • Accumulation theories: H^+, P_i, lactate

■ **Fig. 12.13**

Fatigue Has Multiple Causes

The physiological term **fatigue** describes a reversible condition in which a muscle is no longer able to generate or sustain the expected power output. Fatigue is highly variable. It is influenced by the intensity and duration of the contractile activity, by whether the muscle fiber is using aerobic or anaerobic metabolism, by the composition of the muscle, and by the fitness level of the individual. The study of fatigue is quite complex, and research in this area is complicated by the fact that experiments are done under a wide range of conditions, from "skinned" (sarcolemma removed) single muscle fibers to exercising humans.

Factors that have been proposed to play a role in fatigue are classified into **central fatigue** mechanisms, which arise in the central nervous system, and **peripheral fatigue** mechanisms, which arise anywhere between the neuromuscular junction and the contractile elements of the muscle (■ Fig. 12.13). Most experimental evidence suggests that muscle fatigue arises from excitation-contraction failure in the muscle fiber rather than from failure of control neurons or neuromuscular transmission.

Central fatigue includes subjective feelings of tiredness and a desire to cease activity. Several studies have shown that this psychological fatigue precedes physiological fatigue in the muscles and therefore may be a protective mechanism. Low pH from acid production during ATP hydrolysis is often mentioned as a possible cause of fatigue, and some evidence suggests that acidosis may influence the sensation of fatigue perceived by the brain. However, homeostatic mechanisms for pH balance maintain blood pH at normal levels until exertion is nearly maximal, so pH as a factor in central fatigue probably applies only in cases of maximal exertion.

Neural causes of fatigue could arise either from communication failure at the neuromuscular junction or from failure of the CNS command neurons. For example, if ACh is not synthesized in the axon terminal fast enough to keep up with neuron firing rate, neurotransmitter release at the synapse decreases. Consequently, the muscle end-plate potential fails to reach the threshold value needed to trigger a muscle fiber action potential, resulting in contraction failure. This type of fatigue is associated with some neuromuscular diseases, but it is probably not a factor in normal exercise.

Fatigue within the muscle fiber can occur in any of several sites. In extended submaximal exertion, fatigue is associated with the depletion of muscle glycogen stores. Because most studies show that lack of ATP is not a limiting factor, glycogen depletion may be affecting some other aspect of contraction, such as the release of Ca^{2+} from the sarcoplasmic reticulum.

The cause of fatigue in short-duration maximal exertion seems to be different. One theory is based on the increased levels of inorganic phosphate (P_i) produced when ATP and phosphocreatine are used for energy in the muscle fiber. Elevated cytoplasmic P_i may slow P_i release from myosin and thereby alter the power stroke (see Fig. 12.9 (4)).

Another theory suggests that elevated phosphate levels decrease Ca^{2+} release because the phosphate combines with Ca^{2+} to become calcium phosphate. Some investigators feel that alterations in Ca^{2+} release from the sarcoplasmic reticulum play a major role in fatigue.

Ion imbalances have also been implicated in fatigue. During maximal exercise, K^+ leaves the muscle fiber with each action potential, and as a result K^+ concentrations rise in the extracellular fluid of the t-tubules. The shift in K^+ alters the membrane potential of the muscle fiber. Changes in N^+-K^+-ATPase activity may also be involved. In short, although many different factors have been *associated with* fatigue, the factors that *cause* fatigue are still uncertain.

Concept Check Answer: p. 439

16. If K^+ concentration increases in the extracellular fluid surrounding a cell but does not change significantly in the cell's cytoplasm, the cell membrane ________ (*depolarizes/hyperpolarizes*) and becomes ________ (*more/less*) negative.

Skeletal Muscle Is Classified by Speed and Fatigue Resistance

Skeletal muscle fibers have traditionally been classified on the basis of their speed of contraction and their resistance to fatigue with repeated stimulation. But like so much in physiology, the more scientists learn, the more complicated the picture becomes. Muscles have plasticity and can shift their type depending on their activity. The currently accepted muscle fiber types include **slow-twitch fibers** (also called *ST* or *type I*), **fast-twitch oxidative-glycolytic fibers** (*FOG* or *type IIA*), and **fast-twitch glycolytic fibers** (*FG* or *type IIB*).

Fast-twitch muscle fibers (type II) develop tension two to three times faster than slow-twitch fibers (type I). The speed with which a muscle fiber contracts is determined by the isoform of myosin ATPase present in the fiber's thick filaments. Fast-twitch fibers split ATP more rapidly and can therefore complete multiple contractile cycles more rapidly than slow-twitch fibers. This speed translates into faster tension development in the fast-twitch fibers.

The duration of contraction also varies according to fiber type. Twitch duration is determined largely by how fast the sarcoplasmic reticulum removes Ca^{2+} from the cytosol. As cytosolic Ca^{2+} concentrations fall, Ca^{2+} unbinds from troponin, allowing tropomyosin to move into position to partially block the myosin-binding sites. With the power stroke inhibited in this way, the muscle fiber relaxes.

Fast-twitch fibers pump Ca^{2+} into their sarcoplasmic reticulum more rapidly than slow-twitch fibers do, so fast-twitch fibers have quicker twitches. The twitches in fast-twitch fibers last only about 7.5 msec, making these muscles useful for fine, quick movements, such as playing the piano. Contractions in slow-twitch muscle fibers may last more than 10 times as long. Fast-twitch fibers are used occasionally, but slow-twitch fibers are used almost constantly for maintaining posture, standing, or walking.

The second major difference between muscle fiber types is their ability to resist fatigue. Glycolytic fibers (fast-twitch type IIB) rely primarily on anaerobic glycolysis to produce ATP. However,

RUNNING PROBLEM

Two forms of periodic paralysis exist. One form, called *hypokalemic periodic paralysis*, is characterized by decreased blood levels of K^+ during paralytic episodes. The other form, *hyperkalemic periodic paralysis (hyperKPP)*, is characterized by either normal or increased blood levels of K^+ during episodes. Results of a blood test revealed that Paul has the hyperkalemic form.

Q3: In people with hyperKPP, attacks may occur after a period of exercise (that is, after a period of repeated muscle contractions). What ion is responsible for the repolarization phase of the muscle action potential, and in which direction does this ion move across the muscle fiber membrane? How might this be linked to hyperKPP?

400 413 415 425 428 435

the accumulation of H^+ from ATP hydrolysis contributes to acidosis, a condition implicated in the development of fatigue, as noted previously. As a result, glycolytic fibers fatigue more easily than do oxidative fibers, which do not depend on anaerobic metabolism.

Oxidative fibers rely primarily on oxidative phosphorylation [p. 115] for production of ATP—hence their descriptive name. These fibers, which include slow-twitch fibers and fast-twitch oxidative-glycolytic fibers, have more mitochondria (the site of enzymes for the citric acid cycle and oxidative phosphorylation) than glycolytic fibers do. They also have more blood vessels in their connective tissue to bring oxygen to the cells (■ Fig. 12.14).

The efficiency with which muscle fibers obtain oxygen is a factor in their preferred method of glucose metabolism. Oxygen in the blood must diffuse into the interior of muscle fibers in order to reach the mitochondria. This process is facilitated by the presence of **myoglobin**, a red oxygen-binding pigment with a high affinity for oxygen. This affinity allows myoglobin to act as a transfer molecule, bringing oxygen more rapidly to the interior of the fibers. Because oxidative fibers contain more myoglobin, oxygen diffusion is faster than in glycolytic fibers. Oxidative fibers are described as *red muscle* because large amounts of myoglobin give them their characteristic color.

In addition to myoglobin, oxidative fibers have smaller diameters, so the distance through which oxygen must diffuse before reaching the mitochondria is shorter. Because oxidative fibers have more myoglobin and more capillaries to bring blood to the cells and are smaller in diameter, they maintain a better supply of oxygen and are able to use oxidative phosphorylation for ATP production.

Glycolytic fibers, in contrast, are described as *white muscle* because of their lower myoglobin content. These muscle fibers are also larger in diameter than slow-twitch fibers. The combination of larger size, less myoglobin, and fewer blood vessels means that glycolytic fibers are more likely to run out of oxygen after repeated contractions. Glycolytic fibers therefore rely primarily on anaerobic glycolysis for ATP synthesis and fatigue most rapidly.

Fast-twitch oxidative-glycolytic fibers exhibit properties of both oxidative and glycolytic fibers. They are smaller than fast-twitch glycolytic fibers and use a combination of oxidative and glycolytic metabolism to produce ATP. Because of their intermediate size and the use of oxidative phosphorylation for ATP synthesis, fast-twitch oxidative-glycolytic fibers are more fatigue resistant than their fast-twitch glycolytic cousins. Fast-twitch oxidative-glycolytic fibers, like slow-twitch fibers, are classified as red muscle because of their myoglobin content.

Human muscles are a mixture of fiber types, with the ratio of types varying from muscle to muscle and from one individual to another. For example, who would have more fast-twitch fibers in leg muscles, a marathon runner or a high-jumper?

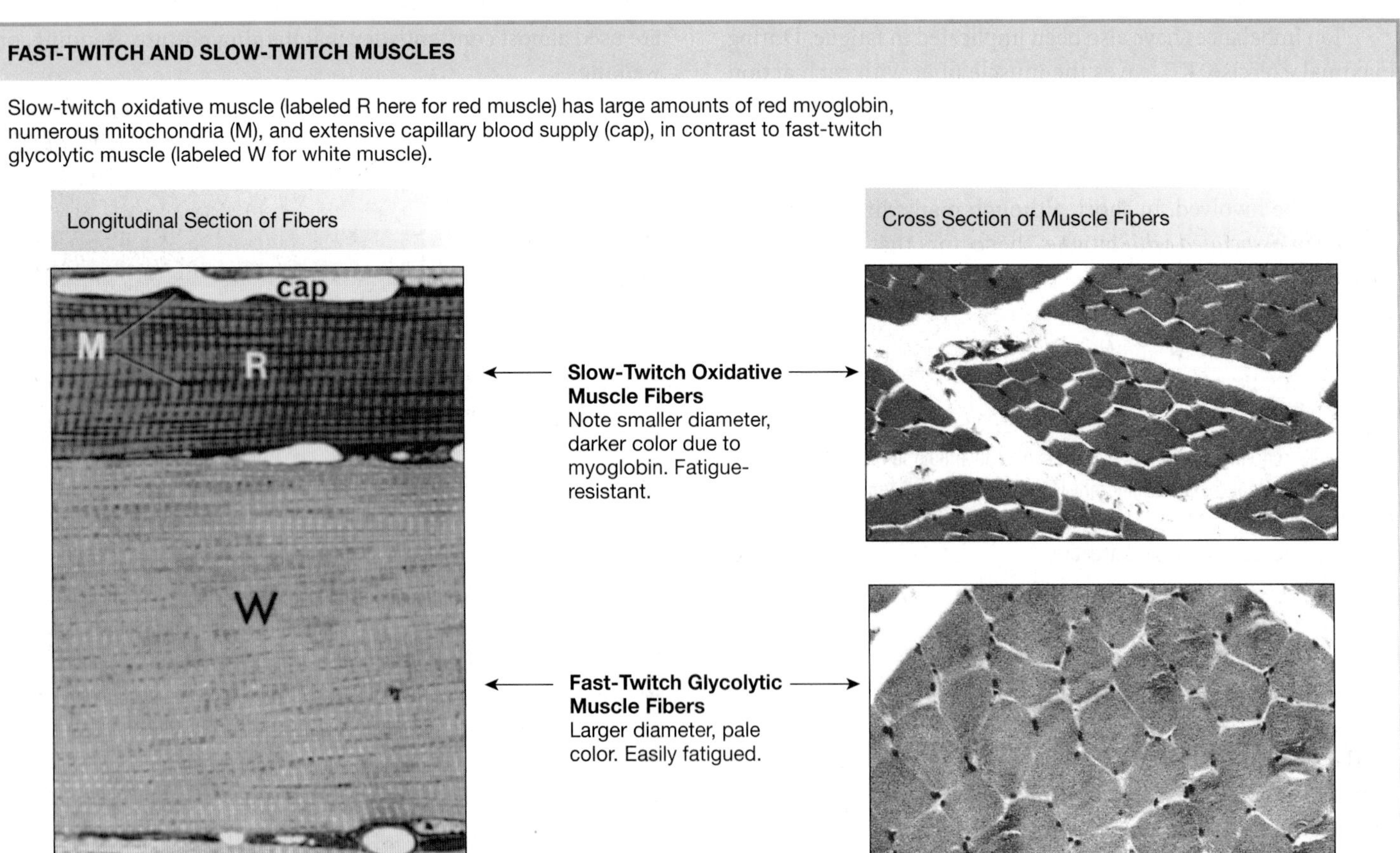
FAST-TWITCH AND SLOW-TWITCH MUSCLES
Slow-twitch oxidative muscle (labeled R here for red muscle) has large amounts of red myoglobin, numerous mitochondria (M), and extensive capillary blood supply (cap), in contrast to fast-twitch glycolytic muscle (labeled W for white muscle).
Longitudinal Section of Fibers
cap
M
R
W
Cross Section of Muscle Fibers
Slow-Twitch Oxidative Muscle Fibers
Note smaller diameter, darker color due to myoglobin. Fatigue-resistant.
Fast-Twitch Glycolytic Muscle Fibers
Larger diameter, pale color. Easily fatigued.

■ **Fig. 12.14**

Table 12.2 Characteristics of Muscle Fiber Types

	Slow-Twitch Oxidative; Red Muscle	Fast-Twitch Oxidative-Glycolytic; Red Muscle	Fast-Twitch Glycolytic; White Muscle
Speed of development of maximum tension	Slowest	Intermediate	Fastest
Myosin ATPase activity	Slow	Fast	Fast
Diameter	Small	Medium	Large
Contraction duration	Longest	Short	Short
Ca^{2+}-ATPase activity in SR	Moderate	High	High
Endurance	Fatigue resistant	Fatigue resistant	Easily fatigued
Use	Most used: posture	Standing, walking	Least used: jumping; quick, fine movements
Metabolism	Oxidative; aerobic	Glycolytic but becomes more oxidative with endurance training	Glycolytic; more anaerobic than fast-twitch oxidative-glycolytic type
Capillary density	High	Medium	Low
Mitochondria	Numerous	Moderate	Few
Color	Dark red (myoglobin)	Red	Pale

Characteristics of the three muscle fiber types are compared in ■ Table 12.2.

Resting Fiber Length Affects Tension

In a muscle fiber, the tension developed during a twitch is a direct reflection of the length of individual sarcomeres before contraction begins (■ Fig. 12.15). Each sarcomere contracts with optimum force if it is at optimum length (neither too long nor too short) before the contraction begins. Fortunately, the normal resting length of skeletal muscles usually ensures that sarcomeres are at optimum length when they begin a contraction.

At the molecular level, sarcomere length reflects the overlap between the thick and thin filaments (Fig. 12.15). The sliding filament theory predicts that *the tension a muscle fiber can generate is directly proportional to the number of crossbridges formed between the thick and thin filaments.* If the fibers start a contraction at a very long sarcomere length, the thick and thin filaments barely overlap and form few crossbridges (Fig. 12.15e). This means that in the initial part of the contraction, the sliding filaments interact only minimally and therefore cannot generate much force.

At the optimum sarcomere length (Fig. 12.15c), the filaments begin contracting with numerous crossbridges between the thick and thin filaments, allowing the fiber to generate optimum force in that twitch. If the sarcomere is shorter than optimum length at the beginning of the contraction (Fig. 12.15b), the thick and thin fibers have too much overlap before the contraction begins. Consequently, the thick filaments can move the thin filaments only a short distance before the thin actin filaments from opposite ends of the sarcomere start to overlap. This overlap prevents crossbridge formation. If the sarcomere is so short that the thick filaments run into the Z disks (Fig. 12.15a), myosin is unable to find new binding sites for crossbridge formation, and tension decreases rapidly. Thus the development of single-twitch tension in a muscle fiber is a passive property that depends on filament overlap and sarcomere length.

Force of Contraction Increases with Summation

Although we have just seen that single-twitch tension is determined by the length of the sarcomere, it is important to note that a single twitch does not represent the maximum force that a muscle fiber can develop. The force generated by the contraction of a single muscle fiber can be increased by increasing the rate (frequency) at which muscle action potentials stimulate the muscle fiber.

LENGTH-TENSION RELATIONSHIPS

Too much or too little overlap of thick and thin filaments in resting muscle results in decreased tension.

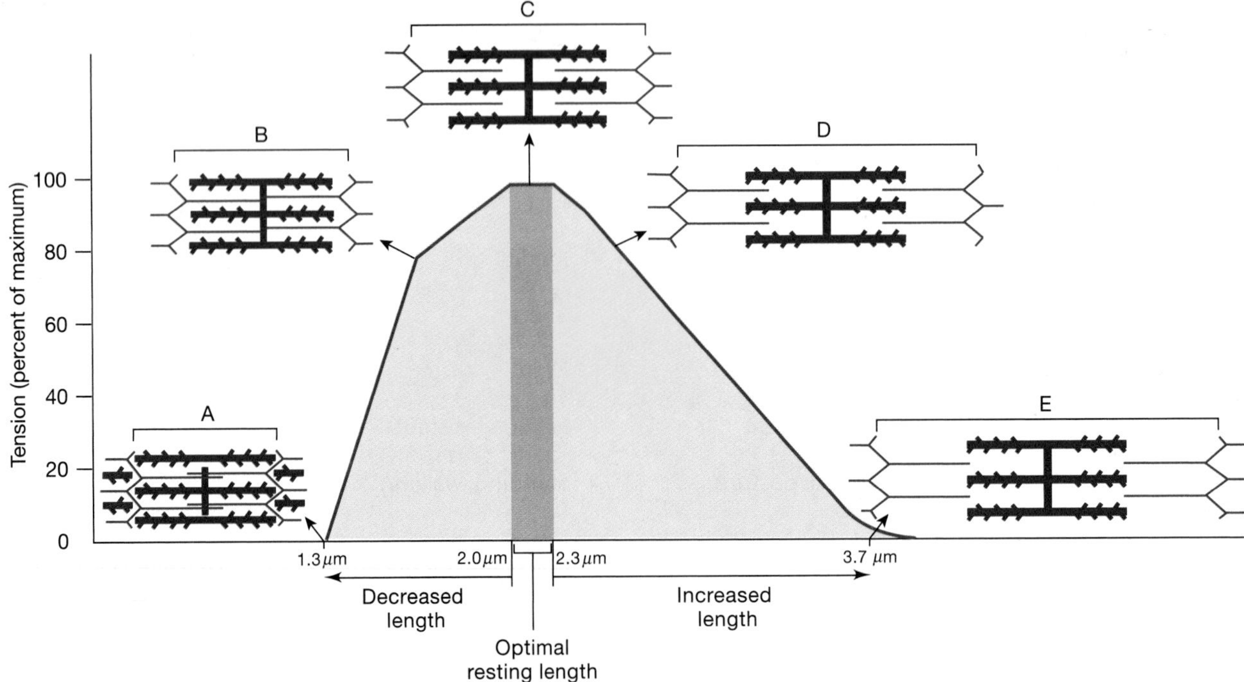

■ **Fig. 12.15** *Adapted from A. M. Gordon et al.,* J Physiol 184: *170–192, 1966.*

A typical muscle action potential lasts between 1 and 3 msec, while the muscle contraction may last 100 msec (see Fig. 12.11). If repeated action potentials are separated by long intervals of time, the muscle fiber has time to relax completely between stimuli (■ Fig. 12.16a). If the interval of time between action potentials is shortened, the muscle fiber does not have time to relax completely between two stimuli, resulting in a more forceful contraction (Fig. 12.16b). This process is known as **summation** and is similar to the temporal summation of graded potentials that takes place in neurons [p. 277].

If action potentials continue to stimulate the muscle fiber repeatedly at short intervals (high frequency), relaxation between contractions diminishes until the muscle fiber achieves a state of maximal contraction known as **tetanus**. There are two types of tetanus. In *incomplete*, or *unfused, tetanus,* the stimulation rate of the muscle fiber is not at a maximum value, and consequently the fiber relaxes slightly between stimuli (Fig. 12.16c). In *complete*, or *fused*, *tetanus,* the stimulation rate is fast enough that the muscle fiber does not have time to relax. Instead, it reaches maximum tension and remains there (Fig. 12.16d).

Thus it is possible to increase the tension developed in a single muscle fiber by changing the rate at which action potentials occur in the fiber. Muscle action potentials are initiated by the somatic motor neuron that controls the muscle fiber.

Concept Check Answers: p. 439

17. Summation in muscle fibers means that the __________ of the fiber increases with repeated action potentials.

18. Temporal summation in neurons means that the __________ of the neuron increases when two depolarizing stimuli occur close together in time.

A Motor Unit Is One Motor Neuron and Its Muscle Fibers

The basic unit of contraction in an intact skeletal muscle is a **motor unit**, composed of a group of muscle fibers that function together and the somatic motor neuron that controls them (■ Fig. 12.17). When the somatic motor neuron fires an action potential, all muscle fibers in the motor unit contract. Note that although one somatic motor neuron innervates multiple fibers, each muscle fiber is innervated by only a single neuron.

The number of muscle fibers in a motor unit varies. In muscles used for fine motor actions, such as the *extraocular* muscles

SUMMATION OF CONTRACTIONS

(a) Single twitches: Muscle relaxes completely between stimuli (▲).

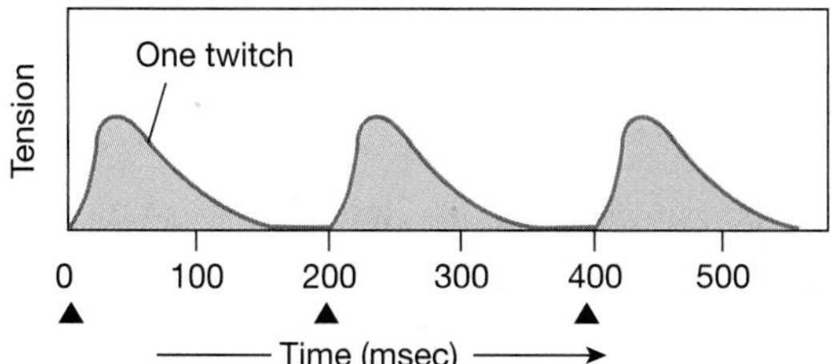

(b) Summation: Stimuli closer together do not allow muscle to relax fully.

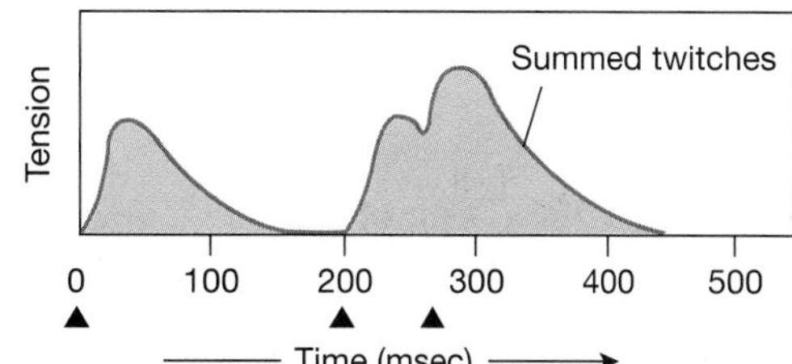

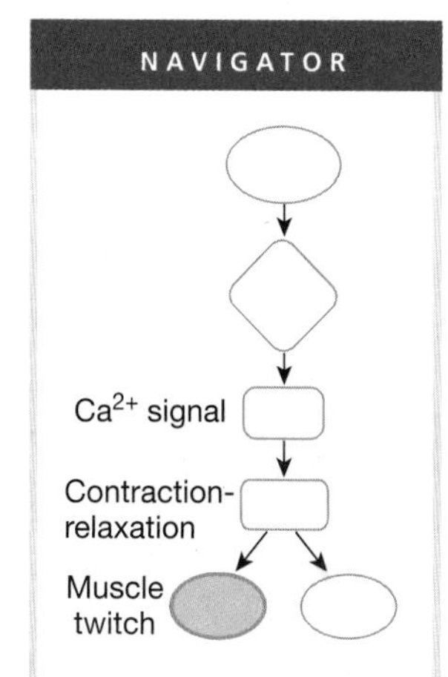

(c) Summation leading to unfused tetanus: Stimuli are far enough apart to allow muscle to relax slightly between stimuli.

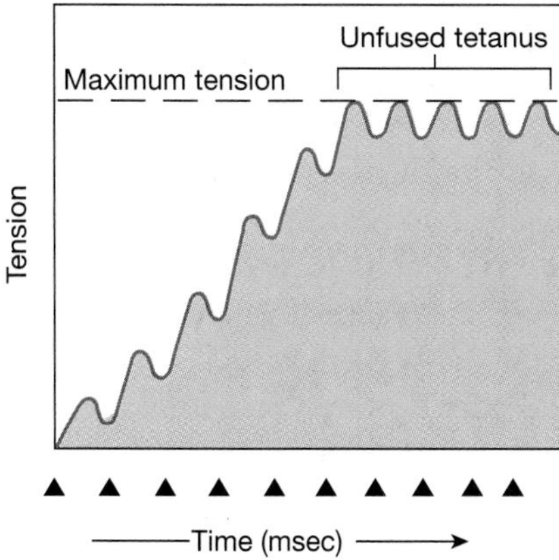

(d) Summation leading to complete tetanus: Muscle reaches steady tension. If muscle fatigues, tension decreases rapidly.

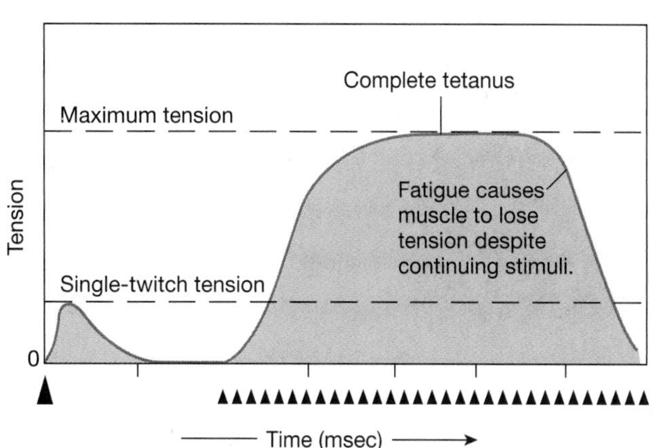

Fig. 12.16

that move the eyes or the muscles of the hand, one motor unit contains as few as three to five muscle fibers. If one such motor unit is activated, only a few fibers contract, and the muscle response is quite small. If additional motor units are activated, the response increases by small increments because only a few more muscle fibers contract with the addition of each motor unit. This arrangement allows fine gradations of movement.

In muscles used for gross motor actions such as standing or walking, each motor unit may contain hundreds or even thousands of muscle fibers. The gastrocnemius muscle in the calf of the leg, for example, has about 2000 muscle fibers in each motor unit. Each time an additional motor unit is activated in these muscles, many more muscle fibers contract, and the muscle response jumps by correspondingly greater increments.

All muscle fibers in a single motor unit are of the same fiber type. For this reason there are fast-twitch motor units and slow-twitch motor units. Which kind of muscle fiber associates with a particular neuron appears to be a function of the neuron. During embryological development, each somatic motor neuron secretes a growth factor that directs the differentiation of all muscle fibers in its motor unit so that they develop into the same fiber type.

Intuitively, it would seem that people who inherit a predominance of one fiber type over another would excel in certain sports. They do, to some extent. Endurance athletes, such as distance runners and cross-country skiers, have a predominance of slow-twitch fibers, whereas sprinters, ice hockey players, and weight lifters tend to have larger percentages of fast-twitch fibers.

Inheritance is not the only determining factor for fiber composition in the body, however, because the metabolic characteristics of muscle fibers have some plasticity. With endurance training, the aerobic capacity of some fast-twitch fibers can be enhanced until they are almost as fatigue-resistant as slow-twitch fibers. Because the conversion occurs only in those

MOTOR UNITS

A motor unit consists of one motor neuron and all the muscle fibers it innervates. A muscle may have many motor units of different types.

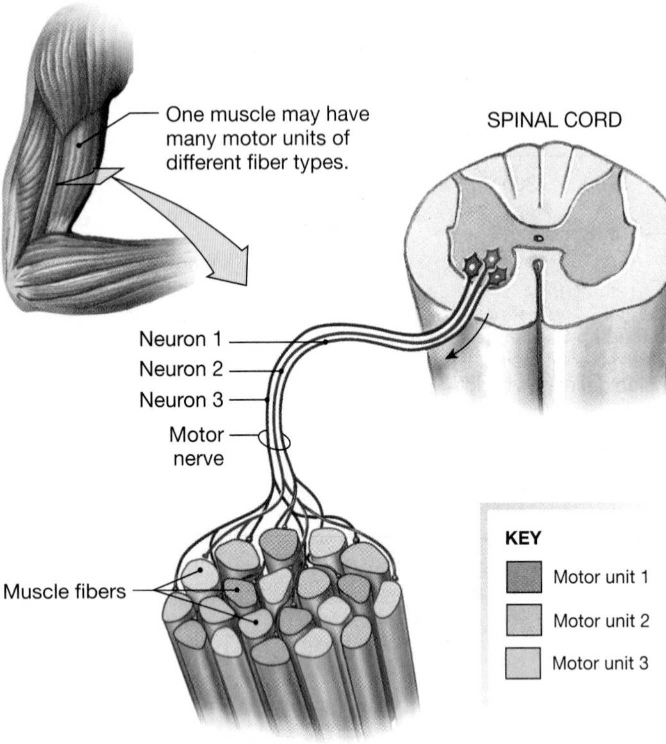

Fig. 12.17

muscles that are being trained, a neuromodulator chemical is probably involved. In addition, endurance training increases the number of capillaries and mitochondria in the muscle tissue, allowing more oxygen-carrying blood to reach the contracting muscle and contributing to the increased aerobic capacity of the muscle fibers.

Concept Check Answer: p. 439

19. Which type of runner would you expect to have more slow-twitch fibers, a sprinter or a marathoner?

Contraction Force Depends on the Types and Numbers of Motor Units

Within a skeletal muscle, each motor unit contracts in an all-or-none manner. How then can muscles create graded contractions of varying force and duration? The answer lies in the fact that muscles are composed of multiple motor units of different types (Fig. 12.17). This diversity allows the muscle to vary contraction by (1) changing the types of motor units that are active or (2) changing the number of motor units that are responding at any one time.

The force of contraction in a skeletal muscle can be increased by recruiting additional motor units. **Recruitment** is controlled by the nervous system and proceeds in a standardized sequence. A weak stimulus directed onto a pool of somatic motor neurons in the central nervous system activates only the neurons with the lowest thresholds [p. 254]. Studies have shown that these low-threshold neurons control fatigue-resistant slow-twitch fibers, which generate minimal force.

As the stimulus onto the motor neuron pool increases in strength, additional motor neurons with higher thresholds begin to fire. These neurons in turn stimulate motor units composed of fatigue-resistant fast-twitch oxidative-glycolytic fibers. Because more motor units (and thus more muscle fibers) are participating in the contraction, greater force is generated in the muscle.

As the stimulus increases to even higher levels, somatic motor neurons with the highest thresholds begin to fire. These neurons stimulate motor units composed of glycolytic fast-twitch fibers. At this point, the muscle contraction is approaching its maximum force. Because of differences in myosin and crossbridge formation, fast-twitch fibers generate more force than slow-twitch fibers do. However, because fast-twitch fibers fatigue more rapidly, it is impossible to hold a muscle contraction at maximum force for an extended period of time. You can demonstrate this by clenching your fist as hard as you can: how long can you hold it before some of the muscle fibers begin to fatigue?

Sustained contractions in a muscle require a continuous train of action potentials from the central nervous system to the muscle. As you learned earlier, however, increasing the stimulation rate of a muscle fiber results in summation of its contractions. If the muscle fiber is easily fatigued, summation leads to fatigue and diminished tension (Fig. 12.16d).

One way the nervous system avoids fatigue in sustained contractions is by **asynchronous recruitment** of motor units. The nervous system modulates the firing rates of the motor neurons so that different motor units take turns maintaining muscle tension. The alternation of active motor units allows some of the motor units to rest between contractions, preventing fatigue.

Asynchronous recruitment prevents fatigue only in submaximal contractions, however. In high-tension, sustained contractions, the individual motor units may reach a state of unfused tetanus, in which the muscle fibers cycle between contraction and partial relaxation. In general, we do not notice this cycling because the different motor units in the muscle are contracting and relaxing at slightly different times. As a result, the contractions and relaxations of the motor units average out and appear to be one smooth contraction. But as different motor units fatigue, we are unable to maintain the same amount of tension in the muscle, and the force of the contraction gradually decreases.

Concept Check Answers: p. 439

20. What is the response of a muscle fiber to an increase in the firing rate of the somatic motor neuron?
21. How does the nervous system increase the force of contraction in a muscle composed of many motor units?

Mechanics of Body Movement

Because one main role of skeletal muscles is to move the body, we now turn to the mechanics of body movement. The term *mechanics* refers to how muscles move loads and how the anatomical relationship between muscles and bones maximizes the work the muscles can do.

Isotonic Contractions Move Loads; Isometric Contractions Create Force Without Movement

When we described the function of muscles earlier in this chapter, we noted that they can create force to generate movement but can also create force without generating movement. You can demonstrate both properties with a pair of heavy weights. Pick up one weight in each hand and then bend your elbows so that the weights touch your shoulders. You have just performed an **isotonic contraction** {*iso*, equal + *teinein*, to stretch}. Any contraction that creates force and moves a load is an isotonic contraction.

When you bent your arms at the elbows and brought the weights to your shoulders, the biceps muscles shortened. Now slowly extend your arms, resisting the gravitational forces pulling the weights down. The biceps muscles are again active, but now you are performing a *lengthening (eccentric) contraction*. Lengthening contractions are thought to contribute most to cellular damage after exercise and to lead to delayed muscle soreness.

If you pick up the weights and hold them stationary in front of you, the muscles of your arms are creating tension (force) to overcome the load of the weights but are not creating movement. Contractions that create force without moving a load are called **isometric** (static) **contractions** {*iso*, equal + *metric*, measurement}. Isotonic and isometric contractions are illustrated in ■ Figure 12.18. To demonstrate an isotonic contraction experimentally, we hang a weight (the load) from the muscle in Figure 12.18a and electrically stimulate the muscle to contract. The muscle contracts, lifting the weight. The graph on the right shows the development of force throughout the contraction.

To demonstrate an isometric contraction experimentally, we attach a heavier weight to the muscle, as shown in Figure 12.18b. When the muscle is stimulated, it develops tension, but the force created is not enough to move the load. In isometric contractions, muscles create force without shortening significantly. For example, when your exercise instructor yells at you to "tighten those glutes," your response is isometric contraction of the gluteal muscles in your buttocks.

How can an isometric contraction create force if the length of the muscle does not change significantly? The elastic elements of the muscle provide the answer. All muscles contain elastic fibers in the tendons and other connective tissues that attach muscles to bone, and in the connective tissue between muscle fibers. In muscle fibers, elastic cytoskeletal proteins occur between the myofibrils and as part of the sarcomere. All of these elastic components behave collectively as if they were connected in series (one after the other) to the contractile elements of the muscle. Consequently, they are often called the **series elastic elements** of the muscle (■ Fig. 12.19).

When the sarcomeres shorten in an isometric contraction, the elastic elements stretch. This stretching of the elastic elements allows the fibers to maintain a relatively constant length even though the sarcomeres are shortening and creating tension (Fig. 12.19 ②). Once the elastic elements have been stretched and the force generated by the sarcomeres equals the load, the muscle shortens in an isotonic contraction and lifts the load.

Bones and Muscles Around Joints Form Levers and Fulcrums

The anatomical arrangement of muscles and bones in the body is directly related to how muscles work. The body uses its bones and joints as levers and fulcrums on which muscles exert force to move or resist a load. A **lever** is a rigid bar that pivots around a point known as the **fulcrum**. In the body, bones form levers, flexible joints form the fulcrums, and muscles attached to bones create force by contracting.

Most lever systems in the body are similar to a fishing pole, like the one shown in ■ Figure 12.20a. In these lever systems, the fulcrum is located at one end of the lever, the load is near the other end of the lever, and force is applied between the fulcrum and the load. This arrangement maximizes the distance and speed with which the lever can move the load but also requires more force than some other lever systems. Let's see how flexion of the forearm illustrates lever system function.

In the lever system of the forearm, the elbow joint acts as the fulcrum around which rotational movement of the forearm (the lever) takes place (Fig. 12.20b). The biceps muscle is attached at its origin at the shoulder and inserts onto the radius

ISOTONIC AND ISOMETRIC CONTRACTIONS

In an isotonic contraction, the muscle contracts, shortens, and creates enough force to move the load.
In an isometric contraction, the muscle contracts but does not shorten. The force created cannot move the load.

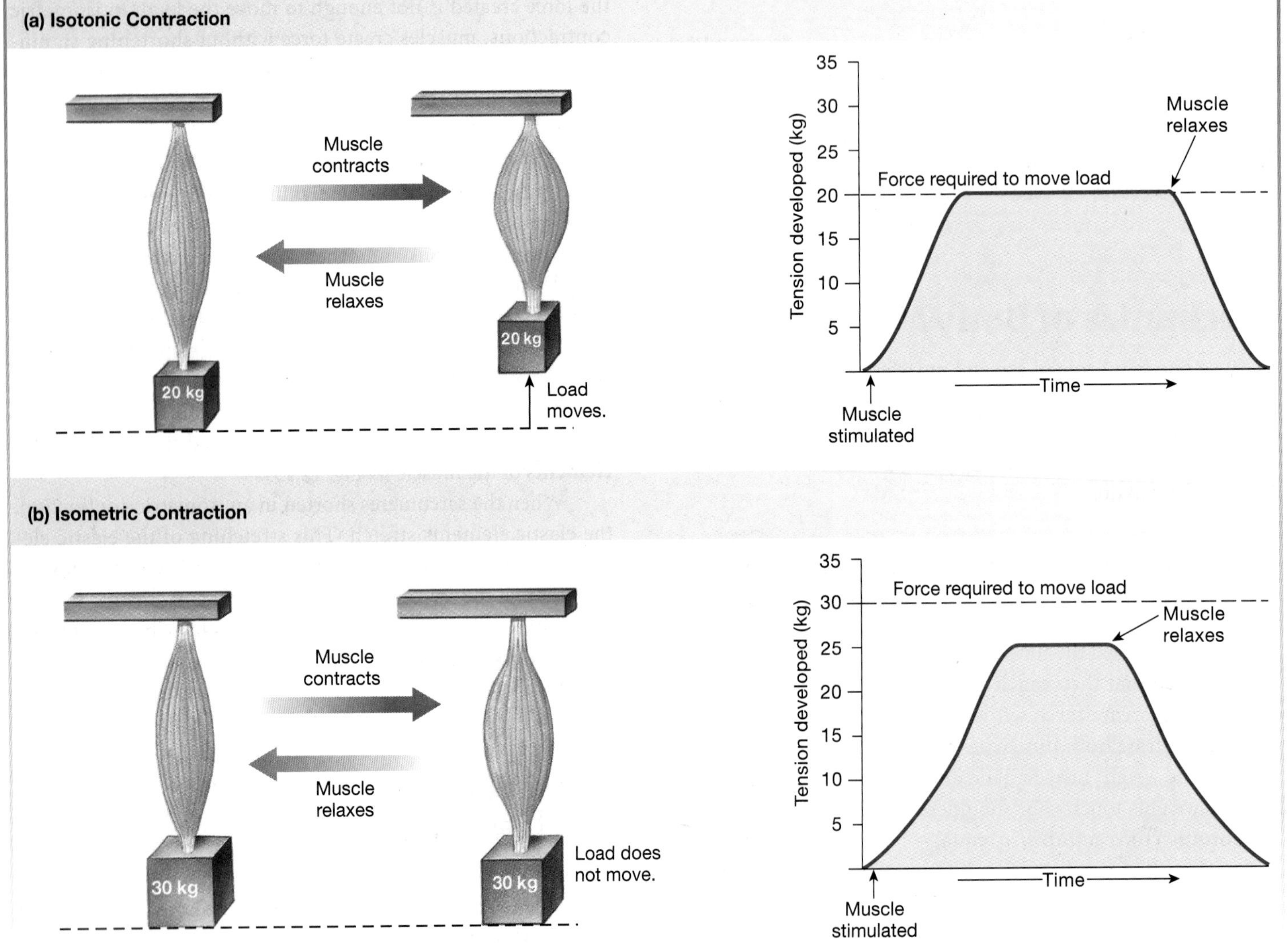

(a) Isotonic Contraction

Muscle contracts
Muscle relaxes
20 kg
20 kg
Load moves.
Tension developed (kg)
Force required to move load
Muscle relaxes
Muscle stimulated
Time

(b) Isometric Contraction

Muscle contracts
Muscle relaxes
30 kg
30 kg
Load does not move.
Tension developed (kg)
Force required to move load
Muscle relaxes
Muscle stimulated
Time

Fig. 12.18

bone of the forearm a few centimeters away from the elbow joint. When the biceps contracts, it creates the upward force F_1 (Fig. 12.20c) as it pulls on the bone. The total rotational force* created by the biceps depends on two things: (1) the force of muscle contraction and (2) the distance between the fulcrum and the point at which the muscle inserts onto the radius.

If the biceps is to hold the forearm stationary and flexed at a 90° angle, the muscle must exert enough upward rotational force to exactly oppose the downward rotational force exerted by gravity on the forearm (Fig. 12.20c). The downward rotational force on the forearm is proportional to the weight of the forearm (F_2) times the distance from the fulcrum to the forearm's center of gravity (the point along the lever at which the forearm load exerts its force). For the arm illustrated in Figure 12.20c, the biceps must exert 6 kg of force to hold the arm at a 90° angle. Because the muscle is not shortening, this is an isometric contraction.

Now what happens if a 7-kg weight is placed in the hand? This weight places an additional load on the lever that is farther from the fulcrum than the forearm's center of gravity. Unless the biceps can create additional upward force to offset the downward force created by the weight, the hand falls. If you know the force exerted by the added weight and its distance from the

*In physics, rotational force is expressed as *torque,* and the force of contraction is expressed in newtons (mass × acceleration due to gravity). For simplicity, we ignore the contribution of gravity in this discussion and use the mass unit "kilograms" for force of contraction.

SERIES ELASTIC ELEMENTS IN MUSCLE

Muscle has both contractile components (sarcomeres, shown here as a gear and ratchet) and elastic components (shown here as a spring).

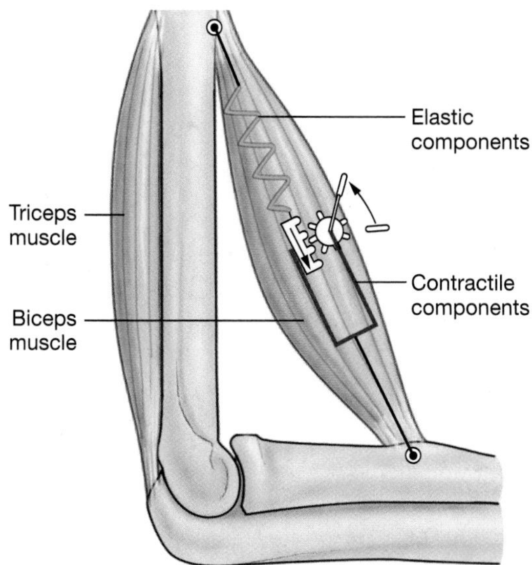

Elastic elements allow isometric contractions.

- In an isometric contraction, sarcomeres shorten, generating force, but elastic elements stretch, allowing muscle length to remain the same.
- In isotonic contractions, sarcomeres shorten more but, because elastic elements are already stretched, the muscles shorten.

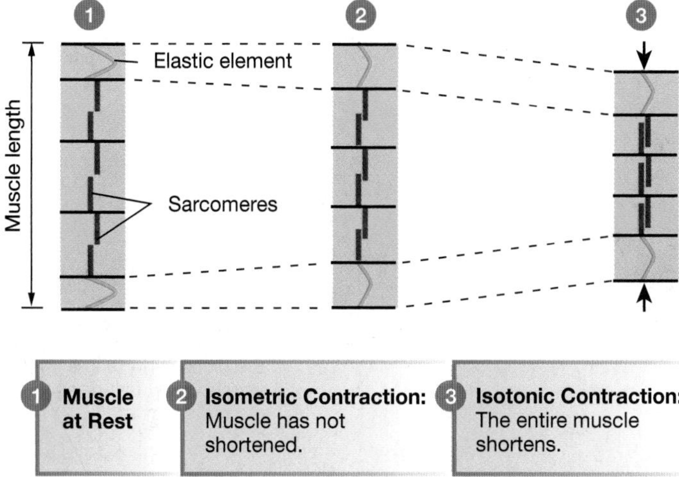

Fig. 12.19

elbow, you can calculate the additional muscle force needed to keep the arm from dropping the 7-kg weight.

What happens to the force required of the biceps to support a weight if the distance between the fulcrum and the muscle insertion point changes? Genetic variability in the insertion point can have a dramatic effect on the force required to move or resist a load. For example, if the biceps in Figure 12.20b inserted 6 cm from the fulcrum instead of 5 cm, it would only need to generate 5 kg of force to offset the weight of the arm. Some studies have shown a correlation between muscle insertion points and success in certain athletic events.

In the example so far, we have assumed that the load is stationary and that the muscle is contracting isometrically. What happens if we want to flex the arm and lift the load? To move the load from its position, the biceps must exert a force that exceeds the force created by the stationary load.

The disadvantage of a lever system in which the fulcrum is positioned near one end of the lever is that the muscle is required to create large amounts of force to move or resist a small load, as we just saw. However, the advantage of this type of lever-fulcrum system is that it maximizes speed and mobility. A small movement of the forearm at the point where the muscle inserts becomes a much larger movement at the hand (Fig. 12.20d). In addition, the two movements occur in the same amount of time, and so the speed of contraction at the insertion point is amplified at the hand. Thus, the lever-fulcrum system of the arm amplifies both the distance the load is moved and the speed at which this movement takes place.

In muscle physiology, the speed with which a muscle contracts depends on the type of muscle fiber (fast-twitch or slow-twitch) and on the load that is being moved. Intuitively, you can see that you can flex your arm much faster with nothing in your hand than you can while holding a 7-kg weight in your hand. The relationship between load and velocity (speed) of contraction in a muscle fiber, determined experimentally, is graphed in ■ Figure 12.21.

Contraction is fastest when the load on the muscle is zero. When the load on the muscle equals the ability of the muscle to create force, the muscle is unable to move the load and the velocity drops to zero. The muscle can still contract, but the contraction becomes isometric instead of isotonic. Because speed is a function of load and muscle fiber type, it cannot be regulated by the body except through recruitment of faster muscle fiber types. However, the arrangement of muscles, bones, and joints allows the body to amplify speed so that regulation at the cellular level becomes less important.

Concept Check Answer: p. 439

22. One study found that many world-class athletes have muscle insertions that are farther from the joint than in the average person. Why would this trait translate into an advantage for a weight lifter?

Muscle Disorders Have Multiple Causes

Dysfunction in skeletal muscles can arise from a problem with the signal from the nervous system, from miscommunication at the neuromuscular junction, or from defects in the muscle. Unfortunately, in many muscle conditions, even the simple ones, we do not fully understand the mechanism of the primary

THE ARM IS A LEVER AND FULCRUM SYSTEM

(a) The lever system of the forearm is like that of a fishing pole. The fulcrum is at one end of the lever and the load is at the other end. Force is applied between the fulcrum and the load.

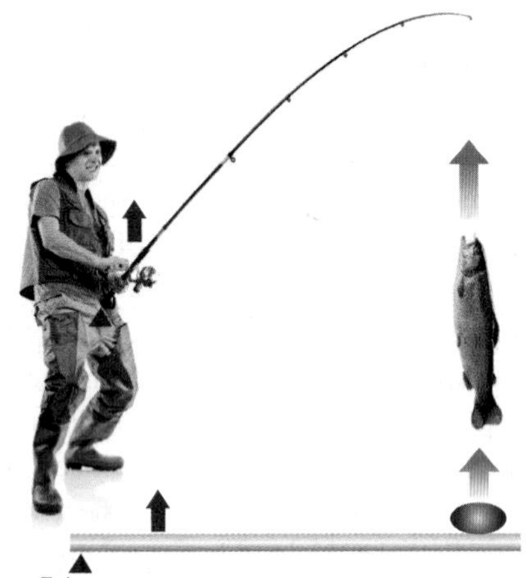

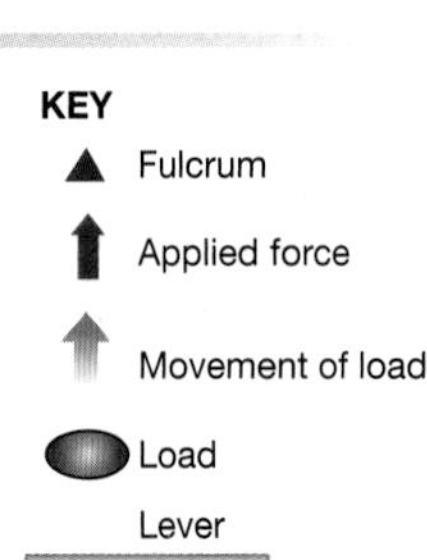

(b) The human forearm acts as a lever. The fulcrum is the elbow joint. The load is gravity acting on the mass of the forearm and hand.

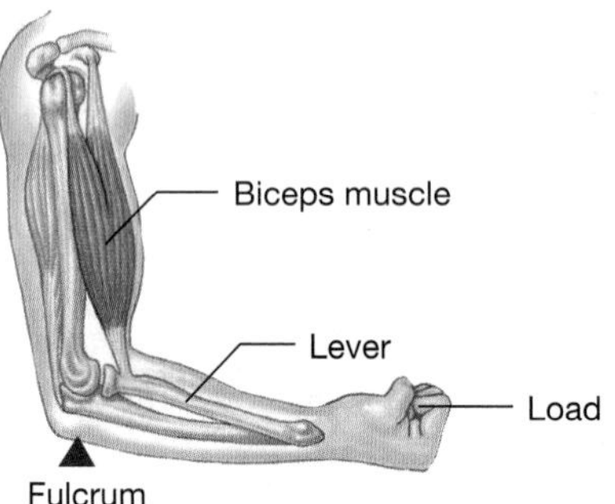

(c) Force calculations

Biceps contraction creates upward force F_1.

The biceps inserts into the lever 5 cm from the fulcrum.

Rotational $\text{force}_{up} \propto$ biceps force $F_1 \times 5$ cm from the fulcrum.

F_1

$F_2 = 2$ kg

5 cm

15 cm

The weight of the forearm exerts a downward force of 2 kg at its center of gravity, which is 15 cm from the fulcrum.

Rotational $\text{force}_{down} \propto$ load $F_2 \times 15$ cm

$\propto 2 \text{ kg} \times 15 \text{ cm}$

To hold the arm stationary at 90 degrees, the rotational force created by the contracting biceps must exactly oppose the downward rotation created by the forearm's weight.

Rotational force_{up} = Rotational force_{down}

Biceps force × 5 cm = 2 kg × 15 cm

$$\text{Biceps force} = \frac{30 \text{ kg·cm}}{5 \text{ cm}}$$

Biceps force = 6 kg

FIGURE QUESTION

How much additional force must the biceps exert to keep from dropping the weight?

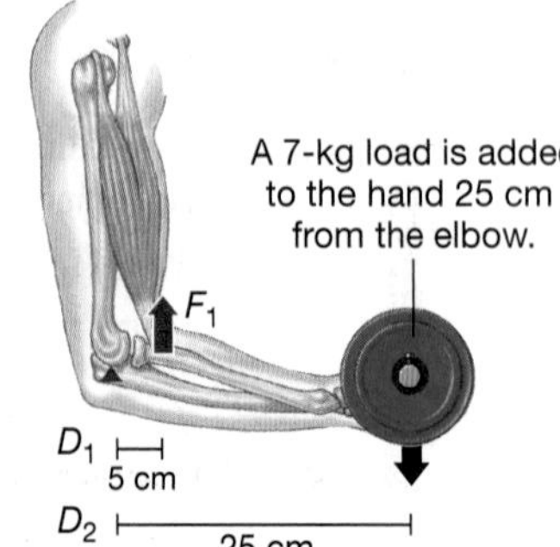

(d) The arm amplifies speed of movement of the load.

Because the insertion of the biceps is close to the fulcrum, a small movement of the biceps becomes a much larger movement of the hand.

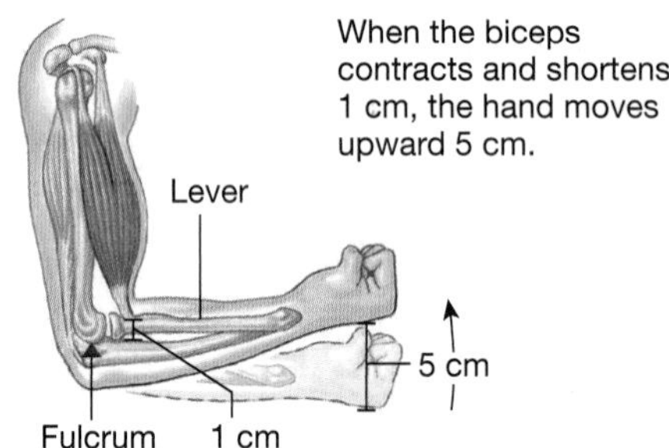

FIGURE QUESTION

If the biceps shortens 1 cm in 1 second, how fast does the hand move upward?

■ Fig. 12.20

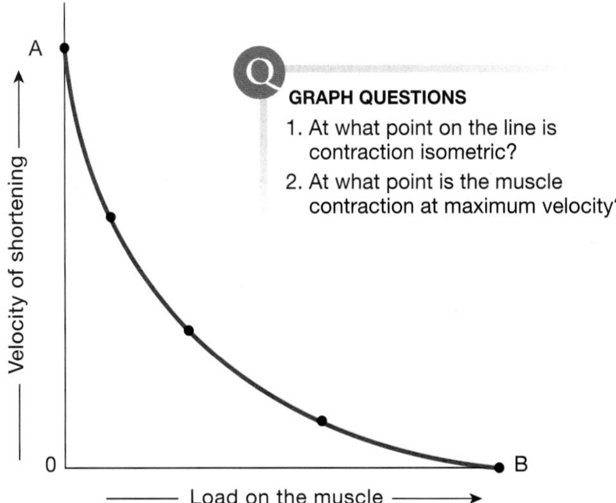

Fig. 12.21

defect. As a result, we can treat the symptoms but may not be able to cure the problem.

One common muscle disorder is a "charley horse," or *muscle cramp*—a sustained painful contraction of skeletal muscles. Many muscle cramps are caused by hyperexcitability of the somatic motor neurons controlling the muscle. As the neuron fires repeatedly, the muscle fibers of its motor unit go into a state of painful sustained contraction. Sometimes muscle cramps can be relieved by forcibly stretching the muscle. Apparently, stretching sends sensory information to the central nervous system that inhibits the somatic motor neuron, relieving the cramp.

The simplest muscle disorders arise from overuse. Most of us have exercised too long or too hard and suffered from fatigue or soreness as a result. With more severe trauma, muscle fibers, the connective tissue sheath, or the union of muscle and tendon may tear.

Disuse of muscles can be as traumatic as overuse. With prolonged inactivity, such as may occur when a limb is immobilized in a cast, the skeletal muscles atrophy. Blood supply to the muscle diminishes, and the muscle fibers get smaller. If activity is resumed in less than a year, the fibers usually regenerate. Atrophy of longer than one year is usually permanent. If the atrophy results from somatic motor neuron dysfunction, therapists now try to maintain muscle function by administering electrical impulses that directly stimulate the muscle fibers.

Acquired disorders that affect the skeletal muscle system include infectious diseases, such as influenza, that lead to weakness and achiness, and poisoning by toxins, such as those produced in botulism (*Clostridium botulinus*) and tetanus (*Clostridium tetani*). Botulinum toxin acts by decreasing the release of acetylcholine from the somatic motor neuron. Clinical investigators have successfully used injections of botulinum toxin as a treatment for writer's cramp, a disabling cramp of the hand that apparently arises as a result of hyperexcitability in the distal portion of the somatic motor neuron. Botox® injections are now widely used for cosmetic wrinkle reduction. Botulinum toxin injected under the skin temporarily paralyzes facial muscles that pull the skin into wrinkles.

Inherited muscular disorders are the most difficult to treat. These conditions include various forms of muscular dystrophy as well as biochemical defects in glycogen and lipid storage. In **Duchenne muscular dystrophy**, the structural protein **dystrophin**, which links actin to proteins in the cell membrane, is absent. In muscle fibers that lack dystrophin, extracellular Ca^{2+} enters the fiber through small tears in the membrane or possibly through stretch-activated channels. Calcium entry activates intracellular enzymes, resulting in breakdown of the fiber components. The major symptom of Duchenne dystrophy is progressive muscle weakness, and patients usually die before age 30 from failure of the respiratory muscles.

McArdle's disease, also known as *myophosphorylase deficiency*, is a condition in which the enzyme that converts glycogen to glucose 6-phosphate is absent in muscles. As a result, muscles lack a usable glycogen energy supply, and exercise tolerance is limited.

One way physiologists are trying to learn more about muscle diseases is by using animal models, such as genetically engineered mice that lack the genes for certain muscle proteins. Researchers are trying to correlate the absence of protein with particular disruptions in function.

12

RUNNING PROBLEM

Paul's doctor explains to Mrs. Leong that the paralytic attacks associated with hyperkalemic periodic paralysis last only a few minutes to a few hours and generally involve only the muscles of the extremities, which become weak and unable to contract (*flaccid paralysis*). "Is there any treatment?" asks Mrs. Leong. The doctor replies that although the inherited condition cannot be cured, attacks may be prevented with drugs. Diuretics, for example, increase the rate at which the body excretes water and ions (including Na^+ and K^+), and these medications have been shown to help prevent attacks of paralysis in people with hyperKPP.

Q4: Draw a map to explain why a Na^+ channel that does not inactivate results in a muscle that cannot contract (flaccid paralysis).

Smooth Muscle

Although skeletal muscle has the most muscle mass in the body, cardiac and smooth muscle are more important in the maintenance of homeostasis. Smooth muscle is challenging to describe because smooth muscles in the body have so much functional variability. There are many ways to categorize the different types of smooth muscle, but we will consider three:

1. **By location.** Smooth muscles with widely differing properties are found throughout the animal kingdom. In humans, smooth muscle can be divided into six major groups: *vascular* (blood vessel walls), *gastrointestinal* (walls of digestive tract and associated organs, such as the gallbladder), *urinary* (walls of bladder and ureters), *respiratory* (airway passages), *reproductive* (uterus in females and other reproductive structures in both females and males), and *ocular* (eye). These muscles have different functions in the body, and their physiology reflects their specialized functions. In contrast, skeletal muscle is relatively uniform throughout the body.
2. **By contraction pattern**. Smooth muscle can be classified by whether it alternates between contraction and relaxation states or whether it is continuously contracted. Muscles that undergo periodic contraction and relaxation cycles are said to be **phasic smooth muscles**. An example would be the wall of the lower esophagus, which contracts only when food passes through it (■ Fig. 12.22a). Some phasic smooth muscles, such as those in the wall of the intestine, cycle rhythmically through contractions alternating with relaxation (Fig. 12.22b).

   Muscles that are continuously contracted are called **tonic smooth muscles** because they are always maintaining some level of muscle tone. The esophageal and urinary bladder **sphincters** {*sphingein,* to close} are examples of tonically contracted muscles that close off the opening to a hollow organ. These sphincters relax when it is necessary to allow material to enter or leave the organ (Fig. 12.22c). The tonic smooth muscle in the walls of some blood vessels maintain an intermediate level of contraction. Under *tonic control* by the nervous system [p. 192], this vascular smooth muscle contracts or relaxes as the situation demands (Fig. 12.22d).
3. **By their communication with neighboring cells.** In some smooth muscles, the cells are electrically connected by gap junctions, and they contract as a coordinated unit. These muscles are called **single-unit smooth muscle**, or *unitary smooth muscle*. In **multi-unit smooth muscle,** the cells are not linked electrically and each muscle cell functions independently.

SMOOTH MUSCLE CONTRACTIONS

(a) A phasic smooth muscle that is usually relaxed. Example: esophagus

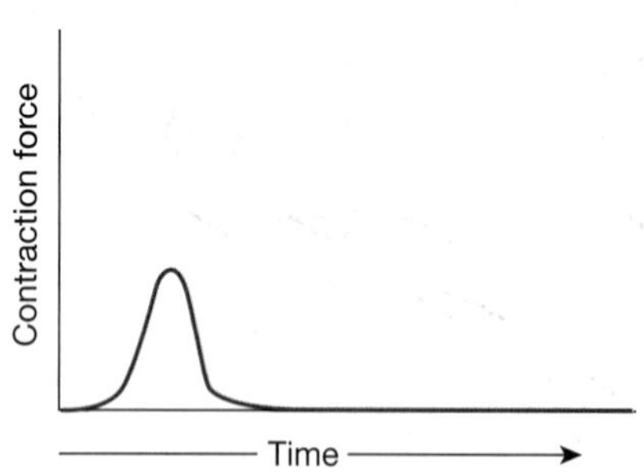

(b) A phasic smooth muscle that cycles between contraction and relaxation. Example: intestine

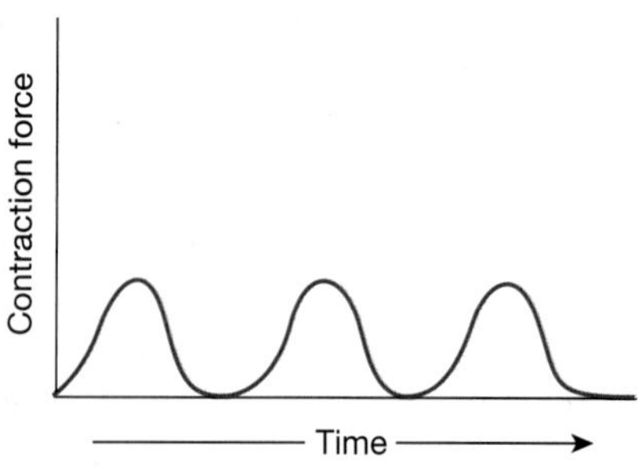

(c) A tonic smooth muscle that is usually contracted. Example: a sphincter that relaxes to allow material to pass.

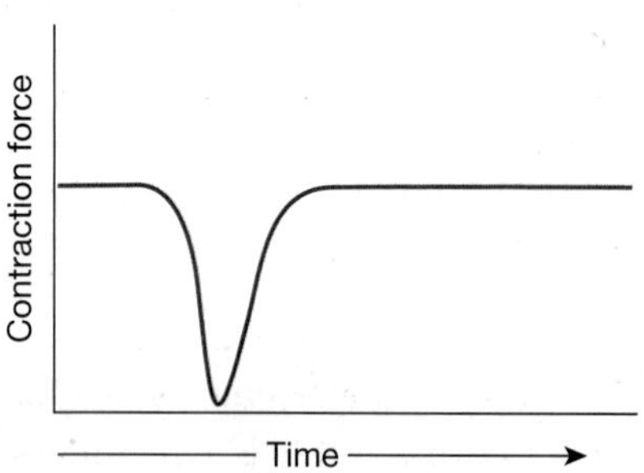

(d) A tonic smooth muscle whose contraction is varied as needed. Example: vascular smooth muscle

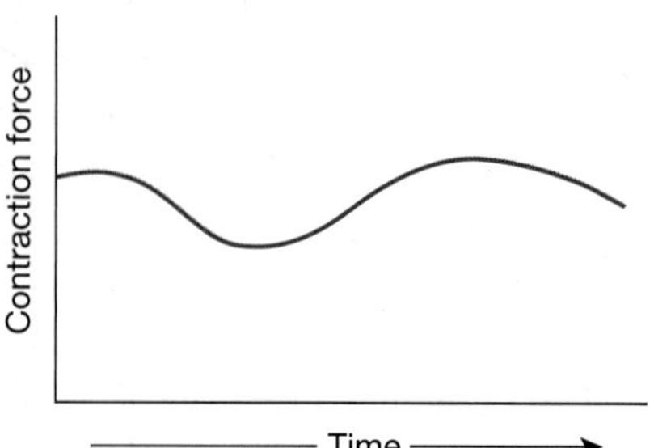

■ **Fig. 12.22**

SMOOTH MUSCLE COORDINATION

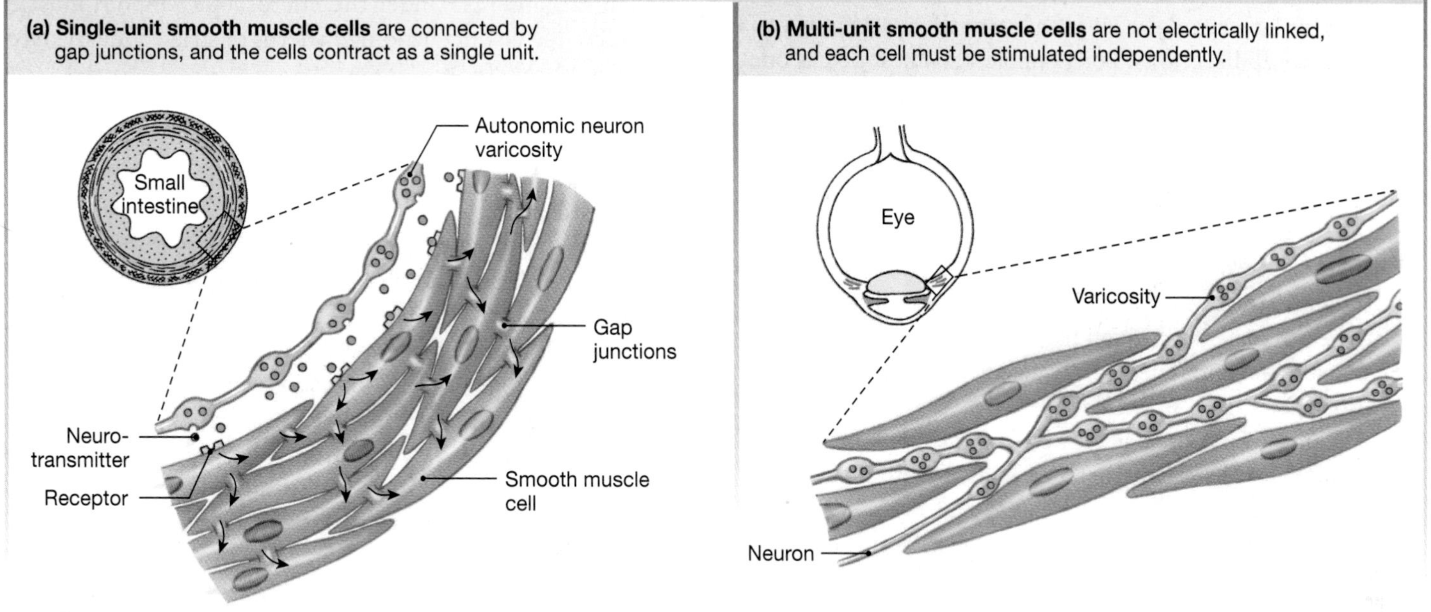

(a) Single-unit smooth muscle cells are connected by gap junctions, and the cells contract as a single unit.

(b) Multi-unit smooth muscle cells are not electrically linked, and each cell must be stimulated independently.

■ **Fig. 12.23**

Most smooth muscle is single-unit smooth muscle. Single-unit smooth muscle is also called **visceral smooth muscle** because it forms the walls of internal organs (viscera), such as the intestinal tract. The fibers of single-unit smooth muscle are connected to one another by gap junctions. An electrical signal in one cell spreads rapidly through the entire sheet of tissue to create a coordinated contraction (■ Fig. 12.23a). Because all fibers contract every time, no reserve units are left to be recruited to increase contraction force. Instead, the amount of Ca^{2+} that enters the cell determines the force of contraction, as you will learn in the discussion that follows.

In multi-unit smooth muscle, the cells are not linked electrically and they must be stimulated independently to contract. Each individual muscle cell is closely associated with an axon terminal or varicosity (Fig. 12.23b). This arrangement allows fine control of contractions in these muscles through selective activation of individual muscle cells. As in skeletal muscle, increasing the force of contraction requires recruitment of additional fibers.

Multi-unit smooth muscle is found in the iris and ciliary body of the eye [p. 361], in part of the male reproductive tract, and in the uterus except just prior to labor and delivery. Interestingly, the multi-unit smooth muscle of the uterus changes and becomes single-unit during the final stages of pregnancy. Genes for synthesis of gap junction connexin proteins turn on, apparently under the influence of pregnancy hormones. The addition of gap junctions to the uterine muscle cells synchronizes electrical signals, allowing the uterine muscle to contract more effectively while expelling the baby.

Because of the variability in smooth muscle types, we introduce only their general features in this chapter. You will learn properties that are specific to a certain type when you study the different organ systems.

12

Concept Check Answer: p. 440

23. What is the difference in how contraction force is varied in multi-unit and single-unit smooth muscle?

Smooth Muscle Is More Variable Than Skeletal Muscle

Two of the principles that you learned in previous sections for skeletal muscle apply to all smooth muscle. First, force is created by actin-myosin crossbridge interaction between sliding filaments. Second, contraction in smooth muscle, as in skeletal and cardiac muscle, is initiated by an increase in free cytosolic Ca^{2+} concentrations. However, in most other ways smooth muscle function is more complex than skeletal muscle function. Let's examine some differences, starting at the organ level and working to the cellular level.

1 **Smooth muscles must operate over a range of lengths.** Smooth muscle is found predominantly in the walls of hollow organs and tubes, many of which expand and contract as they fill and empty. The bladder, which fills with urine, is an example of a distensible organ. Smooth muscles in organs like this must function efficiently over a range of

muscle lengths. In contrast, most skeletal muscles are attached to bone and operate over a narrow range of lengths.

2 **Within an organ, the layers of smooth muscle may run in several directions.** For example, the intestine has one muscle layer that encircles the lumen and a perpendicular layer that runs the length of the intestine. The stomach adds a third layer that is set obliquely to the other two. Contraction in different layers changes the shape of the organ. Sometimes smooth muscles generate force to move material through the lumen of the organ, such as the sequential waves of smooth muscle contraction that move ingested material through the small intestine. In contrast, most skeletal muscles are arranged so that their contraction shortens the muscle.

3 When you compare a single muscle twitch in muscle types, **smooth muscles contract and relax much more slowly** than skeletal or cardiac muscle (■ Fig. 12.24).

4 **Smooth muscle uses less energy to generate and maintain a given amount of force.** Smooth muscles can develop force rapidly but have the ability to slow down their myosin ATPase so that crossbridges cycle slowly as they maintain their force. As a result, their use of ATP is lower than that in striated muscles. Smooth muscle has fewer mitochondria than striated muscles and relies more on glycolysis for its ATP production.

5 **Smooth muscle can sustain contractions for extended periods without fatiguing.** This property allows organs such as the bladder to maintain tension despite a continued load. It also allows some smooth muscles to be tonically contracted and maintain tension most of the time.

6 **Smooth muscles have small, spindle-shaped cells with a single nucleus**, in contrast to the large multinucleated fibers of skeletal muscles.

Duration of Muscle Twitch in the Three Types of Muscle

Smooth muscles are the slowest to contract and to relax.

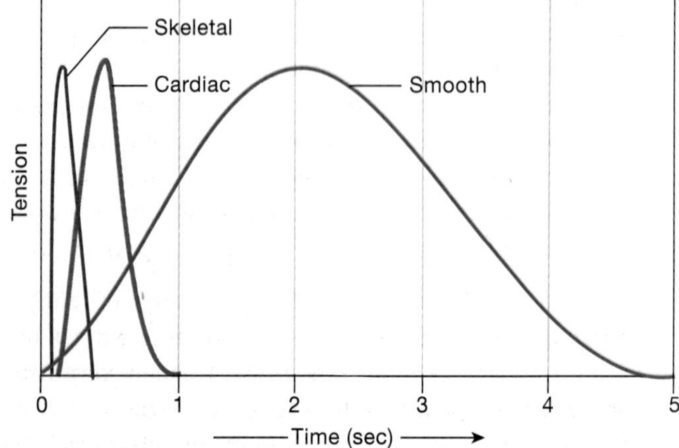

■ **Fig. 12.24**

7 In smooth muscle, **the contractile fibers are not arranged in sarcomeres**. Under the microscope, smooth muscle lacks the distinct banding patterns of striated muscle (see Fig. 12.1c).

8 **Contraction in smooth muscle may be initiated by electrical or chemical signals or both.** Skeletal muscle contraction always begins with an action potential in the muscle fiber.

9 **Smooth muscle is controlled by the autonomic nervous system.** Skeletal muscle is controlled by the somatic motor division of the nervous system.

10 **Smooth muscle lacks specialized receptor regions** such as the motor end plates found in skeletal muscle synapses. Instead, receptors are found all over the cell surface. Neurotransmitter is released from autonomic neuron varicosities [p. 384] close to the surface of the muscle fibers and simply diffuses across the cell surface until it finds a receptor.

11 In smooth muscle, **the Ca^{2+} for contraction comes from the extracellular fluid as well as from the sarcoplasmic reticulum.** In skeletal muscle, the Ca^{2+} comes from the sarcoplasmic reticulum.

12 In smooth muscle, **the Ca^{2+} signal initiates a cascade that ends with phosphorylation of myosin light chains and activation of myosin ATPase.** In skeletal muscle, the Ca^{2+} signal binds to troponin to initiate contraction. (Smooth muscle has no troponin.)

With these points in mind, we will now look at some details of smooth muscle function.

Concept Check Answer: p. 440

24. When the circular muscle layer of the intestine contracts, what happens to the shape of the tube? When the longitudinal layer contracts, what happens to the shape?

RUNNING PROBLEM

Three weeks later, Paul had another attack of paralysis, this time at kindergarten after a game of tag. He was rushed to the hospital and given glucose by mouth. Within minutes, he was able to move his legs and arms and asked for his mother.

Q5: Explain why oral glucose might help bring Paul out of his paralysis. (*Hint:* Glucose stimulates insulin release, and insulin increases Na^+-K^+-ATPase activity. What happens to the extracellular K^+ level when Na^+-K^+-ATPase is more active?)

400 413 415 425 428 435

Smooth Muscle Lacks Sarcomeres

Smooth muscle has the same contractile elements as skeletal muscle—actin and myosin that interact through crossbridges—as well sarcoplasmic reticulum that stores and releases Ca^{2+}. However, details of the structural elements differ in the two muscle types.

Actin and Myosin Actin is more plentiful in smooth muscle than in striated muscle, with an actin-to-myosin ratio of 10–15 to 1, compared with 2–4 to 1 in striated muscle. Smooth muscle actin is associated with tropomyosin, as in skeletal muscle. However, unlike skeletal muscle, smooth muscle lacks troponin.

Smooth muscles have less myosin than skeletal muscle. The less numerous myosin filaments are surrounded by actin filaments and are arranged so that each myosin molecule is in the center of a bundle of 12–15 actin molecules. These contractile units are arranged so that they run parallel to the long axis of the cell.

Myosin filaments in smooth muscle are longer than in skeletal muscle, and the entire surface of the filament is covered by myosin heads (■ Fig. 12.25b). This unique organization enables smooth muscle to stretch more while still maintaining enough overlap to create optimum tension. This is an important property for internal organs, such as the bladder, whose volume varies as it alternately fills and empties.

Smooth muscle cells have an extensive cytoskeleton consisting of intermediate filaments and protein **dense bodies** in the cytoplasm and along the cell membrane. Actin filaments attach to the dense bodies (Fig. 12.25a). Cytoskeleton fibers linking dense bodies to the cell membrane help hold actin in place. Protein fibers in the extracellular matrix tie the smooth muscle cells of a tissue together and transfer force from a contracting cell to its neighbors.

Sarcoplasmic Reticulum The amount of SR in smooth muscle varies from one type of smooth muscle to another. The arrangement of smooth muscle SR is less organized than in skeletal muscle, consisting of a network of tubules that extend from just under the cell membrane into the interior of the cell. There are no t-tubules in smooth muscle, but the SR is closely associated with the membrane invaginations called *caveolae* [p. 157], which apparently participate in cell signaling.

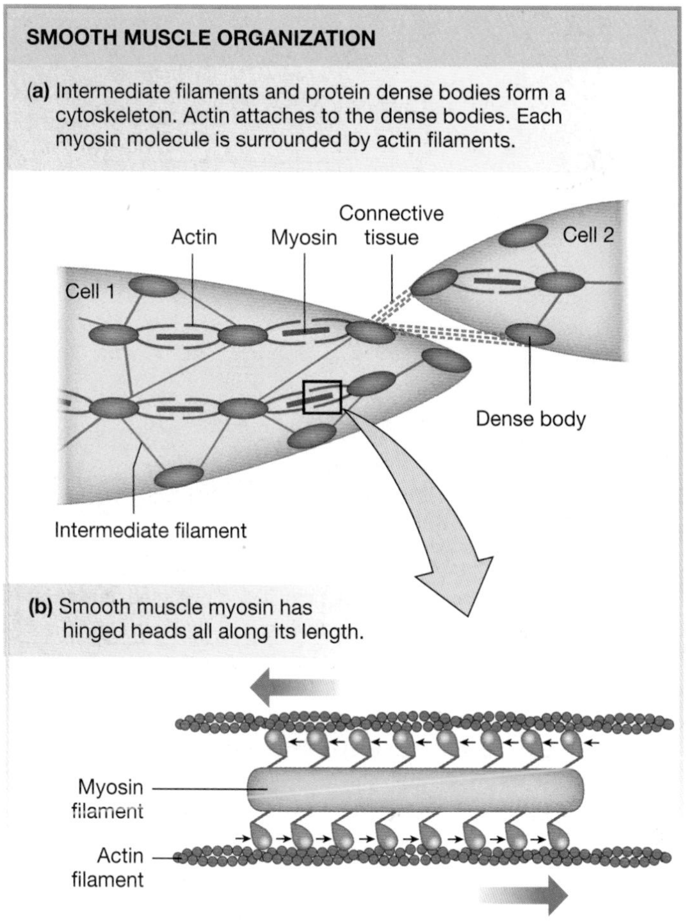

SMOOTH MUSCLE ORGANIZATION

(a) Intermediate filaments and protein dense bodies form a cytoskeleton. Actin attaches to the dense bodies. Each myosin molecule is surrounded by actin filaments.

(b) Smooth muscle myosin has hinged heads all along its length.

■ Fig. 12.25 *Figure courtesy of Marion J. Siegman, Jefferson Medical College*

Concept Check Answers: p. 440

25. The dense bodies that anchor smooth muscle actin are analogous to what structure in a sarcomere? (*Hint:* See Fig. 12.5.)
26. Name two ways smooth muscle myosin differs from skeletal muscle myosin.
27. Name one way actin and its associated proteins differ in skeletal and smooth muscle.

12

Myosin Phosphorylation Controls Contraction

The molecular events of smooth muscle contraction are similar in many ways to those in skeletal muscle, but some important differences exist. Here is a summary of our current understanding of the key points of smooth muscle contraction. In smooth muscle:

1. An increase in cytosolic Ca^{2+} initiates contraction. This Ca^{2+} is released from the sarcoplasmic reticulum but also enters from the extracellular fluid.
2. Ca^{2+} binds to **calmodulin**, a calcium-binding protein found in the cytosol.
3. Ca^{2+} binding to calmodulin is the first step in a cascade that ends in phosphorylation of myosin light chains.
4. Phosphorylation of myosin light chains enhances myosin ATPase activity and results in contraction. Thus, smooth muscle contraction is controlled through myosin-linked regulatory processes rather than through tropomyosin.

We begin our discussion with steps 2–4 because those steps are common to all types of smooth muscle. We then go back and look at the different pathways that create Ca^{2+} signals.

■ Figure 12.26 illustrates the steps of smooth muscle contraction. Contraction begins when cytosolic Ca^{2+} concentrations

■ Fig. 12.26 ESSENTIALS

Smooth Muscle Contraction and Relaxation

Smooth muscle contraction and relaxation are similar to those of skeletal muscle, but differ in several important ways: (1) Ca^{2+} comes from the ECF as well as the sarcoplasmic reticulum, (2) an action potential is not required for Ca^{2+} release, (3) there is no troponin, so Ca^{2+} initiates contraction through a cascade that includes phosphorylation of myosin light chains, and (4) an additional step in smooth muscle relaxation is dephosphorylation of myosin light chains by myosin phosphatase.

Smooth Muscle Contraction

Increased cytosolic calcium is the signal for contraction.

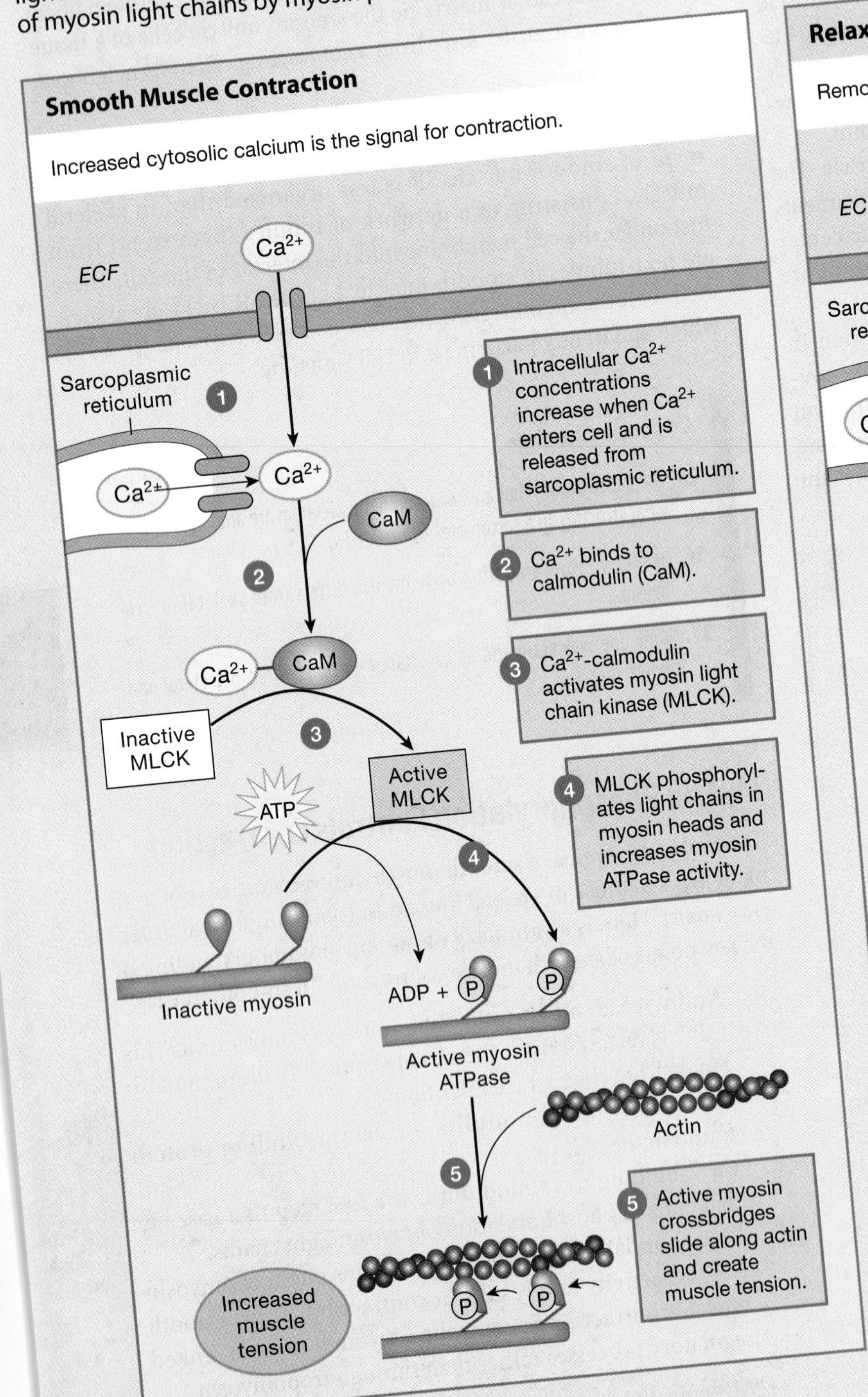

1. Intracellular Ca^{2+} concentrations increase when Ca^{2+} enters cell and is released from sarcoplasmic reticulum.
2. Ca^{2+} binds to calmodulin (CaM).
3. Ca^{2+}-calmodulin activates myosin light chain kinase (MLCK).
4. MLCK phosphorylates light chains in myosin heads and increases myosin ATPase activity.
5. Active myosin crossbridges slide along actin and create muscle tension.

Relaxation in Smooth Muscle

Removal of Ca^{2+} from the cytosol is the first step in relaxation.

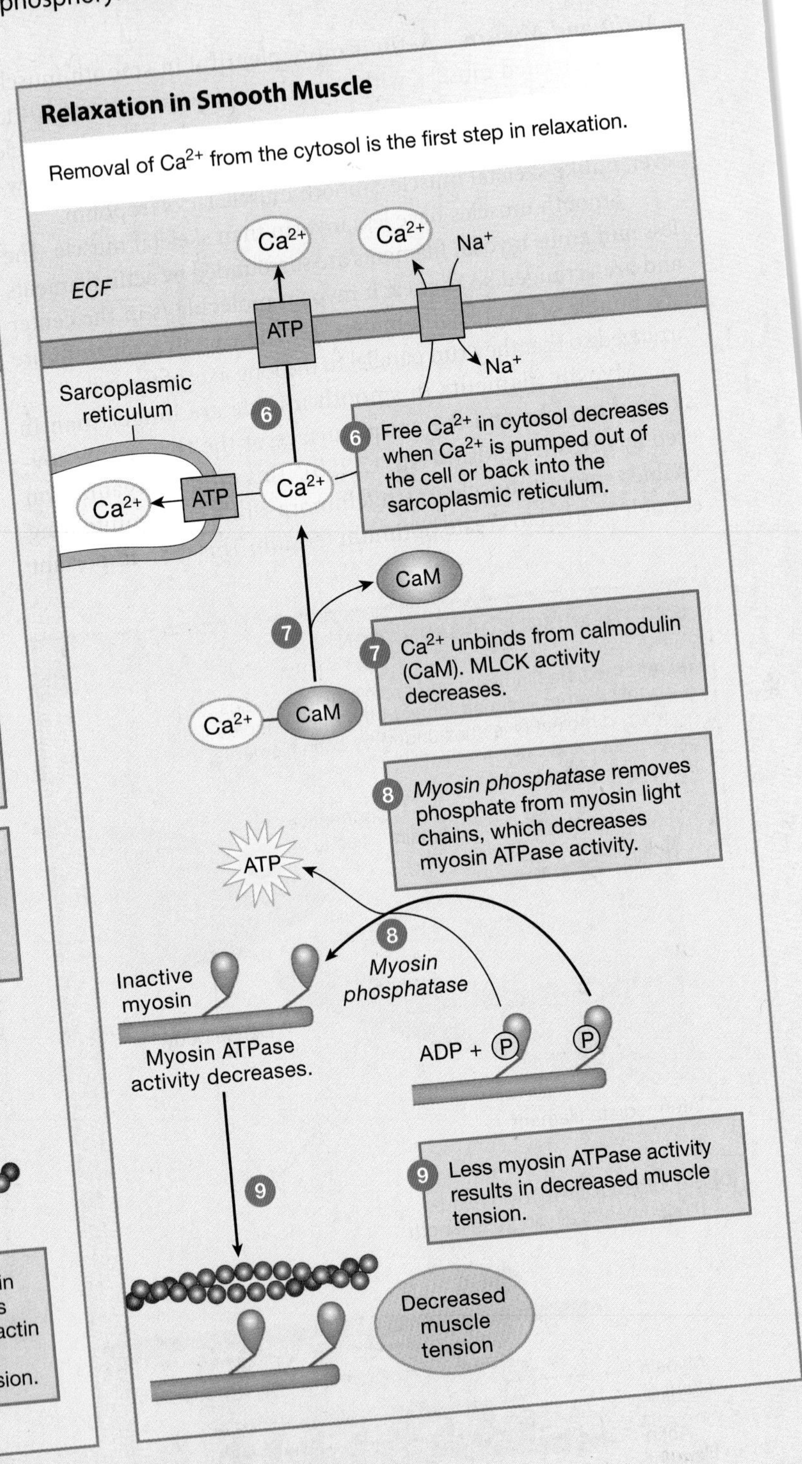

6. Free Ca^{2+} in cytosol decreases when Ca^{2+} is pumped out of the cell or back into the sarcoplasmic reticulum.
7. Ca^{2+} unbinds from calmodulin (CaM). MLCK activity decreases.
8. *Myosin phosphatase* removes phosphate from myosin light chains, which decreases myosin ATPase activity.
9. Less myosin ATPase activity results in decreased muscle tension.

KEY

MLCK = myosin light chain kinase

increase following Ca^{2+} entry from the extracellular fluid and Ca^{2+} release from the sarcoplasmic reticulum (1). The Ca^{2+} ions bind to calmodulin (CaM) (2), obeying the law of mass action [p. 51]. The Ca^{2+}-calmodulin complex then activates an enzyme called **myosin light chain kinase (MLCK)** (3).

At the base of the myosin head is a small regulatory protein chain called a **myosin light chain**. Phosphorylation and dephosphorylation of the myosin light chain control contraction and relaxation in smooth muscle. When Ca^{2+}-calmodulin activates MLCK, the enzyme phosphorylates the myosin light protein chains (4).

Phosphorylation of myosin enhances myosin ATPase activity. When myosin ATPase activity is high, actin binding and crossbridge cycling increase tension in the muscle (5). The myosin ATPase isoform in smooth muscle is much slower that in skeletal muscle, which decreases the rate of crossbridge cycling.

Dephosphorylation of the myosin light chain by the enzyme **myosin light chain phosphatase (MLCP)** decreases myosin ATPase activity. Interestingly, dephosphorylation of myosin does not automatically result in relaxation. Under conditions that we do not fully understand, dephosphorylated myosin may remain attached to actin for a period of time in what is known as a **latch state**. This condition maintains tension in the muscle fiber while consuming little ATP. It is a significant factor in the ability of smooth muscle to sustain contraction without fatiguing.

Relaxation Because dephosphorylation of myosin does not automatically cause relaxation, it is the ratio of MLCK to MLCP activity that determines the contraction state of smooth muscle. MLCP is always active to some degree in smooth muscle, so the activity of MLCK is often the critical factor. As you learned above, MLCK activity depends on Ca^{2+}-calmodulin.

Relaxation in a smooth muscle fiber is a multistep process (Fig. 12.26b). As in skeletal muscle, free Ca^{2+} is removed from the cytosol when Ca^{2+}-ATPase pumps it back into the sarcoplasmic reticulum. In addition, some Ca^{2+} is pumped out of the cell with the help of Ca^{2+}-ATPase and the Na^+-Ca^{2+} exchanger (NCX) [p. 151] (6).

By the law of mass action, a decrease in free cytosolic Ca^{2+} causes Ca^{2+} to unbind from calmodulin (7). In the absence of Ca^{2+}-calmodulin, myosin light chain kinase inactivates. As MLCK becomes less active, myosin ATPase activity decreases. MLCP dephosphorylates myosin, and the muscle relaxes.

MLCP Controls Ca^{2+} Sensitivity

From the discussion above, it would appear that calcium and its regulation of MLCK activity is the primary factor responsible for control of smooth muscle contraction. But chemical signals such as neurotransmitters, hormones, and paracrine molecules alter smooth muscle **Ca^{2+} sensitivity** by modulating myosin light chain phosphatase (MLCP) activity. If MLCK and Ca^{2+}-calmodulin are constant but MLCP activity increases, the MLCK/MLCP ratio shifts so that MLCP dominates. Myosin ATPase dephosphorylates

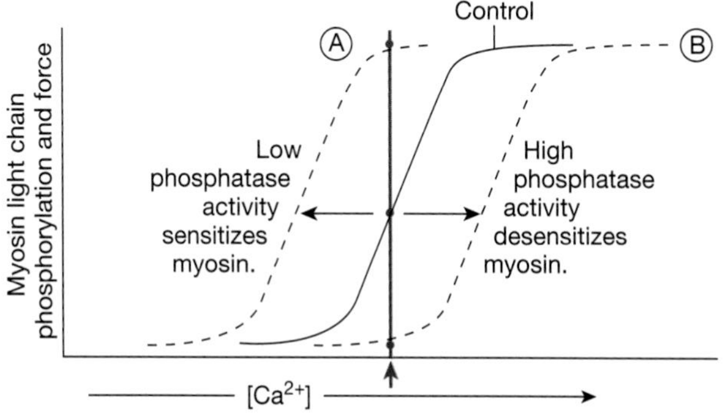

FIGURE QUESTION
At the $[Ca^{2+}]$ indicated by the red arrow, which graph shows increased myosin light chain phosphorylation?

Fig. 12.27

and contraction force decreases, even though the cytosolic Ca^{2+} concentration has not changed (Fig. 12.27). The contraction process is said to be *desensitized* to calcium—the calcium signal is less effective at causing a contraction. Conversely, signal molecules that *decrease* myosin light chain phosphatase activity make the cell *more sensitive* to Ca^{2+}, and contraction force increases even though $[Ca^{2+}]$ has not changed.

Calcium Initiates Smooth Muscle Contraction

We now step back to look in detail at the processes that initiate smooth muscle contraction. Contraction can start with electrical signals—changes in membrane potential—or chemical signals. Contraction caused by electrical signaling is termed *electromechanical coupling*. Contractions initiated by chemical signals without a significant change in membrane potential are called **pharmacomechanical coupling**. Chemical signals may also relax muscle tension without a change in membrane potential. Figure 12.28 is a generalized summary of these pathways.

The Ca^{2+} to initiate contraction comes from two sources: the sarcoplasmic reticulum and the extracellular fluid (Fig. 12.26a). Variable amounts of Ca^{2+} can enter the cytosol from these sources, creating *graded contractions* whose force varies according to the strength of the Ca^{2+} signal.

Sarcoplasmic Ca^{2+} Release The smooth muscle's intracellular Ca^{2+} store is the sarcoplasmic reticulum (SR). SR Ca^{2+} release is mediated both by a ryanodine receptor (RyR) calcium release channel and by an **IP_3-receptor channel**. The RyR channel opens in response to Ca^{2+} entering the cell, a process known as **calcium-induced calcium release (CICR)**. You will learn more about CICR when you study cardiac muscle.

12

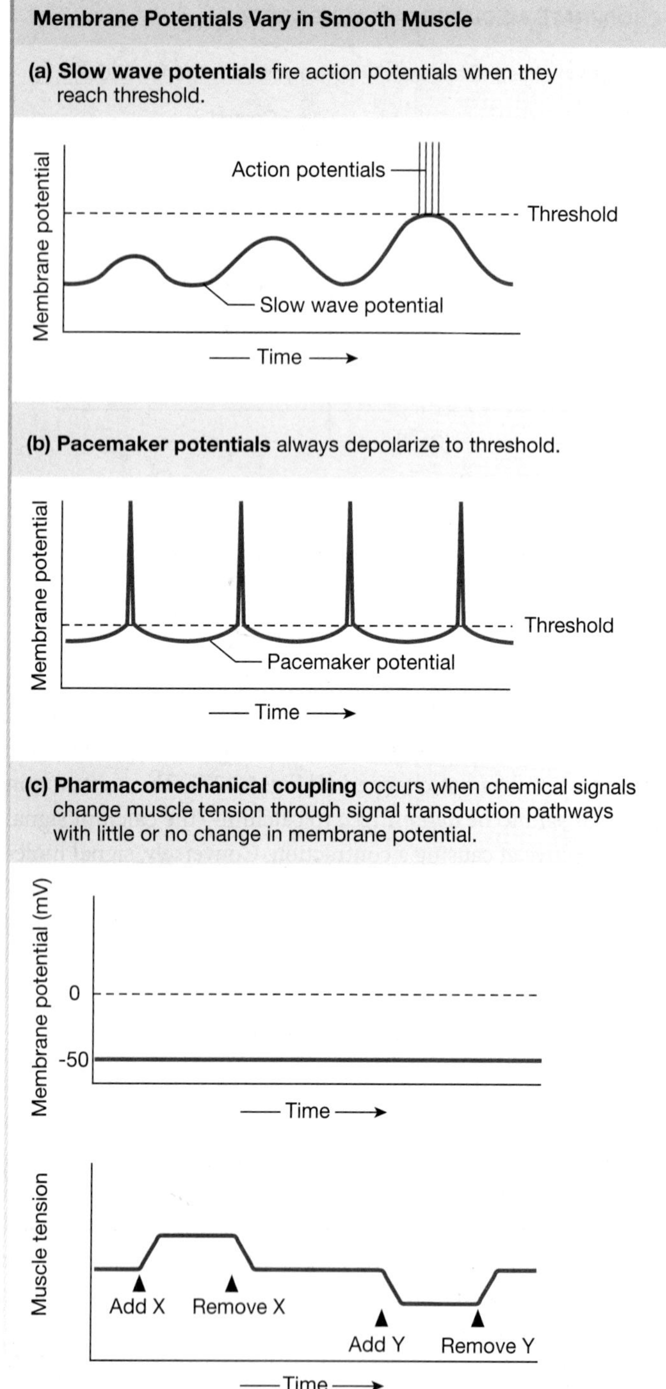
Membrane Potentials Vary in Smooth Muscle
(a) Slow wave potentials fire action potentials when they reach threshold.
Membrane potential
Action potentials
Threshold
Slow wave potential
Time
(b) Pacemaker potentials always depolarize to threshold.
Membrane potential
Threshold
Pacemaker potential
Time
(c) Pharmacomechanical coupling occurs when chemical signals change muscle tension through signal transduction pathways with little or no change in membrane potential.
Membrane potential (mV)
0
-50
Time
Muscle tension
Add X
Remove X
Add Y
Remove Y
Time

Fig. 12.28

The IP_3 channels open when G protein–coupled receptors activate phospholipase C signal transduction pathways [p. 183]. *Inositol trisphosphate* (IP_3) is a second messenger created in that pathway. When IP_3 binds to the SR IP_3-receptor channel, the channel opens and Ca^{2+} flows out of the SR into the cytosol.

Smooth muscle cells have sufficient SR Ca^{2+} stores for contraction. However, because some Ca^{2+} is lost to the ECF through the membrane pumps, the cells must monitor their SR Ca^{2+} stores. The molecular details are still being worked out, but it appears that when SR Ca^{2+} stores become reduced, a protein sensor on the SR membrane communicates with a sarcolemma membrane protein. That protein then opens a set of **store-operated Ca^{2+} channels** to allow more Ca^{2+} into the cell. The Ca^{2+}-ATPase then pumps the Ca^{2+} into the SR to replenish its stores.

Cell Membrane Ca^{2+} Entry Ca^{2+} entry from the extracellular fluid takes place with the help of membrane channels that are voltage-gated, ligand-gated, or mechanically gated [p. 148].

1. Voltage-gated Ca^{2+} channels open in response to a depolarizing stimulus. Action potentials maybe generated in the muscle cell or may enter from neighboring cells via gap junctions. Subthreshold graded potentials may open a few Ca^{2+} channels, allowing small amounts of Ca^{2+} into the cell. This cation entry depolarizes the cell and opens additional voltage-gated Ca^{2+} channels. Sometimes chemical signal molecules open cation channels, and the resulting depolarization opens the Ca^{2+} channels.
2. Ligand-gated Ca^{2+} channels are also known as *receptor-operated calcium channels* or ROCC. These channels open in response to ligand binding and allow enough Ca^{2+} into the cell to induce calcium release from the SR.
3. Stretch-activated channels: Some smooth muscle cells, such as those in blood vessels, contain stretch-activated channels that open when pressure or other force distorts the cell membrane. The exact process is still being worked out, but the cell depolarizes, opening neighboring voltage-gated Ca^{2+} channels. Because contraction in this instance originates from a property of the muscle fiber itself, it is known as a **myogenic contraction**. Myogenic contractions are common in blood vessels that maintain a certain amount of tone at all times.

Although stretch may initiate a contraction, some types of smooth muscle adapt if the muscle cells are stretched for an extended period of time. As the stretch stimulus continues, the Ca^{2+} channels begin to close in a time-dependent fashion. Then, as Ca^{2+} is pumped out of the cell, the muscle relaxes. This adaptation response explains why the bladder develops tension as it fills, then relaxes as it adjusts to the increased volume. (There is a limit to the amount of stretch the muscle can endure, however, and once a critical volume is reached, the urination reflex empties the bladder.)

Concept Check

Answers: p. 440

28. Compare the following aspects of skeletal and smooth muscle contraction:
    (a) signal for crossbridge activation
    (b) source(s) of calcium for the Ca^{2+} signal
    (c) signal that releases Ca^{2+} from the sarcoplasmic reticulum
29. What happens to contraction if a smooth muscle is placed in a saline bath from which all calcium has been removed?
30. Compare Ca^{2+} release channels in skeletal and smooth muscle sarcoplasmic reticulum.

Some Smooth Muscles Have Unstable Membrane Potentials

The role of membrane potentials in smooth muscle contraction is more complex than in skeletal muscle, where contraction always begins in response to an action potential. Smooth muscle exhibits a variety of electrical behaviors: it can hyperpolarize as well as depolarize. Hyperpolarization of the cell decreases the likelihood of contraction. Smooth muscle can also depolarize without firing action potentials. Contraction may take place after an action potential, after a subthreshold graded potential, or without any change in membrane potential.

Many types of smooth muscle display resting membrane potentials that vary between −40 and −80 mV. Cells that exhibit cyclic depolarization and repolarization of their membrane potential are said to have **slow wave potentials** (Fig. 12.28a). Sometimes the cell simply cycles through a series of subthreshold slow waves. However, if the peak of the depolarization reaches threshold, action potentials fire, followed by contraction of the muscle.

Other types of smooth muscle with oscillating membrane potentials have regular depolarizations that always reach threshold and fire an action potential (Fig. 12.28b). These depolarizations are called **pacemaker potentials** because they create regular rhythms of contraction. Pacemaker potentials are found in some cardiac muscles as well as in smooth muscle. Both slow wave and pacemaker potentials are due to ion channels in the cell membrane that spontaneously open and close.

In pharmacomechanical coupling, the membrane potential of the muscle may not change at all. In the next section we consider how this occurs.

Concept Check Answers: p. 440

31. How do pacemaker potentials differ from slow wave potentials?
32. When tetrodotoxin (TTX), a poison that blocks Na^+ channels, is applied to certain types of smooth muscle, it does not alter the spontaneous generation of action potentials. From this observation, what conclusion can you draw about the action potentials of these types of smooth muscle?

Chemical Signals Influence Smooth Muscle Activity

In this section we look at how smooth muscle function is influenced by neurotransmitters, hormones, or paracrines. These chemical signals may be either excitatory or inhibitory, and they modulate contraction by second messenger action at the level of myosin as well as by influencing Ca^{2+} signals (■ Fig. 12.29). One of the interesting properties of smooth muscle is that signal transduction may cause muscle relaxation as well as contraction.

Autonomic Neurotransmitters and Hormones Many smooth muscles are under antagonistic control by both sympathetic and parasympathetic divisions of the autonomic nervous system. Other smooth muscles, such as those found in blood vessels, are controlled by only one of the two autonomic branches. In this type of *tonic control*, the response is graded by increasing or decreasing the amount of neurotransmitter released onto the muscle.

CONTROL OF SMOOTH MUSCLE CONTRACTION

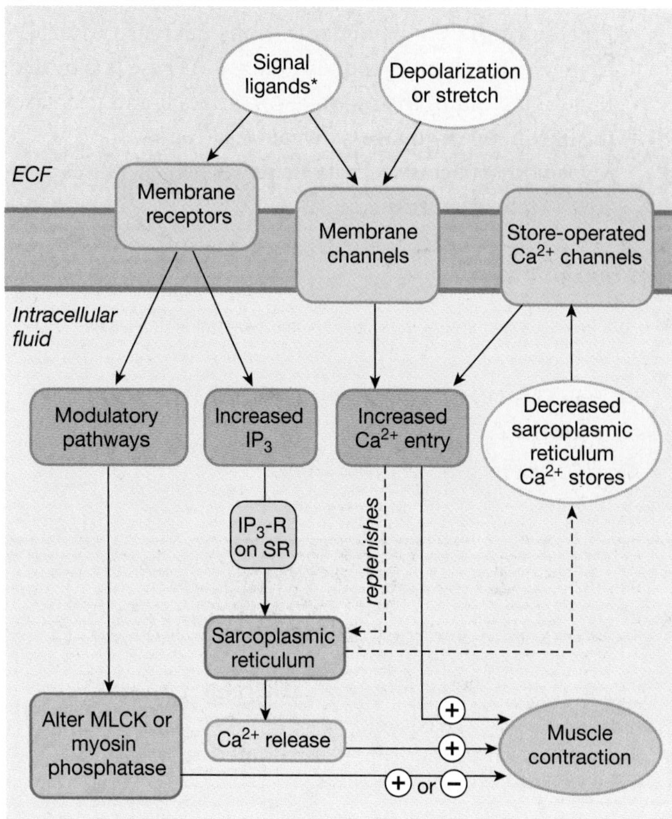

KEY

IP_3-R = IP_3-activated receptor channel

* Ligands include norepinephrine, ACh, other neurotransmitters, hormones, and paracrines.

■ **Fig. 12.29**

A chemical signal can have different effects in different tissues, depending on the receptor type to which it binds [p. 189]. For this reason, it is important to specify the signal molecule and its receptor and subtype when describing the control of a tissue. For example, the sympathetic neurohormone epinephrine causes smooth muscle contraction when it binds to α-adrenergic receptors but relaxation when it binds to β_2-adrenergic receptors.

Most smooth muscle neurotransmitters and hormones bind to G protein–linked receptors. The second messenger pathways then determine the muscle response: IP_3 triggers contraction and cAMP promotes relaxation.

Pathways that increase IP_3 cause contraction several ways:

- IP_3 opens IP_3 channels on the SR to release Ca^{2+}.
- Diacylglycerol (DAG), another product of the phospholipase C signal pathway, indirectly inhibits myosin phosphatase activity. Increasing the MLCK/MLCP ratio promotes crossbridge activity and muscle tension.

12

Signals that increase cAMP production cause muscle relaxation through the following mechanisms:

- Free cytosolic Ca^{2+} concentrations decrease when IP_3 channels are inhibited and the SR Ca^{2+}-ATPase is activated.
- K^+ leaking out of the cell hyperpolarizes it and decreases the likelihood of voltage-activated Ca^{2+} entry.
- Myosin phosphatase activity increases, which causes a decrease in muscle tension.

Concept Check Answers: p. 440

33. How can a neuron alter the amount of neurotransmitter it releases? [*Hint:* See Fig. 8.21, p. 274.]

34. Explain how hyperpolarization decreases the likelihood of contraction in smooth muscle.

35. What causes relaxation in skeletal muscle?

Paracrines Locally released paracrine signals can also alter smooth muscle contraction. For example, asthma is a condition in which smooth muscle of the airways constricts in response to histamine release. This constriction can be reversed by the administration of epinephrine, a neurohormone that relaxes smooth muscle and dilates the airway. Note from this example that not all physiological responses are adaptive or favorable to the body: constriction of the airways triggered during an asthma attack, if left untreated, can be fatal.

Another important paracrine that affects smooth muscle contraction is *nitric oxide* [p. 187]. This gas is synthesized by the endothelial lining of blood vessels and relaxes adjacent smooth muscle that regulates the diameter of the blood vessels. For many years, the identity of this *endothelium-derived relaxing factor*, or EDRF, eluded scientists even though its presence could be demonstrated experimentally. We know

Comparison of the Three Muscle Types — Table 12.3

	Skeletal	Smooth	Cardiac
Appearance under light microscope	Striated	Smooth	Striated
Fiber arrangement	Sarcomeres	No sarcomeres	Sarcomeres
Location	Attached to bones; a few sphincters close off hollow organs	Forms the walls of hollow organs and tubes; some sphincters	Heart muscle
Tissue morphology	Multinucleate; large, cylindrical fibers	Uninucleate; small spindle-shaped fibers	Uninucleate; shorter branching fibers
Internal structure	T-tubule and sarcoplasmic reticulum	No t-tubules; sarcoplasmic reticulum	T-tubule and sarcoplasmic reticulum
Fiber proteins	Actin, myosin; troponin and tropomyosin	Actin, myosin; tropomyosin	Actin, myosin; troponin and tropomyosin
Control	• Ca^{2+} and troponin • Fibers independent of one another	• Ca^{2+} and calmodulin • Some fibers electrically linked via gap junctions; others independent	• Ca^{2+} and troponin • Fibers electrically linked via gap junctions
Contraction speed	Fastest	Slowest	Intermediate
Contraction force of single fiber twitch	Not graded	Graded	Graded
Initiation of contraction	Requires ACh from motor neuron	Stretch, chemical signals. Can be autorhythmic	Autorhythmic
Neural control of contraction	Somatic motor neuron	Autonomic neurons	Autonomic neurons
Hormonal influence on contraction	None	Multiple hormones	Epinephrine

now that EDRF is nitric oxide, an important paracrine in many systems of the body.

Because several different signals might reach a muscle fiber simultaneously, smooth muscle fibers must act as integrating centers. For example, sometimes blood vessels receive contradictory messages from two sources: one message signals for contraction, and the other for relaxation. The smooth muscle fibers must integrate the two signals and execute an appropriate response. The complexity of overlapping signal pathways influencing smooth muscle tone can make the tissue difficult to work with in the laboratory.

Although smooth muscles do not have nearly the mass of skeletal muscles, they play a critical role in body function. You will learn more about smooth muscle physiology as you study the different organ systems.

Cardiac Muscle

Cardiac muscle, the specialized muscle of the heart, has features of both smooth and skeletal muscle (■ Tbl. 12.3). Like skeletal muscle fibers, cardiac muscle fibers are striated and have a sarcomere structure. However, cardiac muscle fibers are shorter than skeletal muscle fibers, may be branched, and have a single nucleus (unlike multinucleate skeletal muscle fibers).

As in single-unit smooth muscle, cardiac muscle fibers are electrically linked to one another. The gap junctions are contained in specialized cell junctions known as *intercalated disks*. Some cardiac muscle, like some smooth muscle, exhibits pacemaker potentials. In addition, cardiac muscle is under sympathetic and parasympathetic control as well as hormonal control. You will learn more about cardiac muscle and how it functions within the heart when you study the cardiovascular system.

RUNNING PROBLEM CONCLUSION

Periodic Paralysis

In this running problem, you were introduced to hyperkalemic periodic paralysis (hyperKPP), a condition caused by a genetic defect in voltage-gated Na^+ channels on muscle cell membranes. The periodic paralyses are a family of related disorders caused by muscle ion channel mutations. To learn more about periodic paralyses, visit the Periodic Paralysis Newsdesk at *www.hkpp.org*. Read the information there to compare the hyperkalemic and hypokalemic forms of the disease. For a more detailed discussion of these two conditions, read GeneReviews on the GeneTests web site (*www.genetests.org*).

Now check your understanding of this running problem by comparing your answers with the information in the following summary table.

Question	Facts	Integration and Analysis
1. When Na^+ channels on the muscle membrane open, which way does Na^+ move?	Na^+ is more concentrated in the ECF than in the ICF, and cells have a negative membrane potential.	The electrochemical gradient causes Na^+ to move into cells.
2. What effect would continued movement of Na^+ have on the membrane potential of muscle fibers?	The resting membrane potential of cells is negative relative to the extracellular fluid.	The influx of positive charge depolarizes the muscle, and it remains depolarized.
3. What ion is responsible for the repolarization phase of the muscle action potential, and in which direction does this ion move across the muscle fiber membrane? How might this be linked to hyperKPP?	In the repolarization phase of the action potential, K^+ leaves the cell.	During repeated contractions, K^+ leaves the muscle fiber, which could contribute to elevated extracellular $[K^+]$ (hyperkalemia).
4. Draw a map to explain why a Na^+ channel that does not inactivate results in a muscle that cannot contract (flaccid paralysis).	During an attack, the Na^+ channels remain open and continuously admit Na^+, and the muscle fiber remains depolarized.	If the muscle fiber is unable to repolarize, it cannot fire additional action potentials. The first action potential causes a twitch, but the muscle then goes into a state of flaccid (uncontracted) paralysis.
5. Explain why oral glucose might help bring Paul out of his paralysis. (*Hint:* What happens to the extracellular K^+ level when Na^+-K^+-ATPase is more active?)	The Na^+-K^+-ATPase moves K^+ into cells and Na^+ out of cells.	Providing glucose to cells triggers insulin release. Insulin increases Na^+-K^+-ATPase activity, which removes Na^+ from the cells and helps them repolarize.

400 413 415 425 428 435

Chapter Summary

Muscles provide an excellent system for studying *structure-function* relationships at all levels, from actin, myosin, and sliding filaments in the cell to muscles pulling on bones and joints. *Mechanical properties* of muscles that influence contraction include elastic components, such as the protein titin and the series elastic elements of the intact muscle. *Compartmentation* is essential to muscle function, as demonstrated by the concentration of Ca^{2+} in the sarcoplasmic reticulum and the key role of Ca^{2+} signals in initiating contraction. The *law of mass action* is at work in the dynamics of Ca^{2+}-calmodulin and Ca^{2+}-troponin binding and unbinding. Muscles also show how *biological energy use* transforms stored energy in ATP's chemical bonds to the movement of motor proteins.

Muscles provide many examples of *communication* and *control* in the body. Communication occurs on a scale as small as electrical signals spreading among smooth muscle cells via gap junctions, or as large as a somatic motor neuron innervating multiple muscle fibers. Skeletal muscles are controlled only by somatic motor neurons, but smooth and cardiac muscle have complex regulation that ranges from neurotransmitters to hormones and paracrines.

1. Muscles generate motion, force, and heat. (p. 399)
2. The three types of muscle are **skeletal muscle, cardiac muscle**, and **smooth muscle**. Skeletal and cardiac muscles are **striated muscles**. (p. 399; Fig. 12.1)
3. Skeletal muscles are controlled by somatic motor neurons. Cardiac and smooth muscle are controlled by autonomic innervation, paracrines, and hormones. Some smooth and cardiac muscles are autorhythmic and contract spontaneously. (p. 399)

Skeletal Muscle

iP Muscular Physiology

4. Skeletal muscles are usually attached to bones by tendons. The **origin** is the end of the muscle attached closest to the trunk or to the more stationary bone. The **insertion** is the more distal or mobile attachment. (p. 400)
5. At a flexible joint, muscle contraction moves the skeleton. **Flexors** bring bones closer together; **extensors** move bones away from each other. Flexor-extensor pairs are examples of **antagonistic muscle groups**. (p. 400; Fig. 12.2)
6. A skeletal muscle is a collection of **muscle fibers**, large cells with many nuclei. (p. 402; Fig. 12.3)
7. **T-tubules** allow action potentials to move rapidly into the interior of the fiber and release calcium from the **sarcoplasmic reticulum**. (p. 404; Fig. 12.4)
8. **Myofibrils** are intracellular bundles of contractile and elastic proteins. **Thick filaments** are made of **myosin. Thin filaments** are made mostly of **actin. Titin** and **nebulin** hold thick and thin filaments in position. (pp. 402, 406; Figs. 12.3, 12.6)
9. Myosin binds to actin, creating **crossbridges** between the thick and thin filaments. (p. 402; Fig. 12.3d)
10. One **sarcomere** is composed of two **Z disks** and the filaments between them. A sarcomere is divided into **I bands** (thin filaments only), an **A band** that runs the length of a thick filament, and a central **H zone** occupied by thick filaments only. The **M line** and Z disks represent attachment sites for myosin and actin, respectively. (p. 405; Fig. 12.5)
11. The force created by a contracting muscle is called **muscle tension**. The **load** is a weight or force that opposes contraction of a muscle. (p. 404)
12. The **sliding filament theory of contraction** states that during contraction, overlapping thick and thin filaments slide past each other in an energy-dependent manner as a result of actin-myosin crossbridge movement. (p. 405; Fig. 12.5d,e)
13. In relaxed muscle, **tropomyosin** partially blocks the myosin-binding site on actin. To initiate contraction, Ca^{2+} binds to **troponin**. This unblocks the myosin-binding sites and allows myosin to complete its power stroke. (p. 407; Fig. 12.8)
14. During relaxation, the sarcoplasmic reticulum uses a **Ca^{2+}-ATPase** to pump Ca^{2+} back into its lumen. (p. 401)
15. Myosin converts energy from ATP into motion. **Myosin ATPase** hydrolyzes ATP to ADP and P_i. (p. 409; Fig. 12.9)
16. When myosin releases P_i, the myosin head moves in the **power stroke**. At the end of the power stroke, myosin releases ADP. The cycle ends in the **rigor state**, with myosin tightly bound to actin. (p. 409; Fig. 12.9)
17. In **excitation-contraction coupling**, a somatic motor neuron releases ACh, which initiates a skeletal muscle action potential that leads to contraction. (p. 411; Fig. 12.10a)
18. Voltage-sensing Ca^{2+} channels called **DHP receptors** in the t-tubules open RyR **Ca^{2+} release channels** in the sarcoplasmic reticulum. (p. 411; Fig. 12.10b)
19. Relaxation occurs when Ca^{2+} is pumped back into the SR by a Ca^{2+}-ATPase. (p. 411; Fig. 12.10c)
20. A single contraction-relaxation cycle is known as a **twitch**. The **latent period** between the end of the muscle action potential and the beginning of muscle tension development represents the time required for Ca^{2+} release and binding to troponin. (p. 412; Fig. 12.11)
21. Muscle fibers store energy for contraction in **phosphocreatine**. Anaerobic metabolism of glucose is a rapid source of ATP but is not

efficient. Aerobic metabolism is very efficient but requires an adequate supply of oxygen to the muscles. (p. 413; Fig. 12.12)

22. **Muscle fatigue** is a reversible condition in which a muscle is no longer able to generate or sustain the expected power output. Fatigue has multiple causes. (p. 414; Fig. 12.13)
23. Skeletal muscle fibers can be classified on the basis of their speed of contraction and resistance to fatigue into **fast-twitch glycolytic fibers, fast-twitch oxidative-glycolytic fibers,** and **slow-twitch (oxidative) fibers.** Oxidative fibers are the most fatigue resistant. (pp. 416, 417; Fig. 12.14; Tbl. 12.2)
24. **Myoglobin** is an oxygen-binding pigment that transfers oxygen to the interior of the muscle fiber. (p. 416)
25. The tension of a skeletal muscle contraction is determined by the length of the sarcomeres before contraction begins. (p. 418; Fig. 12.15)
26. Increasing the stimulus frequency causes summation of twitches with an increase of tension. A state of maximal contraction is known as **tetanus.** (p. 419; Fig. 12.16)
27. A **motor unit** is composed of a group of muscle fibers and the somatic motor neuron that controls them. The number of muscle fibers in a motor unit varies, but all fibers in a single motor unit are of the same fiber type. (p. 420; Fig. 12.17)
28. The force of contraction within a skeletal muscle can be increased by **recruitment** of additional motor units. (p. 420)

Mechanics of Body Movement

29. An **isotonic contraction** creates force as the muscle shortens and moves a load. An **isometric contraction** creates force without moving a load. *Lengthening contractions* create force while the muscle lengthens. (p. 422; Fig. 12.18)
30. Isometric contractions occur because **series elastic elements** allow the fibers to maintain constant length even though the sarcomeres are shortening and creating tension. (p. 423; Fig. 12.19)
31. The body uses its bones and joints as **levers** and **fulcrums.** Most lever-fulcrum systems in the body maximize the distance and speed that a load can be moved but also require that muscles do more work than they would without the lever. (p. 424; Fig. 12.20)
32. Contraction speed is a function of muscle fiber type and load. Contraction is fastest when the load on the muscle is zero. (p. 425; Fig. 12.21)

Smooth Muscle

33. Smooth muscle is slower than skeletal muscle but can sustain contractions for longer without fatiguing. (p. 428; Fig. 12.24)
34. **Phasic muscles** are usually relaxed or cycle through contractions. **Tonic smooth muscle** is usually contracted. (p. 426; Fig. 12.22)
35. **Single-unit smooth muscle** contracts as a single unit when depolarizations pass from cell to cell through gap junctions. In **multi-unit smooth muscle,** individual muscle fibers are stimulated independently. (p. 427; Fig. 12.23)
36. Smooth muscle has less myosin than skeletal muscle. Each myosin is associated with about 12–15 actin molecules. Smooth muscle actin lacks troponin. (p. 429; Fig. 12.25)
37. Smooth muscle sarcoplasmic reticulum has both RyR Ca^{2+} release channels and **IP_3-receptor channels.** Calcium also enters the cell from the extracellular fluid (p. 431)
38. In smooth muscle contraction, Ca^{2+} binds to **calmodulin** and activates **myosin light chain kinase** (MLCK). (p. 430; Fig. 12.26a)
39. MLCK phosphorylates **myosin light protein chains,** which activates myosin ATPase. This allows crossbridge power strokes. (p. 430; Fig. 12.26a)
40. During relaxation, Ca^{2+} is pumped out of the cytosol, and myosin light chains are dephosphorylated by **myosin phosphatase.** (p. 430; Fig. 12.26b)
41. Smooth muscle **calcium sensitivity** can be altered by changing myosin phosphatase activity. (p. 431; Fig. 12.27)
42. In **myogenic contraction,** stretch on the cell depolarizes it and opens membrane Ca^{2+} channels. (p. 432)
43. Unstable membrane potentials in smooth muscle take the form of either **slow wave potentials** or **pacemaker potentials.** (p. 432; Fig. 12.28a, b)
44. In **pharmacomechanical coupling,** smooth muscle contraction initiated by chemical signals can take place without a significant change in membrane potential. (p. 432; Fig. 12.28c)
45. Smooth muscle contraction is influenced by sympathetic and parasympathetic neurons and a variety of hormones and paracrine signals. (p. 433; Fig. 12.29)

Cardiac Muscle

46. Cardiac muscle fibers are striated, have a single nucleus, and are electrically linked through gap junctions. Cardiac muscle shares features with both skeletal and smooth muscle. (p. 434; Tbl. 12.3)

12

Questions

Answers: p. A-1

Level One Reviewing Facts and Terms

1. The three types of muscle tissue found in the human body are ________, ________, and ________. Which type is attached to the bones, enabling it to control body movement?
2. Which two muscle types are striated?
3. Which type of muscle tissue is controlled only by somatic motor neurons?
4. Which of the following statement(s) is (are) true about skeletal muscles?
   (a) They constitute about 60% of a person's total body weight.
   (b) They position and move the skeleton.
   (c) The insertion of the muscle is more distal or mobile than the origin.
   (d) They are often paired into antagonistic muscle groups called flexors and extensors.
5. Arrange the following skeletal muscle components in order, from outermost to innermost: sarcolemma, connective tissue sheath, thick and thin filaments, myofibrils.
6. The modified endoplasmic reticulum of skeletal muscle is called the ________. Its role is to sequester ________ ions.
7. T-tubules allow ________ to move to the interior of the muscle fiber.
8. List six proteins that make up the myofibrils. Which protein creates the power stroke for contraction?
9. List the letters used to label the elements of a sarcomere. Which band has a Z disk in the middle? Which is the darkest band? Why?

Which element forms the boundaries of a sarcomere? Name the line that divides the A band in half. What is the function of this line?

10. Briefly explain the functions of titin and nebulin.
11. During contraction, the _________ band remains a constant length. This band is composed primarily of _________ molecules. Which components approach each other during contraction?
12. Explain the sliding filament theory.
13. Explain the roles of troponin, tropomyosin, and Ca^{2+} in skeletal muscle contraction.
14. Which neurotransmitter is released by somatic motor neurons?
15. What is the motor end plate, and what kinds of receptors are found there? Explain how neurotransmitter binding to these receptors creates an action potential.
16. Match the following characteristics with the appropriate type(s) of muscle.

(a) has the largest diameter	1. fast-twitch glycolytic fibers
(b) uses anaerobic metabolism, thus fatigues quickly	2. fast-twitch oxidative-glycolytic fibers
(c) has the most blood vessels	3. slow-twitch oxidative fibers
(d) has some myoglobin	
(e) is used for quick, fine movements	
(f) is also called red muscle	
(g) uses a combination of oxidative and glycolytic metabolism	
(h) has the most mitochondria	

17. A single contraction-relaxation cycle in a skeletal muscle fiber is known as a(n) _________.
18. List the steps of skeletal muscle contraction that require ATP.
19. The basic unit of contraction in an intact skeletal muscle is the _________. The force of contraction within a skeletal muscle is increased by _________ additional motor units.
20. The two functional types of smooth muscle are _________ and _________.

Level Two Reviewing Concepts

21. Make a map of muscle fiber structure using the following terms. Add terms if you like.

- actin
- Ca^{2+}
- cell
- cell membrane
- contractile protein
- crossbridges
- cytoplasm
- elastic protein
- glycogen
- mitochondria
- muscle fiber
- myosin
- nucleus
- regulatory protein
- sarcolemma
- sarcoplasm
- sarcoplasmic reticulum
- titin
- tropomyosin
- troponin
- t-tubule

22. How does an action potential in a muscle fiber trigger a Ca^{2+} signal inside the fiber?
23. Muscle fibers depend on a continuous supply of ATP. How do the fibers in the different types of muscle generate ATP?
24. Define muscle fatigue. Summarize factors that could play a role in its development. How can muscle fibers adapt to resist fatigue?
25. Explain how you vary the strength and effort made by your muscles in picking up a pencil versus picking up a full gallon container of milk.
26. Compare and contrast the following in skeletal and smooth muscle:
    (a) cellular anatomy
    (b) neural and chemical control of contraction
27. Arrange the following terms to create a map of skeletal muscle excitation, contraction, and relaxation. Terms may be used more than once. Add terms if you like.

- acetylcholine
- ACh receptor
- actin
- action potential
- ADP
- ATP
- axon terminal
- Ca^{2+}
- myosin
- Na^+
- neuromuscular junction
- P_i
- power stroke
- relaxation
- rigor state
- Ca^{2+}-ATPase
- calcium-release channels
- contraction
- crossbridge
- DHP receptor
- end-plate potential
- exocytosis
- motor end plate
- sarcoplasmic reticulum
- somatic motor neuron
- tropomyosin
- troponin
- t-tubules
- voltage-gated Ca^{2+} channels

28. What is the role of the sarcoplasmic reticulum in muscular contraction? How can smooth muscle contract when it has so little sarcoplasmic reticulum?
29. Compare and contrast:
    (a) fast-twitch oxidative-glycolytic, fast-twitch glycolytic, and slow-twitch muscle fibers
    (b) a twitch and tetanus
    (c) action potentials in motor neurons and action potentials in skeletal muscles
    (d) temporal summation in motor neurons and summation in skeletal muscles
    (e) isotonic contraction and isometric contraction
    (f) slow-wave and pacemaker potentials
    (g) the source and role of Ca^{2+} in skeletal and smooth muscle contraction
30. Explain the different factors that influence Ca^{2+} entry and release in smooth muscle fibers.

Level Three Problem Solving

31. One way that scientists study muscles is to put them into a state of rigor by removing ATP. In this condition, actin and myosin are strongly linked but unable to move. On the basis of what you know about muscle contraction, predict what would happen to these muscles in a state of rigor if you (a) added ATP but no free calcium ions; (b) added ATP with a substantial concentration of calcium ions.
32. When curare, a South American Indian arrow poison, is placed on a nerve-muscle preparation, the muscle does not contract when the nerve is stimulated, even though neurotransmitter is still being released from the nerve. Give all possible explanations for the action of curare that you can think of.
33. On the basis of what you have learned about muscle fiber types and metabolism, predict what variations in structure you would find among these athletes:
    (a) a 7-foot, 2-inch-tall, 325-pound basketball player
    (b) a 5-foot, 10-inch-tall, 180-pound steer wrestler
    (c) a 5-foot, 7-inch-tall, 130-pound female figure skater
    (d) a 4-foot, 11-inch-tall, 89-pound female gymnast

Level Four Quantitative Problems

34. Look at the following graph, created from data published in "Effect of ambient temperature on human skeletal muscle metabolism during fatiguing submaximal exercise," *J Appl Physiol* 86(3): 902–908, 1999. What hypotheses might you develop about the cause(s) of muscle fatigue based on these data?

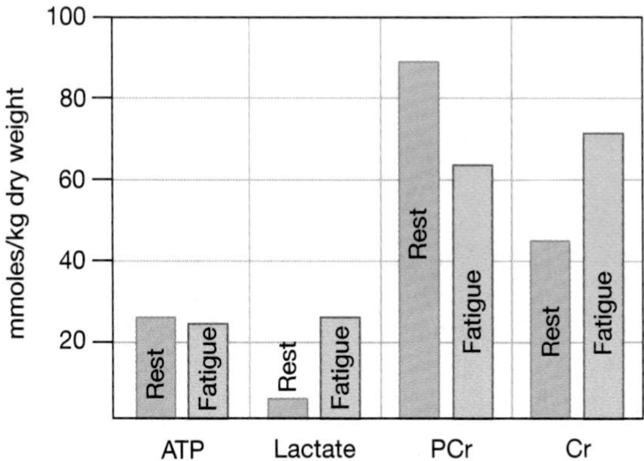

Muscle metabolites in resting muscle and after cycling exercise to fatigue

35. Use the arm in Figure 12.20c to answer the following questions.
    (a) How much force would a biceps muscle inserted 4 cm from the fulcrum need to exert to hold the arm stationary at a 90° angle? How does this force compare with the force needed when the insertion point is 5 cm from the fulcrum?
    (b) Suppose a 7-kg weight band is placed around the wrist 20 cm from the fulcrum. How much force does the biceps inserted 5 cm from the fulcrum need to exert to hold the arm stationary at a 90° angle? How does this force compare with the force needed to keep the arm horizontal in the situation shown in Figure 12.20c, with the same weight in the hand (25 cm from the fulcrum)?

Answers

Answers to Concept Check Questions

Page 400

1. Some examples are biceps/triceps in the upper arm; hamstring (flexor)/quadriceps (extensor) in the upper leg; tibialis anterior (flexor)/gastrocnemius (extensor) for foot movement at the ankle.

Page 404

2. Ends of the A bands are darkest because they are where the thick and thin filaments overlap.
3. T-tubules allow action potentials to travel from the surface of the muscle fiber to its interior.
4. The banding pattern of organized filaments in the sarcomere forms striations in the muscle.

Page 406

5. A neuromuscular junction consists of axon terminals from one somatic motor neuron, the synaptic cleft, and the motor end plate on the muscle fiber.
6. The chemical signal at a neuromuscular junction is acetylcholine.

Page 408

7. Each myosin molecule has binding sites for ATP and actin.
8. F-actin is a polymer filament of actin made from globular G-actin molecules.
9. Enzymes that hydrolyze ATP are ATPases.

Page 410

10. Titin is an elastic fiber in the sarcomere.
11. The crossbridges do not all unlink at one time, so while some myosin heads are free and swiveling, others are still tightly bound.

Page 413

12. The release of myosin heads from actin requires ATP binding. Energy from ATP is required for the power stroke. Relaxation does not directly require ATP, but relaxation cannot occur unless Ca^{2+} is pumped back into the sarcoplasmic reticulum using a Ca^{2+}-ATPase.
13. The events of the latent period include creation of the muscle action potential, release of Ca^{2+} from the sarcoplasmic reticulum, and diffusion of Ca^{2+} to the contractile filaments.

Page 414

14. *Creatine* is the substrate, and *kinase* tells you that this enzyme phosphorylates the substrate.
15. Because creatine kinase catalyzes the reaction in both directions, the relative concentrations of the reactants and products determine the direction of the reaction. The reaction obeys the law of mass action and goes to equilibrium.

Page 415

16. Increasing extracellular K^+ causes the cell to depolarize and become less negative.

Page 418

17. Tension
18. Strength of the graded potential

Page 420

19. A marathoner probably has more slow-twitch muscle fibers, and a sprinter probably has more fast-twitch muscle fibers.

Page 421

20. Increased motor neuron firing rate causes summation in a muscle fiber, which increases the force of contraction.
21. The nervous system increases the force of contraction by recruiting additional motor units.

Page 423

22. If the muscle insertion point is farther from the joint, the leverage is better and a contraction creates more rotational force.

12

Page 427

23. Multi-unit smooth muscle increases force by recruiting additional muscle fibers; single-unit smooth muscle increases force by increasing Ca^{2+} entry.

Page 428

24. Contraction of the circular layer decreases the diameter of a tube. Contraction of the longitudinal layer shortens the tube.

Page 429

25. Dense bodies are analogous to Z disks.
26. Smooth muscle myosin is longer and has heads the entire length of the filament.
27. Smooth muscle actin lacks troponin.

Page 432

28. (a) Skeletal muscle: Ca^{2+} binds to troponin. Smooth muscle: myosin phosphorylated. (b) Skeletal muscles: all Ca^{2+} comes from the sarcoplasmic reticulum. Smooth muscle: Ca^{2+} from both SR and ECF. (c) Skeletal muscle: depolarization signal. Smooth muscle: IP_3 signal.
29. Without ECF Ca^{2+}, contraction decreases because smooth muscle depends on ECF Ca^{2+} for contraction.
30. Skeletal muscle Ca^{2+}-release (RyR) channels are mechanically linked to DHP receptors. Smooth muscle also has Ca^{2+}-release channels that are activated by IP_3.

Page 433

31. Pacemaker potentials always reach threshold and create regular rhythms of contraction. Slow wave potentials are variable in magnitude and may not reach threshold each time.
32. The depolarization phase of the action potentials must not be due to Na^+ entry. In these muscles, depolarization is due to Ca^{2+} entry.

Page 434

33. Increased frequency of action potentials in the neuron increases neurotransmitter release.
34. Many Ca^{2+} channels open with depolarization; therefore, hyperpolarization decreases the likelihood that these channels open. The presence of Ca^{2+} is necessary for contraction.
35. Relaxation in skeletal muscle occurs when troponin releases Ca^{2+} and tropomyosin moves back to block actin's binding site for myosin.

Answers to Figure and Graph Questions

Page 412

Figure 12.11: Both neuronal and muscle action potentials are due to Na^+ entering the fiber during depolarization and K^+ leaving during repolarization. The neuronal channel for Na^+ entry is a voltage-gated Na^+ channel, but the muscle channel for Na^+ entry is the acetylcholine-gated monovalent cation channel.

Page 413

Figure 12.20: (c) Biceps force $\times$ 5 cm = 7 kg $\times$ 25 cm = 35 kg (additional force). (d) The hand moves upward at a speed of 5 cm/sec.

Page 425

Figure 12.21: Contraction is isometric at B because at this point muscle does not shorten. Maximum velocity is at A, where the load on the muscle is zero.

Page 431

Figure 12.27: Graph A. Phosphorylation increases myosin ATPase activity and crossbridge formation.

Integrative Physiology I: Control of Body Movement

Extracting signals directly from the brain to directly control robotic devices has been a science fiction theme that seems destined to become fact.

—Dr. Eberhard E. Fetz, *Science News* 156: 142, 8/28/99

Background Basics

Each dot of a microarray represents one gene. Genes that are active show up in bright colors.

RUNNING PROBLEM

Tetanus

"She hasn't been able to talk to us. We're afraid she may have had a stroke." That is how her neighbors described 77-year-old Cecile Evans when they brought her to the emergency room. But when a neurological examination revealed no problems other than Mrs. Evans's inability to open her mouth and stiffness in her neck, emergency room physician Dr. Doris Ling began to consider other diagnoses. She noticed some scratches healing on Mrs. Evans's arms and legs and asked the neighbors if they knew what had caused them. "Oh, yes. She told us a few days ago that her dog jumped up and knocked her against the barbed wire fence." At that point, Dr. Ling realized she was probably dealing with her first case of tetanus.

Think of a baseball pitcher standing on the mound. As he looks at the first batter, he receives sensory information from multiple sources: the sound of the crowd, the sight of the batter and the catcher, the smell of grass, the feel of the ball in his hand, and the alignment of his body as he begins his windup. Sensory receptors code this information and send it to the central nervous system (CNS), where it is integrated.

The pitcher acts consciously on some of the information: he decides to throw a fastball. But he processes other information at the subconscious level and acts on it without conscious thought. As he thinks about starting his motion, for instance, he shifts his weight to offset the impending movement of his arm. The integration of sensory information into an involuntary response is the hallmark of a *reflex* [p. 15].

Neural Reflexes

All neural reflexes begin with a stimulus that activates a sensory receptor. The sensor sends information in the form of action potentials through sensory afferent neurons to the CNS [p. 239]. The CNS is the integrating center that evaluates all incoming information and selects an appropriate response. It then initiates action potentials in efferent neurons to direct the response of muscles and glands—the targets.

A key feature of many reflex pathways is *negative feedback* [p. 17]. Feedback signals from muscle and joint receptors keep the CNS continuously informed of changing body position. Some reflexes have a *feedforward* component that allows the body to anticipate a stimulus and begin the response [p. 18]. Bracing yourself in anticipation of a collision is an example of a feedforward response.

Neural Reflex Pathways Can Be Classified in Different Ways

Reflex pathways in the nervous system consist of chains or networks of neurons that link sensory receptors to muscles or glands. Neural reflexes can be classified in several ways (■ Tbl. 13.1):

1 *By the efferent division of the nervous system that controls the response.* Reflexes that involve somatic motor neurons and skeletal muscles are known as **somatic reflexes**. Reflexes whose responses are controlled by autonomic neurons are called **autonomic reflexes**.

2 *By the CNS location where the reflex is integrated.* **Spinal reflexes** are integrated in the spinal cord. These reflexes may be modulated by higher input from the brain, but they can occur without that input. Reflexes integrated in the brain are called **cranial reflexes**.

3 *By whether the reflex is innate or learned.* Many reflexes are **innate**; in other words, we are born with them, and they are genetically determined. One example is the knee jerk, or patellar tendon reflex: when the patellar tendon at the lower edge of the kneecap is stretched with a tap from a reflex hammer, the lower leg kicks out. Other reflexes are acquired through experience [p. 316]. The example of Pavlov's dogs salivating upon hearing a bell is the classic example of a **learned reflex**, also referred to as a **conditioned reflex**.

4 *By the number of neurons in the reflex pathway.* The simplest reflex is a **monosynaptic reflex**, named for the single synapse between the two neurons in the pathway: a sensory afferent neuron (often just called *a sensory afferent*) and an efferent somatic motor neuron (■ Fig. 13.1a). These

Table 13.1 Classification of Neural Reflexes

Neural Reflexes Can Be Classified by:

1. **Efferent division that controls the effector**
   a. Somatic motor neurons control skeletal muscles.
   b. Autonomic neurons control smooth and cardiac muscle, glands, and adipose tissue.

2. **Integrating region within the central nervous system**
   a. Spinal reflexes do not require input from the brain.
   b. Cranial reflexes are integrated within the brain.

3. **Time at which the reflex develops**
   a. Innate (inborn) reflexes are genetically determined.
   b. Learned (conditioned) reflexes are acquired through experience.

4. **The number of neurons in the reflex pathway**
   a. Monosynaptic reflexes have only two neurons: one afferent (sensory) and one efferent. Only somatic motor reflexes can be monosynaptic.
   b. Polysynaptic reflexes include one or more interneurons between the afferent and efferent neurons. All autonomic reflexes are polysynaptic because they have three neurons: one afferent and two efferent.

■ Fig. 13.1 *ESSENTIALS*

Neural Reflexes

SKELETAL MUSCLE REFLEXES

(a) A **monosynaptic reflex** has a single synapse between the afferent and efferent neurons.

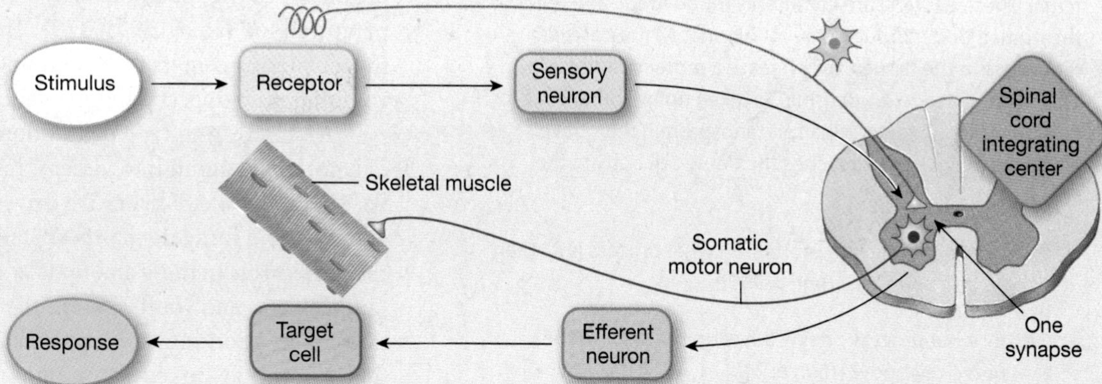

(b) Polysynaptic reflexes have two or more synapses. This somatic motor reflex has both synapses in the CNS.

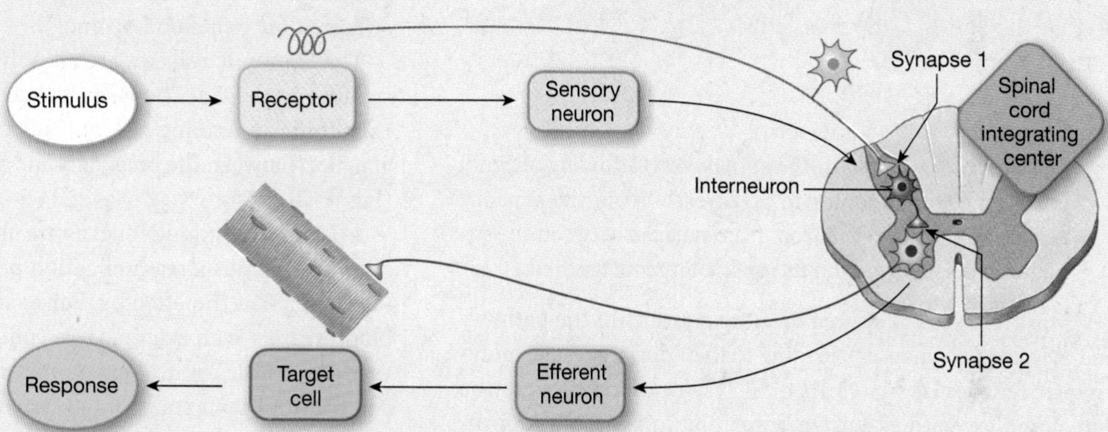

AUTONOMIC REFLEXES

(c) All autonomic reflexes are polysynaptic, with at least one synapse in the CNS and another in the autonomic ganglion.

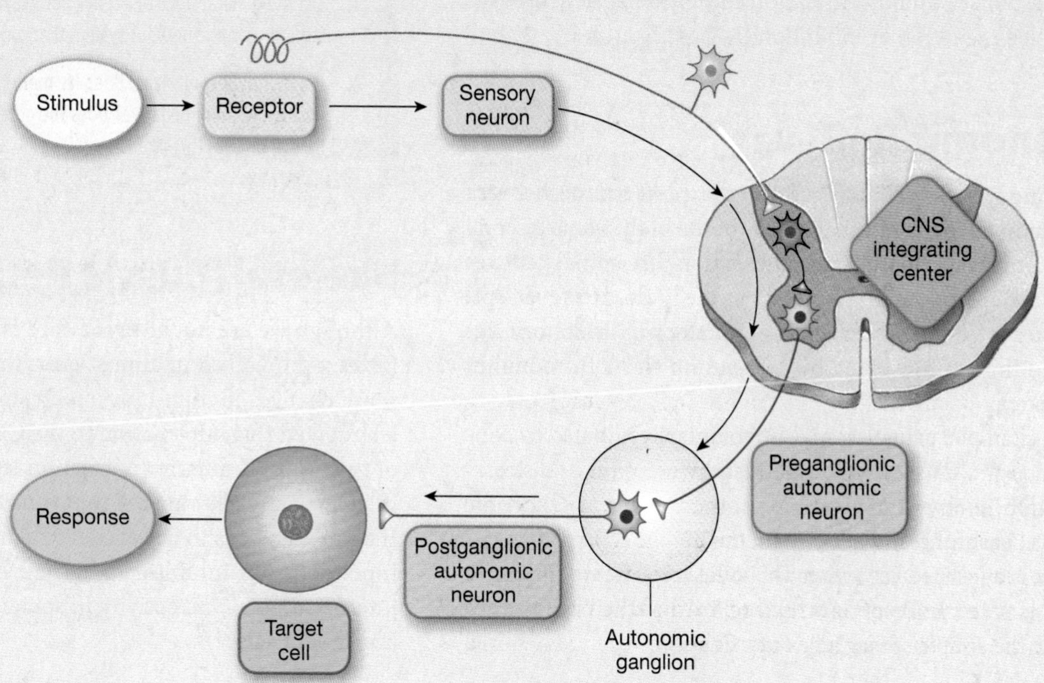

RUNNING PROBLEM

Tetanus {*tetanus,* a muscle spasm}, also known as lockjaw, is a devastating disease caused by the bacterium *Clostridium tetani.* These bacteria are commonly found in soil and enter the human body through a cut or wound. As the bacteria reproduce in the tissues, they release a protein neurotoxin. This toxin, called *tetanospasmin,* is taken up by somatic motor neurons at the axon terminals. Tetanospasmin then travels along the axons until it reaches the nerve cell body in the spinal cord.

Q1: a. Tetanospasmin is a protein. By what process is it taken up into neurons? [*Hint:* p. 155]

b. By what process does it travel up the axon to the nerve cell body? [*Hint:* p. 243]

two neurons synapse in the spinal cord, allowing a signal initiated at the receptor to go directly from the sensory neuron to the motor neuron. (The synapse between the somatic motor neuron and its muscle target is ignored.)

Most reflexes have three or more neurons in the pathway (and at least two synapses), leading to their designation as **polysynaptic reflexes** (■ Fig. 13.1b, c). Polysynaptic reflexes may be quite complex, with extensive branching in the CNS to form networks involving multiple interneurons. *Divergence* of pathways allows a single stimulus to affect multiple targets [p. 274]. *Convergence* integrates the input from multiple sources to modify the response. The modification in polysynaptic pathways may involve excitation or inhibition [p. 274].

Autonomic Reflexes

Autonomic reflexes are also known as *visceral reflexes* because they often involve the internal organs of the body. Some visceral reflexes, such as urination and defecation, are spinal reflexes that can take place without input from the brain. However, spinal reflexes are often modulated by excitatory or inhibitory signals from the brain, carried by descending tracts from higher brain centers.

For example, urination may be voluntarily initiated by conscious thought. Or it may be inhibited by emotion or a stressful situation, such as the presence of other people (a syndrome known as "bashful bladder"). Often, the higher control of a spinal reflex is a learned response. The toilet training we master as toddlers is an example of a learned reflex that the CNS uses to modulate the simple spinal reflex of urination.

Other autonomic reflexes are integrated in the brain, primarily in the hypothalamus, thalamus, and brain stem. These regions contain centers that coordinate body functions needed to maintain homeostasis, such as heart rate, blood pressure, breathing, eating, water balance, and maintenance of body temperature [see Fig. 11.2, p. 379]. The brain stem also contains the integrating centers for autonomic reflexes such as salivating, vomiting, sneezing, coughing, swallowing, and gagging.

An interesting type of autonomic reflex is the conversion of emotional stimuli into visceral responses. The limbic system [p. 305]—the site of primitive drives such as sex, fear, rage, aggression, and hunger—has been called the "visceral brain" because of its role in these emotionally driven reflexes. We speak of "gut feelings" and "butterflies in the stomach"—all transformations of emotion into somatic sensation and visceral function. Other emotion-linked autonomic reflexes include urination, defecation, blushing, blanching, and *piloerection,* in which tiny muscles in the hair follicles pull the shaft of the hair erect ("I was so scared my hair stood on end!").

Autonomic reflexes are all polysynaptic, with at least one synapse in the CNS between the sensory neuron and the preganglionic autonomic neuron, and an additional synapse in the ganglion between the preganglionic and postganglionic neurons (Fig. 13.1c).

Many autonomic reflexes are characterized by *tonic activity,* a continuous stream of action potentials that creates ongoing activity in the effector. For example, the tonic control of blood vessels is an example of a continuously active autonomic reflex [p. 192]. You will encounter many autonomic reflexes as you continue your study of the systems of the body.

Concept Check Answers: p. 461

1. List the general steps of a reflex pathway, including the anatomical structures in the nervous system that correspond to each step.
2. If a cell hyperpolarizes, does its membrane potential become more positive or more negative? Does the potential move closer to threshold or farther from threshold?

Skeletal Muscle Reflexes

Although we are not always aware of them, skeletal muscle reflexes are involved in almost everything we do. Receptors that sense changes in joint movements, muscle tension, and muscle length feed this information to the CNS, which responds in one of two ways. If muscle contraction is the appropriate response, the CNS activates somatic motor neurons to the muscle fibers. If a muscle needs to be relaxed to achieve the response, sensory input activates inhibitory interneurons in the CNS, and these interneurons *inhibit* activity in somatic motor neurons controlling the muscle.

Recall that excitation of somatic motor neurons always causes contraction in skeletal muscle [p. 391]. There is no inhibitory neuron that synapses on skeletal muscles to cause them to relax. Instead, *relaxation results from the absence of excitatory input by the somatic motor neuron.* Inhibition and excitation of somatic motor neurons and their associated skeletal muscles must occur at synapses within the CNS.

Skeletal muscle reflexes have the following components:

1. *Sensory receptors,* known as **proprioceptors**, are located in skeletal muscles, joint capsules, and ligaments. Proprioceptors monitor the position of our limbs in space, our movements, and the effort we exert in lifting objects. The input signal from proprioceptors goes to the CNS through sensory neurons.
2. *The central nervous system* integrates the input signal using networks and pathways of *excitatory and inhibitory interneurons.* In a reflex, sensory information is integrated and acted on subconsciously. However, some sensory information may be integrated in the cerebral cortex and become perception, and some reflexes can be modulated by conscious input.
3. *Somatic motor neurons* carry the output signal. The somatic motor neurons that innervate skeletal muscle contractile fibers are called **alpha motor neurons** (■ Fig. 13.2a).

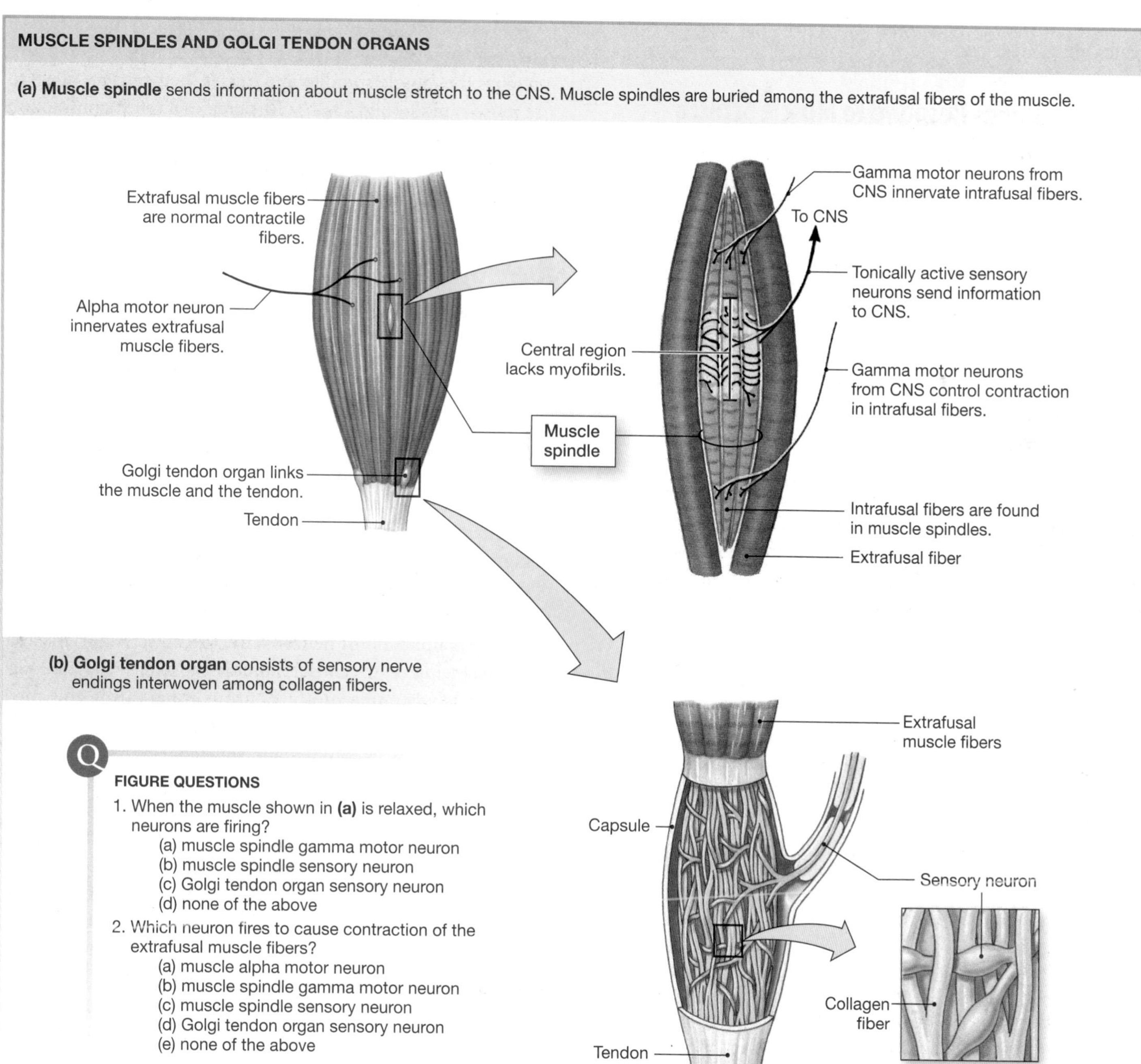
MUSCLE SPINDLES AND GOLGI TENDON ORGANS
(a) Muscle spindle sends information about muscle stretch to the CNS. Muscle spindles are buried among the extrafusal fibers of the muscle.
Extrafusal muscle fibers are normal contractile fibers.
Alpha motor neuron innervates extrafusal muscle fibers.
Golgi tendon organ links the muscle and the tendon.
Tendon
Central region lacks myofibrils.
Muscle spindle
Gamma motor neurons from CNS innervate intrafusal fibers.
To CNS
Tonically active sensory neurons send information to CNS.
Gamma motor neurons from CNS control contraction in intrafusal fibers.
Intrafusal fibers are found in muscle spindles.
Extrafusal fiber
(b) Golgi tendon organ consists of sensory nerve endings interwoven among collagen fibers.
Extrafusal muscle fibers
Capsule
Sensory neuron
Collagen fiber
Tendon
FIGURE QUESTIONS
1. When the muscle shown in (a) is relaxed, which neurons are firing?
(a) muscle spindle gamma motor neuron
(b) muscle spindle sensory neuron
(c) Golgi tendon organ sensory neuron
(d) none of the above
2. Which neuron fires to cause contraction of the extrafusal muscle fibers?
(a) muscle alpha motor neuron
(b) muscle spindle gamma motor neuron
(c) muscle spindle sensory neuron
(d) Golgi tendon organ sensory neuron
(e) none of the above

■ **Fig. 13.2**

4 The effectors are contractile skeletal muscle fibers, also known as **extrafusal muscle fibers**. Action potentials in alpha motor neurons cause extrafusal fibers to contract.

Three types of proprioceptors are found in the body: muscle spindles, Golgi tendon organs, and joint receptors. *Joint receptors* are found in the capsules and ligaments around joints in the body. They are stimulated by mechanical distortion that accompanies changes in the relative positioning of bones linked by flexible joints. Sensory information from joint receptors is integrated primarily in the cerebellum.

In the next two sections we examine the function of muscle spindles and Golgi tendon organs, two interesting and unique receptors. These receptors lie inside skeletal muscles and sense changes in muscle length and tension. Their sensory output activates muscle reflexes.

Muscle Spindles Respond to Muscle Stretch

Muscle spindles are stretch receptors that send information to the spinal cord and brain about muscle length and changes in muscle length. They are small, elongated structures scattered among and arranged parallel to the contractile extrafusal muscle fibers (Fig. 13.2a). With the exception of one muscle in the jaw, every skeletal muscle in the body has many muscle spindles. For example, a small muscle in the index finger of a newborn human has on average about 50 spindles.

Each muscle spindle consists of a connective tissue *capsule* that encloses a group of small muscle fibers known as **intrafusal fibers** {*intra-*, within + *fusus*, spindle}. Intrafusal muscle fibers are modified so that the ends are contractile but the central region lacks myofibrils (Fig. 13.2). The contractile ends of the intrafusal fibers have their own innervation from **gamma motor neurons**. The noncontractile central region of each intrafusal fiber is wrapped by sensory nerve endings that are stimulated by stretch. These sensory neurons project to the spinal cord and synapse directly on alpha motor neurons innervating the muscle in which the spindles lie.

When a muscle is at its resting length, the central region of each muscle spindle is stretched enough to activate the sensory fibers (■ Fig. 13.3a). As a result, sensory neurons from the spindles are tonically active, sending a steady stream of action potentials to the CNS. Because of this tonic activity, even a muscle at rest maintains a certain level of tension, known as **muscle tone**.

Muscle spindles are anchored in parallel to the extrafusal muscle fibers. Any movement that increases muscle length also stretches the muscle spindles and causes their sensory fibers to fire more rapidly (Fig. 13.3b). This creates a reflex contraction of the muscle, which prevents damage from overstretching. The reflex pathway in which muscle stretch initiates a contraction response is known as a **stretch reflex**.

Concept Check

Answer: p. 461

3. Using the standard steps of a reflex pathway (stimulus, receptor, and so forth), draw a reflex map of the stretch reflex.

Muscle stretch activates muscle spindles, but what happens to spindle activity when a resting muscle contracts and shortens? You might predict that the release of tension on the center of the intrafusal fibers in the absence of gamma motor neuron activity would cause the spindle afferents to slow their firing rate, as shown in ■ Figure 13.4b. However, the presence of gamma motor neurons in a normal muscle keeps the muscle spindles active, no matter what the muscle length is.

When alpha motor neurons fire, the muscle shortens and releases tension on the muscle spindle capsule (Fig. 13.4a). Simultaneously, gamma motor neurons innervating the contractile ends of the muscle spindle fire, which causes the ends of intrafusal fibers to contract and shorten. Contraction of the spindle ends lengthens the central region of the spindle and maintains stretch on the sensory nerve endings. As a result, the spindle remains active even when the muscle contracts. Excitation of gamma motor neurons and alpha motor neurons at the same time is a process known as **alpha-gamma coactivation**.

An example of how muscle spindles work during a stretch reflex is shown in ■ Figure 13.5 a–c. You can demonstrate this yourself with an unsuspecting friend. Have your friend stand with eyes closed, one arm extended with the elbow at 90°, and the hand palm up. Place a small book or other flat weight in the outstretched hand and watch the arm muscles contract to compensate for the added weight.

CLINICAL FOCUS

Reflexes and Muscle Tone

Clinicians use reflexes to investigate the condition of the nervous system and the muscles. For a reflex to be normal, there must be normal conduction through all neurons in the pathway, normal synaptic transmission at the neuromuscular junction, and normal muscle contraction. A reflex that is absent, abnormally slow, or greater than normal (hyperactive) suggests the presence of a pathology. Interestingly, not all abnormal reflexes are caused by neuromuscular disorders. For example, slowed relaxation of the ankle flexion reflex suggests hypothyroidism. (The cellular mechanism linking low thyroid to slow reflexes is not known.) Besides testing reflexes, clinicians assess muscle tone. Even when relaxed and at rest, muscles have a certain resistance to stretch that is the result of continuous (tonic) output by alpha motor neurons. The absence of muscle tone or increased muscle resistance to being stretched by an examiner (increased tone) indicates a problem with the pathways that control muscle contraction.

THE STRETCH REFLEX

(a) Spindles are tonically active and firing even when muscle is relaxed.

Sensory neuron endings

Intrafusal fibers of muscle spindle

Sensory neuron

Alpha motor neuron

Spinal cord

1 Extrafusal muscle fibers at resting length

2 Sensory neuron is tonically active.

3 Spinal cord integrates function.

4 Alpha motor neurons to extrafusal fibers receive tonic input from muscle spindles.

5 Extrafusal fibers maintain a certain level of tension even at rest.

(b) Muscle stretch can trigger a stretch reflex. When muscles stretch and lengthen, muscle spindle sensory afferent neurons fire more. The reflex response is muscle contraction to prevent damage from over-stretching.

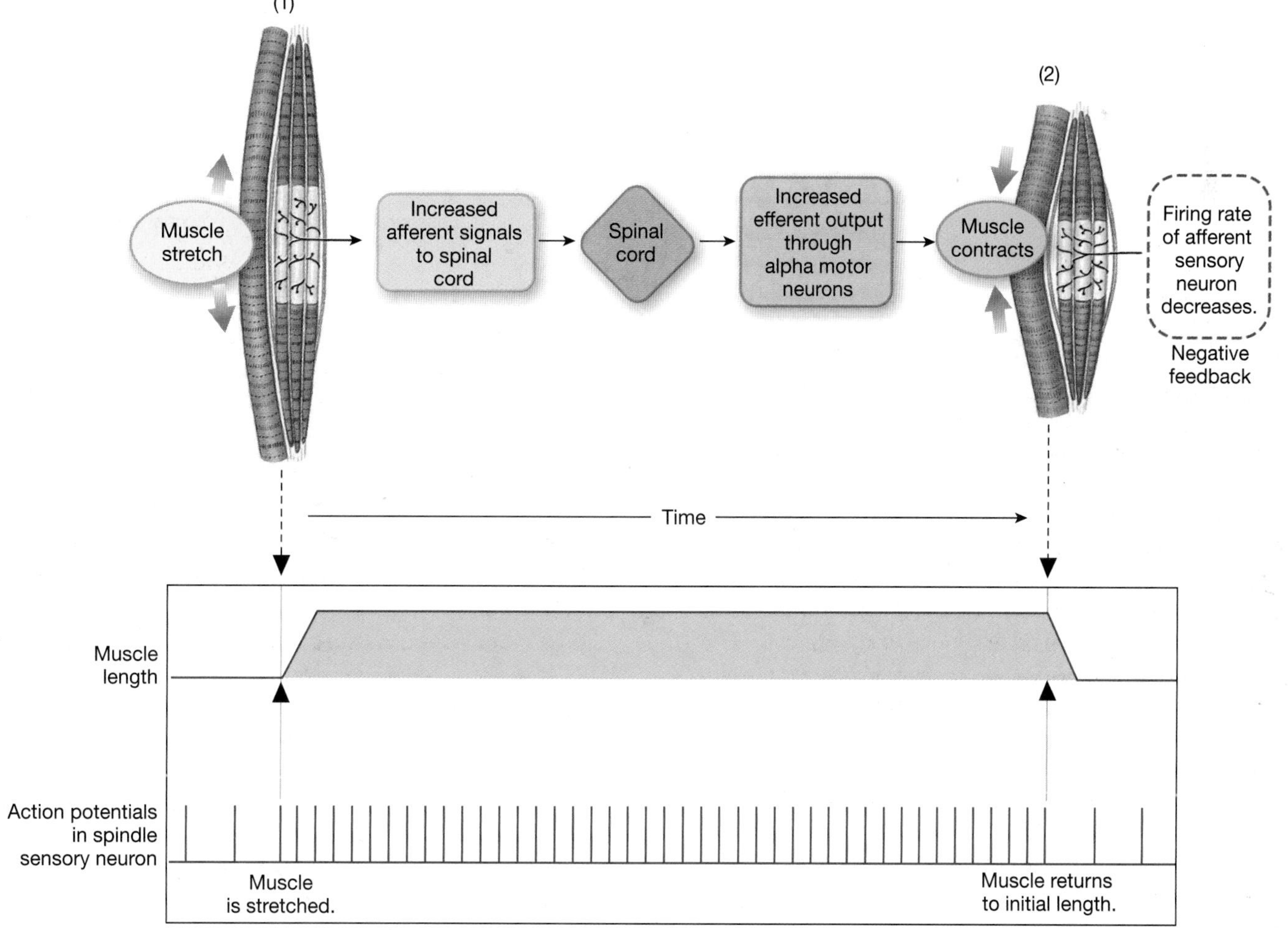

■ **Fig. 13.3**

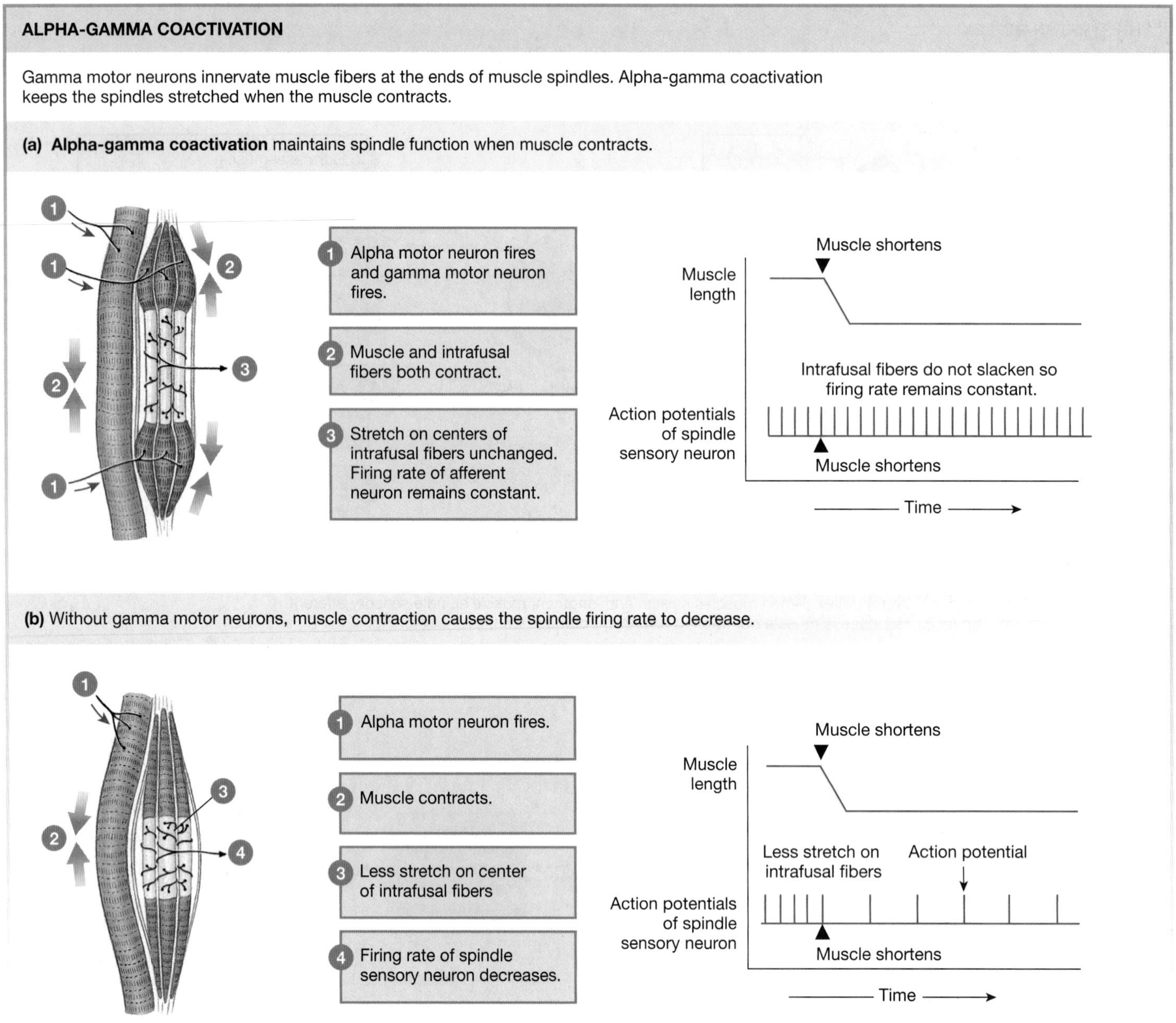

ALPHA-GAMMA COACTIVATION

Gamma motor neurons innervate muscle fibers at the ends of muscle spindles. Alpha-gamma coactivation keeps the spindles stretched when the muscle contracts.

(a) Alpha-gamma coactivation maintains spindle function when muscle contracts.

1 Alpha motor neuron fires and gamma motor neuron fires.
2 Muscle and intrafusal fibers both contract.
3 Stretch on centers of intrafusal fibers unchanged. Firing rate of afferent neuron remains constant.
Muscle length
Muscle shortens
Intrafusal fibers do not slacken so firing rate remains constant.
Action potentials of spindle sensory neuron
Muscle shortens
Time

(b) Without gamma motor neurons, muscle contraction causes the spindle firing rate to decrease.

1 Alpha motor neuron fires.
2 Muscle contracts.
3 Less stretch on center of intrafusal fibers
4 Firing rate of spindle sensory neuron decreases.
Muscle length
Muscle shortens
Less stretch on intrafusal fibers
Action potential
Action potentials of spindle sensory neuron
Muscle shortens
Time

Fig. 13.4

Now suddenly drop a heavier load, such as another book, onto the subject's hand. The added weight will send the hand downward, stretching the biceps muscle and activating its muscle spindles. Sensory input into the spinal cord then activates the alpha motor neurons of the biceps muscle. The biceps will contract, bringing the arm back to its original position.

Golgi Tendon Organs Respond to Muscle Tension

A second type of muscle proprioceptor is the **Golgi tendon organ** (Fig. 13.2b). These receptors are found at the junction of tendons and muscle fibers, placing them in series with the muscle fibers. Golgi tendon organs respond primarily to muscle tension created during an isometric contraction and are relatively insensitive to muscle stretch. Golgi tendon reflexes cause relaxation, the opposite of the reflex contraction caused by muscle spindle reflexes.

Golgi tendon organs are composed of free nerve endings that wind between collagen fibers inside a connective tissue capsule (Fig. 13.2b). When a muscle contracts, its tendons act as an elastic component during the isometric phase of the contraction [p. 421]. Contraction pulls collagen fibers within the Golgi tendon organ tight, pinching sensory endings of the afferent neurons and causing them to fire.

Afferent input from activation of the Golgi tendon organ excites *inhibitory* interneurons in the spinal cord. The interneurons inhibit alpha motor neurons innervating the muscle, and muscle contraction decreases or ceases. Under

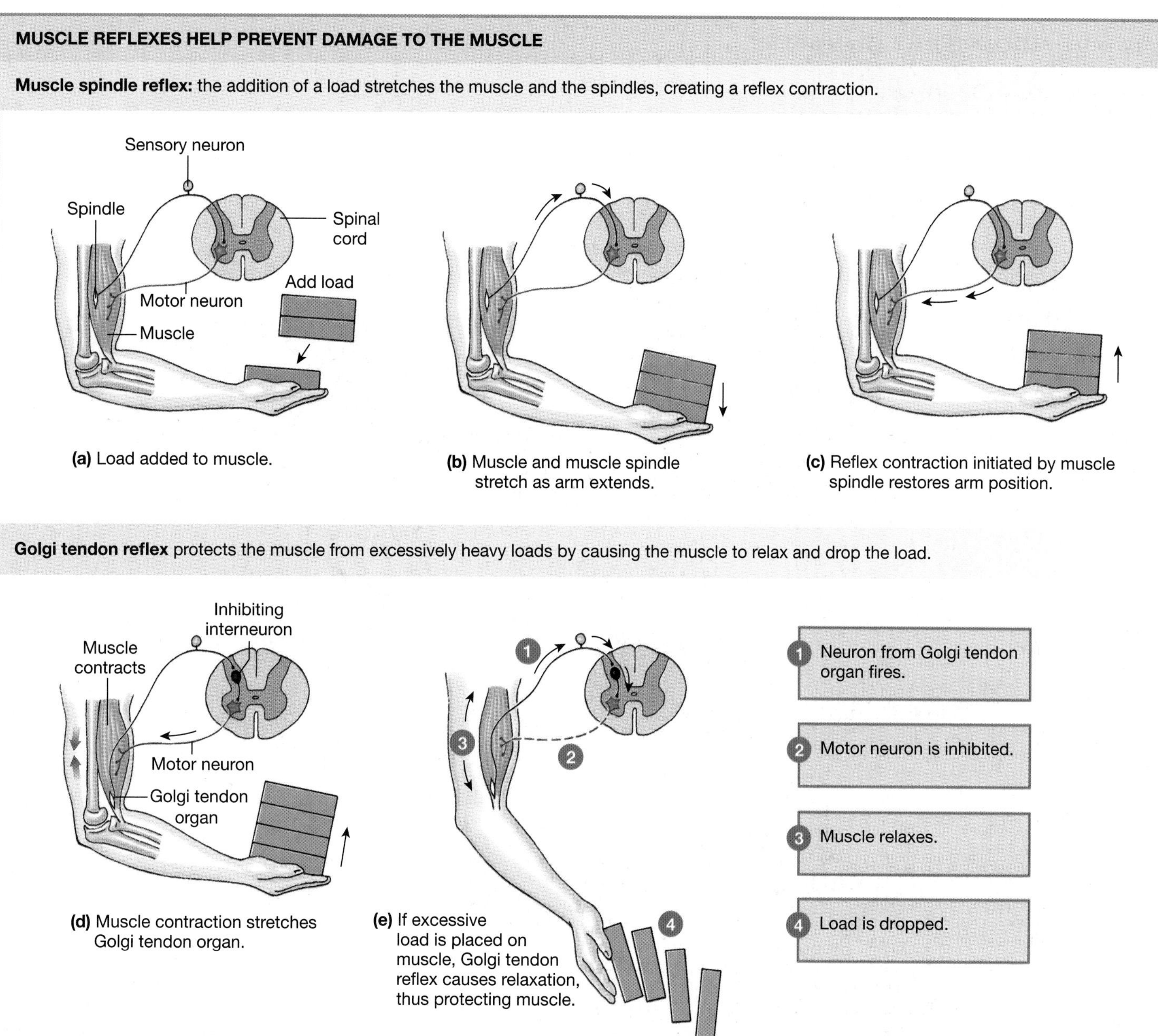
MUSCLE REFLEXES HELP PREVENT DAMAGE TO THE MUSCLE
Muscle spindle reflex: the addition of a load stretches the muscle and the spindles, creating a reflex contraction.
Sensory neuron
Spindle
Spinal cord
Add load
Motor neuron
Muscle
(a) Load added to muscle.
(b) Muscle and muscle spindle stretch as arm extends.
(c) Reflex contraction initiated by muscle spindle restores arm position.
Golgi tendon reflex protects the muscle from excessively heavy loads by causing the muscle to relax and drop the load.
Inhibiting interneuron
Muscle contracts
Motor neuron
Golgi tendon organ
(d) Muscle contraction stretches Golgi tendon organ.
(e) If excessive load is placed on muscle, Golgi tendon reflex causes relaxation, thus protecting muscle.
1 Neuron from Golgi tendon organ fires.
2 Motor neuron is inhibited.
3 Muscle relaxes.
4 Load is dropped.

Fig. 13.5

most circumstances, this reflex slows muscle contraction as the force of contraction increases. In other instances, the Golgi tendon organs prevent excessive contraction that might injure the muscle.

Think back to the example of books placed on the outstretched hand. If supporting the added weight requires more tension than the muscle can develop, the Golgi tendon organ will respond as muscle tension nears its maximum. The Golgi tendon organ triggers reflex *inhibition* of the biceps motor neurons, causing the biceps to relax and the arm to fall. The person then drops the added weight before the muscle fibers can be damaged (Fig. 13.5d, e). Golgi tendon organ input is an important source of inhibition to alpha motor neurons.

Concept Check Answer: p. 461

4. Using the standard steps of a reflex pathway, create a map showing alpha-gamma coactivation and the Golgi tendon reflex. Begin with the stimulus "Alpha motor neuron fires."

Stretch Reflexes and Reciprocal Inhibition Control Movement Around a Joint

Movement around most flexible joints in the body is controlled by groups of synergistic and antagonistic muscles that act in a coordinated fashion. Sensory neurons from muscle receptors

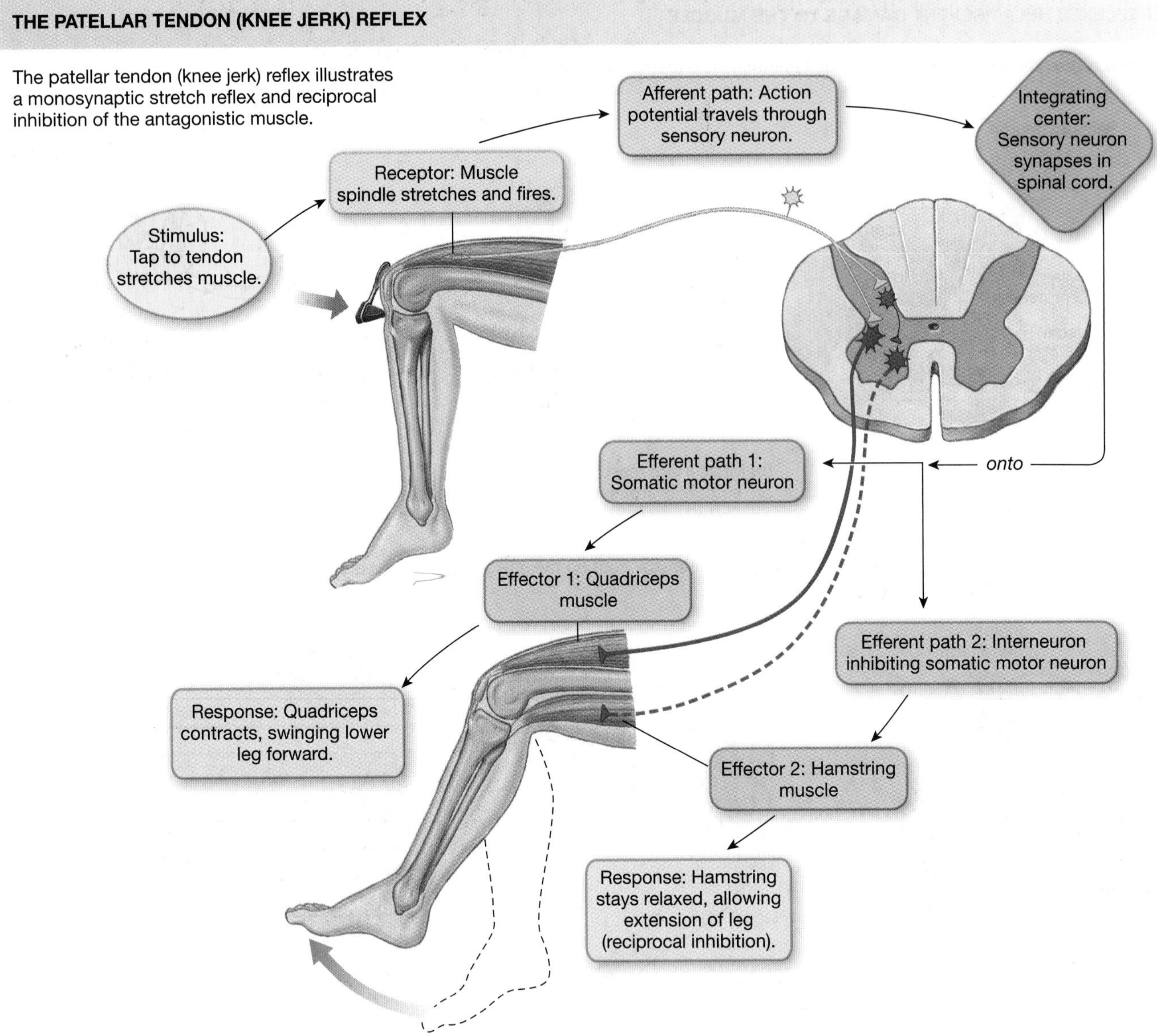
THE PATELLAR TENDON (KNEE JERK) REFLEX
The patellar tendon (knee jerk) reflex illustrates a monosynaptic stretch reflex and reciprocal inhibition of the antagonistic muscle.
Afferent path: Action potential travels through sensory neuron.
Integrating center: Sensory neuron synapses in spinal cord.
Receptor: Muscle spindle stretches and fires.
Stimulus: Tap to tendon stretches muscle.
Efferent path 1: Somatic motor neuron
onto
Effector 1: Quadriceps muscle
Efferent path 2: Interneuron inhibiting somatic motor neuron
Response: Quadriceps contracts, swinging lower leg forward.
Effector 2: Hamstring muscle
Response: Hamstring stays relaxed, allowing extension of leg (reciprocal inhibition).

■ **Fig. 13.6**

and efferent motor neurons that control the muscle are linked by diverging and converging pathways of interneurons within the spinal cord. The collection of pathways controlling a single joint is known as a **myotatic unit** {*myo-*, muscle + *tasis*, stretching}.

The simplest reflex in a myotatic unit is the **monosynaptic stretch reflex,** which involves only two neurons: the sensory neuron from the muscle spindle and the somatic motor neuron to the muscle. The patellar tendon reflex is an example of a monosynaptic stretch reflex (■ Fig. 13.6).

To demonstrate this reflex, a person sits on the edge of a table so that the lower leg hangs relaxed. When the patellar tendon below the kneecap is tapped with a small rubber hammer, the tap stretches the quadriceps muscle, which runs up the front of the thigh. This stretching activates muscle spindles and sends action potentials via the sensory fibers to the spinal cord. The sensory neurons synapse directly onto the motor neurons that control contraction of the quadriceps muscle (a monosynaptic reflex). Excitation of the motor neurons causes motor units in the quadriceps to contract, and the lower leg swings forward.

For muscle contraction to extend the leg, the antagonistic flexor muscles must relax (**reciprocal inhibition**). In the leg, this requires relaxation of the hamstring muscles running up the back of the thigh. The single stimulus of the tap to the tendon accomplishes both contraction of the quadriceps muscle and reciprocal inhibition of the hamstrings. The sensory fibers branch upon entering the spinal cord. Some of the branches activate motor neurons innervating the quadriceps, while the other branches synapse on inhibitory interneurons. The inhibitory interneurons suppress activity in the motor neurons controlling the hamstrings (a polysynaptic reflex). The result is a relaxation of the hamstrings that allows contraction of the quadriceps to proceed unopposed.

RUNNING PROBLEM

Once in the spinal cord, tetanospasmin is released from the motor neuron. It then selectively blocks neurotransmitter release at inhibitory synapses. Patients with tetanus experience muscle spasms that begin in the jaw and may eventually affect the entire body. When the extremities become involved, the arms and legs may go into painful, rigid spasms.

Q2: Using the reflex pathways diagrammed in Figures 13.6 and 13.7, explain why inhibition of inhibitory interneurons might result in uncontrollable muscle spasms.

Flexion Reflexes Pull Limbs away from Painful Stimuli

Flexion reflexes are polysynaptic reflex pathways that cause an arm or leg to be pulled away from a noxious stimulus, such as a pinprick or a hot stove. These reflexes, like the reciprocal inhibition reflex just described, rely on divergent pathways in the spinal cord. ■ Figure 13.7 uses the example of stepping on a tack to illustrate a flexion reflex.

When the foot contacts the point of the tack, nociceptors (pain receptors) in the foot send sensory information to the spinal cord. Here the signal diverges, activating multiple excitatory interneurons. Some of these interneurons excite alpha motor neurons, leading to contraction of the flexor muscles of the stimulated limb. Other interneurons simultaneously activate inhibitory interneurons

THE CROSSED EXTENSOR REFLEX

A flexion reflex in one limb causes extension in the opposite limb. The coordination of reflexes with postural adjustments is essential for maintaining balance.

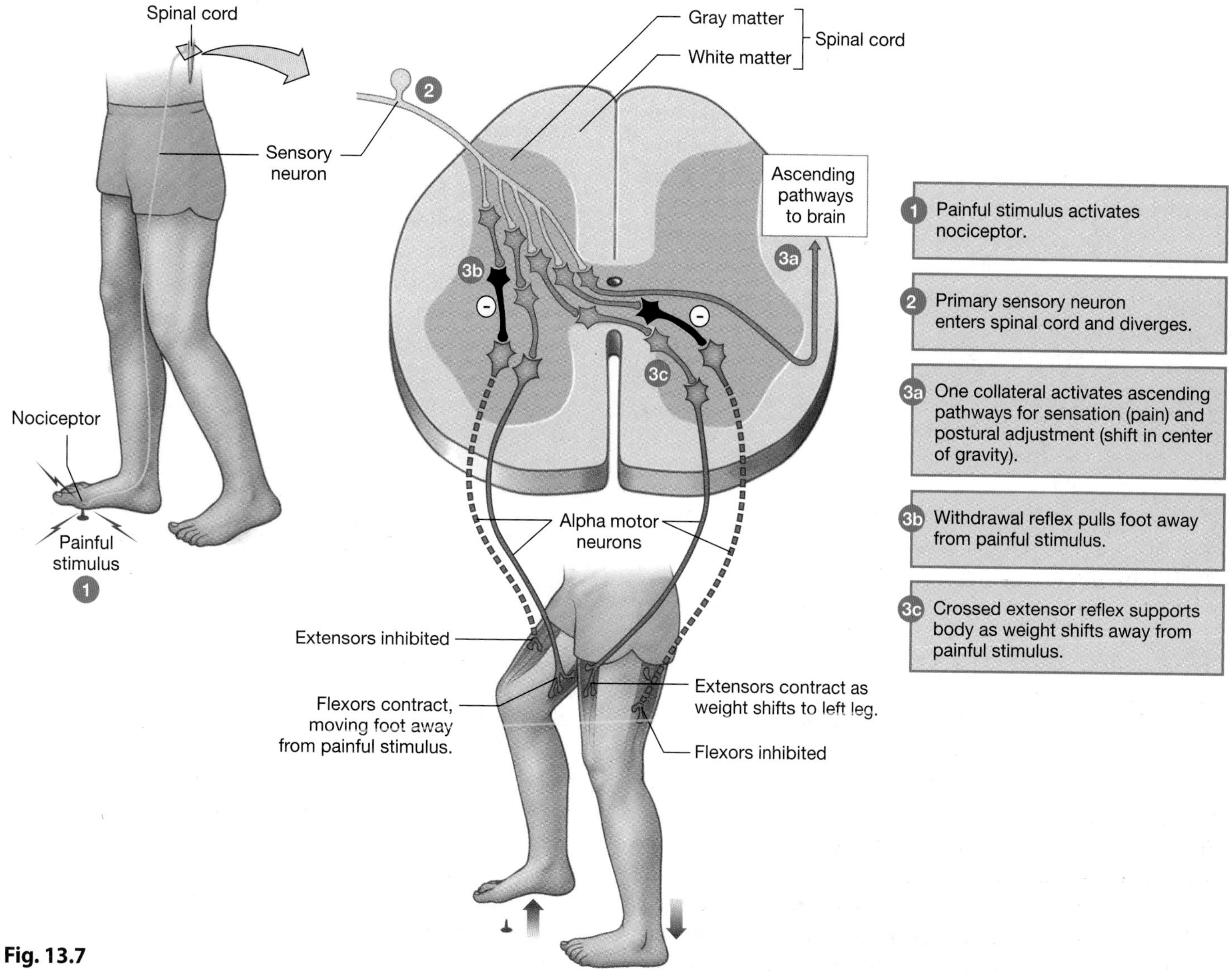

Spinal cord
Sensory neuron
Nociceptor
Painful stimulus
Gray matter
White matter
Spinal cord
Ascending pathways to brain
Alpha motor neurons
Extensors inhibited
Flexors contract, moving foot away from painful stimulus.
Extensors contract as weight shifts to left leg.
Flexors inhibited

1. Painful stimulus activates nociceptor.
2. Primary sensory neuron enters spinal cord and diverges.
3a. One collateral activates ascending pathways for sensation (pain) and postural adjustment (shift in center of gravity).
3b. Withdrawal reflex pulls foot away from painful stimulus.
3c. Crossed extensor reflex supports body as weight shifts away from painful stimulus.

■ **Fig. 13.7**

that cause relaxation of the antagonistic muscle groups. Because of this reciprocal inhibition, the limb is flexed, withdrawing it from the painful stimulus. This type of reflex requires more time than a stretch reflex (such as the knee jerk reflex) because it is a polysynaptic rather than a monosynaptic reflex.

Concept Check Answer: p. 461

5. Draw a reflex map of the flexion reflex initiated by a painful stimulus to the sole of a foot.

Flexion reflexes, particularly in the legs, are usually accompanied by the **crossed extensor reflex**. The crossed extensor reflex is a postural reflex that helps maintain balance when one foot is lifted from the ground. The quick withdrawal of the right foot from a painful stimulus (a tack) is matched by extension of the left leg so that it can support the sudden shift in weight (Fig. 13.7). The extensors contract in the supporting left leg and relax in the withdrawing right leg, while the opposite occurs in the flexor muscles.

Note in Figure 13.7 how the one sensory neuron synapses on multiple interneurons. Divergence of the sensory signal permits a single stimulus to control two sets of antagonistic muscle groups as well as to send sensory information to the brain. This type of complex reflex with multiple neuron interactions is more typical of our reflexes than the simple monosynaptic knee jerk stretch reflex.

In the next section we look at how the CNS controls movements that range from involuntary reflexes to complex, voluntary movement patterns such as dancing, throwing a ball, or playing a musical instrument.

Concept Check Answers: p. 461

6. Add the crossed extensor reflex in the supporting leg to the map you created in Concept Check 5.
7. As you pick up a heavy weight, which of the following are active in your biceps muscle: alpha motor neurons, gamma motor neurons, muscle spindle afferents, Golgi tendon organ afferent neurons?
8. What distinguishes a stretch reflex from a crossed extensor reflex?

The Integrated Control of Body Movement

Most of us never think about how our body translates thoughts into action. Even the simplest movement requires proper timing so that antagonistic and synergistic muscle groups contract in the appropriate sequence and to the appropriate degree. In addition, the body must continuously adjust its position to compensate for differences between the intended movement and

RUNNING PROBLEM

Dr. Ling admits Mrs. Evans to the intensive care unit. There Mrs. Evans is given tetanus antitoxin to deactivate any toxin that has not yet entered motor neurons. She also receives penicillin, an antibiotic that kills the bacteria, and drugs to help relax her muscles. Despite these treatments, by the third day Mrs. Evans is having difficulty breathing because of spasms in her chest muscles. Dr. Ling calls in the chief of anesthesiology to administer metocurine, a drug similar to curare. Curare and metocurine induce temporary paralysis of muscles by binding to ACh receptors on the motor end plate. Patients receiving metocurine must be placed on respirators that breathe for them. For people with tetanus, however, metocurine can temporarily halt the muscle spasms and allow the body to recover.

Q3: a. Why does the binding of metocurine to ACh receptors on the motor end plate induce muscle paralysis? (*Hint*: what is the function of ACh in synaptic transmission?)

b. Is metocurine an agonist or an antagonist of ACh?

the actual one. For example, the baseball pitcher steps off the mound to field a ground ball but in doing so slips on a wet patch of grass. His brain quickly compensates for the unexpected change in position through reflex muscle activity, and he stays on his feet to intercept the ball.

Skeletal muscles cannot communicate with one another directly, and so they send messages to the CNS, allowing the integrating centers to take charge and direct movement. Most body movements are highly integrated, coordinated responses that require input from multiple regions of the brain. Let's examine a few of the CNS integrating centers that are responsible for control of body movement.

Movement Can Be Classified as Reflex, Voluntary, or Rhythmic

Movement can be loosely classified into three categories: reflex movement, voluntary movement, and rhythmic movement (■ Tbl. 13.2). **Reflex movements** are the least complex and are integrated primarily in the spinal cord (for example, see the knee jerk reflex in Fig. 13.6). However, like other spinal reflexes, reflex movements can be modulated by input from higher brain centers. In addition, the sensory input that initiates reflex movements, such as the input from muscle spindles and Golgi tendon organs, goes to the brain and participates in the coordination of voluntary movements and postural reflexes.

Postural reflexes help us maintain body position as we stand or move through space. These reflexes are integrated in

Types of Movement — Table 13.2

	Reflex	Voluntary	Rhythmic
Stimulus that initiates movement	Primarily external via sensory receptors; minimally voluntary	External stimuli or at will	Initiation and termination voluntary
Example	Knee jerk, cough, postural reflexes	Playing piano	Walking, running
Complexity	Least complex; integrated at level of spinal cord or brain stem with higher center modulation	Most complex; integrated in cerebral cortex	Intermediate complexity; integrated in spinal cord with higher center input required
Comments	Inherent, rapid	Learned movements that improve with practice; once learned, may become subconscious ("muscle memory")	Spinal circuits act as pattern generators; activation of these pathways requires input from brain stem

the brain stem. They require continuous sensory input from visual and vestibular (inner ear) sensory systems and from the muscles themselves. Muscle, tendon, and joint receptors provide information about *proprioception,* the positions of various body parts relative to one another. You can tell if your arm is bent even when your eyes are closed because these receptors provide information about body position to the brain.

Information from the vestibular apparatus of the ear and visual cues help us maintain our position in space. For example, we use the horizon to tell us our spatial orientation relative to the ground. In the absence of visual cues, we rely on tactile input. People trying to move in a dark room instinctively reach for a wall or piece of furniture to help orient themselves. Without visual and tactile cues, our orientation skills may fail. The lack of cues is what makes flying airplanes in clouds or fog impossible without instruments. The effect of gravity on the vestibular system is such a weak input when compared with visual or tactile cues that pilots may find themselves flying upside down relative to the ground.

Voluntary movements are the most complex type of movement. They require integration at the cerebral cortex, and they can be initiated at will without external stimuli. Learned voluntary movements improve with practice, and some even become involuntary, like reflexes. Think about learning to ride a bicycle. It may have been difficult at first but once you learned to pedal smoothly and to keep your balance, the movements became automatic. "Muscle memory" is the name dancers and athletes give the ability of the unconscious brain to reproduce voluntary, learned movements and positions.

Rhythmic movements, such as walking or running, are a combination of reflex movements and voluntary movements. Rhythmic movements are initiated and terminated by input from the cerebral cortex, but once activated, networks of CNS interneurons called **central pattern generators** (CPGs) maintain the spontaneous repetitive activity. Changes in rhythmic activity, such as changing from walking to skipping, are also initiated by input from the cerebral cortex.

As an analogy, think of a battery-operated bunny. When the switch is thrown to "on," the bunny begins to hop. It continues its repetitive hopping until someone turns it off (or until the battery runs down). In humans, rhythmic movements controlled by central pattern generators include locomotion and the unconscious rhythm of quiet breathing.

An animal paralyzed by a spinal cord injury is unable to walk because damage to descending pathways blocks the "start walking" signal from the brain to the legs' motor neurons in the spinal cord. However, these paralyzed animals can walk if they are supported on a moving treadmill and given an electrical stimulus to activate the spinal CPG governing that motion. As the treadmill moves the animal's legs, the CPG, reinforced by sensory signals from muscle spindles, drives contraction of the leg muscles.

The ability of central pattern generators to sustain rhythmic movement without continued sensory input has proved important for research on spinal cord injuries. Researchers are trying to take advantage of CPGs and rhythmic reflexes in people with spinal cord injuries by artificially stimulating portions of the spinal cord to restore movement to formerly paralyzed limbs.

The distinctions among reflex, voluntary, and rhythmic movements are not always clear-cut. The precision of voluntary movements improves with practice, but so does that of some reflexes. Voluntary movements, once learned, can become reflexive. In addition, most voluntary movements require continuous input from postural reflexes. **Feedforward reflexes** allow the body to prepare for a voluntary movement, and feedback mechanisms are used to create a smooth, continuous motion. Coordination of movement requires cooperation from many parts of the brain.

Table 13.3

Neural Control of Movement

Location	Role	Receives Input from:	Sends Integrative Output to:
Spinal cord	Spinal reflexes; locomotor pattern generators	Sensory receptors and brain	Brain stem, cerebellum, thalamus/cerebral cortex
Brain stem	Posture, hand and eye movements	Cerebellum, visual and vestibular sensory receptors	Spinal cord
Motor areas of	Planning and coordinating complex movement	Thalamus	Brain stem, spinal cord (corticospinal tract), cerebellum, basal ganglia
Cerebellum	Monitors output signals from motor areas and adjusts movements	Spinal cord (sensory), cerebral cortex (commands)	Brain stem, cerebral cortex (Note: All output is inhibitory.)
Thalamus	Contains relay nuclei that modulate and pass messages to cerebral cortex	Basal ganglia, cerebellum, spinal cord	Cerebral cortex
Basal nuclei	Motor planning	Cerebral cortex	Cerebral cortex, brain stem

The CNS Integrates Movement

Three levels of the nervous system control movement: (1) the spinal cord, which integrates spinal reflexes and contains central pattern generators; (2) the brain stem and cerebellum, which control postural reflexes and hand and eye movements; and (3) the cerebral cortex and basal ganglia [p. 304], which are responsible for voluntary movements. The thalamus relays and modifies signals as they pass from the spinal cord, basal ganglia, and cerebellum to the cerebral cortex (■Tbl. 13.3).

Reflex movements do not require input from the cerebral cortex. Proprioceptors such as muscle spindles, Golgi tendon organs, and joint capsule receptors provide information to the spinal cord, brain stem, and cerebellum (■Fig. 13.8). The brain stem is in charge of postural reflexes and hand and eye movements. It also gets commands from the cerebellum, the part of the brain responsible for "fine-tuning" movement. The result is reflex movement. However, some sensory information is sent through ascending pathways to sensory areas of the cortex, where it can be used to plan voluntary movements.

Voluntary movements require coordination between the cerebral cortex, cerebellum, and basal ganglia. The control of voluntary movement can be divided into three steps: (1) decision-making and planning, (2) initiating the movement, and (3) executing the movement (■Fig. 13.9). The cerebral cortex plays a key role in the first two steps. Behaviors such as movement require knowledge of the body's position in space (where am I?), a decision on what movement should be executed (what shall I do?),

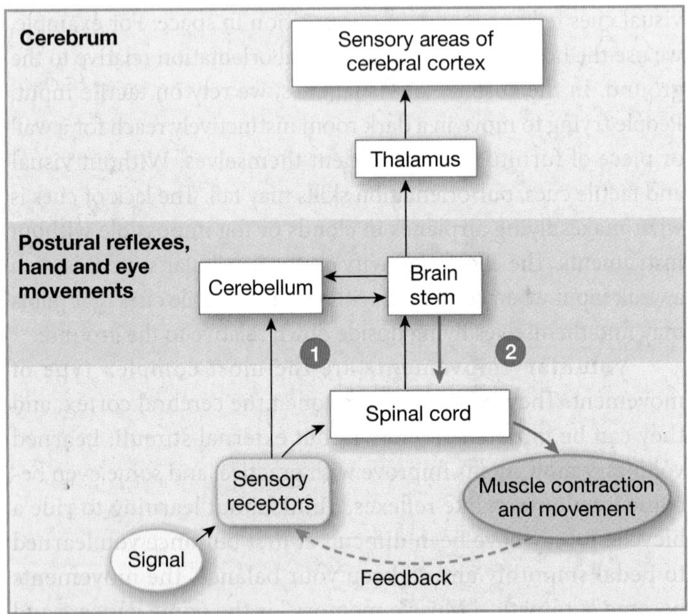
INTEGRATION OF MUSCLE REFLEXES
Cerebrum
Sensory areas of cerebral cortex
Thalamus
Postural reflexes, hand and eye movements
Cerebellum
Brain stem
1
2
Spinal cord
Sensory receptors
Muscle contraction and movement
Signal
Feedback
1 Sensory input (→) from receptors goes to spinal cord, cerebral cortex, and cerebellum. Signals from the vestibular apparatus go directly to the cerebellum.
2 Postural and spinal reflexes do not require integration in the cortex. Output signals (→) initiate movement without higher input.

■ Fig. 13.8

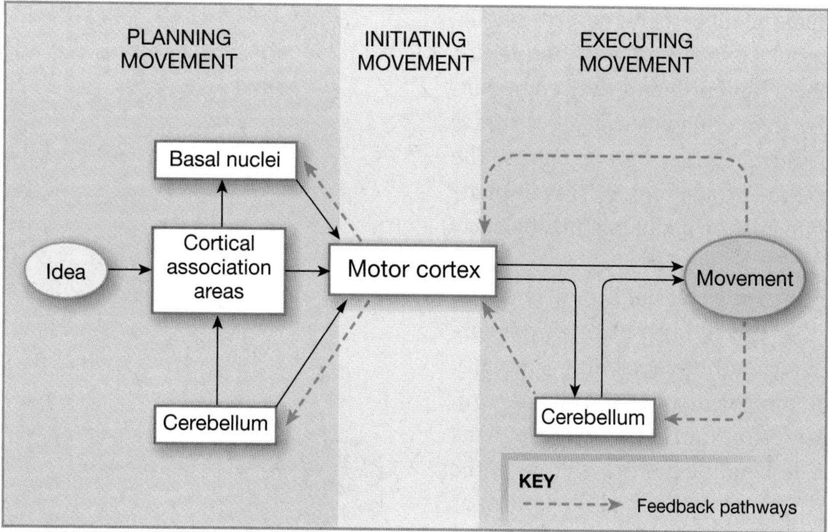

■ **Fig. 13.9**

a plan for executing the movement (how shall I do it?), and the ability to hold the plan in memory long enough to carry it out (now, what was I just doing?). As with reflex movements, sensory feedback is used to continuously refine the process.

Let's return to our baseball pitcher and trace the process as he decides whether to throw a fastball or a slow curve. Standing out on the mound, the pitcher is acutely aware of his surroundings: the other players on the field, the batter in the box, and the dirt beneath his feet. With the help of visual and somatosensory input to the sensory areas of the cortex, he is aware of his body position as he steadies himself for the pitch (■ Fig. 13.10 ①). Deciding which type of pitch to throw and anticipating the consequences

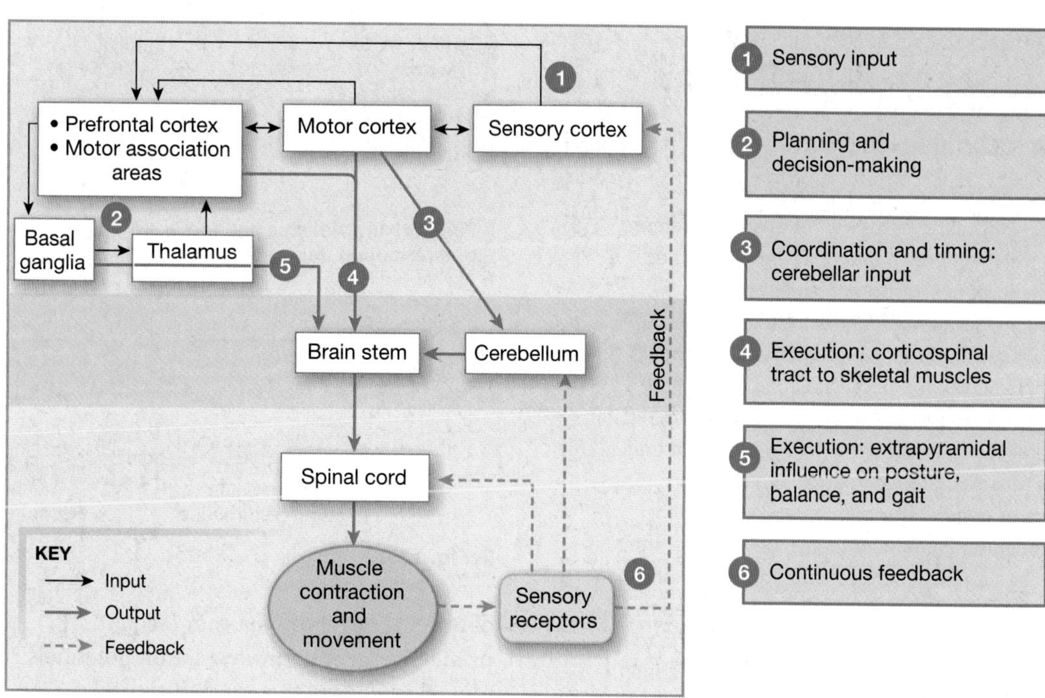

■ **Fig. 13.10**

occupy many pathways in his prefrontal cortex and association areas (2). These pathways loop down through the basal ganglia and thalamus for modulation before cycling back to the cortex.

Once the pitcher makes the decision to throw a fastball, the motor cortex takes charge of organizing the execution of this complex movement. To initiate the movement, descending information travels from the motor association areas and motor cortex to the brain stem, the spinal cord, and the cerebellum (3)–(4). The cerebellum assists in making postural adjustments by integrating feedback from peripheral sensory receptors. The basal ganglia, which assisted the cortical motor areas in planning the pitch, also provide information about posture, balance, and gait to the brain stem (5).

The pitcher's decision to throw a fastball now is translated into action potentials that travel down through the **corticospinal tract**, a group of interneurons controlling voluntary movement that run from the motor cortex to the spinal cord, where they synapse directly onto somatic motor neurons (■ Fig. 13.11). Most of these descending pathways cross to the opposite side of the body in a region of the medulla known as the *pyramids*. Consequently, this pathway is sometimes called the *pyramidal tract*.

Neurons from the basal ganglia [p. 304] also influence body movement. These neurons have multiple synapses in the CNS and make up what is sometimes called the *extrapyramidal tract* or the *extrapyramidal system*. It was once believed that the pyramidal and extrapyramidal pathways were separate systems, but we now know that they interact and are not as distinct in their function as was once believed.

As the pitcher begins the pitch, *feedforward postural reflexes* adjust the body position, shifting weight slightly in anticipation of the changes about to occur (■ Fig. 13.12). Through the appropriate divergent pathways, action potentials race to the somatic motor neurons that control the muscles used for pitching: some are excited, others are inhibited. The neural circuitry allows

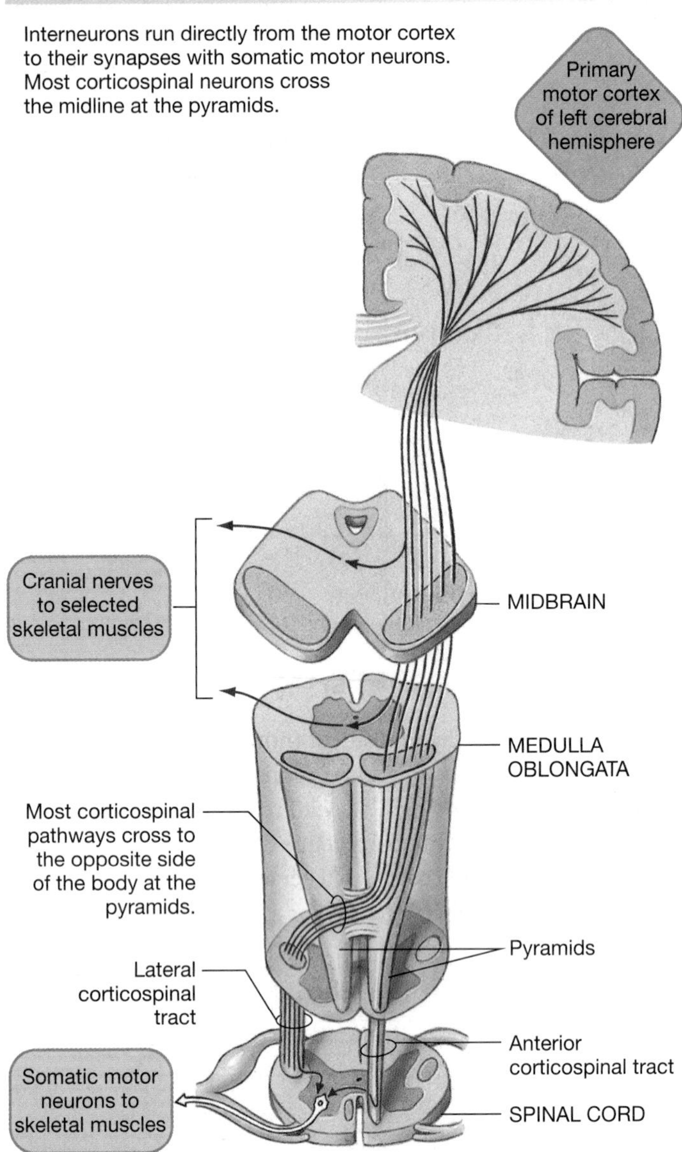

■ **Fig. 13.11**

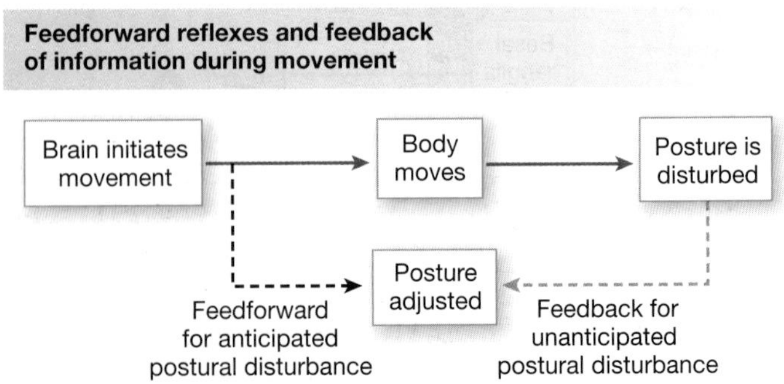

■ **Fig. 13.12**

EMERGING CONCEPTS

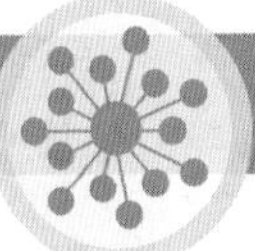

Visualization Techniques in Sports

Researchers now believe that *presynaptic facilitation,* in which modulatory input increases neurotransmitter release, is the physiological mechanism that underlies the success of visualization techniques in sports. Visualization, also known as *guided imagery,* enables athletes to maximize their performance by "psyching" themselves, picturing in their minds the perfect vault or the perfect fastball. By pathways that we still do not understand, the mental image conjured up by the cerebral cortex is translated into signals that find their way to the muscles. Guided imagery is also being used in medicine as *adjunct* (supplementary) therapy for cancer treatment and pain management. The ability of the conscious brain to alter physiological function is only one example of the many fascinating connections between the higher brain and the body. To learn more about this, go to *http://sportsmedicine.about.com* and search for *visualization*.

precise control over antagonistic muscle groups as the pitcher flexes and retracts his right arm. His weight shifts onto his right foot as his right arm moves back.

Each of these movements activates sensory receptors that feed information back to the spinal cord, brain stem, and cerebellum, initiating postural reflexes. These reflexes adjust his body position so that the pitcher does not lose his balance and fall over backward. Finally, he releases the ball, catching his balance on the follow-through—another example of postural reflexes mediated through sensory feedback. His head stays erect, and his eyes track the ball as it reaches the batter. Whack! Home run. As the pitcher's eyes follow the ball and he evaluates the result of his pitch, his brain is preparing for the next batter, hoping to use what it has learned from these pitches to improve those to come.

Symptoms of Parkinson's Disease Reflect Basal Ganglia Function

Our understanding of the role of the basal ganglia in the control of movement has been slow to develop because, for many years, animal experiments yielded little information. Randomly destroying portions of the basal ganglia did not appear to affect research animals. However, research focusing on **Parkinson's disease** (Parkinsonism) in humans has been more fruitful. From studying patients with Parkinson's, scientists have learned that the basal ganglia play a role in cognitive function and memory as well as in the coordination of movement.

Parkinson's disease is a progressive neurological disorder characterized by abnormal movements, speech difficulties, and cognitive changes. These signs and symptoms are associated with loss of neurons in the basal ganglia that release the neurotransmitter dopamine. One abnormal sign that most Parkinson patients have is tremors in the hands, arms, and legs, particularly at rest. In addition, they have difficulty initiating movement and walk slowly with stooped posture and shuffling gait. They lose facial expression, fail to blink (the reptilian stare), and may develop depression, sleep disturbances, and personality changes.

The cause of Parkinson's disease is usually not known and appears to be a combination of environmental factors and genetic susceptibility. However, a few years ago, a number of young drug users were diagnosed with Parkinsonism. Their disease was traced to the use of homemade heroin containing a toxic contaminant that destroyed *dopaminergic* (dopamine-secreting) neurons. This contaminant has been isolated and now enables researchers to induce Parkinson's disease in experimental animals so that we have an animal model on which to test new treatments.

The primary current treatment for Parkinson's is administration of drugs designed to enhance dopamine activity in the brain. Dopamine cannot cross the blood-brain barrier, so patients take *L-dopa*, a precursor of dopamine that crosses the blood-brain barrier, then is metabolized to dopamine. Other drug treatments include dopamine agonists and inhibitors of enzymes that break down dopamine, such as MAO [p. 385]. In severe cases, selected parts of the brain may be destroyed to reduce tremors and rigidity.

Experimental treatments include transplants of dopamine-secreting neurons. Proponents of stem cell research feel that Parkinson's may be one of the conditions that would benefit from the transplant of stem cells into affected brains. For more information on Parkinson's treatments, see *www.parkinson.org*, the National Parkinson Foundation.

RUNNING PROBLEM

Four weeks later, Mrs. Evans is ready to go home, completely recovered and showing no signs of lingering effects. Once she could talk, Mrs. Evans, who was born on the farm where she still lived, was able to tell Dr. Ling that she had never had immunization shots for tetanus or any other diseases. "Well, that made you one of only a handful of people in the United States who will develop tetanus this year," Dr. Ling told her. "You've been given your first two tetanus shots here in the hospital. Be sure to come back in six months for the last one so that this won't happen again." Because of national immunization programs begun in the 1950s, tetanus is now a rare disease in the United States. However, in developing countries without immunization programs, tetanus is still a common and serious condition.

Q4: On the basis of what you know about who receives immunization shots in the United States, predict the age and background of people who are most likely to develop tetanus this year.

Control of Movement in Visceral Muscles

Movement created by contracting smooth and cardiac muscles is very different from that created by skeletal muscles, in large part because smooth and cardiac muscle are not attached to bone. In the internal organs, or viscera, muscle contraction usually changes the shape of an organ, narrowing the lumen of a hollow organ or shortening the length of a tube. In many hollow internal organs, muscle contraction pushes material through the lumen of the organ: the heart pumps blood, the digestive tract moves food, the uterus expels a baby.

Visceral muscle contraction is often reflexively controlled by the autonomic nervous system, but not always. Some types of smooth and cardiac muscle are capable of generating their own action potentials, independent of an external signal. Both the heart and digestive tract have spontaneously depolarizing muscle fibers (often called *pacemakers*) that give rise to regular, rhythmic contractions.

Reflex control of visceral smooth muscle varies from that of skeletal muscle. Skeletal muscles are controlled only by the nervous system, but in many types of visceral muscle, hormones are important in regulating contraction. In addition, some visceral muscle cells are connected to one another by gap junctions that allow electrical signals to pass directly from cell to cell.

Because smooth and cardiac muscle have such a variety of control mechanisms, we will discuss their control as we cover the appropriate organ system for each type of muscle.

RUNNING PROBLEM CONCLUSION

Tetanus

In this running problem, you learned about the tetanus toxin tetanospasmin, a potent poison made by the bacterium *Clostridium tetani*. As little as 175 billionths of a gram (175 nanograms) can be fatal to a 70-kg human. Both tetanus toxin and botulinum toxin cause paralysis, but tetanus is a rigid (contracted muscle) paralysis, while botulism is a *flaccid* (relaxed muscle) paralysis. To learn more about tetanus, visit the web site of the U.S. Centers for Disease Control and Prevention (*www.cdc.gov*). Now check your understanding of this running problem by comparing your answers with the information in the summary table.

Question	Facts	Integration and Analysis
1a By what process is tetanospasmin taken up into neurons?	Tetanospasmin is a protein.	Proteins are too large to cross cell membranes by mediated transport. Therefore, tetanospasmin must be taken up by endocytosis [p. 155].
1b By what process does tetanospasmin travel up the axon to the nerve cell body?	Substances move from the axon terminal to the cell body by retrograde axonal transport [p. 243].	Tetanospasmin is taken up by endocytosis, so it will be contained in endocytotic vesicles. These vesicles "walk" along microtubules through retrograde axonal transport.
2 Using the reflex pathways diagrammed in Figures 13.6 and 13.7, explain why inhibition of inhibitory interneurons might result in uncontrollable muscle spasms.	Muscles often occur in antagonistic pairs. When one muscle is contracting, its antagonist must be inhibited.	If the inhibitory interneurons are not functioning, both sets of antagonistic muscles can contract at the same time. This would lead to muscle spasms and rigidity because the bones attached to the muscles would be unable to move in any direction.
3a Why does the binding of metocurine to ACh receptors on the motor end plate induce muscle paralysis?	ACh is the somatic motor neuron neurotransmitter that initiates skeletal muscle contraction.	If metocurine binds to ACh receptors, it prevents ACh from binding. Without ACh binding, the muscle fiber will not depolarize and cannot contract, resulting in paralysis.
3b Is metocurine an agonist or an antagonist of ACh?	Agonists mimic the effects of a substance; antagonists block the effects of a substance.	Metocurine blocks ACh action; therefore, it is an antagonist.
4 On the basis of what you know about who receives immunization shots in the United States, predict the age and background of people who are most likely to develop tetanus this year.	Immunizations are required for all children of school age. This practice has been in effect since about the 1950s. In addition, most people who suffer puncture wounds or dirty wounds receive tetanus booster shots when they are treated for those wounds.	Most cases of tetanus in the United States will occur in people over the age of 60 who have never been immunized, in immigrants (particularly migrant workers), and in newborn infants. Another source of the disease is contaminated heroin; injection of the drug under the skin may cause tetanus in drug users who do not receive tetanus booster shots.

442 444 451 452 457 458

Chapter Summary

How many times have you heard people say, "I did it without thinking"? In effect, they were saying that their action was a reflex response. There are many ways to control the functions of muscles and glands of the body, but a neural reflex is the simplest and the fastest.

This chapter discusses how the *nervous system* controls body movement. Postural and spinal reflexes follow the basic pattern of a reflex: sensory input is integrated in the CNS, then acted on when an output signal goes to skeletal muscles. Voluntary movements do not require sensory input to be initiated, but they integrate sensory feedback to ensure smooth execution.

Neural Reflexes

1. A neural reflex consists of the following elements: stimulus, receptor, sensory neurons, integrating center, efferent neurons, effectors (muscles and glands), and response. (p. 442)
2. Neural reflexes can be classified in several ways. **Somatic reflexes** involve somatic motor neurons and skeletal muscles. **Autonomic** (or **visceral**) **reflexes** are controlled by autonomic neurons. (p. 442; Tbl. 13.1)
3. **Spinal reflexes** are integrated in the spinal cord. **Cranial reflexes** are integrated in the brain. (p. 442)
4. Many reflexes are innate. Others are acquired through experience. (p. 442)
5. The simplest reflex pathway is a **monosynaptic reflex** with only two neurons. **Polysynaptic reflexes** have three or more neurons in the pathway. (p. 443; Fig. 13.1)

Autonomic Reflexes

6. Some autonomic reflexes are spinal reflexes that are modulated by input from the brain. Other reflexes needed to maintain homeostasis are integrated in the brain, primarily in the hypothalamus, thalamus, and brain stem. (p. 444)
7. Autonomic reflexes are all polysynaptic, and many are characterized by tonic activity. (p. 443; Fig. 13.1c)

Skeletal Muscle Reflexes

8. Skeletal muscle relaxation must be controlled by the CNS because somatic motor neurons always cause contraction in skeletal muscle. (p. 444)
9. The normal contractile fibers of a muscle are called **extrafusal muscle fibers**. Their contraction is controlled by **alpha motor neurons**. (p. 445; Fig. 13.2)
10. **Muscle spindles** send information about muscle length to the CNS. These receptors consist of **intrafusal fibers** with sensory neurons wrapped around the noncontractile center. **Gamma motor neurons** innervate the contractile ends of the intrafusal fibers. (p. 445; Fig. 13.2)
11. Muscle spindles are tonically active stretch receptors. Their output creates tonic contraction of extrafusal muscle fibers. Because of this tonic activity, a muscle at rest maintains a certain level of tension, known as **muscle tone**. (p. 447; Fig. 13.3a)
12. If a muscle stretches, the intrafusal fibers of its spindles stretch and initiate reflex contraction of the muscle. The contraction prevents damage from overstretching. This reflex pathway is known as a **stretch reflex**. (p. 447; Fig. 13.3b)
13. When a muscle contracts, **alpha-gamma coactivation** ensures that its muscle spindle remains active. Activation of gamma motor neurons causes contraction of the ends of the intrafusal fibers. This contraction lengthens the central region of the intrafusal fibers and maintains stretch on the sensory nerve endings. (p. 448; Fig. 13.4)
14. **Golgi tendon organs** are found at the junction of the tendons and muscle fibers. They consist of free nerve endings that wind between collagen fibers. Golgi tendon organs respond to muscle contraction by causing a reflexive relaxation. (p. 445, 449; Figs. 13.2, 13.5)
15. The synergistic and antagonistic muscles that control a single joint are known as a **myotatic unit**. When one set of muscles in a myotatic unit contracts, the antagonistic muscles must relax through a reflex known as **reciprocal inhibition**. (p. 450; Fig. 13.6)
16. **Flexion reflexes** are polysynaptic reflexes that cause an arm or leg to be pulled away from a painful stimulus. Flexion reflexes that occur in the legs are usually accompanied by the **crossed extensor reflex**, a postural reflex that helps maintain balance when one foot is lifted from the ground. (p. 451, Fig. 13.7)
17. **Central pattern generators** are networks of neurons in the CNS that can produce rhythmic motor movements without sensory feedback or higher brain commands. (p. 453)

The Integrated Control of Body Movement

18. Movement can be loosely classified into three categories: reflex movement, voluntary movement, and rhythmic movement. (p. 452; Tbl. 13.2)
19. **Reflex movements** are integrated primarily in the spinal cord. **Postural reflexes** are integrated in the brain stem. (p. 454; Fig. 13.8, Tbl. 13.3)
20. **Voluntary movements** are integrated in the cerebral cortex and can be initiated at will. Learned voluntary movements improve with practice and may even become involuntary, like reflexes. (p. 455; Fig. 13.10)

21. **Rhythmic movements,** such as walking, are a combination of reflexes and voluntary movements. Rhythmic movements can be sustained by central pattern generators. (p. 453)
22. Most signals for voluntary movement travel from cortex to spinal cord through the **corticospinal tract**. Signals from the **basal ganglia** also influence movement through extrapyramidal pathways. (p. 456; Fig. 13.11)
23. **Feedforward reflexes** allow the body to prepare for a voluntary movement; feedback mechanisms are used to create a smooth, continuous motion. (p. 456; Fig. 13.12)

Control of Movement in Visceral Muscles

24. Contraction in smooth and cardiac muscles may occur spontaneously or may be controlled by hormones or by the autonomic division of the nervous system. (p. 457)

Questions

Answers: p. A-1

Level One Reviewing Facts and Terms

1. All neural reflexes begin with a(n) ________ that activates a receptor.
2. Somatic reflexes involve ________ muscles; ________ (or visceral) reflexes are controlled by autonomic neurons.
3. The pathway pattern that brings information from many neurons into a smaller number of neurons is known as ________.
4. When the axon terminal of a modulatory neuron (cell M) terminates close to the axon terminal of a presynaptic cell (cell P) and decreases the amount of neurotransmitter released by cell P, the resulting type of modulation is called ________. [*Hint:* See p. 277.]
5. Autonomic reflexes are also called ________ reflexes. Why?
6. Some autonomic reflexes are spinal reflexes; others are integrated in the brain. List some examples of each.
7. Which part of the brain transforms emotions into somatic sensation and visceral function? List three autonomic reflexes that are linked to emotions.
8. How many synapses occur in the simplest autonomic reflexes? Where do the synapses occur?
9. List the three types of sensory receptors that convey information for muscle reflexes.
10. Because of tonic activity in neurons, a resting muscle maintains a low level of tension known as ________.
11. Stretching a skeletal muscle causes sensory neurons to (increase/decrease) their rate of firing, causing the muscle to contract, thereby relieving the stretch. Why is this a useful reflex?
12. Match the structure to all correct statements about it.

(a) muscle spindle (b) Golgi tendon organ (c) joint capsule mechanoreceptor	1. is strictly a sensory receptor 2. has sensory neurons that send information to the CNS 3. is associated with two types of motor neurons 4. conveys information about the relative positioning of bones 5. is innervated by gamma motor neurons 6. modulates activity in alpha motor neurons

13. The Golgi tendon organ responds to both ________ and ________, although ________ elicits the stronger response. Its activation (increases/decreases) muscle contraction via the ________ neuron.
14. The simplest reflex requires a minimum of how many neurons? How many synapses? Give an example.
15. List and differentiate the three categories of movement. Give an example of each.

Level Two Reviewing Concepts

16. What is the purpose of alpha-gamma coactivation? Explain how it occurs.
17. Modulatory neuron M synapses on the axon terminal of neuron P, just before P synapses with the effector organ. If M is an inhibitory neuron, what happens to neurotransmitter release by P? What effect does M's neurotransmitter have on the postsynaptic membrane potential of P? (*Hint:* Draw this pathway.)
18. At your last physical, your physician checked your patellar tendon reflex by tapping just below your knee while you sat quietly on the edge of the table. (a) What was she checking when she did this test? (b) What would happen if you were worried about falling off the table and were very tense? Where does this additional input to the efferent motor neurons originate? Are these modulatory neurons causing EPSPs or IPSPs [p. 277] at the spinal motor neuron? (c) Your physician notices that you are tense and asks you to count backward from 100 by 3's while she repeats the test. Why would carrying out this counting task enhance your reflex?

Level Three Problem Solving

19. There are several theories about how presynaptic inhibition works at the cellular level. Use what you have learned about membrane potentials and synaptic transmission to explain how each of the following mechanisms would result in presynaptic inhibition:
    (a) Voltage-gated Ca^{2+} channels in axon terminal are inhibited.
    (b) Cl^- channels in axon terminal open.
    (c) K^+ channels in axon terminal open.
20. Andy is working on improving his golf swing. He must watch the ball, swing the club back and then forward, twist his hips, straighten his left arm, then complete the follow-through, where the club arcs in front of him. Which parts of the brain are involved in adjusting how hard he hits the ball, keeping all his body parts moving correctly, watching the ball, and then repeating these actions once he has verified that this swing is successful?
21. It's Halloween, and you are walking through the scariest haunted house around. As you turn a corner and enter the dungeon, a skeleton reaches out and grabs your arm. You let out a scream. Your heart rate quickens, and you feel the hairs on your arm stand on end. (a) What has just happened to you? (b) Where in the brain is fear processed? What are the functions of this part of the brain? Which branch (somatic or autonomic) of the motor output does it control? What are the target organs for this response? (c) How is it possible for your hair to stand on end when hair is made of proteins that do not contract? [*Hint:* See p. 91.] Given that the autonomic nervous system is mediating this reflex response, which type of tissue do you expect to find attached to hair follicles?
22. Using what you have learned about tetanus and botulinum toxins, make a table to compare the two. In what ways are tetanus and botulinum toxin similar? How are they different?

Answers

Answers to Concept Check Questions

Page 444

1. Sensor (sensory receptor), input signal (sensory afferent neuron), integrating center (central nervous system), output signal (autonomic or somatic motor neuron), targets (muscles, glands, some adipose tissue).
2. Upon hyperpolarization, the membrane potential becomes more negative and moves farther from threshold.

Page 446

3. Your map of a stretch reflex should match the components shown in Figure 13.3b.

Page 449

4. Your map of alpha-gamma coactivation should match the steps in Figure 13.4a. The stimulus of muscle contraction is the same for the Golgi tendon reflex, but your map should then branch to show the steps in Figure 13.5d and e.

Page 452

5. Your flexion reflex map should match the steps shown for the knee jerk in Figure 13.6, with the added contraction of hip flexor muscles in addition to the quadriceps.

Page 452

6. The initial steps of the crossed extensor reflex are the same as those of the flexion reflex until the CNS. There the crossed extensor reflex follows the diagram shown in Figure 13.7, step 3c.
7. When you pick up a weight, alpha and gamma neurons, spindle afferents, and Golgi tendon organ afferents are all active.
8. A stretch reflex is initiated by stretch and causes a reflex contraction. A crossed extensor reflex is a postural reflex initiated by withdrawal from a painful stimulus; the extensor muscles contract, but the corresponding flexors are inhibited.

Answers to Figure Questions

Page 445

Figure 13.2: 1. b. 2. (a) Firing of the alpha motor neuron causes extrafusal fibers to contract. However, the tonic activity of (c) spindle afferents will activate the alpha motor neuron.

13

14 Cardiovascular Physiology

Only in the 17th century did the brain displace the heart as the controller of our actions.

—Mary A. B. Brazier, *A History of Neurophysiology in the 19th Century*, 1988

Background Basics

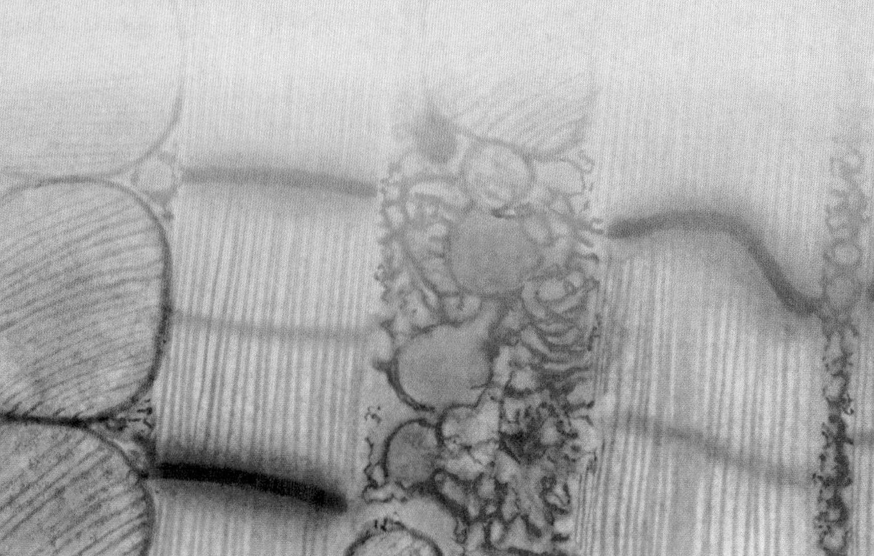

Colored electron micrograph of cardiac muscle. Oval mitochondria lie between pink striated muscle fibers.

In the classic movie *Indiana Jones and the Temple of Doom,* the evil priest reaches into the chest of a sacrificial victim and pulls out his heart, still beating. This act was not dreamed up by some Hollywood scriptwriter—it was taken from rituals of the ancient Mayans, who documented this grisly practice in their carvings and paintings. The heart has been an object of fascination for centuries, but how can this workhorse muscle, which pumps 7200 liters of blood a day, keep beating outside the body? Before we can answer that question, we must first consider the role of hearts in circulatory systems.

As life evolved, simple one-celled organisms began to band together, first into cooperative colonies and then into multicelled organisms. In most multicellular animals, only the surface layer of cells is in direct contact with the environment. This body plan presents a problem because diffusion slows as distance increases [p. 141]. For example, oxygen consumption in the interior cells of larger animals exceeds the rate at which oxygen can diffuse from the body surface.

One solution to overcome slow diffusion was the evolutionary development of circulatory systems that move fluid between the body's surface and its deepest parts. In simple animals, muscular activity creates fluid flow when the animal moves. More complex animals have muscular pumps called hearts to circulate internal fluid.

In the most efficient circulatory systems, the heart pumps blood through a closed system of vessels. This one-way circuit steers the blood along a specific route and ensures systematic distribution of gases, nutrients, signal molecules, and wastes. A circulatory system comprising a heart, blood vessels, and blood is known as a **cardiovascular system** {*kardia,* heart + *vasculum,* little vessel}.

Although the idea of a closed cardiovascular system that cycles blood in an endless loop seems intuitive to us today, it has not always been so. **Capillaries** {*capillus,* hair}, the microscopic vessels where blood exchanges material with the interstitial fluid, were not discovered until Marcello Malpighi, an Italian anatomist, observed them through a microscope in the middle of the seventeenth century. At that time, European medicine was still heavily influenced by the ancient belief that the cardiovascular system distributed both blood and air.

Blood was thought to be made in the liver and distributed throughout the body in the veins. Air went from the lungs to the heart, where it was digested and picked up "vital spirits." From the heart, air was distributed to the tissues through vessels called arteries. Anomalies—such as the fact that a cut artery squirted blood rather than air—were ingeniously explained by unseen links between arteries and veins that opened upon injury.

According to this model of the circulatory system, the tissues consumed all blood delivered to them, and the liver had to synthesize new blood continuously. It took the calculations of William Harvey (1578–1657), court physician to King Charles I of England, to show that the weight of blood pumped by the heart in a single hour exceeds the weight of the entire body! Once it became obvious that the liver could not make blood as rapidly as the heart pumped it, Harvey looked for an anatomical route that would allow the blood to recirculate rather than be consumed in the tissues. He showed that valves in the heart and veins created a one-way flow of blood, and that veins carried blood back to the heart, not out to the limbs. He also showed that blood entering the right side of the heart had to go to the lungs before it could go to the left side of the heart.

These studies created a furor among Harvey's contemporaries, leading Harvey to say in a huff that no one under the age of 40 could understand his conclusions. Ultimately, Harvey's work became the foundation of modern cardiovascular physiology. Today, we understand the structure of the cardiovascular system at microscopic and molecular levels that Harvey never dreamed existed. Yet some things have not changed. Even now, with our sophisticated technology, we are searching for "spirits" in the blood, although today we call them by such names as *hormone* and *cytokine.*

RUNNING PROBLEM

Myocardial Infarction

At 9:06 A.M., the blood clot that had silently formed in Walter Parker's left anterior descending coronary artery made its sinister presence known. The 53-year-old advertising executive had arrived at the Dallas Convention Center feeling fine, but suddenly a dull ache started in the center of his chest, and he became nauseated. At first he brushed it off as the after-effects of the convention banquet the night before. When the chest pain persisted, however, he thought of his family history of heart disease and took an aspirin, remembering a radio ad that said to do this if you were having symptoms of a heart attack. Walter then made his way to the Center's Aid Station. "I'm not feeling very well," he told the emergency medical technician (EMT). The EMT, on hearing Walter's symptoms and seeing his pale, sweaty face, immediately thought of a heart attack. "Let's get you over to the hospital and get this checked out."

463 467 477 485 486 495 501

Overview of the Cardiovascular System

In the simplest terms, a cardiovascular system is a series of tubes (the blood vessels) filled with fluid (blood) and connected to a pump (the heart). Pressure generated in the heart propels blood through the system continuously. The blood picks up oxygen at the lungs and nutrients in the intestine and then delivers these substances to the body's cells while simultaneously removing cellular wastes and heat for excretion. In addition, the cardiovascular system plays an important role in cell-to-cell communication and in defending the body against foreign invaders. This chapter focuses on an overview of the cardiovascular system and on the heart as a pump. You will learn about the properties of the blood vessels and the homeostatic controls that regulate blood flow and blood pressure later.

The Cardiovascular System Transports Materials throughout the Body

The primary function of the cardiovascular system is the transport of materials to and from all parts of the body. Substances transported by the cardiovascular system can be divided into (1) nutrients, water, and gases that enter the body from the external environment, (2) materials that move from cell to cell within the body, and (3) wastes that the cells eliminate (■ Tbl. 14.1).

Oxygen enters the body at the exchange surface of the lungs. Nutrients and water are absorbed across the intestinal epithelium. Once all these materials are in the blood, the cardiovascular system distributes them. A steady supply of oxygen for the cells is particularly important because many cells deprived of oxygen become irreparably damaged within a short period of time. For example, about 5–10 seconds after blood flow to the brain is stopped, a person loses consciousness. If oxygen delivery stops for 5–10 minutes, permanent brain damage results. Neurons of the brain have a very high rate of oxygen consumption and cannot meet their metabolic need for ATP by using anaerobic pathways, which have low yields of ATP/glucose [p. 117]. Because of the brain's sensitivity to *hypoxia* {*hypo-*, low + *oxia,* oxygen}, homeostatic controls do everything possible to maintain cerebral blood flow, even if it means depriving other cells of oxygen.

Cell-to-cell communication is a key function of the cardiovascular system. For example, hormones secreted by endocrine glands travel in the blood to their targets. Blood also carries nutrients, such as glucose from the liver and fatty acids from adipose tissue, to metabolically active cells. Finally, the defense team of white blood cells and antibodies patrols the circulation to intercept foreign invaders.

The cardiovascular system also picks up carbon dioxide and metabolic wastes released by cells and transports them to the lungs and kidneys for excretion. Some waste products are transported to the liver for processing before they are excreted in the urine or feces. Heat also circulates through the blood, moving from the body core to the surface, where it dissipates.

Table 14.1 Transport in the Cardiovascular System

Substance Moved	From	To
Materials entering the body		
Oxygen	Lungs	All cells
Nutrients and water	Intestinal tract	All cells
Materials moved from cell to cell		
Wastes	Some cells	Liver for processing
Immune cells, antibodies, clotting proteins	Present in blood continuously	Available to any cell that needs them
Hormones	Endocrine cells	Target cells
Stored nutrients	Liver and adipose tissue	All cells
Materials leaving the body		
Metabolic wastes	All cells	Kidneys
Heat	All cells	Skin
Carbon dioxide	All cells	Lungs

The Cardiovascular System Consists of the Heart, Blood Vessels, and Blood

The cardiovascular system is composed of the heart, the blood vessels (also known as the *vasculature*), and the cells and plasma of the blood. Blood vessels that carry blood away from the heart are called **arteries**. Blood vessels that return blood to the heart are called **veins**.

As blood moves through the cardiovascular system, a system of valves in the heart and veins ensures that the blood flows in one direction only. Like the turnstiles at an amusement park, the valves keep blood from reversing its direction of flow. ■ Figure 14.1 is a schematic diagram that shows these components and the route that blood follows through the body. Notice in this illustration, as well as in most other diagrams of the heart, that the right side of the heart is on the left side of the page, which means that the heart is labeled as if you were viewing the heart of a person facing you.

The heart is divided by a central wall, or **septum**, into left and right halves. Each half functions as an independent pump that consists of an **atrium** {*atrium,* central room; plural *atria*} and a **ventricle** {*ventriculus,* belly}. The atrium receives blood returning to the heart from the blood vessels, and the ventricle pumps blood out into the blood vessels. The right side of the heart receives blood from the tissues and sends it to the lungs

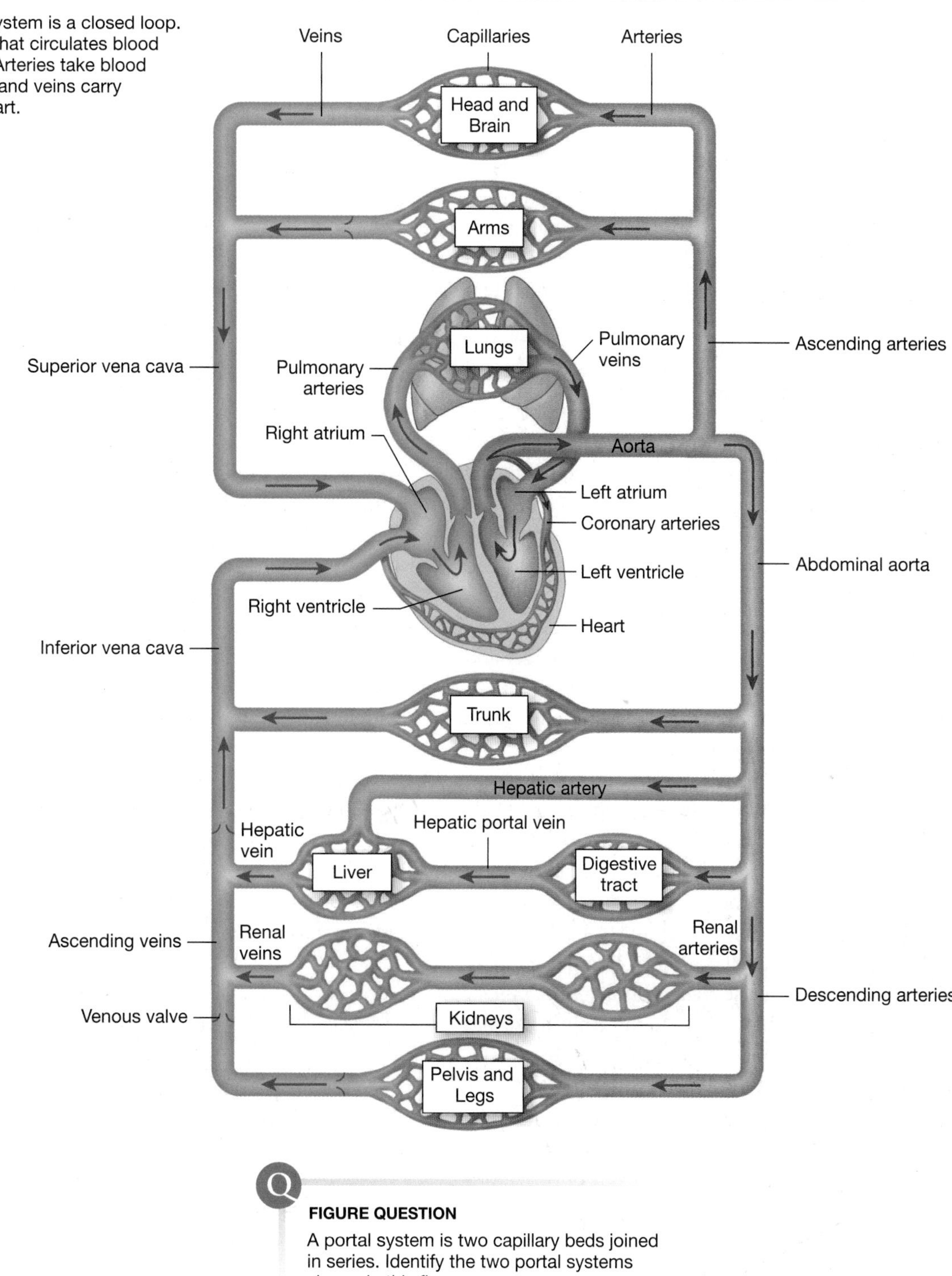

Fig. 14.1

for oxygenation. The left side of the heart receives newly oxygenated blood from the lungs and pumps it to tissues throughout the body.

Starting in the right atrium in Figure 14.1, trace the path taken by blood as it flows through the cardiovascular system. Note that blood in the right side of the heart is colored blue. This is a convention used to show blood from which the tissues have extracted oxygen. Although this blood is often described as *deoxygenated,* it is not completely devoid of oxygen. It simply has less oxygen than blood going from the lungs to the tissues.

In living people, well-oxygenated blood is bright red, and low-oxygen blood is a darker red. Under some conditions, low-oxygen blood can impart a bluish color to certain areas of the skin, such as around the mouth and under the fingernails. This condition, known as *cyanosis* {*kyanos,* dark blue}, is the reason blue is used in drawings to indicate blood with lower oxygen content.

From the right atrium, blood flows into the right ventricle of the heart. From there it is pumped through the **pulmonary arteries** {*pulmo,* lung} to the lungs, where it is oxygenated. Note the color change from blue to red in Figure 14.1, indicating higher oxygen content after the blood leaves the lungs. From the lungs, blood travels to the left side of the heart through the **pulmonary veins**. The blood vessels that go from the right ventricle to the lungs and back to the left atrium are known collectively as the **pulmonary circulation**.

Blood from the lungs enters the heart at the left atrium and passes into the left ventricle. Blood pumped out of the left ventricle enters the large artery known as the **aorta**. The aorta branches into a series of smaller and smaller arteries that finally lead into networks of capillaries. Notice at the top of Figure 14.1 the color change from red to blue as the blood passes through the capillaries, indicating that oxygen has left the blood and diffused into the tissues.

After leaving the capillaries, blood flows into the venous side of the circulation, moving from small veins into larger and larger veins. The veins from the upper part of the body join to form the **superior vena cava**. Those from the lower part of the body form the **inferior vena cava**. The two *venae cavae* empty into the right atrium. The blood vessels that carry blood from the left side of the heart to the tissues and back to the right side of the heart are collectively known as the **systemic circulation**.

Return to Figure 14.1 and follow the divisions of the aorta after it leaves the left ventricle. The first branch represents the *coronary arteries,* which nourish the heart muscle itself. Blood from these arteries flows into capillaries, then into the *coronary veins,* which empty directly into the right atrium at the *coronary sinus.* Ascending branches of the aorta go to the arms, head, and brain. The abdominal aorta supplies blood to the trunk, the legs, and the internal organs such as liver (*hepatic artery*), digestive tract, and the kidneys (*renal arteries*).

Notice two special arrangements of the circulation. One is the blood supply to the digestive tract and liver. Both regions receive well-oxygenated blood through their own arteries, but, in addition, blood leaving the digestive tract goes directly to the liver by means of the *hepatic portal vein.* The liver is an important site for nutrient processing and plays a major role in the detoxifying foreign substances. Most nutrients absorbed in the intestine are routed directly to the liver, allowing that organ to process material before it is released into the general circulation. The two capillary beds of the digestive tract and liver, joined by the hepatic portal vein, are an example of a *portal system.*

A second portal system occurs in the kidneys, where two capillary beds are connected in series. A third portal system, discussed earlier but not shown here, is the hypothalamic-hypophyseal portal system, which connects the hypothalamus and the anterior pituitary [p. 223].

Concept Check Answers: p. 506

1. A cardiovascular system has what three major components?
2. What is the difference between (a) the pulmonary and systemic circulations, (b) an artery and a vein, (c) an atrium and a ventricle?

Pressure, Volume, Flow, and Resistance

If you ask people why blood flows through the cardiovascular system, many of them respond, "So that oxygen and nutrients can get to all parts of the body." This is true, but it is a teleological answer, one that describes the purpose of blood flow. In physiology, we are also concerned with how blood flows—in other words, with the mechanisms or forces that create blood flow.

A simple mechanistic answer to "Why does blood flow?" is that liquids and gases flow down **pressure gradients** (**ΔP**) from regions of higher pressure to regions of lower pressure. For this reason, blood can flow in the cardiovascular system only if one region develops higher pressure than other regions.

In humans, the heart creates high pressure when it contracts. Blood flows out of the heart (the region of highest pressure) into the closed loop of blood vessels (a region of lower pressure). As blood moves through the system, pressure is lost because of friction between the fluid and the blood vessel walls. Consequently, pressure falls continuously as blood moves farther from the heart (■ Fig. 14.2). The highest pressure in the vessels of the cardiovascular system is found in the aorta and systemic arteries as they

Blood Flows Down a Pressure Gradient.

The mean blood pressure of the systemic circulation ranges from a high of 93 mm Hg (millimeters of mercury) in the aorta to a low of a few mm Hg in the venae cavae.

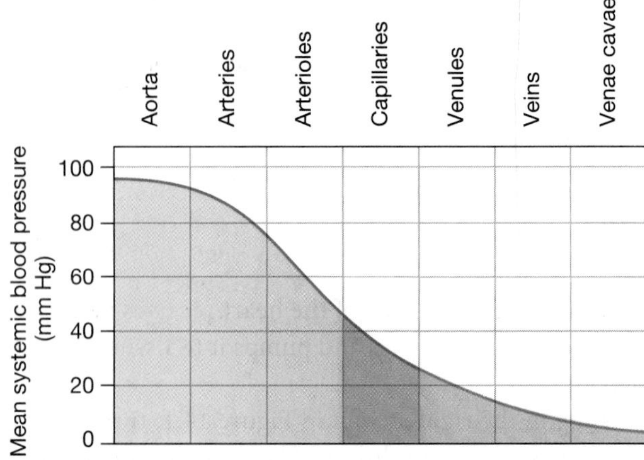

■ **Fig. 14.2**

receive blood from the left ventricle. The lowest pressure is in the venae cavae, just before they empty into the right atrium.

Now let's review the laws of physics that explain the interaction of pressure, volume, flow, and resistance in the cardiovascular system. Many of these principles apply broadly to the flow of all types of liquids and gases, including the flow of air in the respiratory system. However, in this chapter we focus on blood flow and its relevance to the function of the heart.

The Pressure of Fluid in Motion Decreases over Distance

Pressure in a fluid is the force exerted by the fluid on its container. In the heart and blood vessels, pressure is commonly measured in *millimeters of mercury* (mm Hg), where one millimeter of mercury is equivalent to the hydrostatic pressure exerted by a 1-mm-high column of mercury on an area of 1 cm^2. Some physiological literature reports pressures in *torr* (1 torr = 1 mm Hg) or in *centimeters of water:* 1 cm H_2O = 0.74 mm Hg.

If fluid is not moving, the pressure it exerts is called **hydrostatic pressure** (■ Fig. 14.3a), and force is exerted equally in all directions. For example, a column of fluid in a tube exerts hydrostatic pressure on the floor and sides of the tube.

In a system in which fluid is flowing, pressure falls over distance as energy is lost because of friction (Fig. 14.3b). In addition, the pressure exerted by moving fluid has two components: a dynamic, flowing component that represents the kinetic energy of the system, and a lateral component that represents the hydrostatic pressure (potential energy) exerted on the walls of the system. Pressure within our cardiovascular system is usually called hydrostatic pressure even though it is a system in which fluid is in motion. Some textbooks are beginning to replace the term *hydrostatic pressure* with the term *hydraulic pressure.* Hydraulics is the study of fluid in motion.

RUNNING PROBLEM

When people speak of a "heart attack," they are actually referring to a clot that stops the blood supply to part of the heart, creating a condition known as *ischemia* {*ischien,* to suppress + *emia,* blood}. In medical terms, a heart attack is called a *myocardial infarction* (MI), referring to an area of heart muscle that is dying because of a lack of blood supply. The clot in Walter's coronary artery had restricted blood flow to part of his left ventricle, and its cells were beginning to die from lack of oxygen. When someone has a heart attack, immediate medical intervention is critical. In the ambulance on the way to the emergency room, the EMT gave Walter oxygen, hooked him up to a heart monitor, and started an intravenous (IV) injection of normal (isotonic) saline. With an IV line in place, other drugs could be given rapidly if Walter's condition should suddenly worsen.

Q1: Why did the EMT give Walter oxygen?

Q2: What effect would the injection of isotonic saline have on Walter's extracellular fluid volume? On his intracellular fluid volume? On his total body osmolarity? [*Hint:* p. 134]

463 467 477 485 486 495 501

Pressure Changes in Liquids without a Change in Volume

If the walls of a fluid-filled container contract, the pressure exerted on the fluid in the container increases. You can demonstrate this principle by filling a balloon with water and squeezing the water balloon in your hand. Water is minimally compressible, and so the pressure you apply to the balloon is transmitted throughout the fluid. As you squeeze, higher pressure in the fluid causes parts of the balloon to bulge. If the pressure becomes high enough, the stress on the balloon causes it to pop. The water volume inside the balloon did not change, but the pressure in the fluid increased.

In the human heart, contraction of the blood-filled ventricles is similar to squeezing a water balloon: pressure created by the contracting muscle is transferred to the blood. This high-pressure blood then flows out of the ventricle and into the blood vessels, displacing lower-pressure blood already in the vessels. The pressure created in the ventricles is called the **driving pressure** because it is the force that drives blood through the blood vessels.

When the walls of a fluid-filled container expand, the pressure exerted on the fluid decreases. For this reason, when the heart relaxes and expands, pressure in the fluid-filled chambers falls.

Pressure changes can also take place in the blood vessels. If blood vessels dilate, blood pressure inside them falls. If blood vessels constrict, blood pressure increases. Volume changes of the blood vessels and heart are major factors that influence blood pressure in the cardiovascular system.

Blood Flows from Higher Pressure to Lower Pressure

As stated earlier, blood flow through the cardiovascular system requires a pressure gradient. This pressure gradient is analogous to the difference in pressure between two ends of a tube through which fluid flows (Fig. 14.3c). Flow through the tube is directly proportional to ($\propto$) the pressure gradient (ΔP):

$$\text{Flow} \propto \Delta P \qquad (1)$$

Where $\Delta P = P_1 - P_2$. This relationship says that the higher the pressure gradient, the greater the fluid flow.

The Physics of Fluid Flow

Pressure in Static and Flowing Fluids

(a) Hydrostatic pressure is the pressure exerted on the walls of the container by the fluid within the container. Hydrostatic pressure is proportional to the height of the water column.

(b) Once fluid begins to flow through the system, pressure falls with distance as energy is lost because of friction. This is the situation in the cardiovascular system.

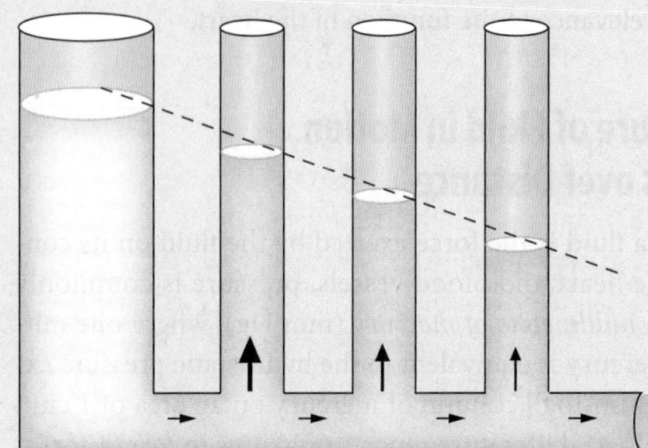

Fluid flow through a tube depends on the pressure gradient.

(c) Fluid flows only if there is a positive pressure gradient (ΔP).

Higher P —— Flow ——> Lower P

Flow →

P_1 P_2

$P_1 - P_2 = \Delta P$

This tube has no pressure gradient, so no flow.

100 mm Hg 100 mm Hg

$\Delta P = 0$, so no flow

(d) Flow depends on the pressure gradient (ΔP), not on the absolute pressure (P). ΔP is equal in these tubes so flow is the same.

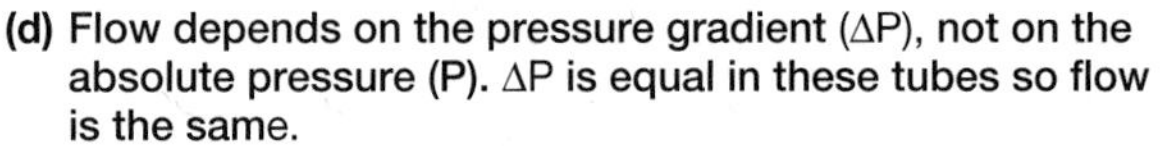

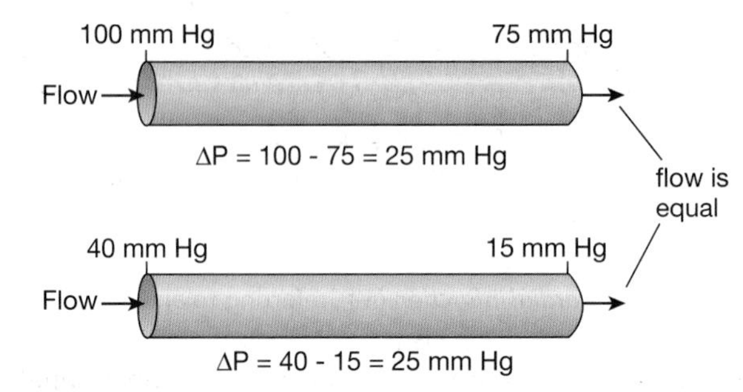

As the radius of a tube decreases, the resistance to flow increases.

(e)

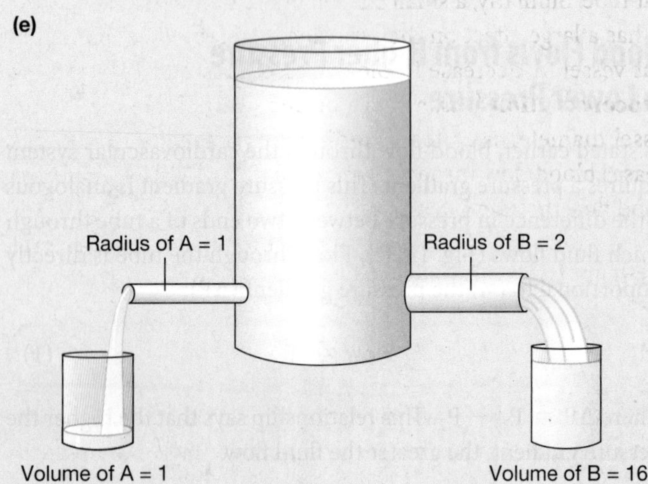

Resistance $\propto \frac{1}{\text{radius}^4}$	
Tube A	Tube B
$R \propto \frac{1}{1^4}$	$R \propto \frac{1}{2^4}$
$R \propto 1$	$R \propto \frac{1}{16}$

Flow $\propto \frac{1}{\text{resistance}}$	
Tube A	Tube B
Flow $\propto \frac{1}{1}$	Flow $\propto \frac{1}{\frac{1}{16}}$
Flow $\propto 1$	Flow $\propto 16$

FIGURE QUESTION

If the radius of A changes to 3, the flow through A will be about ______ times the flow through B.

A pressure gradient is not the same thing as the absolute pressure in the system. For example, the tube in Figure 14.3c has an absolute pressure of 100 mm Hg at each end. However, because there is no pressure gradient between the two ends of the tube, there is no flow through the tube.

On the other hand, two identical tubes can have very different absolute pressures but the same flow. The top tube in Figure 14.3d has a hydrostatic pressure of 100 mm Hg at one end and 75 mm Hg at the other end, which means that the pressure gradient between the ends of the tube is 25 mm Hg. The identical bottom tube has a hydrostatic pressure of 40 mm Hg at one end and 15 mm Hg at the other end. This tube has lower absolute pressure all along its length but the same pressure gradient as the top tube: 25 mm Hg. Because the pressure difference in the two tubes is identical, fluid flow through the tubes is the same.

Resistance Opposes Flow

In an ideal system, a substance in motion would remain in motion. However, no system is ideal because all movement creates friction. Just as a ball rolled across the ground loses energy to friction, blood flowing through blood vessels encounters friction from the walls of the vessels and from cells within the blood rubbing against one another as they flow.

The tendency of the cardiovascular system to oppose blood flow is called the system's **resistance** to flow. Resistance (R) is a term that most of us understand from everyday life. We speak of people being resistant to change or taking the path of least resistance. This concept translates well to the cardiovascular system because blood flow also takes the path of least resistance. An increase in the resistance of a blood vessel results in a decrease in the flow through that vessel. We can express that relationship as

$$\text{Flow} \propto 1/R \quad (2)$$

This expression says that flow is inversely proportional to resistance: if resistance increases, flow decreases; and if resistance decreases, flow increases.

What parameters determine resistance? For fluid flowing through a tube, resistance is influenced by three components: the radius of the tube (r), the length of the tube (L), and the **viscosity** (thickness) of the fluid (η, the Greek letter eta). The following equation, derived by the French physician Jean Leonard Marie Poiseuille and known as **Poiseuille's law**, shows the relationship of these factors:

$$R = 8L\eta/\pi r^4 \quad (3)$$

Because the value of $8/\pi$ is a constant, this factor can be removed from the equation, and the relationship can be rewritten as

$$R \propto L\eta/r^4 \quad (4)$$

This expression says that (1) the resistance to fluid flow offered by a tube increases as the length of the tube increases, (2) resistance increases as the viscosity of the fluid increases, but (3) resistance decreases as the tube's radius increases.

To remember these relationships, think of drinking through a straw. You do not need to suck as hard on a short straw as on a long one (the resistance offered by the straw increases with length). Drinking water through a straw is easier than drinking a thick milkshake (resistance increases with viscosity). And drinking the milkshake through a fat straw is much easier than through a skinny cocktail straw (resistance increases as radius decreases).

How significant are tube length, fluid viscosity, and tube radius to blood flow in a normal individual? The length of the systemic circulation is determined by the anatomy of the system and is essentially constant. Blood viscosity is determined by the ratio of red blood cells to plasma and by how much protein is in the plasma. Normally, viscosity is constant, and small changes in either length or viscosity have little effect on resistance. This leaves changes in the radius of the blood vessels as the main variable that affects resistance in the systemic circulation.

Let's return to the example of the straw and the milkshake to illustrate how changes in radius affect resistance. If we assume that the length of the straw and the viscosity of the milkshake do not change, this system is similar to the cardiovascular system—the radius of the tube has the greatest effect on resistance. If we consider only resistance (R) and radius (r) from equation 4, the relationship between resistance and radius can be expressed as

$$R \propto 1/r^4 \quad (5)$$

If the skinny straw has a radius of 1, its resistance is proportional to $1/1^4$, or 1. If the fat straw has a radius of 2, the resistance it offers is $1/2^4$, or 1/16th, that of the skinny straw (Fig. 14.3e). Because flow is inversely proportional to resistance, flow increases 16-fold when the radius doubles.

As you can see from this example, a small change in the radius of a tube has a large effect on the flow of a fluid through that tube. Similarly, a small change in the radius of a blood vessel has a large effect on the resistance to blood flow offered by that vessel. A decrease in blood vessel diameter is known as **vasoconstriction** {*vas*, a vessel or duct}. An increase in blood vessel diameter is called **vasodilation**. Vasoconstriction decreases blood flow through a vessel, and vasodilation increases blood flow through a vessel.

In summary, by combining equations 1 and 2, we get the equation

$$\text{Flow} \propto \Delta P/R \quad (6)$$

which, translated into words, says that the flow of blood in the cardiovascular system is directly proportional to the pressure gradient in the system, and inversely proportional to the resistance of the system to flow. If the pressure gradient remains constant, then flow varies inversely with resistance.

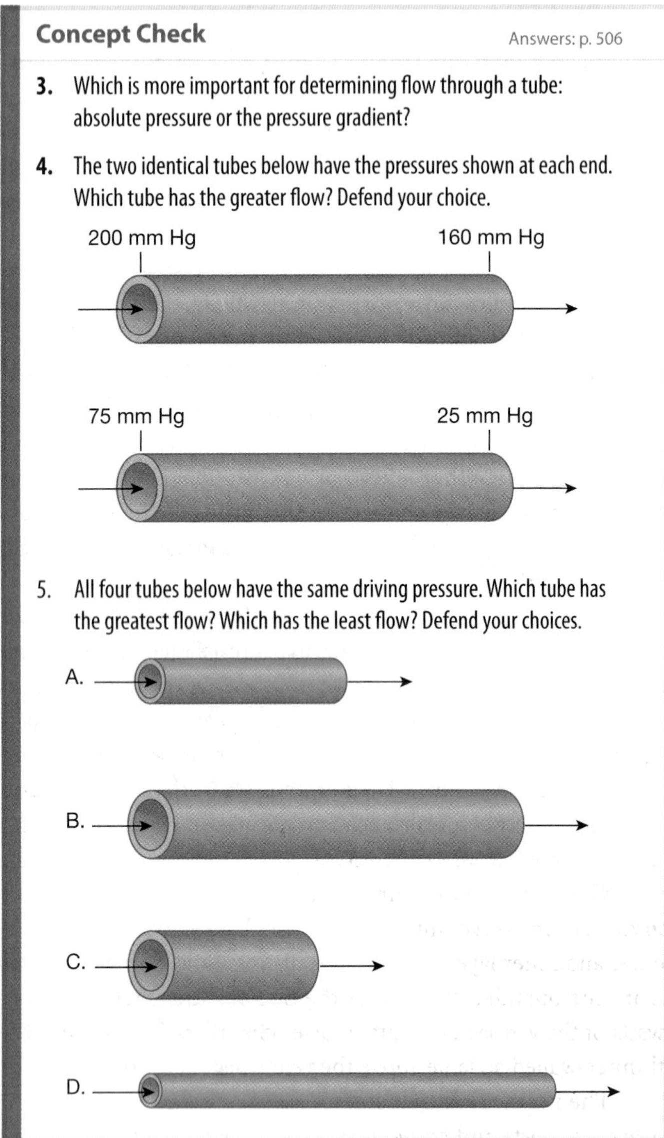

Concept Check Answers: p. 506

3. Which is more important for determining flow through a tube: absolute pressure or the pressure gradient?
4. The two identical tubes below have the pressures shown at each end. Which tube has the greater flow? Defend your choice.

5. All four tubes below have the same driving pressure. Which tube has the greatest flow? Which has the least flow? Defend your choices.

Velocity Depends on the Flow Rate and the Cross-Sectional Area

The word *flow* is sometimes used imprecisely in cardiovascular physiology, leading to confusion. Flow usually means **flow rate**, the volume of blood that passes a given point in the system per unit time. In the circulation, flow is expressed in either liters per minute (L/min) or milliliters per minute (mL/min). For instance, blood flow through the aorta of a 70-kg man at rest is about 5 L/min.

Flow rate should not be confused with **velocity of flow** (or simply *velocity*), the distance a fixed volume of blood travels in a given period of time. Velocity is a measure of *how fast* blood flows past a point. In contrast, flow rate measures *how much* (volume) blood flows past a point in a given period of time. For example, look through the open door at the hallway outside your classroom. The number of people passing the door in one minute is the flow rate of people through the hallway. How quickly those people are walking past the door is their velocity.

The relationship between velocity of flow (v), flow rate (Q), and cross-sectional area of the tube (A) is expressed by the equation

$$v = Q/A \qquad (7)$$

which says that the velocity of flow through a tube equals the flow rate divided by the tube's cross-sectional area. In a tube of fixed diameter (and thus fixed cross-sectional area), velocity is directly related to flow rate. In a tube of variable diameter, if the flow rate is constant, velocity varies inversely with the diameter. In other words, velocity is faster in narrow sections, and slower in wider sections.

■ Figure 14.4 shows how the velocity of flow varies as the cross-sectional area of the tube changes. The vessel in the figure has variable width, from narrow, with a cross-sectional area of 1 cm^2, to wide, with a cross-sectional area of 12 cm^2. The flow rate is identical along the length of the vessel: 12 cm^3 per minute (1 cm^3 = 1 cubic centimeter (cc) = 1 mL). This flow rate means that in one minute, 12 cm^3 of fluid flows past point X in the narrow section, and 12 cm^3 of fluid flows past point Y in the wide section.

But *how fast* does the fluid need to flow to accomplish that rate? According to equation 7, the velocity of flow at point X is 12 cm/min, but at point Y it is only 1 cm/min. As you can see, fluid flows more rapidly through narrow sections of a tube than through wide sections.

To see this principle in action, watch a leaf as it floats down a stream. Where the stream is narrow, the leaf moves rapidly, carried by the fast velocity of the water. In sections where the stream widens into a pool, the velocity of the water decreases and the leaf meanders more slowly.

In this chapter and the next, we apply the physics of fluid flow to the cardiovascular system. The heart generates pressure when it contracts and pumps blood into the arterial side of the circulation. Arteries act as a pressure reservoir during the heart's relaxation phase, maintaining the *mean arterial pressure* (MAP) that is the primary driving force for blood flow. Mean arterial pressure is influenced by two parameters: *cardiac output* (the volume of blood the heart pumps per minute) and *peripheral resistance* (the resistance of the blood vessels to blood flow through them):

$$\text{Mean arterial pressure} \propto \text{cardiac output} \times \text{peripheral resistance}$$

We will return to a discussion of peripheral resistance and blood flow later. In the remainder of this chapter, we examine heart function and the parameters that influence cardiac output.

Flow rate is not the same as velocity of flow.

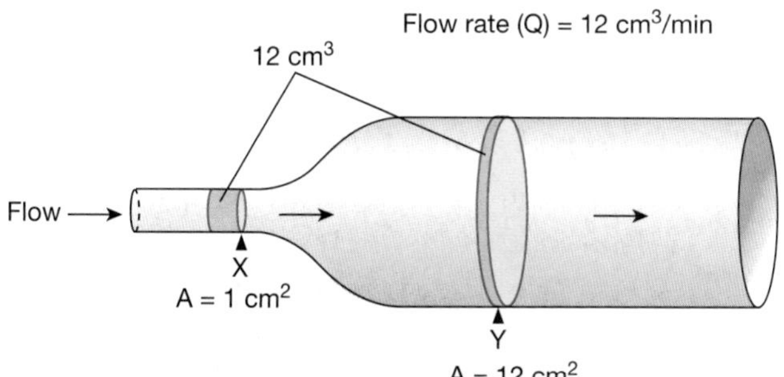

The narrower the vessel, the faster the velocity of flow.

Velocity (v) = $\frac{\text{Flow rate (Q)}}{\text{Cross-sectional area (A)}}$

At point X	At point Y
$v = \frac{12\ cm^3/min}{1\ cm^2}$	$v = \frac{12\ cm^3/min}{12\ cm^2}$
v = 12 cm/min	v = 1 cm/min

FIGURE QUESTION

If the cross-sectional area of this pipe is 3 cm^2, what is the velocity of the flow?

■ Fig. 14.4

> **Concept Check** — Answer: p. 506
>
> **6.** Two canals in Amsterdam are identical in size, but the water flows faster through one than through the other. Which canal has the higher flow rate?

Cardiac Muscle and the Heart

To ancient civilizations, the heart was more than a pump—it was *the seat of the mind.* When ancient Egyptians mummified their dead, they removed most of the viscera but left the heart in place so that the gods could weigh it as an indicator of the owner's worthiness. Aristotle characterized the heart as the most important organ of the body, as well as *the seat of intelligence.* We can still find evidence of these ancient beliefs in modern expressions such as "heartfelt emotions." The link between the heart and mind is one that is still explored today as scientists study the effects of stress and depression on the development of cardiovascular disease.

The heart is the workhorse of the body, a muscle that contracts continually, resting only in the milliseconds-long pause between beats. By one estimate, in one minute the heart performs work equivalent to lifting a 5-pound weight up 1 foot. The energy demands of this work require a continuous supply of nutrients and oxygen to the heart muscle.

The Heart Has Four Chambers

The heart is a muscular organ, about the size of a fist. It lies in the center of the *thoracic cavity* (see Anatomy Summary, ■ Fig. 14.5a, b, c). The pointed *apex* of the heart angles down to the left side of the body, while the broader *base* lies just behind the breastbone, or *sternum.* Because we usually associate the word *base* with the bottom, remember that the base of a cone is the broad end, and the apex is the pointed end. Think of the heart as an inverted cone with apex down and base up. Within the thoracic cavity, the heart lies on the ventral side, sandwiched between the two lungs, with its apex resting on the diaphragm (Fig. 14.5c).

The heart is encased in a tough membranous sac, the **pericardium** {*peri,* around + *kardia,* heart} (Fig.14.5d, e). A thin layer of clear pericardial fluid inside the pericardium lubricates the external surface of the heart as it beats within the sac. Inflammation of the pericardium (*pericarditis*) may reduce this lubrication to the point that the heart rubs against the pericardium, creating a sound known as a *friction rub.*

The heart itself is composed mostly of cardiac muscle, or **myocardium** {*myo,* muscle + *kardia,* heart}, covered by thin outer and inner layers of epithelium and connective tissue. Seen from the outside, the bulk of the heart is the thick muscular walls of the ventricles, the two lower chambers (Fig. 14.5f). The thinner-walled atria lie above the ventricles.

The major blood vessels all emerge from the base of the heart. The aorta and *pulmonary trunk* (artery) direct blood from the heart to the tissues and lungs, respectively. The venae cavae and pulmonary veins return blood to the heart (■ Tbl. 14.2 on page 474). When the heart is viewed from the front (anterior view), as in Figure14.5f, the pulmonary veins are hidden behind the other major blood vessels. Running across the surface of the ventricles are shallow grooves containing the **coronary arteries** and **coronary veins,** which supply blood to the heart muscle.

The relationship between the atria and ventricles can be seen in a cross-sectional view of the heart (Fig. 14.5g). As noted earlier, the left and right sides of the heart are separated by the interventricular septum, so that blood on one side does not mix with blood on the other side. Although blood flow in the left heart is separated from flow in the right heart, the two sides contract in a coordinated fashion. First the atria contract together, then the ventricles contract together.

The Heart

(a) The heart lies in the center of the thorax.

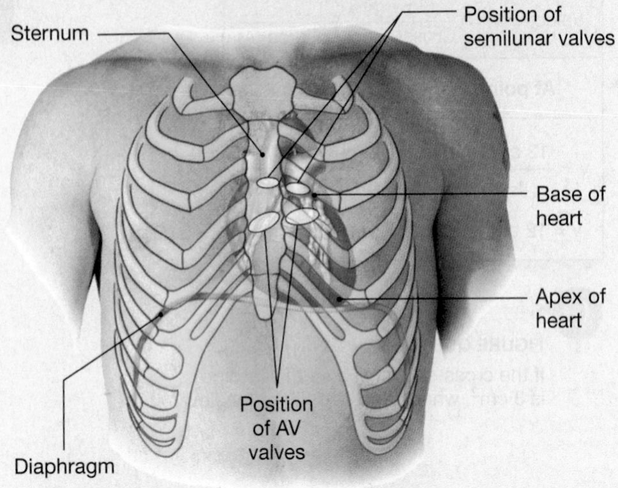

(b) Vessels that carry well-oxygenated blood are red; those with less well-oxygenated blood are blue.

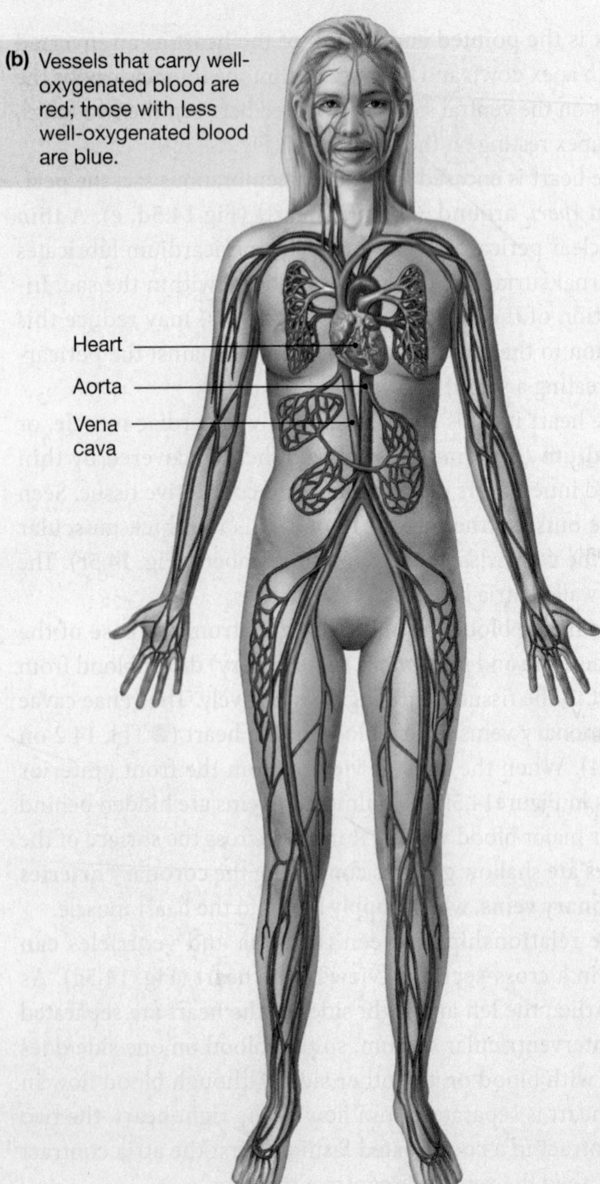

Anatomy of the Thoracic Cavity

(c) The heart is on the ventral side of the thoracic cavity, sandwiched between the lungs.

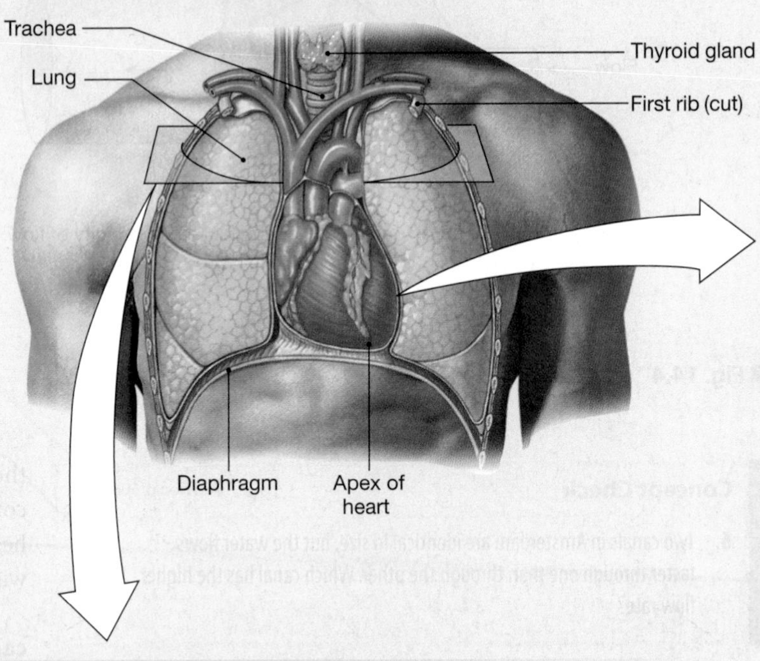

(d) Superior view of transverse plane in (b)

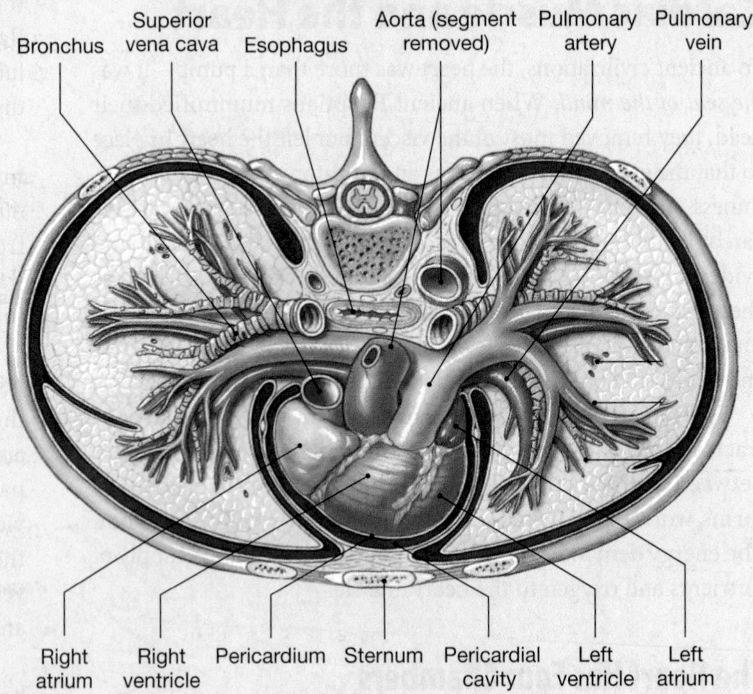

Structure of the Heart

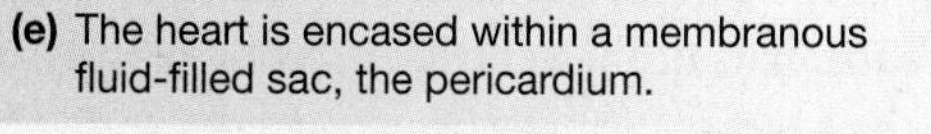

(e) The heart is encased within a membranous fluid-filled sac, the pericardium.

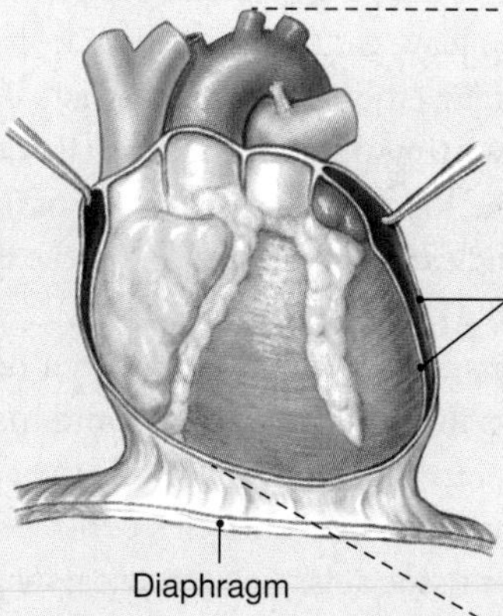

Pericardium

Diaphragm

(f) The ventricles occupy the bulk of the heart. The arteries and veins all attach to the base of the heart.

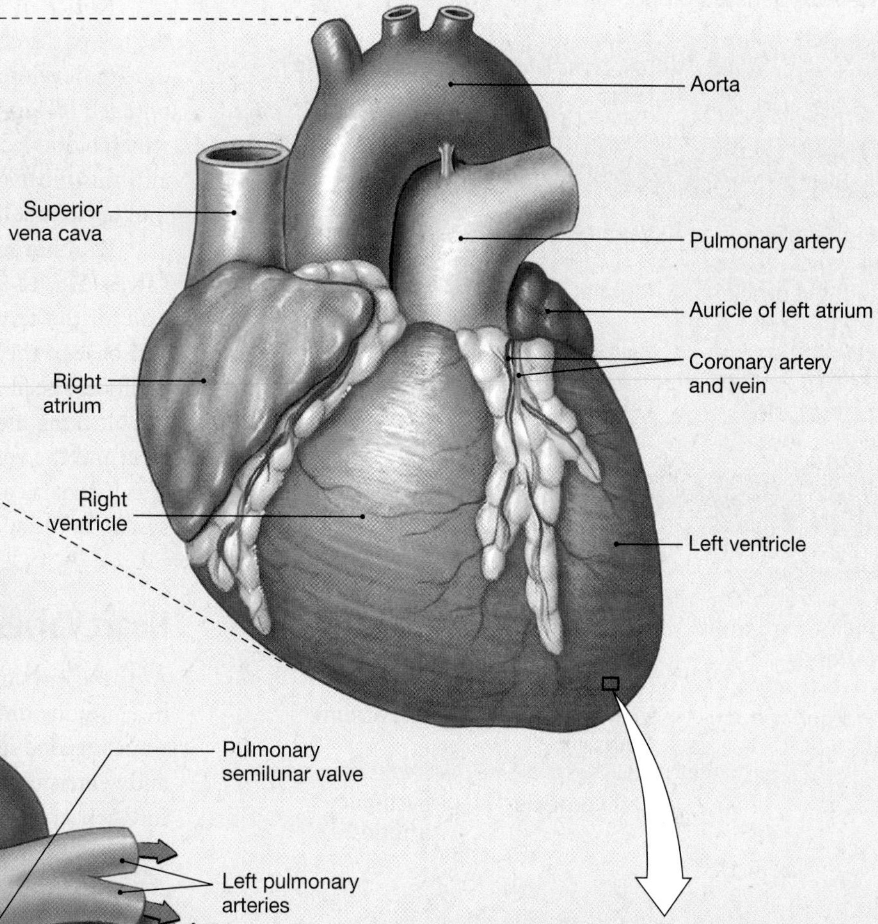

Aorta

Superior vena cava

Pulmonary artery

Auricle of left atrium

Coronary artery and vein

Right atrium

Right ventricle

Left ventricle

(g) One-way flow through the heart is ensured by two sets of valves.

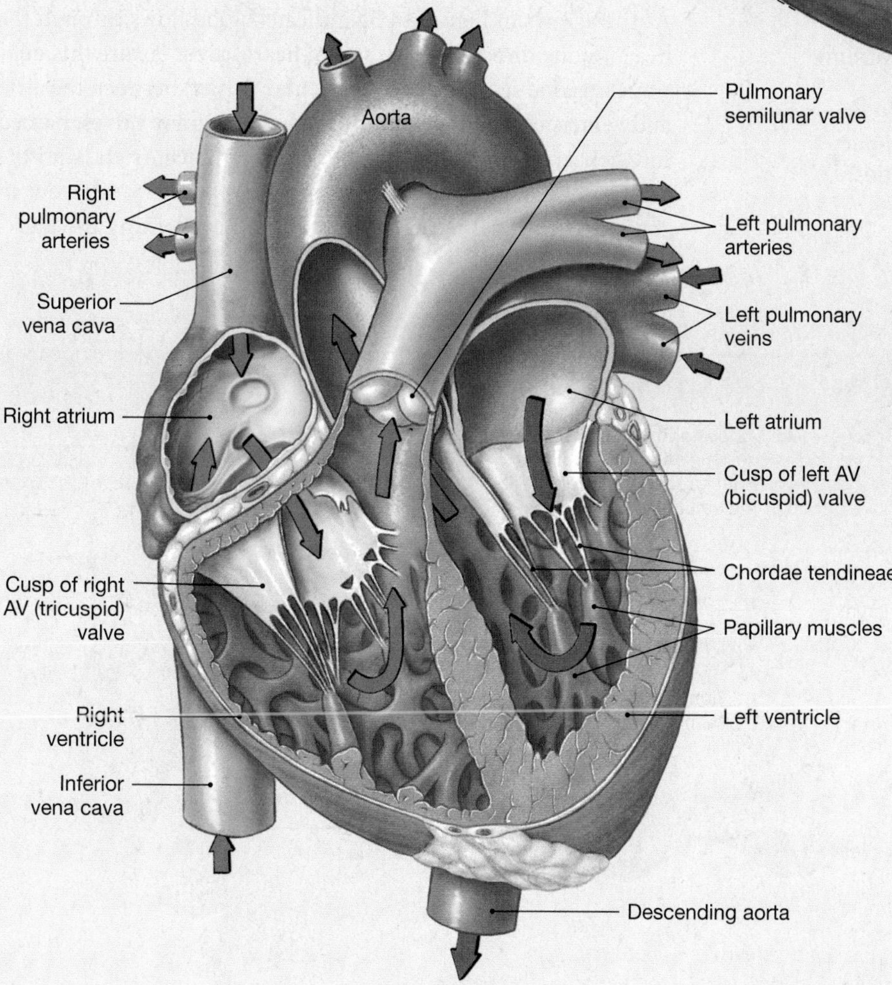

Aorta

Pulmonary semilunar valve

Right pulmonary arteries

Left pulmonary arteries

Superior vena cava

Left pulmonary veins

Right atrium

Left atrium

Cusp of left AV (bicuspid) valve

Chordae tendineae

Cusp of right AV (tricuspid) valve

Papillary muscles

Right ventricle

Left ventricle

Inferior vena cava

Descending aorta

(h) Myocardial muscle cells are branched, have a single nucleus, and are attached to each other by specialized junctions known as intercalated disks.

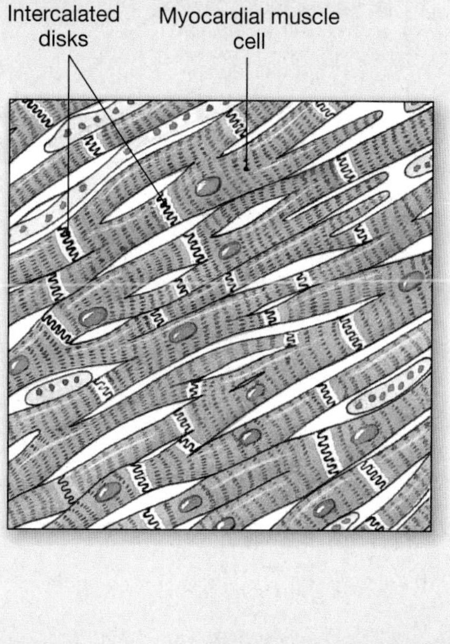

Intercalated disks

Myocardial muscle cell

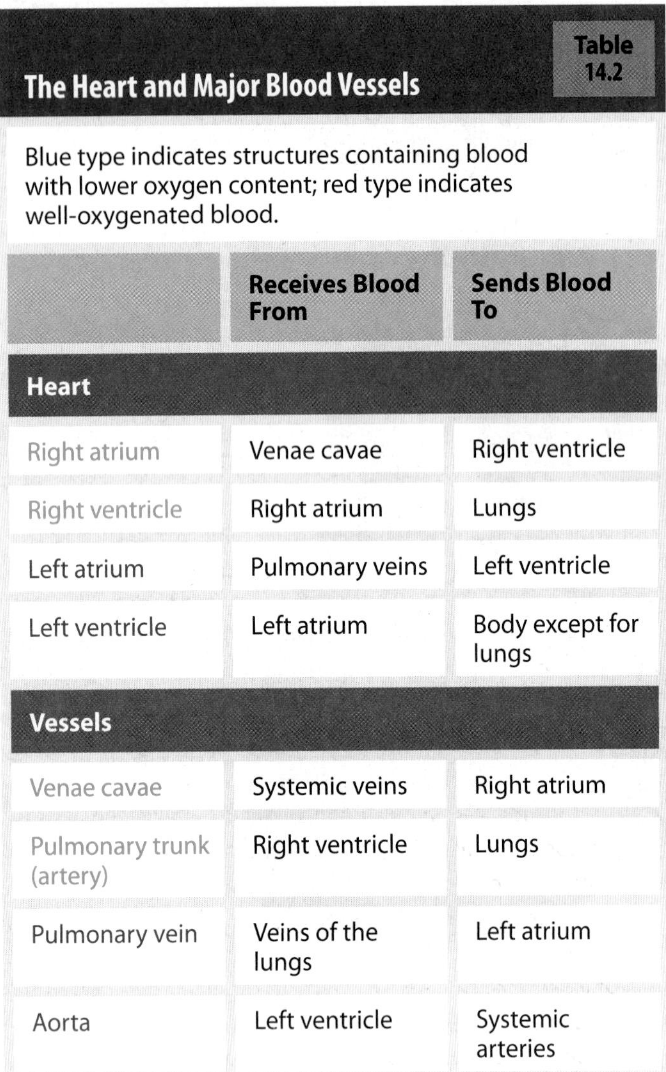

Table 14.2 The Heart and Major Blood Vessels

Blue type indicates structures containing blood with lower oxygen content; red type indicates well-oxygenated blood.

	Receives Blood From	Sends Blood To
Heart		
Right atrium	Venae cavae	Right ventricle
Right ventricle	Right atrium	Lungs
Left atrium	Pulmonary veins	Left ventricle
Left ventricle	Left atrium	Body except for lungs
Vessels		
Venae cavae	Systemic veins	Right atrium
Pulmonary trunk (artery)	Right ventricle	Lungs
Pulmonary vein	Veins of the lungs	Left atrium
Aorta	Left ventricle	Systemic arteries

Blood flows from veins into the atria and from there through one-way valves into the ventricles, the pumping chambers. Blood leaves the heart via the pulmonary trunk from the right ventricle and via the aorta from the left ventricle. A second set of valves guards the exits of the ventricles so that blood cannot flow back into the heart once it has been ejected.

Notice in Figure 14.5g that blood enters each ventricle at the top of the chamber but also leaves at the top. This is because during development, the tubular embryonic heart twists back on itself (■ Fig. 14.6b). This twisting puts the arteries (through which blood leaves) close to the top of the ventricles. Functionally, this means that the ventricles must contract from the bottom up so that blood is squeezed out of the top.

Four fibrous connective tissue rings surround the four heart valves (Fig. 14.5g). These rings form both the origin and insertion for the cardiac muscle, an arrangement that pulls the apex and base of the heart together when the ventricles contract. In addition, the fibrous connective tissue acts as an electrical insulator, blocking most transmission of electrical signals between the atria and the ventricles. This arrangement ensures that the electrical signals can be directed through a specialized conduction system to the apex of the heart for the bottom-to-top contraction.

Heart Valves Ensure One-Way Flow in the Heart

As the arrows in Figure 14.5g indicate, blood flows through the heart in one direction. Two sets of heart valves ensure this one-way flow: one set (the **atrioventricular valves**) between the atria and ventricles, and the second set (the **semilunar valves**, named for their crescent-moon shape) between the ventricles and the arteries. Although the two sets of valves are very different in structure, they serve the same function: preventing the backward flow of blood.

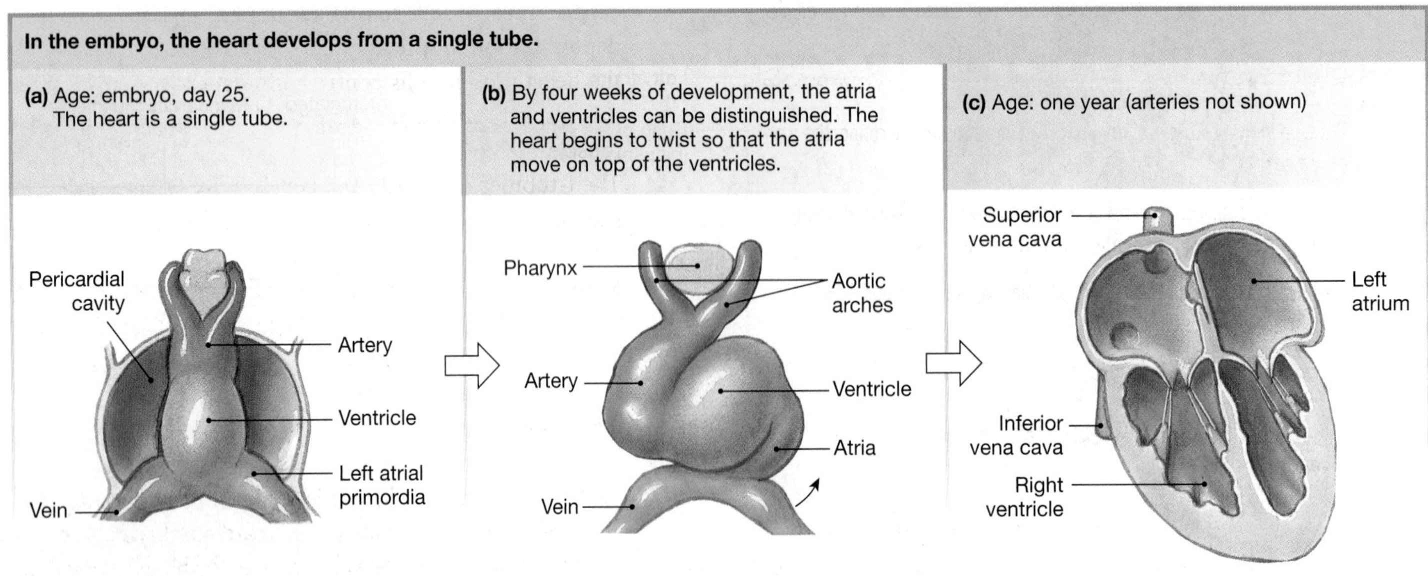
In the embryo, the heart develops from a single tube.
(a) Age: embryo, day 25. The heart is a single tube.
Pericardial cavity
Artery
Ventricle
Left atrial primordia
Vein
(b) By four weeks of development, the atria and ventricles can be distinguished. The heart begins to twist so that the atria move on top of the ventricles.
Pharynx
Aortic arches
Artery
Ventricle
Atria
Vein
(c) Age: one year (arteries not shown)
Superior vena cava
Left atrium
Inferior vena cava
Right ventricle

■ **Fig. 14.6**

The opening between each atrium and its ventricle is guarded by an atrioventricular (AV) valve (Fig. 14.5g). The AV valve is formed from thin flaps of tissue joined at the base to a connective tissue ring. The flaps are slightly thickened at the edge and connect on the ventricular side to collagenous tendons, the **chordae tendineae** (■ Fig. 14.7a, c).

Most of the chordae fasten to the edges of the valve flaps. The opposite ends of the chordae are tethered to moundlike extensions of ventricular muscle known as the **papillary muscles** {*papilla,* nipple}. These muscles provide stability for the chordae, but they cannot actively open and close the AV valves. The valves move passively when flowing blood pushes on them.

When a ventricle contracts, blood pushes against the bottom side of its AV valve and forces it upward into a closed position (Fig. 14.7a). The chordae tendineae prevent the valve from being pushed back into the atrium, just as the struts on an umbrella keep the umbrella from turning inside out in a high wind. Occasionally, the chordae fail, and the valve is pushed back into the atrium during ventricular contraction, an abnormal condition known as *prolapse*.

The two AV valves are not identical. The valve that separates the right atrium and right ventricle has three flaps and is called the **tricuspid valve** {*cuspis,* point} (Fig. 14.7b). The valve between the left atrium and left ventricle has only two flaps and is called the **bicuspid valve**. The bicuspid is also called the **mitral valve** because of its resemblance to the tall headdress, known as a miter, worn by popes and bishops. You can match AV valves to the proper side of the heart by remembering that the Right Side has the Tricuspid (R-S-T).

The semilunar valves separate the ventricles from the major arteries. The **aortic valve** is between the left ventricle and the aorta, and the **pulmonary valve** lies between the right ventricle and the pulmonary trunk. Each semilunar valve has three cuplike leaflets that snap closed when blood attempting to flow back into the ventricles fills them (Fig. 14.7c, d). Because of their shape, the semilunar valves do not need connective tendons as the AV valves do.

Concept Check Answers: p. 506

7. What prevents electrical signals from passing through the connective tissue in the heart?
8. Trace a drop of blood from the superior vena cava to the aorta, naming all structures the drop encounters along its route.
9. What is the function of the AV valves? What happens to blood flow if one of these valves fails?

Cardiac Muscle Cells Contract without Innervation

The bulk of the heart is composed of cardiac muscle cells, or myocardium. Most cardiac muscle is contractile, but about 1% of the myocardial cells are specialized to generate action potentials spontaneously. These cells account for a unique property of the heart: its ability to contract without any outside signal. As mentioned in the introduction to this chapter, records tell us of Spanish explorers in the New World witnessing human sacrifices in which hearts torn from the chests of living victims continued to beat for minutes. The heart can contract without a connection to other parts of the body because the signal for contraction is *myogenic,* originating within the heart muscle itself.

The signal for myocardial contraction comes not from the nervous system but from specialized myocardial cells known as **autorhythmic cells**. The autorhythmic cells are also called **pacemakers** because they set the rate of the heartbeat. Myocardial autorhythmic cells are anatomically distinct from contractile cells: autorhythmic cells are smaller and contain few contractile fibers. Because they do not have organized sarcomeres, autorhythmic cells do not contribute to the contractile force of the heart.

Contractile cells are typical striated muscle, however, with contractile fibers organized into sarcomeres [p. 404]. Cardiac muscle differs in significant ways from skeletal muscle and shares some properties with smooth muscle:

1. Cardiac muscle fibers are much smaller than skeletal muscle fibers and usually have a single nucleus per fiber.
2. Individual cardiac muscle cells branch and join neighboring cells end-to-end to create a complex network (Figs. 14.5h, ■ 14.8b). The cell junctions, known as **intercalated disks** {*inter-,* between + *calare,* to proclaim}, consist of interdigitated membranes. Intercalated disks have two components: *desmosomes* [p. 80] and gap junctions [p. 78]. Desmosomes are strong connections that tie adjacent cells together, allowing force created in one cell to be transferred to the adjacent cell.
3. *Gap junctions* in the intercalated disks electrically connect cardiac muscle cells to one another. They allow waves of depolarization to spread rapidly from cell to cell, so that all the heart muscle cells contract almost simultaneously. In this respect, cardiac muscle resembles single-unit smooth muscle.
4. The t-tubules of myocardial cells are larger than those of skeletal muscle, and they branch inside the myocardial cells.
5. Myocardial sarcoplasmic reticulum is smaller than that of skeletal muscle, reflecting the fact that cardiac muscle depends in part on extracellular Ca^{2+} to initiate contraction. In this respect, cardiac muscle resembles smooth muscle.
6. Mitochondria occupy about one-third the cell volume of a cardiac contractile fiber, a reflection of the high energy demand of these cells. By one estimate, cardiac muscle consumes 70–80% of the oxygen delivered to it by the blood, more than twice the amount extracted by other cells in the body.

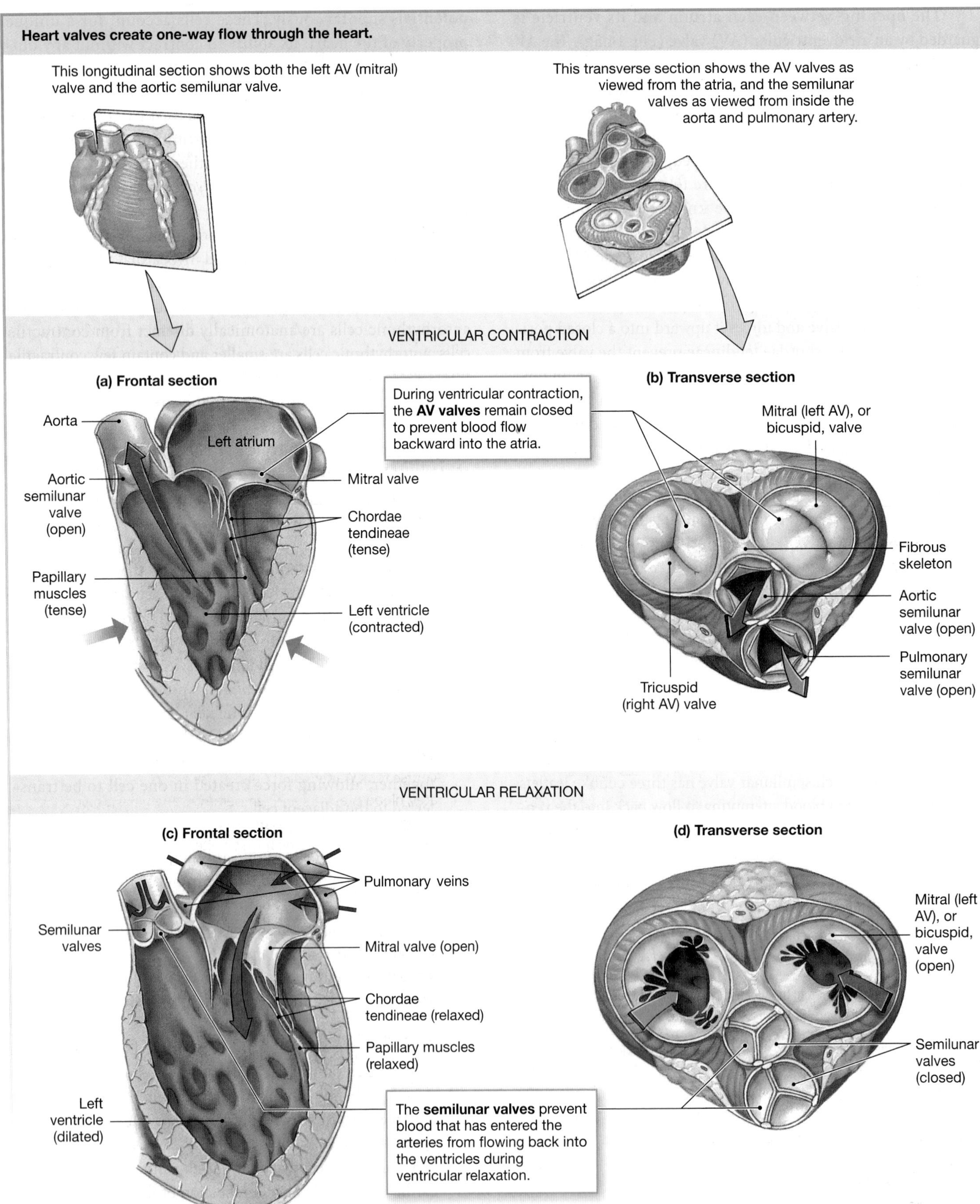
Heart valves create one-way flow through the heart.
This longitudinal section shows both the left AV (mitral) valve and the aortic semilunar valve.
This transverse section shows the AV valves as viewed from the atria, and the semilunar valves as viewed from inside the aorta and pulmonary artery.
VENTRICULAR CONTRACTION
(a) Frontal section
Aorta
Left atrium
Aortic semilunar valve (open)
Papillary muscles (tense)
During ventricular contraction, the AV valves remain closed to prevent blood flow backward into the atria.
Mitral valve
Chordae tendineae (tense)
Left ventricle (contracted)
(b) Transverse section
Mitral (left AV), or bicuspid, valve
Fibrous skeleton
Aortic semilunar valve (open)
Pulmonary semilunar valve (open)
Tricuspid (right AV) valve
VENTRICULAR RELAXATION
(c) Frontal section
Pulmonary veins
Semilunar valves
Mitral valve (open)
Chordae tendineae (relaxed)
Papillary muscles (relaxed)
Left ventricle (dilated)
The semilunar valves prevent blood that has entered the arteries from flowing back into the ventricles during ventricular relaxation.
(d) Transverse section
Mitral (left AV), or bicuspid, valve (open)
Semilunar valves (closed)

Fig. 14.7

CARDIAC MUSCLE

(a) The spiral arrangement of ventricular muscle allows ventricular contraction to squeeze the blood upward from the apex of the heart.

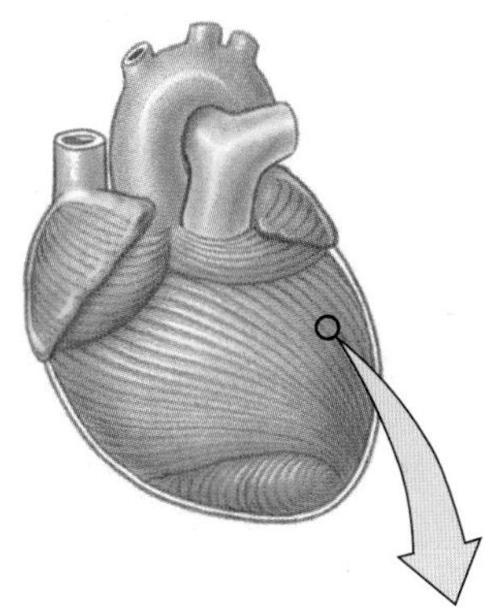

(b) Intercalated disks contain desmosomes that transfer force from cell to cell, and gap junctions that allow electrical signals to pass rapidly from cell to cell.

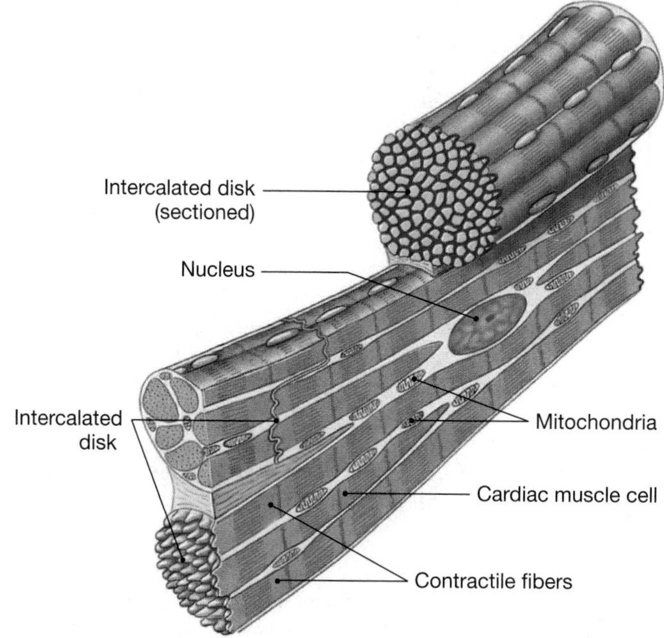

■ **Fig. 14.8**

During periods of increased activity, the heart uses almost all the oxygen brought to it by the coronary arteries. As a result, the only way to get more oxygen to exercising heart muscle is to increase the blood flow. Reduced myocardial blood flow from narrowing of a coronary vessel by a clot or fatty deposit can damage or even kill myocardial cells.

See Table 12.3, p. 434, for a summary comparison of the three muscle types.

Calcium Entry Is a Feature of Cardiac EC Coupling

In skeletal muscle, acetylcholine from a somatic motor neuron causes a skeletal muscle action potential to begin excitation-contraction coupling (EC coupling) [p. 406]. In cardiac muscle, an action potential also initiates EC coupling, but the action potential originates spontaneously in the heart's pacemaker cells and spreads into the contractile cells through gap junctions. Other aspects of cardiac EC coupling repeat processes you encountered in skeletal and smooth muscle contraction.

■ Figure 14.9 illustrates EC coupling and relaxation in cardiac muscle. An action potential that enters a contractile cell moves across the sarcolemma and into the t-tubules (1), where it opens voltage-gated L-type Ca^{2+} channels in the cell membrane (2). Ca^{2+} enters the cell and opens *ryanodine receptor Ca^{2+} release channels (RyR)* in the sarcoplasmic reticulum (3).

When the RyR channels open, stored Ca^{2+} flows out of the sarcoplasmic reticulum and into the cytosol (4), creating a Ca^{2+} "spark" that can be seen using special biochemical methods [p. 188]. Multiple sparks from different RyR channels sum to create a Ca^{2+} signal (5). This process of EC coupling in cardiac muscle is also called **Ca^{2+}-induced Ca^{2+} release** (CICR) because the myocardial RyR channels open in response to Ca^{2+} binding.

RUNNING PROBLEM

When Walter arrived at the University of Texas Southwestern Medical Center emergency room, one of the first tasks was to determine whether he had actually had a heart attack. A nurse took Walter's vital signs (pulse and breathing rates, blood pressure, and temperature) and he was given nitroglycerin to dilate his coronary blood vessels [p. 188]. The doctor was pleased to hear that Walter had taken an aspirin to decrease blood clotting. A technician then drew blood for enzyme assays to determine the level of cardiac creatine kinase (CK-MB) in Walter's blood. When heart muscle cells die, they release various enzymes such as creatine kinase that serve as markers of a heart attack. A second tube of blood was sent for an assay of its troponin I level. Troponin I (TnI) is a good indicator of heart damage following a heart attack.

Q3: A related form of creatine kinase, CK-MM, is found in skeletal muscle. What are related forms of an enzyme called? [*Hint:* p. 105]

Q4: What is troponin, and why would elevated blood levels of troponin indicate heart damage? [*Hint:* p. 407]

463 467 477 485 486 495 501

EC COUPLING IN CARDIAC MUSCLE

This figure shows the cellular events leading to contraction and relaxation in a cardiac contractile cell.

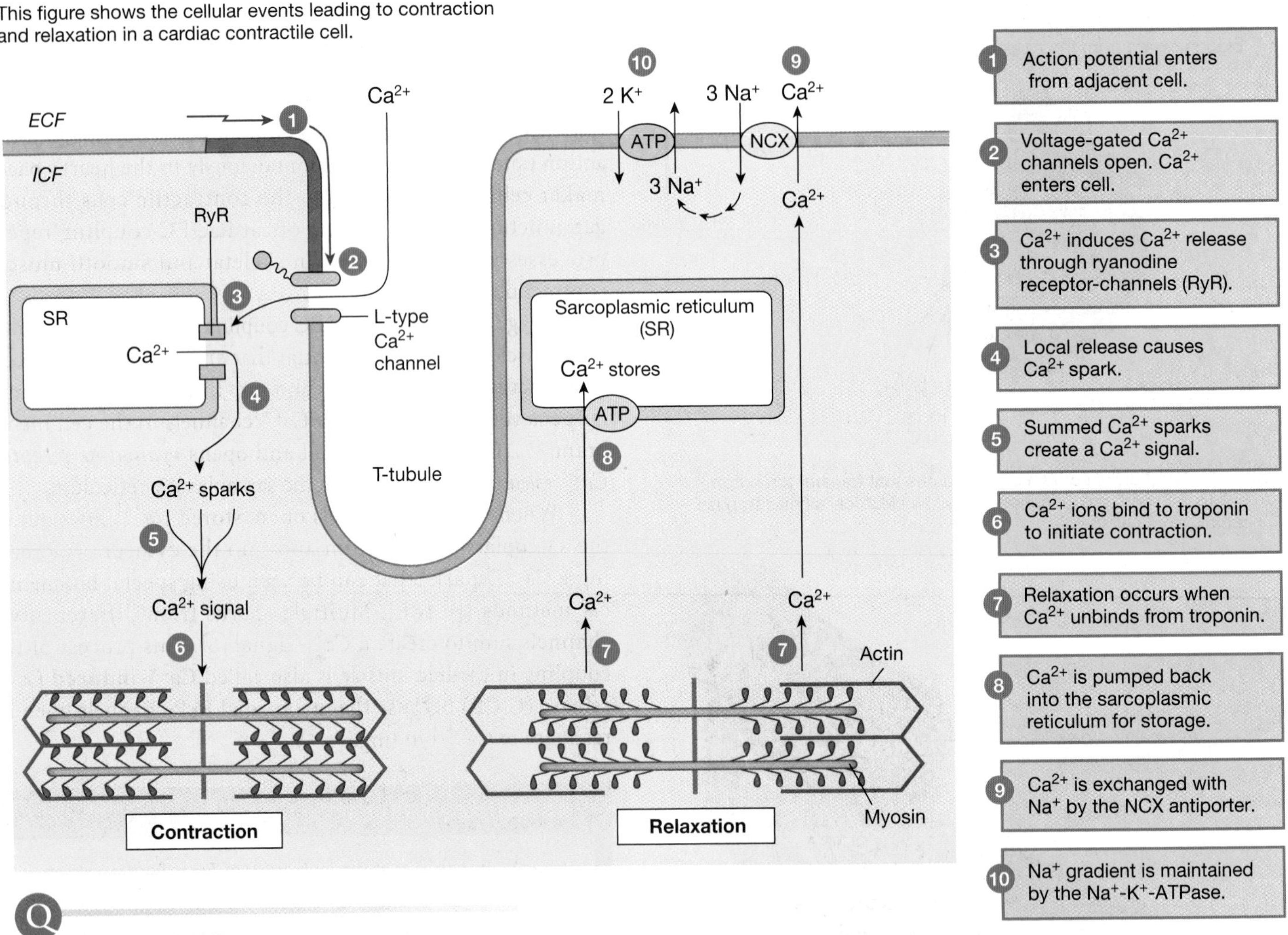

ECF
ICF
Ca2+
RyR
SR
Ca2+
L-type Ca2+ channel
T-tubule
Ca2+ sparks
Ca2+ signal
Contraction
2 K+
3 Na+
Ca2+
ATP
NCX
3 Na+
Ca2+
Sarcoplasmic reticulum (SR)
Ca2+ stores
ATP
Ca2+
Ca2+
Actin
Myosin
Relaxation
1 Action potential enters from adjacent cell.
2 Voltage-gated Ca2+ channels open. Ca2+ enters cell.
3 Ca2+ induces Ca2+ release through ryanodine receptor-channels (RyR).
4 Local release causes Ca2+ spark.
5 Summed Ca2+ sparks create a Ca2+ signal.
6 Ca2+ ions bind to troponin to initiate contraction.
7 Relaxation occurs when Ca2+ unbinds from troponin.
8 Ca2+ is pumped back into the sarcoplasmic reticulum for storage.
9 Ca2+ is exchanged with Na+ by the NCX antiporter.
10 Na+ gradient is maintained by the Na+-K+-ATPase.

FIGURE QUESTION

Using the numbered steps, compare the events shown to EC coupling in skeletal and smooth muscle [see Figs.12.10 and 12.26].

Fig. 14.9

Calcium released from the sarcoplasmic reticulum provides about 90% of the Ca^{2+} needed for muscle contraction, with the remaining 10% entering the cell from the extracellular fluid. Calcium diffuses through the cytosol to the contractile elements, where the ions bind to troponin and initiate the cycle of crossbridge formation and movement (6). Contraction takes place by the same type of sliding filament movement that occurs in skeletal muscle [p. 406].

Relaxation in cardiac muscle is generally similar to that in skeletal muscle. As cytoplasmic Ca^{2+} concentrations decrease, Ca^{2+} unbinds from troponin, myosin releases actin, and the contractile filaments slide back to their relaxed position (7). As in skeletal muscle, Ca^{2+} is transported back into the sarcoplasmic reticulum with the help of a Ca^{2+}-ATPase (8). However, in cardiac muscle Ca^{2+} is also removed from the cell in exchange for Na^+ via the *Na^+-Ca^{2+} exchanger* (NCX) (9). Each Ca^{2+} moves out of the cell against its electrochemical gradient in exchange for 3 Na^+ entering the cell down their electrochemical gradient. Sodium that enters the cell during this transfer is removed by the Na^+-K^+-ATPase (10).

Concept Check

Answers: p. 506

10. Compare the receptors and channels involved in cardiac EC coupling to those found in skeletal muscle EC coupling. [Hint: p. 406]

11. If a myocardial contractile cell is placed in interstitial fluid and depolarized, the cell contracts. If Ca^{2+} is removed from the fluid surrounding the myocardial cell and the cell is depolarized, it does not contract. If the experiment is repeated with a skeletal muscle fiber, the skeletal muscle contracts when depolarized, whether or not Ca^{2+} is present in the surrounding fluid. What conclusion can you draw from the results of this experiment?

Cardiac Muscle Contraction Can Be Graded

A key property of cardiac muscle cells is the ability of a single muscle fiber to execute *graded contractions,* in which the fiber varies the amount of force it generates. (Recall that in skeletal muscle, contraction in a single fiber is all-or-none at any given fiber length.) The force generated by cardiac muscle is proportional to the number of crossbridges that are active. The number of active crossbridges is determined by how much Ca^{2+} is bound to troponin.

If cytosolic Ca^{2+} concentrations are low, some crossbridges are not activated and contraction force is small. If additional Ca^{2+} enters the cell from the extracellular fluid, more Ca^{2+} is released from the sarcoplasmic reticulum. This additional Ca^{2+} binds to troponin, enhancing the ability of myosin to form crossbridges with actin and creating additional force.

Another factor that affects the force of contraction in cardiac muscle is the sarcomere length at the beginning of contraction. In the intact heart, stretch on the individual fibers is a function of how much blood is in the chambers of the heart. The relationship between force and ventricular volume is an important property of cardiac function, and we discuss it in detail later in this chapter.

Concept Check Answer: p. 506

12. A drug that blocks all Ca^{2+} channels in the myocardial cell membrane is placed in the solution around the cell. What happens to the force of contraction in that cell?

Myocardial Action Potentials Vary

Cardiac muscle, like skeletal muscle and neurons, is an excitable tissue with the ability to generate action potentials. Each of the two types of cardiac muscle cells has a distinctive action potential. In both types, Ca^{2+} plays an important role in the action potential, in contrast to the action potentials of skeletal muscle and neurons.

Myocardial Contractile Cells The action potentials of myocardial contractile cells are similar in several ways to those of neurons and skeletal muscle [p. 254]. The rapid depolarization phase of the action potential is the result of Na^+ entry, and the steep repolarization phase is due to K^+ leaving the cell (■ Fig. 14.10). The main difference between the action potential of the myocardial contractile cell and those of skeletal muscle fibers and neurons is that the myocardial cell has a longer action potential due to Ca^{2+} entry. Let's take a look at these longer action potentials. By convention, the action potential phases start with zero.

Phase 4: resting membrane potential. Myocardial contractile cells have a stable resting potential of about −90 mV.

Phase 0: depolarization. When a wave of depolarization moves into a contractile cell through gap junctions, the membrane potential becomes more positive. Voltage-gated Na^+ channels open, allowing Na^+ to enter the cell and rapidly depolarize it. The membrane potential reaches about +20 mV before the Na^+ channels close. These are double-gated Na^+ channels, similar to the voltage-gated Na^+ channels of the axon [p. 257].

ACTION POTENTIAL OF A CARDIAC CONTRACTILE CELL

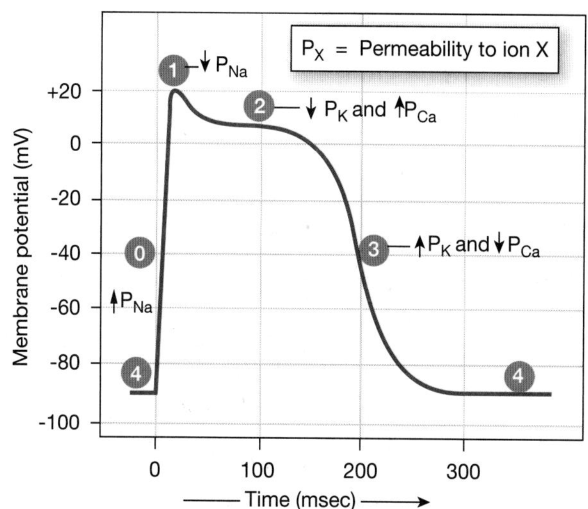

Phase*	Membrane channels
0	Na^+ channels open
1	Na^+ channels close
2	Ca^{2+} channels open; fast K^+ channels close
3	Ca^{2+} channels close; slow K^+ channels open
4	Resting potential

*The phase numbers are a convention.

FIGURE QUESTION

Compare ion movement during this action potential to ion movement of a neuron's action potential [Fig. 8.9, p. 256].

■ **Fig. 14.10**

Phase 1: initial repolarization. When the Na^+ channels close, the cell begins to repolarize as K^+ leaves through open K^+ channels.

Phase 2: the plateau. The initial repolarization is very brief. The action potential then flattens into a plateau as the result of two events: a decrease in K^+ permeability and an increase in Ca^{2+} permeability. Voltage-gated Ca^{2+} channels activated by depolarization have been slowly opening during phases 0 and 1. When they finally open, Ca^{2+} enters the cell. At the same time, some "fast" K^+ channels close. The combination of Ca^{2+} influx and decreased K^+ efflux causes the action potential to flatten out into a plateau.

Phase 3: rapid repolarization. The plateau ends when Ca^{2+} channels close and K^+ permeability increases once more. The "slow" K^+ channels responsible for this phase are similar to

REFRACTORY PERIODS AND SUMMATION

Summation in skeletal muscle leads to tetanus, which would be fatal if it happened in the heart.

Cardiac Muscle

(a) Cardiac muscle fiber: The refractory period lasts almost as long as the entire muscle twitch.

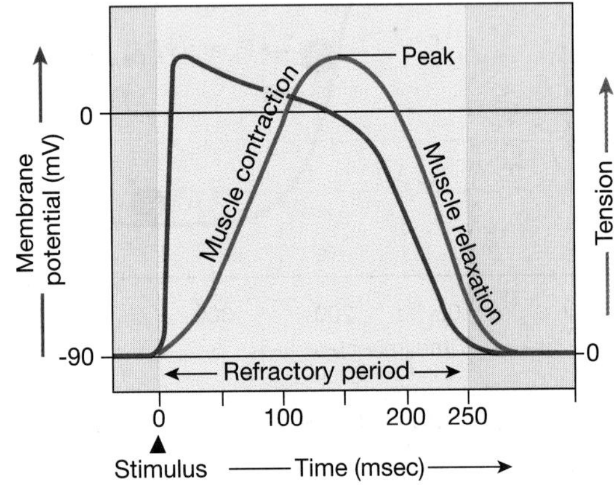

(b) Long refractory period in a cardiac muscle prevents tetanus.

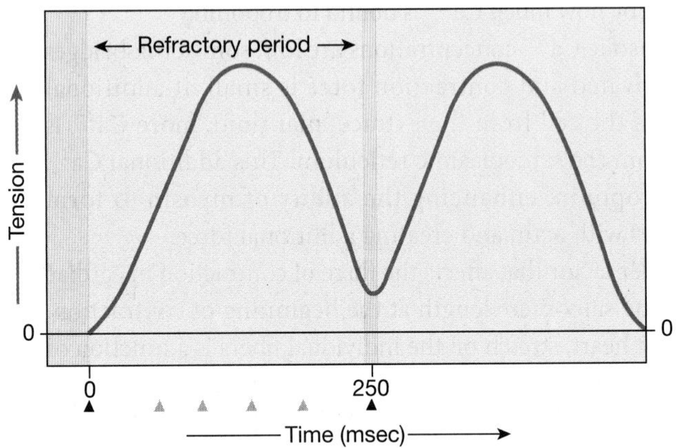

Skeletal Muscle

(c) Skeletal muscle fast-twitch fiber: The refractory period (yellow) is very short compared with the amount of time required for the development of tension.

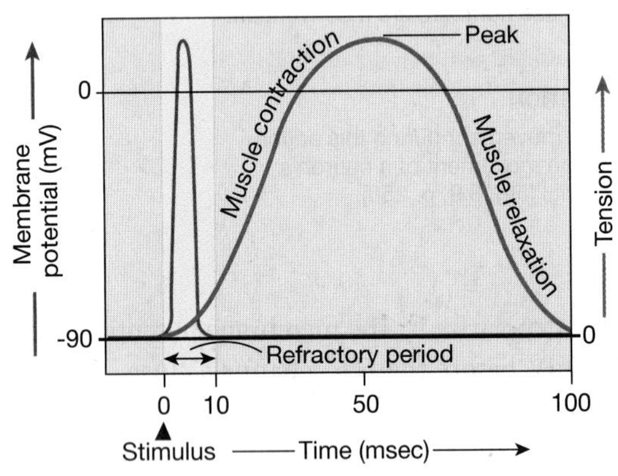

(d) Skeletal muscles that are stimulated repeatedly will exhibit summation and tetanus (action potentials not shown).

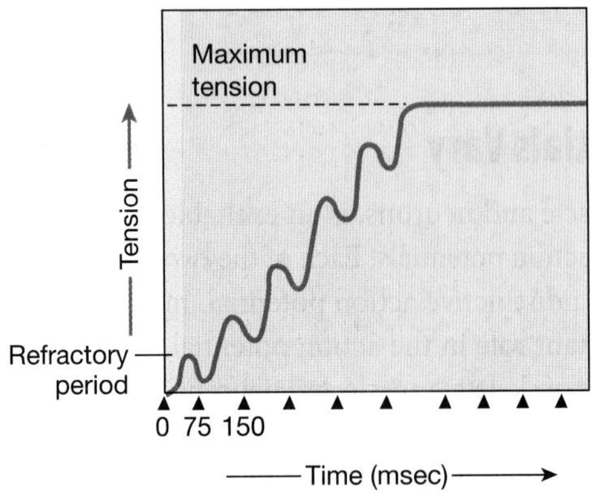

KEY

▲ = Stimulus for action potential

— = Action potential (mV)

— = Muscle tension

Fig. 14.11

those in the neuron: they are activated by depolarization but are slow to open. When the slow K^+ channels open, K^+ exits rapidly, returning the cell to its resting potential (phase 4).

The influx of Ca^{2+} during phase 2 lengthens the total duration of a myocardial action potential. A typical action potential in a neuron or skeletal muscle fiber lasts between 1 and 5 msec. In a contractile myocardial cell, the action potential typically lasts 200 msec or more.

The longer myocardial action potential helps prevent the sustained contraction called tetanus. Prevention of tetanus in the heart is important because cardiac muscles must relax between contractions so the ventricles can fill with blood. To understand how a longer action potential prevents tetanus, let's compare the relationship between action potentials, refractory periods [p. 258], and contraction in skeletal and cardiac muscle cells (■ Fig. 14.11).

As you may recall, the *refractory period* is the time following an action potential during which a normal stimulus cannot trigger a second action potential. In cardiac muscle, the long action potential (red curve) means the refractory period (yellow background) and the contraction (blue curve) end almost simultaneously (Fig. 14.11a). By the time a second action potential can take place, the myocardial cell has almost completely relaxed. Consequently, no summation occurs (Fig. 14.11b).

In contrast, the skeletal muscle action potential and refractory period are ending just as contraction begins (Fig. 14.11c).

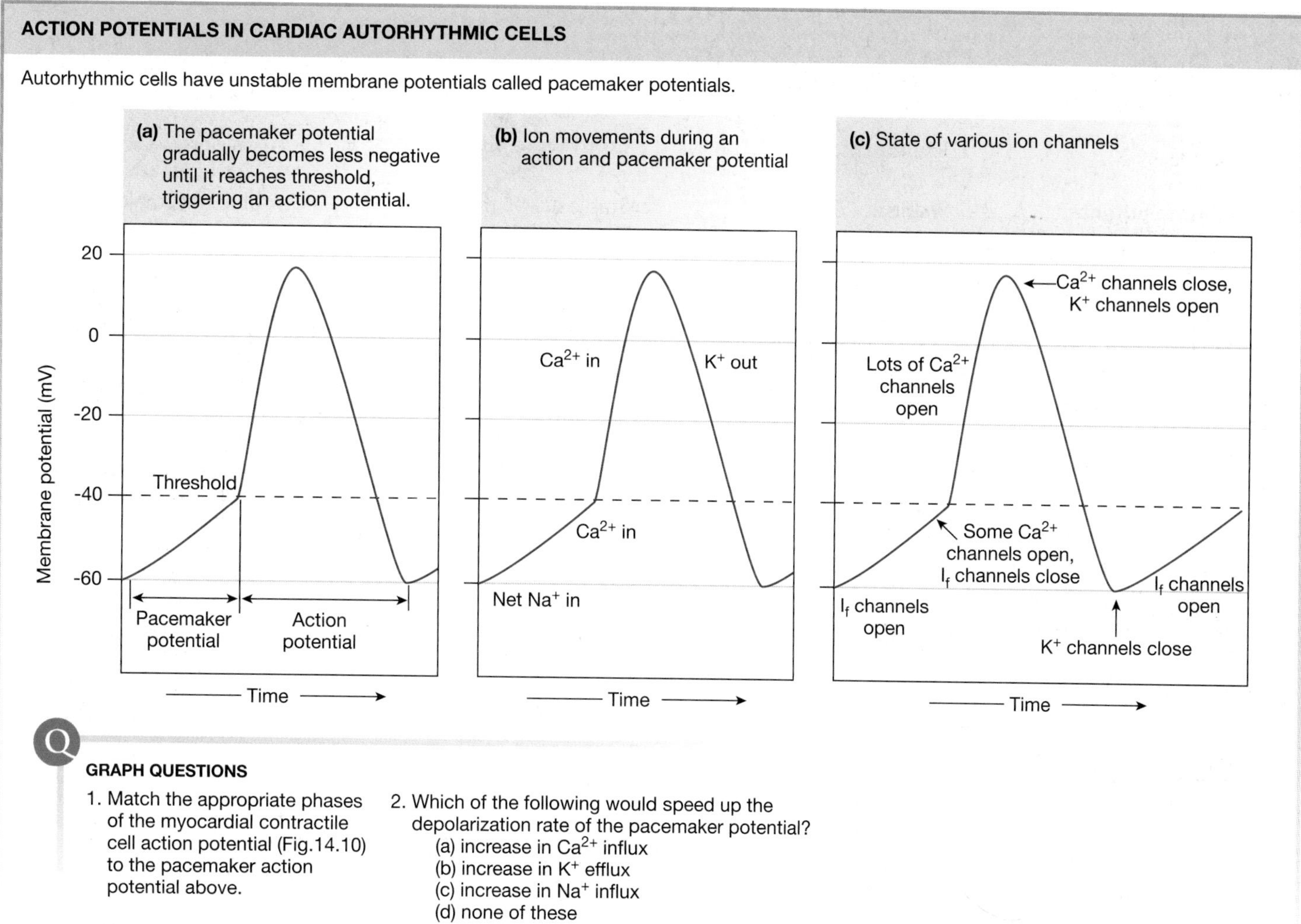

ACTION POTENTIALS IN CARDIAC AUTORHYTHMIC CELLS
Autorhythmic cells have unstable membrane potentials called pacemaker potentials.
(a) The pacemaker potential gradually becomes less negative until it reaches threshold, triggering an action potential.
(b) Ion movements during an action and pacemaker potential
(c) State of various ion channels
Membrane potential (mV)
20
0
-20
-40
-60
Threshold
Pacemaker potential
Action potential
Time
Ca2+ in
K+ out
Ca2+ in
Net Na+ in
Ca2+ channels close, K+ channels open
Lots of Ca2+ channels open
Some Ca2+ channels open, If channels close
If channels open
If channels open
K+ channels close
GRAPH QUESTIONS
1. Match the appropriate phases of the myocardial contractile cell action potential (Fig.14.10) to the pacemaker action potential above.
2. Which of the following would speed up the depolarization rate of the pacemaker potential?
(a) increase in Ca2+ influx
(b) increase in K+ efflux
(c) increase in Na+ influx
(d) none of these

Fig. 14.12

For this reason, a second action potential fired immediately after the refractory period causes summation of the contractions (Fig. 14.11d). If a series of action potentials occurs in rapid succession, the sustained contraction known as tetanus results.

> **Concept Check** Answers: p. 506
>
> **13.** Which ions moving in what directions cause the depolarization and repolarization phases of a neuronal action potential?
>
> **14.** At the molecular level, what is happening during the refractory period in neurons and muscle fibers?
>
> **15.** Lidocaine is a molecule that blocks the action of voltage-gated cardiac Na^+ channels. What happens to the action potential of a myocardial contractile cell if lidocaine is applied to the cell?

Myocardial Autorhythmic Cells What gives myocardial autorhythmic cells their unique ability to generate action potentials spontaneously in the absence of input from the nervous system? This ability results from their unstable membrane potential, which starts at −60 mV and slowly drifts upward toward threshold (Fig. 14.12a). This unstable membrane potential is called a **pacemaker potential** rather than a resting membrane potential because it never "rests" at a constant value. Whenever a pacemaker potential depolarizes to threshold, the autorhythmic cell fires an action potential.

What causes the membrane potential of these cells to be unstable? Our current understanding is that the autorhythmic cells contain channels that are different from the channels of other excitable tissues. When the cell membrane potential is −60 mV, **I_f channels** that are permeable to both K^+ and Na^+ open (Fig. 14.12c). These channels are called I_f channels because they allow current (I) to flow and because of their unusual properties. The researchers who first described the ion current through these channels initially did not understand its behavior and named it *funny* current—hence the subscript *f*. The I_f channels belong to the family of *HCN channels*, or *hyperpolarization-activated cyclic nucleotide-gated channels*. Other members of the HCN family are found in neurons.

When I_f channels open at negative membrane potentials, Na^+ influx exceeds K^+ efflux. (This is similar to what happens at the neuromuscular junction when nonspecific cation channels open [p. 391].) The net influx of positive charge slowly

Table 14.3 Comparison of Action Potentials in Cardiac and Skeletal Muscle

	Skeletal Muscle	Contractile Myocardium	Autorhythmic Myocardium
Membrane potential	Stable at −70 mV	Stable at −90 mv	Unstable pacemaker potential; usually starts at −60 mV
Events leading to threshold potential	Net Na^+ entry through ACh-operated channels	Depolarization enters via gap junctions	Net Na^+ entry through I_f channels; reinforced by Ca^{2+} entry
Rising phase of action potential	Na^+ entry	Na^+ entry	Ca^{2+} entry
Repolarization phase	Rapid; caused by K^+ efflux	Extended plateau caused by Ca^{2+} entry; rapid phase caused by K^+ efflux	Rapid; caused by K^+ efflux
Hyperpolarization	Due to excessive K^+ efflux at high K^+ permeability. When K^+ channels close; leak of K^+ and Na^+ restores potential to resting state	None; resting potential is −90 mV, the equilibrium potential for K^+	Normally none; when repolarization hits −60 mV, the I_f channels open again. ACh can hyperpolarize the cell.
Duration of action potential	Short: 1–2 msec	Extended: 200+ msec	Variable; generally 150 + msec
Refractory period	Generally brief	Long because resetting of Na^+ channel gates delayed until end of action potential	Not significant in normal function

depolarizes the autorhythmic cell (Fig. 14.12b). As the membrane potential becomes more positive, the I_f channels gradually close and one set of Ca^{2+} channels opens. The resulting influx of Ca^{2+} continues the depolarization, and the membrane potential moves steadily toward threshold.

When the membrane potential reaches threshold, a different set of voltage-gated Ca^{2+} channels open. Calcium rushes into the cell, creating the steep depolarization phase of the action potential. Note that this process is different from that in other excitable cells, in which the depolarization phase is due to the opening of voltage-gated Na^+ channels.

When the Ca^{2+} channels close at the peak of the action potential, slow K^+ channels have opened (Fig. 14.12c). The repolarization phase of the autorhythmic action potential is due to the resultant efflux of K^+ (Fig. 14.12b). This phase is similar to repolarization in other types of excitable cells.

The speed with which pacemaker cells depolarize determines the rate at which the heart contracts (the heart rate). The interval between action potentials can be modified by altering the permeability of the autorhythmic cells to different ions, which in turn changes the duration of the pacemaker potential. This topic is discussed in detail at the end of the chapter.

■ Table 14.3 compares action potentials of the two types of myocardial muscle with those of skeletal muscle. Next we look at how action potentials of autorhythmic cells spread throughout the heart to coordinate contraction.

Concept Check

Answers: p. 506

16. What does increasing K^+ permeability do to the membrane potential of the cell?
17. A new cardiac drug called *ivabradine* selectively blocks I_f channels in the heart. What effect would it have on heart rate and for what medical condition might it be used?
18. Do you think that the Ca^{2+} channels in autorhythmic cells are the same as the Ca^{2+} channels in contractile cells? Defend your answer.
19. What happens to the action potential of a myocardial autorhythmic cell if tetrodotoxin, which blocks voltage-gated Na^+ channels, is applied to the cell?
20. In an experiment, the *vagus nerve*, which carries parasympathetic signals to the heart, was cut. The investigators noticed that heart rate increased. What can you conclude about the vagal neurons that innervate the heart?

The Heart As a Pump

We now turn from single myocardial cells to the intact heart. How can one tiny noncontractile autorhythmic cell cause the entire heart to beat? And why do those doctors on TV shows shock patients with electric paddles when their hearts malfunction? You're about to learn the answers to these questions.

Electrical Signals Coordinate Contraction

A simple way to think of the heart is to imagine a group of people around a stalled car. One person can push on the car, but it's not likely to move very far unless everyone pushes together. In the same way, individual myocardial cells must depolarize and contract in a coordinated fashion if the heart is to create enough force to circulate the blood.

Electrical communication in the heart begins with an action potential in an autorhythmic cell. The depolarization spreads rapidly to adjacent cells through gap junctions in the intercalated disks (■ Fig. 14.13). The depolarization wave is followed by a wave of contraction that passes across the atria, then moves into the ventricles.

The depolarization begins in the **sinoatrial node (SA node),** autorhythmic cells in the right atrium that serve as the main pacemaker of the heart (■ Fig. 14.14). The depolarization wave then spreads rapidly through a specialized conducting system of noncontractile autorhythmic fibers. A branched **internodal pathway** connects the SA node to the **atrioventricular node (AV node)**, a group of autorhythmic cells near the floor of the right atrium.

From the AV node, the depolarization moves into the ventricles. **Purkinje fibers**, specialized conducting cells, transmit electrical signals very rapidly down the **atrioventricular bundle (AV bundle,** also called the **bundle of His** ("hiss")) in the ventricular septum. A short way down the septum, the AV bundle fibers divide into left and right **bundle branches**. The bundle branch fibers continue downward to the apex of the heart, where they divide into smaller Purkinje fibers that spread outward among the contractile cells.

The electrical signal for contraction begins when the SA node fires an action potential and the depolarization spreads to adjacent cells through gap junctions (Fig. 14.14 (1)). Electrical conduction is rapid through the internodal conducting pathways (2) but slower through the contractile cells of the atria (3).

As action potentials spread across the atria, they encounter the fibrous skeleton of the heart at the junction of the atria and ventricles. This barricade prevents the transfer of electrical signals from the atria to the ventricles. Consequently, the AV node is the only pathway through which action potentials can reach the contractile fibers of the ventricles.

The electrical signal passes from the AV node through the AV bundle and bundle branches to the apex of the heart (Fig. 14.14 (4)). The Purkinje fibers transmit impulses very rapidly, with speeds up to 4 m/sec, so that all contractile cells in the apex contract nearly simultaneously (5).

Why is it necessary to direct the electrical signals through the AV node? Why not allow them to spread downward from the atria? The answer lies in the fact that blood is pumped out of the ventricles through openings at the top of the chambers (see Fig. 14.7a). If electrical signals from the atria were conducted directly into the ventricles, the ventricles would start contracting at the top. Then blood would be squeezed downward and would become trapped in the bottom of the ventricles

ELECTRICAL CONDUCTION IN MYOCARDIAL CELLS

Autorhythmic cells spontaneously fire action potentials. Depolarizations of the autorhythmic cells then spread rapidly to adjacent contractile cells through gap junctions.

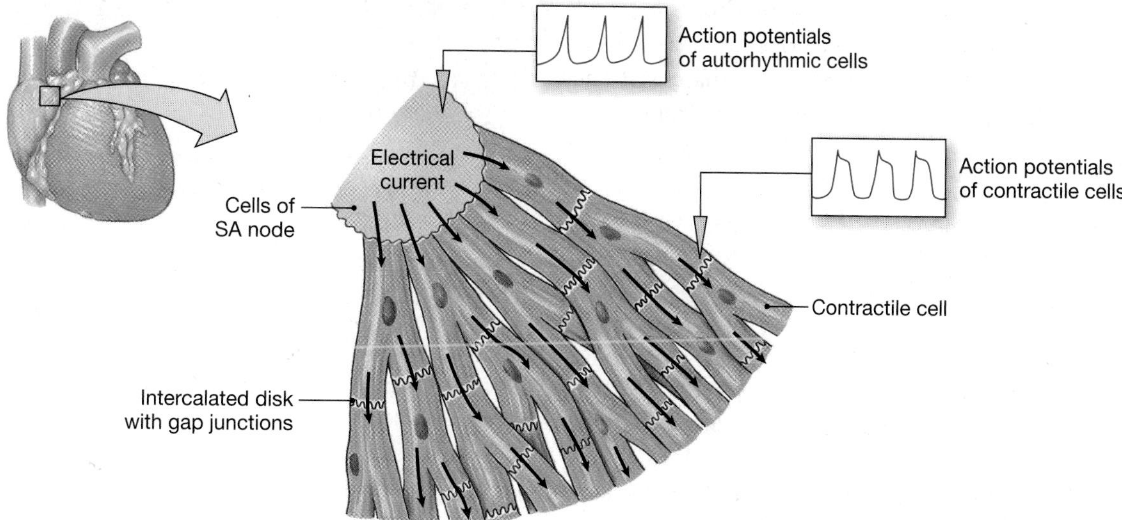

■ **Fig. 14.13**

THE CONDUCTING SYSTEM OF THE HEART

Electrical signaling begins in the SA node.

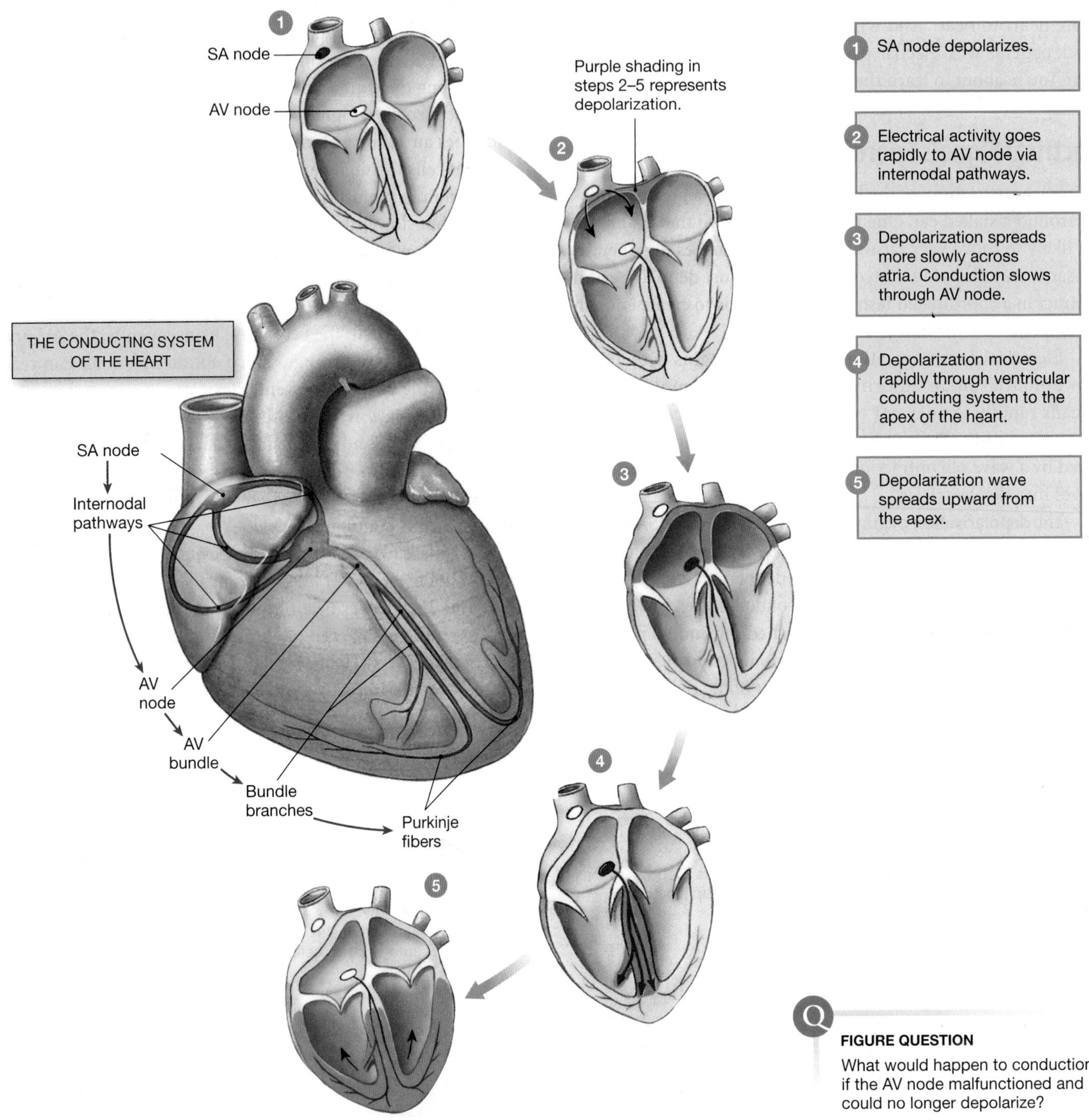
1
SA node
AV node
Purple shading in steps 2–5 represents depolarization.
2
3
4
5
1 SA node depolarizes.
2 Electrical activity goes rapidly to AV node via internodal pathways.
3 Depolarization spreads more slowly across atria. Conduction slows through AV node.
4 Depolarization moves rapidly through ventricular conducting system to the apex of the heart.
5 Depolarization wave spreads upward from the apex.
THE CONDUCTING SYSTEM OF THE HEART
SA node
Internodal pathways
AV node
AV bundle
Bundle branches
Purkinje fibers
Q
FIGURE QUESTION
What would happen to conduction if the AV node malfunctioned and could no longer depolarize?

Fig. 14.14

(think of squeezing a toothpaste tube at the top). The apex-to-base contraction squeezes blood toward the arterial openings at the base of the heart.

The ejection of blood from the ventricles is aided by the spiral arrangement of the muscles in the walls (see Fig. 14.8a). As these muscles contract, they pull the apex and base of the heart closer together, squeezing blood out the openings at the top of the ventricles.

A second function of the AV node is to slow down the transmission of action potentials slightly. This delay allows the atria to complete their contraction before ventricular contraction begins. **AV node delay** is accomplished by slower conduction of signals

RUNNING PROBLEM

The results of creatine kinase and troponin I assays may not come back from the laboratory for an hour. If a coronary artery were blocked, damage to the heart muscle could be severe by that time. In Walter's case, an electrocardiogram (ECG) showed an abnormal pattern of electrical activity. "He's definitely had an MI," said the ER physician, referring to a myocardial infarction, or heart attack. "Let's start him on a beta blocker and tPA." tPA (short for *tissue plasminogen activator*) activates plasminogen, a substance that is produced in the body and dissolves blood clots. Given within a couple of hours of a heart attack, tPA can help dissolve blood clots that are blocking blood flow to the heart muscle. This will help limit the extent of ischemic damage.

Q5: How do electrical signals move from cell to cell in the myocardium?

Q6: What happens to contraction in a myocardial contractile cell if a wave of depolarization passing through the heart bypasses it?

Q7: A beta blocker is an antagonist to β_1-adrenergic receptors. What will this drug do to Walter's heart rate? Why is that response helpful following a heart attack?

through the nodal cells. Action potentials here move at only 1/20 the rate of action potentials in the atrial internodal pathway.

Pacemakers Set the Heart Rate

The cells of the SA node set the pace of the heartbeat. Other cells in the conducting system, such as the AV node and the Purkinje fibers, have unstable resting potentials and can also act as pacemakers under some conditions. However, because their rhythm is slower than that of the SA node, they do not usually have a chance to set the heartbeat. The Purkinje fibers, for example, can spontaneously fire action potentials, but their firing rate is very slow, between 25 and 40 beats per minute.

Why does the fastest pacemaker determine the pace of the heartbeat? Consider the following analogy. A group of people are playing "follow the leader" as they walk. Initially, everyone is walking at a different pace—some fast, some slow. When the game starts, everyone must match his or her pace to the pace of the person who is walking the fastest. The fastest person in the group is the SA node, walking at 70 steps per minute. Everyone else in the group (autorhythmic and contractile cells) sees that the SA node is fastest, and so they pick up their pace and follow the leader. In the heart, the cue to follow the leader is the electrical signal sent from the SA node to the other cells.

14

Now suppose the SA node gets tired and drops out of the group. The role of leader defaults to the next fastest person, the AV node, who is walking at a rate of 50 steps per minute. The group slows to match the pace of the AV node, but everyone is still following the fastest walker.

What happens if the group divides? Suppose that when they reach a corner, the AV node leader goes left but a renegade Purkinje fiber decides to go right. Those people who follow the AV node continue to walk at 50 steps per minute, but the people who follow the Purkinje fiber slow down to match his pace of 35 steps per minute. Now there are two leaders, each walking at a different pace.

In the heart, the SA node is the fastest pacemaker and normally sets the heart rate. If this node is damaged and cannot function, one of the slower pacemakers in the heart takes over. Heart rate then matches the rate of the new pacemaker. It is even possible for different parts of the heart to follow different pacemakers, just as the walking group split at the corner.

In a condition known as *complete heart block,* the conduction of electrical signals from the atria to the ventricles through the AV node is disrupted. The SA node fires at its rate of 70 beats per minute, but those signals never reach the ventricles. So the ventricles coordinate with their fastest pacemaker. Because ventricular autorhythmic cells discharge only about 35 times a minute, the rate at which the ventricles contract is much slower than the rate at which the atria contract. If ventricular contraction is too slow to maintain adequate blood flow, it may be necessary

CLINICAL FOCUS

Fibrillation

Coordination of myocardial contraction is essential for normal cardiac function. In extreme cases in which the myocardial cells contract in a disorganized manner, a condition known as *fibrillation* results. Atrial fibrillation is a common condition, often without symptoms, that can lead to serious consequences (such as stroke) if not treated. Ventricular fibrillation, on the other hand, is an immediately life-threatening emergency because without coordinated contraction of the muscle fibers, the ventricles cannot pump enough blood to supply adequate oxygen to the brain. One way to correct this problem is to administer an electrical shock to the heart. The shock creates a depolarization that triggers action potentials in all cells simultaneously, coordinating them again. You have probably seen this procedure on television hospital shows, when a doctor places flat paddles on the patient's chest and tells everyone to stand back ("Clear!") while the paddles pass an electrical current through the body.

for the heart's rhythm to be set artificially by a surgically implanted mechanical pacemaker. These battery-powered devices artificially stimulate the heart at a predetermined rate.

Concept Check Answers: p. 506

21. Name two functions of the AV node. What is the purpose of AV node delay?
22. Where is the SA node located?
23. Occasionally an ectopic pacemaker {*ektopos*, out of place} develops in part of the heart's conducting system. What happens to heart rate if an ectopic atrial pacemaker depolarizes at a rate of 120 times per minute?

The Electrocardiogram Reflects Electrical Activity

At the end of the nineteenth century, physiologists discovered that they could place electrodes on the skin's surface and record the electrical activity of the heart. It is possible to use surface electrodes to record internal electrical activity because salt solutions, such as our NaCl-based extracellular fluid, are good conductors of electricity. These recordings, called **electrocardiograms** (ECGs or sometimes EKGs—from the Greek word *kardia*, meaning *heart*) show the summed electrical activity generated by all cells of the heart (■ Fig. 14.15a).

The first human electrocardiogram was recorded in 1887, but the procedure was not refined for clinical use until the first years of the twentieth century. The father of the modern ECG was a Dutch physiologist named Walter Einthoven. He named the parts of the ECG as we know them today and created "Einthoven's triangle," a hypothetical triangle created around the heart when electrodes are placed on both arms and the left leg (Fig. 14.15b). The sides of the triangle are numbered to correspond with the three *leads* ("leeds"), or pairs of electrodes, used for a recording.

An ECG is recorded from one lead at a time. One electrode acts as the positive electrode of a lead, and a second electrode acts as the negative electrode of the lead. (The third electrode is inactive). For example, in lead I, the left arm electrode is designated as positive and the right arm electrode is designated as negative. When an electrical wave moving through the heart is directed toward the positive electrode, the ECG wave goes up from the baseline (Fig, 14.15d). If net charge movement through the heart is toward the negative electrode, the wave points downward.

An ECG is not the same as a single action potential (Fig. 14.15e). An action potential is one electrical event in a single cell, recorded using an intracellular electrode. The ECG is an extracellular recording that represents the sum of multiple action potentials taking place in many heart muscle cells. In addition, the amplitudes of action potential and ECG recordings are very different. A ventricular action potential has a voltage change of 110 mV, for example, but the ECG signal has an amplitude of only 1 mV by the time it reaches the surface of the body.

There are two major components of an ECG: waves and segments (Fig. 14.15f). *Waves* appear as deflections above or below the baseline. *Segments* are sections of baseline between two waves. *Intervals* are combinations of waves and segments. Different waves of the ECG reflect depolarization or repolarization of the atria and ventricles.

Three major waves can be seen on a normal ECG recorded from lead I (Fig. 14.15f). The first wave is the **P wave,** which corresponds to depolarization of the atria. The next trio of waves, the **QRS complex**, represents the progressive wave of ventricular depolarization. The final wave, the **T wave**, represents the repolarization of the ventricles. Atrial repolarization is not represented by a special wave but is incorporated into the QRS complex.

One thing many people find confusing is that you cannot tell if an ECG recording represents depolarization or repolarization simply by looking at the shape of the waves relative to the baseline. For example, the P wave represents atrial depolarization and the T wave represents ventricular repolarization, but both the P wave and the T wave are deflections above the baseline. This is very different from the intracellular recordings of neurons and muscle fibers, in which an upward deflection always represents depolarization [Fig. 5.26, p. 166]. Remember that the direction of the ECG trace reflects only the direction of the current flow relative to the axis of the lead. Some waves even change direction in different leads.

Now let's follow an ECG through a single contraction-relaxation cycle, otherwise known as a **cardiac cycle** (■ Fig. 14.16 on page 490). Because depolarization initiates muscle contraction, the *electrical events* (waves) of an ECG can be associated with contraction or relaxation (collectively referred to as the *mechanical events* in the heart). The mechanical events of the cardiac cycle lag slightly behind the electrical signals, just as the contraction of a single cardiac muscle cell follows its action potential (see Fig. 14.11a).

RUNNING PROBLEM

The electrocardiogram indicated that Walter suffered a myocardial infarction, resulting from blockage of blood vessels nourishing the left ventricle. The exact location of the damage depends on which artery and which branch or branches have become occluded.

Q8: If the ventricle of the heart is damaged, in which wave or waves of the electrocardiogram would you expect to see abnormal changes?

463 467 477 485 **486** 495 501

The cardiac cycle begins with both atria and ventricles at rest. The ECG begins with atrial depolarization. Atrial contraction starts during the latter part of the P wave and continues during the P-R segment. During the P-R segment, the electrical signal is slowing down as it passes through the AV node and AV bundle (AV node delay).

Ventricular contraction begins just after the Q wave and continues through the T wave. The ventricles are repolarizing during the T wave, which is followed by ventricular relaxation. During the T-P segment the heart is electrically quiet.

An important point to remember is that an ECG is an electrical "view" of a three-dimensional object. This is one reason we use multiple leads to assess heart function. Think of looking at an automobile. From the air, it looks like a rectangle, but from the side and front it has different shapes. Not everything that you see from the front of the car can be seen from its side, and vice versa. In the same way, the leads of an ECG provide different electrical "views" and give information about different regions of the heart.

A 12-lead ECG is now the standard for clinical use. It is recorded using various combinations of the three limb electrodes plus another 6 electrodes placed on the chest and trunk. The additional leads provide detailed information about electrical conduction in the heart. Electrocardiograms are important diagnostic tools in medicine because they are quick, painless, and noninvasive (that is, do not puncture the skin).

An ECG provides information on heart rate and rhythm, conduction velocity, and even the condition of tissues in the heart. Thus, although obtaining an ECG is simple, interpreting some of its subtleties can be quite complicated. The interpretation of an ECG begins with the following questions (Fig. 14.15g).

1. *What is the heart rate?* Heart rate is normally timed either from the beginning of one P wave to the beginning of the next P wave or from the peak of one R wave to the peak of the next R wave. A normal resting heart rate is 60–100 beats per minute, although trained athletes often have slower heart rates at rest. A faster-than-normal rate is known as *tachycardia,* and a slower-than-normal rate is called *bradycardia* {*tachys,* swift; *bradys,* slow}.
2. *Is the rhythm of the heartbeat regular (that is, occurs at regular intervals) or irregular?* An irregular rhythm, or *arrhythmia* {*a-*, without + rhythm}, can result from a benign extra beat or from more serious conditions such as atrial fibrillation, in which the SA node has lost control of the pacemaking.
3. *Are all normal waves present in recognizable form?* After determining heart rate and rhythm, the next step in analyzing an ECG is to look at the individual waves. To help your analysis, you might want to write the letters above the P, R, and T waves.
4. *Is there one QRS complex for each P wave? If yes, is the P-R segment constant in length?* If not, a problem with conduction of signals through the AV node may exist. In heart block (the conduction problem mentioned earlier), action potentials from the SA node sometimes fail to be transmitted through the AV node to the ventricles. In these conditions, one or more P waves may occur without initiating a QRS complex. In the most severe (third-degree) form of heart block, the atria depolarize regularly at one pace while the ventricles contract at a much slower pace (Fig. 14.15h (2)).

The more difficult aspects of interpreting an ECG include looking for subtle changes, such as alterations in the shape, timing, or duration of various waves or segments. An experienced clinician can find signs pointing to changes in conduction velocity, enlargement of the heart, or tissue damage resulting from periods of ischemia. An amazing number of conclusions can be drawn about heart function simply by looking at alterations in the heart's electrical activity as recorded on an ECG.

Cardiac arrhythmias are a family of cardiac pathologies that range from benign to those with potentially fatal consequences. Arrhythmias are electrical problems that arise during the generation or conduction of action potentials through the heart, and they can usually be seen on an ECG. Some arrhythmias are "dropped beats" that result when the ventricles do not get their usual signal to contract. Other arrhythmias, such as *premature ventricular contractions* (*PVCs*), are extra beats that occur when an autorhythmic cell other than the SA node jumps in and fires an action potential out of sequence.

One interesting heart condition that can be observed on an ECG is *long QT syndrome* (*LQTS*), named for the change in the QT interval. LQTS has several forms. Some are inherited channelopathies, in which mutations occur in myocardial Na^+ or K^+ channels [p. 251]. In another form of LQTS, the ion channels are normal but the protein *ankyrin-B* that anchors the channels to the cell membrane is defective.

Iatrogenic (physician-caused) forms of LQTS can occur as a side effect of taking certain medications. One well-publicized incident occurred in the 1990s when patients took a non-sedating antihistamine called terfenadine (Seldane®) that binds to K^+ repolarization channels. After at least eight deaths were attributed to the drug, the U.S. Food and Drug Administration removed Seldane from the market.

The Heart Contracts and Relaxes during a Cardiac Cycle

Each cardiac cycle has two phases: **diastole**, the time during which cardiac muscle relaxes, and **systole**, the time during which the muscle contracts {*diastole,* dilation; *systole,* contraction}. Because the atria and ventricles do not contract and relax at the same time, we discuss atrial and ventricular events separately.

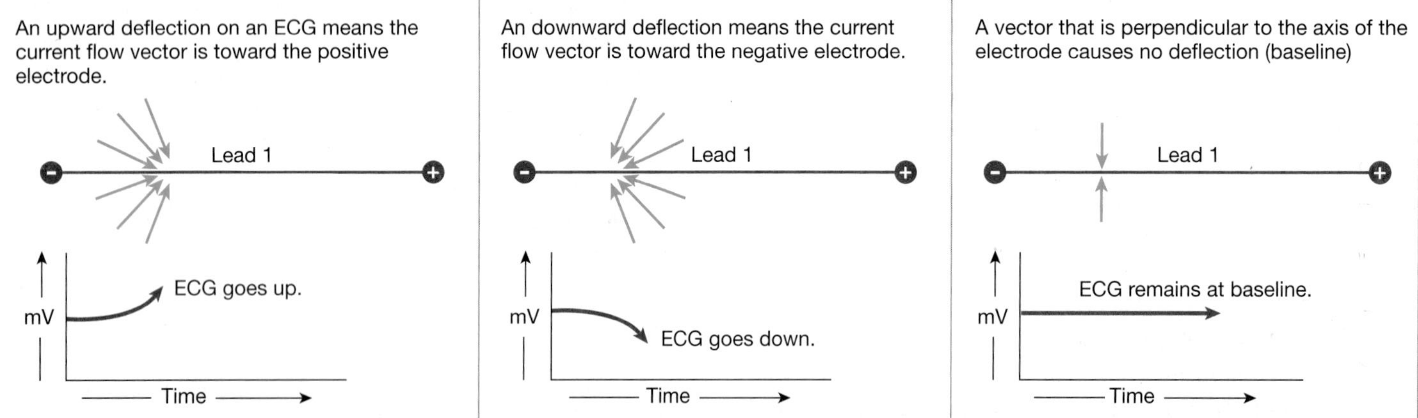
An upward deflection on an ECG means the current flow vector is toward the positive electrode.
Lead 1
mV
ECG goes up.
Time
An downward deflection means the current flow vector is toward the negative electrode.
Lead 1
mV
ECG goes down.
Time
A vector that is perpendicular to the axis of the electrode causes no deflection (baseline)
Lead 1
mV
ECG remains at baseline.
Time

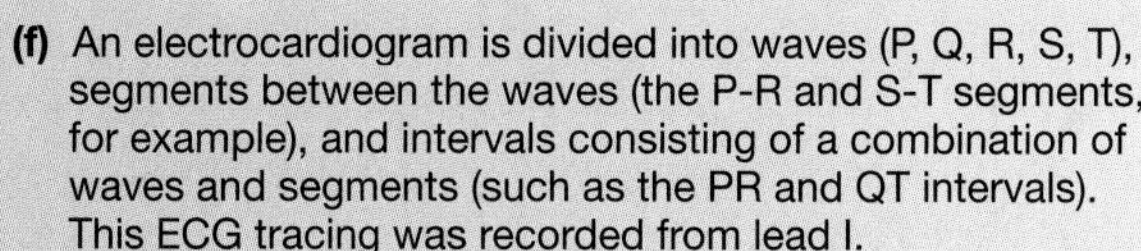

(f) An electrocardiogram is divided into waves (P, Q, R, S, T), segments between the waves (the P-R and S-T segments, for example), and intervals consisting of a combination of waves and segments (such as the PR and QT intervals). This ECG tracing was recorded from lead I.

P wave: atrial depolarization

P-R segment: conduction through AV node and AV bundle

QRS complex: ventricular depolarization

T wave: ventricular repolarization

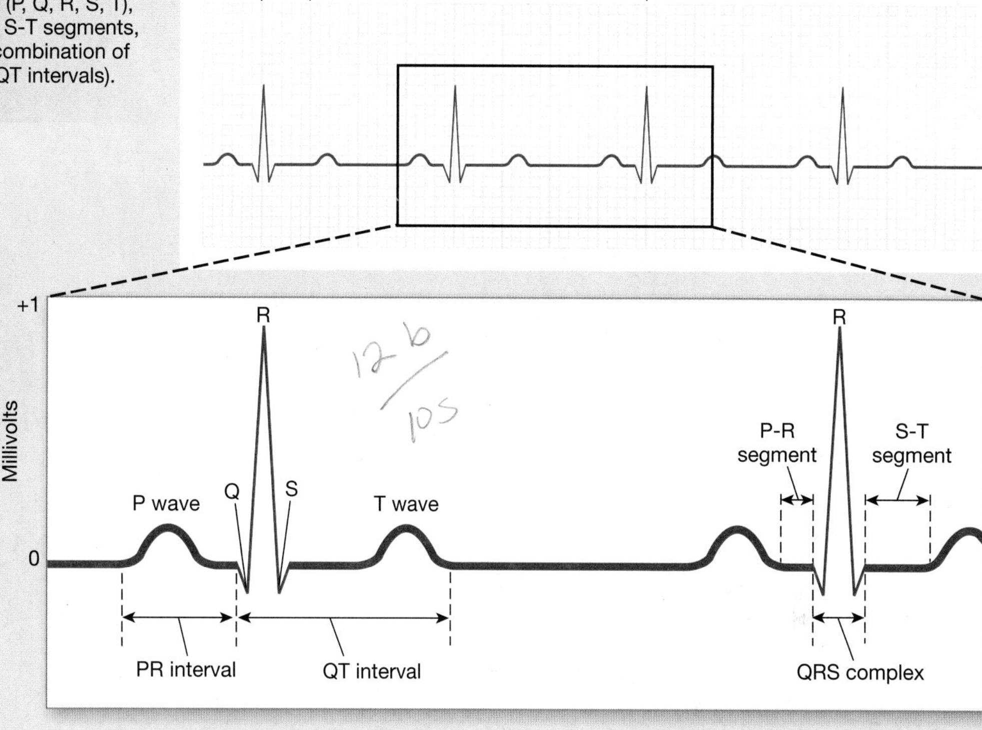

FIGURE QUESTION

1. If the ECG records at a speed of 25 mm/sec, what is the heart rate of the person? (1 little square = 1 mm)

(g)

QUESTIONS TO ASK WHEN ANALYZING ECG TRACINGS:

1. What is the rate? Is it within the normal range of 60–100 beats per minute?
2. Is the rhythm regular?
3. Are all normal waves present in recognizable form?
4. Is there one QRS complex for each P wave? If yes, is the P-R segment constant in length? If there is not one QRS complex for each P wave, count the heart rate using the P waves, then count it according to the R waves. Are the rates the same? Which wave would agree with the pulse felt at the wrist?

FIGURE QUESTIONS

2. Three abnormal ECGs are shown at right. Study them and see if you can relate the ECG changes to disruption of the normal electrical conduction pattern in the heart.
3. Identify the waves on the ECG in part (5). Look at the pattern of their occurrence and describe what has happened to electrical conduction in the heart.

(h) Normal and abnormal ECGs. All tracings represent 10-sec recordings.

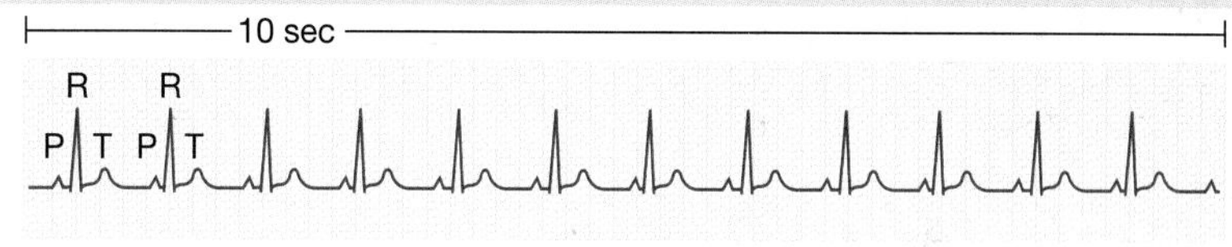

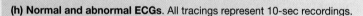

(1) Normal ECG

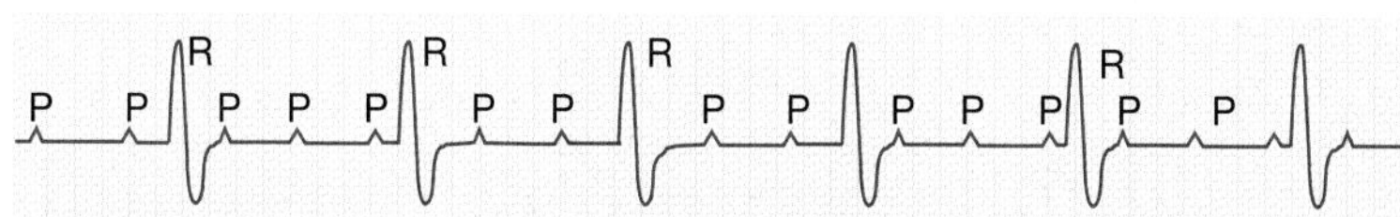

(2) Third-degree block

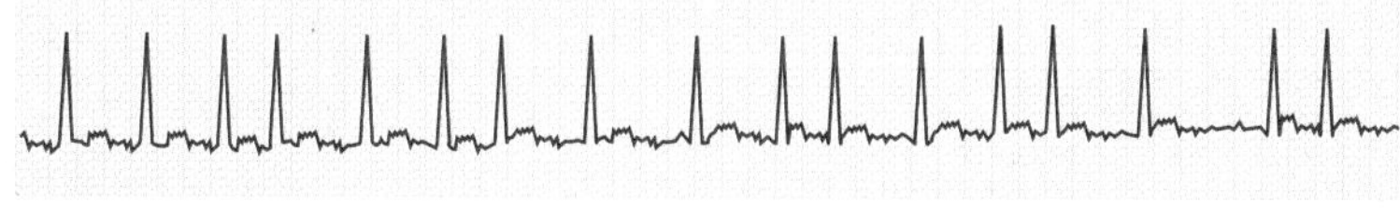

(3) Atrial fibrillation

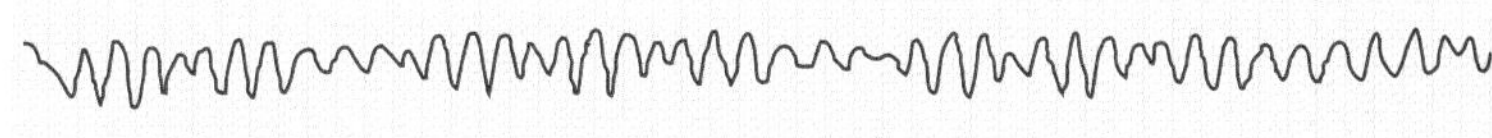

(4) Ventricular fibrillation

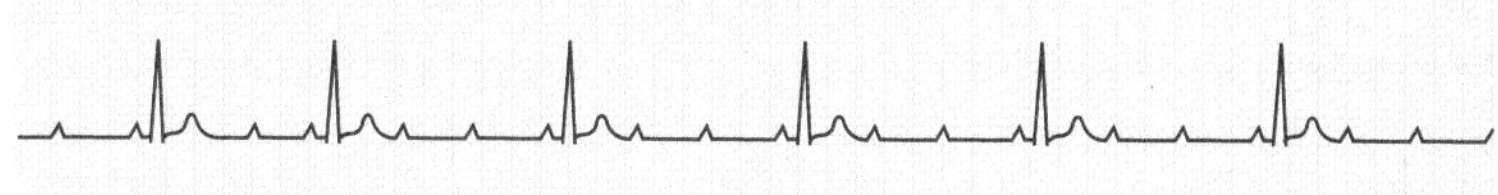

(5) Analyze this abnormal ECG.

CORRELATION BETWEEN AN ECG AND ELECTRICAL EVENTS IN THE HEART

The figure shows the correspondence between electrical events in the ECG and depolarizing (purple) and repolarizing (peach) regions of the heart.

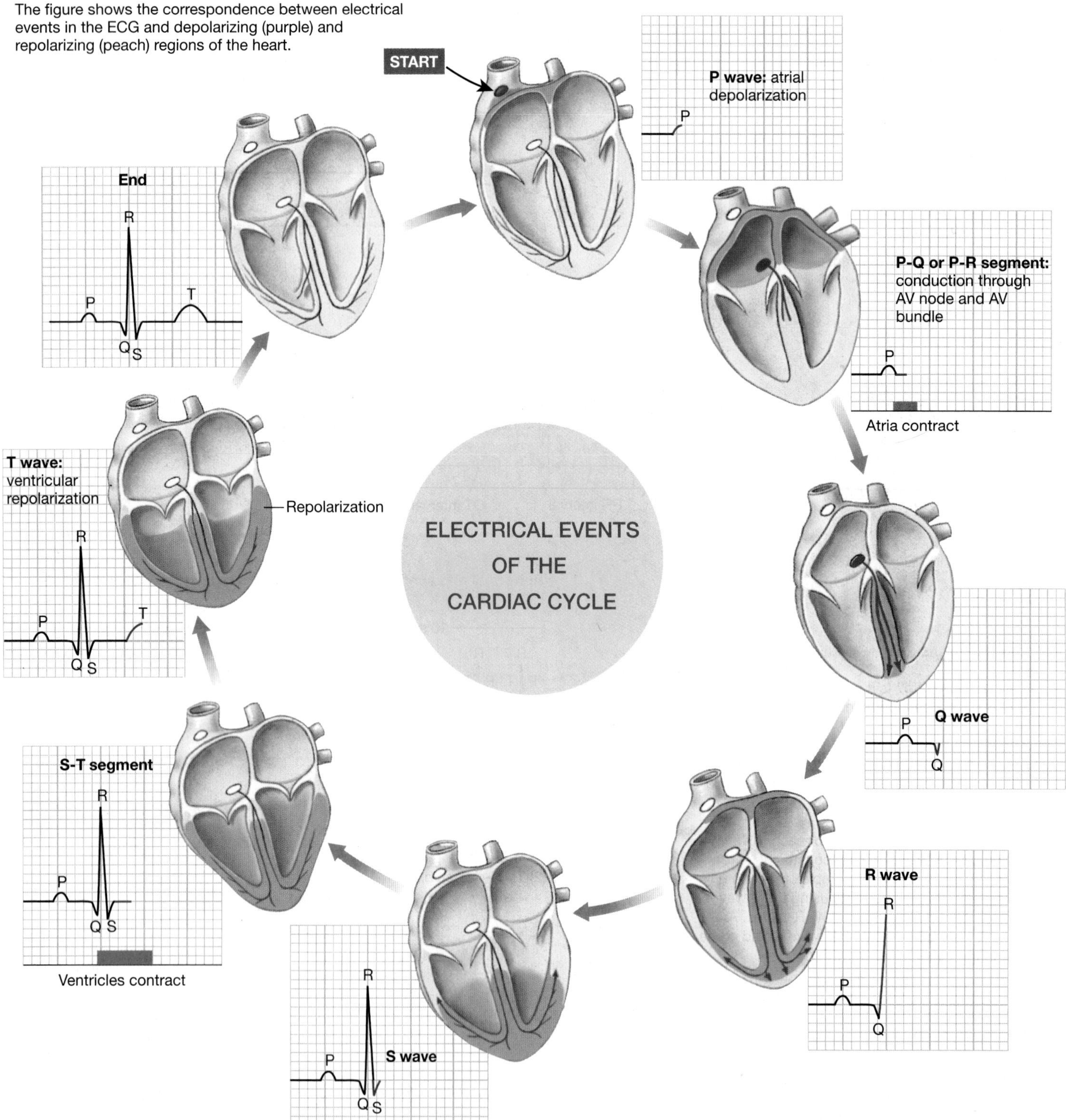

■ Fig. 14.16

In thinking about blood flow during the cardiac cycle, remember that blood flows from an area of higher pressure to one of lower pressure, and that contraction increases pressure while relaxation decreases pressure. In this discussion, we divide the cardiac cycle into the five phases shown in ■ Figure 14.17a:

1 **The heart at rest: atrial and ventricular diastole**. We enter the cardiac cycle at the brief moment during which both the atria and the ventricles are relaxing. The atria are filling with blood from the veins, and the ventricles have just completed a contraction. As the ventricles relax,

MECHANICAL EVENTS OF THE CARDIAC CYCLE

(a) The heart cycles between contraction (systole) and relaxation (diastole).

START

1. **Late diastole—**both sets of chambers are relaxed and ventricles fill passively.

2. **Atrial systole—**atrial contraction forces a small amount of additional blood into ventricles.

3. **Isovolumic ventricular contraction—**first phase of ventricular contraction pushes AV valves closed but does not create enough pressure to open semilunar valves.

4. **Ventricular ejection—**as ventricular pressure rises and exceeds pressure in the arteries, the semilunar valves open and blood is ejected.

5. **Isovolumic ventricular relaxation—**as ventricles relax; pressure in ventricles falls, blood flows back into cusps of semilunar valves and snaps them closed.

Ventricular diastole
Atrial systole
S_1
S_2
Atrial diastole
Ventricular systole

(b) Left ventricular pressure-volume changes during one cardiac cycle. This pressure-volume curve represents one cardiac cycle. Moving around the curve from A to B, C, D and back to A represents time passing as the heart fills with blood, then contracts.

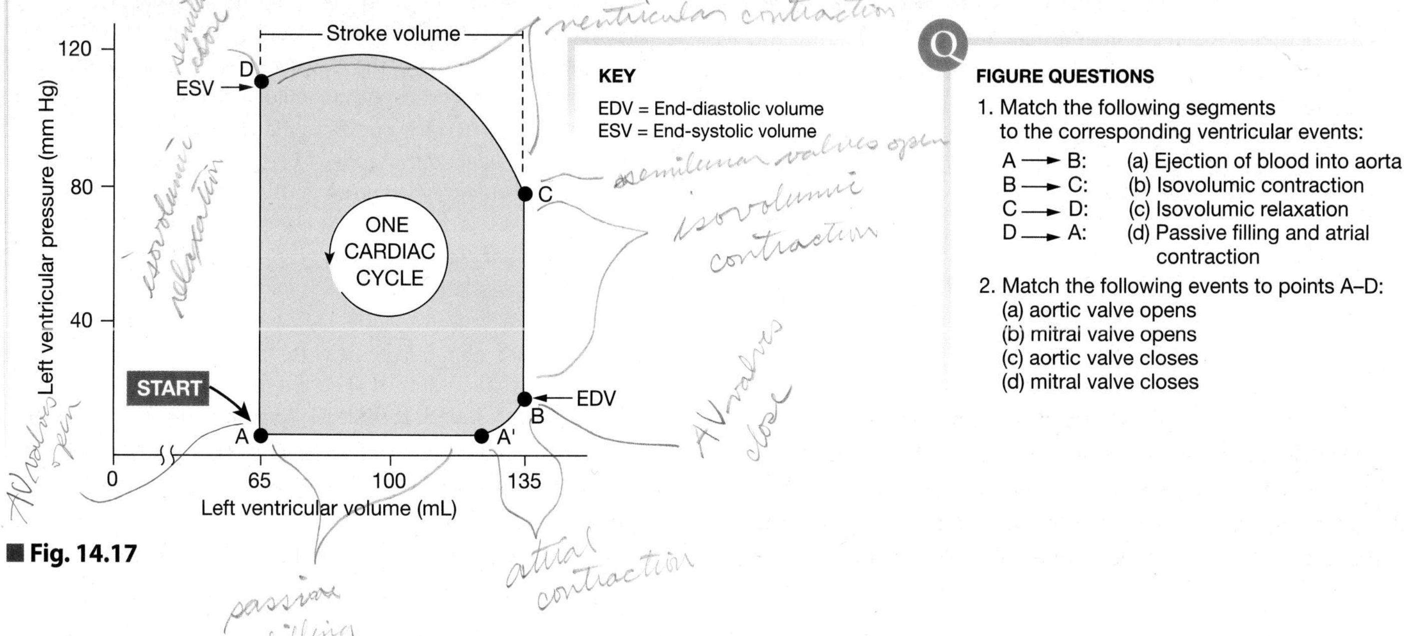

FIGURE QUESTIONS

1. Match the following segments to the corresponding ventricular events:

A → B:	(a) Ejection of blood into aorta
B → C:	(b) Isovolumic contraction
C → D:	(c) Isovolumic relaxation
D → A:	(d) Passive filling and atrial contraction

2. Match the following events to points A–D:
   (a) aortic valve opens
   (b) mitral valve opens
   (c) aortic valve closes
   (d) mitral valve closes

Fig. 14.17

the AV valves between the atria and ventricles open. Blood flows by gravity from the atria into the ventricles. The relaxing ventricles expand to accommodate the entering blood.

Concept Check Answer: p. 506

24. During atrial filling, is pressure in the atrium higher or lower than pressure in the venae cavae?

2 **Completion of ventricular filling: atrial systole.** Although most blood enters the ventricles while the atria are relaxed, the last 20% of filling is accomplished when the atria contract and push blood into the ventricles. (This applies to a normal person at rest. When heart rate increases, as during exercise, atrial contraction plays a greater role in ventricular filling). Atrial systole, or contraction, begins following the wave of depolarization that sweeps across the atria. The pressure increase that accompanies contraction pushes blood into the ventricles.

A small amount of blood is forced backward into the veins because there are no one-way valves to block backward flow, although the openings of the veins do narrow during contraction. This retrograde movement of blood back into the veins may be observed as a pulse in the jugular vein of a normal person who is lying with the head and chest elevated about 30°. (Look in the hollow formed where the sternocleidomastoid muscle runs under the clavicle.) An observable jugular pulse higher on the neck of a person sitting upright is a sign that pressure in the right atrium is higher than normal.

3 **Early ventricular contraction and the first heart sound.** As the atria are contracting, the depolarization wave is moving slowly through the conducting cells of the AV node, then rapidly down the Purkinje fibers to the apex of the heart. Ventricular systole begins there as spiral bands of muscle squeeze the blood upward toward the base. Blood pushing against the underside of the AV valves forces them closed so that blood cannot flow back into the atria. Vibrations following closure of the AV valves create the **first heart sound, S_1**, the "lub" of "lub-dup."

With both sets of AV and semilunar valves closed, blood in the ventricles has nowhere to go. Nevertheless, the ventricles continue to contract, squeezing on the blood in the same way that you might squeeze a water balloon in your hand. This is similar to an isometric contraction, in which muscle fibers create force without movement [p. 421]. To return to the toothpaste tube analogy, it is like squeezing the tube with the cap on: high pressure develops within the tube, but the toothpaste has nowhere to go. This phase is called **isovolumic ventricular contraction** {*iso-*, equal}, to underscore the fact that the volume of blood in the ventricle is not changing.

While the ventricles begin to contract, the atrial muscle fibers are repolarizing and relaxing. When atrial pressure falls below that in the veins, blood flows from the veins into the atria again. Closure of the AV valves isolates the upper and lower cardiac chambers, meaning that atrial filling is independent of events taking place in the ventricles.

4 **The heart pumps: ventricular ejection.** As the ventricles contract, they generate enough pressure to open the semilunar valves and push blood into the arteries. The pressure created by ventricular contraction becomes the driving force for blood flow. High-pressure blood is forced into the arteries, displacing the low-pressure blood that fills them and pushing it farther into the vasculature. During this phase, the AV valves remain closed and the atria continue to fill.

5 **Ventricular relaxation and the second heart sound.** At the end of ventricular ejection, the ventricles begin to repolarize and relax. As they do so, ventricular pressure decreases. Once ventricular pressure falls below the pressure in the arteries, blood starts to flow backward into the heart. This backflow of blood fills the cuplike cusps of the semilunar valves, forcing them together into the closed position. The vibrations created by semilunar valve closure are the **second heart sound, S_2**, the "dup" of "lub-dup."

Once the semilunar valves close, the ventricles again become sealed chambers. The AV valves remain closed because ventricular pressure, although falling, is still higher than atrial

CLINICAL FOCUS

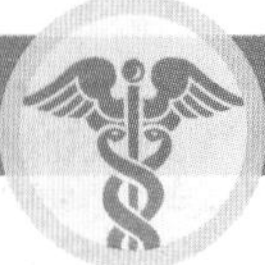

Gallops, Clicks, and Murmurs

The simplest direct assessment of heart function consists of listening to the heart through the chest wall, a process known as **auscultation** {*auscultare*, to listen to} that has been practiced since ancient times. In its simplest form, auscultation is done by placing an ear against the chest. Today, however, it is usually performed by listening through a stethoscope placed against the chest and the back. Normally, there are two audible heart sounds. The first ("lub") is associated with closure of the AV valves. The second ("dup") is associated with closure of the semilunar valves.

Two additional heart sounds can be recorded with very sensitive electronic stethoscopes. The third heart sound is caused by turbulent blood flow into the ventricles during ventricular filling, and the fourth sound is associated with turbulence during atrial contraction. In certain abnormal conditions, these latter two sounds may become audible through a regular stethoscope. They are called gallops because their timing puts them close to one of the normal heart sounds: "lub—dup-dup," or "lub-lub—dup." Other abnormal heart sounds include clicking, caused by abnormal movement of one of the valves, and murmurs, caused by the "whoosh" of blood leaking through an incompletely closed or excessively narrowed (*stenotic*) valve.

pressure. This period is called **isovolumic ventricular relaxation** because the volume of blood in the ventricles is not changing.

When ventricular relaxation causes ventricular pressure to become less than atrial pressure, the AV valves open. Blood that has been accumulating in the atria during ventricular contraction rushes into the ventricles. The cardiac cycle has begun again.

Concept Check Answers: p. 506

25. Which chamber—atrium or ventricle—has higher pressure during the following phases of the cardiac cycle? (a) ventricular ejection, (b) isovolumic ventricular relaxation, (c) atrial and ventricular diastole, (d) isovolumic ventricular contraction

26. *Murmurs* are abnormal heart sounds caused either by blood forced through a narrowed valve opening or by backward flow (regurgitation) through a valve that has not closed completely. *Valvular stenosis* {*stenos*, narrow} may be an inherited condition or may result from inflammation or other disease processes. At which step(s) in the cardiac cycle (Fig. 14.17a) would you expect to hear a murmur caused by the following pathologies? (a) aortic valvular stenosis, (b) mitral valve regurgitation, (c) aortic valve regurgitation

Pressure-Volume Curves Represent One Cardiac Cycle

Another way to describe the cardiac cycle is with a pressure-volume graph, shown in Figure 14.17b. This figure represents the changes in volume (x-axis) and pressure (y-axis) that occur during one cardiac cycle.

The flow of blood through the heart is governed by the same principle that governs the flow of all liquids and gases: flow proceeds from areas of higher pressure to areas of lower pressure. When the heart contracts, the pressure increases and blood flows out of the heart into areas of lower pressure. Figure 14.17b represents pressure and volume changes in the left ventricle, which sends blood into the systemic circulation. The left side of the heart creates higher pressures than the right side, which sends blood through the shorter pulmonary circuit.

The cycle begins at point A. The ventricle has completed a contraction and contains the minimum amount of blood that it will hold during the cycle. It has relaxed, and its pressure is also at its minimum value. Blood is flowing into the atrium from the pulmonary veins.

Once pressure in the atrium exceeds pressure in the ventricle, the mitral valve between the atrium and ventricle opens (Fig. 14.17b, point A). Atrial blood now flows into the ventricle, increasing its volume (point A to point B). As blood flows in, the relaxing ventricle expands to accommodate the entering blood. Consequently, the volume of the ventricle increases, but the pressure in the ventricle goes up very little.

The last portion of ventricular filling is completed by atrial contraction (point A′ to B). The ventricle now contains the maximum volume of blood that it will hold during this cardiac cycle (point B). Because maximum filling occurs at the end of ventricular relaxation (diastole), this volume is called the **end-diastolic volume (EDV)**. In a 70-kg man at rest, end-diastolic volume is about 135 mL, but this value varies under different conditions. During periods of very high heart rate, for instance, when the ventricle does not have time to fill completely between beats, the end-diastolic value may be less than 135 mL.

When ventricular contraction begins, the mitral valve closes. With both the AV valve and the semilunar valve closed, blood in the ventricle has nowhere to go. Nevertheless, the ventricle continues to contract, causing the pressure in this chamber to increase rapidly during isovolumic contraction (B → C in Fig. 14.17b). Once ventricular pressure exceeds the pressure in the aorta, the aortic valve opens (point C). Pressure continues to increase as the ventricle contracts further, but ventricular volume decreases as blood is pushed out into the aorta (C → D).

The heart does not empty itself completely of blood each time the ventricle contracts. The amount of blood left in the ventricle at the end of contraction is known as the **end-systolic volume (ESV)**. The ESV (point D) is the minimum amount of blood the ventricle contains during one cycle. An average ESV value in a person at rest is 65 mL, meaning that nearly half of the 135 mL that was in the ventricle at the start of the contraction is still there at the end of the contraction.

At the end of each ventricular contraction, the ventricle begins to relax. As it does so, ventricular pressure decreases. Once pressure in the ventricle falls below aortic pressure, the semilunar valve closes, and the ventricle again becomes a sealed chamber. The remainder of relaxation occurs without a change in blood volume, and so this phase is called *isovolumic relaxation* (Fig. 14.17b, D → A). When ventricular pressure finally falls to the point at which atrial pressure exceeds ventricular pressure, the mitral valve opens and the cycle begins again.

The electrical and mechanical events of the cardiac cycle are summarized together in ■ Figure 14.18, known as a Wiggers diagram after the physiologist who first created it.

Concept Check Answers: p. 506

27. In Figure 14.17a, at what points in the cycle do EDV and ESV occur?

28. On the Wiggers diagram in Figure 14.18, match the following events to the lettered boxes: (a) end-diastolic volume, (b) aortic valve opens, (c) mitral valve opens, (d) aortic valve closes, (e) mitral valve closes, (f) end-systolic volume

29. Why does atrial pressure increase just to the right of point C in Figure 14.18? Why does it decrease during the initial part of ventricular systole, then increase? Why does it decrease to the right of point D?

30. Why does ventricular pressure shoot up suddenly at point C in Figure 14.18?

THE WIGGERS DIAGRAM

This diagram follows left heart and aortic pressures, left ventricular volume, and the ECG through one cardiac cycle. The boxed letters refer to Concept Checks 28–30.

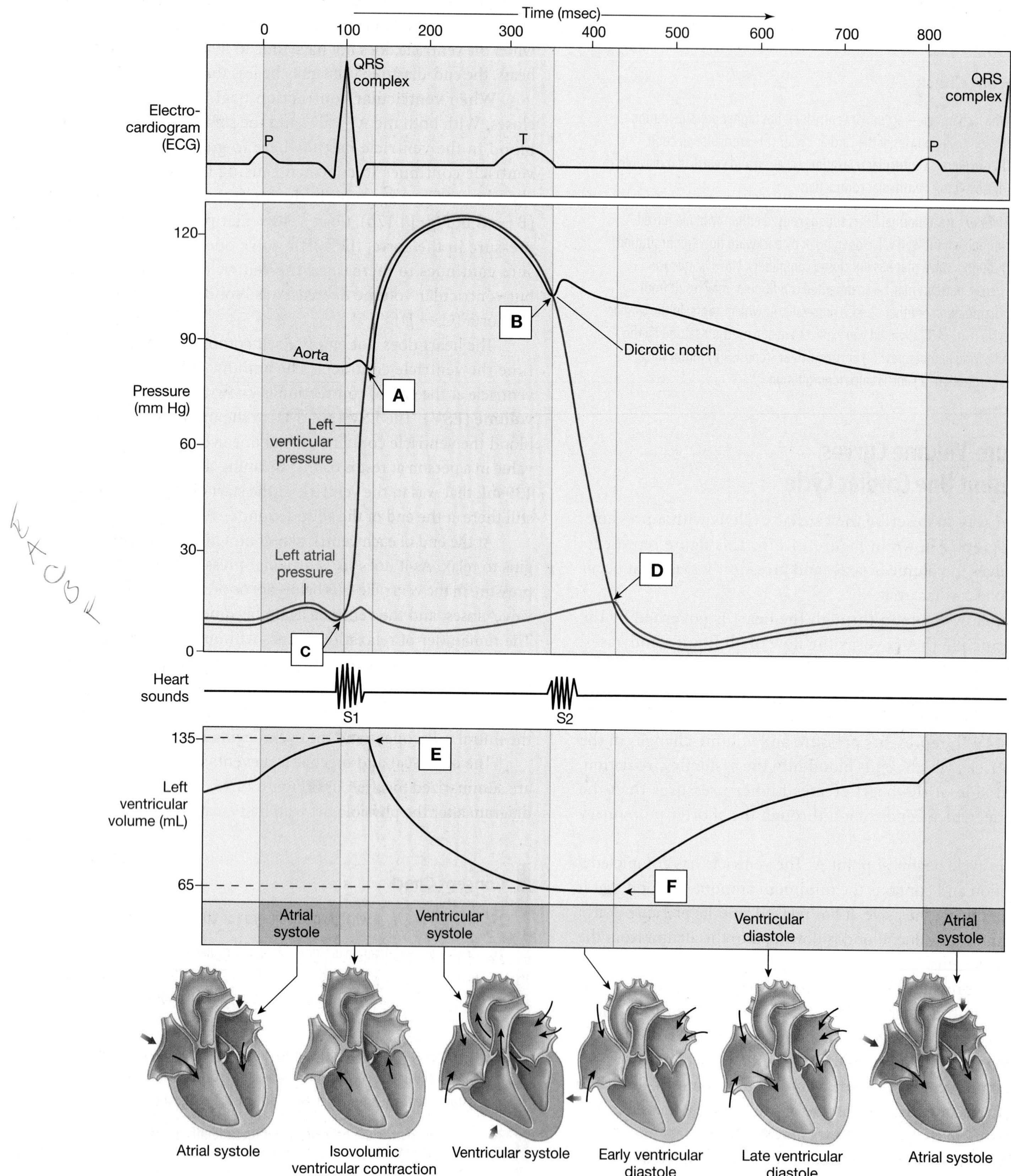

■ **Fig. 14.18**

Stroke Volume Is the Volume of Blood Pumped per Contraction

What is the purpose of blood remaining in the ventricles at the end of each contraction? For one thing, the resting end-systolic volume of 65 mL provides a safety margin. With a more forceful contraction, the heart can decrease its ESV, sending additional blood to the tissues. Like many organs of the body, the heart does not usually work "all out."

Stroke volume is the amount of blood pumped by one ventricle during a contraction. It is measured in milliliters per beat and can be calculated as follows:

Volume of blood before contraction –
volume of blood after contraction = stroke volume

$$\text{EDV} - \text{ESV} = \text{stroke volume}$$

For the average contraction in a person at rest:

$$135\ \text{mL} - 65\ \text{mL} = 70\ \text{mL, the normal stroke volume}$$

Stroke volume is not constant and can increase to as much as 100 mL during exercise. Stroke volume, like heart rate, is regulated by mechanisms we discuss later in this chapter.

Cardiac Output Is a Measure of Cardiac Performance

How can we assess the effectiveness of the heart as a pump? One way is to measure **cardiac output** (CO), the volume of blood pumped by one ventricle in a given period of time. Because all blood that leaves the heart flows through the tissues, cardiac output is an indicator of total blood flow through the body. However, cardiac output does not tell us how blood is distributed to various tissues. That aspect of blood flow is regulated at the tissue level.

Cardiac output (CO) can be calculated by multiplying heart rate (beats per minute) by stroke volume (mL per beat, or per contraction):

$$\text{Cardiac output} = \text{heart rate} \times \text{stroke volume}$$

For an average resting heart rate of 72 beats per minute and a stroke volume of 70 mL per beat, we have

$$\begin{aligned} \text{CO} &= 72\ \text{beats/min} \times 70\ \text{mL/beat} \\ &= 5040\ \text{mL/min (or approx. 5 L/min)} \end{aligned}$$

Average total blood volume is about 5 liters. This means that, at rest, one side of the heart pumps all the blood in the body through it in only one minute!

Normally, cardiac output is the same for both ventricles. However, if one side of the heart begins to fail for some reason and is unable to pump efficiently, cardiac output becomes mismatched. In that situation, blood pools in the circulation behind the weaker side of the heart.

RUNNING PROBLEM

Walter was in the cardiac care unit by 1:00 P.M., where the cardiologist visited him. "We need to keep an eye on you here for the next few days. There is a possibility that the damage from your heart attack could cause an irregular heartbeat." Once Walter was stable, he would have a coronary angiogram, a procedure in which an opaque dye visible on X-rays shows where coronary artery lumens have narrowed from atherosclerotic plaques. Depending on the results of that test, the physician might recommend either balloon angioplasty, in which a tube passed into the coronary artery is inflated to open up the blockage, or coronary bypass surgery, in which veins from other parts of the body are grafted onto the heart arteries to provide bypass channels around blocked regions.

Q9: If Walter's heart attack has damaged the muscle of his left ventricle, what do you predict will happen to his cardiac output?

During exercise, cardiac output may increase to 30–35 L/min. Homeostatic changes in cardiac output are accomplished by varying the heart rate, the stroke volume, or both. Both local and reflex mechanisms can alter cardiac output, as you will see in the sections that follow.

Concept Check Answer: p. 507

31. If the stroke volume of the left ventricle is 250 mL/beat and the stroke volume of the right ventricle is 251 mL/beat, what happens to the relative distribution of blood between the systemic and pulmonary circulation after 10 beats?

The Autonomic Division Modulates Heart Rate

An average resting heart rate in an adult is about 70 beats/minute (bpm). The normal range is highly variable, however. Trained athletes may have resting heart rates of 50 bpm or less, while someone who is excited or anxious may have a rate of 125 bpm or higher. Children have higher average heart rates than adults. Heart rate is initiated by autorhythmic cells in the SA node, but it is modulated by neural and hormonal input.

The sympathetic and parasympathetic branches of the autonomic division influence heart rate through antagonistic control (■ Fig. 14.19). Parasympathetic activity slows heart rate, while sympathetic activity speeds it up.

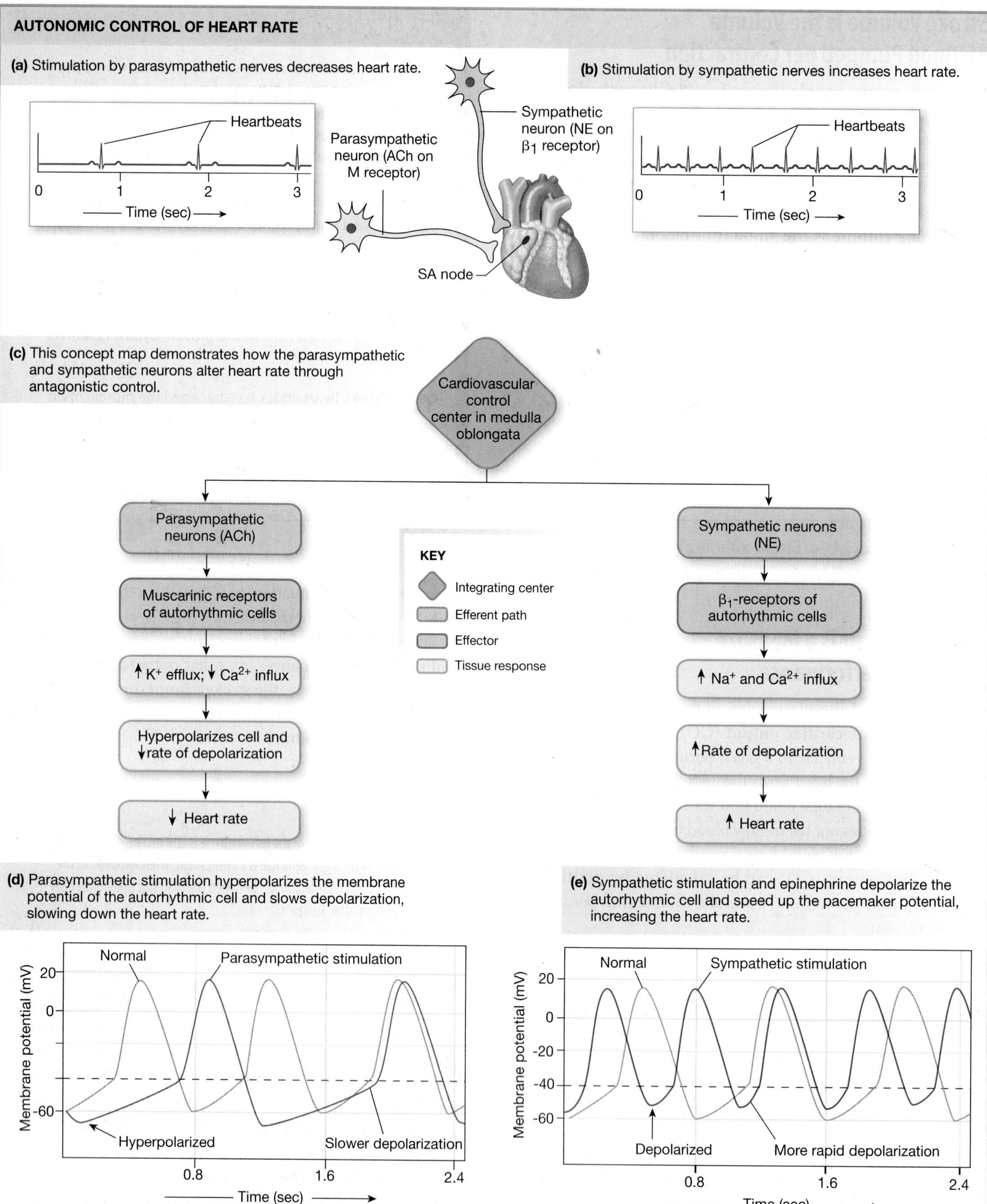
AUTONOMIC CONTROL OF HEART RATE
(a) Stimulation by parasympathetic nerves decreases heart rate.
Heartbeats
0
1
2
3
Time (sec)
Parasympathetic neuron (ACh on M receptor)
Sympathetic neuron (NE on β1 receptor)
SA node
(b) Stimulation by sympathetic nerves increases heart rate.
Heartbeats
0
1
2
3
Time (sec)
(c) This concept map demonstrates how the parasympathetic and sympathetic neurons alter heart rate through antagonistic control.
Cardiovascular control center in medulla oblongata
Parasympathetic neurons (ACh)
Muscarinic receptors of autorhythmic cells
↑ K+ efflux; ↓ Ca2+ influx
Hyperpolarizes cell and ↓rate of depolarization
↓ Heart rate
KEY
Integrating center
Efferent path
Effector
Tissue response
Sympathetic neurons (NE)
β1-receptors of autorhythmic cells
↑ Na+ and Ca2+ influx
↑Rate of depolarization
↑ Heart rate
(d) Parasympathetic stimulation hyperpolarizes the membrane potential of the autorhythmic cell and slows depolarization, slowing down the heart rate.
Normal
Parasympathetic stimulation
Membrane potential (mV)
20
0
-60
Hyperpolarized
Slower depolarization
0.8
1.6
2.4
Time (sec)
(e) Sympathetic stimulation and epinephrine depolarize the autorhythmic cell and speed up the pacemaker potential, increasing the heart rate.
Normal
Sympathetic stimulation
Membrane potential (mV)
20
0
-20
-40
-60
Depolarized
More rapid depolarization
0.8
1.6
2.4
Time (sec)

Fig. 14.19

Parasympathetic Control The parasympathetic neurotransmitter acetylcholine (ACh) slows heart rate. Acetylcholine activates muscarinic cholinergic receptors that influence K^+ and Ca^{2+} channels in the pacemaker cell (Fig. 14.19c). Potassium permeability increases, hyperpolarizing the cell so that the pacemaker potential begins at a more negative value (Fig. 14.19d). At the same time, Ca^{2+} permeability of the pacemaker decreases. Decreased Ca^{2+} permeability slows the rate at which the pacemaker potential depolarizes. The combination of the two effects causes the cell to take longer to reach threshold, delaying the onset of the action potential in the pacemaker and slowing the heart rate.

Sympathetic Control Sympathetic stimulation of pacemaker cells speeds up heart rate (Fig. 14.19b). The catecholamines norepinephrine (from sympathetic neurons) and epinephrine (from the adrenal medulla) increase ion flow through both I_f and Ca^{2+} channels. More rapid cation entry speeds up the rate of the pacemaker depolarization, causing the cell to reach threshold faster and increasing the rate of action potential firing (Fig. 14.19e). When the pacemaker fires action potentials more rapidly, heart rate increases.

Catecholamines exert their effect by binding to and activating β_1-adrenergic receptors on the autorhythmic cells. The β_1-receptors use a cAMP second messenger system to alter the transport properties of the ion channels. In the case of the I_f channels, which are cyclic nucleotide-gated channels, cAMP itself is the messenger. When cAMP binds to open I_f channels, they remain open longer. Increased permeability to Na^+ and Ca^{2+} during the pacemaker potential phase speeds up depolarization and heart rate.

Tonic Control Normally, tonic control of heart rate is dominated by the parasympathetic branch. This control can be shown experimentally by blocking all autonomic input to the heart. When all sympathetic and parasympathetic input is blocked, the spontaneous depolarization rate of the SA node is 90–100 times per minute. To achieve a resting heart rate of 70 beats per minute, tonic parasympathetic activity must slow the intrinsic rate down from 90 beats per minute.

An increase in heart rate can be achieved in two ways. The simplest method for increasing rate is to decrease parasympathetic activity. As parasympathetic influence is withdrawn from the autorhythmic cells, they resume their intrinsic rate of depolarization, and heart rate increases to 90–100 beats per minute. Alternatively, sympathetic input is required to increase heart rate above the intrinsic rate. Norepinephrine (or epinephrine) on β_1-receptors speeds up the depolarization rate of the autorhythmic cells and increases heart rate.

Both autonomic branches also alter the rate of conduction through the AV node. Acetylcholine slows the conduction of action potentials through the AV node, thereby increasing AV node delay. In contrast, the catecholamines epinephrine and norepinephrine enhance conduction of action potentials through the AV node and through the conducting system.

Multiple Factors Influence Stroke Volume

Stroke volume, the volume of blood pumped per ventricle per contraction, is directly related to the force generated by cardiac muscle during a contraction. Normally, as contraction force increases, stroke volume increases. In the isolated heart, the force of ventricular contraction is affected by two parameters: the length of muscle fibers at the beginning of contraction and the contractility of the heart. The volume of blood in the ventricle at the beginning of contraction (the end-diastolic volume) determines the length of the muscle. **Contractility** is the intrinsic ability of a cardiac muscle fiber to contract at any given fiber length and is a function of Ca^{2+} interaction with the contractile filaments.

Length-Tension Relationships and the Frank-Starling Law of the Heart In striated muscles, the force created by a muscle fiber is directly related to the length of the sarcomere, as indicated by the initial length of the muscle fiber [p. 417]. The longer the muscle fiber and sarcomere when a contraction begins, the greater the tension developed, up to a maximum (■ Fig. 14.20a).

The length-tension relationship observed in isolated muscles can also be seen in the intact heart: as stretch of the ventricular wall increases, so does the stroke volume (Fig. 14.20b). If additional blood flows into the ventricles, the muscle fibers stretch, then contract more forcefully, ejecting more blood. The degree of myocardial stretch before contraction begins is called the **preload** on the heart because this stretch represents the load placed on cardiac muscles before they contract.

This relationship between stretch and force in the intact heart was first described by a German physiologist, Otto Frank. A British physiologist, Ernest Starling, then expanded on Frank's work. Starling attached an isolated heart-lung preparation from a dog to a reservoir so that he could regulate the amount of blood returning to the heart. He found that in the absence of any nervous or hormonal control, the heart pumped all the blood that returned to it.

The relationship between stretch and force in the intact heart is plotted on a *Starling curve* (Fig. 14.20b). The x-axis represents the end-diastolic volume. This volume is a measure of stretch in the ventricles, which in turn determines sarcomere length. The y-axis of the Starling curve represents the stroke volume and is an indicator of the force of contraction.

The graph shows that stroke volume is proportional to EDV. As additional blood enters the heart, the heart contracts more forcefully and ejects more blood. This relationship is known as the **Frank-Starling law of the heart**. It means that within physiological limits, the heart pumps all the blood that returns to it.

Stroke Volume and Venous Return According to the Frank-Starling law, stroke volume increases as end-diastolic volume increases. End-diastolic volume is normally determined by **venous return**, the amount of blood that enters the heart from the venous circulation. Three factors affect venous return:

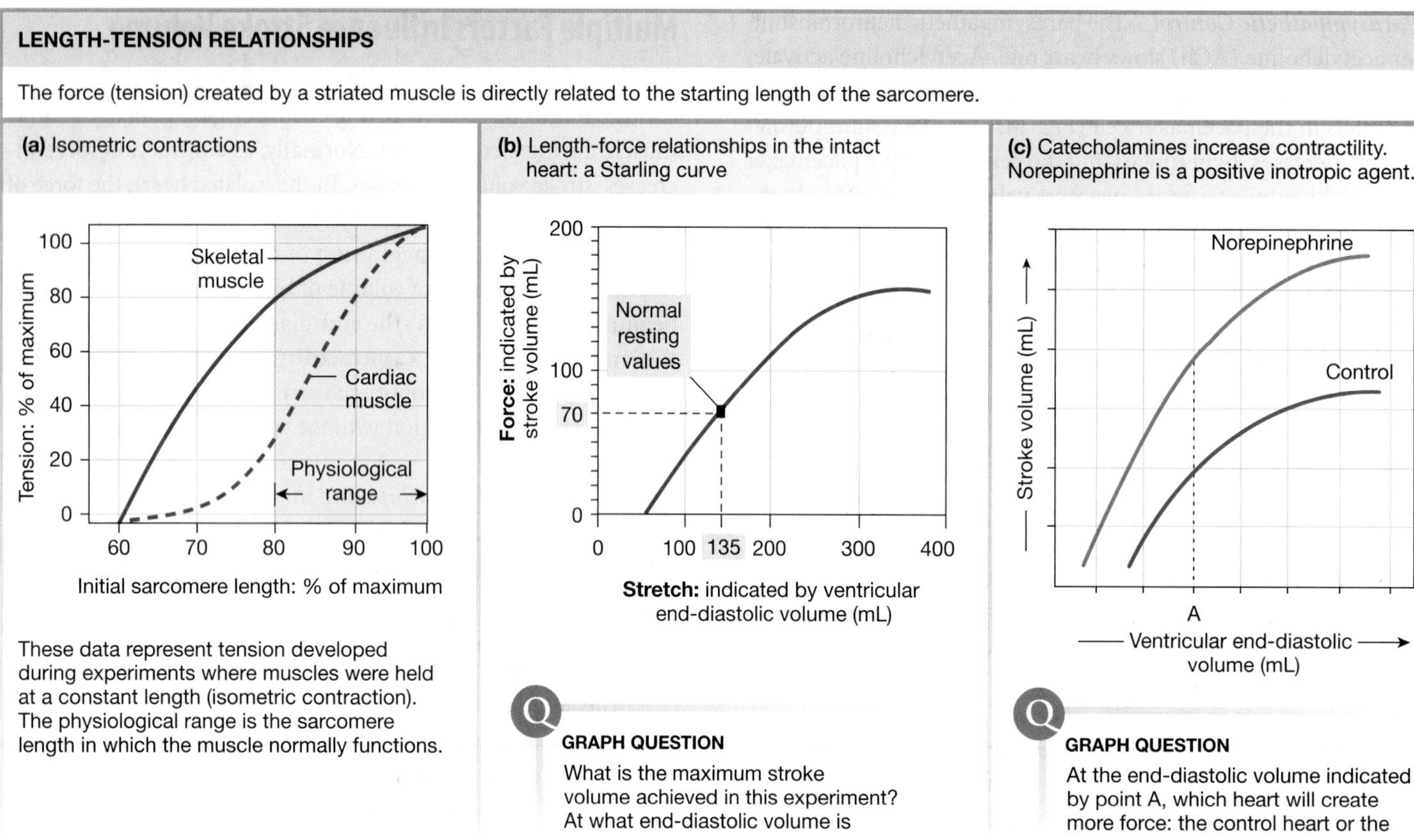
LENGTH-TENSION RELATIONSHIPS
The force (tension) created by a striated muscle is directly related to the starting length of the sarcomere.
(a) Isometric contractions
Skeletal muscle
Cardiac muscle
Physiological range
Tension: % of maximum
Initial sarcomere length: % of maximum
These data represent tension developed during experiments where muscles were held at a constant length (isometric contraction). The physiological range is the sarcomere length in which the muscle normally functions.
(b) Length-force relationships in the intact heart: a Starling curve
Normal resting values
Force: indicated by stroke volume (mL)
Stretch: indicated by ventricular end-diastolic volume (mL)
GRAPH QUESTION
What is the maximum stroke volume achieved in this experiment? At what end-diastolic volume is maximum stroke volume first achieved?
(c) Catecholamines increase contractility. Norepinephrine is a positive inotropic agent.
Norepinephrine
Control
Stroke volume (mL)
A
Ventricular end-diastolic volume (mL)
GRAPH QUESTION
At the end-diastolic volume indicated by point A, which heart will create more force: the control heart or the heart under the influence of norepinephrine?

Fig. 14.20

(1) contraction or compression of veins returning blood to the heart (the skeletal muscle pump), (2) pressure changes in the abdomen and thorax during breathing (the respiratory pump), and (3) sympathetic innervation of veins.

Skeletal muscle pump is the name given to skeletal muscle contractions that squeeze veins (particularly in the legs), compressing them and pushing blood toward the heart. During exercise that involves the lower extremities, the skeletal muscle pump helps return blood to the heart. During periods of sitting or standing motionless, the skeletal muscle pump does not assist venous return.

The **respiratory pump** is created by movement of the thorax during inspiration (breathing in). As the chest expands and the diaphragm moves toward the abdomen, the thoracic cavity enlarges and develops a subatmospheric pressure. This low pressure decreases pressure in the inferior vena cava as it passes through the thorax, which helps draw more blood into the vena cava from veins in the abdomen. The respiratory pump is aided by the higher pressure placed on the outside of abdominal veins when the abdominal contents are compressed during inspiration. The combination of increased pressure in the abdominal veins and decreased pressure in thoracic veins enhances venous return during inspiration.

Constriction of veins by sympathetic activity is the third factor that affects venous return. When the veins constrict, their volume decreases, squeezing more blood out of them and into the heart. With a larger ventricular volume at the beginning of the next contraction, the ventricle contracts more forcefully, sending the blood out into the arterial side of the circulation. In this manner, sympathetic innervation of veins allows the body to redistribute some venous blood to the arterial side of the circulation.

Contractility Is Controlled by the Nervous and Endocrine Systems

Any chemical that affects contractility is called an **inotropic agent** {*ino*, fiber}, and its influence is called an **inotropic effect.** If a chemical increases the force of contraction, it is said to have a positive inotropic effect. For example, the catecholamines epinephrine and norepinephrine and drugs such as digitalis enhance contractility and are therefore considered to have a positive inotropic effect. Chemicals with negative inotropic effects decrease contractility.

Figure 14.20c illustrates a normal Starling curve (the control curve) along with a curve showing how the stroke volume changes with increased contractility due to norepinephrine. Note that contractility is distinct from the length-tension relationship. A muscle can remain at one length (for example, the

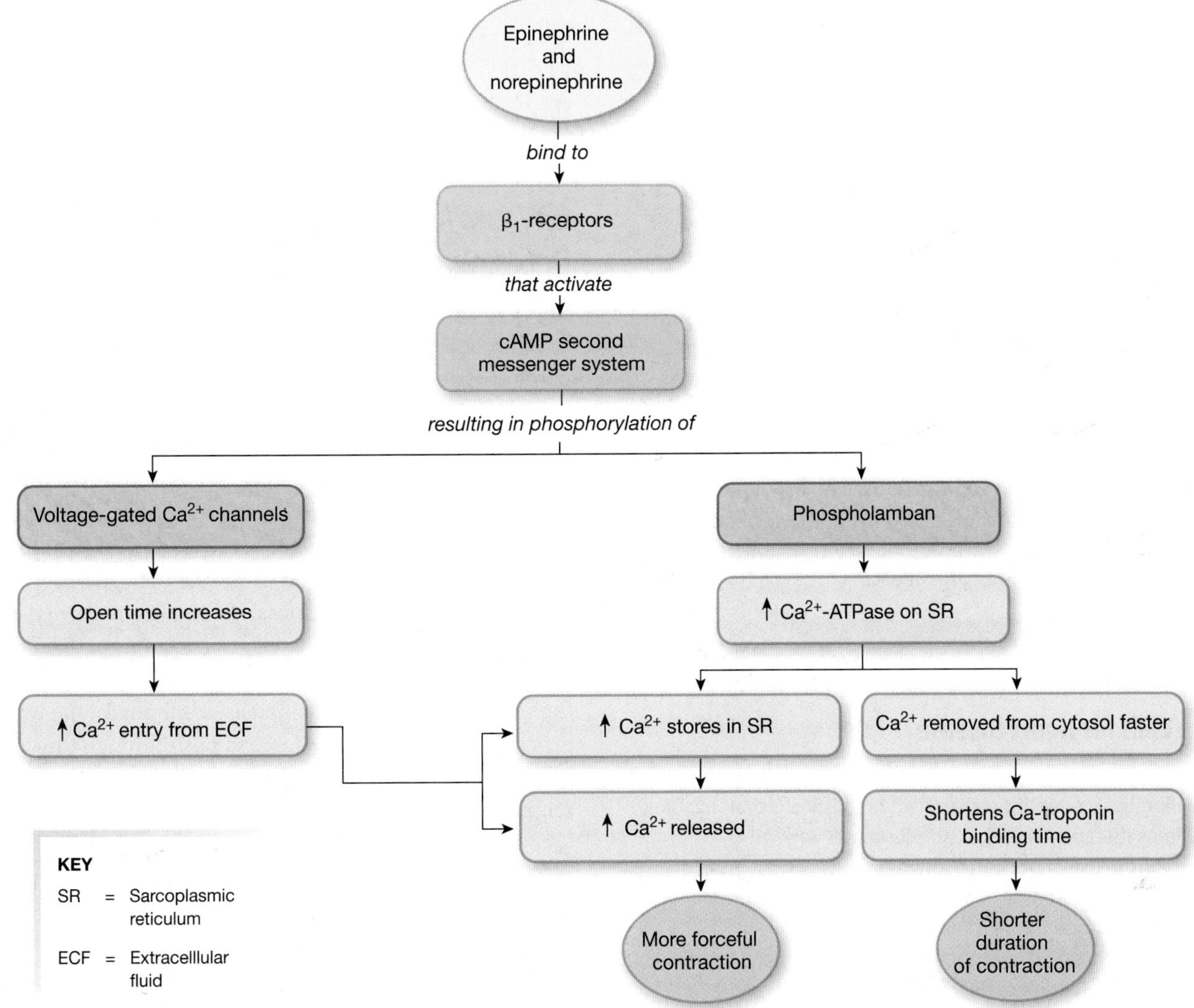

■ **Fig. 14.21**

end-diastolic volume marked A in Figure 14.20c) but show increased contractility. Contractility increases as the amount of calcium available for contraction increases. Contractility was once considered to be distinct from changes in force resulting from variations in muscle (sarcomere) length. However, it now appears that increasing sarcomere length also makes cardiac muscle more sensitive to Ca^{2+}, thus linking contractility to muscle length.

The mechanism by which catecholamines increase Ca^{2+} entry and storage and exert their positive inotropic effect is mapped in ■ Figure 14.21. The signal molecules bind to and activate β_1-adrenergic receptors [p. 385] on the contractile myocardial cell membrane. Activated β_1-receptors use a cyclic AMP second messenger system to phosphorylate specific intracellular proteins [p. 183]. Phosphorylation of voltage-gated Ca^{2+} channels increases the probability that they will open and stay open longer. More open channels allow more Ca^{2+} to enter the cell.

The catecholamines increase Ca^{2+} storage through the use of a regulatory protein called **phospholamban** (Fig. 14.21). Phosphorylation of phospholamban enhances Ca^{2+}-ATPase activity in the sarcoplasmic reticulum. The ATPase concentrates Ca^{2+} in the sarcoplasmic reticulum, making more Ca^{2+} available for calcium-induced calcium release. Because more cytosolic Ca^{2+} means more active crossbridges, and because the force of contraction is proportional to the number of active crossbridges, the net result of catecholamine stimulation is a stronger contraction.

In addition to increasing the force of cardiac contraction, catecholamines also shorten the duration of contraction. The enhanced Ca^{2+}-ATPase speeds up removal of Ca^{2+} from the cytosol. This in turn shortens the time that Ca^{2+} is bound to troponin and decreases the active time of the myosin crossbridges. The muscle twitch is therefore briefer.

A different mechanism that enhances contractility can be triggered by the administration of cardiac glycosides, a class of molecules first discovered in the plant *Digitalis purpurea* (purple foxglove). Cardiac glycosides include digitoxin and the related compound *ouabain*, a molecule used to inhibit sodium transport in research studies. Glycosides increase contractility by slowing Ca^{2+} removal from the cytosol (in contrast to the catecholamines just discussed, which speed up Ca^{2+} removal). This mechanism is a pharmacological effect and does not occur in the absence of the drug.

Cardiac glycosides have been used since the eighteenth century as a remedy for *heart failure*, a pathological condition in which the heart is unable to contract forcefully. These highly toxic drugs depress Na^+-K^+-ATPase activity in all cells, not just those of the heart. With depressed Na^+-K^+-ATPase activity, Na^+ builds up in the cytosol, and the concentration gradient for Na^+ across the cell membrane diminishes. This in turn decreases the potential energy available for indirect active transport [p. 150].

In the myocardial cell, cardiac glycosides decrease the cell's ability to remove Ca^{2+} by means of the Na^+-Ca^{2+} exchanger. The resultant increase in cytosolic Ca^{2+} causes more forceful myocardial contractions.

Concept Check Answer: p. 507

32. Using the myocardial cell in Figure 14.9 as a model, draw a contractile cell and show how catecholamines increase myocardial contractility.

EDV and Arterial Blood Pressure Determine Afterload

Many of the experiments that uncovered the relationship between myocardial stretch and contractile force were conducted using isolated hearts. In the intact animal, ventricular force must be used to overcome the resistance created by blood filling the arterial system. The combined load of EDV and arterial resistance during ventricular contraction is known as **afterload**.

As an analogy, think of waiters carrying trays of food through a swinging door. A tray is a load equivalent to blood in the ventricles at the beginning of contraction. The door is an additional load that the waiter must push against to leave the kitchen. Normally this additional load is relatively minor. If someone decides to play a prank, however, and piles furniture against the dining room side of the door (increased afterload), the waiter must expend considerably more force to push through the door. Similarly, ventricular contraction must push a load of blood through a semilunar valve and out into the blood-filled arteries.

Increased afterload is found in several pathological situations, including elevated arterial blood pressure and loss of stretchability (*compliance*) in the aorta. To maintain constant stroke volume when afterload increases, the ventricle must increase its force of contraction, which then increases the muscle's need for oxygen and ATP production. If increased afterload becomes a chronic situation, the myocardial cells hypertrophy, resulting in increased thickness of the ventricular wall.

Clinically, arterial blood pressure is often used as an indirect indicator of afterload. Other aspects of ventricular function can be assessed noninvasively by echocardiography, an ultrasound procedure in which sound waves are reflected off heart tissue. A common functional index derived from this procedure is the **ejection fraction**, or percentage of EDV ejected with one contraction (stroke volume/EDV). Using our standard values for the 70-kg man, ejection fraction at rest is 70 mL/135 mL, or 52%. If stroke volume increases to 100 mL with exercise, the ejection fraction increases to 74%.

Concept Check Answer: p. 507

33. A person's aortic valve opening has become constricted, creating a condition known as *aortic stenosis*. Which ventricle is affected by this change? What happens to the afterload on this ventricle?

EMERGING CONCEPTS

Stem Cells for Heart Disease

One of the interesting (and scary) aspects of translating basic scientific research into medicine is that sometimes therapies work but no one knows why. An example is the use of bone marrow stem cells to treat heart disease. After a heart attack, portions of the myocardium may be so damaged from lack of oxygen that they can no longer contract and contribute to cardiac function. A therapy that could replace dead and damaged cells and restore function would be a dream come true. In 2001 a group of researchers reported that bone marrow stem cells injected into mice with damaged hearts differentiated into new myocardial cells. This dramatic result prompted rapid translation of the basic research into human clinical trials. By 2008 there were more than 251 clinical trials looking at whether stem cell injections could help impaired cardiac function. Early results indicated that some patients were exhibiting small (about 5%) functional improvement. At the same time, however, scientists reported that they had been unable to duplicate the 2001 findings that bone marrow stem cells differentiate into myocardial cells. At this time, no one has an explanation for the mechanism(s) by which stem cell injections improve function. Theories with supporting evidence include stem cell secretion of growth factors that prevent tissue death or scarring or that cause new blood vessels to grow into damaged areas. One study is looking at whether skeletal muscle stem cells (*myoblasts*) can differentiate into functional cardiac muscle.

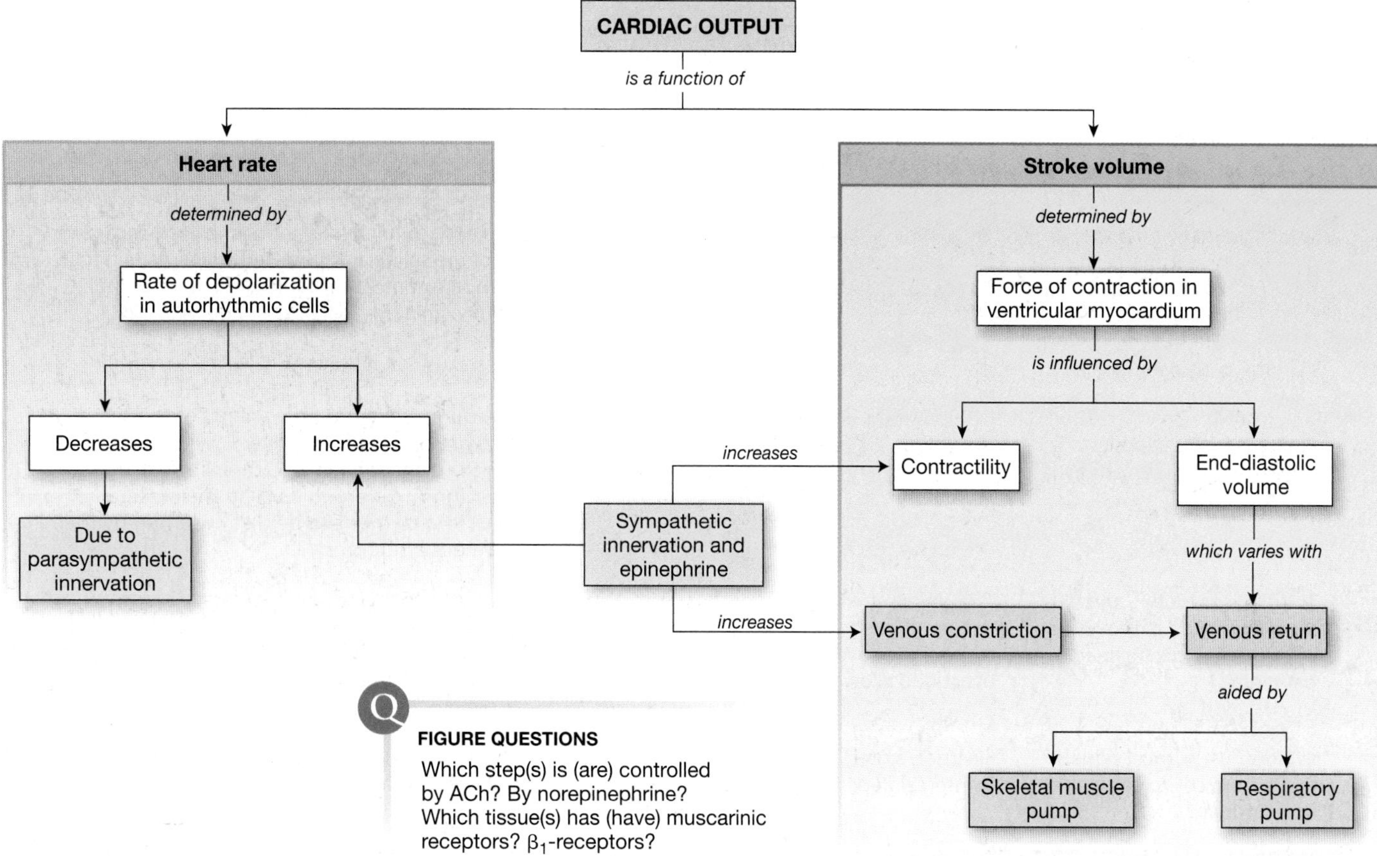

■ **Fig. 14.22**

The factors that determine cardiac output are summarized in ■ Figure 14.22. Cardiac output varies with both heart rate and stroke volume. Heart rate is modulated by the autonomic division of the nervous system and by epinephrine. The determination of stroke volume is more complex because stroke volume is a function of an intrinsic myocardial response to stretch (the length-tension relationship of the Frank-Starling law) interacting with adrenergically mediated changes in contractility. Venous return is a major determinant of end-diastolic volume and stretch.

The heart is a complex organ with many parts that can malfunction. Next, we examine how cardiac output plays a key role in blood flow through the circulation. You will learn about high blood pressure and atherosclerosis, and how these conditions can cause the heart to fail in its role as a pump.

RUNNING PROBLEM CONCLUSION

Myocardial Infarction

Walter's angiogram showed two small blocked arteries, which were opened by balloon angioplasty. He returned home with instructions from his doctor for modifying his lifestyle to include a better diet, regular exercise, and no cigarette smoking. As part of his follow-up, Walter had a *myocardial perfusion imaging* test, in which he was administered radioactive thallium. The distribution of thallium throughout the heart is an indicator of blood flow to the heart muscle.

In this running problem, you learned about some current techniques for diagnosing and treating heart attacks. Walter's symptoms are the classic ones, but many women have symptoms that are different. For more information on heart attack symptoms and other cardiovascular diseases, visit *www.americanheart.org*, the American Heart Association web site. Check your understanding of this physiology by comparing your answers with the information in the summary table.

RUNNING PROBLEM CONCLUSION *(continued)*

Question	Facts	Integration and Analysis
1. Why did the EMT give Walter oxygen?	The EMT suspects that Walter has had a heart attack. Blood flow and oxygen supply to the heart muscle may be blocked.	If the heart is not pumping effectively, the brain may not receive adequate oxygen. Administration of oxygen increases the amount of oxygen that reaches both the heart and the brain.
2. What effect would the injection of isotonic saline have on Walter's extracellular fluid volume? On his intracellular fluid volume? On his total body osmolarity?	An isotonic solution is one that does not change cell volume [p. 134]. Isotonic saline (NaCl) is isosmotic to the body.	The extracellular volume will increase because all of the saline administered will remain in that compartment. Intracellular volume and total body osmolarity will not change.
3. A related form of creatine kinase is found in skeletal muscle. What are related forms of an enzyme called?	Related forms of an enzyme are called isozymes.	Although isozymes are variants of the same enzymes, their activity may vary under different conditions, and their structures are slightly different. Cardiac and skeletal muscle isozymes can be distinguished by their different structures.
4. What is troponin, and why would elevated blood levels of troponin indicate heart damage?	Troponin is the regulatory protein bound to tropomyosin [p. 407]. Ca^{2+} binding to troponin uncovers the myosin-binding site of actin to allow contraction.	Troponin is part of the contractile apparatus of the muscle cell. If troponin escapes from the cell and enters the blood, this is an indication that the cell either has been damaged or is dead.
5. How do electrical signals move from cell to cell in the myocardium?	Electrical signals pass through gap junctions in intercalated disks [p. 175].	The cells of the heart are electrically linked by gap junctions.
6. What happens to contraction in a myocardial contractile cell if a wave of depolarization passing through the heart bypasses it?	Depolarization in a muscle cell is the signal for contraction.	If a myocardial cell is not depolarized, it will not contract. Failure to contract creates a nonfunctioning region of heart muscle and impairs the pumping function of the heart.
7. What will a beta blocker do to Walter's heart rate? Why is that response helpful following a heart attack?	A beta blocker is an antagonist to β_1-adrenergic receptors. Activation of β_1-receptors increases heart rate.	A beta blocker therefore decreases heart rate and lowers oxygen demand. Cells that need less oxygen are less likely to die if their blood supply is diminished.
8. If the ventricle of the heart is damaged, in which wave or waves of the electrocardiogram would you expect to see abnormal changes?	The P wave represents atrial depolarization. The QRS complex and T wave represent ventricular depolarization and repolarization, respectively.	The QRS complex and the T wave are most likely to show changes after a heart attack. Changes indicative of myocardial damage include enlargement of the Q wave, shifting of the S-T segment off the baseline (elevated or depressed), and inversion of the T wave.
9. If Walter's heart attack has damaged the muscle of his left ventricle, what do you predict will happen to his left cardiac output?	Cardiac output equals stroke volume times heart rate.	If the ventricular myocardium has been weakened, stroke volume may decrease. Decreased stroke volume in turn decreases cardiac output.

463 467 477 485 486 495 501

Chapter Summary

The cardiovascular system exemplifies many of the basic themes in physiology. Blood flows through vessels as a result of high pressure created during ventricular contraction (*mass flow*). The circulation of blood provides an essential route for *cell-to-cell communication,* particularly for hormones and other chemical signals. Myocardial contraction, like contraction in skeletal and smooth muscle, demonstrates the importance of *molecular interactions, biological energy use,* and the *mechanical properties* of cells and tissues. This chapter also introduced the *control systems* for cardiovascular physiology, a theme that will be expanded in the next chapter.

Overview of the Cardiovascular System

iP Cardiovascular—Anatomy Review: The Heart

1. The human **cardiovascular system** consists of a **heart** that pumps **blood** through a closed system of **blood vessels.** (p. 465; Fig. 14.1)
2. The primary function of the cardiovascular system is the transport of nutrients, water, gases, wastes, and chemical signals to and from all parts of the body. (p. 464; Tbl. 14.1)
3. Blood vessels that carry blood away from the heart are called **arteries.** Blood vessels that return blood to the heart are called **veins. Valves** in the heart and veins ensure unidirectional blood flow. (p. 465; Fig. 14.1)
4. The heart is divided into two **atria** and two **ventricles.** (p. 465; Fig. 14.1)
5. The **pulmonary circulation** goes from the right side of the heart to the lungs and back to the heart. The **systemic circulation** goes from the left side of the heart to the tissues and back to the heart. (p. 465; Fig. 14.1)

Pressure, Volume, Flow, and Resistance

6. Blood flows down a **pressure gradient** (ΔP), from the highest pressure in the **aorta** and arteries to the lowest pressure in the **venae cavae** and **pulmonary veins.** (p. 466; Fig. 14.2)
7. In a system in which fluid is flowing, pressure decreases over distance. (p. 468; Fig. 14.3)
8. The pressure created when the ventricles contract is called the **driving pressure** for blood flow. (p. 467)
9. **Resistance** of a fluid flowing through a tube increases as the length of the tube and the **viscosity** (thickness) of the fluid increase, and as the radius of the tube decreases. Of these three factors, radius has the greatest effect on resistance. (p. 469)
10. If resistance increases, flow rate decreases. If resistance decreases, flow rate increases. (p. 468; Fig. 14.3)
11. Fluid flow through a tube is proportional to the pressure gradient (ΔP). A pressure gradient is not the same thing as the absolute pressure in the system. (p. 468; Fig. 14.3)
12. **Flow rate** is the volume of blood that passes one point in the system per unit time. (p. 470)
13. **Velocity** is the distance a volume of blood travels in a given period of time. At a constant flow rate, the velocity of flow through a small tube is faster than the velocity through a larger tube. (p. 471; Fig. 14.4)

Cardiac Muscle and the Heart

iP Cardiovascular: Cardiac Action Potential

14. The heart is composed mostly of cardiac muscle, or **myocardium.** Most cardiac muscle is typical striated muscle. (p. 473; Fig. 14.5h)
15. The signal for contraction originates in **autorhythmic cells** in the heart. Autorhythmic cells are noncontractile myocardium. (p. 475)
16. Myocardial cells are linked to one another by **intercalated disks** that contain gap junctions. The junctions allow depolarization to spread rapidly from cell to cell. (p. 477; Fig. 14.8)
17. In contractile cell excitation-contraction coupling, an action potential opens Ca^{2+} channels. Ca^{2+} entry into the cell triggers the release of additional Ca^{2+} from the sarcoplasmic reticulum through **calcium-induced calcium release.** (p. 478; Fig. 14.9)
18. The force of cardiac muscle contraction can be graded according to how much Ca^{2+} enters the cell. (p. 478)
19. The action potentials of myocardial contractile cells have a rapid depolarization phase created by Na^+ influx, and a steep repolarization phase due to K^+ efflux. The action potential also has a plateau phase created by Ca^{2+} influx. (p. 479; Fig. 14.10)
20. Autorhythmic myocardial cells have an unstable membrane potential called a **pacemaker potential.** The pacemaker potential is due to **I_f channels** that allow net influx of positive charge. (p. 481; Fig. 14.12)
21. The steep depolarization phase of the autorhythmic cell action potential is caused by Ca^{2+} influx. The repolarization phase is due to K^+ efflux. (p. 481; Fig. 14.12)

The Heart As a Pump

iP Cardiovascular: Intrinsic Conduction System

22. Action potentials originate at the **sinoatrial node** (SA node) and spread rapidly from cell to cell in the heart. Action potentials are followed by a wave of contraction. (p. 484; Fig. 14.14)

23. The electrical signal moves from the SA node through the **internodal pathway** to the **atrioventricular node** (AV node), then into the **AV bundle**, **bundle branches**, terminal **Purkinje fibers**, and myocardial contractile cells. (p. 484; Fig. 14.14)
24. The SA node sets the pace of the heartbeat. If the SA node malfunctions, other autorhythmic cells in the AV node or ventricles take control of heart rate. (p. 483)
25. An **electrocardiogram** (ECG) is a surface recording of the electrical activity of the heart. The **P wave** represents atrial depolarization. The **QRS complex** represents ventricular depolarization. The **T wave** represents ventricular repolarization. Atrial repolarization is incorporated in the QRS complex. (p. 488; Fig. 14.15)
26. An ECG provides information on heart rate and rhythm, conduction velocity, and the condition of cardiac tissues. (p. 486)

iP Cardiovascular: The Cardiac Cycle

27. One **cardiac cycle** includes one cycle of contraction and relaxation. **Systole** is the contraction phase; **diastole** is the relaxation phase. (p. 491; Fig. 14.17)
28. Most blood enters the ventricles while the atria are relaxed. Only 20% of ventricular filling at rest is due to atrial contraction. (p. 491)
29. The **AV valves** prevent backflow of blood into the atria. Vibrations following closure of the AV valves create the **first heart sound**. (pp. 491, 494; Fig. 14.7, 14.18)
30. During **isovolumic ventricular contraction**, the ventricular blood volume does not change, but pressure rises. When ventricular pressure exceeds arterial pressure, the **semilunar valves** open, and blood is ejected into the arteries. (p. 494; Fig. 14.18)
31. When the ventricles relax and ventricular pressure falls, the semilunar valves close, creating the **second heart sound**. (p. 494; Fig. 14.18)
32. The amount of blood pumped by one ventricle during one contraction is known as the **stroke volume**. (p. 495)

iP Cardiovascular: Cardiac Output

33. **Cardiac output** is the volume of blood pumped per ventricle per unit time. It is equal to heart rate times stroke volume. The average cardiac output at rest is 5 L/min. (p. 495)
34. Homeostatic changes in cardiac output are accomplished by varying heart rate, stroke volume, or both. (p. 501; Fig. 14.22)
35. Parasympathetic activity slows heart rate; sympathetic activity speeds it up. Norepinephrine and epinephrine act on β_1-receptors to speed up the rate of the pacemaker depolarization. Acetylcholine activates muscarinic receptors to hyperpolarize the pacemakers. (p. 496; Fig. 14.19)
36. The longer a muscle fiber is when it begins to contract, the greater the force of contraction. The **Frank-Starling law of the heart** says that an increase in **end-diastolic volume** results in a greater stroke volume. (p. 498; Fig. 14.20)
37. Epinephrine and norepinephrine increase the force of myocardial contraction when they bind to β_1-adrenergic receptors. They also shorten the duration of cardiac contraction. (p. 499; Fig. 14.21)
38. End-diastolic volume and **preload** are determined by **venous return**. Venous return is affected by skeletal muscle contractions, the respiratory pump, and constriction of veins by sympathetic activity. (p. 497)
39. **Contractility** of the heart is enhanced by catecholamines and certain drugs. Chemicals that alter contractility are said to have an **inotropic effect**. (p. 498; Fig. 14.20c)
40. **Afterload** is the load placed on the ventricle as it contracts. Afterload reflects the preload and the effort required to push the blood out into the arterial system. Mean arterial pressure is a clinical indicator of afterload. (p. 500)
41. **Ejection fraction**, the percent of EDV ejected with one contraction (stroke volume/EDV), is one measure for evaluating ventricular function. (p. 500)

Questions

Answers: p. A-1

Level One Reviewing Facts and Terms

1. What contributions to understanding the cardiovascular system did each of the following people make?
   (a) William Harvey
   (b) Otto Frank and Ernest Starling
   (c) Marcello Malpighi
2. List three functions of the cardiovascular system.
3. Put the following structures in the order in which blood passes through them, starting and ending with the left ventricle:
   (a) left ventricle
   (b) systemic veins
   (c) pulmonary circulation
   (d) systemic arteries
   (e) aorta
   (f) right ventricle
4. The primary factor causing blood to flow through the body is a(n) ________ gradient. In humans, the value of this gradient is highest at the ________ and in the ________. It is lowest in the ________. In a system in which fluid is flowing, pressure decreases over distance because of ________.
5. If vasodilation occurs in a blood vessel, pressure (increases/decreases).
6. The specialized cell junctions between myocardial cells are called ________. These areas contain ________ that allow rapid conduction of electrical signals.
7. Trace an action potential from the SA node through the conducting system of the heart.
8. Distinguish between the two members of each of the following pairs:
   (a) end-systolic volume and end-diastolic volume
   (b) sympathetic and parasympathetic control of heart rate
   (c) diastole and systole
   (d) systemic and pulmonary circulation
   (e) AV node and SA node

9. Match the descriptions with the correct anatomic term(s). Not all terms are used and terms may be used more than once. Give a definition for the unused terms.

(a) tough membranous sac that encases the heart	1. aorta
(b) valve between ventricle and a main artery	2. apex
(c) a vessel that carries blood away from the heart	3. artery
(d) lower chamber of the heart	4. atria
(e) valve between left atrium and left ventricle	5. atrium
(f) primary artery of the systemic circulation	6. AV valve
(g) muscular layer of the heart	7. base
(h) narrow end of the heart; points downward	8. bicuspid valve
(i) valve with papillary muscles	9. endothelium
(j) the upper chambers of the heart	10. myocardium
	11. pericardium
	12. semilunar valve
	13. tricuspid valve
	14. ventricle

10. What events cause the two principal heart sounds?
11. What is the proper term for each of the following?
    (a) number of heart contractions per minute
    (b) volume of blood in the ventricle before the heart contracts
    (c) volume of blood that enters the aorta with each contraction
    (d) volume of blood that leaves the heart in one minute
    (e) volume of blood in the entire body

Level Two Reviewing Concepts

12. Concept maps:
    (a) Create a map showing blood flow through the heart and body. Label as many structures as you can.
    (b) Create a map for control of cardiac output using the following terms. You may add additional terms.

• ACh	• heart rate
• adrenal medulla	• length-tension relationship
• autorhythmic cells	• muscarinic receptor
• β_1-receptor	• norepinephrine
• Ca^{2+}	• parasympathetic neurons
• Ca^{2+}-induced Ca^{2+} release	• respiratory pump
• cardiac output	• skeletal muscle pump
• contractile myocardium	• stroke volume
• contractility	• sympathetic neurons
• force of contraction	• venous return

13. List the events of the cardiac cycle in sequence, beginning with atrial and ventricular diastole. Note when valves open and close. Describe what happens to pressure and blood flow in each chamber at each step of the cycle.
14. Compare and contrast the structure of a cardiac muscle cell with that of a skeletal muscle cell. What unique properties of cardiac muscle are essential to its function?
15. Explain why contractions in cardiac muscle cannot sum or exhibit tetanus.
16. Correlate the waves of an ECG with mechanical events in the atria and ventricles. Why are there only three electrical events but four mechanical events?
17. Match the following ion movements with the appropriate phrase. More than one ion movement may apply to a single phrase. Some choices may not be used.

(a) slow rising phase of autorhythmic cells	1. K^+ from ECF to ICF
(b) plateau phase of contractile cells	2. K^+ from ICF to ECF
(c) rapid rising phase of contractile cells	3. Na^+ from ECF to ICF
(d) rapid rising phase of autorhythmic cells	4. Na^+ from ICF to ECF
(e) rapid falling phase of contractile cells	5. Ca^{2+} from ECF to ICF
(f) falling phase of autorhythmic cells	6. Ca^{2+} from ICF to ECF
(g) cardiac muscle contraction	
(h) cardiac muscle relaxation	

18. List and briefly explain four types of information that an ECG provides about the heart.
19. Define inotropic effect. Name two drugs that have a positive inotropic effect on the heart.

Level Three Problem Solving

20. Two drugs used to reduce cardiac output are calcium channel blockers and beta (receptor) blockers. What effect do these drugs have on the heart that explains how they decrease cardiac output?
21. Police Captain Jeffers has suffered a myocardial infarction.
    (a) Explain to his (nonmedically oriented) family what has happened to his heart.
    (b) When you analyzed his ECG, you referred to several different leads, such as lead I and lead III. What are leads?
    (c) Why is it possible to record an ECG on the body surface without direct access to the heart?
22. What might cause a longer-than-normal PR interval in an ECG?
23. The following paragraph is a summary of a newspaper article:

    A new treatment for atrial fibrillation due to an excessively rapid rate at the SA node involves a high-voltage electrical pulse administered to the AV node to destroy its autorhythmic cells. A ventricular pacemaker is then implanted in the patient.

    Briefly explain the physiological rationale for this treatment. Why is a rapid atrial depolarization rate dangerous? Why is the AV node destroyed in this procedure? Why must a pacemaker be implanted?

Level Four Quantitative Problems

24. Police Captain Jeffers in question 21 has an ejection fraction (SV divided by EDV) of only 25%. His stroke volume is 40 mL/beat, and his heart rate is 100 beats/min. What are his EDV, ESV, and CO? Show your calculations.
25. If 1 cm water = 0.74 mm Hg:
    (a) Convert a pressure of 120 mm Hg to cm H_2O.
    (b) Convert a pressure of 90 cm H_2O to mm Hg.
26. Calculate cardiac output if stroke volume is 65 mL/beat and heart rate is 80 beats/min.
27. Calculate end-systolic volume if end-diastolic volume is 150 mL and stroke volume is 65 mL/beat.

28. A person has a total blood volume of 5 L. Of this total, assume that 4 L is contained in the systemic circulation and 1 L is in the pulmonary circulation. If the person has a cardiac output of 5 L/min, how long will it take (a) for a drop of blood leaving the left ventricle to return to the left ventricle and (b) for a drop of blood to go from the right ventricle to the left ventricle?

Answers

Answers to Concept Check Questions

Page 466

1. A cardiovascular system has tubes (vessels), fluid (blood), and a pump (heart).
2. (a) The pulmonary circulation takes blood to and from the lungs; the systemic circulation takes blood to and from the rest of the body. (b) An artery carries blood away from the heart; a vein carries blood to the heart. (c) An atrium is an upper heart chamber that receives blood entering the heart; a ventricle is a lower heart chamber that pumps blood out of the heart.

Page 470

3. The pressure gradient is more important.
4. The bottom tube has the greater flow because it has the larger pressure gradient (50 mm Hg versus 40 mm Hg for the top tube).
5. Tube C has the highest flow because it has the largest radius of the four tubes (less resistance) and the shorter length (less resistance). (Tube B has the same radius as tube C but a longer length and therefore offers greater resistance to flow). Tube D, with the greatest resistance due to longer length and narrow radius, has the lowest flow.

Page 471

6. If the canals are identical in size and therefore in cross-sectional area A, the canal with the higher velocity of flow v has the higher flow rate Q. (From equation 7, $Q = v \times A$).

Page 475

7. Connective tissue is not excitable and is therefore unable to conduct action potentials.
8. Superior vena cava → right atrium → tricuspid (right AV) valve → right ventricle → pulmonary (right semilunar) valve → pulmonary trunk → pulmonary vein → left atrium → mitral (bicuspid, left AV) valve → left ventricle → aortic (left semilunar) valve → aorta
9. The AV valves prevent backward flow of blood. If one fails, blood leaks back into the atrium.

Page 478

10. Skeletal muscle L-type Ca^{2+} channels (also called DHP receptors) are mechanically linked to the RyR Ca^{2+} release channels of the sarcoplasmic reticulum. Myocardial L-type Ca^{2+} channels open to allow Ca^{2+} into the cell. In both muscles, sarcolemma Ca^{2+} channels are associated with RyR Ca^{2+} release channels on the SR.
11. From this experiment, it is possible to conclude that myocardial cells require extracellular calcium for contraction but skeletal muscle cells do not.

Page 479

12. If all calcium channels in the muscle cell membrane are blocked, there will be no contraction. If only some are blocked, the force of contraction will be smaller than the force created with all channels open.

Page 481

13. Na^+ influx causes neuronal depolarization, and K^+ efflux causes neuronal repolarization.
14. The refractory period represents the time required for the Na^+ channel gates to reset (activation gate closes, inactivation gate opens).
15. If cardiac Na^+ channels are completely blocked with lidocaine, the cell will not depolarize and therefore will not contract. Partial blockade will decrease electrical conduction.

Page 482

16. Increasing K^+ permeability hyperpolarizes the membrane potential.
17. Ivabradine slows heart rate and is used to lower abnormally high heart rates.
18. The Ca^{2+} channels in autorhythmic cells are not the same as those in contractile cells. Autorhythmic Ca^{2+} channels open rapidly when the membrane potential reaches about −50 mV and close when it reaches about +20 mV. The Ca^{2+} channels in contractile cells are slower and do not open until the membrane has depolarized fully.
19. If tetrodotoxin is applied to a myocardial autorhythmic cell, nothing will happen because there are no voltage-gated Na^+ channels in the cell.
20. Cutting the vagus nerve caused heart rate to increase, so the parasympathetic fibers in the nerve must slow heart rate.

Page 486

21. The AV node conducts action potentials from atria to ventricles. It also slows down the speed at which those action potentials are conducted, allowing atrial contraction to end before ventricular contraction begins.
22. The SA node is in the upper right atrium.
23. The fastest pacemaker sets the heart rate, so the heart rate increases to 120 beats/min.

Page 492

24. The atrium has lower pressure than the venae cavae.

Page 493

25. (a) ventricle, (b) ventricle, (c) atrium, (d) ventricle
26. (a) ventricular ejection, (b) isovolumic ventricular contraction and ventricular ejection (c) from isovolumic ventricular relaxation until ventricular contraction begins again
27. EDV occurs in step 3, and ESV occurs in step 5.
28. (a) E, (b) A, (c) D, (d) B, (e) C, (f) F
29. Atrial pressure increases because pressure on the mitral valve pushes the valve back into the atrium, decreasing atrial volume. Atrial pressure decreases during the initial part of ventricular systole as the atrium relaxes. The pressure then increases as the atrium fills with blood. Atrial pressure begins to decrease at point D, when the mitral valve opens and blood flows down into the ventricles.
30. Ventricular pressure shoots up when the ventricles contract on a fixed volume of blood.

Page 495

31. After 10 beats, the pulmonary circulation will have gained 10 mL of blood and the systemic circulation will have lost 10 mL.

Page 500

32. Your drawing should show a β_1-receptor on the cell membrane activating intracellular cAMP, which should have an arrow drawn to Ca^{2+} channels on the sarcoplasmic reticulum. Open channels should be shown increasing cytoplasmic Ca^{2+}. A second arrow should go from cAMP to Ca^{2+}-ATPase on the SR and the cell membrane, showing increased uptake in the SR and increased removal of Ca^{2+} from the cell.
33. The aortic valve is found between the left ventricle and the aorta. A stenotic aortic valve would increase afterload on the ventricle.

Answers to Figure and Graph Questions

Page 465

Figure 14.1: The two portal systems are in the GI tract and in the kidneys, with two capillary beds connected in series for each portal system.

Page 468

Figure 14.3: If radius = 3, R = 1/81 and flow = 1/81, which is about 5 × flow through B.

Page 471

Figure 14.4: If A = 3, $v = 4$ cm/min.

Page 478

Figure 14.9: Smooth and cardiac muscle are the same except where indicated. (1) Multi-unit smooth muscle and skeletal muscle require neurotransmitters to initiate the action potential. (2) No significant Ca^{2+} entry in skeletal muscle. (3) No CICR in skeletal muscle. (4) Ca^{2+} leaves the SR in all types. (5) Calcium signal in all types. (6)-(7) Smooth muscle lacks troponin. Skeletal muscle is similar to cardiac. (8) Same in all types. (9) NCX lacking in skeletal muscle. (10) Same in all types.

Page 479

Figure 14.10: The only difference is Ca^{2+} entry during the plateau phase.

Page 481

Figure 14.12: 1. Phase 2 (the plateau) of the contractile cell action potential has no equivalent in the autorhythmic cell action potential. Phase 4 is approximately equivalent to the pacemaker potential. Both action potentials have rising phases, peaks, and falling phases. 2. (a) and (c)

Page 484

Figure 14.14: If the AV node could not depolarize, there would be no conduction of electrical activity into the ventricles. Ventricular pacemakers would take over.

Page 488

Figure 14.15: 1. The heart rate is either 75 beats/min or 80 beats/min, depending on how you calculate it. If you use the data from one R peak to the next, the time interval between the two peaks is 0.8 sec; therefore, 1 beat/0.8 sec × 60 sec/1 min = 75 beats/min. However, it is more accurate to estimate rate by using several seconds of the ECG tracing rather than one RR interval because beat-to-beat intervals may vary. If you start counting at the first R wave on the top graph and go right for 3 sec, there are 4 beats in that time period, which means 4 beats/3 sec × 60 sec/1 min = 80 beats/min. 2. In (2), notice that there is no regular association between the P waves and the QRS complexes (the P-R segment varies in length). Notice also that not every P wave has an associated QRS complex. Both P waves and QRS complexes appear at regular intervals, but the atrial rate (P waves) is faster than the ventricular rate (QRS complexes). The QRS complexes are not their usual shape, and the T wave is absent because the ventricular depolarization is not following its normal path. In (3), there are identifiable R waves but no P waves. In (4), there are no recognizable waves at all, indicating that the depolarizations are not following the normal conduction path. 3. Starting at left, the waves are P, P, QRS, T, P, P, QRS, T, P, P, P, and so on. Each P wave that is not followed by a QRS wave suggests an intermittent conduction block at the AV node.

Page 491

Figure 14.17: 1. (a) C → D, (b) B → C, (c) D → A, (d) A → B. 2. (a) C, (b) A, (c) D, (d) B.

Page 498

Figure 14.20: (b) Maximum stroke volume is about 160 mL/beat, first achieved when end-diastolic volume is about 330 mL. (c) At point A, the heart under the influence of norepinephrine has a larger stroke volume and is therefore creating more force.

Page 501

Figure 14.22: Heart rate is the only parameter controlled by ACh. Heart rate and contractility are both controlled by norepinephrine. The SA node has muscarinic receptors. The SA node and contractile myocardium have β_1-receptors.

15 Blood Flow and the Control of Blood Pressure

Since 1900, CVD (cardiovascular disease) has been the No. 1 killer in the United States every year but 1918.

—American Heart Association, *Heart Disease and Stroke Statistics*

Background Basics

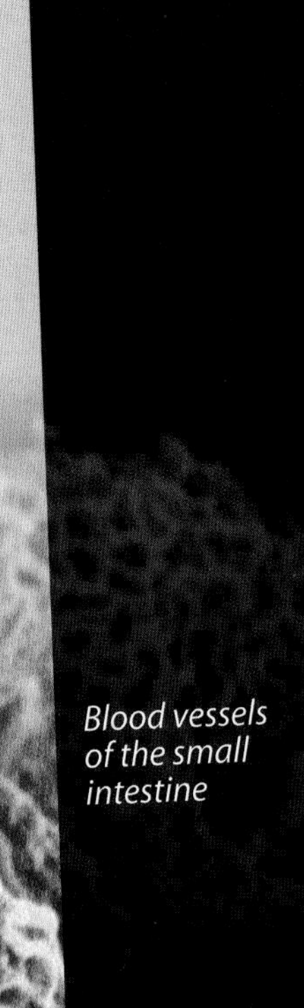

Blood vessels of the small intestine

Anthony was sure he was going to be a physician, until the day in physiology laboratory they studied blood types. When the lancet pierced his fingertip and he saw the drop of bright red blood well up, the room started to spin, and then everything went black. He awoke, much embarrassed, to the sight of his classmates and the teacher bending over him.

Anthony suffered an attack of *vasovagal syncope* (syncope = fainting), a benign and common emotional reaction to blood, hypodermic needles, or other upsetting sights. Normally, homeostatic regulation of the cardiovascular system maintains blood flow, or *perfusion*, to the heart and brain. In vasovagal syncope, signals from the nervous system cause a sudden decrease in blood pressure, and the individual faints from lack of oxygen to the brain. In this chapter you will learn how the heart and blood vessels work together most of the time to prevent such problems.

A simplified model of the cardiovascular system (■ Fig. 15.1) illustrates the key points we discuss in this chapter. This model shows the heart as two separate pumps, with the

RUNNING PROBLEM

Essential Hypertension

"Doc, I'm as healthy as a horse," says Kurt English, age 56, during his long-overdue annual physical examination. "I don't want to waste your time. Let's get this over with." But to Dr. Arthur Cortez, Kurt does not appear to be the picture of health: he is about 30 pounds overweight. When Dr. Cortez asks about his diet, Kurt replies, "Well, I like to eat." Exercise? "Who has the time?" replies Kurt. Dr. Cortez wraps a blood pressure cuff around Kurt's arm and takes a reading. "Your blood pressure is 164 over 100," says Dr. Cortez. "We'll take it again in 15 minutes. If it's still high, we'll need to discuss it further." Kurt stares at his doctor, flabbergasted. "But how can my blood pressure be too high? I feel fine!" he protests.

509 515 519 522 533 537

15

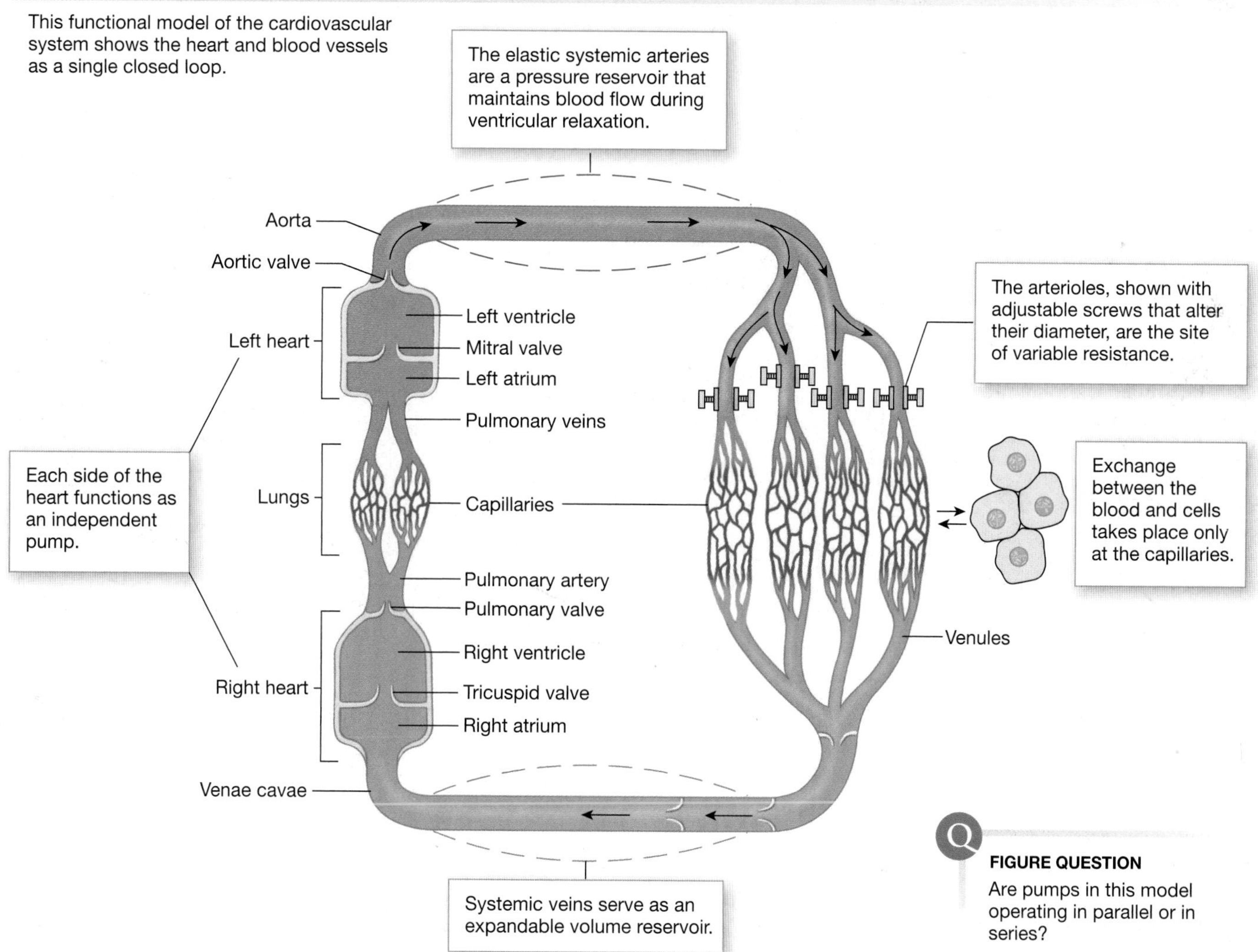
FUNCTIONAL MODEL OF THE CARDIOVASCULAR SYSTEM
This functional model of the cardiovascular system shows the heart and blood vessels as a single closed loop.
The elastic systemic arteries are a pressure reservoir that maintains blood flow during ventricular relaxation.
Aorta
Aortic valve
Left heart
Left ventricle
Mitral valve
Left atrium
Pulmonary veins
Each side of the heart functions as an independent pump.
Lungs
Capillaries
Pulmonary artery
Pulmonary valve
Right heart
Right ventricle
Tricuspid valve
Right atrium
Venae cavae
The arterioles, shown with adjustable screws that alter their diameter, are the site of variable resistance.
Exchange between the blood and cells takes place only at the capillaries.
Venules
Systemic veins serve as an expandable volume reservoir.
Q
FIGURE QUESTION
Are pumps in this model operating in parallel or in series?

■ **Fig. 15.1**

right heart pumping blood to the lungs and back to the left heart. The left heart then pumps blood through the rest of the body and back to the right heart.

Blood leaving the left heart enters systemic arteries, shown here as an expandable, elastic region. Pressure produced by contraction of the left ventricle is stored in the elastic walls of arteries and slowly released through *elastic recoil.* This mechanism maintains a continuous driving pressure for blood flow during ventricular relaxation. For this reason, the arteries are known as the *pressure reservoir* {*reservare,* to retain} of the circulatory system.

Downstream from the arteries, small vessels called **arterioles** create a high-resistance outlet for arterial blood flow. Arterioles direct distribution of blood flow to individual tissues by selectively constricting and dilating, so they are known as the site of *variable resistance.* Arteriolar diameter is regulated both by local factors, such as tissue oxygen concentrations, and by the autonomic nervous system and hormones.

When blood flows into the capillaries, their leaky epithelium allows exchange of materials between the plasma, the interstitial fluid, and the cells of the body. At the distal end of the capillaries, blood flows into the venous side of the circulation. The veins act as a *volume reservoir* from which blood can be sent to the arterial side of the circulation if blood pressure falls too low. From the veins, blood flows back to the right heart.

Total blood flow through any level of the circulation is equal to cardiac output. For example, if cardiac output is 5 L/min, blood flow through all the systemic capillaries is 5 L/min. In the same manner, blood flow through the pulmonary side of the circulation is equal to blood flow through the systemic circulation.

BLOOD VESSEL STRUCTURE

The walls of blood vessels vary in diameter and composition. The bars show the relative proportions of the different tissues. The endothelium and its underlying elastic tissue together form the tunica intima. (Adapted from A.C. Burton, *Physiol Rev* 34: 619–642, 1954).

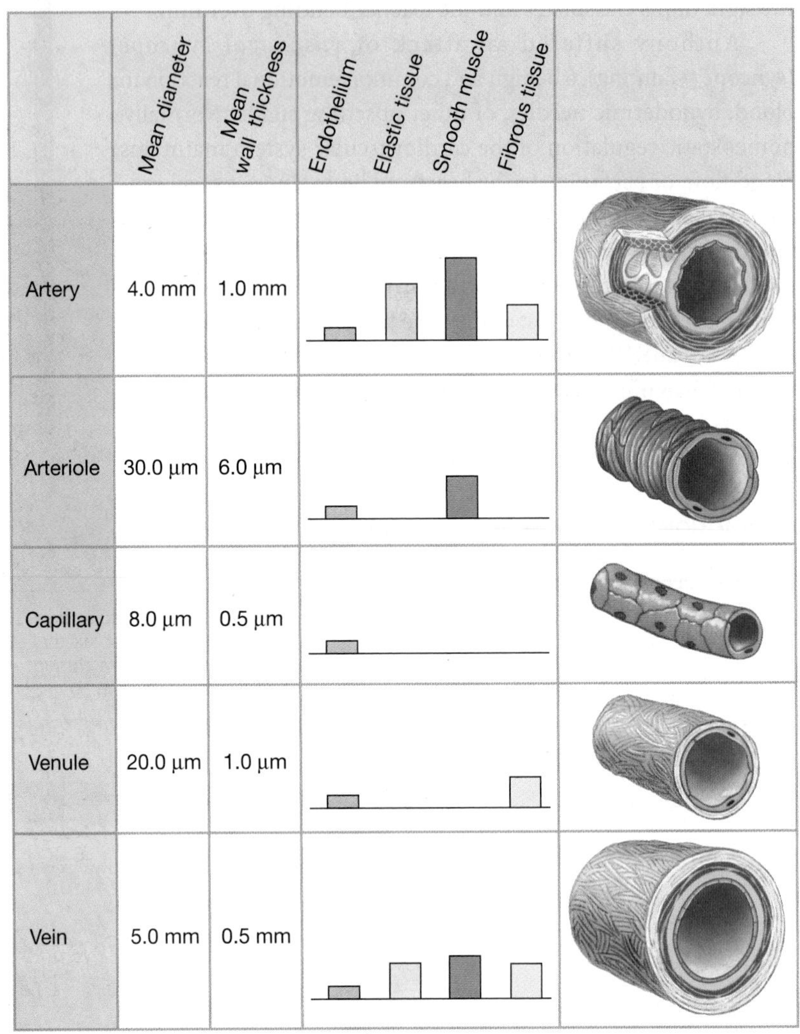

Mean diameter
Mean wall thickness
Endothelium
Elastic tissue
Smooth muscle
Fibrous tissue
Artery
4.0 mm
1.0 mm
Arteriole
30.0 µm
6.0 µm
Capillary
8.0 µm
0.5 µm
Venule
20.0 µm
1.0 µm
Vein
5.0 mm
0.5 mm

Fig. 15.2

The Blood Vessels

The walls of blood vessels are composed of layers of smooth muscle, elastic connective tissue, and fibrous connective tissue (Fig. 15.2). The inner lining of all blood vessels is a thin layer of **endothelium,** a type of epithelium. For years, the endothelium was thought to be simply a passive barrier. However, we now know that endothelial cells secrete many paracrines and play important roles in the regulation of blood pressure, blood vessel growth, and absorption of materials. Some scientists have even proposed that endothelium be considered a separate physiological organ system.

In most vessels, layers of connective tissue and smooth muscle surround the endothelium. The endothelium and its adjacent elastic connective tissue together make up the *tunica intima,* usually called simply the *intima* {*intimus,* innermost}. The thickness of the smooth muscle–connective tissue layers surrounding the intima varies in different vessels. The descriptions that follow apply to the vessels of the systemic circulation, although those of the pulmonary circulation are generally similar.

Blood Vessels Contain Vascular Smooth Muscle

The smooth muscle of blood vessels is known as **vascular smooth muscle.** Most blood vessels contain smooth muscle, arranged in either circular or spiral layers. *Vasoconstriction* narrows the diameter of the vessel lumen, and *vasodilation* widens it.

In most blood vessels, smooth muscle cells maintain a state of partial contraction at all times, creating the condition known as *muscle tone* [p. 446]. Contraction of smooth muscle, like that of cardiac muscle, depends on the entry of Ca^{2+} from the extracellular fluid through Ca^{2+} channels [p. 431]. A variety of chemicals, including neurotransmitters, hormones, and paracrines, influences vascular smooth muscle tone. Many vasoactive paracrines are secreted either by endothelial cells lining blood vessels or by tissues surrounding the vessels.

Arteries and Arterioles Carry Blood Away from the Heart

The aorta and major arteries are characterized by walls that are both stiff and springy. Arteries have a thick smooth muscle layer and large amounts of elastic and fibrous connective tissue (Fig. 15.2). Because of the stiffness of the fibrous tissue, substantial amounts of energy are required to stretch the walls of an artery outward, but that energy can be stored by the stretched elastic fibers and released through elastic recoil.

The arteries and arterioles are characterized by a divergent {*divergere,* bend apart} pattern of blood flow. As major arteries divide into smaller and smaller arteries, the character of the wall changes, becoming less elastic and more muscular. The walls of arterioles contain several layers of smooth muscle that contract and relax under the influence of various chemical signals.

Arterioles, along with capillaries and small postcapillary vessels called venules, form the *microcirculation*. Regulation of blood flow through the microcirculation is an active area of physiological research.

Some arterioles branch into vessels known as **metarterioles** {*meta-,* beyond} (■ Fig. 15.3). True arterioles have a continuous smooth muscle layer, but the wall of a metarteriole is only partially surrounded by smooth muscle. Blood flowing through metarterioles can take one of two paths. If muscle rings called **precapillary sphincters** {*sphingein,* to hold tight} are relaxed, blood flowing into a metarteriole is directed into adjoining capillary beds (Fig. 15.3b).

If the precapillary sphincters are constricted, metarteriole blood bypasses the capillaries and goes directly to the venous circulation (Fig. 15.3c). In addition, metarterioles allow white blood cells to go directly from the arterial to the venous circulation. Capillaries are barely large enough to let red blood cells through, much less white blood cells, which are twice as large.

Exchange Takes Place in the Capillaries

Capillaries are the smallest vessels in the cardiovascular system. They and the postcapillary venules are the site of exchange between the blood and the interstitial fluid. To facilitate

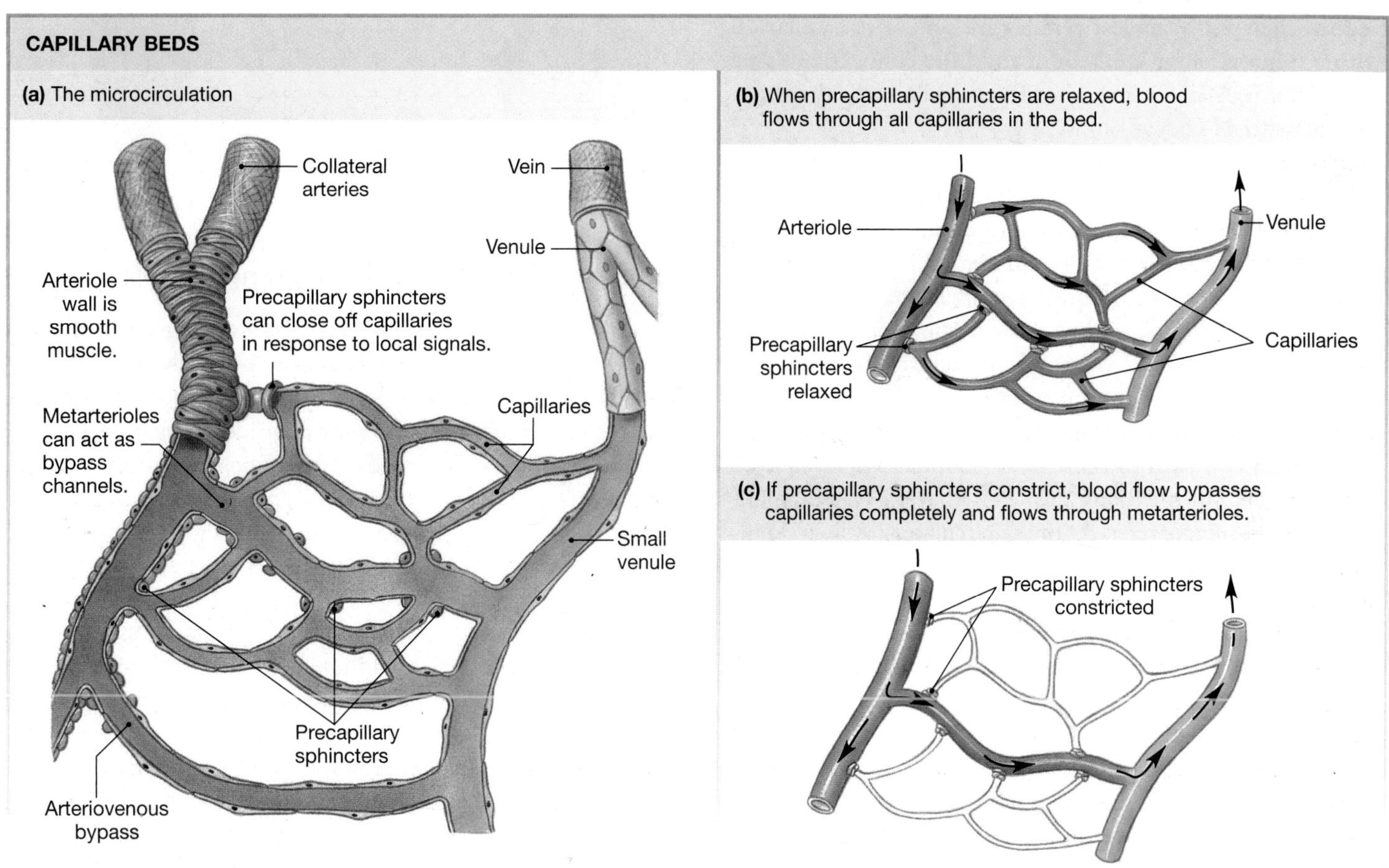
CAPILLARY BEDS
(a) The microcirculation
Collateral arteries
Vein
Venule
Arteriole wall is smooth muscle.
Precapillary sphincters can close off capillaries in response to local signals.
Metarterioles can act as bypass channels.
Capillaries
Small venule
Precapillary sphincters
Arteriovenous bypass
(b) When precapillary sphincters are relaxed, blood flows through all capillaries in the bed.
Arteriole
Venule
Precapillary sphincters relaxed
Capillaries
(c) If precapillary sphincters constrict, blood flow bypasses capillaries completely and flows through metarterioles.
Precapillary sphincters constricted

■ **Fig. 15.3**

exchange of materials, capillaries lack smooth muscle and elastic or fibrous tissue reinforcement (Fig. 15.2). Instead, their walls consist of a flat layer of endothelium, one cell thick, supported on an acellular matrix called the *basal lamina* (basement membrane) [p. 82].

Many capillaries are closely associated with cells known as **pericytes** {*peri-*, around}. In most tissues, these highly branched contractile cells surround the capillaries, forming a mesh-like outer layer between the capillary endothelium and the interstitial fluid. Pericytes contribute to the "tightness" of capillary permeability: the more pericytes, the less leaky the capillary endothelium. Cerebral capillaries, for example, are surrounded by pericytes and glial cells, and have tight junctions that create the *blood-brain barrier* [p. 296].

Pericytes secrete factors that influence capillary growth, and they can differentiate to become new endothelial or smooth muscle cells. Loss of pericytes around capillaries of the retina is a hallmark of the disease *diabetic retinopathy,* a leading cause of blindness. Scientists are now trying to determine whether pericyte loss is a cause or consequence of the retinopathy.

Blood Flow Converges in the Venules and Veins

Blood flows from the capillaries into small vessels called **venules**. The very smallest venules are similar to capillaries, with a thin exchange epithelium and little connective tissue (Fig. 15.2). They are distinguished from capillaries by their convergent pattern of flow.

Smooth muscle begins to appear in the walls of larger venules. From venules, blood flows into veins that become larger in diameter as they travel toward the heart. Finally, the largest veins, the venae cavae, empty into the right atrium. To assist venous flow, some veins have internal one-way valves (■ Fig. 15.4). These valves, like those in the heart, ensure that blood passing the valve cannot flow backward. Once blood reaches the vena cava, there are no valves.

Veins are more numerous than arteries and have a larger diameter. As a result of their large volume, the veins hold more than half of the blood in the circulatory system, making them the *volume reservoir* of the circulatory system. Veins lie closer to the surface of the body than arteries, forming the bluish blood vessels that you see running just under the skin. Veins have thinner walls than arteries, with less elastic tissue. As a result, they expand easily when they fill with blood.

When you have blood drawn from your arm (*venipuncture*), the technician uses a tourniquet to exert pressure on the blood vessels. Blood flow into the arm through deep high-pressure arteries is not affected, but pressure exerted by the tourniquet stops outflow through the low-pressure veins. As a result, blood collects in the surface veins, making them stand out against the underlying muscle tissue.

Valves ensure one-way flow in veins.

Valves in the veins prevent backflow of blood.

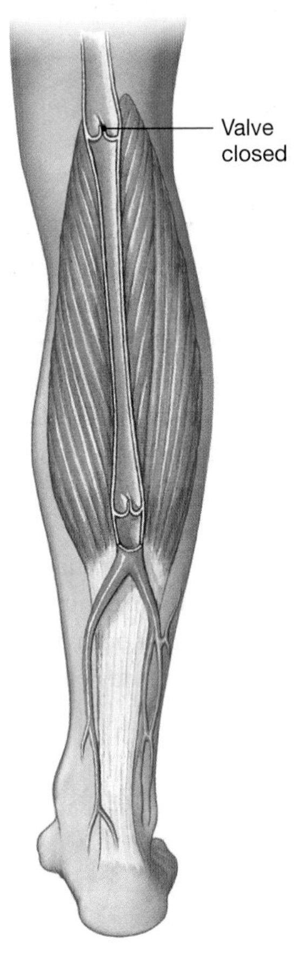

When the skeletal muscles compress the veins, they force blood toward the heart (the skeletal muscle pump).

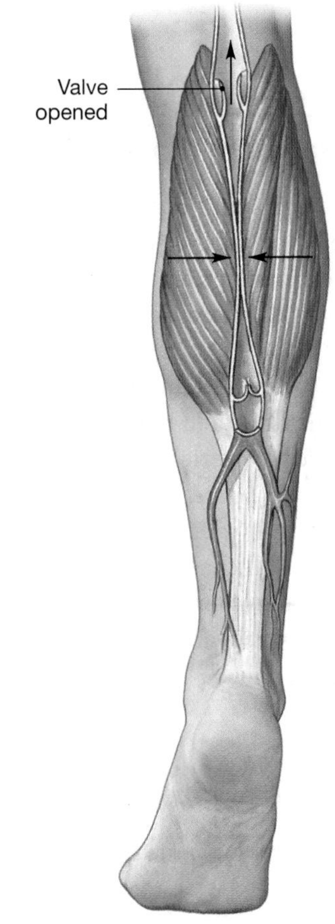

■ **Fig. 15.4**

Angiogenesis Creates New Blood Vessels

One topic of great interest to researchers is **angiogenesis** {*angeion,* vessel + *gignesthai,* to beget}, the process by which new blood vessels develop, especially after birth. In children, blood vessel growth is necessary for normal development. In adults, angiogenesis takes place as wounds heal and as the uterine lining grows after menstruation. Angiogenesis also occurs with endurance exercise training, enhancing blood flow to the heart muscle and to skeletal muscles.

The growth of malignant tumors is a disease state that requires angiogenesis. As cancer cells invade tissues and multiply, they instruct the host tissue to develop new blood vessels to feed the growing tumor. Without these new vessels, the interior cells of a cancerous mass would be unable to get adequate oxygen and nutrients, and would die.

From studies of normal blood vessels and tumor cells, scientists learned that angiogenesis is controlled by a balance of angiogenic and antiangiogenic cytokines. A number of related growth factors, including *vascular endothelial growth factor* (VEGF) and *fibroblast growth factor* (FGF), promote angiogenesis.

These growth factors are *mitogens,* meaning they promote mitosis, or cell division. They are normally produced by smooth muscle cells and pericytes.

Cytokines that inhibit angiogenesis include *angiostatin,* made from the blood protein plasminogen, and *endostatin* {*stasis,* a state of standing still}. Scientists are currently testing these cytokines for treating cancer, to see if they can block angiogenesis and literally starve tumors to death.

In contrast, **coronary heart disease,** also known as *coronary artery disease,* is a condition in which blood flow to the myocardium is decreased by fatty deposits that narrow the lumen of the coronary arteries. In some individuals, new blood vessels develop spontaneously and form *collateral circulation* that supplements flow through the partially blocked artery. Researchers are testing angiogenic cytokines to see if they can duplicate this natural process and induce angiogenesis to replace *occluded* vessels {*occludere,* to close up}.

Blood Pressure

The force that creates blood flow through the cardiovascular system is ventricular contraction [p. 490]. As blood under pressure is ejected from the left ventricle, the aorta and arteries expand to accommodate it (■ Fig. 15.5a). When the ventricle relaxes and the semilunar valve closes, the elastic arterial walls recoil, propelling the blood forward into smaller arteries and arterioles (Fig. 15.5b). By sustaining the *driving pressure* for blood flow during ventricular relaxation, the arteries keep blood flowing continuously through the blood vessels.

Blood flow obeys the rules of fluid flow [p. 467]. Flow is directly proportional to the pressure gradient between any two points, and inversely proportional to the resistance of the vessels to flow (■ Tbl. 15.1). Unless otherwise noted, the discussion that follows is restricted to the events that take place in the systemic circuit. You will learn about pulmonary blood flow when you study the respiratory system.

Blood Pressure Is Highest in Arteries and Lowest in Veins

Blood pressure is highest in the arteries and decreases continuously as blood flows through the circulatory system (■ Fig. 15.6). The decrease in pressure occurs because energy is lost as a result of the resistance to flow offered by the vessels. Resistance to blood flow also results from friction between the blood cells.

In the systemic circulation, the highest pressure occurs in the aorta and results from pressure created by the left ventricle. Aortic

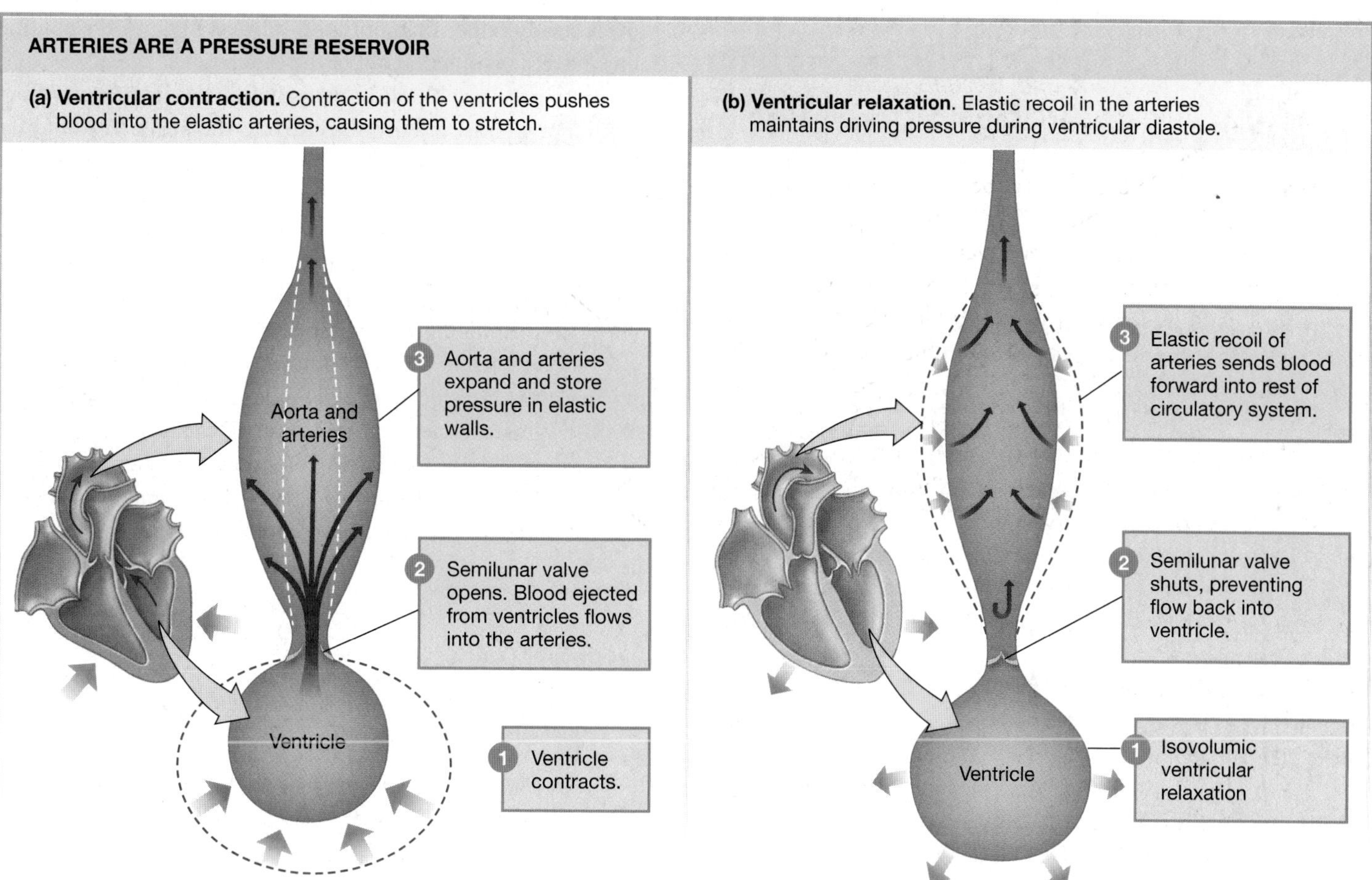

■ **Fig. 15.5**

15

Pressure, Flow, and Resistance in the Cardiovascular System	Table 15.1
Flow ∝ ΔP/R	
1. Blood flows if a pressure gradient (ΔP) is present.	
2. Blood flows from areas of higher pressure to areas of lower pressure.	
3. Blood flow is opposed by the resistance (R) of the system.	
4. Three factors affect resistance: radius of the blood vessels, length of the blood vessels, and viscosity of the blood [p. 469].	
5. Flow is usually expressed in either liters or milliliters per minute (L/min or mL/min).	
6. Velocity of flow is usually expressed in either centimeters per minute (cm/min) or millimeters per second (mm/sec).	
7. The primary determinant of velocity (when flow rate is constant) is the total cross-sectional area of the vessel(s).	

SYSTEMIC CIRCULATION PRESSURES

Pressure waves created by ventricular contraction travel into the blood vessels. Pressure in the arterial side of the circulation cycles but the pressure waves diminish in amplitude with distance and disappear at the capillaries.

Pulse pressure = Systolic pressure minus diastolic pressure

Mean arterial pressure = Diastolic pressure + 1/3 (pulse pressure)

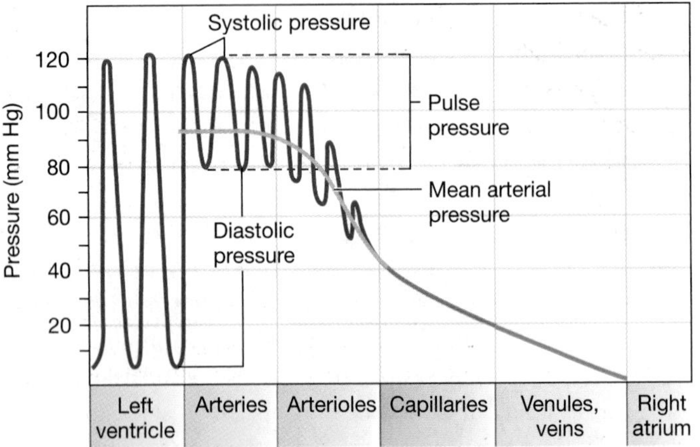

Fig. 15.6

pressure reaches an average high of 120 mm Hg during ventricular systole (**systolic pressure**), then falls steadily to a low of 80 mm Hg during ventricular diastole (**diastolic pressure**). Notice that pressure in the ventricle falls to only a few mm Hg as the ventricle relaxes, but diastolic pressure in the large arteries remains relatively high. The high diastolic pressure in arteries reflects the ability of those vessels to capture and store energy in their elastic walls.

The rapid pressure increase that occurs when the left ventricle pushes blood into the aorta can be felt as a **pulse**, or pressure wave, transmitted through the fluid-filled arteries. The pressure wave travels about 10 times faster than the blood itself. Even so, a pulse felt in the arm is occurring slightly after the ventricular contraction that created the wave.

The amplitude of the pressure wave decreases over distance because of friction, and the wave finally disappears at the capillaries (Fig. 15.6). **Pulse pressure**, a measure of the strength of the pressure wave, is defined as systolic pressure minus diastolic pressure:

$$\text{Systolic pressure} - \text{diastolic pressure} = \text{pulse pressure}$$

For example, in the aorta:

$$120 \text{ mm Hg} - 80 \text{ mm Hg} = 40 \text{ mm Hg pressure}$$

By the time blood reaches the veins, pressure has fallen because of friction, and a pressure wave no longer exists. Venous blood flow is steady rather than pulsatile, pushed along by the continuous movement of blood out of the capillaries.

Low-pressure blood in veins below the heart must flow "uphill," or against gravity, to return to the heart. Try holding your arm straight down without moving for several minutes and notice how the veins in the back of your hand begin to stand out as they fill with blood. (This effect may be more evident in older people, whose subcutaneous connective tissue has lost elasticity). Then raise your hand so that gravity assists the venous flow and watch the bulging veins disappear.

Blood return to the heart, known as *venous return,* is aided by valves, the *skeletal muscle pump*, and the *respiratory pump* [p. 498]. When muscles such as those in the calf of the leg contract, they compress the veins, forcing blood upward past the valves. While your hand is hanging down, try clenching and unclenching your fist to see the effect muscle contraction has on distention of the veins.

Concept Check Answers: p. 542

1. Would you expect to find valves in the veins leading from the brain to the heart? Defend your answer.
2. If you checked the pulse in a person's carotid artery and left wrist at the same time, would the pressure waves occur simultaneously? Explain.
3. Who has the higher pulse pressure, someone with blood pressure of 90/60 or someone with blood pressure of 130/95?

Arterial Blood Pressure Reflects the Driving Pressure for Blood Flow

Arterial blood pressure, or simply "blood pressure," reflects the driving pressure created by the pumping action of the heart. Because ventricular pressure is difficult to measure, it is customary to assume that arterial blood pressure reflects ventricular pressure. Because arterial pressure is pulsatile, we use a single value—the **mean arterial pressure** (MAP)—to represent driving pressure. MAP is represented graphically in Fig. 15.6.

Mean arterial pressure is estimated as diastolic pressure plus one-third of pulse pressure:

$$\text{MAP} = \text{diastolic P} + 1/3\,(\text{systolic P} - \text{diastolic P})$$

For a person whose systolic pressure is 120 and diastolic pressure is 80:

$$\begin{aligned}\text{MAP} &= 80\text{ mm Hg} + 1/3\,(120 - 80\text{ mm Hg})\\ &= 93\text{ mm Hg}\end{aligned}$$

Mean arterial pressure is closer to diastolic pressure than to systolic pressure because diastole lasts twice as long as systole.

Abnormally high or low arterial blood pressure can be indicative of a problem in the cardiovascular system. If blood pressure falls too low (*hypotension*), the driving force for blood flow is unable to overcome opposition by gravity. In this instance, blood flow and oxygen supply to the brain are impaired, and the person may become dizzy or faint.

On the other hand, if blood pressure is chronically elevated (a condition known as *hypertension*, or high blood pressure), high pressure on the walls of blood vessels may cause weakened areas to rupture and bleed into the tissues. If a rupture occurs in the brain, it is called a *cerebral hemorrhage* and may cause the loss of neurological function commonly called a *stroke*. If a weakened area ruptures in a major artery, such as the descending aorta, rapid blood loss into the abdominal cavity causes blood pressure to fall below the critical minimum. Without prompt treatment, rupture of a major artery is fatal.

Concept Check

Answers: p. 542

4. The formula given for calculating MAP applies to a typical resting heart rate of 60–80 beats/min. If heart rate increases, would the contribution of systolic pressure to mean arterial pressure decrease or increase, and would MAP decrease or increase?

5. Peter's systolic pressure is 112 mm Hg, and his diastolic pressure is 68 mm Hg (written 112/68). What is his pulse pressure? His mean arterial pressure?

RUNNING PROBLEM

Kurt's second blood pressure reading is 158/98. Dr. Cortez asks him to take his blood pressure at home daily for two weeks and then return to the doctor's office. When Kurt comes back with his diary, the story is the same: his blood pressure continues to average 160/100. After running some tests, Dr. Cortez concludes that Kurt is one of approximately 50 million adult Americans with high blood pressure, also called hypertension. If not controlled, hypertension can lead to heart failure, stroke, and kidney failure.

Q1: Why are people with high blood pressure at greater risk for having a hemorrhagic (or bleeding) stroke?

15

Blood Pressure Is Estimated by Sphygmomanometry

We estimate arterial blood pressure in the radial artery of the arm using a *sphygmomanometer*, an instrument consisting of an inflatable cuff and a pressure gauge {*sphygmus*, pulse + *manometer*, an instrument for measuring pressure of a fluid}. The cuff encircles the upper arm and is inflated until it exerts pressure higher than the systolic pressure driving arterial blood. When cuff pressure exceeds arterial pressure, blood flow into the lower arm stops (■ Fig. 15.7a).

Now pressure on the cuff is gradually released. When cuff pressure falls below systolic arterial blood pressure, blood begins to flow again. As blood squeezes through the still-compressed artery, a thumping noise called a **Korotkoff sound** can be heard with each pressure wave (Fig. 15.7b). Once the cuff pressure no longer compresses the artery, the sounds disappear (Fig. 15.7c).

The pressure at which a Korotkoff sound is first heard represents the highest pressure in the artery and is recorded as the systolic pressure. The point at which the Korotkoff sounds disappear is the lowest pressure in the artery and is recorded as the diastolic pressure. By convention, blood pressure is written as systolic pressure over diastolic pressure.

For years the "average" value for blood pressure has been stated as 120/80. Like many average physiological values, however, these numbers are subject to wide variability, both from one person to another and within a single individual from moment to moment. A systolic pressure that is consistently over 140 mm Hg at rest, or a diastolic pressure that is chronically over 90 mm Hg, is considered a sign of hypertension in an otherwise healthy person.

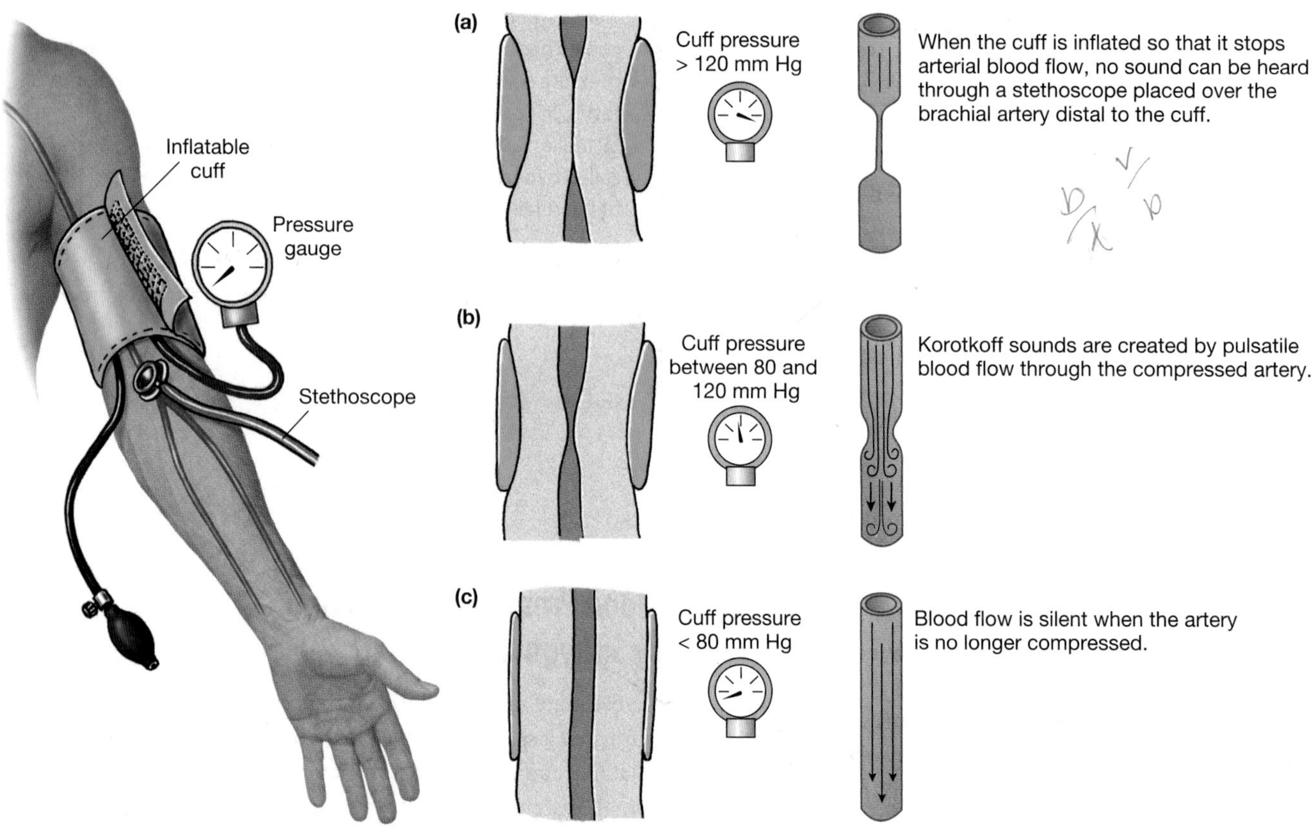
SPHYGMOMANOMETRY
Arterial blood pressure is measured with a sphygmomanometer (an inflatable cuff plus a pressure gauge) and a stethoscope. The inflation pressure shown is for a person whose blood pressure is 120/80.
Inflatable cuff
Pressure gauge
Stethoscope
(a)
Cuff pressure > 120 mm Hg
When the cuff is inflated so that it stops arterial blood flow, no sound can be heard through a stethoscope placed over the brachial artery distal to the cuff.
(b)
Cuff pressure between 80 and 120 mm Hg
Korotkoff sounds are created by pulsatile blood flow through the compressed artery.
(c)
Cuff pressure < 80 mm Hg
Blood flow is silent when the artery is no longer compressed.

■ **Fig. 15.7**

Furthermore, the guidelines published in the 2003 JNC 7 Report* now recommend that individuals maintain their blood pressure *below* 120/80. Persons whose systolic pressure is consistently in the range of 120–139 or whose diastolic pressure is in the range of 80–89 are now considered to be prehypertensive and should be counseled on lifestyle modification strategies to reduce their blood pressure.

Cardiac Output and Peripheral Resistance Determine Mean Arterial Pressure

Mean arterial pressure is the driving force for blood flow, but what determines mean arterial pressure? Arterial pressure is a balance between blood flow into the arteries and blood flow out of the arteries. If flow in exceeds flow out, blood collects in the arteries, and mean arterial pressure increases. If flow out exceeds flow in, mean arterial pressure falls.

Blood flow into the aorta is equal to the cardiac output of the left ventricle. Blood flow out of the arteries is influenced primarily by **peripheral resistance,** defined as the resistance to flow offered by the arterioles (■ Fig. 15.8a). Mean arterial pressure (MAP) then is proportional to cardiac output (CO) times resistance (R) of the arterioles:

$$MAP \propto CO \times R_{arterioles}$$

Let's consider how this works. If cardiac output increases, the heart pumps more blood into the arteries per unit time. If resistance to blood flow out of the arteries does not change, flow into the arteries is greater than flow out, blood volume in the arteries increases, and arterial blood pressure increases.

In another example, suppose cardiac output remains unchanged but peripheral resistance increases. Flow into arteries is unchanged, but flow out is decreased. Blood again accumulates in the arteries, and the arterial pressure again increases. Most cases of hypertension are believed to be caused

*Seventh Report of the Joint National Committee on Prevention, Detection, Evaluation, and Treatment of High Blood Pressure, National Institutes of Health. *www.nhlbi.nih.gov/guidelines/hypertension*. JNC 8 will be published in Spring 2012.

■ Fig. 15.8 *ESSENTIALS*

Mean Arterial Pressure

(a) Mean arterial pressure (MAP) is a function of cardiac output and resistance in the arterioles (peripheral resistance). MAP illustrates mass balance: the volume of blood in the arteries is determined by input (cardiac output) and flow out (altered by changing peripheral resistance). As arterial volume increases, pressure increases. In this model, the ventricle is represented by a syringe. The variable diameter of the arterioles is represented by adjustable screws.

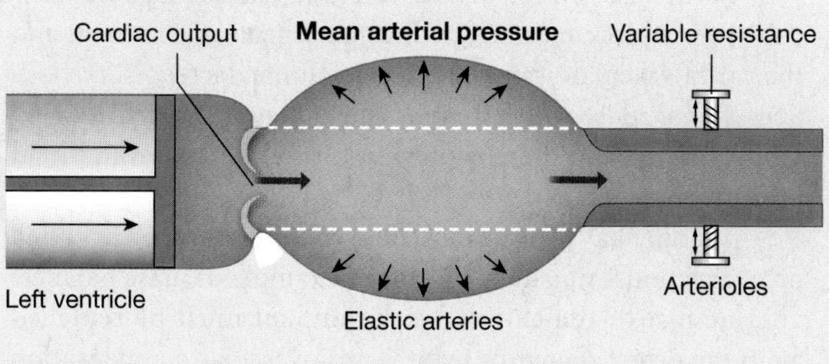

Mean arterial pressure α cardiac output × resistance

FIGURE QUESTIONS

1. If arterioles constrict, what happens to blood flow out of the arteries? What happens to MAP?
2. If cardiac output decreases, what happens to arterial blood volume? What happens to MAP?
3. If veins constrict, what happens to blood volume in the veins? What happens to volume in the arteries and to MAP?

(b) Factors that influence mean arterial pressure

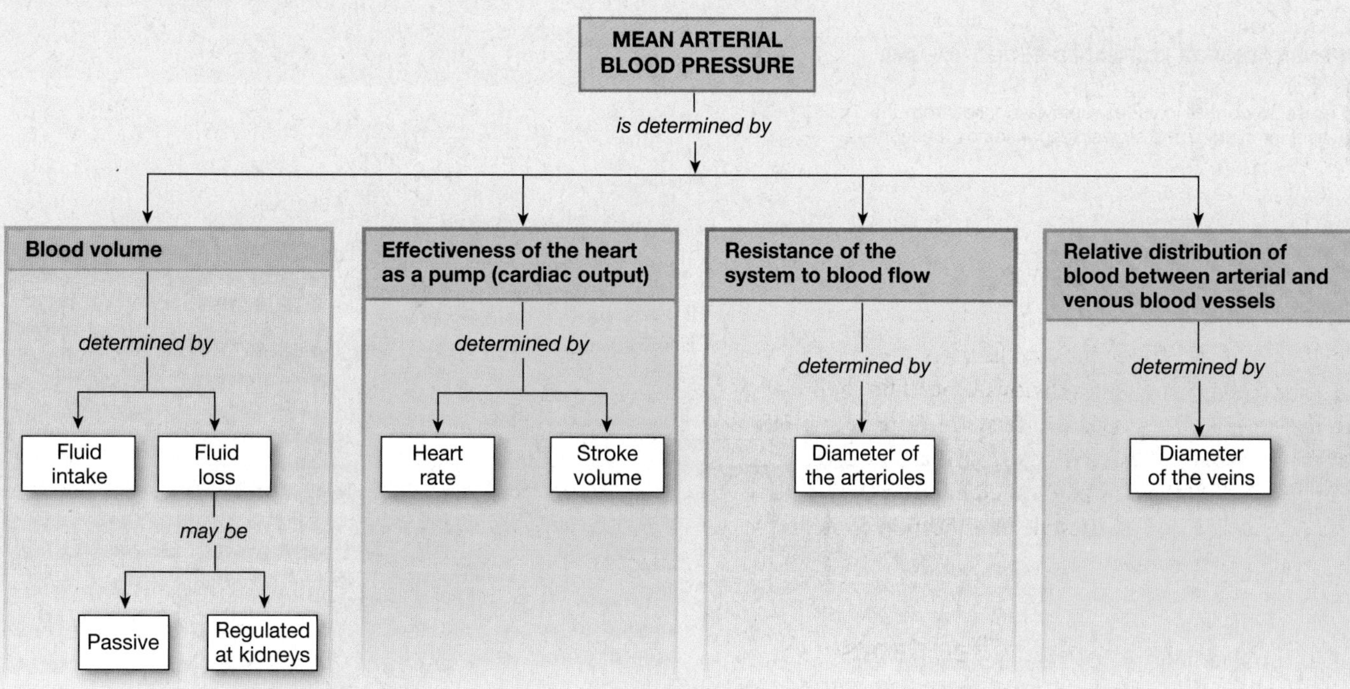

by increased peripheral resistance without changes in cardiac output.

Two additional factors can influence arterial blood pressure: the distribution of blood in the systemic circulation and the total blood volume. The relative distribution of blood between the arterial and venous sides of the circulation can be an important factor in maintaining arterial blood pressure. Arteries are low-volume vessels that usually contain only about 11% of total blood volume at any one time. Veins, in contrast, are high-volume vessels that hold about 60% of the circulating blood volume at any one time.

The veins act as a *volume reservoir* for the circulatory system, holding blood that can be redistributed to the arteries if needed. If arterial blood pressure falls, increased sympathetic activity constricts veins, decreasing their holding capacity. Venous return sends blood to the heart, which according to the Frank-Starling law of the heart, pumps all the venous return out to the systemic side of the circulation [p. 497]. Thus, constriction of the veins redistributes blood to the arterial side of the circulation and raises mean arterial pressure.

Changes in Blood Volume Affect Blood Pressure

Although the volume of the blood in the circulation is usually relatively constant, changes in blood volume can affect arterial blood pressure (Fig. 15.8b). If blood volume increases, blood pressure increases. When blood volume decreases, blood pressure decreases.

To understand the relationship between blood volume and pressure, think of the circulatory system as an elastic balloon filled with water. If only a small amount of water is in the balloon, little pressure is exerted on the walls, and the balloon is soft and flabby. As more water is added to the balloon, more pressure is exerted on the elastic walls. If you fill a balloon close to the bursting point, you risk popping the balloon. The best way to reduce this pressure is to remove some of the water.

Small increases in blood volume occur throughout the day due to ingestion of food and liquids, but these increases usually do not create long-lasting changes in blood pressure because of homeostatic compensations. Adjustments for increased blood volume are primarily the responsibility of the kidneys. If blood volume increases, the kidneys restore normal volume by excreting excess water in the urine (■ Fig. 15.9).

Compensation for decreased blood volume is more difficult and requires an integrated response from the kidneys and the cardiovascular system. If blood volume decreases, *the kidneys cannot restore the lost fluid*. The kidneys can only *conserve* blood volume and thereby prevent further decreases in blood pressure.

The only way to restore lost fluid volume is through drinking or intravenous infusions. This is an example of mass balance: volume lost to the external environment must be replaced from the external environment. Cardiovascular compensation for decreased blood volume includes vasoconstriction and increased sympathetic stimulation of the heart to increase cardiac output [Fig. 14.22, p. 501]. However, there are limits to the effectiveness of cardiovascular compensation—if fluid loss

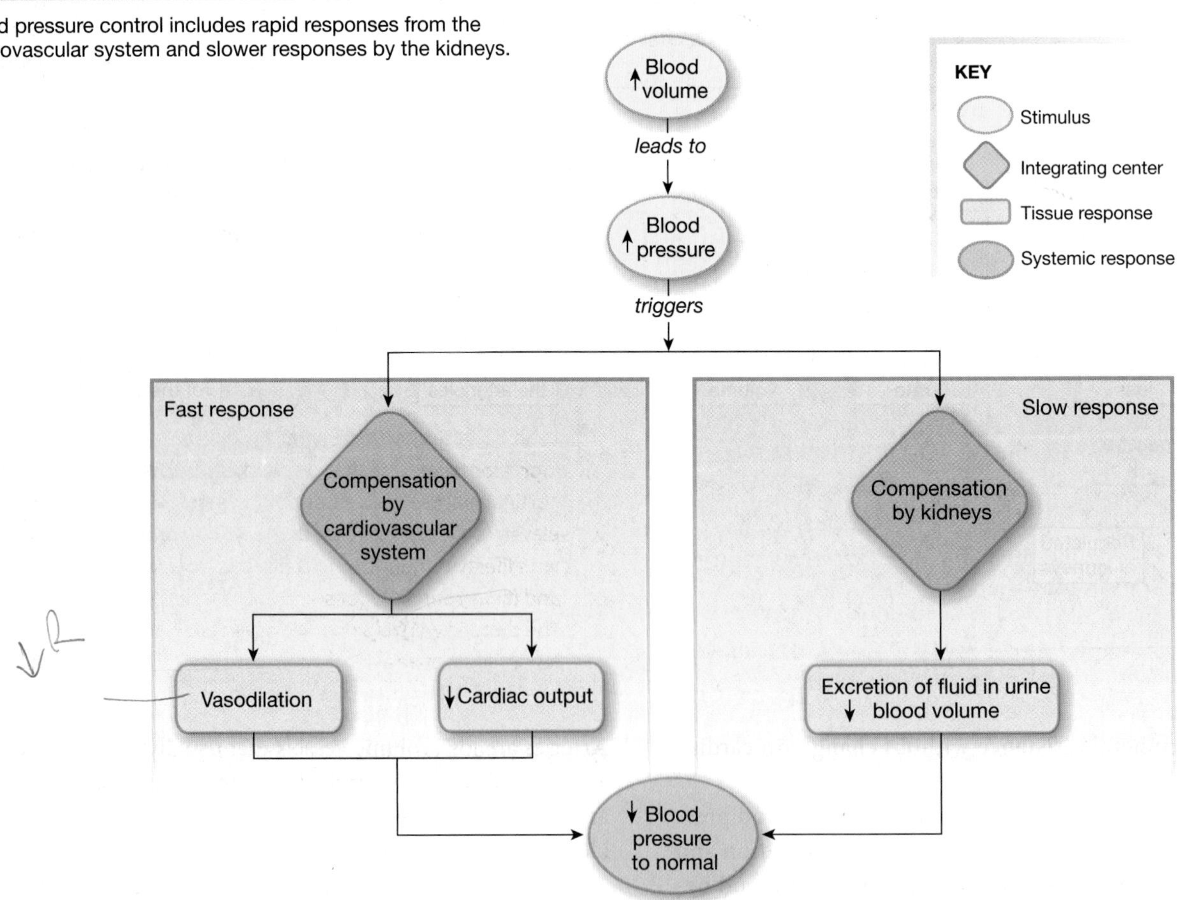

■ **Fig. 15.9**

CLINICAL FOCUS

Shock

Shock is a broad term that refers to generalized, severe circulatory failure. Shock can arise from multiple causes: failure of the heart to maintain normal cardiac output (*cardiogenic shock*), decreased circulating blood volume (*hypovolemic shock*), bacterial toxins (*septic shock*), and miscellaneous causes, such as the massive immune reactions that cause *anaphylactic shock*. No matter what the cause, the results are similar: low cardiac output and falling peripheral blood pressure. When tissue perfusion can no longer keep up with tissue oxygen demand, the cells begin to sustain damage from inadequate oxygen and from the buildup of metabolic wastes. Once this damage occurs, a positive feedback cycle begins. The shock becomes progressively worse until it becomes irreversible, and the patient dies. The management of shock includes administration of oxygen, fluids, and norepinephrine, which stimulates vasoconstriction and increases cardiac output. If the shock arises from a cause that is treatable, such as a bacterial infection, measures must also be taken to remove the precipitating cause.

is too great, the body cannot maintain adequate blood pressure. Typical events that might cause significant changes in blood volume include dehydration, hemorrhage, and ingestion of a large quantity of fluid.

Figure 15.8b summarizes the four key factors that influence mean arterial blood pressure.

Resistance in the Arterioles

Peripheral resistance is one of the two main factors influencing blood pressure. According to Poiseuille's Law [p. 469], resistance to blood flow (R) is directly proportional to the length of the tubing through which the fluid flows (L) and to the viscosity (η) of the fluid, and inversely proportional to the fourth power of the tubing radius (r):

$$R \propto L\,\eta/r^4$$

Normally the length of the systemic circulation and the blood's viscosity are relatively constant. That leaves only the radius of the blood vessels as the primary resistance to blood flow:

$$R \propto 1/r^4$$

The arterioles are the main site of variable resistance in the systemic circulation and contribute more than 60% of the total resistance to flow in the system. Resistance in arterioles is variable because of the large amounts of smooth muscle in the arteriolar walls. When the smooth muscle contracts or relaxes, the radius of the arterioles changes.

Arteriolar resistance is influenced by both local and systemic control mechanisms:

1. *Local control of arteriolar resistance* matches tissue blood flow to the metabolic needs of the tissue. In the heart and skeletal muscle, these local controls often take precedence over reflex control by the central nervous system.
2. *Sympathetic reflexes* mediated by the CNS maintain mean arterial pressure and govern blood distribution for certain homeostatic needs, such as temperature regulation.
3. *Hormones*—particularly those that regulate salt and water excretion by the kidneys—influence blood pressure by acting directly on the arterioles and by altering autonomic reflex control.

■ Table 15.2 lists significant chemicals that mediate arteriolar resistance by producing vasoconstriction or vasodilation. In the following sections we look at some factors that influence blood flow at the tissue level.

Myogenic Autoregulation Automatically Adjusts Blood Flow

Vascular smooth muscle has the ability to regulate its own state of contraction, a process called **myogenic autoregulation**. In the absence of autoregulation, an increase in blood pressure increases blood flow through an arteriole. However, when smooth muscle fibers in the wall of the arteriole stretch because of increased blood pressure, the arteriole constricts. This vasoconstriction increases the resistance offered by the arteriole, automatically decreasing blood flow through the vessel. With

RUNNING PROBLEM

Most hypertension is *essential hypertension,* which means high blood pressure that cannot be attributed to any particular cause. "Since your blood pressure is only mildly elevated," Dr. Cortez tells Kurt, "let's see if we can control it with lifestyle changes and a diuretic. You need to reduce salt and fat in your diet, get some exercise, and lose some weight. The diuretic will help your kidneys get rid of excess fluid." "Looks like you're asking me to turn over a whole new leaf," says Kurt. "I'll try it."

Q2: What is the rationale for reducing salt intake and taking a diuretic to control hypertension? (Hint: Salt causes water retention.)

509 515 519 522 533 537

Table 15.2 Chemicals Mediating Vasoconstriction and Vasodilation

Chemical	Physiological Role	Source	Type
Vasoconstriction			
Norepinephrine (α-receptors)	Baroreceptor reflex	Sympathetic neurons	Neurotransmitter
Serotonin	Platelet aggregation, smooth muscle contraction	Neurons, digestive tract, platelets	Paracrine, neurotransmitter
Endothelin	Local control of blood flow	Vascular endothelium	Paracrine
Vasopressin	Increases blood pressure in hemorrhage	Posterior pituitary	Neurohormone
Angiotensin II	Increases blood pressure	Plasma hormone	Hormone
Vasodilation			
Epinephrine (β_2-receptors)	Increase blood flow to skeletal muscle, heart, liver	Adrenal medulla	Neurohormone
Acetylcholine	Erection reflex (indirectly through NO production)	Parasympathetic neurons	Neurotransmitter
Nitric oxide (NO)	Local control of blood flow	Endothelium	Paracrine
Bradykinin (via NO)	Increases blood flow	Multiple tissues	Paracrine
Adenosine	Increases blood flow to match metabolism	Hypoxic cells	Paracrine
$\downarrow O_2$, $\uparrow CO_2$, $\uparrow H^+$, $\uparrow K^+$	Increase blood flow to match metabolism	Cell metabolism	Paracrine
Histamine	Increases blood flow	Mast cells	Paracrine
Natriuretic peptides (example—ANP)	Reduce blood pressure	Atrial myocardium, brain	Hormone, neurotransmitter
Vasoactive intestinal peptide	Digestive secretion, relax smooth muscle	Neurons	Neurotransmitter, neurohormone

this simple and direct response to pressure, arterioles have limited ability to regulate their own blood flow.

How does myogenic autoregulation work at the cellular level? When vascular smooth muscle cells in arterioles are stretched, mechanically gated channels in the muscle membrane open. Cation entry depolarizes the cell. The depolarization opens voltage-gated Ca^{2+} channels, and Ca^{2+} flows into the cell down its electrochemical gradient. Calcium entering the cell combines with calmodulin and activates myosin light chain kinase [p. 431]. MLCK in turn increases myosin ATPase activity and crossbridge activity, resulting in contraction.

Paracrines Alter Vascular Smooth Muscle Contraction

Local control is an important strategy by which individual tissues regulate their own blood supply. In a tissue, blood flow into individual capillaries can be regulated by the precapillary sphincters described earlier in the chapter. When these small bands of smooth muscle at metarteriole-capillary junctions constrict, they restrict blood flow into the capillaries (see Fig. 15.3). When the sphincters dilate, blood flow into the capillaries increases. This mechanism provides an additional site for local control of blood flow.

Local regulation also takes place by changing arteriolar resistance in a tissue. This is accomplished by paracrines (including the gases O_2, CO_2, and NO) secreted by the vascular endothelium or by cells to which the arterioles are supplying blood (Tbl. 15.2).

The concentrations of many paracrines change as cells become more or less metabolically active. For example, if aerobic metabolism increases, tissue O_2 levels decrease while CO_2 production goes up. Both low O_2 and high CO_2 dilate arterioles. This vasodilation increases blood flow into the tissue, bringing additional O_2 to meet the increased metabolic demand and removing waste CO_2 (■ Fig. 15.10a). The process in which an increase in blood flow accompanies an increase in metabolic activity is known as **active hyperemia** {*hyper-*, above normal + *(h)aimia*, blood}.

If blood flow to a tissue is occluded {*occludere*, to close up} for a few seconds to a few minutes, O_2 levels fall and metabolic paracrines such as CO_2 and H^+ accumulate in the interstitial fluid. Local *hypoxia* {*hypo-*, low + *oxia*, oxygen} causes endothelial cells to synthesize the vasodilator nitric oxide. When blood flow to the tissue resumes, the increased concentrations of NO, CO_2, and other paracrines immediately trigger significant vasodilation. As the vasodilators are metabolized or washed away by the restored tissue blood flow, the radius of the arteriole gradually returns to normal. An increase in tissue blood flow following a period of low perfusion (blood flow) is known as **reactive hyperemia** (Fig. 15.10b).

Nitric oxide is probably best known for its role in the male erection reflex, and drugs used to treat erectile dysfunction prolong NO activity. Decreases in endogenous NO activity are suspected to play a role in other medical conditions, including hypertension and *preeclampsia*, the elevated blood pressure that sometimes occurs during pregnancy.

Another vasodilator paracrine is the nucleotide **adenosine**. If oxygen consumption in heart muscle exceeds the rate at which oxygen is supplied by the blood, myocardial hypoxia results. In response to low tissue oxygen, the myocardial cells release adenosine. Adenosine dilates coronary arterioles in an attempt to bring additional blood flow into the muscle.

Not all vasoactive paracrines reflect changes in metabolism. For example, *kinins* and *histamine* are potent vasodilators that play a role in inflammation. *Serotonin* (5-HT), previously mentioned as a CNS neurotransmitter [p. 310], is also a vasoconstricting paracrine released by activated platelets. When damaged blood vessels activate platelets, the subsequent serotonin-mediated vasoconstriction helps slow blood loss.

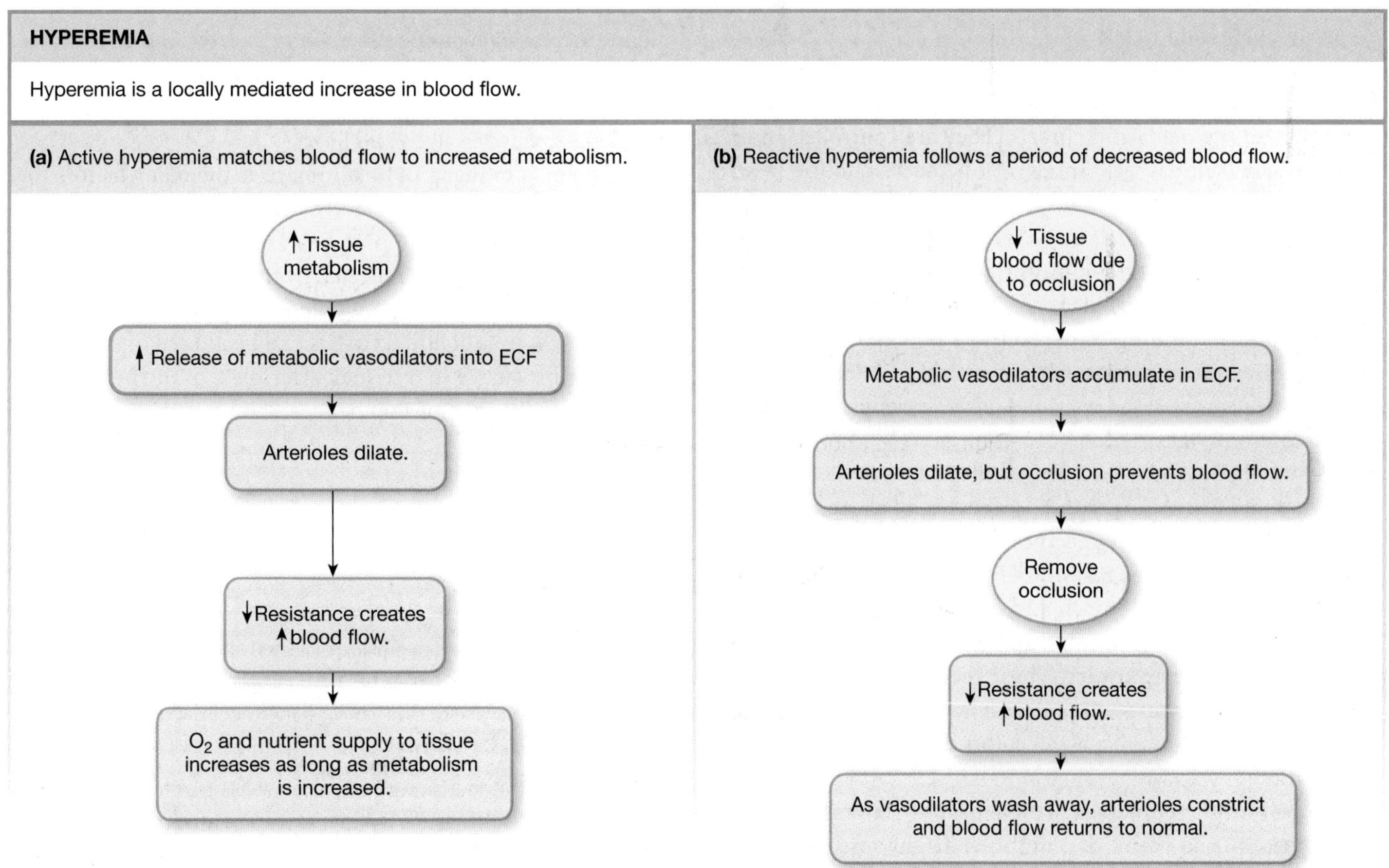
HYPEREMIA
Hyperemia is a locally mediated increase in blood flow.
(a) Active hyperemia matches blood flow to increased metabolism.
↑ Tissue metabolism
↑ Release of metabolic vasodilators into ECF
Arterioles dilate.
↓Resistance creates ↑blood flow.
O_2 and nutrient supply to tissue increases as long as metabolism is increased.
(b) Reactive hyperemia follows a period of decreased blood flow.
↓ Tissue blood flow due to occlusion
Metabolic vasodilators accumulate in ECF.
Arterioles dilate, but occlusion prevents blood flow.
Remove occlusion
↓Resistance creates ↑blood flow.
As vasodilators wash away, arterioles constrict and blood flow returns to normal.

■ **Fig. 15.10**

Serotonin agonists called triptans (for example, *sumatriptan*) are drugs that bind to 5-HT_1 receptors and cause vasoconstriction. These drugs are used to treat migraine headaches, which are caused by inappropriate cerebral vasodilation.

Concept Check Answers: p. 542

6. Resistance to blood flow is determined *primarily* by which? (a) blood viscosity, (b) blood volume, (c) cardiac output, (d) blood vessel diameter, or (e) blood pressure gradient (ΔP)
7. The extracellular fluid concentration of K^+ increases in exercising skeletal muscles. What effect does this increase in K^+ have on blood flow in the muscles?

RUNNING PROBLEM

After two months, Kurt returns to the doctor's office for a checkup. He has lost five pounds and is walking at least a mile daily, but his blood pressure has not changed. "I swear, I'm trying to do better," says Kurt, "but it's difficult." Because lifestyle changes and the diuretic have not lowered Kurt's blood pressure, Dr. Cortez adds an antihypertensive drug. "This drug, called an ACE inhibitor, blocks production of a chemical called angiotensin II, a powerful vasoconstrictor. This medication should bring your blood pressure back to a normal value."

Q3: Why would blocking the action of a vasoconstrictor lower blood pressure?

509 515 519 **522** 533 537

The Sympathetic Branch Controls Most Vascular Smooth Muscle

Smooth muscle contraction in arterioles is regulated by neural and hormonal signals in addition to locally produced paracrines. Among the hormones with significant vasoactive properties are *atrial natriuretic peptide* and *angiotensin II (ANG II)*. These hormones also have significant effects on the kidney's excretion of ions and water.

Most systemic arterioles are innervated by sympathetic neurons. A notable exception is arterioles involved in the erection reflex of the penis and clitoris. They are controlled indirectly by parasympathetic innervation. Acetylcholine from parasympathetic neurons causes paracrine release of nitric oxide, resulting in vasodilation.

Tonic discharge of norepinephrine from sympathetic neurons helps maintain myogenic tone of arterioles (■ Fig. 15.11a). Norepinephrine binding to α-receptors on vascular smooth muscle causes vasoconstriction. If sympathetic release of norepinephrine decreases, the arterioles dilate. If sympathetic stimulation increases, the arterioles constrict.

Epinephrine from the adrenal medulla travels through the blood and also binds with α-receptors, reinforcing vasoconstriction. However, α-receptors have a lower affinity for epinephrine and do not respond as strongly to it as they do to norepinephrine [p. 386]. In addition, epinephrine binds to β_2-receptors, found on vascular smooth muscle of heart, liver, and skeletal muscle arterioles. These receptors are not innervated and therefore respond primarily to circulating epinephrine. Activation of vascular β_2-receptors by epinephrine causes vasodilation.

One way to remember which tissues' arterioles have β_2-receptors is to think of a fight-or-flight response to a stressful event [p. 378]. This response includes a generalized increase in sympathetic activity, along with the release of epinephrine. Blood vessels that have β_2-receptors respond to epinephrine by vasodilating. Such β_2-mediated vasodilation enhances blood flow to the heart, skeletal muscles, and liver, tissues that are active during the fight-or-flight response. (The liver produces glucose for muscle contraction.) During fight or flight, increased sympathetic activity at arteriolar α-receptors causes vasoconstriction. The increase in resistance diverts blood from nonessential organs, such as the gastrointestinal tract, to the skeletal muscles, liver, and heart.

The map in Fig. 15.11b summarizes the many factors that influence blood flow in the body. The pressure to drive blood flow is created by the pumping heart and captured by the arterial pressure reservoir, as reflected by the mean arterial pressure. Flow through the body as a whole is equal to the cardiac output, but flow to individual tissues can be altered by selectively changing resistance in a tissue's arterioles. In the next section we consider the relationship between blood flow and arteriolar resistance.

Concept Check Answers: p. 542

8. What happens when epinephrine combines with β_1-receptors in the heart? With β_2-receptors in the heart? (*Hint:* "in the heart" is vague. The heart has multiple tissue types. Which heart tissues possess the different types of β-receptors? [p. 501]
9. Skeletal muscle arterioles have both α- and β-receptors on their smooth muscle. Epinephrine can bind to both. Will these arterioles constrict or dilate in response to epinephrine? Explain.

RESISTANCE AND FLOW

(a) Tonic control of arteriolar diameter

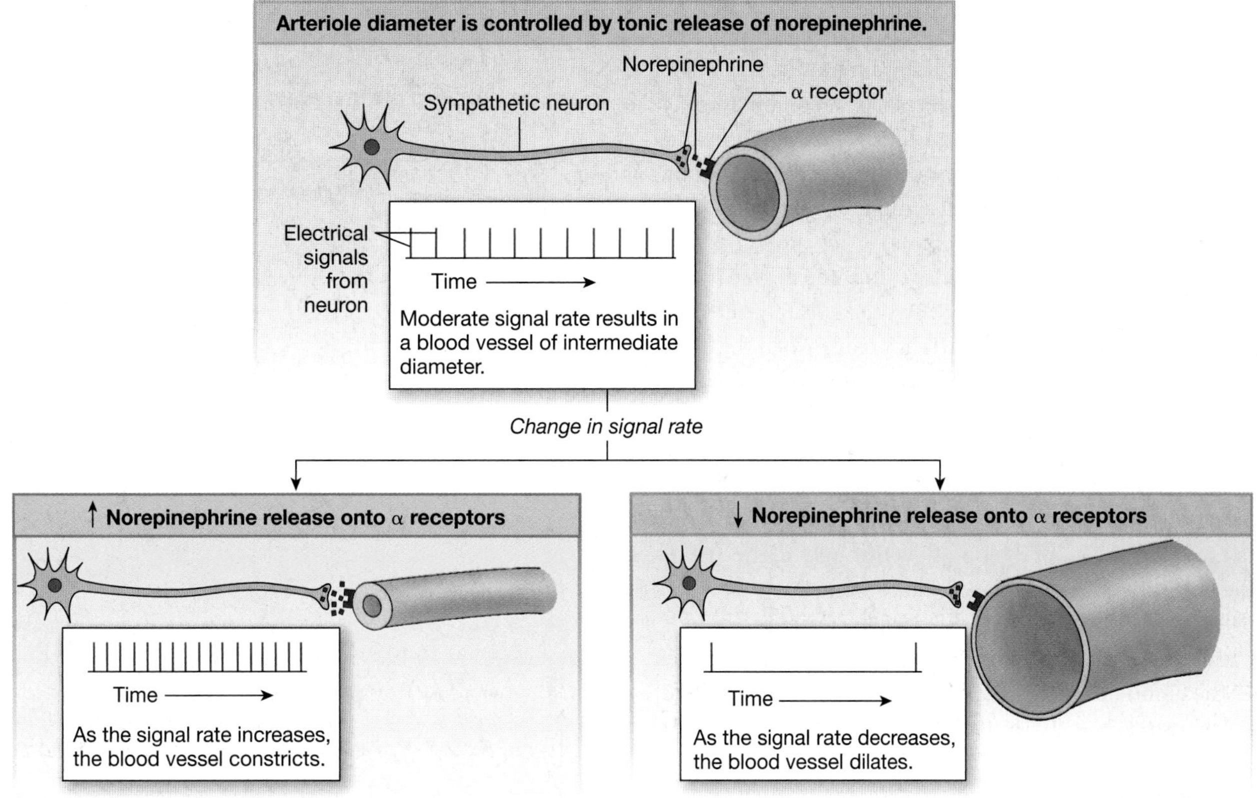

(b) Factors influencing peripheral blood flow

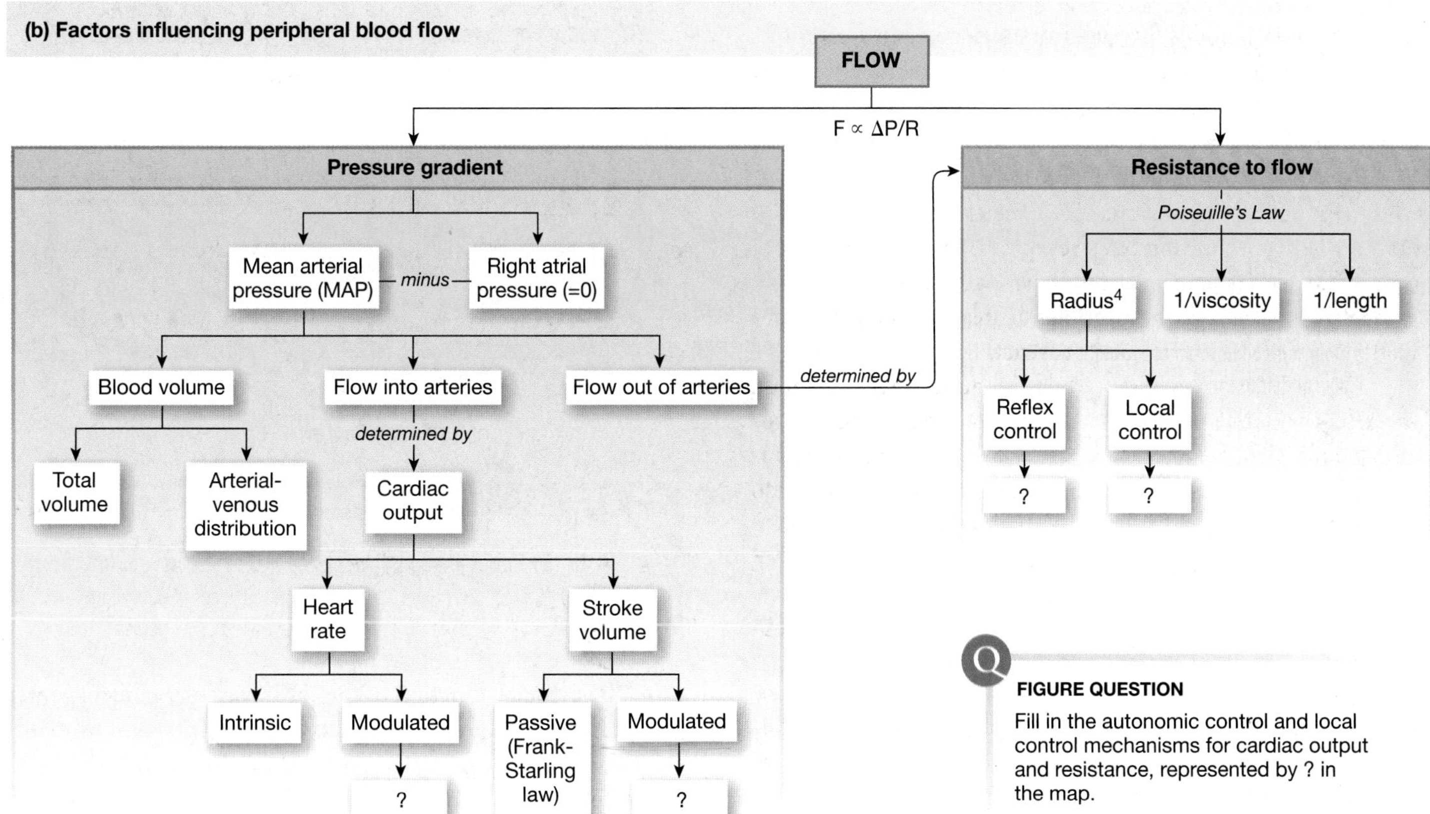

FIGURE QUESTION

Fill in the autonomic control and local control mechanisms for cardiac output and resistance, represented by ? in the map.

Fig. 15.11

Distribution of Blood to the Tissues

The nervous system's ability to selectively alter blood flow to organs is an important aspect of cardiovascular regulation. The distribution of systemic blood varies according to the metabolic needs of individual organs and is governed by a combination of local control mechanisms and homeostatic reflexes. For example, skeletal muscles at rest receive about 20% of cardiac output. During exercise, when the muscles use more oxygen and nutrients, they receive as much as 85%.

Blood flow to individual organs is set to some degree by the number and size of arteries feeding the organ. ■ Figure 15.12 shows how blood is distributed to various organs when the body is at rest. Usually, more than two-thirds of the cardiac output is routed to the digestive tract, liver, muscles, and kidneys.

Variations in blood flow to individual tissues are possible because the arterioles in the body are arranged in parallel. In other words, all arterioles receive blood at the same time from the aorta (see Fig. 15.1). Total blood flow through *all* the arterioles of the body always equals the cardiac output.

However, the flow through individual arterioles depends on their resistance (R). The higher the resistance in an arteriole, the lower the blood flow through it. If an arteriole constricts and resistance increases, blood flow through that arteriole decreases (Fig. ■ 15.13):

$$\text{Flow}_{\text{arteriole}} \propto 1/R_{\text{arteriole}}$$

In other words, blood is diverted from high-resistance arterioles to lower-resistance arterioles. You might say that blood traveling through the arterioles takes the path of least resistance. We will return to this subject after we look at the control mechanisms that govern blood flow and blood pressure.

Concept Check Answer: p. 542

10. Use Fig. 15.12 to answer these questions. (a) Which tissue has the highest blood flow per unit weight? (b) Which tissue has the least blood flow, regardless of weight?

DISTRIBUTION OF BLOOD IN THE BODY AT REST

Blood flow to the major organs is represented in three ways: as a percentage of total flow, as volume per 100 grams of tissue per minute, and as an absolute rate of flow (in L/min).

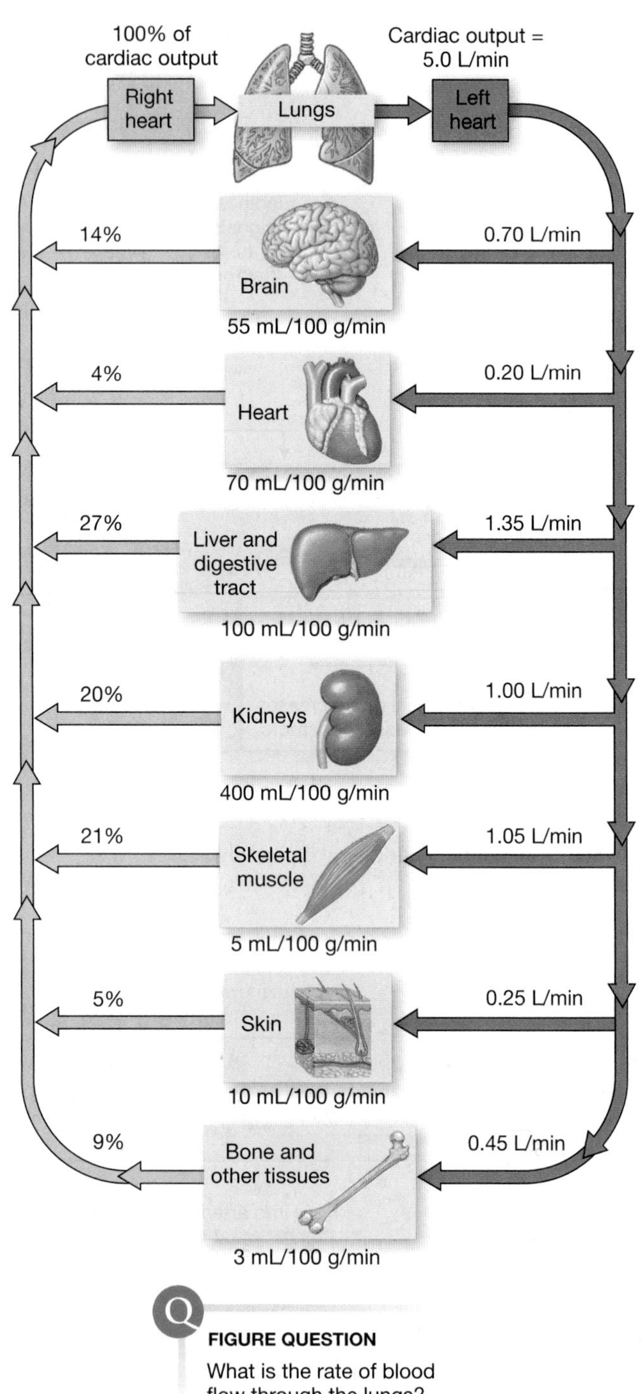

FIGURE QUESTION
What is the rate of blood flow through the lungs?

■ **Fig. 15.12**

Regulation of Cardiovascular Function

The central nervous system coordinates the reflex control of blood pressure and distribution of blood to the tissues. The main integrating center is in the medulla oblongata. Because of the complexity of the neural networks involved in cardiovascular control, we will simplify this discussion and refer to this medullary network as the **cardiovascular control center** (CVCC).

The primary function of the cardiovascular control center is to ensure adequate blood flow to the brain and heart by maintaining sufficient mean arterial pressure. However, the CVCC also receives input from other parts of the brain, and it has the ability to alter function in a few organs or tissues while leaving

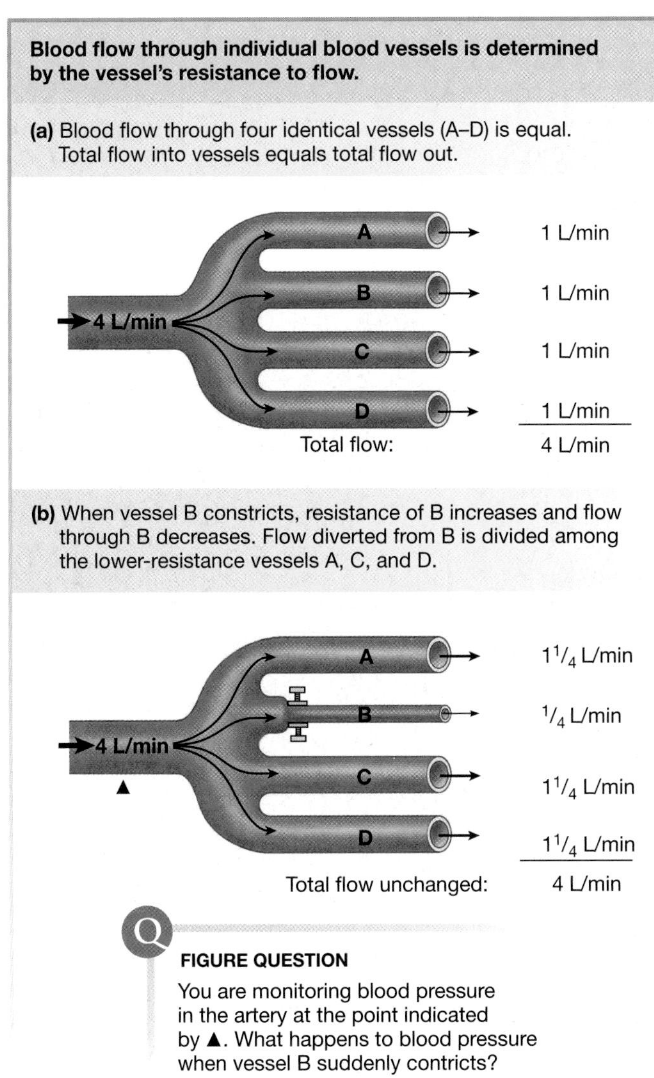
Blood flow through individual blood vessels is determined by the vessel's resistance to flow.
(a) Blood flow through four identical vessels (A–D) is equal. Total flow into vessels equals total flow out.
4 L/min
A
B
C
D
1 L/min
1 L/min
1 L/min
1 L/min
Total flow:
4 L/min
(b) When vessel B constricts, resistance of B increases and flow through B decreases. Flow diverted from B is divided among the lower-resistance vessels A, C, and D.
4 L/min
1 1/4 L/min
1/4 L/min
1 1/4 L/min
1 1/4 L/min
Total flow unchanged:
4 L/min
FIGURE QUESTION
You are monitoring blood pressure in the artery at the point indicated by ▲. What happens to blood pressure when vessel B suddenly contricts?

■ Fig. 15.13

others unaffected. For example, thermoregulatory centers in the hypothalamus communicate with the CVCC to alter blood flow to the skin. Brain-gut communication following a meal increases blood flow to the intestinal tract. Reflex control of blood flow to selected tissues changes mean arterial pressure, so the CVCC is constantly monitoring and adjusting its output as required to maintain homeostasis.

The Baroreceptor Reflex Controls Blood Pressure

The primary reflex pathway for homeostatic control of mean arterial blood pressure is the **baroreceptor reflex**. The components of the reflex are illustrated in ■ Figure 15.14a. Stretch-sensitive mechanoreceptors known as **baroreceptors** are located in the walls of the carotid arteries and aorta, where they can monitor the pressure of blood flowing to the brain (carotid baroreceptors) and to the body (aortic baroreceptors).

The carotid and aortic baroreceptors are tonically active stretch receptors that fire action potentials continuously at normal blood pressures. When increased blood pressure in the arteries stretches the baroreceptor membrane, the firing rate of the receptor increases. If blood pressure falls, the firing rate of the receptor decreases.

If blood pressure changes, the frequency of action potentials traveling from the baroreceptors to the medullary cardiovascular control center changes. The CVCC integrates the sensory input and initiates an appropriate response. The response of the baroreceptor reflex is quite rapid: changes in cardiac output and peripheral resistance occur within two heartbeats of the stimulus.

Output signals from the cardiovascular control center are carried by both sympathetic and parasympathetic autonomic neurons. Peripheral resistance is under tonic sympathetic control, with increased sympathetic discharge causing vasoconstriction.

Heart function is regulated by antagonistic control. Increased sympathetic activity increases heart rate at the SA node, shortens conduction time through the AV node, and enhances the force of myocardial contraction. Increased parasympathetic activity slows heart rate but has only a small effect on ventricular contraction.

The baroreceptor reflex in response to increased blood pressure is mapped in Fig. 15.14b. Baroreceptors increase their firing rate as blood pressure increases, activating the medullary cardiovascular control center. In response, the cardiovascular control center increases parasympathetic activity and decreases sympathetic activity to slow down the heart and dilate arterioles.

When heart rate falls, cardiac output falls. In the vasculature, decreased sympathetic activity causes dilation of arterioles, lowering their resistance and allowing more blood to flow out of the arteries. Because mean arterial pressure is directly proportional to cardiac output and peripheral resistance (MAP $\propto$ CO $\times$ R), the combination of reduced cardiac output and decreased peripheral resistance lowers the mean arterial blood pressure.

It is important to remember that the baroreceptor reflex is functioning all the time, not just with dramatic disturbances in blood pressure, and that it is not an all-or-none response. A change in blood pressure can result in a change in both cardiac output and peripheral resistance or a change in only one of the two variables. Let's look at an example.

For this example, we will use the schematic diagram in ■ Figure 15.15, which combines the concepts introduced in Figures 15.8 and 15.13. In this model there are four sets of variable resistance arterioles (A–D) whose diameters can be independently controlled by local or reflex control mechanisms. Baroreceptors in the arteries monitor mean arterial pressure and communicate with the medullary cardiovascular control center.

■ Fig. 15.14 *ESSENTIALS*

Cardiovascular Control

The intrinsic rate of the heartbeat is modulated by sympathetic and parasympathetic neurons. Blood vessel diameter is under tonic control by the sympathetic division.

(a) CNS control of the heart and blood vessels

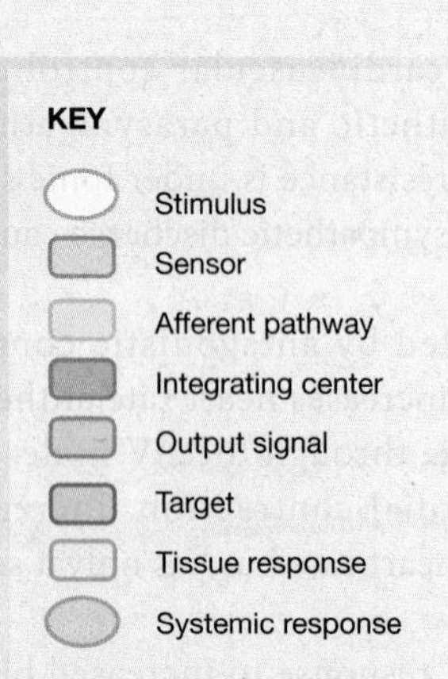

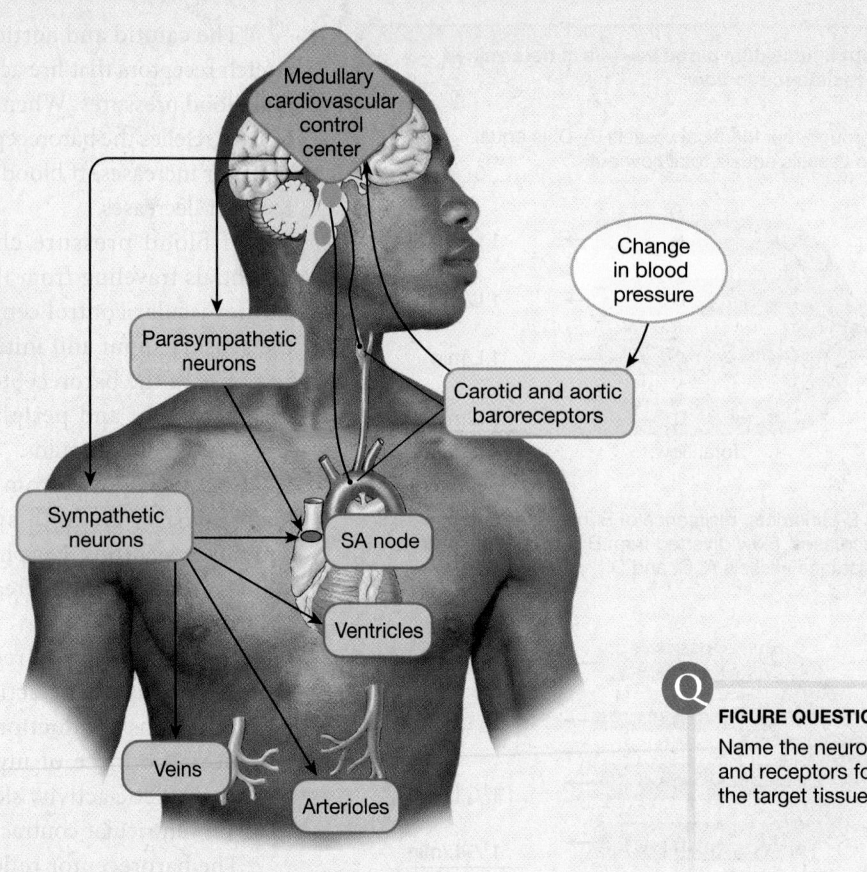

FIGURE QUESTION

Name the neurotransmitters and receptors for each of the target tissues.

(b) The baroreceptor reflex

This map shows the reflex response to an increase in mean arterial pressure.

↑ Blood pressure → ↑ Firing of baroreceptors in carotid arteries and aorta → Sensory neurons → Cardiovascular control center in medulla oblongata

- ↓ Sympathetic output (*less NE released*)
  - *α-receptor* → Arteriolar smooth muscle → Vasodilation → ↓ Peripheral resistance
  - *β₁-receptor* → Ventricular myocardium → ↓ Force of contraction → ↓ Cardiac output
  - *β₁-receptor* → SA node → ↓ Heart rate → ↓ Cardiac output
- ↑ Parasympathetic output (*more ACh on muscarinic receptor*) → SA node → ↓ Heart rate → ↓ Cardiac output

↓ Peripheral resistance + ↓ Cardiac output → ↓ Blood pressure

Negative feedback (⊖) from ↓ Blood pressure to firing of baroreceptors

INTEGRATION OF RESISTANCE CHANGES AND CARDIAC OUTPUT

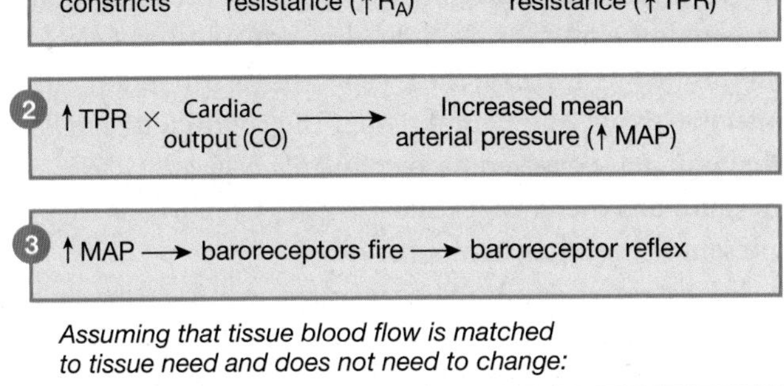

Fig. 15.15

Suppose arteriole set A constricts because of local control mechanisms. Vasoconstriction increases resistance in A and decreases flow through A. Total peripheral resistance (TPR) across all the arterioles also increases. Using the relationship MAP $\propto$ CO $\times$ TPR, an increase in total resistance results in an increase in mean arterial pressure. The arterial baroreceptors sense the increase in MAP and activate the baroreceptor reflex.

Output from the cardiovascular control center can alter either cardiac output, arteriolar resistance, or both. In this instance, we can assume that blood flow in arteriole sets A–D now matches tissue needs and should remain constant. That means the CVCC should not change resistance in the arterioles. The only option left to decrease MAP is to decrease cardiac output. So the efferent signals from the CVCC decrease cardiac output, which in turn brings mean arterial pressure down. Blood pressure homeostasis is restored. In this example, the output signal of the baroreceptor reflex altered cardiac output but did not change peripheral resistance.

Concept Check Answer: p. 542

11. Baroreceptors have stretch-sensitive ion channels in their cell membrane. Increased pressure stretches the receptor cell membrane, opens the channels, and initiates action potentials. What ion probably flows through these channels and in which direction (into or out of the cell)?

Orthostatic Hypotension Triggers the Baroreceptor Reflex

The baroreceptor reflex functions every morning when you get out of bed. When you are lying flat, gravitational forces are distributed evenly up and down the length of your body, and blood is distributed evenly throughout the circulation. When you stand up, gravity causes blood to pool in the lower extremities. This pooling creates an instantaneous decrease in venous return. As a result, less blood is in the ventricles at the beginning of the next contraction. Cardiac output falls from 5 L/min to 3 L/min, causing arterial blood pressure to decrease. This decrease in blood pressure upon standing is known as *orthostatic hypotension* {*orthos,* upright + *statikos,* to stand}.

Orthostatic hypotension normally triggers the baroreceptor reflex. The combination of increased cardiac output and increased peripheral resistance increases mean arterial pressure and brings it back up to normal within two heartbeats. The skeletal muscle pump also contributes to the recovery by enhancing venous return when abdominal and leg muscles contract to maintain an upright position.

The baroreceptor reflex is not always effective, however. For example, during extended bed rest or in the zero-gravity conditions of space flights, blood from the lower extremities is distributed evenly throughout the body rather than pooled in the lower extremities. This even distribution raises arterial pressure, triggering the kidneys to excrete what is perceived as excess fluid. Over the course of three days, excretion of water leads to a 12% decrease in blood volume.

When the person finally gets out of bed or returns to earth, gravity again causes blood to pool in the legs. Orthostatic hypotension occurs, and the baroreceptors attempt to compensate. In this instance, however, the cardiovascular system is unable to restore normal pressure because of the loss of blood volume. As a result, the individual may become dizzy or even faint from reduced delivery of oxygen to the brain.

Concept Check Answer: p. 542

12. Use the map in Fig. 15.14b to map the reflex response to orthostatic hypotension.

Other Systems Influence Cardiovascular Function

Cardiovascular function can be modulated by input from peripheral receptors other than the baroreceptors. For example, arterial chemoreceptors activated by low blood oxygen levels increase cardiac output. The cardiovascular control center also has reciprocal communication with centers in the medulla that control breathing.

The integration of function between the respiratory and circulatory systems is adaptive. If tissues require more oxygen, it is supplied by the cardiovascular system working in tandem with the respiratory system. Consequently, increases in breathing rate are usually accompanied by increases in cardiac output.

Blood pressure is also subject to modulation by higher brain centers, such as the hypothalamus and cerebral cortex. The hypothalamus mediates vascular responses involved in body temperature regulation and for the fight-or-flight response. Learned and emotional responses may originate in the cerebral cortex and be expressed by cardiovascular responses such as blushing and fainting.

One such reflex is *vasovagal syncope,* which may be triggered in some people by the sight of blood or a hypodermic needle. (Recall Anthony's experience at the beginning of this chapter.) In this pathway, increased parasympathetic activity and decreased sympathetic activity slow heart rate and cause widespread vasodilation. Cardiac output and peripheral resistance both decrease, triggering a precipitous drop in blood pressure. With insufficient blood to the brain, the individual faints.

Regulation of blood pressure in the cardiovascular system is closely tied to regulation of body fluid balance by the kidneys. Certain hormones secreted from the heart act on the kidneys, while hormones secreted from the kidneys act on the heart and blood vessels. Together, the heart and kidneys play a major role in maintaining homeostasis of body fluids, an excellent example of the integration of organ system function.

Concept Check Answer: p. 542

13. In the movie *Jurassic Park,* Dr. Ian Malcolm must flee from the *T. rex.* Draw a reflex map showing the cardiovascular response to his fight-or-flight situation. Remember that fight-or-flight causes epinephrine secretion as well as output from the cardiovascular control center. (*Hints:* What is the stimulus? Fear is integrated in the limbic system.)

Exchange at the Capillaries

The transport of materials around the body is only part of the function of the cardiovascular system. Once blood reaches the capillaries, the plasma and the cells exchange materials across the thin capillary walls. Most cells are located within 0.1 mm of the nearest capillary, and diffusion over this short distance proceeds rapidly.

The capillary density in any given tissue is directly related to the metabolic activity of the tissue's cells. Tissues with a higher metabolic rate require more oxygen and nutrients. Those tissues have more capillaries per unit area. Subcutaneous tissue and cartilage have the lowest capillary density. Muscles and glands have the highest. By one estimate, the adult human body has about 50,000 miles of capillaries, with a total exchange surface area of more than 6300 m^2, nearly the surface area of two football fields.

Capillaries have the thinnest walls of all the blood vessels, composed of a single layer of flattened endothelial cells supported on a basal lamina (Fig. 15.2). The diameter of a capillary is barely larger than that of a red blood cell, forcing the RBCs to pass through in single file. Cell junctions between the endothelial cells vary from tissue to tissue and help determine the "leakiness" of the capillary.

The most common capillaries are **continuous capillaries,** whose endothelial cells are joined to one another with leaky junctions (■ Fig. 15.16a). These capillaries are found in muscle, connective tissue, and neural tissue. The continuous capillaries of the brain have evolved to form the *blood-brain barrier*, with tight junctions that protect neural tissue from toxins that may be present in the bloodstream [p. 296].

Fenestrated capillaries {*fenestra,* window} have large pores (*fenestrae*) that allow high volumes of fluid to pass rapidly between the plasma and interstitial fluid (Fig. 15.16b). These capillaries are found primarily in the kidney and the intestine, where they are associated with absorptive transporting epithelia.

Three tissues—the bone marrow, the liver, and the spleen—do not have typical capillaries. Instead they have modified vessels called **sinusoids** that are as much as five times wider than a capillary. The sinusoid endothelium has fenestrations, and there may be gaps between the cells as well. Sinusoids are found in locations where blood cells and plasma proteins need to cross the endothelium to enter the blood. [Figure 16.4c, Focus on Bone Marrow, shows blood cells leaving the bone marrow by squeezing between endothelial cells.] In the liver, the sinusoidal endothelium lacks a basal lamina, which allows even more free exchange between plasma and interstitial fluid.

Velocity of Blood Flow Is Lowest in the Capillaries

The rate at which blood flows through the capillaries plays a role in the efficiency of exchange between the blood and the interstitial fluid. At a constant flow rate, velocity of flow is higher in a smaller diameter tube than in a larger one [p. 470]. From this, you might conclude that blood moves very rapidly through the capillaries because they are the smallest blood vessels. However, the primary determinant for velocity is not the diameter of an individual capillary but the *total cross-sectional area* of *all* the capillaries.

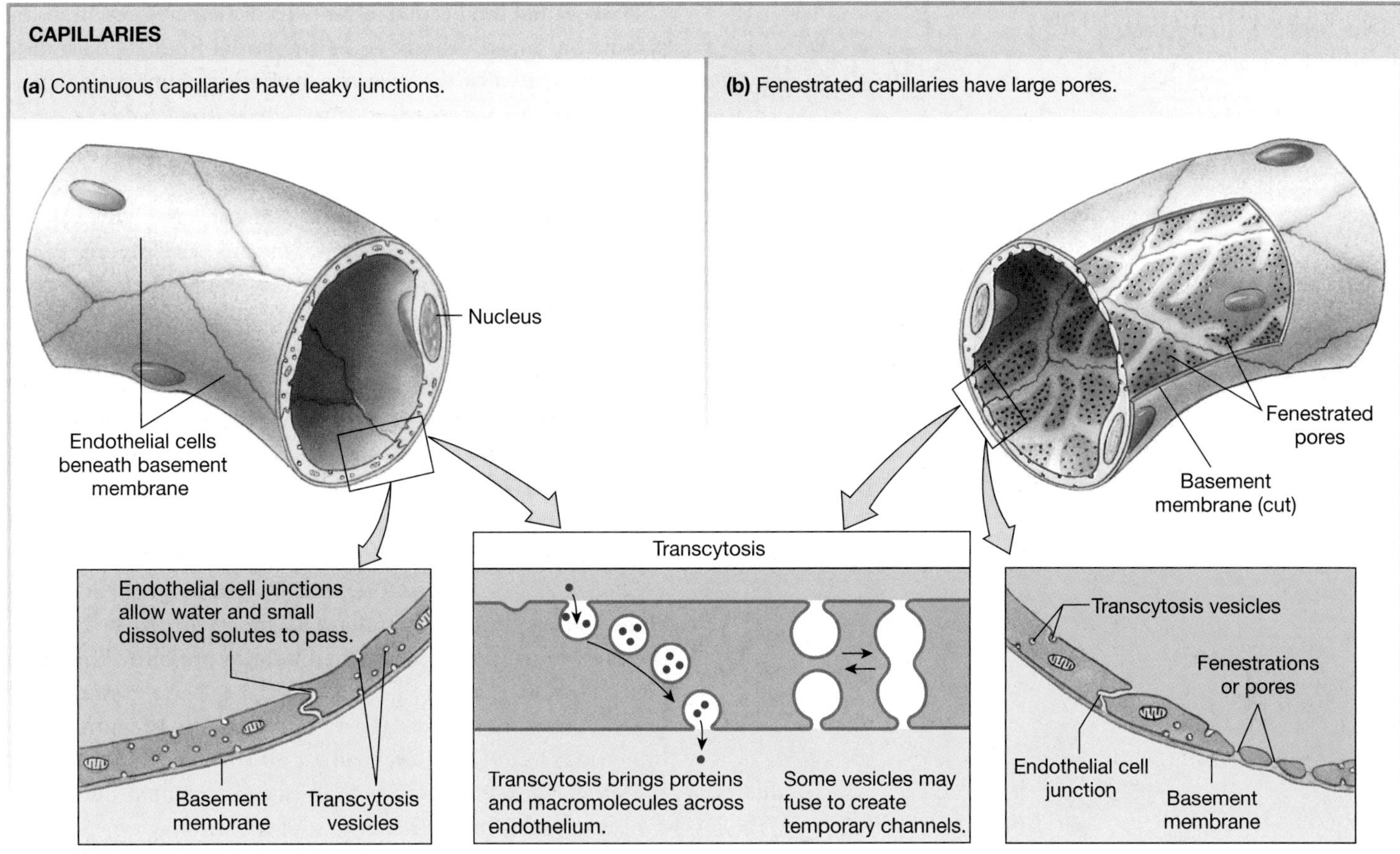
CAPILLARIES
(a) Continuous capillaries have leaky junctions.
(b) Fenestrated capillaries have large pores.
Nucleus
Endothelial cells beneath basement membrane
Fenestrated pores
Basement membrane (cut)
Transcytosis
Endothelial cell junctions allow water and small dissolved solutes to pass.
Basement membrane
Transcytosis vesicles
Transcytosis brings proteins and macromolecules across endothelium.
Some vesicles may fuse to create temporary channels.
Transcytosis vesicles
Fenestrations or pores
Endothelial cell junction
Basement membrane

■ **Fig. 15.16**

What is total cross-sectional area? Imagine circles representing cross sections of all the capillaries placed edge to edge, and you have it. For the capillaries, those circles would cover an area much larger than the total cross-sectional areas of all the arteries and veins combined. Therefore, because total cross-sectional area of the capillaries is so large, the velocity of flow through them is low.

■ Figure 15.17 compares cross-sectional areas of different parts of the systemic circulation with the velocity of blood flow in each part. The fastest flow is in the relatively small-diameter arterial system. The slowest flow is in the capillaries and venules, which collectively have the largest cross-sectional area. The low velocity of flow through capillaries is a useful characteristic that allows enough time for diffusion to go to equilibrium [p. 141].

Most Capillary Exchange Takes Place by Diffusion and Transcytosis

Exchange between the plasma and interstitial fluid takes place either by movement between endothelial cells (the *paracellular pathway*) or by movement through the cells (*endothelial transport*). Smaller dissolved solutes and gases move by diffusion between or through the cells, depending on their lipid solubility [p. 141]. Larger solutes and proteins move mostly by vesicular transport [p. 155].

The diffusion rate for dissolved solutes is determined primarily by the concentration gradient between the plasma and the interstitial fluid. Oxygen and carbon dioxide diffuse freely across the thin endothelium. Their plasma concentrations reach equilibrium with the interstitial fluid and cells by the time blood reaches the venous end of the capillary. In capillaries with leaky cell junctions, most small dissolved solutes can diffuse freely between the cells or through the fenestrae.

In continuous capillaries, blood cells and most plasma proteins are unable to pass through the junctions between endothelial cells. However, we know that proteins do move from plasma to interstitial fluid and vice versa. In most capillaries, larger molecules (including selected proteins) are transported across the endothelium by *transcytosis* [p. 160]. The endothelial cell surface appears dotted with numerous *caveolae* and noncoated pits that become vesicles for transcytosis. It appears that in some capillaries, chains of vesicles fuse to create open channels that extend across the endothelial cell (Fig. 15.16).

Capillary Filtration and Absorption Take Place by Bulk Flow

A third form of capillary exchange is bulk flow into and out of the capillary. **Bulk flow** refers to the mass movement of fluid as the result of hydrostatic or osmotic pressure gradients. If the direction of bulk flow is into the capillary, the fluid movement is called **absorption**. If the direction of flow is out of the capillary,

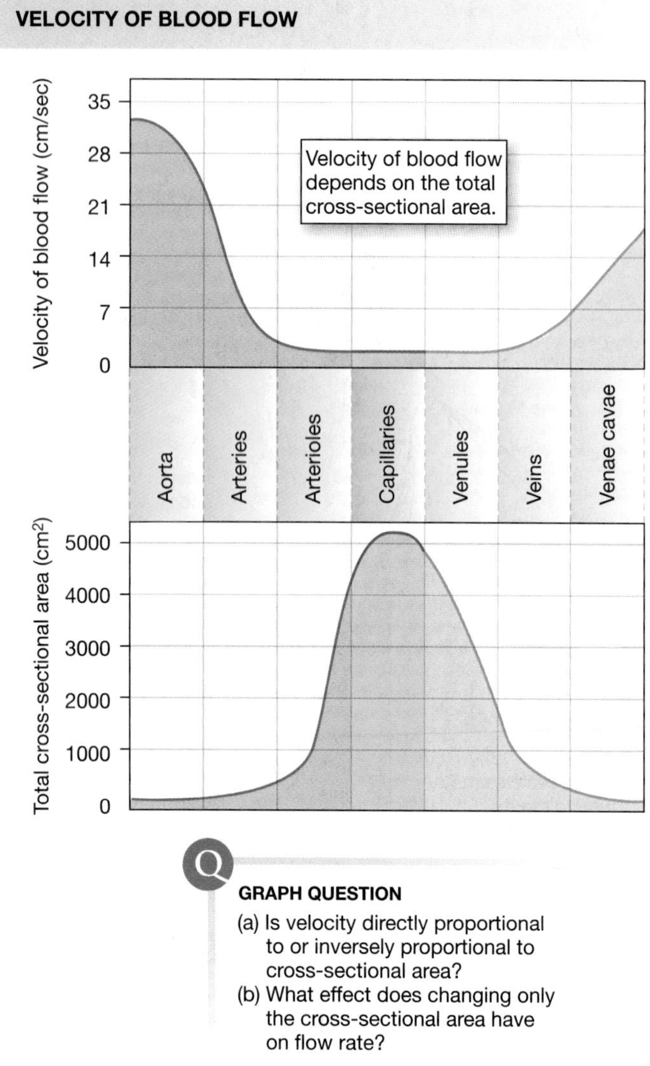

■ **Fig. 15.17**

the fluid movement is known as **filtration**. Capillary filtration is caused by hydrostatic pressure that forces fluid out of the capillary through leaky cell junctions. As an analogy, think of garden "soaker" hoses whose perforated walls allow water to ooze out.

Most capillaries show a transition from net filtration at the arterial end to net absorption at the venous end. There are some exceptions to this rule, though. Capillaries in part of the kidney filter fluid along their entire length, for instance, and some capillaries in the intestine are only absorptive, picking up digested nutrients that have been transported into the interstitial fluid from the lumen of the intestine.

Two forces regulate bulk flow in the capillaries. One is hydrostatic pressure, the lateral pressure component of blood flow that pushes fluid out through the capillary pores [p. 467], and the other is osmotic pressure [p. 133]. These forces are sometimes called *Starling forces,* after the English physiologist E. H. Starling, who first described them (the same Starling as in the Frank-Starling law of the heart).

Osmotic pressure is determined by solute concentration of a compartment. The main solute difference between plasma and interstitial fluid is due to proteins, which are present in the plasma but mostly absent from interstitial fluid. The osmotic pressure created by the presence of these proteins is known as **colloid osmotic pressure** (π), also called *oncotic pressure.*

Colloid osmotic pressure is *not* equivalent to the total osmotic pressure in a capillary. It is simply a measure of the osmotic pressure created by proteins. Because the capillary endothelium is freely permeable to ions and other solutes in the plasma and interstitial fluid, these other solutes do not contribute to the osmotic gradient.

Colloid osmotic pressure is higher in the plasma (π_{cap} = 25 mm Hg) than in the interstitial fluid (π_{IF} = 0 mm Hg) Therefore, the osmotic gradient favors water movement by osmosis from the interstitial fluid into the plasma, represented by the red arrows in ■ Figure 15.18b. For the purposes of our discussion, colloid osmotic pressure is constant along the length of the capillary, at $\pi = 25$ mm Hg

Capillary hydrostatic pressure (P_H), by contrast, decreases along the length of the capillary as energy is lost to friction. Average values for capillary hydrostatic pressure, shown in Fig. 15.18b, are 32 mm Hg at the arterial end of a capillary and 15 mm Hg at the venous end. The hydrostatic pressure of the interstitial fluid (P_{IF}) is very low, and so we consider it to be essentially zero. This means that water movement due to hydrostatic pressure is directed out of the capillary, as denoted by the blue arrows in Fig. 15.18b, with the pressure gradient decreasing from the arterial end to the venous end.

If we assume that the interstitial hydrostatic and colloid osmotic pressures are zero, as discussed above, then the net pressure driving fluid flow across the capillary is determined by the difference between the hydrostatic pressure (P_H) and the colloid osmotic pressure (π):

$$\text{Net pressure} = P_H - \pi$$

A positive value for the net pressure indicates net filtration and a negative value indicates netabsorption.

Using the hydrostatic and oncotic pressure values given in Fig. 15.18b, we can calculate the following values at the arterial end of a capillary:

$$\text{Net pressure} = P_H\ (32\text{ mm Hg}) - \pi\ (25\text{ mm Hg}) \\ = 7\text{ mm Hg}$$

At the arterial end, P_H is greater than π, so the net pressure is 7 mm Hg of filtration pressure.

At the venous end, where capillary hydrostatic pressure is less:

$$\text{Net pressure}_{\text{venous end}} = (15\text{ mm Hg} - 25\text{ mm Hg}) \\ = -10\text{ mm Hg}$$

At the venous end, π is greater than P_H; the net pressure is −10 mm Hg, favoring absorption. (A negative net pressure indicates absorption.)

15

CAPILLARY FLUID EXCHANGE

(a) A net average of 3 L/day of fluid filters out of the capillaries. The excess water and solutes that filter out of the capillary are picked up by the lymph vessels and returned to the circulation.

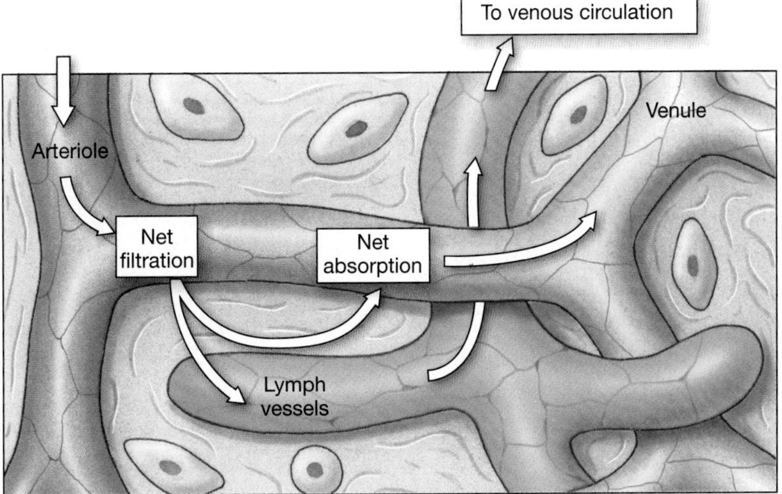

(b) Filtration in systemic capillaries

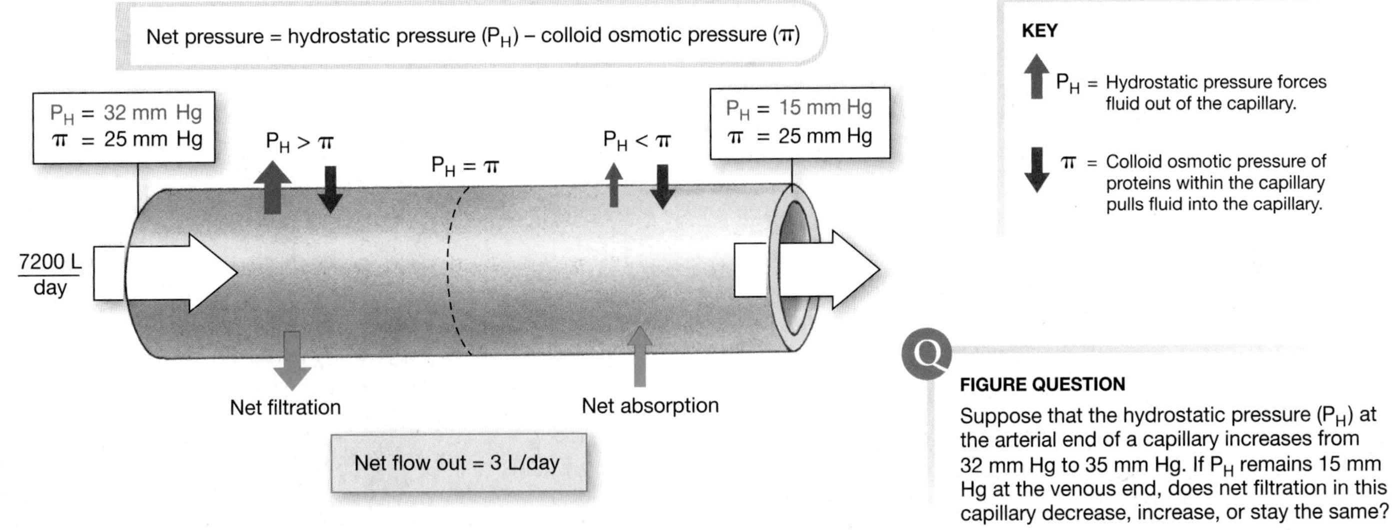

FIGURE QUESTION

Suppose that the hydrostatic pressure (P_H) at the arterial end of a capillary increases from 32 mm Hg to 35 mm Hg. If P_H remains 15 mm Hg at the venous end, does net filtration in this capillary decrease, increase, or stay the same?

Fig. 15.18

Fluid movement down the length of a capillary is shown in Fig. 15.18b. There is net filtration at the arterial end and net absorption at the venous end. If the point at which filtration equals absorption occurred in the middle of the capillary, there would be no net movement of fluid. All volume that was filtered at the arterial end would be absorbed at the venous end. However, filtration is usually greater than absorption, resulting in bulk flow of fluid out of the capillary into the interstitial space.

By most estimates, that bulk flow amounts to about 3 liters per day, which is the equivalent of the entire plasma volume! If this filtered fluid could not be returned to the plasma, the blood would turn into a sludge of blood cells and proteins. Restoring fluid lost from the capillaries to the circulatory system is one of the functions of the lymphatic system, which we discuss next.

Concept Check Answers: p. 542–543

14. A man with liver disease loses the ability to synthesize plasma proteins. What happens to the colloid osmotic pressure of his blood? What happens to the balance between filtration and absorption in his capillaries?
15. Why did this discussion refer to the colloid osmotic pressure of the plasma rather than the osmolarity of the plasma?

The Lymphatic System

The vessels of the lymphatic system interact with three other physiological systems: the cardiovascular system, the digestive system, and the immune system. Functions of the lymphatic system include (1) returning fluid and proteins filtered out of the capillaries to the circulatory system, (2) picking up fat absorbed at the small intestine and transferring it to the circulatory system, and (3) serving as a filter to help capture and destroy foreign pathogens. In this discussion we focus on the role of the lymphatic system in fluid transport.

The lymphatic system allows the one-way movement of interstitial fluid from the tissues into the circulation. Blind-end lymph vessels (*lymph capillaries*) lie close to all blood capillaries except those in the kidney and central nervous system (Fig. 15.18a). The smallest lymph vessels are composed of a single layer of flattened endothelium that is even thinner than the capillary endothelium.

The walls of these tiny lymph vessels are anchored to the surrounding connective tissue by fibers that hold the thin-walled vessels open. Large gaps between cells allow fluid, interstitial proteins, and particulate matter such as bacteria to be swept into the lymph vessels, also called lymphatics, by bulk flow. Once inside the lymphatics, this clear fluid is called simply **lymph**.

Lymph vessels in the tissues join one another to form larger lymphatic vessels that progressively increase in size (■ Fig. 15.19). These vessels have a system of semilunar valves, similar to valves in the venous circulation. The largest lymph ducts empty into the venous circulation just under the collarbones, where the left and right subclavian veins join the internal jugular veins. At intervals along the way, vessels enter **lymph nodes,** bean-shaped nodules of tissue with a fibrous outer capsule and an internal collection of immunologically active cells, including lymphocytes and macrophages.

The lymphatic system has no single pump like the heart. Lymph flow depends primarily on waves of contraction of smooth muscle in the walls of the larger lymph vessels. Flow is aided by contractile fibers in the endothelial cells, by the one-way valves, and by external compression created by skeletal muscles.

The skeletal muscle pump plays a significant role in lymph flow, as you know if you have ever injured a wrist or ankle. An immobilized limb frequently swells from the accumulation of fluid in the interstitial space, a condition known as **edema** {*oidema*, swelling}. Patients with edema in an injured limb are told to elevate the limb above the level of the heart so that gravity can assist lymph flow back to the blood.

An important reason for returning filtered fluid to the circulation is the recycling of plasma proteins. The body must maintain a low protein concentration in the interstitial fluid because colloid osmotic pressure is the only significant force that opposes capillary hydrostatic pressure. If proteins move from the plasma to the interstitial fluid, the osmotic pressure gradient that opposes filtration decreases. With less opposition to capillary hydrostatic pressure, additional fluid moves into the interstitial space.

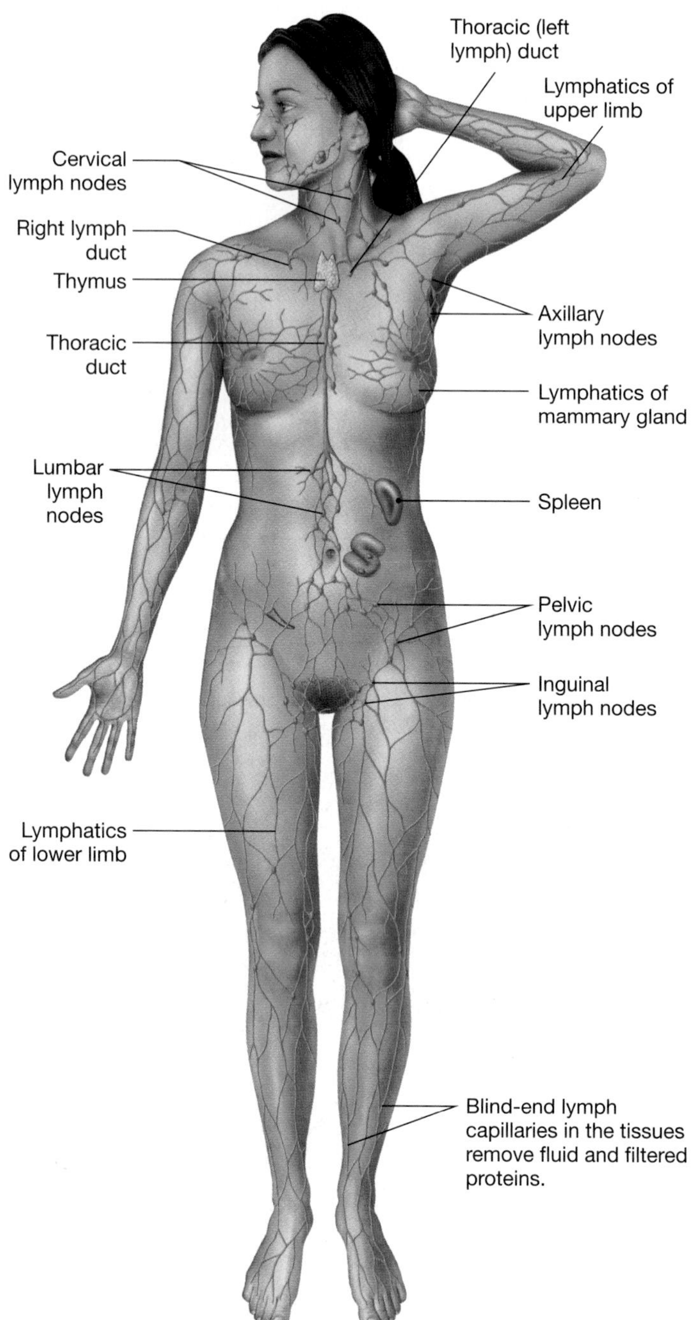

■ **Fig. 15.19**

Inflammation is an example of a situation in which the balance of colloid osmotic and hydrostatic pressures is disrupted. Histamine released in the inflammatory response makes capillary walls leakier and allows proteins to escape from the plasma into the interstitial fluid. The local swelling that accompanies a region of inflammation is an example of edema caused by redistribution of proteins from the plasma to the interstitial fluid.

RUNNING PROBLEM

Another few weeks go by, and Kurt again returns to Dr. Cortez for a checkup. Kurt's blood pressure is finally closer to the normal range and has been averaging 135/87. "But, Doc, can you give me something for this dry, hacking cough I've been having? I don't feel bad, but it's driving me nuts." Dr. Cortez explains that a dry cough is an occasional side effect of taking ACE inhibitors. "It is more of a nuisance than anything else, but let's change your medicine. I'd like to try you on a calcium channel blocker instead of the ACE inhibitor."

Q4: How do calcium channel blockers lower blood pressure?

Edema Results from Alterations in Capillary Exchange

Edema is a sign that normal exchange between the circulatory system and the lymphatics has been disrupted. Edema usually arises from one of two causes: (1) inadequate drainage of lymph or (2) blood capillary filtration that greatly exceeds capillary absorption.

Inadequate lymph drainage occurs with obstruction of the lymphatic system, particularly at the lymph nodes. Parasites, cancer, or fibrotic tissue growth caused by therapeutic radiation can block the movement of lymph through the system. For example, *elephantiasis* is a chronic condition marked by gross enlargement of the legs and lower appendages when parasites block the lymph vessels. Lymph drainage may also be impaired if lymph nodes are removed during surgery, a common procedure in the diagnosis and treatment of cancer.

Factors that disrupt the normal balance between capillary filtration and absorption include:

1. *An increase in capillary hydrostatic pressure.* Increased hydrostatic pressure is usually indicative of elevated venous pressure. An increase in arterial pressure is generally not noticeable at the capillaries because of autoregulation of pressure in the arterioles.

   One common cause of increased venous pressure is *heart failure,* a condition in which one ventricle loses pumping power and can no longer pump all the blood sent to it by the other ventricle. For example, if the right ventricle begins to fail but the left ventricle maintains its cardiac output, blood accumulates in the systemic circulation. Blood pressure rises first in the right atrium, then in the veins and capillaries draining into the right side of the heart. When capillary hydrostatic pressure increases, filtration greatly exceeds absorption, leading to edema.
2. *A decrease in plasma protein concentration.* Plasma protein concentrations may decrease as a result of severe malnutrition or liver failure. The liver is the main site for plasma protein synthesis.
3. *An increase in interstitial proteins.* As discussed earlier, excessive leakage of proteins out of the blood decreases the colloid osmotic pressure gradient and increases net capillary filtration.

On occasion, changes in the balance between filtration and absorption help the body maintain homeostasis. For example, if arterial blood pressure falls, capillary hydrostatic pressure also decreases. This change increases fluid absorption. If blood pressure falls low enough, there is net absorption in the capillaries rather than net filtration. This passive mechanism helps maintain blood volume in situations in which blood pressure is very low, such as hemorrhage or severe dehydration.

Concept Check Answers: p. 543

16. If the left ventricle fails to pump normally, blood backs up into what set of blood vessels? Where would you expect edema to occur?
17. Malnourished children who have inadequate protein in their diet often have grotesquely swollen bellies. This condition, which can be described as edema of the abdomen, is called *ascites* (■ Fig. 15.20). Use the information you have just learned about capillary filtration to explain why malnutrition causes ascites.

ASCITES

This 1960s photo from a Nigerian refugee camp shows ascites (abdominal edema) in a child with protein malnutrition, or kwashiorkor.

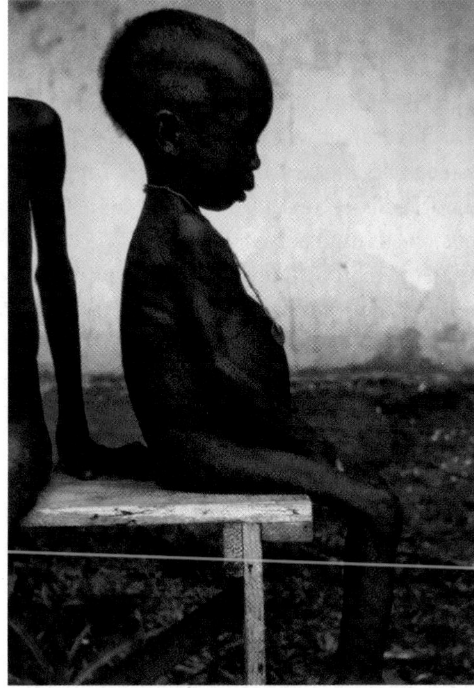

■ **Fig. 15.20**

15

Cardiovascular Disease

Disorders of the heart and blood vessels, such as heart attacks and strokes, play a role in more than half of all deaths in the United States. The American Heart Association predicted that by 2030 over 40% of the U.S. population will have cardiovascular disease. The direct medical costs for these people are expected to triple, to more than $800 billion. The prevalence of cardiovascular disease is reflected in the tremendous amount of research being done worldwide. The scientific investigations range from large-scale clinical studies that track cardiovascular disease in thousands of people, such as the Framingham (Massachusetts) Heart Study, to experiments at the cellular and molecular levels.

Much of the research at the cellular and molecular levels is designed to expand our understanding of both normal and abnormal function in the heart and blood vessels. Scientists are studying a virtual alphabet soup of transporters and regulators. Some of these molecules, such as adenosine, endothelin, vascular endothelial growth factor (VEGF), phospholamban, and nitric oxide, you have studied here.

As we increase our knowledge of cardiovascular function, we also begin to understand the actions of drugs that have been used for centuries. A classic example is the cardiac glycoside *digitalis* [p. 500], whose mechanism of action was explained when scientists discovered the role of Na^+-K^+-ATPase. It is a sobering thought to realize that for many therapeutic drugs, we know *what* they do without fully understanding *how* they do it.

CLINICAL FOCUS

Diabetes and Cardiovascular Disease

Having diabetes is one of the major risk factors for developing cardiovascular disease, and almost two-thirds of people with diabetes will die from cardiovascular problems. In diabetes, cells that cannot use glucose turn to fats and proteins for their energy. The body breaks down fat into fatty acids [p. 33] and dumps them into the blood. Plasma cholesterol levels are also elevated. When LDL-C remains in the blood, the excess is ingested by macrophages, starting a series of events that lead to atherosclerosis. Because of the pivotal role that LDL-C plays in atherosclerosis, many forms of therapy, ranging from dietary modification and exercise to drugs, are aimed at lowering LDL-C levels. Left untreated, blockage of small and medium-sized blood vessels in the lower extremities can lead to loss of sensation and *gangrene* (tissue death) in the feet. Atherosclerosis in larger vessels causes heart attacks and strokes. To learn more about diabetes and the increased risk of cardiovascular disease, visit the web sites of the American Diabetes Association (*www.diabetes.org*) and the American Heart Association (*www.americanheart.org*).

Risk Factors Include Smoking and Obesity

Conducting and interpreting research on humans is a complicated endeavor in part because of the difficulty of designing well-controlled experiments [p. 21]. The economic and social importance of cardiovascular disease (CVD) makes it the focus of many studies each year as researchers try to improve treatments and prediction algorithms. (An *algorithm* is a set of rules or a sequence of steps used to solve a problem). We can predict the likelihood that a person will develop cardiovascular disease during his or her lifetime by examining the various risk factors that the person possesses. The list of risk factors described here is the result of following the medical histories of thousands of people for many years in studies such as the Framingham Heart Study. As more data become available, additional risk factors may be added.

Risk factors are generally divided into those over which the person has no control and those that can be controlled. Medical intervention is aimed at reducing risk from the controllable factors. The risk factors that cannot be controlled include sex, age, and a family history of early cardiovascular disease. As noted earlier in the chapter, *coronary heart disease* (CHD) is a form of cardiovascular disease in which the coronary arteries become blocked by cholesterol deposits and blood clots. Up until middle age, men have a 3–4 times higher risk of developing CHD than do women. After age 55, when most women have entered menopause, the death rate from CHD equalizes in men and women. In general, the risk of coronary heart disease increases as people age. Heredity also plays an important role. If a person has one or more close relatives with this condition, his or her risk is elevated.

Risk factors that can be controlled include cigarette smoking, obesity, sedentary lifestyle, and untreated hypertension. In the United States, smoking-related illnesses are the primary preventable cause of death, followed by conditions related to overweight and obesity. Physical inactivity and obesity have been steadily increasing in the United States since 1991, and currently nearly 70% of U.S. adults are either overweight or obese.

Two risk factors for cardiovascular disease—diabetes mellitus and elevated blood lipids—have both an uncontrollable genetic component and a modifiable lifestyle component. Diabetes mellitus is a metabolic disorder that puts a person at risk for developing coronary heart disease by contributing to the development of **atherosclerosis** ("hardening of the arteries"), in which fatty deposits form inside arterial blood vessels. Elevated serum cholesterol and triglycerides also lead to atherosclerosis. The increasing prevalence of these risk factors has created an epidemic in the United States, with one in every 3.4 deaths in 2009 attributed to all forms of cardiovascular disease.

Atherosclerosis Is an Inflammatory Process

Coronary heart disease accounts for the majority of cardiovascular disease deaths and is the single largest killer of Americans, both men and women. Let's look at the underlying cause of this disease: atherosclerosis.

The role of elevated blood cholesterol in the development of atherosclerosis is well established. Cholesterol, like other lipids, is not very soluble in aqueous solutions, such as the plasma. Therefore, when cholesterol in the diet is absorbed from the digestive tract, it combines with lipoproteins to make it more soluble. Clinicians generally are concerned with two of these: **high-density lipoprotein-cholesterol (HDL-C)** complexes and **low-density lipoprotein-cholesterol (LDL-C)** complexes. HDL-C is the more desirable form of blood cholesterol because high levels of HDL are associated with lower risk of heart attacks. (Memory aid: "H" in HDL stands for "healthy.")

LDL-C is sometimes called "bad" cholesterol because elevated plasma LDL-C levels are associated with coronary heart disease. (Remember this by associating "L" with "lethal.") Normal levels of LDL-C are not bad, however, because LDL is necessary for cholesterol transport into cells. LDL-C's binding site—a protein called **apoB**—combines with an LDL receptor found in clathrin-coated pits on the cell membrane, and the receptor-LDL-C complex is brought into the cell by endocytosis. The LDL receptor recycles to the cell membrane, and the endosome fuses with a lysosome. LDL-C's proteins are digested to amino acids, and the freed cholesterol is used to make cell membranes or steroid hormones.

Although LDL is needed for cellular uptake of cholesterol, excess levels of plasma LDL-C lead to atherosclerosis (■ Fig. 15.21). Endothelial cells lining the arteries transport LDL-C into the extracellular space so that it accumulates just under the intima (1). There, white blood cells called macrophages ingest cholesterol and other lipids to become lipid-filled *foam cells* (2). Cytokines released by the macrophages promote smooth muscle cell division (3). This early-stage *lesion* {*laesio*, injury} is called a *fatty streak*.

As the condition progresses, the lipid core grows, and smooth muscle cells reproduce, forming bulging *plaques* that protrude into the lumen of the artery (4). In the advanced stages of atherosclerosis, the plaques develop hard, calcified regions and fibrous collagen caps (5)–(7). The mechanism by which calcium carbonate is deposited is still being investigated.

Scientists once believed that the occlusion (blockage) of coronary blood vessels by large plaques that triggered blood clots was the primary cause of heart attacks, but that model has been revised. The new model indicates that blood clot formation on plaques is more dependent on the structure of a plaque than on its size. Atherosclerosis is now considered to be an inflammatory process in which macrophages release enzymes that convert stable plaques to vulnerable plaques (8). *Stable plaques* have thick fibrous caps that separate the lipid core from the blood and do not activate platelets. *Vulnerable plaques* have thin fibrous caps that are more likely to rupture, exposing collagen and activating platelets that initiate a blood clot (*thrombus*) (9).

If a clot blocks blood flow to the heart muscle, a heart attack, or *myocardial infarction*, results. Blocked blood flow in a coronary artery cuts off the oxygen supply to myocardial cells supplied by that artery. The oxygen-starved cells must then rely on anaerobic metabolism [p. 117], which produces lactate. As ATP production declines, the contractile cells are unable to pump Ca^+ out of the cell.

The unusually high Ca^+ concentration in the cytosol closes gap junctions in the damaged cells. Closure electrically isolates the damaged cells so that they no longer contract, and it forces action potentials to find an alternate route from cell to cell. If the damaged area of myocardium is large, the disruption can lead to an irregular heartbeat (*arrhythmia*) and potentially result in cardiac arrest or death.

EMERGING CONCEPTS

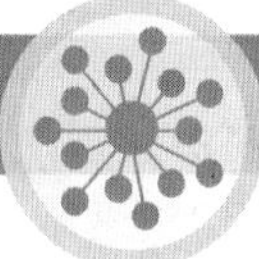

Inflammatory Markers for Cardiovascular Disease

In clinical studies, it is sometimes difficult to determine whether a factor that has a positive correlation with a disease functions in a cause-effect relationship or represents a simple association. For example, two factors associated with higher incidence of heart disease are C-reactive protein and homocysteine. *C-reactive protein* (CRP) is a molecule involved in the body's response to inflammation. In one study, women who had elevated blood CRP levels were more than twice as likely to have a serious cardiovascular problem as women with low CRP. Does this finding mean that CRP is causing cardiovascular disease? Or could it simply be a marker that can be used clinically to predict who is more likely to develop cardiovascular complications, such as a heart attack or stroke?

Similarly, elevated homocysteine levels are associated with an increased incidence of CVD. (*Homocysteine* is an amino acid that takes part in a complicated metabolic pathway that also requires folate and vitamin B_{12} as cofactors). Should physicians routinely measure homocysteine along with cholesterol? Currently there is little clinical evidence to show that reducing either CRP or homocysteine decreases a person's risk of developing CVD. If these two markers are not indicators for *modifiable* risk factors, should a patient's insurance be asked to pay for the tests used to detect them?

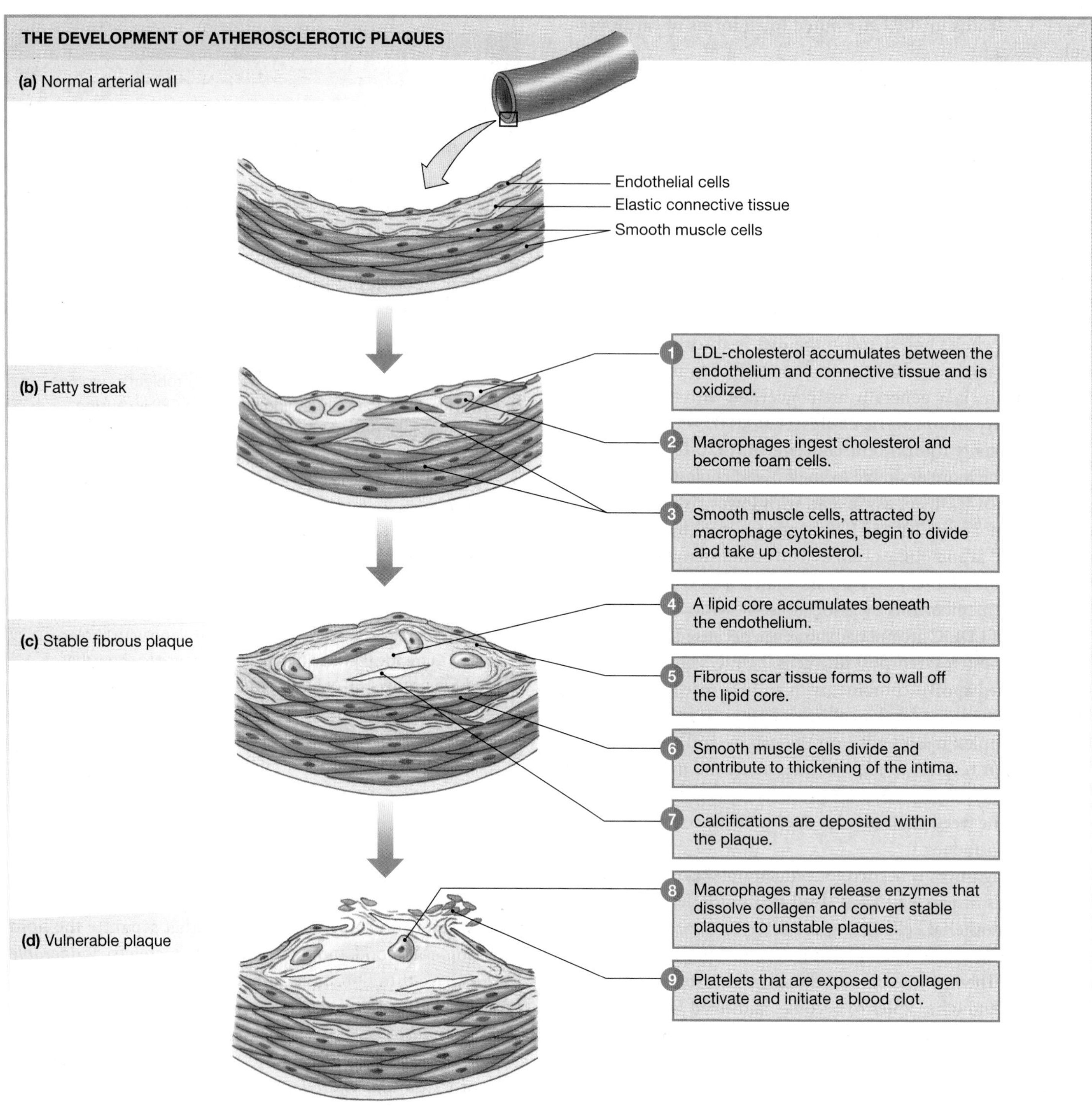

THE DEVELOPMENT OF ATHEROSCLEROTIC PLAQUES
(a) Normal arterial wall
Endothelial cells
Elastic connective tissue
Smooth muscle cells
(b) Fatty streak
1 LDL-cholesterol accumulates between the endothelium and connective tissue and is oxidized.
2 Macrophages ingest cholesterol and become foam cells.
3 Smooth muscle cells, attracted by macrophage cytokines, begin to divide and take up cholesterol.
(c) Stable fibrous plaque
4 A lipid core accumulates beneath the endothelium.
5 Fibrous scar tissue forms to wall off the lipid core.
6 Smooth muscle cells divide and contribute to thickening of the intima.
7 Calcifications are deposited within the plaque.
(d) Vulnerable plaque
8 Macrophages may release enzymes that dissolve collagen and convert stable plaques to unstable plaques.
9 Platelets that are exposed to collagen activate and initiate a blood clot.

■ Fig. 15.21

Hypertension Represents a Failure of Homeostasis

One controllable risk factor for cardiovascular disease is hypertension—chronically elevated blood pressure, with systolic pressures greater than 130–140 mm Hg or diastolic pressures greater than 80–90 mm Hg. Hypertension is a common disease in the United States and is one of the most common reasons for visits to physicians and for the use of prescription drugs. High blood pressure is associated with increasing risk of CVD: the risk doubles for each 20/10 mm Hg increase in blood pressure over a baseline value of 115/75 (■ Fig. 15.22).

More than 90% of all patients with hypertension are considered to have *essential* (or *primary*) *hypertension*, with no clear-cut cause other than heredity. Cardiac output is usually normal in these people, and their elevated blood pressure appears to be associated with increased peripheral resistance. Some investigators have speculated that the increased resistance may be due to a lack of nitric

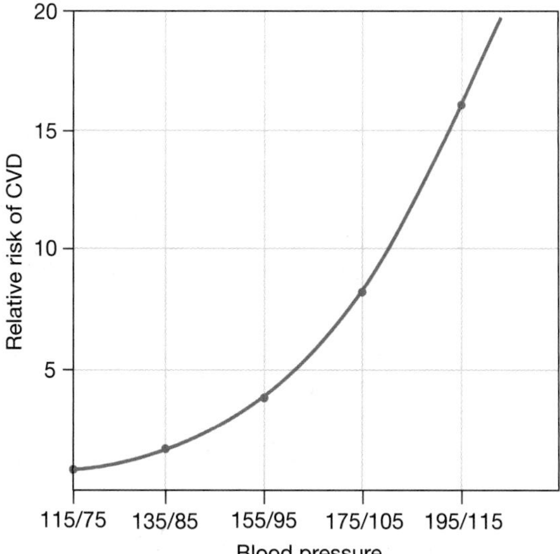

Fig. 15.22

oxide, the locally produced vasodilator formed by endothelial cells in the arterioles. In the remaining 5–10% of hypertensive cases, the cause is more apparent, and the hypertension is considered to be secondary to an underlying pathology. For instance, the cause might be an endocrine disorder that causes fluid retention.

A key feature of hypertension from all causes is adaptation of the carotid and aortic baroreceptors to higher pressure, with subsequent down-regulation of their activity. Without input from the baroreceptors, the cardiovascular control center interprets the high blood pressure as "normal," and no reflex reduction of pressure occurs.

Hypertension is a risk factor for atherosclerosis because high pressure in the arteries damages the endothelial lining of the vessels and promotes the formation of atherosclerotic plaques. In addition, high arterial blood pressure puts additional strain on the heart by increasing afterload [p. 500]. When resistance in the arterioles is high, the myocardium must work harder to push the blood into the arteries.

Amazingly, stroke volume in hypertensive patients remains constant up to a mean blood pressure of about 200 mm Hg, despite the increasing amount of work that the ventricle must perform as blood pressure increases. The cardiac muscle of the left ventricle responds to chronic high systemic resistance in the same way that skeletal muscle responds to a weight-lifting routine. The heart muscle *hypertrophies,* increasing the size and strength of the muscle fibers.

However, if resistance remains high over time, the heart muscle cannot meet the work load and begins to fail: cardiac output by the left ventricle decreases. If cardiac output of the right heart remains normal while the output from the left side decreases, fluid collects in the lungs, creating *pulmonary edema.* At this point, a detrimental positive feedback loop begins. Oxygen exchange in the lungs diminishes because of the pulmonary edema, leading to less oxygen in the blood. Lack of oxygen for aerobic metabolism further weakens the heart, and its pumping effectiveness diminishes even more. Unless treated, this condition, known as *congestive heart failure,* eventually leads to death.

Many of the treatments for hypertension have their basis in the cardiovascular physiology you have learned. For example, calcium entry into vascular smooth muscle and cardiac muscle can be decreased by a class of drugs known as *calcium channel blockers.* These drugs bind to Ca^{+} channel proteins, making it less likely that the channels will open in response to depolarization. With less Ca^{2+} entry, vascular smooth muscle dilates, while in the heart the depolarization rate of the SA node and the force of contraction decrease.

Vascular smooth muscle is more sensitive than cardiac muscle to certain classes of calcium channel blockers, and it is possible to get vasodilation at drug doses that are low enough to have no effect on heart rate. Other tissues with Ca^{+} channels, such as neurons, are only minimally affected by calcium channel blockers because their Ca^{+} channels are of a different subtype.

Other drugs used to treat hypertension include diuretics, which decrease blood volume, and beta-blocking drugs that target β_1-receptors and decrease catecholamine stimulation of cardiac output. Two other groups of antihypertensive drugs, the ACE inhibitors and the angiotensin receptor blockers, act by decreasing the activity of angiotensin, a powerful vasoconstrictor substance. You will learn more about angiotensin when you study the integrated control of blood pressure by the cardiovascular and renal systems. In the future, we may be seeing new treatments for hypertension that are based on other aspects of the molecular physiology of the heart and blood vessels.

RUNNING PROBLEM CONCLUSION

Essential Hypertension

Kurt remained on the calcium channel blocker and diuretic, and after several months his cough went away and his blood pressure stabilized at 130/85—a significant improvement. Kurt's new diet also brought his total blood cholesterol down below 200 mg/dL plasma. By improving two of his controllable risk factors, Kurt decreased his

RUNNING PROBLEM CONCLUSION *(continued)*

chances of having a heart attack. To learn more about hypertension and some of the therapies currently used to treat it, visit the web site of the American Heart Association (*www.americanheart.org*). Now check your understanding of this running problem by comparing your answers with the information in the summary table.

Question	Facts	Integration and Analysis
1. Why are people with high blood pressure at greater risk for having a hemorrhagic (or bleeding) stroke?	High blood pressure exerts force on the walls of the blood vessels.	If an area of blood vessel wall is weakened or damaged, high blood pressure may cause that area to rupture, allowing blood to leak out of the vessel into the surrounding tissues.
2. What is the rationale for reducing salt intake and taking a diuretic to control hypertension?	Salt causes water retention. Diuretics increase renal fluid excretion.	Blood pressure increases if the circulating blood volume increases. By restricting salt in the diet, a person can decrease retention of fluid in the extracellular compartment, which includes the plasma. Diuretics also help decrease blood volume.
3. Why would blocking the action of a vasoconstrictor lower blood pressure?	Blood pressure is determined by cardiac output and peripheral resistance.	Resistance is inversely proportional to the radius of the blood vessels. Therefore, if blood vessels dilate as a result of blocking a vasoconstrictor, resistance and blood pressure decrease.
4. How do calcium channel blockers lower blood pressure?	Calcium entry from the extracellular fluid plays an important role in both smooth muscle and cardiac muscle contraction.	Blocking Ca^{2+} entry through Ca^{2+} channels decreases the force of cardiac contraction and decreases the contractility of vascular smooth muscle. Both of these effects lower blood pressure.

509 515 519 522 533 537

Chapter Summary

Blood flow through the cardiovascular system is an excellent example of *mass flow* in the body. Cardiac contraction creates high pressure in the ventricles, and this pressure drives blood through the vessels of the systemic and pulmonary circuits, speeding up cell-to-cell *communication.* Resistance to flow is regulated by *local and reflex control mechanisms* that act on arteriolar smooth muscle and help match tissue perfusion to tissue needs. The *homeostatic* baroreceptor reflex monitors arterial pressure to ensure adequate perfusion of the brain and heart. Capillary *exchange of material* between the plasma and interstitial fluid *compartments* uses several transport mechanisms, including diffusion, transcytosis, and bulk flow.

1. Homeostatic regulation of the cardiovascular system is aimed at maintaining adequate blood flow to the brain and heart. (p. 510)
2. Total blood flow at any level of the circulation is equal to the cardiac output. (p. 510)

The Blood Vessels

iP Cardiovascular—Anatomy Review: Blood Vessel Structure & Function

3. Blood vessels are composed of layers of smooth muscle, elastic and fibrous connective tissue, and **endothelium**. (p. 510; Fig. 15.2)
4. **Vascular smooth muscle** maintains a state of muscle tone. (p. 510)
5. The walls of the aorta and major arteries are both stiff and springy. This property allows them to absorb energy and release it through elastic recoil. (p. 511)
6. **Metarterioles** regulate blood flow through capillaries and allow white blood cells to go directly from arterioles to the venous circulation. Blood flow into individual capillaries can be regulated by **precapillary sphincters**. (p. 511; Fig. 15.3)
7. Capillaries and postcapillary **venules** are the site of exchange between blood and interstitial fluid. (p. 512)
8. Veins hold more than half of the blood in the circulatory system. Veins have thinner walls with less elastic tissue than arteries, so veins expand easily when they fill with blood. (p. 512)
9. **Angiogenesis** is the process by which new blood vessels grow and develop, especially after birth. (p. 512)

Blood Pressure

iP Cardiovascular: Measuring Blood Pressure

10. The ventricles create high pressure that is the driving force for blood flow. The aorta and arteries act as a pressure reservoir during ventricular relaxation. (p. 513; Fig. 15.5)
11. Blood pressure is highest in the arteries and decreases as blood flows through the circulatory system. At rest, desirable **systolic pressure** is 120 mm Hg or less, and desirable **diastolic pressure** is 80 mm Hg or less. (p. 514; Fig. 15.6)
12. Pressure created by the ventricles can be felt as a **pulse** in the arteries. **Pulse pressure** equals systolic pressure minus diastolic pressure. (p. 514)
13. Blood flow against gravity in the veins is assisted by one-way valves and by the respiratory and skeletal muscle pumps. (p. 512; Fig. 15.4)
14. Arterial blood pressure is indicative of the driving pressure for blood flow. **Mean arterial pressure** (MAP) is defined as diastolic pressure + 1/3 (systolic pressure – diastolic pressure). (p. 515)
15. Arterial blood pressure is usually measured with a sphygmomanometer. Blood squeezing through a compressed brachial artery makes **Korotkoff sounds**. (p. 516; Fig. 15.7)
16. Arterial pressure is a balance between cardiac output and the resistance to blood flow offered by the arterioles (**peripheral resistance**). (p. 517; Fig. 15.8)
17. If blood volume increases, blood pressure increases. If blood volume decreases, blood pressure decreases. (p. 518; Fig. 15.9)
18. Venous blood volume can be shifted to the arteries if arterial blood pressure falls. (p. 509; Fig. 15.1)

Resistance in the Arterioles

iP Cardiovascular: Factors That Affect Blood Pressure

19. The arterioles are the main site of variable resistance in the systemic circulation. A small change in the radius of an arteriole creates a large change in resistance: $R \propto 1/r^4$. (p. 519)
20. Arterioles regulate their own blood flow through **myogenic autoregulation**. Vasoconstriction increases the resistance offered by an arteriole and decreases the blood flow through the arteriole. (p. 519)
21. Arteriolar resistance is influenced by local control mechanisms that match tissue blood flow to the metabolic needs of the tissue. Vasodilator paracrines include nitric oxide, H^+, K^+, CO_2, prostaglandins, adenosine, and histamine. Low O_2 causes vasodilation. Endothelins are powerful vasoconstrictors. (p. 520; Tbl. 15.2)
22. **Active hyperemia** is a process in which increased blood flow accompanies increased metabolic activity. **Reactive hyperemia** is an increase in tissue blood flow following a period of low perfusion. (p. 521; Fig. 15.10)
23. Most systemic arterioles are under tonic sympathetic control. Norepinephrine causes vasoconstriction. Decreased sympathetic stimulation causes vasodilation. (p. 522)
24. Epinephrine binds to arteriolar α-receptors and causes vasoconstriction. Epinephrine on β_2-receptors, found in the arterioles of the heart, liver, and skeletal muscle, causes vasodilation. (p. 522)

Distribution of Blood to the Tissues

25. Changing the resistance of the arterioles affects mean arterial pressure and alters blood flow through the arteriole. (p. 527; Fig. 15.15)
26. The flow through individual arterioles depends on their resistance. The higher the resistance in an arteriole, the lower the blood flow in that arteriole: $Flow_{arteriole} \propto 1/R_{arteriole}$. (p. 524)

Regulation of Cardiovascular Function

iP Cardiovascular: Blood Pressure Regulation

27. The reflex control of blood pressure resides in the medulla oblongata. **Baroreceptors** in the carotid artery and the aorta monitor arterial blood pressure and trigger the **baroreceptor reflex**. (p. 526; Fig. 15.14)
28. Efferent output from the medullary **cardiovascular control center** goes to the heart and arterioles. Increased sympathetic activity increases heart rate and force of contraction. Increased parasympathetic activity slows heart rate. Increased sympathetic discharge at the arterioles causes vasoconstriction. There is no significant parasympathetic control of arterioles. (p. 524)
29. Cardiovascular function can be modulated by input from higher brain centers and from the respiratory control center of the medulla. (p. 525)
30. The baroreceptor reflex functions each time a person stands up. The decrease in blood pressure upon standing is known as orthostatic hypotension. (p. 525)

Exchange at the Capillaries

iP Cardiovascular: Autoregulation and Capillary Dynamics

31. Exchange of materials between the blood and the interstitial fluid occurs primarily by diffusion. (p. 528)

15

32. **Continuous capillaries** have leaky junctions between cells but also transport material using transcytosis. Continuous capillaries with tight junctions form the blood-brain barrier. (p. 529; Fig. 15.16)
33. **Fenestrated capillaries** have pores that allow large volumes of fluid to pass rapidly. (p. 529; Fig. 15.16)
34. The velocity of blood flow through the capillaries is slow, allowing diffusion to go to equilibrium. (p. 530; Fig. 15.17)
35. The mass movement of fluid between the blood and the interstitial fluid is **bulk flow**. Fluid movement is called **filtration** if the direction of flow is out of the capillary and **absorption** if the flow is directed into the capillary. (p. 531; Fig. 15.18)
36. The osmotic pressure difference between plasma and interstitial fluid due to the presence of plasma proteins is the **colloid osmotic pressure**. (p. 530)

The Lymphatic System

iP Fluids & Electrolytes: Electrolyte Homeostasis, Edema

37. About 3 liters of fluid filter out of the capillaries each day. The lymphatic system returns this fluid to the circulatory system. (p. 532; Fig. 15.19)
38. Lymph capillaries accumulate fluid, interstitial proteins, and particulate matter by bulk flow. Lymph flow depends on smooth muscle in vessel walls, one-way valves, and the skeletal muscle pump. (p. 532)
39. The condition in which excess fluid accumulates in the interstitial space is called **edema**. Factors that disrupt the normal balance between capillary filtration and absorption cause edema. (p. 533)

Cardiovascular Disease

40. **Cardiovascular disease** is the leading cause of death in the United States. Risk factors predict the likelihood that a person will develop cardiovascular disease during her or his lifetime. (p. 534)
41. **Atherosclerosis** is an inflammatory condition in which fatty deposits called plaques develop in arteries. If plaques are unstable, they may block the arteries by triggering blood clots. (p. 536; Fig. 15.21)
42. Hypertension is a significant risk factor for the development of cardiovascular disease. (p. 537; Fig. 15.22)

Questions

Answers: p. A-1

Level One Reviewing Facts and Terms

1. The first priority of blood pressure homeostasis is to maintain adequate perfusion to which two organs?
2. Match the types of systemic blood vessels with the terms that describe them. Each vessel type may have more than one match, and matching items may be used more than once.

(a) arterioles	1. store pressure generated by the heart
(b) arteries	2. have walls that are both stiff and elastic
(c) capillaries	3. carry low-oxygen blood
(d) veins	4. have thin walls of exchange epithelium
(e) venules	5. act as a volume reservoir
	6. their diameter can be altered by neural input
	7. blood flow slowest through these vessels
	8. have lowest blood pressure
	9. are the main site of variable resistance

3. List the four tissue components of blood vessel walls, in order from inner lining to outer covering. Briefly describe the importance of each tissue.
4. Blood flow to individual tissues is regulated by selective vasoconstriction and vasodilation of which vessels?
5. Aortic pressure reaches a typical high value of ________ (give both numeric value and units) during ________, or contraction of the heart. As the heart relaxes during the event called ________, aortic pressure declines to a typical low value of ________. This blood pressure reading would be written as ________/________.
6. The rapid pressure increase that occurs when the left ventricle pushes blood into the aorta can be felt as a pressure wave, or ________. What is the equation used to calculate the strength of this pressure wave?
7. List the factors that aid venous return to the heart.
8. What is hypertension, and why is it a threat to a person's health?
9. When measuring a person's blood pressure, at what point in the procedure are you likely to hear Korotkoff sounds?
10. List three paracrines that cause vasodilation. What is the source of each one? In addition to paracrines, list two other ways to control smooth muscle contraction in arterioles.
11. What is hyperemia? How does active hyperemia differ from reactive hyperemia?
12. Most systemic arterioles are innervated by the ________ branch of the nervous system. Increased sympathetic input will have what effect on arteriole diameter?
13. Match each event in the left column with all appropriate neurotransmitter(s) *and* receptor(s) from the list on the right.

(a) vasoconstriction of intestinal arterioles	1. norepinephrine
	2. epinephrine
(b) vasodilation of coronary arterioles	3. acetylcholine
	4. β_1-receptor
(c) increased heart rate	5. α-receptor
(d) decreased heart rate	6. β_2-receptor
(e) vasoconstriction of coronary arterioles	7. nicotinic receptor
	8. muscarinic receptor

14. Which organs receive more than two-thirds of the cardiac output at rest? Which organs have the highest flow of blood on a per unit weight basis?
15. By looking at the density of capillaries in a tissue, you can make assumptions about what property of the tissue? Which tissue has the lowest capillary density? Which tissue has the highest?
16. What type of transport is used to move each of the following substances across the capillary endothelium?
    (a) oxygen
    (b) proteins
    (c) glucose
    (d) water

17. With which three physiological systems do the vessels of the lymphatic system interact?
18. Define edema. List some ways in which it can arise.
19. Define the following terms and explain their significance to cardiovascular physiology.
    (a) perfusion
    (b) colloid osmotic pressure
    (c) vasoconstriction
    (d) angiogenesis
    (e) metarterioles
    (f) pericytes
20. The two major lipoprotein carriers of cholesterol are ________ and ________. Which type is bad when present in the body in elevated amounts?

Level Two Reviewing Concepts

21. **Concept map:** Map all the following factors that influence mean arterial pressure. You may add terms.

• aorta	• parasympathetic neuron
• arteriole	• peripheral resistance
• baroreceptor	• SA node
• blood volume	• sensory neuron
• cardiac output	• stroke volume
• carotid artery	• sympathetic neuron
• contractility	• vein
• heart rate	• venous return
• medulla oblongata	• ventricle

22. Compare and contrast the following sets of terms:
    (a) lymphatic capillaries and systemic capillaries
    (b) roles of the sympathetic and parasympathetic branches in blood pressure control
    (c) lymph and blood
    (d) continuous capillaries and fenestrated capillaries
    (e) hydrostatic pressure and colloid osmotic pressure in systemic capillaries
23. Calcium channel blockers prevent Ca^{2+} movement through Ca^{2+} channels. Explain two ways this action lowers blood pressure. Why are neurons and other cells unaffected by these drugs?
24. Define myogenic autoregulation. What mechanisms have been proposed to explain it?
25. Left ventricular failure may be accompanied by edema, shortness of breath, and increased venous pressure. Explain how these signs and symptoms develop.

Level Three Problem Solving

26. Robert is a 52-year-old nonsmoker. He weighs 180 lbs and stands 5'9" tall, and his blood pressure averaged 145/95 on three successive visits to his doctor's office. His father, grandfather, and uncle all had heart attacks in their early 50s, and his mother died of a stroke at the age of 71.
    (a) Identify Robert's risk factors for coronary heart disease.
    (b) Does Robert have hypertension? Explain.
    (c) Robert's doctor prescribes a drug called a beta blocker. Explain the mechanism by which a beta-receptor-blocking drug may help lower blood pressure.
27. The following figure is a schematic representation of the systemic circulation. Use it to help answer the following questions. (CO = cardiac output, MAP = mean arterial pressure).
    (a) If resistance in vessels 1 and 2 increases because of the presence of local paracrines but cardiac output is unchanged, what happens to MAP? What happens to flow through vessels 1 and 2? Through vessels 3 and 4?
    (b) Homeostatic compensation occurs within seconds. Draw a reflex map to explain the compensation (stimulus, receptor, and so on).
    (c) When vessel 1 constricts, what happens to the filtration pressure in the capillaries downstream from that arteriole?

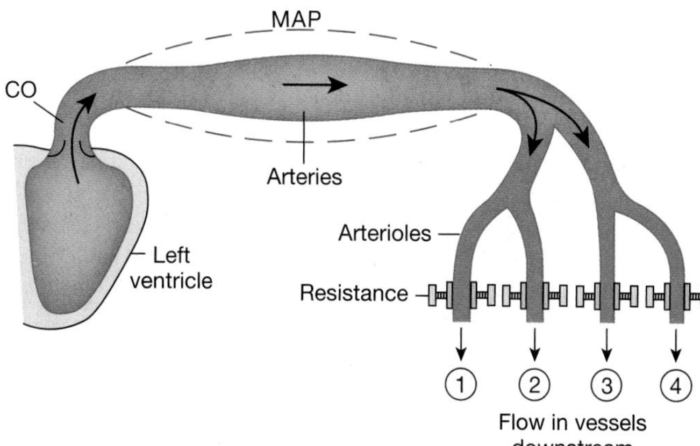

28. The following graphs are recordings of contractions in an isolated frog heart. The intact frog heart is innervated by sympathetic neurons that increase heart rate and by parasympathetic neurons that decrease heart rate. Based on these four graphs, what conclusion can you draw about the mechanism of action of atropine? (Atropine does not cross the cell membrane.)

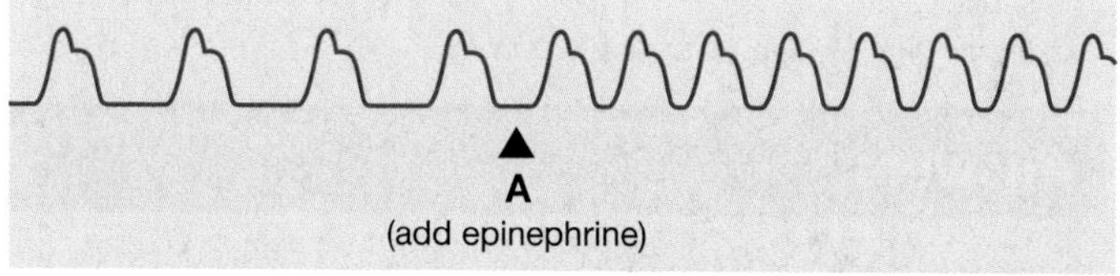

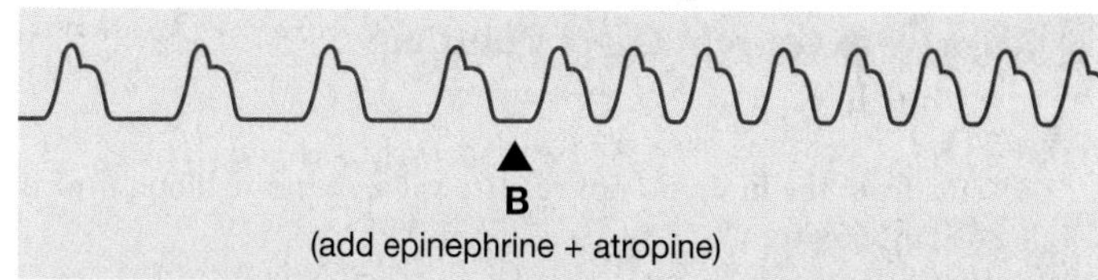

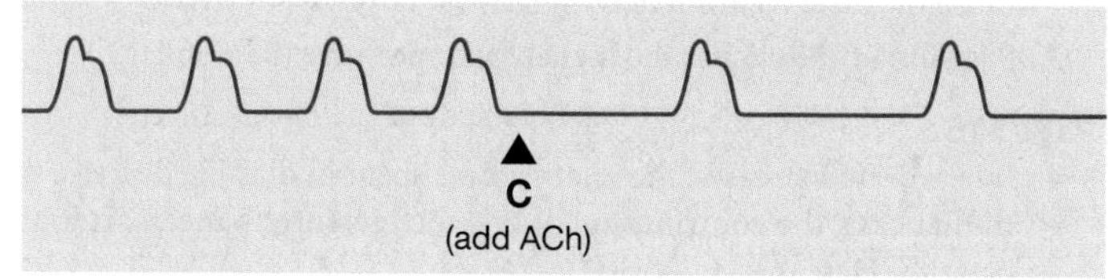

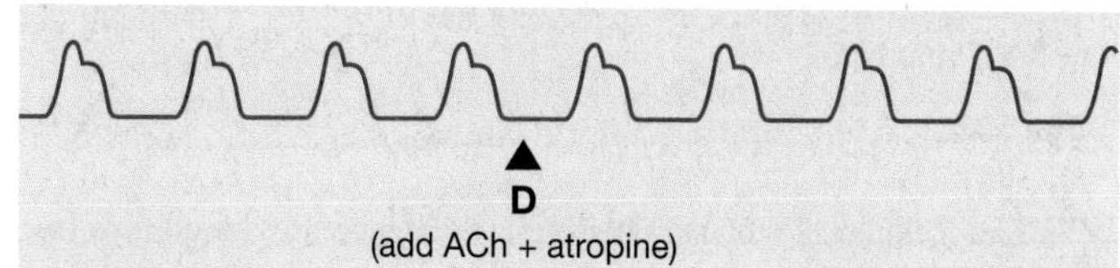

29. Draw a reflex map that explains Anthony's vasovagal syncope at the sight of blood. Include all the steps of the reflex, and explain whether autonomic pathways are being stimulated or inhibited.
30. A physiologist placed a section of excised arteriole in a perfusion chamber containing saline. When the oxygen content of the saline perfusing (flowing through) the arteriole was reduced, the arteriole

dilated. In a follow-up experiment, she used an isolated piece of arteriolar smooth muscle that had been stripped away from the other layers of the arteriole wall. When the oxygen content of the saline was reduced as in the first experiment, the isolated muscle showed no response. What do these two experiments suggest about how low oxygen exerts local control over arterioles?

31. In advanced atherosclerosis, calcified plaques cause the normally elastic aorta and arteries to become stiff and noncompliant. (a) What effect does this change in the aorta have on afterload? (b) If cardiac output remains unchanged, what happens to peripheral resistance and mean arterial pressure?

32. During fetal development, most blood in the pulmonary artery bypasses the lungs and goes into the aorta by way of a channel called the *ductus arteriosus*. Normally this fetal bypass channel closes during the first day after birth, but each year about 4000 babies in the United States maintain a *patent* (open) ductus arteriosus and require surgery to close the channel.
    (a) Use this information to draw an anatomical diagram showing blood flow in an infant with a patent ductus arteriosus.
    (b) In the fetus, why does most blood bypass the lungs?
    (c) If the systemic side of the circulatory system is longer than the pulmonary side, which circuit has the greater resistance?
    (d) If flow is equal in the pulmonary and systemic circulations, which side of the heart must generate more pressure to overcome resistance?
    (e) Use your answer to (d) to figure out which way blood will flow through a patent ductus arteriosus.

Level Four Quantitative Problems

33. Using the appropriate equation, mathematically explain what happens to blood flow if the *diameter* of a blood vessel increases from 2 mm to 4 mm.

34. Duplicate the calculations that led William Harvey to believe that blood circulated in a closed loop:
    (a) Take your resting pulse.
    (b) Assume that your heart at rest pumps 70 mL/beat, and that 1 mL of blood weighs one gram. Calculate how long it would take your heart to pump your weight in blood. (2.2 pounds = 1 kilogram)

35. Calculate the mean arterial pressure (MAP) and pulse pressure for a person with a blood pressure of 115/73.

36. According to the Fick principle, the oxygen consumption rate of an organ is equal to the blood flow through that organ times the amount of oxygen extracted from the blood as it flows through the organ:

    Oxygen consumption rate = blood flow × (arterial oxygen content – venous oxygen content)

    $$(\text{mL } O_2 \text{ consumed/min}) = (\text{mL blood/min} \times \text{mL } O_2/\text{mL blood})$$

    A woman has a total body oxygen consumption rate of 250 mL/min. The oxygen content of blood in her aorta is 200 mL O_2/L blood, the oxygen content of her pulmonary artery blood is 160 mL O_2/L blood. What is her cardiac output?

37. Beau has an average daily heart rate of 75 beats per minute. If his net capillary filtration rate is 3.24 L/day, how much fluid filters from his capillary with each beat of his heart?

Answers

Answers to Concept Check Questions

Page 514

1. Veins from the brain do not require valves because blood flow is aided by gravity.
2. The carotid wave would arrive slightly ahead of the wrist wave because the distance from heart to carotid artery is shorter.
3. Pressure of 130/95 has the higher pulse pressure (35 mm Hg).

Page 515

4. If heart rate increases, the relative time spent in diastole decreases. In that case, the contribution of systolic pressure to mean arterial pressure increases, and MAP increases.
5. Pulse pressure is $112 - 68 = 44$ mm Hg. MAP is $68 + 1/3\,(44) = 82.7$ mm Hg.

Page 522

6. (d)
7. Extracellular K^+ dilates arterioles, which increases blood flow (see Tbl. 15.2).

Page 522

8. Epinephrine binding to myocardial β_1-receptors increases heart rate and force of contraction. Epinephrine binding to β_2-receptors on heart arterioles causes vasodilation.
9. α-Receptors have lower affinity for epinephrine than β_2-receptors, so the β_2-receptors dominate and arterioles dilate.

Page 524

10. (a) The kidney has the highest blood flow per unit weight. (b) The heart has the lowest total blood flow.

Page 527

11. The most likely ion is Na^+ moving into the receptor cell.

Page 527

12. This map should look exactly like Fig. 15.14b except that the directions of the arrows is reversed.

Page 528

13. Stimulus: sight, sound, and smell of the *T. rex*. Receptors: eyes, ears, and nose. Integrating center: cerebral cortex, with descending pathways through the limbic system. Divergent pathways go to the cardiovascular control center, which increases sympathetic output to heart and arterioles. A second descending spinal pathway goes to the adrenal medulla, which releases epinephrine. Epinephrine on β_2-receptors of liver, heart, and skeletal muscle arterioles causes vasodilation of those arterioles. Norepinephrine onto α-receptors in other arterioles causes vasoconstriction. Both catecholamines increase heart rate and force of contraction.

Page 531

14. Loss of plasma proteins will decrease colloid osmotic pressure. As a result, hydrostatic pressure will have a greater effect in the filtration-absorption balance, and filtration will increase.
15. Using osmotic pressure rather than osmolarity allows a direct comparison between absorption pressure and filtration pressure, both of which are expressed in mm Hg.

Page 533

16. If the left ventricle fails, blood backs up into the left atrium and pulmonary veins, and then into lung capillaries. Edema in the lungs is known as *pulmonary edema*.
17. Low-protein diets result in a low concentration of plasma proteins. Capillary absorption is reduced while filtration remains constant, resulting in edema and ascites.

Answers to Figure and Graph Questions

Page 509

Fig. 15.1: The pumps are arranged in series (one after the other).

Page 517

Fig. 15.8: 1. Flow decreases and MAP increases. 2. Volume and MAP decrease. 3. Venous volume decreases, arterial volume increases, and MAP increases.

Page 523

Fig. 15.11: Sympathetic innervation and epinephrine increase heart rate and stroke volume; parasympathetic innervation decreases heart rate. Sympathetic input causes vasoconstriction but epinephrine causes vasodilation in selected arterioles. For paracrine factors that influence arteriolar diameter, see Table 15.2.

Page 524

Fig. 15.12: Blood flow through the lungs is 5 L/min.

Page 525

Fig. 15.13: Blood pressure upstream increases.

Page 526

Fig. 15.14a: SA node has muscarinic cholinergic receptors for ACh and β_1-receptors for catecholamines. Ventricles have β_1-receptors for catecholamines. Arterioles and veins have α-receptors for norepinephrine.

Page 530

Fig. 15.17: (a) Velocity of flow is inversely proportional to area: as area increases, velocity decreases. (b) Changing only cross-sectional area has no effect on flow rate because flow rate is determined by cardiac output.

Page 531

Fig. 15.18: Net filtration will increase as a result of the increased hydrostatic pressure.

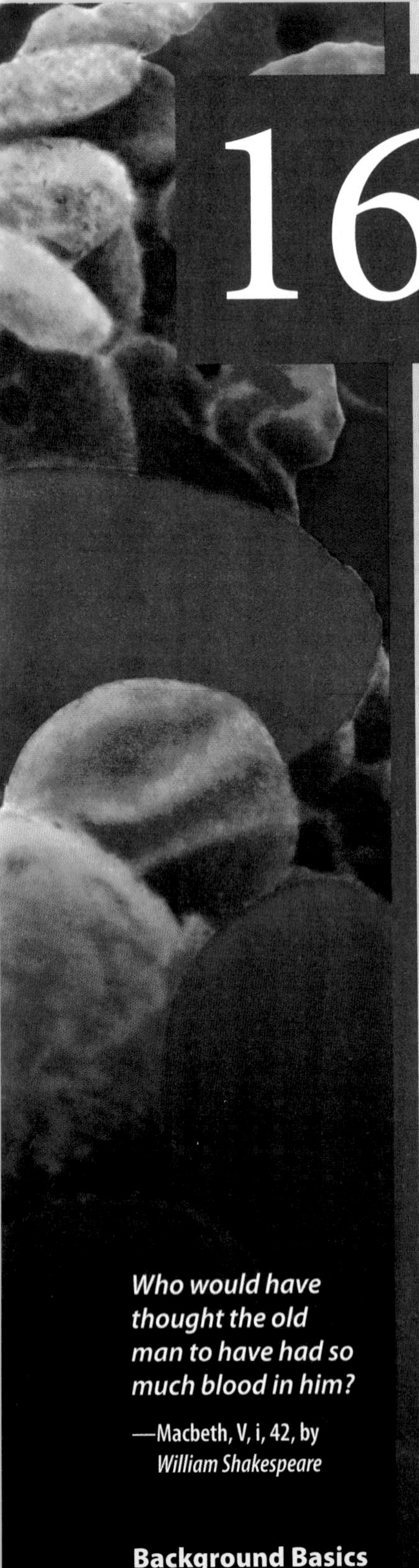

16 Blood

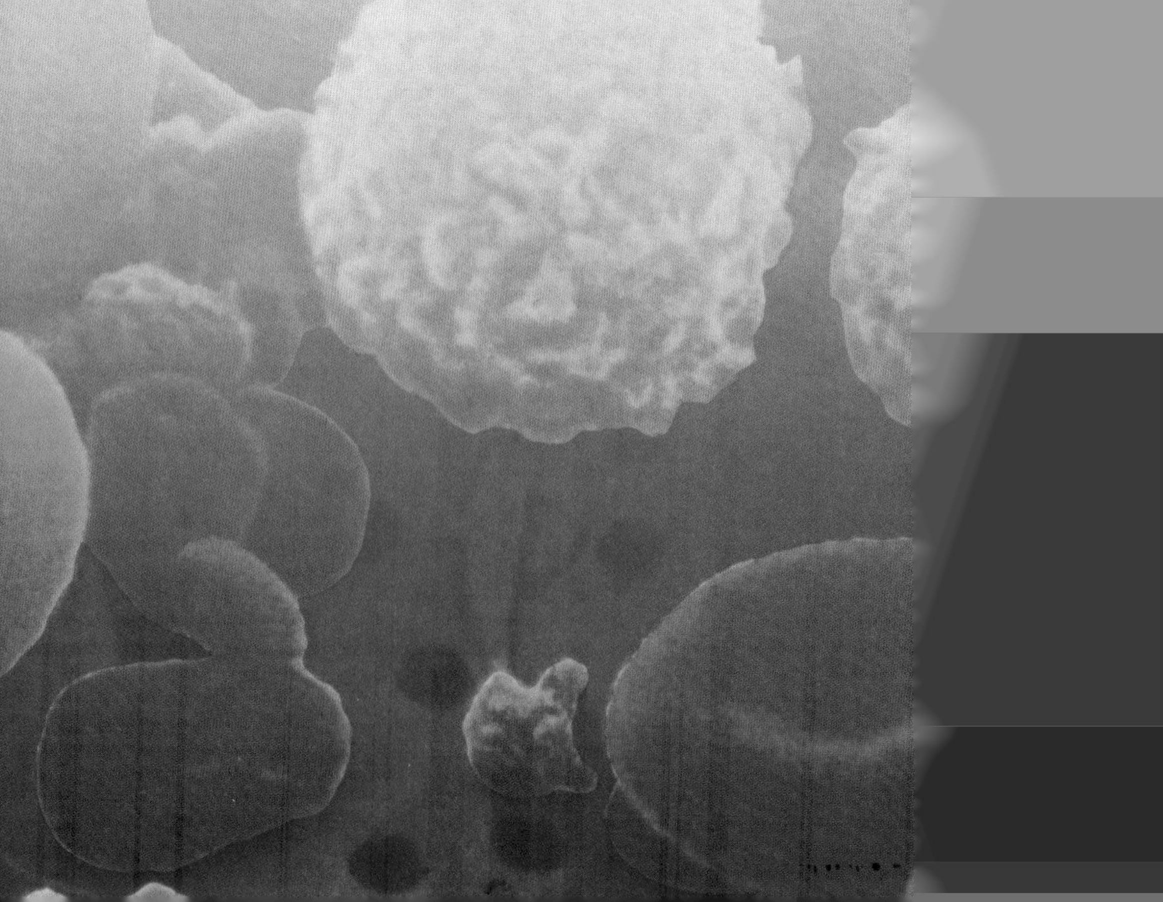

Who would have thought the old man to have had so much blood in him?

—Macbeth, V, i, 42, by *William Shakespeare*

Background Basics

85 Connective tissue

155 Phagocytosis

180 Second messenger cascade

469 Viscosity and resistance

87 Collagen

74 Cell organelles

177 Cytokines

Blood, the fluid that circulates in the cardiovascular system, has occupied a prominent place throughout history as an almost mystical fluid. Humans undoubtedly had made the association between blood and life by the time they began to fashion tools and hunt animals. A wounded animal that lost blood would weaken and die if the blood loss was severe enough. The logical conclusion was that blood was necessary for existence. This observation eventually led to the term *lifeblood,* meaning anything essential for existence.

Ancient Chinese physicians linked blood to energy flow in the body. They wrote about the circulation of blood through the heart and blood vessels long before William Harvey described it in seventeenth-century Europe. In China, changes in blood flow were used as diagnostic clues to illness. Chinese physicians were expected to recognize some 50 variations in the pulse. Because blood was considered a vital fluid to be conserved and maintained, bleeding patients to cure disease was not a standard form of treatment.

In contrast, Western civilizations came to believe that disease-causing evil spirits circulated in the blood. The way to remove these spirits was to remove the blood containing them. Because blood was recognized as an essential fluid, however, bloodletting had to be done judiciously. Veins were opened with knives or sharp instruments (*venesection*), or blood-sucking leeches were applied to the skin. In ancient India, people believed that leeches could distinguish between healthy and infected blood.

There is no written evidence that venesection was practiced in ancient Egypt, but the work carried out by Galen of Pergamum in the second century influenced Western medicine for nearly 2000 years. This early Greek physician advocated bleeding as treatment for many disorders. The location, timing, and frequency of the bleeding depended on the condition, and the physician was instructed to remove enough blood to bring the patient to the point of fainting. Over the years, this practice undoubtedly killed more people than it cured.

What is even more remarkable is the fact that as late as 1923, an American medical textbook advocated bleeding for treating certain infectious diseases, such as pneumonia! Now that we better understand the importance of blood in the immune response, it is doubtful that modern medicine will ever again turn to blood removal as a nonspecific means of treating disease. It is still used, however, for selected *hematological disorders* {*haima,* blood}.

RUNNING PROBLEM

Blood Doping in Athletes

Athletes spend hundreds of hours training, trying to build their endurance. For Johann Muehlegg, a cross-country skier at the 2002 Salt Lake City Winter Olympics, it appeared that his training had paid off when he captured three gold medals. On the last day of the Games, however, Olympics officials expelled Muehlegg and stripped him of his gold medal in the 50-kilometer classical race. The reason? Muehlegg had tested positive for a performance-enhancing chemical that increased the oxygen-carrying capacity of his blood. Officials claimed Muehlegg's endurance in the grueling race was the result of blood doping, not training.

545 554 557 560 565

16

Plasma and the Cellular Elements of Blood

What is this remarkable fluid that flows through the circulatory system? Blood makes up one-fourth of the extracellular fluid, the internal environment that bathes cells and acts as a buffer between cells and the external environment. Blood is the circulating portion of the extracellular fluid, responsible for carrying material from one part of the body to another.

Total blood volume in a 70-kg man is equal to about 7% of his total body weight, or 0.07×70 kg = 4.9 kg. Thus, if we assume that 1 kg of blood occupies a volume of 1 liter, a 70-kg man has about 5 liters of blood. Of this volume, about 2 liters is composed of blood cells, while the remaining 3 liters is composed of plasma, the fluid portion of the blood. The 58-kilogram "Reference Woman" [p. 132] has about 4 L total blood volume.

In this chapter we present an overview of the components of blood and the functions of plasma, red blood cells, and platelets. You will learn more about hemoglobin when you study oxygen transport in the blood, and more about white blood cells and blood types when you study the immune system.

Plasma Is Extracellular Matrix

Plasma is the fluid matrix of the blood, within which cellular elements are suspended (■ Fig. 16.1). Water is the main component of plasma, accounting for about 92% of its weight. Proteins account for another 7%. The remaining 1% is dissolved organic molecules (amino acids, glucose, lipids, and nitrogenous wastes), ions (Na^+, K^+, Cl^-, H^+, Ca^{2+}, and HCO_3^-), trace elements and vitamins, and dissolved oxygen (O_2) and carbon dioxide (CO_2).

Plasma is identical in composition to interstitial fluid except for the presence of **plasma proteins**. **Albumins** are the most prevalent type of protein in the plasma, making up about 60% of the total. Albumins and nine other proteins—including *globulins,* the clotting protein *fibrinogen,* and the iron-transporting protein *transferrin*—make up more than 90% of all plasma proteins. The liver makes most plasma proteins and secretes them into the blood. Some globulins, known as *immunoglobulins* or

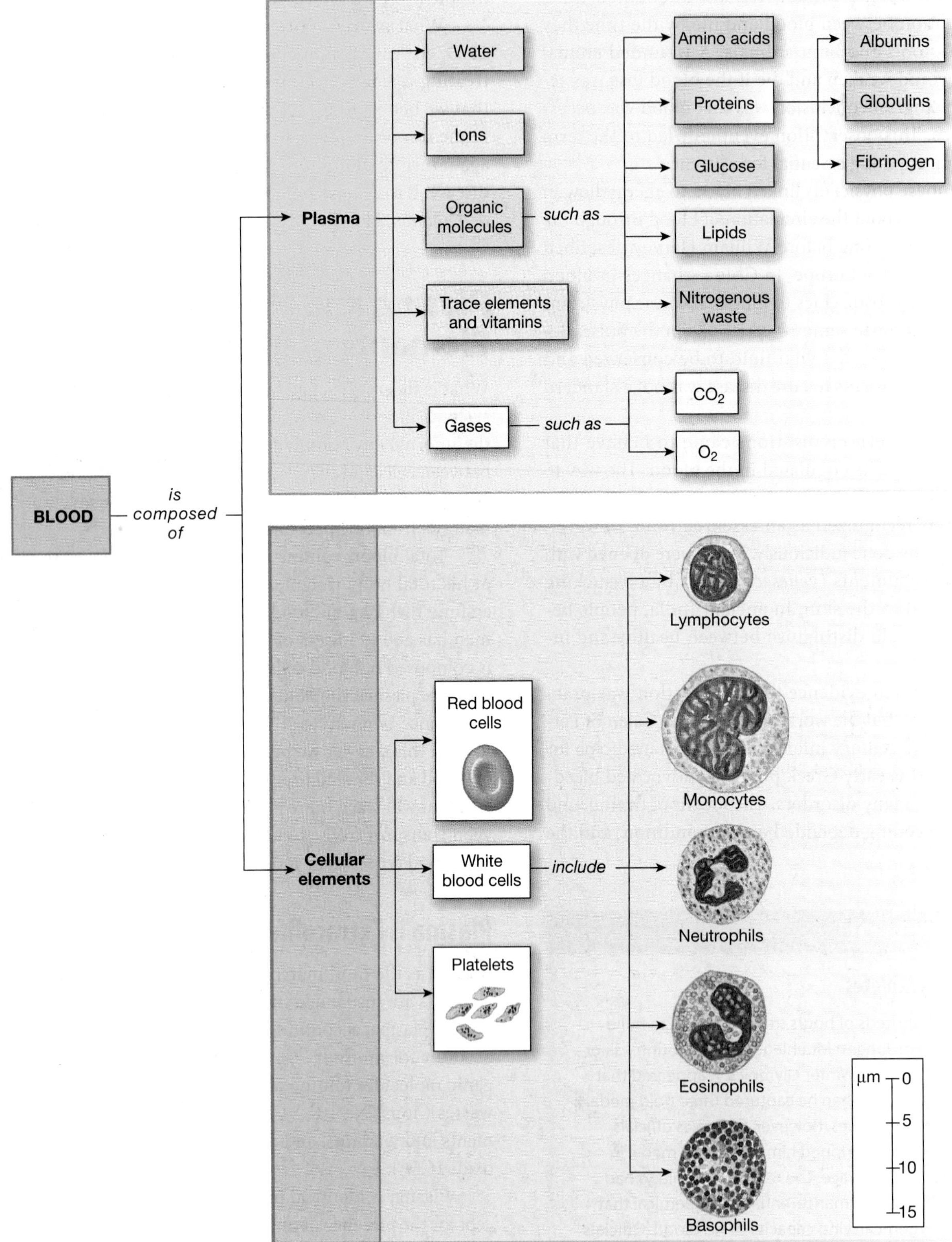
Blood consists of plasma and cellular elements.
BLOOD
is composed of
Plasma
Water
Ions
Organic molecules
such as
Amino acids
Proteins
Glucose
Lipids
Nitrogenous waste
Albumins
Globulins
Fibrinogen
Trace elements and vitamins
Gases
such as
CO_2
O_2
Cellular elements
Red blood cells
White blood cells
include
Platelets
Lymphocytes
Monocytes
Neutrophils
Eosinophils
Basophils
µm
0
5
10
15

Fig. 16.1

Table 16.1 Functions of Plasma Proteins

Name	Source	Function
Albumins (multiple types)	Liver	Major contributors to colloid osmotic pressure of plasma; carriers for various substances
Globulins (multiple types)	Liver and lymphoid tissue	Clotting factors, enzymes, antibodies, carriers for various substances
Fibrinogen	Liver	Forms fibrin threads essential to blood clotting
Transferrin	Liver and other tissues	Iron transport

antibodies, are synthesized and secreted by specialized blood cells rather than by the liver.

The presence of proteins in the plasma makes the osmotic pressure of the blood higher than that of the interstitial fluid. This osmotic gradient tends to pull water from the interstitial fluid into the capillaries and offset filtration out of the capillaries created by blood pressure [p. 529].

Plasma proteins participate in many functions, including blood clotting and defense against foreign invaders. In addition, they act as carriers for steroid hormones, cholesterol, drugs, and certain ions such as iron (Fe^{2+}). Finally, some plasma proteins act as hormones or as extracellular enzymes. ■ Table 16.1 summarizes the functions of plasma proteins.

Cellular Elements Include RBCs, WBCs, and Platelets

Three main cellular elements are found in blood (Fig. 16.1): **red blood cells** (RBCs), also called **erythrocytes** {*erythros,* red}; **white blood cells** (WBCs), also called **leukocytes** {*leukos,* white}; and **platelets** or *thrombocytes* {*thrombo-,* lump, clot}. White blood cells are the only fully functional cells in the circulation. Red blood cells have lost their nuclei by the time they enter the bloodstream, and platelets, which also lack a nucleus, are cell fragments that have split off a relatively large parent cell known as a **megakaryocyte** {*mega,* extremely large + *karyon,* kernel + *-cyte,* cell}.

Red blood cells play a key role in transporting oxygen from lungs to tissues, and carbon dioxide from tissues to lungs. Platelets are instrumental in *coagulation,* the process by which blood clots prevent blood loss in damaged vessels. White blood cells play a key role in the body's immune responses, defending the body against foreign invaders, such as parasites, bacteria, and viruses. Although most white blood cells circulate through the body in the blood, their work is usually carried out in the tissues rather than in the circulatory system.

Blood contains five types of mature white blood cells: (1) **lymphocytes**, (2) **monocytes**, (3) **neutrophils**, (4) **eosinophils**, and (5) **basophils**. Monocytes that leave the circulation and enter the tissues develop into **macrophages**. Tissue basophils are called **mast cells**.

The types of white blood cells may be grouped according to common morphological or functional characteristics. Neutrophils, monocytes, and macrophages are collectively known as **phagocytes** because they can engulf and ingest foreign particles such as bacteria (phagocytosis) [p. 155]. Lymphocytes are sometimes called **immunocytes** because they are responsible for specific immune responses directed against invaders. Basophils, eosinophils, and neutrophils are called **granulocytes** because they contain cytoplasmic inclusions that give them a granular appearance.

Concept Check Answers: p. 567

1. Name the five types of leukocytes.
2. Why do we say that erythrocytes and platelets are not fully functional cells?
3. On the basis of what you have learned about the origin and role of plasma proteins, explain why patients with advanced liver degeneration frequently suffer from edema [p. 533].

Blood Cell Production

Where do these different blood cells come from? They are all descendants of a single precursor cell type known as the *pluripotent hematopoietic stem cell* (■ Fig. 16.2). This cell type is found primarily in **bone marrow**, a soft tissue that fills the hollow center of bones. Pluripotent stem cells have the remarkable ability to develop into many different cell types. As they specialize,

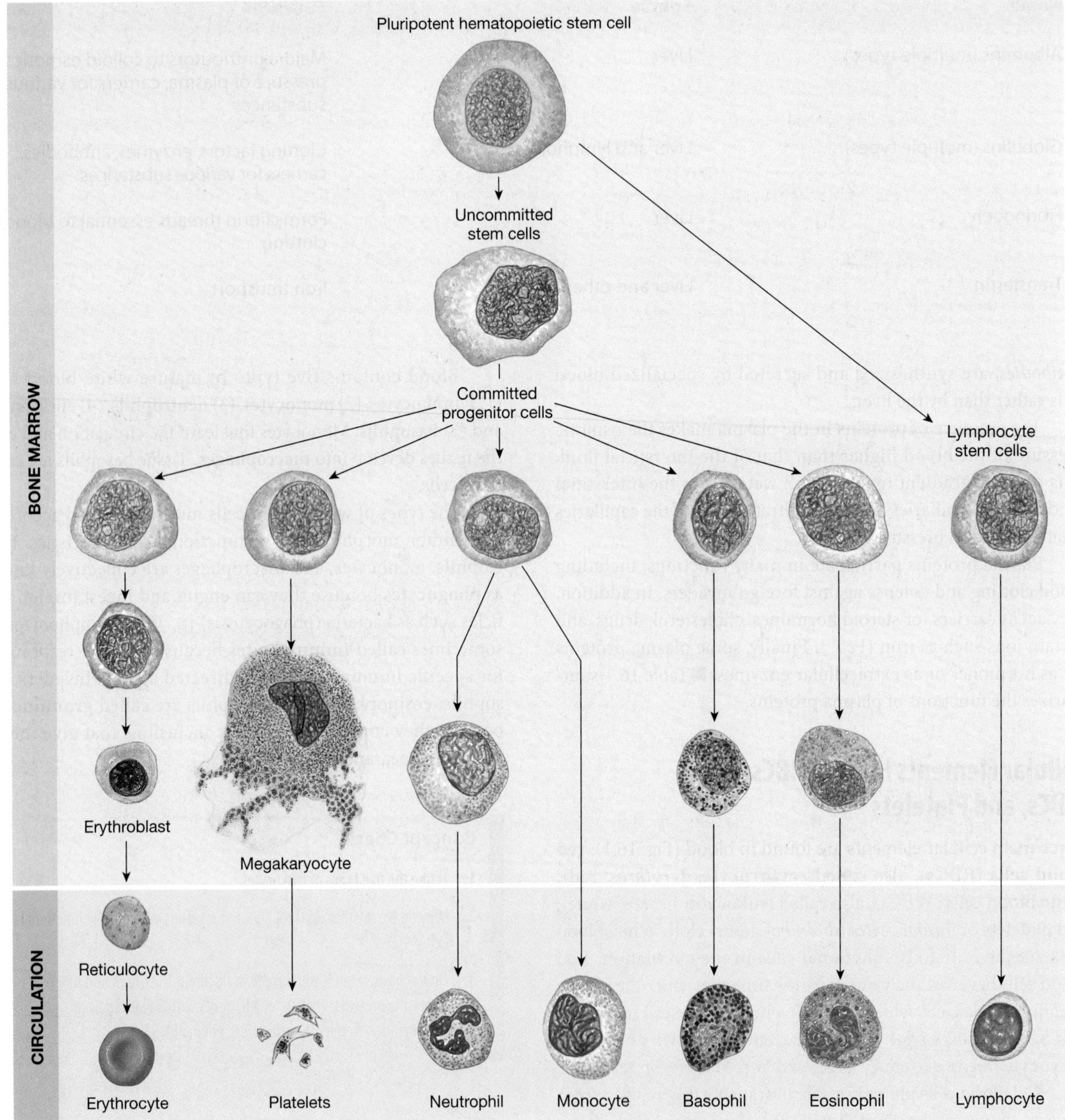

Fig. 16.2

they narrow their possible fates. First they become *uncommitted stem cells,* then *progenitor cells* that are committed to developing into one or perhaps two cell types. Progenitor cells differentiate into red blood cells, lymphocytes, other white blood cells, and megakaryocytes, the parent cells of platelets. It is estimated that only about one out of every 100,000 cells in the bone marrow is an uncommitted stem cell, making it difficult to isolate and study these cells.

In recent years scientists have been working to isolate and grow uncommitted hematopoietic stem cells to use as replacements in patients whose own stem cells have been killed by cancer chemotherapy. Originally scientists obtained these stem cells from bone marrow or peripheral blood. Now umbilical cord blood, collected at birth, has been found to be a rich source of hematopoietic stem cells that can be used for transplants in patients with hematological diseases such as leukemia. Public and private cord blood banking programs are active in the United States and Europe, and the American National Marrow Donor Program Registry now includes genetic marker information from banked cord blood to help patients find stem cell matches. Currently researchers are working on techniques for culturing cord blood to increase the number of stem cells in each unit.

Blood Cells Are Produced in the Bone Marrow

Hematopoiesis {*haima*, blood + *poiesis*, formation}, the synthesis of blood cells, begins early in embryonic development and continues throughout a person's life. In about the third week of fetal development, specialized cells in the yolk sac of the embryo form clusters. Some of these cell clusters are destined to become the endothelial lining of blood vessels, while others become blood cells. The common embryological origin of the endothelium and blood cells perhaps explains why many cytokines that control hematopoiesis are released by the vascular endothelium.

As the embryo develops, blood cell production spreads from the yolk sac to the liver, spleen, and bone marrow. By birth, the liver and spleen no longer produce blood cells. Hematopoiesis continues in the marrow of all the bones of the skeleton until age five. As the child continues to age, the active regions of marrow decrease. In adults, the only areas producing blood cells are the pelvis, spine, ribs, cranium, and proximal ends of long bones.

Active bone marrow is red because it contains **hemoglobin**, the oxygen-binding protein of red blood cells. Inactive marrow is yellow because of an abundance of adipocytes (fat cells). (You can see the difference between red and yellow marrow the next time you look at bony cuts of meat in the grocery store.) Although blood synthesis in adults is limited, the liver, spleen, and inactive (yellow) regions of marrow can resume blood cell production in times of need.

In the regions of marrow that are actively producing blood cells, about 25% of the developing cells are red blood cells, while 75% are destined to become white blood cells. The life span of white blood cells is considerably shorter than that of red blood cells, and so WBCs must be replaced more frequently. For example, neutrophils have a six-hour half-life, and the body must make more than *100 million* neutrophils each day in order to replace those that die. Red blood cells, on the other hand, live for nearly four months in the circulation.

Hematopoiesis Is Controlled by Cytokines

What controls the production and development of blood cells? The chemical factors known as cytokines are responsible. Cytokines are peptides or proteins released from one cell that affect the growth or activity of another cell [p. 177]. Newly discovered cytokines are often called *factors* and given a modifier that describes their actions: growth factor, differentiating factor, trophic factor.

Some of the best-known cytokines in hematopoiesis are the *colony-stimulating factors*, molecules made by endothelial cells and white blood cells. Others are the **interleukins** {*inter-*, between + *leuko*, white}, such as IL-3. The name *interleukin* was first given to cytokines released by one white blood cell to act on another white blood cell. Numbered interleukin names, such as interleukin-3, are given to cytokines once their amino acid sequences have been identified. Interleukins also play important roles in the immune system.

Another hematopoietic cytokine is *erythropoietin*, which controls red blood cell synthesis. Erythropoietin is usually called a hormone, but technically it fits the definition of a cytokine because it is made on demand rather than stored in vesicles.

■ Table 16.2 lists a few of the many cytokines linked to hematopoiesis. The role cytokines play in blood cell production is so complicated that one review on this topic was titled "Regulation of hematopoiesis in a sea of chemokine family members with a plethora of redundant activities"!* Because of the complexity of the subject, we give only an overview of the key hematopoietic cytokines.

*H. E. Broxmeyer, H. E. and C. H. Kim, *Exp Hematol* 27(7): 1113–1123, 1999, July.

Cytokines Involved in Hematopoiesis — Table 16.2

Name	Sites of Production	Influences Growth or Differentiation of
Erythropoietin (EPO)	Kidney cells primarily	Red blood cells
Thrombopoietin (TPO)	Liver primarily	Megakaryocytes
Colony-stimulating factors, interleukins, stem cell factor	Endothelium and fibroblasts of bone marrow, leukocytes	All types of blood cells; mobilizes hematopoietic stem cells

16

Colony-Stimulating Factors Regulate Leukopoiesis

Colony-stimulating factors (CSFs) were identified and named for their ability to stimulate the growth of leukocyte colonies in culture. These cytokines, made by endothelial cells, marrow fibroblasts, and white blood cells, regulate leukocyte production and development, or **leukopoiesis**. CSFs induce both cell division (mitosis) and cell maturation in stem cells. Once a leukocyte matures, it loses its ability to undergo mitosis.

One fascinating aspect of leukopoiesis is that production of new white blood cells is regulated in part by existing white blood cells. This form of control allows leukocyte development to be very specific and tailored to the body's needs. When the body's defense system is called on to fight off foreign invaders, both the absolute number of white blood cells and the relative proportions of the different types of white blood cells in the circulation change.

Clinicians often rely on a *differential white cell count* to help them arrive at a diagnosis (■ Fig. 16.3). For example, a person with a bacterial infection usually has a high total number of white blood cells in the blood, with an increased percentage that are neutrophils. Cytokines released by active white blood cells fighting the bacterial infection stimulate the production of additional neutrophils and monocytes. A person with a viral infection may have a high, normal, or low total white cell count but often shows an increase in the percentage of lymphocytes. The complex process by which leukocyte production is matched to need is still not completely understood and is an active area of research.

Scientists are working to create a model for the control of leukopoiesis so that they can develop effective treatments

THE BLOOD COUNT

This table lists the normal ranges of values.

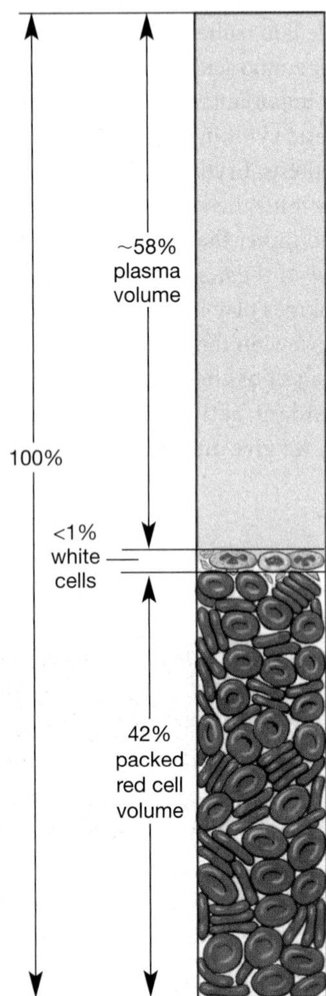

	MALES	FEMALES
Hematocrit		
Hematocrit is the percentage of total blood volume that is occupied by packed (centrifuged) red blood cells.	40–54%	37–47%
Hemoglobin (g Hb/dL* whole blood)		
The hemoglobin value reflects the oxygen-carrying capacity of red blood cells. (*1 deciliter (dL) = 100 mL)	14–17	12–16
Red cell count (cells/μL)		
A machine counts erythrocytes as they stream through a beam of light.	$4.5–6.5 \times 10^3$	$3.9–5.6 \times 10^3$
Total white count (cells/μL)		
A total white cell count includes all types of leukocytes but does not distinguish between them.	$4–11 \times 10^3$	$4–11 \times 10^3$
Differential white cell count		
The differential white cell count presents estimates of the relative proportions of the five types of leukocytes in a thin blood smear stained with biological dyes.		
Neutrophils	50–70%	50–70%
Eosinophils	1–4%	1–4%
Basophils	<1%	<1%
Lymphocytes	20–40%	20–40%
Monocytes	2–8%	2–8%
Platelets (per μL)		
Platelet count is suggestive of the blood's ability to clot.	$150–450 \times 10^3$	$150–450 \times 10^3$

■ **Fig. 16.3**

for diseases characterized by either a lack or an excess of white blood cells. The *leukemias* are a group of diseases characterized by the abnormal growth and development of white blood cells. In *neutropenias* {*penia*, poverty}, patients have too few white blood cells and are unable to fight off bacterial and viral infections. Researchers hope to find better treatments for both leukemias and neutropenias by unlocking the secrets of how the body regulates cell growth and division.

Thrombopoietin Regulates Platelet Production

Thrombopoietin (TPO) is a glycoprotein that regulates the growth and maturation of megakaryocytes, the parent cells of platelets. (Recall that *thrombocyte* is an alternative name for *platelet*). TPO is produced primarily in the liver but is also present in the kidney. This cytokine was first described in 1958, but its gene was not cloned until 1994. Within a year, TPO was widely available to researchers through the use of recombinant DNA techniques, and there has been an explosion of papers describing its effects on megakaryocytes and platelet production. Scientists still do not understand everything about the basic biology of thrombopoiesis, but TPO synthesis is a huge step forward in the search for answers.

Erythropoietin Regulates RBC Production

Red blood cell production (**erythropoiesis**) is controlled by the glycoprotein **erythropoietin (EPO)**, assisted by several cytokines. Erythropoietin is made primarily in the kidneys of adults. The stimulus for EPO synthesis and release is *hypoxia*, low oxygen levels in the tissues. Hypoxia stimulates production of a transcription factor called *hypoxia-inducible factor 1* (HIF-1), which turns on the EPO gene to increase EPO synthesis. This pathway, like other endocrine pathways, helps the body maintain homeostasis. By stimulating the synthesis of red blood cells, EPO puts more hemoglobin into the circulation to carry oxygen.

The existence of a hormone controlling red blood cell production was first suggested in the 1950s, but two decades passed before scientists succeeded in purifying the substance. One reason for the delay is that EPO is made on demand and not stored, as in an endocrine cell. It took scientists another nine years to identify the amino acid sequence of EPO and to isolate and clone the gene for it. However, an incredible leap was made after the EPO gene was isolated: only two years later, the hormone was produced by recombinant DNA technology and put into clinical use.

In recent years physicians have been able to prescribe not only genetically engineered EPO, such as epoetin, but also several colony-stimulating factors (sargramostim and filgrastim) that stimulate white blood cell synthesis. Cancer patients in whom hematopoiesis has been suppressed by chemotherapy have benefited from injections of these hematopoietic hormones, but in 2007 the Food and Drug Administration issued new dosing instructions and warnings about an increased risk of blood clots in patients taking erythropoiesis-stimulating agents. Scientists are currently monitoring the safety of CSFs to ensure that they do not increase the likelihood of developing hematological diseases.

Concept Check Answer: p. 567

4. Name the cytokine(s) that regulate(s) growth and maturation in (a) erythrocytes, (b) leukocytes, and (c) megakaryocytes.

Red Blood Cells

Erythrocytes are the most abundant cell type in the blood. A microliter of blood contains about 5 million red blood cells, compared with only 4000–11,000 white blood cells and 150,000–450,000 platelets. The primary function of red blood cells is to facilitate oxygen transport from the lungs to cells, and carbon dioxide transport from cells to lungs.

The ratio of red blood cells to plasma is indicated clinically by the **hematocrit** and is expressed as a percentage of the total blood volume (Fig. 16.3). Hematocrit is determined by drawing a blood sample into a narrow capillary tube and spinning it in a centrifuge so that the heavier red blood cells go to the bottom of the sealed tube, leaving the thin "buffy layer" of lighter white blood cells and platelets in the middle, and plasma on top.

The column of *packed red cells* is measured, and the hematocrit value is reported as a percentage of the total sample volume. The normal range of hematocrit is 40–54% for a man and 37–47% for a woman. This test provides a rapid and inexpensive way to estimate a person's red cell count because blood for a hematocrit can be collected by simply sticking a finger.

Mature RBCs Lack a Nucleus

In the bone marrow, committed progenitor cells differentiate through several stages into large, nucleated *erythroblasts*. As erythroblasts mature, the nucleus condenses and the cell shrinks in diameter from 20 μm to about 7 μm. In the last stage before maturation, the nucleus is pinched off and phagocytized by bone marrow macrophages. At the same time, other membranous organelles (such as mitochondria) break down and disappear. The final immature cell form, called a *reticulocyte*, leaves the marrow and enters the circulation, where it matures into an erythrocyte in about 24 hours (■ Fig. 16.4c).

Mature mammalian red blood cells are biconcave disks, shaped much like jelly doughnuts with the filling squeezed out of

Bone Marrow

(a) The bone marrow, hidden within the bones of the skeleton, is easily overlooked as a tissue, although collectively it is nearly the size and weight of the liver!

(b) Marrow is a highly vascular tissue, filled with blood sinuses, widened regions lined with epithelium.

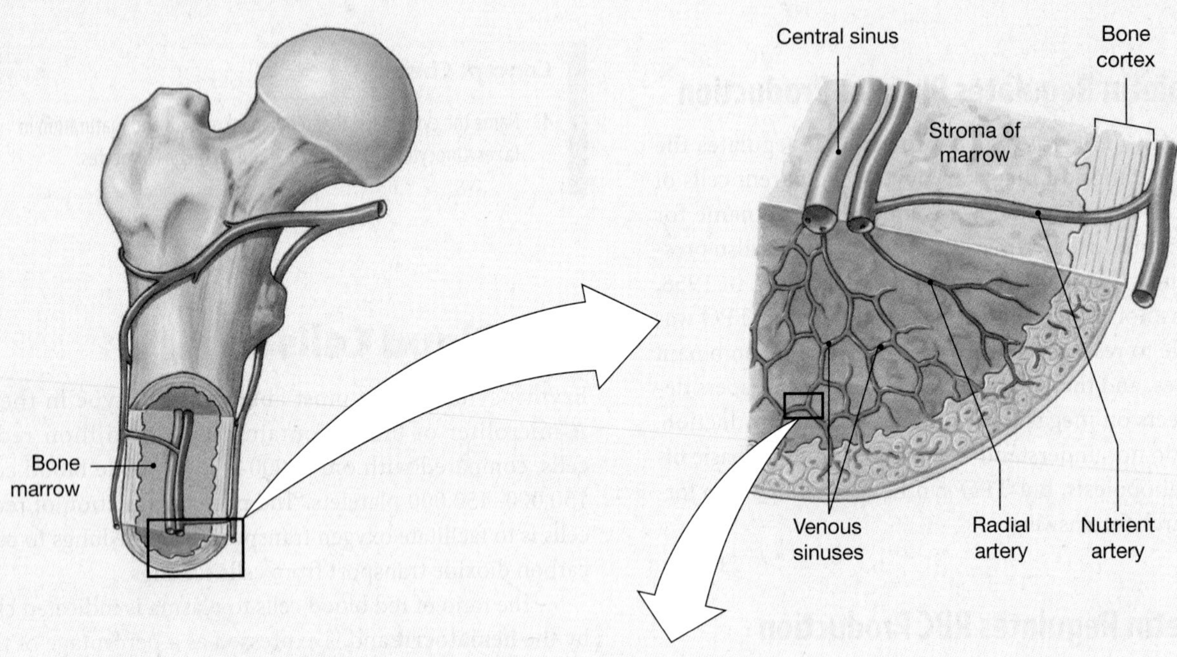

(c) Bone marrow consists of blood cells in different stages of development and supporting tissue known as the **stroma** {mattress}.

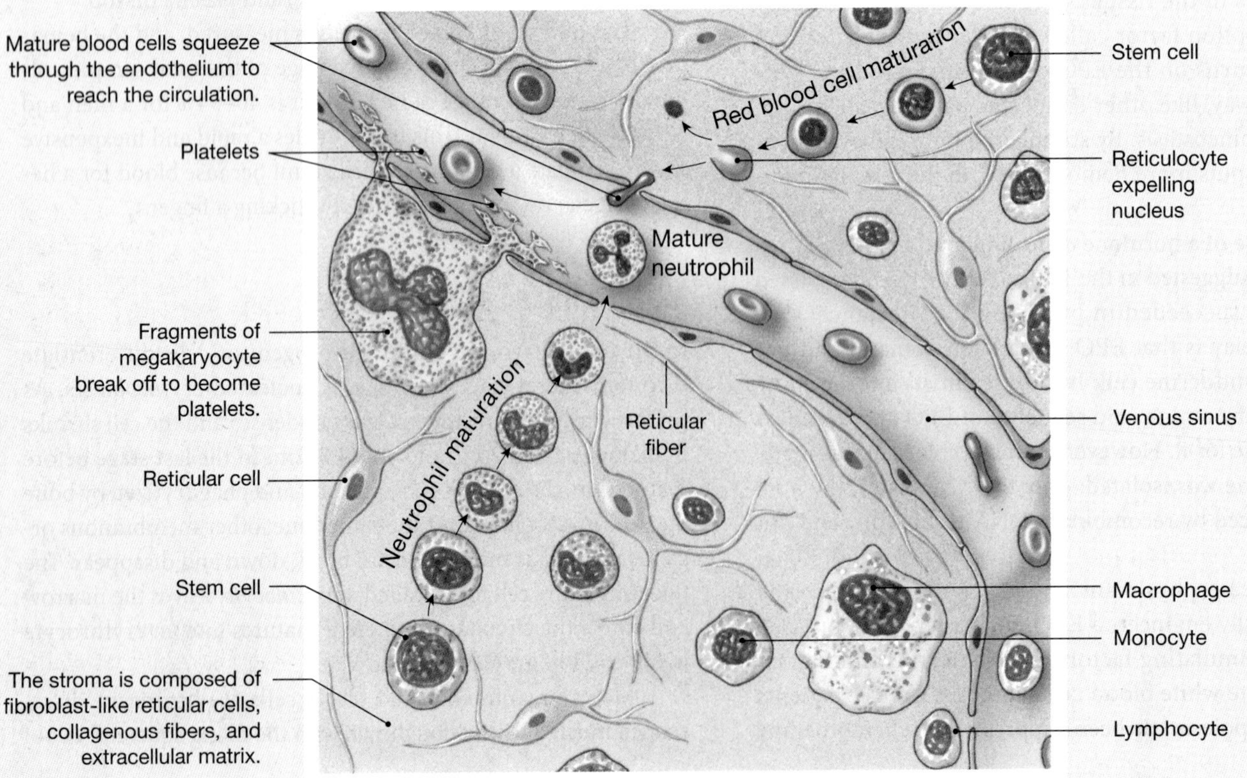

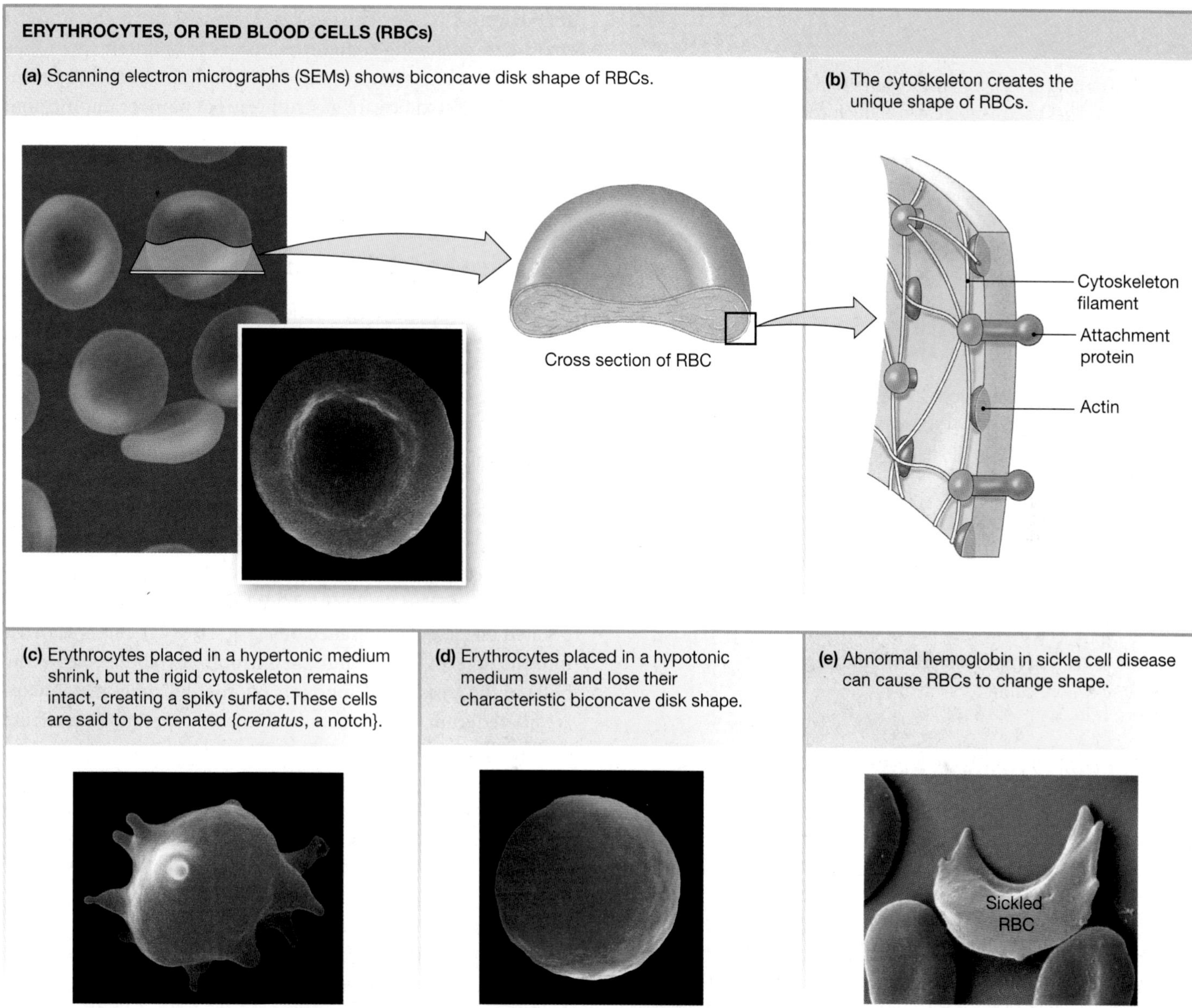
ERYTHROCYTES, OR RED BLOOD CELLS (RBCs)
(a) Scanning electron micrographs (SEMs) shows biconcave disk shape of RBCs.
Cross section of RBC
(b) The cytoskeleton creates the unique shape of RBCs.
Cytoskeleton filament
Attachment protein
Actin
(c) Erythrocytes placed in a hypertonic medium shrink, but the rigid cytoskeleton remains intact, creating a spiky surface.These cells are said to be crenated {crenatus, a notch}.
(d) Erythrocytes placed in a hypotonic medium swell and lose their characteristic biconcave disk shape.
(e) Abnormal hemoglobin in sickle cell disease can cause RBCs to change shape.
Sickled RBC

■ **Fig. 16.5**

the middle (■ Fig. 16.5a). They are simple membranous "bags" filled with enzymes and hemoglobin. Because red blood cells contain no mitochondria, they cannot carry out aerobic metabolism. Glycolysis is their primary source of ATP. Without a nucleus and endoplasmic reticulum to carry out protein synthesis, erythrocytes are unable to make new enzymes or to renew membrane components. This inability leads to an increasing loss of membrane flexibility, making older cells more fragile and likely to rupture.

The biconcave shape of the red blood cell is one of its most distinguishing features. The membrane is held in place by a complex cytoskeleton composed of filaments linked to transmembrane attachment proteins (Fig. 16.5b). Despite the cytoskeleton, red cells are remarkably flexible, like a partially filled water balloon that can compress into various shapes. This flexibility allows erythrocytes to change shape as they squeeze through the narrow capillaries of the circulation.

The disk-like structure of red blood cells also allows them to modify their shape in response to osmotic changes in the blood. In hypertonic media, red blood cells shrink up and develop a spiky surface when the membrane pulls tight against the cytoskeleton (Fig. 16.5c). An erythrocyte placed in a slightly hypotonic medium [p. 134] swells and forms a sphere without disruption of its membrane integrity (Fig. 16.5d).

The **morphology** {*morphe,* form} of red blood cells can provide clues to the presence of disease. Sometimes the cells lose their flattened disk shape and become spherical (*spherocytosis*), a shape similar to that of the cell in hypotonic medium. In sickle cell anemia, the cells are shaped like a sickle or crescent moon (Fig. 16.5e). In some disease states, the size of red blood cells—the **mean red cell volume** (MCV)—may be either abnormally large or abnormally small. For example, red blood cells can be abnormally small, or *microcytic*, in iron-deficiency anemia. If they are

RUNNING PROBLEM

Blood doping to increase the oxygen-carrying capacity of the blood has been a problem in endurance sports for more than 30 years. The first sign that Muehlegg might be cheating by this method was the result of a simple blood test for hemoglobin and a hematocrit, taken several hours before his 50-kilometer race. Muehlegg's blood hemoglobin level registered above 17.5 g/dL. A repeat test was within acceptable limits, however, and Muehlegg was allowed to race.

Q1: What is the normal range for Muehlegg's hemoglobin (Fig. 16.3)?

Q2: Olympic officials also tested Muehlegg's hematocrit. With blood doping, would you expect a hematocrit value to be lower or higher than normal?

pale due to lack of red hemoglobin, they are described as *hypochromic* {*chroma*, color}.

Hemoglobin Synthesis Requires Iron

Hemoglobin, the main component of red blood cells, is best known for its role in oxygen transport. Hemoglobin (Hb) is a large, complex protein with four globular protein chains, each of which is wrapped around an iron-containing *heme group* (■ Fig. 16.6a). There are several isoforms of **globin** proteins in hemoglobin. The most common isoforms are designated *alpha* (α), *beta* (β), *gamma* (γ), and *delta* (δ), depending on the structure of the chain. Most adult hemoglobin (designated *HbA*) has two alpha chains and two beta chains, as shown. However, a small portion of adult hemoglobin (about 2.5%) has two alpha chains and two delta chains (HbA_2).

The four **heme groups** in a hemoglobin molecule are identical. Each consists of a carbon-hydrogen-nitrogen *porphyrin ring* with an iron atom (Fe) in the center (Fig. 16.6b). About 70% of the iron in the body is found in the heme groups of hemoglobin. Consequently, hemoglobin synthesis requires an adequate supply of iron in the diet (Fig. 16.6c (1)). Most dietary iron comes from red meat, beans, spinach, and iron-fortified bread.

Iron is absorbed in the small intestine by active transport (Fig. 16.6c (2)). A carrier protein called **transferrin** binds iron and transports it in the blood (3). The bone marrow takes up iron and uses it to make the heme group of hemoglobin for developing red blood cells (4).

Iron ingested in amounts greater than needed for hemoglobin synthesis is stored, mostly in the liver, as the molecule **ferritin** and its derivatives (9). Excess iron in the body is toxic, and poisoning sometimes occurs in children when they ingest too many vitamin pills containing iron. Initial symptoms of iron toxicity are gastrointestinal pain, cramping, and internal bleeding, which occurs as iron corrodes the digestive epithelium. Subsequent problems include liver failure, which can be fatal.

RBCs Live About Four Months

Red blood cells in the circulation live for about 120 ± 20 days. Increasingly fragile older erythrocytes may rupture as they try to squeeze through narrow capillaries, or they may be engulfed by scavenging macrophages as they pass through the spleen (Fig. 16.6 (6)). Many components of the destroyed cells are recycled. Amino acids from the globin chains of hemoglobin are incorporated into new proteins, and some iron from the heme groups is reused to make new heme groups.

The spleen and liver convert remnants of the heme groups to a colored pigment called **bilirubin**. Bilirubin is carried by plasma albumin to the liver, where it is metabolized and incorporated into a secretion called **bile** (Fig. 16.6 (8)). Bile is secreted into the digestive tract, and the bilirubin metabolites leave the body in the feces. Small amounts of other bilirubin metabolites are filtered from the blood in the kidneys, where they contribute to the yellow color of urine (7).

In some circumstances, bilirubin levels in the blood become elevated (*hyperbilirubinemia*). This condition, known as **jaundice**, causes the skin and whites of the eyes to take on a yellow cast. The accumulation of bilirubin can occur from several different causes. Newborns whose fetal hemoglobin is being broken down and replaced with adult hemoglobin are particularly susceptible to bilirubin toxicity, so doctors monitor babies for jaundice in the first weeks of life. Another common cause of

CLINICAL FOCUS: DIABETES

Hemoglobin and Hyperglycemia

One of the goals of diabetes treatment is to keep blood glucose concentrations as close to normal as possible, but how can a clinician tell if a patient has been doing this? One way is to analyze the patient's hemoglobin. Glucose in the plasma binds covalently to hemoglobin, producing a glycohemoglobin known as **hemoglobin A_{1C}** ("A-one-C"). The amount of hemoglobin A_{1C} in the plasma is directly related to hemoglobin's exposure to glucose over the preceding 8–12 weeks. By using this assay, a clinician can monitor long-term fluctuations in blood glucose levels and adjust a diabetic patient's therapy appropriately.

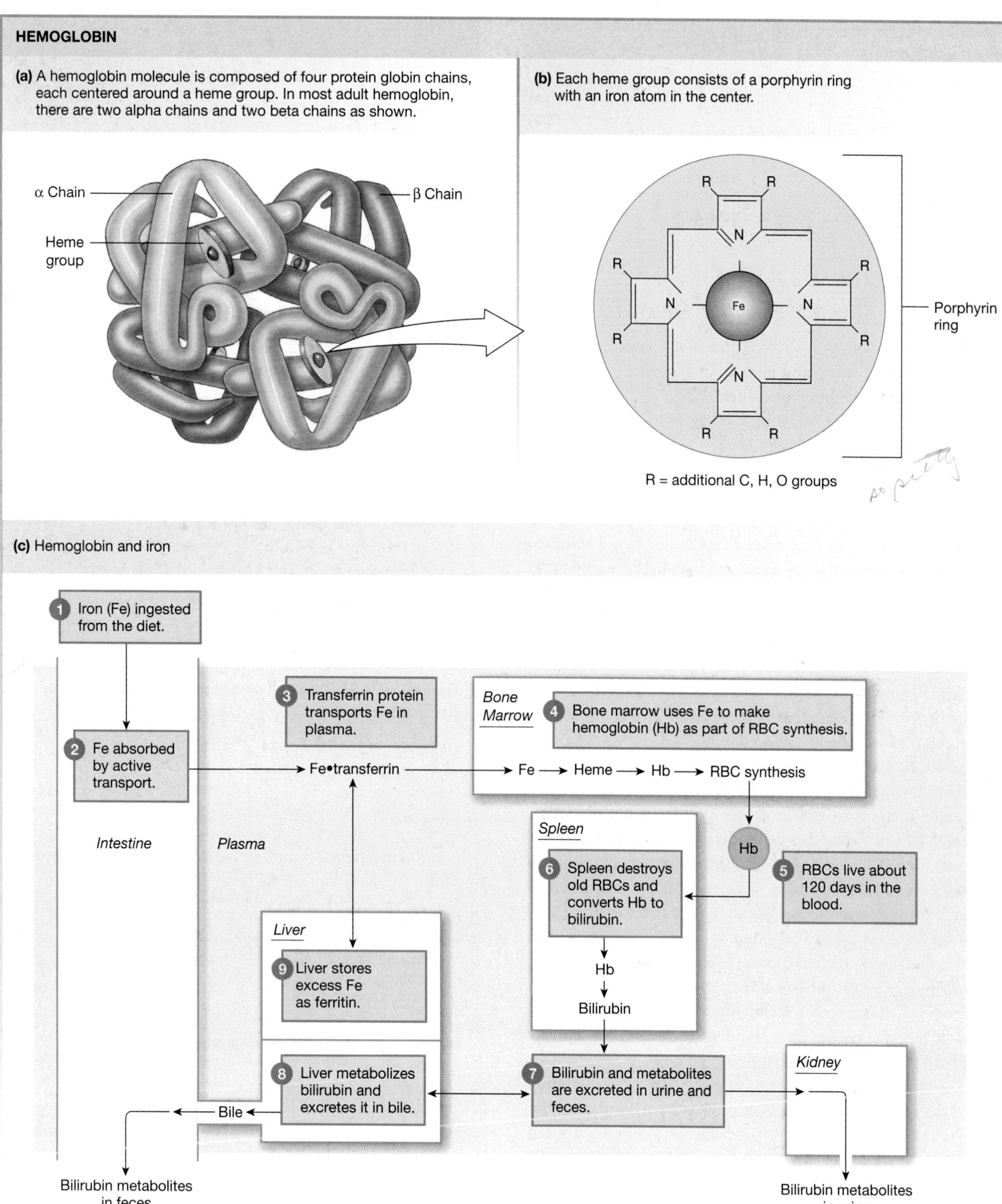
HEMOGLOBIN
(a) A hemoglobin molecule is composed of four protein globin chains, each centered around a heme group. In most adult hemoglobin, there are two alpha chains and two beta chains as shown.
α Chain
β Chain
Heme group
(b) Each heme group consists of a porphyrin ring with an iron atom in the center.
R
N
Fe
Porphyrin ring
R = additional C, H, O groups
(c) Hemoglobin and iron
1 Iron (Fe) ingested from the diet.
2 Fe absorbed by active transport.
3 Transferrin protein transports Fe in plasma.
Fe•transferrin
Bone Marrow
4 Bone marrow uses Fe to make hemoglobin (Hb) as part of RBC synthesis.
Fe → Heme → Hb → RBC synthesis
Intestine
Plasma
Spleen
6 Spleen destroys old RBCs and converts Hb to bilirubin.
Hb
Bilirubin
5 RBCs live about 120 days in the blood.
Liver
9 Liver stores excess Fe as ferritin.
8 Liver metabolizes bilirubin and excretes it in bile.
Bile
7 Bilirubin and metabolites are excreted in urine and feces.
Kidney
Bilirubin metabolites in feces
Bilirubin metabolites in urine

■ **Fig. 16.6**

jaundice is liver disease, in which the liver is unable to process or excrete bilirubin.

Concept Check Answers: p. 567

5. Distinguish between (a) heme and hemoglobin, and (b) ferritin and transferrin.
6. Is bile an endocrine secretion or an exocrine secretion?

RBC Disorders Decrease Oxygen Transport

Because hemoglobin plays a critical role in oxygen transport, the red blood cell count and hemoglobin content of the body are important. If hemoglobin content is too low—a condition known as **anemia**—the blood cannot transport enough oxygen to the tissues. People with anemia are usually tired and weak, especially during exercise. The major causes of anemia are summarized in ■ Table 16.3.

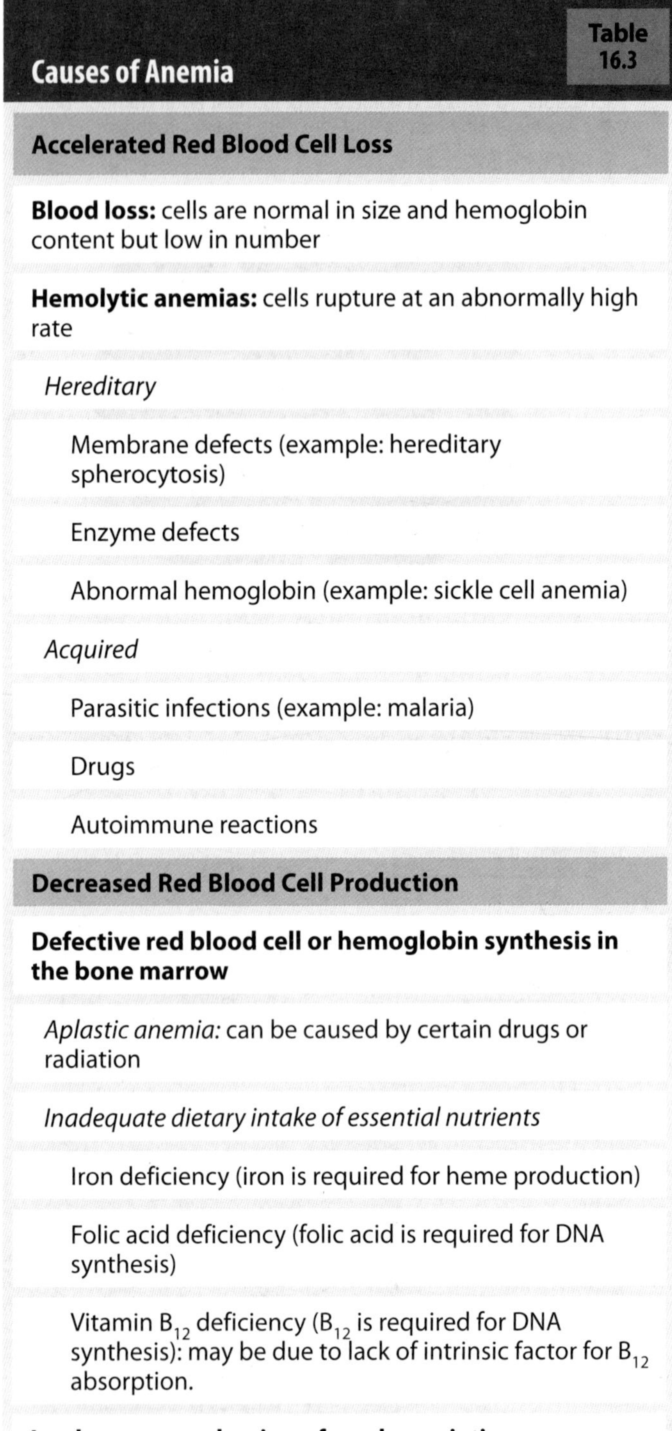

Table 16.3 Causes of Anemia

Causes of Anemia
Accelerated Red Blood Cell Loss
Blood loss: cells are normal in size and hemoglobin content but low in number
Hemolytic anemias: cells rupture at an abnormally high rate
Hereditary
Membrane defects (example: hereditary spherocytosis)
Enzyme defects
Abnormal hemoglobin (example: sickle cell anemia)
Acquired
Parasitic infections (example: malaria)
Drugs
Autoimmune reactions
Decreased Red Blood Cell Production
Defective red blood cell or hemoglobin synthesis in the bone marrow
Aplastic anemia: can be caused by certain drugs or radiation
Inadequate dietary intake of essential nutrients
Iron deficiency (iron is required for heme production)
Folic acid deficiency (folic acid is required for DNA synthesis)
Vitamin B_{12} deficiency (B_{12} is required for DNA synthesis): may be due to lack of intrinsic factor for B_{12} absorption.
Inadequate production of erythropoietin

In the *hemolytic anemias* {*lysis,* rupture}, the rate of red blood cell destruction exceeds the rate of red blood cell production. The hemolytic anemias are usually hereditary defects in which the body makes fragile cells. For example, in *hereditary spherocytosis,* the erythrocyte cytoskeleton does not link properly because of defective or deficient cytoskeletal proteins. Consequently, the cells are shaped more like spheres than like biconcave disks. This disruption in the cytoskeleton results in red blood cells that rupture easily and are unable to withstand osmotic changes as well as normal cells can. Several of the hemolytic anemias are acquired diseases, as indicated in Table 16.3.

Some anemias are the result of abnormal hemoglobin molecules. *Sickle cell disease* is a genetic defect in which glutamate, the sixth amino acid in the 146–amino acid beta chain of hemoglobin, is replaced by valine. The result is abnormal hemoglobin (a form referred to as *HbS*) that crystallizes when it gives up its oxygen. This crystallization pulls the red blood cells into a sickle shape, like a crescent moon (Fig. 16.5e). The sickled cells become tangled with other sickled cells as they pass through the smallest blood vessels, causing the cells to jam up and block blood flow to the tissues. This blockage creates tissue damage and pain from hypoxia.

One treatment for sickle cell disease is the administration of *hydroxyurea,* a compound that inhibits DNA synthesis. Hydroxyurea alters bone marrow function so that immature red blood cells produce the fetal form of hemoglobin (*HbF*) instead of adult hemoglobin. HbF interferes with the crystallization of hemoglobin, so that HbS no longer forms and the red blood cells no longer sickle. In addition, some studies show improvements in sickle cell symptoms before HbF levels increase. One theory of why this happens is based on the finding that hydroxyurea is metabolized to nitric oxide (NO), which causes vasodilation. Inhaled nitric oxide is now being tested as a treatment for sickle cell disease symptoms.

Other anemias result from the failure of the bone marrow to make adequate amounts of hemoglobin. One of the most common examples of an anemia that results from insufficient hemoglobin synthesis is *iron-deficiency anemia.* If iron loss by the body exceeds iron intake, the marrow does not have adequate iron to make heme groups, and hemoglobin synthesis slows.

People with iron-deficiency anemia have either a low red blood cell count (reflected in a low hematocrit) or low hemoglobin content in their blood. Their red blood cells are often smaller than usual (*microcytic* red blood cells), and the lower hemoglobin content may cause the cells to be paler than normal, in which case they are described as being *hypochromic* {*hypo-*, below normal; *chrom-*, color}. Women who menstruate are likely to suffer from iron-deficiency anemia because of iron loss in menstrual blood.

Although the anemias are common, it is also possible to have too many red blood cells. *Polycythemia vera* {*vera*, true} is a stem cell dysfunction that produces too many blood cells, white as well as red. These patients may have hematocrits as high as 60–70% (normal is 37–54%). The increased number of cells causes the blood to become more viscous and thus more resistant to flow through the circulatory system [p. 469].

In *relative polycythemia*, the person's red blood cell number is normal, but the hematocrit is elevated because of low plasma volume. This might occur with dehydration, for example. The opposite problem can also occur. If an athlete overhydrates, the hematocrit may decrease temporarily because of increased plasma volume. In both of these situations, there is no actual pathology involving the red blood cells.

Concept Check Answer: p. 567

7. A person who goes from sea level to a city that is 5000 feet above sea level begins to show an increased hematocrit within days. Draw the reflex pathway that links the hypoxia of high altitude to increased red blood cell production.

RUNNING PROBLEM

In its earliest form, blood doping was accomplished by blood transfusions, which increased the athlete's oxygen-carrying capacity. One hallmark of a recent blood transfusion is elevated hemoglobin and hematocrit levels. Muehlegg claimed that his elevated hemoglobin was a result of his special diet and of dehydration from diarrhea he had suffered the night before.

Q3: Explain how diarrhea could cause a temporarily elevated hematocrit.

Q4: How might Muehlegg quickly reduce his hematocrit without removing red blood cells?

Platelets and Coagulation

Because of its fluid nature, blood flows freely throughout the circulatory system. However, if there is a break in the "piping" of the system, blood will be lost unless steps are taken. One of the challenges for the body is to plug holes in damaged blood vessels while still maintaining blood flow through the vessel.

It would be simple to block off a damaged blood vessel completely, like putting a barricade across a street full of potholes. However, just as shopkeepers on that street lose business if traffic is blocked, cells downstream from the point of injury die from lack of oxygen and nutrients if the vessel is completely blocked. The body's task is to allow blood flow through the vessel while simultaneously repairing the damaged wall.

This challenge is complicated by the fact that blood in the system is under pressure. If the repair "patch" is too weak, it is blown out by the blood pressure. For this reason, stopping blood loss involves several steps. First, the pressure in the vessel must be decreased long enough to create a secure mechanical seal in the form of a blood clot. Once the clot is in place and blood loss has been stopped, the body's repair mechanisms can take over. Then, as the wound heals, enzymes gradually dissolve the clot while scavenger white blood cells ingest and destroy the debris.

Platelets Are Small Fragments of Cells

As noted earlier, platelets are cell fragments produced in the bone marrow from huge cells called megakaryocytes. Megakaryocytes develop their formidable size by undergoing mitosis up to seven times without undergoing nuclear or cytoplasmic division. The result is a *polyploid* cell with a lobed nucleus (■ Fig. 16.7).

The outer edges of marrow megakaryocytes extend through the endothelium into the lumen of marrow blood sinuses, where the cytoplasmic extensions fragment into disk-like platelets (Fig. 16.4c). Platelets are smaller than red blood cells, are colorless, and have no nucleus. Their cytoplasm contains mitochondria, smooth endoplasmic reticulum, and many granules filled with clotting proteins and cytokines.

The typical life span of a platelet is about 10 days. Platelets are always present in the blood, but they are not active unless damage occurs to the walls of the circulatory system.

Hemostasis Prevents Blood Loss from Damaged Vessels

Hemostasis {*haima*, blood + *stasis*, stoppage} is the process of keeping blood within a damaged blood vessel (■ Fig. 16.8). (The opposite of hemostasis is *hemorrhage* {*-rrhagia*, abnormal flow}.)

MEGAKARYOCYTES AND PLATELETS

(a) Megakaryocytes are giant cells with multiple copies of DNA in the nucleus.

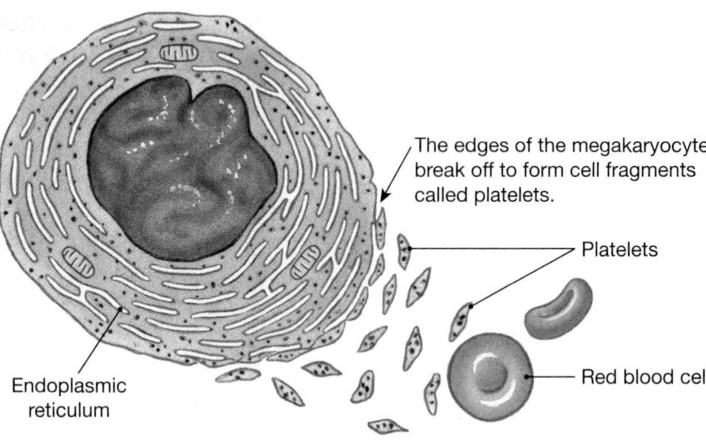

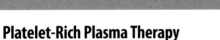

(b) Inactive platelets are small disk-like cell fragments.

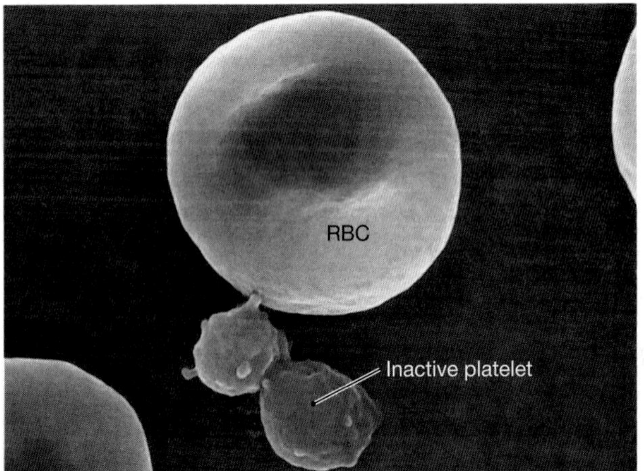

(c) Activated platelets (shown enlarged) develop a spiky outer surface and adhere to each other.

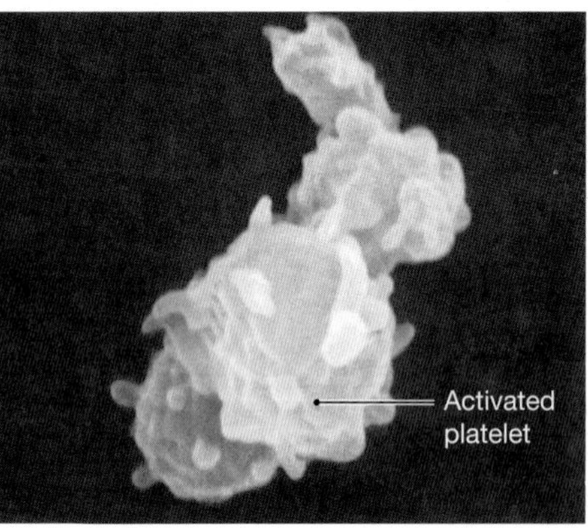

Fig. 16.7

EMERGING CONCEPTS

Platelet-Rich Plasma Therapy

Professional sports are big business around the world, so any treatment that will speed healing in an injured athlete makes news. That's what happened in 2010 when *platelet-rich plasma (PRP) therapy* became the buzz on the Internet after Tiger Woods, the golfer, announced he had used RPR to aid his recovery from knee surgery. Tendons and ligaments have minimal blood supply and are notoriously slow to heal. So how can platelets, whose main function is the first step in clot formation, help connective tissues repair themselves? The answer lies inside the platelets. These tiny cell fragments are filled with vesicles containing a wide variety of cytokines and growth factors, and it is these chemicals that are believed to promote healing. At this time PRP therapy is still considered experimental, however, and evidence for PRP efficacy from well-designed, placebo-controlled studies is lacking.

Hemostasis has three major steps: (1) vasoconstriction, (2) temporary blockage of a break by a platelet plug, and (3) *coagulation*, the formation of a clot that seals the hole until tissues are repaired.

The first step in hemostasis is immediate constriction of damaged vessels caused by vasoconstrictive paracrines released by the endothelium. Vasoconstriction temporarily decreases blood flow and pressure within the vessel. When you put pressure on a bleeding wound, you also decrease flow within the damaged vessel.

Vasoconstriction is rapidly followed by the second step, mechanical blockage of the hole by a loose **platelet plug**. Plug formation begins with **platelet adhesion,** when platelets *adhere* or stick to exposed collagen in the damaged area. The adhered platelets activate, releasing cytokines into the area around the injury. These platelet factors reinforce local vasoconstriction and activate more platelets, which *aggregate* or stick to one another to form a loose platelet plug.

HEMOSTASIS AND TISSUE REPAIR

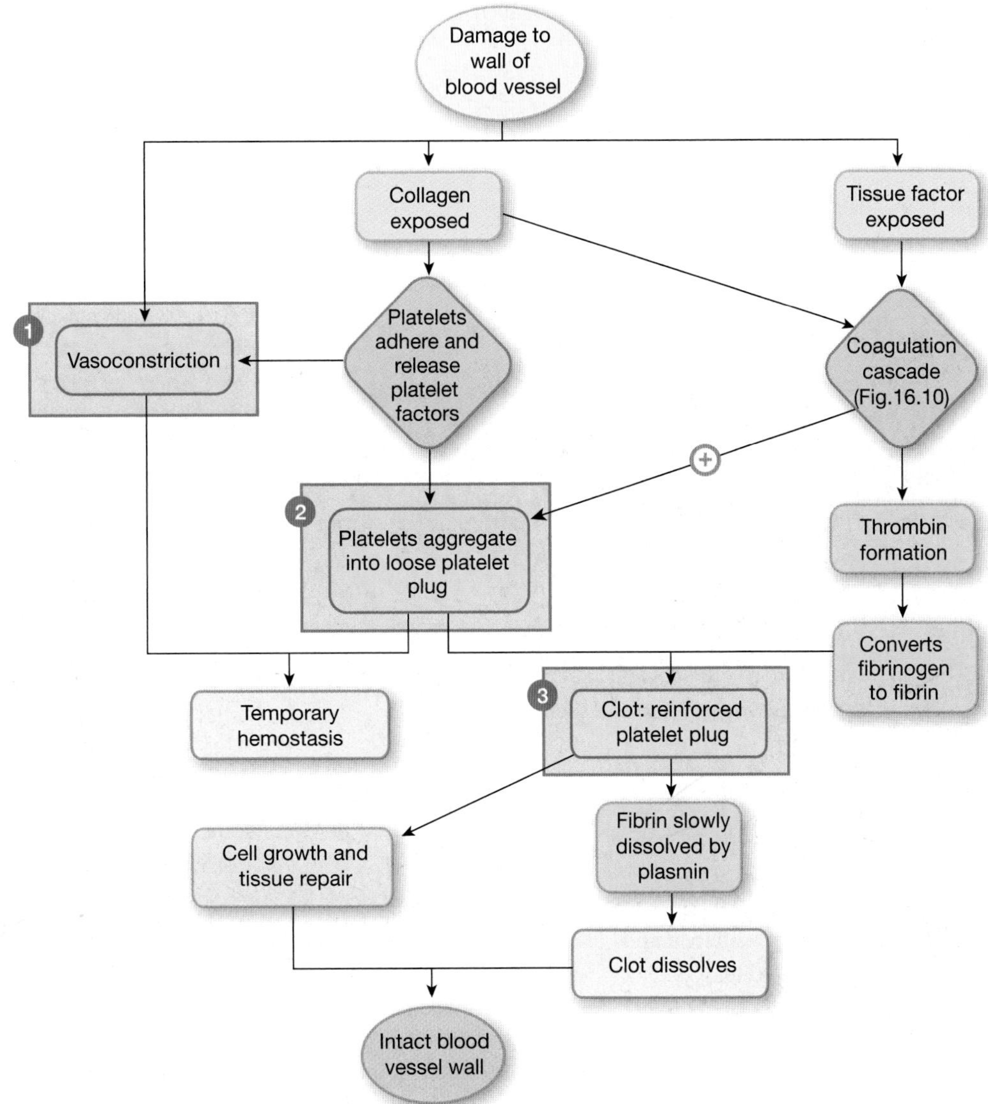

■ **Fig. 16.8**

Simultaneously, exposed collagen and **tissue factor** (a protein-phospholipid mixture) initiate the third step, the formation of a *fibrin* protein mesh that stabilizes the platelet plug to form a **clot**. Fibrin is the end product of a series of enzymatic reactions known as the **coagulation cascade**. Some chemical factors involved in the coagulation cascade also promote platelet adhesion and aggregation in the damaged region. Eventually, as the damaged vessel repairs itself, the clot retracts when fibrin is slowly dissolved by the enzyme *plasmin*.

The body must maintain the proper balance during hemostasis. Too little hemostasis allows excessive bleeding; too much creates a **thrombus**, a blood clot that adheres to the undamaged wall of a blood vessel {*thrombos*, a clot or lump}. A large thrombus can block the lumen of the vessel and stop blood flow.

Hemostasis seems straightforward, but unanswered questions remain at the cellular and molecular levels. Because inappropriate blood clotting plays an important role in strokes and heart attacks, this area of research is very active. Research has led to the development and use of "clot busters," enzymes that can dissolve clots in arteries after heart attacks and strokes.

A detailed study of hemostasis involves many chemical factors, some of which play multiple roles and have multiple names. For this reason, learning about hemostasis can be especially challenging. For example, some factors participate in both platelet plug formation and coagulation, and one factor in the cascade activates enzymes for both clot formation and clot dissolution. Because of the complexity of the coagulation cascade, we discuss only a few aspects of hemostasis in additional detail.

PLATELET PLUG FORMATION

Platelets will not adhere to intact endothelium. Damage triggers platelet plug formation where collagen has been exposed.

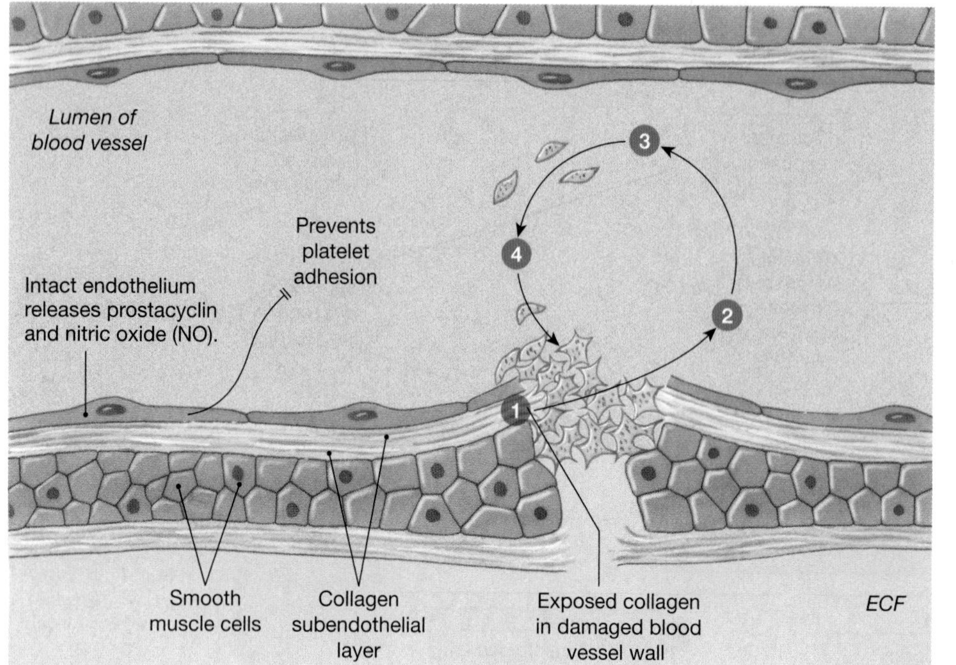

1 Exposed collagen binds and activates platelets.

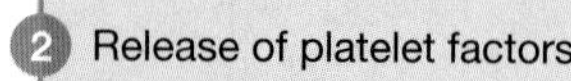
2 Release of platelet factors

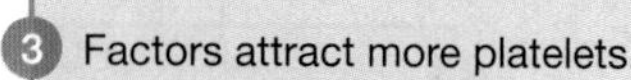
3 Factors attract more platelets.

4 Platelets aggregate into platelet plug.

■ **Fig. 16.9**

Platelet Activation Begins the Clotting Process

When a blood vessel wall is first damaged, exposed collagen and chemicals from endothelial cells activate platelets (■ Fig. 16.9 ①). Normally, the blood vessel's endothelium separates the collagenous matrix fibers from the circulating blood. But when the vessel is damaged, collagen is exposed, and platelets rapidly begin to adhere to it.

Platelets adhere to collagen with the help of *integrins*, membrane receptor proteins that are linked to the cytoskeleton [p. 80]. Binding activates platelets so that they release the contents of their intracellular granules, including *serotonin* (5-hydroxytryptamine), ADP, and **platelet-activating factor (PAF)**. PAF sets up a positive feedback loop by activating more platelets. It also initiates pathways that convert platelet membrane phospholipids into **thromboxane A2** [p. 189]. Serotonin and thromboxane A2 are vasoconstrictors. They also contribute to platelet aggregation, along with ADP and PAF (■ Tbl. 16.4). The net result is a growing platelet plug that seals the damaged vessel wall.

If platelet aggregation is a positive feedback event, what prevents the platelet plug from continuing to form and spreading beyond the site of injury to other areas of the vessel wall? The answer lies in the fact that platelets do not adhere to normal endothelium. Intact vascular endothelial cells convert their membrane lipids into **prostacyclin**, an eicosanoid [p. 33] that

RUNNING PROBLEM

After the 1984 Olympics Games, where U.S. cyclists reportedly suffered bad side effects following blood transfusions, the International Olympic Committee and other organizations banned blood doping. Then recombinant human EPO (rhEPO) became available in the late 1980s, and athletes started injecting the drug to increase their body's red blood cell production. Subsequently the biotechnology firm Amgen created a longer-acting derivative of EPO named darbepoietin. Athletes using rhEPO and darbepoietin hoped to escape detection by using these natural hormones, but sports organizations worked with scientists to develop methods for detection.

Q5: Endogenous EPO, rhEPO, and darbepoietin all induce red blood cell synthesis, but they can be distinguished from one another when a urine sample is tested by electrophoresis [p. 106]. Explain how three hormones made from the same gene can all be active yet different enough from one another to be detectable in the laboratory.

Q6: One hallmark of illegal EPO use is elevated reticulocytes in the blood. Why would this suggest greater-than-normal EPO activity?

545 554 557 560 565

Factors Involved in Platelet Function

Table 16.4

Chemical Factor	Source	Activated by or Released in Response to	Role in Platelet Plug Formation	Other Roles and Comments
Collagen	Subendothelial extracellular matrix	Injury exposes platelets to collagen	Binds platelets to begin platelet plug	N/A
von Willebrand factor (vWF)	Endothelium, megakaryocytes	Exposure to collagen	Links platelets to collagen	Deficiency or defect causes prolonged bleeding
Serotonin	Secretory vesicles of platelets	Platelet activation	Platelet aggregation	Vasoconstrictor
Adenosine diphosphate (ADP)	Platelet mitochondria	Platelet activation, thrombin	Platelet aggregation	N/A
Platelet-activating factor (PAF)	Platelets, neutrophils, monocytes	Platelet activation	Platelet aggregation	Plays role in inflammation; increases capillary permeability
Thromboxane A2	Phospholipids in platelet membranes	Platelet-activating factor	Platelet aggregation	Vasoconstrictor; eicosanoid
Platelet-derived growth factor (PDGF)	Platelets	Platelet activation	N/A	Promotes wound healing by attracting fibroblasts and smooth muscle cells

blocks platelet adhesion and aggregation (Fig. 16.9). Nitric oxide released by normal, intact endothelium also inhibits platelets from adhering. The combination of platelet attraction to the injury site and repulsion from the normal vessel wall creates a localized response that limits the platelet plug to the area of damage.

Coagulation Converts a Platelet Plug into a Clot

The third major step in hemostasis, coagulation, is a complex process in which fluid blood forms a gelatinous clot. Coagulation is divided into two pathways that eventually merge into one (■ Fig. 16.10). An **intrinsic pathway** (yellow) begins when damage to the tissue exposes collagen. The intrinsic pathway uses proteins already present in the plasma. Collagen activates the first enzyme, factor XII, to begin the cascade. An **extrinsic pathway** (blue) starts when damaged tissues expose tissue factor, also called *tissue thromboplastin* or factor III. Tissue factor activates factor VII to begin the extrinsic pathway. The two pathways unite at the **common pathway** (green) to create **thrombin**, the enzyme that converts **fibrinogen** into insoluble **fibrin** polymers. These fibrin fibers become part of the clot.

Coagulation was initially regarded as a cascade similar to second messenger cascades of signal transduction [p. 180]. At each step an enzyme converts an inactive precursor into an active enzyme, often with the help of Ca^{2+}, membrane phospholipids, or additional factors. We now know, however, that the process is more than a simple cascade. Factors in the intrinsic and extrinsic pathways interact with each other, making coagulation a network rather than a simple cascade. In addition, several positive feedback loops sustain the cascade until one or more of the participating plasma proteins are completely consumed.

The final step of coagulation is the conversion of fibrinogen into fibrin, a reaction catalyzed by the enzyme *thrombin* (■ Fig. 16.11a). The fibrin fibers weave through the platelet plug and trap red blood cells within their mesh (Fig. 16.11b). Active factor XIII converts fibrin into a cross-linked polymer that stabilizes the clot.

THE COAGULATION CASCADE

Inactive plasma proteins (white boxes) are converted into active enzymes in each step of the pathway.

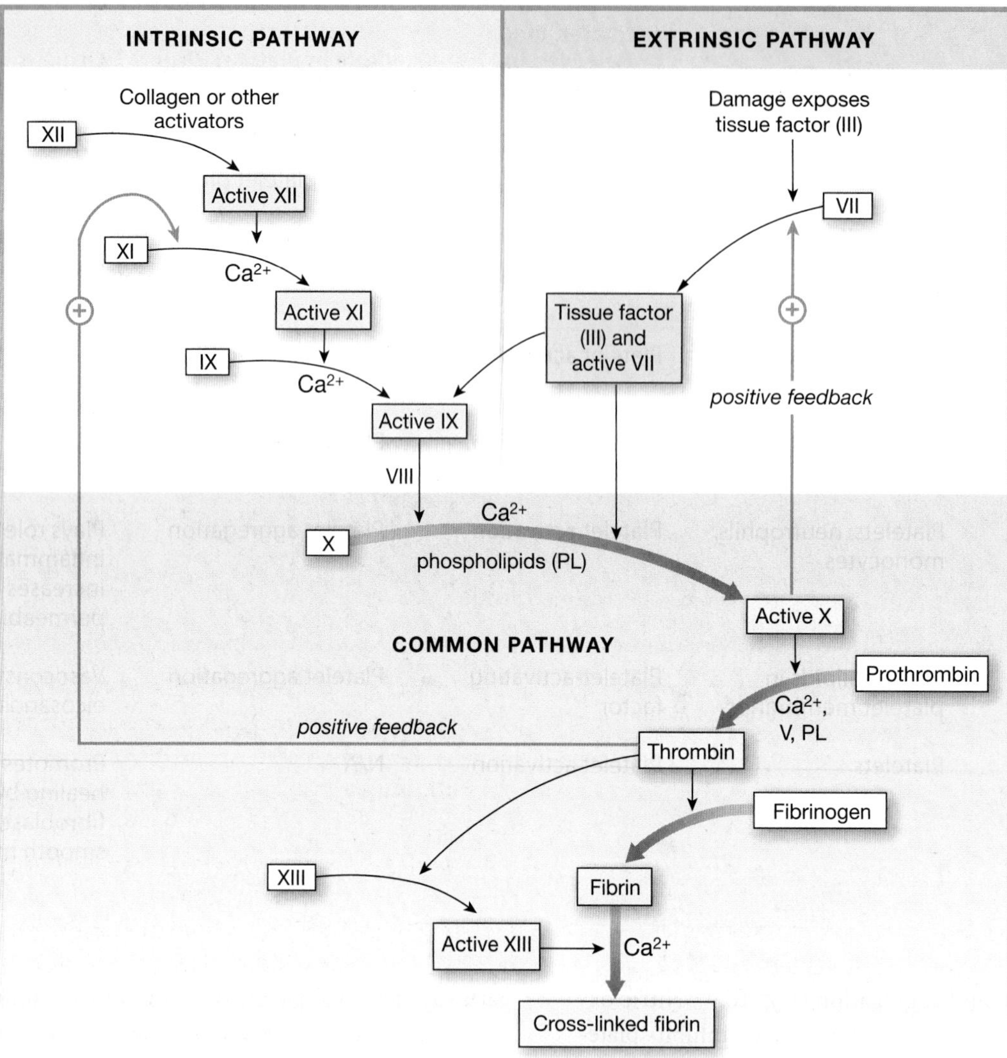

■ **Fig. 16.10**

Clots are only a temporary fix. As the damaged vessel wall slowly repairs itself, the clot disintegrates when fibrin is broken into fragments by the enzyme **plasmin** (Fig. 16.11a). An inactive form of plasmin, **plasminogen**, is part of the clot. After coagulation, thrombin, a factor in the coagulation cascade, works with a second factor called **tissue plasminogen activator (tPA)** to convert inactive plasminogen into plasmin. Plasmin then breaks down fibrin, a process known as **fibrinolysis**.

The large number of factors involved in coagulation and the fact that a single factor may have many different names can be confusing (■ Tbl. 16.5). Scientists assigned numbers to the coagulation factors, but the factors are not numbered in the order in which they participate in the coagulation cascade. Instead, they are numbered according to the order in which they were discovered.

Concept Check Answer: p. 567

8. In Figure 16.8, which box corresponds to the beginning of the intrinsic pathway of coagulation? Which corresponds to the beginning of the extrinsic pathway? To the beginning of the common pathway?

Anticoagulants Prevent Coagulation

Once coagulation begins, what keeps it from continuing until the entire circulation has clotted? Two mechanisms limit the extent of blood clotting within a vessel: (1) inhibition of platelet adhesion and (2) inhibition of the coagulation cascade and

COAGULATION AND FIBRINOLYSIS

(a) Conversion of fibrinogen into fibrin, and subsequent fibrinolysis

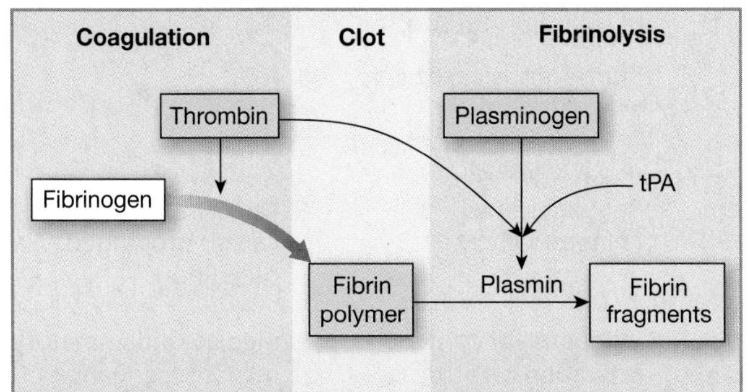

(b) Red blood cells are trapped in the fibrin mesh of a clot.

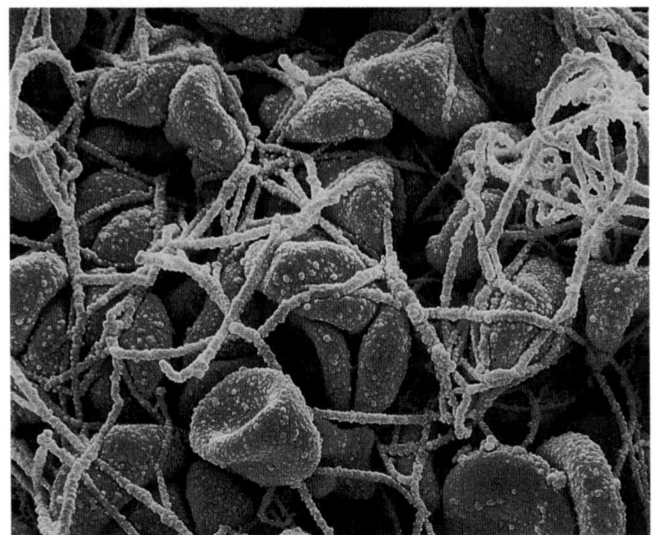

■ **Fig. 16.11**

fibrin production (■ Tbl. 16.6). As mentioned earlier, factors such as prostacyclin in the blood vessel endothelium and plasma ensure that the platelet plug is restricted to the area of damage (see left side of Fig. 16.9).

In addition, endothelial cells release chemicals known as **anticoagulants**, which prevent coagulation from taking place. Most act by blocking one or more of the reactions in the coagulation cascade. The body produces two anticoagulants, **heparin** and **antithrombin III**, which work together to block active factors IX, X, XI, and XII. **Protein C**, another anticoagulant in the body, inhibits clotting factors V and VIII.

The discovery of the factors controlling coagulation and fibrinolysis was an important step in developing treatments for many diseases related to coagulation problems. For example, heart attacks, more properly called *myocardial infarctions* (MIs), occur when a coronary blood vessel is blocked by a blood clot. Unless the blockage is removed promptly, the tissue will die or be severely damaged. One option for dissolving blood clots is to use fibrinolytic drugs—such as *streptokinase* (from bacteria) and tissue plasminogen activator (tPA)—to dissolve the clots. These drugs are now being combined with *antiplatelet agents* to prevent further platelet plug and clot formation. Some antiplatelet agents act as antagonists to platelet integrin receptors and prevent platelets from adhering to collagen.

Acetylsalicylic acid (aspirin) is an agent that prevents platelet plug formation. It acts by inhibiting the COX enzymes [p. 189] that promote synthesis of the platelet activator thromboxane A2. People who are at risk of developing small blood clots are sometimes told to take one aspirin every other day "to thin the blood." The aspirin does not actually make the blood less viscous, but it does prevent clots from forming by blocking platelet aggregation. Aspirin is now given routinely as emergency treatment for a suspected heart attack.

16

Anticoagulant drugs may be prescribed for people who are in danger of forming small blood clots that could block off critical vessels in the brain, heart, or lungs. The *coumarin anticoagulants,* such as *warfarin* (Coumadin®), block the action of vitamin K, a cofactor [p. 52] in the synthesis of clotting factors II (thrombin), VII, IX, and X. These anticoagulants were discovered when cattle that developed severe bleeding problems were found to have been eating spoiled sweet clover.

When blood samples are drawn into glass tubes, clotting takes place very rapidly unless the tube contains an anticoagulant. Several of the anticoagulants used for this purpose remove free Ca^{2+} from the plasma. Calcium is an essential clotting factor, so with no Ca^{2+}, no coagulation can occur. In the living body, however, plasma Ca^{2+} levels never decrease to levels that interfere with coagulation.

Several inherited diseases affect the coagulation process. Patients with coagulation disorders bruise easily. In severe forms, spontaneous bleeding may occur throughout the body. Bleeding into the joints and muscles can be painful and disabling. If bleeding occurs in the brain, it can be fatal.

The best-known coagulation disorder is **hemophilia**, a name given to several diseases in which one of the factors in the coagulation cascade is either defective or lacking. Hemophilia A, a factor VIII deficiency, is the most common form, occurring in about 80% of all cases. This disease is a recessive sex-linked trait that usually affects only males.

One exciting development in the treatment of hemophilia was a report on the first patients to be given gene therapy for hemophilia B, a deficiency in clotting factor IX. Patients injected with a virus engineered to carry the gene for factor IX started to produce some of the factor on their own, reducing their need for expensive injections of artificial factor IX. To learn more about these clinical trials and the latest treatments for hemophilia, visit the National Hemophilia Foundation web site at *www.hemophilia.org.*

Factors Involved in Coagulation

Table 16.5

Chemical Factor	Source	Activated by or Released in Response to	Role in Coagulation	Other Roles and Comments
Collagen	Subendothelial extracellular matrix	Injury that exposes collagen to plasma clotting factors	Starts intrinsic pathway	N/A
von Willebrand factor (vWF)	Endothelium, megakaryocytes	Exposure to collagen	Regulates level of factor VIII	Deficiency or defect causes prolonged bleeding
Kininogen and kallikrein	Liver and plasma	Cofactors normally present in plasma pathway	Cofactors for contact activation of intrinsic pathway	Mediate inflammatory response; enhance fibrinolysis
Tissue factor (tissue thromboplastin or factor III)	Most cells except platelets	Damage to tissue	Starts extrinsic pathway	N/A
Prothrombin and thrombin (factor II)	Liver and plasma	Platelet lipids, Ca^{2+} and factor V	Fibrin production	N/A
Fibrinogen and fibrin (factor I)	Liver and plasma	Thrombin	Form insoluble fibers that stabilize platelet plug	N/A
Fibrin-stabilizing factor (XIII)	Liver, megakaryocytes	Platelets	Cross-links fibrin polymers to make stable mesh	N/A
Ca^{2+} (factor IV)	Plasma ions	N/A	Required for several steps of coagulation cascade	Never a limiting factor
Vitamin K	Diet	N/A	Needed for synthesis of factors II, VII, IX, X	N/A

Endogenous Factors Involved in Fibrinolysis and Anticoagulation

Table 16.6

Chemical Factor	Source	Activated by or Released in Response to	Role in Anticoagulation or Fibrinolysis	Other Roles and Comments
Plasminogen and plasmin	Liver and plasma	tPA and thrombin	Dissolves fibrin and fibrinogen	N/A
Tissue plasminogen activator (tPA)	Many tissues	Normally present; levels increase with stress, protein C	Activates plasminogen	Recombinant tPA used clinically to dissolve clots
Antithrombin III	Liver and plasma	N/A	Anticoagulant; blocks factors IX, X, XI, XII, thrombin, kallikrein	Facilitated by heparin; no effect on thrombin despite name
Prostacyclin (prostaglandin I, or PGI_2)	Endothelial cells	N/A	Blocks platelet aggregation	Vasodilator

RUNNING PROBLEM CONCLUSION

Blood Doping in Athletes

Johann Muehlegg's elevated hemoglobin and hematocrit prior to his 50-km race meant an automatic urine drug test following the race. At the time of the 2002 Olympics, athletes knew that there was a urine test for EPO, but they were not aware that the same test could detect darbepoietin. Both of Muehlegg's urine samples tested positive for darbepoietin, and he was stripped of his 50-km gold metal. The International Olympic Committee tested other athletes for rhEPO at the 2002 Salt Lake City Winter Olympics and had more than 100 positive results. Despite official prohibitions, blood doping in endurance sports remains a major problem.

Now check your understanding of the physiology behind blood doping by comparing your answers with the information in the following table.

Question	Facts	Integration and Analysis
1. What is the normal range for Muehlegg's hemoglobin?	Normal hemoglobin range for males is 14–17 g/dL whole blood.	N/A
2. With blood doping, would you expect a hematocrit value to be lower or higher than normal?	Hematocrit is the percent of a blood sample volume that is packed red blood cells. A primary function of red blood cells is to carry oxygen.	Blood doping is done to increase oxygen-carrying capacity; therefore, the athlete would want more blood cells. This would mean a higher hematocrit.
3. Explain how diarrhea could cause a temporarily elevated hematocrit.	Diarrhea causes dehydration, which is loss of fluid volume. Plasma is the fluid component of blood.	If the total volume of red cells is unchanged but plasma volume decreases with dehydration, the hematocrit will increase.
4. How might Muehlegg quickly reduce his hematocrit without removing red blood cells?	Hematocrit = RBC volume/total blood volume (total blood volume = plasma volume + cell volume)	If plasma volume increases, hematocrit will decrease even though red cell volume does not change. By drinking fluids, Muehlegg could increase his plasma volume quickly.
5. Explain how endogenous EPO, rhEPO, and darbepoietin made from the same gene can all be active yet different enough from one another to be detectable in the laboratory.	Activity depends on the protein binding to the receptor's binding site. Post-translational modification allows proteins from the same gene to be altered so that they are different from one another.	The three hormones have sites that bind to and activate the EPO receptor, but they have different sizes or charges that cause them to separate during electrophoresis. For example, the glycosylation pattern [p. 32] of rhEPO is different from the pattern in endogenous EPO.
6. One hallmark of illegal EPO use is elevated reticulocytes in the blood. Why would this suggest greater-than-normal EPO activity?	Reticulocytes are the final immature stage of red blood cell development. Maturation usually takes place in the marrow.	If red blood cell development becomes more rapid, more reticulocytes may be released into the blood before they have time to mature.

545 554 557 560 565

Chapter Summary

Blood is an interesting tissue, with blood cells and cell fragments suspended in a liquid matrix—the plasma—that forms one of the two extracellular *compartments*. *Exchange* between the plasma and interstitial fluid takes place only in the capillaries. *Bulk flow* of blood through the body depends on the pressure gradient created by the heart. At the same time, high pressure in the blood vessels poses a danger should the wall of a vessel rupture. Collectively, the cellular and protein components of blood serve as a functional unit that provides protection against hemorrhage. Blood cells are also essential for oxygen transport and defense, as you will learn in later chapters.

Plasma and the Cellular Elements of Blood

1. Blood is the circulating portion of the extracellular fluid. (p. 545)
2. **Plasma** is composed mostly of water, with dissolved proteins, organic molecules, ions, and dissolved gases. (p. 546; Fig. 16.1)
3. The **plasma proteins** include **albumins, globulins**, and the clotting protein **fibrinogen**. They function in blood clotting, defense, and as hormones, enzymes, or carriers for different substances. (p. 545)
4. The cellular elements of blood are **red blood cells (erythrocytes), white blood cells (leukocytes)**, and **platelets**. Platelets are fragments of cells called **megakaryocytes**. (p. 546; Fig. 16.1)
5. Blood contains five types of white blood cells: (1) **lymphocytes**, (2) **monocytes**, (3) **neutrophils**, (4) **eosinophils**, and (5) **basophils**. (p. 546; Fig. 16.1)

Blood Cell Production

6. All blood cells develop from a pluripotent hematopoietic stem cell. (p. 548; Fig. 16.2)
7. **Hematopoiesis** begins early in embryonic development and continues throughout a person's life. Most hematopoiesis takes place in the bone marrow. (p. 552; Fig. 16.4)
8. **Colony-stimulating factors** and other cytokines control white blood cell production. **Thrombopoietin** regulates the growth and maturation of megakaryocytes. Red blood cell production is regulated primarily by **erythropoietin**. (p. 551)

Red Blood Cells

9. Mature mammalian red blood cells are biconcave disks lacking a nucleus. They contain hemoglobin, a red oxygen-carrying pigment. (p. 553; Fig. 16.5)
10. Hemoglobin synthesis requires iron in the diet. Iron is transported in the blood on **transferrin** and stored mostly in the liver, on **ferritin**. (p. 555; Fig. 16.6)
11. When hemoglobin is broken down, some heme groups are converted into **bilirubin**, which is incorporated into **bile** and excreted. Elevated bilirubin concentrations in the blood cause **jaundice**. (p. 555; Fig. 16.6)

Platelets and Coagulation

12. Platelets are cell fragments filled with granules containing clotting proteins and cytokines. Platelets are activated by damage to vascular endothelium. (p. 558; Fig. 16.7)
13. **Hemostasis** begins with vasoconstriction and the formation of a **platelet plug**. (p. 559; Fig. 16.8)
14. Exposed collagen triggers **platelet adhesion** and **platelet aggregation**. The platelet plug is converted into a clot when reinforced by **fibrin**. (p. 559; Fig. 16.8)
15. In the last step of the **coagulation cascade**, fibrin is made from **fibrinogen** through the action of **thrombin**. (p. 562; Fig. 16.10)
15. As the damaged vessel is repaired, **plasmin** trapped in the platelet plug dissolves fibrin (**fibrinolysis**) and breaks down the clot. (p. 563; Fig. 16.11)
16. Platelet plugs are restricted to the site of injury by **prostacyclin** in the membrane of intact vascular endothelium. **Anticoagulants** limit the extent of blood clotting within a vessel. (p. 560; Fig. 16.9)

Questions

Answers: p. A-1

Level One Reviewing Facts and Terms

1. The fluid portion of the blood, called ________, is composed mainly of ________.
2. List the three types of plasma proteins. Name at least one function of each type. Which type is most prevalent in the body?
3. List the cellular elements found in blood, and name at least one function of each.
4. Blood cell production is called ________ When and where does it occur?
5. What role do colony-stimulating factors, cytokines, and interleukins play in blood cell production? How are these chemical signal molecules different? Give two examples of each.
6. List the technical terms for production of red blood cells, production of platelets, and production of white blood cells.
7. The hormone that directs red blood cell synthesis is called ________. Where is it produced, and what is the stimulus for its production?
8. How are the terms *hematocrit* and *packed red cells* related? What are normal hematocrit values for men and women?
9. Distinguish between an erythroblast and an erythrocyte. Give three distinct characteristics of erythrocytes.
10. Which chemical element in the diet is important for hemoglobin synthesis?
11. Define the following terms and explain their significance in hematology.

(a) jaundice	(c) transferrin
(b) anemia	(d) hemophilia

12. Chemicals that prevent blood clotting from occurring are called ________.

Level Two Reviewing Concepts

13. **Concept maps:** Combine each list of terms into a map. You may add other terms.

List 1

- ADP
- collagen
- integrins
- membrane
- phospholipids
- platelet-activating factor
- platelet activation
- platelet adhesion
- platelet aggregation
- platelet plug
- positive feedback
- serotonin
- thromboxane A2
- vasoconstriction

List 2

- clot
- coagulation
- fibrin
- fibrinogen
- fibrinolysis
- infarct
- plasmin
- plasminogen
- polymer
- thrombin

List 3

- bile
- bilirubin
- bone marrow
- erythropoietin
- ferritin
- globin
- hematocrit
- heme
- hemoglobin
- intestine
- iron
- liver
- reticulocyte
- transferrin

14. Distinguish between the intrinsic, extrinsic, and common pathways of the coagulation cascade.
15. Once platelets are activated to aggregate, what factors halt their activity?

Level Three Problem Solving

16. Rachel is undergoing chemotherapy for breast cancer. She has blood cell counts at regular intervals, with these results:

	Normal Count (cells) / μL	Patient Count 10 Days Post-Chemotherapy	Patient Count 20 Days Post-Chemotherapy
WBC	$4–11 \times 10^3$	2.6	4.9
RBC	$3.9–5.6 \times 12^6$	3.85	4.2
Platelets	$150–450 \times 10^3$	133	151

At the time of the 10-day test, Jen (the nurse) notes that Rachel, although pale and complaining of being tired, does not have any bruises on her skin. Jen tells Rachel to eat foods high in protein, take a multivitamin tablet containing iron, and stay home and away from crowds as much as possible. How are Jen's observations and recommendations related to the results of the 10-day and 20-day blood tests?

17. Hemochromatosis is an inherited condition in which the body absorbs iron excessively, resulting in an elevated total body load of iron.
    (a) What plasma protein would you expect to be elevated in this disease?
    (b) Which organ(s) would you expect to show damage in this disease?
    (c) Can you think of a simple treatment that could decrease the body's overload of iron in hemochromatosis?
18. Erythropoietin (EPO) was first isolated from the urine of anemic patients who had high circulating levels of the hormone. Although these patients had high concentrations of EPO, they were unable to produce adequate amounts of hemoglobin or red cells. Give some possible reasons why the patients' EPO was unable to correct their anemia.

Level Four Quantitative Problems

19. If we estimate that total blood volume is 7% of body weight, calculate the total blood volume in a 200-lb man and in a 130-lb woman (2.2 lb/kg). What are their plasma volumes if the man's hematocrit is 52% and the woman's hematocrit is 41%?
20. The total blood volume of an average person is 7% of total body weight. Using this figure and the fact that 1 kg of blood occupies a volume of about 1 liter, figure the total erythrocyte volume of a 50-kg woman with a hematocrit of 40%.

Answers

Answers to Concept Check Questions

Page 547

1. The five types of leukocytes are lymphocytes, monocytes/macrophages, basophils/mast cells, neutrophils, and eosinophils.
2. Erythrocytes and platelets lack nuclei, which would make them unable to carry out protein synthesis.
3. Liver degeneration reduces the total plasma protein concentration, which reduces the osmotic pressure in the capillaries. This decrease in osmotic pressure increases net capillary filtration, and edema results.

Page 551

4. (a) erythropoietin (EPO), (b) colony-stimulating factors (CSFs), (c) thrombopoietin (TPO).

Page 556

5. (a) Heme is an iron-containing subunit of a hemoglobin molecule. (b) Ferritin is the liver protein that stores iron. Transferrin is the plasma protein that transports iron in the blood.
6. Bile is an exocrine secretion because it is secreted into the intestine.

Page 557

7. Low atmospheric oxygen at high altitude → low arterial oxygen → sensed by kidney cells → secrete erythropoietin → acts on bone marrow → increased production of red blood cells.

Page 562

8. The intrinsic pathway starts at the gold "Collagen exposed" box, the extrinsic pathway starts at the gold "Tissue factor exposed" box, and the common pathway begins at the red "Coagulation cascade" diamond.

16

17 Mechanics of Breathing

This being of mine, whatever it really is, consists of a little flesh, a little breath, and the part which governs.

—Marcus Aurelius Antoninus (C.E. 121–180)

Background Basics

Colored x-ray of the lung showing the branching airways.

Imagine covering the playing surface of a racquetball court (about 75 m^2) with thin plastic wrap, then crumpling up the wrap and stuffing it into a 3-liter soft drink bottle. Impossible? Maybe so, if you use plastic wrap and a drink bottle. But the lungs of a 70-kg man have a gas exchange surface the size of that plastic wrap, compressed into a volume that is less than that of the bottle. This tremendous surface area for gas exchange is needed to supply the trillions of cells in the body with adequate amounts of oxygen.

Aerobic metabolism in cells depends on a steady supply of oxygen and nutrients from the environment, coupled with the removal of carbon dioxide. In very small aquatic animals, simple diffusion across the body surface meets these needs. Distance limits diffusion rate, however, so most multicelled animals require specialized respiratory organs associated with a circulatory system. Respiratory organs take a variety of forms, but all possess a large surface area compressed into a small space.

Besides needing a large exchange surface, humans and other terrestrial animals face an additional physiological challenge: dehydration. The exchange surface must be thin and moist to allow gases to pass from air into solution, and yet at the same time it must be protected from drying out as a result of exposure to air. Some terrestrial animals, such as the slug (a shell-less snail), meet the challenge of dehydration with behavioral adaptations that restrict them to humid environments and nighttime activities.

A more common solution is anatomical: an internalized respiratory epithelium. Human lungs are enclosed in the chest cavity to control their contact with the outside air. Internalization creates a humid environment for the exchange of gases with the blood and protects the delicate exchange surface from damage.

Internalized lungs create another challenge, however: how to move air between the atmosphere and an exchange surface deep within the body. Air flow requires a muscular pump to create pressure gradients. More complex respiratory systems therefore consist of two separate components: a muscle-driven pump and a thin, moist exchange surface. In humans, the pump is the musculoskeletal structure of the thorax [p. 64]. The lungs themselves consist of the exchange epithelium and associated blood vessels.

The four primary functions of the respiratory system are:

1. *Exchange of gases between the atmosphere and the blood.* The body brings in O_2 for distribution to the tissues and eliminates CO_2 waste produced by metabolism.
2. *Homeostatic regulation of body pH.* The lungs can alter body pH by selectively retaining or excreting CO_2.
3. *Protection from inhaled pathogens and irritating substances.* Like all other epithelia that contact the external environment, the respiratory epithelium is well supplied with defense mechanisms to trap and destroy potentially harmful substances before they can enter the body.
4. *Vocalization.* Air moving across the vocal cords creates vibrations used for speech, singing, and other forms of communication.

(5) Smell

In addition to serving these functions, the respiratory system is also a significant source of water loss and heat loss from the body. These losses must be balanced using homeostatic compensations.

In this chapter you will learn how the respiratory system carries out these functions by exchanging air between the environment and the interior air spaces of the lungs. This exchange is the *bulk flow* of air, and it follows many of the same principles that govern the bulk flow of blood through the cardiovascular system:

1. Flow takes place from regions of higher pressure to regions of lower pressure.
2. A muscular pump creates pressure gradients.
3. Resistance to air flow is influenced primarily by the diameter of the tubes through which the air is flowing.

Air and blood are both fluids. The primary difference between air flow in the respiratory system and blood flow in the circulatory system is that air is a less viscous, compressible mixture of gases while blood is a noncompressible liquid.

RUNNING PROBLEM

Emphysema

You could hear her whistling, wheezing breathing preceding her down the hall. "Diagnosis: COPD," reads Edna Wilson's patient chart. COPD—chronic obstructive pulmonary disease—is the name given to diseases in which air exchange is impaired by narrowing of the lower airways. Most people with COPD have emphysema or chronic bronchitis or a combination of the two. Individuals in whom chronic bronchitis predominates are sometimes called "blue bloaters," owing to the bluish tinge of their skin (from low blood oxygen levels) and a tendency to be overweight. In contrast, patients with emphysema have been nicknamed "pink puffers." They tend to be thin, have normal (pink) skin coloration, and often breathe out through pursed lips, which helps open their airways. More than 12 million people in the United States have COPD. Its most common cause is smoking, and most people can avoid the disease simply by not smoking. Unfortunately, Edna has been a heavy smoker for 35 of her 47 years.

569 576 582 584 587 594

The Respiratory System

The word *respiration* has several meanings in physiology (■ Fig. 17.1). **Cellular respiration** refers to the intracellular reaction of oxygen with organic molecules to produce carbon

EXTERNAL RESPIRATION

The respiratory and circulatory systems coordinate to move oxygen and CO_2 between the atmosphere and the cells.

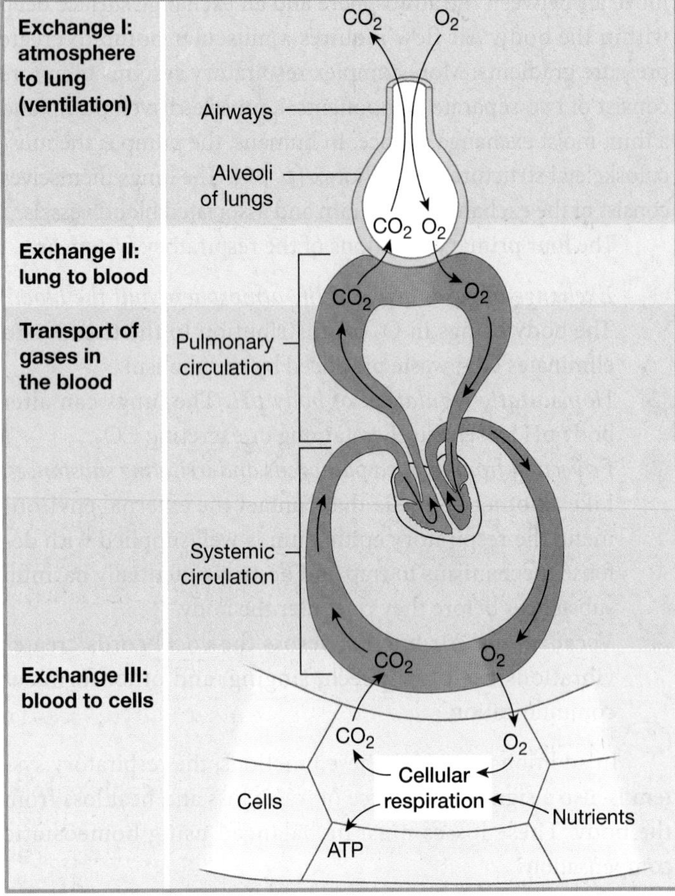

■ **Fig. 17.1**

dioxide, water, and energy in the form of ATP [p. 111]. **External respiration,** the topic of this chapter and the next, is the movement of gases between the environment and the body's cells. External respiration can be subdivided into four integrated processes, illustrated in Figure 17.1:

1. *The exchange of air between the atmosphere and the lungs.* This process is known as **ventilation**, or breathing. **Inspiration** (inhalation) is the movement of air into the lungs. **Expiration** (exhalation) is the movement of air out of the lungs. The mechanisms by which ventilation takes place are collectively called the *mechanics of breathing.*
2. *The exchange of O_2 and CO_2 between the lungs and the blood.*
3. *The transport of O_2 and CO_2 by the blood.*
4. *The exchange of gases between blood and the cells.*

External respiration requires coordination between the respiratory and cardiovascular systems. The **respiratory system** consists of structures involved in ventilation and gas exchange (■ Fig. 17.2 on pages 572–573):

1. The **conducting system** of passages, or **airways**, that lead from the external environment to the exchange surface of the lungs.
2. The **alveoli** (singular **alveolus**) {*alveus,* a concave vessel}, a series of interconnected sacs and their associated *pulmonary capillaries.* These structures form the exchange surface, where oxygen moves from inhaled air to the blood, and carbon dioxide moves from the blood to air that is about to be exhaled.
3. The bones and muscles of the thorax (chest cavity) and abdomen that assist in ventilation.

The respiratory system can be divided into two parts. The **upper respiratory tract** consists of the mouth, nasal cavity, pharynx, and larynx. The **lower respiratory tract** consists of the trachea, two primary bronchi {*bronchos,* windpipe; singular—bronchus}, their branches, and the lungs. The lower tract is also known as the *thoracic portion* of the respiratory system because it is enclosed in the thorax.

Bones and Muscles of the Thorax Surround the Lungs

The thorax is bounded by the bones of the spine and rib cage and their associated muscles. Together the bones and muscles are called the *thoracic cage.* The ribs and spine (the *chest wall*) form the sides and top of the cage. A dome-shaped sheet of skeletal muscle, the **diaphragm**, forms the floor (Fig. 17.2b).

Two sets of **intercostal muscles**, internal and external, connect the 12 pairs of ribs (Fig. 17.2a). Additional muscles, the **sternocleidomastoids** and the **scalenes**, run from the head and neck to the sternum and first two ribs.

Functionally, the thorax is a sealed container filled with three membranous bags, or sacs. One, the *pericardial sac,* contains the heart. The other two bags, the **pleural sacs**, each surround a lung {*pleura,* rib or side}. The esophagus and thoracic blood vessels and nerves pass between the pleural sacs (Fig. 17.2c).

Pleural Sacs Enclose the Lungs

The **lungs** (Fig. 17.2b, d) consist of light, spongy tissue whose volume is occupied mostly by air-filled spaces. These irregular cone-shaped organs nearly fill the thoracic cavity, with their bases resting on the curved diaphragm. Semi-rigid conducting airways—the bronchi—connect the lungs to the main airway, the trachea.

Each lung is surrounded by a double-walled pleural sac whose membranes line the inside of the thorax and cover the outer surface of the lungs (■ Fig. 17.3 on page 574). Each *pleural membrane,* or **pleura**, contains several layers of elastic connective tissue and numerous capillaries. The opposing layers of pleural membrane are held together by a thin film of **pleural fluid** whose total volume is only about 25–30 mL in a 70-kg man. The result is similar to an air-filled balloon (the lung) surrounded by a water-filled balloon (the pleural sac). Most illustrations exaggerate the volume of the pleural fluid, but you can appreciate its thinness if you imagine spreading 25 mL of water evenly over the surface of a 3-liter soft drink bottle.

Pleural fluid serves several purposes. First, it creates a moist, slippery surface so that the opposing membranes can slide across one another as the lungs move within the thorax. Second, it holds the lungs tight against the thoracic wall. To visualize this arrangement, think of two panes of glass stuck together by a thin film of water. You can slide the panes back and forth across each other, but you cannot pull them apart because of the cohesiveness of the water [p. 39]. A similar fluid bond between the two pleural membranes makes the lungs "stick" to the thoracic cage and holds them stretched in a partially inflated state, even at rest.

Airways Connect Lungs to the External Environment

Air enters the upper respiratory tract through the mouth and nose and passes into the **pharynx**, a common passageway for food, liquids, and air {*pharynx,* throat}. From the pharynx, air flows through the **larynx** into the **trachea**, or windpipe (Fig. 17.2b). The larynx contains the **vocal cords**, connective tissue bands that vibrate and tighten to create sound when air moves past them.

The trachea is a semiflexible tube held open by 15 to 20 C-shaped cartilage rings. It extends down into the thorax, where it branches (division 1) into a pair of **primary bronchi**, one *bronchus* to each lung (Fig. 17.2b). Within the lungs, the bronchi branch repeatedly (divisions 2–11) into progressively smaller bronchi (Fig. 17.2e). Like the trachea, the bronchi are semi-rigid tubes supported by cartilage.

Within the lungs, the smallest bronchi branch to become **bronchioles**, small collapsible passageways with walls of smooth muscle. The bronchioles continue branching (divisions 12–23) until the *respiratory bronchioles* form a transition between the airways and the exchange epithelium of the lung.

The diameter of the airways becomes progressively smaller from the trachea to the bronchioles, but as the individual airways get narrower, their numbers increase geometrically (■ Fig. 17.4 on page 574). As a result, the total cross-sectional area increases with each division of the airways. Total cross-sectional area is lowest in the upper respiratory tract and greatest in the bronchioles, analogous to the increase in cross-sectional area that occurs from the aorta to the capillaries in the circulatory system [p. 470].

Concept Check

Answers: p. 598

1. What is the difference between cellular respiration and external respiration?
2. Name the components of the upper respiratory tract and those of the lower respiratory tract.
3. Based on the total cross-sectional area of different airways, where is the velocity of air flow highest and lowest?
4. Give two functions of pleural fluid.
5. Name the components (including muscles) of the thoracic cage. List the contents of the thorax.
6. Which air passages of the respiratory system are collapsible?

17

The Airways Warm, Humidify, and Filter Inspired Air

During breathing, the upper airways and the bronchi do more than simply serve as passageways for air. They play an important role in conditioning air before it reaches the alveoli. Conditioning has three components:

1. *Warming* air to body temperature (37 °C), so that core body temperature does not change and alveoli are not damaged by cold air;
2. *Adding water vapor* until the air reaches 100% humidity, so that the moist exchange epithelium does not dry out; and
3. *Filtering out foreign material,* so that viruses, bacteria, and inorganic particles do not reach the alveoli.

Inhaled air is warmed by the body's heat and moistened by water evaporating from the mucosal lining of the airways. Under normal circumstances, by the time air reaches the trachea, it has been conditioned to 100% humidity and 37 °C. Breathing through the mouth is not nearly as effective at warming and moistening air as breathing through the nose. If you exercise outdoors in very cold weather, you may be familiar with the ache in your chest that results from breathing cold air through your mouth.

Air is filtered both in the trachea and in the bronchi. These airways are lined with ciliated epithelium whose cilia are bathed in a watery saline layer (■ Fig. 17.5 on page 575). The saline is produced by epithelial cells when Cl^- secreted into the lumen by apical anion channels draws Na^+ into the lumen through the paracellular pathway (Fig. 17.5c). Movement of solute from the ECF to the lumen creates an osmotic gradient, and water follows the ions into the airways. The *CFTR channel*, whose malfunction causes cystic fibrosis, is one of the anion channels found on the apical surface of this epithelium [p. 148].

A sticky layer of mucus floats over the cilia to trap most inhaled particles larger than 2 μm. The mucus layer is secreted

The Lungs and Thoracic Cavity

(a) Muscles of the thorax, neck, and abdomen create the force to move air during breathing.

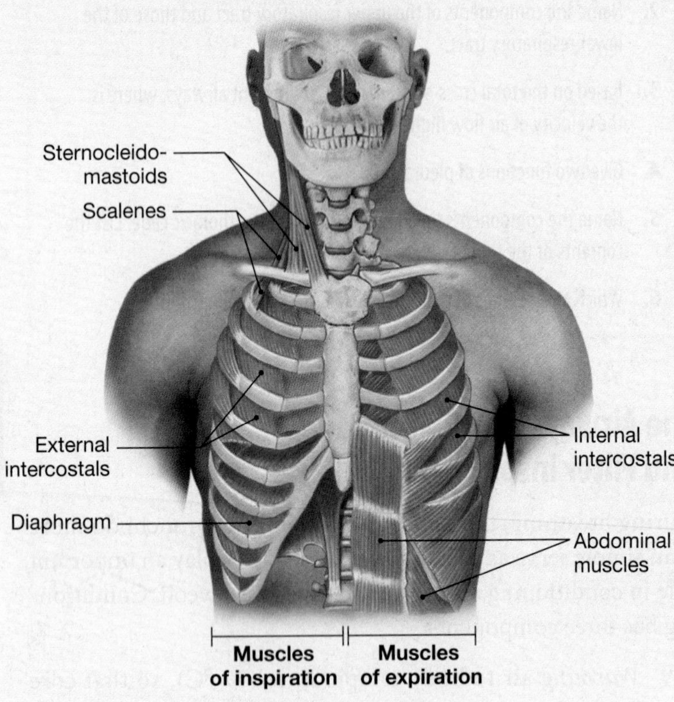

(b) The respiratory system is divided into upper and lower regions.

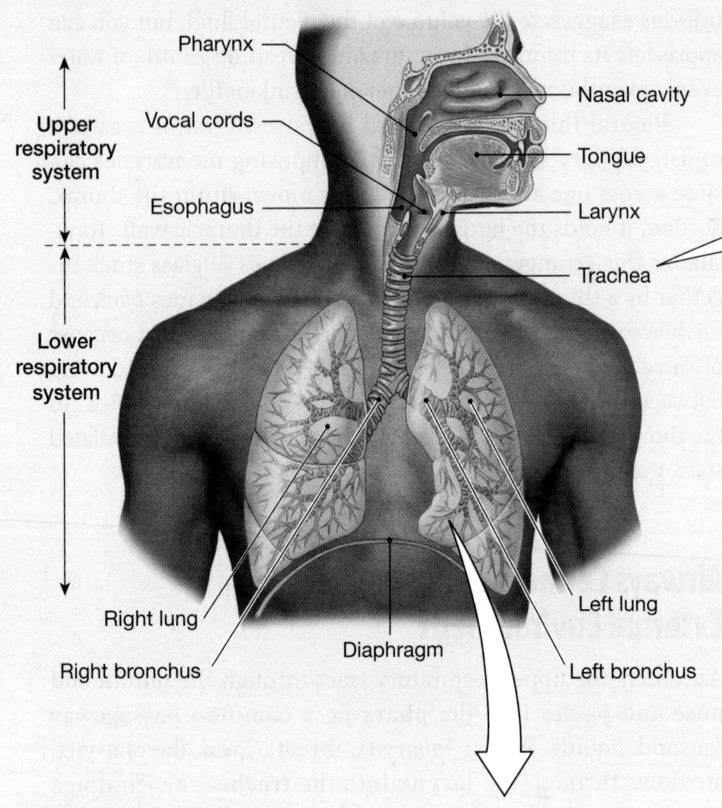

(c) Sectional view of chest. Each lung is enclosed in two pleural membranes. The esophagus and aorta pass through the thorax between the pleural sacs.

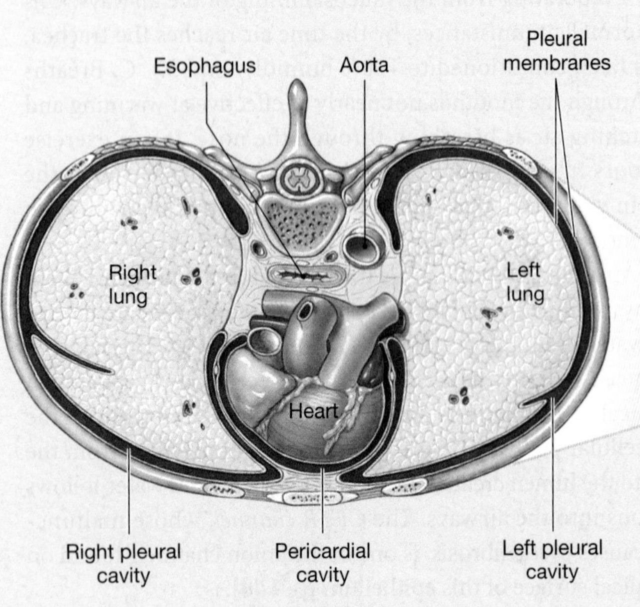

(d) On external view, the right lung is divided into three lobes, and the left lung is divided into two lobes.

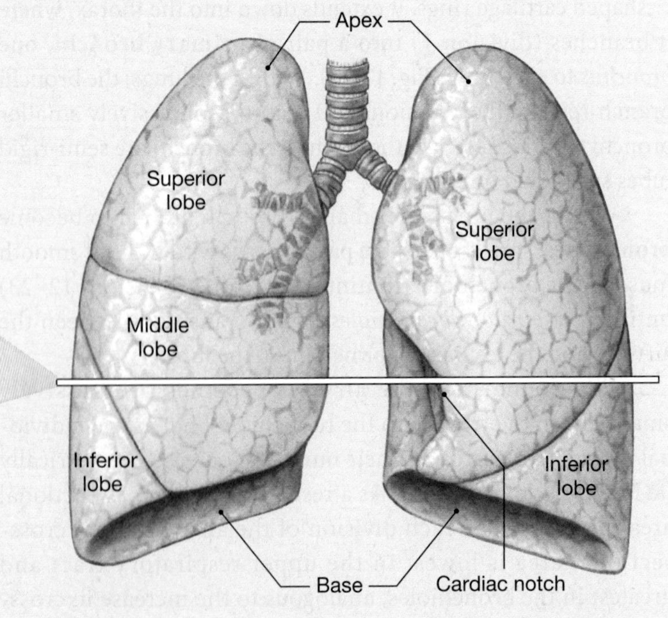

The Bronchi and Alveoli

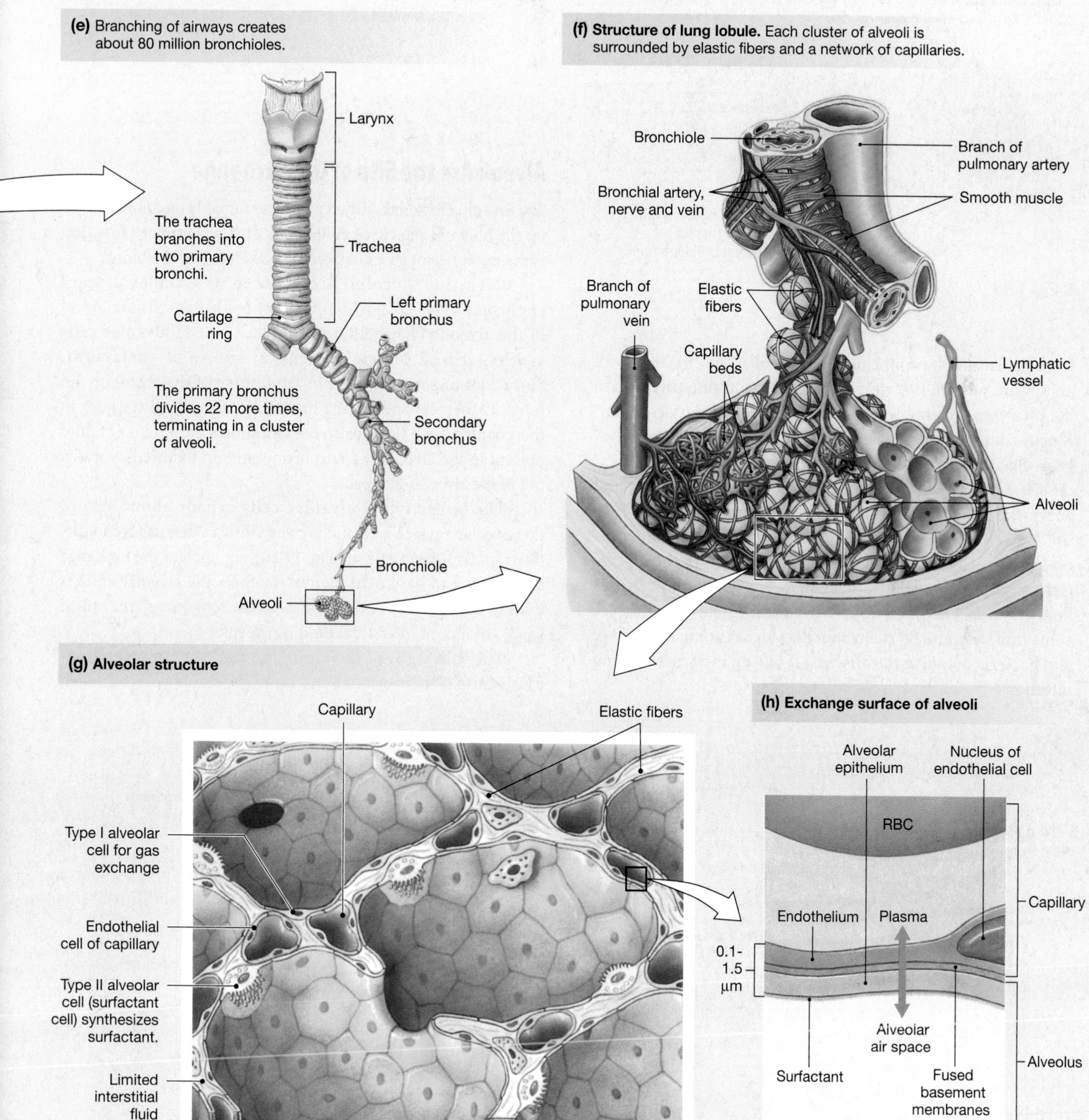

(e) Branching of airways creates about 80 million bronchioles.
Larynx
The trachea branches into two primary bronchi.
Trachea
Cartilage ring
Left primary bronchus
The primary bronchus divides 22 more times, terminating in a cluster of alveoli.
Secondary bronchus
Bronchiole
Alveoli
(f) Structure of lung lobule. Each cluster of alveoli is surrounded by elastic fibers and a network of capillaries.
Bronchiole
Branch of pulmonary artery
Bronchial artery, nerve and vein
Smooth muscle
Branch of pulmonary vein
Elastic fibers
Capillary beds
Lymphatic vessel
Alveoli
(g) Alveolar structure
Capillary
Elastic fibers
Type I alveolar cell for gas exchange
Endothelial cell of capillary
Type II alveolar cell (surfactant cell) synthesizes surfactant.
Limited interstitial fluid
Alveolar macrophage ingests foreign material.
(h) Exchange surface of alveoli
Alveolar epithelium
Nucleus of endothelial cell
RBC
Capillary
Endothelium
Plasma
0.1-1.5 µm
Alveolar air space
Surfactant
Fused basement membranes
Alveolus

Blue arrow represents gas exchange between alveolar air space and the plasma.

THE PLEURAL SAC

The pleural sac forms a double membrane surrounding the lung, similar to a fluid-filled balloon surrounding an air-filled balloon.

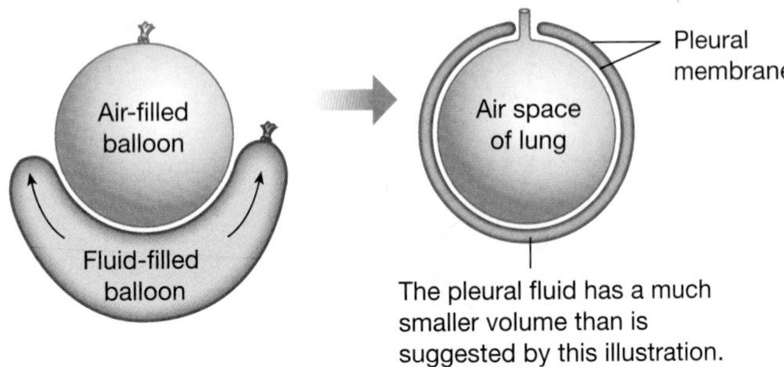

Air-filled balloon
Fluid-filled balloon
Air space of lung
Pleural membrane
The pleural fluid has a much smaller volume than is suggested by this illustration.

■ Fig. 17.3

by *goblet cells* in the epithelium (Fig. 17.5b). The cilia beat with an upward motion that moves the mucus continuously toward the pharynx, creating what is called the *mucociliary escalator*. Mucus contains *immunoglobulins* that can disable many pathogens. Once mucus reaches the pharynx, it can be spit out (*expectorated*) or swallowed. For swallowed mucus, stomach acid and enzymes destroy any remaining microorganisms.

Secretion of the watery saline layer beneath the mucus is essential for a functional mucociliary escalator. In the disease *cystic fibrosis,* for example, inadequate ion secretion decreases fluid movement in the airways. Without the saline layer, cilia become trapped in thick, sticky mucus. Mucus cannot be cleared, and bacteria colonize the airways, resulting in recurrent lung infections.

Concept Check Answer: p. 598

7. Cigarette smoking paralyzes cilia in the airways and increases mucus production. Why would these effects cause smokers to develop a cough?

Alveoli Are the Site of Gas Exchange

The alveoli, clustered at the ends of terminal bronchioles, make up the bulk of lung tissue (Fig. 17.2f, g). Their primary function is the exchange of gases between themselves and the blood.

Each tiny alveolus is composed of a single layer of epithelium (Fig. 17.2g). Two types of epithelial cells are found in the alveoli. The smaller but thicker **type II alveolar cells** synthesize and secrete a chemical known as **surfactant.** Surfactant mixes with the thin fluid lining of the alveoli to aid lungs as they expand during breathing, as you will see later in this chapter. Type II cells also help minimize the amount of fluid present in the alveoli by transporting solutes, followed by water, out of the alveolar air space.

The larger **type I alveolar cells** occupy about 95% of the alveolar surface area and are very thin so that gases can diffuse rapidly through them (Fig. 17.2h). In much of the exchange area, a layer of basement membrane fuses the alveolar epithelium to the capillary endothelium. In the remaining area only a small amount of interstitial fluid is present.

The thin walls of the alveoli do not contain muscle because muscle fibers would block rapid gas exchange. As a result,

BRANCHING OF THE AIRWAYS

	Name	Division	Diameter (mm)	How many?	Cross-sectional area (cm^2)
Conducting system	Trachea	0	15-22	1	2.5
	Primary bronchi	1	10-15	2	↓
	Smaller bronchi ↓	2	1-10	4 ↓	
		3			
		4			
		5			
		6-11		1 x 10^4	
	Bronchioles	12-23	0.5-1	2 x 10^4 ↓ 8 x 10^7	100 ↓ 5 x 10^3
Exchange surface	Alveoli	24	0.3	3-6 x 10^8	>1 x 10^6

■ Fig. 17.4

AIRWAY EPITHELIUM

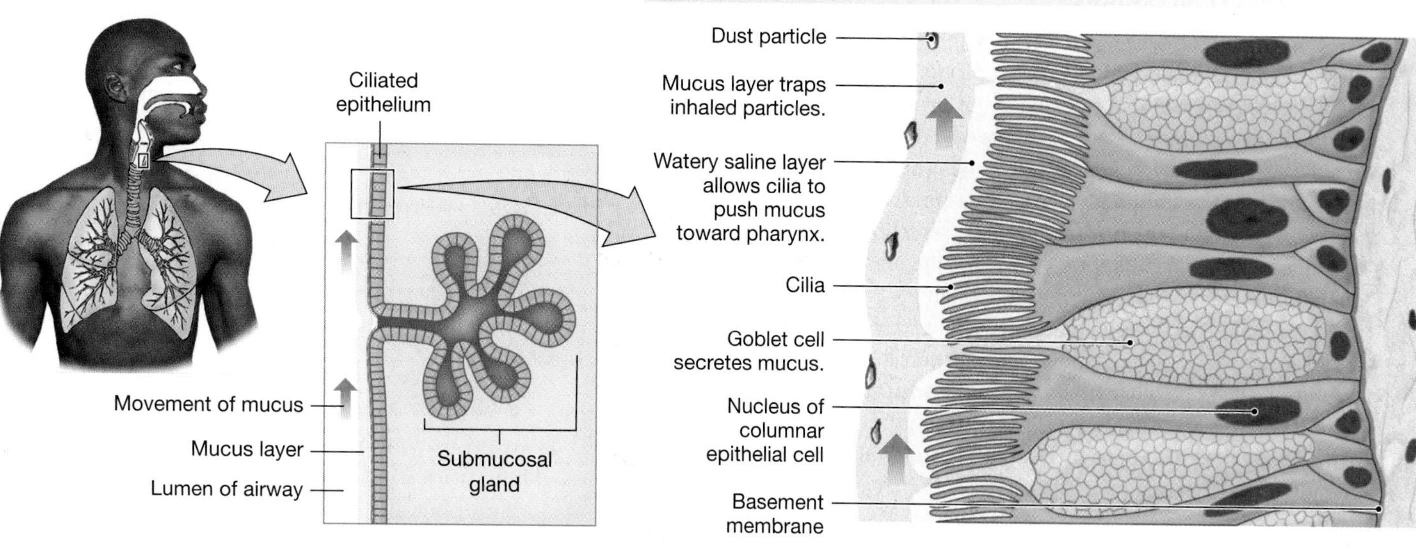
(a) Epithelial cells lining the airways and submucosal glands secrete saline and mucus.
Ciliated epithelium
Movement of mucus
Mucus layer
Lumen of airway
Submucosal gland
(b) Cilia move the mucus layer toward the pharynx, removing trapped pathogens and particulate matter.
Dust particle
Mucus layer traps inhaled particles.
Watery saline layer allows cilia to push mucus toward pharynx.
Cilia
Goblet cell secretes mucus.
Nucleus of columnar epithelial cell
Basement membrane

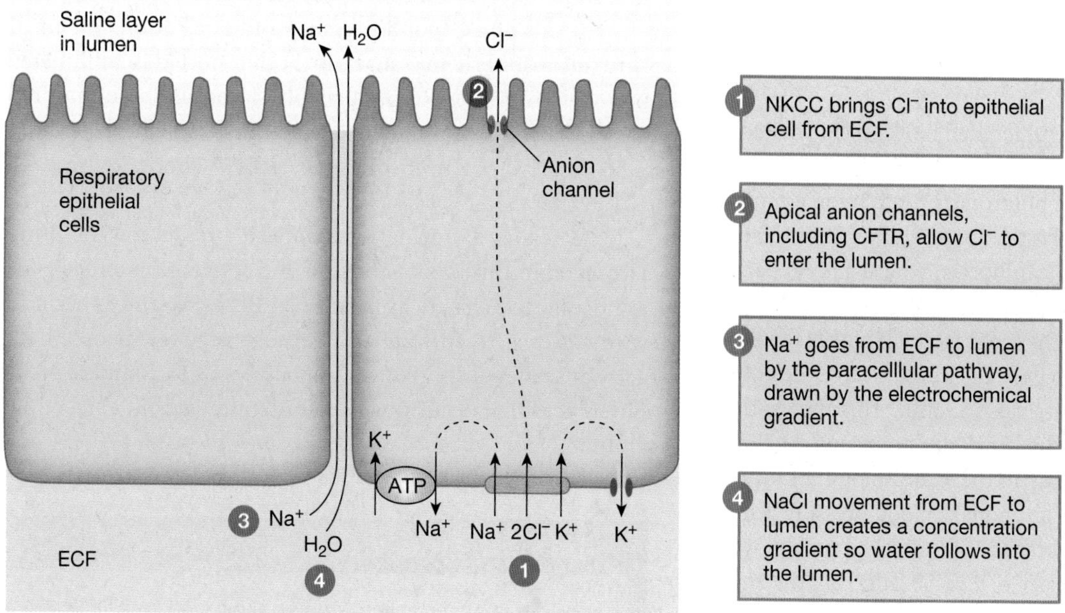
(c) One model of saline secretion by airway epithelial cells
Saline layer in lumen
Na⁺ H₂O
Cl⁻
Anion channel
Respiratory epithelial cells
K⁺
ATP
Na⁺
Na⁺ 2Cl⁻ K⁺
K⁺
ECF
1 NKCC brings Cl⁻ into epithelial cell from ECF.
2 Apical anion channels, including CFTR, allow Cl⁻ to enter the lumen.
3 Na⁺ goes from ECF to lumen by the paracelllular pathway, drawn by the electrochemical gradient.
4 NaCl movement from ECF to lumen creates a concentration gradient so water follows into the lumen.

Fig. 17.5

lung tissue itself cannot contract. However, connective tissue between the alveolar epithelial cells contains many elastin and collagen fibers that create elastic recoil when lung tissue is stretched.

The close association of the alveoli with an extensive network of capillaries demonstrates the intimate link between the respiratory and cardiovascular systems. Blood vessels fill 80–90% of the space between alveoli, forming an almost continuous "sheet" of blood in close contact with the air-filled alveoli. The proximity of capillary blood to alveolar air is essential for the rapid exchange of gases.

Pulmonary Circulation Is High-Flow, Low-Pressure

The pulmonary circulation begins with the pulmonary trunk, which receives low-oxygen blood from the right ventricle. The pulmonary trunk divides into two pulmonary arteries, one to

CLINICAL FOCUS

Congestive Heart Failure

When is a lung problem not a lung problem? The answer: when it's really a heart problem. Congestive heart failure (CHF) is an excellent example of the interrelationships among body systems and demonstrates how disruptions in one system can have a domino effect in the others. The primary symptoms of heart failure are shortness of breath (*dyspnea*), wheezing during breathing, and sometimes a productive cough that may be pinkish from the presence of blood. Congestive heart failure arises when the right heart is a more effective pump than the left heart. When blood accumulates in the pulmonary circulation, increased volume increases pulmonary blood pressure and capillary hydrostatic pressure. Capillary filtration exceeds the ability of the lymph system to drain interstitial fluid, resulting in pulmonary edema. Treatment of CHF includes increasing urinary output, which brings yet another organ system into the picture. By current estimates, about 5 million Americans suffer from CHF. To learn more about this condition, visit the American Heart Association web site (*www.americanheart.org*) or MedlinePlus, published by the National Institutes of Health (*www.nlm.nih.gov/medlineplus/heartfailure.html*).

each lung [Fig. 14.1, p. 465]. Oxygenated blood from the lungs returns to the left atrium via the pulmonary veins.

At any given moment, the pulmonary circulation contains about 0.5 liter of blood, or 10% of total blood volume. About 75 mL of this amount is found in the capillaries, where gas exchange takes place, with the remainder in pulmonary arteries and veins. The rate of blood flow through the lungs is much higher than the rate in other tissues [p. 466] because the lungs receive the entire cardiac output of the right ventricle: 5 L/min. This means that as much blood flows through the lungs in one minute as flows through the entire rest of the body in the same amount of time!

Despite the high flow rate, pulmonary blood pressure is low. Pulmonary arterial pressure averages 25/8 mm Hg, much lower than the average systemic pressure of 120/80 mm Hg. The right ventricle does not have to pump as forcefully to create blood flow through the lungs because resistance of the pulmonary circulation is low. This low resistance can be attributed to the shorter total length of pulmonary blood vessels and to the distensibility and large total cross-sectional area of pulmonary arterioles.

Normally, the net hydrostatic pressure filtering fluid out of a pulmonary capillary into the interstitial space is low because of low mean blood pressure [p. 529]. The lymphatic system efficiently removes filtered fluid, and lung interstitial fluid volume is usually minimal. As a result, the distance between the alveolar air space and the capillary endothelium is short, and gases diffuse rapidly between them.

Concept Check Answers: p. 598

8. Is blood flow through the pulmonary trunk greater than, less than, or equal to blood flow through the aorta?
9. A person has left ventricular failure but normal right ventricular function. As a result, blood pools in the pulmonary circulation, doubling pulmonary capillary hydrostatic pressure. What happens to net fluid flow across the walls of the pulmonary capillaries?
10. Calculate the mean pressure in a person whose pulmonary arterial pressure is 25/8 mm Hg. [p. 515]

Gas Laws

Respiratory air flow is very similar in many respects to blood flow in the cardiovascular system because both air and blood are fluids. Their primary difference is that blood is a non-compressible liquid but air is a compressible mixture of gases. ■ Figure 17.6 summarizes the laws that govern the behavior of gases in air and provide the basis for the exchange of air between the external environment and the alveoli. We will consider the gas laws that govern the solubility of gases in solution when we talk about oxygen transport in blood.

In this book, blood pressure and environmental air pressure (**atmospheric pressure**) are both reported in millimeters of mercury (mm Hg). Respiratory physiologists sometimes report gas pressures in centimeters of water instead, where 1 mm Hg = 1.36 cm H_2O, or in kiloPascals (kPa), where 760 mm Hg = 101.325 kPa.

At sea level, normal atmospheric pressure is 760 mm Hg. However, in this book we follow the convention of designating atmospheric pressure as 0 mm Hg. Because atmospheric pressure varies with altitude and because very few people live exactly at sea level, this convention allows us to compare pressure differences that occur during ventilation without correcting for altitude.

RUNNING PROBLEM

Edna has not been able to stop smoking, and her COPD is a combination of emphysema and bronchitis. Patients with chronic bronchitis have excessive mucus production and exhibit general inflammation of the entire respiratory tract. The mucus narrows the airways and makes breathing difficult.

Q1: What does narrowing of the airways do to airway resistance? [Hint: Poiseuille's law, p. 469]

569 576 582 584 587 594

Gas Laws

This figure summarizes the rules that govern the behavior of gases in air. These rules provide the basis for the exchange of air between the external environment and the alveoli.

(a) The ideal gas equation

$PV = nRT$ Where P is pressure, V is volume, n is the moles of gas, T is absolute temperature, and R is the universal gas constant, 8.3145 j/mol × K

In the human body we can assume that the number of moles and temperature are constant. Removing the constants leaves the following equation:

$V = 1/P$ This relationship says that if the volume of gas increases, the pressure decreases, and vice versa.

(b) Boyle's Law

Boyle's law also expresses this inverse relationship between pressure and volume.

$P_1V_1 = P_2V_2$ For example, the container on the left is 1 L (V_1) and has a pressure of 100 mm Hg (P_1).

What happens to the pressure when the volume decreases to 0.5 L?

100 mm Hg × 1 L = P_2 × 0.5 L

200 mm Hg = P_2

The pressure has increased ×2.

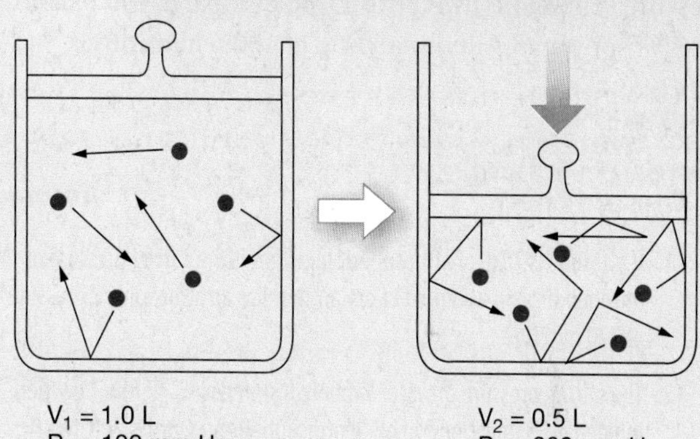

The Ideal Gas law and Boyle's law apply to all gases or mixtures of gases.

(c) Dalton's Law

Dalton's law says that the total pressure of a mixture of gases is the sum of the pressures of the individual gases. The pressure of an individual gas in a mixture is known as the **partial pressure** of the gas (P_{gas}).

For example, at sea level, atmospheric pressure (P_{atm}) is 760 mm Hg, and oxygen is 21% of the atmosphere. What is the partial pressure of oxygen (P_{O_2})?

To find the partial pressure of any one gas in a sample of dry air, multiply the atmospheric pressure (P_{atm}) by the gas's relative contribution (%) to P_{atm}:

Partial pressure of a gas = P_{atm} × % of gas in atmosphere

P_{O_2} = 760 mm Hg x 21% P_{O_2}

= 760 mm x 0.21 = 160 mm Hg

The partial pressure of oxygen (P_{O_2}) in dry air at sea level is 160 mm Hg.

The pressure exerted by an individual gas is determined only by its relative abundance in the mixture and is independent of the molecular size or mass of the gas.

In humid air, water vapor "dilutes" the contribution of other gases to the total pressure.

Partial Pressures (P_{gas}) of Atmospheric Gases at 760 mm Hg

Gas and its percentage in air	P_{gas} in dry 25 °C air	P_{gas} in 25 °C air, 100% humidity	P_{gas} in 37 °C air, 100% humidity
Oxygen (O_2) 21%	160 mm Hg	155 mm Hg	150 mm Hg
Carbon dioxide (CO_2) 0.03%	0.25 mm Hg	0.24 mm Hg	0.235 mm Hg
Water vapor	0 mm Hg	24 mm Hg	47 mm Hg

To calculate the partial pressure of a gas in humid air, you must first subtract the water vapor pressure from the total pressure. At 100% humidity and 25° C, water vapor pressure (P_{H_2O}) is 24 mm Hg.

P_{gas} in humid air = (P_{atm} – P_{H_2O}) × % of gas

P_{O_2} = (760 – 24) x 21% = 155 mm Hg

Air Is a Mixture of Gases

The atmosphere surrounding the earth is a mixture of gases and water vapor. **Dalton's law** states that the total pressure exerted by a mixture of gases is the sum of the pressures exerted by the individual gases (Fig. 17.6c). For example, in dry air at an atmospheric pressure of 760 mm Hg, 78% of the total pressure is due to N_2, 21% to O_2, and so on.

In respiratory physiology, we are concerned not only with total atmospheric pressure but also with the individual pressures of oxygen and carbon dioxide. The pressure of a single gas in a mixture is known as its **partial pressure** (P_{gas}). The pressure exerted by an individual gas is determined only by its relative abundance in the mixture and is independent of the molecular size or mass of the gas.

The partial pressures of gases in air vary slightly depending on how much water vapor is in the air because the pressure of water vapor "dilutes" the contribution of other gases to the total pressure. The table in Figure 17.6c compares the partial pressures of some gases in dry air and at 100% humidity.

Concept Check Answers: p. 598

11. If nitrogen is 78% of atmospheric air, what is the partial pressure of nitrogen (P_{N_2}) in a sample of dry air that has an atmospheric pressure of 720 mm Hg?
12. The partial pressure of water vapor in inspired air is 47 mm Hg when inhaled air is fully humidified. If atmospheric pressure is 700 mm Hg and oxygen is 21% of the atmosphere at 0% humidity, what is the P_{O_2} of fully humidified air?

Gases Move Down Pressure Gradients

Air flow occurs whenever there is a pressure gradient. Bulk flow of air, like blood flow, is directed from areas of higher pressure to areas of lower pressure. Meteorologists predict the weather by knowing that areas of high atmospheric pressure move in to replace areas of low pressure. In ventilation, bulk flow of air down pressure gradients explains how air is exchanged between the external environment and the lungs. Movement of the thorax during breathing creates alternating conditions of high and low pressure in the lungs.

Diffusion of gases down concentration (partial pressure) gradients applies to single gases. For example, oxygen moves from areas of higher oxygen partial pressure (P_{O_2}) to areas of lower oxygen partial pressure. Diffusion of individual gases is important in the exchange of oxygen and carbon dioxide between alveoli and blood and from blood to cells.

Boyle's Law Describes Pressure-Volume Relationships

The pressure exerted by a gas or mixture of gases in a sealed container is created by the collisions of moving gas molecules with the walls of the container and with each other. If the size of the container is reduced, the collisions between the gas molecules and the walls become more frequent, and the pressure rises (Fig. 17.6b). This relationship between pressure and volume was first noted by Robert Boyle in the 1600s and can be expressed by the equation of **Boyle's law** of gases:

$$P_1V_1 = P_2V_2$$

where P represents pressure and V represents volume.

Boyle's law states that if the volume of a gas is reduced, the pressure increases. If the volume increases, the pressure decreases.

In the respiratory system, changes in the volume of the chest cavity during ventilation cause pressure gradients that create air flow. When chest volume increases, alveolar pressure falls, and air flows into the respiratory system. When the chest volume decreases, alveolar pressure increases, and air flows out into the atmosphere. This movement of air is bulk flow because the entire gas mixture is moving rather than merely one or two of the gases in the air.

Ventilation

This bulk flow exchange of air between the atmosphere and the alveoli is ventilation, or breathing (Fig. 17.1). A single **respiratory cycle** consists of an inspiration followed by an expiration.

Lung Volumes Change During Ventilation

Physiologists and clinicians assess a person's pulmonary function by measuring how much air the person moves during quiet breathing, then with maximum effort. These **pulmonary function tests** use a **spirometer**, an instrument that measures the volume of air moved with each breath (■ Fig. 17.7a). (Most spirometers in clinical use today are small computerized machines rather than the traditional wet spirometer illustrated here.)

When a subject is attached to the traditional spirometer through a mouthpiece and the subject's nose is clipped closed, the subject's respiratory tract and the spirometer form a closed system. When the subject breathes in, air moves from the spirometer into the lungs, and the recording pen, which traces a graph on a rotating cylinder, moves up. When the subject exhales, air moves from the lungs back into the spirometer, and the pen moves down.

Lung Volumes The air moved during breathing can be divided into four lung volumes: (1) tidal volume, (2) inspiratory reserve

PULMONARY FUNCTION TESTS

(a) The Spirometer

This figure shows a traditional wet spirometer. The subject inserts a mouthpiece that is attached to an inverted bell filled with air or oxygen. The volume of the bell and the volume of the subject's respiratory tract create a closed system because the bell is suspended in water.

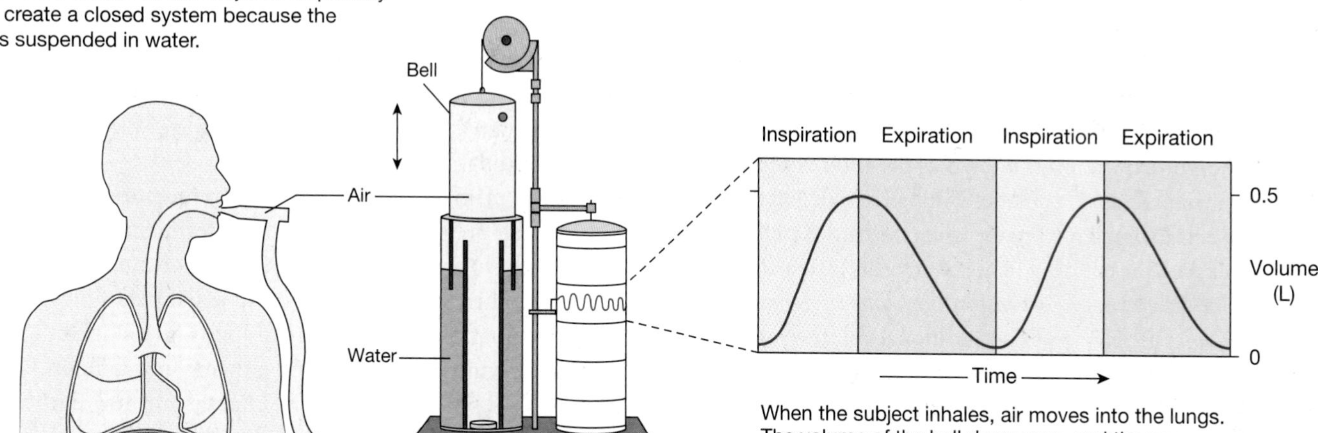

When the subject inhales, air moves into the lungs. The volume of the bell decreases, and the pen rises on the tracing.

(b) Lung Volumes and Capacities

The four lung volumes

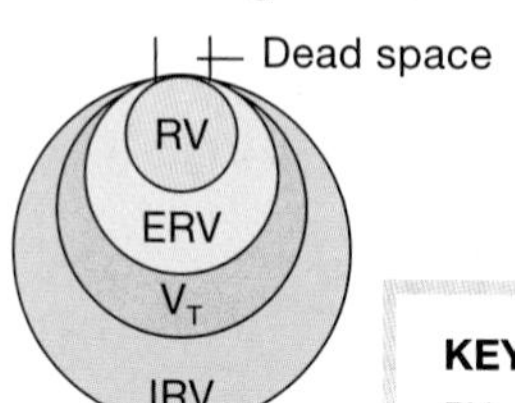

KEY

RV = Residual volume
ERV = Expiratory reserve volume
V_T = Tidal volume
IRV = Inspiratory reserve volume

Capacities are sums of 2 or more volumes.

Inspiratory capacity = V_T + IRV
Vital capacity = V_T + IRV + ERV
Total lung capacity = V_T + IRV + ERV + RV
Functional residual capacity = ERV + RV

Pulmonary Volumes and Capacities*

		Males	Females
Vital capacity	IRV	3000	1900
	V_T	500	500
	ERV	1100	700
Residual volume		1200	1100
Total lung capacity		5800 mL	4200 mL

A spirometer tracing showing lung volumes and capacities.

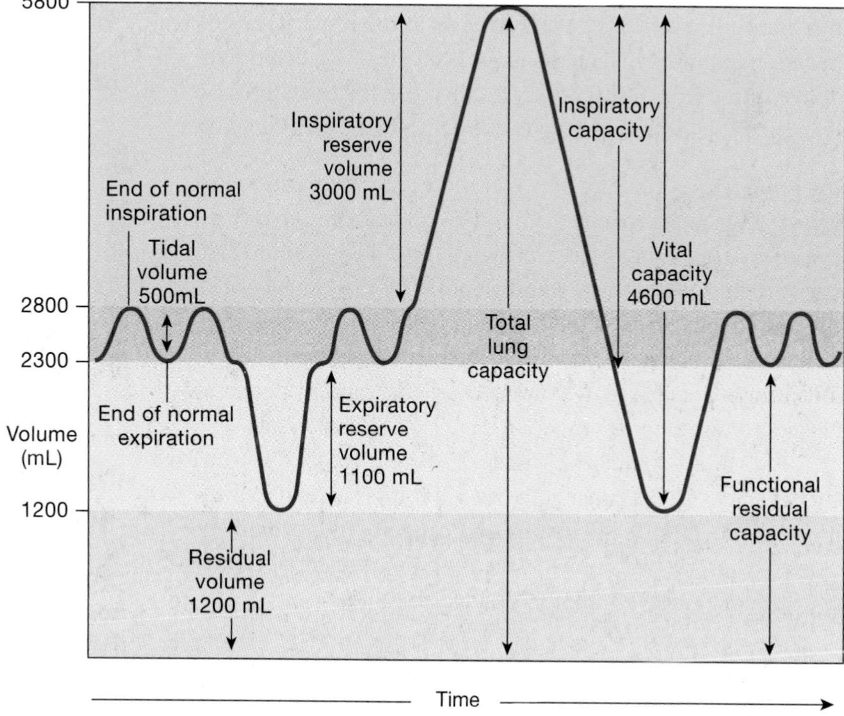

*Pulmonary volumes are given for a normal 70-kg man or a 50-kg woman, 28 years old.

Fig. 17.7

volume, (3) expiratory reserve volume, and (4) residual volume. The numerical values used on the graph in Figure 17.7b represent average volumes for a 70-kg man. The volumes for women are typically less, as shown in Figure 17.7b. Lung volumes vary considerably with age, sex, height, and weight, so clinicians use algorithms based on those parameters to predict lung volumes. (An *algorithm* is an equation or series of steps used to solve a problem.)

Each of the following paragraphs begins with the instructions you would be given if you were being tested for these volumes.

"Breathe quietly." The volume of air that moves during a single inspiration or expiration is known as the **tidal volume** (V_T). Average tidal volume during quiet breathing is about 500 mL. (It is hard for subjects to breathe normally when they are thinking about their breathing, so the clinician may not give this instruction.)

"Now, at the end of a quiet inspiration, take in as much additional air as you possibly can." The additional volume you inspire above the tidal volume represents your **inspiratory reserve volume** (IRV). In a 70-kg man, this volume is about 3000 mL, a sixfold increase over the normal tidal volume.

"Now stop at the end of a normal exhalation, then exhale as much air as you possibly can." The amount of air forcefully exhaled after the end of a normal expiration is the **expiratory reserve volume** (ERV), which averages about 1100 mL.

The fourth volume cannot be measured directly. Even if you blow out as much air as you can, air still remains in the lungs and the airways. The volume of air in the respiratory system after maximal exhalation—about 1200 mL—is called the **residual volume** (RV). Most of this residual volume exists because the lungs are held stretched against the ribs by the pleural fluid.

Lung Capacities The sum of two or more lung volumes is called a capacity. The **vital capacity** (VC) is the sum of the inspiratory reserve volume, expiratory reserve volume, and tidal volume. Vital capacity represents the maximum amount of air that can be voluntarily moved into or out of the respiratory system with one breath. To measure vital capacity, you would instruct the person to take in as much air as possible, then blow it all out. Vital capacity decreases with age as muscles weaken and the lungs become less elastic.

Vital capacity plus the residual volume yields the **total lung capacity** (TLC). Other capacities of importance in pulmonary medicine include the **inspiratory capacity** (tidal volume + inspiratory reserve volume) and the **functional residual capacity** (expiratory reserve volume + residual volume).

Concept Check Answers: p. 598

13. How are lung volumes related to lung capacities?
14. Which lung volume cannot be measured directly?
15. If vital capacity decreases with age but total lung capacity does not change, which lung volume must be changing? In which direction?
16. As inhaled air becomes humidified passing down the airways, what happens to the P_{O_2} of the air?

During Ventilation, Air Flows Because of Pressure Gradients

Breathing is an active process that requires muscle contraction. Air flows into the lungs because of pressure gradients created by a pump, just as blood flows because of the pumping action of the heart. In the respiratory system, muscles of the thoracic cage and diaphragm function as the pump because most lung tissue is thin exchange epithelium. When these muscles contract, the lungs expand, held to the inside of the chest wall by the pleural fluid.

The primary muscles involved in quiet breathing (breathing at rest) are the diaphragm, the external intercostals, and the scalenes. During forced breathing, other muscles of the chest and abdomen may be recruited to assist. Examples of physiological situations in which breathing is forced include exercise, playing a wind instrument, and blowing up a balloon.

As we noted earlier in the chapter, air flow in the respiratory tract obeys the same rule as blood flow:

$$\text{Flow} \propto \Delta P/R$$

This equation means that (1) air flows in response to a pressure gradient (ΔP) and (2) flow decreases as the resistance (R) of the system to flow increases. Before we discuss resistance, let's consider how the respiratory system creates a pressure gradient. The pressure-volume relationships of Boyle's law provide the basis for pulmonary ventilation.

Concept Check Answers: p. 598

17. Compare the direction of air movement during one respiratory cycle with the direction of blood flow during one cardiac cycle.
18. Explain the relationship between the lungs, the pleural membranes, the pleural fluid, and the thoracic cage.

Inspiration Occurs When Alveolar Pressure Decreases

For air to move into the alveoli, pressure inside the lungs must become lower than atmospheric pressure. According to Boyle's law, an increase in volume will create a decrease in pressure. During inspiration, thoracic volume increases when certain skeletal muscles of the rib cage and diaphragm contract.

When the diaphragm contracts, it drops down toward the abdomen. In quiet breathing, the diaphragm moves about 1.5 cm, increasing thoracic volume (■ Fig. 17.8b). Contraction of the diaphragm causes between 60% and 75% of the inspiratory volume change during normal quiet breathing.

Movement of the rib cage creates the remaining 25–40% of the volume change. During inhalation, the external intercostal

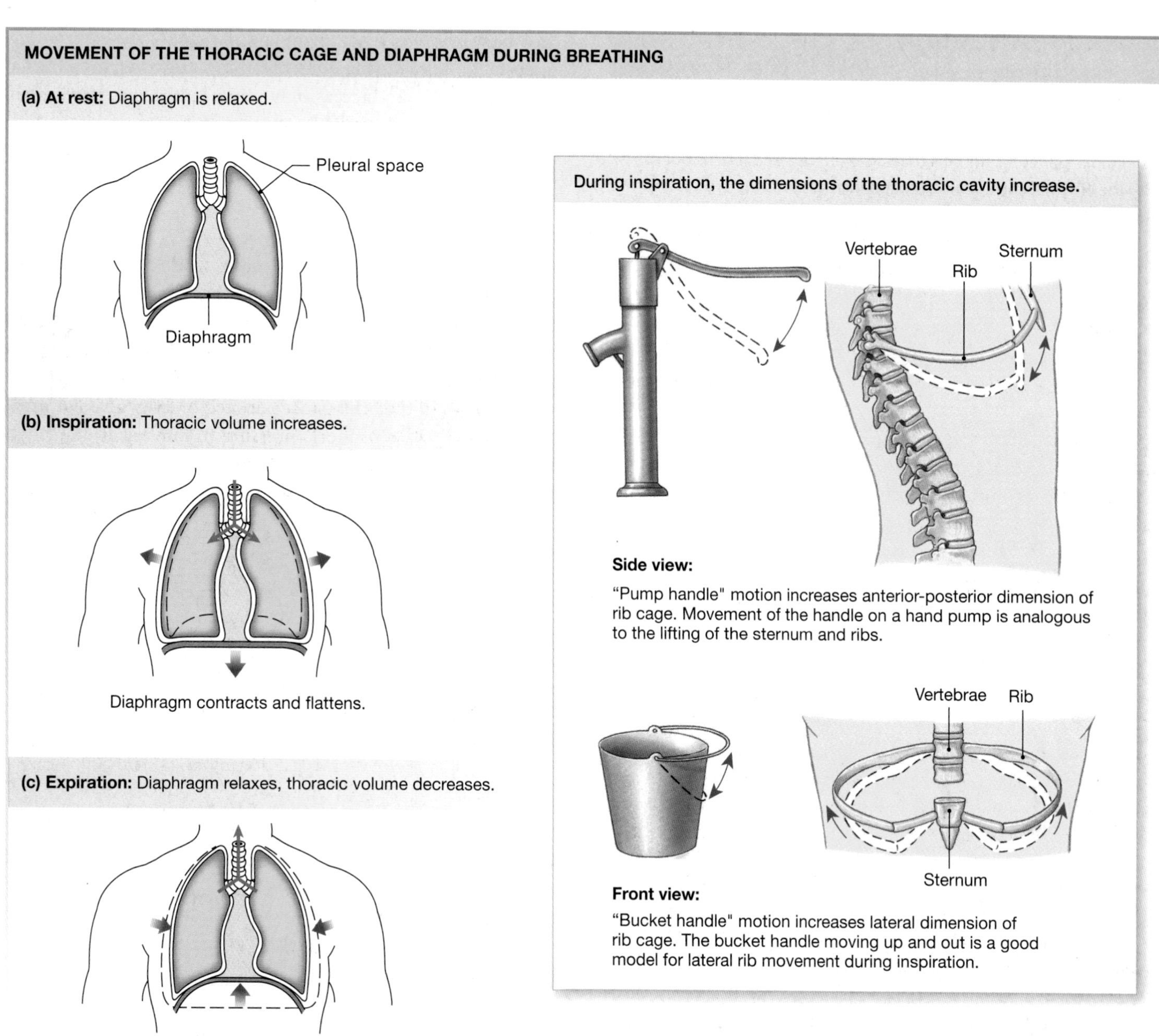
MOVEMENT OF THE THORACIC CAGE AND DIAPHRAGM DURING BREATHING
(a) At rest: Diaphragm is relaxed.
Pleural space
Diaphragm
(b) Inspiration: Thoracic volume increases.
Diaphragm contracts and flattens.
(c) Expiration: Diaphragm relaxes, thoracic volume decreases.
During inspiration, the dimensions of the thoracic cavity increase.
Vertebrae
Rib
Sternum
Side view:
"Pump handle" motion increases anterior-posterior dimension of rib cage. Movement of the handle on a hand pump is analogous to the lifting of the sternum and ribs.
Vertebrae
Rib
Sternum
Front view:
"Bucket handle" motion increases lateral dimension of rib cage. The bucket handle moving up and out is a good model for lateral rib movement during inspiration.

Fig. 17.8

and scalene muscles (see Fig. 17.2a) contract and pull the ribs upward and out (Fig. 17.8b). Rib movement during inspiration has been likened to a pump handle lifting up and away from the pump (the ribs moving up and away from the spine) and to the movement of a bucket handle as it lifts away from the side of a bucket (ribs moving outward in a lateral direction). The combination of these two movements broadens the rib cage in all directions. As thoracic volume increases, pressure decreases, and air flows into the lungs.

For many years, quiet breathing was attributed solely to the action of the diaphragm and the external intercostal muscles. It was thought that the scalenes and sternocleidomastoid muscles were active only during deep breathing. In recent years, however, studies have changed our understanding of how these accessory muscles contribute to quiet breathing.

If an individual's scalenes are paralyzed, inspiration is achieved primarily by contraction of the diaphragm. Observation of patients with neuromuscular disorders has revealed that although the contracting diaphragm increases thoracic volume by moving toward the abdominal cavity, it also tends to pull the lower ribs inward, working against inspiration. In normal individuals, we know that the lower ribs move up and out during inspiration rather than inward. The fact that there is no up-and-out rib motion in patients with paralyzed scalenes tells us that normally the scalenes must be contributing to inspiration by lifting the sternum and upper ribs.

RUNNING PROBLEM

Edna's COPD began with chronic bronchitis and a morning cough that produced lots of mucus (*phlegm*). Cigarette smoke paralyzes the cilia that sweep debris and mucus out of the airways, and smoke irritation increases mucus production in the airway. Without functional cilia, mucus and debris pool in the airways, leading to a chronic cough. Eventually, smokers may begin to develop emphysema in addition to their bronchitis.

Q2: Why do people with chronic bronchitis have a higher-than-normal rate of respiratory infections?

New evidence also downplays the role of the external intercostal muscles during quiet breathing. However, the external intercostals play an increasingly important role as respiratory activity increases. Because the exact contribution of external intercostals and scalenes varies depending on the type of breathing, we group these muscles together and simply call them the *inspiratory muscles*.

Now let's see how alveolar pressure (P_A) changes during a single inspiration. Follow the graphs in ■ Figure 17.9 as you read through the process. Remember that atmospheric pressure is assigned a value of 0 mm Hg. Negative numbers designate subatmospheric pressures, and positive numbers denote higher-than-atmospheric pressures.

Time 0. In the brief pause between breaths, alveolar pressure is equal to atmospheric pressure (0 mm Hg at point A_1). When pressures are equal, there is no air flow.

PRESSURE CHANGES DURING QUIET BREATHING

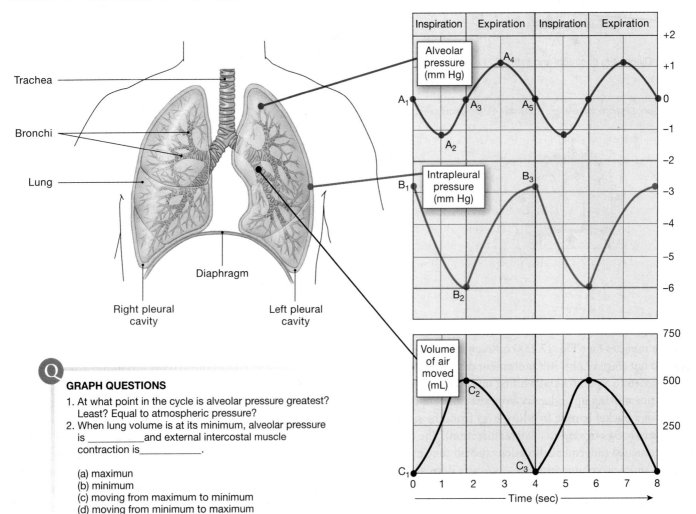

GRAPH QUESTIONS

1. At what point in the cycle is alveolar pressure greatest? Least? Equal to atmospheric pressure?
2. When lung volume is at its minimum, alveolar pressure is ___________and external intercostal muscle contraction is___________.

   (a) maximun
   (b) minimum
   (c) moving from maximum to minimum
   (d) moving from minimum to maximum

3. What is this person's ventilation rate?

Normally expiration takes 2–3 times longer than inspiration (not shown to scale on this idealized graph).

■ **Fig. 17.9**

Time 0–2 sec: Inspiration. As inspiration begins, inspiratory muscles contract, and thoracic volume increases. With the increase in volume, alveolar pressure falls about 1 mm Hg below atmospheric pressure (−1 mm Hg, point A_2), and air flows into the alveoli (point C_1 to point C_2). Because the thoracic volume changes faster than air can flow, alveolar pressure reaches its lowest value about halfway through inspiration (point A_2).

As air continues to flow into the alveoli, pressure increases until the thoracic cage stops expanding, just before the end of inspiration. Air movement continues for a fraction of a second longer, until pressure inside the lungs equalizes with atmospheric pressure (point A_3). At the end of inspiration, lung volume is at its maximum for the respiratory cycle (point C_2), and alveolar pressure is equal to atmospheric pressure.

You can demonstrate this phenomenon by taking a deep breath and stopping the movement of your chest at the end of inspiration. (Do not "hold your breath" because doing so closes the opening of the pharynx and prevents air flow.) If you do this correctly, you notice that air flow stops after you freeze the inspiratory movement. This exercise shows that at the end of inspiration, alveolar pressure is equal to atmospheric pressure.

Expiration Occurs When Alveolar Pressure Increases

At the end of inspiration, impulses from somatic motor neurons to the inspiratory muscles cease, and the muscles relax. Elastic recoil of the lungs and thoracic cage returns the diaphragm and rib cage to their original relaxed positions, just as a stretched elastic waistband recoils when released. Because expiration during quiet breathing involves passive elastic recoil rather than active muscle contraction, it is called passive expiration.

Time 2–4 sec: expiration. As lung and thoracic volumes decrease during expiration, air pressure in the lungs increases, reaching a maximum of about 1 mm Hg above atmospheric pressure (Fig. 17.9, point A_4). Alveolar pressure is now higher than atmospheric pressure, so air flow reverses and air moves out of the lungs.

Time 4 sec. At the end of expiration, air movement ceases when alveolar pressure is again equal to atmospheric pressure (point A_5). Lung volume reaches its minimum for the respiratory cycle (point C_3). At this point, the respiratory cycle has ended and is ready to begin again with the next breath.

The pressure differences shown in Figure 17.9 apply to quiet breathing. During exercise or forced heavy breathing, these values become proportionately larger. **Active expiration** occurs during voluntary exhalations and when ventilation exceeds 30–40 breaths per minute. (Normal resting ventilation rate is 12–20 breaths per minute for an adult.) Active expiration uses the internal intercostal muscles and the abdominal muscles (see Fig. 17.2a), which are not used during inspiration. These muscles are collectively called the *expiratory muscles*.

The internal intercostal muscles line the inside of the rib cage. When they contract, they pull the ribs inward, reducing the volume of the thoracic cavity. To feel this action, place your hands on your rib cage. Forcefully blow as much air out of your lungs as you can, noting the movement of your hands as you do so.

The internal and external intercostals function as antagonistic muscle groups [p. 400] to alter the position and volume of the rib cage during ventilation. The diaphragm, however, has no antagonistic muscles. Instead, abdominal muscles contract during active expiration to supplement the activity of the internal intercostals. Abdominal contraction pulls the lower rib cage inward and decreases abdominal volume, actions that displace the intestines and liver upward. The displaced viscera push the diaphragm up into the thoracic cavity and passively decrease chest volume even more. The action of abdominal muscles during forced expiration is why aerobics instructors tell you to blow air out as you lift your head and shoulders during abdominal "crunches." The active process of blowing air out helps contract the abdominals, the muscles you are trying to strengthen.

17

Any neuromuscular disease that weakens skeletal muscles or damages their motor neurons can adversely affect ventilation. With decreased ventilation, less fresh air enters the lungs. In addition, loss of the ability to cough increases the risk of pneumonia and other infections. Examples of diseases that affect the motor control of ventilation include *myasthenia gravis* [p. 269], an illness in which acetylcholine receptors of the motor end plates of skeletal muscles are destroyed, and *polio* (poliomyelitis), a viral illness that paralyzes skeletal muscles.

Concept Check

Answers: p. 598

19. Scarlett O'Hara is trying to squeeze herself into a corset with an 18-inch waist. Will she be more successful by taking a deep breath and holding it or by blowing all the air out of her lungs? Why?
20. Why would loss of the ability to cough increase the risk of respiratory infections? (*Hint:* What does coughing do to mucus in the airways?)

Intrapleural Pressure Changes During Ventilation

Ventilation requires that the lungs, which are unable to expand and contract on their own, move in association with the expansion and relaxation of the thorax. As we noted earlier in this chapter, the lungs are enclosed in the fluid-filled pleural sac. The surface of the lungs is covered by the *visceral pleura*, and the portion of the sac that lines the thoracic cavity is called the *parietal pleura* {*paries*, wall}. Cohesive forces exerted by the fluid between the two pleural membranes cause the stretchable lung to adhere to the thoracic cage. When the thoracic cage moves during breathing, the lungs move with it.

RUNNING PROBLEM

Emphysema is characterized by a loss of elastin, the elastic fibers that help the alveoli recoil during expiration. Elastin is destroyed by *elastase*, an enzyme released by alveolar macrophages, which must work overtime in smokers to rid the lungs of irritants. People with emphysema have more difficulty exhaling than inhaling. Their alveoli have lost elastic recoil, which makes expiration—normally a passive process—require conscious effort.

Q3: Name the muscles that patients with emphysema use to exhale actively.

569 576 582 **584** 587 594

The intrapleural pressure in the fluid between the pleural membranes is normally subatmospheric. This subatmospheric pressure arises during fetal development, when the thoracic cage with its associated pleural membrane grows more rapidly than the lung with its associated pleural membrane. The two pleural membranes are held together by the pleural fluid bond, so the elastic lungs are forced to stretch to conform to the larger volume of the thoracic cavity. At the same time, however, elastic recoil of the lungs creates an inwardly directed force that tries to pull the lungs away from the chest wall (■ Fig. 17.10a). The combination of the outward pull of the thoracic cage and inward recoil of the elastic lungs creates a subatmospheric intrapleural pressure of about −3 mm Hg.

You can create a similar situation by half-filling a syringe with water and capping it with a plugged-up needle. At this point, the pressure inside the barrel is equal to atmospheric pressure. Now hold the syringe barrel (the chest wall) in one hand while you try to withdraw the plunger (the elastic lung pulling away from the chest wall). As you pull on the plunger, the volume inside the barrel increases slightly, but the cohesive forces between the water molecules cause the water to resist expansion. The pressure in the barrel, which was initially equal to atmospheric pressure, decreases slightly as you pull on the plunger. If you release the plunger, it snaps back to its resting position, restoring atmospheric pressure inside the syringe.

What happens to subatmospheric intrapleural pressure if an opening is made between the sealed pleural cavity and the atmosphere? A knife thrust between the ribs, a broken rib that punctures the pleural membrane, or any other event that opens

SUBATMOSPHERIC PRESSURE IN THE PLEURAL CAVITY HELPS KEEP THE LUNGS INFLATED

(a) In the normal lung at rest, pleural fluid keeps the lung adhered to the chest wall.

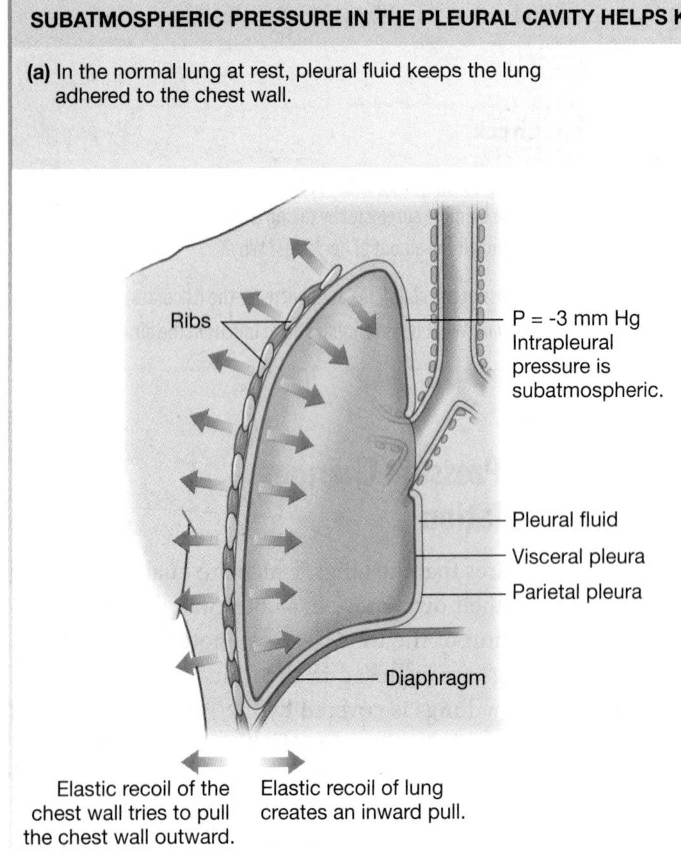
Ribs
P = -3 mm Hg
Intrapleural pressure is subatmospheric.
Pleural fluid
Visceral pleura
Parietal pleura
Diaphragm
Elastic recoil of the chest wall tries to pull the chest wall outward.
Elastic recoil of lung creates an inward pull.

(b) Pneumothorax. If the sealed pleural cavity is opened to the atmosphere, air flows in. The bond holding the lung to the chest wall is broken, and the lung collapses, creating a pneumothorax (air in the thorax).

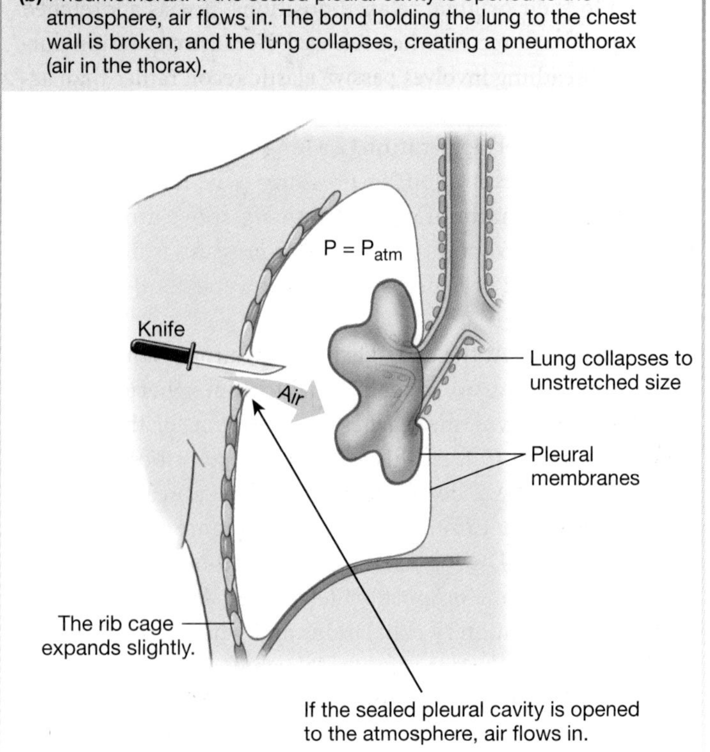
P = P_{atm}
Knife
Air
Lung collapses to unstretched size
Pleural membranes
The rib cage expands slightly.
If the sealed pleural cavity is opened to the atmosphere, air flows in.

■ **Fig. 17.10**

the pleural cavity to the atmosphere allows air to flow down its pressure gradient into the cavity, just as air enters when you break the seal on a vacuum-packed can.

Air in the pleural cavity breaks the fluid bond holding the lung to the chest wall. The chest wall expands outward while the elastic lung collapses to an unstretched state, like a deflated balloon (Fig. 17.10b). This condition, called **pneumothorax** {*pneuma*, air + *thorax*, chest}, results in a collapsed lung that is unable to function normally. Pneumothorax can also occur spontaneously if a congenital *bleb*, a weakened section of lung tissue, ruptures, allowing air from inside the lung to enter the pleural cavity.

Correction of a pneumothorax has two components: removing as much air from the pleural cavity as possible with a suction pump, and sealing the hole to prevent more air from entering. Any air remaining in the cavity is gradually absorbed into the blood, restoring the pleural fluid bond and reinflating the lung.

Pressures in the pleural fluid vary during a respiratory cycle. At the beginning of inspiration, intrapleural pressure is about −3 mm Hg (Fig. 17.9, point B1). As inspiration proceeds, the pleural membranes and lungs follow the expanding thoracic cage because of the pleural fluid bond, but the elastic lung tissue resists being stretched. The lungs attempt to pull farther away from the chest wall, causing the intrapleural pressure to become even more negative (Fig. 17.9, point B_2).

Because this process is difficult to visualize, let's return to the analogy of the water-filled syringe with the plugged-up needle. You can pull the plunger out a small distance without much effort, but the cohesiveness of the water makes it difficult to pull the plunger out any farther. The increased amount of work you do trying to pull the plunger out is paralleled by the work your inspiratory muscles must do when they contract during inspiration. The bigger the breath, the more work is required to stretch the elastic lung.

By the end of a quiet inspiration, when the lungs are fully expanded, intrapleural pressure falls to around −6 mm Hg (Fig. 17.9, point B_2). During exercise or other powerful inspirations, intrapleural pressure may reach −8 mm Hg or lower.

During expiration, the thoracic cage returns to its resting position. The lungs are released from their stretched position, and the intrapleural pressure returns to its normal value of about −3 mm Hg (point B_3). Notice that intrapleural pressure never equilibrates with atmospheric pressure because the pleural cavity is a closed compartment.

Pressure gradients required for air flow are created by the work of skeletal muscle contraction. Normally, about 3–5% of the body's energy expenditure is used for quiet breathing. During exercise, the energy required for breathing increases substantially. The two factors that have the greatest influence on the amount of work needed for breathing are the stretchability of the lungs and the resistance of the airways to air flow.

Concept Check Answers: p. 598

21. A person has periodic spastic contractions of the diaphragm, otherwise known as hiccups. What happens to intrapleural and alveolar pressures when a person hiccups?
22. A stabbing victim is brought to the emergency room with a knife wound between the ribs on the left side of his chest. What has probably happened to his left lung? To his right lung? Why does the left side of his rib cage seem larger than the right side?

Lung Compliance and Elastance May Change in Disease States

Adequate ventilation depends on the ability of the lungs to expand normally. Most of the work of breathing goes into overcoming the resistance of the elastic lungs and the thoracic cage to stretching. Clinically, the ability of the lung to stretch is called **compliance**.

Compliance refers to the amount of force that must be exerted in a body to deform it. In the lung, we can express compliance as the change of volume (V) that results from a given force or pressure (P) exerted on the lung: $\Delta V/\Delta P$. A high-compliance lung stretches easily, just as a compliant person is easy to persuade. A low-compliance lung requires more force from the inspiratory muscles to stretch it.

Compliance is the reciprocal of **elastance** (elastic recoil), the ability to resist being deformed. Elastance also refers to the ability of a body to return to its original shape when a deforming force is removed. A lung that stretches easily (high compliance) has probably lost its elastic tissue and will not return to its resting volume when the stretching force is released (low elastance). You may have experienced something like this with old gym shorts. After many washings the elastic waistband is easy to stretch (high compliance) but lacking in elastance, making it impossible for the shorts to stay up around your waist.

Analogous problems occur in the respiratory system. For example, as noted in the Running Problem, emphysema is a disease in which elastin fibers normally found in lung tissue are destroyed. Destruction of elastin results in lungs that exhibit high compliance and stretch easily during inspiration. However, these lungs also have decreased elastance, so they do not recoil to their resting position during expiration.

To understand the importance of elastic recoil to expiration, think of an inflated balloon and an inflated plastic bag. The balloon is similar to the normal lung. Its elastic walls squeeze on the air inside the balloon, thereby increasing the internal air pressure. When the neck of the balloon is opened to the atmosphere, elastic recoil causes air to flow out of the balloon. The inflated plastic bag, on the other hand, is like the lung of an individual with emphysema. It has high compliance and is easily inflated, but it has little elastic recoil. If the inflated plastic bag is opened to the atmosphere, most of the air remains inside the bag.

A decrease in lung compliance affects ventilation because more work must be expended to stretch a stiff lung. Pathological conditions in which compliance is reduced are called **restrictive lung diseases**. In these conditions, the energy expenditure required to stretch less-compliant lungs can far exceed the normal work of breathing. Two common causes of decreased compliance are inelastic scar tissue formed in *fibrotic lung diseases,* and inadequate alveolar production of surfactant, a chemical that facilitates lung expansion.

Pulmonary **fibrosis** is characterized by the development of stiff, fibrous scar tissue that restricts lung inflation. In *idiopathic* pulmonary fibrosis {*idios*, one's own}, the cause is unknown. Other forms of fibrotic lung disease result from chronic inhalation of fine particulate matter, such as asbestos and silicon, that escapes the mucus lining the airways and reaches the alveoli. Wandering alveolar macrophages (see Fig. 17.2g) then ingest the inhaled particulate matter. If the particles are organic, the macrophages can digest them with lysosomal enzymes. However, if the particles cannot be digested or if they accumulate in large numbers, an inflammatory process ensues. The macrophages then secrete growth factors that stimulate fibroblasts in the lung's connective tissue to produce inelastic collagen. Pulmonary fibrosis cannot be reversed.

Surfactant Decreases the Work of Breathing

For years, physiologists assumed that elastin and other elastic fibers were the primary source of resistance to stretch in the lung. However, studies comparing the work required to expand air-filled and saline-filled lungs showed that air-filled lungs are much harder to inflate. From this result, researchers concluded that lung tissue itself contributes less to resistance than once thought. Some other property of the normal air-filled lung, a property not present in the saline-filled lung, must create most of the resistance to stretch.

This property is the surface tension [p. 39] created by the thin fluid layer between the alveolar cells and the air. At any air-fluid interface, the surface of the fluid is under tension, like a thin membrane being stretched. When the fluid is water, surface tension arises because of the hydrogen bonds between water molecules. The water molecules on the fluid's surface are attracted to other water molecules beside and beneath them but are not attracted to gases in the air at the air-fluid interface.

Alveolar surface tension is similar to the surface tension that exists in a spherical bubble, even though alveoli are not perfect spheres. The surface tension created by the thin film of fluid is directed toward the center of the bubble and creates pressure in the interior of the bubble. The **law of LaPlace** is an expression of this pressure. It states that the pressure (P) inside a bubble formed by a fluid film is a function of two factors: the surface tension of the fluid (T) and the radius of the bubble (r). This relationship is expressed by the equation

$$P = 2T/r$$

Notice in ■ Figure 17.11a that if two bubbles have different diameters but are formed by fluids with the same surface tension, the pressure inside the smaller bubble is greater than that inside the larger bubble.

How does this apply to the lung? In physiology, we can equate the bubble to a fluid-lined alveolus (although alveoli are not perfect spheres). The fluid lining all the alveoli creates surface tension. If the surface tension (T) of the fluid were the

LAW OF LaPLACE

(a) The two bubbles shown have the same surface tension (T). According to the Law of LaPlace, pressure is greater in the smaller bubble.

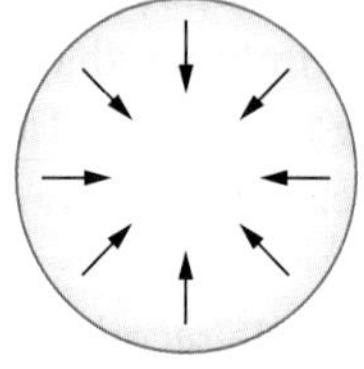

Larger bubble
r = 2
T = 3
P = (2 × 3)/2
P = 3

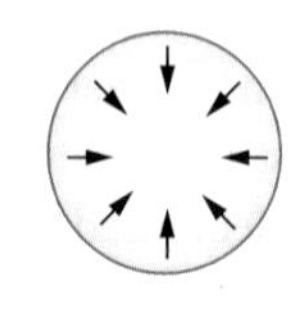

Smaller bubble
r = 1
T = 3
P = (2 × 3)/1
P = 6

Law of LaPlace

P = 2T/r

P = pressure
T = surface tension
r = radius

According to the law of LaPlace, if two bubbles have the same surface tension, the smaller bubble will have higher pressure.

(b) Surfactant (●) reduces surface tension (T). In the lungs, smaller alveoli have more surfactant, which equalizes the pressure between large and small alveoli.

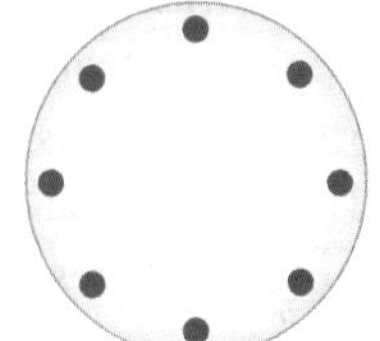

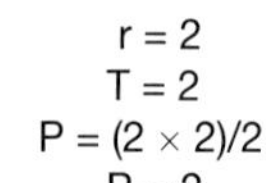

r = 2
T = 2
P = (2 × 2)/2
P = 2

More surfactant decreases surface tension.

r = 1
T = 1
P = (2 × 1)/1
P = 2

■ **Fig. 17.11**

same in small and large alveoli, small alveoli would have higher inwardly directed pressure than larger alveoli, and increased resistance to stretch. As a result, more work would be needed to expand smaller alveoli.

Normally, however, our lungs secrete a surfactant that reduces surface tension. Surfactants ("*surf*ace *act*ive *agents*") are molecules that disrupt cohesive forces between water molecules by substituting themselves for water at the surface. For example, that product you add to your dishwasher to aid in the rinse cycle is a surfactant that keeps the rinse water from beading up on the dishes (and forming spots when the water beads dry). In the lungs, surfactant decreases surface tension of the alveolar fluid and thereby decreases resistance of the lung to stretch.

Surfactant is more concentrated in smaller alveoli, making their surface tension less than that in larger alveoli (Fig. 17.11b). Lower surface tension helps equalize the pressure among alveoli of different sizes and makes it easier to inflate the smaller alveoli. With lower surface tension, the work needed to expand the alveoli with each breath is greatly reduced. Human surfactant is a mixture containing proteins and phospholipids, such as *dipalmitoylphosphatidylcholine*, which are secreted into the alveolar air space by type II alveolar cells (see Fig. 17.2g).

Normally, surfactant synthesis begins about the twenty-fifth week of fetal development under the influence of various hormones. Production usually reaches adequate levels by the thirty-fourth week (about six weeks before normal delivery). Babies who are born prematurely without adequate concentrations of surfactant in their alveoli develop *newborn respiratory distress syndrome (NRDS)*. In addition to having "stiff" (low-compliance) lungs, NRDS babies also have alveoli that collapse each time they exhale. These infants must use a tremendous amount of energy to expand their collapsed lungs with each breath. Unless treatment is initiated rapidly, about 50% of these infants die. In the past, all physicians could do for NRDS babies was administer oxygen. Today, however, the prognosis for NRDS babies is much better. Amniotic fluid can be sampled to assess whether or not the fetal lungs are producing adequate amounts of surfactant. If they are not, and if delivery cannot be delayed, NRDS babies can be treated with aerosol administration of artificial surfactant until the lungs mature enough to produce their own. The current treatment also includes artificial ventilation that forces air into the lungs (*positive-pressure ventilation*) and keeps the alveoli open.

RUNNING PROBLEM

Edna has been experiencing shortness of breath while exercising, so her physician runs some tests, including measuring Edna's lung volumes with spirometry. Part of the test is a forced expiratory volume. With her lungs filled to their maximum with air, Edna is told to blow out as fast and as forcefully as she can. The volume of air that Edna expels in the first second of the test (the forced expiratory volume in one second, or FEV_1) is lower than normal because in COPD, airway resistance is increased. Another test the physician orders is a complete blood count (CBC). The results of this test show that Edna has higher-than-normal red blood cell count and hematocrit [p. 551].

Q4: When Edna fills her lungs maximally, the volume of air in her lungs is known as the ________ capacity. When she exhales all the air she can, the volume of air left in her lungs is the ________.

Q5: Why are Edna's RBC count and hematocrit increased? (*Hint:* Because of Edna's COPD, her arterial P_{O_2} is low.)

569 576 582 584 **587** 594

Airway Diameter Determines Airway Resistance

The other factor besides compliance that influences the work of breathing is the resistance of the respiratory system to air flow. Resistance in the respiratory system is similar in many ways to resistance in the cardiovascular system [p. 469]. Three parameters contribute to resistance (R): the system's length (L), the viscosity of the substance flowing through the system (η), and the radius of the tubes in the system (r). As with flow in the cardiovascular system, Poiseuille's law relates these factors to one another:

$$R \propto L\eta/r^4$$

Because the length of the respiratory system is constant, we can ignore L in the equation. The viscosity of air is almost constant, although you may have noticed that it feels harder to breathe in a sauna filled with steam than in a room with normal humidity. Water droplets in the steam increase the viscosity of the steamy air, thereby increasing its resistance to flow. Viscosity also changes slightly with atmospheric pressure, decreasing as pressure decreases. A person at high altitude may feel less resistance to air flow than a person at sea level. Despite these exceptions, viscosity plays a very small role in resistance to air flow.

Length and viscosity are essentially constant for the respiratory system. As a result, the radius (or diameter) of the airways becomes the primary determinant of airway resistance. Normally, however, the work needed to overcome resistance of the airways to air flow is much less than the work needed to overcome the resistance of the lungs and thoracic cage to stretch.

Nearly 90% of airway resistance normally can be attributed to the trachea and bronchi, rigid structures with the smallest total cross-sectional area. Because these structures are supported by cartilage and bone, their diameters normally do not change, and their resistance to air flow is constant. However, accumulation of mucus from allergies or infections can dramatically increase resistance. If you have ever tried breathing through your nose when you have a cold, you can appreciate how the narrowing of an upper airway limits air flow!

The bronchioles normally do not contribute significantly to airway resistance because their total cross-sectional area is about 2000 times that of the trachea. Because the bronchioles are collapsible tubes, however, a decrease in their diameter can suddenly turn them into a significant source of airway resistance. **Bronchoconstriction** increases resistance to air flow and decreases the amount of fresh air that reaches the alveoli.

Bronchioles, like arterioles, are subject to reflex control by the nervous system and by hormones. However, most minute-to-minute changes in bronchiolar diameter occur in response to paracrines. Carbon dioxide in the airways is the primary paracrine that affects bronchiolar diameter. Increased CO_2 in expired air relaxes bronchiolar smooth muscle and causes **bronchodilation**.

Histamine is a paracrine that acts as a powerful bronchoconstrictor. This chemical is released by *mast cells* [p. 547] in response to either tissue damage or allergic reactions. In severe allergic reactions, large amounts of histamine may lead to widespread bronchoconstriction and difficult breathing. Immediate medical treatment in these patients is imperative.

The primary neural control of bronchioles comes from parasympathetic neurons that cause bronchoconstriction, a reflex designed to protect the lower respiratory tract from inhaled irritants. There is no significant sympathetic innervation of the bronchioles in humans. However, smooth muscle in the bronchioles is well supplied with β_2-receptors that respond to epinephrine. Stimulation of β_2-receptors relaxes airway smooth muscle and results in bronchodilation. This reflex is used therapeutically in the treatment of asthma and various allergic reactions characterized by histamine release and bronchoconstriction. ■ Table 17.1 summarizes the factors that alter airway resistance.

Concept Check

Answers: p. 598

23. In a normal person, which contributes more to the work of breathing: airway resistance or lung and chest wall elastance?

24. Coal miners who spend years inhaling fine coal dust have much of their alveolar surface area covered with scarlike tissue. What happens to their lung compliance as a result?

25. How does the work required for breathing change when surfactant is not present in the lungs?

26. A cancerous lung tumor has grown into the walls of a group of bronchioles, narrowing their lumens. What has happened to the resistance to air flow in these bronchioles?

27. Name the neurotransmitter and receptor for parasympathetic bronchoconstriction.

Rate and Depth of Breathing Determine the Efficiency of Breathing

You may recall that the efficiency of the heart is measured by the cardiac output, which is calculated by multiplying heart rate by stroke volume. Likewise, we can estimate the effectiveness of ventilation by calculating **total pulmonary ventilation,** the volume of air moved into and out of the lungs each minute (■ Fig. 17.12a). Total pulmonary ventilation, also known as the *minute volume*, is calculated as follows:

$$\text{Total pulmonary ventilation} = \text{ventilation rate} \times \text{tidal volume}$$

Table 17.1 Factors That Affect Airway Resistance

Factor	Affected by	Mediated by
Length of the system	Constant; not a factor	
Viscosity of air	Usually constant; humidity and altitude may alter slightly	
Diameter of airways		
Upper airways	Physical obstruction	Mucus and other factors
Bronchioles	Bronchoconstriction	Parasympathetic neurons (muscarinic receptors), histamine, leukotrienes
	Bronchodilation	Carbon dioxide, epinephrine (β_2-receptors)

Fig. 17.12 *ESSENTIALS*

Ventilation

(a) Total pulmonary ventilation is greater than alveolar ventilation because of dead space.

Total pulmonary ventilation

Total pulmonary ventilation = ventilation rate × tidal volume (V_T)

For example: 12 breaths/min × 500 mL/breath = 6000 mL/min

Alveolar ventilation

Alveolar ventilation is a better indication of how much fresh air reaches the alveoli. Fresh air remaining in the dead space does not get to the alveoli.

Alveolar ventilation = ventilation rate × (V_T – dead space volume V_D)

If dead space is 150 mL: 12 breaths/min × (500 – 150 mL) = 4200 mL/min

(b) Because the conducting airways do not exchange gases with the blood, they are known as **anatomic dead space.**

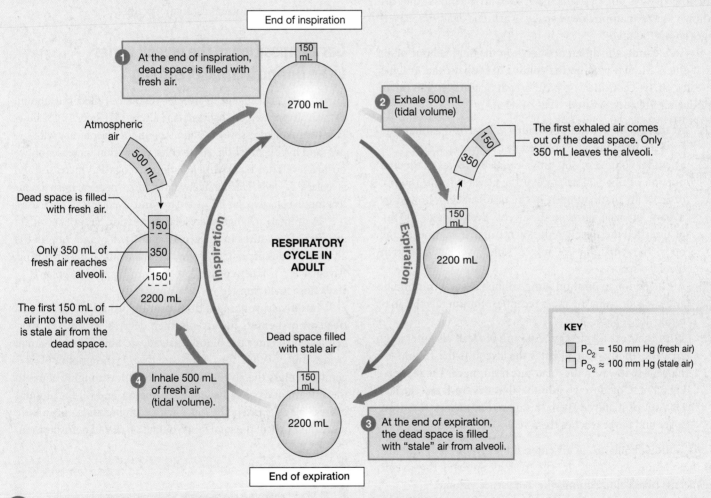

FIGURE QUESTION

Complete this table showing the effects of breathing pattern on alveolar ventilation. Assume dead space volume is 150 mL. Which pattern is the most efficient?

Tidal volume (mL)	Ventilation rate (breaths/min)	Total pulmonary ventilation (mL/min)	Fresh air to alveoli (mL)	Alveolar ventilation (mL/min)
500 (normal)	12 (normal)	6000	350	4200
300 (shallow)	20 (rapid)			
750 (deep)	8 (slow)			

The normal ventilation rate for an adult is 12–20 breaths (br) per minute. Using the average tidal volume (500 mL) and the slowest ventilation rate, we get:

Total pulmonary ventilation =
12 br/min × 500 mL/br = 6000 mL/min = 6 L/min

Total pulmonary ventilation represents the physical movement of air into and out of the respiratory tract, but is it a good indicator of how much fresh air reaches the alveolar exchange surface? Not necessarily.

Some air that enters the respiratory system does not reach the alveoli because part of every breath remains in the conducting airways, such as the trachea and bronchi. Because the conducting airways do not exchange gases with the blood, they are known as the **anatomic dead space.** Anatomic dead space averages about 150 mL.

To illustrate the difference between the total volume of air that enters the airways and the volume of fresh air that reaches the alveoli, let's consider a typical breath that moves 500 mL of air during a respiratory cycle (Fig. 17.12b).

1. At the end of an inspiration, lung volume is maximal, and fresh air from the atmosphere fills the dead space.
2. The tidal volume of 500 mL is exhaled. However, the first portion of this 500 mL to exit the airways is the 150 mL of fresh air that had been in the dead space, followed by 350 mL of "stale" air from the alveoli. Even though 500 mL of air exited the alveoli, only 350 mL of that volume left the body. The remaining 150 mL of "stale" alveolar air stays in the dead space.
3. At the end of expiration, lung volume is at its minimum, and stale air from the most recent expiration fills the anatomic dead space.
4. With the next inspiration, 500 mL of fresh air enters the airways. The first air to enter the alveoli is the 150 mL of stale air that was in the anatomic dead space. The remaining 350 mL of air to go into the alveoli is fresh air. The last 150 mL of inspired fresh air again remains in the dead space and never reaches the alveoli.

Thus, although 500 mL of air entered the alveoli, only 350 mL of that volume was fresh air. The fresh air entering the alveoli equals the tidal volume minus the dead space volume.

Because a significant portion of inspired air never reaches an exchange surface, a more accurate indicator of ventilation efficiency is **alveolar ventilation**, the amount of fresh air that reaches the alveoli each minute. Alveolar ventilation is calculated by multiplying ventilation rate by the volume of fresh air that reaches the alveoli:

Alveolar ventilation =
ventilation rate × (tidal volume – dead space)

Using the same ventilation rate and tidal volume as before, and a dead space of 150 mL, then

Alveolar ventilation =
12 br/min × (500 – 150 mL/br) = 4200 mL/min

Thus, at 12 breaths per minute, the alveolar ventilation is 4.2 L/min. Although 6 L/min of fresh air enters the respiratory system, only 4.2 L reaches the alveoli.

Alveolar ventilation can be drastically affected by changes in the rate or depth of breathing, as you can calculate using the figure question in Figure 17.12. **Maximum voluntary ventilation**, which involves breathing as deeply and quickly as possible, may increase total pulmonary ventilation to as much as 170 L/min. ■ Table 17.2 describes various patterns of ventilation, and ■ Table 17.3 gives normal ventilation values.

Gas Composition in the Alveoli Varies Little During Normal Breathing

How much can a change in alveolar ventilation affect the amount of fresh air and oxygen that reach the alveoli? ■ Figure 17.13 shows how the partial pressures P_{O_2} and P_{CO_2} in the alveoli vary with hyper- and hypoventilation. As alveolar ventilation increases above normal levels during hyperventilation, alveolar P_{O_2} increases, and alveolar P_{CO_2} falls. During hypoventilation, when less fresh air enters the alveoli, alveolar P_{O_2} decreases and alveolar P_{CO_2} increases.

A dramatic change in alveolar ventilation pattern can affect gas partial pressures in the alveoli, but the P_{O_2} and P_{CO_2} in the alveoli change surprisingly little during normal quiet breathing. Alveolar P_{O_2} is fairly constant at 100 mm Hg, and alveolar P_{CO_2} stays close to 40 mm Hg.

Intuitively, you might think that P_{O_2} would increase when fresh air first enters the alveoli, then decrease steadily as oxygen leaves to enter the blood. Instead, we find only very small swings in P_{O_2}. Why? The reasons are that (1) the amount of oxygen that enters the alveoli with each breath is roughly equal to the amount of oxygen that enters the blood, and (2) the amount of fresh air that enters the lungs with each breath is only a little more than 10% of the total lung volume at the end of inspiration.

Concept Check Answers: p. 598

28. If a person increases his tidal volume, what would happen to his alveolar P_{O_2}?

29. If his breathing rate increases, what would happen to his alveolar P_{O_2}?

Ventilation and Alveolar Blood Flow Are Matched

Moving oxygen from the atmosphere to the alveolar exchange surface is only the first step in external respiration. Next, gas exchange must occur across the alveolar-capillary interface. Finally, blood flow (*perfusion*) past the alveoli must be high enough to

Types and Patterns of Ventilation

Table 17.2

Name	Description	Examples
Eupnea	Normal quiet breathing	
Hyperpnea	Increased respiratory rate and/or volume in response to increased metabolism	Exercise
Hyperventilation	Increased respiratory rate and/or volume without increased metabolism	Emotional hyperventilation; blowing up a balloon
Hypoventilation	Decreased alveolar ventilation	Shallow breathing; asthma; restrictive lung disease
Tachypnea	Rapid breathing; usually increased respiratory rate with decreased depth	Panting
Dyspnea	Difficulty breathing (a subjective feeling sometimes described as "air hunger")	Various pathologies or hard exercise
Apnea	Cessation of breathing	Voluntary breath-holding; depression of CNS control centers

Normal Ventilation Values in Pulmonary Medicine

Table 17.3

Total pulmonary ventilation	6 L/min
Total alveolar ventilation	4.2 L/min
Maximum voluntary ventilation	125–170 L/min
Respiration rate	12–20 breaths/min

pick up the available oxygen. Matching the ventilation rate into groups of alveoli with blood flow past those alveoli is a two-part process involving local regulation of both air flow and blood flow.

Alterations in pulmonary blood flow depend almost exclusively on properties of the capillaries and on such local factors as the concentrations of oxygen and carbon dioxide in the lung tissue. Capillaries in the lungs are unusual because they are collapsible. If the pressure of blood flowing through the capillaries falls below a certain point, the capillaries close off, diverting blood to pulmonary capillary beds in which blood pressure is higher.

In a person at rest, some capillary beds in the apex (top) of the lung are closed off because of low hydrostatic pressure. Capillary beds at the base of the lung have higher hydrostatic pressure because of gravity and thus remain open. Consequently, blood flow is diverted toward the base of the lung. During exercise, when blood pressure rises, the closed apical capillary beds open, ensuring that the increased cardiac output can be fully oxygenated as it passes through the lungs. The ability of the lungs to recruit additional capillary beds during exercise is an example of the reserve capacity of the body.

At the local level, the body attempts to match air flow and blood flow in each section of the lung by regulating the diameters of the arterioles and bronchioles. Bronchiolar diameter is mediated primarily by CO_2 levels in exhaled air passing through them (■ Fig. 17.14). An increase in the P_{CO_2} of expired air causes bronchioles to dilate. A decrease in the P_{CO_2} of expired air causes bronchioles to constrict.

Although there is some autonomic innervation of pulmonary arterioles, there is apparently little neural control of pulmonary blood flow. The resistance of pulmonary arterioles to blood flow is regulated primarily by the oxygen content of the interstitial fluid around the arteriole. If ventilation of alveoli in one area of the lung is diminished, as shown in Figure 17.14b, the P_{O_2} in that area decreases, and the arterioles respond by constricting, as shown in Figure 17.14c. This local vasoconstriction is adaptive because it diverts blood away from the under-ventilated region to better-ventilated parts of the lung.

Note that constriction of pulmonary arterioles in response to low P_{O_2} is the opposite of what occurs in the systemic circulation [p. 521]. In the systemic circulation, a decrease in the P_{O_2} of a tissue causes local arterioles to dilate, delivering more oxygen-carrying blood to those tissues that are consuming oxygen. In the lungs, blood is picking up oxygen, so it does not make sense to send more blood to an area with low tissue P_{O_2} due to poor ventilation.

Another important point must be noted here. Local control mechanisms are not effective regulators of air and blood flow under all circumstances. If blood flow is blocked in one pulmonary artery, or if air flow is blocked at the level of the larger

ALVEOLAR GASES

As alveolar ventilation increases, alveolar P_{O_2} increases and P_{CO_2} decreases. The opposite occurs as alveolar ventilation decreases.

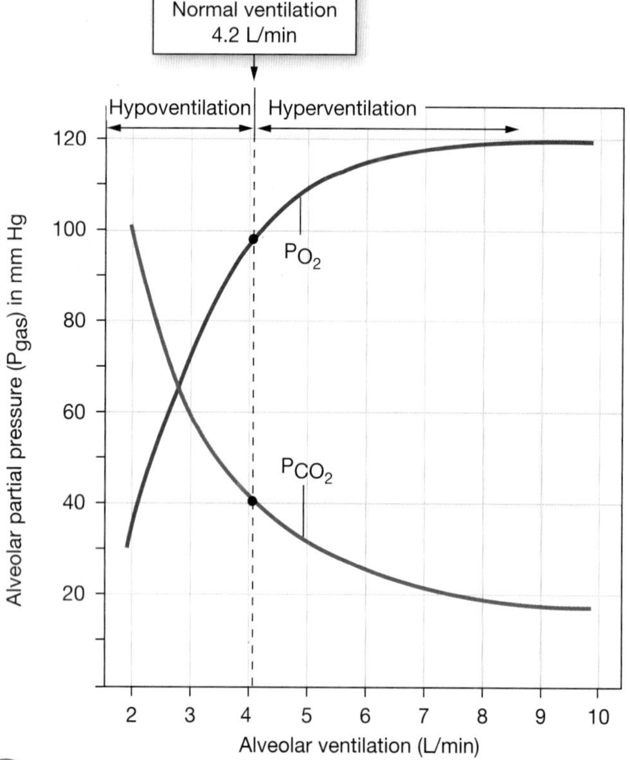

GRAPH QUESTION
What are the maximum alveolar P_{O_2} and minimum P_{CO_2} shown in this graph?

Fig. 17.13

airways, local responses that shunt air or blood to other parts of the lung are ineffective because in these cases no part of the lung has normal ventilation or perfusion.

Concept Check Answer: p. 598

30. If a lung tumor decreases blood flow in one small section of the lung to a minimum, what happens to P_{O_2} in the alveoli in that section and in the surrounding interstitial fluid? What happens to P_{CO_2} in that section? What is the compensatory response of the bronchioles in the affected section? Will the compensation bring ventilation in the affected section of the lung back to normal? Explain.

Auscultation and Spirometry Assess Pulmonary Function

Most pulmonary function tests are relatively simple to perform. Auscultation of breath sounds is an important diagnostic technique in pulmonary medicine, just as auscultation of heart sounds is an important technique in cardiovascular diagnosis [p. 492]. Breath sounds are more complicated to interpret than heart sounds, however, because breath sounds have a wider range of normal variation.

Normally, breath sounds are distributed evenly over the lungs and resemble a quiet "whoosh" made by flowing air. When air flow is reduced, such as in pneumothorax, breath sounds may be either diminished or absent. Abnormal sounds include various squeaks, pops, wheezes, and bubbling sounds caused by fluid and secretions in the airways or alveoli. Inflammation of the pleural membrane results in a crackling or grating sound known as a *friction rub*. It is caused by swollen, inflamed pleural membranes rubbing against each other, and it disappears when fluid again separates them.

Diseases in which air flow is diminished because of increased airway resistance are known as **obstructive lung diseases**. When patients with obstructive lower airway diseases are asked to exhale forcefully, air whistling through the narrowed airways creates a wheezing sound that can be heard even without a stethoscope. Depending on the severity of the disease, the bronchioles may even collapse and close off before a forced expiration is completed, reducing both the amount and rate of air flow as measured by a spirometer.

Obstructive lung diseases include asthma, obstructive sleep apnea, emphysema, and chronic bronchitis. The latter two are sometimes called *chronic obstructive pulmonary disease* (COPD) because of their ongoing, or chronic, nature. *Obstructive sleep apnea* {*apnoia*, breathless} results from obstruction of the upper airway, often due to abnormal relaxation of the muscles of the pharynx and tongue that increases airway resistance during inspiration.

Asthma is an inflammatory condition, often associated with allergies, that is characterized by bronchoconstriction and airway edema. Asthma can be triggered by exercise (exercise-induced asthma) or by rapid changes in the temperature or humidity of inspired air. Asthmatic patients complain of "air hunger" and difficulty breathing, or *dyspnea*. The severity of asthma attacks ranges from mild to life threatening. Studies of asthma at the cellular level show that a variety of chemical signals may be responsible for inducing asthmatic bronchoconstriction. Among these are acetylcholine, histamine, substance P (a neuropeptide), and leukotrienes secreted by mast cells, macrophages, and eosinophils. *Leukotrienes* are lipid-like bronchoconstrictors that are released during the inflammatory response. Asthma is treated with inhaled and oral medications that include β_2-adrenergic agonists, anti-inflammatory drugs, and leukotriene antagonists.

Concept Check Answers: p. 598

31. Restrictive lung diseases decrease lung compliance. How will inspiratory reserve volume change in patients with a restrictive lung disease?
32. Chronic obstructive lung disease causes patients to lose the ability to exhale fully. How does residual volume change in these patients?

This completes our discussion of the mechanics of ventilation. In the next chapter, we shift focus from the bulk flow of air to the diffusion and transport of oxygen and carbon dioxide as they travel between the air spaces of the alveoli and the cells of the body.

Local control mechanisms attempt to match ventilation and perfusion.

(a) Normally perfusion of blood past alveoli is matched to alveolar ventilation to maximize gas exchange.

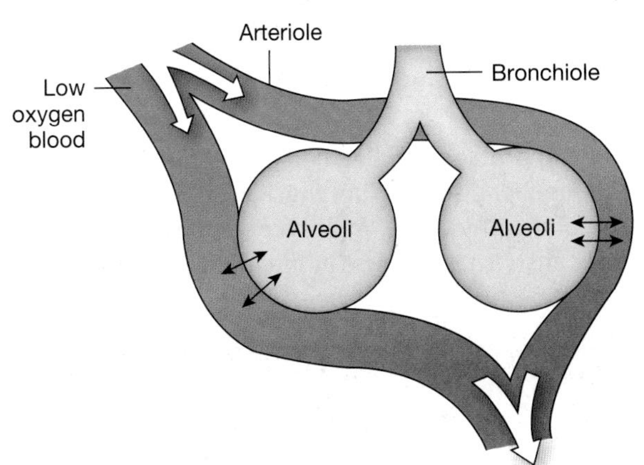

Arteriole
Bronchiole
Low oxygen blood
Alveoli
Alveoli

(b) Ventilation-perfusion mismatch caused by under-ventilated alveoli.

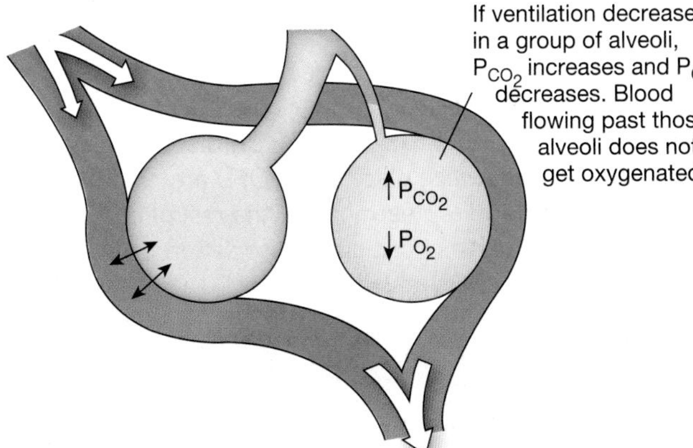

If ventilation decreases in a group of alveoli, P_{CO_2} increases and P_{O_2} decreases. Blood flowing past those alveoli does not get oxygenated.
↑P_{CO_2}
↓P_{O_2}

(c) Local control mechanisms try to keep ventilation and perfusion matched.

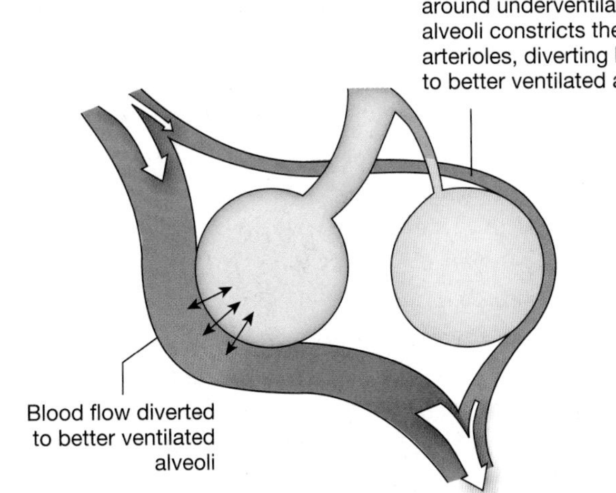

Decreased tissue P_{O_2} around underventilated alveoli constricts their arterioles, diverting blood to better ventilated alveoli.
Blood flow diverted to better ventilated alveoli

(d) Bronchiole diameter is mediated primarily by CO_2 levels in exhaled air passing through them.

Local Control of Arterioles and Bronchioles by Oxygen and Carbon Dioxide

Gas composition	Bronchioles	Pulmonary arteries	Systemic arteries
P_{CO_2} increases	Dilate	(Constrict)*	Dilate
P_{CO_2} decreases	Constrict	(Dilate)	Constrict
P_{O_2} increases	(Constrict)	(Dilate)	Constrict
P_{O_2} decreases	(Dilate)	Constrict	Dilate

* Parentheses indicate weak responses.

FIGURE QUESTIONS

A blood clot prevents gas exchange in a group of alveoli.

1. What happens to tissue and alveolar gases?
2. What do bronchioles and arterioles do in response?

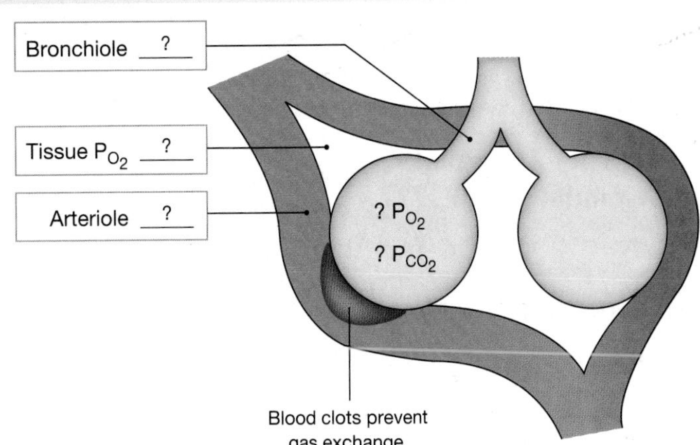

Bronchiole ?
Tissue P_{O_2} ?
Arteriole ?
? P_{O_2}
? P_{CO_2}
Blood clots prevent gas exchange.

Fig. 17.14

RUNNING PROBLEM CONCLUSION

Emphysema

Edna leaves the office with prescriptions for a mucus-thinning drug, a bronchodilator, and anti-inflammatory drugs to keep her airways as open as possible. She has agreed to try to stop smoking once more and also has a prescription and brochures for that. Unfortunately, the lung changes that take place with COPD are not reversible, and Edna will require treatment for the rest of her life. According to the American Lung Association (*www.lungusa.org*), COPD is the fourth leading cause of death in the United States and costs more than $30 billion per year in direct medical costs and indirect costs such as lost wages.

In this running problem you learned about chronic obstructive pulmonary disease. Now check your understanding of the physiology in the problem by comparing your answers with those in the following table.

Question	Facts	Integration and Analysis
1. What does narrowing of the airways do to the resistance airways offer to air flow?	The relationship between tube radius and resistance is the same for air flow as for blood flow: as radius decreases, resistance increases [p. 469].	When resistance increases, the body must use more energy to create air flow.
2. Why do people with chronic bronchitis have a higher-than-normal rate of respiratory infections?	Cigarette smoke paralyzes the cilia that sweep debris and mucus out of the airways. Without the action of cilia, mucus and trapped particles pool in the airways.	Bacteria trapped in the mucus can multiply and cause respiratory infections.
3. Name the muscles that patients with emphysema use to exhale actively.	Normal passive expiration depends on elastic recoil of muscles and elastic tissue in the lungs.	Forceful expiration involves the internal intercostal muscles and the abdominal muscles.
4. When Edna fills her lungs maximally, the volume of air in her lungs is known as the _________ capacity. When she exhales all the air she can, the volume of air left in her lungs is the _________.	The maximum volume of air in the lungs is the *total lung capacity*. Air left in the lungs after maximal exhalation is the *residual volume*.	N/A
5. Why are Edna's RBC count and hematocrit increased?	Because of Edna's COPD, her arterial P_{O_2} is low. The major stimulus for red blood cell synthesis is hypoxia.	Low arterial oxygen levels trigger EPO release, which increases the synthesis of red blood cells [p. 551]. More RBCs provide more binding sites for oxygen transport.

569 576 582 584 587 594

Chapter Summary

Air flow into and out of the lungs is another example of the principle of *mass flow*. Like blood flow, air flow is bulk flow that requires a pump to create a pressure gradient and that encounters resistance, primarily from changes in the diameter of the tubes through which it flows. The *mechanical properties* of the pleural sacs and elastic recoil in the chest wall and lung tissue are essential for normal ventilation.

1. Aerobic metabolism in living cells consumes oxygen and produces carbon dioxide. (p. 569)
2. Gas exchange requires a large, thin, moist exchange surface; a pump to move air; and a circulatory system to transport gases to the cells. (p. 569)
3. Respiratory system functions include gas exchange, pH regulation, vocalization, and protection from foreign substances. (p. 569)

The Respiratory System

iP Respiratory: Anatomy Review

4. **Cellular respiration** refers to cellular metabolism that consumes oxygen. **External respiration** is the exchange of gases between the atmosphere and cells of the body. It includes ventilation, gas exchange at the lung and cells, and transport of gases in the blood. **Ventilation** is the movement of air into and out of the lungs. (p. 570; Fig. 17.1)
5. The **respiratory system** consists of anatomical structures involved in ventilation and gas exchange. (p. 570)
6. The **upper respiratory tract** includes the mouth, nasal cavity, **pharynx**, and **larynx**. The **lower respiratory tract** includes the **trachea**, **bronchi**, **bronchioles**, and exchange surfaces of the **alveoli**. (p. 572; Fig. 17.2b)
7. The thoracic cage is bounded by the ribs, spine, and **diaphragm**. Two sets of **intercostal muscles** connect the ribs. (p. 572; Fig. 17.2a)
8. Each **lung** is contained within a double-walled **pleural sac** that contains a small quantity of **pleural fluid**. (p. 572, 574; Figs. 17.2c, 17.3)
9. The two **primary bronchi** enter the lungs. Each primary bronchus divides into progressively smaller bronchi and finally into collapsible **bronchioles**. (p. 572, 574; Figs. 17.2e, 17.4)
10. The upper respiratory system filters, warms, and humidifies inhaled air. (p. 572)
11. The alveoli consist mostly of thin-walled **type I alveolar cells** for gas exchange. **Type II alveolar cells** produce surfactant. A network of capillaries surrounds each alveolus. (p. 573; Fig. 17.2f, g)
12. Blood flow through the lungs equals cardiac output. Resistance to blood flow in the pulmonary circulation is low. Pulmonary arterial pressure averages 25/8 mm Hg. (p. 575)

Gas Laws

iP Respiratory: Pulmonary Ventilation

13. The total pressure of a mixture of gases is the sum of the pressures of the individual gases in the mixture (**Dalton's law**). **Partial pressure** is the pressure contributed by a single gas in a mixture. (p. 577; Fig. 17.6)
14. Bulk flow of air occurs down pressure gradients, as does the movement of any individual gas making up the air. (p. 576)
15. **Boyle's law** states that as the volume available to a gas increases, the gas pressure decreases. The body creates pressure gradients by changing thoracic volume. (p. 577; Fig. 17.6b)

Ventilation

iP Respiratory: Pulmonary Ventilation

16. A single **respiratory cycle** consists of an inspiration and an expiration. (p. 578)
17. **Tidal volume** is the amount of air taken in during a single normal inspiration. **Vital capacity** is tidal volume plus **expiratory** and **inspiratory reserve volumes**. Air volume in the lungs at the end of maximal expiration is the **residual volume**. (p. 579; Fig. 17.7b)
18. Air flow in the respiratory system is directly proportional to the pressure gradient, and inversely related to the resistance to air flow offered by the airways. (p. 580)
19. During **inspiration**, **alveolar pressure** decreases, and air flows into the lungs. Inspiration requires contraction of the inspiratory muscles and the diaphragm. (p. 582; Fig. 17.9)
20. **Expiration** is usually passive, resulting from elastic recoil of the lungs. (p. 580)
21. **Active expiration** requires contraction of the internal intercostal and abdominal muscles. (p. 583)
22. **Intrapleural pressures** are subatmospheric because the pleural cavity is a sealed compartment. (p. 582, 584; Figs. 17.9, 17.10)
23. **Compliance** is a measure of the ease with which the chest wall and lungs expand. Loss of compliance increases the work of breathing. **Elastance** is the ability of a stretched lung to return to its normal volume. (p. 585)
24. **Surfactant** decreases surface tension in the fluid lining the alveoli. Reduced surface tension prevents smaller alveoli from collapsing and also makes it easier to inflate the lungs. (p. 586; Fig. 17.11)
25. The diameter of the bronchioles determines how much resistance they offer to air flow. (p. 586)
26. Increased CO_2 in expired air dilates bronchioles. Parasympathetic neurons cause **bronchoconstriction** in response to irritant stimuli. There is no significant sympathetic innervation of bronchioles, but epinephrine causes **bronchodilation**. (p. 588; Tbl. 17.1)
27. **Total pulmonary ventilation** = tidal volume × ventilation rate. **Alveolar ventilation** = ventilation rate × (tidal volume − dead space volume). (p. 589; Fig. 17.12a)
28. Alveolar gas composition changes very little during a normal respiratory cycle. **Hyperventilation** increases alveolar P_{O_2} and decreases alveolar P_{CO_2}. **Hypoventilation** has the opposite effect. (p. 592; Fig. 17.13)
29. Local mechanisms match air flow and blood flow around the alveoli. Increased levels of CO_2 dilate bronchioles, and decreased O_2 constricts pulmonary arterioles. (p. 593; Fig. 17.14)

Questions

Answers: p. A-1

Level One Reviewing Facts and Terms

1. List four functions of the respiratory system.
2. Give two definitions for the word *respiration*.
3. Which sets of muscles are used for normal quiet inspiration? For normal quiet expiration? For active expiration? What kind(s) of muscles are the different respiratory muscles (skeletal, cardiac, or smooth)?
4. What is the function of pleural fluid?
5. Name the anatomical structures that an oxygen molecule passes on its way from the atmosphere to the blood.
6. Diagram the structure of an alveolus, and state the function of each part. How are capillaries associated with an alveolus?
7. Trace the path of the pulmonary circulation. About how much blood is found here at any given moment? What is a typical arterial blood pressure for the pulmonary circuit, and how does this pressure compare with that of the systemic circulation?
8. What happens to inspired air as it is conditioned during its passage through the airways?
9. During inspiration, most of the thoracic volume change is the result of movement of the ________.
10. Describe the changes in alveolar and intrapleural pressure during one respiratory cycle.
11. What is the function of surfactants in general? In the respiratory system?
12. Of the three factors that contribute to the resistance of air flow through a tube, which plays the largest role in changing resistance in the human respiratory system?
13. Match the following items with their correct effect on the bronchioles:

(a) histamine	1. bronchoconstriction
(b) epinephrine	2. bronchodilation
(c) acetylcholine	3. no effect
(d) increased P_{CO_2}	

14. Refer to the spirogram in the figure below:

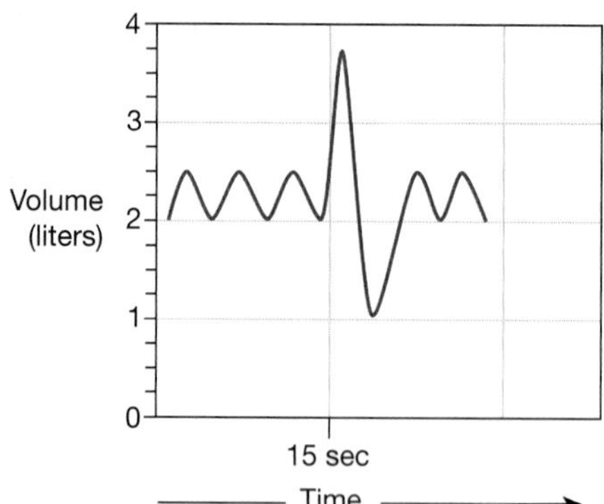

(a) Label tidal volume (V_T), inspiratory and expiratory reserve volumes (IRV and ERV), residual volume (RV), vital capacity (VC), total lung capacity (TLC).
(b) What is the value of each of the volumes and capacities you labeled?
(c) What is this person's ventilation rate?

Level Two Reviewing Concepts

15. Compare and contrast the terms in each of the following sets:
    (a) compliance and elastance
    (b) inspiration, expiration, and ventilation
    (c) intrapleural pressure and alveolar pressure
    (d) total pulmonary ventilation and alveolar ventilation
    (e) type I and type II alveolar cells
    (f) pulmonary circulation and systemic circulation
16. List the major paracrines and neurotransmitters that cause bronchoconstriction and bronchodilation. What receptors do they act through? (muscarinic, nicotinic, α, β_1, β_2)
17. Compile the following terms into a map of ventilation. Use up arrows, down arrows, greater than symbols (>), and less than symbols (<) as modifiers. You may add other terms.

• abdominal muscles	• inspiratory muscles
• air flow	• internal intercostals
• contract	• P_A
• diaphragm	• P_{atm}
• expiratory muscles	• $P_{intrapleural}$
• external intercostals	• quiet breathing
• forced breathing	• relax
• in, out, from, to	• scalenes

18. Decide whether each of the following parameters will increase, decrease, or not change in the situations given.
    (a) airway resistance with bronchodilation
    (b) intrapleural pressure during inspiration
    (c) air flow with bronchoconstriction
    (d) bronchiolar diameter with increased P_{CO_2}
    (e) tidal volume with decreased compliance
    (f) alveolar pressure during expiration
19. Define the following terms: pneumothorax, spirometer, auscultation, hypoventilation, bronchoconstriction, minute volume, partial pressure of a gas.
20. The cartoon coyote is blowing up a balloon in another attempt to catch the roadrunner. He first breathes in as much air as he can, then blows out all he can into the balloon.
    (a) The volume of air in the balloon is equal to the ________ of the coyote's lungs. This volume can be measured directly by measuring the balloon volume or by adding which respiratory volumes together?
    (b) In 10 years, when the coyote is still chasing the roadrunner, will he still be able to put as much air into the balloon in one breath? Explain.
21. Match the descriptions to the appropriate phase(s) of ventilation:

(a) usually depend(s) on elastic recoil	1. inspiration
(b) is/are easier when lung compliance decreases	2. expiration
(c) is/are driven mainly by positive intrapleural pressure generated by muscular contraction	3. both inspiration and expiration
(d) is usually an active process requiring smooth muscle contraction	4. neither

22. Draw and label a graph showing the P_{O_2} of air in the primary bronchi during one respiratory cycle. (*Hint:* What parameter goes on each axis?)
23. Lung compliance increases but chest wall compliance decreases as we age. In the absence of other changes, would the following parameters increase, decrease, or not change as compliance decreases?
    (a) work required for breathing
    (b) ease with which lungs inflate
    (c) lung elastance
    (d) airway resistance during inspiration
24. Will pulmonary surfactant increase, decrease, or not change the following?
    (a) work required for breathing
    (b) lung compliance
    (c) surface tension in the alveoli

Level Three Problem Solving

25. Assume a normal female has a resting tidal volume of 400 mL, a respiratory rate of 13 breaths/min, and an anatomic dead space of 125 mL. When she exercises, which of the following scenarios would be most efficient for increasing her oxygen delivery to the lungs?
    (a) increase respiratory rate to 20 breaths/min but have no change in tidal volume
    (b) increase tidal volume to 550 mL but have no change in respiratory rate
    (c) increase tidal volume to 500 mL and respiratory rate to 15 breaths/min

    Which of these scenarios is most likely to occur during exercise in real life?
26. A 30-year-old computer programmer has had asthma for 15 years. When she lies down at night, she has spells of wheezing and coughing. Over the years, she has found that she can breathe better if she sleeps sitting nearly upright. Upon examination, her doctor finds that she has an enlarged thorax. Her lungs are overinflated on X-ray. Here are the results of her examination and pulmonary function tests. Use the normal values and abbreviations in Figure 17.8 to help answer the questions.

    Ventilation rate: 16 breaths/min
    Tidal volume: 600 mL
    ERV: 1000 mL
    RV: 3500 mL
    Inspiratory capacity: 1800 mL
    Vital capacity: 2800 mL
    Functional residual capacity: 4500 mL
    TLC: 6300 mL
    After she is given a bronchodilator, her vital capacity increased to 3650 mL.

    (a) What is her minute volume?
    (b) Explain the change in vital capacity with bronchodilators.
    (c) Which other values are abnormal? Can you explain why they might be, given her history and findings?

Level Four Quantitative Problems

27. A container of gas with a movable piston has a volume of 500 mL and a pressure of 60 mm Hg. The piston is moved, and the new pressure is 150 mm Hg. What is the new volume of the container?
28. You have a mixture of gases in dry air, with an atmospheric pressure of 760 mm Hg. Calculate the partial pressure of each gas if the composition of the air is:
    (a) 21% oxygen, 78% nitrogen, 0.3% carbon dioxide
    (b) 40% oxygen, 13% nitrogen, 45% carbon dioxide, 2% hydrogen
    (c) 10% oxygen, 15% nitrogen, 1% argon, 25% carbon dioxide
29. Li is a tiny woman, with a tidal volume of 400 mL and a respiratory rate of 12 breaths per minute at rest. What is her total pulmonary ventilation? Just before a physiology exam, her ventilation increases to 18 breaths per minute from nervousness. Now what is her total pulmonary ventilation? Assuming her anatomic dead space is 120 mL, what is her alveolar ventilation in each case?
30. You collected the following data on your classmate Neelesh:

    Minute volume = 5004 mL/min
    Respiratory rate = 3 breaths/15 sec
    Vital capacity = 4800 mL
    Expiratory reserve volume = 1000 mL
    What are Neelesh's tidal volume and inspiratory reserve volume?
31. Use the figure below to help solve this problem. A spirometer with a volume of 1 liter (V_1) is filled with a mixture of oxygen and helium, with the helium concentration being 4 g/L (C_1). Helium does not move from the lungs into the blood or from the blood into the lungs. A subject is told to blow out all the air he possibly can. Once he finishes that exhalation, his lung volume is V_2. He then puts the spirometer tube in his mouth and breathes quietly for several breaths. At the end of that time, the helium is evenly dispersed in the spirometer and the subject's lungs. A measurement shows the new concentration of helium is 1.9 g/L. What was the subject's lung volume at the start of the experiment? (*Hint:* $C_1V_1 = C_2V_2$)

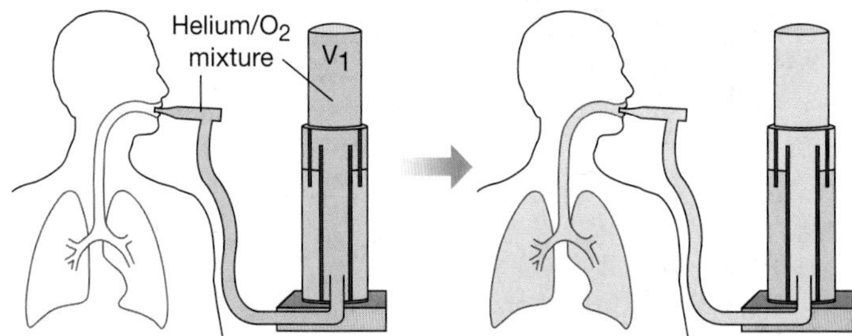

32. The graph shows one lung under two different conditions, A and B. What does this graph show? (a) the effect of lung volume on pressure, or (b) the effect of pressure on lung volume? In which condition does the lung have higher compliance, or is compliance the same in the two situations?

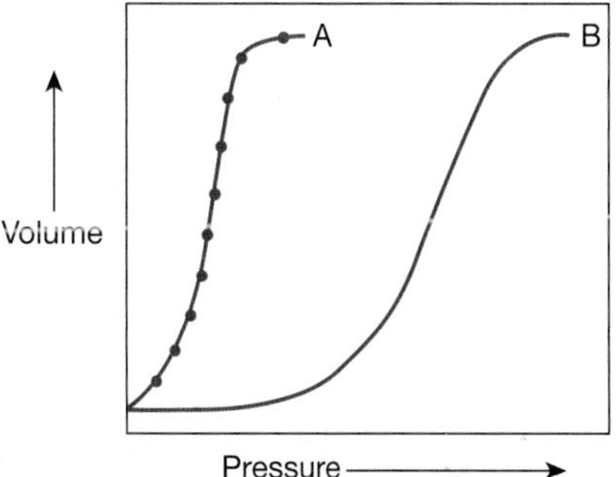

17

Answers

Answers to Concept Check Questions

Page 571

1. Cellular respiration is intracellular and uses O_2 and organic substrates to produce ATP. External respiration is exchange and transport of gases between the atmosphere and cells.
2. The upper respiratory tract includes the mouth, nasal cavity, pharynx, and larynx. The lower respiratory tract includes the trachea, bronchi, bronchioles, and exchange surface of lungs.
3. Velocity is highest in the trachea and lowest in the bronchioles.
4. Pleural fluid reduces friction and holds the lungs tight against the chest wall.
5. The thoracic cage consists of the rib cage with intercostal muscles, spinal (vertebral) column, and diaphragm. The thorax contains two lungs in pleural sacs, the heart and pericardial sac, esophagus, and major blood vessels.
6. The bronchioles are collapsible.

Page 574

7. If cilia cannot move mucus, the mucus collecting in the airways triggers a cough reflex to clear out the mucus.

Page 576

8. Blood flow is approximately equal in the pulmonary trunk and aorta. (Normally some venous blood leaving the bronchi, pleura, and part of the heart bypasses the pulmonary circulation and drains directly into the left side of the heart. This is called an anatomic shunt.)
9. Increased hydrostatic pressure causes greater net filtration out of capillaries and may result in pulmonary edema.
10. Mean pressure $= 8$ mm Hg $+ 1/3(25 - 8)$ mm Hg $= 8 + 17/3$ mm Hg $= 13.7$ mm Hg.

Page 578

11. 720 mm Hg $\times$ 0.78 = 562 mm Hg
12. 700 mm Hg − 47 mm Hg = 653 mm Hg $\times$ 21% = 137.1 mm Hg P_{O_2}

Page 580

13. Lung capacities are the sum of two or more lung volumes.
14. Residual volume cannot be measured directly.
15. If aging individuals have reduced vital capacity while total lung capacity does not change, then residual volume must increase.
16. As air becomes humidified, the P_{O_2} decreases.

Page 580

17. Air flow reverses direction during a respiratory cycle, but blood flows in a loop and never reverses direction.
18. See Figures 17.2c and 17.3. The lungs are enclosed in a pleural sac. One pleural membrane attaches to the lung, and the other lines the thoracic cage. Pleural fluid fills the pleural sac.

Page 583

19. Scarlett will be more successful if she exhales deeply, as this will decrease her thoracic volume and will pull her lower rib cage inward.
20. Inability to cough decreases the ability to expel the potentially harmful material trapped in airway mucus.

Page 585

21. A hiccup causes a rapid decrease in both intrapleural pressure and alveolar pressure.
22. The knife wound would collapse the left lung if the knife punctured the pleural membrane. Loss of adhesion between the lung and chest wall would release the inward pressure exerted on the chest wall, and the rib cage would expand outward. The right side would be unaffected as the right lung is contained in its own pleural sac.

Page 588

23. Normally, lung and chest wall elastance contribute more to the work of breathing.
24. Scar tissue reduces lung compliance.
25. Without surfactant, the work of breathing increases.
26. When bronchiolar diameter decreases, resistance increases.
27. Neurotransmitter is acetylcholine, and receptor is muscarinic.

Page 590

28. Increased tidal volume increases alveolar P_{O_2}.
29. Increased breathing rate increases alveolar P_{O_2}. Increasing breathing rate or tidal volume increases alveolar ventilation.

Page 592

30. P_{O_2} in alveoli in the affected section will increase because O_2 is not leaving the alveoli. P_{CO_2} will decrease because new CO_2 is not entering the alveoli from the blood. Bronchioles constrict when P_{CO_2} decreases (see Fig. 17.14), shunting air to areas of the lung with better blood flow. This compensation cannot restore normal ventilation in this section of lung, and local control is insufficient to maintain homeostasis.

Page 592

31. Inspiratory reserve volume decreases.
32. Residual volume increases in patients who cannot fully exhale.

Answers to Figure and Graph Questions

Page 582

Figure 17.9: 1. Alveolar pressure is greatest in the middle of expiration and least in the middle of inspiration. It is equal to atmospheric pressure at the beginning and end of inspiration and expiration. 2. When lung volume is at its minimum, alveolar pressure is (c) moving from maximum to minimum and external intercostal muscle contraction is (b) minimal. 3. 2 breaths/8 sec = ? breaths/60 sec = 15 breaths/min.

Page 589

Figure 17.12: Shallow and rapid: total pulmonary ventilation = 6000 mL/min, 150 mL fresh air, alveolar ventilation = 3000 mL/min. Slow and deep: total pulmonary ventilation = 6000 mL/min, 600 mL fresh air, alveolar ventilation = 4800 mL/min. Slow and deep is the most efficient.

Page 592

Figure 17.13: Alveolar P_{O_2} goes to 120 mm Hg and P_{CO_2} falls to about 19 mm Hg.

Page 593

Figure 17.14: 1. Alveolar P_{O_2} increases and P_{CO_2} decreases in the affected alveoli. Local tissue P_{O_2} increases. 2. This constricts local arterioles, which then shunts blood to better-perfused sections of lung. Bronchioles constrict to divert air to better-perfused alveoli.

18 Gas Exchange and Transport

The successful ascent of Everest without supplementary oxygen is one of the great sagas of the 20th century.

—John B. West, *Climbing with O's*, NOVA Online (*www.pbs.org*)

Background Basics

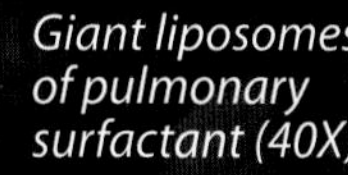

Giant liposomes of pulmonary surfactant (40X)

RUNNING PROBLEM

High Altitude

In 1981 a group of 20 physiologists, physicians, and climbers, supported by 42 Sherpa assistants, formed the American Medical Research Expedition to Mt. Everest. The purpose of the expedition was to study human physiology at extreme altitudes, starting with the base camp at 5400 m (18,000 ft) and continuing on to the summit at 8850 m (over 29,000 ft). From the work of these scientists and others, we now have a good picture of the physiology of high-altitude acclimatization.

The book *Into Thin Air* by Jon Krakauer chronicles an ill-fated trek to the top of Mt. Everest. To reach the summit of Mt. Everest, climbers must pass through the "death zone" located at about 8000 meters (over 26,000 ft). Of the thousands of people who have attempted the summit, only about 2000 have been successful, and more than 185 have died. What are the physiological challenges of climbing Mt. Everest (8850 m or 29,035 ft), and why did it take so many years before humans successfully reached the top? The lack of oxygen at high altitude is part of the answer.

The mechanics of breathing includes the events that create bulk flow of air into and out of the lungs. In this chapter we focus on the two gases most significant to human physiology, oxygen and carbon dioxide, and look at how they move between alveolar air spaces and the cells of the body. The process can be divided into two components: the exchange of gases between compartments, which requires diffusion across cell membranes, and the transport of gases in the blood. ■ Figure 18.1 presents an overview of the topics that we cover in this chapter.

If the diffusion of gases between alveoli and blood is significantly impaired, or if oxygen transport in the blood is inadequate, **hypoxia** (a state of too little oxygen) results. Hypoxia frequently (but not always!) goes hand in hand with **hypercapnia**, elevated concentrations of carbon dioxide. These two conditions are clinical signs, not diseases, and clinicians must gather additional information to pinpoint their cause. ■ Table 18.1 lists several types of hypoxia and some typical causes.

PULMONARY GAS EXCHANGE AND TRANSPORT

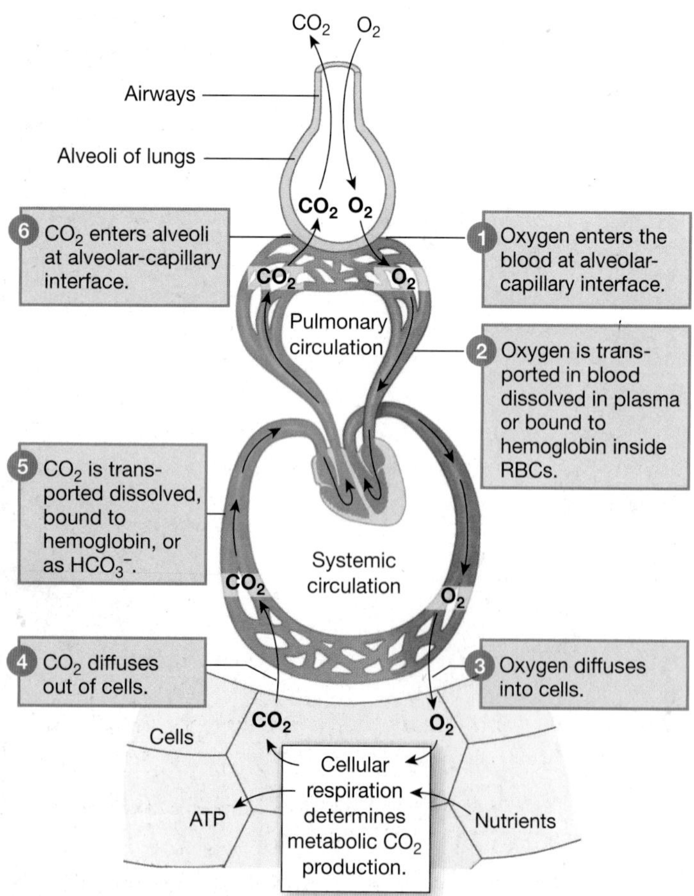

■ **Fig. 18.1**

To avoid hypoxia and hypercapnia, the body uses sensors that monitor arterial blood composition. These sensors respond to three regulated variables:

1. *Oxygen*. Arterial oxygen delivery to the cells must be adequate to support aerobic respiration and ATP production.
2. *Carbon dioxide* (CO_2) is produced as a waste product during the citric acid cycle [p. 114]. Excretion of CO_2 by the lungs is important for two reasons: high levels of CO_2 are a central nervous system depressant, and elevated CO_2 causes a state of acidosis (low pH) through the following reaction: $CO_2 + H_2O \rightleftharpoons H_2CO_3 \rightleftharpoons H^+ + HCO_3^-$.
3. *pH*. Maintaining pH homeostasis is critical to prevent denaturation of proteins [p. 54]. The respiratory system monitors plasma pH and uses changes in ventilation to alter pH. This process is discussed later along with renal contributions to pH homeostasis.

The normal values for these three parameters are given in ■ Table 18.2. In this chapter we will consider the mechanisms by which oxygen and CO_2 move from the lungs to the cells and back again.

Gas Exchange in the Lungs and Tissues

Breathing is the bulk flow of air into and out of the lungs. Once air reaches the alveoli, individual gases such as oxygen and CO_2 diffuse from the alveolar air space into the blood. Recall that diffusion is movement of a molecule from a region of higher concentration to one of lower concentration [p. 141].

Classification of Hypoxias

Table 18.1

Type	Definition	Typical Causes
Hypoxic hypoxia	Low arterial P_{O2}	High altitude; alveolar hypoventilation; decreased lung diffusion capacity; abnormal ventilation-perfusion ratio
Anemic hypoxia	Decreased total amount of O_2 bound to hemoglobin	Blood loss; anemia (low [Hb] or altered HbO_2 binding); carbon monoxide poisoning
Ischemic hypoxia	Reduced blood flow	Heart failure (whole-body hypoxia); shock (peripheral hypoxia); thrombosis (hypoxia in a single organ)
Histotoxic hypoxia	Failure of cells to use O_2 because cells have been poisoned	Cyanide and other metabolic poisons

18

Normal Blood Values in Pulmonary Medicine

Table 18.2

	Arterial	Venous
P_{O2}	95 mm Hg (85–100)	40 mm Hg
P_{CO2}	40 mm Hg (35–45)	46 mm Hg
pH	7.4 (7.38–7.42)	7.37

When we think of concentrations of solutions, units such as moles/liter and milliosmoles/liter come to mind. However, respiratory physiologists commonly express plasma gas concentrations in partial pressures to establish whether there is a concentration gradient between the alveoli and the blood. Gases move from regions of higher partial pressure to regions of lower partial pressure.

■ Figure 18.2 shows the partial pressures of oxygen and CO_2 in air, the alveoli, and inside the body. Normal alveolar P_{O_2} at sea level is about 100 mm Hg. The P_{O_2} of "deoxygenated" venous blood arriving at the lungs is 40 mm Hg. Oxygen therefore diffuses down its partial pressure (concentration) gradient from the alveoli into the capillaries. Diffusion goes to equilibrium, and the P_{O_2} of arterial blood leaving the lungs is the same as in the alveoli: 100 mm Hg.

When arterial blood reaches tissue capillaries, the gradient is reversed. Cells are continuously using oxygen for oxidative phosphorylation [p.115]. In the cells of a person at rest, intracellular P_{O_2} averages 40 mm Hg. Arterial blood arriving at the cells has a P_{O_2} of 100 mm Hg. Because P_{O_2} is lower in the cells, oxygen diffuses down its partial pressure gradient from plasma into cells. Once again, diffusion goes to equilibrium. As a result, venous blood has the same P_{O_2} as the cells it just passed.

Conversely, P_{CO_2} is higher in tissues than in systemic capillary blood because of CO_2 production during metabolism (Fig. 18.2). Cellular P_{CO_2} in a person at rest is about 46 mm Hg, compared to an arterial plasma P_{CO_2} of 40 mm Hg. The gradient causes CO_2 to diffuse out of cells into the capillaries. Diffusion goes to equilibrium, and systemic venous blood averages a P_{CO_2} of 46 mm Hg.

At the pulmonary capillaries, the process reverses. Venous blood bringing waste CO_2 from the cells has a P_{CO_2} of 46 mm Hg. Alveolar P_{CO_2} is 40 mm Hg. Because P_{CO_2} is higher in the plasma, CO_2 moves from the capillaries into the alveoli. By the time blood leaves the alveoli, it has a P_{CO_2} of 40 mm Hg, identical to the P_{CO_2} of the alveoli.

In the sections that follow we will consider some of the other factors that affect the transfer of gases between the alveoli and the body's cells.

Concept Check

Answers: p. 625

1. Cellular metabolism review: which of the following three metabolic pathways—glycolysis, the citric acid cycle, and the electron transport system—is *directly* associated with (a) O_2 consumption and with (b) CO_2 production?
2. Why doesn't the movement of oxygen from the alveoli to the plasma decrease the P_{O2} of the alveoli? [*Hint*: p. 590]
3. If nitrogen is 78% of atmospheric air, what is the partial pressure of this gas when the dry atmospheric pressure is 720 mm Hg?

GASES DIFFUSE DOWN CONCENTRATION GRADIENTS

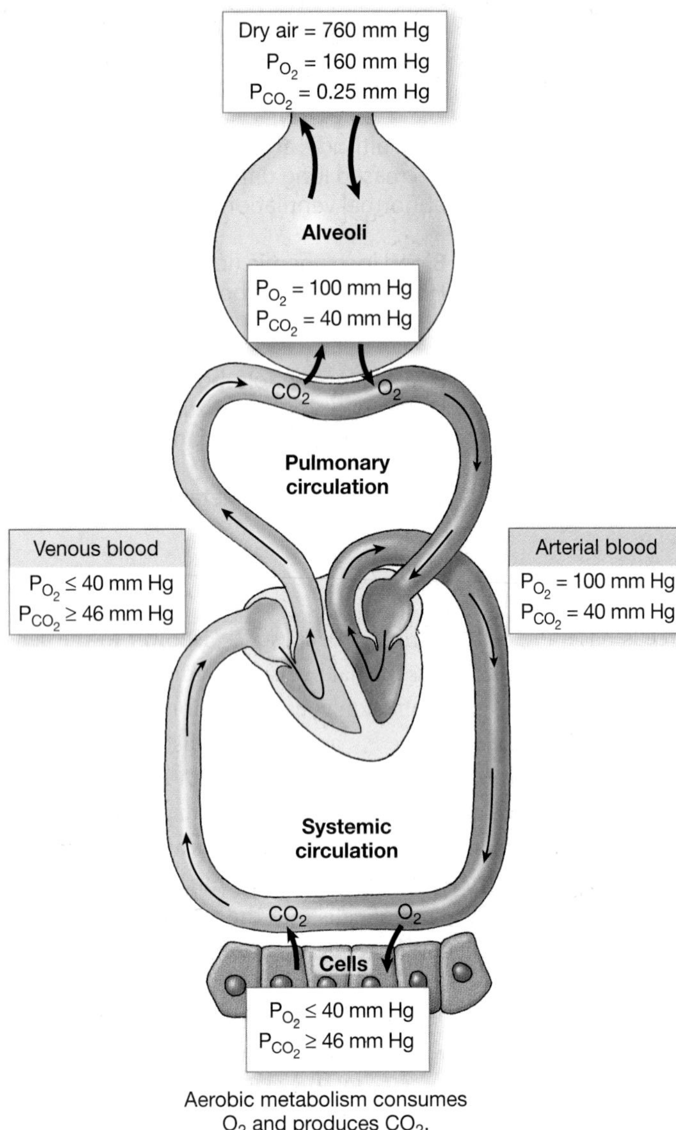

■ **Fig. 18.2**

Lower Alveolar P_{O_2} Decreases Oxygen Uptake

Many variables influence the efficiency of alveolar gas exchange and determine whether arterial blood gases are normal (■ Fig. 18.3a). First, adequate oxygen must reach the alveoli. A decrease in alveolar P_{O_2} means that less oxygen is available to enter the blood. There can also be problems with the transfer of gases between the alveoli and pulmonary capillaries. Finally, blood flow, or *perfusion,* of the alveoli must be adequate [p. 590]. If something impairs blood flow to the lung, then the body is unable to acquire the oxygen it needs. Let's look in more detail at these factors.

There are two possible causes of low alveolar P_{O_2}: either (1) the inspired air has low oxygen content or (2) alveolar ventilation [p. 590], is inadequate.

RUNNING PROBLEM

Hypoxia is the primary problem that people experience when ascending to high altitude. High altitude is considered anything above 1500 m (5000 ft), but most pathological responses to altitude occur above 2500 m (about 8000 ft). By one estimate, 25% of people arriving at 2590 m will experience some form of altitude sickness.

Q1: If water vapor contributes 47 mm Hg to the pressure of fully humidified air, what is the P_{O_2} of inspired air reaching the alveoli at 2500 m, where dry atmospheric pressure is 542 mm Hg? How does this value for P_{O_2} compare with that of fully humidified air at sea level?

Composition of the Inspired Air The first requirement for adequate oxygen delivery to the tissues is adequate oxygen intake from the atmosphere. The main factor that affects atmospheric oxygen content is altitude. The partial pressure of oxygen in air decreases along with total atmospheric pressure as you move from sea level (where normal atmospheric pressure is 760 mm Hg) to higher altitudes.

For example, Denver, 1609 m above sea level, has an atmospheric pressure of about 628 mm Hg. The P_{O_2} of dry air in Denver is 132 mm Hg, down from 160 mm Hg at sea level. For fully humidified atmospheric air reaching the alveoli, the P_{O_2} is even lower: P_{atm} 628 mm Hg$-P_{H_2O}$ 47 mm Hg)$=$581 mm Hg $\times$ 21% $= P_{O_2}$ of 122 mm Hg, down from 150 mm Hg at sea level. Notice that water vapor pressure is the same no matter what the altitude, making its contribution to total pressure in the lungs more important as you go higher.

Alveolar Ventilation Unless a person is traveling, altitude remains constant. If the composition of inspired air is normal but alveolar P_{O_2} is low, then the problem must lie with alveolar ventilation. Low alveolar ventilation is also known as *hypoventilation* and is characterized by lower-than-normal volumes of fresh air entering the alveoli. Pathological changes that can result in alveolar hypoventilation (Fig. 18.3c) include decreased lung compliance [p. 585], increased airway resistance [p. 587], or CNS depression that slows ventilation rate and decreases depth. Common causes of CNS depression in young people include alcohol poisoning and drug overdoses.

Concept Check Answers: p. 625

4. At the summit of Mt. Everest, an altitude of 8850 m, atmospheric pressure is only 250 mm Hg. What is the P_{O_2} of dry atmospheric air atop Everest? If water vapor added to inhaled air at the summit has a partial pressure of 47 mm Hg, what is the P_{O_2} of the inhaled air when it reaches the alveoli?

GAS EXCHANGE IN THE ALVEOLI

(a) Alveolar gas exchange

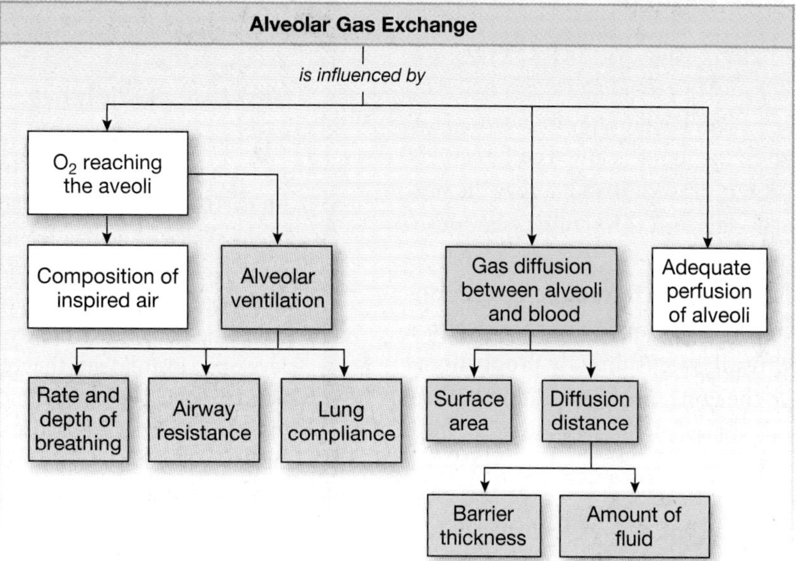

(b) Cells form a diffusion barrier between lung and blood.

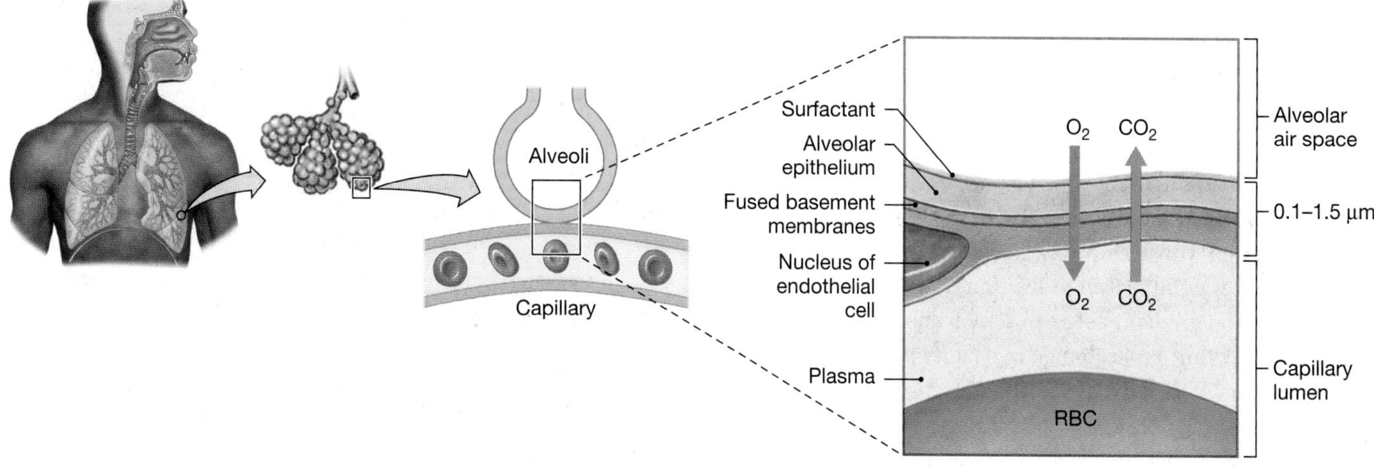

(c) Pathologies that cause hypoxia

Diffusion ∝ surface area × barrier permeability/distance[2]

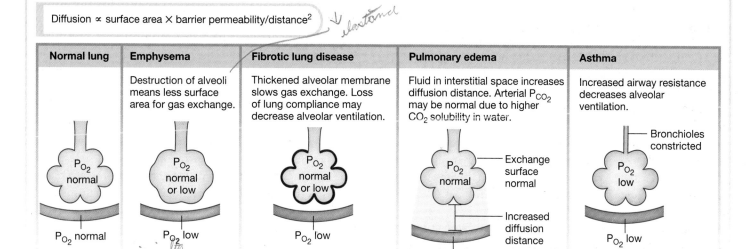

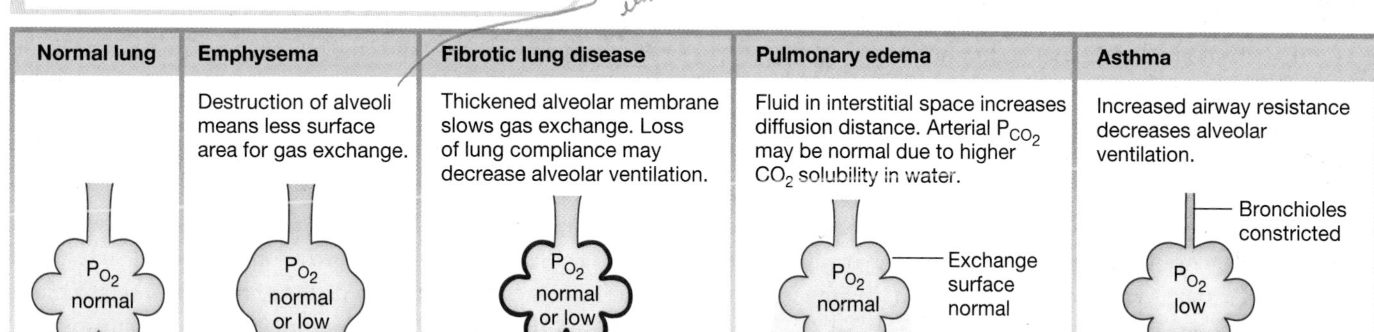

Normal lung
Emphysema
Destruction of alveoli means less surface area for gas exchange.
Fibrotic lung disease
Thickened alveolar membrane slows gas exchange. Loss of lung compliance may decrease alveolar ventilation.
Pulmonary edema
Fluid in interstitial space increases diffusion distance. Arterial P_{CO_2} may be normal due to higher CO_2 solubility in water.
Asthma
Increased airway resistance decreases alveolar ventilation.
P_{O_2} normal
P_{O_2} normal
P_{O_2} normal or low
P_{O_2} low
P_{O_2} normal or low
P_{O_2} low
P_{O_2} normal
Exchange surface normal
Increased diffusion distance
P_{O_2} low
Bronchioles constricted
P_{O_2} low
P_{O_2} low

Fig. 18.3

Diffusion Problems Cause Hypoxia

If hypoxia is not caused by hypoventilation, then the problem usually lies with some aspect of gas exchange between alveoli and blood. In these situations, alveolar P_{O_2} may be normal, but the P_{O_2} of arterial blood leaving the lungs is low. The transfer of oxygen from alveoli to blood requires diffusion across the barrier created by type I alveolar cells and the capillary endothelium (Fig. 18.3b).

The exchange of oxygen and carbon dioxide across this diffusion barrier obeys the same rules as simple diffusion across a membrane [p. 142]. The diffusion rate is directly proportional to the available surface area, the concentration gradient of the gas, and the permeability of the barrier:

$$\text{Diffusion rate} \propto \text{surface area} \times \text{concentration gradient} \times \text{barrier permeability}$$

From the general rules for diffusion, we can add a fourth factor: *diffusion distance.* Diffusion is inversely proportional to the square of the distance or, in simpler terms—diffusion is most rapid over short distances [p. 141]

$$\text{Diffusion rate} \propto 1/\text{distance}^2$$

Under most circumstances, diffusion distance, surface area, and barrier permeability in the body are constants and are maximized to facilitate diffusion. Gas exchange in the lungs is rapid, blood flow through pulmonary capillaries is slow, and diffusion reaches equilibrium in less than 1 second. This leaves the concentration gradient between alveoli and blood as the primary factor affecting gas exchange in healthy people.

The factors of surface area, diffusion distance, and membrane permeability do come into play with various diseases. Pathological changes that adversely affect gas exchange include (1) a decrease in the amount of alveolar surface area available for gas exchange, (2) an increase in the thickness of the alveolar-capillary exchange barrier, and (3) an increase in the diffusion distance between the alveolar air space and the blood.

Surface Area Physical loss of alveolar surface area can have devastating effects in *emphysema*, a degenerative lung disease most often caused by cigarette smoking (Fig. 18.3c). The irritating effect of smoke chemicals and tar in the alveoli activates alveolar macrophages that release elastase and other proteolytic enzymes. These enzymes destroy the elastic fibers of the lung [p. 87] and induce apoptosis of cells, breaking down the walls of the alveoli. The result is a high-compliance/low-elastic recoil lung with fewer and larger alveoli and less surface area for gas exchange.

Diffusion Barrier Permeability Pathological changes in the alveolar-capillary diffusion barrier may alter its properties and slow gas exchange. For example, in fibrotic lung diseases, scar tissue thickens the alveolar wall (Fig. 18.3c). Diffusion of gases through this scar tissue is much slower than normal. However, because the lungs have a built-in reserve capacity, one-third of the exchange epithelium must be incapacitated before arterial P_{O_2} falls significantly.

Diffusion Distance Normally the pulmonary diffusion distance is small because the alveolar and endothelial cells are thin and there is little or no interstitial fluid between the two cell layers (Fig. 18.3b). However, in certain pathological states, excess fluid increases the diffusion distance between the alveolar air space and the blood. Fluid accumulation may occur inside the alveoli or in the interstitial compartment between the alveolar epithelium and the capillary.

In **pulmonary edema,** accumulation of interstitial fluid increases the diffusion distance and slows gas exchange (Fig. 18.3c).

BIOTECHNOLOGY

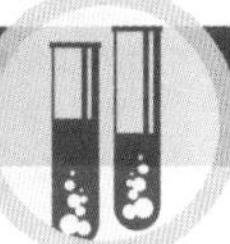

The Pulse Oximeter

One important clinical indicator of the effectiveness of gas exchange in the lungs is the concentration of oxygen in arterial blood. Obtaining an arterial blood sample is difficult for the clinician and painful for the patient because it means finding an accessible artery. (Most blood is drawn from superficial veins rather than from arteries, which lie deeper within the body). Over the years, however, scientists have developed instruments that quickly and painlessly measure blood oxygen levels through the surface of the skin on a finger or earlobe. One such instrument, the *pulse oximeter*, clips onto the skin and in seconds gives a digital reading of arterial hemoglobin saturation. The oximeter works by measuring light absorbance of the tissue at two wavelengths. Another instrument, the *transcutaneous oxygen sensor*, measures dissolved oxygen using a variant of traditional gas-measuring electrodes. Both methods have limitations but are popular because they provide a rapid, noninvasive means of estimating arterial oxygen content.

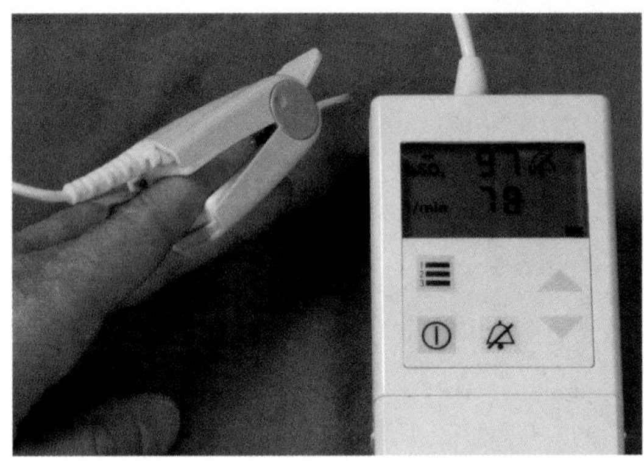

Normally, only small amounts of interstitial fluid are present in the lungs, the result of low pulmonary blood pressure and effective lymph drainage. However, if pulmonary blood pressure rises for some reason, such as left ventricular failure or mitral valve dysfunction, the normal filtration/reabsorption balance at the capillary is disrupted [Fig. 15.18, p. 531].

When capillary hydrostatic pressure increases, more fluid filters out of the capillary. If filtration increases too much, the lymphatics are unable to remove all the fluid, and excess accumulates in the pulmonary interstitial space, creating pulmonary edema. In severe cases, if edema exceeds the tissue's ability to retain it, fluid leaks from the interstitial space into the alveolar air space, flooding the alveoli. Normally the inside of the alveoli is a moist surface lined by a very thin (about 2–5 μm) layer of fluid with surfactant (see Fig. 18.3b). With alveolar flooding, this fluid layer can become much thicker and seriously impair gas exchange. Alveolar flooding can also occur with leakage when alveolar epithelium is damaged, such as from inflammation or inhaling toxic gases. If hypoxia due to alveolar fluid accumulation is severe and cannot be corrected by oxygen therapy, the condition may be called *adult respiratory distress syndrome* or ARDS.

RUNNING PROBLEM

Acute mountain sickness is the mildest illness caused by altitude hypoxia. The primary symptom is a headache that may be accompanied by dizziness, nausea, fatigue, or confusion. More severe illnesses are *high-altitude pulmonary edema* (HAPE) and *high-altitude cerebral edema.* HAPE is the major cause of death from altitude sickness. It is characterized by high pulmonary arterial pressure, extreme shortness of breath, and sometimes a productive cough yielding a pink, frothy fluid. Treatment is immediate relocation to lower altitude and administration of oxygen.

Q2: Why would someone with HAPE be short of breath?

Q3: Based on what you learned about the mechanisms for matching ventilation and perfusion in the lung [p. 590], can you explain why patients with HAPE have elevated pulmonary arterial blood pressure?

Concept Check Answers: p. 625

5. Why would left ventricular failure or mitral valve dysfunction cause elevated pulmonary blood pressure?
6. If alveolar ventilation increases, what happens to arterial P_{O_2}? To arterial P_{CO_2}? To venous P_{O_2} and P_{CO_2}? Explain your answers.

Gas Solubility Affects Diffusion

A final factor that can affect gas exchange in the alveoli is the solubility of the gas. The movement of gas molecules from air into a liquid is directly proportional to three factors: (1) the pressure gradient of the gas, (2) the solubility of the gas in the liquid, and (3) temperature. Because temperature is relatively constant in mammals, we can ignore its contribution in this discussion.

When a gas is placed in contact with water and there is a pressure gradient, gas molecules move from one phase to the other. If gas pressure is higher in the water than in the gaseous phase, then gas molecules leave the water. If gas pressure is higher in the gaseous phase than in water, then the gas dissolves into the water.

For example, consider a container of water exposed to air with a P_{O_2} of 100 mm Hg (■ Fig. 18.4a). Initially, the water has no oxygen dissolved in it (water $P_{O_2} = 0$ mm Hg). As the air stays in contact with the water, some of the moving oxygen molecules in the air diffuse into the water and dissolve (Fig. 18.4b). This process continues until equilibrium is reached. At equilibrium (Fig. 18.4c), the movement of oxygen from the air into the water is equal to the movement of oxygen from the water back into the air.

We refer to the concentration of oxygen dissolved in the water at any given P_{O_2} as the *partial pressure of the gas in solution.* In our example, therefore, if the air has a P_{O_2} of 100 mm Hg, at equilibrium the water also has a P_{O_2} of 100 mm Hg.

Note that this does *not* mean that the concentration of oxygen is the same in the air and in the water! The concentration of dissolved oxygen also depends on the *solubility* of oxygen in water. The ease with which a gas dissolves in a liquid is the **solubility** of the gas in that liquid. If a gas is very soluble, large numbers of gas molecules go into solution at a low gas partial pressure. With less soluble gases, even a high partial pressure may cause only a few molecules of the gas to dissolve in the liquid.

For example, when P_{O_2} is 100 mm Hg in both the air and the water, air contains 5.2 mmol O_2/L air, but water contains only 0.15 mmol O_2/L water (Fig. 18.4c). As you can see, oxygen is not very soluble in water and, by extension, in any aqueous solution. Its low solubility was a driving force for the evolution of oxygen-carrying molecules in the aqueous solution we call blood.

Now compare oxygen solubility with CO_2 solubility (Fig. 18.4d). Carbon dioxide is 20 times more soluble in water than oxygen is. At a P_{CO_2} of 100 mm Hg, the CO_2 concentration in air is 5.2 mmol CO_2/L air, and its concentration in water is 3.0 mmol/L water. So although P_{O_2} and P_{CO_2} are both 100 mm Hg in the water, the amount of each gas that dissolves in the water is very different.

Why is solubility important in physiology? The answer is that oxygen's low solubility in aqueous solutions means that very little oxygen can be carried dissolved in plasma. Its low solubility

GASES IN SOLUTION

When temperature remains constant, the amount of a gas that dissolves in a liquid depends on both the solubility of the gas in the liquid and the partial pressure of the gas.

Oxygen solubility

(a) Initial state: no O_2 in solution

(b) Oxygen dissolves.

(c) At equilibrium, P_{O_2} in air and water are equal. Low O_2 solubility means concentrations are not equal.

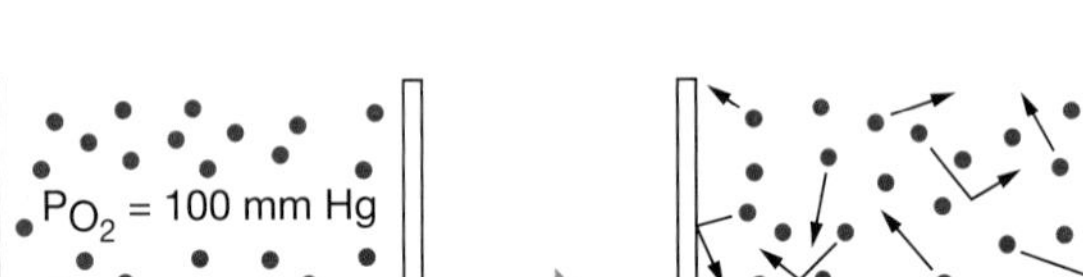

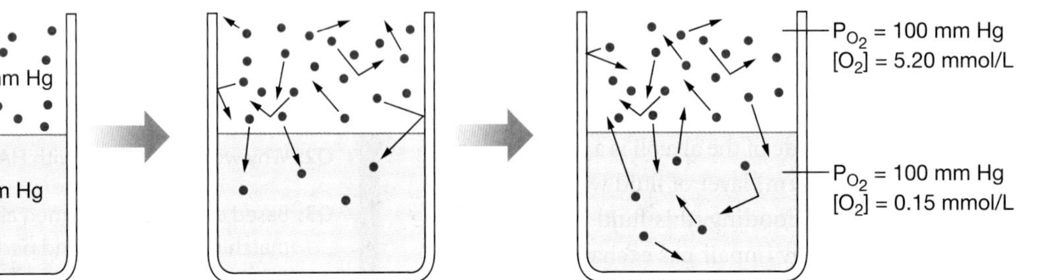

CO_2 solubility

(d) When CO_2 is at equilibrium at the same partial pressure (100 mm Hg), more CO_2 dissolves.

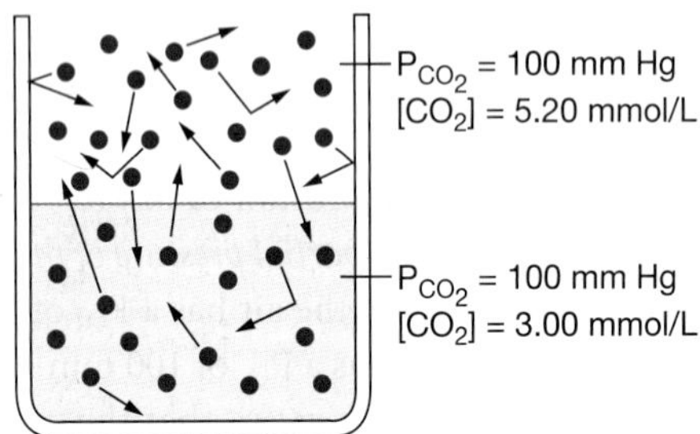

FIGURE QUESTION

Physiologists also express dissolved gases in blood using the following equation:

$[Gas]_{diss} = \alpha\ [P_{gas}]$

α for oxygen is (0.03 mL O_2/L blood)/mm Hg P_{O_2}

α for CO_2 is (0.7 mL CO_2/L blood)/mm Hg P_{CO_2}

If arterial blood has a P_{O_2} of 95 mm Hg and a P_{CO_2} of 40 mm Hg, what are the oxygen and CO_2 concentrations (in mL gas/L blood)?

■ Fig. 18.4

also means oxygen is slower to cross the increased diffusion distance present in pulmonary edema. Diffusion of oxygen into alveolar capillaries does not have time to come to equilibrium before the blood has left the capillaries. The result is decreased arterial P_{O_2} even though alveolar P_{O_2} may be normal.

Carbon dioxide, in contrast, is relatively soluble in body fluids, so increased diffusion distance may not significantly affect CO_2 exchange. In some cases of pulmonary edema, arterial P_{O_2} is low but arterial P_{CO_2} is normal because of the different solubilities of the two gases.

Concept Check Answers: p. 625

7. True or false? Plasma with a P_{O_2} of 40 mm Hg and a P_{CO_2} of 40 mm Hg has the same concentrations of oxygen and carbon dioxide.
8. A saline solution is exposed to a mixture of nitrogen gas and hydrogen gas in which $P_{H_2} = P_{N_2}$. What information do you need to predict whether equal amounts of H_2 and N_2 dissolve in the solution?

Gas Transport in the Blood

Now that we have described how gases enter and leave the capillaries, we turn our attention to oxygen and carbon dioxide transport in the blood. Gases that enter the capillaries first dissolve in the plasma. But dissolved gases play only a small part in providing the cells with oxygen. The red blood cells, or *erythrocytes*, have a critical role in ensuring that gas transport between lung and cells is adequate to meet cell needs. Without hemoglobin in the red blood cells, the blood would be unable to transport sufficient oxygen to sustain life (■ Fig. 18.5).

Oxygen transport in the circulation and oxygen consumption by tissues are excellent ways to illustrate the general principles of mass flow and mass balance. *Mass flow* [p. 13] is defined as amount of x moving per minute, where mass flow = concentration × volume flow. We can calculate the mass flow of oxygen traveling from lungs to the cells by using the oxygen content of the arterial blood × cardiac output.

OXYGEN TRANSPORT

More than 98% of the oxygen in blood is bound to hemoglobin in red blood cells, and less than 2% is dissolved in plasma.

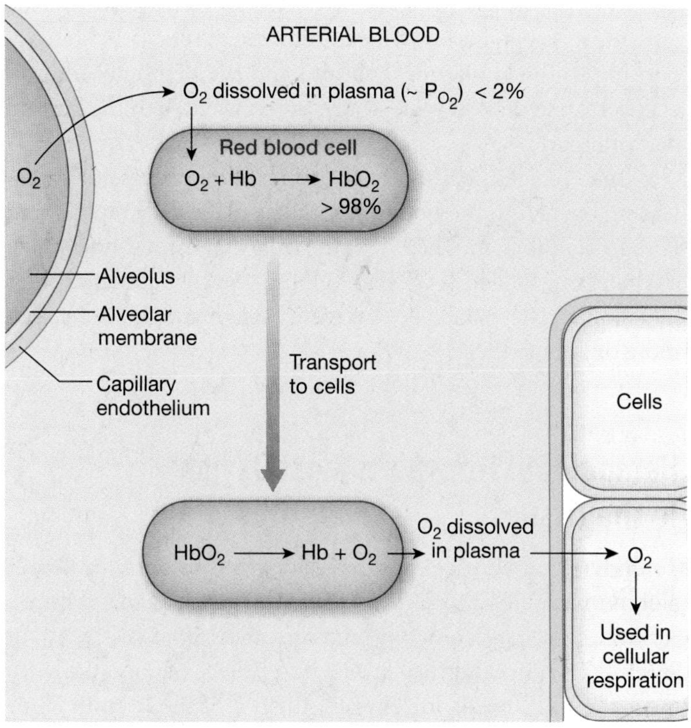

FIGURE QUESTION

How many cell membranes will O_2 cross in its passage between the airspace of the alveolus and binding to hemoglobin?

■ **Fig. 18.5**

If arterial blood contains, on average, 200 mL O_2/L blood and the cardiac output is 5 L/min:

$$\text{mL } O_2/\text{min to cells} = 200 \text{ mL } O_2/\text{L blood} \times 5 \text{ L blood/min}$$
$$= 1000 \text{ mL } O_2/\text{min delivered to tissues}$$

If we know the mass flow of oxygen in the venous blood leaving the cells, we can use the principle of mass balance [p. 11] to calculate the uptake and consumption of oxygen by the cells (■ Fig. 18.6):

$$\text{Arterial } O_2 \text{ transport} - \text{cell use of } O_2 = \text{venous } O_2 \text{ transport}$$

where oxygen transport is mass flow, mL O_2 being transported per minute. This equation rearranges to:

$$\text{Arterial } O_2 \text{ transport} - \text{venous } O_2 \text{ transport} = \text{cell use of } O_2$$

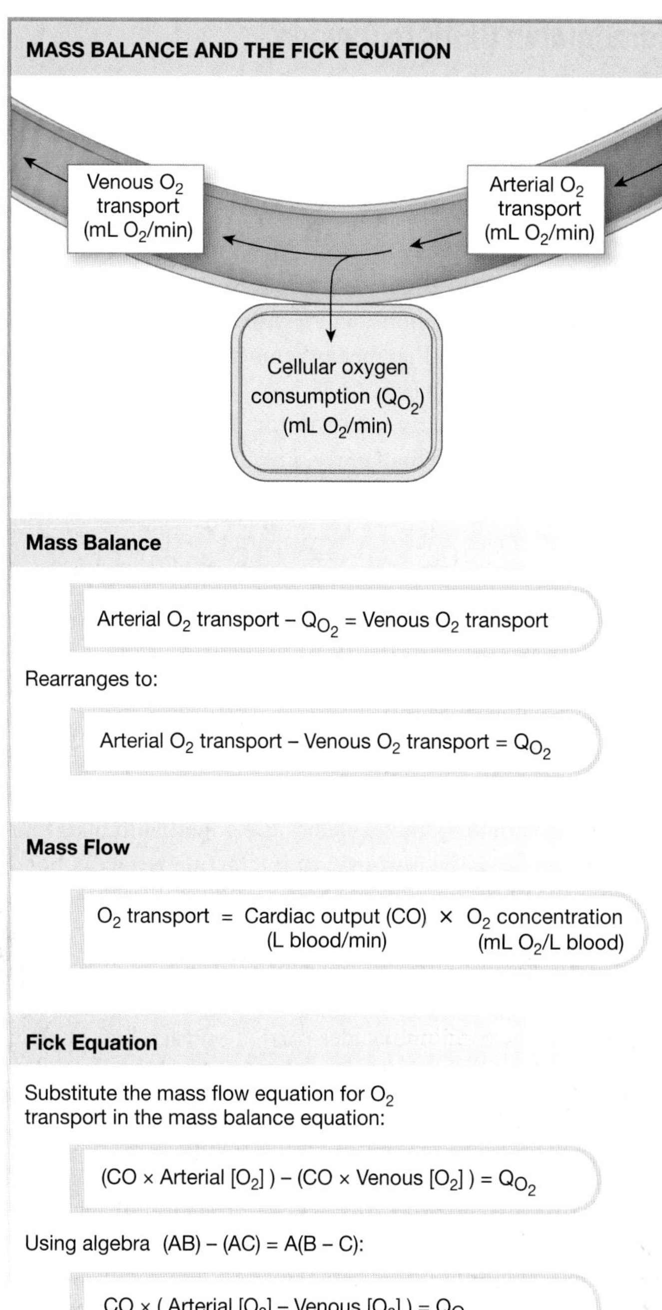

■ **Fig. 18.6**

Adolph Fick, the nineteenth-century physiologist who derived Fick's law of diffusion, combined the mass flow and mass balance equations above to relate oxygen consumption (Q_{O_2}), cardiac output (CO), and blood oxygen content. The result is the **Fick equation**:

$$Q_{O_2} = \text{CO} \times (\text{arterial oxygen content} - \text{venous oxygen content})$$

The Fick equation can be used to estimate cardiac output or oxygen consumption, assuming that arterial and venous blood gases can be measured.

Hemoglobin Binds to Oxygen

Oxygen transport in the blood has two components: the oxygen that is dissolved in the plasma (the P_{O_2}) and oxygen bound to hemoglobin (Hb). In other words:

$$\text{Total blood } O_2 \text{ content} = \text{dissolved } O_2 + O_2 \text{ bound to Hb}$$

As you learned in the previous section, oxygen is only slightly soluble in aqueous solutions, and less than 2% of all oxygen in the blood is dissolved. That means hemoglobin transports more than 98% of our oxygen (Fig. 18.5).

Hemoglobin, the oxygen-binding protein that gives red blood cells their color, binds reversibly to oxygen, as summarized in the equation

$$Hb + O_2 \rightleftharpoons HbO_2$$

Why is hemoglobin an effective oxygen carrier? The answer lies in its molecular structure. Hemoglobin (Hb) is a tetramer with four globular protein chains (*globins*), each centered around an iron-containing *heme* group [p. 554]. The central iron atom of each heme group can bind reversibly with one oxygen molecule. With four heme groups per hemoglobin molecule, one hemoglobin molecule has the potential to bind four oxygen molecules. The iron-oxygen interaction is a weak bond that can be easily broken without altering either the hemoglobin or the oxygen.

Hemoglobin bound to oxygen is known as **oxyhemoglobin**, abbreviated HbO_2. It would be more accurate to show the number of oxygen molecules carried on each hemoglobin molecule—$Hb(O_2)_{1-4}$—but we use the simpler abbreviation because the number of bound oxygen molecules varies from one hemoglobin molecule to another.

Oxygen Binding Obeys the Law of Mass Action

The hemoglobin binding reaction $Hb + O_2 \rightleftharpoons HbO_2$ obeys the law of mass action [p. 51]. As the concentration of free O_2 increases, more oxygen binds to hemoglobin and the equation shifts to the right, producing more HbO_2. If the concentration of O_2 decreases, the equation shifts to the left. Hemoglobin releases oxygen and the amount of oxyhemoglobin decreases.

In the blood, the free oxygen available to bind to hemoglobin is dissolved oxygen, indicated by the P_{O_2} of plasma (Fig. 18.5). In the pulmonary capillaries, oxygen from the alveoli dissolves in plasma. Dissolved O_2 then diffuses into the red blood cells, where it can binds to hemoglobin. The hemoglobin acts like a sponge, soaking up oxygen from the plasma until the reaction $Hb + O_2 \rightleftharpoons HbO_2$ reaches equilibrium.

The transfer of oxygen from alveolar air to plasma to red blood cells and onto hemoglobin occurs so rapidly that blood in the pulmonary capillaries normally picks up as much oxygen as the P_{O_2} of the plasma and the number of red blood cells permit.

Once arterial blood reaches the tissues, the exchange process that took place in the lungs reverses. Dissolved oxygen diffuses out of systemic capillaries into cells, and the resultant decrease in plasma P_{O_2} disturbs the equilibrium of the oxygen-hemoglobin binding reaction by removing oxygen from the left side of the equation. The equilibrium shifts to the left according to the law of mass action, and the hemoglobin molecules release their oxygen stores, as represented in the bottom half of Figure 18.5.

Like oxygen loading at the lungs, this process of transferring oxygen to the body's cells takes place very rapidly and goes to equilibrium. The P_{O_2} of the cells determines how much oxygen is unloaded from hemoglobin. As cells increase their metabolic activity, their P_{O_2} decreases, and hemoglobin releases more oxygen to them.

Hemoglobin Transports Most Oxygen to the Tissues

To understand why we must have adequate amounts of hemoglobin in our blood to survive, consider the following example. Assume that a person's oxygen consumption at rest is about 250 mL O_2/min and the cardiac output is 5 L blood/min. To meet the cells' needs for oxygen, the 5 L of blood/min coming to the tissues would need to contain at least 250 mL O_2, or 50 mL O_2/L blood.

The low solubility of oxygen means that only 3 mL of O_2 will dissolve in the plasma fraction of 1 liter of arterial blood (■ Fig. 18.7a). The dissolved oxygen delivery to the cells is

$$3 \text{ mL } O_2/\text{L blood} \times 5 \text{ L blood/min} = 15 \text{ mL } O_2/\text{min}$$

The cells use at least 50 mL O_2/min, so the small amount of oxygen that dissolves in plasma cannot meet the needs of the tissues at rest.

Now let's consider the difference in oxygen delivery if hemoglobin is available. At normal hemoglobin levels, red blood cells carry about 197 mL O_2/L blood (Fig. 18.7b).

$$\begin{aligned}\text{Total blood } O_2 \text{ content} &= \text{dissolved } O_2 + O_2 \text{ bound to Hb}\\ &= 3 \text{ mL } O_2/\text{L blood} + 197 \text{ mL } HbO_2/\text{L blood}\\ &= 200 \text{ mL } O_2/\text{L blood}\end{aligned}$$

If cardiac output remains 5 L/min, hemoglobin-assisted oxygen delivery to cells is 1000 mL/min:

$$200 \text{ mL } O_2/\text{L blood} \times 5 \text{ L blood/min} = 1000 \text{ mL } O_2/\text{min}$$

This is four times the oxygen consumption needed by the tissues at rest. The extra serves as a reserve for times when oxygen demand increases, such as with exercise.

HEMOGLOBIN INCREASES OXYGEN TRANSPORT

(a) Oxygen transport in blood without hemoglobin. Alveolar P_{O_2} = arterial P_{O_2}

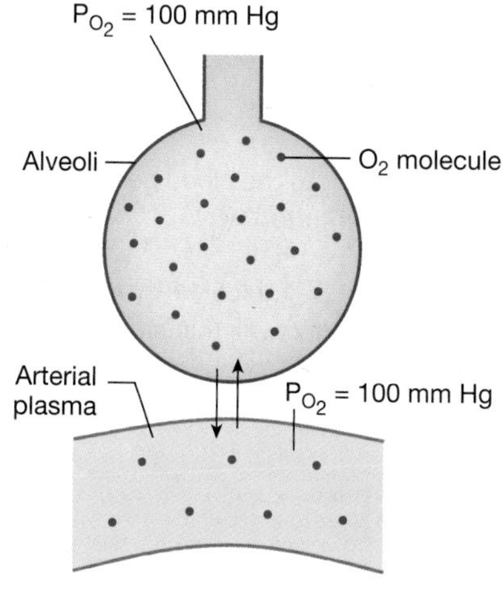

Oxygen dissolves in plasma.

O_2 content of plasma	= 3 mL O_2/L blood
O_2 content of red blood cells	= 0
Total O_2 carrying capacity	3 mL O_2/L blood

(b) Oxygen transport at normal P_{O_2} in blood with hemoglobin

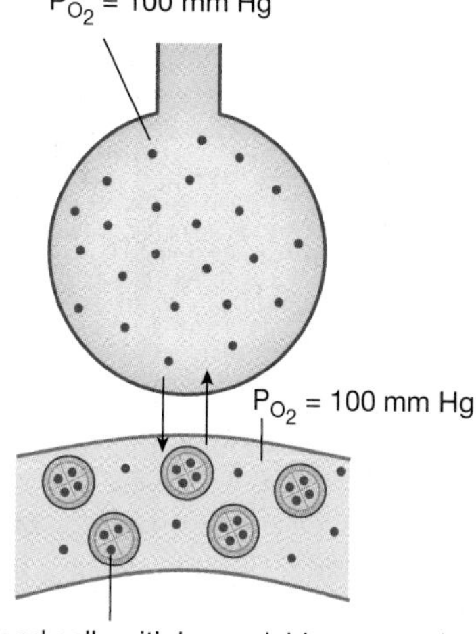

Red blood cells with hemoglobin are carrying 98% of their maximum load of oxygen.

O_2 content of plasma =	3 mL O_2/L blood
O_2 content of red blood cells	= 197 mL O_2/L blood
Total O_2 carrying capacity	200 mL O_2/L blood

(c) Oxygen transport at reduced P_{O_2} in blood with hemoglobin

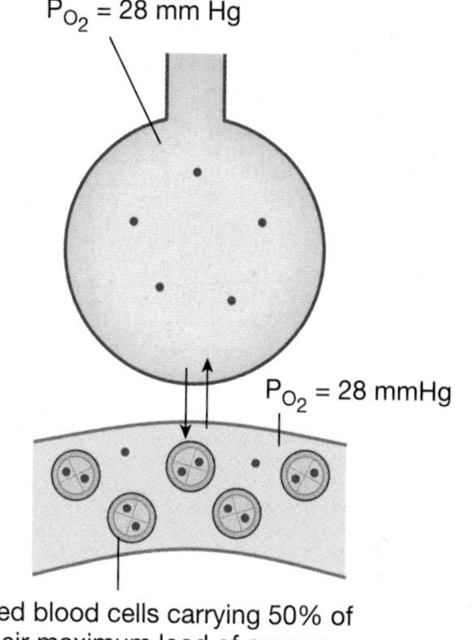

Red blood cells carrying 50% of their maximum load of oxygen.

O_2 content of plasma =	0.8 mL O_2/L blood
O_2 content of red blood cells	= 99.5 mL O_2/L blood
Total O_2 carrying capacity	100.3 mL O_2/L blood

■ **Fig. 18.7**

EMERGING CONCEPTS

Blood Substitutes

Physiologists have been attempting to find a substitute for blood ever since 1878, when an intrepid physician named T. Gaillard Thomas transfused a patient with whole milk in place of blood. (It helped but the patient died anyway.) Although milk seems an unlikely replacement for blood, it has two important properties: proteins to provide colloid osmotic pressure and molecules (emulsified lipids) capable of binding to oxygen. In the development of hemoglobin substitutes, oxygen transport is the most difficult property to mimic. A hemoglobin solution would seem to be the obvious answer, but hemoglobin that is not compartmentalized in red blood cells behaves differently than hemoglobin that is compartmentalized. Investigators are making progress by polymerizing hemoglobin into larger, more stable molecules and loading these hemoglobin polymers into phospholipid liposomes [p. 66]. Perfluorocarbon emulsions are also being tested as oxygen carriers. To learn more about this research, read "Physiological properties of blood substitutes," *News Physiol Sci* 16(1): 38–41, 2001 Feb (*http://nips.physiology.org*).

P_{O_2} Determines Oxygen-Hb Binding

The amount of oxygen that binds to hemoglobin depends on two factors: (1) the P_{O_2} in the plasma surrounding the red blood cells and (2) the number of potential Hb binding sites available in the red blood cells (■ Fig. 18.8). Plasma P_{O_2} is the primary factor determining what percentage of the available hemoglobin binding sites are occupied by oxygen, known as *the percent saturation of hemoglobin*. As you learned in previous sections, arterial P_{O_2} is established by (1) the composition of inspired air, (2) the alveolar ventilation rate, and (3) the efficiency of gas exchange from alveoli to blood. Figure 18.7c shows what happens to O_2 transport when P_{O_2} decreases.

The total number of oxygen-binding sites depends on the number of hemoglobin molecules in red blood cells. Clinically, this number can be estimated either by counting the red blood cells and quantifying the amount of hemoglobin per red blood cell (*mean corpuscular hemoglobin*) or by determining the blood hemoglobin content (g Hb/dL whole blood). Any pathological condition that decreases the amount of hemoglobin in the cells or the number of red blood cells adversely affects the blood's oxygen-transporting capacity.

People who have lost large amounts of blood need to replace hemoglobin for oxygen transport. A blood transfusion is

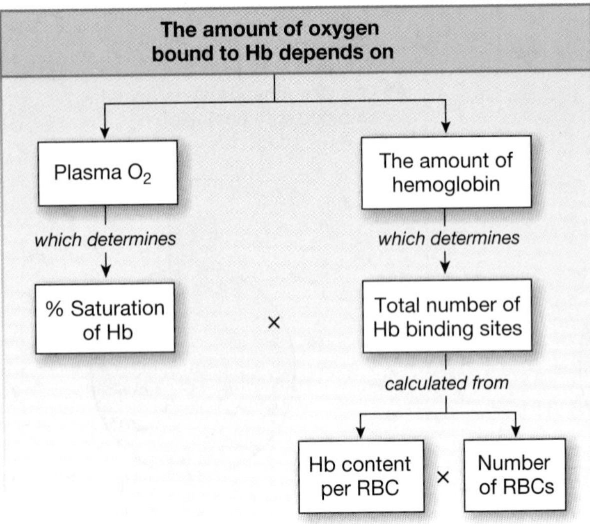

Fig. 18.8

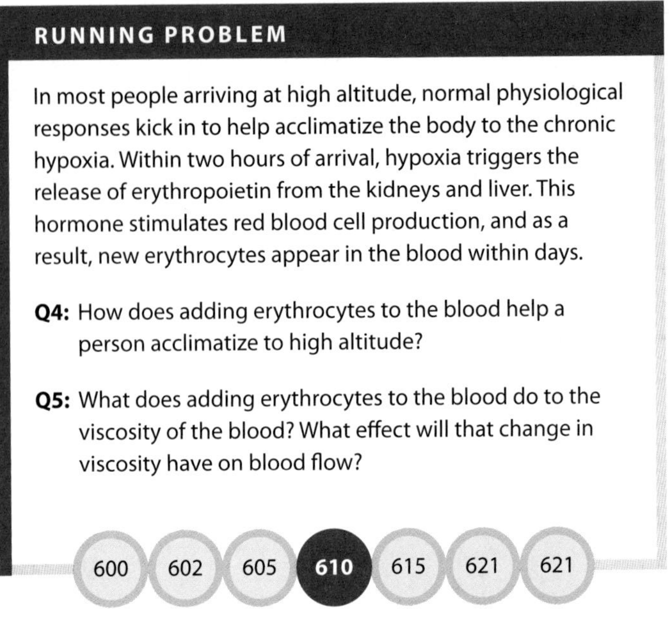

RUNNING PROBLEM

In most people arriving at high altitude, normal physiological responses kick in to help acclimatize the body to the chronic hypoxia. Within two hours of arrival, hypoxia triggers the release of erythropoietin from the kidneys and liver. This hormone stimulates red blood cell production, and as a result, new erythrocytes appear in the blood within days.

Q4: How does adding erythrocytes to the blood help a person acclimatize to high altitude?

Q5: What does adding erythrocytes to the blood do to the viscosity of the blood? What effect will that change in viscosity have on blood flow?

600 602 605 610 615 621 621

the ideal replacement for blood loss, but in emergencies this is not always possible. Saline infusions can replace lost blood volume, but saline (like plasma) cannot transport sufficient quantities of oxygen to support cellular respiration. Faced with this problem, researchers are currently testing artificial oxygen carriers to replace hemoglobin. In times of large-scale disasters, these hemoglobin substitutes would eliminate the need to identify a patient's blood type before giving transfusions.

Oxygen Binding Is Expressed As a Percentage

As you just learned, the amount of oxygen bound to hemoglobin at any given P_{O_2} is expressed as the **percent saturation of hemoglobin**, where

$$(\text{Amount of } O_2 \text{ bound/maximum that could be bound}) \times 100 = \text{percent saturation of hemoglobin}$$

If all binding sites on all hemoglobin molecules are occupied by oxygen molecules, the blood is 100% oxygenated, or *saturated* with oxygen. If half the available binding sites are carrying oxygen, the hemoglobin is 50% saturated, and so on.

The relationship between plasma P_{O_2} and percent saturation of hemoglobin can be explained with the following analogy. The hemoglobin molecules carrying oxygen are like students moving books from an old library to a new one. Each student (a hemoglobin molecule) can carry a maximum of four books (100% saturation). The librarian in charge controls how many books (O_2 molecules) each student will carry, just as plasma P_{O_2} determines the percent saturation of hemoglobin.

The total number of books being carried depends on the number of available students, just as the amount of oxygen delivered to the tissues depends on the number of available hemoglobin molecules. For example, if there are 100 students, and the librarian gives each of them four books (100% saturation), then 400 books are carried to the new library. If the librarian gives three books to each student (decreased plasma P_{O_2}), then only 300 books go to the new library, even though each student could carry four. (Students carrying only three of a possible four books correspond to 75% saturation of hemoglobin.) If the librarian is handing out four books per student but only 50 students show up (fewer hemoglobin molecules), then only 200 books get to the new library, even though the students are taking the maximum number of books they can carry.

The physical relationship between P_{O_2} and how much oxygen binds to hemoglobin can be studied *in vitro*. Researchers expose samples of hemoglobin to various P_{O_2} levels and quantitatively determine the amount of oxygen that binds. **Oxyhemoglobin saturation curves**, such as the ones shown in Figure 18.9, are the result of these *in vitro* binding studies. (These curves are also called *dissociation curves*.)

The shape of the $Hb \cdot O_2$ saturation curve reflects the properties of the hemoglobin molecule and its affinity for oxygen. If you look at the curve, you find that at normal alveolar and arterial P_{O_2} (100 mm Hg), 98% of the hemoglobin is bound to oxygen (Fig. 18.9a). In other words, as blood passes through the lungs under normal conditions, hemoglobin picks up nearly the maximum amount of oxygen that it can carry.

Notice that the curve is nearly flat at P_{O_2} levels higher than 100 mm Hg (that is, the slope approaches zero). At P_{O_2} above 100 mm Hg, even large changes in P_{O_2} cause only minor changes in percent saturation. In fact, hemoglobin is not 100% saturated until the P_{O_2} reaches nearly 650 mm Hg, a partial pressure far higher than anything we encounter in everyday life.

The flattening of the saturation curve at higher P_{O_2} also means that alveolar P_{O_2} can fall a good bit below 100 mm Hg

Fig. 18.9 ESSENTIALS

Oxygen-hemoglobin Binding Curves

Binding properties of adult and fetal hemoglobin

(a) The oxyhemoglobin saturation curve is determined *in vitro* in the laboratory.

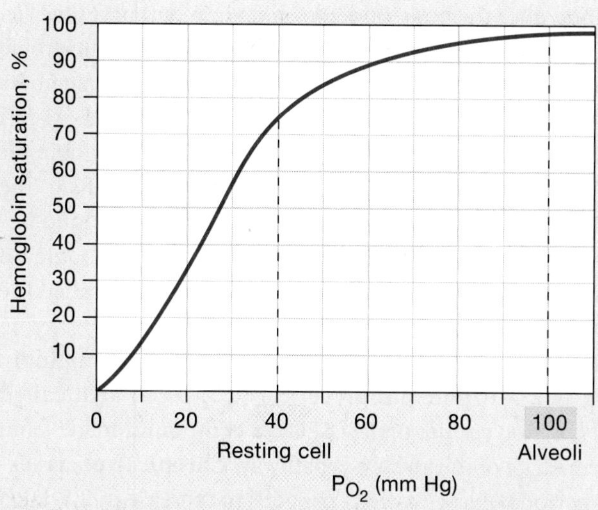

(b) Maternal and fetal hemoglobin have different oxygen-binding properties.

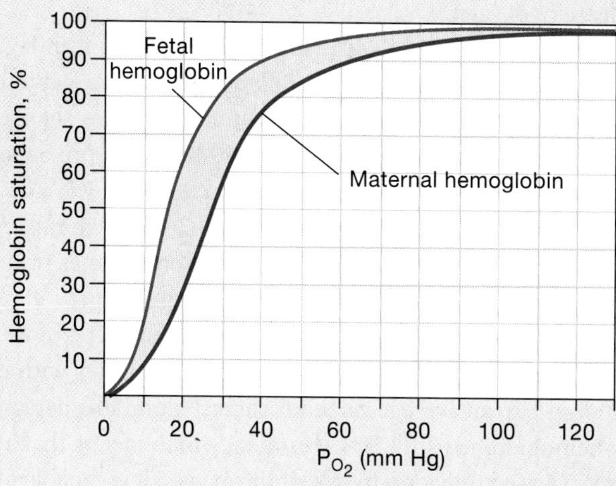

Physical factors alter hemoglobin's affinity for oxygen

(c) Effect of pH

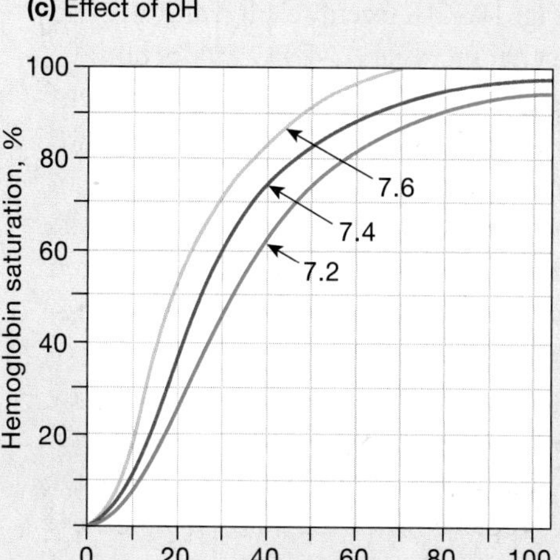

(d) Effect of temperature

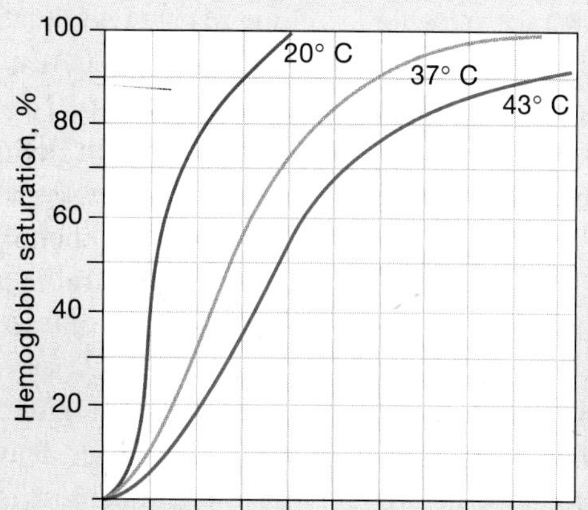

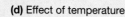

(e) Effect of P_{CO_2}

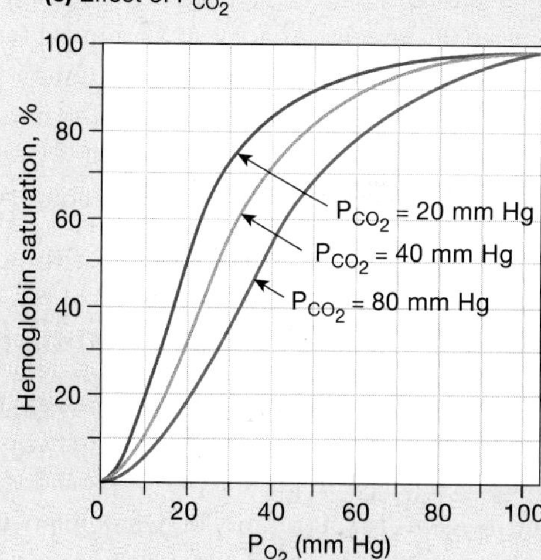

(f) Effect of the metabolic compound 2,3-DPG

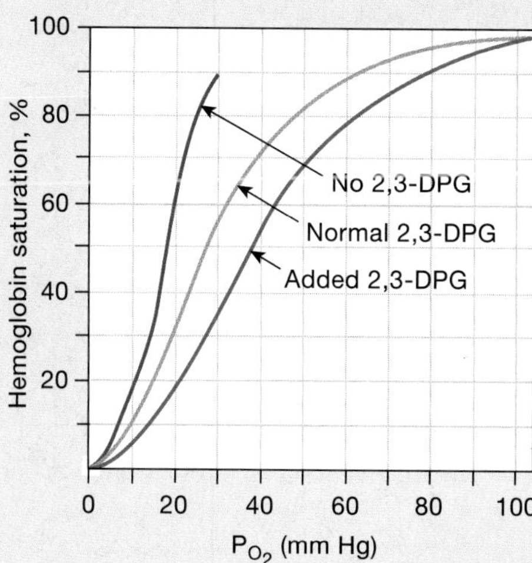

GRAPH QUESTIONS

1. For the graph in (a):
   (a) When the P_{O_2} is 20 mm Hg, what is the percent O_2 saturation of hemoglobin?
   (b) At what P_{O_2} is hemoglobin 50% saturated with O_2?
2. At a P_{O_2} of 20 mm Hg, how much more oxygen is released at an exercising muscle cell whose pH is 7.2 than at a cell with a pH of 7.4?
3. What happens to oxygen release when the exercising muscle cell warms up?
4. Blood stored in blood banks loses its normal content of 2,3-DPG. Is this good or bad? Explain.
5. Because of incomplete gas exchange across the thick membranes of the placenta, hemoglobin in fetal blood leaving the placenta is 80% saturated with oxygen. What is the P_{O_2} of that placental blood?
6. Blood in the vena cava of the fetus has a P_{O_2} around 10 mm Hg. What is the percent O_2 saturation of maternal hemoglobin at the same P_{O_2}?

without significantly lowering hemoglobin saturation. As long as P_{O_2} in the alveoli (and thus in the pulmonary capillaries) stays above 60 mm Hg, hemoglobin is more than 90% saturated and maintains near-normal levels of oxygen transport. However, once P_{O_2} falls below 60 mm Hg, the curve becomes steeper. The steep slope means that a small decrease in P_{O_2} causes a relatively large release of oxygen.

For example, if P_{O_2} falls from 100 mm Hg to 60 mm Hg, the percent saturation of hemoglobin goes from 98% to about 90%, a decrease of 8%. This is equivalent to a saturation change of 2% for each 10 mm Hg change. If P_{O_2} falls further, from 60 to 40 mm Hg, the percent saturation goes from 90% to 75%, a decrease of 7.5% for each 10 mm Hg. In the 40–20 mm Hg range, the curve is even steeper. Hemoglobin saturation declines from 75% to 35%, a change of 20% for each 10 mm Hg change.

What is the physiological significance of the shape of the saturation curve? In blood leaving systemic capillaries with a P_{O_2} of 40 mm Hg (an average value for venous blood in a person at rest), hemoglobin is still 75% saturated, which means that at the cells it released only one-fourth of the oxygen it is capable of carrying. The oxygen that remains bound serves as a reservoir that cells can draw on if metabolism increases.

When metabolically active tissues use additional oxygen, their cellular P_{O_2} decreases, and hemoglobin releases additional oxygen at the cells. At a P_{O_2} of 20 mm Hg (an average value for exercising muscle), hemoglobin saturation falls to about 35%. With this 20 mm Hg decrease in P_{O_2} (40 mm Hg to 20 mm Hg), hemoglobin releases an additional 40% of the oxygen it is capable of carrying. This is another example of the built-in reserve capacity of the body.

Several Factors Affect Oxygen-Hb Binding

Any factor that changes the conformation of the hemoglobin protein may affect its ability to bind oxygen. In humans, physiological changes in plasma pH, P_{CO_2}, and temperature all alter the oxygen-binding affinity of hemoglobin. Changes in binding affinity are reflected by changes in the shape of the HbO_2 saturation curve.

Increased temperature, increased P_{CO_2}, or decreased pH decrease the affinity of hemoglobin for oxygen and shift the oxygen-hemoglobin saturation curve to the right (Fig. 18.9c–e). When these factors change in the opposite direction, binding affinity increases, and the curve shifts to the left. Notice that when the curve shifts in either direction, the changes are much more pronounced in the steep part of the curve. Physiologically, this means that oxygen binding at the lungs (in the 90–100 mm Hg P_{O_2} range) is not greatly affected, but oxygen delivery at the tissues (in the 20–40 mm Hg range) is significantly altered.

Let's examine one situation, the affinity shift that takes place when pH decreases from 7.4 (normal) to 7.2 (more acidic). (The normal range for blood pH is 7.38–7.42, but a pH of 7.2 is compatible with life). Look at the graph in Figure 18.c. At a P_{O_2} of 40 mm Hg (equivalent to a resting cell) and pH of 7.4, hemoglobin is about 75% saturated. At the same P_{O_2}, if the pH falls to 7.2, the percent saturation decreases to about 62%. This means that hemoglobin molecules release 13% more oxygen at pH 7.2 than they do at pH 7.4.

When does the body undergo shifts in blood pH? One situation is with maximal exertion that pushes cells into anaerobic metabolism. Anaerobic metabolism in exercising muscle fibers releases H^+ into the cytoplasm and extracellular fluid. As H^+ concentrations increase, pH falls, the affinity of hemoglobin for oxygen decreases, and the HbO_2 saturation curve shifts to the right. More oxygen is released at the tissues as the blood becomes more acidic (pH decreases). A shift in the hemoglobin saturation curve that results from a change in pH is called the **Bohr effect**.

An additional factor that affects oxygen-hemoglobin binding is **2,3-diphosphoglycerate** (2,3-DPG; also called *2,3-bisphosphoglycerate* or 2,3-BPG), a compound made from an intermediate of the glycolysis pathway. **Chronic hypoxia** (extended periods of low oxygen) triggers an increase in 2,3-DPG production in red blood cells. Increased levels of 2,3-DPG lower the binding affinity of hemoglobin and shift the HbO_2 saturation curve to the right (Fig. 18.9f). Ascent to high altitude and anemia are two situations that increase 2,3-DPG production.

Changes in hemoglobin's structure also change its oxygen-binding affinity. For example, *fetal hemoglobin* (*HbF*) has two gamma protein chains in place of the two beta chains found in adult hemoglobin. The presence of gamma chains enhances the ability of fetal hemoglobin to bind oxygen in the low-oxygen environment of the placenta. The altered binding affinity is reflected by the different shape of the fetal HbO_2 saturation curve (Fig. 18.9b). At any given placental P_{O_2}, oxygen released by maternal hemoglobin is picked up by the higher-affinity fetal hemoglobin for delivery to the developing fetus. Shortly after birth, fetal hemoglobin is replaced with the adult form as new red blood cells are made.

■ Figure 18.10 summarizes all the factors that influence the total oxygen content of arterial blood.

Concept Check

Answers: p. 625

9. Can a person breathing 100% oxygen at sea level achieve 100% saturation of her hemoglobin?
10. What effect does hyperventilation have on the percent saturation of arterial hemoglobin? [*Hint*: Fig. 17.13, p. 592]
11. A muscle that is actively contracting may have a cellular P_{O_2} of 25 mm Hg. What happens to oxygen binding to hemoglobin at this low P_{O_2}? What is the P_{O_2} of the venous blood leaving the active muscle?

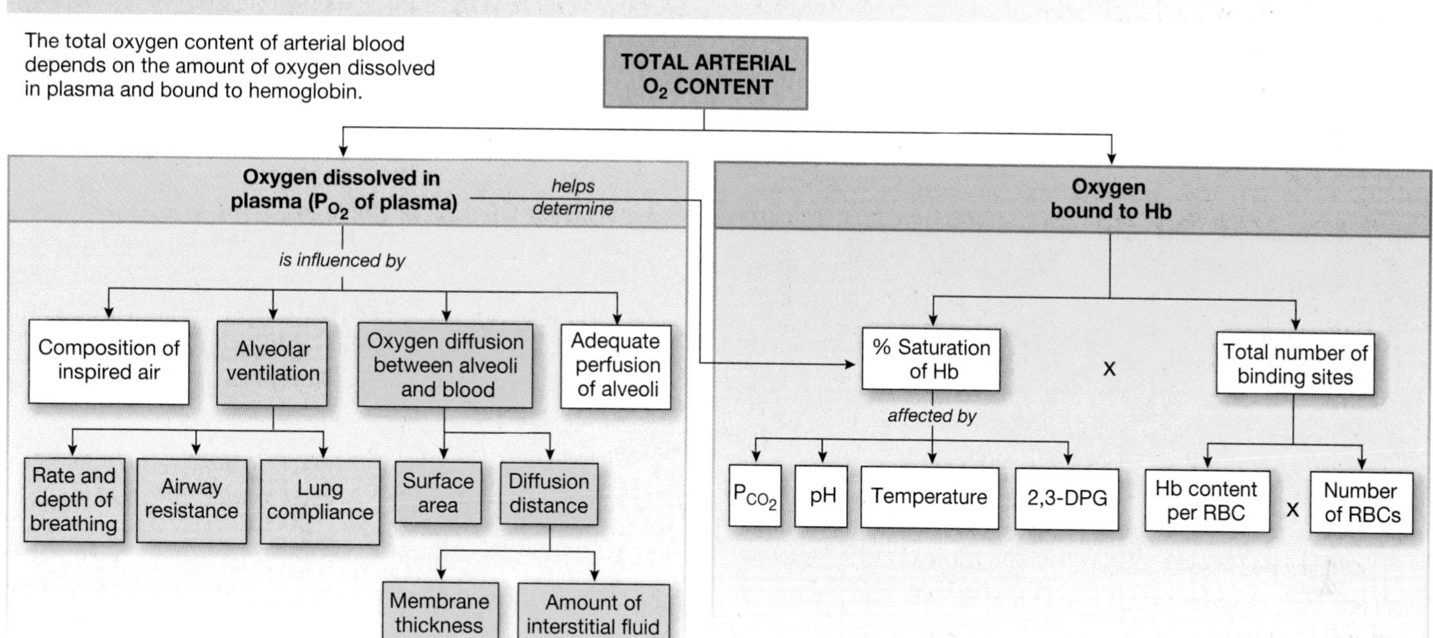

■ **Fig. 18.10**

18

Carbon Dioxide Is Transported in Three Ways

Gas transport in the blood includes carbon dioxide removal from the cells as well as oxygen delivery to cells, and hemoglobin also plays an important role in CO_2 transport. Carbon dioxide is a by-product of cellular respiration [p. 114]. It is more soluble in body fluids than oxygen is, but the cells produce far more CO_2 than can dissolve in the plasma. Only about 7% of the CO_2 carried by venous blood is dissolved in the blood. The remaining 93% diffuses into red blood cells, where 70% is converted to bicarbonate ion, as explained below, and 23% binds to hemoglobin ($HbCO_2$). ■ Figure 18.11 summarizes these three mechanisms of carbon dioxide transport in the blood.

Why is removing CO_2 from the body so important? First, elevated P_{CO_2} (*hypercapnia*) causes the pH disturbance known as *acidosis*. Extremes of pH interfere with hydrogen bonding of molecules and can denature proteins [p. 54]. Abnormally high P_{CO_2} levels also depress central nervous system function, causing confusion, coma, or even death. For these reasons, CO_2 is a potentially toxic waste product that must be removed by the lungs.

CO_2 and Bicarbonate Ions As we just noted, about 70% of the CO_2 that enters the blood is transported to the lungs as bicarbonate ions (HCO_3^-) dissolved in the plasma. The conversion of CO_2 to HCO_3^- serves two purposes: (1) it provides an additional means of CO_2 transport from cells to lungs, and (2) HCO_3^- is available to act as a buffer for metabolic acids [p. 49], thereby helping stabilize the body's pH.

How does CO_2 turn into HCO_3^-? The rapid conversion depends on the presence of **carbonic anhydrase (CA)**, an enzyme found concentrated in red blood cells. Let's see how this happens. Dissolved CO_2 in the plasma diffuses into red blood cells, where it may react with water in the presence of carbonic anhydrase to form *carbonic acid* (H_2CO_3, top portion of Fig. 18.11). Carbonic acid then dissociates into a hydrogen ion and a bicarbonate ion:

$$CO_2 + H_2O \underset{}{\overset{\text{Carbonic anhydrase}}{\rightleftharpoons}} \underset{\text{Carbonic acid}}{H_2CO_3} \rightleftharpoons H^+ + HCO_3^-$$

Because carbonic acid dissociates readily, we sometimes ignore the intermediate step and summarize the reaction as:

$$CO_2 + H_2O \rightleftharpoons H^+ + HCO_3^-$$

This reaction is reversible. The rate in either direction depends on the relative concentrations of the substrates and obeys the law of mass action.

The conversion of carbon dioxide to H^+ and HCO_3^- continues until equilibrium is reached. (Water is always in excess in the body, so water concentration plays no role in the dynamic equilibrium of this reaction.) To keep the reaction going, the products (H^+ and HCO_3^-) must be removed from the cytoplasm of the red blood cell. If the product concentrations are kept low, the reaction cannot reach equilibrium. Carbon dioxide continues to move out of plasma into the red blood cells, which in turn allows more CO_2 to diffuse out of tissues into the blood.

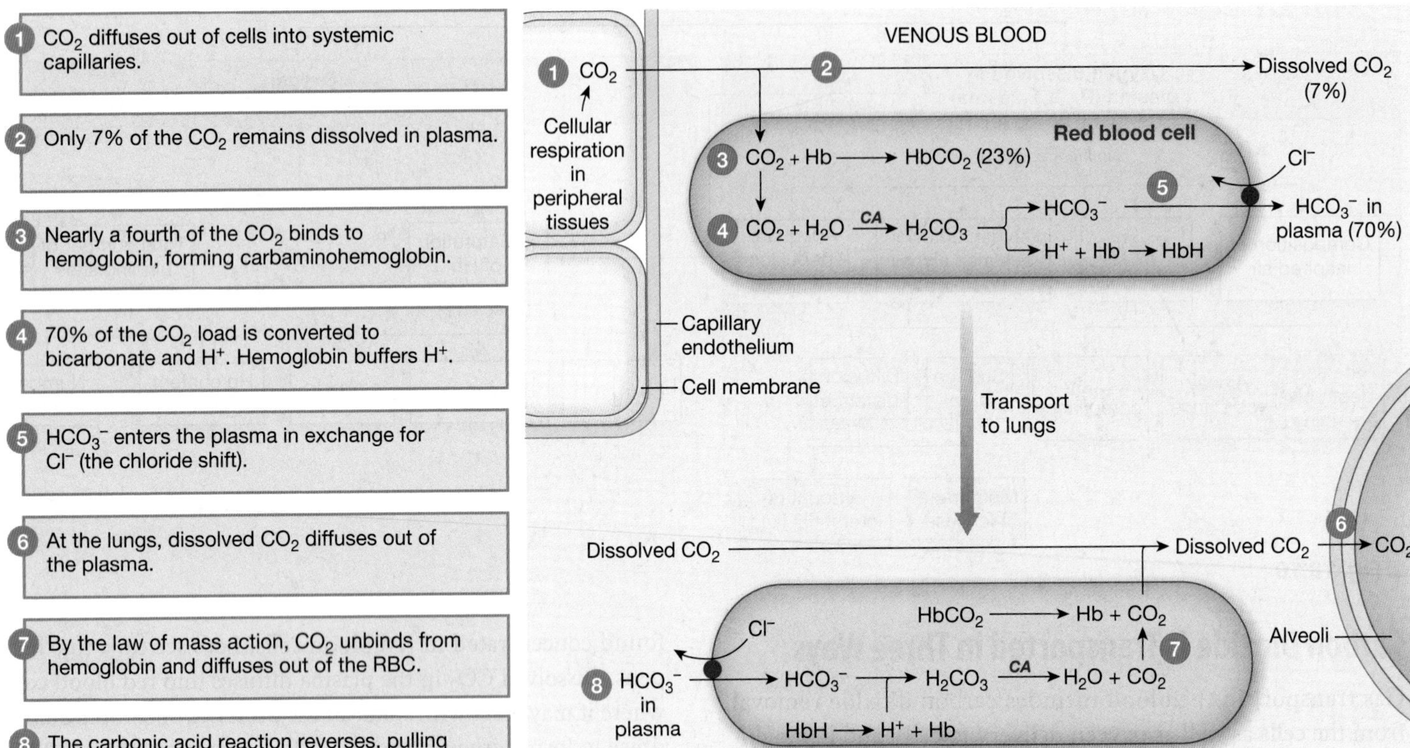

Fig. 18.11

Two separate mechanisms remove free H^+ and HCO_3^-. In the first, bicarbonate leaves the red blood cell on an antiport protein [p. 148]. This transport process, known as the **chloride shift**, exchanges HCO_3^- for Cl^-. The anion exchange maintains the cell's electrical neutrality. The transfer of HCO_3^- into the plasma makes this buffer available to moderate pH changes caused by the production of metabolic acids. Bicarbonate is the most important extracellular buffer in the body.

Hemoglobin and H⁺ The second mechanism removes free H^+ from the red blood cell cytoplasm. Hemoglobin within the red blood cell acts as a buffer and binds hydrogen ions in the reaction

$$H^+ + Hb \rightleftharpoons HbH$$

Hemoglobin's buffering of H^+ is an important step that prevents large changes in the body's pH. If blood P_{CO_2} is elevated much above normal, the hemoglobin buffer cannot soak up all the H^+ produced from the reaction of CO_2 and water. In those cases, excess H^+ accumulates in the plasma, causing the condition known as **respiratory acidosis**. You will learn more about the role of the respiratory system in maintaining pH homeostasis when you study acid-base balance.

Hemoglobin and CO₂ Although most carbon dioxide that enters red blood cells is converted to bicarbonate ions, about 23% of the CO_2 in venous blood binds directly to hemoglobin. At the cells, when oxygen leaves its binding sites on the hemoglobin molecule, CO_2 binds with free hemoglobin at exposed amino groups ($-NH_2$), forming **carbaminohemoglobin:**

$$CO_2 + Hb \rightleftharpoons HbCO_2 \text{ (carbaminohemoglobin)}$$

The presence of CO_2 and H^+ facilitates formation of carbaminohemoglobin because both these factors decrease hemoglobin's binding affinity for oxygen (see Fig. 18.9).

RUNNING PROBLEM

The usual homeostatic response to high-altitude hypoxia is hyperventilation, which begins on arrival. Hyperventilation enhances alveolar ventilation, but this may not help elevate arterial P_{O_2} levels significantly when atmospheric P_{O_2} is low. However, hyperventilation does lower plasma P_{CO_2}.

Q6: What happens to plasma pH during hyperventilation? (*Hint*: Apply the law of mass action to figure out what happens to the balance between CO_2 and $H^+ + HCO_3^-$).

Q7: How does this change in pH affect oxygen binding at the lungs when P_{O_2} is decreased? How does it affect unloading of oxygen at the cells?

CO_2 Removal at the Lungs When venous blood reaches the lungs, the processes that took place in the systemic capillaries reverse (bottom portion of Fig. 18.11). The P_{CO_2} of the alveoli is lower than that of venous blood in the pulmonary capillaries. Therefore, CO_2 diffuses down its pressure gradient—in other words, out of plasma into the alveoli—and the plasma P_{CO_2} begins to fall.

The decrease in plasma P_{CO_2} allows dissolved CO_2 to diffuse out of the red blood cells. As CO_2 levels in the red blood cells decrease, the equilibrium of the CO_2-HCO_3^- reaction is disturbed, shifting toward production of more CO_2. Removal of CO_2 causes H^+ to leave the hemoglobin molecules, and the chloride shift reverses: Cl^- returns to the plasma in exchange for HCO_3^- moving back into the red blood cells. The HCO_3^- and newly released H^+ re-form into carbonic acid, which is then converted into water and CO_2. This CO_2 is then free to diffuse out of the red blood cell and into the alveoli.

■ Figure 18.12 shows the combined transport of CO_2 and O_2 in the blood. At the alveoli, O_2 diffuses down its pressure gradient, moving from the alveoli into the plasma and then from the plasma into the red blood cells. Hemoglobin binds to O_2, increasing the amount of oxygen that can be transported to the cells.

At the cells, the process reverses. Because P_{O_2} is lower in cells than in the arterial blood, O_2 diffuses from the plasma into the cells. The decrease in plasma P_{O_2} causes hemoglobin to release O_2, making additional oxygen available to enter cells.

Carbon dioxide from aerobic metabolism simultaneously leaves cells and enters the blood, dissolving in the plasma. From there, CO_2 enters red blood cells, where most is converted to HCO_3^- and H^+. The HCO_3^- is returned to the plasma in exchange for a Cl^- while the H^+ binds to hemoglobin. A fraction of the CO_2 that enters red blood cells also binds directly to hemoglobin. At the lungs, the process reverses as CO_2 diffuses out of the pulmonary capillaries and into the alveoli.

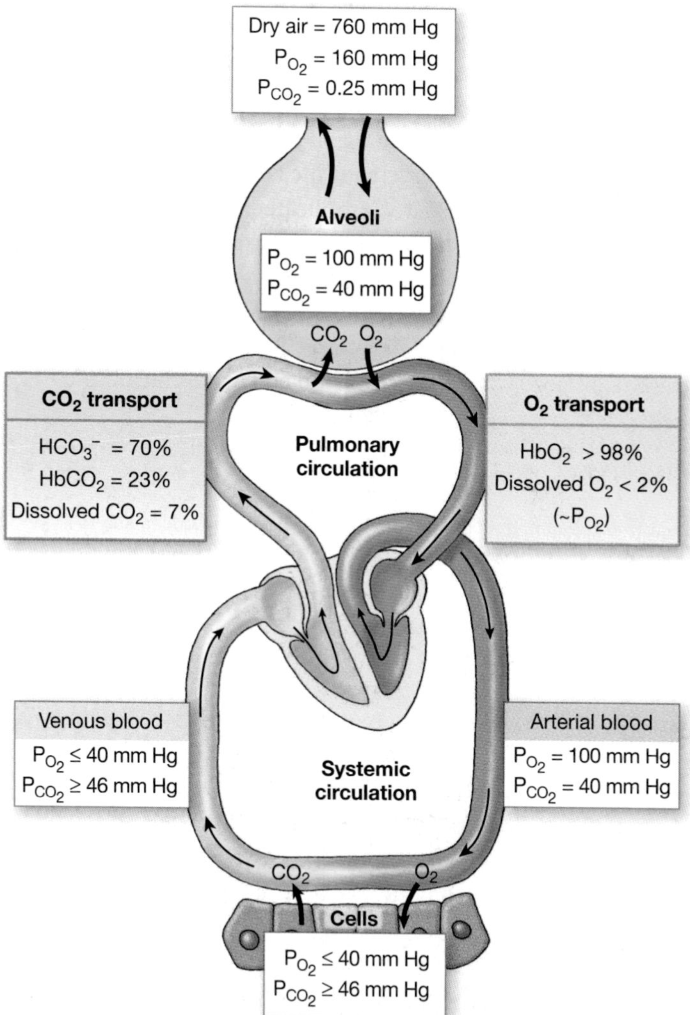

■ **Fig. 18.12**

To understand fully how the respiratory system coordinates delivery of oxygen to the lungs with transport of oxygen in the circulation, we now consider the central nervous system control of ventilation.

Concept Check Answer: p. 625

12. How would an obstruction of the airways affect alveolar ventilation, arterial P_{CO_2}, and the body's pH?

Regulation of Ventilation

Breathing is a rhythmic process that usually occurs without conscious thought or awareness. In that respect, it resembles the rhythmic beating of the heart. However, skeletal muscles, unlike autorhythmic cardiac muscles, are not able to contract spontaneously. Instead, skeletal muscle contraction must be initiated by somatic motor neurons, which in turn are controlled by the central nervous system.

In the respiratory system, contraction of the diaphragm and other muscles is initiated by a spontaneously firing network of neurons in the brain stem (■ Fig. 18.13). Breathing occurs automatically throughout a person's life but can also be controlled voluntarily, up to a point. Complicated synaptic interactions between neurons in the network create the rhythmic cycles of inspiration and expiration, influenced continuously by sensory input, especially that from chemoreceptors for CO_2, O_2, and H^+. Ventilation pattern depends in large part on the levels of those three substances in the arterial blood and extracellular fluid.

The neural control of breathing is one of the few "black boxes" left in systems-level physiology. Although we know the major regions of the brain stem that are involved, the details remain elusive and controversial. The brain stem network that controls breathing behaves like a *central pattern generator* [p. 453], with intrinsic rhythmic activity that probably arises from *pacemaker neurons* with unstable membrane potentials.

Some of our understanding of how ventilation is controlled has come from observing patients with brain damage. Other information has come from animal experiments in which neural connections between major parts of the brain stem are severed, or sections of brain are studied in isolation. Research on CNS respiratory control is difficult because of the complexity of the neural network and its anatomical location,

THE REFLEX CONTROL OF VENTILATION

Central and peripheral chemoreceptors monitor blood gases and pH. Control networks in the brain stem regulate activity in somatic motor neurons leading to respiratory muscles.

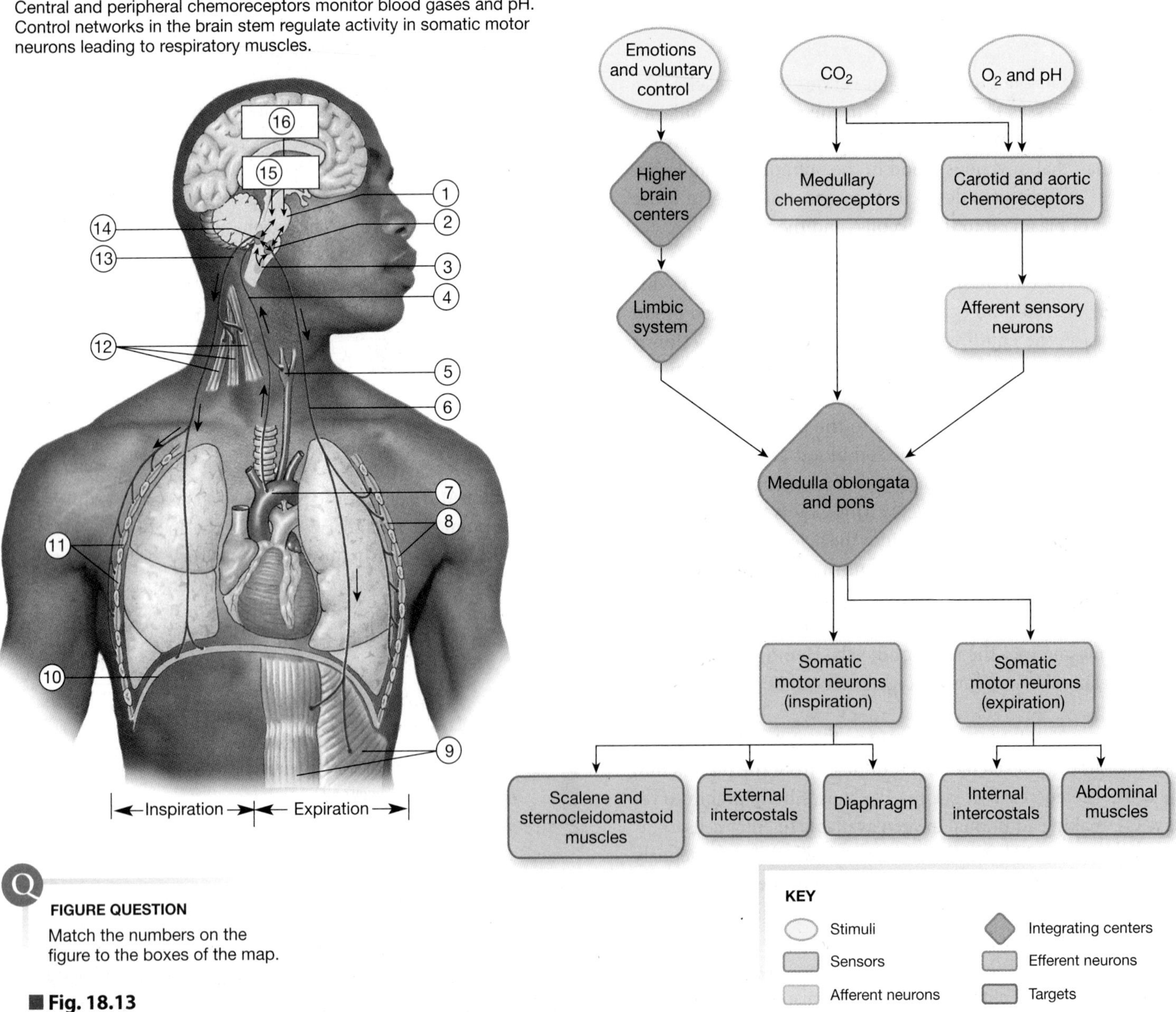

FIGURE QUESTION

Match the numbers on the figure to the boxes of the map.

■ **Fig. 18.13**

but in recent years scientists have developed better techniques for studying the system.

The details that follow represent a contemporary model for the control of ventilation. Although some parts of the model are well supported with experimental evidence, other aspects are still under investigation. This model states that:

1 Respiratory neurons in the medulla control inspiratory and expiratory muscles.
2 Neurons in the pons integrate sensory information and interact with medullary neurons to influence ventilation.
3 The rhythmic pattern of breathing arises from a neural network with spontaneously discharging neurons.
4 Ventilation is subject to continuous modulation by various chemoreceptor- and mechanoreceptor-linked reflexes and by higher brain centers.

Neurons in the Medulla Control Breathing

Classic descriptions of how the brain controls ventilation divided the brain stem into various control centers. More recent descriptions, however, are less specific about assigning function to particular "centers" and instead look at complex interactions between neurons in a network. We know that respiratory neurons are concentrated bilaterally in two areas of the medulla oblongata. ■ Figure 18.14 shows these areas on the left side of the brain stem. One area called the **nucleus tractus solitarius** (NTS) contains the **dorsal respiratory group** (DRG) of neurons that control mostly muscles of inspiration. Output from the DRG goes via the **phrenic nerves** to the diaphragm and via the **intercostal nerves** to the intercostal muscles. In addition, the NTS receives sensory information from peripheral chemo- and mechanoreceptors through the *vagus* and *glossopharyngeal nerves* (cranial nerves X and IX).

Respiratory neurons in the pons receive sensory information from the DRG and in turn influence the initiation and termination of inspiration. The **pontine respiratory groups** (previously called the pneumotaxic center) and other pontine neurons provide tonic input to the medullary networks to help coordinate a smooth respiratory rhythm.

The **ventral respiratory group** (VRG) of the medulla has multiple regions with different functions. One area known as the **pre-Bötzinger complex** contains spontaneously firing neurons that may act as the basic pacemaker for the respiratory rhythm. Other areas control muscles used for active expiration or for greater-than-normal inspiration, such as occurs during vigorous exercise. In addition, nerve fibers from the VRG innervate muscles of the larynx, pharynx, and tongue to keep the upper airways open during breathing. Inappropriate relaxation of these muscles during sleep contributes to *obstructive sleep apnea*, a sleeping disorder associated with snoring and excessive daytime sleepiness.

The integrated action of the respiratory control networks can be seen by monitoring electrical activity in the phrenic

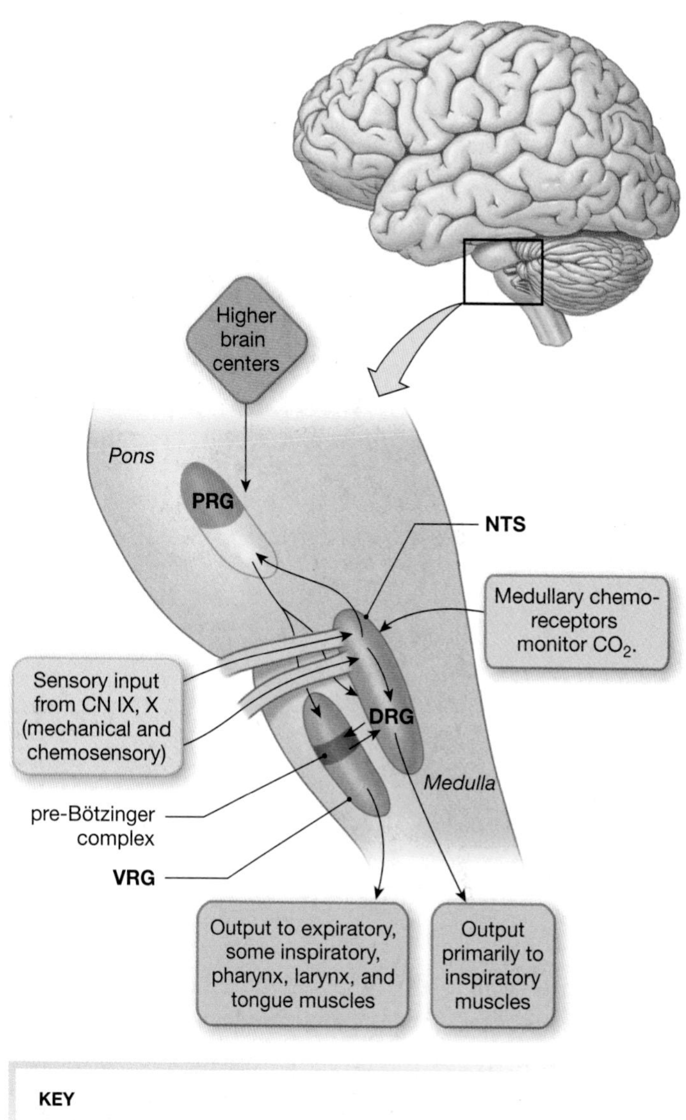

■ **Fig. 18.14**

nerve and other motor nerves (■ Fig. 18.15). During quiet breathing, a pacemaker initiates each cycle, and inspiratory neurons gradually increase stimulation of the inspiratory muscles. This increase is sometimes called *ramping* because of the shape of the graph of inspiratory neuron activity. A few inspiratory neurons fire to begin the ramp. The firing of these neurons recruits other inspiratory neurons to fire in an apparent positive feedback loop. As more neurons fire, more skeletal muscle fibers are recruited. The rib cage expands smoothly as the diaphragm contracts.

At the end of inspiration, the inspiratory neurons abruptly stop firing, and the respiratory muscles relax. Over the next few seconds, passive expiration occurs because of elastic recoil of the inspiratory muscles and elastic lung tissue. However, some

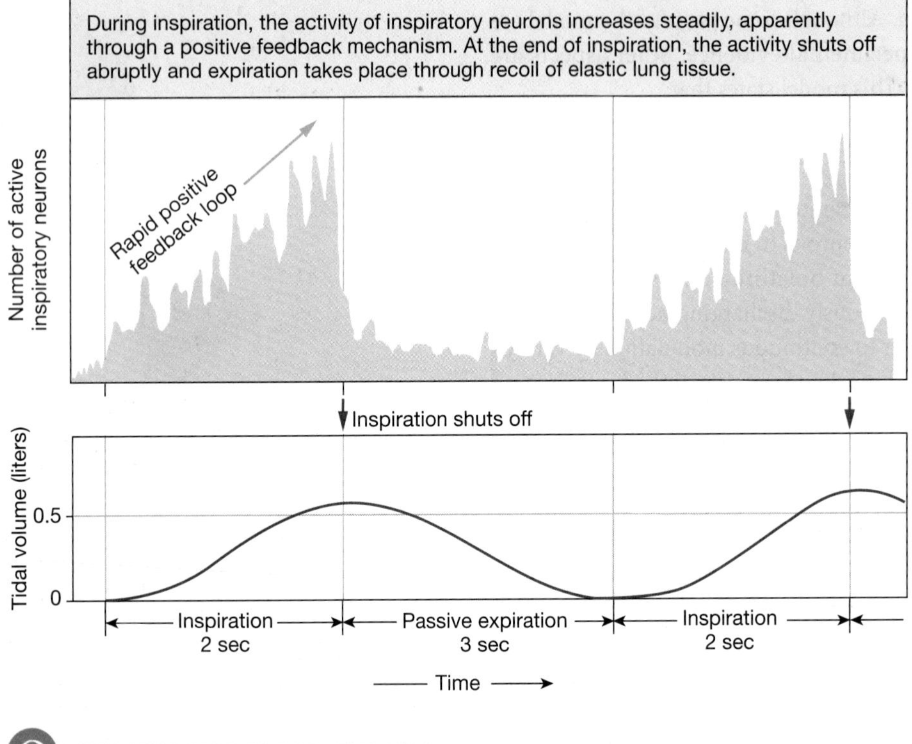
NEURAL ACTIVITY DURING QUIET BREATHING
During inspiration, the activity of inspiratory neurons increases steadily, apparently through a positive feedback mechanism. At the end of inspiration, the activity shuts off abruptly and expiration takes place through recoil of elastic lung tissue.
Number of active inspiratory neurons
Rapid positive feedback loop
Inspiration shuts off
Tidal volume (liters)
0.5
0
Inspiration 2 sec
Passive expiration 3 sec
Inspiration 2 sec
Time

GRAPH QUESTION
What is the ventilation rate of the person in this example?

Fig. 18.15

motor neuron activity can be observed during passive expiration, suggesting that perhaps muscles in the upper airways contract to slow the flow of air out of the respiratory system.

Many neurons of the VRG remain inactive during quiet respiration. They function primarily during forced breathing, when inspiratory movements are exaggerated, and during active expiration. In forced breathing, increased activity of inspiratory neurons stimulates accessory muscles, such as the sternocleidomastoids. Contraction of these accessory muscles enhances expansion of the thorax by raising the sternum and upper ribs.

With active expiration, expiratory neurons from the VRG activate the internal intercostal and abdominal muscles. There seems to be some communication between inspiratory and expiratory neurons, as inspiratory neurons are inhibited during active expiration.

Carbon Dioxide, Oxygen, and pH Influence Ventilation

Sensory input from central and peripheral chemoreceptors modifies the rhythmicity of the control network to help maintain blood gas homeostasis. Carbon dioxide is the primary stimulus for changes in ventilation. Oxygen and plasma pH play lesser roles.

The chemoreceptors for oxygen and carbon dioxide are strategically associated with the arterial circulation. If too little oxygen is present in arterial blood destined for the brain and other tissues, the rate and depth of breathing increase. If the rate of CO_2 production by the cells exceeds the rate of CO_2 removal by the lungs, arterial P_{CO_2} increases, and ventilation is intensified to match CO_2 removal to production. These homeostatic reflexes operate constantly, keeping arterial P_{O_2} and P_{CO_2} within a narrow range.

Peripheral chemoreceptors located in the carotid and aortic arteries sense changes in the P_{O_2}, pH, and P_{CO_2} of the plasma (Fig. 18.13). These **carotid** and **aortic bodies** are close to the locations of the baroreceptors involved in reflex control of blood pressure [p. 525]. **Central chemoreceptors** in the brain respond to changes in the concentration of CO_2 in the cerebrospinal fluid. These central receptors lie on the ventral surface of the medulla, close to neurons involved in respiratory control.

Peripheral Chemoreceptors When specialized **glomus cells** {*glomus*, a ball-shaped mass} in the carotid and aortic bodies are activated by a decrease in P_{O_2} or pH or by an increase in P_{CO_2}, they trigger a reflex increase in ventilation. Under most normal circumstances, oxygen is not an important factor in modulating ventilation because arterial P_{O_2} must fall to less than 60 mm Hg before ventilation is stimulated. This large decrease in P_{O_2} is equivalent to ascending to an altitude of 3000 m. (For reference, Denver is located at an altitude of 1609 m). However, any condition that reduces plasma pH or increases P_{CO_2} will activate the carotid and aortic glomus cells and increase ventilation.

The details of glomus cell function remain to be worked out, but the basic mechanism by which these chemoreceptors respond to low oxygen is similar to the mechanism you learned for insulin release by pancreatic beta cells [p. 167] or taste transduction in taste buds [p. 344].

In all three examples, a stimulus inactivates K^+ channels, causing the receptor cell to depolarize (■ Fig. 18.16). Depolarization opens voltage-gated Ca^{2+} channels, and Ca^{2+} entry causes exocytosis of neurotransmitter onto the sensory neuron. In the carotid and aortic bodies, neurotransmitters initiate action potentials in sensory neurons leading to the brain stem respiratory networks, signaling them to increase ventilation.

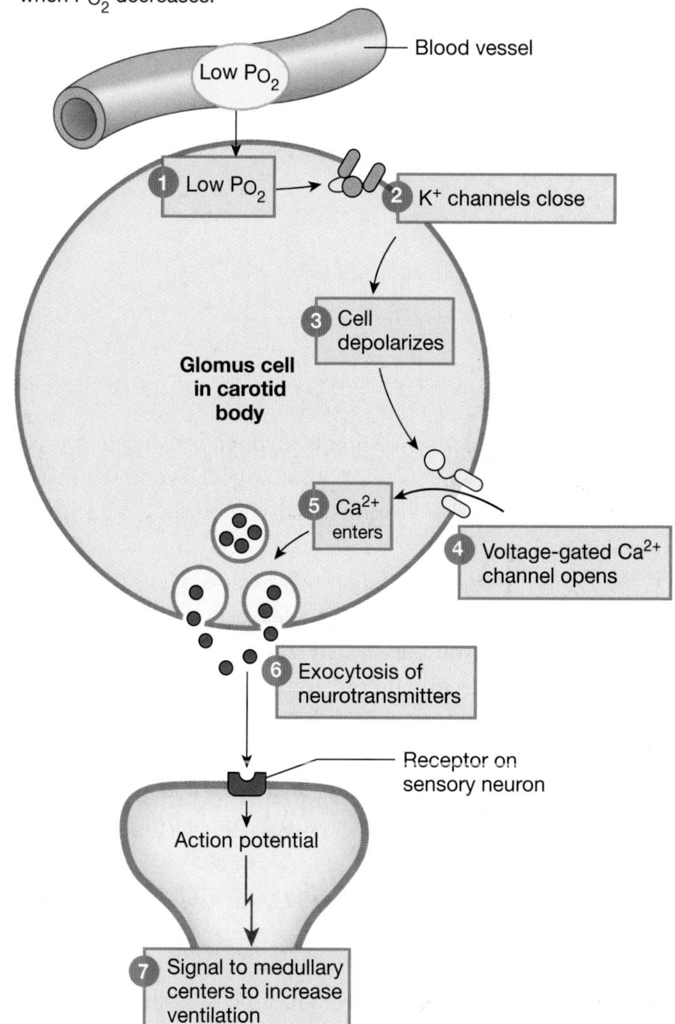

■ **Fig. 18.16**

Because the peripheral chemoreceptors respond only to dramatic changes in arterial P_{O_2}, arterial oxygen concentrations do not play a role in the everyday regulation of ventilation. However, unusual physiological conditions, such as ascending to high altitude, and pathological conditions, such as chronic obstructive pulmonary disease (COPD), can reduce arterial P_{O_2} to levels that are low enough to activate the peripheral chemoreceptors.

Central Chemoreceptors The most important chemical controller of ventilation is carbon dioxide, mediated both through the peripheral chemoreceptors just discussed and through central chemoreceptors located in the medulla (■ Fig. 18.17). These receptors set the respiratory pace, providing continuous input into the control network. When arterial P_{CO_2} increases, CO_2 crosses the blood-brain barrier and activates the central chemoreceptors. These receptors signal the control network to increase the rate and depth of ventilation, thereby enhancing alveolar ventilation and removing CO_2 from the blood.

Although we say that the central chemoreceptors monitor CO_2, they actually respond to pH changes in the cerebrospinal fluid (CSF). Carbon dioxide that diffuses across the blood-brain barrier into the CSF is converted to carbonic acid, which dissociates to bicarbonate and H^+. Experiments indicate that the H^+ produced by this reaction is what initiates the chemoreceptor reflex, rather than the increased level of CO_2.

Note, however, that pH changes in the plasma *do not* usually influence the central chemoreceptors directly. Although plasma P_{CO_2} enters the CSF readily, plasma H^+ crosses the blood-brain barrier very slowly and therefore has little direct effect on the central chemoreceptors.

When plasma P_{CO_2} increases, the chemoreceptors initially respond strongly by increasing ventilation. However, if P_{CO_2} remains elevated for several days, ventilation falls back toward normal rates as the chemoreceptor response adapts by mechanisms that are not clear. Fortunately for people with chronic lung diseases, the response of peripheral chemoreceptor to low arterial P_{O_2} remains intact over time, even though the central chemoreceptor response adapts to high P_{CO_2}.

In some situations, low P_{O_2} becomes the primary chemical stimulus for ventilation. For example, patients with severe chronic lung disease, such as COPD, have chronic hypercapnia and hypoxia. Their arterial P_{CO_2} may rise to 50–55 mm Hg (normal is 35–45) while their P_{O_2} falls to 45–50 mm Hg (normal 75–100). Because these levels are chronic, the chemoreceptor response adapts to the elevated P_{CO_2}. Most of the chemical stimulus for ventilation in this situation then comes from low P_{O_2}, sensed by the carotid and aortic chemoreceptors. If these patients are given too much oxygen, they may stop breathing because their chemical stimulus for ventilation is eliminated.

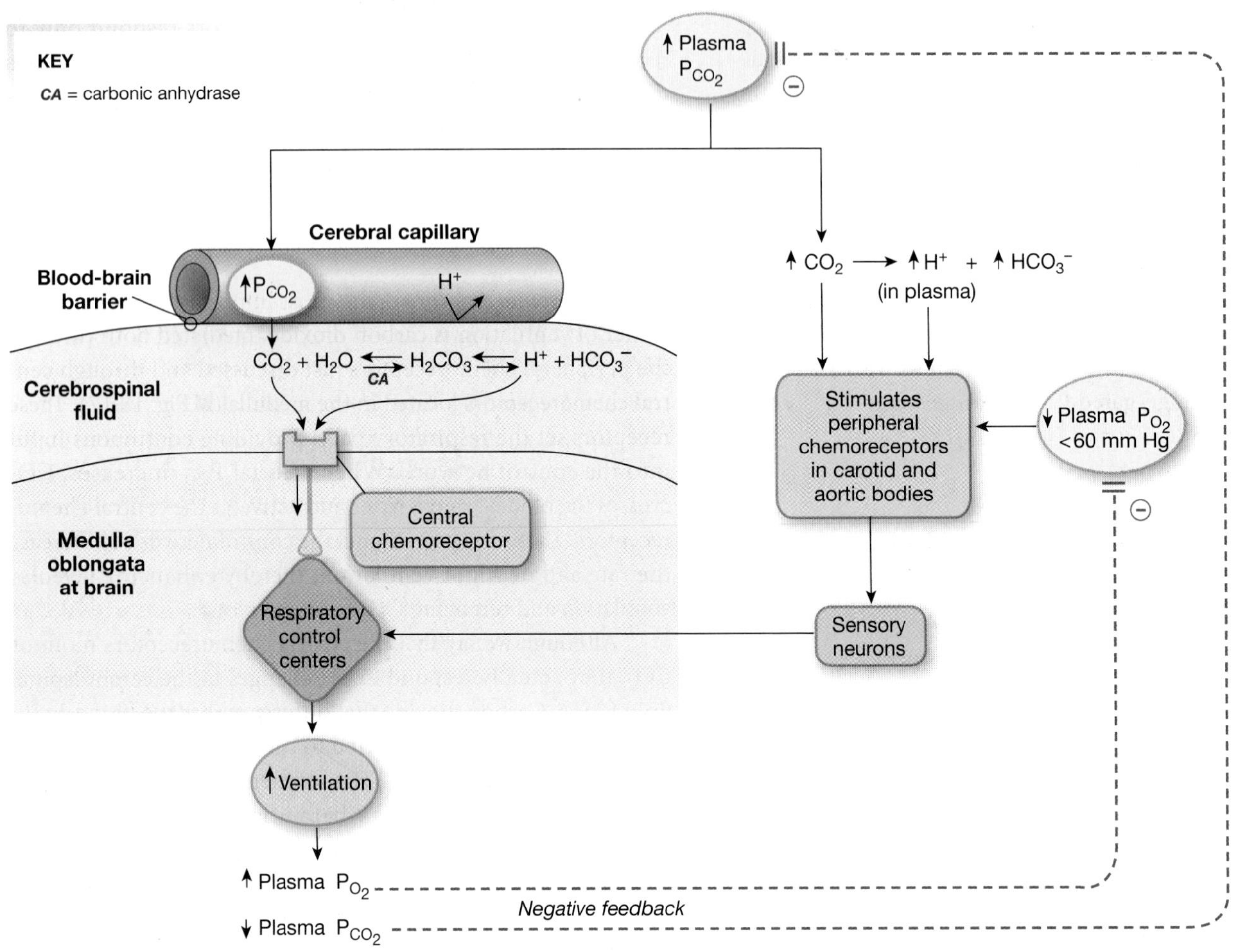

■ **Fig. 18.17**

The central chemoreceptors respond to decreases in arterial P_{CO_2} as well as to increases. If alveolar P_{CO_2} falls, as it might during hyperventilation, plasma P_{CO_2} and cerebrospinal fluid P_{CO_2} follow suit. As a result, central chemoreceptor activity declines, and the control network slows the ventilation rate. When ventilation decreases, carbon dioxide begins to accumulate in alveoli and the plasma. Eventually, the arterial P_{CO_2} rises above the threshold level for the chemoreceptors. At that point, the receptors fire, and the control network again increases ventilation.

Protective Reflexes Guard the Lungs

In addition to the chemoreceptor reflexes that help regulate ventilation, the body has protective reflexes that respond to physical injury or irritation of the respiratory tract and to overinflation of the lungs. The major protective reflex is *bronchoconstriction,* mediated through parasympathetic neurons that innervate bronchiolar smooth muscle. Inhaled particles or noxious gases stimulate **irritant receptors** in the airway mucosa. The irritant receptors send signals through sensory neurons to integrating centers in the CNS that trigger bronchoconstriction. Protective reflex responses also include coughing and sneezing.

The *Hering-Breuer inflation reflex* was first described in the late 1800s in anesthetized dogs. In these animals, if tidal volume exceeded a certain volume, stretch receptors in the lung signaled the brain stem to terminate inspiration. However, this reflex is difficult to demonstrate in adult humans and does not operate during quiet breathing and mild exertion. Studies on human infants, however, suggest that the Hering-Breuer inflation reflex may play a role in limiting their ventilation volumes.

Higher Brain Centers Affect Patterns of Ventilation

Conscious and unconscious thought processes also affect respiratory activities. Higher centers in the hypothalamus and cerebrum can alter the activity of the brain stem control network to change ventilation rate and depth. Voluntary control of ventilation falls into this category. Higher brain center control is not a *requirement* for ventilation, however. Even if the brain stem above the pons is severely damaged, essentially normal respiratory cycles continue.

Respiration can also be affected by stimulation of portions of the limbic system. For this reason, emotional and autonomic activities such as fear and excitement may affect the pace and depth of respiration. In some of these situations, the neural pathway goes directly to the somatic motor neurons, bypassing the control network in the brain stem.

Although we can temporarily alter our respiratory performance, we cannot override the chemoreceptor reflexes. Holding your breath is a good example. You can hold your breath voluntarily only until elevated P_{CO_2} in the blood and cerebrospinal fluid activates the chemoreceptor reflex, forcing you to inhale.

Small children having temper tantrums sometimes attempt to manipulate parents by threatening to hold their breath until they die. However, the chemoreceptor reflexes make it impossible for the children to carry out that threat. Extremely strong-willed children can continue holding their breath until they turn blue and pass out from hypoxia, but once they are unconscious, normal breathing automatically resumes.

RUNNING PROBLEM

The hyperventilation response to hypoxia creates a peculiar breathing pattern called *periodic breathing*, in which the person goes through a 10–15-second period of breath-holding followed by a short period of hyperventilation. Periodic breathing occurs most often during sleep.

Q8: Based on your understanding of how the body controls ventilation, why do you think periodic breathing occurs most often during sleep?

Breathing is intimately linked to cardiovascular function. The integrating centers for both functions are located in the brain stem, and interneurons project between the two networks, allowing signaling back and forth. The cardiovascular, respiratory, and renal systems all work together to maintain fluid and acid-base homeostasis, as you will see.

RUNNING PROBLEM CONCLUSION

High Altitude

On May 29, 1953, Edmund Hillary and Tenzing Norgay of the British Everest Expedition were the first humans to reach the summit of Mt. Everest. They carried supplemental oxygen with them, as it was believed that this feat was impossible without it. In 1978, however, Reinhold Messner and Peter Habeler achieved the "impossible." On May 8, they struggled to the summit using sheer willpower and no extra oxygen. In Messner's words, "I am nothing more than a single narrow gasping lung, floating over the mists and summits." Learn more about these Everest expeditions by doing a Google search for Hillary Everest or Messner Everest.

To learn more about different types of mountain sickness, see the International Society for Mountain Medicine (*www.ismmed.org/np_altitude_tutorial.htm*); "High altitude medicine," *Am Fam Physician* 1998 Apr. 15 (*www.aafp.org/afp/980415ap/harris.html*); and "High-altitude pulmonary edema" (*www.emedicine.com/MED/topic1956.htm*).

In this running problem you learned about normal and abnormal responses to high altitude. Check your understanding of the physiology behind this respiratory challenge by comparing your answers with the information in the following table.

Question	Facts	Integration and Analysis
1. What is the P_{O_2} of inspired air reaching the alveoli when dry atmospheric pressure is 542 mm Hg? How does this value for P_{O_2} compare with the P_{O_2} value for fully humidified air at sea level?	Water vapor contributes a partial pressure of 47 mm Hg to fully humidified air. Oxygen is 21% of dry air. Normal atmospheric pressure at sea level is 760 mm Hg.	Correction for water vapor: $542 - 47 = 495$ mm Hg $\times$ 21% P_{O_2} $= 104$ mm Hg P_{O_2}. In humidified air at sea level, $P_{O_2} = 150$ mm Hg.
2. Why would someone with HAPE be short of breath?	Pulmonary edema increases the diffusion distance for oxygen.	Slower oxygen diffusion means less oxygen reaching the blood, which worsens the normal hypoxia of altitude.

RUNNING PROBLEM CONCLUSION *(continued)*

Question	Facts	Integration and Analysis
3. Based on mechanisms for matching ventilation and perfusion in the lung, why do patients with HAPE have elevated pulmonary arterial blood pressure?	Low oxygen levels constrict pulmonary arterioles.	Constriction of pulmonary arterioles causes blood to collect in the pulmonary arteries behind the constriction. This increases pulmonary arterial blood pressure.
4. How does adding erythrocytes to the blood help a person acclimatize to high altitude?	98% of arterial oxygen is carried bound to hemoglobin.	Additional hemoglobin increases the oxygen-carrying capacity of the blood.
5. What does adding erythrocytes to the blood do to the viscosity of the blood? What effect will that change in viscosity have on blood flow?	Adding cells increases blood viscosity.	According to Poiseuille's law, increased viscosity increases resistance to flow, so blood flow will decrease.
6. What happens to plasma pH during hyperventilation?	Apply the law of mass action to the equation $CO_2 + H_2O \rightleftharpoons H^+ + HCO_3^-$.	The amount of CO_2 in the plasma decreases during hyperventilation, which means the equation shifts to the left. This shift decreases H^+, which increases pH (alkalosis).
7. How does this change in pH affect oxygen binding at the lungs when P_{O_2} is decreased? How does it affect unloading of oxygen at the cells?	See Figure 18.9c.	The left shift of the curve means that, at any given P_{O_2}, more O_2 binds to hemoglobin. Less O_2 will unbind at the tissues for a given P_{O_2}, but P_{O_2} in the cells is probably lower than normal, and consequently there may be no change in unloading.
8. Why do you think periodic breathing occurs most often during sleep?	Periodic breathing alternates periods of breath-holding (apnea) and hyperventilation.	An awake person is more likely to make a conscious effort to breathe during the breath-holding spells, eliminating the cycle of periodic breathing.

Test your understanding with:

- Practice Tests
- Running Problem Quizzes
- A&PFlix™ Animations
- PhysioEx™ Lab Simulations
- Interactive Physiology Animations

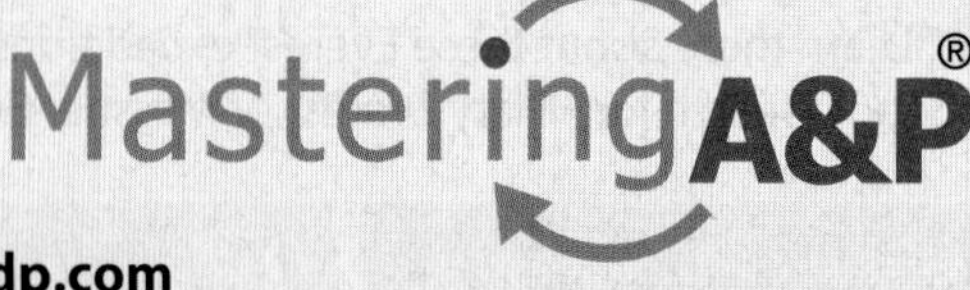

www.masteringaandp.com

Chapter Summary

In this chapter, you learned why climbing Mt. Everest is such a respiratory challenge for the human body, and why people with emphysema experience the same respiratory challenges at sea level. The exchange and transport of oxygen and carbon dioxide in the body illustrate the *mass flow* of gases along concentration gradients. *Homeostasis* of these blood gases demonstrates *mass balance*: the concentration in the blood varies according to what enters or leaves at the lungs and tissues. The *law of mass action* governs the chemical reactions through which hemoglobin binds O_2, and carbonic anhydrase catalyzes the conversion of CO_2 and water to carbonic acid.

Gas Exchange in the Lungs and Tissues

iP Respiratory: Gas Exchange

1. Normal alveolar and arterial P_{O_2} is about 100 mm Hg. Normal alveolar and arterial P_{CO_2} is about 40 mm Hg. Normal venous P_{O_2} is 40 mm Hg, and normal venous P_{CO_2} is 46 mm Hg. (p. 602; Fig. 18.2)
2. Body sensors monitor blood oxygen, CO_2, and pH in an effort to avoid **hypoxia** and **hypercapnia.** (p. 604)
3. Both the composition of inspired air and the effectiveness of alveolar ventilation affect alveolar P_{O_2}. (p. 605)
4. Changes in alveolar surface area, in diffusion barrier thickness, and in fluid distance between the alveoli and pulmonary capillaries can all affect gas exchange efficiency and arterial P_{O_2}. (p. 603; Fig. 18.3)
5. The amount of a gas that dissolves in a liquid is proportional to the partial pressure of the gas and to the **solubility** of the gas in the liquid. Carbon dioxide is 20 times more soluble in aqueous solutions than oxygen is. (p. 606; Fig. 18.4)

Gas Transport in the Blood

iP Respiratory: Gas Transport

6. Gas transport demonstrates mass flow and mass balance. The **Fick equation** relates blood oxygen content, cardiac output, and tissue oxygen consumption. (p. 607; Fig. 18.6)
7. Oxygen is transported dissolved in plasma (<2%) and bound to hemoglobin (>98%). (p. 607; Fig. 18.5)
8. The P_{O_2} of plasma determines how much oxygen binds to hemoglobin. (p. 610; Fig. 18.8)
9. Oxygen-hemoglobin binding is influenced by pH, temperature, and **2,3-diphosphoglycerate** (2,3-DPG). (p. 611; Fig. 18.9)
10. Venous blood carries 7% of its carbon dioxide dissolved in plasma, 23% as **carbaminohemoglobin**, and 70% as bicarbonate ion in the plasma. (p. 614; Fig. 18.11)
11. **Carbonic anhydrase** in red blood cells converts CO_2 to carbonic acid, which dissociates into H^+ and HCO_3^-. The H^+ then binds to hemoglobin, and HCO_3^- enters the plasma using the **chloride shift**. (p. 613)

Regulation of Ventilation

iP Respiratory: Control of Respiration

12. Respiratory control resides in networks of neurons in the medulla and pons, influenced by input from central and peripheral sensory receptors and higher brain centers. (p. 616; Fig. 18.13)
13. The medullary **dorsal respiratory group** (DRG) contains mostly inspiratory neurons that control somatic motor neurons to the diaphragm. The **ventral respiratory group** (VRG) includes the **pre-Bötzinger complex** with its apparent pacemakers as well as neurons for inspiration and active expiration. (p. 617; Fig. 18.14)
14. **Peripheral chemoreceptors** in the carotid and aortic bodies monitor P_{O_2}, P_{CO_2}, and pH. P_{O_2} below 60 mm Hg triggers an increase in ventilation. (p. 620; Fig. 18.17)
15. Carbon dioxide is the primary stimulus for changes in ventilation. Chemoreceptors in the medulla and carotid bodies respond to changes in P_{CO_2}. (p. 620; Fig. 18.17)
16. Protective reflexes monitored by peripheral receptors prevent injury to the lungs from inhaled irritants. (p. 620)
17. Conscious and unconscious thought processes can affect respiratory activity. (p. 620)

Questions

Answers: p. A-1

Level One Reviewing Facts and Terms

1. List five factors that influence the diffusion of gases between alveolus and blood.
2. More than ________% of the oxygen in arterial blood is transported bound to hemoglobin. How is the remaining oxygen transported to the cells?
3. Name four factors that influence the amount of oxygen that binds to hemoglobin. Which of these four factors is the most important?
4. Describe the structure of a hemoglobin molecule. What chemical element is essential for hemoglobin synthesis?
5. The networks for control of ventilation are found in the ________ and ________ of the brain. What do the dorsal and ventral respiratory groups of neurons control? What is a central pattern generator?
6. Describe the chemoreceptors that influence ventilation. What chemical is the most important controller of ventilation?
7. Describe the protective reflexes of the respiratory system.
8. What causes the exchange of oxygen and carbon dioxide between alveoli and blood or between blood and cells?
9. List five possible physical changes that could result in less oxygen reaching the arterial blood.

Level Two Reviewing Concepts

10. Concept map: Construct a map of gas transport using the following terms. You may add other terms.

• alveoli	• hemoglobin saturation
• arterial blood	• oxyhemoglobin
• carbaminohemoglobin	• P_{CO_2}
• carbonic anhydrase	• plasma
• chloride shift	• P_{O_2}
• dissolved CO_2	• pressure gradient
• dissolved O_2	• red blood cell
• hemoglobin	• venous blood

11. In respiratory physiology, it is customary to talk of the P_{O_2} of the plasma. Why is this not the most accurate way to describe the oxygen content of blood?
12. Compare and contrast the following pairs of concepts:
    (a) transport of O_2 and CO_2 in arterial blood
    (b) partial pressure and concentration of a gas dissolved in a liquid
13. Does HbO_2 binding increase, decrease, or not change with decreased pH?
14. Define hypoxia, COPD, and hypercapnia.
15. Why did oxygen-transporting molecules evolve in animals?
16. Draw and label the following graphs:
    (a) the effect of ventilation on arterial P_{O2}
    (b) the effect of arterial P_{CO_2} on ventilation
17. As the P_{O_2} of plasma increases:
    (a) what happens to the amount of oxygen that dissolves in plasma?
    (b) what happens to the amount of oxygen that binds to hemoglobin?

18. If a person is anemic and has a lower-than-normal level of hemoglobin in her red blood cells, what is her arterial P_{O_2} compared to normal?
19. Create reflex pathways (stimulus, receptor, afferent path, and so on) for the chemical control of ventilation, starting with the following stimuli:
    (a) Increased arterial P_{CO_2}
    (b) Arterial P_{O_2} = 55 mm Hg

    Be as specific as possible regarding anatomical locations. Where known, include neurotransmitters and their receptors.

Level Three Problem Solving

20. Marco tries to hide at the bottom of a swimming hole by breathing in and out through two feet of garden hose, which greatly increases his anatomic dead space. What happens to the following parameters in his arterial blood, and why?
    (a) P_{CO_2}
    (b) P_{O_2}
    (c) bicarbonate ion
    (d) pH
21. Which person carries more oxygen in his blood?
    (a) one with Hb of 15 g/dL and arterial P_{O_2} of 80 mm Hg
    (b) one with Hb of 12 g/dL and arterial P_{O_2} of 100 mm Hg
22. What would happen to each of the following parameters in a person suffering from pulmonary edema?
    (a) arterial P_{O_2}
    (b) arterial hemoglobin saturation
    (c) alveolar ventilation
23. In early research on the control of rhythmic breathing, scientists made the following observations. What hypotheses might the researchers have formulated from each observation?
    (a) *Observation.* If the brain stem is severed below the medulla, all respiratory movement ceases.
    (b) *Observation.* If the brain stem is severed above the level of the pons, ventilation is normal.
    (c) *Observation.* If the medulla is completely separated from the pons and higher brain centers, ventilation becomes irregular but a pattern of inspiration/expiration remains.
24. A hospitalized patient with severe chronic obstructive lung disease has a P_{CO_2} of 55 mm Hg and a P_{O_2} of 50 mm Hg. To elevate his blood oxygen, he is given pure oxygen through a nasal tube. The patient immediately stops breathing. Explain why this might occur.
25. You are a physiologist on a space flight to a distant planet. You find intelligent humanoid creatures inhabiting the planet, and they willingly submit to your tests. Some of the data you have collected are described below.

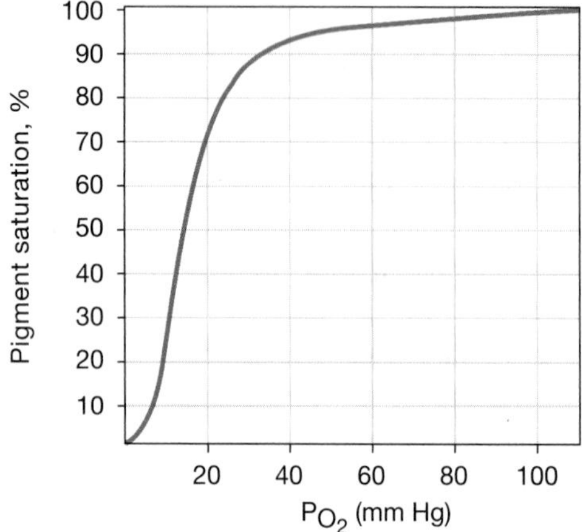

The graph above shows the oxygen saturation curve for the oxygen-carrying molecule in the blood of the humanoid named Bzork. Bzork's normal alveolar P_{O_2} is 85 mm Hg. His normal cell P_{O_2} is 20 mm Hg, but it drops to 10 mm Hg with exercise.

    (a) What is the percent saturation for Bzork's oxygen-carrying molecule in blood at the alveoli? In blood at an exercising cell?
    (b) Based on the graph above, what conclusions can you draw about Bzork's oxygen requirements during normal activity and during exercise?
26. The next experiment on Bzork involves his ventilatory response to different conditions. The data from that experiment are graphed below. Interpret the results of experiments A and C.

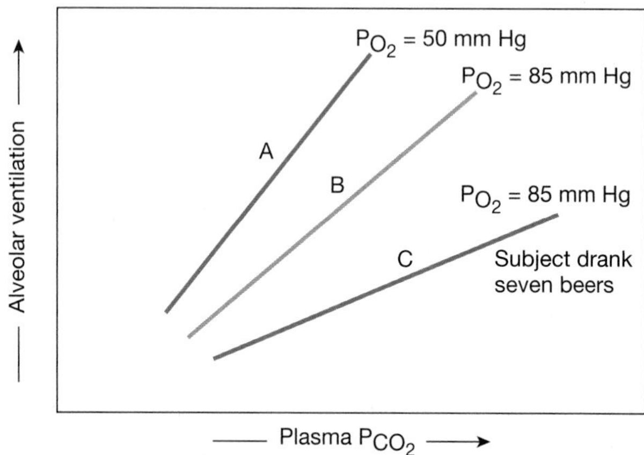

27. The alveolar epithelium is an absorptive epithelium and is able to transport ions from the fluid lining of alveoli into the interstitial space, creating an osmotic gradient for water to follow. Draw an alveolar epithelium and label apical and basolateral surfaces, the airspace, and interstitial fluid. Arrange the following proteins on the cell membrane so that the epithelium absorbs sodium and water: aquaporins, Na^+-K^+-ATPase, epithelial Na^+ channel (ENaC). (Remember: Na^+ concentrations are higher in the ECF than in the ICF.)

Level Four Quantitative Problems

28. You are given the following information on a patient.

    Blood volume = 5.2 liters
    Hematocrit = 47%
    Hemoglobin concentration = 12 g/dL whole blood
    Total amount of oxygen carried in blood = 1015 mL
    Arterial plasma P_{O_2} = 100 mm Hg

    You know that when plasma P_{O_2} is 100 mm Hg, plasma contains 0.3 mL O_2/dL, and that hemoglobin is 98% saturated. Each hemoglobin molecule can bind to a maximum of four molecules of oxygen. Using this information, calculate the maximum oxygen-carrying capacity of hemoglobin (100% saturated). Units will be mL O_2/g Hb.
29. Adolph Fick, the nineteenth-century physiologist who derived Fick's law of diffusion, also developed the Fick equation that relates oxygen consumption, cardiac output, and blood oxygen content:

    O_2 consumption = cardiac output × (arterial oxygen content − venous oxgen content)

    A person has a cardiac output of 4.5 L/min, an arterial oxygen content of 105 mL O_2/L blood, and a vena cava oxygen content of 50 mL O_2/L blood. What is this person's oxygen consumption?
30. Describe what happens to the oxygen-hemoglobin saturation curve in Figure 18.9a when blood hemoglobin falls from 15 g/dL blood to 10 g/dL blood.

Answers

Answers to Concept Check Questions

Page 601

1. (a) electron transport system (b) citric acid cycle
2. The P_{O_2} of the alveoli is constantly being replenished by fresh air. [p. 590]
3. 720 mm Hg × 0.78 N_2 = 561.6 mm Hg

Page 602

4. Air is 21% oxygen. Therefore, for dry air on Everest, $P_{O_2} = 0.21 \times 250$ mm Hg = 53 mm Hg. Correction for P_{H_2O}: P_{O_2} = (250 mm Hg − 47 mm Hg) × 21 = (203 mm Hg) × 0.21 = 43 mm Hg.

Page 605

5. Blood pools in the lungs because the left heart is unable to pump all the blood coming into it from the lungs. Increased blood volume in the lungs increases pulmonary blood pressure.
6. When alveolar ventilation increases, arterial P_{O_2} increases because more fresh air enters the alveoli. Arterial P_{CO_2} decreases because the low P_{CO_2} of fresh air dilutes alveolar P_{CO_2}. The CO_2 pressure gradient between venous blood and the alveoli increases, causing more CO_2 to leave the blood. Venous P_{O_2} and P_{CO_2} do not change because these pressures are determined by oxygen consumption and CO_2 production in the cells.

Page 606

7. False. Plasma is essentially water, and Figure 18.4 shows that CO_2 is more soluble in water than O_2.
8. The other factor that affects how much of each gas dissolves in the saline solution is the solubility of the gas in that solution.

Page 612

9. Yes. Hemoglobin reaches 100% saturation at 650 mm Hg. At sea level, atmospheric pressure is 760 mm Hg, and if the "atmosphere" is 100% oxygen, then P_{O_2} is 760 mm Hg.
10. The flatness at the top of the P_{O_2} curve tells you that hyperventilation causes only a minimal increase in percent saturation of arterial Hb.
11. As the P_{O_2} falls, more oxygen is released. The P_{O_2} of venous blood leaving the muscle is 25 mm Hg, same as the P_{O_2} of the muscle.

Page 615

12. An airway obstruction would decrease alveolar ventilation and increase arterial P_{CO_2}. Elevated arterial P_{CO_2} would increase the H^+ concentration in the arterial blood and decrease pH.

Answers to Figure and Graph Questions

Page 606

Figure 18.4: Oxygen is 2.85 mL/L blood and CO_2 is 28 mL/L blood.

Page 607

Figure 18.5: O_2 crosses five cell membranes: two of the alveolar cell, two of the capillary endothelium, and one of the red blood cell.

Page 611

Figure 18.9: 1. (a) When P_{O_2} is 20 mm Hg, Hb saturation is 34%. (b) Hemoglobin is 50% saturated with oxygen at a P_{O_2} of 28 mm Hg. 2. When pH falls from 7.4 to 7.2, Hb saturation decreases by 13%, from about 37% saturation to 24%. 3. When an exercising muscle cell warms up, Hb releases more oxygen. 4. Loss of 2,3-DPG is not good because then hemoglobin binds more tightly to oxygen at the P_{O_2} values found in cells. 5. The P_{O_2} of placental blood is about 28 mm Hg. 6. At a P_{O_2} of 10 mm Hg, maternal blood is only about 8% saturated with oxygen.

Page 616

Figure 18.13: 1. pons; 2. ventral respiratory group; 3. medullary chemoreceptor; 4. sensory neuron; 5. carotid chemoreceptor; 6. somatic motor neuron (expiration); 7. aortic chemoreceptor; 8. internal intercostals; 9. abdominal muscles; 10. diaphragm; 11. external intercostals; 12. scalenes and sternocleidomastoids; 13. somatic motor neuron (inspiration); 14. dorsal respiratory group; 15. limbic system; 16. higher brain centers (emotions and voluntary control)

Page 618

Figure 18.15: One breath takes 5 seconds, so there are 12 breaths/min.

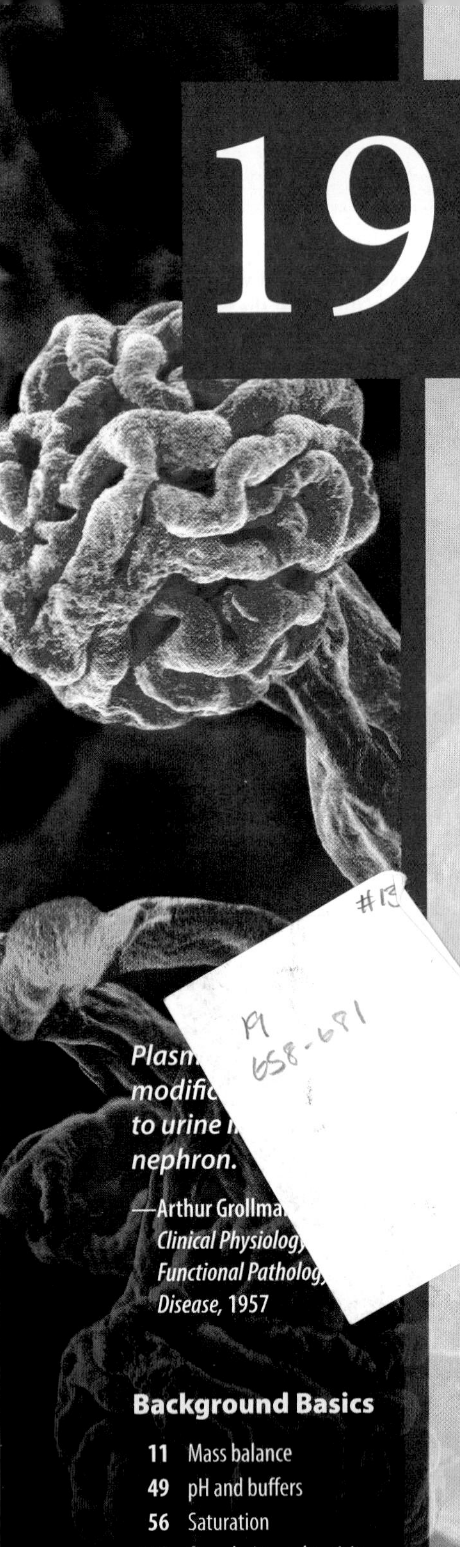

19 The Kidneys

Plasm... modific... to urine i... nephron.

—Arthur Grollma... *Clinical Physiolog... Functional Patholog... Disease*, 1957

Background Basics

11 Mass balance
49 pH and buffers
56 Saturation
133 Osmolarity and tonicity
144 Membrane transport
153 Competition
82 Transporting epithelium
157 Epithelial transport
160 Transcytosis
529 Capillary filtration
528 Fenestrated capillaries
521 Autoregulation of vascular resistance
413 Phosphocreatine

About C.E. 100, Aretaeus the Cappadocian wrote, "Diabetes is a wonderful affection, not very frequent among men, being a melting down of the flesh and limbs into urine. . . . The patients never stop making water {urinating}, but the flow is incessant, as if from the opening of aqueducts." Physicians have known since ancient times that **urine,** the fluid waste produced by the kidneys, reflects the functioning of the body. To aid them in their diagnosis of illness, they even carried special flasks for the collection and inspection of patients' urine.

The first step in examining a urine sample is to determine its color. Is it dark yellow (concentrated), pale straw (dilute), red (indicating the presence of blood), or black (indicating the presence of hemoglobin metabolites)? One form of malaria was called *blackwater fever* because metabolized hemoglobin from the abnormal breakdown of red blood cells turned victims' urine black or dark red.

Physicians also inspected urine samples for clarity, froth (indicating abnormal presence of proteins), smell, and even taste. Physicians who did not want to taste the urine themselves would allow their students the "privilege" of tasting it for them. A physician without students might expose insects to the urine and study their reaction.

Probably the most famous example of using urine for diagnosis was the taste test for diabetes mellitus, historically known as the *honey-urine disease*. Diabetes is an endocrine disorder characterized by the presence of glucose in the urine. The urine of diabetics tasted sweet and attracted insects, making the diagnosis clear.

Today we have much more sophisticated tests for glucose in the urine, but the first step of a *urinalysis* is still to examine the color, clarity, and odor of the urine. In this chapter you will learn why we can tell so much about how the body is functioning by what is present in the urine.

RUNNING PROBLEM

Gout

Michael, 43, had spent the last two days on the sofa, suffering from a relentless throbbing pain in his left big toe. When the pain began, Michael thought he had a mild sprain or perhaps the beginnings of arthritis. Then the pain intensified, and the toe joint became hot and red. Michael finally hobbled into his doctor's office, feeling a little silly about his problem. On hearing his symptoms and looking at the toe, the doctor seemed to know instantly what was wrong. "Looks to me like you have gout," said Dr. Garcia.

627 629 642 646 651 651

Functions of the Kidneys

If you ask people on the street, "What is the most important function of the kidney?" they are likely to say, "The removal of wastes." Actually, the most important function of the kidney is the homeostatic regulation of the water and ion content of the blood, also called *salt and water balance* or *fluid and electrolyte balance*. Waste removal is important, but disturbances in blood volume or ion levels cause serious medical problems before the accumulation of metabolic wastes reaches toxic levels.

The kidneys maintain normal blood concentrations of ions and water by balancing intake of those substances with their excretion in the urine, obeying the principle of *mass balance* [p. 11]. We can divide kidney function into six general areas:

1. **Regulation of extracellular fluid volume and blood pressure.** When extracellular fluid volume decreases, blood pressure also decreases [p. 516]. If ECF volume and blood pressure fall too low, the body cannot maintain adequate blood flow to the brain and other essential organs. The kidneys work in an integrated fashion with the cardiovascular system to ensure that blood pressure and tissue perfusion remain within an acceptable range.
2. **Regulation of osmolarity.** The body integrates kidney function with behavioral drives, such as thirst, to maintain blood osmolarity at a value close to 290 mOsM. We examine the reflex pathways for regulation of ECF volume and osmolarity later.
3. **Maintenance of ion balance.** The kidneys keep concentrations of key ions within a normal range by balancing dietary intake with urinary loss. Sodium (Na^+) is the major ion involved in the regulation of extracellular fluid volume and osmolarity. Potassium (K^+) and calcium (Ca^{2+}) concentrations are also closely regulated.
4. **Homeostatic regulation of pH.** The pH of plasma is normally kept within a narrow range. If extracellular fluid becomes too acidic, the kidneys remove H^+ and conserve bicarbonate ions (HCO_3^-), which act as a buffer [p. 43]. Conversely, when extracellular fluid becomes too alkaline, the kidneys remove HCO_3^- and conserve H^+. The kidneys play a significant role in pH homeostasis, but they do not correct pH disturbances as rapidly as the lungs do.
5. **Excretion of wastes.** The kidneys remove metabolic waste products and foreign substances, such as drugs and environmental toxins. Metabolic wastes include creatinine from muscle metabolism [p. 413] and the nitrogenous wastes *urea* and *uric acid*. A metabolite of hemoglobin called *urobilinogen* gives urine its characteristic yellow color. Hormones are another endogenous substance the kidneys clear from the blood. Examples of foreign substances that the kidneys actively remove include the artificial sweetener *saccharin* and the anion *benzoate,* part of

the preservative *potassium benzoate*, which you ingest each time you drink a diet soft drink.

6 **Production of hormones.** Although the kidneys are not endocrine glands, they play important roles in three endocrine pathways. Kidney cells synthesize *erythropoietin*, the cytokine/hormone that regulates red blood cell synthesis [p. 551]. They also release *renin*, an enzyme that regulates the production of hormones involved in sodium balance and blood pressure homeostasis. Renal enzymes help convert vitamin D3 into a hormone that regulates Ca^{2+} balance.

The kidneys, like many other organs in the body, have a tremendous reserve capacity. By most estimates, you must lose nearly three-fourths of your kidney function before homeostasis begins to be affected. Many people function perfectly normally with only one kidney, including the one person in 1000 born with only one kidney (the other fails to develop during gestation) or those people who donate a kidney for transplantation.

Concept Check Answers: p. 655

1. Ion regulation is a key feature of kidney function. What happens to the resting membrane potential of a neuron if extracellular K^+ levels decrease? [p. 265]
2. What happens to the force of cardiac contraction if plasma Ca^{2+} levels decrease substantially? [p. 477]

CLINICAL FOCUS

Urinary Tract Infections

Because of the shorter length of the female urethra and its proximity to bacteria leaving the large intestine, women are more prone than men to develop bacterial infections of the bladder and kidneys, or **urinary tract infections** (UTIs). The most common cause of UTIs is the bacterium *Escherichia coli*, a normal inhabitant of the human large intestine. *E. coli* is not harmful while restricted to the lumen of the large intestine, but it is pathogenic {*patho-*, disease + *-genic*, causing} if it gets into the urethra. The most common symptoms of a UTI are pain or burning during urination and increased frequency of urination. A urine sample from a patient with a UTI often contains many red and white blood cells, neither of which is commonly found in normal urine. UTIs are treated with antibiotics.

Anatomy of the Urinary System

The **urinary system** is composed of the kidneys and accessory structures (■ Fig. 19.1a). The study of kidney function is called **renal physiology,** from the Latin word *renes,* meaning "kidneys."

The Urinary System Consists of Kidneys, Ureters, Bladder, and Urethra

Let's begin by following the route a drop of water takes on its way from plasma to excretion in the urine. In the first step of urine production, water and solutes move from plasma into the hollow tubules (*nephrons*) that make up the bulk of the paired **kidneys.** These tubules modify the composition of the fluid as it passes through. The modified fluid leaves the kidney and passes into a hollow tube called a **ureter.** There are two ureters, one leading from each kidney to the **urinary bladder.** The bladder expands and fills with urine until, by reflex action, it contracts and expels urine through a single tube, the **urethra.**

The urethra in males exits the body through the shaft of the penis. In females, the urethral opening is found anterior to the openings of the vagina and anus. *Micturition,* or urination, is the process by which urine is excreted.

The kidneys are the site of urine formation. They lie on either side of the spine at the level of the eleventh and twelfth ribs, just above the waist (Fig. 19.1b). Although they are below the diaphragm, they are technically outside the abdominal cavity, sandwiched between the membranous **peritoneum,** which lines the abdomen, and the bones and muscles of the back. Because of their location behind the peritoneal cavity, the kidneys are sometimes described as being *retroperitoneal* {*retro-*, behind}.

The concave surface of each kidney faces the spine. The renal blood vessels, nerves, lymphatics, and ureters all emerge from this surface. **Renal arteries,** which branch off the abdominal aorta, supply blood to the kidneys. **Renal veins** carry blood from the kidneys to the inferior vena cava.

At any given time, the kidneys receive 20–25% of the cardiac output, even though they constitute only 0.4% of total body weight (4.5–6 ounces each). This high rate of blood flow through the kidneys is critical to renal function.

The Nephron Is the Functional Unit of the Kidney

A cross section through a kidney shows that the interior is arranged in two layers: an outer **cortex** and inner **medulla** (Fig. 19.1c). The layers are formed by the organized arrangement of microscopic tubules called **nephrons.** About 80% of the nephrons in a kidney are almost completely contained within the cortex (*cortical* nephrons), but the other 20%—called *juxtamedullary* nephrons {*juxta-*, beside}—dip down into the medulla (Fig. 19.1f, h).

The nephron is the functional unit of the kidney. (A *functional unit* is the smallest structure that can perform all the functions of an organ.) Each of the 1 million nephrons in a kidney is divided into sections (Fig. 19.1i), and each section is closely associated with specialized blood vessels (Fig. 19.1g, h).

Vascular Elements of the Kidney Blood enters the kidney through the renal artery before flowing into smaller arteries and then into arterioles in the cortex (Fig. 19.1d, e). At this point, the arrangement of blood vessels turns into a portal system, one of three in the body [p. 223]. Blood flows from the **afferent arteriole** into a ball-like network of capillaries known as the **glomerulus** {*glomus,* a ball-shaped mass; plural *glomeruli*} (Fig. 19.1g, j).

Blood leaving the glomerulus flows into an **efferent arteriole,** then into a second set of capillaries, the **peritubular capillaries** {*peri-*, around} that surround the tubule (Fig. 19.1g). In juxtamedullary nephrons, the long peritubular capillaries that dip into the medulla are called the **vasa recta** (Fig. 19.1h). Finally, renal capillaries join to form venules and small veins, conducting blood out of the kidney through the renal vein.

The function of the renal portal system is first to filter fluid out of the blood and into the lumen of the nephron at the glomerular capillaries, then to *reabsorb* fluid from the tubule back into the blood at the peritubular capillaries. The forces behind fluid movement in the renal portal system are similar to those that govern filtration of water and molecules out of systemic capillaries in other tissues.

Concept Check Answers: p. 655

3. If net filtration out of glomerular capillaries occurs, then you know that capillary hydrostatic pressure must be (*greater than/less than/equal to*) capillary colloid osmotic pressure. [p. 530]
4. If net reabsorption into peritubular capillaries occurs, then capillary hydrostatic pressure must be (*greater than/less than/ equal to*) the capillary colloid osmotic pressure.

Tubular Elements of the Kidney The nephron begins with a hollow, ball-like structure called **Bowman's capsule** that surrounds the glomerulus (Fig. 19.1i). The endothelium of the glomerulus is fused to the epithelium of Bowman's capsule so that fluid filtering out of the capillaries passes directly into the lumen of the tubule. The combination of glomerulus and Bowman's capsule is called the **renal corpuscle.**

From Bowman's capsule, filtered fluid flows into the **proximal tubule** {*proximal,* close or near}, then into the **loop of Henle,** a hairpin-shaped segment that dips down toward the medulla and then back up. The loop of Henle is divided into two limbs, a thin **descending limb** and an **ascending limb** with thin and thick segments. The fluid then passes into the **distal tubule** {*distal,* distant or far}. The distal tubules of up to eight nephrons drain into a single larger tube called the **collecting duct.** (The distal tubule and its collecting duct together form the **distal nephron.**) Collecting ducts pass from the cortex through the medulla and drain into the **renal pelvis** (Fig. 19.1c). From the renal pelvis, the filtered and modified fluid, now called **urine,** flows into the ureter on its way to excretion.

RUNNING PROBLEM

Gout is a metabolic disease characterized by high blood concentrations of uric acid (*hyperuricemia*). If uric acid concentrations reach a critical level (7.5–8 mg/dL), monosodium urate precipitates out of solution and forms crystals in peripheral joints, particularly in the feet, ankles, and knees. These crystals trigger an inflammatory reaction and cause periodic attacks of excruciating pain. Uric acid crystals may also form kidney stones in the renal pelvis.

Q1. Trace the route followed by these kidney stones when they are excreted.

Q2. Name the anion formed when uric acid dissociates.

Notice in Figure 19.1g how the nephron twists and folds back on itself so that the final part of the ascending limb of the loop of Henle passes between the afferent and efferent arterioles. This region is known as the **juxtaglomerular apparatus.** The proximity of the ascending limb and the arterioles allows paracrine communication between the two structures, a key feature of kidney autoregulation. Because the twisted configuration of the nephron makes it difficult to follow fluid flow, we unfold the nephron in many of the remaining figures in this chapter so that fluid flows from left to right across the figure, as in Figure 19.1i.

19

Overview of Kidney Function

Imagine drinking a 12-ounce soft drink every three minutes around the clock: by the end of 24 hours, you would have consumed the equivalent of 90 two-liter bottles. The thought of putting 180 liters of liquid into your intestinal tract is staggering, but that is how much plasma moves into the nephrons every day! But because the average volume of urine leaving the kidneys is only 1.5 L/day, more than 99% of the fluid that enters nephrons must find its way back into the blood, or the body would rapidly dehydrate.

Kidneys Filter, Reabsorb, and Secrete

Three basic processes take place in the nephron: filtration, reabsorption, and secretion (■ Fig. 19.2). **Filtration** is the movement of fluid from blood into the lumen of the nephron. Filtration takes place only in the renal corpuscle, where the walls of glomerular capillaries and Bowman's capsule are modified to allow bulk flow of fluid.

The Urinary System

(a) The urinary system

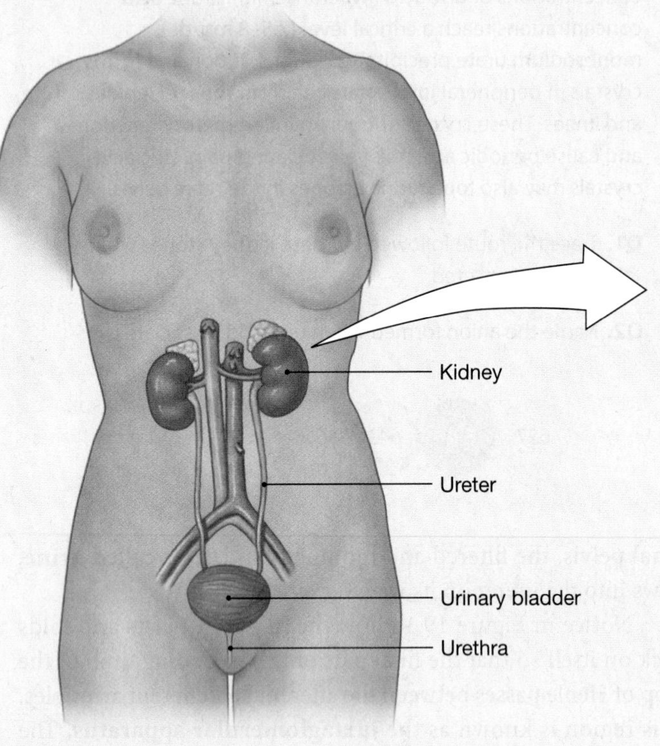

(b) The kidneys are located retroperitoneally at the level of the lower ribs.

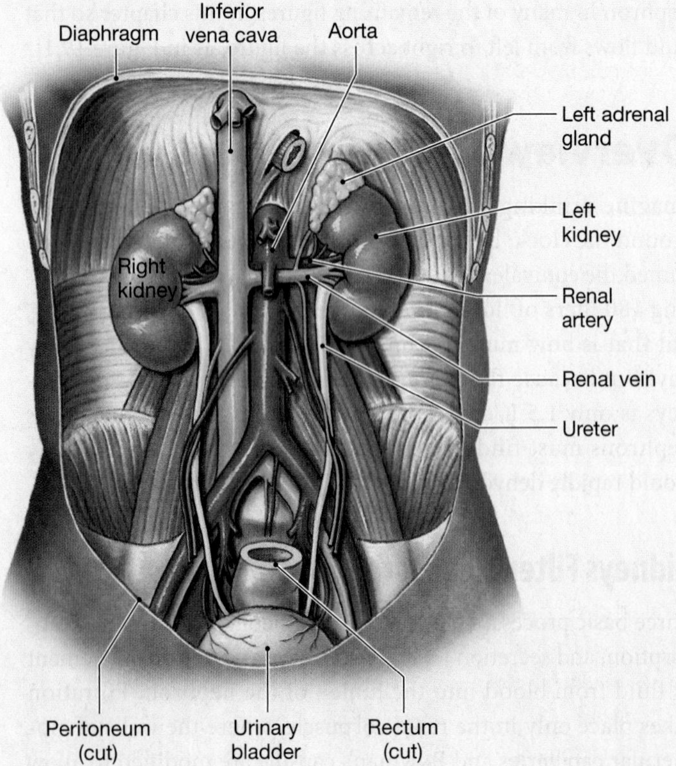

Structure of the Kidney

(c) In cross section, the kidney is divided into an outer cortex and an inner medulla. Urine leaving the nephrons flows into the renal pelvis prior to passing through the ureter into the bladder.

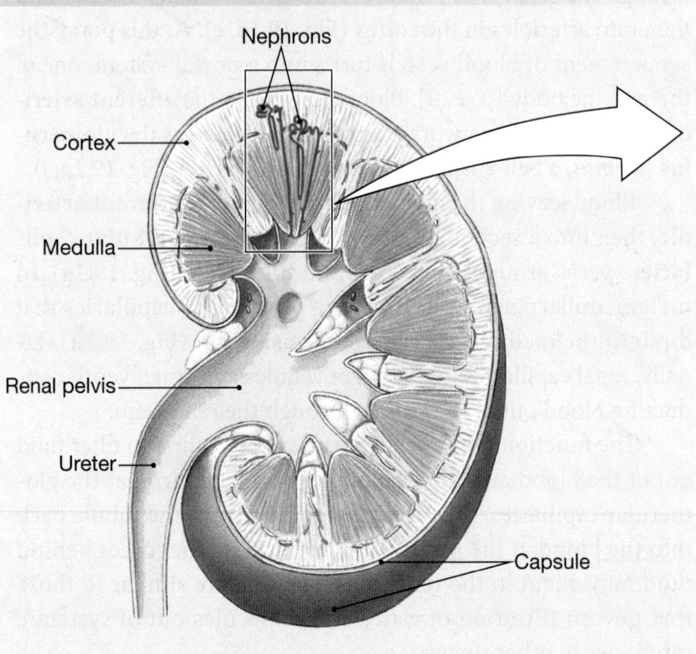

(d) Renal arteries take blood to the cortex.

(e) Afferent arterioles and glomeruli are all found in the cortex.

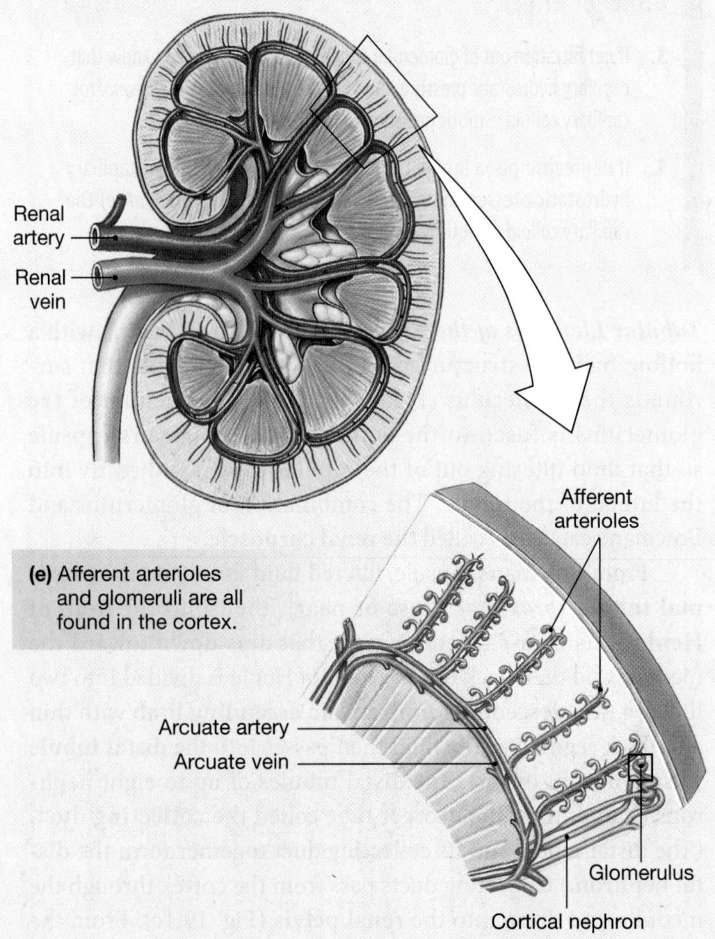

Structure of the Nephron

(f) Some nephrons dip deep into the medulla.

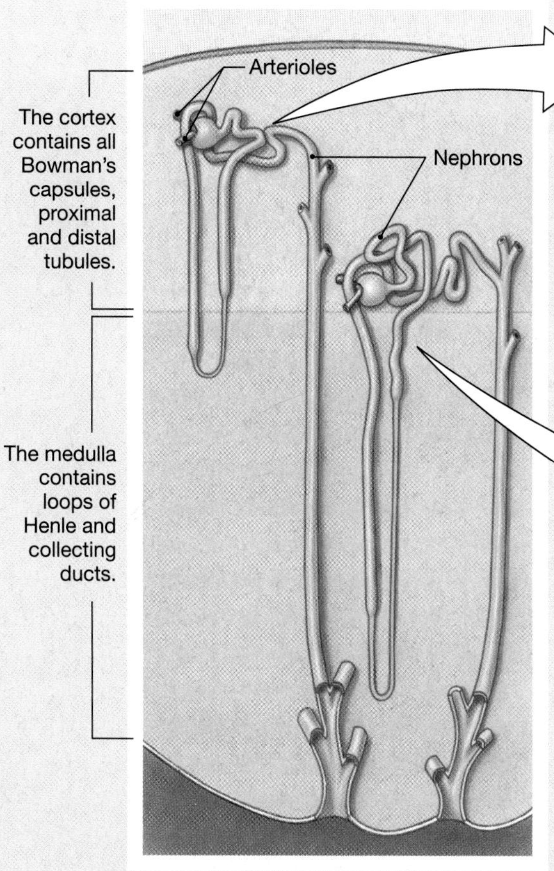

Arterioles
Nephrons
The cortex contains all Bowman's capsules, proximal and distal tubules.
The medulla contains loops of Henle and collecting ducts.

(g) One nephron has two arterioles and two sets of capillaries that form a portal system.

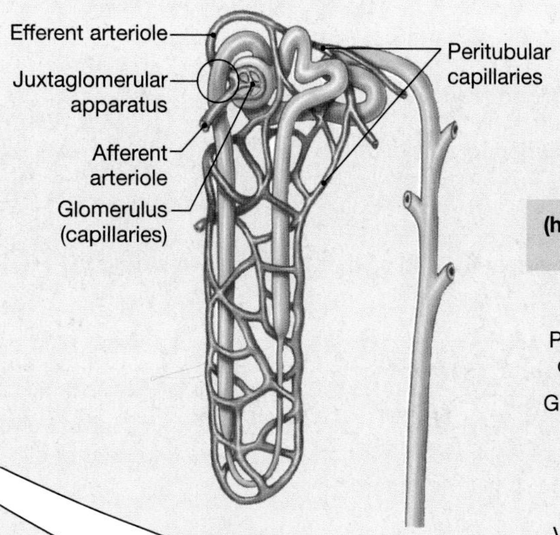

Efferent arteriole
Juxtaglomerular apparatus
Afferent arteriole
Glomerulus (capillaries)
Peritubular capillaries

(h) Juxtamedullary nephron with vasa recta

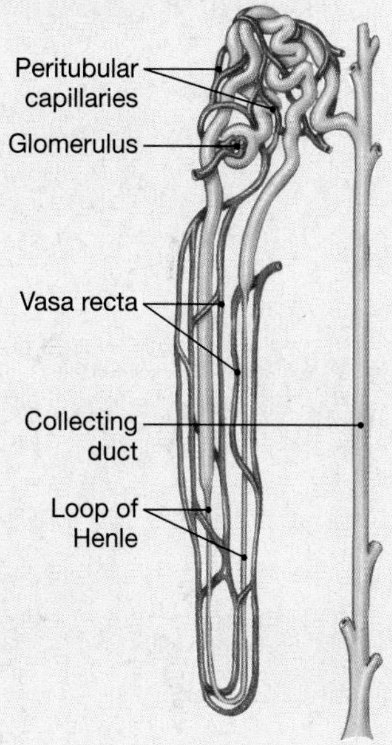

Peritubular capillaries
Glomerulus
Vasa recta
Collecting duct
Loop of Henle

(i) Parts of a nephron. In this view the nephron has been untwisted so that flow goes left to right. Compare with the nephrons in (f).

Bowman's capsule
Proximal tubule
Distal tubule
Descending limb of loop begins
Ascending limb of loop ends
Collecting duct
Descending limb
Ascending limb
Loop of Henle
To bladder

(j) The capillaries of the glomerulus form a ball-like mass.

Glomerulus
Cut edge of nephron tubule

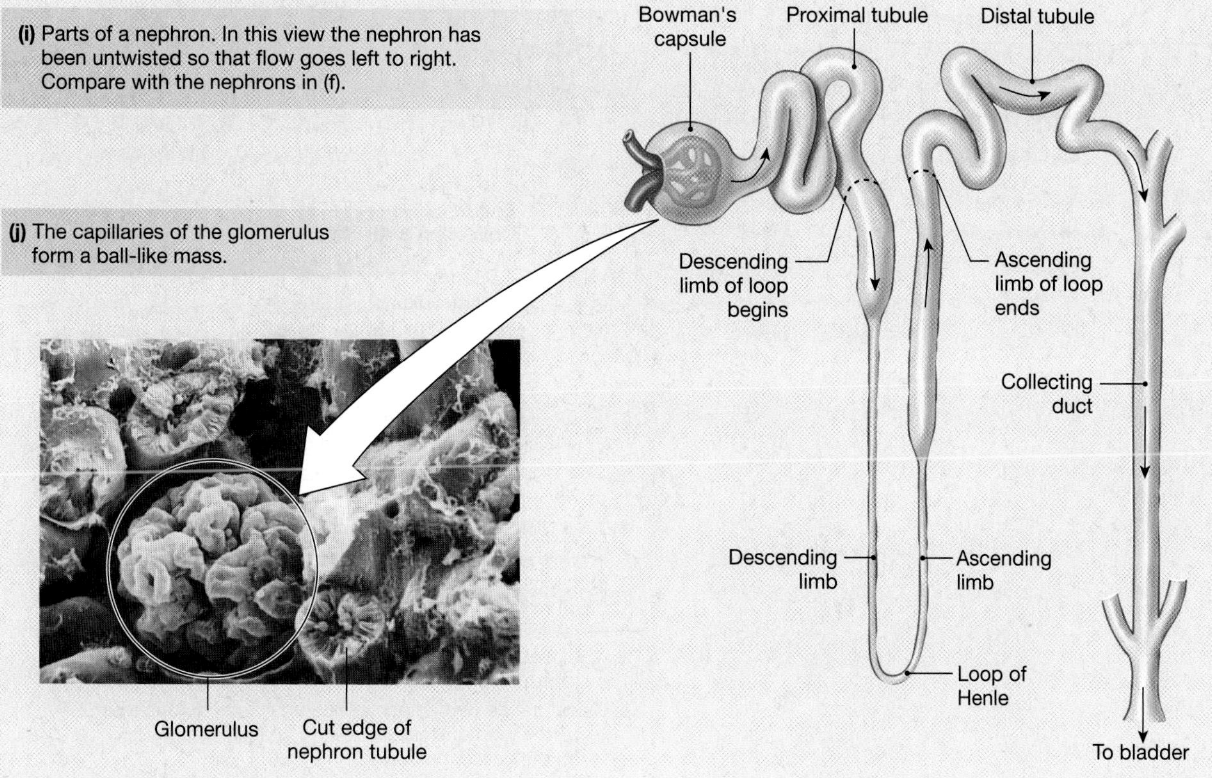

■ Fig. 19.2 *ESSENTIALS*

Nephron Function

The four processes of the nephron are:

- F = **Filtration**: movement from blood to lumen
- R = **Reabsorption**: from lumen to blood
- S = **Secretion**: from blood to lumen
- E = **Excretion**: from lumen to outside the body

This model nephron has been untwisted so that fluid flows left to right.

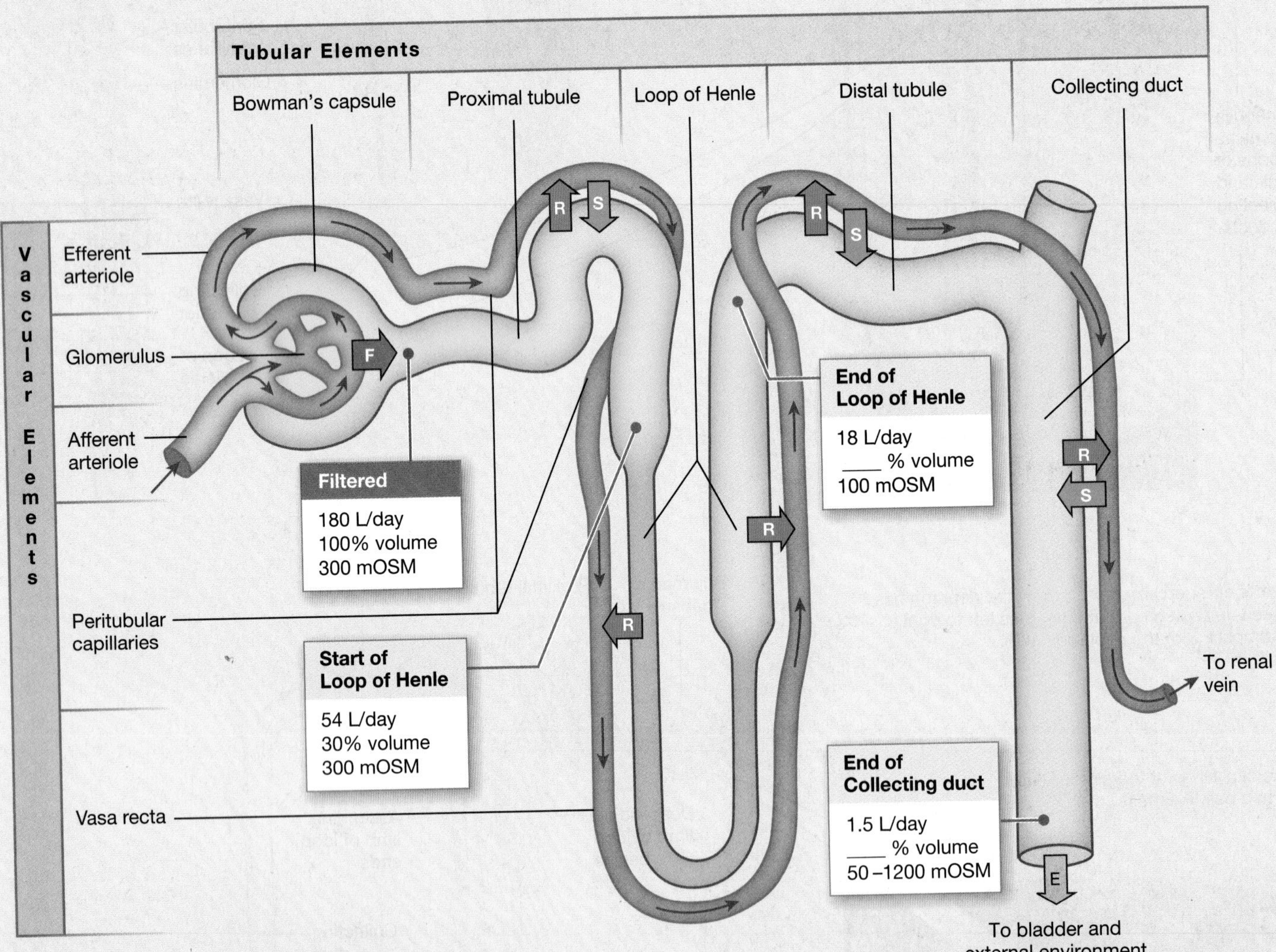

FIGURE QUESTIONS

1. In which segments of the nephron do the following processes take place:
   (a) filtration
   (b) reabsorption
   (c) secretion
   (d) excretion
2. Calculate the percentage of filtered volume that leaves
   (a) the loop of Henle
   (b) the collecting duct

Once the filtered fluid, called *filtrate*, passes into the lumen of the nephron, it becomes part of the body's external environment, just as substances in the lumen of the intestinal tract are part of the external environment [Tbl. 1.1, p. 4]. For this reason, anything that filters into the nephron is destined for **excretion**, removal in the urine, unless it is reabsorbed into the body.

After filtrate leaves Bowman's capsule, it is modified by reabsorption and secretion. **Reabsorption** is the process of moving substances in the filtrate from the lumen of the tubule back into the blood flowing through peritubular capillaries. **Secretion** removes selected molecules from the blood and adds them to the filtrate in the tubule lumen. Although secretion and glomerular filtration both move substances from blood into the tubule, secretion is a more selective process that usually uses membrane proteins to move molecules across the tubule epithelium.

The Nephron Modifies Fluid Volume and Osmolarity

Now let's follow some filtrate through the nephron to learn what happens to it in the various segments (Fig. 19.2). The 180 liters of fluid that filters into Bowman's capsule each day are almost identical in composition to plasma and nearly isosmotic—about 300 mOsM [p. 133]. As this filtrate flows through the proximal tubule, about 70% of its volume is reabsorbed, leaving 54 liters in the lumen.

Reabsorption occurs when proximal tubule cells transport solutes out of the lumen, and water follows by osmosis. Filtrate leaving the proximal tubule has the same osmolarity as filtrate that entered. For this reason, we say that the primary function of the proximal tubule is the reabsorption of isosmotic fluid.

Filtrate leaving the proximal tubule passes into the loop of Henle, the primary site for creating dilute urine. As filtrate passes through the loop, proportionately more solute is reabsorbed than water, and the filtrate becomes hyposmotic relative to the plasma. By the time filtrate flows out of the loop, it averages 100 mOsM, and its volume has fallen from 54 L/day to about 18 L/day. Most of the volume originally filtered into Bowman's capsule has been reabsorbed into the capillaries.

From the loop of Henle, filtrate passes into the distal tubule and the collecting duct. In these two segments, the fine regulation of salt and water balance takes place under the control of several hormones. Reabsorption and (to a lesser extent) secretion determine the final composition of the filtrate. By the end of the collecting duct, the filtrate has a volume of 1.5 L/day and an osmolarity that can range from 50 mOsM to 1200 mOsM. The final volume and osmolarity of urine depend on the body's need to conserve or excrete water and solute.

A word of caution here: it is very easy to confuse *secretion* with *excretion*. Try to remember the origins of the two prefixes. *Se-* means *apart,* as in to separate something from its source. In the nephron, secreted solutes are moved from plasma to tubule lumen. *Ex-* means *out,* or *away,* as in out of or away from the body. Excretion refers to the removal of a substance from the body. Besides the kidneys, other organs that carry out excretory processes include the lungs (CO_2) and intestines (undigested food, bilirubin).

Figure 19.2 summarizes filtration, reabsorption, secretion, and excretion. Filtration takes place in the renal corpuscle as fluid moves from the capillaries of the glomerulus into Bowman's capsule. Reabsorption and secretion occur along the remainder of the tubule, transferring materials between the lumen and the peritubular capillaries. The quantity and composition of the substances being reabsorbed and secreted vary in different segments of the nephron. Filtrate that remains in the lumen at the end of the nephron is excreted as urine.

The amount of any substance excreted in the urine reflects how that substance was handled during its passage through the nephron (■ Fig. 19.3). The amount excreted is equal to the amount filtered into the tubule, minus the amount reabsorbed into the blood, plus the amount secreted into the tubule lumen:

$$\text{Amount filtered} - \text{amount reabsorbed} + \text{amount secreted} = \text{amount excreted}$$

This equation is a useful way to think about renal handling of solutes. In the following sections, we look in more detail at the important processes of filtration, reabsorption, secretion, and excretion.

19

The urinary excretion of a substance depends on its filtration, reabsorption, and secretion.

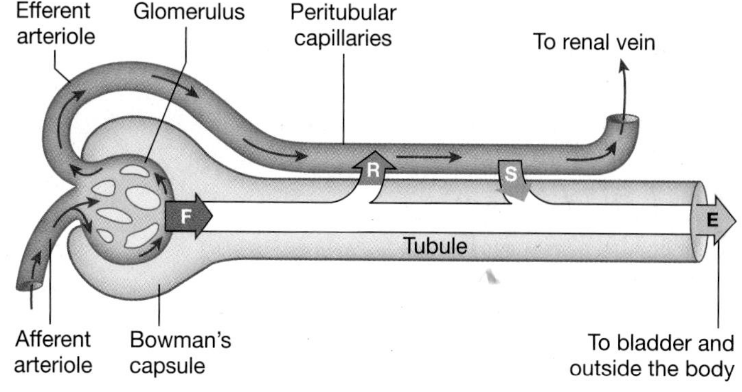

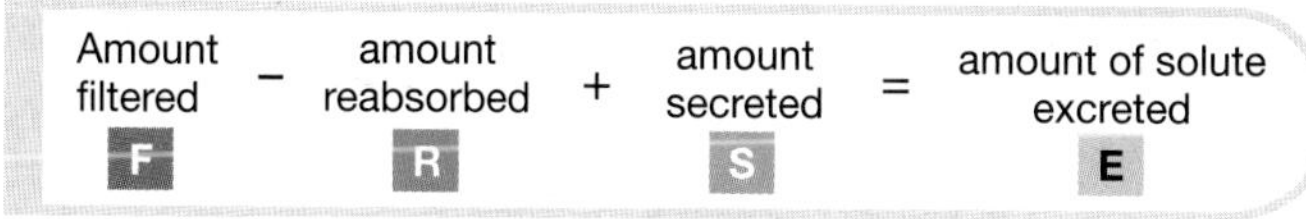

FIGURE QUESTION

A person filters 720 millimoles of K^+ in a day and secrets 43 millimoles. She excretes 79 millimoles in her urine. What happened to the rest of the K^+ and how much was it?

■ **Fig. 19.3**

Concept Check Answers: p. 655

5. Name one way in which filtration and secretion are alike. Name one way in which they differ.
6. A water molecule enters the renal corpuscle from the blood and ends up in the urine. Name all the anatomical structures that the molecule passes through on its trip to the outside world.
7. What would happen to the body if filtration continued at a normal rate but reabsorption dropped to half the normal rate?

Filtration

The filtration of plasma into the kidney tubule is the first step in urine formation. This relatively nonspecific process creates a filtrate whose composition is like that of plasma minus most of the plasma proteins. Under normal conditions, blood cells remain in the capillary, so that the filtrate is composed of water and dissolved solutes.

When you visualize plasma filtering out of the glomerular capillaries, it is easy to imagine that all the plasma in the capillary moves into Bowman's capsule. However, filtration of all the plasma would leave behind a sludge of blood cells and proteins that could not flow out of the glomerulus. Instead, only about one-fifth of the plasma that flows through the kidneys filters into the nephrons. The remaining four-fifths of the plasma, along with most plasma proteins and blood cells, flows into the peritubular capillaries (■ Fig. 19.4). The percentage of total plasma volume that filters into the tubule is called the **filtration fraction.**

The Renal Corpuscle Contains Filtration Barriers

Filtration takes place in the renal corpuscle (■ Fig. 19.5), which consists of the glomerular capillaries surrounded by Bowman's capsule. Substances leaving the plasma must pass through three *filtration barriers* before entering the tubule lumen: the glomerular capillary endothelium, a basal lamina (basement membrane), and the epithelium of Bowman's capsule (Fig. 19.5d). The details of how these filtration barriers function are still under investigation.

The first barrier is the capillary endothelium. Glomerular capillaries are *fenestrated capillaries* [p. 528] with large pores that allow most components of the plasma to filter through the endothelium. The pores are small enough, however, to prevent blood cells from leaving the capillary. The negatively charged proteins on the pore surfaces also help repel negatively charged plasma proteins.

Glomerular **mesangial cells** lie between and around the glomerular capillaries (Fig. 19.5c). Mesangial cells have cytoplasmic bundles of actin-like filaments that enable them to contract and alter blood flow through the capillaries. In addition, mesangial cells secrete cytokines associated with immune and inflammatory processes. Disruptions of mesangial cell function have been linked to several disease processes in the kidney.

The second filtration barrier is the **basal lamina,** an acellular layer of extracellular matrix that separates the capillary endothelium from the epithelial lining of Bowman's capsule (Fig. 19.5d). The basal lamina consists of negatively charged glycoproteins, collagen, and other proteins. The lamina acts like a coarse sieve, excluding most plasma proteins from the fluid that filters through it.

THE FILTRATION FRACTION

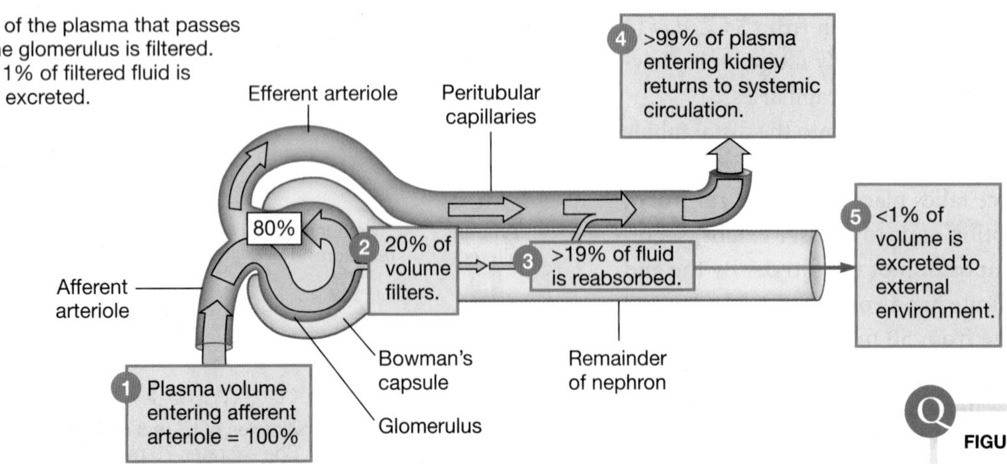

FIGURE QUESTION

If 120 mL of plasma filter each minute and the filtration fraction is 20%, what is the daily renal plasma flow?

■ **Fig. 19.4**

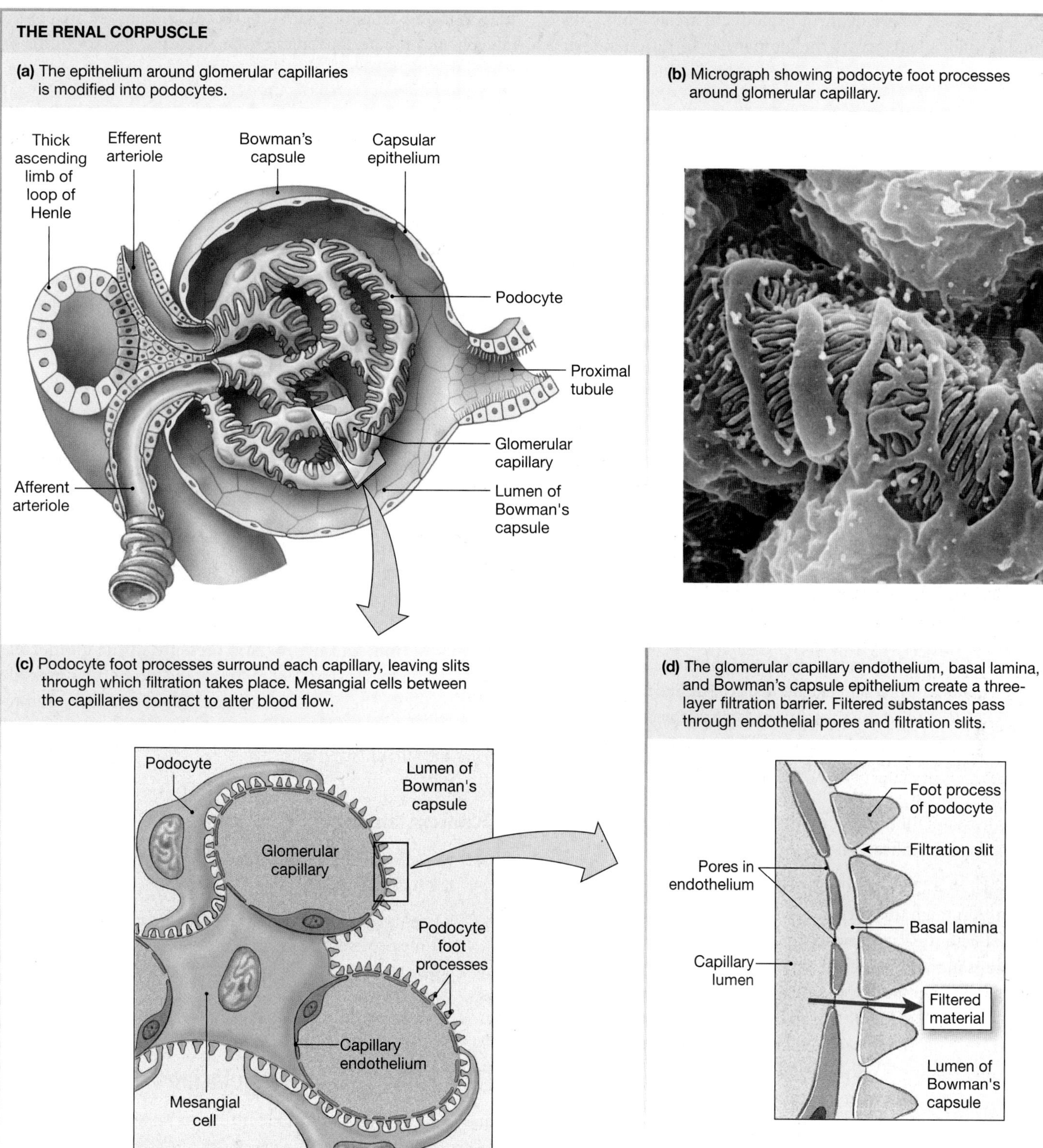
THE RENAL CORPUSCLE
(a) The epithelium around glomerular capillaries is modified into podocytes.
Thick ascending limb of loop of Henle
Efferent arteriole
Bowman's capsule
Capsular epithelium
Podocyte
Proximal tubule
Glomerular capillary
Lumen of Bowman's capsule
Afferent arteriole
(b) Micrograph showing podocyte foot processes around glomerular capillary.
(c) Podocyte foot processes surround each capillary, leaving slits through which filtration takes place. Mesangial cells between the capillaries contract to alter blood flow.
Podocyte
Lumen of Bowman's capsule
Glomerular capillary
Podocyte foot processes
Capillary endothelium
Mesangial cell
(d) The glomerular capillary endothelium, basal lamina, and Bowman's capsule epithelium create a three-layer filtration barrier. Filtered substances pass through endothelial pores and filtration slits.
Foot process of podocyte
Filtration slit
Pores in endothelium
Basal lamina
Capillary lumen
Filtered material
Lumen of Bowman's capsule

Fig. 19.5

The third filtration barrier is the epithelium of Bowman's capsule. The portion of the capsule epithelium that surrounds each glomerular capillary consists of specialized cells called **podocytes** {*podos,* foot} (Fig. 19.5c). Podocytes have long cytoplasmic extensions called **foot processes** that extend from the main cell body (Fig. 19.5a, b).

Foot processes wrap around the glomerular capillaries and interlace with one another, leaving narrow **filtration slits** closed by a semiporous membrane. The filtration slit membrane contains several unique proteins, including *nephrin* and *podocin*. These proteins were discovered by investigators looking for the gene mutations responsible for two congenital kidney diseases.

In these diseases, where nephrin or podocin are absent or abnormal, proteins leak across the glomerular filtration barrier into the urine.

Capillary Pressure Causes Filtration

What drives filtration across the walls of the glomerular capillaries? The process is similar in many ways to filtration of fluid out of systemic capillaries [p. 529]. The three pressures that influence glomerular filtration—capillary blood pressure, capillary colloid osmotic pressure, and capsule fluid pressure—are summarized in ■ Figure 19.6a.

1 The *hydrostatic pressure* (P_H) of blood flowing through the glomerular capillaries forces fluid through the leaky endothelium. Capillary blood pressure averages 55 mm Hg and favors filtration into Bowman's capsule. Although pressure declines along the length of the capillaries, it remains higher than the opposing pressures. Consequently, filtration takes place along nearly the entire length of the glomerular capillaries.

2 The *colloid osmotic pressure* (π) inside glomerular capillaries is higher than that of the fluid in Bowman's capsule. This pressure gradient is due to the presence of proteins in the plasma. The osmotic pressure gradient averages 30 mm Hg and favors fluid movement back into the capillaries.

3 Bowman's capsule is an enclosed space (unlike the interstitial fluid), and so the presence of fluid in the capsule creates a hydrostatic *fluid pressure* (P_{fluid}) that opposes fluid movement into the capsule. Fluid filtering out of the capillaries must displace the fluid already in the capsule lumen. Hydrostatic fluid pressure in the capsule averages 15 mm Hg, opposing filtration.

The net driving force is 10 mm Hg in the direction favoring filtration. Although this pressure may not seem very high, when combined with the very leaky nature of the fenestrated capillaries, it results in rapid fluid filtration into the tubules.

The volume of fluid that filters into Bowman's capsule per unit time is the **glomerular filtration rate (GFR).** Average GFR is 125 mL/min, or 180 L/day, an incredible rate considering that total plasma volume is only about 3 liters. This rate means that the kidneys filter the entire plasma volume 60 times a day, or 2.5 times every hour. If most of the filtrate were not reabsorbed during its passage through the nephron, we would run out of plasma in only 24 minutes of filtration!

GFR is influenced by two factors: the net filtration pressure just described and the filtration coefficient. Filtration pressure is determined primarily by renal blood flow and blood pressure. The **filtration coefficient** has two components: the surface area of the glomerular capillaries available for filtration and the permeability of interface between the capillary and Bowman's capsule. In this respect, glomerular filtration is similar to gas exchange at the alveoli, where the rate of gas exchange depends on partial pressure differences, the surface area of the alveoli, and the permeability of the alveolar-capillary diffusion barrier [p. 604].

GFR Is Relatively Constant

Blood pressure provides the hydrostatic pressure that drives glomerular filtration. Therefore, it might seem reasonable to assume that if blood pressure increased, GFR would increase, and if blood pressure fell, GFR would decrease. That is not usually the case, however. Instead, GFR is remarkably constant over a wide range of blood pressures. As long as mean arterial blood pressure remains between 80 mm Hg and 180 mm Hg, GFR averages 180 L/day (Fig. 19.6b).

GFR is controlled primarily by regulation of blood flow through the renal arterioles. If the overall resistance of the renal arterioles increases, renal blood flow decreases, and blood is diverted to other organs [Fig. 15.13, p. 525]. The effect of increased resistance on GFR, however, depends on *where* the resistance change takes place.

If resistance increases in the *afferent* arteriole (Fig. 19.6d), hydrostatic pressure decreases on the glomerular side of the constriction. This translates into a decrease in GFR. If resistance increases in the *efferent* arteriole, blood "dams up" in front of the constriction, and hydrostatic pressure in the glomerular capillaries increases (Fig. 19.8e). Increased glomerular pressure increases GFR. The opposite changes occur with decreased resistance in the afferent or efferent arterioles. Most regulation occurs at the afferent arteriole.

Concept Check Answers: p. 655

8. Why is the osmotic pressure of plasma in efferent arterioles higher than that in afferent arterioles?
9. If a hypertensive person's blood pressure is 143/107 mm Hg and mean arterial pressure is diastolic pressure + ⅓ the pulse pressure, what is this person's mean arterial pressure? What is this person's GFR according to Figure 19.6b?

GFR Is Subject to Autoregulation

Autoregulation of glomerular filtration rate is a local control process in which the kidney maintains a relatively constant GFR in the face of normal fluctuations in blood pressure. One important function of GFR autoregulation is to protect the filtration barriers from high blood pressures that might damage them. We do not completely understand the autoregulation process, but several mechanisms are at work. The **myogenic response** is the intrinsic ability of vascular smooth muscle to respond to pressure changes. **Tubuloglomerular feedback** is a paracrine signaling mechanism through which changes in fluid flow through the loop of Henle influence GFR.

■ Fig. 19.6 *ESSENTIALS*

Glomerular Filtration Rate

Filtration out of glomerular capillaries is similar to filtration in other systemic capillaries. Filtration pressure depends on hydrostatic pressure, and is opposed by colloid osmotic pressure and capsule fluid pressure.

(a) Calculating glomerular filtration pressure

$P_H - \pi - P_{fluid}$ = net filtration pressure

55 – 30 – 15 = 10mm Hg

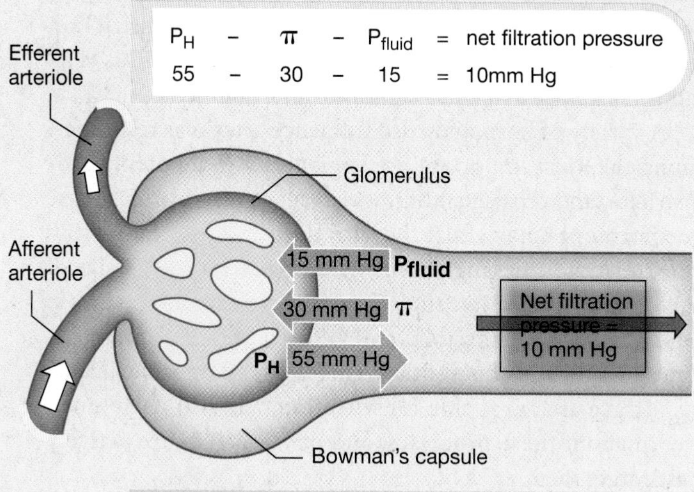

KEY

P_H = Hydrostatic pressure (blood pressure)

π = Colloid osmotic pressure gradient due to proteins in plasma but not in Bowman's capsule

P_{fluid} = Fluid pressure created by fluid in Bowman's capsule

(b) Autoregulation of glomerular filtration rate takes place over a wide range of blood pressures.

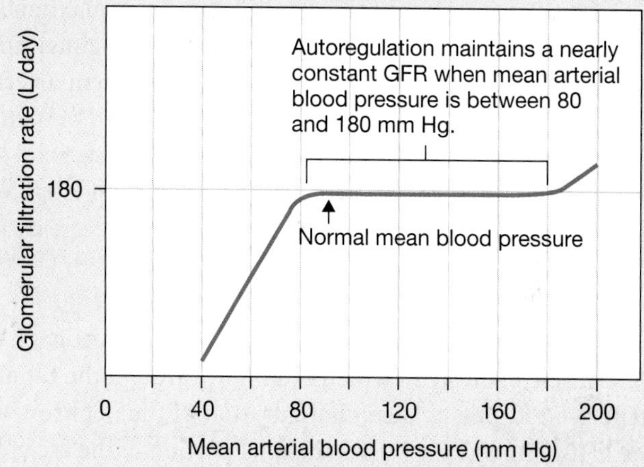

(c) Resistance changes in renal arterioles alter renal blood flow and GFR.

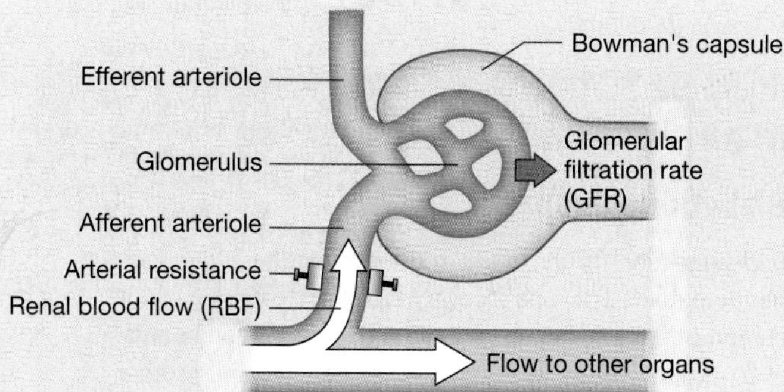

(d) Vasoconstriction of the afferent arteriole increases resistance and decreases renal blood flow, capillary blood pressure (P_H), and GFR.

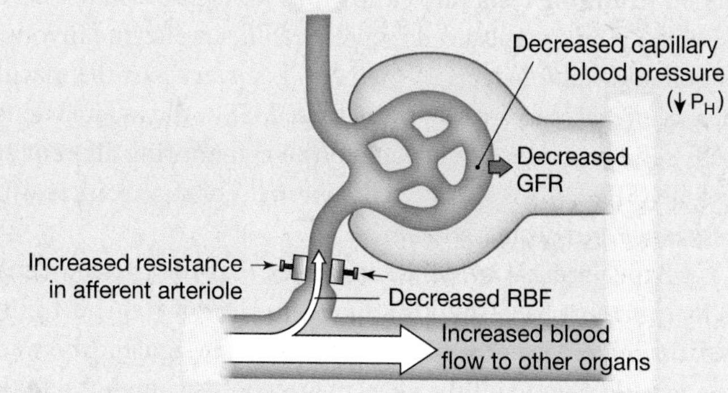

(e) Increased resistance of efferent arteriole decreases renal blood flow but increases P_H and GFR.

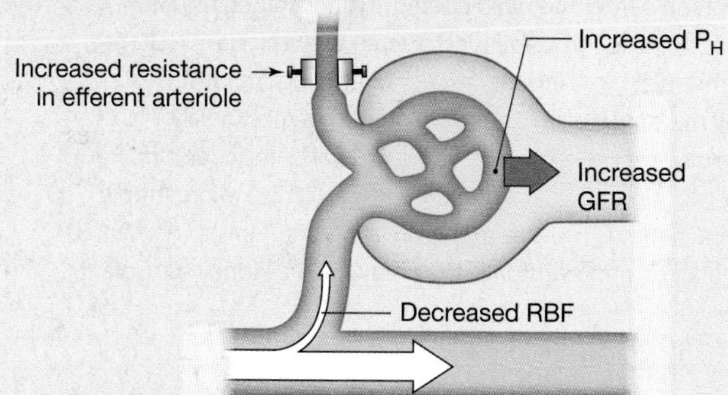

FIGURE QUESTION

What happens to capillary blood pressure, GFR, and RBF when the afferent arteriole dilates?

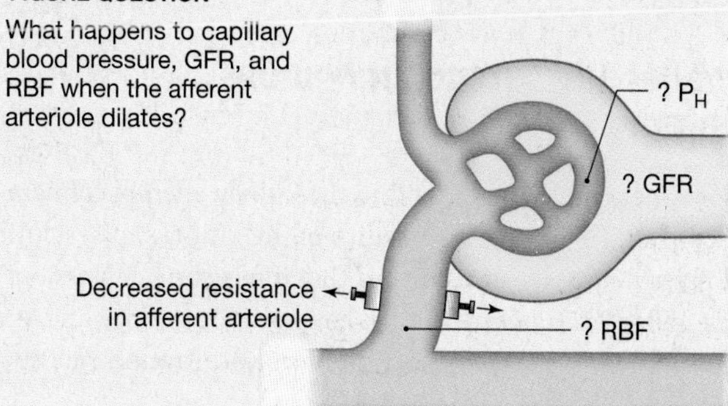

Myogenic Response The myogenic response of afferent arterioles is similar to autoregulation in other systemic arterioles. When smooth muscle in the arteriole wall stretches because of increased blood pressure, stretch-sensitive ion channels open, and the muscle cells depolarize. Depolarization opens voltage-gated Ca^{2+} channels, and the vascular smooth muscle contracts [p. 519]. Vasoconstriction increases resistance to flow, and so blood flow through the arteriole diminishes. The decrease in blood flow decreases filtration pressure in the glomerulus.

If blood pressure decreases, the tonic level of arteriolar contraction disappears, and the arteriole becomes maximally dilated. However, vasodilation is not as effective at maintaining GFR as vasoconstriction because normally the afferent arteriole is fairly relaxed. Consequently, when mean blood pressure drops below 80 mm Hg, GFR decreases. This decrease is adaptive in the sense that if less plasma is filtered, the potential for fluid loss in the urine is decreased. In other words, a decrease in GFR helps the body conserve blood volume.

Tubuloglomerular Feedback Tubuloglomerular feedback is a local control pathway in which fluid flow through the tubule influences GFR. The twisted configuration of the nephron, as shown in ■ Figure 19.7a, causes the final portion of the ascending limb of the loop of Henle to pass between the afferent and efferent arterioles. The tubule and arteriolar walls are modified in the regions where they contact each other and together form the *juxtaglomerular apparatus*.

The modified portion of the tubule epithelium is a plaque of cells called the **macula densa** (Fig. 19.7b). The adjacent wall of the afferent arteriole has specialized smooth muscle cells called **granular cells** (also known as *juxtaglomerular cells* or *JG cells*). The granular cells secrete *renin*, an enzyme involved in salt and water balance. When NaCl delivery past the macula densa increases as a result of increased GFR, the macula densa cells send a paracrine message to the neighboring afferent arteriole (Fig. 19.7c). The afferent arteriole constricts, increasing resistance and decreasing GFR.

Experimental evidence indicates that the macula densa cells transport NaCl, and that increases in salt transport initiate tubuloglomerular feedback. The paracrine signaling between the macula densa and the afferent arteriole is complex, and the details are still being worked out. Experiments show that multiple paracrine signals, including ATP, adenosine, and nitric oxide, pass from the macula densa to the arteriole.

Hormones and Autonomic Neurons Also Influence GFR

Although local mechanisms within the kidney attempt to maintain a constant GFR, the importance of the kidneys in systemic blood pressure homeostasis means that integrating centers outside the kidney can override local controls. Hormones and the autonomic nervous system alter glomerular filtration rate two ways: by changing resistance in the arterioles and by altering the filtration coefficient.

Neural control of GFR is mediated by sympathetic neurons that innervate both the afferent and efferent arterioles. Sympathetic innervation of α-receptors on vascular smooth muscle causes vasoconstriction [p. 522]. If sympathetic activity is moderate, there is little effect on GFR. If systemic blood pressure drops sharply, however, as occurs with hemorrhage or severe dehydration, sympathetically induced vasoconstriction of the arterioles decreases GFR and renal blood flow. This is an adaptive response that helps conserve fluid volume.

A variety of hormones also influence arteriolar resistance. Among the most important are *angiotensin II*, a potent vasoconstrictor, and prostaglandins, which act as vasodilators. These same hormones may affect the filtration coefficient by acting on podocytes or mesangial cells. Podocytes change the size of the glomerular filtration slits. If the slits widen, more surface area is available for filtration, and GFR increases. Contraction of mesangial cells apparently changes the glomerular capillary surface area available for filtration. We still have much to learn about these processes, and physiologists are actively investigating them.

Concept Check Answers: p. 655

10. If systemic blood pressure remains constant but the afferent arteriole of a nephron constricts, what happens to renal blood flow and GFR in that nephron?
11. A person with cirrhosis of the liver has lower-than-normal levels of plasma proteins and consequently a higher-than-normal GFR. Explain why a decrease in plasma protein concentration causes an increase in GFR.

EMERGING CONCEPTS: DIABETES

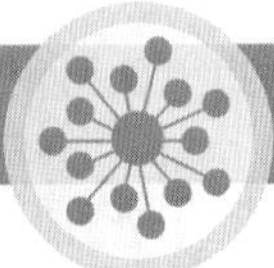

Diabetic Nephropathy

End-stage renal failure, in which kidney function has deteriorated beyond recovery, is a life-threatening complication in 30–40% of people with type 1 diabetes and in 10–20% of those with type 2 diabetes. As with many other complications of diabetes, the exact causes of renal failure are not clear. Diabetic nephropathy usually begins with an increase in glomerular filtration. This is followed by the appearance of proteins in the urine (*proteinuria*), an indication that the normal filtration barrier has been altered. In later stages, filtration rates decline. This stage is associated with thickening of the glomerular basal lamina and changes in both podocytes and mesangial cells. Abnormal growth of mesangial cells compresses the glomerular capillaries and impedes blood flow, contributing to the decrease in glomerular filtration. At this point, patients must have their kidney function supplemented by dialysis, and eventually they may need a kidney transplant.

THE JUXTAGLOMERULAR APPARATUS

The juxtaglomerular apparatus consists of macula densa and granular cells. Paracrine signaling between the nephron and afferent arteriole influences GFR.

(a) The nephron loops back on itself so that the ascending limb of the loop of Henle passes between the afferent and efferent arterioles.

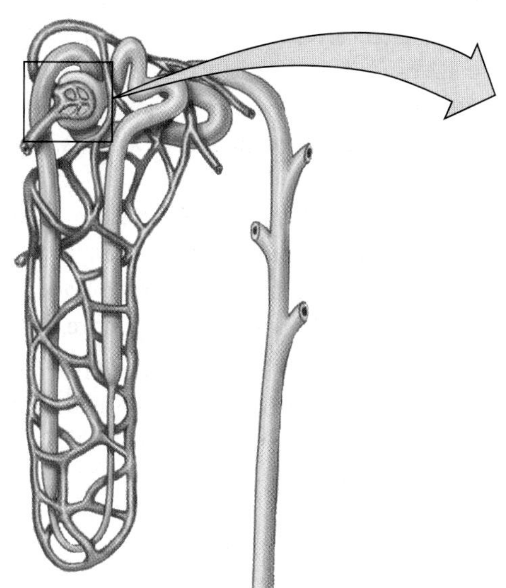

(b) The macula densa cells sense distal tubule flow and release paracrines that affect afferent arteriole diameter.

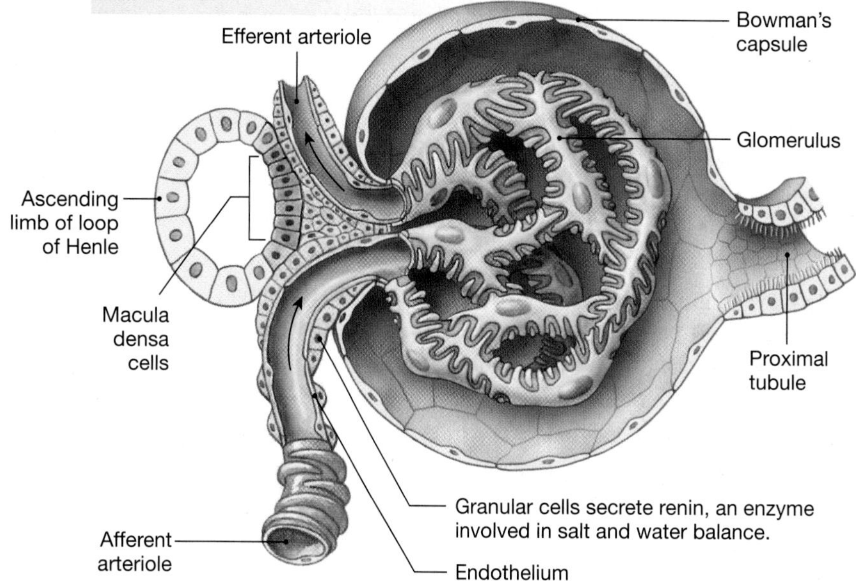

(c) Tubuloglomerular feedback helps GFR autoregulation.

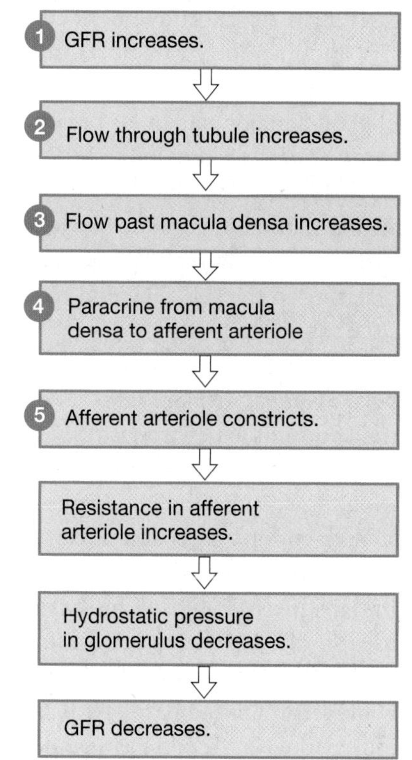

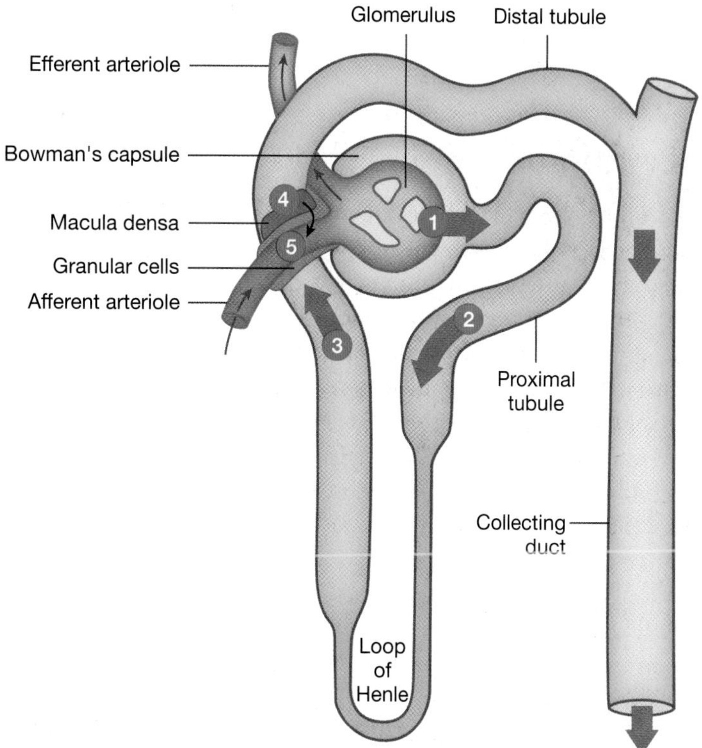

■ **Fig. 19.7**

Reabsorption

Each day, 180 liters of filtered fluid pass from the glomerular capillaries into the tubules, yet only about 1.5 liters are excreted in the urine. Thus more than 99% of the fluid entering the tubules must be reabsorbed into the blood as filtrate moves through the nephrons. Most of this reabsorption takes place in the proximal tubule, with a smaller amount of reabsorption in the distal segments of the nephrons. Regulated reabsorption in the distal nephron allows the kidneys to return ions and water to the plasma selectively—as needed to maintain homeostasis.

One question you might be asking is, "Why bother to filter 180 L/day and then reabsorb 99% of it? Why not simply filter and excrete the 1% that needs to be eliminated?" There are two reasons. First, many foreign substances are filtered into the tubule but not reabsorbed into the blood. The high daily filtration rate helps clear such substances from the plasma very rapidly.

Once a substance filters into the lumen of Bowman's capsule, it is no longer part of the body's internal environment. The lumen of the nephron is external environment, and anything in the filtrate is destined to leave the body in the urine unless there is a tubule mechanism for reclaiming it. Many small nutrients, such as glucose and citric acid cycle intermediates, are filtered, but the proximal tubule very efficiently reabsorbs them.

Second, filtering ions and water into the tubule simplifies their regulation. If a portion of filtrate that reaches the distal nephron is not needed to maintain homeostasis, it passes into the urine. With a high GFR, this excretion can occur quite rapidly. However, if the ions and water are needed, they are reabsorbed.

Reabsorption May Be Active or Passive

Reabsorption of water and solutes from the tubule lumen to the extracellular fluid depends on active transport. The filtrate flowing out of Bowman's capsule into the proximal tubule has the same solute concentrations as extracellular fluid. To move solute out of the lumen, the tubule cells must therefore use active transport to create concentration or electrochemical gradients. Water osmotically follows solutes as they are reabsorbed.

■ Figure 19.8a is an overview of renal reabsorption. Active transport of Na^+ from the tubule lumen to the extracellular fluid creates a transepithelial electrical gradient in which the lumen is more negative than the ECF. Anions then follow the positively charged Na^+ out of the lumen. The removal of Na^+ and anions from lumen to ECF dilutes the luminal fluid and increases the concentration of the ECF, so water leaves the tubule by osmosis.

The loss of volume from the lumen increases the concentration of solutes (including K^+, Ca^{2+}, and urea) left behind in the filtrate: the same amount of solute in a smaller volume equals higher solute concentration. Once luminal solute concentrations are higher than solute concentrations in the extracellular fluid, the solutes diffuse out of the lumen if the epithelium of the tubule is permeable to them.

Reabsorption involves both epithelial transport and paracellular transport. In **epithelial transport** (also called *transcellular transport*), substances cross the apical and basolateral membranes of the tubule epithelial cell [p. 158] to reach the interstitial fluid. In the **paracellular pathway,** substances pass through the cell-cell junction between two adjacent cells. Which route a solute takes depends on the permeability of the epithelial junctions and on the electrochemical gradient for the solute.

For solutes that move by epithelial transport, their concentration or electrochemical gradients determine their transport mechanisms. Solutes moving down their gradient use open leak channels or facilitated diffusion carriers to cross the cell membrane. Molecules that need to be pushed against their gradient are moved by either primary or indirect (usually secondary) active transport. Sodium is directly or indirectly involved in many instances of both passive and active transport.

Active Transport of Sodium The active reabsorption of Na^+ is the primary driving force for most renal reabsorption. As noted earlier, filtrate entering the proximal tubule is similar in ion composition to plasma, with a higher Na^+ concentration than is found in cells. Thus Na^+ in the filtrate can enter tubule cells passively by moving down its electrochemical gradient (Fig. 19.8b). Apical movement of Na^+ uses a variety of symport and antiport transport proteins [p. 148] or open leak channels. In the proximal tubule, the *Na^+-H^+ exchanger (NHE)* plays a major role in Na^+ reabsorption, as does the apical *epithelial Na^+ channel (ENaC,* pronounced ee-knack). Once inside a tubule cell, Na^+ is actively transported out across the basolateral membrane in exchange for K^+ by the Na^+-K^+-ATPase. A basolateral K^+ leak channel prevents K^+ from accumulating in the cell. The end result is Na^+ reabsorption across the tubule epithelium.

Secondary Active Transport: Symport with Sodium Sodium-linked secondary active transport in the nephron is responsible for the reabsorption of many substances, including glucose, amino acids, ions, and various organic metabolites. Figure 19.8c shows one example: Na^+-dependent glucose reabsorption across the proximal tubule epithelium [Fig. 5.21, p. 536]. The apical membrane contains the *Na^+-glucose cotransporter* (SGLT) that brings glucose into the cytoplasm against its concentration gradient by harnessing the energy of Na^+ moving down its electrochemical gradient. On the basolateral side of the cell, Na^+ is pumped out by the Na^+-K^+-ATPase, while glucose diffuses out with the aid of a facilitated diffusion GLUT transporter.

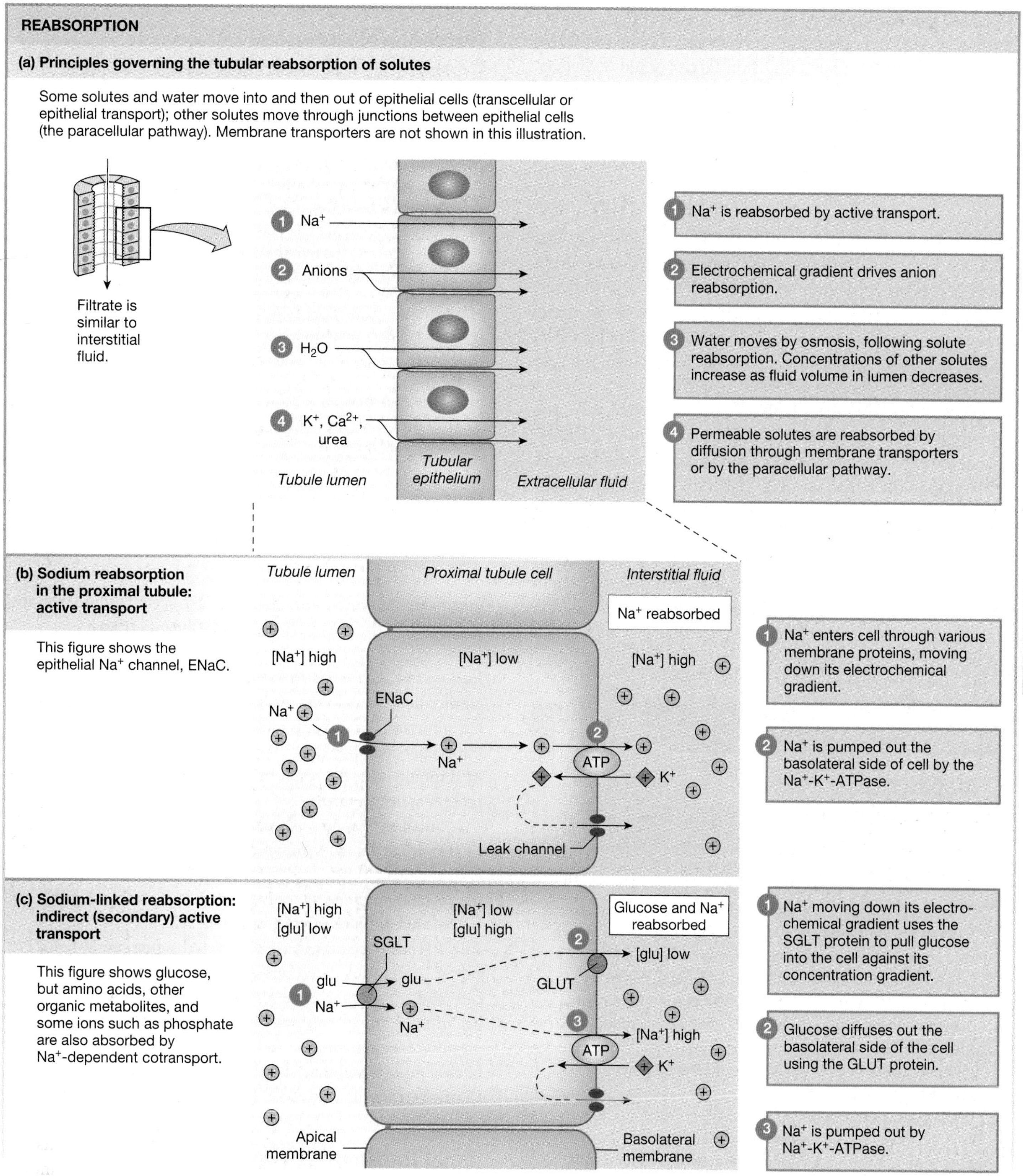
REABSORPTION
(a) Principles governing the tubular reabsorption of solutes
Some solutes and water move into and then out of epithelial cells (transcellular or epithelial transport); other solutes move through junctions between epithelial cells (the paracellular pathway). Membrane transporters are not shown in this illustration.
Filtrate is similar to interstitial fluid.
1 Na+
2 Anions
3 H2O
4 K+, Ca2+, urea
Tubule lumen
Tubular epithelium
Extracellular fluid
1 Na+ is reabsorbed by active transport.
2 Electrochemical gradient drives anion reabsorption.
3 Water moves by osmosis, following solute reabsorption. Concentrations of other solutes increase as fluid volume in lumen decreases.
4 Permeable solutes are reabsorbed by diffusion through membrane transporters or by the paracellular pathway.
(b) Sodium reabsorption in the proximal tubule: active transport
This figure shows the epithelial Na+ channel, ENaC.
Tubule lumen
Proximal tubule cell
Interstitial fluid
Na+ reabsorbed
[Na+] high
[Na+] low
[Na+] high
ENaC
Na+
Na+
ATP
K+
Leak channel
1 Na+ enters cell through various membrane proteins, moving down its electrochemical gradient.
2 Na+ is pumped out the basolateral side of cell by the Na+-K+-ATPase.
(c) Sodium-linked reabsorption: indirect (secondary) active transport
This figure shows glucose, but amino acids, other organic metabolites, and some ions such as phosphate are also absorbed by Na+-dependent cotransport.
[Na+] high
[glu] low
[Na+] low
[glu] high
Glucose and Na+ reabsorbed
SGLT
glu
glu
Na+
Na+
GLUT
[glu] low
[Na+] high
ATP
K+
Apical membrane
Basolateral membrane
1 Na+ moving down its electrochemical gradient uses the SGLT protein to pull glucose into the cell against its concentration gradient.
2 Glucose diffuses out the basolateral side of the cell using the GLUT protein.
3 Na+ is pumped out by Na+-K+-ATPase.

Fig. 19.8

The same basic pattern holds for many other molecules absorbed by Na^+-dependent transport: an apical symport protein and a basolateral facilitated diffusion carrier or ion exchanger. Other molecules that are reabsorbed by similar mechanisms include amino acids, lactate, citric acid cycle intermediates such as citrate and α-ketoglutarate (αKG), and ions such as phosphate and sulfate. A few of the apical transporters use H^+ in place of Na^+.

Passive Reabsorption: Urea The nitrogenous waste product urea has no active transporters in the proximal tubule but can move across the epithelium by diffusion if there is a urea concentration gradient. Initially, urea concentrations in the filtrate and extracellular fluid are equal. However, the active transport of Na^+ and other solutes in the proximal tubule creates a urea concentration gradient by the following process.

When Na^+ and other solutes are reabsorbed from the proximal tubule, the transfer of osmotically active particles makes the extracellular fluid more concentrated than the filtrate remaining in the lumen (see Fig. 19.8a). In response to the osmotic gradient, water moves by osmosis across the epithelium. Up to this point, no urea molecules have moved out of the lumen because there has been no urea concentration gradient.

When water is reabsorbed, the concentration of urea in the lumen increases—the same amount of urea is contained in a smaller volume. Once a concentration gradient for urea exists, urea moves out of the lumen into the extracellular fluid by transport through the cells or by the paracellular pathway.

RUNNING PROBLEM

Uric acid, the molecule that causes gout, is a normal product of purine metabolism. Increased uric acid production may be associated with cell and tissue breakdown, or it may occur as a result of inherited enzyme defects. Plasma urate, the anionic form of uric acid, filters freely into Bowman's capsule but is almost totally reabsorbed in the first part of the proximal tubule. The middle section of the proximal tubule then secretes about half of the reabsorbed urate back into the lumen, and the terminal section of the proximal tubule again reabsorbs some of it. The end result is net secretion.

Q3. Purines are part of which category of biomolecules? Using that information, explain why uric acid levels in the blood go up when cell breakdown increases.

Q4. Based on what you have learned about uric acid and urate, predict two ways a person may develop hyperuricemia.

Endocytosis: Plasma Proteins Filtration of plasma at the glomerulus normally leaves most plasma proteins in the blood, but some smaller proteins and peptides can pass through the filtration barrier. Most filtered proteins are reabsorbed in the proximal tubule, with the result that normally only trace amounts of protein appear in urine.

Small as they are, filtered proteins are too large to be reabsorbed by carriers or through channels. Instead they enter proximal tubule cells by receptor-mediated endocytosis [p. 155] at the apical membrane. Once in the cells, the proteins are digested in lysosomes and released as amino acids.

Recently scientists have discovered a membrane-bound receptor protein in coated pits on the surface of proximal tubule cells and some other tissues in the body. The protein is a member of the LDL-receptor family [p. 157] and has been named *megalin*. Megalin appears to be responsible for reabsorption of filtered protein and may also play a role in cellular uptake of carrier-bound steroid hormones and lipid-soluble vitamins.

BIOTECHNOLOGY

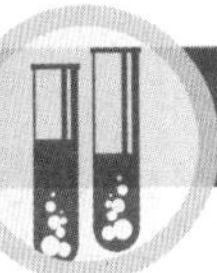

Artificial Kidneys

Many people with severe renal disease depend on *dialysis*, a medical procedure that either supplements or completely replaces their kidney function. Imagine trying to make a machine or develop a procedure that performs the functions of the kidney. What would it have to do? Dialysis is based on diffusion through a semipermeable membrane. Solutes and water pass from a patient's extracellular fluid across the membrane into a dialysis fluid. *Hemodialysis* routes blood from the arm past a membrane in an external dialysis machine. This technique requires attachment to the machine for 3–5 hours three days a week and is used for more severe cases of renal failure. *Peritoneal dialysis* is also called *continuous ambulatory peritoneal dialysis* (CAPD) because it takes place while the patient moves about during daily activities. In CAPD, dialysis fluid is injected into the peritoneal cavity, where it accumulates waste products from the blood for 4–6 hours before being drained out. For more information about dialysis, see the web site for the National Institute of Diabetes and Digestive and Kidney Diseases (*www.niddk.nih.gov*) and search for *dialysis*.

Renal Transport Can Reach Saturation

Most transport in the nephron uses membrane proteins and exhibits the three characteristics of mediated transport: saturation, specificity, and competition [p. 56].

Saturation refers to the maximum rate of transport that occurs when all available carriers are occupied by (are saturated with) substrate. At substrate concentrations below the saturation point, transport rate is directly related to substrate concentration (■ Fig. 19.9). At substrate concentrations equal to or above the saturation point, transport occurs at a maximum rate. The transport rate at saturation is the **transport maximum,** or T_m [p. 154].

Glucose reabsorption in the nephron is an excellent example of the consequences of saturation. At normal plasma glucose concentrations, all glucose that enters the nephron is reabsorbed before it reaches the end of the proximal tubule. The tubule epithelium is well supplied with carriers to capture glucose as the filtrate flows past.

But what happens if blood glucose concentrations become excessive, as they do in diabetes mellitus? In that case, glucose is filtered faster than the carriers can reabsorb it. The carriers become saturated and are unable to reabsorb all the glucose that flows through the tubule. As a result, some glucose escapes reabsorption and is excreted in the urine.

Consider the following analogy. Assume that the carriers are like seats on a train at Disney World. Instead of boarding the stationary train from a stationary platform, passengers step onto a moving sidewalk that rolls them past the train. As the passengers see an open seat, they grab it. However, if more people are allowed onto the moving sidewalk than there are seats in the train, some people will not find seats. And because the sidewalk is moving people past the train toward an exit, they cannot wait for the next train. Instead, they end up being transported out the exit.

Glucose molecules entering Bowman's capsule in the filtrate are like passengers stepping onto the moving sidewalk. To be reabsorbed, each glucose molecule must bind to a transporter as the filtrate flows through the proximal tubule. If only a few glucose molecules enter the tubule at a time, each one can find a free transporter and be reabsorbed, just as a small number of people on the moving sidewalk all find seats on the train. However, if glucose molecules filter into the tubule faster than the glucose carriers can transport them, some glucose remains in the lumen and is excreted in the urine.

■ Figure 19.10 is a graphic representation of glucose handling by the kidney. Figure 19.10a shows that the filtration rate of glucose from plasma into Bowman's capsule is proportional to the plasma concentration of glucose. Because filtration does not exhibit saturation, the graph continues infinitely in a straight line: the filtrate glucose concentration is always equal to the plasma glucose concentration.

Figure 19.10b plots the reabsorption rate of glucose in the proximal tubule against the plasma concentration of glucose. Reabsorption exhibits a maximum transport rate (T_m) when the carriers reach saturation. Notice that normal plasma glucose concentrations are well below the saturation point.

Figure 19.10c plots the excretion rate of glucose in relation to the plasma concentration of glucose. Remember that excretion equal filtration minus reabsorption ($E = F - R$). When plasma glucose concentrations are low enough that 100% of the filtered glucose is reabsorbed, no glucose is excreted. Once the carriers reach saturation, glucose excretion begins. The plasma concentration at which glucose first appears in the urine is called the **renal threshold** for glucose.

Figure 19.10d is a composite graph that compares filtration, reabsorption, and excretion of glucose. Recall from our earlier discussion that

Amount excreted = amount filtered − amount reabsorbed + amount secreted

For glucose, which is not secreted, the equation can be rewritten as

Glucose excreted = glucose filtered − glucose reabsorbed

Under normal conditions, all filtered glucose is reabsorbed. In other words, filtration is equal to reabsorption.

Notice in Figure 19.10d that the lines representing filtration and reabsorption are identical up to the plasma glucose concentration that equals the renal threshold. If filtration equals reabsorption, the algebraic difference between the two is zero, and there is no excretion. Once the renal threshold is reached, filtration begins to exceed reabsorption. Notice on the graph

19

SATURATION OF MEDIATED TRANSPORT

The transport rate of a sustance is proportional to the plasma concentration of the substance, up to the point at which transporters become saturated. Once saturation occurs, transport rate reaches a maximum. The plasma concentration of substrate at which the transport maximum occurs is called the renal threshold.

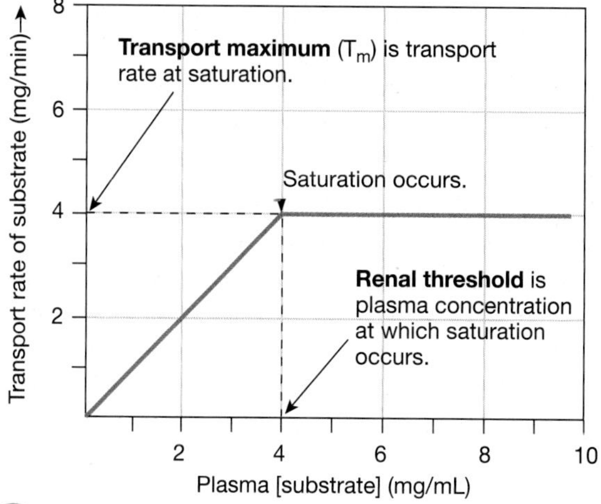

GRAPH QUESTION

What is the transport rate at the following plasma substrate concentrations: 3 mg/mL, 5 mg/mL, 8 mg/mL? At what plasma substrate concentration is the transport rate 2 mg/min?

■ **Fig. 19.9**

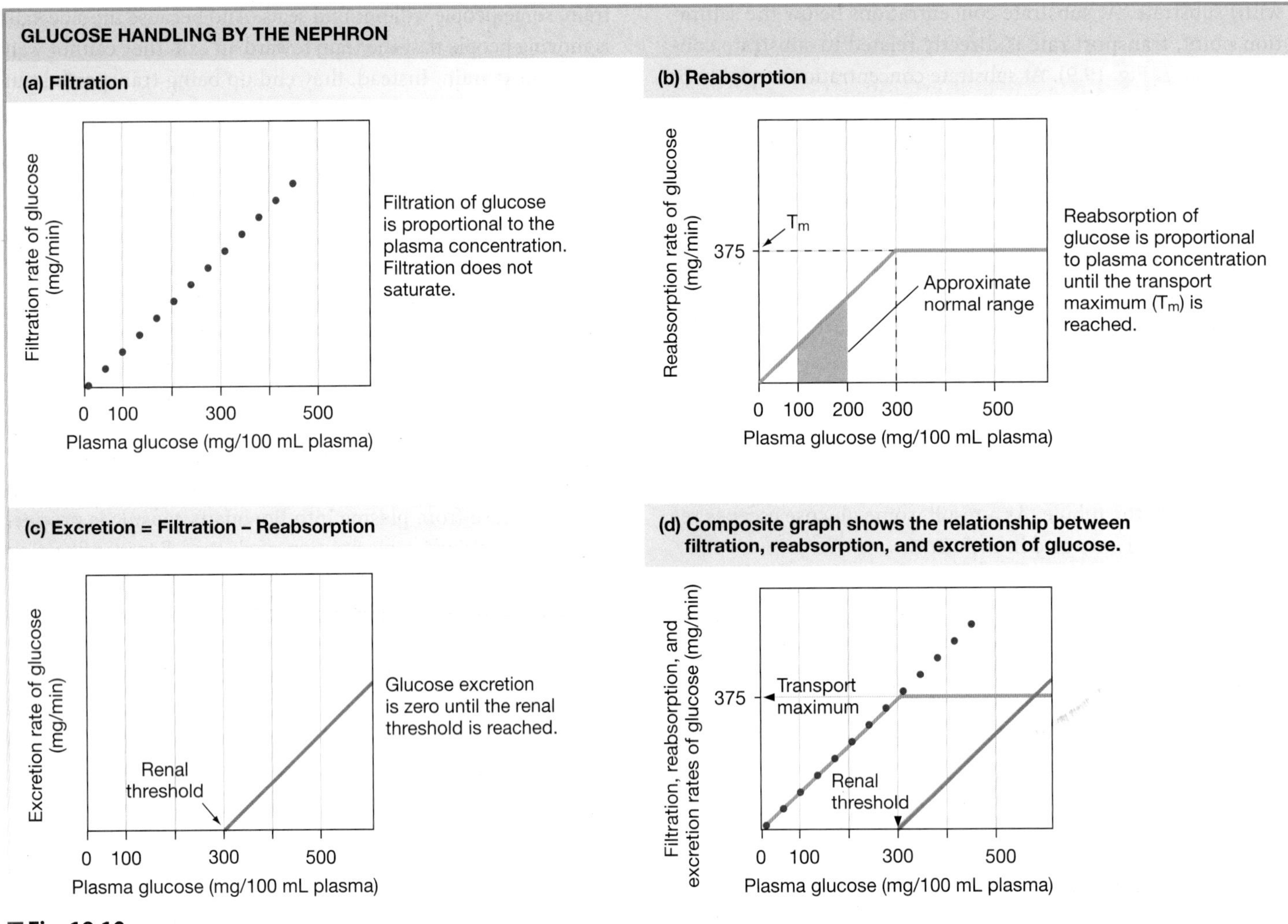

■ **Fig. 19.10**

that the filtration and reabsorption lines diverge at this point. The difference between the filtration line and the reabsorption line represents the excretion rate:

$$\text{Excretion} = \underset{\text{(increasing)}}{\text{filtration}} - \underset{\text{(constant)}}{\text{reabsorption}}$$

Excretion of glucose in the urine is called **glucosuria** or **glycosuria** {*-uria,* in the urine} and usually indicates an elevated blood glucose concentration. Rarely, glucose appears in the urine even though the blood glucose concentrations are normal. This situation is due to a genetic disorder in which the nephron does not make enough carriers.

Peritubular Capillary Pressures Favor Reabsorption

The reabsorption we have just discussed refers to the movement of solutes and water from the tubule lumen to the interstitial fluid. How does that reabsorbed fluid then get into the capillary? The answer is that the hydrostatic pressure that exists along the entire length of the peritubular capillaries is less than the colloid osmotic pressure, so the net pressure gradient favors reabsorption (■ Fig. 19.11). The peritubular capillaries have an average hydrostatic pressure of 10 mm Hg (in contrast to the glomerular capillaries, where hydrostatic pressure averages 55 mm Hg). Colloid osmotic pressure, which favors movement of fluid into the capillaries, is 30 mm Hg. As a result, the pressure gradient in peritubular capillaries is 20 mm Hg, favoring the absorption of fluid into the capillaries. Fluid that is reabsorbed passes from the peritubular capillaries to the venous circulation and returns to the heart.

Secretion

Secretion is the transfer of molecules from extracellular fluid into the lumen of the nephron (see Fig. 19.2). Secretion, like reabsorption, depends mostly on membrane transport systems. The secretion of K^+ and H^+ by the distal nephron is important in the homeostatic regulation of those ions. In addition, many organic compounds are secreted. These compounds include

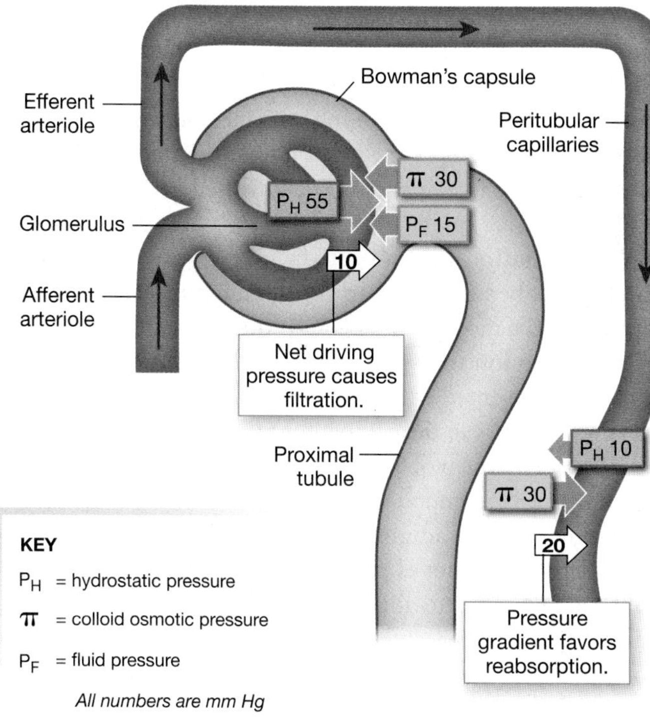

FIGURE QUESTION

Why is the hydrostatic pressure so much lower in the peritubular capillaries than in the glomerulus?

Fig. 19.11

both metabolites produced in the body and substances brought into the body, or *xenobiotics*.

Secretion enables the nephron to enhance excretion of a substance. If a substance is filtered and not reabsorbed, it is excreted very efficiently. If, however, the substance is filtered into the tubule, not reabsorbed, *and* then more of it is secreted into the tubule from the peritubular capillaries, excretion is even more efficient.

Secretion is an active process because it requires moving substrates against their concentration gradients. Most organic compounds are secreted across the proximal tubule epithelium into the lumen by indirect active transport. Let's look at how the tubule handles the secretion of organic anions (■ Fig. 19.12).

The transporters responsible for organic solute excretion have broad specificity. For example, the **organic anion transporter** (OAT) family, shown in this figure, is able to transport a wide variety of endogenous and exogenous anions, ranging from bile salts to benzoate used as a preservative in soft drinks, salicylate from aspirin, and the artificial sweetener saccharine. Secretion of organic anions on the OAT is an example of *tertiary* active transport, where the use of energy from ATP is two steps removed from the OAT. Let's see how this works.

In the first step of the process, which is direct active transport, the proximal tubule cell uses ATP to maintain the low intracellular concentration of Na^+. In the second step, the Na^+ gradient is then used to concentrate a *dicarboxylate* inside the tubule cell, using a *Na^+-dicarboxylate cotransporter* called the NaDC. The NaDC is found on both apical and basolateral membranes in the proximal tubule.

Dicarboxylates are the anion form of dicarboxylic acids, which have two carboxyl (−COOH) groups. Most of the citric acid cycle intermediates, such as citrate, oxaloacetate, and

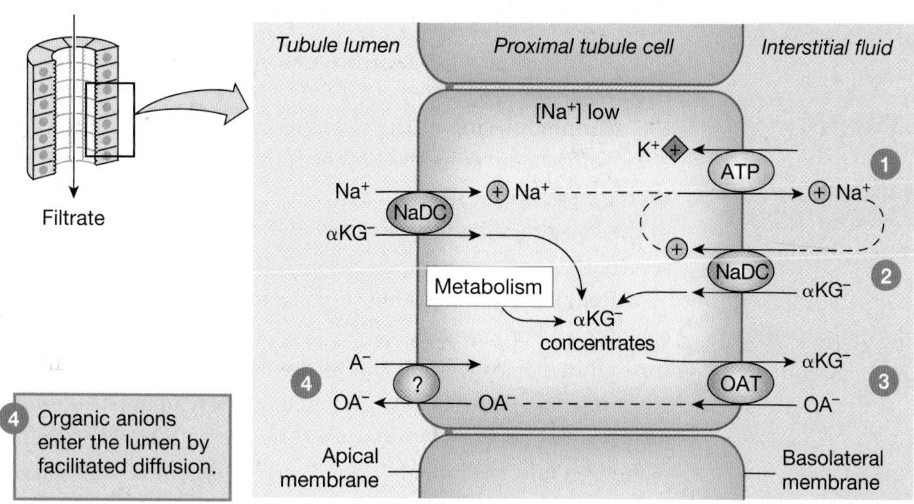

Fig. 19.12

α-ketoglutarate (αKG), are dicarboxylates. Figure 19.12 shows αKG as the dicarboxylate.

The concentration of dicarboxylate inside the tubule cell drives the third step of organic anion secretion. The OAT is an indirect active transporter that uses the dicarboxylate moving out of the cell down it concentration gradient to move an organic anion against its gradient into the cell. In the final step, once the organic anion is concentrated inside the tubule cell, it can use facilitated diffusion to enter the lumen. The apical transporters have not been identified but appear to be anion exchangers.

Competition Decreases Penicillin Secretion

The broad specificity of the organic anion transporters means that different substrates must compete for the transporter binding sites [p. 153]. An interesting and important example of an organic molecule secreted by the OAT is the antibiotic *penicillin*. Many people today take antibiotics for granted, but until the early decades of the twentieth century, infections were a leading cause of death.

In 1928, Alexander Fleming discovered a substance in the bread mold *Penicillium* that retarded the growth of bacteria. But the antibiotic was difficult to isolate, so it did not become available for clinical use until the late 1930s. During World War II, penicillin made a major difference in the number of deaths and amputations caused by infected wounds. The only means of producing penicillin, however, was to isolate it from bread mold, and supplies were limited.

Demand for the drug was heightened by the fact that kidney tubules secrete penicillin. Renal secretion is so efficient at clearing foreign molecules from the blood that within three to four hours after a dose of penicillin has been administered, about 80% has been excreted in the urine. During the war, the drug was in such short supply that it was common procedure to collect the urine from patients being treated with penicillin so that the antibiotic could be isolated and reused.

This solution was not satisfactory, however, and so researchers looked for a way to slow penicillin secretion. They hoped to find a molecule that could compete with penicillin for the organic anion transporter responsible for secretion. That way, when presented with both drugs, the OAT carrier would bind preferentially to the competitor and secrete it, leaving penicillin behind in the blood.

A synthetic compound named *probenecid* was the answer. When probenecid is administered concurrently with penicillin, the transporter removes probenecid preferentially, prolonging the activity of penicillin in the body. Once mass-produced synthetic penicillin became available and supply was no longer a problem, the medical use of probenecid declined.

RUNNING PROBLEM

Michael found it amazing that a metabolic problem could lead to pain in his big toe. "How do we treat gout?" he asked. Dr. Garcia explained that the treatment includes anti-inflammatory agents, lots of water, and avoidance of alcohol, which can trigger gout attacks. "In addition, I would like to put you on a uricosuric agent, like probenecid, which will enhance renal excretion of urate," replied Dr. Garcia. "By enhancing excretion, we can reduce uric acid levels in your blood and thus provide relief." Michael agreed to try these measures.

Q5. Urate is reabsorbed by some proximal tubule cells and secreted by others using membrane transporters, one on the apical membrane and one on the basolateral membrane. Could the same transporters be used by cells that reabsorb urate and cells that secrete it? Defend your reasoning.

Q6. Uricosuric agents, like urate, are organic acids. Given this fact, explain how uricosuric agents might enhance excretion of urate.

627 629 642 **646** 651 651

Excretion

Urine output is the result of all the processes that take place in the kidney. By the time fluid reaches the end of the nephron, it bears little resemblance to the filtrate that started in Bowman's capsule. Glucose, amino acids, and useful metabolites are gone, having been reabsorbed into the blood, and organic wastes are more concentrated. The concentrations of ions and water in the urine are highly variable, depending on the state of the body.

Although excretion tells us what the body is eliminating, excretion by itself cannot tell us the details of renal function. Recall that for any substance,

$$\text{Excretion} = \text{filtration} - \text{reabsorption} + \text{secretion}$$

Simply looking at the excretion rate of a substance tells us nothing about how the kidney handled that substance. The excretion rate of a substance depends on (1) the filtration rate of the substance and (2) whether the substance is reabsorbed, secreted, or both, as it passes through the tubule.

Renal handling of a substance and GFR are often of clinical interest. For example, clinicians use information about a person's glomerular filtration rate as an indicator of overall kidney function. And pharmaceutical companies developing drugs must provide the Food and Drug Administration with complete information on how the human kidney handles each new compound.

But how can investigators dealing with living humans assess filtration, reabsorption, and secretion at the level of the individual nephron? They have no way to do this directly because the kidneys are not easily accessible and the nephrons are microscopic. Scientists therefore had to develop a technique that would allow them to assess renal function using only analysis of the urine and blood. To do this, they apply the concept of clearance.

Clearance Is a Noninvasive Way to Measure GFR

Clearance of a solute is the rate at which that solute disappears from the body by excretion or by metabolism [p. 13]. The general equation for clearance is:

$$\text{Clearance of } X = \frac{\text{excretion rate of } X \text{ (mg/min)}}{[X]_{\text{plasma}} \text{ (mg/mL plasma)}}$$

where clearance is mL plasma cleared of X per minute. Notice that the units for clearance are mL plasma and time. Substance *X* does not appear anywhere in the clearance units.

For any solute that is cleared only by renal excretion, clearance is expressed as the volume of plasma passing through the kidneys that has been totally cleared of that solute in a given period of time. Because this is such an indirect way to think of excretion (how much blood has been cleared of *X* rather than how much *X* has been excreted), clearance is often a very difficult concept to grasp.

Before we jump into the mathematical expression of clearance, let's look at an example that shows how clearance relates to kidney function. For our example, we use **inulin,** a polysaccharide isolated from the tuberous roots of a variety of plants. (Inulin is not the same as *insulin,* the protein hormone that regulates glucose metabolism.) Scientists discovered from experiments with isolated nephrons that inulin injected into the plasma filters freely into the nephron. As it passes through the kidney tubule, inulin is neither reabsorbed nor secreted. In other words, 100% of the inulin that filters into the tubule is excreted.

How does this relate to clearance? To answer this question, take a look at ■ Figure 19.13, which assumes that 100% of a filtered volume of plasma is reabsorbed. (This is not too far off the actual value, which is more than 99%.) In Figure 19.13a, inulin has been injected so that its plasma concentration is 4 inulin molecules per 100 mL plasma. If GFR is 100 mL plasma filtered per minute, we can calculate the filtration rate, or *filtered load,* of inulin using the equation

$$\begin{aligned}\text{Filtered load of } X &= [X]_{\text{plasma}} \times \text{GFR}\\ \text{Filtered load of inulin} &= (4 \text{ inulin}/100 \text{ mL plasma})\\ &\quad \times 100 \text{ mL plasma filtered/min}\\ &= 4 \text{ inulin/min filtered}\end{aligned}$$

As the filtered inulin and the filtered plasma pass along the nephron, all the plasma is reabsorbed, but all the inulin remains in the tubule. The reabsorbed plasma contains no inulin, so we say it has been totally *cleared* of inulin. The *inulin clearance* therefore is 100 mL of plasma cleared/min. At the same time, the excretion rate of inulin is 4 inulin molecules excreted per minute.

What good is this information? For one thing, we can use it to calculate the glomerular filtration rate. Notice from Figure 19.13a that inulin clearance (100 mL plasma cleared/min) is equal to the GFR (100 mL plasma filtered/min). Thus, *for any substance that is freely filtered but neither reabsorbed nor secreted, its clearance is equal to GFR.*

Now let's show mathematically that inulin clearance is equal to GFR. We already know that

$$\text{Filtered load of } X = [X]_{\text{plasma}} \times \text{GFR} \qquad (1)$$

We also know that 100% of the inulin that filters into the tubule is excreted. In other words:

$$\text{Filtered load of inulin} = \text{excretion rate of inulin} \qquad (2)$$

Because of this equality, we can substitute excretion rate for filtered load in equation (1) by using algebra (if A = B and A = C, then B = C):

$$\text{Excretion rate of inulin} = [\text{inulin}]_{\text{plasma}} \times \text{GFR} \qquad (3)$$

This equation can be rearranged to read

$$\text{GFR} = \frac{\text{excretion rate of inulin}}{[\text{inulin}]_{\text{plasma}}} \qquad (4)$$

It turns out that the right side of this equation is identical to the clearance equation for inulin. Thus the general equation for the clearance of any substance *X* (mL plasma cleared/min) is

$$\text{Clearance of } X = \frac{\text{excretion rate of } X \text{ (mg/min)}}{[X]_{\text{plasma}} \text{ (mg/mL plasma)}} \qquad (5)$$

For inulin:

$$\text{Inulin clearance} = \frac{\text{excretion rate of inulin}}{[\text{inulin}]_{\text{plasma}}} \qquad (6)$$

The right sides of equations (4) and (6) are identical, so by using algebra again, we can say that:

$$\text{GFR} = \text{inulin clearance} \qquad (7)$$

So why is this important? For one thing, you have just learned how we can measure GFR in a living human by taking only blood and urine samples. Try the example in Concept Check 12 to see if you understand the preceding discussion. ■ Table 19.1 is a summary table of equations you will find useful for renal physiology.

19

Renal Clearance

These figures show the relationship between clearance and excretion. Each figure represents the events taking place in one minute. For simplicity, 100% of the filtered volume is assumed to be reabsorbed.

(a) Inulin clearance is equal to GFR.

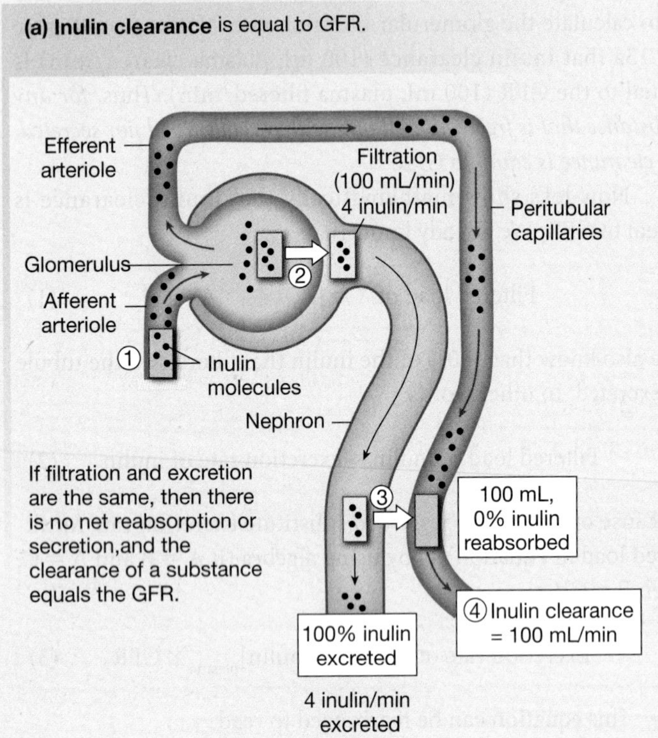

If filtration and excretion are the same, then there is no net reabsorption or secretion,and the clearance of a substance equals the GFR.

(b) Glucose clearance: Normally all glucose that filters is reabsorbed.

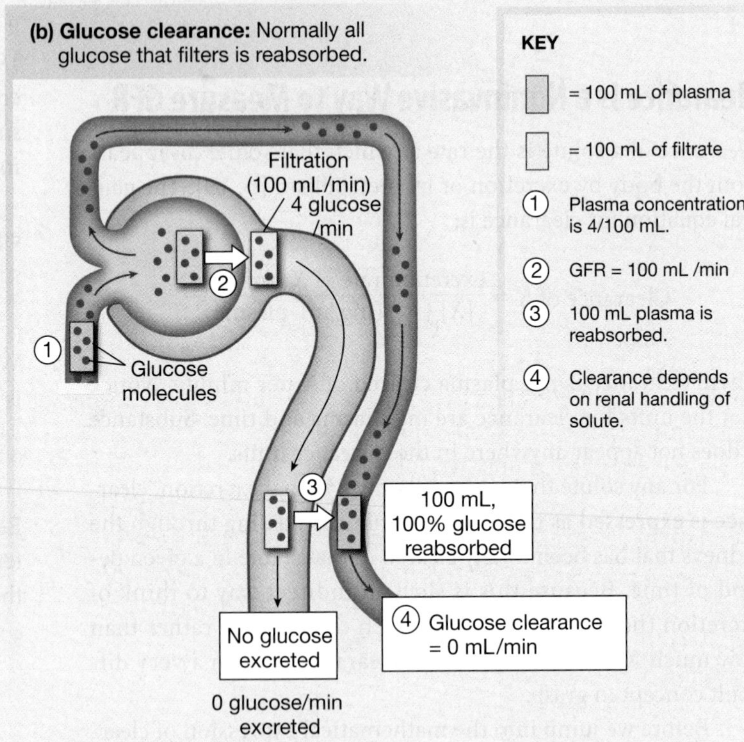

(c) Urea clearance is an example of net reabsorption. If filtration is greater than excretion, there is net reabsorption.

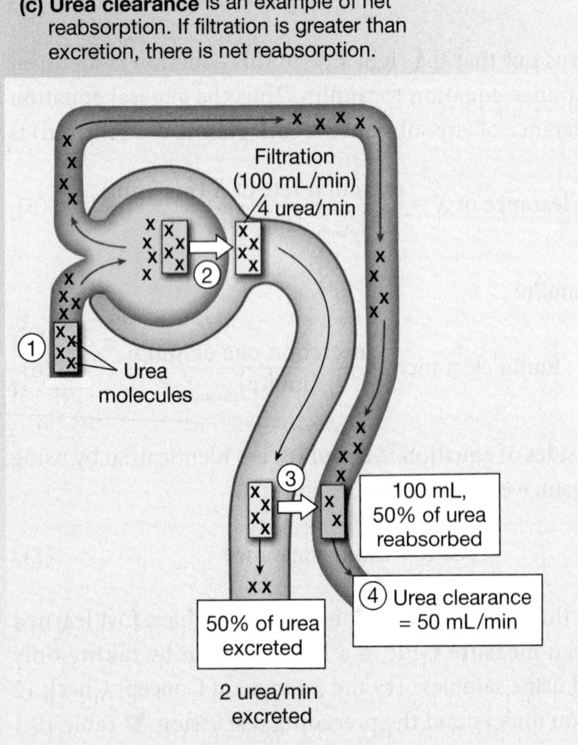

If clearance of a substance is less than GFR, there is net reabsorption.

(d) Penicillin clearance is an example of net secretion. If excretion is greater than filtration, there is net secretion.

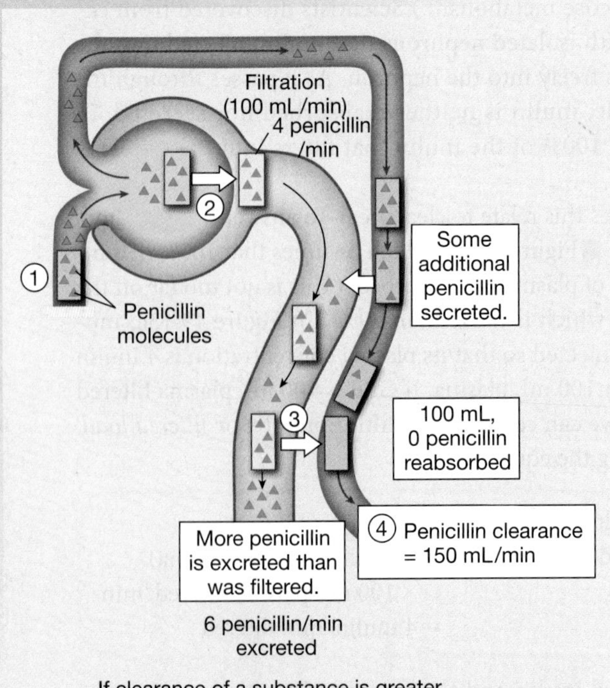

If clearance of a substance is greater than GFR, there is net secretion.

Table 19.1

Useful Equations in Renal Physiology

- Excretion = Filtration − Reabsorption + Secretion
- Filtration of $X = [X]_{plasma} \times \text{GFR}$
- Clearance of $X = \dfrac{\text{excretion rate of } X \text{ (mg/min)}}{[X]_{plasma} \text{ (mg/mL plasma)}}$
- When $[X]_{plasma}$ = renal threshold for X, then reabsorption of X = transport maximum for X.

Inulin is not practical for routine clinical applications because it does not occur naturally in the body and must be administered by continuous intravenous infusion. As a result, inulin use is restricted to research. Unfortunately, no substance that occurs naturally in the human body is handled by the kidney exactly the way inulin is handled.

In clinical settings, physicians use creatinine to estimate GFR. **Creatinine** is a breakdown product of phosphocreatine, an energy-storage compound found primarily in muscles [p. 413]. It is constantly produced by the body and need not be administered. Normally, the production and breakdown rates of phosphocreatine are relatively constant, and the plasma concentration of creatinine does not vary much.

Although creatinine is always present in the plasma and is easy to measure, it is not the perfect molecule for estimating GFR because a small amount is secreted into the urine. However, the amount secreted is small enough that, in most people, *creatinine clearance* is routinely used to estimate GFR.

Concept Check Answer: p. 655

12. If plasma creatinine = 1.8 mg/100 mL plasma, urine creatinine = 1.5 mg/mL urine, and urine volume is 1100 mL in 24 hours, what is the creatinine clearance? What is GFR?

Clearance Helps Us Determine Renal Handling

Once we know a person's GFR, we can determine how the kidney handles any solute by measuring the solute's plasma concentration and its excretion rate. If we assume that the solute is freely filtered at the glomerulus, we know from equation (1) that

$$\text{Filtered load of } X = [X]_{plasma} \times \text{GFR}$$

By comparing the filtered load of the solute with its excretion rate, we can tell how the nephron handled that substance (Fig. 19.13). For example, if less of the substance appears in the urine than was filtered, net reabsorption occurred (excreted = filtered − reabsorbed). If more of the substance appears in the urine than was filtered, there must have been net secretion of the substance into the lumen (excreted = filtered + secreted). If the same amount of the substance is filtered and excreted, then the substance is handled like inulin—neither reabsorbed nor secreted. Let's look at some examples.

Suppose that glucose is present in the plasma at 100 mg glucose/dL plasma, and GFR is calculated from creatinine clearance to be 125 mL plasma/min. For these values, equation (1) tells us that

$$\text{Filtered load of glucose} = (100 \text{ mg glucose}/100 \text{ mL plasma}) \times 125 \text{ mL plasma/min}$$

$$\text{Filtered load of glucose} = 125 \text{ mg glucose/min}$$

There is no glucose in this person's urine, however: glucose excretion is zero. Because glucose was filtered at a rate of 125 mg/min but excreted at a rate of 0 mg/min, it must have been totally reabsorbed.

Clearance values can also be used to determine how the nephron handles a filtered solute. In this method, researchers calculate creatinine or inulin clearance, then compare the clearance of the solute being investigated with the creatinine or inulin clearance. If clearance of the solute is less than the inulin clearance, the solute has been reabsorbed. If the clearance of the solute is higher than the inulin clearance, additional solute has been secreted into the urine. More plasma was cleared of the solute than was filtered, so the additional solute must have been removed from the plasma by secretion.

Figure 19.13 shows filtration, excretion, and clearance of three molecules: glucose, urea, and penicillin. All solutes have the same concentration in the blood entering the glomerulus: 4 molecules/100 mL plasma. GFR is 100 mL/min, and we assume for simplicity that the entire 100 mL of plasma filtered into the tubule is reabsorbed.

For any solute, its clearance reflects how the kidney tubule handles it. For example, 100% of the glucose that filters is reabsorbed, and glucose clearance is zero (Fig. 19.13b). On the other hand, urea is partially reabsorbed; four molecules filter, but two are reabsorbed and two are excreted (Fig. 19.13c). Consequently, urea clearance is 50 mL plasma per minute. Urea and glucose clearance are both less than the inulin clearance of 100 mL/min, which tells you that urea and glucose have been reabsorbed.

As you learned earlier, penicillin is filtered, not reabsorbed, and additional penicillin molecules are secreted from plasma in the peritubular capillaries. In Figure 19.13d, four penicillin are filtered, but six are excreted. An extra 50 mL of plasma have been cleared of penicillin in addition to the original 100 mL that were filtered. The penicillin clearance therefore is 150 mL plasma cleared per minute. Penicillin clearance is greater than the inulin clearance of 100 mL/min, which tells you that net secretion of penicillin occurs.

Note that a comparison of clearance values tells you only the *net* handling of the solute. It does not tell you if a molecule is both reabsorbed and secreted. For example, nearly all K^+ filtered is reabsorbed in the proximal tubule and loop of Henle, and then a small amount is secreted back into the tubule lumen at the distal nephron. On the basis of K^+ clearance, it appears that only reabsorption occurred.

Clearance calculations are relatively simple because all you need to know are the urine excretion rates and the plasma concentrations for any solute of interest, and both values are easily obtained. If you also know either inulin or creatinine clearance, then you can determine the renal handling of any compound.

Micturition

Once filtrate leaves the collecting ducts, it can no longer be modified, and its composition does not change. The filtrate, now called urine, flows into the renal pelvis and then down the *ureter* to the bladder with the help of rhythmic smooth muscle contractions. The bladder is a hollow organ whose walls contain well-developed layers of smooth muscle. In the bladder, urine is stored until released in the process known as urination, voiding, or more formally, **micturition** {*micturire*, to desire to urinate}.

The bladder can expand to hold a volume of about 500 mL. The neck of the bladder is continuous with the *urethra*, a single tube through which urine passes to reach the external environment. The opening between the bladder and urethra is closed by two rings of muscle called *sphincters* (■ Fig. 19.14a).

The **internal sphincter** is a continuation of the bladder wall and consists of smooth muscle. Its normal tone keeps it contracted. The **external sphincter** is a ring of skeletal muscle controlled by somatic motor neurons. Tonic stimulation from the central nervous system maintains contraction of the external sphincter except during urination.

Micturition is a simple spinal reflex that is subject to both conscious and unconscious control from higher brain centers. As the bladder fills with urine and its walls expand, stretch receptors send signals via sensory neurons to the spinal cord (Fig. 19.14b). There the information is integrated and

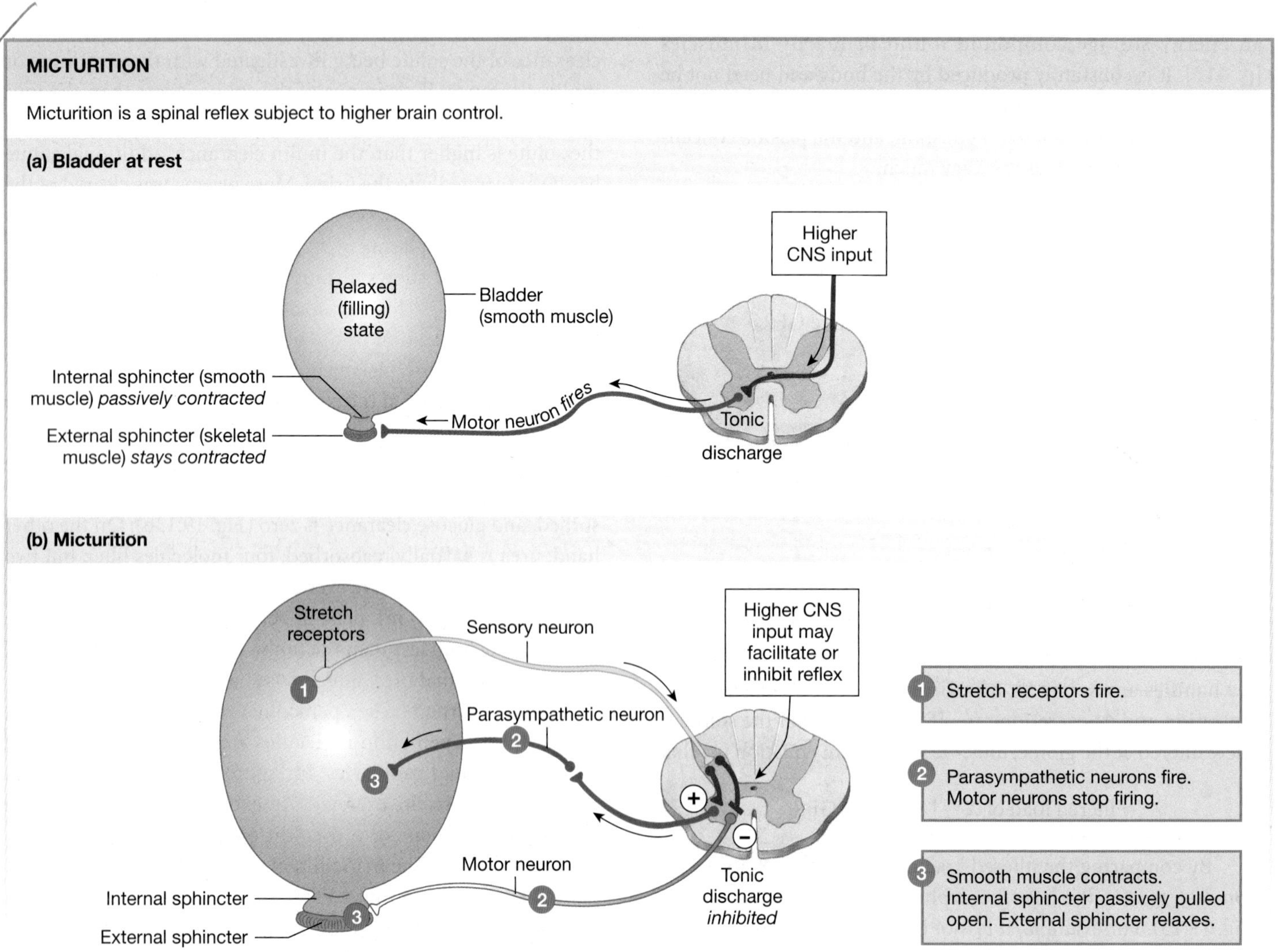

■ **Fig. 19.14**

RUNNING PROBLEM

Three weeks later, Michael was back in Dr. Garcia's office. The anti-inflammatory drugs and probenecid had eliminated the pain in his toe, but last night he had gone to the hospital with a very painful kidney stone. "We'll have to wait until the analysis comes back," said Dr. Garcia, "but I will guess that it is a uric acid stone. Did you drink as much water as I told you to?" Sheepishly, Michael admitted that he had good intentions but could never find the time at work to drink much water. "You have to drink enough water while on this drug to produce 3 liters or more of urine a day. That's more than three quarts. Otherwise, you may end up with another uric acid kidney stone." Michael remembered how painful the kidney stone was and agreed that this time he would follow instructions to the letter.

Q7. Explain why not drinking enough water while taking uricosuric agents may cause uric acid crystals to form kidney stones in the urinary tract.

transferred to two sets of neurons. The stimulus of a full bladder excites parasympathetic neurons leading to the smooth muscle in the bladder wall. The smooth muscle contracts, increasing the pressure on the bladder contents. Simultaneously, somatic motor neurons leading to the external sphincter are inhibited.

Contraction of the bladder occurs in a wave that pushes urine downward toward the urethra. Pressure exerted by the urine forces the internal sphincter open while the external sphincter relaxes. Urine passes into the urethra and out of the body, aided by gravity.

This simple micturition reflex occurs primarily in infants who have not yet been toilet trained. A person who has been toilet trained acquires a learned reflex that keeps the micturition reflex inhibited until she or he consciously desires to urinate. The learned reflex involves additional sensory fibers in the bladder that signal the degree of fullness. Centers in the brain stem and cerebral cortex receive that information and override the basic micturition reflex by directly inhibiting the parasympathetic fibers and by reinforcing contraction of the external sphincter. When an appropriate time to urinate arrives, those same centers remove the inhibition and facilitate the reflex by inhibiting contraction of the external sphincter.

In addition to conscious control of urination, various subconscious factors can affect the micturition reflex. "Bashful bladder" is a condition in which a person is unable to urinate in the presence of other people despite the conscious intent to do so. The sound of running water facilitates micturition and is often used to help patients urinate if the urethra is irritated from insertion of a *catheter*, a tube inserted into the bladder to drain it passively.

RUNNING PROBLEM CONCLUSION

Gout

In this running problem, you learned that gout, which often presents as a debilitating pain in the big toe, is a metabolic problem whose cause and treatment can be linked to kidney function. Urate handling by the kidney is a complex process because urate is both secreted and reabsorbed in different segments of the proximal tubule. Scientists have now identified three different but related transport proteins that are involved in the process: the *organic anion transporter* (OAT), which exchanges two anions in an electrically neutral exchange; *urate transporter* 1 (URAT1), which is also an anion exchanger but with high specificity for urate; and *urate transporter* (UAT), an electrogenic uniport urate transporter. The arrangement of these transport proteins on the polarized cell membrane determines whether the cell reabsorbs or secretes urate.

Gout is one of the oldest known diseases and for many years was considered a "rich man's" disease caused by too much rich food and drink. Thomas Jefferson and Benjamin Franklin both suffered from gout. To learn more about its causes, symptoms, and treatments, go to the Mayo Clinic's health information pages (*www.mayoclinic.com*) and search for *gout*. Check your understanding of this running problem by comparing your answers against the information in the summary table.

Question	Facts	Integration and Analysis
1. Trace the route followed by kidney stones when they are excreted.	Kidney stones often form in the renal pelvis.	From the renal pelvis, a stone passes down the ureter, into the urinary bladder, then into the urethra and out of the body.
2. Name the anion formed when uric acid dissociates.	The suffix *–ate* is used to identify the anion of organic acids.	The anion of uric acid is urate.

RUNNING PROBLEM CONCLUSION *(continued)*

Question	Facts	Integration and Analysis
3. Purines are part of which category of biomolecules? Using that information, explain why uric acid levels in the blood go up when cell breakdown increases.	Purines include adenine and guanine, which are components of DNA, RNA, and ATP [p. 36]. When a cell dies, nuclear DNA and other chemical components are broken down.	Degradation of the cell's DNA, RNA, and ATP increases purine production, which in turn increases uric acid production.
4. Based on what you have learned about uric acid and urate, predict two ways a person may develop hyperuricemia.	Hyperuricemia is a disturbance of mass balance. Uric acid is made from purines. Urate is filtered by the kidneys with net secretion.	Hyperuricemia results either from overproduction of uric acid or from a defect in the renal excretion of urate.
5. Could the same transporters be used by cells that reabsorb urate and cells that secrete it? Defend your reasoning.	Some transporters move substrates in one direction only but others are reversible. Assume one urate transporter brings urate into the cell and another takes it out.	You could use the same two transporters if you reverse their positions on the apical and basolateral membranes. Cells reabsorbing urate would bring it in on the apical side and move it out on the basolateral. Cells secreting urate would reverse this pattern.
6. Uricosuric agents, like urate, are organic acids. With that information, explain how uricosuric agents might enhance excretion of urate.	Mediated transport exhibits competition, in which related molecules compete for one transporter. Usually, one molecule binds preferentially and therefore inhibits transport of the second molecule [p. 153].	Uricosuric agents are organic anions, so they may compete with urate for the proximal tubule organic anion transporter. Preferential binding of the uricosuric agents would block urate access to the OAT, leaving urate in the lumen and increasing its excretion.
7. Explain why not drinking enough water while taking uricosuric agents may cause uric acid stones to form in the urinary tract.	Uric acid stones form when uric acid concentrations exceed a critical level and crystals precipitate.	If a person drinks large volumes of water, the excess water will be excreted by the kidneys. Large amounts of water dilute the urine, thereby preventing the high concentrations of uric acid needed for stone formation.

627 629 642 646 651 651

Chapter Summary

The urinary system, like the lungs, uses the principle of *mass balance* to maintain homeostasis. The components of urine are constantly changing and reflect the kidney's functions of regulating ions and water and removing wastes. One of the body's three *portal systems*—each of which includes two capillary beds—is found in the kidney. Filtration occurs in the first capillary bed and reabsorption in the second. The *pressure-flow-resistance* relationship you encountered in the cardiovascular and pulmonary systems also plays a role in glomerular filtration and urinary excretion. *Compartmentation* is illustrated by the movement of water and solutes between the internal and external environments as filtrate is modified along the nephron. Reabsorption and secretion of solutes depend on *molecular interactions* and on the *movement of molecules across membranes* of the tubule cells.

Functions of the Kidneys

1. The kidneys regulate extracellular fluid volume, blood pressure, and osmolarity; maintain ion balance; regulate pH; excrete wastes and foreign substances; and participate in endocrine pathways. (p. 627)

Anatomy of the Urinary System

iP Urinary System: Glomerular Filtration

2. The **urinary system** is composed of two kidneys, two ureters, a bladder, and a urethra. (p. 630; Fig. 19.1a)
3. Each **kidney** has about 1 million microscopic **nephrons.** In cross section, a kidney is arranged into an outer **cortex** and inner **medulla.** (p. 630; Fig. 19.1c)
4. Renal blood flow goes from **afferent arteriole** to **glomerulus** to **efferent arteriole** to **peritubular capillaries.** The **vasa recta** capillaries dip into the medulla. (p. 631; Fig. 19.1g, h, j)
5. Fluid filters from the glomerulus into **Bowman's capsule.** From there, it flows through the **proximal tubule, loop of Henle, distal tubule,** and **collecting duct,** then drains into the **renal pelvis. Urine** flows through the **ureter** to the **urinary bladder.** (p. 630; Fig. 19.1b, c, i)

Overview of Kidney Function

6. **Filtration** is the movement of fluid from plasma into Bowman's capsule. **Reabsorption** is the movement of filtered materials from tubule to blood. **Secretion** is the movement of selected molecules from blood to tubule. (p. 632; Fig. 19.2)
7. Average urine volume is 1.5 L/day. Osmolarity varies between 50 and 1200 mOsM. (p. 632; Fig. 19.2)
8. The amount of a solute excreted equals the amount filtered minus the amount reabsorbed plus the amount secreted. (p. 633; Fig. 19.3)

Filtration

iP Urinary System: Glomerular Filtration

9. One-fifth of renal plasma flow filters into the tubule lumen. The percentage of total plasma volume that filters is called the **filtration fraction.** (p. 634; Fig. 19.4)
10. Bowman's capsule epithelium has specialized cells called **podocytes** that wrap around the glomerular capillaries and create **filtration slits. Mesangial cells** are associated with the glomerular capillaries. (p. 635; Fig. 19.5a, c)
11. Filtered solutes must pass first through glomerular capillary endothelium, then through a **basal lamina,** and finally through Bowman's capsule epithelium before reaching the lumen of Bowman's capsule. (p. 635; Fig. 19.5d)
12. Filtration allows most components of plasma to enter the tubule but excludes blood cells and most plasma proteins. (p. 634)
13. Hydrostatic pressure in glomerular capillaries averages 55 mm Hg, favoring filtration. Opposing filtration are colloid osmotic pressure of 30 mm Hg and hydrostatic capsule fluid pressure averaging 15 mm Hg. The net driving force is 10 mm Hg, favoring filtration. (p. 637; Fig. 19.6)
14. The **glomerular filtration rate** (GFR) is the amount of fluid that filters into Bowman's capsule per unit time. Average GFR is 125 mL/min, or 180 L/day. (p. 636)
15. Hydrostatic pressure in glomerular capillaries can be altered by changing resistance in the afferent and efferent arterioles. (p. 637; Fig. 19.6c)
16. Autoregulation of glomerular filtration is accomplished by a **myogenic response** of vascular smooth muscle in response to pressure changes and by **tubuloglomerular feedback.** When fluid flow through the distal tubule increases, the **macula densa** cells send a paracrine signal to the afferent arteriole, which constricts. (p. 639; Fig. 19.7c)
17. Reflex control of GFR is mediated through systemic signals, such as hormones, and through the autonomic nervous system. (p. 638)

Reabsorption

iP Urinary System: Early Filtrate Processing

18. Most reabsorption takes place in the proximal tubule. Finely regulated reabsorption takes place in the more distal segments of the nephron. (p. 640)
19. The active transport of Na^+ and other solutes creates concentration gradients for passive reabsorption of urea and other solutes. (p. 641; Fig. 19.8a)
20. Most reabsorption involves transepithelial transport, but some solutes and water are reabsorbed by the paracellular pathway. (p. 640)
21. Glucose, amino acids, ions, and various organic metabolites are reabsorbed by Na^+-linked secondary active transport. (p. 641; Fig. 19.8c)
22. Most renal transport is mediated by membrane proteins and exhibits saturation, specificity, and competition. The **transport maximum** Tm is the transport rate at saturation. (p. 643; Fig. 19.9)
23. The **renal threshold** is the plasma concentration at which a substance first appears in the urine. (p. 643; Fig. 19.9)
24. Peritubular capillaries reabsorb fluid along their entire length. (p. 645; Fig. 19.11)

19

Secretion

25. Secretion enhances excretion by removing solutes from the peritubular capillaries. K^+, H^+, and a variety of organic compounds are secreted. (p. 645; Fig. 19.12)
26. Molecules that compete for renal carriers slow the secretion of a molecule. (p. 646)

Excretion

27. The excretion rate of a solute depends on (1) its filtered load and (2) whether it is reabsorbed or secreted as it passes through the nephron. (p. 646)
28. **Clearance** describes how many milliliters of plasma passing through the kidneys have been totally cleared of a solute in a given period of time. (p. 647)
29. **Inulin** clearance is equal to GFR. In clinical settings, **creatinine** is used to measure GFR. (p. 648; Fig. 19.13)
30. Clearance can be used to determine how the nephron handles a solute filtered into it. (p. 648; Fig. 19.13)

Micturition

31. The external sphincter of the bladder is skeletal muscle that is tonically contracted except during urination. (p. 650; Fig. 19.14)
32. Micturition is a simple spinal reflex subject to conscious and unconscious control. (p. 650)
33. Parasympathetic neurons cause contraction of the smooth muscle in the bladder wall. Somatic motor neurons leading to the external sphincter are simultaneously inhibited. (p. 651)

Questions

Answers: p. A-1

Level One Reviewing Facts and Terms

1. List and explain the significance of the five characteristics of urine that can be found by physical examination.
2. List and explain the six major kidney functions.
3. At any given time, what percentage of cardiac output goes to the kidneys?
4. List the major structures of the urinary system in their anatomical sequence, from the kidneys to the urine leaving the body. Describe the function of each structure.
5. Arrange the following structures in the order that a drop of water entering the nephron would encounter them:
   (a) afferent arteriole
   (b) Bowman's capsule
   (c) collecting duct
   (d) distal tubule
   (e) glomerulus
   (f) loop of Henle
   (g) proximal tubule
   (h) renal pelvis
6. Name the three filtration barriers that solutes must cross as they move from plasma to the lumen of Bowman's capsule. What components of blood are usually excluded by these layers?
7. What force(s) promote(s) glomerular filtration? What force(s) oppose(s) it? What is meant by the term *net driving force*?
8. What does the abbreviation GFR stand for? What is a typical numerical value for GFR in milliliters per minute? In liters per day?
9. Identify the following structures, then explain their significance in renal physiology:
   (a) juxtaglomerular apparatus
   (b) macula densa
   (c) mesangial cells
   (d) podocytes
   (e) sphincters in the bladder
   (f) renal cortex
10. In which segment of the nephron does most reabsorption take place? When a molecule or ion is reabsorbed from the lumen of the nephron, where does it go? If a solute is filtered and not reabsorbed from the tubule, where does it go?
11. Match each of the following substances with its mode(s) of transport in proximal tubule reabsorption.

(a) Na^+	1. simple diffusion
(b) glucose	2. primary active transport
(c) urea	3. indirect active transport
(d) plasma proteins	4. facilitated diffusion
(e) water	5. movement through open channels
	6. endocytosis
	7. paracellular movement

12. List three solutes secreted into the tubule lumen.
13. What solute that is normally present in the body is used to estimate GFR in humans?
14. What is micturition?

Level Two Reviewing Concepts

15. Map the following terms. You may add terms if you like.

• α-receptor	• glomerulus
• afferent arteriole	• JG cells
• autoregulation	• macula densa
• basal lamina	• mesangial cell
• Bowman's capsule	• myogenic autoregulation
• capillary blood pressure	• norepinephrine
• capsule fluid pressure	• paracrine
• colloid osmotic pressure	• plasma proteins
• efferent arteriole	• podocyte
• endothelium	• resistance
• epithelium	• vasoconstriction
• GFR	

16. Define, compare, and contrast the items in the following sets of terms:
   (a) filtration, secretion, and excretion
   (b) saturation, transport maximum, and renal threshold
   (c) probenecid, creatinine, inulin, and penicillin
   (d) clearance, excretion, and glomerular filtration rate
17. What are the advantages of a kidney that filters a large volume of fluid and then reabsorbs 99% of it?
18. If the afferent arteriole of a nephron constricts, what happens to GFR in that nephron? If the efferent arteriole of a nephron constricts, what happens to GFR in that nephron? Assume that no autoregulation takes place.
19. Diagram the micturition reflex. How is this reflex altered by toilet training? How do higher brain centers influence micturition?
20. Antimuscarinic drugs are the accepted treatment for an overactive bladder. Explain why they work for this condition.

Level Three Problem Solving

21. You have been asked to study kidney function in a new species of rodent found in the Amazonian jungle. You isolate some nephrons and expose them to inulin. The following graph shows the results of your studies. (a) How is the rodent nephron handling inulin? Is inulin filtered? Is it excreted? Is there net inulin reabsorption? Is there net secretion? (b) On the graph, accurately draw a line indicating the net reabsorption or secretion. (*Hint:* excretion = filtration − reabsorption + secretion)

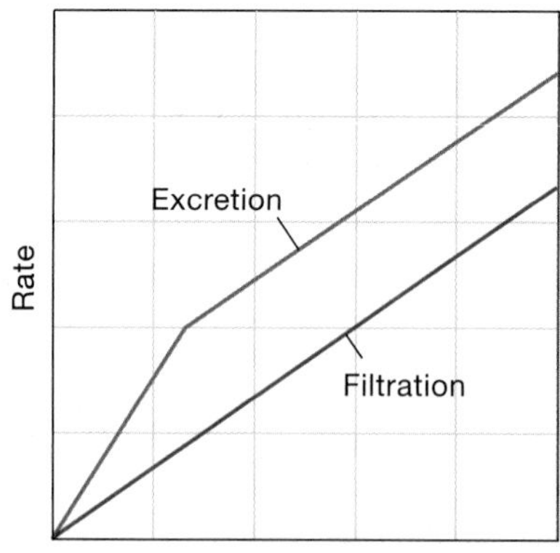

22. Draw a section of renal tubule epithelium showing three cells joined by cell junctions. Label the apical and basolateral membranes, the tubule lumen, and the extracellular fluid. Use the following written description of proximal tubule processes to draw a model cell.

    The proximal tubule cells contain carbonic anhydrase, which promotes the conversion of CO_2 and water to carbonic acid. Carbonic acid then dissociates to H^+ and HCO_3^-. Sodium is reabsorbed by an apical Na^+-H^+ antiporter and a basolateral Na^+-K^+-ATPase. Chloride is passively reabsorbed by movement through the paracellular pathway. Bicarbonate produced in the cytoplasm leaves the cell on a basolateral Na^+-HCO_3^- symporter.

23. Read the box on hemodialysis on p. 642 and see if you can create a model system that would work for dialysis. Draw two compartments (one to represent blood and one to represent dialysis fluid) separated by a semipermeable membrane. In the blood compartment, list normal extracellular fluid solutes and their concentrations (see the table with normal values of blood components inside the back cover of this book). What will happen to the concentrations of these solutes during kidney failure? Which of these solutes should you put in the dialysis fluid, and what should their concentrations be? (*Hint:* Do you want diffusion into the dialysis fluid, out of the dialysis fluid, or no net movement?) How would you change the dialysis fluid if the patient was retaining too much water?

Level Four Quantitative Problems

24. Darlene weighs 50 kg. Assume that her total blood volume is 8% of her body weight, that her heart pumps her total blood volume once a minute, and that her renal blood flow is 25% of her cardiac output. Calculate the volume of blood that flows through Darlene's kidneys each minute.

25. Dwight was competing for a spot on the Olympic equestrian team. As his horse, Nitro, cleared a jump, the footing gave way, causing the horse to somersault, landing on Dwight and crushing him. The doctors feared kidney damage and ran several tests. Dwight's serum creatinine level was 2 mg/100 mL. His 24-hour urine specimen had a volume of 1 L and a creatinine concentration of 20 mg/mL. A second specimen taken over the next 24 hours had the same serum creatinine value and urine volume, but a urine creatinine concentration of 4 mg/mL. How many milligrams of creatinine are in each specimen? What is Dwight's creatinine clearance in each test? What is his GFR? Evaluate these results and comment on Dwight's kidney function.

26. You are a physiologist taking part in an archeological expedition to search for Atlantis. One of the deep-sea submersibles has come back with a mermaid, and you are taking a series of samples from her. You have determined that her GFR is 250 mL/min and that her kidneys reabsorb glucose with a transport maximum of 50 mg/min. What is her renal threshold for glucose? When her plasma concentration of glucose is 15 mg/mL, what is its glucose clearance?

27. If 140 liters of plasma are filtered in a day, and the filtration fraction is 20%:
    (a) What is the renal plasma flow?
    (b) If this person has a hematocrit of 30% [p. 551], what is the renal blood flow?
    (c) If renal blood flow is 20% of this person's cardiac output, what is her cardiac output in L/min?

Answers

Answers to Concept Check Questions

Page 628

1. If extracellular K^+ decreases, more K^+ leaves the neuron, and the membrane potential hyperpolarizes (becomes more negative, increases).
2. If plasma Ca^{2+} decreases, the force of contraction decreases.

Page 629

3. When net filtration out of the glomerular capillaries occurs, the capillary hydrostatic pressure must be *greater than* the capillary colloid osmotic pressure.
4. When net reabsorption into the peritubular capillaries occurs, the capillary hydrostatic pressure must be *less than* the capillary colloid osmotic pressure.

Page 634

5. Filtration and secretion both represent movement from the extracellular fluid into the lumen. Filtration takes place only at Bowman's capsule; secretion takes place all along the rest of the tubule.
6. Glomerulus → Bowman's capsule → proximal tubule → loop of Henle → distal tubule → collecting duct → renal pelvis → ureter → urinary bladder → urethra.
7. If reabsorption decreases to half the normal rate, the body would run out of plasma in under an hour.

Page 636

8. Osmotic pressure is higher in efferent arterioles because fluid volume is decreased there, leaving the same amount of protein in a smaller volume.
9. This person's mean arterial pressure is 119 mm Hg. This person's GFR is 180 L/day.

Page 638

10. If the afferent arteriole constricts, the resistance in that arteriole increases, and blood flow through that arteriole is diverted to lower-resistance arterioles. GFR decreases in the nephron whose arteriole constricted.
11. The primary driving force for GFR is blood pressure opposed by fluid pressure in Bowman's capsule and colloid osmotic pressure due to plasma proteins (Fig. 19.6). With fewer plasma proteins, the plasma has lower-than-normal colloid osmotic pressure. With less colloid osmotic pressure opposing GFR, GFR increases.

Page 649

12. Creatinine clearance = creatinine excretion rate/[creatinine]$_{plasma}$ = (1.5 mg creatinine/mL urine × 1.1 L urine/day)/1.8 mg creatinine/100 mL plasma. Creatinine clearance is about 92 L/day, and GFR is equal to creatinine clearance.

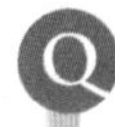

Answers to Figure and Graph Questions

Page 632

Figure 19.2: 1. (a) Bowman's capsule, (b) proximal tubule, loop of Henle, distal tubule, collecting duct, (c) proximal tubule, distal tubule, collecting duct, (d) collecting duct. 2. (a) 18/180 = 10%, 1.5/180 = 0.8%

Page 633

Figure 19.3: E = F − R + S. 79 mmol/day = 720 − R + 43. R = 684 mmol reabsorbed per day.

Page 634

Figure 19.4: 120 mL/min × 1440 min/day = 172,800 mL/day filtered = 172.8 L. 172.8 L = 20% of plasma flow. Plasma flow = 864 L/day.

Page 637

Figure 19.6: Capillary blood pressure, GFR, and renal blood flow all increase.

Page 643

Figure 19.9: The transport rate at 3 mg/mL is 3 mg/min; at 5 and 8 mg/mL, it is 4 mg/min. The transport rate is 2 mg/min at a plasma concentration of 2 mg/mL.

Page 645

Figure 19.11: Pressure is lower because blood flowing out of the glomerulus loses pressure as it moves along the peritubular capillaries.

20 Integrative Physiology II: Fluid and Electrolyte Balance

At a 10% loss of body fluid, the patient will show signs of confusion, distress, and hallucinations and at 20%, death will occur.

—Poul Astrup, in *Salt and Water in Culture and Medicine*, 1993

Background Basics

Each dot of a microarray represents one gene. Genes that are active show up in bright colors.

The American businesswoman in Tokyo finished her workout and stopped at the snack bar of the fitness club to ask for a sports drink. The attendant handed her a bottle labeled "Pocari Sweat®." Although the thought of drinking sweat is not very appealing, the physiological basis for the name is sound.

During exercise, the body secretes sweat, a dilute solution of water and ions, particularly Na^+, K^+, and Cl^-. To maintain homeostasis, the body must replace any substances it has lost to the external environment. For this reason, the replacement fluid a person consumes after exercise should resemble sweat.

In this chapter, we explore how humans maintain salt and water balance, also known as fluid and electrolyte balance. The homeostatic control mechanisms for fluid and electrolyte balance in the body are aimed at maintaining four parameters: fluid volume, osmolarity, the concentrations of individual ions, and pH.

Fluid and Electrolyte Homeostasis

The human body is in a state of constant flux. Over the course of a day we ingest about 2 liters of fluid that contains 6–15 grams of NaCl. In addition, we take in varying amounts of other electrolytes, including K^+, H^+, Ca^{2+}, HCO_3^-, and phosphate ions (HPO_4^{2-}). The body's task is to maintain *mass balance* [p. 11]: what comes in must be excreted if the body does not need it.

The body has several routes for excreting ions and water. The kidneys are the primary route for water loss and for removal of many ions. Under normal conditions, small amounts of both water and ions are lost in the feces and sweat as well. In addition, the lungs lose water and help remove H^+ and HCO_3^- by excreting CO_2.

RUNNING PROBLEM

Hyponatremia

Lauren was competing in her first ironman distance triathlon, a 140.6-mile race consisting of 2.4 miles of swimming, 112 miles of cycling, and 26.2 miles of running. At mile 22 of the run, approximately 16 hours after starting the race, she collapsed. After being admitted to the medical tent, Lauren complained of nausea, a headache, and general fatigue. The medical staff noted that Lauren's face and clothing were covered in white crystals. When they weighed her and compared that value with her pre-race weight recorded at registration, they realized Lauren had gained 2 kg during the race.

658 660 673 676 683 690

Although physiological mechanisms that maintain fluid and electrolyte balance are important, behavioral mechanisms also play an essential role. *Thirst* is critical because drinking is the only normal way to replace lost water. *Salt appetite* is a behavior that leads people and animals to seek and ingest salt (sodium chloride, NaCl).

Why are we concerned with homeostasis of these substances? Water and Na^+ are associated with extracellular fluid volume and osmolarity. Disturbances in K^+ balance can cause serious problems with cardiac and muscle function by disrupting the membrane potential of excitable cells. Ca^{2+} is involved in a variety of body processes, from exocytosis and muscle contraction to bone formation and blood clotting, and H^+ and HCO_3^- are the ions whose balance determines body pH.

ECF Osmolarity Affects Cell Volume

Why is maintaining osmolarity so important to the body? The answer lies in the fact that water crosses most cell membranes freely. If the osmolarity of the extracellular fluid changes, water moves into or out of cells and changes intracellular volume. If ECF osmolarity decreases as a result of excess water intake, water moves into the cells and they swell. If ECF osmolarity increases as a result of salt intake, water moves out of the cells and they shrink. Cell volume is so important that many cells have independent mechanisms for maintaining it.

For example, renal tubule cells in the medulla of the kidney are constantly exposed to high extracellular fluid osmolarity, yet these cells maintain normal cell volume. They do so by synthesizing organic solutes as needed to make their intracellular osmolarity match that of the medullary interstitial fluid. The organic solutes used to raise intracellular osmolarity include sugar alcohols and certain amino acids. Other cells in the body regulate their volume by changing their ionic composition.

In a few instances, changes in cell volume are believed to act as signals that initiate certain cellular responses. For example, swelling of liver cells activates protein and glycogen synthesis, and shrinkage of these cells causes protein and glycogen breakdown. In many cases, however, inappropriate changes in cell volume—either shrinking or swelling—impair cell function. The brain, encased in the rigid skull, is particularly vulnerable to damage from swelling. In general, maintenance of ECF osmolarity within a normal range is essential to maintain cell volume homeostasis.

Multiple Systems Integrate Fluid and Electrolyte Balance

The process of fluid and electrolyte balance is truly integrative because it involves the respiratory and cardiovascular systems in addition to renal and behavioral responses. Adjustments made by the lungs and cardiovascular system are primarily under neural control and can be made quite rapidly. Homeostatic compensation by the kidneys occurs more slowly because the kidneys are primarily under endocrine and neuroendocrine control. For example, small changes in blood pressure that result from increases or decreases in blood volume are quickly corrected by the cardiovascular control center in the brain [p. 524]. If volume changes are persistent or of large magnitude, the kidneys step in to help maintain homeostasis.

■ Figure 20.1 summarizes the integrated response of the body to changes in blood volume and blood pressure. Signals from carotid and aortic baroreceptors and atrial volume receptors initiate a quick neural response mediated through the cardiovascular control center and a slower response elicited from the kidneys. In addition, low blood pressure stimulates thirst. In both situations, renal function integrates with the cardiovascular system to keep blood pressure within a normal range.

Because of the overlap in their functions, a change made by one system—whether renal or cardiovascular—is likely to have consequences that affect the other. Endocrine pathways initiated by the kidneys have direct effects on the cardiovascular system, for instance, and hormones released by myocardial cells act on the kidneys. Sympathetic pathways from the cardiovascular control center affect not only cardiac output and vasoconstriction but also glomerular filtration and hormone release by the kidneys.

In this way the maintenance of blood pressure, blood volume, and ECF osmolarity forms a network of interwoven control pathways. This integration of function in multiple systems is one of the more difficult concepts in physiology, but it is also one of the most exciting areas of medicine and physiological research.

Water Balance

Water is the most abundant molecule in the body, constituting about 50% of total body weight in females ages 17 to 39, and 60% of total body weight in males of the same age group.

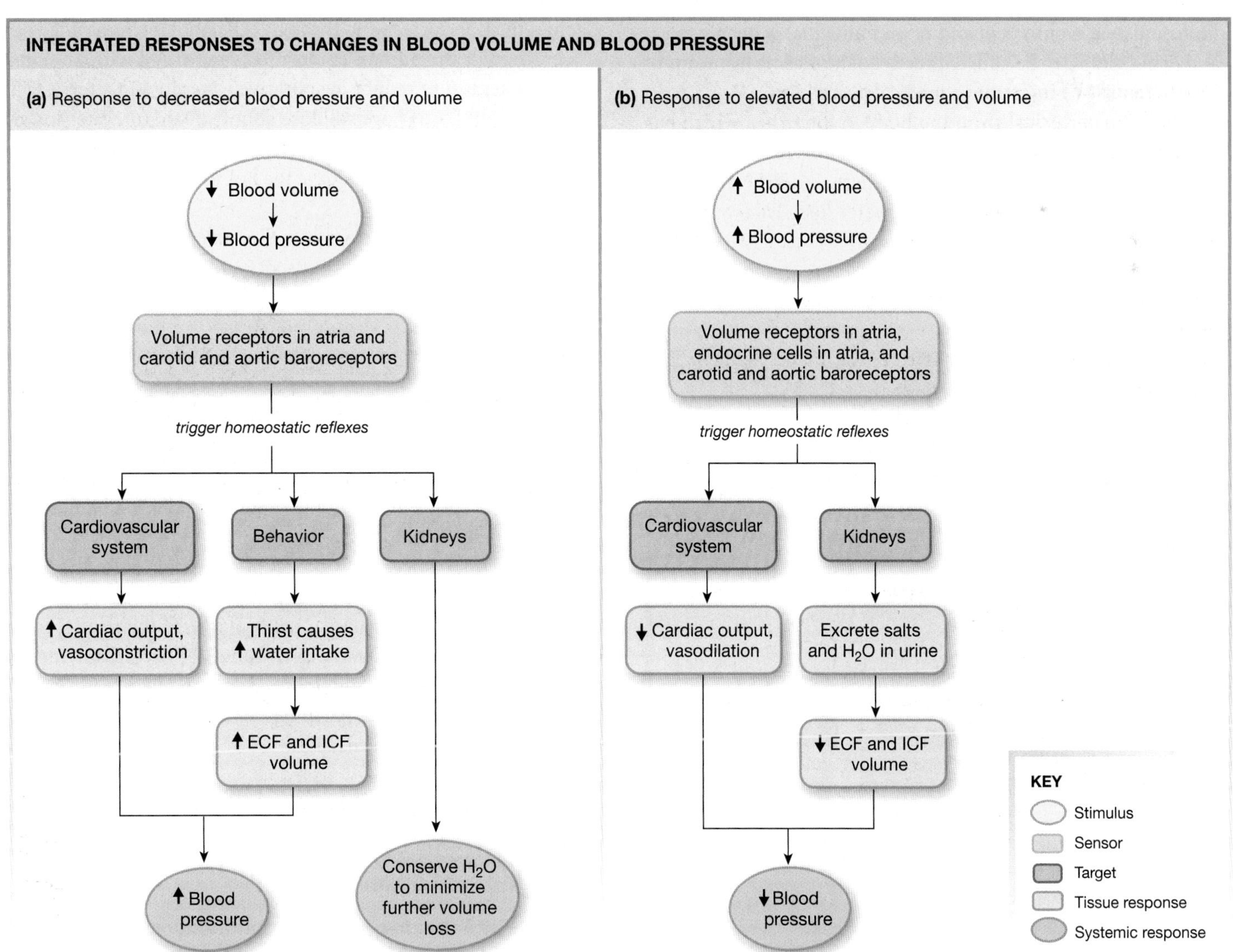

■ **Fig. 20.1**

A 60-kg (132-lb) woman contains about 30 liters of body water, and the "standard" 70-kg man contains about 42 liters. Two-thirds of his water (about 28 liters) is inside the cells, about 3 liters are in the plasma, and the remaining 11 liters are in the interstitial fluid [Fig. 5.1, p. 131].

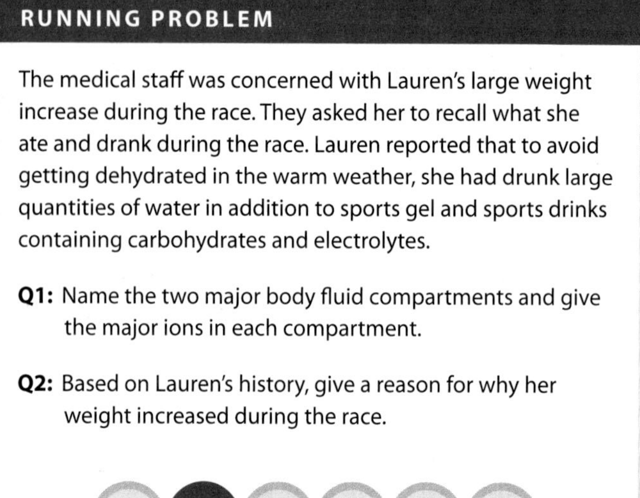
RUNNING PROBLEM

The medical staff was concerned with Lauren's large weight increase during the race. They asked her to recall what she ate and drank during the race. Lauren reported that to avoid getting dehydrated in the warm weather, she had drunk large quantities of water in addition to sports gel and sports drinks containing carbohydrates and electrolytes.

Q1: Name the two major body fluid compartments and give the major ions in each compartment.

Q2: Based on Lauren's history, give a reason for why her weight increased during the race.

658 660 673 676 683 690

Daily Water Intake and Excretion Are Balanced

To maintain a constant volume of water in the body, we must take in the same amount of water that we excrete: intake must equal output. There are multiple avenues for daily water gain and loss (■ Fig. 20.2). On average, an adult ingests a little more than 2 liters of water in food and drink in a day. Normal metabolism, especially aerobic respiration (glucose + $O_2 \rightarrow CO_2 + H_2O$), adds about 0.3 liter of water, bringing the total daily intake to approximately 2.5 liters.

Notice that the only means by which water normally enters the body from the external environment is by absorption through the digestive tract. Unlike some animals, we cannot absorb significant amounts of water directly through our skin. If fluids must be rapidly replaced or an individual is unable to eat and drink, fluid can be added directly to the plasma by means of **intravenous (IV) injection,** a medical procedure.

Most water is lost from the body in the urine, which has a daily volume of about 1.5 liters (Fig. 20.2). A small volume of water (about 100 mL) is lost in the feces. Additionally, water leaves the body through **insensible water loss.** This water loss, called *insensible* because we are not normally aware of it, occurs across the skin surface and through exhalation of humidified air. Even though the human epidermis is modified with an outer layer of keratin to reduce evaporative water loss in a terrestrial environment [p. 91], we still lose about 900 mL of water insensibly each day. Thus the 2.5 liters of water we take in are balanced by the 2.5 liters that leave the body. Only water loss in the urine can be regulated.

WATER BALANCE IN THE BODY

Water gain
2.2 L/day Food and drink
0.3 L/day Metabolism
2.5 L/day Totals

Water loss
Skin, Lungs: Insensible water loss 0.9 L/day
Urine 1.5 L/day
Feces 0.1 L/day
2.5 L/day

Intake 2.2 L/day + Metabolic production 0.3 L/day − Output 2.5 L/day = 0

■ **Fig. 20.2**

Although urine is normally the major route of water loss, in certain situations other routes of water loss can become significant. Excessive sweating is one example. Another way in which water is lost is through *diarrhea*, a condition that can pose a major threat to the maintenance of water balance, particularly in infants.

Pathological water loss disrupts homeostasis in two ways. Volume depletion of the extracellular compartment decreases blood pressure. If blood pressure cannot be maintained through homeostatic compensations, the tissues do not get adequate oxygen. Also, if the fluid lost is hyposmotic to the body (as is the case in excessive sweating), the solutes left behind in the body raise osmolarity, potentially disrupting cell function.

Normally, water balance takes place automatically. Salty food makes us thirsty. Drinking 42 ounces of a soft drink means an extra trip to the bathroom. Salt and water balance is a subtle process that we are only peripherally aware of, like breathing and the beating of the heart.

Now that we have discussed *why* regulation of osmolarity is important, let's see *how* the body accomplishes that goal.

The Kidneys Conserve Water

■ Figure 20.3 summarizes the role of the kidneys in water balance. The bottom line is that the kidneys can remove excess fluid by excreting it in the urine, but the kidneys cannot replace lost

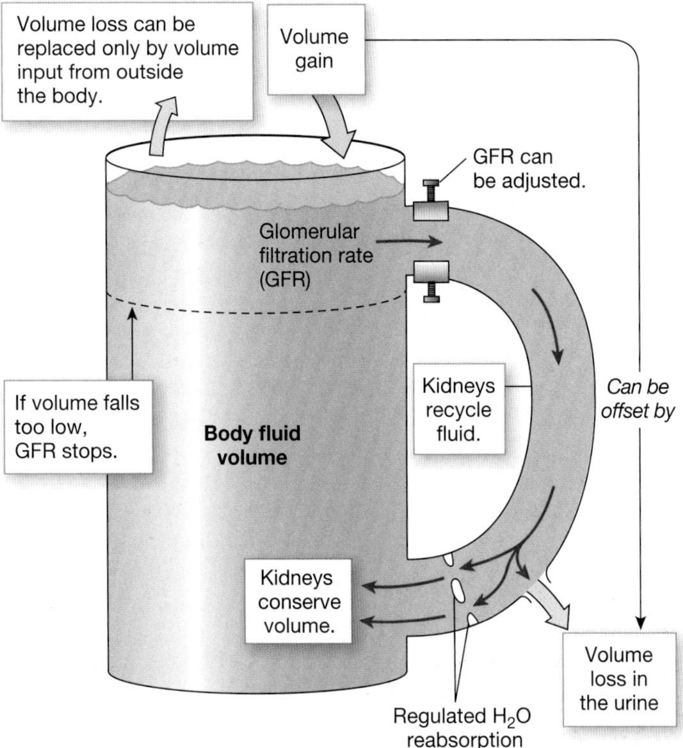
THE KIDNEYS CONSERVE VOLUME
Kidneys cannot restore lost volume. They only conserve fluid.
Volume loss can be replaced only by volume input from outside the body.
Volume gain
GFR can be adjusted.
Glomerular filtration rate (GFR)
If volume falls too low, GFR stops.
Body fluid volume
Kidneys recycle fluid.
Can be offset by
Kidneys conserve volume.
Volume loss in the urine
Regulated H_2O reabsorption

■ **Fig. 20.3**

volume. Volume lost to the environment must be replaced from the environment.

The mug represents the body, and its hollow handle represents the kidneys, where body fluid filters into the nephrons. Once fluid filters, it is in the outside environment. Unless it is reabsorbed, it will go into the urine. The volume that leaves can be regulated, as indicated by the little gates at the bottom of the handle.

The normal range for fluid volume in the mug lies between the dashed line and the open top. Fluid in the mug enters the handle (equivalent to being filtered into the kidney) and cycles back into the body of the mug to maintain the mug's volume. If fluid is added to the mug and threatens to overflow, the extra fluid is allowed to drain out of the handle (comparable to excess water excreted in urine). If a small volume is lost from the mug, fluid still flows through the handle, but fluid loss from the handle is turned off to prevent additional fluid loss.

The only way to replace lost fluid is to add water from a source outside the mug. Translating this model to the body underscores the fact that *the kidneys cannot replenish lost water: all they can do is conserve it.* And as shown in the mug model, if fluid loss is severe and volume falls below the dashed line, fluid no longer flows into the handle, just as a major fall in blood volume and blood pressure shuts down renal filtration.

The Renal Medulla Creates Concentrated Urine

The concentration, or osmolarity, of urine is a measure of how much water is excreted by the kidneys. When maintenance of homeostasis requires eliminating excess water, the kidneys produce copious amounts of dilute urine with an osmolarity as low as 50 mOsM. Removal of excess water in urine is known as **diuresis** {*diourein,* to pass in urine}. Drugs that promote the excretion of urine are called *diuretics*. In contrast, if the kidneys need to conserve water, the urine becomes quite concentrated. Specialized mechanisms in the medulla of the kidney allow urine to be up to four times as concentrated as the blood (1200 mOsM versus the blood's 300 mOsM).

The kidneys control urine concentration by varying the amounts of water and Na^+ reabsorbed in the distal nephron (distal tubule and collecting duct). To produce dilute urine, the kidney must reabsorb solute without allowing water to follow by osmosis. This means that the apical tubule cell membranes must not be permeable to water. On the other hand, if urine is to become concentrated, the nephron must be able to reabsorb water but leave solute in the tubule lumen.

Mechanistically, it seems simple enough to create an epithelium that transports solutes but is impermeable to water (dilute urine)—simply remove all water pores on the apical cell membrane. But mechanistically it seems much more difficult to create concentrated urine. How can the kidney reabsorb water without first reabsorbing solute? At one time, scientists speculated that water was actively transported on carriers, just as Na^+ and other ions are. However, once scientists developed micropuncture techniques for sampling fluid inside kidney tubules, they discovered that water is reabsorbed only by osmosis through water pores (*aquaporins*).

The mechanism for absorbing water without solute turned out to be simple: make the collecting duct cells and interstitial fluid surrounding them more concentrated than the fluid flowing into the tubule. Then, if the tubule cells have water pores, water can be absorbed from the lumen without first reabsorbing solute. This is indeed the situation in the kidney. Through an unusual arrangement of blood vessels and renal tubules, which we discuss later, the renal medulla maintains a high osmotic concentration in its cells and interstitial fluid. This high *medullary interstitial osmolarity* allows urine to be concentrated as it flows through the collecting duct.

Let's follow some filtered fluid through a nephron to see where these changes in osmolarity take place (■ Fig. 20.4). The renal cortex has an interstitial osmolarity of about 300 mOsM. Reabsorption in the proximal tubule is isosmotic [p. 633], and filtrate entering the loop of Henle has an osmolarity of about 300 mOsM (Fig. 20.4 ①).

As the nephrons dip into the medulla, the interstitial osmolarity steadily increases until it reaches about 1200 mOsM where the collecting ducts empty into the renal pelvis [Fig. 19.1c, p. 630]. Fluid passing through the descending limb

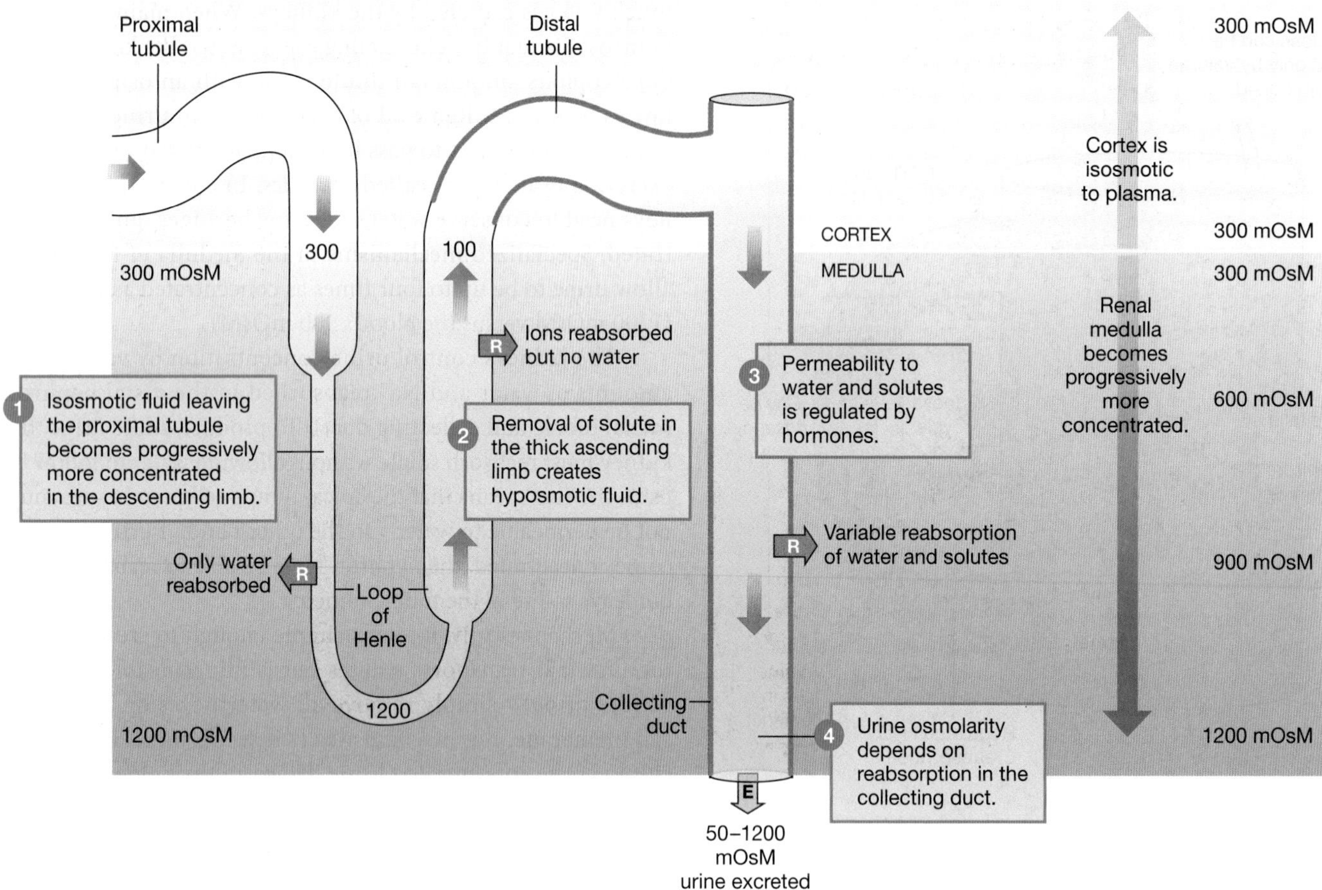
OSMOLARITY CHANGES THROUGH THE NEPHRON
Proximal tubule
Distal tubule
300 mOsM
Cortex is isosmotic to plasma.
300
100
300 mOsM
CORTEX
MEDULLA
300 mOsM
300 mOsM
Ions reabsorbed but no water
Permeability to water and solutes is regulated by hormones.
Renal medulla becomes progressively more concentrated.
Isosmotic fluid leaving the proximal tubule becomes progressively more concentrated in the descending limb.
Removal of solute in the thick ascending limb creates hyposmotic fluid.
600 mOsM
Variable reabsorption of water and solutes
Only water reabsorbed
Loop of Henle
900 mOsM
1200
Collecting duct
1200 mOsM
Urine osmolarity depends on reabsorption in the collecting duct.
1200 mOsM
50–1200 mOsM urine excreted

Fig. 20.4

of the loop loses water to the interstitium. Tubule fluid at the bottom of the loop will be of the same osmolarity as in the medulla.

In the ascending limb, the permeability of the tubule wall changes. The cells in the thick portion of the ascending limb of the loop have apical surfaces (facing the tubule lumen) that are impermeable to water. These cells do transport ions out of the tubule lumen (Fig. 20.4 (2)), but in this part of the nephron, solute movement is not followed by water movement. The reabsorption of solute without water decreases the concentration of the tubule fluid. Fluid leaving the loop of Henle therefore is hyposmotic, with an osmolarity of around 100 mOsM. The loop of Henle is the primary site where the kidney creates hyposmotic fluid.

Once hyposmotic fluid leaves the loop of Henle, it passes into the distal nephron. Here the water permeability of the tubule cells is variable and under hormonal control (Fig. 20.4 (3)). When the apical membrane of distal nephron cells is not permeable to water, water cannot leave the tubule, and the filtrate remains dilute. A small amount of additional solute can be reabsorbed as fluid passes along the collecting duct, making the filtrate even more dilute. When this happens, the concentration of urine can be as low as 50 mOsM (Fig. 20.4 (4)).

CLINICAL FOCUS: DIABETES

Osmotic Diuresis

The primary sign of diabetes mellitus is an elevated blood glucose concentration. In untreated diabetics, if blood glucose levels exceed the renal threshold for glucose reabsorption [p. 643], glucose is excreted in the urine. This may not seem like a big deal, but any additional solute that remains in the lumen forces additional water to be excreted, causing *osmotic diuresis*. Suppose, for example, that the nephrons must excrete 300 milliosmoles of NaCl. If the urine is maximally concentrated at 1200 mOsM, the NaCl is excreted in a volume of 0.25 L. However, if the NaCl is joined by 300 milliosmoles of glucose that must be excreted, the volume of urine doubles, to 0.5 L. Osmotic diuresis in untreated diabetics (primarily type 1) causes *polyuria* (excessive urination) and *polydipsia* (excessive thirst) {*dipsios*, thirsty} as a result of dehydration and high plasma osmolarity.

On the other hand, if the body needs to conserve water by reabsorbing it, the tubule epithelium in the distal nephron must become permeable to water. Under hormonal control, the cells insert water pores into their apical membranes. Once water can enter the epithelial cells, osmosis draws water out of the less-concentrated lumen and into the more concentrated interstitial fluid. At maximal water permeability, removal of water from the tubule leaves behind concentrated urine with an osmolarity that can be as high as 1200 mOsM (Fig. 20.4 (4)).

Water reabsorption in the kidneys conserves water and can decrease body osmolarity to some degree when coupled with excretion of solute in the urine. But remember that the kidney's homeostatic mechanisms can do nothing to restore lost fluid volume. Only the ingestion or infusion of water can replace water that has been lost.

Vasopressin Controls Water Reabsorption

How do the distal tubule and collecting duct cells alter their permeability to water? The process involves adding or removing water pores in the apical membrane under the direction of the posterior pituitary hormone **vasopressin** [p. 219]. In most mammals, the nine-amino-acid peptide contains the amino acid arginine, so vasopressin is called *arginine vasopressin* or *AVP.* Because vasopressin causes the body to retain water, it is also known as *antidiuretic hormone (ADH).*

When vasopressin acts on target cells, the collecting duct epithelium becomes permeable to water, allowing water to move out of the lumen (■ Fig. 20.5a). The water moves by osmosis because solute concentration in the cells and interstitial fluid of the renal medulla is higher than that of fluid in the tubule. In the absence of vasopressin, the collecting duct is impermeable to water (Fig. 20.5b). Although a concentration gradient is present across the epithelium, water remains in the tubule, producing dilute urine.

The water permeability of the collecting duct is not an all-or-none phenomenon, as the previous paragraph might suggest. Permeability is variable, depending on how much vasopressin is present. The graded effect of vasopressin allows the body to match urine concentration closely to the body's needs.

Vasopressin and Aquaporins Most membranes in the body are freely permeable to water. What makes the cells of the distal nephron different? The answer lies with the *water pores* found in these cells. Water pores are **aquaporins,** a family of membrane channels with at least 10 different isoforms that occur in mammalian tissues. The kidney has multiple isoforms of aquaporins, including *aquaporin-2* (AQP2), the water channel regulated by vasopressin.

AQP2 in a collecting duct cell may be found in two locations: on the apical membrane facing the tubule lumen and in the membrane of cytoplasmic storage vesicles (Fig. 20.5c). (Two other isoforms of aquaporins are present in the basolateral membrane, but they are not regulated by vasopressin.) When vasopressin levels and, consequently, collecting duct water permeability are low, the collecting duct cell has few water pores in its apical membrane and stores its AQP2 water pores in cytoplasmic storage vesicles.

When vasopressin arrives at the collecting duct, it binds to its *V2 receptors* on the basolateral side of the cell (step 1 in Fig. 20.5c). Binding activates a G-protein/cAMP second messenger system [p. 183]. Subsequent phosphorylation of intracellular proteins causes the AQP2 vesicles to move to the apical membrane and fuse with it. Exocytosis inserts the AQP2 water pores into the apical membrane. Now the cell is permeable to water. This process, in which parts of the cell membrane are alternately added by exocytosis and withdrawn by endocytosis, is known as **membrane recycling** [Fig. 5.19, p. 156].

Concept Check Answers: p. 694

1. Does the apical membrane of a collecting duct cell have more water pores when vasopressin is present or when it is absent?
2. If vasopressin secretion is suppressed, will urine be dilute or concentrated?

Blood Volume and Osmolarity Activate Osmoreceptors

What stimuli control vasopressin secretion? There are three: plasma osmolarity, blood volume, and blood pressure (■ Fig. 20.6). The most potent stimulus for vasopressin release is an increase in plasma osmolarity. Osmolarity is monitored by **osmoreceptors,** stretch-sensitive neurons that increase their firing rate as osmolarity increases. Our current model indicates that when the osmoreceptors shrink, cation channels linked to actin filaments open, depolarizing the cell.

20

The primary osmoreceptors for vasopressin release are in the hypothalamus. When plasma osmolarity is below the threshold value of 280 mOsM, the osmoreceptors do not fire, and vasopressin release from the pituitary ceases (Fig. 20.6b). If plasma osmolarity rises above 280 mOsM, the osmoreceptors stimulate release of vasopressin.

Decreases in blood pressure and blood volume are less powerful stimuli for vasopressin release. The primary receptors for decreased volume are stretch-sensitive receptors in the atria. Blood pressure is monitored by the same carotid and aortic baroreceptors that initiate cardiovascular responses [p. 525]. When blood pressure or blood volume is low, these receptors signal the hypothalamus to secrete vasopressin and conserve fluid.

In adults, vasopressin secretion also shows a circadian rhythm, with increased secretion during the overnight hours. As a result of this increase, less urine is produced overnight than during the day, and the first urine excreted in the morning is

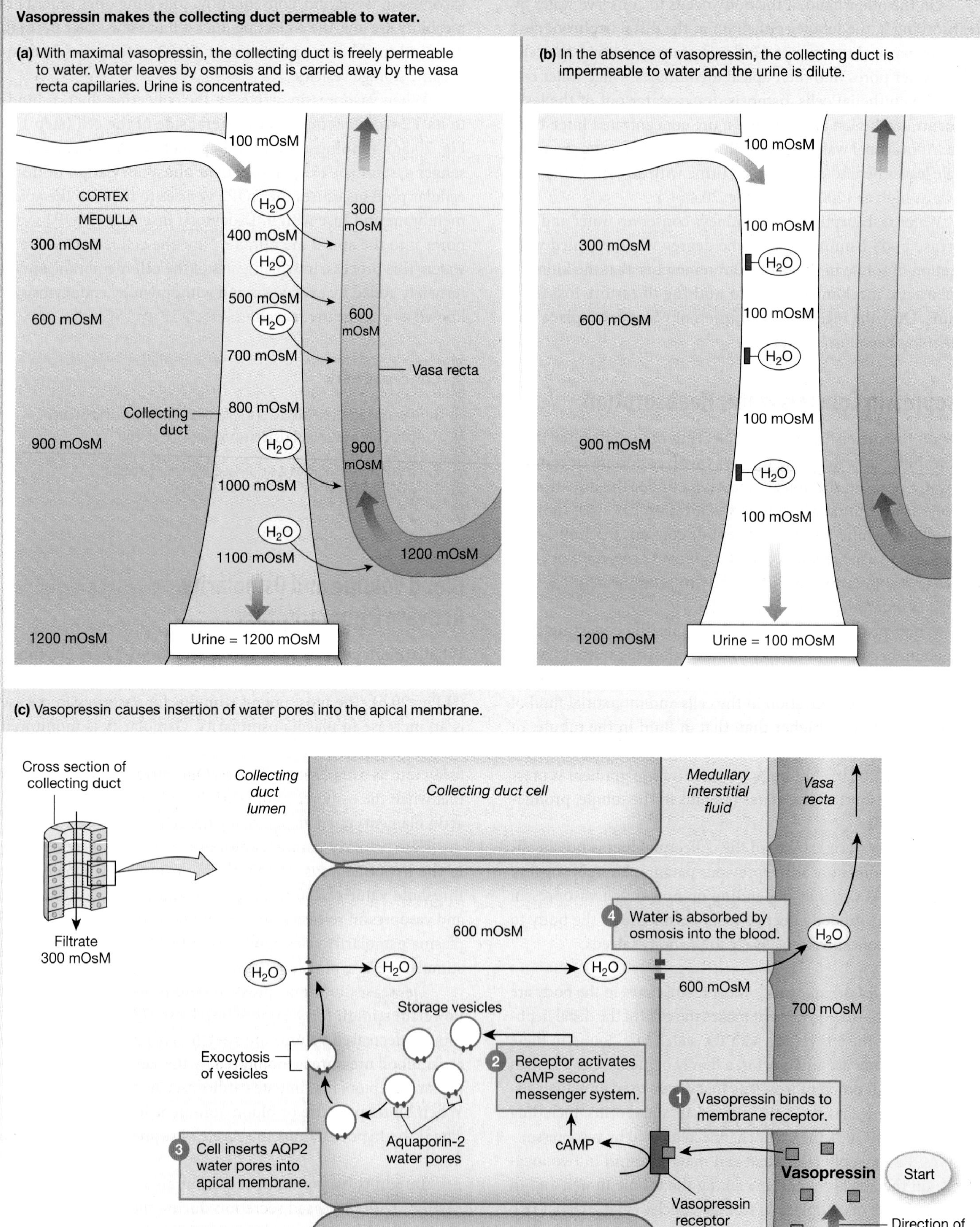

Vasopressin makes the collecting duct permeable to water.
(a) With maximal vasopressin, the collecting duct is freely permeable to water. Water leaves by osmosis and is carried away by the vasa recta capillaries. Urine is concentrated.
100 mOsM
CORTEX
MEDULLA
300 mOsM
H2O
300 mOsM
400 mOsM
500 mOsM
600 mOsM
600 mOsM
700 mOsM
Vasa recta
Collecting duct
800 mOsM
900 mOsM
900 mOsM
1000 mOsM
1100 mOsM
1200 mOsM
1200 mOsM
Urine = 1200 mOsM
(b) In the absence of vasopressin, the collecting duct is impermeable to water and the urine is dilute.
100 mOsM
100 mOsM
300 mOsM
100 mOsM
600 mOsM
100 mOsM
900 mOsM
100 mOsM
1200 mOsM
Urine = 100 mOsM
(c) Vasopressin causes insertion of water pores into the apical membrane.
Cross section of collecting duct
Filtrate 300 mOsM
Collecting duct lumen
Collecting duct cell
Medullary interstitial fluid
Vasa recta
4 Water is absorbed by osmosis into the blood.
600 mOsM
600 mOsM
700 mOsM
Storage vesicles
Exocytosis of vesicles
2 Receptor activates cAMP second messenger system.
1 Vasopressin binds to membrane receptor.
3 Cell inserts AQP2 water pores into apical membrane.
Aquaporin-2 water pores
cAMP
Vasopressin
Start
Vasopressin receptor
Direction of blood flow

■ **Fig. 20.5**

Vasopressin

High osmolarity or low blood pressure cause vasopressin release.

(a) Control of vasopressin secretion

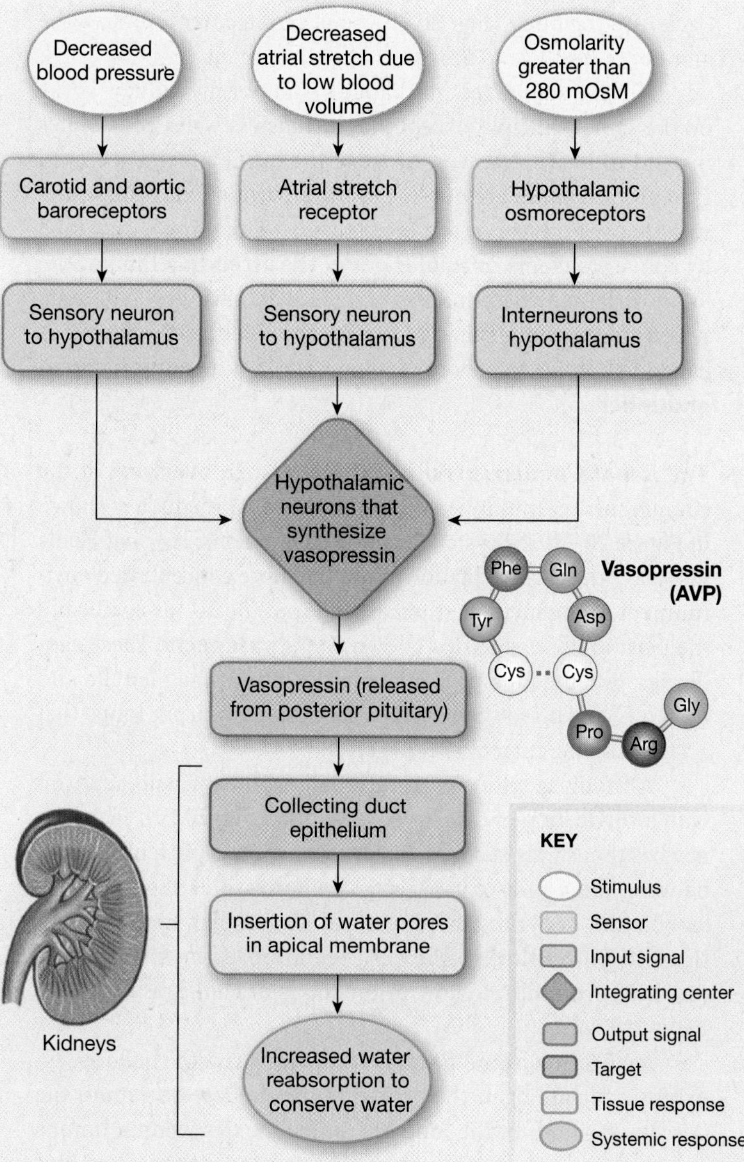

HYPOTHALAMUS

1. AVP is made and packaged in cell body of neuron.
2. Vesicles are transported down the cell.
3. Vesicles containing AVP are stored in posterior pituitary.
4. AVP is released into blood.

Posterior pituitary

Vein

ARGININE VASOPRESSIN (AVP), Antidiuretic hormone (ADH)	
Origin	Hypothalamic neurons. Released from posterior pituitary
Chemical nature	9-amino acid peptide
Transport in the circulation	Dissolved in plasma
Half-life	15 min
Factors affecting release	↑ Osmolarity (hypothalamic osmoreceptors) ↓ Blood pressure or volume (carotid, aortic, atrial receptors)
Target cells or tissues	Renal collecting duct
Receptor/second messenger	V2 receptor/cAMP
Tissue action	Increases renal water reabsorption
Action at cellular-molecular level	Inserts AQP water pores in apical membrane

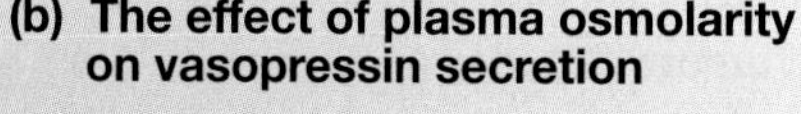

(b) The effect of plasma osmolarity on vasopressin secretion

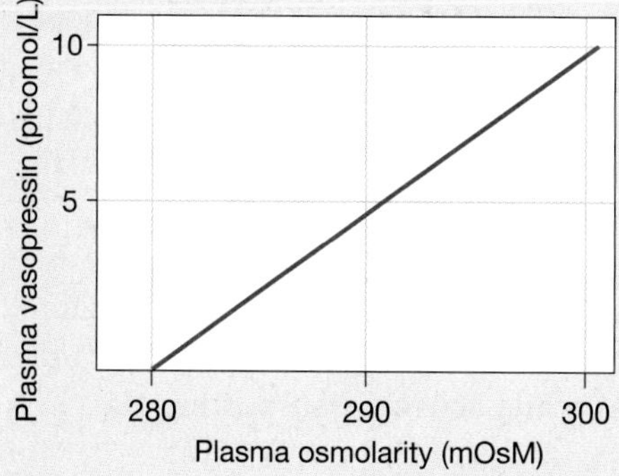

FIGURE QUESTIONS

1. What is the threshold osmolarity for vasopressin release?
2. What signal in the AVP neuron triggers exocytosis of AVP-containing vesicles?

more concentrated. One theory for the cause of bed-wetting, or **nocturnal enuresis,** in children is that these children have a developmental delay in the normal pattern of increased vasopressin secretion at night. With less vasopressin, the children's urine output stays elevated, causing the bladder to fill to its maximum capacity and empty spontaneously during sleep. Many of these children can be successfully treated with a nasal spray of *desmopressin*, a vasopressin derivative, administered at bedtime.

Concept Check Answers: p. 694

3. A scientist monitoring the activity of osmoreceptors notices that infusion of hyperosmotic saline (NaCl) causes increased firing of the osmoreceptors. Infusion of hyperosmotic urea (a penetrating solute) [p. 134] had no effect on the firing rate. If osmoreceptors fire only when cell volume decreases, explain why hyperosmotic urea did not affect them.
4. If vasopressin increases water reabsorption by the nephron, would vasopressin secretion be increased or decreased with dehydration?
5. Experiments suggest that there are peripheral osmoreceptors in the lumen of the upper digestive tract and in the hepatic portal vein [Fig. 14.1, p. 465]. What is the adaptive significance of osmoreceptors in these locations?

The Loop of Henle Is a Countercurrent Multiplier

Vasopressin is the signal for water reabsorption out of the nephron tubule, but the key to the kidney's ability to produce concentrated urine is the high osmolarity of the medullary *interstitium* (interstitial fluid compartment of the kidney). Without it, there would be no concentration gradient for osmotic movement of water out of the collecting duct. What creates this high ECF osmolarity? And why isn't the interstitial fluid osmolarity reduced as water is reabsorbed from the collecting duct and descending limb of the loop of Henle (see Fig. 20.4)? The answers to these questions lie in the anatomical arrangement of the loop of Henle and its associated blood vessels, the vasa recta. Together, these structures form a *countercurrent exchange system.*

Countercurrent Exchange Systems **Countercurrent exchange systems** require arterial and venous blood vessels that pass very close to each other, with their fluid flow moving in opposite directions (the name *countercurrent* reflects the fact that the two flows run *counter to* each other). This anatomical arrangement allows the passive transfer of heat or molecules from one vessel to the other. Because the countercurrent heat exchanger is easier to understand, we first examine how it works and then apply the same principle to the kidney.

The countercurrent heat exchanger in mammals and birds evolved to reduce heat loss from flippers, tails, and other limbs that are poorly insulated and have a high surface-area-to-volume ratio. Without a heat exchanger, warm blood flowing from the body core into the limb would easily lose heat to the surrounding environment (■ Fig. 20.7a). With a countercurrent heat exchanger, warm arterial blood entering the limb transfers its heat to cooler venous blood flowing from the tip of the limb back into the body (Fig. 20.7b). This arrangement reduces the amount of heat lost to the external environment.

The countercurrent exchange system of the kidney works on the same principle, except that it transfers water and solutes instead of heat. However, because the kidney forms a closed system, the solutes are not lost to the environment. Instead, the solutes concentrate in the interstitium. This process is aided by active transport of solutes out of the ascending limb of the loop of Henle, which makes the ECF osmolarity even greater. A countercurrent exchange system in which exchange is enhanced by active transport of solutes is called a **countercurrent multiplier**.

The Renal Countercurrent Multiplier An overview of the countercurrent multiplier system in the renal medulla is shown in Figure 20.7c. The system has two components: loops of Henle that leave the cortex, dip down into the more concentrated environment of the medulla, then ascend into the cortex again, and the peritubular capillaries known as the **vasa recta**. These capillaries, like the loop of Henle, dip down into the medulla and then go back up to the cortex, also forming hairpin loops that act as a countercurrent exchanger.

Although textbooks traditionally show a single nephron with a single loop of capillary (as we do in Fig. 20.7c), each kidney has thousands of collecting ducts and loops of Henle packed between thousands of vasa recta capillaries, blurring the direct association between a nephron and its vascular supply. Functionally, blood flow in the vasa recta moves in the opposite direction from filtrate flow in the loops of Henle, as shown in Figure 20.7c.

Let's follow some fluid as it moves through the loop. Isosmotic filtrate from the proximal tubule first flows into the descending limb of the loop of Henle. The descending limb is permeable to water but does not transport ions. As the loop dips into the medulla, water moves by osmosis from the descending limb into the progressively more concentrated interstitial fluid, leaving solutes behind in the tubule lumen.

The filtrate becomes progressively more concentrated as it moves deeper into the medulla. At the tips of the longest loops of Henle, the filtrate reaches a concentration of 1200 mOsM. Filtrate in shorter loops (which do not extend into the most concentrated regions of the medulla) does not reach such a high concentration.

When the fluid flow reverses direction and enters the ascending limb of the loop, the properties of the tubule epithelium change. The tubule epithelium in this segment of the nephron is impermeable to water while actively transporting Na^+, K^+

COUNTERCURRENT MECHANISMS

A countercurrent heat exchanger

(a) If blood vessels are not close to each other, heat is dissipated to the external environment.

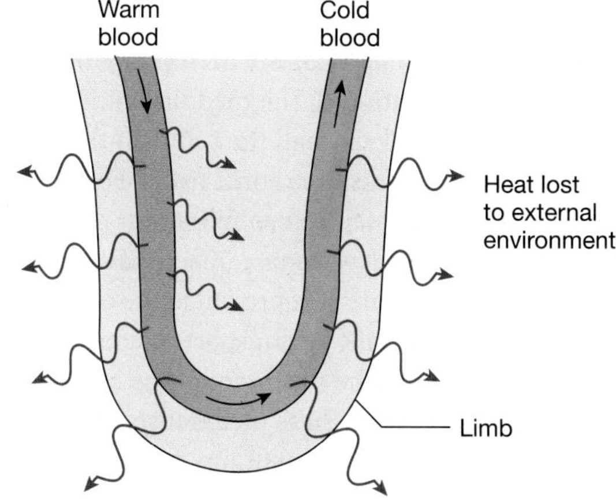

(b) Countercurrent heat exchanger allows warm blood entering the limb to transfer heat directly to blood flowing back into the body.

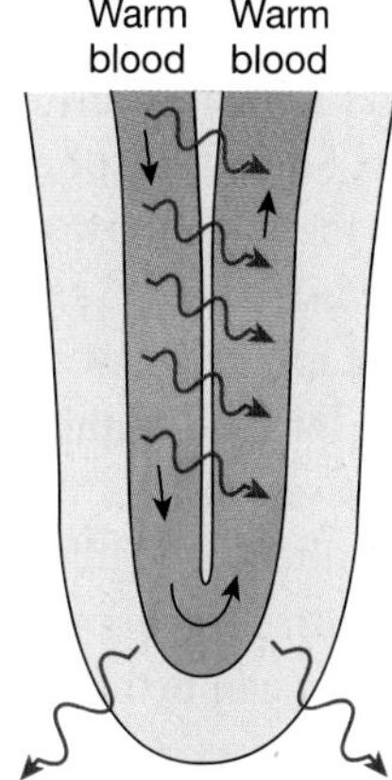

(c) Countercurrent exchange in the vasa recta

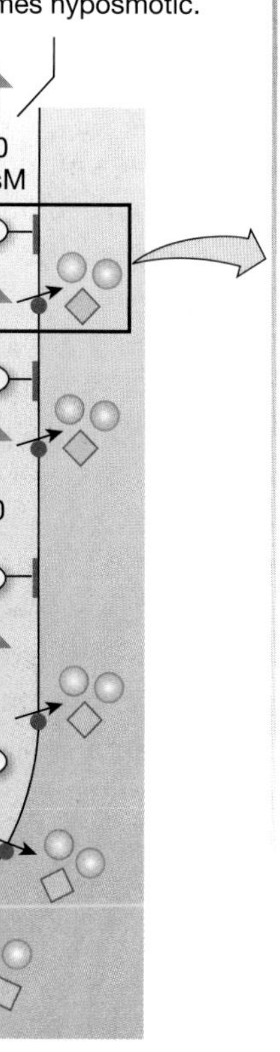

KEY
H_2O = ○ K^+ = ◇
Cl^- = ● Na^+ = ●

(d) The apical surface of the ascending limb is not permeable to water. Active reabsorption of ions in this region creates a dilute filtrate in the lumen.

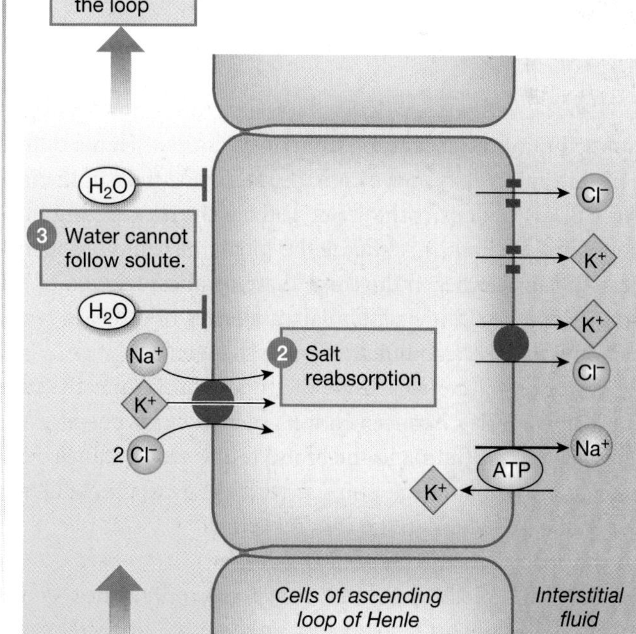

Fig. 20.7

20

and Cl^- out of the tubule into the interstitial fluid. The loss of solute from the lumen causes the filtrate osmolarity to decrease steadily, from 1200 mOsM at the bottom of the loop to 100 mOsM at the point where the ascending limb leaves the medulla and enters the cortex. The net result of the countercurrent multiplier in the kidney is to produce hyperosmotic interstitial fluid in the medulla and hyposmotic filtrate leaving the loop of Henle.

Normally, about 25% of all Na^+ and K^+ reabsorption takes place in the ascending limb of the loop. Some transporters responsible for active ion reabsorption in the thick portion of the ascending limb are shown in Figure 20.7d. The *NKCC symporter* uses energy stored in the Na^+ concentration gradient to transport Na^+, K^+, and 2 Cl^- from the lumen into the epithelial cells of the ascending limb. The Na^+-K^+-ATPase removes Na^+ from the cells on the basolateral side of the epithelium, while K^+ and Cl^- leave the cells together on a cotransport protein or through open channels. NKCC-mediated transport can be inhibited by drugs known as "loop diuretics," such as *furosemide* (Lasix).

Concept Check Answers: p. 694

6. Explain why patients taking a loop diuretic that inhibits solute reabsorption excrete greater-than-normal volumes of urine.
7. Loop diuretics that inhibit the NKCC symporter are sometimes called "potassium-wasting" diuretics. Explain why people who are on loop diuretics must increase their dietary K^+ intake.

The Vasa Recta Removes Water It is easy to see how transport of solute out of the ascending limb of the loop of Henle dilutes the filtrate and helps concentrate the interstitial fluid in the medulla. Still, why doesn't the water leaving the descending limb of the loop (see Fig. 20.7c) *dilute* the interstitial fluid of the medulla? The answer lies in the close anatomical association of the loop of Henle and the peritubular capillaries of the vasa recta, which functions as a countercurrent exchanger.

Water or solutes that leave the tubule move into the vasa recta if an osmotic or concentration gradient exists between the medullary interstitium and the blood in the vasa recta. For example, assume that at the point at which the vasa recta enters the medulla, the blood in the vasa recta is 300 mOsM, isosmotic with the cortex. As the blood flows deeper into the medulla, it loses water and picks up solutes transported out of the ascending limb of the loop of Henle, carrying these solutes farther into the medulla. By the time the blood reaches the bottom of the vasa recta loop, it has a high osmolarity, similar to that of the surrounding interstitial fluid (1200 mOsM).

Then, as blood in the vasa recta flows back toward the cortex, the high plasma osmolarity attracts the water that is being lost from the descending limb, as Figure 20.7c shows. The movement of this water into the vasa recta decreases the osmolarity of the blood while simultaneously preventing the water from diluting the concentrated medullary interstitial fluid.

The end result of this arrangement is that blood flowing through the vasa recta removes the water reabsorbed from the loop of Henle. Without the vasa recta, water moving out of the descending limb of the loop of Henle would eventually dilute the medullary interstitium. The vasa recta thus plays an important part in keeping the medullary solute concentration high.

Urea Increases the Osmolarity of the Medullary Interstitium The high solute concentration in the medullary interstitium is only partly due to NaCl. Nearly half the solute in this compartment is urea. Where does this urea come from? For many years scientists thought urea crossed cell membranes only by passive transport. However, in recent years researchers have learned that membrane transporters for urea are present in the collecting duct and loops of Henle. One family of transporters consists of facilitated diffusion carriers, and the other family has Na^+-dependent secondary active transporters. These urea transporters help concentrate urea in the medullary interstitium, where it contributes to the high interstitial osmolarity.

Sodium Balance and ECF Volume

With an average American diet, we ingest a lot of NaCl—about 9 grams per day. This is about 2 teaspoons of salt, or 155 milliosmoles of Na^+ and 155 milliosmoles of Cl^-. Let's see what would happen to our bodies if the kidneys could not get rid of this Na^+.

Our normal plasma Na^+ concentration, measured from a venous blood sample, is 135–145 milliosmoles Na^+ per liter of plasma. Because Na^+ distributes freely between plasma and interstitial fluid, this value also represents our ECF Na^+ concentration. If we add 155 milliosmoles of Na^+ to the ECF, how much water would we have to add to keep the ECF Na^+ concentration at 140 mOsM? One form of an equation asking this question is

$$\begin{aligned} 155 \text{ mosmol}/x \text{ liters} &= 140 \text{ mosmol/liter} \\ x &= 1.1 \text{ liters} \end{aligned}$$

We would have to add more than a liter of water to the ECF to compensate for the addition of the Na^+. Normal ECF volume is about 14 liters, and so that increase in volume would represent about an 8% gain! Imagine what that volume increase would do to blood pressure.

Suppose, however, that instead of adding water to keep plasma concentrations constant, we add the NaCl but don't drink any water. What happens to osmolarity now? If we assume that normal total body osmolarity is 300 mOsM and that the volume of fluid in the body is 42 L, the addition of 155 milliosmoles of Na^+ and 155 milliosmoles of Cl^- would increase total body osmolarity to 307 mOsM*—a substantial increase.

*(155 mosmol Na^+ + 155 mosmol Cl^-)/42 L = 7.4 mosmol/L added; 300 mosmol/L initial + 7.4 mosmol/L added = 307 mOsM final

In addition, because NaCl is a nonpenetrating solute, it would stay in the ECF. Higher osmolarity in the ECF would draw water from the cells, shrinking them and disrupting normal cell function.

Fortunately, our homeostatic mechanisms usually maintain mass balance: anything extra that comes into the body is excreted. ■ Figure 20.8 shows a generalized homeostatic pathway for sodium balance in response to salt ingestion. Here's how it works:

The addition of NaCl to the body raises osmolarity. This stimulus triggers two responses: vasopressin secretion and thirst. Vasopressin release causes the kidneys to conserve water (by reabsorbing water from the filtrate) and concentrate the urine. Thirst prompts us to drink water or other fluids. The increased fluid intake decreases osmolarity, but the combination of salt and water intake increases both ECF volume and blood pressure. These increases then trigger another series of control pathways, which bring ECF volume, blood pressure, and total-body osmolarity back into the normal range by excreting extra salt and water.

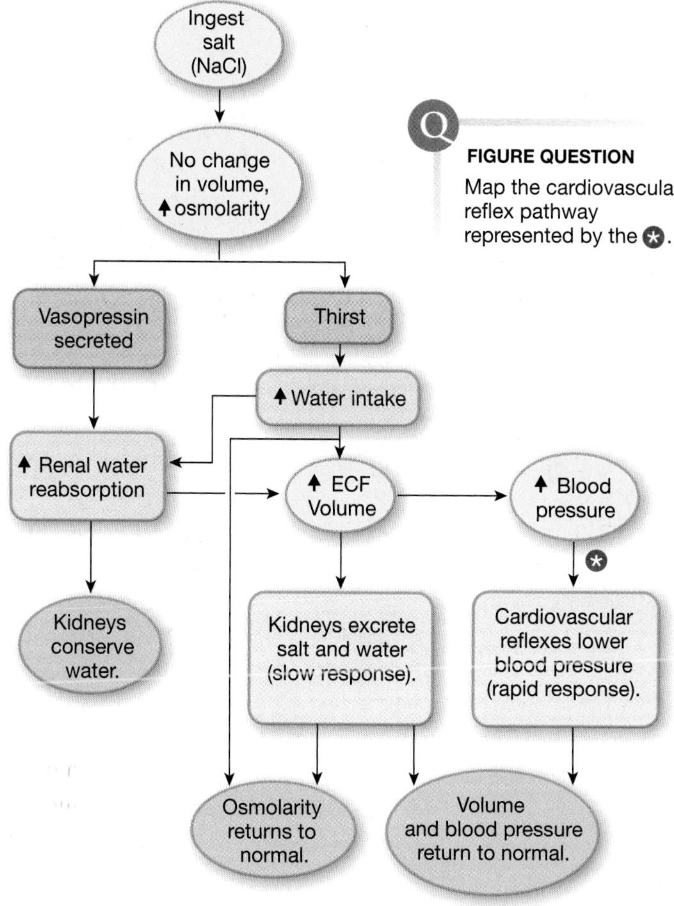
HOMEOSTATIC RESPONSES TO SALT INGESTION
Ingest salt (NaCl)
No change in volume, ↑ osmolarity
FIGURE QUESTION
Map the cardiovascular reflex pathway represented by the ✪.
Vasopressin secreted
Thirst
↑ Water intake
↑ Renal water reabsorption
↑ ECF Volume
↑ Blood pressure
Kidneys conserve water.
Kidneys excrete salt and water (slow response).
Cardiovascular reflexes lower blood pressure (rapid response).
Osmolarity returns to normal.
Volume and blood pressure return to normal.

■ Fig. 20.8

The kidneys are responsible for most Na^+ excretion, and normally only a small amount of Na^+ leaves the body in feces and perspiration. However, in situations such as vomiting, diarrhea, and heavy sweating, we may lose significant amounts of Na^+ and Cl^- through nonrenal routes.

Although we speak of ingesting and losing salt (NaCl), only renal Na^+ absorption is regulated. And actually, the stimuli that set the Na^+ balance pathway in motion are more closely tied to blood volume and blood pressure than to Na^+ levels. Chloride movement usually follows Na^+ movement, either indirectly via the electrochemical gradient created by Na^+ transport or directly via membrane transporters such as the NKCC transporter of the loop of Henle or the Na^+-Cl^- symporter of the distal tubule.

Aldosterone Controls Sodium Balance

The regulation of blood Na^+ levels takes place through one of the most complicated endocrine pathways of the body. The reabsorption of Na^+ in the distal tubules and collecting ducts of the kidney is regulated by the steroid hormone **aldosterone**: the more aldosterone, the more Na^+ reabsorption. Because one target of aldosterone is increased activity of the Na^+-K^+-ATPase, aldosterone also causes K^+ secretion (■ Fig. 20.9).

Aldosterone is a steroid hormone synthesized in the adrenal cortex, the outer portion of the adrenal gland that sits atop each kidney [p. 214]. Like other steroid hormones, aldosterone is secreted into the blood and transported on a protein carrier to its target.

The primary site of aldosterone action is the last third of the distal tubule and the portion of the collecting duct that runs through the kidney cortex (the *cortical collecting duct*). The primary target of aldosterone is **principal cells,** or **P cells** (Fig. 20.9b). Principal cells are arranged much like other polarized transporting epithelial cells, with Na^+-K^+-ATPase pumps on the basolateral membrane, and various channels and transporters on the apical membrane [p. 155]. In principal cells, the apical membranes contain leak channels for Na^+ (called ENaC, for epithelial Na^+ channel) and for K^+ (called ROMK, for renal outer medulla K^+ channel).

Aldosterone enters P cells by simple diffusion. Once inside, it combines with a cytoplasmic receptor (Fig. 20.9b ①). In the early response phase, apical Na^+ and K^+ channels increase their open time under the influence of an as-yet-unidentified signal molecule. As intracellular Na^+ levels rise, the Na^+-K^+-ATPase pump speeds up, transporting cytoplasmic Na^+ into the ECF and bringing K^+ from the ECF into the P cell. The net result is a rapid increase in Na^+ reabsorption and K^+ secretion that does not require the synthesis of new channel or ATPase proteins. In the slower phase of aldosterone action, newly synthesized channels and pumps are inserted into epithelial cell membranes (Fig. 20.9b ③).

20

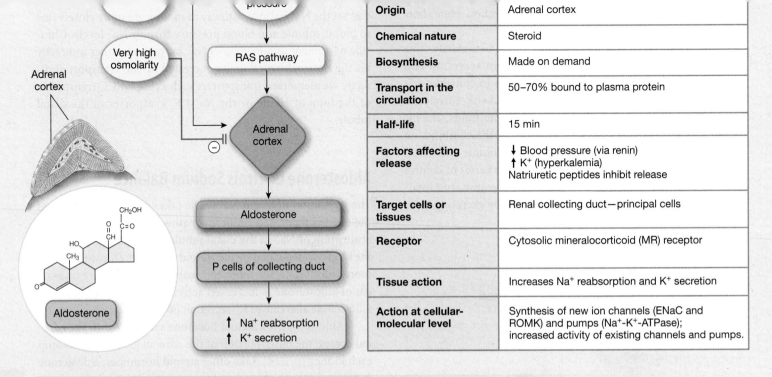

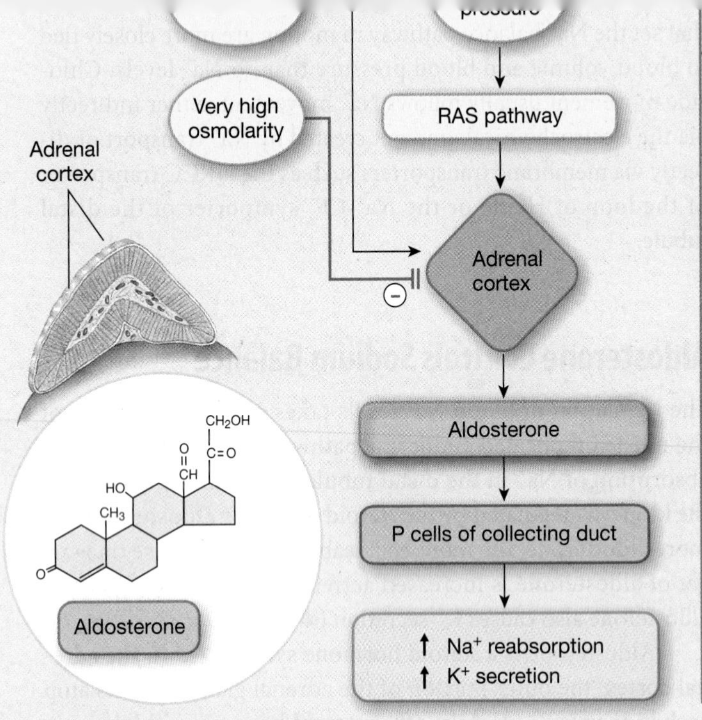

Origin	Adrenal cortex
Chemical nature	Steroid
Biosynthesis	Made on demand
Transport in the circulation	50–70% bound to plasma protein
Half-life	15 min
Factors affecting release	↓ Blood pressure (via renin) ↑ K^+ (hyperkalemia) Natriuretic peptides inhibit release
Target cells or tissues	Renal collecting duct—principal cells
Receptor	Cytosolic mineralocorticoid (MR) receptor
Tissue action	Increases Na^+ reabsorption and K^+ secretion
Action at cellular-molecular level	Synthesis of new ion channels (ENaC and ROMK) and pumps (Na^+-K^+-ATPase); increased activity of existing channels and pumps.

(b) Aldosterone acts on principal cells.

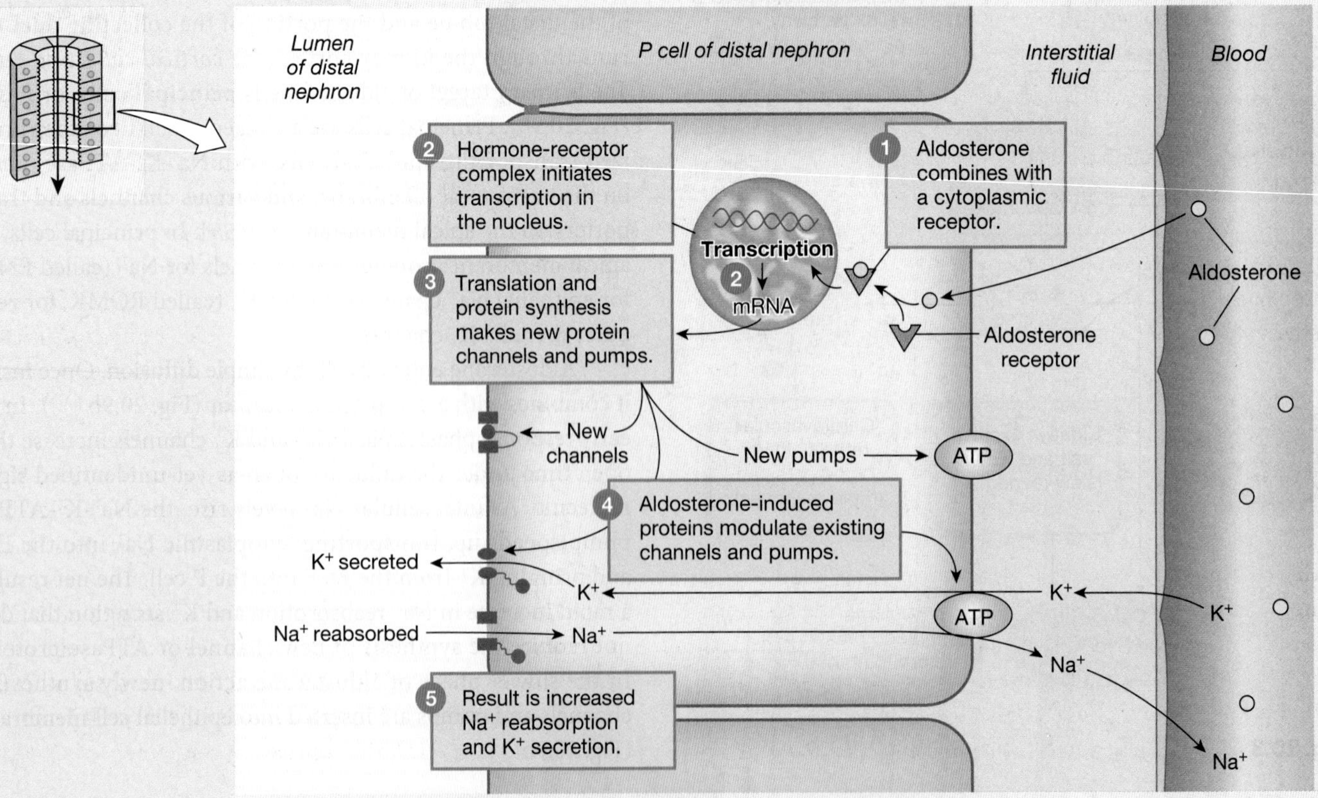

Note that Na^+ and water reabsorption are separately regulated in the distal nephron. Water does not automatically follow Na^+ reabsorption: vasopressin must be present to make the distal-nephron epithelium permeable to water. In contrast, Na^+ reabsorption in the proximal tubule is automatically followed by water reabsorption because the proximal tubule epithelium is always freely permeable to water.

Concept Check Answers: p. 694

8. In Figure 20.9b, what forces cause Na^+ and K^+ to cross the apical membrane?
9. If a person experiences hyperkalemia, what happens to resting membrane potential and the excitability of neurons and the myocardium?

Low Blood Pressure Stimulates Aldosterone Secretion

What controls physiological aldosterone secretion from the adrenal cortex? There are two primary stimuli: increased extracellular K^+ concentration and decreased blood pressure (Fig. 20.9a). Elevated K^+ concentrations act directly on the adrenal cortex in a reflex that protects the body from hyperkalemia. Decreased blood pressure initiates a complex pathway that results in release of a hormone, **angiotensin II,** that stimulates aldosterone secretion in most situations.

Two additional factors modulate aldosterone release in pathological states: an increase in ECF osmolarity acts directly on adrenal cortex cells to inhibit aldosterone secretion during dehydration, and an abnormally large (10–20 mEq/L) decrease in plasma Na^+ can directly stimulate aldosterone secretion.

The Renin-Angiotensin Pathway Angiotensin II (ANG II) is the usual signal controlling aldosterone release from the adrenal cortex. ANG II is one component of the **renin-angiotensin system (RAS),** a complex, multistep pathway for maintaining blood pressure. The RAS pathway begins when juxtaglomerular granular cells in the afferent arterioles of a nephron [p. 638] secrete an enzyme called **renin** (■ Fig. 20.10). Renin converts an inactive plasma protein, **angiotensinogen,** into **angiotensin I (ANG I).** (The suffix *-ogen* indicates an inactive precursor.) When ANG I in the blood encounters an enzyme called **angiotensin-converting enzyme (ACE),** ANG I is converted into ANG II.

This conversion was originally thought to take place only in the lungs, but ACE is now known to occur on the endothelium of blood vessels throughout the body. When ANG II in the blood reaches the adrenal gland, it causes synthesis and release of aldosterone. Finally, at the distal nephron, aldosterone initiates the intracellular reactions that cause the tubule to reabsorb Na^+.

The stimuli that begin the RAS pathway are all related either directly or indirectly to low blood pressure (Fig. 20.10):

1. The *granular cells* are directly sensitive to blood pressure. They respond to low blood pressure in renal arterioles by secreting renin.
2. *Sympathetic neurons,* activated by the cardiovascular control center when blood pressure decreases, terminate on the granular cells and stimulate renin secretion.
3. *Paracrine feedback*—from the macula densa in the distal tubule to the granular cells—stimulates renin release [p. 636]. When fluid flow through the distal tubule is relatively high, the macula densa cells release paracrines, which inhibit renin release. When fluid flow in the distal tubule decreases, macula densa cells signal the granular cells to secrete renin.

Sodium reabsorption does not directly raise low blood pressure, but retention of Na^+ increases osmolarity, which stimulates thirst. Fluid intake when the person drinks more water increases ECF volume (see Fig. 20.8). When blood volume increases, blood pressure also increases.

The effects of the RAS pathway are not limited to aldosterone release, however. Angiotensin II is a remarkable hormone with additional effects directed at raising blood pressure. These actions make ANG II an important hormone in its own right, not merely an intermediate step in the aldosterone control pathway.

ANG II Has Many Effects

Angiotensin II has significant effects on fluid balance and blood pressure beyond stimulating aldosterone secretion, underscoring the integrated functions of the renal and cardiovascular systems. ANG II increases blood pressure both directly and indirectly through four additional pathways (Fig. 20.10):

1. *ANG II increases vasopressin secretion.* ANG II receptors in the hypothalamus initiate this reflex. Fluid retention in the kidney under the influence of vasopressin helps conserve blood volume, thereby maintaining blood pressure.
2. *ANG II stimulates thirst.* Fluid ingestion is a behavioral response that expands blood volume and raises blood pressure.
3. *ANG II is one of the most potent vasoconstrictors* known in humans. Vasoconstriction causes blood pressure to increase without a change in blood volume.
4. *Activation of ANG II receptors in the cardiovascular control center increases sympathetic output to the heart and blood vessels.* Sympathetic stimulation increases cardiac output and vasoconstriction, both of which increase blood pressure.
5. *ANG II increases proximal tubule Na^+ reabsorption.* ANG II stimulates an apical transporter, the Na^+-H^+ exchanger (NHE). Sodium reabsorption in the proximal tubule is followed by water reabsorption, so the net effect is reabsorption of isosmotic fluid, conserving volume.

The Renin-angiotensin System (RAS)

This map outlines the control of aldosterone secretion as well as the blood pressure–raising effects of ANG II. The pathway begins when decreased blood pressure stimulates renin secretion.

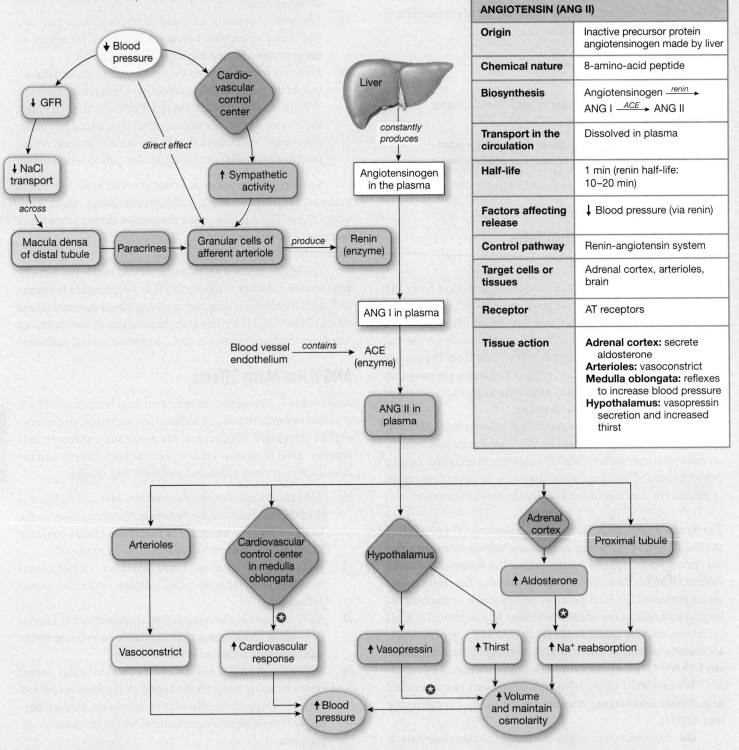

ANGIOTENSIN (ANG II)	
Origin	Inactive precursor protein angiotensinogen made by liver
Chemical nature	8-amino-acid peptide
Biosynthesis	Angiotensinogen $\xrightarrow{renin}$ ANG I $\xrightarrow{ACE}$ ANG II
Transport in the circulation	Dissolved in plasma
Half-life	1 min (renin half-life: 10–20 min)
Factors affecting release	↓ Blood pressure (via renin)
Control pathway	Renin-angiotensin system
Target cells or tissues	Adrenal cortex, arterioles, brain
Receptor	AT receptors
Tissue action	**Adrenal cortex:** secrete aldosterone **Arterioles:** vasoconstrict **Medulla oblongata:** reflexes to increase blood pressure **Hypothalamus:** vasopressin secretion and increased thirst

↓ Blood pressure

↓ GFR

Cardio-vascular control center

direct effect

↓ NaCl transport

across

↑ Sympathetic activity

Macula densa of distal tubule

Paracrines

Granular cells of afferent arteriole

produce

Renin (enzyme)

Liver

constantly produces

Angiotensinogen in the plasma

ANG I in plasma

Blood vessel endothelium

contains

ACE (enzyme)

ANG II in plasma

Arterioles

Cardiovascular control center in medulla oblongata

Hypothalamus

Adrenal cortex

Proximal tubule

↑ Aldosterone

Vasoconstrict

↑ Cardiovascular response

↑ Vasopressin

↑ Thirst

↑ Na^+ reabsorption

↑ Blood pressure

↑ Volume and maintain osmolarity

FIGURE QUESTION

Add efferent pathways and/or targets to the pathways marked with a ✪.

Once these blood pressure–raising effects of ANG II became known, it was not surprising that pharmaceutical companies started looking for drugs to block ANG II. Their research produced a new class of antihypertensive drugs called *ACE inhibitors*. These drugs block the ACE-mediated conversion of ANG I to ANG II, thereby helping to relax blood vessels and lower blood pressure. Less ANG II also means less aldosterone release, a decrease in Na^+ reabsorption and, ultimately, a decrease in ECF volume. All these responses contribute to lowering blood pressure.

However, the ACE inhibitors have side effects in some patients. ACE inactivates a cytokine called *bradykinin*. When ACE is inhibited by drugs, bradykinin levels increase, and in some patients this creates a dry, hacking cough. One solution was the development of drugs called *angiotensin receptor blockers (ARBs)*, which block the blood pressure–raising effects of ANG II at target cells by binding to *AT_1 receptors*. Recently another new class of drugs, *direct renin inhibitors*, was approved. Direct renin inhibitors decrease the plasma activity of renin, which in turn blocks production of ANG I and inhibits the entire RAS pathway.

Concept Check Answers: p. 694

10. A man comes to the doctor with high blood pressure. Tests show that he also has elevated plasma renin levels and atherosclerotic plaques that have nearly blocked blood flow through his renal arteries. How does decreased blood flow in his renal arteries cause elevated renin levels?
11. Map the pathways through which elevated renin causes high blood pressure in the man mentioned in Concept Check 10.
12. Why is it more effective to put ACE in the pulmonary vasculature than in the systemic vasculature?

ANP Promotes Na⁺ and Water Excretion

Once it was known that aldosterone and vasopressin increase Na^+ and water reabsorption, scientists speculated that other hormones might cause Na^+ loss, or **natriuresis** {*natrium*, sodium + *ourein*, to urinate} and water loss (diuresis) in the urine. If found, these hormones might be used clinically to lower blood volume and blood pressure in patients with essential hypertension [p. 536]. During years of searching, however, evidence for the other hormones was not forthcoming.

Then, in 1981, a group of Canadian researchers found that injections of homogenized rat atria caused rapid but short-lived excretion of Na^+ and water in the rats' urine. They hoped they had found the missing hormone, one whose activity would complement that of aldosterone and vasopressin. As it turned out, they had discovered the first *natriuretic peptide* (NP), one member of a family of hormones that appear to be endogenous RAS antagonists (■ Fig. 20.11).

Atrial natriuretic peptide (ANP; also known as *atriopeptin*) is a peptide hormone produced in specialized myocardial cells primarily in the atria of the heart. ANP is synthesized as part of a large prohormone that is cleaved into several active hormone fragments [p. 212]. A related hormone, **brain natriuretic peptide** (BNP), is synthesized by ventricular myocardial cells and certain brain neurons. Both natriuretic peptides are released by the heart when myocardial cells stretch more than normal. The natriuretic peptides bind to membrane receptor-enzymes that work through a cGMP second messenger system.

ANP is the more important signal molecule in normal physiology. ANP is released when increased blood volume causes increased atrial stretch. At the systemic level, ANP enhances Na^+ and water excretion to decrease blood volume. ANP acts at multiple sites. In the kidney it increases GFR by dilating the afferent arterioles, and it directly decreases Na^+ reabsorption in the collecting duct.

Natriuretic peptides also act indirectly to increase Na^+ and water excretion by suppressing the release of renin, aldosterone, and vasopressin (Fig. 20.11), actions that reinforce the natriuretic-diuretic effect. In addition, natriuretic peptides act directly on the cardiovascular control center of the medulla to lower blood pressure.

Brain natriuretic peptide (BNP) is now recognized as an important biological marker for heart failure because production of this substance increases with ventricular dilation and increased ventricular pressure. Hospital emergency departments now use BNP levels to distinguish dyspnea (difficulty breathing) in heart failure from other causes. BNP levels are also used as an independent predictor of heart failure and sudden death from cardiac arrhythmias.

RUNNING PROBLEM

The medical staff analyzed Lauren's blood for electrolyte concentrations. Her serum Na^+ concentration was 124 mEq/L. The normal range is 135–145 mEq/L. Lauren's diagnosis was **hyponatremia** {*hypo-*, below + *natri-*, sodium + *-emia*, blood}, defined as a serum Na^+ concentration below 135 mEq/L. Hyponatremia induced by the consumption of large quantities of low-sodium or sodium-free fluid, which is what happened in Lauren's case, is sometimes called *dilutional hyponatremia*.

Q3: Which body fluid compartment is being diluted in dilutional hyponatremia?

Q4: One way to estimate body osmolarity is to double the plasma Na^+ concentration. Estimate Lauren's osmolarity and explain what effect the dilutional hyponatremia has on her cells.

Q5: In dilutional hyponatremia, the medical personnel are most concerned about which organ or tissue?

658 660 673 676 683 690

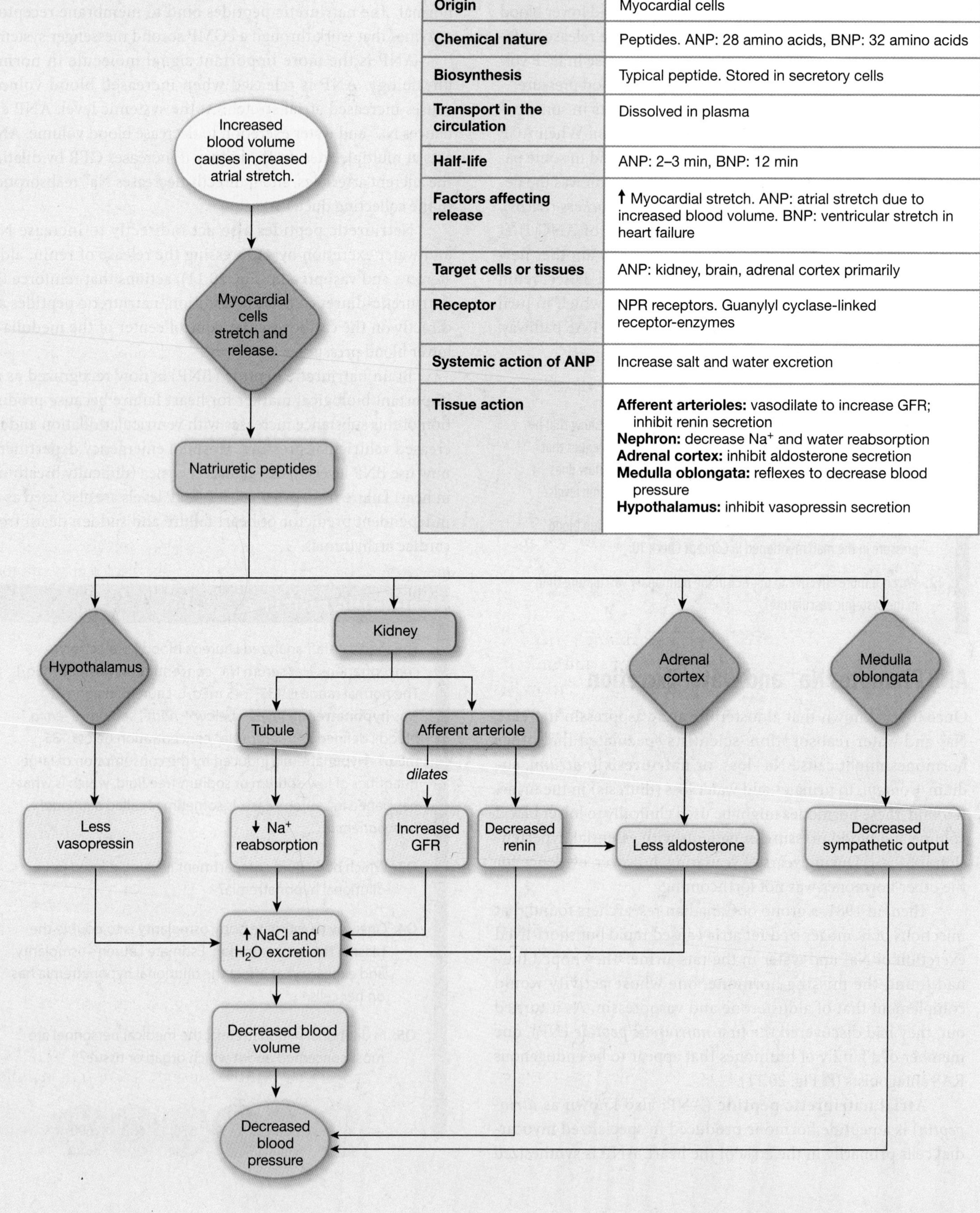

Origin	Myocardial cells
Chemical nature	Peptides. ANP: 28 amino acids, BNP: 32 amino acids
Biosynthesis	Typical peptide. Stored in secretory cells
Transport in the circulation	Dissolved in plasma
Half-life	ANP: 2–3 min, BNP: 12 min
Factors affecting release	↑ Myocardial stretch. ANP: atrial stretch due to increased blood volume. BNP: ventricular stretch in heart failure
Target cells or tissues	ANP: kidney, brain, adrenal cortex primarily
Receptor	NPR receptors. Guanylyl cyclase-linked receptor-enzymes
Systemic action of ANP	Increase salt and water excretion
Tissue action	**Afferent arterioles:** vasodilate to increase GFR; inhibit renin secretion **Nephron:** decrease Na^+ and water reabsorption **Adrenal cortex:** inhibit aldosterone secretion **Medulla oblongata:** reflexes to decrease blood pressure **Hypothalamus:** inhibit vasopressin secretion

Potassium Balance

Aldosterone (but not other factors in the RAS pathway) also plays a critical role in potassium homeostasis. Only about 2% of the body's K^+ load is in the ECF, but regulatory mechanisms keep plasma K^+ concentrations within a narrow range (3.5–5 meq/L). Under normal conditions, mass balance matches K^+ excretion to K^+ ingestion. If intake exceeds excretion and plasma K^+ goes up, aldosterone is released into the blood through the direct effect of hyperkalemia on the adrenal cortex. Aldosterone acting on distal-nephron P cells keeps the cells' apical ion channels open longer and speeds up the Na^+-K^+-ATPase pump, enhancing renal excretion of K^+.

The regulation of body potassium levels is essential to maintaining a state of well-being. Changes in extracellular K^+ concentration affect the resting membrane potential of all cells [Fig. 8.17, p. 265]. If plasma (and ECF) K^+ concentrations decrease (*hypokalemia*), the concentration gradient between the cell and the ECF becomes larger, more K^+ leaves the cell, and the resting membrane potential becomes more negative. If ECF K^+ concentrations increase (*hyperkalemia*), the concentration gradient decreases and more K^+ remains in the cell, depolarizing it. (Remember that when plasma K^+ concentrations change, anions such as Cl^- are also added to or subtracted from the ECF in a 1:1 ratio, maintaining overall electrical neutrality.)

Because of the effect of plasma K^+ on excitable tissues, such as the heart, clinicians are always concerned about keeping plasma K^+ within its normal range. If K^+ falls below 3 meq/L or rises above 6 meq/L, the excitable tissues of muscle and nerve begin to show altered function. For example, hypokalemia causes muscle weakness because it is more difficult for hyperpolarized neurons and muscles to fire action potentials. The danger in this condition lies in the failure of respiratory muscles and the heart. Fortunately, skeletal muscle weakness is usually significant enough to lead patients to seek treatment before cardiac problems occur. Mild hypokalemia may be corrected by oral intake of K^+ supplements and K^+-rich foods, such as orange juice and bananas.

Hyperkalemia is a more dangerous potassium disturbance because in this case depolarization of excitable tissues makes them more excitable initially. Subsequently, the cells are unable to repolarize fully and actually become *less* excitable. In this state, they have action potentials that are either smaller than normal or nonexistent. Cardiac muscle excitability affected by changes in plasma K^+ can lead to life-threatening cardiac arrhythmias.

Disturbances in K^+ balance may result from kidney disease, eating disorders, loss of K^+ in diarrhea, or the use of certain types of diuretics that prevent the kidneys from fully reabsorbing K^+. Inappropriate correction of dehydration can also create K^+ imbalance. Consider a golfer playing a round of golf when the temperature was above 100 °F. He was aware of the risk of dehydration, so he drank lots of water to replace fluid lost through sweating. The replacement of lost sweat with pure water kept his ECF volume normal but dropped his total blood osmolarity and his K^+ and Na^+ concentrations. He was unable to finish the round of golf because of muscle weakness, and he required medical attention that included ion replacement therapy. A more suitable replacement fluid would have been one of the sports drinks that include salt and K^+.

Potassium balance is also closely tied to acid-base balance, as you will learn in the final section of this chapter. Correction of a pH disturbance requires close attention to plasma K^+ levels. Similarly, correction of K^+ imbalance may alter body pH.

Behavioral Mechanisms in Salt and Water Balance

Although neural, neuroendocrine, and endocrine reflexes play key roles in salt and water homeostasis, behavioral responses are critical in restoring the normal state, especially when ECF volume decreases or osmolarity increases. Drinking water is normally the only way to restore lost water, and eating salt is the only way to raise the body's Na^+ content. Both behaviors are essential for normal salt and water balance. Clinicians must recognize the absence of these behaviors in patients who are unconscious or otherwise unable to obey behavioral urges, and must adjust treatment accordingly. The study of the biological basis for behaviors, including drinking and eating, is a field known as *physiological psychology*.

20

Drinking Replaces Fluid Loss

Thirst is one of the most powerful urges known in humans. In 1952, the Swedish physiologist Bengt Andersson showed that stimulating certain regions of the hypothalamus triggered drinking behavior. This discovery led to the identification of hypothalamic osmoreceptors that initiate drinking when body osmolarity rises above 280 mOsM. This is an example of a behavior initiated by an internal stimulus.

It is interesting to note that although increased osmolarity triggers thirst, the act of drinking is sufficient to relieve thirst. The ingested water need not be absorbed in order for thirst to be quenched. As-yet-unidentified receptors in the mouth and pharynx (*oropharynx receptors*) respond to cold water by decreasing thirst and decreasing vasopressin release even though plasma osmolarity remains high. This oropharynx reflex is one reason surgery patients are allowed to suck on ice chips: the ice alleviates their thirst without putting significant amounts of fluid into the digestive system.

A similar reflex exists in camels. When led to water, they drink just enough to replenish their water deficit. Oropharynx receptors apparently act as a feedforward "metering" system that helps prevent wide swings in osmolarity by matching water intake to water need.

In humans, cultural rituals complicate the thirst reflex. For example, we may drink during social events, whether or not we are thirsty. As a result, our bodies must be capable of eliminating fluid ingested in excess of our physiological needs.

Concept Check Answer: p. 694

13. Incorporate the thirst reflex into Figure 20.8.

RUNNING PROBLEM

During exercise in the heat, sweating rate and sweat composition are quite variable among athletes and depend partly on how acclimatized the individual is to the heat. Sweat fluid losses can range from less than 0.6 L/h to more than 2.5 L/h, and sweat Na^+ concentrations can range from less than 20 mEq/L to more than 90 mEq/L. The white salt crystals noted on Lauren's face and clothing suggest that she is a "salty sweater" who probably lost a large amount of salt during the race. Follow-up testing revealed that Lauren's sweat Na^+ concentration was 70 mEq/L.

Q6: Assuming a sweating rate of 1 L/hr, how much Na^+ did Lauren lose during the 16-hour race?

Q7: Total body water for a 60-kg female is approximately 30 L, and her ECF volume is 10 L. Based on the information given in the problem so far, calculate how much fluid Lauren probably ingested during the race.

Low Na^+ Stimulates Salt Appetite

Thirst is not the only urge associated with fluid balance. **Salt appetite** is a craving for salty foods that occurs when plasma Na^+ concentrations drop. It can be observed in deer and cattle attracted to salt blocks or naturally occurring salt licks. In humans, salt appetite is linked to aldosterone and angiotensin, hormones that regulate Na^+ balance. The centers for salt appetite are in the hypothalamus close to the center for thirst.

Avoidance Behaviors Help Prevent Dehydration

Other behaviors play a role in fluid balance by preventing or promoting dehydration. Desert animals avoid the heat of the day and become active only at night, when environmental temperatures fall and humidity rises. Humans, especially now that we have air conditioning, are not always so wise.

The midday nap, or *siesta,* is a cultural adaptation in tropical countries that keeps people indoors during the hottest part of the day, thereby helping prevent dehydration and overheating. In the United States, we have abandoned this civilized custom and are active continuously during daylight hours, even when the temperature soars during summer in the South and Southwest. Fortunately, our homeostatic mechanisms usually keep us out of trouble.

Integrated Control of Volume and Osmolarity

The body uses an integrated response to correct disruptions of salt and water balance. The cardiovascular system responds to changes in blood volume, and the kidneys respond to changes in blood volume or osmolarity. Maintaining homeostasis throughout the day is a continuous process in which the amounts of salt and water in the body shift, according to whether you just drank a soft drink or sweated through an aerobics class.

In that respect, maintaining fluid balance is like driving a car down the highway and making small adjustments to keep the car in the center of the lane. However, just as exciting movies feature wild car chases, not sedate driving, the exciting part of fluid homeostasis is the body's response to crisis situations, such as severe dehydration or hemorrhage. In this section we examine challenges to salt and water balance.

Osmolarity and Volume Can Change Independently

Normally, volume and osmolarity are homeostatically maintained within an acceptable range. Under some circumstances, however, fluid loss exceeds fluid gain or vice versa, and the body goes out of balance. Common pathways for fluid loss include excessive sweating, vomiting, diarrhea, and hemorrhage. All these situations may require medical intervention. In contrast, fluid gain is seldom a medical emergency, unless it is addition of water that decreases osmolarity below an acceptable range.

Volume and osmolarity of the ECF can each have three possible states: normal, increased, or decreased. The relation of volume and osmolarity changes can be represented by the matrix in ■ Figure 20.12. The center box represents the normal state, and the surrounding boxes represent the most common examples of the variations from normal.

In all cases, the appropriate homeostatic compensation for the change acts according to the principle of mass balance: whatever fluid and solute were added to the body must be removed,

DISTURBANCES IN VOLUME AND OSMOLARITY

Volume \ Osmolarity	Decrease	No change	Increase
Increase	Drinking large amount of water	Ingestion of isotonic saline	Ingestion of hypertonic saline
No change	Replacement of sweat loss with plain water	Normal volume and osmolarity	Eating salt without drinking water
Decrease	Incomplete compensation for dehydration	Hemorrhage	Dehydration (e.g., sweat loss or diarrhea)

■ **Fig. 20.12**

or whatever was lost must be replaced. However, perfect compensation is not always possible. Let's begin at the upper right corner of Figure 20.12 and move right to left across each row.

1 *Increased volume, increased osmolarity.* A state of increased volume and increased osmolarity might occur if you ate salty food and drank liquids at the same time, such as popcorn and a soft drink at the movies. The net result could be ingestion of hypertonic saline that increases ECF volume and osmolarity. The appropriate homeostatic response is excretion of hypertonic urine. For homeostasis to be maintained, the osmolarity and volume of the urinary output must match the salt and water input from the popcorn and soft drink.

2 *Increased volume, no change in osmolarity.* Moving one cell to the left in the top row, we see that if the proportion of salt and water in ingested food is equivalent to an isotonic NaCl solution, volume increases but osmolarity does not change. The appropriate response is excretion of isotonic urine whose volume equals that of the ingested fluid.

3 *Increased volume, decreased osmolarity.* This situation would occur if you drank pure water without ingesting any solute. The goal here would be to excrete very dilute urine to maximize water loss while conserving salts. However, because our kidneys cannot excrete pure water, there is always some loss of solute in the urine. In this situation, urinary output cannot exactly match input, and so compensation is imperfect.

4 *No change in volume, increased osmolarity.* This disturbance (middle row, right cell) might occur if you ate salted popcorn without drinking anything. The ingestion of salt without water increases ECF osmolarity and causes some water to shift from cells to the ECF. The homeostatic response is intense thirst, which prompts ingestion of water to dilute the added solute. The kidneys help by creating highly concentrated urine of minimal volume, conserving water while removing excess NaCl. Once water is ingested, the disturbance becomes that described in situation 1 or situation 2.

5 *No change in volume, decreased osmolarity.* This scenario (middle row, left cell) might occur when a person who is dehydrated replaces lost fluid with pure water, like the golfer described earlier. The decreased volume resulting from the dehydration is corrected, but the replacement fluid has no solutes to replace those lost. Consequently, a new imbalance is created.

This situation led to the development of electrolyte-containing sports beverages. If people working out in hot weather replace lost sweat with pure water, they may restore volume but run the risk of diluting plasma K^+ and Na^+ concentrations to dangerously low levels (*hypokalemia* and *hyponatremia*, respectively).

6 *Decreased volume, increased osmolarity.* Dehydration is a common cause of this disturbance (bottom row, right cell). Dehydration has multiple causes. During prolonged heavy exercise, water loss from the lungs can double while sweat loss may increase from 0.1 liter to as much as 5 liters! Because the fluid secreted by sweat glands is hyposmotic, the fluid left behind in the body becomes hyperosmotic.

Diarrhea {*diarhein,* to flow through}, excessively watery feces, is a pathological condition involving major water and solute loss, this time from the digestive tract. In both sweating and diarrhea, if too much fluid is lost from the circulatory system, blood volume decreases to the point that the heart can no longer pump blood effectively to the brain. In addition, cell shrinkage caused by increased osmolarity disrupts cell function.

7 *Decreased volume, no change in osmolarity.* This situation (bottom row, middle cell) occurs with hemorrhage. Blood loss represents the loss of isosmotic fluid from the extracellular compartment, similar to scooping a cup of seawater out of a large bucketful. If a blood transfusion is not immediately available, the best replacement solution is one that is isosmotic and remains in the ECF, such as isotonic NaCl.

8 *Decreased volume, decreased osmolarity.* This situation (bottom row, left cell) might also result from incomplete compensation of dehydration, but it is uncommon.

Dehydration Triggers Homeostatic Responses

To understand the body's integrated response to changes in volume and osmolarity, you must first have a clear idea of which pathways become active in response to various stimuli. ■ Table 20.1

20

Table 20.1 Responses Triggered by Changes in Volume, Blood Pressure, and Osmolarity

Stimulus	Organ or Tissue Involved	Response(s)	Figure(s)
DECREASED BLOOD PRESSURE/VOLUME			
Direct effects			
	Granular cells	Renin secretion	20.10
	Glomerulus	Decreased GFR	19.6, 20.10
Reflexes			
Carotid and aortic baroreceptors	Cardiovascular control center	Increased sympathetic output, decreased parasympathetic output	15.14b, 20.10
Carotid and aortic baroreceptors	Hypothalamus	Thirst stimulation	20.1a
Carotid and aortic baroreceptors	Hypothalamus	Vasopressin secretion	20.6
Atrial volume receptors	Hypothalamus	Thirst stimulation	20.1a
Atrial volume receptors	Hypothalamus	Vasopressin secretion	20.6
INCREASED BLOOD PRESSURE			
Direct effects			
	Glomerulus	Increased GFR (transient)	19.6, 19.7
	Myocardial cells	Natriuretic peptide secretion	20.11
Reflexes			
Carotid and aortic baroreceptors	Cardiovascular control center	Decreased sympathetic output, increased parasympathetic output	15.14b
Carotid and aortic baroreceptors	Hypothalamus	Thirst inhibition	
Carotid and aortic baroreceptors	Hypothalamus	Vasopressin inhibition	
Atrial volume receptors	Hypothalamus	Thirst inhibition	
Atrial volume receptors	Hypothalamus	Vasopressin inhibition	
INCREASED OSMOLARITY			
Direct effects			
Pathological dehydration	Adrenal cortex	Decreased aldosterone secretion	20.13

Table 20.1 Responses Triggered by Changes in Volume, Blood Pressure, and Osmolarity (Continued)

Stimulus	Organ or Tissue Involved	Response(s)	Figure(s)
Reflexes			
Osmoreceptors	Hypothalamus	Thirst stimulation	20.8
Osmoreceptors	Hypothalamus	Vasopressin secretion	20.6
DECREASED OSMOLARITY			
Direct effects			
Pathological hyponatremia	Adrenal cortex	Increased aldosterone secretion	
Reflexes			
Osmoreceptors	Hypothalamus	Decreased vasopressin secretion	

is a summary of the many pathways involved in the homeostasis of salt and water balance. For details of individual pathways, refer to the figures cited in Table 20.1.

The homeostatic response to severe dehydration is an excellent example of how the body works to maintain blood volume and cell volume in the face of decreased volume and increased osmolarity. It also illustrates the role of neural and endocrine integrating centers. In severe dehydration, the adrenal cortex receives two opposing signals. One says, "Secrete aldosterone"; the other says, "Do not secrete aldosterone." The body has multiple mechanisms for dealing with diminished blood volume, but high ECF osmolarity causes cells to shrink and presents a more immediate threat to well-being. Thus, faced with decreased volume and increased osmolarity, the adrenal cortex does not secrete aldosterone. (If secreted, aldosterone would cause Na^+ reabsorption, which could worsen the already-high osmolarity associated with dehydration.)

In severe dehydration, compensatory mechanisms are aimed at restoring normal blood pressure, ECF volume, and osmolarity by (1) conserving fluid to prevent additional loss, (2) triggering cardiovascular reflexes to increase blood pressure, and (3) stimulating thirst so that normal fluid volume and osmolarity can be restored. ■ Figure 20.13 maps the interwoven nature of these responses. This figure is complex and intimidating at first glance, so let's discuss it step by step.

At the top of the map (in yellow) are the two stimuli caused by dehydration: decreased blood volume/pressure, and increased osmolarity. Decreased ECF volume causes decreased blood pressure. Blood pressure acts both directly and as a stimulus for several reflex pathways that are mediated through the carotid and aortic baroreceptors and the pressure-sensitive granular cells. Decreased volume is sensed by the atrial volume receptors.

1. *The carotid and aortic baroreceptors signal the cardiovascular control center (CVCC) to raise blood pressure.* Sympathetic output from the CVCC increases while parasympathetic output decreases.
   (a) Heart rate goes up as control of the SA node shifts from predominantly parasympathetic to sympathetic.
   (b) The force of ventricular contraction also increases under sympathetic stimulation. The increased force of contraction combines with increased heart rate to increase cardiac output.
   (c) Simultaneously, sympathetic input causes arteriolar vasoconstriction, increasing peripheral resistance.
   (d) Sympathetic vasoconstriction of afferent arterioles in the kidneys decreases GFR, helping conserve fluid.
   (e) Increased sympathetic activity at the granular cells of the kidneys increases renin secretion.
2. *Decreased peripheral blood pressure directly decreases GFR.* A lower GFR conserves ECF volume by filtering less fluid into the nephron.
3. *Paracrine feedback causes the granular cells to release renin.* Lower GFR decreases fluid flow past the macula densa. This triggers renin release.
4. *Granular cells respond to decreased blood pressure by releasing renin.* The combination of decreased blood pressure, increased sympathetic input onto granular cells, and

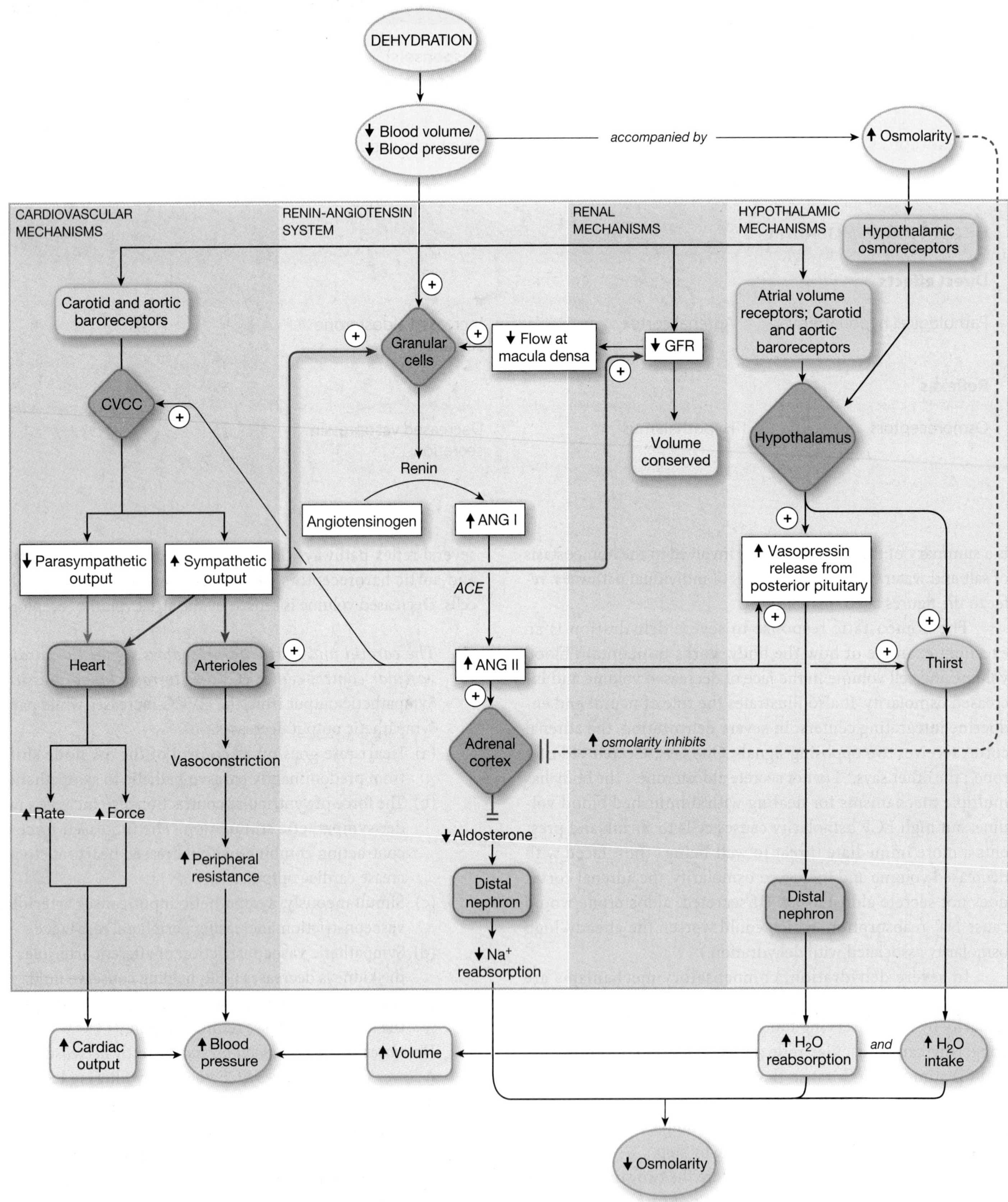
HOMEOSTATIC COMPENSATION FOR SEVERE DEHYDRATION
DEHYDRATION
↓ Blood volume/ ↓ Blood pressure
accompanied by
↑ Osmolarity
CARDIOVASCULAR MECHANISMS
RENIN-ANGIOTENSIN SYSTEM
RENAL MECHANISMS
HYPOTHALAMIC MECHANISMS
Hypothalamic osmoreceptors
Carotid and aortic baroreceptors
Granular cells
↓ Flow at macula densa
↓ GFR
Atrial volume receptors; Carotid and aortic baroreceptors
CVCC
Volume conserved
Hypothalamus
Renin
Angiotensinogen
↑ ANG I
↓ Parasympathetic output
↑ Sympathetic output
ACE
↑ Vasopressin release from posterior pituitary
Heart
Arterioles
↑ ANG II
Thirst
Adrenal cortex
↑ osmolarity inhibits
Vasoconstriction
↑ Rate
↑ Force
↓ Aldosterone
↑ Peripheral resistance
Distal nephron
Distal nephron
↓ Na^+ reabsorption
↑ Cardiac output
↑ Blood pressure
↑ Volume
↑ H_2O reabsorption
and
↑ H_2O intake
↓ Osmolarity

■ **Fig. 20.13**

signals from the macula densa stimulates renin release and ensures increased production of ANG II.

5 *Decreased blood pressure, decreased blood volume, increased osmolarity, and increased ANG II production all stimulate vasopressin and the thirst centers of the hypothalamus.*

The redundancy in the control pathways ensures that all four main compensatory mechanisms are activated: cardiovascular responses, ANG II, vasopressin, and thirst.

1 *Cardiovascular responses* combine increased cardiac output and increased peripheral resistance to raise blood pressure. Note, however, that this increase in blood pressure does *not necessarily* mean that blood pressure returns to normal. If dehydration is severe, compensation may be incomplete, and blood pressure may remain below normal.

2 *Angiotensin II* has a variety of effects aimed at raising blood pressure, including stimulation of thirst, vasopressin release, direct vasoconstriction, and reinforcement of cardiovascular control center output. ANG II also reaches the adrenal cortex and attempts to stimulate aldosterone release. In dehydration, however, Na^+ reabsorption worsens the already high osmolarity. Consequently, high osmolarity at the adrenal cortex directly inhibits aldosterone release, blocking the action of ANG II. The RAS pathway in dehydration produces the beneficial blood pressure–enhancing effects of ANG II while avoiding the detrimental effects of Na^+ reabsorption. This is a beautiful example of integrated function.

3 *Vasopressin* increases the water permeability of the renal collecting ducts, allowing water reabsorption to conserve fluid. Without fluid replacement, however, vasopressin cannot bring volume and osmolarity back to normal.

4 *Oral (or intravenous) intake of water* in response to thirst is the only mechanism for replacing lost fluid volume and for restoring ECF osmolarity to normal.

The net result of all four mechanisms is (1) restoration of volume by water conservation and fluid intake, (2) maintenance of blood pressure through increased blood volume, increased cardiac output, and vasoconstriction, and (3) restoration of normal osmolarity by decreased Na^+ reabsorption and increased water reabsorption and intake.

Using the pathways listed in Table 20.1 and Figure 20.13 as a model, try to create reflex maps for the seven other disturbances of volume and osmolarity shown in Figure 20.12.

Acid-Base Balance

Acid-base balance (also called pH homeostasis) is one of the essential functions of the body. The pH of a solution is a measure of its H^+ concentration [p. 43]. The H^+ concentration of normal arterial plasma sample is 0.00004 meq/L, minute compared with the concentrations of other ions. (For example, the plasma concentration of Na^+ is about 135 meq/L.)

Because the body's H^+ concentration is so low, it is commonly expressed on a logarithmic pH scale of 0–14, in which a pH of 7.0 is neutral (neither acidic nor basic). If the pH of a solution is below 7.0, the solution has an H^+ concentration greater than 1×10^{-7} M and is considered acidic. If the pH is above 7.0, the solution has an H^+ concentration lower than 1×10^{-7} M and is considered alkaline (basic).

The normal pH of the body is 7.40, slightly alkaline. A change of 1 pH unit represents a 10-fold change in H^+ concentration. To review the concept of pH and the logarithmic scale on which it is based, see Appendix B.

pH Changes Can Denature Proteins

The normal pH range of plasma is 7.38–7.42. Extracellular pH usually reflects intracellular pH, and vice versa. Because monitoring intracellular conditions is difficult, plasma values are used clinically as an indicator of ECF and whole body pH. Body fluids that are "outside" the body's internal environment, such as those in the lumen of the gastrointestinal tract or kidney tubule, can have a pH that far exceeds the normal range. Acidic secretions in the stomach, for instance, may create a gastric pH as low as 1, and the pH of urine varies between 4.5 and 8.5, depending on the body's need to excrete H^+ or HCO_3^-.

The concentration of H^+ in the body is closely regulated. Intracellular proteins, such as enzymes and membrane channels, are particularly sensitive to pH because the function of these proteins depends on their three-dimensional shape. Changes in H^+ concentration alter the tertiary structure of protei
teracting with hydrogen bonds in the molecules, disru
proteins' three-dimensional structures and activities [p

Abnormal pH may significantly affect the activi
nervous system. If pH is too low—the condition k
acidosis—neurons become less excitable, and CNS de
results. Patients become confused and disoriented, then
a coma. If CNS depression progresses, the respirator
cease to function, causing death.

If pH is too high—the condition known as **alk**
neurons become hyperexcitable, firing action potentia
slightest signal. This condition shows up first as sensory changes, such as numbness or tingling, then as muscle twitches. If alkalosis is severe, muscle twitches turn into sustained contractions (*tetanus*) that paralyze respiratory muscles.

Disturbances of acid-base balance are associated with disturbances in K^+ balance. This is partly due to a renal transporter that moves K^+ and H^+ ions in an antiport fashion. In acidosis, the kidneys excrete H^+ and reabsorb K^+ using an *H^+-K^+-ATPase*. In alkalosis, the kidneys reabsorb H^+ and excrete K^+. Potassium imbalance usually shows up as disturbances in excitable tissues, especially the heart.

Acids and Bases in the Body Come from Many Sources

In day-to-day functioning, the body is challenged by intake and production of acids more than bases. Hydrogen ions come both from food and from internal metabolism. Maintaining mass balance requires that acid intake and production be balanced by acid excretion. Hydrogen balance in the body is summarized in ■ Figure 20.14.

Acid Input Many metabolic intermediates and foods are organic acids that ionize and contribute H^+ to body fluids.* Examples of organic acids include amino acids, fatty acids, intermediates in the citric acid cycle, and lactate produced by anaerobic metabolism. Metabolic production of organic acids each day generates a significant amount of H^+ that must be excreted to maintain mass balance.

Under extraordinary circumstances, metabolic organic acid production can increase significantly and create a crisis. For example, severe anaerobic conditions, such as circulatory collapse, produce so much lactate that normal homeostatic mechanisms cannot keep pace, resulting in a state of *lactic acidosis.* In diabetes mellitus, abnormal metabolism of fats and amino acids creates strong acids known as **ketoacids**. These acids cause a state of metabolic acidosis known as *ketoacidosis.*

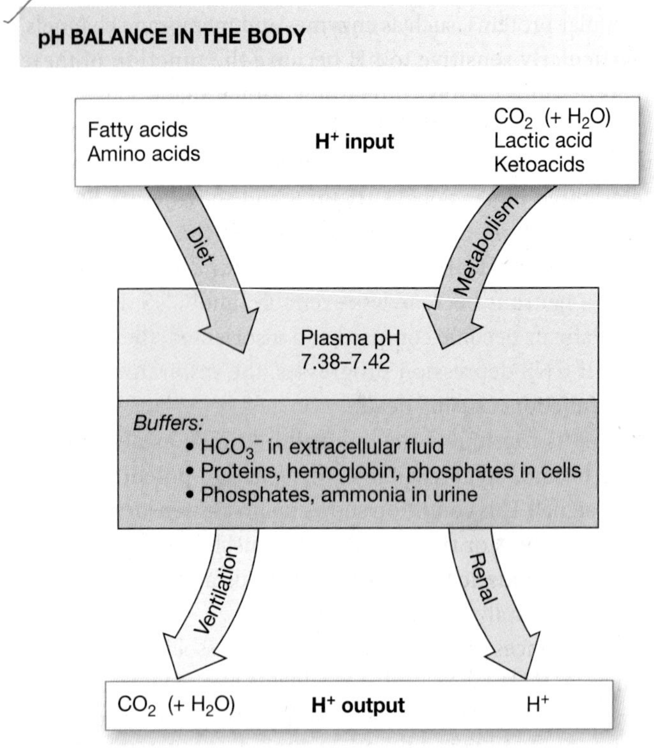

pH BALANCE IN THE BODY

Fatty acids
Amino acids

H^+ input

CO_2 (+ H_2O)
Lactic acid
Ketoacids

Diet

Metabolism

Plasma pH
7.38–7.42

Buffers:
- HCO_3^- in extracellular fluid
- Proteins, hemoglobin, phosphates in cells
- Phosphates, ammonia in urine

Ventilation

Renal

CO_2 (+ H_2O)

H^+ output

H^+

■ Fig. 20.14

*The anion forms of many organic acids end with the suffix *–ate,* such as pyruvate and lactate.

The biggest source of acid on a daily basis is the production of CO_2 during aerobic respiration. Carbon dioxide is not an acid because it does not contain any hydrogen atoms. However, CO_2 from respiration combines with water to form carbonic acid (H_2CO_3), which dissociates into H^+ and bicarbonate ion, HCO_3^-.

$$CO_2 + H_2O \rightleftharpoons H_2CO_3 \rightleftharpoons H^+ + HCO_3^-$$

This reaction takes place in all cells and in the plasma, but at a slow rate. However, in certain cells of the body, the reaction proceeds very rapidly because of the presence of large amounts of *carbonic anhydrase* [p. 613]. This enzyme catalyzes the conversion of CO_2 and H_2O to H^+ and HCO_3^-.

The production of H^+ from CO_2 and H_2O is the single biggest source of acid input under normal conditions. By some estimates, CO_2 from resting metabolism produces 12,500 meq of H^+ each day. If this amount of acid were placed in a volume of water equal to the plasma volume, it would create an H^+ concentration of 4167 meq/L, over one hundred million (10^8) times as concentrated as the normal plasma H^+ concentration of 0.00004 meq/L!

These numbers show that CO_2 from aerobic respiration has the potential to affect pH in the body dramatically. Fortunately, homeostatic mechanisms normally prevent CO_2 from accumulating in the body.

Base Input Acid-base physiology focuses on acids for good reasons. First, our diet and metabolism have few significant sources of bases. Some fruits and vegetables contain anions that metabolize to HCO_3^-, but the influence of these foods is far outweighed by the contribution of acidic fruits, amino acids, and fatty acids. Second, acid-base disturbances due to excess acid are more common than those due to excess base. For these reasons, the body expends far more resources removing excess acid.

pH Homeostasis Depends on Buffers, Lungs, and Kidneys

How does the body cope with minute-to-minute changes in pH? There are three mechanisms: (1) buffers, (2) ventilation, and (3) renal regulation of H^+ and HCO_3^-. Buffers are the first line of defense, always present and waiting to prevent wide swings in pH. Ventilation, the second line of defense, is a rapid, reflexively controlled response that can take care of 75% of most pH disturbances. The final line of defense lies with the kidneys. They are slower than buffers or the lungs but are very effective at coping with any remaining pH disturbance under normal conditions. Usually these three mechanisms help the body balance acid so effectively that normal body pH varies only slightly. Let's take a closer look at each of them.

RUNNING PROBLEM

The human body attempts to maintain fluid and sodium balance via several hormonal mechanisms. During exercise sessions, increased sympathetic output causes increased production of aldosterone and vasopressin, which promote the retention of Na^+ and water by the kidneys.

Q8: What would you expect to happen to vasopressin and aldosterone production in response to dilutional hyponatremia?

Buffer Systems Include Proteins, Phosphate Ions, and HCO_3^-

A buffer is a molecule that moderates but does not prevent changes in pH by combining with or releasing H^+ [p. 49]. In the absence of buffers, the addition of acid to a solution causes a sharp change in pH. In the presence of a buffer, the pH change is moderated or may even be unnoticeable. Because acid production is the major challenge to pH homeostasis, most physiological buffers combine with H^+.

Buffers are found both within cells and in the plasma. Intracellular buffers include cellular proteins, phosphate ions (HPO_4^{2-}), and hemoglobin. Hemoglobin in red blood cells buffers the H^+ produced by the reaction of CO_2 with H_2O [Fig. 18.11, p. 614].

Each H^+ ion buffered by hemoglobin leaves a matching bicarbonate ion inside the red blood cell. This HCO_3^- can then leave the red blood cell in exchange for plasma Cl^-, the *chloride shift* [p. 614].

The large amounts of plasma bicarbonate produced from metabolic CO_2 create the most important extracellular buffer system of the body. Plasma HCO_3^- concentration averages 24 meq/L, which is approximately 600,000 times as concentrated as plasma H^+. Although H^+ and HCO_3^- are created in a 1:1 ratio from CO_2 and H_2O, intracellular buffering of H^+ by hemoglobin is a major reason the two ions do not appear in the plasma in the same concentration. The HCO_3^- in plasma is then available to buffer H^+ from nonrespiratory sources, such as metabolism.

The relationship between CO_2, HCO_3^-, and H^+ in the plasma is expressed by the equation we just looked at:

$$CO_2 + H_2O \rightleftharpoons \underset{\text{carbonic acid}}{H_2CO_3} \rightleftharpoons H^+ + HCO_3^- \quad (1)$$

According to the law of mass action, any change in the amount of CO_2, H^+, or HCO_3^- in the reaction solution causes the reaction to shift until a new equilibrium is reached. (Water is always in excess in the body and does not contribute to the reaction equilibrium.) For example, if CO_2 increases (red), the equation shifts to the right, creating one additional H^+ and one additional HCO_3^- from each CO_2 and water:

$$\uparrow CO_2 + H_2O \rightarrow H_2CO_3 \rightarrow \uparrow H^+ + \uparrow HCO_3^- \quad (2)$$

Once a new equilibrium is reached, both H^+ and HCO_3^- levels have increased. The addition of H^+ makes the solution more acidic and therefore lowers its pH. In this reaction, it does not matter that a HCO_3^- buffer molecule has also been produced because HCO_3^- acts as a buffer only when it binds to H^+ and becomes carbonic acid.

Now suppose H^+ (red) is added to the plasma from some metabolic source, such as lactic acid:

$$CO_2 + H_2O \rightleftharpoons H_2CO_3 \rightleftharpoons \uparrow H^+ + HCO_3^- \quad (3)$$

In this case, plasma HCO_3^- *can* act as a buffer by combining with some of the added H^+ until the reaction reaches a new equilibrium state. The increase in H^+ shifts the equation to the left:

$$CO_2 + H_2O \leftarrow H_2CO_3 \leftarrow \uparrow H^+ + HCO_3^- \quad (4)$$

Converting some of the added H^+ and bicarbonate buffer to carbonic acid means that at equilibrium, H^+ is still elevated, but not as much as it was initially. The concentration of HCO_3^- is decreased because some has been used as a buffer. The buffered H^+ is converted to CO_2 and H_2O, increasing the amounts of both. At equilibrium, the reaction looks like this:

$$\uparrow CO_2 + \uparrow H_2O \rightleftharpoons H_2CO_3 \rightleftharpoons \uparrow H^+ + \downarrow HCO_3^- \quad (5)$$

The law of mass action is a useful way to think about the relationship between changes in the concentrations of H^+, HCO_3^-, and CO_2, as long as you remember certain qualifications. First, a change in HCO_3^- concentration (as indicated in reaction 5) may not show up clinically as a HCO_3^- concentration outside the normal range. This is because HCO_3^- is 600,000 times more concentrated in the plasma than H^+ is. If both H^+ and HCO_3^- are added to the plasma, you may observe changes in pH but not in HCO_3^- concentration because so much bicarbonate was present initially. Both H^+ and HCO_3^- experience an *absolute* increase in concentration, but because so many HCO_3^- were in the plasma to begin with, the *relative increase* in HCO_3^- goes unnoticed.

As an analogy, think of two football teams playing in a stadium packed with 80,000 fans. If 10 more players (H^+) run out onto the field, everyone notices. But if 10 people (HCO_3^-) come into the stands at the same time, no one pays any attention because there were already so many people watching the game that 10 more make no significant difference.

The relationship between pH, HCO_3^- concentration in mM, and dissolved CO_2 concentration is expressed mathematically by the **Henderson-Hasselbalch equation**. One variant of the equation that is more useful in clinical medicine uses P_{CO_2} instead of dissolved CO_2 concentration:

$$pH = 6.1 + \log[HCO_3^-]/0.03 \times P_{CO_2}$$

If you know a patient's P_{CO2} and plasma bicarbonate concentration, you can predict the plasma pH.

The second qualification for the law of mass action is that when the reaction shifts to the left and increases plasma CO_2, a nearly instantaneous increase in ventilation takes place in a normal person. If extra CO_2 is ventilated off, arterial P_{CO_2} may remain normal or even fall below normal as a result of hyperventilation.

Ventilation Can Compensate for pH Disturbances

The increase in ventilation just described is a *respiratory compensation* for acidosis. Ventilation and acid-base status are intimately linked, as shown by the equation

$$CO_2 + H_2O \rightleftharpoons H_2CO_3 \rightleftharpoons H^+ + HCO_3^-$$

Changes in ventilation can correct disturbances in acid-base balance, but they can also cause them. Because of the dynamic equilibrium between CO_2 and H^+, any change in plasma P_{CO_2} affects both H^+ and HCO_3^- content of the blood.

For example, if a person hypoventilates and P_{CO_2} increases (red), the equation shifts to the right. More carbonic acid is formed, and H^+ goes up, creating a more *acidotic* state:

$$\uparrow CO_2 + H_2O \rightarrow H_2CO_3 \rightarrow \uparrow H^+ + \uparrow HCO_3^- \qquad (6)$$

On the other hand, if a person hyperventilates, blowing off CO_2 and thereby decreasing the plasma P_{CO_2} (red), the equation shifts to the left, which means that H^+ combines with HCO_3^- and becomes carbonic acid, thereby decreasing the H^+ concentration. Lower H^+ means an increase in pH:

$$\downarrow CO_2 + H_2O \leftarrow H_2CO_3 \leftarrow \downarrow H^+ + \downarrow HCO_3^- \qquad (7)$$

In these two examples, you can see that a change in P_{CO_2} affects the H^+ concentration and therefore the pH of the plasma. The body uses ventilation as a method for adjusting pH only if a stimulus associated with pH triggers the reflex response. Two stimuli can do so: H^+ and CO_2.

Ventilation is affected directly by plasma H^+ levels through carotid and aortic chemoreceptors (■ Fig. 20.15). These chemoreceptors are located in the aorta and carotid arteries along with oxygen sensors and blood pressure sensors [p. 618]. An increase in plasma H^+ stimulates the chemoreceptors, which in turn signal the medullary respiratory control centers to increase ventilation. Increased alveolar ventilation allows the lungs to excrete more CO_2 and convert H^+ to carbonic acid.

The central chemoreceptors of the medulla oblongata cannot respond directly to changes in plasma pH because H^+ does not cross the blood-brain barrier. However, changes in pH change P_{CO_2}, and CO_2 stimulates the central chemoreceptors [Fig. 18.17, p. 620]. Dual control of ventilation through the central and peripheral chemoreceptors helps the body respond rapidly to changes in either pH or plasma CO_2.

Concept Check Answer: p. 694

14. In equation 6, the amount of HCO_3^- present is increased at equilibrium. Why doesn't this HCO_3^- act as a buffer and prevent acidosis from occurring?

Kidneys Use Ammonia and Phosphate Buffers

The kidneys take care of the 25% of compensation that the lungs cannot handle. They alter pH two ways: (1) directly, by excreting or reabsorbing H^+ and (2) indirectly, by changing the rate at which HCO_3^- buffer is reabsorbed or excreted.

In acidosis, the kidney secretes H^+ into the tubule lumen using direct and indirect active transport (■ Fig. 20.16). Ammonia from amino acids and phosphate ions (HPO_4^{2-}) in the kidney act as buffers, trapping large amounts of H^+ as NH_4^+ and $H_2PO_4^-$. These buffers allow more H^+ to be excreted. Phosphate ions are present in filtrate and combine with H^+ secreted into the nephron lumen:

$$HPO_4^{2-} + H^+ \rightleftharpoons H_2PO_4^-$$

Even with these buffers, urine can become quite acidic, down to a pH of about 4.5. While H^+ is being excreted, the kidneys make new HCO_3^- from CO_2 and H_2O. The HCO_3^- is reabsorbed into the blood to act as a buffer and increase pH.

In alkalosis, the kidney reverses the general process just described for acidosis, excreting HCO_3^- and reabsorbing H^+ in an effort to bring pH back into the normal range. Renal compensations are slower than respiratory compensations, and their effect on pH may not be noticed for 24–48 hours. However, once activated, renal compensations effectively handle all but severe acid-base disturbances.

The cellular mechanisms for renal handling of H^+ and HCO_3^- resemble transport processes in other epithelia. However, these mechanisms involve some membrane transporters that you have not encountered before:

1 The **apical Na^+-H^+ exchanger (NHE)** is an indirect active transporter that brings Na^+ into the epithelial cell in exchange for moving H^+ against its concentration gradient into the lumen. This transporter also plays a role in proximal tubule Na^+ reabsorption.

RESPIRATORY COMPENSATION FOR METABOLIC ACIDOSIS

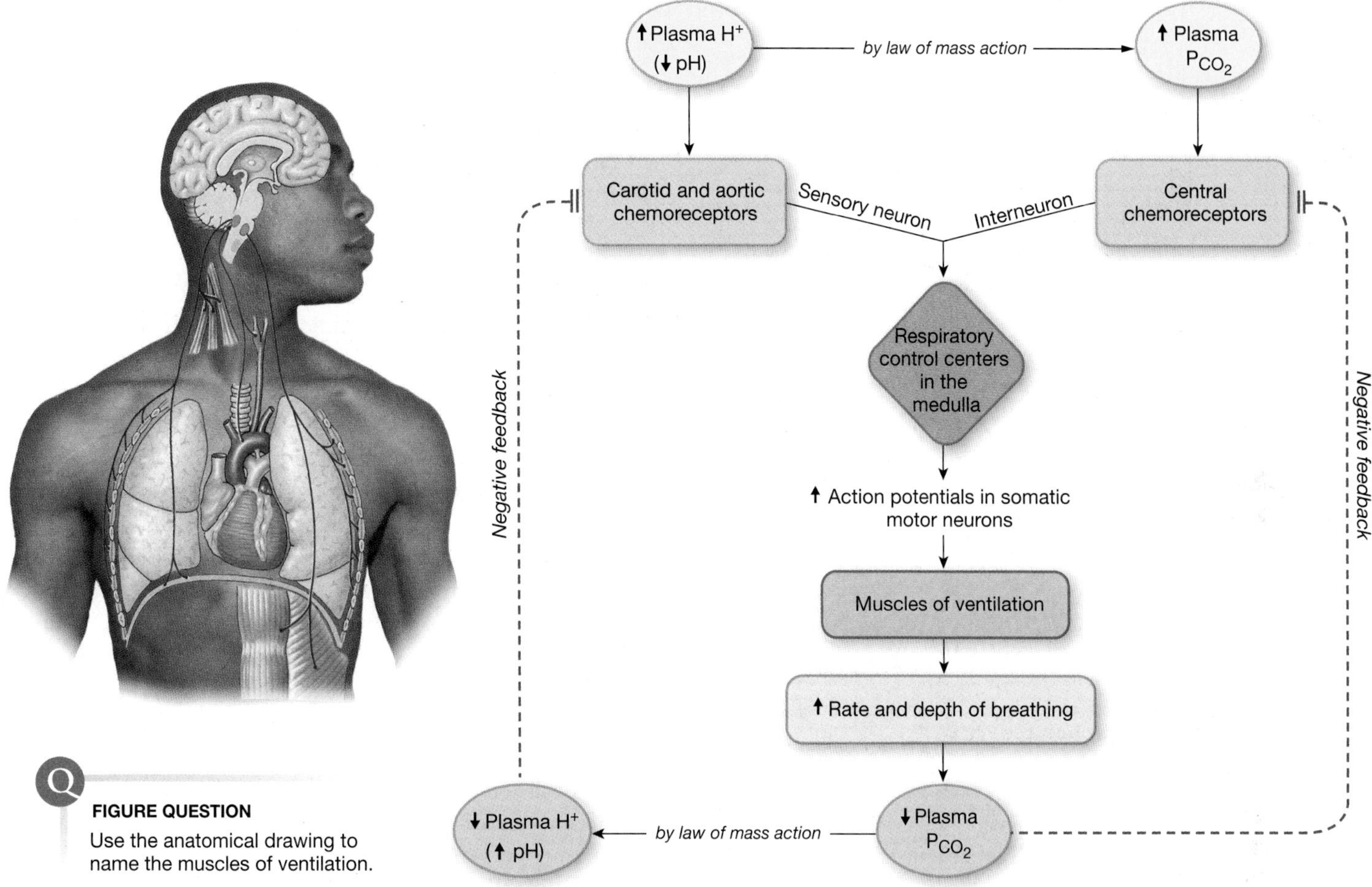

FIGURE QUESTION
Use the anatomical drawing to name the muscles of ventilation.

Fig. 20.15

2 The **basolateral Na^+-HCO_3^- symport** moves Na^+ and HCO_3^- out of the epithelial cell and into the interstitial fluid. This indirect active transporter couples the energy of HCO_3^- diffusing down its concentration gradient to the uphill movement of Na^+ from the cell to the ECF.

3 The **H^+-ATPase** uses energy from ATP to acidify the urine, pushing H^+ against its concentration gradient into the lumen of the distal nephron.

4 The **H^+-K^+-ATPase** puts H^+ into the urine in exchange for reabsorbed K^+. This exchange contributes to the potassium imbalance that sometimes accompanies acid-base disturbances.

5 A **Na^+-NH_4^+ antiport** moves NH_4^+ from the cell to the lumen in exchange for Na^+.

In addition to these transporters, the renal tubule also uses the ubiquitous Na^+-K^+-ATPase and the same HCO_3^--Cl^- antiport protein that is responsible for the chloride shift in red blood cells.

The Proximal Tubule Secretes H^+ and Reabsorbs HCO_3^-

The amount of bicarbonate ion the kidneys filter each day is equivalent to the bicarbonate in a pound of baking soda ($NaHCO_3$)! Most of this HCO_3^- must be reabsorbed to maintain the body's buffer capacity. The proximal tubule reabsorbs most filtered HCO_3^- by indirect methods because there is no apical membrane transporter to bring HCO_3^- into the tubule cell.

■ Figure 20.17a shows the two pathways by which bicarbonate is reabsorbed in the proximal tubule. (The numbers in the following lists correspond to the steps shown in the figure.) By following this illustration, you will see how the transporters listed in the previous section function together.

The first pathway converts filtered HCO_3^- into CO_2, then back into HCO_3^-, which is reabsorbed:

1 H^+ is secreted from the proximal tubule cell into the lumen in exchange for filtered Na^+, which moves from the

OVERVIEW OF RENAL COMPENSATION FOR ACIDOSIS

The kidney secretes H^+, which is buffered in the urine by ammonia and phosphate ions. It reabsorbs bicarbonate to act as an extracellular buffer.

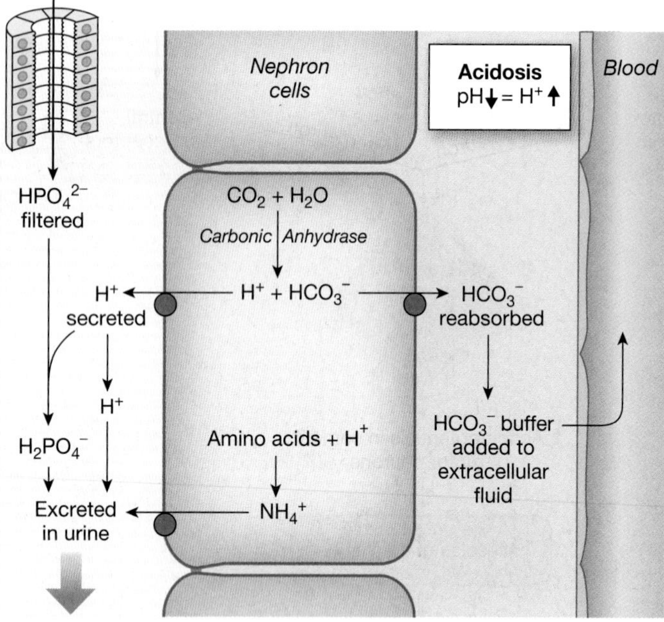

The transporters shown here are generic membrane proteins. For specific transporters involved, see Fig. 20.17.

Fig. 20.16

lumen into the tubule cell. This exchange takes place using the NHE.

2 The secreted H^+ combines with filtered HCO_3^- to form CO_2 in the lumen. This reaction is facilitated by carbonic anhydrase that is bound to the luminal membrane of the tubule cells.

3 The newly formed CO_2 diffuses from the lumen into the tubule cell.

4 In the cytoplasm CO_2 combines with water to form H_2CO_3, which dissociates to H^+ and HCO_3^-.

5 The H^+ created in step 4 can be secreted into the lumen again, replacing the H^+ that combined with filtered HCO_3^- in step 2. It can combine with another filtered bicarbonate or be buffered by filtered phosphate ion and excreted.

6 The HCO_3^- created in step 3 is transported out of the cell on the basolateral side of the proximal tubule cell by the HCO_3^--Na^+ symporter.

The net result of this process is reabsorption of filtered Na^+ and HCO_3^- and secretion of H^+.

A second way to reabsorb bicarbonate and excrete H^+ comes from metabolism of the amino acid glutamine:

7 Glutamine in the proximal tubule cell loses its two amino groups, which become ammonia (NH_3). The ammonia buffers H^+ to become ammonium ion.

The ammonium ion is transported into the lumen in exchange for Na^+. The α-ketoglutarate (αKG) molecule made from deamination of glutamine is metabolized further to HCO_3^-, which is transported into the blood along with Na^+.

This pathway also reabsorbs filtered Na^+ and HCO_3^- and secretes H^+, but buffered this time by ammonia. The net result of both pathways is secretion of acid (H^+) and reabsorption of buffer in the form of sodium bicarbonate—baking soda, $NaHCO_3$.

The Distal Nephron Controls Acid Excretion

The distal nephron plays a significant role in the fine regulation of acid-base balance. The **intercalated cells** (or **I cells**) interspersed among the principal cells are responsible for acid-base regulation.

Intercalated cells are characterized by high concentrations of carbonic anhydrase in their cytoplasm. This enzyme allows them to rapidly convert CO_2 and water into H^+ and HCO_3^-. The H^+ ions are pumped out of the intercalated cell either by the H^+-ATPase or by an ATPase that exchanges one H^+ for one K^+. Bicarbonate leaves the cell by means of the HCO_3^--Cl^- antiport exchanger.

There are two types of intercalated cells, and their transporters are found on different faces of the epithelial cell. During periods of acidosis, type A intercalated cells secrete H^+ and reabsorb bicarbonate. During periods of alkalosis, type B intercalated cells secrete HCO_3^- and reabsorb H^+.

Figure 20.17b shows how type A intercalated cells work during acidosis, secreting H^+ and reabsorbing HCO_3^-. The process is similar to H^+ secretion in the proximal tubule except for the specific H^+ transporters. The distal nephron uses apical H^+-ATPase and H^+-K^+-ATPase rather than the Na^+-H^+ antiport protein found in the proximal tubule.

During alkalosis, when the H^+ concentration of the body is too low, H^+ is reabsorbed and HCO_3^- buffer is excreted in the urine (Fig. 20.17c). Once again, the ions are formed from H_2O and CO_2. Hydrogen ions are reabsorbed by transport into the ECF on the basolateral side of the cell, and HCO_3^- is secreted into the lumen. The polarity of the two types of I cells is reversed, with the same transport processes taking place, but on the opposite sides of the cell.

The H^+-K^+-ATPase of the distal nephron helps create parallel disturbances of acid-base balance and K^+ balance. In acidosis, when plasma H^+ is high, the kidney secretes H^+ and reabsorbs K^+. For this reason, acidosis is often accompanied by hyperkalemia. (Other nonrenal events also contribute to elevated ECF K^+ concentrations in acidosis.) The reverse is true for alkalosis, when blood H^+ levels are low. The mechanism that allows the distal nephron to reabsorb H^+ simultaneously causes it to secrete K^+, with the result that alkalosis goes hand in hand with hypokalemia.

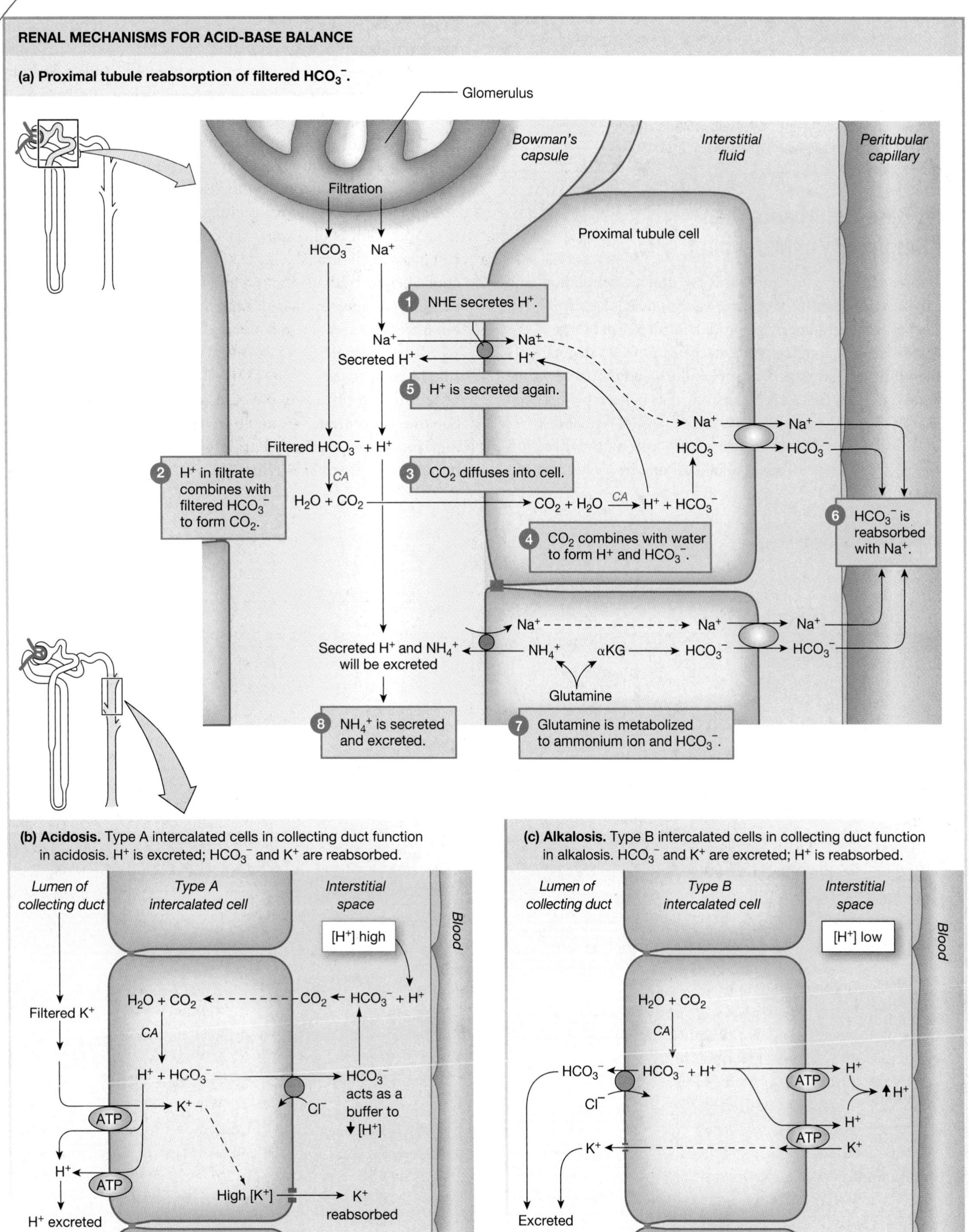

RENAL MECHANISMS FOR ACID-BASE BALANCE
(a) Proximal tubule reabsorption of filtered HCO_3^-.
Glomerulus
Bowman's capsule
Interstitial fluid
Peritubular capillary
Filtration
HCO_3^-
Na^+
Proximal tubule cell
1 NHE secretes H^+.
Na^+
Secreted H^+
Na^+
H^+
5 H^+ is secreted again.
Na^+
HCO_3^-
Na^+
HCO_3^-
Filtered HCO_3^- + H^+
2 H^+ in filtrate combines with filtered HCO_3^- to form CO_2.
CA
3 CO_2 diffuses into cell.
$H_2O + CO_2$
$CO_2 + H_2O$ CA $H^+ + HCO_3^-$
6 HCO_3^- is reabsorbed with Na^+.
4 CO_2 combines with water to form H^+ and HCO_3^-.
Na^+
Na^+
Na^+
Secreted H^+ and NH_4^+ will be excreted
NH_4^+
αKG
HCO_3^-
HCO_3^-
Glutamine
8 NH_4^+ is secreted and excreted.
7 Glutamine is metabolized to ammonium ion and HCO_3^-.
(b) Acidosis. Type A intercalated cells in collecting duct function in acidosis. H^+ is excreted; HCO_3^- and K^+ are reabsorbed.
Lumen of collecting duct
Type A intercalated cell
Interstitial space
Blood
$[H^+]$ high
$H_2O + CO_2$
CO_2
$HCO_3^- + H^+$
Filtered K^+
CA
$H^+ + HCO_3^-$
HCO_3^- acts as a buffer to ↓$[H^+]$
K^+
Cl^-
ATP
H^+
ATP
High $[K^+]$
K^+ reabsorbed
H^+ excreted in urine
(c) Alkalosis. Type B intercalated cells in collecting duct function in alkalosis. HCO_3^- and K^+ are excreted; H^+ is reabsorbed.
Lumen of collecting duct
Type B intercalated cell
Interstitial space
Blood
$[H^+]$ low
$H_2O + CO_2$
CA
HCO_3^-
$HCO_3^- + H^+$
H^+
ATP
↑H^+
Cl^-
H^+
ATP
K^+
K^+
Excreted in urine

■ **Fig. 20.17**

Concept Check Answers: p. 694

15. Why is ATP required for H^+ secretion by the H^+-K^+ transporter but not for the Na^+-H^+ exchanger?

16. In hypokalemia, the intercalated cells of the distal nephron reabsorb K^+ from the tubule lumen. What happens to blood pH as a result?

Acid-Base Disturbances May Be Respiratory or Metabolic

The three compensatory mechanisms (buffers, ventilation, and renal excretion) take care of most variations in plasma pH. But under some circumstances, the production or loss of H^+ or HCO_3^- is so extreme that compensatory mechanisms fail to maintain pH homeostasis. In these states, the pH of the blood moves out of the normal range of 7.38–7.42. If the body fails to keep pH between 7.00 and 7.70, acidosis or alkalosis can be fatal (■ Fig. 20.18).

Acid-base problems are classified both by the direction of the pH change (acidosis or alkalosis) and by the underlying cause (metabolic or respiratory). Changes in P_{CO_2} resulting from hyperventilation or hypoventilation cause pH to shift. These disturbances are said to be of respiratory origin. If the pH problem arises from acids or bases of non-CO_2 origin, the problem is said to be a metabolic problem.

Note that by the time an acid-base disturbance becomes evident as a change in plasma pH, the body's buffers are ineffectual. The loss of buffering ability leaves the body with only two options: respiratory compensation or renal compensation. And if the problem is of respiratory origin, only one homeostatic compensation is available—the kidneys. If the problem is of metabolic origin, both respiratory and renal mechanisms can compensate. Compensation can bring the pH back closer to normal but may not correct the disturbance completely (Fig. 20.18).

The combination of an initial pH disturbance and the resultant compensatory changes is one factor that makes analysis of acid-base disorders in the clinical setting so difficult. In this book we concentrate on simple scenarios with a single underlying cause. Changes that occur in the four simple acid-base disturbances are listed in ■ Table 20.2.

ACID-BASE DISTURBANCES

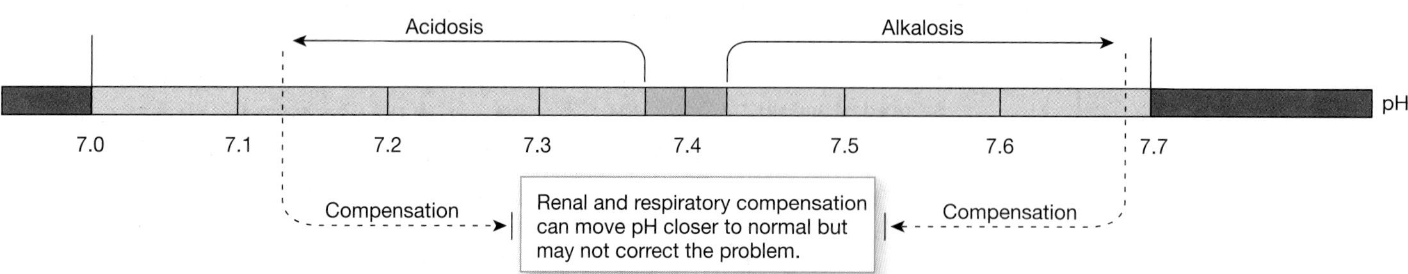

■ **Fig. 20.18**

Table 20.2 Plasma P_{CO_2}, Ions, and pH in Acid-Base Disturbances

Disturbance	P_{CO_2}	H^+	pH	HCO_3^-
Acidosis				
Respiratory	↑	↑	↓	↑
Metabolic	Normal* or ↓	↑	↓	↓
Alkalosis				
Respiratory	↓	↓	↑	↓
Metabolic	Normal* or ↑	↓	↑	↑

* These values are different from what you would expect from the law of mass action because almost instantaneous respiratory compensation keeps P_{CO_2} from changing significantly.

Respiratory Acidosis A state of respiratory acidosis occurs when alveolar hypoventilation results in CO_2 retention and elevated plasma P_{CO_2}. Some situations in which this occurs are respiratory depression due to drugs (including alcohol), increased airway resistance in asthma, impaired gas exchange in fibrosis or severe pneumonia, and muscle weakness in muscular dystrophy and other muscle diseases. The most common cause of respiratory acidosis is *chronic obstructive pulmonary disease* (COPD), which includes emphysema. In emphysema inadequate alveolar ventilation is compounded by loss of alveolar exchange area.

No matter what the cause of respiratory acidosis, plasma CO_2 levels increase (red), leading to elevated H^+ and HCO_3^-:

$$\uparrow CO_2 + H_2O \rightarrow H_2CO_3 \rightarrow \uparrow H^+ + \uparrow HCO_3^- \qquad (8)$$

The hallmark of respiratory acidosis is decreased pH with elevated bicarbonate levels (Tbl. 20.2). Because the problem is of respiratory origin, the body cannot carry out respiratory compensation. (However, depending on the problem, mechanical ventilation can sometimes be used to assist breathing.)

Any compensation for respiratory acidosis must occur through renal mechanisms that excrete H^+ and reabsorb HCO_3^-. The excretion of H^+ raises plasma pH. Reabsorption of HCO_3^- provides additional buffer that combines with H^+, lowering the H^+ concentration and therefore raising the pH.

In chronic obstructive pulmonary disease, renal compensation mechanisms for acidosis can moderate the pH change, but they may not be able to return the pH to its normal range. If you look at pH and HCO_3^- levels in patients with compensated respiratory acidosis, you find that both those values are closer to normal than they were when the acidosis was at its worst.

Metabolic Acidosis Metabolic acidosis is a disturbance of mass balance that occurs when the dietary and metabolic input of H^+ exceeds H^+ excretion. Metabolic causes of acidosis include *lactic acidosis*, which is a result of anaerobic metabolism, and *ketoacidosis*, which results from excessive breakdown of fats or certain amino acids. The metabolic pathway that produces ketoacids is associated with type 1 diabetes mellitus and with low-carbohydrate diets, like the Atkins diet [Chapter 22]. Ingested substances that cause metabolic acidosis include methanol, aspirin, and ethylene glycol (antifreeze).

Metabolic acidosis is expressed by the equation

$$\uparrow CO_2 + H_2O \leftarrow H_2CO_3 \leftarrow \uparrow H^+ + \downarrow HCO_3^- \qquad (9)$$

Hydrogen ion concentration increases (red) because of the H^+ contributed by the metabolic acids. This increase shifts the equilibrium represented in the equation to the left, increasing CO_2 levels and using up HCO_3^- buffer.

Metabolic acidosis can also occur if the body loses HCO_3^-. The most common cause of bicarbonate loss is diarrhea, during which HCO_3^- is lost from the intestines. The pancreas produces HCO_3^- from CO_2 and H_2O by a mechanism similar to the renal mechanism illustrated in Figure 20.16. The H^+ made at the same time is released into the blood. Normally, the HCO_3^- is released into the small intestine, then reabsorbed into the blood, buffering the H^+. However, if a person is experiencing diarrhea, HCO_3^- is not reabsorbed, and a state of acidosis may result.

Whether HCO_3^- concentration is elevated or decreased is an important criterion for distinguishing metabolic acidosis from respiratory acidosis (Tbl. 20.2).

You would think from looking at equation 9 that metabolic acidosis would be accompanied by elevated P_{CO_2}. However, unless the individual also has a lung disease, respiratory compensation takes place almost instantaneously. Both elevated CO_2 and elevated H^+ stimulate ventilation through the pathways described earlier. As a result, P_{CO_2} decreases to normal or even below-normal levels due to hyperventilation.

Uncompensated metabolic acidosis is rarely seen clinically. Actually, a common sign of metabolic acidosis is hyperventilation, evidence of respiratory compensation occurring in response to the acidosis.

The renal compensations discussed for respiratory acidosis also take place in metabolic acidosis: secretion of H^+ and reabsorption of HCO_3^-. Renal compensations take several days to reach full effectiveness, and so they are not usually seen in recent-onset (acute) disturbances.

Respiratory Alkalosis States of alkalosis are much less common than acidotic conditions. Respiratory alkalosis occurs as a result of hyperventilation, when alveolar ventilation increases without a matching increase in metabolic CO_2 production. Consequently, plasma P_{CO_2} falls (red), and alkalosis results when the equation shifts to the left:

$$\downarrow CO_2 + H_2O \leftarrow H_2CO_3 \leftarrow \downarrow H^+ + \downarrow HCO_3^- \qquad (10)$$

The decrease in CO_2 shifts the equilibrium to the left, and both plasma H^+ and plasma HCO_3^- decrease. Low plasma HCO_3^- levels in alkalosis indicate a respiratory disorder.

The primary clinical cause of respiratory alkalosis is excessive artificial ventilation. Fortunately, this condition is easily corrected by adjusting the ventilator. The most common physiological cause of respiratory alkalosis is hysterical hyperventilation caused by anxiety. When this is the cause, the neurological symptoms caused by alkalosis can be partially reversed by having the patient breathe into a paper bag. In doing so, the patient rebreathes exhaled CO_2, a process that raises arterial P_{CO_2} and corrects the problem.

Because this alkalosis has respiratory cause, the only compensation available to the body is renal. Filtered bicarbonate, which if reabsorbed could act as a buffer and increase pH even more, is not reabsorbed in the proximal tubule and is secreted in the distal nephron. The combination of HCO_3^- excretion and H^+ reabsorption in the distal nephron decreases the body's HCO_3^- and increases its H^+, both of which help correct the alkalosis.

Metabolic Alkalosis Metabolic alkalosis has two common causes: excessive vomiting of acidic stomach contents and excessive ingestion of bicarbonate-containing antacids. In both cases, the resulting alkalosis reduces H^+ concentration (red):

$$\downarrow CO_2 + H_2O \rightarrow H_2CO_3 \rightarrow \downarrow H^+ + \uparrow HCO_3^- \quad (11)$$

The decrease in H^+ shifts the equilibrium to the right, meaning that carbon dioxide (P_{CO_2}) decreases and HCO_3^- goes up.

Just as in metabolic acidosis, respiratory compensation for metabolic alkalosis is rapid. The increase in pH and drop in P_{CO_2} depress ventilation. Hypoventilation means the body retains CO_2, raising the P_{CO_2} and creating more H^+ and HCO_3^-. This respiratory compensation helps correct the pH problem but elevates HCO_3^- levels even more. However, this respiratory compensation is limited because hypoventilation also causes hypoxia. Once the arterial P_{O_2} drops below 60 mm Hg, hypoventilation ceases.

The renal response to metabolic alkalosis is the same as that for respiratory alkalosis: HCO_3^- is excreted and H^+ is reabsorbed.

This chapter has used fluid balance and acid-base balance to illustrate functional integration in the cardiovascular, respiratory, and renal systems. Changes in body fluid volume, reflected by changes in blood pressure, trigger both cardiovascular and renal homeostatic responses. Disturbances of acid-base balance are met with compensatory responses from both the respiratory and renal systems. Because of the interwoven responsibilities of these three systems, a disturbance in one system is likely to cause disturbances in the other two. Recognition of this fact is an important aspect of treatment for many clinical conditions.

RUNNING PROBLEM CONCLUSION

Hyponatremia

In acute cases of dilutional hyponatremia such as Lauren's, the treatment goal is to correct the body's depleted Na^+ load and raise the plasma osmolarity to reduce cerebral swelling. The physicians in the emergency medical tent started a slow intravenous drip of 3% saline and restricted Lauren's oral fluid intake. Over the course of several hours, the combination of Na^+ intake and excretion of dilute urine returned Lauren's plasma Na^+ to normal levels.

Hyponatremia has numerous causes, including inappropriate secretion of antidiuretic hormone (a condition known as SIADH, which stands for **s**yndrome of **i**nappropriate **a**nti**d**iuretic **h**ormone secretion). To learn more about medical causes of hyponatremia, Google *hyponatremia.* To learn more about exercise-associated hyponatremia, visit the Gatorade Sports Science Institute at *www.gssiweb.com.*

Question	Facts	Integration and Analysis
1. Name the two major body fluid compartments and give the major ions in each compartment.	The major compartments are the intracellular fluid (ICF) and extracellular fluid (ECF) compartments. The primary ICF ion is K^+, and the major ECF ions are Na^+ and Cl^-.	N/A*
2. Based on Lauren's history, give a reason for why her weight increased during the race.	Lauren reported drinking lots of water and sports drinks. One liter of pure water has a mass of 1 kg.	Lauren's fluid intake was greater than her fluid loss from sweating. A 2-kg increase in body weight means she drank an excess of about 2 L.
3. Which body fluid compartment is being diluted in dilutional hyponatremia?	Ingested water distributes itself throughout the ECF and ICF. Sodium is one of the major extracellular cations.	Lauren consumed a large amount of Na-free fluid and therefore diluted her Na^+ stores. However, the body compartments are in osmotic equilibrium so both ECF and ICF have lower osmolarities.
4. One way to estimate osmolarity is to double the plasma Na^+ concentration. Estimate Lauren's osmolarity and explain what effect the dilutional hyponatremia has on her cells.	Lauren's plasma Na^+ is 124 mEq/L. For Na^+, 1 mEq = 1 milliosmole. Doubling this value tells you that Lauren's estimated plasma osmolarity is 248 mOsM. Water distributes to maintain osmotic equilibrium.	At the start of the race, Lauren's cells were at 280 mOsM. The water she ingested distributed to maintain osmotic equilibrium, so water entered the ICF from the ECF, resulting in cell swelling.

RUNNING PROBLEM CONCLUSION *(continued)*

Question	Facts	Integration and Analysis
5. In dilutional hyponatremia, the medical personnel are most concerned about which organ or tissue?	All cells in Lauren's body swell as a result of excess water ingestion. The brain is encased in the rigid skull.	The bony skull restricts the swelling of brain tissue, causing neurological symptoms, including confusion, headache, and loss of coordination. With lower Na^+ concentrations, death can result.
6. Assuming a sweating rate of 1.0 L/hr, how much Na^+ did Lauren lose during the 16-hour race?	1.0 L sweat lost/hr $\times$ 16 hr $\times$ 70 mEq Na^+/L sweat = 1120 mEq Na^+ lost during the16-hour race.	N/A*
7. Total body water for a 60-kg female is approximately 30 L, and her ECF volume is 10 L. Based on the information given so far, how much fluid did Lauren ingest during the race?	From the sweating rate given in question 6, you know that Lauren lost 16 liters of sweat during the race. You also know that she gained 2 kg in weight. One liter of water weighs 1 kg.	Lauren must have ingested at least 18 liters of fluid. You have no information on other routes of fluid loss, such as urine and insensible water lost during breathing.
8. What would you expect to happen to vasopressin and aldosterone production in response to dilutional hyponatremia?	Vasopressin secretion is inhibited by a decrease in osmolarity. The usual stimuli for renin or aldosterone release are low blood pressure and hyperkalemia.	Vasopressin secretion decreases with hyponatremia. The usual stimuli for aldosterone secretion are absent, but a pathological decrease in plasma Na^+ of 10 mEq/L can stimulate the adrenal cortex to secrete aldosterone. Thus, Lauren's plasma Na^+ may be low enough to increase her aldosterone secretion.

*N/A = not applicable

This problem was developed by Matt Pahnke while he was a kinesiology graduate student at the University of Texas.

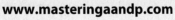

Chapter Summary

Homeostasis of body fluid volume, electrolytes, and pH follows the principle of *mass balance:* to maintain constant amount of a substance in the body, any intake or production must be offset by metabolism or excretion. The *control systems* that regulate these parameters are among the most complicated reflexes of the body because of the overlapping functions of the kidneys, lungs, and cardiovascular system. At the cellular level, however, the *movement of molecules across membranes* follows familiar patterns, as transfer of water and solutes from one *compartment* to another depends on osmosis, diffusion, and protein-mediated transport.

Fluid and Electrolyte Homeostasis

1. The renal, respiratory, and cardiovascular systems control fluid and electrolyte balance. Behaviors such as drinking also play an important role. (p. 659; Fig. 20.1)
2. Pulmonary and cardiovascular compensations are more rapid than renal compensation. (p. 658)

Water Balance

iP Urinary: Early Filtrate Processing

3. Most water intake comes from food and drink. The largest water loss is 1.5 liters/day in urine. Smaller amounts are lost in feces, by evaporation from skin, and in exhaled humidified air. (p. 660; Fig. 20.2)
4. Renal water reabsorption conserves water but cannot restore water lost from the body. (p. 661; Fig. 20.3)
5. To produce dilute urine, the nephron must reabsorb solute without reabsorbing water. To concentrate urine, the nephron must reabsorb water without reabsorbing solute. (p. 661)
6. Filtrate leaving the ascending limb of the loop of Henle is dilute. The final concentration of urine depends on the water permeability of the collecting duct. (p. 662; Fig. 20.4)
7. The hypothalamic hormone **vasopressin** controls collecting duct permeability to water in a graded fashion. When vasopressin is absent, water permeability is nearly zero. (p. 664; Fig. 20.5a,b)
8. Vasopressin causes distal nephron cells to insert **aquaporin** water pores in their apical membrane. (p. 664; Fig. 20.5c)
9. An increase in ECF osmolarity or a decrease in blood pressure stimulates vasopressin release from the posterior pituitary. Osmolarity is monitored by hypothalamic **osmoreceptors.** Blood pressure and blood volume are sensed by receptors in the carotid and aortic bodies, and in the atria, respectively. (p. 665; Fig. 20.6)
10. The loop of Henle is a **countercurrent multiplier** that creates high osmolarity in the medullary interstitial fluid by actively transporting Na^+, Cl^-, and K^+ out of the nephron. This high medullary osmolarity is necessary for formation of concentrated urine as filtrate flows through the collecting duct. (p. 667; Fig. 20.7)
11. The **vasa recta** capillaries is a countercurrent exchanger that carries away water leaving the tubule so that the water does not dilute the medullary interstitium. (p. 667; Fig. 20.7)
12. Urea contributes to the high osmolarity in the renal medulla. (p. 668)

Sodium Balance and ECF Volume

iP Urinary: Late Filtrate Processing

13. The total amount of Na^+ in the body is a primary determinant of ECF volume. (p. 669; Fig. 20.8)
14. The steroid hormone **aldosterone** increases Na^+ reabsorption and K^+ secretion. (p. 670; Fig. 20.9a)
15. Aldosterone acts on **principal cells** (P cells) of the distal nephron. This hormone enhances Na^+-K^+-ATPase activity and increases open time of Na^+ and K^+ leak channels. It also stimulates the synthesis of new pumps and channels. (p. 670; Fig. 20.9b)
16. Aldosterone secretion can be controlled directly at the adrenal cortex. Increased ECF K^+ stimulates aldosterone secretion, but very high ECF osmolarity inhibits it. (p. 670; Fig. 20.9)
17. Aldosterone secretion is also stimulated by **angiotensin II.** Granular cells in the kidney secrete **renin,** which converts **angiotensinogen** in the blood to **angiotensin I. Angiotensin-converting enzyme** (ACE) converts ANG I to ANG II. (p. 672; Fig. 20.10)
18. The stimuli for the release of renin are related either directly or indirectly to low blood pressure. (p. 672; Fig. 20.10)
19. ANG II has additional effects that raise blood pressure, including increased vasopressin secretion, stimulation of thirst, vasoconstriction, and activation of the cardiovascular control center. (p. 672; Fig. 20.10)
20. **Atrial natriuretic peptide** (ANP) and **brain natriuretic peptide** (BNP) enhance Na^+ excretion and urinary water loss by increasing GFR, inhibiting tubular reabsorption of NaCl, and inhibiting the release of renin, aldosterone, and vasopressin. (p. 674; Fig. 20.11)

Potassium Balance

21. Potassium homeostasis keeps plasma K^+ concentrations in a narrow range. **Hyperkalemia** and **hypokalemia** cause problems with excitable tissues, especially the heart. (p. 675)

Behavioral Mechanisms in Salt and Water Balance

22. Thirst is triggered by hypothalamic osmoreceptors and relieved by drinking. (p. 675)
23. **Salt appetite** is triggered by aldosterone and angiotensin. (p. 676)

Integrated Control of Volume and Osmolarity

iP Fluids & Electrolytes: Water Homeostasis

24. Homeostatic compensations for changes in salt and water balance follow the law of mass balance. Fluid and solute added to the body must be removed, and fluid and solute lost from the body must be replaced. However, perfect compensation is not always possible. (p. 678; Tbl. 20.1)

Acid-Base Balance

iP Fluids & Electrolytes: Acid/Base Homeostasis

25. The body's pH is closely regulated because pH affects intracellular proteins, such as enzymes and membrane channels. (p. 681)
26. Acid intake from foods and acid production by the body's metabolic processes are the biggest challenge to body pH. The most significant source of acid is CO_2 from respiration, which combines with water to form carbonic acid (H_2CO_3). (p. 682; Fig. 20.14)
27. The body copes with changes in pH by using buffers, ventilation, and renal secretion or reabsorption of H^+ and HCO_3^-. (p. 682; Fig. 20.14)
28. Bicarbonate produced from CO_2 is the most important extracellular buffer of the body. Bicarbonate buffers organic acids produced by metabolism. (p. 683)
29. Ventilation can correct disturbances in acid-base balance because changes in plasma P_{CO_2} affect both the H^+ content and the HCO_3^- content of the blood. An increase in P_{CO_2} stimulates central chemoreceptors. An increase in plasma H^+ stimulates carotid and aortic chemoreceptors. Increased ventilation excretes CO_2 and decreases plasma H^+. (p. 685; Fig. 20.15)
30. In **acidosis,** the kidneys secrete H^+ and reabsorb HCO_3^-. (p. 686, 687; Figs. 20.16, 20.17b)
31. In **alkalosis,** the kidneys secrete HCO_3^- and reabsorb H^+. (p. 687; Fig. 20.17c)
32. **Intercalated cells** in the collecting duct are responsible for the fine regulation of acid-base balance. (p. 687; Fig. 20.17b, c)

Questions

Answers: p. A-1

Level One Reviewing Facts and Terms

1. What is an electrolyte? Name five electrolytes whose concentrations must be regulated by the body.
2. List five organs and four hormones important in maintaining fluid and electrolyte balance.
3. Compare the routes by which water enters the body with the routes by which the body loses water.
4. List the receptors that regulate osmolarity, blood volume, blood pressure, ventilation, and pH. Where are they located, what stimulates them, and what compensatory mechanisms are triggered by them?
5. How do the two limbs of the loop of Henle differ in their permeability to water? What makes this difference in permeability possible?
6. Which ion is a primary determinant of ECF volume? Which ion is the determinant of extracellular pH?
7. What happens to the resting membrane potential of excitable cells when plasma K^+ concentrations decrease? Which organ is most likely to be affected by changes in K^+ concentration?
8. Appetite for what two substances is important in regulating fluid volume and osmolarity?
9. Write out the words for the following abbreviations: ADH, ANP, ACE, ANG II, JG apparatus, P cell, I cell.
10. Make a list of all the different membrane transporters in the kidney. For each transporter, tell (a) which section(s) of the nephron contain(s) the transporter; (b) whether the transporter is on the apical membrane only, on the basolateral membrane only, or on both; (c) whether it participates in reabsorption only, in secretion only, or in both.
11. List and briefly explain three reasons why monitoring and regulating ECF pH are important. What three mechanisms does the body use to cope with changing pH?
12. Which is more likely to accumulate in the body, acids or bases? List some sources of each.
13. What is a buffer? List three intracellular buffers. Name the primary extracellular buffer.
14. Name two ways the kidneys alter plasma pH. Which compounds serve as urinary buffers?
15. Write the equation that shows how CO_2 is related to pH. What enzyme increases the rate of this reaction? Name two cell types that possess high concentrations of this enzyme.
16. When ventilation increases, what happens to arterial P_{CO_2}? To plasma pH? To plasma H^+ concentration?

Level Two Reviewing Concepts

17. Concept map: Map the homeostatic reflexes that occur in response to each of the following situations:
    (a) decreased blood volume, normal blood osmolarity
    (b) increased blood volume, increased blood osmolarity
    (c) normal blood volume, increased blood osmolarity
18. Figures 20.15 and 20.17b show the respiratory and renal compensations for acidosis. Draw similar maps for alkalosis.
19. Explain how the loop of Henle and vasa recta work together to create dilute renal filtrate.
20. Diagram the mechanism by which vasopressin alters the composition of urine.
21. Make a table that specifies the following for each substance listed: hormone or enzyme? steroid or peptide? produced by which cell or tissue? target cell or tissue? target has what response?
    (a) ANP
    (b) aldosterone
    (c) renin
    (d) ANG II
    (e) vasopressin
    (f) angiotensin-converting enzyme
22. Name the four main compensatory mechanisms for restoring low blood pressure to normal. Why do you think there are so many homeostatic pathways for raising low blood pressure?
23. Compare and contrast the terms in each set:
    (a) principal cells and intercalated cells
    (b) renin, ANG II, aldosterone, ACE
    (c) respiratory acidosis and metabolic acidosis, including causes and compensations
    (d) water reabsorption in proximal tubule, distal tubule, and ascending limb of the loop of Henle
    (e) respiratory alkalosis and metabolic alkalosis, including causes and compensations
24. The interstitial fluid in contact with the basolateral side of collecting duct cells has an extremely high osmolarity, and yet the cells do not shrivel up. How can they maintain normal cell volume in the face of such high ECF osmolarity?

Level Three Problem Solving

25. A 45-year-old man visiting from out of town arrives at the emergency room having an asthma attack caused by pollen.
    (a) Blood drawn before treatment showed the following: $HCO_3^- =$ 30 meq/L (normal: 24), $P_{CO_2} = 70$ mm Hg, pH = 7.24. What is the man's acid-base state? Is this an acute or a chronic situation?
    (b) The man was treated and made a complete recovery. Over the next ten years he continued to smoke a pack of cigarettes a day, and a year ago his family doctor diagnosed chronic obstructive pulmonary disease (emphysema). The man's most recent blood test showed the following: $HCO_3^- = 45$ meq/L, $P_{CO_2} = 85$ mm Hg, pH = 7.34. What is the man's acid-base state now? Is this an acute or a chronic situation?
    (c) Explain why in his second illness his plasma bicarbonate level and P_{CO_2} are higher than in the first illness but his pH is closer to normal.
26. The U.S. Food and Drug Administration recently approved a new class of drugs called *vasopressin receptor antagonists*. Predict the effect these drugs would have on renal function and describe some clinical situations or diseases in which these drugs might be useful.
27. Karen has bulimia, in which she induces vomiting to avoid weight gain. When the doctor sees her, her weight is 89 lb and her respiration rate is 6 breaths/min (normal 12). Her blood HCO_3^- is 62 meq/L (normal: 24–29), arterial blood pH is 7.61, and P_{CO_2} is 61 mm Hg.
    (a) What is her acid-base condition called?
    (b) Explain why her plasma bicarbonate level is so high.
    (c) Why is she hypoventilating? What effect does this have on the pH and total oxygen content of her blood? Explain your answers.
28. Hannah, a 31-year-old woman, decided to have colonic irrigation, a procedure during which large volumes of distilled water were infused into her rectum. During the treatment she absorbed 3000 mL

of water. About 12 hours later, her roommate found her in convulsions and took her to the emergency room. Her blood pressure was 140/90, her plasma Na^+ concentration was 106 meq/L (normal: 135 meq/L), and her plasma osmolarity was 270 mOsM. In a concept map or flow chart, diagram all the homeostatic responses her body was using to attempt compensation for the changes in blood pressure and osmolarity.

Level Four Quantitative Problems

29. The **Henderson-Hasselbalch equation** is a mathematical expression of the relationship between pH, HCO_3^- concentration, and dissolved CO_2 concentration. One variant of the equation uses P_{CO_2} instead of dissolved CO_2 concentration:

$$pH = 6.1 + \log[HCO_3^-] / 0.03 \times P_{CO_2}$$

(a) If arterial blood has a P_{CO_2} of 40 mm Hg and its HCO_3^- concentration is 24 mM, what is its pH? (Use a log table or calculator with a logarithmic function capability.)
(b) What is the pH of venous blood with the same HCO_3^- concentration but a P_{CO_2} of 46 mm Hg?

30. In extreme dehydration, urine can reach a concentration of 1400 mOsM. If the minimum amount of waste solute that a person must excrete daily is about 600 milliosmoles, what is the minimum urine volume that is excreted in one day?

31. Hyperglycemia in a diabetic patient leads to osmotic diuresis and dehydration. Given the following information, answer the questions.

Plasma glucose = 400 mg/dL
Normal urine flow = 1 L per day
GFR = 130 mL/min
Normal urine osmolarity = 300 mOsM
Glucose T_m = 400 mg/min
Molecular mass of glucose = 180 dalton
Renal plasma flow = 500 mL/min

(a) How much glucose filters into the nephron each minute?
(b) How much glucose is reabsorbed each minute?
(c) How much glucose is excreted in the urine each day?
(d) Assuming that dehydration causes maximal vasopressin secretion and allows the urine to concentrate to 1200 mOsM, how much additional urine does this diabetic patient excrete in a day?

Answers

Answers to Concept Check Questions

Page 663

1. Apical membranes have more water pores when vasopressin is present.
2. If vasopressin secretion is suppressed, the urine is dilute.

Page 666

3. Hyperosmotic NaCl is hypertonic and causes the osmoreceptors to shrink, but hyperosmotic urea is hypotonic and causes them to swell. Because only cell shrinkage causes firing, osmoreceptors exposed to urea do not fire.
4. Because vasopressin enhances water reabsorption, vasopressin levels would increase with dehydration.
5. Osmoreceptors in the lumen of the digestive tract and hepatic portal vein would sense high-osmolarity food or drink that has been ingested and absorbed, before it is in the general circulation. This would allow an anticipatory, or feed-forward, secretion of vasopressin to conserve body water.

Page 668

6. Solutes that remain in the lumen when the NKCC symporter is inhibited force water to remain in the lumen with them because urine can be concentrated only to 1200 mOsM. Thus each 12 milliosmoles of un-reabsorbed solute "holds" an additional 10 mL of water in the urine.
7. Diuretics that inhibit the NKCC symporter leave K^+ in the tubule lumen, where it is likely to be excreted, thus increasing urinary K^+ loss.

Page 671

8. Na^+ and K^+ are moving down their electrochemical gradients.
9. In hyperkalemia, resting membrane potential depolarizes. Excitable tissues fire one action potential but are unable to repolarize to fire a second one.

Page 673

10. Atherosclerotic plaques block blood flow, which decreases GFR and decreases pressure in the afferent arteriole. Both events are stimuli for renin release.
11. Renin secretion begins a cascade that produces ANG II. This powerful hormone then causes vasoconstriction, acts on the medullary cardiovascular control center to increase blood pressure, increases production of vasopressin and aldosterone, and increases thirst, resulting in drinking and an increased fluid volume in the body. All these responses may contribute to increased blood pressure.
12. All blood passes through the pulmonary blood vessels with each circuit. Unless ACE was in every systemic blood vessel, some blood might not be exposed to ACE every time.

Page 676

13. On the left side of Figure 20.8, interneurons also lead from hypothalamic osmoreceptors to the hypothalamic thirst centers.

Page 684

14. The bicarbonate level increases as the reaction shifts to the right as a result of added CO_2. Once a new equilibrium state is achieved, bicarbonate cannot act as a buffer because the system is at equilibrium.

Page 688

15. In the distal nephron, both K^+ and H^+ are being moved against their concentration gradients, which requires ATP. In the proximal tubule, Na^+ is moving down its concentration gradient, providing the energy to push H^+ against its gradient.
16. When intercalated cells reabsorb K^+, they secrete H^+, and therefore blood pH increases.

Answers to Figure Questions

Page 665

Figure 20.6: 1. Threshold is 280 mOsM. 2. An action potential arriving at the axon terminal initiates exocytosis.

Page 669

Figure 20.8: See Figure 15.14b, p. 526.

Page 672

Figure 20.10: See Figure 15.14b, p. 526, for the cardiovascular pathway, Figure 20.9b for the target cell involved in aldosterone action, and Figure 20.5c for the target cell involved in vasopressin action.

Page 685

Figure 20.15: The muscles of inspiration are the diaphragm, the external intercostals, the scalenes, and sternocleidomastoid. Muscles of expiration are the abdominals and internal intercostals.

21 The Digestive System

Give me a good digestion, Lord, and also something to digest.

—Anonymous, *A Pilgrim's Grace*

Background Basics

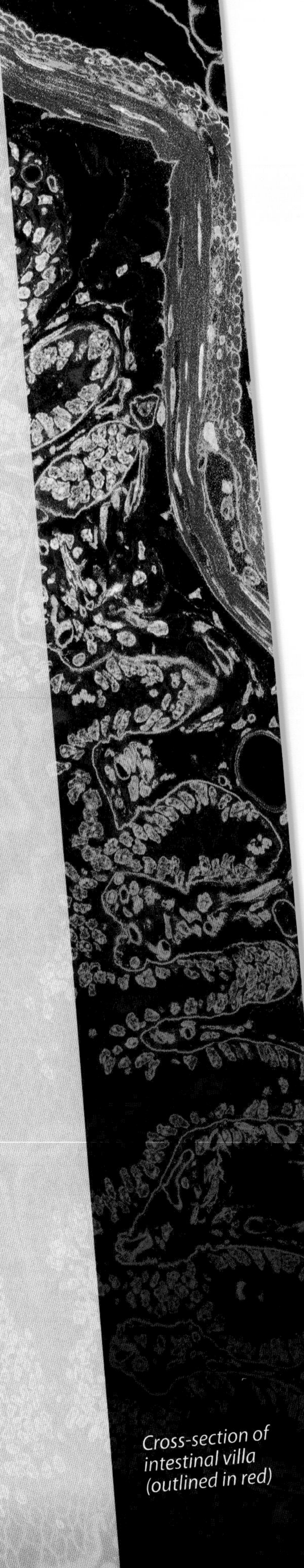

Cross-section of intestinal villa (outlined in red)

A shotgun wound to the stomach seems an unlikely beginning to the scientific study of digestive processes. But in 1822 at Fort Mackinac, a young Canadian trapper named Alexis St. Martin narrowly escaped death when a gun discharged three feet from him, tearing open his chest and abdomen and leaving a hole in his stomach wall. U.S. Army surgeon William Beaumont attended to St. Martin and nursed him back to health over the next two years.

The gaping wound over the stomach failed to heal properly, leaving a *fistula,* or opening, into the lumen. St. Martin was destitute and unable to care for himself, so Beaumont "retained St. Martin in his family for the special purpose of making physiological experiments." In a legal document, St. Martin even agreed to "obey, suffer, and comply with all reasonable and proper experiments of the said William [Beaumont] in relation to . . . the exhibiting . . . of his said stomach and the power and properties . . . and states of the contents thereof."

Beaumont's observations on digestion and on the state of St. Martin's stomach under various conditions created a sensation. In 1832, just before Beaumont's observations were published, the nature of gastric juice {*gaster*, stomach} and digestion in the stomach was a subject of much debate. Beaumont's careful observations went far toward solving the mystery. Like physicians of old who tasted urine when making a diagnosis, Beaumont tasted the mucous lining of the stomach and the gastric juices. He described them both as "saltish," but mucus was not at all acidic, and gastric fluid was very acidic. Beaumont collected copious amounts of gastric fluid through the fistula, and in controlled experiments he confirmed that gastric fluid digested meat, using a combination of hydrochloric acid and another active factor now known to be the enzyme pepsin.

RUNNING PROBLEM

Cholera in Haiti

Brooke was looking for a way to spend her 2011 winter break, so she volunteered to join a disaster relief team going to Haiti. Upon her arrival in the earthquake-devastated country, Brooke was appalled to see the living conditions. Many people were still living in tents with little or no running water and sanitation. To make matters worse, in October 2010 the World Health Organization (WHO) issued a global outbreak alert for a cholera epidemic. *Vibrio cholerae*, the cholera bacterium, causes secretory diarrhea and vomiting in people who consume contaminated food or water. There had been no cholera in Haiti for nearly a hundred years, but since the earthquake, over 73,000 cases and 259 deaths had already been reported.

697 699 708 715 722 731

These observations and others about motility and digestion in the stomach became the foundation of what we know about digestive physiology. Although research today is conducted more at the cellular and molecular level, researchers still create surgical fistulas in experimental animals to observe and sample the contents of the digestive tract.

Why is the digestive system—also referred to as the **gastrointestinal system** {*intestinus,* internal}—of such great interest? The reason is that gastrointestinal diseases today account for nearly one-tenth of the money spent on health care. Many of these conditions, such as heartburn, indigestion, gas, and constipation, are troublesome rather than major health risks, but their significance should not be underestimated. Go into any drugstore and look at the number of over-the-counter medications for digestive disorders to get a feel for the impact digestive diseases have on our society. In this chapter we examine the gastrointestinal system and the remarkable way in which it transforms the food we eat into nutrients for the body's use.

Digestive Function and Processes

The **gastrointestinal tract**, or **GI tract**, is a long tube passing through the body. The tube has muscular walls lined with epithelium and is closed off by a skeletal-muscle sphincter at each end. Because the GI tract opens to the outside world, the lumen and its contents are actually part of the external environment. (Think of a hole passing through the center of a bead.) [Tbl. 1.1, p. 4] The primary function of the GI tract is to move nutrients, water, and electrolytes from the external environment into the body's internal environment.

The food we eat is mostly in the form of macromolecules, such as proteins and complex carbohydrates, so our digestive systems must secrete powerful enzymes to digest food into molecules that are small enough to be absorbed into the body. At the same time, however, these enzymes must not digest the cells of the GI tract itself (*autodigestion*). If protective mechanisms against autodigestion fail, we may develop raw patches known as *peptic ulcers* {*peptos,* digested} on the walls of the GI tract.

Another challenge the digestive system faces daily is mass balance: matching fluid input with output (■ Fig. 21.1). People ingest about 2 liters of fluid a day. In addition, exocrine glands and cells secrete 7 liters or so of enzymes, mucus, electrolytes, and water into the lumen of the GI tract. That volume
fluid is the equivalent of one-sixth of the body's total l
(42 liters), or more than twice the plasma volume of 3
it must be reabsorbed or the body would rapidly deh

Normally reabsorption is very efficient, and
100 mL of fluid is lost in the feces. However, vomiti
arrhea (excessively watery feces) can become an
when GI secretions that would normally be reabsor
to the environment. In severe cases, this fluid loss c

MASS BALANCE IN THE DIGESTIVE SYSTEM

To maintain homeostasis, the volume of fluid entering the GI tract by intake or secretion must equal the volume leaving the lumen.

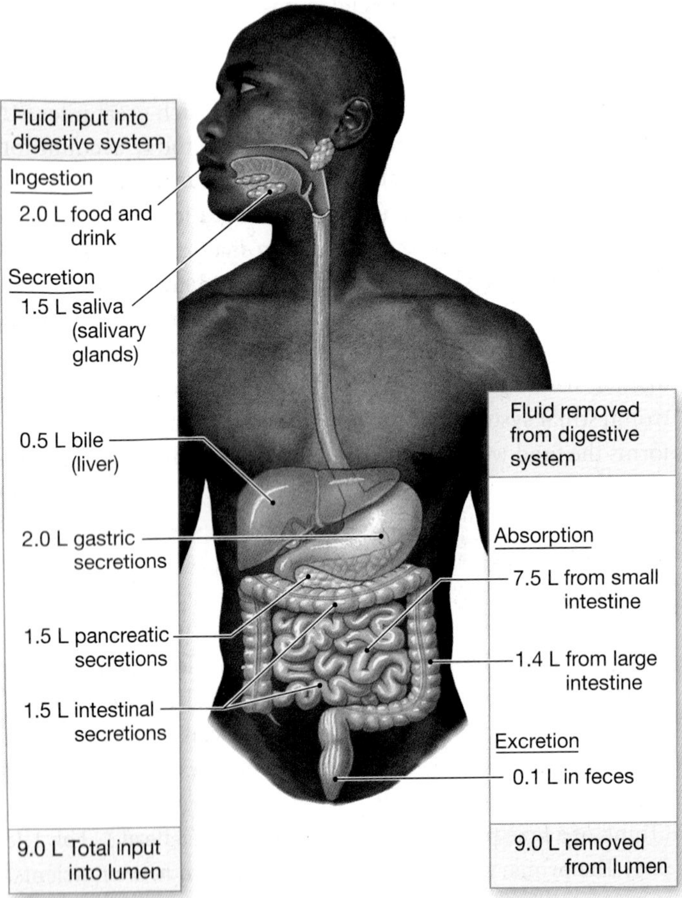

Fig. 21.1

FOUR PROCESSES OF THE DIGESTIVE SYSTEM

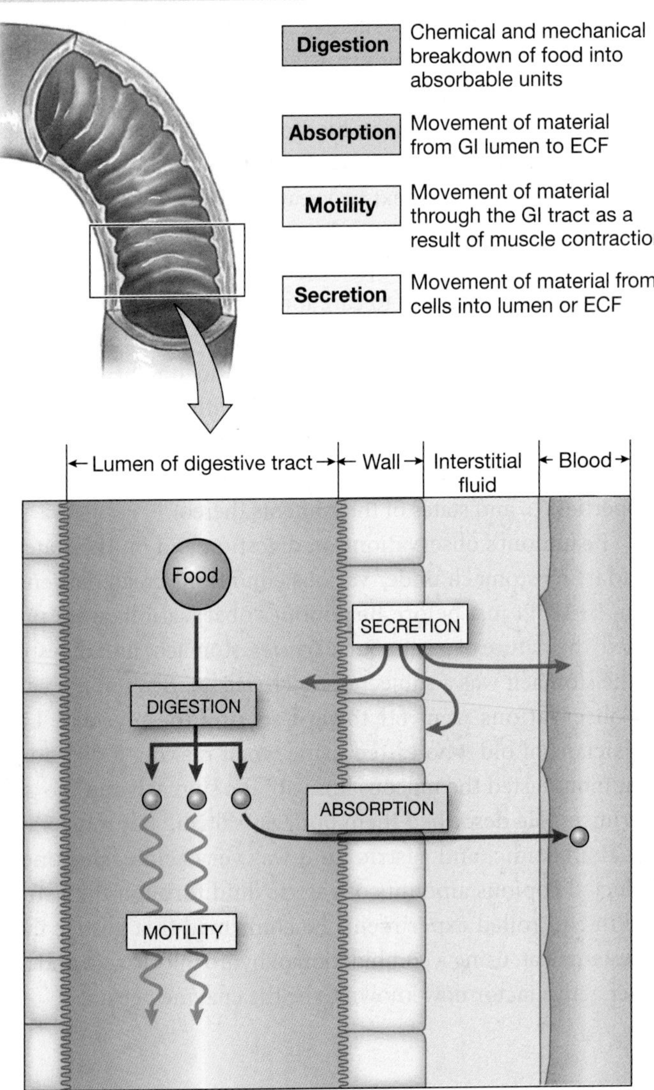

Fig. 21.2

extracellular fluid volume to the point that the circulatory system is unable to maintain adequate blood pressure.

A final challenge the digestive system faces is repelling foreign invaders. It is counterintuitive, but the largest area of contact between the internal environment and the outside world is in the lumen of the digestive system. As a result, the GI tract, with a total surface area about the size of a tennis court, faces daily conflict between the need to absorb water and nutrients, and the need to keep bacteria, viruses, and other pathogens from entering the body. To this end, the transporting epithelium of the GI tract is assisted by an array of physiological defense mechanisms, including mucus, digestive enzymes, acid, and the largest collection of lymphoid tissue in the body, the **gut-associated lymphoid tissue** (GALT). By one estimate, 80% of all lymphocytes [p. 547] in the body are found in the small intestine.

The human body meets these sometimes conflicting physiological challenges by coordinating the four basic processes of the digestive system: digestion, absorption, motility, and secretion (Fig. 21.2). **Digestion** is the chemical and mechanical breakdown of foods into smaller units that can be taken across the intestinal epithelium into the body. **Absorption** is the active or passive transfer of substances from the lumen of the GI tract to the extracellular fluid. **Motility** {*movere,* move + *tillis,* characterized by} is movement of material in the GI tract as a result of muscle contraction. **Secretion** refers to the transepithelial transfer of water and ions from the ECF to the digestive tract lumen as well as to the release of substances synthesized by GI epithelial cells.

Scientists used to believe that nutrient absorption was not regulated and that "what you eat is what you get." Now, however, evidence indicates that some nutrient absorption changes in response to long-term environmental changes. Motility and secretion are continuously regulated to maximize the availability of absorbable material. Motility is regulated because if food moves through the system too rapidly, there is not enough time for everything in the lumen to be digested and absorbed. Secretion is regulated because if digestive enzymes are not secreted in adequate amounts, food in the GI tract cannot be broken down into an absorbable form.

When digested nutrients have been absorbed and have reached the body's cells, cellular metabolism directs their use or storage. Some of the same chemical signal molecules that alter digestive motility and secretion also participate in the control of metabolism, providing an integrating link between the two steps.

Anatomy of the Digestive System

The digestive system begins with the oral cavity (mouth and pharynx), which serves as a receptacle for food (■ Fig. 21.3a). In the oral cavity the first stages of digestion begin with chewing and the secretion of saliva by three pairs of **salivary glands:** *sublingual glands* under the tongue, *submandibular glands* under the mandible (jawbone), and *parotid glands* lying near the hinge of the jaw (Fig. 21.3b).

Once swallowed, food moves into the GI tract. At intervals along the tract, rings of muscle function as *sphincters* to separate the tube into segments with distinct functions. Food moves through the tract propelled by waves of muscle contraction. Along the way, secretions are added to the food by secretory epithelium, the liver, and the pancreas, creating a soupy mixture known as **chyme.**

Digestion takes place primarily in the lumen of the tube. The products of digestion are absorbed across the epithelium and pass into the extracellular compartment. From there, they move into the blood or lymph for distribution throughout the body. Any waste remaining in the lumen at the end of the GI tract leaves the body through the opening known as the *anus.*

Concept Check

Answers: p. 735

1. Define digestion.
2. What is the difference between digestion and metabolism [p. 109]?
3. What is the difference between absorption and secretion?

The Digestive System Is a Tube

When you swallow a piece of food, it passes into the **esophagus,** a narrow tube that travels through the thorax to the abdomen (Fig. 21.3a). The esophageal walls are skeletal muscle initially but transition to smooth muscle about two-thirds of the way down the length. Just below the diaphragm, the esophagus ends at the **stomach,** a baglike organ that can hold as much as 2 liters of food and fluid when fully (if uncomfortably) expanded.

The stomach is divided into three sections: the upper **fundus,** the central **body,** and the lower **antrum** (Fig. 21.3c). The stomach continues digestion that began in the mouth by mixing food with acid and enzymes to create chyme. The **pylorus** {gatekeeper} or opening between the stomach and the **small intestine** is guarded by the **pyloric valve.** This thickened band of smooth muscle relaxes to allow only small amounts of chyme into the small intestine at any one time.

The stomach acts as an intermediary between the behavioral act of eating and the physiological events of digestion and absorption in the intestine. Integrated signals and feedback loops between the intestine and stomach regulate the rate at which chyme enters the duodenum. This ensures that the intestine is not overwhelmed with more than it can digest and absorb.

Most digestion takes place in the small intestine, which is also divided into three sections: the **duodenum** (the first 25 cm), **jejunum,** and **ileum** (the latter two together are about 260 cm long). Digestion is carried out by intestinal enzymes, aided by exocrine secretions from two *accessory glandular organs:* the pancreas and the liver (Fig. 21.3a). Secretions from these two organs enter the initial section of the duodenum through ducts. A tonically contracted sphincter (the *sphincter of Oddi*) keeps pancreatic fluid and bile from entering the small intestine except during a meal.

Digestion is essentially completed in the small intestine, and nearly all digested nutrients and secreted fluids are absorbed there, leaving about 1.5 liters of chyme per day to pass into the **large intestine** (Fig. 21.3a). In the **colon**—the proximal section of the large intestine—watery chyme is converted into semisolid **feces** {*faeces,* dregs} as water and electrolytes are absorbed out of the chyme and into the ECF.

When feces are propelled into the terminal section of the large intestine, known as the **rectum,** distension of the rectal wall triggers a *defecation reflex.* Feces leave the GI tract through the

RUNNING PROBLEM

Facing a cholera epidemic in the country, members of the relief team were apprehensive. A worker from the U.S. Centers for Disease Control (CDC) spoke to the group about proper precautions. He warned them to be careful about what they ate and drank, and to wash their hands often. Then, about five days into her trip, Brooke had a bout of copious and watery diarrhea, which she initially attributed to the emotional stress of the relief work. But when she developed dizziness and a rapid heartbeat, she went to the medical tent. There she was diagnosed with dehydration from cholera-induced diarrhea.

Q1: Given Brooke's watery diarrhea, what would you expect her ECF volume to be?

Q2: Why was Brooke experiencing a rapid heartbeat?

697 699 708 715 722 731

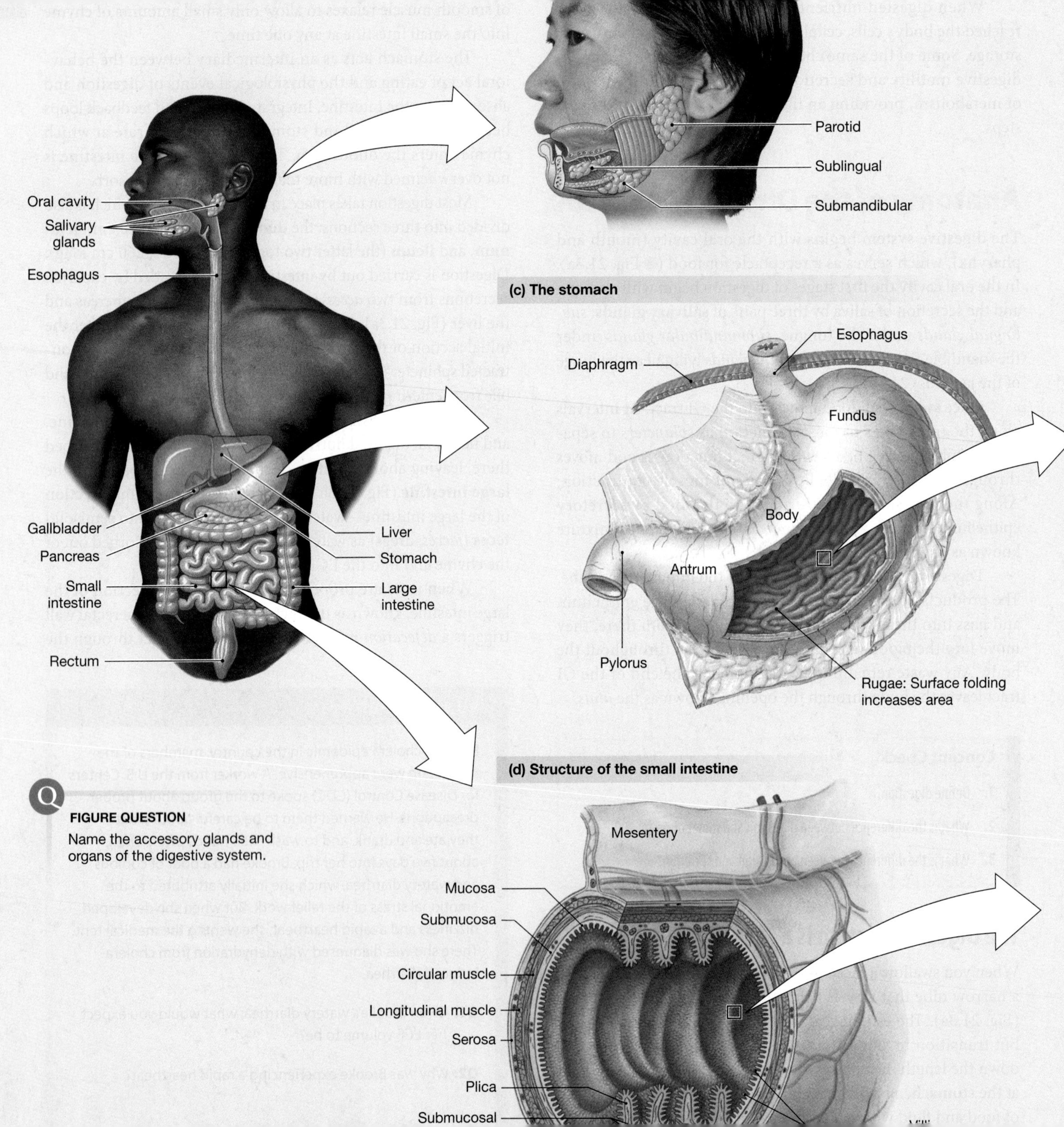

FIGURE QUESTION

Name the accessory glands and organs of the digestive system.

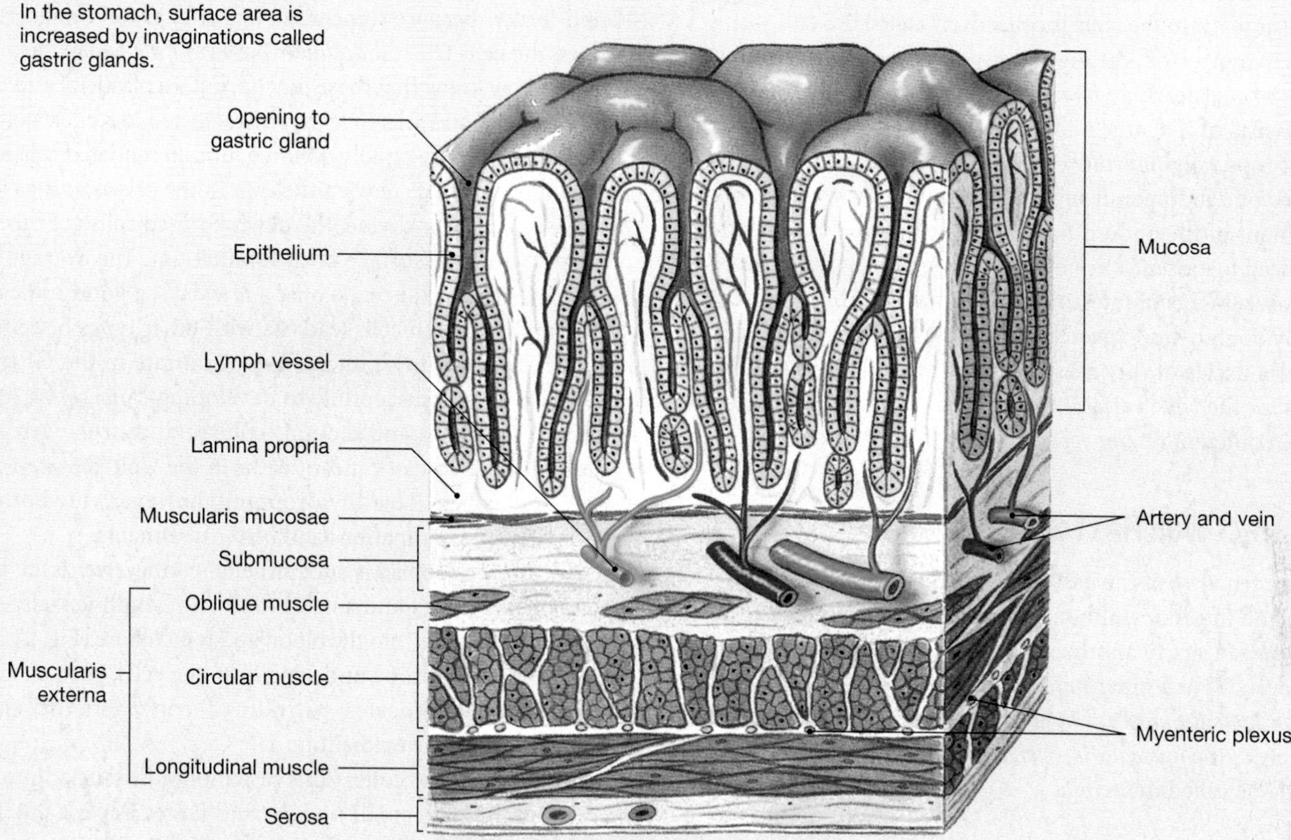
(e) Sectional view of the stomach
In the stomach, surface area is increased by invaginations called gastric glands.
Opening to gastric gland
Epithelium
Lymph vessel
Lamina propria
Muscularis mucosae
Submucosa
Oblique muscle
Muscularis externa
Circular muscle
Longitudinal muscle
Serosa
Mucosa
Artery and vein
Myenteric plexus

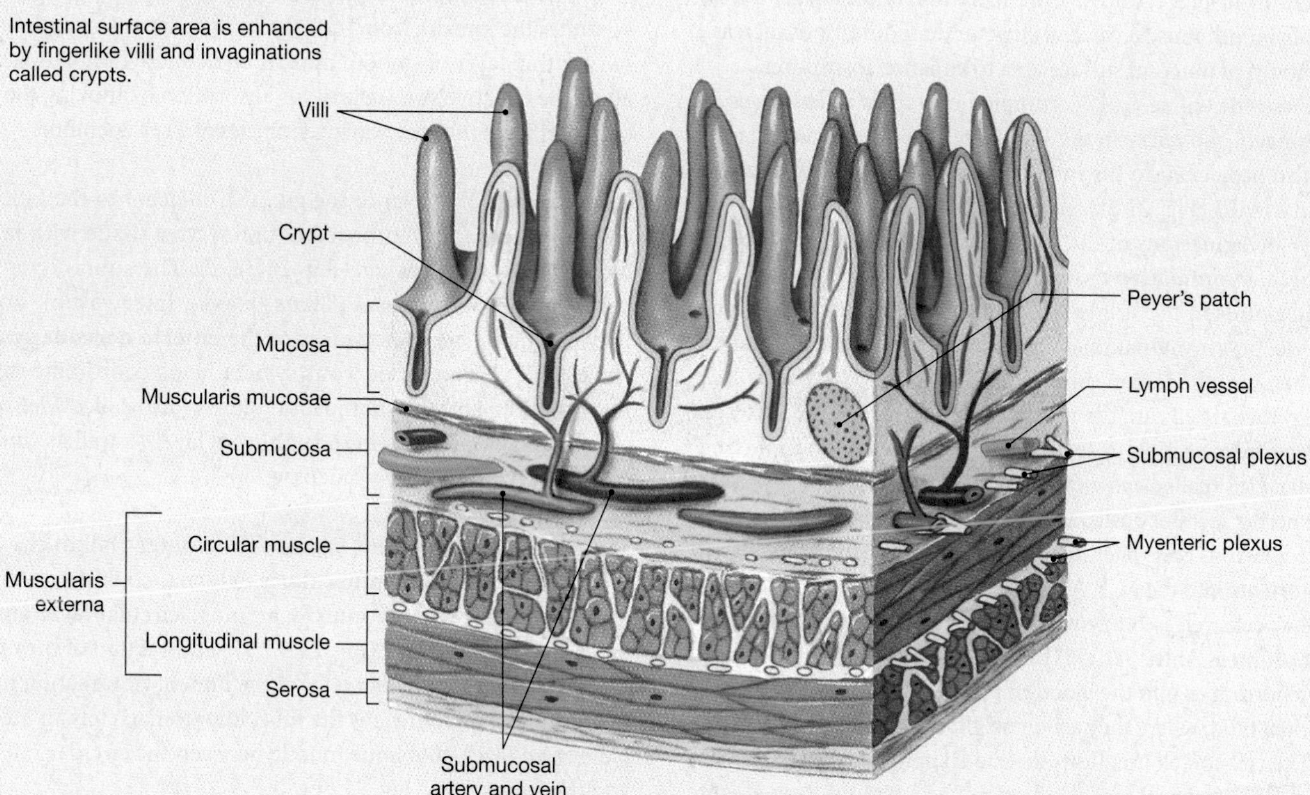
(f) Sectional view of the small intestine
Intestinal surface area is enhanced by fingerlike villi and invaginations called crypts.
Villi
Crypt
Mucosa
Muscularis mucosae
Submucosa
Muscularis externa
Circular muscle
Longitudinal muscle
Serosa
Submucosal artery and vein
Peyer's patch
Lymph vessel
Submucosal plexus
Myenteric plexus

anus, with its **external anal sphincter** of skeletal muscle, which is under voluntary control. The portion of the GI tract running from the stomach to the anus is collectively called the **gut.**

In a living person, the digestive system from mouth to anus is about 450 cm (nearly 15 ft) long! Of this length, 395 cm (about 13 ft) consists of the large and small intestines. Try to imagine 13 feet of rope ranging from 1 to 3 inches in diameter all coiled up inside your abdomen from the belly button down. The tight arrangement of the abdominal organs helps explain why you feel the need to loosen your belt after consuming a large meal.

Measurements of intestinal length made during autopsies are nearly double those given here because after death, the longitudinal muscles of the intestinal tract relax. This relaxation accounts for the wide variation in intestinal length you may encounter in different references.

The GI Tract Wall Has Four Layers

The basic structure of the gastrointestinal wall is similar in the stomach and intestines, although variations exist from one section of the GI tract to another (Fig. 21.3e, f). The wall consists of four layers: (1) an inner *mucosa* facing the lumen, (2) a layer known as the *submucosa,* (3) layers of smooth muscle known collectively as the *muscularis externa,* and (4) a covering of connective tissue called the *serosa.*

Mucosa The **mucosa,** the inner lining of the gastrointestinal tract, is created from (1) a single layer of epithelial cells; (2) the **lamina propria,** subepithelial connective tissue that holds the epithelium in place; and (3) the **muscularis mucosae,** a thin layer of smooth muscle. Several structural modifications increase the amount of mucosal surface area to enhance absorption.

First, the entire wall is crumpled into folds called *rugae* in the stomach, and *plicae* in the small intestine. The intestinal mucosa also projects into the lumen in small fingerlike extensions known as **villi** (Fig. 21.3f). Additional surface area is added by tubular invaginations of the surface that extend down into the supporting connective tissue. These invaginations are called **gastric glands** in the stomach and **crypts** in the intestine. Some of the deepest invaginations form secretory **submucosal glands** that open into the lumen through ducts.

Epithelial cells are the most variable feature of the GI tract, changing from section to section. The cells include transporting epithelial cells (called *enterocytes* in the small intestine), endocrine and exocrine secretory cells, and stem cells. Transporting epithelial cells secrete ions and water into the lumen, and absorb ions, water, and nutrients into the ECF. At the *mucosal* (apical) surface [p. 82], secretory cells release enzymes, mucus, and paracrine molecules into the lumen. At the *serosal* (basolateral) surface, secretory cells secrete hormones into the blood or paracrine messengers into the interstitial fluid, where they act on neighboring cells.

The cell-to-cell junctions that tie GI epithelial cells together vary [p. 78]. In the stomach and colon, the junctions form a tight barrier so that little can pass between the cells. In the small intestine, junctions are not as tight. This intestinal epithelium is considered "leaky" because some water and solutes can be absorbed between the cells (the *paracellular pathway*) instead of through them. We now know that these junctions have plasticity and that their "tightness" and selectivity can be regulated to some extent.

GI *stem cells* are rapidly dividing, undifferentiated cells that continuously produce new epithelium in the crypts and gastric glands. As stem cells divide, the newly formed cells are pushed toward the luminal surface of the epithelium. The average life span of a GI epithelial cell is only a few days, a good indicator of the rough life such cells lead. As with other types of epithelium, the rapid turnover and cell division rate in the GI tract make these organs susceptible to developing cancers. In 2010, cancers of the colon and rectum (colorectal cancer) were the second leading cause of cancer deaths in the United States. The death rate has been steadily dropping, however, due both to more screening examinations and better treatments.

The *lamina propria* is subepithelial connective tissue that contains nerve fibers and small blood and lymph vessels. Absorbed nutrients pass into the blood and lymph here (Fig. 21.3e). This layer also contains wandering immune cells, such as macrophages and lymphocytes, patrolling for invaders that enter through breaks in the epithelium.

In the intestine, collections of lymphoid tissue adjoining the epithelium form small *nodules* and larger **Peyer's patches** that create visible bumps in the mucosa (Fig. 21.3f). These lymphoid aggregations are a major part of the gut-associated lymphoid tissue (GALT).

The third region of the mucosa, the *muscularis mucosae,* separates the mucosa from the submucosa. The muscularis mucosae is a thin layer of smooth muscle, and contraction of this layer alters the effective surface area for absorption by moving the villi back and forth, like the waving tentacles of a sea anemone.

Submucosa The layer of the gut wall adjacent to the mucosa, the **submucosa,** is composed of connective tissue with larger blood and lymph vessels (Fig. 21.3e, f). The submucosa also contains the **submucosal plexus** {*plexus,* interwoven}, one of the two major nerve networks of the **enteric nervous system** [p. 239]. The enteric nervous system helps coordinate digestive function, and the submucosal plexus (also called *Meissner's plexus*) innervates cells in the epithelial layer as well as smooth muscle of the muscularis mucosae.

Muscularis Externa and Serosa The outer wall of the gastrointestinal tract, the **muscularis externa,** consists primarily of two layers of smooth muscle: an inner circular layer and an outer longitudinal layer (Fig. 21.3d, f). Contraction of the circular layer decreases the diameter of the lumen. Contraction of the longitudinal layer shortens the tube. The stomach has an incomplete third layer of oblique muscle between the circular muscles and the submucosa (Fig. 21.3e).

The second nerve network of the enteric nervous system, the **myenteric plexus** {*myo-*, muscle + *enteron,* intestine}, lies between the longitudinal and circular muscle layers. The myenteric plexus (also called *Auerbach's plexus*) controls and coordinates the motor activity of the muscularis externa.

The outer covering of the entire digestive tract, the **serosa,** is a connective tissue membrane that is a continuation of the **peritoneal membrane** (*peritoneum*) lining the abdominal cavity [p. 63]. The peritoneum also forms sheets of **mesentery** that hold the intestines in place so that they do not become tangled as they move.

Now let's take a brief look at the four processes of motility, secretion, digestion, and absorption. Gastrointestinal physiology is a rapidly expanding field, and this textbook does not attempt to be all inclusive. Instead, it focuses on selected broad aspects of digestive physiology.

Concept Check Answers: p. 735

4. Is the lumen of the digestive tract on the apical or basolateral side of the intestinal epithelium? On the serosal or mucosal side?
5. Name the four layers of the GI tract wall, starting at the lumen and moving out.
6. Name the structures a piece of food passes through as it travels from mouth to anus.
7. Why is the digestive system associated with the largest collection of lymphoid tissue in the body?

Motility

Motility in the gastrointestinal tract serves two purposes: moving food from the mouth to the anus and mechanically mixing food to break it into uniformly small particles. This mixing maximizes exposure of the particles to digestive enzymes by increasing particle surface area. Gastrointestinal motility is determined by the properties of the tract's smooth muscle and modified by chemical input from nerves, hormones, and paracrine signals.

GI Smooth Muscle Exhibits Different Patterns of Contraction

Muscle contractions in the gastrointestinal tract occur in three general patterns (■ Fig. 21.4). Between meals, when the tract is largely empty, a series of contractions begins in the stomach and passes slowly from section to section, each series taking about 90 minutes to reach the large intestine. This pattern, known as the **migrating motor complex,** is a "housekeeping" function that sweeps food remnants and bacteria out of the upper GI tract and into the large intestine (Fig. 21.4c).

Muscle contractions during and following a meal fall into one of two other patterns. **Peristalsis** {*peri-*, surrounding + *stalsis,* contraction} is progressive waves of contraction that move from one section of the GI tract to the next, just like the human "waves" that ripple around a football stadium or basketball arena. In peristalsis, circular muscles contract just behind a mass, or **bolus,** of food (Fig. 21.4a). This contraction pushes the bolus forward into a *receiving segment,* where the circular muscles are relaxed. The receiving segment then contracts, continuing the forward movement. Peristaltic contractions push a bolus forward at speeds between 2 and 25 cm/sec.

Peristalsis in the esophagus propels material from pharynx to stomach. Peristalsis contributes to food mixing in the stomach, but in normal digestion intestinal peristaltic waves are limited to short distances. Hormones, paracrine signals, and the autonomic nervous system influence peristalsis in all regions of the GI tract.

In **segmental contractions,** short (1–5 cm) segments of intestine alternately contract and relax (Fig. 21.4b). In the contracting segments, circular muscles contract while longitudinal muscles relax. These contractions may occur randomly along the intestine or at regular intervals. Alternating segmental contractions churn the intestinal contents, mixing them and keeping them in contact with the absorptive epithelium. When segments contract sequentially, in an oral-to-aboral direction {*ab-*, away}, digesting material is propelled short distances.

Motility disorders are among the more common gastrointestinal problems. They range from esophageal spasms and delayed gastric (stomach) emptying to constipation and diarrhea. *Irritable bowel syndrome* is a chronic functional disorder characterized by altered bowel habits and abdominal pain.

GI Smooth Muscle Contracts Spontaneously

Most of the gastrointestinal tract is composed of single-unit smooth muscle, with groups of cells electrically connected by gap junctions [p. 175] to create contracting segments. Different regions exhibit different types of contraction. **Tonic contractions**

CLINICAL FOCUS: DIABETES

Delayed Gastric Emptying

Diabetes mellitus has an impact on almost every organ system. The digestive tract is not exempt. One problem that plagues more than a third of all diabetics is *gastroparesis,* also called delayed gastric emptying. In these patients, the migrating motor complex is absent between meals, and the stomach empties very slowly after meals. Many patients suffer nausea and vomiting as a result. The causes of diabetic gastroparesis are unclear, but recent studies of animal models and human patients show loss or dysfunction of the interstitial cells of Cajal, which serve as pacemakers and as a link between GI smooth muscle and the enteric and autonomic nervous systems. Adopting the cardiac model of an external pacemaker, scientists are now testing an implantable gastric pacemaker to promote gastric motility in diabetic patients with severe gastroparesis.

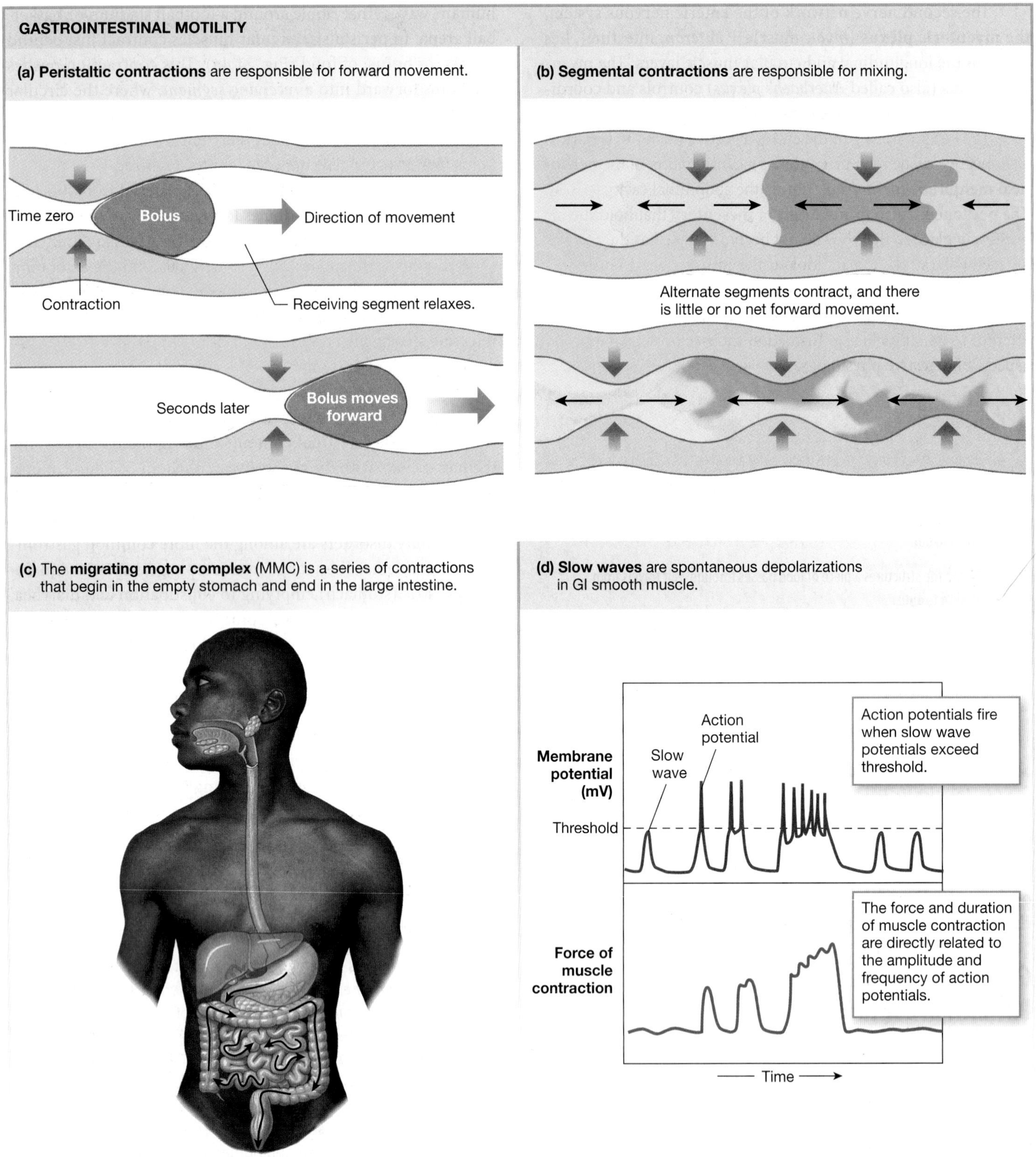

GASTROINTESTINAL MOTILITY

(a) **Peristaltic contractions** are responsible for forward movement.
Time zero
Bolus
Direction of movement
Contraction
Receiving segment relaxes.
Seconds later
Bolus moves forward
(b) **Segmental contractions** are responsible for mixing.
Alternate segments contract, and there is little or no net forward movement.
(c) The **migrating motor complex** (MMC) is a series of contractions that begin in the empty stomach and end in the large intestine.
(d) **Slow waves** are spontaneous depolarizations in GI smooth muscle.
Membrane potential (mV)
Threshold
Slow wave
Action potential
Action potentials fire when slow wave potentials exceed threshold.
Force of muscle contraction
The force and duration of muscle contraction are directly related to the amplitude and frequency of action potentials.
Time

Fig. 21.4

that are sustained for minutes or hours occur in some smooth muscle sphincters and in the anterior portion of the stomach. **Phasic contractions,** with contraction-relaxation cycles lasting only a few seconds, occur in the posterior region of the stomach and in the small intestine.

Cycles of smooth muscle contraction and relaxation are associated with spontaneous cycles of depolarization and repolarization known as **slow wave potentials** (or simply *slow waves*). Slow wave potentials differ from myocardial pacemaker potentials in that the GI waves have a much slower rate (3–12 waves/min

GI versus 60–90 waves/min myocardial) and do not reach threshold with each cycle (Fig. 21.4d).

A slow wave that does not reach threshold does not cause a contraction in the muscle fiber. The likelihood of a slow wave firing an action potential depends on input from the enteric nervous system. When a slow wave potential does reach threshold, voltage-gated Ca^{2+} channels in the muscle fiber open, Ca^{2+} enters, and the cell fires one or more action potentials. The depolarization phase of the action potential, like that in myocardial autorhythmic cells, is the result of Ca^{2+} entry into the cell. In addition, Ca^{2+} entry initiates muscle contraction [p. 431].

Contraction of smooth muscle, like that of cardiac muscle, is graded according to the amount of Ca^{2+} that enters the fiber. The longer the duration of the slow wave, the more action potentials fire, and the greater the contraction force in the muscle. Similarly, the longer the duration of the slow wave, the longer the duration of contraction. Both amplitude and duration can be modified by neurotransmitters, hormones, or paracrine molecules.

Slow wave frequency varies by region of the digestive tract, ranging from 3 waves/min in the stomach to 12 waves/min in the duodenum. Current research indicates that slow waves originate in a network of cells called the **interstitial cells of Cajal** (named for the Spanish neuroanatomist Santiago Ramón y Cajal), or ICCs. These modified smooth muscle cells lie between smooth muscle layers and the intrinsic nerve plexuses, and they may act as an intermediary between the neurons and smooth muscle.

It appears that ICCs function as the pacemakers for slow wave activity in different regions of the GI tract. Slow waves that begin spontaneously in ICCs spread to adjacent smooth muscle layers through gap junctions. Just as in the cardiac conducting system, the fastest pacemaker in a group of ICCs sets the pace for the entire group [p. 481]. The observation that ICCs seem to coordinate GI motility now has researchers working to establish a link between ICCs and functional bowel disorders, such as irritable bowel syndrome and chronic constipation.

Concept Check Answer: p. 735

8. Why are some sphincters of the digestive system tonically contracted?

Secretion

In a typical day, 9 liters of fluid pass through the lumen of an adult's gastrointestinal tract—equal to the contents of three 3-liter soft drink bottles! Only about 2 liters of that volume enter the GI system through the mouth. The remaining 7 liters of fluid come from body water secreted along with enzymes and mucus (see Fig. 21.1). About half of the secreted fluid comes from accessory organs and glands such as the salivary glands, pancreas, and liver. The remaining 3.5 liters are secreted by epithelial cells of the digestive tract itself.

The Digestive System Secretes Ions and Water

A large portion of the 7 liters of fluid secreted by the digestive system each day is composed of water and ions, particularly Na^+, K^+, Cl^-, HCO_3^-, and H^+. The ions are first secreted into the lumen of the tract, then reabsorbed. Water follows osmotic gradients created by the transfer of solutes from one side of the epithelium to the other. Water moves through the epithelial cells via membrane channels or between cells (the paracellular pathway).

Gastrointestinal epithelial cells, like those in the kidney, have distinct apical and basolateral membranes. Each cell surface contains proteins for active transport, facilitated diffusion, and ion movement through channels. The arrangement of channels and transporters on the apical and basolateral membranes determines the direction of movement of solutes and water across the epithelium. In some parts of the digestive tract, Na^+ and water may move through leaky junctions between cells.

Many of the membrane transporters of the GI tract are similar to those of the renal tubule. The basolateral membrane contains the ubiquitous Na^+-K^+-ATPase. Cotransporters include the Na^+-K^+-$2Cl^-$ symporter (NKCC), Cl^--HCO_3^- exchangers, the Na^+-H^+ exchanger (NHE), and H^+-K^+-ATPase. Ion channels include the apical Na^+ channel ENaC, K^+ channels, and Cl^- channels, such as the gated Cl^- channel known as the **cystic fibrosis transmembrane conductance regulator,** or **CFTR chloride channel.** Defects in CFTR channel structure or function cause the disease *cystic fibrosis*.

The sections that follow describe our current models of how cells arrange these membrane proteins to create acid secretion by the stomach, bicarbonate secretion by the pancreas and duodenum, and isotonic NaCl secretion by the intestines and salivary glands.

Acid Secretion **Parietal cells** deep in the gastric glands secrete hydrochloric acid into the lumen of the stomach. Acid secretion in the stomach averages 1–3 liters per day and can create a luminal pH as low as 1. The cytoplasmic pH of the parietal cells is about 7.2, which means the cells are pumping H^+ against a gradient that is 2.5 *million* times more concentrated in the lumen.

The parietal cell pathway for acid secretion is depicted in ■ Figure 21.5a. The process begins when H^+ from water inside the parietal cell is pumped into the stomach lumen by an H^+-K^+-ATPase in exchange for K^+ entering the cell. Cl^- then follows H^+ through open chloride channels, resulting in net secretion of HCl by the cell. (The molecular identity of this chloride channel is uncertain.)

While acid is being secreted into the lumen, bicarbonate made from CO_2 and the OH^- from water is absorbed into the blood. The buffering action of HCO_3^- makes blood leaving the stomach less acidic, creating an "*alkaline tide*" that can be measured as a meal is being digested.

Bicarbonate Secretion Bicarbonate secretion into the duodenum neutralizes acid entering from the stomach. A small amount of bicarbonate is secreted by duodenal cells, but most

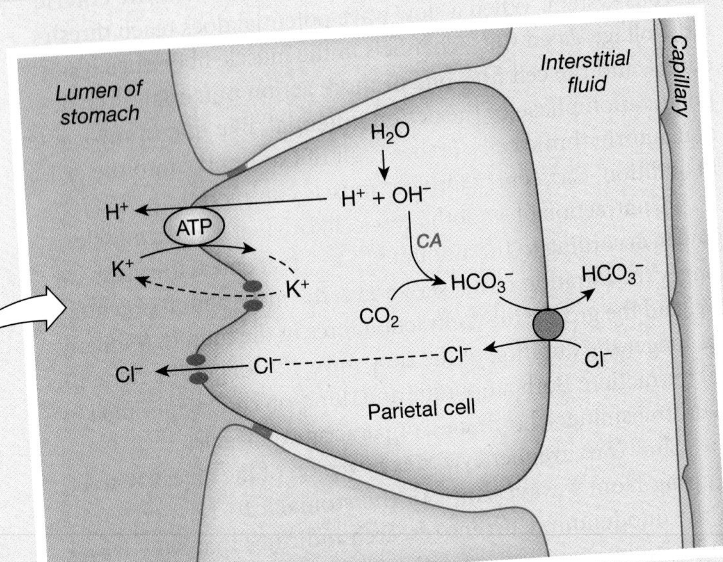

Lumen of stomach
Interstitial fluid
Capillary
H_2O
$H^+ + OH^-$
H^+
ATP
CA
K^+
K^+
HCO_3^-
HCO_3^-
CO_2
Cl^-
Cl^-
Cl^-
Cl^-
Parietal cell

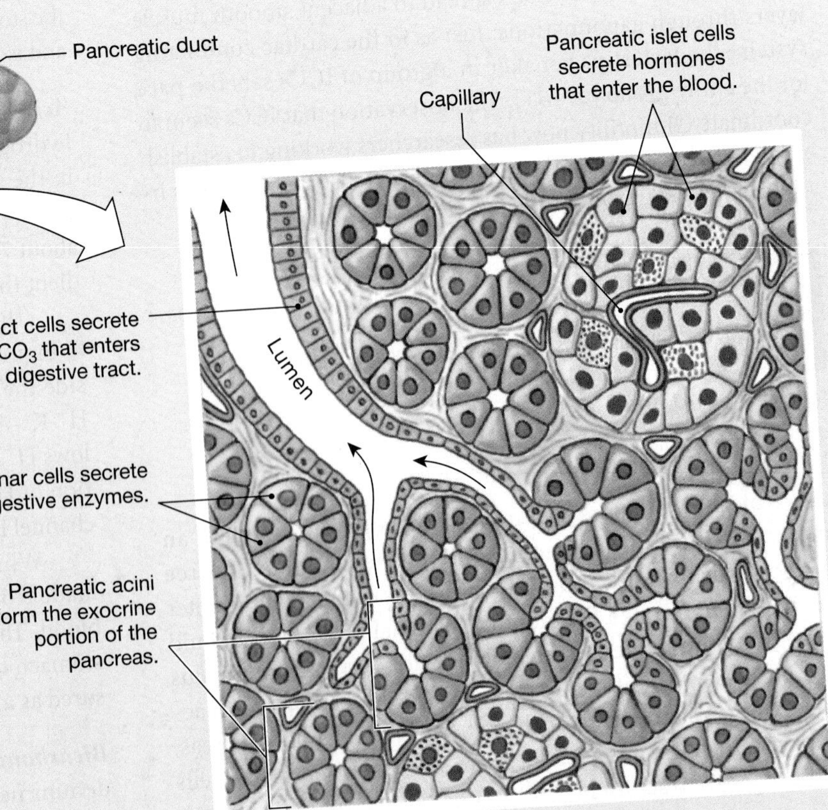

Pancreatic duct
Capillary
Pancreatic islet cells secrete hormones that enter the blood.
ct cells secrete
CO_3 that enters
digestive tract.
Lumen
nar cells secrete
gestive enzymes.
Pancreatic acini
orm the exocrine
portion of the
pancreas.

Intestinal and Pancreatic Secretion

Leaky junctions allow paracellular movement of ions and water. Chloride enters cells by indirect active transport and leaves the apical side through a CFTR channel. Negative Cl^- in the lumen attracts Na^+ by the paracellular pathway. Water follows. Cells that produce bicarbonate have high concentrations of carbonic anhydrase (CA).

(c) Cl^- secretion by intestinal and colonic crypt cells and salivary gland acini

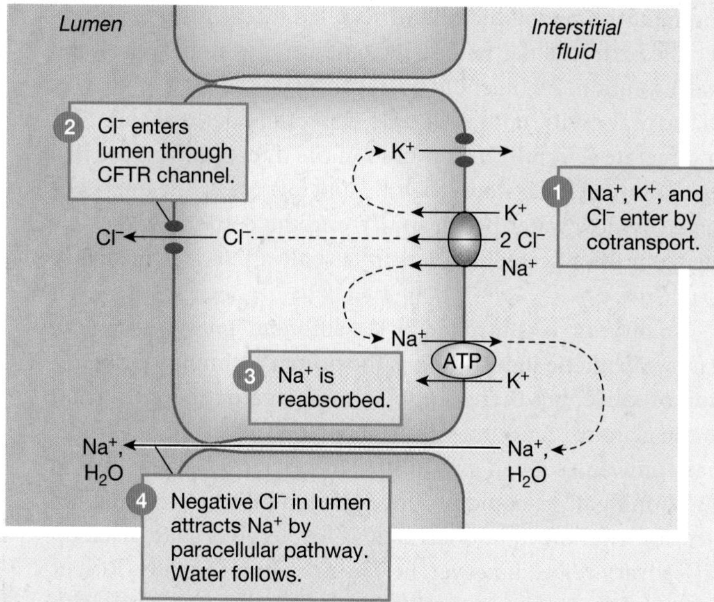

(d) Bicarbonate secretion in the pancreas and duodenum

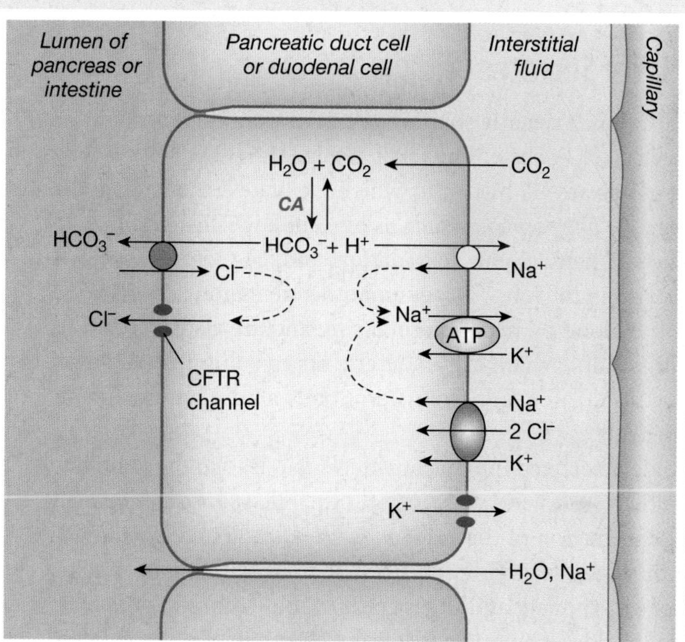

comes from the pancreas, which secretes a watery solution of $NaHCO_3$. The exocrine portion of the pancreas consists of lobules called **acini** {*acinus,* grape or berry} that open into ducts whose lumens are part of the body's external environment (Fig. 21.5b). The *acinar cells* secrete digestive enzymes, and the duct cells secrete the $NaHCO_3$ solution. The pancreas also secretes hormones from *islet cells* tucked among the acinar cells.

Bicarbonate production requires high levels of the enzyme *carbonic anhydrase,* levels similar to those found in renal tubule cells and red blood cells [pp. 682, 613]. Bicarbonate produced from CO_2 and water is secreted by an apical Cl^--HCO_3^- exchanger (Fig. 21.5d). Chloride enters the cell on a basolateral NKCC cotransporter and leaves via an apical CFTR channel. Luminal Cl^- then re-enters the cell in exchange for HCO_3^- entering the lumen. Hydrogen ions produced along with bicarbonate leave the cell on basolateral Na^+-H^+ exchangers. The H^+ thus reabsorbed into the intestinal circulation helps balance HCO_3^- put into the blood when parietal cells secrete H^+ into the stomach (see Fig. 21.5a).

Sodium and water movement in these tissues is a passive process, driven by electrochemical and osmotic gradients. The net movement of negative ions from the ECF to the lumen attracts Na^+, which moves down its electrochemical gradient through leaky junctions between the cells. The secretion of Na^+ and HCO_3^- into the lumen creates an osmotic gradient, and water follows by osmosis. The net result is secretion of a watery sodium bicarbonate solution.

In cystic fibrosis, an inherited defect causes the CFTR channel to be defective or absent. As a result, secretion of Cl^- and fluid ceases, but goblet cells [p. 84] continue to secrete mucus, resulting in thickened mucus. In the digestive system, the thick mucus clogs small pancreatic ducts and prevents digestive enzyme secretion into the intestine. In airways of the respiratory system, where the CFTR channel is also found, failure to secrete fluid clogs the mucociliary escalator [p. 574] with thick mucus, leading to recurrent lung infections.

NaCl Secretion Crypt cells in the small intestine and colon secrete an isotonic NaCl solution that mixes with mucus secreted by goblet cells to help lubricate the contents of the gut. The active step is Cl^- secretion similar to that just described for pancreatic cells (Fig. 21.5c). Chloride from the ECF enters cells via NKCC transporters, then exits into the lumen via apical CFTR channels. Na^+ and water follow along the paracellular pathway, with the end result being secretion of isotonic saline solution. The same model is used to explain isotonic NaCl secretion in the acini of the salivary glands.

Digestive Enzymes Are Secreted into the Lumen

Digestive enzymes are secreted either by exocrine glands (salivary glands and the pancreas) or by epithelial cells in the mucosa of the stomach and small intestine. Enzymes are proteins,

RUNNING PROBLEM

A hallmark of *Vibrio cholerae* infection is profuse, dilute diarrhea sometimes said to resemble "rice water." The toxin secreted by *Vibrio cholerae* is a protein complex with six subunits. The toxin binds to intestinal cells, and the A subunit is taken into the enterocytes by endocytosis. Once inside the enterocyte, the toxin turns on adenylyl cyclase, which then produces cAMP continuously. Because the CFTR channel of the enterocyte is a cAMP-gated channel, the effect of cholera toxin is to open the CFTR channels and keep them open.

Q3: Why would continuously open enterocyte CFTR channels cause secretory diarrhea and dehydration in humans?

which means that they are synthesized on the rough endoplasmic reticulum, packaged by the Golgi complex into secretory vesicles, and then stored in the cell until needed. On demand, they are released by exocytosis [p. 157]. Many intestinal enzymes are not released free into the lumen but remain bound to the apical membranes of intestinal cells, anchored by transmembrane protein "stalks" or lipid anchors [p. 68].

Some digestive enzymes are secreted in an inactive *proenzyme* form known collectively as *zymogens*. Zymogens must be activated in the GI lumen before they can carry out digestion [p. 52]. This late activation allows enzymes to be stockpiled in the cells that make them without damaging those cells. Zymogen names often have the suffix *-ogen* added to the enzyme name, such as *pepsinogen*.

The control pathways for enzyme release vary but include a variety of neural, hormonal, and paracrine signals. Usually, stimulation of parasympathetic neurons in the vagus nerve enhances enzyme secretion.

Specialized Cells Secrete Mucus

Mucus is a viscous secretion composed primarily of glycoproteins collectively called **mucins**. The primary functions of mucus are to form a protective coating over the GI mucosa and to lubricate the contents of the gut. Mucus is made in specialized exocrine cell called *mucous cells* in the stomach, *serous cells* in salivary glands, and *goblet cells* in the intestine [Fig. 3.10d, p. 83]. Goblet cells make up between 10% and 24% of the intestinal cell population.

The signals for mucus release include parasympathetic innervation, a variety of neuropeptides found in the enteric nervous system, and cytokines from immunocytes. Parasitic infections and inflammatory processes in the gut also cause substantial increases in mucus secretion as the body attempts to fortify its protective barrier.

Saliva Is an Exocrine Secretion

Saliva is a complex hyposmotic fluid secreted by the salivary glands of the oral cavity. The salivary glands, like the exocrine pancreas, are organized into acini and ducts (Fig. 21.5b). The components of saliva include water, ions, mucus, and proteins such as enzymes and immunoglobulins.

The ionic composition of saliva is determined in two epithelial transport steps. The salivary glands are exocrine glands, with a secretory epithelium that opens to the outside environment through a duct. Fluid secreted by the acinar cells resembles extracellular fluid in its ionic composition: an isotonic NaCl solution. As this fluid passes through the duct on its way to the oral cavity, epithelial cells along the duct reabsorb Na^+ and secrete K^+ until the ion ratio in the duct fluid is more like that of intracellular fluid (high K^+ and low Na^+). The duct cells have very low water permeability, and the net removal of solute from the secreted fluid results in saliva that is hyposmotic to plasma.

Salivation is controlled by the autonomic nervous system. Parasympathetic innervation is the primary stimulus for secretion of saliva, but there is also some sympathetic innervation to the glands. In ancient China, a person suspected of a crime was sometimes given a mouthful of dry rice to chew during questioning. If he could produce enough saliva to moisten the rice and swallow it, he went free. If his nervous state dried up his salivary reflex, however, he was pronounced guilty. Recent research has confirmed that stress, such as that associated with lying or anxiety from being questioned, decreases salivary secretion.

The Liver Secretes Bile

Bile is a nonenzymatic solution secreted from **hepatocytes,** or liver cells (see Focus on the Liver, ■ Fig. 21.6). The key components of bile are (1) **bile salts,** which facilitate enzymatic fat digestion, (2) *bile pigments,* such as bilirubin, which are the waste products of hemoglobin degradation, and (3) *cholesterol,* which is excreted in the feces. Drugs and other xenobiotics are cleared from the blood by hepatic processing and are also excreted in bile. Bile salts, which act as detergents to solubilize fats during digestion, are made from steroid **bile acids** combined with amino acids.

Bile is secreted into hepatic ducts that lead to the **gallbladder,** which stores and concentrates the bile solution. During a meal, contraction of the gallbladder sends bile into the duodenum through the **common bile duct,** along with a watery solution of bicarbonate and digestive enzymes from the pancreas. The gallbladder is an organ that is not essential for normal digestion, and if the duct becomes blocked by hard deposits known as gallstones, the gallbladder can be removed without creating long-term problems.

Fig. 21.6 *FOCUS ON ...*

The Liver

(a) The liver is the largest of the internal organs, weighing about 1.5 kg (3.3 lb) in an adult. It lies just under the diaphragm, toward the right side of the body.

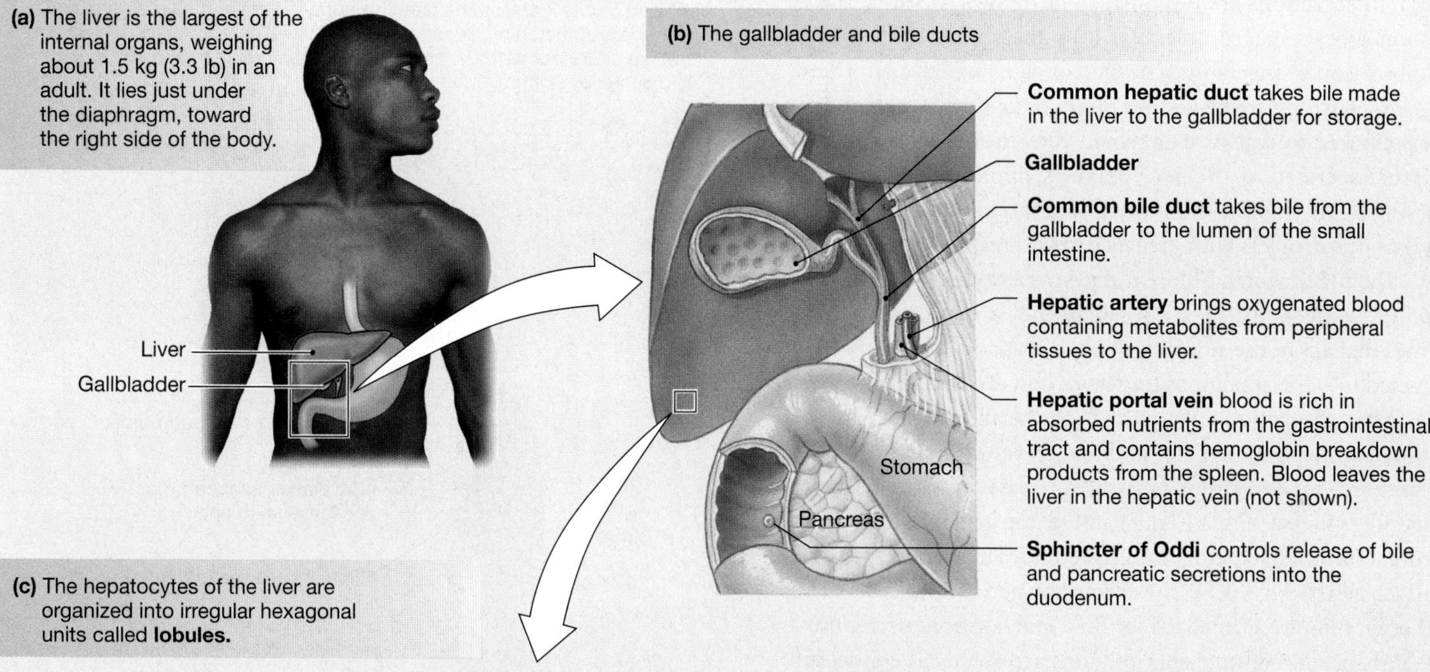

(b) The gallbladder and bile ducts

Common hepatic duct takes bile made in the liver to the gallbladder for storage.

Gallbladder

Common bile duct takes bile from the gallbladder to the lumen of the small intestine.

Hepatic artery brings oxygenated blood containing metabolites from peripheral tissues to the liver.

Hepatic portal vein blood is rich in absorbed nutrients from the gastrointestinal tract and contains hemoglobin breakdown products from the spleen. Blood leaves the liver in the hepatic vein (not shown).

Sphincter of Oddi controls release of bile and pancreatic secretions into the duodenum.

(c) The hepatocytes of the liver are organized into irregular hexagonal units called **lobules.**

Bile canaliculi

Sinusoid

Hepatic artery

Hepatic portal vein

Bile ductule

Hepatocytes are liver cells. About 70% of the surface area of each hepatocyte faces the sinusoids, maximizing the exchange between the blood and the cells.

Each lobule is centered around a central vein that drains blood into the hepatic vein.

Along its periphery, a lobule is associated with branches of the hepatic portal vein and hepatic artery.

These vessels branch among the hepatocytes, forming **sinusoids** into which the blood flows.

The **bile canaliculi** are small channels into which bile is secreted. The canaliculi coalesce into bile ductules that run through the liver alongside the portal veins.

(d) Blood entering the liver brings nutrients and foreign substances from the digestive tract, bilirubin from hemoglobin breakdown, and metabolites from peripheral tissues of the body. In turn, the liver excretes some of these in the bile and stores or metabolizes others. Some of the liver's products are wastes to be excreted by the kidney; others are essential nutrients, such as glucose. In addition, the liver synthesizes an assortment of plasma proteins.

Absorbed from gastrointestinal tract
- Bilirubin
- Nutrients
- Drugs
- Foreign substances

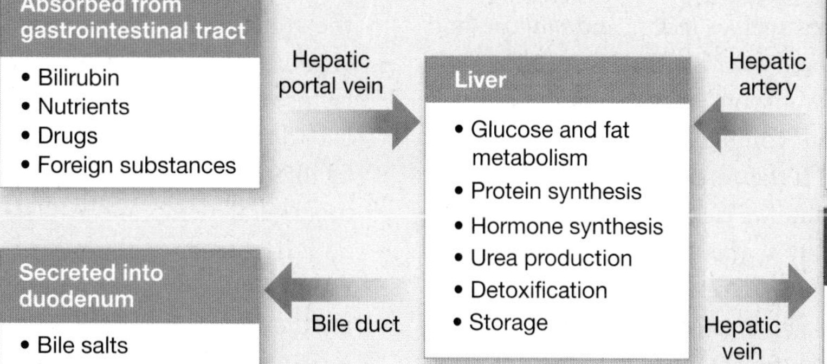

Metabolites and drugs from peripheral tissues
- Bilirubin
- Metabolites of hormones and drugs
- Nutrients

Secreted into duodenum
- Bile salts
- Bilirubin
- Water, ions
- Phospholipids

Metabolites to peripheral tissues
- Glucose
- Plasma proteins: Albumin, clotting factors, angiotensinogen
- Urea
- Vitamin D, somatomedins
- Metabolites for excretion

Digestion and Absorption

Most GI secretions are designed to facilitate digestion. The GI system digests macromolecules into absorbable units using a combination of mechanical and enzymatic breakdown. Chewing and churning create smaller pieces of food with more surface area exposed to digestive enzymes. Bile, the complex chemical mixture secreted by the liver, serves a similar purpose by dispersing lipids (more commonly called *fats* in digestive physiology) as fine droplets with greater surface area.

The pH at which different digestive enzymes function best [p. 106] reflects the location where they are most active. Enzymes that act in the stomach work well at acidic pH, and those secreted into the small intestine work best at alkaline pH.

Most absorption takes place in the small intestine, with additional absorption of water and ions in the large intestine. The surface area for absorption is greatly increased by the presence of fingerlike villi (see Fig. 21.3d, f) and by the **brush border** on the luminal surface of enterocytes, created from numerous microvilli on each cell (Fig. 21.7). The crypts also add to surface area, but the crypt cells are specialized for fluid and hormone secretion.

Absorption of nutrients and ions across the GI epithelium, like secretion, uses many of the same transport proteins as the kidney tubule. Once absorbed, most nutrients enter capillaries within the villi. The exception is fats, which mostly enter lymph vessels called **lacteals.**

Digestion and absorption are not directly regulated except in a few instances. Instead, they are influenced primarily by motility and secretion in the digestive tract, the two processes that in turn are regulated by hormones, the nervous system, and local control mechanisms.

THE VILLUS AND A CRYPT IN THE SMALL INTESTINE

Villi and crypts increase the effective surface area of the small intestine. Stem cells in the crypts produce new epithelial cells to replace those that die or are damaged. Most absorption occurs along the villi. Most fluid secretion occurs in the crypts.

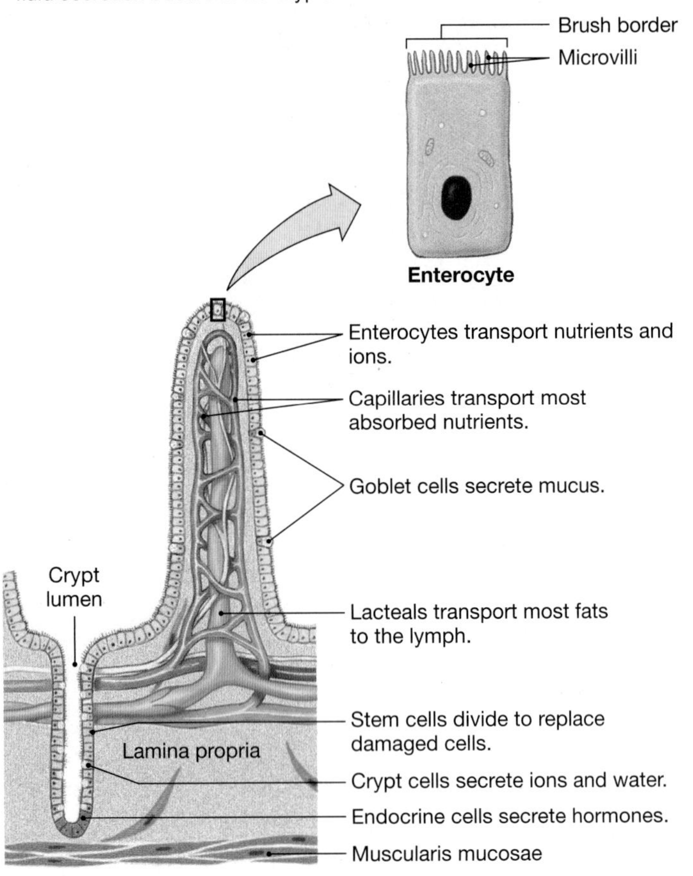
Brush border
Microvilli
Enterocyte
Enterocytes transport nutrients and ions.
Capillaries transport most absorbed nutrients.
Goblet cells secrete mucus.
Crypt lumen
Lacteals transport most fats to the lymph.
Stem cells divide to replace damaged cells.
Lamina propria
Crypt cells secrete ions and water.
Endocrine cells secrete hormones.
Muscularis mucosae

Fig. 21.7

Carbohydrates Are Absorbed As Monosaccharides

About half the calories the average American ingests are in the form of carbohydrates, mainly *starch* and *sucrose* (table sugar). Other dietary carbohydrates include the glucose polymers *glycogen* and *cellulose,* disaccharides such as *lactose* and *maltose,* and the monosaccharides *glucose* and *fructose* [Fig. 2.2, p. 34].

Intestinal carbohydrate transport is restricted to monosaccharides, which means that all complex carbohydrates and disaccharides must be digested if they are to be absorbed. We are unable to digest cellulose because we lack the necessary enzymes. As a result, the cellulose in plant matter becomes what is known as dietary *fiber* or *roughage* and is excreted undigested. Similarly, *sucralose* (Splenda®), the artificial sweetener made from sucrose, cannot be digested because chlorine atoms substituted for three hydroxyl groups block enzymatic digestion of the sugar derivative.

The complex carbohydrates we can digest are starch and glycogen (Fig. 21.8a). The enzyme **amylase** breaks long glucose polymers into smaller glucose chains and into the disaccharide maltose. Maltose and other disaccharides are broken down by intestinal brush-border enzymes known as **disaccharidases** (maltase, sucrase, and lactase). The end products of carbohydrate digestion are glucose, galactose, and fructose.

Intestinal glucose and galactose absorption uses transporters identical to those found in the renal proximal tubule: the apical Na^+-glucose SGLT symporter and the basolateral GLUT2 transporter (Fig. 21.8b). These transporters move galactose as well as glucose. Fructose absorption, however, is not Na^+-dependent. Fructose moves across the apical membrane by facilitated diffusion on the GLUT5 transporter and across the basolateral membrane by GLUT2 [p. 149].

If glucose is the major metabolic substrate for aerobic respiration, why don't enterocytes use the glucose they absorb for their own metabolism? How can these cells keep intracellular glucose concentrations high so that facilitated diffusion moves glucose into the extracellular space? The metabolism of enterocytes (and proximal tubule cells) apparently differs from that of most other cells in that these transporting cells do not use glucose as their preferred energy source. Current studies indicate that these cells use the amino acid glutamine as their

■ Fig. 21.8 *ESSENTIALS*

Digestion and Absorption: Carbohydrates and Proteins

CARBOHYDRATES

Most carbohydrates in our diets are disaccharides and complex carbohydrates. Cellulose is not digestible. All other carbohydrates must be digested to monosaccharides before they can be absorbed.

(a) Carbohydrates break down into monosaccharides.

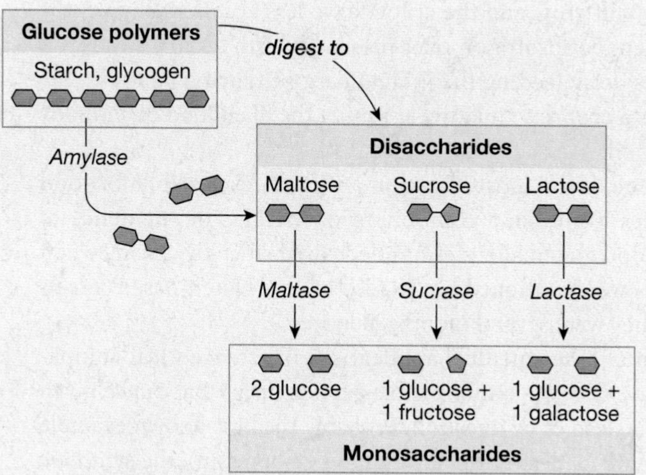

(b) Carbohydrate absorption in the small intestine

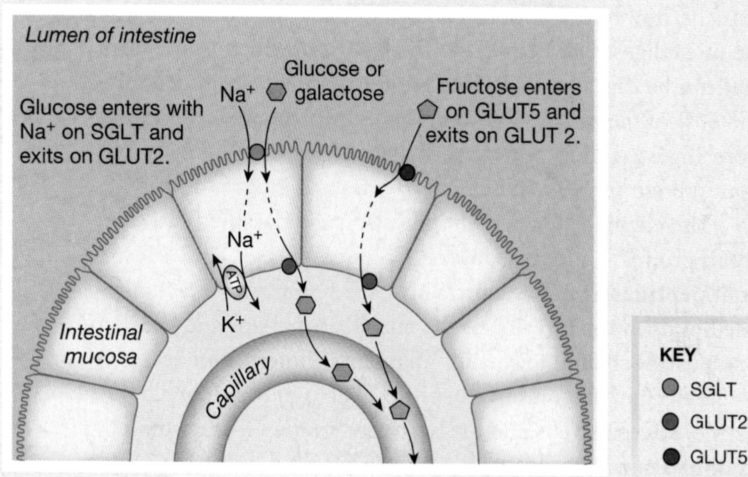

PROTEINS

Proteins are chains of amino acids.

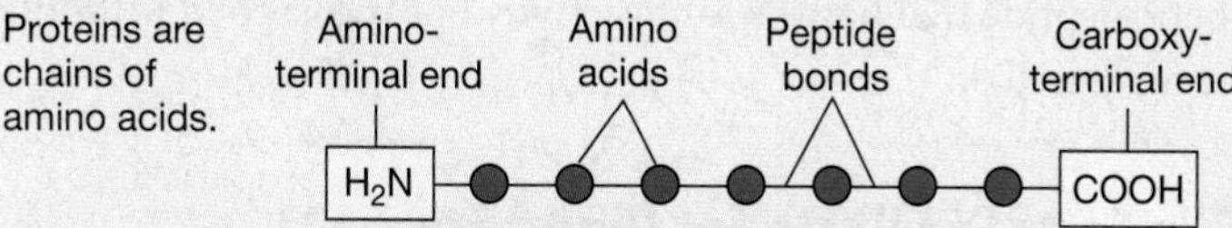

(c) Enzymes for protein digestion

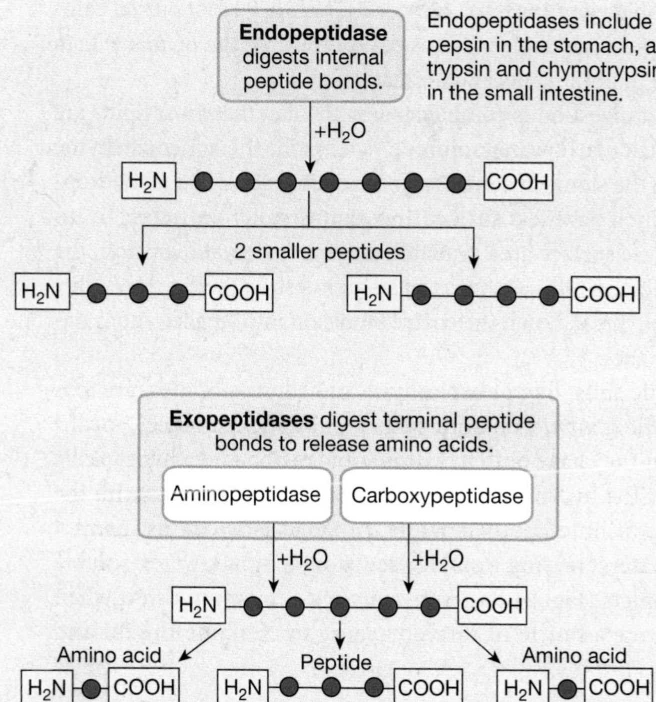

(d) Peptide absorption

After digestion, proteins are absorbed mostly as free amino acids. A few di- and tripeptides are absorbed. Some peptides larger than tripeptides can be absorbed by transcytosis.

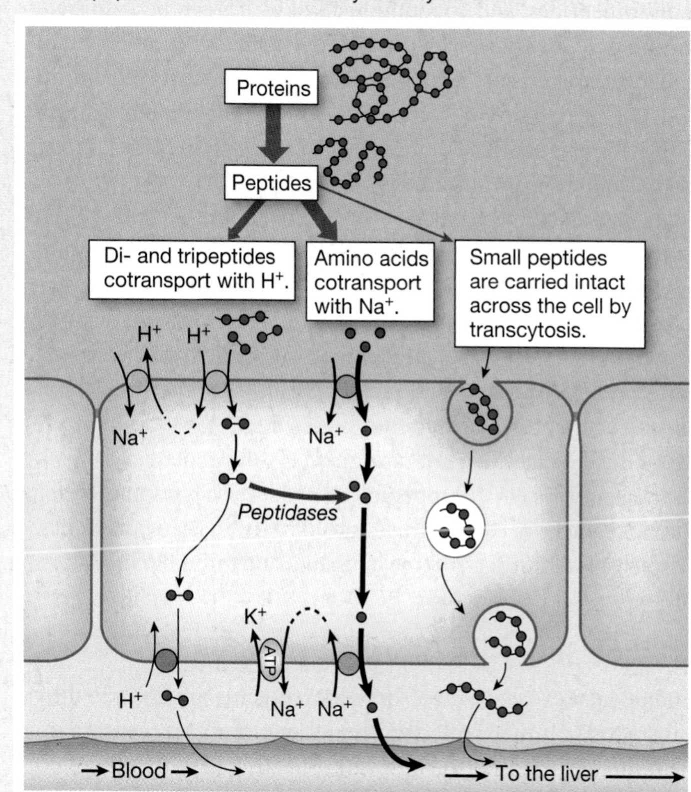

main source of energy, thus allowing absorbed glucose to pass unchanged into the bloodstream.

Proteins Are Digested into Small Peptides and Amino Acids

Unlike carbohydrates, which are ingested in forms ranging from simple to complex, most ingested proteins are polypeptides or larger [Fig. 2.3, p. 35]. Not all proteins are equally digestible by humans, however. Plant proteins are the least digestible. Among the most digestible is egg protein, 85–90% of which is in a form that can be digested and absorbed. Surprisingly, between 30% and 60% of the protein found in the intestinal lumen comes not from ingested food but from the sloughing of dead cells and from protein secretions such as enzymes and mucus.

The enzymes for protein digestion are classified into two broad groups: endopeptidases and exopeptidases (Fig. 21.8c). **Endopeptidases,** more commonly called *proteases,* attack peptide bonds in the interior of the amino acid chain and break a long peptide chain into smaller fragments. Proteases are secreted as inactive proenzymes from epithelial cells in the stomach, intestine, and pancreas and are activated in the GI tract lumen. Examples of proteases include **pepsin** secreted in the stomach, and **trypsin** and **chymotrypsin** secreted by the pancreas.

Exopeptidases release single amino acids from peptides by chopping them off the ends, one at a time (Fig. 21.8c). The most important digestive exopeptidases are two isozymes of *carboxypeptidase* secreted by the pancreas. *Aminopeptidases* play a lesser role in digestion.

The primary products of protein digestion are free amino acids, dipeptides, and tripeptides, all of which can be absorbed. Amino acid structure is so variable that multiple amino acid transport systems are found in the intestine. Most free amino acids are carried by Na^+-dependent cotransport proteins similar to those in the proximal tubule of the kidney (Fig. 21.8d). A few amino acid transporters are H^+-dependent.

Dipeptides and tripeptides are carried into the mucosal cell on the oligopeptide transporter *PepT1* that uses H^+-dependent cotransport (Fig. 21.8d). Once inside the epithelial cell, the *oligopeptides* {*oligos,* little} have two possible fates. Most are digested by cytoplasmic peptidases into amino acids, which are then transported across the basolateral membrane and into the circulation. Those oligopeptides that are not digested are transported intact across the basolateral membrane on an H^+-dependent exchanger. The transport system that moves oligopeptides also is responsible for intestinal uptake of certain drugs, including beta-lactam antibiotics, angiotensin-converting enzyme inhibitors, and thrombin inhibitors.

Some Larger Peptides Can Be Absorbed Intact

Some peptides larger than three amino acids are absorbed by transcytosis [p. 160] after binding to membrane receptors on the luminal surface of the intestine. The discovery that ingested proteins can be absorbed as small peptides has implications in medicine because these peptides may act as *antigens,* substances that stimulate antibody formation and result in allergic reactions. Consequently, the intestinal absorption of peptides may be a significant factor in the development of food allergies and food intolerances.

In newborns, peptide absorption takes place primarily in intestinal crypt cells (Fig. 21.7). At birth, intestinal villi are very small, so the crypts are well exposed to the luminal contents. As the villi grow and the crypts have less access to chyme, the high peptide absorption rates present at birth decline steadily. If parents delay feeding the infant allergy-inducing peptides, the gut has a chance to mature, lessening the likelihood of antibody formation.

One of the most common antigens responsible for food allergies is gluten, a component of wheat. The incidence of childhood gluten allergies has decreased since the 1970s, when parents were cautioned not to feed infants gluten-based cereals until they were several months old.

In another medical application, pharmaceutical companies have developed indigestible peptide drugs that can be given orally instead of by injection. Probably the best-known example is DDAVP (1-deamino-8-D-arginine vasopressin), the synthetic analog of vasopressin. If the natural hormone vasopressin is ingested, it is digested rather than absorbed intact. By changing the structure of the hormone slightly, scientists created a synthetic peptide that has the same activity but is absorbed without being digested.

Bile Salts Facilitate Fat Digestion

Fats and related molecules in the Western diet include triglycerides, cholesterol, phospholipids, long-chain fatty acids, and the fat-soluble vitamins [Fig. 2.1, p. 33]. Nearly 90% of our fat calories come from triglycerides because they are the primary form of lipid in both plants and animals.

Fat digestion is complicated by the fact that most lipids are not particularly water soluble. As a result, the aqueous chyme leaving the stomach contains a coarse emulsion of large fat droplets, which have less surface area than smaller particles. To increase the surface area available for enzymatic fat digestion, the liver secretes bile salts into the small intestine (■ Fig. 21.9). Bile salts help break down the coarse emulsion into smaller, more stable particles.

Bile salts, like phospholipids of cell membranes, are *amphipathic* {*amphi-,* on both sides + *pathos,* experience}, meaning that they have both a hydrophobic region and a hydrophilic region. The hydrophobic regions of bile salts associate with the surface of lipid droplets while the polar side chains interact with water, creating a stable emulsion of small, water-soluble fat droplets (Fig. 21.9a). You can see a similar emulsion when you shake a bottle of salad dressing to combine the oil and aqueous layers.

Fig. 21.9 *ESSENTIALS*

Digestion and Absorption: Fats

Most lipids are hydrophobic and must be emulsified to facilitate digestion in the aqueous environment of the intestine.

(a) Bile salts coat lipids to make emulsions.

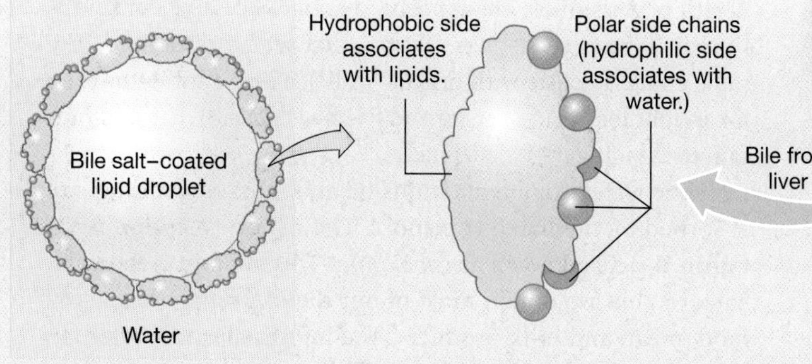

(b) Micelles are small disks with bile salts, phospholipids, fatty acids, cholesterol, and mono- and diglycerides.

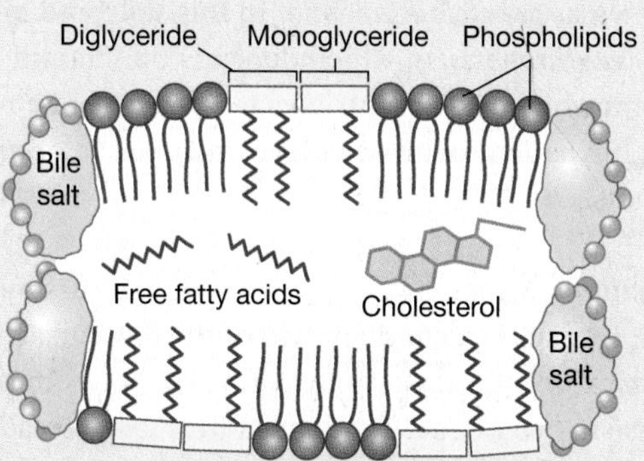

(c) Lipase and colipase digest triglycerides.

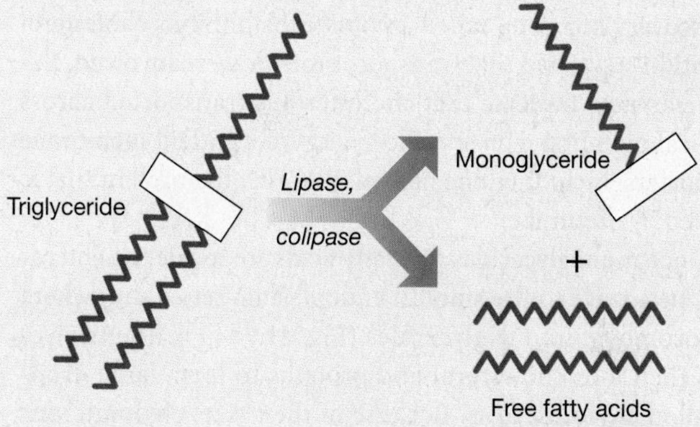

(d) Fat digestion and absorption

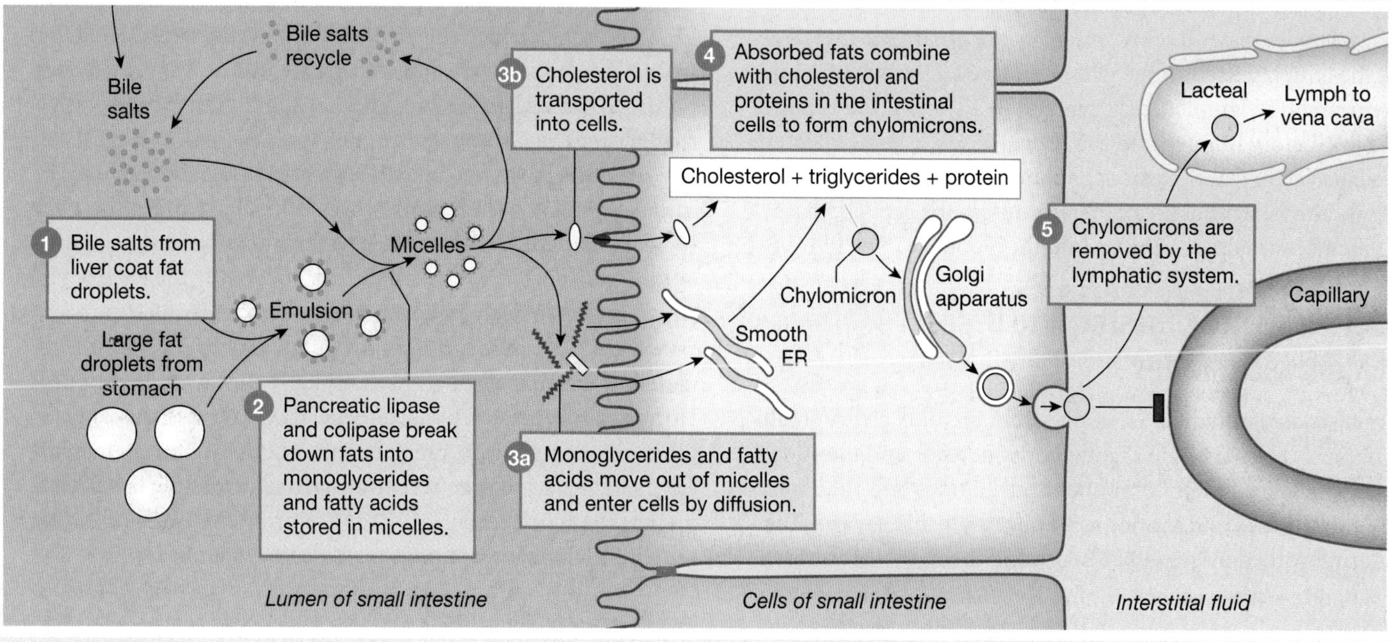

Enzymatic fat digestion is carried out by **lipases,** enzymes that remove two fatty acids from each triglyceride molecule. The result is one monoglyceride and two free fatty acids (Fig. 21.9c). The bile salt coating of the intestinal emulsion complicates digestion, however, because lipase is unable to penetrate the bile salts. For this reason, fat digestion also requires **colipase,** a protein cofactor secreted by the pancreas. Colipase displaces some bile salts, allowing lipase access to fats inside the bile salt coating. Phospholipids are digested by pancreatic *phospholipase*. Free cholesterol need not be digested before being absorbed.

As enzymatic and mechanical digestion proceed, fatty acids, bile salts, monoglycerides, phospholipids, and cholesterol form small disk-shaped **micelles** (Fig. 21.9b) [p. 66]. Micelles then enter the unstirred aqueous layer close to the absorptive cells (*enterocytes*) lining the small intestine lumen.

Because fats are lipophilic, many are absorbed primarily by simple diffusion. Fatty acids and monoglycerides move out of their micelles and diffuse across the apical membrane into the epithelial cells (Fig. 21.9d). Initially scientists believed that cholesterol also diffused across the enterocyte membrane, but the discovery of a drug called *ezetimibe* that inhibits cholesterol absorption suggested that transport proteins were involved. Experiments now indicate that cholesterol is transported across the apical membrane on specific, energy-dependent membrane transporters, including one named *NPC1L1*, the protein that is inhibited by ezetimibe.

Once monoglycerides and fatty acids are inside the enterocytes, they move to the smooth endoplasmic reticulum, where they recombine into triglycerides (Fig. 21.9d (4)). The triglycerides then join cholesterol and proteins to form large droplets called **chylomicrons**. Because of their size, chylomicrons must be packaged into secretory vesicles and leave the cell by exocytosis.

The large size of chylomicrons also prevents them from crossing the basement membrane to enter capillaries (Fig. 21.9 (5)). Instead, chylomicrons are absorbed into *lacteals*, the lymph vessels of the villi (see Fig. 21.7). Chylomicrons pass through the lymphatic system and finally enter the venous blood just before it flows into the heart [p. 532]. Some shorter fatty acids (10 or fewer carbons) are not assembled into chylomicrons. These fatty acids can therefore cross the capillary basement membrane and go directly into the blood.

Nucleic Acids Are Digested into Bases and Monosaccharides

The nucleic acid polymers DNA and RNA are only a very small part of most diets. They are digested by pancreatic and intestinal enzymes, first into their component nucleotides and then into nitrogenous bases and monosaccharides [Fig. 2.4, p. 36]. The bases are absorbed by active transport, and the monosaccharides are absorbed by facilitated diffusion and secondary active transport, as other simple sugars are.

The Intestine Absorbs Vitamins and Minerals

In general, the fat-soluble vitamins (A, D, E, and K) are absorbed in the small intestine along with fats—one reason that health professionals are concerned about excessive consumption of "fake fats," such as Olestra, that are not absorbed. The same concern exists with orlistat (Alli®), a lipase inhibitor used for weight loss, and users are advised to take a daily multivitamin to avoid vitamin deficiencies.

The water-soluble vitamins (C and most B vitamins) are absorbed by mediated transport. The major exception is **vitamin** B_{12}, also known as *cobalamin*. This vitamin is made by bacteria, but we obtain most of our dietary supply from seafood, meat, and milk products. The intestinal transporter for B_{12} is found only in the ileum and recognizes B_{12} only when the vitamin is complexed with a protein called **intrinsic factor,** secreted by the same gastric parietal cells that secrete acid. In the absence of intrinsic factor, vitamin B_{12} deficiency causes the condition known as *pernicious anemia*. In this state, red blood cell synthesis (*erythropoiesis*), which depends on vitamin B_{12}, is severely diminished. Lack of intrinsic factor cannot be remedied directly, but patients with pernicious anemia can be given vitamin B_{12} shots.

Iron and Calcium Mineral absorption usually occurs by active transport. Iron and calcium are two of the few substances whose intestinal absorption is actively regulated. For both minerals, a decrease in body concentrations of the mineral leads to enhanced uptake at the intestine.

Dietary iron is ingested as heme iron [p. 554] in meat and as ionized iron in some plant products. Heme iron is absorbed by an apical transporter on the enterocyte (Fig. 21.10a). Ionized iron (Fe^{2+}) is actively absorbed by apical cotransport with H^+ on a protein called the *divalent metal transporter 1* (*DMT1*). Inside the cell, enzymes convert heme iron to Fe^{2+}, and both pools of ionized iron leave the cell on a transporter called *ferroportin*.

Iron uptake by the body is regulated by a peptide hormone called *hepcidin*. When body stores of iron are high, the liver secretes hepcidin, which binds to ferroportin. The hepcidin-bound transporter is targeted for destruction by the enterocyte, which results in decreased iron uptake across the intestine.

Most Ca^{2+} absorption in the gut occurs by passive, unregulated movement through paracellular pathways (Fig. 21.10b). Hormonally regulated transepithelial Ca^{2+} transport takes place in the duodenum. Calcium enters the enterocyte through apical Ca^{2+} channels and is actively transported across the basolateral membrane by either a Ca^{2+}-ATPase or by the Na^+-Ca^{2+} antiporter. Calcium absorption is regulated by vitamin D_3.

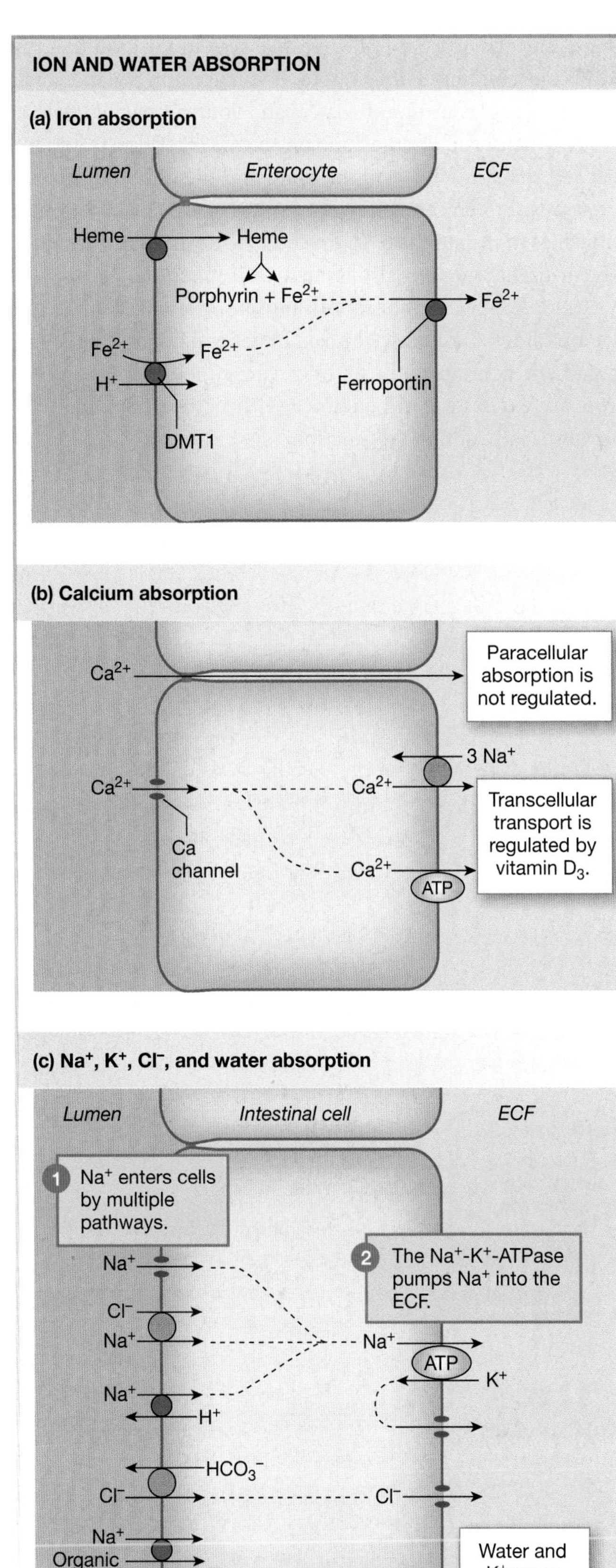

Fig. 21.10

RUNNING PROBLEM

Rehydrating people with cholera is the key to their survival. Most patients who develop cholera can be treated successfully with oral rehydration salts. But in about 5% of patients, the dehydration caused by cholera-induced diarrhea can be severe. If left untreated, these patients can die from circulatory collapse as soon as 18 hours after infection. Because Brooke's blood pressure was so low, the medical personnel decided that she needed intravenous (IV) fluids to restore her volume.

Q4: Recipes for oral rehydration therapy usually include sugar (sucrose) and table salt. Explain how the sugar enhances intestinal absorption of Na^+.

Q5: Which type of IV solution would you select for Brooke, and why? Your choices are normal (isotonic) saline, half-normal saline, and 5% dextrose in water (D-5-W).

The Intestine Absorbs Ions and Water

Most water absorption takes place in the small intestine, with an additional 0.5 L per day absorbed in the colon. As with intestinal fluid secretion, water follows osmotic gradients created by solute movement. Enterocytes in the small intestine and **colonocytes,** the epithelial cells on the luminal surface of the colon, absorb Na^+ using three membrane proteins (Fig. 21.10c): apical Na^+ channels, a Na^+-Cl^- symporter, and the Na^+-H^+ exchanger (NHE). In the small intestine, a significant fraction of Na^+ absorption also takes place through Na^+-dependent organic solute uptake, such as the SGLT and Na^+-amino acid transporters (see Fig. 21.8).

On the basolateral side of both enterocytes and colonocytes, the primary transporter for Na^+ is Na^+-K^+-ATPase. Chloride uptake uses an apical Cl^--HCO_3^- exchanger and a basolateral Cl^- channel. Potassium and water absorption in the intestine occur by the paracellular pathway.

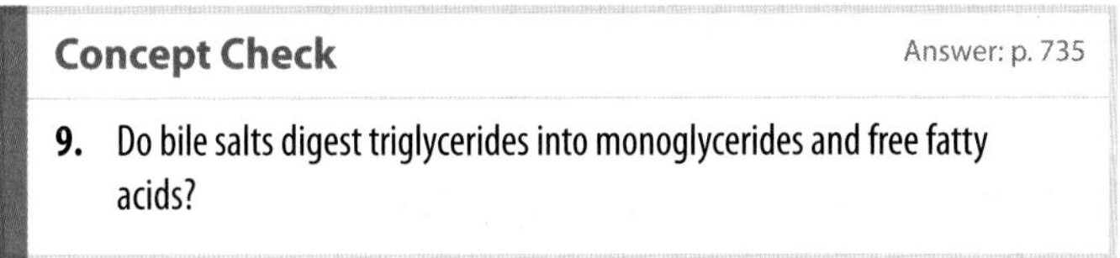

Concept Check

Answer: p. 735

9. Do bile salts digest triglycerides into monoglycerides and free fatty acids?

Regulation of GI Function

The digestive system has remarkably complex regulation of motility and secretion. The various control mechanisms, with neural, endocrine, and local components, include the following pathways.

Long Reflexes Integrated in the CNS A classic neural reflex begins with a stimulus transmitted along a sensory neuron to the CNS, where the stimulus is integrated and acted on. In the digestive system, some classic reflexes originate with sensory receptors in the GI tract, but others originate outside the digestive system (gray arrows in ■ Fig. 21.11). No matter where they originate, digestive reflexes integrated in the CNS are called **long reflexes**.

Long reflexes that originate completely outside the digestive system include feedforward reflexes [p. 18] and emotional reflexes. These reflexes are called **cephalic reflexes** because they originate in the brain {*cephalicus,* head}. *Feedforward reflexes* begin with stimuli such as the sight, smell, sound, or thought of food, and they prepare the digestive system for food that the brain is anticipating. For example, if you are hungry and smell dinner cooking, your mouth waters and your stomach growls.

The influence of emotions on the GI tract illustrates another reflexive link between the brain and the digestive system. GI responses to emotions range from traveler's constipation to "butterflies in the stomach" to psychologically induced diarrhea. Fight-or-flight reactions also influence GI function.

In long reflexes, the smooth muscle and glands of the GI tract are under autonomic control. In general, we say that parasympathetic neurons to the GI tract, carried mostly in the vagus nerve, are excitatory and enhance GI functions. Sympathetic neurons usually inhibit GI function.

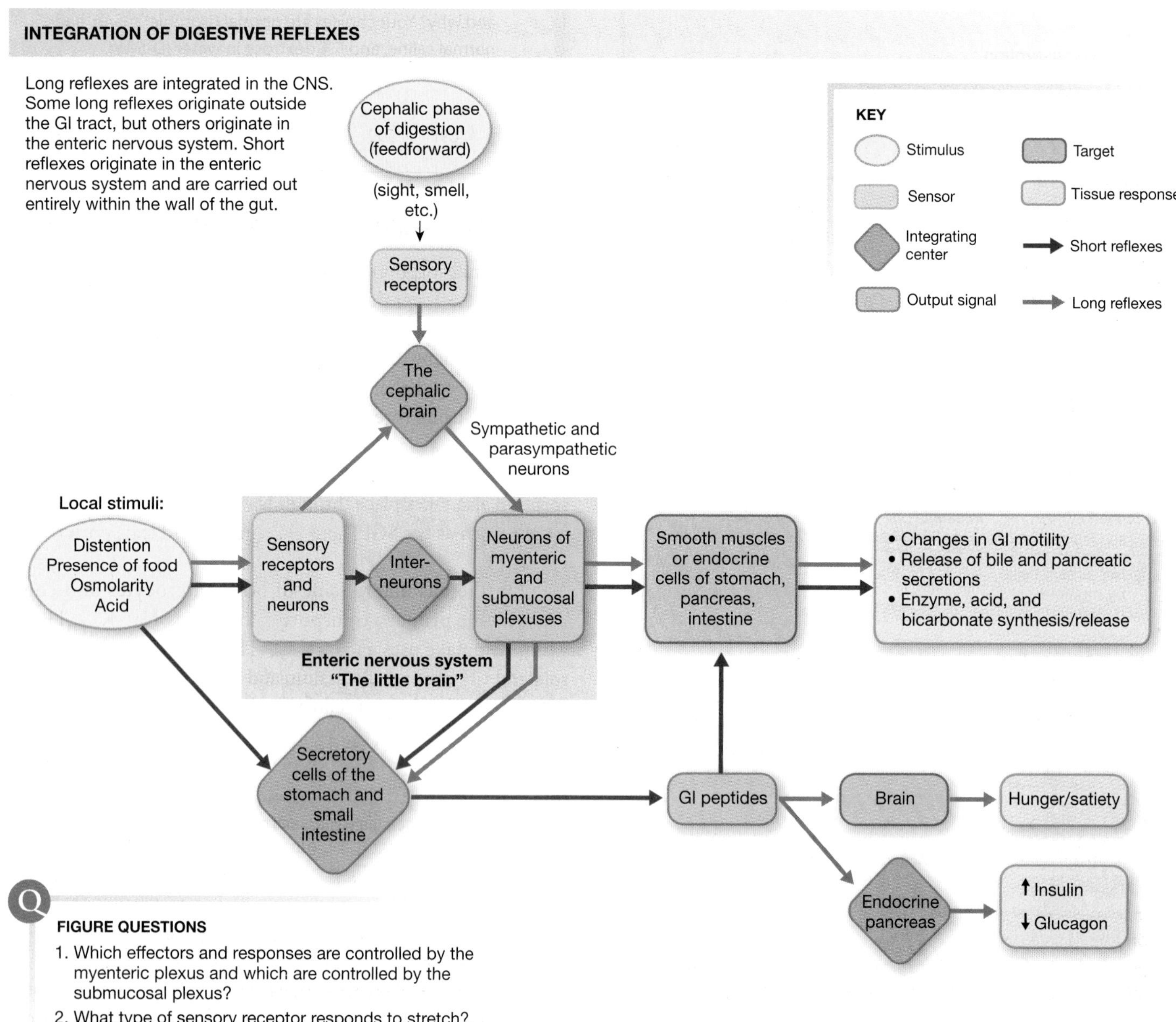

■ **Fig. 21.11**

Short Reflexes Integrated in the Enteric Nervous System Neural control of the GI tract does not rely strictly on the CNS. Instead, the enteric nerve plexus in the gut wall acts as a "little brain," allowing local reflexes to begin, be integrated, and end completely in the GI tract (red arrows in Fig. 21.11). Reflexes that originate within the enteric nervous system (ENS) and are integrated there without outside input are called **short reflexes.** Although the ENS can work in isolation, it also coordinates function with autonomic neurons bringing signals from the CNS.

The processes controlled by the enteric nervous system are related to motility, secretion, and growth. The submucosal plexus contains sensory neurons that receive signals from the lumen of the gut. The ENS network integrates the sensory information, then initiates responses through submucosal neurons that control secretion by GI epithelial cells. The ENS works through neurons in the myenteric plexus to influence motility.

Reflexes Involving GI Peptides Peptides secreted by cells of the digestive tract may act as hormones or paracrine signals. Some of the peptide signal molecules involved in digestive regulation were first identified in other body systems. Because their names have nothing to do with their function in the gastrointestinal system, learning the terminology can be a challenge.

GI hormones, like all hormones, are secreted into the blood. They act on the GI tract, on accessory organs such as the pancreas, and on more distant targets, such as the brain. Paracrine molecules are secreted either into the lumen of the GI tract or into the extracellular fluid. Luminal paracrine signals combine with receptors on the apical membrane of the luminal epithelium to elicit a response. Paracrine molecules in the ECF act locally, on cells close to where they were secreted.

EMERGING CONCEPTS

Taste Receptors in the Gut

Scientists have known for years that the GI tract has the ability to sense and respond specifically and differentially to the composition of a meal. Fats and proteins do not stimulate the same endocrine and exocrine responses as a meal of pure carbohydrate. But how does the gut "know" what's in a meal? Traditional sensory receptors, such as osmoreceptors and stretch receptors, are not tuned to respond to biomolecules. New research indicates that epithelial cells in the gut, especially some of the endocrine cells, express the same G protein–coupled receptors and the taste-linked G protein gustducin as taste buds [p. 344]. Researchers using knockout mice and cultured cell lines are now trying to establish the functional link between gut "taste receptors" and physiological responses to food.

In the digestive system, GI peptides excite or inhibit motility and secretion (Fig. 21.11). Motility effects include altered peristaltic activity, contraction of the gallbladder for bile release, and regulated gastric emptying to maximize digestion and absorption. Secretory processes influenced by GI peptides include both endocrine and exocrine functions, such as insulin secretion from the endocrine pancreas.

Some of the most interesting pathways for GI peptides involve the brain. For example, in experimental studies the GI hormone cholecystokinin (CCK) was found to enhance *satiety,* the feeling that hunger has been satisfied. However, these findings are complicated by the fact that CCK is also manufactured by neurons and functions as a neurotransmitter in the brain. Another GI peptide, *ghrelin,* is secreted by the stomach and acts on the brain to increase food intake.

The Enteric Nervous System Can Act Independently

The ENS was first recognized more than a century ago, when scientists noted that an increase in GI intraluminal pressure caused a reflex wave of peristaltic contraction to sweep along sections of isolated intestine removed from the body. What they observed was the ability of the ENS to carry out a reflex independent of control by the CNS.

In this respect, the ENS is much like the nerve networks of jellyfish and sea anemones (phylum Cnidaria) [p. 289]. You might have seen sea anemones being fed at an aquarium. As a piece of shrimp or fish drifts close to the tentacles, they begin to wave, picking up chemical "odors" through the water. Once the food contacts the tentacles, it is directed toward the mouth, passed from one tentacle to another until it disappears into the digestive cavity.

This purposeful reflex is accomplished without a brain, eyes, or a nose. The anemone's nervous system consists of a network composed of sensory neurons, interneurons, and efferent neurons that control the muscles and secretory cells of the anemone's body. The neurons of the network are linked in a way that allows them to integrate information and act on it. In the same way that an anemone captures its food, the ENS receives stimuli and acts on them. For this reason, the ENS is able to function autonomously, independent of efferent signals from the CNS.

Anatomically and functionally, the ENS shares many features with the CNS:

1. *Intrinsic neurons.* The **intrinsic neurons** of the two nerve plexuses of the digestive tract are those neurons that lie completely within the wall of the gut, just as interneurons are completely contained within the CNS. Autonomic neurons that bring signals from the CNS to the digestive system are called **extrinsic neurons.**
2. *Neurotransmitters and neuromodulators.* ENS neurons release more than 30 neurotransmitters and neuromodulators, most of which are identical to molecules found in the

brain. These neurotransmitters are sometimes called *non-adrenergic, noncholinergic* to distinguish them from the traditional autonomic neurotransmitters norepinephrine and acetylcholine. Among the best known neurotransmitters and neuromodulators are serotonin, vasoactive intestinal peptide, and nitric oxide.

3 *Glial support cells.* The glial cells of neurons within the ENS are more similar to astroglia of the brain than to Schwann cells of the peripheral nervous system.

4 *Diffusion barrier.* The capillaries that surround ganglia in the ENS are not very permeable and create a diffusion barrier that is similar to the blood-brain barrier of cerebral blood vessels.

5 *Integrating center.* As noted earlier, reflexes that originate in the GI tract can be integrated and acted on without neural signals leaving the ENS. For this reason, the neuron network of the ENS is its own integrating center, much like the brain and spinal cord.

It was once thought that if we could explain how the ENS integrates simple behaviors, we could use the system as a model for CNS function. But studying ENS function is difficult because enteric reflexes have no discrete command center. Instead, in an interesting twist, GI physiologists are applying information gleaned from studies of the brain and spinal cord to investigate ENS function. The complex interactions between the enteric and central nervous systems, the endocrine system, and the immune system promise to provide scientists with questions to investigate for many years to come.

GI Peptides Include Hormones, Neuropeptides, and Cytokines

The hormones of the gastrointestinal tract occupy an interesting place in the history of endocrinology. In 1902, two English physiologists, W. M. Bayliss and E. H. Starling, discovered that acidic chyme entering the small intestine from the stomach caused the release of pancreatic juices even when all nerves to the pancreas were cut. Because the only communication remaining between intestine and pancreas was the blood supply that ran between them, Bayliss and Starling postulated the existence of some blood-borne (*humoral*) factor released by the intestine.

When duodenal extracts applied directly to the pancreas stimulated secretion, they knew they were dealing with a chemical produced by the duodenum. They named the substance *secretin*. Starling further proposed that the general name *hormone*, from the Greek word meaning "I excite," be given to all humoral agents that act at a site distant from their release.

In 1905, J. S. Edkins postulated the existence of a gastric hormone that stimulated gastric acid secretion. It took more than 30 years for researchers to isolate a relatively pure extract of the gastric hormone, and it was 1964 before the hormone, named *gastrin*, was finally purified.

Why was research on the digestive hormones so slow to develop? A major reason is that GI hormones are secreted by isolated endocrine cells scattered among other cells of the mucosal epithelium. At one time, the only way to obtain these hormones was to make a crude extract of the entire epithelium, a procedure that also liberated digestive enzymes and paracrine molecules made in adjacent cells. For this reason, it was very difficult to tell whether the physiological effect elicited by the extract came from one hormone, from more than one hormone, or from a paracrine signal such as histamine.

Although researchers have now sequenced more than 30 peptides from the GI mucosa, only some of them are widely accepted as hormones. A few peptides have well-defined paracrine effects, but most fall into a long list of candidate hormones. In addition, we know of nonpeptide regulatory molecules, such as histamine, that function as paracrine signals. Because of the uncertainty associated with the field, we restrict our focus to the major regulatory molecules.

The gastrointestinal hormones are usually divided into three families. All the members of a family have similar amino acid sequences, and in some cases there is overlap in their ability to bind to receptors.

The *gastrin family* includes the hormones **gastrin** and **cholecystokinin** (CCK), plus several variants of each. Their structural similarity means that both gastrin and CCK can bind to and activate the CCKB receptor found on parietal cells.

The *secretin family* includes **secretin**; **vasoactive intestinal peptide** (VIP), a neurocrine [p. 177] molecule; and **GIP**, a hormone known originally as *gastric inhibitory peptide* because it inhibited gastric acid secretion in early experiments. Subsequent studies, however, indicated that GIP administered in lower physiological doses does not block acid secretion. Researchers proposed a new name with the same initials—**glucose-dependent insulinotropic peptide**—that more accurately describes the hormone's action: it stimulates insulin release in response to glucose in the intestinal lumen. However, for the most part *gastric inhibitory peptide* has remained the preferred name.

Another member of the secretin family is the hormone **glucagon-like peptide-1** (GLP-1), which also plays an important role in glucose homeostasis. GIP and GLP-1 act together as feedforward signals for insulin release, as you will learn when you study the endocrine pancreas [Chapter 22].

The third family of peptides contains those that do not fit into the other two families. The primary member of this group is the hormone **motilin**. Increases in motilin secretion are associated with the migrating motor complex. The sources, targets, and effects of the major GI hormones are summarized in Table 21.1.

In the remainder of this chapter we integrate mobility, secretion, digestion, and absorption as we follow food passing

Table 21.1 The Digestive Hormones

	Stimulus for Release	Primary Target(s)	Primary Effect(s)	Other Information
Stomach				
Gastrin (G cells)	Peptides and amino acids; neural reflexes	ECL cells and parietal cells	Stimulates gastric acid secretion and mucosal growth.	Somatostatin inhibits release.
Intestine				
Cholecystokinin (CCK)	Fatty acids and some amino acids	Gallbladder, pancreas, stomach	- Stimulates gallbladder contraction and pancreatic enzyme secretion. - Inhibits gastric emptying and acid secretion.	Promotes satiety. Some effects may be due to CCK as a neurotransmitter.
Secretin	Acid in small intestine	Pancreas, stomach	- Stimulates HCO_3^- secretion. - Inhibits gastric emptying and acid secretion.	
Motilin	Fasting: periodic release every 1.5–2 hours	Gastric and intestinal smooth muscle	Stimulates migrating motor complex.	Inhibited by eating a meal.
Gastric inhibitory peptide (GIP)	Glucose, fatty acids, and amino acids in small intestine	Beta cells of pancreas	- Stimulates insulin release (feedforward mechanism) - Inhibits gastric emptying and acid secretion.	
Glucagon-like peptide-1 (GLP-1)	Mixed meal that includes carbohydrates or fats in the lumen	Endocrine pancreas	- Stimulates insulin release. - Inhibits glucagon release and gastric function.	Promotes satiety.

through the GI tract. ■ Figure 21.12 is a summary of the main events that occur in each section of the GI tract. Food processing is traditionally divided into three phases: a cephalic phase, a gastric phase, and an intestinal phase.

Integrated Function: The Cephalic Phase

Digestive processes in the body begin before food ever enters the mouth. Simply smelling, seeing, or even *thinking* about food can make our mouths water and our stomachs rumble. These long reflexes that begin in the brain create a feedforward response known as the **cephalic phase** of digestion (■ Fig. 21.13).

Anticipatory stimuli and the stimulus of food in the oral cavity activate neurons in the medulla oblongata. The medulla in turn sends an efferent signal through autonomic neurons to the salivary glands, and through the vagus nerve to the enteric nervous system. In response to these signals, the stomach, intestine, and accessory glandular organs begin secretion and increase motility in anticipation of the food to come.

Chemical and Mechanical Digestion Begins in the Mouth

When food first enters the mouth, it is met by a flood of the secretion we call *saliva*. Salivary secretion is under autonomic control and can be triggered by multiple stimuli, including the sight, smell, touch, and even thought of food. The water and mucus in saliva soften and lubricate food to make it easier to swallow. You can appreciate this function if you've ever tried to swallow a dry soda cracker without chewing it thoroughly. Saliva also dissolves food so that we can taste it.

Chemical digestion begins with the secretion of *salivary amylase*. Amylase breaks starch into maltose after the enzyme is activated by Cl^- in saliva. If you chew on an unsalted soda cracker for a long time, you may be able to detect the conversion of the cracker's flour starch to maltose, which is sweeter.

OVERVIEW OF MOTILITY, SECRETION, DIGESTION, AND ABSORPTION

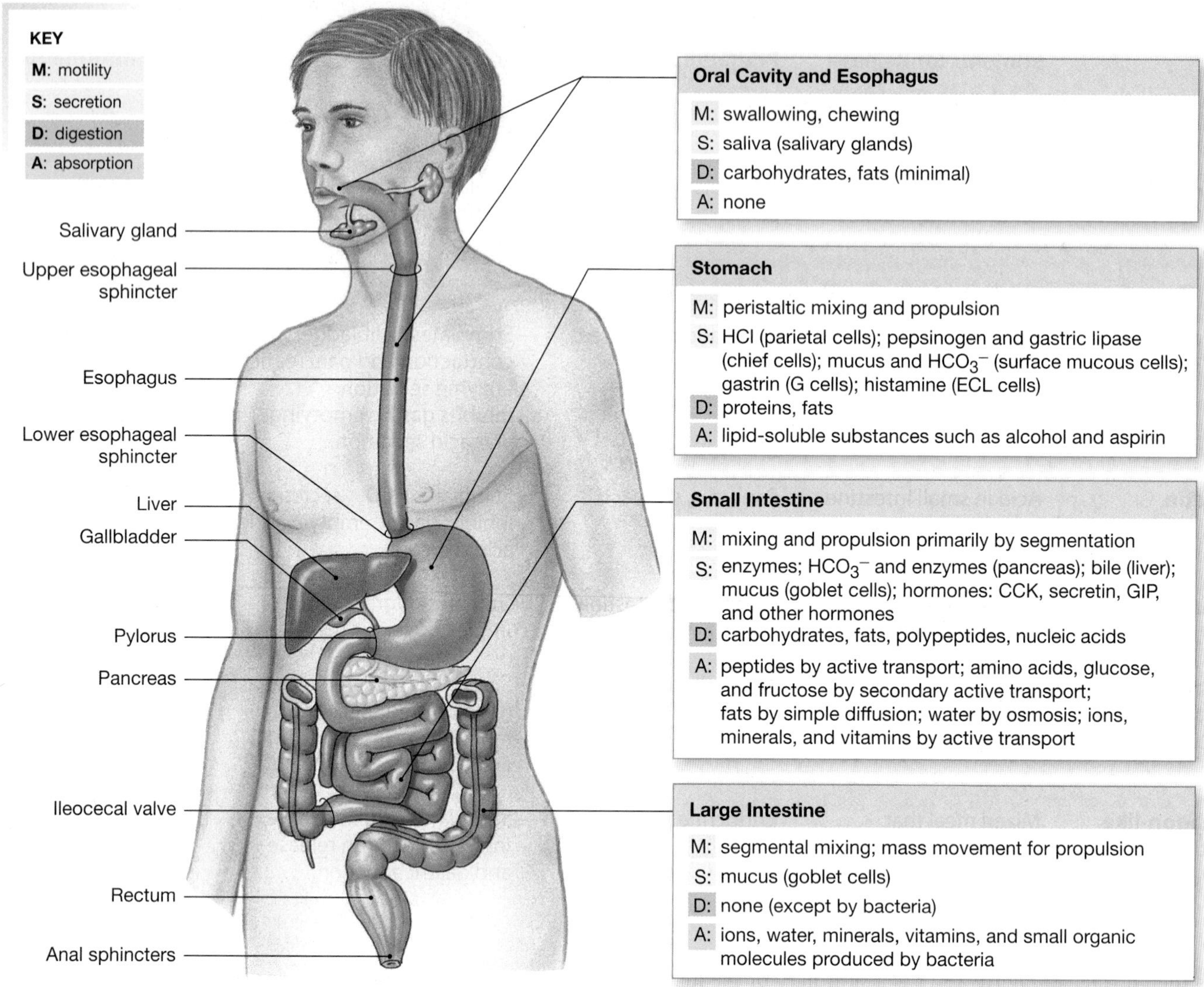

Fig. 21.12

The final function of saliva is protection. *Lysozyme* is an antibacterial salivary enzyme, and salivary *immunoglobulins* disable bacteria and viruses. In addition, saliva helps wash the teeth and keep the tongue free of food particles.

Mechanical digestion of food begins in the oral cavity with chewing. The lips, tongue, and teeth all contribute to the **mastication** {*masticare*, to chew} of food, creating a softened, moistened mass (*bolus*) that can be easily swallowed.

Swallowing Moves Food from Mouth to Stomach

Swallowing, or **deglutition** {*glutire*, to swallow}, is a reflex action that pushes a bolus of food or liquid into the esophagus (Fig. 21.14). The stimulus for swallowing is pressure created when the tongue pushes the bolus against the soft palate and the back of the mouth. Pressure from the bolus activates sensory afferents running through the glossopharyngeal nerve (cranial nerve IX) to a swallowing center in the medulla oblongata. Output from the swallowing center consists of somatic motor neurons that control the skeletal muscles of pharynx and upper esophagus as well as autonomic neurons that act on the lower portions of the esophagus.

As the reflex begins, the soft palate elevates to close off the nasopharynx. Muscle contractions move the larynx up and forward, which helps close off the trachea and open the upper esophageal sphincter. As the bolus moves down toward the esophagus, the **epiglottis** folds down, completing closure of the upper airway and preventing food and liquid from entering the airways. At the same time, respiration is briefly inhibited.

LONG AND SHORT REFLEXES OF THE CEPHALIC AND GASTRIC PHASES OF DIGESTION

The sight, smell, and taste of food initiate long reflexes that prepare the stomach for the arrival of food.

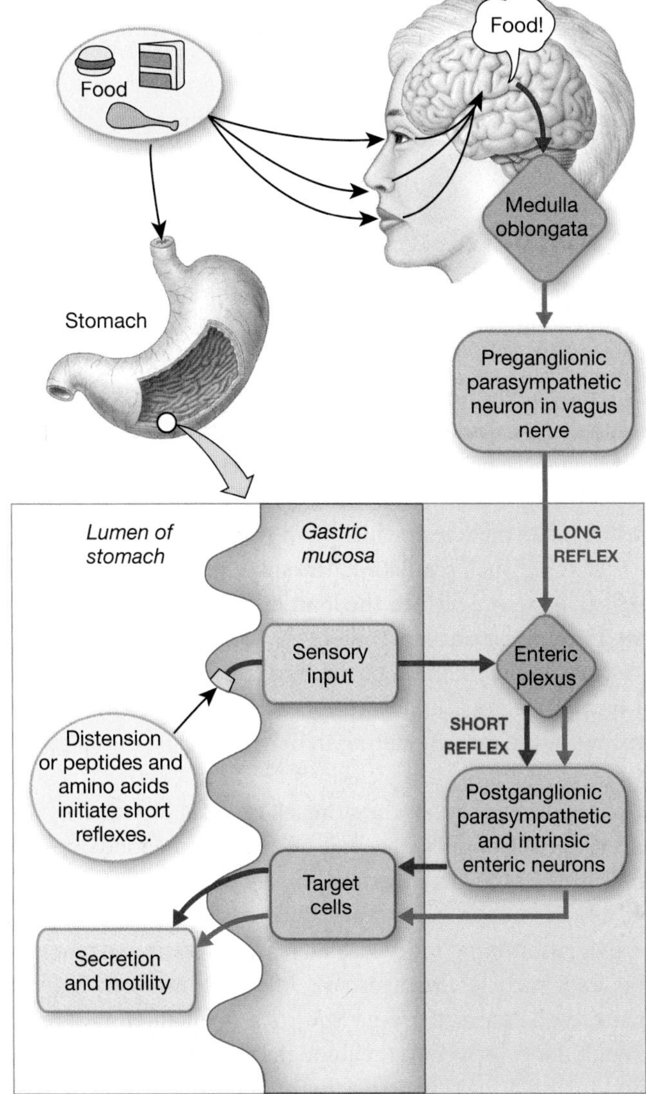

Fig. 21.13

DEGLUTITION: The Swallowing Reflex

Swallowing is integrated in the medulla oblongata. Sensory afferents in cranial nerve IX and somatic motor and automomic neurons mediate the reflex.

1 Tongue pushes bolus against soft palate and back of mouth, triggering swallowing reflex.

Soft palate elevates, closing off the nasopharynx.
Hard palate
Tongue
Bolus
Epiglottis
Glottis
Larynx moves up and forward.
Tonically contracted upper esophageal sphincter

2 Breathing is inhibited as the bolus passes the closed airway.

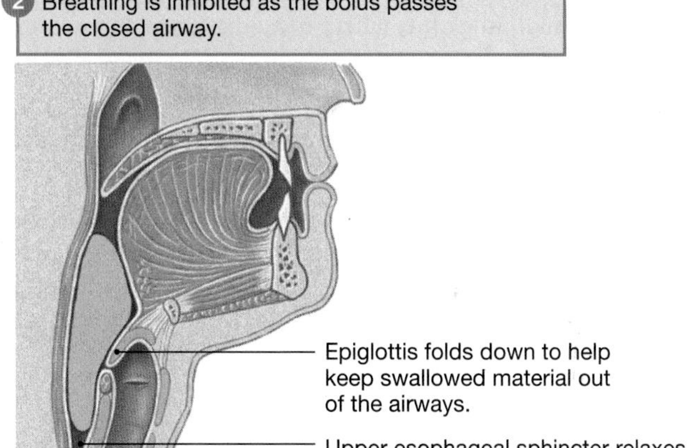

3 Food moves downward into the esophagus, propelled by peristaltic waves and aided by gravity.

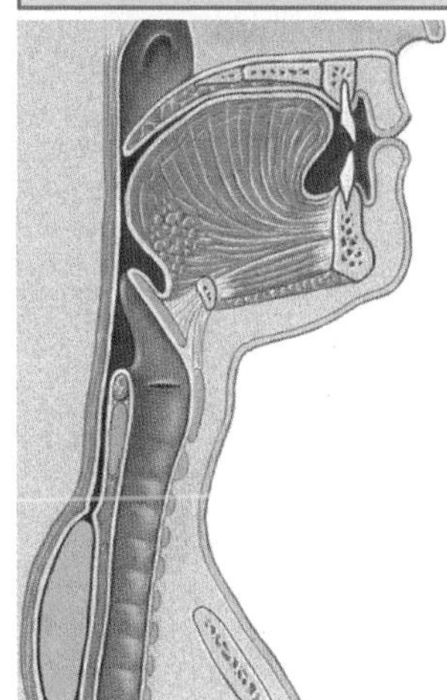

Fig. 21.14

When the bolus reaches the esophagus, the upper esophageal sphincter relaxes. Waves of peristaltic contractions then push the bolus toward the stomach, aided by gravity. Gravity is not required, however, as you know if you have ever participated in the party trick of swallowing while standing on your head.

The lower end of the esophagus lies just below the diaphragm and is separated from the stomach by the lower esophageal sphincter. This area is not a true sphincter but a region of relatively high muscle tension that acts as a barrier between the esophagus and the stomach. When food is swallowed, the tension relaxes, allowing the bolus to pass into the stomach.

If the lower esophageal sphincter does not stay contracted, gastric acid and pepsin can irritate the lining of the esophagus, leading to the pain and irritation of *gastroesophageal reflux* {*re-*, backward + *fluxus*, flow}, more commonly called heartburn. During the inspiratory phase of breathing, when the intrapleural pressure falls, the walls of the esophagus expand [p. 584]. This expansion creates subatmospheric pressure in the esophageal lumen and can suck acidic contents out of the stomach if the sphincter is relaxed. The churning action of the stomach when filled with food can also squirt acid back into the esophagus if the sphincter is not fully contracted. *Gastroesophageal reflux disorder* or GERD is one the most common digestive disorders in American society.

Integrated Function: The Gastric Phase

About 3.5 liters of food, drink, and saliva enter the fundus of the stomach each day. The stomach has three general functions:

1. **Storage.** The stomach stores food and regulates its passage into the small intestine, where most digestion and absorption take place.
2. **Digestion.** The stomach chemically and mechanically digests food into the soupy mixture of uniformly small particles called chyme.
3. **Protection.** The stomach protects the body by destroying many of the bacteria and other pathogens that are swallowed with food or trapped in airway mucus. At the same time, the stomach must protect itself from being damaged by its own secretions.

RUNNING PROBLEM

Brooke, who had always been healthy, was baffled. How could she have contracted cholera? But after discussing the methods of transmission with her healthcare providers, she realized that she hadn't been as careful about consuming only bottled water as she should have been. One of the doctors noticed that Brooke's medical history form listed Nexium® (esomeprazole) among her current medications. "You know, taking Nexium might have also contributed to your contracting cholera."

Q6: Esomeprazole is a proton pump inhibitor (PPI). For what symptom or condition might Brooke have been taking this drug?

Q7: Why might taking a protein pump inhibitor like esomeprazole have increased Brooke's chances of contracting cholera?

697 699 708 715 722 731

Before food even arrives, digestive activity in the stomach begins with the long **vagal reflex** of the cephalic phase (Fig. 21.13). Then, once food enters the stomach, stimuli in the gastric lumen initiate a series of short reflexes that constitute the **gastric phase** of digestion.

In gastric phase reflexes, distension of the stomach and the presence of peptides or amino acids in the lumen activate endocrine cells and enteric neurons. Hormones, neurocrine secretions, and paracrine molecules then influence motility and secretion.

The Stomach Stores Food

When food arrives from the esophagus, the stomach relaxes and expands to hold the increased volume. This neurally mediated reflex is called *receptive relaxation*. The upper half of the stomach remains relatively quiet, holding food until it is ready to be digested. The storage function of the stomach is perhaps the least obvious aspect of digestion. However, whenever we ingest more than we need from a nutritional standpoint, the stomach must regulate the rate at which food enters the small intestine.

Without such regulation, the small intestine would not be able to digest and absorb the load presented to it, and significant amounts of unabsorbed chyme would pass into the large intestine. The epithelium of the large intestine is not designed for large-scale nutrient absorption, so most of the chyme would pass out in the feces, resulting in diarrhea. This "dumping syndrome" is one of the less pleasant side effects of surgery that removes portions of either the stomach or small intestine.

While the upper stomach is quietly holding food, the lower stomach is busy with digestion. In the distal half of the stomach, a series of peristaltic waves pushes the food down toward the pylorus, mixing food with acid and digestive enzymes. As large food particles are digested to the more uniform texture of chyme, each contractile wave squirts a small amount of chyme through the pylorus into the duodenum. Enhanced gastric motility during a meal is primarily under neural control and is stimulated by distension of the stomach.

The Stomach Secretes Acid and Enzymes

When food comes into the mouth, the feedforward cephalic vagal reflex begins secretion in the stomach. The various secretions, their stimuli for release, and their functions are summarized in ■ Figure 21.15 and described below.

Acid Secretion **Parietal cells** deep in the gastric glands secrete **gastric acid** (HCl). Gastric acid has multiple functions. It activates pepsin, the enzyme that digests proteins, and it *denatures* the proteins [p. 54] by breaking disulfide and hydrogen bonds that hold the protein in its tertiary structure. Unfolding protein chains make the peptide bonds between amino acids accessible to pepsin. Gastric acid helps kill

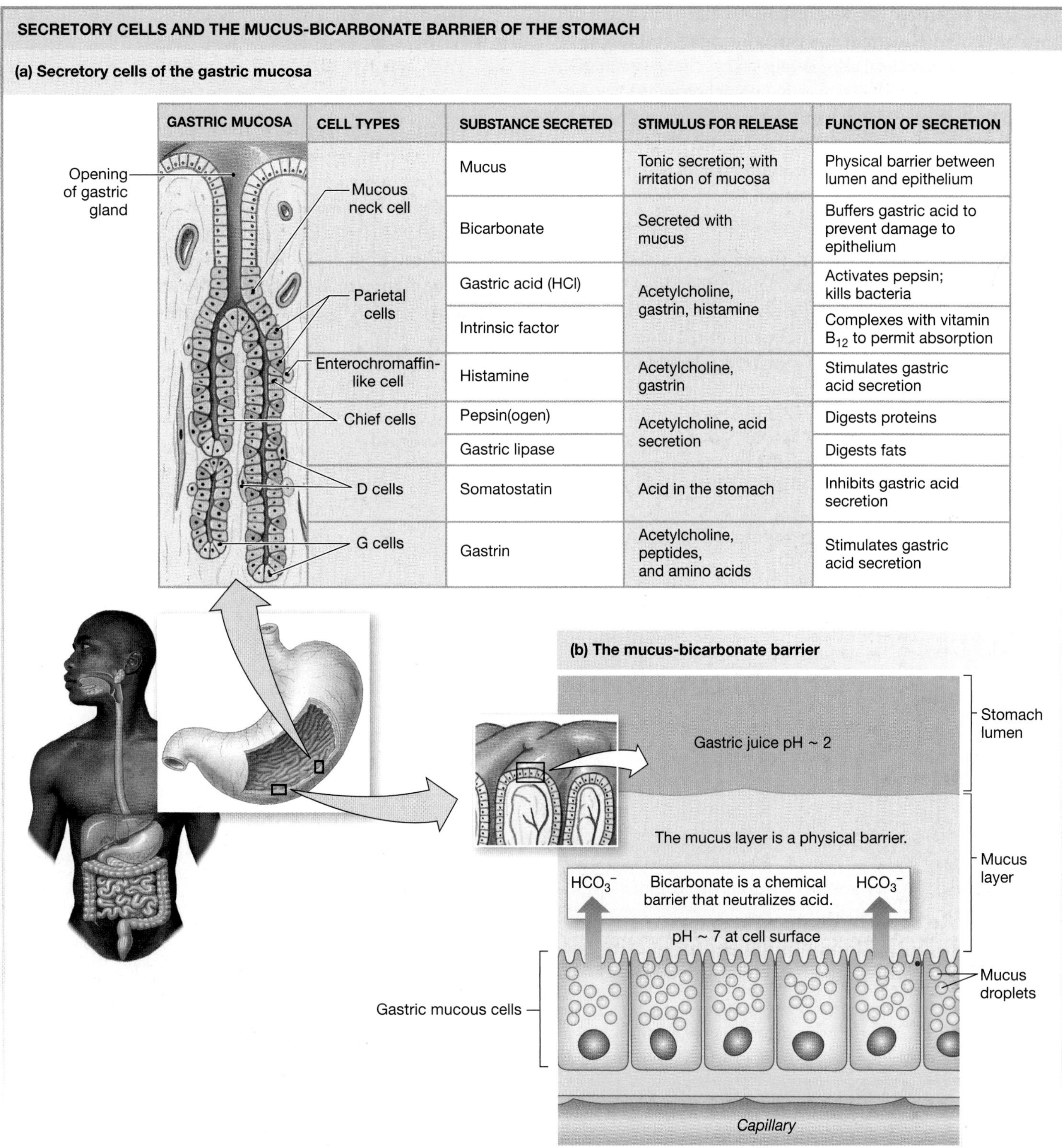

GASTRIC MUCOSA	CELL TYPES	SUBSTANCE SECRETED	STIMULUS FOR RELEASE	FUNCTION OF SECRETION
	Mucous neck cell	Mucus	Tonic secretion; with irritation of mucosa	Physical barrier between lumen and epithelium
		Bicarbonate	Secreted with mucus	Buffers gastric acid to prevent damage to epithelium
	Parietal cells	Gastric acid (HCl)	Acetylcholine, gastrin, histamine	Activates pepsin; kills bacteria
		Intrinsic factor		Complexes with vitamin B_{12} to permit absorption
	Enterochromaffin-like cell	Histamine	Acetylcholine, gastrin	Stimulates gastric acid secretion
	Chief cells	Pepsin(ogen)	Acetylcholine, acid secretion	Digests proteins
		Gastric lipase		Digests fats
	D cells	Somatostatin	Acid in the stomach	Inhibits gastric acid secretion
	G cells	Gastrin	Acetylcholine, peptides, and amino acids	Stimulates gastric acid secretion

Fig. 21.15

bacteria and other ingested microorganisms, and it inactivates salivary amylase, which stops carbohydrate digestion that began in the mouth.

Enzyme Secretion **Chief cells** in the gastric glands secrete the inactive enzyme pepsinogen. Pepsinogen is cleaved to active pepsin in the lumen of the stomach by the action of H^+. Pepsin is an endopeptidase that carries out the initial digestion of proteins. It is particularly effective on collagen and therefore plays an important role in digesting meat.

Gastric lipase is co-secreted with pepsin. However, less than 10% of fat digestion takes place in the stomach.

Paracrine Secretion **Enterochromaffin-like** (ECL) **cells** secrete histamine. Histamine is a paracrine signal that promotes acid secretion by parietal cells. **D cells** secrete somatostatin (SS), also known as growth hormone-inhibiting hormone for its action on the pituitary gland. As a GI paracrine molecule, somatostatin is the primary negative feedback signal for gastric phase secretion. It shuts down acid secretion directly and indirectly and also inhibits pepsinogen secretion.

Hormone Secretion **G cells,** found deep in the gastric glands, secrete the hormone gastrin into the blood. Gastrin release is stimulated by the presence of amino acids and peptides in the stomach, by distension of the stomach, and by neural reflexes mediated by **gastrin-releasing peptide.** Coffee (even if decaffeinated) also stimulates gastrin release—one reason people with excess acid secretion syndromes are advised to avoid coffee.

The coordinated function of gastric secretion is illustrated in Figure 21.16:

1. In a cephalic reflex, parasympathetic neurons from the vagus nerve stimulate G cells to release gastrin into the blood (2). The presence of amino acids or peptides in the lumen triggers a short reflex for gastrin release.
2. Gastrin in turn promotes acid release, both directly and indirectly by stimulating histamine release.
3. Histamine is released from ECL cells in response to gastrin and acetylcholine from the enteric nervous system. Histamine diffuses to its target, the parietal cells, and stimulates acid secretion by combining with H_2 *receptors* on parietal cells.
4. Acid in the stomach lumen stimulates pepsinogen release from chief cells through a short reflex (3). In the lumen, acid converts pepsinogen into pepsin, and protein digestion begins.

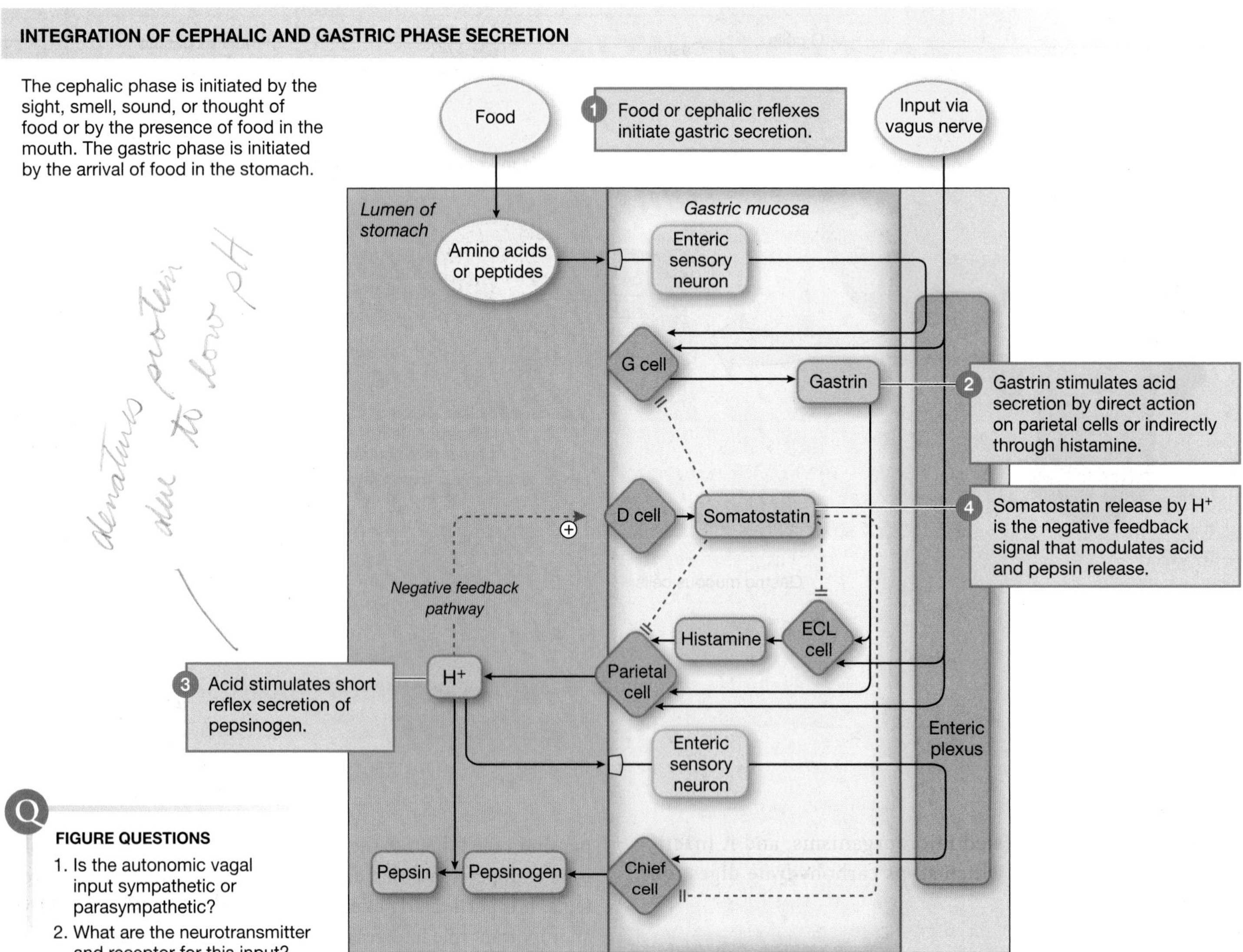
INTEGRATION OF CEPHALIC AND GASTRIC PHASE SECRETION
The cephalic phase is initiated by the sight, smell, sound, or thought of food or by the presence of food in the mouth. The gastric phase is initiated by the arrival of food in the stomach.
Food
1 Food or cephalic reflexes initiate gastric secretion.
Input via vagus nerve
Lumen of stomach
Gastric mucosa
Amino acids or peptides
Enteric sensory neuron
G cell
Gastrin
2 Gastrin stimulates acid secretion by direct action on parietal cells or indirectly through histamine.
D cell
Somatostatin
4 Somatostatin release by H+ is the negative feedback signal that modulates acid and pepsin release.
Negative feedback pathway
Histamine
ECL cell
Parietal cell
H+
3 Acid stimulates short reflex secretion of pepsinogen.
Enteric plexus
Enteric sensory neuron
Pepsin
Pepsinogen
Chief cell
denatures protein due to low pH

FIGURE QUESTIONS

1. Is the autonomic vagal input sympathetic or parasympathetic?
2. What are the neurotransmitter and receptor for this input?

Fig. 21.16

5 Acid also triggers somatostatin release from D cells (4). Somatostatin acts via negative feedback to inhibit secretion of gastric acid, gastrin, histamine, and pepsinogen.

The Stomach Balances Digestion and Protection

Under normal conditions, the gastric mucosa protects itself from acid and enzymes by a mucus-bicarbonate barrier. **Mucous cells** in the neck of gastric glands secrete both substances. The mucus forms a physical barrier, and the bicarbonate creates a chemical buffer barrier underlying the mucus (Fig. 21.15b). Researchers using microelectrodes have shown that the bicarbonate layer just above the cell surface in the stomach has a pH that is close to 7, even when the pH in the lumen is highly acidic at pH 2. Mucus secretion is increased when the stomach is irritated, such as by the ingestion of aspirin (acetylsalicylic acid) or alcohol.

Even the mucus-bicarbonate barrier can fail at times. In *Zollinger-Ellison syndrome,* patients secrete excessive levels of gastrin, usually from gastrin-secreting tumors in the pancreas. As a result, hyperacidity in the stomach overwhelms the normal protective mechanisms and causes a peptic ulcer. In peptic ulcers, acid and pepsin destroy the mucosa, creating holes that extend into the submucosa and muscularis of the stomach and duodenum. *Acid reflux* into the esophagus can erode the mucosal layer there as well.

Excess acid secretion is an uncommon cause of peptic ulcers. By far the most common causes are nonsteroidal anti-inflammatory drugs (NSAIDs), such as aspirin, and *Helicobacter pylori,* a bacterium that creates inflammation of the gastric mucosa.

For many years the primary therapy for excess acid secretion, or *dyspepsia,* was the ingestion of *antacids,* agents that neutralize acid in the gastric lumen. But as molecular biologists explored the mechanism for acid secretion by parietal cells, the potential for new therapies became obvious. Today we have two classes of drugs to fight hyperacidity: H_2 receptor antagonists (cimetidine and ranitidine, for example) that block histamine action, and *proton pump inhibitors* (PPIs), which block H^+-K^+-ATPase (omeprazole and lansoprazole, for example).

Integrated Function: The Intestinal Phase

The net result of the gastric phase is the digestion of proteins in the stomach by pepsin; the formation of chyme by the action of pepsin, acid, and peristaltic contractions; and the controlled entry of chyme into the small intestine, where further digestion and absorption can take place. Once chyme enters the small intestine, the **intestinal phase** of digestion begins. Sensors in the intestine trigger a series of reflexes that feed back to regulate the delivery rate of chyme from the stomach, and feed forward to promote digestion, motility, and utilization of nutrients.

The feedback signals to the stomach are both neural and hormonal (Fig. 21.17a):

1 Chyme in the intestine activates *the enteric nervous system,* which then slows gastric motility and secretion. In addition, three hormones reinforce the motility feedback signal: secretin, cholecystokinin (CCK), and gastric inhibitory peptide (GIP) (see Tbl. 21.1).

2 *Secretin* is released by the presence of acidic chyme in the duodenum. Secretin inhibits acid production and gastric motility, slowing gastric emptying. In addition, secretin stimulates production of pancreatic HCO_3^- to neutralize the acidic chyme that has entered the intestine.

3 *CCK* is secreted into the bloodstream if a meal contains fats. CCK also slows gastric motility and acid secretion. Because fat digestion proceeds more slowly than either protein or carbohydrate digestion, it is crucial that the stomach allow only small amounts of fat into the intestine at one time.

4 The *incretin hormones* GIP and glucagon-like peptide-1 (GLP-1) are released if the meal contains carbohydrates. Both hormones feed forward to promote insulin release by the endocrine pancreas, allowing cells to prepare for glucose that is about to be absorbed. They also slow the entry of food into the intestine by decreasing gastric motility and acid secretion.

5 The mixture of acid, enzymes, and digested food in chyme usually forms a hyperosmotic solution. Osmoreceptors in the intestine wall are sensitive to the osmolarity of the entering chyme. When stimulated by high osmolarity, the receptors inhibit gastric emptying in a reflex mediated by some unknown blood-borne substance.

Once food enters the small intestine, it is slowly propelled forward by a combination of segmental and peristaltic contractions. These actions mix chyme with enzymes and expose digested nutrients to the mucosa for absorption. Forward movement of chyme through the intestine must be slow enough to allow digestion and absorption to go to completion. Parasympathetic innervation, gastrin, and CCK all promote intestinal motility; sympathetic innervation inhibits it.

Bicarbonate Neutralizes Gastric Acid

About 5.5 liters of food, fluid, and secretions enter the small intestine each day, and about 3.5 liters of hepatic, pancreatic, and intestinal secretions are added there, making a total input of 9 liters into the lumen (see Fig. 21.1). The added secretions include bicarbonate, mucus, bile, and digestive enzymes.

1 *Bicarbonate secretion* into the small intestine neutralizes the highly acidic chyme that enters from the stomach. Most bicarbonate comes from the pancreas and is released in response to neural stimuli and secretin.

THE INTESTINAL PHASE

(a) Integration of gastric and intestinal phases

Chyme moving into the duodenum triggers neural and endocrine reflexes that

1. Initiate enzyme and bicarbonate secretion;
2. Feed back to slow gastric digestion and emptying;
3. Feed forward to start insulin secretion.

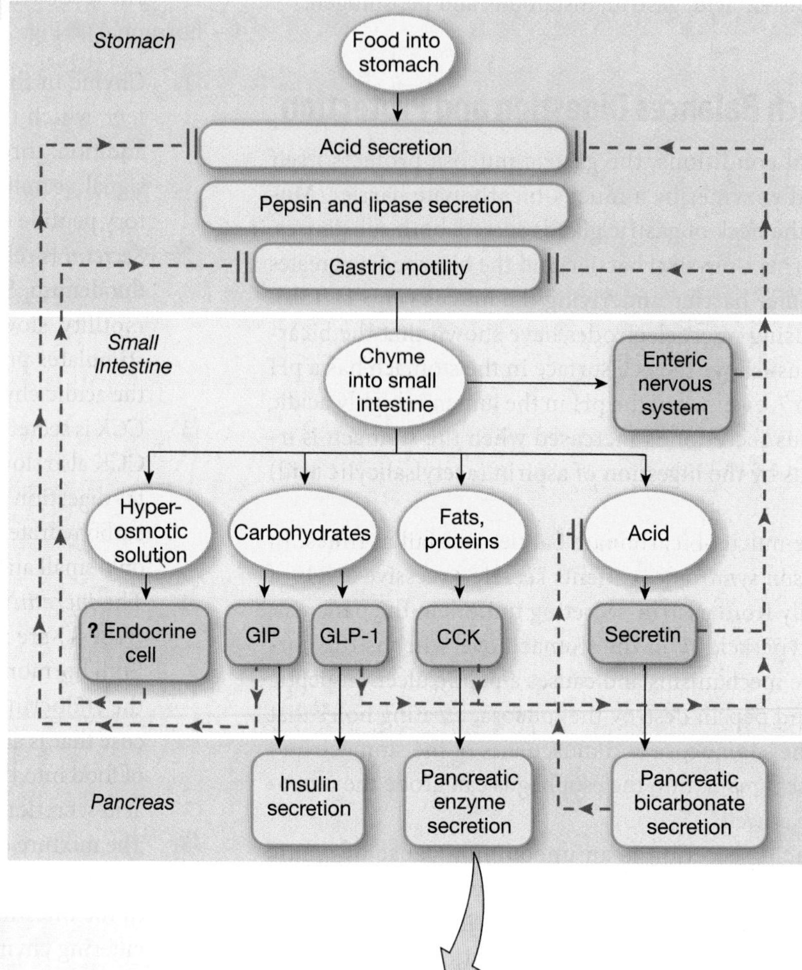

(b) Activation of pancreatic zymogens

Inactive enzymes secreted by the pancreas are activated in a cascade. Trypsinogen is activated to trypsin by brush border enteropeptidase, and trypsin then activates other pancreatic enzymes.

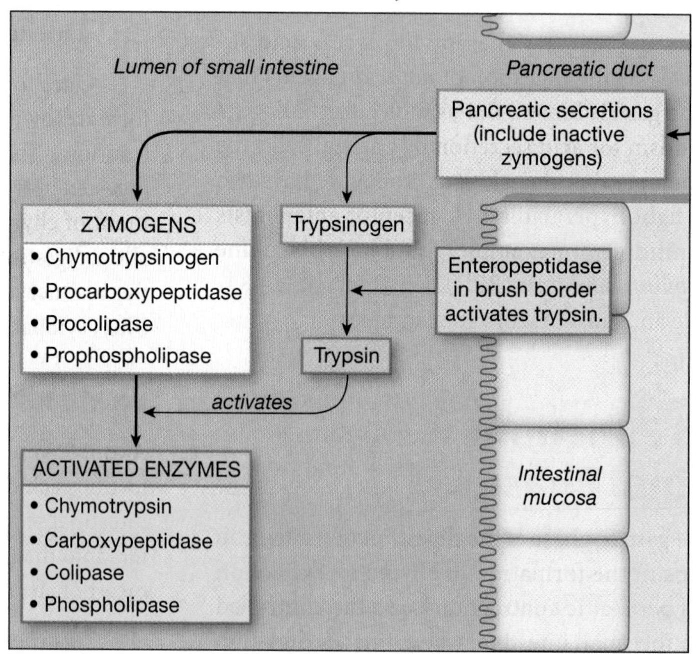

Fig. 21.17

2 *Mucus* from intestinal goblet cells protects the epithelium and lubricates the gut's contents.

3 *Bile* release into the intestine occurs when CCK stimulates gallbladder contraction following ingestion of fats. Bile salts are not altered during fat digestion. When they reach the terminal section of the small intestine (the ileum), they encounter cells that transport them back into the circulation. From there, bile salts return to the liver, are taken back up into the hepatocytes, and re-secreted. This recirculation of bile salts is essential to fat digestion because the body's pool of bile salts must cycle from two to five times for each meal. Bilirubin and other wastes secreted in bile are not reabsorbed and pass into the large intestine for excretion.

4 *Digestive enzymes* are produced by the intestinal epithelium and acinar cells of the exocrine pancreas. The brush border enzymes, which include peptidases, disaccharidases, and a protease called **enteropeptidase** (previously called *enterokinase*), are anchored to the luminal enterocyte membrane and are not swept out of the small intestine as chyme is propelled forward.

The signals for pancreatic enzyme release include distension of the small intestine, the presence of food in the intestine, neural signals, and the hormone CCK. Pancreatic enzymes enter the intestine in a watery fluid that also contains bicarbonate.

Most pancreatic enzymes are secreted as zymogens that must be activated upon arrival in the intestine. This activation process is a cascade that begins when brush border enteropeptidase converts inactive trypsinogen to active trypsin (Fig. 21.17b). Trypsin then converts the other pancreatic zymogens to their active forms.

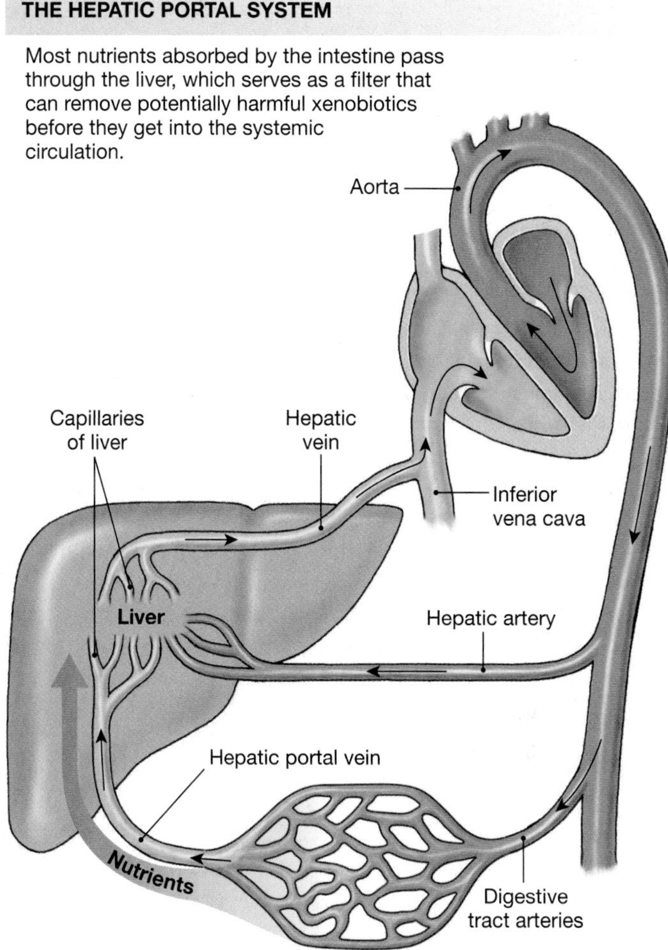

Fig. 21.18

Most Fluid Is Absorbed in the Small Intestine

Of the 9 liters that enter the small intestine daily, most (7.5 liters) is reabsorbed there. The transport of organic nutrients and ions, which takes place mostly in the duodenum and jejunum, creates an osmotic gradient for water absorption. Most absorbed nutrients move into capillaries in the villi and from there into the *hepatic portal system* [p. 466]. This specialized region of the circulation has two sets of capillary beds: one that picks up absorbed nutrients at the intestine, and another that delivers the nutrients directly to the liver (Fig. 21.18). Most digested fats go into the lymphatic system rather than into the blood.

The delivery of absorbed materials directly to the liver underscores the importance of that organ as a biological filter. Hepatocytes contain a variety of enzymes, such as the *cytochrome P450* isozymes, that metabolize drugs and xenobiotics and clear them from the bloodstream before they reach the systemic circulation. Hepatic clearance is one reason a drug administered orally must often be given in higher doses than the same drug administered by IV infusion.

Most Digestion Occurs in the Small Intestine

Chyme entering the small intestine has undergone relatively little chemical digestion. Protein digestion started in the stomach, but pepsin entering the small intestine is soon inactivated at the higher intestinal pH. Pancreatic proteases and at least 17 additional proteases and peptidases in the brush border continue protein digestion, creating small peptides and free amino acids that can be absorbed.

Carbohydrate digestion in the small intestine finishes converting digestible polysaccharides and disaccharides into monosaccharides that can be absorbed. Pancreatic amylase continues to digest starch into maltose. Maltose and disaccharides from food, such as sucrose and lactose (milk sugar), are digested by the appropriate brush border disaccharidase to their absorbable end products: glucose, galactose, and fructose.

Fats enter the small intestine in the form of a coarse emulsion. In the duodenum, bile salts coat the fat droplets to stabilize them so that digestion can be carried out by pancreatic lipase. Cholesterol and fatty acids are absorbed as described earlier in this chapter.

CLINICAL FOCUS

Lactose Intolerance

Lactose, or milk sugar, is a disaccharide composed of glucose and galactose. Ingested lactose must be digested before it can be absorbed, a task accomplished by the intestinal brush border enzyme *lactase*. Generally, lactase is found only in juvenile mammals, except in some humans of European descent. Those people inherit a dominant gene that allows them to produce lactase after childhood. Scientists believe the lactase gene provided a selective advantage to their ancestors, who developed a culture in which milk and milk products played an important role. In non-Western cultures, in which dairy products are not part of the diet after weaning, most adults lack the gene and synthesize less intestinal lactase. Decreased lactase activity is associated with a condition known as *lactose intolerance*. If a person with lactose intolerance drinks milk or eats dairy products, diarrhea may result. In addition, bacteria in the large intestine ferment lactose to gas and organic acids, leading to bloating and flatulence. The simplest remedy is to remove milk products from the diet, although milk predigested with lactase is available.

Concept Check Answers: p. 735

10. Bile acids are reabsorbed in the distal intestine by an apical sodium-dependent bile acid transporter (ASBT) and a basolateral organic anion transporter (OAT). Draw one enterocyte. Label the lumen, ECF, and basolateral and apical sides. Diagram bile acid reabsorption as described.
11. List the actions of CCK on the digestive system, and explain how these functions coordinate to promote fat digestion and absorption.

The Large Intestine Concentrates Waste

By the end of the ileum, only about 1.5 liters of unabsorbed chyme remain. The colon absorbs most of this volume so that normally only about 0.1 liter of water is lost daily in feces. Chyme enters the large intestine through the **ileocecal valve.** This is a tonically contracted region of muscularis that narrows the opening between the ileum and the **cecum,** the initial section of the large intestine (■ Fig. 21.19). The ileocecal valve relaxes each time a peristaltic wave reaches it. It also relaxes when food leaves the stomach as part of the *gastroileal reflex.*

The large intestine has seven regions. The cecum is a dead-end pouch with the *appendix*, a small fingerlike projection, at its ventral end. Material moves from the cecum upward through the **ascending colon,** horizontally across the body through the **transverse colon,** then down through the **descending colon** and **sigmoid colon** {*sigmoeides,* shaped like a sigma, Σ}. The **rectum** is the short (about 12 cm) terminal section of the large intestine. It is separated from the external environment by the **anus,** an opening closed by two sphincters, an internal smooth muscle sphincter and an external skeletal muscle sphincter.

The wall of the colon differs from that of the small intestine in that the muscularis of the large intestine has an inner circular layer but a discontinuous longitudinal muscle layer concentrated into three bands called the **tenia coli**. Contractions of the tenia pull the wall into bulging pockets called **haustra** {*haustrum*, bucket or scoop}.

The mucosa of the colon has two regions, like that of the small intestine. The luminal surface lacks villi and appears smooth. It is composed of colonocytes and mucus-secreting goblet cells. The crypts contain stem cells that divide to produce new epithelium, as well as goblet cells, endocrine cells, and maturing colonocytes.

Motility in the Large Intestine Chyme that enters the colon continues to be mixed by segmental contractions. Forward movement is minimal during mixing contractions and depends primarily on a unique colonic contraction known as **mass movement.** A wave of contraction decreases the diameter of a segment of colon and sends a substantial bolus of material forward. These contractions occur 3–4 times a day and are associated with eating and distension of the stomach through the *gastrocolic reflex*. Mass movement is responsible for the sudden distension of the rectum that triggers defecation.

The **defecation reflex** removes undigested feces from the body. Defecation resembles urination in that it is a spinal reflex triggered by distension of the organ wall. The movement of fecal material into the normally empty rectum triggers the reflex. Smooth muscle of the **internal anal sphincter** relaxes, and peristaltic contractions in the rectum push material toward the anus. At the same time, the external anal sphincter, which is under voluntary control, is consciously relaxed if the situation is appropriate. Defecation is often aided by conscious abdominal contractions and forced expiratory movements against a closed glottis (the *Valsalva maneuver*).

Defecation, like urination, is subject to emotional influence. Stress may increase intestinal motility and cause psychosomatic diarrhea in some individuals but may decrease motility and cause *constipation* in others. When feces are retained in the colon, either through consciously ignoring a defecation reflex or through decreased motility, continued water absorption creates hard, dry feces that are difficult to expel. One treatment for constipation is glycerin suppositories, small bullet-shaped wads that are inserted through the anus into the rectum. Glycerin attracts water and helps soften the feces to promote defecation.

Digestion and Absorption in the Large Intestine According to the traditional view of the large intestine, no significant digestion of organic molecules takes place there. However, in recent

ANATOMY OF THE LARGE INTESTINE

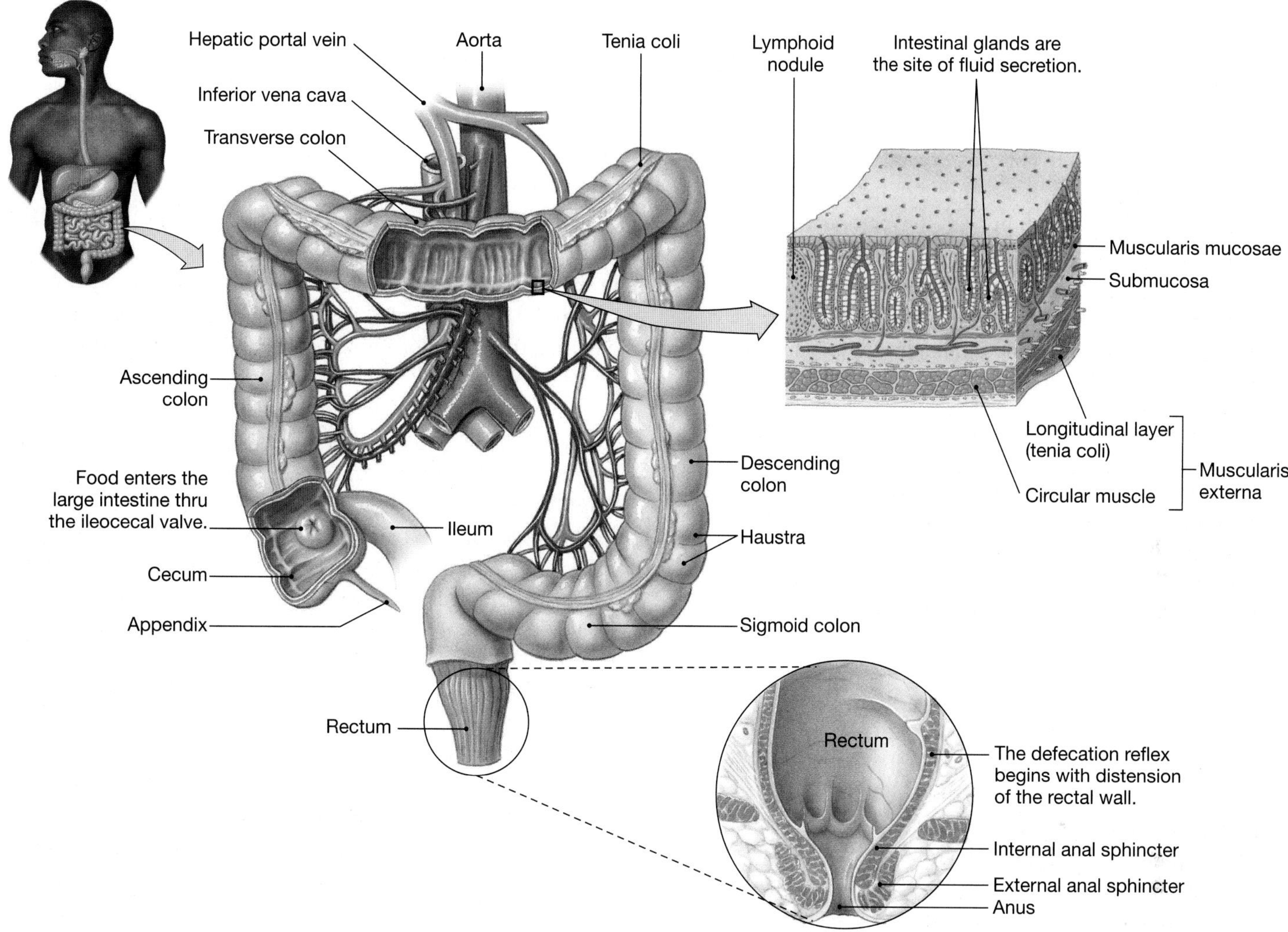

Fig. 21.19

years this view has been revised. We now know that the numerous bacteria inhabiting the colon break down significant amounts of undigested complex carbohydrates and proteins through fermentation. The end products include lactate and short-chain fatty acids, such as butyric acid. Several of these products are lipophilic and can be absorbed by simple diffusion. The fatty acids, for example, are used by colonocytes as their preferred energy substrate.

Colonic bacteria also produce significant amounts of absorbable vitamins, especially vitamin K. Intestinal gases, such as hydrogen sulfide, that escape from the gastrointestinal tract are a less useful product. Some starchy foods, such as beans, are notorious for their tendency to produce intestinal gas (**flatus**).

Diarrhea Can Cause Dehydration

Diarrhea is a pathological state in which intestinal secretion of fluid is not balanced by absorption, resulting in watery stools. Diarrhea occurs if normal intestinal water absorption mechanisms are disrupted or if there are unabsorbed osmotically active solutes that "hold" water in the lumen. Substances that cause *osmotic diarrhea* include undigested lactose and sorbitol, a sugar alcohol from plants. Sorbitol is used as an "artificial" sweetener in some chewing gums and in foods made for diabetics. Another unabsorbed solute that can cause osmotic diarrhea, intestinal cramping, and gas is Olestra, the "fake fat" made from vegetable oil and sugar.

In clinical settings, patients who need to have their bowels cleaned out before surgery or other procedures are often given four liters of an isotonic solution of polyethylene glycol and electrolytes to drink. Because polyethylene glycol cannot be absorbed, a large volume of unabsorbed solution passes into the colon, where it triggers copious diarrhea that removes all solid waste from the GI tract.

Secretory diarrheas occur when bacterial toxins, such as cholera toxin from *Vibrio cholerae* and *Escherichia coli* enterotoxin, enhance colonic Cl^- secretion (see Fig. 21.5c). When excessive fluid secretion is coupled with increased motility, diarrhea results. Secretory diarrhea in response to intestinal infection can be viewed as adaptive because it helps flush pathogens out of the lumen. However, it also has the potential to cause dehydration if fluid loss is excessive.

The World Health Organization estimates that in developing countries, 4 million people die from diarrhea each year. In the United States, diarrhea in children causes about 200,000 hospitalizations a year. Oral replacement fluids for treatment of diarrheal salt and water loss can prevent the *morbidity* (illness) and *mortality* (death) associated with diarrhea. Oral rehydration solutions usually contain glucose or sucrose as well as Na^+, K^+, and Cl^- because the inclusion of a sugar enhances Na^+ absorption. If dehydration is severe, intravenous fluid therapy may be necessary.

Concept Check Answer: p. 735

12. In secretory diarrhea, epithelial cells in the intestinal villi may be damaged or may slough off. In these cases, would it be better to use an oral rehydration solution containing glucose or one containing sucrose? Explain your reasoning.

Immune Functions of the GI Tract

As you learned at the beginning of the chapter, the GI tract is the largest immune organ in the body. Its luminal surface is continuously exposed to disease-causing organisms, and the immune cells of the GALT must prevent these pathogens from entering the body through delicate absorptive tissues. The first lines of defense are the enzymes and immunoglobulins in saliva and the highly acidic environment of the stomach. If pathogens or toxic materials make it into the small intestine, sensory receptors and the immune cells of the GALT respond. Two common responses are diarrhea, just described, and vomiting.

M Cells Sample Gut Contents

The immune system of the intestinal mucosa consists of immune cells scattered throughout the mucosa, clusters of immune cells in Peyer's patches (see Fig. 21.3f), and specialized epithelial cells called **M cells** that overlie the Peyer's patches. The M cells provide information about the contents of the lumen to the immune cells of the GALT.

The microvilli of M cells are fewer in number and more widely spaced than in the typical intestinal cell. The apical surface of M cells contains clathrin-coated pits [p. 155] with embedded membrane receptors. When antigens bind to these receptors, the M cell uses transcytosis to transport them to its basolateral membrane, where they are released into the interstitial fluid. Macrophages and lymphocytes [p. 547] are waiting in the extracellular compartment for the M cell to present them with antigens.

If the antigens are substances that threaten the body, the immune cells swing into action. They secrete cytokines to attract additional immune cells that can attack the invaders and cytokines that trigger an inflammatory response. A third response to cytokines is increased secretion of Cl^-, fluid, and mucus to flush the invaders from the GI tract.

In *inflammatory bowel diseases* (such as ulcerative colitis and Crohn's disease), the immune response is triggered inappropriately by the normal contents of the gut. One apparently successful experimental therapy for these diseases involves blocking the action of cytokines released by the gut-associated lymphoid tissues.

How certain pathogenic bacteria cross the barrier created by the intestinal epithelium has puzzled scientists for years. The discovery of M cells may provide the answer. It appears that some bacteria, such as *Salmonella* and *Shigella*, have evolved surface molecules that bind to M cell receptors. The M cells then obligingly transport the bacteria across the epithelial barrier and deposit them inside the body, where the immune system immediately reacts. Both bacteria cause diarrhea, and *Salmonella* also causes fever and vomiting.

Vomiting Is a Protective Reflex

Vomiting, or *emesis*, the forceful expulsion of gastric and duodenal contents from the mouth, is a protective reflex that removes toxic materials from the GI tract before they can be absorbed. However, excessive or prolonged vomiting, with its loss of gastric acid, can cause metabolic alkalosis [p. 690].

The vomiting reflex is coordinated through a vomiting center in the medulla. The reflex begins with stimulation of sensory receptors and is often (but not always) accompanied by nausea. A variety of stimuli from all over the body can trigger vomiting. They include chemicals in the blood, such as cytokines and certain drugs; pain; and disturbed equilibrium, such as occurs in a moving car or rocking boat. Tickling the back of the pharynx can also induce vomiting.

Efferent signals from the vomiting center initiate a wave of reverse peristalsis that begins in the small intestine and moves upward. The motility wave is aided by abdominal contraction that increases intra-abdominal pressure. The stomach relaxes so that the increased pressure forces gastric and intestinal contents back into the esophagus and out of the mouth.

During vomiting, respiration is inhibited. The epiglottis and soft palate close off the trachea and nasopharynx to prevent the *vomitus* from being inhaled (*aspirated*). Should acid or small food particles get into the airways, they could damage the respiratory system and cause *aspiration pneumonia*.

RUNNING PROBLEM CONCLUSION

Cholera in Haiti

Vibrio cholerae, the bacterium that causes cholera, was first identified in India in the 1800s. It has caused seven worldwide epidemics in the years since. About 75% of people who become infected with *V. cholerae* have no symptoms, but the remaining 25% develop potentially fatal diarrhea. The gut immune systems in most people overcome the infection within about a week. But until that happens, even asymptomatic people shed the bacteria in their feces, which contributes to the spread of the disease. In Haiti, plagued by inadequate sanitation and an earthquake-damaged water supply, cholera spread rapidly. By May 2011, nearly 400,000 cases and more than 5000 deaths had been reported. Genetic analysis of the *Vibrio cholera* strain in Haiti suggests that the bacterium was accidentally brought to the island by asymptomatic United Nations peacekeepers from Asia.

To learn more about cholera in Haiti, see the CDC (*www.cdc.gov*) and WHO (*www.who.int*) web sites. If you plan to travel to Haiti or any place with a declared cholera epidemic, visit *www.cdc.gov/travel* and review proper guidelines and procedures on avoiding contact with this potentially lethal bacterium. Now check your understanding of this running problem by comparing your answers to the information in the following summary table.

Question	Facts	Integration and Analysis
1. What would you expect her ECF volume to be?	Fluid in diarrhea is secreted from the body into the lumen of the GI tract.	Loss of fluid from the body would decrease ECF volume.
2. Why was Brooke experiencing a rapid heartbeat?	Loss of ECF volume with the diarrhea decreased Brooke's blood pressure.	Decreased blood pressure triggered a baroreceptor reflex [p. 525]. Increased sympathetic and decreased parasympathetic output to the SA node resulted in a faster heart rate.
3. Why would continuously open enterocyte CFTR channels cause secretory diarrhea and dehydration in humans?	Chloride leaves enterocytes by the CFTR channel. Na^+ and water follow by the paracellular pathway. See Figure 21.5c.	A continuously open CFTR channel means increased secretion of NaCl and water into the lumen, which leads to watery diarrhea. The salt and water come from the ECF, and their loss causes dehydration.
4. Recipes for oral rehydration therapy usually include sugar (sucrose) and table salt. Explain how the sugar enhances intestinal absorption of Na^+.	Sucrose digests to glucose and fructose. Glucose is absorbed by Na^+-dependent indirect active transport on the SGLT.	Na^+ uptake by the SGLT provides an additional pathway for Na^+ absorption and speeds the replenishing of fluid loss.
5. Which type of IV solution would you select for Brooke and why? Your choices are normal (isotonic) saline, half-normal saline, and 5% dextrose in water (D-5-W).	Chloride secretion by enterocytes causes Na^+ and water to follow, with the net result being secretion of isotonic fluid. The replacement fluid should match the fluid loss as closely as possible.	Normal saline (isosmotic) approximates the fluid lost in cholera diarrhea. Half-normal saline would dilute the body's osmolarity. D-5-W is not acceptable because it is equivalent to giving pure water and would not replace the lost NaCl.
6. Esomeprazole is a proton pump inhibitor (PPI). For what symptom or condition might Brooke have been taking this drug?	"Proton pump" is another name for an ATP-dependent H^+ transporter. Stomach acid is secreted by H^+-K^+-ATPase.	A proton pump inhibitor would decrease stomach acid, so Brooke may have been taking the PPI for heartburn or gastroesophageal reflex disorder (GERD).
7. Why might taking a protein pump inhibitor like esomeprazole have increased Brooke's chances of contracting cholera?	Proton pump inhibitors decrease the acidity in the stomach. Low gastric pH is one of the body's defense mechanisms.	In a less acidic stomach environment, more cholera bacteria might survive passage through the stomach to the small intestine, where they could infect the enterocytes.

This problem was written by Claire Conroy when she was an undergraduate Nutritional Sciences/Pre-Physical Therapy student at the University of Texas at Austin.

697 699 708 715 722 731

Chapter Summary

The digestive system, like the renal system, plays a key role in *mass balance* in the body. Most material that enters the system, whether by mouth or by secretion, is absorbed before it reaches the end of the GI tract. In pathologies such as diarrhea, in which absorption and secretion are unbalanced, the loss of material through the GI tract can seriously disrupt *homeostasis*. Absorption and secretion in the GI tract provide numerous examples of *movement across membranes*, and most transport processes follow patterns you have encountered in the kidney and other systems. Finally, regulation of GI tract function illustrates the complex interactions that take place between endocrine and neural *control systems* and the immune system.

Digestive Function and Processes

1. The GI tract moves nutrients, water, and electrolytes from the external environment to the internal environment. (p. 697)
2. About 2 liters of fluid per day enter the GI tract through the mouth. Another 7 liters of water, ions, and proteins are secreted by the body. Nearly all of this volume is reabsorbed. (p. 698; Fig. 21.1)
3. The four processes of the digestive system are digestion, absorption, motility, and secretion. **Digestion** is chemical and mechanical breakdown of foods into absorbable units. **Absorption** is transfer of substances from the lumen of the GI tract to the ECF. **Motility** is movement of material through the GI tract. **Secretion** refers to the transfer of fluid and electrolytes from ECF to lumen or the release of substances from cells. (p. 698; Fig. 21.2)
4. The GI tract contains the largest collection of lymphoid tissue in the body, the **gut-associated lymphoid tissue (GALT)**. (p. 698)

Anatomy of the Digestive System

iP GI: Anatomy Review

5. **Chyme** is a soupy substance created as ingested food is broken down by mechanical and chemical digestion. (p. 699)
6. Food entering the digestive system passes through the mouth, pharynx, **esophagus, stomach (fundus, body, antrum), small intestine (duodenum, jejunum, ileum), large intestine (colon, rectum)**, and **anus.** (p. 700; Fig. 21.3a)
7. The **salivary glands, pancreas**, and **liver** add exocrine secretions containing enzymes and mucus to the lumen. (p. 700; Fig. 21.3a)
8. The wall of the GI tract consists of four layers: mucosa, submucosa, muscle layers, and serosa. (p. 700; Fig. 21.3d)
9. The **mucosa** faces the lumen and consists of epithelium, the **lamina propria**, and the **muscularis mucosae**. The lamina propria contains immune cells. Small **villi** and invaginations increase the surface area. (p. 701; Fig. 21.3e, f)
10. The **submucosa** contains blood vessels and lymph vessels and the **submucosal plexus** of the **enteric nervous system**. (p. 701; Fig. 21.3f)
11. The **muscularis externa** consists of a layer of circular muscle and a layer of longitudinal muscle. The **myenteric plexus** lies between these two muscle layers. (p. 701; Fig. 21.3e, f)
12. The **serosa** is the outer layer of the GI tract wall. It is a connective tissue membrane that is a continuation of the peritoneal membrane. (p. 700; Fig. 21.3d)

Motility

iP GI: Motility

13. Motility moves food from mouth to anus and mechanically mixes food. (p. 703)
14. Between meals, the **migrating motor complex** moves food remnants from the upper GI tract to the lower regions. (p. 703)
15. **Peristaltic contractions** are progressive waves of contraction that occur mainly in the esophagus. (p. 704; Fig. 21.4a)
16. **Segmental contractions** are primarily mixing contractions. (p. 704; Fig. 21.4b)
17. GI motility is mediated by the enteric nervous system and modulated by hormones, paracrine signals, and neuropeptides. (p. 703)
18. GI smooth muscle cells are electrically connected by gap junctions. Some segments are **tonically contracted,** but others exhibit **phasic contractions**. (p. 704)
19. Intestinal muscle exhibits spontaneous **slow wave potentials**. When a slow wave reaches threshold, it fires action potentials and contracts. (p. 704; Fig. 21.4d)
20. Slow waves originate in the **interstitial cells of Cajal**. (p. 705)

Secretion

iP GI: Secretion

21. **Parietal cells** in gastric glands secrete hydrochloric acid. (p. 706; Fig. 21.5a)
22. The pancreas secretes a watery $NaHCO_3$ solution from duct cells. (p. 706; Fig. 21.5b, d)
23. Intestinal cells secrete Cl^- using the **CFTR chloride channel**. Water and Na^+ follow passively down osmotic and electrochemical gradients. (p. 706; Fig. 21.5c)
24. Digestive enzymes are secreted by exocrine glands or by gastric and intestinal epithelium. (p. 707)

25. Mucus from **mucous cells** and **goblet cells** forms a protective coating and lubricates contents of the gut. (p. 708)
26. **Saliva** is an exocrine secretion that contains water, ions, mucus, and proteins. Salivation is under autonomic control. (p. 708)
27. Bile made by **hepatocytes** contains **bile salts**, bilirubin, and cholesterol. Bile is stored and concentrated in the gallbladder (p. 709; Fig. 21.6)

Digestion and Absorption

iP GI: Digestion and Absorption

28. Digestion combines mechanical and enzymatic breakdown of food. Digestion is not directly regulated but depends on secretion and motility. (p. 710)
29. Most nutrient absorption takes place in the small intestine. The large intestine absorbs water and ions. (p. 710)
30. **Amylase** digests starch to maltose. **Disaccharidases** digest disaccharides to monosaccharides. (p. 711; Fig. 21.8a)
31. Glucose absorption uses the SGLT Na^+-glucose symporter and GLUT2 transporter. Fructose uses the GLUT5 and GLUT2 transporters. (p. 711; Fig. 21.8b)
32. **Endopeptidases** (also called proteases) break proteins into smaller peptides. **Exopeptidases** remove amino acids from peptides. (p. 711; Fig. 21.8c)
33. Amino acids are absorbed via Na^+- or H^+-dependent cotransport. Dipeptides and tripeptides are absorbed via H^+-dependent cotransport. Some larger peptides are absorbed intact via transcytosis. (p. 711; Fig. 21.8d)
34. Fat digestion requires the enzyme **lipase** and the cofactor **colipase**. (p. 713; Fig. 21.9)
35. Fat digestion is facilitated by bile salts, which emulsify fats. As enzymatic and mechanical digestion proceed, fat droplets form **micelles**. (p. 713; Fig. 21.9b, d)
36. Fat absorption occurs primarily by simple diffusion. Cholesterol is actively transported. (p. 713; Fig. 21.9d)
37. Monoglycerides and fatty acids reassemble into triglycerides in the intestinal cell, then combine with cholesterol and proteins into **chylomicrons**. Chylomicrons are absorbed into the lymph. (p. 713; Fig. 21.9d)
38. Nucleic acids are digested and absorbed as nitrogenous bases and monosaccharides. (p. 714)
39. Fat-soluble vitamins are absorbed along with fats. Water-soluble vitamins are absorbed by mediated transport. Vitamin B_{12} absorption requires **intrinsic factor** secreted by the stomach. (p. 714)
40. Mineral absorption usually occurs via active transport. Some calcium moves by the paracellular pathway. Ions and water move by the paracellular pathway as well as by membrane proteins. (p. 715; Fig. 21.10)

Regulation of GI Function

iP GI: Control of the Digestive System

41. **Short reflexes** originate in the **enteric nervous system** and are integrated there. **Long reflexes** are integrated in the CNS. Long reflexes may originate in the ENS or outside it. (p. 716; Fig. 21.11)
42. Generally, parasympathetic innervation is excitatory for GI function, and sympathetic innervation is inhibitory. (p. 716)
43. GI peptides excite or inhibit motility and secretion. Most stimuli for GI peptide secretion arise from the ingestion of food. (p. 716; Fig. 21.11)
44. GI paracrine molecules, such as histamine, interact with hormones and neural reflexes. (p. 717)
45. The enteric nervous system is called "the little brain" because it can integrate information without input from the CNS. **Intrinsic neurons** lie completely within the ENS. (p. 717)
46. Digestive hormones are divided into the gastrin family (**gastrin, cholecystokinin**), secretin family (**secretin, gastric inhibitory peptide, glucagon-like peptide-1**), and hormones that do not fit into either of those two families (**motilin**). (p. 719; Tbl. 21.1)

The Cephalic Phase

47. In the **cephalic phase** of digestion, the sight, smell, or taste of food initiates GI reflexes. (p. 721; Fig. 21.13)
48. Mechanical digestion begins with chewing, or **mastication**. Saliva moistens and lubricates food. Salivary amylase digests carbohydrates. (p. 720)
49. Swallowing, or **deglutition**, is a reflex integrated by a medullary center. (p. 721; Fig. 21.14)

The Gastric Phase

50. The stomach stores food, begins protein and fat digestion, and protects the body from swallowed pathogens. (p. 722)
51. The stomach secretes mucus and bicarbonate from **mucous cells**, gastric acid from parietal cells, pepsinogen from **chief cells**, somatostatin from **D cells**, histamine from **ECL cells**, and gastrin from **G cells**. (p. 723; Fig. 21.15)
52. Gastric function is integrated with the cephalic and intestinal phases of digestion. (p. 724; Fig. 21.16)

The Intestinal Phase

53. Acid in the intestine, CCK, and secretin delay gastric emptying. (p. 726; Fig. 21.17a)
54. The pancreas secretes bicarbonate to neutralize gastric acid. (p. 725)
55. Intestinal enzymes are part of the **brush border**. Most pancreatic enzymes are secreted as zymogens that need to be activated. (p. 726; Fig. 21.17b)
56. Most absorption takes place in the small intestine. (p. 727)
57. Digestion and absorption of nutrients are completed in the small intestine. Most absorbed nutrients go directly to the liver via the hepatic portal system before entering the systemic circulation. (p. 727)
58. The large intestine concentrates the 1.5 liters of chyme that enter it daily. (p. 728)
59. Undigested material in the colon moves forward by **mass movement**. The **defecation reflex**, a spinal reflex subject to higher control, is triggered by sudden distension of the rectum. (p. 728)
60. Colonic bacteria use fermentation to digest organic material. (p. 729)
61. Cells of the colon can both absorb and secrete fluid. Excessive fluid secretion or decreased absorption causes diarrhea. (p. 729)

Immune Functions of the GI Tract

62. Protective mechanisms of the GI tract include acid and mucus production, vomiting, and diarrhea. (p. 730)
63. **M cells** sample gut contents and present antigens to cells of the GALT. (p. 730)
64. **Vomiting** is a protective reflex integrated in the medulla. (p. 730)

Questions

Answers: p. A-1

Level One Reviewing Facts and Terms

1. Define the four basic processes of the digestive system and give an example of each.
2. For most nutrients, the processes ________ and ________ are not regulated, whereas ________ and ________ are continuously regulated. Why do you think these differences exist? Defend your answer.
3. Match each of the following descriptions with the appropriate term(s):

(a) chyme is produced here	1. colon
(b) organ where most digestion occurs	2. stomach
(c) initial section of small intestine	3. small intestine
(d) this adds exocrine secretions to duodenum via a duct	4. duodenum
(e) sphincter between stomach and intestine	5. ileum
(f) enzymes produced here	6. jejunum
(g) distension of its walls triggers the defecation reflex	7. pancreas
	8. pylorus
	9. rectum
	10. liver

4. List the four layers of the GI tract walls. What type of tissue predominates in each layer?
5. Describe the functional types of epithelium lining the stomach and intestines.
6. What are Peyer's patches? M cells of the intestine?
7. What purposes does motility serve in the gastrointestinal tract? Which types of tissue contribute to gut motility? Which types of contraction do the tissues undergo?
8. What is a zymogen? What is a proenzyme? List two examples of each.
9. Match each of the following cells with the product(s) it secretes. Items may be used more than once.

(a) parietal cells	1. enzymes
(b) goblet cells	2. histamine
(c) brush border cells	3. mucus
(d) pancreatic c ells	4. pepsinogen
(e) D cells	5. gastrin
(f) ECL cells	6. somatostatin
(g) chief cells	7. HCO_3^-
(h) G cells	8. HCl
	9. intrinsic factor

10. How does each of the following factors affect digestion? Briefly explain how and where each factor exerts its effects.
    (a) emulsification
    (b) neural activity
    (c) pH
    (d) size of food particles
11. Most digested nutrients are absorbed into the ________ of the ________ system, delivering nutrients to the ________ (organ). However, digested fats go into the ________ system because intestinal capillaries have a(n) ________ around them that most lipids are unable to cross.
12. What is the enteric nervous system, and what is its function?
13. What are short reflexes? What types of responses do they regulate? What is meant by the term *long reflex*?
14. What role do paracrines play in digestion? Give specific examples.

Level Two Reviewing Concepts

15. Mapping. Map 1: List the three major groups of biomolecules across the top of a large piece of paper. Down the left side of the paper write *mouth, stomach, small intestine, absorption.* For each biomolecule in each location, fill in the enzymes that digest the biomolecule, the products of digestion for each enzyme, and the location and mechanisms by which these products are absorbed.

    Map 2: Create a diagram or map using the following terms related to iron absorption:

DMT1	ferroportin	Fe^{2+}
endocytosis	heme iron	liver
enterocyte	hepcidin	

16. Define, compare, and contrast the following pairs or sets of terms:
    (a) mastication, deglutition
    (b) microvilli, villi
    (c) peristalsis, segmentation, migrating motor complex, mass movements, defecation, vomiting, diarrhea
    (d) chyme, feces
    (e) short reflexes, long reflexes
    (f) submucosal plexus, myenteric plexus, enteric nervous system, vagus nerve
    (g) cephalic, gastric, and intestinal phases of digestion
17. (a) Diagram the mechanisms by which Na^+, K^+, and Cl^- are reabsorbed from the intestine.
    (b) Diagram the mechanisms by which H^+ and HCO_3^- are secreted into the lumen.
18. Compare the enteric nervous system with the cephalic brain. Give some specific examples of neurotransmitters, neuromodulators, and supporting cells in the two.
19. List and briefly describe the actions of the members of each of the three groups of GI hormones.
20. Explain how H_2-receptor antagonists and proton pump inhibitors decrease gastric acid secretion.

Level Three Problem Solving

21. In the disease state called *hemochromatosis,* the hormone hepcidin is either absent or not functional. Use your understanding of iron homeostasis to predict what would happen to intestinal iron uptake and plasma levels of iron in this disease.
22. Erica's baby, Justin, has had a severe bout of diarrhea and is now dehydrated. Is his blood more likely to be acidotic or alkalotic? Why?
23. Mary Littlefeather arrives in her physician's office complaining of severe, steady pain in the upper right quadrant of her abdomen. The pain began shortly after she ate a meal of fried chicken, French fries, and peas. Lab tests and an ultrasound reveal the presence of gallstones in the common bile duct running from the liver, gallbladder, and pancreas into the small intestine.

(a) Why was Mary's pain precipitated by the meal she ate?
(b) Which of the following processes will be affected by the gallstones: micelle formation in the intestine, carbohydrate digestion in the intestine, protein absorption in the intestine. Explain your reasoning.

24. Using what you have learned about epithelial transport in this chapter and the renal chapters, draw a picture of the salivary duct cells and lumen. Arrange membrane channels and transporters on the apical and basolateral membranes so that the duct cell absorbs Na^+ and secretes K^+. With neural stimulation, the flow rate of saliva can increase from 0.4 mL/min to 2 mL/min. What do you think happens to the Na^+ and K^+ content of saliva at the higher flow rate?

Level Four Quantitative Problems

25. Intestinal transport of the amino acid analog MIT (monoiodotyrosine) can be studied using the "everted sac" preparation. A length of intestine is turned inside out, filled with a solution containing MIT, tied at both ends, and then placed in a bath containing nutrients, salts, and an equal concentration of MIT. Changes in the concentration of MIT are monitored in the bath (mucosal or apical side of the inverted intestine), in the intestinal cells (tissue), and within the sac (serosal or basolateral side of the intestine) over a 240-minute period. The results are displayed in the graph shown here. (Data from Nathans *et al.*, *Biochimica et Biophysica Acta* 41: 271–282, 1960)

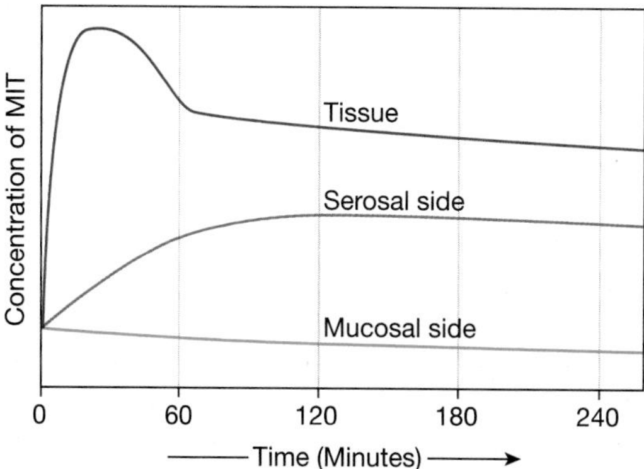

(a) Based on the data shown, is the transepithelial transport of MIT a passive process or an active process?
(b) Which way does MIT move: (1) apical to tissue to basolateral, or (2) basolateral to tissue to apical? Is this movement absorption or secretion?
(c) Is transport across the apical membrane active or passive? Explain your reasoning.
(d) Is transport across the basolateral membrane active or passive? Explain your reasoning.

Answers

Answers to Concept Check Questions

Page 699

1. Digestion is the chemical and mechanical breakdown of food into absorbable units.
2. Digestion takes place in the GI tract lumen, which is external to the body; metabolism takes place in the body's internal environment.
3. Absorption moves material from the GI lumen into the ECF; secretion moves substances from the cells or the ECF into the lumen.

Page 703

4. The lumen of the digestive tract is on the apical or mucosal side of the intestinal epithelium.
5. The four layers are mucosa, submucosa, muscularis externa, and serosa.
6. Mouth → pharynx → esophagus → stomach (fundus, body, antrum) → small intestine (duodenum, jejunum, ileum) → large intestine (colon, rectum) → anus
7. Because the GI tract has a large, vulnerable surface area facing the external environment, it needs the immune cells of lymphoid tissue to combat potential invaders.

Page 705

8. Some sphincters are tonically contracted to close off the GI tract from the outside world and to keep material from passing freely from one section of the tract to another.

Page 715

9. Bile salts do not digest triglycerides. They simply emulsify them into small particles so that lipase can digest them.

Page 728

10. The ASBT faces the lumen and brings Na^+ and bile acid into the enterocyte together. The OAT transports bile acid by itself out of the enterocyte and into the ECF.
11. CCK slows gastric motility and acid secretion, promotes release of pancreatic enzymes and bile, and promotes intestinal motility. Bile from the gallbladder and pancreatic lipase and colipase are necessary for efficient fat digestion. Fat digestion takes longer than digestion of other foodstuffs, and slowing gastric motility gives the intestine more time to digest fats. Decreasing gastric acid secretion also delays movement of chyme from stomach to intestine. Promoting intestinal motility ensures that bile salts can emulsify fats and that micelles will get adequate exposure to the brush border, where they are absorbed.

Page 730

12. Damage to epithelial cells means that brush border sucrase may be less effective or absent. In these cases, sucrose digestion would be impaired, so it would be better to use a solution containing glucose because this sugar need not be digested before being absorbed.

Answers to Figure Questions

Page 700

Figure 21.3: Glands are the salivary glands. Organs are liver, gallbladder, and pancreas.

Page 716

Figure 21.11: 1. The myenteric plexus controls smooth muscle contraction; the submucosal plexus controls smooth muscle contraction and endocrine and exocrine secretions by secretory cells. 2. Stretch receptors respond to stretch, osmoreceptors to osmolarity, and chemoreceptors to products of digestion.

Page 724

Figure 21.16: 1. The vagal input is parasympathetic, which stimulates digestion. 2. Neurotransmitter is ACh, and receptor is muscarinic.

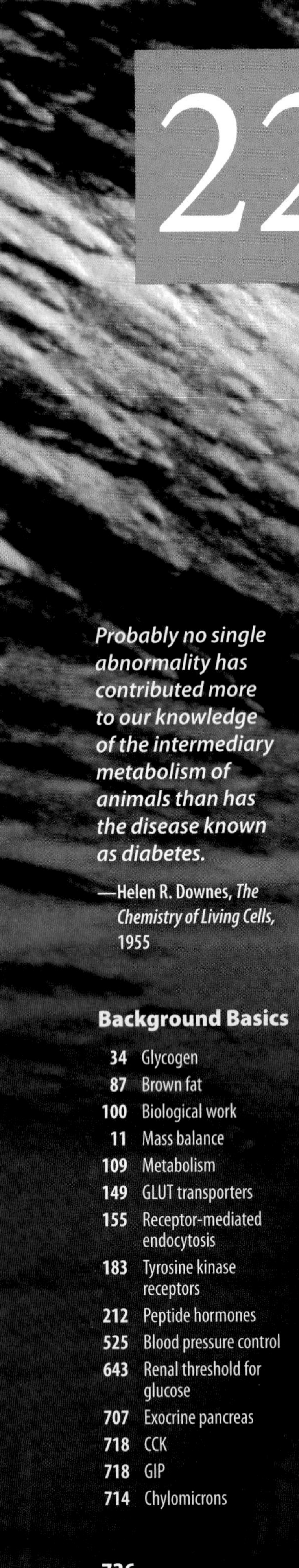

22 Metabolism and Energy Balance

Probably no single abnormality has contributed more to our knowledge of the intermediary metabolism of animals than has the disease known as diabetes.

—Helen R. Downes, *The Chemistry of Living Cells*, 1955

Background Basics

Vitamin C crystals

Magazine covers at grocery store checkout stands reveal a lot about Americans. Headlines screaming "Lose 10 pounds in a week without dieting" or "CCK: the hormone that makes you thin" vie for attention with glossy photographs of fat-laden, high-calorie desserts dripping with chocolate and whipped cream. As one magazine article put it, we are a nation obsessed with staying trim and with eating—two mutually exclusive occupations. But what determines when, what, and how much we eat? The factors influencing food intake are an area of intense research because the act of eating is the main point at which our bodies exert control over energy input.

Appetite and Satiety

The control of food intake is a complex process. The digestive system does not regulate energy intake, so we must depend on behavioral mechanisms, such as hunger and **satiety** {*satis*, enough}, to tell us when and how much to eat. Psychological and social aspects of eating, such as parents who say "Clean your plate," complicate the physiological control of food intake. As a result, we still do not fully understand what governs when, what, and how much we eat. What follows is an overview of this increasingly complex and constantly changing field.

Our current model for behavioral regulation of food intake is based on two hypothalamic centers [Fig. 11.3, p. 379]: a **feeding center** that is tonically active and a **satiety center** that stops food intake by inhibiting the feeding center. Output signals from these centers cause changes in eating behavior and create sensations of hunger and fullness. Animals whose feeding centers are destroyed cease to eat. If the satiety centers are destroyed, animals overeat and become obese {*obesus*, plump or fat}.

RUNNING PROBLEM

Eating Disorders

Sara Inman and Nicole Baker had been best friends growing up but had not seen each other for several semesters. When Sara heard that Nicole was in the hospital, she went to visit but wasn't prepared for what she saw. Nicole, who had always been thin, now was so emaciated that Sara hardly recognized her. Nicole had been taken to the emergency room when she fainted in a ballet class. When Dr. Ayani saw Nicole, her concern about her weight was as great as her concern about her broken wrist. At 5′6″, Nicole weighed only 95 pounds (normal healthy range for that height is 118–155 pounds). Dr. Ayani suspected an eating disorder, probably anorexia nervosa, and she ordered blood tests to confirm her suspicion.

737 739 741 744 765 770

Studies using transgenic and knockout mice show that control of these hypothalamic centers is complex. Higher brain centers, including the cerebral cortex and limbic system, provide input to the hypothalamus. Many different chemical signals influence food intake and satiety, including neuropeptides, "brain-gut" hormones secreted by the GI tract, and chemical signals called **adipocytokines**, which are secreted by adipose tissue.

There are two classic theories for regulation of food intake: the glucostatic theory and the lipostatic theory. Current evidence indicates that these two classic theories are too simple to be the only models, however. The **glucostatic theory** states that glucose metabolism by hypothalamic centers regulates food intake. When blood glucose concentrations decrease, the satiety center is suppressed, and the feeding center is dominant. When glucose metabolism increases, the satiety center inhibits the feeding center.

The **lipostatic theory** of energy balance proposes that a signal from the body's fat stores to the brain modulates eating behavior so that the body maintains a particular weight. If fat stores are increased, eating decreases. In times of starvation, eating increases. Obesity results from disruption of this pathway.

The 1994 discovery of **leptin** {*leptos*, thin}, a protein hormone synthesized in adipocytes, provided evidence for the lipostatic theory. Leptin acts as a negative-feedback signal between adipose tissue and the brain. As fat stores increase, adipose cells secrete more leptin, and food intake decreases.

Leptin is synthesized in adipocytes under control of the *obese (ob) gene*. Mice that lack the *ob* gene (and therefore lack leptin) become obese, as do mice with defective leptin receptors. However, these findings did not translate well to humans, as only a small percentage of obese humans are leptin deficient. The majority of them have *elevated* leptin levels. As scientists have learned, leptin is only part of the story.

A key signal molecule is **neuropeptide Y** (NPY), a brain neurotransmitter that seems to be the stimulus for food intake. In normal-weight animals, leptin inhibits NPY in a negative feedback pathway (■ Fig. 22.1). Other neuropeptides, hormones, and adipocytokines also influence NPY, leptin release by adipocytes, and the hypothalamic centers controlling food intake.

For example, the peptide **ghrelin** is secreted by the stomach during fasting and increases hunger when infused into human subjects. Other peptides, such as the hormones CCK and GLP-1, are released by the gut during a meal and help decrease hunger. Many of these appetite-regulating peptides have functions in addition to control of food intake. Ghrelin promotes release of growth hormone, for instance. The brain peptides called *orexins* appear to play a role in wakefulness and sleep. Our understanding of how all these factors interact is incomplete, and many scientists are studying this topic.

Appetite and eating are also influenced by sensory input through the nervous system. The simple acts of swallowing and chewing food help create a sensation of fullness. The sight, smell, and taste of food can either stimulate or suppress appetite.

Complex Chemical Signaling Controls Food Intake

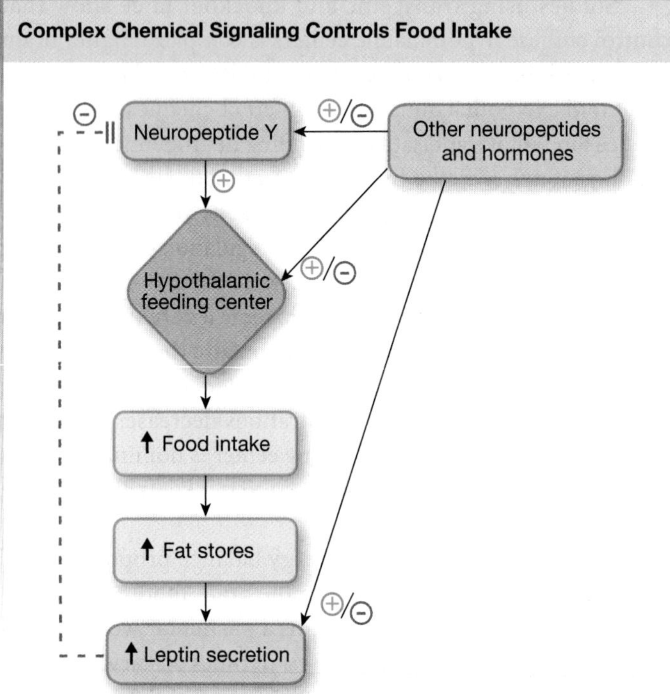

Other Key Peptides That Modulate Food Intake	
Peptide	**Source**
Increase food intake	
Ghrelin	Stomach
Orexins (also called hypocretins)	Hypothalamus
Decrease food intake	
CCK	Small intestine, neurons
Glucagon-like peptide-1 (GLP-1)	Intestines
PYY	Intestines
Corticotropin-releasing hormone (CRH)	Hypothalamus
α-Melanocyte-stimulating hormone (α-MSH)	Hypothalamus
CART (cocaine-and amphetamine-regulated transcript)	Hypothalamus

Fig. 22.1

BIOTECHNOLOGY

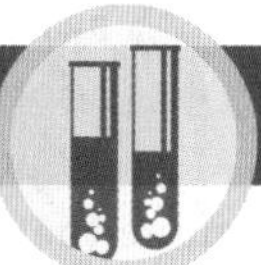

Discovering Peptides: Research in Reverse

In the early days of molecular biology, scientists collected tissues that contained active peptides, isolated and purified the peptides, and then analyzed their amino acid sequence. Now, as we enter the age of proteomics, investigators are using reverse techniques to discover new proteins in the body. One group of researchers, for example, found the hunger-inducing *orexin* (or *hypocretin*) peptides by isolating mRNA expressed in a particular region of the hypothalamus. The investigators then used this mRNA to create the amino acid sequence of the prepropeptide. At the same time, a different group of scientists also discovered orexin peptides by working backwards from an "orphan" G protein–coupled receptor [p. 183] to find its peptide ligand. The endogenous hunger hormone *ghrelin* was discovered by a similar method. Pharmacologists testing synthetic peptides to stimulate the release of growth hormone found that their peptides were binding to a previously unknown receptor, and from that receptor they discovered ghrelin.

In one interesting study researchers tried to determine whether chocolate craving is attributable to psychological factors or to physiological stimuli, such as chemicals in the chocolate.* Subjects were given either dark chocolate, white chocolate (which contains none of the pharmacological agents of cocoa), cocoa capsules, or placebo capsules. The researchers found that white chocolate was the best substitute for the real thing, which suggests that taste and aroma play a significant role in satisfying chocolate craving.

Psychological factors, such as stress, can also play a significant role in regulating food intake. In another study** researchers found that subjects who imagined eating 30 M&M's one at a time ate fewer real M&M's than subjects who thought about eating only 3 M&M's. A repeat of the experiment using cheese cubes instead of M&M's had the same result.

The eating disorder *anorexia nervosa* has both psychological and physiological components, which complicates its treatment. And the concept of *appetite* is closely linked to the psychology of eating, which may explain why dieters who crave smooth, cold ice cream cannot be satisfied by a crunchy carrot stick. A considerable amount of money is directed to research on eating behaviors.

*W. Michener and P. Rozin. Pharmacological versus sensory factors in the satiation of chocolate craving. *Physiol. Behav.* 56(3): 419–422, 1994.

**C. K. Morewedge *et al.* Thought for food: Imagined consumption reduces actual consumption. *Science* 330: 1530–1533, 2010.

RUNNING PROBLEM

Sara and Nicole first met in ballet class at age 11. They were both serious about dance and worked hard to maintain the perfect thin ballet-dancer's body. Both girls occasionally took diet pills or laxatives, and it was almost a contest to see who could eat the least food. Sara was always impressed with Nicole's willpower, but then again, Nicole was a perfectionist. Two years ago, Sara found herself less interested in dance and went away to college. Nicole was accepted into a prestigious ballet company and remained focused on her dancing—and on her body image. Nicole's regimented diet became stricter, and if she felt she'd eaten too much, she would make herself throw up. Her weight kept dropping, but each time she looked in the mirror, she saw a fat girl looking back.

Q1: If you measured Nicole's leptin level, what would you expect to find?

Q2: Would you expect Nicole to have elevated or depressed levels of neuropeptide Y?

737 739 741 744 765 770

Concept Check Answers: p. 774

1. Explain the roles of the satiety and feeding centers. Where are they located?
2. Studies show that most obese humans have elevated leptin levels in their blood. Based on your understanding of endocrine disorders [p. 227], propose some reasons why leptin is not decreasing food intake in these people.

Energy Balance

Once food has been digested and absorbed, the body's chemical reactions—known collectively as metabolism—determine what happens to the nutrients in the food. Are they destined to burn off as heat? Become muscle? Or turn into extra pounds that make it difficult to zip up blue jeans? In this section we examine energy balance in the body.

Energy Input Equals Energy Output

The *first law of thermodynamics* [p. 102] states that the total amount of energy in the universe is constant. By extension, this statement means that all energy that goes into a biological system, such as the human body, can be accounted for (Fig. 22.2). In the body, most stored energy is contained in the chemical bonds of molecules.

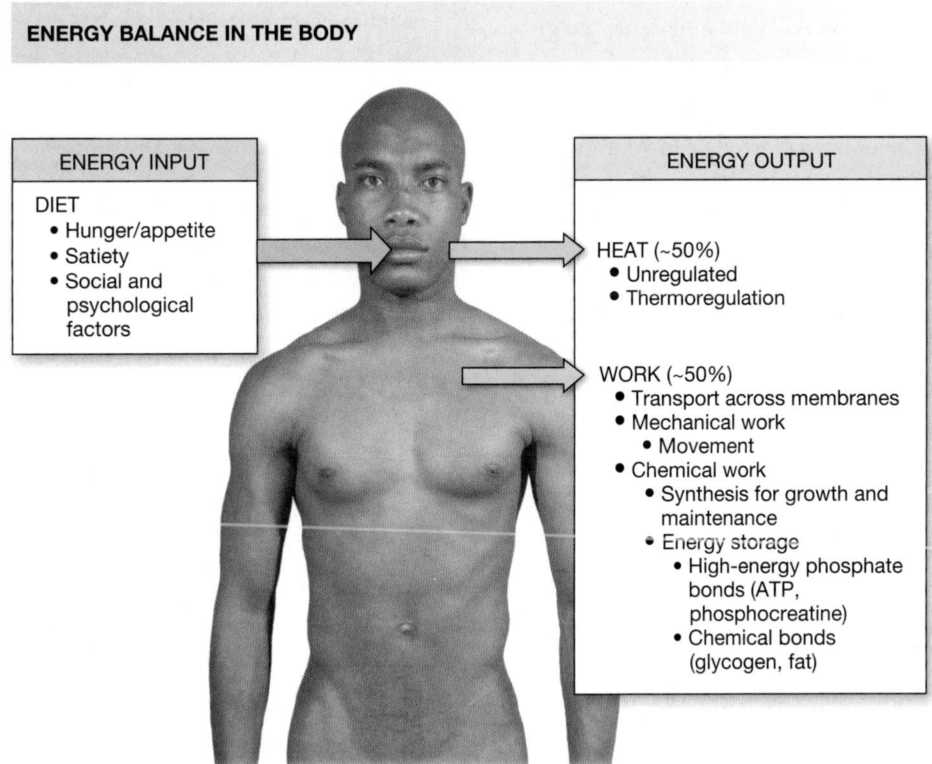

Fig. 22.2

We can apply the concept of mass balance to energy balance: changes to the body's energy stores result from the difference between the energy put into the body and the energy used.

$$\text{Total body energy} = \text{energy stored} + \text{energy intake} - \text{energy output} \quad (1)$$

Energy intake for humans consists of energy in the nutrients we eat, digest, and absorb. *Energy output* is a combination of work performed and energy returned to the environment as heat:

$$\text{Energy output} = \text{work} + \text{heat} \quad (2)$$

In the human body, at least half the energy released in chemical reactions is lost to the environment as unregulated "waste" heat.

The work in equation (2) takes one of three forms [p. 100]:

1. **Transport work** moves molecules from one side of a membrane to the other. Transport processes bring materials into and out of the body and transfer them between compartments.
2. **Mechanical work** uses intracellular fibers and filaments to create movement. This form of work includes external work, such as movement created by skeletal muscle contraction, and internal work, such as the movement of cytoplasmic vesicles and the pumping of the heart.
3. **Chemical work** is used for growth, maintenance, and storage of information and energy. Chemical work in the body can be subdivided into synthesis and storage. Storage includes both *short-term energy storage* in high-energy phosphate compounds such as ATP and *long-term energy storage* in the chemical bonds of glycogen and fat.

Most of this energy-consuming work in the body is not under conscious control. The only way people can voluntarily increase energy *output* is through body movement, such as walking and exercise. People can control their energy *intake*, however, by watching what they eat.

Although energy balance is a very simple concept, it is a difficult one for people to accept. Behavior modifications, such as eating less and exercising more, are among the most frequent instructions healthcare professionals give to patients. These instructions are among the most difficult for people to follow, and patient compliance is low.

CLINICAL FOCUS

Estimating Fat—The Body Mass Index

One way to estimate the total amount of energy stored in the body is a person's body weight. Although it is a simplistic approach, we say that when energy intake exceeds energy output, a person gains weight. If energy use exceeds dietary energy input, the body taps into its energy stores and the person loses weight. One current assessment for healthy weight is a measurement known as the **body mass index**, or BMI. To calculate your BMI:

$$\text{weight (lb)} \times 703/\text{height}^2\text{ (in)} = \text{BMI}$$
$$\text{weight (kg)}/\text{height}^2\text{(m)} = \text{BMI}$$

A BMI between 18.5 and 24.9 is considered normal weight. Less than 18.5 is underweight, and more than 24.5 is overweight. A BMI over 30 indicates obesity. Higher risk for a number of diseases, including diabetes, heart disease, and high blood pressure, correlate with higher BMI values. The BMI calculation does not distinguish between fat mass and muscle mass, however, and heavily muscled athletes, such as football players, may have a BMI that seems unhealthy. Muscle tissue weighs more than fat, which explains the discrepancy. BMI calculations also do not allow for differences due to age, gender, and ethnicity. For example, data indicate that Asians with BMIs in the normal weight range may still be at higher risk of certain diseases. Researchers have suggested that a fat mass index (fat mass / height^2) is a better health indicator than BMI.

Oxygen Consumption Reflects Energy Use

To compile an energy balance sheet for the human body, we must estimate both the energy content of food (energy intake) and the energy expenditure from heat loss and various types of work (energy output). The most direct way to measure the energy content of food is by **direct calorimetry.** In this procedure, food is burned in an instrument called a *bomb calorimeter,* and the heat released is trapped and measured.

The heat released is a direct measure of the energy content of the burned food and is usually measured in kilocalories (kcal). One **kilocalorie** (kcal) is the amount of heat needed to raise the temperature of 1 liter of water by 1 °C. A kilocalorie is the same as a *Calorie* (with a capital C). Although direct calorimetry is a quick way of measuring the total energy content of food, the *metabolic* energy content of food is slightly less because most foods cannot be fully digested and absorbed.

The caloric content of any food can be calculated by multiplying the number of grams of each component by its metabolic energy content. The metabolic energy content of proteins and carbohydrates is 4 kcal/g. Fats contain more than twice as much energy—9 kcal/g. For example, a plain bagel contains:

2 g of fat	2 g fat × 9 kcal/g fat =	18 kcal
7 g of protein	7 g protein × 4 kcal/g protein =	28 kcal
38 g of carbohydrates	38 g carbohydrate × 4 kcal/g CHO =	152 kcal
	Total calories =	198 kcal

RUNNING PROBLEM

When Nicole's blood test results came back, Dr. Ayani immediately wrote orders to start an intravenous infusion and heart monitoring. The laboratory report showed plasma potassium of 2.5 mEq/L (normal: 3.5–5.0 mEq/L), plasma HCO_3^- of 40 mEq/L (normal: 24–29), and plasma pH of 7.52 (normal: 7.38–7.42). Dr. Ayani admitted Nicole to the hospital for further treatment and evaluation, hoping to convince her that she needed help for her anorexia. Anorexia, meaning "no appetite," can have both physiological and psychological origins.

Q3: What is Nicole's K^+ disturbance called? What effect does it have on the resting membrane potential of her cells?

Q4: Why does Dr. Ayani want to monitor Nicole's cardiac function?

Q5: Based on her clinical values, what is Nicole's acid-base status?

737 739 741 744 765 770

In the United States, you can find the energy content for various foods on the *Nutrition Facts* label of food packages.

Estimating an individual's energy expenditure, or **metabolic rate,** is more complex than figuring the caloric content of ingested food. According to the law of mass balance, a person's caloric intake minus heat production is the energy used for chemical, mechanical, and transport work. Heat released by the body can be measured by enclosing a person in a sealed compartment. Practically speaking, however, measuring total body heat release is not a very easy way to measure energy use.

Probably the most common method for estimating metabolic rate is to measure a person's **oxygen consumption,** the rate at which the body consumes oxygen as it metabolizes nutrients. Recall [p. 111] that metabolism of glucose to trap energy in the bonds of ATP is most efficient in the presence of adequate oxygen:

$$C_6H_{12}O_6 + O_2 + ADP + P_i \rightarrow CO_2 + H_2O + ATP + \text{heat}$$

Studies have shown that oxygen consumption for different foods is relatively constant at a rate of 1 liter of oxygen consumed for each 4.5–5 kcal of energy released from the food being metabolized. The measurement of oxygen consumption is one form of **indirect calorimetry**.

Another method of estimating metabolic rate is to measure carbon dioxide production, either alone or in combination with oxygen consumption. The equation above shows that aerobic metabolism consumes O_2 and produces CO_2. However, the ratio of CO_2 produced to O_2 consumed varies with the composition of the diet. This ratio of CO_2 produced to O_2 consumed is known as the **respiratory quotient (RQ)** or the **respiratory exchange ratio (RER)**. RQ varies from a high of 1.0 for a pure carbohydrate diet to 0.8 for pure protein and 0.7 for pure fat. The average American diet has an RQ of about 0.82.

Metabolic rate is calculated by multiplying oxygen consumption by the number of kilocalories metabolized per liter of oxygen consumed:

$$\text{Metabolic rate (kcal/day)} = \text{L } O_2 \text{ consumed/day} \times \text{kcal/L } O_2$$

A mixed diet with an RQ of 0.8 requires one liter of O_2 for each 4.80 kcal metabolized. For a 70-kg male whose resting oxygen consumption is 430 L/day, this means:

$$\text{Resting metabolic rate} = 430 \text{ L } O_2\text{/day} \times 4.80 \text{ kcal/L } O_2 = 2064 \text{ kcal/day}$$

Many Factors Influence Metabolic Rate

Whether measured by O_2 consumption or by CO_2 production, metabolic rate can be highly variable from one person to another or from day to day in a single individual. An individual's lowest metabolic rate is considered the **basal metabolic rate (BMR)**. In reality, metabolic rate would be lowest when an individual is asleep. However, because measuring the BMR of a sleeping person is difficult, metabolic rate is often measured after a 12-hour fast in a person who is awake but resting (a **resting metabolic rate, RMR**).

Other factors that affect metabolic rate in humans include age, sex, amount of lean muscle mass, activity level, diet, hormones, and genetics.

1. **Age and sex**. Adult males have an average BMR of 1.0 kcal per hour per kilogram of body weight. Adult females have a lower rate than males: 0.9 kcal/hr/kg. The difference arises because women have a higher percentage of adipose tissue and less lean muscle mass. Metabolic rates in both sexes decline with age. Some of this decline is due to decreases in lean muscle mass.
2. **Amount of lean muscle mass**. Muscle has higher oxygen consumption than adipose tissue, even at rest. (Most of the volume of an adipose tissue cell is occupied by metabolically inactive lipid droplets.) This is one reason weight loss advice often includes weight training in addition to aerobic exercise. Weight training adds muscle mass to the body, which increases basal metabolic rate and results in more calories being burned at rest.
3. **Activity level**. Physical activity and muscle contraction increase metabolic rate over the basal rate. Sitting and lying down consume relatively little energy. Competitive

22

rowing and cycling are among the activities that expend the most energy.

4 **Diet**. Resting metabolic rate increases after a meal, a phenomenon termed **diet-induced thermogenesis**. In other words, there is an energetic cost to the digestion and assimilation of food. Diet-induced thermogenesis is related to the type and amount of food ingested. Fats cause relatively little diet-induced thermogenesis, and proteins increase heat production the most. This phenomenon may support the claim of some nutritionists that eating a calorie of fat is different from eating a calorie of protein, although they contain the same amount of energy when measured by direct calorimetry.

5 **Hormones.** Basal metabolic rate is increased by thyroid hormones and by catecholamines (epinephrine and norepinephrine). Some of the peptides that regulate food intake also appear to influence metabolism.

6 **Genetics.** The effect of inherited traits on energy balance can be observed in the variety of normal body types. Some people have very efficient metabolism that converts food energy into energy stored in adipose tissue with little heat loss, while others can eat large amounts of food and never gain weight because their metabolism is less efficient.

Of the factors affecting metabolic rate, a person can voluntarily control only two: energy intake (how much food is eaten) and level of physical activity. If a person's activity includes strength training, which increases lean muscle mass, resting metabolic rate goes up. The addition of lean muscle mass to the body creates additional energy use, which in turn decreases the number of calories that go into storage.

Energy Is Stored in Fat and Glycogen

A person's daily energy requirement, expressed as caloric intake, varies with the needs and activity of the body. For example, during the 2008 Olympics, swimming champion Michael Phelps consumed more than 12,000 kcal per day. On the other hand, a woman engaged in normal activities may require only 2000 kcal/day.

Suppose that our woman's energy requirement could be met by ingesting only glucose. Glucose has an energy content of 4 kcal/g, which means that to get 2000 kcal, she would have to consume 500 g, or 1.1 pounds, of glucose each day. Our bodies cannot absorb crystalline glucose, however, so those 500 g of glucose would have to be dissolved in water. If the glucose were made as an isosmotic 5% solution, the 500 g would have to be dissolved in 10 liters of water—a substantial volume to drink in a day!

Fortunately, we do not usually ingest glucose as our primary fuel. Proteins, complex carbohydrates, and fats also provide energy. The glucose polymer glycogen is a more compact form of energy than an equal number of individual glucose molecules. Glycogen also requires less water for hydration.

For this reason, our cells convert glucose to glycogen for storage. Normally we keep about 100 g of glycogen in the liver and 200 g in skeletal muscles. But even this 300 g of glycogen can provide only enough energy for 10 to 15 hours. The brain alone requires 150 g of glucose per day.

Consequently, the body keeps most of its energy reserves in compact, high-energy fat molecules. One gram of fat has 9 kcal, more than twice the energy content of an equal amount of carbohydrate or protein. This feature makes adipose tissue very efficient at storing large amounts of energy in minimal space. Metabolically, however, the energy in fat is harder to access, and the metabolism of fats is slower than that of carbohydrates.

Concept Check Answers: p. 774

3. Name seven factors that can influence a person's metabolic rate.
4. Why does the body store most of its extra energy in fat and not in glycogen?
5. Complete and balance the following equation for aerobic metabolism of one glucose molecule:

$$C_6H_{12}O_6 + O_2 \rightarrow ? + ?$$

6. What is the RQ for the balanced equation in Concept Check 5?

Metabolism

Metabolism is the sum of all chemical reactions in the body. The reactions making up these pathways (1) extract energy from nutrients, (2) use energy for work, and (3) store excess energy so that it can be used later. Metabolic pathways that synthesize large molecules from smaller ones are called **anabolic pathways** {*ana-*, completion + *metabole,* change}. Those that break large molecules into smaller ones are called **catabolic** pathways {*cata-*, down or back}.

The classification of a pathway is its net result, not what happens in any individual step of the pathway. For example, in the first step of *glycolysis* [p. 113], glucose gains a phosphate to become a larger molecule, glucose 6-phosphate. This single reaction is anabolic, but by the end of glycolysis the initial 6-carbon glucose molecule has been converted to two 3-carbon pyruvate molecules, which makes glycolysis a catabolic pathway.

In the human body, we divide metabolism into two states. The period of time following a meal, when the products of digestion are being absorbed, used, and stored, is called the **fed state** or the **absorptive state**. This is an anabolic state in which the energy of nutrient biomolecules is transferred to high-energy compounds or stored in the chemical bonds of other molecules.

Once nutrients from a recent meal are no longer in the bloodstream and available for use by the tissues, the body enters what is called the **fasted state** or the **postabsorptive state**. As the pool of available nutrients in the blood decreases, the body taps into its stored reserves. The fasted state is catabolic because cells break down large molecules into smaller molecules. The energy released by breaking chemical bonds of large molecules is used to do work.

Ingested Energy May Be Used or Stored

The biomolecules we ingest are destined to meet one of three fates:

1. **Energy**. Biomolecules can be metabolized immediately, with the energy released from broken chemical bonds trapped in ATP, phosphocreatine, and other high-energy compounds. This energy can then be used to do mechanical work.
2. **Synthesis**. Biomolecules entering cells can be used to synthesize basic components needed for growth and maintenance of cells and tissues.
3. **Storage**. If the amount of food ingested exceeds the body's requirements for energy and synthesis, the excess energy goes into storage in the bonds of glycogen and fat. Storage makes energy available for times of fasting.

The fate of an absorbed biomolecule depends on whether it is a carbohydrate, protein, or fat. ■ Figure 22.3 is a schematic diagram that follows these biomolecules from the diet into the three *nutrient pools* of the body: the free fatty acid pool, the glucose pool, and the amino acid pool. **Nutrient pools** are nutrients that are available for immediate use. They are located primarily in the plasma.

Free fatty acids form the primary pool of fats in the blood. They can be used as an energy source by many tissues but are also easily stored as fat (*triglycerides*) in adipose tissue.

Carbohydrates are absorbed mostly as glucose. Plasma glucose concentration is the most closely regulated of the three nutrient pools because glucose is the only fuel the brain can metabolize, except in times of starvation. Notice the locations of the exit "pipes" on the glucose pool in Figure 22.3. If the glucose pool falls below a certain level, only the brain has access to glucose. This conservation measure ensures that the brain has an adequate energy supply. Just as the circulatory system gives priority to supplying oxygen to the brain, metabolism also gives priority to the brain.

22

NUTRIENT POOLS AND METABOLISM

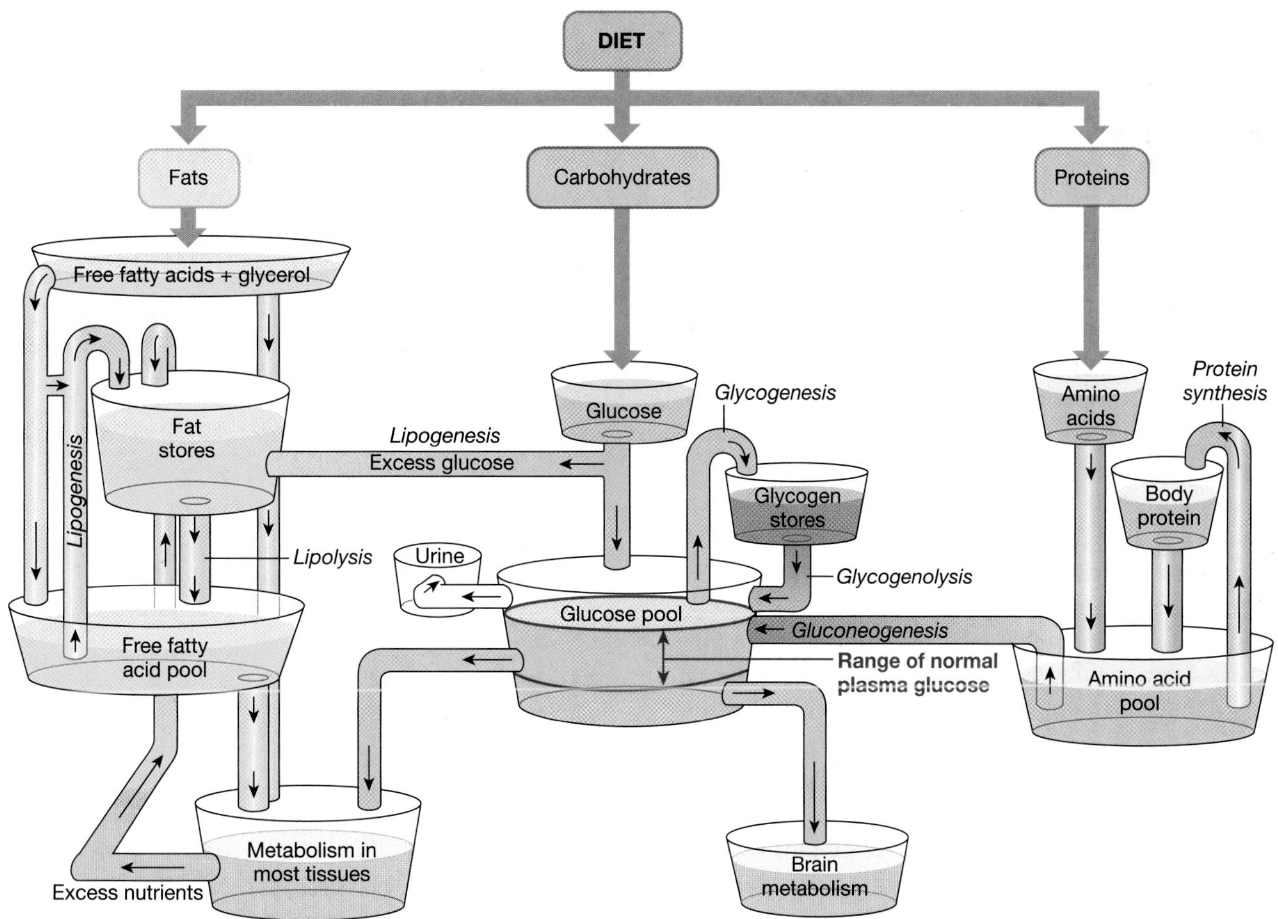

■ **Fig. 22.3** Adapted from L. L. Langley, *Homeostasis* (New York: Reinhold, 1965).

RUNNING PROBLEM

When Nicole was admitted to the hospital, her blood pressure was 80/50, and her pulse was a weak and irregular 90 beats per minute. She weighed less than 85% of the minimal healthy weight for a woman of her height and age. She had an intense fear of gaining weight, even though she was underweight. Her menstrual periods were irregular, she had just suffered a fractured wrist from a fall that normally shouldn't have caused a fracture, and her hair was thinning. When Dr. Ayani questioned Nicole, she admitted that she had been feeling weak during dance rehearsals and had been having difficulty concentrating at times.

Q6: Based on what you know about heart rate and blood pressure, speculate on why Nicole has low blood pressure with a rapid pulse.

Q7: Would you expect Nicole's renin and aldosterone levels to be normal, elevated, or depressed? How might these levels relate to her K^+ disturbance?

Q8: Give some possible reasons Nicole had been feeling weak during dance rehearsals.

If the body's glucose pool is within the normal range, most tissues use glucose for their energy source. Excess glucose goes into storage as glycogen. The synthesis of glycogen from glucose is known as **glycogenesis.** Glycogen stores are limited, however, and additional excess glucose is converted to fat by **lipogenesis**.

If plasma glucose concentrations decrease, the body converts glycogen to glucose through **glycogenolysis**. The body maintains plasma glucose concentrations within a narrow range by balancing oxidative metabolism, glycogenesis, glycogenolysis, and lipogenesis.

If homeostasis fails and plasma glucose exceeds a critical level, as occurs in diabetes mellitus, excess glucose is excreted in the urine. Glucose excretion takes place only when the *renal threshold* for glucose reabsorption is exceeded [p. 643].

The amino acid pool of the body is used primarily for protein synthesis. However, if glucose intake is low, amino acids can be converted into glucose through the pathways known as **gluconeogenesis.** This word literally means "the birth (*genesis*) of new (*neo*) glucose" and refers to the synthesis of glucose from a noncarbohydrate precursor.

Amino acids are the main source for glucose through the gluconeogenesis pathways, but glycerol from triglycerides can also be used. Both gluconeogenesis and glycogenolysis are important backup sources for glucose during periods of fasting.

Enzymes Control the Direction of Metabolism

How does the body control the shift of metabolism between the fed state and the fasted state? One key feature of metabolic regulation is the use of different enzymes to catalyze forward and reverse reactions. This dual control, sometimes called *push-pull control,* allows close regulation of a reaction's direction.

Figure 22.4 shows how push-pull control can regulate the flow of nutrients through metabolic pathways. In Figure 22.4a, enzyme 1 catalyzes the reaction A $\rightarrow$ B, and enzyme 2 catalyzes the reverse reaction, B $\rightarrow$ A. When the activity of the two enzymes is roughly equal, as soon as A is converted into B, B is converted back into A. Turnover of the two substrates is rapid, but there is no net production of either A or B.

To alter the net direction of the reaction, the enzyme activity must change. Enzymes are proteins that bind ligands, so their activity can be modulated [p. 52]. Most modulation of metabolic enzymes is controlled by hormones.

Figure 22.4b represents the series of reactions through which glucose becomes glycogen. During the fed state, the pancreatic hormone insulin stimulates the enzymes promoting glycogenesis and inhibits the enzymes for glycogenolysis. The net result is glycogen synthesis from glucose.

The reverse pattern is shown in Figure 22.4c. In the fasted state, *glucagon*, another pancreatic hormone, is dominant.

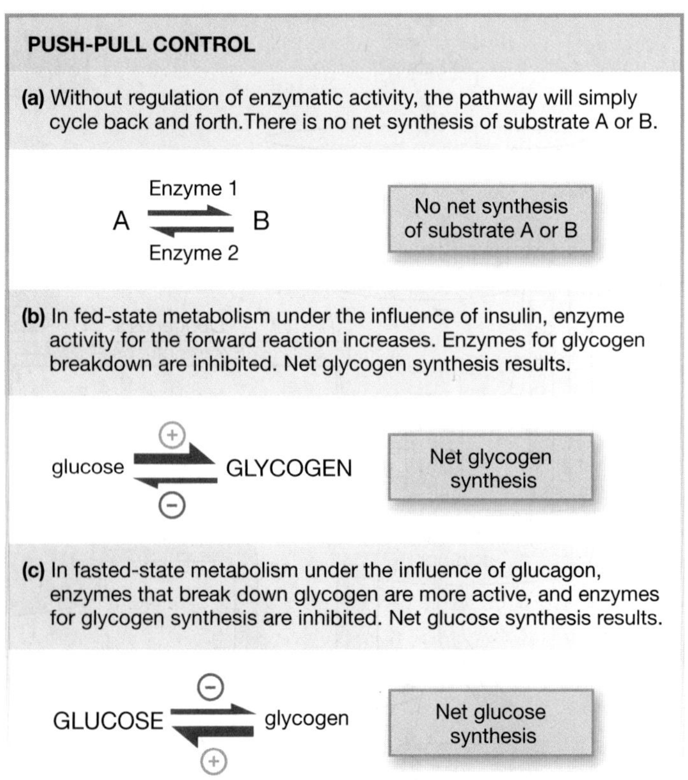

Fig. 22.4 ***In push-pull control, different enzymes catalyze forward and reverse reactions.***

Glucagon stimulates the enzymes of glycogenolysis while inhibiting the enzymes for glycogenesis. The net result is glucose synthesis from glycogen.

Fed-State Metabolism

The fed state following ingestion of nutrients is anabolic: absorbed nutrients are being used for energy, synthesis, and storage. The table at the bottom of ■ Figure 22.5 summarizes the fates of carbohydrates, proteins, and fats in the fed state. In the sections that follow, we examine some of these pathways.

Carbohydrates Make ATP

The most important biochemical pathways for energy production are summarized in Figure 22.5. This figure does not include all of the metabolic intermediates in each pathway [see Chapter 4, p. 113 for detailed pathways]. Instead, it emphasizes the points at which different pathways intersect, because these intersections are often key points at which metabolism is controlled.

Glucose is the primary substrate for ATP production. Glucose absorbed from the intestine enters the hepatic portal vein and is taken directly to the liver, where about 30% of all ingested glucose is metabolized. The remaining 70% continues in the bloodstream for distribution to the brain, muscles, and other organs and tissues.

Glucose moves from interstitial fluid into cells by membrane GLUT transporters [p. 149]. Most glucose absorbed from a meal goes immediately into glycolysis and the citric acid cycle to make ATP. Some glucose is used by the liver for lipoprotein synthesis. Glucose that is not required for energy and synthesis is stored either as glycogen or fat. The body's ability to store glycogen is limited, so most excess glucose is converted to triglycerides and stored in adipose tissue.

Glucose Storage Glycogen, a large polysaccharide, is the main storage form of glucose in the body. Glycogen is a glucose polymer, created by linking many individual glucose molecules together into a branching chain [see Fig. 2.2, p. 34]. A single glycogen particle in the cytoplasm may contain as many as 55,000 linked glucose molecules! Glycogen granules occur as insoluble inclusions in the cytosol of cells [p. 72].

Glycogen is found in all cells of the body, but the liver and skeletal muscles contain especially high concentrations. Glycogen in skeletal muscles provides a ready energy source for muscle contraction. Glycogen in the liver acts as the main source of glucose for the body in periods between meals (the fasted state). It is estimated that the liver keeps about a four-hour supply of glucose stored as glycogen.

Concept Check Answer: p. 774

7. Are GLUT transporters active or passive transporters?

Amino Acids Make Proteins

Most amino acids absorbed from a meal go to the tissues for protein synthesis [p. 117]. Like glucose, amino acids are taken first to the liver by the hepatic portal system. The liver uses them to synthesize lipoproteins and plasma proteins, such as albumin, clotting factors, and angiotensinogen.

Amino acids not taken up by the liver are used by cells to create structural or functional proteins, such as cytoskeletal elements, enzymes, and hormones. Amino acids are also incorporated into nonprotein molecules, such as amine hormones and neurotransmitters.

If glucose intake is low, amino acids can be used for energy, as described in the next section on fasted-state metabolism. However, if more protein is ingested than is needed for synthesis and energy expenditures, excess amino acids are converted to fat. Some bodybuilders spend large amounts of money on amino-acid supplements advertised to build bigger muscles. But these amino acids do not automatically go into protein synthesis. When amino acid intake exceeds the body's need for protein synthesis, excess amino acids are burned for energy or stored as fat.

Fats Store Energy

Most ingested fats are assembled inside intestinal epithelial cells into lipoprotein and lipid complexes called *chylomicrons* [p. 714]. Chylomicrons leave the intestine and enter the venous circulation via the lymphatic vessels (■ Fig. 22.6a). Chylomicrons consist of cholesterol, triglycerides, phospholipids, and lipid-binding proteins called **apoproteins**, or *apolipoproteins* {*apo-*, derived from}. Once these lipid complexes begin to circulate through the blood, the enzyme **lipoprotein lipase** bound to the capillary endothelium of muscles and adipose tissue converts the triglycerides to free fatty acids and glycerol. These molecules may be used for energy by most cells or reassembled into triglycerides for storage in adipose tissue.

Chylomicron remnants that remain in the circulation are taken up and metabolized by the liver (Fig. 22.6a). Cholesterol from the remnants joins the liver's pool of lipids. If cholesterol is in excess, some may be converted to bile salts and excreted in the bile. The remaining cholesterol is added to newly synthesized cholesterol and fatty acids, and packaged into lipoprotein complexes for secretion into the blood.

The lipoprotein complexes that re-enter the blood contain varying amounts of triglycerides, phospholipids, cholesterol, and apoproteins. The more protein a complex contains, the heavier it is, with plasma lipoprotein complexes ranging from *very-low-density lipoprotein* (VLDL) to high-density lipoprotein (HDL). The combination of lipids with proteins makes cholesterol more soluble in plasma, but the complexes are unable to diffuse through cell membranes. Instead, they must be brought into cells by *receptor-mediated endocytosis* [p. 155].

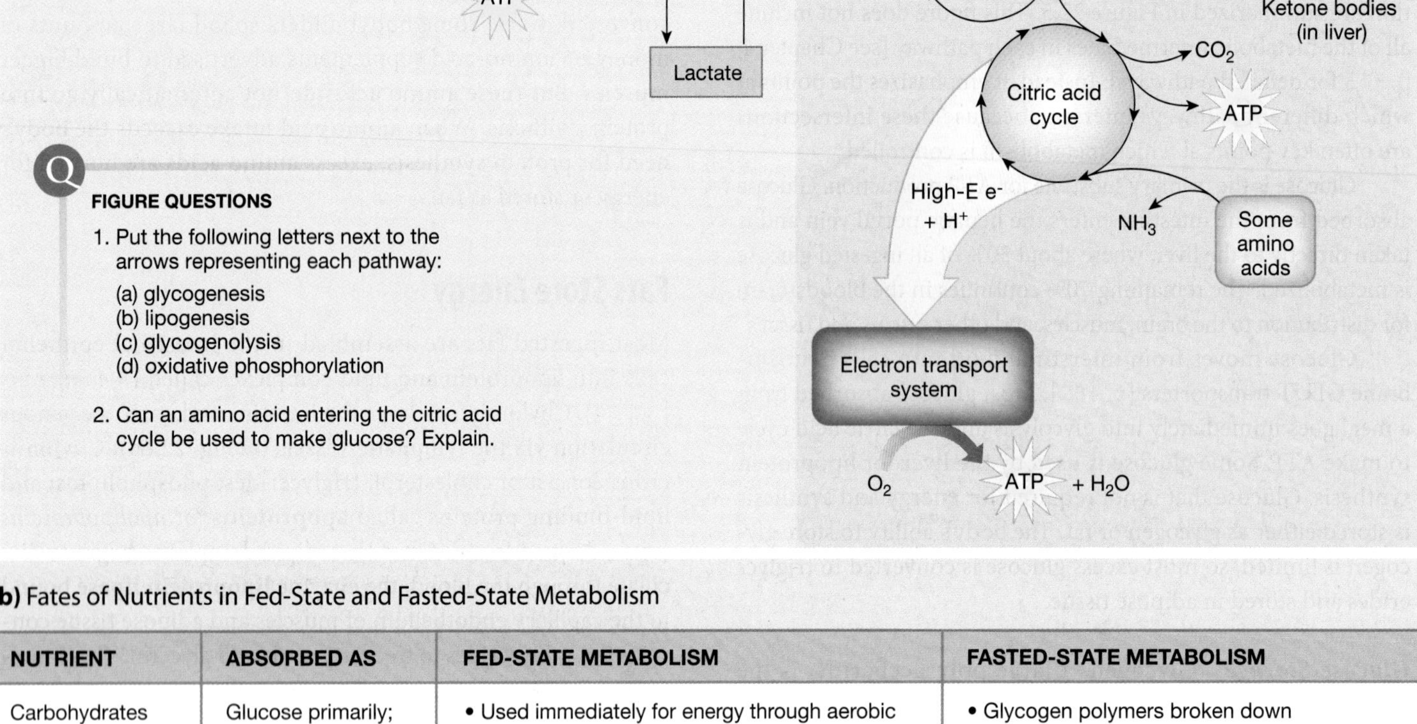

FIGURE QUESTIONS

1. Put the following letters next to the arrows representing each pathway:
   (a) glycogenesis
   (b) lipogenesis
   (c) glycogenolysis
   (d) oxidative phosphorylation
2. Can an amino acid entering the citric acid cycle be used to make glucose? Explain.

(b) Fates of Nutrients in Fed-State and Fasted-State Metabolism

NUTRIENT	ABSORBED AS	FED-STATE METABOLISM	FASTED-STATE METABOLISM
Carbohydrates	Glucose primarily; also fructose and galactose	• Used immediately for energy through aerobic pathways* (*glycolysis* and *citric acid cycle*) • Used for lipoprotein synthesis in liver • Stored as glycogen in liver and muscle (*glycogenesis*) • Excess converted to fat and stored in adipose tissue (*lipogenesis*)	• Glycogen polymers broken down (*glycogenolysis*) to glucose in liver and kidney or to glucose 6-phosphate for use in glycolysis
Proteins	Amino acids primarily plus some small peptides	• Most amino acids go to tissues for protein synthesis* • If needed for energy, amino acids converted in liver to intermediates for aerobic metabolism (*deamination*) • Excess converted to fat and stored in adipose tissue (*lipogenesis*)	• Proteins broken down into amino acids • Amino acids deaminated in liver for ATP production or used to make glucose (*gluconeogenesis*)
Fats	Fatty acids, triglycerides and cholesterol	• Stored as triglycerides primarily in the liver and adipose tissue* (*lipogenesis*) • Cholesterol used for steroid synthesis or as a membrane component • Fatty acids used for lipoprotein and eicosanoid synthesis	• Triglycerides broken down into fatty acids and glycerol (*lipolysis*) • Fatty acids used for ATP production through aerobic pathways (β-*oxidation*)

Fat Synthesis

(a) Transport and Fate of Dietary Fats

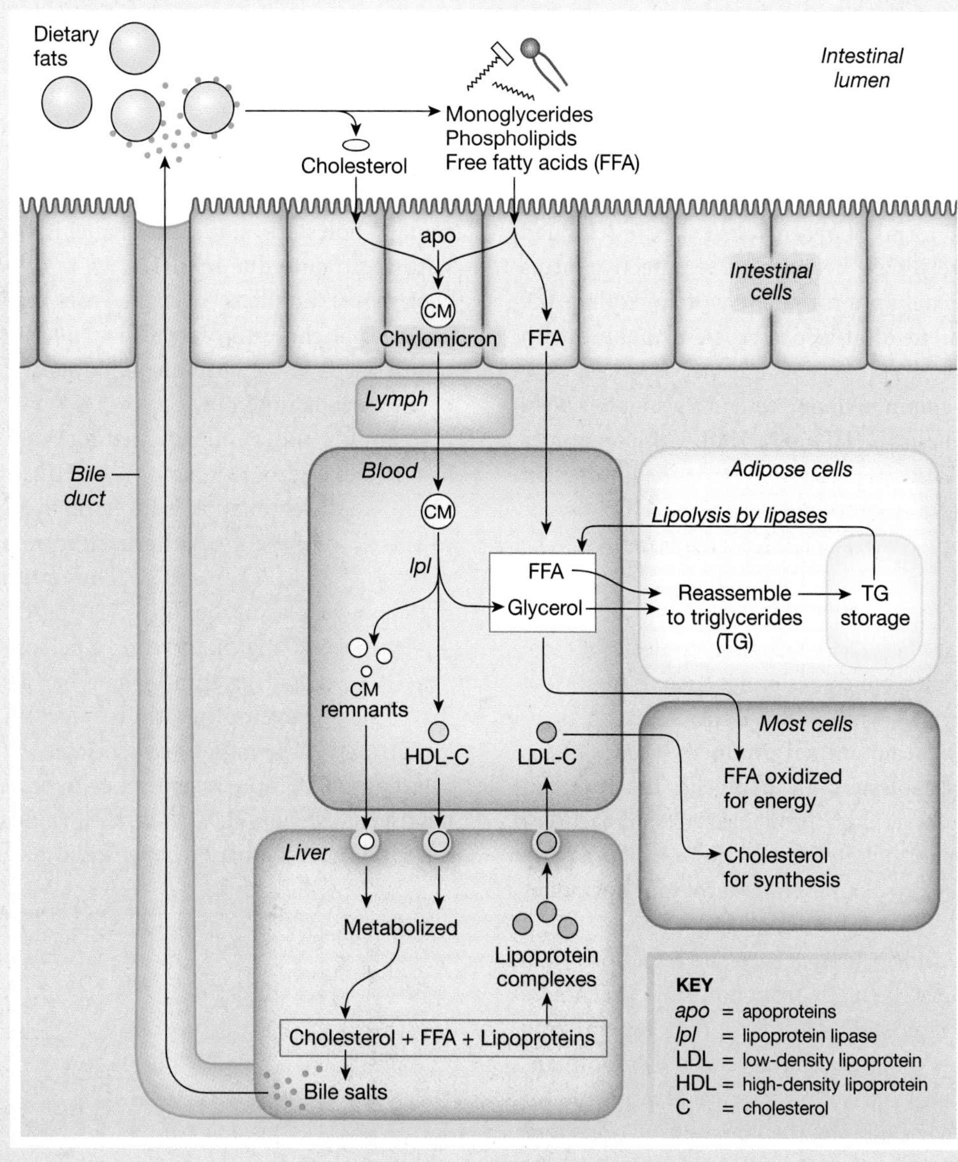

(b) Triglyceride and Cholesterol Synthesis from Glucose

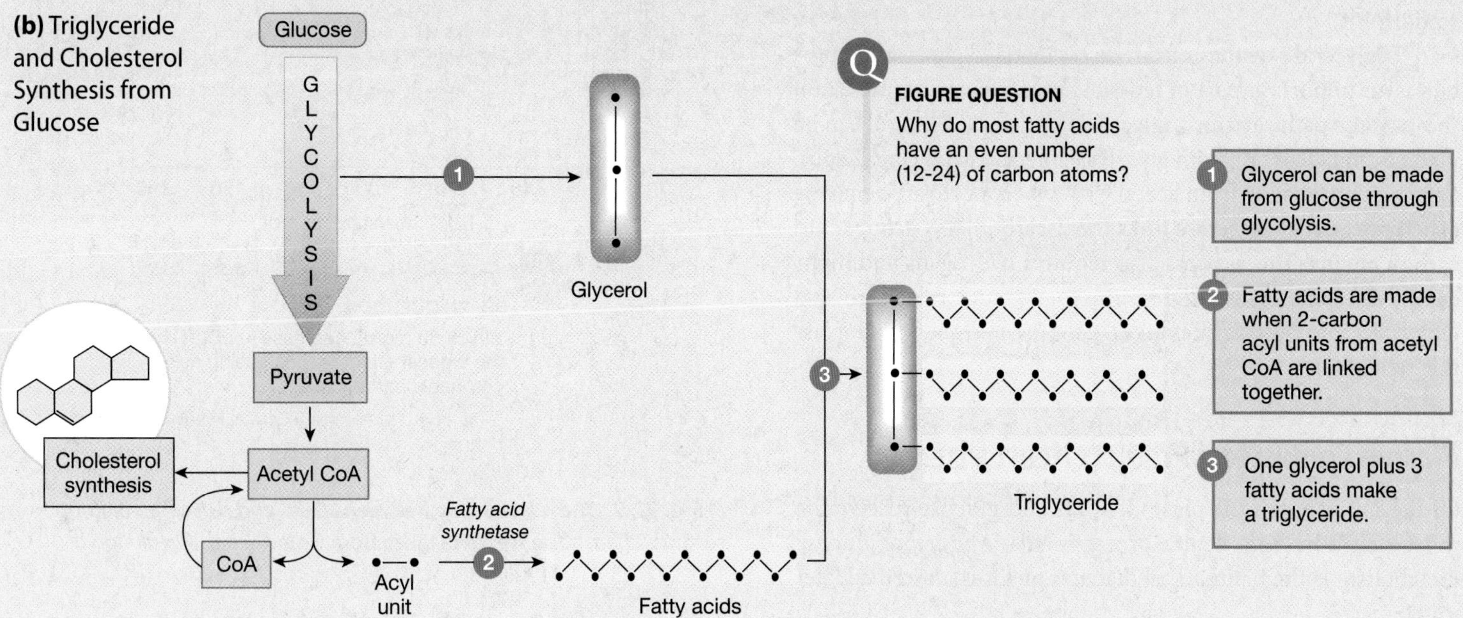

The apoproteins in the complexes have specific membrane receptors in different tissues.

Most lipoprotein in the blood is *low-density lipoprotein-cholesterol* [p. 535]. LDL-C is sometimes known as the "lethal cholesterol" because elevated concentrations of plasma LDL-C are associated with the development of atherosclerosis [p. 535]. LDL-C complexes contain *apoprotein B* (apoB), which combines with receptors that bring LDL-C into most cells of the body. Several inherited forms of *hypercholesterolemia* (elevated plasma cholesterol levels) have been linked to defective forms of apoB. These abnormal apoproteins may help explain the accelerated development of atherosclerosis in people with hypercholesterolemia.

The second most common lipoprotein in the blood is *high-density lipoprotein-cholesterol* (HDL-C). HDL-C is sometimes called the "healthy cholesterol" because HDL is the lipoprotein involved in cholesterol transport out of the plasma. HDL-C contains *apoprotein A* (apoA), which facilitates cholesterol uptake by the liver and other tissues.

Lipid Synthesis Most people get sufficient cholesterol from animal products in the diet, but cholesterol is such an important molecule that the body will synthesize it if the diet is deficient. Even vegetarians who eat no animal products (vegans) have substantial amounts of cholesterol in their cells. The body can make cholesterol from acetyl CoA through a series of reactions. Once the ring structure of cholesterol is synthesized, it is a fairly simple matter for the cell to change cholesterol into hormones and other steroids.

Other fats needed for cell structure and function, such as phospholipids, can also be made from non-lipid precursors during the fed state. Lipids are so diverse that generalizing about their synthesis is difficult. Enzymes in the smooth endoplasmic reticulum and cytosol of cells are responsible for most lipid synthesis. For example, the phosphorylation steps that turn triglycerides into phospholipids take place in the smooth ER.

Triglyceride synthesis from excess ingested glucose and protein is an important part of fed-state metabolism. Figure 22.6b shows some pathways for triglyceride synthesis. Glycerol can be made from glucose or from glycolysis intermediates (Fig. 22.6b). Fatty acids are made from acetyl CoA when a cytosolic enzyme called *fatty acid synthetase* links the 2-carbon acyl groups into carbon chains. This process also requires hydrogens and high-energy electrons from NADPH. The combination of glycerol and fatty acids into triglycerides takes place in the smooth endoplasmic reticulum.

Plasma Cholesterol Predicts Heart Disease

Of the nutrients in the plasma, lipids and glucose receive the most attention from health professionals. Abnormal glucose metabolism is the hallmark of diabetes mellitus, described later in this chapter. Abnormal plasma lipids are used as predictors of atherosclerosis and coronary heart disease (CHD) [p. 534].

Tests to measure blood lipids and assess cardiovascular risk range from simple but less accurate finger-stick blood samples to expensive tests on venous blood that look at all sizes of lipoproteins, from VLDL through HDL. As more epidemiological and treatment data are gathered, experts continue to redefine desirable lipid values. The U.S. National Cholesterol Education Panel issued guidelines in 2004, with an update due in 2012 (*www.nhlbi.nih.gov/guidelines/cholesterol*). Emphasis over the years has shifted from concern about total cholesterol levels (< 200 mg/dL of plasma is recommended) to the absolute amounts and relative proportions of the various subtypes.

Some studies indicate that elevated LDL-C is the single largest cholesterol risk factor for CHD (Fig. 22.7). Oxidized LDL-C in the circulation is taken up by macrophages and leads to the development of atherosclerotic plaques inside blood vessels [Fig. 15.22, p. 536]. The optimum LDL-C value is < 100 mg/dL plasma, but up to 130 mg/dL is acceptable. LDL-C higher than 130 mg/dL increases a person's risk for CHD.

The HDL-C level in plasma has also been used to predict risk of developing atherosclerosis. Like high LDL-C, low HDL-C (< 40 mg/dL) is associated with a higher risk of developing CHD. More recently, health professionals have started looking at the *non-HDL cholesterol value* (total cholesterol minus HDL-C) as perhaps a better indicator of CHD risk.

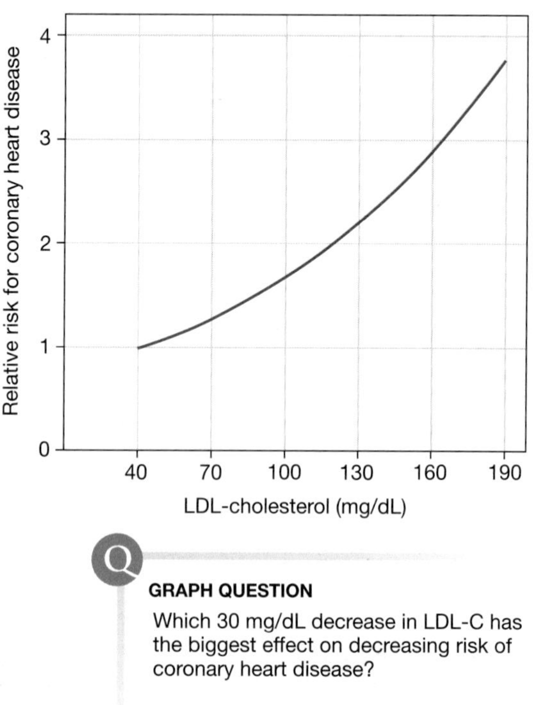

Fig. 22.7 ***The relationship between LDL-C and risk of developing coronary heart disease.*** Data taken from Grundy *et al.*, *Circulation* 110: 227–239, 2004 July 13.

Lifestyle modifications (improved diet, smoking cessation, and exercise) can be very effective in improving lipid profiles but can be difficult for patients to implement and sustain. All the pharmacological therapies developed to treat elevated cholesterol target some aspect of cholesterol absorption and metabolism, underscoring the principle of mass balance in cholesterol homeostasis. Decreasing cholesterol uptake or synthesis and increasing cholesterol clearance by metabolism or excretion are effective mechanisms for reducing the amount of cholesterol in the body.

The drugs known as *bile acid sequestrants* {*sequestrare*, to put in the hands of a trustee} bind to bile acids in the intestinal lumen, preventing their reabsorption and increasing cholesterol excretion [p. 713]. In the absence of recycled bile salts, the liver increases bile acid synthesis from cholesterol, thereby decreasing plasma cholesterol when hepatic LDL receptors import cholesterol from the blood. A drug-free way to accomplish the same result is to eat more soluble fiber—indigestible fiber found in oatmeal, for example. Soluble fiber traps bile salts and increases their excretion in the feces.

Another lipid-lowering drug, *ezetimibe*, inhibits intestinal cholesterol transport [p. 713]. A different strategy for reducing intestinal cholesterol uptake is adding plant *sterols* (steroid-alcohols) and *stanols* (saturated sterols) to the diet by eating nuts, seeds, cereals, and vegetable oils. Sterols and stanols displace cholesterol in chylomicrons, decreasing cholesterol absorption.

The remaining lipid-lowering drugs all affect cholesterol metabolism in the liver. The drugs called *statins* inhibit the enzyme *HMG CoA reductase*, which mediates cholesterol synthesis in hepatocytes. The *fibrates*, which stimulate a transcription factor called *PPARα* (pronounced *p-par-alpha*), and *niacin* (vitamin B_3, nicotinic acid) decrease LDL-C and increase HDL-C by mechanisms that are not well understood.

Concept Check Answers: p. 774

8. What is a dL (as in mg/dL)?
9. Use your understanding of digestive physiology to predict the common side effects of taking bile acid sequestrants and ezetimibe.

Fasted-State Metabolism

Once all nutrients from a meal have been digested, absorbed, and distributed to various cells, plasma concentrations of glucose begin to fall. This is the signal for the body to shift from fed (absorptive) state to fasted (postabsorptive) state metabolism. Metabolism is under the control of hormones whose goal is to maintain blood glucose homeostasis so that the brain and neurons have adequate fuel.

Glucose homeostasis is achieved through catabolic pathways that convert glycogen, proteins, and fats into intermediates that can be used to make either glucose or ATP. Using proteins or fats for ATP production spares plasma glucose for use by the brain. ■ Figure 22.8 summarizes the catabolic pathways of the fasted state in different organs.

22

Glycogen Converts to Glucose

The most readily available source of glucose for plasma glucose homeostasis is the body's glycogen store, most of which is in the liver (Fig. 22.8). Liver glycogen can provide enough glucose to meet 4–5 hours of the body's energy needs.

In glycogenolysis, glycogen is broken down to glucose or glucose 6-phosphate (■ Fig. 22.9). Most glycogen is converted directly to glucose 6-phosphate in a reaction that splits a glucose molecule from the polymer with the help of an inorganic phosphate from the cytosol. Only about 10% of stored glycogen is hydrolyzed to plain glucose molecules.

CLINICAL FOCUS

Antioxidants Protect the Body

One of the daily rituals of childhood is taking your chewable vitamin. Many vitamins are coenzymes [p. 106] and necessary in small amounts for metabolic reactions. Some, such as vitamins C and E, also act as antioxidants. We hear antioxidants promoted everywhere—in cosmetics as well as foods. Many antioxidants occur naturally in fruits and vegetables. **Antioxidants** are molecules that prevent damage to our cells by taking or giving up an electron without becoming free radicals. So what are free radicals, and why are they bad?

Free radicals are unstable molecules with an unpaired electron. Electrons in an atom are most stable in pairs, so free radicals, with their one unpaired electron, look for an electron they can "steal" from another atom. This creates a chain reaction of free radical production that can disrupt normal cell function. Free radicals are thought to contribute to aging and to the development of certain diseases, such as some cancers.

Free radicals are created during normal metabolism or through exposure to radiation, both natural, such as from the sun, and manufactured, such as from microwave ovens and television sets. One common free radical is the *superoxide* ion, $\cdot O_2^-$. The body constantly manufactures this free radical during normal metabolism whenever a neutral oxygen molecule (O_2) gains an extra electron (represented by placing a • in front of the molecule). Antioxidants can absorb the extra electron from superoxide, thereby stopping the destructive chain of free radical formation.

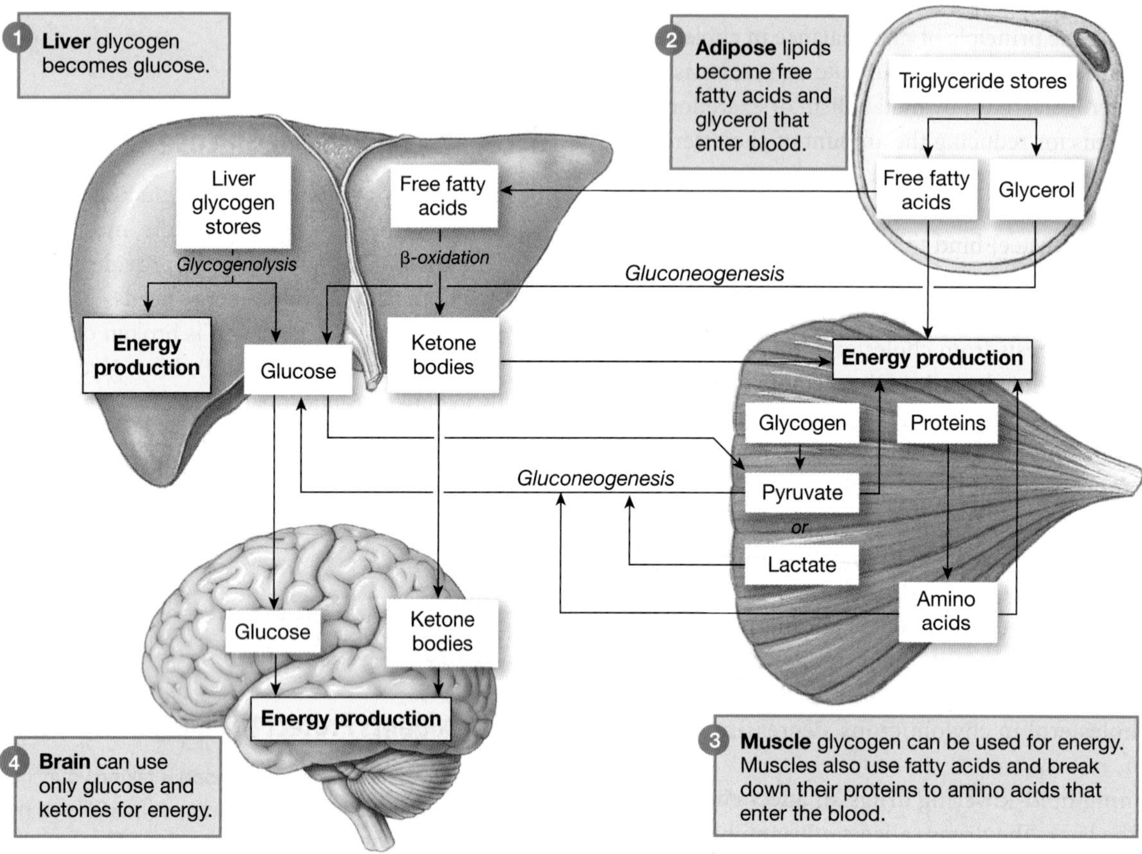
FASTED-STATE METABOLISM
Fasted-state metabolism must maintain plasma glucose homeostasis for the brain.
1 Liver glycogen becomes glucose.
Liver glycogen stores
Glycogenolysis
Energy production
Glucose
Free fatty acids
β-oxidation
Ketone bodies
2 Adipose lipids become free fatty acids and glycerol that enter blood.
Triglyceride stores
Free fatty acids
Glycerol
Gluconeogenesis
Energy production
Glycogen
Proteins
Pyruvate
or
Lactate
Amino acids
Gluconeogenesis
Glucose
Ketone bodies
Energy production
3 Muscle glycogen can be used for energy. Muscles also use fatty acids and break down their proteins to amino acids that enter the blood.
4 Brain can use only glucose and ketones for energy.

Fig. 22.8

In the fasted state, skeletal muscle glycogen can be metabolized to glucose, but not directly (Fig. 22.8). Muscle cells, like most other cells, lack the enzyme that makes glucose from glucose 6-phosphate. Consequently, glucose 6-phosphate produced from glycogenolysis in skeletal muscle is metabolized to either pyruvate (aerobic conditions) or lactate (anaerobic conditions). Pyruvate and lactate are then transported to the liver, which uses them to make glucose via gluconeogenesis.

Proteins Can Be Used to Make ATP

The free amino acid pool of the body normally provides substrate for ATP production during the fasted state. If the fasted state goes on for an extended time, however, muscle proteins can be broken down to provide energy. The first step in protein catabolism is the digestion of a protein into smaller polypeptides by enzymes called *proteases* (*endopeptidases*) [p. 712]. Then enzymes known as *exopeptidases* break the peptide bonds at the ends of the smaller polypeptide, freeing individual amino acids.

Amino acids can be converted into intermediates that enter either glycolysis or the citric acid cycle (Fig. 22.5). The first step in this conversion is **deamination**, the removal of the amino group from the amino acid (Fig. 22.10a). Deamination creates an ammonia molecule and an organic acid. Some of the organic acids created this way are pyruvate, acetyl CoA, and several intermediates of the citric acid cycle. The organic acids can then enter the pathways of aerobic metabolism for ATP production.

The *ammonia* (NH_3) molecules created during deamination rapidly pick up hydrogen ions (H^+) to become *ammonium ions* (NH_4^+), as indicated in Figure 22.10b. Both ammonia and ammonium ions are toxic, however, so liver cells rapidly convert them into **urea** (CH_4N_2O). Urea is the main nitrogenous waste of the body and is excreted by the kidneys.

If glycogen stores run low and plasma glucose homeostasis is threatened, proteins can also be used to make glucose. In the liver, amino acids or pyruvate made from amino acids

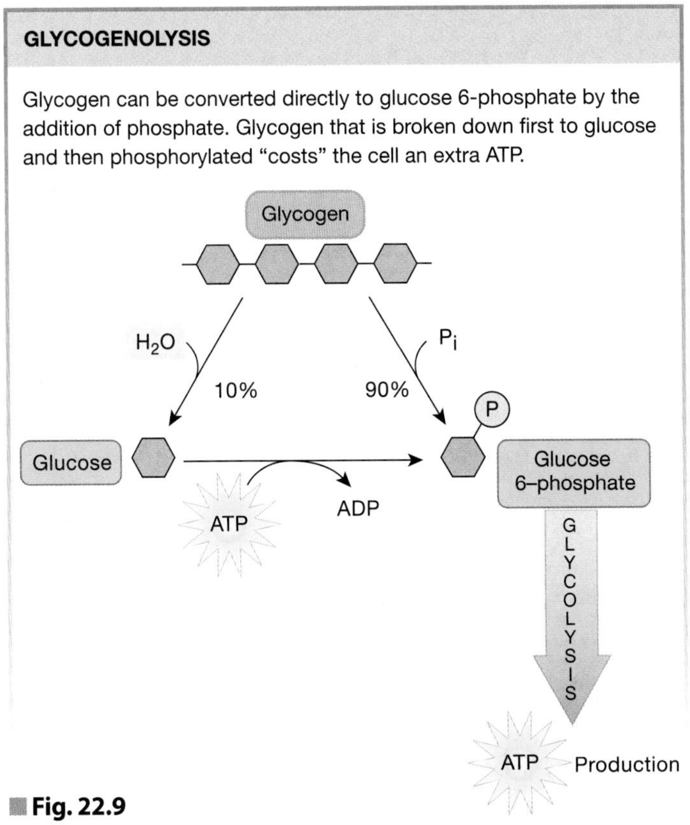

GLYCOGENOLYSIS

Glycogen can be converted directly to glucose 6-phosphate by the addition of phosphate. Glycogen that is broken down first to glucose and then phosphorylated "costs" the cell an extra ATP.

Glycogen
H_2O
P_i
10%
90%
Glucose
P
Glucose 6-phosphate
ATP
ADP
GLYCOLYSIS
ATP Production

■ **Fig. 22.9**

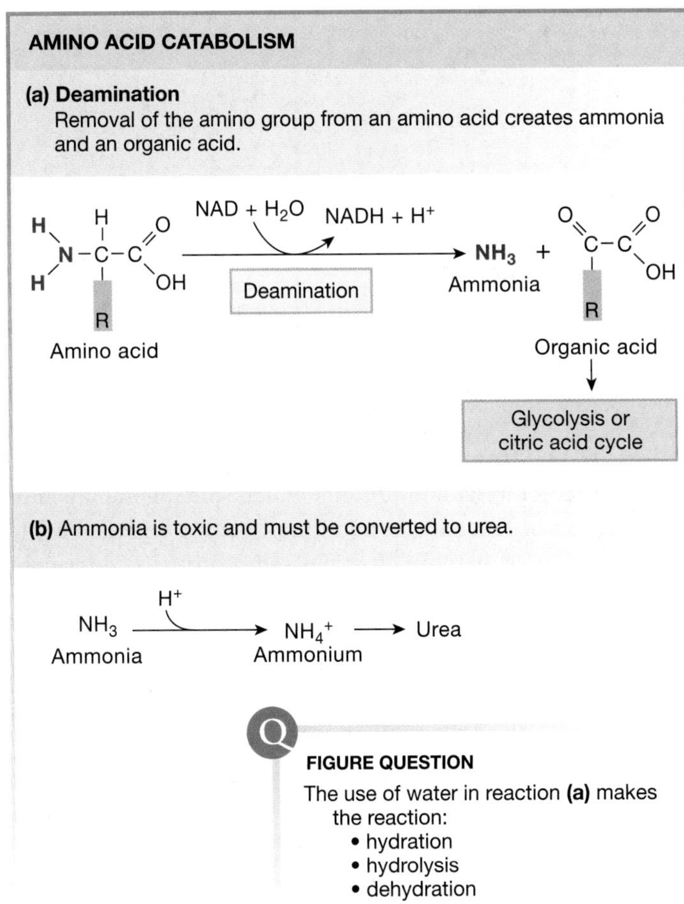

■ **Fig. 22.10**

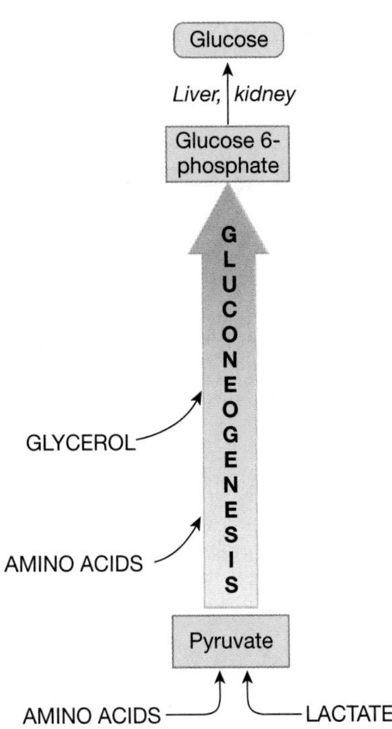

■ **Fig. 22.11**

enter the glycolysis pathway (■ Fig. 22.11). The pathway then runs backward to create glucose 6-phosphate and glucose (gluconeogenesis).

Lipids Store More Energy than Glucose or Protein

Lipids are the primary fuel-storage molecule of the body because they have higher energy content than proteins or carbohydrates. When the fasted-state body needs to use stored energy, *lipases* break down triglycerides into glycerol and fatty acids through a series of reactions collectively known as **lipolysis** (■ Fig. 22.12). Glycerol feeds into glycolysis about halfway through the pathway (2). It then goes through the same reactions as glucose from that point on.

The long carbon chains of fatty acids are more difficult to turn into ATP. Most fatty acids must first be transported from the cytosol into the mitochondrial matrix (Fig. 22.12). There they are slowly disassembled as 2-carbon units are chopped off the end of the chain, one unit at a time, in a process called **beta-oxidation** or β-oxidation.

In most cells, the 2-carbon units from fatty acids are converted into acetyl CoA, whose 2-carbon acyl units feed directly into the citric acid cycle (Fig. 22.12 (4)). Because many acetyl CoA molecules can be produced from a single fatty acid, lipids contain 9 kcal of stored energy per gram, compared with 4 kcal per gram for proteins and carbohydrates.

In the fasted state, adipose tissue releases fatty acids and glycerol into the blood (Fig. 22.8 (2)). Glycerol goes to the liver

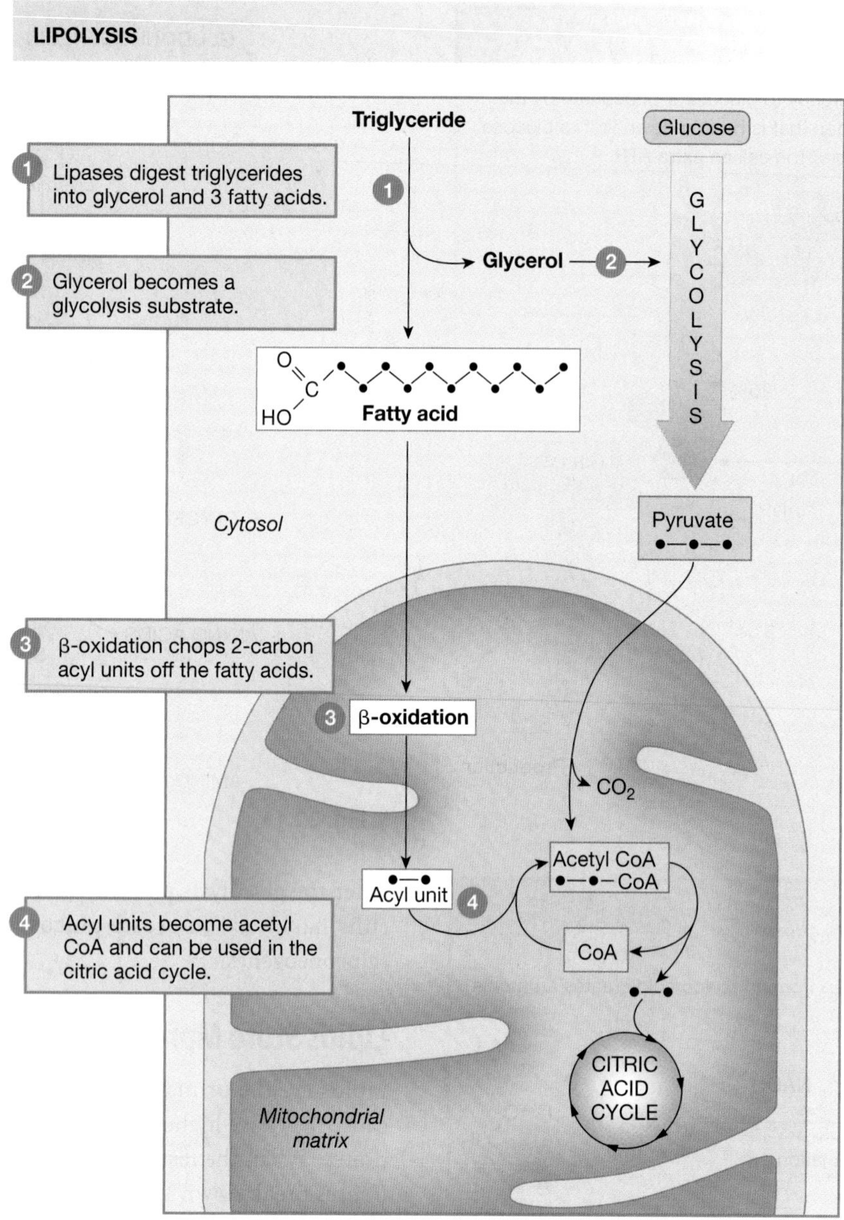

LIPOLYSIS
Triglyceride
Glucose
1 Lipases digest triglycerides into glycerol and 3 fatty acids.
2 Glycerol becomes a glycolysis substrate.
Glycerol
GLYCOLYSIS
Fatty acid
HO
Cytosol
Pyruvate
3 β-oxidation chops 2-carbon acyl units off the fatty acids.
β-oxidation
CO2
Acetyl CoA
CoA
Acyl unit
4 Acyl units become acetyl CoA and can be used in the citric acid cycle.
CoA
CITRIC ACID CYCLE
Mitochondrial matrix

Fig. 22.12

and can be converted to glucose through gluconeogenesis. The fatty acids are taken up by many tissues and used for energy production.

In the liver, if fatty acid breakdown produces acetyl CoA faster than the citric acid cycle can metabolize it, the excess acyl units become **ketone bodies** (often called simply *ketones* in physiology and medicine). Ketone bodies become a significant source of energy for the brain in cases of prolonged starvation, when glucose is low (Fig. 22.8). Ketone bodies enter the blood, creating a state of *ketosis*. The breath of people in ketosis has a fruity odor caused by acetone, a volatile ketone whose odor you might recognize from nail polish remover.

Low-carbohydrate diets, such as the Atkins and South Beach diet plans, are *ketogenic* because most of their calories come from fats. The diets are very low in carbohydrate and high in fat and protein, which shifts metabolism to β-oxidation of fats and production of ketone bodies. People who go on these diets are often pleased by initial rapid weight loss, but it is due to glycogen breakdown and water loss, not body fat reduction.

Ketogenic diets have been used successfully to treat children who have epilepsy that is not responding fully to drug therapy. For reasons that we do not understand, maintaining a state of ketosis in these children decreases the incidence of seizures. But ketogenic diets can also be dangerous. Ketone bodies such as *acetoacetic acid* and *β-hydroxybutyric acid* are strong metabolic acids that can seriously disrupt the body's pH balance and lead to a state of *ketoacidosis*, a

THE PANCREAS

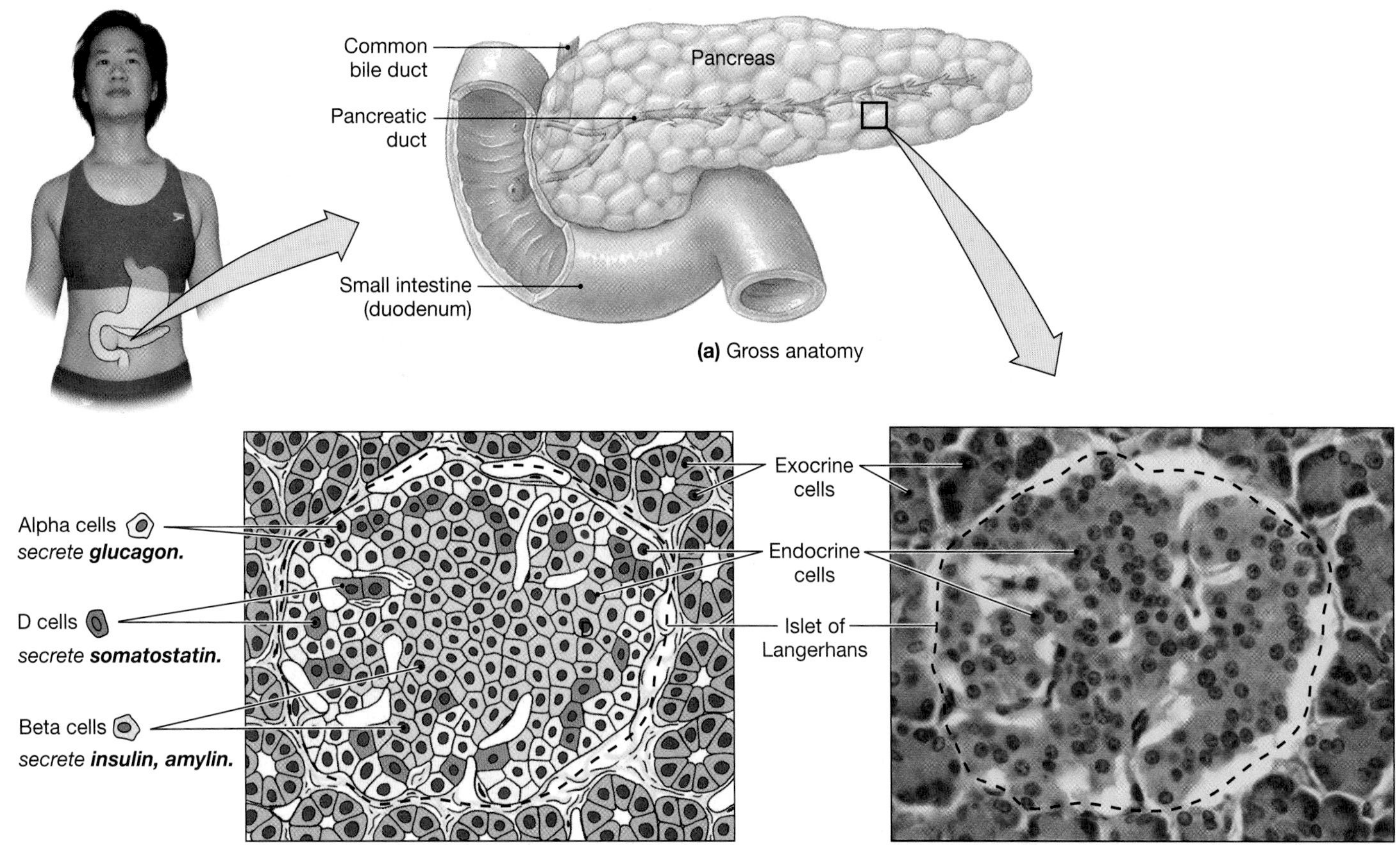

(b) Cells of the islets of Langerhans, which constitute the endocrine pancreas

Fig. 22.13 Part (b) based on data from Orci and Unger, *Lancet* 2: 1243–1244, 1975.

metabolic acidosis [p. 681]. Among the other risks associated with ketogenic diets are dehydration, electrolyte loss, inadequate intake of calcium and vitamins, gout, and kidney problems.

With this summary of metabolic pathways as background, we now turn to the endocrine and neural regulation of metabolism.

Concept Check Answers: p. 774

10. What is the difference between glycogenesis and gluconeogenesis?
11. When amino acids are used for energy, which pathways in Figure 22.5 do they follow?
12. Cholesterol is soluble in lipids, so why does plasma cholesterol need the help of a membrane transporter to enter cells?

Homeostatic Control of Metabolism

The endocrine system has primary responsibility for metabolic regulation, although the nervous system does have some influence, particularly in terms of governing food intake. Hour-to-hour regulation depends primarily on the ratio of insulin to glucagon, two hormones secreted by endocrine cells of the pancreas. Both hormones have short half-lives and must be continuously secreted if they are to have a sustained effect.

The Pancreas Secretes Insulin and Glucagon

In 1869, the German anatomist Paul Langerhans described small clusters of cells—now known as the **islets of Langerhans**—scattered throughout the body of the pancreas (Fig. 22.13b). Most pancreatic tissue is devoted to the production and exocrine secretion of digestive enzymes and bicarbonate [Figure 21.5b, p. 706] but Langerhans had found the pancreas's endocrine cells, which make up less than 2% of the total mass. The islets of Langerhans contain four distinct cell types, each associated with secretion of one or more peptide hormones.

Nearly three-quarters of the islet cells are **beta cells**, which produce *insulin* and a peptide called *amylin*. Another 20% are **alpha cells**, which secrete **glucagon**. Most of the remaining cells are *somatostatin*-secreting **D cells**. A few rare cells called *PP cells* (or *F cells*) produce *pancreatic polypeptide*.

Like all endocrine glands, the islets are closely associated with capillaries into which the hormones are released. Both sympathetic and parasympathetic neurons terminate on the islets, providing a means by which the nervous system can influence metabolism.

The Insulin-to-Glucagon Ratio Regulates Metabolism

As noted earlier, insulin and glucagon act in antagonistic fashion to keep plasma glucose concentrations within an acceptable range. Both hormones are present in the blood most of the time. The ratio of the two hormones determines which hormone dominates.

In the fed state, when the body is absorbing nutrients, insulin dominates, and the body undergoes net anabolism (■ Fig. 22.14a). Ingested glucose is used for energy production, and any excess is stored as glycogen or fat. Amino acids go primarily to protein synthesis.

In the fasted state, metabolic regulation prevents low plasma glucose concentrations (*hypoglycemia*). When glucagon predominates, the liver uses glycogen and nonglucose intermediates to synthesize glucose for release into the blood (Fig. 22.14b).

Figure 22.14c shows plasma glucose, glucagon, and insulin concentrations before and after a meal. In a normal person, fasting plasma glucose is maintained around 90 mg/dL of plasma, insulin secretion is low, and plasma glucagon levels are relatively high. After absorption of nutrients from a meal, plasma glucose rises. The increase in blood glucose inhibits glucagon secretion and stimulates insulin release. Insulin in turn promotes glucose transfer into cells. As a result, plasma glucose concentrations begin to fall back toward fasting levels shortly after each meal. Insulin secretion decreases along with the glucose, and glucagon secretion slowly begins to increase.

Insulin Is the Dominant Hormone of the Fed State

Insulin is a typical peptide hormone (■ Tbl. 22.1). It is synthesized as an inactive prohormone and activated prior to secretion [Figure 7.3c, p. 213]. Glucose is an important stimulus for

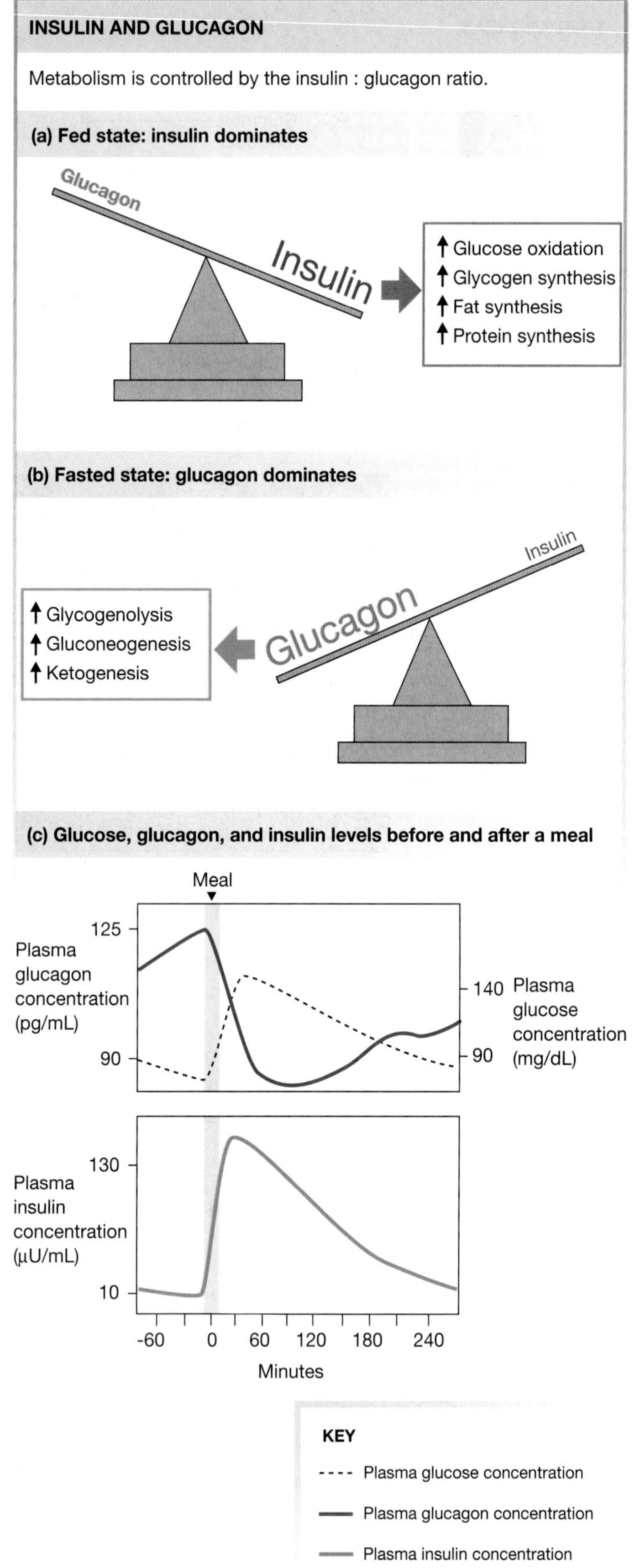

■ **Fig. 22.14** Part (c) adapted from Unger, *New Engl J Med* 285: 443–449, 1971.

Insulin	Table 22.1
Cell of origin	Beta cells of pancreas
Chemical nature	51-amino acid peptide
Biosynthesis	Typical peptide
Transport in the circulation	Dissolved in plasma
Half-life	5 minutes
Factors affecting release	Plasma [glucose] > 100 mg/dL; ↑ blood amino acids; GLP-1 (feedforward reflex). Parasympathetic activity amplifies. Sympathetic activity inhibits.
Target cells or tissues	Liver, muscle, and adipose tissue primarily; brain, kidney, and intestine not insulin dependent
Target receptor	Membrane receptor with tyrosine kinase activity; pathways with insulin-receptor substrates
Whole body or tissue action	↓ Plasma [glucose] by ↑ transport into cells or ↑ metabolic use of glucose
Action at cellular level	↑ Glycogen synthesis; ↑ aerobic metabolism of glucose; ↑ protein and triglyceride synthesis
Action at molecular level	Inserts GLUT4 transporters in muscle and adipose cells; alters enzyme activity. Complex signal transduction pathways
Feedback regulation	↓ Plasma [glucose] shuts off insulin release
Other information	Growth hormone and cortisol are antagonistic

insulin secretion, but additional factors may enhance, amplify, or inhibit secretion (■ Fig. 22.15).

1 **Increased plasma glucose.** A major stimulus for insulin release is plasma glucose concentrations greater than 100 mg/dL. Glucose absorbed from the small intestine reaches pancreatic beta cells, where it is taken up by GLUT2 transporters [Figure 5.27b, p. 167]. With more glucose available as substrate, ATP production increases and ATP-gated K^+ channels close. The cell depolarizes, voltage-gated Ca^{2+} channels open, and Ca^{2+} entry initiates exocytosis of insulin.

2 **Increased plasma amino acids.** Increased plasma amino acid concentrations following a meal also trigger insulin secretion.

3 **Feedforward effects of GI hormones.** Recently it has been shown that as much as 50% of insulin secretion is stimulated by the hormone *glucagon-like peptide-1* (GLP-1). GLP-1 and GIP (gastric inhibitory peptide) are *incretin* hormones produced by cells of the ileum and jejunum in response to nutrient ingestion. The incretins travel through the circulation to pancreatic beta cells and may reach them even before the first glucose is absorbed. The anticipatory release of insulin in response to these hormones prevents a sudden surge in plasma glucose concentrations when the meal is absorbed. Other GI hormones, such as CCK and gastrin, amplify insulin secretion.

4 **Parasympathetic activity.** Parasympathetic activity to the GI tract and pancreas increases during and following a meal. Parasympathetic input to beta cells stimulates insulin secretion.

5 **Sympathetic activity.** Insulin secretion is inhibited by sympathetic neurons. In times of stress, sympathetic input to the endocrine pancreas increases, reinforced by catecholamine release from the adrenal medulla (■ Tbl. 22.2). Epinephrine and norepinephrine inhibit insulin secretion and switch metabolism to gluconeogenesis to provide extra fuel for the nervous system and skeletal muscles.

Insulin Promotes Anabolism

Like other peptide hormones, insulin combines with a membrane receptor on its target cells (■ Fig. 22.16). The insulin receptor has *tyrosine kinase* activity, which initiates complex intracellular cascades whose details are still not completely understood. The activated insulin receptor phosphorylates proteins that include a group known as the **insulin-receptor substrates** (IRS). These proteins act through complicated pathways to influence transport and cellular metabolism. The enzymes that regulate metabolic pathways may be inhibited or activated directly, or their synthesis may be influenced indirectly through transcription factors.

The primary target tissues for insulin are the liver, adipose tissue, and skeletal muscles (Fig. 22.15). The usual target cell response is increased glucose metabolism. In some target tissues, insulin also regulates the GLUT transporters. Other tissues, including the brain and transporting epithelia of the kidney and intestine, are insulin-independent—they do not require insulin for glucose uptake and metabolism.

Insulin lowers plasma glucose in the following four ways:

1 **Insulin increases glucose transport into most, but not all, insulin-sensitive cells.** Adipose tissue and resting skeletal muscle require insulin for glucose uptake

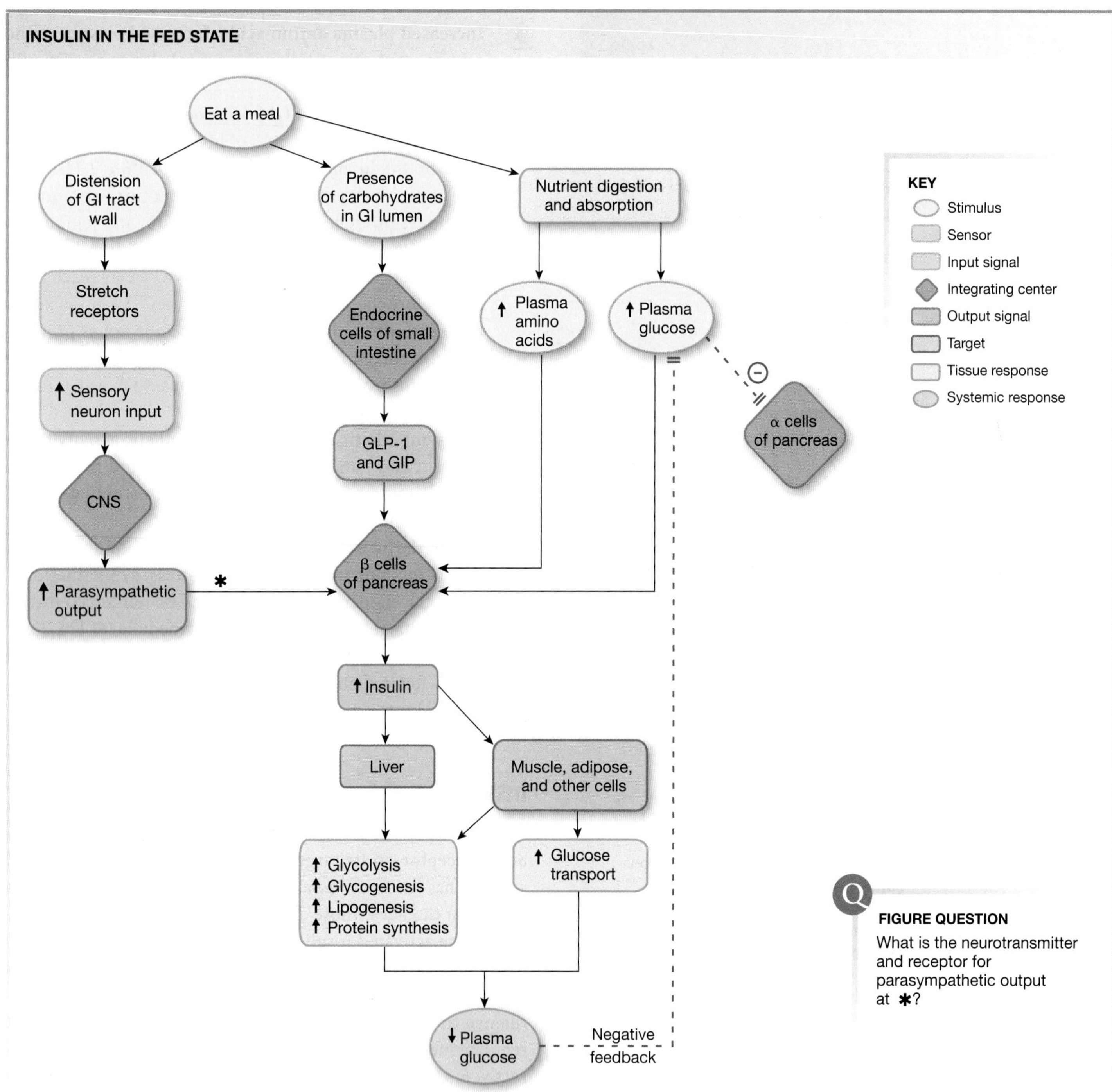

Fig. 22.15

(Fig. 22.17a). Without insulin, their GLUT4 transporters are withdrawn from the membrane and stored in cytoplasmic vesicles—another example of membrane recycling. When insulin binds to the receptor and activates it, the resulting signal transduction cascade causes the vesicles to move to the cell membrane and insert the GLUT4 transporters by exocytosis (Fig. 22.17b). The cells then take up glucose from the interstitial fluid by facilitated diffusion.

Curiously, *exercising* skeletal muscle is not dependent on insulin activity for its glucose uptake. When muscles contract, GLUT4 transporters are inserted into the membrane even in the absence of insulin, and glucose uptake increases. The intracellular signal for this is complex but appears to involve Ca^{2+} as well as a variety of intracellular proteins.

Glucose transport in liver cells (*hepatocytes*) is not *directly* insulin dependent but is influenced by the presence

22

Table 22.2 Adrenal Catecholamines (Epinephrine and Norepinephrine)

Origin	Adrenal medulla (90% epinephrine and 10% norepinephrine)
Chemical nature	Amines made from tyrosine
Biosynthesis	Typical peptide
Transport in the circulation	Some bound to sulfate
Half-life	2 minutes
Factors affecting release	Primarily fight-or-flight reaction through CNS and autonomic nervous system; hypoglycemia
Target cells or tissues	Mainly neurons, pancreatic endocrine cells, heart, blood vessels, adipose tissue
Target receptor	G protein–coupled receptors; α and β subtypes
Second messenger	cAMP for α_2 and all β receptors; IP_3 for α_1 receptors
Whole body or tissue action	↑ Plasma [glucose]; activate fight-or-flight and stress reactions; ↑ glucagon and ↓ insulin secretion
Onset and duration of action	Rapid and brief

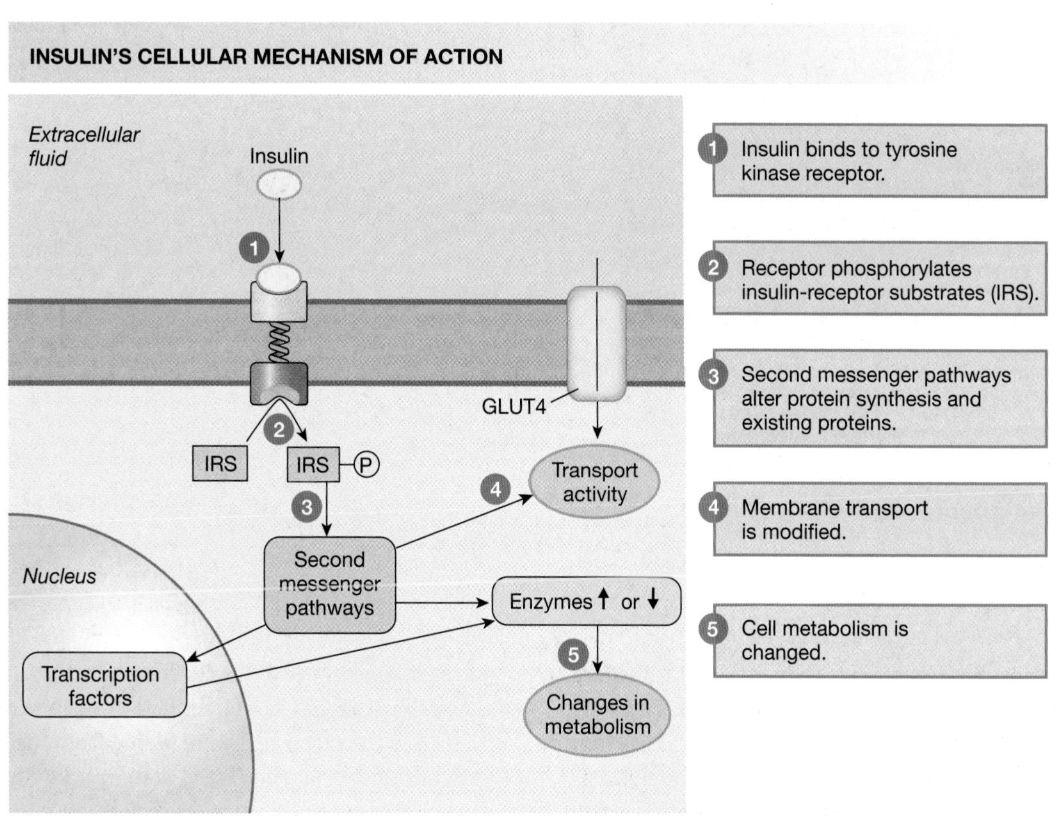
INSULIN'S CELLULAR MECHANISM OF ACTION
Extracellular fluid
Insulin
GLUT4
IRS
IRS-P
Transport activity
Second messenger pathways
Nucleus
Enzymes ↑ or ↓
Transcription factors
Changes in metabolism
1 Insulin binds to tyrosine kinase receptor.
2 Receptor phosphorylates insulin-receptor substrates (IRS).
3 Second messenger pathways alter protein synthesis and existing proteins.
4 Membrane transport is modified.
5 Cell metabolism is changed.

Fig. 22.16

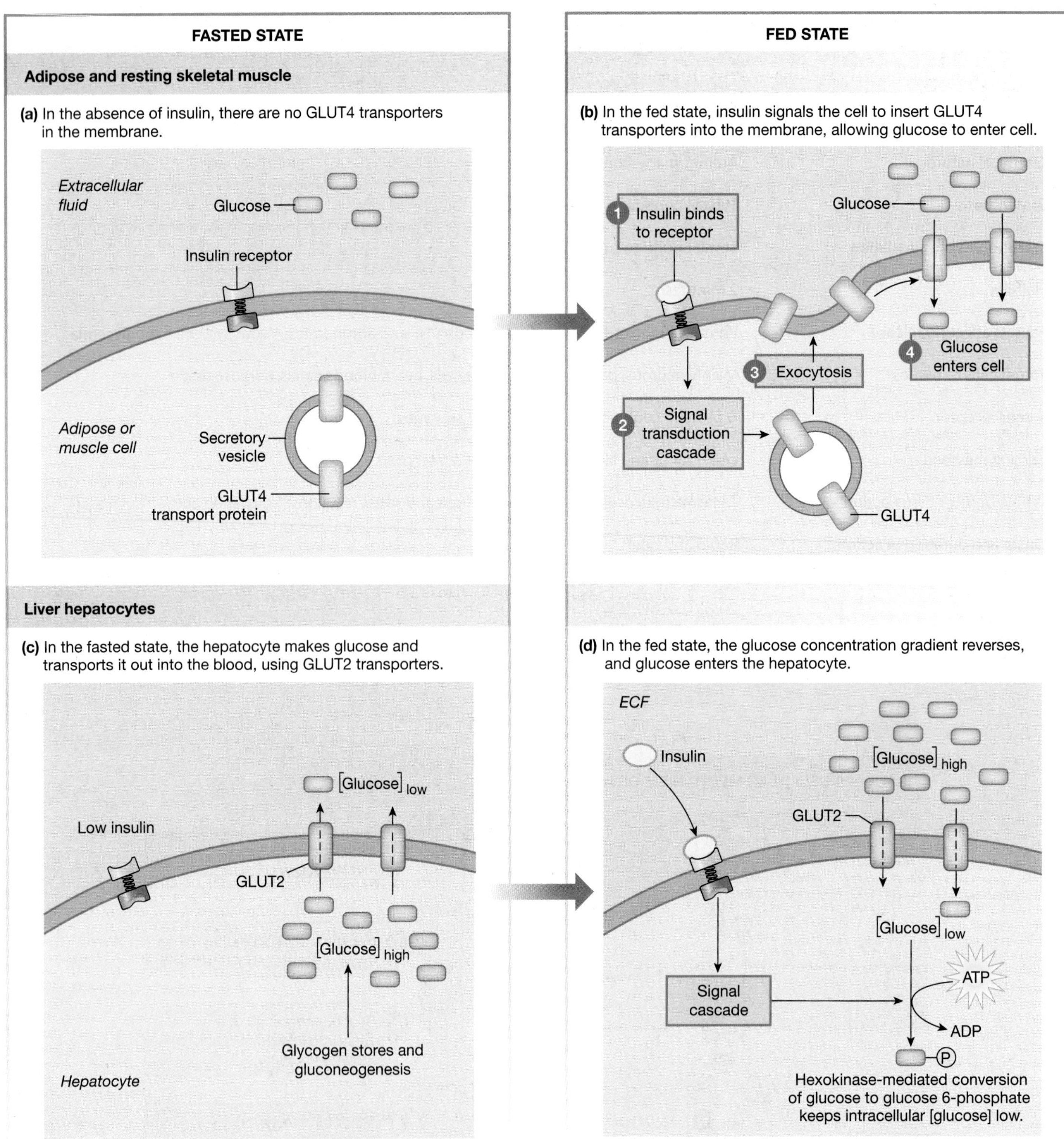
FASTED STATE
FED STATE
Adipose and resting skeletal muscle
(a) In the absence of insulin, there are no GLUT4 transporters in the membrane.
Extracellular fluid
Glucose
Insulin receptor
Adipose or muscle cell
Secretory vesicle
GLUT4 transport protein
(b) In the fed state, insulin signals the cell to insert GLUT4 transporters into the membrane, allowing glucose to enter cell.
1 Insulin binds to receptor
Glucose
2 Signal transduction cascade
3 Exocytosis
4 Glucose enters cell
GLUT4
Liver hepatocytes
(c) In the fasted state, the hepatocyte makes glucose and transports it out into the blood, using GLUT2 transporters.
Low insulin
[Glucose] low
GLUT2
[Glucose] high
Glycogen stores and gluconeogenesis
Hepatocyte
(d) In the fed state, the glucose concentration gradient reverses, and glucose enters the hepatocyte.
ECF
Insulin
[Glucose] high
GLUT2
[Glucose] low
ATP
ADP
Signal cascade
P
Hexokinase-mediated conversion of glucose to glucose 6-phosphate keeps intracellular [glucose] low.

Fig. 22.17 ***Glucose transport in the fed and fasted states***

or absence of insulin. Hepatocytes have GLUT2 transporters always present in the cell membrane. In the fasted state, when insulin levels are low, glucose moves *out* of the liver and into the blood to help maintain glucose homeostasis. In this process (Fig. 22.17c), hepatocytes are converting glycogen stores and amino acids to glucose. Newly formed glucose moves down its concentration gradient out of the cell using GLUT2 facilitated diffusion transporters. If the GLUT transporters were pulled from the hepatocyte membrane during the fasted state, as they are in muscle and adipose tissue, glucose would have no way to leave the hepatocytes.

In the fed state (Fig. 22.17d), insulin activates *hexokinase,* the enzyme that phosphorylates glucose to glucose 6-phosphate. This phosphorylation reaction keeps free intracellular glucose concentrations low relative to plasma concentrations [Fig. 5.13, p. 150]. Now glucose diffuses into the hepatocyte on the GLUT2 transporter operating in the reverse direction.

2 **Insulin enhances cellular utilization and storage of glucose.** Insulin activates enzymes for glucose utilization (*glycolysis*), and for glycogen synthesis (*glycogenesis*). Insulin simultaneously inhibits enzymes for glycogen breakdown (*glycogenolysis*), glucose synthesis (*gluconeogenesis*), and fat breakdown (*lipolysis*) to ensure that metabolism moves in the anabolic direction. If more glucose has been ingested than is needed for energy and synthesis, the excess is made into glycogen or fatty acids.

3 **Insulin enhances utilization of amino acids.** Insulin activates enzymes for protein synthesis and inhibits enzymes that promote protein breakdown. If a meal includes protein, amino acids in the ingested food are used for protein synthesis by both the liver and muscle. Excess amino acids are converted to fatty acids.

4 **Insulin promotes fat synthesis.** Insulin inhibits β-oxidation of fatty acids and promotes conversion of excess glucose or amino acids into triglycerides (*lipogenesis*). Excess triglyceride is stored as lipid droplets in adipose tissue.

In summary, insulin is an anabolic hormone because it promotes glycogen, protein, and fat synthesis. When insulin is absent or deficient, cells go into catabolic metabolism.

Concept Check Answers: p. 774

13. What are the primary target tissues for insulin?
14. Why are glucose metabolism and glucose transport independent of insulin in renal and intestinal epithelium and in neurons?
15. What is the advantage to the body of inhibiting insulin release during a sympathetically mediated fight-or-flight response?

Glucagon Is Dominant in the Fasted State

Glucagon, secreted by pancreatic alpha cells, is generally antagonistic to insulin in its effects on metabolism (Tbl. 22.3). When plasma glucose concentrations decrease after a meal, insulin secretion slows, and the effects of glucagon on tissue metabolism take on greater significance (Fig. 22.14c). As noted earlier, it is the ratio of insulin to glucagon that determines the direction of metabolism rather than an absolute amount of either hormone.

Table 22.3 Glucagon

Cell of origin	Alpha cells of pancreas
Chemical nature	29-amino acid peptide
Biosynthesis	Typical peptide
Transport in the circulation	Dissolved in plasma
Half-life	4–6 minutes
Factors affecting release	Enhanced secretion when plasma [glucose] < 65–70 mg/dL; ↑ blood amino acids
Target cells or tissues	Liver primarily
Target receptor/second messenger	G protein–coupled receptor linked to cAMP
Whole body or tissue action	↑ Plasma [glucose] by glycogenolysis and gluconeogenesis; ↑ lipolysis leads to ketogenesis in liver
Action at molecular level	Alters existing enzymes and stimulates synthesis of new enzymes
Feedback regulation	↑ Plasma [glucose] shuts off glucagon secretion
Other information	Member of secretin family (along with VIP, GIP, and GLP-1)

The function of glucagon is to prevent hypoglycemia, so the primary stimulus for glucagon release is low plasma glucose concentration (Fig. 22.18). When plasma glucose falls below 100 mg/dL, glucagon secretion rises dramatically. At glucose concentrations above 100 mg/dL, when insulin is being secreted, glucagon secretion is inhibited and remains at a low but relatively constant level. The strong relationship between insulin secretion and glucagon inhibition has led to speculation that alpha cells are regulated by some factor linked to insulin rather than by plasma glucose concentrations directly.

The liver is glucagon's primary target tissue (Fig. 22.18). Glucagon stimulates glycogenolysis and gluconeogenesis to increase glucose output. It is estimated that during an overnight fast, 75% of the glucose produced by the liver comes from glycogen stores, and the remaining 25% from gluconeogenesis.

Glucagon release is also stimulated by plasma amino acids. This pathway prevents hypoglycemia after ingestion of a pure protein meal. Let's see how hypoglycemia might occur in the absence of glucagon.

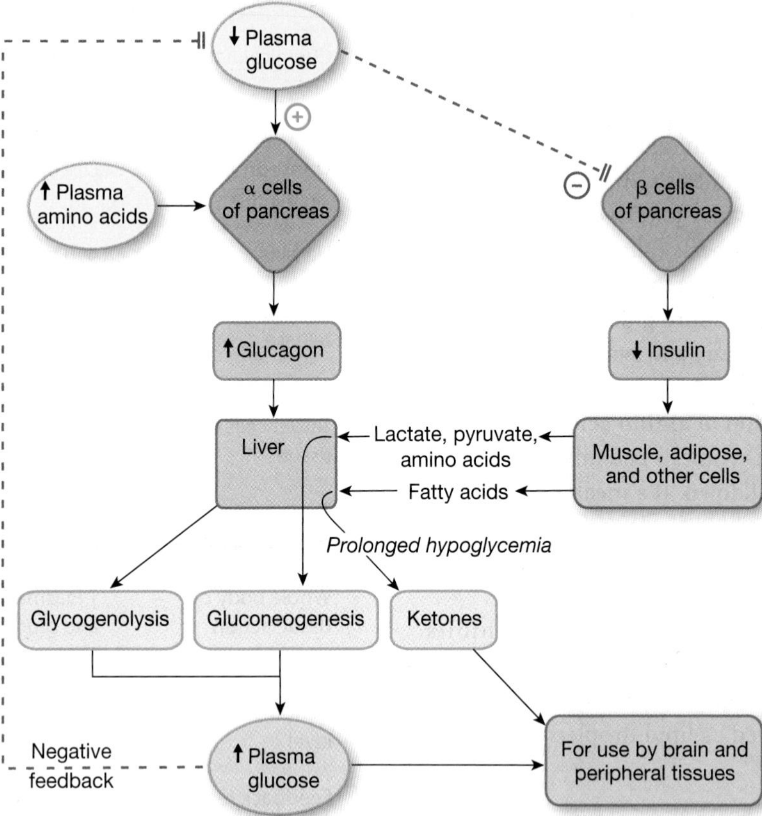

Fig. 22.18

If a meal contains protein but no carbohydrate, amino acids absorbed from the food cause insulin secretion. Even though no glucose has been absorbed, insulin-stimulated glucose uptake increases, and plasma glucose concentrations fall. Unless something counteracts this process, the brain's fuel supply is threatened by hypoglycemia.

Co-secretion of glucagon in this situation prevents hypoglycemia by stimulating hepatic glucose output. As a result, although only amino acids were ingested, both glucose and amino acids are made available to peripheral tissues.

Diabetes Mellitus Is a Family of Diseases

The most common pathology of the pancreatic endocrine system is the family of metabolic disorders known as **diabetes mellitus.** Diabetes is characterized by abnormally elevated plasma glucose concentrations (*hyperglycemia*) resulting from inadequate insulin secretion, abnormal target cell responsiveness [p. 227], or both. Chronic hyperglycemia and its associated metabolic abnormalities cause the many complications of diabetes, including damage to blood vessels, eyes, kidneys, and the nervous system.

Diabetes is reaching epidemic proportions in the United States. In 2011 the U.S. Centers for Disease Control estimated that more than 25 million people in the U.S. (8.3% of the population) had diabetes and that another 79 million (26%) had pre-diabetes. Pre-diabetes is a condition that will likely become diabetes if those people do not alter their eating and exercise habits. Experts attribute the cause of the epidemic to our sedentary lifestyle, ample food, and overweight and obesity, which affect more than 50% of the population.

Diabetes has been known to affect humans since ancient times, and written accounts of the disorder highlight the calamitous consequences of insulin deficiency. Aretaeus the Cappadocian (C.E. 81–138) wrote of the "wonderful"* nature of this disease that consists of the "melting down of the flesh . . . into urine," accompanied by terrible thirst that cannot be quenched. The copious production of glucose-laden urine gave the disease its name. *Diabetes* refers to the flow of fluid through a siphon, and *mellitus* comes from the word for honey. In the Middle Ages, diabetes was known as "the pissing evil."

*In the sense of "causing wonder."

The severe type of diabetes described by Aretaeus is **type 1 diabetes mellitus.** It is a condition of insulin deficiency resulting from beta cell destruction. Type 1 diabetes is most commonly an *autoimmune disease* in which the body fails to recognize the beta cells as "self" and destroys them with antibodies and white blood cells.

The other major variant of diabetes mellitus is **type 2 diabetes.** This type of diabetes is also known as *insulin-resistant diabetes* because in most patients, insulin levels in the blood are normal or even elevated initially. Later in the disease process, many type 2 diabetics become insulin deficient and require insulin injections. Type 2 diabetes is actually a whole family of diseases with a variety of causes.

Diabetes is diagnosed by testing blood glucose. The first indication that someone may have diabetes or pre-diabetes is often seen in the results of routine health checkup blood tests. These tests are done after the person has fasted for at least 8 hours. A fasting plasma glucose concentration between 100 and 125 mg/dL indicates pre-diabetes, and a fasting value greater than 125 mg/dL is diagnostic for diabetes.

Another test for diabetes is the 2-hour oral *glucose tolerance test* (Fig. 22.19). First, the person's fasting plasma glucose concentration is determined (time 0). Then the person consumes a special drink containing 75 g of glucose dissolved in water. Plasma glucose is measured every 30 minutes for 2 hours.

Normal individuals show a slight increase in plasma glucose concentration soon after drinking glucose, but the level rapidly returns to normal with insulin secretion. In diabetic patients, however, fasting plasma glucose concentrations are above normal and go even higher as glucose is absorbed into the body. In diabetes, plasma glucose remains elevated above 200 mg/dL after 2 hours. This slow response indicates that cells are not taking up and metabolizing glucose normally. Pre-diabetic patients show an intermediate response, with 2-hour plasma glucose concentrations in the 140–199 mg/dL range.

Elevated fasting glucose levels and abnormal oral glucose tolerance tests simply show that the body's response to an ingested glucose load is not normal. The tests cannot distinguish between problems with insulin synthesis, insulin release, or the responsiveness of target tissues to insulin.

DIAGNOSING DIABETES

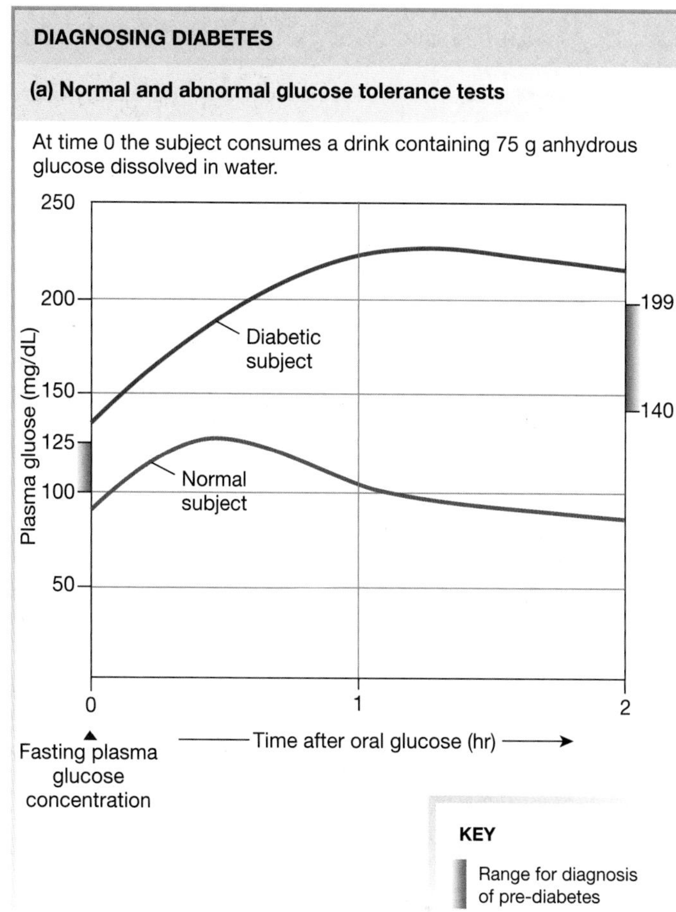

(b) Diagnostic criteria for diabetes

	Fasting blood glucose	After 2-hour oral glucose tolerance test
Normal	< 100 mg/dL (< 5.6 mM)	< 140 mg/dL (<7.8 mM)
Pre-diabetes	100–125 mg/dL (5.6–6.9 mM)	140–199 mg/dL (7.8–11 mM)
Diabetes	> 125 mg/dL (> 6.9 mM)	> 199 mg/dL (> 11 mM)

Fig. 22.19

Concept Check Answer: p. 774

16. For the oral glucose tolerance test, could you use table sugar or a sugary soft drink and expect the same results? Explain.

Type 1 Diabetics Are Prone to Ketoacidosis

Type 1 diabetes is a complex disorder whose onset in genetically susceptible individuals is sometimes preceded by a viral infection. Many type 1 diabetics develop their disease in childhood, giving rise to the old name *juvenile-onset diabetes*. About 10% of all diabetics have type 1 diabetes.

Because individuals with type 1 diabetes are insulin deficient, the only treatment is insulin injections. Until the arrival of genetic engineering, most pharmaceutical insulin came from swine, cow, and sheep pancreases. However, once the gene for human insulin was cloned, biotechnology companies began to manufacture artificial human insulin for therapeutic use. In addition, scientists are developing techniques for implanting encapsulated beta cells in the body, in the hope that individuals with type 1 diabetes will no longer need to rely on regular insulin injections.

The events that follow ingestion of food in an insulin-deficient diabetic create a picture of what happens to metabolism in the absence of insulin (Fig. 22.20). They also show the integrative

ACUTE PATHOPHYSIOLOGY OF TYPE 1 DIABETES MELLITUS

Untreated type 1 diabetes is marked by tissue breakdown, glucosuria, polyuria, polydipsia, polyphagia, and metabolic ketoacidosis.

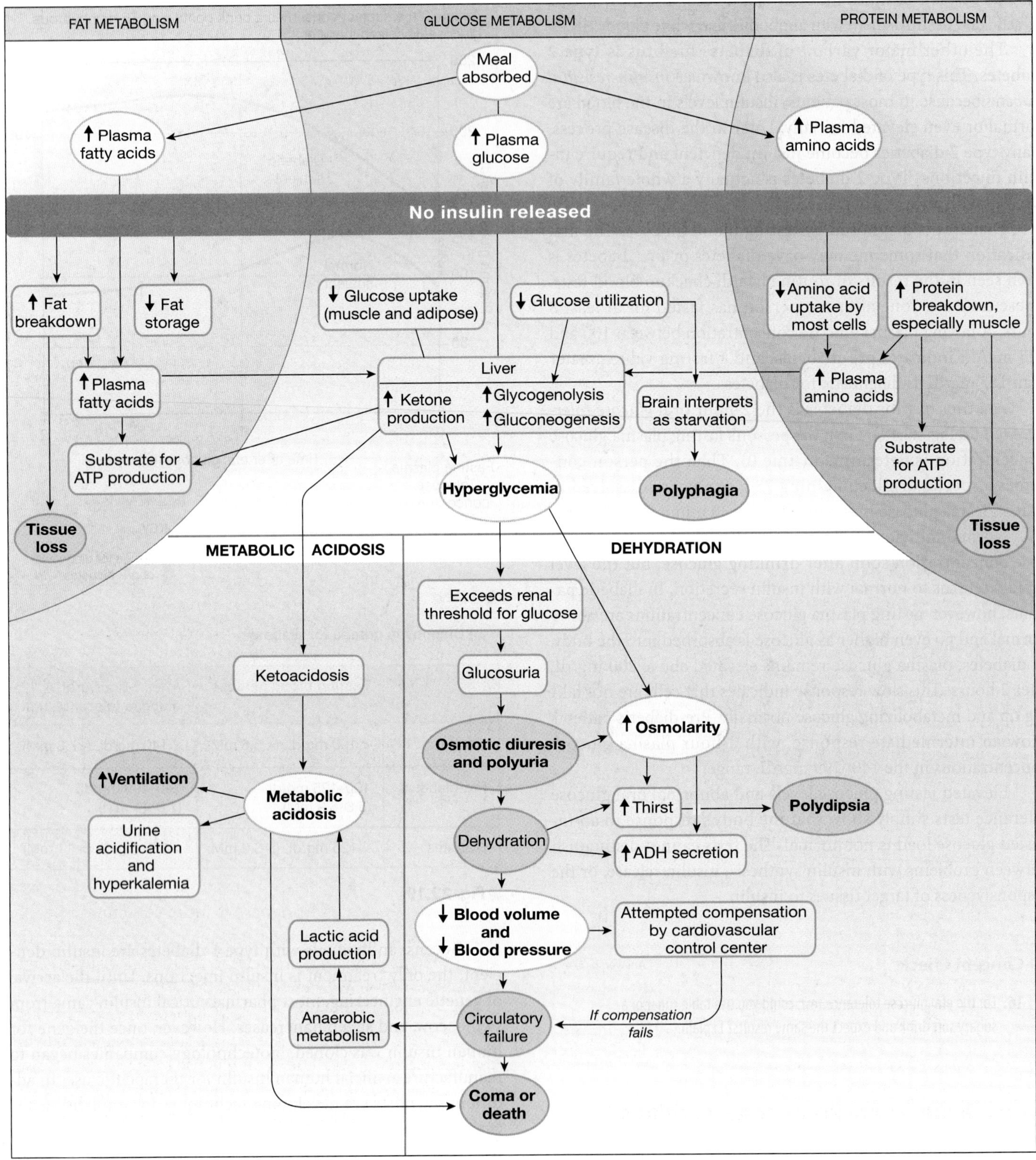

Fig. 22.20

nature of physiology because the problems that arise from abnormal metabolism affect nearly every organ system of the body.

Following a meal, nutrient absorption by the intestine proceeds normally because this process is insulin independent. However, nutrient uptake from the blood and cellular metabolism in many tissues are insulin dependent and therefore severely diminished in the absence of insulin. Lacking nutrients to metabolize, cells go into fasted-state metabolism:

1. *Protein metabolism.* Without glucose for energy or amino acids for protein synthesis, muscles break down their proteins to provide a substrate for ATP production. Amino acids are also converted to pyruvate and lactate, which leave the muscles and are transported to the liver.
2. *Fat metabolism.* Adipose tissue in fasted-state metabolism breaks down its fat stores. Fatty acids enter the blood for transport to the liver. The liver uses β-oxidation to break down fatty acids. However, this organ is limited in its ability to send fatty acids through the citric acid cycle, and the excess fatty acids are converted to ketones.

   Ketone bodies re-enter the circulation and can be used by other tissues (such as muscle and brain) for ATP synthesis. (The breakdown of muscle and adipose tissue in the absence of insulin leads to tissue loss and the "melting down of the flesh" described by Aretaeus.) However, ketones are also metabolic acids, creating a state of ketoacidosis (see 7 below).
3. *Glucose metabolism.* In the absence of insulin, glucose remains in the blood, causing hyperglycemia. The liver, unable to metabolize this glucose, initiates fasted-state pathways of glycogenolysis and gluconeogenesis. These pathways produce *additional* glucose from glycogen, amino acids, and glycerol. When the liver dumps this glucose into the blood, hyperglycemia worsens.

   Diabetic hyperglycemia will increase the osmolarity of the blood and create a *hyperglycemic hyperosmolar state.* Plasma glucose may be as high as 600–1200 mg/dL and total osmolarity ranges from 330–380 mOsM. The high osmolarity will trigger vasopressin (ADH) secretion and thirst in an effort to conserve water and return osmolarity back to the normal range [p. 663].
4. *Brain metabolism.* Tissues that are not insulin dependent, such as most neurons in the brain, carry on metabolism as usual. However, neurons in the brain's satiety center *are* insulin sensitive. Therefore, in the absence of insulin, the satiety center is unable to take up plasma glucose. The center perceives the absence of intracellular glucose as starvation and allows the feeding center to increase food intake. The result is *polyphagia* (excessive eating), a classic symptom associated with untreated type 1 diabetes mellitus.
5. *Osmotic diuresis and polyuria.* If the hyperglycemia of diabetes causes plasma glucose concentrations to exceed the *renal threshold* for glucose, glucose reabsorption in the proximal tubule of the kidney becomes saturated [p. 643]. As a result, some filtered glucose is not reabsorbed, and it is excreted in the urine (*glucosuria*).

   The presence of additional solute in the collecting duct lumen causes less water to be reabsorbed and more to be excreted [see Chapter 20, question 31, p. 694]. This creates large volumes of urine (*polyuria*) and, if unchecked, results in dehydration. The loss of water in the urine due to unreabsorbed solutes is known as **osmotic diuresis.**
6. *Dehydration.* Dehydration caused by osmotic diuresis leads to decreased circulating blood volume and decreased blood pressure. Low blood pressure triggers homeostatic mechanisms for raising blood pressure, including secretion of vasopressin, thirst that causes constant drinking (*polydipsia*), and cardiovascular compensations [Fig. 20.13, p. 680].
7. *Metabolic acidosis.* Metabolic acidosis in diabetes has two potential sources: anaerobic metabolism and ketone body production. The primary cause of metabolic acidosis in type 1 diabetics is the production of acidic ketone bodies by the liver. Patients in *diabetic ketoacidosis* exhibit the signs of metabolic acidosis: increased ventilation, acidification of the urine, and hyperkalemia [p. 681].

Tissues may also go into anaerobic glycolysis (which creates lactate) if low blood pressure decreases blood flow to the point that oxygen delivery to peripheral tissues becomes inadequate. Lactate leaves the cells and enters the blood, contributing to a state of metabolic acidosis. If untreated, the combination of ketoacidosis and hypoxia from circulatory collapse can cause coma and even death. The treatment for a patient in diabetic ketoacidosis is insulin replacement, accompanied by fluid and electrolyte therapy to replenish lost volume and ions.

Type 2 Diabetics Often Have Elevated Insulin Levels

Type 2 diabetics account for 90% of all diabetics. A significant genetic predisposition to develop the disease exists among certain ethnic groups. For example, about 25% of Hispanics over age 45 have diabetes. The disease is more common in people over the age of 40, but there is growing concern about the increased diagnosis of type 2 diabetes in children and adolescents. About 80% of type 2 diabetics are obese.

A common hallmark of type 2 diabetes is **insulin resistance**, demonstrated by the delayed response to an ingested glucose load seen in the 2-hour oral glucose tolerance test. Some type 2 diabetics have both resistance to insulin action and decreased insulin secretion. Others have normal-to-high insulin secretion but decreased target cell responsiveness.

In addition, although type 2 diabetics are hyperglycemic, they often have elevated glucagon levels as well. This seems contradictory until you realize that the pancreatic alpha cells, like muscle and adipose cells, require insulin for glucose uptake.

This means that in diabetes, the alpha cells do not take up glucose, which prompts them to secrete glucagon. Glucagon then contributes to hyperglycemia by promoting glycogenolysis and gluconeogenesis.

In type 2 diabetes, the acute symptoms of the disease are not nearly as severe as in type 1 because insulin is usually present, and the cells, although resistant to insulin's action, are able to carry out a certain amount of glucose metabolism. The liver, for example, does not have to turn to ketone production. As a result, ketosis is uncommon in type 2 diabetes.

Nevertheless, overall metabolism is not normal in type 2 diabetes, and patients with this condition develop a variety of physiological problems because of abnormal glucose and fat metabolism. Complications of type 2 diabetes include atherosclerosis, neurological changes, renal failure, and blindness from diabetic retinopathy. As many as 70% of type 2 diabetics die from cardiovascular disease.

Because many people with type 2 diabetes are asymptomatic when diagnosed, they can be very difficult to treat. People who come in for their yearly checkup feeling fine, only to be told that they have diabetes, can be reluctant to make dramatic lifestyle changes when they do not feel sick. Unfortunately, by the time diabetic symptoms appear, damage to tissues and organs is well under way. Patient compliance at that point can slow the progress of the disease but cannot reverse the pathological changes. The goal of treatment is to correct hyperglycemia to prevent the complications described above.

The first therapy recommended for most type 2 diabetics and pre-diabetics, and for those individuals at high risk of developing the disease, is to exercise and lose weight. For some patients, simply losing weight eliminates their insulin resistance. Exercise decreases hyperglycemia because exercising skeletal muscle does not require insulin for glucose uptake.

Drugs used to treat type 2 diabetes may (1) stimulate beta-cell secretion of insulin, (2) slow the digestion or absorption of carbohydrates in the intestine, (3) inhibit hepatic glucose output, or (4) make target tissues more responsive to insulin (■ Tbl. 22.4). Many of the newest anti-diabetic drugs mimic endogenous hormones. For example, *pramlintide* is an analog of **amylin**, a peptide hormone that is co-secreted with insulin. Amylin helps regulate glucose homeostasis following a meal by slowing digestion and absorption of carbohydrates. Amylin also decreases food intake by a central effect on appetite, and it decreases secretion of glucagon.

Other hormone-based therapies recently approved by the FDA are incretin *mimetics* (agonists). *Exendin-4* (Byetta®) is a GLP-1 mimetic derived from a compound found in the venomous saliva of gila monsters. Exendin-4 has four primary effects: it increases insulin production, decreases production of glucagon, slows digestion, and increases satiety. It has also been associated with weight loss.

In normal physiology, the combined actions of amylin, GIP, and GLP-1 create a self-regulating cycle for glucose absorption and fed-state glucose metabolism. Glucose in the intestine

Drugs for Treating Diabetes

Table 22.4

Drug Class	Effect	Mechanism of Action
Sulfonylureas and meglitinides	Stimulate insulin secretion	Close beta cell K_{ATP} channels and depolarize the cell
α-Glucosidase inhibitors	Decrease intestinal glucose uptake	Block intestinal enzymes that digest complex carbohydrates
Biguanides (e.g., metformin)	Reduce plasma glucose by decreasing hepatic gluconeogenesis	Unclear
PPAR activators ("glitazones")	Increase gene transcription for proteins that promote glucose utilization and fatty acid metabolism	Activate PPARγ, a nuclear receptor activator
Amylin analogs (pramlintide)	Reduce plasma glucose	Delay gastric emptying, suppress glucagon secretion, and promote satiety
Incretin (GLP-1) analogs (exendin-4)	Reduce plasma glucose and induce weight loss	Stimulate insulin secretion, reduce glucagon secretion, delay gastric emptying, and promote satiety
DPP4 inhibitors (sitagliptin)	Increase insulin secretion and decrease gastric emptying	Inhibit dipeptidyl peptidase-4, which breaks down GLP-1 and GIP

following a meal causes feedforward release of GIP and GLP-1 (Fig. 22.15). The two incretins travel through the circulation to the pancreas, where they initiate insulin and amylin secretion. Amylin then acts on the GI tract to slow the rate at which food enters the intestine, while insulin acts on target tissues to promote glucose uptake and utilization.

Metabolic Syndrome Links Diabetes and Cardiovascular Disease

Clinicians have known for years that people who are overweight are prone to develop type 2 diabetes, atherosclerosis, and high blood pressure. The combination of these three conditions has been formalized into a diagnosis called **metabolic syndrome**, which highlights the integrative nature of metabolic pathways. People with metabolic syndrome meet at least three of the following five criteria: central (visceral) obesity, blood pressure $\geq$ 130/85 mm Hg, fasting plasma glucose $\geq$ 110 mg/dL, elevated fasting plasma triglyceride levels, and low plasma HDL-C levels. *Central obesity* is defined as a waist circumference greater than 40″ in men and 35″ in women. Women who have an apple-shaped body (widest at the waist) are more prone to developing metabolic syndrome than women who have a pear-shaped body (widest at the hips).

The association between obesity, diabetes, and cardiovascular disease illustrates the fundamental disturbances in cellular metabolism that occur with obesity. One common mechanism known to play a role in both glucose metabolism and lipid metabolism involves the family of nuclear receptors called *peroxisome proliferator-activated receptors* (PPARs, pronounced *p-pars*). Lipids and lipid-derived molecules bind to PPARs, which then turn on a variety of genes. The PPAR subtype called PPARγ (*p-par-gamma*) has been linked to adipocyte differentiation, type 2 diabetes, and foam cells, the endothelial macrophages that have ingested oxidized cholesterol. PPARα, mentioned earlier in the discussion of cholesterol metabolism, is important in hepatic cholesterol metabolism. The PPARs may be important clues to the link between obesity, type 2 diabetes, and atherosclerosis that has evaded scientists for so long.

Concept Check

Answers: p. 774

17. Why must insulin be administered as a shot and not as an oral pill?
18. Patients admitted to the hospital with acute diabetic ketoacidosis and dehydration are given insulin and fluids that contain K^+ and other ions. The acidosis is usually accompanied by hyperkalemia, so why is K^+ included in the rehydration fluids? (*Hint:* dehydrated patients may have a high *concentration* of K^+, but their total body fluid volume is low.)
19. In 2006 the FDA approved sitagliptin (Januvia®), a DPP4 inhibitor. This drug blocks action of the enzyme *dipeptidyl peptidase-4*, which breaks down GLP-1 and GIP. Explain how sitagliptin is helpful in treating diabetes.

RUNNING PROBLEM

The medical staff and Nicole's family decided to talk to her about her weight and eating habits. The disorder Dr. Ayani suspects, **anorexia nervosa** (AN), can have serious physiological consequences. As a result, AN has the highest death rate of any psychiatric disorder, and the mortality rate of young women ages 15–24 with AN is 12 times greater than that of the general population. The most common causes of death are cardiac arrest, electrolyte imbalance, and suicide. AN reportedly affects as many as 3% of females in industrialized nations at some point in their lifetime. (While 90% of AN cases are female, the number of male cases is increasing.) Successful treatment of AN includes providing nutrition, psychotherapy, and family therapy. Current research is investigating the usefulness of neuropeptide Y and other brain peptides in treating anorexia.

Q9: Why might an NPY agonist help in cases of anorexia?

737 739 741 744 765 770

Regulation of Body Temperature

Type 2 diabetes is an excellent example of the link between body weight and metabolism. The development of obesity may be linked to the efficiency with which the body converts food energy into cell and tissue components. According to one theory, people who are more efficient in transferring energy from food to fat are the ones who put on weight. In contrast, people who are less metabolically efficient can eat the same number of calories and not gain weight because more food energy is released as heat. Much of what we know about the regulation of energy balance comes from studies on body temperature regulation.

Body Temperature Balances Heat Production, Gain, and Loss

Temperature regulation in humans is linked to metabolic heat production (*thermogenesis*). Humans are **homeothermic** animals, which means our bodies regulate internal temperature within a relatively narrow range. Average body temperature is 37 °C (98.6 °F), with a normal range of 35.5–37.7 °C (96–99.9 °F).

These values are subject to considerable variation, both among individuals and throughout the day in a single individual. The site at which temperature is measured also makes a difference because core body temperature may be higher than temperature at the skin surface. Oral temperatures are about 0.5 °C lower than rectal temperatures.

Several factors affect body temperature in a given individual. Body temperature increases with exercise or after a meal (because of *diet-induced thermogenesis*). Temperature also cycles throughout the day: the lowest (basal) body temperature occurs in the early morning, and the highest occurs in the early evening. Women of reproductive age also exhibit a monthly temperature cycle: basal body temperatures are about 0.5 °C higher in the second half of the menstrual cycle (after ovulation) than before ovulation.

Heat Gain and Loss Are Balanced Temperature balance in the body, like energy balance, depends on a dynamic equilibrium between heat input and heat output (■ Fig. 22.21). Heat input has two components: *internal heat production*, which includes heat from normal metabolism and heat released during muscle contraction, and *external heat input* from the environment through either *radiation* or *conduction*.

All objects with a temperature above absolute zero give off radiant energy (radiation) with infrared or visible wavelengths. This energy can be absorbed by other objects and constitutes **radiant heat gain** for those objects. You absorb radiant energy each time you sit in the sun or in front of a fire. **Conductive heat gain** is the transfer of heat between objects that are in contact with each other, such as the skin and a heating pad or the skin and hot water.

We lose heat from the body in four ways: conduction, radiation, convection, and evaporation. **Conductive heat loss** is the loss of body heat to a cooler object that is touching the body, such as an icepack or a cold stone bench. **Radiant heat loss** from the human body is estimated to account for nearly half the heat lost from a person at rest in a normal room. *Thermography* is a diagnostic imaging technique that measures radiant heat loss. Some cancerous tumors can be visually identified because they have higher metabolic activity and give off more heat than surrounding tissues.

Radiant and conductive heat loss is enhanced by **convective heat loss**, the process in which heat is carried away by warm air rising from the body's surface. Convective air currents are created wherever a temperature difference in the air exists: hot air rises and is replaced by cooler air. Convection helps move warmed air away from the skin's surface. Clothing, which traps air and prevents convective air currents, helps retain heat close to the body.

The fourth type of heat loss from the body is **evaporative heat loss**, which takes place as water evaporates at the skin's surface and in the respiratory tract. The conversion of water from the liquid state to the gaseous state requires the input of substantial amounts of heat energy. When water on the body evaporates, it removes heat from the body.

You can demonstrate the effect of evaporative cooling by wetting one arm and letting the water evaporate. As it dries, the wet arm feels much cooler than the rest of your body because heat is being drawn from the arm to vaporize the water.

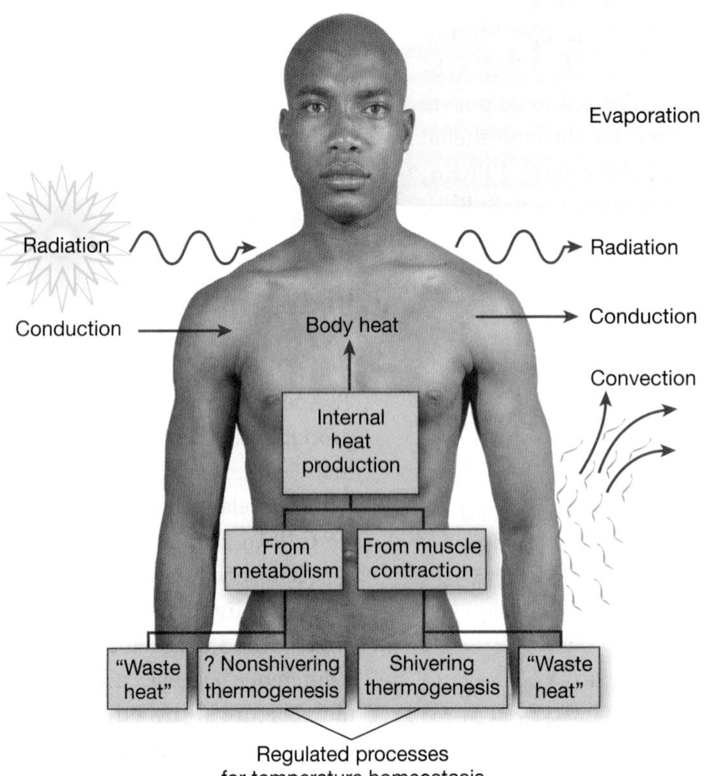

■ **Fig. 22.21** ***Heat balance in the body***

Similarly, the half-liter of water vapor that leaves the body through the lungs and skin each day takes with it a significant amount of body heat. Evaporative heat loss is affected by the humidity of surrounding air: less evaporation occurs at higher humidities.

Conductive, convective, and evaporative heat loss from the body is enhanced by the *bulk flow* of air across the body, such as air moved across the skin by a fan or breeze. The effect of wind on body temperature regulation in the winter is described by the *wind chill factor,* a combination of absolute environmental temperature and the effect of convective heat loss.

Body Temperature Is Homeostatically Regulated

The human body is usually warmer than its environment and therefore loses heat. However, normal metabolism generates enough heat to maintain body temperature when the environmental temperature stays between 27.8–30 °C (82–86 °F). This range is known as the **thermoneutral zone**.

In temperatures above the thermoneutral zone, the body has a net gain of heat because heat production exceeds heat loss. Below the thermoneutral zone, heat loss exceeds heat production. In both cases, the body must use homeostatic compensation to maintain a constant internal temperature.

A human without clothing can thermoregulate at environmental air temperatures between 10–55 °C (50–131°F). Because we are seldom exposed to the higher end of that temperature range, the main physiological challenge in thermal regulation is posed by cold environments.

Humans have been described by some physiologists as tropical animals because we are genetically adapted for life in warm climates. But we have retained a certain amount of genetic flexibility, and the physiological mechanisms by which we thermoregulate have some ability to adapt to changing conditions.

The autonomic control of body temperature regulation is generally considered a function of *thermoregulatory centers* in the hypothalamus. Sensory neurons known as **thermoreceptors** are located peripherally in the skin and centrally in the anterior hypothalamus. These sensors monitor skin temperature and core body temperature, respectively, and send that information to the thermoregulatory center. The hypothalamic "thermostat" then compares the input signals with the desired temperature setpoint and coordinates an appropriate physiological response to raise or lower the core temperature (■ Fig. 22.22). Heat loss from the body is promoted by dilation of blood vessels in the skin and by sweating. Heat gain is generated by shivering and possibly by nonshivering thermogenesis.

Alterations in Cutaneous Blood Flow Conserve or Release Heat Heat loss across the skin surface is regulated by controlling blood flow in *cutaneous* {*cutis,* skin} blood vessels (vessels near the skin's surface). These blood vessels can pick up heat from the environment by convection and transfer it to the body core, or they can lose heat to the surrounding air. Blood flow through cutaneous blood vessels varies from close to zero when heat must be conserved, to nearly one-third of cardiac output when heat must be released to the environment. Local control influences cutaneous blood flow to a certain degree, possibly through vasodilators produced by the vascular endothelium. However, neural regulation is the primary determining factor.

22

Most arterioles in the body are under tonic sympathetic adrenergic control [p. 522]. If core body temperature falls, the hypothalamus selectively activates sympathetic neurons innervating cutaneous arterioles. The arterioles constrict, increasing their resistance to blood flow and diverting blood to lower-resistance blood vessels in the interior of the body. This response keeps warmer core blood away from the cooler skin surface, thereby reducing heat loss.

In warm temperatures, the opposite happens: cutaneous arterioles dilate to increase blood flow near the skin surface and enhance heat loss. Only a small fraction of vasodilation results from the withdrawal of tonic sympathetic input. Active cutaneous vasodilation is mediated through **sympathetic cholinergic neurons**, specialized neurons that co-secrete acetylcholine and other molecules. Some proposed mediators of active vasodilation include nitric oxide, substance P, histamine, and prostaglandins. It remains unclear which vasodilator substances are most important in the thermoregulatory response.

Concept Check Answers: p. 774

20. What neurotransmitter and neurotransmitter receptor mediate cutaneous vasoconstriction?

21. What observations might have prompted researchers who discovered sympathetic neurons that secrete ACh to classify them as sympathetic rather than parasympathetic? [*Hint:* p. 382]

Sweat Contributes to Heat Loss Surface heat loss is enhanced by the evaporation of sweat. By some estimates the human integument has 2–3 million sweat glands. The highest concentrations are found on the forehead, scalp, axillae (armpits), palms of the hands, and soles of the feet.

Sweat glands are made of transporting epithelium. Cells deep in the gland secrete an isotonic solution similar to interstitial fluid. As the fluid travels through the duct to the skin, NaCl is reabsorbed, resulting in hypotonic sweat. A typical value for sweat production is 1.5 L/hr. With acclimatization to hot weather, some people sweat at rates of 4–6 L/hr. However, they can maintain this high rate only for short periods unless they are drinking to replace lost fluid volume. Sweat production is regulated by cholinergic sympathetic neurons.

Cooling by evaporative heat loss depends on the evaporation of water from sweat on the skin's surface. Because water evaporates rapidly in dry environments but slowly or not at all in humid ones, the body's ability to withstand high temperatures

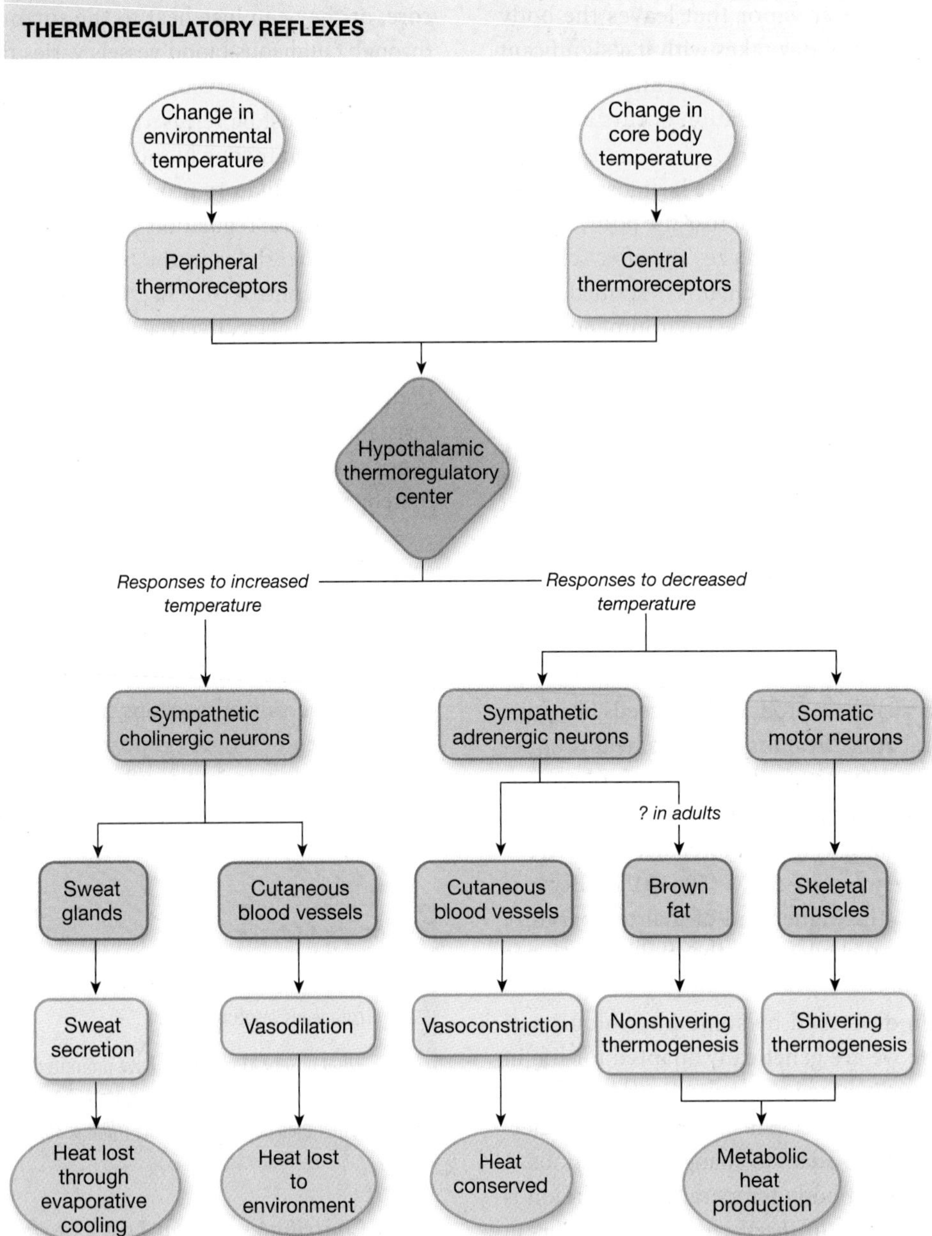

Fig. 22.22

is directly related to the relative humidity of the air. Meteorologists report the combination of heat and humidity as the *heat index* or *humidex*. Air moving across a sweaty skin surface enhances evaporation even with high humidity, which is one reason fans are useful in hot weather.

Movement and Metabolism Produce Heat

Heat production by the body falls into two broad categories: (1) unregulated heat production from voluntary muscle contraction and normal metabolic pathways, and (2) regulated heat production for maintaining temperature homeostasis in low environmental temperatures. Regulated heat production is further divided into shivering thermogenesis and nonshivering thermogenesis (Fig. 22.21).

In **shivering thermogenesis**, the body uses shivering (rhythmic tremors caused by skeletal muscle contraction) to generate heat. Signals from the hypothalamic thermoregulatory center initiate these skeletal muscle tremors. Shivering muscle generates five to six times as much heat as resting muscle. Shivering can be partially suppressed by voluntary control.

Nonshivering thermogenesis is metabolic heat production by means other than shivering. In laboratory animals such as the rat, cold exposure significantly increases heat production in brown fat [p. 87]. The mechanism for brown fat heat production is *mitochondrial uncoupling*. In this process, energy flowing through the electron transport system [p. 115] is released as heat rather than being trapped in ATP. Mitochondrial uncoupling in response to cold is promoted by thyroid hormones and by increased sympathetic activity on β_3 adrenergic receptors.

The importance of nonshivering thermogenesis in adult humans is unclear but becoming a topic of increasing interest. Humans are born with significant amounts of brown fat, found primarily in the *interscapular* area between the shoulder blades. In newborns, nonshivering thermogenesis in this brown fat contributes significantly to raising and maintaining body temperature. As children age, white fat gradually replaces most brown fat. Recently, however, imaging studies used for cancer diagnosis showed that some adult humans still have active brown fat. Scientists are now investigating further because increasing brown fat activity or amount might be one way to help people burn calories as heat instead of storing them as fat. See the January 29, 2011 issue of the *New England Journal of Medicine* for several articles on this topic.

The body's responses to high and low temperatures are summarized in ■ Figure 22.23. In cold environments, the body tries to reduce heat loss while increasing internal heat production. In hot temperatures, the opposite is true. Notice from Figure 22.23 that voluntary behavioral responses play a significant role in temperature regulation. We reduce activity during hot weather, thereby decreasing muscle heat production. In cold weather, we put on extra clothing, tuck our hands in our armpits, or curl up in a ball to slow heat loss.

The Body's Thermostat Can Be Reset

22

Variations in body temperature regulation can be either physiological or pathological. Examples of physiological variation include the circadian rhythm of body temperature mentioned earlier, menstrual cycle variations, postmenopausal hot flashes, and fever. These processes share a common mechanism: resetting of the hypothalamic thermostat.

Hot flashes appear to be transient decreases in the thermostat's setpoint caused by the absence of estrogen. When the setpoint is lower, a room temperature that had previously been comfortable suddenly feels too hot. This discomfort triggers the usual thermoregulatory responses to heat, including sweating and cutaneous vasodilation, which leads to flushing of the skin.

For many years *fever* was thought to be a pathological response to infection, but it is now considered part of the body's normal immune response. Toxins from bacteria and other pathogens trigger the release of chemicals known as **pyrogens** {*pyr,* fire} from various immunocytes. Pyrogens are fever-producing cytokines that also have many other effects.

Experimentally, some interleukins (IL-1, IL-6), some interferons, and tumor necrosis factor have all been shown to induce

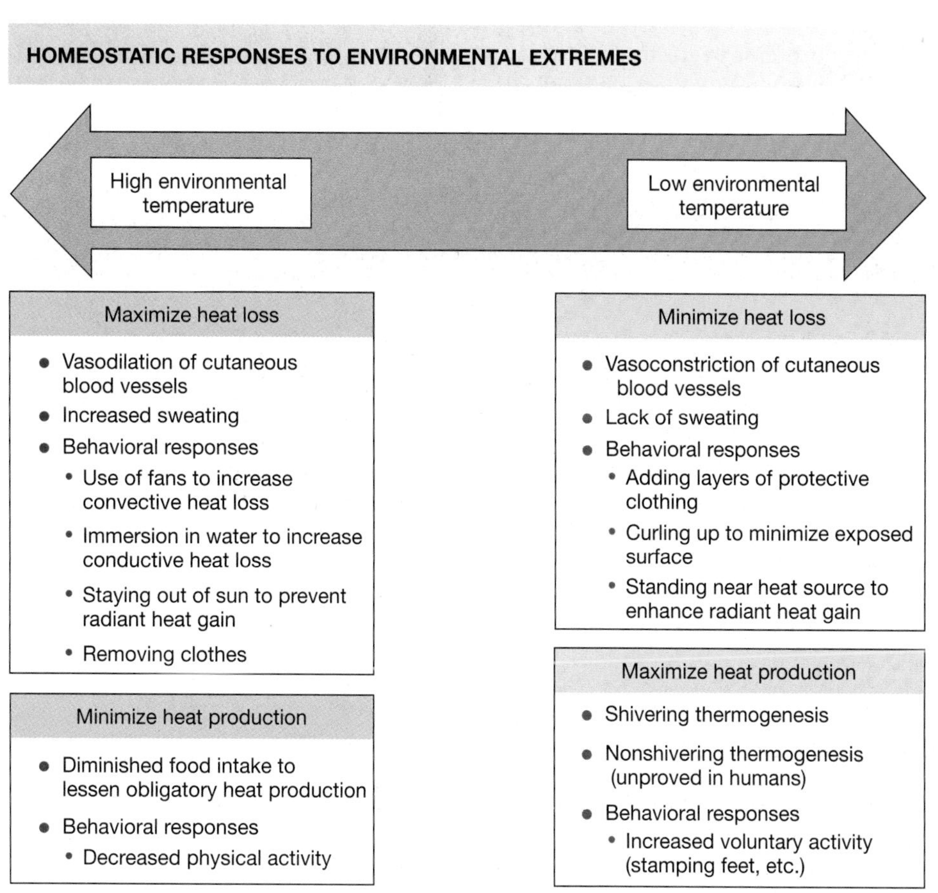

■ **Fig. 22.23**

fever. They do so by resetting the hypothalamic thermostat to a higher setpoint. Normal room temperature feels too cold, and the patient begins to shiver, creating additional heat. Pyrogens may also increase nonshivering thermogenesis, causing body temperature to rise.

The adaptive significance of fever is still unclear, but it seems to enhance the activity of white blood cells involved in the immune response. For this reason, some people question whether patients with a fever should be given aspirin and other fever-reducing drugs simply for the sake of comfort. High fever can be dangerous, however, as a fever of 41 °C (106 °F) for more than a brief period causes brain damage.

Pathological conditions in which body temperature strays outside the normal range include different states of *hyperthermia* and *hypothermia.* Heat exhaustion and heat stroke are the most common forms of **hyperthermia**, a condition in which body temperature rises to abnormally high values. **Heat exhaustion** is marked by severe dehydration and core body temperatures of 37.5–39 °C (99.5–102.2 °F). Patients may experience muscle cramps, nausea, and headache. They are usually pale and sweating profusely. Heat exhaustion often occurs in people who are physically active in hot, humid climates to which they are not acclimatized. It also occurs in the elderly, whose ability to thermoregulate is diminished.

Heat stroke is a more severe form of hyperthermia, with higher core body temperatures. The skin is usually flushed and dry. Immediate and rapid cooling of these patients is important, as enzymes and other proteins begin to denature at temperatures above 41 °C (106 °F). Mortality in heat stroke is nearly 50%.

Malignant hyperthermia, in which body temperature becomes abnormally elevated, is a genetically linked condition. A defective Ca^{2+} channel in skeletal muscle releases too much Ca^{2+} into the cytoplasm. As cell transporters work to move the Ca^{2+} back into mitochondria and the sarcoplasmic reticulum, the heat released from ATP hydrolysis substantially raises body temperature. Some investigators have suggested that a mild version of this process plays a role in nonshivering thermogenesis in mammals.

Hypothermia, a condition in which body temperature falls abnormally low, is also a dangerous condition. As core body temperature falls, enzymatic reactions slow, and the person loses consciousness. When metabolism slows, oxygen consumption also decreases.

Victims of drowning in cold water can sometimes be revived without brain damage if they have gone into a state of hypothermia. This observation led to the development of induced hypothermia for certain surgical procedures, such as heart surgery. The patient is cooled to 21–24 °C (70–75 °F) so that tissue oxygen demand can be met by artificial oxygenation of the blood as it passes through a bypass pump. After surgery is complete, the patient is gradually rewarmed.

Concept Check Answers: p. 774

22. Why must a water bed be heated to allow a person to sleep on it comfortably?
23. Will a person who is exercising outside overheat faster when the air humidity is low or when it is high?

RUNNING PROBLEM CONCLUSION

Eating Disorders

Nicole finally agreed to undergo counseling and enter a treatment program for anorexia nervosa. She was lucky—her wrist would heal, and her medical complications could have been much worse. After seeing Nicole and discussing her anorexia, Sara realized that she also needed to see a counselor. Even though she was no longer dancing, Sara still used diet pills, diuretics, and laxatives when she became uncomfortable with her weight, and she had started binging and purging—eating much more than normal when she was stressed and then forcing herself to vomit to avoid gaining any weight. These are the behavioral patterns of *bulimia nervosa* (BN), a condition that is as serious as AN and that affects an estimated 4% of females. Its physiological effects and treatments are similar to those for AN.

To learn more about anorexia and bulimia, and for help finding a support group, see the National Association of Anorexia Nervosa and Associated Disorders web site at *www.anad.org* or *www.nationaleatingdisorders.org.*

Question	Facts	Integration and Analysis
1. If you measured Nicole's leptin level, what would you expect to find?	Leptin is a hormone secreted by adipose tissue.	Nicole has little adipose tissue, so she would have a low leptin level.

RUNNING PROBLEM CONCLUSION *(continued)*

Question	Facts	Integration and Analysis
2. Would you expect Nicole to have elevated or depressed levels of neuropeptide Y?	NPY is inhibited by leptin. NPY stimulates feeding centers.	Because her leptin level is low, you might predict that NPY would be elevated and feeding stimulated. However, the feeding center is affected by other factors besides NPY (Fig. 22.1). Brain studies of anorexic patients show high levels of CRH, which opposes NPY and depresses feeding.
3. What is Nicole's K^+ disturbance called? What effect does it have on the resting membrane potential of her cells?	Nicole's K^+ is 2.5 mEq/L, and normal is 3.5–5 mEq/L.	Low plasma K^+ is called hypokalemia. Hypokalemia causes the membrane potential to hyperpolarize [p. 265].
4. Why does Dr. Ayani want to monitor Nicole's cardiac function?	Cardiac muscle is an excitable tissue whose activity depends on changes in membrane potential.	Hypokalemia can alter the membrane potential of cardiac autorhythmic and contractile cells and cause a potentially fatal cardiac arrhythmia.
5. Based on her clinical values, what is Nicole's acid-base status?	Nicole's pH is 7.52, and her plasma HCO_3^- is elevated at 40 mEq/L.	Normal pH is 7.38–7.42, so she is in alkalosis. Her elevated HCO_3^- indicates a metabolic alkalosis. The cause is probably induced vomiting and loss of HCl from her stomach.
6. Based on what you know about heart rate and blood pressure, speculate on why Nicole has low blood pressure with a rapid pulse.	Her blood pressure is 80/50 (low), and her pulse is 90 (high).	Normally, increasing the heart rate would increase blood pressure. In this case, the increased pulse is a compensatory attempt to raise her low blood pressure. The low blood pressure probably results from dehydration.
7. Would you expect Nicole's renin and aldosterone levels to be normal, elevated, or depressed? How might these levels relate to her K^+ disturbance?	All the primary stimuli for renin secretion are associated with low blood pressure. Renin begins the RAAS pathway that stimulates aldosterone secretion.	Because Nicole's blood pressure is low, you would expect elevated renin and aldosterone levels. Aldosterone promotes renal K^+ secretion, which would lower her body load of K^+. She probably also has low dietary K^+ intake, which contributes to her hypokalemia.
8. Give some possible reasons Nicole had been feeling weak during dance rehearsals.	In fasted-state metabolism, the body breaks down skeletal muscle.	Loss of skeletal muscle proteins, hypokalemia, and possibly hypoglycemia could all be causes of Nicole's weakness.
9. Why might an NPY agonist help in cases of anorexia?	NPY stimulates the feeding center.	An NPY agonist might stimulate the feeding center and help Nicole want to eat.

737 739 741 744 765 770

Test your understanding with:

- Practice Tests
- Running Problem Quizzes
- A&PFlix™ Animations
- PhysioEx™ Lab Simulations
- Interactive Physiology Animations

www.masteringaandp.com

Chapter Summary

Energy balance in the body means that the body's energy intake equals its energy output. The same balance principle applies to metabolism and body temperature. The amount of nutrient in each of the body's nutrient pools depends on intake and output. Glucose *homeostasis* is one of the most important goals of regulated metabolism, for without adequate glucose, the brain is unable to function. Flow of material through the biochemical pathways of metabolism depends on the *molecular interactions* of substrates and enzymes.

Appetite and Satiety

1. The hypothalamus contains a tonically active **feeding center** and a **satiety center** that inhibits the feeding center. (p. 737)
2. Blood glucose concentrations (the **glucostatic theory**) and body fat content (the **lipostatic theory**) influence food intake. (p. 737)
3. Food intake is influenced by a variety of peptides, including **leptin, neuropeptide Y,** and **ghrelin.** (p. 738; Fig. 22.1)

Energy Balance

4. To maintain a constant amount of energy in the body, energy intake must equal energy output. (p. 739; Fig. 22.2)
5. The body uses energy for transport, movement, and chemical work. About half of energy used is released as heat. (p. 740)
6. **Direct calorimetry** measures the energy content of food. (p. 740)
7. The body's **oxygen consumption rate** is the most common method of estimating energy expenditure. (p. 741)
8. The **respiratory quotient** (RQ) or **respiratory exchange ratio** (RER) is the ratio of CO_2 produced to O_2 consumed. RQ varies with diet. (p. 741)
9. **Basal metabolic rate** (BMR) is an individual's lowest metabolic rate. Metabolic rate (kcal/day) = L O_2 consumed/day × kcal/L O_2. (p. 741)
10. **Diet-induced thermogenesis** is an increase in heat production after eating. (p. 742)
11. Glycogen and fat are the two primary forms of energy storage in the human body. (p. 742)

Metabolism

12. **Metabolism** is all the chemical reactions that extract, use, or store energy. (p. 743, 746; Figs. 22.3, 22.5)
13. **Anabolic pathways** synthesize small molecules into larger ones. **Catabolic pathways** break large molecules into smaller ones. (p. 742)
14. Metabolism is divided into the **fed** (absorptive) **state** and **fasted** (postabsorptive) **state**. The fed state is anabolic; the fasted state is catabolic. (p. 742)
15. **Glycogenesis** is glycogen synthesis. (p. 746; Fig. 22.5)
16. Ingested fats enter the circulation as chylomicrons. **Lipoprotein lipase** removes triglycerides, leaving chylomicron remnants to be taken up and metabolized by the liver. (p. 747; Fig. 22.6)
17. The liver secretes lipoprotein complexes, such as LDL-C. **Apoproteins A** and **B** are the ligands for receptor-mediated endocytosis of lipoprotein complexes. (p. 747; Fig. 22.6)
18. Elevated blood LDL-C and low blood HDL-C are risk factors for coronary heart disease. Therapies for lowering cholesterol decrease cholesterol uptake or synthesis or increase cholesterol clearance. (p. 748)
19. The function of fasted-state metabolism is to maintain adequate plasma glucose concentrations because glucose is normally the only fuel that the brain can metabolize. (p. 750; Fig. 22.8)
20. **Glycogenolysis** is glycogen breakdown. **Gluconeogenesis** is glucose synthesis from noncarbohydrate precursors, especially amino acids. (p. 751; Figs. 22.9, 22.11)
21. In the fasted state, the liver produces glucose from glycogen and amino acids. **Beta oxidation** of fatty acids forms acidic **ketone bodies**. (p. 750; Fig. 22.8)

Homeostatic Control of Metabolism

22. Hour-to-hour metabolic regulation depends on the ratio of insulin to glucagon. Insulin dominates the fed state and decreases plasma glucose. Glucagon dominates the fasted state and increases plasma glucose. (p. 754; Fig. 22.14)
23. The **islets of Langerhans** secrete insulin and amylin from beta cells, glucagon from alpha cells, and somatostatin from D cells. (p. 753; Fig. 22.13)
24. Increased plasma glucose and amino acid levels stimulate insulin secretion. GI hormones and parasympathetic input amplify it. Sympathetic signals inhibit insulin secretion. (p. 756; Fig. 22.15)
25. Insulin binds to a tyrosine kinase receptor and activates multiple **insulin-receptor substrates**. (p. 757; Fig. 22.16)
26. Major insulin target tissues are the liver, adipose tissue, and skeletal muscles. Some tissues are insulin independent. (p. 757)
27. Insulin increases glucose transport into muscle and adipose tissue, as well as glucose utilization and storage of glucose and fat. (p. 758; Fig. 22.17)
28. **Glucagon** stimulates glycogenolysis and gluconeogenesis. (p. 760; Fig. 22.18)
29. **Diabetes mellitus** is a family of disorders marked by abnormal secretion or activity of insulin that causes hyperglycemia. In **type 1 diabetes**, pancreatic beta cells are destroyed by antibodies. In **type 2 diabetes**, target tissues fail to respond normally to insulin. (pp. 760–761)
30. Type 1 diabetes is marked by catabolism of muscle and adipose tissue, glucosuria, polyuria, and metabolic ketoacidosis. Type 2 diabetics have less acute symptoms. In both types, complications include atherosclerosis, neurological changes, and problems with the eyes and kidneys. (p. 762; Fig. 22.20)
31. **Metabolic syndrome** is a condition in which people have central obesity, elevated fasting glucose levels, and elevated lipids. These people are at high risk for developing cardiovascular disease. (p. 765)

Regulation of Body Temperature

32. Body temperature homeostasis is controlled by the hypothalamus. (p. 765)
33. Heat loss from the body takes place by radiation, conduction, convection, and evaporation. Heat loss is promoted by cutaneous vasodilation and sweating. (p. 766, 768; Figs. 22.21 and 22.22)
34. Heat is generated by **shivering thermogenesis** and by **nonshivering thermogenesis**. (p. 766; Fig. 22.21)

Questions

Answers: p. A-1

Level One Reviewing Facts and Terms

1. Define metabolic, anabolic, and catabolic pathways.
2. List and briefly explain the three forms of biological work.
3. Define a kilocalorie. What is direct calorimetry?
4. What is the respiratory quotient (RQ)? What is a typical RQ value for an American diet?
5. Define basal metabolic rate (BMR). Under what conditions is it measured? Why does the average BMR differ in adult males and females? List at least four factors other than sex that may affect BMR in humans.
6. What are the three general fates of biomolecules in the body?
7. What are the main differences between metabolism in the absorptive and postabsorptive states?
8. What is a nutrient pool? What are the three primary nutrient pools of the body?
9. What is the primary goal of fasted-state metabolism?
10. In what forms is excess energy stored in the body?
11. What are the three possible fates for ingested proteins? For ingested fats?
12. Name the two hormones that regulate glucose metabolism, and explain what effect each hormone has on blood glucose concentrations.
13. Which noncarbohydrate molecules can be made into glucose? What are the pathways called through which these molecules are converted to glucose?
14. Under what circumstances are ketone bodies formed? From what biomolecule are ketone bodies formed? How are they used by the body, and why is their formation potentially dangerous?
15. Name two stimuli that increase insulin secretion, and one stimulus that inhibits insulin secretion.
16. What are the two types of diabetes mellitus? How do their causes and basic symptoms differ?
17. What factors release glucagon? What organ is the primary target of glucagon? What effect(s) do(es) glucagon produce?
18. Define the following terms and explain their physiological significance:
    - (a) lipoprotein lipase
    - (b) amylin
    - (c) ghrelin
    - (d) neuropeptide Y
    - (e) apoprotein
    - (f) leptin
    - (g) osmotic diuresis
    - (h) insulin resistance
19. What effect does insulin have on:
    - (a) glycolysis
    - (b) gluconeogenesis
    - (c) glycogenesis
    - (d) lipogenesis
    - (e) protein synthesis

Level Two Reviewing Concepts

20. Map: Draw a map that compares the fed state and the fasted state. For each state, compare metabolism in skeletal muscles, the brain, adipose tissue, and the liver. Indicate which hormones are active in each stage and at what points they exert their influence.
21. Examine the graphs of insulin and glucagon secretion in Figure 22.14c. Why have some researchers concluded that the ratio of these two hormones determines whether glucose is stored or removed from storage?
22. Define, compare, and contrast or relate the terms in each of the following sets:
    - (a) glucose, glycogenolysis, glycogenesis, gluconeogenesis, glucagon, glycolysis
    - (b) shivering thermogenesis, nonshivering thermogenesis, diet-induced thermogenesis
    - (c) lipoproteins, chylomicrons, cholesterol, HDL-C, LDL-C, apoproteins
    - (d) direct and indirect calorimetry
    - (e) conductive heat loss, radiant heat loss, convective heat loss, evaporative heat loss
    - (f) absorptive and postabsorptive states
23. Describe (or map) the physiological events that lead to the following signs or symptoms in a type 1 diabetic:
    - (a) hyperglycemia
    - (b) glucosuria
    - (c) polyuria
    - (d) ketosis
    - (e) dehydration
    - (f) severe thirst
24. Both insulin and glucagon are released following ingestion of a protein meal that raises plasma amino acid levels. Why is the secretion of both hormones necessary?
25. Explain the current theory of the control of food intake. Use the following terms in your explanation: hypothalamus, feeding center, satiety center, appetite, leptin, NPY, neuropeptides.
26. Compare human thermoregulation in hot environments and cold environments.

Level Three Problem Solving

27. Scott is a bodybuilder who consumes large amounts of amino acid supplements in the belief that they will increase his muscle mass. He believes that the amino acids he consumes are stored in his body until he needs them. Is Scott correct? Explain.
28. Draw and label a graph showing the effect of insulin secretion on plasma glucose concentration.
29. One of the debates in fluid therapy for diabetic ketoacidosis (DKA) is whether to administer bicarbonate (bicarb). Although it is generally accepted that bicarb should be given if the patient's blood pH is < 7.1 (life-threatening), most authorities do not give bicarb otherwise. One reason for not administering bicarb relates to the oxygen-binding capacity of hemoglobin. In DKA, patients have low levels of 2,3-DPG [p. 612]. When acidosis is corrected rapidly, 2,3-DPG is much slower to recover and may take 24 or more hours to return to normal.

    Draw and label a graph of the normal oxygen-dissociation curve [p. 610]. *Briefly* explain and *draw lines on the same graph to show:*
    - (a) what happens to oxygen release during DKA as a result of acidosis and low 2,3-DPG levels.
    - (b) what happens to oxygen release when the metabolic acidosis is rapidly corrected with bicarbonate.

30. The signal molecules involved in active cutaneous vasodilation are unclear but it is known that sympathetic cholinergic neurons are involved. In one experiment* scientists used botulinum toxin [p. 425] to block release of chemicals from the sympathetic axon terminal. When they did this, the vasodilation response disappeared. In the next experiment, they applied atropine, a muscarinic receptor antagonist and observed that some but not all of the vasodilation response disappeared. What conclusion could they draw from these two experiments?

*DL Kellogg Jr. *et al.* Cutaneous active vasodilation in humans is mediated by cholinergic nerve cotransmission. *Circ Res 77: 1222–1228, 1995.*

Level Four Quantitative Problems

31. One way to estimate obesity is to calculate a person's body mass index (BMI). A body mass index greater than 30 is considered a sign of obesity. To calculate your BMI, divide your body weight in kilograms by the square of your height in meters: kg/m^2. (To convert your weight from pounds to kilograms, use the conversion factor 1 kg/2.2 lb. To convert your height from inches to meters, use the factor 1 m/39.24 in.)
32. What is the calorie content of a serving of spaghetti and meatballs that contains 6 g fat, 30 g carbohydrate, and 8 g protein? What percentage of the calories comes from fat?

Answers

Answers to Concept Check Questions

Page 739

1. The feeding center causes an animal to eat, and the satiety center causes an animal to cease eating. Both centers are located in the hypothalamus.
2. The problem might involve abnormal tissue responsiveness—a target cell with no leptin receptors or defective receptors. There might also be a problem with leptin's signal transduction/second messenger pathway.

Page 742

3. The seven factors are age, sex, lean muscle mass, activity, diet, hormones, and genetics.
4. One g of fat contains more than twice the energy of 1 g of glycogen.
5. $C_6H_{12}O_6 + 6\,O_2 \rightarrow 6\,CO_2 + 6\,H_2O$
6. $RQ = CO_2/O_2 = 6/6 = 1$

Page 745

7. GLUT transporters are passive facilitated diffusion transporters.

Page 749

8. dL is the abbreviation for deciliter, or 1/10th of a liter (100 mL).
9. Bile acid sequestrants and ezetimibe leave bile salts and cholesterol in the intestinal lumen to be excreted, so possible side effects are loose, fatty feces and inadequate absorption of fat-soluble vitamins.

Page 453

10. Glycogenesis is glycogen synthesis; gluconeogenesis is synthesis of glucose from amino acids or glycerol.
11. Amino acids used for energy become pyruvate or enter the citric acid cycle.
12. Plasma cholesterol is bound to a carrier protein and can't diffuse across the cell membrane.

Page 759

13. The primary target tissues for insulin are liver, skeletal muscle, and adipose tissue.
14. If glucose uptake depended on insulin, the intestine, kidney tubule, and neurons could not absorb glucose in the fasted state. Neurons use glucose exclusively for metabolism and must always be able to take it up.
15. During fight-or-flight, skeletal muscles need glucose for energy. Inhibiting insulin secretion causes the liver to release glucose into the blood and prevents adipose cells from taking it up, making more glucose available for exercising muscle, which does not require insulin for glucose uptake.

Page 761

16. No, you would not get the same result because you would not be ingesting the same amount of glucose. Table sugar is sucrose: half glucose and half fructose. Most soft drinks are sweetened with high-fructose corn syrup.

Page 765

17. Insulin is a protein and is digested if administered orally.
18. Although dehydrated patients may have elevated plasma K^+ concentrations, their total *amount* of K^+ is below normal. If fluid volume is restored to normal with no K^+ added, a below-normal K^+ concentration results.
19. Sitagliptin inhibits DPP4, the enzyme that breaks down GLP-1 and GIP. Prolonging the action of these two gut hormones enhances insulin release and slows digestion, which gives cells time to take up and use absorbed glucose.

Page 767

20. Norepinephrine binds to α-receptors to elicit vasoconstriction.
21. Researchers probably classified the neurons as sympathetic because of where they leave the spinal cord.

Page 770

22. Water cooler than body temperature draws heat away from the body through conductive heat transfer. If this loss exceeds the body's heat production, the person feels cold.
23. A person exercising in a humid environment loses the benefit of evaporative cooling and is likely to overheat faster.

Answers to Figure and Graph Questions

Page 746

Figure 22.5: 1. (a) next to left arrow from G-6-P to glycogen, (b) next to arrow going from acetyl CoA to fatty acids, (c) next to right arrow from glycogen to G-6-P, (d) with electron transport system. 2. No, amino acids entering the citric acid cycle cannot be used to make glucose because the step from pyruvate to acetyl CoA is not reversible.

Page 748

Figure 22.7: The decrease from 190 to 160 mg/dL has the greatest effect.

Page 751

Figure 22.10: hydrolysis

Page 756

Figure 22.15: acetylcholine on muscarinic receptor

23 Endocrine Control of Growth and Metabolism

Disorders of hormone action will be more common causes of endocrinopathy than states of hormone deficiency and excess combined.

—Jean D. Wilson, *Endocrinology: Survival as a Discipline in the 21st Century?* Annu Rev Physiol 62: 947-950, 2000

Background Basics

Cells embedded in the nonliving matrix of bone

In 1998 Mark McGwire made news when he hit 70 home runs, surpassing the single-season home run record Roger Maris established in 1961. McGwire also created a firestorm of controversy when he admitted to taking *androstenedione*, a performance-enhancing steroid prohormone banned by the International Olympic Committee and other groups but not by professional baseball. As a result of the controversy, Congress passed the Anabolic Steroids Act of 2004, which made androstenedione and some other steroid prohormones controlled substances available only by prescription.

What is this prohormone, and why is it so controversial? You will learn more about androstenedione in this chapter as we discuss the hormones that play a role in long-term regulation of metabolism and growth. In normal individuals, these hormones can be difficult to study because their effects are subtle and their interactions with one another complex. As a result, much of what we know about endocrinology comes from studying pathological conditions in which a hormone is either oversecreted or undersecreted. In recent years, however, advances in molecular biology and the use of transgenic animal models have enabled scientists to learn more about hormone action at the cellular level.

BIOTECHNOLOGY

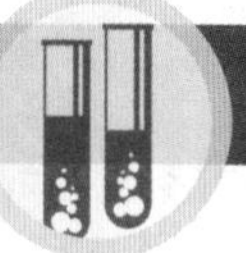

Mutant Mouse Models

The use of animal models to study human diseases has become a valuable part of biomedical research. The mouse genome sequencing effort tells us that 99% of the approximately 30,000 mouse genes have direct human *homologs* (equivalents). This means that in many cases we can use the mouse to understand what genes, and the proteins they encode, do in healthy and sick humans. Sometimes, natural mutations produce animal diseases that resemble human diseases. Two examples of such mutants are the twitcher mouse, in which normal myelin degenerates owing to an inherited metabolic problem, and the wobbler mouse, in which somatic motor neurons controlling the limbs die. In other cases, scientists have used biotechnology techniques to create mice that lack specific genes (**knockout mice**) or to breed mice that contain extra genes that were inserted artificially (**transgenic mice**). The mouse is the ideal organism for conducting these experiments. It is small, relatively inexpensive, and has a short lifespan. Additionally, its biological processes are similar to those of humans, and the mouse genome can be easily manipulated by genetic engineering technologies.

To learn more, read the articles in a special Mouse Genome issue of the journal *Nature* (*www.nature.com/nature/mousegenome*). You can obtain additional information about the different knockout and transgenic mice being used for research from the Mouse Genome Informatics databases at The Jackson Laboratory (*www.informatics.jax.org*).

Review of Endocrine Principles

Before we delve into the different hormones, let's do a quick review of some basic principles and patterns of endocrinology.

1. **The hypothalamic-pituitary control system** [p. 221]. Several of the hormones described in this chapter are controlled by hypothalamic and anterior pituitary (*adenohypophyseal*) trophic hormones.
2. **Feedback patterns** [p. 17]. The negative feedback signal for simple endocrine pathways is the systemic response to the hormone. For example, insulin secretion shuts off when blood glucose concentrations decrease. In complex pathways using the hypothalamic-pituitary control system, the feedback signal may be the hormone itself. In pathological states, endocrine cells may not respond appropriately to feedback signals.
3. **Hormone receptors** [p. 212]. Hormone receptors may be on the cell surface or inside the cell.
4. **Cellular responses** [p. 212]. In general, hormone target cells respond by altering existing proteins or by making new proteins. The historical distinctions between the actions of peptide and steroid hormones are no longer valid. Some steroid hormones exert rapid, nongenomic effects, and some peptide hormones alter transcription and translation.
5. **Modulation of target cell response** [p. 191]. The amount of active hormone available to the cell and the number and activity of target cell receptors determine the magnitude

RUNNING PROBLEM

Hyperparathyroidism

"Broken bones, kidney stones, abdominal groans, and psychic moans." Medical students memorize this saying when they learn about *hyperparathyroidism*, a disease in which parathyroid glands (see Fig. 23.12, p. 794) work overtime and produce excess parathyroid hormone (PTH). Dr. Adiaha Spinks suddenly recalls the saying as she examines Prof. Bob Magruder, who has arrived at her office in pain from a kidney stone lodged in his ureter. When questioned about his symptoms, Prof. Magruder also mentions pain in his shin bones, muscle weakness, stomach upset, and a vague feeling of depression. "I thought it was all just the stress of getting my book published," he says. To Dr. Spinks, however, Prof. Magruder's combination of symptoms sounds suspiciously like he might be suffering from hyperparathyroidism.

776 780 784 793 796 797

of target cell response. Cells may up-regulate or down-regulate their receptors to alter their response. Cells that do not have hormone receptors are nonresponsive.

6 **Endocrine pathologies** [p. 227]. Endocrine pathologies result from (a) excess hormone secretion, (b) inadequate hormone secretion, and (c) abnormal target cell response to the hormone. It now appears that failure of the target cell to respond appropriately to its hormone is a major cause of endocrine disorders.

In the following sections we first examine adrenal corticosteroids and thyroid hormones, two groups of hormones that influence long-term metabolism. We then consider the endocrine control of growth.

Adrenal Glucocorticoids

The paired adrenal glands sit on top of the kidneys like little caps (Fig. 23.1). Each adrenal gland, like the pituitary gland, is two embryologically distinct tissues that merged during development. This complex organ secretes multiple hormones, both neurohormones and classic hormones. The *adrenal medulla* occupies a little over a quarter of the inner mass and is composed of modified sympathetic ganglia that secrete catecholamines (mostly epinephrine) to mediate rapid responses in fight-or-flight situations [p. 378]. The *adrenal cortex* forms the outer three-quarters of the gland and secretes a variety of steroid hormones.

The Adrenal Cortex Secretes Steroid Hormones

The adrenal cortex secretes three major types of steroid hormones: aldosterone (sometimes called a *mineralocorticoid* because of its effect on the minerals sodium and potassium) [p. 669], glucocorticoids, and sex hormones. Histologically, the adrenal cortex is divided into three layers, or zones (Fig. 23.1a). The outer *zona glomerulosa* secretes only aldosterone. The inner *zona reticularis* secretes mostly *androgens,* the sex hormones dominant in men. The middle *zona fasciculata* secretes mostly **glucocorticoids,** named for their ability to increase plasma glucose concentrations. **Cortisol** is the main glucocorticoid secreted by the adrenal cortex.

The generalized synthesis pathway for steroid hormones is shown in Figure 23.1b. All steroid hormones begin with cholesterol, which is modified by multiple enzymes to end up as aldosterone, glucocorticoids, or sex steroids (androgens as well as *estrogens* and *progesterone,* the dominant sex hormones in females). The pathways are the same in the adrenal cortex, gonads, and placenta, but what differs from tissue to tissue is the distribution of enzymes that catalyze the different reactions. For example, the enzyme that makes aldosterone is found in only one of the three adrenal cortex zones.

This chapter opened with the story of baseball player Mark McGwire and his controversial use of the supplement androstenedione. Figure 23.1b shows that this prohormone is one intermediate in the synthesis of testosterone and dihydrotestosterone. One androstenedione precursor, *dehydroepiandrosterone* (DHEA), is used as a dietary supplement. In the United States, purchase of DHEA is not regulated, despite the fact that this substance is metabolically converted to androstenedione and testosterone, both controlled substances whose use is widely banned by sports associations.

The close structural similarity among steroid hormones means that the binding sites on their receptors are also similar, leading to *crossover effects* when one steroid binds to the receptor for a related molecule. For example, *mineralocorticoid receptors* (MRs) for aldosterone are found in the distal nephron. MRs also bind and respond to cortisol, which may be 100 times more concentrated in the blood than aldosterone. What is to keep cortisol from binding to an MR and influencing Na^+ and K^+ excretion? It turns out that renal tubule cells with MRs have an enzyme (*11β-hydroxysteroid dehydrogenase*) that converts cortisol to a less active form with low specificity for the MR. By inactivating cortisol, these cells normally prevent crossover effects from cortisol. However, crossover activity and the structural similarities of steroid hormones mean that in many endocrine disorders, patients may experience symptoms related to more than one hormone.

Concept Check Answers: p. 800

1. Name the two parts of the adrenal gland and the major hormones secreted by each part.
2. For what hormones is androstenedione a prohormone? (See Fig. 23.1b.) Why might this prohormone give an athlete an advantage?

Cortisol Secretion Is Controlled by ACTH

The control pathway for cortisol secretion is known as the *hypothalamic-pituitary-adrenal (HPA) pathway* (Fig. 23.2a on page 779). The HPA pathway begins with hypothalamic **corticotropin-releasing hormone** (CRH), which is secreted into the hypothalamic-hypophyseal portal system and transported to the anterior pituitary. CRH stimulates release of **adrenocorticotropic hormone** (ACTH or *corticotropin*) from the anterior pituitary. ACTH in turn acts on the adrenal cortex to promote synthesis and release of cortisol. Cortisol then acts as a negative feedback signal, inhibiting ACTH and CRH secretion.

Cortisol secretion is continuous and has a strong diurnal rhythm (Fig. 23.2c). Secretion normally peaks in the morning and diminishes during the night. Cortisol secretion also increases with stress.

Cortisol is a typical steroid hormone and is synthesized on demand [Fig. 7.5, p. 215]. Once synthesized, it diffuses out of

THE ADRENAL GLAND

(a) The paired adrenal glands sit on top of the kidneys. Each region secretes different hormones.

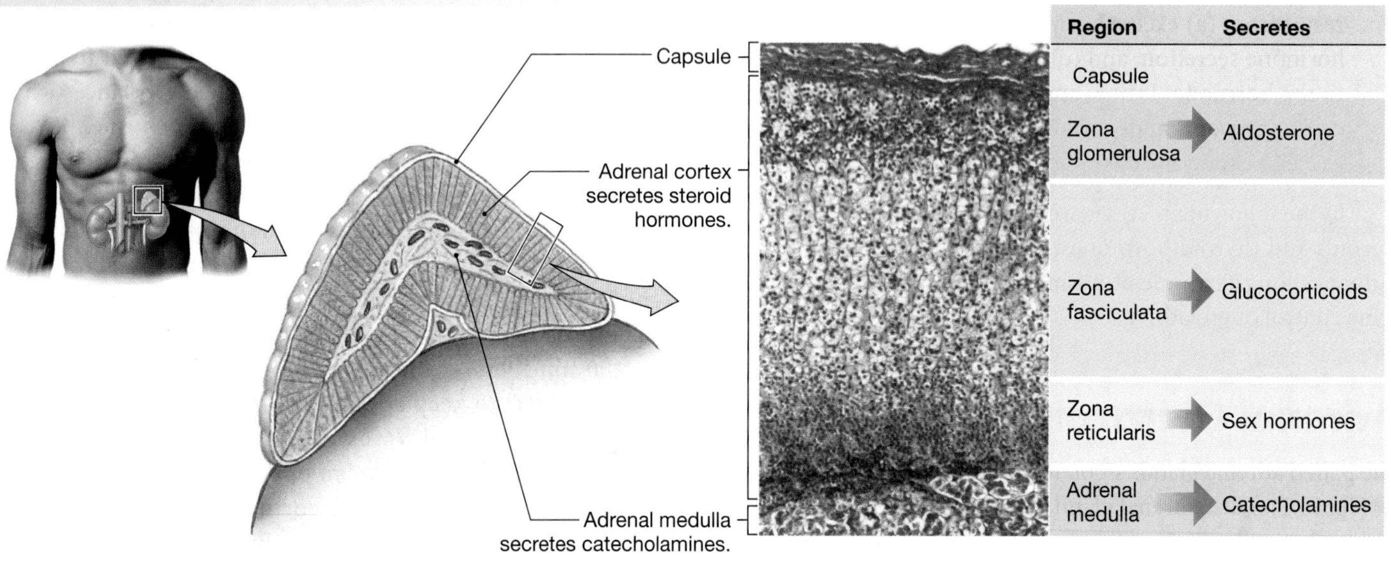

Region	Secretes
Capsule	
Zona glomerulosa	Aldosterone
Zona fasciculata	Glucocorticoids
Zona reticularis	Sex hormones
Adrenal medulla	Catecholamines

(b) Synthesis pathways for steroid hormones

All steroid hormones are synthesized from cholesterol. The blank boxes represent intermediate compounds whose names have been omitted for simplicity. Each step is catalyzed by an enzyme, but only two enzymes are shown in the figure.

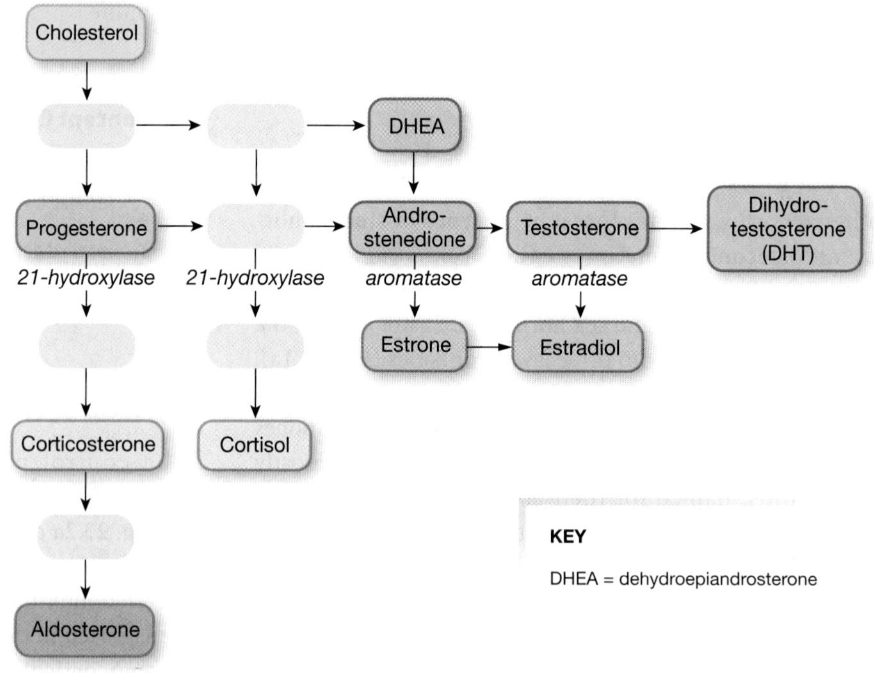

KEY

DHEA = dehydroepiandrosterone

FIGURE QUESTIONS

1. A baby is born with a genetic mutation that results in a deficiency of the enzyme 21-hydroxylase. Based on the role of this enzyme in the pathway illustrated, what symptoms might you predict in the baby?
2. Would men or women have more aromatase activity?

Fig. 23.1

adrenal cells into the plasma, where most of it is transported by a carrier protein, *corticosteroid-binding globulin* (CBG or *transcortin*). Unbound hormone is free to diffuse into target cells.

All nucleated cells of the body have cytoplasmic glucocorticoid receptors. The hormone-receptor complex enters the nucleus, binds to DNA with the aid of a *hormone-response element,* and alters gene expression, transcription, and translation. In general, a tissue's response to glucocorticoid hormones is not evident for 60–90 minutes. However, cortisol's negative feedback effect on ACTH secretion occurs within minutes.

THE HYPOTHALAMIC-PITUITARY-ADRENAL (HPA) PATHWAY

(a) The control of cortisol secretion

Circadian rhythm
Stress
Hypothalamus
CRH
Anterior pituitary
POMC
ACTH
long-loop negative feedback
Adrenal cortex
Cortisol
Immune system
Liver
Muscle
Adipose tissue
Function suppressed
Gluco-neogenesis
Protein catabolism
Lipolysis

(b) The properties of cortisol

CORTISOL	
Origin	Adrenal cortex
Chemical nature	Steroid
Biosynthesis	From cholesterol; made on demand; not stored
Transport in the circulation	On corticosteroid-binding globulin (made in liver)
Half-life	60–90 min
Factors affecting release	Circadian rhythm of tonic secretion; stress enhances release
Control pathway	CRH (hypothalamus) ⟶ ACTH (anterior pituitary) ⟶ cortisol (adrenal cortex)
Target cells or tissues	Most tissues
Target receptor	Intracellular
Whole body or tissue reaction	↑ Plasma [glucose]; ↓ immune activity; permissive for glucagon and catecholamines
Action at cellular level	↑ Gluconeogenesis and glycogenolysis; ↑ protein catabolism. Blocks cytokine production by immune cells
Action at molecular level	Initiates transcription, translation, and new protein synthesis
Feedback regulation	Negative feedback to anterior pituitary and hypothalamus

(c) The circadian rhythm of cortisol secretion

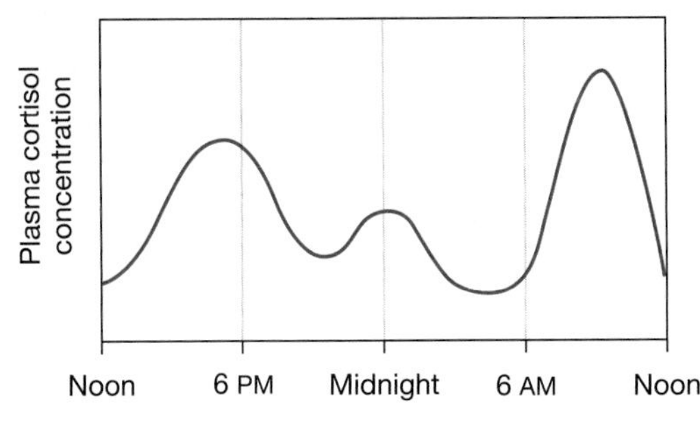

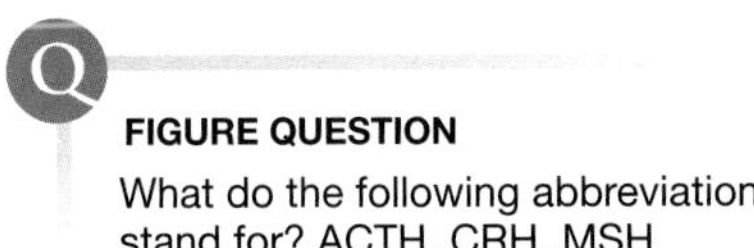

FIGURE QUESTION

What do the following abbreviations stand for? ACTH, CRH, MSH

(d) Post-translational processing of POMC creates a variety of active peptides.

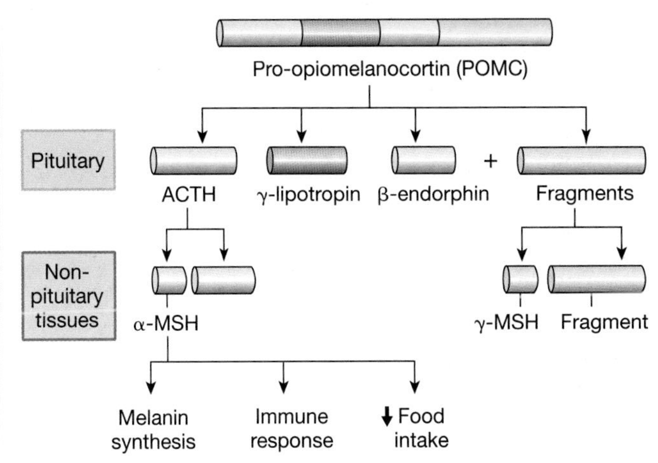

Fig. 23.2

Cortisol Is Essential for Life

Adrenal glucocorticoids are sometimes called the body's stress hormones because of their role in the mediation of long-term stress. Adrenal catecholamines, particularly epinephrine, are responsible for rapid metabolic responses needed in fight-or-flight situations.

Cortisol is essential for life. Animals whose adrenal glands have been removed die if exposed to any significant environmental stress. The most important metabolic effect of cortisol is its protective effect against *hypoglycemia*. When blood glucose decreases, the normal response is secretion of pancreatic glucagon, which promotes gluconeogenesis and glycogen breakdown [p. 744]. In the absence of cortisol, however, glucagon is unable to respond adequately to a hypoglycemic challenge. Because cortisol is required for full glucagon and catecholamine activity, it is said to have a *permissive effect* on those hormones [p. 226].

Cortisol receptors are found in every tissue of the body, but for many targets we do not fully understand the physiological actions of cortisol. However, we can speculate on these actions based on tissue responses to high levels (*pharmacological doses*) of cortisol administered for therapeutic reasons or associated with hypersecretion.

All the metabolic effects of cortisol are directed at preventing hypoglycemia. Overall, cortisol is catabolic (Fig. 23.2a, b).

1 **Cortisol promotes gluconeogenesis** in the liver. Some glucose produced in the liver is released into the blood, and the rest is stored as glycogen. As a result, cortisol increases blood glucose concentrations.

2 **Cortisol causes the breakdown of skeletal muscle proteins** to provide a substrate for gluconeogenesis.

3 **Cortisol enhances lipolysis** so that fatty acids are available to peripheral tissues for energy use. The glycerol from fatty acids can be used for gluconeogenesis.

4 **Cortisol suppresses the immune system** through multiple pathways. This effect is discussed in more detail below.

5 **Cortisol causes negative calcium balance.** Cortisol decreases intestinal Ca^{2+} absorption and increases renal Ca^{2+} excretion, resulting in net Ca^{2+} loss from the body. In addition, cortisol is catabolic in bone tissue, causing net breakdown of calcified bone matrix. As a result, people who take therapeutic cortisol for extended periods have a higher-than-normal incidence of broken bones.

6 **Cortisol influences brain function.** States of cortisol excess or deficiency cause mood changes and memory and learning alterations. Some of these effects may be mediated by hormones in the cortisol release pathway, such as CRH. We discuss this effect of cortisol in more detail below.

RUNNING PROBLEM

Hyperparathyroidism causes breakdown of bone and the release of calcium phosphate into the blood. Elevated plasma Ca^{2+} can affect the function of excitable tissues, such as muscles and neurons. Surprisingly, however, most people with hyperparathyroidism have no symptoms. The condition is usually discovered during blood work performed for a routine health evaluation.

Q1: What role does Ca^{2+} play in the normal functioning of muscles and neurons?

Q2: What is the technical term for "elevated levels of calcium in the blood"? (Use your knowledge of word roots to construct this term.)

776 780 784 793 796 797

Concept Check Answers: p. 800

3. What do the abbreviations HPA and CBG stand for? If there is an alternate name for each term, what is it?
4. You are mountain-biking in Canada and encounter a bear, which chases you up a tree. Is your stress response mediated by cortisol? Explain.
5. The illegal use of anabolic steroids by bodybuilders and athletes periodically receives much attention. Do these illegal steroids include cortisol? Explain.

Cortisol Is a Useful Therapeutic Drug

Cortisol suppresses the immune system by preventing cytokine release and antibody production by white blood cells. It also inhibits the inflammatory response by decreasing leukocyte mobility and migration. These *immunosuppressant effects* of cortisol make it a useful drug for treating a variety of conditions, including bee stings, poison ivy, and pollen allergies. Cortisol also helps prevent rejection of transplanted organs. However, glucocorticoids also have potentially serious side effects because of their metabolic actions. Once *nonsteroidal anti-inflammatory drugs* (NSAIDs), such as ibuprofen, were developed, the use of glucocorticoids for treating minor inflammatory problems was discontinued.

Exogenous administration of glucocorticoids has a negative feedback effect on the anterior pituitary and may shut down ACTH production [Fig. 7.13, p. 227]. Without ACTH stimulation, the adrenal cells that produce cortisol atrophy. For this reason, it is essential that patients taking steroids taper their dose gradually, giving the pituitary and adrenal glands a chance to recover, rather than stopping the drug abruptly.

Cortisol Pathologies Result from Too Much or Too Little Hormone

The most common HPA pathologies result from hormone deficiency and hormone excess. Abnormal tissue responsiveness is an uncommon cause of adrenal steroid disorders.

Hypercortisolism Excess cortisol in the body is called **hypercortisolism**. It can arise from hormone-secreting tumors or from exogenous administration of the hormone. Cortisol therapy with high doses for more than a week has the potential to cause hypercortisolism—also known as *Cushing's syndrome*, after Dr. Harvey Cushing, who first described the condition in 1932.

Most signs of hypercortisolism can be predicted from the normal actions of the hormone. Excess gluconeogenesis causes hyperglycemia, which mimics diabetes. Muscle protein breakdown and lipolysis cause tissue wasting. Paradoxically, excess cortisol deposits extra fat in the trunk and face, perhaps in part because of increased appetite and food intake. The classic appearance of patients with hypercortisolism is thin arms and legs, obesity in the trunk, and a "moon face" with plump cheeks (Fig. 23.3). CNS effects of too much cortisol include initial mood elevation followed by depression, as well as difficulty with learning and memory.

Hypercortisolism has three common causes:

1. **An adrenal tumor that autonomously secretes cortisol.** These tumors are not under the control of pituitary ACTH. This condition is an instance of *primary hypercortisolism* [p. 228].
2. **A pituitary tumor that autonomously secretes ACTH.** Excess ACTH prompts the adrenal gland to oversecrete cortisol (*secondary hypercortisolism*). The tumor does not respond to negative feedback. This condition is also called *Cushing's disease* because it was the actual disease described by Dr. Cushing. (Hypercortisolism from any cause is called Cushing's *syndrome*.)
3. **Iatrogenic (physician-caused) hypercortisolism** occurs secondary to cortisol therapy for some other condition.

Hypocortisolism Hyposecretion pathologies are far less common than Cushing's syndrome. **Addison's disease** is hyposecretion of all adrenal steroid hormones, usually following autoimmune destruction of the adrenal cortex. Hereditary defects in the enzymes needed for adrenal steroid production cause several related syndromes (see the question in Fig. 23.1). These inherited disorders are often marked by excess androgen secretion because substrate that cannot be made into cortisol or aldosterone is converted to androgens. In newborn girls, excess androgens cause masculinization of the external genitalia, a condition called *adrenogenital syndrome*.

Concept Check Answers: p. 800

6. For primary, secondary, and iatrogenic hypercortisolism, indicate whether the ACTH level is normal, higher than normal, or lower than normal.
7. Would someone with Addison's disease have normal, low, or high levels of ACTH in the blood?

CRH and ACTH Have Additional Physiological Functions

In recent years, research interest has shifted away from glucocorticoids to CRH and ACTH, the trophic hormones of the HPA pathway. Both peptides are now known to belong to larger families of related molecules, with multiple receptor types found

HYPERCORTISOLISM (CUSHING'S SYNDROME)

(a) Moon face. A "moon face" with red cheeks is typical in this condition.

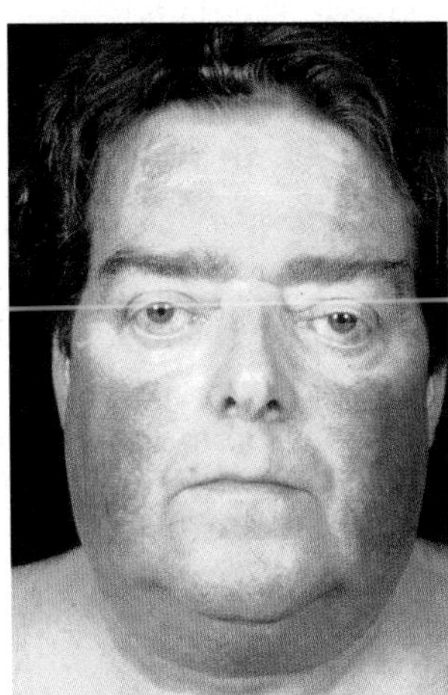

(b) Abdominal fat with striations. Fat also deposits in the trunk. The dark striations result from protein breakdown in the skin.

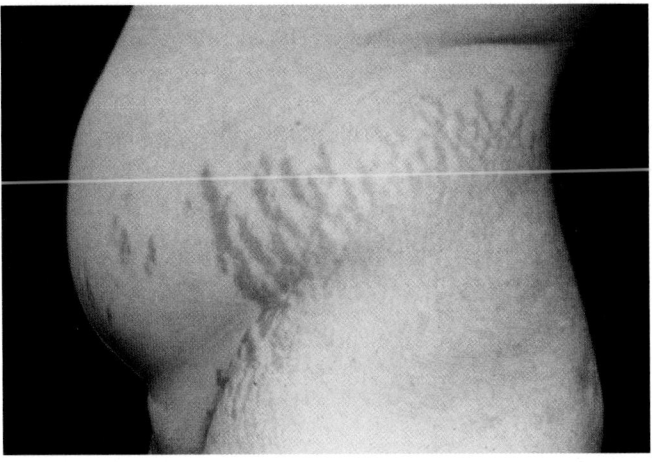

Fig. 23.3

in numerous tissues. Experiments with knockout mice lacking a particular receptor have revealed some of the physiological functions of peptides related to CRH and ACTH.

Two interesting findings of this research are that cytokines secreted by the immune system can stimulate the HPA pathway and that immune cells have receptors for ACTH and CRH. The association between stress and immune function appears to be mediated through CRH and ACTH, and this association provides one explanation for mind-body interactions in which mental state influences physiological function.

CRH Family The CRH family includes CRH and a related brain neuropeptide called *urocortin*. In addition to its involvement in inflammation and the immune response, CRH is known to decrease food intake [Fig. 22.1, p. 738] and has been associated with signals that mark the onset of labor in pregnant women. Additional evidence links CRH to anxiety, depression, and other mood disorders.

POMC and Melanocortins CRH acting on the anterior pituitary stimulates secretion of ACTH. ACTH is synthesized from a large glycoprotein called **pro-opiomelanocortin** (POMC, pronounced *pom-see*). POMC undergoes post-translational processing to produce a variety of biologically active peptides in addition to ACTH (Fig. 23.2d). In the pituitary, POMC products include **β-endorphin,** an endogenous opioid that binds to receptors that block pain perception [p. 269].

Processing of POMC in nonpituitary tissues creates additional peptides, such as *melanocyte-stimulating hormone (MSH)*. **α-MSH** is produced in the brain, where it inhibits food intake, and in the skin, where it acts on **melanocytes**. Melanocytes contain pigments called **melanins** that influence skin color in humans and coat color in rodents (see the Emerging Concepts box).

POMC-producing neurons in the hypothalamus are being studied for their role in food intake, energy balance, and thermoregulation. Recent investigations indicate that the action of nicotine on POMC neurons explains why smoking decreases food intake, for example. Other research suggests that POMC neurons may respond to changes in blood glucose and possibly participate in the glucostat control mechanism influencing food intake [p. 738].

The MSH hormones plus ACTH have been given the family name **melanocortins**. Five *melanocortin receptors* (MCRs) have been identified. MC2R responds only to ACTH and is the adrenal cortex receptor. MC1R is found in skin melanocytes and responds equally to α-MSH and ACTH. When ACTH is elevated in Addison's disease, the action of ACTH on MC1R leads to increased melanin production and the apparent "tan," or skin darkening, characteristic of this disorder.

Concept Check Answer: p. 800

8. What might be the advantage for the body to co-secrete ACTH and β-endorphin?

EMERGING CONCEPTS

Melanocortins and the *Agouti* Mouse

Much of what we are learning about the functions of the hypothalamic-pituitary-adrenal (HPA) pathway comes from research on mutant mice. One such animal is the *agouti* mouse, a strain that resulted from a spontaneous mutation first described in 1905. *Agouti* mice with one mutated gene overproduce a protein (called **agouti protein**) that gives them a characteristic yellow coat. Agouti protein is an antagonist to the melanocortin receptor MC1R, which controls melanin synthesis in hair. Of more interest to physiologists, however, is the fact that *agouti* mice overeat and develop adult-onset obesity, hyperglycemia, and insulin resistance—in other words, these mice are a model for obesity-related type 2 diabetes. Their agouti protein blocks the MC4R receptor in the brain, a receptor that normally depresses feeding behavior. If the receptor is not functional, the animal overeats and becomes obese. This link between melanocortin receptors, eating behavior, and diabetes has opened up a new area of research on treatments to prevent type 2 diabetes.

Thyroid Hormones

The thyroid gland is a butterfly-shaped gland that lies across the trachea at the base of the throat, just below the larynx (■ Fig. 23.4a). It is one of the larger endocrine glands, weighing 15–20 g. The thyroid gland has two distinct endocrine cell types: *C ("clear") cells,* which secrete a calcium-regulating hormone called *calcitonin,* and *follicular cells,* which secrete thyroid hormone. Calcitonin is discussed with calcium homeostasis.

Thyroid Hormones Contain Iodine

Thyroid hormones, like glucocorticoids, have long-term effects on metabolism. Unlike glucocorticoids, however, thyroid hormones are not essential for life. They are essential for normal growth and development in children, however, and infants born with thyroid deficiency will be developmentally delayed unless treated promptly. Because of the importance of thyroid hormones in children, the United States and Canada test all newborns for thyroid deficiency.

Thyroid hormones are amines derived from the amino acid tyrosine, and they are unusual because they contain the element iodine (Fig. 23.4c). Currently, thyroid hormones are the only known use for iodine in the body, although a few other tissues also concentrate this mineral.

Synthesis of thyroid hormones takes place in the thyroid **follicles** (also called *acini*), spherical structures whose walls are

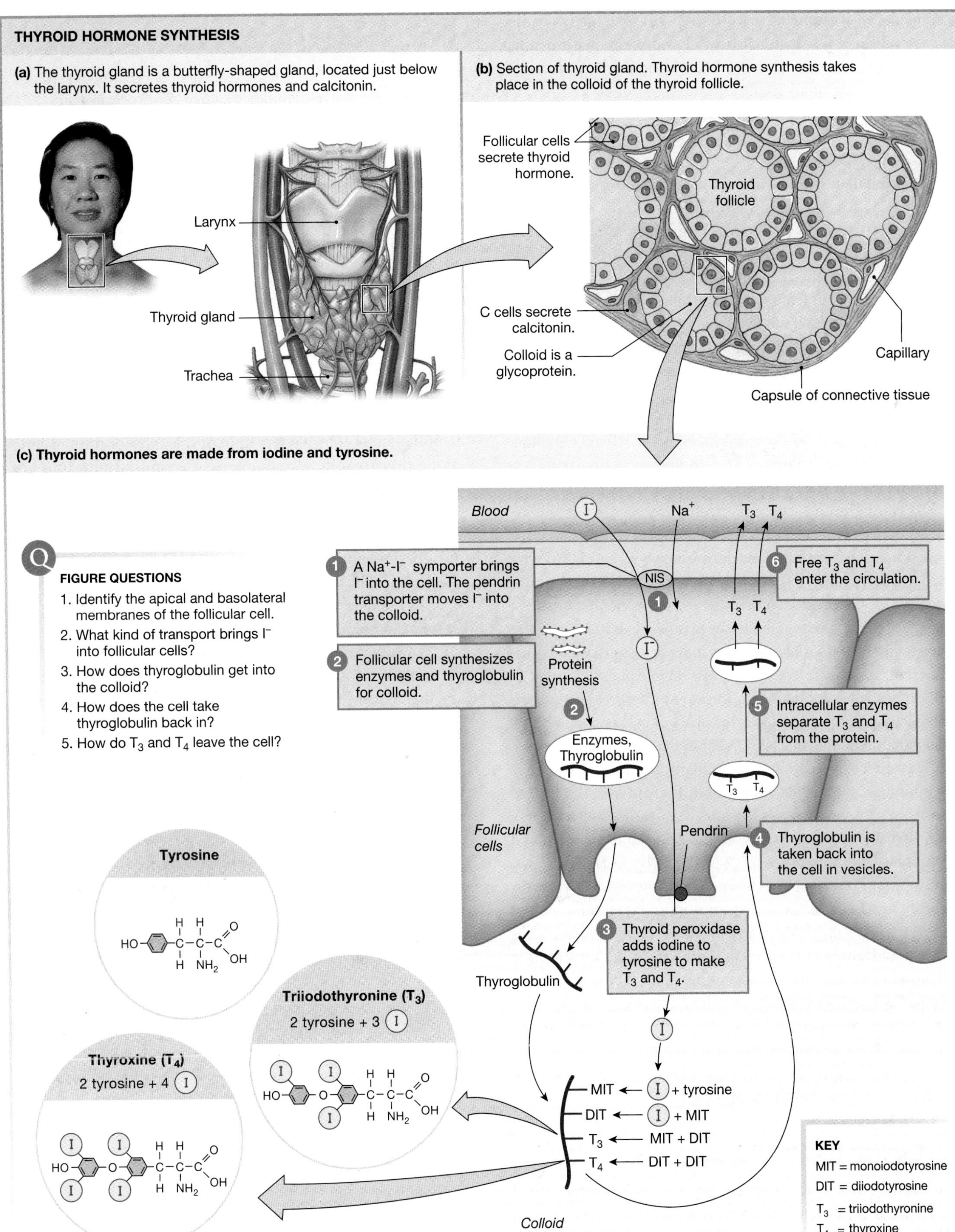
THYROID HORMONE SYNTHESIS
(a) The thyroid gland is a butterfly-shaped gland, located just below the larynx. It secretes thyroid hormones and calcitonin.
Larynx
Thyroid gland
Trachea
(b) Section of thyroid gland. Thyroid hormone synthesis takes place in the colloid of the thyroid follicle.
Follicular cells secrete thyroid hormone.
Thyroid follicle
C cells secrete calcitonin.
Colloid is a glycoprotein.
Capillary
Capsule of connective tissue
(c) Thyroid hormones are made from iodine and tyrosine.
Blood
Na+
T3 T4
1 A Na+-I− symporter brings I− into the cell. The pendrin transporter moves I− into the colloid.
NIS
6 Free T3 and T4 enter the circulation.
2 Follicular cell synthesizes enzymes and thyroglobulin for colloid.
Protein synthesis
Enzymes, Thyroglobulin
5 Intracellular enzymes separate T3 and T4 from the protein.
Follicular cells
Pendrin
4 Thyroglobulin is taken back into the cell in vesicles.
3 Thyroid peroxidase adds iodine to tyrosine to make T3 and T4.
Thyroglobulin
MIT ← I + tyrosine
DIT ← I + MIT
T3 ← MIT + DIT
T4 ← DIT + DIT
Colloid
KEY
MIT = monoiodotyrosine
DIT = diiodotyrosine
T3 = triiodothyronine
T4 = thyroxine
Tyrosine
Triiodothyronine (T3)
2 tyrosine + 3 I
Thyroxine (T4)
2 tyrosine + 4 I
FIGURE QUESTIONS
1. Identify the apical and basolateral membranes of the follicular cell.
2. What kind of transport brings I− into follicular cells?
3. How does thyroglobulin get into the colloid?
4. How does the cell take thyroglobulin back in?
5. How do T3 and T4 leave the cell?

Fig. 23.4

a single layer of epithelial cells (Fig. 23.4b). The hollow center of each follicle is filled with a sticky glycoprotein mixture called **colloid**. The colloid holds a 2–3 month supply of thyroid hormones at any one time.

The follicular cells surrounding the colloid manufacture a glycoprotein called **thyroglobulin** and enzymes for thyroid hormone synthesis (Fig. 23.4c (1)). These proteins are packaged into vesicles, then secreted into the center of the follicle. Follicular cells also actively concentrate dietary iodide, I^-, using the *sodium-iodide symporter* (NIS) (2). I^- transport into the colloid is mediated by an anion transporter known as *pendrin* (SLC26A4).

As I^- enters the colloid, the enzyme *thyroid peroxidase* removes an electron from the iodide ion and adds iodine to tyrosine on the thyroglobulin molecule (3). The addition of one iodine to tyrosine creates **monoiodotyrosine** (MIT). The addition of a second iodine creates **diiodotyrosine** (DIT). MIT and DIT then undergo *coupling reactions*. One MIT and one DIT combine to create the thyroid hormone **triiodothyronine**, or T_3. (Note the change from *tyrosine* to *thyronine* in the name.) Two DIT couple to form **tetraiodothyronine** (T_4, also known as **thyroxine**). At this point, the hormones are still attached to thyroglobulin.

When hormone synthesis is complete, the thyroglobulin–T_3/T_4 complex is taken back into the follicular cells in vesicles (4). There intracellular enzymes free the hormones T_3 and T_4 from the thyroglobulin protein (5). For many years scientists believed that the lipophilic nature of T_3 and T_4 allowed the hormones to diffuse out of the follicular cells and into the plasma, but current evidence indicates that the thyroid hormones move across cell membranes by protein carriers (6). Transporters for thyroid gland export of T_3 and T_4 have not been identified. Target tissue uptake transporters include an amino acid transporter (*MCT8*) and one member of the organic anion transporter (OAT) family.

T_3 and T_4 have limited solubility in plasma because they are lipophilic molecules. As a result, thyroid hormones bind to plasma proteins, such as **thyroid-binding globulin** (TBG). Most thyroid hormone in the plasma is in the form of T_4, and for years it was thought that T_4 was the active hormone. However, we now know that T_3 is three to five times more active biologically, and that it is the active hormone in target cells.

Target cells make about 85% of active T_3 by using enzymes called **deiodinases** to remove an iodine from T_4. Target tissue activation of the hormone adds another layer of control because individual target tissues can alter their exposure to active thyroid hormone by regulating their tissue deiodinase synthesis.

Thyroid receptors, with multiple isoforms, are in the nucleus of target cells. Hormone binding initiates transcription, translation, and synthesis of new proteins.

TSH Controls the Thyroid Gland

The control of thyroid hormone secretion follows the typical hypothalamic-pituitary-peripheral endocrine gland pattern (Fig. 23.5). **Thyrotropin-releasing hormone** (TRH) from the hypothalamus controls secretion of the anterior pituitary hormone **thyrotropin,** also known as **thyroid-stimulating hormone** (TSH). TSH in turns acts on the thyroid gland to promote hormone synthesis. The thyroid hormones normally act as a negative feedback signal to prevent oversecretion.

Thyroid hormones are not essential for life, but they do affect the quality of life if over- or under-secreted. Patients with thyroid excess or deficiency may experience decreased tolerance to heat or cold and mood disturbances, in addition to other symptoms.

The main function of thyroid hormones in adults is to provide substrates for oxidative metabolism. Thyroid hormones are thermogenic [p. 765] and increase oxygen consumption in most tissues. The exact mechanism is unclear but is at least partly related to changes in ion transport across the cell and mitochondrial membranes. Thyroid hormones also interact with other hormones to modulate protein, carbohydrate, and fat metabolism.

In children, thyroid hormones are necessary for full expression of growth hormone, which means they are essential for normal growth and development, especially of the nervous system. In the first few years after birth, myelin and synapse formation requires T_3 and T_4. Cytological studies suggest that thyroid hormones regulate microtubule assembly, which is an essential part of neuronal growth. Thyroid hormone is also necessary for proper bone growth.

The actions of thyroid hormones are most observable in people who secrete too much or too little hormone. Physiological effects that are subtle in normal people often become exaggerated in patients with endocrine disorders.

Thyroid Pathologies Affect Quality of Life

Problems with thyroid hormone secretion can arise either in the thyroid gland or along the control pathway depicted in Figure 23.5. The trophic action of TSH on the thyroid gland

RUNNING PROBLEM

Elevated blood Ca^{2+} leads to high Ca^{2+} concentrations in the kidney filtrate. Calcium-based kidney stones occur when calcium phosphate or calcium oxalate crystals form and aggregate with organic material in the lumen of the kidney tubule. Once Prof. Magruder's kidney stone passes into the urine, Dr. Spinks sends it for a chemical analysis.

Q3: Only free Ca^{2+} in the blood filters into Bowman's capsule at the nephron. A significant portion of plasma Ca^{2+} cannot be filtered. Use what you have learned about filtration at the glomerulus to speculate on why some plasma Ca^{2+} cannot filter [p. 635].

776 780 784 793 796 797

THYROID HORMONE CONTROL PATHWAY

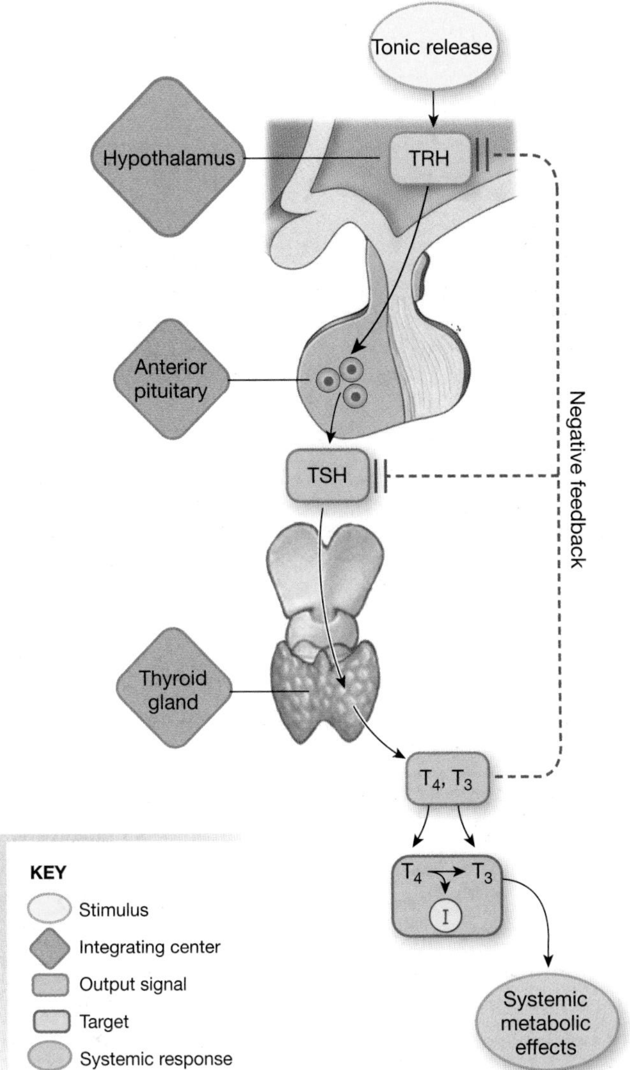

Thyroid Hormones	
Cell of origin	Thyroid follicle cells
Chemical nature	Iodinated amine
Biosynthesis	From iodine and tyrosine. Formed and stored on thyroglobulin in follicle colloid.
Transport in the circulation	Bound to thyroxine-binding globulin and albumins
Half-life	6-7 days for thyroxine (T_4); about 1 day for triiodothyronine (T_3)
Factors affecting release	Tonic release
Control pathway	TRH (hypothalamus) → TSH (anterior pituitary) → T_3 + T_4 (thyroid) → T_4 deiodinates in tissues to form more T_3
Target cells or tissues	Most cells of the body
Target receptor	Nuclear receptor
Whole body or tissue reaction	↑ Oxygen consumption (thermogenesis). Protein catabolism in adults but anabolism in children. Normal development of nervous system
Action at cellular level	Increases activity of metabolic enzymes and Na^+-K^+-ATPase
Action at molecular level	Production of new enzymes
Feedback regulation	T_3 has negative feedback on anterior pituitary and hypothalamus.

Fig. 23.5

causes enlargement, or *hypertrophy,* of follicular cells. In pathological conditions with elevated TSH levels, the thyroid gland will enlarge, a condition known as a **goiter**. A large goiter can weigh hundreds of grams and almost encircle the neck (Fig. 23.6a).

Goiters are the result of excess TSH stimulation of the thyroid gland. Simply knowing that someone has a goiter does not tell you what the pathology is, however. Let's see how both hypothyroidism and hyperthyroidism can be associated with goiter.

Hyperthyroidism A person whose thyroid gland secretes too much hormone suffers from **hyperthyroidism.** Excess thyroid hormone causes changes in metabolism, the nervous system, and the heart.

1 Hyperthyroidism increases oxygen consumption and metabolic heat production. Because of the internal heat generated, these patients have warm, sweaty skin and may complain of being intolerant of heat.

2 Excess thyroid hormone increases protein catabolism and may cause muscle weakness. Patients often report weight loss.

3 The effects of excess thyroid hormone on the nervous system include hyperexcitable reflexes and psychological disturbances ranging from irritability and insomnia to psychosis. The mechanism for psychological disturbances is unclear, but morphological changes in the hippocampus and effects on β-adrenergic receptors have been suggested.

4 Thyroid hormones are known to influence β-adrenergic receptors in the heart, and these effects are exaggerated with hypersecretion. A common sign of hyperthyroidism is rapid heartbeat and increased force of contraction due to up-regulation of β_1-receptors on the myocardium [p. 503].

SIGNS OF THYROID PATHOLOGIES

(a) **Goiter.** Excessive stimulation of the thyroid gland by TSH causes the gland to enlarge (goiter).

(b) **Myxedema.** In hypothyroid individuals, mucopolysaccharide deposits beneath the skin may cause bags under the eyes.

(c) **Exophthalmos.** In hyperthyroid states, excessive deposition of mucopolysaccharides in the bony orbit may cause the bulging eyeball called exophthalmos.

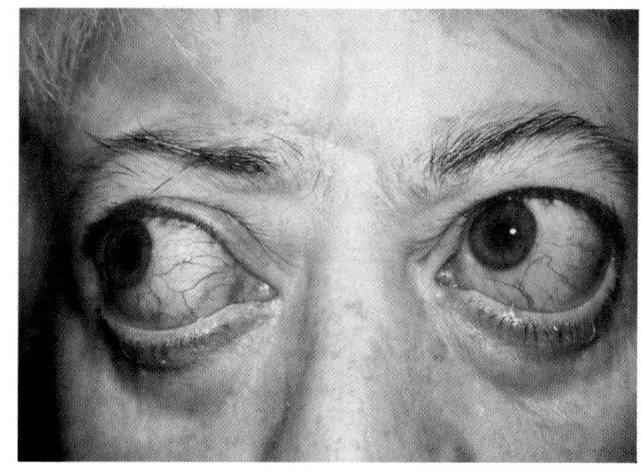

Fig. 23.6

The most common cause of hyperthyroidism is *Graves' disease* (Fig. 23.7a). In this condition, the body produces antibodies called **thyroid-stimulating immunoglobulins**, or **TSI**. These antibodies mimic the action of TSH by combining with and activating TSH receptors on the thyroid gland. The result is goiter, hypersecretion of T_3 and T_4, and symptoms of hormone excess.

Negative feedback by the high levels of T_3 and T_4 shuts down the body's TRH and TSH secretion but does nothing to block the TSH-like activity of TSI on the thyroid gland. Graves' disease is often accompanied by **exophthalmos** (Fig. 23.6c), a bug-eyed appearance caused by immune-mediated enlargement of muscles and tissue in the eye socket. The English comic Marty Feldman was known for his wild-eyed appearance caused by exophthalmos.

Thyroid gland tumors are another cause of primary hyperthyroidism. Secondary hyperthyroidism will occur with pituitary tumors secreting TSH.

Hypothyroidism Hyposecretion of thyroid hormones affects the same systems altered by hyperthyroidism.

1 Decreased thyroid hormone secretion slows metabolic rate and oxygen consumption. Patients become intolerant of cold because they are generating less internal heat.

2 Hypothyroidism decreases protein synthesis. In adults, this causes brittle nails, thinning hair, and dry, thin skin. Hypothyroidism also causes accumulation of *mucopolysaccharides* under the skin. These molecules attract water and cause the puffy appearance of *myxedema* (Fig. 23.6b). Hypothyroid children have slow bone and tissue growth and are shorter than normal for their age.

3 Nervous system changes in adults include slowed reflexes, slow speech and thought processes, and feelings of fatigue. Deficient thyroid hormone secretion in infancy causes **cretinism,** a condition marked by decreased mental capacity.

4 The primary cardiovascular change in hypothyroidism is *bradycardia* (slow heart rate).

Primary hypothyroidism is most commonly caused by a lack of iodine in the diet. Without iodine, the thyroid gland cannot make thyroid hormones (Fig. 23.7b). Low levels of T_3 and T_4 in the blood mean no negative feedback to the hypothalamus and anterior pituitary. In the absence of negative feedback, TSH secretion rises dramatically, and TSH stimulation enlarges the thyroid gland (goiter). Despite hypertrophy, the gland cannot obtain iodine to make hormone, so the patient remains hypothyroid. These patients exhibit the previously described signs of hypothyroidism. The goiter shown in the photograph of Figure 23.6a is probably due to iodine deficiency.

Therapy for thyroid disorders depends on the cause of the problem. Hypothyroidism is treated with oral thyroxine (T_4). Hyperthyroidism can be treated by surgical removal of all or part of the gland, by destruction of thyroid cells with radioactive iodine, or by drugs that block either hormone synthesis (thiourea drugs) or peripheral conversion of T_4 to T_3 (propylthiouracil).

Concept Check Answers: p. 800

9. A woman who had her thyroid gland removed because of cancer was given pills containing only T_4. Why was this less active form of the hormone an effective treatment for her hypothyroidism?

10. Why would excessive production of thyroid hormone, which uncouples mitochondrial ATP production and proton transport [p. 115], cause a person to become intolerant of heat?

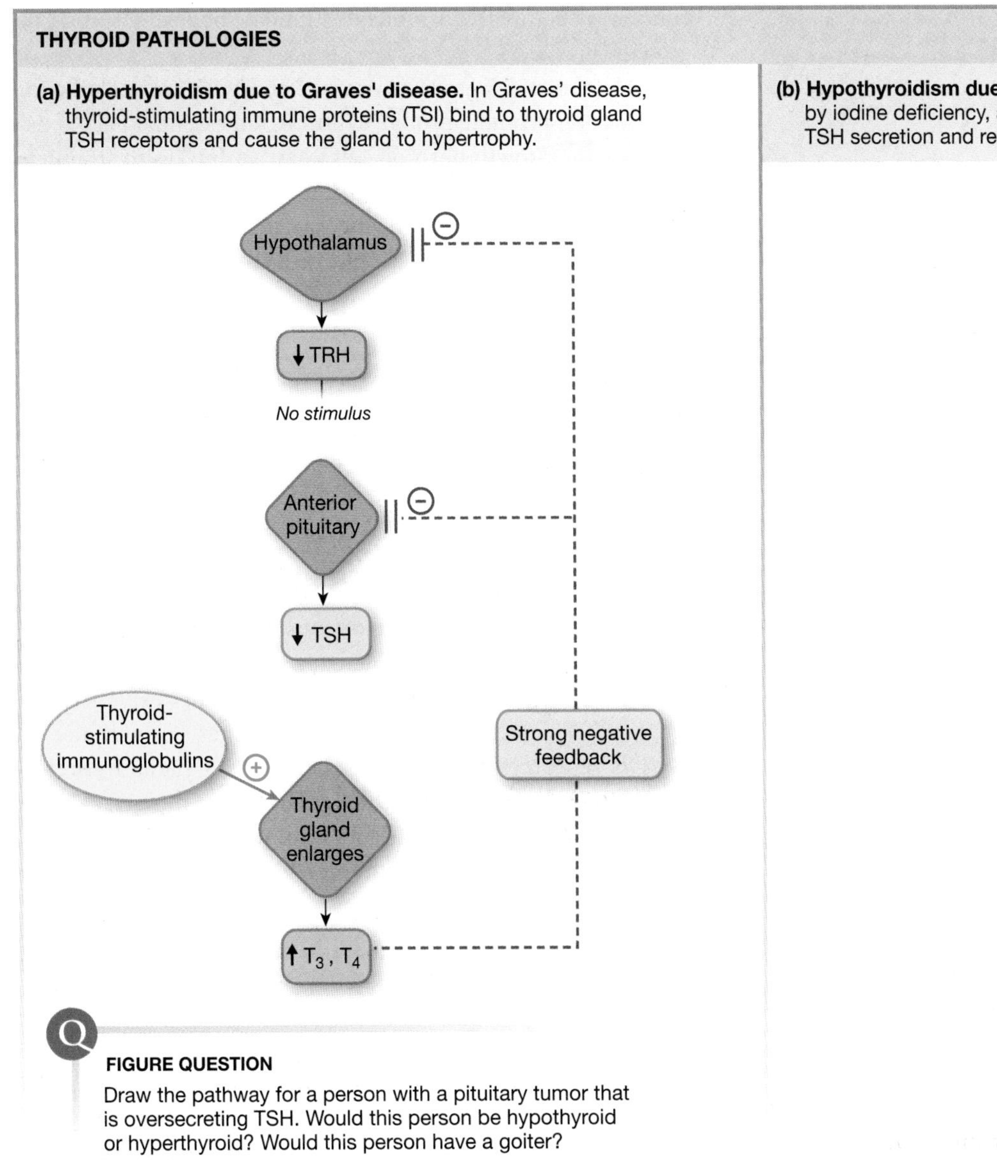

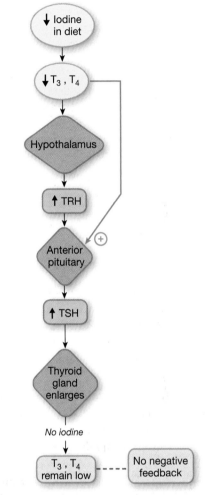

FIGURE QUESTION

Draw the pathway for a person with a pituitary tumor that is oversecreting TSH. Would this person be hypothyroid or hyperthyroid? Would this person have a goiter?

Fig. 23.7

Growth Hormone

Growth in human beings is a continuous process that begins before birth. However, growth rates in children are not steady, with the first two years of life and the adolescent years marked by spurts of rapid growth and development. Normal growth is a complex process that depends on a number of factors:

1. **Growth hormone and other hormones.** Without adequate amounts of growth hormone, children simply fail to grow. Thyroid hormones, insulin, and the sex hormones at puberty also play both direct and permissive roles. A deficiency in any one of these hormones leads to abnormal growth and development.
2. **An adequate diet** that includes protein, sufficient energy (caloric intake), vitamins, and minerals. Many amino acids can be manufactured in the body from other precursors, but essential amino acids must come from dietary sources. Among the minerals, calcium in particular is needed for proper bone formation.
3. **Absence of chronic stress.** Cortisol from the adrenal cortex is released in times of stress and has significant catabolic effects that inhibit growth. Children who are subjected to stressful environments may exhibit a condition known as *failure to thrive* that is marked by abnormally slow growth.
4. **Genetics.** Each human's potential adult size is genetically determined at conception.

Growth Hormone Is Anabolic

Growth hormone (GH or somatotropin [p. 221]) is released throughout life, although its biggest role is in children. Peak GH secretion occurs during the teenage years. The stimuli for growth hormone release are complex and not well

CLINICAL FOCUS

New Growth Charts

When you were growing up, did your family mark your growth each year on a special wall chart? Monitoring growth is an important part of health care for children and adolescents, particularly as we see a growing problem with childhood obesity in the United States. In 2000 the U.S. Centers for Disease Control and Prevention (CDC) issued new growth charts for the first time since 1977. In 2006 they recommended that clinicians use an international chart from the World Health Organization for children under two years of age. The old charts were based on 1929–1979 data from mostly bottle-fed, middle-class white children. We now know that breast-fed babies grow more rapidly than bottle-fed infants in the first two months, then more slowly for the remainder of the first year. We also have data showing that babies in lower socioeconomic groups grow more slowly. The new charts take these differences into account and also include body mass index (BMI) information up to age 20. To see the new charts and learn more about monitoring growth in infants and children, visit the CDC web site at *www.cdc.gov/growthcharts.*

understood, but they include circulating nutrients, stress, and other hormones interacting with a daily rhythm of secretion (Fig. 23.8).

The stimuli for GH secretion are integrated in the hypothalamus, which secretes two neuropeptides into the hypothalamic-hypophyseal portal system: **growth hormone–releasing hormone** (GHRH) and *growth hormone–inhibiting hormone,* better known as *somatostatin* (SS). On a daily basis, pulses of GHRH from the hypothalamus stimulate GH release. In adults, the largest pulse of GH release occurs in the first two hours of sleep. It is speculated that GHRH has sleep-inducing properties, but the role of GH in sleep cycles is unclear.

GH is secreted by cells in the anterior pituitary. It is a typical peptide hormone in most respects, except that nearly half the GH in blood is bound to a plasma **growth hormone–binding protein**. The binding protein protects plasma GH from being filtered into the urine and extends its half-life by 12 minutes. Researchers have hypothesized that genetic determination of binding protein concentration plays a role in determining adult height.

The target tissues for GH include both endocrine and nonendocrine cells. GH acts as a trophic hormone to stimulate secretion of **insulin-like growth factors** (IGFs; formerly called *somatomedins*) from the liver and other tissues. IGFs have a negative feedback effect on growth hormone secretion by acting on the anterior pituitary and on the hypothalamus. IGFs act in concert with growth hormone to stimulate bone and soft tissue growth (Fig. 23.8).

Metabolically, growth hormone and IGFs are anabolic for proteins and promote protein synthesis, an essential part of tissue growth. Growth hormone also acts with IGFs to stimulate bone growth. IGFs are responsible for cartilage growth. GH increases plasma fatty acid and glucose concentrations by promoting fat breakdown and hepatic glucose output.

Concept Check Answer: p. 800

11. Which pituitary hormone in addition to GH has two hypothalamic factors that regulate its release?

Growth Hormone Is Essential for Normal Growth

The disorders that reflect the actions of growth hormone are most obvious in children. Severe growth hormone deficiency in childhood leads to **dwarfism,** which can result from a problem either with growth hormone synthesis or with defective GH receptors. Unfortunately, neither bovine nor porcine growth hormone is effective as replacement therapy, as only primate growth hormone is active in humans. Prior to 1985, when genetically engineered human growth hormone became available, donated human pituitaries harvested at autopsy were the only source of growth hormone. Fortunately, severe growth hormone deficiency is relatively rare. At the opposite extreme, oversecretion of growth hormone in children leads to **giantism**.

Once bone growth stops in late adolescence, growth hormone cannot further increase height. GH and IGFs can continue to act on cartilage and soft tissues, however. Adults with excessive secretion of growth hormone develop a condition known as **acromegaly,** characterized by lengthening of the jaw, coarsening of facial features, and growth of hands and feet (Fig. 23.9). Andre the Giant, a French wrestler who also had a role in the classic movie *The Princess Bride,* exhibited signs of both giantism (he grew to 7′4″ tall) and acromegaly before his death at age 47.

Genetically Engineered hGH Raises Ethical Questions

When genetically engineered human growth hormone (hGH) became available in the mid-1980s, the medical profession was faced with a dilemma. Obviously the hormone should be used to treat children who would otherwise be dwarfs, but what about children with only partial GH deficiency or genetically short children with normal GH secretion levels? This question is complicated by the difficulty of accurately identifying children with partial growth hormone deficiency. And what about

GROWTH HORMONE

Growth hormone control pathway

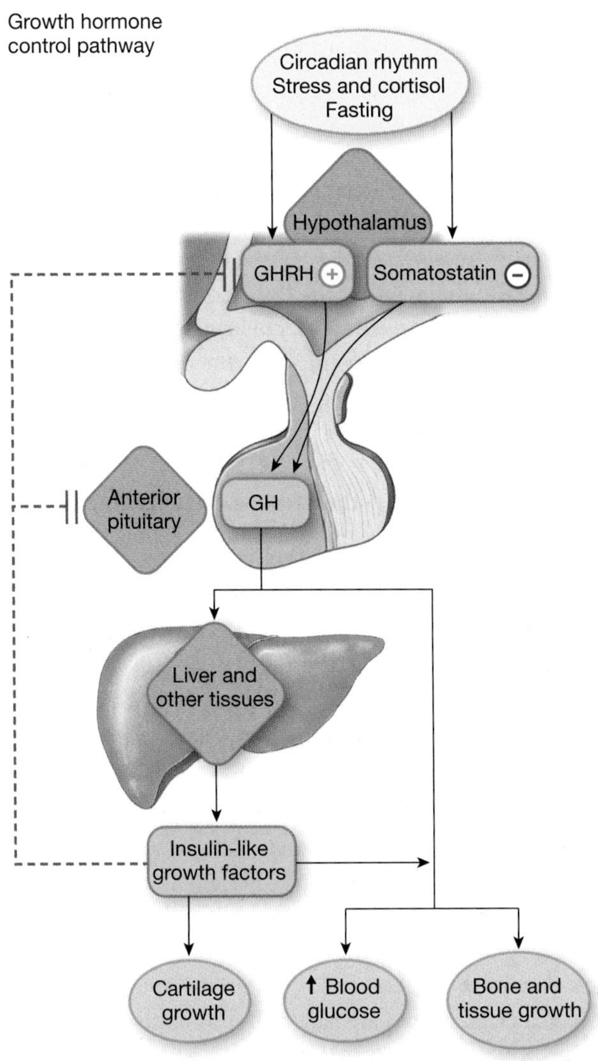

GROWTH HORMONE (hGH)	
Origin	Anterior pituitary
Chemical nature	191-amino acid peptide; several closely related forms
Biosynthesis	Typical peptide
Transport in the circulation	Half is dissolved in plasma, half is bound to a binding protein whose structure is identical to that of the GH receptor
Half-life	18 minutes
Factors affecting release	Circadian rhythm of tonic secretion; influenced by circulating nutrients, stress, and other hormones in a complex fashion
Control pathway	GHRH, somatostatin (hypothalamus) → growth hormone (anterior pituitary)
Target cells or tissues	Trophic on liver for insulin-like growth factor production; also acts directly on many cells
Target receptor	Membrane receptor with tyrosine kinase activity
Whole body or tissue reaction (with IGFs)	Bone and cartilage growth; soft tissue growth; ↑ plasma glucose
Action at cellular level	Receptor linked to kinases that phosphorylate proteins to initiate transcription

Fig. 23.8

ACROMEGALY

Excess growth hormone secretion in adults causes acromegaly, with lengthening of the jaw, coarsening of the features, and growth in hands and feet. These photographs show the progression of acromegaly in one woman.

Age 16

Age 33

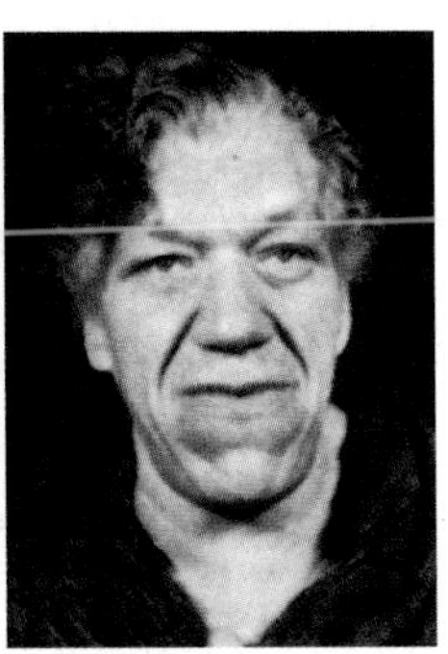
Age 52

Fig. 23.9

children whose parents want them to be taller? Should these healthy children be given the hormone?

In 2003 the U.S. Food and Drug Administration approved use of a recombinant human growth hormone for treating children with non-GH-deficient short stature, defined as being more than 2.25 standard deviations below the mean height for their age and sex. (This means children in the bottom 1% of their age-sex group.) In clinical trials, daily injections of the drug for two years resulted in an average height increase of 1.3" (3.3 cm). According to a 2006 analysis in a pediatric medicine journal, the cost for this treatment was more than $52,000 per inch of height gained. Side effects reported during hGH studies include glucose intolerance [p. 761] and *pancreatitis* (inflammation of the pancreas). Long-term risks associated with hGH treatment are unknown, and parents must be made aware that hGH therapy has the potential to create psychological problems in children if the results are less than optimum.

Tissue and Bone Growth

Growth can be divided into two general areas: soft tissue growth and bone growth. In children, bone growth is usually assessed by measuring height, and tissue growth by measuring weight. Multiple hormones have direct or permissive effects on growth. In addition, we are just beginning to understand how paracrine growth factors interact with classic hormones to influence tissue development and differentiation.

Tissue Growth Requires Hormones and Paracrines

Soft tissue growth requires adequate amounts of growth hormone, thyroid hormone, and insulin. Growth hormone and IGFs are required for tissue protein synthesis and cell division. Under the influence of these hormones, cells undergo both *hypertrophy* (increased cell size) and **hyperplasia** (increased cell number).

Thyroid hormones play a permissive role in growth and contribute directly to nervous system development. At the target tissue level, thyroid hormone interacts synergistically with growth hormone in protein synthesis and nervous system development. Children with untreated hypothyroidism (cretinism) do not grow to normal height even if they secrete normal amounts of growth hormone.

Insulin supports tissue growth by stimulating protein synthesis and providing energy in the form of glucose. Because insulin is permissive for growth hormone, insulin-deficient children fail to grow normally even though they may have normal concentrations of growth and thyroid hormones.

Bone Growth Requires Adequate Dietary Calcium

Bone growth, like soft tissue development, requires the proper hormones and adequate amounts of protein and calcium. Bone contains calcified extracellular matrix formed when calcium phosphate crystals precipitate and attach to a collagenous lattice support. The most common form of calcium phosphate is **hydroxyapatite,** $Ca_{10}(PO_4)_6(OH)_2$.

Although the large amount of inorganic matrix in bone makes some people think of it as nonliving, bone is a dynamic tissue, constantly being formed and broken down, or *resorbed.* Spaces in the collagen-calcium matrix are occupied by living cells that are well supplied with oxygen and nutrients by blood vessels that run through adjacent channels (■ Fig. 23.10).

Bones generally have two layers: an outer layer of dense **compact bone** and an inner layer of spongy **trabecular bone**. In some bones, a central cavity is filled with bone marrow. Compact bone provides strength and is thickest where support is needed (such as in the long bones of the legs) or where muscles attach. Trabecular bone is less sturdy and has open, cell-filled spaces between struts of calcified lattice.

Bones grow when matrix is deposited faster than it is resorbed. Specialized bone-forming cells called **osteoblasts** produce enzymes and **osteoid,** a mixture of collagen and other proteins to which hydroxyapatite binds. Recent research has found two other proteins, *osteocalcin* and *osteonectin,* that appear to aid in deposition of the calcified matrix.

Bone diameter increases when matrix deposits on the outer surface of the bone. Linear growth of long bones occurs at specialized regions called **epiphyseal plates,** located at each end of the bone shaft (*diaphysis*) (Fig. 23.10b). The side of the plate closer to the end (*epiphysis*) of the bone contains continuously dividing columns of **chondrocytes,** collagen-producing cells of cartilage. As the collagen layer thickens, the older cartilage calcifies and older chondrocytes degenerate, leaving spaces that osteoblasts invade. The osteoblasts then lay down bone matrix on top of the cartilage base.

As new bone is added at the ends, the shaft lengthens. Long bone growth continues as long as the epiphyseal plate is active. When osteoblasts complete their work, they revert to a less active form known as **osteocytes.**

Growth of long bone is under the influence of growth hormone and the insulin-like growth factors. In the absence of these hormones, normal bone growth does not occur. Long bone growth is also influenced by steroid sex hormones. The growth spurt of adolescent boys used to be attributed solely to increased androgen production but it now appears that estrogens play a significant role in pubertal bone growth in both sexes.

In all adolescents, the sex hormones eventually inactivate the epiphyseal plate so that long bones no longer grow. Because the epiphyseal plates of various bones close in a regular, ordered sequence, X-rays that show which plates are open and which have closed can be used to calculate a child's "bone age."

Linear bone growth ceases in adults, but bones are dynamic tissues that undergo continual remodeling throughout life. The *resorption* or breakdown of bone is controlled by **osteoclasts,** large, mobile, multinucleate cells derived from hematopoietic stem cells [p. 547]. Osteoclasts are responsible for dissolving bone.

Osteoclasts attach around their periphery to a section of matrix, much like a suction cup (Fig. 23.10c). The central region of the osteoclast secretes hydrochloric acid with the aid of carbonic anhydrase and an H^+-ATPase. Osteoclasts also secrete *protease* enzymes that work at low pH. The combination of acid and enzymes dissolves the calcified hydroxyapatite matrix and its collagen support. Ca^{2+} from hydroxyapatite becomes part of the ionized Ca^{2+} pool and can enter the blood.

Bone mass in the body is another example of mass balance. In children, bone deposition exceeds bone resorption, and bone mass increases. In young adults up to about age 30, deposition and resorption are balanced. From age 30 on, resorption begins to exceed deposition, with concurrent loss of bone from the skeleton. Bone loss and osteoporosis are discussed in more detail at the end of this chapter.

■ Fig. 23.10 *ESSENTIALS*

Bone

(a) Composition of bone

Bone is composed largely of calcified extracellular matrix.

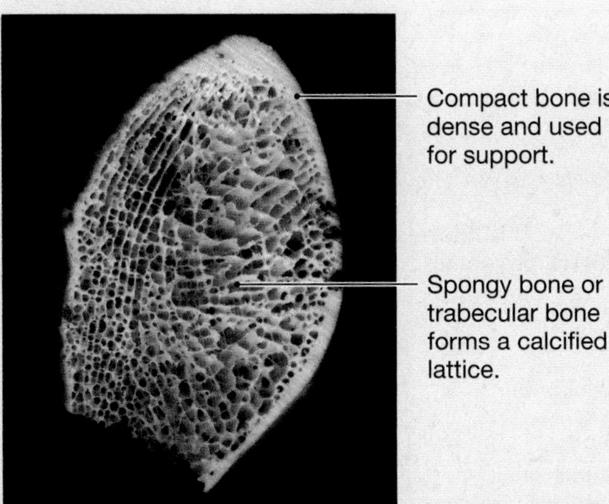

Compact bone is dense and used for support.
Spongy bone or trabecular bone forms a calcified lattice.

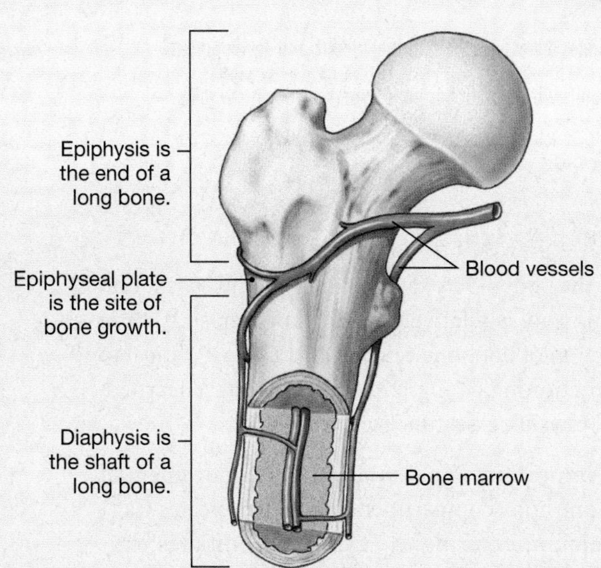

Epiphysis is the end of a long bone.
Blood vessels
Epiphyseal plate is the site of bone growth.
Diaphysis is the shaft of a long bone.
Bone marrow

(b) Bone growth

Chondrocytes form cartilage. Osteoblasts create calcium phosphate crystals to replace cartilage.

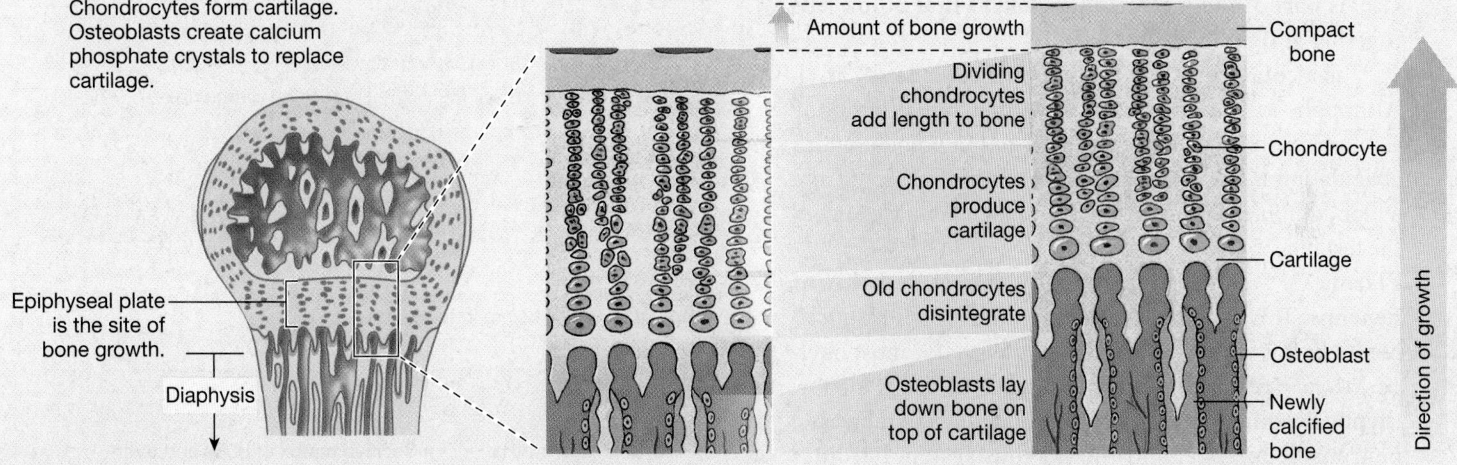

(c) Bone resorption

Osteoclasts are responsible for bone resorption. Osteoclasts secrete acid and enzymes that dissolve calcium phosphate in bone.

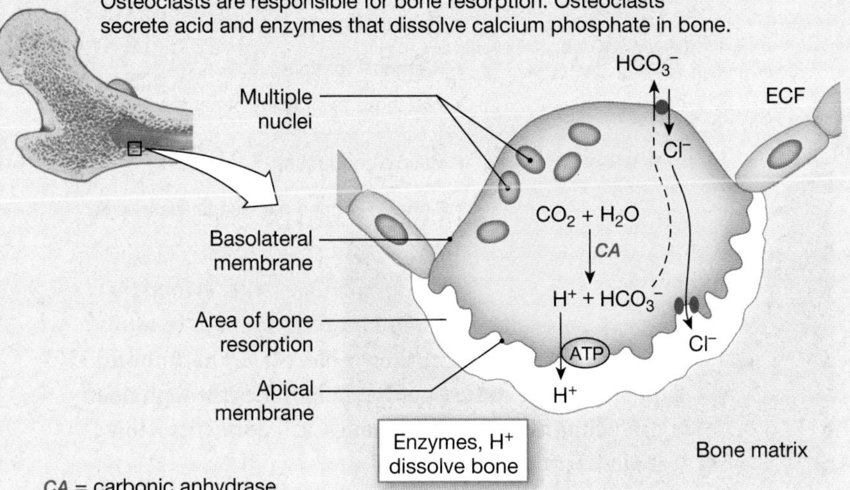

CA = carbonic anhydrase

These photographs dramatically illustrate why people with osteoporosis have a high incidence of bone fractures.

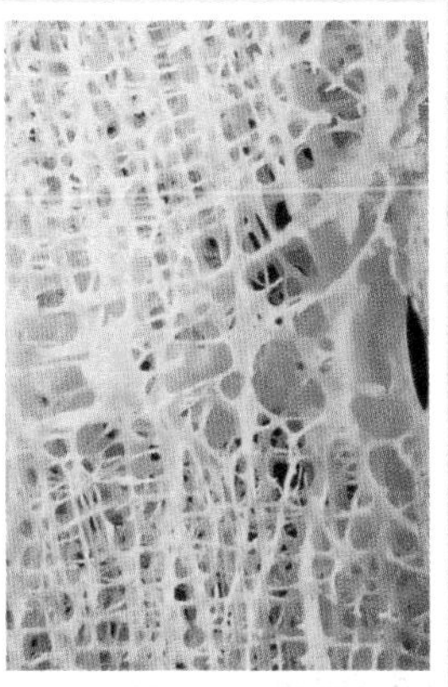

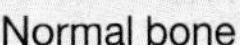

Normal bone

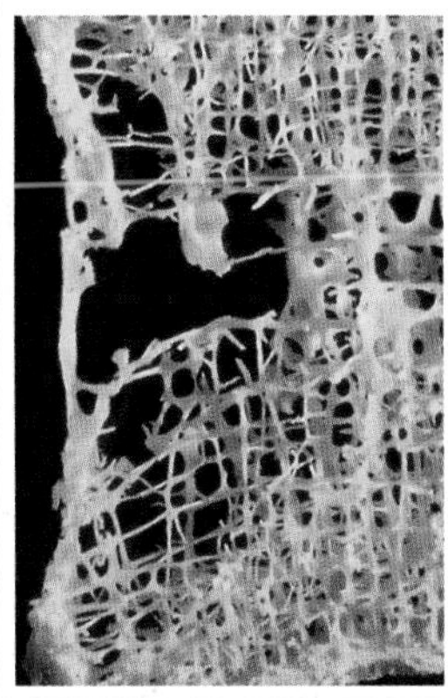

Bone loss in osteoporosis

Concept Check Answers: p. 800

12. Which hormones are essential for normal growth and development?
13. Why don't adults with growth hormone hypersecretion grow taller?

Calcium Balance

Most calcium in the body—99%, or nearly 2.5 pounds—is found in the bones. This pool is relatively stable, however, so it is the body's small fraction of nonbone calcium that is most critical to physiological functioning (■ Fig. 23.11). As you have learned, Ca^{2+} has several physiological functions:

1. **Ca^{2+} is an important signal molecule.** The movement of Ca^{2+} from one body compartment to another creates Ca^{2+} signals. Calcium entering the cytoplasm initiates exocytosis of synaptic and secretory vesicles, contraction in muscle fibers, or altered activity of enzymes and transporters. Removal of Ca^{2+} from the cytoplasm requires active transport.
2. **Ca^{2+} is part of the intercellular cement that holds cells together at tight junctions.**
3. **Ca^{2+} is a cofactor in the coagulation cascade** [p. 561]. Although Ca^{2+} is essential for blood coagulation, body Ca^{2+} concentrations never decrease to the point at which coagulation is inhibited. However, removal of Ca^{2+} from a blood sample will prevent the specimen from clotting in the test tube.
4. **Plasma Ca^{2+} concentrations affect the excitability of neurons.** This function of Ca^{2+} has not been introduced before in this text, but it is the function that is most obvious in Ca^{2+}-related disorders. If plasma Ca^{2+} falls too low (**hypocalcemia**), neuronal permeability to Na^+ increases, neurons depolarize, and the nervous system becomes hyperexcitable. In its most extreme form, hypocalcemia causes sustained contraction (*tetany*) of the respiratory muscles, resulting in asphyxiation. **Hypercalcemia** has the opposite effect, depressing neuromuscular activity.

Plasma Calcium Is Closely Regulated

Because calcium is critical to so many physiological functions, the body's plasma Ca^{2+} concentration is very closely regulated. Calcium homeostasis follows the principle of mass balance:

$$\text{Total body calcium} = \text{intake} - \text{output}$$

1. **Total body** Ca^{2+} is all the calcium in the body, distributed among three compartments (Fig. 23.11):
   a. *Extracellular fluid.* Ionized Ca^{2+} is concentrated in the ECF. In the plasma, nearly half the Ca^{2+} is bound to plasma proteins and other molecules. The unbound Ca^{2+} is free to diffuse across membranes through open Ca^{2+} channels. Total plasma Ca^{2+} concentration is about 2.5 mM.

CALCIUM BALANCE IN THE BODY

(a) To maintain calcium balance, dietary intake should equal Ca^{2+} loss in the urine and feces.

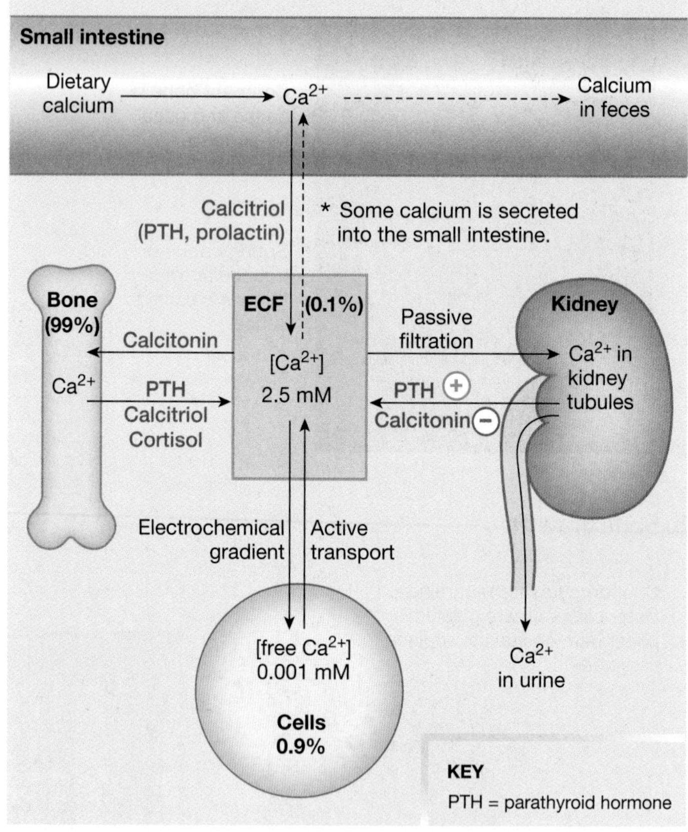

(b) Functions of calcium in the body

Location	Function
Extracellular matrix	• Cacified matrix of bone and teeth
Extracellular fluid Ca^{2+}	• Neurotransmitter release at synapse • Role in myocardial and smooth muscle contraction • Cofactor in coagulation cascade • "Cement" for tight junctions • Influences excitability of neurons
Intracellular Ca^{2+}	• Muscle contraction • Signal in second messenger pathways

■ **Fig. 23.11**

b. *Intracellular Ca^{2+}*. The concentration of free Ca^{2+} in the cytosol is about 0.001 mM. In addition, Ca^{2+} is concentrated inside mitochondria and the sarcoplasmic reticulum. Electrochemical gradients favor movement of Ca^{2+} into the cytosol when Ca^{2+} channels open.

c. *Extracellular matrix (bone)*. Bone is the largest Ca^{2+} reservoir in the body, with most bone Ca^{2+} in the form of hydroxyapatite crystals. Bone Ca^{2+} forms a reservoir that can be tapped to maintain plasma Ca^{2+} homeostasis. Usually only a small fraction of bone Ca^{2+} is ionized and readily exchangeable, and this pool remains in equilibrium with Ca^{2+} in the interstitial fluid.

2 **Intake** is the Ca^{2+} ingested in the diet and absorbed in the small intestine. Only about one-third of ingested Ca^{2+} is absorbed, and unlike organic nutrients, Ca^{2+} absorption is hormonally regulated. Many people do not eat enough Ca^{2+}-containing foods, however, and intake may not match output.

Intestinal calcium absorption is apparently both transcellular and paracellular (between the cells). Transcellular transport is accomplished by entry into the enterocyte through apical Ca^{2+} channels (*TRPV6, also called ECaC*) and exit through basolateral Na^{+}-Ca^{2+} exchanger (NCX) and Ca^{2+}-ATPase transporters.

3 **Output**, or Ca^{2+} loss from the body, occurs primarily through the kidneys, with a small amount excreted in feces. Ionized Ca^{2+} is freely filtered at the glomerulus and then reabsorbed along the length of the nephron. Hormonally regulated reabsorption takes place only in the distal nephron and uses transporters similar to those found in the intestine. There is no paracellular transport in the kidney.

Concept Check Answers: p. 800

14. What does hypercalcemia do to neuronal membrane potential, and why does that effect depress neuromuscular excitability?

15. Draw a picture of a distal tubule cell and label apical and basolateral membranes, lumen, and ECF. Use the description of intestinal Ca^{2+} absorption in list item 2 (Intake) to draw the appropriate transporters.

16. Describe the renal transport of Ca^{2+} from the tubule lumen to the ECF as active, passive, facilitated diffusion, and so on.

Three Hormones Control Calcium Balance

Three hormones regulate the movement of Ca^{2+} between bone, kidney, and intestine: parathyroid hormone, calcitriol (vitamin D_3), and calcitonin (Fig. 23.11). Of these, parathyroid hormone and calcitriol are the most important in adult humans.

The **parathyroid glands,** which secrete parathyroid hormone {*para-*, alongside of}, were discovered in the 1890s by physiologists studying the role of the thyroid gland. These scientists noticed that if they removed all of the thyroid gland from dogs and cats, the animals died in a few days. In contrast, rabbits died only if the little parathyroid "glandules" alongside the thyroid were removed. The scientists then looked for parathyroid glands in dogs and cats and found them tucked away behind the larger thyroid gland. If the parathyroid glands were left behind when the thyroid was surgically removed, the animals lived. The scientists concluded that the parathyroids contained a substance that was essential for life, although the thyroid gland did not. That essential substance was parathyroid hormone.

RUNNING PROBLEM

Prof. Magruder's blood work reveals that his Ca^{2+} level is 12.3 mg/dL plasma (normal: 8.5–10.5 mg/dL). These results support the suspected diagnosis of hyperparathyroidism. "Do you take vitamin D or use a lot of antacids?" Dr. Spinks asks. "Those could raise your blood calcium." Prof. Magruder denies using either substance. "Well, we need one more test before we can say conclusively that you have hyperparathyroidism," Dr. Spinks says.

Q4: What one test could definitively prove that Prof. Magruder has hyperparathyroidism?

776 780 784 793 796 797

Parathyroid Hormone Four small parathyroid glands lie on the dorsal surface of the thyroid gland (Fig. 23.12). They secrete **parathyroid hormone** (PTH, also called *parathormone*), a peptide whose main function is to increase plasma Ca^{2+} concentrations. The stimulus for PTH release is a decrease in plasma Ca^{2+}, monitored by a cell membrane **Ca^{2+}-sensing receptor** (CaSR). The CaSR, a G protein–coupled receptor, was the first membrane receptor identified whose ligand was an ion rather than an organic molecule.

PTH acts on bone, kidney, and intestine to increase plasma Ca^{2+} concentrations (Fig. 23.12). Increased plasma Ca^{2+} acts as negative feedback and shuts off PTH secretion. Parathyroid hormone raises plasma Ca^{2+} in three ways:

1 **PTH mobilizes calcium from bone.** Increased bone resorption by osteoclasts takes about 12 hours to become measurable. Curiously, although osteoclasts are responsible for dissolving the calcified matrix and would be logical targets for PTH, they do not have PTH receptors. Instead, PTH effects are mediated by a collection of paracrines, including *osteoprotegerin* (OPG) and an osteoclast differentiation factor called *RANKL*. These paracrine factors

PARATHYROID GLANDS AND PARATHYROID HORMONE (PTH)

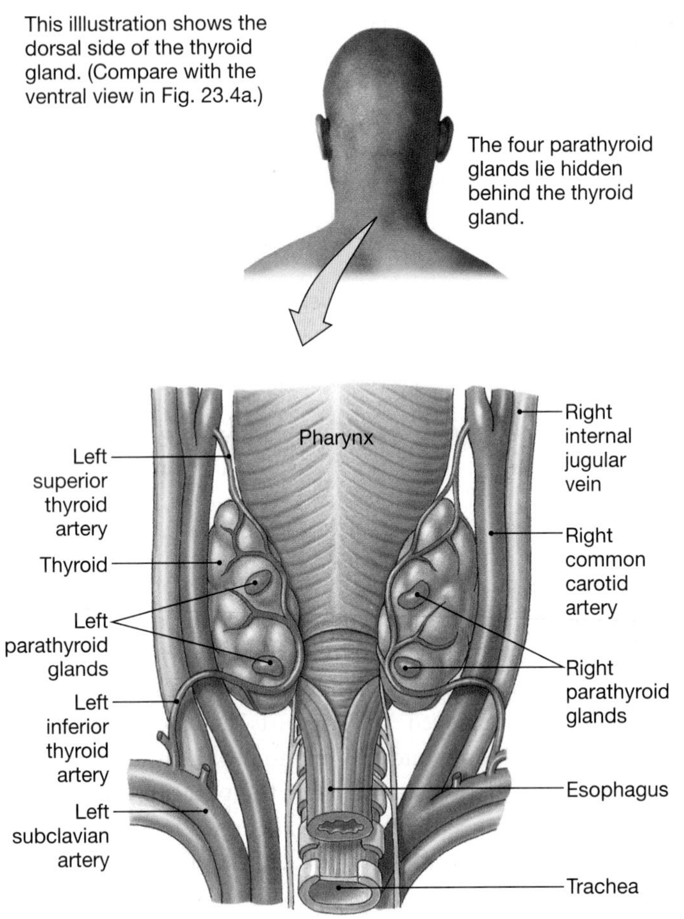

PARATHYROID HORMONE (PTH)	
Origin	Parathyroid glands
Chemical nature	84-amino acid peptide
Biosynthesis	Continuous production, little stored
Transport in the circulation	Dissolved in plasma
Half-life	Less than 20 minutes
Factors affecting release	↓ Plasma Ca^{2+}
Target cells or tissues	Kidney, bone, intestine
Target receptor	Membrane receptor acts via cAMP
Whole body or tissue reaction	↑ Increased plasma Ca^{2+}
Action at cellular level	↑ Vitamin D synthesis; ↑ renal reabsorption of Ca^{2+}; ↑bone resorption
Action at molecular level	Rapidly alters Ca^{2+} transport but also initiates protein synthesis in osteolcasts
Onset of action	2–3 hours for bone, with increased osteoclast activity requiring 1–2 hours. 1–2 days for intestinal absorption. Within minutes for renal transport
Feedback regulation	Negative feedback by ↑ plasma Ca^{2+}
Other information	Osteoclasts have no PTH receptors and are regulated by PTH-induced paracrines. PTH is essential for life. Absence causes hypocalcemic tetany

Fig. 23.12

are receiving intense scrutiny as potential pharmacological agents. In late 2010 a RANKL-inhibitor named denosumab was approved for use as treatment for conditions with excessive bone loss.

2 **PTH enhances renal reabsorption of calcium.** As we mentioned previously, regulated Ca^{2+} reabsorption takes place in the distal nephron. PTH simultaneously enhances renal excretion of phosphate by reducing its reabsorption. The opposing effects of PTH on calcium and phosphate are needed to keep their combined concentrations below a critical level. If the concentrations exceed that level, calcium phosphate crystals form and precipitate out of solution. High concentrations of calcium phosphate in the urine are one cause of kidney stones. Additional aspects of phosphate homeostasis are discussed below.

3 **PTH indirectly increases intestinal absorption of calcium** through its influence on vitamin D_3, a process described below.

Calcitriol Intestinal absorption of calcium is enhanced by the action of a hormone known as **1,25-dihydroxycholecalciferol** ($1,25(OH)_2D_3$, also known as **calcitriol** or vitamin D_3 (Fig. 23.13). The body makes calcitriol from vitamin D that has been obtained through diet or made in the skin by the action of sunlight on precursors made from acetyl CoA. People who live above 37 degrees of latitude north or below 37 degrees south do not get enough sunlight to make adequate vitamin D except in the summer, and they should consider taking vitamin supplements.

Vitamin D is modified in two steps—first in the liver, then in the kidneys—to make vitamin D_3, or calcitriol. Calcitriol is the primary hormone responsible for enhancing Ca^{2+} uptake from the small intestine. In addition, calcitriol facilitates renal reabsorption of Ca^{2+} and helps mobilize Ca^{2+} out of bone.

The production of calcitriol is regulated at the kidney by the action of PTH. Decreased plasma Ca^{2+} increases PTH secretion, which stimulates calcitriol synthesis. Intestinal and

ENDOCRINE CONTROL OF CALCIUM BALANCE

PTH works with calcitriol to promote bone resorption, intestinal Ca^{2+} absorption, and distal nephron Ca^{2+} reabsorption, all of which tend to elevate plasma Ca^{2+} concentrations.

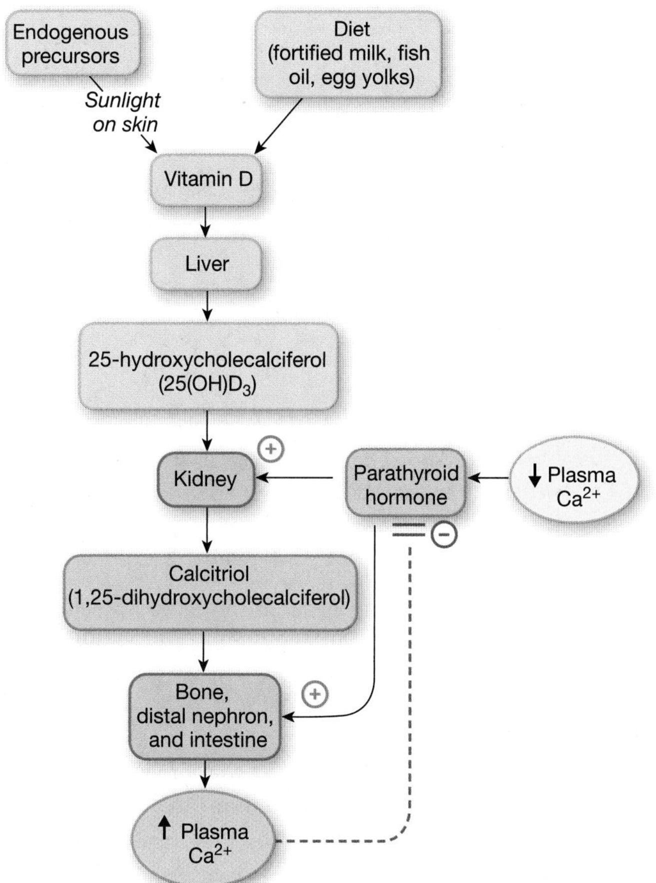

VITAMIN D_3 (CALCITRIOL, 1,25-DIHYDROXYCHOLECALCIFEROL)	
Origin	Complex biosynthesis; see below
Chemical nature	Steroid
Biosynthesis	Vitamin D formed by sunlight on precursor molecules or ingested in food; converted in two steps (liver and kidney) to $1,25(OH)_2D_3$
Transport in the circulation	Bound to plasma protein
Stimulus for synthesis	↓ Ca^{2+}. Indirectly via PTH. Prolactin also stimulates synthesis.
Target cells or tissues	Intestine, bone, and kidney
Target receptor	Nuclear
Whole body or tissue reaction	↑ Plasma Ca^{2+}
Action at molecular level	Stimulates production of calbindin, a Ca^{2+}-binding protein, and of CaSR in parathyroid gland. Associated with intestinal transport by unknown mechanism
Feedback regulation	↑ Plasma Ca^{2+} shuts off PTH secretion

Fig. 23.13

renal absorption of Ca^{2+} raises blood Ca^{2+}, turning off PTH in a negative feedback loop that decreases calcitriol synthesis.

Prolactin, the hormone responsible for milk production in breast-feeding (lactating) women, also stimulates calcitriol synthesis. This action ensures maximal absorption of Ca^{2+} from the diet at a time when metabolic demands for calcium are high.

Calcitonin The third hormone involved with calcium metabolism is **calcitonin**, a peptide produced by the C cells of the thyroid gland (Tbl. 23.1). Its actions are opposite to those of parathyroid hormone. Calcitonin is released when plasma Ca^{2+} increases. Experiments in animals have shown that calcitonin decreases bone resorption and increases renal calcium excretion.

Calcitonin apparently plays only a minor role in daily calcium balance in adult humans. Patients whose thyroid glands have been removed show no disturbance in calcium balance, and people with thyroid tumors that secrete large amounts of calcitonin also show no ill effects.

Calcitonin has been used medically to treat patients with *Paget's disease,* a genetically linked condition in which osteoclasts are overactive and bone is weakened by resorption. Calcitonin in these patients stabilizes the abnormal bone loss, leading scientists to speculate that this hormone is most important during childhood growth, when net bone deposition is needed, and during pregnancy and lactation, when the mother's body must supply calcium for both herself and her child.

Calcium and Phosphate Homeostasis Are Linked

Phosphate homeostasis is closely linked to calcium homeostasis. Phosphate is the second key ingredient in the hydroxyapatite of bone, $Ca_{10}(PO_4)_6(OH)_2$, and most phosphate in the body is found in bone. However, phosphates have other significant physiological roles, including energy transfer and

Table 23.1 Calcitonin

Cell of origin	C cells of thyroid gland (parafollicular cells)
Chemical nature	32-amino acid peptide
Biosynthesis	Typical peptide
Transport in the circulation	Dissolved in plasma
Half-life	<10 minutes
Factors affecting release	↑ Plasma [Ca^{2+}]
Target cells or tissues	Bone and kidney
Target receptor	G protein–coupled membrane receptor
Whole body or tissue action	Prevents bone resorption. Enhances kidney excretion
Action at molecular level	Signal transduction pathways appear to vary during cell cycle
Other information	Experimentally decreases plasma Ca^{2+} but has little apparent physiological effect in adult humans. Possible effect on skeletal development; possible protection of bone Ca^{2+} stores during pregnancy and lactation

RUNNING PROBLEM

The results of Prof. Magruder's last test confirm that he has hyperparathyroidism. He goes on a low-calcium diet, avoiding milk, cheese, and other dairy products, but several months later he returns to the emergency room with another painful kidney stone. Dr. Spinks sends him to an endocrinologist, who recommends surgical removal of the overactive parathyroid glands. "We can't tell which of the parathyroid glands is most active," the specialist says, "and we'd like to leave you with some parathyroid hormone of your own. So I will take out all four glands, but we'll reimplant two of them in the muscle of your forearm. In many patients, the implanted glands secrete just enough PTH to maintain calcium homeostasis. And if they secrete too much PTH, it is much easier to take them out of your arm than do major surgery on your neck again."

Q5: Why can't Prof. Magruder simply take replacement PTH by mouth? (Hint: PTH is a peptide hormone.)

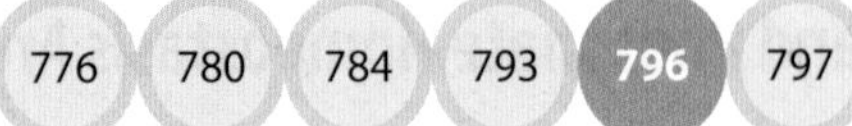

storage in high-energy phosphate bonds, and activation or deactivation of enzymes, transporters, and ion channels through phosphorylation and dephosphorylation. Phosphates also form part of the DNA and RNA backbone.

Phosphate homeostasis parallels that of Ca^{2+}. Phosphate is absorbed in the intestines, filtered and reabsorbed in the kidneys, and divided between bone, ECF, and intracellular compartments. Vitamin D_3 enhances intestinal absorption of phosphate. Renal excretion is affected by both PTH (which promotes phosphate excretion) and vitamin D_3 (which promotes phosphate reabsorption).

Concept Check Answers: p. 800

17. Name two compounds that store energy in high-energy phosphate bonds.
18. What are the differences between a kinase, a phosphatase, and a phosphorylase?

Osteoporosis Is a Disease of Bone Loss

One of the best-known pathologies of bone function is **osteoporosis,** a metabolic disorder in which bone resorption exceeds bone deposition. The result is fragile, weakened bones that are more easily fractured (Fig. 23.10c). Most bone resorption takes place in spongy trabecular bone, particularly in the vertebrae, hips, and wrists.

Osteoporosis is most common in women after menopause, when estrogen concentrations fall. However, older men also develop osteoporosis. Bone loss and small fractures and compression in the spinal column lead to *kyphosis* {hump-back}, the stooped, hunchback appearance that is characteristic of advanced osteoporosis in the elderly. Osteoporosis is a complex disease with genetic and environmental components. Risk factors include small, thin body type; postmenopausal age; smoking; and low dietary Ca^{2+} intake.

For many years estrogen or estrogen/progesterone *hormone replacement therapy* (HRT) was used to prevent osteoporosis. However, estrogen therapy alone increases the risk of endometrial and possibly other cancers, and some studies suggest that combined estrogen/progesterone HRT might increase risk of heart attacks and strokes. A *selective estrogen receptor modulator (SERM)* called raloxifene has been used to treat osteoporosis.

The most effective drugs for preventing or treating osteoporosis act more directly on bone metabolism. They include *bisphosphonates,* which induce osteoclast apoptosis and suppress bone resorption, and *teriparatide*, a PTH derivative, which stimulates formation of new bone. Teriparatide consists of the first 34 amino acids of the 84-amino acid PTH molecule and must be injected rather than taken orally. Currently clinical studies are investigating whether some combination of bisphosphonates and teriparatide is more effective in combating osteoporosis than either drug alone.

To avoid osteoporosis in later years, young women need to maintain adequate dietary calcium intake and perform weight-bearing exercises, such as running or aerobics, which increase bone density. Loss of bone mass begins by age 30, long before people think they are at risk, and many women suffer from low bone mass (*osteopenia*) before they are aware of a problem. Bone mass testing can help with early diagnosis of osteopenia.

RUNNING PROBLEM CONCLUSION

Hyperparathyroidism

Prof. Magruder had the surgery, and the implanted glands produced an adequate amount of PTH. He must have his plasma Ca^{2+} levels checked regularly for the rest of his life to ensure that the glands continue to function adequately. To learn more about hyperparathyroidism, see this article in *The American Family Physician, www.aafp.org/afp/20040115/333.html.*

Check your understanding of this running problem by comparing your answers with the information in the summary table.

Question	Facts	Integration and Analysis
1. What role does Ca^{2+} play in the normal functioning of muscles and neurons?	Calcium triggers neurotransmitter release [p. 270] and uncovers the myosin-binding sites on muscle actin filaments [p. 407].	Muscle weakness in hyperparathyroidism is the opposite of what you would predict from knowing the role of Ca^{2+} in muscles and neurons. However, calcium also affects the Na^+ permeability of neurons, and it is this effect that leads to muscle weakness and CNS effects.
2. What is the technical term for "elevated levels of calcium in the blood"?	Prefix for elevated levels: *hyper-*. Suffix for "in the blood": -emia.	*Hypercalcemia* is the technical term for elevated levels of calcium in the blood.
3. Speculate on why some plasma Ca^{2+} cannot filter into Bowman's capsule.	Filtration at the glomerulus is a selective process that excludes blood cells and most plasma proteins [p. 635].	A significant amount of plasma Ca^{2+} is bound to plasma proteins and therefore cannot filter.
4. What one test could definitively prove that Prof. Magruder has hyperparathyroidism?	Hyperparathyroidism is a condition in which excessive amounts of PTH are secreted.	A test for the amount of PTH in the blood would confirm the diagnosis of hyperparathyroidism.
5. Why can't Prof. Magruder simply take replacement PTH by mouth?	PTH is a peptide hormone.	Ingested peptides are digested by proteolytic enzymes. This means PTH taken orally cannot be absorbed intact into the body and consequently will not be effective.

776 780 784 793 796 797

Chapter Summary

Endocrinology is based on the physiological principles of *homeostasis* and *control systems.* Each hormone has stimuli that initiate its secretion, and feedback signals that modulate its release. *Molecular interactions* and *communication across membranes* are also essential to hormone activity. In many instances, such as calcium and phosphate homeostasis, the principle of *mass balance* is the focus of homeostatic regulation.

Review of Endocrine Principles

1. Basic components of endocrine pathways include hormone receptors, feedback loops, and cellular responses. (p. 776)

Adrenal Glucocorticoids

2. The **adrenal cortex** secretes **glucocorticoids,** sex steroids, and aldosterone. (p. 778; Fig. 23.1)
3. **Cortisol** secretion is controlled by hypothalamic **CRH** and **ACTH** from the pituitary. Cortisol is the feedback signal. Cortisol is a typical steroid hormone in its synthesis, secretion, transport, and action. (p. 779; Fig. 23.2)
4. Cortisol is catabolic and essential for life. It promotes gluconeogenesis, breakdown of skeletal muscle proteins and adipose tissue, Ca^{2+} excretion, and suppression of the immune system. (p. 777)
5. **Hypercortisolism** usually results from a tumor or therapeutic administration of the hormone. **Addison's disease** is hyposecretion of all adrenal steroids. (p. 781)
6. CRH and the **melanocortins** have physiological actions in addition to cortisol release. (p. 779; Fig. 23.2d)

Thyroid Hormones

7. The thyroid **follicle** has a hollow center filled with **colloid** containing **thyroglobulin** and enzymes. (p. 783; Fig. 23.4b)
8. Thyroid hormones are made from tyrosine and iodine. **Tetraiodothyronine** (thyroxine, T_4) is converted in target tissues to the more active hormone **triiodothyronine** (T_3). (p. 783; Fig. 23.4)
9. Thyroid hormones are not essential for life, but they influence metabolic rate as well as protein, carbohydrate, and fat metabolism. (p. 784)
10. Thyroid hormone secretion is controlled by **thyrotropin** (thyroid-stimulating hormone, TSH) and **thyrotropin-releasing hormone** (TRH). (p. 785; Fig. 23.5)

Growth Hormone

11. Normal growth requires growth hormone, thyroid hormones, insulin, and sex hormones at puberty. Growth also requires adequate diet and absence of chronic stress. (p. 787)
12. Growth hormone is secreted by the anterior pituitary and stimulates secretion of **insulin-like growth factors** (**IGFs**) from the liver and other tissues. These hormones promote bone and soft tissue growth. (p. 789; Fig. 23.8)
13. Secretion of growth hormone is controlled by **growth hormone–releasing hormone** (GHRH) and growth hormone–inhibiting hormone (**somatostatin**). (p. 789; Fig. 23.8)

Tissue and Bone Growth

14. **Bone** is composed of **hydroxyapatite** crystals attached to a collagenous support. Bone is a dynamic tissue with living cells. (p. 790)
15. **Osteoblasts** synthesize bone. Long bone growth occurs at **epiphyseal plates,** where **chondrocytes** produce cartilage. (p. 791; Fig. 23.10)

Calcium Balance

iP Fluids & Electrolytes: Electrolyte Homeostasis

16. Calcium acts as an intracellular signal for second messenger pathways, exocytosis, and muscle contraction. It also plays a role in cell junctions, coagulation, and neural function (p. 792)
17. Ca^{2+} homeostasis balances dietary intake, urinary output, and distribution of Ca^{2+} among bone, cells, and the ECF. (p. 792; Fig. 23.11)
18. Decreased plasma Ca^{2+} stimulates **parathyroid hormone** (**PTH**) secretion by the **parathyroid glands.** (p. 794; Fig. 23.12)
19. PTH promotes Ca^{2+} resorption from bone, enhances renal Ca^{2+} reabsorption, and increases intestinal Ca^{2+} absorption through its effect on **calcitriol.** (p. 794; Fig. 23.12)
20. Calcitonin from the thyroid gland plays only a minor role in daily calcium balance in adult humans. (p. 796; Tbl. 23.1)

Questions

Answers: p. A-1

Level One Reviewing Facts and Terms

1. Name the zones of the adrenal cortex and the primary hormones secreted in each zone.
2. For (a) cortisol, (b) growth hormone, (c) parathyroid hormone, and (d) T_3 and T_4, draw the full control pathway and show feedback where appropriate. Do not use abbreviations.

3. List four conditions that are necessary for people to achieve their full growth. Include five specific hormones known to exert an effect on growth.
4. Name the thyroid hormones. Which one has the highest activity? How and where is most of it produced?
5. Define each of the following terms and explain its physiological significance:
   (a) melanocortins
   (b) osteoporosis
   (c) hydroxyapatite
   (d) mineralocorticoid
   (e) trabecular bone
   (f) POMC
   (g) epiphyseal plates
6. List seven functions of calcium in the body.
7. Make a table showing the effects of cortisol, thyroid hormones, growth hormone, insulin, and glucagon on protein, carbohydrate, and lipid metabolism.

Level Two Reviewing Concepts

8. Mapping exercise: Create a reflex map with feedback loops for each of the following situations:
   (a) Hypercortisolism from an adrenal tumor
   (b) Hypercortisolism from a pituitary tumor
   (c) Hyperthyroidism from a hormone-secreting thyroid tumor
   (d) Hypothyroidism from a pituitary problem that decreases TSH synthesis
9. Define, compare, and contrast or relate the terms in each set:
   (a) cortisol, glucocorticoids, ACTH, CRH
   (b) thyroid, C cell, follicle, colloid
   (c) thyroglobulin, tyrosine, iodide, TBG, deiodinase, TSH, TRH
   (d) somatotropin, IGF, GHRH, somatostatin, growth hormone-binding protein
   (e) giantism, acromegaly, dwarfism
   (f) hyperplasia, hypertrophy
   (g) osteoblast, osteoclast, chondrocyte, osteocyte
   (h) vitamin D, calcitriol, 1,25-dihydroxycholecalciferol, calcitonin, estrogen, PTH
10. Based on what you know about the cellular mechanism of action for T_3, would you expect to see tissue response to this hormone within a few minutes or in more than an hour?
11. If average plasma $[Ca^{2+}]$ is 2.5 mmol/L, what is the concentration in mEq/L?
12. Osteoclasts make acid (H^+) from CO_2 and H_2O. They secrete the acid at their apical membrane and put bicarbonate into the ECF. Draw an osteoclast and diagram this process, including enzymes and the appropriate transporters on each membrane. How many different transporters can you think of that could be used to reabsorb bicarbonate?

Level Three Problem Solving

13. Diabetic patients who have surgery, become sick, or are under other physiological stress are told to monitor their blood sugar carefully because they may need to increase their insulin dose temporarily. What is the physiological explanation behind this advice?
14. One diagnostic test to determine the cause of hypercortisolism is a *dexamethasone suppression test.* Dexamethasone blocks secretion of ACTH by the pituitary. The following table shows the results from two patients given a dexamethasone suppression test.

Plasma Cortisol Concentration	Before Test	After Test
Patient A	High	High
Patient B	High	Low

Can you tell from these results where the patients' pathologies originate? Explain for each patient.

15. When blood test results came back last week, someone in the office spilled a cup of coffee on them, smearing the patient names and some of the numbers. One report shows elevated TSH levels, but the thyroid levels are so low they are unreadable. You have three charts waiting for test results on thyroid hormone levels. Your tentative diagnoses, based on physical findings and symptoms, for those three patients are:

    Mr. A: primary hypothyroidism
    Ms. B: primary hyperthyroidism
    Ms. C: secondary hyperthyroidism

    (a) Can you tell whose results are on the smeared report, based on the TSH results and the tentative diagnosis?
    (b) Can you rule out any of the three people based on those same criteria? Explain.
16. The following graph shows the results of a study done in Boston that compared blood vitamin D levels during summer and winter. Boston is located at 42 degrees north latitude, and weak sunlight in winter there does not allow skin synthesis of vitamin D. (Data from *Am J Med* 112: 659–662, 2002 Jun 1)

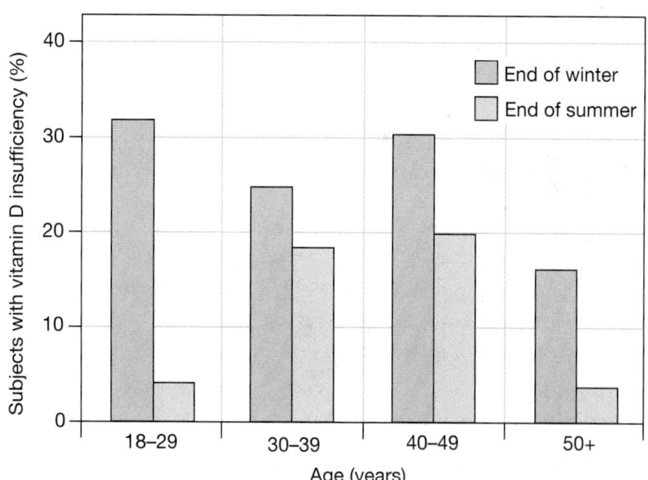

    (a) Summarize the results shown in the graph. How many variables are shown in the graph that you must address in your summary?
    (b) Based on what you know, how could you explain the results of the study?
    (c) Would taking a multivitamin supplement affect the results?

Level Four Quantitative Problems

17. Filterable plasma Ca^{2+} is about 5 mg/L. Assume a man has a GFR of 125 mL of plasma filtered per minute.
    (a) How much Ca^{2+} does he filter in a day?
    (b) Net dietary Ca^{2+} intake is 170 mg/day. To remain in Ca^{2+} balance, how much Ca^{2+} must he excrete?
    (c) What percentage of filtered Ca^{2+} is reabsorbed by the kidney tubule?

Answers

Answers to Concept Check Questions

Page 777

1. The medulla secretes catecholamines (epinephrine, norepinephrine), and the cortex secretes aldosterone, glucocorticoids, and sex hormones.
2. Androstenedione is a prohormone for testosterone. Testosterone is anabolic for skeletal muscle, which might give an athlete a strength advantage.

Page 780

3. HPA = hypothalamic-pituitary-adrenal. CBG = corticosteroid-binding globulin or transcortin.
4. This immediate stress response is too rapid to be mediated by cortisol and must be a fight-or-flight response mediated by the sympathetic nervous system and catecholamines.
5. No, because cortisol is catabolic on muscle proteins.

Page 781

6. Primary and iatrogenic hypercortisolism: ACTH is lower than normal because of negative feedback. Secondary hypercortisolism: ACTH is higher because of the ACTH-secreting tumor.
7. Addison's disease: high ACTH due to reduced corticosteroid production and lack of negative feedback.

Page 782

8. ACTH is secreted during stress. If the stress is a physical one caused by an injury, the endogenous opioid β-endorphin can decrease pain and help the person continue functioning.

Page 786

9. In peripheral tissues T_4 is converted to T_3, which is the more active form of the hormone.
10. When mitochondria are uncoupled, energy normally captured in ATP is released as heat. This raises the person's body temperature and causes heat intolerance.

Page 788

11. Prolactin also has two hypothalamic factors that regulate its release [p. 222].

Page 792

12. Normal growth and development require growth hormone, thyroid hormone, insulin, and insulin-like growth factors.
13. Adults who hypersecrete growth hormone do not grow taller because their epiphyseal plates have closed.

Page 793

14. Hypercalcemia hyperpolarizes the membrane potential, which makes it harder for the neuron to fire an action potential.
15. The figure should resemble Figure 21.10b.
16. The Na^+-Ca^{2+} exchanger is a secondary active transporter, and the Ca^{2+}-ATPase is an active transporter.

Page 796

17. ATP and phosphocreatine store energy in high-energy phosphate bonds.
18. A kinase transfers a phosphate group from one substrate to another. A phosphatase removes a phosphate group, and a phosphorylase adds one.

Answers to Figure Questions

Page 778

Figure 23.1: 1. A baby born with deficient 21-hydroxylase would have low aldosterone and cortisol levels and an excess of sex steroids, particularly androgens. Low cortisol would decrease the child's ability to respond to stress. Excess androgen would cause masculinization in female infants. 2. Women, who synthesize more estrogens, would have more aromatase.

Page 779

Figure 23.2: ACTH = adrenocorticotropic hormone or corticotropin. CRH = corticotropin-releasing hormone. MSH = melanocyte-stimulating hormone.

Page 783

Figure 23.4: 1. The apical membrane faces the colloid, and the basolateral membrane faces the ECF. 2. I^- comes into the cell by secondary active transport (co-transport with Na^+). 3 and 4. Thyroglobulin moves between colloid and cytoplasm by exocytosis and endocytosis. 5. Thyroid hormones leave the cell by an unknown membrane transporter.

Page 787

Figure 23.7: A pituitary tumor hypersecreting TSH would cause hyperthyroidism and an enlarged thyroid gland. The pathway would show decreased TRH resulting from short-loop negative feedback from TSH to the hypothalamus, increased TSH caused by the tumor, elevated thyroid hormones, but no negative feedback from the thyroid hormones to the anterior pituitary because the tumor does not respond to feedback signals.

24 The Immune System

Although, at first sight, the immune system may appear to be autonomous, it is connected by innumerable structural and functional bridges with the nervous system and the endocrine system, so as to constitute a multisystem.

—Branislav D. Jankovic, *Neuroimmunomodulation: The State of the Art*, 1994

Background Basics

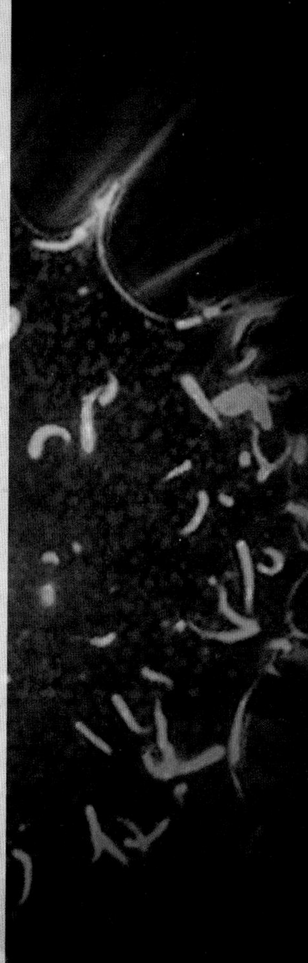

Cell infected with poxvirus (650x)

It seems that since the 1980s plagues of infectious diseases have been increasing in frequency. Some of them, such as tuberculosis and rabies, are old enemies that we thought we had under control. Others, such as the "bird flu," severe acute respiratory syndrome (SARS), monkey pox, and the Hanta and Ebola viruses, seem to be cropping up out of nowhere. The conflict between humans and the invaders that cause disease has always been a fight for survival on both sides. The primary job of our immune system is to recognize "self" from "nonself" and rally defenses when needed.

The body's ability to protect itself from itself as well as from viruses, bacteria, and other disease-causing entities is known as **immunity**, from the Latin word *immunis,* meaning *exempt.* The human immune system consists of the lymphoid tissues of the body, the immune cells, and chemicals (both intracellular and secreted) that coordinate and execute immune functions. Most immune functions depend heavily on cell-to-cell communication, particularly local communication by cytokines and contact-dependent signaling [p. 177].

RUNNING PROBLEM

HPV: To Vaccinate or Not?

Rebecca and her daughter Lizzie arrived at the doctor's office for Lizzie's annual back-to-school checkup. Lizzie is 12 years old and will be in 6th grade in the fall. Halfway through the appointment, Dr. Paul asked Rebecca if Lizzie had started the series of vaccinations to protect her against HPV, the human papillomavirus. "Isn't HPV the virus that causes cervical cancer?" Rebecca asked. When the doctor confirmed that it was, Rebecca said, "No, Lizzie doesn't need that yet. She's only 12 and she's a good girl." "Let's talk about it," Dr. Paul suggested. "The American Academy of Pediatrics, the American Cancer Society, The American Congress of Obstetricians and Gynecologists, and the Centers for Disease Control all recommend this vaccine for girls, and now is the time to start it."

Overview

The key features of the immune system are *specificity* and *memory*. Together these processes enable the body to distinguish "self" from "non-self" and mount a targeted response to specific invaders. The immune system serves three major functions:

1. **It tries to recognize and remove abnormal "self" cells** created when normal cell growth and development go wrong. For example, the diseases we call *cancer* result from abnormal cells that multiply uncontrollably, crowding out normal cells and disrupting function. Scientists believe that cancer cells form on a regular basis but are usually detected and destroyed by the immune system before they get out of control.
2. **It removes dead or damaged cells,** as well as old red blood cells. Scavenger cells of the immune system, such as macrophages, patrol the extracellular compartment, gobbling up and digesting dead or dying cells.
3. **It protects the body from disease-causing invaders** known as **pathogens**. Microorganisms (*microbes*) that act as pathogens include bacteria, viruses, fungi, and one-celled protozoans. Larger pathogens include multicellular **parasites,** such as hookworms and tapeworms. Almost any exogenous molecule or cell has the potential to elicit an immune response. Pollens, chemicals, and foreign bodies are examples of substances to which the body may react. Substances that trigger the body's immune response are called *immunogens*. Immunogens that react with products of the response are known as **antigens**.

Sometimes the body's immune system fails to perform its normal functions. Pathologies of the immune system generally fall into one of three categories: incorrect responses, overactive responses, or lack of response.

1. *Incorrect responses.* If mechanisms for distinguishing self from non-self fail and the immune system attacks the body's normal cells, an *autoimmune disease* results. Type 1 diabetes mellitus, in which proteins made by immune cells destroy pancreatic beta cells, is an example of an autoimmune disease in humans.
2. *Overactive responses.* Allergies are conditions in which the immune system creates a response that is out of proportion to the threat posed by the antigen. In extreme cases, the systemic effects of allergic responses can be life threatening.
3. *Lack of response.* **Immunodeficiency diseases** arise when some component of the immune system fails to work properly. *Primary immunodeficiency* is a family of genetically inherited disorders that range from mild to severe. *Acquired immunodeficiencies* may occur as a result of infection, such as *acquired immunodeficiency syndrome* (AIDS) caused by the *human immunodeficiency virus* (HIV). Acquired immunodeficiencies may also arise as a side effect of drug or radiation therapy, such as those used to treat cancer.

Pathogens of the Human Body

In the United States, our most prevalent infectious diseases are of viral and bacterial origin. Worldwide, parasites are an additional significant public health concern. For example, malaria, a pathogenic protozoan whose life cycle alternates between human and mosquito hosts, was estimated to infect as many as 225 million people in 2009, with nearly 800,000 deaths that year.

Many parasitic organisms, such as the malaria protozoan, are introduced into the body by biting insects. Others enter the body through the digestive tract, brought in by contaminated food and water. Some, such as the fungi that cause valley fever and histoplasmosis, are inhaled. A few, such as the blood fluke *Schistosoma*, burrow through the host's skin. Once in the body, microbes and parasites may enter host cells in an effort to evade the immune response, or they may remain in the extracellular compartment.

Bacteria and Viruses Require Different Defense Mechanisms

Bacteria and viruses differ from each other in several ways. These differences require the body to have a variety of immune responses.

1 **Structure.** Bacteria are cells, with a cell membrane that is usually surrounded by a cell wall (■ Fig. 24.1a). Some *encapsulated* bacteria also produce an additional protective outer layer known as a *capsule*. Viruses are not cells. They

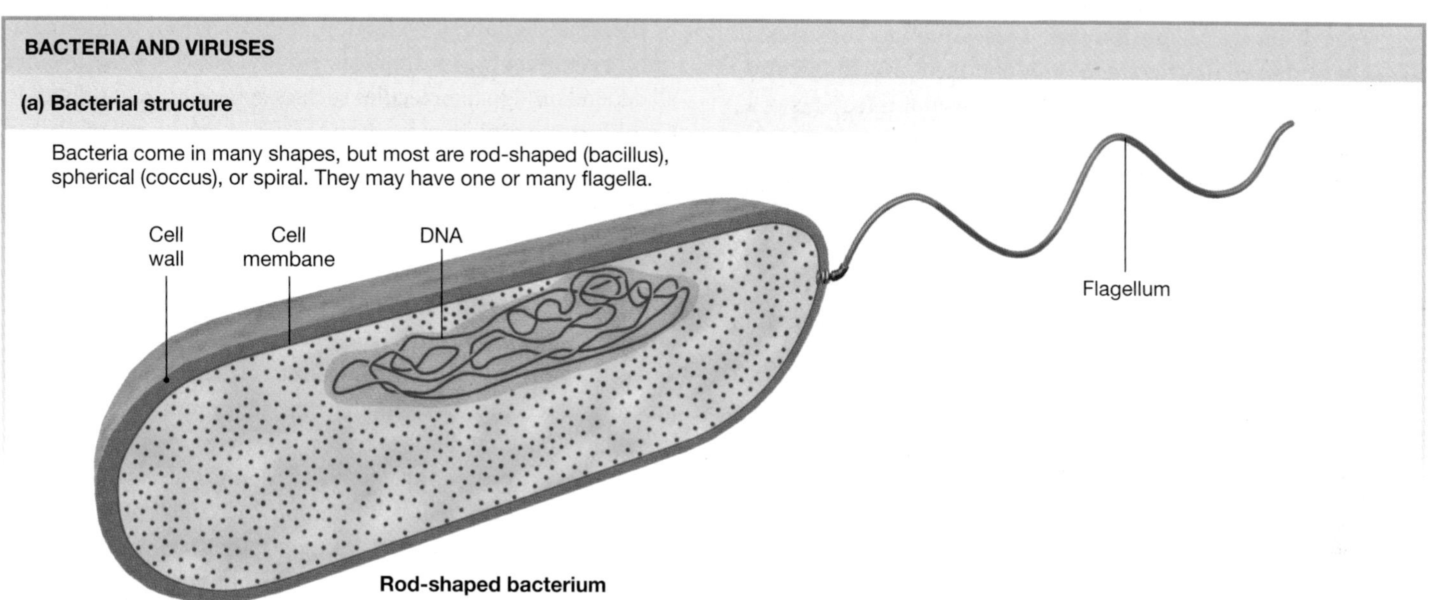

(b) Viral structure

Viruses may be made of DNA or RNA. This shows influenza, an RNA virus.

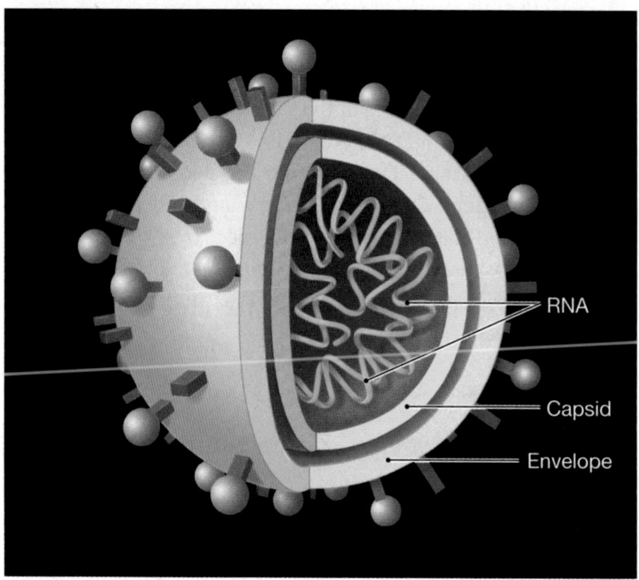

(c) Differences between bacteria and viruses

	BACTERIA	VIRUSES
Structure	Cells. Usually surrounded by cell wall	Not cells. Nucleic acid core with protein coat
Living conditions	Most can survive and reproduce outside a host.	Parasitic. Must have a host cell to reproduce.
Susceptibility to drugs	Most can be killed or inhibited by antibiotics.	Cannot be killed with antibiotics. Some can be inhibited with antiviral drugs.

■ **Fig. 24.1**

consist of nucleic acid (DNA or RNA) enclosed in a coat of viral proteins called a *capsid*. Some viruses add an *envelope* of phospholipid and protein made from the host's cell membrane and incorporate viral proteins into the envelope (Fig. 24.1b).

2 **Living conditions and reproduction.** Most bacteria can survive and reproduce outside a host if they have the required nutrients, temperature, pH, and so on. Viruses *must* use the intracellular machinery of a host cell to replicate. The location of pathogens in both major compartments of the body requires different defense mechanisms for each compartment.

3 **Susceptibility to drugs.** Most bacteria can be killed by the drugs we call **antibiotics**. These drugs act directly on bacteria and destroy them or inhibit their growth. Viruses cannot be killed by antibiotics. A few viral infections can be treated with *antiviral drugs,* which target specific stages of viral replication.

Viruses Can Replicate Only inside Host Cells

The replication cycle of a virus begins when the virus invades the host cell. To cross the human host cell membrane, the virus binds to membrane receptors, triggering endocytosis of the entire virus particle. In an alternative scenario, the virus envelope fuses with the host cell membrane, allowing the core of the virus to enter the cytoplasm.

Once inside the host cell and free of the capsid, the virus's nucleic acid takes over the host cell's resources to make new viral nucleic acid and viral proteins. These components assemble into additional virus particles that are released from the host cell to infect other cells.

Viruses can be released from host cells in one of two ways. (1) The virus causes the host cell to rupture, releasing virus particles into the ECF, or (2) virus particles surround themselves with a layer of host cell membrane and then bud off from the surface of the host cell.

Viruses do other kinds of damage to host cells. In taking over a host cell, they may totally disrupt the cell's metabolism so that the cell dies. Some viruses (*Herpes simplex type 1* and varicella-zoster virus, which cause cold sores and chicken pox, respectively) "hide out" in the host cell and replicate only sporadically. Other viruses incorporate their DNA into the host cell DNA. Viruses with this characteristic include HIV and **oncogenic viruses,** which cause cancer.

Concept Check

Answers: p. 834

1. Explain the differences between a virus and a bacterium.
2. List the sequence of steps by which a virus replicates itself.

The Immune Response

The body has two lines of defense. Physical and chemical barriers, such as skin, mucus, and stomach acid, first try to keep pathogens out of the body's internal environment (■ Fig. 24.2). If this first line of defense fails, then the internal **immune response** takes over. The internal immune response has four basic steps: (1) *detection* and *identification* of the foreign substance, (2) *communication* with other immune cells to rally an organized response, (3) *recruitment* of assistance and *coordination* of the response among all participants, and (4) *destruction* or *suppression* of the invader.

The immune response is distinguished by its extensive use of chemical signaling. Detection, identification, communication, recruitment, coordination, and the attack on the invader all depend on signal molecules such as cytokines and antibodies. **Cytokines** are protein messengers released by one cell that affect the growth or activity of another cell [p. 177]. **Antibodies**, proteins secreted by certain immune cells, bind antigens and make them more visible to the immune system.

The human immune response is generally divided into two categories: nonspecific innate immunity and specific acquired immunity. **Innate immunity** is present from birth {*innatus,* inborn} and is the body's **nonspecific immune response** to invasion. The membrane receptors that mediate innate immunity have broad specificity and allow some immune cells to respond to molecular signals that are both unique and common to pathogenic microorganisms. An example of a common pathogen-specific signal would be certain components of the bacterial cell wall. Because the nonspecific innate response does not target a particular

RUNNING PROBLEM

Dr. Paul explained to Rebecca that there are over 100 known types of human papillomavirus. HPV is a DNA-based virus that can live only in the thin, flat epithelial cells known as *squamous epithelium* {*squama*, a scale}. Squamous epithelium is found on the surface of the skin and in mucous membrane, such as those lining the mouth and genital tract. HPV can cause both ordinary warts on the skin and the growths known as genital warts. More importantly, HPV infections cause nearly all cases of cervical cancer. (The *cervix* is neck of the uterus that opens into the vagina.) Two high-risk strains of HPV, designated types 16 and 18, account for about 70% of cervical cancer cases.

Q1: Why must viruses live and reproduce inside host cells?

Q2: How might you test a person to see if she is infected with HPV?

802 804 816 825 829 830

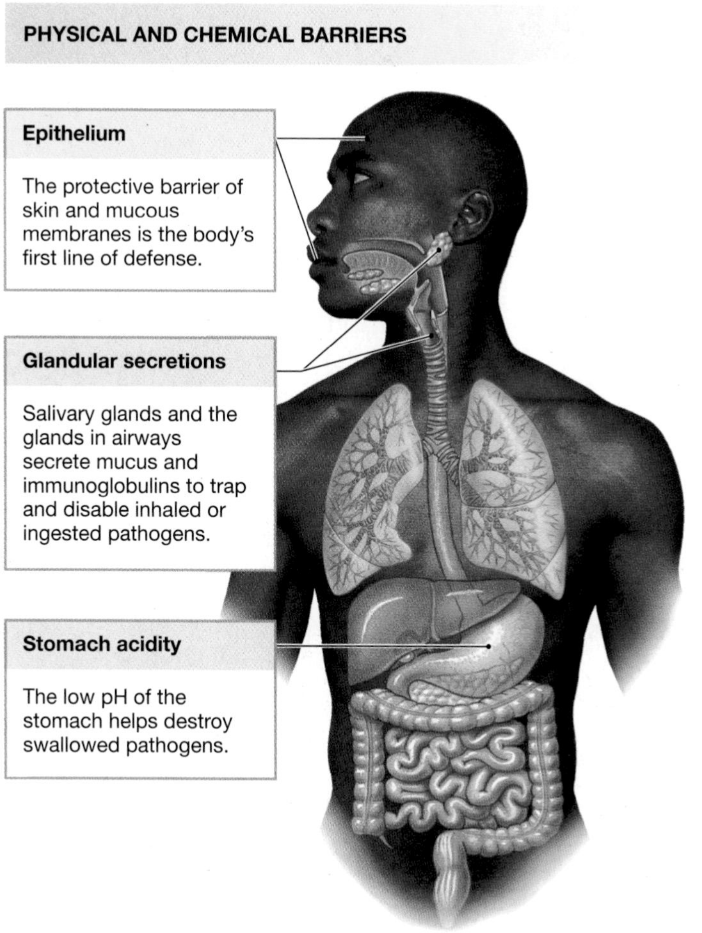

PHYSICAL AND CHEMICAL BARRIERS
Epithelium
The protective barrier of skin and mucous membranes is the body's first line of defense.
Glandular secretions
Salivary glands and the glands in airways secrete mucus and immunoglobulins to trap and disable inhaled or ingested pathogens.
Stomach acidity
The low pH of the stomach helps destroy swallowed pathogens.

Fig. 24.2

pathogen, it begins within minutes to hours. **Inflammation**, apparent on the skin as a red, warm, swollen area, is a hallmark reaction of cytokine-mediated innate immunity.

Acquired immunity (also called *adaptive immunity*) is directed at specific invaders and for this reason is the body's **specific immune response**. The membrane receptors that mediate acquired immunity are highly specific and can distinguish between different pathogens. One characteristic of acquired immunity is that a specific immune response following first exposure to a pathogen may take days. With repeated exposures, however, the immune system "remembers" prior exposure to the pathogen and reacts more rapidly.

Acquired immunity can be divided into cell-mediated immunity and humoral immunity. *Cell-mediated immunity* uses contact-dependent signaling in which an immune cell binds through receptors to its target cell. *Humoral immunity*, also known as *antibody-mediated immunity*, uses the secreted proteins known as antibodies to carry out the immune response. Antibodies combine with foreign substances to make them more visible to the cells of the immune system. (The term *humoral,* referring to the blood, comes from the ancient Hippocratic school of medicine, which classified the body's fluids into four *humors:* blood, phlegm, black bile, and yellow bile.)

As you will see later in this chapter, the nonspecific and specific immune responses overlap. Although we have described them as if they are separate, in reality they are two interconnected parts of a single process. The innate response is the more rapid response, but it does not target a specific invader. It is reinforced by the antigen-specific acquired response, which amplifies the efficacy of the innate response. Communication and coordination among all the different pathways of immunity are vital for maximum protective effect.

Keep in mind that not all invaders can be destroyed by the body's immune system. In some cases the best the body can do is control the damage and keep the invader from spreading. Pathogens that are suppressed by the immune system rather than destroyed include the bacterium that causes tuberculosis, which hides inside macrophages in the lung; the malaria parasite, which hides inside liver cells; and the herpesviruses responsible for outbreaks of cold sores or genital lesions, which hide inside cells in the skin.

Because of the huge numbers of molecules involved in human immunity, and because of the complex interactions between different components of the immune system, the field of immunology is continuously expanding. In this chapter we cover the basics.

Anatomy of the Immune System

The immune system is probably the least anatomically identifiable system of the body because most of it is integrated into the tissues of other organs, such as the skin and gastrointestinal tract. Yet the mass of all immune cells in the body equals the mass of the brain! The immune system has two anatomical components: lymphoid tissues and the cells responsible for the immune response.

Lymphoid Tissues Are Everywhere

Lymphoid tissues are found all over the body (■ Fig. 24.3). The two *primary lymphoid tissues* are the **thymus gland** (see Focus on the Thymus Gland, Fig. 24.10, p. 819) and the **bone marrow** [Fig. 16.4, p. 552], both sites where cells involved in the immune response form and mature. Some types of mature immune cells do not specialize until their first exposure to the pathogen they will fight. These mature but unspecialized immune cells are said to be *naïve* cells {*naïf*, natural}.

In the *secondary lymphoid tissues*, mature immune cells interact with pathogens and initiate a response. Secondary tissues are divided into encapsulated tissues and unencapsulated diffuse lymphoid tissues. The **encapsulated lymphoid tissues** are the **spleen** (■ Fig. 24.4) and the **lymph nodes** (Fig. 24.3). Both spleen and lymph nodes have an outer wall formed from fibrous collagenous capsules. The spleen contains immune cells positioned so that they monitor the blood for foreign invaders.

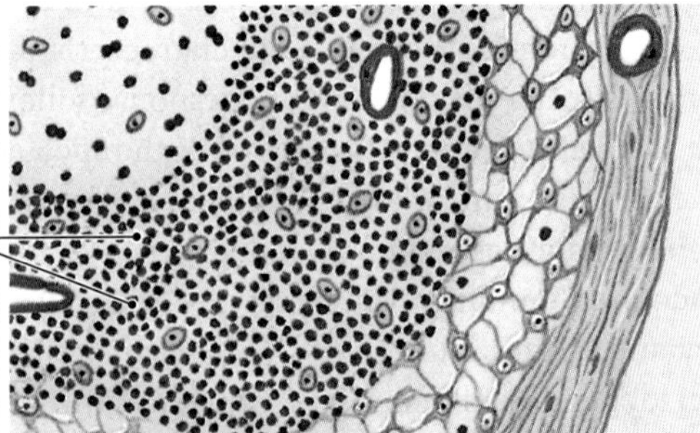

■ Fig. 24.4 *FOCUS ON ...*

The Spleen

The spleen is the largest lymphoid organ in the body, located in the upper left quadrant of the abdomen close to the stomach.

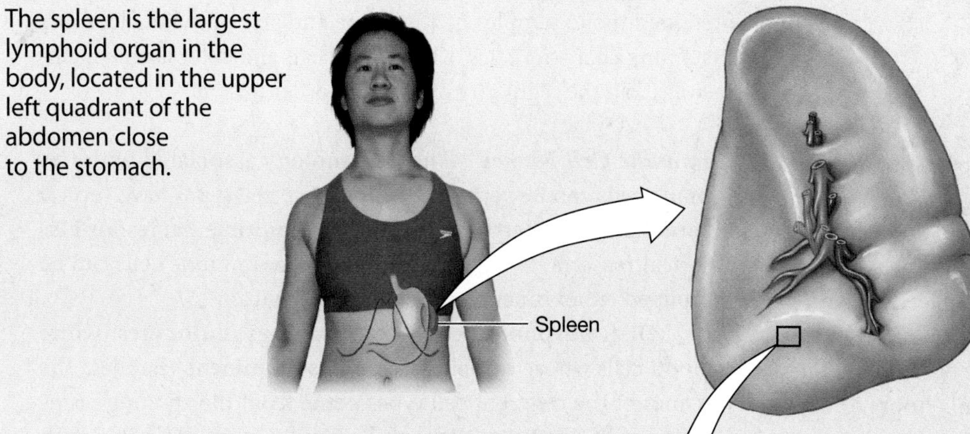

The outer surface of the spleen is a connective tissue capsule that extends into the interior to create an open framework that supports the blood vessels and lymphoid tissue.

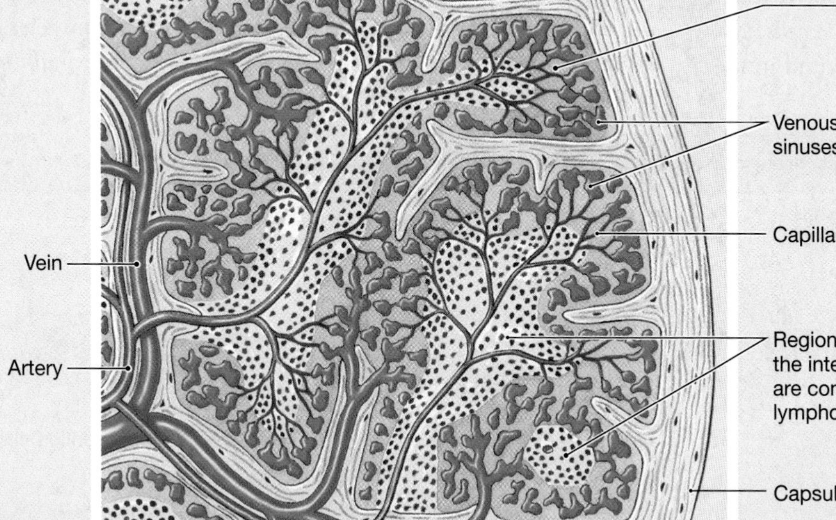

Darker regions of **red pulp** are closely associated with extensive blood vessels and open venous sinuses.The red pulp contains many macrophages that act as a filter by trapping and destroying foreign material circulating in the blood. In addition, the macrophages ingest old, damaged, and abnormal red blood cells, breaking down their hemoglobin molecules into amino acids, iron, and bilirubin that is transported to the liver for excretion.

Regions of **white pulp** resemble the interior of lymph nodes and are composed mainly of lymphocytes.

Phagocytic cells in the spleen also trap and remove aging red blood cells.

The lymph nodes are part of the lymphatic circulation, which is closely associated with capillaries of the cardiovascular system. Recall that blood pressure creates net flow of fluid out of capillaries and into the interstitial space [p. 529]. The filtered fluid, amounting to about 3 L/day, is picked up by lymph capillaries and passes through the encapsulated lymph nodes on its journey back to the heart.

Inside lymph nodes, clusters of immune cells intercept pathogens that have entered the interstitial fluid through breaks in the skin or through mucous membranes (Fig. 24.3). Once these microbes have been swept into the lymph, immune cells in the lymph nodes help prevent their spread throughout the body. You have probably noticed that if you have a sinus infection or a sore throat, the lymph nodes in your neck become swollen. These sore, swollen nodes result from the presence of active immune cells that have collected in the nodes to fight the infection.

The unencapsulated **diffuse lymphoid tissues** are aggregations of immune cells that appear in other organs of the body (Fig. 24.3). They include the **tonsils** at the posterior nasopharynx; the **gut-associated lymphoid tissue** (GALT), which lies just under the epithelium of the esophagus and intestines [p. 688]; and clusters of lymphoid tissue associated with the skin and the respiratory, urinary, and reproductive tracts. In each case, these tissues contain immune cells positioned to intercept invading pathogens before they get into the general circulation. Because of the large surface area of the digestive tract epithelium, some authorities consider the GALT to be the largest immune organ. Anatomically, the immune system is positioned wherever pathogens are most likely to enter the body.

Leukocytes Mediate Immunity

The white blood cells (WBCs), or **leukocytes**, are the primary cells responsible for the immune responses of the body. Most leukocytes are much larger than red blood cells, and they are not nearly as numerous. A microliter (μL) of whole blood contains about 5 million red blood cells but only about 7000 leukocytes.

Although most leukocytes circulate in the blood, they usually leave the capillaries and function *extravascularly* (outside the vessels). Some types of leukocytes can live out in the tissues for several months, but others may live for only hours or days. Leukocytes are divided into six basic types: (1) *eosinophils,* (2) *basophils* in the blood and the related *mast cells* in the tissues, (3) *neutrophils,* (4) *monocytes* and their derivative *macrophages,* (5) *lymphocytes* and their derivative *plasma cells,* and (6) *dendritic cells*. Dendritic cells are not usually found in the blood, and therefore are often excluded from discussion of leukocytes in the blood.

Leukocytes can be distinguished from one another in stained tissue samples by the shape and size of the nucleus, the staining characteristics of the cytoplasm and cytoplasmic inclusions, and the regularity of the cell border.

Immune Cell Names The terminology associated with immune cells can be very confusing. Some cell types have several variants, and others have been given multiple names for historical reasons (■ Fig. 24.5). In addition, immune cells can be grouped either functionally or morphologically.

One morphological group is the **granulocytes**, white blood cells whose cytoplasm contains prominent granules. The names of the different cell types come from the staining properties of the granules. Basophil granules stain dark blue with basic (alkaline) dye, and eosinophil granules stain dark pink with the acidic dye *eosin* {Eos, Greek goddess of the dawn}. In all three types of granulocytes, the granule contents are released from the cell by exocytosis, a process known as *degranulation*.

CELLS OF THE IMMUNE SYSTEM Circulating leukocytes, tissue macrophages, and dendritic cells are the body's immunocytes.

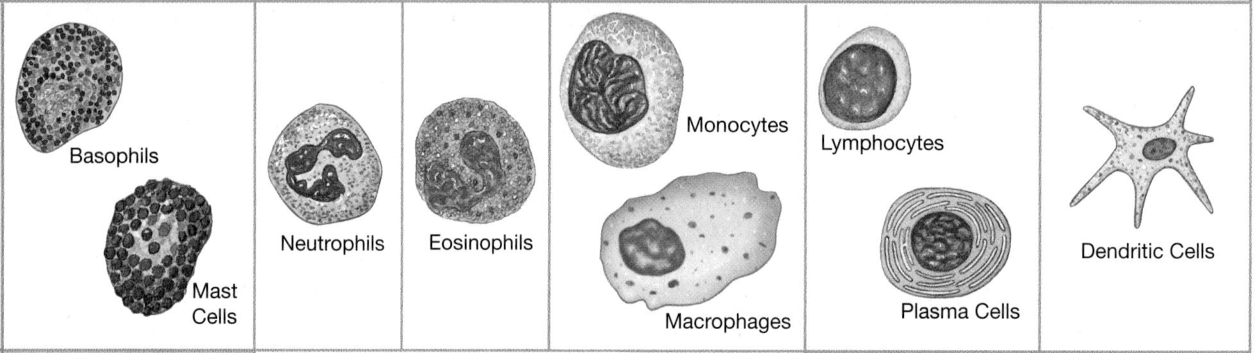

Types of cells	Basophils, Mast Cells	Neutrophils	Eosinophils	Monocytes, Macrophages	Lymphocytes, Plasma Cells	Dendritic Cells
Classifications		Phagocytes	Phagocytes	Phagocytes		
	Granulocytes	Granulocytes	Granulocytes			
			Cytotoxic cells		Cytotoxic cells (some types)	
				Antigen-presenting cells	Antigen-presenting cells	Antigen-presenting cells
% of WBCs in blood	Rare	50–70%	1–3%	1–6%	20–35%	N/A
Subtypes and nicknames		Called "polys" or "segs." Immature forms called "bands" or "stabs."		Called the mononuclear phagocyte system	B lymphocytes Plasma cells Memory cells T lymphocytes Cytotoxic T cells Helper T cells Natural killer cells	Also called Langerhans cells, veiled cells
Primary function(s)	Release chemicals that mediate inflammation and allergic responses	Ingest and destroy invaders	Destroy invaders, particularly antibody-coated parasites	Ingest and destroy invaders. Antigen presentation	Specific responses to invaders, including antibody production	Recognize pathogens and activate other immune cells by antigen presentation

■ **Fig. 24.5**

Neutrophil granules do not stain darkly with standard blood stains and are therefore "neutral."

One functional group of leukocytes is the **phagocytes,** white blood cells that engulf and ingest their targets by phagocytosis [p. 155]. This group includes the neutrophils, eosinophils, macrophages, and monocytes (which are macrophage precursors). A second functional group is the **cytotoxic cells,** so named because they kill the cells they attack. This group includes eosinophils and some types of lymphocytes. A third group is made up of **antigen-presenting cells** (APCs), which display fragments of foreign proteins on their cell surface. This group includes macrophages, monocytes, one type of lymphocyte, and dendritic cells.

The terminology associated with macrophages has changed over the history of histology and immunology. For many years, tissue macrophages were known as the *reticuloendothelial system* and were not associated with white blood cells. To confuse matters, the cells were named when they were first described in different tissues, before they were all identified as macrophages. For this reason, *histiocytes* in skin, *Kupffer cells* in the liver, *osteoclasts* in bone, *microglia* in the brain, and *reticuloendothelial cells* in the spleen are all names for specialized macrophages. The new name for the reticuloendothelial system is the **mononuclear phagocyte system**, a term that refers both to macrophages in the tissues and to their parent monocytes circulating in the blood.

In the sections that follow, we take a closer look at the six basic types of leukocytes.

Basophils Basophils are rare in the circulation but are easily recognized in a stained blood smear by the large, dark blue granules in their cytoplasm. They are very similar to the mast cells of tissues, and both cell types release mediators that contribute to inflammation. The granules of these cells contain histamine, **heparin** (an anticoagulant that inhibits blood clotting), cytokines, and other chemicals involved in allergic and immune responses.

Mast cells are concentrated in the connective tissue of skin, lungs, and the gastrointestinal tract. In these locations, mast cells are ideally situated to intercept pathogens that are inhaled or ingested or that enter through breaks in the epidermis.

Neutrophils **Neutrophils** are phagocytic cells that typically ingest and kill 5–20 bacteria during their short programmed life span of one or two days (■ Fig. 24.6). They are the most abundant white blood cells (50–70% of the total) and are most easily identified by a segmented nucleus made up of three to five lobes connected by thin strands of nuclear material. Because of the segmented nucleus, neutrophils are also called *polymorphonuclear leukocytes* ("polys") and *"segs."* Immature neutrophils, occasionally found in the circulation, can be identified by their horseshoe-shaped nucleus. These immature neutrophils go by the nicknames of *"bands"* and *"stabs."*

Neutrophils, like other blood cells, are formed in the bone marrow and released into the circulation. Most neutrophils remain in the blood but can leave the circulation if attracted to an extravascular site of damage or infection. In addition to ingesting bacteria and foreign particles, neutrophils release a variety of cytokines, including fever-causing *pyrogens* [p. 769] and chemical mediators of the inflammatory response.

Eosinophils **Eosinophils** are easily recognized by the bright pink-staining granules in their cytoplasm. These immune cells are associated with allergic reactions and parasitic diseases. Normally, few eosinophils are found in the peripheral circulation, where they account for only 1–3% of all leukocytes. The life span of an eosinophil in the blood is estimated to be only 6–12 hours.

Most functioning eosinophils are found in the digestive tract, lungs, urinary and genital epithelia, and connective tissue of the skin. These locations reflect their role in defense against parasitic invaders. Eosinophils are known to attach to large antibody-coated parasites, such as the blood fluke *Schistosoma,* and to release substances from their granules that damage or kill the parasites. Because eosinophils kill pathogens, they are classified as cytotoxic cells. Eosinophils also participate in allergic reactions, where they contribute to inflammation and tissue damage by releasing toxic enzymes, oxidative substances, and a protein called *eosinophil-derived neurotoxin*. Although eosinophils are classified as phagocytes because they have been observed ingesting foreign particles *in vitro*, the importance of this behavior in the body is unclear.

Monocytes and Macrophages **Monocytes** are the precursor cells of tissue macrophages. Monocytes are not very common in the blood (1–6% of all white blood cells). By some estimates, they spend only eight hours there in transit from the bone marrow to their permanent positions in the tissues.

Once in the tissues, monocytes enlarge and differentiate into phagocytic **macrophages**. Some tissue macrophages patrol the tissues, creeping along by amoeboid motion. Others find a location and remain fixed in place. In either case, macrophages are the primary scavengers within tissues. They are larger and more effective than neutrophils, ingesting up to 100 bacteria during their life span. Macrophages also remove larger particles, such as old red blood cells and dead neutrophils.

Macrophages play a very important role in the development of acquired immunity because they are antigen-presenting cells. After a macrophage ingests and digests molecular or cellular antigens, it can insert fragments of processed antigen into its membrane so that the antigen fragment becomes as part of surface protein complexes (Fig. 24.6c). In addition, dendritic cells and the lymphocytes known as *B lymphocytes* can also act as antigen-presenting cells, or APCs. When APCs have bits of antigen on their surface, they can bind to and activate other types of immune cells.

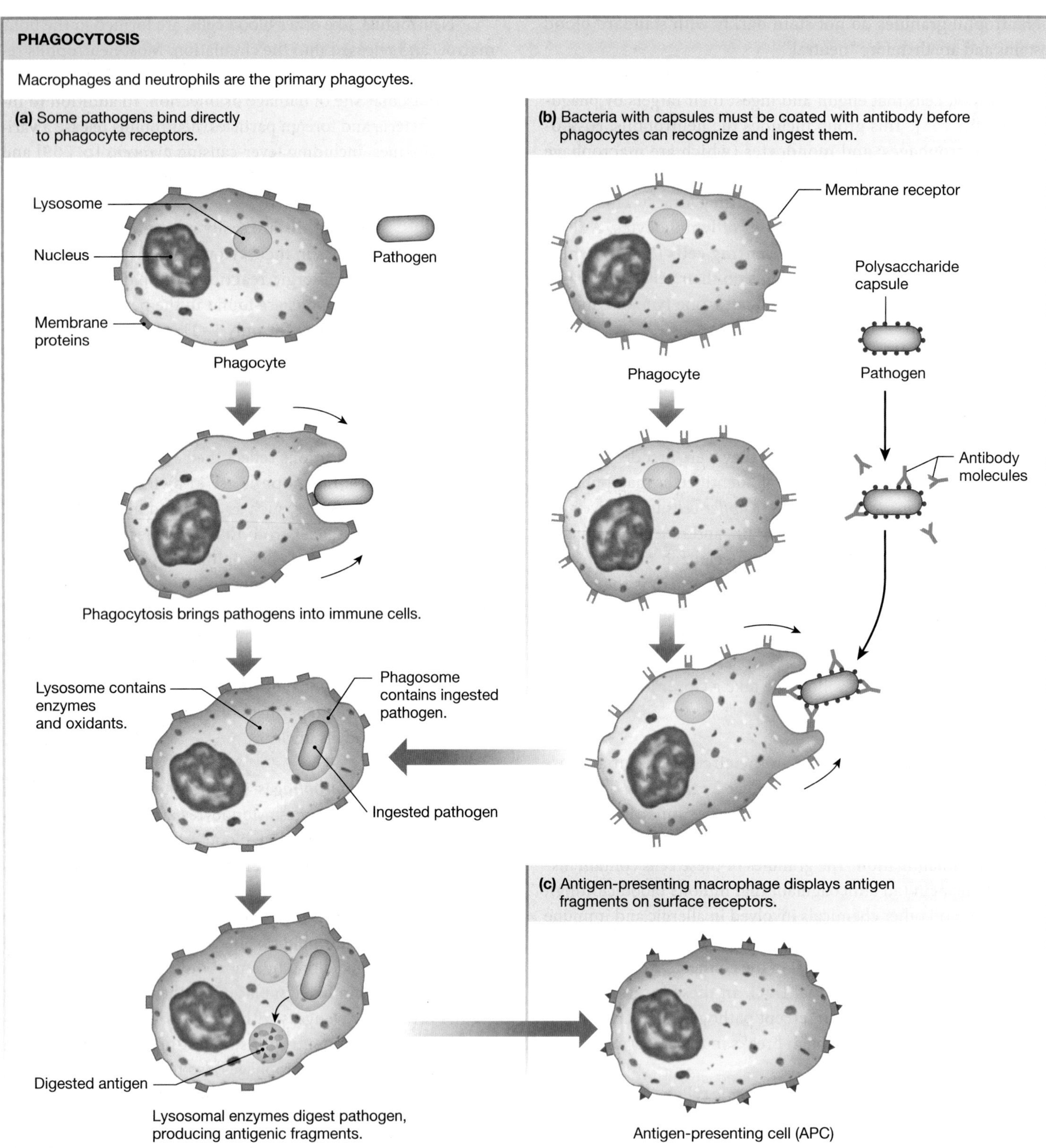
PHAGOCYTOSIS
Macrophages and neutrophils are the primary phagocytes.
(a) Some pathogens bind directly to phagocyte receptors.
Lysosome
Nucleus
Membrane proteins
Pathogen
Phagocyte
Phagocytosis brings pathogens into immune cells.
Lysosome contains enzymes and oxidants.
Phagosome contains ingested pathogen.
Ingested pathogen
Digested antigen
Lysosomal enzymes digest pathogen, producing antigenic fragments.
(b) Bacteria with capsules must be coated with antibody before phagocytes can recognize and ingest them.
Membrane receptor
Polysaccharide capsule
Phagocyte
Pathogen
Antibody molecules
(c) Antigen-presenting macrophage displays antigen fragments on surface receptors.
Antigen-presenting cell (APC)

Fig. 24.6

Lymphocytes **Lymphocytes** and their derivatives are the key cells that mediate the acquired immune response of the body. By one estimate, the adult body contains a trillion lymphocytes at any one time. Only 5% of these are found in the circulation, where they constitute 20–35% of all white blood cells. Most lymphocytes are found in lymphoid tissues, where they are especially likely to encounter invaders.

Although lymphocytes all look alike under the microscope, there are three major sub-types with significant differences in function and specificity, as you will learn later in the chapter.

B lymphocytes and their derivatives are responsible for antibody production and antigen presentation. *T lymphocytes* and *natural killer cells (NK cells)* play important roles in defense against intracellular pathogens, such as viruses.

Dendritic Cells **Dendritic cells** are antigen-presenting cells characterized by long, thin processes that resemble neuronal dendrites. Dendritic cells are found in the skin (where they are called *Langerhans cells*) and in various organs. When dendritic cells recognize and capture antigens, they migrate to secondary lymphoid tissues, such as lymph nodes, where they present the antigens to lymphocytes. Antigen binding activates the lymphocytes.

Innate Immunity: Nonspecific Responses

The body's first line of defense is to exclude pathogens by physical and chemical barriers. If invaders get past those barriers, the innate immune system provides the second line of defense. The innate immune response consists of patrolling and stationary immunocytes that attack and destroy invaders. These immune cells are genetically programmed to respond to a broad range of material that they recognize as foreign, which is why innate immunity is considered nonspecific. Innate immunity either clears the infection or contains it until the acquired immune response is activated.

Barriers Are the Body's First Line of Defense

Physical barriers of the body include the skin [p. 91], the protective mucous linings of the gastrointestinal and genitourinary tracts, and the ciliated epithelium of the respiratory tract (Fig. 24.2). The digestive and respiratory systems are most vulnerable to microbial invasion because these regions have extensive areas of thin epithelium in direct contact with the external environment. In women, the reproductive tract is also vulnerable, but to a lesser degree. The opening to the uterus is normally sealed by a plug of mucus that keeps bacteria in the vagina from ascending into the uterine cavity.

In the respiratory system, inhaled particulate matter is trapped by mucus lining the upper respiratory system. The mucus is then transported upward on the mucociliary escalator to be expelled or swallowed [p. 574]. Swallowed pathogens may be disabled by the acidity of the stomach. In addition, respiratory tract secretions contain **lysozyme**, an enzyme with antibacterial activity. Lysozyme attacks cell wall components of unencapsulated bacteria and breaks them down. However, it cannot digest the capsules of encapsulated bacteria.

Phagocytes Ingest Foreign Material

Pathogens that get past the physical barriers of skin and mucus are dealt with first by the innate immune response. A key element of the innate immune response is the ability of certain leukocytes to recognize molecules that are unique to microorganisms (*pathogen-associated molecular patterns*, or PAMPs) and initiate an appropriate response. PAMPs bind to leukocyte *pattern recognition receptors* (PRRs) that activate the nonspecific immune response. The initial response of these immune cells to invaders is to kill them or ingest them.

First, patrolling and stationary phagocytes are attracted to areas of invasion by chemical signals known as **chemotaxins**. Chemotaxins include bacterial toxins or cell wall components that act as PAMPs. Products of tissue injury, such as fibrin and collagen fragments, may also indicate a location that needs defending. Once on site, activated leukocytes fighting the invaders secrete their own chemotaxic cytokines to bring additional leukocytes to the infection site.

If the pathogen is in a tissue, circulating phagocytes leave the blood (*extravasation*) by squeezing through pores in the capillary endothelium. If an area of infection attracts a large number of phagocytes, the material we call **pus** may form. This thick, whitish to greenish substance is a collection of living and dead neutrophils and macrophages, along with tissue fluid, cell debris, and other remnants of the immune process.

Tissue macrophages and neutrophils are the primary phagocytic cells responsible for defense. Phagocyte membranes contain receptors that recognize many different types of foreign particles, both organic and inorganic. In macrophages, pattern recognition receptors known as **Toll-like receptors** (TLRs) activate the cell to secrete inflammatory cytokines (see below). Phagocytes ingest unencapsulated bacteria, cell fragments, carbon, and asbestos particles, among other materials. They will even ingest tiny polystyrene beads, providing one way that scientists in the laboratory analyze phagocytic activity.

Phagocytosis is a receptor-mediated event, which ensures that only unwanted particles are ingested. In the simplest phagocytic reactions, surface molecules on the pathogen act as ligands that bind directly to PRRs on the phagocyte membrane (Fig. 24.6a). In a sequence reminiscent of a zipper closing, the ligands and receptors combine sequentially, so that the phagocyte surrounds the unwanted foreign particle. The process is aided by actin filaments that push arms of the phagocytic cell around the invader.

The ingested particle ends up in a cytoplasmic vesicle called a **phagosome** (Fig. 24.6). Phagosomes fuse with intracellular lysosomes [p. 75], which contain powerful chemicals that destroy ingested pathogens. Lysosomal contents include enzymes and oxidizing agents, such as hydrogen peroxide (H_2O_2), nitric oxide (NO), and the superoxide anion (O_2^-).

Phagocytes cannot instantly recognize all foreign substances, however, because some pathogens lack markers that react with the phagocyte receptors. For example, certain bacteria have evolved a polysaccharide capsule that masks their surface markers from the host immune system. These encapsulated bacteria are not as quickly recognized by phagocytes and consequently are more pathogenic because they can grow unchecked until the immune system finally recognizes them and makes antibodies against them. Once that happens, the bacteria are "tagged" with a coat of antibodies so that phagocytes recognize them as something to be ingested (Fig. 24.6b).

The antibodies that tag encapsulated bacteria, along with additional plasma proteins, are known collectively as **opsonins** {*opsonin,* to buy provisions}. In the body, opsonins convert unrecognizable particles into "food" for phagocytes. Opsonins act as bridges between pathogens and phagocytes by binding to receptors on the phagocytes.

NK Cells Kill Infected and Tumor Cells

One class of lymphocyte—**natural killer cells**, or **NK cells**—participates in the innate response against viral infections. NK cells act more rapidly than other lymphocytes, responding within hours of a primary viral infection. NK cells recognize virus-infected cells and induce them to commit suicide by *apoptosis* [p. 90] before the virus can replicate. Complete elimination of the virus requires activation of a specific immune response. NK cells also attack some tumor cells.

NK cells and other lymphocytes secrete multiple antiviral cytokines, including **interferons,** which were named for their ability to interfere with viral replication. **Interferon-alpha** (IFN-α) and **interferon-beta** (IFN-β) target host cells and promote synthesis of antiviral proteins to prevent viral replication. **Interferon-gamma** (IFN-γ) activates macrophages and other immune cells.

Cytokines Create the Inflammatory Response

Inflammation is a hallmark reaction of innate immunity. Inflammation has three important roles in fighting infection in damaged tissue: (1) attracting immune cells and chemical mediators to the site, (2) producing a physical barrier to retard the spread of infection, and (3) promoting tissue repair once the infection is under control (a non-immunological function).

The inflammatory response is created when activated tissue macrophages release cytokines. These chemicals attract other immune cells, increase capillary permeability, and cause fever. Immune cells attracted to the site in turn release their own cytokines. Some representative examples of chemicals involved in the innate immune response are described below and listed in ■ Table 24.1, a convenient "scorecard" for following the "players" in these complex processes.

Acute-Phase Proteins In the time immediately following an injury or pathogen invasion (the acute phase), the body responds by increasing the concentration of various plasma proteins. Some of these proteins, produced mostly by the liver, are given the general name of **acute-phase proteins**. They include molecules that act as opsonins by coating pathogens; antiprotease molecules that help prevent tissue damage; and *C-reactive protein* (CRP).

Normally, the levels of acute-phase proteins decline to normal as the immune response proceeds, but in *chronic inflammatory diseases,* such as rheumatoid arthritis, elevated levels of acute-phase proteins may persist. One interesting finding is that increased levels of CRP are associated with increased risk of coronary heart disease. This association exists because atherosclerosis is an inflammatory process that begins when macrophages in blood vessels ingest excess cholesterol and become foam cells [p. 535] that secrete CRP, other cytokines, and growth factors. Elevated CRP is not a specific indicator of atherosclerosis, however, because CRP levels can also be elevated in other acute and chronic inflammatory conditions.

Histamine Histamine is found primarily in the granules of mast cells and basophils, and it is the active molecule that helps initiate the inflammatory response when mast cells degranulate. Histamine's actions bring more leukocytes to the injury site to kill bacteria and remove cellular debris.

Histamine opens pores in capillaries, allowing plasma proteins to escape into the interstitial space. This process causes local edema, or swelling. In addition, histamine dilates blood vessels (vasodilation), increasing blood flow to the area. The result of histamine release is a hot, red, swollen area around a wound or infection site.

Mast cell degranulation is triggered by different cytokines in the immune response. Because mast cells are concentrated under mucous membranes that line the airways and digestive tract, the inhalation or ingestion of certain antigens can trigger histamine release. The resultant edema in the nasal passages leads to one annoying symptom of seasonal pollen allergies: the stuffy nose. Fortunately, pharmacologists have developed a variety of drugs called *antihistamines,* which block the action of histamine at its receptor.

Scientists once believed that histamine was the primary chemical responsible for the bronchoconstriction of asthma and for the severe systemic allergic reaction called *anaphylactic shock* [p. 519]. However, they now know that mast cells release powerful lipid mediators in addition to histamine, including leukotrienes, platelet-activating factor, and prostaglandins. These mediators work with histamine to create the bronchoconstriction and hypotension characteristic of anaphylactic shock.

Interleukins Interleukins are cytokines initially thought to mediate communication among the body's different types of leukocytes. Scientists have since discovered that many different tissues in the body respond to interleukins. **Interleukin-1** (IL-1) is secreted by activated macrophages and other immune cells. Its main role is to mediate the inflammatory response, but it also has widespread systemic effects on immune function and metabolism. IL-1 modulates the immune response by:

1 *Altering blood vessel endothelium* to ease passage of white blood cells and proteins during the inflammatory response.
2 *Stimulating production of acute-phase proteins* by the liver.

Table 24.1 Chemicals of the Immune Response

Functional Classes
Acute phase proteins: Liver proteins that act as opsonins and that enhance the inflammatory response
Chemotaxins: Molecules that attract phagocytes to a site of infection
Cytokines: Proteins released by one cell that affect growth or activity of another cell
Opsonins: Proteins that coat pathogens so that phagocytes recognize and ingest them
Pyrogens: Fever-producing substances
Specific Chemicals and Their Functions
Antibodies (immunoglobulins, gamma globulins): Proteins secreted by B lymphocytes that fight specific invaders
Bradykinin: Stimulates pain receptors; vasodilator
Complement: Plasma and cell membrane proteins that act as opsonins, cytolytic agents, and mediators of inflammation
C-reactive protein: Opsonin that activates complement cascade
Granzymes: Cytotoxic enzymes that initiate apoptosis
Heparin: An anticoagulant
Histamine: Vasodilator and bronchoconstrictor released by mast cells and basophils
Interferons (IFN): Cytokines that inhibit viral replication and modulate the immune response
Interleukins (IL): Cytokines secreted by leukocytes to act primarily on other leukocytes; IL-1 mediates inflammatory response and induces fever
Kinins: Plasma proteins that activate to form bradykinin
Lysozyme: An extracellular enzyme that attacks bacteria
Major histocompatibility complex (MHC): Membrane protein complexes involved in cell recognition
Membrane attack complex: A membrane pore protein made in the complement cascade
Perforin: A membrane pore protein that allows granzymes to enter the cell; made by NK and cytotoxic T cells
Superoxide anion (O_2^-): Powerful oxidant in phagocyte lysosomes
T-cell receptors: T lymphocyte receptors that recognize and bind antigen presented by MHC receptors
Tumor necrosis factor (TNF): Cytokines that promote inflammation and can cause cells to self-destruct through apoptosis

3. *Inducing fever* by acting on the hypothalamic thermostat [p. 767]. IL-1 is a known pyrogen.
4. *Stimulating cytokine and endocrine secretion* by a variety of other cells.

Bradykinin Kinins are a group of inactive plasma proteins that participate in a cascade similar to the coagulation cascade [p. 561]. The end product of the kinin cascade is **bradykinin**, a molecule that has the same vasodilator effects as histamine. Bradykinin also stimulates pain receptors, creating the tenderness associated with inflammation. Pain draws the brain's attention to the injury.

Complement Proteins **Complement** is a collective term for a group of more than 25 plasma proteins and cell membrane proteins. The *complement cascade* is similar to the blood coagulation cascade. The complement proteins are secreted in inactive forms that are activated as the cascade proceeds. Intermediates of the complement cascade act as opsonins, chemical attractants for leukocytes, and agents that cause mast cell degranulation.

The complement cascade terminates with the formation of **membrane attack complex**, a group of lipid-soluble proteins that insert themselves into the cell membranes of pathogens and virus-infected cells and form giant pores (■ Fig. 24.7). These pores allow water and ions to enter the pathogen cells. As a result, the cells swell and lyse.

Concept Check Answer: p. 834

3. How does the action of histamine on capillary permeability result in swelling?

Acquired Immunity: Antigen-Specific Responses

Acquired immune responses are *antigen-specific responses* in which the body recognizes a particular foreign substance and selectively reacts to it. Acquired immunity is mediated primarily by lymphocytes.

Membrane attack complex creates pores in pathogens.

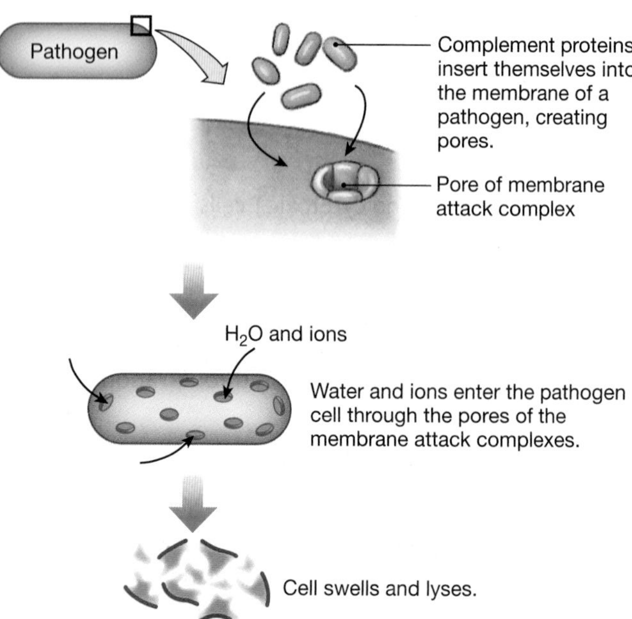
Pathogen
Complement proteins insert themselves into the membrane of a pathogen, creating pores.
Pore of membrane attack complex
H_2O and ions
Water and ions enter the pathogen cell through the pores of the membrane attack complexes.
Cell swells and lyses.

■ **Fig. 24.7**

There are three main types of lymphocytes: B lymphocytes, T lymphocytes, and natural killer (NK) cells. Activated **B lymphocytes** develop into **plasma cells**, which secrete antibodies. Activated **T lymphocytes** develop either into cells that attack and destroy virus-infected cells (**cytotoxic T cells, T_C**) or into cells that regulate other immune cells (**helper T cells, T_H**). NK cells attack and destroy virus-infected cells and tumor cells as part of the innate response discussed previously. All lymphocytes secrete cytokines that act on immune cells, on non-immune cells, and, sometimes, on pathogens.

The process of acquired immunity overlaps with the process of innate immunity. Cytokines released by the inflammatory response attract lymphocytes to the site of an immune reaction. The lymphocytes release additional cytokines that enhance the inflammatory response.

Acquired immunity can be subdivided into active immunity and passive immunity. **Active immunity** occurs when the body is exposed to a pathogen and produces its own antibodies. Active immunity can occur naturally, when a pathogen invades the body, or artificially, as when we are given vaccinations containing dead or disabled pathogens.

Passive immunity occurs when we acquire antibodies made by another organism. The transfer of antibodies from mother to fetus across the placenta is one example. Injections containing antibodies are another. Travelers going abroad may be injected with *gamma globulin* (antibodies extracted from donated human plasma), but this passive immunity lasts only about three months as the injected proteins degrade and are cleared from the circulation.

Lymphocytes Mediate the Acquired Immune Response

On a microscopic level, all lymphocytes look alike. At the molecular level, however, the cell types can be distinguished from one another by their membrane receptors. Each lymphocyte binds only one particular ligand. All lymphocytes that bind that ligand form a group known as a **clone** {*klon*, a twig} (■ Fig. 24.8a).

If each pathogen that entered the body needs a dedicated type of lymphocyte, there must be millions of different types of lymphocytes ready to combat millions of different pathogens. But how can the body store both the number and variety of lymphocytes needed for adequate defense? As it turns out, the immune system keeps only a few of each type of lymphocyte on hand. If the pathogen these cells fight appears, the cells reproduce to provide the numbers needed.

At an individual's birth, each clone of lymphocytes is represented by only a few cells, called **naïve lymphocytes**. Because the small number of cells in each naïve clone is not enough to fight off foreign invaders, the first exposure to an antigen activates the appropriate clone and stimulates it to divide

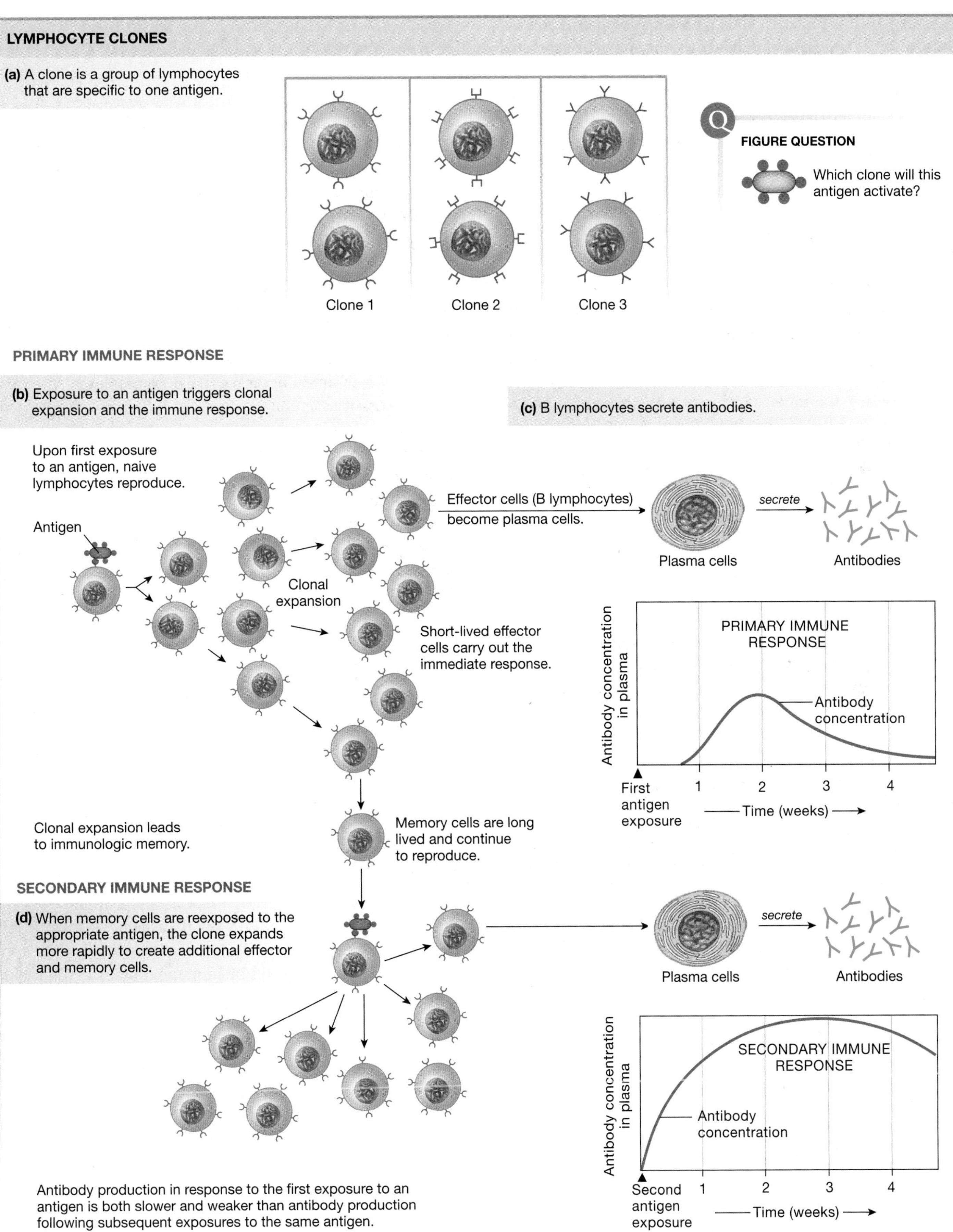
LYMPHOCYTE CLONES
(a) A clone is a group of lymphocytes that are specific to one antigen.
Clone 1
Clone 2
Clone 3
FIGURE QUESTION
Which clone will this antigen activate?
PRIMARY IMMUNE RESPONSE
(b) Exposure to an antigen triggers clonal expansion and the immune response.
Upon first exposure to an antigen, naive lymphocytes reproduce.
Antigen
Clonal expansion
Effector cells (B lymphocytes) become plasma cells.
Short-lived effector cells carry out the immediate response.
Clonal expansion leads to immunologic memory.
Memory cells are long lived and continue to reproduce.
(c) B lymphocytes secrete antibodies.
Plasma cells
secrete
Antibodies
PRIMARY IMMUNE RESPONSE
Antibody concentration
Antibody concentration in plasma
First antigen exposure
1
2
3
4
Time (weeks)
SECONDARY IMMUNE RESPONSE
(d) When memory cells are reexposed to the appropriate antigen, the clone expands more rapidly to create additional effector and memory cells.
Plasma cells
secrete
Antibodies
SECONDARY IMMUNE RESPONSE
Antibody concentration
Antibody concentration in plasma
Second antigen exposure
1
2
3
4
Time (weeks)
Antibody production in response to the first exposure to an antigen is both slower and weaker than antibody production following subsequent exposures to the same antigen.

Fig. 24.8

(Fig. 24.8b). This process, called **clonal expansion**, creates additional cells in the clone. Naïve cells continue to be generated throughout an individual's lifetime.

The newly formed lymphocytes in an expanded clone differentiate into effector cells and memory cells. **Effector cells** carry out the immediate response and then die within a few days. **Memory cells**, in contrast, are long lived and continue reproducing themselves. Second and subsequent exposures to the antigen activate the memory cells and cause rapid clonal expansion, creating a quicker and stronger secondary response to the antigen.

B Lymphocytes Become Plasma Cells and Memory Cells

B lymphocytes (also called **B cells**) develop in the bone marrow. (Memory aid: bone starts with B.) Clones of activated B lymphocytes differentiate primarily into specialized cells that secrete antibodies. The word *antibody* describes what the molecules do: work against foreign bodies. Antibodies are also called **immunoglobulins**, and this alternative name describes what the molecules are: globular proteins that participate in the humoral immune response.

Mature B lymphocytes insert antibody molecules into their cell membranes so that the antibodies become surface receptors marking the members of each clone (Fig. 24.8). When a clone of B lymphocytes responds to antigen exposure, some of the effector cells differentiate into plasma cells. Plasma cells do not have antibody proteins bound in their membranes. Instead, they synthesize and secrete additional antibody molecules at incredible rates, estimated to be as high as 2000 molecules per second! Plasma cell antibodies form *humoral immunity,* the soluble antibodies of the plasma.

After each invader has been successfully repulsed, the short-lived plasma cells die off—it could be dangerous to have them continuing to secrete antibody after the antigen is gone. A few memory cells of the clone remain behind, waiting to respond to the next exposure to the same antigen. Figure 24.8 shows B cell responses following the first and second exposures to an antigen.

The **primary immune response** occurs after the initial exposure. Antibody production by plasma cells is slower and lower in magnitude because the body has not encountered the antigen previously. The **secondary immune response,** which takes place after second and subsequent exposures, is quicker and larger because of memory cells that remain behind after the first exposure. Clonal expansion is enhanced by lymphocytes that carry a molecular memory of the first exposure to the antigen, so antibody production begins sooner and reaches higher concentrations.

The existence of a secondary immune response is what allows vaccinations to be an effective protection from disease. A vaccine contains an altered pathogen that no longer harms the host but that can be recognized as foreign by immune cells. The altered pathogen triggers creation of memory cells specific to that particular pathogen. If a vaccinated person is later infected by the pathogen, memory cells created by the vaccine produce a stronger and more rapid secondary immune response. We are now learning, however, that immunity after immunization may not last for a person's lifetime.

RUNNING PROBLEM

In 2007 there were more than 12,000 cases of cervical cancer in the United States, and more than 4000 women died from the disease. The HPV vaccines were developed to prevent HPV infection and decrease the incidence of cervical cancer. The first vaccine, Gardasil®, protects against four different strains of HPV: 16 and 18, associated with cervical cancer, and types 6 and 11, which cause genital warts. A newer vaccine, Cervarix®, protects against only types 16 and 18. The vaccines are composed of viruslike particles (VLPs), which are proteins from the HPV virus capsid. When injected into a person, VLPs initiate an antibody response that has a protective effect against the types of HPV from which they were derived.

Q3: Antibody production begins with activation of which type of lymphocyte? What type of cell produces antibodies?

Q4: In a normal infection, HPV infects cells in the superficial layers of the skin but does not enter the circulation. What kind of immune cell is mostly likely to encounter HPV in the skin?

Antibodies Are Proteins Secreted by Plasma Cells

Antibodies were among the first aspects of the immune system to be discovered, and traditionally their names—agglutinins, precipitins, hemolysins, and so on—indicated what they do. Today, however, antibodies or immunoglobulins (Ig) are divided into five general classes: IgG, IgA, IgE, IgM, and IgD (pronounced *eye-gee-[letter]*). Antibodies are collectively referred to as **gamma globulins**.

IgGs make up 75% of plasma antibody in adults because they are produced in secondary immune responses. Some maternal IgGs cross the placental membrane and give infants immunity in the first few months of life. Some IgGs activate complement.

IgA antibodies are found in external secretions, such as saliva, tears, intestinal and bronchial mucus, and breast milk, where they bind to pathogens and flag them for phagocytosis if they reach the internal environment.

IgEs target gut parasites and are associated with allergic responses. When mast cell receptors bind with IgEs and antigen, the mast cells degranulate and release chemical mediators, such as histamine.

IgM antibodies are associated with primary immune responses and with the antibodies that react to blood group antigens. IgMs strongly activate complement.

IgD antibody proteins appear on the surface of B lymphocytes along with IgM, but the physiological role of IgDs is unclear.

Antibody Proteins The basic antibody molecule has four polypeptide chains linked into a Y shape (■ Fig. 24.9a). The two sides of the Y are identical, with one **light chain** attached to

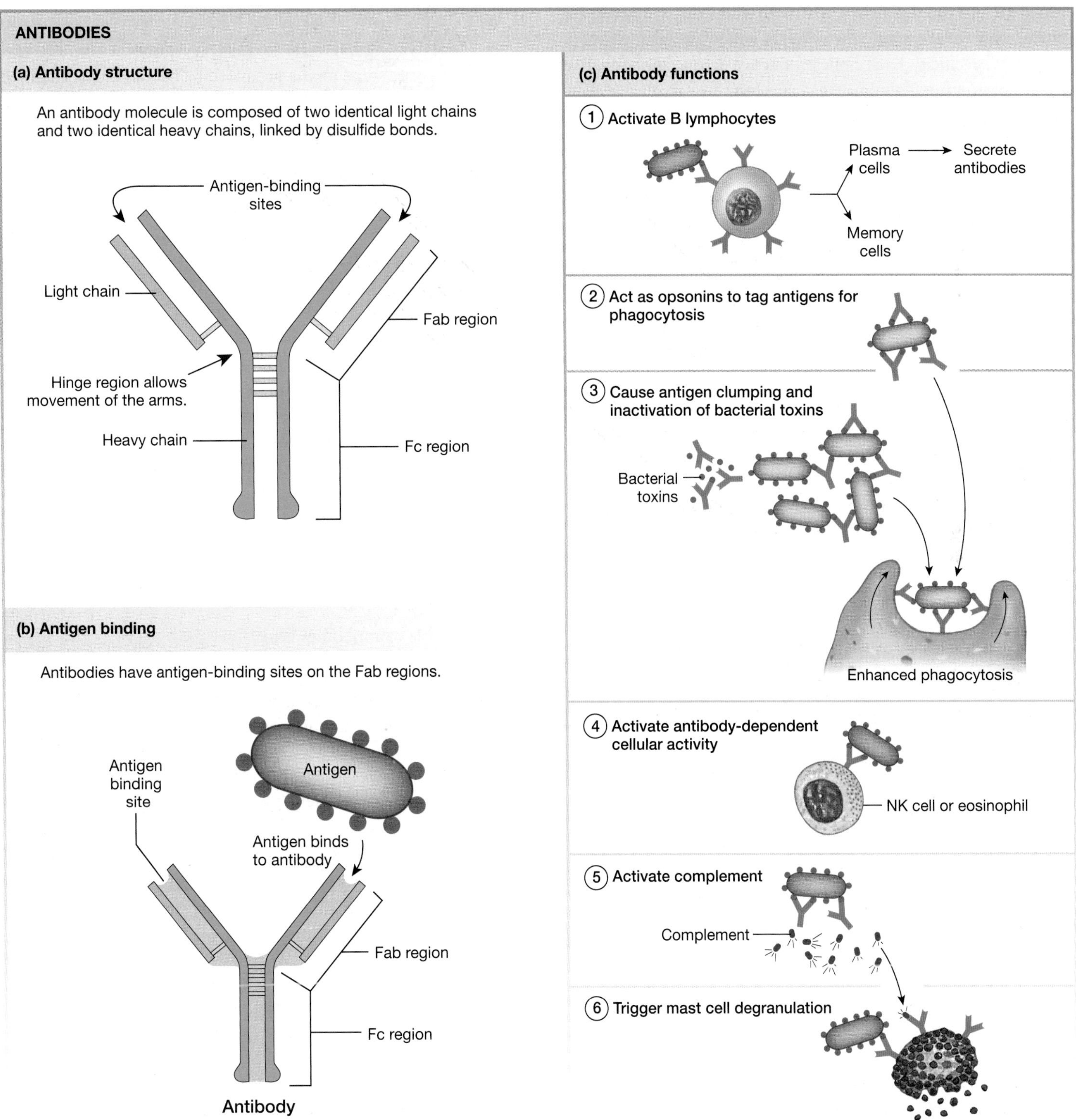
ANTIBODIES
(a) Antibody structure
An antibody molecule is composed of two identical light chains and two identical heavy chains, linked by disulfide bonds.
Antigen-binding sites
Light chain
Fab region
Hinge region allows movement of the arms.
Heavy chain
Fc region
(b) Antigen binding
Antibodies have antigen-binding sites on the Fab regions.
Antigen binding site
Antigen
Antigen binds to antibody
Fab region
Fc region
Antibody
(c) Antibody functions
1 Activate B lymphocytes
Plasma cells
Secrete antibodies
Memory cells
2 Act as opsonins to tag antigens for phagocytosis
3 Cause antigen clumping and inactivation of bacterial toxins
Bacterial toxins
Enhanced phagocytosis
4 Activate antibody-dependent cellular activity
NK cell or eosinophil
5 Activate complement
Complement
6 Trigger mast cell degranulation

■ **Fig. 24.9**

one **heavy chain.** The two arms, or **Fab regions,** form antigen-binding sites that confer the antibody's specificity (Fig. 24.9b).

The stem of the Y-shaped antibody molecule is known as the **Fc region**. The Fc region determines the Ig class to which the antibody belongs. A *hinge region* between the arms and the stem allows flexible positioning of the arms as the antibody binds to the antigen.

In any given antibody molecule, the two light chains are identical and the two heavy chains are identical. However, the chains vary widely among the different antibodies, giving the antibody its specificity. Each clone produces a unique antibody. Two classes of immunoglobulins (IgM and IgA) are secreted as polymers: IgM is made up of five Y-shaped antibody molecules, and IgA has from one to four antibody molecules.

Concept Check Answers: p. 834

4. Because antibodies are proteins, they are too large to cross cell membranes on transport proteins or through channels. How then do IgAs and other antibodies become part of external secretions such as saliva, tears, and mucus?
5. A child is stung by a bee for the first time. Why should the parent be particularly alert when the child is stung a second time?

Antibody Functions Most antibodies are found in the blood, where they make up about 20% of the plasma proteins in a healthy individual. These antibodies are most effective against extracellular pathogens (such as bacteria), some parasites, antigenic macromolecules, and viruses that have not yet invaded their host cells.

Antibody function is summarized in Figure 24.9c. Although antibodies are not toxic and cannot destroy antigens, they help the immune system react to the antigen. Sometimes the antibody binds first to the antigen. Other times immune cells have antibody inserted into their cell membranes, waiting for the antigen to bind to the cell.

First let's consider examples where the antibody binds first to the antigen. The second step is for the Fc region of the antibody-antigen complex to bind to Fc receptors on an immune cell. The presence of a single Fc receptor type eliminates the need for immune cells to have millions of different receptors that recognize different antigens. Instead, with one type of receptor the immune cells are activated by any antibody-bound antigen. If the immune cell is a phagocyte, Fc binding initiates phagocytosis (Fig. 24.9c). To facilitate phagocytosis, antibodies:

1 **Act as opsonins** (2). Soluble antibodies coat antigens to facilitate recognition and phagocytosis by immune cells.

2 **Make antigens clump** (3). Antibody-caused clumping of antigens enhances recognition of the antigen for phagocytosis.

If the immune cell that binds the antibody is an eosinophil or NK cell, Fc binding:

3 **Activates cytotoxic cell responses.** Cytotoxic cells release chemicals that destroy the antibody-bound antigen (4). This nonspecific response of cytotoxic cells to antibody binding is called **antibody-dependent cell-mediated cytotoxicity.**

Antibodies also enhance inflammation by:

4 **Inactivating bacterial toxins** (3). Antibodies bind to and inactivate some toxins produced by bacteria. One example where antibodies neutralize a bacterial toxin is infection by *Corynebacterium diphtheria. C. diphtheria* is the bacterium that causes diphtheria, an upper respiratory infection. In this disease, the bacterial toxin kills host cells, leaving ulcers that have a characteristic grayish membrane. Natural immunity to the disease occurs when the host produces antibodies that disable the toxin. To develop a vaccine for diphtheria, researchers created an inactivated toxin preparation that did not harm living cells. When administered to a person, the vaccine triggers antibody production without causing any symptoms of the disease. As a result, diphtheria has been almost eliminated in countries with good immunization programs.

5 **Activating complement** (5). Antigen-bound antibodies use the Fc end of the antibody molecule to activate complement.

6 **Activating mast cells** (6). Mast cells have IgE antibodies attached to their surface. When antigens or complement proteins bind to IgE, the mast cells degranulate, releasing chemicals that mediate the inflammatory response.

The final role of antibodies is to:

7 **Activate B lymphocytes.** The surface of every B lymphocyte is covered with as many as 100,000 antibody molecules whose Fc ends are inserted into the lymphocyte membrane (Fig. 24.9c (1)). This arrangement leaves the Fab regions of the membrane-bound antibodies available to bind to free-floating extracellular antigens, such as viruses or bacterial toxins, or to bits of antigen on the surface of antigen-presenting cells (Fig. 24.6c). Once antigen is bound, the activated B cells then differentiate into plasma cells and secrete more antibodies. Some B cells differentiate into memory cells to await a subsequent invasion.

T Lymphocytes Use Contact-Dependent Signaling

Antibodies are typically effective only against extracellular pathogens because antibodies can bind only to soluble or exposed antigens. Once a pathogen gets inside a host cell, it can no longer be "seen" by the humoral immune system. Defending the body against intracellular pathogens is the role of cytotoxic T lymphocytes, which carry out **cell-mediated immunity**. In this process, cytotoxic T cells bind to cells that display foreign antigen fragments as part of a *major histocompatibility complex* (MHC) on their surface.

T lymphocytes develop in the thymus gland (■ Fig. 24.10). Immature precursor cells migrate from the bone marrow to the thymus. During the development phase, cells that do not react with "self" form clones and insert **T-cell receptors** into their cell membranes (■ Fig. 24.11a). T-cell receptors are not antibodies like the receptors on B lymphocytes, although the proteins are closely related. T cell receptors can bind only to MHC-antigen complexes on the surface of an antigen-presenting cell. That means that T cells cannot bind to free-floating antigens as B cells do.

MHC and Antigen What is MHC? The **major histocompatibility complexes** are a family of membrane protein complexes encoded by a specific set of genes. (These proteins were named when they were discovered to play a role in rejecting foreign tissue following organ or tissue transplants.) We know now that every nucleated cell of the body has MHC proteins on its membrane.

MHC proteins combine with fragments of antigens that have been digested within the cell. The MHC-antigen complex is then inserted into the cell membrane so that the antigen is visible on the extracellular surface (see Fig. 24.6c for an example). Free antigen in the extracellular fluid cannot bind to unoccupied MHC receptors on the cell surface.

There are two types of MHC molecules. **MHC class I molecules** are found on all nucleated human cells. When viruses and bacteria invade the cell, they are digested into peptide fragments and loaded onto MHC-I "platforms." If a **cytotoxic T cell** (T_C cell) encounters a host cell with foreign antigen fragment on its MHC-I, the T_C cell recognizes the target as either a virus-infected cell or as a tumor cell and kills it to prevent it from reproducing (Fig. 24.11a).

MHC class II molecules are found primarily on the antigen-presenting cells (APCs): macrophages, B lymphocytes, and dendritic cells. When an immune cell engulfs and digests an antigen, the fragments are returned to the immune cell membrane combined with MHC-II proteins. If a **helper T cell** (T_H cell) encounters an APC with a foreign antigen fragment on its MHC-II, the T_H cell responds by secreting cytokines that enhance the immune response.

24

■ **Fig. 24.10** ***FOCUS ON ...***

The Thymus Gland

The thymus gland is a two-lobed organ located in the thorax just above the heart.

The thymus gland reaches its greatest size during adolescence.Then it shrinks and is largely replaced by adipose tissue as a person ages.

During development in the thymus, those cells that would be self-reactive are eliminated. Those that do not react with "self" tissues multiply to form clones.

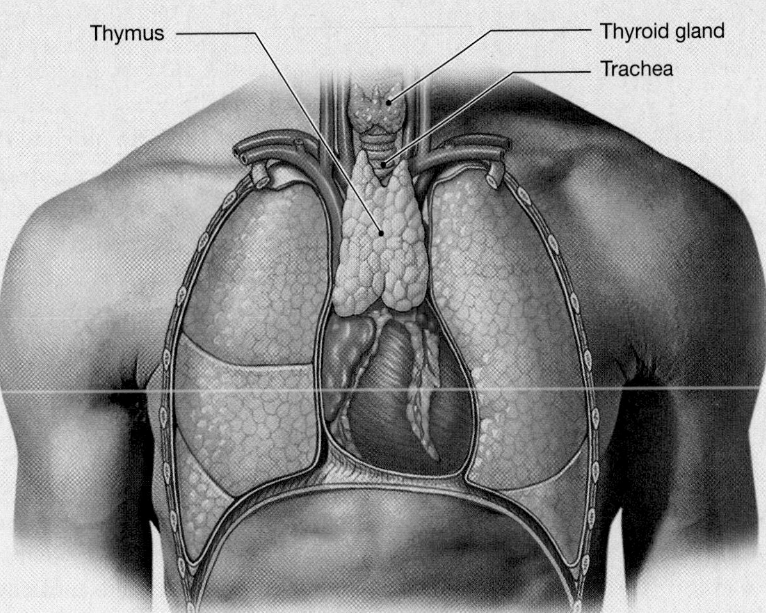

The thymus gland produces:

- T lymphocytes
- Peptides
  thymosin
  thymopoietin
  thymulin

FIGURE QUESTION

New T lymphocyte production in the thymus is low in adults, but the number of T lymphocytes in the blood does not decrease. What conclusion(s) about T lymphocytes can you draw from this information?

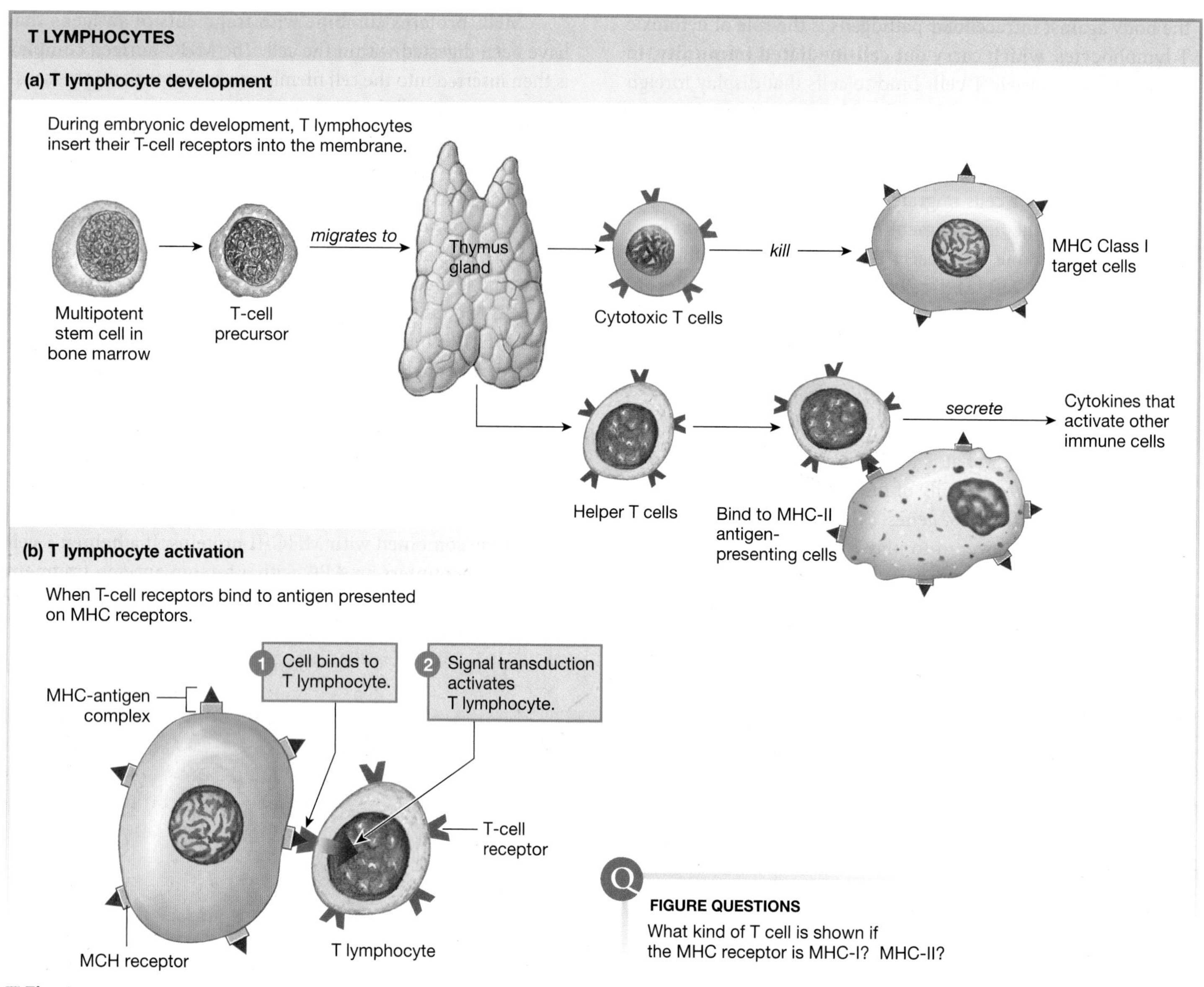
T LYMPHOCYTES
(a) T lymphocyte development
During embryonic development, T lymphocytes insert their T-cell receptors into the membrane.
migrates to
Thymus gland
Multipotent stem cell in bone marrow
T-cell precursor
kill
MHC Class I target cells
Cytotoxic T cells
secrete
Cytokines that activate other immune cells
Helper T cells
Bind to MHC-II antigen-presenting cells
(b) T lymphocyte activation
When T-cell receptors bind to antigen presented on MHC receptors.
1 Cell binds to T lymphocyte.
2 Signal transduction activates T lymphocyte.
MHC-antigen complex
T-cell receptor
MCH receptor
T lymphocyte
FIGURE QUESTIONS
What kind of T cell is shown if the MHC receptor is MHC-I? MHC-II?

Fig. 24.11

All MHC proteins are related, but they vary from person to person because of the huge number of alleles (variants of a gene) people inherit from their parents. There are so many alleles that it is unlikely that any two people other than identical twins inherit exactly the same set. Major histocompatibility complexes are one reason tissues cannot be transplanted from one person to another without first establishing compatibility.

Cytotoxic T Cells Cytotoxic T (T_C) cells attack and destroy cells that display MHC-I-antigen complexes. Although this may seem to be an extreme response, it prevents the reproduction of intracellular invaders such as viruses, some parasites, and some bacteria when the cells infected by these pathogens are targeted for destruction.

How do cytotoxic T cells kill their targets? They do so in two ways. First they can release a cytotoxic pore-forming molecule called **perforin** along with **granzymes**, enzymes related to the digestive enzymes trypsin and chymotrypsin. When granzymes enter the target cell through perforin channels, they activate an enzyme cascade that induces the cell to commit suicide (*apoptosis*). Second, cytotoxic T cells instruct target cells to undergo apoptosis by activating **Fas**, a "death receptor" protein on the target cell membrane that is linked to the enzyme cascade.

Helper T (T_H) cells Helper T (T_H) cells do not directly attack pathogens and infected cells, but they play an essential role in the immune response by secreting cytokines that activate other immune cells. The cytokines secreted by T_H cells include (1) *interferon-gamma* (IFN-γ), which activates macrophages; (2) *interleukins* that activate antibody production and cytotoxic T lymphocytes; (3) *colony-stimulating factors*, which enhance leukocyte production; and (4) *interleukins* that support the actions of mast cells and eosinophils. Helper T cells also bind to B cells and promote their differentiation into plasma cells and memory B cells. HIV, the virus that causes AIDS, preferentially infects and destroys T_H cells, leaving the host unable to respond to pathogens that otherwise could be easily suppressed.

Immune Response Pathways

How does the body respond to different kinds of immune challenges? The details depend on the particular challenge, but the basic pattern is the same. The innate response starts first, and it is reinforced by the more specific acquired response. The two pathways are interconnected, so cooperation and communication are essential. In the following sections we examine the body's responses in four situations: an extracellular bacterial infection, a viral infection, an allergic response to pollen, and the transfusion of incompatible blood.

Bacterial Invasion Causes Inflammation

What happens when bacteria invade? As we've seen, if the passive barricades of the skin and mucous membranes fail, bacteria reach the extracellular fluid. There they usually cause an inflammatory response that represents the combined effects of many cells working to get rid of the invader. If bacteria enter the lymph, infection-fighting takes place in lymph nodes as well.

Inflammation is characterized by a red, swollen warm area that is tender or painful. In addition to the nonspecific inflammatory response, lymphocytes attracted to the area produce antibodies keyed to the specific type of bacterium. The entry of bacteria sets off several interrelated reactions (■ Fig. 24.12):

1 *Activity of the complement system.* Components of the bacterial cell wall activate the complement system. Some products of the complement cascade are chemical signals (*chemotaxins*) that attract leukocytes from the circulation to help fight the infection. Others act as opsonins to enhance phagocytosis.

Complement also causes degranulation of mast cells and basophils. Cytokines secreted by mast cells act as additional chemotaxins, attracting more immune cells. Vasoactive chemicals such as histamine dilate blood vessels and increase capillary permeability. The enhanced blood supply to the site creates the redness and warmth of inflammation. Plasma proteins that escape into the interstitial space pull water with them, leading to tissue edema (swelling).

The complement cascade ends with formation of membrane attack complex molecules that insert themselves into the bacterial wall of unencapsulated bacteria. The subsequent influx of ions and water lyses the bacteria, aided by the enzyme lysozyme. This is a purely chemical process that does not involve immune cells.

2 *Activity of phagocytes.* If the bacteria are not encapsulated, macrophages can begin to ingest the bacteria immediately. But if the bacteria are encapsulated, the capsule hides the bacteria from recognition by the macrophage receptors. Antibodies must coat the capsule before the bacteria can be identified and ingested by phagocytes. Opsonins enhance phagocytosis for bacteria that are not encapsulated. Molecules that act as opsonins include complement, acute-phase proteins, and antibodies.

3 *Role of the acquired immune response.* Some elements of the acquired immune response are called into play in bacterial infections. If antibodies against the bacteria are already present, they enhance the innate response by acting as opsonins and neutralizing bacterial toxins. Antigen-presenting cells that digest the bacteria can then move to secondary lymphoid tissues, where they present antigen to memory cells to initiate more antibody production.

If the infection is a novel one, some of the bacterial antigens activate naïve B cells with the assistance of APC and helper T cells. Antigen-presenting cells ingest the bacteria and present bacterial fragments to helper T cells to activate them. This triggers cytokine secretion from the T_H cells, B cell clonal expansion, antibody production by plasma cells, and formation of memory B and T_H cells.

4 *Initiation of repair.* If the initial wound damaged blood vessels underlying the skin, platelets and the proteins of the coagulation cascade are also recruited to minimize the damage [p. 561]. Once the bacteria are removed by the immune response, the injured site is repaired under the control of growth factors and other cytokines.

Viral Infections Require Intracellular Defense

What happens when viruses invade the body? First, they encounter an extracellular phase of immune response similar to that described for bacteria. In the early stages of a viral infection, innate immune responses and antibodies can help control the invasion of the virus.

Once the viruses enter the host's cells, humoral immunity in the form of antibodies is no longer effective. Cytotoxic T lymphocytes (and to a lesser extent, NK cells) are the main defense against intracellular viruses. When these cells recognize infected host cells, they destroy them.

For years, T cell–mediated immunity and humoral immunity controlled by B lymphocytes were considered independent processes. We now know that the two types of immunity are linked. ■ Figure 24.13 depicts how these two types of lymphocytes coordinate to destroy viruses and virus-infected cells. In this figure, we have assumed prior exposure to the virus and the presence of preexisting antibodies in the circulation. Antibodies can play an important defensive role in the early extracellular stages of a viral infection.

1 Antibodies act as opsonins (1), coating viral particles to make them better targets for antigen-presenting cells such

Fig. 24.12 ESSENTIALS

Immune Responses to Extracellular Bacteria

Bacterial infections cause inflammation and trigger specific immune responses.

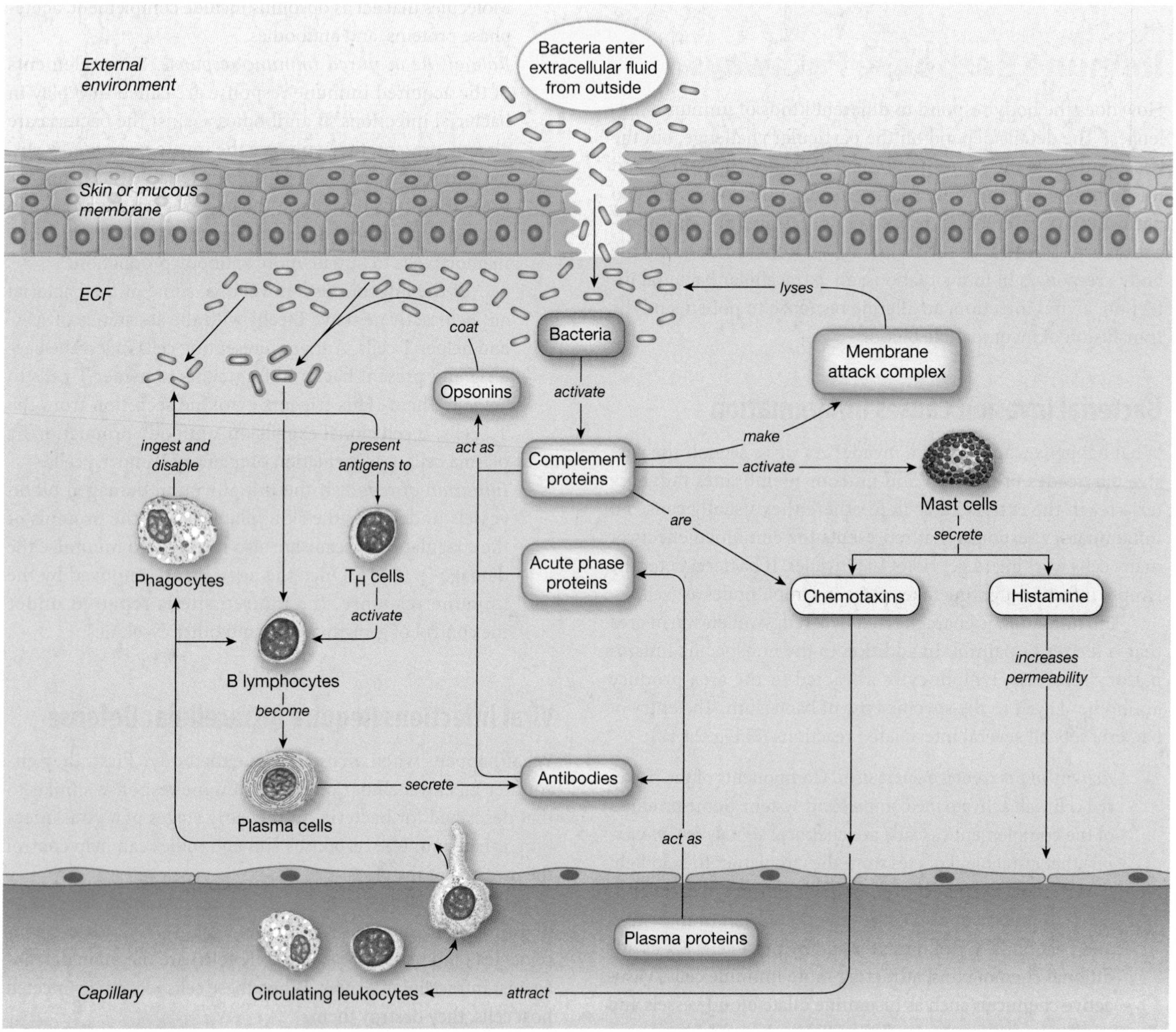

External environment
Bacteria enter extracellular fluid from outside
Skin or mucous membrane
ECF
coat
Bacteria
lyses
Membrane attack complex
Opsonins
activate
make
ingest and disable
present antigens to
act as
Complement proteins
activate
Mast cells
are
secrete
Phagocytes
T_H cells
Acute phase proteins
Chemotaxins
Histamine
activate
B lymphocytes
increases permeability
become
secrete
Antibodies
Plasma cells
act as
Plasma proteins
Capillary
Circulating leukocytes
attract

as macrophages. Antibodies can also bind to virus particles, preventing them from entering their target cells. However, once the virus is inside host cells, antibodies are no longer as effective.

2 Macrophages that ingest viruses insert fragments of viral antigen into MHC-II molecules on their membranes (2). Macrophages also secrete a variety of cytokines. Some of these cytokines initiate the inflammatory response. They produce interferon-α, which causes host cells to make antiviral proteins that keep viruses from replicating. Other macrophage cytokines stimulate NK cells and helper T cells.

3 Helper T cells bind to viral antigen on macrophage MHC-II molecules (3). Activated T_H cells then secrete cytokines to stimulate B lymphocytes and cytotoxic cells.

4 Previous exposure to the virus created memory B lymphocytes with the viral antibody on their surface. This second

■ Fig. 24.13 *ESSENTIALS*

Immune Responses to Viruses

This figure assumes prior exposure to the virus and preexisting antibodies.

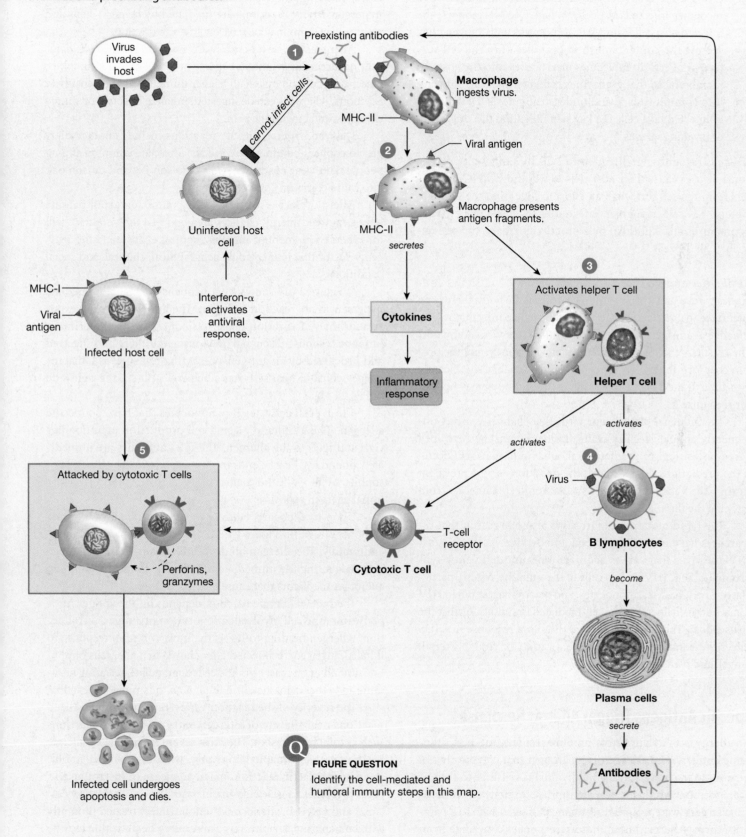

Virus invades host

Preexisting antibodies

1

cannot infect cells

Macrophage ingests virus.

MHC-II

2

Viral antigen

MHC-II

Macrophage presents antigen fragments.

secretes

Uninfected host cell

MHC-I

Viral antigen

Infected host cell

Interferon-α activates antiviral response.

Cytokines

Inflammatory response

3

Activates helper T cell

Helper T cell

activates

activates

4

Virus

B lymphocytes

become

Plasma cells

secrete

Antibodies

5

Attacked by cytotoxic T cells

Perforins, granzymes

T-cell receptor

Cytotoxic T cell

Infected cell undergoes apoptosis and dies.

Q

FIGURE QUESTION

Identify the cell-mediated and humoral immunity steps in this map.

exposure to the virus activates the memory cells and promotes development of plasma cells, resulting in additional antibody production (4).

5 Cytotoxic T cells use viral antigen–MHC-I complexes to recognize infected host cells. When T_C cells recognize the infected host cell from MHC-I receptors with antigen, they secrete the contents of their granules onto the cell surface (5). Perforin molecules insert pores into the host cell membrane so that granzymes can enter the cell, inducing it to commit suicide and undergo apoptosis. Destruction of infected host cells is a key step in halting the replication of invading viruses.

NK cells recognize virally infected cells by a different process. Some viruses cause their host cells to withdraw MHC-I receptors from the cell surface in an effort to hide from the immune system. NK cells recognize infected host cells lacking MHC-I complexes and kill them by a process similar to the one described for T_C cells.

Antibodies and Viruses Although Figure 24.13 shows antibodies from a previous infection protecting the body, there is no guarantee that antibodies produced during one infection will be effective against the next invasion by the same virus. Why not? The reason is that many viruses mutate constantly, and the protein coat forming the primary antigen may change significantly over time. If the protein coat changes, the antibody may no longer recognize it.

The influenza virus is one virus that changes yearly. Consequently, annual vaccines against influenza must be developed based on virologists' predictions of what mutations will occur. If the predictions do not match the mutations or the prevalent strains, the vaccine for that year cannot keep someone from catching the flu.

The rapid mutation of viruses is also one reason that researchers have not yet developed an effective vaccine against HIV, the virus that causes **acquired immunodeficiency syndrome** (AIDS). HIV infects cells of the immune system, particularly T lymphocytes, monocytes, and macrophages. When HIV wipes out the helper T cells, cell-mediated immunity against the virus is lost. The general loss of an immune response in AIDS leaves these patients susceptible to a variety of viral, bacterial, fungal, and parasitic infections.

Specific Antigens Trigger Allergic Responses

An **allergy** is an inflammatory immune response to a nonpathogenic antigen. The **allergen** is an antigen that is typically not harmful to the body. But if an individual is sensitive to the antigen, the body produces an inappropriate immune response, as if the antigen were a more threatening pathogen such as a parasitic worm. Allergic inflammatory responses can range from mild tissue damage to fatal reactions.

The immune response in allergies is called **sensitivity** or **hypersensitivity** to the antigen. **Immediate hypersensitivity reactions** are mediated by antibodies and occur within minutes of exposure to antigens, which are called **allergens**. **Delayed hypersensitivity reactions** are mediated by helper T cells and macrophages and may take several days to develop.

Allergens can be practically any exogenous molecule: naturally occurring or synthetic, organic or inorganic. Certain foods, insect venoms, and pollen all trigger immediate hypersensitivity reactions. Allergens can be ingested, inhaled, injected, or simply come in contact with the skin.

Delayed hypersensitivity reactions include contact allergies to copper and other base metals. These are common among people who wear costume jewelry. Poison ivy and poison oak also cause common contact allergies.

Allergies have a strong genetic component, so if parents have a ragweed allergy, chances are good that their children will too. The development of allergies requires exposure to the allergen, a factor that is affected by geographical, cultural, and social conditions.

■ Figure 24.14 shows what happens during an immediate hypersensitivity reaction to pollen. The initial step—first exposure or the sensitization phase—is equivalent to the primary immune response discussed previously: the allergen is ingested and processed by an antigen-presenting cell such as a macrophage, which in turn activates a helper T cell (see the right side of Fig. 24.13).

The T_H cell activates B lymphocytes that have bound the allergen. This results in plasma cell production of antibodies (IgE and IgG) to the allergen. The IgE antibodies are immediately bound by their Fc ends to the surface of mast cells and basophils. Memory T and memory B cells store the record of the initial allergen exposure.

Upon re-exposure, equivalent to the secondary immune response shown in Figure 24.8d, the body reacts very strongly and rapidly. The allergen binds to IgE already present on mast cells, triggering the immediate release of histamine, cytokines, and other mediators that cause allergic symptoms (Fig. 24.14).

The type of allergic reaction depends on the antigen antibody, immune cell involved, and location of antigen introduction. Allergen binding to IgE is the most common response to inhaled, ingested, or injected allergens. When allergens bind to IgE antibodies on mast cells, the cells degranulate, releasing histamine and other cytokines. The result is an inflammatory reaction.

The severity of the reaction varies, ranging from localized reactions near the site of allergen entry to systemic reactions such as total body rashes. The most severe IgE-mediated allergic reaction is called **anaphylaxis** {*ana-*, without + *phylax*, guard}. In an anaphylactic reaction, massive release of histamine and other cytokines causes widespread vasodilation, circulatory collapse, and severe bronchoconstriction. Unless treated promptly with epinephrine, anaphylaxis can result in death within as little as 20 minutes.

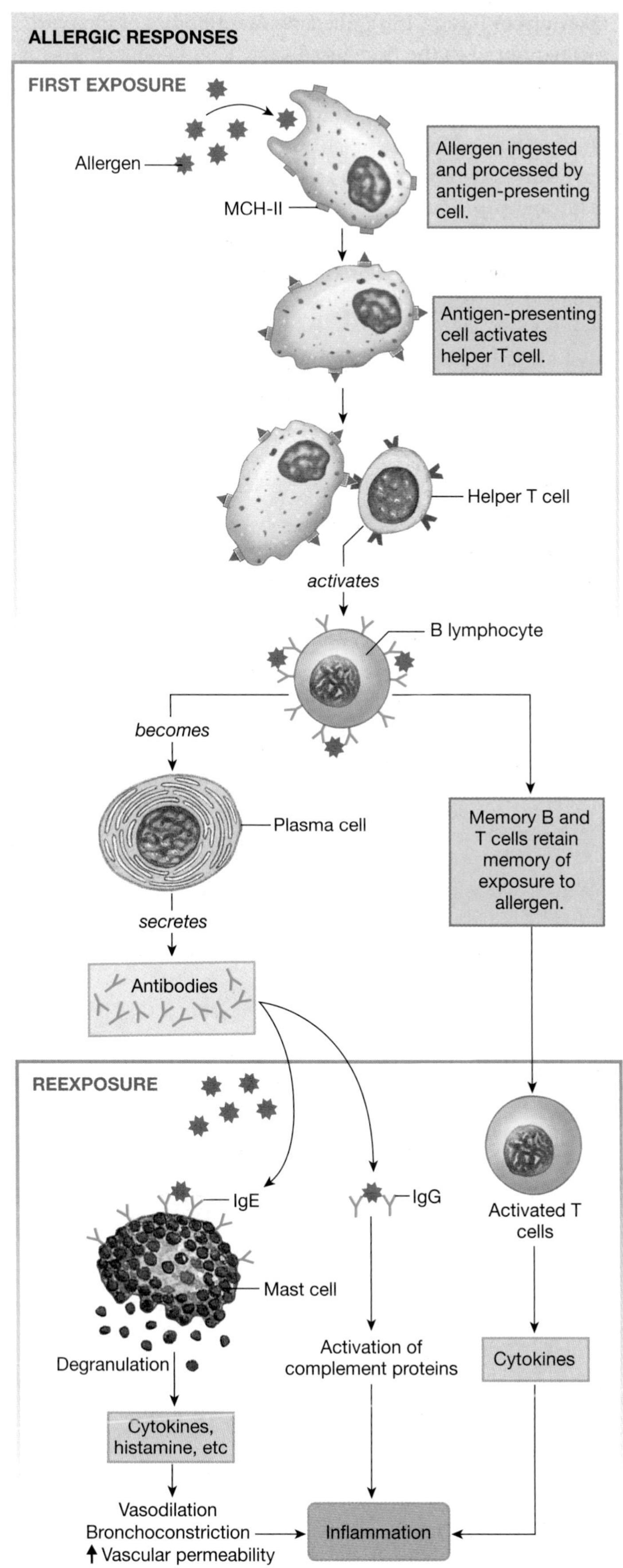

Fig. 24.14

RUNNING PROBLEM

Rebecca cares greatly about the health and well-being of her daughter but is not convinced that her daughter needs the Gardasil immunizations—at least not yet. "Isn't HPV a sexually transmitted disease?" Rebecca asked Dr. Paul. "Why can't Lizzie wait until she is older to get the shots?" Dr. Paul explained that although HPV transmission is usually associated with sexual intercourse, it can be spread by skin-to-skin or hand-to-genital contact. Because the vaccine is a preventive measure rather than a treatment, it is effective only when given before a person has been exposed to HPV.

Q5: Why are the vaccinations ineffective if the person receiving them already has HPV of the type in the vaccine?

802 804 816 825 829 830

MHC Proteins Allow Recognition of Foreign Tissue

As surgeons have developed techniques to transplant organs between human beings or from animals to humans, physicians have had to wrestle with the problems of rejection of the host by the donor tissue (known as *graft versus host*) or donor tissue rejection by the recipient's immune system (*host versus graft*). The ubiquitous major histocompatibility complex proteins (MHC) are the primary tissue antigens that determine whether donated tissue is recognized as foreign by the recipient's immune system.

The MHC proteins are also known as **human leukocyte antigens (HLA)** and are classified according to an international HLA system. A tissue graft or transplanted organ is more likely to be successful if donor and recipient share HLA antigens. Incompatible matches trigger antibody production and activate cytotoxic T cells and T_H cells. Generally it is T cells that are responsible for acute rejection of solid tissue grafts.

One of the most common examples of tissue donation is blood transfusion. Human red blood cell (RBC) membranes contain antigenic proteins and glycoproteins, but RBCs lack the MHC protein markers for recognizing foreign tissue that are found on nucleated cells. In the absence of MHC proteins, two surface proteins—the ABO blood group antigens and the Rh* antigens—become the most important causes of a rejection reaction after a blood transfusion.

The ABO blood groups consist of four blood types created by combinations of two different glycoprotein antigens (designated A and B) found on the membrane of red blood cells

*Rh stands for Rhesus, the type of monkey in which the antigen was first discovered. Rh antigens are also called D antigens.

(■ Fig. 24.15a). Each person inherits two alleles (one from each parent) of three possible ABO alleles: A, B, and O (neither A nor B antigen is produced). Because alleles A and B are dominant to allele O, the various combinations of two alleles produce four blood types: A, B, AB, and O (■ Tbl. 24.2).

Problems with blood transfusions arise because plasma normally contains antibodies to the ABO group antigens. These antibodies are thought to be produced early in life in response to bacterial antigens or food antigens in the gut. The antibodies can be measured in the blood of infants as early as age 3–6 months.

People express antibodies to the red blood cell antigen(s) that they do not possess. For this reason, people with blood type A have anti-B antibodies in their plasma, people with blood type B have anti-A antibodies in their plasma, people with no antigens on their red blood cells (blood type O) have both anti-A and anti-B antibodies, and people with both antigens on their red blood cells (blood type AB) have no antibodies to A or B antigens.

How does the body respond to a transfusion of incompatible blood? If a person with blood type O is mistakenly given a transfusion of type A blood, for example, an immune reaction takes place (Fig. 24.15b). The anti-A antibodies of the type O recipient bind to the transfused type A red blood cells, causing them to clump (*agglutinate*). This reaction is easily seen in a blood sample and forms the basis for the blood-typing tests often performed in student laboratories.

Antibody binding also activates the complement system, resulting in production of membrane attack complexes that cause the transfused cells to swell and leak hemoglobin. Free hemoglobin released into the plasma can result in acute renal failure as the kidneys try to filter the large molecules from the blood. Cross-matching donor and recipient blood is critical prior to giving a blood transfusion.

Concept Check

Answers: p. 834

6. A person with AB blood type is transfused with type O blood. What happens and why?
7. A person with O blood type is transfused with type A blood. What happens? Why?

ABO BLOOD GROUPS

(a)

Blood type	Antigen on red blood cell	Antibodies in plasma
O	RBC No A or B antigens	"Anti-A" and "anti-B"
A	A antigens	"Anti-B"
B	B antigens	"Anti-A"
AB	A and B antigens	None to A or B

(b) A mixture of type O and type A blood

When red blood cells with group A antigens on their membranes are mixed with plasma containing antibodies to group A, the antibodies cause the blood cells to clump, or agglutinate.

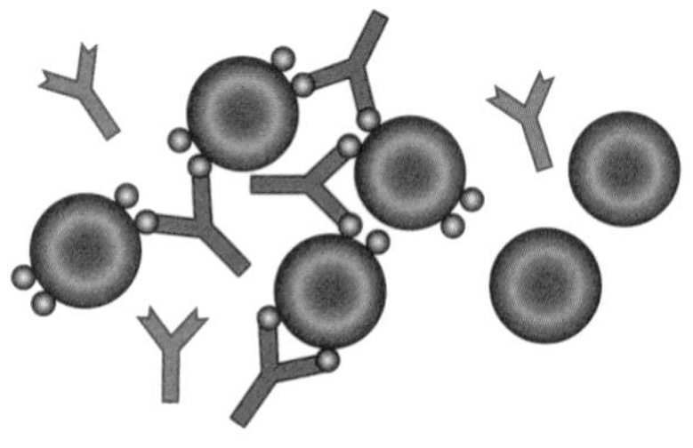

FIGURE QUESTION

Each person inherits one allele for ABO blood groups from each parent. A and B are dominant to O but equal if they occur together (blood type AB). Fill in the table showing combinations of inherited alleles. In the shaded blocks show the blood type that would be expressed.

		Mother		
		A	B	O
Father	A	A A		
		A		
	B			
	O			

■ **Fig. 24.15**

ABO Blood Group Frequencies in the United States Table 24.2

Blood Group	U.S. Population (%)			
	White	Black	Asian	Native American
O	44	49	43	79
A	43	27	27	16
B	9	20	25	4
AB	4	4	5	<1

Some Common Autoimmune Diseases in Humans Table 24.3

Disease	Antibodies Produced Against
Graves' disease (hyperthyroidism)	TSH receptor on thyroid cells
Insulin-dependent diabetes mellitus	Pancreatic beta cell antigens
Multiple sclerosis	Myelin of CNS neurons
Myasthenia gravis	Acetylcholine receptor of motor endplate
Rheumatoid arthritis	Collagen
Systemic lupus erythematosus	Intracellular nucleic acid protein complexes (antinuclear antibodies)
Guillain-Barré syndrome (acute inflammatory demyelinating polyneuropathy)	Myelin of peripheral nerves

The Immune System Must Recognize "Self"

One amazing property of immunity is the ability of the body to distinguish its own cells from foreign cells. The lack of immune response by lymphocytes to cells of the body is known as **self-tolerance** and appears to be due to the elimination of self-reactive lymphocytes in a process known as *clonal deletion*. During development, some lymphocyte clones develop with antibodies that can combine with MHC-self-antigen complexes. The primary lymphoid tissues contain self-antigens that can combine with these self-reactive lymphocytes and eliminate them by inducing apoptosis.

When self-tolerance fails, the body makes antibodies against its own components through T cell–activated B lymphocytes. The body's attack on its own cells leads to **autoimmune diseases** {*auto-*, self}. The antibodies produced in autoimmune diseases are specific against a particular antigen and are usually restricted to a particular organ or tissue type. ■ Table 24.3 lists some common autoimmune diseases of humans.

Why does self-tolerance suddenly fail? We still do not know. We have learned, however, that autoimmune diseases often begin in association with an infection. One potential trigger for autoimmune diseases is foreign antigens that are similar to human antigens. When the body makes antibodies to the foreign antigen, those antibodies have enough cross-reactivity with human tissues to do some damage.

One example of an autoimmune disease is type 1 diabetes mellitus, in which the body makes *islet cell antibodies* that destroy pancreatic beta cells but leave other endocrine cells untouched [p. 761]. Another autoimmune condition is the hyperthyroidism of Graves' disease [p. 786]. The body makes *thyroid-stimulating immunoglobulins* that mimic thyroid-stimulating hormone and cause the thyroid gland to oversecrete hormone.

Severe autoimmune diseases are sometimes treated by administering glucocorticosteroids, such as cortisol and its derivatives. Glucocorticoids depress immune system function by suppressing bone marrow production of leukocytes and decreasing the activity of circulating immune cells. Although helpful in suppressing symptoms of the autoimmune disease, large doses of steroids may trigger the same signs and symptoms as hormone oversecretion, causing patients to develop the central obesity and moon face characteristic of Cushing's syndrome [p. 781].

Immune Surveillance Removes Abnormal Cells

What is the role of the human immune system in protecting the body against cancer? This question holds considerable interest for scientists. One theory, known as **immune surveillance**, proposes that cancerous cells develop on a regular basis, but they are detected and destroyed before they can spread. Some evidence supports that theory, but it does not explain why so many cancerous tumors develop every year, or why people with depressed immune function do not develop cancerous tumors in all types of tissues. However, immune surveillance does appear to recognize and control some virus-associated tumors. In addition, some types of endogenous tumors lack surface antigens, which

BIOTECHNOLOGY

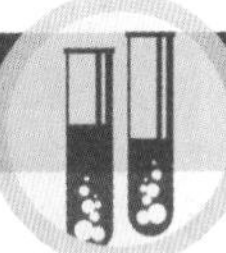

Engineered Antibodies

One goal of cancer treatments is to destroy cancerous cells while minimizing damage to normal body cells. Scientists have used both chemotherapy and radiation for this approach, but with mixed success. One of the newest cancer treatments is based on the genetic engineering of antibodies against tumor surface proteins, such as the growth factor receptor HER2 that occurs in high density on some breast cancer cells. The engineered antibody then binds to its antigen on the tumor cell. In the case of the HER2 receptor, the anti-HER2 antibody blocks the action of a growth factor that the tumor cell needs to reproduce. Other antibodies have been engineered to carry chemotherapy drugs or radioactive molecules that specifically target cancer cells.

One strategy for developing these antibody treatments involves injecting the antigen into mice, which then produce antibodies against the antigen. Murine (mouse) antibodies injected into humans normally would trigger an immune response against the mouse protein, so scientists have used genetic engineering to create "humanized antibodies." In this procedure, the antigen-binding portion of the mouse Fab region is integrated into the Fab region of a human antibody, resulting in a protein that is less likely to be rejected in the human body.

allow NK cells to recognize these cells as abnormal and destroy them. One active area of cancer research is investigating ways to activate the immune system to fight cancerous tumor cells.

Neuro-Endocrine-Immune Interactions

One of the most fascinating and rapidly developing areas in physiology involves the link between the mind and the body. Although serious scientists scoffed at the topic for many years, the interaction of emotions and somatic illnesses has been described for centuries. The Old Testament says, "A merry heart doeth good like a medicine, but a broken spirit drieth the bones" (Proverbs 17:22). Most societies have tales of people who lost the will to live and subsequently died without obvious illness, or of people given up for dead who made remarkable recoveries.

Today, the medical field is beginning to recognize the reality of the mind-body connection. Psychosomatic illnesses and the placebo effect have been accepted for many years. Why should we not investigate the possibility that cancer patients can enhance their immune function by visualizing their immune cells gobbling up the abnormal cancer cells?

The study of brain-immune interactions is now a recognized field known as **neuroimmunomodulation**. Currently the field is still developing, and the results of many studies raise more questions than they answer. In one experiment, for example, mice were repeatedly injected with a chemical that suppressed the activity of their lymphocytes. During each injection, they were also exposed to the odor of camphor, a chemical that does not affect the immune system. After a period of conditioning, the mice were exposed only to the camphor odor. When the researchers checked the mice's lymphocyte function, they found that the odor of camphor had suppressed the activity of the immune cells, just as the chemical suppressant had done previously. The pathways through which this conditioning occurred are still largely unexplained.

What do we know at this point about the relationship between the immune, nervous, and endocrine systems?

1. **The three systems share common signal molecules and common receptors for those molecules.** Chemicals once thought to be exclusive to a single cell or tissue are being discovered all over the body. Immune cells secrete hormones, and leukocyte cytokines are produced by nonimmune cells. For example, lymphocytes secrete thyrotropin (TSH), ACTH, growth hormone, and prolactin, as well as hypothalamic corticotropin-releasing hormone (CRH).

   Receptors for hormones, neurotransmitters, and cytokines are being discovered everywhere. Neurons in the brain have receptors for cytokines produced by immune cells. Natural killer cells have opiate receptors and beta adrenergic receptors. Everywhere we turn, the nervous, endocrine, and immune systems seem to share chemical signal molecules and their receptors.
2. **Hormones and neuropeptides can alter the function of immune cells.** It has been known for years that increased cortisol levels due to stress were associated with decreased antibody production, depressed lymphocyte proliferation, and diminished activity of natural killer cells. Substance P, a neuropeptide, has been shown to induce degranulation of mast cells in the mucosa of the intestines and the respiratory tract. And sympathetic innervation of the bone marrow increases antibody synthesis and production of cytotoxic T cells.
3. **Cytokines from the immune system can affect neuroendocrine function.** Stressors such as bacterial and viral infections or tumors can induce stress responses in the central nervous system through cytokines released by immune cells. Interleukin-1 is probably the best-studied cytokine in this response, but the induction of cortisol release by lymphocyte ACTH is receiving considerable attention.

   It was once believed that cortisol secretion depended on neural signals translated through the hypothalamic CRH–anterior pituitary ACTH pathway. Now it appears that pathogenic stressors can activate the cortisol pathway by causing immune cells to secrete ACTH.

The interaction of the three systems is summarized in the model shown in ■ Figure 24.16a. The nervous, endocrine, and immune systems are closely linked through bidirectional communication carried out using cytokines, hormones, and neuropeptides. The brain is linked to the immune system by autonomic neurons, CNS neuropeptides, and leukocyte cytokines. The nervous system controls endocrine glands by secretion of hypothalamic-releasing hormones—but immune cells secrete some of the same trophic hormones. Hormones from endocrine glands feed back to influence both the nervous and immune systems.

The result is a complex network of chemical signals subject to modulation by outside factors. These outside factors include physical and emotional stimuli integrated through the brain, pathogenic stressors integrated through the immune system, and a variety of miscellaneous factors, including magnetic fields, chemical factors from brown adipose tissue, and melatonin from the pineal gland. It will take years for us to decipher these complex pathways.

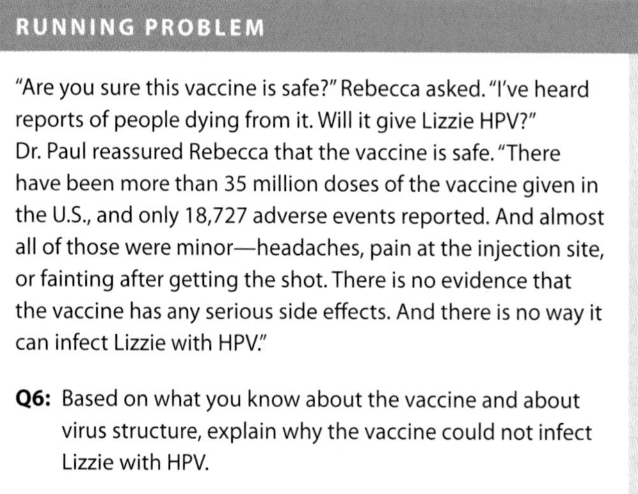

RUNNING PROBLEM

"Are you sure this vaccine is safe?" Rebecca asked. "I've heard reports of people dying from it. Will it give Lizzie HPV?" Dr. Paul reassured Rebecca that the vaccine is safe. "There have been more than 35 million doses of the vaccine given in the U.S., and only 18,727 adverse events reported. And almost all of those were minor—headaches, pain at the injection site, or fainting after getting the shot. There is no evidence that the vaccine has any serious side effects. And there is no way it can infect Lizzie with HPV."

Q6: Based on what you know about the vaccine and about virus structure, explain why the vaccine could not infect Lizzie with HPV.

802 804 816 825 **829** 830

Stress Alters Immune System Function

One area of interest is the link between inability to cope with stress and the development of illnesses. The modern study of stress is attributed to Hans Selye, beginning in 1936. He defined **stress** as nonspecific stimuli that disturb homeostasis and elicit an invariable stress response that he termed the *general adaptation syndrome.*

Selye's stress response consisted of stimulation of the adrenal glands, followed by suppression of the immune system due

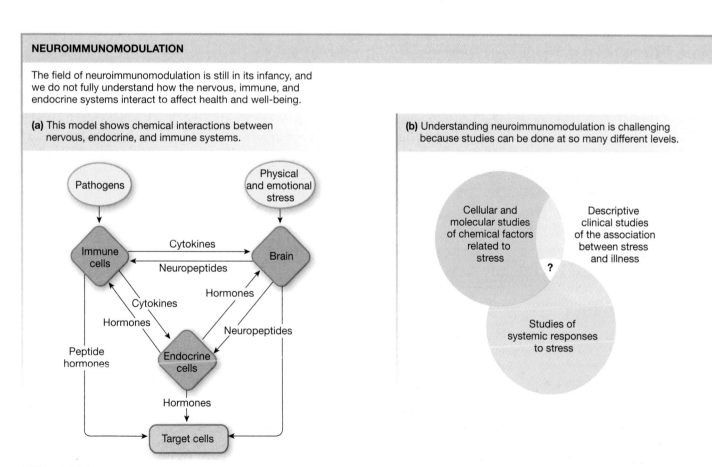

NEUROIMMUNOMODULATION

The field of neuroimmunomodulation is still in its infancy, and we do not fully understand how the nervous, immune, and endocrine systems interact to affect health and well-being.

(a) This model shows chemical interactions between nervous, endocrine, and immune systems.

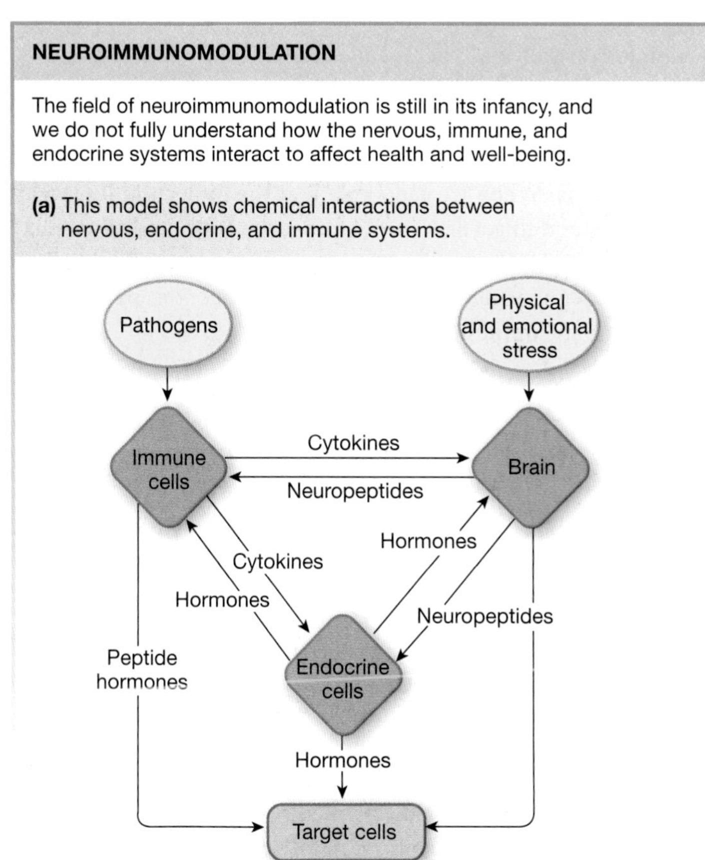

(b) Understanding neuroimmunomodulation is challenging because studies can be done at so many different levels.

Cellular and molecular studies of chemical factors related to stress

Descriptive clinical studies of the association between stress and illness

?

Studies of systemic responses to stress

■ **Fig. 24.16**

to high levels of circulating glucocorticoids [p. 780]. For many years, Selye's experiment was the benchmark for defining stress. However, since the 1970s, the definition of stress and the stress response has broadened.

Stressors—the events or items that create stress—are highly variable and difficult to define in experimental settings. Acute stress is different from chronic stress. A person's reaction to stress is affected by loneliness and by whether the person feels in control of the stressful situation. Many stressors are sensed and interpreted by the brain, leading to modulation of the stressor by experience and expectations. A stressor to one person may not affect another.

Most physical and emotional stressors are integrated through the CNS. The two classic stress responses are (1) the rapid, neurally mediated **fight-or-flight reaction**, and (2) the elevation of adrenal cortisol levels with associated suppression of the immune response. Fight-or-flight is a rapid reaction to acute stress. The cortisol response is a better indicator of chronic or repetitive stress.

One difficulty in studying the body's response to stress is the complexity involved in integrating information from three levels of response (Fig. 24.16b). Scientists have gathered a good deal of information at each of three levels: cellular and molecular factors related to stress, systemic responses to stress, and clinical studies describing the relationships between stress and illness. In some cases, we have linked two different levels together, although in other cases the experimental evidence is scanty or even contradictory. We still do not understand the big picture, the way that all three levels of response overlap.

Experimental progress is slow because many traditional study animals, such as rodents, are not good models for studying stress in humans. At the same time, controlled experiments that create stress in humans are difficult to design and execute for both practical and ethical reasons. Much of our research is observational. One study, for example, followed natural killer cell activity over an 18-month period in two groups of people: spouse-caregivers of Alzheimer's disease patients (a group that reports high psychosocial stress) and a matched group of controls. For a subset of the caregivers, high perceived stress at the beginning of the study was associated with low NK cell activity at the end. However, many variables—including age, gender, exercise, and use of alcohol and prescription drugs—must be taken into account before we can conclude that stress was the causative factor.

Modern Medicine Includes Mind-Body Therapeutics

The role of mind-body interactions in traditional medicine is receiving more attention in recent years. The U.S. National Center for Complementary and Alternative Medicine (*http://nccam.nih.gov*) supports rigorously designed research studies on the efficacy of complementary medical therapies, such as meditation, yoga, hypnosis, visual imagery, and biofeedback. Although the placebo and nocebo effects [p. 21] are widely accepted in the traditional medical community, the role of complementary therapies such as laughter therapy is still being studied. Is laughter really the best medicine, as the old saying goes? Can watching a funny movie help your immune system? Recent studies suggest that it can. In these studies, researchers measured the activity of immune cells in subjects before and after watching a humorous movie or video. In both studies, immune cells became more active. You can learn more about *humor therapy* from the Association for Applied and Therapeutic Humor (*www.aath.org*).

In spite of the difficulties of studying the integration of the immune, endocrine, and nervous systems, scientists are beginning to piece together the intricate tapestry of mind-body interactions. When we can explain at the molecular level why students tend to get sick just before exam time, and why relaxation therapy enhances the immune response, we will be well on the way to understanding the complex connections between the brain and the immune system.

RUNNING PROBLEM CONCLUSION

HPV: To Vaccinate or Not?

Human papillomavirus is the most common sexually transmitted disease in the United States. By some estimates, half to three-fourths of all sexually active people have been infected with HPV. In most people, their natural immune response finds the virus and clears it from the body. But for some people, HPV infections lead to genital warts and cervical cancer. Some opponents of vaccinating adolescents for HPV point to the relatively small number of women who develop cervical cancer. But each year about 3.5 million women have an abnormal Pap test, which shows that cervical epithelial cells have been affected by HPV. The direct medical costs of following these women may be as much a $6 billion a year, and the emotional cost of the abnormal test reports cannot be measured. Most of the time, the abnormal tests revert to normal as the woman's body finds and removes the HPV virus. But scientists, medical practitioners, and public health advocates feel that preventing the infections is a more effective and cost-efficient approach. In 2009, use of the Gardasil vaccine in boys was approved. While cervical

RUNNING PROBLEM CONCLUSION *(continued)*

cancer is not an issue for boys and men, HPV can cause genital warts, anogenital cancers, and various forms of head and throat cancers in males. Not only can the Gardasil vaccine protect them from these illnesses, but it can also help reduce the *spread* of HPV, effectively lowering rates of cervical cancer in women as well. To learn more about HPV, cervical cancer, and the vaccines, visit the web site of the Centers for Disease Control, *www.cdc.gov*.

Now check your understanding of this running problem by comparing your answers with the information in the summary table.

Question	Facts	Integration and Analysis
1. Why must viruses live and reproduce inside host cells?	Viruses are not cells. They are nucleic acids with a protein coat.	Viruses need to use the host cell's resources so the virus can reproduce.
2. How might you test a person to see if she is infected with HPV?	HPV inserts its DNA into the host cell.	Viral DNA will be different from the host DNA. Test host cells to see if viral DNA is present.
3. Antibody production begins with activation of which type of lymphocyte? What type of cell produces antibodies?	B lymphocytes activate and become plasma cells, which make antibodies.	N/A
4. What kind of immune cell is mostly likely to encounter HPV in the skin?	Dendritic cells and macrophages are the primary immune cells of the tissues.	Dendritic cells are found in the skin and are most likely to encounter HPV. These cells are also called Langerhans cells.
5. Why are the vaccinations ineffective if the person receiving them already has HPV of the type in the vaccine?	The HPV vaccines cause the body to create antibodies to the viral capsid proteins.	If a person already has HPV, the virus is inside the cells. Antibodies work in the extracellular compartment and cannot affect viruses inside cells.
6. Based on what you know about the vaccine and about virus structure, explain why the vaccine could not infect Lizzie with HPV.	The vaccine is made of viruslike proteins that are proteins of the virus capsid.	Because the vaccine does not contain any viral DNA, the vaccine can't cause HPV infection.

802 804 816 825 829 830

Chapter Summary

In this chapter you learned how the immune system protects the body from pathogens and how it removes damaged tissue and abnormal cells. An immune response requires detection of the foreign substance, communication with other immune cells, and coordination of the response. These functions depend on *chemical communication* and *molecular interactions* between receptors, antibodies, and antigens. The *compartmentation* of the body affects how the immune system fights pathogens: some pathogens hide inside cells, while others remain in the extracellular compartment.

24

Overview

iP Immune System: Immune System Overview

1. **Immunity** is the ability of the body to protect itself from pathogens. The human immune system consists of lymphoid tissues, immune cells, and cytokines. (p. 802)
2. **Antigens** are substances that trigger an immune response and react with products of that response. (p. 802)

Pathogens of the Human Body

3. Bacteria are cells that can survive and reproduce outside a living host. Antibiotic drugs kill bacteria. (p. 803)
4. Viruses are DNA or RNA enclosed in protective proteins. They invade host cells and reproduce using the host's intracellular machinery. The immune system must destroy infected host cells to kill a virus. (p. 803; Fig. 24.1)

The Immune Response

5. The immune response includes: (1) *detection* and *identification*, (2) *communication* among immune cells, (3) *recruitment* of assistance and *coordination* of the response, and (4) *destruction* or *suppression* of the invader. (p. 804)
6. **Innate immunity** is the rapid, **nonspecific immune response**. **Acquired immunity** is slower and is an **antigen-specific response**. The two types of immunity overlap. (p. 804)

Anatomy of the Immune System

iP Immune System: Anatomy Review

7. Immune cells form and mature in the **thymus gland** and **bone marrow**. The **spleen** and **lymph nodes** are encapsulated lymphoid tissues. Unencapsulated **diffuse lymphoid tissues** include **tonsils** and the **gut-associated lymphoid tissue** (**GALT**). (p. 806, 807; Figs. 24.3, 24.4, 24.10)
8. **Eosinophils** are **cytotoxic cells** that kill parasites. **Basophils** and **mast cells** release histamine, **heparin**, and other cytokines. (p. 808; Fig. 24.5)
9. **Neutrophils** are phagocytic cells that release pyrogens and cytokines that mediate inflammation. (p. 808; Fig. 24.5)
10. **Monocytes** are the precursors of tissue **macrophages**. **Antigen-presenting cells** (**APCs**) display antigen fragments in proteins on their membrane. (p. 808, 810; Figs. 24.5, 24.6)
11. **Lymphocytes** mediate antigen-specific immune responses. (p. 810)
12. **Dendritic cells** capture antigens and present them to lymphocytes. (p. 811)

Innate Immunity: Nonspecific Responses

iP Immune System: Innate Host Defenses

13. Physical and chemical barriers, such as skin and mucus, are the first line of defense. (p. 805; Fig. 24.2)
14. **Chemotaxins** attract immune cells to pathogens. (p. 811)
15. Phagocytosis requires that surface molecules on the pathogen bind to membrane receptors on the phagocyte. **Opsonins** are antibodies or complement proteins that coat foreign material. (p. 810; Fig. 24.6)
16. **Natural killer (NK) cells** kill certain tumor cells and virus-infected cells. When eosinophils and NK cells bind to the Fc stem of antibodies bound to pathogens, they kill the pathogen. This is called **antibody-dependent cell-mediated cytotoxicity**. (p. 817; Fig. 24.9)
17. **Inflammation** results from cytokines released by activated tissue macrophages. (p. 812)
18. Inflammatory cytokines include **acute-phase proteins**, histamine, **interleukins**, **bradykinin**, and **complement proteins**. (p. 812)
19. The complement cascade forms **membrane attack complex**, pore-forming molecules that cause pathogens to rupture. (p. 814; Fig. 24.7)

Acquired Immunity: Antigen-Specific Responses

iP Immune System: Common Characteristics of B and T Lymphocytes

20. Lymphocytes mediate the acquired immune response. Lymphocytes that react to one type of antigen form a **clone**. (p. 815; Fig. 24.8)
21. First exposure to an antigen activates a **naïve lymphocyte** clone and causes it to divide (**clonal expansion**). Newly formed lymphocytes differentiate into **effector cells** or **memory cells**. Additional exposure to the antigen creates a faster and stronger **secondary immune response**. (p.815; Fig. 24.8)
22. Lymphocyte cytokines include **interleukins** and **interferons**, which interfere with viral replication. **Interferon-**γ activates macrophages and other immune cells. (p. 820)
23. **B lymphocytes** develop into **plasma cells** and **memory cells**. **T lymphocytes** and NK cells attack and destroy infected cells. (p. 814)
24. **Antibodies** bind to pathogens. Mature B lymphocytes are covered with antibodies that act as surface receptors. Activated B lymphocytes differentiate into **plasma cells**, which secrete soluble antibodies. (p. 814)
25. The five classes of immunoglobulins (Ig), or antibodies, are IgG, IgA, IgE, IgM, and IgD, collectively known as **gamma globulins**. (p. 816)
26. The antibody molecule is Y shaped, with four polypeptide chains. The arms, or **Fab regions**, contain antigen-binding sites. The stem, or **Fc region**, can bind to immune cells. Some immunoglobulins are polymers. (p. 817; Fig. 24.9)
27. Soluble antibodies act as opsonins, serve as a bridge between antigens and leukocytes, activate complement proteins and mast cells, and disable viruses and toxins. (p. 815; Fig. 24.8)
28. B cell antibodies bind antigens to activate B lymphocytes and initiate the production of additional antibodies. (p. 815; Fig. 24.8)
29. **T lymphocytes** are responsible for **cell-mediated immunity**, in which the lymphocytes bind to target cells using **T-cell receptors**. (p. 820; Fig. 24.11)
30. T lymphocytes react to antigen bound to **major histocompatibility complex** proteins on the target cell. **MHC-I molecules** displaying antigen cause cytotoxic T (T_C) cells to kill the target cell. **MHC-II molecules** with antigen activate helper T (T_H) cells to secrete cytokines. (p. 820; Fig. 24.11)
31. **Cytotoxic T** (T_C) **cells** release **perforin**, a pore-forming molecule that allows **granzymes** to enter the target cell and trigger it into committing suicide (**apoptosis**). (p. 820)
32. **Helper T** (T_H) **cells** secrete cytokines that influence other cells. (p. 820; Fig. 24.11)

Immune Response Pathways

iP Immune System: Humoral Immunity, Cellular Immunity

33. Bacteria usually elicit a nonspecific **inflammatory response**. In addition, lymphocytes produce antibodies keyed to the specific type of bacterium. (p. 822; Fig. 24.12)
34. Innate immune responses and antibodies help control the early stages of a viral infection. Once the viruses enter the host's cells, cytotoxic T and NK cells must destroy infected host cells. (p. 823; Fig. 24.13)
35. An **allergy** is an inflammatory immune response to a nonpathogenic **allergen**. The body gets rid of the allergen through the inflammatory response. (p. 825; Fig. 24.14)
36. **Immediate hypersensitivity reactions** to allergens are mediated by antibodies and occur within minutes of exposure to antigen. **Delayed hypersensitivity reactions** are mediated by T lymphocytes and may take several days to develop. (p. 824)
37. The **major histocompatibility complex** (MHC) proteins found on all nucleated cells determine tissue compatibility. Lymphocytes attack any MHC complexes they do not recognize as self proteins. (p. 825)
38. Human red blood cells lack MHC proteins but contain other antigenic membrane proteins and glycoproteins, such as the ABO and Rh antigens. (p. 826; Fig. 24.15)
39. **Self-tolerance** is the body's ability to recognize its own cells and not produce an immune response. When self-tolerance fails, the body makes antibodies against its own components, creating an **autoimmune disease.** (p. 827)
40. The theory of **immune surveillance** proposes that cancerous cells develop on a regular basis but are detected and destroyed by the immune system before they can spread. (p. 827)

Neuro-Endocrine-Immune Interactions

41. The nervous, endocrine, and immune systems are linked together by signal molecules and receptors. The study of these interactions is a field known as **neuroimmunomodulation.** (p. 829; Fig. 24.16)
42. The link between inability to cope with stress and the development of illnesses is believed to result from neuroimmunomodulation. (p. 830)

24

Questions

Answers: p. A-1

Level One Reviewing Facts and Terms

1. Define immunity. What is meant by memory and specificity in the immune system?
2. Name the anatomical components of the immune system.
3. List and briefly summarize the three main functions of the immune system.
4. Name two ways viruses are released from a host cell. Give three ways in which viruses damage host cells or disrupt their function.
5. Suppose a pathogen has invaded the body. What four steps occur if the body is able to destroy the pathogen? What compromise may be made if the pathogen cannot be destroyed?
6. Define the following terms and explain their significance:
   (a) anaphylaxis
   (b) agglutinate
   (c) extravascular
   (d) degranulation
   (e) acute-phase protein
   (f) clonal expansion
   (g) immune surveillance
7. How are histiocytes, Kupffer cells, osteoclasts, and microglia related?
8. What is the mononuclear phagocytic system, and what role does it play in the immune system?
9. Match each of the following cell types with its description:

(a) lymphocyte (b) neutrophil (c) monocyte (d) dendritic cell (e) eosinophil (f) basophil	1. most abundant leukocyte; phagocytic; lives 1–2 days 2. cytotoxic; is associated with allergic reactions and parasitic infestations 3. precursor of macrophages; relatively rare in blood smears 4. related to mast cells; releases cytokines such as histamine and heparin 5. most of these cells reside in the lymphoid tissues 6. these cells capture antigens, then migrate to the lymphoid tissues

10. Give examples of physical and chemical barriers to infection.
11. Name the three main types of lymphocytes and their subtypes. Explain the functions and interactions of each group.
12. Define self-tolerance. Which cells are responsible? Relate self-tolerance to autoimmune disease.
13. What is meant by the term *neuroimmunomodulation*?
14. Explain the terms *stress, stressor,* and *general adaptation syndrome*.

Level Two Reviewing Concepts

15. **Concept map:** Map the following terms. You may add terms.

• acquired immunity • antibody • antigen • antigen-presenting cell • B lymphocyte • bacteria • basophil • chemotaxin • cytokine • cytotoxic cell • cytotoxic T cell • dendritic cell • eosinophil • Fab region • Fc region	• helper T cell • innate immunity • memory cell • MHC class I • MHC class II • monocyte • neutrophil • NK cell • opsonin • phagocyte • phagosome • plasma cell • T lymphocyte • viruses

16. Why do lymph nodes often swell and become tender or even painful when you are sick?
17. Summarize the effects of histamine, interleukin-1, acute-phase proteins, bradykinin, complement, and interferon-γ. Are these chemicals antagonistic or synergistic to one another?
18. Compare and contrast the terms in the following sets:
    (a) pathogen, microbe, pyrogen, antigen, antibody, antibiotic
    (b) infection, inflammation, allergy, autoimmune disease

(c) virus, bacteria, allergen
(d) chemotaxin, opsonin, cytokine, interleukin, bradykinin, interferon
(e) acquired immunity, innate immunity, humoral immunity, cell-mediated immunity, specific immunity, nonspecific immunity
(f) immediate and delayed hypersensitivity
(g) membrane attack complex, granzymes, perforin

19. Diagram and label the parts of an antibody molecule. What is the significance of each part?
20. Diagram the steps in the process of inflammation.
21. Diagram the steps in the process of fighting viral infections.
22. Diagram the steps in the process of an allergic reaction.

Level Three Problem Solving

23. One ABO blood type is called the "universal donor," and one is called the "universal recipient." Which types are these, and why are they given those names?
24. Maxie Moshpit claims that Snidley Superstar is the father of her child. Maxie has blood type O. Snidley has blood type B. Baby Moshpit has blood type O. Can you rule out Snidley as the father of the baby with only this information? Explain.
25. Every semester around exam time, the number of students visiting the Student Health Center with colds and viral infections increases. The physicians attribute the increased incidence of illness to stress. Can you explain the biological basis for that relationship? What are some other possible explanations?
26. Barbara has rheumatoid arthritis, characterized by painful, swollen joints from inflamed connective tissue. She notices that it flares up when she is tired and stressed. This condition is regarded as an autoimmune disease. Outline the physiological reactions that would lead to this condition.
27. The differential white cell count [Fig. 16.3, p. 550] is an important diagnostic tool in medicine. If a patient's "diff" shows an increase in the number of immature neutrophils ("bands"), do you think the cause is more likely to be a bacterial or a viral infection? If the count shows an increase in eosinophils (*eosinophilia*), is the cause more likely to be a viral infection or a parasitic infection? Explain your reasoning.

Answers

Answers to Concept Check Questions

Page 804

1. Bacteria are cells; viruses are not. Most bacteria can reproduce outside a host; viruses cannot. Bacteria can be killed or inhibited by antibiotics; viruses cannot.
2. Virus enters host cell and uses its viral nucleic acids to tell host cell to make viral nucleic acids and proteins. These assemble into new virus particles, which are then released from the host cell.

Page 814

3. When capillary permeability increases, proteins move from plasma to interstitial fluid. This decreases the colloid osmotic force opposing capillary filtration, and additional fluid accumulates in the interstitial space (swelling or edema).

Page 818

4. Antibodies can be moved across cells by transcytosis or released from cells by exocytosis.
5. The child developed antibodies to bee venom on first exposure, so if the child will have a severe allergic reaction to bee venom, it will occur on the second exposure. Allergic reactions are mediated by IgE. After the first bee sting, IgE antibodies secreted in response to the venom are bound to the surface of mast cells. At the second exposure, bee venom binding to the IgE causes the mast cells to degranulate, resulting in an allergic reaction that may be severe enough to cause anaphylaxis.

Page 826

6. Type O red blood cells have neither A nor B antigens, so recipient antibodies, even if present, would not react with them. Type AB blood cells have both A and B antigens but type O blood has no antibodies. Therefore, no immune reaction will occur.
7. The type O recipient's anti-A antibodies will cause agglutination of the type A blood cells. The donor blood has anti-B antibodies but type O cells have no antigens so there is no reaction to donor antibodies.

Answers to Figure Questions

Page 815

Figure 24.8: The antigen will activate clone 1.

Page 819

Figure 24.10: Either the lymphocytes live for a long time or they can reproduce themselves outside the thymus gland.

Page 820

Figure 24.11: T_C recognize MHC-I antigen and T_H recognize MHC-II antigen.

Page 823

Figure 24.13: Steps 3 and 5 are cell-mediated immunity. Steps 1 and 4 are humoral immunity.

Page 826

Figure 24.15:

	Mother			
		A	B	O
Father	A	AA	AB	AO
		A	AB	A
	B	AB	BB	BO
		AB	B	B
	O	AO	BO	OO
		A	B	O

25 Integrative Physiology III: Exercise

One fascinating aspect of the physiology of the human being at work is that it provides basic information about the nature and the range of the functional capacity of different organ systems.

—Per-Olof Åstrand and Kaare Rodahl, *Textbook of Work Physiology*, 1977

Background Basics

Each dot of a microarray represents one gene. Genes that are active show up in bright colors.

In July 2008 the world watched with great anticipation as Michael Phelps swam his way to a record eight gold medals at the 2008 Olympics. The work done by Phelps and the hundreds of other Olympic athletes from around the world represents one of the most common challenges to body homeostasis: exercise. Exercise comes in many forms. Distance running, swimming, and cycling are examples of a dynamic endurance exercise. Weight lifting and strength training are examples of resistance training. In this chapter we examine dynamic exercise as a challenge to homeostasis that is met with an integrated response from multiple body systems.

In many ways, exercise is the ideal example for teaching physiological integration. Everyone is familiar with it, and unlike high-altitude mountaineering or deep-sea diving, it involves no special environmental conditions. Moreover, exercise is a normal physiological state, not a pathological one—although it can be affected by disease and (if excessive) can result in injury.

In addition to being an excellent teaching example, exercise physiology is a very active area of integrative physiology research. The coordinated functioning of multiple body systems is still not well understood in many instances because of complex interactions between neural and local control mechanisms. Researchers use a combination of animal models and studies with human subjects, including elite athletes, in their quest to explain how the body adapts to the metabolic demands of exercise. Note that this chapter is not intended to be a complete discussion of exercise physiology. For further information, readers should consult an exercise physiology textbook.

Metabolism and Exercise

Exercise begins with skeletal muscle contraction, an active process that requires ATP for energy. But where does the ATP for muscle contraction come from? A small amount is in the muscle fiber when contraction begins (■ Fig. 25.1 ④). As this ATP is used for muscle contraction and transformed into ADP, another phosphate compound, **phosphocreatine** (PCr), transfers energy from its high-energy phosphate bond to ADP. The transfer replenishes the muscle's supply of ATP [p. 413].

Together, the combination of muscle ATP and phosphocreatine is adequate to support only about 15 seconds of intense exercise, such as sprinting or power lifting. Subsequently, muscle fibers must manufacture additional ATP from energy stored in nutrients. Some of these molecules are contained within the muscle fiber itself. Others must be mobilized from the liver and adipose tissue and then transported to muscles through the circulation.

The primary substrates for energy production are carbohydrates and fats. The most efficient production of ATP occurs through aerobic pathways such as the glycolysis-citric acid cycle pathway [p. 111]. If the cell has adequate amounts of oxygen for oxidative phosphorylation, then both glucose and fatty acids can be metabolized to provide ATP (Fig. 25.1 ①, ②).

If the oxygen requirement of a muscle fiber exceeds its oxygen supply, energy production from fatty acids decreases dramatically, and glucose metabolism shifts to anaerobic pathways. In low-oxygen conditions, when the cell lacks oxygen for oxidative phosphorylation, the final product of glycolysis—pyruvate—is converted to lactate instead of being converted to acetyl CoA and entering the citric acid cycle [p. 117]. In general, exercise that depends on anaerobic metabolism cannot be sustained for an extended period. Cells that obtain their ATP by anaerobic metabolism of glucose to lactate are said to be carrying out *glycolytic metabolism*.

Anaerobic metabolism has the advantage of speed, producing ATP 2.5 times as rapidly as aerobic pathways do (■ Fig. 25.2). But this advantage comes with two distinct disadvantages: (1) anaerobic metabolism provides only 2 ATP per glucose, compared with an average of 30–32 ATP per glucose for oxidative metabolism, and (2) anaerobic metabolism contributes to a state of metabolic acidosis by producing H^+. (However, the CO_2 generated during exercise is a more significant source of acid.)

Where does glucose for aerobic and anaerobic ATP production come from? The body has three sources: the plasma glucose pool, intracellular stores of glycogen in muscles and liver, and "new" glucose made in the liver through *gluconeogenesis* [p. 744]. Muscle and liver glycogen stores provide enough energy substrate to release about 2000 kcal (equivalent to about 20 miles of running in the average person), more than adequate for the exercise that most of us do. However, glucose alone cannot provide sufficient ATP for endurance athletes such as marathon runners. To meet their energy demands, they rely on the energy stored in fats.

In reality, aerobic exercise of any duration uses both fatty acids and glucose as substrates for ATP production. About 30 minutes after aerobic exercise begins, the concentration of free fatty acids in the blood increases significantly, indicating

RUNNING PROBLEM

Heat Stroke

"Dangerous heat continues for fifth straight day," reads the morning headline. Had Colleen, 19, taken time to scan the story, she might have avoided the life-threatening health emergency she faced a few hours later. But Colleen overslept and rushed off to field hockey practice. The weather this week had been unbearably hot and humid, just as it had been all summer. The forecast today was for a high of 105 °F, with high humidity. In her haste, Colleen forgot her water bottle. "No big deal," she thought as she hopped on her bike to pedal two miles to the practice field.

836 838 842 843 844 846

OVERVIEW OF MUSCLE METABOLISM

ATP for muscle contraction is continuously produced by aerobic metabolism of glucose and fatty acids. During short bursts of activity, when ATP demand exceeds the rate of aerobic ATP production, aerobic glycolysis produces ATP, lactate, and H^+.

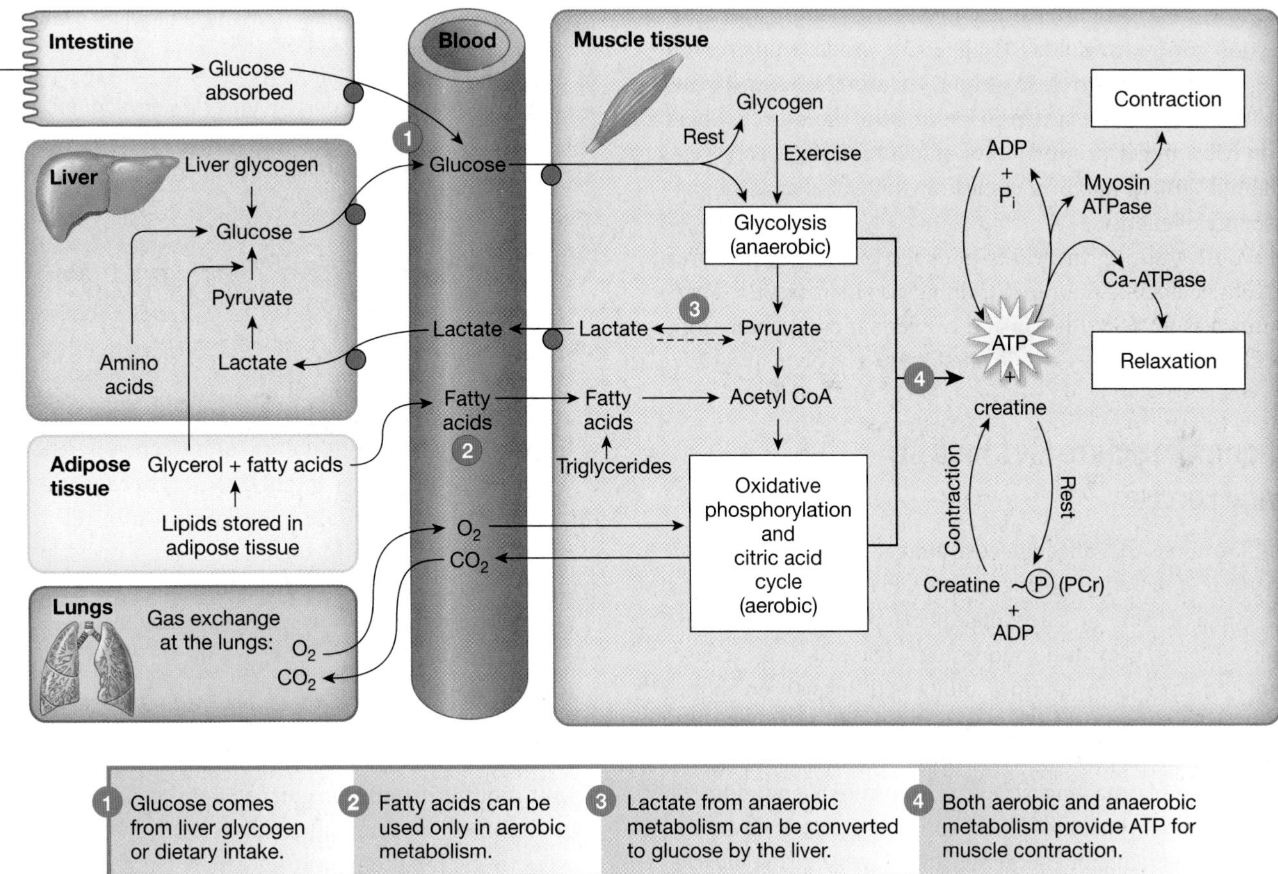

Fig. 25.1

AEROBIC VERSUS ANAEROBIC METABOLISM

Anaerobic metabolism produces ATP 2.5 times faster than aerobic metabolism, but aerobic metabolism can support exercise for hours.

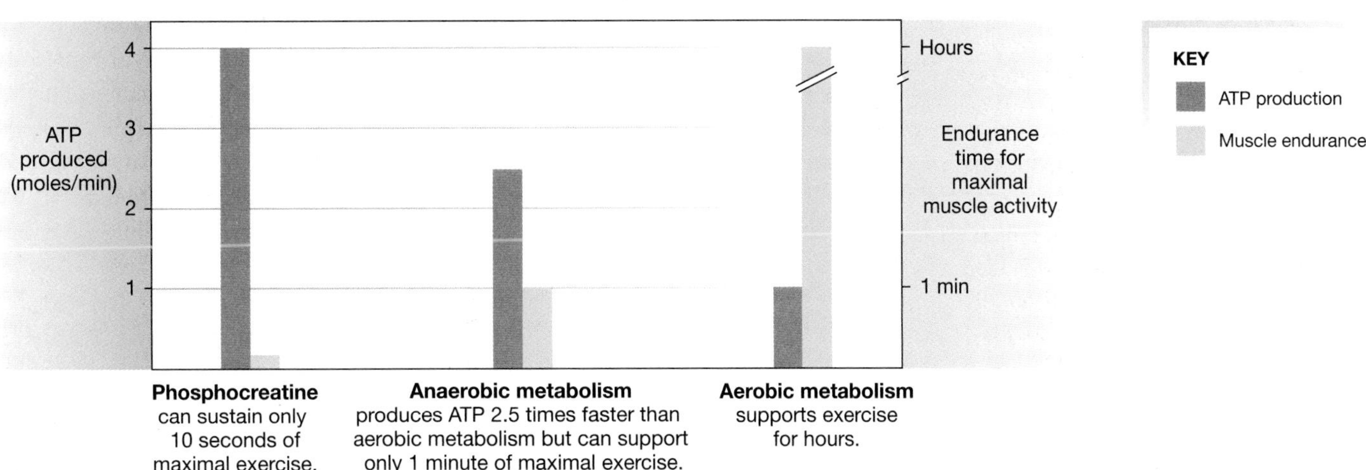

Fig. 25.2

that fats are being mobilized from adipose tissue. However, the breakdown of fatty acids through the process of β-oxidation [p. 751] is slower than glucose metabolism through glycolysis, so muscle fibers use a combination of fatty acids and glucose to meet their energy needs.

At lower exercise intensities, most of the energy for ATP production comes from fats (Fig. 25.3), which is one reason walking is a good way to lose weight. As exercise intensity increases and ATP is consumed more rapidly, the muscle fibers begin to use a larger proportion of glucose. When exercise exceeds about 70% of maximum, carbohydrates become the primary source of energy.

Aerobic training increases both fat and glycogen stores within muscle fibers. Endurance training also increases the activity of enzymes for β-oxidation and converts muscle fibers from fast-twitch glycolytic to fast-twitch oxidative-glycolytic [p. 415].

Hormones Regulate Metabolism During Exercise

Several hormones that affect glucose and fat metabolism change their pattern of secretion during exercise. Plasma concentrations of glucagon, cortisol, the catecholamines (epinephrine and norepinephrine), and growth hormone all increase during exercise. Cortisol and the catecholamines, along with growth hormone, promote the conversion of triglycerides to glycerol and fatty acids. Glucagon, catecholamines, and cortisol also mobilize liver glycogen and raise plasma glucose levels. A hormonal environment that favors the conversion of glycogen into glucose is desirable, because glucose is a major energy substrate for exercising muscle.

ENERGY SUBSTRATE USE DURING EXERCISE

At low-intensity exercise, muscles get more energy from fats than from glucose (CHO). During high-intensity exercise (levels greater than 70% of maximum), glucose becomes the main energy source.

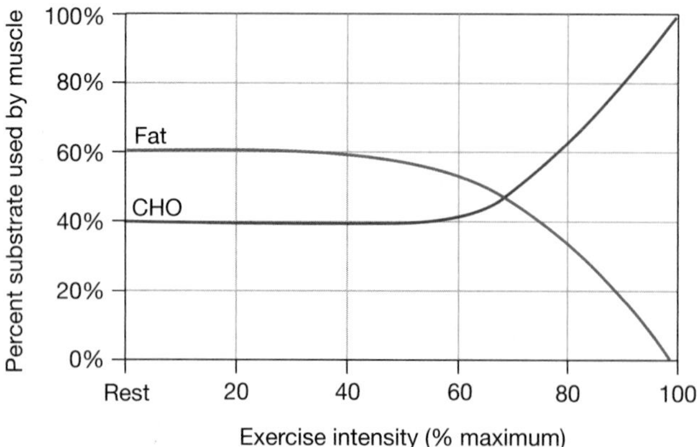

Data from G. A. Brooks and J. Mercier, *J App Physiol* 76: 2253–2261, 1994

Fig. 25.3

RUNNING PROBLEM

By the time Colleen reached the practice field, she was already sweating and her face was flushed. At 9:00 A.M. the air temperature was 80 °F and the humidity was 54%. Colleen took a quick drink of water from the large water container and ran out to the field.

Q1: "Humidity" is the percentage of water vapor present in air. Why is the thermoregulatory mechanism of sweating less efficient in humid environments?

836 838 842 843 844 846

Curiously, although plasma glucose concentrations rise with exercise, the secretion of insulin decreases. This response is contrary to what you might predict, because normally an increase in plasma glucose stimulates insulin release. During exercise, however, insulin secretion is suppressed, probably by sympathetic input onto the beta cells of the pancreas.

What could be the advantage of lower insulin levels during exercise? For one thing, less insulin means that cells other than muscle fibers reduce their glucose uptake, sparing blood glucose for use by muscles. Actively contracting muscle cells, on the other hand, are not affected by low levels of insulin because they do not require insulin for glucose uptake. Contraction stimulates the insulin-independent translocation of GLUT4 transporters to the muscle membrane, increasing glucose uptake in proportion to contractile activity.

Oxygen Consumption Is Related to Exercise Intensity

The activities we call exercise range widely in intensity and duration, from the rapid and relatively brief burst of energy exerted by a sprinter or power lifter, to the sustained effort of a marathoner. Physiologists traditionally quantify the intensity of a period of exercise by measuring oxygen consumption (V_{O_2}). **Oxygen consumption** refers to the fact that oxygen is used up, or consumed, during oxidative phosphorylation, when it combines with hydrogen in the mitochondria to form water [p. 115].

Oxygen consumption is a measure of cellular respiration and is usually measured in liters of oxygen consumed per minute. A person's *maximal rate of oxygen consumption* (V_{O_2max}) is an indicator of the ability to perform endurance exercise. The greater the V_{O_2max}, the greater the person's predicted ability to do work.

A metabolic hallmark of exercise is an increase in oxygen consumption that persists even after the activity ceases (Fig. 25.4). When exercise begins, oxygen consumption

OXYGEN CONSUMPTION AND EXERCISE

Oxygen supply to exercising cells lags behind energy use, creating an oxygen deficit. Excess postexercise oxygen consumption compensates for the oxygen deficit.

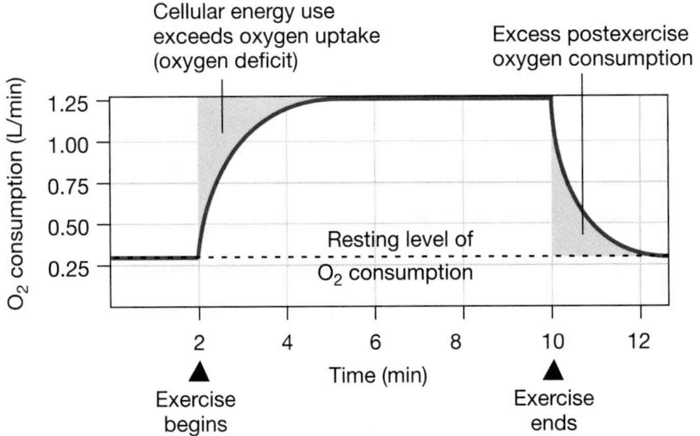

Fig. 25.4

increases so rapidly that it is not immediately matched by the oxygen supplied to the muscles. During this lag time, ATP is provided by muscle ATP reserves, phosphocreatine, and aerobic metabolism supported by oxygen stored on muscle myoglobin and blood hemoglobin [p. 419].

The use of these muscle stores creates an *oxygen deficit* because their replacement requires aerobic metabolism and oxygen uptake. Once exercise stops, oxygen consumption is slow to resume its resting level. The **excess postexercise oxygen consumption** (EPOC; formerly called the oxygen debt) represents oxygen being used to metabolize lactate, restore ATP and phosphocreatine levels, and replenish the oxygen bound to myoglobin. Other factors that play a role in elevating postexercise oxygen consumption include increased body temperature and circulating catecholamines.

Several Factors Limit Exercise

What factors limit a person's exercise capacity? To some extent, the answer depends on the type of exercise. Resistance training such as strength training depends heavily on anaerobic metabolism to meet energy needs. The situation is more complex with aerobic or endurance exercise. Is the limiting factor for aerobic exercise the ability of the exercising muscle to use oxygen efficiently? Or is it the ability of the cardiovascular system to deliver oxygen to the tissues? Or the ability of the respiratory system to provide oxygen to the blood?

One possible limiting factor in exercise is the ability of muscle fibers to obtain and use oxygen. If muscle mitochondria are limited in number, or if they have insufficient oxygen supply, the muscle fibers are unable to produce ATP rapidly. Data suggest that muscle metabolism is not the limiting factor for maximum exercise capacity, but muscle metabolism has been shown to influence submaximal exercise capacity. This finding explains the increase in numbers of muscle mitochondria and capillaries with endurance training.

The question of whether the pulmonary system or the cardiovascular system limits maximal exercise was resolved when research showed that ventilation is only 65% of its maximum when cardiac output has reached 90% of its maximum. From that information, exercise physiologists concluded that the ability of the cardiovascular system to deliver oxygen and nutrients to the muscle at a rate that supports aerobic metabolism is a major factor in determining maximum oxygen consumption. One goal of training is to improve cardiac efficiency.

Next we examine the reflexes that integrate breathing and cardiovascular function during exercise.

Ventilatory Responses to Exercise

Think about what happens to your breathing when you exercise. Exercise is associated with both increased rate and increased depth of breathing, resulting in enhanced alveolar ventilation [p. 590]. **Exercise hyperventilation**, or **hyperpnea,** results from a combination of feedforward signals from central command neurons in the motor cortex and sensory feedback from peripheral receptors.

When exercise begins, mechanoreceptors and proprioreceptors [p. 445] in muscles and joints send information about movement to the motor cortex. Descending pathways from the motor cortex to the respiratory control center of the medulla oblongata then immediately increase ventilation (Fig. 25.5).

VENTILATION AND EXERCISE

Ventilation rate jumps as soon as exercise begins, despite the fact that neither arterial P_{CO_2} nor P_{O_2} has changed. This suggests there is a feedforward component to the ventilatory response.

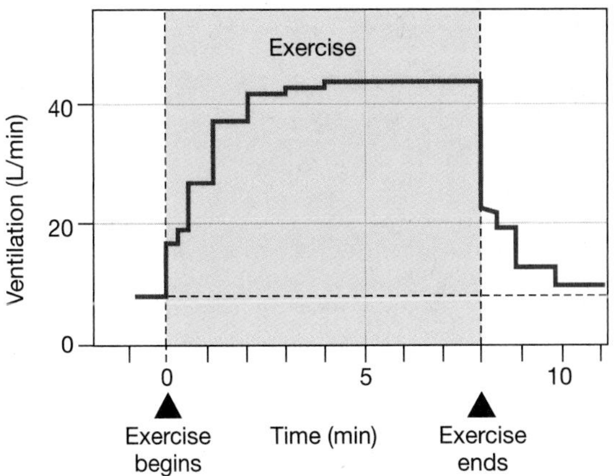

Modified from P. Dejours, *Handbook of Physiology* (Washington, D.C.: American Physiological Society, 1964).

Fig. 25.5

As muscle contraction continues, sensory information feeds back to the respiratory control center to ensure that ventilation and tissue oxygen use remain closely matched. Sensory receptors involved in the secondary response probably include central, carotid, and aortic chemoreceptors that monitor P_{CO_2}, pH, and P_{O_2} [p. 618]; proprioceptors in the joints; and possibly receptors located within the exercising muscle itself. Pulmonary stretch receptors were once thought to play a role, but recipients of heart-lung transplants display a normal ventilatory response to exercise even though the neural connections between lung and brain are absent.

Exercise hyperventilation maintains nearly normal arterial P_{O_2} and P_{CO_2} by steadily increasing alveolar ventilation in proportion to the level of exercise. The compensation is so effective that when arterial P_{O_2}, P_{CO_2}, and pH are monitored during mild to moderate exercise, they show no significant change (Fig. 25.6). This observation means that the once-accepted causes of increased ventilation during mild to moderate exercise—reduced arterial P_{O_2}, elevated arterial P_{CO_2}, and decreased plasma pH—must not be correct. Instead the chemoreceptors or medullary respiratory control center, or both, must be responding to other exercise-induced signals.

Several factors have been postulated to be these signals, including sympathetic input to the carotid body and changes in plasma K^+ concentration. During even mild exercise, extracellular K^+ increases as repeated action potentials in the muscle fibers allow K^+ to move out of cells. Carotid chemoreceptors are known to respond to increased K^+ by increasing ventilation. However, because K^+ concentration changes slowly, this mechanism does not explain the sharp initial rise in ventilation at the onset of activity.

It appears likely that the initial increase in ventilation is caused by sensory input from muscle mechanoreceptors combined with parallel descending pathways from the motor cortex to the respiratory control centers. Once exercise is under way, sensory input keeps ventilation matched to metabolic needs.

BLOOD GASES AND EXERCISE

Arterial blood gases and pH remain steady with submaximal exercise.

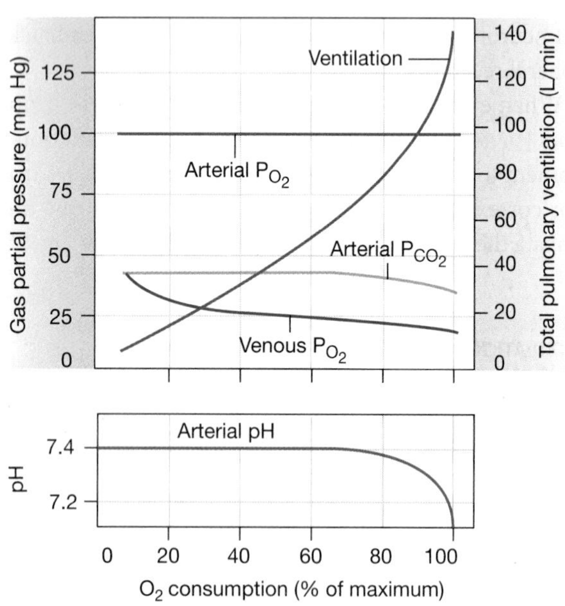

FIGURE QUESTIONS

1. Ventilation increases with exercise. Why doesn't arterial P_{O_2} increase as well?
2. What happens to O_2 delivery to cells with increasing exercise?
3. Why does venous P_{O_2} decrease?
4. Why doesn't arterial P_{CO_2} increase with exercise?
5. Why does arterial P_{CO_2} decrease with maximum exercise?

Adapted from P. O. Astrand, *et al. Textbook of Work Physiology*, 4th ed. New York: McGraw Hill, 2003.

Fig. 25.6

Concept Check Answer: p. 849

1. If venous P_{O_2} decreases as exercise intensity increases, what do you know about the P_{O_2} of muscle cells as exercise intensity increases?

Cardiovascular Responses to Exercise

When exercise begins, mechanosensory input from working limbs combines with descending pathways from the motor cortex to activate the cardiovascular control center in the medulla oblongata. The center responds with sympathetic discharge that increases cardiac output and causes vasoconstriction in many peripheral arterioles.

Cardiac Output Increases During Exercise

During strenuous exercise, cardiac output rises dramatically. In untrained individuals, cardiac output goes up fourfold, from 5 L/min to 20 L/min. In trained athletes, it may go up six to eight times, reaching as much as 40 L/min. Because oxygen delivery by the cardiovascular system is the primary factor determining exercise tolerance, trained athletes are therefore capable of more strenuous exercise than untrained people.

Cardiac output is determined by heart rate and stroke volume:

$$\text{Cardiac output (CO)} = \text{heart rate} \times \text{stroke volume}$$

If the factors that influence heart rate and stroke volume are considered, then

$$CO = (\text{SA node rate} + \text{autonomic nervous system input}) \times (\text{venous return} + \text{force of contraction})$$

Which of these factors has the greatest effect on cardiac output during exercise in a healthy heart? Venous return is enhanced by skeletal muscle contraction and deep inspiratory movements during exericse [p. 498], so it is tempting to postulate that the cardiac muscle fibers simply stretch in response to increased venous return, thereby increasing contractility.

However, overfilling of the ventricles is potentially dangerous, because overstretching may damage the fibers. One factor that counters increased venous return is increased heart rate. If the interval between contractions is shorter, the heart has less time to fill and is less likely to be damaged by excessive stretch.

The initial change in heart rate at the onset of exercise is due to decreased parasympathetic activity at the sinoatrial (SA) node [p. 495]. As cholinergic inhibition lessens, heart rate rises from its resting rate to around 100 beats per minute, the intrinsic pacemaker rate of the SA node. At that point, sympathetic output from the cardiovascular control center escalates.

Sympathetic stimulation has two effects on the heart. First, it increases contractility so that the heart squeezes out more blood per stroke (increased stroke volume). Second, sympathetic innervation increases heart rate so that the heart has less time to relax, protecting it from overfilling. In short, the combination of faster heart rate and greater stroke volume increases cardiac output during exercise.

Muscle Blood Flow Increases During Exercise

At rest, skeletal muscles receive less than a fourth of the cardiac output, or about 1.2 L/min. During exercise, a significant shift in peripheral blood flow takes place because of local and reflex reactions (■ Fig. 25.7). During strenuous exercise in highly trained athletes, the combination of increased cardiac output and vasodilation can increase blood flow through

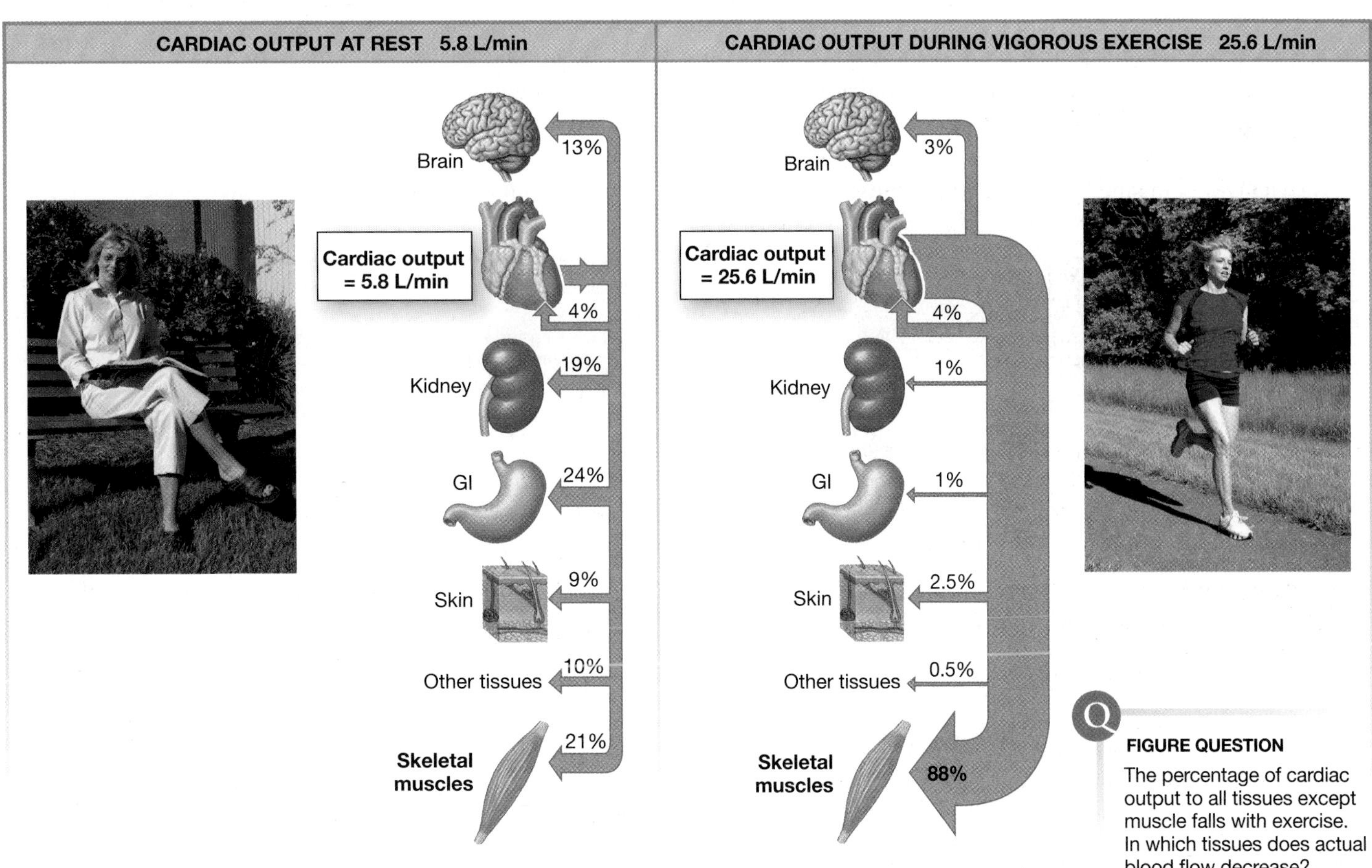

■ **Fig. 25.7**

RUNNING PROBLEM

By 10:00 A.M. the temperature had risen to 93 °F, with 50% humidity. Colleen felt dizzy and nauseated, but she pressed on. She had made the team by the skin of her teeth and felt she had to prove herself to her teammates. She took only a short drink of water during a break and hustled back onto the field. At 10:07 A.M. Colleen collapsed on the field. One of her teammates with training in first aid felt Colleen's skin. It was hot and dry. A call went in to emergency medical services.

Q2: Individuals with a heat emergency called heat exhaustion have cool, moist skin. Hot, dry skin indicates a more serious emergency called heat stroke. Why is heat exhaustion less serious than heat stroke? (Hint: How does skin temperature relate to the body's ability to regulate body temperature?)

836 838 842 843 844 846

exercising muscle to more than 22 L/min! The relative distribution of blood flow to tissues also shifts. About 88% of cardiac output is diverted to the exercising muscle, up from 21% at rest.

The redistribution of blood flow during exercise results from a combination of vasodilation in skeletal muscle arterioles and vasoconstriction in other tissues. At the onset of exercise, sympathetic signals from the cardiovascular control center cause vasoconstriction in peripheral tissues. As muscles become active, changes in the microenvironment of muscle tissue take place: tissue O_2 concentrations decrease, while temperature, CO_2, and acid in the interstitial fluid around muscle fibers increase. All these factors act as paracrines causing local vasodilation that overrides the sympathetic signal for vasoconstriction. The net result is shunting of blood flow from inactive tissues to the exercising muscles, where it is needed.

Blood Pressure Rises Slightly During Exercise

What happens to blood pressure during exercise? Peripheral blood pressure is determined by a combination of cardiac output and peripheral resistance [p. 515]:

$$\text{Mean arterial blood pressure} = \text{cardiac output} \times \text{peripheral resistance}$$

Cardiac output increases during exercise, thereby contributing to increased blood pressure. The changes resulting from peripheral resistance are harder to predict, however, because some peripheral arterioles are constricting while others are dilating.

Skeletal muscle vasodilation decreases peripheral resistance to blood flow. At the same time, sympathetically induced vasoconstriction in nonexercising tissues offsets the vasodilation, but only partially. Consequently, total peripheral resistance to blood flow falls dramatically as exercise commences, reaching a minimum at about 75% of V_{O_2max} (Fig. 25.8a).

If no other compensation occurred, this decrease in peripheral resistance would dramatically lower arterial blood pressure. However, increased cardiac output cancels out decreased peripheral resistance. When blood pressure is monitored during exercise, mean arterial blood pressure actually increases slightly

BLOOD PRESSURE AND EXERCISE

(a) Peripheral resistance decreases due to vasodilation in exercising muscle.

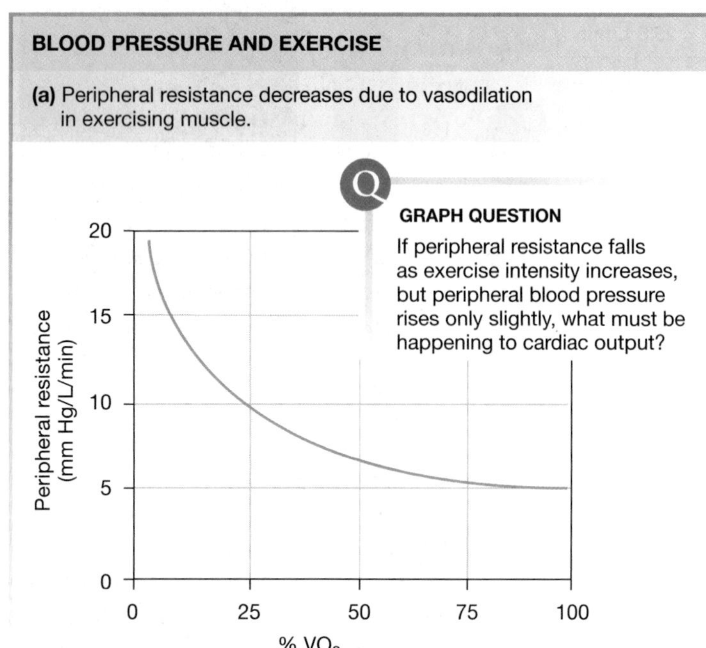

(b) Mean arterial blood pressure rises slightly despite drop in resistance.

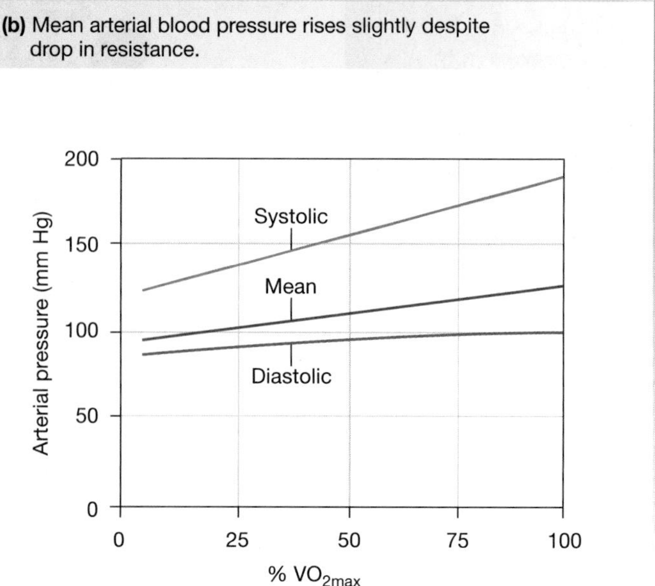

Fig. 25.8

as exercise intensity increases (Fig. 25.8b). The fact that it increases at all, however, suggests that the normal baroreceptor reflexes that control blood pressure are functioning differently during exercise.

Concept Check Answer: p. 849

2. In Figure 25.8b, why does the line for mean blood pressure lie closer to diastolic pressure instead of being evenly centered between systolic and diastolic pressures? (*Hint:* What is the equation for calculating mean blood pressure?)

The Baroreceptor Reflex Adjusts to Exercise

Normally, homeostasis of blood pressure is regulated through peripheral baroreceptors in the carotid and aortic bodies: an increase in blood pressure initiates responses that return blood pressure to normal. But during exercise, blood pressure increases without activating homeostatic compensation. What happens to the normal baroreceptor reflex during exercise?

There are several theories. According to one, signals from the motor cortex during exercise reset the arterial baroreceptor threshold to a higher pressure. Blood pressure can then increase slightly during exercise without triggering the homeostatic counter-regulatory responses.

Another theory suggests that signals in baroreceptor afferent neurons are blocked in the spinal cord by presynaptic inhibition [p. 277] at some point before the afferent neurons synapse with central nervous system neurons. This central inhibition inactivates the baroreceptor reflex during exercise.

A third theory is based on the postulated existence of muscle chemoreceptors that are sensitive to metabolites (probably H^+) produced during strenuous exercise. When stimulated, these chemoreceptors signal the CNS that tissue blood flow is not adequate to remove muscle metabolites or keep the muscle in aerobic metabolism. The chemoreceptor input is reinforced by sensory input from mechanoreceptors in the working limbs. The CNS response to this sensory input is to override the baroreceptor reflex and raise blood pressure to enhance muscle perfusion. The same hypothetical muscle chemoreceptors may play a role in ventilatory responses to exercise.

Feedforward Responses to Exercise

Interestingly, there is a significant *feedforward* element [p. 18] in the physiological responses to exercise. It is easy to explain physiological changes that occur with exercise as reactions to the disruption of homeostasis. However, many of these changes occur in the absence of the normal stimuli or before the stimuli are present. For example, as you may know from your own experience, ventilation rates jump as soon as exercise begins (Fig. 25.5), even though experiments have shown that arterial P_{O_2} and P_{CO_2} do not change (Fig. 25.6).

RUNNING PROBLEM

While they waited for the ambulance, Colleen's coaches began to try to cool her down by misting her with cool water and using a fan to blow air across her body. When the paramedics reached Colleen on the field, they took her blood pressure, and it was extremely low. Her body temperature was 104 °F. The paramedics rushed Colleen to the hospital.

Q3: Why was Colleen's blood pressure so low?

Q4: Why would misting and fanning help lower Colleen's body temperature?

836 838 842 843 844 846

25

How does the feedforward response work? One model says that as exercise begins, proprioceptors in the muscles and joints send information to the motor cortex of the brain. Descending signals from the motor cortex go not only to the exercising muscles but also along parallel pathways to the cardiovascular and respiratory control centers and to the limbic system of the brain.

Output from the limbic system and cardiovascular control center triggers generalized sympathetic discharge. As a result, an immediate slight increase in blood pressure marks the beginning of exercise. Sympathetic discharge causes widespread vasoconstriction, increasing blood pressure. Once exercise has begun, this increase in blood pressure compensates for decreases in blood pressure resulting from muscle vasodilation.

As exercise proceeds, reactive compensations become superimposed on the feedforward changes. For example, when exercise reaches 50% of aerobic capacity, muscle chemoreceptors detect the buildup of H^+, lactate, and other metabolites, and send this information to central command centers in the brain. The command centers then maintain changes in ventilation and circulation that were initiated in a feedforward manner. Thus, the integration of systems in exercise probably involves both common reflex pathways and some unique centrally mediated reflex pathways.

Temperature Regulation During Exercise

As exercise continues, heat released through metabolism creates an additional challenge to homeostasis. Most of the energy released during metabolism is not converted into ATP but instead is released as heat. (Efficiency of energy conversion from organic substrates to ATP is only 20–25%.) With continued exercise, heat production exceeds heat loss, and core body temperature rises. In endurance events, body temperature can reach 40–42 °C (104–108 °F), which we would normally consider a fever.

This rise in body temperature during exercise triggers two thermoregulatory mechanisms: sweating and increased cutaneous blood flow [p. 767]. Both mechanisms help regulate body temperature, but both can also disrupt homeostasis in other ways. While sweating lowers body temperature through evaporative cooling, the loss of fluid from the extracellular compartment can cause dehydration and significantly reduce circulating blood volume. Because sweat is a hypotonic fluid, the extra water loss increases body osmolarity. The combination of decreased ECF volume and increased osmolarity during extended exercise sets in motion the complex homeostatic pathways for overcoming dehydration, including thirst and renal conservation of water [Fig. 20.13, p. 680].

The other thermoregulatory mechanism—increased blood flow to the skin—causes body heat loss to the environment through convection [p. 766]. However, increased sympathetic output during exercise tends to vasoconstrict cutaneous blood vessels, which opposes the thermoregulatory response. The primary control of vasodilation in hairy regions of skin, such as trunk and limbs, during exercise appears to come from a sympathetic vasodilator system. Activation of these acetylcholine-secreting sympathetic neurons as body core temperature rises dilates some cutaneous blood vessels without altering sympathetic vasoconstriction in other body tissues.

Although cutaneous vasodilation is essential for thermoregulation, it can disrupt homeostasis by decreasing peripheral resistance and diverting blood flow from the muscles. In the face of these contradictory demands, the body initially gives preference to thermoregulation. However, if central venous pressure falls below a critical minimum, the body abandons thermoregulation in the interest of maintaining blood flow to the brain.

The degree to which the body can adjust to both demands depends on the type of exercise being performed and its intensity and duration. Strenuous exercise in hot, humid environments can severely impair normal thermoregulatory mechanisms and cause *heat stroke,* a potentially fatal condition. Unless prompt measures are taken to cool the body, core temperatures can go as high as 43 °C (109 °F).

It is possible for the body to adapt to repeated exercise in hot environments, however, through **acclimatization**. In this process, physiological mechanisms shift to fit a change in environmental conditions. As the body adjusts to exercise in the heat, sweating begins sooner and doubles or triples in volume, enhancing evaporative cooling. With acclimatization, sweat also becomes more dilute, as salt is reabsorbed from the sweat glands under the influence of increased aldosterone. Salt loss in an unacclimatized person exercising in the heat may reach 30 g NaCl per day, but that value decreases to as little as 3 g after a month of acclimatization.

Concept Check Answer: p. 849

3. The active vasodilator nerves to the skin secrete ACh but are classified as sympathetic neurons. On what basis were they identified as sympathetic?

Exercise and Health

Physical activity has many positive effects on the human body. The lifestyles of humans have changed dramatically since we were hunter-gatherers, but our bodies still seem to work best with a certain level of physical activity. Several common pathological conditions—including high blood pressure, strokes, and diabetes mellitus—can be improved by physical activity. Even so, developing regular exercise habits is one lifestyle change that many people find difficult to make. In this section we look at the effects exercise has on several common health conditions.

Exercise Lowers the Risk of Cardiovascular Disease

As early as the 1950s, scientists showed that physically active men have a lower rate of heart attacks than do men who lead sedentary lives. These studies started many investigations into the exact relationship between cardiovascular disease and exercise. Scientists have subsequently demonstrated that exercise has positive benefits for both men and women. These benefits include lowering blood pressure, decreasing plasma triglyceride levels, and raising plasma HDL-cholesterol levels. High blood pressure is a major risk factor for strokes, and elevated triglycerides and low HDL-cholesterol levels are associated with development of atherosclerosis and increased risk of heart attack.

Overall, exercise reduces the risk of death or illness from a variety of cardiovascular diseases, although the exact mechanisms by which this occurs are still unclear. Even such mild exercise as walking has significant health benefits that could

RUNNING PROBLEM

When Colleen arrived in the emergency room, she was quickly diagnosed with heat stroke, a life-threatening condition. She was immersed in a tub of cool water and given intravenous fluids.

Q5: What steps should Colleen have taken to avoid heat stroke? (Start at 9:00 A.M., when Colleen set out for the practice field.)

836 838 842 843 844 846

reduce the risk of developing cardiovascular diseases or diabetes and the complications of obesity in the estimated 40 million adult Americans with sedentary lifestyles.

Type 2 Diabetes Mellitus May Improve with Exercise

Regular exercise is now widely accepted as effective in preventing and alleviating type 2 diabetes mellitus and its complications, including microvascular retinopathy [p. 512], diabetic neuropathy [p. 339], and cardiovascular disease [p. 534]. With regular exercise, skeletal muscle fibers up-regulate both the number of GLUT4 glucose transporters and the number of insulin receptors on their membrane. The addition of insulin-independent GLUT4 transporters decreases the muscle's dependence on insulin for glucose uptake. Glucose uptake into the exercising muscle also helps correct the hyperglycemia of diabetes.

Up-regulation of insulin receptors with exercise makes the muscle fibers more sensitive to insulin. A smaller amount of insulin then can achieve a response that previously required more insulin. Because the cells are responding to lower insulin levels, the endocrine pancreas secretes less insulin. This lessens the stress on the pancreas, resulting in a lower incidence of type 2 diabetes mellitus.

Figure 25.9 shows the effects of seven days of exercise on glucose utilization and insulin secretion in men with mild type 2 diabetes. Individuals in the experiment underwent *glucose tolerance tests,* in which they ingested 100 g of glucose after an overnight fast. Their plasma glucose levels were assessed before and for 120 minutes after ingesting the glucose. Simultaneous measurements were made of plasma insulin.

The graph in Figure 25.9a shows glucose tolerance tests in control subjects (blue line) and in the diabetic men before and after exercise (red and green lines, respectively). Figure 25.9b shows concurrent insulin secretion in the three groups. After only seven days of exercise, both the glucose tolerance test and insulin secretion in exercising diabetic subjects had shifted to a pattern that was more like that of the normal control subjects. These results demonstrate the beneficial effect of exercise on glucose transport and metabolism, and support the recommendation that patients with type 2 diabetes maintain a regular exercise program.

Stress and the Immune System May Be Influenced by Exercise

Another health-related topic receiving much attention is the interaction of exercise with the immune system, and "exercise immunology" has become a recognized scientific discipline.

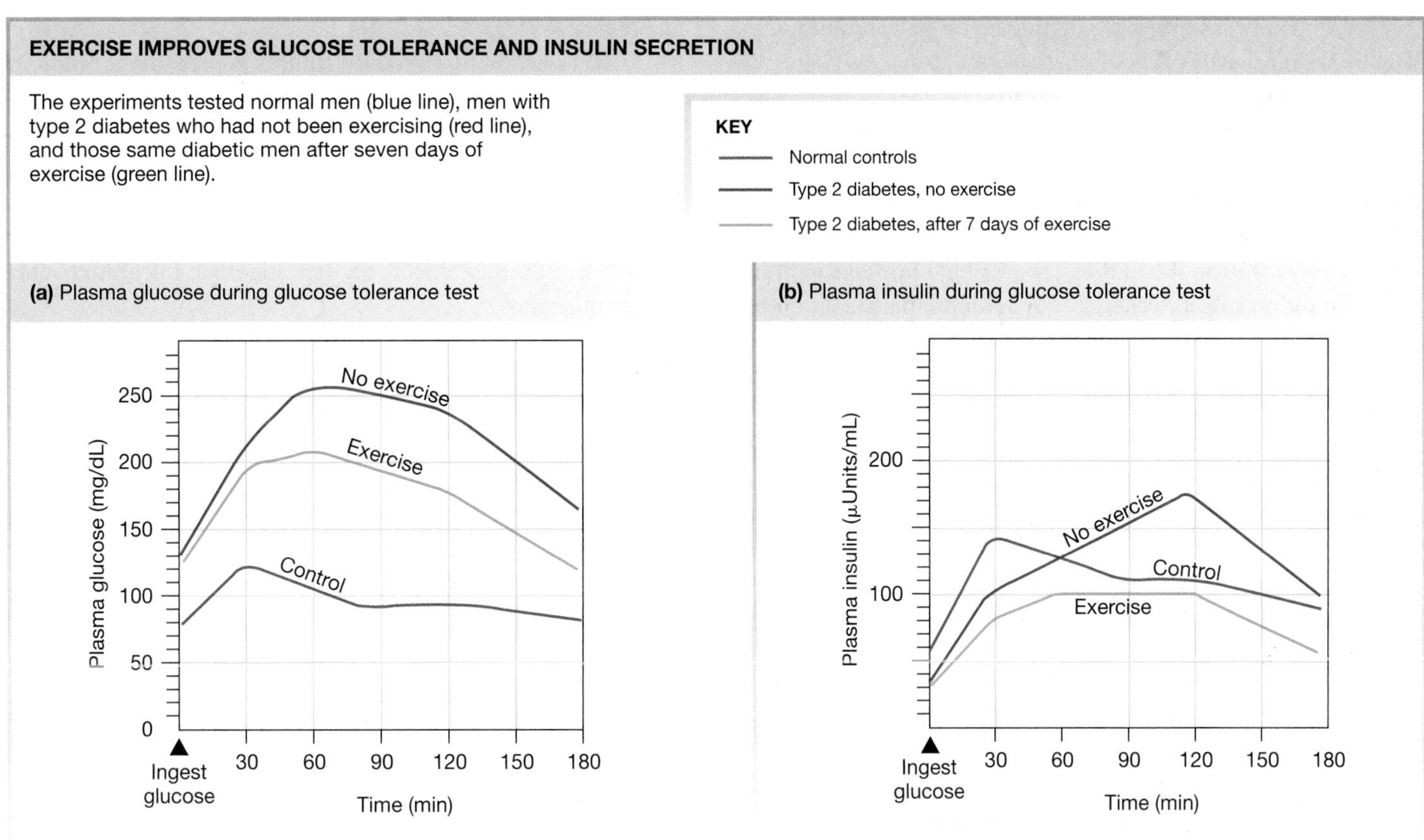

Fig. 25.9

Epidemiological studies looking at large populations of people suggest that exercise is associated with a reduced incidence of disease and with increased longevity. Moreover, many people believe that exercise boosts immunity, prevents cancer, and helps HIV-infected patients combat AIDS.

However, there are few rigidly controlled research studies providing evidence to support those viewpoints. Indeed, other evidence suggests that strenuous exercise is a form of stress that suppresses the immune response. Immune suppression may be due to corticosteroid release, or it may be due to release of interferon-γ during strenuous exercise.

Researchers have proposed that the relationship between exercise and immunity can be represented by a J-shaped curve (Fig. 25.10). People who exercise moderately have slightly more effective immune systems than those who are sedentary, but people who exercise strenuously may experience a decrease in immune function because of the stress of the exercise.

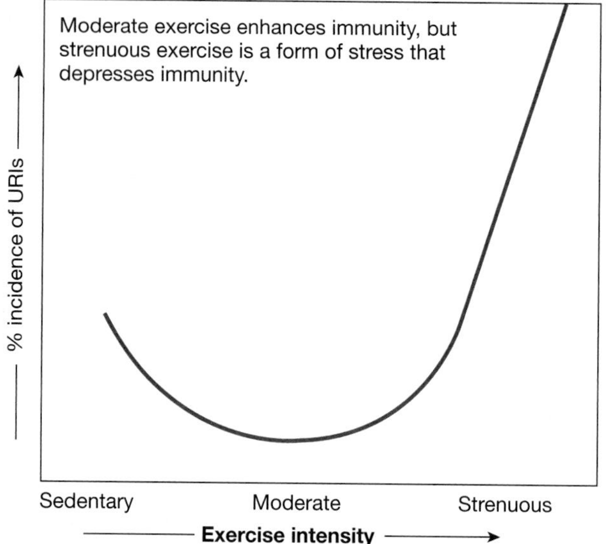

IMMUNE FUNCTION AND EXERCISE
Individuals who exercise in moderation have fewer upper respiratory infections (URIs) than sedentary individuals or those who exercise strenously.
Moderate exercise enhances immunity, but strenuous exercise is a form of stress that depresses immunity.
% incidence of URIs
Sedentary
Moderate
Strenuous
Exercise intensity

Fig. 25.10

Another area of exercise physiology filled with interesting though contradictory results is the effect of exercise on stress, depression, and other psychological parameters. Research has shown an inverse relationship between exercise and depression: people who exercise regularly are significantly less likely to be clinically depressed than are people who do not exercise regularly. Although the association exists, assigning cause and effect to the two parameters is difficult. Are the exercisers less depressed because they exercise? Or do depressed individuals exercise less because they are depressed? What physiological factors are involved?

Many published studies appear to show that regular exercise is effective in reducing depression. But a careful analysis of experimental design suggests that the conclusions of some of those studies may be overstated. The subjects in many of the experiments were being treated concurrently with drugs or psychotherapy, so it is difficult to attribute improvement in their condition solely to exercise. In addition, participation in exercise studies gives subjects a period of social interaction, another factor that might play a role in the reduction of stress and depression.

The assertion that exercise reduces depression has support from studies showing that exercise increases serotonin in the brain. Drugs that enhance serotonin activity, such as the selective serotonin reuptake inhibitors [p. 314], are currently being used to treat depression, and a way to achieve the same result without drugs would be desirable. A number of clinical trials that look at exercise effects on depression and health are currently underway.

RUNNING PROBLEM CONCLUSION

Heat Stroke

After a day of treatment, Colleen recovered enough to be sent home. She was unable to practice with the team for the remainder of the season because victims of heat stroke are more sensitive to high temperature for some time after the episode.

Heat stroke can occur in athletes who overexert themselves in extremely hot weather, but it is also common in elderly individuals, whose thermoregulatory mechanisms are less efficient than those in younger people. To learn more about the symptoms and treatment of heat stroke, try a Google search. Then check your understanding of this problem by comparing your answers to those in the table below.

RUNNING PROBLEM CONCLUSION *(continued)*

Question	Facts	Integration and Analysis
1. Why is the thermoregulatory mechanism of sweating less efficient in humid environments?	"Humidity" is the percentage of water vapor present in air. Thermoregulation in warm environments includes sweating and evaporative cooling.	Evaporation is slower in humid air, so evaporative cooling is less effective in high humidity.
2. Individuals with a heat emergency called heat exhaustion have cool, moist skin. Hot, dry skin indicates a more serious emergency called heat stroke. Why is heat exhaustion less serious than heat stroke?	Sweating helps the body regulate temperature through evaporative cooling. As evaporation occurs, the skin surface cools.	Cool, moist skin indicates that the sweating mechanism is still functioning. Hot, dry skin indicates that the sweating mechanism has failed. Subjects with hot, dry skin are likely to have higher internal temperatures.
3. Why was Colleen's blood pressure so low?	Blood volume decreases when the body loses large amounts of water through sweating.	Colleen had been sweating but not replacing the fluid she lost. This caused her blood volume to decrease, with a corresponding decrease in blood pressure.
4. Why would misting and fanning help lower Colleen's body temperature?	Evaporation of water creates evaporative cooling.	Putting water on Colleen's skin, then using a fan to evaporate the water helps cool the skin.
5. What steps could Colleen have taken to avoid heat stroke? (Start at 9:00 A.M., when Colleen set out for the practice field.)	Heat stroke is caused by dehydration resulting from excessive sweating. Dehydration results in a drop in blood pressure. Peripheral blood vessels constrict in an effort to maintain pressure and blood flow to the brain. Constricted blood vessels in the skin cannot release excess heat. In addition, the sweating response is inhibited to prevent further fluid loss.	Colleen could have avoided heat stroke by: (1) consuming large amounts of fluid (water plus electrolytes) before and during practice; (2) avoiding unnecessary exercise such as bicycling to practice; (3) stopping activity at the first signs of heat emergency (dizziness and nausea); (4) seeking shade and drinking large amounts of fluid to replenish lost fluids; (5) applying cool wet towels or ice to lower body temperature.

836 838 842 843 844 846

Test your understanding with:

- Practice Tests
- Running Problem Quizzes
- A&PFlix™ Animations
- PhysioEx™ Lab Simulations
- Interactive Physiology Animations

MasteringA&P®

Chapter Summary

In this chapter you learned about exercise and the physiological challenges it presents. *Integration* and *coordination* between the body's *physiological control systems* allow the internal environment to remain relatively constant, despite the challenges to *homeostasis* that exercise presents.

Metabolism and Exercise

1. Exercising muscle requires a steady supply of ATP from metabolism or from conversion of phosphocreatine. (p. 837; Fig. 25.1)
2. Carbohydrates and fats are the primary energy substrates. Glucose can be metabolized through both oxidative and anaerobic pathways, but fatty acid metabolism requires oxygen. (p. 837; Fig. 25.1)
3. Anaerobic **glycolytic metabolism** converts glucose to lactate and H^+. Glycolytic metabolism is 2.5 times more rapid than aerobic pathways but is not as efficient at ATP production. (p. 837; Fig. 25.2)
4. Glucagon, cortisol, catecholamines, and growth hormone influence glucose and fatty acid metabolism during exercise. These hormones favor the conversion of glycogen to glucose. (p. 838)
5. Plasma glucose concentrations rise with exercise, but insulin secretion decreases. This response reduces glucose uptake by most cells, making more glucose available for exercising muscle. (p. 838)
6. The intensity of exercise is indicated by **oxygen consumption** (V_{O_2}). A person's maximal rate of oxygen consumption (V_{O_2max}) is an indicator of that person's ability to perform endurance exercise. (p. 838)
7. Oxygen consumption increases rapidly at the onset of exercise. **Excess postexercise oxygen consumption** is due to ongoing metabolism, increased body temperature, and circulating catecholamines. (p. 839; Fig. 25.4)
8. Muscle mitochondria increase in size and number with endurance training. (p. 839)
9. At maximal exertion, the ability of the cardiovascular system to deliver oxygen and nutrients appears to be the primary limiting factor. (p. 839)

Ventilatory Responses to Exercise

10. Exercise hyperventilation results from feedforward signals from the motor cortex and sensory feedback from peripheral sensory receptors. (p. 839; Fig. 25.5)
11. Arterial P_{O_2}, P_{CO_2}, and pH do not change significantly during mild to moderate exercise. (p. 840; Fig. 25.6)

Cardiovascular Responses to Exercise

12. Cardiac output increases with exercise because of increased venous return and sympathetic stimulation of heart rate and contractility. (p. 841; Fig. 25.7)
13. Blood flow through exercising muscle increases dramatically when skeletal muscle arterioles dilate. Arterioles in other tissues constrict. (p. 841; Fig. 25.7)
14. Decreased tissue O_2 and glucose or increased muscle temperature, CO_2, and acid act as paracrine signals and cause local vasodilation. (p. 842)
15. Mean arterial blood pressure increases slightly as exercise intensity increases. The baroreceptors that control blood pressure change their setpoints during exercise. (p. 842; Fig. 25.8)

Feedforward Responses to Exercise

16. When exercise begins, feedforward responses prevent significant disruption of homeostasis. (p. 843)

Temperature Regulation During Exercise

17. Heat released during exercise is dissipated by sweating and increased cutaneous blood flow. (p. 843)

Exercise and Health

18. Physical activity can help prevent or decrease the risk of developing high blood pressure, strokes, and type 2 diabetes mellitus. (p. 844)
19. Studies suggest that serotonin release during exercise may help alleviate depression. (p. 845)

Questions

Answers: p. A-1

Level One Reviewing Facts and Terms

1. Name the two muscle compounds that store energy in the form of high-energy phosphate bonds.
2. The most efficient ATP production is through *aerobic/anaerobic* pathways. When these pathways are being used, then *glucose/fatty acids/ both/neither* can be metabolized to provide ATP.
3. What are the differences between aerobic and anaerobic metabolism?
4. List three sources of glucose that can be metabolized to ATP, either directly or indirectly.
5. List four hormones that promote the conversion of triglycerides into fatty acids. What effects do these hormones have on plasma glucose levels?
6. What is meant by the term oxygen deficit, and how is it related to excessive postexercise oxygen consumption?
7. What organ system is the limiting factor for maximal exertion?
8. In endurance events, body temperature can reach 40–42 °C. What is normal body temperature? What two thermoregulatory mechanisms are triggered by this change in temperature during exercise?

Level Two Reviewing Concepts

9. Concept map: Map the metabolic, cardiovascular, and respiratory changes that occur during exercise. Include the signals to and from the nervous system, and show what specific areas signal and coordinate the exercise response.
10. What causes insulin secretion to decrease during exercise, and why is this decrease adaptive?
11. State two advantages and two disadvantages of anaerobic glycolysis.
12. Compare and contrast each of the terms in the following sets of terms, especially as they relate to exercise:
    (a) ATP, ADP, PCr
    (b) myoglobin, hemoglobin
13. Match the following brain areas with the response(s) that each controls. Brain areas may control one response, more than one response, or none at all. Some responses may be associated with more than one brain area.

(a) pons	1. changes in cardiac output
(b) medulla oblongata	2. vasoconstriction
(c) midbrain	3. exercise hyperventilation
(d) motor cortex	4. increased stroke volume

(e) hypothalamus	5. increased heart rate
(f) cerebellum	6. coordination of skeletal muscle movement
(g) brain not involved (i.e., local control)	

14. Specify whether each of the following parameters stays the same, increases, or decreases when a person becomes better conditioned for athletic activities:
    (a) heart rate during exercise
    (b) resting heart rate
    (c) cardiac output during exercise
    (d) resting cardiac output
    (e) breathing rate during exercise
    (f) blood flow to muscles during exercise
    (g) blood pressure during exercise
    (h) total peripheral resistance during exercise
15. Why doesn't increased venous return during exercise overstretch the heart muscle?
16. Diagram the three theories that explain why the normal baroreceptor reflex is absent during exercise.
17. List and briefly discuss the benefits of a lifestyle that includes regular exercise.
18. Explain how exercise decreases blood glucose in type 2 diabetes mellitus.

Level Three Problem Solving

19. You have decided to manufacture a new sports drink that will help athletes, from football players to gymnasts. List at least four different ingredients you would include in your drink, and indicate why each is important for the athlete.

Level Four Quantitative Problems

20. You are a well-conditioned athlete. At rest, your heart rate is 60 beats per minute and your stroke volume is 70 mL/beat. What is your cardiac output? At one point during exercise, your heart rate goes up to 120 beats/min. Does your cardiac output increase proportionately? Explain.
21. The graph below shows left ventricular pressure-volume curves in one individual. Curve A is the person sitting at rest. Curve B shows the person's cardiac response to mild exercise on a stationary bicycle. Curve C shows the cardiac response during maximum intensity cycling.
    (a) Calculate the stroke volume for each of the curves.
    (b) Given the following cardiac outputs (CO), calculate the heart rates for each condition. $CO_A = 6$ L/min, $CO_B = 10.5$ L/min, $CO_C = 19$ L/min
    (c) Which exercise curve shows an increase in stroke volume due primarily to increased contractility? Which exercise curve shows an increase in stroke volume due primarily to increased venous return?
    (d) Mechanistically, why did the end-diastolic volume in curve C fall back toward the resting value?

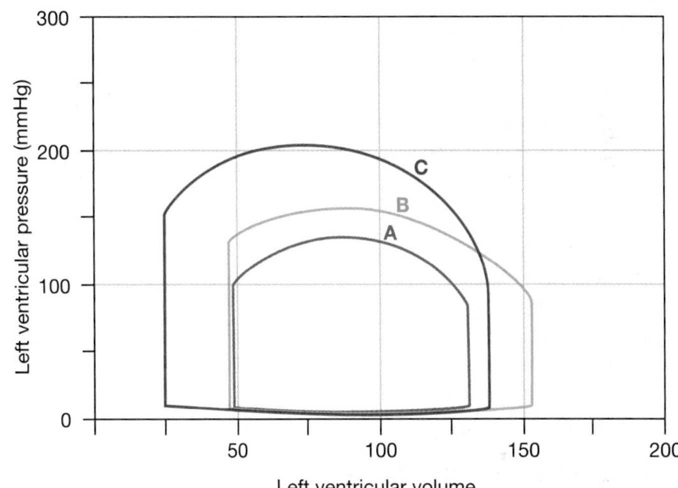

Data from G. D. Plotnick, *et al.*, *Am J Physiol* 251: H1101–H1105, 1986.

Answers

Answers to Concept Check Questions

Page 840

1. If venous P_{O_2} decreases, P_{O_2} in the cells is also decreasing.

Page 843

2. The mean blood pressure line lies closer to the diastolic pressure line because the heart spends more time in diastole than systole.

Page 844

3. The neurons are classified as sympathetic because of where they originate along the spinal cord.

Answers to Figure and Graph Questions

Page 840

Figure 25.6: 1. Arterial P_{O_2} remains constant because pulmonary ventilation is matched to blood flow through the lungs. 2. Although arterial P_{O_2} is constant, oxygen delivery to cells increases due to increased cardiac output (not shown). 3. Venous P_{O_2} drops as exercise increases because cells remove more oxygen from hemoglobin as oxygen consumption increases. 4. Arterial P_{CO_2} does not increase because increased production is matched by increased ventilation. 5. As the person begins to hyperventilate, arterial (and alveolar) P_{CO_2} declines.

Page 841

Figure 25.7: Blood flow to an organ is calculated by multiplying cardiac output (L/min) times the percentage of flow to that organ. When rest and exercise values are compared, actual blood flow decreases only in the kidneys, GI tract, and "other tissues."

Page 842

Figure 25.8: Mean arterial pressure is cardiac output × resistance. If resistance is falling but MAP is increasing, then cardiac output must be increasing.

25

26 Reproduction and Development

Birth, and copulation, and death. That's all the facts when you come to brass tacks.

—T.S. Eliot, *Sweeney Agonistes*

Background Basics

Imagine growing up as a girl, then at the age of 12 or so, finding that your voice is deepening and your genitals are developing into those of a man. This scenario actually happens to a small number of men who have a condition known as *pseudohermaphroditism* {*pseudes*, false + *hermaphrodites*, the dual-sex offspring of Hermes and Aphrodite}. These men have the internal sex organs of a male but inherit a gene that causes a deficiency in one of the male hormones. Consequently, they are born with external genitalia that appear feminine, and they are raised as girls. At **puberty** {*pubertas*, adulthood}, the period when a person makes the transition from being nonreproductive to being reproductive, pseudohermaphrodites begin to secrete more male hormones. As a result, they develop some, but not all, of the characteristics of men. Not surprisingly, a conflict arises: should these individuals change gender or remain female? Most choose to change and continue life as men.

Reproduction is one area of physiology in which we humans like to think of ourselves as significantly advanced over other animals. We mate for pleasure as well as procreation, and women are always sexually receptive (that is, not only during fertile periods). But just how different are we?

Like many other terrestrial animals, humans have internal fertilization that allows motile flagellated sperm to remain in an aqueous environment. To facilitate the process, we have mating and courtship rituals, as do other animals. Development is also internal, within the uterus, which protects the growing embryo from dehydration and cushions it in a layer of fluid.

Humans are *sexually dimorphic* {*di-*, two + *morphos*, form}, meaning that males and females are physically distinct. This distinction is sometimes blurred by dress and hairstyle, but these are cultural acquisitions. Although everyone agrees that male and female humans are physically dimorphic, we are still debating whether we are behaviorally and psychologically dimorphic as well.

RUNNING PROBLEM

Infertility

Peggy and Larry have just about everything to make them happy: successful careers, a loving marriage, a comfortable home. But one thing is missing: after five years of marriage, they have been unable to have a child. Today, Peggy and Larry have their first appointment with Dr. Coddington, an infertility specialist. "Finding the cause of your infertility is going to require some painstaking detective work," Dr. Coddington explains. He will begin his workup of Peggy and Larry by asking detailed questions about their reproductive histories. Based on the answers to these questions, he will then order tests to pinpoint the problem.

851 865 873 874 878 881 886

Sex hormones play a significant role in the behavior of other mammals, acting on adults as well as influencing the brain of the developing embryo. Their role in humans is more controversial. Human fetuses are exposed to sex hormones while in the uterus, but it is unclear how much influence these hormones have on behavior later in life. Does the preference of little girls for dolls and of little boys for toy guns have a biological basis or a cultural basis? We have no answer yet, but growing evidence suggests that at least part of our brain structure is influenced by sex hormones before we ever leave the womb.

In this chapter we address the biology of human reproduction and development. We begin our discussion with gametes that fuse to form the fertilized egg, or **zygote.** As the zygote begins to divide (2-cell stage, 4-cell stage, etc.), it becomes first an **embryo** (weeks 0–8 of development), then a **fetus** (8 weeks until birth).

Sex Determination

The male and female sex organs consist of three sets of structures: the gonads, the internal genitalia, and the external genitalia. **Gonads** {*gonos*, seed} are the organs that produce **gametes** {*gamein*, to marry}, the eggs and sperm that unite to form a new individual. The male gonads are the **testes** (singular *testis*), which produce **sperm** (*spermatozoa*). The female gonads are the **ovaries,** which produce eggs, or **ova** (singular *ovum*). The undifferentiated gonadal cells destined to produce eggs and sperm are called **germ cells.** The **internal genitalia** consist of accessory glands and ducts that connect the gonads with the outside environment. The **external genitalia** include all external reproductive structures. 26

Sexual development is programmed in the human genome. Each nucleated cell of the body except eggs and sperm contains 46 chromosomes. This set of chromosomes is called the *diploid number* because the chromosomes occur in pairs: 22 matched, or *homologous*, pairs of **autosomes** plus one pair of sex **chromosomes** (Fig. 26.1a). The 22 pairs of autosomal chromosomes direct development of the human body form and of variable characteristics such as hair color and blood type. The two sex chromosomes, designated as either X or Y, contain genes that direct development of internal and external sex organs. The X chromosome is larger than the Y chromosome and includes many genes that are missing from the Y chromosome.

Eggs and sperm are *haploid* cells with 23 chromosomes, one from each matched pair and one sex chromosome. When egg and sperm unite, the resulting zygote then contains a unique set of 46 chromosomes, with one chromosome of each matched pair coming from the mother and the other from the father.

Sex Chromosomes Determine Genetic Sex

The sex chromosomes a person inherits determine the genetic sex of that individual. Genetic females are XX, and genetic males are XY (Fig. 26.1b). Females inherit one X chromosome from

HUMAN CHROMOSOMES

(a) Humans have 23 pairs of chromosomes: 22 pairs of autosomes and one pair of sex chromosomes. X and Y chromosomes (lower right) mean that these chromosomes came from a male. The autosomes are arranged in homologous pairs in this figure.

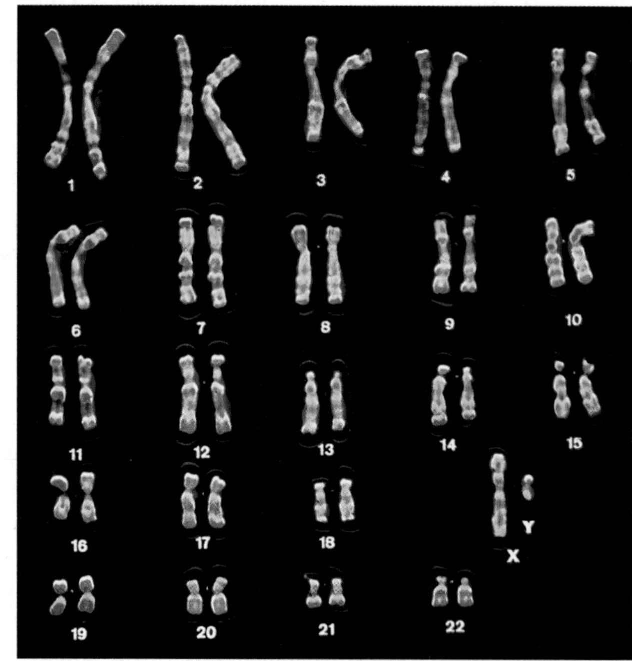

(b) X and Y chromosomes determine sex. Each egg produced by a female (XX) has an X chromosome. Sperm produced by a male (XY) have either an X chromosome or a Y chromosome.

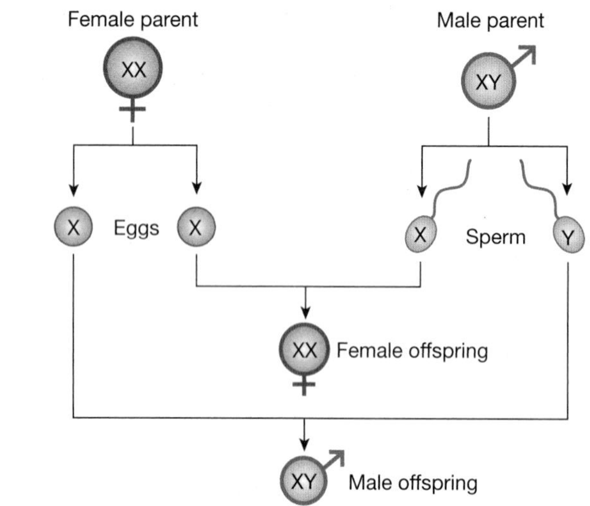

Fig. 26.1

each parent. Males inherit a Y chromosome from the father and an X chromosome from the mother. The Y chromosome is essential for development of the male reproductive organs.

If sex chromosomes are abnormally distributed at fertilization, the presence or absence of a Y chromosome determines whether development proceeds along male or female lines. The presence of a Y chromosome means the embryo will become male, even if the zygote also has multiple X chromosomes. For instance, an XXY zygote will become male. A zygote that inherits only a Y chromosome (YO) will die because the larger X chromosome contains essential genes that are missing from the Y chromosome.

In the absence of a Y chromosome, an embryo will develop into a female. For this reason, a zygote that gets only one X chromosome (XO; Turner's syndrome) will develop into a female. Two X chromosomes are needed for normal female reproductive function, however.

Once the ovaries develop in a female fetus, one X chromosome in each cell of her body inactivates and condenses into a clump of nuclear chromatin known as a *Barr body*. (Barr bodies in females can be seen in stained cheek epithelium.) The selection of the X chromosome that becomes inactive during development is random: some cells will have an active maternal X chromosome and others have an active paternal X chromosome. Because inactivation occurs early in development—before cell division is complete—all cells of a given tissue will usually have the same active X chromosome, either maternal or paternal.

Concept Check Answer: p. 890

1. Name the male and female gonads and gametes.

Sexual Differentiation Occurs Early in Development

The sex of an early embryo is difficult to determine because reproductive structures do not begin to differentiate until the seventh week of development. Before differentiation, the embryonic tissues are considered *bipotential* because they cannot be morphologically identified as male or female.

The bipotential gonad has an outer cortex and an inner medulla (■ Fig. 26.2a on page 854). Under the influence of the appropriate developmental signal (described below), the medulla will develop into a testis. In the absence of that signal, the cortex will differentiate into ovarian tissue.

The bipotential internal genitalia consist of two pairs of accessory ducts: **Wolffian ducts** (mesonephric) derived from the embryonic kidney, and **Müllerian ducts** (paramesonephric ducts). As development proceeds along either male or female lines, one pair of ducts develops while the other degenerates (Fig. 26.2a ②).

The bipotential external genitalia consist of a *genital tubercle, urethral folds, urethral groove,* and *labioscrotal swellings* (Fig. 26.2b). These structures differentiate into the male and female reproductive structures as development progresses.

What directs some single-cell zygotes to become males, and others to become females? Sex determination depends on

CLINICAL FOCUS

X-Linked Inherited Disorders

Normally, a person inherits two copies of the gene for a given trait: one copy from each parent. However, many genes found on the X chromosome, called *X-linked genes*, have no matching gene on the much smaller Y chromosome. Females always get two copies of X-linked genes, so the expression of X-linked traits follows the usual pattern of gene dominance and recession. Males, however, receive only one copy of an X-linked gene—on the X chromosome from their mother—so males *always* exhibit the traits associated with an X-linked gene. If the maternally inherited X-linked gene is defective, male offspring will exhibit the mutation. Among the identified X-linked diseases are Duchenne muscular dystrophy [p. 425], hemophilia [p. 563], and color-blindness.

the presence or absence of the ***sex-determining region of the Y*** *chromosome*, or **SRY gene.** In the presence of a functional SRY gene, the bipotential gonads develop into testes. In the absence of the SRY gene and under the direction of multiple female-specific genes, the gonads develop into ovaries.

Male Embryonic Development The SRY gene produces a protein (*testis-determining factor* or TDF) that binds to DNA and activates additional genes, including SOX9, WT1 (Wilms' tumor protein), and SF1 (steroidogenic factor). The protein products of these genes direct development of the gonadal medulla into a testis (■ Fig. 26.3 on page 856). Note that testicular development does *not* require male sex hormones such as **testosterone.** The developing embryo cannot secrete testosterone until after the gonads differentiate into testes.

Once the testes differentiate, they begin to secrete three hormones that influence development of the male internal and external genitalia. Testicular **Sertoli cells** secrete glycoprotein **anti-Müllerian hormone** (AMH; also called *Müllerian-inhibiting substance*). Testicular **Leydig cells** secrete testosterone and its derivative **dihydrotestosterone (DHT).** These two **androgens** {*andro-*, male} are the dominant steroid hormones in males. Testosterone and DHT both bind to the same *androgen receptor,* but the two ligands elicit different responses.

In the developing fetus, anti-Müllerian hormone causes the embryonic Müllerian ducts to regress (Fig. 26.2a, (2) male). Testosterone converts the Wolffian ducts into male accessory structures: epididymis, vas deferens, and seminal vesicle ((3) male). Later in fetal development, testosterone controls migration of the testes from the abdomen into the *scrotum,* or scrotal sac. The remaining male sex characteristics, such as differentiation of the external genitalia, are controlled primarily by DHT.

The importance of DHT in male development came to light in studies of the male pseudohermaphrodites described in the opening of this chapter. These men inherit a defective gene for 5α-**reductase,** the enzyme that catalyzes the conversion of testosterone to DHT (■ Fig. 26.4 on page 856). Despite normal testosterone secretion, these men have inadequate levels of DHT, and as a result the male external genitalia and prostate gland fail to develop fully during fetal development.

At birth, the infants appear to be female and are raised as such. However, at puberty, the testes again begin to secrete testosterone, causing masculinization of the external genitalia, pubic hair growth (although scanty facial and body hair), and deepening voice. By studying the 5α-reductase defect in these individuals, scientists have been able to separate the effects of testosterone from those of DHT.

Exposure of nongenital tissues to testosterone during embryonic development is known to have masculinizing effects, such as altering the brain's responsiveness to certain hormones. One controversial aspect of the masculinizing effects of testosterone is its influence on human sexual behavior and gender identity. It is well documented that in many nonhuman mammals, adult sexual behavior depends on the absence or presence of testosterone during critical periods of brain development. However, a similar cause-effect relationship has never been proved in humans. In human behavior, it is very difficult to separate biological influences from environmental factors, and it will probably be years before this question is resolved.

Female Embryonic Development In female embryos, which have no SRY gene, the cortex of the bipotential gonad develops into ovarian tissue (Fig. 26.2a (1) female). Research indicates that female development is more complex than originally thought, with multiple genes required for the development of functional ovaries.

In the absence of testicular AMH, the Müllerian ducts develop into the upper portion of the **vagina,** the **uterus,** and the **fallopian tubes, named** after the anatomist Fallopius, who first described them (Fig. 26.2a (3) female). Fallopian tubes are also called **oviducts.** Without testosterone, the Wolffian ducts degenerate (Fig. 26.2a (2) female). Without DHT, the external genitalia take on female characteristics (Fig. 26.2b).

Concept Check

Answers: p. 890

2. Where in a target cell would you expect to find receptors for androgens? Where would you expect to find receptors for AMH?
3. Why was King Henry VIII of England wrong to blame his wives when they were unable to produce a male heir to the throne?
4. Which sex will a zygote become if it inherits only one X chromosome (XO)?
5. If the testes are removed from an early male embryo, why does it develop a uterus and fallopian tubes rather than the normal male accessory structures? Will the embryo have male or female external genitalia? Explain.

■ Fig. 26.2 *ESSENTIALS*

Sexual Development in the Human Embryo

(a) Development of Internal Organs

Bipotential stage: 6 week fetus
The internal reproductive organs have the potential to develop into male or female structures.

Bipotential stage (6 week fetus)

Müllerian duct

Gonad (bipotential)

Wolffian duct

Kidney

Cloacal opening

IF FEMALE:

Gonad (cortex) forms ovary.

Gonad (medulla) regresses.

Wolffian duct regresses (testosterone absent).

Müllerian duct becomes fallopian tube, uterus, cervix, and upper 1/2 of vagina (AMH absent).

IF MALE:

Gonad (cortex) regresses.

Gonad (medulla) forms testis.

Wolffian duct forms epididymis, vas deferens, and seminal vesicle (testosterone present).

Müllerian duct regresses (AMH present).

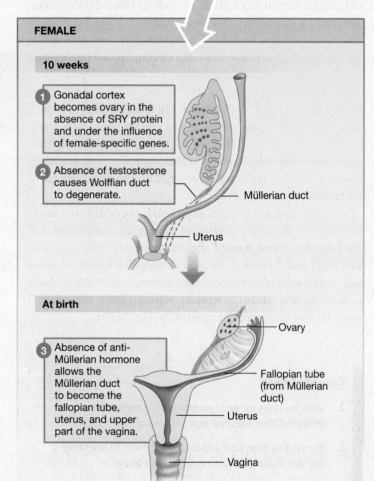

FEMALE

10 weeks

1. Gonadal cortex becomes ovary in the absence of SRY protein and under the influence of female-specific genes.

2. Absence of testosterone causes Wolffian duct to degenerate.

Müllerian duct

Uterus

At birth

3. Absence of anti-Müllerian hormone allows the Müllerian duct to become the fallopian tube, uterus, and upper part of the vagina.

Ovary

Fallopian tube (from Müllerian duct)

Uterus

Vagina

MALE

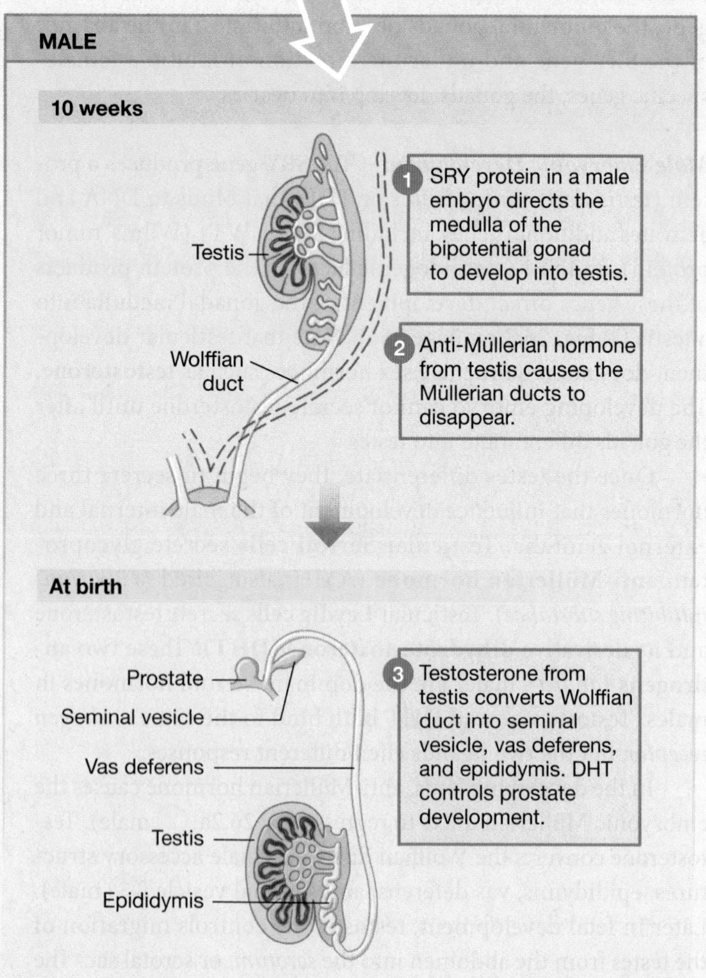

(b) Development of External Genitalia

Bipotential stage: The external genitalia of a 6-week fetus cannot be visually identified as male or female.

Bipotential stage (6 week fetus)

Genital tubercle

Urethral groove

Urethral fold

Labioscrotal swelling

Anus

IF FEMALE:

Genital tubercle forms clitoris.

Urethral folds and grooves form labia minora, opening of vagina and urethra.

Labioscrotal swellings form labia majora.

IF MALE:

Genital tubercle forms glans penis.

Urethral folds and grooves form shaft of penis.

Labioscrotal swellings form shaft of penis and scrotum.

FEMALE

10 weeks

Clitoris

Urethral fold

Labioscrotal swelling

Anus

1 In the absence of androgens, the external genitalia are feminized.

At birth

Labia majora

Labia minora

Clitoris

Urethral opening

Vaginal opening

Anus

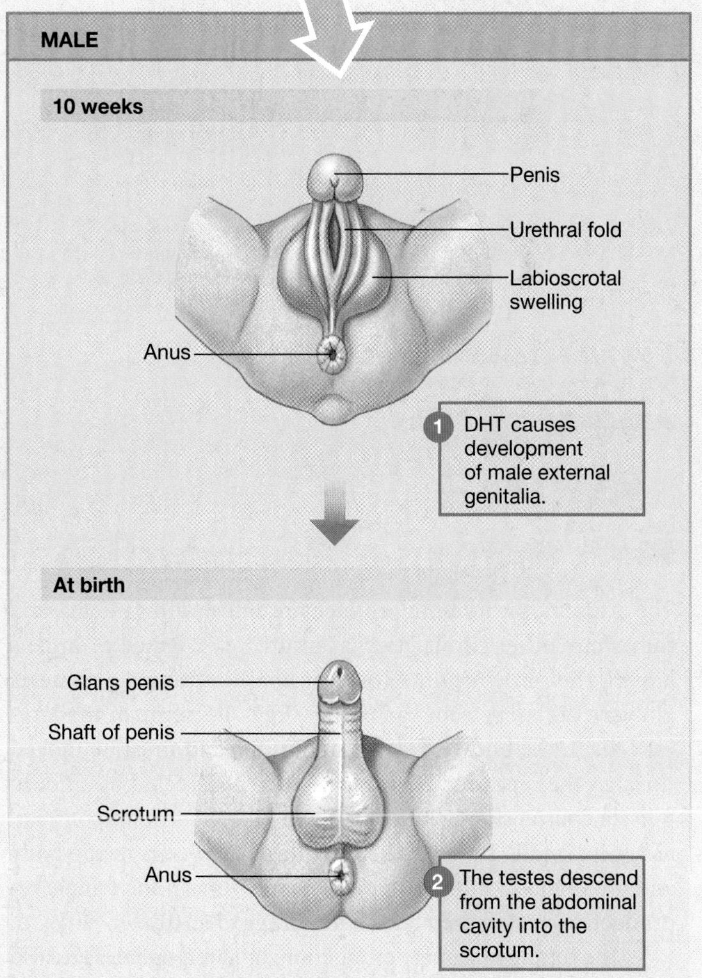

The SRY gene directs male development.

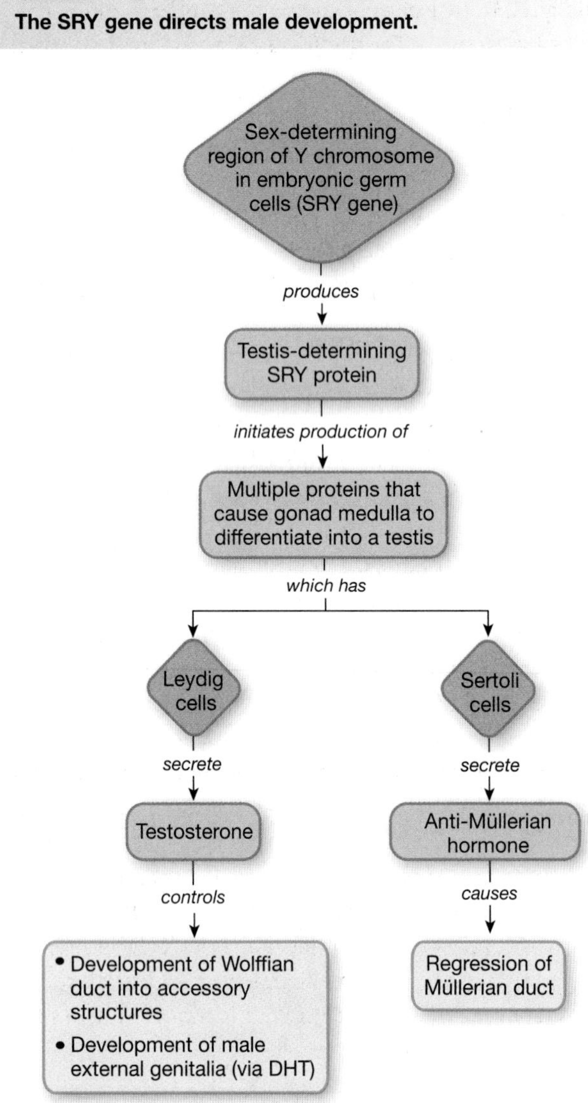

Fig. 26.3

Basic Patterns of Reproduction

The testis and ovary both produce hormones and gametes, and they share other similarities, as might be expected of organs having the same origin. However, male and female gametes are very different from each other. Eggs are some of the largest cells in the body. They are nonmotile and must be moved through the reproductive tract on currents created by smooth muscle contraction or the beating of cilia. Sperm, in contrast, are quite small. They are the only flagellated cells in the body and are highly motile so that they can swim up the female reproductive tract in their search for an egg to fertilize.

The timing of gamete production, or **gametogenesis**, is also very different in males and females. Most evidence indicates that women are born with all the eggs, or **oocytes**, they will ever have, although recent reports suggest there may be stem cells in the ovary. During the reproductive years, eggs mature in a cyclic pattern and

CLINICAL FOCUS

Determining Sex

The first question new parents typically ask about their child is, "Is it a boy or a girl?" Sometimes the answer is not obvious because in approximately 1 in 3000 births, the sex of the child cannot easily be determined. Multiple criteria might be used to establish an individual's sex: genetic, chromosomal, gonadal, morphological, or even psychological characteristics. For example, presence of a Y chromosome with a functional SRY gene could be one criterion for "maleness." However, it is possible for an infant to have a Y chromosome and not appear to be male because of a defect in some aspect of development. Currently there is ongoing debate about how best to decide sex in cases where there is doubt. Traditionally, sex determination has been based on appearance of the external genitalia at birth, but the idea that individuals should be allowed to choose their sex when they become old enough is gaining ground. The sex a person considers himself or herself to be is called the person's *gender identity*. You can read more about causes of ambiguous genitalia and the current criteria used to decide a child's sex in the American Academy of Pediatrics policy statement "Evaluation of the Newborn with Developmental Anomalies of the External Genitalia," *Pediatrics* 106(1): 138–142, 2000 (July) (available online at *http://pediatrics.aappublications.org*).

SYNTHESIS PATHWAYS FOR STEROID HORMONES

The blank boxes represent intermediate compounds whose names have been omitted for simplicity.

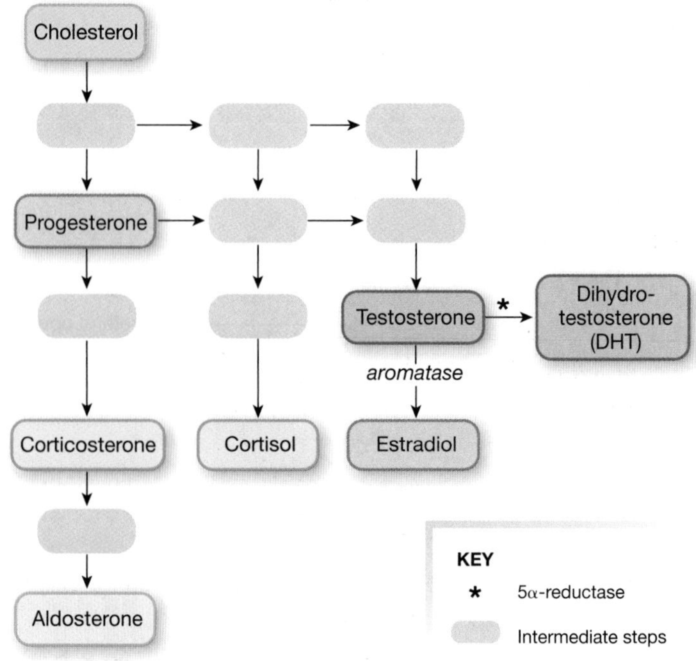

Fig. 26.4

are released from the ovaries roughly once a month. After about 40 years, female reproductive cycles cease (*menopause*).

Men, in contrast, manufacture sperm continuously from the time they reach reproductive maturity. Sperm production and testosterone secretion diminish with age but do not cease as women's reproductive cycles do.

Gametogenesis Begins *in Utero*

■ Figure 26.5 on page 858 compares the male and female patterns of gametogenesis. In both sexes, germ cells in the embryonic gonads first undergo a series of mitotic divisions to increase their numbers (1). After that, the germ cells are ready to undergo **meiosis**, the cell division process that forms gametes.

In the first step of meiosis, the germ cell's DNA replicates until each chromosome is duplicated (46 chromosomes duplicated = 92 chromosomes). The cell, now called a **primary spermatocyte** or **primary oocyte**, contains twice the normal amount of DNA (2). However, cell and chromosomal division do not take place as they do in mitosis. Instead, each duplicated chromosome forms two identical **sister chromatids,** linked together at a region known as the **centromere.** The primary gametes are then ready to undergo meiotic divisions to create four haploid cells.

In the **first meiotic division**, one primary gamete divides into two *secondary gametes* (**secondary spermatocyte** or **secondary oocyte**) (3). Each secondary gamete gets one copy of each duplicated autosome plus one sex chromosome. In the **second meiotic division**, the sister chromatids separate (4). In males, the cells split during the second meiotic division, resulting in two haploid sperm from each secondary spermatocyte.

In females, the second meiotic division creates one egg and one small cell called a *polar body*. What happens after that depends on whether or not the egg is fertilized.

The timing of mitotic and meiotic divisions is very different in males and females. Let's take a closer look at gametogenesis in each sex.

Male Gametogenesis At birth, the testes of a newborn boy have not progressed beyond mitosis and contain only immature germ cells (Fig. 26.5 (1)). After birth, the gonads become *quiescent* (relatively inactive) until puberty, the period in the early teen years when the gonads mature.

At puberty, germ cell mitosis resumes. From that point onward, the germ cells, known as **spermatogonia** (singular *spermatogonium*), have two possible fates. Some continue to undergo mitosis throughout the male's reproductive life. Others are destined to start meiosis and become primary spermatocytes (2).

Each primary spermatocyte creates four sperm. In the first meiotic division, a primary spermatocyte divides into two secondary spermatocytes (3). In the second meiotic division, each secondary spermatocyte divides into two spermatids. Each spermatid has 23 single chromosomes, the haploid number characteristic of a gamete (4). The spermatids then mature into sperm.

Female Gametogenesis In the embryonic ovary, germ cells are called **oögonia** (singular *oögonium*) (Fig. 26.5 (1)). Oögonia complete mitosis and the DNA duplication stage of meiosis by the fifth month of fetal development (2). At this time, germ cell mitosis ceases and no further oocytes can be formed. At birth each ovary contains about half a million primary gametes, or **primary oocytes.**

In the ovary, meiosis does not resume until puberty (3). Each primary oocyte divides into two cells, a large **egg (secondary oocyte)** and a tiny **first polar body**. Despite the size difference, the egg and polar body each contain 23 duplicated chromosomes. This first polar body disintegrates.

Meanwhile, the egg begins the second meiotic division (4). After the sister chromatids separate from each other, meiosis pauses. The final step of meiosis, in which sister chromatids go to separate cells, does not take place unless the egg is fertilized.

The ovary releases the mature egg during a process known as **ovulation**. If the egg is not fertilized, meiosis never goes to completion, and the egg disintegrates or passes out of the body (5). If fertilization by a sperm occurs, the final step of meiosis takes place (6). Half the sister chromatids remain in the fertilized egg (zygote), while the other half are released in a **second polar body.** The second polar body, like the first, degenerates. As a result of meiosis, each primary oocyte gives rise to only one egg.

Gametogenesis in both males and females is under the control of hormones from the brain and from endocrine cells in the gonads. Some of these hormones are identical in males and females, but others are different.

Concept Check

Answers: p. 890

6. At what stage of development is the gamete in a newborn male? In a newborn female?
7. Compare the amount of DNA in the first polar body with the amount of DNA in the second polar body.
8. How many gametes are formed from one primary oocyte? From one primary spermatocyte?

The Brain Directs Reproduction

The reproductive system has some of the most complex control pathways of the body, in which multiple hormones interact in an ever-changing fashion. The pathways that regulate reproduction begin with secretion of peptide hormones by the hypothalamus and anterior pituitary. These trophic hormones control gonadal secretion of the steroid sex hormones, including androgens, **estrogens**, and **progesterone**.

The sex steroids are closely related to one another and arise from the same steroid precursors (Fig. 26.4). Both sexes produce

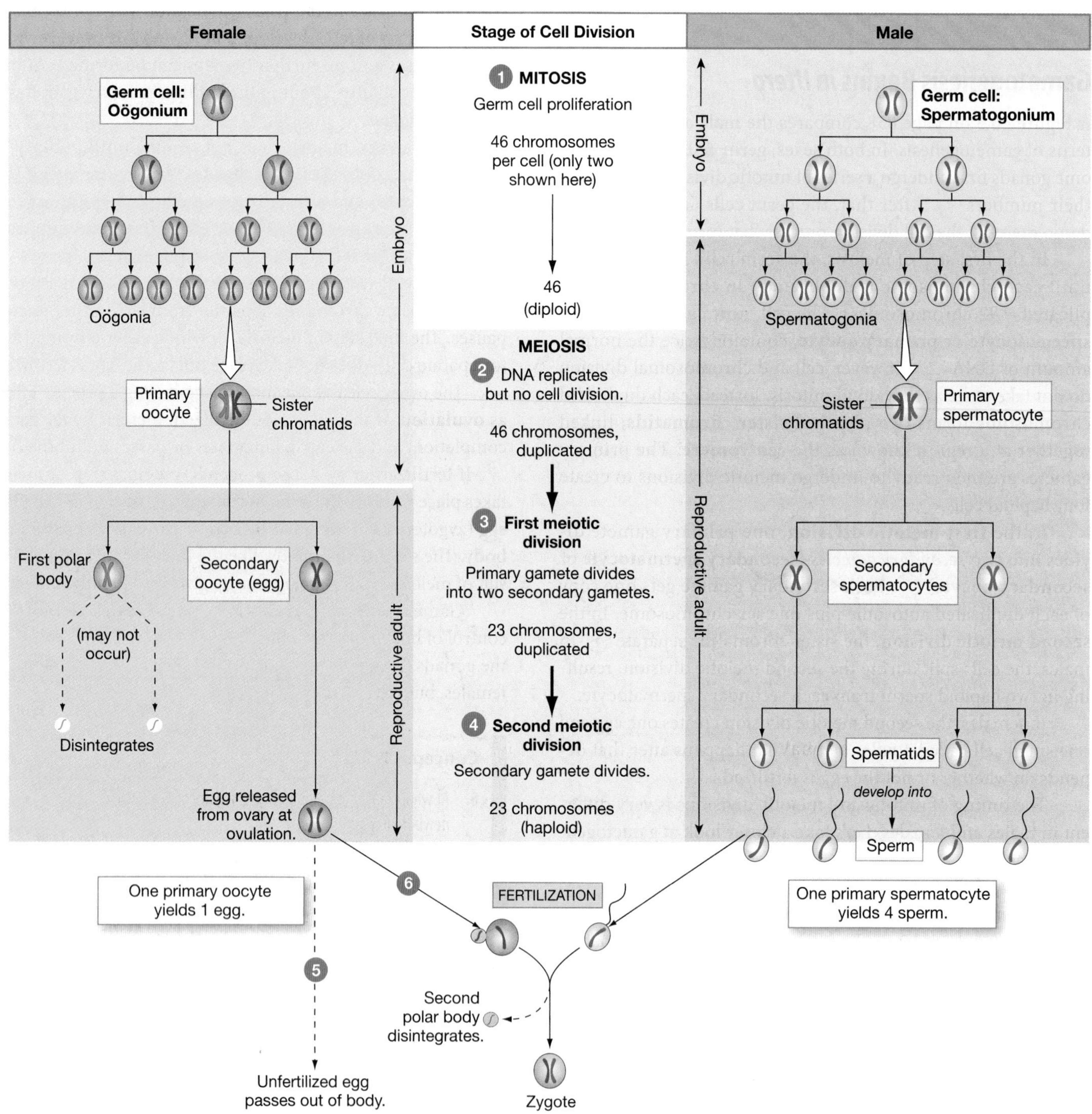

Fig. 26.5

both androgens and estrogens, but androgens predominate in males, and estrogens are dominant in females.

In men, most testosterone is secreted by the testes, but about 5% comes from the adrenal cortex. Testosterone is converted in peripheral tissues to its more potent derivative DHT. Some of the physiological effects attributed to testosterone are actually the result of DHT activity.

Males synthesize some estrogens, but the feminizing effects of these compounds are usually not obvious in males. Both testes and ovaries contain the enzyme **aromatase,** which converts

HORMONAL CONTROL OF REPRODUCTION

(a) In both sexes, the brain controls reproduction through GnRH and pituitary gonadotropins (FSH and LH).

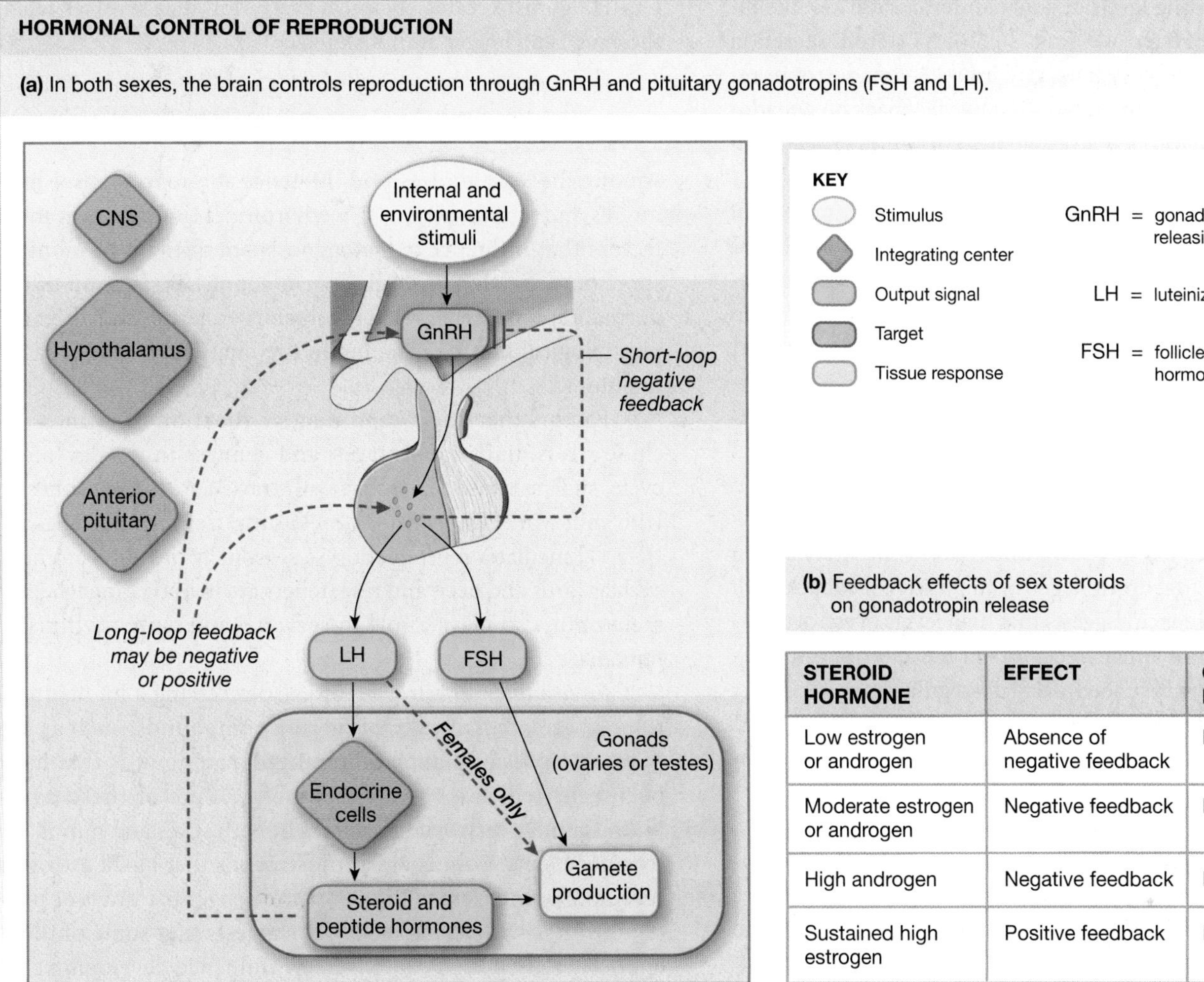

(b) Feedback effects of sex steroids on gonadotropin release

STEROID HORMONE	EFFECT	GONADOTROPIN LEVEL
Low estrogen or androgen	Absence of negative feedback	Increases
Moderate estrogen or androgen	Negative feedback	Decreases
High androgen	Negative feedback	Decreases
Sustained high estrogen	Positive feedback	Increases

Fig. 26.6

26

androgens to estrogens, the female sex hormones. A small amount of estrogen is also made in peripheral tissues.

In women, the ovary produces estrogens (particularly **estradiol** and *estrone*) and *progestins*, particularly progesterone. The ovary and the adrenal cortex produce small amounts of androgens.

Control Pathways The hormonal control of reproduction in both sexes follows the basic hypothalamus-anterior pituitary-peripheral gland pattern (Fig. 26.6). **Gonadotropin-releasing hormone** (GnRH*) from the hypothalamus controls secretion of two anterior pituitary **gonadotropins: follicle-stimulating hormone (FSH)** and **luteinizing hormone (LH).** FSH and LH in turn act on the gonads. FSH, along with steroid sex hormones, is required to initiate and maintain gametogenesis. LH acts primarily on endocrine cells, stimulating production of the steroid sex hormones.

Although primary control of gonadal function arises in the brain, the gonads also influence their own function. Both ovary and testis secrete peptide hormones that feed back to act directly on the pituitary. **Inhibins** inhibit FSH secretion, and related peptides called **activins** stimulate FSH secretion. Activins also promote spermatogenesis, oocyte maturation, and development of the embryonic nervous system. These gonadal peptides are produced in nongonadal tissues as well, and their other functions are still being investigated.

AMH, introduced earlier in the discussion of sexual differentiation during development, is also made by cells of both ovary and testis after birth. The inhibins, activins, and AMH are part of a large family of related growth and differentiation factors known as the *transforming growth factor-β* family.

Feedback Pathways The feedback loops of the reproductive system also become quite complex. The feedback pathways for trophic hormones follow the general patterns for long-loop and short-loop feedback [p. 224]. Gonadal hormones alter secretion of GnRH, FSH, and LH in a long-loop response, and the pituitary gonadotropins inhibit GnRH release from the hypothalamus by a short-loop path (Fig. 26.6a).

*GnRH is sometimes called *luteinizing hormone releasing hormone* (LHRH) because it was first thought to have its primary effect on LH.

When circulating levels of gonadal steroids are low, the pituitary secretes FSH and LH (Fig. 26.6b). As steroid secretion increases, negative feedback usually inhibits gonadotropin release. Androgens always maintain negative feedback on gonadotropin release: as androgen levels go up, FSH and LH secretion decreases.

However, in an unusual twist, higher concentrations of estrogen can exert either positive or negative feedback. Low levels of estrogen have no feedback effect. Moderate concentrations of estrogen have a negative feedback effect. But if estrogen rises rapidly to a threshold level and remains high for at least 36 hours, feedback switches from negative to positive, and gonadotropin release (particularly LH) is *stimulated.* The paradoxical effects of estrogen on gonadotropin release play a significant role in the female reproductive cycle, as you will learn later in this chapter.

Scientists still do not fully understand the mechanism underlying the change from negative to positive feedback with estrogen. Some evidence suggests that high levels of estrogen increase the number of GnRH receptors in the anterior pituitary, making it more sensitive to GnRH (up-regulation of receptors [p. 191]). Other evidence points to estrogen influencing GnRH release by altering the release of a peptide called **kisspeptin** from hypothalamic neurons.

Pulsatile GnRH Release Tonic GnRH release from the hypothalamus occurs in small pulses every 1–3 hours in both males and females. The region of the hypothalamus that contains the GnRH neuron cell bodies has been called a **GnRH pulse generator** because it coordinates the periodic pulsatile secretion of GnRH.

Scientists wondered why tonic GnRH release occurred in pulses rather than in a steady fashion, but several studies have shown the significance of the pulses. Children who suffer from a GnRH deficiency will not mature sexually in the absence of gonadotropin stimulation of the gonads. If treated with steady infusions of GnRH through drug-delivery pumps, these children still fail to mature sexually. But if the pumps are adjusted to deliver GnRH in pulses similar to those that occur naturally, the children will go through puberty. Apparently, steady high levels of GnRH cause down-regulation of the GnRH receptors on gonadotropin cells, making the pituitary unable to respond to GnRH.

This receptor down-regulation is the basis for the therapeutic use of GnRH in treating certain disorders. For example, patients with prostate and breast cancers stimulated by androgens or estrogens may be given GnRH agonists to slow the growth of the cancer cells. It seems paradoxical to give these patients a drug that stimulates secretion of androgens and estrogens, but after a brief increase in FSH and LH, the pituitary becomes insensitive to GnRH. Then FSH and LH secretion decreases, and gonadal output of steroid hormones also falls. In essence, the GnRH agonist creates chemical castration that reverses when the drug is no longer administered.

Environmental Factors Influence Reproduction

Among the least-understood influences on reproductive hormones and gametogenesis are environmental effects. In men, factors that influence gametogenesis are difficult to monitor short of requesting periodic sperm counts. Disruption of the normal reproductive cycle in women is easier to study because physiological uterine bleeding in the menstrual cycle is easily monitored.

Factors that affect reproductive function in women include stress, nutritional status, and changes in the day-night cycle, such as those that occur with travel across time zones or with shift work. The hormone **melatonin** from the pineal gland [p. 231] mediates reproduction in seasonally breeding animals, such as birds and deer, and researchers are investigating whether melatonin also plays a role in seasonal and daily rhythms in humans.

Environmental estrogens are also receiving a lot of attention. These are naturally occurring compounds, such as the *phytoestrogens* of plants, or synthetic compounds that have been released into the environment. Some of these compounds bind to estrogen receptors in both sexes and mimic estrogen's effects. Others are anti-estrogens that block estrogen receptors or interfere with second messenger pathways or protein synthesis. Growing evidence suggests that some of these endocrine disruptors can adversely influence developing embryos and even have their effects passed down to subsequent generations.

Now that you have learned the basic patterns of hormone secretion and gamete development, let's look in detail at the male and female reproductive systems.

Concept Check

Answers: p. 890

9. What does aromatase do?
10. What do the following abbreviations stand for? (Spelling counts!) FSH, DHT, SRY, LH, GnRH, AMH
11. Name the hypothalamic and anterior pituitary hormones that control reproduction.

Male Reproduction

The male reproductive system consists of the testes, the internal genitalia (accessory glands and ducts), and the external genitalia. The external genitalia consist of the **penis** and the **scrotum**, a saclike structure that contains the testes. The

urethra serves as a common passageway for sperm and urine, although not simultaneously. It runs through the ventral aspect of the shaft of the penis (■ Fig. 26.7a) and is surrounded by a spongy column of tissue known as the **corpus spongiosum** {*corpus,* body; plural *corpora*}. The corpus spongiosum and two columns of tissue called the **corpora cavernosa** constitute the erectile tissue of the penis.

The tip of the penis is enlarged into a region called the **glans** that at birth is covered by a layer of skin called the **foreskin,** or **prepuce.** In some cultures, the foreskin is removed surgically in a procedure called **circumcision.** In the United States, this practice goes through cycles of popularity. Proponents of the procedure claim that it is necessary for good hygiene, and they cite evidence suggesting that the incidence of penile cancer, sexually transmitted diseases, and urinary tract infections is lower in circumcised men. Studies from Africa indicate that circumcising heterosexual adult men helps prevent infection with the HIV virus that causes AIDS (acquired immunodeficiency syndrome). Opponents of circumcision claim that subjecting newborn boys to this surgical procedure is unnecessary.

The scrotum is an external sac into which the testes migrate during fetal development. This location outside the abdominal cavity is necessary because normal sperm development requires a temperature that is 2–3 °F lower than core body temperature. Men who have borderline or low sperm counts are advised to switch from jockey-style underwear, which keeps the scrotum close to the body, to boxer shorts, which allow the testes to stay cooler.

The failure of one or both testes to descend is known as **cryptorchidism** {*crypto,* hidden + *orchis,* testicle} and occurs in 1–3% of newborn males. If left alone, about 80% of cryptorchid testes spontaneously descend later. Those that remain in the abdomen through puberty become sterile and are unable to produce sperm.

Although cryptorchid testes lose their sperm-producing potential, they can produce androgens, indicating that hormone production is not as temperature sensitive as sperm production. Because undescended testes are prone to become cancerous, authorities recommend that they be moved to the scrotum with testosterone treatment or, if necessary, surgically.

The male accessory glands and ducts include the **prostate gland**, the **seminal vesicles,** and the **bulbourethral (Cowper's) glands** (Fig. 26.7b). The bulbourethral glands and seminal vesicles empty their secretions into the urethra through ducts. The individual glands of the prostate open directly into the urethral lumen.

The prostate gland is the best known of the three accessory glands because of its medical significance. Cancer of the prostate is the most common form of cancer in men, and *benign prostatic hypertrophy* (enlargement) creates problems for many men after age 50. Because the prostate gland completely encircles the urethra, its enlargement causes difficulty in urinating by narrowing the passageway.

Fetal development of the prostate gland, like that of the external genitalia, is under the control of dihydrotestosterone. Discovery of the role of DHT in prostate growth led to the development of *finasteride,* a 5α-reductase inhibitor that blocks DHT production. This drug was the first nonsurgical treatment for benign prostatic hypertrophy.

The Prostate Cancer Prevention Trial (PCPT) was a placebo-controlled study to see if finasteride could also prevent prostate cancer. Nearly 19,000 men participated, with half of them receiving the drug and half receiving a placebo. The trial was stopped a year early after analysis of the results showed that the risk of developing prostate cancer fell by 25% in the men taking the drug.

Testes Produce Sperm and Hormones

The human testes are paired ovoid structures about 5 cm by 2.5 cm (Fig. 26.7a). The word *testis* means "witness" in Latin, but the reason for its application to male reproductive organs is not clear. Testes are also called *testicles.*

The testes have a tough outer fibrous capsule that encloses masses of coiled **seminiferous tubules** clustered into 250–300 compartments (Fig. 26.7c). Interstitial tissue consisting primarily of blood vessels and the testosterone-producing Leydig cells lies between the tubules (Fig. 26.7e). The seminiferous tubules constitute nearly 80% of the testicular mass in an adult. Each individual tubule is 0.3–1 meter long, and, if stretched out and laid end to end, the entire mass would extend for about the length of two and a half football fields.

The seminiferous tubules leave the testis and join the **epididymis** {*epi-,* upon + *didymos,* twin}, a single duct that forms a tightly coiled cord on the surface of the testicular capsule (Fig. 26.7c). The epididymis becomes the **vas deferens** {*vas,* vessel + *deferre,* to carry away from}, also known as the **ductus deferens.** This duct passes into the abdomen, where it eventually empties into the urethra, the passageway from the urinary bladder to the external environment (see Fig. 26.7a).

Seminiferous Tubules The seminiferous tubules are the site of sperm production and contain two types of cells: spermatogonia in various stages of becoming sperm and **Sertoli cells** (Fig. 26.7d, e). The developing spermatocytes stack in columns from the outer edge of the tubule to the lumen. Between each column is a single Sertoli cell that extends from the outer edge of the tubule to the lumen. Surrounding the outside of the tubule is a basal lamina (Fig. 26.7e) that acts as a barrier, preventing certain large molecules in the interstitial fluid from entering the tubule but allowing testosterone to enter easily.

Adjacent Sertoli cells in a tubule are linked to each other by tight junctions that form an additional barrier between the

The Male Reproductive System

(a) Reproductive anatomy of the male

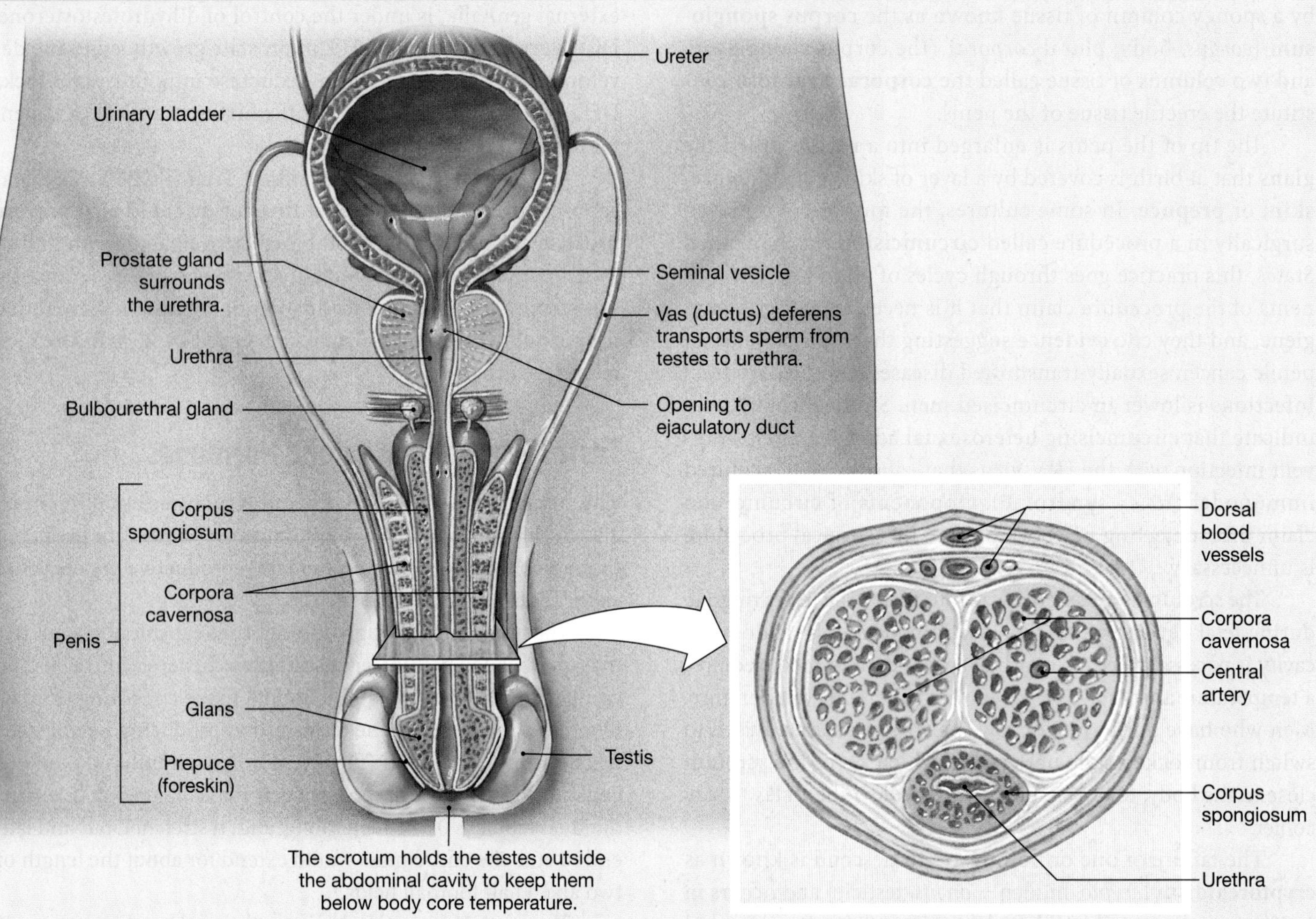

Ureter
Urinary bladder
Prostate gland surrounds the urethra.
Seminal vesicle
Vas (ductus) deferens transports sperm from testes to urethra.
Urethra
Opening to ejaculatory duct
Bulbourethral gland
Corpus spongiosum
Corpora cavernosa
Penis
Glans
Prepuce (foreskin)
Testis
The scrotum holds the testes outside the abdominal cavity to keep them below body core temperature.
Dorsal blood vessels
Corpora cavernosa
Central artery
Corpus spongiosum
Urethra

(b) Lateral view

Ureter
Rectum
Urinary bladder
These accessory glands contribute secretions to semen.
Pubic symphysis (cartilage)
Seminal vesicle
Vas deferens
Prostate gland
Ejaculatory duct
Bulbourethral gland
Urethra
Penis
Epididymis
The testis is the site of sperm and hormone production.
Scrotum

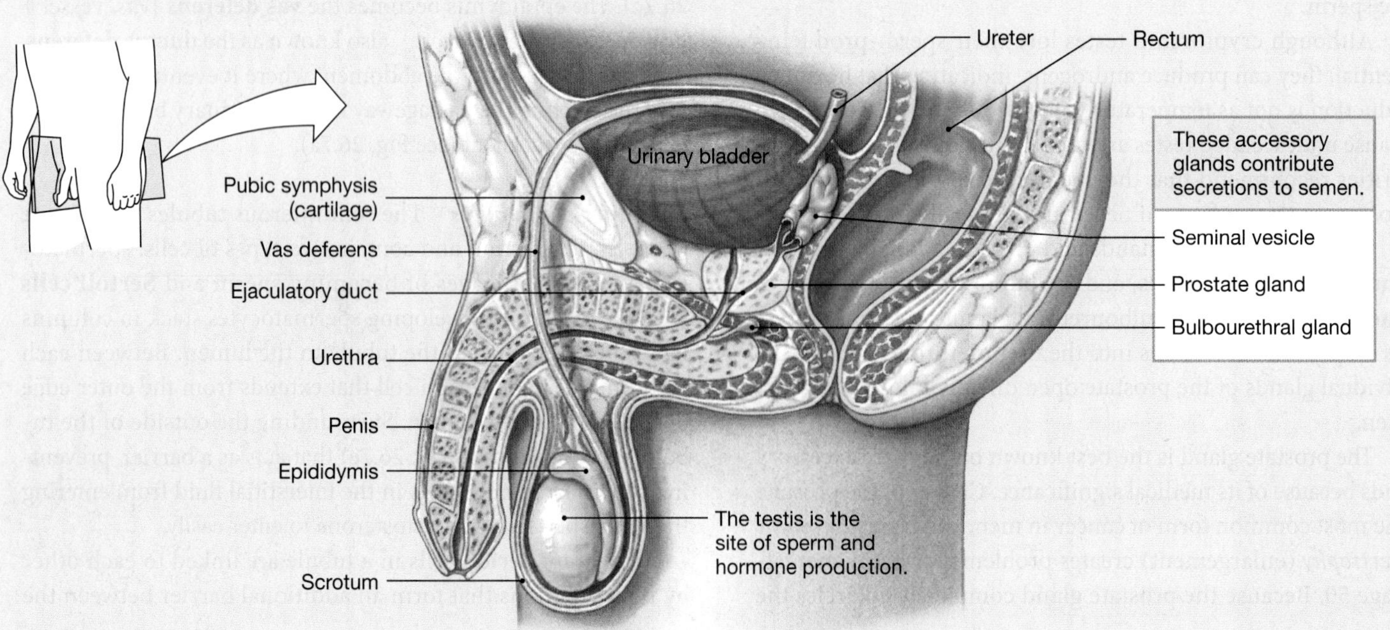

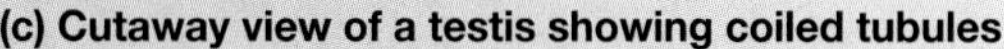

(c) Cutaway view of a testis showing coiled tubules

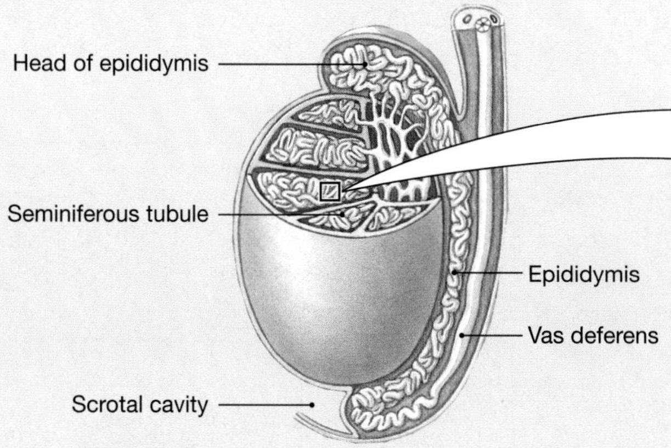

Head of epididymis
Seminiferous tubule
Epididymis
Vas deferens
Scrotal cavity

(d) Cross section of a seminiferous tubule

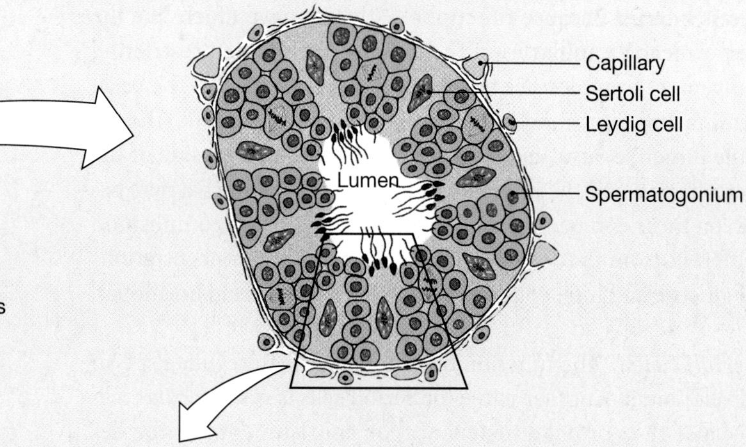

Capillary
Sertoli cell
Leydig cell
Lumen
Spermatogonium

(e) Sperm development

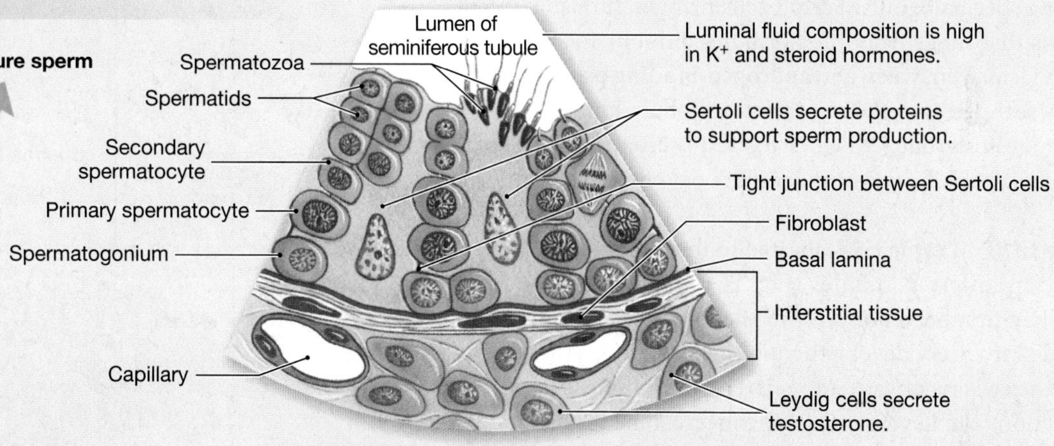

Mature sperm
Germ cells
Lumen of seminiferous tubule
Spermatozoa
Spermatids
Secondary spermatocyte
Primary spermatocyte
Spermatogonium
Capillary
Luminal fluid composition is high in K+ and steroid hormones.
Sertoli cells secrete proteins to support sperm production.
Tight junction between Sertoli cells
Fibroblast
Basal lamina
Interstitial tissue
Leydig cells secrete testosterone.

(f) Semen is composed of sperm and secretions from the accessory glands.

COMPONENT	FUNCTION	SOURCE
Sperm	Gametes	Seminiferous tubules
Mucus	Lubricant	Bulbourethral glands
Water	Provides liquid medium	All accessory glands
Buffers	Neutralize acidic environment of vagina	Prostate, bulbo-urethral glands
Nutrients	Nourish sperm	
Fructose		Seminal vesicles
Citric acid		Prostate
Vitamin C		Seminal vesicles
Carnitine		Epididymis
Enzymes	Clot semen in vagina, then liquefy the clot	Seminal vesicles and prostate
Zinc	Unknown, possible association with fertility	Unknown
Prostaglandins	Smooth muscle contraction; may aid sperm transport	Seminal vesicles

(g) A sperm consists of a head with enzymes and DNA, a long tail, and mitochondria.

Head
Mid piece
Tail (flagellum)
Acrosome contains enzymes to aid fertilization.
Nucleus
Centrioles
Mitochondrial spiral
Microtubules

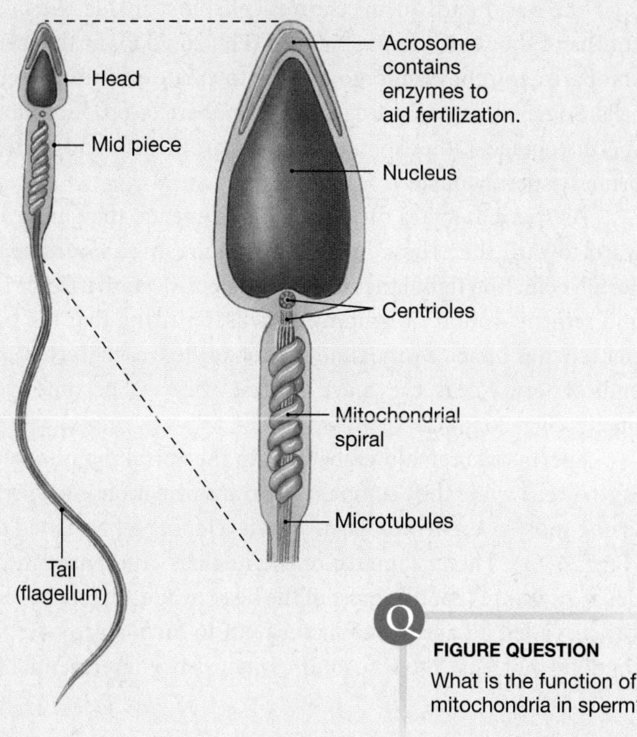

FIGURE QUESTION

What is the function of mitochondria in sperm?

lumen of the tubule and the interstitial fluid outside the tubule. These tight junctions are sometimes called the **blood-testis barrier** because functionally they behave much like the impermeable capillaries of the blood-brain barrier, restricting movement of molecules between two compartments. The basal lamina and tight junctions create three compartments: the tubule lumen, a *basal compartment* on the basolateral side of the Sertoli cells, and the interstitial fluid. Because of the barriers between these compartments, the luminal fluid has a composition different from that of interstitial fluid, with low concentrations of glucose and high concentrations of K^+ and steroid hormones.

Sertoli Cells The function of Sertoli cells is to regulate sperm development. Another name for Sertoli cells is *sustentacular cells* because they provide sustenance, or nourishment, for the developing spermatogonia. Sertoli cells manufacture and secrete proteins that range from the hormones inhibin and activin to growth factors, enzymes, and **androgen-binding protein** (ABP). ABP is secreted into the seminiferous tubule lumen, where it binds to testosterone (■ Fig. 26.8). Testosterone bound to protein is less lipophilic and cannot diffuse out of the tubule lumen.

Leydig Cells Leydig cells, located in the interstitial tissue between seminiferous tubules (Fig. 26.7d,e), secrete testosterone. They first become active in the fetus, when testosterone is needed to direct development of male characteristics. After birth, the cells inactivate. At puberty they resume testosterone production. The Leydig cells also convert some testosterone to estradiol.

Sperm Production Spermatogonia, the germ cells that undergo meiotic division to become sperm, are found clustered near the basal ends of the Sertoli cells, just inside the basal lamina of the seminiferous tubules (Fig. 26.7d,e). In this basal compartment, they undergo mitosis to create additional germ cells. Some of the spermatogonia remain here to produce future spermatogonia. Other spermatogonia start meiosis and become primary spermatocytes.

As spermatocytes differentiate into sperm, they move inward toward the tubule lumen, continuously surrounded by Sertoli cells. The tight junctions of the blood-testis barrier break and reform around the migrating cells, ensuring that the barrier remains intact. By the time spermatocytes reach the luminal ends of Sertoli cells, they have divided twice and become spermatids (Fig. 26.5).

Spermatids remain embedded in the apical membrane of Sertoli cells while they complete the transformation into sperm, losing most of their cytoplasm and developing a flagellated tail (Fig. 26.7g). The chromatin of the nucleus condenses into a dense structure that fills most of the head, while a lysosome-like vesicle called an **acrosome** flattens out to form a cap over the tip of the nucleus. The acrosome contains enzymes essential for fertilization. Mitochondria to produce energy for sperm movement concentrate in the midpiece of the sperm body, along with microtubules that extend into the tail [p. 73]. The result is a small, motile gamete that bears little resemblance to the parent spermatid.

Sperm are released into the lumen of the seminiferous tubule, along with secreted fluid. From there, they are free to move out of the testis. The entire development process—from spermatogonium division until sperm release—takes about 64 days. At any given time, different regions of the tubule contain spermatocytes in different stages of development. The staggering of developmental stages allows sperm production to remain nearly constant at a rate of *200 million* sperm per day. That may sound like an extraordinarily high number, but it is about the number of sperm released in a single ejaculation.

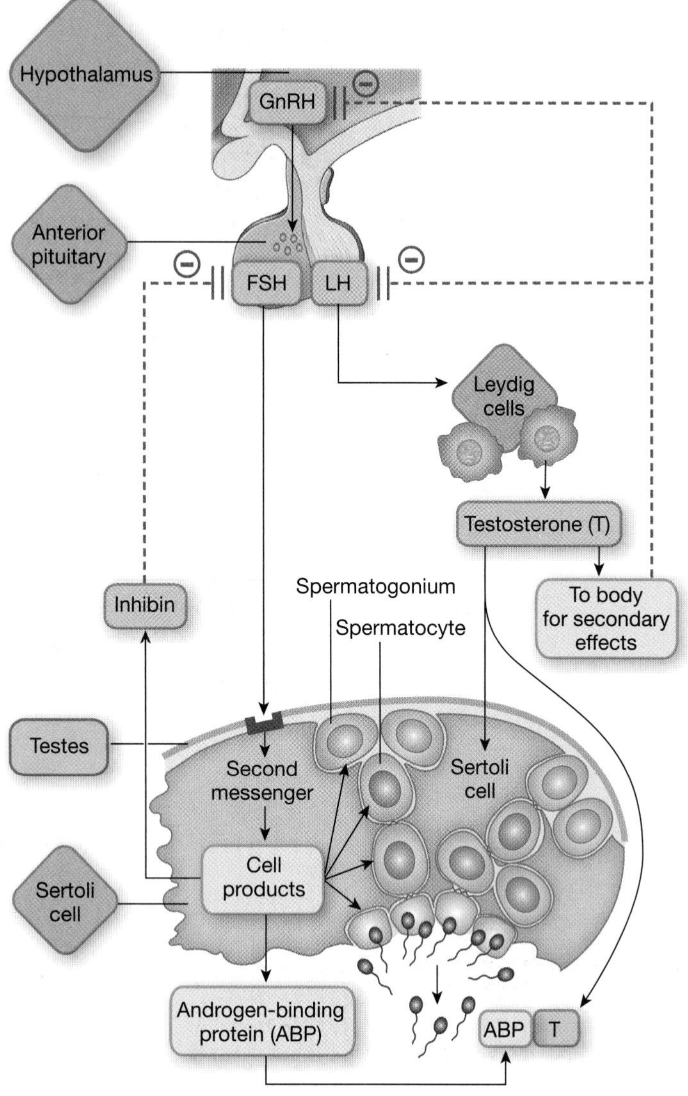

■ **Fig. 26.8**

Sperm just released from Sertoli cells are not yet mature and are incapable of swimming. They are pushed out of the tubule lumen by other sperm and by bulk flow of the fluid secreted by Sertoli cells. Sperm entering the epididymis complete their maturation during the 12 or so days of their transit time, aided by protein secretions from epididymal cells.

Spermatogenesis Requires Gonadotropins and Testosterone

The hormonal control of spermatogenesis follows the general pattern described previously: hypothalamic GnRH promotes release of LH and FSH from the anterior pituitary (Fig. 26.8). FSH and LH in turn stimulate the testes. The gonadotropins were named originally for their effect on the female ovary, but the same names have been retained in the male.

GnRH release is pulsatile, peaking every 1.5 hours, and LH release follows the same pattern. FSH levels are not as obviously related to GnRH secretion because FSH secretion is also influenced by inhibin and activin.

FSH targets Sertoli cells. Unlike oocytes, male germ cells do not have FSH receptors. Instead, FSH stimulates Sertoli synthesis of paracrine molecules needed for spermatogonia mitosis and spermatogenesis. In addition, FSH stimulates production of androgen-binding protein and inhibin.

The primary target of LH is the Leydig cells, which produce testosterone. In turn, testosterone feeds back to inhibit LH and GnRH release. Testosterone is essential for spermatogenesis, but its actions appear to be mediated by Sertoli cells, which have androgen receptors. Spermatocytes lack androgen receptors and cannot respond directly to testosterone.

Spermatogenesis is a very difficult process to study *in vivo* or *in vitro*, and the available animal models may not accurately reflect the situation in the human testis. For these reasons, it may be some time before we can say with certainty how testosterone and FSH regulate spermatogenesis.

RUNNING PROBLEM

Infertility can be caused by problems in either the man or the woman. Sometimes, however, both partners have problems that contribute to their infertility. In general, male infertility is caused by low sperm counts, abnormalities in sperm morphology, or abnormalities in the reproductive structures that carry sperm. Female infertility may be caused by problems in hormonal pathways that govern maturation and release of eggs or by abnormalities of the reproductive structures (cervix, uterus, ovaries, oviducts). Because tests of male fertility are simple to perform, Dr. Coddington first analyzes Larry's sperm. In this test, trained technicians examine a fresh sperm sample under a microscope. They note the shape and motility of the sperm and estimate the concentration of sperm in the sample.

Q1: Name (in order) the male reproductive structures that carry sperm from the testes to the external environment.

Q2: A new technique for the treatment of male infertility involves retrieval of sperm from the epididymis. The retrieved sperm can be used to fertilize an egg, which is then implanted in the uterus. Which causes of male infertility might make this treatment necessary?

851 **865** 873 874 878 881 886

Concept Check Answers: p. 890

12. What do Sertoli cells secrete? What do Leydig cells secrete?
13. Because GnRH agonists cause down-regulation of GnRH receptors, what would be the advantages and disadvantages of using these drugs as a male contraceptive?
14. Name another important lipophilic molecule that binds to protein to make it more soluble in body fluid.

Male Accessory Glands Contribute Secretions to Semen

The male reproductive tract has three accessory glands—bulbourethral glands, seminal vesicles, and prostate—whose primary function is to secrete various fluid mixtures (Fig. 26.7b). When sperm leave the vas deferens during ejaculation, they are joined by these secretions, resulting in a sperm-fluid mixture known as **semen**. About 99% of the volume of semen is fluid added from the accessory glands.

Accessory gland contributions to the composition of semen are listed in Figure 26.7f. Semen provides a liquid medium for delivering sperm. The bulbourethral glands contribute mucus for lubrication and buffers to neutralize the usually acidic environment of the vagina. Seminal vesicles contribute prostaglandins [p. 189] that appear to influence sperm motility and transport in both male and female reproductive tracts. Prostaglandins were originally believed to come from the prostate gland, and the name was well established by the time their true source was discovered. Both the prostate and seminal vesicles contribute nutrients for sperm metabolism.

In addition to providing a medium for sperm, accessory gland secretions help protect the male reproductive tract from pathogens that might ascend the urethra from the external environment. The secretions physically flush out the urethra and supply immunoglobulins, lysozyme, and other compounds with antibacterial action. One interesting component of semen is zinc. Its role in reproduction is unclear, but concentrations of zinc below a certain level are associated with male infertility.

Androgens Influence Secondary Sex Characteristics

Androgens have a number of effects on the body in addition to gametogenesis. These effects are divided into primary and secondary sex characteristics. **Primary sex characteristics** are the internal sexual organs and external genitalia that distinguish males from females. As you have already learned, androgens are responsible for the differentiation of male genitalia during embryonic development and for their growth during puberty.

The **secondary sex characteristics** are other traits that distinguish males from females. The male body shape is sometimes described as an inverted triangle, with broad shoulders and narrow waist and hips. The female body is usually more pear shaped, with broad hips and narrow shoulders. Androgens are responsible for such typically male traits as beard and body hair growth, muscular development, thickening of the vocal chords with subsequent lowering of the voice, and behavioral effects, such as the sex drive, also called **libido** {*libido*, desire, lust].

Androgens are anabolic hormones that promote protein synthesis, which gives them their street name of *anabolic steroids*. The illicit use of these drugs by athletes has been widespread despite possible adverse side effects such as liver tumors, infertility, and excessive aggression (*'roid rage*). One of the more interesting side effects is the apparent addictiveness of anabolic steroids. Withdrawal from the drugs may be associated with behavioral changes that include depression, psychosis, or aggression. These psychiatric disturbances suggest that human brain function can be modulated by sex steroids, just as the brain function of other animals can. Fortunately, many side effects of anabolic steroids are reversible once their use is discontinued.

Concept Check Answer: p. 890

15. Explain why the use of exogenous anabolic steroids might shrink a man's testes and make him temporarily infertile.

Female Reproduction

Female reproduction is an example of a physiological process that is cyclic rather than steady state. The cycles of gamete production in the ovary and the interactions of reproductive hormones and feedback pathways are part of one of the most complex control systems of the human body.

Females Have an Internal Uterus

The female external genitalia are known collectively as either the **vulva** or the **pudendum** {*vulva*, womb; *pudere*, to be ashamed}. They are shown in ■ Figure 26.9c on page 868, the view seen by a healthcare worker who is about to do a pelvic exam or take a Pap smear [p. 63].

Starting at the periphery are the **labia majora** {*labium*, lip}, folds of skin that arise from the same embryonic tissue as the scrotum. Within the labia majora are the **labia minora,** derived from embryonic tissues that in the male give rise to the shaft of the penis (see Fig. 26.2b). The **clitoris** is a small bud of erectile, sensory tissue at the anterior end of the vulva, enclosed by the labia minora and an additional fold of tissue equivalent to the foreskin of the penis.

In females, the urethra opens to the external environment between the clitoris and the vagina {*vagina*, sheath}, the cavity that acts as receptacle for the penis during intercourse. At birth, the external opening of the vagina is partially closed by a thin ring of tissue called the **hymen,** or *maidenhead*. The hymen is external to the vagina, not within it, so the normal use of tampons during menstruation will not rupture the hymen. However, it can be stretched by normal activities such as cycling and horseback riding and therefore is not an accurate indicator of a woman's virginity.

Now let's follow the path of sperm deposited in the vagina during intercourse. To continue into the female reproductive tract, sperm must pass through the narrow opening of the **cervix,** the neck of the uterus that protrudes slightly into the upper end of the vagina (Fig. 26.9a). The cervical canal is lined with mucous glands whose secretions create a barrier between the vagina and uterus.

Sperm that make it through the cervical canal arrive in the lumen of the uterus, or *womb,* a hollow, muscular organ slightly smaller than a woman's clenched fist. The uterus is the structure in which fertilized eggs implant and develop during pregnancy. It is composed of three tissue layers (Fig. 26.9d): a thin outer connective tissue covering, a thick middle layer of smooth muscle known as the **myometrium,** and an inner layer known as the **endometrium** {*metra,* womb}.

The endometrium consists of an epithelium with glands that dip into a connective tissue layer below. The thickness and character of the endometrium vary during the menstrual cycle. Cells of the epithelial lining alternately proliferate and slough off, accompanied by a small amount of bleeding in the process known as **menstruation** {*menstruus,* monthly}.

Sperm swimming upward through the uterus leave its cavity through openings into the two fallopian tubes (Fig. 26.9a). The fallopian tubes are 20–25 cm long and about the diameter of a drinking straw. Their walls have two layers of smooth muscle, longitudinal and circular, similar to the walls of the intestine. A ciliated epithelium lines the inside of the tubes.

Fluid movement created by the cilia and aided by muscular contractions transports an egg along the fallopian tube toward the uterus. If sperm moving up the tube encounter an egg moving down the tube, fertilization may occur. Pathological conditions in which ciliary function is absent are associated with female infertility and with pregnancies in which the embryo implants in the fallopian tube rather than the uterus.

The flared open end of the fallopian tube divides into fingerlike projections called **fimbriae** {*fimbriae,* fringe}. The fimbriae (Fig. 26.9a) are held close to the adjacent ovary by connective tissue, which helps ensure that eggs released from the surface of the ovary will be swept into the tube rather than floating off into the abdominal cavity.

The Ovary Produces Eggs and Hormones

The ovary is an elliptical structure, about 2–4 cm long (Fig. 26.9e). It has an outer connective tissue layer and an inner connective tissue framework known as the **stroma** {*stroma,* mattress}. Most of the ovary consists of a thick outer *cortex* filled with ovarian follicles in various stages of development or decline. The small central *medulla* contains nerves and blood vessels.

The ovary, like the testis, produces both gametes and hormones. As mentioned earlier, about 7 million oögonia in the embryonic ovary develop into half a million primary oocytes. Each primary oocyte is enclosed in a **primary follicle** with a single layer of **granulosa cells** separated by a basement membrane from an outer layer of cells known as the **theca** {*theke,* case or cover}.

A Menstrual Cycle Lasts about One Month

Female humans produce gametes in monthly cycles (average 28 days; normal range 24–35 days). These cycles are commonly called **menstrual cycles** because they are marked by a 3–7 day period of bloody uterine discharge known as the **menses** {*menses,* months}, or **menstruation**. The menstrual cycle can be described by following changes that occur in follicles of the ovary, the **ovarian cycle,** or by following changes in the endometrial lining of the uterus, the **uterine cycle.** ■ Figure 26.10 is a summary figure showing a typical menstrual cycle and its phases.

Notice that the ovarian cycle is divided into three phases:

1. **Follicular phase.** The first part of the ovarian cycle, known as the **follicular phase,** is a period of follicular growth in the ovary. This phase is the most variable in length and lasts from 10 days to 3 weeks.
2. **Ovulation.** Once one or more follicles have ripened, the ovary releases the oocyte(s) during **ovulation**.
3. **Luteal phase.** The phase of the ovarian cycle following ovulation is known as the *postovulatory* or **luteal phase**. The second name comes from the transformation of a ruptured follicle into a **corpus luteum** {*corpus,* body + *luteus,* yellow}, named for its yellow pigment and lipid deposits. The corpus luteum secretes hormones that continue the preparations for pregnancy. If a pregnancy does not occur, the corpus luteum ceases to function after about two weeks, and the ovarian cycle begins again.

The endometrial lining of the uterus also goes through a cycle—the uterine cycle—regulated by ovarian hormones:

1. **Menses.** The beginning of the follicular phase in the ovary corresponds to menstrual bleeding from the uterus.
2. **Proliferative phase.** The latter part of the ovary's follicular phase corresponds to the **proliferative phase** in the uterus, during which the endometrium adds a new layer of cells in anticipation of pregnancy.
3. **Secretory phase.** After ovulation, hormones from the corpus luteum convert the thickened endometrium into a secretory structure. This means that the luteal phase of the ovarian cycle corresponds to the **secretory phase** of the uterine cycle. If no pregnancy occurs, the superficial layers of the secretory endometrium are lost during menstruation as the uterine cycle begins again.

Hormonal Control of the Menstrual Cycle Is Complex

26

The ovarian and uterine cycles are under the primary control of various hormones:

- GnRH from the hypothalamus
- FSH and LH from the anterior pituitary
- Estrogen, progesterone, inhibin, and AMH from the ovary

During the follicular phase, the dominant steroid hormone is estrogen (Fig. 26.10). Ovulation is triggered by surges in LH and FSH. In the luteal phase, progesterone is dominant, although estrogen is still present.

Anti-Müllerian hormone (AMH) was first known for its role in male development, but scientists have discovered that AMH is also produced by ovarian follicles in the first part of the ovarian cycle. AMH apparently acts as a brake to keep too many follicles from developing at one time.

Now let's go through an ovarian cycle in detail.

Early Follicular Phase The first day of menstruation is day 1 of a cycle. This point was chosen to start the cycle because the bleeding of menstruation is an easily observed physical sign. Just before the beginning of each cycle, gonadotropin secretion from the anterior pituitary increases. Under the influence of FSH, several follicles in the ovaries begin to mature (second row of Fig. 26.10 and ■ Fig. 26.11).

As the follicles grow, their granulosa cells (under the influence of FSH) and their thecal cells (under the influence of LH) start to produce steroid hormones (■ Fig. 26.12). Granulosa cells also begin to secrete AMH. This AMH decreases follicle

The Female Reproductive System

(a) Internal reproductive structures

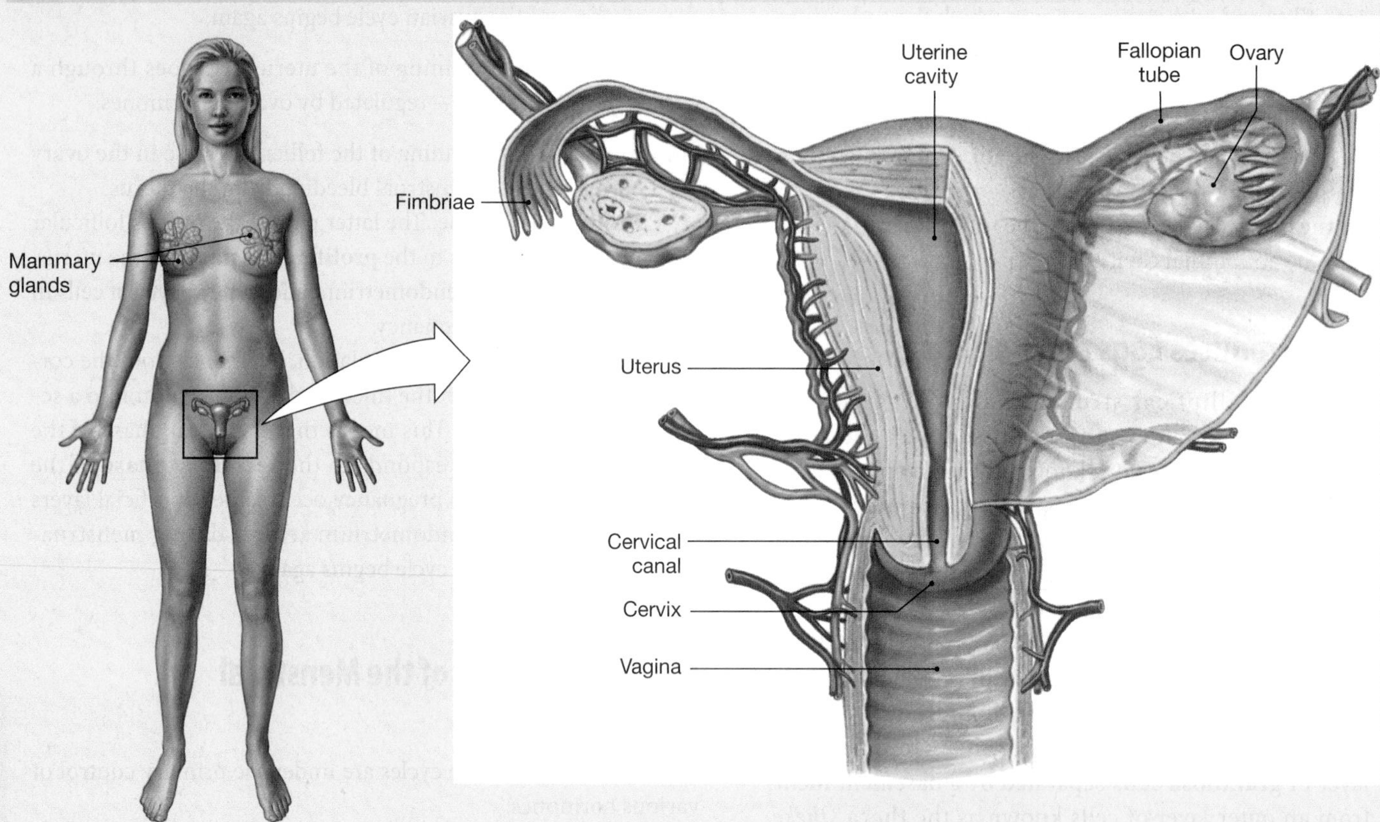

(b) Cross-sectional view of pelvis

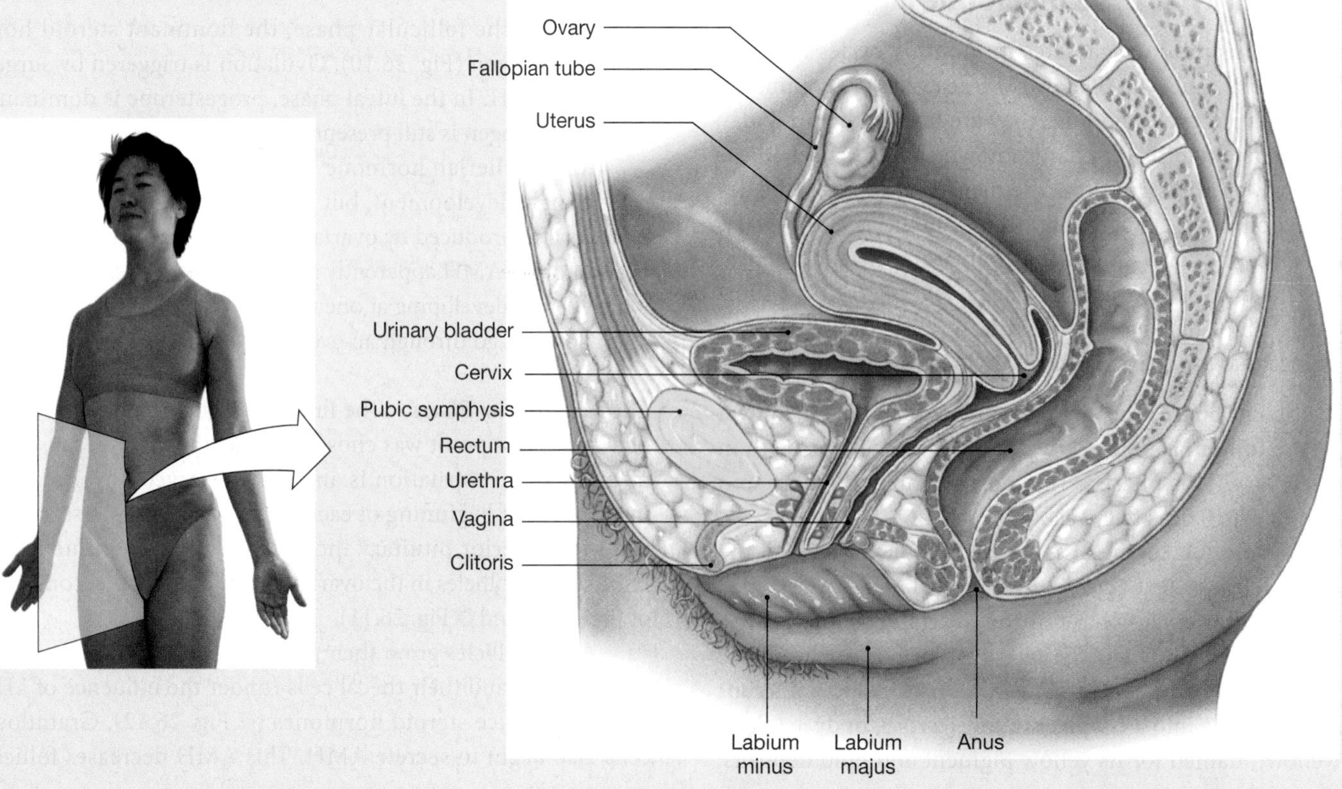

(c) Female external genitalia

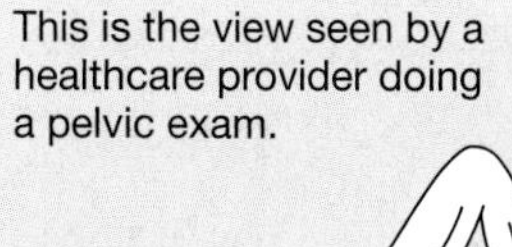

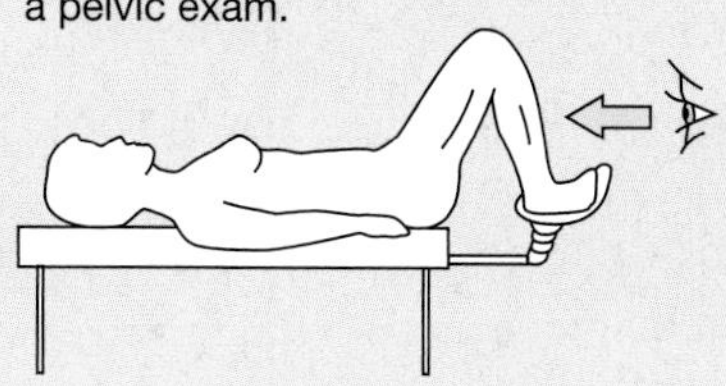

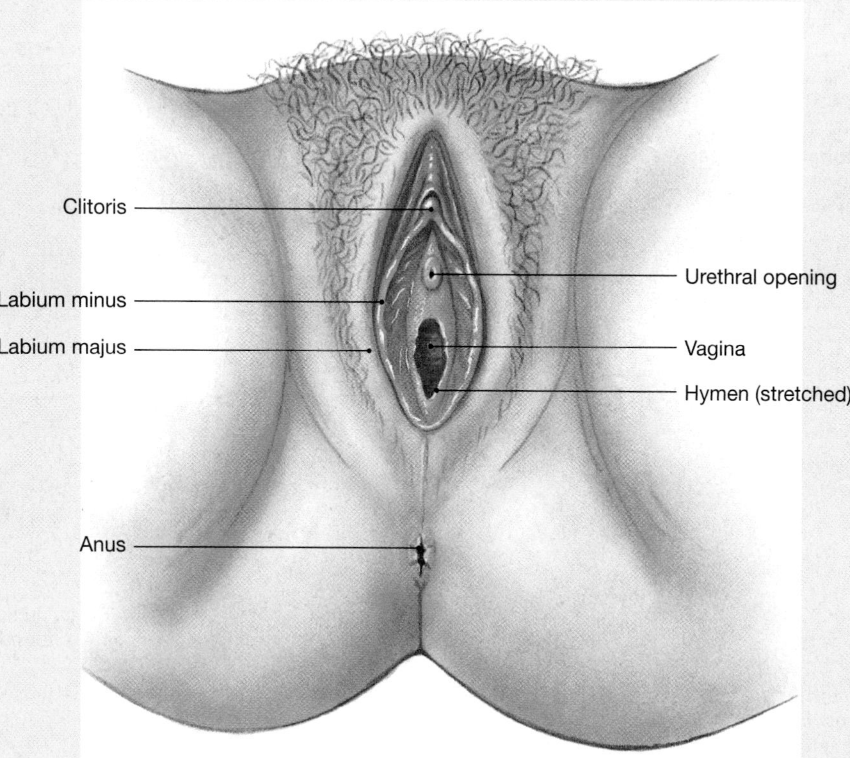

(d) Structure of the uterus

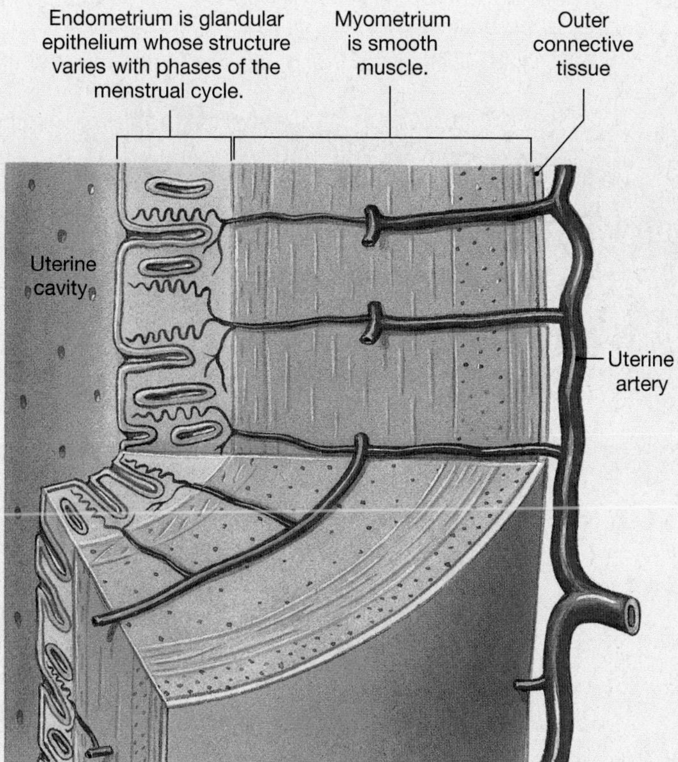

Endometrium is glandular epithelium whose structure varies with phases of the menstrual cycle.
Myometrium is smooth muscle.
Outer connective tissue
Uterine cavity
Uterine artery

(e) Cross section of an ovary, showing all different stages of follicular development.

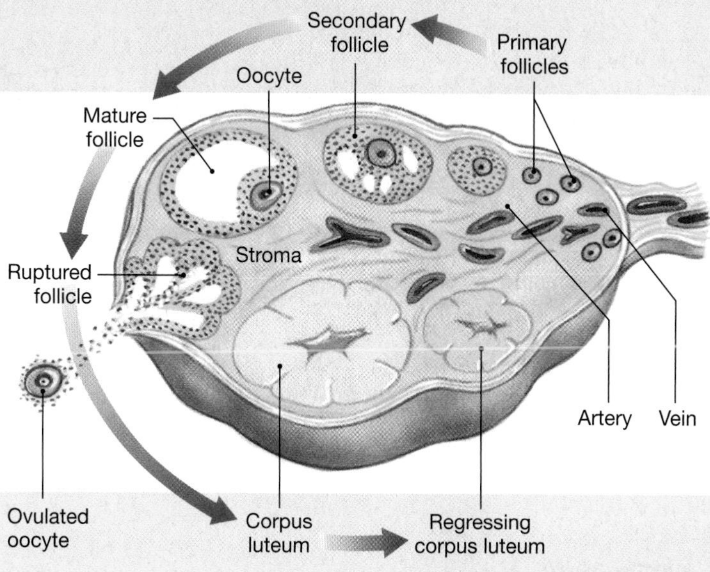

THE MENSTRUAL CYCLE

This 28-day menstrual cycle is divided into phases based on events in the ovary (ovarian cycle) and in the uterus (uterine cycle).

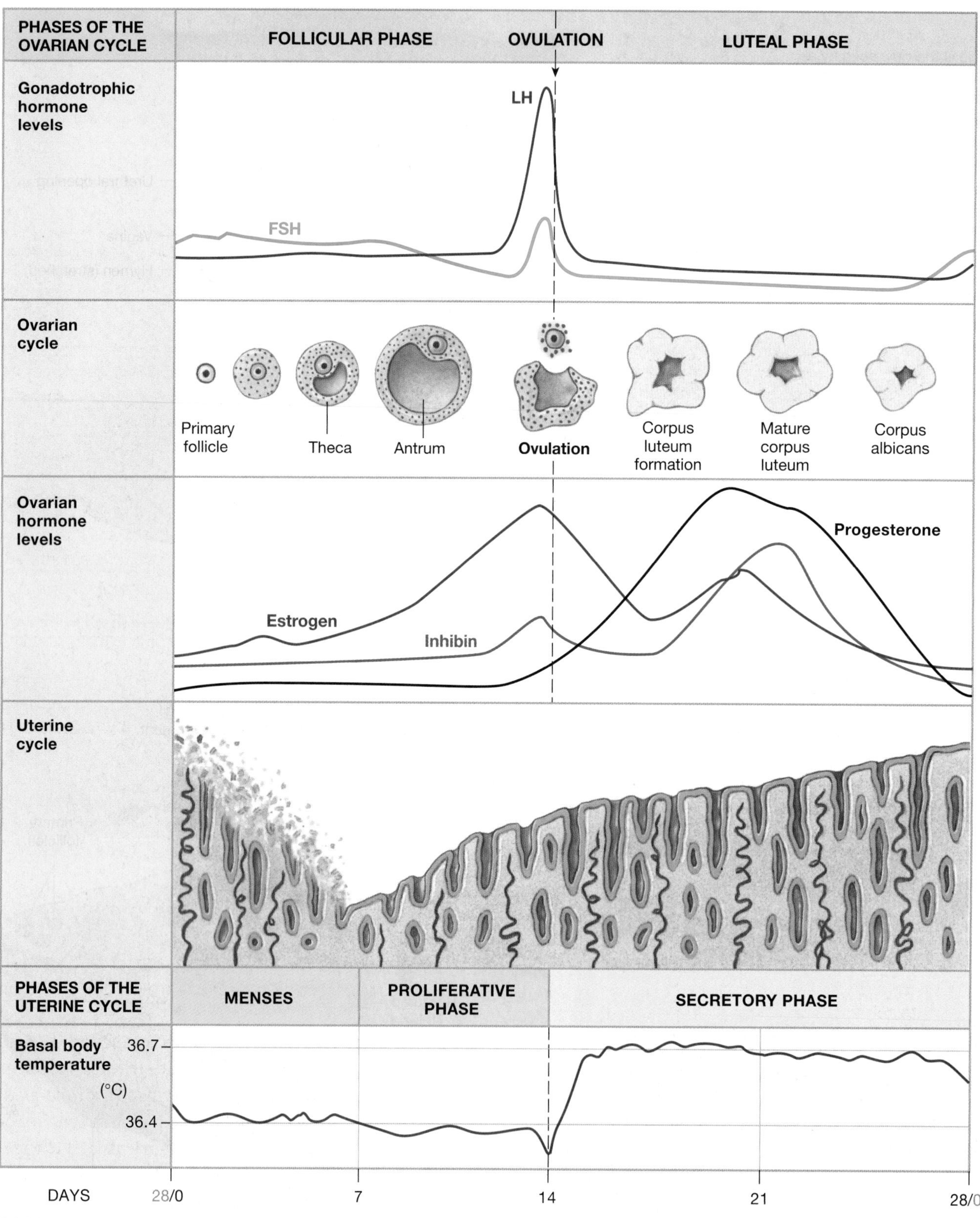

Fig. 26.10

FOLLICULAR DEVELOPMENT

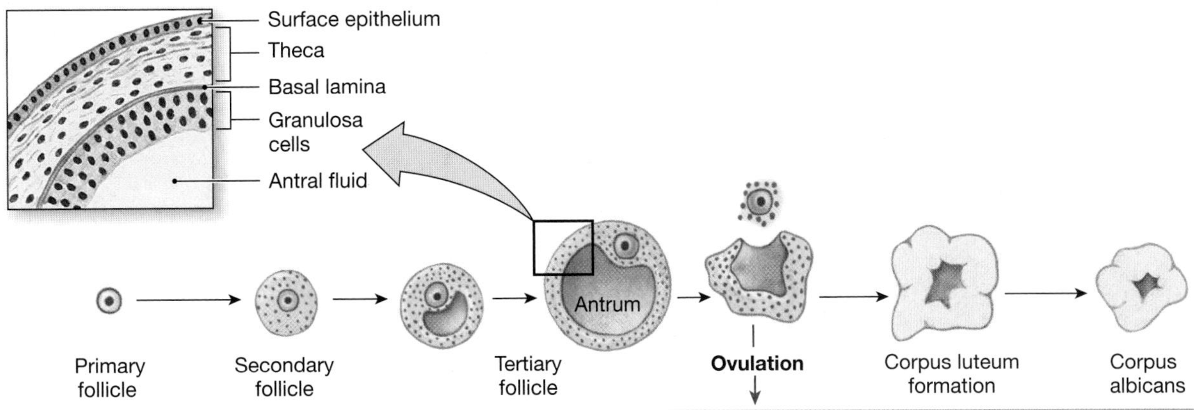

OVARIAN PHASE	BEFORE FSH STIMULATION	EARLY FOLLICULAR PHASE	LATE FOLLICULAR PHASE	LUTEAL PHASE	POST-LUTEAL PHASE
Follicle stage	Primary follicle	Secondary follicle	Tertiary follicle	Corpus luteum	Corpus albicans
Ovum	Primary oocyte	Primary oocyte	Becomes secondary oocyte with division arrested	None	None
Zona pellucida*	Minimal	Increased in width	Present	None	None
Granulosa cells	Single layer	2–6 cell layer	3–4 cell layer	Converted to luteal cells	Cells degenerate
Antrum	None	None	Develops within granulosa layer and fills with fluid; swells to 15–20 mm in diameter	Fills with migrating cells	None
Basal lamina	Separates granulosa and theca	Present	Present	Disappears	None
Theca	Single cell layer plus blood vessels	Single cell layer	*Inner layer:* secretory and small blood vessels *Outer layer:* connective tissue, smooth muscle cells, large blood vessels	Converted to luteal cells	Cells degenerate

*The zona pellucida is a glycoprotein coat that protects the ovum.

Fig. 26.11

sensitivity to FSH, which apparently prevents recruitment of additional primary follicles once one group has started developing. Physicians now use blood AMH levels as an indicator of how many follicles are developing early in a cycle and as a marker for the condition known as *polycystic ovary syndrome (PCOS)*, in which ovarian follicles form fluid-filled cysts.

Thecal cells synthesize androgens that diffuse into the neighboring granulosa cells, where aromatase converts them to estrogens (Fig. 26.12a). Gradually increasing estrogen levels in the circulation have several effects. Estrogen exerts negative feedback on pituitary FSH and LH secretion, which prevents the development of additional follicles in the same cycle. At the same time, estrogen stimulates additional estrogen production by the granulosa cells. This positive feedback loop allows the follicles to continue estrogen production even though FSH and LH levels decrease.

In the uterus, menstruation ends during the early follicular phase (Fig 26.10). Under the influence of estrogen from developing follicles, the endometrium begins to grow, or proliferate. This period is characterized by an increase in cell number and by enhanced blood supply to bring nutrients and oxygen to the thickening endometrium. Estrogen also causes mucous glands of the cervix to produce clear, watery mucus.

Mid to Late Follicular Phase As the follicles enlarge, granulosa cells begin to secrete fluid that collects in a central cavity in

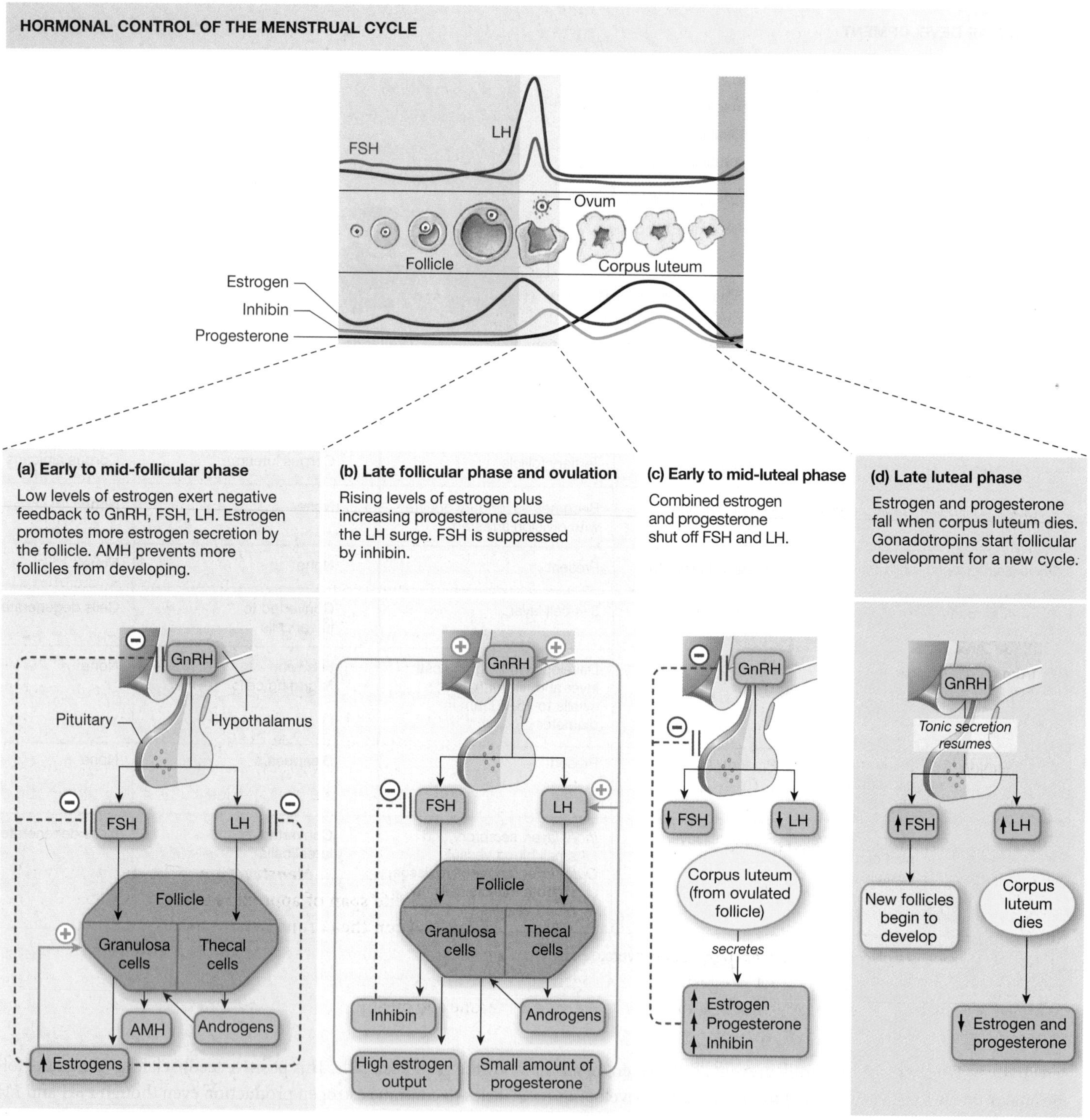

HORMONAL CONTROL OF THE MENSTRUAL CYCLE

(a) Early to mid-follicular phase

Low levels of estrogen exert negative feedback to GnRH, FSH, LH. Estrogen promotes more estrogen secretion by the follicle. AMH prevents more follicles from developing.

(b) Late follicular phase and ovulation

Rising levels of estrogen plus increasing progesterone cause the LH surge. FSH is suppressed by inhibin.

(c) Early to mid-luteal phase

Combined estrogen and progesterone shut off FSH and LH.

(d) Late luteal phase

Estrogen and progesterone fall when corpus luteum dies. Gonadotropins start follicular development for a new cycle.

Fig. 26.12

the follicle known as the **antrum** {*antron*, cave} (Fig. 26.11). Antral fluid contains hormones and enzymes needed for ovulation. At each stage of follicular development, some follicles undergo *atresia* (hormonally regulated cell death). Only a few follicles reach the final stage, and usually only one dominant follicle develops until ovulation.

As the follicular phase nears its end, ovarian estrogen secretion peaks (Fig. 26.12b). By this point of the cycle, only one follicle is still developing. As the follicular phase ends, granulosa cells of the dominant follicle begin to secrete inhibin and progesterone in addition to estrogen. Estrogen, which had exerted a negative feedback effect on GnRH earlier in the follicular phase, changes to positive feedback, leading to a preovulatory GnRH surge.

Immediately before ovulation, the persistently high levels of estrogen, aided by rising levels of progesterone, enhance pituitary responsiveness to GnRH. As a result, LH secretion increases dramatically, a phenomenon known as the *LH surge*. FSH surges also, but to a lesser degree, presumably because it is being suppressed by inhibin and estrogen.

The LH surge is an essential part of ovulation. Without it, the final steps of oocyte maturation cannot take place. Meiosis resumes in the developing follicle with the first meiotic division, which converts the primary oocyte into a secondary oocyte (egg) and a polar body, which is extruded (Fig. 26.5). While this division is taking place, antral fluid collects and the follicle grows to its greatest size, preparing to release the egg.

High levels of estrogen in the late follicular phase prepare the uterus for a possible pregnancy. The endometrium grows to a thickness of 3–4 mm (Fig. 26.10). Just before ovulation, the cervical glands produce copious amounts of thin, stringy mucus to facilitate sperm entry. The stage is set for ovulation.

RUNNING PROBLEM

The results of Larry's sperm analysis are normal. Dr. Coddington is therefore able to rule out sperm abnormalities as a cause of Peggy and Larry's infertility. Peggy is instructed to take her body temperature daily and record the results on a chart. This temperature tracking is intended to determine whether or not she is ovulating. Following ovulation, body temperature rises slightly and remains elevated through the remainder of the menstrual cycle.

Q3: For which causes of female infertility is temperature tracking useful? For which causes is it not useful?

Ovulation About 16–24 hours after LH peaks, ovulation occurs (Fig. 26.10). The mature follicle secretes *collagenase*, an enzyme that dissolves collagen in the connective tissue holding the follicular cells together. The breakdown products of collagen create an inflammatory reaction, attracting leukocytes that secrete prostaglandins into the follicle. The prostaglandins may cause smooth muscle cells in the outer theca to contract, rupturing the follicle wall at its weakest point. Antral fluid spurts out along with the egg, which is surrounded by two to three layers of granulosa cells. The egg is swept into the fallopian tube and carried away to be fertilized or to die.

In addition to promoting follicular rupture, the LH surge causes follicular thecal cells to migrate into the antral space, mingling with the former granulosa cells and filling the cavity. Both cell types then transform into *luteal cells* of the corpus luteum. This process, known as *luteinization,* involves biochemical and morphological changes.

Early to Mid-Luteal Phase After ovulation, newly formed luteal cells accumulate lipid droplets and glycogen granules in their cytoplasm and begin to secrete progesterone. Estrogen synthesis diminishes initially but as the luteal phase progresses, the corpus luteum produces steadily increasing amounts of progesterone and estrogen. Progesterone is the dominant hormone of the luteal phase. Estrogen levels increase but never reach the peak seen before ovulation.

The combination of estrogen and progesterone exerts negative feedback on the hypothalamus and anterior pituitary (Fig. 26.12c). Gonadotropin secretion, further suppressed by luteal inhibin production, remains shut down throughout most of the luteal phase.

Under the influence of progesterone, the endometrium continues its preparation for pregnancy and becomes a secretory structure. Endometrial glands coil, and additional blood vessels grow into the connective tissue layer. Endometrial cells deposit lipids and glycogen in their cytoplasm. These deposits will provide nourishment for a developing embryo while the **placenta,** the fetal-maternal connection, is developing.

Progesterone also causes cervical mucus to thicken. Thicker mucus creates a plug that blocks the cervical opening, preventing bacteria as well as sperm from entering the uterus.

One interesting effect of progesterone is its thermogenic ability. During the luteal phase of an ovulatory cycle, a woman's basal body temperature, taken immediately upon awakening and before getting out of bed, jumps 0.3–0.5 °F and remains elevated until menstruation. Because this change in the temperature setpoint occurs after ovulation, it cannot be used effectively to predict ovulation. However, it is a simple way to assess whether a woman is having ovulatory or *anovulatory* (non-ovulating) cycles.

26

Late Luteal Phase and Menstruation The corpus luteum has an intrinsic life span of approximately 12 days. If pregnancy does not occur, the corpus luteum spontaneously undergoes *apoptosis* [p. 90] to become an inactive structure called a **corpus albicans** {*albus,* white}. As the luteal cells degenerate, progesterone and estrogen production decrease (Fig. 26.12d). This fall removes the negative feedback signal to the pituitary and hypothalamus, so secretion of FSH and LH increases.

Maintenance of a secretory endometrium depends on the presence of progesterone. When the corpus luteum degenerates and hormone production decreases, blood vessels in the surface layer of the endometrium contract. Without oxygen and nutrients, the surface cells die. About two days after the corpus luteum ceases to function, or 14 days after ovulation, the endometrium begins to slough its surface layer, and menstruation begins.

Menstrual discharge from the uterus totals about 40 mL of blood and 35 mL of serous fluid and cellular debris. There are usually few clots of blood in the menstrual flow because of the presence of *plasmin* [p. 562], which breaks up clots.

Menstruation continues for 3–7 days, well into the follicular phase of the next ovulatory cycle.

Hormones Influence Female Secondary Sex Characteristics

Estrogens control the development of primary sex characteristics in females, just as androgens control them in males. Estrogens also control the most prominent female secondary sex traits: breast development and the female pattern of fat distribution (hips and upper thighs). Other female secondary sex characteristics, however, are governed by androgens produced in the adrenal cortex. Pubic and axillary (armpit) hair growth and libido (sex drive) are under the control of adrenal androgens.

Concept Check Answers: p. 890

16. Name the phases of the ovarian cycle and the corresponding phases of the uterine cycle.
17. What side effects would you predict in female athletes who take anabolic steroids to build muscles?
18. Aromatase converts testosterone to estrogen. What would happen to the ovarian cycle of a woman given an aromatase inhibitor?
19. On what day of the menstrual cycle will a woman with the following cycle lengths ovulate?

    (a) 28 days (b) 23 days (c) 31 days

Procreation

Reproduction throughout the animal kingdom is marked by species-specific behaviors designed to ensure that egg and sperm meet. For aquatic animals that release gametes into the water, coordinated timing is everything. Interaction between males and females of these species may be limited to chemical communication by pheromones.

In terrestrial vertebrates, internal fertilization requires interactive behaviors and specialized adaptations of the genitalia. For example, the female must have an internal receptacle for sperm (the vagina in humans), and the male must possess an organ (the penis in humans) that can place sperm in the receptacle. The human penis is *flaccid* (soft and limp) in its resting state, not capable of penetrating the narrow opening of the vagina. In the male sex act, the penis first stiffens and enlarges during **erection,** and then releases sperm from the ducts of the reproductive tract during **ejaculation**. Without these events, fertilization cannot take place.

The Human Sexual Response Has Four Phases

The human sex act—also known as sexual intercourse, copulation, or **coitus** {*coitio*, a coming together}—is highly variable in some ways and highly stereotypical in other ways.

Human sexual response in both sexes is divided into four phases: (1) excitement, (2) plateau, (3) orgasm, and (4) resolution. In the excitement phase, various erotic stimuli prepare the genitalia for the act of copulation. For the male, excitement involves erection of the penis. For the female, it includes erection of the clitoris and vaginal lubrication. In both sexes, erection is a state of vasocongestion in which arterial blood flow into spongy erectile tissue exceeds venous outflow.

Erotic stimuli include sexually arousing tactile stimuli as well as psychological stimuli. Because the latter vary widely among individuals and among cultures, what is erotic to one person or in one culture may be considered disgusting by another individual or in another culture. Regions of the body that possess receptors for sexually arousing tactile stimuli are called **erogenous zones** and include the genitalia as well as the lips, tongue, nipples, and ear lobes.

In the plateau phase, changes that started during excitement intensify and peak in an **orgasm** (climax). In both sexes, orgasm is a series of muscular contractions accompanied by intense pleasurable sensations and increased blood pressure, heart rate, and respiration rate. In females, the uterus and walls of the vagina contract. In males, the contractions usually result in the ejaculation of semen from the penis. Female orgasm is not required for pregnancy.

The final phase of the sexual response is resolution, a period during which the physiological parameters that changed in the first three phases slowly return to normal.

The Male Sex Act Includes Erection and Ejaculation

A key element to successful copulation is the ability of the male to achieve and sustain an erection. Sexual excitement from either tactile or psychological stimuli triggers the **erection reflex,**

RUNNING PROBLEM

Results of the temperature tracking for several months reveal that Peggy is ovulating regularly. Dr. Coddington therefore believes that her ovaries are functioning normally. Other possible causes for this couple's infertility include abnormalities in Peggy's cervix, fallopian tubes, or uterus. Dr. Coddington next decides to order a postcoital test. In this test, the couple is instructed to have intercourse 12 hours before the physician visit. Cervical mucus is then analyzed. This test will also analyze the interaction between sperm and mucus.

Q4: What abnormalities in the cervix, fallopian tubes, and uterus could cause infertility?

851 865 873 874 878 881 886

a spinal reflex that is subject to control from higher centers in the brain. The urination and defecation reflexes are similar types of reflexes [pp. 650, 728].

In its simplest form, the erection reflex begins with tactile stimuli sensed by mechanoreceptors in the glans penis or other erogenous zones (■ Fig. 26.13). Sensory neurons signal the spinal integration center, which inhibits vasoconstrictive sympathetic input on penile arterioles. Simultaneously, nitric oxide produced by increased parasympathetic input actively dilates the penile arterioles. As arterial blood flows into the open spaces of the erectile tissue, it passively compresses the veins and traps blood. The erectile tissue becomes engorged, stiffening and lengthening the penis within 5–10 seconds.

The climax of the male sexual act coincides with emission and ejaculation. **Emission** is the movement of sperm out of the vas deferens and into the urethra, where they are joined by secretions from the accessory glands to make semen. The average semen volume is 3 mL (range 2–6 mL), of which less than 10% is sperm.

During ejaculation, semen in the urethra is expelled to the exterior by a series of rapid muscular contractions accompanied by sensations of intense pleasure—the orgasm. A sphincter at the base of the bladder contracts to prevent sperm from entering the bladder and urine from joining the semen.

Both erection and ejaculation can occur in the absence of mechanical stimulation. Sexually arousing thoughts, sights, sounds, emotions, and dreams can all initiate sexual arousal and even lead to orgasm in both men and women. In addition, nonsexual penile erection accompanies rapid eye movement (REM) sleep.

THE ERECTION REFLEX

Erection can take place without input from higher brain centers. It can also be stimulated (and inhibited) by descending pathways from the cerebral cortex. Spontaneous erections occur during REM sleep.

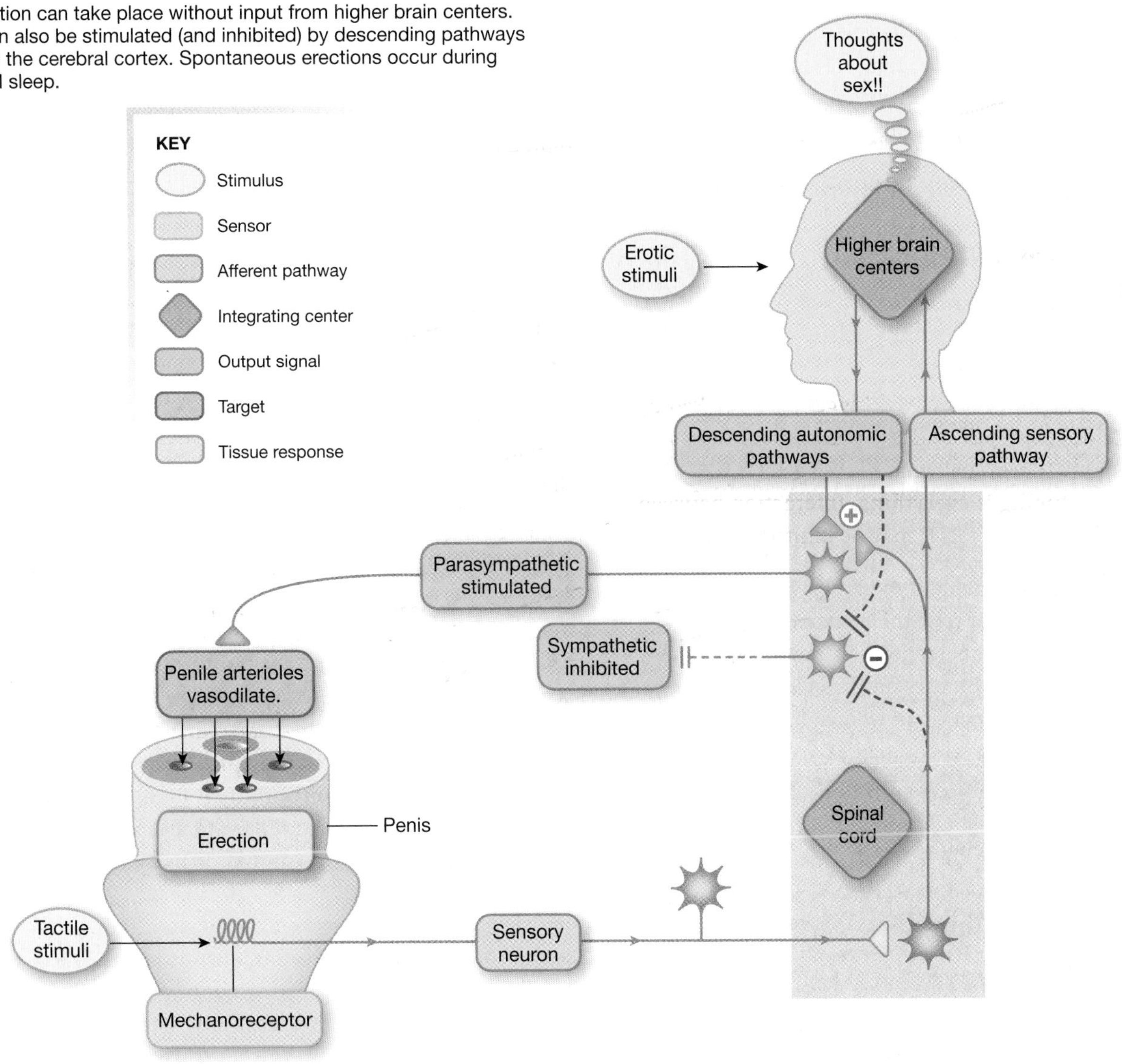

■ **Fig. 26.13**

Sexual Dysfunction Affects Males and Females

The inability to achieve or sustain a penile erection is known as **erectile dysfunction** (ED) or *impotence*. Erectile dysfunction is a matter of global concern because inability to achieve and sustain an erection disrupts the sex act for both men and women. Organic (physiological and anatomical) causes of ED include neural and hormonal problems, vascular insufficiency, and drug-induced ED. A variety of psychological causes can also contribute to ED.

Alcohol inhibits sexual performance in both men and women, as noted by Shakespeare in *Macbeth* (II, iii). When Macduff asks, "What three things does drink especially provoke?" the porter answers, "Marry, sir, nose-painting, sleep, and urine. Lechery, sir, it provokes and unprovokes: it provokes the desire, but it takes away the performance." Several antidepressant drugs list loss of libido among their side effects.

Erectile dysfunction in men over age 40 is now considered a marker for cardiovascular disease and atherosclerosis, and sometimes ED is the first clinical sign of these conditions. Erections occur when neurotransmitters from pelvic nerves increase endothelial production of nitric oxide (NO), which increases cGMP and results in vasodilation of penile arterioles. Endothelial dysfunction and failure to produce adequate NO occur in atherosclerosis and diabetes mellitus, making ED an early manifestation of vascular pathology.

In 1998 the U.S. Food and Drug Administration (FDA) approved sildenafil (Viagra®) for the treatment of erectile dysfunction. Sildenafil and similar drugs in the same class prolong the effects of nitric oxide by blocking *phosphodiesterase-5* (PDE-5), the enzyme that degrades cGMP. Clinical trials have shown that phosphodiesterase inhibitors are very effective in correcting ED but are not without side effects. The U.S. Federal Aviation Administration issued an order that pilots should not take sildenafil within six hours of flying because 3% of men report impaired color vision (a blue or greenish haze). This impairment occurs because sildenafil also inhibits an enzyme in the retina.

When the FDA approved PDE-5 inhibitors for male erectile dysfunction, women wondered if the drug, which promotes the erection reflex, would improve their sexual response. Although women do have clitoral erections, the female sexual response is more complicated. Studies on the efficacy of PDE-5 inhibitors for orgasmic dysfunction in women have had mixed results. Instead, pharmaceutical companies are testing other drugs for female sexual dysfunction. The most promising candidates, now in late clinical trials, are based on testosterone, the androgen that creates libido in both sexes.

Contraceptives Are Designed to Prevent Pregnancy

One disadvantage of sexual intercourse for pleasure rather than reproduction is the possibility of an unplanned pregnancy. On average, 85% of young women who have sexual intercourse without using any form of birth control will get pregnant within a year. Many women, however, get pregnant after just a single unprotected encounter. Couples who hope to avoid unwanted pregnancies generally use some form of birth control, or **contraception.**

Contraceptive practices fall into several broad groups. **Abstinence**, the total avoidance of sexual intercourse, is the surest method to avoid pregnancy (and sexually transmitted diseases). Some couples practice abstinence only during times of suspected fertility calculated using *fertility-awareness methods* of birth control.

Sterilization is the most effective contraceptive method for sexually active people, but it is a surgical procedure and is not easily reversed. Female sterilization is called **tubal ligation**. It consists of tying off and cutting the fallopian tubes. A woman with a tubal ligation still ovulates, but the eggs remain in the abdomen. The male form of sterilization is the **vasectomy**, in which the vas deferens is tied and clipped. Sperm are still made in the seminiferous tubules, but because they cannot leave the reproductive tract, they are reabsorbed.

Interventional methods of contraception include (1) barrier methods, which prevent union of eggs and sperm; (2) methods that prevent implantation of the fertilized egg; and (3) hormonal treatments that decrease or stop gamete production. The efficacy of interventional contraceptives depends in part on how consistently and correctly they are used (■ Tbl. 26.1).

Barrier Methods Contraceptive methods based on chemical or physical barriers are among the earliest recorded means of birth control. Once people made the association between pregnancy and semen, they concocted a variety of physical barriers and *spermicides* {*cida*, killer} to kill sperm. An ancient Egyptian papyrus with the earliest known references to birth control describes the use of vaginal plugs made of leaves, feathers, figs, and alum held together with crocodile and elephant dung. Sea sponges soaked in vinegar and disks of oiled silk have also been used at one time or another. In subsequent centuries women used douches of garlic, turpentine, and rose petals to rinse the vagina after intercourse. As you can imagine, many of these methods also caused vaginal or uterine infections.

Modern versions of the female barrier include the **diaphragm,** introduced into the United States in 1916. These rubber domes and a smaller version called a *cervical cap* are usually filled with a spermicidal cream, then inserted into the top of the vagina so they cover the cervix. One advantage to the diaphragm is that it is nonhormonal. When used properly and regularly, diaphragms are highly effective (97–99%). However, they are not always used because they must be inserted close to the time of intercourse, and consequently about 20% of women who depend on diaphragms for contraception are pregnant within the first year. Another female barrier contraceptive that was recently reintroduced is the **contraceptive sponge,** which contains a spermicidal chemical.

Table 26.1 Efficacy of Various Contraceptive Methods

Method	Pregnancy Rate With Typical Use*
No contraception	85%
Spermicides	29%
Abstinence during times of predicted fertility	25%
Diaphragm, cervical cap, or sponge	16–32%†
Oral contraceptive pills	8%
Intrauterine devices (IUDs)	< 1%
Contraceptive hormone injection	< 1%
Male condom	15%
Female condom	21%
Sterilization	< 1%

*Rates reflect unintentional pregnancies in the first year of using the method. Data are from *www.contraceptivetechnology.org/table.html* (Accessed 7/22/11).

†Lower rates are in women who have never delivered a child.

The male barrier contraceptive is the **condom**, a closed sheath that fits closely over the penis to catch ejaculated semen. Males have used condoms made from animal bladders and intestines for centuries. Condoms lost popularity when oral contraceptives came into widespread use in the 1960s and 1970s, but in recent years they have regained favor because they combine pregnancy protection with protection from many sexually transmitted diseases. However, latex condoms may cause allergic reactions, and there is evidence that HIV can pass through pores in some condoms currently produced. A female version of the condom is also commercially available. It covers the cervix and completely lines the vagina, providing more protection from sexually transmitted diseases.

Implantation Prevention Some contraceptive methods do not prevent fertilization but do keep a fertilized egg from establishing itself in the endometrium. They include **intrauterine devices (IUDs)** as well as chemicals that change the properties of the endometrium. IUDs are copper-wrapped plastic devices that are inserted into the uterine cavity, where they create a mild inflammatory reaction that prevents implantation. They have low failure rates (0.5% per year) but side effects that range from pain and bleeding to infertility caused by pelvic inflammatory disease and blockage of the fallopian tubes.

Hormonal Treatments Techniques for decreasing gamete production depend on altering the hormonal milieu of the body. In centuries past, women would eat or drink various plant concoctions for contraception. Some of these substances actually worked because the plants contained estrogen-like compounds. Modern pharmacology has improved on this method, and now women can choose between oral contraceptive pills, injections lasting three months, or a vaginal contraceptive ring (NuvaRing®).

The **oral contraceptives**, also known as *birth control pills*, first became available in 1960. They rely on various combinations of estrogen and progesterone that inhibit gonadotropin secretion from the pituitary. Without adequate FSH and LH, ovulation is suppressed. In addition, progesterones in the contraceptive pills thicken the cervical mucus and help prevent sperm penetration. These hormonal methods of contraception are highly effective when used correctly but also carry some risks, including an increased incidence of blood clots and strokes, especially in women who smoke.

Development of a male hormonal contraceptive has been slow because of undesirable side effects. Contraceptives that block testosterone secretion or action are also likely to decrease the male libido or even cause impotence. Both side effects are unacceptable to men who would be most interested in using the contraceptive. Some early male oral contraceptives irreversibly suppressed sperm production, which was also unacceptable. It now appears, however, that a combination of oral progestin to suppress sperm production plus injected testosterone to maintain libido is a promising candidate for a male hormonal contraceptive.

Contraceptive vaccines are based on antibodies against various components of the male and female reproductive systems, such as antisperm or antiovum antibodies. These contraceptives can be administered as shots. However, clinical trials of human vaccines have been disappointing and vaccines may not be a practical contraceptive for humans.

Infertility Is the Inability to Conceive

While some couples are trying to prevent pregnancy, others are spending thousands of dollars trying to get pregnant. *Infertility* is the inability of a couple to conceive a child after a year of unprotected intercourse. For years, infertile couples had no choice but adoption if they wanted to have a child, but incredible strides have been made in this field since the 1970s. As a result, many infertile couples today are able to have children.

Infertility can arise from a problem in the male, the female, or both. Male infertility usually results from a low sperm count or an abnormally high number of defective sperm. Female infertility can be mechanical (blocked fallopian tubes or other structural problems) or hormonal, leading to decreased or absent ovulation. One problem involving both partners is that the woman may produce antibodies to her partner's sperm. In addition, not all pregnancies go to a successful conclusion. By some estimates, as many as a third of all pregnancies spontaneously terminate—many within the first weeks, before the woman is even aware that she was pregnant.

Some of the most dramatic advances have been made in the field of **assisted reproductive technology** (ART), strategies in which both sperm and eggs are manipulated. For *in vitro* fertilization, a woman's ovaries are manipulated with hormones to ovulate multiple eggs at one time. The eggs are collected surgically and fertilized outside the body. The developing embryos are then placed in the woman's uterus, which has been primed for pregnancy by hormonal therapy. Because of the expense and complicated nature of the procedure, multiple embryos are usually placed in the uterus at one time, which may result in multiple births. *In vitro* fertilization has allowed some infertile couples to have children, with a 2009 success rate in the United States averaging 31%. Success varies significantly with age, ranging from 41% for women younger than 35 to 12% for women older than 40.

Pregnancy and Parturition

Now let's return to a recently ovulated egg and some sperm deposited in the vagina and follow them through fertilization, pregnancy, and **parturition**, the birth process.

Fertilization Requires Capacitation

Once an egg is released from the ruptured follicle, it is swept into the fallopian tube by beating cilia. Meanwhile, sperm deposited in the vagina must go through their final maturation step, **capacitation**, which enables the sperm to swim rapidly and fertilize an egg. The process apparently involves the reorganization of molecules in the outer membrane of the sperm head.

RUNNING PROBLEM

Analysis of Peggy's postcoital cervical mucus shows that sperm are present but not moving. Dr. Coddington explains that it is likely that Peggy's cervical mucus contains antibodies that destroy Larry's sperm.

Q5: Speculate on how this kind of infertility problem might be treated.

851 865 873 874 878 881 886

Normally, capacitation takes place in the female reproductive tract, which presents a problem for *in vitro* fertilization. Those sperm must be artificially capacitated by placing them in physiological saline supplemented with human serum. Much of what we know about human fertilization has come from infertility research aimed at improving the success rate of *in vitro* fertilization.

Fertilization of an egg by a sperm is the result of a chance encounter, possibly aided by chemical attractants produced by the egg. An egg can be fertilized for only about 12–24 hours after ovulation. Sperm in the female reproductive tract remain viable for 5–6 days.

Fertilization normally takes place in the distal part of the fallopian tube. Of the millions of sperm in a single ejaculation, only about 100 reach this point. To fertilize the egg, a sperm must penetrate both an outer layer of loosely connected granulosa cells (the *corona radiata*) and a protective glycoprotein coat called the **zona pellucida** (■ Fig. 26.14b). To get past these barriers, capacitated sperm release powerful enzymes from the acrosome in the sperm head, a process known as the **acrosomal reaction**. The enzymes dissolve cell junctions and the zona pellucida, allowing the sperm to wiggle their way toward the egg.

The first sperm to reach the egg quickly finds sperm-binding receptors on the oocyte membrane and fuses its membrane to the egg membrane (Fig. 26.14c). The fused section of membrane opens, and the sperm nucleus sinks into the egg's cytoplasm. Fusion of the egg and sperm membranes signals the egg to resume meiosis and complete its second division. The final meiotic division creates a second polar body, which is ejected. At this point, the 23 chromosomes of the sperm join the 23 chromosomes of the egg, creating a zygote nucleus with a full set of genetic material.

The fusion of sperm and oocyte membrane triggers a chemical reaction called the **cortical reaction**. Membrane-bound **cortical granules** in the peripheral cytoplasm of the egg release their contents into the space just outside the egg membrane. These chemicals rapidly alter the membrane and surrounding zona pellucida so that additional sperm cannot penetrate or bind. The cortical reaction prevents **polyspermy**, in which more than one sperm fertilizes an egg.

Once an egg is fertilized and becomes a zygote, it begins mitosis as it slowly makes its way along the fallopian tube to the uterus, where it will settle for the remainder of the **gestation** period {*gestare*, to carry in the womb}.

The Developing Embryo Implants in the Endometrium

The dividing embryo takes four or five days to move through the fallopian tube into the uterine cavity (Fig. 26.14d). Under the influence of progesterone, smooth muscle of the tube relaxes, and transport proceeds slowly. By the time the developing

■ Fig. 26.14 *ESSENTIALS*

Fertilization

Fertilization must occur within 24 hours of ovulation.

(a) This photograph shows the tremendous difference in the sizes of human sperm and egg.

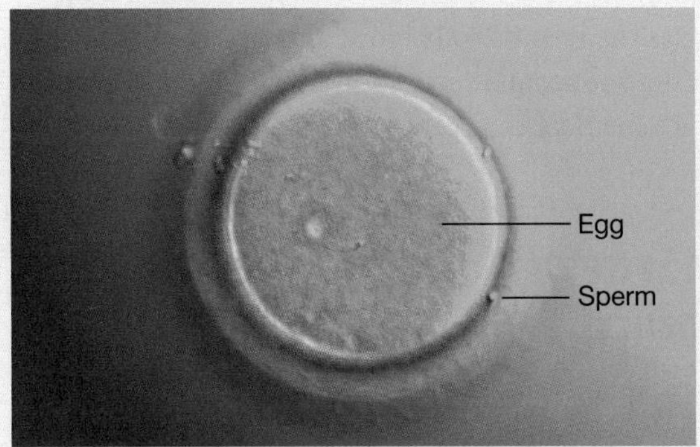

(b) Capacitated sperm release enzymes from their acrosomes in order to penetrate the cells and zona pellucida surrounding the egg.

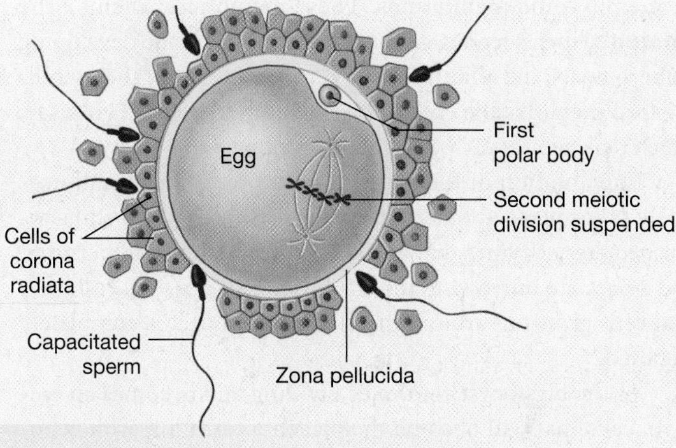

(c) The first sperm to fuse with the egg fertilizes it.

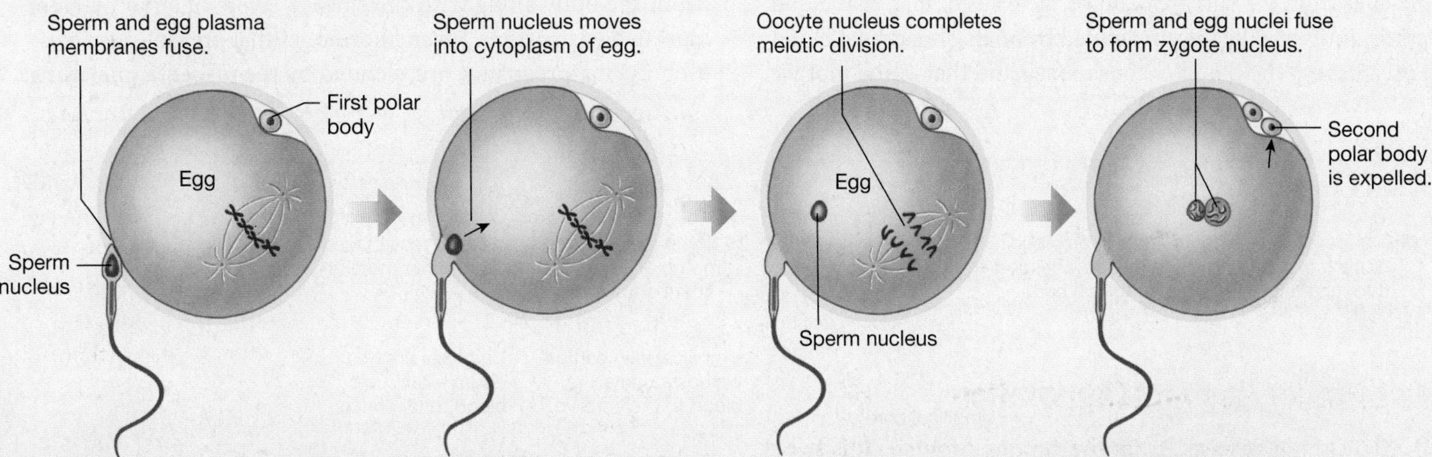

(d) Timing of ovulation, fertilization, and implantation

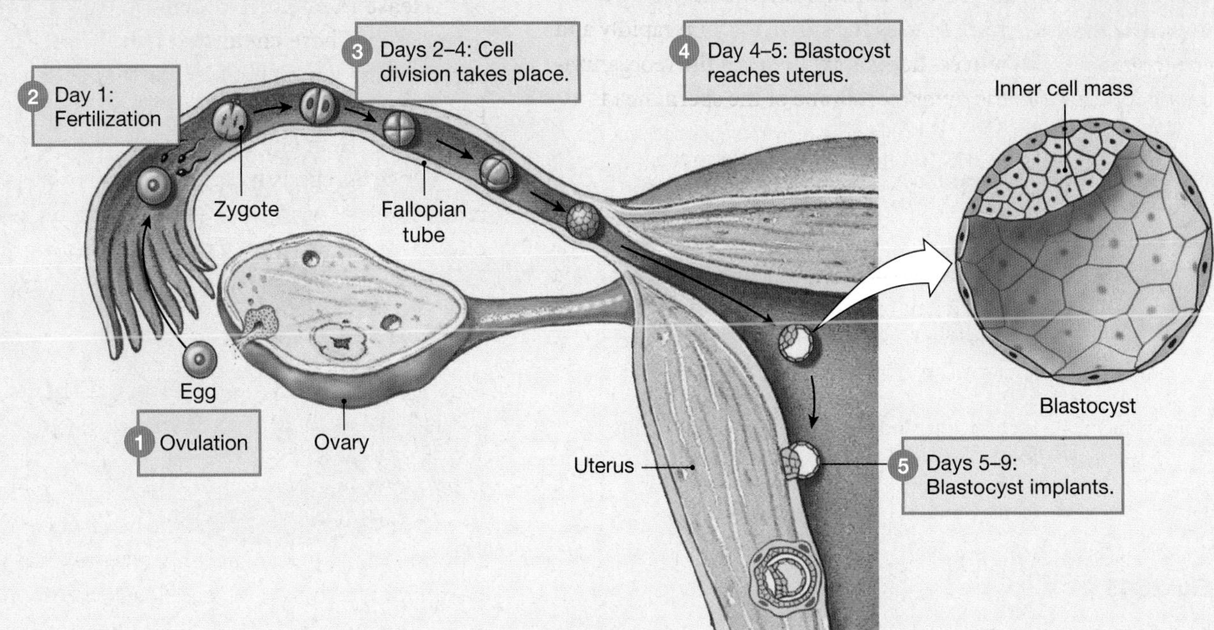

embryo reaches the uterus, it consists of a hollow ball of about 100 cells called a **blastocyst**.

Some of the outer layer of blastocyst cells will become the **chorion,** an *extraembryonic membrane* that will enclose the embryo and form the placenta (■ Fig. 26.15a). The inner cell mass of the blastocyst will develop into the embryo and into other extraembryonic membranes. These membranes include the **amnion,** which secretes *amniotic fluid* in which the developing embryo floats; the **allantois,** which becomes part of the umbilical cord that links the embryo to the mother; and the **yolk sac,** which degenerates early in human development.

Implantation of the blastocyst into the uterine wall normally takes place about 7 days after fertilization. The blastocyst secretes enzymes that allow it to invade the endometrium, like a parasite burrowing into its host. As it does so, endometrial cells grow out around the blastocyst until it is completely engulfed.

As the blastocyst continues dividing and becomes an embryo, cells that will become the placenta form fingerlike **chorionic villi** that penetrate into the vascularized endometrium. Enzymes from the villi break down the walls of maternal blood vessels until the villi are surrounded by pools of maternal blood (Fig. 26.15b). The blood of the embryo and that of the mother do not mix, but nutrients, gases, and wastes exchange across the membranes of the villi. Many of these substances move by simple diffusion, but some, such as maternal antibodies, must be transported across the membrane.

The placenta continues to grow during pregnancy until, by delivery, it is about 20 cm in diameter (the size of a small dinner plate). The placenta receives as much as 10% of the total maternal cardiac output. The tremendous blood flow to the placenta is one reason sudden, abnormal separation of the placenta from the uterine wall is a medical emergency.

The Placenta Secretes Hormones During Pregnancy

As the blastocyst implants in the uterine wall and the placenta begins to form, the corpus luteum is nearing the end of its preprogrammed 12-day life span. Unless the developing embryo sends a hormonal signal, the corpus luteum disintegrates, progesterone and estrogen levels drop, and the embryo is flushed from the body along with the surface layers of endometrium during menstruation. Several hormones that prevent menstruation during pregnancy are secreted by the placenta, including

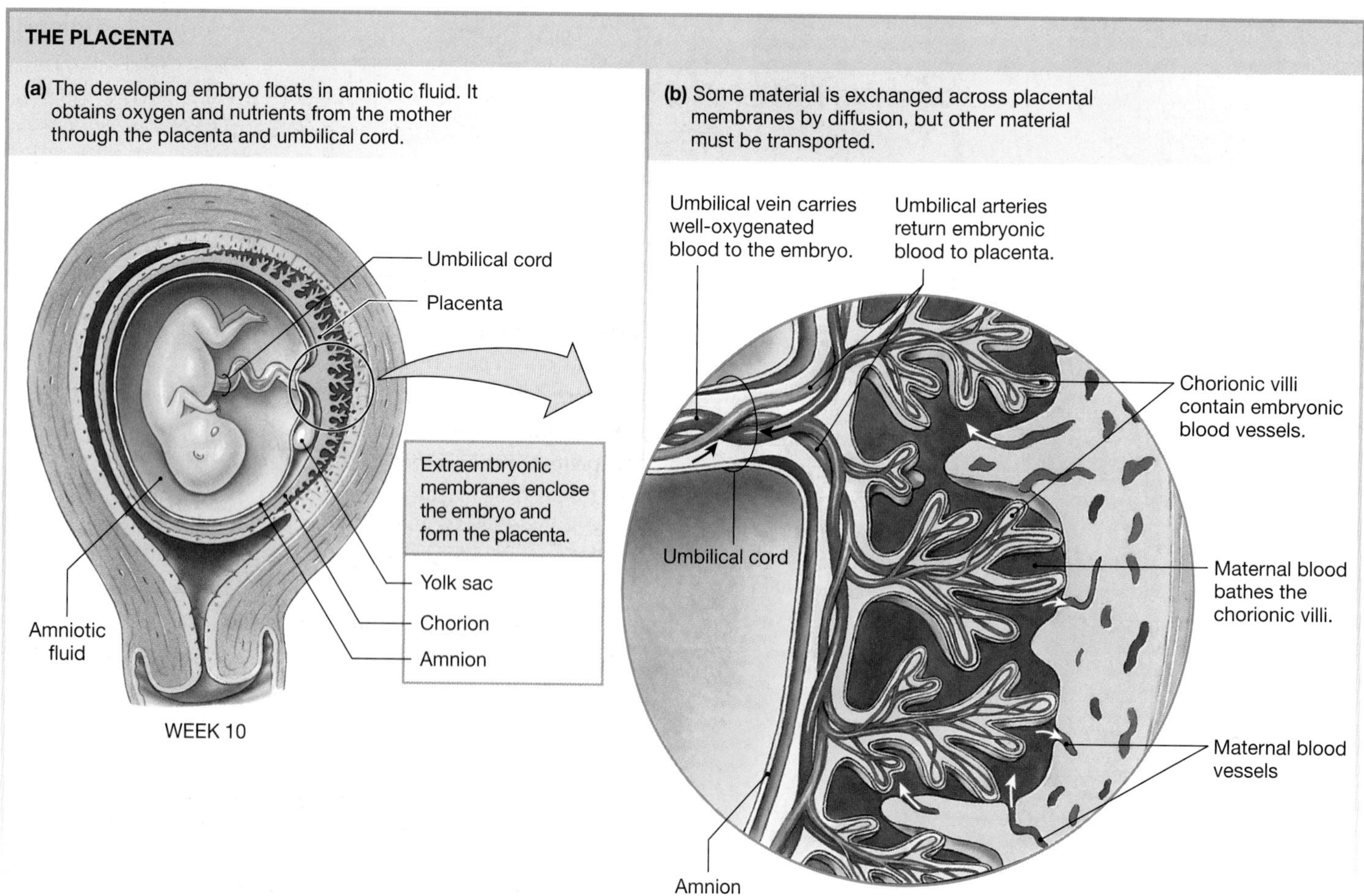
THE PLACENTA
(a) The developing embryo floats in amniotic fluid. It obtains oxygen and nutrients from the mother through the placenta and umbilical cord.
Umbilical cord
Placenta
Extraembryonic membranes enclose the embryo and form the placenta.
Yolk sac
Chorion
Amnion
Amniotic fluid
WEEK 10
(b) Some material is exchanged across placental membranes by diffusion, but other material must be transported.
Umbilical vein carries well-oxygenated blood to the embryo.
Umbilical arteries return embryonic blood to placenta.
Chorionic villi contain embryonic blood vessels.
Umbilical cord
Maternal blood bathes the chorionic villi.
Maternal blood vessels
Amnion

■ **Fig. 26.15**

human chorionic gonadotropin, human placental lactogen, estrogen, and progesterone.

Human Chorionic Gonadotropin The corpus luteum remains active during early pregnancy because of **human chorionic gonadotropin** (hCG), a peptide hormone secreted by the chorionic villi and developing placenta. Human chorionic gonadotropin is structurally related to LH, and it binds to LH receptors. Under the influence of hCG, the corpus luteum keeps producing progesterone to keep the endometrium intact.

By the seventh week of development, however, the placenta has taken over progesterone production, and the corpus luteum is no longer needed. At that point, it finally degenerates. Human chorionic gonadotropin production by the placenta peaks at three months of development, then diminishes.

A second function of hCG is stimulation of testosterone production by the developing testes in male fetuses. As you learned in the opening sections of this chapter, fetal testosterone and its metabolite DHT are essential for expression of male characteristics and for descent of the testes into the scrotum before birth.

Human chorionic gonadotropin is the chemical detected by pregnancy tests. Because hCG can induce ovulation in rabbits, years ago rabbits were used for pregnancy testing. If a woman suspected she was pregnant, her urine was injected into a rabbit. The rabbit's ovaries were then inspected for signs of ovulation. It took several days for women to learn the results of this test. Today, with modern biochemical techniques, women can perform their own pregnancy tests in a few minutes in the privacy of their home.

RUNNING PROBLEM

Assisted reproductive technologies (ART) are one treatment option currently available to infertile couples. All ART techniques involve either artificially stimulating the ovaries to produce eggs or using an egg from an egg donor. The eggs are harvested surgically and are fertilized *in vitro*. The zygote may be placed in the fallopian tube immediately or may be allowed to develop into an early embryo before being placed into the uterus. A different technique used to overcome infertility is *intrauterine insemination*. In this procedure, sperm that have been washed to remove antigenic material are introduced into the uterus through a tube inserted through the cervix so that fertilization takes place *in vivo*.

Q6: Based on the results of their infertility workup, which intervention—ART or intrauterine insemination—should be recommended for Peggy and Larry? Why?

851 865 873 874 878 881 886

Human Placental Lactogen (hPL) Another peptide hormone produced by the placenta is **human placental lactogen** (hPL), also known as *human chorionic somatomammotropin (hCS)*. This hormone, structurally related to growth hormone and prolactin, was initially believed to be necessary for breast development during pregnancy and for milk production **(lactation)**. Although hPL probably does contribute to lactation, women who do not make hPL during pregnancy because of a genetic defect still have adequate breast development and milk production.

A second role for hPL is alteration of the mother's glucose and fatty acid metabolism to support fetal growth. Maternal glucose moves across the membranes of the placenta by facilitated diffusion and enters the fetal circulation. During pregnancy, about 4% of women develop *gestational diabetes mellitus*, with elevated blood glucose levels caused by insulin resistance, similar to type 2 diabetes. After delivery, glucose metabolism in most of these women returns to normal, but these mothers and their babies are at higher risk of developing type 2 diabetes later in life.

Estrogen and Progesterone Estrogen and progesterone are produced continuously during pregnancy, first by the corpus luteum under the influence of hCG and then by the placenta. With high circulating levels of these steroid hormones, feedback suppression of the pituitary continues throughout pregnancy, preventing another set of follicles from beginning development.

During pregnancy, estrogen contributes to the development of the milk-secreting ducts of the breasts. Progesterone is essential for maintaining the endometrium and in addition helps suppress uterine contractions. The placenta makes a variety of other hormones, including inhibin and prorenin, but the function of most of them remains unclear.

Pregnancy Ends with Labor and Delivery

Parturition normally occurs between the 38th and 40th week of gestation. What triggers this process? For many years, researchers developed animal models of the signals that initiate parturition, only to discover recently that there are no good non-primate models that apply to humans. Parturition begins with **labor,** the rhythmic contractions of the uterus that push the fetus out into the world. Signals that initiate these contractions could begin with either the mother or the fetus, or they could be a combination of signals from both.

In many nonhuman mammals, a decrease in estrogen and progesterone levels marks the beginning of parturition. A decrease in progesterone levels is logical, as progesterone inhibits uterine contractions. In humans, however, levels of these hormones do not decrease until labor is well under way.

Another possible labor trigger is oxytocin, the peptide hormone that causes uterine muscle contraction. As a pregnancy nears full term, the number of uterine oxytocin receptors increases. However, studies have shown that oxytocin secretion does not increase until after labor begins. Synthetic oxytocin is often used to induce labor in pregnant women, but it is not always effective. Apparently, the start of labor requires something more than adequate amounts of oxytocin.

Another possibility is that the fetus somehow signals that it has completed development. One theory supported by clinical evidence is that corticotropin-releasing hormone (CRH) secreted by the placenta is the signal to begin labor. (CRH is also a hypothalamic releasing factor that controls release of ACTH from the anterior pituitary.) In the weeks prior to delivery, maternal blood CRH levels increase rapidly. In addition, women with elevated CRH levels as early as 15 weeks of gestation are more likely to go into premature labor.

Although we do not know for certain what initiates parturition, we do understand the sequence of events. In the days prior to the onset of active labor, the cervix softens ("ripens") and ligaments holding the pelvic bones together loosen as enzymes destabilize collagen in the connective tissue. The control of these processes is not clear and may be due to estrogen or the peptide hormone **relaxin,** which is secreted by ovaries and the placenta.

Once the contractions of labor begin, a positive feedback loop consisting of mechanical and hormonal factors is set into motion. The fetus is normally oriented head down (■ Fig. 26.16a). At the beginning of labor it repositions itself lower in the abdomen ("the baby has dropped") and begins to push on the softened cervix (Fig. 26.16b).

Cervical stretch triggers uterine contractions that move in a wave from the top of the uterus down, pushing the fetus farther into the pelvis. The lower portion of the uterus stays relaxed, and the cervix stretches and dilates. Cervical stretch starts a positive feedback cycle of escalating contractions (Fig. 26.16d). The contractions are reinforced by secretion of oxytocin from the posterior pituitary [p. 221], with continued cervical stretch reinforcing oxytocin secretion.

Prostaglandins are produced in the uterus in response to oxytocin and CRH secretion. Prostaglandins are very effective at causing uterine muscle contractions at any time. They are the primary cause of menstrual cramps and have been used to induce abortion in early pregnancy. During labor and delivery, prostaglandins reinforce the uterine contractions induced by oxytocin (Fig. 26.16d).

As the contractions of labor intensify, the fetus moves down though the vagina and out into the world (Fig. 26.16c), still attached to the placenta. The placenta then detaches from the uterine wall and is expelled a short time later. Uterine contractions clamp the maternal blood vessels and help prevent excessive bleeding, although typically the mother loses about 240 mL of blood.

The Mammary Glands Secrete Milk During Lactation

A newborn has lost its source of maternal nourishment through the placenta and must rely on an external source of food instead. Primates, who normally have only one or two offspring at a time, have two functional mammary glands. A mammary gland is composed of 15–20 milk-secreting lobes (■ Fig. 26.17a on page 884). Each lobe branches into lobules, and the lobules terminate in secretory structures called alveoli or acini. Each alveolus is composed of secretory epithelium that secretes into a duct, similar to the exocrine secretions of the pancreas [Fig. 21.5b, p. 706]. Each alveolus is surrounded by contractile *myoepithelium*. Interestingly, the mammary gland epithelium is closely related to the secretory epithelium of sweat glands, so milk secretion and sweat secretion share some common features.

During puberty, the breasts begin to develop under the influence of estrogen. The milk ducts grow and branch, and fat is deposited behind the glandular tissue. During pregnancy, the glands develop further under the direction of estrogen, growth hormone, and cortisol. The final development step also requires progesterone, which converts the duct epithelium into a secretory structure. This process is similar to progesterone's effect on the uterus, in which progesterone makes the endometrium into a secretory tissue during the luteal phase.

Although estrogen and progesterone stimulate mammary development, they inhibit secretion of milk. Milk production is stimulated by prolactin from the anterior pituitary [p. 223]. Prolactin is an unusual pituitary hormone in that its secretion is primarily controlled by **prolactin-inhibiting hormone** (PIH) from the hypothalamus. Good evidence suggests that PIH is actually *dopamine*, an amine neurohormone related to epinephrine and norepinephrine [p. 216].

During the later stages of pregnancy, PIH secretion falls, and prolactin reaches levels 10 or more times those found in nonpregnant women. Prior to delivery, when estrogen and progesterone are also high, the mammary glands produce only small amounts of a thin, low-fat secretion called **colostrum**. After delivery, when estrogen and progesterone decrease, the glands produce greater amounts of milk that contains 4% fat and substantial amounts of calcium. Proteins in colostrum and milk include maternal immunoglobulins, secreted into the duct and absorbed by the infant's intestinal epithelium [p. 712]. This process transfers some of the mother's immunity to the infant during its first weeks of life.

Suckling, the mechanical stimulus of the infant nursing at the breast, reinforces the inhibition of PIH begun in the last weeks of pregnancy (Fig. 26.17b). In the absence of PIH, prolactin secretion increases, resulting in milk production. Pregnancy is not a requirement for lactation, and some women who have adopted babies have been successful in breast-feeding.

The ejection of milk from the glands, known as the **let-down reflex,** requires the presence of oxytocin from the posterior pituitary. Oxytocin initiates smooth muscle contraction in the uterus and breasts. In the *postpartum* (after delivery) uterus, oxytocin-induced contractions help return the uterus to its pre-pregnancy size.

PARTURITION: THE BIRTH PROCESS

(a) Fully developed fetus. As labor begins, the fetus is normally head down in the uterus.

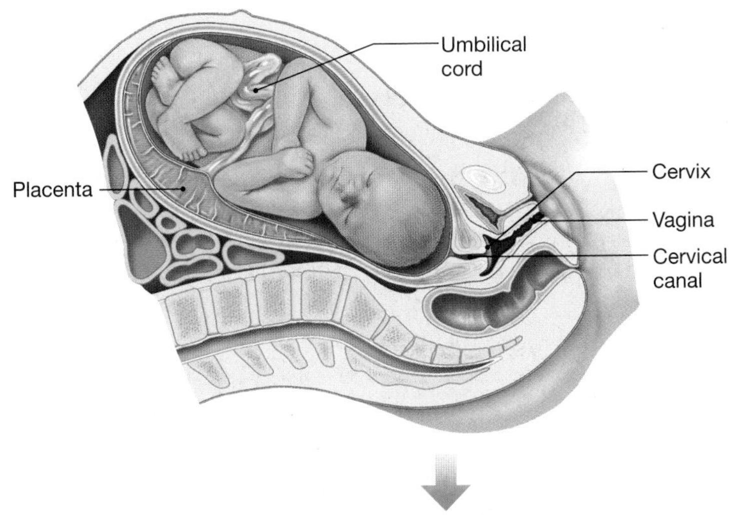

(b) Cervical dilation. Uterine contractions push the head against the softened cervix, stretching and dilating it.

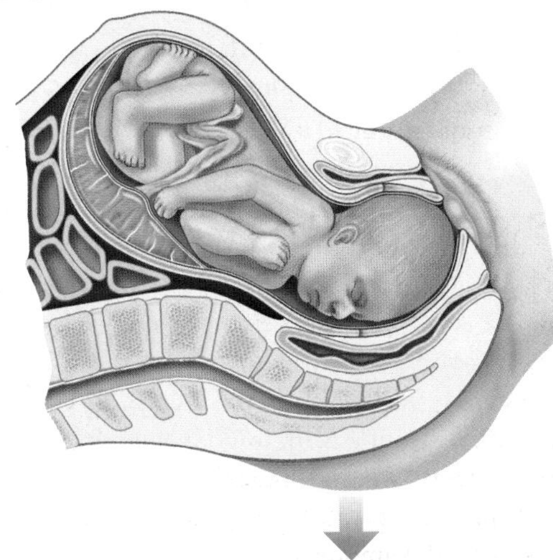

(c) Delivery. Once the cervix is fully dilated and stretched, the uterine contractions push the fetus out through the vagina.

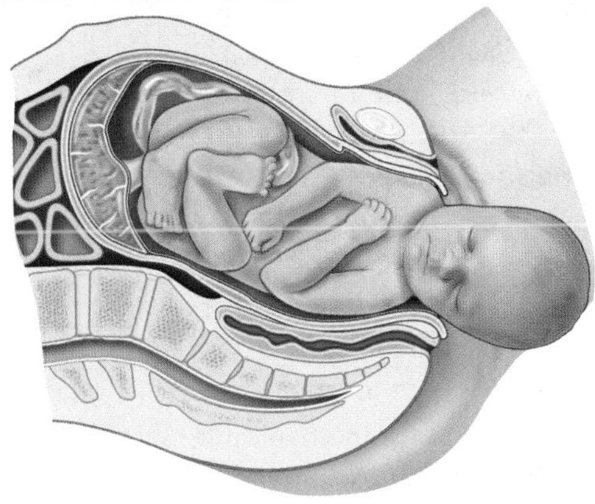

(d) The process of labor is controlled by a positive feedback loop that ends with delivery.

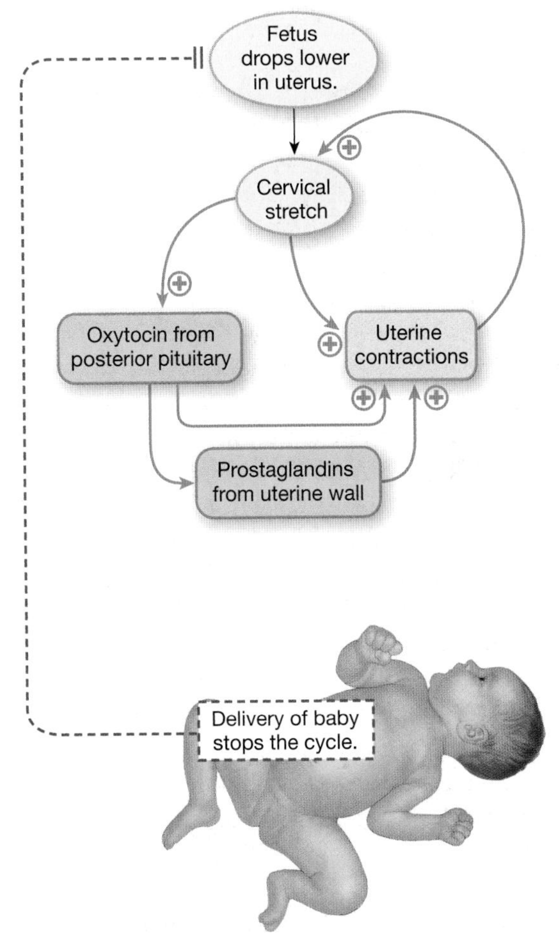

Fig. 26.16

26

LACTATION

(a) Mammary glands

Epithelial cells of the mammary glands secrete milk into the ducts of the gland. Contraction of the myoepithelium forces fluid out of the ducts through openings in the nipple.

(b) The hormonal control of milk secretion and release

Prolactin controls milk secretion, and oxytocin causes smooth muscle contraction to eject milk.

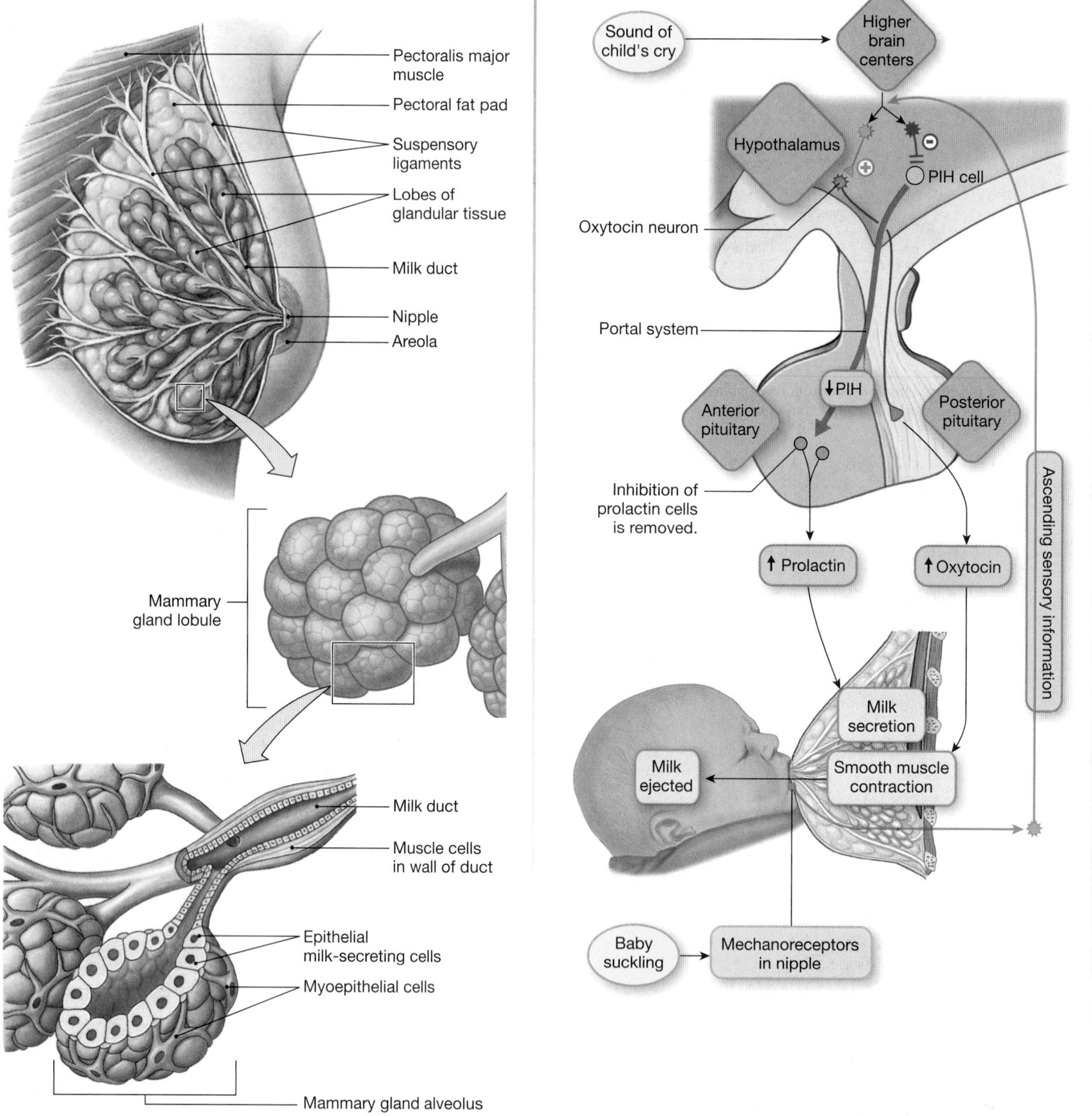

Fig. 26.17

In the lactating breast, oxytocin causes contraction of myoepithelial cells surrounding the alveoli and in the walls of the ducts. Contraction creates high pressure in the ducts that sends the milk squirting into the infant's mouth. Although prolactin release requires the mechanical stimulus of suckling, oxytocin release can be stimulated by various cerebral stimuli, including the thought of the child. Many nursing mothers experience inappropriate milk release triggered by hearing someone else's child cry.

Prolactin Has Other Physiological Roles

Although we discussed prolactin in the context of nursing mothers, all non-nursing women and men have tonic prolactin secretion that exhibits a diurnal cycle, peaking during sleep. Prolactin is related to growth hormone and plays a role in other reproductive and nonreproductive processes. For example, prolactin is synthesized in the uterine endometrium during normal menstrual cycles. Male knockout mice who lack prolactin or a prolactin receptor have decreased fertility.

Some interesting research has established a role for prolactin in neuroimmunomodulation [p. 828]. Both prolactin and growth hormone appear to be necessary for normal differentiation of T lymphocytes in the thymus gland, an observation supported by impaired immune function in animals with *hypoprolactinemia*. In contrast, several autoimmune diseases, including multiple sclerosis, systemic lupus erythematosus, and autoimmune thyroiditis, have been linked to elevated levels of prolactin.

Growth and Aging

The reproductive years begin with the events surrounding puberty and end with decreasing gonadal hormone production.

Puberty Marks the Beginning of the Reproductive Years

In girls, the onset of puberty is marked by budding breasts and the first menstrual period, called **menarche,** a time of ritual significance in many cultures. In the United States, the average age at menarche is 12 years (normal range is considered 8 to 13 years).

In boys, the onset of puberty is more subtle. The signs include growth and maturation of the external genitalia; development of secondary sex characteristics, such as pubic and facial hair and lowering of voice pitch; change in body shape; and growth in height. The age range for male puberty is 9 to 14 years.

Puberty requires maturation of the hypothalamic-pituitary control pathway. Before puberty, the child has low levels of both steroid sex hormones and gonadotropins. Because low sex hormone levels normally enhance gonadotropin release, the combination of low steroids and low gonadotropins indicates that the hypothalamus and pituitary are not yet sensitive to steroid levels in the blood.

At puberty, the hypothalamic GnRH-secreting neurons increase their pulsatile secretion of GnRH, which in turn increases gonadotropin release. The signals responsible for the onset of puberty are complex, but several of them appear to be mediated by the hypothalamic neuropeptide *kisspeptin*. One theory holds that the genetically programmed maturation of hypothalamic neurons initiates puberty. We know that puberty has a genetic basis because inherited patterns of maturation are common. If a woman did not start her menstrual periods until she was 16, for example, it is likely that her daughters will also have late menarche.

The adipose tissue hormone *leptin* [p. 737] also contributes to the onset of puberty. Undernourished women with little adipose tissue and low leptin levels often stop having menstrual periods (*amenorrhea*), and knockout mice without leptin are infertile. Presumably improved nutrition over the last century increased individuals' prepubertal fat stores and leptin secretion, which could interact with other factors to initiate puberty.

Menopause and Andropause Are a Consequence of Aging

In nineteenth century America, many people died of acute illnesses while still reproductively active. Now modern medicine has overcome most acute illnesses, and most of us will live well past the time we are likely to have children.

Women's reproductive cycles stop completely at the time known as **menopause**. The physiology of menopause has been well studied. After about 40 years of menstrual cycles, a woman's periods become irregular (*perimenopause*) and finally cease. The cessation of reproductive cycles is due not to the pituitary but to the ovaries, which can no longer respond to gonadotropins. In the absence of negative feedback, gonadotropin levels increase dramatically in an effort to stimulate the ovaries into maturing more follicles.

26

The absence of estrogen in postmenopausal women leads to symptoms of varying severity. These may include hot flashes [p. 769], atrophy of genitalia and breasts, and osteoporosis as calcium is lost from bones [p. 796]. Hormone replacement therapy (HRT) for women in menopause traditionally consists of estrogen or a combination of estrogen and progesterone. This treatment has become controversial, however, because of some studies that suggest that HRT risks outweigh its benefits.

A newer drug therapy for menopause uses *selective estrogen receptor modulators* (SERMs). These drugs bind with different affinities to the two estrogen receptor subtypes, which allows the drugs to mimic the beneficial effects of estrogen on bone while avoiding the potentially detrimental effects on breasts and uterus.

In men, testosterone production decreases with age, and about half of men over the age of 50 have symptoms of **andropause,** a term coined as the counterpart to menopause. The existence of physiological andropause in men is still controversial because the physical and psychological symptoms of aging in men are not clearly linked to a decline in testosterone. Many men remain reproductively active as they age, and it is not uncommon for men in their fifties or sixties to have children with younger women. Postmenopausal women also remain sexually active, although not reproductively active, and some report a more fulfilling sex life once the fear of unwanted pregnancy has been removed.

RUNNING PROBLEM CONCLUSION

Infertility

Peggy and Larry used home ovulation tests that measured Peggy's LH levels to determine when she was ovulating. On the day her LH surged, Peggy went to her doctor's office, where a washed sperm sample was inserted into her uterus through a small tube. After the third cycle Peggy became pregnant and later gave birth to a little girl.

In this running problem you learned how the cause of infertility is diagnosed in a typical couple. To learn more about infertility, see literature from the American Society for Reproductive Medicine at *www.asrm.org* or go to Medline Plus (*www.nlm.nih.gov/medlineplus*) and look under Health Topics. The Centers for Disease Control keep statistics on the success of assisted reproductive technologies (ART) on their web site (*www.cdc.gov*).

Now test your understanding of the running problem by checking your answers against the information in this summary table.

Question	Facts	Integration and Analysis
1. Name (in order) the male reproductive structures that carry sperm from the testes to the external environment.	The male reproductive structures include the testes, accessory glandular organs, a series of ducts, and the external genitalia.	Sperm leaving the testes pass into the epididymis, then into the vas deferens, and finally exit the body via the urethra.
2. Which causes of male infertility might make retrieval of sperm from the epididymis necessary?	The epididymis is the first duct the sperm enter upon leaving the testes.	If the infertility problem is due to blockage or congenital defects in the vas deferens or urethra, removal of sperm from the epididymis might be useful. If the problem is caused by low sperm count or abnormal sperm morphology, this technique would probably not be useful.
3. For which causes of female infertility is temperature tracking useful? For which causes is it not useful?	Basal body temperature rises slightly following ovulation.	Temperature tracking is a useful way to tell if a woman is ovulating, but it cannot reveal structural problems in the female reproductive tract.
4. What abnormalities in the cervix, fallopian tubes, and uterus could cause infertility?	The cervix, fallopian tubes, and uterus are hollow structures through which sperm must pass.	Any blockage of these organs resulting from disease or congenital defects would prevent normal movement of sperm and cause infertility. Hormonal problems might cause the endometrium to develop incompletely, preventing implantation of the embryo.
5. Speculate on how infertility due to cervical mucus antibodies against sperm might be treated.	Antibodies in the cervical mucus react with antigenic material in the semen or on the sperm, causing the sperm to become immobile.	If the antigenic material can be removed from the semen, this might help the problem of sperm immobilization. A semen sample can be washed to remove nonsperm components. If the antigens are part of the sperm, this method will not work.
6. Should ART or intrauterine insemination be recommended for Peggy and Larry? Why?	ART is used when ovulation is abnormal. Intrauterine insemination is used when ovulation is normal.	Intrauterine insemination can be used to overcome cervical factors, as this technique bypasses the cervix. Because Peggy can ovulate, she and Larry should try intrauterine insemination first.

851 865 873 874 878 881 886

Chapter Summary

In this chapter you learned how the human species perpetuates itself through reproduction. The reproductive system has some of the most complex *control systems* of the body, in which multiple hormones interact in an ever-changing fashion. *Homeostasis* in the adult reproductive system is anything but steady state, particularly during the female menstrual cycle, when the *feedback effects* of estrogen change from negative to positive and back again. An example of *positive feedback* occurs with oxytocin secretion during labor and delivery. The testis provides a nice example of *compartmentation*, with the lumen of the seminiferous tubules, where sperm develop, isolated from the rest of the extracellular compartment.

Sex Determination

1. The sex organs consist of **gonads, internal genitalia,** and **external genitalia.** (p. 851)
2. **Testes** produce **sperm. Ovaries** produce eggs, or **ova.** Embryonic cells that will produce **gametes** (eggs and sperm) are called **germ cells.** (p. 851)
3. Humans have 46 chromosomes. (p. 852; Fig. 26.1a)
4. The genetic sex of an individual depends on the **sex chromosomes:** females are XX, and males are XY. In the absence of a Y chromosome, an embryo will develop into a female. (p. 852; Fig. 26.1b)
5. The **SRY gene** on the Y chromosomes produces SRY protein, a *testis-determining factor* that converts the bipotential gonad into a testis. In the absence of SRY protein, the gonad becomes an ovary. (p. 852)
6. Testicular **Sertoli cells** secrete **anti-Müllerian hormone** (AMH), which causes the **Müllerian ducts** to regress. **Leydig cells** secrete **testosterone,** which converts **Wolffian ducts** to male accessory structures. **Dihydrotestosterone** (DHT) promotes development of the prostate gland and external genitalia. (p. 854; Fig. 26.2)
7. Absence of testosterone and AMH causes Müllerian ducts to develop into **fallopian tubes (oviducts), uterus,** and **vagina.** In females, the Wolffian ducts regress. (p. 854; Fig. 26.2)

Basic Patterns of Reproduction

8. **Gametogenesis** begins with mitotic divisions of **spermatogonia** and **oögonia.** The first step of meiosis creates **primary spermatocytes** and **primary oocytes.** The first meiotic division creates two identical **secondary spermatocytes** in males or a large secondary oocyte (egg) and a tiny **first polar body** in females. (p. 858; Fig. 26.5)
9. The second meiotic division in males creates haploid **spermatids** that mature into sperm. In females, the second meiotic division does not take place unless the egg is fertilized. (p. 858; Fig. 26.5)
10. In both sexes, **gonadotropin-releasing hormone** (GnRH) controls the secretion of **follicle-stimulating hormone** (FSH) and **luteinizing hormone** (LH) from the anterior pituitary. FSH and steroid sex hormones regulate gametogenesis in gonadal gamete-producing cells. LH stimulates production of steroid sex hormones. (p. 859; Fig. 26.6)
11. The steroid sex hormones include **androgens, estrogens,** and **progesterone. Aromatase** converts androgens to estrogens. **Inhibin** inhibits secretion of FSH, and **activin** stimulates FSH secretion. (p. 858)
12. Gonadal steroids generally suppress secretion of GnRH, FSH, and LH. However, if estrogen rises rapidly above a threshold level for at least 36 hours, its feedback changes to positive and stimulates gonadotropin release. (p. 860)
13. After puberty, tonic GnRH release occurs in small pulses every 1–3 hours from a region of the hypothalamus called a **pulse generator.** (p. 860)

Male Reproduction

14. The **corpus spongiosum** and **corpora cavernosa** make up the erectile tissue of the penis. The **glans** is covered by the **foreskin.** The urethra runs though the **penis.** (p. 862; Fig. 26.7)
15. The testes migrate into the scrotum during fetal development. Failure of one or both testes to descend is known as **cryptorchidism.** (p. 861)
16. The testes consist of **seminiferous tubules** and interstitial tissue containing blood vessels and Leydig cells. The seminiferous tubules join the **epididymis,** which becomes the **vas deferens.** The vas deferens empties into the urethra. (p. 862; Fig. 26.7b)
17. A seminiferous tubule contains spermatogonia, spermatocytes, and Sertoli cells. Tight junctions between Sertoli cells form a **blood-testis barrier.** (p. 863; Fig. 26.7d, e)
18. Spermatogonia in the tubule undergo meiosis, becoming primary spermatocytes, spermatids, and finally sperm in about 64 days. (p. 863; Fig. 26.7e)
19. Sertoli cells regulate sperm development. They also produce inhibin, activin, growth factors, enzymes, and **androgen-binding protein.** (p. 864; Fig. 26.8)
20. Leydig cells produce 95% of a male's testosterone. The other 5% comes from the adrenal cortex. (p. 864)
21. FSH stimulates Sertoli cell production of androgen-binding protein, inhibin, and paracrine molecules. Leydig cells produce testosterone under the direction of LH. (p. 864; Fig. 26.8)
22. The **prostate gland, seminal vesicles,** and **bulbourethral glands** secrete the fluid component of **semen.** (p. 865)
23. **Primary sex characteristics** are the internal sexual organs and external genitalia. **Secondary sex characteristics** are other features of the body, such as body shape. (p. 866)

Female Reproduction

24. Female external genitalia, called the **vulva** or **pudendum,** are the **labia majora, labia minora,** and **clitoris.** The urethra opening is between the clitoris and the vagina. (p. 868; Fig. 26.9)
25. The uterine tissue layers are outer connective tissue, **myometrium,** and **endometrium.** (p. 869; Fig. 26.9d)
26. Fallopian tubes are lined with ciliated epithelium. The bulk of an ovary consists of ovarian follicles. (p. 869; Fig. 26.9e)
27. Eggs are produced in monthly **menstrual cycles.** (p. 870; Fig. 26.10)
28. In the **ovarian cycle,** the **follicular phase** is a period of follicular growth. **Ovulation** is the release of an egg from its follicle. In the **luteal phase,** the ruptured follicle becomes a **corpus luteum.** (p. 870; Fig. 26.10)
29. The **menses** begin the **uterine cycle.** This is followed by a **proliferative phase,** with endometrial thickening. Following ovulation, the endometrium goes into a **secretory phase.** (p. 870; Fig. 26.10)

30. Follicular **granulosa cells** secrete estrogen. As the follicular phase ends, a surge in LH is necessary for oocyte maturation. (p. 870; Fig. 26.10)
31. The corpus luteum secretes progesterone and some estrogen, which exert negative feedback on the hypothalamus-anterior pituitary. (p. 872; Fig. 26.12)
32. Estrogens and androgen control primary and secondary sex characteristics in females. (p. 874)

Procreation

33. The human sex act is divided into four phases; (1) excitement, (2) plateau, (3) orgasm, and (4) resolution. (p. 874)
34. The male **erection reflex** is a spinal reflex that can be influenced by higher brain centers. Parasympathetic input mediated by nitric oxide actively vasodilates the penile arterioles. (p. 875; Fig. 26.13)
35. **Emission** is the movement of sperm out of the vas deferens and into the urethra. **Ejaculation** is the expulsion of semen to the external environment. (p. 875)
36. Contraceptive methods include **abstinence, barrier methods, implantation prevention,** and **hormonal treatments.** (p. 876)
37. Infertility can arise from a problem in the male, the female, or both. *In vitro* fertilization has allowed some infertile couples to have children. (p. 877)

Pregnancy and Parturition

38. Sperm must go through **capacitation** before they can fertilize an egg. (p. 878)
39. Fertilization normally takes place in the fallopian tube. Capacitated sperm release acrosomal enzymes (the **acrosomal reaction**) to dissolve cell junctions and the **zona pellucida** of the egg. The first sperm to reach the egg fertilizes it. (p. 879; Fig. 26.14)
40. Fusion of egg and sperm membranes initiates a **cortical reaction** that prevents **polyspermy.** (p. 878)
41. The developing embryo is a hollow **blastocyst** when it reaches the uterus. Once the blastocyst implants, it develops extraembryonic membranes. (p. 879, 880; Figs. 26.14d and 26.15)
42. The **chorionic villi** of the placenta are surrounded by pools of maternal blood where nutrients, gases, and wastes are exchanged between mother and embryo. (p. 880; Fig. 26.15)
43. The corpus luteum remains active during early pregnancy because of **human chorionic gonadotropin** (hCG) produced by the developing embryo. (p. 881)
44. The placenta secretes hCG, estrogen, progesterone, and **human placental lactogen.** This last hormone plays a role in maternal metabolism. (p. 881)
45. Estrogen during pregnancy contributes to development of milk-secreting ducts in the breasts. Progesterone is essential for maintaining the endometrium and, along with **relaxin,** helps suppress uterine contractions. (p. 882)
46. **Parturition** normally occurs between the 38th and 40th week of gestation. It begins with **labor** and ends with delivery of the fetus and placenta. A positive feedback loop of oxytocin secretion causes uterine muscle contraction. (p. 883; Fig. 26.16)
47. Following delivery, the mammary glands produce milk under the influence of prolactin. Milk is released during nursing by oxytocin, causing mammary gland myoepithelial cells to contract. (p. 884; Fig. 26.17)
48. Prolactin plays a role in immune function in both sexes. (p. 885)

Growth and Aging

49. In girls, puberty begins with **menarche,** the first menstrual period, at age 8–13 years. The age range for the onset of puberty in boys is 9 to 14 years. (p. 885)
50. The cessation of reproductive cycles in women is known as the **menopause.** With increasing age, some men exhibit symptoms of testosterone deficiency. (p. 885)

Questions

Answers: p. A-1

The Physiology Place

Level One Reviewing Facts and Terms

1. Match each of the following items with all the terms it applies to:

(a) X or Y	1. chromosomes other than sex chromosomes
(b) inactivated X chromosome	2. fertilized egg
(c) XX	3. sperm or ova
(d) XY	4. sex chromosomes
(e) XX or XY	5. germ cells
(f) autosomes	6. male chromosomes
	7. female chromosomes
	8. Barr body

2. The Y chromosome contains a region for male sex determination that is known as the ________ gene.
3. List the functions of the gonads. How do the products of gonadal function differ in males and females?
4. Trace the anatomical routes to the external environment followed by a newly formed sperm and by an ovulated egg. Name all structures the gametes pass through on their journey.
5. Define each of the following terms and describe its significance to reproductive physiology:
   (a) aromatase
   (b) blood-testis barrier
   (c) androgen-binding protein
   (d) first polar body
   (e) acrosome
6. Decide whether each of the following statements is true or false, and defend your answer.
   (a) All testosterone is produced in the testes.
   (b) Each sex hormone is produced only by members of one sex.
   (c) Anabolic steroid use appears to be addictive, and withdrawal symptoms include psychological disturbances.
   (d) High levels of estrogen in the late follicular phase help prepare the uterus for menstruation.
   (e) Progesterone is the dominant hormone of the luteal phase of the ovarian cycle.
7. What is semen? What are its main components, and where are they produced?
8. List and give a specific example of the various methods of contraception. Which is/are most effective? Least effective?

Level Two Reviewing Concepts

9. **Concept maps:** Map the following groups of terms. You may add terms.

List 1	List 2
• AMH • DHT • Leydig cells • Müllerian ducts • Sertoli cells • sperm • spermatids • spermatocytes • spermatogonia • SRY • testosterone • Wolffian ducts	• antrum • corpus luteum • endometrium • follicle • granulosa cells • myometrium • ovum • thecal cells

10. Diagram the hormonal control of gametogenesis in males.
11. Diagram the menstrual cycle, distinguishing between the ovarian cycle and the uterine cycle. Include all relevant hormones.
12. Why are X-linked traits exhibited more frequently by males than females?
13. Define and relate each of the following terms in each group:
    (a) gamete, zygote, germ cell, embryo, fetus
    (b) coitus, erection, ejaculation, orgasm, emission, erogenous zones
    (c) capacitation, zona pellucida, acrosomal reaction, cortical reaction, cortical granules
    (d) puberty, menarche, menopause, andropause
14. Compare the actions of each of the following hormones in males and females:
    (a) FSH
    (b) inhibin
    (c) activin
    (d) GnRH
    (e) LH
    (f) DHT
    (g) estrogen
    (h) testosterone
    (i) progesterone
15. Compare and contrast the events of the four phases of sexual intercourse in males and in females.
16. Discuss the roles of each of the following hormones in pregnancy, labor and delivery, and mammary gland development and lactation:
    (a) human chorionic gonadotropin
    (b) luteinizing hormone
    (c) human placental lactogen
    (d) estrogen
    (e) progesterone
    (f) relaxin
    (g) prolactin

Level Three Problem Solving

17. Down syndrome is a chromosomal defect known as "trisomy" (three copies instead of two) of chromosome 21. The extra chromosome usually comes from the mother. Speculate what causes trisomy, using what you have learned about the events surrounding fertilization.
18. Sometimes the follicle fails to rupture at ovulation, even though it appears to have gone through all stages of development. This condition results in benign ovarian cysts, and the unruptured follicles can be palpated as bumps on the surface of the ovary. If the cysts persist, symptoms of this condition often mimic pregnancy, with missed menstrual periods and tender breasts. Explain how these symptoms occur, using diagrams as needed.
19. An XY individual inherits a mutation that results in completely nonfunctional androgen receptors.
    (a) Is this person genetically male or female?
    (b) Will this person have functional ovaries, functional testes, or incompletely developed or nonfunctional gonads?
    (c) Will this person have Wolffian ducts or their derivatives? Müllerian ducts or their derivatives?
    (d) Will this person have the external appearance of a male or a female?
20. The babies of mothers with gestational diabetes mellitus tend to weigh more at birth. They are also at risk of developing hypoglycemia immediately following birth. Use what you have learned about diabetes and insulin to explain these two observations. *Hint:* these babies have normal insulin responses.

Level Four Quantitative Problems

21. The following graph shows the results of an experiment in which normal men were given testosterone over a period of months (indicated by the beige bar from A to E). Control values of hormones were measured prior to the start of the experiment. From time B to time C, the men were also given FSH. From time D to time E, they were also given LH. Based on the information given, answer the following questions.

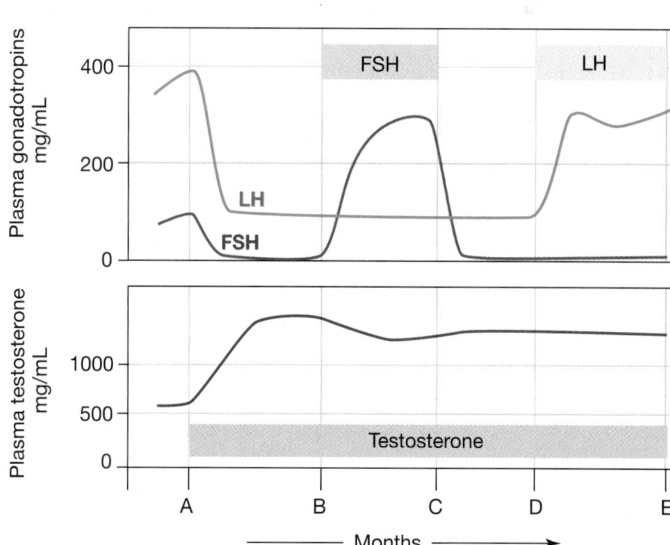

    (a) Why did testosterone level increase beginning at point A?
    (b) Why did LH and FSH levels decrease beginning at point A?
    (c) Predict what happened to the men's sperm production in the A–B interval, the B–C interval, and the D–E interval.

Answers

Answers to Concept Check Questions

Page 852

1. Female gonad: ovary; female gamete: egg, or ovum. Male gonad: testis; male gametes: sperm.

Page 853

2. Primary androgen receptors are in the cytoplasm or nucleus of the target cell. AMH has membrane receptors.
3. The male parent donates the chromosome that determines sex of the zygote; therefore, the wives were not at fault.
4. An XO fetus will be a female because she lacks a Y chromosome.
5. Lack of AMH from the testes allows Müllerian ducts to develop into uterus and fallopian tubes. External genitalia will be female because there is no DHT for development of male genitalia.

Page 858

6. A newborn male's gametes are spermatogonia; a newborn female's gametes are primary oocytes.
7. The first polar body has twice as much DNA as the second polar body.
8. Each primary oocyte forms one egg; each primary spermatocyte forms four sperm.

Page 860

9. Aromatase converts testosterone to estradiol.
10. FSH = follicle-stimulating hormone, DHT = dihydrotestosterone, SRY = sex-determining region of Y chromosome, LH = luteinizing hormone, GnRH = gonadotropin-releasing hormone, AMH = anti-Müllerian hormone.
11. Hypothalamic GnRH, and FSH and LH from the anterior pituitary, control reproduction.

Page 865

12. Sertoli cells secrete inhibin, activin, androgen-binding protein, enzymes, and growth factors. Leydig cells secrete testosterone.
13. The advantage is that GnRH agonists decrease FSH and LH, so the testes stop producing sperm. The disadvantage is that the testes also stop producing testosterone, which causes decreased sex drive.
14. Cholesterol and steroid hormones such as cortisol are examples of lipophilic molecules that bind to protein carriers.

Page 866

15. Exogenous anabolic steroids (androgens) shut down FSH and LH secretion. In response, the testes shrink and stop producing sperm.

Page 874

16. Ovarian cycle: follicular phase, ovulation, and luteal phase. The menses and proliferative phases of the uterine cycle correspond to the follicular phase and ovulation; the secretory uterine phase corresponds to the luteal phase.
17. Women who take anabolic steroids may experience growth of facial and body hair, deepening of the voice, increased libido, and irregular menstrual cycles.
18. A woman given an aromatase inhibitor would have decreased estrogen production.
19. Ovulation occurs about 14 days before the end of the cycle, which would be (a) day 14, (b) day 9, or (c) day 17.

Answers to Figure Questions

Page 862

Figure 26.7: Mitochondria produce ATP to power the flagellum.

Answers to Review Questions

Chapter 1

Level One Reviewing Facts and Terms

1. The normal functioning of a living organism. Anatomy is the study of structure.
2. See Fig. 1.1.
3. See Tbl. 1.1.
4. Physiology integrates body function across all levels of organization and emphasizes the coordinated function of body systems.
5. The maintenance of internal stability. Examples: body temperature and water balance.
6. Homeostasis and control systems; structure-function relationships; biological energy; communication.
7. Stimulus, sensor, input signal, integrating center, output signal, target, response.
8. Circadian rhythm.

Level Two Reviewing Concepts

9. Maps are highly individual. Evaluate your map by comparing it to some done by classmates or to ask your instructor for comments.
10. (a) Tissues—collections of cells that carry out related functions. Organs—collections of tissues that form structural and functional units. (b) x-axis—independent variable; y-axis—dependent variable. (c) Independent variable is manipulated to change the dependent variable. (d) Teleological—functional approach, the "why" of a system. Mechanistic approach—physiological mechanisms, the "how" of a system. (e) Internal environment—extracellular fluid; external environment—the world outside the body. (f) Blind study: subjects do not know the treatment they are receiving. Double-blind study: neither subjects nor experimenters know which treatment is the active one. Crossover study: each subject serves as both control and experimental. (g) Sensors receive signals. Targets respond to signals.
11. Nasal and oral cavities, external ear, lacrimal ducts, sweat, sebaceous, and mammary gland ducts, lumens of esophagus, stomach, small and large intestines, ducts of the salivary glands, pancreas, liver and gall bladder, urinary tract organs, reproductive organs, respiratory organs.
12. Coordinate: endocrine and nervous systems. Protection: integumentary, digestive, cardiovascular, and immune systems. Exchange with external environment: respiratory exchanges gases; digestive system takes in nutrients; digestive and urinary eliminate waste products. Integumentary loses water and solutes.
13. Negative feedback—feedback signal turns response loop off; helps maintain homeostasis. Positive feedback—feedback keeps the response loop going; makes a change bigger. Feedforward—starts response loop before the stimulus does; minimizes change.

Level Three Problem Solving

14. (a) incorrect mechanistic answer (b) correct teleological answer (c) correct teleological answer (d) correct mechanistic answer
15. Other problems: requirement of an aqueous environment for fertilization (internal fertilization in mammals; many other terrestrial animals return to water to breed); aqueous environment for embryonic development (eggs in birds, some reptiles and insects; internal development in mammals, some reptiles, and insects); physical support (exoskeletons in insects, internal skeletons in vertebrates)

Level Four Quantitative Problems

16. (a) independent—time; dependent—body length (b) There was no control. (c) Should be a line graph with time in days on x-axis and body length on y-axis. (d) Growth slowest from days 0–3 and most rapid for days 6–9 and days 18–21.
17. (a) independent solution concentration; dependent volume change (b) The volume measurements before soaking provide a baseline but there is no control. (c) A scatter plot with best-fit line would allow you to estimate volume change at intermediate salt concentrations, such as 5%.
18. (a) scatter plot (b) Is there a relationship between midarm muscle circumference and aerobic fitness? (c) There appears to be no relationship between midarm muscle circumference and aerobic fitness.
19. (a) There is no "correct" answer. For peer critiques of the study, see *New England Journal of Medicine* 347(2):132–33 and 137–39, 2002, July 11. (b) The subjects believed that the surgery had helped (a placebo effect) or other interventions, such as physical therapy, helped. (c) The study is directly applicable to a limited population: male veterans, under age 76, predominantly white, with osteoarthritis or degenerative joint disease. (d) blind study (e) The investigators were trying to determine whether a placebo effect could account for post-surgical improvement.

Chapter 2

Level One Reviewing Facts and Terms

1. Proteins (collagen, hemoglobin, enzymes); carbohydrates (glucose, sucrose); lipids (cholesterol, phospholipids); and nucleic acids (ATP, DNA, RNA).
2. False. All biomolecules are organic molecules, but the reverse is not true.
3. *molecule*
4. One carbon atom needs to share four electrons to fill its outer shell; therefore it will form four covalent bonds.
5. covalent; polar; nonpolar
6. Oxygen and nitrogen strongly attract electrons and tend to form polar bonds.
7. Table sugar dissolves easily, so it is polar. Vegetable oil does not dissolve in water, so it is nonpolar.
8. *anion, cation*
9. pH = H^+ concentration. pH $<$ 7 is *acidic*. pH $>$ 7 is *basic* or *alkaline*.
10. *buffer*
11. *lipoproteins; glycoproteins*
12. *ligand*
13. (a) 4, (b) 3, (c) 2
14. *cofactor*
15. *denatured*

Level Two Reviewing Concepts

16. Check your map with your instructor or your fellow students. Maps will vary.
17. 10^{-3} M = pH 3; acidic. 10^{-10} M = pH 10; basic.
18. ATP: usable energy in a high-energy bond. DNA stores genetic information. RNA translates genetic information into proteins. cAMP: transfer of signals into cells. NAD and FAD transfer energy.
19. Isoforms are structurally similar, with similar functions but differing affinities for ligands. They may function best under different conditions.
20. (a) 4, 5 (b) 3 (c) 2, 1

Level Three Problem Solving

21. Nucleotides contain all of the elements listed in the right ratio. Carbohydrates have a C:H:O ratio of 1:2:1, so alien does not have enough H. Fats are mostly C and H with little O (not enough H and too much O). Proteins do not have P and have less N relative to C.
22. More CO_2 means more H^+ by the law of mass action. More H^+ means a decrease in pH.

Level Four Quantitative Problems

23. 0.9% = 0.9 g/100 mL. Dissolve 9 g NaCl in water to yield 1 L of solution.
24. (a) 6.02×10^{23} molecules of NaCl. (b) 1000 millimoles. (c) 1 equivalent. (d) 5.85% solution.

25. 10% glucose = 10 g/100 mL or 20 g/200 mL solution. Molarity: 10 g/100 mL = 100 g/L × 1 mole/180 g = 0.556 moles/L or 556 millimoles/L (556 mM). 500 mL of 10% glucose would have 50 g glucose × 1 mole/180 g = 278 millimoles glucose.
26. Myoglobin has a higher affinity for O_2 because at lower oxygen concentrations, myoglobin binds more O_2 than hemoglobin does.

Chapter 3

Level One Reviewing Facts and Terms

1. Create a barrier between cell and ECF; regulate exchange of material between cell and ECF; transfer information between the cell and other cells; provide structural support.
2. *phospholipids; proteins; carbohydrates*
3. phospholipids and proteins
4. Inclusions: particles of insoluble material, such as glycogen and ribosomes. Organelles, such as mitochondria and Golgi apparatus, are separated from cytosol by membranes.
5. A flexible, changeable, three-dimensional scaffold of actin, microfilaments, intermediate filaments, and microtubules. Functions: mechanical strength; stabilize position of organelles; transport material; link cells together; movement.
6. (a) 2, (b) 3, (c) 1, (d) 4
7. *serous secretions; mucous secretions*
8. (a) 3, (b) 5, (c) 4, (d) 1, (e) 2
9. very acidic conditions
10. *endocrine*
11. connective tissue (tendons that hold muscles to bones); epithelium (skin); neural tissue (the brain); and muscular tissue (heart and skeletal muscles)
12. *skin*
13. (a) 1, (b) 1, (c) 4, (d) 3, (e) 4, (f) 4, (g) 4, (h) 1, (i) 1
14. sweat glands—sweat; apocrine glands—waxy or milky secretions; sebaceous glands—a mixture of lipids
15. mitochondrial matrix—the internal compartment; tissue matrix—noncellular material found outside cells

Level Two Reviewing Concepts

16. Anchoring junctions (skin)—allow twisting and stretching of tissue. Tight junctions (epithelia)—prevent movement of materials between cells. Gap junctions (some muscles)—allow material to pass directly from cytoplasm of one cell to another.
17. Rough ER is where proteins are made, so pancreatic cells would have more.
18. Vesicles—membranous spheres. Examples: lysosomes, peroxisomes, secretory vesicles.
19. Stratified has many cell layers for protection; simple epithelium only has one layer.
20. Map: See Figure 3.2.
21. See Figure 3.10e. Tight junctions prevent movement of material between cells; leaky junctions allow some material to pass between cells.
22. *intracellular fluid; interstitial fluid; plasma.* Interstitial fluid and plasma are ECF.
23. Cholesterol molecules fill space between phospholipid tails.
24. Bone is rigid due to calcification; cartilage is firm but elastic. Bones are the primary support structure for the body; cartilage forms the ear, nose, larynx, and spine and helps hold bones together at the joints.
25. (a) lumen—hollow inside of an organ or tube; wall—cell layer. (b) cytoplasm—everything inside the cell except the nucleus; cytosol—semi-gelatinous, intracellular fluid. (c) myosin—motor protein filament; keratin—structural protein fiber.
26. Apoptosis—it is a normal part of development.
27. (a) cell junctions: 1 (gap junctions), 2 (tight junction proteins), 4 (strength of desmosomes) (b) cell membrane: 1 (receptors), 2 (enzymes), 3 (barrier), 4 (fluidity), 5 (ATP-dependent transporters) (c) cytoskeleton: 2 (microtubules direct movement), 4 (strength), 5 (ATP required for actin-myosin interaction) (d) organelles: 2 (mRNA binds to ribosomes), 3 (membrane-bounded organelles), 5 (ATP-dependent processes) (e) cilia: 2 (microtubules and dynein), 4 (flexibility), 5 (ATP-dependent movement)
28. The matrix can be dissolved and re-assembled.

Level Three Problem Solving

29. Cilia sweep mucus and particles up and out of the airways. When they fail, inhaled pathogens are more likely to reach the lungs, resulting in infections, inflammation, or cancer. The smoker's cough removes the mucus that would normally be swept away by the cilia.
30. Many epithelia are vulnerable to damage and need to be replaced frequently. Cells undergoing frequent mitosis are more likely to develop abnormal cell division.
31. MMPs are enzymes that dissolve the extracellular matrix, so blocking them might inhibit tissue growth and repair.

Chapter 4

Level One Reviewing Facts and Terms

1. Transport work (moving substances across membranes); chemical work (making proteins); and mechanical work (muscle contraction).
2. Potential energy = stored energy; kinetic energy = energy of motion.
3. First Law: there is a fixed amount of energy in the universe. Second Law: without input of energy, an open system will become progressively less organized.
4. *metabolism*
5. *substrates; rate*
6. *enzymes; decreasing*
7. 1 (d), 2 (a), 3 (f), 4 (c)
8. *-ase*
9. *coenzymes; vitamins*
10. *reduced; oxidized*
11. *dehydration; hydrolysis*
12. *deamination; transamination*
13. *catabolic; anabolic.* Kilocalories.
14. *feedback inhibition*
15. H^+ transported into the inner compartment stores energy in a concentration gradient. When the ions move back across the membrane, the released energy is trapped in the high-energy bond of ATP.
16. NADH and $FADH_2$.

Level Two Reviewing Concepts

17. Map 1: start with Fig. 4.11. Map 2: use Figs. 4.19, 4.20, and 4.21.
18. Perform work, transfer to another molecule, or be released as heat.
19. 1 (b), 2 (a), 3 (b), 4 (a), 5 (c), 6 (c) or (a)
20. When inactive, they cannot harm the cell if accidentally released.
21. Aerobic breakdown = 30–32 ATP; anaerobic breakdown = 2 ATP. Anaerobic is faster and does not require oxygen, but energy yield is lower.
22. Transcription: synthesis of RNA from the sense strand of DNA. Translation: conversion of information coded in mRNA into a string of amino acids.
23. Anticodons are part of tRNA. Amino acids attach to tRNA.
24. Chemical bond energy is potential energy.
25. If the reaction requires ATP, the activation energy must be large compared to a reaction that does not require ATP.

Level Three Problem Solving

26. mRNA: GCGAUGUUCAGUGCAUGGCAUUGC. Amino acids: Methionine (AUG), serine (UCA), cysteine (UGC), histadine (CAU)

Level Four Quantitative Problems

27. Exergonic
28. 149 amino acids

Chapter 5

Level One Reviewing Facts and Terms

1. Structural proteins (link cell to matrix); transporters (water channels); receptors (hormone receptors); and enzymes (intestinal digestive enzymes)

2. Active: requires direct or indirect use of energy. Passive: uses energy stored in a concentration gradient.
3. Passive: simple and facilitated diffusion, osmosis. Active: phagocytosis, exocytosis, and endocytosis.
4. Greater concentration gradient, smaller distance, higher temperature, and smaller molecular size
5. Simple diffusion, protein-mediated transport, or vesicular transport
6. *symport; antiport; uniport*
7. *primary (direct)* and *secondary (indirect)*
8. *penetrating; nonpenetrating*
9. (d), (a), (b), (c).
10. Osmolarity: concentration of osmotically active particles, expressed as osmol/L or milliosmoles per liter
11. Hypotonic: net influx of water into the cell at equilibrium. Hypertonic: net water loss at equilibrium. Tonicity is determined by relative concentrations of nonpenetrating solutes in cell versus solution.
12. (a) 2, 4; (b) 1, 6; (c) 2, 3; (d) 2, 5
13. (1) Like charges repel; opposites attract. (2) Every positive ion has a matching negative ion. (3) Energy must be used to separate ions or electrons and protons. (4) Conductors allow ions to move through them, while insulators keep ions separated.
14. (a) 7; (b) 1, 7; (c) 6; (d) 5; (e) 8; (f) 3; (g) 2
15. *equilibrium potential*
16. *conductor; insulator*

Level Two Reviewing Concepts

17. Use Figs. 5.5, 5.8, 5.10, 5.21, and 5.24 to create your map.
18. See Fig. 5.1c and d.
19. Lipid solubility, so that a molecule can pass through the lipid core of the membrane. Diffusion is slower for larger or heavier molecules and faster when there is more membrane surface area.
20. Specificity: enzyme or transporter works on one molecule or class of molecules. Competition: similar substrates can compete for the protein binding site. Saturation: rate reaches a maximum when all binding sites are filled. GLUT is specific for hexose sugars. If glucose and fructose are both present, they compete for GLUT binding sites. If enough sugar is present, transport saturates.
21. (a) *hypotonic* (b) *into the cells*
22. Active transport. Must use energy to go from a state of equilibrium to one of disequilibrium.
23. (a) *hyperosmotic* (convert molarity to osmolarity). (b) True. Water moves from *B* to *A*.
24. Chemical gradient = concentration gradient. Electrical gradient = separation of electrical charge. Electrochemical gradient includes both concentration and electrical gradients.

Level Three Problem Solving

25. Apical side: Na^+ leak channels but no water pores. Basolateral side: Na^+-K^+-ATPase and K^+ leak channels. May also have water channels.
26. Insulin could increase the number or affinity of GLUT proteins or could act on cell metabolism to keep the intracellular glucose concentration low.
27. Both enzymes and transporters are proteins that bind ligands at a specific binding site. Enzymes alter their substrates. Transporters move substrates unchanged across a membrane.
28. Must convert units from mM to mOsM. (a) hyperosmotic, isotonic; (b) hyposmotic, hypotonic; (c) isosmotic, hypotonic; (d) hyperosmotic, isotonic; (e) hyperosmotic, hypotonic
29. Sugars are added to proteins inside the organelle/vesicle, therefore will face the ECF after being inserted into the membrane.

Level Four Quantitative Problems

30. 296 mOsM
31. (a) ICF = 29.5 L; interstitium = 9.8 L. (b) Total solute = 12.432 osmoles; ECF = 3.7 osmoles; ICF = 8.732 osmoles; plasma = 0.799 osmoles.
32. 154 mOsM
33. (a) increases; (b) decreases; (c) increase; (d) decrease
34. Simple diffusion (a). Cannot be active transport because $\text{concentration}_{in}$ never exceeds $\text{concentration}_{out}$.

Chapter 6

Level One Reviewing Facts and Terms

1. Neurons and blood
2. Nervous and endocrine systems
3. Chemical (available to all cells) and electrical
4. *receptor, targets (effectors)* or *proteins*
5. (a) *adenylyl cyclase,* (b) *guanylyl cyclase,* (c) *phospholipase C*
6. *phosphate, ATP*
7. Central: located within the central nervous system. Peripheral: found outside the CNS
8. *nucleus, cytosol, cell membrane*
9. *decreased*
10. It may down-regulate receptor number or decrease receptor affinity for the substrate.
11. *opposite*

Level Two Reviewing Concepts

12. (a) *Gap junctions* connect two cells using protein channels called *connexons,* made from *connexin* subunits. (b) All are chemicals secreted into the ECF. *Paracrines* act on nearby cells; *autocrines* act on the cell that secretes them. *Cytokines* are peptide autocrine and paracrine signals or *hormones*. *Neurocrines* are chemicals secreted by neurons. (c) *Agonists* have the same action as another molecule; *antagonists* act to oppose the action of another molecule. (d) *Transduction:* a signal molecule transfers information from ECF to the cytoplasm. *Cascade:* a series of steps. *Amplification:* one signal molecule creates a larger signal.
13. Ligand-gated channels (ATP-gated K^+ channel); integrin receptors (platelet receptors); receptor enzymes (tyrosine kinase receptor); G-protein–coupled receptors (adenylyl cyclase/cAMP-linked receptors).
14. The father of American physiology. (1) The nervous system keeps body functions within normal limits. (2) Some functions have tonic control rather than on-off control. (3) Some signals act in opposition to each other. (4) Cell response depends on the cell's receptor for a signal.
15. Input signal: information transmission from stimulus to integrating center (sensory nerve). Integrating center: cell or cells that receive information, decide whether and how it should be acted upon, and send a signal to initiate a response (the brain). Output signal: electrical or chemical signals that take information from integrating center to target (nerve or hormone). Response: what target cell does to react to the stimulus (pull hand away from hot stove). Sensor: cell that perceives the stimulus (temperature receptor). Stimulus: change that begins a response (touching a hot stove). Target: the cell or tissue that carries out the response (muscle).
16. Neural control is faster than endocrine and better for short-acting responses. Endocrine can affect widely separated tissues with a single signal and better for long-acting responses.
17. (a) negative, (b) positive, (c) negative, (d) negative
18. (a) tissues that respond to glucagon, such as liver, (b) breast, (c) bladder, (d) sweat glands
19. (a) pancreatic endocrine cells that secrete glucagon, (b)–(d) nervous system

Level Three Problem Solving

20. (a) stimulus = decrease in body temperature to decrease; sensor = temperature receptors; input = sensory neurons; integrating center = CNS; output = efferent neurons; targets = muscles used to pull up afghan; response = afghan conserves heat (b) stimulus = smell of sticky buns; sensor = odor receptors in the nose; input = sensory neurons; integrating center = CNS; output = efferent neurons; target = skeletal muscles; response = walk to bakery, buy buns, and eat

21. (a) *antagonistic* (b) Neurotransmitters act on nearby cells (paracrine action). Neurohormones act on distant targets. (c) Epinephrine is secreted in larger amounts because it will be diluted by the blood volume before reaching its target.

Level Four Quantitative Problems

22. (a) amplification and a cascade (b) (1000×4000) or 4,000,000 GMP

Chapter 7

Level One Reviewing Facts and Terms

1. *endocrinology*
2. Alter the rate of enzymatic reactions, control transport of molecules into and out of cells, or change gene expression and protein synthesis in target cells.
3. See Figure 7.2.
4. (a) 4, (b) 5, (c) 1, (d) 2, (e) 3
5. (d) , (b) , (c) , (a)
6. *blood; distant target; very low*
7. the time required for half a dose of hormone to disappear from the blood
8. *kidneys* and *liver; urine* and *bile*
9. *factor*
10. Peptides—three or more amino acids; example: insulin. Steroids—derived from cholesterol; example: estrogen. Amino acid–derived—made from single amino acids; example: thyroid hormone
11. (a) peptide (b) peptide (c) steroid (d) peptide (e) peptide (f) steroid (g) peptide (h) all classes (i) steroid (j) steroid
12. Steroid hormones usually initiate new protein synthesis, which takes time. Peptides modify existing proteins.
13. *transcription* factor; *genes; proteins*
14. *cell membrane*
15. *tryptophan; tyrosine*
16. *trophic*
17. *negative feedback*
18. synthesized by and secreted from neurons
19. oxytocin and vasopressin, both peptide neurohormones
20. The portal system is composed of hypothalamic capillaries that take up hormones and deliver them directly to capillaries in the anterior pituitary. The direct connection allows very small amounts of hypothalamic hormone to control the anterior pituitary endocrine cells.
21. See Figure 7.9.
22. Long-loop—hormone from peripheral endocrine gland turns off pituitary and hypothalamic hormone secretion. Short-loop—anterior pituitary hormone turns off hypothalamus.
23. *synergism; permissiveness; antagonistic*

Level Two Reviewing Concepts

24. (a) Paracrines—local; cytokines—local or long distance; hormones—long distance. Cytokines—peptides; hormones—peptides, steroids, or amines. Cytokines—made on demand; peptides—made in advance and stored. (b) Primary pathology arises in the last endocrine gland of the pathway. Secondary pathology arises in a gland secreting a trophic hormone. (c) Hypersecretion—too much hormone; hyposecretion—too little hormone. (d) Both secrete peptide hormones. Anterior pituitary gland—true endocrine gland; posterior pituitary—neural tissue.
25. See Table 7.1.
26. Use Figure. 7.3 for List 1 and Figures. 7.8 and 7.9 for List 2.

Level Three Problem Solving

27. The meanings do not change significantly. Enzymes, hormone receptors, transport proteins, and receptors are all proteins that bind ligands.
28. Patient A—cortisol hypersecretion results from ACTH hypersecretion. When dexamethasone suppresses ACTH secretion, the adrenal gland is no longer stimulated. Cortisol secretion decreases as a result. Patient B—problem in the adrenal gland. His normal negative feedback pathways do not operate, and the adrenal gland continues oversecreting cortisol even though ACTH secretion has been suppressed by dexamethasone.
29. (a) See Fig. 26.11. (b) Both LH and testosterone needed for gamete formation. Testosterone does not directly suppress gamete formation, but it does have a negative feedback effect and shuts off LH secretion. LH is needed for gamete production, so its absence would suppress gamete synthesis.

Level Four Quantitative Problems

30. Half-life is 3 hours.
31. (a) Group A (b) Group B (c) Group A
32. x-axis—plasma glucose; y-axis—insulin secretion. As X increases, Y increases.

Chapter 8

Level One Reviewing Facts and Terms

1. Sensory afferents carry messages from sensory receptors to CNS. Their cell bodies are located close to the CNS. Interneurons are completely contained within the CNS and are often extensively branched. Efferents carry signals from the CNS to effectors. They have short, branched dendrites and long axons.
2. *skeletal muscles; autonomic*
3. *sympathetic* or *parasympathetic*
4. (a) 3, (b) 1, (c) 2, (d) 5, (e) 4
5. neurons and glial cells
6. See Figs. 8.2 and 8.3.
7. (c). Answer (b) is only partly correct because not all axonal transport uses microtubules and not all substances moved will be secreted.
8. (a) 1, 4; (b) 2, 3, 5, 6
9. (a) 3, (b) 1, (c) 4, 6; (d) 2, (e) 5, 6; (f) 5
10. Na^+ channels (voltage-gated along axon; any type of gating on dendrites); voltage-gated K^+ channels along axon; voltage-gated Ca^{2+} channels in axon terminal; chemically gated Cl^- channels
11. (e) – (b) – (d) – (a) – (c)
12. (b) and (d)
13. (a) K^+, Na^+, Na^+, (b) Na^+, (c) K^+, (d) Na^+, (e) K^+
14. Insulating membranes around neurons that prevent current leak
15. larger axon diameter and the presence of myelin
16. enzymatic degradation, reabsorption, and diffusion
17. See Figs. 8.9, 8.10, and 8.12.

Level Two Reviewing Concepts

18. See Figs. 8.1 and 8.5.
19. (d)
20. See Tbl. 8.4.
21. (f) – (c) – (g) – (e) – (b) – (k) – (c) – (a) – (h) – (j) – (i) – (d)
22. (a) depolarize, (b) hyperpolarize, (c) repolarize, (d) depolarize
23. (a) depolarize, (b) hyperpolarize, (c) hyperpolarize, (d) depolarize
24. (a) Threshold signals trigger action potentials. Suprathreshold also trigger action potentials, but subthreshold do not unless summed. Action potentials are all-or-none events. Overshoot—portion of the action potential above 0 mV. Undershoot—after-hyperpolarization portion of the action potential. (b) Graded potentials may be depolarizing or hyperpolarizing. Graded potential in a postsynaptic cell is an EPSP if depolarizing and an IPSP if hyperpolarizing. (c) No stimulus can trigger another action potential during the absolute refractory period, but a suprathreshold stimulus can trigger an action potential during the relative refractory period. (d) See answer to question 1. (e) Sensory are afferents; all others are efferents. (f) Fast synaptic potentials result from neurotransmitters altering ion channel gating, occur rapidly, and are short-lived. Slow synaptic potentials are mediated through second messengers, may involve protein modification, and last longer. (g) Temporal

summation—multiple stimuli arrive at the trigger zone close together in time. Spatial summation—multiple stimuli from different locations arrive simultaneously at the trigger zone. (h) Divergence—a single neuron branches and its collaterals synapse on multiple targets. Convergence—many presynaptic neurons provide input to a smaller number of postsynaptic neurons.

25. Strength is coded by the frequency of action potentials; duration is coded by the duration of a train of repeated action potentials.
26. (b)

Level Three Problem Solving

27. All the necessary synapses have not yet been made between neurons or between neurons and effectors.
28. Inactivation gates also respond to depolarization, but they close more slowly than the activation gates open, allowing ions to flow for a short period of time.
29. (b), (d), and (h)
30. (a) thermal, (b) chemical, (c) chemical, (d) chemical, (e) chemical, (f) mechanical
31. Unmyelinated axons have many ion channels, so more ions cross during an action potential and must be returned to their original compartments by the Na^+-K^+-ATPase, using energy from ATP.

Level Four Quantitative Problems

32. (a) -80 mV (b) $+63$ mV (c) -86 mV (d) -73 mV
33. (a) $(12 \times 2\text{ mV} = 24) + (3 \times -3\text{ mV} = -9) =$ signal strength $= 15$. $V_m = -70 + 15 = -55$. Threshold is -50, so no action potential. (V_m must be equal to or more positive than threshold.) (b) Signal $= +13$. $V_m = -57$. Action potential will fire. (c) Signal $= +19$. $V_m = -51$. No action potential.

Chapter 9

Level One Reviewing Facts and Terms

1. *plasticity*
2. *cognitive; affective*
3. *cerebrum*
4. *skull; vertebral column*
5. From the bones inward: dura mater, arachnoid membrane, pia mater
6. Buoyancy reduces brain's weight; cushion between the brain and bone; chemical protection by creating a closely regulated ECF for brain cells. CSF is secreted by the choroid plexus.
7. (a) H^+ is higher in CSF, (b) Na^+ is the same in CSF and plasma, (c) K^+ is lower in CSF
8. *glucose; hypoglycemia; oxygen*; 15%
9. (a) 5 (b) 7 (c) 9 (d) 3 (e) 1 (f) 2 (g) 6 (h) 8 (i) 4
10. Capillaries that are less leaky due to tight junctions between endothelial cells. Function—regulate substances allowed into brain tissue.
11. Gray matter: nerve cell bodies, dendrites, and axon terminals. Forms nuclei or layers in the brain and spinal cord. Information passes from neuron to neuron. White matter: mostly myelinated axons; tracts carry information up and down the spinal cord.
12. (a) Sensory areas—perception. (b) Motor cortex—movement. (c) Association areas integrate information and direct voluntary behavior.
13. Asymmetrical distribution of function between the two lobes of the cerebrum. Left brain—language and verbal functions; right brain—spatial skills.
14. See Table 9.1. Mnemonic: **O**h **O**nce **O**ne **T**akes **T**he **A**natomy **F**inal, **V**ery **G**ood **V**acations **S**ound **H**eavenly.
15. REM (rapid eye movement) sleep—when most dreaming takes place. Rapid, low-amplitude EEG waves, flaccid paralysis, and depression of homeostatic function. Slow-wave (deep) sleep—high amplitude, low frequency EEG waves and unconscious body movements.
16. Homeostasis of body temperature and osmolarity, reproduction functions, hunger and satiety, and cardiovascular function. Emotional input from the limbic system.
17. *amygdala*
18. Associative and nonassociative. Habituation—a person responds less and less to a repeated stimulus; sensitization—an enhanced response to a dangerous or unpleasant stimulus.
19. Broca's area and Wernicke's area

Level Two Reviewing Concepts

20. Include information from Table 9.1 and Figures 9.3, 9.4, and 9.6.
21. Secreted into the ventricles and flows into the subarachnoid space around the brain and spinal cord before being reabsorbed by the cerebral arachnoid membrane.
22. sensory system, behavioral state system, and cognitive system
23. Wernicke's area—understand language; Broca's area—produce language.
24. (a) Diffuse modulatory—attention, motivation, wakefulness, memory, motor control, mood, and metabolism. Reticular formation—arousal and sleep, muscle tone, breathing, blood pressure, and pain. Reticular activating system—helps maintain consciousness. Limbic system—links higher cognitive functions with more primitive emotions such as fear. (b) Short-term memory—disappears unless consolidated; long-term memory—stored for recall. Long-term includes reflexive, or unconscious, memory, and declarative, or conscious, memory. (c) Nuclei—clusters of nerve cell bodies in the CNS; ganglia—clusters of nerve cell bodies outside the CNS. (d) Tracts—bundles of axons within the CNS. Nerves—bundles of axons outside the CNS. Horns—extensions of spinal cord gray matter that connect to peripheral nerves. Nerve fibers—bundles of axons. Roots—branches of peripheral nerves that enter or exit the spinal cord.
25. Primary somatic sensory cortex—parietal lobe. Visual cortex—processes information from eyes. Auditory cortex—processes information from ears. Olfactory cortex—processes information from nose. Frontal lobe motor cortices—control skeletal muscle movements. Association areas—integrate sensory information into perception.
26. (a) Lower frequency: wave peaks farther apart. (b) Larger amplitude: taller peaks. (c) Higher frequency: peaks closer together.
27. Drives increase arousal, initiate goal-oriented behavior, and coordinate disparate behaviors to achieve goals.
28. new synapses and changes in effectiveness of synaptic transmission

Level Three Problem Solving

29. Expressive aphasia—could understand people but unable to communicate in any way that made sense. Speech centers are in the left brain. If music centers are in the right brain, then perhaps information from Wernicke's area can be integrated by the right brain so that Mr. Anderson can musically string together words so that they make sense.
30. Learning probably occurred, but need not be translated into behavioral responses. The participants who didn't buckle their seat belts learned that wearing seat belts was important but did not consider this knowledge important enough to act on.
31. Sleep-deprived dogs are producing a substance that induces sleep. Controls: putting CSF from normal dogs into sleep-deprived dogs, CSF from normal dogs into normal dogs, and CSF from sleep-deprived dogs into other sleep-deprived dogs.
32. (a) No, other information that should be taken into consideration include genetics, age, and general health. (b) The application of this study would be limited to women of similar age, background, and health. Other factors you might be interested in would include the ethnicity of the participants, factors as listed in (a), and geographical location.
33. Decreasing ECF osmolality by drinking a lot of water causes water to move into cells. The brain is enclosed in the bony cranium and has limited room in which to expand. If pressure inside the skull rises because of brain swelling, seizures will result.

Chapter 10

Level One Reviewing Facts and Terms

1. Carry information from sensory receptors to the CNS.
2. The ability to tell where our body is in space and to sense the relative locations of different body parts.

3. A sensor and a sensory neuron. Could be one cell or two.
4. Mechanoreceptors—pressure, sound, stretch, etc. Chemoreceptors—specific chemicals. Photoreceptors–photons of light. Thermoreceptors—heat and cold.
5. *sensory field*
6. (a) 3, (b) 2, (c) 1, 2, (d) 2, 3, (e) 4
7. *transduction; adequate stimulus; threshold*
8. *Receptor potentials* are graded potentials.
9. Adequate stimulus—form of energy to which a receptor is most sensitive.
10. *cortex*. Exceptions—olfaction and hearing.
11. Sensory neurons surrounding a sensory field are inhibited, which enhances contrast between the stimulus and surrounding areas.
12. Tonic receptors, such as for heat, adapt slowly and respond to stimuli that need to be constantly monitored. Phasic receptors adapt rapidly and stop responding unless the stimulus changes. An example is smell.
13. *referred*
14. Sweet and umami indicate nutritious foods, and bitter may contain toxins. Salty (Na^+) and sour (H^+) ions are related to body osmolarity and pH, respectively.
15. Sound waves per second—*hertz (Hz)*. Loudness—a function of the *wave amplitude* and measured in *decibels (dB)*. Range of hearing: *20–20,000 Hz*. Most acute hearing: *1000–3000 Hz.*
16. Basilar membrane. Spatial coding—association of wave frequencies with different areas of the membrane.
17. (a)
18. Signals from cochlea to *medulla*, with collaterals to *reticular formation* and *cerebellum*. Synapses in *midbrain* and *thalamus* before projecting to *auditory cortex* in the *cerebrum*.
19. *Semicircular canals*—rotation; *otolith organs*—linear forces.
20. (b), (a), (d), (c), (e)
21. *Red, blue,* and *green; cones; color-blindness*
22. Rods and cones (photoreceptors), bipolar cells, ganglion cells, horizontal cells, and amacrine cells. Photoreceptors transduce light energy. Remaining cells carry out signal processing.

Level Two Reviewing Concepts

23. (a) Special senses have receptors localized in the head. Somatic senses have receptors located all over the body. (b) See Figure 10.10. (c) Sharp pain—small, myelinated *Aδ* fibers. Dull pain—small, unmyelinated C fibers. (d) Conductive loss: sound cannot be transmitted through the external or middle ear. Sensorineural loss: inner ear is damaged. Central hearing loss: auditory pathways are damaged. (e) Minimal convergence of retinal neurons in the fovea results in the sharpest vision. Minimal convergence of primary somatic sensory neurons creates smaller receptive fields, and two-point discrimination is better. Regions with more convergence have less acute vision or poor two-point discrimination.
24. Seven distinct areas: 1, 2, 3, 1+2, 1+3, 2+3, and 1+2+3.
25. Ascending pathways for pain go to the limbic system (emotional distress) and hypothalamus (nausea and vomiting).
26. Olfactory receptors—olfactory bulb—secondary sensory neuron—higher-order neurons—olfactory cortex, with parallel pathways to amygdala and hippocampus. G_{olf}—G protein of olfactory receptors.
27. Bitter, sweet, and umami: membrane receptors on type II receptor cells, with different G protein–linked receptors and signal transduction pathways for each ligand. Salt ions (Na^+) apparently enter type 1 support cells through ion channels. H^+ may enter presynaptic cells through channels or bind to a membrane receptor.
28. (a), (g), (j), (h), (c), (e), (i), (b), (f), (d)
29. See Figure 10.22.
30. The lens changes shape due to contraction/relaxation of the ciliary muscles. Loss of this reflex—presbyopia.
31. Presbyopia—loss of accommodation due to stiffening of the lens with age. Myopia or near-sightedness—longer-than-normal distance between lens and retina; hyperopia or far-sightedness—shorter-than-normal distance. Color-blindness—defective cones.
32. Intensity—action potential frequency. Duration—duration of a train of action potentials.
33. See Table 10.1 and the section for each special sense.
34. Start with Figure 10.25 and the basic components of vision. Work in details and related terms from the text.

Level Three Problem Solving

35. Testing touch-pressure, mediated through free nerve endings and Merkel receptors. Feeling only one probe means both needles are within the same receptive field.
36. Walk a straight line, stand on one leg with the eyes closed, count backward by 3s.
37. Test hearing first. If children cannot hear well, they cannot imitate speech.
38. Absence of the consensual reflex upon stimulating the left eye suggests damage to the left retina and/or to the left optic nerve.
39. To dilate: a sympathetic agonist (a) or something that blocks muscarinic receptors (b). To constrict: a cholinergic agonist (c), a nicotinic agonist (e), or an anticholinesterase (d), which prevents breakdown of ACh.
40. Circular muscles form a ring on the inner part of the iris, surrounding the pupil. When these muscles contract, the pupil gets smaller. The radial muscles extend from the outer edge of the iris to the circular muscles. When the radial muscles contract, they pull on the relaxed circular muscles and expand the diameter of the pupil (dilation).
41. Loss of rods explains loss of night vision.

Level Four Quantitative Problems

42. (a) 0.02 m (b) $1/0.02\text{ m} = 1/0.3048 + 1/Q$. $Q = 21.4$ mm, so lens must become rounder to make F smaller.

Chapter 11

Level One Reviewing Facts and Terms

1. Somatic motor—skeletal muscles. Autonomic—smooth and cardiac muscle, glands, some adipose tissue.
2. *Visceral* nervous system because it controls internal organs (viscera) and functions such as heart rate and digestion.
3. Sympathetic and parasympathetic divisions. Sympathetic neurons exit the spinal cord in the thoracic and lumbar regions; ganglia are close to the spinal cord. Parasympathetic exit from the brain stem or sacral region; ganglia on or close to their targets. Sympathetic—fight-or-flight; parasympathetic—rest-and-digest.
4. adrenal medulla
5. *Cholinergic*—acetylcholine; *adrenergic* or *noradrenergic*—norepinephrine.
6. Diffuse away from the synapse, broken down by enzymes in the synapse, taken back into the presynaptic neuron, or bind to a membrane receptor.
7. *monoamine oxidase, MAO*
8. excitatory, single neuron, and synapse with skeletal muscle
9. enzyme that breaks down ACh
10. nicotinic cholinergic receptors

Level Two Reviewing Concepts

11. Divergence allows one signal to affect multiple targets.
12. (a) Neuroeffector junction—distal ends of autonomic axons, anywhere there is a varicosity. Neuromuscular junction—axon terminals of the somatic motor neuron. (b) Alpha and beta adrenergic; nicotinic and muscarinic cholinergic. Nicotinic—on skeletal muscle and postganglionic autonomic neurons. Adrenergic and muscarinic receptors—autonomic targets.
13. Use Figures 11.9 and 11.10 to create this map.

14. (a) Autonomic ganglia—nerve cell bodies of postganglionic autonomic neurons. CNS nuclei—nerve cell bodies in the brain and spinal cord. (b) Both have true endocrine tissue and neuroendocrine tissue. (c) Boutons—ends of axons; varicosities—strung out along the ends of autonomic neurons.
15. (a) 1, 2 (b) 3 (c) 4 (d) 3
16. (d), (e)

Level Three Problem Solving

17. The electrochemical gradient for Na^+ is greater than that for K^+.
18. (a) endocytosis (b) parasympathetic autonomic (c) acetylcholine
19. Skeletal muscles would become paralyzed. Monkey could not flee.

Level Four Quantitative Problems

20. Increased 1991–1997, then began to decrease. Little change from 2003 to 2005. (b) Most likely—white females. Least likely—black females.

Chapter 12

Level One Reviewing Facts and Terms

1. *smooth, cardiac, skeletal.* Skeletal are attached to bones.
2. Cardiac and skeletal muscle
3. Skeletal muscle
4. (a) false (b) true (c) true (d) true
5. Connective tissue, sarcolemma, myofibrils, thick and thin filaments
6. *sarcoplasmic reticulum; Ca^{2+}* ions
7. *action potentials*
8. Actin, myosin, troponin, tropomyosin, titin, and nebulin. Myosin produces the power stroke.
9. Z disk—ends of a sarcomere. I band—Z disk in the middle. A band (thick filaments)—darkest; H zone—lighter region of A band. M line divides A band in half; thick filaments link to each other.
10. They keep actin and myosin in alignment. Titin helps stretched muscles return to resting length.
11. *A band; myosin.* Z disks approach each other.
12. Contraction occurs when thin and thick filaments slide past each other as myosin binds to actin, swivels, and pulls actin toward the center of the sarcomere.
13. Ca^{2+} binds to troponin, which repositions tropomyosin, uncovering actin's myosin-binding sites.
14. Acetylcholine
15. The region of a muscle fiber where the synapse occurs. Contains ACh receptors. Influx of Na^+ through ACh receptor-channels depolarizes muscle.
16. Fast-twitch glycolytic fibers: a, b, e; Fast-twitch oxidative-glycolytic fibers: d, f, g; Slow-twitch oxidative fibers: c, d, f, h
17. *twitch*
18. ATP binding—myosin dissociates from actin. ATP hydrolysis—myosin head swings and binds to a new actin. Release of P_i initiates the power stroke.
19. *motor unit; recruitment*
20. *single-unit* (visceral) and *multi-unit*

Level Two Reviewing Concepts

21. Use Figures 12.3 to 12.6.
22. Action potential activates DHP receptors that open SR Ca^{2+} channels.
23. Generate ATP by energy transfer from phosphocreatine. Oxidative fibers use oxygen to make ATP from glucose and fatty acids; glycolytic fibers get ATP primarily from anaerobic glycolysis.
24. Fatigue—a reversible state in which a muscle can no longer generate or sustain the expected force. May involve changes in ion concentrations, depletion of nutrients, or excitation-contraction coupling. Increase size and number of mitochondria or increase blood supply.
25. The body uses different types of motor units and recruits different numbers of motor units. Small movements use motor units with fewer muscle fibers; gross movements use motor units with more fibers.
26. See Table 12.3.
27. Use Figures 12.8 to 12.10.
28. Stores and releases Ca^{2+} on command. Smooth muscle uses Ca^{2+} from the ECF.
29. (a) Fast-twitch oxidative-glycolytic—smaller, some myoglobin, use both oxidative and glycolytic metabolism, more fatigue-resistant. Fast-twitch glycolytic fibers—largest, rely primarily on anaerobic glycolysis, least fatigue-resistant. Slow-twitch—develop tension more slowly, maintain tension longer, the most fatigue-resistant, depend primarily on oxidative phosphorylation, more mitochondria, greater vascularity, large amounts of myoglobin, smallest in diameter. (b) twitch—a single contraction-relaxation cycle. Tetanus—contraction with little to no relaxation. (c) Both result from inward Na^+ current and outward K^+ current through voltage-gated channels. Motor neuron action potential triggers ACh release. Muscle action potential triggers Ca^{2+} release from the sarcoplasmic reticulum. (d) Motor neuron temporal summation determines whether or not the neuron fires an action potential. Muscle cell summation increases force of contraction. (e) Isotonic contraction moves a load. Isometric contraction creates tension without moving a load. (f) Slow-wave potentials—cycles of depolarization and repolarization in smooth muscle cells. Pacemaker potentials—repetitive depolarizations to threshold in some smooth muscle and cardiac muscle. (g) Skeletal muscle—sarcoplasmic reticulum. Smooth muscle—ECF and sarcoplasmic reticulum.
30. Ca^{2+} release from smooth muscle SR uses RyR and IP_3-activated channels. Influx from ECF uses mechanically, chemically, or voltage-gated channels.

Level Three Problem Solving

31. (a) Adding ATP allows crossbridges to detach. If insufficient Ca^{2+} is available, the muscle will relax. (b) With ATP and Ca^{2+}, the muscle will continue in the contraction cycle until it is completely contracted.
32. Curare must interfere with a process that follows ACh release: diffusion of ACh across the synaptic cleft, ACh binding to receptors, and opening of the receptor-channel. Curare binds to the ACh receptor and stops the channel from opening.
33. Muscle length is related to bone length. Assuming these athletes are lean, differences in weight are correlated with muscle strength, so heavier athletes should have stronger muscles. More important factors are the relative endurance and strength required for a given sport. Any given muscle will have a combination of three fiber types, with the exact ratios depending upon genetics and specific type of athletic training. (a) Basketball: endurance and strength. Leg muscles—fast-twitch glycolytic fibers, to generate strength, and fast-twitch oxidative, for endurance. The arm and shoulder muscles—fast-twitch glycolytic, because shooting requires fast and precise contraction. (b) Steer wrestler: great strength but less endurance. Fast-twitch glycolytic fibers. (c) Figure skaters: strength and endurance. Trunk muscles—slow-twitch oxidative fibers for endurance. Leg muscles—fast-twitch oxidative, for moving across the ice, and fast-twitch glycolytic, for powering jumps. (d) Gymnastics—great strength in arms and legs, and great endurance in trunk and limb muscles. Arm and leg muscles—fast-twitch glycolytic fibers. Limb and trunk muscles—slow-twitch oxidative fibers.

Level Four Quantitative Problems

34. The data suggest lactate accumulation or loss of PCr. Find the original paper at *http://jap.physiology.org*
35. (a) 7.5 kg of force, a 125% increase. (b) an additional 28 kg of force. This is less than if the weight is placed in the hand.

Chapter 13

Level One Reviewing Facts and Terms

1. *stimulus*
2. *skeletal; autonomic*
3. *convergence*
4. *presynaptic inhibition*
5. *visceral* reflexes because many of them involve internal organs (the viscera)

6. Spinal reflexes: urination and defecation. Cranial reflexes: control of heart rate, blood pressure, and body temperature.
7. Limbic system. Emotional reflexes: blushing, heart rate, gastrointestinal function
8. Two neuron-neuron synapses in the spinal cord and the autonomic ganglion, and one neuron-target synapse.
9. Golgi tendon organ, the muscle spindle, and joint capsule mechanoreceptors
10. *tone*
11. *increase*. This reflex prevents damage from overstretching.
12. (a) 2, 3, 5, 6 (b) 1, 2, 6 (c) 1, 2, 4
13. *stretch; contraction; contraction; decreases; alpha motor* neuron
14. Two neurons and one synapse between them (monosynaptic). The knee jerk (patellar tendon) reflex is an example.
15. Reflex movements, such as the knee jerk, can be integrated in the spinal cord. Voluntary movements, such as playing the piano, and rhythmic movements, such as walking, must involve the brain. Reflex movements are involuntary; the initiation, modulation, and termination of rhythmic movements are voluntary.

Level Two Reviewing Concepts

16. Alpha-gamma coactivation allows muscle spindles to continue functioning when the muscle contracts. When the muscle contracts, the ends of the spindles also contract to maintain stretch on the central portion of the spindle.
17. Neurotransmitter release will decrease when M's neurotransmitter hyperpolarizes P.
18. (a) Assessing the components that regulate limb movement, including quadriceps muscle, the nerves that control it, and the area of the spinal cord where the reflex integrates. (b) The reflex would probably be less apparent. The origin of this inhibition is the primary motor cortex. The inhibitory cells will produce IPSPs in the spinal motor neuron. (c) If the brain is distracted by some other task, the inhibitory signals will presumably stop.

Level Three Problem Solving

19. (a) Prevents Ca^{2+}-activated transmitter release. (b) Cell hyperpolarizes and voltage-gated Ca^{2+} channels in terminal will not open. (c) Same as (b).
20. See Figures 13.10 and 13.12. Parts of the brain include the brain stem, cerebellum, basal ganglia, thalamus, cerebral cortex (visual cortex, association areas, motor cortex).
21. (a) Fright activates the sympathetic nervous system fight-or-flight response. (b) Limbic system processes fear. Other functions include regulating drives such as sex, rage, aggression, and hunger, and reflexes including urination, defecation, and blushing. Limbic system influences autonomic motor output. Heart, blood vessels, respiratory muscles, smooth muscle, and glands are some of the target organs involved. (c) Smooth muscles attach to the base of each hair and pull them upright. [See Focus on Skin, p. 91.]
22. Both toxins are produced by bacteria of the Clostridium genus. *Clostridium tetani* enter the body through a cut. *Clostridium botulini* enter the body through ingestion. Both toxins produce skeletal muscle paralysis. Tetanus toxin inhibits secretion of glycine from interneurons that normally inhibit somatic motor neurons. This releases the neurons from inhibition, so they trigger prolonged contractions in skeletal muscles, or spastic paralysis. Botulinum toxin blocks secretion of acetylcholine from somatic motor neurons, so skeletal muscles cannot contract, which is flaccid paralysis.

Chapter 14

Level One Reviewing Facts and Terms

1. (a) first European to describe the closed circulatory system (b) described the relationship between ventricular muscle stretch and force of contraction (c) described capillaries
2. transport of materials entering and leaving the body, defense, and cell-to-cell communication
3. a—e—d—b—f—c—a
4. *pressure; left ventricle; aorta; right atrium; friction*
5. *decreases*
6. *intercalated disks; gap junctions*
7. SA node to internodal pathways to AV node to bundle of His (left and right branches) to Purkinje fibers to ventricular myocardium
8. (a) ESV—volume of blood in ventricle at end of contraction; EDV—volume of blood in the ventricle at beginning of contraction (b) Sympathetic increases heart rate; parasympathetic decreases heart rate. (c) Diastole = relaxation; systole = contraction (d) Pulmonary goes to the lungs; systemic goes to rest of body. (e) SA node is the (atrial) pacemaker; AV node transmits signals from atria to ventricles.
9. (a) 11, (b) 12, (c) 3, (d) 14, (e) 8, (f) 1, (g) 10, (h) 2, (i) 6, (j) 4
10. Vibrations from AV closure cause the "lub" sound and from semilunar valve closure cause the "dup" sound.
11. (a) heart rate (b) end-diastolic volume (c) stroke volume (d) cardiac output (e) blood volume

Level Two Reviewing Concepts

12. (a) Refer to Figure 14.1. (b) Use Figures 14.19 and 14.22 as starting point for a map.
13. See Figures 14.17 and 14.18.
14. See Table 12.3. Cardiac muscle has strong cell-to-cell junctions, gap junctions for electrical conduction, and the modification of some muscle cells into autorhythmic cells.
15. The long refractory period prevents a new action potential until the heart muscle has relaxed.
16. See Figure 14.16. Atrial relaxation and ventricular contraction overlap during the QRS complex.
17. (a) 3; 5 in the last part, (b) 5, (c) 3, (d) 5, (e) 2, (f) 2, (g) 5, (h) 6
18. Heart rate, heart rhythm (regular or irregular), conduction velocity, and the electrical condition of heart tissue. An ECG does not give any direct information about force of contraction.
19. An effect on force of contraction. Norepinephrine and cardiac glycosides

Level Three Problem Solving

20. Calcium channel blockers slow heart rate by blocking Ca^{2+} entry and decrease force of contraction by decreasing Ca^{2+}-induced Ca^{2+} release. Beta blockers decrease effect of norepinephrine and epinephrine, preventing increased heart rate and force of contraction.
21. (a) His heart muscle has been damaged by lack of oxygen and the cells are unable to contract as strongly. Thus, less blood is being pumped out of the ventricle each time the heart contracts. (b) Leads are recording electrodes placed on the surface of the body to measure electrical activity. (c) Leads are effective because electricity is conducted through body fluids to the skin surface.
22. A conduction problem at the AV node or in the ventricular conduction system might cause a long PR interval.
23. Destroying the AV node will prevent rapid atrial signals from being passed to the ventricles. A ventricular pacemaker is implanted so that the ventricles have an electrical signal telling them to contract at an appropriate rate. Rapid atrial depolarization rate is dangerous because if the rate is too fast, only some action potentials will initiate contractions due to the refractory period of muscle. This can cause an arrhythmia.

Level Four Quantitative Problems

24. $SV/EDV = 0.25$. If $SV = 40$ mL, $EDV = 160$ mL. $SV = EDV - ESV$, so $ESV = 120$ mL. $CO = HR \times SV = 4$ L/min.
25. (a) 162.2 cm H_2O (b) 66.6 mm Hg
26. 5200 mL/min or 5.2 L/min
27. 85 mL
28. (a) 1 min (b) 12 sec

Chapter 15

Level One Reviewing Facts and Terms

1. brain and heart
2. (a) 6, 9; (b) 1, 2; (c) 4, 7; (d) 3, 5, 6, 8; (e) 3, 4

3. endothelium (capillary exchange and paracrine secretion); elastic tissue (recoil); smooth muscle (contraction); fibrous connective tissue (resistance to stretch).
4. arterioles
5. *120 mm Hg; systole; diastole; 80 mm Hg; 120/80*
6. *pulse.* Pulse pressure = systolic pressure − diastolic pressure
7. One-way valves in the veins, skeletal muscle pump, and low pressure in the thorax during breathing
8. Elevated blood pressure can cause a weakened blood vessel to rupture and bleed.
9. Korotkoff sounds occur when cuff pressure is lower than systolic pressure but higher than diastolic pressure.
10. See Table 15.2. Sympathetic neurons (α-receptors) vasoconstrict, and epinephrine on β_2-receptors in certain organs vasodilates.
11. A region of increased blood flow. Active—increased blood flow is in response to an increase in metabolism. Reactive—increase in flow follows a period of decreased blood flow.
12. *Sympathetic* innervation causes vasoconstriction.
13. (a) 1, 5; (b) 2, 6; (c) 1, 2, 4; (d) 3, 8; (e) none of the above
14. Digestive tract, liver, kidneys, and skeletal muscles. Kidneys have the highest flow on a per unit weight basis.
15. Capillary density is proportional to the tissue's metabolic rate. Cartilage—lowest; muscles and glands—highest.
16. (a) diffusion (b) diffusion or transcytosis (c) facilitated diffusion (d) osmosis
17. immune, circulatory, and digestive systems
18. Edema is excess fluid in the interstitial space. Causes include lower capillary oncotic pressure due to decreased plasma proteins or blockage of the lymphatic vessels by a tumor or other pathology.
19. (a) blood flow though a tissue (b) the contribution of plasma proteins to the osmotic pressure of the plasma (c) a decrease in blood vessel diameter (d) growth of new blood vessels, especially capillaries, into a tissue (e) small vessels between arterioles and venules that can act as bypass channels (f) cells surrounding the capillary endothelium that regulate capillary leakiness
20. *HDL* and *LDL*. LDL-C is harmful in elevated amounts.

Level Two Reviewing Concepts

21. Use Figure 15.8 as the starting point.
22. (a) Pores of lymphatic capillaries are larger. Lymphatic capillaries have contractile fibers to help fluid flow; systemic capillaries depend on systemic blood pressure for flow. (b) Sympathetic division raises blood pressure by increasing cardiac output and causing vasoconstriction. Parasympathetic division can decrease heart rate. (c) Lymph fluid is similar to blood plasma minus most plasma proteins. Blood also has nearly half its volume occupied by blood cells. (d) Continuous capillaries have smaller pores and regulate the movement of substances better than fenestrated capillaries. Fenestrated can open large gaps to allow proteins and blood cells to pass. (e) Hydrostatic pressure forces fluid out of capillaries; colloid osmotic pressure of plasma proteins draws fluid into capillaries.
23. Preventing Ca^{2+} entry decreases ability of cardiac and smooth muscles to contract. Decreasing Ca^{2+} entry into autorhythmic cells decreases heart rate. Neurons and other cells are unaffected because they have types of calcium channels not affected by the drugs.
24. The ability of vascular smooth muscle to regulate its own contraction. Probably results from Ca^{2+} influx when the muscle is stretched.
25. Left ventricular failure causes blood to pool in the lungs, increasing pulmonary capillary hydrostatic pressure. This may cause pulmonary edema and shortness of breath when oxygen has trouble diffusing into the body. Blood backing up into the systemic circulation increases venous pressure.

Level Three Problem Solving

26. (a) Uncontrollable: male, middle-aged, family history of cardiovascular disease on both sides of his family. Controllable: elevated blood pressure. (b) Yes, because blood pressure >140 or diastolic pressure >90 on several occasions. It would be useful to confirm that this was not "white coat hypertension" by having him take his blood pressure for a week or so at locations away from the doctor's office, such as at a drug store. (c) Beta blockers block β_1-receptors in the heart, thus lowering cardiac output and MAP.
27. (a) MAP increases, flow through vessels 1 and 2 decreases, flow through 3 and 4 increases. (b) Pressure increase → arterial baroreceptor → cardiovascular control center → arteriolar vasodilation and decreased CO → decreased pressure (c) decreases
28. sight of blood → cerebral cortex → CVCC in the medulla oblongata → increased parasympathetic and decreased sympathetic output → decreased heart rate and vasodilation → decreased blood pressure
29. Cells (endothelium) in the intact wall detect changes in oxygen and communicate these changes to the smooth muscle.
30. Atropine is an ACh antagonist, possibly by binding to an ACh receptor.
31. (a) increases (b) resistance increases and pressure increases
32. (a) In Figure 14.1 draw a connection from pulmonary artery to aorta. In Figure 14.5f you can see a remnant of the closed ductus as a small ligament connecting the aorta and pulmonary artery. (b) The lungs are not functioning. (c) Systemic (d) Left side (e) From the aorta into the pulmonary artery

Level Four Quantitative Problems

33. increases 16-fold
34. Answers will vary. For a 50-kg individual with a resting pulse of 70 bpm, she will pump her weight in blood in about 10 minutes.
35. MAP = 87 mm Hg; pulse pressure = 42 mm Hg
36. 250 mL oxygen/min = CO × (200 − 160 mL oxygen/L blood). CO = 6.25 L/min
37. 75 beats/min × 1440 min/day = 108,000 beats/day. 3240 mL filtered/day × day/108,000 beats = 0.03 mL/beat.

Chapter 16

Level One Reviewing Facts and Terms

1. *plasma; water*
2. albumins (most prevalent), globulins, and fibrinogen. Functions: Table 16.1.
3. erythrocytes (transport O_2 and CO_2); leukocytes or white blood cells (defense); platelets (clotting)
4. *hematopoiesis.* Embryo—yolk sac, liver, spleen, and bone marrow. At birth—restricted to the bone marrow. By adulthood—only in axial skeleton and proximal ends of long bones.
5. Colony-stimulating factors stimulate hematopoiesis. Cytokines are released by one cell to act on another cell. Interleukins are cytokines released by leukocytes to act on other leukocytes. All influence growth and differentiation of blood cells. Examples: see Table 16.2.
6. RBC—erythropoiesis; WBC—leukopoiesis; platelets—thrombopoiesis
7. *erythropoietin.* Produced primarily in the kidney in response to low oxygen.
8. Hematocrit—% total blood volume occupied by packed (centrifuged) red cells. Men: 40–54%; women: 37–47%.
9. Erythroblast is an immature, large, nucleated precursor of the erythrocyte. Characteristics: biconcave disk shape, no nucleus, and red color due to hemoglobin.
10. iron
11. (a) yellow color to the skin due to elevated bilirubin (b) low level of hemoglobin (c) plasma protein that acts as a carrier for iron (d) inherited defects of the coagulation cascade, resulting in decreased clotting ability
12. *anticoagulants*

Level Two Reviewing Concepts

13. List 1: see Figures 16.8 and 16.9 and Table 16.5. List 2: see Figure 16.11. List 3: see Figure 16.6.
14. Intrinsic pathway—exposed collagen and other triggers activate factor XII. Extrinsic pathway—damaged tissue exposes tissue factor (III), which activates factor VII. The two pathways unite at the common pathway to initiate the formation of thrombin. See Figure 16.10.

15. Activated platelets cannot stick to undamaged regions of endothelium that release prostacyclin and nitric oxide (NO).

Level Three Problem Solving

16. Rachel is pale and tired because she is anemic. Bruising is a sign that platelet count is low. Proteins and vitamins promote hemoglobin synthesis and the production of new blood cell components. Iron is also necessary for hemoglobin synthesis. Avoid crowds to prevent being exposed to infections because her WBC count and ability to fight infection are decreased. By day 20, all blood counts are back into the low-normal range.
17. (a) transferrin (b) the liver, which stores iron (c) withdraw blood. This illustrates mass balance: if input exceeds output, restore body load by increasing output.
18. Some other factor essential for RBC synthesis, such as iron, folic acid, or vitamin B_{12}, must be lacking.

Level Four Quantitative Problems

19. 200-lb man: blood 6.4 L and plasma about 3.1 L. 130-lb woman: blood 4.1 L and plasma about 2.4 L.
20. Blood volume is 3.5 L, and total erythrocyte volume is 1.4 L.

Chapter 17

Level One Reviewing Facts and Terms

1. gas exchange, vocalization, pH regulation, and protection
2. Cellular respiration—oxygen and nutrients are used for energy production. External respiration—gas exchange between atmosphere and cells.
3. Quiet inspiration—external intercostals, scalenes, and diaphragm. Quiet expiration—no significant muscle contraction. Active expiration—internal intercostals and abdominal muscles. These are all skeletal muscles.
4. Lubrication between lungs and internal thoracic surface
5. Nose and mouth, pharynx, larynx, trachea, main bronchus, secondary bronchi, bronchioles, epithelium of the alveoli, interstitial fluid, and capillary endothelium
6. See Figure 17.2g and h. Type I—gas exchange; type II—surfactant. Macrophages ingest foreign material. Capillary endothelium is almost fused to the alveolar epithelium, and the space between alveoli is almost filled with capillaries.
7. Right ventricle to pulmonary trunk, to left and right pulmonary arteries, smaller arteries, arterioles, capillaries, venules, small veins, pulmonary veins, left atrium. Contains about 0.5 L of blood. Arterial pressure is 25/8, compared with 120/80 for systemic.
8. Warmed, humidified, and cleaned (filtered)
9. *diaphragm*
10. See Figure 17.9.
11. Surfactant decreases surface tension of water and makes it easier for lungs to inflate and stay inflated.
12. radius of the airways
13. (a) 1 (b) 2 (c) 1 (d) 2
14. (a) See Figure 17.7. (b) $V_T = 0.5$ L, IRV = 1.25 L, ERV = 1.0 L. (c) 3 breaths/15 sec × 60 sec/min = 12 br/min

Level Two Reviewing Concepts

15. (a) Compliance—ability to deform in response to force; elastance—ability to resume original shape after deforming force has been removed. (b) Ventilation—air exchange between atmosphere and lungs. Inspiration—air movement into lungs. Expiration—air movement out of lungs. (c) Intrapleural pressure—always subatmospheric (except during forced expiration, when it may become positive); alveolar pressures vary from subatmospheric to above atmospheric. (d) Total pulmonary ventilation—volume of air entering or leaving airways in a given period of time. Alveolar ventilation—volume of air entering or leaving alveoli in a given period of time. (e) Type I—thin cells for gas exchange; Type II—synthesize and secrete surfactant. (f) Pulmonary—from right heart to lung and back to left atrium. Systemic—left heart to most tissues and back to right atrium.
16. Bronchoconstrictors: histamine, leukotrienes, acetylcholine (muscarinic); bronchodilators: carbon dioxide, epinephrine (β_2)
17. See Figs. 17.8 and 17.9.
18. (a) decrease (b) decrease (c) decrease (d) increase (e) decrease (f) increase
19. Pneumothorax—air in the pleural cavity. Spirometer—device used to measure ventilation. Auscultation—listening for body sounds. Hypoventilation—decreased pulmonary ventilation. Bronchoconstriction—decrease in bronchiole radius. Minute volume—total pulmonary ventilation. Partial pressure of gas—portion of total pressure in a mixture of gases that is contributed by a specific gas.
20. (a) *vital capacity*. Sum of tidal volume and expiratory and inspiratory reserve volumes. (b) No, because lung function decreases with age as elastance and compliance diminish.
21. (a) 2, (b) 2, (c) 4, (d) 4
22. *x*-axis—time; *y*-axis—P_{O_2}. During inspiration, the P_{O_2} of the primary bronchi will increase, as fresh air (P_{O_2} = 160 mm Hg) replaces the stale air (P_{O_2} = 100 mm Hg). During expiration, the P_{O_2} will decrease, as the oxygen-depleted air exits the alveoli. The curve will vary from 100 mm Hg to 160 mm Hg.
23. (a) Work increases. (b) Lungs inflate more easily. (c) Elastance decreases. (d) Airway resistance is not affected.
24. (a) decrease (b) increase (c) decrease

Level Three Problem Solving

25. Resting alveolar ventilation = 3575 mL/min. Exercising: (a) 5500 mL/min (b) 5525 mL/min (c) 5625 mL/min. Increasing both rate and depth has the largest effect and is what would happen in real life.
26. (a) 9600 mL/min. (b) Dilating bronchioles reduces airway resistance. The patient is able to force more air out of the lungs on expiration, which increases her ERV and decreases her RV. (c) Respiratory rate is normal, but lung volumes are abnormal. Her high RV is confirmed by the X-ray. In obstructive lung diseases such as asthma, the bronchioles collapse on expiration, trapping air in the lungs and resulting in hyperinflation. Her low IRV accounts for most of the low vital capacity and is to be expected in someone with asthma, where the lungs are already overinflated at the beginning of inspiration. Her higher tidal volume may be the result of the effort she must exert to breathe.

Level Four Quantitative Problems

27. $P_1V_1 = P_2V_2$. New volume = 200 mL
28. (a) O_2 = 160 mm Hg, nitrogen = 593 mm Hg, CO_2 = 2.3 mm Hg. (b) O_2 = 304 mm Hg, nitrogen = 99 mm Hg, CO_2 = 342 mm Hg, H_2 = 15 mm Hg. (c) O_2 = 76 mm Hg, nitrogen = 114 mm Hg, argon = 8 mm Hg, CO_2 = 190 mm Hg.
29. Total pulmonary ventilation = 4800 mL/min. Before an exam, ventilation is 7200 mL/min. Alveolar ventilation is 3360 mL/min (at rest) and 5040 mL/min (before exam).
30. Tidal volume = 417 mL/breath. IRV = 3383 mL
31. Lung volume is 1.1 L. (Did you forget to subtract the volume of the spirometer?)
32. (b). The lung in A has the highest compliance.

Chapter 18

Level One Reviewing Facts and Terms

1. Pressure gradients, solubility in water, alveolar capillary perfusion, blood pH, temperature.
2. 98%. Remainder is dissolved in plasma.
3. P_{O_2}, temperature, pH, and the amount of hemoglobin available for binding (most important).
4. Four globular protein chains, each wrapped around a central heme group with iron.

5. *medulla* and *pons*. Dorsal—neurons for inspiration; ventral—neurons for inspiration and active expiration. Central pattern generator—group of neurons that interact spontaneously to control rhythmic contraction of certain muscle groups.
6. Medullary chemoreceptors increase ventilation when P_{CO_2} increases. Carotid and aortic chemoreceptors respond to P_{CO_2}, pH, and $P_{O_2} < 60$ mm Hg. P_{CO_2} is most important.
7. They include irritant-mediated bronchoconstriction and the cough reflex.
8. Partial pressure gradients
9. Decreased atmospheric P_{O_2}, decreased alveolar perfusion, loss of hemoglobin, increased thickness of respiratory membrane, decreased respiratory surface area, increased diffusion distance.

Level Two Reviewing Concepts

10. Start with Figure 18.10.
11. Most oxygen is bound to hemoglobin, not dissolved in the plasma.
12. (a) Most O_2 is transported bound to hemoglobin, but most CO_2 is converted to bicarbonate. (b) Concentration is amount of gas per volume of solution, measured in units such as moles per liter. While solution partial pressure and concentration are proportional, concentration is affected by the gas solubility, and therefore is not the same as partial pressure.
13. decrease
14. Hypoxia—low oxygen inside cells. COPD—chronic obstructive pulmonary disease (includes chronic bronchitis and emphysema). Hypercapnia—elevated CO_2.
15. Oxygen is not very soluble in water, and the metabolic requirement for oxygen in most multicellular animals would not be met without an oxygen-transport molecule.
16. (a) *x*-axis—ventilation in L/min; *y*-axis—arterial P_{O_2}, in mm Hg. See Figure 18.9. (b) *x*-axis—arterial P_{CO_2} in mm Hg; *y*-axis—ventilation in L/min. As arterial P_{CO_2} increases, ventilation increases. There is a maximum ventilation rate, and the slope of the curve decreases as it approaches this maximum.
17. (a) increases (b) increases
18. Normal, because P_{O_2} depends on the P_{O_2} of the alveoli, not on how much Hb is available for oxygen transport.
19. (a) See Figure 18.17. (b) See Figure 18.13.

Level Three Problem Solving

20. Increased dead space decreases alveolar ventilation. (a) increases (b) decreases (c) increases (d) decreases
21. Person (a) has slightly reduced dissolved O_2 but at $P_{O_2} = 80$, Hb saturation is still about 95%. If oxygen content is 197 mL O_2/L at $P_{O_2} = 100$ and 98% saturation, then oxygen content at $P_{O_2} = 80$ mm Hg and 95% saturation is 190 mL O_2/L blood $(197 \times (0.95/0.98))$, with Hb constant. Person (b) has reduced hemoglobin of 12 g/dL, but it is still 98% saturated. So, oxygen content would be 157.6 mL O_2/L blood $(197 \times (12/15))$. The increased P_{O_2} did not compensate for the decreased hemoglobin content.
22. (a) decrease (b) decrease (c) decrease
23. (a) Respiratory movements originate above the level of the cut, which could include any area of the brain. (b) Ventilation depends upon signals from the medulla and/or pons. (c) Respiratory rhythm is controlled by the medulla alone, but other important aspects of respiration depend upon signals originating in the pons or higher.
24. With chronic elevated P_{CO_2}, the chemoreceptor response adapts, and CO_2 is no longer a chemical drive for ventilation. The primary chemical signal for ventilation becomes low oxygen (below 60 mm Hg). Thus, when the patient is given O_2, there is no chemical drive for ventilation, and the patient stops breathing.
25. (a) Alveoli—96%; exercising cell—23% (b) At rest Bzork only uses about 20% of the oxygen that his hemoglobin can carry. With exercise, his oxygen consumption increases, and his hemoglobin releases more than 3/4 of the oxygen it can carry.
26. All three lines show that as P_{CO_2} increases, ventilation increases. Line A shows that a decrease in P_{CO_2} potentiates this increase in ventilation (when compared to line B). Line C shows that ingestion of alcohol lessens the effect of increasing P_{CO_2} on ventilation. Because alcohol is a CNS-depressant, we can hypothesize that the pathway that links increased P_{CO_2} and increased ventilation is integrated in the CNS.
27. Apical—faces airspace; basolateral—faces interstitial fluid. Apical side has ENaC and aquaporins; basolateral side has aquaporins and Na^+-K^+-ATPase. Na^+ enters the cell through ENaC, then is pumped out the basolateral side. (Cl^- follows to maintain electrical neutrality.) Translocation of NaCl allows water to follow by osmosis.

Level Four Quantitative Problems

28. 1.65 mL O_2/gm Hb
29. 247.5 mL O_2/min
30. Nothing. The percent saturation of Hb is unchanged at any given P_{O_2}. However, with less Hb available, less oxygen will be transported.

Chapter 19

Level One Reviewing Facts and Terms

1. Color (concentration), odor (infection or excreted substances), clarity (presence of cells), taste (presence of glucose), and froth (presence of proteins)
2. Regulation of extracellular fluid volume (to maintain adequate blood pressure), regulation of osmolarity, maintenance of ion balance (neuron function), regulation of pH (proteins denature if pH not maintained), excretion of wastes and foreign substances (to prevent toxic effects), and production of hormones (that regulate RBC synthesis, Ca^{2+} and Na^+ balance).
3. 20–25%
4. Nephrons through ureters to urinary bladder (storage), leaving through the urethra
5. (a), (e), (b), (g), (f), (d), (c), (h)
6. Glomerular capillary endothelium, basal lamina, and epithelium of Bowman's capsule. Blood cells and most plasma proteins are excluded.
7. Capillary hydrostatic pressure promotes filtration. Fluid pressure in Bowman's capsule and colloid osmotic (oncotic) pressure of plasma oppose it. Net driving force is the sum of these pressures.
8. GFR—glomerular filtration rate. 125 mL/min or 180 L/day.
9. (a) Found where distal tubule passes between afferent and efferent arterioles. Composed of macula densa cells in the distal tubule and granular cells in arteriole wall. (b) Macula densa paracrine signals control autoregulation of GFR and renin secretion. (c) Alter the size of filtration slits. (d) Specialized epithelial cells that surround glomerular capillaries. Changes in slit size alter GFR. (e) An internal smooth muscle sphincter that is passively contracted and an external skeletal muscle sphincter that is tonically (actively) contracted. (f) Outer layer of the kidney that contains renal corpuscles, proximal and distal tubules, and parts of the loop of Henle and collecting ducts.
10. 70% occurs in the proximal tubule. Reabsorbed molecules go into the peritubular capillaries and the systemic venous circulation. If filtered and not reabsorbed, a molecule is excreted in the urine.
11. (a) 2, 3, 5; (b) 3, 4; (c) 4, 7; (d) 6; (e) 5, 7
12. penicillin, K^+, and H^+
13. creatinine
14. urination

Level Two Reviewing Concepts

15. Use Figures 19.5, 19.6, and 19.7.
16. (a) Filtration and secretion both move material from blood to tubule lumen, but filtration is a bulk flow process while secretion is a selective process. Excretion is also bulk flow but involves movement from the kidney lumen to the outside world. (b) Saturation—all transporter binding sites are occupied by ligand. Transport maximum—the maximum rate at which carriers are saturated by substrate. Renal threshold—plasma concentration at which saturation occurs. (c) Creatinine and inulin—compounds used to determine GFR. Penicillin and probenecid—xenobiotics that are secreted. (d) Clearance—rate at which plasma is cleared of a substance (mL plasma cleared of substance X/min). GFR—filtration rate of plasma (mL plasma filtered/min). Excretion—removal of urine, mL urine/min.

17. Allows rapid removal of foreign substances that are filtered but not reabsorbed.
18. Afferent arteriole constricts, GFR decreases. Efferent arteriole constricts, GFR increases.
19. See Figure 19.14. Toilet training allows higher brain centers to inhibit the reflex until an appropriate time. Higher brain centers can also initiate the reflex.
20. Bladder smooth muscle contracts under parasympathetic control, so blocking muscarinic receptors decreases bladder contraction.

Level Three Problem Solving

21. (a) Inulin is filtered, secreted, and excreted. No evidence for reabsorption is presented. (b) The line indicating net secretion will be close to the filtration line until the slope changes, after which the secretion line is horizontal (no further increase in rate due to saturation).
22. See Figure 19.8. Place transporters as described. Cl^- moves between the cells.
23. Dialysis fluid should resemble plasma without waste substances, such as urea. This will allow diffusion of solutes and water from the blood into the dialysis fluid, but diffusion will stop at the desired concentration. To remove excess water from the blood, you can make the dialysis fluid more concentrated than plasma.

Level Four Quantitative Problems

24. 1 L/min
25. First specimen clearance = 1000 L plasma/day. Normally creatinine clearance = GFR. However, this value is not at all realistic for GFR (normal average is 180 L/day). The repeat test has 4000 mg of creatinine and gives a clearance of 200 L/day, which is within normal limits. The abnormal values on the first test were probably a laboratory error. Dwight's kidney function is normal.
26. For any solute that filters: plasma concentration $\times$ GFR = filtration rate. At the transport maximum: filtration rate = reabsorption rate of T_m. By substitution: plasma concentration $\times$ GFR = T_m. The renal threshold represents the plasma concentration at which the transporters are working at their maximum (T_m). By substitution: renal threshold $\times$ GFR = T_m. Mermaid's GFR is 250 mL/min and T_m is 50 mg/min, so renal threshold is 0.2 mg glucose/mL plasma. Clearance = excretion rate/plasma concentration. At 15 mg glucose/mL plasma, 3750 mg/min filter and 50 mg/min reabsorb, so 3700 mg/min are excreted.
27. (a) 140 L/day is 20% of renal plasma flow (RBF), so plasma flow is 700 L/day. (b) Hematocrit is percent of blood occupied by packed red blood cells; the remainder (70%) is plasma. 700 L/day is 70% of RBF, so RBF is 1000 L/day. (c) If RBF is 20% of cardiac output (CO), then CO = 5000 L/day or 3.47 L/min.

Chapter 20

Level One Reviewing Facts and Terms

1. Electrolytes are ions, which can conduct electric current through a solution. Examples: Na^+, K^+, Ca^{2+}, H^+, HPO_4^{2-} and HCO_3^-.
2. Organs: kidneys, lungs, heart, blood vessels, digestive tract. Hormones: vasopressin (ADH), aldosterone, atrial natriuretic peptides (ANP), RAS pathway.
3. Entry: ingested and a small amount from metabolism. Loss: exhaled air, evaporation and perspiration from skin, excreted by kidneys, and in feces.
4. See Table 20.1 and Figure 20.15.
5. Descending limb: permeable to water but lacks transporters for salts. Ascending limb: impermeable to water but reabsorbs NaCl.
6. ECF volume—Na^+; pH—H^+
7. More K^+ leaves the cell, and membrane potential becomes more negative (hyperpolarizes). The heart is most likely to be affected.
8. Salt and water
9. ADH = antidiuretic hormone; ANP = atrial natriuretic peptide; ACE = andiotensin-converting enzyme; ANG II = angiotensin II; JG (apparatus) = juxtaglomerular; P cell = principal cell; I cell = intercalated cell.
10. Use Figures 19.8, 19.12, 20.5b, , 20.7d, 20.9, and 20.17.
11. pH alters protein structure (enzyme activity, membrane transporters, neural function). Buffers, renal and respiratory compensation.
12. Acids from CO_2, metabolism, and food are more likely. Sources of bases include some foods.
13. A molecule that moderates changes in pH. Intracellular: proteins, HPO_4^{2-}, and hemoglobin. Extracellular: HCO_3^-.
14. Kidneys excrete or reabsorb H^+ or HCO_3^-. Ammonia and phosphates.
15. $CO_2 + H_2O \rightleftharpoons H_2CO_3 \rightleftharpoons H^+ + HCO_3^-$. Carbonic anhydrase. High in renal tubule cells and RBCs.
16. Arterial P_{CO_2} decreases, pH increases, and plasma H^+ concentration decreases.

Level Two Reviewing Concepts

17. Use the information in Table 20.1 and compile multiple pathways into a single map similar to Figure 20.13. Include all steps of the reflex.
18. Combine information from Figures 20.15 and 20.17c.
19. See Figure 20.7.
20. See Figure 20.6.
21. (a) ANP—peptide from atrial myocardial cells. Causes Na^+ and water excretion; inhibits ADH secretion. (b) Aldosterone—steroid from adrenal cortex. Increases distal nephron Na^+ reabsorption and K^+ excretion. (c) Renin—enzyme from JG cells. Converts plasma angiotensinogen to ANG I. (d) ANG II—peptide hormone made from ANG I. Increases blood pressure by actions on arterioles, brain, and adrenal cortex. (e) Vasopressin—hypothalamic peptide. Increases distal nephron water reabsorption. (f) ACE—enzyme on vascular endothelium. Converts ANG I to ANG II.
22. Vasoconstriction, increased cardiac output, water conservation by kidneys, and increased thirst. If blood pressure falls too low, oxygen supply to the brain will decrease, resulting in damage or death.
23. (a) Both are in the distal nephron. P cells are associated with aldosterone-mediated Na^+ reabsorption; I cells are involved with acid-base regulation. (b) All are parts of the RAS system. Renin and ACE—enzymes; ANG II and aldosterone—hormones. See Figure 20.10. (c) In both, body pH falls below 7.38. Respiratory—results from CO_2 retention (from any number of causes); metabolic—results from excessive production of metabolic acids. Respiratory compensation—renal H^+ excretion and HCO_3^- retention. Metabolic compensation—increased ventilation, renal H^+ excretion, and HCO_3^- retention. Respiratory—arterial P_{CO_2} is elevated; metabolic—P_{CO_2} usually decreased. (d) Proximal tubule—not regulated; distal nephron—regulated by vasopressin. Ascending limb—impermeable to water. (e) Both—pH goes above 7.42. Metabolic—may be caused by excessive ingestion of bicarbonate-containing antacids or vomiting; respiratory—hyperventilation. Metabolic compensation—decrease ventilation, decreased renal H^+ excretion, increased HCO_3^- excretion. Respiratory compensation—decreased renal H^+ excretion, increased HCO_3^- excretion.
24. The cells concentrate organic solutes to increase their internal osmolarity.

Level Three Problem Solving

25. (a) Acute respiratory acidosis (b) Chronic respiratory acidosis (c) Renal compensation has increased his pH by H^+ excretion and HCO_3^- reabsorption. His P_{CO_2} is elevated because of his emphysema.
26. These drugs decrease ADH-mediated water reabsorption. Useful in people who secrete too much vasopressin (SIADH, or syndrome of inappropriate ADH secretion) or in hyponatremia, such as the woman in this chapter's Running Problem.
27. (a) Metabolic alkalosis, partially compensated. (b) After vomiting acid (H^+), her body was left with HCO_3^-. (c) Hypoventilation increases P_{CO_2}, HCO_3^-, and H^+. Increased H^+ decreases pH (compensation). Hypoventilation also decreases arterial P_{O_2} and decreases the total oxygen content of blood (see Fig. 17.13).
28. Blood pressure is high, plasma Na^+ and osmolarity are low. Use Table 20.1 to select reflex pathways for the map.

Level Four Quantitative Problems

29. (a) $pH = 6.1 + \log[24/(0.03 \times 40)] = 7.40$ (b) 7.34
30. 428.6 mL. $(600 \text{ mosmol}/? \text{ L} = 1400 \text{ mosmol/L})$
31. (a) 400 mg glucose/100 mL × 130 mL/min = 520 mg glucose/min filters. (b) Can reabsorb up to T_m so 400 mg/min reabsorbed. (c) Excreted = filtered − reabsorbed = 120 mg/min × 1440 min/day = 172.8 g/day excreted. (d) Convert grams to milliosmoles: 172.8 g × mole/180 g × 1000 mosmol/mole = 960 mosmol glucose excreted/day. Concentration = amount/volume. 1200 mosmol/L = 960 mosmol/? liters. Will require 0.8 L additional volume.

Chapter 21

Level One Reviewing Facts and Terms

1. Digestion—chemical or mechanical breakdown of nutrients (proteins). Absorption—transport from lumen to ECF (water). Secretion—transport from ECF to lumen (enzymes). Motility—movement of material through the digestive tract.
2. *absorption* and *digestion; secretion* and *motility*. By not regulating absorption and digestion, the body ensures that it will always absorb the maximum available nutrients.
3. (a) 2; (b) 3; (c) 4; (d) 7, 10; (e) 8; (f) 2, 3, 7; (g) 9
4. Layers (lumen outward): mucosa (epithelium, connective tissue, and smooth muscle), submucosa (connective tissue), musculature (smooth muscle), serosa (connective tissue).
5. Secretory epithelium (endocrine and exocrine) lines the stomach; absorptive epithelium with a few secretory cells lines the intestines.
6. Peyer's patches—nodes of lymphoid tissue. M cells—epithelial cells that transfer information from gut lumen to Peyer's patches.
7. Moves food through the GI tract and helps mix food with secretions. Results from contraction of longitudinal and circular muscle layers to create propulsive peristaltic movements or mixing segmental movements.
8. An inactive digestive proenzyme. Must have a segment of protein chain removed to activate. Examples: pepsinogen-pepsin, trypsinogen-trypsin.
9. (a) 8, 9; (b) 3; (c) 1, 3, 7; (d) 1, 7; (e) 6; (f) 2; (g) 4; (h) 5
10. (a) Increases surface area for enzymes to work; stomach and small intestine. (b) Motility and secretion along the length of the digestive tract. (c) Acidic pH in stomach helps break down food and digest microorganisms. (d) Size determines the surface area upon which enzymes can act.
11. *capillaries; hepatic portal system; liver; lymphatic; basement membrane (basal lamina)*
12. ENS: network of neurons within the GI tract that can sense a stimulus, integrate information, and create an appropriate response without integration or input from the CNS. Also interacts with the CNS through sensory and autonomic neurons.
13. Short reflexes—mediated entirely within the ENS; regulate secretion and motility. Long reflexes—GI reflexes integrated in the CNS.
14. Paracrines help mediate secretion and motility. Examples: serotonin (5-HT) and histamine.

Level Two Reviewing Concepts

15. Map 1: use Figures 21.8 and 21.9 and add details. Map 2: Use Figure 21.10.
16. (a) Mastication—chewing; deglutition—swallowing. (b) Villi—folds of intestine; microvilli—folds of cell membrane. Both increase surface area. (c) All patterns of GI muscle contraction. Migrating motor complex—move material from stomach to large intestine between meals. Peristalsis—progressive waves of contraction. Segmental contraction—contraction and relaxation of short intestinal segments. Mass movements—push material into rectum, triggering defecation. Vomiting—forceful expulsion of GI contents from the mouth. Diarrhea—excessive amounts of watery stool. (d) Chyme—semidigested food and secretions, produced in the stomach. Feces—solid waste material that remains after digestion and absorption are complete; produced in the large intestine. (e) Short reflexes—integrated within the ENS. Long reflexes—integrated within the CNS. (f) Submucosal plexus—ENS in the submucosal layer. Myenteric plexus—ENS that lies between muscle layers of the GI tract wall. Vagus nerve—carries sensory and efferent signals between the brain and ENS. (g) Cephalic phase—digestive reflexes triggered by stimuli received in the brain. Gastric phase—short reflexes that begin with food entering the stomach. Intestinal phase—begins when chyme enters the small intestine.
17. (a) See Figures 21.10 and 21.8. (b) See Figure 21.5.
18. Both use similar neurotransmitters and neuromodulators (serotonin, VIP, NO). Enteric support cells similar to CNS astroglia. GI capillaries not very permeable, like blood-brain barrier. Both act as integrating centers.
19. See Table 21.1 for specific hormones.
20. See Figures 21.5a and 21.16.

Level Three Problem Solving

21. Hepcidin causes enterocytes to destroy ferroportin transporters. If hepcidin is absent or not functional, intestinal iron uptake cannot be down-regulated when iron levels become too high, so these patients have elevated plasma levels of iron.
22. Severe diarrhea → loss of small intestine HCO_3^- → metabolic acidosis
23. (a) Ingestion of a fatty meal triggers contraction of the gallbladder to release bile salts, but the blocked bile duct prevented bile secretion, causing pain. (b) Micelle formation—decreased due to lack of bile salts. Carbohydrate digestion—decreased because pancreatic secretions with amylase not able to pass blockage. Protein absorption—decreased slightly because of low pancreatic secretion; however, brush border enzymes also digest protein, so digestion does not stop completely when the bile duct is blocked. Therefore, some digested proteins will be absorbed.
24. Apical membrane has Na^+ and K^+ leak channels. Basolateral membrane has the Na^+-K^+-ATPase. At high flow, saliva has more Na^+ and less K^+.

Level Four Quantitative Problems

25. (a) MIT started out with equal concentrations in both solutions, but by the end of the experiment MIT was more concentrated on the serosal side. Therefore, MIT must be moving by active transport. (b) MIT moves apical to basolateral, which is absorption. (c) Transport across the apical membrane goes from bath into tissue. Tissue MIT is more concentrated than bath. Therefore, this must be active transport. (d) Transport across the basolateral membrane goes from tissue into the sac. Tissue MIT is more concentrated than sac fluid, so this must be passive transport.

Chapter 22

Level One Reviewing Facts and Terms

1. Metabolic—all pathways for synthesis or energy production, use, or storage. Anabolic—primarily synthetic; catabolic—break down large molecules into smaller ones.
2. Transport (moving molecules across membranes), mechanical work (movement of muscles), chemical work (protein synthesis).
3. The amount of heat required to raise the temperature of 1 L water by 1°C. In direct calorimetry, food is burned to see how much energy it contains.
4. The ratio of CO_2 produced to O_2 used in cellular metabolism. Typical RQ is 0.82.
5. BMR—an individual's lowest metabolic rate, measured at rest after sleep and a 12-hour fast. Higher in adult males because females have more adipose tissue with a lower respiration rate. Factors that affect BMR: age, physical activity, lean muscle mass, diet, hormones, and genetics.
6. Broken down for energy, used for synthesis, or stored.
7. Absorptive state—anabolic reactions and nutrient storage. Postabsorptive—mobilizes stored nutrients for energy and synthesis.
8. A group of nutrients (glucose, free fatty acids, and amino acids), mostly in the blood, available for cell use.
9. To maintain adequate glucose supply for the brain.
10. Glycogen and adipose tissue fat
11. Proteins: protein synthesis, energy, and conversion to fat for storage. Fats: lipid synthesis, energy, and storage as fats.
12. Insulin decreases blood glucose and glucagon increases it.

13. Amino acids and glycerol. Gluconeogenesis.
14. Excessive breakdown of fatty acids, as occurs in starvation. Can be burned as fuel by neurons and other tissues. Many ketone bodies are strong acids and can cause metabolic acidosis.
15. Increased plasma glucose or amino acids and parasympathetic input stimulate, sympathetic input inhibits.
16. Type 1: absolute lack of insulin. Type 2: cells do not respond normally to insulin. Both: elevated fasting blood glucose levels. Type 1: body uses fats and proteins for fuel. Type 2: not as severe because the cells can use some glucose.
17. Low plasma glucose or increased plasma amino acids stimulate. Primary target—liver, which increases glycogenolysis and gluconeogenesis.
18. (a) Capillary endothelium enzyme that converts triglycerides into free fatty acids and monoglycerides. (b) Co-secreted with insulin; slows gastric emptying and gastric acid secretion. (c) "hunger hormone" secreted by the stomach. (d) Hypothalamic peptide that increases food intake. (e) protein components of lipoproteins. Apoprotein B on LDL-C facilitates transport into most cells. (f) "Satiety hormone" produced by adipocytes. (g) Loss of water in the urine due to high amounts of urine solutes. Hyperglycemia causes dehydration through osmotic diuresis. (h) Target cells fail to respond normally to insulin.
19. (a) stimulates (b) inhibits (c) stimulates (d) stimulates (e) stimulates

Level Two Reviewing Concepts

20. Use Figures 22.5 and 22.8. Try different colors for each organ or hormone.
21. Glucagon and insulin cycle according to food intake, but both hormones are always present in some amount. So it appears that the ratio rather than an absolute amount of hormone determines the direction of metabolism.
22. (a) Glucose—monosaccharide. Glycogenolysis—glycogen breakdown. Glycogenesis—glycogen production from glucose. Gluconeogenesis—glucose synthesis from amino acids and fats. Glucagon—hormone that increases plasma glucose. Glycolysis—first pathway in glucose metabolism for ATP production. (b) Thermogenesis—heat production by cells. Shivering thermogenesis—muscle twitches produce heat as a by-product. Nonshivering thermogenesis occurs in all cells. Diet-induced thermogenesis—heat generated by digestive and anabolic reactions during the absorptive state. (c) Lipoproteins—transport molecules. Chylomicrons—lipoprotein complexes assembled in intestinal epithelium and absorbed into lymphatic system. Cholesterol—steroid component of cell membranes and precursor to steroid hormones. HDL-C—takes cholesterol into liver cells, where it is metabolized or excreted. LDL-C—elevated concentrations are associated with atherosclerosis. Apoprotein—protein component of lipoproteins. (d) Calorimetry—measurement of energy content and a means of determining metabolic rate. Direct calorimetry—measuring heat production when food is burned. Indirect calorimetry—measures oxygen consumption or CO_2 production. (e) Conductive heat loss—loss of body heat to a cooler object. Radiant heat loss—loss from production of infrared electromagnetic waves. Convective heat loss—upward movement of warm air and its replacement by cooler air. Evaporative heat loss—heat lost when water evaporates. (f) Absorptive state—following a meal, when anabolism exceeds catabolism. Postabsorptive state—catabolism exceeds anabolism.
23. (a) Hyperglycemia results from the lack of insulin production and failure of cells to take up and use glucose. (b) Glucosuria results when filtered glucose exceeds the kidney's capacity to reabsorb it. (c) Polyuria results from osmotic diuresis caused by glucosuria. (d) Ketosis results from increased fatty acid metabolism. (e) Dehydration is a consequence of polyuria due to osmotic diuresis. (f) Severe thirst is a consequence of dehydration.
24. If a person ingests a pure protein meal and only insulin is released, blood glucose concentrations might fall too low. Glucagon co-secretion ensures that blood glucose remains within normal levels.
25. See Figure 22.1. The satiety center inhibits the feeding center.
26. See Figures 22.22 and 22.23.

Level Three Problem Solving

27. Amino acids in excess of what's needed for protein synthesis are stored as glycogen or fat.
28. As insulin secretion (x-axis) increases, plasma glucose (y-axis) decreases.
29. (a) See Figure 18.9 on p. 611. Acidosis shifts the curve to the right. Low DPG, however, would shift the curve to the left like Figure 18.9f. The net effect of both conditions simultaneously would be close to normal oxygen binding. (b) As pH approaches normal, the curve shifts back to the left. With DPG still low, the curve would be between the left shift for low DPG and normal. Oxygen release after treatment would therefore be less than normal.
30. They concluded that some other neurotransmitter besides acetylcholine (which binds to muscarinic receptors) is involved in the vasodilation reflex.

Level Four Quantitative Problems

31. Answers vary. A 64-inch woman weighing 50 kg has a BMI of 19.
32. Fat: 6 g × 9 kcal/g = 54 kcal. Carbohydrate: 30 g × 4 kcal/g = 120 kcal. Protein: 8 g × 4 kcal/g = 32 kcal. Total = 206 kcal. 54/206 = 26% of calories from fat.

Chapter 23

Level One Reviewing Facts and Terms

1. Zona glomerulosa (aldosterone), zona fasciculata (glucocorticoids), zona reticularis (sex steroids, primarily androgens).
2. (a) corticotropin releasing hormone (hypothalamus) → adrenocorticotropic hormone (anterior pituitary) → cortisol (adrenal cortex) feeds back to inhibit secretion of both CRH and ACTH. (b) growth hormone releasing hormone and somatostatin (hypothalamus) → growth hormone (anterior pituitary) (c) decreased blood Ca^{2+} → parathyroid hormone (parathyroid glands) → increases blood Ca^{2+} by increasing reabsorption of bone, among other effects → negative feedback inhibits secretion of PTH. (d) Thyrotropin releasing hormone (hypothalamus) → thyroid stimulating hormone (thyrotropin) (anterior pituitary) → triiodothyronine (T_3) and thyroxine (T_4) (thyroid gland) → negative feedback to hypothalamus and anterior pituitary
3. Conditions: adequate diet, absence of chronic stress, and adequate amounts of thyroid and growth hormones. Other important hormones: insulin, IGFs (somatomedins), and sex hormones at puberty.
4. Triiodothyronine (T_3) and tetraiodothyronine (T_4 or thyroxine). T_3 is the more active; most of it is made from T_4 in peripheral tissues.
5. (a) Include ACTH (cortisol secretion) and MSH (not significant in humans). (b) Bone loss that occurs when bone reabsorption exceeds bone deposition. (c) The inorganic portion of bone matrix, mostly calcium salts. (d) Steroid hormones that regulate minerals, i.e., aldosterone. (e) Spongy bone, with an open latticework. (f) Pro-opiomelanocortin, inactive precursor to ACTH and other molecules. (g) Growth zones in long bones, comprised of cartilage.
6. Functions: blood clotting, cardiac muscle excitability and contraction, skeletal and smooth muscle contraction, second messenger systems, exocytosis, tight junctions, strength of bones and teeth.
7. In this table, A indicates anabolism, C indicates catabolism. CHO = *carbohydrate*

HORMONE	PROTEIN	CHO	FAT
Cortisol	C (skeletal muscle)	C	C
Thyroid	A (children), C (adults)	C	C
GH	A	C	C
Insulin	A	A	A
Glucagon	C	C	C

Level Two Reviewing Concepts

8. (a) See Figure 7.14a, p. 229. (b) See Figure 7.14b, p. 229. (c) Substitute thyroid pathway hormones into Figure 7.14a. (d) TRH high, TSH low, thyroid hormones low.
9. (a) Hypothalamic CRH stimulates anterior pituitary secretion of ACTH, which stimulates adrenal cortex (zona fasciculata) secretion of glucocorticoids such as cortisol. (b) Thyroid gland follicle cells secrete colloid from which thyroid hormones are produced; C cells secrete calcitonin. (c) Thyroid hormone synthesis is controlled by TSH, whose release is controlled by TRH.

In the thyroid gland, tyrosine and iodine combine on thyroglobulin to make thyroid hormones. Thyroid-binding globulin (TBG) carries thyroid hormones in the blood. Target cell deiodinase removes iodine from T_4 to make T_3. (d) Growth hormone releasing hormone (GHRH) stimulates anterior pituitary secretion of growth hormone (GH or somatotropin). Somatostatin (GHIH) inhibits production of GH. Growth hormone binding protein binds about half the GH in the blood. Insulin-like growth factors (IGFs) from the liver act with GH to promote growth. (e) Dwarfism results from severe GH deficiency in childhood. Giantism results from hypersecretion of GH during childhood. Acromegaly is lengthening of jaw and growth in hands and feet, caused by hypersecretion of GH in adults. (f) Hyperplasia—increased cell number. Hypertrophy—increased cell size. (g) Osteoblasts—bone cells that secrete organic bone matrix. Osteocytes—inactive form of osteoblasts. Chondrocytes—cartilage cells. Osteoclasts—bone-destroying cells. (h) PTH increases blood Ca^{2+} by stimulating bone and renal reabsorption, and intestinal absorption of Ca^{2+}. Calcitriol (1,25-dihydroxycholecalciferol) is a vitamin D derivative that mediates PTH effect on intestinal absorption of Ca^{2+}. Calcitonin decreases bone reabsorption of Ca^{2+}. Estrogen promotes bone deposition.

10. Thyroid hormones have intracellular receptors, so you expect 60–90 minute onset of action. However, effects on metabolic rate are apparent within a few minutes and are thought to be related to changes in ion transport across cell and mitochondrial membranes.
11. Equivalent = ion's molarity × the number of charges/ion. Ca^{2+}—2.5 mmoles × 2 = 5 mEq.
12. See Figure 23.10c. The cell uses carbonic anhydrase to make H^+ from $CO_2 + H_2O$. Apical membrane: H^+-ATPase secretes H^+; basolateral membrane secretes HCO_3^- with Cl^--HCO_3^- antiporter. Could also be Na^+-HCO_3^- symporter.

Level Three Problem Solving

13. Physiological stress stimulates secretion of cortisol, which increases blood glucose. Increased insulin opposes this effect.
14. Normal response: dexamethasone → ACTH suppression → decrease in cortisol. Patient A: no response to dexamethasone suggests there is adrenal hypersecretion that is insensitive to ACTH. Patient B: dexamethasone decreases cortisol, suggesting that the problem is in the pituitary.
15. Mr. A—elevated TSH. Ms. B—low TSH. Ms. C—elevated TSH. (a) Not possible to determine if the lab slip has the results of Mr. A or Ms. C without knowing the thyroid hormone levels. (b) Ms. B can be ruled out, because her TSH would be low if the tentative diagnosis is correct.
16. (a) People in all age groups showed vitamin D insufficiency at the end of winter. Deficiency was most pronounced in the 18–29 age group and least pronounced in the 50+ age group. At the end of summer, fewer subjects were deficient in vitamin D. Variables: season when blood collected, age group, and % of people with vitamin D insufficiency. (b) Energy from the sun is required for precursors in the skin to be converted to vitamin D. Days are shorter in the winter, and at northern latitudes like Boston, people spend less time outside during the winter. This explains the difference between the two seasons. Fewer than half the people tested were deficient, however, suggesting that most people consumed enough vitamin D. The biggest seasonal difference was in the 18–29 age group, who probably spent more time outside in the summer than members of other groups. (c) Taking multivitamin supplements containing vitamin D should reduce vitamin D insufficiency.

Level Four Quantitative Problems

17. (a) 5 mg Ca^{2+}/L plasma × 125 mL plasma filtered/min × 1440 min/day = 900 mg Ca^{2+} filtered/day (b) To remain in Ca^{2+} balance, he must excrete 170 mg/day. (c) 900 mg filtered−170 mg excreted = 730 mg reabsorbed. 730/900 = 81%.

Chapter 24

Level One Reviewing Facts and Terms

1. The body's ability to defend itself against disease-causing pathogens. Memory—immune cells remember prior exposure to an antigen and create a stronger immune response. Specificity—antibodies that target specific antigens.
2. Thymus gland, bone marrow, spleen, lymph nodes, and diffuse lymphoid tissues
3. Protect the body against foreign pathogens; remove dead or damaged tissues and cells; recognize and remove abnormal "self" cells.
4. Viruses bud off the host cell or kill and rupture it. Viruses damage host cells by killing them, taking over their metabolism, or causing them to become cancerous and reproduce uncontrollably.
5. Detect the pathogen, recognize it as foreign, organize a response, and recruit assistance from other cells. If a pathogen cannot be destroyed, it may be suppressed.
6. (a) Severe IgE-mediated allergic reaction with widespread vasodilation, circulatory collapse, and bronchoconstriction. (b) To clump together. When blood cells are exposed to an antibody, an antibody-antigen reaction may cause the blood cells to agglutinate. (c) Outside the blood vessels. Many immune reactions are extravascular. (d) Release of cytoplasmic granule chemicals into the ECF. (e) Opsonins that coat pathogens; released in the early stages of injury or infection. (f) One cell of a clone divides to make many identical cells. (g) Ability of the immune system to find and destroy abnormal cells (especially cancerous).
7. They are all names given to specialized tissue macrophages before scientitsts recognized that they were the same cell type.
8. It consists of monocytes and macrophages, which ingest and destroy invaders and abnormal cells
9. (a) 5, (b) 1, (c) 3, (d) 6, (e) 2, (f) 4
10. Physical: skin, mucous membranes, respiratory mucociliary escalator. Chemical: lysozymes, opsonins, enzymes, and antibodies.
11. B lymphocytes secrete antibodies; T lymphocytes and NK cells kill infected cells. T lymphocytes bind to antigen presented by MHC complexes; NK cells can also bind to antibodies coating foreign cells.
12. The ability of the body's immune system to ignore the body's own cells. Occurs because T lymphocytes that react with "self" cells are eliminated by clonal deletion. If self-tolerance fails, the body makes antibodies against itself (autoimmune disease).
13. The ability of the nervous system to influence immune function, either positively or negatively.
14. Stress—nonspecific stimulus that disturbs homeostasis. Stressor—stimulus that causes stress. General adaptation syndrome—stress response that includes activation of the adrenal glands (fight-or-flight response by adrenal medulla and cortisol secretion by the cortex).

Level Two Reviewing Concepts

15. Use figures and tables of the chapter to create the map.
16. When lymph nodes trap bacteria, a localized inflammatory response (swelling and cytokine activation of nociceptors) results in swollen, sore nodes.
17. Histamine—opens pores in capillaries so immune cells and proteins can leave the blood. IL-1—increases capillary permeability, stimulates acute phase proteins, causes fever. Acute phase proteins—opsonins. Bradykinin—vasodilator; stimulates pain receptors. Complement—opsonins, chemotaxins, histamine release, membrane attack complex. Interferon-γ—activates macrophages. These molecules work together so they are not antagonistic. If their combined effect is greater than the sum of their individual effects, they are synergistic.
18. (a) Pathogen—any organism that causes disease. Microbes—microscopic organisms, pathogens or not. Pyrogens—fever-causing chemicals. Antigens—substances that trigger an immune response and react with products of the response. Antibodies—disease-fighting chemicals produced by the body. Antibiotics—drugs that destroy bacteria and fungi. (b) Infection—illness caused by pathogens, especially viruses or bacteria. Inflammation—nonspecific response to cell damage or invaders, including nonpathogens such as a splinter. Allergy—inflammatory response to a nonpathogenic invader, such as plant pollen. Autoimmune disease—body creates antibodies to its own cells. (c) Allergens—nonpathogenic substances that create allergic reactions. Bacteria—cellular organisms. Viruses—acellular parasites that must invade the host's cells to reproduce. (d) Chemotaxins—chemicals that attract immune cells. Cytokines—peptides made on demand and secreted for action

on other cells. Opsonins—proteins that coat and tag foreign material so that it can be recognized by the immune system. Interleukins—cytokines initially thought to act only on leukocytes. Interferons—lymphocyte cytokines that aid in the immune response. Bradykinin—paracrine vasodilator. (e) Innate immunity—nonspecific, present from birth; acquired—directed at specific invaders. Acquired can be divided into cell-mediated and humoral (antibodies). (f) Immediate hypersensitivity response—mediated by antibodies; occurs within minutes of exposure to allergen. Delayed—may take several days to develop; mediated by helper T cells and macrophages. (g) All chemicals of the immune response. Membrane attack complex and perforin are membrane pore proteins. Perforins allow granzymes (cytotoxic enzymes) to enter the cell.

19. See Figure 24.9. Fc region—determines antibody class; Fab region—antigen-binding sites that confer the antibody's specificity.
20. See Figure 24.12.
21. See Figure 24.13.
22. See Figure 24.14.

Level Three Problem Solving

23. Type O—universal donor because these RBCs lack A or B surface antigens and do not trigger an immune response. Type AB—universal recipient because these RBCs have both A and B antigens and no A or B antibodies.
24. Maxie and baby are both OO. Snidley could be either BB or BO. Baby received an O gene from Maxie, and could have received the other O gene from Snidely. Thus, it is possible that Snidley is the father of Maxie's baby.
25. Emotional stress → increases cortisol secretion → immune system suppression. Also likely that students are spending more time inside and having closer contact with fellow students.
26. Barbara's immune cells recognize her connective tissue as an antigen, and attack it. Autoimmune diseases often begin in association with an infection and are thought to represent cross-reactivity of antibodies that developed because of the infection. Stress suppresses the immune system.
27. Increase in neutrophils—bacterial infection because neutrophils eat bacteria. Increase in eosinophils—parasitic infection because eosinophils kill parasites.

Chapter 25

Level One Reviewing Facts and Terms

1. ATP and phosphocreatine
2. *aerobic; both glucose and fatty acids*
3. Aerobic metabolism: requires O_2; glucose goes through glycolysis and citric acid cycle; produces 30–32 ATP/glucose through oxidative phosphorylation. Anaerobic: no O_2 used; glucose undergoes glycolysis to lactate; produces only 2 ATP/glucose.
4. Glycogen, plasma glucose, glucose produced through gluconeogenesis
5. Cortisol, growth hormone, epinephrine, and norepinephrine all increase plasma glucose.
6. At the beginning of exercise, muscle ATP use exceeds aerobic ATP production, so cellular stores of ATP are used. This creates an oxygen deficit reflected by increased oxygen consumption after exercise ceases.
7. Cardiovascular system
8. Normal—37 °C. Sweating and cutaneous vasodilation

Level Two Reviewing Concepts

9. Look for figures in Chapters 4, 15, 17, 18, 23, and 25.
10. Sympathetic input on pancreatic beta cells decreases insulin secretion. Less insulin → liver produces glucose; insulin-sensitive tissues do not take up blood glucose → blood glucose available for brain and exercising muscle (glucose uptake does not require insulin).
11. Advantages: fast; uses readily available glucose. Disadvantages: low ATP yield per glucose; contributes to metabolic acidosis.
12. (a) ATP—energy for muscle contraction. ADP—accepts high-energy phosphate from PCr and becomes ATP. (b) Myoglobin—muscle O_2-binding protein that aids O_2 diffusion from blood to mitochondria. Hemoglobin—RBC O_2-binding pigment that transports O_2 from lungs to cells.
13. (a) 3; (b) 1, 2, 3, 4, 5; (c) 1, 2, 4, 5, 6; (d) 6; (e) no match; (f) 6; (g) 1 (venous return), 4
14. (a) increases (b) decreases (c) increases (d) increases (e) increases (f) increases (g) stays the same (h) decreases
15. Increased heart rate shortens filling time and helps offset increased end diastolic volume that might be expected from increased venous return.
16. (1) the baroreceptor reflex resets to a higher setpoint, (2) afferent signals from the baroreceptors are being blocked in transit up the spinal cord, or (3) chemo- and mechanoreceptor input from exercising tissues overrides the baroreceptor input.
17. Regular exercise lowers risk of heart attacks, lowers blood pressure, creates better lipid profiles, and lowers risk of developing type 2 diabetes.
18. Exercising muscle does not require insulin for glucose uptake, so regular exercise can help keep blood glucose levels normal.

Level Three Problem Solving

19. Water, NaCl, and K^+ to replace fluid and ions lost in sweat plus a carbohydrate that is easily absorbed and metabolized to form ATP

Level Four Quantitative Problems

20. CO = 60 beats/minute × 70 mL/beat = 4200 mL/min. If heart rate doubles, CO goes to 8400 mL/min, or doubles also.
21. Values are approximate. (a) A: 128 − 48 = 88 mL, B: 155 − 47 = 108 mL, C = 138 − 24 = 114 mL. (b) CO = HR × SV. A: 6000 mL/min = 88 mL/beat × SV. SV = 68 bpm. B = 97 bpm. C = 167 bpm. (c) Curve C shows increased contractility and curve B shows increased venous return. (d) As heart rate sped up to 167 bpm, there was less time for ventricular filling and the end-diastolic volume decreased.

Chapter 26

Level One Reviewing Facts and Terms

1. (a) 3, 4, 5; (b) 8; (c) 2, 7; (d) 2, 6; (e) 2; (f) 1
2. *SRY*
3. Gonads produce gametes and secrete sex hormones. Female gamete—egg (ovum); male—sperm. Female gonadal hormones—estrogen, progesterone, androgens, and inhibin; male—androgens and inhibin.
4. Newly formed sperm: seminiferous tubule → epididymis → ductus (vas) deferens → ejaculatory duct (passing the seminal vesicles, prostate gland, and bulbourethral glands) → urethra. Ovulated egg: fallopian tube → uterine cavity → cervix → vagina
5. (a) converts androgens to estrogens (b) tight junctions that prevent free movement of substances between blood and seminiferous tubule lumen (c) Sertoli cell protein secreted into seminiferous tubule lumen, where it binds and concentrates androgens (d) formed by the first meiotic division of a primary oocyte; disintegrates and has no function (e) lysosome-like structure in the head of sperm; contains enzymes essential for fertilization
6. (a) False. Some is produced in the adrenal glands of both sexes. (b) False. Both sexes produce them. (c) True. (d) False. High levels of late follicular estrogen help prepare the uterus for implantation of a fertilized ovum. (e) True.
7. A sperm-fluid mixture made mostly by the accessory glands. See Figure 26.7f for components.
8. The most effective contraception is abstinence. Least effective forms rely on avoiding intercourse during times when the female thinks she might be fertile.

Level Two Reviewing Concepts

9. List 1: use Figures 26.2, 26.3, and 26.4. List 2: use Figures 26.9, 26.10, and 26.12.
10. See Figure 26.8.

11. See Figures 26.10 and 26.12.
12. Males have one Y chromosome, which often does not have a gene to match one found on the X chromosome. Thus, a male may inherit a recessive X trait and will exhibit it, while a female who inherits the same recessive trait will not exhibit it if her second X chromosome has the dominant gene for the trait.
13. (a) Gamete—eggs and sperm. Germ cell—cell that will become a gamete. Zygote—formed from the fusion of egg and sperm; undergoes mitosis to become an embryo. In 8th week of pregnancy, embryo becomes a fetus. (b) Coitus—intercourse. Erection—stiffening and enlargement of the penis. Male orgasm—sperm move into the urethra during emission, then out of the body in semen during ejaculation. Erogenous zones—portions of the body with receptors for sexually arousing stimuli. (c) Capacitation—sperm maturation necessary before it can fertilize an egg. Zona pellucida—protective glycoprotein coat around the ovum. Acrosomal reaction—enzymes help sperm penetrate the zona pellucida. Cortical reaction—granules in egg cytoplasm release their contents at fertilization to change the egg membrane properties. (d) Puberty—time of sexually maturation. Menarche—the first menstrual period. Menopause—female reproductive cycles cease. Andropause—male counterpart to menopause.
14. (a) FSH—stimulates gamete production in both sexes. (b) Inhibin—inhibits FSH secretion. (c) Activin—stimulates FSH secretion. (d) GnRH—stimulates release of FSH and LH. (e) LH—stimulates gonadal sex hormone production; in females, also necessary for gamete maturation. (f) DHT—testosterone metabolite responsible for fetal development of male genitalia. (g) Estrogen—present in both sexes but dominant in females; female gamete formation and some secondary characteristics. (h) Testosterone in males—gamete formation. Both sexes—some secondary sex traits such as hair growth. (i) Progesterone—females only; helps prepare the uterus for pregnancy.
15. The four phases are similar in both sexes. Excitement—penis and clitoris become erect due to increased blood flow. The vagina secretes fluids for lubrication. In male orgasm, ejaculation takes place, while in female orgasm the uterus and vaginal walls contract.
16. (a) hCG—keeps the corpus luteum from dying. (b) LH—no direct role in pregnancy. (c) HPL—regulation of maternal metabolism during pregnancy. (d) Estrogen—breast development; negative feedback signal to prevent new follicles from developing. (e) Progesterone—maintenance of the uterine lining; prevents uterine contractions; mammary gland development. (f) Relaxin—prevents uterine contractions. (g) Prolactin—PIH levels decrease so that prolactin levels will increase, allowing milk production.

Level Three Problem Solving

17. Normally after fertilization the second polar body, containing a haploid set of chromosomes, is released from the zygote. If all or some of the second polar body chromosomes are retained, the embryo will have three copies of a chromosome instead of just two.
18. If the unovulated cysts continue to secrete estrogen and do not develop into corpora lutea, the uterine lining will continue to grow and the breasts will develop, just as during pregnancy.
19. (a) male (b) nonfunctional testes (c) no ducts of either type (d) female
20. During pregnancy the mother's blood glucose is available to the fetus, which metabolizes the extra energy and gains weight. The fetus also up-regulates insulin secretion to handle the glucose coming across the placenta. After birth, when insulin is still high but glucose drops to normal, the baby may become hypoglycemic.

Level Four Quantitative Problems

21. (a) Because it was being administered to the subjects. (b) Negative feedback by testosterone. (c) Sperm production decreased in the A–B interval because FSH and LH decreased. It increased toward the end of the B–C interval because FSH allowed sperm production to resume. Sperm production did not increase significantly during the D–E interval.

Richard D. Hill and Daniel Biller
University of Texas

Introduction

This appendix discusses selected aspects of **biophysics,** the study of physics as it applies to biological systems. Because living systems are in a continual exchange of force and energy, it is necessary to define these important concepts. According to the seventeenth-century scientist Sir Isaac Newton, a body at rest tends to stay at rest, and a body in motion tends to continue moving in a straight line unless the body is acted upon by some force (Newton's First Law).

Newton further defined **force** as an influence, measurable in both intensity and direction, that operates on a body in such a manner as to produce an alteration of its state of rest or motion. Put another way, force gives **energy** to a quantity, or mass, thereby enabling it to do work. In general, a driving force multiplied by a quantity yields energy, or work. For example:

$$\text{force} \times \text{distance} = \text{work}$$

Energy exists in two general forms: kinetic energy and potential energy. **Kinetic energy** {*kinein,* to move} is the energy possessed by a mass in motion. **Potential energy** is energy possessed by a mass because of its position. Kinetic energy (KE) is equal to one-half the mass (m) of a body in motion multiplied by the square of the velocity (v) of the body:

$$KE = \tfrac{1}{2}\, mv^2$$

Potential energy (PE) is equal to the mass (m) of a body multiplied by acceleration due to gravity (g) times the height (h) of the body above the earth's surface:

$$PE = mgh \quad \text{where } g = 10\ \text{m/s}^2$$

Both kinetic and potential energy are measured in joules.

Basic Units of Measurement

For physical concepts to be useful in scientific endeavors, they must be measurable and should be expressed in standard units of measurement. Some fundamental units of measure include the following:

Length (l): Length is measured in meters (m).
Time (t): Time is measured in seconds (s).
Mass (m): Mass is measured in kilograms (kg), and is defined as the weight of a body in a gravitational field.
Temperature (T): Absolute temperature is measured on the Kelvin (K) scale,
where K = degrees Celsius (°C) + 273.15
and °C = (degrees Fahrenheit − 32)/1.8

Electric current (I): Electric current is measured in amperes (A).
Amount of substance (n): The amount of a substance is measured in moles (mol).

Using these fundamental units of measure, we can now establish standard units for physical concepts (● Tbl. B.1). Although these are the standard units for these concepts at this time, they are not the only units ever used to describe them. For instance, force can also be measured in dynes, energy can be measured in calories, pressure can be measured in torr or mm Hg, and power can be measured in horsepower. However, all of these units can be converted into a standard unit counterpart, and vice versa.

The remainder of this appendix discusses some biologically relevant applications of physical concepts. This discussion includes topics such as bioelectrical principles, osmotic principles, and behaviors of gases and liquids relevant to living organisms.

Bioelectrical Principles

Living systems are composed of different molecules, many of which exist in a charged state. Cells are filled with charged particles such as proteins and organic acids, and ions are in continual flux across the cell membrane. Therefore, electrical forces are important to life.

When molecules gain or lose electrons, they develop positive or negative charges. A basic principle of electricity is that opposite charges attract and like charges repel. A force must act on a charged particle (a mass) to bring about changes in its position.

Standard Units for Physical Concepts — Table B-1

Measured Concept	Standard (SI*) Unit	Mathematical Derivation/ Definition
Force	Newton (N)	$1\ \text{N} = 1\ \text{kg} \cdot \text{m/s}^2$
Energy/Work/ Heat	Joule (J)	$1\ \text{J} = 1\ \text{N} \cdot \text{m}$
Power	Watt (W)	$1\ \text{W} = 1\ \text{J/s}$
Electrical charge	Coulomb (C)	$1\ \text{C} = 1\ \text{A} \cdot \text{s}$
Potential	Volt (V)	$1\ \text{V} = 1\ \text{J/C}$
Resistance	Ohm (Ω)	$1\ \Omega = 1\ \text{V/A}$
Capacitance	Farad (F)	$1\ \text{F} = 1\ \text{C/V}$
Pressure	Pascal (Pa)	$1\ \text{Pa} = 1\ \text{N/m}^2$

* SI = Système International d'Unités

Therefore, there must be a force acting on charged particles to cause attraction or repulsion, and this electrical force can be measured.

Electrical force increases as the strength (number) of charges increases, and it decreases as the distance between the charges increases (● Fig. B.1). This observation has been called **Coulomb's law**, and can be written:

$$F = q_1 q_2 / \varepsilon d^2$$

where q_1 and q_2 are the electrical charges (coulombs), d is the distance between the charges (meters), ε is the dielectric constant, and F is the force of attraction or repulsion, depending on the type of charge on the particles.

When opposite charges are separated, a force acts over a distance to draw them together. As the charges move together, work is being done by the charged particles and energy is being released. Conversely, to separate the united charges, energy must be added and work done. If charges are separated and kept apart, they have the potential to do work. This electrical potential is called **voltage.** Voltage is measured in **volts (V).**

If electrical charges are separated and there is a potential difference between them, then the force between the charges allows electrons to flow. Electron flow is called an electric **current.** The **Faraday constant (F)** is an expression of the electrical charge carried by one mole of electrons and is equal to 96,485 coulombs/mole.

The amount of current that flows depends on the nature of the material between the charges. If a material hinders electron flow, then it is said to offer **resistance (R),** measured in ohms. Current is inversely proportional to resistance, such that current decreases as resistance increases. If a material offers high resistance to electron flow, then that material is called an **insulator.** If resistance is low, and current flows relatively freely, then the material is called a **conductor.** Current, voltage, and resistance are related by **Ohm's law,** which states:

$$V = IR$$

where V = potential difference in volts
I = current in amperes
R = resistance in ohms

In biological systems, pure water is not a good conductor, but water containing dissolved NaCl is a good conductor because ions provide charges to carry the current. In biological membranes, the lipids have few or no charged groups, so they offer high resistance to current flow across them. Thus, different cells can have different electrical properties depending on their membrane lipid composition and the permeability of their membranes to ions.

ELECTRICAL FORCE

If you separate two opposite charges, there will be an electric force between them.

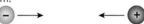

If you increase the number of charges that are separated, the force increases.

If you increase the distance between the charges, the force decreases.

If charges are separated by some distance *d*, they have the potential to do work. This electrical potential is called voltage.

If separated charges are allowed to move together, they do work and energy is released. The amount of work done depends on the number of particles and the distance between them.

To separate the charged particles, energy must be put into the system and work is done.

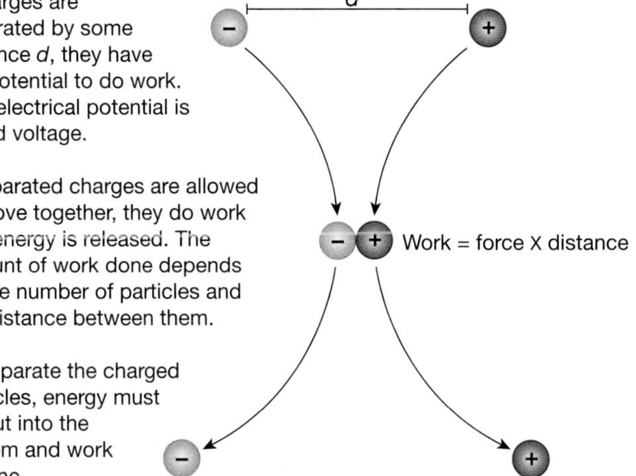

● **Fig. B-1**

Osmotic Principles

Freezing point, vapor pressure, boiling point, and osmotic pressure are properties of solutions collectively called **colligative properties.** These properties depend on the number of solute particles present in a solution. **Osmotic pressure** is the force that drives the diffusion of water across a membrane. Because there are no solutes in pure water, it has no osmotic pressure. However, if one adds a solute like NaCl, the greater the concentration (c) of a solute dissolved in water, the greater the osmotic pressure. The osmotic pressure (π) varies directly with the concentration of solute (number of particles (n) per volume (V)):

$$\pi = (n/V)RT$$

$$\pi = cRT$$

where R is the ideal gas constant ($8.3145\ \text{joules}/K \cdot mol$) and T is the absolute temperature in Kelvin. Osmotic pressure can be measured by determining the mechanical pressure that must be applied to a solution so that osmosis ceases.

Water balance in the body is under the control of osmotic pressure gradients (concentration gradients). Most cell membranes allow water to pass freely, primarily through open channels. To control the movement of water, the body either removes these channels from the membrane or control solute movement that creates concentration gradients.

Relevant Behaviors of Gases and Liquids

The respiratory and circulatory systems of the human body obey the physical laws that govern the behavior of gases and liquids. This section discusses some of the important laws that govern these behaviors and how our body systems utilize these laws.

Gases

The **ideal gas law** states:

$$PV = nRT$$

where P = pressure of gases in the system
V = volume of the system
n = number of moles in gas
T = temperature
R = ideal gas constant (8.3145 J/K mol)

If n and T are kept constant for all pressures and volumes in a system of gases, then any two pressures and volumes in that system are related by Boyle's Law,

$$P_1V_1 = P_2V_2$$

where P represents pressure and V represents volume.

This principle is relevant to the human lungs because the concentration of gas in the lungs is relatively equal to that in the atmosphere. In addition, body temperature is maintained at a constant temperature by homeostatic mechanisms. Therefore, if the volume of the lungs is changed, then the pressure in the lungs changes inversely. For example, an increase in pressure causes a decrease in volume, and vice versa.

Liquids

Fluid pressure (or hydrostatic pressure) is the pressure exerted by a fluid on a real or hypothetical body. In other words, the pressure exists whether or not there is a body submerged in the fluid. Fluid exerts a pressure (P) on an object submerged in it at a certain depth from the surface (h). **Pascal's law** allows us to find the fluid pressure at a specified depth for any given fluid. It states:

$$P = \rho gh$$

where P = fluid pressure (measured in pascals, Pa)
ρ = density of the fluid
g = acceleration due to gravity ($10\ \text{m/s}^2$)
h = depth below the surface of the fluid

Fluid pressure is unrelated to the shape of the container in which the fluid is situated.

Review of Logarithms

Understanding logarithms ("logs") is important in biology because of the definition of pH:

$$\text{pH} = -\log_{10}[\text{H}^+]$$

This equation is read as "pH is equal to the negative log of the hydrogen ion concentration." But what is a logarithm?

A logarithm is the exponent to which you would have to raise the base (10) to get the number in which you are interested. For example, to get the number 100, you would have to square the base (10):

$$10^2 = 100$$

The base 10 was raised to the second power; therefore, the log of 100 is 2:

$$\log 100 = 2$$

Some other simple examples include:

$10^1 = 10$ The log of 10 is 1.

$10^0 = 1$ The log of 1 is 0.

$10^{-1} = 0.1$ The log of 0.1 is -1.

What about numbers that fall between the powers of 10? If the log of 10 is 1 and the log of 100 is 2, the log of 70 would be between 1 and 2. The actual value can be looked up on a log table or ascertained with most calculators.

To calculate pH, you need to know another rule of logs that says:

$$-\log x = \log(1/x)$$

and a rule of exponents that says:

$$1/10^x = 10^{-x}$$

Suppose you have a solution whose hydrogen ion concentration $[\text{H}^+]$ is 10^{-7} mEq/L. What is the pH of this solution?

$$\text{pH} = -\log[\text{H}^+]$$

$$\text{pH} = -\log[10^{-7}]$$

Using the rule of logs, this can be rewritten as

$$\text{pH} = \log(1/10^{-7})$$

Using the rule of exponents, this can be rewritten as

$$\text{pH} = \log 10^7$$

The log of 10^7 is 7, so the solution has a pH of 7.

Natural logarithms (ln) are logs in the base e. The mathematical constant e is approximately equal to 2.7183.

Genetics

Appendix C

Richard D. Hill
University of Texas

What is DNA?

Deoxyribonucleic acid (DNA) is the macromolecule that stores all information a cell needs to survive and reproduce. DNA and its cousin RNA belong to a group of macromolecules called **nucleic acids**. [For illustrations of nucleic acid structure, see Figure 2.4 on page 36.] Nucleic acids are polymers made from monomers {*mono*-, one} called **nucleotides**. Each nucleotide consists of a *nucleoside* (a pentose, or 5-carbon, sugar covalently bound to a nitrogenous base) and a phosphoric acid with at least one phosphate group.

In humans, many millions of nucleotides join together to form a DNA molecule. Eukaryotic DNA is commonly in the form of a double helix that looks like a twisted ladder or twisted zipper. The two sugar-phosphate sides, or backbones, are the same for every DNA molecule, but the sequence of the nucleotides is unique for each individual organism.

Functions of DNA

Cells use the information stored in DNA to build their structural and functional components. DNA also provides the basis for inheritance when DNA is passed from parent to offspring. The union of these concepts about DNA allows us to devise a working definition of a gene. A **gene** is a segment of DNA that codes for the synthesis messenger RNA (mRNA) to make proteins. Genes also act as a unit of inheritance that can be transmitted from generation to generation. The external appearance (**phenotype**) of an organism is determined to a large extent by the genes it inherits (**genotype**). Thus, one can begin to see how variation at the DNA level can cause variation at the level of the entire organism. These concepts form the basis of **genetics** and evolutionary theory.

DNA's primary function in most cells is to initiate the synthesis of proteins needed for cell structure or function. The information coded in DNA is first *transcribed* into mRNA. mRNA leaves the cell nucleus and enters the cytoplasm, where its code is *translated* into proteins. The second key function of DNA is its ability act as a unit of inheritance when transmitted across generations.

Before we discuss DNA as a unit of inheritance, let's explain a few terms you need to know. A **chromosome** is one complete molecule of DNA. Each chromosome contains many DNA sequences that act as genes. Every gene comes in variants called **alleles.** Interactions between the cell products of alleles determine how that gene will be expressed in the phenotype of an individual.

Somatic cells {*soma*, body} are those cells that make up the majority of the body (e.g., a skin cell, a liver cell); they are not directly involved with passing on genetic information to future generations. Each somatic cell in a human contains two alleles of each gene, one allele inherited from each parent. For this reason, human somatic cells are called **diploid** ("two chromosome sets"), meaning that they have two complete sets of all their chromosomes.

In contrast, **germ cells** pass genetic information directly to the next generation. In human males, the germ cells are the **spermatozoa** (sperm), and in human females, the germ cells are the **oocytes** (eggs). Human germ cells are called **haploid** ("half of the chromosome sets") because each germ cell only contains one chromosome set, which is equal to half of the chromosomes in somatic cells. When a human male germ cell joins with a human female germ cell, the result is a fertilized egg (zygote) containing the diploid number of chromosomes. If this zygote eventually develops into a healthy adult, that adult will have diploid somatic cells and haploid germ cells.

Cell Division

Cells alternate between periods of cell growth and cell division. There are two types of cell division: mitosis and meiosis. **Mitosis** is cell division by somatic cells that results in two daughter cells identical to the parent cell. Each daughter cell has a diploid set of chromosomes. **Meiosis**, in contrast, is cell division that results in four daughter cells, each with a haploid set of chromosomes. After meiosis, the daughter cells develop into germ cells, or eggs and sperm.

Cells that are not dividing are said to be in **interphase**. Interphase is divided into three stages: $\mathbf{G_1}$, a period of cell growth, protein synthesis, and organelle production; **S,** the period during which DNA is replicated in preparation for cell division; and $\mathbf{G_2}$, a period of protein synthesis and final preparations for cell division (● Fig. C.1). During interphase, the DNA in the nucleus is not visible under the light microscope without dyes because it is uncoiled and diffuse. However, as a cell prepares for division, it condenses all its DNA to form more manageable packages. Each eukaryotic DNA molecule has millions of nucleotides which, if laid end-to-end, could stretch out to about 6 cm. If this DNA molecule did not coil tightly and condense for cell division, imagine how difficult moving it around during cell division would be.

There is a hierarchy of DNA packaging in the cell (● Fig. C.2a). Each chromosome begins with a linear molecule of DNA about 2 nm in diameter. Then proteins called **histones** associate with the DNA to form **nucleosomes**, which consist of histones wrapped in DNA. A series of nucleosomes creates a fiber about 10 nm in diameter that looks like "beads on a string." The beaded string can twist into a **chromatin** fiber about 30 nm in diameter, with about 6 nucleosomes per turn. When cells get ready to divide, their chromatin fibers then coil even more to form the **chromosome fiber** (about 700 nm in diameter). Once DNA is in this state of condensed packaging, the cell is ready for division.

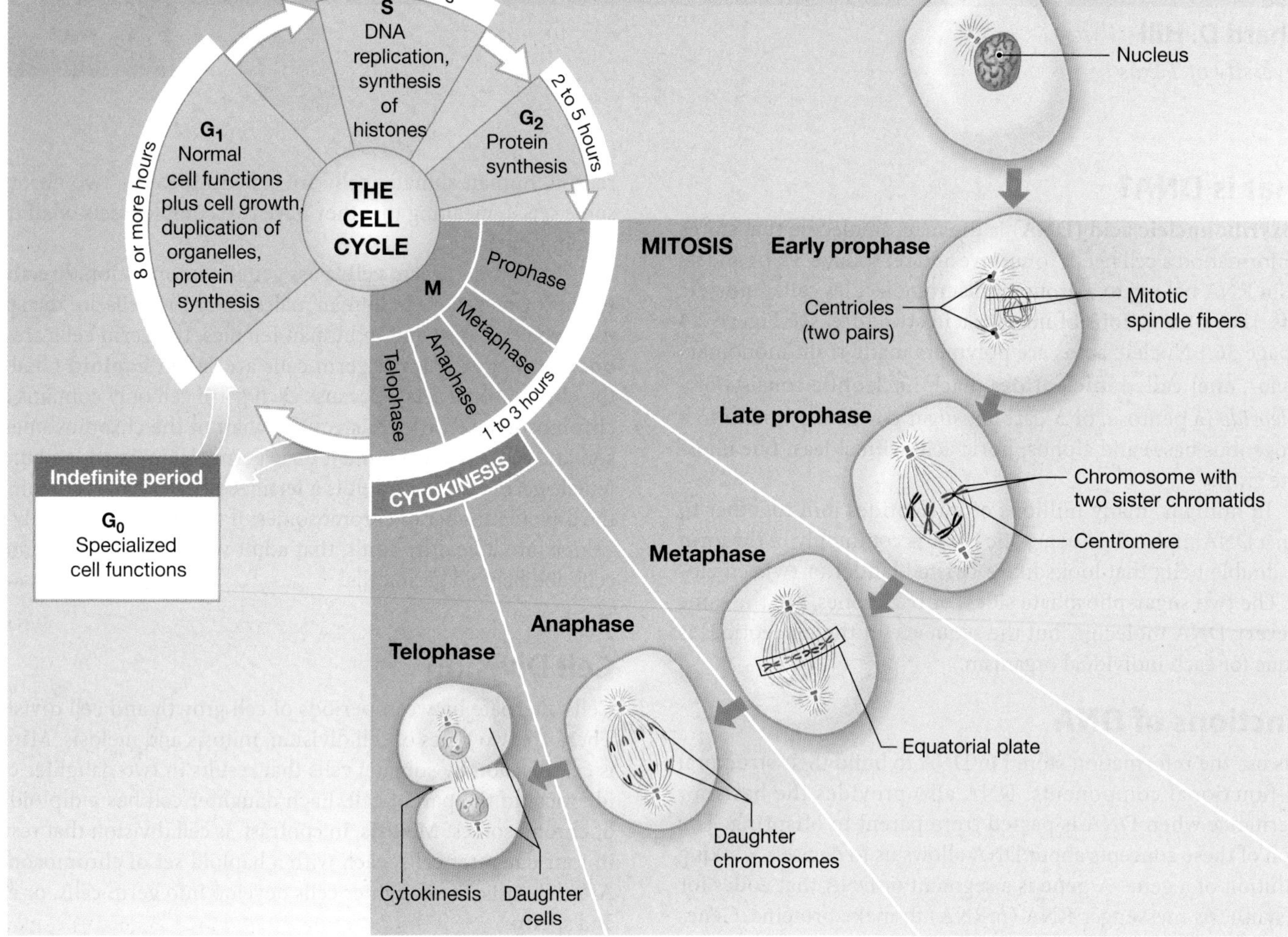

Fig. C.1

Mitosis Creates Two Identical Daughter Cells

As stated earlier, mitosis is the division of a somatic cell that results in two diploid daughter cells. The DNA of the parent cell first duplicates itself into two complete sets of chromosomes. One set of chromosomes then goes to each daughter cell, and the daughter cells separate.

The four main steps of mitosis are **prophase, metaphase, anaphase**, and **telophase** (Fig. C.1). The entire somatic cell cycle can be remembered by the acronym, IPMAT, in which the "I" stands for interphase and the other letters stand for the steps of mitosis that follow.

Prophase

During prophase, chromatin becomes condensed and microscopically visible as duplicate chromosomes. The duplicated chromosomes form **sister chromatids**, which are joined to each other at the **centromere**. The cell's centriole pair duplicates and the two centriole pairs move to opposing ends of the cell. The **mitotic spindle**, composed of microtubules, assembles between the centriole pairs. The nuclear membrane begins to break down and disappears by the end of prophase.

Metaphase

In metaphase, mitotic spindle fibers extending from the centrioles attach to the centromere of each chromosome. The 46 chromosomes, each consisting of a pair of sister chromatids, line up at the "equator" of the cell.

Anaphase

During anaphase, the spindle fibers pull the sister chromatids apart, so that an identical copy of each chromosome moves toward each pole of the cell. By the end of anaphase, an identical set of 46 chromosomes is present at each pole. At this point, the cell has a total of 92 chromosomes, double the diploid number.

Telophase

The actual division of the parent cell into two daughter cells takes place during telophase. In **cytokinesis**, the cytoplasm divides when an actin contractile ring tightens at the midline of the cell. The result is two separate daughter cells, each with a full diploid set of chromosomes. The spindle fibers disintegrate, nuclear envelopes form around the chromosomes in each cell, and the chromatin returns to its loosely coiled state.

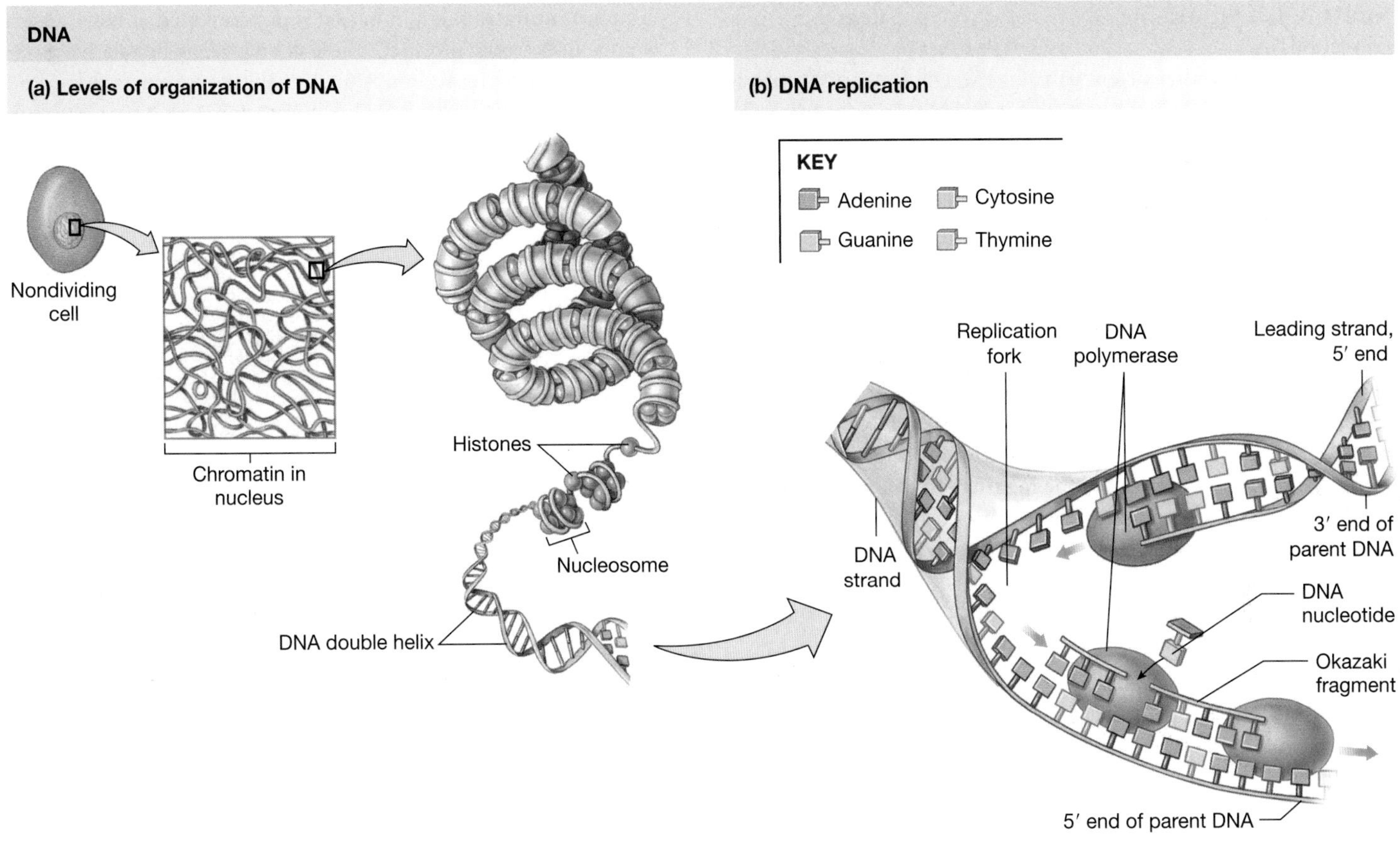

Fig. C.2

DNA Replication

The information stored in DNA is encoded in the nucleotide sequence of the molecule. When nucleotides link together, the phosphate group of one nucleotide bonds covalently to the sugar group of the adjacent nucleotide. The end of the polymer that has an unbound sugar is called the 3′ ("three prime") end. The end of the polymer with the unbound phosphate is called the 5′ end. A DNA molecule has four types of nucleotides, distinguished by their nitrogenous bases.

The nitrogenous bases in nucleic acids are classified as either **purines** or **pyrimidines**. The purine bases are **guanine (G)** and **adenine (A)**. The pyrimidine bases are **cytosine (C)**, and either **thymine (T)**, found in DNA only, or **uracil (U)**, found in RNA only. To remember which DNA bases are pyrimidines, look at the first syllable. The word "pyrimidine" and names of the DNA pyrimidine bases all have a "y" in the first syllable.

The "rungs" of the DNA double helix are created when the nitrogenous bases on one DNA strand form hydrogen bonds with nitrogenous bases on the adjoining DNA strand. This phenomenon is called **base-pairing**. The base-pairing rules are as follows:

1. Purines pair only with pyrimidines.
2. Guanine (G) bonds with cytosine (C) in both DNA and RNA.
3. Adenine (A) bonds with thymine (T) in DNA or with uracil (U) in RNA.

The two strands of DNA are bound in **antiparallel** orientation, so that the 3′ end of one strand is bound to the 5′ end of the second strand. This organization has important implications for DNA replication.

DNA Replication Is Semi-Conservative

To be transmitted from one generation to the next, DNA must be replicated. Furthermore, the process of replication must be accurate and fast enough for a living system. The base-pairing rules for nitrogenous bases provide a means for making an appropriate replication system.

In DNA replication, special proteins unzip the DNA double helix and build new DNA by pairing new nucleotide molecules to the two existing DNA strands. The result of this replication is two double-stranded DNA molecules, such that each DNA molecule contains one DNA strand from the template and one newly synthesized DNA strand. This form of replication is called **semi-conservative replication**.

Replication of DNA is bidirectional. A portion of DNA that is "unzipped" and has enzymes performing replication is called a **replication fork** (Fig. C.2b). Replication begins at many points (**origins of replication**), and it continues along both parent strands simultaneously until all the replication forks join.

Nucleotides bond together to form new strands of DNA with the help of an enzyme called **DNA polymerase**. DNA polymerase can add nucleotides only to the 3′ end of a growing strand of DNA. For this reason, DNA is said to replicate in a 5′ to 3′ direction.

The antiparallel orientation of the DNA strands and the directionality of DNA polymerase force replication into two different modes: **leading strand replication** and **lagging strand replication**. The DNA polymerase can replicate continuously along only one parent strand of DNA: the parent strand in the 3′ to 5′ orientation. The DNA replicated continuously is called the **leading strand**.

The DNA replication along the other parent strand is discontinuous because of the strand's 5′ to 3′ orientation. DNA replication on this strand occurs in short fragments called **Okazaki fragments** that are synthesized in the direction away from the replication fork. Another enzyme known as **DNA ligase** later connects these fragments into a continuous strand. The DNA replicated in this way is called the **lagging strand**. Because the 5′ ends of the lagging strand of DNA cannot be replicated by DNA polymerase, a specialized enzyme called **telomerase** has arisen to replicate the 5′ ends.

Much of the accuracy of DNA replication comes from base pairing, but on occasion, mistakes in replication happen. However, several quality control mechanisms are in place to keep the error rate at 1 error/10^9 to 10^{12} base pairs. **Genome** (the entire amount of DNA in an organism) sizes in eukaryotes range from 10^9 to 10^{11} base pairs per genome, so this error rate is low enough to prevent many lethal mutations, yet still allows genetic variation to arise.

Mutations Change the Sequence of DNA

Over the course of a lifetime, there are countless opportunities for mistakes to arise in the replication of DNA. A change in a DNA sequence, such as the addition, substitution, or deletion of a base, is a **point mutation**. If a mutation is not corrected, it may cause a change in the gene product. These changes may be relatively minor, or they may result in dysfunctional gene products that could kill the cell or the organism. Only rarely does a mutation result in a beneficial change in a gene product. Fortunately, our cells contain enzymes that detect and repair damage to DNA.

Some mutations are caused by **mutagens**, factors that increase the rate of mutation. Various chemicals, ionizing radiation such as X-rays and atomic radiation, ultraviolet light, and other factors can behave as mutagens. Mutagens either alter the base code of DNA or interfere with repair enzymes, thereby promoting mutation.

Mutations that occur in body cells are called **somatic mutations**. Somatic mutations are perpetuated in the somatic cells of an individual, but they are not passed on to subsequent generations. However, **germ-line mutations** can also occur. Because these mutations arise in the germ cells of an individual, they are passed on through gametes to future generations.

Oncogenes and Cancer

Proto-oncogenes are normal genes in the genome of an organism that code primarily for protein products that regulate cell growth, cell division, and cell adhesion. Mutations in these proto-oncogenes give rise to **oncogenes** {*onkos*,a mass}, genes that induce uncontrolled cell proliferation and the condition known as **cancer.** The mutations in proto-oncogenes that give rise to cancer-causing oncogenes are often the result of viral activity.

Photo Credits

1 **CO:** Alfred Pasieka/Photo Researchers, Inc.

2 **CO:** Michael W. Davidson/Photo Reseachers, Inc.; **2.6:** Ohio State University; **2.7a:** Charles D. Winters/Photo Researchers, Inc.; **2.7b:** Sergio Ponomarev/Shutterstock.com; **2.7c:** Katarina Pantelic/ iStockphoto.com; **2.7d:** Lisa F. Young/Shutterstock.com.

3 **CO:** J. W. Shuler/Photo Researchers, Inc.; **3.3:** Dr. J. M. Edwardson; **3.5:** Prof. P. Motta/Dept. of Anatomy/University, "La Sapienza," Rome/ Science Photo Library/Custom Medical Stock Photo; **3.8(L):** Don W. Fawcett/Photo Researchers, Inc.; **3.8(R):** Prof. H. Wartenberg/Dr. H. Jastrow's EM Atlas, (222.drjastrow.de). **3.9c:** Robert B. Tallitsch; **3.10b:** Robert B. Tallitsch; **3.10c–d:** Todd Derksen; **3.13a:** Ward's Natural Science Establishment; **3.13b–e:** Robert B. Tallitsch; **3.14:** SPL/ Photo Researchers; **3.15:** Courtesy of Elizabeth A. Abel, M.D., from the Leonard C. Winograd Memorial Slide Collection, Stanford University School of Medicine.

4 **CO:** Stefan Eberhard.

5 **CO:** Marian Miller; **5.4:** Darren Hubley/iStockphoto.com.

6 **CO:** Alila Sao Mai /Shutterstock.com, **Page 188:** Russ Bishop/ Alamy.

7 **CO:** CNRI/Photo Researchers, Inc.; **7.1:** Dee U. Silverthorn.

8 **CO:** C. J. Guerin, PhD, MRC Toxicology Unit/Photo Researchers, Inc.; **8.4:** Courtesy of Dr. Timothy Gomez, University of Wisconsin-Madison; **8.15:** Todd Derksen; **8.18:** Don W. Fawcett/Photo Researchers, Inc.; **8.22:** David Becker/Photo Researchers, Inc.

9 **CO:** Steve Gschmeissner/Photo Researchers, Inc.; **Page 291:** Dr. Tamily Weissman, Department of Molecular and Cellular Beology, Harvard University, Cambridge, Massachusetts; **9.20:** Marcus E. Raichle, M.D., Washington University School of Medicine.

10 **CO:** Dr. David Furness/Wellcome Images; **10.14:** Todd Derksen; **10.25b:** Webvision, John Moran Eye Center, University of Utah; **10.27:** Gene Chutka/iStockphoto.com.

11 **CO:** Kent Wood/Photo Researchers, Inc.

12 **CO:** Biology Media/Photo Researchers, Inc.; **12.1a–c:** Robert B. Tallitsch; **12.14(L):** Comack, D. (ed): *Ham's Histology,* 9th ed. Philadelphia: J. B. Lippincott, 1987. By Permission. **12.14(R):** Frederic H. Martini; **12.20:** Ljupco/iStockphoto.

13 **CO:** Alila Sao Mai/Shutterstock.com.

14 **CO:** Quest/Photo Researchers, Inc.

15 **CO:** Susumu Nishinaga/Photo Researchers, Inc.; **15.20:** Dr. Lyle Conrad/CDC.

16 **CO:** National Cancer Institute/Photo Researchers, Inc.; **16.5a, c–d:** David M. Phillips/Photo Researchers, Inc.; **16.5e:** Oliver Meckes & Nicole Ottawa/Photo Researchers, Inc.; **16.7b–c:** Todd Derksen; **16.11b:** Susumu Nishinaga/Photo Researchers, Inc.

17 **CO:** Biophoto Associates/Photo Researchers, Inc.

18 **CO:** Jorge Bernardino de la Serna, MEMPHYS-Center for Biomembrane Physics, University of Southern Denmark; **Page 604:** Photomick/iStockphoto.com.

19 **CO:** Susumu Nishinaga/Photo Researchers, Inc.; **19.1f:** Todd Derksen; **19.5b:** Todd Derksen.

20 **CO:** Alila Sao Mai/Shutterstock.com.

21 **CO:** Thomas Deerinck, NCMIR/Photo Researchers, Inc.

22 **CO:** Oliver Sun Kim/iStockphoto.com; **22.13b:** Ward's Natural Science Establishment.

23 **CO:** Terry Freeman, Ph.D.; **23.3a:** Biophoto Associates/Science Source/Photo Researchers, Inc.; **23.3b:** Biophoto Associates/Photo Researchers, Inc.; **23.6a:** Alison Wright/Photo Researchers, Inc.; **23.6b:** Braden Gunem/Alamy; **23.6c:** Ralph Eagle/Science Source/Photo Researchers, Inc.; **23.9(all):** Reprinted by permission from the publisher from American Journal of Medicine 20:133, 1956. Copyright © 1956 by Excepta Medica, Inc.; **23.10a:** MedImage/Photo Researchers, Inc.; **23.10c:** Dr. Michael Klein/Peter Arnold, Inc./PhotoLibrary.

24 **CO:** Dr. Dan Kalman/Photo Researchers, Inc.

25 **CO:** Alila Sao Mai/Shutterstock.com

26 **CO:** Image Courtesy of R&D Systems; **26.1:** CNRI/Science Photo Library/Photo Researchers, Inc.; **26.14a:** Densey Clyne/Photolibrary.

Cover: Cell Signaling Technology. Image reproduced courtesy of Cell Signaling Technology, Inc. (www.cellsignal.com)

Glossary/Index

NOTE: Page numbers followed by a *t* refer to tables. Page numbers followed by an *f* refer to figures.

B

C

D

E

ear canal, 346, 347f, 349f
eardrum, 346, 352f. *See also* tympanic membrane
early follicular phase, 867, 871
early ventricular diastole, 494f
eating. *See* food intake
eating behavior, 379f
eating disorders, 737, 739, 741, 744, 765, 770. *See also* anorexia nervosa
E-C coupling. *See* excitation-contraction coupling
ECF. *See* extracellular fluid
ECG. *See* electrocardiogram
ECL cell. *See* enterochromaffin-like cell
E. coli. *See Escherichia coli*
ecology, 3f
ectohormone Signal molecules secreted to the external environment (Ch 7), 209
ED. *See* erectile dysfunction
edema The accumulation of fluid in the interstitial space (Ch 15), 532–533
EDRF. *See* endothelial-derived relaxing factor
EDV. *See* end-diastolic volume
EEG. *See* electroencephalogram
effector The cell or tissue that carries out the homeostatic response (Ch 6, 14), 195, 199f, 496f
effector cell, 815f, 816
effector organ, 383f
efferent, 298f, 302t
efferent arteriole, 629, 631f–634f, 637f, 639f, 645f, 648f
efferent division, (11), 240f, 377, 442, 442t
efferent lymph vessel, 806f
efferent neuron A peripheral neuron that carries signals from the central nervous system to the target cells (Ch 6, 7, 8, 9, 11, 13), 197f, 199f, 218f, 239, 240f–241f, 298f, 380, 442t, 443f
efferent output, 447f
efferent pathway Outgoing signal that goes from the integrating center to an effector (Ch 13, 14), 450f, 496f
efficiency, 101
efflux, 257
egg, 852f, 858f, 867, 879f. *See also* ovum
egg (secondary oocyte), 857
eicosanoid Modified 20-carbon fatty acids that act as regulators of physiological functions (Ch 2, 6, 7), 33f, 41f, 177, 189, 209, 561t
Einstein, Albert, 141
Einthoven's triangle The triangle formed by the three lead electrodes of the simple ECG (Ch 14), 488f
E_{ion}. *See* equilibrium potential
E_{ion}, 165, 249t
ejaculation Semen in the urethra is expelled to the exterior (Ch 11, 26), 382f–383f, 874–875
ejaculatory duct, 862f
ejection fraction, 500
EKG. *See* electrocardiogram
elastance Ability of a stretched substance to return to its unstretched state (Ch 1, 3, 17), 9, 87, 585
elastase, 584
elastic arteries, 513f, 517f
elastic component, 423f
elastic connective tissue, 536f
elastic fiber, 573f
elastic recoil, 510, 513f, 584f
elastic systemic arteries, 509f
elastic tissue, 510f
elastin A coiled, wavy protein that returns to its original length after being stretched (Ch 3), 86f, 87
electrical activity, 261, 311f, 486, 488f
electrical charge, 161, 162f
electrical communication, 177
electrical conduction in myocardial cells, 483f
electrical current, 255f, 483f
electrical disequilibrium, 130
electrical event, 486, 490f
electrical gradient Uneven distribution of electrical change, especially across a membrane (Ch 5), 142, 162
electrical signal, 175, 185f–187f, 195f, 248–265, 272f, 428, 523f. *See also* action potential; phototransduction
  ear, 348, 349f
  eye, 365–366
  heart, 477f, 483–484, 484f
electrical synapse Synapse where electrical signals pass directly from cell to cell through gap junctions (Ch 8), 266
electricity, 161
electrocardiogram (ECG) A recording of the summed electrical events of the cardiac cycle (Ch 14), 486, 488f–490f, 494f
electrochemical gradient The combined concentration and electrical gradients for an ion (Ch 5, 8, 17, 19, 23), 142, 163, 251, 575f, 641f, 792f
electrodes, 163, 412f, 488f
electroencephalogram (EEG), 309, 311f
electroencephalography, 309
electrogenic pump, 165
electrolyte An ion (Ch 19, 20), 627, 658–659. *See also* ion electrolyte balance, *specific electrolyte*
electromechanical coupling, 431
electron, 38, 40f
electron molecular bonds, 42f
electron transport system (ETS), 111, 112f, 115f–116f, 746f
electrophoresis, 106
electrostatic attraction, 39
element The simplest kind of matter, such as oxygen and carbon (Ch 2), 40f
embryo, 474f, 851, 878–879, 880f
embryological development. *See also specific structure*
embryonic blood, 880f
embryonic blood vessels, 880f
embryonic development, 292f, 820f
embryonic germ cells (SRY gene), 856f
emergent property Some property of a system that cannot be predicted from the simple sum of its parts (Ch 1, 4, 8, 9), 2, 99, 238, 289
emesis Vomiting (Ch 21), 730
emission Movement of sperm from vas deferens to the urethra (Ch 26), 875
emotion, 300f, 303t, 305f, 312–313, 313f, 616f
emotional hyperventilation, 591t
emotional states, 314
emotional stress, 829f
emphysema, 576
emphysema Lung disease characterized by loss of elastance and alveolar surface area (Ch 5, 16, 17, 18), 569, 582, 584, 587, 594, 603f, 604
emulsion Small droplets suspended in a liquid (Ch 21), 713f
E_{Na}. *See* equilibrium potential
ENaC (epithelial sodium channel), 670f
encapsulated lymphoid tissue Lymph nodes and the spleen (Ch 24), 805, 806f
enclosed nerve ending, 327f
end-diastolic volume (EDV) The maximum volume of blood that the ventricles hold during a cardiac cycle (Ch 14), 493, 500–507, 500f
endergonic reaction A reaction that requires net input of energy from an outside source (Ch 4), 103, 103f, 104, 104f
endocrine cell, 240f, 464t, 659f, 710f, 716f, 726f, 753f, 756f, 829f, 859f
  hormones, 211
  hormone release, 217
  hypothalamic anterior pituitary pathway, 222
  pituitary gland, 220
endocrine control, 197f, **775–800**, 777–790, 792–796, 795f
endocrine gland A ductless gland or single cell that secretes a hormone (Ch 3, 7, 8, 11), 85, 85f, 225f, 240f, 268t, 387t, 390f. *See also* endocrine system
endocrine pathology. *See* endocrine system; endocrine system, disorders of
endocrine pathway, 225f
endocrine reflex, 15, 198t, 201t, 217, 218f, 219, 222f
endocrine system The cells and tissues of the body that secrete hormones (Ch 1, 6, 7, 9, 11, 14, 21, 22, 23, 24), 4, 4t, 176f, 498–499, 829f. *See also endocrine gland*; *specific gland*
  autonomic responses, 313f, 760f
  disorders of, 227–230, 777
  glands, 208
  growth, 776–797
  integrating center, 197f, 199f
  introduction to, (Ch. 7), 206
  and metabolism, 776–797
  pancreas, 303t, 383f, 706f, 716f, 719t, 753f
  principles, review of, 776
  principles review, 776–777
endocrine system sensor-integrating center, 197f
endocrinology, 207. *See also* endocrine system
endocytosis, 140f, 155, 156f, 160f, 642
endogenous factor, 564t
endogenous hormone, 227f
endogenous precursor, 795f
endolymph High K^+, low Na^+ fluid that fills the cochlear duct of the ear (Ch 10), 349, 355f
endolymphatic hydrops, 331
endometrium The secretory inner lining of the uterus (Ch 26), 866, 869f, 878–879
endopeptidase An enzyme that attacks peptide bonds in the interior of an amino acid chain (Ch 21, 22), 711f, 712, 750
endoplasmic reticulum (ER) A network of interconnected membrane tubes in the cytoplasm; site of protein and lipid synthesis (Ch 7, 12, 16), 71f, 75, 213f, 401t, 558f
endorphin, 269, 779f
endosome Vesicle formed by endocytosis (Ch 5), 156
endostatin, 513
endothelial cell, 296f, 529f, 536f, 564t, 573f
endothelial cell junction, 529f
endothelial-derived relaxing factor (EDRF) Nitric oxide released by endothelial cells; relaxes vascular smooth muscle, 188. *See also* nitric oxide endothelial
endothelial lining, 295f
endothelin, 520t
endothelium Layer of thin epithelial cells that line the lumen of the heart and blood vessels (Ch 3, 15, 16, 17, 19), 82, 510, 510f, 561t, 564t, 573f, 639f
  atherosclerotic plaques, 536f
  bone marrow, 552f
  capillary exchange, 529f
  hematopoiesis, 549t
  platelet plug formation, 560f
  vasoconstriction, 520t
endothelium-derived relaxing factor (EDRF), 434
end-plate potential (EPP) Depolarization at the motor end plate due to acetylcholine (Ch 12), 410
end-product inhibition, 110. *See also* feedback inhibition
end-systolic volume (ESV) The amount of blood left in the ventricle at the end of contraction (Ch 14), 493
energy, (4), 9, 9t, 162f, 363f
  activation and reactions, 103, 107
  and ATP transfer, 111
  in biological systems, 100–102, 125
  biological transport in the body requires, 169
  body, balance in the, 739f
  capture and transfer, 36f
  cells and, 99–100
  cellular metabolism, 98–123
  chemical reactions, 102–104
  conversions, 101
  and enzymes, 105–107
  fat and glycogen, storage in, 742
  fats store, 745–747
  input, 739, 739f
  intake, 740
  kinetic and potential, 101
  lipid storage, 751

H

ivabradine, 482
IV fluid therapy. *See* intravenous (IV) injection

J

JAK kinase, 183
Janus family tyrosine kinase, 183
jaundice A yellow tint to the skin and sclera due to excessive levels of bilirubin (Ch 16), 554
"java jolt", 312
jejunum The middle section of the small intestine (Ch 21), 699
JG cells (granular), 638
joint, 449–450
joint receptor Sensory receptors that send information about the relative positioning of bones linked by flexible joints (Ch 13), 446
juvenile-onset diabetes, 761. *See also* type 1 diabetes mellitus
juxtaglomerular (JG) apparatus Region where the distal tubule of the nephron passes between afferent and efferent arterioles (Ch 19), 629, 631*f*, 639*f*
juxtamedullary nephrons, 628

K

K^+. *See* potassium ion
kallikrein, 564*t*
K_{ATP} channel. *See* ATP-gated K^+ channel
kcal. *See* kilocalorie
K_d. *See* dissociation constant
K_{eq}. *See* equilibrium constant
keratin Insoluble protein prevalent in hair and nails (Ch 2, 3), 47, 72, 79*f*, 84
ketoacidosis A state of acidosis that results from excessive ketone production (Ch 20, 22), 682, 689, 752, 761–762, 762*f*, 763
ketoacids, 682, 682*f*
ketogenesis, 754*f*, 759*t*
ketogenic, 752
ketone bodies, 746*f*, 750*f*, 752
ketone production, 762*f*
ketones, 752, 760*f*
ketosis, 752
key, 22*f*
kidney, 341*f*, 464*t*, 465*f*, 524*f*, 555*f*, **628–649**, 630*f*, 746*f*, 751*f*, 755*t*
- adrenal gland, 778*f*
- ammonia and phosphate buffers, use of, 684–685
- calcium balance, 792*f*, 794*f*–796*f*
- conserve volume, 661*f*
- epithelial transport, 153*f*, 158*f*–159*f*
- excretion, 647–649, 796*f*
- filter, 629–632
- filtration, 634–639
- H^+ secretion, 686*f*
- hormones, 211*f*
- natriuretic peptides, 674*f*
- pH homeostasis depends on, 682
- pituitary gland, 221*f*,
- reabsorption, 629–632, 640–645
- salt and water excretion, 669*f*
- secretion, 629–632, 646
- structure of the, 630*f*
- tubular elements, 629
- urinary system, 628, 630*f*
- vasoconstrictor, 517*f*, 518, 518*f*
- water balance, 659*f*, 660–661
- water conservation, 660, 669*f*

kidney cell, 549*t*
kidney function, 627–629, 633
killer T cell. *See* cytotoxic T cell
kilocalorie (kcal or Calorie) Amount of energy needed to raise the temperature of 1 liter of water by 1° C (Ch 4, 22), 109, 740
kinase An enzyme that adds a phosphate group to the substrate (Ch 4, 23), 107–108, 108*t*, 789*f*
kinesin A motor protein (Ch 3), 74
kinetic energy The energy of motion (Ch 4), 101, 101*f*
kinin, 521, 813, 813*t*
kinocilium, 354
kiss-and-run pathway Secretion in which the secretory vesicle fuses transiently with the membrane, then pulls away (Ch 8), 271
kisspeptin, 860, 885
knee jerk, 453*t*
knockout mouse A mouse whose DNA has been altered to remove ("knock out") a specific gene (Ch 23), 776
knowledge, 314
Korotkoff sound, 515, 516*f*
Krebs, Hans A., 111
Krebs cycle, 111. *See also* citric acid cycle
kwashiokor, 533*f*
kyphosis, 797

L

labeled line coding The 1:1 association of a sensory receptor with a sensation (Ch 10), 331
labia majora Outer lips of the vulva (Ch 26), 855*f*, 866
labia minora Small inner lips of the vulva (Ch 26), 855*f*, 866
labioscrotal swelling, 852, 855*f*
labium majus, 868*f*–869*f*
labium minus, 868*f*–869*f*
labor, 881, 883*f*
labor and delivery, 881–882
lack of response, 802
lacrimal apparatus Tear ducts and glands (Ch 10), 357
lacrimal gland, 357*f*
lactase Enzyme that breaks down the milk sugar lactose (Ch 21), 710, 711*f*, 728
lactate, 116*f*–117*f*, 414*f*, 746*f*, 750*f*–751*f*, 760*f*, 837*f*
lactate accumulation, 414*f*
lactate dehydrogenase (LDH), 105, 105*t*, 117
lactation Milk production by the mammary gland (Ch 23, 26), 796*f*, 881–884, 884*f*
lacteal A fingerlike projection of the lymph system that extends into the villi of the intestine (Ch 21), 710*f*, 713*f*, 714
lactic acid The end product of anaerobic glycolysis (Ch 20), 682*f*
lactic acidosis, 682, 689
lactic acid production, 762*f*
lactose Milk sugar (Ch 2, 21), 34*f*, 710, 711*f*
lactose intolerance, 728
lamina propria Subepithelial connective tissue that holds the epithelium in place (Ch 10, 21), 343*f*, 701*f*, 702, 710*f*
laminin Insoluble protein fiber in extracellular matrix (Ch 3, 8), 78, 79*f*, 244
Langerhans cell Alternate name for dendritic cell (Ch 24), 808*f*, 811
language, 317
language processing, 318*f*
LaPlace, law of Pressure of a fluid sphere equals 2 times the surface tension of the fluid divided by the radius of the sphere (Ch 17), 586, 586*f*
large intestine The terminal portion of the intestine (Ch 21), 698*f*, 699, 700*f*, 704*f*, 720*f*, 728, 729*f*
larynx The "voice box" that contains vocal cords (Ch 8, 17, 21, 23), 571, 572*f*–573*f*, 617*f*, 721*f*, 783*f*
latch state, 431
late diastole, 491*f*
latent period Delay between the muscle action potential and beginning of muscle tension that represents the time required for Ca^{2+} release and binding to troponin (Ch 12), 412, 412*f*
lateral corticospinal tract, 456*f*
lateral dimension of rib cage, 581*f*
lateral geniculate body/nucleus Nucleus in the thalamus where optic fibers synapse with neurons going to the visual cortex (Ch 9, 10), 293, 358, 360*f*, 370*f*
lateral horn, 298*f*
lateral inhibition Process in which sensory neurons close to a stimulus are inhibited to intensify the perception of the stimulus (Ch 10), 332, 332*f*
lateral neurons, 332*f*
lateral ventricles, 293, 295*f*, 304*f*
late ventricular diastole, 494*f*
law of conservation of electrical charge The body is electrically neutral (Ch 5), 161
law of conservation of energy The total amount of energy in the universe is constant. Also called the first law of thermodynamics (Ch 4), 102
law of LaPlace. *See* LaPlace, law of, 585, 585*f*
law of mass action For a reaction at equilibrium, the ratio of substrates to products is always the same (Ch 2, 18, 20), 50–51, 51*f*, 608, 614*f*, 685*f*
law of mass balance If the amount of a substance in the body remains constant, any gain must be offset by an equal loss (Ch 1, 18), 11, 12*f*, 608
LDH. *See* lactate dehydrogenase
LDL-cholesterol. *See* low-density lipoprotein (LDL)
L-dopa Dopamine precursor that can cross the blood-brain barrier (Ch 9, 14), 297, 457
leads, 486, 488*f*
leak channel Ion channels that spend most of their time in an open state (Ch 5), 147
leaky epithelia Epithelia that allow movement between the cells (Ch 3)
leaky junctions, 529*f*, 707*f*
learned reflex, 442, 442*t*
learning, 300*f*, 305*f*, 310*f*, 314
Le Châtelier's principle, 51. *See also* law of mass action
left atrium Chamber of the heart that receives blood from the lungs (Ch 14, 15), 465*f*, 472*f*–474*f*, 474*t*, 476*f*, 509*f*
left ventricle Chamber of the heart that pumps blood to the systemic circulation (Ch 14, 15), 465*f*, 472*f*–473*f*, 474*t*, 476*f*, 509*f*, 514*f*, 517*f*, 527*f*
legs, 465*f*, 488*f*
length constant, 251
lengthening (eccentric) contraction, 421
length-force relationship, 498*f*
length-tension relationship, 418*f*, 497, 498*f*
lens Portion of the eye that focuses light upon the retina (Ch 10), 358–360, 359*f*, 362*f*, 364*f*
lens curvature, 359*f*
leptin Protein hormone from adipocytes that acts as a satiety factor (Ch 22, 26), 737, 738*f*, 885
lesion, 535
let-down reflex Neuroendocrine reflex that triggers oxytocin release and ejection of milk from the mammary gland (Ch 26), 882
leukemia, 551
leukocyte White blood cells that defend the body against foreign invaders (Ch 16, 24), 547, 550*f*, 808–810, 808*f*, 813*t*
leukopoiesis, 550
leukotriene Eicosanoid signal molecule; plays a role in the etiology of asthma (Ch 6, 17), 189, 189*f*, 588*t*, 592
levels of organization, 2, 3*f*, 64*f*
levers, 421–422
lever system, 424*f*
Leydig cell Testicular cells that secrete testosterone (Ch 26), 853, 856*f*, 863*f*, 864, 864*f*
LH. *See* luteinizing hormone
LHRH. *See* luteinizing hormone releasing hormone
libido Sex drive (Ch 26), 866
lifeblood, 545
ligament Connective tissue that connects one bone to another (Ch 3, 10), 87, 362*f*
ligand The molecule that binds to a protein (Ch 2, 5, 6, 10), 50, 51*f*, 55*f*, 146*f*, 156*f*, 178, 189–190, 190*f*, 195*f*, 345*f*
ligand-gated ion channel Synonym: chemically gated ion channel (Ch 6), 179
light, 360*f*, 362*f*, 364*f*, 365–366, 366*f*, 368*f*
- absorption, 367*f*
- bipolar cell, 369*f*
- transduction, 366*f*. *See also* phototransduction

light chain (myosin) Small protein chains that make up part of the smooth muscle myosin head (Ch 12, 24), 401, 817, 817*f*

light-off (OFF bipolar cells), 368
light-on (ON bipolar cells), 368
light rays, 362*f*
limb, 451, 667*f*
limbic system Region of the cerebrum that acts as the link between higher cognitive functions and more primitive emotional responses (Ch 9, 10, 18), 300*f*, 303*t*, 304*f*, 305, 305*f*, 313*f*, 343*f*, 616*f*
line graph, 23*f*
lipase Enzyme that digests lipids (Ch 4, 21, 22), 108, 108*t*, 713*f*, 714, 747*f*, 751, 752*f*
lipase secretion, 726*f*
lipid Synonym: fats (Ch 2, 3, 6, 8, 16, 21, 22), 33*f*, 41*f*, 66, 66*t*, 69, 188–189, 269, 546*f*, 713*f*, 751, 837*f*
lipid-anchored protein, 68
lipid bilayer, 67*f*, 140*f*
lipid core, 536*f*
lipid-derived, 182*f*
lipid-derived second messengers, 183
lipid layer, 142
lipid layer composition, 144*f*
lipid raft, 68, 68*f*
lipid-related molecule, 33*f*, 41*f*
lipid solubility, 144*f*
lipid-soluble paracrine, 189*f*
lipid-soluble substance, 720*f*
lipogenesis, 743*f*, 744, 746*f*, 756*f*, 759
lipolysis Lipid breakdown (Ch 22, 23), 743*f*, 746*f*–747*f*, 751, 752*f*, 759, 759*t*, 779*f*
lipophilic molecules Molecules that can diffuse through cell membranes (Ch 5), 142
lipophilic signal molecule, 178, 179*f*
lipophobic molecules Molecules that cannot diffuse through the phospholipid bilayer (Ch 5, 6), 142, 179, 183
lipoprotein Protein combined with a lipid (Ch 2), 32, 41*f*
lipoprotein lipase Enzyme that digests lipoproteins (Ch 22), 745, 747*f*
liposome Spherical structures with an exterior composed of a phospholipid bilayer, leaving a hollow center with an aqueous core (Ch 3), 66, 67*f*, 68
lipostatic theory Control of food intake is based on a set point for body weight that is set by adipocytes (Ch 22), 737
lipotropin, 779*f*
lipoxygenase Enzyme that converts arachidonic acid to leukotrienes (Ch 6), 189, 189*f*
liver, 211*f*, 222*f*, 224*f*, 341*f*, 382*f*, 524*f*, 547*f*, 549*t*, 555*f*, 564*t*, 672*f*, 698*f*, 700*f*, 713*f*
  bile secretion, 708–709, 709*f*
  calcium balance, 795*f*
  cardiovascular system, 464*t*, 465*f*
  digestion, 720*f*
  gluconeogenesis, 779*f*
  growth hormone, 789*f*
  hepatic portal system, 727*f*
  metabolism, 746*f*–747*f*, 751*f*, 755*t*, 756*f*, 759*t*, 760*f*, 762*f*
  muscle contraction, 520*t*, 522
liver bile, 708
liver cell, 709*f*
liver disease, 105*t*
liver glycogen, 750*f*, 837*f*
liver hepatocyte, 758*f*
liver protein, 813*t*
living conditions, 803–804
load A weight or force that opposes contraction of a muscle (Ch 12, 13), 12, 404, 424*f*, 449*f*
load-velocity relationship, 425*f*
lobes of glandular tissue, 884*f*
lobule, 709*f*
local communication, 175, 176*f*, 177
local control Homeostatic control that takes place strictly at the tissue or cell by using paracrine or autocrine signals (Ch 1, 15), 15, 15*f*, 523*f*
local control mechanisms, 593*f*
local current flow A wave of electrical current that spreads throughout the cytoplasm (Ch 8), 252, 262*f*
local signal, 177, 511*f*
locomotor pattern generator, 454*t*
locus coeruleus, 310*f*
long-distance communication, 175, 176*f*, 177
long-distance pathway, 194
long-distance signaling, 15, 177
longitudinal layer (tenai coli), 729*f*
longitudinal muscle, 700*f*–701*f*
longitudinal studies, 24
long-loop feedback, 859*f*
long-loop negative feedback Negative feedback from a peripheral endocrine gland hormone to the hypothalamus and anterior pituitary (Ch 3, 7), 224, 779*f*
long QT syndrome (LQTS) (Ch 8, 14), 251, 487
long reflex A GI reflex that is integrated in the CNS rather than in the enteric nervous system (Ch 21), 716, 716*f*, 721*f*
long-term depression (LTD), 280
long-term energy storage, 740
long-term memory, 315*f*, 316, 316*t*
long-term potentiation (LTP) Physical changes in a synapse that allow the response of the postsynaptic cell to a constant stimulus to be enhanced (Ch 8), 280
loop of Henle Portion of the renal tubule that creates dilute urine and sets up the conditions needed to make concentrated urine (Ch 19, 20), 629, 631*f*–632*f*, 633, 634*f*, 639*f*, 662*f*, 666–668, 667*f*, 669
loose connective tissue Elastic connective tissues that underlie skin and provide support for small glands (Ch 3), 86*f*, 87, 88*f*
loose platelet plug, 559*f*
loudness, 347
Lou Gehrig's disease. *See* amyotrophic lateral sclerosis (ALS)
low blood pressure, 665*f*, 671
low current flow, 261*f*
low-density lipoprotein (LDL) The "bad" protein carrier for plasma cholesterol (Ch 5, 15), 157, 536*f*
low-density lipoprotein-cholesterol (LDL-C), 535, 747*f*, 748, 748*f*
Lower, Richard, 219
lower esophageal sphincter, 720*f*
lower eyelid, 357*f*
lower respiratory system, 572*f*
lower respiratory tract, 570
low frequency, 352*f*
low oxygen blood, 593*f*
LQTS. *See* long QT syndrome
LTD. *See* long-term depression
LTP. *See* long-term potentiation
L-type calcium channel, 410, 478*f*
lumbar, 298, 389*t*
lumbar lymph node, 532*f*
lumbar puncture, 296
lumbar spinal nerve, 294*f*
lumen The cavity of a hollow tube or organ (Ch 3, 5, 6, 9, 17, 20, 21), 63–65, 77*f*, 153*f*, 158*f*–159*f*, 192*t*, 292*f*, 575*f*, 667*f*, 698*f*, 715*f*, 723*f*
  of blood vessel, 560*f*
  of collecting duct, 687*f*
  digestive enzymes secretion, 707
  of digestive tract, 698*f*
  of distal nephron, 670*f*
  fats in the, 719*t*
  of intestine, 711*f*
  pancreas, 706*f*
  of seminiferous tubule, 863*f*
  of the small intestine, 709*f*
  stomach, 706*f*
  of stomach, 721*f*, 724*f*
luminal fluid, 863*f*
lung(s) Organs where gases are exchanged with the blood (Ch 11, 14, 15, 17, 18, 20), 382*f*–383*f*, 464*t*, 465*f*, 472*f*, 474*t*, 509*f*, 524*f*, 570–571, 572*f*, 582*f*, 620, 682
  alveoli, 600*f*
  base, 572*f*
  cardiac notch, 572*f*
  CO_2 diffusion, 614*f*
  collapse, 584*f*
  in disease state, 585
  exchange of air between the atmosphere and the, 570
  external respiration, 570
  gas exchange in the, 602–607, 837*f*
  hypoxia, pathologies that cause, 603*f*
  inferior lobe, 572*f*
  inflation, 326*t*
  lobule, 573*f*
  middle lobe, 572*f*
  pH homeostasis depends on, 682
  pleural sacs, 574*f*
  protective reflexes guard the, 620
  pulmonary function tests, 579*f*
  at rest, 584*f*
  restrictive lung disease, 591*t*
  subatmospheric pressure in the pleural cavity, 584*f*
  superior lobe, 572*f*
  surface tension (T), 586*f*
  water loss, 660*f*
lung capacity Sums of two or more lung volumes (Ch 17), 580
lung compliance, 585, 603*f*, 613*f*
lung elastance, 585, 618*f*
lung volume, 578, 579*f*, 580
luteal cell, 871*f*, 873
luteal phase The portion of the menstrual cycle following ovulation, when the corpus luteum produces estrogen and progesterone (Ch 26), 867, 870*f*, 871, 871*f*–872*f*, 873
luteinization Conversion of the follicle to a corpus luteum (Ch 26), 873
luteinizing hormone (LH) Anterior pituitary hormone that acts on the gonads to influence hormone production (Ch 7, 26), 211*f*, 221, 222*f*, 223, 859, 859*f*, 864*f*, 872*f*
luteinizing hormone releasing hormone, 859n
luteinizing hormone surge, 873
lymph The fluid within the lymphatic system that moves from the tissues to the venous side of the systemic circulation (Ch 15, 21, 22), 532, 710*f*, 747*f*
lymphatics of lower limb, 532*f*
lymphatics of mammary gland, 532*f*
lymphatics of upper limbs, 532*f*
lymphatic system, 4, 532*f*, 713*f*, 806*f*
lymphatic vessel, 573*f*
lymph capillary Small vessels of the lymph system (Ch 15), 532
lymph duct, 532*f*
lymph fluid, 532*f*
lymph node Collections of immune cells that monitor the lymph for pathogens (Ch 15, 24), 532, 805–807, 806*f*
lymphocyte A white blood cell responsible primarily for the acquired immune response (Ch 7, 16, 24), 211*f*, 546*f*, 547, 548*f*, 550*f*, 552*f*, 807*f*, 808, 808*f*, 811, 814–815, 815*f*. *See also* B lymphocyte; T lymphocyte
lymphocyte clone, 815*f*, 816
lymphocyte stem cell, 548*f*
lymphoid nodule, 729*f*
lymphoid organ, 807*f*
lymphoid tissues The tissues of the immune system, including the thymus gland, bone marrow, lymph nodes, and spleen (Ch 11, 16, 24), 383*f*, 547*f*, 805–807, 807*f*. *See also* lymphatic system
lymph vessel, 531*f*, 701*f*, 806*f*
lysosomal enzyme, 810*f*
lysosomal storage disease, 76
lysosome, 63, 70*f*, 75–76, 77*f*, 156*f*, 243*f*, 810*f*
lysozyme Antibacterial enzyme found in respiratory tract secretions and tears (Ch 21, 24), 720, 811, 813, 813*t*

M

O

P

parasympathetic control, 497
parasympathetic division, 385t
parasympathetic innervation, 500f
parasympathetic nerve, 193f, 496f
parasympathetic nervous system, 378
parasympathetic neuron, 496f, 520t, 526f, 588t, 650f, 716f
parasympathetic output, 678t, 680f, 756f
parasympathetic pathway, 383f, 386, 390f
parasympathetic response, 383f
parasympathetic stimulated, 875f
parathormone, 793. *See also* parathyroid hormone
parathyroid gland, 192t, 211f, 793, 794f–795f
parathyroid hormone (parathormone, or PTH) Hormone from the parathyroid glands that increases plasma Ca^{2+} concentration (Ch 7, 23), 211f, 217, 218f, 792f, 793–794, 794f–795f
paraventricular, 219
paravertebral, 391
parietal bone, 301f
parietal cells Cells of the stomach that secrete hydrochloric acid (Ch 21), 705, 706f, 719t, 722, 723f–724f
parietal lobe, 300f, 304, 306f
parietal pleura, 583, 584f
Parkinson's disease, 297, 457
parotid, 700f
parotid salivary gland secretion, 302t
partial pressure The pressure of a single gas (Ch 18), 609
partial pressure (Pgas), 577f, 578, 605
partition, 142
parturition The birth process (Ch 26), 878–878, 883f
parvocellular ganglion cell (P cell), 370
passive, 141
passive filtration, 792f
passive immunity, 814
passive process, 141
passive reabsorption, 640–641, 640, 642
passive transport Movement across a membrane that does not depend on an outside source of energy (Ch 5), 140
patellar tendon (knee jerk) reflex, 450f
pathalogical hyponatremia, 679t
pathogen Any substance capable of causing disease (Ch 24), 802–804, 805f–806f, 810f, 814f, 829f
pathogen-associated molecular patterns (PAMPs), 811
pathogen cell, 814f
pathological dehydration, 678
pathological hyponatremia, 679
pathological pain, 339
pathologies, 591t
pathophysiology, 11
pathways Network of interconnected chemical reactions formed by the enzymatic reactions of metabolism (Ch 9, 10, 22), 299, 755t
  collateral, 360f
  pupillary reflexes, 360f
  vision, 360f
pattern recognition receptor (PRR), 811
patterns of reproduction, 856–861
Pavlov, 315
P_{cap}. *See also* capillary hydrostatic pressure
P cell (1) Parvocellular ganglion cells of the retina that transmit information about color, form, and texture (Ch 10, 20), 370, 669, 670f
Pco_2, 592f–593f, 601t, 602f, 613f, 685f
  hemoglobin's affinity for oxygen, effect of, 611f
PCOS. *See* polycystic ovary syndrome
PDE-5. *See* phosphodiesterase-5
PDGF. *See* platelet-derived growth factor
pectoral fat pad, 884f
pectoralis major muscle, 884f
pelvic cavity, 64f
pelvic lymph node, 532f
pelvic nerve, 382f
pelvis, 63, 465f, 868f
pendrin (SLC26A4), 783f, 784
pendrin transporter, 783f
penetrating solute A solute that freely crosses the cell membrane (Ch 5), 135
Penfield, W., 337f
penicillin, 646
penicillin clearance, 648f
penicillin molecules, 648f
penicillin secretion, 646
Penicillium, 54, 646
penile arterioles vasodilate, 875f
penis, 382f, 855f, 860, 862f, 875f
pentose, 34f
pepsinogen The inactive form of pepsin (Ch 21), 708, 720f, 723f–724f
pepsin release, 724f
pepsin secretion, 726f
PepT1, 712
peptic ulcer, 697
peptidase Enzyme that breaks up peptides into smaller peptides or amino acids (Ch 2, 4), 50, 108, 108t
peptide A chain of 2-9 nine amino acids (Ch 2, 8, 20, 21, 22, 23, 24), 35f, 243f, 269, 674f, 712, 719t, 721f, 723f–724f, 738, 738f, 746f, 755t, 757t, 759t, 779f, 789f, 796f, 819f
  absorption, 711f, 712
  by active transport, 720f
  fragment, 213f
peptide bond Bond formed between carboxyl group of one amino acid and amino group of another amino acid (Ch 2, 4, 21), 35f, 122, 711f
peptide hormone Any hormone made of amino acids, including peptides, proteins, and glycoproteins (Ch 7, 24, 26), 211f, 212t, 214f, 829f, 859f
  cellular mechanism of action of, 214
  membrane receptors and signal transduction for, 214f
  synthesis, storage, and release, 212
  synthesis and processing, 213f
  transport in the blood and half-life of, 214
percent saturation of hemoglobin, 609–610
percent solution Solution concentration expressed as parts of solute per 100 parts of total solution (Ch 2), 45f
perception, 300f, 306f, 308, 308f
perceptual threshold The level of stimulus intensity necessary for awareness (Ch 10), 330
perforin Pore-forming protein secreted by immune cells (Ch 24), 813, 813t, 820, 823f
perfusion Blood flow to lung tissues (Ch 15, 17, 18), 509, 590, 593f, 602. *See also* blood flow
pericardial cavity, 472f, 474f, 572f
pericardial membrane The connective tissue membrane of the pericardium (Ch 3), 65
pericardial sac, 63, 64f, 570
pericarditis Inflammation of the pericardial sac (Ch 14), 471
pericardium A tough membranous sac that encloses the heart (Ch 14), 471, 472f–473f
pericytes Cells that form a meshlike outer layer between the capillary endothelium and the interstitial fluid (Ch 15), 512
perilymph Fluid within the vestibular and tympanic ducts of the cochlea (Ch 10), 349
periodic breathing, 621
periodic paralysis, 400, 413, 415, 425, 435
periodic table of the elements, 40f, inside cover
peripheral blood flow, 523f
peripheral blood pressure, 679
peripheral chemoreceptor Chemoreceptors not found in the CNS (Ch 18), 616f, 618, 620f
peripheral component, 390f
peripheral fatigue, 414, 414f
peripheral ganglia, 389t
peripheral nerve, 827t
peripheral nervous system (PNS) All neurons that lie completely or partially outside the central nervous system (Ch 8, 9), 239, 240f, 246f–247f, 292f
  efferent division of the, 240f
  sensory division of the, 240f
peripheral neuron injury, 248f
peripheral protein Proteins attached to membrane-spanning proteins or to the polar regions of membrane phospholipids. (Ch 3, 5), 68, 145f
peripheral receptor Sensory receptors that are not located in or close to the brain (Ch 6), 195f
peripheral resistance Resistance to blood flow created primarily by the arterioles (Ch 14, 15, 20, 25), 470, 516, 517f, 526f, 680f, 842f
peripheral sensory receptor, 195f
peripheral thermoreceptor, 768f
peripheral tissue, 709f, 760f
peristalsis Waves of contraction that move along the gastrointestinal tract (Ch 21), 703
peristaltic contraction, 704f
peristaltic mixing, 720f
peristaltic propulsion, 720f
peritoneal membrane Lines the inside of abdominal cavity (Ch 3, 21), 65, 703
peritoneum A membrane that lines the abdomen (Ch 3, 19, 21), 63, 628, 630f, 703
peritubular capillaries, 629, 631f–634f, 645f, 648f, 687f
peritubular capillary pressure, 644–645
permeability to ion, 479f
permeable, 140
permissive effect, 780
permissive hormone, 226
permissiveness One hormone cannot exert its effects fully unless a second hormone is present (Ch 7), 226
pernicious anemia, 714
peroxisome Storage vesicles that contain enzymes to degrade long-chain fatty acids and potentially toxic foreign molecules (Ch 3), 70f, 76
peroxisome profilerator-activated receptors (PPARs), 765
personality, 318
PET scan, 318f
Peyer's patch Bump of lymphoid tissue visible in the mucosa of the GI tract (Ch 21), 701f, 702
P_{fluid}. *See* fluid pressure
P_{gas}. *See* partial pressure
PGI_2. *See* prostaglandin I
P_H. *See* hydrostatic pressure
pH A measure of the concentration of H^+; pH^+ log $[H^+]$ (Ch 2, 4, 10, 17, 18, 20, 24), 47, 48f, 54t, 55f, 106f, 326t, 569, 600, 601t, 613f, 618
  in acid-base disturbances, 688t
  balance in the body, 682f
  central and peripheral chemoreceptors monitor blood gases and, 616f
  denature proteins, changes can, 681
  disturbances, 684
  hemoglobin's affinity for oxygen, effect of, 611f
  homeostasis, 682. *See also* acid-base balance
  renal and respiratory compensation, 688f
  of the stomach, 805f
  ventilation, influence on, 618–619
phagocyte Immune cell that ingests material by phagocytosis (Ch 5, 16, 24), 155, 547, 808, 808f, 809, 810f, 811, 813t, 821, 822f
phagocyte lysosome, 813t
phagocyte receptor, 810f
phagocytosis The process by which a cell engulfs a particle into a vesicle by using the cytoskeleton to push the membrane around the particle (Ch 5, 24), 140f, 155, 155f, 810–811, 810f, 817f
phagosome The vesicle formed around ingested material during phagocytosis; site of digestion (Ch 5, 24), 155, 810f, 811–812
phantom limb pain, 331
pharmacogenomics, 3
pharmacological doses, 780
pharmacomechanical coupling Contraction that occurs in smooth muscle as a result of a ligand binding; not accompanied by a change in membrane potential (Ch 12), 431, 432f

S

transverse tubule (t-tubules) Invaginations of the muscle fiber membrane, associated with the sarcoplasmic reticulum (Ch 12), 401
TRH. *See* thyrotropin-releasing hormone
triacylglycerol. *See also* triglyceride
triad One t-tubule with its flanking terminal cisternae (Ch 12), 401, 404*f*
tricarboxylic acid cycle. *See* citric acid cycle
tricep, 400*f*, 423*f*
tricep brachii, 400
tricuspid (right AV) valve, 476*f*, 509*f*
tricuspid valve The right AV valve of the heart (Ch 14), 475
tricyclic antidepressant, 314
trigeminal, 302*t*
trigeminal nerve, 346
trigger zone The region of the axon where graded potentials are integrated and an action potential begins if the signal is above threshold (Ch 8, 10), 252*t*, 253*f*, 254, 255*f*, 258, 262*f*, 274*f*, 278*f*–279*f*, 333*f*
triglyceride Lipid composed of one glycerol and three fatty acids. Synonym: triacylglycerol (Ch 2, 21, 22, 25), 33*f*, 41*f*, 713*f*, 743, 746*f*–747*f*, 750*f*, 752*f*, 837*f*
triglyceride synthesis, 755*t*
triiodothyronine (T3) Most active form of thyroid hormone; produced mostly in peripheral tissues from T4 (Ch 7, 23), 211*f*, 217*f*, 783*f*, 784
tRNA. *See* transfer RNA
trochlear, 302*t*
trophic, 245, 789*f*
trophic hormone Any hormone that controls the secretion of another hormone (Ch 7, 9), 211*f*, 221, 222*f*, 225*f*, 303*t*
tropic hormone. *See* trophic hormone
- tropin (suffix), 223
tropomyosin A regulatory protein that blocks the myosin-binding site on actin (Ch 12), 401, 402*f*–403*f*, 407, 407*f*, 409*f*, 411*f*, 434*t*
troponin Complex of three proteins associated with tropomyosin (Ch 6, 12, 14), 187, 401, 402*f*–403*f*, 407, 407*f*, 411*f*, 430*f*, 434*t*, 478*f*
troponin C, 408
$TRPV_1$ channel. *See* transient receptor potential V_1 ($TRPV_1$) channel
trunk, 465*f*
trypsin Enzyme that digests proteins (Ch 21), 711*f*, 712, 726*f*
trypsinogen Inactive form of trypsin (Ch 3, 21), 52, 726*f*
tryptophan Amino acid from which the hormone melatonin is made (Ch 7), 231*f*
TSH receptor, 827
TSH (thyroid-stimulating hormone). *See* thyrotropin
TSI. *See* thyroid-stimulating immunoglobulin
T-tubule, 401, 402*f*–404*f*, 411*f*, 434*t*, 478*f*. *See also* transverse tubule
TTX. *See* tetrodotoxin
tubal ligation, 876
tubular element, 632*f*
tubular epithelium, 641*f*
tubular reabsorption, 641*f*
tubule, 633*f*, 674*f*
tubule lumen, 641*f*
tubulin, 72
tubuloglomerular feedback The process by which changes in fluid flow through the distal tubule influence glomerular filtration rate (Ch 19), 636, 638, 639*f*
tumor. *See also* cancer
tumor cell, 811
tumor necrosis factor (TNF), 813, 813*t*
tunica intima, 510
tuning fork, 348*f*
T wave ECG wave that represents ventricular repolarization (Ch 14), 486, 489*f*–490*f*
twitch A single contraction/relaxation cycle in a muscle fiber (Ch 12), 406, 412
two-point discrimination test, 329, 329*f*
2010 Advanced Placement Biology Curriculum (College Board), 9*t*
tympanic duct (perilymph), 349, 349*f*–350*f*
tympanic membrane, 346, 347*f*, 349*f*
type 1. *See* slow-twitch fibers
type 1 diabetes mellitus, 178, 761–762, 762*f*, 827, 827*t*
type 2 diabetes mellitus, 18, 24, 761, 763–764, 845
type I alveolar cell Thin alveolar cells for gas exchange (Ch 17), 573*f*
type IIA. *See* fast-twitch oxidative-glycolytic fibers
type II alveolar cell Alveolar cells that synthesize and secrete surfactant (Ch 17), 573*f*, 574
type IIB. *See* fast-twitch glycolytic fibers
type III taste cells, 344
type II taste cell, 344
type I support cell, 345*f*
tyrosine Amino acid that is the basis for thyroid hormones and the catecholamines (Ch 7, 11, 22, 23), 217*f*, 384*f*, 385*t*, 757*t*, 783*f*, 784–785, 785*f*
tyrosine derivative, 212*t*
tyrosine kinase (TK) Membrane enzyme that adds a phosphate group to the tyrosine residue of a cytoplasmic protein, enhancing or inhibiting its activity (Ch 6, 7, 22), 183, 186*f*, 214*f*, 755, 755*t*
tyrosine kinase activity, 789*f*
tyrosine kinase receptor, 183*f*, 757*f*

U

ubiquitin Protein that tags molecules for destruction by proteasomes (Ch 4), 124
ultra-short-loop feedback, 224
umami The taste sensation triggered by glutamate and associated with nutritious food (Ch 10), 344, 345*f*
umbilical arteries, 880*f*
umbilical cord, 880*f*, 883*f*
umbilical vein, 880*f*
unbound hormone, 215*f*
undershoot, 257
under-ventilated alveoli, 593*f*
unfused tetanus, 418, 419*f*
uninucleate, 434*t*
uniport carrier A membrane transport protein that moves only one kind of molecule (Ch 5), 147*f*, 148
unitary smooth muscle, 426. *See also* visceral smooth muscle
United States National Institutes of Health, 175
universal gas constant, 577*f*
unmyelinated axon, 263*f*, 327*f*
unmyelinated axon membrane, 247*f*
unmyelinated fiber, 339*f*
unmyelinated nerve ending, 338*f*
unregulated heat, 739*f*
unsaturated fatty acid Fatty acid with one or more double bonds between carbons (Ch 2), 33*f*
unstable plaque, 536*f*
up and down-regulation, 191
upper esophageal sphincter, 720*f*–721*f*
upper eyelid, 357*f*
upper respiratory infections (URIs), 846*f*
upper respiratory system, 572*f*
upper respiratory tract, 570
up-regulation Increase in protein number or binding affinity that increases the response of the target cell (Ch 2, 6), 56, 191
uracil, 36*f*–37*f*
urate, 646
urate transporter (UAT), 651
urate transporter 1 (URAT1), 651
urea Nitrogenous waste product produced from amino groups (Ch 4, 19, 20, 21), 627, 642, 668, 709*f*, 750, 751*f*
urea and medullary interstitium, 668
urea clearance, 648*f*
urea molecule, 648*f*
urea production, 709*f*
ureter Tube that links a kidney to the urinary bladder (Ch 10, 19, 26), 341*f*, 628, 630*f*, 862*f*
urethra Single tube that drains urine from the bladder to the external environment (Ch 19, 26), 628, 630*f*, 855*f*, 861, 862*f*, 868*f*
urethral fold, 852, 855*f*
urethral groove, 852, 855*f*
urethral opening, 855*f*, 869*f*
uric acid Nitrogenous waste product (Ch 19), 642
uricosuric, 646
urinalysis, 627
urinary, 426
urinary bladder, 382*f*–383*f*, 628, 630*f*, 632*f*–633*f*, 650*f*, 862*f*, 868*f*
urinary bladder control, 379*f*
urinary excretion, 633*f*
urinary system The kidneys, bladder, and accessory structures (Ch 1, 3, 19), 3, 4*t*, 81*f*, 628–629, 630*f*. *See also specific structure*
urinary tract infections (UTI), 628, 792–793
urination. *See* micturition
urine Fluid waste product produced by the kidneys (Ch 11, 15, 16, 19, 20, 22, 23), 382*f*, 518*f*, 555*f*, 627, 659*f*–660*f*, 661–662, 661*f*, 664*f*, 682*f*, 686*f*–687*f*, 743*f*, 792*f*
urine acidification, 762*f*
urine osmolarity, 662*f*
urobilinogen, 627
urocortin, 782
U.S. Food and Drug Administration (FDA), 876
uterine artery, 869*f*
uterine cavity, 868*f*–869*f*
uterine contraction, 18, 18*f*, 883*f*
uterine cycle, 867, 870*f*
uterine muscle, 211*f*
uterine wall, 883*f*
uterus, 211*f*, 221*f*, 382*f*–383*f*, 853, 854*f*, 866, 868*f*–870*f*, 879*f*, 883*f*
utricle One of the otolith organs of the vestibular apparatus (Ch 10), 354, 355*f*
UV rays, 363*f*

V

V2 receptor, 663, 665*f*
vagal, 196
vagal reflex, 722
vagina, 853, 854*f*–855*f*, 868*f*–869*f*, 883*f*
vaginal opening, 855*f*
vagotomy Operation that severs the vagus nerve (Ch 11), 381
vagus, 196, 302*t*
vagus nerve Cranial nerve that carries sensory information and efferent signals to many internal organs including the heart and GI tract (Ch 9, 11, 14, 21), 301, 381, 382*f*, 482, 721*f*, 724*f*
Valsalva maneuver Abdominal contraction and forced expiratory movement against a closed glottis (Ch 21), 728
valve, heart, 512*f*. *See also specific valve*
valvular stenosis, 493
van der Waals force Weak attractive force that occurs between two polar molecules or a polar molecule and an ion (Ch 2), 42*f*, 43
vanilloid receptor, 337
variability, 21
variable resistance, 510, 517*f*
varicosity Swollen regions along autonomic axons that store and release neurotransmitter (Ch 8, 11), 242, 384*f*, 389*t*, 390*f*, 427*f*
vasa recta Peritubular capillaries in the kidney that dip into the medulla and then go back up to the cortex, forming hairpin loops (Ch 19, 20), 629, 631*f*–632*f*, 664*f*, 666, 667*f*
vascular, 426
vascular elements of the kidney, 629
vascular endothelial growth factor (VEGF) Growth factors that regulate angiogenesis (Ch 15), 512
vascular endothelium, 520*t*
vascular permeability, 825*f*
vascular smooth muscle The smooth muscle of blood vessels (Ch 12, 15), 426*f*, 510, 520, 522
vascular tissue, 552*f*

W

X

Y

Z

Anatomical Positions of the Body

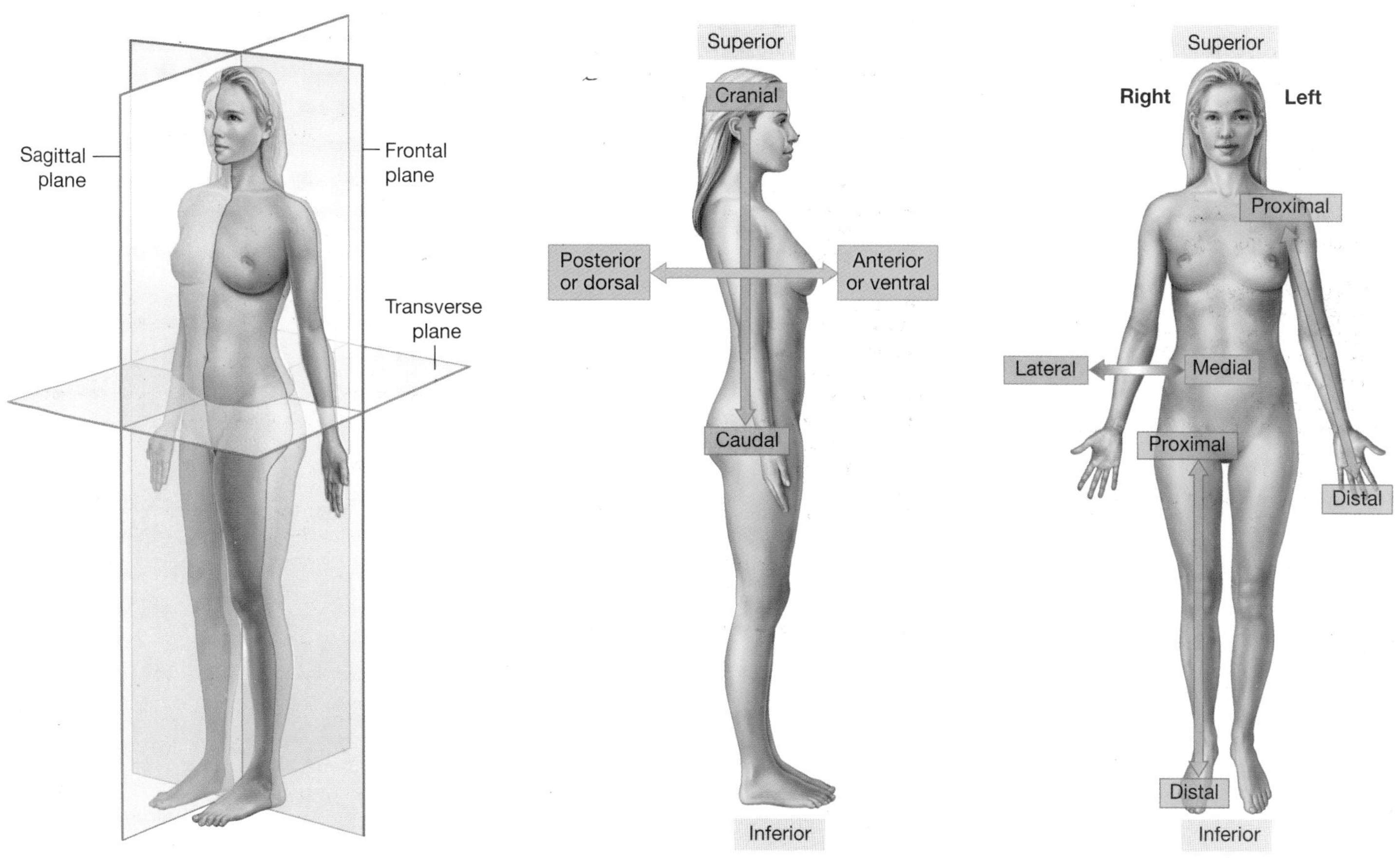

(a) Sectional planes

(b) Body directions

Anterior	(situated in front of): in humans, toward the front of the body (see VENTRAL).
Posterior	(situated behind): in humans, toward the back of the body (see DORSAL).
Medial	(middle, as in *median strip*): located nearer to the midline of the body (the line that divides the body into mirror-image halves)
Lateral	(side, as in a *football lateral*): located toward the sides of the body
Distal	(distant): farther away from the point of reference or from the center of the body
Proximal	(closer, as in *proximity*): closer to the center of the body
Superior	(higher): located toward the head or the upper part of the body
Inferior	(lower): located away from the head or from the upper part of the body
Prone:	lying on the stomach, face downward
Supine:	lying on the back, face up
Dorsal:	refers to the back of the body
Ventral:	refers to the front of the body
Ipsilateral:	on the same side as
Contralateral:	on the opposite side from

Measurements and Conversions

PREFIXES

deci-	(d)	1/10	0.1	1×10^{-1}
centi-	(c)	1/100	0.01	1×10^{-2}
milli-	(m)	1/1000	0.001	1×10^{-3}
micro-	(μ)	1/1,000,000	0.000001	1×10^{-6}
nano-	(n)	1/1,000,000,000	0.000000001	1×10^{-9}
pico-	(p)	1/1,000,000,000,000	0.000000000001	1×10^{-12}
kilo-	(k)		1000.	1×10^{3}

METRIC SYSTEM

1 meter (m)	=	100 centimeters (cm)	=	1000 millimeters (mm)
1 centimeter (cm)	=	10 millimeters (mm)	=	0.01 meter (m)
1 millimeter (mm)	=	1000 micrometers (μm; also called micron, μ)		
1 angstrom (Å)	=	1/10,000 micrometer	=	1×10^{-7} millimeters
1 liter (L)	=	1000 milliliters (mL)		
1 deciliter (dL)	=	100 milliliters (mL)	=	0.1 liter (L)
1 cubic centimeter (cc)	=	1 milliliter (mL)		
1 milliliter (mL)	=	1000 microliters (μL)		
1 kilogram (kg)	=	1000 grams (g)		
1 gram (g)	=	1000 milligrams (mg)		
1 milligram (mg)	=	1000 micrograms (μg)		

CONVERSIONS

1 yard (yd)	= 0.92 meter
1 inch (in)	= 2.54 centimeters
1 meter	= 1.09 yards
1 centimeter	= 0.39 inch
1 liquid quart (qt)	= 946 milliliters
1 fluid ounce (oz)	= 8 fluid drams = 29.57 milliliters (mL)
1 liter	= 1.05 liquid quarts
1 pound (lb)	= 453.6 grams
1 kilogram	= 2.2 pounds

TEMPERATURE

FREEZING

0 degrees Celsius (°C)	=	32 degrees Fahrenheit (°F)	=	273 Kelvin (K)

To convert degrees Celsius (°C) to degrees Fahrenheit (°F): (°C × 9/5) + 32

To convert degrees Fahrenheit (°F) to degrees Celsius (°C): (°F − 32) × 5/9

NORMAL VALUES OF BLOOD COMPONENTS

SUBSTANCE OR PARAMETER	NORMAL RANGE	MEASURED
Calcium (Ca^{2+})	4.3–5.3 meq/L	Serum
Chloride (Cl^{-})	100–108 meq/L	Serum
Potassium (K^{+})	3.5–5.0 meq/L	Serum
Sodium (Na^{+})	135–145 meq/L	Serum
pH	7.35–7.45	Whole blood
P_{O_2}	75–100 mm Hg	Arterial blood
P_{CO_2}	34–45 mm Hg	Arterial blood
Osmolality	280–296 mosmol/kg water	Serum
Glucose, fasting	70–110 mg/dL	Plasma
Creatinine	0.6–1.5 mg/dL	Serum
Protein, total	6.0–8.0 g/dL	Serum

Modified from W. R Ganong, *Review of Medical Physiology* (Norwalk: Appleton & Lange). 1995.